FOR REFERENCE

Do Not Take From This Room

Encyclopedia of the Solar System

Image compilation: S. Pieth, DLR.

Encyclopedia of the Solar System

Third Edition

Edited by

Tilman Spohn
Institute of Planetary Research
German Aerospace Centre (DLR)
Berlin, Germany

Doris Breuer
Institute of Planetary Research
German Aerospace Centre (DLR)
Berlin, Germany

Torrence V. Johnson
Jet Propulsion Laboratory, NASA
Pasadena, CA
USA

Waubonsee Community College
Aurora Campus
18 S. River Street
Aurora, IL 60506

ELSEVIER

AMSTERDAM • BOSTON • HEIDELBERG • LONDON • NEW YORK • OXFORD
PARIS • SAN DIEGO • SAN FRANCISCO • SINGAPORE • SYDNEY • TOKYO

Elsevier
Radarweg 29, PO Box 211, 1000 AE Amsterdam, Netherlands
The Boulevard, Langford Lane, Kidlington, Oxford, OX5 1GB, UK
225 Wyman Street, Waltham, MA 02451, USA

First edition 1999
Second edition 2007
Third edition 2014

Copyright © 2014, 2007, 1999 Elsevier Inc. All rights reserved.

No part of this publication may be reproduced, stored in a retrieval system or transmitted in any form or by any means electronic, mechanical, photocopying, recording or otherwise without the prior written permission of the publisher

Permissions may be sought directly from Elsevier's Science & Technology Rights Department in Oxford, UK: phone (+44) (0) 1865 843830; fax (+44) (0) 1865 853333; email: permissions@elsevier.com. Alternatively you can submit your request online by visiting the Elsevier web site at http://elsevier.com/locate/permissions, and selecting Obtaining permission to use Elsevier material

Notice
No responsibility is assumed by the publisher for any injury and/or damage to persons or property as a matter of products liability, negligence or otherwise, or from any use or operation of any methods, products, instructions or ideas contained in the material herein

Library of Congress Cataloging-in-Publication Data
A catalog record for this book is available from the Library of Congress

Encyclopedia of the solar system / edited by Tilman Spohn, Doris Breuer and Torrence Johnson. – Third edition.
   pages cm
   Includes bibliographical references and index.
   Summary: "This book is filled with the knowledge about our solar system that resulted from all this exploration, whether by spacecraft or by telescopes both in space and earth-bound. All of this new knowledge is based on discoveries made in the interim by scientist-explorers who have followed their inborn human imperative to explore and to understand. Many old mysteries, misunderstandings, and fears that existed 50 years ago about what lay beyond the Earth have been eliminated. We now know the major features of the landscape in our cosmic backyard and can look forward to the adventure, excitement, and new knowledge that will result from more in-depth exploration by today's spacecraft, such as those actually exploring the surface of these faraway places, including the Huygens Titan lander and the Mars Exploration rovers, doing things that were unimaginable before the Space Age began. The Encyclopedia of the Solar System is filled with images, illustrations, and charts to aid in understanding. Every object in the solar system is covered by at least one chapter. Other chapters are devoted to the relationships among the objects in the solar system and with the galaxy beyond. The processes that operate on solar system objects, in their atmospheres, on their surfaces, in their interiors, and interactions with space itself are all described in detail. There are chapters on how we explore and learn about the solar system and about the investigations used to make new discoveries. And there are chapters on the history of solar system exploration and the missions that have carried out this enterprise. All written by an international set of world-class scientists using rigorous yet easy-to-understand prose"–Provided by publisher.
   ISBN 978-0-12-415845-0
   1. Solar system–Encyclopedias. I. Spohn, T. (Tilman), editor of compilation. II. Breuer, Doris, 1965- editor of compilation. III. Johnson, T. V. (Torrence V.), editor of compilation.
   QB501.E53 2014
   523.203–dc23
                                        2014002257

British Library Cataloguing in Publication Data
A catalogue record for this book is available from the British Library

ISBN: 978-0-12-415845-0

For information on all Elsevier publications
visit our web site at store.elsevier.com

Printed and bound in China
14 15 16 17 18   10 9 8 7 6 5 4 3 2 1

Contents

Foreword   ix
Preface to the Third Edition   xi
Preface to the Second Edition   xiii
Preface to the First Edition   xv
About the Editors   xvii
Contributors   xix

Part I
The Solar System

1. The Solar System and Its Place in the Galaxy   3
   Paul R. Weissman

2. The Origin of the Solar System   29
   John E. Chambers and Alex N. Halliday

3. Solar System Dynamics: Regular and Chaotic Motion   55
   Jack J. Lissauer and Carl D. Murray

Part II
Fundamental Planetary Processes and Properties

4. Planetary Impacts   83
   Richard A.F. Grieve, Gordon R. Osinski, and Livio L. Tornabene

5. Planetary Volcanism   101
   Lionel Wilson

6. Magnetic Field Generation in Planets   121
   Sabine Stanley

7. Planetary Magnetospheres   137
   Margaret Galland Kivelson and Fran Bagenal

8. Rotation of Planets   159
   Véronique Dehant and Tim Van Hoolst

9. Evolution of Planetary Interiors   185
   Nicola Tosi, Doris Breuer, and Tilman Spohn

10. Astrobiology   209
    Christopher P. McKay and Wanda L. Davis

Part III
The Sun

11. The Sun   235
    Markus J. Aschwanden

12. The Solar Wind   261
    J.T. Gosling

Part IV
Earthlike Planets

13. Mercury   283
    Scott L. Murchie, Ronald J. Vervack, Jr., Carolyn M. Ernst, and Robert G. Strom

14. Venus: Atmosphere   305
    Fredric W. Taylor and Donald M. Hunten

15. **Venus: Surface and Interior** 323
    Suzanne E. Smrekar, Ellen R. Stofan, and Nils Mueller

16. **Mars Atmosphere: History and Surface Interactions** 343
    David C. Catling

17. **Mars: Surface and Interior** 359
    Michael H. Carr and James F. Bell, III

18. **Interior Structure and Evolution of Mars** 379
    Tim Van Hoolst and Attilio Rivoldini

19. **Mars: Landing Site Geology, Mineralogy, and Geochemistry** 397
    Matthew P. Golombek and Harry Y. McSween

Part V
Earth and Moon as Planets

20. **Earth as a Planet: Atmosphere and Oceans** 423
    Adam P. Showman and Timothy E. Dowling

21. **Earth as a Planet: Surface and Interior** 445
    David Pieri and Adam Dziewonski

22. **Space Weather** 479
    J.G. Luhmann and S.C. Solomon

23. **The Moon** 493
    Harald Hiesinger and Ralf Jaumann

24. **Interior of the Moon** 539
    Renee C. Weber

25. **Lunar Exploration** 555
    Ian A. Crawford, Katherine H. Joy, and Mahesh Anand

Part VI
Asteroids, Dust and Comets

26. **Main-Belt Asteroids** 583
    Daniel T. Britt, Guy Consolmagno, S.J., and Larry Lebofsky

27. **Near-Earth Objects** 603
    Alan W. Harris, Line Drube, Lucy A. McFadden, and Richard P. Binzel

28. **Meteorites** 625
    Michael E. Lipschutz and Ludolf Schultz

29. **Dust in the Solar System** 657
    Harald Krüger and Eberhard Grün

30. **Physics and Chemistry of Comets** 683
    John C. Brandt

31. **Comet Populations and Cometary Dynamics** 705
    Harold F. Levison and Luke Dones

Part VII
Giant Planets and their Satellites

32. **Atmospheres of the Giant Planets** 723
    Robert A. West

33. **Interiors of the Giant Planets** 743
    Mark S. Marley and Jonathan J. Fortney

34. **Planetary Satellites** 759
    Bonnie J. Buratti and Peter C. Thomas

35. **Io: The Volcanic Moon** 779
    Rosaly M.C. Lopes

36. **Europa** 793
    Louise M. Prockter and Robert T. Pappalardo

37. Ganymede and Callisto 813
    Geoffrey Collins and Torrence V. Johnson

38. Titan 831
    Athena Coustenis

39. Enceladus 851
    Francis Nimmo and Carolyn Porco

40. Triton 861
    William B. McKinnon and Randolph L. Kirk

41. Planetary Rings 883
    Matthew S. Tiscareno and Matthew M. Hedman

Part VIII
Beyond the Planets

42. Pluto 909
    S. Alan Stern

43. Kuiper Belt: Dynamics 925
    Alessandro Morbidelli and Harold F. Levison

44. Kuiper Belt Objects: Physical Studies 941
    Stephen C. Tegler

45. Extrasolar Planets 957
    Michael Endl

Part IX
Exploring the Solar System

46. Strategies of Modern Solar System Exploration 981
    Berndt Feuerbacher and Bernhard Hufenbach

47. A History of Solar System Studies 999
    David Leverington

48. X-rays in the Solar System 1019
    Anil Bhardwaj, Carey M. Lisse, and Konrad Dennerl

49. The Solar System at Ultraviolet Wavelengths 1047
    Amanda R. Hendrix, Robert M. Nelson, and Deborah L. Domingue

50. Infrared Views of the Solar System from Space 1073
    Mark V. Sykes

51. New Generation Ground-Based Optical/Infrared Telescopes 1089
    Alan T. Tokunaga

52. The Solar System at Radio Wavelengths 1107
    Imke de Pater and William S. Kurth

53. Planetary Radar 1133
    Catherine D. Neish and Lynn M. Carter

54. Remote Sensing of Chemical Elements Using Nuclear Spectroscopy 1161
    Thomas H. Prettyman

55. Probing the Interiors of Planets with Geophysical Tools 1185
    W. Bruce Banerdt, Véronique Dehant, Robert Grimm, Matthias Grott, Philippe Lognonné, and Suzann E. Smrekar

56. Planetary Exploration Missions 1205
    James D. Burke

57. **Exploration and Analysis of Planetary Shape and Topography Using Stereophotogrammetry** 1223

Jürgen Oberst, Klaus Gwinner and Frank Preusker

Appendix 1235
Glossary 1261
Index 1283

Foreword*

The solar system has become humankind's new backyard. It is the playground of robotic planetary spacecraft that has surveyed just about every corner of this vast expanse in space. Nowadays, every schoolchild knows what even the farthest planets look like. Fifty years ago, these places could only be imagined, and traveling to them was the realm of fiction. In just this short time in the long history of the human species we have leapt off the surface of our home planet and sent robotic extensions of our eyes, ears, noses, arms, and legs to the far reaches of the solar system and beyond.

In the early twentieth century, we were using airplanes to extend our reach to the last unexplored surface regions of our own planet. Now 100 years later, at the beginning of the twenty-first century, we are using spacecraft to extend our reach from the innermost planet Mercury to the outmost planet Neptune, and we have a spacecraft on the way to Pluto and the Kuiper Belt. Today, there are telescopes beyond imagination 100 or even 50 years ago that can image Pluto and detect planets around other stars! Now, Sol's planets can say "we are not alone"; there are objects just like us elsewhere in the universe. As humanity's space technology improves, perhaps in the next 100 years or so human beings also may be able to say "we are not alone."

When I was a kid more than 50 years ago, I was thrilled by the paintings of Chesley Bonestell and others who put their imagination on canvas to show us what it might be like "out there." Werner Von Braun's *Collier's* magazine articles of 1952–1954 superbly illustrated how we would go to the Moon and Mars using new rocket technologies. Reading those fabulous articles crystallized thoughts in my young mind about what to do with my life. I wanted to be part of the adventure to find out what these places were like. Not so long after the *Collier's* articles appeared, we did go to the Moon, and pretty much as illustrated, although perhaps not in such a grand manner. We have not sent humans to Mars—at least we have not yet—but we have sent our robots to Mars and to just about every other place in the solar system as well.

This book is filled with the knowledge about our solar system that resulted from all this exploration, whether by spacecraft or by telescopes both in space and earth-bound. It could not have been written 50 years ago as almost everything in this Encyclopedia was unknown back then. All of this new knowledge is based on discoveries made in the interim by scientist-explorers who have followed their inborn human imperative to explore and to understand. Many old mysteries, misunderstandings, and fears that existed 50 years ago about what lay beyond the Earth have been eliminated.

We now know the major features of the landscape in our cosmic backyard and can look forward to the adventure, excitement, and new knowledge that will result from more in-depth exploration by today's spacecraft, such as those actually exploring the surface of these faraway places, including the *Huygens Titan* lander, the *Mars Exploration* and the Curiosity rovers, doing things that were unimaginable before the Space Age began.

The *Encyclopedia of the Solar System* is filled with images, illustrations, and charts to aid in understanding. Every object in the solar system is covered by at least one chapter. Other chapters are devoted to the relationships among the objects in the solar system and with the galaxy beyond. The processes that operate on solar system objects, in their atmospheres, on their surfaces, in their interiors, and interactions with space itself are all described in detail. There are chapters on how we explore and learn about the solar system and about the investigations used to make new discoveries. And there are chapters on the history of solar system exploration and the missions that have carried out this enterprise. All written by an international set of world-class scientists using rigorous yet easy-to-understand prose.

Everything you want to know about the solar system is here. This is your highway to the solar system. It is as much fun exploring this Encyclopedia as all the exploration it took to get the information that it contains. Let your fingers be the spacecraft as you thumb through this book visiting all the planets, moons, and other small objects in the solar system. Experience what it is like to look at our solar system with ultraviolet eyes, infrared eyes, radio eyes, and radar eyes.

---

* This foreword to the second edition has been editorially updated to be included in the present edition.

It has been almost 15 years since the first edition. The exploration of space has continued at a rapid pace since then, and many missions have flown in the interim. New discoveries are being made all the time. This third edition will catch you up on all that has happened since the previous editions, including several new chapters based on information from our latest missions.

I invite you to enjoy a virtual exploration of the solar system by flipping through the pages in this volume. This book deserves a place in any academic setting and wherever there is a need to understand the cosmos beyond our home planet. It is the perfect solar system reference book, lavishly illustrated and well written. The editors and authors have done a magnificent job.

We live in a wonderful time of exploration and discovery. Here is your window to the adventure.

Wesley T. Huntress
Geophysical Laboratory,
Carnegie Institution of Washington,
Washington, D.C.

Preface to the Third Edition

The known is finite, the unknown infinite; intellectually we stand on an islet in the midst of an illimitable ocean of inexplicability. Our business in every generation is to reclaim a little more land.
                                                     Thomas Henry Huxley

It is now 15 years since the first publication of the Encyclopedia of the Solar System and 8 years since the second and revised edition. The book has been an undebated jewel in every library of books in solar system science and a great success with readers.

When Elsevier approached us to prepare a third edition, with a largely new editorship we thought hard on how we would proceed. Elsevier had left us to decide whether we wanted a completely new concept or to simply update the existing book. We finally settled on a concept that tried to further evolve an outstandingly successful work.

The past decade has seen an increasing importance of geophysical tools for the exploration of planets. In addition, our theoretical knowledge of the inner workings of terrestrial planets has substantially increased. We have acknowledged this by adding a chapter on geophysical exploration tools, in general, and on exploration of the Moon and on potential landing sites on Mars. We further added a chapter on rotation of the planets and using its observation to constrain models of the interior of terrestrial planets. Chapters on the interiors of Mars and the Moon—the two planets we know best—have been added as well as theoretical chapters pertaining to the inner workings of terrestrial planets—the generation of their magnetic fields and the relation between their thermal evolution, convection in the interior and their tectonics.

High resolution and stereo imaging is another novel tool of planetary exploration that we serve by adding a chapter. For the outer solar system we have added a chapter on Enceladus. Finally, we have complemented the suite of chapters dealing with the history of space exploration with a chapter describing the strategies that the space fairing nations have jointly developed in the *International Space Exploration Initiative* to take us from robotic exploration to human exploration to permanent human outposts.

Most of the authors of the previous edition have stayed on the team. They have worked meticulously to bring their chapters up-to-date, reflecting our current state of knowledge. A few authors have been unable to contribute in which case we have found new (co)authors or—in just a couple of cases—have reprinted the chapters after editorial updates.

It is sad to note that Conway Leovy, author of the Mars atmosphere chapter for the previous editions and Don Hunten, author of the Venus atmosphere chapter have passed away. David Catling was so kind to provide us with a newly written chapter on the Martian atmosphere while Fred Taylor updated Don's chapter on Venus.

There have been significant advances in our knowledge, many related to new missions (compare the table in the appendix). Since 2006, when most of the chapters of the previous edition went to press, six missions have been launched to the sun, among them, a Russian mission and a French microsatellite. Unfortunately, the Russian satellite failed. The French microsatellite is an example of a new tool to explore the solar system, dedicated small to very small and affordable missions. China, India, and Japan have sent their own missions to the Moon in these years in addition to three NASA missions, totaling eight new missions altogether. The Moon continues to be the reachable target of great scientific interest. But even the Apollo data and samples remain valuable as the new discovery of water in lunar rocks and the seismic confirmation of the core and the discovery of its layering show.

Among the new missions since 2006 is Messenger, a NASA orbiter mission that revolutionized our knowledge of Mercury, the innermost planet and one of the two mostly unexplored places up to then. (The other being Pluto to which the New Horizon mission is on its way.) Another new mission is Venus Express, the first European Venus orbiter. In the inner solar system where the earthlike planets are located, Mars continues to be the prime target next to the Moon with NASA continuing to launch a mission at almost every opportunity, not the least because of Mars' astrobiological potential. Here, exploration has proceeded along the classical exploration path where flybys are followed by orbiters, orbiters by landers, and landers by rovers. Robotic exploration would culminate with sample return, planned for the next decade only to be topped by human landing and exploration.

To the outer solar system, New Horizons has been launched in 2006 to visit Pluto and the Kuiper Belt for the first time after almost a decade-long journey. The mission shares this long travel time with Rosetta, the ESA mission to orbit and land in the fall of 2014 on comet nucleus Churyumov–Gerasimenko. A new mission to Jupiter, Juno, has been launched by NASA in 2011 to revisit the king of the planets. Missions to small bodies launched in recent years have been ESA's Rosetta mission, NASA's Dawn mission to explore Vesta and Ceres, and Japan's innovative Hayabusa mission that has brought back samples from Asteroid Itokawa.

Earthbound and space telescopes have pushed the frontiers of planetary sciences beyond the solar system. NASA's Kepler and the European Corot missions together with earthbound telescopes have increased the count of confirmed planets to almost 1800, about an order magnitude more than were known when the previous edition went to print. Three thousand planet candidates have been identified many of which, unfortunately, will never be confirmed by backup observation in the foreseeable future because of their large orbital distances and periods. The first nearly earth-sized planet in the habitable zone has been discovered by Kepler just recently bringing renewed interest to the science of Astrobiology. The habitable zone is defined as the range of orbital distance from the central star in which temperature on a planet in equilibrium with stellar radiation would allow liquid water to exist on the surface. We may speculate that the next edition of this book will perhaps report on the discovery of biosignatures in spectroscopic data from an extrasolar planet.

Other missions have continued their work in orbit or on the surfaces of planets. Among these are NASA's Mars Exploration Rover Spirit, ESA's Mars Express, and NASA's Cassini mission to Saturn all three of which have celebrated their 10 years anniversary at their target planet at the time of this writing.

The missions we cited above just like the missions that preceded them have been extremely helpful to "reclaim a little more land" as Thomas Henry Huxley has put it. More missions are on the horizon such as NASA's InSight mission, a geophysical station with a seismometer and a heat flow probe from Europe, ESA's JUICE mission to the Jovian moons, and ESA's PLATO and NASA's TESS exoplanet telescopes.

We, the editors are deeply thankful to our outstanding colleagues who authored the chapters in this book and to about as many friends and colleagues who gave us their time and thoroughly reviewed the chapters.

We are equally indebted to the people at Elsevier who helped the project along over the past 4 years. John Fedor has sewed the first seeds to get the new edition under way. Katy Morrisey, Jill Cetel, and Louisa Hutchins have been our Editorial Project Managers at Elsevier and Poulouse Joseph and Paul Chandramohan have succeeded each other as Project Managers at Elsevier Book Productions in Chennai, India overseeing the proof composition and corrections.

With the editors of the previous edition, we share the "hope that this Encyclopedia will help you, the reader, appreciate this ongoing process of discovery and change as much as we do."

Tilman Spohn
Torrence V. Johnson
Doris Breuer
May 13, 2014

Preface to the Second Edition

Knowledge is not static. Science is a process, not a product. Some of what is presented in this volume will inevitably be out of date by the time you read it.

From the Preface to the first edition, 1999.

Written on the eve of the new millennium, the statement above was our acknowledgment that we cannot simply 'freeze' our knowledge of the solar system we inhabit; we box it up and display it like a collection of rare butterflies in the nineteenth-century "cabinet of curiosities." Rather our goal was to provide our readers with an introduction to understanding the solar system as an interacting system, shaped by its place in the universe, its history, and the chemical and physical processes that operate from the extreme pressures and temperatures of the Sun's interior to the frigid realm of the Oort cloud. We aimed to provide a work that was useful to students, professionals, and serious amateurs at a variety of levels, containing both detailed technical material and clear expositions of general principles and findings. With the help of our extremely talented colleagues who agreed to author the chapters, we humbly believe we achieved at least some of these ambitious goals.

How to decide when to update a work whose subject matter is in a constant, exuberant state of flux? It is difficult. Waiting for our knowledge of the solar system to be "complete" was deemed impractical, since our thesis is that this will never happen. Picking an anniversary date (30 years since this, or 50 years after that) seemed arbitrary. We compromised on taking an informal inventory of major events and advances in knowledge since that last edition whenever we got together at conferences and meetings. When we realized that virtually every chapter in the first edition needed major revisions and that new chapters would be called for to properly reflect new material, we decided to undertake the task of preparing a second edition with the encouragement and help from our friends and colleagues at Academic Press.

Consider how much has happened in the relatively short time since the first edition, published in 1999. An international fleet of spacecraft is now in place around Mars and two rovers are roaming its surface, with more to follow. *Galileo* ended its mission of discovery at Jupiter with a spectacular fiery plunge into the giant planet's atmosphere. We have reached out and touched one comet with the *Deep Impact* mission and brought back precious fragments from another with *Stardust*. *Cassini* is sending back incredible data from the Saturn system and the *Huygens* probe descended to the surface of the giant, smog-shrouded moon Titan, revealing an eerily earthlike landscape carved by methane rains. *NEAR* and *Hayabusa* each orbited and then touched down on the surface of near-earth asteroids Eros and Itokawa, respectively. Scientists on the earth are continually improving the capabilities of telescopes and instruments, while laboratory studies and advances in theory improve our ability to synthesize and understand the vast amounts of new data being returned.

What you have before you is far more than a minor tweak to add a few new items to a table here or a figure there. It is a complete revamping of the Encyclopedia to reflect the solar system as we understand it today. We have attempted to capture the excitement and breadth of all this new material in the layout of the new edition. The authors of existing chapters were eager to update them to reflect our current state of knowledge, and many new authors have been added to bring fresh perspectives to the work. To all of those authors who contributed to the second edition and to the army of reviewers who carefully checked each chapter, we offer our sincere thanks and gratitude.

The organization of the chapters remains based on the logic of combining individual surveys of objects and planets, reviews of common elements and processes, and discussions of the latest techniques used to observe the solar system. Within this context you will find old acquaintances and many new friends. The sections on our own home planet have been revised and a new chapter on the Sun–Earth connection added to reflect our growing understanding of the intimate relationship between our star and conditions here on Earth. The treatment of Mars has been updated and a new chapter included incorporating the knowledge gained from the rovers *Spirit* and *Opportunity* and new orbital exploration of the red planet. Galileo's remarkable discovery of evidence for subsurface oceans on the icy Galilean satellites is treated fully in new chapters devoted to Europa and to Ganymede and Callisto. New information from the *Deep Impact* mission and the *Stardust* sample return is included as well. We continue to find out more and more about the denizens of the most distant reaches of the solar system, and have expanded the discussion of the Kuiper belt with a new chapter on physical properties. The area of observational techniques and instrumentation has been expanded to include chapters covering the X-ray portion of the spectrum, new generation telescopes, and remote chemical analysis.

Finally, nothing exemplifies the dynamic character of our knowledge than the area of extrasolar planets, which completes the volume. In the first edition the chapter on extrasolar planets contained a section entitled, "What is a Planet?" which concluded with this: "The reader is cautioned that these definitions are not uniformly accepted." The chapter included a table of 19 objects cautiously labeled "Discovered Substellar Companions." As this work goes to press, more than 200 extrasolar planets are known, many in multiplanet systems, with more being discovered everyday. And at the 2006 General Assembly of the International Astronomical Union, the question of the definition of "planet" was still being hotly debated. The current IAU definition is discussed in the introductory chapter by one of us (PRW) and other views concerning the status of Pluto may be found in the chapter on that body.

In addition to the energy and hard work of all of our authors, this edition of the Encyclopedia is greatly enhanced by the vision and talents of our friends at Academic Press. Specifically, we wish to thank Jennifer Helé, our Publishing Editor, who oversaw the project and learned the hard truth that herding scientists and herding cats are the same thing. Jennifer was the task master who made us realize that we could not just keep adding exciting new results to the volume, but one day had to stop and actually publish it. Francine Ribeau was our very able Marketing Manager and Deena Burgess, our Publishing Services Manager in the U.K., handled all of the last minute loose ends and made certain that the book was published without a hitch yet on a very tight schedule. Frank Cynar was our Publishing Editor for the first edition and for the beginning of the second, assisted by Gail Rice who was the Developmental Editor early on for the second edition. At Techbooks, Frank Scott was the Project Manager who oversaw all the final chapter and figure submissions and proof checking. Finally, also at Techbooks, was Carol Field, our Developmental Editor, simply known as Fabulous Carol, who seemed to work 30-h days for more than a year to see the volume through to fruition, while still finding time to get married in the midst of it all. This Encyclopedia would not exist without the tireless efforts of all of these extremely talented and dedicated individuals. To all of them we offer our eternal thanks.

Extensive use of color and new graphic designs have made the Encyclopedia even more beautiful and enhanced its readability while at the same time allowing the authors to display their information more effectively. The Encyclopedia before you is the result of all these efforts and we sincerely hope you will enjoy reading it as much as we enjoyed the process of compiling it.

Which brings us back to the quotation at the start of the Preface. We sincerely hope that this edition of the Encyclopedia will indeed also be out of date by the time you read it. The *New Horizons* spacecraft is on its way to the Pluto/Charon system, *MESSENGER* is on its way to Mercury, *Rosetta* is en route to a rendezvous with periodic comet Churyumov–Gerasimenko, new spacecraft are probing Venus and Mars, many nations are refocusing on exploration of the Moon, plans are being laid to study the deep interior of Jupiter and return to Europa, while the results from the Saturn system, Titan, and Enceladus have sparked a multitude of ideas for future exploration. We hope this Encyclopedia will help you, the reader, appreciate and enjoy this ongoing process of discovery and change as much as we do.

Lucy-Ann McFadden,
Paul R. Weissman,
Torrence V. Johnson
November 1, 2006

Preface to the First Edition

This is what hydrogen atoms can accomplish after four billion years of evolution.

Carl Sagan, Cosmos, 1981.

The quote above comes from the final episode of the public television series "Cosmos," which was created by Carl Sagan and several colleagues in 1981. Carl was describing the incredible accomplishments of the scientists and engineers who made the Voyager 1 and 2 missions to Jupiter and Saturn possible. But he just as easily could have been describing the chapters in this book.

This Encyclopedia is the product of the many scientists, engineers, technicians, and managers who produced the spacecraft missions which have explored our solar system over the past four decades. It is our attempt to provide to you, the reader, a comprehensive view of all we have learned in that 40 years of exploration and discovery. But we cannot take credit for this work. It is the product of the efforts of thousands of very talented and hardworking individuals in a score of countries who have contributed to that exploration. And it includes not only those involved directly in space missions, but also the many ground-based telescopic observers (both professional and amateur), laboratory scientists, theorists, and computer specialists who have contributed to creating that body of knowledge called solar system science. To all of these individuals, we say thank you.

Our goal in creating this Encyclopedia is to provide an integrated view of all we have learned about the solar system, at a level that is useful to the advanced amateur or student, to teachers, to nonsolar system astronomers, and to professionals in other scientific and technical fields. What we present here is an introduction to the many different specialties that constitute solar system science, written by the world's leading experts in each field. A reader can start at the beginning and follow the course we have laid out, or delve into the volume at almost any point and pursue his or her own personal interests. If the reader wishes to go further, the lists of recommended reading at the end of each article provide the next step in learning about any of the subjects covered.

Our approach is to have the reader understand the solar system not only as a collection of individual and distinct bodies, but also as an integrated, interacting system, shaped by its initial conditions and by a variety of physical and chemical processes. The Encyclopedia begins with an overview chapter which describes the general features of the solar system and its relationship to the Milky Way galaxy, followed by a chapter on the origin of the system. Next we proceed from the Sun outward. We present the terrestrial planets (Mercury, Venus, Earth, and Mars) individually with separate chapters on their atmospheres and satellites (where they exist). For the giant planets (Jupiter, Saturn, Uranus, and Neptune) our focus shifts to common areas of scientific knowledge: atmospheres, interiors, satellites, rings, and magnetospheres. In addition, we have singled out three amazing satellites for individual chapters: Io, Titan, and Triton. Next is a chapter on the planetary system's most distant outpost, Pluto, and its icy satellite, Charon. From there we move into discussing the small bodies of the solar system: comets, asteroids, meteorites, and dust. Having looked at the individual members of the solar system, we next describe the different view of those members at a variety of wavelengths outside the normal visual region. From there we consider the important processes that have played such an important role in the formation and evolution of the system: celestial dynamics, chaos, impacts, and volcanism. Last, we look at three topics which are as much in our future as in our past: life on other planets, space exploration missions, and the search for planets around other stars.

A volume like this one does not come into being without the efforts of a great number of very dedicated people. We express our appreciation to the more than 50 colleagues who wrote chapters, sharing their expertise with you, the reader. In addition to providing chapters that captured the excitement of their individual fields, the authors have endured revisions, rewrites, endless questions, and unforeseen delays. For all of these we offer our humble apologies. To ensure the quality and accuracy of each contribution, at least two independent reviewers critiqued each chapter. The peer review process maintains its integrity through the anonymity of the reviewers. Although we cannot acknowledge them by name, we thank all the reviewers for their time and their conscientious efforts.

We are also deeply indebted to the team at Academic Press. Our Executive Editor, Frank Cynar, worked tirelessly with us to conceptualize and execute the encyclopedia, while allowing us to maintain the highest intellectual and scientific standards. We thank him for his patience and perseverance in seeing this volume through to completion. Frank's assistants, Daniela Dell'Orco, Della Grayson, Linda McAleer, Cathleen Ryan, and Suzanne Walters, kept the entire process moving and attended to the myriad of details and questions that arise with such a large and complex volume. Advice and valuable guidance came from Academic Press' director of major reference works, Chris Morris. Lori Asbury masterfully oversaw the production and copyediting. To all of the people at Academic Press, we give our sincere thanks.

Knowledge is not static. Science is a process, not a product. Some of what is presented in this volume will inevitably be out of date by the time you read it. New discoveries seem to come everyday from our colleagues using earth-based and orbiting telescopes, and from the flotilla of new small spacecraft that are out there adding to our store of knowledge about the solar system. In this spirit we hope that you, the reader, will benefit from the knowledge and understanding compiled in the following pages. The new millennium will surely add to the legacy presented herein, and we will all be the better for it. Enjoy, wonder, and keep watching the sky.

**Paul R. Weissman,
Lucy-Ann McFadden,
Torrence V. Johnson**

About the Editors

Tilman Spohn is director of the Institute of Planetary Research of the German Aerospace Center (DLR) in Berlin, Germany and Professor of Planetology at the Westfälische Wilhelms-University Münster, Germany. He specializes in the Thermodynamics of Planetary Interiors and in physical problems of Astrobiology and has written 140 papers for scientific journals and books. He is a Principal Investigator (PI) for MUPUS on the Rosetta Lander Philae, for BELA on BepiColombo, and the instrument PI for HP3 on the InSight mission. Spohn has served as member and chairperson of ESA scientific working and advisory groups and is presently a member of ESA's HISPAC. He is the chairman of the science committee of the international Space Science Institute and has served as editor for Earth and Planetary Science Letters, Reviews of Geophysics, and the Treatise on Geophysics. He is the recipient of the 2013 EGU Runcorn-Florensky Medal and a fellow of the American Geophysical Union.

Doris Breuer is Head of the Department of Planetary Physics at the Institute of Planetary Research of the German Aerospace Center (DLR) in Berlin, Germany and an Associate Professor at the Institute de Physique du Globe de Paris, France. She specializes in the interior dynamics, thermo-chemical evolution and interior structure of terrestrial bodies. She has published over 60 articles in refereed journals, has been coinvestigator on ESA's BepiColombo mission and has served on several ESA science definition teams and on the ESA Solar System Working Group.

Torrence V. Johnson is a specialist on icy satellites in the solar system. He has written over 130 papers for scientific journals. He received a PhD in planetary science from the California Institute of Technology and is currently a Senior Research Scientist at the Jet Propulsion Laboratory. Johnson was on the *Voyager* camera team during its exploration of the outer solar system and was the Project Scientist for the *Galileo* mission. He is currently an active investigator on the *Cassini* mission exploring the Saturn system. He is the recipient of two NASA Exceptional Scientific Achievement Medals and the NASA Outstanding Leadership Medal and has an honorary doctorate from the University of Padua, where Galileo made his first observations of the solar system.

Contributors

Mahesh Anand Department of Physical Sciences, The Open University, UK

Markus J. Aschwanden Lockheed Martin ATC Solar and Astrophysics Laboratory, Palo Alto, CA, USA

Fran Bagenal Department of Astrophysical & Planetary Sciences, Laboratory for Atmospheric & Space Physics, University of Colorado, Boulder, Boulder, CO, USA

W. Bruce Banerdt Jet Propulsion Laboratory, California Institute of Technology, Pasadena, CA, USA

James F. Bell, III School of Earth and Space Exploration, Arizona State University, Tempe, AZ, USA

Anil Bhardwaj Space Physics Laboratory, Vikram Sarabhai Space Centre, Trivandrum, Kerala, India

Richard P. Binzel Department of Earth, Atmospheric and Planetary Sciences, Massachusetts Institute of Technology, Cambridge, MA 02139, USA

John C. Brandt Department of Astronomy, University of Washington, Seattle, Washington, USA

Doris Breuer Institute for Planetary Research, German Aerospace Center (DLR), Berlin, Germany

Daniel T. Britt University of Central Florida, Orlando, FL, USA

Bonnie J. Buratti Jet Propulsion Laboratory, California Institute of Technology, Pasadena, CA, USA

James D. Burke The Planetary Society, Pasadena, CA, USA

Michael H. Carr U. S. Geological Survey, Menlo Park, CA, USA

Lynn M. Carter Planetary Geodynamics Laboratory; NASA Goddard Space Flight Center, Greenbelt, MD, USA

David C. Catling University of Washington, Dept. of Earth and Space Sciences/Astrobiology Program, Seattle, WA, USA

John E. Chambers Department of Terrestrial Magnetism, Carnegie Institution of Washington, Washington, DC, USA

Geoffrey Collins Physics and Astronomy Dept., Wheaton College, Norton, Massachuse, USA

Athena Coustenis LESIA - Observatoire de Paris, CNRS, UPMC Univ. Paris 06, Univ. Paris-Diderot — Meudon, France

Ian A. Crawford Department of Earth and Planetary Sciences, Birkbeck College, University of London, London, UK

Wanda L. Davis Space Science Division, NASA Ames Research Center, Moffett Field, CA, USA

Véronique Dehant Royal Observatory of Belgium, Brussels, Belgium

Konrad Dennerl Max-Planck-Institut für extraterrestrische Physik, Garching, Germany

Imke de Pater Astronomy Department, University of California, Berkeley, CA, USA; Faculty of Aerospace Engineering, Delft University of Technology, Delft, NL; SRON Netherlands Institute for Space Research, Utrecht, The Netherlands

Deborah L. Domingue Planetary Science Institute, Tucson, AZ, USA

Luke Dones Southwest Research Institute, Boulder, CO, USA

Timothy E. Dowling Department of Physics and Astronomy, University of Louisville, Louisville, KY, USA

Line Drube German Aerospace Center (DLR), Institute of Planetary Research, 12489 Berlin, Germany

Adam M. Dziewonski Department of Earth and Planetary Sciences, Harvard University, Cambridge, MA, USA

Michael Endl McDonald Observatory, University of Texas at Austin, Austin, TX, USA

Carolyn M. Ernst Space Department, Johns Hopkins University Applied Physics Laboratory, Laurel, MD, USA

Berndt Feuerbacher German Aerospace Center, Cologne, Germany (ret.)

Jonathan J. Fortney Department of Astronomy and Astrophysics, University of California, Santa Cruz, CA, USA

Matthew P. Golombek Jet Propulsion Laboratory, California Institute of Technology, Pasadena, CA, USA

J.T. Gosling Laboratory for Atmospheric and Space Physics, University of Colorado, Boulder, CO, USA

Richard A.F. Grieve Centre for Planetary Science and Exploration, University of Western Ontario, London, ON, Canada

Robert Grimm Southwest Research Institute, Boulder, CO, USA

Matthias Grott German Aerospace Center (DLR), Institute of Planetary Research, Berlin, Germany

Eberhard Grün Max-Planck-Institut für Kernphysik, Heidelberg, Germany and LASP, University of Colorado, Boulder, CO, USA

Guy Consolmagno, S. J. Specola Vaticana, Vatican City State

Klaus Gwinner Planetary Geodesy Department, German Aerospace Center, Institute of Planetary Research, Berlin, Germany

Alex N. Halliday Department of Earth Sciences, University of Oxford, Oxford, UK

Alan W. Harris German Aerospace Center (DLR), Institute of Planetary Research, 12489 Berlin, Germany

Matthew M. Hedman Center for Radiophysics and Space Research, Cornell University, Ithaca, New York, USA

Amanda R. Hendrix Planetary Science Institute, Tucson, AZ, USA

Harald Hiesinger Institut für Planetologie, Westfälische Wilhelms-Universität, Münster, Germany

Bernhard Hufenbach European Space Agency, Noordwijk, The Netherlands

Donald M. Hunten University of Arizona, AZ, USA

Ralf Jaumann Deutsches Zentrum für Luft- und Raumfahrt (DLR), Berlin, Germany; Freie Universität Berlin, Institut für Geologische Wissenschaften, Berlin, Germany

Torrence V. Johnson Jet Propulsion Laboratory, California Institute of Technology, Pasadena, California, USA

Katherine H. Joy School of Earth, Atmospheric and Environmental Sciences, University of Manchester, UK

Randolph L. Kirk U.S. Geological Survey, Flagstaff, AZ, USA

Margaret Galland Kivelson Department of Earth & Space Sciences, University of California, Los Angeles, CA, USA and Department of Atmospheric, Oceanic, and Space Sciences, University of Michigan, Ann Arbor, MI, USA

Harald Krüger Max-Planck-Institut für Sonnensystemforschung, Göttingen, Germany

William S. Kurth Department of Physics and Astronomy, University of Iowa, Iowa City, IA, USA

Larry Lebofsky Planetary Science Institute, Tucson, AZ, USA

David Leverington Stoke Lacy, Herefordshire, United Kingdom

Harold F. Levison Southwest Research Institute, Boulder, CO, USA

Michael E. Lipschutz Purdue University, West Lafayette, IN, USA (Professor Emeritus)

Jack J. Lissauer Space Science & Astrobiology Division, NASA Ames Research Center Moffett Field, CA, USA

Carey M. Lisse Applied Physics Laboratory, Johns Hopkins University, Laurel, Maryland

Philippe Lognonné Institut de Physique du Globe de Paris, Paris, France

Rosaly M.C. Lopes Jet Propulsion Laboratory, California Institute of Technology, Pasadena, CA, USA

J.G. Luhmann Space Sciences Laboratory, University of California, Berkeley, CA, USA

Mark S. Marley Space Science Division, NASA Ames Research Center, Moffett Field, California, USA

Lucy A. McFadden Planetary Systems Laboratory, NASA Goddard Space Flight Center, Greenbelt, MD 20771 USA

Christopher P. McKay Space Science Division, NASA Ames Research Center, Moffett Field, CA, USA

William B. McKinnon Department of Earth and Planetary Sciences and McDonnell Center for the Space Sciences, Washington University, Saint Louis, MO, USA

Harry Y. McSween Department of Earth & Planetary Sciences, University of Tennessee, Knoxville, TN, USA

Alessandro Morbidelli Observatoire de la Côte d'Azur, Nice, France

Nils Mueller German Aerospace Center (DLR), Institute of Planetary Research, Berlin, Germany

Scott L. Murchie Space Department, Johns Hopkins University Applied Physics Laboratory, Laurel, MD, USA

Carl D. Murray Astronomy Unit, Queen Mary University of London, London, England, UK

Contributors

Catherine D. Neish Department of Physics and Space Sciences; Florida Institute of Technology, Melbourne, FL, USA

Robert M. Nelson Planetary Science Institute, Tucson, AZ, USA

Francis Nimmo Dept. Earth and Planetary Sciences, University of California Santa Cruz, CA, USA

Jürgen Oberst Planetary Geodesy Department, German Aerospace Center, Institute of Planetary Research, Berlin, Germany

Gordon R. Osinski Centre for Planetary Science and Exploration, University of Western Ontario, London, ON, Canada

Robert T. Pappalardo Jet Propulsion Laboratory, California Institute of Technology, Pasadena, CA, USA

David C. Pieri Jet Propulsion Laboratory, California Institute of Technology, Pasadena, CA, USA

Carolyn Porco CICLOPS, Space Science Institute, Boulder, CO, USA

Thomas H. Prettyman Planetary Science Institute, Tucson, AZ, USA

Frank Preusker Planetary Geodesy Department, German Aerospace Center, Institute of Planetary Research, Berlin, Germany

Louise M. Prockter Johns Hopkins University Applied Physics Laboratory, Laurel, MD, USA

Attilio Rivoldini Royal Observatory of Belgium, Brussels, Belgium

Ludolf Schultz Max-Planck-Institut für Chemie, Mainz, Germany (retired)

Adam P. Showman Department of Planetary Sciences, Lunar and Planetary Laboratory, University of Arizona, Tucson, AZ, USA

Suzanne E. Smrekar Earth and Space Sciences, Jet Propulsion Laboratory/Caltech, Pasadena, CA, USA

Sue Smrekar Jet Propulsion Laboratory, California Institute of Technology, Pasadena, CA

S.C. Solomon High Altitude Observatory, National Center for Atmospheric Research, Boulder, CO, USA

Tilman Spohn Institute for Planetary Research, German Aerospace Center (DLR), Berlin, Germany

Sabine Stanley Department of Physics, University of Toronto, Toronto, ON, Canada

S. Alan Stern Space Science and Engineering Division, Southwest Research Institute, Boulder, CO, USA

Ellen R. Stofan National Aeronautics and Space Administration, Headquarters, Washington, DC, USA

Robert G. Strom Department of Planetary Sciences, University of Arizona, Tucson, AZ, USA

Mark V. Sykes Planetary Science Institute, Tucson, AZ, USA

Fredric W. Taylor University of Oxford, Oxford, UK

Stephen C. Tegler Northern Arizona University, Flagstaff, Arizona, USA

Peter C. Thomas Department of Astronomy, Center for Radiophysics & Space Research, Cornell University, Ithaca, NY, USA

Matthew S. Tiscareno Center for Radiophysics and Space Research, Cornell University, Ithaca, NY, USA

Alan T. Tokunaga Institute for Astronomy, University of Hawaii, Honolulu, HI, USA

Livio L. Tornabene Centre for Planetary Science and Exploration, University of Western Ontario, London, ON, Canada

Nicola Tosi Institute for Planetary Research, German Aerospace Center (DLR), Berlin, Germany

Tim Van Hoolst Instituut voor Sterrenkunde, KU Leuven, Celestijnenlaan 200D, B-3001 Leuven, Belgium; Royal Observatory of Belgium, Brussels, Belgium

Ronald J. Vervack, Jr. Space Department, Johns Hopkins University Applied Physics Laboratory, Laurel, MD, USA

Renee C. Weber NASA Marshall Space Flight Center, Huntsville, AL, USA

Paul R. Weissman Jet Propulsion Laboratory, California Institute of Technology, Pasadena, CA, USA

Robert A. West Jet Propulsion Laboratory, California Institute of Technology, Pasadena, California, USA

Lionel Wilson Lancaster Environment Centre, Lancaster University, Lancaster, UK

Part I

The Solar System

Chapter 1

The Solar System and Its Place in the Galaxy

Paul R. Weissman
Jet Propulsion Laboratory, California Institute of Technology, Pasadena, CA, USA

Chapter Outline

1. Introduction 3
2. The Definition of a Planet 4
3. The Architecture of the Solar System 5
   3.1. Dynamics 5
   3.2. Nature and Composition 9
   3.3. Satellites, Rings, and Things 15
   3.4. The Solar Wind and the Heliosphere 20
4. The Origin of the Solar System 22
5. The Solar System's Place in the Galaxy 24
6. The Fate of the Solar System 27
7. Concluding Remarks 28
Bibliography 28

1. INTRODUCTION

The origins of modern astronomy lie with the study of our solar system. When ancient humans first gazed at the skies, they recognized the same patterns of fixed stars rotating over their heads each night. They identified these fixed patterns, now called constellations, with familiar objects or animals, or with stories from their mythologies and their culture. But along with the fixed stars, there were a few bright points of light that moved each night, slowly following similar paths through a belt of constellations around the sky. The Sun and Moon also appeared to move through the same belt of constellations. These wandering objects were the **planets** of our solar system. Indeed, the name "planet" derives from the Latin *planeta*, meaning wanderer.

The ancients recognized five planets that they could see with their naked eyes. We now know that the solar system consists of eight planets, at least five **dwarf planets**, plus a myriad of smaller objects: satellites, asteroids, comets, rings, and dust. Discoveries of new objects and new classes of objects are continuing even today. Thus, our view of the solar system is constantly changing and evolving as new data and new theories to explain (or anticipate) the data become available.

The solar system we see today is the result of the complex interaction of physical, chemical, and dynamical processes that have shaped the planets and other bodies. By studying each of the planets and other bodies individually as well as collectively, we seek to gain an understanding of those processes and the steps that led to the current solar system. Many of those processes operated most intensely early in the solar system's history, as the Sun and planets formed from an interstellar cloud of dust and gas, 4.567 billion years ago. The first billion years of the solar system's history was a violent period as the planets cleared their orbital zones of much of the leftover debris from the process of planet formation, flinging small bodies into planet-crossing, and often planet-impacting, **orbits** or out to interstellar space. In comparison, the present-day solar system is a much quieter place, although many of these processes continue today on a lesser scale.

Our knowledge of the solar system has exploded in the past five decades as interplanetary exploration spacecraft have provided close-up views of all the planets, as well as of a diverse collection of satellites, rings, asteroids, and comets. Earth-orbiting telescopes have provided an unprecedented view of the solar system, often at wavelengths not accessible from the Earth's surface. Ground-based observations have also continued to produce exciting new discoveries through the application of a variety of new technologies such as charge-coupled device cameras, infrared detector arrays, adaptive optics, and powerful planetary radars. Theoretical studies have also contributed significantly to our understanding of the solar system, largely through the use of advanced computer codes and high-speed, dedicated computers. Serendipity has also played an important role in many new discoveries.

Along with this increased knowledge have come numerous additional questions as we attempt to explain the complexity and diversity that we observe on each newly encountered world. The increased spatial and spectral resolution of the observations, along with in situ measurements of atmospheres, surface materials, and **magnetospheres**, have revealed that each body is unique, the result of a different combination of the physical, chemical, and dynamical processes that formed and shaped it. Also, each body's formation zone (i.e. distance from the Sun) and the different initial solar nebula composition at that distance play an important role. Yet, at the same time, there are broad systematic trends and similarities that are clues to the collective history of the solar system.

We have now begun an exciting new age of discovery with the detection of numerous planet-sized bodies around nearby stars. Although the properties and placement of many of these extrasolar planets appear to be very different from those in our solar system, they are likely the prelude to the discovery of planetary systems that may more closely resemble our own.

We may also be on the brink of discovering evidence for life on other planets, in particular, Mars. There is an ongoing debate as to whether biogenic materials have been discovered in meteorites that were blasted off the surface of Mars and have found their way to the Earth. Although still very controversial, this finding, if confirmed, would have profound implications for the existence of life elsewhere in the solar system and the galaxy.

The goal of this chapter is to provide the reader with an introduction to the solar system. It seeks to provide a broad overview of the solar system and its constituent parts, to note the location of the solar system in the galaxy and to describe the local galactic environment. Detailed discussions of each of the bodies that make up the solar system, as well as the processes that have shaped those bodies and the techniques for observing the planetary system, are provided in the following chapters of this book. The reader is referred to those chapters for more detailed discussions of each of the topics introduced herein.

Some brief notes about planetary nomenclature will be useful. The names of the planets are all taken from Greek and Roman mythology (with the exception of the Earth, which is named for a goddess from Norse mythology), as are the names of their satellites, with the exception of the Moon and the Uranian satellites, the latter being named after Shakespearean characters. The Earth is occasionally referred to as Terra and the Moon as Luna, each the Latin version of their names. The naming system for planetary rings is different at each planet and includes descriptive names of the structures (at Jupiter), letters of the Roman alphabet (at Saturn), Greek letters and Arabic numerals (at Uranus), and the names of scientists associated with the discovery of Neptune (at Neptune).

Asteroids were initially named after women in Greek and Roman mythology. As their numbers have increased, asteroids have been named after the family members of the discoverers, after observatories, universities, cities, provinces, historical figures, scientists, writers, artists, literary figures, and, in at least one case, the astronomer's cat. Initial discoveries of asteroids are designated by the year of their discovery and a letter/number code. Once the orbits of the asteroids are firmly established, they are given official numbers in the asteroid catalog: about 632,000 asteroids have been discovered and 385,000 asteroids have been numbered (as of January 2014). The discoverer(s) of an asteroid are given the privilege of suggesting its name, if done so within 10 years from when it was officially numbered.

Comets are generally named for their discoverers, although in a few well-known cases such as comets Halley and Encke, they are named for the individuals who first computed their orbits and linked several apparitions. Because some astronomers have discovered more than one short-period comet, a number is added at the end of the name in order to differentiate them, although this system is not applied to long-period comets. Comets are also designated by the year of their discovery and a letter code (a recently abandoned system used lowercase Roman letters and Roman numerals in place of the letter codes). The naming of newly discovered comets, asteroids, and satellites, as well as surface features on solar system bodies, is overseen by several working groups of the International Astronomical Union (IAU).

2. THE DEFINITION OF A PLANET

No formal definition of a planet existed until very recently. Originally, the ancients recognized five planets that could be seen with the naked eye: Mercury, Venus, Mars, Jupiter and Saturn, plus the Earth. Two more **giant planets**, Uranus and Neptune, were discovered telescopically in 1781 and 1846, respectively.

The largest asteroid, Ceres, was discovered in 1801 in an orbit between Mars and Jupiter and was hailed as a new planet because it fit into Bode's law (see discussion later in this chapter). However, it was soon recognized that Ceres was much smaller than any of the known planets. As more and more asteroids were discovered in similar orbits between Mars and Jupiter, it became evident that Ceres was simply the largest body of a huge swarm of bodies between Mars and Jupiter that we now call the Asteroid Belt. A new term was coined, "**minor planet**", to describe these bodies.

Searches for planets beyond Neptune continued and culminated in the discovery of Pluto in 1930. As with Ceres, it was soon suspected that Pluto was much smaller than any of the neighboring giant planets. Later, measurements of Pluto's diameter by stellar occultations showed that it was also smaller than any of the **terrestrial planets**, in fact, even smaller than the Earth's Moon. As a result, Pluto's status as a planet was called into question.

In the 1980s, dynamical calculations suggested the existence of a belt of many small objects in orbits beyond Neptune, left over from the formation of the solar system. In the early 1990s the first of these objects, 1992 QB_1, was discovered at a distance of 40.9 **astronomical units** (AU). More discoveries followed and over 1500 bodies have now been found in the trans-Neptunian region (as of September 2013). They are collectively known as the **Kuiper belt**.

The existence of the Kuiper belt suggested that Pluto, like Ceres in the asteroid belt, was simply the largest body among a huge swarm of bodies beyond Neptune, again calling Pluto's status into question. Then came the discovery of 136199 Eris (2003 UB_{313}), a Kuiper belt object (KBO) in a distant orbit, which turned out to be comparable in size to Pluto and somewhat more massive.

In response, the IAU, the governing body for astronomers worldwide, formed a committee to create a formal definition of a planet. The definition was presented at the IAU's triennial gathering in Prague in 2006, where it was revised several times by the astronomers at the meeting. Eventually the IAU voted and passed a resolution that defined a planet.

That resolution states that a planet must have three qualities: (1) it must be round, indicating its interior is in hydrostatic equilibrium; (2) it must orbit the Sun; and (3) it must have gravitationally cleared its zone of other debris. The last requirement means that a planet must be massive enough to be gravitationally dominant in its zone in the solar system. Any round body orbiting the Sun that fails condition (3) is labeled a "dwarf planet" by the IAU.

This outcome left the solar system with the eight major planets discovered through 1846, and reclassified Ceres, Pluto, and Eris as dwarf planets. Two other KBOs, 136108 Haumea and 136472 Makemake, have also been added to that list. Other large objects in the asteroid and Kuiper belts may be added to the list of dwarf planets if observations show that they too are large and round.

There are weaknesses in the definition, particularly in condition (3), which may be modified by an IAU committee tasked with improving the definition. However, the likelihood of the definition being changed sufficiently to again classify Pluto as a planet is small.

The IAU has a somewhat different definition for planets discovered around other stars, known as "extrasolar" planets. At some point the two definitions need to be reconciled. See the chapter on Extra-Solar Planets for more discussion of this matter.

3. THE ARCHITECTURE OF THE SOLAR SYSTEM

The solar system consists of the Sun at its center, eight planets, five dwarf planets, 173 known natural **satellites** (or moons) of planets (as of September 2013), four ring systems, approximately 1 million asteroids (greater than 1 km in diameter), perhaps a trillion comets (greater than 1 km in diameter), the **solar wind**, and a large cloud of interplanetary dust. The arrangement and nature of all these bodies are the result of physical and dynamical processes during their origin and subsequent evolution, and their complex interactions with one another.

At the center of the solar system is the Sun, a rather ordinary, **main sequence** star. The Sun is classified spectrally as a G2V dwarf, which means that it emits the bulk of its radiation in the visible region of the spectrum, peaking at yellow-green wavelengths. The Sun contains 99.86% of the mass in the solar system, but only about 0.5% of the angular momentum. The low angular momentum of the Sun results from the transfer of momentum to the accretion disk surrounding the Sun during the formation of the planetary system, and to a slow spin down due to angular momentum being carried away by the solar wind.

The Sun is composed of hydrogen (70% by mass), helium (28%), and heavier elements (2%). The Sun produces energy through nuclear fusion at its center, hydrogen atoms combining to form helium and releasing energy that eventually makes its way to the Sun's surface as visible sunlight. The central temperature of the Sun where fusion takes place is 15.7 million K, while the temperature at the visible surface, the photosphere, is ~5800 K. The Sun has an outer atmosphere called the corona, which is only visible during solar eclipses, or through the use of specially designed telescopes called coronagraphs.

A star like the Sun is believed to have a typical lifetime of 9–10 billion years on the main sequence. The present age of the Sun (and the entire solar system) is estimated to be 4.567 billion years, so it is about halfway through its nominal lifetime. The age estimate comes from radioisotope dating of meteorites, as well as from theories of stellar evolution.

3.1. Dynamics

The planets all orbit the Sun in roughly the same plane, known as the **ecliptic** (the plane of the Earth's orbit), and in the same direction, counterclockwise as viewed from the north ecliptic pole. Because of gravitational torques from the other planets, the ecliptic is not inertially fixed in space, and so dynamicists often use the invariable plane, which is

TABLE 1.1 Orbits of the Planets[1] and Dwarf Planets

Name	Semimajor Axis (AU)	Eccentricity	Inclination (°)	Period (years)
Mercury	0.38710	0.205631	7.0049	0.2408
Venus	0.72333	0.006773	3.3947	0.6152
Earth	1.00000	0.016710	0.0000	1.0000
Mars	1.52366	0.093412	1.8506	1.8808
Ceres[2]	2.7665	0.078375	10.5834	4.601
Jupiter	5.20336	0.048393	1.3053	11.862
Saturn	9.53707	0.054151	2.4845	29.457
Uranus	19.1913	0.047168	0.7699	84.018
Neptune	30.0690	0.008586	1.7692	164.78
Pluto[2]	39.4817	0.248808	17.1417	248.4
Haumea[2]	43.0127	0.196577	28.1976	282.1
Makemake[2]	45.4904	0.161420	29.0130	306.82
Eris[2]	68.1461	0.432439	43.7408	562.55

[1] Mean ecliptic and equinox J2000, Epoch: January 1, 2000.
[2] Dwarf planet.

the plane defined by the summed angular momentum vectors of all the planets.

To first order, the motion of any body about the Sun is governed by Kepler's Laws of Planetary Motion. These laws state that (1) each planet moves about the Sun in an orbit that is an ellipse, with the Sun at one focus of the ellipse; (2) the straight line joining a planet and the Sun sweeps out equal areas in space in equal intervals of time; and (3) the squares of the sidereal periods of the planets are in direct proportion to the cubes of the semimajor axes of their orbits. The laws of planetary motion, first set down by J. Kepler in 1609 and 1619, are easily shown to be the result of the inverse square law of gravity with the Sun as the central body, and the conservation of angular momentum and energy. Parameters for the orbits of the eight planets and five dwarf planets are listed in Table 1.1.

Because the planets themselves have finite masses, they exert small gravitational tugs on one another, which cause their orbits to depart from perfect ellipses. The major effects of these long-term or "secular" perturbations are to cause the **perihelion** direction of each orbit to precess (rotate counterclockwise) in space, and the line of nodes (the intersection between the planet's orbital plane and the ecliptic plane) of each orbit to regress (rotate clockwise). Additional effects include slow oscillations in the **eccentricity** and **inclination** of each orbit, and the inclination of the planet's rotation pole to the planet's orbit plane (called the obliquity). For the Earth, these orbital oscillations have periods of 19,000–100,000 years. They have been identified with long-term variations in the Earth's climate, known as Milankovitch cycles, although the linking physical mechanism is not well understood.

Relativistic effects also play a small but detectable role. They are most evident in the precession of the perihelion of the orbit of Mercury, the planet deepest in the Sun's gravitational potential well. General relativity adds 43 arcsec/century to the precession rate of Mercury's orbit, which is 574 arcsec/century. Prior to Einstein's theory of general relativity in 1916, it was thought that the excess in the precession rate of Mercury was due to a planet orbiting interior to it. This hypothetical planet was given the name Vulcan, and extensive searches were conducted for it, primarily during solar eclipses. No planet was detected.

A more successful search for a new planet occurred in 1846. Two celestial mechanicians, U. J. J. Leverrier and J. G. Adams, independently used the observed deviations of Uranus from its predicted orbit to successfully predict the existence and position of Neptune. Neptune was found by J. G. Galle on September 23, 1846, using Leverrier's prediction.

More complex dynamical interactions are also possible, in particular when the orbital period of one body is a small integer ratio of another's orbital period. This is known as a "mean-motion resonance" and can have dramatic effects. For example, Pluto is locked in a 2:3 mean-motion resonance with Neptune, and although the orbits of the two bodies cross in space, the resonance prevents them from ever coming within 14 AU of each other. Also, when two bodies have identical perihelion precession rates or nodal regression rates, they are said to be in a "secular resonance", and similarly interesting dynamical effects can result. In many cases, mean-motion and secular resonances can lead to chaotic motion, driving a body onto a planet-crossing orbit, which will then lead to it being dynamically scattered among the planets and eventually either ejected from the solar system, or impacted on the Sun or a planet. In other cases, such as Pluto and some asteroids, the mean-motion resonance is actually a stabilizing factor for the orbit.

Chaos has become a very exciting topic in solar system dynamics in the past 25 years and has been able to explain many features of the planetary system that were not previously understood. It should be noted that the dynamical definition of chaos is not always the same as the general dictionary definition. In celestial mechanics, the term "chaos" is applied to describe systems that are not perfectly predictable over time. That is, small variations in the initial conditions, or the inability to specify the initial conditions precisely, will lead to a growing error in predictions of the long-term behavior of the system. If the error grows exponentially, then the system is said to be chaotic. However, the chaotic zone, the allowed area in phase space over which an orbit may vary, may still be quite constrained. Thus, although studies have found that the orbits of the planets are chaotic, this does not mean that Jupiter may one day become Earth-crossing, or vice versa. It means that the precise position of the Earth or Jupiter in their orbits is not predictable over very long periods. Because this happens for all the planets, the long-term **secular perturbations** of the planets on one another are also not perfectly predictable and can vary.

On the other hand, chaos can result in some extreme changes in orbits, with sudden increases in eccentricity that can throw small bodies onto planet-crossing orbits. One well-recognized case occurs near mean-motion resonances in the asteroid belt, which causes small asteroids to be thrown onto Earth-crossing orbits, allowing for the delivery of **meteoroids** to the Earth.

The natural satellites of the planets and their ring systems (where they exist) are governed by the same dynamical laws of motion. Most major satellites and all ring systems are deep within their planets' gravitational potential wells and so they move, to first order, on Keplerian ellipses. The Sun, planets, and other satellites all act as perturbers on the satellite and ring particle orbits. Additionally, the equatorial bulges of the planets, caused by the planets' rotation, act as perturbers on the orbits. Finally, the satellites raise tides on the planets (and vice versa), and these result in yet another dynamical effect, causing the planets to transfer rotational angular momentum to the satellite orbits in the case of direct or prograde orbits (satellites in retrograde orbits lose angular momentum). As a result, satellites may slowly move away from their planets into larger orbits (or into smaller orbits in the case of retrograde satellites).

The mutual gravitational interactions can be quite complex, particularly in multisatellite systems. For example, the three innermost Galilean satellites of Jupiter (so named because they were discovered by Galileo in 1610)—Io, Europa, and Ganymede—are locked in a 4:2:1 mean-motion resonance with one another. In other words, Ganymede's orbital period is twice that of Europa and four times that of Io. At the same time, the other Jovian satellites (primarily Callisto), the Sun, and Jupiter's oblateness perturb the orbits, forcing them to be slightly eccentric and inclined to one another, while the tidal interaction with Jupiter forces the orbits to evolve outward. These competing dynamical processes result in considerable energy deposition in the satellites, which manifests itself as volcanic activity on Io, as a possible subsurface ocean on Europa, and as past tectonic activity on Ganymede.

This illustrates an important point in understanding the solar system. The bodies in the solar system do not exist as independent, isolated entities, with no physical interactions between them. Even these "action-at-a-distance" gravitational interactions can lead to profound physical and chemical changes in the bodies involved. To understand the solar system as a whole, one must recognize and understand the processes that were involved in its formation and its subsequent evolution, and that continue to act today.

An interesting feature of the planetary orbits is their regular spacing. This is described by Bode's law, first discovered by J. B. Titius in 1766 and brought to prominence by J. E. Bode in 1772. The law states that the semimajor axes of the planets in **astronomical units** can be roughly approximated by taking the sequence 0, 3, 6, 12, 24,..., adding 4, and dividing by 10. The values for Bode's law and the actual semimajor axes of the planets and two dwarf planets are listed in Table 1.2. It can be seen that the law works very well for the planets as far as Uranus, but it then breaks down. It also predicts a planet between Mars and Jupiter, the current location of the asteroid belt. Yet Bode's law predates the discovery of the first asteroid by 35 years, as well as the discovery of Uranus by 15 years.

The reason why Bode's law works so well is not understood. H. Levison has recently suggested that, at least for the giant planets, it is a result of their spacing themselves at distances where they are equally likely to scatter a smaller body inward or outward to the next planet in either direction.

However, it has also been argued that Bode's law may just be a case of numerology and not reflect any real physical principle at all. Since Bode's law was formulated after the semimajor axes of the first six planets were known, Titius and Bode were free to fit the form of the equation to the known data.

Computer-based dynamical simulations have shown that the spacing of the planets is such that a body placed in a

TABLE 1.2 Bode's Law: $a_1 = 0.4$, $a_n = 0.3 \times 2^{n-2} + 0.4$

Planet	Semimajor Axis (AU)	n	Bode's Law
Mercury	0.387	1	0.4
Venus	0.723	2	0.7
Earth	1.000	3	1.0
Mars	1.524	4	1.6
Ceres[1]	2.767	5	2.8
Jupiter	5.203	6	5.2
Saturn	9.537	7	10.0
Uranus	19.19	8	19.6
Neptune	30.07	9	38.8
Pluto[1]	39.48	10	77.2

[1]Dwarf planet.

circular orbit between any pair of neighboring planets will likely be dynamically unstable. It will not survive over the history of the solar system unless protected by some dynamical mechanism such as a mean-motion resonance with one of the planets. Over the history of the solar system, the planets have generally cleared their zones of smaller bodies through gravitational scattering. The larger planets, in particular Jupiter and Saturn, are capable of throwing small bodies onto hyperbolic orbits, allowing the objects to escape to interstellar space. In the course of doing this, the planets themselves "migrate" moving either closer or farther from the Sun as a result of the angular momentum exchange with many smaller bodies.

Thus, the comets and asteroids we now see in planet-crossing orbits must have been introduced into the planetary system relatively recently from storage locations either outside the planetary system, or in protected, dynamically stable reservoirs. Because of its position at one of the Bode's law locations, the asteroid belt is a relatively stable reservoir. However, the asteroid belt's proximity to Jupiter's substantial gravitational influence results in some highly complex dynamics. Mean-motion and secular resonances, as well as mutual collisions, act to remove objects from the asteroid belt and throw them into planet-crossing orbits. The failure of a major planet to grow in the asteroid belt is generally attributed to the gravitational effects of Jupiter disrupting the slow growth by accretion of a planet-sized body in the neighboring asteroid belt region. As Jupiter gravitationally perturbed the orbits in the asteroid belt, collision velocities increased and the collisions changed from accretionary to disruptive.

It is generally believed that comets originated as icy **planetesimals** in the outer regions of the **solar nebula**, at the orbit of Jupiter and beyond. Those protocomets with orbits between the giant planets were gravitationally ejected, mostly to interstellar space. However, a fraction of the protocomets, about 4%, were flung into distant but still bound orbits; the Sun's gravitational sphere of influence extends $\sim 2 \times 10^5$ AU, or about 1 **parsec (pc = 206,264.8 AU)**. These orbits were sufficiently distant from the Sun that they were perturbed by random passing stars and by the tidal perturbation from the galactic disk. The stellar and galactic perturbations raised the perihelia of the comet orbits out of the planetary region. Additionally, the stellar perturbations randomized the inclinations of the comet orbits, forming a spherical cloud of comets around the planetary system and extending halfway to the nearest stars. This region is now called the **Oort cloud**, after J. H. Oort who first suggested its existence in 1950.

The current population of the Oort cloud is estimated to be about 2×10^{12} comets, with a total mass of about one Earth mass of material. Between 20 and 50% of the Oort cloud population is in a dynamically active shell between 10^4 and 2×10^5 AU from the Sun. Comets in this shell are perturbed by random passing stars and the galactic tide. The perturbations can change the perihelion distances of comets, sending them back into the planetary region where they are observed as long-period comets (those with orbital periods greater than 200 years). Interior to the shell is a dense inner Oort cloud that contains 50–80% of the comets, extending as close as 1000 AU from the Sun. The inner Oort cloud is not dynamically active. It is too close to the Sun to be significantly perturbed by external perturbers, unless the latter come very close, such as stars passing through the Oort cloud.

A second reservoir of comets is the Kuiper belt beyond the orbit of Neptune, named after G. P. Kuiper who in 1951 was one of the first to suggest its existence. Because no large planet grew beyond Neptune, there was no body to scatter away the icy planetesimals formed in that region. (The failure of a large planet to grow beyond Neptune is generally attributed to the increasing timescale for planetary accretion and the decreasing density of solar nebula materials with increasing **heliocentric** distance.) This belt of remnant planetesimals may terminate at ~ 50 AU or may extend out several 100 AU from the Sun, analogous to the disks of dust that have been discovered around main sequence stars such as Vega and Beta Pictoris (Figure 1.1).

The Kuiper belt actually consists of two different dynamical populations. The classical Kuiper belt is the population in low-inclination, low-eccentricity orbits beyond Neptune. Some of this population, including Pluto, is trapped in mean-motion resonances with Neptune at both the 3:2 and 2:1 resonances. The second population is objects in more eccentric and inclined orbits, typically with larger semimajor

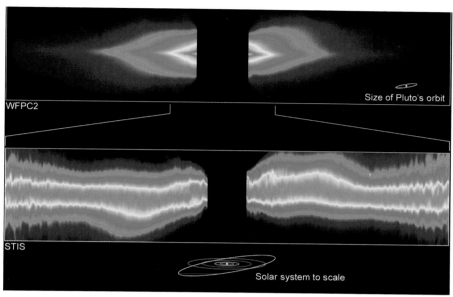

FIGURE 1.1 False-color images of the dust disk around the star Beta Pictoris, discovered by the *Infrared Astronomical Satellite* in 1983. The disk is viewed nearly edge on and is over 900 AU in diameter. The gaps in the center of each image are where the central star image has been removed. The top image shows the full disk as imaged with the Wide Field Planetary Camera 2 (WFPC2) onboard the *Hubble Space Telescope* (*HST*). The lower image shows the inner disk as viewed by the Space Telescope Imaging Spectrograph (STIS) instrument on HST. The orbits of the outer planets of our solar system, including the dwarf planet Pluto, are shown to scale for comparison. There is evidence of a warping of the Beta Pic disk, possibly caused by perturbations from a passing star. Infrared data show that the disk does not extend all the way in to the star, but that it has an inner edge at about 30 AU from Beta Pic. The disk interior to that distance may have been swept up by the accretion of planets in the nebula around the star. This disk is a possible analog for the Kuiper belt around our own solar system.

axes, called the **scattered disk**. These latter objects all have perihelia relatively close to Neptune's orbit, such that they continue to gravitationally interact with Neptune.

The Kuiper belt may contain many tens of Earth masses of comets, although the mass within 50 AU is currently estimated as ~ 0.1 Earth mass. A slow gravitational erosion of comets from the Kuiper belt, in particular from the scattered disk, due to the perturbing effect of Neptune, causes these comets to "leak" into the planetary region. Eventually, some fraction of the comets evolves due to gravitational scattering by the giant planets into the terrestrial planets region where they are observed as short-period comets. Short-period comets from the Kuiper belt are often called Jupiter-family or ecliptic comets because most are in orbits that can have close encounters with Jupiter, and are also in orbits with low inclinations, close to the ecliptic plane. Based on the observed number of ecliptic comets, the number of comets in the Kuiper belt between 30 and 50 AU has been estimated at $\sim 10^9$ objects larger than 1 km diameter, with a roughly equal number in the scattered disk. Current studies suggest that the Kuiper belt has been collisionally eroded out to a distance of ~ 100 AU from the Sun, but that considerably more mass may still exist in orbits beyond that distance.

Although gravity is the dominant force in determining the motion of bodies in the solar system, other forces do come into play in special cases. Dust grains produced by asteroid collisions or liberated from the sublimating icy surfaces of comet nuclei are small enough to be affected by radiation pressure forces. For submicron grains, radiation pressure from sunlight is sufficient to blow the grains out of the solar system. For larger grains, radiation pressure causes the grains to depart from Keplerian orbits. Radiation effects can also cause centimeter-sized grains to slowly spiral in toward the Sun through the Poynting–Robertson effect, and meter- to kilometer-sized bodies to slowly spiral either inward or outward due to the Yarkovsky effect.

Electromagnetic forces play a role in planetary magnetospheres where ions are trapped and spiral back and forth along magnetic field lines, and in cometary Type I plasma tails where ions are accelerated away from the cometary coma by the solar wind. Dust grains trapped in planetary magnetospheres and in interplanetary space also respond to electromagnetic forces, although to a lesser extent than ions because of their much lower charge-to-mass ratios.

3.2. Nature and Composition

The solar nebula, the cloud of dust and gas out of which the planetary system formed, almost certainly exhibited a strong temperature gradient with heliocentric distance, hottest near the forming proto-Sun at its center, and cooler as one moved outward through the planetary region. This

TABLE 1.3 Physical Parameters for the Sun, Planets, and Dwarf Planets

Name	Mass (kg)	Equatorial Radius (km)	Density (g/cm^3)	Rotation Period	Obliquity (o)	Escape Velocity (km/s)
Sun	1.989×10^{30}	695,508	1.41	25.4–35. d	7.25	617.7
Mercury	3.302×10^{23}	2440	5.43	56.646 d	0	4.25
Venus	4.869×10^{24}	6052	5.24	243.018 d	177.33	10.36
Earth	5.974×10^{24}	6378	5.52	23.934 h	23.45	11.18
Mars	6.419×10^{23}	3397	3.94	24.623 h	25.19	5.02
Ceres[1]	9.47×10^{20}	474	2.1	9.075 h	3	0.52
Jupiter	1.899×10^{27}	71,492	1.33	9.925 h	3.08	59.54
Saturn	5.685×10^{26}	60,268	0.70	10.656 h	26.73	35.49
Uranus	8.662×10^{25}	25,559	1.30	17.24 h	97.92	21.26
Neptune	1.028×10^{26}	24,764	1.76	16.11 h	28.80	23.53
Pluto[1]	1.314×10^{22}	1151	2.0	6.387 d	119.6	1.23
Haumea[1]	4.006×10^{21}	575–718	2.6–3.3	3.915 h		0.84
Makemake[1]	3×10^{21}	715	1.7	7.771 h		0.74
Eris[1]	1.67×10^{22}	1163	2.52	25.9 h		1.38

[1]Dwarf planet.

temperature gradient is reflected in the compositional arrangement of the planets and their satellites vs heliocentric distance. Parts of the gradient are also preserved in the asteroid belt between Mars and Jupiter and possibly in the Kuiper belt beyond Neptune.

Physical parameters for the planets and dwarf planets are given in Table 1.3. The planets fall into two major compositional groups. The terrestrial or Earth-like planets are Mercury, Venus, Earth, and Mars and are shown in Figure 1.2. The terrestrial planets are characterized by predominantly silicate compositions with iron cores. Gravitational potential energy heated the terrestrial planets as they formed resulting in them melting and then chemically differentiating. Their volatile content, i.e. atmospheres and oceans, may have accreted directly with the solid matter or may have been added later by asteroid and comet bombardment. Also, the modest masses of the terrestrial planets and their closeness to the Sun did not allow them to capture and retain gas directly from the solar nebula. The terrestrial planets all have solid surfaces that are modified to varying degrees by both cratering and internal processes (tectonics, weather, etc.).

Mercury is the most heavily cratered planet because it has no appreciable atmosphere to protect it from impacts or weather to erode the cratered terrain, and also because encounter velocities with Mercury are very high that close to the Sun. Additionally, tectonic processes on Mercury appear to have played a role in modifying its surface, which is partially covered by lava flows, like the Earth's Moon. Mars is next in the degree of cratering, in large part because of its proximity to the asteroid belt. Also, Mars' thin atmosphere affords little protection against impactors However, Mars also displays substantial volcanic and tectonic features, and evidence of erosion by wind and flowing water, the latter presumably having occurred early in the planet's history.

The surface of Venus is dominated by a wide variety of volcanic terrains. The degree of cratering on Venus is less than that on Mercury or Mars for two reasons: (1) Venus' thick CO_2 atmosphere (surface pressure = 93 bar) breaks up smaller asteroids and comets before they can reach the surface and (2) vulcanism on the planet has covered over the older craters on the planet surface. The surface of Venus is estimated to be 300–600 million years in age.

The Earth's surface is dominated by plate tectonics, in which large plates of the crust can move about the planet, and whose motions are reflected in such features as mountain ranges (where plates collide) and volcanic zones (where one plate dives under another). The Earth is the only planet with the right combination of atmospheric surface pressure and temperature to permit liquid water on its surface, and some 70% of the planet is covered by oceans.

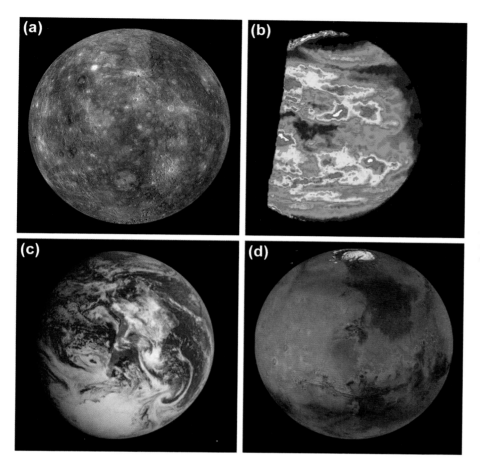

FIGURE 1.2 The four terrestrial planets. The heavily cratered surface of Mercury as imaged by the *MErcury Surface Space ENvironment GEochemistry and Ranging (MESSENGER)* spacecraft in 2011 (a); false-color image of clouds on the nightside of Venus, backlit by the intense infrared radiation from the planet's hot surface, as seen by the *Galileo* Near-Infrared Mapping Spectrometer instrument in 1990 (b); South America and Antarctica as imaged by the *Galileo* camera during a gravity assist flyby of the Earth in 1990 (c); Mars with its icy polar cap and Valles Marineris (at lower center in the image), a 3000-km-long canyon on Mars as imaged by the *Viking 1* orbiter in 1980 (d). The planets are not shown to scale; see Table 1.3 for their dimensions.

Craters on the Earth are rapidly erased by its active geology and weather, although the atmosphere only provides protection against very modest size impactors, on the order of 60 m diameter or less. Still, 181 impact craters or their remnants have been found on the Earth's surface or under its oceans.

The terrestrial planets each have substantially different atmospheres. Mercury has a tenuous atmosphere arising from its interaction with the solar wind. Hydrogen and helium ions are captured directly from the solar wind, while oxygen, sodium, and potassium are likely the product of sputtering at the surface. In contrast, Venus has a dense CO_2 atmosphere with a surface pressure 93 times the pressure at the Earth's surface. Nitrogen is also present in the Venus atmosphere at a few percent relative to CO_2. The dense atmosphere results in a massive greenhouse on the planet, heating the surface to a mean temperature of 735 K. The middle and upper atmosphere contain thick clouds composed of H_2SO_4 and H_2O, which shroud the surface from view. However, thermal radiation from the surface does penetrate the clouds, making it possible to view surface features through infrared "windows".

The Earth's atmosphere is unique because of its large abundance of free oxygen, which is normally tied up in oxidized surface materials on other planets. The reason for this unusual state is the presence of life on the planet, which traps and buries CO_2 as carbonates and also converts the CO_2 to free oxygen. Still, the bulk of the Earth's atmosphere is nitrogen (78%), with oxygen making up 21% and argon about 1%. The water vapor content of the atmosphere varies from about 1% to 4%. Various lines of evidence suggest that the composition of the Earth's atmosphere has evolved considerably over the history of the solar system and that the original atmosphere was denser than the present-day atmosphere and dominated by CO_2.

Mars has a relatively modest CO_2 atmosphere with a mean surface pressure of only 6 mbar. The atmosphere also contains a few percent of N_2 and argon. Mineralogic and isotopic evidence and geologic features suggest that the past atmosphere of Mars may have been much denser and warmer, allowing liquid water to flow across the surface in massive floods.

The volatiles in the terrestrial planets' atmospheres and in the Earth's oceans may have been contained in hydrated

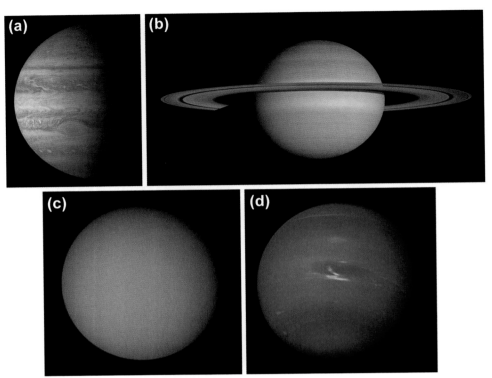

FIGURE 1.3 The giant planets. The complex, belted atmosphere of Jupiter with the Great Red Spot at the lower center, as imaged by the *Cassini* spacecraft during its gravity-assist flyby in 2000 (a); Saturn, its beautiful ring system, and its largest satellite Titan at lower left, as seen by *Cassini* in 2008 (b); the featureless atmosphere of Uranus, obscured by a high-altitude methane haze, as imaged by *Voyager 2* in 1986 (c); several large storm systems and a banded structure, similar to that of Jupiter, in Neptune's atmosphere, as imaged by *Voyager 2* in 1989 (d). The planets are not shown to scale; see Table 1.3 for their dimensions.

minerals in the planetesimals that originally formed the planets, and/or may have been added later due to asteroid and comet bombardment as the planets dynamically cleared their individual zones of leftover planetesimals. It appears most likely that all these reservoirs contributed some fraction of the volatiles on the terrestrial planets.

The giant or Jupiter-like planets are Jupiter, Saturn, Uranus, and Neptune and are shown in Figure 1.3. The giant planets are also referred to as the gas giants. They are characterized by low mean densities and thick hydrogen—helium atmospheres, presumably captured directly from the solar nebula during the formation of these planets. The composition of the giant planets is similar to that of the Sun, although more enriched in heavier elements. Because of their primarily gaseous composition and their high internal temperatures and pressures, the giant planets do not have solid surfaces. However, they may each have silicate—iron cores of several to tens of Earth masses of material.

Because they formed at heliocentric distances where ices could condense, the giant planets may have initially had a much greater local density of solid material to grow from. This may, in fact, have allowed them to form before the terrestrial planets interior to them. Studies of the dissipation of nebula dust disks around nearby solar-type protostars suggest that the timescale for the formation of giant planets is on the order of 10 million years or less. This is very rapid as compared with the ~100 million year timescale currently estimated for the formation of the terrestrial planets (although questions have now been raised as to the correctness of that accretionary timescale). Additionally, the higher uncompressed densities of Uranus and Neptune (0.5 g/cm^3) vs those of Jupiter and Saturn (0.3 g/cm^3) suggest that the outer two giant planets contain a significantly lower fraction of gas captured from the nebula. This may mean that the outer pair formed later than the inner two giant planets, consistent with the increasing timescale for planetary accretion at larger heliocentric distances.

Because of their heliocentric arrangement, the terrestrial and giant planets are occasionally called the inner and outer planets, respectively, although sometimes the term "inner planets" is used only to denote Mercury and Venus, the planets interior to the Earth's orbit.

There are currently five recognized dwarf planets, described below.

Ceres, discovered in 1801, is the largest body in the asteroid belt and the only main belt object classified as a dwarf planet. It has a surface composition and density

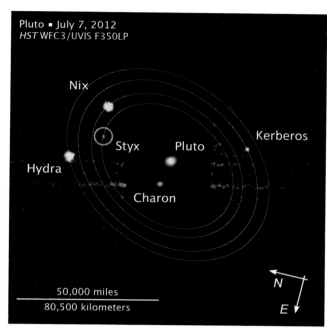

FIGURE 1.4 Hubble Space Telescope (HST) image of the dwarf planet Pluto (center) with its large moon Charon (just below and to the left of Pluto), and the four small satellites discovered with HST. The images of Pluto and Charon have been deliberately reduced in brightness so that the smaller satellites can be seen. A NASA spacecraft mission, *New Horizons*, was launched in 2006 and will fly by Pluto and Charon in 2015. *Courtesy of NASA and the Space Telescope Science Institute.*

similar to carbonaceous chondrite meteorites. This is a primitive class of meteorites that shows only limited processing during and since formation. Water frost has also been detected on the surface of Ceres. Because of its large size, the interior of Ceres is likely differentiated. A National Aeronautics and Space Administration (NASA) mission, Dawn, recently visited the large asteroid Vesta and is now on its way to Ceres, arriving in 2015.

Pluto, discovered in 1930, is the first object discovered in the Kuiper belt. It is classified as a dwarf planet, and has five satellites, the largest being Charon, which is about half the size of Pluto, Both are shown in Figure 1.4 along with the four smaller satellites. Pluto bears a strong resemblance to Triton, Neptune's large icy satellite (which is slightly larger than Pluto) and to other large icy objects in the Kuiper belt beyond the orbit of Neptune. Pluto has a thin, extended atmosphere, probably methane and nitrogen, which is slowly escaping because of Pluto's low gravity. This puts it in a somewhat intermediate state between a freely outflowing cometary coma and a bound planetary atmosphere. Spectroscopic evidence shows that methane frost covers much of the surface of Pluto, whereas its largest satellite Charon appears to be covered with water frost. Nitrogen frost has also been detected on Pluto. The density of Pluto is ~ 2 g/cm^3, suggesting that the rocky component of the dwarf planet accounts for about 70% of its total mass.

The Pluto–Charon system is fully tidally evolved. This means that Pluto and Charon each rotate with the same period, 6.38723 days, which is also the revolution period of Charon in its orbit. As a result, Pluto and Charon always show the same faces to each other. It is suspected that the Pluto–Charon system was formed by a giant impact between two large KBOs.

Haumea, discovered in 2004, is in an orbit that ranges between 35 and 51 AU from the Sun, and inclined 28° to the ecliptic. It may be trapped in a 7:12 mean-motion resonance with Neptune. Spectra show that Haumea is covered with a layer of crystalline ice, much like Pluto's satellite Charon. Haumea has two known satellites: Hi'iaka and Namaka.

Makemake, discovered in 2005, has an orbit that ranges between 38 and 53 AU with an inclination of 28°. Spectra of Makemake show the presence of methane ices on the surface, similar to Pluto. No satellites have been detected around Makemake.

The dwarf planet Eris was discovered in 2005 and is a scattered disk object in a distant orbit that ranges from 37.8 to 97.5 AU from the Sun, with an inclination of 43°. It is comparable in size to Pluto, has a somewhat higher bulk density, and also displays evidence for methane frost on its surface. Eris has one satellite, Dysnomia.

There has been considerable speculation as to the existence of a major planet beyond Neptune, often dubbed "Planet X". The search program that found Pluto in 1930 was continued for many years afterward but failed to detect any other distant objects, even though the limiting magnitude was considerably fainter than Pluto's visual magnitude of ~ 13.5. Other searches have been carried out, most notably by the *Infrared Astronomical Satellite (IRAS)* in 1983–1984. An automated algorithm was used to search for a distant planet in the *IRAS* data; it successfully "discovered" Neptune, but nothing else. More recently, the WISE (Widefield Infrared Survey Explorer) spacecraft surveyed the infrared sky in four wavelengths in 2010–2011 with much higher sensitivity than IRAS. Although nothing was found, analysis of the WISE data is continuing. As noted above, telescopic searches for KBOs have found objects comparable to Pluto in size, but none significantly larger.

Gravitational analyses of the orbits of Uranus and Neptune show no evidence of an additional perturber at greater heliocentric distances. Studies of the trajectories of the *Pioneer 10* and *11* and *Voyager 1* and *2* spacecraft have also yielded negative results. Analyses of the spacecraft trajectories do provide an upper limit on the unaccounted mass within the orbit of Neptune of $< 3 \times 10^{-6}$ solar masses ($M_\odot$), equal to about one Earth mass.

The compositional gradient in the solar system is perhaps best visible in the asteroid belt, whose members range from silicate-rich bodies in the inner belt (inside of ~ 2.6 AU), to volatile-rich carbonaceous bodies in the outer main belt (out to about 3.3 AU). (See Figure 1.5.)

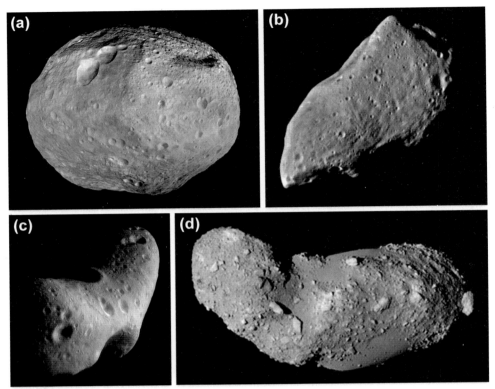

FIGURE 1.5 A sampling of main belt and near-Earth asteroids. 4 Vesta, the third largest asteroid (a); 951 Gaspra, the first asteroid encountered by an interplanetary spacecraft (b); 433 Eros, an Earth-approaching asteroid (c); and 25143 Itokawa, an Earth-crossing asteroid (d). All these asteroids, with the exception of Vesta, are stony types; Vesta has a surface resembling basaltic lava flows on the Earth. Most of the asteroids exhibit heavily cratered surfaces, but Itokawa is an exception, appearing to be a complete rubble pile. Vesta has dimensions of 573 × 557 × 446 km, Gaspra is 18 × 10 × 9 km, Eros is 34 × 11 × 11 km, and Itokawa is only 550 × 300 × 260 m. Vesta was imaged by the Dawn spacecraft while in orbit in 2011–2012, Gaspra was imaged by the *Galileo* spacecraft while it was en route to Jupiter, in 1991, Eros was imaged by the *NEAR (Near Earth Asteroid Rendezvous)* in 2000, and Itokawa by the *Hayabusa* spacecraft while in orbit in 2005. *Courtesy of NASA/JPL and JAXA.*

There also exist thermally processed asteroids, such as Vesta, whose surface material resembles a basaltic lava flow, and iron–nickel objects, presumably the differentiated cores of larger asteroids that were subsequently disrupted by collisions. The thermal gradient that processed the asteroids appears to be very steep and likely cannot be explained simply by the individual distances of these bodies from the forming proto-Sun. Rather, various special mechanisms such as magnetic induction, short-lived radioisotopes, or massive solar flares have been invoked to explain the heating event that so strongly processed the inner third of the asteroid belt.

The largest asteroid is Ceres, now classified as a dwarf planet, at a mean distance of 2.77 AU from the Sun. Ceres was the first asteroid discovered, by G. Piazzi on January 1, 1801. Ceres is 948 km in diameter, rotates in 9.075 h, and appears to have a surface composition similar to that of carbonaceous chondrite meteorites. The second largest asteroid is Pallas, also a carbonaceous type with a diameter of 532 km. Pallas is also at 2.77 AU, but its orbit has an unusually large inclination of 34.8°. Over 385,000 asteroids have had their orbits accurately determined and have been given official numbers in the asteroid catalog (as of January 2014). Another 247,300 asteroids have been observed well enough to obtain preliminary orbits, 130,200 of them at more than one opposition. Note that these numbers include all objects nominally classified as asteroids: main belt, near-Earth, Trojans, Centaurs, and KBOs (including Pluto, Eris, Haumea, and Makemake).

As a result of the large number of objects in the asteroid belt, impacts and collisions are frequent. Several "families" of asteroids have been identified by their closely grouped orbital elements and are likely fragments of larger asteroids that collided. Spectroscopic studies have shown that the members of these families often have very similar surface compositions, further evidence that they are related. The largest asteroids such as Ceres, Pallas, and Vesta are likely too large to be disrupted by impacts, but most of the smaller asteroids have probably been collisionally processed. Increasing evidence suggests that many asteroids may be "rubble piles", that is, asteroids that have been broken up but not dispersed by previous collisions and that now form a single but poorly consolidated body.

Beyond the main asteroid belt there exist small groups of asteroids locked in dynamical resonances with Jupiter. These include the Hildas at the 3:2 mean-motion resonance, the Thule group at the 4:3 resonance, and the Trojans, which are in a 1:1 mean-motion resonance with Jupiter. The effect of the resonances is to prevent these asteroids from making close approaches to Jupiter, even though many of the asteroids are in Jupiter-crossing orbits.

The Trojans are particularly interesting. They are essentially in the same orbit as Jupiter, but they librate about points 60° ahead and 60° behind the planet in its orbit, known as the Lagrange L_4 and L_5 points. These are pseudostable points in the three-body problem (Sun—Jupiter—asteroid) where bodies can remain dynamically stable for extended periods of time. Some estimates have placed the total number of objects in the Jupiter L_4 and L_5 Trojan swarms as equivalent to the population of the main asteroid belt. Trojan-type 1:1 librators have also been found for the Earth (one), Mars (three), Uranus (one), and Neptune (nine). Interestingly, the Saturnian satellites Dione and Tethys also have small satellites locked in Trojan-type librations in their respective orbits.

Much of what we know about the asteroid belt and about the early history of the solar system comes from meteorites recovered on the Earth. It appears that the asteroid belt is the source of almost all recovered meteorites. A modest number of meteorites that are from the Moon and from Mars, presumably blasted off of those bodies by asteroid and/or comet impacts, have been found. Cometary meteoroids are thought to be too fragile to survive atmospheric entry. In addition, cometary meteoroids typically encounter the Earth at higher velocities than asteroidal debris and thus are more likely to fragment and burn up during atmospheric entry. However, we may have cometary meteorites in our sample collections and simply not yet be knowledgeable enough to recognize them.

Recovered meteorites are roughly equally split between silicate and carbonaceous types, with a few percent being iron—nickel meteorites. The most primitive meteorites (i.e. the meteorites which appear to show the least processing in the solar nebula) are the volatile-rich carbonaceous chondrites. However, even these meteorites show evidence of some thermal processing and aqueous alteration (i.e. processing in the presence of liquid water). Study of carbonaceous and ordinary (silicate) chondrites provides significant information on the composition of the original solar nebula, on the physical and chemical processes operating in the solar nebula, and on the chronology of the early solar system.

The other major group of primitive bodies in the solar system is the comets. Because comets formed farther from the Sun than the asteroids, in colder environments, they contain a significant fraction of volatile ices. Water ice is the dominant and most stable volatile. Typical comets also contain modest amounts of CO, CO_2, CH_4, NH_3, H_2CO, and CH_3OH, most likely in the form of ices, but possibly also contained within complex organic molecules and/or in clathrate hydrates. Organics make up a significant fraction of the cometary nucleus, as well as silicate grains. F. Whipple described this icy conglomerate mix as "a dirty snowball", although the term "frozen mudball" may be more appropriate since the comets are more than 60% organics and silicates. It appears that the composition of comets is very similar to the condensed (solid) grains and ices observed in dense interstellar cloud cores where new stars are forming, with little or no evidence of processing in the solar nebula. Thus, comets appear to be the most primitive bodies in the solar system. As a result, the study of comets is extremely valuable for learning about the origin of the planetary system and the conditions in the solar nebula 4.567 billion years ago.

Five cometary nuclei—periodic comets Halley, Borrelly, Wild 2, Tempel 1, and Hartley 2—have been encountered by interplanetary spacecraft and imaged (Figure 1.6). These irregular nuclei range from about 2 to 12 km in mean diameter and have low albedos, only 3—4%. The nuclei exhibit a variety of complex surface morphologies unlike any other bodies in the solar system. It has been suggested that cometary nuclei are weakly bound conglomerations of smaller dirty snowballs, assembled at low velocity and low temperature in the giant planets region (and beyond in the Kuiper belt) of the solar nebula. Thus, comets may be "primordial rubble piles", in some ways similar to the asteroids. Recent studies have suggested that cometary nuclei, like the asteroids, may have undergone intense collisional evolution, either while resident in the Kuiper belt or in the giant planets region prior to their dynamical ejection to the Oort cloud.

Subtle and not-so-subtle differences in cometary compositions have been observed. However, it is not entirely clear if these differences are intrinsic or due to the physical evolution of cometary surfaces over many close approaches to the Sun. Because the comets that originated among the giant planets have all been ejected to the Oort cloud or to interstellar space, the compositional spectrum resulting from the heliocentric thermal profile is not spatially preserved as it has been in the asteroid belt. Although comets in the classical Kuiper belt are likely located close to their formation distances, physical studies of these distant objects are still in an early stage. There is an observed compositional trend, but it is associated with orbital eccentricity and inclination, rather than **semimajor axis**.

3.3. Satellites, Rings, and Things

The natural satellites of the planets, listed in the appendix to this volume, show as much diversity as the planets they orbit (see Figure 1.7). Among the terrestrial planets, the

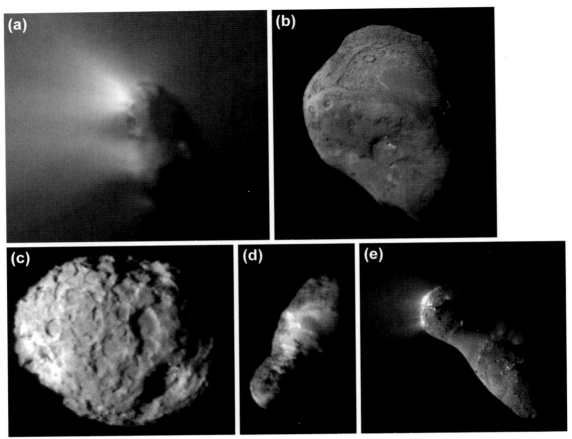

FIGURE 1.6 Five cometary nuclei imaged by flyby spacecraft. 1P/Halley in 1986 (Giotto, a), 9P/Tempel 1 in 2005 (*Deep Impact*, b), 81P/Wild 2 in 2004 (*Stardust*, c), 19P/Borrelly in 2001 (*Deep Space 1*, d), and 103P/Hartley 2 (*EPOXI*, e). The nuclei show considerable diversity both in shape and in surface topography. The Halley nucleus has dimensions of 15 × 8 km, the Tempel 1 nucleus is 7.6 × 4.9 km, the Wild 2 nucleus is 5.2 × 4.0 km, the Borrelly nucleus is 8 × 3.2 km, and the Harley 2 nucleus is 2.3 × 1.2 km. The Halley and Hartley 2 images show bright dust jets emanating from active areas on the nucleus surface. The other three nuclei were also active during their respective flybys but the activity was too faint to show in these images. *Courtesy of ESA and NASA/JPL.*

only known satellites are the Earth's Moon and the two small moons of Mars, Phobos and Deimos. The Earth's Moon is unusual in that it is so large relative to its primary. The Moon has a silicate composition similar to the Earth's mantle and a small iron core.

It is now widely believed that the Moon formed as a result of a collision between the proto-Earth and another protoplanet about the size of Mars, late in the accretion of the terrestrial planets. Such "giant impacts" are now recognized as being capable of explaining many of the features of the solar system, such as the unusually high density of Mercury and the large obliquities of several of the planetary rotation axes. In the case of the Earth, the collision with another protoplanet resulted in the cores of the two planets merging, while a fraction of the mantles of both bodies was thrown into orbit around the Earth where some of the material reaccreted to form the Moon. The tidal interaction between the Earth and the Moon then slowly evolved the orbit of the Moon outward to its present position, at the same time slowing the rotation of both the Earth and the Moon. The giant impact hypothesis is capable of explaining many of the features of the Earth–Moon system, including the similarity in composition between the Moon and the Earth's mantle, the lack of a significant iron core within the Moon, and the high angular momentum of the Earth–Moon system.

Like most large natural satellites, the Moon has tidally evolved to where its rotation period matches its revolution period in its orbit. This is known as synchronous rotation. It results in the Moon showing the same face to the Earth at all times, although there are small departures from this because of the eccentricity and inclination of the Moon's orbit.

The Moon's surface displays a record of the intense bombardment all the planets have undergone over the history of the solar system. Returned lunar samples have been age-dated based on decay of long-lived radioisotopes. This has allowed the determination of a chronology of lunar bombardment by comparing the sample ages with the crater counts on the lunar plains where the samples were collected. The lunar plains, or maria, are the result of

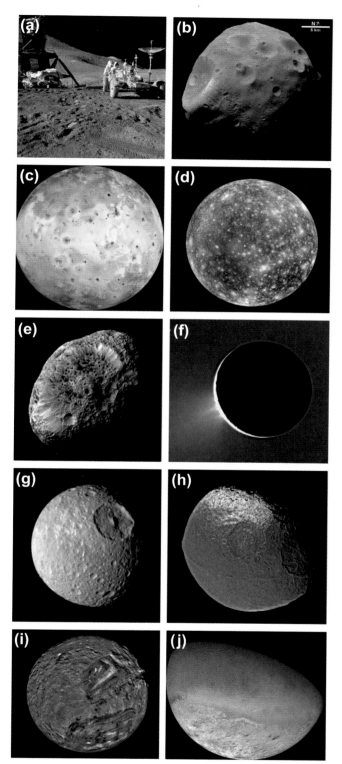

massive eruptions of lava during the first billion years of the Moon's history. The revealed chronology shows that the Moon experienced a massive bombardment between 4.2 and 3.8 billion years ago, known as the Late Heavy Bombardment. This time period is relatively late as compared with the 100–200 million years required to form the terrestrial planets and to clear the orbital zones of most interplanetary debris. Similarities in crater size distributions on the Moon, Mercury, and Mars suggest that the Late Heavy Bombardment swept over all the terrestrial planets. Recent explanations for the Late Heavy Bombardment have focused on the possibility that it came from the clearing of the outer planets' zones of their cometary debris. However, the detailed dynamical calculations of the timescales for that process are still being determined.

Like almost all other satellites in the solar system, the Moon has no substantial atmosphere. There is a transient atmosphere due to helium atoms in the solar wind striking the lunar surface and being captured. Argon has been detected escaping from surface rocks and being temporarily cold-trapped during the lunar night. Also, sodium and potassium have been detected, likely the result of sputtering of surface materials due to solar wind particles, as on Mercury. Water ice has been detected in craters at the Moon's south pole, but in very limited quantities.

Unlike the Earth's Moon, the two natural satellites of Mars are both small, irregular bodies, and in orbits relatively close to the planet. In fact, Phobos, the larger and closer satellite, orbits Mars faster than the planet rotates. Both of the Martian satellites have surface compositions that appear to be similar to carbonaceous chondrites. This has resulted in speculation that the satellites are captured asteroids. A problem with this hypothesis is that Mars is located close to the inner edge of the asteroid belt, where thermally processed silicate asteroids dominate the asteroid population, and where carbonaceous asteroids are relatively rare. Also, both satellites are located very close to the planet and in near-circular orbits, which is unusual for captured objects.

In contrast to the satellites of the terrestrial planets, the satellites of the giant planets are numerous and are arranged in complex systems. Jupiter has four major satellites, easily

FIGURE 1.7 A sampling of satellites in the solar system. The dusty surface of the Earth's Moon, still the only other celestial body visited by humans (a); Phobos, the larger of Mars' two moons showing the large crater Stickney at left (b); the innermost Galilean satellite, Io, displays active vulcanism on its sulfur-rich surface (c); the outermost Galilean satellite, Callisto, displays a heavily cratered surface, likely dating back to the origin of the solar system (d); one of Saturn's smaller satellites, Hyperion, is irregularly shaped, in chaotic rotation, and displays a very unusual surface morphology (e); Saturn's satellite Enceladus is one of several in the solar system that has active geysers on its surface (f); another small Saturnian satellite, Mimas, displays an immense impact crater on one hemisphere (g); Saturn's satellite Iapetus is black on one hemisphere and white on the other, and has a high ridge circling it at the equator (h); Uranus' outermost major satellite, Miranda, has a complex surface morphology suggesting that the satellite was disrupted and reaccreted (i); Neptune's one large satellite, Triton, displays a mix of icy terrains and ice vulcanism (j). The satellites are not shown to scale.

visible in small telescopes from Earth, and 63 known lesser satellites. The discovery of the four major satellites by Galileo in 1610, now known as the Galilean satellites, was one of the early confirmations of the Copernican theory of a heliocentric solar system. The innermost Galilean satellite, Io, is about the same size as the Earth's Moon and has active vulcanism on its surface as a result of Jupiter's tidal perturbation and the gravitational interaction with Europa and Ganymede (see Section 3.1). The next satellite outward is Europa, somewhat smaller than Io, which appears to have a thin ice crust overlying a possible liquid water ocean, also the result of tidal heating by Jupiter and the satellite—satellite gravitational interactions. Estimates of the age of the surface of Europa, based on counting impact craters, are very young, suggesting that the thin ice crust may repeatedly break up and reform. The next satellite outward from Jupiter is Ganymede, the largest satellite in the solar system, even larger than the planet Mercury. Ganymede is another icy satellite and shows evidence of tectonic activity and of being partially resurfaced at some time(s) in its past. The final Galilean satellite is Callisto, another icy satellite that appears to preserve an impact record of comets and asteroids dating back to the origin of the solar system. As previously noted, the orbits of the inner three Galilean satellites are locked into a 4:2:1 mean-motion resonance.

The lesser satellites of Jupiter include four within the orbit of Io, and 59 at very large distance from the planet. The latter are mostly in retrograde orbits, which suggests that they are likely captured comets and asteroids. The orbital parameters of many of these satellites fall into several tightly associated groups. This suggests that each group consists of fragments of a larger object that was disrupted, most likely by a collision with another asteroid or comet. Possibly, the collision occurred within the gravitational sphere of Jupiter, which then could have led to the dynamical capture of some of the fragments.

All the close-orbiting Jovian satellites (out to the orbit of Callisto) appear to be in synchronous rotation with Jupiter. However, rotation periods have been determined for two of the outer satellites, Himalia and Elara, and these are approximately 8 and 12 h, respectively, much shorter than their $\sim$250-day periods of revolution about the planet.

Saturn's satellite system is very different from Jupiter's in that it contains only one large satellite, Titan, comparable in size to the Galilean satellites, seven intermediate-sized satellites, and 54 smaller satellites. Titan is the only satellite in the solar system with a substantial atmosphere. Clouds of organic compounds in its atmosphere prevent easy viewing of the surface of that moon, although the *Cassini* spacecraft has had success in viewing the surface at infrared and radar wavelengths. The atmosphere is primarily nitrogen and also contains methane and possibly argon. The surface temperature on Titan has been measured at 94 K, and the surface pressure is 1.5 bar. *Cassini* radar imaging has revealed a complex surface morphology on Titan that includes rivers, lakes, and possible cryovulcanism.

The intermediate satellites of Saturn all appear to have icy compositions and have undergone substantial processing, possibly as a result of tidal heating and also due to collisions. Orbital resonances exist between several pairs of satellites, and most are in synchronous rotation with Saturn. An interesting exception is Hyperion, which is a highly nonspherical body and which appears to be in chaotic rotation. Another moon, Enceladus, has a ring of material in its orbit that likely has come from geysers discovered at the icy satellite's south pole. Two other satellites, Dione and Tethys, have two companion satellites each, in the same orbit, which oscillate about the Trojan-libration points for the Saturn—Dione and Saturn—Tethys systems, respectively. Yet another particularly interesting satellite of Saturn is Iapetus, which is dark on one hemisphere and bright on the other and has a narrow ridge circling the satellite at its equator. The dark material appears to be a coating on the satellite's leading hemisphere that is suspected of coming from Phoebe. The equatorial ridge is believed to be a remnant from a time when the satellite was warmer and larger, but this is by no means certain.

Saturn has one very distant, intermediate-sized satellite, Phoebe, which is in a retrograde orbit and which is suspected of being a captured, early solar system planetesimal, albeit a very large one. Phoebe is not in synchronous rotation, but rather has a rotation period of about 10 h. The 54 known small satellites of Saturn include 11 embedded in or immediately adjacent to the planet's ring system, 4 Trojan-type librators, and 39 in distant orbits. As with Jupiter, the majority of these distant objects are in retrograde orbits and some are in groups, which suggests that they too are collisional fragments.

The Uranian system consists of five intermediate-sized satellites and 22 smaller ones. Again, these are all icy bodies. These satellites also exhibit evidence of past heating and possible tectonic activity. The satellite Miranda is particularly unusual in that it exhibits a wide variety of complex terrains. It has been suggested that Miranda, and possibly many other icy satellites, were collisionally disrupted at some time in their history, and the debris then reaccreted in orbit to form the currently observed satellites, but preserved some of the older surface morphology. Such disruption/reaccretion phases may have even reoccurred on several occasions for some of the satellites over the history of the solar system. Of the smaller Uranian satellites, 13 are embedded in the ring system and nine are in distant, mostly retrograde orbits. Again, these are likely captured objects.

Neptune's satellite system consists of one large icy satellite and 13 smaller ones. Triton is somewhat larger than Pluto and is unusual in that it is in a retrograde orbit. As a result, the tidal interaction with Neptune is causing the

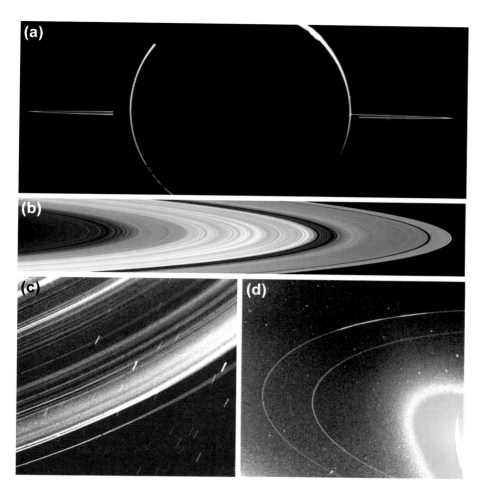

FIGURE 1.8 The ring systems of the giant planets. Jupiter's single ring photographed in forward scattered light, while the *Galileo* spacecraft was in eclipse behind the giant planet: the lit circle is sunlight filtering through the atmosphere of Jupiter (a); Saturn's rings break up into hundreds of ringlets when viewed at high resolution, as in this *Cassini* mosaic (b); Uranus' system of narrow rings as viewed in forward scattered light by *Voyager 2* as it passed behind the planet (c); two of Neptune's rings showing the unusual azimuthal concentrations, as photographed by *Voyager 2* as it passed behind the planet; the greatly overexposed crescent of Neptune is visible at lower right in the image (d). The ring systems are not shown to scale.

satellite's orbit to decay, and eventually Triton will be torn apart by the planet's gravity when it passes within the **Roche limit**. The retrograde orbit is often cited as evidence that Triton must have been captured from interplanetary space and did not actually form in orbit around the planet. Despite its tremendous distance from the Sun, Triton's icy surface displays a number of unusual terrain types that strongly suggest thermal processing and possibly even current activity. The *Voyager 2* spacecraft photographed what appeared to be plumes from "ice volcanoes" on Triton.

Neptune has one intermediate-sized satellite, Nereid, in a distant and eccentric orbit. The lesser satellites of Neptune include six that are either in or adjacent to the ring system and five in distant orbits, three of which are retrograde.

In addition to their satellite systems, all the giant planets have ring systems (Figure 1.8). As with the satellite systems, each ring system is distinctly different from its neighbors. Jupiter has a single ring at 1.72−1.81 planetary radii, discovered by the Voyager 1 spacecraft. The ring has several components, related to the four small satellites in or close to the ring. The micron-sized ring particles appear to be material sputtered off the embedded satellites.

Saturn has an immense, broad ring system extending between 1.11 and 2.27 planetary radii, easily seen in a small telescope from the Earth. The ring system consists of three major rings, known as A, B, and C ordered from the outside in toward the planet, a diffuse ring labeled D inside the C ring and extending down almost to the top of the Saturnian atmosphere, and several other narrow, individual rings.

Closer examination by the *Voyager* spacecraft revealed that the A, B, and C rings were each composed of thousands of individual ringlets. This complex structure is the result of mean-motion resonances with the many Saturnian satellites, as well as with small satellites embedded within the rings themselves. Some of the small satellites act as gravitational "shepherds", focusing the ring particles into narrow ringlets. Additional narrow and diffuse rings are located outside the main ring system.

The Uranian ring system was discovered accidentally in 1977 during observation of a stellar occultation by Uranus. A symmetric pattern of five narrow dips in the stellar signal was seen on both sides of the planet. Later observations of other stellar occultations found an additional five narrow rings. *Voyager 2* detected several more, fainter, diffuse rings and provided detailed imaging of the entire ring system.

The success with finding Uranus' rings led to similar searches for a ring system around Neptune using stellar occultations. Rings were detected but were not always symmetric about the planet, suggesting gaps in the rings. Subsequent *Voyager 2* imaging revealed large azimuthal concentrations of material in one of the six detected rings.

All of the ring systems are within the Roche limits of their respective planets, at distances where tidal forces from the planet will disrupt any solid body, unless it is small enough and strong enough to be held together by its own material strength. This has led to the general belief that the rings are disrupted satellites, or possibly material that could never successfully form into satellites. Ring particles have typical sizes ranging from micron-sized dust to meter-sized objects and appear to be made primarily of icy materials, although in some cases contaminated with carbonaceous materials. Jupiter's ring is an exception because it appears to be composed of carbonaceous and silicate materials, with no ice.

Another component of the solar system is the zodiacal dust cloud, a huge, continuous cloud of fine dust extending throughout the planetary region and generally concentrated toward the ecliptic plane. The cloud consists of dust grains liberated from comets as the nucleus ices sublimate and from collisions between asteroids. Comets are estimated to account for about two-thirds of the total material in the **zodiacal cloud**, with asteroid collisions providing the rest. Dynamical processes tend to spread the dust uniformly around the Sun, although some structure is visible as a result of the most recent asteroid collisions. These structures, or bands as they are also known, are each associated with specific asteroid collisional families.

Dust particles will typically burn up due to friction with the atmosphere when they encounter the Earth, appearing as visible meteors. However, particles less than about 50 μm in radius have sufficiently large area-to-mass ratios that they can be decelerated high in the atmosphere at an altitude of about 100 km and can radiate away the energy generated by friction without vaporizing the particles. These particles then settle slowly through the atmosphere and are eventually incorporated into terrestrial sediments. In the 1970s, NASA began experimenting with collecting interplanetary dust particles (IDPs, also known as Brownlee particles because of the pioneering work of D. Brownlee) using high-altitude U2 reconnaissance aircraft. Terrestrial sources of particulates in the stratosphere are rare and consist largely of volcanic aerosols and aluminum oxide particles from solid rocket fuel exhausts, each of which are readily distinguishable from extraterrestrial materials.

The composition of the IDPs reflects the range of source bodies that produce them and include ordinary and carbonaceous chondritic material and suspected cometary

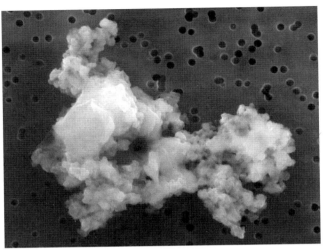

FIGURE 1.9 A suspected cometary interplanetary dust particle (IDP). The IDP is a highly porous, apparently random collection of submicron silicate grains embedded in a carbonaceous matrix. This particle is ~10 μm across. The voids in the IDP may have once been filled with cometary ices.

particles. Because the degree of heating during atmospheric deceleration is a function of the encounter velocity, recovered IDPs are strongly biased toward asteroidal particles from the main belt, which approach the Earth in lower eccentricity orbits. Nevertheless, suspected cometary particles are included in the IDPs. The cometary IDPs show a random, "botryoidal" (cluster-of-grapes) arrangement of submicron silicate grains similar in size to interstellar dust grains, intimately mixed in a carbonaceous matrix. Voids in cometary IDPs may have once been filled by cometary ices. In 2006, the *Stardust* spacecraft returned samples of cometary dust collected during a flyby of comet Wild 2; these are providing an important comparison with the IDPs collected by high-flying aircraft. An example of a suspected cometary IDP is shown in Figure 1.9.

Extraterrestrial particulates are also collected on the Earth in Antarctic ice cores, in melt ponds in Greenland, and as millimeter-sized silicate and nickel−iron melt products in ocean sediments. The IDP component in terrestrial sediments can be determined by measuring the abundance of ^{3}He. ^{3}He has normal abundances in terrestrial materials of 10^{-6} or less. The ^{3}He is implanted in the IDP grains during their exposure to the solar wind. Using this technique, one can look for variations in the infall rate of extraterrestrial particulates over time, and such variations are seen, sometimes correlated with impact events on the Earth.

3.4. The Solar Wind and the Heliosphere

A largely unseen part of the solar system is the solar wind, an ionized plasma that streams continuously into space from the Sun. The solar wind is composed primarily of protons (hydrogen nuclei) and electrons with some alpha particles (helium nuclei) and trace amounts of heavier ions. It is accelerated to supersonic speed in the solar corona and streams outward at a typical velocity of 400 km/s. The solar wind is highly variable, changing with both the solar rotation period of ~25 days and with the 22-year solar cycle, as well as on much more rapid timescales. As the solar wind expands outward, it carries the solar magnetic field with it in a spiral pattern caused by the rotation of the Sun. The solar wind was first inferred in the early 1950s by L. Biermann based on observations of cometary Type I plasma tails. The theory of the supersonic solar wind was first described by E. N. Parker in 1958, and the solar wind itself was detected in 1962 by the *Explorer 10* spacecraft in Earth orbit, and the *Mariner 2* spacecraft while en route to a flyby of Venus.

The solar wind interaction with the planets and the other bodies in the solar system is also highly variable, depending primarily on whether or not the body has its own intrinsic magnetic field. For bodies without a magnetic field, such as Venus and the Moon, the solar wind impinges directly on the top of the atmosphere or on the solid surface, respectively. For bodies like the Earth or Jupiter, which do have magnetic fields, the field acts as a barrier and deflects the solar wind around it. Because the solar wind is expanding at supersonic speeds, a shock wave, or bow shock, develops at the interface between the interplanetary solar wind and the planetary magnetosphere. The planetary magnetospheres can be quite large, extending out ~12 planetary radii upstream (sunward) of the Earth, and 50–100 radii sunward of Jupiter. Solar wind ions can leak into the planetary magnetospheres near the poles, and these can result in visible aurora, which have been observed on the Earth, Jupiter (Figure 1.10), and Saturn. As it flows past the planet, the interaction of the solar wind with the planetary magnetospheres results in huge magnetotail structures that often extend over interplanetary distances.

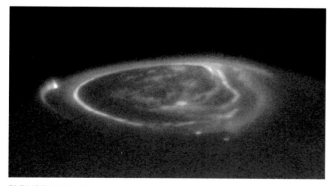

FIGURE 1.10 The auroral ring over the north polar region of Jupiter, as imaged by the *Hubble Space Telescope*. Several of the bright spots correspond to flux tube "footprints", magnetic field lines that connect the planet to the Galilean satellites.

All the giant planets, as well as the Earth, have substantial magnetic fields and thus planetary magnetospheres. Mercury has a weak magnetic field, but Venus has no detectable field. Mars has a patchy field, indicative of a past magnetic field at some point in the planet's history, but it has no organized magnetic field at this time. The *Galileo* spacecraft detected a magnetic field associated with Ganymede, the largest of the Galilean satellites. However, no magnetic field was detected for Europa or Callisto. The Earth's Moon has no magnetic field.

The most visible manifestation of the solar wind is cometary plasma tails, which result when the evolving gases in the cometary comae are ionized by sunlight and by charge exchange with the solar wind and then accelerated by the solar magnetic field. The ions stream away from the cometary comae at high velocity in the antisunward direction. Structures in the tail are visible as a result of fluorescence by CO^+ and other ions, although the most abundant ion in the plasma tails is H_2O^+.

At some distance from the Sun, far beyond the orbits of the planets, the solar wind reaches a point where the ram pressure from the wind is equal to the external pressure from the local interstellar wind flowing past the solar system. A termination shock develops upstream of that point, and the solar wind will be decelerated from supersonic to subsonic. *Voyager 1* detected the termination shock at 94 AU in 2004 and Voyager 2 detected it at 84 AU in 2007. Beyond this distance is a region called the heliosheath, still dominated by the subsonic solar plasma and extending out another 15–25 AU. The outer boundary of this region is known as the heliopause and defines the limit between solar system-dominated plasma and the interstellar wind. It is not currently known if the flow of interstellar medium past the solar system is supersonic or subsonic. If it is supersonic, then there must additionally be a bow shock beyond the heliopause, where the interstellar medium encounters the obstacle presented by the **heliosphere**. A diagram of the major features of the heliosphere is shown in Figure 1.11.

The *Voyager 1* spacecraft crossed the heliopause in August 2012 and is now in interstellar space. It is 126.9 AU from the Sun and continues to move outward at 3.6 AU/year (as of February 2014). *The Voyager 2* spacecraft continues to study the outermost region of the heliosphere, known as the "heliosheath" and is expected to cross the heliopause in 2017. Voyager 2 is currently 104.0 AU from the Sun and is moving outward at 3.3 AU/year. The *Voyager* 1 and 2 spacecraft are expected to continue to send measurements at least until the year 2020, when they will be at about 148 and 129 AU from the Sun, respectively. To many planetary scientists, the heliopause defines the boundary of the solar system because it marks the changeover from the solar wind to an interstellar medium-dominated space. However, as already noted, the Sun's gravitational sphere of influence extends out much farther,

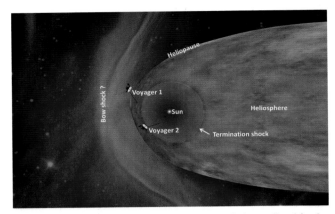

FIGURE 1.11 Artist's concept of the major boundaries predicted for the heliosphere and the locations of the two *Voyager* spacecraft. *Voyager 1* crossed the termination shock in 2004 at 94 AU from the Sun, and *Voyager 2* crossed it in 2007 at 84 AU. Voyage 1 crossed the heliopause and entered interstellar space in August 2012 at 122 AU. *Courtesy NASA/JPL-Caltech.*

to $\sim 2 \times 10^5$ AU (~ 1 pc), and there are bodies in orbit around the Sun at those distances. These include the Kuiper belt and scattered disk, which may each extend out to $\sim 10^3$ AU (possibly even farther for the scattered disk), and the Oort comet cloud which is populated to the limits of the Sun's gravitational field.

4. THE ORIGIN OF THE SOLAR SYSTEM

Our knowledge of the origin of the Sun and the planetary system comes from two sources: study of the solar system itself and study of star formation in nearby giant molecular clouds. The two sources are radically different. In the case of the solar system, we have an abundance of detailed information on the planets, their satellites, and numerous small bodies. But the solar system we see today is highly evolved and has undergone massive changes since it first condensed from the natal interstellar cloud. We must learn to recognize which qualities reflect that often violent evolution and which truly record conditions at the time of solar system formation.

In contrast, when studying even the closest star-forming regions (which are about 140 pc from the Sun), we are handicapped by a lack of adequate resolution and detail. In addition, we are forced to take a "snapshot" view of many young stars at different stages in their formation, and from that attempt to generate a time-ordered sequence of those different stages and processes involved. When we observe the formation of other stars, we also need to recognize that some of the observed processes or events may not be applicable to the formation of our own Sun and planetary system.

Still, a coherent picture has emerged of the major events and processes in the formation of the solar system. That picture assumes that the Sun is a typical star and that it formed in a similar way to many of the low-mass protostars we see today.

The birthplace of stars is giant molecular clouds in the galaxy. These huge clouds of molecular hydrogen have masses of $10^5-10^6\,M_\odot$. Within these clouds are denser regions or cores where star formation actually takes place. Some process, perhaps the shock wave from a nearby supernova, triggers the gravitational collapse of a cloud core. Material falls toward the center of the core under its own self-gravity and a massive object begins to grow at the center of the cloud. Heated by the gravitational potential energy of the infalling matter, the object becomes self-luminous and is then described as a protostar. Although central pressures and temperatures are not yet high enough to ignite nuclear fusion, the protostar begins to heat the growing nebula around it. The timescale of the infall of the cloud material for a solar mass cloud is about 10^6 years.

The infalling cloud material consists of both gas and dust. The gas is mostly hydrogen (75% by mass) and helium (22%). The dust (2%) is a mix of interstellar grains, including silicates, organics, and condensed ices. A popular model suggests that the silicate grains are coated with icy organic mantles. As the dust grains fall inward, they experience a pressure from the increasing density of gas toward the center of the nebula. This slows and even halts the inward radial component of their motion. However, the dust grains can still move vertically with respect to the central plane of the nebula, as defined by the rotational angular momentum vector of the original cloud core. As a result, the grains settle toward the central plane.

As the grains settle, they begin to collide with one another. The grains stick and quickly grow from microscopic to macroscopic objects, perhaps meters in size (initial agglomerations of grains may look very much like the suspected cometary IDP in Figure 1.9). This process continues and even increases as the grains reach the denser environment at the central plane of the nebula. The meter-sized bodies grow to kilometer-sized bodies and the kilometer-sized bodies grow to 100 km-sized bodies. These bodies are known as planetesimals. As a planetesimal begins to acquire significant mass, its cross-section for accretion grows beyond its physical cross-section because it is now capable of gravitationally deflecting smaller planetesimals toward it. These larger planetesimals then "run away" from the others, growing at an ever increasing rate.

The actual process is far more complex than described here, and there are many details of this scenario that still need to be worked out. For example, the role of turbulence in the nebula is not well quantified. Turbulence would tend to slow or even prevent the accretion of grains into larger objects. Also, the role of electrostatic and magnetic effects in the nebula is not understood.

Nevertheless, it appears that accretion in the central plane of the solar nebula can account for the growth of

FIGURE 1.12 Artist's concept of the accretion disk in the solar nebula, showing dust, orbiting planetesimals, and the proto-Sun at the center. *Painting by William Hartmann.*

planets from interstellar grains. An artist's concept of the accretion disk in the solar nebula is shown in Figure 1.12. In the inner region of the solar nebula, close to the forming Sun, the higher temperatures would vaporize icy and organic grains, leaving only silicate grains to form the planetesimals, which eventually merged to form the terrestrial planets. At larger distances where the nebula was cooler, organic and icy grains would condense, and these would combine with the silicates to form the cores of the giant planets. Because the total mass of ice and organics may have been several times the mass of silicates, the cores of the giant planets may actually have grown faster than the terrestrial planets interior to them.

At some point, the growing cores of the giant planets became sufficiently massive to begin capturing hydrogen and helium directly from the nebula gas. Because of the lower temperatures in the outer planets zone, the giant planets were able to retain the gas and continue to grow even larger. The terrestrial planets close to the Sun may have acquired some nebula gas, but probably they could not hold on to it at their higher temperatures.

Observations of protostars in nearby molecular clouds have found substantial evidence for accretionary disks and gas nebulae surrounding these stars. The relative ages of these protostars can be estimated by comparing their luminosity and color with theoretical predictions of their location in the Hertzsprung—Russell diagram. One of the more interesting observations is that the nebula dust and gas around solar mass protostars seem to dissipate after about 10^7 years. It appears that the nebula and dust may be swept away by mass outflows, essentially superpowerful solar winds, from the protostars. If the Sun formed similarly to the protostars we see today, then these observations set strong limits on the likely formation times of Jupiter and Saturn.

An interesting process that must have occurred during the late stages of planetary accretion is "giant impacts", i.e. collisions between very large protoplanetary objects. As noted in Section 3.3, a giant impact between a Mars-size protoplanet and the proto-Earth is now the accepted explanation for the origin of the Earth's Moon. Although it was previously thought that such giant impacts were low-probability events, they are now recognized to be a natural consequence of the final stages of planetary accretion.

Another interesting process late in the accretion of the planets is the clearing of debris from the planetary zones. At some point in the growth of the planets, their gravitational spheres of influence grew sufficiently large that an encounter with a planetesimal would more likely lead to the planetesimal being gravitationally scattered into a different orbit, rather than an actual collision. This would be particularly true for the massive giant planets, both because of their stronger gravitational fields and because of their larger distances from the Sun.

Because it is just as likely that a planet will scatter objects inward as outward, the clearing of the planetary zones resulted in planetesimals being flung throughout the solar system and in a massive bombardment of all planets and satellites. Many planetesimals were also flung out of the planetary system to interstellar space or to distant orbits in the Oort cloud. Although the terrestrial planets are too small to eject objects out of the solar system, they can scatter objects to Jupiter-crossing orbits where Jupiter will quickly dispose of them in about 10^6 years or less.

The clearing of the planetary zones has several interesting consequences. The dynamical interaction between the planets and the remaining planetesimals results in an exchange of angular momentum. Computer-based dynamical simulations have shown that this causes the semimajor axes of the planets to migrate. In general, Saturn, Uranus, and Neptune are expected to first move inward and then later outward as the ejection of material progresses. Jupiter, which ejects the most material because of its huge mass, migrates inward but by only a few tenths of an astronomical unit.

This migration of the giant planets has significant consequences for the populations of small bodies in the planetary region. As the planets move, the locations of their mean-motion and secular resonances will move with them. This will result in some small bodies being captured into resonances while others will be thrown into chaotic orbits, leading to their eventual ejection from the system or possibly to impacts on the planets and the Sun. The radial migration of the giant planets has been invoked both in the clearing of the outer regions of the main asteroid belt and the inner regions of the Kuiper belt.

Another consequence of the clearing of the planetary zones is that rocky planetesimals formed in the terrestrial planets zone will be scattered throughout the giant planets region, and vice versa, for icy planetesimals formed in the outer planets zone. The bombardment of the terrestrial planets by icy planetesimals is of particular interest, both as an explanation for the Late Heavy Bombardment and as a means of delivering the volatile reservoirs of the terrestrial planets. Isotopic studies suggest that some fraction of the water in the Earth's oceans may have come from comets and/or volatile-rich asteroids, although not all of it. Also, the discovery of an asteroidal-appearing object, 1996 PW, on a long-period comet orbit has provided evidence that asteroids may indeed have been ejected to the Oort cloud, where they may make up 1—3% of the population there.

5. THE SOLAR SYSTEM'S PLACE IN THE GALAXY

The Milky Way galaxy is classified as a barred spiral with loosely wound arms, SBc in the Hubble catalog of galaxies. It consists of four major structures: the galactic disk, the central bar, the halo, and the corona (Figure 1.13 and 1.14). As the name implies, the disk is a highly flattened, rotating structure about 15—25 kpc in radius and about 0.5—1.3 kpc thick, depending on which population of stars is used to trace the disk. Note that galactic distances are measured in parsecs and kiloparsecs (1000 pc), where a parsec is defined as the distance where a star would have a parallax of 1 arc-second as viewed from the Earth's orbit. A parsec is equivalent to 206,264.8 AU, or 3.26 light years.

The galactic disk contains 100—400 million relatively young stars and interstellar clouds, arranged in a multiarm spiral structure. At the center of the disk is the bar, a prolate spheroid about 3 kpc in radius in the plane of the disk, and with a radius of about 1.5 kpc perpendicular to the disk. The bar rotates more slowly than the disk and consists largely of densely packed older stars and interstellar clouds. It does not display spiral structure. At the center of the bar is the galactic nucleus, a complex region only 4—5 pc across (see Figure 1.15), which appears to have a super massive black hole at its center. The mass of the central black hole has been estimated at ~ 4 million $M_\odot$.

The halo surrounds both of these structures, extending ~ 30 kpc from the galactic center. The halo has an oblate spheroid shape and contains older stars and globular clusters of stars. The corona appears to be a yet more distant halo extending 60—100 kpc and consists of ionized gas and dark matter, unobservable except for the effect it has on the dynamics of observable bodies in the galaxy. The corona may be several times more massive than the other three galactic components combined.

FIGURE 1.13 NGC 1300, a barred spiral galaxy in the constellation Eridanus, as photographed by the *Hubble Space Telescope*. NGC 1300 is about ~ 34 kpc in diameter and is 18.7 Mpc from our galaxy. The Milky Way galaxy may appear similar to this.

FIGURE 1.14 The spiral structure of the Milky Way galaxy as inferred from the positions of HII regions (clouds of ionized hydrogen) in the galaxy. The Sun and solar system are located at the lower center, as indicated by the ⊙ symbol at the center of the grid lines.

in that direction, which is the view of the central bulge and bar.

The disk is not perfectly flat; there is evidence for warping in the outer reaches of the disk, between 15 and 25 kpc. The warp may be the result of gravitational perturbations due to encounters with other galaxies and/or with the Magellanic Clouds, two nearby, irregular dwarf galaxies that appear to be in orbit around the Milky Way. In addition, the Milky Way's central bar appears to be tilted relative to the plane of the galactic disk. The nonspherical shape of the bar and the tilt have important implications for understanding stellar dynamics and the long-term evolution of the galaxy.

Stars in the galactic disk have different characteristic velocities as a function of their stellar classification, and hence age. Low-mass older stars, like the Sun, have relatively high random velocities and, as a result, can move farther out of the galactic plane. Younger, more massive stars have lower mean velocities and thus smaller scale heights above and below the plane. Giant molecular clouds, the birthplace of stars, also have low mean velocities and thus are confined to regions relatively close to the galactic plane. The galactic disk rotates clockwise as viewed from "galactic north", at a relatively constant velocity of ∼220 km/s. This motion is distinctly non-Keplerian, the result of the nonspherical mass distribution in the disk. The rotation velocity for a circular galactic orbit in the galactic plane defines the Local Standard of Rest (LSR). The LSR is then used as the reference frame for describing local stellar dynamics.

The galactic disk is visible in the night sky as the Milky Way, a bright band of light extending across the celestial sphere. When examined with a small telescope, the Milky Way is resolved into thousands or even millions of individual stars and numerous nebulae and star clusters. The direction to the center of the galaxy is in the constellation Sagittarius (best seen from the southern hemisphere in June), and the disk appears visibly wider

The Sun and the solar system are located approximately 8.5 kpc from the galactic center (although some estimates put it closer at ∼7 kpc or farther at 8.7 kpc), and 5–30 pc above the central plane of the galactic disk. The Sun and the

FIGURE 1.15 The center of the Milky Way galaxy as imaged in the infrared by the Spitzer Space Telescope. The colors in the image are not real: older cool stars are blue, dust features lit up by large hot stars are shown in a reddish hue, and the bright white spot in the middle marks the site of Sagittarius A, the supermassive black hole at the center of the Galaxy. The imaged area is about 273 pc wide and 196 pc high.

solar system are moving at approximately 17–22 km/s relative to the LSR. The Sun's velocity vector is currently directed toward a point in the constellation of Hercules, approximately at right ascension 18 h 0 m, and declination +30°, known as the solar apex. Because of this motion relative to the LSR, the solar system's galactic orbit is not circular. The Sun and planets move in a quasielliptical orbit between about 8.4 and 9.7 kpc from the galactic center, with a period of revolution of about 225–250 million years. The solar system is currently close to and moving inward toward "perigalacticon", the point in the orbit closest to the galactic center. In addition, the solar system moves perpendicular to the galactic plane in a harmonic fashion, with an estimated period of 52–74 million years, and an amplitude of ±49–93 pc out of the galactic plane. (The uncertainties in the estimates of the period and amplitude of the motion are caused by the uncertainty in the amount of dark matter in the galactic disk.) The Sun and planets passed through the galactic plane about 2 to 3 million years ago, moving "northward."

The Sun and solar system are located at the inner edge of one of the spiral arms of the galaxy, known as the Orion or local arm, although also called the "Orion spur". Nearby spiral structures can be traced by constructing a three-dimensional map of stars, star clusters, and interstellar clouds in the solar neighborhood. Two well-defined neighboring structures are the Perseus arm, farther from the galactic center than the local arm, and the Sagittarius arm, toward the galactic center. The arms are about 0.5 kpc wide, and the spacing between the spiral arms is ~1.2–1.6 kpc.

The Sun's velocity relative to the LSR is low as compared with other G-type stars, which have typical velocities of 40–45 km/s relative to the LSR. Stars are accelerated by encounters with giant molecular clouds in the galactic disk. Thus, older stars can be accelerated to higher mean velocities, as noted earlier. The reason(s) for the Sun's low velocity is not known. Velocity-altering encounters with giant molecular clouds occur with a typical frequency of once every 300–500 million years.

The local density of stars in the solar neighborhood is about $0.11/pc^3$, although many of the stars are in binary or multiple star systems. The local density of binary and multiple star systems is $0.086/pc^3$. Most of these are low-mass stars, less massive and less luminous than the Sun. The star nearest to the solar system is Proxima Centauri, which is a low-mass ($M \simeq 0.1\, M_\odot$), distant companion to Alpha Centauri, which itself is a double star system of two close-orbiting solar-type stars. Proxima Centauri is currently about 1.3 pc from the Sun and about 0.06 pc (1.35×10^4 AU) from the Alpha Centauri pair it is orbiting. The second nearest star is Barnard's star, a fast-moving red dwarf at a distance of 1.83 pc. The brightest star within 5 pc of the Sun is Sirius, an A1 star ($M \simeq 2\, M_\odot$) about 2.6 pc away. Sirius is also a double star, with a faint, white dwarf companion. The stars in the solar neighborhood are shown in Figure 1.16.

The Sun's motion relative to the LSR, as well as the random velocities of the stars in the solar neighborhood, will

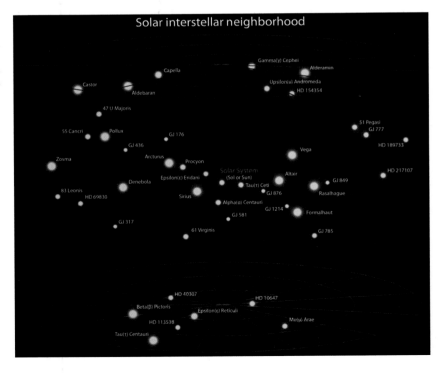

FIGURE 1.16 A three-dimensional representation of the stars in the solar neighborhood. The size of the dot representing each star denotes its relative brightness. The four circles represent radii of 1, 2, 3, and 4 light years, respectively. A light year is the distance that a photon of light travels in a year; one light year is equal to 0.307 pc.

occasionally result in close encounters between the Sun and other stars. Using the value above for the density of stars in the solar neighborhood, one can predict that ~ 12 star systems (single or multiple stars) will pass within 1 pc of the Sun per million years. The total number of stellar encounters scales as the square of the encounter distance. This rate has been confirmed in part by data from the *Hipparcos* astrometry satellite, which measured the distances and proper motions of $\sim 118,000$ stars, and which was used to reconstruct the trajectories of stars in the solar neighborhood.

Based on this rate, the closest stellar approach over the lifetime of the solar system would be expected to be at ~ 900 AU. Such an encounter would result in a major perturbation of the Oort cloud and would eject many comets to interstellar space. It would also send a shower of comets into the planetary region, raising the impact rate on the planets for a period of about 2—3 million years, and having other effects that may be detectable in the stratigraphic record on the Earth or on other planets. A stellar encounter at 900 AU could also have a substantial perturbative effect on the orbits of comets in the Kuiper belt and scattered disk and would likely disrupt the outer regions of those populations. Obviously, the effect that any such stellar passage will have is a strong function of the mass and velocity of the passing star.

Because the Sun likely formed in a star cluster, and because the Sun will move through denser regions of the galactic disk (in particular, the spiral arms), the encounter rate mentioned above is likely a lower limit and was higher at times in the past. That also means that the closest stellar encounters may have been even closer than 900 to the planetary system.

The advent of space-based astronomy, primarily through Earth-orbiting ultraviolet and X-ray telescopes, has made it possible to study the local interstellar medium surrounding the solar system. The structure of the local interstellar medium has turned out to be quite complex. The solar system appears to be on the edge of an expanding bubble of hot plasma about 120 pc in radius, which appears to have originated from multiple supernovae explosions in the Scorpius—Centaurus OB association. The Sco-Cen association is a nearby star-forming region that contains many young, high-mass O- and B-type stars. Such stars have relatively short lifetimes and end their lives in massive supernova explosions, before collapsing into black holes. The expanding shells of hot gas blown off the stars in the supernova explosions are able to "sweep" material before them, leaving a low-density "bubble" of hot plasma.

Within this bubble, known as the Local Bubble, the solar system is at this time within a small interstellar cloud, perhaps 2—5 pc across, known as the Local Interstellar Cloud. That cloud is apparently a fragment of the expanding shells of gas from the supernova explosions, and there appear to be a number of such clouds within the local solar neighborhood.

6. THE FATE OF THE SOLAR SYSTEM

Stars like the Sun are expected to have lifetimes on the main sequence of about 10^{10} years. The main sequence lifetime refers to the time period during which the star produces energy through hydrogen fusion in its core. As the hydrogen fuel in the core is slowly depleted over time, the core contracts to maintain the internal pressure. This raises the central temperature and as a result, the rate of nuclear fusion also increases and the star slowly brightens. Thus, temperatures throughout the solar system will slowly increase over time. Presumably, this slow brightening has already been going on since the formation of the Sun and solar system.

A $1\text{-}M_\odot$ star like the Sun is expected to run out of hydrogen at its core in about 10^{10} years. As the production of energy declines, the core again contracts. The rising internal temperature and pressure are then able to ignite hydrogen burning in a shell surrounding the depleted core. The hydrogen burning in the shell heats the surrounding mass of the star and causes it to expand. The radius of the star increases and the surface temperature drops. The luminosity of the star increases dramatically, and it becomes a red giant. Eventually the star reaches a brightness about 10^3 times more luminous than the present-day Sun, a surface temperature of 3000 K, and a radius of 100—200 solar radii. One hundred solar radii is equal to 0.46 AU, larger than the orbit of Mercury. Two hundred radii is just within the orbit of the Earth. Thus, Mercury and likely Venus will be incorporated into the outer shell of the red giant Sun and will be vaporized.

The increased solar luminosity during the red giant phase will result in a fivefold rise in temperatures throughout the solar system. At the Earth's orbit this temperature increase will vaporize the oceans and roast the planet at a temperature on the order of ~ 1400 K or more. At Jupiter's orbit it will melt the icy Galilean satellites and cook them at a more modest temperature of about 600 K, about the same as current noontime temperatures on the surface of Mercury. Typical temperatures at the orbit of Neptune will be about the same as they are today at the orbit of the Earth. Comets in the inner portion of the Kuiper belt will be warmed sufficiently to produce visible comae.

The lowered gravity at the surface of the greatly expanded Sun will result in a substantially increased solar wind, and the Sun will slowly lose mass from its outer envelope. Meanwhile, the core of the Sun will continue to contract until the central temperature and pressure are great enough to ignite helium burning in the core. During this time, hydrogen burning continues in a shell around the core. Helium burning continues during the red giant phase until the helium in the core is also exhausted. The star again contracts, and this permits helium burning to ignite in a shell around the core. This is an unstable situation, and the star can undergo successive contractions and reignition pulses, during which it will blow off part or all of its outer envelope into space. These huge mass ejections produce an expanding nebula around the star, known as a planetary nebula (because it looks somewhat like the disk of a giant planet through a telescope). For a star with the mass of the Sun, the entire red giant phase lasts about 7×10^8 years.

As the Sun loses mass in this fashion, the orbits of the surviving planets will slowly spiral outward. This will also be true for comets in the Kuiper belt and Oort cloud. The gravitational sphere of influence of the Sun will shrink as a result of the Sun's decreasing mass, so comets will be lost to interstellar space at the outer limits of the Oort cloud.

As a red giant star loses mass, its core continues to contract. However, for an initially 1-$M_\odot$ star like the Sun, the central pressure and temperature cannot rise sufficiently to ignite carbon burning in the core, the next phase in nuclear fusion. With no way of producing additional energy other than gravitational contraction, the luminosity of the star plunges. The star continues to contract and cool, until the contraction is halted by degenerate electron pressure in the superdense core. At this point, the mass of the star has been reduced to about 70% of its original mass and the diameter is about the same as the present-day Earth. Such a star is known as a white dwarf. The remnants of the previously roasted planets will be plunged into a deep freeze as the luminosity of the white dwarf slowly declines.

The white dwarf star will continue to cool over a period of about 10^9 years, to the point where its luminosity drops below detectable levels. Such a star is referred to as a black dwarf. A nonluminous star is obviously very difficult to detect. There is some suggestion that they may have been found through an observing technique known as microlensing events. Dark stars provide one of the possible explanations for the dark matter in the galaxy.

7. CONCLUDING REMARKS

This chapter has provided an introduction to the solar system and its varied members, viewing them as components of a large and complex system. Each of them (the Sun, the planets, their satellites, the comets and asteroids, etc.) is also a fascinating world in its own right. The ensuing chapters provide more detailed descriptions of each of these members of the solar system, as well as descriptions of important physical and dynamical processes, discussions of some of the more advanced ways we study the solar system, the search for life elsewhere in the solar system, and finally, the search for planetary systems around other stars.

BIBLIOGRAPHY

Lewis, J. S. (2004). *Physics and chemistry of the solar system* (2nd ed.). San Diego: Elsevier Academic Press.

Beatty, J. K., Petersen, C. C., & Chaikin, A. (Eds.). (1999). *The new solar system* (4th ed.). Cambridge, MA: Sky Publishing Corp.

de Pater, I., & Lissauer, J. J. (2010). *Planetary science* (2nd ed.). Cambridge University Press.

Lissauer, J. J., & de Pater, I. (2013). *Fundamental planetary science*. Cambridge University Press.

von Steiger, R., Lallement, R., & Lee, M. A. (Eds.). (1996). *The heliosphere in the local interstellar medium*. Dordrecht, The Netherlands: Kluwer.

Sparke, L. S., & Gallagher, J. S. (2000). *Galaxies in the Universe: An introduction*. Cambridge, UK: Cambridge University Press.

Chapter 2

The Origin of the Solar System

John E. Chambers
Department of Terrestrial Magnetism, Carnegie Institution of Washington, Washington, DC, USA

Alex N. Halliday
Department of Earth Sciences, University of Oxford, Oxford, UK

Chapter Outline

1. Introduction — 29
2. Star Formation and Protoplanetary Disks — 30
3. Meteorites and the Origin of the Solar System — 33
4. Nucleosynthesis and Short-lived Isotopes — 38
5. Early Stages of Planetary Growth — 41
6. Formation of Terrestrial Planets — 43
7. The Asteroid Belt — 48
8. Growth of Gas and Ice Giant Planets — 49
9. Planetary Satellites — 51
10. Extrasolar Planets — 52
11. Summary and Future Prospects — 53
Bibliography — 54

1. INTRODUCTION

The origin of the solar system has long been a fascinating subject posing difficult questions of deep significance. It takes one to the heart of the question of our origins, of how we came to be here and why our surroundings look the way they do. Unfortunately, we currently lack a self-consistent model for the origin of the solar system and other planetary systems. The early stages of planet formation are obscure and we have only a modest understanding of how much the orbits of planets change during and after their **formation**. At present, we cannot say whether **terrestrial planets** similar to the Earth are commonplace or highly unusual. Nor do we understand where the water came from that makes our planet habitable.

In the face of such uncertainty, one might ask whether we will ever understand how planetary systems form. In fact, the last 10 years have seen rapid progress in almost every area of planetary science, and our understanding of the origin of the solar system and other planetary systems has improved greatly as a result. Planetary science today is as exciting as it has been at any time since the Apollo landings on the Moon, and the coming decade looks set to continue this trend.

Some key recent developments are

1. Two decades ago, the first planet orbiting another Sun-like star was discovered. Since then, hundreds of new planets have been discovered using ground-based telescopes, and several thousand planetary candidates have been identified by the space-based Kepler mission. Most of the first planets to be found appear to be gas giants similar to Jupiter and Saturn. Recently, many smaller planets have been found, and at least some of these may be akin to terrestrial planets like Earth.
2. In the last 10 years there have been a number of highly successful space missions to other bodies in the solar system, including Mercury, Mars, and several asteroids and comets, as well as the ongoing Cassini mission to Saturn. Information and images returned from these missions have transformed our view of these objects, while spacecraft have recently obtained samples of an asteroid, a comet, and particles from the **solar wind**. All this information is greatly enhancing our understanding of the origin and evolution of the solar system.
3. The discovery that one can physically separate and analyze stardust—presolar grains that can be extracted from **meteorites** and that formed in the envelopes of other stars, has meant that scientists can for the first

time test decades of theory on how stars work. The parallel development of methods for extracting isotopic information at the submicron scale has opened up a new window to the information stored in such grains.

4. The development of multiple collector inductively coupled plasma mass spectrometry has made it possible to use new isotopic systems for determining the mechanisms and timescales for the growth of bodies early in the solar system.
5. Our theoretical understanding of planet formation has advanced substantially in several areas, including new models for the rapid growth of giant planets, a better understanding of the physical and chemical evolution of protoplanetary disks, and the growing realization of the ways in which planets can migrate substantially during and after their formation.
6. Powerful new computer codes and **equations of state** have been developed recently, which make it possible to make realistic, high-resolution simulations of collisions between planet-sized bodies. These developments are greatly improving the realism of models for planetary growth, and may offer the solution to some long-standing puzzles about the origin of Mercury, the Moon, and asteroids.

Today, the formation of the solar system is being studied using three complementary approaches.

- Astronomical observations of protoplanetary disks around young stars are providing valuable information about probable conditions during the early history of the solar system and the timescales involved in planet formation. The discovery of new planets orbiting other stars is adding to the astonishing diversity of possible planetary systems, and providing additional tests for theories of how planetary systems form.
- Physical, chemical, and isotopic analysis of meteorites and samples returned by space missions is generating important information about the formation and evolution of objects in the solar system and their constituent materials. This field of *cosmochemistry* has taken off in several important new directions in recent years, including the determination of timescales involved in the formation of the terrestrial planets and asteroids, and constraints on the origin of the materials that make up the Solar System.
- Theoretical calculations and numerical simulations are being used to examine every stage in the formation of the solar system. These provide valuable insights into the complex interplay of physical and chemical processes involved, and help to fill in some of the gaps when astronomical and cosmochemical data are unavailable.

In this chapter we will describe what we currently know about how the solar system formed, and highlight some of the main areas of uncertainty that await future discoveries.

2. STAR FORMATION AND PROTOPLANETARY DISKS

The solar system formed 4.5–4.6 billion years ago by collapse of a portion of a **molecular cloud** composed of gas and dust, rather like the Eagle or Orion Nebulae. Some of the stardust from that ancient nebula has now been isolated from primitive meteorites. Their isotopic compositions are vastly different from those of our own solar system and provide fingerprints of nearby stars that preceded our Sun. These include red giants, asymptotic giant branch (AGB) stars, supernovae and novae. From studying modern molecular clouds it has also become clear that stars like our Sun can form in significant numbers in close proximity to each other. Such observation also provide clues as to how own solar system formed because they have provided us with images of circumstellar disks—the environments in which planetary objects are born.

Observations from infrared telescopes such as the Spitzer Space Telescope have shown that many young stars give off more infrared radiation than would be expected for blackbodies of the same size. This *infrared excess* comes from micron-sized grains of dust orbiting the star in an *optically thick* (opaque) disk. Dark, dusty disks can be seen with the Hubble Space Telescope surrounding some young stars in the Orion Nebula (Figure 2.1). These disks have been dubbed *proplyds*, short for *protoplanetary disks*. It is thought that protoplanetary disks are mostly composed of gas, especially hydrogen and helium, and in a few cases this gas has been detected, although gas is generally much harder to see than dust. The fraction of stars having a massive disk declines with stellar age, and large infrared excesses are rarely seen in stars older than 10 Myrs. In some cases, such as the disk surrounding the

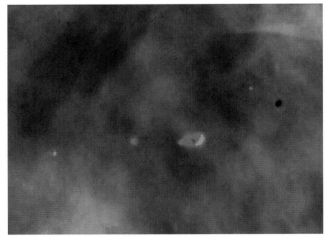

FIGURE 2.1 Proplyds are young stellar objects embedded in an optically dense envelope of gas and dust. The objects shown here are from the Orion Nebula.

Chapter | 2 The Origin of the Solar System

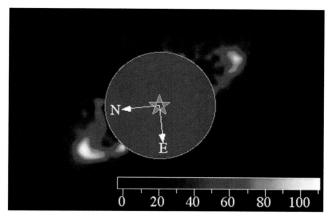

FIGURE 2.2 The circumstellar disk surrounding HR 4796A as revealed by interferometry measurements of the infrared excess. Note the area close into the star swept clear of dust, which has presumably been incorporated into planetary objects.

star HR 4796A, there are signs that the inner portion of a disk has been cleared of dust (Figure 2.2), perhaps due to the presence of one or more planets.

Roughly half of stars up to a few hundred million years old have low-mass, optically thin (nearly transparent) disks containing some dust but apparently little or no gas. In a few cases, such as the star Beta Pictoris, the disk can be seen at visible wavelengths if the glare from the star itself is blocked. Dust grains in these disks will be quickly accelerated out of the system by the pressure of radiation from the central star, or destroyed by high-speed collisions with other grains. Any primordial dust should have been removed on a timescale that is short compared to the age of the star. For this reason, the dust in these disks is thought to be second-generation material formed by collisions between asteroids or sublimation from comets orbiting these stars in more massive analogues of the **Kuiper belt** in our own solar system. These are often referred to as *debris disks* since asteroids and comets are presumed to be debris left over from planet formation. In a few cases, such as Beta Pictoris, a planet has been discovered orbiting the same star, reinforcing the link between disks and planetary systems.

In the solar system, the planets all orbit the Sun in the same direction, and their orbits are very roughly coplanar. This suggests the solar system originated from a disk-shaped region of material referred to as the ***solar nebula***, an idea going back more than two centuries to Kant and later Laplace. The discovery of disks of gas and dust around many young stars provides strong support for this idea, and implies that planet formation is associated with the formation of stars themselves. Stars typically form in clusters of a few hundred to a few thousand objects in dense regions of the interstellar medium called *molecular clouds* (see Figure 2.3). The gas in molecular clouds is cold (roughly 10 K) and dense compared to that in other regions of space (roughly 10^4 atoms per cubic centimeter) but still much more tenuous than the gas in a typical laboratory "vacuum". Stars in these clusters are typically separated by

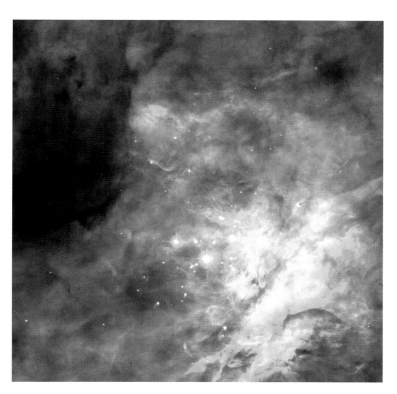

FIGURE 2.3 This Hubble Space Telescope image of the Orion Nebula shows molecular clouds of gas and dust illuminated by radiation from young stars. Some early stars appear shrouded in dusty disks (see Figure 2.1). Scientists think that our solar system formed by collapse of a portion of a similar kind of molecular cloud leading to formation of a new star embedded in a dusty disk. How that collapse occurred is unclear. It may have been triggered by a shock wave carrying material being shed from another star such as an AGB star or supernova.

about 0.1 **parsecs** (0.3 light-years), much less than the distance between stars in the Sun's neighborhood.

It is unclear precisely what causes the densest portions of a molecular cloud (called *molecular cloud cores*) to collapse to form stars. It may be that contraction of a cloud core is inevitable sooner or later due to the gravitational attraction of material in the core, or an external event may cause the *triggered collapse* of a core. The original triggered collapse theory was based on the sequencing found in the ages of stars in close proximity to one another in molecular clouds. This suggests that the formation and evolution of some stars triggered the formation of additional stars in neighboring regions of the cloud. However, several other triggering mechanisms are possible, such as the impact of energetic radiation and gas ejected from other newly formed stars, the effects of a nearby, pulsating AGB star, or a shock wave from the supernova explosion of a massive star.

Gas in molecular cloud cores is typically moving. When a core collapses, the gas has too much **angular momentum** for all the material to form a single, isolated star. In many cases a binary star system forms instead. In others cases, a single protostar forms (called a *T Tauri* star or *pre-main sequence star*), while a significant fraction of the gas goes into orbit about the star forming a disk that is typically 100 **astronomical units (AU)** in diameter. Temperatures in T Tauri stars are initially too low for nuclear reactions to take place. However, T Tauri stars are much brighter than older stars like the Sun due to the release of gravitational energy as the star contracts. The initial collapse of a molecular cloud core takes roughly 10^5 years, and material continues to fall onto both the star and its disk until the surrounding molecular cloud core is depleted.

The spectra of T Tauri stars contain strong ultraviolet and visible emission lines caused by hot gas falling onto the star. This provides evidence that disks lose mass over time as material moves inward through the disk and onto the star, a process called *viscous accretion*. This process provides one reason why older stars do not have disks, the other reason being planet formation itself. Estimated disk accretion rates range from 10^{-6} to 10^{-9} solar masses per year. The mechanism responsible for viscous accretion is unclear. A promising candidate is *magnetorotational instability* (MRI), in which partially ionized gas in the disk becomes coupled to the local magnetic field. Because stars rotate, the magnetic field sweeps around rapidly, increasing the orbital velocity of material that couples strongly to it and moving it outward. Friction causes the remaining material to move inward. As a result, a disk loses mass to its star and spreads outward over time. This kind of disk evolution explains why the planets currently contain only 0.1% of the mass in the solar system but have retained more than 99% of its angular momentum. MRI requires a certain fraction of the gas to be ionized, and it may not be effective in all portions of a disk, creating so-called dead zones where material flows inward more slowly and the gas becomes denser. Disks are also eroded over time by photoevaporation. In this process, gas is accelerated when atoms absorb ultraviolet photons from the central star or nearby, energetic stars, until the gas is moving fast enough to escape into interstellar space.

T Tauri stars often have jets of material moving rapidly away from the star perpendicular to the plane of the disk. These jets are powered by the inward accretion of material through the disk coupled with the rotating magnetic field. Outward flowing winds also arise from the inner portions of a disk. T Tauri stars are strong emitters of X-rays, generating fluxes up to 10^4 times greater than that of the Sun during the strongest solar **flares**. Careful sampling of large populations of young solar mass stars in the Orion Nebula shows that this is normal behavior in young stars. This energetic flare activity is strongest in the first million years and declines at later times, persisting for up to 10^8 years. From this it has been concluded that the young Sun generated 10^5 times as many energetic protons as today. It is thought that reactions between these protons and material in the disk may have provided some of the short-lived isotopes whose daughter products are seen today in meteorites, although the formation of most of these isotopes predate that of the solar system (see Section 4).

The minimum mass of material that passed through the solar nebula can be estimated from the total mass of the planets, asteroids, and comets in the solar system. However, all these objects are depleted in hydrogen and helium relative to the Sun. Ninety percent of the mass of the terrestrial planets is made up of oxygen, magnesium, silicon, and iron (Figure 2.4), and while Jupiter and Saturn are mostly composed of hydrogen and helium, they are enriched in the heavier elements compared to the Sun. When the missing hydrogen and helium is added, the *minimum-mass solar nebula* (MMSN) turns out to be 1–2% of the Sun's mass. The major uncertainties in this

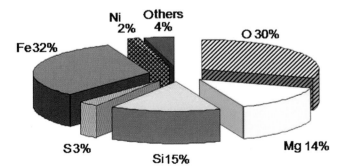

FIGURE 2.4 Pie chart showing the bulk composition of the Earth. Most of the iron (Fe), nickel (Ni), and sulfur (S) are in Earth's core, while the silicate Earth mostly contains magnesium (Mg), silicon (Si), and oxygen (O) together with some iron.

number come from the fact that the interior compositions of the giant planets and the initial mass of the Kuiper belt are poorly known. Not all this mass necessarily existed in the nebula at the same time, but it must have been present at some point. Current theoretical models predict that planet formation is an inefficient process, with some mass falling into the Sun or being ejected into interstellar space, so the solar nebula was probably more massive than the MMSN.

Gas in the solar nebula became hotter as it viscously accreted toward the Sun, releasing gravitational energy and absorbing sunlight. The presence of large amounts of dust meant the inner portions of the nebula were optically thick to infrared radiation so these regions held on to much of this heat. Numerical disk models show that temperatures probably exceeded 1500 K in the terrestrial-planet forming region early in the disk's history. Viscous heating mainly took place at the disk midplane where most of the mass was concentrated. The surfaces of the disk would have been much cooler. The amount of energy generated by viscous accretion declined rapidly with distance from the Sun. In the outer nebula, solar irradiation was the more important effect. Protoplanetary disks are thought to be *flared*, so that their vertical thickness grows more rapidly than their radius, As a result, the surface layers are always irradiated by the central star. For this reason, the surface layers of the outer solar nebula may have been warmer than the midplane.

The nebula cooled over time as the viscous accretion rate declined and dust was swept up by larger bodies, reducing the **optical depth**. In the inner nebula, cooling was probably rapid. Models show that at the midplane at 1 AU, the temperature probably fell to about 300 K after 10^5 years. Because the energy generated by viscous accretion and solar irradiation declined with distance from the Sun, disk temperatures also declined with heliocentric distance. At some distance from the Sun, temperatures became low enough for water ice to form, a location referred to as the *ice line*. Initially, the ice line may have been 5—6 AU from the Sun, but it moved inward over time as the nebula cooled. Some asteroids contain **hydrated** minerals formed by reactions between water ice and dry rock. This suggests water ice was present when these asteroids formed, in which case the ice line would have been no more than 2—3 AU from the Sun at the time.

Meter-sized icy bodies drifted rapidly inward through the solar nebula due to **gas drag** (see Section 5). When these objects crossed the ice line they would have evaporated, depositing water vapor in the nebular gas. As a result, the inner nebula probably became more oxidizing over time as the level of oxygen from water increased. When the flux of drifting particles dwindled, the inner nebula may have become chemically reducing again, as water vapor diffused outward across the ice line, froze to form ice, and became incorporated into growing planets.

3. METEORITES AND THE ORIGIN OF THE SOLAR SYSTEM

Much of the above is based on theory and observations of other stars. To find out how our own solar system formed it is necessary to study *meteorites and* **interplanetary dust particles** (**IDPs**). These are fragments of rock and metal from other bodies in the solar system that have fallen to Earth and survived passage through its atmosphere. Meteorites and IDPs tend to have broadly similar compositions, and the difference is mainly one of size. IDPs are much smaller of the two, typically 10—100 μm in diameter, while meteorites can range up to several meters in size. Most such objects are quite unlike any objects formed on Earth. Therefore, we cannot readily link them to natural present-day processes as earth scientists do when unraveling past geological history. Yet the approaches that are used are in some respects very similar. The research that is conducted on meteorites and IDPs is dominated by two fields: petrography and **geochemistry**. Petrography is the detailed examination of mineralogical and textural features. Geochemistry uses the isotopic and chemical compositions. This combined approach to these fascinating archives has provided a vast amount of information on our Sun and solar system and how they formed. We know about the stars and events that predated formation of the Sun, the nature of the material from which the planets were built, the solar nebula, the timescales for planetary accretion, and the interior workings and geological histories of other planets. Not only these, meteorites provide an essential frame of reference for understanding how our own planet Earth formed and **differentiated**.

The geochemistry of meteorites and IDPs provides evidence that the Sun's protoplanetary disk as well as the planets it seeded had a composition that was similar in some respects to that of the Sun itself (Figure 2.5). In other respects, however, it is clear the disk was a highly modified residuum that generated a vast range of planetary compositions. The composition of the Sun can be estimated from the depths of lines associated with each element in the Sun's spectra (although this is problematic for the lightest elements and the noble gases). Today, the Sun contains almost 99.9% of the total mass of the solar system. A sizable fraction of this material passed through the solar nebula at some point, which tells us that the composition of the original nebula would have been similar to that of the Sun today. The challenge is therefore to explain how it is possible that a disk that formed gas giant objects like Jupiter and Saturn with compositions like the Sun, also generated rocky terrestrial planets like the Earth (Figure 2.4).

Most meteorites are thought to come from *parent bodies* in the main asteroid belt that formed during the first few million years of the solar system. As a result, these objects

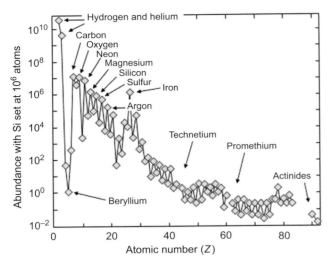

FIGURE 2.5 The abundances of elements in our Sun and solar system is estimated from the spectroscopic determination of the composition of the Sun and the laboratory analysis of primitive meteorites called carbonaceous chondrites—thought to represent unprocessed dust and other solid debris from the circumsolar disk. To compare the abundances of different elements it is customary to scale the elements relative to 1 million atoms of silicon. The pattern provides powerful clues to how the various elements were created. See text for details. *Based on a figure in Broecker, W. S. How to build a habitable planet with kind permission.*

carry a record of processes that occurred in the solar nebula during the formation of the planets. In a few cases, the trajectories of falling meteorites have been used to establish that they arrived on orbits coming from the asteroid belt. Most other meteorites are deduced to come from asteroids based on their age and composition. IDPs are thought to come from both asteroids and comets. A few meteorites did not originate in the asteroid belt. The young ages and abundances of the noble gases trapped inside the Shergottite-Nakhlite-Chassignite (SNC) meteorites suggest they come from Mars. Roughly a 100 SNC meteorites have been found to date, and a comparable number of *lunar meteorites* from the Moon are also known.

The Earth is currently accumulating meteoritic material at the rate of about 5×10^7 kg per year. At this rate it would take more than 10^{17} years to obtain the Earth's current mass of 5.97×10^{24} kg, which is much longer than the age of the universe. While it is thought that the Earth did form as the result of the accumulation of smaller bodies, it is clear that the rate of impacts was much higher while the planets were forming than it is today.

Broadly speaking meteorites can be divided into three types: **chondrites, achondrites**, and irons, which can be distinguished as follows:

1. **Chondrites** are mixtures of grains from submicron-sized dust to millimeter- to centimeter-sized particles of rock and metal, apparently assembled in the solar nebula. Most elements in chondrites are present in broadly similar ratios to those in the Sun, with the exception of carbon, nitrogen, hydrogen, and the noble gases, which are all highly depleted. For this reason chondrites have long been viewed as representative of the dust and debris in the circumstellar disk from which the planets formed. So, for example, **refractory** elements that would have resided in solid phases in the circumstellar disk have chondritic (and therefore solar) relative proportions in the Earth, even though the **volatile** elements are vastly depleted. The nonmetallic components of chondrites are mostly **silicates** such as olivine and pyroxene. *Chondrules* are a major component of most chondrites (see Figure 2.6). These are roughly millimeter-sized rounded beads of rock that formed by melting, either partially or completely. Their mineral-grain textures suggest they cooled over a period of a few hours, presumably in the nebula, with the heating possibly caused by passage through shock waves in the nebular gas. Some chondrules are thought to have formed later in collisions between planetary objects. Most chondrites also contain *calcium–aluminum-rich inclusions* (CAIs, see Figure 2.7), which have chemical compositions similar to those predicted for objects that condensed from a gas of roughly solar composition at very high temperatures. It is possible that CAIs formed in the very innermost regions of the solar nebula close to the Sun. Dating based on radioactive isotopes suggest that CAIs are the oldest surviving materials to have formed in the solar system. CAIs in the Efremovka chondrite are 4.5673 ± 0.0002 Ga old based on the $^{235/238}$U-$^{207/206}$Pb system, and this date is often used to define the canonical start to the solar system. The oldest chondrules appear to have

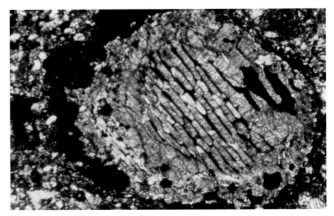

FIGURE 2.6 Chondrules are spherical objects, sometimes partly flattened and composed of **mafic** silicate minerals, metals, and oxides. They are thought to form by sudden (flash) heating in the solar nebula. Some formed as much as 2 million years after the start of the solar system. *Photograph courtesy of Drs M. Grady and S. Russell and the Natural History Museum, London.*

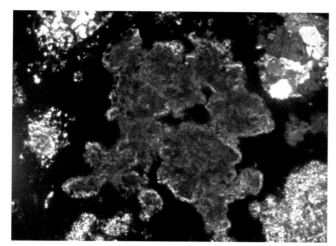

FIGURE 2.7 Calcium–aluminum refractory inclusions are found in chondrite meteorites and are thought to be the earliest objects that formed within our solar system. They have a chemical composition consistent with condensation from a hot gas of solar composition. How they formed exactly is unclear but some have suggested they were produced close in to the Sun and then scattered across the disk. *Photograph courtesy of Drs M. Grady and S. Russell and the Natural History Museum, London.*

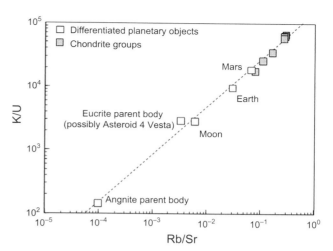

FIGURE 2.9 Comparison between the K/U and Rb/Sr ratios of the Earth and other differentiated objects compared with chondrites. The alkali elements K and Rb are both relatively volatile compared with U and Sr, which are refractory. Therefore, these trace element ratios provide an indication of the degree of volatile element depletion in inner solar system differentiated planets relative to chondrites, which are relatively primitive. It can be seen that the differentiated objects are more depleted in moderately volatile elements than are chondrites. *Based on a figure that first appeared in Halliday, A. N., & Porcelli, D. (2001).*

formed at about the same time, but most chondrules are 1–3 million years younger than this (Figure 2.8). The space between the chondrules and CAIs in chondrites is filled with fine-grained dust called *matrix*. Most chondrites are variably depleted in moderately volatile elements like potassium (K) and rubidium (Rb) (Figure 2.9). This depletion is more a feature of the chondrules and CAIs than the matrix. Chondrites are subdivided into groups of like objects thought to come originally from the same parent body. Currently, about 15 groups are firmly established, half of which are collectively referred to as *carbonaceous chondrites*.

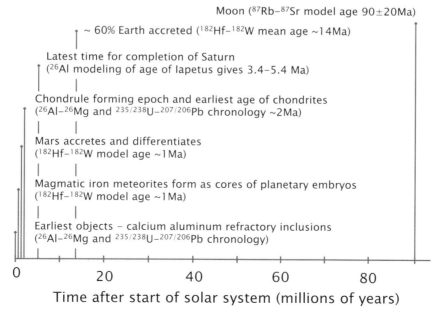

FIGURE 2.8 The current best estimates for the timescales over which very early inner solar system objects and the terrestrial planets formed. The approximated mean life of accretion is the time taken to achieve 63% growth at exponentially decreasing rates of growth. *Based on a figure that first appeared in Halliday, A. N., & Kleine, T. (2006).*

These tend to be richer in highly volatile elements such as carbon and nitrogen compared to other chondrites, although as with all meteorites these elements are less abundant than they are in the Sun. *Ordinary chondrites* are more depleted in certain volatile elements than carbonaceous chondrites, and are largely made of silicates and metal grains. *Enstatite chondrites* are similar but highly reduced. Chondrules are absent from the most primitive, volatile-rich group of carbonaceous chondrites (the CI group), either because their parent body formed entirely from matrixlike material or because chondrule structures have been erased by subsequent reactions with water in the parent body. Chondrites also contain *presolar grains* that are submicron grains that are highly anomalous isotopically and have compositions that match those predicted to form by condensation in the outer envelopes of various stars. These represent a remarkable source of information on stellar **nucleosynthesis** and can be used to test theoretical models.

2. **Achondrites** are silicate-rich mafic and ultramafic igneous rocks not too dissimilar from those forming on Earth but with slightly different chemistry and isotopic compositions. They clearly represent the near-surface rocks of planets and asteroids that have melted and differentiated. A few achondrites come from asteroids that appear to have undergone only partial differentiation. In principle, it is possible to group achondrites and distinguish which planet or asteroid they came from. The oxygen isotopic composition of a meteorite is particularly useful in this respect. Isotopically, oxygen is extremely heterogeneous in the solar system, and planets that formed in different parts of the nebula seem to have specific oxygen isotope compositions. This makes it possible to link all the Martian meteorites together, for example (Figure 2.10). These meteorites are specifically linked to Mars because nearly all of them are too young to have formed on any asteroid; they had to come from an object that was large enough to be geologically active in the recent past. This was confirmed by a very close match between the composition of the atmosphere measured with the Viking lander and that measured in fluids trapped in alteration products in Martian meteorites. In fact, Martian meteorites provide an astonishing archive of information into how Mars formed and evolved as discussed in Section 6. To date, only one asteroidal source has been positively identified: Vesta, whose spectrum and orbital location strongly suggest it is the source of the howardite, eucrite, and diogenite (HED) meteorites.

3. **Irons** (see Figure 2.11) are largely composed of iron, nickel (about 10% by mass), and sulfides, together with

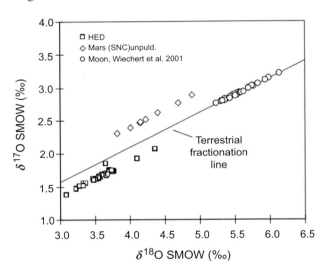

FIGURE 2.10 The oxygen isotopic composition of the components in chondrites, in particular CAIs, is highly heterogeneous for reasons that are unclear. (The figure shows deviations in parts per thousand relative to Earth's oceans or SMOW - standard mean ocean water.) The net result of this variability is that different planets possess distinct oxygen isotopic compositions that define an individual mass **fractionation** lines as shown here for eucrites, howardites, and diogenites, which come from Vesta and Martian (SNC) meteorites, thought to come from Mars. The Moon is thought to have formed from the debris produced in a giant impact between the proto-Earth when 90% formed and an impacting Mars-sized planet sometimes named "Theia". The fact that the data for lunar samples are collinear with the terrestrial fractionation line could mean that the Moon formed from the Earth, or the planet from which it was created was formed at the same heliocentric distance, or it could mean that the silicate reservoirs of the two planets homogenized during the impact process, for example, by mixing in a vapor cloud from which lunar material condensed. *From Halliday, A. N. (2003).*

FIGURE 2.11 Iron meteorites are the most abundant kind of meteorite found because they are distinctive and survive long after other kinds of meteorite are destroyed by weathering. In contrast chondrites are the most abundant class of meteorite observed to fall. Some iron meteorites are thought to represent disrupted fragments of planetesimal cores. Others appear to have formed at low pressures, probably as metal-rich pools formed from impacts on asteroids. The Henbury meteorite shown here is a type IIIAB magmatic iron that fell near Alice Springs, Australia, about 5000 years ago. The texture shown on the sawn face are Windmanstatten patterns formed by slow cooling, consistent with an origin from a core located deep within a meteorite parent body. *Photograph courtesy of Drs M. Grady and S. Russell and the Natural History Museum, London.*

other elements that have a chemical affinity for iron, called *siderophile* elements. Like chondrites, irons can be grouped according to their likely parent body, and several dozen groups or unique irons have been found. The textures of mineral grains in iron meteorites have been used to estimate how quickly their parent bodies cooled, and thus the depth at which they formed. It appears that most irons are samples of metallic **cores** of small asteroidal parent bodies, 10–100 km in radius. These appear to have formed very early, probably within a million years of CAIs, when there was considerable heat available from decay of short-lived radioactive nuclides (see Section 6). Others appear to have formed by impact melting at the surface of asteroids and these are later. A rare class of *stony-iron* meteorites (amounting to about 5% of all nonchondritic meteorites) called *pallasites* contain an intricate mixture of metal and silicate (Figure 2.12). It is thought these come from the core—**mantle** boundary regions of differentiated asteroids that broke up during a collision.

Note that there are no clear examples of mantle material within meteorite collections. The isotopic compositions of some elements in irons reveal that they have been exposed to cosmic rays for long periods—up to hundreds of millions of years. This means their parent bodies broke up a long time ago. Because they are extremely hard they survived the collisions that destroyed their parent body as well as any subsequent impacts. In contrast fragments of mantle material (as with samples excavated by volcanoes on the Earth) are extremely friable and are more easily disrupted by collisions.

Survivability is also an issue for meteorites entering Earth's atmosphere and being recovered in recognizable form. Chondrites and achondrites are mainly composed of silicates that undergo physical and chemical alterations on the surface of Earth more rapidly than the material in iron meteorites. Furthermore, iron meteorites are highly distinctive, so they are easier to recognize than stony meteorites. For this reason, most meteorites found on the ground are irons, whereas most meteorites that are seen to fall from the sky (referred to as *falls*) are actually chondrites. Most falls are ordinary chondrites, which probably reflects the fact that they survive passage through the atmosphere better than the weaker carbonaceous chondrites. The parent bodies of ordinary chondrites may also have orbits in the asteroid belt that favor their delivery to Earth. IDPs are less prone to destruction during passage through the atmosphere than meteorites so they probably provide a less biased sample of the true population of interplanetary material. Most IDPs are compositionally similar to carbonaceous rather than ordinary chondrites and this suggests that the asteroid belt is dominated by carbonaceous-chondrite like material.

Mass spectrometric measurements on meteorites and lunar samples provide evidence that the isotopes of most elements are present in similar proportions in the Earth, Moon, Mars, and the asteroids. The isotopes of elements heavier than hydrogen and helium were made by nucleosynthesis in stars which generate extremely anomalous isotopic compositions compared to the solar system. Since the solar nebula probably formed from material from a variety of sources, the observed isotopic homogeneity was originally interpreted as indicative that the inner solar nebula was very hot and planetary material condensed from a ~2000 K gas of solar composition. However, a variety of observations including the preservation of presolar grains in chondrites suggest that the starting point of planet formation was cold dust and gas. This homogeneity is therefore nowadays interpreted as indicating that the inner nebula was initially turbulent, allowing dust to become thoroughly mixed. CAIs sometimes contain nucleosynthetic isotopic anomalies. This suggests that CAIs sampled varied proportions of the isotopes of the elements before they became homogenized in the swirling disk. With improved mass spectrometric measurements evidence has been accumulating for small differences in isotopic composition in some elements between certain meteorites and those of the Earth and Moon. This area of study that searches for nucleosynthetic isotopic heterogeneity in the solar system is ongoing and is now providing a method for tracking the provenance of different portions of the disk.

However, oxygen and the noble gases are very different in this respect. Extreme isotopic variations have been found for these elements. The different oxygen and noble gas isotope ratios provide evidence of mixing between compositions of dust and those of volatile (gaseous) components. Some of this mixing may have arisen later when

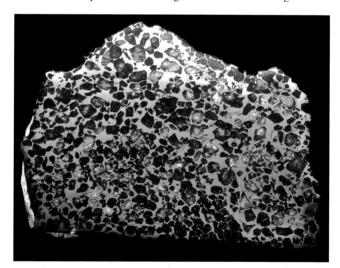

FIGURE 2.12 The pallasite Esquel is a mixture of silicate (olivine) and iron metal that may have formed at a planetary core/mantle boundary. *Photograph courtesy of Drs M. Grady and S. Russell and the Natural History Museum, London.*

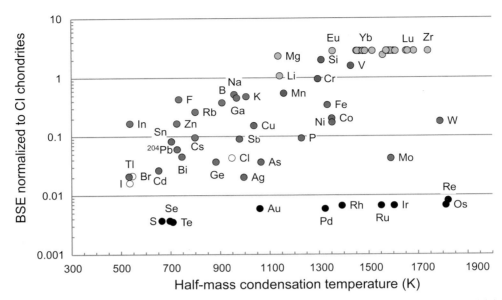

FIGURE 2.13 The estimated composition of the silicate portion of the Earth as a function of the calculated temperature at which half the mass of the element would have condensed. The concentrations of the various elements are normalized to the average composition of the solid matter in the disk as represented by CI carbonaceous chondrites. White: atmophile elements; yellow: refractory **lithophile** elements; blue: moderately volatile lithophile elements; green: slightly siderophile elements; red: moderately siderophile and chalcophile elements; black: highly siderophile elements. It can be seen that refractory lithophile elements are enriched relative to CI concentrations. This is because of core formation and volatile losses compared with CI chondrites. The moderately volatile lithophile elements like K are depleted because of loss of volatiles. Siderophile and chalcophile elements are depleted by core formation. However, the pattern of depletion is not as strong as expected given the ease with which these elements should enter the core. The explanation is that there was addition of a late veneer of chondritic material to the silicate Earth after core formation. *From Halliday, A. N. (2003).*

the nebula cooled, possibly because large amounts of isotopically distinct material are thought to have arrived from the outer nebula in the form of water ice. There are also possibilities for generating some of the heterogeneity in oxygen by irradiation within the solar nebula itself. Samples of the solar wind obtained by the Genesis space mission suggest that the oxygen isotope composition of the solar nebula changed over time as the first stages of planet formation took place.

The terrestrial planets and asteroids are not just depleted in nebular gas relative to the Sun. They are also very depleted in *moderately volatile elements* (elements such as lead, potassium, and rubidium that condense at temperatures in the range 700–1350 K) (Figure 2.9 and 2.13). In chondritic meteorites, the degree of depletion becomes larger as an element's condensation temperature decreases. It was long assumed that this is the result of the loss of gas from a hot nebula before it cooled. For example, by the time temperatures became cool enough for lead to condense, much of the lead had already accreted onto the Sun as a gas. However, it is clear that moderately volatile elements are depleted in chondrites at least in part because they contain CAIs and chondrules that lost volatiles by evaporation during heating events. The least depleted chondrites (CI carbonaceous chondrites) contain no CAIs or chondrules. Another mechanism for losing moderately volatile elements is planetary collisions. Energetic collisions between large bodies would have generated high temperatures and could have caused further loss of moderately volatile elements. For this reason, the terrestrial planets have compositions that differ from one another and also from chondritic meteorites. The Moon is highly depleted in moderately volatile elements (Figure 2.9) and is thought to be the product of such an energetic planetary collision.

4. NUCLEOSYNTHESIS AND SHORT-LIVED ISOTOPES

With the exception of hydrogen and helium, the elements we see in the solar system were mainly made by nuclear reactions in the interiors of other stars, a process called *stellar nucleosynthesis*. If one examines Figure 2.5, seven rather striking features stand out.

- The estimated abundances of the elements in the Sun and the solar nebula span a huge range of 13 orders of magnitude. For this reason they are most easily compared by plotting on a log scale of relative abundance such that the number of atoms of Si is 10^6.
- Hydrogen and helium are by far the most abundant elements in the Sun, as they are elsewhere in the

universe. These two elements were made from subatomic particles shortly after the Big Bang.
- The abundances of the heavier elements generally decrease with increasing atomic number. This is because most of the elements are themselves formed from lighter elements by stellar nucleosyntheis.
- Iron is about 1000 times more abundant than its neighbors in the periodic table because the binding energy of an atomic nucleus is highest for iron. This provided enhanced stability for iron nuclei during nucleosynthesis.
- Lithium, beryllium, and boron are all relatively underabundant compared to other light elements because they are unstable in stellar interiors.
- A saw-toothed variability is superimposed on the overall trend reflecting the relatively high stability of even-numbered isotopes compared to odd-numbered ones.
- All the elements in the periodic table are present in the solar system except those with no long-lived or stable isotopes, viz. technetium (Tc), promethium (Pm), and the transuranic elements.

Elements lighter than iron can be made by nuclear fusion because the process of combining two nuclei to make a heavier nucleus releases energy for elements up to and including iron. Fusion provides the main source of energy in stars, and is activated when the central pressure exceeds a critical threshold, i.e. when a star reaches a certain mass. Larger stars exert more pressure on their cores such that fusion reactions proceed more quickly. Massive stars shine more brightly than small stars, and have shorter lifetimes as a result. When a star has converted all the hydrogen in its core to helium, nuclear reactions will cease if the star is small, or proceed to the next fusion cycle such as the conversion of helium to carbon if the star is sufficiently massive to drive this reaction. Lithium, beryllium, and boron are unstable at the temperatures and pressures of stellar interiors, and they are rapidly consumed. Small amounts of these elements are made by spallation reactions from heavier elements by irradiation in the outer portions of stars.

Nearly all nuclides heavier than iron have to be made by **neutron** irradiation because their synthesis via fusion would consume energy. Nuclear reactions in stars generate large numbers of neutrons, and these neutrons are readily absorbed by atoms since they are not repelled by the nuclei's electrical charge. Neutron addition continues until an unstable isotope is made that decays to an isotope of another element, which then receives more neutrons until another unstable nuclide is made and so forth. These are *s-process* isotopes (produced by the *slow* but continuous production of neutrons in stars). However, some of these isotopes cannot be made simply by adding a neutron to a stable nuclide because there is no stable isotope with a suitable mass. Such nuclides are instead created with a very high flux of neutrons such that unstable nuclides produced by neutron irradiation receive additional neutrons before they have time to decay, jumping the gap to very heavy nuclides. These are *r-process* isotopes (produced by a *rapid* burst of neutrons). Such extremely high fluxes of neutrons are generated in supernova explosions and in particular in the cores of very large stars (e.g. 25 solar masses).

The composition of the Sun and solar system represents the cumulative ~8 billion year previous history of such stellar processes in this portion of the galaxy prior to collapse of the solar nebula (Figure 2.14). It is unknown how constant these processes were. However, the isotopes of some elements in meteorites provide evidence that stellar nucleosynthesis was still going on just prior to the formation of the solar nebula. In fact the formation of the solar system may have been triggered by material being ejected from a massive star as it was exploding, seeding the solar nebula with freshly synthesized nuclides.

Chondrites show evidence that they once contained short-lived radioactive isotopes probably produced in massive stars shortly before the solar system formed. As already pointed out, most stable isotopes are present in the same ratios in the Earth, the Moon, Mars, and different groups of meteorites, which argues that material in the solar nebula was thoroughly mixed at an early stage. However, a few isotopes such as ^{26}Mg are heterogeneously distributed in chondrites. In most cases, these isotopes are

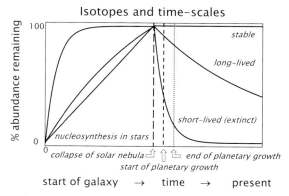

FIGURE 2.14 Most solar system nuclides heavier than hydrogen and helium were produced in stars over the history of our galaxy. This schematic figure shows the difference between nuclides that are stable, those that have very long half-lives (such as ^{238}U used for determining the ages of geological events and the solar system itself), and those that have short half-lives of $<10^8$ years, assuming all were produced at a constant rate through the history of the galaxy. The short-lived nuclides decay very fast and provide crucial insights into the timescales of events, including planet formation, immediately following their incorporation into the solar nebula.

the daughter products of short-lived isotopes. In other words, the excess ^{26}Mg comes from the radioactive decay of ^{26}Al. Every atom of ^{26}Al decays to a daughter atom of ^{26}Mg; therefore

$$\left(^{26}\text{Mg}\right)_{\text{today}} = \left(^{26}\text{Mg}\right)_{\text{original}} + \left(^{26}\text{Al}\right)_{\text{original}} \quad (2.1)$$

Because it is easier to measure these effects using isotopic ratios rather than absolute numbers of atoms we divide by another isotope of Mg:

$$\left(\frac{^{26}\text{Mg}}{^{24}\text{Mg}}\right)_{\text{today}} = \left(\frac{^{26}\text{Mg}}{^{24}\text{Mg}}\right)_{\text{original}} + \left(\frac{^{26}\text{Al}}{^{24}\text{Mg}}\right)_{\text{original}} \quad (2.2)$$

However, the ^{26}Al is no longer extant and so cannot be measured. For this reason we convert Eqn (2.2) to a form that includes a monitor of the amount of ^{26}Al that would have been present determined from the amount of Al today. Aluminum has only one stable nuclide ^{27}Al. Hence, Eqn (2.2) becomes

$$\left(\frac{^{26}\text{Mg}}{^{24}\text{Mg}}\right)_{\text{today}} = \left(\frac{^{26}\text{Mg}}{^{24}\text{Mg}}\right)_{\text{original}} + \left\{\left(\frac{^{26}\text{Al}}{^{27}\text{Al}}\right)_{\text{original}} \times \left(\frac{^{27}\text{Al}}{^{24}\text{Mg}}\right)_{\text{today}}\right\} \quad (2.3)$$

which represents the equation for a straight line (Figure 2.15). A plot of ^{26}Mg/^{24}Mg against ^{27}Al/^{24}Mg for a suite of cogenetic samples or minerals (such as different minerals in the same rock) will define a straight line the slope of which gives the ^{26}Al/^{27}Al at the time the object formed. This can be related in time to the start of the solar system with Soddy and Rutherford's equation for radioactive decay:

$$\left(\frac{^{26}\text{Al}}{^{27}\text{Al}}\right)_{\text{original}} = \left(\frac{^{26}\text{Al}}{^{27}\text{Al}}\right)_{\text{BSSI}} \times e^{-\lambda t} \quad (2.4)$$

in which BSSI is the bulk solar system initial ratio, λ is the decay constant (or probability of decay in unit time), and t is the time that elapsed since the start of the solar system. Using this method and the assumed initial ratio for $(^{26}\text{Al}/^{27}\text{Al})_{\text{BSSI}}$ of $\sim 5 \times 10^{-5}$ (Table 2.1) it has been possible to demonstrate that many chondrules formed 1–3 million years after CAIs.

Over the past 40 years scientists have found evidence that more than 15 short-lived isotopes existed early in the solar system. The main ones are listed in Table 2.1. However, evidence even exists for ^{7}Be with a half-life of just 53 days.

These short-lived isotopes can be broken down into three types on the basis of their origin in the solar nebula:

1. The Sun and the other stars in its cluster inherited a mixture of isotopes from their parent molecular cloud that built up over time from a range of stellar sources.
2. Some short-lived isotopes were probably injected into the Sun's molecular cloud core or the solar nebula itself from at least one nearby star, possibly a supernova.
3. It is likely that a limited number of short-lived isotopes were also generated in the innermost regions of the solar nebula when material was bombarded with energetic particles from the Sun.

Determining the origin of a particular isotope and the timing of its production is often difficult. Isotopes with half-lives of less than 10^6 years must have come from a source close to the solar nebula in order to have survived, while isotopes with longer half-lives may have come from further away as well. Irradiation in the solar nebula could have produced a variety of light isotopes but the relative importance of local production versus external sources is still unclear. Formation in the nebula appears to be a promising source for Be isotopes. However, if all the ^{26}Al had formed this way it seems likely that some of the other isotopes, especially ^{41}Ca, would have been more abundant than they actually were. In fact, there is mounting evidence that many of the short-lived isotopes were uniformly distributed in the solar system, which is hard to explain if they formed in a localized region close to the Sun.

Some of the heavier short-lived isotopes that existed in the early solar system (e.g. ^{107}Pd and ^{129}I) can only be produced in large amounts in a massive star. For example, a large flux of neutrons is required to produce ^{129}I and this is achievable during the enormously energetic death throws of a massive star undergoing a type II supernova explosion. Many of the isotopic ratios in Table 2.1 are similar, lying in

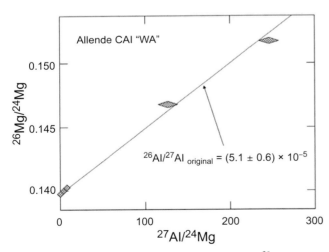

FIGURE 2.15 The decay of a short-lived nuclide such as ^{26}Al generates excess ^{26}Mg in proportion to the elemental ratio Al/Mg. The data here were produced for a CAI from the Allende meteorite. The slope of the line corresponds to the ^{26}Al/^{27}Al at the time of formation of the object. See text for discussion. *Based on a figure in Lee, T., Papanastassiou, D. A., & Wasserburg, G. J. (1976).*

TABLE 2.1 A List of Short-Lived Isotopes Thought to have Existed in the Early Solar System Based on the Distribution of their Decay Products in Meteorites.

Radio-Nuclide	Half-Life (Myrs)	Ratio	Initial Ratio	Daughter
^{10}Be	1.5	^{10}Be/^{9}Be	1×10^{-3}	^{10}B
^{26}Al	0.71	^{26}Al/^{27}Al	5×10^{-5}	^{26}Mg
^{36}Cl	0.30	^{36}Cl/^{35}Cl	2×10^{-5}	^{36}S, ^{36}Ar
^{41}Ca	0.10	^{41}Ca/^{40}Ca	1×10^{-8}	^{41}K
^{53}Mn	3.7	^{53}Mn/^{55}Mn	6×10^{-6}	^{53}Cr
^{60}Fe	2.6	^{60}Fe/^{56}Fe	1×10^{-6}	^{60}Ni
^{92}Nb	35	^{92}Nb/^{93}Nb	3×10^{-5}	^{92}Zr
^{107}Pd	6.5	^{107}Pd/^{110}Pd	9×10^{-5}	^{107}Ag
^{129}I	15.7	^{129}I/^{127}I	1×10^{-4}	^{129}Xe
^{135}Cs	2.3	^{135}Cs/^{133}Cs	5×10^{-4}	^{135}Ba
^{146}Sm	68	^{146}Sm/^{144}Sm	0.008	^{142}Nd
^{182}Hf	8.9	^{182}Hf/^{180}Hf	1×10^{-4}	^{182}W
^{205}Pb	17	^{205}Pb/^{204}Pb	1×10^{-3}	^{205}Tl
^{244}Pu	80	^{244}Pu/^{238}U	7×10^{-3}	131, 132, 134, 136Xe
^{247}Cm	16	^{247}Pu/^{235}U	2×10^{-3}	^{235}U

the range 10^{-6} to 10^{-4} for isotopes with half-lives of 0.7×10^6 to 30×10^6 years. This is as expected if all these isotopes were synthesized in roughly similar proportions just prior to the start of the solar system. Many of these isotopes have initial abundances similar to those that would be formed by an AGB star. However, models for AGB stars do not predict the amounts of ^{53}Mn and ^{182}Hf that once existed. In fact, ^{182}Hf (half-life = 8.9 Myrs) requires a large flux of neutrons of the kind produced in the supernova explosion of a much larger star. It is possible that more than one kind of nucleosynthetic process gave rise to the short-lived isotopes in the early solar system. At present is seems likely that a nearby supernova was involved because the abundance of ^{60}Fe, which has a fairly short half-life, is too high to be explained by alternative sources. Some isotopes that may have been present have yet to be found, including ^{126}Sn with a half-life of 0.3 Myrs. This is an *r-process* isotope that should have been present in the early solar system if a supernova occurred nearby. While modeling these processes is complex, it appears that the supernova explosion of a 25-solar-mass star may explain the correct relative abundances of many of the short-lived isotopes, including ^{182}Hf, provided that roughly 5 solar masses of material was left behind in the form of a supernova remnant or a black hole.

Supernovas are sufficiently energetic that they could tear apart a molecular cloud core rather than cause it to collapse. Shocks waves with a velocity of at least 20–45 km/s are capable of triggering collapse but if the velocity exceeds ~100 km/s, a molecular cloud core will be shredded instead. If the supernova was sufficiently far away the shock wave would have slowed by the time it reached the molecular cloud core. However, the supernova cannot have been more than a few tens of parsecs away otherwise ^{41}Ca (with a half-life of only 0.104 Myrs) would have decayed before it reached the solar nebula. The former presence of ^{41}Ca in CAIs may provide the best constraint on the time between nucleosynthesis of the short-lived isotopes and their incorporation into the solar system. To do this, it will be necessary to ascertain the particular stellar source(s) that gave rise to these isotopes, so that the initial amount of ^{41}Ca can be calculated.

5. EARLY STAGES OF PLANETARY GROWTH

Dust grains are a relatively minor constituent of protoplanetary disks, but they represent the starting point for the formation of rocky planets like Earth, and probably also gas-rich planets like Jupiter. These grains are small, typically 1 μm in diameter or less. In a microgravity environment, electrostatic forces dominate interactions between such grains, and these forces help grains stick together when they collide. Charge transfer during grain collisions can lead to the formation of grain dipoles that align with one another, forming aggregates up to several centimeters in size. Freshly deposited frost surfaces also make grains stickier, and increase the ability of grain aggregates to hold together during subsequent collisions.

Laboratory experiments show that low-velocity collisions between grains tend to result in sticking, while faster collisions often cause grains to rebound. Irregularly shaped micron-sized grains often stick to one another at collision speeds of up to tens of meters per second. Fluffy aggregates may stick more readily than compact solids as some of the energy of impact goes into compaction. However, the primary components of chondritic meteorites are compact chondrules so further compaction cannot have played a big role in the formation of their parent bodies. In general, sticking forces scale with the surface area of an object, while collisional energy scales with mass and hence volume. As a result, growth becomes more difficult, and break up becomes more likely, as aggregates become larger. It is possible that early growth in the solar nebula took place mainly as the result of large objects sweeping up smaller ones. This idea is supported by recent experiments that have found that small dust aggregates tend to embed themselves in larger ones if they collide at speeds above about 10 m/s.

Dust grains, grain aggregates, and chondrules would have been closely coupled to the motion of gas in the solar nebula. The smallest particles were mainly affected by Brownian motion—collisions with individual gas molecules, which caused the particles to move with respect to one another, leading to collisions. Particles also settled slowly toward the disk's midplane due to the vertical component of the Sun's gravitational field. Settling was opposed by gas drag so that each particle fell at its terminal velocity:

$$v_z = -\left(\frac{\rho}{\rho_{gas}}\right)\left(\frac{v_{kep}}{c_s}\right)\left(\frac{rz}{a^2}\right)v_{kep} \qquad (2.5)$$

where r and ρ are the radius and density of the particle, ρ_{gas} is the gas density, a is the orbital distance from the Sun, z is the height above the disk midplane, and c_s is the sound speed in the gas. Here v_{kep} is the speed of a solid body moving on a circular orbit, called the **Keplerian** velocity:

$$v_{kep} = \sqrt{\frac{GM_{sun}}{a}} \qquad (2.6)$$

where M_{sun} is the mass of the Sun. Large particles fell faster than small ones, sweeping up material as they went, increasing their vertical speed further. Calculations show that micron-sized particles would grow and reach the midplane in about 10^3-10^4 orbital periods if these were the only processes operating.

If the gas was turbulent, particles would have become coupled to turbulent eddies due to gas drag. Particles of a given size were coupled most strongly to eddies whose turnover (rotation) time was similar to the particle's stopping time, given by

$$t_s = \frac{\rho r}{\rho_{gas} c_s} \qquad (2.7)$$

Meter-sized particles would have coupled to the largest eddies, with turnover times comparable to the orbital period P. In a strongly turbulent nebula, meter-sized particles would have collided with one another and with smaller particles at high speeds, typically tens of meters per second.

Gas pressure in the nebula generally decreased with distance from the Sun. This means gas orbited the Sun more slowly than solid bodies, which moved at the Keplerian velocity. Large solid bodies thus experienced a headwind of up to 100 m/s. The resulting gas drag removed angular momentum from solid bodies, causing them to undergo *radial drift* toward the Sun. Small particles with $t_s \ll P$ drifted slowly at terminal velocity. Very large objects with $t_s \gg P$ were only weakly affected by gas drag and also drifted slowly. Drift rates were highest for meter-sized bodies with $t_s \approx P$ (see Figure 2.16), and these drifted inward at rates of 1 AU every few hundred years. Rapid inward drift meant that these bodies collided with smaller particles at high speeds. Rapid drift also meant that meter-sized objects had very short lifetimes, and many were probably lost when they reached the hot innermost regions of the nebula and vaporized.

The short drift lifetimes and high collision speeds experienced by meter-sized particles have led some researchers to conclude that particle growth stalled at this

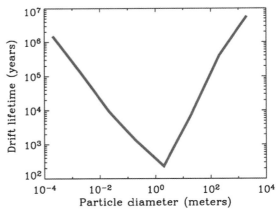

FIGURE 2.16 The lifetime of solid particles orbiting at 1 AU from the Sun in the minimum-mass solar nebula when the particles drift inward due to gas drag. Drift rates are fastest for meter-sized particles, which are lost in a few hundred years unless they rapidly grow larger.

size because particles were destroyed as fast as they formed. This is often referred to as the *meter-sized barrier*. This remains an open question, however, due to a shortage of experimental data regarding the physics of collisions in a microgravity environment, and uncertainty about the level of turbulence in the solar nebula. It is possible that a small population of lucky objects successfully passed through the meter-sized barrier because they never experienced a destructive, high-speed collision.

Bodies larger than 1 km generally took a long time to drift inward due to gas drag. These objects were also large enough to have appreciable gravitational fields, making them better able to hold on to fragments generated in collisions. For these reasons, growth became easier once bodies became this large. Much effort has been devoted to seeing whether kilometer-sized bodies could have formed directly, avoiding the difficulties associated with the meter-sized barrier. *Gravitational instability* (GI) offers a possible way to do this. If the level of turbulence in the nebula was very low, solid particles would have settled close to the nebula midplane increasing their local concentration. If enough particles became concentrated in one place, their combined gravitational attraction would render the configuration unstable, allowing the region to become gravitationally bound and collapse. If the particles were then able to contract enough to form a single solid body, the resulting object would be roughly 1–10 km in radius. Such an object is called a *planetesimal*.

GI faces severe obstacles, however. As solid particles accumulated near the nebula midplane, they would have begun to drag gas around the Sun at Keplerian speeds, while gas above and below the midplane continued to travel at sub-Keplerian speeds. The velocity difference between the layers generated turbulence, puffing up the particle layer until a balance between vertical sedimentation and turbulence was reached. This balance may have prevented particle concentrations from becoming high enough for GI to occur. Calculations suggest that the solid-to-gas ratio in a vertical column of nebula material had to become roughly unity before GI would take place. This means that the concentration of solid material had to become enhanced by one to two orders of magnitude compared to the nebula as a whole. If a region of the disk did start to undergo GI, it would only contract to form a planetesimal if the relative velocities of the particles in that region became low enough. Turbulence and radial drift both lead to large relative velocities between particles and may have rendered GI ineffective.

The presence of turbulence may not have been entirely detrimental to growth. Numerical simulations show that chondrule-size particles would be strongly concentrated in stagnant regions in a turbulent nebula, a process called *turbulent concentration*. Larger, roughly meter-sized particles would have been concentrated by a second process called the streaming instability. These highly mobile particles tended to accumulate at temporary, high-pressure zones in the turbulent gas. As particles accumulated they began to shield one another from the headwind, slowing their inward drift and allowing more particles to accumulate at the same location. Each of these mechanisms provides a possible route to planetesimal formation in turbulent protoplanetary disks.

The difficulties associated with both the meter-sized barrier and GI mean that the question of how planetesimals formed remains open for now. However, the fact that roughly half of young stars have debris disks of dust thought to come from asteroids and comets implies that growth of large solid bodies occurs in many protoplanetary disks, even if the mechanism remains obscure.

6. FORMATION OF TERRESTRIAL PLANETS

The growth of bodies beyond 1 km in size is better understood than planetesimal formation itself. Gravitational interactions and collisions between pairs of planetesimals dominate the evolution from this point onward. A key factor in determining the rate of growth is **gravitational focusing**. The probability that two planetesimals will collide during a close approach depends on their cross-sectional area multiplied by a *gravitational focusing factor* F_g:

$$F_g = 1 + \frac{v_{esc}^2}{v_{rel}^2} \qquad (2.8)$$

where v_{rel} is the planetesimals' relative velocity, and v_{esc} is the escape velocity from a planetesimal, given by

$$v_{esc} = \sqrt{\frac{2GM}{r}} \qquad (2.9)$$

where M and r are the planetesimal's mass and radius, respectively. When planetesimals pass each other slowly, there is time for their mutual gravitational attraction to focus their trajectories toward each other, so F_g is large, and the chance of a collision is high. Fast-moving bodies typically do not collide unless they are traveling directly toward each other because $F_g \cong 1$ in this case. The relative velocities of planetesimals depend on their orbits about the Sun. Objects with similar orbits are the most likely to collide with each other. In particular, planetesimals moving on nearly circular, coplanar orbits have high collision probabilities while ones with highly **inclined, eccentric** (elliptical) orbits do not.

Most close encounters between planetesimals did not lead to a collision, but bodies often passed close enough for their mutual gravitational tug to change their orbits. Statistical studies show that after many such close encounters, high-mass bodies tend to acquire circular, coplanar orbits, while low-mass bodies are perturbed onto eccentric, inclined orbits.

This is called *dynamical friction*, and is analogous to the equipartition of kinetic energy between molecules in a gas. Dynamical friction means that on average, the largest bodies in a particular region experience the strongest gravitational focusing and therefore they grow the fastest (Figure 2.17). This state of affairs is called **runaway growth** for obvious reasons. Most planetesimals remained small, while a few objects, called **planetary embryos**, grew much larger.

Runaway growth continued as long as interactions between planetesimals determined their orbital distribution. However, once embryos became large enough, gravitational perturbations from these objects came to dominate the motion of the smaller planetesimals. This transition took place when

$$M_{emb}\Sigma_{emb} > M_{plan}\Sigma_{plan}$$

where M_{emb} and M_{plan} are the mass of a typical embryo and planetesimal, respectively, and Σ_{emb} and Σ_{plan} are the surface densities of the embryos and planetesimals.

The evolution now entered a new phase called *oligarchic growth*. The relative velocities of planetesimals were determined by a balance between perturbations from nearby embryos and damping due to gas drag. Embryos continued to grow faster than planetesimals, but growth was no longer unrestrained. Large embryos stirred up nearby planetesimals more than small embryos did, weakening gravitational focusing and slowing growth. As a result, neighboring embryos tended to grow at similar rates. Embryos spaced themselves apart at regular radial intervals, with each one staking out an annular region of influence in the nebula called a *feeding zone*.

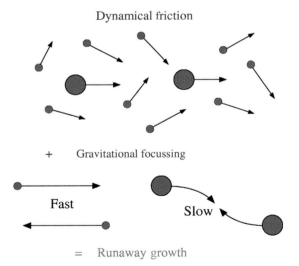

FIGURE 2.17 Runaway growth of a few large planetesimals takes place due to a combination of dynamical friction (which gives large planetesimals circular and coplanar orbits), and gravitational focusing (which increases the chance of a collision between bodies moving on similar orbits).

As embryos became larger, they perturbed planetesimals onto highly inclined and eccentric orbits. The planetesimals began to collide with one another at high speeds, causing fragmentation and breakup. A huge number of subkilometer-sized collision fragments were generated, together with a second generation of fine dust particles. Gas drag operates efficiently on small fragments, so their orbits rapidly became almost circular and coplanar. As a result, many fragments were quickly swept up by embryos, increasing the embryos' growth rates still further.

Numerical calculations show that embryo feeding zones were typically about 10 **Hill radii** in width, where the Hill radius of an embryo with mass M and orbital radius a is given by

$$r_h = a\left(\frac{M}{3M_{sun}}\right)^{1/3} \quad (2.10)$$

If an embryo were to accrete all the solid material in its feeding zone it would stop growing when its mass reached a value called the **isolation mass**, given by

$$M_{iso} \cong \left(\frac{8b^3\pi^3\Sigma_{solid}^3 a^6}{3M_{sun}}\right)^{1/2} \quad (2.11)$$

where Σ is the surface (column) density of solid material in that region of the disk, and $b \approx 10$ is the width of a feeding zone in Hill radii. The surface density in the Sun's protoplanetary nebula is not known precisely, but for plausible values, the isolation masses would have been about 0.1 Earth masses at 1 AU, and around 10 Earth masses in the outer solar system. Calculations suggest that bodies approached their isolation mass in the inner solar system roughly 10^5 years after planetesimals first appeared in large numbers. Growth was slower in the outer solar system, but bodies were probably nearing their isolation mass at 5 AU after 10^6 years.

Large embryos significantly perturbed nearby gas in the nebula forming spiral waves. Gas passing through these waves had a higher density than that in the surrounding region. Gravitational interactions between an embryo and its spiral waves transferred angular momentum between them. For conditions likely to exist in the solar nebula, the net result was that each embryo lost angular momentum and migrated inward toward the Sun. This is called *type-I migration*. In an isothermal (uniform temperature) disk, the migration rate is proportional to an embryo's mass M and the local surface density of gas Σ_{gas} and is given by

$$\frac{da}{dt} \approx -4\left(\frac{M}{M_{sun}}\right)\left(\frac{\Sigma_{gas}a^2}{M_{sun}}\right)\left(\frac{v_{kep}}{c_s}\right)^2 v_{kep} \quad (2.12)$$

where c_s is the sound speed in the gas and v_{kep} is the orbital velocity of a body moving on a circular, Keplerian orbit. Type-I migration became important once embryos grew to about 0.1 Earth masses. Migration rates can be uncomfortably fast, with a 10 Earth mass body at 5 AU migrating into the Sun in 10^5 years in a minimum-mass nebula. It is possible that many objects migrated all the way into the Sun and were lost in this way, and the question of how other bodies survived migration is one of the great unresolved questions of planet formation at present.

Type-I migration rates are modified when radiative transfer within the disk is taken into account. Migration is especially likely to change in regions where there is a discontinuity in the disk such as the ice line or the edge of a dead zone. At these locations, inward migration can slow substantially or even change direction. As a result, there may be particular locations in a disk that are preferred for planet formation since embryos at these points do not migrate, or survive for long enough to outlive the disk.

Oligarchic growth in the inner solar system ended when embryos had swept up roughly half of the solid material. However, these embryos were still an order of magnitude less massive than Earth. Further collisions were necessary to form planets the size of Earth and Venus. With the removal of most of the planetesimals, dynamical friction weakened. As a result, interactions between embryos caused their orbits to become more inclined and eccentric. The embryos' gravitational focusing factors became small and this greatly reduced the collision rate. As a result, the last stage of planet formation was prolonged, and the Earth may have taken 100 million years to finish growing.

Embryos underwent numerous close encounters with one another before colliding. Each encounter changed an embryo's orbit, with the result that embryos moved considerable distances radially in the nebula. Numerical calculations show that the orbital evolution must have been highly chaotic (Figure 2.18). As a result, it is impossible to predict the precise characteristics of a planetary system based on observations of typical protoplanetary disks. Other stars with nebulas similar to the Sun may have formed terrestrial planets that are very different from those in the solar system.

The radial motions of embryos partially erased any chemical gradients that existed in the nebula during the early stages of planet formation. Mixing cannot have been complete, however, since Mars and Earth have distinct compositions. Mars is richer in the more volatile rock-forming elements, and the two planets have distinct oxygen isotope mixtures. Unfortunately, we have no confirmed samples of Mercury and Venus, so we know little about their composition. Mercury is known to have an unexpectedly high density, suggesting it has a large iron-rich core and a small mantle. This probably does not reflect compositional differences in the solar nebula since there is no known reason why iron-rich materials would preferentially form closer to the Sun than silicate materials. A more likely explanation is that Mercury suffered a near-catastrophic impact after it had differentiated, and this stripped away much of the silicate mantle. Mercury's location close to the Sun made it especially vulnerable in this respect since orbital velocities and hence impact speeds are highest close to the Sun.

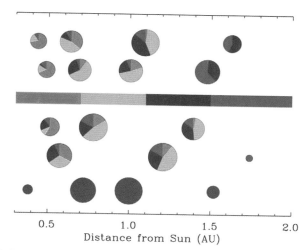

FIGURE 2.18 Four artificial planetary systems generated by numerical simulations of planetary accretion. Each horizontal row of symbols represents one planetary system, with symbol radius proportional to planetary radius, with the largest objects similar in size to the Earth. The shaded segments show the composition of each planet in terms of material that originated in four different portions of the nebula. Planets in these simulations typically contain material from many regions of the nebula. The row of gray symbols shows the terrestrial planets of the solar system for comparison.

Earth and Venus are probably composites of 10 or more embryos so their chemical and isotopic compositions represent averages over a fairly large region of the inner solar system. Mars and Mercury are sufficiently small that they may be individual embryos that did not grow much beyond the oligarchic growth stage. It is currently a mystery why Earth and Venus continued to grow while Mars did not. It may be that Mars formed in a low-density region of the nebula or that all other embryos were removed from that region without colliding with Mars.

The *Grand Tack model* offers one possible explanation for why there was relatively little solid material in the region that gave rise to Mars, and even less in the asteroid belt. This model is based on numerical simulations that show that Jupiter would have migrated inward through the solar nebula until Saturn formed, at which point Jupiter would have migrated outward. If Jupiter migrated inward to about 1.5 AU from the Sun, and then moved outward again to its current location, the planet's gravity would have pulled most planetesimals and planetary embryos out of the region that now contains Mars and the asteroid belt, scattering these objects into the Sun or out of the solar system. Computer simulations show the Grand Tack model can explain many observed features of the solar system as a result, but the model remains unconfirmed at the present time.

As embryos grew larger, their temperatures increased due to kinetic energy released during impacts and the decay of radioactive isotopes in their interiors. Short-lived isotopes such as ^{26}Al and ^{60}Fe, with half-lives of 0.7 and 2.6 Myrs, respectively (Table 2.1), were particularly powerful heat sources early in the solar system. Bodies more than a few kilometers in radius would have melted if they formed within the first 2 million years, when the short-lived isotopes were still abundant. Embryos that melted also differentiated, with iron and siderophile elements sinking to the center to form a core, while lighter silicates formed a mantle closer to the surface.

The abundances of the most highly siderophile elements (such as platinum and osmium) in Earth's mantle are higher than one would expect to find after the planet differentiated since most siderophile material should have been extracted into the core. The amount of the platinum in the core is sufficient to cover Earth's surface to a depth of about a meter. However, even that which is residual in Earth's mantle and which provides our platinum and gold jewelry, is much more than expected unless it was added after core formation had ceased. The most likely explanation for these high abundances therefore is that Earth continued to acquire some material after its core and mantle had finished separating. This *late veneer* may amount to almost 1% of the total mass of the planet.

The degree to which this late veneer also provided Earth with some of its inventory of volatiles, namely, water, carbon, nitrogen, sulfur, and the noble gases, is uncertain; some argue that the major portion was accreted earlier. Even the amount of water that Earth contains is debated. Earth's oceans contain about 0.03% of the planet's total mass. At least as much water exists in the mantle and some think there could be an order of magnitude more. The present amounts of water and other volatiles, even if they were better quantified, may not reflect the original situation. Earth may have also suffered removal of volatiles as it grew from energetic collisions. There would also have been some dissipation of dissociated hydrogen to space. Finally, reactions with iron could have led to segregation to the core of hydrogen, carbon, nitrogen, and sulfur, just like the platinum group elements.

Temperatures at 1 AU are currently too high for water ice to condense, and this was probably also true for most of the history of the solar nebula (pressures were always too low for liquid water to condense). As a result, Earth probably received most of its water as the result of collisions with other embryos or planetesimals that contained water ice or hydrated minerals in their interiors. Planetesimals similar to modern comets almost certainly delivered some water to Earth. However, a typical comet has a probability of only about one in a million of colliding with Earth, so it is unlikely that comets provided the bulk of the planet's water. The deuterium to hydrogen (D/H) ratio seen in most comets is twice that of Earth's oceans, which suggests these comets supplied at most about 10% of Earth's water. To date, one comet has been observed with a D/H ratio similar to Earth, so it is possible that Earth acquired a substantial amount of water from a subpopulation of comets.

Planetesimals from the asteroid belt are another possible source of water. Carbonaceous chondrites are especially promising since they contain up to 10% water by mass in the form of hydrated silicates, and this water would be released upon impact with the Earth. Calculations suggest that if the early asteroid belt was several orders of magnitude more massive than today, it could have supplied the bulk of Earth's water. From the current meteorite record it seems unlikely that this water could have been delivered after core formation like the late veneer of highly siderophile elements. There are a number of reasons for this but a compelling case comes from the fact that carbonaceous chondrites and Earth's mantle have different osmium isotope ratios. As a result, the delivery of water to Earth and its acquisition of a late veneer were separate processes that occurred at different times in its history.

The origin of Earth's atmospheric constituents is also somewhat uncertain. When the solar nebula was still present, planetary embryos probably had thick atmospheres mostly composed of hydrogen and helium captured from the nebula. Most of this atmosphere was lost subsequently by **hydrodynamic escape** as hydrogen atoms were accelerated to escape velocity by ultraviolet radiation from the Sun, dragging other gases along with them. Much of Earth's current atmosphere was probably outgassed from the mantle at a later stage. Some noble gases currently escaping from Earth's interior are similar to those found in the Sun, which suggests they may have been captured into Earth's mantle from the nebula or were trapped in bodies that later collided with Earth. Most of the xenon produced by radioactive decay of ^{244}Pu and ^{129}I (Table 2.1) has been lost, which implies that Earth's atmosphere was still being eroded 100 Myr after the start of the solar system, possibly by impacts.

Radioactive isotopes can be used to place constraints on the timing of planet formation. The hafnium–tungsten system is particularly useful in this respect since the parent nuclide ^{182}Hf is lithophile (tending to reside in silicate mantles) while the daughter nuclide ^{182}W is siderophile (tending to combine with iron during core formation) (Figure 2.19). Isotopic data can be used in a variety of ways to define a timescale for planetary accretion. The simplest method uses a model age calculation, which corresponds to the calculated time when an object or sample would have needed to form from a simple average solar system reservoir, as represented by chondrites, in order generate its

Chapter | 2 The Origin of the Solar System

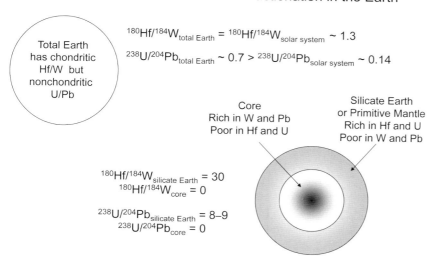

FIGURE 2.19 Hafnium—tungsten chronometry provides insights into the rates and mechanisms of formation of the solar system whereas U-Pb chronometry provides us with an absolute age of the solar system. In both cases the radioactive parent/radiogenic daughter element ratio is fractionated by core formation, an early planetary process. It is this fractionation that is being dated. The Hf/W ratio of the total Earth is chondritic (average solar system) because Hf and W are both refractory elements. The U/Pb ratio of the Earth is enhanced relative to average solar system because approximately >80% of the Pb was lost by volatilization or incomplete condensation mainly at an early stage of the development of the circumstellar disk. The fractionation within the Earth for Hf/W and U/Pb is similar. In both cases the parent (Hf or U) prefers to reside in the silicate portion of the Earth. In both cases the daughter (W or Pb) prefers to reside in the core.

isotopic composition. For the ^{182}Hf-^{182}W system this time is given as

$$t_{CHUR} = \frac{1}{\lambda} \ln \left[\left(\frac{^{182}Hf}{^{180}Hf}\right)_{BSSI} \times \left(\frac{\left(\frac{^{182}W}{^{184}W}\right)_{SAMPLE} - \left(\frac{^{182}W}{^{184}W}\right)_{CHONDRITES}}{\left(\frac{^{180}Hf}{^{184}W}\right)_{SAMPLE} - \left(\frac{^{180}Hf}{^{184}W}\right)_{CHONDRITES}} \right) \right] \quad (2.13)$$

where t_{CHUR} is the time of separation from a CHondritic Uniform Reservoir, $\lambda = (\ln 2/\text{half-life})$ is the decay constant for ^{182}Hf (0.078 per million years), and $(^{182}Hf/^{180}Hf)_{BSSI}$ is the BSSI ratio of ^{182}Hf to ^{180}Hf. Tungsten-182 excesses have been found in Earth, Mars, and the HED meteorites, which are thought to come from asteroid Vesta, indicating that all these bodies differentiated while some ^{182}Hf was still present. Iron meteorites, which come from the cores of differentiated planetesimals, have low Hf/W ratios and are deficient in ^{182}W. This means these planetesimals must have formed at a very early stage before most of the ^{182}Hf had decayed. New, very precise ^{182}Hf-^{182}W chronometry has shown that some of these objects formed within the first 2 million years of the solar system (Figure 2.8).

New modeling of the latest ^{182}Hf-^{182}W data for Martian meteorites also provides evidence that Mars grew and started differentiating within about 1 million years of the start of the solar system. This short timescale is consistent with runaway growth described above. So far, isotopic data for other silicate objects has not been so readily explicable in terms of very rapid growth. However, asteroid Vesta certainly formed within about 3 million years of the start of the solar system (Figure 2.8).

The existence of meteorites from differentiated asteroids suggests that core formation began early and this is confirmed by ^{182}Hf-^{182}W chronometry. Therefore, most planetary embryos would have been differentiated when they collided with one another. Although Mars grew extremely rapidly, Earth does not appear to have reached its current size until the giant impact that was associated with the formation of the Moon (see Section 8). ^{182}Hf-^{182}W chronometry for lunar samples shows that this took place more than 30 Myrs after the start of the solar system. There is other evidence that this could have been as late as 100 Myrs and it has long been recognized that the formation of the Moon probably happened near the end of Earth's accretion, and this is consistent with the results of Moon-forming impact simulations. This is also consistent with the W isotopic composition of the silicate Earth itself (Figure 2.20). This shows that the Earth accreted at least half of its mass within the first 30 Myrs of the solar system. However, the data are fully consistent with the final stage of accretion being around the time of the Moon-forming impact. Because the Earth accreted over a protracted period rather than in a single event it is the simplest to model the W isotope data in terms of an exponentially decreasing rate of growth (Figure 2.20).

$$F = 1 - e^{-(1/\tau) \times t} \quad (2.14)$$

where F is the mass fraction of the Earth that has accumulated, τ is the mean life for accretion in Myrs (Figure 2.20), and t is time in Myrs. This is consistent with the kinds of curves produced by the late George Wetherill who modeled the growth of the terrestrial planets using

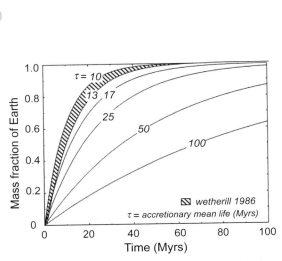

FIGURE 2.20 The mean life of accretion of the Earth (τ) is the inverse of the time constant for exponentially decreasing oligarchic growth from stochastic collisions between planetary embryos and planets. The growth curves corresponding to several such mean lives are shown including the one that most closely matches the calculation made by the late George Wetherill based on Monte Carlo simulations. The mean life determined from tungsten isotopes (Figure 2.8) is in excellent agreement with Wetherill's predictions.

Monte Carlo simulations. The W isotope data are consistent with a mean life of between 10 and 15 Myrs depending on the exact parameters used. This is fully consistent with the timescales proposed by Wetherill. From these protracted timescales it is clear that Earth took much longer to approach its current size than Mars or Vesta, which probably formed from different mechanisms (Figure 2.8).

7. THE ASTEROID BELT

The asteroid belt currently contains only enough material to make a planet 2000 times less massive than Earth, even though the spatial extent of the belt is huge. It seems likely that this region once contained much more mass than it does today. A smooth interpolation of the amount of solid material needed to form the inner planets and the gas giants would place about 2 Earth masses in the asteroid belt. Even if most of this mass was lost at an early stage, the surface density of solid material must have been at least 100 times higher than it is today in order to grow bodies the size of Ceres and Vesta (roughly 900 and 500 km in diameter, respectively) in only a few million years.

Several regions of the asteroid belt contain clusters of asteroids with similar orbits and similar spectral features, suggesting they are made of the same material. These clusters are fragments from the collisional breakup of larger asteroids. There are relatively few of these asteroid families, which implies that catastrophic collisions are quite rare. This suggests the asteroid belt has contained relatively little mass for most of its history. The spectrum of asteroid Vesta, located 2.4 AU from the Sun, shows that it has a basaltic crust. The HED meteorites, which probably come from Vesta, show this crust formed only a few million years after the solar system, according to several isotopic systems. The survival of Vesta's crust suggests the impact rate in the belt has never been much higher than today since the crust formed. For these reasons, it is thought that most of the asteroid belt's original mass was removed at a very early stage by a dynamical process rather than by collisional erosion.

The asteroid belt currently contains a number of *orbital resonances* associated with the giant planets. Resonances occur when either the orbital period or **precession** period of an asteroid has a simple ratio with the corresponding period for one of the planets. Many resonances induce large changes in orbital eccentricity, causing asteroids to fall into the Sun, or to come close to Jupiter, leading to close encounters and ejection from the solar system. For this reason, there are very few asteroids that orbit the Sun twice every time Jupiter orbits the Sun once, for example. When the nebular gas was still present, small asteroids moving on eccentric orbits would have drifted inward rapidly due to gas drag. After the giant planets had formed, a combination of resonances and gas drag may have transferred most objects smaller than a few hundred kilometers from the asteroid belt into the terrestrial-planet region. Larger planetary embryos would not have drifted very far. However, once oligarchic growth ceased, embryos began to gravitationally scatter one another across the belt. Numerical simulations show that most or all of these bodies would eventually enter a resonance and be removed, leaving an asteroid belt greatly depleted in mass and containing no objects bigger than Ceres. The timescale for the depletion of the belt depends sensitively on the orbital eccentricities of the giant planets at the time, which are poorly known. The belt may have been cleared in only a few million years, but it may have required as much as several hundred million years if the giant planets had nearly circular orbits. Finally, as we saw earlier, the asteroid belt may have lost much of its original mass due to the migration of Jupiter while the solar nebula was still present.

The albedos and spectral features of asteroids vary widely from one body to another, but clear trends are apparent as one moves across the asteroid belt. *S-type* asteroids, which generally lie in the inner asteroid belt, appear to be more thermally processed than the *C-type* asteroids that dominate the middle belt. These may include the parent bodies of ordinary and carbonaceous chondrites, respectively. C-types in turn seem more processed than the *P-type* asteroids that mostly lie in the outer belt. These differences may reflect differences in the composition of solid materials in different parts of the nebula, or differences in the time at which asteroids formed. Ordinary and enstatite chondrites, which probably come from the inner asteroid belt, tend to be dry, while carbonaceous chondrites from the middle and outer belt contain up to 10% water by mass in the form of hydrated minerals. This suggests that temperatures were cold enough in the outer asteroid belt for water ice to form and become incorporated into asteroids where it reacted with dry rock. Temperatures were apparently too high for water ice to condense in the inner asteroid belt. It is possible that some of the objects currently in the asteroid belt formed elsewhere. For example, it has been proposed that many of the parent bodies of the iron meteorites, and possibly Vesta, formed in the terrestrial-planet region and were later gravitationally scattered outward to their current orbits.

Iron meteorites from the cores of melted asteroids are common, whereas meteorites from the mantles of these asteroids are rarely seen. This suggests that a substantial amount of collisional erosion took place at an early stage, with only the strong, iron-rich cores of many bodies surviving. A number of other meteorites also show signs that their parent asteroids experienced violent collisions early in their history. Chondrites presumably formed somewhat later than the differentiated asteroids, when the main radioactive heat sources had mostly decayed. Chondrites are mostly composed of chondrules, which typically formed 1–3 Myr after CAIs. Chondrite parent bodies cannot be older than the youngest chondrules they contain, so they must have formed several Myr after the start of the solar system. For this reason, it appears that the early stages of planet formation were prolonged in the asteroid belt. While chondrites have experienced some degree of thermal processing, their late formation meant that their parent bodies never grew hot enough to melt, which has allowed chondrules, CAIs, and matrix grains to survive.

8. GROWTH OF GAS AND ICE GIANT PLANETS

Jupiter and Saturn are mostly composed of hydrogen and helium. These elements do not form solids or liquids at temperatures and pressures found in protoplanetary disks, so they must have been gravitationally captured from the gaseous component of the solar nebula. Observations of young stars indicate that protoplanetary disks survive for only a few million years, and this sets an upper limit for the amount of time required to form giant planets. Uranus and Neptune also contain significant amounts of hydrogen and helium (somewhere in the range 3–25%), and so they probably also formed quickly, before the solar nebula dispersed.

Jupiter and Saturn also contain elements heavier than helium and they are enriched in these elements compared to the Sun. The gravitational field of Saturn strongly suggests it has a core of dense material at its center, containing roughly 1/5 of the planet's total mass. Jupiter may also have a dense core containing a few Earth masses of material. The interior structure of Jupiter remains quite uncertain since we lack adequate equations of state for the behavior of hydrogen at the very high pressures found in the planet's interior. The upper atmospheres of both planets are enriched in elements such as carbon, nitrogen, sulfur, and argon, compared to the Sun. It is thought likely that these enrichments extend deep into the planets' interiors but this remains uncertain.

Giant planets may form directly by the contraction and collapse of gravitationally unstable regions of a protoplanetary disk. This *disk instability* is analogous to the gravitational instabilities that may have formed planetesimals, but instead the instability takes place in nebula gas rather than the solid component of the disk. Instabilities will occur if the *Toomre stability criterion* Q becomes close to or lower than 1, where

$$Q = \frac{M_{\text{sun}} c_s}{\Sigma \pi a^2 v_{\text{kep}}} \quad (2.15)$$

where v_{kep} is the Keplerian velocity, c_s is the sound speed, and Σ is the local surface density of gas in the disk. Gas in an unstable region quickly becomes much denser than the surrounding material. Disk instability requires high surface densities and low sound speeds (cold gas), so it is most likely to occur in the outer regions of a massive protoplanetary disk. Numerical calculations suggest instabilities will occur beyond about 5 AU in a nebula a few times more massive than the MMSN. What happens to an unstable region depends on how quickly the gas cools as it contracts, and this is the subject of much debate. If the gas remains hot, the dense regions will quickly become sheared out and destroyed by the differential rotation of the disk. If cooling is efficient, simulations show that gravitationally bound clumps will form in a few hundred years, and these may ultimately contract to form giant planets. Initially, such planets would be homogeneous and have the same composition as the nebula. Their structure and composition may change subsequently due to gravitational settling of heavier elements to the center and capture of rocky or icy bodies such as comets.

The evidence for dense cores at the centers of Jupiter and Saturn suggests to many scientists that giant planets form by *core accretion* rather than disk instability. In this model, the early stages of giant-planet formation mirror the growth of rocky planets, beginning with the formation of planetesimals, followed by runaway and oligarchic growth. However, planetary embryos would have grown larger in the outer solar system for two reasons. First, feeding zones here are larger because the Sun's gravity is weaker, so each embryo gravitationally holds sway over a larger region of the nebula. Second, temperatures here were cold enough for volatile materials such as tars, water ice, and other ices to condense, so more solid material was available to build large embryos.

In the outer solar system, bodies roughly 10 times more massive than Earth would have formed via oligarchic growth in a million years, provided the disk was a few times more massive than the MMSN. Bodies that grew larger than Mars would have captured substantial atmospheres of gas from the nebula. Such atmospheres remain in equilibrium due to a balance between an embryo's gravity and an outward pressure gradient. However, there is a *critical core mass* above which an embryo can no longer support a static atmosphere. Above this limit, the atmosphere begins to collapse onto the planet forming a massive gas *envelope* that increases in mass over time as more gas is captured from the nebula. As gas falls toward the planet it heats up as gravitational potential energy is released. The rate at which a planet grows depends on how fast this heat can be radiated away. The critical core mass depends on the **opacity** of the envelope and the rate at which planetesimals collide with the core, but calculations suggest it is in the range 3–20 Earth masses, possibly less if the envelope is enriched in heavy elements. The growth of the envelope is slow at first, but speeds up rapidly once an embryo reaches 20–30 Earth masses. Numerical simulations show that Jupiter-mass planets can form this way in 1–5 million years. Such planets are mostly composed of hydrogen-rich nebular gas, and are enriched in heavier elements due to the presence of a solid core. As with the disk instability, the planet's envelope may be further enriched in heavy elements by collisions with comets.

Measurements by the Galileo spacecraft showed that Jupiter's upper atmosphere is enriched in carbon, nitrogen, sulfur, and the noble gases argon, krypton, and xenon by factors of two to three compared to the Sun. If these enrichments are typical of Jupiter's envelope as a whole, it suggests the planet captured a huge number of comets. Argon can be trapped in cometary ices but only if these ices form at temperatures below about 30 K. Temperatures at Jupiter's current distance from the Sun were probably quite a lot higher than this. This suggests either that the comets came from colder regions of the nebula or that Jupiter itself migrated inward over a large distance. However, the fact that relatively refractory elements such as sulfur are present in the same enrichment as the noble gases suggests these elements may all have been captured as gases from the nebula along with hydrogen and helium. If so, Jupiter's envelope must be nonhomogeneous, with the lower layers depleted in heavy elements, perhaps due to exclusion from high-pressure phases of hydrogen, while the upper layers are enriched.

It is unclear why Jupiter and Saturn stopped growing when they reached their current masses. These planets are sufficiently massive that they would continue to grow very rapidly if a supply of gas was available nearby. It is possible, but unlikely, that they stopped growing because the nebula happened to disperse at this point. A more likely explanation is that the growth of these planets slowed because they each became massive enough to clear an annular gap in the nebula around their orbit. Gap clearing happens when a planet's Hill radius becomes comparable to the vertical thickness of the gas disk, which would have been the case for Jupiter. Gas orbiting a little further from the Sun than Jupiter would have been sped up by the planet's gravitational pull, moving the gas away from the Sun. Gas orbiting closer to the Sun than Jupiter was slowed down, causing it to move inward. These forces open up a gap in the disk around Jupiter's orbit, balancing viscous forces that would cause gas to flow back into the gap. Numerical simulations show that generally gaps are not cleared completely, and some gas continues to cross a gap and accrete onto a planet. However, the accretion rate declines as a planet becomes more massive. Saturn's growth may have been truncated because the combined gravity of Jupiter and Saturn cleared a gap in the disk around both planets.

Uranus and Neptune are referred to as *ice-giant planets* since they contain large amounts of materials such as water and methane that form ices at low temperatures. They contain some hydrogen and helium, but they did not acquire the huge gaseous envelopes that Jupiter and Saturn possess. This suggests the nebula gas had largely dispersed in the region where Uranus and Neptune were forming before they became massive enough to undergo rapid gas accretion. This may be because they formed in the outer regions of the protoplanetary disk, where embryo growth rates were slowest. It is also possible that the nebula dispersed more quickly in some regions than others. In particular, the outer regions of the nebula may have disappeared at an early stage as the gas escaped the solar system due to photoevaporation by ultraviolet radiation.

The presence of a gap modifies planetary migration. Planets massive enough to open a gap still generate spiral density waves in the gas beyond the gap, but these waves are located further away from the planet as a result, so migration is slower. As a planet with a gap migrates inward, gas tends to pile up at the inner edge of the gap, and become rarified at the outer edge, slowing migration as a result. The migration of the planet now becomes tied to the inward viscous accretion of the gas toward the star. The planet, its gap, and the nebular gas all move inward at the same rate, given by

$$\frac{da}{dt} = -1.5\alpha\left(\frac{c_s}{v_{\text{kep}}}\right)^2 v_{\text{kep}} \quad (2.16)$$

where $\alpha = \nu v_{\text{kep}}/(ac_s^2)$ and ν is the viscosity of the nebular gas. This is called **type-II migration**. Type-II migration slows when a planet's mass becomes comparable to that of the nebula, and migration ceases as the nebular gas disperses. Migration can also be modified by the presence of additional giant planets as we saw earlier when discussing the Grand Tack model.

Giant planets in the solar system experienced another kind of migration as they interacted gravitationally with planetesimals moving on orbits between the giant planets and in the primordial Kuiper belt. One consequence of this process was the formation of the **Oort cloud** of comets. Once Jupiter approached its current mass, many planetesimals that came close to the planet would have been flung far beyond the outer edge of the protoplanetary disk. Some were ejected from the solar system altogether, but others remained weakly bound to the Sun. Over time, gravitational interactions with molecular clouds, other nearby stars, and the galactic disk circularized the orbits of these objects so they no longer passed through the planetary system. Many of these objects are still present orbiting far from the Sun in the Oort cloud. The ultimate source of angular momentum for these objects came at the expense of Jupiter's orbit, which shrank accordingly. Saturn, Uranus, and Neptune ejected some planetesimals, but they also perturbed many objects inward, which were then ejected by Jupiter. As a result, Saturn, Uranus, and Neptune probably moved outward rather than inward.

As Neptune migrated outward it interacted dynamically with the primordial Kuiper belt of comets orbiting in the very outer region of the nebula. Some of these comets were ejected from the solar system or perturbed inward toward Jupiter. Others were perturbed onto highly eccentric orbits with periods of hundreds or thousands of years, and now form the *Scattered Disk*, a region that extends out beyond the Kuiper belt but whose objects are gradually being removed by close encounters with Neptune. A sizable fraction of the objects in the region beyond Neptune were trapped in external mean-motion resonances and migrated outward with the planet. Pluto, currently located in the 3:2 mean-motion resonance with Neptune, probably represents one of these objects.

As the giant planets migrated, it is possible that they passed through orbital resonances with one another. Such a resonance crossing may have had a profound impact on every part of the solar system, the combined effects of which have come to be known as the *Nice model* after the French city in which it was developed. In particular, if Jupiter and Saturn passed through a mean-motion resonance, the orbital eccentricities of all four giant planets would have increased substantially. The orbits of the ice giants would have penetrated deeply into the primordial Kuiper belt, gravitationally scattering large numbers of objects onto unstable orbits crossing those of the other planets. Many comets would have been perturbed into the inner solar system as a result. In addition, the changing orbits of the giant planets would have perturbed many main-belt asteroids into unstable orbits also leading to a flux of asteroids into orbits crossing the inner planets. Currently, it is unclear whether Jupiter and Saturn passed through a resonance, or when this may have happened. It has been proposed that passage through a resonance was responsible for the late heavy bombardment of the inner planets, which occurred 600–700 million years after the start of the solar system and left a clear record of impacts on the Moon, Mars, and Mercury.

9. PLANETARY SATELLITES

Earth's moon possesses a number of unusual features. It has a low density compared to the inner planets and it has only a very small core. The Moon is depleted in volatile materials such as water. In addition, the Earth–Moon system has a large amount of angular momentum per unit mass. If they were combined into a single body the object would rotate once every 4 h! All these features can be understood if the Moon formed as the result of an oblique impact between Earth and another large, differentiated body, sometimes referred to as *Theia*, late in Earth's formation. Theia is the Greek Titaness goddess who was the mother of Selene, the goddess of the Moon.

Numerical simulations of this giant impact show that much of Theia's core would have sunk through Earth's mantle to coalesce with Earth's core. Molten and vaporized mantle material from both bodies was ejected outward. Gravitational torques from the highly nonspherical distribution of matter during the collision gave some of this mantle material enough angular momentum to go into orbit about Earth. This material quickly formed into a disk, from which the Moon accreted. Certain features of the Moon's composition are very similar to those of the Earth, which means that (1) Theia was formed from similar material, (2) the resulting vapor and debris that condensed to form the Moon totally equilibrated with the outer portions of the Earth, or (3) the Moon is mostly composed of material from Earth rather than Theia, although most numerical simulations tend to find that the opposite is true in this case.

The impact released huge amounts of energy, heating the disk sufficiently that many volatile materials escaped. As a result, the Moon formed mostly from volatile-depleted mantle materials, explaining its current composition. The simulations suggest Theia probably had a mass similar to Mars, which has roughly 1/10 the mass of Earth. We know little about Theia's composition except that, like Mars, it seems to have been rich in geochemical volatile elements such as rubidium compared to Earth (Figure 2.9). The Earth and the Moon have identical oxygen isotope characteristics (Figure 2.10). It was once thought that this meant Earth and Theia had a similar isotopic composition, but this similarity now appears to be the result of exchange of material between the Earth and protolunar disk while the Moon was forming. The similarity may also mean that the Moon was mostly formed from material ejected from Earth's mantle rather than the impactor.

The satellites of the giant planets are much smaller relative to their parent planet than the Moon is compared to the Earth. The Moon is roughly 1/80 of the mass of the Earth, whereas the satellite systems of Jupiter, Saturn, and Uranus each contain about 1/10,000 of the mass of their respective planet. The satellites of the giant planets can be divided into two classes with different properties. Those close to their parent planet tend to have nearly circular orbits in the same plane as the planet's equator and orbiting in the same direction as the planet spins. These are referred to as *regular satellites*. Satellites orbiting further from the planet tend to have highly inclined and eccentric orbits and these are called *irregular satellites* as a result. The regular satellites tend to be larger and include the **Galilean satellites** of Jupiter and Saturn's largest satellite Titan.

The orbits of the regular satellites suggest they formed from *circumplanetary disks* orbiting each planet like miniature versions of the solar nebula, while the irregular satellites are thought to have been captured later. Large satellites would have moved rapidly inward through a massive circumplanetary disk due to type-I migration, on a timescale that was short compared to the lifetime of the solar nebula. For this reason, it is possible that multiple generations of satellites formed, with the satellites we see today being the last to form. It is also possible that the circumplanetary disks had very low masses containing much less gas than the solar nebula itself. Solid material would have slowly accumulated in these *gas-starved disks*, while the gas quickly passed through the disk and accreted onto the planet. Large satellites would have formed slowly as a result, limiting the degree to which they were heated by impacts. This idea is consistent with the fact that three of the Galilean satellites of Jupiter have retained volatile materials such as water ice, while Callisto never grew hot enough to differentiate. Orbital resonances involving two or more satellites are common. For example, the inner three Galilean satellites Io, Europa, and Ganymede have orbital periods in the ratio 1:2:4, and resonances are common in the Saturnian satellite system. This contrasts with the absence of resonances between the major planets. The ubiquity of satellite resonances suggests many of the satellites migrated considerable distances during or after their formation, becoming captured in a resonance en route. Some resonances may have arisen as the growing satellites migrated inward through their planet's accretion disk. Others could have arisen later as tidal interactions between a planet and its satellite caused the satellites to move outward at different rates.

The Neptunian satellite system is different from those of the other giant planets, having relatively few moons with most mass contained in a single large satellite Triton, which is larger than Pluto. Triton is unusual in that its orbit is **retrograde**, unlike all the other large satellites in the solar system. This suggests it was captured rather than forming in situ. Several capture mechanisms have been proposed, but most are low-probability events, which makes them unlikely to explain the origin of Triton. A more plausible idea is that Triton was once part of a binary planet like the Pluto–Charon system, orbiting around the Sun. During a close encounter with Neptune, the binary components were parted. Triton's companion remained in orbit about the Sun, taking with it enough kinetic energy to leave Triton in a bound orbit about Neptune. Triton's orbit would have been highly eccentric initially, but tidal interactions with Neptune caused its orbit to shrink and become more circular over time. As Triton's orbit shrank it would have disturbed the orbits of smaller satellites orbiting Neptune, leading to their destruction by mutual collisions. This is presumably the reason for the paucity of regular satellites orbiting Neptune today.

10. EXTRASOLAR PLANETS

At the time of writing about 800 confirmed planets are known orbiting stars other than the Sun, and several thousand additional candidates await verification. These are referred to as extrasolar planets or exoplanets. Many of these objects have been found using the Doppler radial velocity technique. This makes use of the fact that the gravitational pull of a planet causes its star to move in an ellipse with the same period as the orbital period of the planet. As the star moves toward and away from the observer, lines in its spectra are alternately blue- and redshifted by the Doppler effect, indicating the planet's presence. Current levels of precision allow the detection of gas-giant planets and ice giants, as well as Earth-mass planets that orbit close to their star. The planet's orbital period P can be readily identified from the radial velocity variation. The mean radius of the planet's orbit a can then be found using Kepler's third law if the star's mass M_* is known:

$$a^3 = \frac{P^2 G M_*}{4\pi^2} \qquad (2.17)$$

Unfortunately, the Doppler method determines only one component of the star's velocity, so the orientation of the orbital plane is not known in general. This means one can obtain only a lower limit on the planet's mass. For randomly oriented orbits, however, the true mass of the planet is most likely to lie within 30% of its minimum value.

Many other extrasolar planets have been detected when they transit across the face of their star, typically causing the star to dim by a small amount for a few hours. Only a small fraction of extrasolar planets generate a transit since their orbital plane must be almost edge on as seen from the Earth. However, the space-based Kepler mission has surveyed more than one hundred thousand stars, with the result that several thousand possible planets have been found. These await confirmation by ground-based observers using other techniques.

When a planet is observed using both the Doppler and transit methods, its true mass can be obtained since the orientation of the orbital plane is known. If the stellar radius is also known, the degree of dimming yields the planet's radius and hence its density. The densities of large extrasolar planets observed this way are generally comparable to that of Jupiter and substantially lower than that of Earth. This suggests these planets are composed mainly of gas rather than rock or ice. In one case, hydrogen has been detected escaping from an extrasolar planet. Recently, a number of objects have been found with masses below 10 Earth masses, and it is plausible that these are more akin to terrestrial planets or water-rich worlds with no analogue in the solar system.

Stars with known giant planets tend to have high *metallicities*, that is, they are enriched in elements heavier than helium compared to most stars in the Sun's neighborhood (Figure 2.21). (The Sun also has a high metallicity.) The meaning of this correlation is hotly debated, but it is consistent with the formation of giant planets via core accretion (see Section 8). When a star has a high metallicity, its disk will contain large amounts of the elements needed to form a solid core, promoting rapid growth and increasing the likelihood that a gas giant can form before the gas disk disperses.

Both the Doppler velocity and transit techniques are biased toward finding massive planets since these generate a stronger signal. Both are also biased toward detecting planets lying close to their star. In the case of transits, the probability of suitable orbital alignment declines with increasing orbital distance, while for the Doppler velocity method, one generally needs to observe a planet for at least a full orbital period to obtain a firm detection. Despite these biases, it is clear that at least 20% of Sun-like stars have planets and this fraction may be much higher. The fraction of planets with a given mass increases as the planetary mass grows smaller, despite the strong observational bias working in the opposite direction. Roughly 10% of known extrasolar planets have orbital periods of only a few days, which implies their orbits are several times smaller than Mercury's orbit about the Sun. These planets are often referred to as **hot Jupiters** or *hot Neptunes* due to their likely high temperatures. Theoretical models of planet formation suggest it is unlikely that planets will form this close to a star. Instead, it is thought that these planets formed at larger distances and moved inward due to type-I and/or type-II migration. Alternatively, they may have been scattered onto highly eccentric orbits following close encounters with other planets in the same system. In this case, subsequent tidal interactions with the star will circularize a planet's orbit and cause the orbit to shrink.

More than 100 stars are known to have two or more planets. In a sizable fraction of these cases, the planets are involved in orbital resonances where either the ratio of the orbital periods or precession periods of two planets is close to the ratio of two integers, such as 2:1. This state of affairs has a low probability of occurring by chance, which suggests these planets have been captured into a resonance when the orbits of one or both planets migrated inward. Several of the planetary systems found by the Kepler mission consist of multiple low-mass planets lying close to their star. It seems likely that these planets or their building blocks must have migrated to their current location from more outlying regions of their protoplanetary disk.

11. SUMMARY AND FUTURE PROSPECTS

Thanks to improvements in isotopic chronology we now know the timescales over which the Earth, Moon, Mars, and some asteroids formed. Terrestrial-planet accretion started soon after the solar system formed, leading to the growth of some Mars-sized and smaller objects within the first million years or so. This early accretionary phase was accompanied by widespread melting due to heat generated by short-lived isotopes and the formation of planetary cores. The Moon formed relatively late, at least 30 Myrs after the start of the solar system. This was the last major event in Earth's formation. These isotopic timescales are consistent with theoretical models that predict rapid runaway and oligarchic growth at early times, to form asteroid- to Mars-sized bodies within a million years, while predicting that Earth took tens of millions of years to grow to its final size.

The presence in Earth's mantle of nonnegligible amounts of siderophile elements such as platinum and osmium argues that roughly 1% of Earth's mass arrived after its core had finished forming. For some time it has been postulated that Earth formed in a very dry

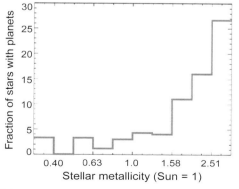

FIGURE 2.21 The fraction of stars that have planets as a function of the stellar metallicity (the abundance of elements heavier than helium compared to the Sun). Here, the iron to hydrogen ratio relative to the Sun is used a proxy for metallicity.

environment and that its water was delivered along with these siderophile elements in a late veneer. This now appears unlikely given the composition of Earth's mantle. Instead, Earth probably acquired its water earlier, perhaps from carbonaceous-chondrite-like asteroids, before core formation was complete. This implies that the planet held onto much of its water during the giant impact that led to the formation of the Moon.

It now seems that chondrites, the most primitive meteorites in our collection both physically and chemically, actually formed at a rather late stage, after the parent bodies of the iron meteorites had formed. Chondrites escaped melting because the potent heat sources ^{26}Al and ^{60}Fe had largely decayed by that point. For a long time it has been thought that chondrites, or something similar, provided the basic building blocks of Earth and the other terrestrial planets, but it now seems that the parent bodies of the iron meteorites provide a better analogue in this respect. Currently, we do not have good dynamical or cosmochemical models for how chondrites and their constituents formed. Chondrules, CAIs, matrix grains, and presolar grains all survived in the nebula for several million years, undergoing different degrees of thermal processing, and then were collected together into large bodies. The refractory CAIs may have formed close to the Sun prior to being scattered across the disk, perhaps by turbulent motions within the gas. Supporting evidence for this hypothesis comes from the recent discovery of high-temperature condensates in samples from comet Wild 2 returned by the Stardust mission. Where chondrules formed remains unclear, but these objects would have been highly mobile as long as nebular gas was present, and they may have drifted radially over large distances.

The origin of giant planets remains a subject of debate, but the observed correlation between stellar metallicity and the presence of giant planets, and the recent discovery of a Saturn-mass extrasolar planet that appears to have a very massive core, lend weight to the core accretion model. Recent simulations using plausible envelope opacities have found that giant planets can form within the typical lifetime of a protoplanetary disk, overcoming a long-standing obstacle for core accretion. It is becoming apparent that planetary migration is an important feature in the formation and early evolution of planetary systems. This presumably explains the fact that extrasolar planets are seen to orbit their stars at a wide range of distances. Planets also migrate when they clear away residual planetesimals. This may have led to a dramatic episode early in the history of the solar system associated with the late heavy bombardment of comets and asteroids onto the Moon and inner planets.

It is impressive to look back on the past two decades of discovery in planetary science partly because the breakthroughs have involved so many diverse areas of research. Technology has been a key driver, be it in the form of more powerful computers, mass spectrometers, instrumentation for planetary missions, or new telescopes and detectors. The near future looks equally exciting. The Atacama Large Millimeter Array promises to transform our knowledge of protoplanetary disks with very high spatial resolution able to observe features as small as 1 AU in size, and sufficient sensitivity to detect many new molecules including organic materials. Space missions will continue to expand our survey of the solar system, such as the New Horizons and Juno probes en route to Pluto and Jupiter, respectively, and the Rosetta spacecraft heading for comet Churyumov-Gerasimenko. In addition to National Aeronautics and Space Administration and European Space Agency, space agencies in Japan, China, and India are also becoming active players in space exploration. The Doppler radial velocity and transit techniques continue to be refined and are set to expand the catalogue of known extrasolar planets. The relatively new microlensing technique is opening up the possibility of finding Earth-mass planets. With luck, the ongoing Kepler mission should finally answer the question of whether Earth-sized planets are common or relatively rare. Here on Earth, continuing analysis of dust samples from comet Wild 2 returned by the Stardust mission, and solar wind samples from the Genesis mission, will enhance our understanding of the cosmochemical evolution of the solar system. New isotopic measurement techniques and a new generation of dynamic secondary ion mass spectrometers or ion probes are sure to generate exciting discoveries at a rapid pace. All in all, we have much to look forward to.

BIBLIOGRAPHY

Halliday, A. N. (2003). The origin and earliest history of the Earth. In A. M. Davies (Series Ed.) & H. D. Holland, & K. K. Turekian (Vol. Eds.), *Meteorites, comets and planets: Vol. 1. Treatise of Geochemistry* (2nd ed.). (pp. 509–557). Oxford: Elsevier-Pergamon.

Halliday, A. N., & Porcelli, D. (2001). In search of lost planets – the paleocosmochemistry of the inner solar system. *Earth and Planetary Science Letters, 192*, 545–559.

Halliday, A. N., & Kleine, T. (2006). Meteorites and the timing, mechanisms and conditions of terrestrial planet accretion and early differentiation. In D. Lauretta, L. Leshin, & H. MacSween (Eds.), *Meteorites and the early solar system II* (pp. 775–801). Univ. Arizona Press.

Lee, T., Papanastassiou, D. A., & Wasserburg, G. J. (1976). *Astrophysical Journal, 211*, L107.

Lewis, J. S. (2004). *Physics and chemistry of the solar system*. Academic Press.

de Pater, I., & Lissauer, J. J. (2001). *Planetary sciences*. Cambridge University Press.

Reipurth, B., Jewitt, D., & Keil, K. (2006). *Protostars and planets V*. University of Arizona Press.

Chapter 3

Solar System Dynamics: Regular and Chaotic Motion

Jack J. Lissauer
Space Science & Astrobiology Division, NASA Ames Research Center Moffett Field, CA, USA

Carl D. Murray
Astronomy Unit, Queen Mary University of London, London, England, UK

Chapter Outline

1. Introduction: Keplerian Motion — 55
   1.1. Kepler's Laws of Planetary Motion — 56
   1.2. Elliptical Motion, Orbital Elements, and the Orbit in Space — 56
2. The Two-Body Problem — 57
   2.1. Newton's Laws of Motion and the Universal Law of Gravitation — 57
   2.2. Reduction to the One-Body Case — 57
   2.3. Energy, Circular Velocity, and Escape Velocity — 58
   2.4. Orbital Elements: Elliptical, Parabolic, and Hyperbolic Orbits — 58
3. Planetary Perturbations and the Orbits of Small Bodies — 58
   3.1. Perturbed Keplerian Motion and Resonances — 59
   3.2. Examples of Resonances: Lagrangian Points and Tadpole and Horseshoe Orbits — 60
      3.2.1. Horseshoe and Tadpole Orbits — 60
      3.2.2. Hill Sphere — 60
   3.3. Examples of Resonances: Ring Particles and Shepherding — 61
4. Chaotic Motion — 63
   4.1. Concepts of Chaos — 63
   4.2. The Three-Body Problem as a Paradigm — 63
      4.2.1. Regular Orbits — 64
      4.2.2. Chaotic Orbits — 65
      4.2.3. Location of Regular and Chaotic Regions — 67
5. Orbital Evolution of Minor Bodies — 68
   5.1. Asteroids — 68
   5.2. Meteorites — 70
   5.3. Comets — 70
   5.4. Small Satellites and Rings — 71
6. Long-Term Stability of Planetary Orbits — 71
   6.1. The N-Body Problem — 71
   6.2. Stability of the Solar System — 72
7. Dissipative Forces and the Orbits of Small Bodies — 72
   7.1. Radiation Force (Micron-Sized Particles) — 72
   7.2. Poynting—Robertson Drag (Centimeter-Sized Grains) — 73
   7.3. Yarkovsky Effect (Meter-Sized Objects) — 73
   7.4. Gas Drag — 73
   7.5. Tidal Interactions and Planetary Satellites — 74
   7.6. Tidal Evolution and Resonances — 75
8. Chaotic Rotation — 76
   8.1. Spin—Orbit Resonance — 76
   8.2. Hyperion — 77
   8.3. Other Satellites — 77
   8.4. Chaotic Obliquity — 78
9. Epilog — 78
Bibliography — 78

1. INTRODUCTION: KEPLERIAN MOTION

The study of the motion of celestial bodies within our solar system has played a key role in the broader development of classical mechanics. In 1687, Isaac Newton published his *Principia*, in which he presented a unified theory of the motion of bodies in the heavens and on the Earth. Newtonian physics has been proved to provide a remarkably good description of a multitude of phenomena on a wide range of length scales. Many of the mathematical tools developed over the centuries to analyze planetary motions in the Newtonian framework have found applications for terrestrial phenomena. Deviations of the orbit of Uranus from that predicted by **Newton's Laws** led to the discovery of the planet Neptune. However, Newtonian gravity is only an approximation to Einstein's general theory of relativity.

This was used to explain deviations of Mercury's orbit that could not be accounted for by Newtonian physics. But general relativistic corrections to planetary motions are quite small, so this chapter concentrates on the rich and varied effects of Newtonian gravitation, together with briefer descriptions of nongravitational forces that affect the motions of some objects in the solar system.

Newton showed that the motion of two spherically symmetric bodies resulting from their mutual gravitational attraction is described by simple conic sections (see Section 2.4). However, the introduction of additional gravitating bodies produces a rich variety of dynamical phenomena, even though the basic interactions between pairs of objects can be straightforwardly described. Even few-body systems governed by apparently simple nonlinear interactions can display remarkably complex behavior, which has come to be known collectively as chaos. The concept of **deterministic chaos**, now known to play a major role in weather patterns on the Earth, was first conceived in connection with planetary motions (by Poincaré, in the late nineteenth century). On sufficiently long timescales, the apparently regular orbital motion of many bodies in the solar system can exhibit symptoms of this chaotic behavior.

An object in the solar system exhibits chaotic behavior in its orbit or rotation if the motion is sensitively dependent on the starting conditions, such that small changes in its initial state produce different final states. Examples of **chaotic motion** in the solar system include the rotation of the Saturnian satellite Hyperion, the orbital evolution of numerous asteroids and comets, and the orbit of Pluto. Numerical investigations suggest that the motion of the planetary system as a whole is chaotic, although there are currently no signs of any gross instability in the orbits of the planets. Chaotic motion has probably played an important role in determining the dynamical structure of the solar system, particularly in its early history.

In this chapter, the basic orbital properties of solar system objects (planets, moons, minor bodies, and dust) and their mutual interactions are described. Several examples of important dynamical processes that occur in the solar system are provided and groundwork is laid for describing some of the phenomena that are discussed in more detail in other chapters of this book.

1.1. Kepler's Laws of Planetary Motion

By analyzing Tycho Brahe's careful observations of the orbits of the planets, Johannes Kepler deduced the following three laws of planetary motion:

1. All planets move along elliptical paths with the Sun at one focus. The heliocentric distance r (i.e. the planet's distance from the Sun) can be expressed as

$$r = \frac{a(1 - e^2)}{1 + e \cos f}, \qquad (3.1)$$

where a is the semimajor axis (average of the minimum and maximum heliocentric distances) and e (the eccentricity of the orbit) $= (1 - b^2/a^2)^{1/2}$, where $2b$ is the minor axis of an ellipse. The true anomaly, f, is the angle between the planet's perihelion (closest heliocentric distance) and its instantaneous position (Figure 3.1).

2. A line connecting a planet and the Sun sweeps out equal areas ΔA in equal periods of time Δt:

$$\frac{\Delta A}{\Delta t} = \text{constant}. \qquad (3.2)$$

Note that the value of this constant differs from one planet to the next.

3. The square of a planet's orbital period P about the Sun (in years) is equal to the cube of its semimajor axis a (in AU):

$$P^2 = a^3. \qquad (3.3)$$

1.2. Elliptical Motion, Orbital Elements, and the Orbit in Space

The Sun contains more than 99.8% of the mass of the known solar system. The gravitational force exerted by a body is proportional to its mass (Eqn (3.5)), so to an excellent first approximation, the motion of the planets and many other bodies can be regarded as being solely due to the influence of a fixed central pointlike mass. For bound objects like the planets, which cannot go arbitrarily far from the Sun, the general solution for the orbit is the ellipse described by Eqn (3.1). The orbital plane, although fixed in

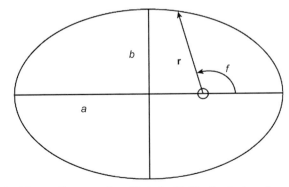

FIGURE 3.1 Geometry of an elliptical orbit. The Sun is at one focus and the vector **r** denotes the instantaneous heliocentric location of the planet (i.e. r is the planet's distance from the Sun). a is the semimajor axis (average heliocentric distance) and b is the semiminor axis of the ellipse. The true anomaly, f, is the angle between the planet's perihelion (closest heliocentric distance) and its instantaneous position.

space, can be arbitrarily oriented with respect to whatever reference plane is chosen (such as Earth's orbital plane about the Sun, which is called the **ecliptic**, or the equator of the primary). The inclination, i, of the orbital plane is the angle between the reference plane and the orbital plane and can range from $0°$ to $180°$. Conventionally, if the orbital angular momentum of the body is aligned with the rotational angular momentum of the primary[1] (or, for heliocentric orbits, with the orbital angular momentum of the Earth), then the inclination is defined to be in the $0°-90°$ range and the orbit is said to be prograde. Bodies traveling in the opposite direction are defined to have inclinations from $90°$ to $180°$ and are said to be on retrograde orbits. The two planes intersect in a line called the line of nodes and the orbit pierces the reference plane at two locations— one as the body passes upward through the plane (the ascending node) and one as it descends (the descending node). A fixed direction in the reference plane is chosen and the angle to the direction of the orbit's ascending node is called the longitude of the ascending node, Ω. Finally, the angle between the line to the ascending node and the line to the direction of **periapse** (perihelion for orbits about the Sun, perigee for orbits about the Earth) is called the argument of periapse, ω. An additional angle, the longitude of periapse $\bar{\omega} = \omega + \Omega$ is sometimes used in place of ω. The six orbital elements $a, e, i, \Omega, \omega, M$, and f uniquely specify the location of the object in space (Figure 3.2). The first three quantities (a, e, and i) are often referred to as the principal orbital elements, as they describe the orbit's size, shape, and tilt, respectively.

2. THE TWO-BODY PROBLEM

In this section, the general solution to the problem of the motion of two objects under the effects of their mutual gravitational interaction is discussed.

2.1. Newton's Laws of Motion and the Universal Law of Gravitation

Although Kepler's laws were originally found from careful observation of planetary motion, they were subsequently shown to be derivable from Newton's laws of motion together with his universal law of gravity. Consider a body of mass m_1 at instantaneous location $\mathbf{r_1}$ with instantaneous velocity $\mathbf{v_1} = d\mathbf{r}_1/dt$ and hence momentum $\mathbf{p}_1 = m_1\mathbf{v}_1$. The acceleration $d\mathbf{v}_1/dt$ produced by a net force $\mathbf{F}_1$ is given by Newton's second law of motion:

$$\mathbf{F}_1 = \frac{d(m_1 \mathbf{v}_1)}{dt}. \quad (3.4)$$

Newton's universal law of gravity states that a second body of mass m_2 at position $\mathbf{r}_2$ exerts an attractive force on the first body given by

$$\mathbf{F}_1 = -\frac{Gm_1m_2}{r_{12}^3}\mathbf{r}_{12} = -\frac{Gm_1m_2}{r_{12}^2}\hat{\mathbf{r}}_{12}, \quad (3.5)$$

where $\mathbf{r}_{12} = \mathbf{r}_1 - \mathbf{r}_2$ is the location of particle 1 with respect to particle 2, $\hat{\mathbf{r}}_{12}$ is the unit vector in the direction of $\mathbf{r}_{12}$, and G is the gravitational constant. Newton's third law states that for every action there is an equal and opposite reaction; thus, the force on each object of a pair is equal in magnitude but opposite in direction. These facts are used to reduce the two-body problem to an equivalent one-body case in the next subsection.

2.2. Reduction to the One-Body Case

From the foregoing discussion of Newton's laws, and the two-body problem, the force exerted by body 1 *on* body 2 is

$$\frac{d(m_2\mathbf{v}_2)}{dt} = \mathbf{F}_2 = -\mathbf{F}_1 = \frac{Gm_1m_2}{r_{12}^3}\mathbf{r}_{12} = \frac{Gm_1m_2}{r_{12}^2}\hat{\mathbf{r}}_{12} \quad (3.6)$$

Thus, from Eqns (3.4) and (3.6)

$$\frac{d(m_1\mathbf{v}_1 + m_2\mathbf{v}_2)}{dt} = \mathbf{F}_1 + \mathbf{F}_2 = 0. \quad (3.7)$$

This is of course a statement that the total linear momentum of the system is conserved, which means that the center of mass of the system moves with constant velocity.

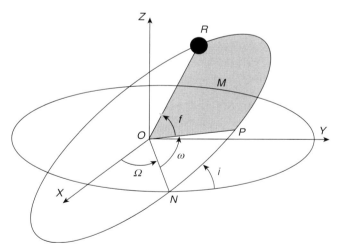

FIGURE 3.2 Geometry of an orbit in three dimensions. The Sun is at one focus of the ellipse (O), and the planet is instantaneously at location R. The location of the perihelion of the orbit is P. The intersection of the orbital plane ($X-Y$) and the reference plane is along the line ON (where N is the ascending node). The various angles shown are described in the text. The mean anomaly M is an angle proportional to the area OPR swept out by the radius vector OR (Kepler's second law).

---

1. That is, the dot product of the two angular momenta is nonnegative.

Multiplying Eqn (3.6) by m_1 and Eqn (3.5) by m_2 and subtracting, the equation for the relative motion of the bodies can be cast in the form

$$\mu_r \frac{d^2 \mathbf{r}_{12}}{dt^2} = \mu_r \frac{d^2(\mathbf{r}_1 - \mathbf{r}_2)}{dt^2} = -\frac{G\mu_r M}{r_{12}^3}\mathbf{r}_{12}, \quad (3.8)$$

where $\mu_r \equiv m_1 m_2/(m_1+m_2)$ is called the reduced mass and $M \equiv m_1 + m_2$ is the total mass. Thus, the relative motion is completely equivalent to that of a particle of reduced mass μ_r orbiting *a fixed* central mass M. For known masses, specifying the elements of the relative orbit and the positions and velocities of the center of mass is completely equivalent to specifying the positions and velocities of both bodies. A detailed solution of the equation of motion (Eqn 3.8) is discussed in any elementary text on orbital mechanics and in most general classical mechanics books. In the remainder of Section 2, a few key results are given.

2.3. Energy, Circular Velocity, and Escape Velocity

The centripetal force necessary to keep an object of mass μ_r in a circular orbit of radius r with speed v_c is $\mu_r v_c^2/r$. Equating this to the gravitational force exerted by the central body of mass M, the circular velocity is

$$v_c = \sqrt{\frac{GM}{r}}. \quad (3.9)$$

Thus the orbital period (the time to move once around the circle) is

$$P = 2\pi r/v_c = 2\pi\sqrt{\frac{r^3}{GM}}. \quad (3.10)$$

The total (kinetic plus potential) energy E of the system is a conserved quantity:

$$E = T + V = \frac{1}{2}\mu_r v^2 - \frac{GM\mu_r}{r}, \quad (3.11)$$

where the first term on the right is the kinetic energy of the system, T, and the second term is the potential energy of the system, V. If $E < 0$, the absolute value of the potential energy of the system is larger than its kinetic energy, and the system is bound. The body will orbit the central mass on an elliptical path. If $E > 0$, the kinetic energy is larger than the absolute value of the potential energy, and the system is unbound. The relative orbit is then described mathematically as a hyperbola. If $E = 0$, the kinetic and potential energies are equal in magnitude, and the relative orbit is a parabola. By setting the total energy equal to zero, the escape velocity at any separation can be calculated:

$$v_e = \sqrt{\frac{2GM}{r}} = \sqrt{2}v_c. \quad (3.12)$$

For circular orbits it is easy to show (using Eqns (3.9) and (3.11)) that both the kinetic energy and the total energy of the system are equal in magnitude to half the potential energy:

$$T = -\frac{1}{2}V, \quad (3.13)$$

$$E = -\frac{GM\mu_r}{2r}. \quad (3.14)$$

For an elliptical orbit, Eqn (3.14) holds if the radius r is replaced by the semimajor axis a:

$$E = -\frac{GM\mu_r}{2a}. \quad (3.15)$$

Similarly, for an elliptical orbit, Eqn (3.10) becomes Newton's generalization of Kepler's third law:

$$P^2 = \frac{4\pi^2 a^3}{G(m_1 + m_2)}. \quad (3.16)$$

It can be shown that Kepler's second law follows immediately from the conservation of angular momentum, $\mathbf{L}$:

$$\frac{d\mathbf{L}}{dt} = \frac{d(\mu_r \mathbf{r} \times \mathbf{v})}{dt} = 0. \quad (3.17)$$

2.4. Orbital Elements: Elliptical, Parabolic, and Hyperbolic Orbits

As noted earlier, the relative orbit in the two-body problem is either an ellipse, a parabola, or a hyperbola depending on whether the energy is negative, zero, or positive, respectively. These curves are known collectively as conic sections and the generalization of Eqn (3.1) is

$$r = \frac{p}{1 + e\cos f}, \quad (3.18)$$

where r and f have the same meaning as in Eqn (3.1), e is the generalized **eccentricity**, and p, the semilatus rectum, is a conserved quantity that depends on the initial conditions. For an ellipse, $p = a(1-e^2)$, as in Eqn (3.1). For a parabola, $e = 1$ and $p = 2q$, where q is the pericentric separation (distance of closest approach). For a hyperbola, $e > 1$ and $p = q(1+e)$, where q is again the pericentric separation. For all orbits, the three orientation angles i, Ω, and ω are defined as in the elliptical case.

3. PLANETARY PERTURBATIONS AND THE ORBITS OF SMALL BODIES

Gravity is not restricted to interactions between the Sun and the planets or individual planets and their satellites, but rather all bodies feel the gravitational force of one another.

Within the solar system, one body typically produces the dominant force on any given body, and the resultant motion can be thought of as a Keplerian orbit about a primary, subject to small perturbations by other bodies. In this section, some important examples of the effects of these perturbations on the orbital motion are considered.

Classically, much of the discussion of the evolution of orbits in the solar system used perturbation theory as its foundation. Essentially, the method involves writing the equations of motion as the sum of a part that describes the independent Keplerian motion of the bodies about the Sun plus a part (called the disturbing function) that contains terms due to the pairwise interactions among the planets and minor bodies and the indirect terms associated with the backreaction of the planets on the Sun. In general, one can then expand the disturbing function in terms of the small parameters of the problem (such as the ratio of the planetary masses to the solar mass, the eccentricities and inclinations, etc.), as well as the other orbital elements of the bodies, including the mean longitudes (i.e. the location of the bodies in their orbits), and attempt to solve the resulting equations for the time dependence of the orbital elements.

3.1. Perturbed Keplerian Motion and Resonances

Although perturbations on a body's orbit are often small, they cannot always be ignored. They must be included in short-term calculations if high accuracy is required, for example, for predicting when an object passes in front of a star (stellar occultation) or targeting spacecraft. Most long-term perturbations are periodic in nature, their directions oscillating with the relative longitudes of the bodies or with some more complicated function of the bodies' orbital elements.

Small perturbations can produce large effects if the forcing frequency is commensurate or nearly commensurate with the natural frequency of oscillation of the responding elements. Under such circumstances, perturbations add coherently, and the effects of many small tugs can build up over time to create a large-amplitude, long-period response. This is an example of resonance forcing, which occurs in a wide range of physical systems.

An elementary example of resonance forcing is given by the simple one-dimensional harmonic oscillator, for which the equation of motion is

$$m \frac{d^2 x}{dt^2} + m\Gamma^2 x = F_0 \cos \varphi t. \qquad (3.19)$$

In Eqn (3.19), m is the mass of the oscillating particle, F_0 is the amplitude of the driving force, Γ is the natural frequency of the oscillator, and φ is the forcing or resonance frequency. The solution to Eqn (3.19) is

$$x = x_0 \cos \varphi t + A \cos \Gamma t + B \sin \Gamma t, \qquad (3.20a)$$

where

$$x_0 \equiv \frac{F_0}{m(\Gamma^2 - \varphi^2)}, \qquad (3.20b)$$

and A and B are constants determined by the initial conditions. Note that if $\varphi \approx \Gamma$, a large-amplitude, long-period response can occur even if F_0 is small. Moreover, if $\varphi = \Gamma$, this solution to Eqn (3.19) is invalid. In this case, the solution is given by

$$x = \frac{F_0}{2m\Gamma} t \sin \Gamma t + A \cos \Gamma t + B \sin \Gamma t. \qquad (3.21)$$

The t in front of the first term on the right-hand side of Eqn (3.21) leads to **secular** growth. Often this linear growth is moderated by the effects of nonlinear terms that are not included in the simple example provided here. However, some perturbations have a secular component.

Nearly exact orbital commensurabilities exist at many places in the solar system. Io orbits Jupiter twice as frequently as Europa does, which in turn orbits Jupiter twice as frequently as Ganymede does. Conjunctions (at which the bodies have the same longitude) always occur at the same position of Io's orbit (its perijove). How can such commensurabilities exist? After all, the probability of randomly picking a rational from the real number line is 0, and the number of small integer ratios is infinitely smaller still! The answer lies in the fact that orbital resonances may be held in place as stable locks, which result from nonlinear effects not represented in the foregoing simple mathematical example. For example, differential tidal recession (see Section 7.5) brings moons into resonance, and nonlinear interactions among the moons can keep them there. Other examples of resonance locks include the Hilda asteroids, the Trojan asteroids, Neptune–Pluto, and the pairs of moons about Saturn, Mimas–Tethys and Enceladus–Dione.

Resonant perturbation can also force material into highly eccentric orbits that may lead to collisions with other bodies; this is thought to be the dominant mechanism for clearing the Kirkwood gaps in the asteroid belt (see Section 5.1). Spiral density waves can propagate away from resonant locations in a self-gravitating particle disk perturbed by an orbiting satellite. Density waves are seen at many resonances in Saturn's rings; they explain most of the structure seen in Saturn's A ring. The vertical analogs of density waves, bending waves, are caused by resonant perturbations perpendicular to the ring plane due to a satellite in an orbit that is inclined to the ring. Spiral bending waves excited by the moons Mimas and Titan have been seen in Saturn's rings. In the next few subsections, these manifestations of resonance effects that do not explicitly involve chaos are discussed. Chaotic motion produced by resonant forcing is discussed later in the chapter.

3.2. Examples of Resonances: Lagrangian Points and Tadpole and Horseshoe Orbits

Many features of the orbits considered in this section can be understood by examining an idealized system in which two massive (but typically of unequal mass) bodies move in circular orbits about their common center of mass. If a third body is introduced that is much less massive than either of the first two, its motion can be followed by assuming that its gravitational force has no effect on the orbits of the other bodies. By considering the motion in a frame corotating with the massive pair (so that the pair remain fixed on a line that can be taken to be the x-axis), Lagrange found that there are five points where particles placed at rest would feel no net force in the rotating frame. Three of the so-called **Lagrange points** (L_1, L_2, and L_3) lie along a line joining the two masses m_1 and m_2. The other two Lagrange points (L_4 and L_5) form equilateral triangles with the two massive bodies.

Particles displaced slightly from the first three Lagrangian points will continue to move away and hence these locations are unstable. The triangular Lagrangian points are potential energy maxima, which are stable for sufficiently large primary to secondary mass ratio due to the Coriolis force. Provided that the most massive body has at least 25 times the mass of the secondary (which is the case for all known examples in the solar system larger than the Pluto–Charon system), the Lagrangian points L_4 and L_5 are stable points. Thus, a particle at L_4 or L_5 that is perturbed slightly will start to "orbit" these points in the rotating coordinate system. Lagrangian points L_4 and L_5 are important in the solar system. For example, the Trojan asteroids in Jupiter's Lagrangian points and both Neptune and Mars confine their own Trojans. There are also small moons in the triangular Lagrangian points of Tethys and Dione, in the Saturnian system. The L_4 and L_5 points in the Earth–Moon system have been suggested as possible locations for space stations.

3.2.1. Horseshoe and Tadpole Orbits

Consider a moon on a circular orbit about a planet. Figure 3.3 shows some important dynamical features in the frame corotating with the moon. All five Lagrangian points are indicated in the picture. A particle just interior to the moon's orbit has a higher angular velocity than the moon in the stationary frame and thus moves with respect to the moon in the direction of corotation. A particle just outside the moon's orbit has a smaller angular velocity and moves away from the moon in the opposite direction. When the outer particle approaches the moon, the particle is slowed down (loses angular momentum) and, provided the initial difference in semimajor axis is not too large, the particle drops to an orbit lower than that of the moon. The particle then recedes in the forward direction. Similarly, the particle at the lower orbit is

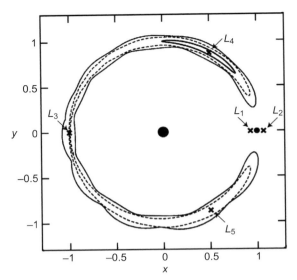

FIGURE 3.3 Diagram showing the five Lagrangian equilibrium points (denoted by crosses) and three representative orbits near these points for the circular restricted three-body problem. In this example, the secondary's mass is 0.001 times the total mass. The coordinate frame has its origin at the barycenter and corotates with the pair of bodies, thereby keeping the primary (large solid circle) and secondary (small solid circle) fixed on the x-axis. Tadpole orbits remain near one or the other of the L_4 and L_5 points. An example is shown near the L_4 point on the diagram. Horseshoe orbits enclose all three of L_3, L_4, and L_5 but do not reach L_1 or L_2. The outermost orbit on the diagram illustrates this behavior. There is a critical curve dividing tadpole and horseshoe orbits that encloses L_4 and L_5 and passes through L_3. A horseshoe orbit near this dividing line is shown as the dashed curve in the diagram.

accelerated as it catches up with the moon, resulting in an outward motion toward the higher, slower orbit. Orbits like these encircle the L_3, L_4, and L_5 points and are called **horseshoe orbits**. Saturn's small moons Janus and Epimetheus execute just such a dance, changing orbits every 4 years.

Since the Lagrangian points L_4 and L_5 are stable, material can librate about these points individually: such orbits are called **tadpole orbits**. The tadpole **libration** width at L_4 and L_5 is roughly equal to $(m/M)^{1/2}r$, and the horseshoe width is $(m/M)^{1/3}r$, where M is the mass of the planet, m the mass of the satellite, and r the distance between the two objects. For a planet of Saturn's mass, $M = 5.7 \times 10^{29}$ g, and a typical small moon of mass $m = 10^{20}$ g (e.g. an object with a 30-km radius, with density of ~ 1 g/cm^3), at a distance of 2.5 Saturnian radii, the tadpole libration half-width is about 3 km and the horseshoe half-width is about 60 km.

3.2.2. Hill Sphere

The approximate limit to a planet's gravitational dominance is given by the extent of its **Hill sphere**,

$$R_H = \left[\frac{m}{3(M+m)}\right]^{1/3} a, \qquad (3.22)$$

where m is the mass of the planet and M is the Sun's mass. A test body located at the boundary of a planet's Hill sphere is subjected to a gravitational force from the planet that is comparable in magnitude to the tidal difference between the force of the Sun on the planet and that on the test body. The Hill sphere essentially stretches out to the L_1 point and is roughly the limit of the Roche lobe (maximum extent of an object held together by gravity alone) of a body with $m \ll M$. Planetocentric orbits that are stable over long periods are those well within the boundary of a planet's Hill sphere; the overwhelming majority of natural satellites lie in this region. The trajectories of the outermost planetary satellites, which lie closest to the boundary of the Hill sphere, show large variations in planetocentric orbital paths (Figure 3.4). Stable heliocentric orbits are those that are always well outside the Hill sphere of any planet.

3.3. Examples of Resonances: Ring Particles and Shepherding

In the discussions in Section 2, the gravitational force produced by a spherically symmetric body was described. In this section, the effects of deviations from spherical symmetry must be included when computing the force. This is most conveniently done by introducing the gravitational potential $\Phi(r)$, which is defined such that the acceleration $d^2\mathbf{r}/dt^2$ of a particle in the gravitational field is

$$d^2\mathbf{r}/dt^2 = \nabla\Phi. \quad (3.23)$$

In empty space, the Newtonian gravitational potential $\Phi(r)$ always satisfies Laplace's equation

$$\nabla^2 \Phi = 0. \quad (3.24)$$

Most planets are very nearly axisymmetric, with the major departure from sphericity being due to a rotationally induced equatorial bulge. Thus, the gravitational potential can be expanded in terms of Legendre polynomials instead of the complete spherical harmonic expansion, which would be required for the potential of a body of arbitrary shape:

$$\Phi(r,\phi,\theta) = -\frac{Gm}{r}\left[1 - \sum_{n=2}^{\infty} J_n P_n(\cos\theta)(R/r)^n\right]. \quad (3.25)$$

This equation uses standard spherical coordinates, so that θ is the angle between the planet's symmetry axis and the vector to the particle. The terms $P_n(\cos\theta)$ are the Legendre polynomials and J_n are the gravitational moments determined by the planet's mass distribution. If the planet's mass distribution is symmetrical about the planet's equator,

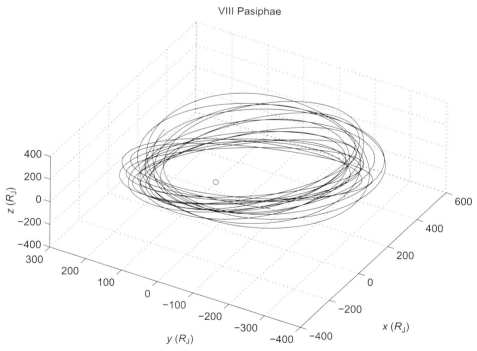

FIGURE 3.4 The orbit of J VIII Pasiphae, a distant retrograde satellite of Jupiter, is shown in a nonrotating coordinate system with Jupiter at the origin (open circle). The satellite was integrated as a massless test particle in the context of the circular restricted three-body problem for approximately 38 years. The unit of distance is Jupiter's radius, R_J. During the course of this integration, the distance to Jupiter varied from 122 to 548 R_J. Note how the large solar perturbations produce significant deviations from a Keplerian orbit. *Figure reprinted with permission from Jose Alvarellos (1996). "Orbital Stability of Distant Satellites of Jovian planets," M.Sc. thesis, San Jose State University.*

J_n are zero for odd n. For large bodies, J_2 is generally substantially larger than the other gravitational moments.

Consider a particle in Saturn's rings, which revolves around the planet in a circular orbit in the equatorial plane ($\theta = 90°$) at a distance r from the center of the planet. The centripetal force must be provided by the radial component of the planet's gravitational force (see Eqn (3.9)), so the particle's angular velocity n satisfies

$$rn^2(r) = \left[\frac{\partial \Phi}{\partial r}\right]_{\theta = 90°}. \tag{3.26}$$

If the particle suffers an infinitesimal displacement from its circular orbit, it will oscillate freely in the horizontal and vertical directions about the reference circular orbit with radial (epicyclic) frequency $\kappa(r)$ and vertical frequency $\mu(r)$, respectively, given by

$$\kappa^2(r) = r^{-3}\frac{d}{dr}\left[(r^2 n)^2\right], \tag{3.27}$$

$$\mu^2(r) = \left[\frac{\partial^2 \Phi}{\partial z^2}\right]_{z=0}. \tag{3.28}$$

From Eqns (3.24)–(3.28), the following relation is found between the three frequencies for a particle in the equatorial plane:

$$\mu^2 = 2n^2 - \kappa^2. \tag{3.29}$$

For a perfectly spherically symmetric planet, $\mu = k = n$. Since Saturn and the other ringed planets are oblate, μ is slightly higher and k is slightly lower than the orbital frequency n.

Using Eqns (3.24)–(3.29), one can show that the orbital and epicyclic frequencies can be written as

$$n^2 = \frac{GM}{r^3}\left[1 + \frac{3}{2}J_2\left(\frac{R}{r}\right)^2 - \frac{15}{8}J_4\left(\frac{R}{r}\right)^4 \right.$$
$$\left. + \frac{35}{16}J_6\left(\frac{R}{r}\right)^6 + ...\right], \tag{3.30}$$

$$\kappa^2 = \frac{GM}{r^3}\left[1 - \frac{3}{2}J_2\left(\frac{R}{r}\right)^2 + \frac{45}{8}J_4\left(\frac{R}{r}\right)^4 \right.$$
$$\left. - \frac{175}{16}J_6\left(\frac{R}{r}\right)^6 + ...\right], \tag{3.31}$$

$$\mu^2 = \frac{GM}{r^3}\left[1 + \frac{9}{2}J_2\left(\frac{R}{r}\right)^2 - \frac{75}{8}J_4\left(\frac{R}{r}\right)^4 \right.$$
$$\left. + \frac{245}{16}J_6\left(\frac{R}{r}\right)^6 + ...\right]. \tag{3.32}$$

Thus, for a particle orbit that is nearly equatorial, the oblateness of a planet causes the line of periapse to precess and the line of nodes to regress.

Resonances occur where the radial (or vertical) frequency of the ring particles is equal to the frequency of a component of a satellite's horizontal (or vertical) forcing, as experienced in the rotating frame of the particle. In this case, the resonating particle is always near the same phase in its radial (or vertical) oscillation when it experiences a particular phase of the satellite's forcing. This situation enables continued coherent "kicks" from the satellite to build up the particle's radial (or vertical) motion, and significant forced oscillations may thus result. The location and strengths of resonances with any given satellite can be determined by decomposing the gravitational potential of the satellite's effect on the ring particle into its Fourier components. The disturbance frequency, $\bar{\omega}$, can be written as the sum of integer multiples of the satellite's angular, vertical, and radial frequencies:

$$\bar{\omega} = jn_s + k\mu_s + \ell\kappa_s, \tag{3.33}$$

where the azimuthal symmetry number, j, is a nonnegative integer, and k and ℓ are integers, with k being even for horizontal forcing and odd for vertical forcing. The subscript s refers to the satellite. A particle placed at distance $r = r_L$ will undergo horizontal (Lindblad) resonance if r_L satisfies

$$\bar{\omega} - jn(r_L) = \pm\kappa(r_L). \tag{3.34}$$

It will undergo vertical resonance if its radial position, r_v, satisfies

$$\bar{\omega} - jn(r_L) = \pm\mu(r_v). \tag{3.35}$$

When Eqn (3.34) is valid for the lower (upper) sign, r_L is referred to as the inner (outer) Lindblad or horizontal resonance. The distance r_v is called an inner (outer) vertical resonance if Eqn (3.35) is valid for the lower (upper) sign. Since all of Saturn's large satellites orbit the planet well outside the main ring system, the satellite's angular frequency n_s is less than the angular frequency of the particle and inner resonances are more important than the outer ones. When $j \neq 1$, the approximation $\mu \approx n \approx k$ may be used to obtain the ratio

$$\frac{n(r_{L,v})}{n_s} = \frac{j+k+\ell}{j-1}. \tag{3.36}$$

The notation $(j+k+\ell)/(j-1)$ or $(j+k+\ell):(j-1)$ is commonly used to identify a given resonance.

The strength of the forcing by the satellite depends, to lowest order, on the satellite's eccentricity, e, and inclination, i, as $e^{|\ell|}(\sin i)^{|k|}$. The strongest horizontal resonances have $k = \ell = 0$, and are of the form $j:(j-1)$. The strongest vertical resonances have $k = 1$, $\ell = 0$, and are of the form

$(j+1):(j-1)$. The location and strengths of such orbital resonances can be calculated from known satellite masses and orbital parameters and Saturn's gravity field. Most strong resonances in the Saturnian system lie in the outer A ring, near the orbits of the moons responsible for them. If $n = \mu = k$, the locations of the horizontal and vertical resonances would be identical: $r_L = r_v$. However, owing to Saturn's oblateness, $\mu > n > k$ and the positions r_L and r_v do not coincide, i.e. $r_v < r_L$. A detailed discussion of spiral density waves, spiral bending waves, and gaps at resonances produced by moons is presented elsewhere in this encyclopedia. (See Planetary Rings.)

4. CHAOTIC MOTION

4.1. Concepts of Chaos

In the nineteenth century, Henri Poincaré studied the mathematics of the circular restricted **three-body problem**. In this problem, one mass (the secondary) moves in a fixed, circular orbit about a central mass (the primary), while a massless (test) particle moves under the gravitational effect of both masses but does not perturb their orbits. From this work, Poincaré realized that despite the simplicity of the equations of motion, some solutions to the problem exhibit complicated behavior.

Poincaré's work in celestial mechanics provided the framework for the modern theory of nonlinear dynamics and ultimately led to a deeper understanding of the phenomenon of chaos, whereby dynamical systems described by simple equations can give rise to unpredictable behavior. The whole question of whether or not a given system is stable to sufficiently small perturbations is the basis of the Kolmogorov-Arnol'd-Moser theory, which has its origins in the work of Poincaré.

One characteristic of chaotic motion is that small changes in the starting conditions can produce vastly different final outcomes. Since all measurements of positions and velocities of objects in the solar system have finite accuracy, relatively small uncertainties in the initial state of the system can lead to large errors in the final state for initial conditions that lie in chaotic regions in **phase space**.

This is an example of what has become known as the "butterfly effect", first mentioned in the context of chaotic weather systems. It has been suggested that under the right conditions, a small atmospheric disturbance (such as the flapping of a butterfly's wings) in one part of the world could ultimately lead to a hurricane in another part of the world.

The changes in an orbit that reveal it to be chaotic may occur very rapidly, for example, during a close approach to the planet, or very slowly as perturbations accumulate over millions or even billions of years. Although there have been a number of significant mathematical advances in the study of nonlinear dynamics since Poincaré's time, the digital computer has been proved to be the most important tool in investigating chaotic motion in the solar system. This is particularly true in studies of the gravitational interaction of all the planets, where there are few analytical results.

4.2. The Three-Body Problem as a Paradigm

The characteristics of chaotic motion are common to a wide variety of dynamical systems. In the context of the solar system, the general properties are best described by considering the planar circular restricted three-body problem. This idealization consists of a (massless) test particle and two bodies of masses m_1 and m_2 moving in circular orbits about their common center of mass at constant separation, with all bodies moving in the same plane. The test particle is attracted to each mass under the influence of the inverse square law of force given in Eqn (3.5). In Eqn (3.16), a is the constant separation of the two masses and $n = 2\pi/P$ is their constant angular velocity about the center of mass. Using x and y as components of the position vector of the test particle referred to the center of mass of the system (Figure 3.5), the equations of motion of the particle in a reference frame rotating at angular velocity n are

$$\ddot{x} - 2n\dot{y} - n^2 x = -G\left(m_1 \frac{x+\mu_2}{r_1^3} - m_2 \frac{x-\mu_1}{r_2^3}\right). \quad (3.37)$$

$$\ddot{y} + 2n\dot{x} - n^2 y = -G\left(\frac{m_1}{r_1^3} + \frac{m_2}{r_2^3}\right) y, \quad (3.38)$$

where $\mu_1 \equiv m_1 a/(m_1 + m_2)$ and $\mu_2 \equiv m_2 a/(m_1 + m_2)$ are constants and

$$r_1^2 = (x + \mu_2)^2 + y^2, \quad (3.39)$$

$$r_2^2 = (x - \mu_1)^2 + y^2, \quad (3.40)$$

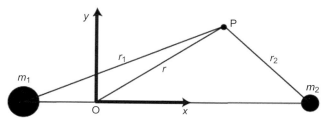

FIGURE 3.5 The rotating coordinate system used in the circular restricted three-body problem. The masses are at a fixed distance from one another and this is taken to be the unit of length. The position and velocity vectors of the test particle (at point P) are referred to the center of mass of the system at O.

where r_1 and r_2 are the distances of the test particle from the masses m_1 and m_2, respectively.

These two second-order, coupled, nonlinear differential equations can be solved numerically provided the initial position (x_0, y_0) and velocity $(\dot{x}_0, \dot{y}_0)$ of the particle are known. Therefore the system is deterministic, and at any given time, the orbital elements of the particle (such as its semimajor axis and eccentricity) can be calculated from its initial position and velocity.

The region of space open to the test particle may be constrained by the existence of a constant of the motion called the Jacobi constant, C, given by

$$C = n^2(x^2 + y^2) + 2G\left(\frac{m_1}{r_1} + \frac{m_2}{r_2}\right) - \dot{x}^2 - \dot{y}^2. \quad (3.41)$$

The values of (x_0, y_0) and $(\dot{x}_0, \dot{y}_0)$ fix the value of C for the system, and this value is preserved for all subsequent motion. At any instant, the particle is at some position on the two-dimensional (x, y) plane. However, since the actual orbit is also determined by the components of the velocity $(\dot{x}, \dot{y})$, the particle can also be thought of as being at a particular position in a four-dimensional (4D) $(x, y, \dot{x}, \dot{y})$ phase space. Note that the use of four dimensions rather than the customary two is simply a means of representing the position *and* the velocity of the particle at a particular instant in time; the particle's motion is always restricted to the x–y plane. The existence of the Jacobi constant implies that the particle is not free to wander over the entire 4D phase space, but rather that its motion is restricted to the three-dimensional "surface" defined by Eqn (3.41). This has an important consequence for studying the evolution of orbits in the problem.

The usual method is to solve the equations of motion; convert $x, y, \dot{x}$, and $\dot{y}$ into orbital elements such as semimajor axis, eccentricity, longitude of periapse, and mean longitude; and then plot the variation of these quantities as a function of time. However, another method is to produce a **surface of section**, also called a Poincaré map. This makes use of the fact that the orbit is always subject to Eqn (3.41), where C is determined by the initial position and velocity. Therefore, if any three of the four quantities $x, y, \dot{x}$, and $\dot{y}$ are known, the fourth can always be determined by solving Eqn (3.41). One common surface of section that can be obtained for the planar circular restricted three-body problem is a plot of values of x and $\dot{x}$ whenever $y = 0$ and $\dot{y}$ is positive. The actual value of $\dot{y}$ can always be determined uniquely from Eqn (3.41), and so the two-dimensional $(x, \dot{x})$ plot implicitly contains all the information about the particle's location in the 4D phase space. Although surfaces of section make it more difficult to study the evolution of the orbital elements, they have the advantage of revealing the characteristic motion of the particle (regular or chaotic) and a number of orbits can be displayed on the same diagram.

As an illustration of the different types of orbits that can arise, the results of integrating a number of orbits using a mass $m_2/(m_1 + m_2) = 10^{-3}$ and Jacobi constant $C = 3.07$ are described next. In each case, the particle was started with the initial longitude of periapse $\bar{\omega}_0 = 0$ and initial mean longitude $\lambda_0 = 0$. This corresponds to $\dot{x} = 0$ and $y = 0$. Since the chosen mass ratio is comparable to that of the Sun–Jupiter system, and Jupiter's eccentricity is small, this will be used as a good approximation to the motion of fictitious asteroids moving around the Sun under the effect of gravitational perturbations from Jupiter. The asteroid is assumed to be moving in the same plane as Jupiter's orbit.

4.2.1. Regular Orbits

The first asteroid has starting values $x = 0.55$, $y = 0$, $\dot{x} = 0$, with $\dot{y} = 0.9290$ determined from the solution of Eqn (3.41). Here a set of dimensionless coordinates are used in which $n = 1$, $G = 1$, and $m_1 + m_2 = 1$. In these units, the orbit of m_2 is a circle at distance $a = 1$ with uniform speed $v = 1$. The corresponding initial values of the heliocentric semimajor axis and eccentricity are $a_0 = 0.6944$ and $e_0 = 0.2065$. Since the semimajor axis of Jupiter's orbit is 5.202 AU, this value of a_0 would correspond to an asteroid at 3.612 AU.

Figure 3.6 shows the evolution of e as a function of time. The plot shows regular behavior with the eccentricity varying from 0.206 to 0.248 over the course of the integration. In fact, an asteroid at this location would be close to an orbit–orbit resonance with Jupiter, where the ratio of the orbital period of the asteroid, T, to Jupiter's period, T_J, is close to a rational number. From **Kepler's third law** of planetary motion, $T^2 \propto a^3$. In this case, $T/T_J = (a/a_J)^{3/2} = 0.564 \approx 4/7$ and the asteroid orbit is close to a 7:4

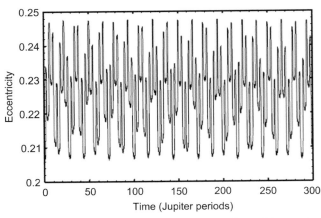

FIGURE 3.6 The eccentricity as a function of time for an object moving in a regular orbit near the 7:4 resonance with Jupiter. The plot was obtained by solving the circular restricted three-body problem numerically using initial values of 0.6944 and 0.2065 for the semimajor axis and eccentricity, respectively. The corresponding position and velocity in the rotating frame were $x_0 = 0.55$, $y_0 = 0$, $\dot{x} = 0$, and $\dot{y} = 0.9290$.

Chapter | 3 Solar System Dynamics: Regular and Chaotic Motion

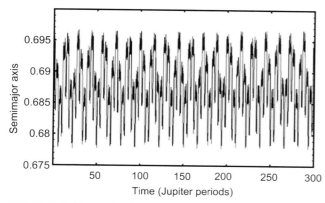

FIGURE 3.7 The semimajor axis as a function of time for an object using the same starting conditions as in Figure 3.6. The units of the semimajor axis are such that Jupiter's semimajor axis (5.202 AU) is taken to be unity.

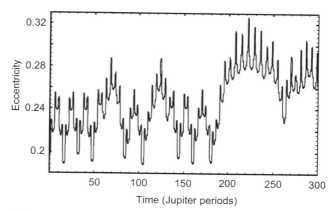

FIGURE 3.9 The eccentricity as a function of time for an object moving in a chaotic orbit starting just outside the 7:4 resonance with Jupiter. The plot was obtained by solving the circular restricted three-body problem numerically using initial values of 0.6984 and 0.1967 for the semimajor axis and eccentricity, respectively. The corresponding position and velocity in the rotating frame were $x_0 = 0.56$, $y_0 = 0$, $\dot{x}_0 = 0$, and $\dot{y} = 0.8998$.

resonance with Jupiter. Figure 3.7 shows the variation of the semimajor axis of the asteroid, a over the same time interval as shown in Figure 3.6. Although the changes in a are correlated with those in e, they are smaller in amplitude and a appears to oscillate about the location of the exact resonance at $a = (4/7)^{2/3} \approx 0.689$. An asteroid in resonance experiences enhanced gravitational perturbations from Jupiter, which can cause regular variations in its orbital elements. The extent of these variations depends on the asteroid's location within the resonance, which is, in turn, determined by the starting conditions.

The equations of motion can be integrated with the same starting conditions to generate a surface of section by plotting the values of x and $\dot{x}$ whenever $y = 0$ with $\dot{y} > 0$ (Figure 3.8). The pattern of three distorted curves or "islands" that emerges is a characteristic of resonant motion when displayed in such plots. If a resonance is of the form $(p+q){:}p$, where p and q are integers, then q is said to be the order of the resonance. The number of islands seen in a surface-of-section plot of a given resonant trajectory is equal to q. In this case, $p = 4$, $q = 3$, and three islands are visible.

The center of each island would correspond to a starting condition that placed the asteroid at exact resonance where the variation in e and a would be minimal. Such points are said to be fixed points of the Poincaré map. If the starting location was moved farther away from the center, the subsequent variations in e and a would get larger, until eventually some starting values would lead to trajectories that were not in resonant motion.

4.2.2. Chaotic Orbits

Figures 3.9 and 3.10 show the plots of e and a as a function of time for an asteroid orbit with starting values $x_0 = 0.56$,

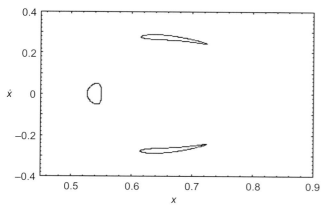

FIGURE 3.8 A surface-of-section plot for the same (regular) orbit shown in Figures 3.6 and 3.7. The 2000 points were generated by plotting the values of x and $\dot{x}$, whenever $y = 0$ with positive $\dot{y}$. The three "islands" in the plot are due to the third-order 7:4 resonance.

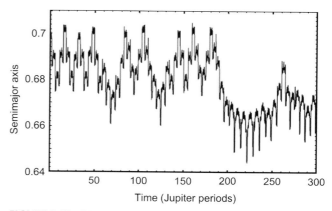

FIGURE 3.10 The semimajor axis as a function of time for an object using the same starting conditions as in Figure 3.9. The units of the semimajor axis are such that Jupiter's semimajor axis (5.202 AU) is taken to be unity.

$y_0 = 0$, and $\dot{x}_0 = 0$, and $\dot{y}$ determined from Eqn (3.41) with $C = 3.07$. The corresponding orbital elements are $a_0 = 0.6984$ and $e_0 = 0.1967$. These values are only slightly different from those used earlier, indeed the initial behavior of the plots is quite similar to that seen in Figures 3.6 and 3.7. However, subsequent variations in e and a are strikingly different. The eccentricity varies from 0.188 to 0.328 in an irregular manner, and the value of a is not always close to the value associated with exact resonance. This is an example of a chaotic trajectory where the variations in the orbital elements have no obvious periodic or quasi-periodic structure. The anticorrelation of a and e can be explained in terms of the Jacobi constant.

The identification of this orbit as chaotic becomes apparent from a study of its surface of section (Figure 3.11). Clearly, this orbit covers a much larger region of phase space than the previous example. Furthermore, the orbit does not lie on a smooth curve, but is beginning to fill an area of the phase space. The points also help to define a number of empty regions, three of which are clearly associated with the 7:4 resonance seen in the regular trajectory. There is also a tendency for the points to "stick" near the edges of the islands; this gives the impression of regular motion for short periods.

Chaotic orbits have the additional characteristic that they are sensitively dependent on initial conditions. This is illustrated in Figure 3.12, where the variation in e as a function of time is shown for two trajectories; the first corresponds to Figure 3.9 (where $x_0 = 0.56$) and the second has $x_0 = 0.56001$. The initial value of $\dot{y}$ was chosen so that the same value of C was obtained. Although both trajectories show comparable initial variations in e, after 60 Jupiter periods it is clear that the orbits have drifted apart.

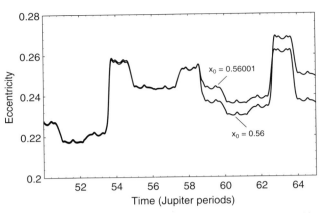

FIGURE 3.12 The variation in the eccentricity for two chaotic orbits starting close to one another. One plot is part of Figure 3.9 using the chaotic orbit that started with $x_0 = 0.56$ and the other is for an orbit with $x_0 = 0.56001$. Although the divergence of the two orbits is exponential, the effect becomes noticeable only after 60 Jupiter periods.

Such a divergence would not occur for nearby orbits in a regular part of the phase space.

The rate of divergence of nearby trajectories in such numerical experiments can be quantified by monitoring the evolution of two orbits that started close together. In a dynamical system such as the three-body problem, there are a number of quantities called the **Lyapunov characteristic exponents**. Measurement of the local divergence of nearby trajectories leads to an estimate of the largest of these exponents, and this can be used to determine whether or not the system is chaotic. If two orbits are separated in phase space by a distance d_0 at time t_0, and d is their separation at time t, then the orbit is chaotic if

$$d = d_0 \exp \gamma (t - t_0), \qquad (3.42)$$

where γ is a positive quantity equal to the maximum Lyapunov characteristic exponent. However, in practice the Lyapunov characteristic exponents can only be derived analytically for a few idealized systems. For practical problems in the solar system, γ can be estimated from the results of a numerical integration by writing

$$\gamma = \lim_{t \to \infty} \frac{\ln(d/d_0)}{t - t_0} \qquad (3.43)$$

and monitoring the behavior of γ with time. A plot of γ as a function of time on a log–log scale reveals a striking difference between regular and chaotic trajectories. For regular orbits, $d \approx d_0$ and a log–log plot has a slope of -1. However, if the orbit is chaotic, then γ tends to a constant nonzero value. This method may not always work because γ is defined only in the limit as $t \to \infty$ and sometimes chaotic orbits may give the appearance of being regular orbits for long periods by sticking close to the edges of the islands, such as those visible in Figure 3.8.

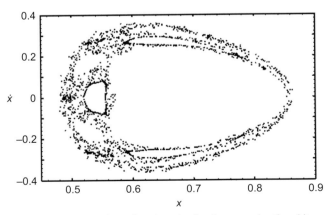

FIGURE 3.11 A surface-of-section plot for the same chaotic orbit as shown in Figures 3.9 and 3.10. The 2000 points were generated by plotting the values of x and $\dot{x}$ whenever $y = 0$ with positive $\dot{y}$. The points are distributed over a much wider region of the $(x, \dot{x})$ plane than the points for the regular orbit shown in Figure 3.8 and they help to define the edges of the regular regions associated with the 7:4 and other resonances.

If the nearby trajectory drifts too far from the original one, then γ is no longer a measure of the local divergence of the orbits. To overcome this problem, it helps to rescale the separation of the nearby trajectory at fixed intervals. Figure 3.13 shows log γ as a function of log t calculated using this method for the regular and chaotic orbits described here. This leads to an estimate of $\gamma = 10^{-0.77}$(Jupiter periods)$^{-1}$ for the maximum Lyapunov characteristic exponent of the chaotic orbit. The corresponding Lyapunov time is given by $1/\gamma$, or in this case ~ 6 Jupiter periods. This indicates that for this starting condition the chaotic nature of the orbit quickly becomes apparent.

It is important to realize that a chaotic orbit is not necessarily unbounded. The maximum Lyapunov characteristic exponent concerns local divergence and provides no information about the global stability of the trajectory. The phrase "wandering on a leash" is an apt description of objects on bounded chaotic orbits—the motion is contained but yet chaotic at the same time. Another consideration is that numerical explorations of chaotic systems have many pitfalls both in how the physical system is modeled and whether or not the model provides an accurate portrayal of the real system.

4.2.3. Location of Regular and Chaotic Regions

The extent of the chaotic regions of the phase space of a dynamical system can depend on a number of factors. In the case of the circular restricted three-body problem, the critical quantities are the values of the Jacobi constant and the mass ratio μ_2. In Figures 3.14 and 3.15, 10 trajectories are shown for each of two different values of the Jacobi constant. In the first case (Figure 3.14), the value is

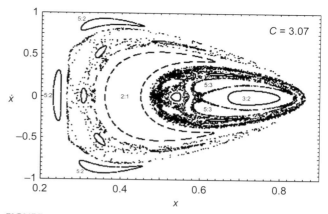

FIGURE 3.14 Representative surface-of-section plots for $x_0 = 0.25$, 0.29, 0.3, 0.45, 0.475, 0.5, 0.55, 0.56, 0.6, and 0.8 with $\dot{x}_0 = 0$, $y_0 = 0$, and Jacobi constant $C = 3.07$. Each trajectory was followed for a minimum of 500 crossing points. The plot uses the points shown in Figures 3.8 and 3.11 (although the scales are different), as well as points from other regular and chaotic orbits. The major resonances are identified.

$C = 3.07$ (the same as the value used in Figures 3.8 and 3.11), whereas in Figure 3.15 it is $C = 3.13$. It is clear that the extent of the chaos is reduced in Figure 3.15. The value of C in the circular restricted problem determines how close the asteroid can get to Jupiter. Larger values of C correspond to orbits with greater minimum distances from Jupiter. For the case $\mu_2 = 0.001$ and $C > 3.04$, it is impossible for their orbits to intersect, although the perturbations can still be significant.

Close inspection of the **separatrices** in Figures 3.14 and 3.15 reveals that they consist of chaotic regions with regular regions on either side. As the value of the Jacobi constant decreases, the extent of the chaotic separatrices

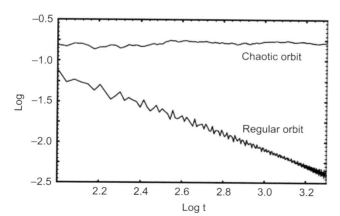

FIGURE 3.13 The evolution of the quantity γ (defined in Eqn (3.43)) as a function of time (in Jupiter periods) for a regular ($x_0 = 0.55$) and a chaotic ($x_0 = 0.56$) orbit. In this log–log plot, the regular orbit shows a characteristic slope of -1 with no indication of log γ tending toward a finite value. However, in the case of the chaotic orbit, log γ tends to a limiting value close to -0.77.

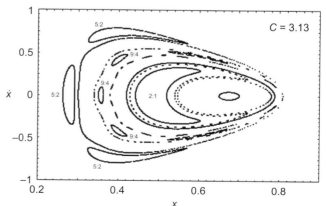

FIGURE 3.15 Representative surface-of-section plots for $x_0 = 0.262$, 0.3, 0.34, 0.35, 0.38, 0.42, 0.52, 0.54, 0.7, and 0.78 with $\dot{x}_0 = 0$, $y_0 = 0$, and Jacobi constant $C = 3.13$. Each trajectory was followed for a minimum of 500 crossing points. It is clear from a comparison with Figure 3.14 that the phase space is more regular; chaotic orbits still exist for this value of C, but they are more difficult to find. The major resonances are identified.

increases until the regular curves separating adjacent resonances are broken down and neighboring chaotic regions begin to merge. This can be thought of as the overlap of adjacent resonances giving rise to chaotic motion. It is this process that permits chaotic orbits to explore regions of the phase space that are inaccessible to the regular orbits. In the context of the Sun—Jupiter—asteroid problem, this observation implies that asteroids in certain orbits are capable of large excursions in their orbital elements.

5. ORBITAL EVOLUTION OF MINOR BODIES

5.1. Asteroids

With more than 380,000 accurately determined orbits and one major perturber (the planet Jupiter), the asteroids provide a natural laboratory in which to study the consequences of regular and chaotic motion. Using suitable approximations, asteroid motion can be studied analytically in some special cases. However, it is frequently necessary to resort to numerical integration. (*See* Main-Belt Asteroids.)

Investigations have shown that a number of asteroids have orbits that result in close approaches to planets. Of particular interest are asteroids such as 433 Eros, 1038 Ganymed, and 4179 Toutatis, because they are on orbits that bring them close to the Earth. One of the most striking examples of the butterfly effect (see Section 4.1) in the context of orbital evolution is the orbit of asteroid 2060 Chiron, which has a **perihelion** inside Saturn's orbit and an **aphelion** close to Uranus's orbit. Numerical integrations based on the best available orbital elements show that it is impossible to determine Chiron's past or future orbit with any degree of certainty since it frequently suffers close approaches to Saturn and Uranus. In such circumstances, the outcome strongly depends on the initial conditions as well as the accuracy of the numerical method. These are the characteristic signs of a chaotic orbit. By integrating several orbits with initial conditions close to the nominal values, it is possible to carry out a statistical analysis of the orbital evolution. Studies suggest that there is a one in eight chance that Saturn will eject Chiron from the solar system on a hyperbolic orbit, while there is a seven in eight chance that it will evolve toward the inner solar system and come under strong perturbations from Jupiter. Telescopic observations of a faint coma surrounding Chiron imply that it is a comet rather than an asteroid; perhaps its future orbit will resemble that of a short-period comet of the Jupiter family.

Numerical studies of the orbital evolution of planet-crossing asteroids under the effects of perturbations from all the planets have shown a remarkable complexity of motion for some objects. For example, the Earth-crossing asteroid 1620 Geographos gets trapped temporarily in a number of resonances with the Earth in the course of its chaotic evolution (Figure 3.16).

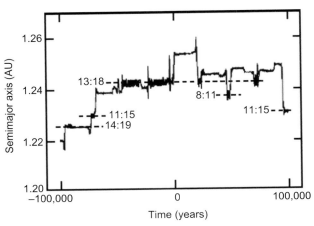

FIGURE 3.16 A plot of the semimajor axis of the near-Earth asteroid 1620 Geographos over a backward and forward integration of 100,000 years starting in 1986. Under perturbations from the planets, Geographos moves in a chaotic orbit and gets temporarily trapped in a number of high-order, orbit—orbit resonances (indicated in the diagram) with the Earth. *The data are taken from a numerical study of planet-crossing asteroids undertaken by A. Milani and coworkers. Courtesy of Academic Press.*

A histogram of the number distribution of asteroid orbits in semimajor axis (Figure 3.17) shows that apart from a clustering of asteroids near Jupiter's semimajor axis at 5.2 AU, there is an absence of objects within 0.75 AU of the orbit of Jupiter. The objects with the same orbital distance (semimajor axis) as Jupiter are the Trojan asteroids (Section 3.2), which librate about the L_4 and L_5 triangular

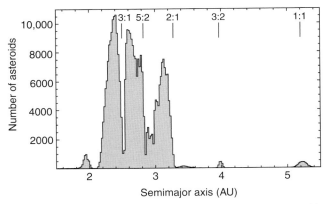

FIGURE 3.17 A histogram of the distribution of the numbered asteroids (as of August 2012) with semimajor axis together with the locations of the major Jovian resonances. Most objects lie in the main belt between 2.0 and 3.3 AU, where the outer edge is defined by the location of the 2:1 resonance with Jupiter. The width of each bin is 0.02 AU. Apart from gaps (the Kirkwood gaps) at the 3:1, 5:2, 2:1, and other resonances in the main belt, there are small concentrations of asteroids at the 3:2 and 1:1 resonances (the Hilda and Trojan groups, respectively). Note that observational biases result in overrepresentation of asteroids orbiting near the inner edge of the asteroid belt and underrepresentation of distant asteroids.

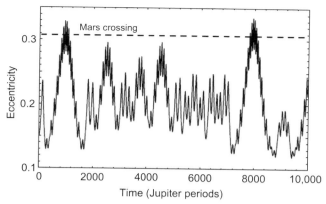

FIGURE 3.18 The chaotic evolution of the eccentricity of a fictitious asteroid at the 3:1 resonance with Jupiter. The orbit was integrated using an algebraic mapping technique developed by J. Wisdom. The line close to $e = 0.3$ denotes the value of the asteroid's eccentricity above which it would cross the orbit of Mars. It is thought that the 3:1 Kirkwood gap was created when asteroids in chaotic zones at the 3:1 resonance reached high eccentricities and were removed by direct encounters with Mars, Earth, or Venus.

Lagrangian points located $\sim 60°$ ahead of and behind Jupiter.

The cleared region near Jupiter's orbit can be understood in terms of chaotic motion due to the overlap of adjacent resonances. In the context of the Sun–Jupiter–asteroid restricted three-body problem, the perturber (Jupiter) has an infinite sequence of first-order resonances that lie closer together as its semimajor axis is approached. For example, the 2:1, 3:2, 4:3, and 5:4 resonances with Jupiter lie at 3.3, 4.0, 4.3, and 4.5 AU, respectively. Since each $(p+1){:}p$ resonance (where p is a positive integer) has a finite width in semimajor axis that is almost independent of p, adjacent resonances will always overlap for some value of p greater than a critical value, p_{crit}. This value is given by

$$p_{\text{crit}} \approx 0.51 \left(\frac{m}{m+M} \right)^{-2/7} \qquad (3.44)$$

where, in this case, m is the mass of Jupiter and M is the mass of the Sun. This equation can be used to predict that resonance overlap and chaotic motion should occur for p values greater than 4; this corresponds to a semimajor axis near 4.5 AU. Therefore, chaos may have played a significant role in the depletion of the outer asteroid belt.

The histogram in Figure 3.17 also shows a number of regions in the main belt where there are few asteroids. The gaps at 2.5 and 3.3 AU were first detected in 1867 by Daniel Kirkwood using a total sample of fewer than 100 asteroids; these are now known as the Kirkwood gaps. Their locations coincide with prominent Jovian resonances (indicated in Figure 3.17), and this led to the hypothesis that they were created by the gravitational effect of Jupiter on asteroids that had orbited at these semimajor axes. The exact removal mechanism was unclear until the 1980s, when several numerical and analytical studies showed that the central regions of these resonances contained large chaotic zones.

The Kirkwood gaps cannot be understood using the model of the circular restricted three-body problem described in Section 4.2. The eccentricity of Jupiter's orbit, although small (0.048), plays a crucial role in producing the large chaotic zones that help to determine the orbital evolution of asteroids. On timescales of several hundreds of thousands of years, the mutual perturbations of the planets act to change their orbital elements and Jupiter's eccentricity can vary from 0.025 to 0.061. An asteroid in the chaotic zone at the 3:1 resonance would undergo large, essentially unpredictable changes in its orbital elements. In particular, the eccentricity of the asteroid could become large enough for it to cross the orbit of Mars. This is illustrated in Figure 3.18 for a fictitious asteroid with an initial eccentricity of 0.15 moving in a chaotic region of the phase space at the 3:1 resonance. Although the asteroid can have periods of relatively low eccentricity, there are large deviations and e can reach values in excess of 0.3. Taking the eccentricity of Mars's orbit to be its maximum value of 0.14, this implies that there will be times when the orbits could intersect (Figure 3.19). In this case, the asteroid orbit would be unstable, since it is likely to either impact the surface of Mars or suffer a close approach that would drastically alter its semimajor axis. Although Jupiter provides the perturbations, it is Mars, Earth, or Venus that ultimately removes the asteroids from the 3:1 resonance.

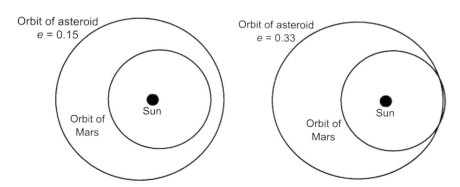

FIGURE 3.19 The effect of an increase in the orbital eccentricity of an asteroid at the 3:1 Jovian resonance on the closest approach between the asteroid and Mars. For $e = 0.15$, the orbits do not cross. However, for $e = 0.33$, a typical maximum value for asteroids in chaotic orbits, there is a clear intersection of the orbits and the asteroid could have a close encounter with Mars (eccentricity 0.14, its maximum value).

Figure 3.20 shows the excellent correspondence between the distribution of asteroids close to the 3:1 resonance and the maximum extent of the chaotic region determined from numerical experiments.

The situation is less clear for other resonances, although there is good evidence for large chaotic zones at the 2:1 and 5:2 resonances. In the outer part of the main belt, large changes in eccentricity will cause the asteroid to cross the orbit of Jupiter before it gets close to Mars. There may also be perturbing effects from other planets. In fact, it is now known that **secular resonances** have an important role to play in the clearing of the Kirkwood gaps, including the one at the 3:1 resonance. Once again, chaos is involved. Studies of asteroid motion at the 3:2 Jovian resonance indicate that the motion is regular, at least for low values of the eccentricity. This may help to explain why there is a local concentration of asteroids (the Hilda group) at this resonance, whereas others are associated with an absence of material.

Since the dynamical structure of the asteroid belt has been determined by the perturbative effects of nearby planets, it seems likely that the original population was much larger and more widely dispersed. Therefore, the current distribution of asteroids may represent objects that are either recent collision products or that have survived in relatively stable orbits over the age of the solar system. Asteroids can also undergo orbital evolution due to nongravitational forces such as the **Yarkovsky effect**.

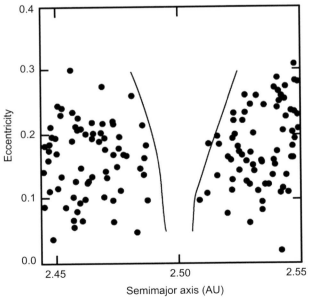

FIGURE 3.20 The eccentricity and semimajor axes of asteroids in the vicinity of the 3:1 Jovian resonance; the Kirkwood gap is centered close to 2.5 AU. The two curves denote the maximum extent of the chaotic zone determined from numerical experiments, and there is excellent agreement between these lines and the edges of the 3:1 gap.

5.2. Meteorites

Most meteorites are thought to be the fragments of material produced from collisions in the asteroid belt, and the reflectance properties of certain meteorites are known to be similar to those of common types of asteroids. Since most collisions take place in the asteroid belt, the fragments have to evolve into Earth-crossing orbits before they can hit the Earth and be collected as samples.

An estimate of the time taken for a given meteorite to reach the Earth after the collisional event that produced it can be obtained from a measure of its cosmic ray exposure age. Prior to the collisions, the fragment may have been well below the surface of a much larger body, and as such it would have been shielded from all but the most energetic cosmic rays. However, after a collision, the exposed fragment would be subjected to cosmic ray bombardment in interplanetary space. A detailed analysis of meteorite samples allows these exposure ages to be measured.

In the case of one common class of meteorites called the ordinary chondrites, the cosmic ray exposure ages are typically less than 20 million years and the samples show little evidence of having been exposed to high pressure, or "shocking". Prior to the application of chaos theory to the origin of the Kirkwood gaps, there was no plausible mechanism that could explain delivery to the Earth within the exposure age constraints and without shocking. However, small increments in the velocity of the fragments as a result of the initial collision or orbital changes due to the **Yarkovsky effect** could easily cause them to enter a chaotic zone near a given resonance. (*See* Meteorites.)

Numerical integrations of such orbits near the 3:1 resonance showed that it was possible for them to achieve eccentricities large enough for them to cross the orbit of the Earth. This result complemented previous research that had established that this part of the asteroid belt was a source region for the ordinary chondrites. In order to obtain agreement between theory and observations, other perturbations such as the **Yarkovsky effect** need to be included.

5.3. Comets

Typical cometary orbits have large eccentricities and therefore planet-crossing trajectories are commonplace. Many comets are thought to originate in the Oort cloud at several tens of thousands of arbitrary units from the Sun; another reservoir of comets, known as the Kuiper belt, exists just beyond the orbit of Neptune. Those that have been detected from the Earth are classified as long period (most of which have made single apparitions and have periods >200 y), Halley-type (with orbital periods of $20-200$ y) or Jupiter family (with orbital periods <20 y). All comets with orbital periods of less than $\sim 10^3$ y have experienced a close approach to Jupiter or one of the other giant planets. By their very nature, the orbits of comets are chaotic, since the outcome of any planetary encounter will be sensitively dependent on the initial conditions.

Studies of the orbital evolution of the short-period comet D/Lexell highlight the possible effects of close approaches. A numerical integration has shown that prior to 1767 it was a short-period comet with a semimajor axis of 4.4 AU and an eccentricity of 0.35. In 1767 and 1779, it suffered close approaches to Jupiter. The first encounter placed it on a trajectory which brought it into the inner solar system and close (0.0146 AU) to the Earth, leading to its discovery and its only apparition in 1770, whereas the second was at a distance of ~3 Jovian radii. This changed its semimajor axis to 45 AU with an eccentricity of 0.88.

A more recent example is the orbital history of comet D/Shoemaker–Levy 9 prior to its spectacular collision with Jupiter in 1994. Orbit computations suggest that the comet was captured by Jupiter at some time during a 9-year interval centered on 1929. Prior to its capture, it is likely that it was orbiting in the outer part of the asteroid belt close to the 3:2 resonance with Jupiter or between Jupiter and Saturn close to the 2:3 resonance with Jupiter. However, the chaotic nature of its orbit means that it is impossible to derive a more accurate history unless prediscovery images of the comet are obtained. (See Physics And Chemistry Of Comets; Cometary Dynamics.)

5.4. Small Satellites and Rings

Chaos is also involved in the dynamics of a satellite embedded in a planetary ring system. The processes differ from those discussed in Section 3.1, because there is a near-continuous supply of ring material and direct scattering by the perturber is now important. In this case, the key quantity is the Hill's sphere of the satellite. Ring particles on near-circular orbits passing close to the satellite exhibit chaotic behavior due to the significant perturbations they receive at close approach. This causes them to collide with surrounding ring material, thereby forming a gap. Studies have shown that for small satellites, the expression for the width of the cleared gap is

$$W \approx 0.44 \left(\frac{m_2}{m_1}\right)^{2/7} a \qquad (3.45)$$

where m_2 and a are the mass and semimajor axis of the satellite, respectively, and m_1 is the mass of the planet. Thus, an icy satellite with a radius of 10 km and a density of 1 g/cm^3 orbiting in Saturn's A ring at a radial distance of 135,000 km would create a gap approximately 140 km wide.

Since such a gap is wider than the satellite that creates it, this provides an indirect method for the detection of small satellites in ring systems. There are two prominent gaps in Saturn's A ring: the ~35-km-wide Keeler gap at 136,530 km and the 325-km-wide Encke gap at 133,570 km. The predicted radii of the icy satellites required to produce these gaps are ~2.5 and ~24 km, respectively. In 1991, an analysis of *Voyager* images by M. Showalter revealed a small satellite, Pan, with a radius of ~10 km orbiting in the Encke gap. In 2005, the moon Daphnis of radius ~3–4 km was discovered in the Keeler gap by the *Cassini* spacecraft. *Voyager 2* images of the dust rings of Uranus show pronounced gaps at certain locations. Although most of the proposed shepherding satellites needed to maintain the narrow rings have yet to be discovered, these gaps may provide indirect evidence of their orbital locations.

6. LONG-TERM STABILITY OF PLANETARY ORBITS

6.1. The N-Body Problem

The entire solar system can be approximated by a system of eight planets orbiting the Sun. In a center-of-mass frame, the vector equation of motion for planet i moving under the Newtonian gravitational effect of the Sun and the eight planets is given by

$$\ddot{\mathbf{r}} = G \sum_{j=0}^{8} \mathbf{m}_j \frac{\mathbf{r}_j - \mathbf{r}_i}{r_{ij}^3} (j \neq i) \qquad (3.46)$$

where $\mathbf{r}_i$ and $\mathbf{m}_i$ are the position vector and mass of planet i ($i = 1, 2, ..., 8$), respectively; $\mathbf{r}_{ij} = \mathbf{r}_j - \mathbf{r}_i$; and the subscript 0 refers to the Sun. These are the equations of the N-body problem for the case where $N = 9$, and although they have a surprisingly simple form, they have no general, analytical solution. However, as in the case of the three-body problem, it is possible to tackle this problem mathematically by making some simplifying assumptions.

Provided the eccentricities and inclinations of the N bodies are small and there are no resonant interactions between the planets, it is possible to derive an analytical solution that describes the evolution of all the eccentricities, inclinations, perihelia, and nodes of the planets. This solution, called Laplace–Lagrange secular perturbation theory, gives no information about the longitudinal location of the planets, yet it demonstrates that there are long-period variations in the planetary orbital elements that arise from mutual perturbations. The secular periods involved are typically tens or hundreds of thousands of years, and the evolving system always exhibits regular behavior. In the case of Earth's orbit, such periods may be correlated with climatic change, and large variations in the eccentricity of Mars are thought to have had important consequences for its climate.

In the early nineteenth century, Pierre-Simon de Laplace claimed that he had demonstrated the long-term stability of the solar system using the results of his secular perturbation theory. Although the actual planetary system violates some of the assumed conditions (e.g. Jupiter and Saturn are close to a 5:2 resonance), the Laplace–Lagrange theory can be modified to account for some of these effects. However, such analytical approaches always involve the neglect of potentially important interactions between planets. The problem becomes even more difficult when the possibility of near-resonances between some of the **secular** periods of the system is considered. However, nowadays it is always possible to carry out numerical investigations of long-term stability.

6.2. Stability of the Solar System

Numerical integrations show that the orbits of the planets are chaotic with a timescale for exponential divergence of 4 or 5 million years. The effect is most apparent in the orbits of the inner planets. Despite this chaos, gross changes in planetary orbits are unlikely on astrophysically important timescales. But in 1% of systems integrated forward for 5 billion years, Mercury and Venus suffer a close approach, which can lead to Mercury colliding with another planet or the Sun or being ejected from the solar system.

The chaotic divergence seen in all long-term integrations implies that the accuracy of the deterministic equations of celestial mechanics to predict the future positions of the planets will always be limited by the accuracy with which their orbits can be measured. For example, if the position of Earth along its orbit is known to within 1 cm today, then the exponential propagation of errors that is characteristic of chaotic motion implies that we have no knowledge of Earth's orbital position 200 million years in the future.

The situation is even less predictable when the gravitational influence of smaller bodies is accounted for. Asteroids exert small perturbations on the orbits of the major planets. These perturbations can be accounted for and do not adversely affect the precision to which planetary orbits can be simulated on timescales of tens of millions of years. However, unlike the major planets, asteroids suffer close approaches to one another. Close approaches between the two largest and most massive asteroids, 1 Ceres and 4 Vesta, led to exponential growth in uncertainty for backward integrations of planetary orbits with doubling times of $<10^6$ years more than 50–60 million years ago.

Pluto's orbit is chaotic, partly as a result of its 3:2 resonance with the planet Neptune, although the perturbing effects of other planets are also important. Despite the fact that the timescale for exponential divergence of nearby trajectories (the inverse of the Lyapunov exponent) is about 20 million years, no study has shown evidence for Pluto leaving the resonance.

7. DISSIPATIVE FORCES AND THE ORBITS OF SMALL BODIES

The previous sections describe the gravitational interactions between the Sun, planets, and moons. Solar radiation has been ignored, but this is an important force for small particles in the solar system. Three effects can be distinguished: (1) the radiation pressure, which pushes particles primarily outward from the Sun (micron-sized dust); (2) the **Poynting–Robertson drag**, which causes centimeter-sized particles to spiral inward toward the Sun; and (3) the **Yarkovsky effect**, which changes the orbits of meter- to kilometer-sized objects owing to uneven temperature distributions at their surfaces. These effects can be significant as they can lead to **secular** changes in orbital angular momentum and energy. Each of these effects is discussed in the next three subsections and then the effect of gas drag is examined. In the final subsection, the influence of tidal interactions is discussed; this effect (in contrast to the other dissipative effects described in this section) is most important for larger bodies such as moons and planets. (See Solar System Dust.)

7.1. Radiation Force (Micron-Sized Particles)

The Sun's radiation exerts a force, F_r, on all other bodies of the solar system. The magnitude of this force is

$$F_r = \frac{LA}{4\pi c r^2} Q_{pr}, \qquad (3.47)$$

where A is the particle's geometric cross-section, L is the solar luminosity, c is the speed of light, r is the heliocentric distance, and Q_{pr} is the radiation pressure coefficient, which is equal to unity for a perfectly absorbing particle and is of order unity unless the particle is small compared to the wavelength of the radiation. The parameter β is defined as the ratio between the forces due to the radiation pressure and the Sun's gravity:

$$\beta \equiv \frac{F_r}{F_g} = 5.7 \times 10^{-5} \frac{Q_{pr}}{\rho R}, \qquad (3.48)$$

where the radius, R, and the density, ρ, of the particle are in c.g.s. units. Note that β is independent of heliocentric distance and that the solar radiation force is important only for micron- and submicron-sized particles. Using the parameter β, a more general expression for the effective gravitational attraction can be written:

$$F_{geff} = \frac{-(1-\beta)GmM}{r^2}, \qquad (3.49)$$

that is, the small particles "see" a Sun of mass $(1-\beta)M$. It is clear that small particles with $\beta > 1$ are repelled by the Sun and thus quickly escape the solar system, unless they are gravitationally bound to one of the planets. Dust that is released from bodies traveling on circular orbits at the Keplerian velocity is ejected from the solar system if $\beta > 0.5$.

The importance of solar radiation pressure can be seen, for example, in comets. Cometary dust is pushed in the antisolar direction by the Sun's radiation pressure. The dust tails are curved because the particles' velocity decreases as they move farther from the Sun, due to conservation of angular momentum. (*See* Cometary Dynamics; Physics And Chemistry Of Comets.)

7.2. Poynting–Robertson Drag (Centimeter-Sized Grains)

A small particle in orbit around the Sun absorbs solar radiation and reradiates the energy isotropically in its own frame. The particle thereby preferentially radiates (and loses momentum) in the forward direction in the inertial frame of the Sun. This leads to a decrease in the particle's energy and angular momentum and causes dust in bound orbits to spiral sunward. This effect is called the Poynting–Robertson drag.

The net force on a rapidly rotating dust grain is given by

$$F_{\text{rad}} \approx \frac{L Q_{\text{pr}} A}{4\pi c r^2} \left[\left(1 - \frac{2v_r}{c}\right) \hat{r} - \frac{v_\theta}{c} \hat{\theta} \right]. \quad (3.50)$$

The first term in Eqn (3.50) is that due to radiation pressure and the second and third terms (those involving the velocity of the particle) represent the Poynting–Robertson drag.

From this discussion, it is clear that small-sized dust grains in the interplanetary medium are removed: (sub) micron-sized grains are blown out of the solar system, whereas larger particles spiral inward toward the Sun. Typical decay times (in years) for circular orbits are given by

$$\tau_{\text{P-R}} \approx 400 \frac{r^2}{\beta}, \quad (3.51)$$

with the distance r in AU.

Particles that produce the bulk of the zodiacal light (at infrared and visible wavelengths) are between 20 and 200 μm, so their lifetimes in the Earth orbit are on the order of 10^5 y, which is much less than the age of the solar system. Sources for the dust grains are comets as well as the asteroid belt, where numerous collisions occur between countless small asteroids.

7.3. Yarkovsky Effect (Meter-Sized Objects)

Consider a rotating body heated by the Sun. Because of thermal inertia, the afternoon hemisphere is typically warmer than the morning hemisphere, by an amount $\Delta T \ll T$. Let us assume that the temperature of the morning hemisphere is $T - \Delta T/2$, and that of the evening hemisphere is $T + \Delta T/2$. The radiation reaction on a surface element dA, normal to its surface, is d$F = 2\sigma T^4 \text{d}A/3c$. For a spherical particle of radius R, the Yarkovski force in the orbit plane due to the excess emission on the evening side is

$$F_Y = \frac{8}{3}\pi R^2 \frac{\sigma T^4}{c} \frac{\Delta T}{T} \cos\psi, \quad (3.52)$$

where σ is the Stefan–Boltzmann constant and ψ is the particle's **obliquity**, that is, the angle between its rotation axis and orbit pole. The reaction force is positive for an object that rotates in the prograde direction, $0 < \psi < 90°$, and negative for an object with retrograde rotation, $90° < \psi < 180°$. In the latter case, the force enhances the Poynting–Robertson drag.

The Yarkovsky force is important for bodies ranging in size from meters to several kilometers. Asymmetric outgassing from comets produces a nongravitational force similar in form to the Yarkovski force. (*See* Cometary Dynamics.)

7.4. Gas Drag

Although interplanetary space generally can be considered an excellent vacuum, there are certain situations in planetary dynamics where interactions with gas can significantly alter the motion of solid particles. Two prominent examples of this process are planetesimal interactions with the gaseous component of the protoplanetary disk during the formation of the solar system and orbital decay of ring particles as a result of drag caused by extended planetary atmospheres.

In the laboratory, gas drag slows solid objects down until their positions remain fixed relative to the gas. In the planetary dynamics case, the situation is more complicated. For example, a body on a circular orbit about a planet loses mechanical energy as a result of drag with a static atmosphere, but this energy loss leads to a decrease in the semimajor axis of the orbit, which implies that the body actually speeds up! Other, more intuitive effects of gas drag are the damping of eccentricities and, in the case where there is a preferred plane in which the gas density is the greatest, the damping of inclinations relative to this plane.

Objects whose dimensions are larger than the mean free path of the gas molecules experience Stokes drag,

$$F_D = -\frac{C_D A \rho v^2}{2}, \quad (3.53)$$

where v is the relative velocity of the gas and the body, ρ is the gas density, A is the projected surface area of the body, and C_D is a dimensionless drag coefficient, which is of order unity unless the **Reynolds number** is very small. Smaller bodies are subject to Epstein drag,

$$F_D = -A\rho v v' \quad (3.54)$$

where v' is the mean thermal velocity of the gas. Note that as the drag force is proportional to surface area and the gravitational force is proportional to volume (for constant particle density), gas drag is usually most important for the dynamics of small bodies.

The gaseous component of the protoplanetary disk in the early solar system is thought to have been partially supported against the gravity of the Sun by a negative pressure gradient in the radial direction. Thus, less centripetal force was required to complete the balance, and consequently the gas orbited less rapidly than the Keplerian velocity. The "effective gravity" felt by the gas is

$$g_{\text{eff}} = -\frac{GM_S}{r^2} - (1/\rho)\frac{dp}{dr}. \qquad (3.55)$$

To maintain a circular orbit, the effective gravity must be balanced by centripetal acceleration, rn^2. For estimated protoplanetary disk parameters, the gas rotated $\sim 0.5\%$ slower than the Keplerian speed.

Large particles moving at (nearly) the Keplerian speed thus encountered a headwind, which removed part of their angular momentum and caused them to spiral inward toward the Sun. Inward drift was greatest for midsized particles, which have large ratios of surface area to mass yet still orbit with nearly Keplerian velocities. The effect diminishes for very small particles, which are so strongly coupled to the gas that the headwind they encounter is very slow. Peak rates of inward drift occur for particles that collide with roughly their own mass of gas in one orbital period. Meter-sized bodies in the inner solar nebula drift inward at a rate of up to 10^6 km/y! Thus, the material that survives to form the planets must complete the transition from centimeter to kilometer size rather quickly, unless it is confined to a thin dust-dominated subdisk in which the gas is dragged along at essentially the Keplerian velocity.

Drag induced by a planetary atmosphere is even more effective for a given density, as atmospheres are almost entirely pressure supported, so the relative velocity between the gas and particles is high. As atmospheric densities drop rapidly with height, particles decay slowly at first, but as they reach lower altitudes, their decay can become very rapid. Gas drag is the principal cause of orbital decay of artificial satellites in low Earth orbit.

7.5. Tidal Interactions and Planetary Satellites

Tidal forces are important to many aspects of the structure and evolution of planetary bodies:

1. On short timescales, temporal variations in tides (as seen in the frame rotating with the body under consideration) cause stresses that can move fluids with respect to more rigid parts of the planet (e.g. the familiar ocean tides) and even cause seismic disturbances (although the evidence that the Moon causes some earthquakes is weak and disputable, it is clear that the tides raised by the Earth are a major cause of moonquakes).
2. On long timescales, tides cause changes in the orbital and spin properties of planets and moons. Tides also determine the equilibrium shape of a body located near any massive body; note that many materials that behave as solids on human timescales are effectively fluids on very long geological timescales (e.g. Earth's mantle).

The gravitational attraction of the Moon and Earth on each other causes tidal bulges that rise in a direction close to the line joining the centers of the two bodies. Particles on the nearside of the body experience gravitational forces from the other body that exceed the centrifugal force of the mutual orbit, whereas particles on the far side experience gravitational forces that are less than the centripetal forces needed for motion in a circle. It is the gradient of the gravitational force across the body that gives rise to the double tidal bulge.

The Moon spins once per orbit, so the same face of the Moon always points toward the Earth and the Moon is always elongated in that direction. Earth, however, rotates much faster than the Earth–Moon orbital period. Thus, different parts of the Earth point toward the Moon and are tidally stretched. If the Earth was perfectly fluid, the tidal bulges would respond immediately to the varying force, but the finite response time of Earth's figure causes the tidal bulge to lag behind, at the point on the Earth where the Moon was overhead slightly earlier. Since the Earth rotates faster than the Moon orbits, this "tidal lag" on the Earth leads the position of the Moon in inertial space. As a result, the tidal bulge of the Earth accelerates the Moon in its orbit. This causes the Moon to slowly spiral outward. The Moon slows down Earth's rotation by pulling back on the tidal bulge, so the angular momentum in the system is conserved. This same phenomenon has caused most, if not all, major moons to be in synchronous rotation: the rotation and orbital periods of these bodies are equal. In the case of the Pluto–Charon system, the entire system is locked in a synchronous rotation and revolution of 6.4 days. Satellites in retrograde orbits (e.g. Triton) or satellites whose orbital periods are less than the planet's rotation period (e.g. Phobos) spiral inward toward the planet as a result of tidal forces.

Mercury orbits the Sun in 88 days and rotates around its axis in 59 days, a 3:2 spin-orbit resonance. Hence, at every perihelion, one of two locations is pointed at the Sun: the subsolar longitude is either 0° or 180°. This configuration is stable because Mercury has both a large orbital eccentricity and a significant permanent deformation that is aligned with the solar direction at perihelion. Indeed, at 0° longitude, there is a large impact crater, Caloris Planitia, which may be the cause of the permanent deformation.

3. Under special circumstances, strong tides can have significant effects on the physical structure of bodies. Generally, the strongest tidal forces felt by solar system bodies (other than Sun-grazing or planet-grazing comets) are those caused by planets on their closest satellites. Near a planet, tides are so strong that they rip a fluid (or weakly aggregated solid) body apart. In such a region, large moons are unstable, and even small moons, which could be held together by material strength, are unable to accrete because of tides. The boundary of this region is known as **Roche's limit**. Inside Roche's limit, solid material remains in the form of small bodies and rings are found instead of large moons.

The closer a moon is to a planet, the stronger is the tidal force to which it is subjected. Let us consider Roche's limit for a spherical satellite in synchronous rotation at a distance r from a planet. This is the distance at which a loose particle on an equatorial subplanet point just remains gravitationally bound to the satellite. At the center of the satellite of mass m and radius R_s, a particle would be in equilibrium and so

$$\frac{GM}{r^2} = n^2 r, \quad (3.56)$$

where $M (\gg m)$ is the mass of the planet. However, at the equator, the particle will experience (1) an excess gravitational or centrifugal force due to the planet, (2) a centrifugal force due to rotation, and (3) a gravitational force due to the satellite. If the equatorial particle is *just* in equilibrium, these forces will balance and

$$-\frac{d}{dr}\left(\frac{GM}{r^2}\right) R_s + n^2 r = \frac{Gm}{R_s^2}. \quad (3.57)$$

In this case, Roche's limit r_Roche is given by

$$r_\text{Roche} = 3^{1/3} \left(\frac{\rho_\text{planet}}{\rho_s}\right)^{1/3} R_\text{planet}. \quad (3.58)$$

where ρ_planet and ρ_s are the densities of the planet and satellite, respectively, and R_planet is the planetary radius. When a fluid moon is considered and flattening of the object due to the tidal distortion is taken into account, the correct result for a liquid moon (no internal strength) is

$$r_\text{Roche} = 2.456 \left(\frac{\rho_\text{planet}}{\rho_s}\right)^{1/3} R_\text{planet}. \quad (3.59)$$

Most bodies have some internal strength, which allows bodies with sizes $\lesssim 100$ km to be stable somewhat inside Roche's limit. Mars's satellite Phobos is well inside Roche's limit; it is subjected to a tidal force equivalent to that in Saturn's B ring.

4. Internal stresses caused by variations in tides on a body in an eccentric orbit or not rotating synchronously with its orbital period can result in significant tidal heating of some bodies, most notably in Jupiter's moon Io. If no other forces were present, this would lead to a decay of Io's orbital eccentricity. In analogy to the Earth–Moon system, the tide raised on Jupiter by Io will cause Io to spiral outward and its orbital eccentricity to decrease. However, there exists a 2:1 mean-motion resonant lock between Io and Europa. Io passes on some of the orbital energy and angular momentum that it receives from Jupiter to Europa, and Io's eccentricity is increased as a result of this transfer. This forced eccentricity maintains a high tidal dissipation rate and large internal heating in Io, which displays itself in the form of active volcanism. [See Io]

7.6. Tidal Evolution and Resonances

Objects in prograde orbits that lie outside the **synchronous orbit** can evolve outward at different rates, so there may have been occasions in the past when pairs of satellites evolved toward an **orbit–orbit resonance**. The outcome of such a resonant encounter depends on the direction from which the resonance is approached. For example, capture into resonance is possible only if the satellites are approaching one another. If the satellites are receding, then capture is not possible but the resonance passage can lead to an increase in the eccentricity and inclination. In certain circumstances, it is possible to study the process using a simple mathematical model. However, this model breaks down near the chaotic separatrices of resonances and in regions of resonance overlap.

It is likely that the major satellites of Jupiter, Saturn, and Uranus have undergone significant tidal evolution and that the numerous resonances in the Jovian and Saturnian systems are a result of resonant capture. The absence of orbit–orbit resonances among the major moons in the Uranian system is thought to be related to the fact that the oblateness of Uranus is significantly less than that of Jupiter or Saturn. In these circumstances, there can be large chaotic regions associated with resonances and stable capture may be impossible. However, temporary capture into some resonances can produce large changes in eccentricity or inclination. For example, the Uranian satellite Miranda has an anomalously large inclination of $4°$, which is thought to be the result of a chaotic passage through the 3:1 resonance with Umbriel at some time in its orbital history. Under tidal forces, a satellite's eccentricity is reduced on a shorter timescale than its inclination, and Miranda's current inclination agrees with estimates derived from a chaotic evolution. (*See* Planetary Satellites.)

8. CHAOTIC ROTATION

8.1. Spin–Orbit Resonance

One of the dissipative effects of the tide raised on a natural satellite by a planet is to cause the satellite to evolve toward a state of synchronous rotation, where the rotational period of the satellite is approximately equal to its orbital period. Such a state is one example of a spin–orbit resonance, where the ratio of the spin period to the orbital period is close to a rational number. The time needed for a near-spherical satellite to achieve this state depends on its mass and orbital distance from the planet. Small, distant satellites take a longer time to evolve into the synchronous state than large satellites that orbit close to the planet. Observations by spacecraft and ground-based instruments suggest that most regular satellites are in the synchronous spin state, in agreement with theoretical predictions.

The lowest energy state of a satellite in synchronous rotation has the moon's longest axis pointing in the approximate direction of the planet–satellite line. Let θ denote the angle between the long axis and the planet–satellite line in the planar case of a rotating satellite (Figure 3.21). The variation of θ with time can be described by equating the time variation of the rotational angular momentum with the restoring torque. The resulting differential equation is

$$\ddot{\theta} + \frac{\omega_0^2}{2r^3}\sin 2(\theta - f) = 0, \qquad (3.60)$$

where ω_0 is a function of the principal moments of inertia of the satellite, r is the radial distance of the satellite from the planet, and f is the true anomaly (or angular position) of the satellite in its orbit. The radius is an implicit function of time and is related to the true anomaly by the equation

$$r = \frac{a(1-e^2)}{1 + e\cos f}, \qquad (3.61)$$

where a and e are the constant semimajor axis and the eccentricity of the satellite's orbit, respectively, and the orbit is taken to be fixed in space.

Equation (3.60) defines a deterministic system where the initial values of θ and $\dot{\theta}$ determine the subsequent rotation of the satellite. Since θ and $\dot{\theta}$ define a unique spin position of the satellite, a surface-of-section plot of $(\theta, \dot{\theta})$ once every orbital period, say at every periapse passage, produces a picture of the phase space. Figure 3.22 shows the resulting surface-of-section plots for a number of starting conditions using $e = 0.1$ and $\omega_0 = 0.2$. The chosen values of ω_0 and e are larger than those that are typical for natural satellites, but they serve to illustrate the structure of the surface of section; large values of e are unusual since tidal forces also act to damp eccentricity. The surface of section shows large, regular regions surrounding narrow islands associated with the 1:2, 1:1, 3:2, 2:1, and 5:2 spin–orbit resonances at $\dot{\theta} = 0.5, 1, 1.5, 2$, and 2.5, respectively. The largest island is associated with the strong 1:1 resonance and, although other spin states are possible, most regular satellites, including the Earth's Moon, are observed to be in this state. Note the presence of diffuse collections of points associated with small chaotic regions at the separatrices of the resonances. These are particularly obvious at the 1:1 spin–orbit state at $\theta = \pi/2$, $\dot{\theta} = 1$. Although this is a completely different dynamical system compared to the circular restricted three-body problem,

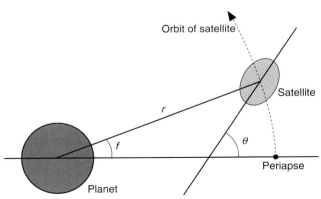

FIGURE 3.21 The geometry used to define the orientation of a satellite in orbit about a planet. The planet–satellite line makes an angle f (the true anomaly) with a reference line, which is taken to be the periapse direction of the satellite's orbit. The orientation angle, θ, of the satellite is the angle between its long axis and the reference direction.

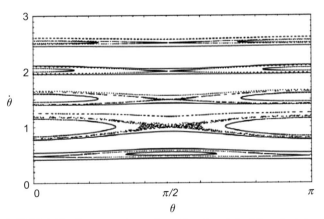

FIGURE 3.22 Representative surface-of-section plots of the orientation angle, θ, and its time derivative, $\dot{\theta}$, obtained from the numerical solution of Eqn (3.60) using $e = 0.1$ and $\omega_0 = 0.2$. The values of θ and $\dot{\theta}$ were obtained at every periapse passage of the satellite. Four starting conditions were integrated for each of the 1:2, 1:1, 3:2, 2:1, and 5:2 spin–orbit resonances in order to illustrate motion inside, at the separatrix, and on either sides of each resonance. The thickest "island" is associated with the strong 1:1 spin–orbit state $\theta = 1$, whereas the thinnest is associated with the weak 5:2 resonance at $\theta = 2.5$.

there are distinct similarities in the types of behavior visible in Figure 3.22 and parts of Figures 3.14 and 3.15.

In the case of near-spherical objects, it is possible to investigate the dynamics of spin—orbit coupling using analytical techniques. The sizes of the islands shown in Figure 3.22 can be estimated by expanding the second term in Eqn (3.60) and isolating the terms that will dominate at each resonance. Using such a method, each resonance can be treated in isolation and the gravitational effects of nearby resonances can be neglected. However, if a satellite is distinctly nonspherical, ω_0 can be large and this approximation is no longer valid. In such cases, it is necessary to investigate the motion of the satellite using numerical techniques.

8.2. Hyperion

Hyperion is a satellite of Saturn that has an unusual shape (Figure 3.23). It has a mean radius of 135 km, an orbital eccentricity of 0.1, a semimajor axis of 24.55 Saturn radii, and a corresponding orbital period of 21.3 days. Such a small object at this distance from Saturn has a large tidal despinning timescale, but the unusual shape implies an estimated value of $\omega_0 = 0.89$.

The surface of section for a *single* trajectory is shown in Figure 3.24 using the same scale as Figure 3.22. It is clear that there is a large chaotic zone that encompasses most of the spin—orbit resonances. The islands associated with the synchronous and other resonances survive but in a much reduced form. Although this calculation assumes that Hyperion's spin axis remains perpendicular to its orbital plane, studies have shown that the satellite should also be undergoing a tumbling motion, such that its axis of rotation is not fixed in space.

Voyager observations of Hyperion indicated a spin period of 13 days, which suggested that the satellite was not in synchronous rotation. However, the standard techniques that are used to determine the period are not applicable if it varies on a timescale that is short compared with the timespan of the observations. In principle, the rotational period can be deduced from ground-based observations by looking for periodicities in plots of the brightness of the object as a function of time (the light curve of the object). The results of one such study for Hyperion are shown in Figure 3.25. Since there is no recognizable periodicity, the light curve is consistent with that of an object undergoing chaotic rotation. Hyperion is the first natural satellite that has been observed to have a chaotic spin state, and results from *Cassini* images confirm this result. Observations and numerical studies of Hyperion's rotation in three dimensions have shown that its spin axis does not point in a fixed direction. Therefore, the satellite also undergoes a tumbling motion in addition to its chaotic rotation.

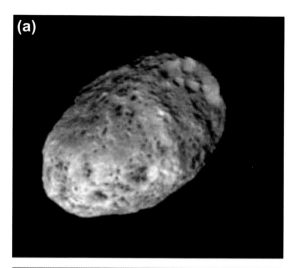

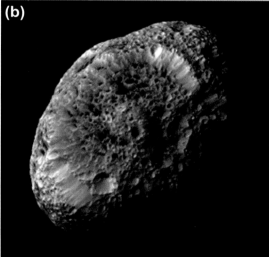

FIGURE 3.23 Two *Cassini* images of the Saturnian satellite Hyperion show the unusual shape of the satellite, which is one cause of its chaotic rotation. Panel (a) is a true color image, while panel (b) uses false color and has better resolution because it was obtained at closer range. *Courtesy of NASA/JPL/Space Science Institute.*

The dynamics of Hyperion's motion is complicated by the fact that it is in a 4:3 orbit—orbit resonance with the larger Saturnian satellite Titan. Although tides act to decrease the eccentricities of satellite orbits, Hyperion's eccentricity is maintained at 0.104 by means of the resonance. Titan effectively forces Hyperion to have this large value of e and so the apparently regular orbital motion inside the resonance results, in part, in the extent of the chaos in its rotational motion. (*See* Planetary Satellites.)

8.3. Other Satellites

Although there is no evidence that other natural satellites are undergoing chaotic rotation at the present time, it is

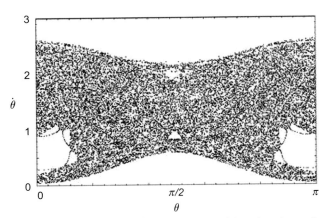

FIGURE 3.24 A single surface-of-section plot of the orientation angle, θ, and its time derivative, $\dot{\theta}$, obtained from the numerical solution of Eqn (3.60) using the values $e = 0.1$ and $\omega_0 = 0.89$, which are appropriate for Hyperion. The points cover a much larger region of the phase space than any of those shown in Figure 3.22, and although there are some remaining islands of stability, most of the phase space is chaotic.

possible that several irregularly shaped regular satellites did experience chaotic rotation at some time in their histories. In particular, since satellites have to cross chaotic separatrices before capture into synchronous rotation can occur, they must have experienced some episode of chaotic rotation. This may also have occurred if the satellite suffered a large impact that affected its rotation. Such episodes could have induced significant internal heating and resurfacing events in some satellites. The Martian moon Phobos and the Uranian moon Miranda have been mentioned as possible candidates for this process. If this happened early in the history of the solar system, then the evidence may well have been obliterated by subsequent cratering events. (*See* Planetary Satellites.)

8.4. Chaotic Obliquity

The fact that a planet is not a perfect sphere means that it experiences additional perturbing effects due to the gravitational forces exerted by its satellites and the Sun, and these can cause long-term evolution in its obliquity (the angle between the planet's equator and its orbit plane). Numerical investigations have shown that chaotic changes in obliquity are particularly common in the inner solar system. For example, it is now known that the stabilizing effect of the Moon results in a variation of $\pm 1.3°$ in Earth's obliquity around a mean value of $23.3°$. Without the Moon, Earth's obliquity would undergo large, chaotic variations. In the case of Mars, there is no stabilizing factor and the obliquity varies chaotically from $0°$ to $60°$ on a timescale of 50 million years. Therefore, an understanding of the long-term changes in a planet's climate can be achieved only by an appreciation of the role of chaos in its dynamical evolution.

9. EPILOG

It is clear that nonlinear dynamics has provided us with a deeper understanding of the dynamical processes that have helped to shape the solar system. Chaotic motion is a natural consequence of even the simplest systems of three or more interacting bodies. The realization that chaos has played a fundamental role in the dynamical evolution of the solar system came about because of contemporary and complementary advances in mathematical techniques and digital computers. This coincided with an explosion in our knowledge of the solar system and its major and minor members. Understanding how a random system of planets, satellites, ring and dust particles, asteroids, and comets interacts and evolves under a variety of chaotic processes and timescales ultimately means that this knowledge can be used to trace the history and predict the fate of other planetary systems.

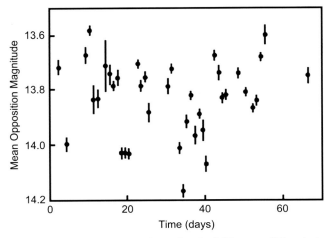

FIGURE 3.25 Ground-based observations by J. Klavetter of Hyperion's light curve obtained over 13 weeks (4.5 orbital periods) in 1987. The fact that there is no obvious curve through the data points is convincing evidence that the rotation of Hyperion is chaotic. *Courtesy of the American Astronomical Society.*

BIBLIOGRAPHY

Burns, J. A. (1987). The motion of interplanetary dust. In M. Fulchignoni, & L. Kresak (Eds.), *The evolution of the small bodies of the solar system* (pp. 252–275). Bologna, Italy: Soc. Italiana di Fisica.

Danby, J. M. A. (1992). *Fundamentals of celestial mechanics*. Richmond, Virginia: Willmann-Bell.

Diacu, F., & Holmes, P. (1996). *Celestial encounters. The origins of chaos and stability*. Princeton, NJ: Princeton Univ. Press.

Duncan, M., & Quinn, T. (1993). The long-term dynamical evolution of the solar system. *Annual Review of Astronomy and Astrophysics, 31*, 265–295.

Ferraz-Mello, S. (Ed.). (1992). *Chaos, resonance and collective dynamical phenomena in the solar system*. Dordrecht, Holland: Kluwer.

Laskar, J. (1994). Large-scale chaos in the solar system. *Astronomy and Astrophysics, 287*, L9–L12.

Laskar, J., Gastineau, M., Delisle, J.-B., Farrés, A., & Fienga, A. (2011). Strong chaos induced by close encounters with Ceres and Vesta. *Astronomy and Astrophysics, 532*, L4.

Lichtenberg, A. J., & Lieberman, M. A. (1992) (2nd ed.). *Regular and chaotic dynamics* (Vol. 38). New York: Springer-Verlag. Applied Mathematical Sciences.

Lissauer, J. J. (1993). Planet formation. *Annual Review of Astronomy and Astrophysics, 31*, 129–174.

Morbidelli, A. (2002). *Modern celestial mechanics*. London: Taylor & Francis. Out of print. Downloadable from: http://www/oca.eu/morby/, 368 pp.

Murray, C. D., & Dermott, S. F. (1999). *Solar system dynamics*. Cambridge: Cambridge University Press.

Peale, S. J. (1976). Orbital resonances in the solar system. *Annual Review of Astronomy and Astrophysics, 14*, 215–246.

Peterson, I. (1993). *Newton's clock. Chaos in the solar system*. New York: W. H. Freeman.

Part II

Fundamental Planetary Processes and Properties

Chapter 4

Planetary Impacts

Richard A.F. Grieve, Gordon R. Osinski and Livio L. Tornabene
Centre for Planetary Science and Exploration, University of Western Ontario, London, ON, Canada

Chapter Outline

1. Impact Craters 83
   1.1. Crater Shape 83
   1.2. Crater Dimensions 88
2. Impact Processes 89
   2.1. Crater Formation 89
   2.2. Changes in the Target Rocks 92
      2.2.1. Solid Effects 92
      2.2.2. Melting 94
3. Impacts and Planetary Evolution 94
   3.1. Impact Origin of Earth's Moon 94
   3.2. Early Crustal Evolution 95
   3.3. Biosphere Evolution 95
4. Impacts as Planetary Probes 97
   4.1. Water and Ices 97
   4.2. Spectral Composition 98
   4.3. Morphologic and Geologic 99
Bibliography 99

1. IMPACT CRATERS

1.1. Crater Shape

On bodies that have no atmosphere, such as the Moon, even the smallest pieces of interplanetary material can produce impact craters down to micrometer-sized cavities on individual mineral grains. On larger bodies, atmospheric passage results in aerodynamic resistant forces, which decelerate incoming bodies and break up weaker ones. On Earth, for example, impacting bodies with masses below 10^4 g can lose up to 90% of their velocity during atmospheric penetration and the resultant impact pit is only slightly larger than the projectile itself. Atmospheric effects on larger incoming masses, however, are less severe, and the body impacts with relatively undiminished velocity, producing a crater that is considerably larger than the impacting body.

The processes accompanying such events are rooted in the physics of impact, with the differences in response among the various planets largely being due to differences in the properties of the planetary bodies (e.g. surface gravity, atmospheric density, and target composition and strength). The basic shape of virtually all impact craters is a depression with an upraised rim. With increasing diameter, impact craters become proportionately shallower, with respect to their diameter, i.e. the depth to diameter ratio decreases. They also develop more complicated rims and floors, including the appearance of central topographic peaks and interior rings. It should be noted that not all impact craters are circular in plan view. For example, the rims of some terrestrial, Martian and Venusian impact craters have straight line segments, reflecting preimpact inhomogeneities and structural features in the target rocks.

There are three major subdivisions in shape: simple craters, complex craters, and impact basins. Simple impact structures have the form of a bowl-shaped depression with an upraised rim (Figure 4.1(A)). An overturned flap of ejected target materials exists on the rim, and the exposed rim, walls, and floor define the apparent crater. Observations at terrestrial impact craters reveal that a lens of brecciated target material, roughly parabolic in cross section, exists beneath the floor of this apparent crater (Figure 4.2). This breccia lens is a mixture of different materials in heterogeneous targets, with fractured blocks set in a finer-grained matrix. These are **allochthonous** materials, having been moved into their present position by the cratering process. Beneath the breccia lens, relatively in-place, or para-utochthonous, fractured target materials define the walls and floor of what is known as the true crater (Figure 4.2). In the case of terrestrial simple craters, the depth to the base of the

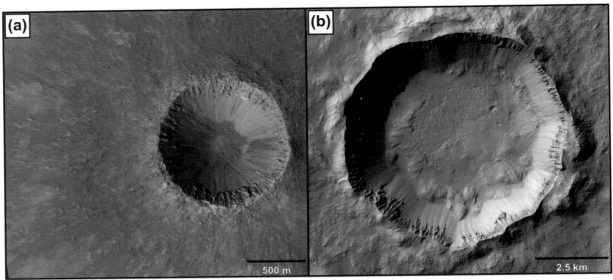

FIGURE 4.1 (a) Winslow Crater on Mars is an example of a fresh simple crater. It is ~1 km in diameter and large blocks ejected late in the cratering process can be seen around the rim area. The ejecta can be differentiated into continuous and discontinuous ejecta, which appear as separate "fingers" and "braids" (herringbone pattern) ~1.5–2 crater radii from the rim. (b) Noord Crater is a fresh transitional crater ~8 km in diameter. It has a relatively flat floor and shows extensive slumping in the form of blocks, where the crater wall meets the crater floor. Note that there is no sign of a central structure, typical of bona fide complex craters (Mars Reconnaissance Orbiter).

breccia lens (i.e. the base of the true crater) is roughly twice the depth to the top of the breccia lens (i.e. the floor of the apparent crater). With increasing diameter, simple craters display signs of wall and rim collapse (Figure 4.1(B)), as they evolve into complex craters. The diameter at which this transition takes place varies between planetary bodies and is, to a first approximation, an inverse function of planetary gravity. Other variables, such as target strength, and possibly projectile type, and impact angle and velocity, play a role and the transition actually occurs over a small range in diameter. For example, the transition between simple and complex craters occurs in the 15–25 km diameter range on the Moon. The effect of target strength is most readily apparent on Earth, where complex craters can occur at diameters as small as 2 km in sedimentary target rocks, but do not occur until diameters of 4 km, or greater, in stronger, crystalline target rocks.

Complex craters are highly modified structures with respect to their final form, compared to simple craters. A typical complex crater is characterized by a central topographic feature (e.g. a peak, pit or some combination thereof), a broad, flat floor, and a terraced, inwardly slumped rim area (Figure 4.3(A)). Observations at terrestrial complex craters show that the flat floor consists of a sheet of **impact melt** rock and/or **polymict** breccia (Figure 4.4). The central region is structurally complex and, is most commonly occupied by the central peak, which is the topographic manifestation of a much broader and extensive volume of uplifted fractured and faulted **parautochthonous** rocks that originate from the target beneath the crater (Figure 4.4).

With increasing diameter, a fragmentary ring of interior peaks appears (Figure 4.3(B)), marking the beginning of the morphologic transition from craters to basins. While a single interior ring is required to define a basin, they can be subdivided further into central-peak basins, with both a peak and ring; peak ring basins (Figure 4.5), with a single ring; and multiring basins, with two or more interior rings (Figure 4.6). The transition from central-peak basins to peak-ring basins to multiring basins also represents a sequence with increasing diameter. As with the simple to complex crater transition, there is a small amount of overlap in basin shape near transition diameters.

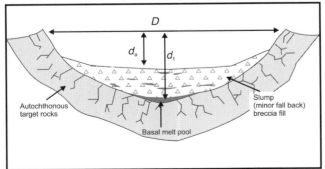

FIGURE 4.2 Schematic cross-section of a simple crater, based on terrestrial observations, D is rim diameter and d_a and d_t are apparent and true depth, respectively. See text for details.

Chapter | 4 Planetary Impacts

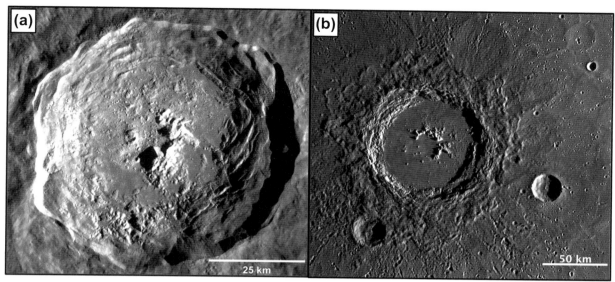

FIGURE 4.3 (a) Jackson Crater is a well-preserved complex crater on the lunar far-side. It is 72 km in diameter and shows extensive terraces and slumps on the crater walls, stepping down to a flat floor with central peak. Smooth, dark deposits of what is likely impact melt rocks are evident on the crater floor, terraces and near-rim ejecta (Lunar Reconnaissance Orbiter). (b) Hokusai Crater is transitional between a central peak crater and a peak ring basin. It is 100 km in diameter and displays continuous (lobate) and discontinuous (braided or rayed) ejecta, as well as secondary craters, on the surrounding terrain. Hokusai is, in fact, the most extensively rayed crater on Mercury, with rays extending over 1000 km from the rim (Messenger).

Ejected target material surrounds impact craters and can be subdivided into continuous and discontinuous ejecta facies (Figures 4.1(A) and 4.3(B)). The continuous deposits are those closest to the crater, being thickest at the rim crest. In the case of simple craters, the net effect of the ejection process is to invert the stratigraphy at the rim, which may continue into the continuous ejecta depending on target heterogeneity (i.e. the deepest materials are deposited near the rim, and the shallowest are most distal). However, as the distance from the crater rim increases, the ejecta are emplaced at higher velocities and, therefore, land with higher kinetic energies, resulting in the mixing of ejecta with local surface material. Thus, at increasing distance from the crater, the final ejecta blanket on the ground includes increasing amounts of local materials. Secondary crater fields, resulting from the impact of larger, coherent blocks and clods of ejecta, surround fresh craters and are particularly evident on bodies that lack or have thin atmospheres, such as the Moon, Mercury (Figure 4.3(B)), and Mars. On the Moon and Mercury, they are often associated with typically bright or high-albedo "rays" that define an overall radial pattern to the primary crater (See; Mercury, The Moon). Two principal processes have been suggested to explain the rays. The first is a compositional effect, where the ejecta are chemically different from the material on which it is deposited. While this most often results in rays that are brighter than the surrounding material, the reverse can also occur. The second effect is a consequence of "maturity" due to prolonged exposure to "space weathering" agents like radiation and micrometeoroid bombardment on surface materials (See Main-Belt Asteroids). Fresher material excavated by an impact and deposited in the rays is generally brighter than the more mature material of the deposition surface: however, this contrasts with the rays on Mars, which are most recognizable in thermal images. More recently, numerical simulations have suggested the rays (at least, on airless bodies) could be the result of the interaction of impact-induced **shock waves** and preexisting surface depressions.

Many Martian craters display examples of apparently fluidized ejecta (Figure 4.7). They have been called

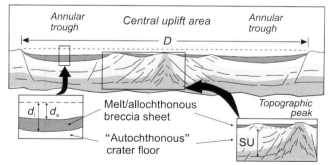

FIGURE 4.4 Schematic cross-section of a complex crater, based on terrestrial data. Notation as in Figure 4.2, with SU corresponding to the structural uplift and D_{cp} to the diameter of the central uplift area. Note the preservation of upper beds (different shades of gray) in the outer portion of the crater floor, indicating excavation was limited to the central area. See text for details.

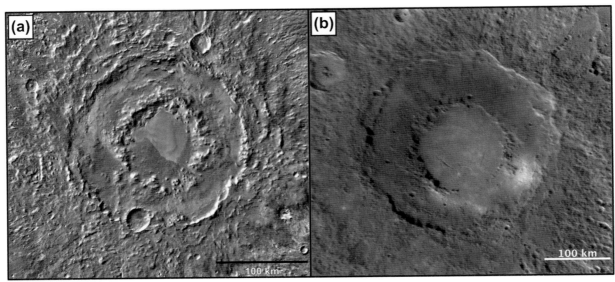

FIGURE 4.5 (a) The 200 km diameter peak ring basin Lowell on Mars (Mars Odyssey). An extensive deposit interior to the inner ring is likely dunes or related peri-glacial (near-surface rock–ice interactions) that occurred well after the formation of the basin. (b) The approximately 290 km peak ring basin Rachmaninoff on Mercury. The dark, smooth fractured deposits interior to the inner ring of this basin may represent extensive deposits rich in impact melt that were emplaced during crater formation (Messenger).

"fluidized–ejecta", "rampart", or "pedestal" craters, where their ejecta deposits indicate emplacement as a ground-hugging flow. Most hypotheses on the origin of these features invoke the presence of ground ice (or water),

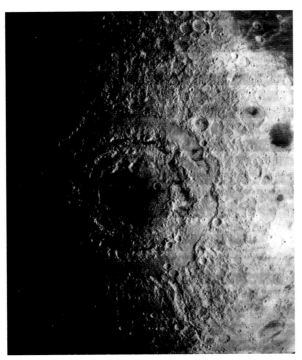

FIGURE 4.6 At a diameter of ∼1000 km, as defined by the outer ring, the Cordillera mountains, Orientale is the youngest and best-preserved lunar multiring basin (Lunar Orbiter Images).

which, upon heating by impact, is incorporated into the ejecta in either liquid or vapor form. This, then, provides lubrication for the mobilized material (See: Mars: Surface and Interior).

On Venus, radar data indicate that impact craters more than 15–20 km in diameter exhibit central peaks and/or peak rings (Figure 4.8) and appear, for the most part, to be similar to complex craters and basins on the other terrestrial planets. Many of the craters smaller than 15 km, however, have rugged, multiple floors or occur as crater clusters (See: Venus: Surface and Interior). This is attributed to the effects of the dense atmosphere of Venus (surface pressure of ∼90 bar), which effectively crushes and breaks up smaller impacting bodies, so that they result in clusters of relatively shallow craters. Also due to atmospheric effects, there is a deficit in the number of expected craters with diameters up to 35 km, and there are no craters smaller than 3 km in diameter on Venus. In principle, this atmospheric effect on small impacting bodies occurs on Earth. Due to its less dense atmosphere, however, the fragments remain relatively close together in the terrestrial case and the net effect is similar to the impact of a coherent impacting body.

In many cases, craters on Venus have ejecta deposits out to greater distances than expected from simple ballistic emplacement and the distal deposits are clearly lobate (Figure 4.8). These deposits likely owe their origin to entrainment by the dense atmosphere and/or the high portion of **impact melt** that would be produced on a high gravity and high surface temperature planet, such as Venus. Another unusual feature on Venus is radar-dark zones surrounding some craters that can extend three to four

Chapter | 4 Planetary Impacts

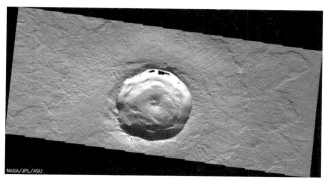

FIGURE 4.7 The 11-km diameter Martian crater Steinheim is a type-example of a complex crater, it has a small central peak and simple terraced walls, and what is referred to as a Double-Layered Ejecta (DLE) crater. The ejecta can be discriminated into outer and inner layer. Both appear to be fluidized material with distinctive lobate margins. The outer layer appears to be overlain by the inner, which is less extensive and displays radial linear features (Mars Odyssey).

crater diameters from the crater center (Figure 4.8). They are believed to be due to the modification of the surface roughness by the atmospheric shock wave produced by the impacting body. Small crater clusters have dark haloes and dark circular areas with no central crater form have been observed. In these latter cases, the impacting body did not survive atmospheric passage, but the accompanying atmospheric shock wave had sufficient energy to interact with the surface to create a dark, radar-smooth area (See Venus: Surface and Interior). The situation is somewhat analogous to the 1908 Tunguska event, when a relatively small body exploded over Siberia at an altitude of ~10 km, and the resultant atmospheric pressure wave leveled some 2000 km^2 of forest. Most recently, on 15 February, 2013, a meteoritic body, with an estimated original mass of 10,000 t, entered the Earth's atmosphere on a shallow trajectory over Russia. It exploded in an air burst at a height of 23.3 km. Damage to infrastructure in the nearby city of Chelyabinsk due to the atmospheric shock wave resulted in the injury of some 1500 people but no fatalities. It is believed to have been the largest meteoritic object to enter the Earth's atmosphere since the 1908 Tunguska event.

Remarkable ring structures occur on the Galilean satellites of Jupiter, Callisto, and Ganymede (See: Ganymede and Callisto). The largest is the 4000-km feature Valhalla on Callisto (Figure 4.9), which consists of a bright central area up to 800 km in diameter, surrounded by a darker terrain with bright ridges 20–30 km apart. This zone is about 300 km wide and gives way to an outer zone with **graben** or rift-like features 50–100 km apart. These (very) multiring basins are generally considered to be of impact origin, but with the actual impact crater confined to the central area. In one working hypothesis, the exterior rings are formed as a result of the original crater puncturing the outer, strong shell, or lithosphere, of these bodies. This permitted the weaker, underlying layer, the asthenosphere, to flow toward the crater, setting up stresses that led to fracturing and the formation of circumscribing scarps and graben.

On Callisto and Ganymede, there is also a unique class of impact craters that no longer have an obvious crater form but appear as bright, or high-albedo, spots on the surfaces

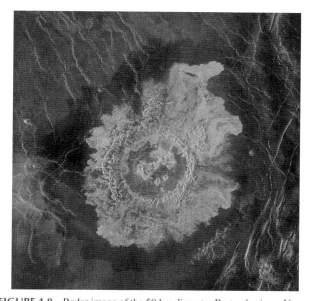

FIGURE 4.8 Radar image of the 50 km diameter Barton basin on Venus, exhibiting a discontinuous peak ring. Barton is close to the lower diameter limit for peak ring basins on Venus. Note the long run-out of lobate ejecta in the north, northeast of the image, suggesting a high degree of flow. The crater and ejecta are surrounded by terrain with a radar dark signature. See text for details (Magellan).

FIGURE 4.9 The Valhalla multiring basin on Callisto. The overall structure may be as large as 4000 km in diameter but only the central bright area is believed to formed directly by impact. See text for details (Galileo).

of these bodies. These are known as palimpsests and are believed to have begun as complex craters but have had their topography relaxed by the slow, viscous creep of the target's icy crust over time. Palimpsests are old impact features and may have been formed when the icy satellites were young and relatively warm, with a thin crust possibly incapable of retaining significant topography.

Impact craters on icy satellites display a wide range of morphologies, some of which have no counterpart on rocky bodies. On these icy satellites, most craters larger than 25 km have a central pit or central dome, rather than a central peak. Pit and dome craters are shallower than other craters of comparable size, and it has been suggested that the pits are due to the formation of slushy or fluid material by impact melting and the domes are due to uplift of the centers of the craters as a result of layers in the crust with different mechanical properties. The fact that some craters on these icy bodies are anomalous has been ascribed to a velocity effect, as higher impact velocities result in greater melting of the target, or to changes in the mechanical behavior of the crust and its response to impact with time. Interpretations of the origin of the various anomalous crater forms on the icy satellites, however, are generally not well constrained.

The 2011–2012 Dawn mission to the asteroid Vesta imaged the largest complex central peak crater in the solar system (See: Main Belt Asteroids). It is the 505 km diameter Rheasilvia structure, which is centered on the south pole. The central peak rises some 23 km above the crater floor, making it the tallest mountain in the solar system. The fact that such a large impact crater has a complex crater form, as opposed to ringed basin form as observed on major planetary bodies, is due to the low gravity of Vesta and serves to dramatically illustrate the effect of the variation in planetary gravity on final crater form with crater size. The Rheasilvia impact is believed to have excavated some 1% of Vesta's mass and is the most likely source of the Howardite–Eucrite–Diogenite (HED) group of differentiated meteorites (See: Meteorites).

1.2. Crater Dimensions

The depth-diameter relations for craters on the terrestrial or silicate planets are given in Table 4.1. Relations are in the form $d = aD^b$ where d is apparent depth, D is rim crest diameter, and units are in kilometers. Other relations involving parameters such as rim height, rim width, central peak diameter, and peak height can be found in the literature. Due to low rate of crater-modifying process, such as erosion, and the abundance of high-resolution data from the Apollo missions, morphometric relations for fresh impact craters were well defined for the Moon. Recent planetary missions have produced laser altimetry data and have, for example, resulted in digital terrain models of all lunar craters with D > 30 km.

TABLE 4.1 Apparent Depth–Diameter Relations for Craters on the Terrestrial Planets

Planetary body	Exponent (b)	Coefficient (a)	Gravity (cm^{-2})
Simple Craters			
Moon	1.010	0.196	162
Mars	1.019	0.204	372
Mercury	0.98	0.18	378
Earth	1.06	0.13	981
Complex Central Peak Craters			
Moon	0.301	1.044	162
Mars	0.25	0.53	372
Mercury	0.415	0.492	378
Venus	0.30	0.40	891
Sedimentary	0.12	0.30	981
Crystalline	0.15	0.43	981

Simple craters have similar apparent depth–diameter relationships on all the terrestrial planets (Table 4.1). At first glance, terrestrial craters appear to be shallower than their planetary counterparts. Compared to the other terrestrial planets, erosion is most severe on Earth, and crater rims and floors are rapidly affected by erosion and subsequent deposition, respectively. Few terrestrial craters have well-preserved rims, and it is common to measure terrestrial crater depths with respect to the ground surface, which is known and is assumed to erode more slowly. In the case of other planetary bodies, depths have been measured most often by the shadow that the rim casts on the crater floor, although some recent planetary missions have produced laser or stereo-derived altimetry data. That is, the topographic measure is a relative one between the rim crests and the floor. Thus, the measurements of depth for Earth and for other planetary bodies differ. For the very few cases in which the rim is well preserved in terrestrial craters, depths from the top of the rim to the crater floor are comparable to those of similar-sized simple craters on the other terrestrial planets.

Unlike simple craters, the depths of complex craters with respect to their diameters do vary between the terrestrial planets (Table 4.1). While the sense of variation is that increasing planetary gravity shallows final crater depths, this is not a strict relationship. For example, Martian complex craters are shallower than equivalent-sized Mercurian craters (Table 4.1), even though the surface gravities of the two planets are very similar. This is probably a function of differences between target materials, with the trapped volatiles and relatively abundant sedimentary deposits making Mars' surface, in general, a weaker target. Mars has also evidence of wind and water processes, which will reduce crater-related topography by erosion and sedimentary infilling. The secondary effect of target strength is also well illustrated by the observation that terrestrial complex craters in sedimentary targets are shallower than those in crystalline targets (Table 4.1). Target effects are also apparent in the recent Lunar Orbiter Laser Altimeter (LOLA) topographic data, with "young" complex craters in the lunar highlands being generally deeper and having higher central peaks than equivalent sized complex craters on the lunar mare.

Data from the Galileo mission indicates that depth—diameter relationships for craters on the icy satellites Callisto, Europa, and Ganymede have the same general trends as those on the rocky terrestrial planets. Interestingly, the depth—diameter relationship for simple craters is equivalent to that on the terrestrial planets. Although the surface gravities of these icy satellites is only 13—14% of that of the Earth, the transition diameter to complex crater forms occurs at ~3 km, similar to that on the Earth. This may be a reflection of the extreme differences in material properties between icy and rocky worlds. There are also inflections and changes in the slopes of the depth—diameter relationships for the complex craters, with a progressive reduction in absolute depth at diameters larger than the inflection diameter. These anomalous characteristics of the depth—diameter relationship have been attributed to changes in the physical behavior of the crust with depth and the presence of subsurface oceans (See Europa; Ganymede and Callisto).

2. IMPACT PROCESSES

The extremely brief timescales and extremely high energies, velocities, pressures, and temperatures that accompany impact are not encountered, as a group, in other geologic processes and make studying impact processes inherently difficult. Small-scale impacts can be produced in the laboratory by firing projectiles at high velocity (generally below about 8 km/s) at various targets. Some insights can also be gained from observations of high-energy, including nuclear, explosions. "Hydrocode" numerical models have been used to simulate impact crater formation. The planetary impact record also provides constraints on the process. The terrestrial record is an important source of ground-truth data, especially with regard to the subsurface nature and spatial relations at impact craters, and the effects of impact on rocks.

When an interplanetary body impacts a planetary surface, it transfers most of its kinetic energy to the target. The energy released in the impact of a 1 kg body with a velocity of approximately 2 km/s is equivalent to that in 1 kg of high explosives. The energy density of impacting interplanetary bodies is even higher, however, as the mean impact velocity on the terrestrial planets for asteroidal bodies ranges from ~12 km/s for Mars to over ~25 km/s for Mercury. The impact velocity of comets is even higher. Long-period comets (those with orbital periods greater than 200 years) have an average impact velocity with Earth of ~55 km/s; whereas, short-period comets have a somewhat lower average impact velocity (See Comet Populations and Cometary Dynamics).

2.1. Crater Formation

On impact, a shock wave propagates back into the impacting body and also into the target. The latter shock wave compresses and heats the target, while accelerating the target material (Figure 4.10). The direction of this acceleration is perpendicular to the shock front, which is roughly hemispherical, so material is accelerated downward and outward. As a state of stress cannot be maintained at a free surface, such as the original ground surface or the edges and rear of the impacting body, a series of secondary release or "rarefaction" waves are generated, which bring the shock-compressed materials back to ambient pressure. As the rarefaction wave interacts with the target material, it alters the direction of the material set in motion by the shock wave, changing some of the outward and downward motions in the relatively near-surface materials to outward and upward, leading to the ejection of material and the growth of a cavity. Directly below the impacting body, however, the two wave fronts are more nearly parallel, and material is still driven downward (Figure 4.10).

These motions define the **cratering flow-field** and a cavity grows by a combination of upward ejection and downward displacement of target materials. This "transient cavity" reaches its maximum depth before its maximum radial dimensions, but it is usually depicted in illustrations at its maximum growth in all directions (Figure 4.10). At this point, it is parabolic in cross section and, at least for the terrestrial case, has a depth-to-diameter ratio of about 1:3. As simple craters throughout the solar system appear to have similar depth-diameter ratios, the 1:3 ratio for the transient cavity can probably be treated as universal.

An asteroidal body of density 3 g/cm^3 impacting crystalline target rocks at 25 km/s will generate initial shock velocities in the target faster than 20 km/s, with corresponding velocities over 10 km/s for the materials set in motion by the shock wave. The shock wave pressure decays with propagated distance and there is a decay in the strength of the cratering flow-field with distance, until it finally ceases to be able to displace target materials and the

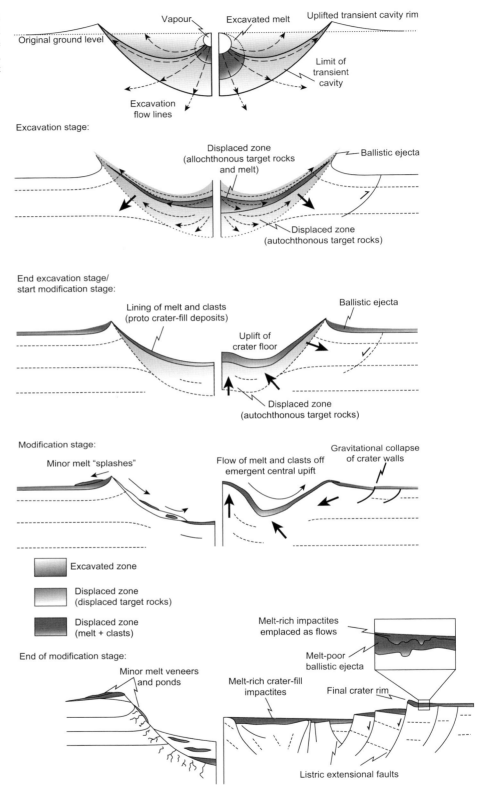

FIGURE 4.10 Schematic model of the formation of simple (left) and complex (right) craters for a typical impact. In the modification stage of complex craters, the arrows labeled "a" to "c" to represent a time sequence. See text for details.

formation of the transient cavity stops Transient-cavity growth is an extremely rapid event. For example, the formation of a 2.5 km diameter transient cavity will take only about 10 s on Earth.

The cratering process is sometimes divided into stages: initial contact and compression, excavation, and modification. In reality, however, it is a continuum with different volumes of the target undergoing different stages of the cratering process at the same time (Figure 4.10). As the excavation stage draws to a close, the direction of movement of target material changes from outward to inward, as the unstable transient cavity collapses to a final topographic form more in equilibrium with gravity. This is the modification stage, with collapse ranging from landslides on the cavity walls of the smaller simple craters to complete collapse and modification of the transient cavity, involving the uplift of the center and collapse of the rim area to form central peaks and terraced, structural rims in larger complex craters (Figure 4.10).

The interior breccia lens of a typical simple crater is the result of this collapse. As the cratering flow comes to an end, the fractured and oversteepened cavity walls become unstable and collapse inward, carrying with them a lining of shocked and melted debris (Figure 4.10). The inward-collapsing walls undergo more fracturing and mixing, eventually coming to rest as the bowl-shaped breccia lens of mixed unshocked and shocked target materials that partially fill simple craters (Figure 4.10). The collapse of the walls increases the rim diameter, such that the final crater diameter in the terrestrial environment is about 20% larger than that of the transient cavity. This is offset by the shallowing of the cavity accompanying production of the breccia lens, with the final apparent crater being about half the depth of the original transient cavity (Figure 4.10). The collapse process is rapid and probably takes place on timescales comparable to those of transient-cavity formation.

Much of our understanding of complex-crater formation comes from observations at terrestrial craters, where it has been possible to trace the movement of beds to show that central peaks are the result of the uplift of rocks from depth (Figure 4.4). Shocked target rocks, analogous to those found in the floors of terrestrial simple craters, constitute the central peak at the centers of complex structures, with the central structure representing the uplifted floor of the original transient cavity. The amount of uplift determined from terrestrial data corresponds to a value of approximately one tenth of the final rim—crest diameter. Further observations at terrestrial complex craters indicate excavation is also limited to the central area and that the transient cavity diameter was about 50—65% of the diameter of the final crater. Radially beyond this, original near-surface units are preserved in the down-dropped annular floor. The rim area is a series of fault terraces, progressively stepping down to the floor (Figure 4.3(A)).

Although models for the formation of complex craters are less constrained than those of simple craters, there is a general consensus that, in their initial stages, complex craters were not unlike simple craters. At complex craters, however, the downward displacements in the transient cavity floor observed in simple craters are not locked in and the cavity floor rebounds upward (Figure 4.10). As the maximum depth of the transient cavity is reached before the cavity's maximum diameter, it is likely that this rebound and reversal of the flow-field in the center of a complex crater occurs while the diameter of the transient cavity is still growing by excavation (Figure 4.10). With the upward movement of material in the transient cavity's floor, the entire rim area of the transient cavity collapses downward and inward (Figure 4.10), greatly enlarging the crater's diameter compared to that of the transient cavity. There have been a number of reconstructions of large lunar craters, in which the terraces are restored to their original, preimpact positions, resulting in estimated transient cavity diameters of about 60% of the final rim—crest diameter. It is clear that uplift and collapse, during the modification stage at complex craters, is extremely rapid and also takes place on time scales comparable to those of transient cavity formation. During the modification stage, the target materials behave as if they were temporally very weak, with the mechanical properties of "normal" fractured rock being restored on final crater formation. A number of mechanisms, including "thermal softening" and "acoustic fluidization", by which strong vibrations cause the rock debris to behave as a fluid, have been suggested as mechanisms to produce the required weakening of the target materials.

There is less of a consensus on the formation of rings within impact basins. The most popular hypothesis for central peak basins is based on the results of modeling; namely, that the rings represent uplifted material in excess of what can be accommodated in a central peak (Figure 4.10). This may explain the occurrence of both peaks and rings in central peak basins but offers little explanation for the absence of peaks and the occurrence of only rings in peak ring and multi-ring basins. A number of analogies have been drawn with the formation of "craters" in liquids and semiconsolidated materials such as muds, where the initial uplifted peak of material has no strength and collapses completely, sometimes oscillating up and down several times. At some time in the formation of ringed basins, however, the target rocks must regain their strength, so as to preserve the interior rings. An alternative explanation is that the uplift process proceeds, as in central peak craters, but the uplifted material in the very center is essentially fluid due to impact melting. In large impact events, the depth of impact melting may reach and even exceed the depth of the transient cavity floor. When the transient cavity is uplifted in such events, the central, melted part has no strength and, therefore, cannot form a positive topographic feature, such as a central peak. Only rings from the unmelted portion of the uplifted transient cavity floor can form some distance out from the center. This is one working hypothesis for the formation of peak ring basins on Mercury (Figure 4.3(B)), which has the largest population and population per area of peak ring structures amongst all the terrestrial planets.

One of the principal characteristics of impact events is the formation and emplacement of ejecta deposits. Recent observations suggest that ejecta may be emplaced as a multistage process. The generation of the continuous ejecta blanket occurs during the excavation stage of cratering, via conventional ballistic ejection, followed by more minor radial surface flow. Most recently, it has been hypnotized that this is followed by the late-stage emplacement of more melt-rich, ground-hugging flows, during the terminal stages of excavation and the modification stage of crater formation (Figure 4.10). Ejecta deposited in this latter is relatively minor in terms of total volume and is influenced by several factors, most importantly planetary gravity, surface temperature and the physical properties of the target rocks.

2.2. Changes in the Target Rocks

The target rocks are initially highly compressed by the passage of the shock wave, transformed into high-density phases, and then rapidly decompressed by the rarefaction wave. As a result, they do not recover fully to their pre-shock state but are of slightly lower density, with the nature of their constituent minerals changed. The collective term for these shock-induced changes in minerals and rocks is **shock metamorphism**. Shock metamorphic effects are found naturally in many lunar samples and meteorites and at terrestrial impact craters. They have also been produced in nuclear explosions and in the laboratory, through shock recovery experiments. No other geologic process is capable of producing the extremely high transient pressures and temperatures required for shock metamorphism and it is diagnostic of impact.

Metamorphism of rocks normally occurs in planetary bodies as a consequence of thermal and tectonic events originating within the planet. The maximum pressures and temperatures recorded in surface rocks by such metamorphic events in planetary crusts are generally on the order of 1 GPa (10 kb) and 1000 °C. During shock metamorphism, materials are compressed from their initial to "shocked" state along straight Rayleigh lines in pressure−volume space. The locus of shocked states in pressure−volume space defines so-called "Hugoniot curves", which differ for individual geologic materials. Shock metamorphic effects do not appear until the material has exceeded its "**Hugoniot elastic limit (HEL)**", which is on the order of 5−10 GPa for most minerals but slightly higher for complex geologic materials with fractures, etc. This is the pressure−volume point beyond which the shocked material no longer deforms elastically and permanent changes are recorded on recovery from shock compression.

The peak pressures generated on impact control the upper limit of shock metamorphism. These vary with the type of impacting body and target material but are, principally, a function of impact velocity, reaching into the hundreds to thousands of GPa. For example, the peak pressure generated when a stony asteroidal body impacts crystalline rock at 15 km/s is over 300 GPa, not much less than the pressure at the center of the Earth ($\sim$390 GPa). Shock metamorphism is also characterized by strain rates that are orders of magnitude higher than those produced by internal geologic processes. For example, the duration of regional metamorphism associated with tectonism on Earth is generally considered to be in the millions of years. In contrast, the peak strains associated with the formation of a crater 20 km in diameter are attained in less than a second.

2.2.1. Solid Effects

At pressures below the HEL, minerals and rocks respond to shock with brittle deformation, which is manifested as fracturing, shattering, and brecciation. Such features are generally not readily distinguished from those produced by endogenic geologic processes, such as tectonism. There is, however, a unique, brittle, shock-metamorphic effect, which results in the development of unusual, striated, and horse-tailed conical fractures, known as shatter cones (Figure 4.11). Shatter cones are best developed at relatively low shock pressures (5−10 GPa) and in fine-grained, structurally homogeneous rocks, such as carbonates, quartzites, and basalts.

Apart from shatter cones, all other diagnostic shock effects are microscopic in character. The most obvious are **planar deformation features** and **diaplectic glasses**. Planar deformation features are intensely deformed, are a few micrometers wide, and are arranged in parallel sets (Figure 4.11). They are best known from the common silicate minerals, quartz and feldspar, for which shock-recovery experiments has calibrated the onset shock pressures for particular crystal orientations. They develop initially at $\sim$10 GPa and continue to 20−30 GPa in crystalline rocks. The increasing effects of shock pressure are mirrored by changes in X-ray characteristics, indicative of the increasing breakdown of the internal crystal structure of individual minerals to smaller and smaller domains.

By shock pressures of $\sim$30−40 GPa, quartz and feldspar are converted to diaplectic (from the Greek, "to strike") glass in crystalline rocks. These are solid-state glasses, with no evidence of flow, that exhibit the same outline as the original crystal. For this reason, they are

FIGURE 4.11 Some shock metamorphic effects at terrestrial impact structures. (a) Shatter cones in basalt at the Slate Islands structure, Canada. (b) Photomicrograph of planar deformation features in quartz from the Mistastin structure, Canada. Width of field of view is 0.5 mm, crossed polars. (c) Hand samples of shocked target rocks from the Wanapitei structure, Canada, which are beginning to melt to form mixed glasses and to vesiculate or froth. (d) Outcrop of coherent impact melt rock some 80 m high, with columnar cooling joints, at the Mistastin structure, Canada.

sometimes referred to as thetamorphic (from the Greek, "same shape") glasses. The variety produced from plagioclase is known as maskelynite and was originally discovered in the Shergotty meteorite in 1872. The thermodynamics of shock processes are highly irreversible, so the pressure-volume work that is done during shock compression is not fully recovered upon decompression. This residual work is manifested as waste heat and, as a result, shock pressures of 40–50 GPa are sufficient to initiate melting in some minerals (Figure 4.11). For example, feldspar grains show incipient melting and flow at shock pressures of $\sim$45 GPa. It is important to note that in porous and potentially volatile-rich sedimentary rocks, the pressures required for the formation of shock features are substantially less than for dense nonporous crystalline rocks. For example, diaplectic quartz glass in sandstones begins to form at pressures as low as $\sim$5.5 GPa and, between $\sim$10 and 20 GPa, almost complete conversion of quartz to diaplectic glass has been observed.

Regardless of the target, melting tends initially to be mineral specific, favoring mineral phases with the highest compressibilities and to be concentrated at grain boundaries, where pressures and temperatures are enhanced by reflections and refractions of the shock wave. In detail, as the shock wave travels through multicomponent systems, such as rocks, it becomes a complex system of multiple reflected and refracted local shock fronts, which may result in the localization of particular shock metamorphic phenomena. The effects of the complex interactions of shock reflections and refractions on melting are most obvious when comparing the pressures required to melt particulate materials, such as those that make up the lunar regolith (see The Moon), and solid rock of similar composition. Shock recovery experiments indicate that intergranular melts can occur at pressures as low as 30 GPa in particulate basaltic material, compared to 45 GPa necessary for the onset of melting of solid basalt.

Most minerals undergo transitions to dense, high-pressure phases during shock compression. Little is known, however, about the mineralogy of the high-pressure phases, as they generally revert to their low-pressure forms during decompression. Nevertheless, metastable high-pressure phases are sometime preserved, as either high-pressure **polymorphs** of preexisting low-pressure phases or high-pressure assemblages due to mineral breakdown. Some known high-pressure phases, such as diamond from carbon or stishovite from quartz (SiO_2), form during shock compression. Others, such as coesite (SiO_2), form by reversion of such minerals during pressure release. Several high-pressure phases that have been noted in shocked meteorites, however, are relatively rare at terrestrial craters. This may be due to postshock thermal effects, which are sufficiently prolonged at a large impact crater to inhibit preservation of metastable phases.

2.2.2. Melting

The waste heat trapped in shocked rocks is sufficient to result in whole-rock melting above shock pressures of ~60 GPa for crystalline rocks and ~30–35 GPa for sandstones. Thus, relatively close to the impact point, a volume of the target rocks is melted and can even be vaporized (Figure 4.10). Ultimately, these liquids cool to form impact melt rocks. These occur as glassy bodies in ejecta and breccias, as dikes in the crater floor, as pools and lenses within the breccia lenses of simple craters (Figures 4.2 and 4.10), and as annular sheets surrounding the central structures and lining the floors of complex craters and basins (Figures 4.4, 4.5(B), and 4.11). Some terrestrial impact melt rocks were initially misidentified as having a volcanic origin. In general, however, impact melt rocks are compositionally distinct from volcanic rocks. They have compositions determined by a mixture of the compositions of the target rocks, in contrast to volcanic rocks that have compositions determined by internal partial melting of more mafic and refractory progenitors, within the planetary body's mantle or crust.

Impact melt rocks can also contain shocked and unshocked fragments of rocks and minerals. During the cratering event, as the melt is driven down into the expanding transient cavity (Figure 4.10), it overtakes and incorporates less-shocked materials such as clasts, ranging in size from small grains to large blocks. Impact melt rocks that cool quickly generally contain large fractions of clasts, while those that cool more slowly show evidence of melting and resorption of the clastic debris, which is possible because impact melts are initially a superheated mixture of liquid melt and vapor. This is another characteristic that sets impact melt rocks apart from volcanic rocks, which are generally erupted at their melting temperature and no higher.

3. IMPACTS AND PLANETARY EVOLUTION

Impacts are fundamental to the origin and evolution of the Solar System. The current working hypothesis for early Solar System history is that, at an early stage, solid material from the preplanetary disk formed a large number of kilometer-sized planetesmals through collisions or impacts (See: The Origin of The Solar System). This planetesmal assemblage evolved through additional impacts to form a small number of larger planetary embryos. Whether a planetesmal grows or erodes through impacts depends on the impact velocity of the collision. Retention of colliding planetesmal material requires the impact velocity to be less than three to five times the escape velocity of the target planetesmal.

As the impact flux has varied through geologic time, so has the potential for impact to act as an evolutionary agent. The ancient highland crust of the Moon records almost the complete record of cratering since its formation. Crater counts combined with isotopic ages on returned lunar samples have established an estimate of the cratering rate on the Moon and its variation with time. Terrestrial data have been used to extend knowledge of the cratering rate, at least in the Earth–Moon system, to more recent geologic time. The lunar data are generally interpreted as indicating an exponential decrease in the rate until ~4.0 billion years (Ga) ago, a slower decline for an additional billion years, and a relatively constant rate, within a factor of two, since ~3.0 Ga ago. The actual rate before ~4.0 Ga ago is imprecisely known, as there is the question of whether the ancient lunar highlands reflect all of the craters that were produced (i.e. a production population) or only those that have not been obliterated by subsequent impacts (i.e. an equilibrium population). Thus, it is possible that the oldest lunar surfaces give only a minimum estimate of the ancient cratering rate. Similarly, there is some question as to whether the largest recorded events, represented by the major multiring basins on the Moon, occurred over the relatively short time period of 4.2–3.8 Ga ago (the "called lunar cataclysm") or were spread more evenly with time (See The Moon).

3.1. Impact Origin of Earth's Moon

The impacts of the greatest magnitude dominate the cumulative effects of the much more abundant smaller impacts in terms of affecting planetary evolution. In the case of Earth, this would be the massive impact that likely produced the Moon. Earth is unique among the terrestrial planets in having a large satellite and the origin of the Moon has always presented a problem. The suggestion that the Moon formed from a massive impact with Earth was originally proposed some 35 years ago, but, with the development of complex numerical calculations and more efficient computers, it has been possible more recently to model such an event. Most models involve the oblique impact of a Mars-sized object with the proto-Earth, which produces an Earth-orbiting disk of impact-produced vapor and debris, consisting mostly of mantle material from Earth and the impacting body. This disk, depleted in volatiles and enriched in refractory elements, would cool, condense, and accrete to form the Moon

(See The Moon). In the computer simulations, very little material from the iron core of the impacting body goes into the accretionary disk, accounting for the low iron and, ultimately, the small core of the Moon. In addition to the formation of the Moon, the effects of such a massive impact on the earliest Earth itself would have been extremely severe, leading to massive remelting of the Earth and loss of any existing atmosphere. The current atmospheres of Venus, Mars and Earth are significantly different, although these planets are generally similar in bulk composition and density. Reasoned modeling suggests that these planets suffered different atmospheric loss and delivery of volatiles by different populations of impactors (asteroids and comets) early in their evolution. In other words, the assumption that these planets had a similar initial volatile budget may be, at least, an oversimplification.

3.2. Early Crustal Evolution

Following planetary formation, the subsequent high rate of bombardment by the remaining "tail" of accretionary debris is recorded on the Moon and the other terrestrial planets and the icy satellites of the outer solar system that have preserved some portion of their earliest crust. Due to the age of its early crust, the relatively large number of space missions, and the availability of samples, the Moon is the source of most interpretations of the effects of such an early, high flux. In the case of the Moon, a minimum of 6000 craters with diameters greater than 20 km are believed to have been formed during this early period. In addition, ~45 impacts produced basins, ranging in diameter from Bailly at 300 km, through the South Pole–Aitken Basin at 2600 km, to the putative Procellarum Basin at 3500 km, the existence of which is still debated. The results of the Apollo missions demonstrate clearly the dominance of impact in the nature of the samples from the lunar highlands. Over 90% of the returned samples from the highlands are impact rock units, with 30–50% of the hand-sized samples being impact melt rocks. The dominance of impact as a process for change is also reflected in the age of the lunar highland samples. The bulk of the near-surface rocks, which are impact products, are in the range of 3.8–4.0 Ga old. Only a few pristine, igneous rocks from the early lunar crust, with ages >3.9 Ga, occur in the Apollo collection. Computer simulations indicate that the cumulative thickness of materials ejected from major craters in the lunar highlands is 2–10 km. Beneath this, the crust is believed to be brecciated and fractured by impacts to a depth of 20–25 km.

The large multiring basins define the major topographic features of the Moon. For example, the topography associated with the Orientale Basin (Figure 4.6), the youngest multiring basin at ~3.8 Ga and, therefore, the basin with the least topographic relaxation, is over 8 km, somewhat less than Mt. Everest at ~9 km. The impact energies released in the formation of impact basins in the 1000 km size range are on the order of $10^{27}-10^{28}$ J, one to 10 million times the present annual output of internal energy of Earth. The volume of crust melted in a basin-forming event of this size is on the order of a 1×10^6 km^3. Although the majority of crater ejecta is generally confined to within ~2.5 diameters of the source crater, this still represents essentially hemispheric redistribution of materials in the case of an Orientale-sized impact on the Moon.

Following formation, these impact basins localized subsequent endogenic geologic activity in the form of tectonism and volcanism. A consequence of such a large impact is the uplift of originally deep-seated isotherms and the subsequent tectonic evolution of the basin, and its immediate environs, is then a function of the gradual loss of this thermal anomaly, which could take as long as a billion years to dissipate completely. Cooling leads to stresses, crustal fracturing, and basin subsidence. In addition to thermal subsidence, the basins may be loaded by later mare volcanism, leading to further subsidence and stress.

All the terrestrial planets experienced the formation of large impact basins early in their histories. Neither Earth nor Venus, however, retains any record of this massive bombardment, so the cumulative effect of such a bombardment on the Earth is unknown. Basin-sized impacts will have also affected any existing atmosphere, hydrosphere, and potential biosphere. For example, the impact on the early Earth of a body in the 500 km size range, similar to the present day asteroids Pallas and Vesta would be sufficient to evaporate the world's present oceans, if only 25% of the impact energy were used in vaporizing the water. Such an event would have effectively sterilized the surface of Earth. The planet would have been enveloped by an atmosphere of hot rock and water vapor that would radiate heat downward onto the surface, with an effective temperature of a few 1000 degrees. It would take thousands of years for the water-saturated atmosphere to rain out and reform the oceans. Models of impact's potential to frustrate early development of life on Earth indicate that life could have survived in a deep marine setting at 4.2–4.0 Ga, but smaller impacts would continue to make the surface inhospitable until ~4.0–3.8 Ga.

3.3. Biosphere Evolution

Evidence from the Earth–Moon system suggests that the cratering rate had essentially stabilized to something approaching a constant value by 3.0 Ga. Although major basin-forming impacts were no longer occurring, there were still occasional impacts resulting in craters in the size range of a few 100 km. The terrestrial record contains remnants of the Sudbury, Canada, and Vredefort, South Africa, structures, which have estimated original crater diameters of ~250 km and ~300 km, respectively, and ages of ~2 Ga. Events of this size are unlikely to have caused significant long-term changes in the solid geosphere, but they likely affected the biosphere of Earth. In addition to these actual Precambrian impact craters, a number of anomalous spherule beds with ages ranging from ~2.0 to 3.5 Ga. have been discovered relatively recently in Australia and South Africa. Geochemical and physical evidence (shocked quartz) indicate an impact origin for some of these beds; at present, however, their source craters are unknown. If, as indicated, one of these spherule beds in Australia is temporally correlated to one in South Africa, its spatial extent would be in excess of 32,000 km^2.

At present, the only case of a direct physical and chemical link between a large impact event and changes in the biostratigraphic record is at the "Cretaceous—Paleogene boundary", which occurred ~65 million years (Ma) ago. The worldwide physical evidence for impact includes: shock-produced, microscopic planar deformation features in quartz and other minerals; the occurrence of stishovite (a high-pressure polymorph of quartz) and impact diamonds; high-temperature minerals believed to be vapor condensates; and various, generally altered, impact-melt spherules. The chemical evidence consists primarily of a geochemical anomaly, indicative of an admixture of meteoritic material. In undisturbed North American sections, which were laid down in swamps and pools on land, the boundary consists of two units: a lower one, linked to ballistic ejecta, and an upper one, linked to atmospheric dispersal in the impact fireball and subsequent fallout over a period of time. This fireball layer occurs worldwide, but the ejecta horizon is known only in North America.

The Cretaceous—Paleogene boundary marks a mass extinction in the biostratigraphic record of the Earth. Originally, it was thought that dust in the atmosphere from the impact led to global darkening, the cessation of photosynthesis, and cooling. Other potential killing mechanisms have been suggested. Soot, for example, has also been identified in boundary deposits, and its origin has been ascribed to globally dispersed wildfires. Soot in the atmosphere may have enhanced or even overwhelmed the effects produced by global dust clouds. Recently, increasing emphasis has been placed on understanding the effects of vaporized and melted ejecta on the atmosphere. Models of the thermal radiation produced by the ballistic re-entry of ejecta condensed from the vapor and melt plume of the impact indicate the occurrence of a thermal-radiation pulse on Earth's surface. The pattern of survival of land animals 65 Ma ago is in general agreement with the concept that this intense thermal pulse was the first global blow to the biosphere.

Although the record in the Cretaceous—Paleogene boundary deposits is consistent with the occurrence of a major impact, it is clear that many of the details of the potential killing mechanism(s) and the associated mass extinction are not fully known. The "killer crater" has been identified as the ~180 km diameter structure, known as Chicxulub, buried under ~1 km of sediments on the Yucatan peninsula, Mexico. Variations in the concentration and size of shocked quartz grains and the thickness of the boundary deposits, particularly the ejecta layer, pointed toward a source crater in Central America. Shocked minerals have been found in deposits both interior and exterior to the Chicxulub structure, as have impact melt rocks, with an isotopic age of 65 Ma.

Chicxulub may hold the clue to potential extinction mechanisms. The target rocks include beds of anhydrite ($CaSO_4$), and model calculations for the Chicxulub impact indicate that the SO_2 released would have sent anywhere between 30 billion and 300 billion tons of sulfuric acid into the atmosphere, depending on the exact impact conditions. Studies have shown that the lowering of temperatures following large volcanic eruptions is mainly due to sulfuric acid aerosols. Models, using both the upper and lower estimates of the mass of sulfuric acid created by the Chicxulub impact, lead to a calculated drop in global temperature of several degrees Celsius. The sulfuric acid would eventually return to Earth as acid rain, which would cause the acidification of the upper ocean and potentially lead to marine extinctions. In addition, impact heating of nitrogen and oxygen in the atmosphere would produce NO_x gases that would affect the ozone layer and, thus, the amount of ultraviolet radiation reaching the Earth's surface. Like the sulfur-bearing aerosols, these gases would react with water in the atmosphere to form nitric acid, which would result in additional acid rains.

The frequency of Chicxulub-size events on Earth is on the order of one every ~100 Ma. Smaller, but still significant, impacts occur on shorter timescales and could affect the terrestrial climate and biosphere to varying degrees. Some model calculations suggest that dust injected into the atmosphere from the formation of impact craters as small as 20 km could produce global light reductions and temperature disruptions. Such impacts occur on Earth with a frequency of approximately two or three every million years but are not likely to have a serious affect upon the biosphere. The most fragile component of the present environment, however, is human civilization, which is highly dependent on an organized and technologically complex infrastructure for its survival. Though we seldom think of civilization in terms of millions of years, there is little doubt that if civilization lasts long enough, it could suffer severely or even be destroyed by an impact event.

Impacts can occur on historical timescales. For example, the Tunguska event in Russia in 1908 was due to the atmospheric explosion of a relatively small body at an altitude of ~10 km. The energy released, based on that required to produce the observed seismic disturbances, has been estimated as being equivalent to the explosion of ~10 megatons of Trinitrotoluene (TNT). Although the air blast resulted in the devastation of ~2000 km² of Siberian forest, there was no loss of human life. Events such as Tunguska occur on timescales of a 1000 of years. Fortunately, 70% of the Earth's surface is ocean and most of the land surface is not densely populated. Such oceanic impacts, however, could result in devastating tsunami waves in coastal areas.

In addition to the aforementioned deleterious effects of meteorite impacts, it has become apparent over the past decade that impact events produce several beneficial effects with respect to microbial life. Most importantly, impact events are now known to produce several habitats that are highly conducive to life and that were not present before the impact event. Major habitats include (1) impact-generated hydrothermal systems, which could provide habitats for thermophilic and hyperthermophilic microorganisms, (2) impact-processed crystalline rocks, which have increased porosity and translucence compared to unshocked materials, improving microbial colonization, (3) impact glasses, which, similar to volcanic glasses, provide an excellent readily available source of bioessential elements, and (4) impact crater lakes, which form protected sedimentary basins with various niches and that increase the preservation potential of fossils and organic material. Thus, impact craters, once formed on Early Earth, and, by analogy on Mars and other planets, may have represented prime sites that served as protected niches, where life could have survived and evolved and, more speculatively, perhaps originated.

Of these habitats, the one that has received most attention is the impact-generated hydrothermal system. This derives from the longstanding suggestion that hydrothermal systems might have provided habitats or "cradles" for the origin and evolution of early life on Earth and possibly other planets, such as Mars. This is consistent with the most ancient organisms in the terrestrial tree of life being thermophilic (optimum growth temperatures >50 °C) or hyperthermophilic (optimum growth temperatures >80 °C) in nature. Studies of a number of impact structures on Earth suggest that most impact events that result in the formation of complex impact craters (i.e. >2–4 and >5–10 km diameter on Earth and Mars, respectively) are potentially capable of generating a hydrothermal system. Studies of the Haughton impact structure in the Canadian Arctic suggest that there are six main locations within and around impact craters where impact-generated hydrothermal deposits can form: (1) crater-fill impact melt rocks and melt-bearing breccias; (2) interior of central uplifts; (3) outer margin of central uplifts; (4) impact ejecta deposits; (5) crater rim region; and (6) post-impact crater lake sediments. The question of whether impact-generated hydrothermal systems form in craters elsewhere in the Solar System remains open, however, in 2010, such evidence was presented from the Toro Crater on Mars.

It has become apparent over the past couple of decades that impact events have profoundly affected the evolution of life on Earth and may have also influenced life's origins. There is also the outstanding question of the potential transfer of life from another planet to Earth through impact events. Experiments have shown that certain organisms can survive the impact process and the harsh conditions of space, at least for the time span that these experiments were conducted, which is obviously limited by the human life-span and research careers. Whether life could have been ejected, survived the potential several millennia journey through Space, survive the impact on Earth and then have the ability to colonize this planet, remains conjecture at this time.

4. IMPACTS AS PLANETARY PROBES

Impacts serve to probe the nature of the subsurface of planetary bodies through the processes of excavation, exposure and uplift. For example, in terrestrial complex structures lithologies exposed in the central uplift originally from depths approximately one tenth of the final rim diameter.

4.1. Water and Ices

As noted earlier, it has been known since the late 1970s that many Martian impact craters possess, "lobate", "rampart", "layered", or "fluidized" ejecta (Figure 4.7). According to the most recent Martian crater database, which contains 79,723 craters ≥3 km in diameter, approximately 50% of all Martian craters possess clearly discernable ejecta blankets and, of these, over 40% possess layered ejecta morphologies. Although the formation of layered ejecta morphologies continues to be debated, it is generally accepted that subsurface volatiles (water-ice) played a role in the formation of these unique ejecta morphologies. In addition to layered ejecta, evidence of deeper subsurface volatiles has been recently bolstered by the discovery of a globally wide-spread crater related pitted deposit (Figure 4.12) observed in 205 craters ranging from 1 to 150 km in diameter. Interestingly, crater-related pitted deposits have also been observed associated with the freshest craters on the asteroid Vesta, and are consistent with the former Martian work that suggests that there may be volatile-rich materials beneath the surface of this large differentiated asteroid. There is also direct evidence of shallow subsurface ice on Mars has been brought to light by small impact craters (10s of meters scale; Figure 4.13). High-resolution images indicate that meter-to-decameter scale impact craters occur frequently on Mars (~20 craters over a 7 year period). Subsequent observations from High Resolution Imaging Science Experiment

FIGURE 4.12 Crater-related pitted deposits on the floor of Zumba Crater, Mars. These pits were possibly formed from the release of volatiles from ices that were mixed with superheated impact melt (Mars Reconnaissance Orbiter). Note pitted deposits are also observed in Noord Crater (Figure 4.1(B)).

(HiRISE) with coordinated Compact Reconnaissance Imaging Spectrometer for Mars (CRISM) of five newly formed craters within the mid-latitude regions of Mars revealed water–ice excavated from just meters below the surface.

Historically, rocks returned from the Moon were composed entirely of anhydrous minerals and, thus, believed to be completely "bone-dry". This was typically considered a testament to the energetic and violent planet-scale impact origin of the Moon. However, recent spectral data of the surface of the Moon indicates the presence of adsorbed water molecules and/or hydroxyl (OH^-) associated with lunar surface materials. Based on a few assumptions, the hydrated proportion likely amounts to ~0.1% of these surface materials. Interestingly, the hydration spectral signature has been noted to correlate with the ejecta blankets of several fresh lunar craters, suggesting that some of these relatively water-rich materials can retain their water just beneath the surface, where they are protected by interactions with cosmic rays and the solar wind.

The concept of an impact as a natural subsurface probe was utilized in a recent lunar mission designed to create an artificial impact and observe in real-time the ejected plume of impact-liberated subsurface materials. The Lunar Crater Observation and Sensing Satellite detected an estimated ~6% by mass of water in the top few meters of the lunar regolith that was ejected by a crater formed from the impact of a spent Centaur rocket stage into a crater near the lunar south pole.

4.2. Spectral Composition

Several recent studies of craters that take advantage of crater as probes emphasize distinct compositions between the exposed subsurface materials and the surrounding surface around impact craters. From a survey of over 100 lunar central peaks in complex craters with Clementine multispectral images, some general conclusions can be drawn regarding the upper and lower lunar crust. These include: (1) that the lunar crust is extremely anorthositic, consistent with the "magma ocean" model (See The Moon), (2) crustal composition gradually increases in mafic content with depth, although mafic compositions are generally rare

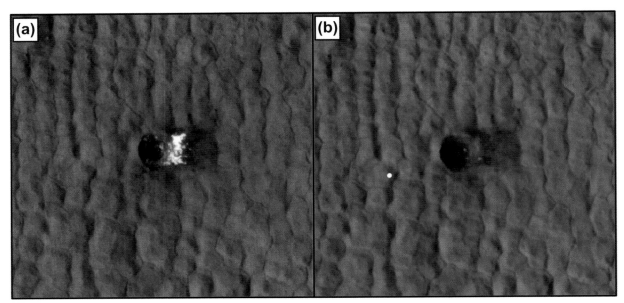

FIGURE 4.13 (a) A recently formed crater with abundant excavated and ejected white materials proposed to be subsurface water–ice. (b) This image taken almost one full Martian year later, shows that the white material has disappeared, consistent with a water–ice interpretation (Mars Reconnaissance Orbiter).

in central peaks, and (3) the lower crust is more compositionally diverse than the highlands. The strong anorthositic character of the lunar crust, as observed through central peaks, was confirmed by the Selene multiband imager, which identified some peaks that approached 100% pure anorthosite. Recent results are generally consistent with the Clementine spectral images of central peaks, with the exception of the discovery of a rare occurrence of Mg-spinel-rich deposits associated with the central peak of Theophilus Crater. As Theophilus lies on the rim of the Nectaris Basin, and with the only other known occurrence of Mg-spinel-rich also being associated with another large impact basin, it has been suggested that these materials may originate from deep within the lunar interior.

The Thermal Emission Imaging System (THEMIS) was the first subkilometer-scale spectral instrument flown in Mars orbit that could effectively be used to detect crater-scale spectral contrasts. Although a great deal of the surface of Mars is spectrally obscured by surface dust, including many crater central peaks, a few THEMIS observations of crater-related deposits or outcrops show higher concentrations of mafic minerals (pyroxene and olivine), with respect to their surrounding surfaces (Figures). A THEMIS-based study also discovered an undetermined high-silica spectral component associated with some crater deposits in the northwestern part of Syrtis Major. Once interpreted to represent exposed granitoids, these crater-related deposits have since been observed to contain altered silicate phases, including smectites, hydrated glasses and zeolites. As such, they are now interpreted to phases altered prior to impact or via multiple impact alteration pathways, including hydrothermalism. Additional craters associated with hydrated silicates are now being observed by the Observatoire pour la Mineralogie, l'Eau, les Glaces et l'Activitie' (OMEGA) and CRISM across Mars. Two independent global-scale classifications of alteration phases using OMEGA and CRISM data indicate that ~70% of the total occurrences of known alteration phases on Mars are associated with impact craters. With the exception of small simple craters, the origin of altered phases within complex craters, especially associated with crater central peaks, is difficult to constrain as large impact events generate a hydrothermal system that typically dominates the central uplift region. Detailed studies of multiple data sets are required to distinguish between these different origins.

4.3. Morphologic and Geologic

High-resolution visible images from recent planetary missions have been documenting exquisite geologic and structural details of craters. For example, images from HiRISE and the Context Camera on the Mars Reconnaissance Orbiter have revealed exposed bedrock in central uplifts of complex craters. The exposed parautochthonous bedrock can be divided into three outcrop-scale textural classes: (1) layered, (2) massive and fractured and (3) brecciated. Of these three classes, the layered bedrock class is the most strongly correlated with specific surface geologic units on Mars. Central peaks with layers occur predominately in units interpreted as extensive and voluminous lava deposits or plains (e.g., Tharsis and the surrounding regions). These uplifts have layered megablocks that are consistent with sampled stratigraphic sections of cyclic volcanism (i.e. alternating lava flows and pyroclastics). Although not surprising, the application of resolving, identifying and mapping bedrock characteristics in crater central uplifts on planetary bodies, other than Earth, is a new and novel approach for aiding planetary scientists to determine the geologic history of the surface and subsurface and to elucidate various aspects of central uplift formation. How different target rocks are exposed and deformed are particularly informative with respect to the impact process. For example, it has been hypothesized that the final orientation of layers that are faulted and rotated during central uplifts show a relationship to the impactor trajectory. Breccia injection dikes and impact melt deposits associated with crater central uplifts, which are now resolvable in HiRISE images of Mars, are also providing important clues with respect to the impact process.

BIBLIOGRAPHY

French, B. M. (1998). *Traces of catastrophe: A handbook of shock-metamorphic effects in terrestrial meteorite impact structures*. Lunar and Planetary Institute Contribution 954. Houston, USA: Lunar and Planetary Institute.

Melosh, H. J. (1989). *"Impact cratering: A geologic process*. New York, USA: Oxford Univ. Press.

Osinski, G. R., & Pierazzo, E. (Eds.). (2012). *Impact cratering: Processes and products*. Oxford, UK: Blackwell.

Spudis, P. D. (1993). *The geology of multi-ring basins: The moon and other planets*. Cambridge, UK: Cambridge Univ. Press.

Chapter 5

Planetary Volcanism

Lionel Wilson
Lancaster Environment Centre, Lancaster University, Lancaster, UK

Chapter Outline

1. Summary of Planetary Volcanic Features   101
　1.1. Earth   101
　1.2. The Moon   103
　1.3. Mars   104
　1.4. Venus   105
　1.5. Mercury   106
　1.6. Io   106
　1.7. The icy Satellites: Cryo-Volcanism   107
　1.8. The Differentiated Asteroids   108
2. Classification of Eruptive Processes   108
3. Effusive Eruptions and Lava Flows   109
4. Explosive Eruptions   111
　4.1. Basic Considerations   111
　4.2. Strombolian Activity   111
　4.3. Vulcanian Activity   112
　4.4. Hawaiian Activity   114
　4.5. Plinian Activity   115
　4.6. Phreato-Magmatic Activity   116
　4.7. Dispersal of Pyroclasts into a Vacuum   117
5. Inferences about Planetary Interiors   117
Bibliography   119

1. SUMMARY OF PLANETARY VOLCANIC FEATURES

1.1. Earth

The ~70% of Earth's surface represented by the **crust** forming the floors of the oceans consists of volcanic **rocks** generally erupted within the last 300 Ma (million years) from long lines of **volcanoes** located along ridges near the centers of ocean basins. This geologically youthful age stimulated the development of the theory of **plate tectonics**, which explained the locations and distributions of volcanoes over Earth's surface. Midocean ridge volcanoes erupt magmas called **basalt**s that are relatively metal-rich and silica- and **volatile**-poor, and these volcanoes mark the constructional margins of Earth's rigid crustal plates. Basalts are the products of the partial melting of **mantle** rocks due to decreasing pressure at the tops of **convection** cells in which temperature variations cause the solid mantle to deform and flow on very long timescales. Basalt compositions are closely related to the bulk composition of the mantle, which makes up most of Earth's volume outside the iron-dominated core. The volcanic edifices produced by ocean-floor volcanism consist mainly of relatively fluid (low-**viscosity**) lava flows with lengths from a few kilometers to a few tens of kilometers. Lava flows erupted along the midocean ridges simply add to the topography of the edges of the growing plates as the plates move slowly (~10 mm/year) away from the ridge crest (See Earth as a Planet: Atmosphere and Oceans; Earth as a Planet: Surface and Interior).

Lavas erupted from vents located some distance away from the ridge crest build roughly symmetrical edifices with convex-upward shapes described as **shield** volcanoes (having relatively shallow flank slopes) or domes (having relatively steeper flanks). Some of these vent systems are not related to the spreading ridges at all, but instead mark the locations of "hot spots" in the underlying mantle, unusually vigorous plumes of mantle material feeding magmas through the overlying plate. Because the plate moves over the hot spot, a chain of shield volcanoes can be built up in this way, marking the trace of the relative motion. The largest shield volcanoes on Earth form such a line of volcanoes, the Hawaiian Islands, and the two largest of these edifices, Mauna Loa and Mauna Kea, rise ~10 km above the ocean floor and have basal diameters of about 200 km.

Eruptive activity on shield volcanoes tends to be concentrated either at the summit or along linear or arcuate zones radiating away from the summit, called rift zones. The low viscosity of the basaltic magmas released in

Hawaiian-style eruptions on these volcanoes (Figure 5.1) allows the lava flows produced to travel relatively great distances (a few tens of km), and is what gives shield volcanoes their characteristic wide, low profiles. It is very common for a long-lived reservoir of magma, a magma chamber, to exist at a depth of a few to several kilometers below the summit. This reservoir, which is roughly equant in shape and may be up to 1–3 km in diameter, intermittently feeds surface eruptions, either when magma ascends vertically in the volcano summit region or when magma flows laterally in a subsurface fracture called a dike, which most commonly follows an established rift zone, to erupt at some distance from the summit. In many cases, magma fails to reach the surface and instead freezes within its fracture, forming an **intrusion**. The summit reservoir is fed, probably episodically, from partial melt zones in the mantle beneath. Rare but important events in which a large volume of magma leaves such a reservoir lead to the collapse of the rocks overlying it, and a characteristically steep-sided crater called a **caldera** is formed, with a width similar to that of the underlying reservoir.

Volcanoes erupting relatively silica-rich and volatile-rich magma (andesite or, less commonly, rhyolite) mark the destructive margins of plates, where the plates bend downward to be subducted into the interior and at least partly remelted. These volcanoes tend to form an arcuate pattern (called an island arc when the volcanoes rise from the sea floor) marking the trace on the surface of the zone where the melting is taking place, at depths on the order of 100–150 km. The andesitic magmas thus produced are the products of melting of a mixture of subducted ocean floor basalt, sedimentary material that had been washed onto the ocean floor from the continents (which are themselves an older, silica-rich product of the chemical **differentiation** of Earth), seawater trapped in the sediments, and the primary mantle materials into which the plates are subducted. Thus, andesites are much less representative of the current composition of the mantle. Andesite magmas are rich in volatiles (mainly water, carbon dioxide, and sulfur compounds), and their high silica contents give them high viscosities, making it hard for gas bubbles to escape. As a result, andesitic volcanoes often erupt explosively in Vulcanian-style eruptions, producing localized **pyroclastic** deposits with a wide range of grain sizes; alternatively, they produce relatively viscous lava flows that travel only short distances (a few kilometers) from the vent. The combination of short flows and localized ash deposits tends to produce steep-sided, roughly conical volcanic edifices.

When large bodies of very silica-and volatile-rich magma (rhyolite) accumulate—in subduction zones or, in some cases, where hot spots exist under continental areas, leading to extensive melting of the continental crustal rocks—the potential exists for very large-scale explosive eruptions to occur, in which finely fragmented magma is blasted at high speed from the vent to form a convecting eruption cloud, called a Plinian cloud, in the atmosphere. These clouds may reach heights up to 50 km, and pyroclastic fragments fall from them to create a characteristic deposit spreading downwind from the vent (Figure 5.2). Under certain circumstances, the cloud cannot convect in a stable fashion and collapses to form a fountain-like structure over the vent, which feeds a series of pyroclastic flows—mixtures of incandescent pyroclastic fragments, volcanic gas, and entrained air—that can travel for at least tens of kilometers from the vent at speeds in excess of

FIGURE 5.1 A Hawaiian style lava fountain feeding a lava flow and building a cinder cone (Pu'u 'O'o on the flank of Kilauea volcano in Hawaii). Steaming ground is visible marking the location of the axis of the rift zone along which a dike propagated laterally to feed the vent. *Photograph by P.J. Mouginis-Mark.*

FIGURE 5.2 Plinian air-fall deposits mantle pre-existing topography with relatively uniform layers, as seen near the middle of this photograph of the wall of the caldera formed in the ~3600-year old eruption of Santorini in the eastern Mediterranean. *Photograph by L. Wilson.*

100 m/s, eventually coming to rest to form a rock body called ignimbrite. These fall and flow deposits may be so widespread around the vent that no appreciable volcanic edifice is recognizable; however, there may still be a caldera, or at least a depression, at the vent site due to the collapse of the surface rocks to replace the large volume of material erupted from depth.

It should be clear from the foregoing descriptions that the distribution of the various types of volcano and characteristic volcanic activity seen on Earth are intimately linked with the processes of plate tectonics. A major finding to emerge from the exploration of the Solar System over the last 30 years is that this type of large-scale tectonism is currently confined to the Earth and may never have been active on any of the other bodies. Virtually all of the major volcanic features that we see elsewhere can be related to the eruption of mantle melts similar to those associated with the midocean ridges and oceanic hotspots on Earth. However, differences between the physical environments (acceleration due to gravity, atmospheric conditions) of the other planets and Earth lead to significant differences in the details of the eruption processes and the deposits and volcanic edifices formed.

1.2. The Moon

Analyses of the samples collected from the Moon by the Apollo missions in the 1970s showed that there were two major rock types on the lunar surface. The relatively bright rocks forming the old, heavily cratered highlands of the Moon are a primitive crust that formed about 4.5 Ga (billion years) ago by the accumulation of solid minerals at the cooling top of a possibly 300 km thick melted layer referred to as a magma ocean. This early crust was extensively modified, mainly prior to about 3.9 Ga ago, by the impacts of **comets** and **asteroid**s with a wide range of sizes to form impact craters and basins. Some of the larger craters and basins (the **mare** basins) were later flooded episodically by extensive lava flows, many more than 100 km long, to form the darker rocks visible on the lunar surface (See The Moon, Planetary Impacts).

Radiometric dating of samples from lava flow units showed that these mare lavas were mostly erupted between 3 and 4 Ga ago, forming extensive, relatively flat deposits inside large basins. Individual flow units, or at least groups of flows, can commonly be distinguished using multispectral remote-sensing imagery on the basis of their differing chemical compositions, which give them differing reflectivities in the visible and near-infrared parts of the spectrum. In composition these lavas are basaltic, and their detailed mineralogy shows that they are the products of partial melting of the lunar mantle at depths between 150 and more than 400 km, the depth of origin increasing with time as the lunar interior cooled. Melting experiments on samples, supported by theoretical calculations based on their mineralogies, show that these lavas were extremely fluid (i.e. had very low viscosities, at least a factor of 3–10 less than those of typical basalts on Earth). This allowed them to travel for great distances, often more than 100 km (Figure 5.3) from their vents; it also meant that they had a tendency to flow back into, and cover up, their vents at the ends of the eruptions. Even so, it is clear from the flow directions that the vents were mainly near the edges of the interiors of the basins that the flows occupy. Many vents were probably associated with the arcuate rilles found in similar positions. These are curved **graben**s, trench-like depressions parallel to the edges of the basins formed as parts of the crust sink between pairs of parallel faults caused by tension. This marginal tension, due to the weight of the lava ponded in the middle of the basin, makes it easier for cracks filled with magma to reach the surface in these places.

A second class of lunar volcanic features associated with the edges of large basins is the sinuous rilles. These are meandering depressions, commonly hundreds of meters wide, tens of meters deep, and tens of kilometers long, which occur mainly within the mare basalts. Some are discontinuous, giving the impression of an underground tube that has been partly revealed by partial collapse of its roof, and these are almost certainly the equivalent of lava tube systems (lava flows whose top surface has completely solidified) on Earth. Other sinuous rilles are continuous open channels all along their length; these generally have origins in source

FIGURE 5.3 Lava flows in S.W. Mare Imbrium on the Moon. The source vents are off the image to the lower left and the ~300 km long flows extend down a gentle slope toward the center of the mare basin beyond the upper right edge of the frame. *NASA* Apollo *photograph.*

depressions two or three times wider than the rille itself, and become narrower and shallower with increasing downslope distance from the source. These sinuous rilles appear to have been caused by long-duration lava flows that were very turbulent, i.e. the hot interior was being constantly mixed with the cooler top and bottom of the flow. As a result the flows were able to heat up the preexisting surface until some of its minerals melted, allowing material to be carried away and an eroded channel to form.

In contrast to the lava flows and lava channels, two types of pyroclastic deposit are recognized on the Moon. There are numerous deposits called dark mantles, often roughly circular and up to at least 200 km in diameter, where the fragmental lunar surface regolith is less reflective than usual, and spectroscopic evidence shows that it contains a component of small volcanic particles in addition to the locally derived rock fragments. The centers of these regions are commonly near the edges of mare basins, suggesting that the dark mantle deposits were produced by the same source vents as the lava flows. Chemical analyses of the Apollo lava samples show that the Moon's mantle is severely depleted in common volatiles like water and carbon dioxide due to its hot origin (See The Moon) and suggest that the main gas released from mare lava vents was carbon monoxide, produced in amounts up to a few 100 parts per million by weight as a result of a chemical reaction between free carbon and metal oxides, mainly iron oxide, in the magma as it neared the surface.

Several small, dark, fragmental deposits occur on the floor of the old, 90-km-diameter impact crater Alphonsus. These patches, called dark haloes, extend for a few kilometers from the rims of subdued craters that are centered on, and elongated along, linear fault-bounded depressions (called linear rilles) on the crater floor. It is inferred that these are the sites of less energetic volcanic explosions.

Localized volcanic constructs such as shield volcanoes and domes are rare on the Moon, though more than 200 low, shield-like features with diameters mainly in the range 3–10 km are found in the Marius region within Oceanus Procellarum, in northeast Mare Tranquillitatis, and in the region between the craters Kepler and Copernicus. Conspicuously absent are edifices with substantial summit calderas. This implies that large, shallow magma reservoirs are very rare, almost certainly a consequence of the difficulty with which dense magmas rising from the mantle penetrate the low-density lunar crust. However, a few collapse pits with diameters up to 3 km do occur, located near the tops of domes or aligned along linear rilles.

1.3. Mars

About half of the surface of Mars consists of an ancient crust containing impact craters and basins. Spectroscopic evidence from orbiting spacecraft suggests that it is composed mainly of volcanic rocks. The other half of the planet consists of relatively young, flat, lower-lying, plains-forming units that are a mixture of wind-blown sediments, lava flows, and rock debris washed into the lowlands by episodes of water release from beneath the surface. Combining orbital observations with analyses made by the several probes that have landed successfully on the surface suggests that most of the magmas erupted on Mars are basalts or basaltic andesites (See Mars: Surface and Interior, Mars Site Geology and Geochemistry).

The most obvious volcanic features on Mars are five extremely large ($\sim$600 km diameter, heights up to $\sim$20 km) shield volcanoes (Olympus Mons, Ascraeus Mons, Pavonis Mons, Arsia Mons and Alba Mons) with the same general morphology as basaltic shield volcanoes found on Earth (Figure 5.4). These volcanoes are surrounded by overlapping lava aprons that collectively form a huge volcanic rise. A second but smaller rise contains the volcanoes Elysium Mons, Hecates Tholus and Albor Tholus. There are also about 20 smaller shield volcanoes on Mars. Counts of small impact craters seen in high-resolution images from orbiting spacecraft show that the ages of the lava flow units on the volcanoes range from more than 3 Ga to less than $\sim$50 Ma. Complex systems of nested and intersecting calderas are found on the larger shields, implying protracted evolution of the internal plumbing of each volcano, typified by cycles of activity in which a volcano is sporadically active for $\sim$1 Ma and then dormant for $\sim$50 Ma. Individual caldera depressions are up to at least 30 km in diameter, much larger in size than any found on Earth, and imply the presence of very large

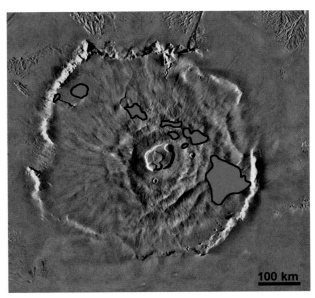

FIGURE 5.4 The Olympus Mons shield volcano on Mars with the Hawaiian islands superimposed for scale. *NASA image with overlay by P.J. Mouginis-Mark.*

shallow magma reservoirs during the active parts of the volcanic cycles. The large size of these reservoirs, like that of the volcanoes themselves, is partly due to the low acceleration due to gravity on Mars and partly due to the absence of plate tectonics, which means that a mantle hot spot builds a single large volcano, rather than a chain of small volcanoes as on Earth. The availability of large volumes of melt in the mantle beneath some of the largest shield volcanoes has apparently led to the production of giant swarms of dikes, propagating radially away from the volcanic centers for more than 2000 km in some cases. The locations of these dikes are indicated by the presence of narrow (1–5 km wide) grabens where the intrusions have exerted extensional stresses on the crust.

Most shields appear to have flanks dominated by lava flows, many more than 100 km long. However, there are examples of sinuous channels like the sinuous rilles on the Moon, presumably caused by hot, turbulent, high-speed lavas melting the ground over which they flow. Some of the older and more eroded edifices, like Tyrrhenus Mons and Hadriacus Mons, appear to contain high proportions of relatively weak, presumably pyroclastic, rocks. There is a hint, from the relative ages of the volcanoes and the stratigraphic positions of the mechanically weaker layers within them, that pyroclastic eruptions were commoner in the early part of Mars's history. It is also possible that some of the plains-forming units, generally interpreted as weathered lava flows, in fact consist of pyroclastic fall or flow deposits. Not all of the sources of these deposits have been identified with certainty — there are a few dozen massifs, each tens of km in size, in the Martian highlands that have been tentatively interpreted as ancient degraded volcanoes.

1.4. Venus

Because of its dense, optically opaque atmosphere, the only detailed synoptic imaging of the Venus surface comes from orbiting satellite-based radar systems. Despite the differences between optical and radar images (radar is sensitive to both the dielectric constant and the roughness of the surface on a scale similar to the radar wavelength), numerous kinds of volcanic features have been unambiguously detected on Venus. Large parts of the planet are covered with plains-forming lava flows, having well-defined lobate edges and showing the clear control of topography on their direction of movement (Figure 5.5). The lengths (which can be up to several hundred kilometers) and thicknesses (generally significantly less than 30 m, since they are not resolvable in the radar altimetry data) of these flows suggest that they are basaltic in composition. This interpretation is supported by the (admittedly small) amounts of major-element chemical data obtained from six Soviet probes that soft-landed on the Venus surface. Some areas show concentrations of

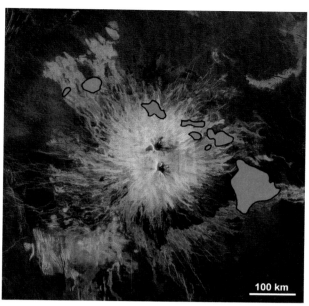

FIGURE 5.5 A variety of radar-bright lava flows radiate from the summit area down the flanks of the shield volcano Sapas Mons on Venus. The Hawaiian islands are superimposed for scale. *NASA* Magellan *image with overlay by P.J. Mouginis-Mark.*

particularly long flows called fluctūs (Latin for floods) and incised channels called canali. Most of the lava plains, judging by the numbers of superimposed impact craters, were emplaced within the last ~700 Ma (See Venus: Surface and Interior).

Many areas within the plains and within other geological units contain groupings (dozens to hundreds) of small volcanic shields or domes, from less than one to several kilometers in diameter. At least 500 such shield fields have been identified. Some of the individual volcanoes have small summit depressions, apparently due to magma withdrawal and collapse, and others are seen to feed lava flows. Quite distinct from these presumably basaltic shields and domes is a class of larger, steep-sided domes with diameters of a few tens of kilometers and heights up to ~1 km. The surface morphologies of these domes suggest that most were emplaced in a single episode, and current theoretical modeling shows that their height-to-width ratio is similar to that expected for highly viscous silicic (perhaps rhyolitic) lavas on Earth.

Many much larger volcanic constructs occur on Venus. About 300 of these are classed as intermediate volcanoes and have a variety of morphologies, not all including extensive lava flows. A further 150, with diameters between 100 and about 600 km, are classed as large volcanoes. These are generally broad shield volcanoes, with extensive systems of lava flows and heights above the surrounding plains of up to 3 km.

Summit calderas are quite common on the volcanoes, ranging in size from a few kilometers to a few tens of kilometers. There are two particularly large volcano-related depressions, called Sacajawea and Colette, located on the upland plateau Lakshmi Planum. With diameters of ~200 km and depths of ~2 km, these features appear to represent the downward sagging of the crust over some unusually deep-seated site of magma withdrawal.

Finally, there are a series of large, roughly circular features on Venus, which, though intimately linked with the large-scale tectonic **stresses** acting on the crust (they range from a few hundred to a few thousand kilometers in diameter), also have very strong volcanic associations. These are the **coronae**, novae, and arachnoids. Though defined in terms of the morphology of circumferential, moat-like depressions and radial fracture systems, these features commonly contain small volcanic edifices (shields or domes), small calderas, or lava flows, the latter often apparently fed from elongate vents coincident with the distal parts of radial fractures. In such cases, it seems extremely likely that the main feature is underlain by some kind of magma reservoir, which feeds the more distant eruption sites via lateral dike systems.

1.5. Mercury

Much of the surface of Mercury is a heavily cratered ancient terrain like that of the Moon and parts of Mars. However, data from the MESSENGER spacecraft orbiting Mercury have confirmed suspicions from the earlier Mariner 10 flyby mission that the relatively smooth plains-forming units dispersed among the craters (the **intercrater plains**) are vast areas of basaltic lava flows (Figure 5.6). As in the case of the Moon, these lavas appear to be characterized by high eruption rates, great travel distances, and the ability to commonly drown their own vents at the end of the eruption, making the identification of source areas difficult.

Localized explosive volcanic activity is indicated by the presence of more than 30 rimless depressions up to ~40 km in diameter surrounded by spectrally distinctive deposits that grade progressively into the surrounding terrain. The lateral extents of these deposits, up to ~30 km, imply up to ~1 wt% of volatiles in the erupting magmas. The nature of the volatiles is uncertain, but a high abundance of sulfur is present in crustal rocks. Additional evidence for volcanic activity includes several irregularly shaped, rimless, steep-sided pits up to ~30 km in diameter that commonly occur on the floors of impact craters. These may indicate collapse after withdrawal of shallow bodies of intruded magma or the decomposition of magma-linked metal sulfide deposits due to the high daytime temperatures (See Mercury).

1.6. Io

The bulk density of Io suggests that it has a silicate composition, similar to that of the inner, Earth-like planets. Io and the Earth's Moon also have similar sizes and masses, and it might therefore be expected by analogy with the Moon that any volcanic activity on Io would have been confined to the first 1 or 2 billion years of its life. However, as the innermost satellite of the gas-giant Jupiter, Io is subjected to strong tidal forces. An orbital period resonance driven by the mutual gravitational interactions of Io, Europa, and Ganymede causes the orbit of Io to be slightly elliptical. This, coupled with the fact that Io rotates synchronously (i.e. the orbital period is the same as that of the axial rotation), means that the interior is subjected to periodic tidal flexing. The inelastic part of this deformation generates heat on a scale that far outweighs any remaining heat of formation or heat from the decay of naturally radioactive elements. As a result, Io is currently the most volcanically active body in the solar system. At any one time there may be up to a dozen erupting vents. Roughly half of these produce lava flows from fissure vents (Figure 5.7) associated with calderas located at the centers of very low shield-like features, and half produce umbrella-shaped eruption clouds into which gases and small pyroclasts are being ejected at speeds of up to 1000 m/s to reach heights up to 300 km (Figure 5.8) (See Io).

The main gases detected in the eruption clouds are sulfur and sulfur dioxide, and much of the surface is coated with highly colored deposits of sulfur and sulfur compounds that have been degassed from the interior over solar

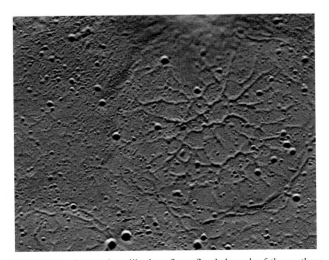

FIGURE 5.6 Long, sheet-like lava flows flooded much of the northern hemisphere of Mercury early in its geologic history. Here, near 82° N, 51° W, lava almost completely buried an old impact crater ~60 km in diameter. Tension due to cooling-induced shrinkage of the deep lava in the center of the crater produced a characteristic pattern of cracks. *NASA MESSENGER image.*

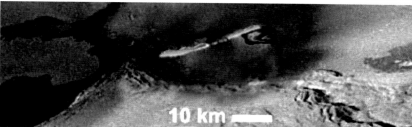

FIGURE 5.7 The upper part of the figure shows the chain of calderas called Tvashtar Catena on Io, with a fissure eruption in progress. The high temperature of the lava overloaded the spacecraft imaging system causing "bleeding" of data values down vertical lines of the image. Using a lower resolution image taken later, the appearance of the eruption was reconstructed as shown in the lower part of the figure. *NASA* Galileo *images.*

system history and are now concentrated in the near-surface layers. However, it seems very likely, based on the presence of hot lava flows and the fluid dynamic and thermodynamic analysis of the eruption clouds, that the underlying cause of the activity is the ascent of very hot basaltic magmas from the mantle. The average eruption rate on Io is so great that the materials forming the surface layers at any one time are buried to depths of order 30 km in only a few million years. This rapid subsidence makes the geothermal gradient very nonlinear, and volatiles like sulfur and sulfur dioxide do not melt until depths of ~20 km area reached. At these depths, these volatiles can be entrained into magma in dikes propagating up from the deeper mantle. The magma may be inherently quite volatile poor, but the addition of the buried volatiles makes it anomalously volatile rich and drives the extremely explosive eruptions. Most of the volatiles condense as they expand and cool and, along with the silicate pyroclasts, eventually fall back to the surface, where their subsequent burial provides the materials to drive future explosive eruptions.

1.7. The icy Satellites: Cryo-Volcanism

Many of the satellites of the gas-giant planets have bulk densities indicating that their interiors are mixtures of silicate rocks and the **ice** of common volatiles, mainly water. On some of these bodies (e.g. Jupiter's satellites Ganymede and Europa, Uranus's satellite Ariel, Neptune's satellite Triton, and Saturn's large satellite Titan), flow-like features are seen, reminiscent of very viscous lava flows (See: Icy Satellites).

However, there is no spectroscopic evidence for silicate magmas having been erupted on these bodies, and the flow-like features have forced us to recognize that there is a more general definition of volcanism than that employed so far. Volcanism is the generation of partial melts from the internal materials of a body and the transport out onto the surface of some fraction of those melts. In the ice-rich bodies, it is the generation of liquid water from solid ice that mimics the partial melting of rocks, in the process called **cryo-volcanism**.

The ability of this water to erupt at the surface is influenced by its content of volatiles like ammonia and methane. Since the surface temperatures of most of these satellites are very much less than the freezing temperature of water (even when the freezing point is lowered by the presence of compounds like ammonia), and since they do not have appreciable atmospheres (except Titan), the fate of

FIGURE 5.8 An explosive eruption plume on Io. The great height of the plume, more than 200 km, implies that magma is mixing with and evaporating volatile materials (sulfur or sulfur dioxide) in the crust through which it erupts. *NASA* New Horizons *image.*

any liquid water erupting at the surface is complex. Cooling will produce ice crystals at all boundaries of the flow and, being less dense than liquid water, these crystals will rise toward the flow surface. Because of the negligible external pressure, evaporation (boiling) will take place within the upper few hundred millimeters of the flow. The vapor produced will freeze as it expands, to settle out as a frost or snow on the surrounding surface. The boiling process extracts heat from the liquid and adds to the rate of ice crystal formation. If enough ice crystals collect at the surface of a flow, they will impede the boiling process, and if a stable ice raft several 100-mm-thick forms, it will suppress further boiling. Thus, if it is thick enough, a liquid water flow may be able to travel a significant distance from its eruption site. It is even possible that solid ice may form flow-like features, in essentially the same way that glaciers flow on Earth, though the very low temperatures will make the timescales much longer.

If liquid water produced below the surface of an icy satellite contains a large enough amount of volatiles it will erupt explosively at high speed in what, near the vent, is the equivalent of a Plinian eruption. The expanding volatiles could cause the eruption cloud to spread sideways (like the umbrella-shaped plumes on Io) and disperse the water droplets, rapidly freezing to snow, over a wide area. If the eruption speed is high enough and the parent body small enough some of the snow may be ejected with escape velocity. Data from the Cassini spacecraft provided graphic evidence for this process occurring near the south pole of Saturn's small satellite Enceladus. The orbit of Enceladus is very close to the brightest of Saturn's many rings, the E ring, which appears to be composed of particles of ice. It now seems clear that these are derived directly from Enceladus, having been ejected fast enough to escape from the satellite but not from Saturn itself (See: Planetary Rings). The volatiles driving the water release include nitrogen, methane and carbon dioxide with traces of propane, ethane, acetylene and ammonia.

1.8. The Differentiated Asteroids

The meteorites that fall to the Earth's surface are fragments ejected from the surfaces of asteroids during mutual collisions. Most meteorites are pieces of silicate rock and, whereas many contain minerals consistent with them never having been strongly heated, the mineralogy of others can only be explained if they are either solidified samples of what was once magma or pieces of what was once a mantle that partially melted and then cooled again after magma was removed from it. Additionally, some meteorites are pieces of a nickel-iron-sulfur alloy that was once molten but subsequently cooled slowly. Taken together these observations imply that some asteroids went through a process of extensive chemical differentiation by melting to form a crust, mantle and core. The trace element composition of the meteorites from these differentiated asteroids shows that they were heated by the radioactive decay of a group of short-half-life isotopes that were present at the time the Solar System formed, the most important of which was ^{26}Al which has a half life of ~ 0.75 Ma. Thus all of the heating, melting and differentiation must have taken place within an interval of only a few million years. Yet during this brief period, asteroids as small as 100 km in diameter were undergoing the same patterns of mantle melting, melt rise to the surface, and explosive and effusive eruptions that would only start to occur on Earth, Mars and Venus many tens of millions of years later.

Earth-based spectroscopic evidence, now supported by remote-sensing measurements from the orbiting Dawn spacecraft, very strongly suggests that the asteroid 4 Vesta is the parent body of one group of volcanically-generated surface, crust and upper-mantle rocks, the Howardite—Eucrite—Diogenite group of meteorites. Unfortunately Vesta has been so modified by impact cratering that no obvious volcanic features like lava flows are visible in the Dawn images. We have not yet identified any other differentiated asteroids with such certainty, but know from their compositions that the Aubite and Ureilite meteorites are rocks from the mantles of two different asteroids on which violently explosive eruptions ejected magma that should have become their crustal rocks into space at escape velocity. Acapulcoite and Lodranite meteorites are rocks from the shallow crust or upper mantle of a body that produced rather small amounts of gas during mantle melting, so that in these meteorites we see gas bubbles trapped in what was once magma traveling through fractures toward the surface. Finally the nickel-iron meteorites cluster into many tens of chemically-similar groups implying that at least this number of differentiated asteroids once existed but have since been largely fragmented in mutual collisions. The importance of these meteorites is that they give us copious samples of the very deep interiors of their parent bodies as well as the surfaces; such deep samples will not be available for a very long time for Venus, Mars and Mercury and are rare even for the Earth and Moon (See: Meteorites; Asteroids).

2. CLASSIFICATION OF ERUPTIVE PROCESSES

Volcanic eruption styles on Earth were traditionally classified mainly in terms of the observed composition and dispersal of the eruption products. Over the last 30 years it has been realized that they might be more systematically classified in terms of the physics of the processes involved. This has the advantage that a similar system can be adopted for all planetary bodies, automatically taking account of the ways in which local environmental factors (especially surface gravity and atmospheric pressure) lead to differences in the morphology of the deposits of the same process occurring on different planets.

Eruptive processes are classified as either explosive or effusive. An effusive eruption is one in which lava spreads steadily away from a vent to form one or more lava flows, whereas in explosive eruptions magma emerging through the vent is torn apart, as a result of the coalescence of expanding gas bubbles, into clots of liquid that are widely dispersed. The clots cool while in flight above the ground and may be partly or completely solid by the time they land to form a layer of pyroclasts. There is some ambiguity concerning this basic distinction between effusive and explosive activity, because many lava flows form from the coalescence, near the vent, of large clots of liquid that have been disrupted by gas expansion but that have not been thrown high enough or far enough to cool appreciably. Thus some, especially Hawaiian-style, eruptions have both an explosive and an effusive component at the same time.

There is also ambiguity about the use of the word explosive in a volcanic context. Conventionally, an explosion involves the sudden release of a quantity of material that has been confined in some way at a high pressure. Most often the expansion of trapped gas drives the explosion process. In volcanology, the term explosive is used not only for this kind of abrupt release of pressurized material, but also for any eruption in which magma is torn apart into pyroclasts that are accelerated by gas expansion, even if the magma is being erupted in a steady stream over a long time period. Eruption styles falling into the first category include Strombolian, Vulcanian, and phreato-magmatic activity, whereas those falling into the second include Hawaiian and Plinian activity. All of these styles are discussed in detail below.

3. EFFUSIVE ERUPTIONS AND LAVA FLOWS

Whatever the complications associated with prior gas loss, an effusive eruption is regarded as taking place once lava leaves the vicinity of a vent as a continuous liquid flow. The morphology of a lava flow, both while it is moving and after it has come to rest as a solid rock body, is an important source of information about the rheology (the deformation properties) of the lava, which is determined largely by its chemical composition, and about the rate at which the lava is being delivered to the surface through the vent. Because lava flows basically similar to those seen on Earth are so well exposed on Mars, Venus, Mercury, the Moon and Io, a great deal of effort has been made to understand lava emplacement mechanisms.

In general, lava contains some proportion of solid crystals of various minerals and also gas bubbles. Above a certain temperature called the liquidus temperature, all the crystals will have melted, and the lava will be completely liquid. Under these circumstances, lavas containing less than about 20% by volume of gas bubbles will have almost perfectly Newtonian rheologies, which means that the rate at which the lava deforms, the **strain** rate, is directly proportional to the stress applied to it under all conditions. This constant ratio of the stress to the strain rate is called the Newtonian viscosity of the lava. At temperatures below the liquidus but above the **solidus** (the temperature at which all the components of the lava form completely solid minerals), the lava in general contains both gas bubbles and crystals and has a non-Newtonian rheology. The ratio of stress to strain rate is now a function of the stress, and is called the apparent viscosity. At high crystal or bubble contents, the lava may develop a nonzero strength, called the yield strength, which must be exceeded by the stress before any flowage of the lava can occur. The simplest kind of non-Newtonian rheology is one in which the increase in stress, after the yield strength is exceeded, is proportional to the increase in strain rate: the ratio of the two is then called the Bingham viscosity and the lava is described as a Bingham plastic.

The earliest theoretical models of lava flows treated them as Newtonian fluids. Such a fluid released on an inclined plane will spread both downslope and sideways indefinitely (unless surface tension stops it, a negligible factor on the scale of lava flows). Some lavas are channeled by preexisting topography, and so it is understandable that they have not spread sideways. However, others clearly stop spreading sideways even when there are no topographic obstacles, and quickly establish a pattern in which lava moves downhill in a central channel between a pair of stationary banks called levées. Also, lavas do not flow downhill indefinitely once the magma supply from the vent ceases: they commonly stop moving quite soon afterward, often while the front of the flow is on ground with an appreciable slope and almost all of the cooling lava is still at least partly liquid. Also, liquid lava present in a channel at the end of an eruption does not drain completely out of the channel: a significant thickness of lava is left in the channel floor. These observations led to the suggestion that no lavas are Newtonian, and attempts were made to model flows as the simplest non-Newtonian fluids, Bingham plastics.

The basis of these models is the idea that the finite thickness of the levées or flow front can be used to determine the yield strength of the lava and that the flow speed in the central channel can be used to give its apparent, and hence Bingham, viscosity. Multiplying the central channel width by its depth and the mean lava flow speed gives the volume flux (the volume per second) being erupted from the vent. Laboratory experiments were used to develop these ideas, and they have been applied by numerous workers to field observation of moving flows on Earth and to images of ancient flows on other planets. For flows on Earth it is possible to deduce all of the parameters just listed; for ancient flow deposits one can obtain the yield strength unambiguously, but only the product of the viscosity and volume flux can be determined.

There is a possible alternative way to estimate the volume flux if it can be assumed that the flow unit being examined has come to rest because of cooling. An empirical relationship has been established for cooling-limited flows on Earth between the effusion rate from the vent and the length of a flow unit, its thickness, and the width of its active channel. If a flow is treated as cooling-limited when in fact it was not (the alternative being that it was volume-limited, meaning that it came to rest because the magma supply from the vent ceased at the end of the eruption), the effusion rate will inevitably be an underestimate by an unknown amount. Cooling-limited flows can sometimes be recognized because they have breakouts from their sides where lava was forced to form a new flow unit when the original flow front came to rest.

Lava rheologies and effusion rates have been estimated in this way for lava flows on Mars, the Moon, and Venus. It should be born in mind, when assessing these published estimates, that a major failing of simple models like the Bingham model is that they assign the same rheological properties to all of the material in a flow. However, lava that has resided in a stationary levée near the vent for a long period will have suffered vastly more cooling than the fresh lava emerging from the vent and will have very different properties. More elaborate models have been evolved since the earliest work, including some that apply to broadly spreading lava lobes that do not have a well-defined levée-channel structure, and others that treat the levées and central channel as separate materials with differing rheologies, but no model yet accounts for all of the factors controlling lava flow emplacement. With this caution, the rheological properties found suggest that essentially all of the lavas studied so far on the other planets have properties similar to those of basaltic to intermediate (basaltic–andesite) lavas on Earth. Many of these lavas have lengths up to several hundred kilometers, to be compared with basaltic flow lengths up to a few tens of kilometers on Earth in geologically recent times, and this implies that they were erupted at much higher volume fluxes than is now common on Earth. There is a possibility, however, that some of these flow lengths have been overestimated. If a flow comes to rest so that its surface cools, but the eruption that fed it continues and forms other flow units alongside it, a breakout may eventually occur at the front of the original flow. A new flow unit is fed through the interior of the old flow, and the cooled top of the old flow, which has now become a lava tube, acts as an excellent insulator. As a result, the breakout flow can form a new unit almost as long as the original flow, and a large, complex compound flow field may eventually form in this way. Unless spacecraft images of the area have sufficiently high resolution for the compound nature of the flows to be recognized, the total length of the group of flows will be interpreted as the length of a single flow, and the effusion rate will be greatly overestimated.

There are, however, certain volcanic features on the Moon and Mars that may be less ambiguous indicators of high effusion rates: the sinuous rilles. The geometric properties of these meandering channels—widths and depths that decrease away from the source, lengths of tens to a few hundred kilometers—are consistent with the channels being produced by very fluid lava erupted at a very high volume flux for a long time. The turbulent motion of the initial flow, meandering downhill away from the vent, led to efficient heating of the ground on which it flowed, and it can be shown theoretically that both mechanical and thermal erosion of the ground surface are expected to have occurred on a timescale from weeks to months. The flow, typically ~ 10 m deep and moving at ~ 10 m/s, slowly subsided into the much deeper channel that it was excavating. Beyond a certain distance, the lava cooled to the point where it could no longer erode the ground, and it continued as an ordinary surface lava flow. The volume eruption rates deduced from the longer sinuous rille channel lengths are very similar to those found for the longest conventional lava flow units. Modeling studies show that the turbulence leading to efficient thermal erosion was probably encouraged by a combination of unusually steep slope and unusually low lava viscosity. A few sinuous channels associated with lava plains are visible on Venus, but the lengths of some of the Venus channels are several to 10 times as great as those seen on the Moon and Mars. It is not yet clear if the thermal erosion process is capable of explaining these channels by the eruption of low-viscosity basalts, or whether some more exotic volcanic fluid (or some other process) was involved.

There are numerous uncertainties in using the foregoing relationships to estimate lava eruption conditions. Thus, there have been many studies of the way heat is transported out of lava flows, taking account of the porosity of the lava generated by gas bubbles, the effects of deep cracks extending inward from the lava surface, and the external environmental conditions—the ability of the planetary atmosphere to remove heat lost by the flow by conduction and convection, and by radiation (whether or not an atmosphere is present). However, none of these has yet dealt in sufficient detail with turbulent flows, or with the fact that cooling must make the rheological properties of a lava flow a function of distance inward from its outer surface, so that any bulk properties estimated in the ways described earlier can only be approximations to the detailed behavior of the interior of the lava flow. There is clearly some feedback between the way a flow advances and its internal pattern of shear stresses. For example, lava flows on Earth have two basic surface textures. Basaltic flows erupted at low effusion rates or while still hot near their vents have smooth, folded surfaces with a texture called pahoehoe (a Hawaiian word), the result of plastic stretching of the outer skin as the lava advances; at higher effusion rates, or at lower temperatures farther from the vent, the surface fractures in a more brittle fashion to produce a very rough texture called 'a'ā. A similar but coarser, rough, blocky texture is seen on the surfaces of more andesitic flows. Because there is a possibility of relating effusion rate and composition to the surface roughness of a flow in this way, there is a growing interest in obtaining relatively high resolution radar images of planetary surfaces (and Earth's surface) in which, as in the Magellan images of Venus, the returned signal intensity is a function of the small-scale roughness.

4. EXPLOSIVE ERUPTIONS

4.1. Basic Considerations

Magmas ascending from the mantle on Earth commonly contain volatiles, mainly water and carbon dioxide together with sulfur compounds and halogens. All of these have solubilities in the melt that are both pressure and temperature dependent. The temperature of a melt does not change greatly if it ascends rapidly enough toward the surface, but the pressure to which it is subjected changes enormously. As a result, the magma generally becomes saturated in one or more of the volatile compounds before it reaches the surface. Only a small degree of supersaturation is needed before the magma begins to exsolve the appropriate volatile mixture into gas bubbles, especially if the magma contains unmelted crystals on which bubbles can nucleate. As a magma ascends to shallower levels, existing bubbles grow by decompression and new ones form. It is found empirically that once the volume fraction of the magma occupied by the bubbles exceeds some value in the range 65–80%, the foam-like fluid can no longer deform fast enough in response to the shear stresses applied to it and as a result disintegrates into a mixture of released gas and entrained clots and droplets that form the pyroclasts. The eruption is then, by definition, explosive. The pyroclasts have a range of sizes dictated by the viscosity of the magmatic liquid, in turn a function of its composition and temperature, the rate at which the decompression is taking place, essentially proportional to the rise speed of the magma, and the rate at which the magma is being sheared, a function of its rise speed and the conduit width.

It is not a trivial matter for the volume fraction of gas in a magma to become large enough to cause disruption into pyroclasts. The lowest pressure to which a magma is ever exposed is the planetary surface atmospheric pressure. On Venus this ranges from about 10 MPa in lowland plains to about 4 MPa at the tops of the highest volcanoes; on Earth it is about 0.1 MPa at sea-level (and 30% less on high volcanoes) but much higher, up to 60 MPa, on the deep ocean floor; on Mars it ranges from about 500 Pa at the mean planetary radius to about 50 Pa at the tops of the highest volcanoes; and it is essentially zero on the Moon, Mercury and Io. If the magma volatile content is small enough, then even at atmospheric pressure too little gas will be exsolved to cause magma fragmentation. Using the solubilities of common volatiles in magmas, calculations show that explosive eruptions can occur on Earth as long as the water content exceeds 0.07 wt% in a basalt. On Mars the critical level is 0.01 wt%. On Venus, however, a basalt would have to contain about 2 wt% water before explosive activity could occur, even at highland sites; this is greater than is common in basalts on Earth, and leads to the suggestion that explosive activity may never happen on Venus, at least at lowland sites, or may happen only when some process leads to the local concentration of volatiles within a magma. Examples of this are discussed later. Finally, the negligible atmospheric pressures on the Moon, Mercury and Io mean that miniscule amounts of magmatic volatiles can in principle cause some kind of explosive activity there.

The above discussion assumes that released magmatic volatiles are the only source of explosive activity. However, many Vulcanian and all phreato-magmatic explosive eruptions involve interaction of erupting magma with solid or liquid volatiles already present at the surface (almost always water or ice on Earth and Mars; mainly sulfur compounds on Io). The total weight fraction of gas in the eruption products in such cases will depend on the detailed nature of the interaction as well as the composition and inherent volatile content of the magma; this is a critical factor in understanding the very explosive activity on Io.

4.2. Strombolian Activity

Strombolian eruptions, named for the style of activity common on the Italian volcanic island Stromboli, are an excellent example of how the rise speed, gas content, and viscosity of a magma are critical in determining the style of explosive activity that occurs. While the magma as a whole is ascending through a fracture in the planetary crust, bubbles of exsolved gas are rising through the liquid at a finite speed determined by the liquid viscosity and the bubble sizes. If the magma rise speed is negligible, for example, when magma is trapped in a shallow reservoir or a shallow intrusion, and if its viscosity is low, as in the case of a basalt, there may be enough time for gas bubbles to rise completely through the magma and escape into overlying fractures that convey the gas to the surface, where it escapes or is added to the atmosphere if there is one. Subsequent eruption of the residual liquid will be essentially perfectly effusive. If a low-viscosity magma is rising to the surface at a slow enough speed, most of the gas will still escape as bubbles rise to the liquid surface and burst. Because relatively large bubbles (those that nucleated first and have decompressed most) will rise through the liquid faster than very small bubbles, it is common in basalts for large bubbles to overtake and coalesce with small ones. The even larger bubbles produced in this way rise even faster and overtake additional smaller bubbles. A runaway situation can develop in which a single large bubble completely fills the diameter of the vent system apart from a thin film of magma lining the walls of the fracture. In extreme cases the bubble may have a much greater vertical extent than its width, in which case it is called a slug of gas. As this body of gas emerges at the surface of the slowly rising liquid magma column, it bursts, and a discrete layer of magma forming the upper "skin" of the bubble or slug disintegrates into clots and droplets up to tens of cm in size. These are blown outward by the expanding gas (Figure 5.9; see also color insert). The pyroclasts produced accumulate around the vent to form a cinder cone that can be up to several tens of meters in size. The time interval between the emergence of successive bubbles or slugs from a vent may range from seconds to at least minutes, making this a distinctly intermittent type of explosive activity. If the largest rising gas bubble does not completely fill the vent, continuous overflow of a lava lake in the vent may take place to form one or more lava flows at the same time that intermittent explosive activity is occurring, resulting in a simultaneously effusive and explosive eruption.

A second method of producing gas slugs has been suggested for some Strombolian eruptions on Earth, in which gas bubbles form during convection in a body of magma beneath the surface and drift upward to accumulate into a layer of foam at the top of the magma body. When the vertical extent of the foam layer exceeds a critical value it collapses. Liquid magma drains from between the bubbles, which coalesce into a large gas pocket that can now rise through any available fracture to the surface. The argument is that if a fracture had been already present, the high effective viscosity of the foam would have inhibited its rise into the fracture, whereas the viscosity of the pure gas is low enough to allow this to occur. If a fracture was not already present, the changing stresses due to the foam collapse may be able to create one.

As long as any volatiles are exsolved from a low-viscosity magma rising sufficiently slowly to the surface, some kind of Strombolian explosive activity, however feeble, should occur at the vent on any planet, even at the high pressures on Venus or on Earth's ocean floors (where there is now evidence for such activity from submersible vehicles). Strombolian eruptions commonly involve excess pressures in the bursting bubbles of only a few tenths of a MPa, so that the amount of gas expansion that drives the dispersal of pyroclasts is small. Pyroclast ranges in air on Earth can be several tens to at most a few hundred meters, and ranges would be much smaller in submarine Strombolian events on the Earth's ocean floor or on Venus because of the higher ambient pressure. Strombolian eruptions on Mars would eject pyroclasts to distances about five times greater than on Earth because of the lower gravity and atmospheric pressure; as a result the deposits formed would have a 25-fold lower relief than on Earth, and perhaps as a result no examples have yet been unambiguously identified in spacecraft images.

4.3. Vulcanian Activity

In a slowly rising viscous magma, it is relatively difficult for gas bubbles to escape from the melt. Particularly if the magma stalls as a shallow intrusion, slow diffusion of gas and rise of bubbles in the liquid concentrate gas in the upper part of the intrusion, and the gas pressure in this region rises. The pressure rise is greatly enhanced if any volatiles existing near the surface (groundwater on Earth; ground ice on Mars; sulfur or sulfur dioxide on Io) are evaporated. Eventually the rocks overlying the zone of high pressure break under the stress and the rapid expansion of the trapped gas drives a sudden, discrete explosion in which fragments of the overlying rock and disrupted magma are scattered around the vent: this is called Vulcanian activity

FIGURE 5.9 Jets of hot gas and entrained incandescent basaltic pyroclasts ejected from a transient Strombolian explosion on the volcano Stromboli. *Photograph by P.J. Mouginis-Mark.*

FIGURE 5.10 A dense cloud of large and small pyroclasts and gas ejected to a height of a few hundred meters in a transient Vulcanian explosion by the volcano Ngauruhoe in New Zealand. *Image courtesy of the University of Colorado in Boulder, Colorado, and the National Oceanic and Atmospheric Administration, National Geophysical Data Center.*

(Figure 5.10), named for the Italian volcanic island Vulcano. Again, as long as any volatiles are released from the magma or are present in the near-surface layers of the planet, activity of this kind can occur. Several Vulcanian events on Earth involving fairly viscous magmas have been analyzed in enough detail to provide estimates of typical pressures and gas concentrations. Bombs approaching a meter in size thrown as far as 5 km imply pressures as high as a few MPa in regions that are tens of meters in size. The gas mass fractions in the explosion products can be up to 10%.

On Mars, with the same initial conditions, the lower atmospheric pressure would cause much more gas expansion to accelerate the ejected fragments, and the lower atmospheric density would exert much less drag on them; also the lower gravity would allow them to travel farther for a given initial velocity. The result is that the largest clasts could travel up to 50 km. This means that the roughly circular deposit from a localized, point-source explosion would be spread over an area 100 times greater than on Earth, being on average 100 times thinner. Apart from the possibility that the pattern of small craters produced by the impact of the largest boulders on the surface might be recognized, such a deposit, with almost no vertical relief and having very little influence on the preexisting surface, would almost certainly go unnoticed in even the highest-resolution spacecraft images, and indeed no such features have yet been identified. However, if the explosion involves a larger, more complex, and especially elongate vent structure, there would not be such large differences. In the Elysium region of Mars a large, water-carved channel, Hrad Vallis, has a complex elongate source depression that appears to have been excavated by a Vulcanian explosion when a dike injected a sill into the ice-rich permafrost of the cryosphere—the outer few kilometers of the crust which is so cold that any H_2O must be present as ice. As heat from the sill magma melted the ice and boiled the resulting water in the cryosphere, violent expansion of the vapor forced intimate mixing of magma and lumps of cryosphere, encouraging ever more vapor production. Soon all of the cryosphere above the sill was thrown out in what is called a fuel—coolant interaction (here the fuel is the magma and the coolant is the ice) to produce a deposit extending about 35 km on either side of the 150 km-long depression. Residual heat from the magma melted the remaining ice in the shattered cryosphere rocks so that for a while, until it froze again, there was liquid water present to form a characteristic "muddy" appearance in the deposit (Figure 5.11).

A Vulcanian explosion on Venus would also be very different from its equivalent on Earth. In this case, the high atmospheric pressure would tend to suppress gas

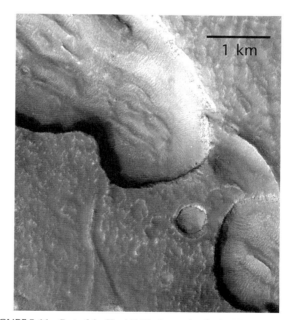

FIGURE 5.11 Part of the Hrad Vallis depression in the Elysium Planitia area of Mars. The depression is surrounded by a "muddy" deposit and is interpreted to have formed when a volcanic sill intruded the cryosphere—cold rocks containing ice. The ice melted and the water flash-boiled to produce a "fuel-coolant" variety of Vulcanian explosion that excavated the depression and threw out a mixture of magma clots and cryosphere rocks. The hot magma melted ice in the cryosphere rocks to produce the mud. *NASA Mars Global Surveyor image.*

expansion, producing a low initial velocity in the ejecta, and the atmospheric drag would also be high. Pyroclasts that would have reached a range of 5 km on Earth would travel less than 200 m on Venus. On the one hand, this should concentrate the eruption products around the vent and make the deposit more obvious; however, the resolution of the best radar images from Magellan is only ~75 m, and so such a deposit would represent only three or four adjacent pixels, which again would probably not be recognized.

On the Moon a number of Vulcanian explosion products are seen. The dark halo craters on the floor of the impact crater Alphonsus have ejecta deposits with ranges up to 5 km. Since the Moon has no atmosphere, the preceding arguments suggest that lunar Vulcanian explosions should eject material to very great ranges. However, the Alphonsus event seems to have involved the intrusion of basaltic magma into the ~10 m thick layer of fragmental material forming the regolith in this area, and the strength of the resulting mixture of partly welded regolith and chilled basalt was quite low. Thus only a small amount of pressure buildup occurred before the retaining rock layer fractured. As a result, the initial speeds of ejected pyroclasts were low and their ranges were unusually small.

4.4. Hawaiian Activity

In some cases, especially where low viscosity basaltic magma travels laterally in dikes at shallow depth, enough gas bubble coalescence and bubble rise occurs for much of the gas to be lost into cracks in the rocks above the dike. Magma then emerges from the vent as a lava flow. However, when basaltic magmas rise mainly vertically at appreciable rates (more than about 1 m/s), some gas bubble coalescence occurs but little gas is lost, and the magma is released at the vent in a nearly continuously explosive manner. A lava fountain, more commonly called a fire fountain, forms over the vent, consisting of pyroclastic clots and droplets of liquid entrained in a magmatic gas stream that fluctuates in its upward velocity on a timescale of a few seconds. The largest clots of liquid, up to tens of cm in size, rise some way up the fountain and fall back around the vent to coalesce into a lava pond that overflows to feed a lava flow—the effusive part of the eruption—whereas smaller clasts travel to greater heights in the fountain. Some of the intermediate-sized pyroclasts cool as they fall from the outer parts of the fountain and collect around the lava pond in the vent to build up a roughly conical edifice called an ash cone, cinder cone, or scoria cone, the term used depending on the sizes of the pyroclasts involved, ash being smallest. Such pyroclastic cones are commonly asymmetric owing to the influence of the prevailing wind.

Atmospheric gases are entrained into the edge of the fire fountain and heated by contact with the hot pyroclasts and mixing with the hot magmatic gas. In this way, a convecting gas cloud is formed over the upper part of the fountain, entraining the smallest pyroclasts so that they take part fully in the convective motion. The whole cloud spreads downwind and cools, and eventually the pyroclasts are released again to form a layer on the ground, the smallest particles being deposited at the greatest distances from the vent. This whole process, involving formation of lava flows and pyroclastic deposits at the same time, is called Hawaiian eruptive activity (Figure 5.12). This style of activity should certainly have occurred on Mars, but may be suppressed in basaltic magmas on Venus by the high atmospheric pressure, especially in lowland areas, unless, as noted earlier, magma volatile contents are several times higher than is common on Earth.

Figure 5.13 shows qualitatively how the combination of erupting mass flux and magma gas content in a Hawaiian eruption on Earth determines the nature and size of the possible products: a liquid lava pond at the vent that directly feeds lava flows; a pile of slightly cooled pyroclasts accumulating fast enough to weld together and form a "rootless" lava flow; a cone in which almost all of the pyroclasts are welded together; or a cone formed from pyroclasts that have had time to cool while in flight so that none, or only a few, weld on landing. Theoretical analyses based on the trends seen in Figure 5.13 confirm that hot lava ponds around vents on Earth are expected to be no more than a few tens of meters wide even at very high mass eruption rates. On the Moon, the greater gas expansion due to the lack of an atmosphere causes very thorough disruption of the magma (even at the low gas contents implied by analysis of the Apollo samples) and

FIGURE 5.12 A Hawaiian eruption from the Pu'u 'O'o vent in Hawaii showing a convecting cloud of gas and small particles in the atmosphere above the 300 m high lava fountain (commonly termed fire fountain) of coarser basaltic pyroclasts. *Photograph by P.J. Mouginis-Mark.*

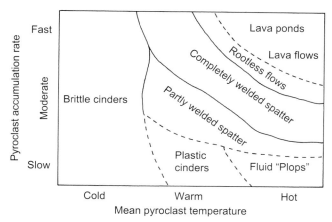

FIGURE 5.13 Schematic indication of the relative influences of the pyroclast temperature on reaching the ground and the volume eruption rate of magma, and hence the pyroclast accumulation rate, on the nature of the pyroclastic material produced in explosive eruptions. *Reprinted from Figure 5.5 in Head and Wilson (1989), with kind permission of Elsevier Science, NL, Sara Burgerhartstraat 25, 1055 KV Amsterdam, The Netherlands.*

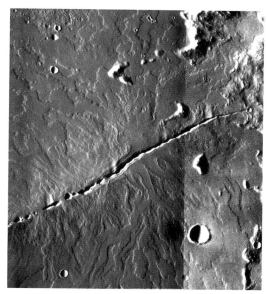

FIGURE 5.14 Mosaic of two images showing a fissure vent near Jovis Tholus volcano on Mars. The eruption produced spatter ramparts and multiple lava flow lobes, probably of basaltic composition. The area shown is 24 km wide. *NASA Mars Odyssey images.*

gives the released volcanic gas a high speed. This, together with the lower gravity, allows greater dispersal of pyroclasts of all sizes, and provides an explanation of the 100–300 km wide dark mantle deposits as the products of extreme dispersal of the smallest, 30–100 µm sized particles.

Nevertheless, it appears that hot lava ponds up to ~5 km in diameter could have formed around basaltic vents on the Moon if the eruption rates were high enough—as high as those postulated to explain the long lava flows and sinuous rilles. The motion of the lava in such ponds would have been thoroughly turbulent, thus encouraging thermal erosion of the base of the pond, and this explains why the circular to oval depressions seen surrounding the sources of many sinuous rilles have just these sizes. Similar calculations for the Mars environment show that, as long as eruption rates are high enough, the atmospheric pressure and gravity are low enough on Mars to allow similar hot lava source ponds to have formed there, again in agreement with the observed sizes of depressions of this type that are seen.

Some noticeable differences occur when Hawaiian eruptions take place from very elongate fissure vents. Instead of a roughly circular pyroclastic cone containing a lava pond feeding one main lava flow, a pair of roughly parallel ridges forms, one on either side of the fissure, called spatter ramparts. Along the parts of the fissure where the eruption rate is highest, pyroclasts may coalesce as they land to form lava flows, so that there are gaps in the ramparts from which the flows spread out. A striking example of this has been found on Mars (Figure 5.14).

4.5. Plinian Activity

In the case of a basaltic magma very rich in volatiles, or (much more commonly on Earth) in the case of a volatile-rich andesitic or rhyolitic magma, fragmentation in a steadily erupting magma is very efficient, and most of the pyroclasts formed are small enough to be entrained by the gas stream. Furthermore, the speed of the mixture emerging from the vent, which is proportional to the square root of the amount of gas exsolved from the magma, will be much higher (perhaps up to 500 m/s) than in the case of a basaltic Hawaiian eruption (where speeds are commonly less than 100 m/s). The fire fountain in the vent now entrains so much atmospheric gas that it develops into a very strongly convecting eruption cloud in which the heat content of the pyroclasts is converted in the buoyancy of the entrained gas. The resulting cloud rises to a height that is proportional to the fourth root of the magma eruption rate (and hence the heat supply rate). Such clouds may reach heights of several tens of kilometers on Earth. Only the very coarsest pyroclasts fall out near the vent, and almost all of the erupted material is dispersed over a wide area from the higher parts of the eruption cloud (Figure 5.15). This activity is termed Plinian, after Pliny's description of the A.D. 79 eruption of Vesuvius.

Not all eruptions of this type produce stable convection clouds. If the vent is too wide or the eruption speed of the magma is too low, insufficient atmospheric gas is entrained to provide the necessary buoyancy for convection, and a

FIGURE 5.15 The Plinian phase of the explosive eruption of Pinatubo volcano in 1991. A dense cloud of large and small pyroclasts and gases is ejected at high speed from the vent and entrains and heats the surrounding air. Convection then drives the resulting cloud to a height of tens of kilometers, where it drifts downwind, progressively releasing the entrained pyroclasts. *Photograph credit: U.S. Geological Survey.*

collapsed fountain forms over the vent, feeding large pyroclastic density currents or smaller, more episodic pyroclastic surges.

Mars is the obvious place other than Earth to look for explosive eruption products: the low atmospheric pressure encourages explosive eruptions to occur and the atmospheric density is high enough to allow convecting eruption clouds to form, at least up to ~20 km. Stable eruption clouds much higher than this cannot form on Mars because the atmosphere becomes too thin to provide the required amount of entrained gas. Nevertheless, the smaller sizes expected for pyroclasts on Mars than Earth mean that winds can transport particles for great distances. Very extensive friable layered deposits are seen in the Arabia Terra and Terra Meridiani areas and in the Medusae Fossae formation. Computations combining eruption cloud formation models with global atmospheric circulation models have shown that large explosive eruptions from the major volcanoes are readily able to explain these deposits.

Although the large magma gas contents needed suggest that large-scale, steady (Plinian) explosive eruptions are rare on Venus, it is possible to calculate the heights to which their eruption clouds would rise. The high density and temperature of the atmosphere lead to rise heights about a factor of two smaller than on Earth for the same eruption rate, and very large (at least a few tens of meters) clasts may be transported into near-vent deposits. At distances greater than a few kilometers from the vent, pyroclastic fall deposits will not be very different from those on Earth. A few examples of elongate markings on the Venus surface have been proposed as fall deposits, but no detailed analysis of them has yet been carried out.

The conditions that cause a steady explosive eruption to generate pyroclastic density currents instead of feeding a stable, convecting eruption cloud are fairly well understood. If the eruption rate exceeds a critical value (which increases with increasing gas content of the mixture emerging through the vent and decreases with increasing vent diameter), stable convection is not possible whatever the nature of the atmosphere. Since pyroclastic density current formation is linked to high eruption rate and, in general, to high eruption speed, which will encourage a great travel distance, it would not be surprising if such large-scale pyroclastic deposits distributed radially around a vent were the products of high discharge rate eruptions of gas-rich magmas. Many of the flanking deposits of some Martian volcanoes, especially Tyrrhenus Mons and Hadriacus Mons, may have been produced in this way.

Short-lived or intermittent explosive eruptions (e.g. Vulcanian explosions, phreato-magmatic explosions, or events in which a gas-rich, high-viscosity lava flow or dome disintegrates into released gas and pyroclasts as a result of excessive gas pressure) can also produce small-scale pyroclastic density currents. Because these are shorter-lived and have characteristically different grain size distributions, they are called surges. The least well understood aspect of these phenomena is the way in which the magmatic material interacts with the overlying atmosphere. As a result, it is currently almost impossible to predict in detail what the results of this kind of activity on Mars or Venus would look like. Such deposits, by the nature of the way they are generated, would not be very voluminous, however, and so would be spread very thinly, and might not be recognized if they were able to travel far from the vent.

4.6. Phreato-Magmatic Activity

Some types of eruption on Earth are controlled by the vigorous interaction of magma with surface or shallow subsurface water. If an intrusion into water-rich ground causes steam explosions, these are called phreatic events (from the Greek word for a well). If some magma also reaches the surface, the term used is phreato-magmatic, as distinct from normal, purely magmatic eruptions. When the equivalents of Strombolian or Hawaiian explosive events take place from eruption sites located in shallow water, they lead to much greater fragmentation of the magma than usual because of the thermal stresses induced as pyroclasts are chilled by contact with the water. This activity is usually called Surtseyan, named after an eruption that formed the island of Surtsey off the south coast of Iceland. A much more vigorous and long-lived eruption under similar circumstances leads to a pyroclastic fall deposit similar to that of a Plinian event, but again involving greater fragmentation of magma: the result is called phreato-Plinian activity. Since the word phreatic does not specifically refer to water as the nonmagmatic volatile involved in these kinds of explosive eruption, it seems safe to apply these terms, as appropriate, to the various kinds of interactions between magma and liquid sulfur or sulfur dioxide forming the plumes currently seen on Io. These eruptions appear to involve about 30% by weight volatiles mixed with the magma; these proportions are close to the optimum for converting the heat of the magma to kinetic energy of the explosion products. Phreatic and phreato-magmatic eruptions should also have occurred on Mars in the distant past if, as many suspect, the atmospheric pressure was high enough to allow liquid water to exist on the surface.

4.7. Dispersal of Pyroclasts into a Vacuum

The conditions in the region above the vent in an explosive eruption on a planet with an appreciable atmosphere (e.g. Venus, Earth, and Mars) are very different from those when the atmospheric pressure is very small (much less than about 1 Pa), as on the Moon, Mercury and Io (and differentiated asteroids in the distant past). If the mass of atmospheric gas displaced from the region occupied by the eruption products after the magmatic gas has decompressed to the local pressure is much less than the mass of the magmatic gas, convecting eruption clouds cannot form in eruptions that would have been classed as Hawaiian or Plinian on Earth. In the region immediately above the vent, the gas expansion involves a series of shock waves. Relatively large pyroclasts will pass through these shocks with only minor deviations in their trajectories, but intermediate-sized particles may follow very complex paths, and few studies have yet been made of these conditions. The magmatic gas eventually expands radially into space, accelerating to reach a limiting velocity that depends on its initial temperature. As the density of the gas decreases, its ability to exert a drag force on pyroclasts also decreases. On bodies the size of the Moon, even the smallest particles eventually decoupled from the gas and fell back to the planetary surface, though in gas-rich eruptions on asteroids these particles were commonly ejected into space.

These are the conditions that led to the formation of the ancient dark mantle deposits on the Moon, with ultimate gas speeds on the order of 500 m/s, leading to ranges up to 150 km for small pyroclasts 30–100 μm in size. They are also the conditions that exist now in the eruption plumes on Io, though with an added complication. The driving volatiles in the Io plumes appear to be mainly sulfur and sulfur dioxide, evaporated from the solid or liquid state by intimate mixing with rising basaltic magma in what are effectively phreato-magmatic eruptions. The Io plume heights imply gas speeds just above the vent of ∼1000 m/s, and these speeds are consistent with the plume materials being roughly equal mixtures of basaltic pyroclasts and evaporated surface volatiles. As the gas phase expands to very low pressures, both sulfur and sulfur dioxide will condense, forming small solid particles that rain back onto the surface along with the silicate pyroclasts to be recycled again in future eruptions.

A final point concerns pyroclastic eruptions on the smallest atmosphereless bodies, the asteroids. Basaltic partial melts formed within these bodies were erupted at the surface at speeds that depended on the released volatile content. This is estimated to have been as much as 0.2–0.3 wt%, leading to speeds up to 150 m/s. These speeds are greater than the escape velocities from asteroids with diameters less than about 200 km, and so instead of falling back to the surface, pyroclasts would have been expelled into space, eventually to spiral into the Sun. This process explains the otherwise puzzling fact that we have many meteorites (e.g. the aubrites and ureilites) representing samples of the residual material left in the mantle of at least two asteroids after partial melting events, but have only a tiny number of meteorites from these asteroids containing grains with the expected partial melt composition.

5. INFERENCES ABOUT PLANETARY INTERIORS

The presence of the collapse depressions called calderas at or near the summits of many volcanoes on Earth, Mars, Venus, and Io suggests that it is common on all of these bodies for large volumes of magma to accumulate in reservoirs at relatively shallow depths. Theories of magma accumulation suggest that the magma in these reservoirs must have an internal pressure greater than the stress produced in the surrounding rocks by the weight of the overlying crust. This excess pressure may be due to the formation of bubbles by gas exsolution, or to the fact that heat loss from the magma to its cooler surroundings causes the growth of crystals that are less dense than the magmatic liquid and so occupy a larger volume. Most commonly, a pressure increase leads to fracturing of the wall of the reservoir and to the propagation of a magma-filled crack, called a dike, as an intrusion into the surrounding rocks. If the dike reaches the surface, an eruption occurs, and removal of magma from the reservoir allows the wall rocks to relax inward elastically as the pressure decreases. If magma does not reach the surface, the dike propagates underground until either the magma within it chills and comes to rest, or the pressure within the reservoir falls to the point where there is no longer a great enough stress at the dike tip for rock fracturing to continue.

Under certain circumstances, an unusually large volume of magma may be removed from a shallow reservoir, reducing the internal pressure beyond the point where the reservoir walls behave elastically. Collapse of the overlying rocks may then occur to fill the potential void left by the magma, and a caldera (or, on a smaller scale, a pit crater) will form. The circumstances causing large-volume eruptions on Earth include the rapid eruption to the surface immediately above the reservoir of large volumes of low-density, gas-rich silicic (rhyolitic) magma, and the drainage of magma through extensive lateral dike systems extending along rift zones to distant flank eruption sites on basaltic volcanoes. This latter process appears to have been associated with caldera formation on Kilauea volcano in Hawaii, and it is tempting to speculate that the very large calderas on some of the Martian basaltic shield volcanoes (especially Pavonis Mons and Arsia Mons) are directly associated with the large-volume eruptions seen on their flanks.

The size of a caldera must be related to the volume of the underlying magma reservoir, or more exactly to the volume of magma removed from it in the caldera-forming event. If the reservoir is shallow enough, the diameter of the caldera is probably similar to that of the reservoir. Diameters from 1 to 3 km are common on basaltic volcanoes on Earth and on Venus, with depths up to a few hundred meters implying magma volumes less than about 10 km^3. In contrast, caldera diameters up to at least 30 km occur on several volcanoes on Mars and, coupled with caldera depths up to 3 km, imply volumes ranging up to as much as 10,000 km^3. The stresses implied by the patterns of fractures on the floors and near the edges of some of these Martian calderas suggest that the reservoirs beneath them are centered on depths on the order of 10–15 km, about three to four times greater than the depths of basaltic reservoirs on Earth. The simplest models of the internal structures of volcanoes suggest that, due to the progressive closing of cavities in rocks as the pressure increases, the density of the rocks forming a volcanic edifice should increase, at first quickly and then more slowly, with depth. Rising magma from deep partial melt zones may stall when its density is similar to that of the rocks around it, so that it is neither positively nor negatively buoyant, and a reservoir may develop in this way. Since the pressure at a given depth inside a volcano is proportional to the acceleration due to gravity, and since on Mars this is about three times less than that on Earth or Venus, the finding that Martian magma reservoirs are centered three to four times deeper than on Earth is not surprising. However, these simple models do not address the reason for the Martian calderas being very much wider than any of those on Earth or many of those on Venus. On Io we see some caldera-like structures, not necessarily associated with obvious volcanic edifices, that are even wider (but not deeper) than those on Mars, though we have too little information about the internal structure of Io's crust to interpret this observation unambiguously. Clearly, much is still not understood about the formation and stability of shallow magma bodies.

Evidence for significant shallow magma storage is very rare on the Moon. The large volumes observed for the great majority of eruptions in the later part of lunar volcanic history, and the high effusion rates inferred for them, imply that almost all of the eruptions took place directly from large bodies of magma stored at very great depth—at least at the base of the crust and possibly in partial melting zones in the lunar mantle. Not all the dikes propagating up from these depths will have reached the surface, however, and some shallow dike intrusions almost certainly exist. Recent work suggests that many of the linear rilles on the Moon represent the surface deformation resulting from the emplacement of such dikes, having thicknesses of at least 100 m, horizontal and vertical extents of ~100 km, and tops extending to within 1 or 2 km of the surface. Minor volcanic activity associated with some of these features, as in the case of Rima Hyginus, would then be the result of gas loss and small-scale magma redistribution as the main body of the dike cooled.

The emplacement of very large dike systems extending most or all of the way from mantle magma source zones to the surface is not confined to the Moon. It has long been assumed that such structures must have existed to feed the high-volume basaltic lava flow sequences called flood basalts that occur on Earth every few tens of millions of years. These kinds of feature are probably closely related to the systems of giant dikes, tens to hundreds of meters wide and traceable laterally for many hundreds to more than 1000 km, that are found exposed in very ancient rocks on the Earth. The radial patterns of these ancient dike swarms suggest that they are associated with major areas of mantle upwelling and partial melting, with magma migrating vertically above the mantle plume to depths of a few tens of kilometers and then traveling laterally to form the longest dikes. Some of the radial surface fracture patterns associated with the novae and coronae on Venus are almost certainly similar features that have been formed more recently in that planet's geologic history. On Mars the systems of linear grabens, some of which show evidence of localized eruptive vents, extending radially from large shield volcanoes such as Arsia Mons, also bear witness to the presence of long-lived mantle upwellings generating giant dike swarms. It seems that there may be a great deal of similarity between the processes taking place in the mantles of all of the Earth-like planets; it is the near-surface conditions, probably strongly influenced by the current presence of its oceans, that drive the plate tectonic processes distinguishing the Earth from its neighbors.

BIBLIOGRAPHY

Davies, A. G. (2007). *Volcanism on Io: A comparison with Earth.* Cambridge, UK: Cambridge University Press.

Gilbert, J. S., & Sparks, R. S. J. (Eds.). (1998). *The physics of explosive volcanic eruptions.* Geological Society of London. Special Publication 145.

Head, J. W., & Wilson, L. (1989). Basaltic pyroclastic eruptions: influence of gas-release patterns and volume fluxes on fountain structure, and the formation of cinder cones, spatter cones, rootless flows, lava ponds and lava flows. *Journal of Volcanology and Geothermal Research, 37,* 261–271.

Houghton, B., Rymer, H., Stix, J., McNutt, S., & Sigurdsson, H. (Eds.). (1999). *The encyclopedia of volcanoes.* San Diego, USA: Academic Press.

Lane, S. J., & Gilbert, J. S. (Eds.). (2008). *Fluid motions in volcanic conduits: A source of seismic and acoustic signals.* Geological Society of London. Special Publication 307.

Lopes, R., & Gregg, T. K. P. (Eds.). (2004). *Volcanic worlds: Volcanism in the solar system.* New York, USA: Praxis Press.

Manga, M., & Ventura, G. (Eds.). (2005). *Kinematics and dynamics of lava flows.* Geological Society of America (Special Paper).

Marti, J., & Ernst, G. G. J. (Eds.). (2008). *Volcanoes and the environment.* Cambridge, UK: Cambridge Univ. Press.

Parfitt, E. A., & Wilson, L. (2008). *Fundamentals of physical volcanology.* Oxford, UK: Blackwell Publishing Ltd.

Schmincke, H.-U. (2004). *Volcanism* (2nd ed.). Berlin, Heidelberg, Germany: Springer-Verlag.

Wilson, L. (2009). Volcanism in the solar system. *Nature Geoscience, 2*(6), 389–397. http://dx.doi.org/10.1038/NGEO529.

Zimbelman, J. R., & Gregg, T. K. P. (Eds.). (2000). *Environmental effects on volcanic eruptions: From deep oceans to deep space.* New York, USA: Kluwer Academic/Plenum Publishing.

Chapter 6

Magnetic Field Generation in Planets

Sabine Stanley

Department of Physics, University of Toronto, Toronto, ON, Canada

Chapter Outline

1. Planetary Magnetic Field Observations — 121
  1.1. Sources of Observed Magnetic Fields — 121
  1.2. Spatial Characteristics of Dynamo-Generated Fields — 122
  1.3. Temporal Characteristics of Observed Magnetic Fields — 125
2. The Dynamo Mechanism — 125
  2.1. What is a Dynamo? — 125
  2.2. Necessary Conditions for a Dynamo — 126
  2.3. Dynamo Generation Regions in Planets — 126
3. The Standard Planetary Dynamo — 127
  3.1. Driving Forces — 127
  3.2. Fluid Motions in Dynamo Regions — 127
  3.3. Generation Mechanisms — 127
  3.4. Beyond the Standard Dynamo — 129
4. Simulations and Experiments — 129
  4.1. Numerical Dynamo Simulations — 129
  4.2. Dynamo Experiments — 131
5. Planetary Dynamos — 131
  5.1. Earth — 131
  5.2. Mercury — 133
  5.3. Jupiter — 133
  5.4. Saturn — 133
  5.5. Uranus and Neptune — 134
  5.6. Ganymede — 134
  5.7. Ancient Moon — 134
  5.8. Ancient Mars — 134
  5.9. Small Bodies — 135
  5.10. Extrasolar Planets — 135
  5.11. Planetary Bodies Lacking Dynamos — 135
6. Conclusions and Future Prospects — 135
Bibliography — 135

1. PLANETARY MAGNETIC FIELD OBSERVATIONS

The Earth's magnetic field has been used for navigation since at least the eleventh century AD, but it was not until the seventeenth century that the source of this magnetic field was attributed to the Earth's interior. Namely, in 1600, William Gilbert published "De Magnete" in which he described his experiments involving magnetic measurements of a lodestone sphere. He concluded that "*Globus terrae sid magneticus & magnes*" which can be loosely translated as "The Earth is a great magnet". Further work by Gellibrand, Halley, Gauss, and others established that the Earth's field was predominantly axially dipolar, but varied in time (see Kono, 2007 for a nice overview of the history of geomagnetism).

It was not until the mid-twentieth century that magnetic fields of other planets were observed. The first observation was indirectly of Jupiter's magnetic field in the 1950s through its radio emissions. These emissions result from interactions of the **solar wind** with the planet's **magnetosphere** (*see chapter on Planetary Magnetospheres*). The radio emissions were intense enough and in an appropriate bandwidth such that they could be detected from the Earth. Discovery of other planetary magnetic fields awaited visits by planetary spacecraft missions with magnetometers. Table 6.1 provides highlights of magnetic missions to planetary bodies and their main magnetic discoveries.

1.1. Sources of Observed Magnetic Fields

Observed planetary magnetic fields can result from a variety of processes. For example, external sources that generate observed planetary magnetic fields include Venus' ionospheric currents, Earth's magnetospheric currents, and **electromagnetic induction** in the saltwater oceans of Europa, Ganymede, and Callisto due to

TABLE 6.1 Magnetic Field Highlights From Planetary Missions

Planet	Spacecraft	Date	Magnetic Highlights
Mercury	Mariner 10	1974–1975	• Detection of dynamo-generated magnetic field
	MESSENGER	2008–	• Field characterized by weak dipole moment, large dipole offset, and small dipole tilt
Venus	Mariners 2,5,10	1962,1967,1974	• No dynamo-generated magnetic field detected
	Venera 4,9,10	1967,1975,1975	• Detection of small induced magnetosphere due to solar wind–ionosphere interaction
	Pioneer Venus Orbiter	1979–1981	
	Magellan	1990–1994	
	Venus Express	2006–2014	
Earth	Magsat	1979–1980	• Mapping of dynamo-generated and crustal magnetic fields to high resolution
	Orsted	1999–	
	Champ	2000–2010	• Detailed observations of magnetic field secular variation
	SAC-c	2000–2004	
Moon	Lunar Prospector	1998–1999	• No dynamo-generated magnetic field detected
			• Detection of localized crustal magnetic fields
Mars	Mars Global Surveyor	1997–2007	• Detection of strong localized crustal magnetic fields
Jupiter	Pioneer 10,11	1973,1974	• Detection of dynamo-generated magnetic field, similar in morphology to Earth's field
	Voyager I, II	1979	
	Galileo	1995–2003	
Galilean moons	Galileo	1995–2003	• Detection of Ganymede's dynamo-generated magnetic field
			• Detection of induced magnetic fields in global subsurface oceans of Europa, Ganymede, and Callisto
Saturn	Pioneer 11	1979	• Detection of dynamo-generated magnetic field
	Voyager I, II	1980,1981	• No nonaxisymmetric field components detected, upper limit on dipole tilt: 0.06°
	Cassini	2004–2017	
Titan	Cassini	2004–2017	• No dynamo-generated magnetic field detected
Uranus	Voyager II	1986	• Detection of dynamo-generated magnetic fields
Neptune		1989	• Field characterized by significant nondipolar, nonaxisymmetric components

Jupiter's time varying field. In contrast, internal sources include remanent magnetization in crustal rocks on Mars, Earth, and the Moon or self-sustaining **dynamos** in the deep interiors of planets. This latter process will be the main focus of this chapter.

1.2. Spatial Characteristics of Dynamo-Generated Fields

Dynamo-generated magnetic fields are typically distinguished from other sources by their spatial and temporal characteristics. Dynamo-generated fields are global in structure and vary on timescales related to the fluid motions in the planetary interior. Figure 6.1 shows the radial component of the magnetic field at the surface of the planets with actively generated dynamos. Jupiter's moon Ganymede also has a dynamo-generated field, but the data can only constrain the dipole moment. Hence a map of its surface magnetic field is not included in the figure. The planetary fields likely contain much smaller scale structure that has not yet been resolvable by available data; however, based on the large-scale fields, major similarities and differences between planetary magnetic fields are obvious.

The surface magnetic fields of Mercury, Earth, Jupiter, Saturn, and possibly Ganymede are dominated by their axial dipolar components. In contrast, fields of Uranus and Neptune do not show this dominance and instead, higher order multipoles and nonaxisymmetric components are as prominent as the axial dipole component. In terms of secondary features, one also notices that Saturn's observed field is purely axisymmetric (i.e. there is no variation in the zonal direction) and Mercury's field has a fairly large northward offset between its geographic equator and magnetic equator compared to the other planets.

In order to analyze the spectral components of the field more quantitatively, the surface magnetic field can be represented using surface spherical harmonics. By assuming that observations are made in a current-free region (i.e. that current $\vec{J} = 0$), Ampere's law implies

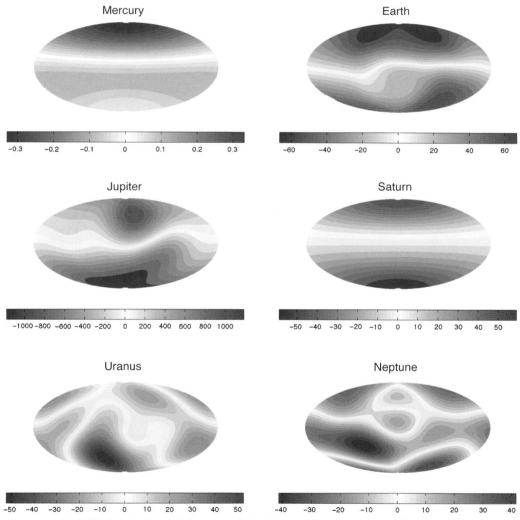

FIGURE 6.1 Surface radial magnetic field of planets with active dynamos. Units are μT.

that the magnetic field $\vec{B}$ is solenoidal (i.e. curl free) and hence that it can be written as the gradient of a scalar potential V:

$$\nabla \times \vec{B} = \mu_0 \vec{J} = 0 \Rightarrow \vec{B} = -\nabla V$$

(where μ_0 is the magnetic permeability of free space). Combining this with Gauss' law for magnetism yields the result that the magnetic scalar potential is the solution to Laplace's equation:

$$\nabla \cdot \vec{B} = 0 \Rightarrow \nabla^2 V = 0.$$

Ignoring any external field sources, the potential can be written in spherical coordinates as:

$$V(r,\theta,\phi) = a \sum_{l=1}^{\infty} \left(\frac{a}{r}\right)^{l+1} \sum_{m=0}^{l} \left[g_l^m \cos(m\phi) + h_l^m \sin(m\phi)\right] P_l^m(\cos\theta)$$

where l and m are the spherical harmonic degree and order, respectively; r is the radius; ϕ is the longitude; θ is the colatitude; a is the planetary radius; and P_l^m are the associated Legendre polynomials. In this expansion, axisymmetric terms are given by $m=0$ terms, and the successive multipoles are determined by l. For example, $l=1$, 2, and 3 represent the dipole, quadrupole, and octupole components respectively. The amplitudes of each harmonic are given by their respective **Gauss coefficients** g_l^m and h_l^m.

Perhaps the easiest way to visualize the similarity in spectral content of the different planetary magnetic fields is through power spectra. Defining the power in each degree and order using the mean-square field intensity:

$$p(l,m,r) = (l+1)\left(\frac{a}{r}\right)^{(2l+4)}\left[\left(g_l^m\right)^2 + \left(h_l^m\right)^2\right]$$

the power in each degree is found by summing over all orders and the power in each order can be found by summing over all degrees. Figure 6.2 plots the power as a function of degree and order for the planets in Figure 6.1. Since the purpose of this plot is to compare between the planets, only the lowest degrees (i.e. largest length scales) are plotted (maximum degree up to three). Observational data for Earth provide spectra to much higher degree, a recent model can be found in Finlay et al. (2010).

The equations above for the magnetic scalar potential and power depend on the distance from the source region. It is common practice to calculate the Gauss coefficients and power spectra at the respective planetary surface radius because we are limited to making observations outside the planet. However, if our goal is to compare and contrast planetary magnetic fields, then it is more appropriate to choose the dynamo source region radius since this removes

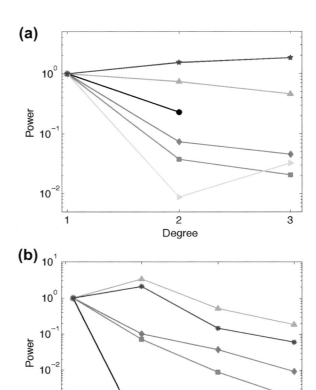

FIGURE 6.2 Surface magnetic power spectra for Mercury (black circles), Earth (green squares), Jupiter (red diamonds), Saturn (yellow side triangles), Uranus (cyan up triangles), and Neptune (blue stars). (a) Power vs degree and (b) power vs order. Saturn's purely axisymmetric field has only the $m=0$ component visible in plot (b). For Mercury, only data up to $l=2$, $m=1$ was used. For each planet, the power is normalized to the dipole power in (a) and to the axisymmetric power in (b). This figure therefore does not demonstrate the relative intensity of the different fields.

the arbitrary differences in distance between the surface and the dynamo source regions for the planets.

There is an inherent danger in extrapolating the field deeper in the planet due to four factors. First, the smaller scale fields will increase in power much faster than the larger scale fields. Since the smaller scale fields are the least resolved at the surface, this can result in significant errors in the extrapolation. Second, to extrapolate using the potential field expansion in Gauss coefficients given above, the region between the surface and the top of the dynamo source region must be an insulator. Any significant electrical conductivity will introduce errors into the field extrapolation. Third, any sources of magnetism between the surface and dynamo source region (e.g. crustal magnetism in terrestrial planets) needs to be accounted for if we are only interested in the dynamo-generated field. Fourth, the radius of the top of the

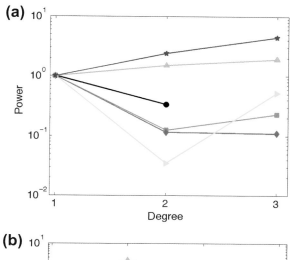

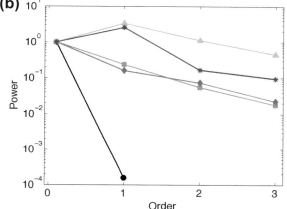

FIGURE 6.3 Same as Figure 6.2 except that the power is calculated at the top of the dynamo source regions (radii given in Table 6.2).

Figure 6.3. The radii of the dynamo source regions used for the figure are given in Table 6.2.

Figure 6.3 demonstrates that the fields at the top of the dynamo source region are not as dipolar dominated as the surface fields, but Uranus and Neptune are the only planets for which the dipole ($l = 1$) and axisymmetric ($m = 0$) components are not the largest contributors to the spectra. The planet whose magnetic field spectrum most resembles that of Earth is Jupiter. Of the planets with a dominant axial dipole, Mercury's field seems to have a relatively large quadrupole, whereas Saturn appears to have a relatively large octupole. Based on current data, both Saturn and Mercury have little to no spectral contributions from nonaxial terms.

1.3. Temporal Characteristics of Observed Magnetic Fields

Dynamo-generated fields are expected to display a myriad of temporal behavior reflecting the temporal nature of the fluid motions generating the fields. Variations in Earth's observed field are discussed in Section 5.1. There is little information about the temporal behavior of other planetary magnetic fields due to a lack of magnetic data resolution, both temporally and spatially. No undisputed **magnetic secular variation** has been observed for Jupiter, Saturn, or Mercury, the three planets that have been visited by multiple magnetic missions with a sufficient time interval to carry out a secular variation study. However, we do expect these fields to exhibit secular variation based on our observations of Earth's magnetic field as well as the solar magnetic field, also generated by a dynamo.

2. THE DYNAMO MECHANISM

2.1. What is a Dynamo?

A dynamo is the process by which mechanical energy is converted to electromagnetic energy through induction. In

dynamo source region is not well known for all the planets. This is especially a problem for the giant planets since they experience a gradual increase in conductivity with depth without a significant compositional change.

Being aware of these limitations, we tentatively plot the power spectra at the top of the dynamo source regions in

TABLE 6.2 Properties of Planetary Dynamo Source Regions[1]

	Mercury	Earth	Jupiter	Ganymede	Saturn	Uranus	Neptune
R_P(km)	2440	6371	69,911	2634	60,268	25,559	24,764
R_D(km)	2030	3486	~55900	~800	~30100	~17900	~19800
Composition	Iron	Iron	Metallic hydrogen	Iron	Metallic hydrogen	Ionic water	Ionic water
σ(S/m)	~10^6	~10^6	~5×10^5	~10^6	~5×10^5	~10^4	~10^4
Re_M	$O(10^3)$	$O(10^3)$	$O(10^4)$	$O(10^2-10^3)$	$O(10^4)$	$O(10^4)$	$O(10^4)$

R_P is the planetary radius; R_D is the radius of the dynamo source region; "~" in front of numbers are meant to convey that these values are not well constrained
[1]Only the dominant constituent of the composition relevant for dynamo action is given. To estimate Re_M, a fluid velocity and length scale had to be assumed. R_D was chosen for the length scale, and for simplicity, an Earth-like estimate for the planetary velocities (5×10^{-4} m/s) was chosen. another approach would be to use scaling laws (e.g. Christensen, 2010) to estimate velocities from various force balances.

planets, the mechanical energy is due to fluid motions and the resulting electromagnetic energy produces the observed planetary magnetic fields. The main equation governing dynamo action is the **magnetic induction equation**:

$$\frac{\partial \vec{B}}{\partial t} = \nabla \times (\vec{v} \times \vec{B}) + \eta \nabla^2 \vec{B}$$

which can be derived from Maxwell's equations and Ohm's law in the **magnetohydrodynamic limit**.

In this equation, the time variation of magnetic field $\vec{B}$ is the result of: (1) the interaction of velocity fields $\vec{v}$ and magnetic fields (represented in the first term on the right-hand side of the equation which we will call the "induction" term) and (2) the diffusion of the field through **Ohmic dissipation** (represented by the second term on the right-hand side of the equation). The magnetic diffusivity $\eta = (\sigma\mu)^{-1}$ is inversely proportional to the electrical conductivity σ and the magnetic permeability μ.

2.2. Necessary Conditions for a Dynamo

In order for a planet to have a dynamo-generated magnetic field, it must contain an electrically conducting fluid region undergoing motions to generate induction. These conditions are easily discernible by examining the magnetic induction equation, i.e. nonzero σ and $\vec{v}$ are required for a nonzero induction term.

However, there are further necessary conditions for the vigor and morphology of the motions that result from the fact that the magnetic field must not decay away due to Ohmic dissipation. A common measure of the required vigor of motions comes from the critical magnetic Reynolds number condition. In order for dynamo action to be sustainable, the induction term must be larger than the diffusion term in the magnetic induction equation. By using some characteristic velocity V and length scale L to represent the magnitude of terms on the right-hand side of the magnetic induction equation, the ratio of the induction to diffusion terms is given by the **magnetic Reynolds number**, Re_M:

$$Re_M = \frac{|\nabla \times (\vec{v} \times \vec{B})|}{|\eta \nabla^2 \vec{B}|} \approx \frac{VL}{\eta}$$

This number must be larger than some critical value Re_M^c in order for dynamo action to occur. Lower bounds can be placed on this critical value using analytic techniques (see Jones (2008) for some common ones). Depending on the choice of characteristic length and velocity scales, the bounds are typically around π to π^2; however, these bounds do not take into account the required complexity of the fluid motions (see below). Investigations of Re_M^c in numerical simulations of dynamos give values around 20–50.

It is believed that active planetary dynamos have Re_M much greater than the critical value. For example, using the secular variation of the field as a characteristic velocity and the core radius as a characteristic length scale results in $Re_M = O(10^3)$ for the Earth.

In addition to the vigor of convection, there are also necessary conditions on the morphology of the velocity field. Antidynamo theorems demonstrate that the flow in a spherical geometry must have a radial component. This rules out some standard fluid motions from being dynamo-capable. For example, in spherical coordinates (r,θ,ϕ), differential rotation of the form $\vec{v}(r,\theta,\phi) = v(s)\hat{\phi}$ (where $s = r\sin\theta$ is the cylindrical radial coordinate and $\hat{\phi}$ is the longitudinal direction) cannot produce a dynamo alone. Similarly, solid body rotation due to the rotation of the planet alone cannot generate a dynamo, although it has an important influence on the flow morphology. Analytic expressions for flows capable of generating a dynamo have been found (see Jones (2008) for a review); however the minimum sufficient conditions for a dynamo are not currently known.

2.3. Dynamo Generation Regions in Planets

The basic necessity for a planetary dynamo is a fluid electrically conducting region in the planet. In terrestrial planets, this region is the liquid layer of the iron-rich core and hence the conductivity is metallic. Although the giant planets likely also contain deep rocky layers with iron-rich cores, these are not the source of the observed magnetic fields for these planets. This is because other materials with good conductivity are undergoing motions in a much larger fraction of these planetary interiors closer to their surfaces.

Jupiter and Saturn, composed predominantly of hydrogen, possess extreme temperatures and pressures in their interiors. Hydrogen under these extreme conditions can metallize and hence produce a good electrical conductor. Even at pressures somewhat lower than the transition to a metallic state, hydrogen can be an effective semiconductor, and if velocities are large enough in this region, the conductivity may be sufficient to generate a dynamo. The approximate radii at which these transitions occur are given in Table 6.2 along with other properties of the dynamo source regions in planets.

In Uranus and Neptune, the hydrogen-rich layer does not extend to high enough pressures to metallize. Instead, the large water-rich portion of these planets reaches pressures and temperatures allowing for the dissociation of molecules and hence a significant ionic conductivity. Although ionic conductivities are not as large as metallic conductivities, the length scales and likely velocities in these regions still result in highly supercritical magnetic Reynolds numbers for these bodies.

3. THE STANDARD PLANETARY DYNAMO

In this section, we discuss the "canonical" planetary dynamo. Perhaps not coincidentally, this canonical dynamo is considered a good representation of Earth's dynamo. As spacecraft missions have provided details of other planetary magnetic fields, it has become obvious that this standard picture is not applicable to all planets and that the differences between planetary magnetic fields must be explained by considering each planet's dynamo region properties more carefully. In Section 3.4, we consider some of the details beyond this standard picture.

In the canonical planetary dynamo, fluid motions are generated in an electrically conducting spherical shell surrounding a solid inner core. The physical properties of the dynamo region (e.g. thermal and electrical conductivities and viscosity) are assumed to be constants.

3.1. Driving Forces

The fluid motions required for dynamo action must have a power source. The most commonly accepted source is gravitational potential energy release due to cooling of the planet. Planetary formation results in significant amounts of heat trapped in planetary interiors. The planets then slowly cool over time. In the simplest picture, this results in an unstable thermal stratification in the planetary dynamo region where hotter (and hence less dense) material lies below colder (and hence more dense) material. Above a critical temperature difference across the dynamo region, convection will occur in which the hotter (i.e. more buoyant) material is transported outward.

In addition to heat of formation, other sources of buoyancy in the dynamo region may include:

1. *Radiogenic heat sources*: If significant concentrations of radiogenic elements are in the dynamo region, then heat from radioactive decay can contribute to the thermal energy in the core and drive convection.

2. *Compositional convection*: Planetary cooling can result in the generation of compositional variations that can also lead to buoyancy differences and hence convection. For example, as the Earth's core cools, the solid inner core freezes out, releasing a light element-rich fluid at the base of the liquid core. Since this fluid contains less iron than the surrounding fluid, it is less dense and hence buoyant. In Mercury and Ganymede, details of the iron–sulfur system suggest that an iron-rich solid (called "iron snow") may condense out at midlayers or at the top of the cores resulting in more dense material at outer radii and hence a negative buoyancy. This can also drive convection.

The rate at which a planet can cool is ultimately determined by how much heat can be removed from the outer layers. In the terrestrial planets, the metallic cores are surrounded by rocky mantles with very different material properties than the cores. Heat transfer in the mantle layer is at a very different pace than that in the core and the amount of heat that can be removed from the core is ultimately determined by how much heat can be transferred through the mantle. Details of the mantle structure are therefore important in determining the cooling properties of the cores of terrestrial planets.

Convection is believed to be the most likely source of motions generating dynamos in planetary cores for two reasons. First, the abundance of heat in most planetary dynamo regions makes this power source capable of generating motions for long times. Second, the form of convection in rapidly rotating fluids results in flow morphologies that are very conducive to dynamo action. These morphologies will be explored in Section 3.2.

3.2. Fluid Motions in Dynamo Regions

Dynamo generation regions are spherical shells undergoing rapid rotation. This results in specific flow morphologies:

1. *Convective flows*: Due to the **Taylor–Proudman theorem**, flows in rapidly rotating, low-viscosity fluid show much smaller variation in the axial direction compared to the cylindrical radial and azimuthal directions. Since convection aims to move buoyant parcels outward, the combination of outward motions and rapid rotation results in columnar motions where fluid in entire vertical spans move outward. Rapid rotation is therefore very good at organizing convective fluid motions on a large scale (see Figure 6.4).

2. *Meridional flows*: These flows are approximately parallel to the rotation axis inside the convection columns caused, for example, by geometric effects due to the boundary curvature.

3. *Zonal flows*: Flows in the azimuthal (i.e. longitudinal or zonal) direction are typically generated by (a) **thermal winds** due to latitudinal variations in buoyancy, (b) **magnetic winds** due to latitudinal variations in magnetic fields, or (c) **Reynolds stresses** due to correlations in small-scale velocity fields. These flows result in differential rotation in the dynamo generation region.

3.3. Generation Mechanisms

Fluid motions must possess a certain amount of complexity in order to generate a dynamo. As discussed in Section 2.2, the flows must be vigorous enough, but must also meet

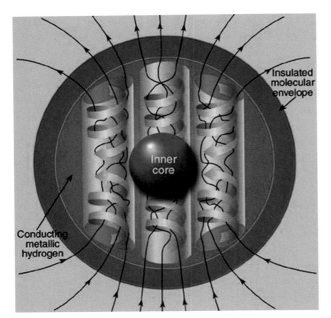

FIGURE 6.4 Sketch of typical convective fluid motions in planetary dynamo regions that are conducive to dynamo action. The geometry and composition in the figure are specific to Jupiter, but the depicted motions could be applied to other planets. Helical motions in Taylor columns are depicted with yellow arrows. Magnetic field lines are in black. *Image courtesy of NASA/GSFC.*

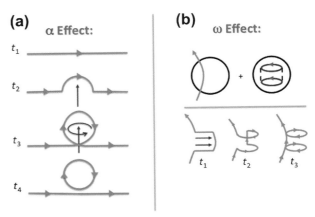

FIGURE 6.5 Sketch of canonical dynamo generation mechanisms. Magnetic field lines are red and velocity fields are blue. (a) In the α-effect, a helical flow stretches (t_2) then twists (t_3) magnetic field. A small amount of diffusion at the twisting point (t_4) can then generate a magnetic field loop orthogonal to the original field. (b) In the ω-effect, a large-scale differential rotation takes a poloidal field line and stretches it in the zonal direction (t_1–t_3) generating toroidal magnetic loops from the poloidal field line. The black circle indicates the outer boundary of the dynamo region. *Adapted from Merrill et al. (1998).*

morphology requirements. There are two categories of motions which are likely to occur in planetary cores that work very well in combination to generate a dynamo: (1) helical flows and (2) differential rotation.

To discuss magnetic field generation mechanisms, it is common to use a poloidal/toroidal decomposition of the field. This is possible because magnetic fields obey Gauss' law for magnetism and so they can be fully represented with two scalar functions:

$$\nabla \cdot \vec{B} = 0 \Rightarrow \vec{B} = \vec{B}_T + \vec{B}_P$$
$$= \nabla \times (T\vec{r}) + \nabla \times (\nabla \times (P\vec{r}))$$

where $\vec{B}_T$ and $\vec{B}_P$ are orthogonal "**toroidal field**" and "**poloidal field**" components and T and P are the toroidal and poloidal scalar functions respectively. Note from the equation above that toroidal field has no radial component, whereas poloidal field generally has components in all three directions. In simplest terms, magnetic field generation can be envisioned as the result of helical flows and differential rotation acting upon some initial toroidal and poloidal fields to generate new fields. If this can be done in such a way as to reproduce the initial fields before their Ohmic decay, then the dynamo is self-sustaining. Heuristic depictions of these generation mechanisms are shown in Figure 6.5.

Helical flows are a natural result of convection in planetary dynamo regions due to their spherical geometry and rapid rotation. Figure 6.5(a) depicts helical flow stretching and twisting a magnetic field line to generate field with an orthogonal component to the original field. The amount of helicity is measured through the helicity parameter $h = \vec{v} \cdot \vec{w}$, where $\vec{w} = \nabla \times \vec{v}$ is the vorticity in the fluid. If this flow acts upon a toroidal magnetic field, then it can generate poloidal field from it. Similarly, if this flow acts upon a poloidal magnetic field, then it can generate toroidal field from it.

The process of generating magnetic fields through helical motions is sometimes called the "**α-effect**". This name is borrowed from mean field dynamo theory, popular in studies of astrophysical dynamos, but has a slightly different connotation in planetary dynamo theory. Specifically, the α-effect in mean field theory represents the combined effect of turbulent microscopic motions, whereas in planetary dynamo theory, the helical motions described above can be macroscopic in scale and laminar (as opposed to turbulent). In general terms, one can consider the α-effect as being the result of motions with helicity.

Differential rotation refers to the shearing motion of zonal flows. As discussed in Section 2.2, these motions alone cannot generate a dynamo; however, they can be used in combination with other motions to very effectively generate magnetic field. Figure 6.5(b) depicts cylindrical differential rotation acting on a poloidal magnetic field. The differential rotation shears the magnetic field line generating a magnetic field component orthogonal to the original magnetic field. The strength of the generated toroidal magnetic field will depend on the shear and the amplitude of the poloidal field where the shear is strong.

The process of magnetic field generation through differential rotation is sometimes called the "**ω-effect**". Here the connotation is similar to that in mean field theory as it describes the effect of large-scale zonal flows on stretching magnetic fields.

The dynamo generation cycle can then be envisioned in the following way: If we initially have a poloidal magnetic field, then either differential rotation or helical flows can act to generate toroidal magnetic field from it. Once this toroidal field is created, then helical flows can act to generate poloidal magnetic field from it. In this way, we are able to regenerate the field we started with and the cycle of field generation can continue in a self-sustained manner.

There is debate as to whether the ω-effect is an important contributor to magnetic field generation in planets. It is possible to generate a dynamo solely through the α-effect and hence the ω-effect is not needed in principle. However, there are mechanisms to generate strong differential rotation in planets, for example, through thermal winds or Reynolds stresses. Since we cannot observe fluid motions deep in planets directly, investigations of the importance of these processes are mainly carried out through computational and laboratory experiments (see Section 4).

3.4. Beyond the Standard Dynamo

Additions to the standard dynamo appear to be necessary to explain unique features of specific planetary dynamos. These include:

1. *Alternative driving mechanisms*: In addition to convection, other driving mechanisms for fluid motions have been proposed and may be relevant for specific planets. Motions due to precession, tides, and boundary driving have all been invoked as possibilities. Although the basic flows due to these forcings are purely toroidal and laminar, and hence, not good at generating dynamos, instabilities of the motions due to these forcings have been shown to be capable of dynamo action. These mechanisms may be important at certain times in planetary evolution when convection is not capable of driving a dynamo. For example, the relatively late lunar dynamo may have been the result of precession or boundary forcings.

2. *Stably stratified layers*: The entire electrically conducting fluid regions may not be convecting either because of compositional or thermal stratification. For example, data and models suggest that an outer thin layer of Earth's core may be stably stratified. Thermal evolution models for Uranus and Neptune also suggest that the deepest water-rich regions of these planets are stably stratified. The presence of **helium rain** in Jupiter and Saturn may also result in stably stratified layers.

3. *Dynamo region geometry*: The dynamo source region is a spherical shell as opposed to a full sphere in most planets. The thickness of the spherical shell has important implications for the convective motions. For example, in Earth, the solid inner core is relatively small and may not affect the location of convection columns, but in a planetary body with a larger inner core (and hence a thinner convecting shell), the fluid motions must allow for the inner core boundary.

4. *Radially-varying physical properties*: The material properties of the fluid, such as the electrical and thermal conductivities, kinematic viscosity, and density are pressure and temperature dependent. The variation of these properties as a function of depth in the dynamo source region may have important implications for dynamo generation. This is likely to be more important for the giant planets than the terrestrial planets since the more massive giant planets experience larger ranges of pressure and temperature in their dynamo source regions.

5. *Laterally-varying boundary conditions:* Conditions at dynamo region boundaries may not be homogeneous. For example, in the Earth, laterally varying heat flux at the core-mantle boundary may have significant influence on the dynamo.

6. *Influence of external fields*: If a source external to the planetary body generates significant magnetic fields in the dynamo source region, these can affect the dynamo. For example, Jupiter's magnetic field is relatively strong at Ganymede and hence may influence Ganymede's dynamo generation. Similarly, magnetospheric currents at Mercury may generate magnetic fields that can be appreciable in Mercury's core.

4. SIMULATIONS AND EXPERIMENTS

In combination with planetary magnetic field observations, properties of planetary dynamos are investigated through numerical simulations and laboratory experiments. It is currently not feasible to accurately represent the parameter regime of planetary interiors with simulations and experiments; however, they are useful tools in investigating mechanisms and force balances in planetary dynamos.

4.1. Numerical Dynamo Simulations

Dynamo simulations computationally solve the governing equations for planetary dynamos. Current models include some of the additions discussed in Section 3.4, but here we will describe the equations solved for the simplest standard planetary dynamo model: that of a Boussinesq, electrically conducting fluid driven by thermal convection in a spherical rotating shell with a linear gravity profile. Equations solved numerically are then:

1. Momentum equation:

$$\frac{\partial \vec{v}}{\partial t} + \vec{v} \cdot \nabla \vec{v} + 2\vec{\Omega} \times \vec{v}$$
$$= -\nabla \tilde{P} - \alpha T \frac{g_0}{r_0} \vec{r} + \frac{1}{\rho_o} \vec{J} \times \vec{B} + \nu \nabla^2 \vec{v}$$

2. Magnetic induction equation:

$$\frac{\partial \vec{B}}{\partial t} = \nabla \times (\vec{v} \times \vec{B}) + \eta \nabla^2 \vec{B}$$

3. Energy equation:

$$\frac{\partial T}{\partial t} + \vec{v} \cdot \nabla T = \kappa \nabla^2 T + Q$$

where $\vec{\Omega}$ is the angular velocity of the body, $\tilde{P}$ is the modified pressure, ρ_o is the background constant density, α is the thermal expansion coefficient, T is the temperature, g_0 is the gravitational acceleration at the top of the dynamo source region (i.e. at r_0), ν is the kinematic viscosity, κ is the thermal diffusivity, and Q is the volumetric heat source.

Typically, these equations are nondimensionalized using characteristic scales for the dimensional variables. There are a variety of ways to carry out the nondimensionalization and we offer one example here. Choosing the dynamo region shell thickness D as a length scale, the magnetic diffusion time $\tau = D^2/\eta$ as the timescale, a **magnetostrophic balance** estimate $B = (2\Omega\mu_0\rho_o\eta)^{1/2}$ as the magnetic field scale, the superadiabatic temperature difference ΔT across D as the temperature scale, and using Ampere's law to represent the current density in terms of the magnetic field, the nondimensional equations can be written in the form:

$$\frac{E}{P_m}\left(\frac{\partial \vec{v}}{\partial t} + \vec{v} \cdot \nabla \vec{v}\right) + 2\vec{\Omega} \times \vec{v}$$
$$= -\nabla \tilde{P} - Ra_{th} T \vec{r} + \vec{J} \times \vec{B} + E\nabla^2 \vec{v}$$

$$\frac{\partial \vec{B}}{\partial t} = \nabla \times (\vec{v} \times \vec{B}) + \nabla^2 \vec{B}$$

$$\frac{\partial T}{\partial t} + \vec{v} \cdot \nabla T = \frac{P_m}{P_r}\nabla^2 T + Q$$

where all variables are now dimensionless. The nondimensional groupings of physical parameters that govern the system of equations are:

1. Ekman number:

$$E = \frac{\nu}{2\Omega D^2}$$

2. Magnetic Prandtl number:

$$P_m = \frac{\nu}{\eta}$$

3. Modified Rayleigh number:

$$Ra_{th} = \frac{\alpha \Delta T g_0 D^2}{2\Omega \eta r_0}$$

4. Prandtl number:

$$P_r = \frac{\nu}{\kappa}$$

Due to numerical constraints, simulations are not able to work in the appropriate parameter regime for planetary dynamo regions. Specifically, the Ekman and magnetic Prandtl numbers are much larger in simulations than in planets, and the Rayleigh number is probably much smaller in simulations than in planets (although it is not well constrained). However, insights into mechanisms and force balances are used to extrapolate results from simulations to planetary conditions using scaling laws, but it is unclear whether these derived scaling laws hold over the many orders of magnitude of extrapolation needed to bridge the gap between simulations and planets.

Although dynamo simulations work in parameter regimes far from that of planets, they are capable of reproducing many salient features of observed planetary magnetic fields. For example, models for Earth's dynamo can produce axially dipolar-dominated fields, smaller scale spectral features, reversals, and other secular variation features that have been observed. This may be coincidental but may also be the result of the fact that dynamo models are producing accurate force balances even if the quantitative values for the parameters are not correct. For example, although the Ekman number is ~ 10 orders of magnitude too large in simulations, the value is still very small ($\sim O(10^{-5})$) indicating that viscous forces are much weaker than Coriolis forces in the models, as we would expect them to be in planetary dynamo regions.

Numerical simulations for other planets are generated by including some of the additions discussed in Section 3.4, where applicable. These are discussed in detail in Section 5 for the respective planets.

The ultimate goal of numerical simulations is to use them to understand the processes occurring in planetary dynamos. Numerical modelers are using the most advanced computational resources available to push parameters as close as possible to planetary values, and the models will only improve in the future. Other efforts include covering wide ranges of parameter space to develop scaling laws for various observable characteristics. For example, Christensen et al. (2009) demonstrated that a scaling law developed from simulations for the strength of the magnetic field seems to work well in predicting planetary field strengths as well as some stellar magnetic field strengths.

Other numerical methods aim to simplify the models to work in more challenging parameter regimes. For example, "quasi-geostrophic" models solve the equations governing fluid motions solely in the equatorial plane and then use constraints from rapid rotation to infer the motions outside of this plane. These types of models provide valuable insight into fluid motions at more extreme parameter values than possible in fully three-dimensional self-consistent models.

4.2. Dynamo Experiments

Building a laboratory experiment that can generate a self-sustaining dynamo is incredibly challenging. The main reason being that the small length scales of experiments result in small magnetic Reynolds numbers (i.e. below critical values) unless velocities are made extremely large. Typical experiments use liquid metals such as sodium or gallium. Early experiments, such as the Karlsruhe and Riga dynamos, used a series of pipes to create flow morphologies that are known analytically to be conducive to dynamo action. They demonstrated the growth of magnetic field intensity; and hence dynamo action; however, they are somewhat nonplanetary-like in geometry.

Present day experiments aim to generate dynamos in more homogenous geometries (e.g. in cylindrical or spherical tanks). In these experiments, the flows are not as constrained as the pipe flows and instead, fluid motions are generated through propellers or boundary differential rotation (see Figure 6.6 for some examples). Generating a laboratory dynamo using convective motions (i.e. through buoyancy) is currently not feasible. In addition, experiments face difficulties in mimicking the radial form of the gravity force in planets.

These experiments have provided important insights into the role of turbulence in helping and hindering dynamos and also provide a means to investigate different regions of parameter space from numerical simulations. For a nice review of dynamo experiments, see Lathrop and Forest (2011).

5. PLANETARY DYNAMOS

Here we outline the major features of planetary dynamos in a comparative fashion. Information comes from magnetic field and other spacecraft observations, as well as theoretical and experimental studies of planetary interior properties. We briefly discuss results from numerical dynamo simulations, but for a deeper review of planetary dynamo simulations, see Stanley and Glatzmaier (2010).

5.1. Earth

The Earth's magnetic field (also known as the geomagnetic field) is dominated by its axial dipole component. The nondipolar component of the field includes two strong normal polarity flux spot pairs, particularly evident in the northern hemisphere over Canada and Russia. Based on paleomagnetic records, these spots appear to be long-lived and it has been suggested that they are features associated with convection columns in the core. These convection columns might be relatively stationary due to thermal influences from the mantle convection morphology at the core-mantle boundary. In addition to these flux spots, the Earth also has intermittent normal and reverse flux patches in equatorial regions, some of which drift westward in time. Figure 6.7 shows the radial component of the Earth's magnetic field at the core-mantle boundary.

Records from satellites such as Magsat and Oersted, ground observatories, and ship logs have provided fairly detailed records of the geomagnetic field over the past 400 years. In addition, paleomagnetic records from crustal rocks on the sea floor provide data on magnetic field reversals over the past ~ 180 Myrs (i.e. up to the age of the oldest seafloor). Prior to this, we rely on paleomagnetic fields in continental rocks.

These data demonstrate the time variability of the geomagnetic field. Reversals occur sporadically, on average, every half a million years. In recent times, equatorial flux spots drift with a speed of approximately $0.2°$/year and the north geomagnetic pole meanders about the geographic pole at about 10 km/year.

The geomagnetic field is generated in the Earth's iron-rich core. The fluid outer core surrounds a solid inner

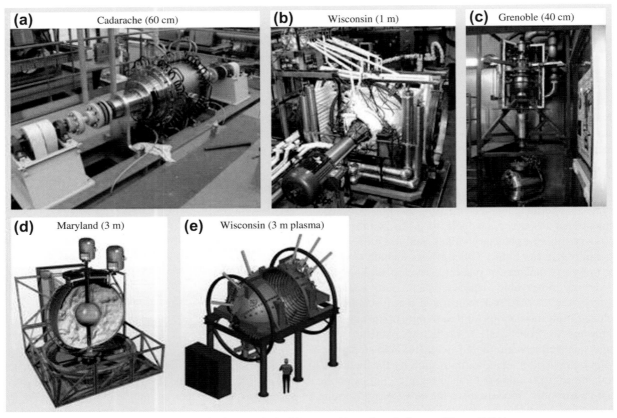

FIGURE 6.6 Magnetic dynamos in the laboratory. *Figure from Lathrop and Forest (2011).*

core that is enriched in iron compared to the outer core. The inner core grows in time as the Earth cools and thermal evolution models suggest that it is ∼1 billion years old. As it grows, it expels light elements resulting in a source of compositional buoyancy at the base of the outer core, in addition to the thermal buoyancy available from core cooling, latent heat, and possibly, radiogenic elements. Paleomagnetic records suggest that the Earth's field is at least 3.4 billion years old, implying that the geodynamo was active before the solid inner core began to grow.

Earth is the only planet with evidence of a solid inner core. Although the Earth's inner core is relatively small

FIGURE 6.7 Radial component of the Earth's magnetic field at the core-mantle boundary. *Figure from Jones (2011).*

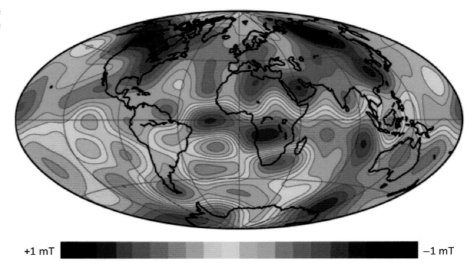

+1 mT　　　　　　　　　　　　　　　　　−1 mT

(approximately 1/3 the radius of the total core), it may provide stability to the axial dipole field since magnetic field lines threading the solid inner core are electromagnetically frozen into it and can only vary on the inner core's magnetic diffusion timescale ($O(10^3)$ years) as opposed to the faster outer core's convective timescale ($O(10^2)$ years).

5.2. Mercury

Mercury's magnetic field was first observed by the Mariner 10 mission during flybys of the planet in the mid-1970s. Data from two of the flybys found a weak dipole moment. More recent data from the MESSENGER mission has confirmed the Mariner 10 results and has also provided more detailed constraints on the low-degree spectral components of the field.

Prior to the Mariner 10 mission, it was considered unlikely that Mercury would have a dynamo based on thermal evolution models for such a small body. Essentially, since smaller planets cool faster, it was suggested that Mercury's iron core should have fully solidified by the present day making dynamo action impossible. To explain the observed global magnetic field, most likely the result of a dynamo, researchers suggested that the core must also contain light elements such as sulfur or silicon in addition to iron in order to depress the melting temperature and hence reduce the speed of solid inner core growth. Radar measurements from Earth, as well as MESSENGER data, have independently confirmed that the Mercury core contains a liquid layer. The size of the inner core is unknown and depends strongly on the fraction of light elements in the core.

Specific features of Mercury's magnetic field are difficult to explain with a "standard" dynamo model. First, the dipole field is about two to three orders of magnitude weaker than that expected by scaling laws that estimate a dynamo-generated field's strength. Second, Mercury's axial dipole offset (a measure of its axial quadrupole component) is fairly large, whereas its dipole tilt (a measure of its nonaxial dipole component) is fairly small. This combination is difficult to produce with a standard dynamo.

Several numerical dynamo models have attempted to explain the weak dipole intensity by appealing to additions to the standard dynamo model (like those in Section 3.4). For example, numerical models with stably stratified layers either in the outer region of the core, or at mid-depth, can produce weak dipole fields, as can models with a relatively large or relatively small solid inner core. In addition, models that appeal to feedback between external magnetospheric fields and the core field produce weaker observed fields. There has not been much investigation of producing models with the combination of a large dipole offset and small dipole tilt due to the freshness of this data, but it is likely that they will appear in the near future.

5.3. Jupiter

Magnetic field data for Jupiter has come primarily from the Voyager I and II and Galileo missions. Jupiter's field is similar to Earth's field in morphology, being axially dipolar dominated with a dipole tilt of approximately 10° and a large-scale spectral structure similar to Earth's field (see Figure 6.3).

The field is generated in the electrically conducting hydrogen region of the planet which extends out to about 0.8–0.9 Jupiter radii. Dynamo simulations for Jupiter generally include radially varying physical properties such as density and electrical conductivity since the dynamo generation region extends through many pressure-scale heights. Although the electrical conductivity increases with depth, the deepest layers are also the most dense and hence, experience the slowest fluid velocities. It is therefore possible that dynamo generation is limited to the outermost layers of the dynamo region where the combination of fluid velocities and electrical conductivity produce the most appreciable magnetic Reynolds numbers. Data from the upcoming Juno mission are expected to provide a significant improvement to the resolution of the field, which should allow testing of dynamo region geometry.

5.4. Saturn

Saturn's magnetic field data has come from Voyager I and II in the 1970s and the 1980s and more recently from Cassini since 2004. Like Jupiter, the field is dominated by its axial dipole, but Saturn is unique in the lack of any observed nonaxisymmetric field components. In addition, the octupole component is larger than the quadrupole component suggesting a preference for odd harmonics in the field. Because the axial octupole Gauss coefficient (g_3^0) has the same sign as the axial dipole Gauss coefficient (g_1^0), the field is concentrated in the polar regions compared to equatorial regions. This is the opposite of what is observed in Earth's or Jupiter's field today, although standard numerical dynamo simulations suggest that nondipolar components can vary significantly in time. Hence, this may just be the result of capturing the magnetic field in an untypical configuration or point toward a different dynamo mechanism.

The observation of a purely axisymmetric field is problematic for a standard dynamo explanation. First, **Cowling's theorem** demonstrates that a perfectly axisymmetric magnetic field cannot be generated by a dynamo. Second, no other planetary magnetic field demonstrates the same amount of axisymmetry as Saturn (although future data from Mercury and Ganymede may alter this statement). Third, standard dynamo models cannot reproduce this level of axisymmetry in the observed field.

Like Jupiter, Saturn's dynamo region is its hydrogen-rich metallic layer. Saturn may differ from Jupiter by the presence of a helium rain layer at pressures and temperatures where hydrogen becomes metallic. Quantum mechanical simulations have demonstrated that at these characteristic pressures and temperature in Saturn, helium may become immiscible in hydrogen and as it separates from the mixture it will be negatively buoyant compared to the hydrogen and hence "rain out". At deeper pressure, helium may then become miscible again (for a review, see McMahon et al., 2012). It is possible that such a layer also exists in Jupiter, but based on thermal evolution calculations, this layer should be much thicker in Saturn than Jupiter and hence, this may be a decent mechanism to explain the difference between these planetary magnetic fields.

The leading theory explaining Saturn's axisymmetric field involves this layer: If Saturn's dynamo is surrounded by this stably stratified electrically conducting layer, and if differential rotation exists in this layer, then the non-axisymmetric field may be preferentially attenuated through the electromagnetic skin effect resulting in a surface field with much more axisymmetry than would occur without such a layer. Dynamo simulations for Saturn explore the effects of stably stratified layers on axi-symmetrizing the field.

5.5. Uranus and Neptune

The magnetic field data of Uranus and Neptune come from single flybys of the planets by the Voyager II mission in the 1980s. The data revealed that, unlike the other planets, these ice giants' magnetic fields were not dominated by their axial dipoles and instead, contained roughly equal contributions from different field harmonics.

Standard numerical dynamo models typically do not produce nonaxially dipolar-dominated fields, except in isolated regions of parameter space or for very large buoyancy forcing relative to Coriolis forcing. Although these may be viable explanations for these planets' field morphologies, these models do not explain another observation of the ice giants: in addition to the anomalous magnetic fields, data also demonstrate that these planets have low intrinsic heat flows.

The dynamo source regions in the ice giants are the water-rich layers at depths such that a significant ionic conductivity results ($\sim 0.7-0.8$ planetary radii). Thermal evolution simulations suggest that the low heat flows are explained if the ice giant interiors are not fully convective, and hence that the inner regions of these ice layers are stably stratified to convection. Numerical dynamo simulations that incorporate a stably stratified interior fluid region below a relatively thin convective shell (where the dynamo generation occurs) can reproduce the magnetic field observations for the ice giants.

5.6. Ganymede

Ganymede's magnetic field was discovered by the Galileo mission in the mid-1990s. Although magnetic induction signatures were found for Europa, Ganymede, and Callisto, resulting from currents generated in salt-water oceans in these bodies, Ganymede was the sole Galilean satellite to demonstrate a self-sustained dynamo-generated field. There is little data available for the spectrum of the field aside from a dipole moment.

Like Mercury, Ganymede's small size suggests that there must be a significant fraction of light elements such as sulfur in its core to keep it liquid at present day. It is unknown whether Ganymede has a solid inner core, but the dynamo may have a compositional driving source if the liquid core is in a regime where it freezes at the outer boundary, releasing negatively buoyant iron-rich fluid, rather than freezing at the inner boundary like in the Earth's core. Dynamo studies including different buoyancy source distributions intended to mimic these solidification processes have been carried out.

5.7. Ancient Moon

The lunar magnetic field was mapped by the Lunar Prospector mission, but this field is due to **remanent magnetization** in the lunar crust rather than an active dynamo. Paleomagnetic data from lunar samples indicate that the field was most likely due to a dynamo that was active from at least 4.2 to 3.5 billion years ago.

The driving source for the lunar dynamo is unclear since the small size of the core suggests that thermal convection would not provide enough energy to drive a dynamo for such a long time after formation. Therefore, alternative mechanisms such as precession or boundary forcing due to oblique impacts have been suggested.

5.8. Ancient Mars

The Martian magnetic field was studied by the Mars Global Surveyor mission in the 1990s. Similar to the Moon, the Martian magnetic field is due to remanent magnetization in the crustal rocks. The magnetizing field was most likely due to a dynamo active in early Martian history, before ~ 3.9 billion years. The crustal field displays a correlation with the hemispheric crustal dichotomy (*see Chapter on Mars: Surface and Interior*) with the southern hemisphere containing more intense fields than the northern hemisphere.

The magnetic dichotomy may be due to postdynamo crustal reworking that preferentially removed magnetism from crust in the northern hemisphere, or it may be due to the morphology of the magnetizing field while the rocks were forming. For example, mechanisms suggested to explain the hemispheric crustal dichotomy include degree-one mantle circulation or a large glancing impact in the northern hemisphere. Both these mechanisms could result in hemispheric thermal variations at Mars' core-mantle boundary. Numerical dynamo simulations have demonstrated that these thermal variations can result in hemispheric dynamos, where the field is much stronger in the southern hemisphere than in the northern hemisphere. It is therefore possible that the difference in crustal magnetization between the hemispheres is the result of a difference in the intensity of the magnetizing field between the hemispheres.

5.9. Small Bodies

Paleomagnetic studies of classes of meteorites such as the **Angrites**, some carbonaceous chondrites and the **HED (Howardite-Eucrite-Diogenite) meteorites** demonstrate that small bodies such as planetesimals and asteroids may have possessed dynamos in the early solar system. Although it is difficult to sustain a convective driving force for a long time in such small bodies, the presence of appreciable heat sources in the early solar system, such as radiogenic elements Al^{26} and Fe^{60}, provide enough driving power for a short time (around 10 million years) to allow iron core formation and then to generate a dynamo. Scaling studies demonstrate that the surface fields are strong enough to magnetize the crusts and explain the observed meteorite magnetism (Weiss et al., 2010).

5.10. Extrasolar Planets

Although one would expect some extrasolar planets to have active dynamo-generated magnetic fields, no extrasolar planetary magnetic field has been unambiguously discovered. However there are potential detection methods that could be feasible in the near future. One possibility of discovery is through radio emissions from the interaction of stellar winds with planetary magnetic fields (similar to the radio emissions observable from Jupiter in our own solar system). New advanced radio antennae such as the Low-Frequency Array might be capable of such detection.

5.11. Planetary Bodies Lacking Dynamos

The only planet in our solar system with no evidence of past or present dynamo action is Venus. This is somewhat unexpected considering the similarity between Venus' and Earth's interior structures. The difference cannot be explained by Venus' slower rotation. The most likely explanation for a lack of dynamo action today in Venus is that the core is not cooling vigorously enough to generate strong convection. This may be due to the fact that Venus experiences a different mode of mantle convection than the Earth, possibly due to a lack of water in Venus' mantle. Venus appears to experiences sluggish, rigid lid convection with episodic large-scale overturn. During the sluggish stage, the mantle may not remove enough heat from the core to generate core convection and hence no dynamo generation occurs.

An interesting question remains as to whether a dynamo onsets during episodic overturning events. Unfortunately, the surface temperatures are above the Curie temperature for most crustal rocks and therefore, it is unlikely that the surface rocks contain appreciable remanent magnetization from the last overturning episode, which occurred approximately 700 million years ago.

Aside from the Moon and Ganymede, no other satellites or small planetary bodies have observed magnetic fields (expect for those inferred from meteorites discussed in the previous section).

6. CONCLUSIONS AND FUTURE PROSPECTS

The past four decades have provided a wealth of new data from satellite missions on planetary magnetic fields. In combination with numerical simulations and laboratory experiments over the past two decades, new insights on planetary magnetic field generation have resulted. The near future promises some exciting advances. First, new data from planned and active magnetic missions such as Cassini (at Saturn until 2017), JUNO (to Jupiter), Juice (to the Jupiter system), BepiColombo (to Mercury), and Swarm (to Earth) will provide great improvements in data resolution and possibly allow study of secular variation of planetary magnetic fields other than Earth. Second, ongoing improvements to computational models and hardware, in addition to laboratory dynamo experiments, will improve our knowledge of the fluid dynamics and magnetohydrodynamics of planetary dynamo regions. Third, new paleomagnetic techniques and data sets will add to our knowledge of past planetary magnetic fields. Finally, planetary magnetic field generation is not an isolated process, and so improvements in our knowledge of planetary interior composition, structure, and dynamics will also be used to provide greater constraints on planetary magnetic fields.

BIBLIOGRAPHY

Breuer, D., Labrosse, S., & Spohn, T. (2010). Thermal evolution and magnetic field generation in terrestrial planets and satellites. *Space Science Reviews, 152*, 449–500.

Christensen, U. R., Holzwarth, V., & Reiners, A. (2009). Energy flux determines magnetic field strength of planets and stars. *Nature, 457*, 167–169.

Christensen, U. R. (2010). Dynamo scaling laws and applications to the planets. *Space Science Reviews, 152*, 565–590.

Finlay, C. C., Maus, S., Beggan, C. D., Bondar, T. N., Chambodut, A., Chernova, T. A., et al. (2010). International geomagnetic reference field: the eleventh generation. *Geophysical Journal International, 183*, 1216–1230.

Jackson, A., & Finlay, C. C. (2007). *Geomagnetic secular variation and its applications to the core*. In G. Schubert (Ed.), *Treatise of Geophysics* (Vol. 5); (pp. 147–193).

Jones, C. A. (2008). Course 2: dynamo theory. In P. Cardin, & L. F. Cugliandolo (Eds.), *Dynamos* (pp. 45–135).

Jones, C. A. (2011). Planetary magnetic fields and fluid dynamos. *Annual Review of Fluid Mechanics, 43*, 583–614.

Kono, M. (2007). *Geomagnetism in perspective*. In G. Schubert (Ed.), *Treatise of Geophysics* (Vol. 5); (pp. 1–31).

Lathrop, D. P., & Forest, C. B. (July 2011). Magnetic dynamos in the lab. *Physics Today*, 40–45.

McMahon, J. M., Morales, M. A., Pierleoni, C., & Ceperley, D. M. (2012). The properties of hydrogen and helium under extreme conditions. *Reviews of Modern Physics, 84*, 1607–1653.

Merrill, R. T., McElhinny, M. W., & McFadden, P. L. (1998). *Magnetic field of the Earth: Paleomagnetism, the core, and the deep mantle*. San Diego, USA: Academic Press.

Stanley, S., & Glatzmaier, G. (2010). Dynamo models for planets other than Earth. *Space Science Reviews, 152*, 617–649.

Weiss, B. P., Gattacceca, J., Stanley, S., Rochette, P., & Christensen, U. R. (2010). Paleomagnetic records of meteorites and early planetesimal differentiation. *Space Science Reviews, 152*, 341–490.

Chapter 7

Planetary Magnetospheres

Margaret Galland Kivelson
Department of Earth & Space Sciences, University of California, Los Angeles, CA, USA and Department of Atmospheric, Oceanic, and Space Sciences, University of Michigan, Ann Arbor, MI, USA

Fran Bagenal
Department of Astrophysical & Planetary Sciences, Laboratory for Atmospheric & Space Physics, University of Colorado, Boulder, Boulder, CO, USA

Chapter Outline

1. What is a Magnetosphere? 137
2. Types of Magnetospheres 138
   2.1. The Heliosphere 138
   2.2. Magnetospheres of the Unmagnetized Planets 140
   2.3. Interactions of the Solar Wind with Asteroids, Comets, and Pluto 142
   2.4. Magnetospheres of Magnetized Planets 144
3. Planetary Magnetic Fields 144
4. Magnetospheric Plasmas 146
   4.1. Sources of Magnetospheric Plasmas 146
   4.2. Energetic Particles 148
5. Dynamics 150
6. Interactions with Moons 154
7. Conclusions 156
Bibliography 157

1. WHAT IS A MAGNETOSPHERE?

The term magnetosphere was coined by T. Gold in 1959 to describe the region above the **ionosphere** in which the magnetic field of the Earth controls the motions of charged particles. The magnetic field traps low-energy charged particles and forms the Van Allen belts, torus-shaped regions in which high-energy ions and electrons (tens of keV and higher) drift around the Earth. The control of charged particles by the planetary magnetic field extends many Earth radii into space but finally terminates near 10 Earth radii in the direction toward the Sun. At this distance, the magnetosphere is confined by a low-density magnetized plasma called the **solar wind** that flows radially outward from the Sun at supersonic speeds. (Plasmas are highly ionized gases composed of electrically charged particles in equal proportions of positive charge on ions and negative charge on electrons whose properties are dominated by their electromagnetic interactions.) Qualitatively, a planetary magnetosphere is the volume of space from which the solar wind is excluded by a planet's magnetic field. (A schematic illustration of the terrestrial magnetosphere is given in Figure 7.1, which shows how the solar wind is diverted around the *magnetopause*, a surface that surrounds the volume containing the Earth, its distorted magnetic field, and the plasma trapped within that field.) This qualitative definition is far from precise. Most of the time, solar wind plasma is not totally excluded from the region that we call the magnetosphere. Some solar wind plasma finds its way in and indeed many important dynamical phenomena give clear evidence of intermittent direct links between the solar wind and the plasmas governed by a planet's magnetic field. Moreover, unmagnetized planets in the flowing solar wind carve out cavities whose properties are sufficiently similar to those of true magnetospheres to allow us to include them in this discussion. Moons embedded in the flowing plasma of a planetary magnetosphere create interaction regions resembling those that surround unmagnetized planets. If a moon is sufficiently strongly magnetized, it may carve out a true magnetosphere completely contained within the magnetosphere of the planet.

Magnetospheric phenomena are of both theoretical and phenomenological interest. Theory has benefited from the data collected in the vast plasma laboratory of space in which different planetary environments provide the analogue of different laboratory conditions. Furthermore, magnetospheric plasma interactions are important to diverse elements of planetary science. For example, plasma trapped in a planetary magnetic field can interact

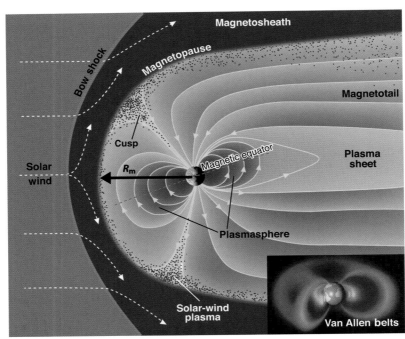

FIGURE 7.1 Schematic illustration of the Earth's magnetosphere. The Earth's magnetic field lines are shown as modified by the interaction with the solar wind. The solar wind, whose flow speed exceeds the speeds at which perturbations of the field and the plasma flow directions can propagate in the plasma, is incident from the left. The pressure exerted by the Earth's magnetic field excludes the solar wind. The boundary of the magnetospheric cavity is called the magnetopause, its nose distance (black arrow) being R_m. Sunward (upstream) of the magnetopause, a standing bow shock slows the incident flow, and the perturbed solar wind plasma between the bow shock and the magnetopause is called the magnetosheath. Antisunward (downstream) of the Earth, the magnetic field lines stretch out to form the magnetotail. In the northern portion of the magnetotail, field lines point generally sunward, while in the southern portion, the orientation reverses. These regions are referred to as the northern and southern lobes, and they are separated by a sheet of electrical current flowing generally dawn to dusk across the near-equatorial magnetotail in the plasmasheet. Low-energy plasma diffusing up from the ionosphere is found close to Earth in a region called the plasmasphere whose boundary is the plasmapause. The dots show the entry of magnetosheath plasma that originated in the solar wind into the magnetosphere, particularly in the polar cusp regions. Inset is a diagram showing the three-dimensional structure of the Van Allen belts of energetic particles that are trapped in the magnetic field and drift around the Earth. Source: The New Solar System (eds. Kelly Beatty et al.), CUP/Sky Publishing. *Credit: Steve Bartlett; Inset: Don Davis.*

strongly with the planet's atmosphere, heating the upper layers, generating neutral winds, ionizing the neutral gases and affecting the ionospheric flow. Energetic ions and electrons that precipitate into the atmosphere can modify atmospheric chemistry. Interaction with plasma particles can contribute to the isotopic fractionation of a planetary atmosphere over the lifetime of a planet. Impacts of energetic charged particles on the surfaces of planets and moons can modify surface properties, changing their albedos and spectral properties. The motions of charged dust grains in a planet's environment are subject to both electrodynamic and gravitational forces; recent studies of dusty plasmas show that the former have been critical in determining the role and behavior of dust in the solar nebula as well as being significant in parts of the present-day solar system.

In Section 2, the different types of magnetospheres and related interaction regions are introduced. Section 3 presents the properties of observed planetary magnetic fields and discusses the mechanisms that produce such fields.

Section 4 reviews the properties of plasmas contained within magnetospheres, describing their distribution, their sources, and some of the currents that they carry. Section 5 covers magnetospheric dynamics, both steady and "stormy". Section 6 addresses the interactions of moons with planetary plasmas. Section 7 concludes the chapter with remarks on plans for future space exploration.

2. TYPES OF MAGNETOSPHERES

2.1. The Heliosphere

The solar system is dominated by the Sun, which forms its own magnetosphere referred to as the *heliosphere*. (See The Sun.) The size and structure of the heliosphere are governed by the motion of the Sun relative to the local interstellar medium, the density of the interstellar plasma, and the pressure exerted on its surroundings by the outflowing solar wind that originates in the solar corona. (See The Solar Wind.) The corona is a highly ionized gas, so hot that it can

Chapter | 7 Planetary Magnetospheres

escape the Sun's immense gravitational field and flow outward at supersonic speeds. Through much of the heliosphere, the solar wind speed is not only supersonic but also much greater than the *Alfvén speed* $(v_A = \mathbf{B}/(\mu_0\rho)^{1/2})$, the speed at which rotational perturbations of the magnetic field propagate along the magnetic field in a magnetized plasma. (Here $\mathbf{B}$ is the magnetic field magnitude, μ_0 is the magnetic permeability of vacuum, and ρ is the mass density of the plasma.)

The solar wind is threaded by magnetic field lines that map back to the Sun. A useful and picturesque description of the field contained within a plasma relies on the idea that if the conductivity of a plasma is sufficiently large, the magnetic field is frozen into the plasma and field lines can be traced from their source by following the motion of the plasma to which it is frozen. Because the roots of the field lines remain linked to the rotating Sun (the Sun rotates about its axis with a period of approximately 25 days), the field lines twist in the form of an Archimedean spiral as illustrated in Figure 7.2. In the direction of the Sun's motion relative to the interstellar plasma, the outflow is terminated by the forces exerted by the interstellar plasma. Elsewhere the flow is diverted within the boundary of the heliosphere. Thus, the Sun and the solar wind are (largely) confined within the heliospheric cavity; the heliosphere is the biggest of the solar system magnetospheres.

Our knowledge of the heliosphere beyond the orbits of the giant planets was for decades principally theoretical, but data acquired by *Voyager 1* and *2* since their last planetary encounters in 1989 have provided important evidence in situ of the structure of the outer heliosphere. The solar wind density continues to decrease as the inverse square of the distance from the Sun; as the plasma becomes sufficiently tenuous, the pressure of the interstellar plasma impedes its further expansion. The solar wind slows down abruptly across a shock (referred to as the termination shock) before reaching the **heliopause**, the boundary that separates the solar wind from the interstellar plasma. (The different plasma regimes are schematically illustrated in Figure 7.3.)

Voyager 1 encountered the termination shock on December 16, 2004, at a distance of 94 AU (AU is an

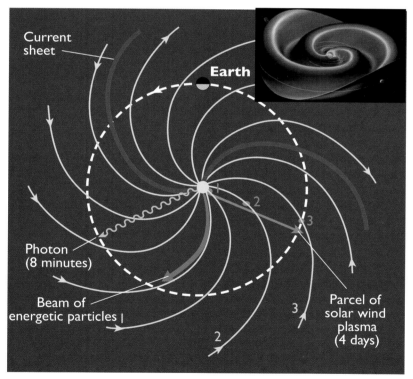

FIGURE 7.2 The magnetic field of the Sun is carried by the solar wind away from the Sun and winds into a spiral. The heliospheric current sheet (colored magenta in the inset three-dimensional diagram) separates magnetic fields of opposite polarities and is warped into a "ballerina skirt" by combined effects of the Sun's spin and the tilt of the magnetic field. The main diagram (two-dimensional projection) shows a cut through the inner heliosphere in the ecliptic plane (the plane of Earth's orbit); the radial flow of the solar wind and the rotation of the Sun combine to twist the solar magnetic field (yellow lines) into a spiral. A parcel of solar wind plasma (traveling radially at an average speed of 400 km/s) takes about 4 days to travel from the Sun to Earth's orbit at 1 AU. The dots and magnetic field lines labeled 1, 2, and 3 represent snapshots during this journey. Energetic particles emitted from the Sun travel much faster than the bulk solar wind and reach the Earth in minutes to hours. Traveling at the speed of light, solar photons reach the Earth in 8 min. *Credit: Van Allen and Bagenal (1999).*

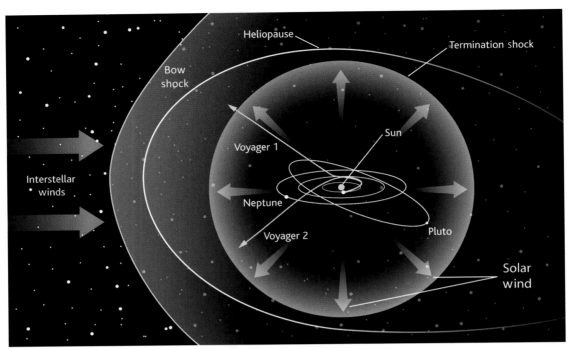

FIGURE 7.3 Schematic illustration of the heliosphere. The direction of plasma flow in the local interstellar medium relative to the Sun is indicated, and the boundary between solar wind plasma and interstellar plasma is identified as the heliopause. A broad internal shock, referred to as the termination shock, is shown within the heliopause. Such a shock, needed to slow the outflow of the supersonic solar wind inside of the heliopause, is a new feature in this type of magnetosphere. Beyond the heliopause, the interstellar flow is diverted around the heliosphere and a shock that slows and diverts flow may or may not exist. *Credit: Fisk, 2005.*

astronomical unit, equal to the mean radius of Earth's orbit or about 1.5×10^8 km) from the Sun and entered the heliosheath, the boundary layer between the termination shock and the heliopause. The encounter with the termination shock had long been anticipated as an opportunity to identify the processes that accelerate a distinct class of cosmic rays, referred to as anomalous cosmic rays (ACRs). ACRs are extremely energetic singly charged ions (energies of the order of 10 MeV/nucleon) produced by ionization of interstellar neutrals. The mechanism that accelerates them to high energy is not established. Some models propose that these particles are ionized and accelerated near the termination shock. Although the *Voyager* data show no sign of a change in the energy spectrum or the intensity of the flux across the termination shock the connection of ACRs to the shock itself may be nonlocal, which could reconcile the observations with the theory. However, at this time there is not full understanding of the mechanism that produces ACRs.

Various sorts of electromagnetic waves and plasma waves have been interpreted as coming from the termination shock or the heliopause. Bursts of radio emissions that do not weaken with distance from known sources within the solar systems were observed intermittently by *Voyager* between 1983 and 2004. They are thought to be emissions generated when an interplanetary shock propagating outward from the Sun reaches the heliopause. Plasma waves driven by electron beams generated at the termination shock and propagating inward along the spiral field lines of the solar wind were also identified. As *Voyager* continues its journey out of the solar system, it will encounter the heliopause and enter the interstellar plasma beyond. Although the schematic heliosphere of Figure 7.3 suggests that beyond the heliopause, there is a region of interstellar wind, there is increasing evidence that the upstream flow may be submagnetosonic, in which case no shock develops. With the Voyager spacecraft continuing to provide data, direct evidence of the properties of the local interstellar medium will be beamed back to earth.

2.2. Magnetospheres of the Unmagnetized Planets

Earth has a planetary magnetic field that has long been used as a guide by such travelers as scouts and sea voyagers. However, not all of the planets are magnetized. Table 7.1 summarizes some key properties of some of the planets including their surface magnetic field strengths. The planetary magnetic field of Mars is extremely small, and the

Chapter | 7 Planetary Magnetospheres

TABLE 7.1 Properties of the Solar Wind and Scales of Planetary Magnetospheres

	Mercury	Venus	Earth	Mars	Jupiter	Saturn	Uranus	Neptune	Pluto
Distance, a_{planet} (AU)[1]	0.31–0.47	0.723	1^2	1.524	5.2	9.5	19	30	30–50
Solar wind density (amu/cm^3)[2]	35–80	16	8	3.5	0.3	0.1	0.02	0.008	0.008–0.003
Radius, R_P (km)	2439	6051	6373	3390	71,398	60,330	25,559	24,764	1153
Surface magnetic field, $\mathbf{B}_0$ (nT)	195	–	30,600	–	430,000	21,400	22,800	13,200	Unknown
R_{MP} (R_{planet})[3]	1.4–1.6 R_M	–	10 R_E	–	46 R_J	20 R	25 R_U	24 R_N	
Observed size of magnetosphere (Rplanet)	1.5 R_M	–	8–12 R_E	–	63–93 R_J	22–27 R_S	18 R_U	23–26 R_N	Unknown
Observed size of magnetosphere (km)	3.6×10^3	–	7×10^4	–	7×10^6	1×10^6	5×10^5	6×10^5	

[1] $1\,AU = 1.5 \times 10^8$ km.
[2] The density of the solar wind fluctuates by about a factor of 5 about typical values of $\rho_{SW} \sim [(8\ amu/cm^3)/a_{planet}^2]$.
[3] Magnetopause nose distance, R_{MP} is calculated using $R_{MP} = (\mathbf{B}_0^2/2\mu_0\rho u^2)^{1/6}$ for typical solar wind conditions of ρ_{sw} given above and $u \sim 400$ km/s. For outer planet magnetospheres, this is usually an underestimate of the actual distance (Kivelson & Russell, 1995).

planetary magnetic field of Venus is nonexistent. (See Mars and Venus: Surface and Interior.) The nature of the interaction between an unmagnetized planet and the supersonic solar wind is determined principally by the electrical conductivity of the body. If conducting paths exist across the planet's interior or ionosphere, then electric currents flow through the body and into the solar wind where they create forces that slow and divert the incident flow. The diverted solar wind flows around a region that is similar to a planetary magnetosphere. Mars and Venus have ionospheres that provide the required conducting paths. The barrier that separates planetary plasma at these planets from solar wind plasma is referred to as an *ionopause* in analogy to the magnetopause of a magnetized planet. Earth's Moon, with no ionosphere and a very low-conductivity surface, does not deflect the bulk of the solar wind incident on it. Instead, the solar wind runs directly into the surface, where it is absorbed. (See The Moon.) The absorption leaves the region immediately downstream of the Moon in the flowing plasma (the wake) devoid of plasma, but the void fills in as solar wind plasma flows toward the center of the wake. The different types of interaction are illustrated in Figure 7.4.

The magnetic structure surrounding Mars and Venus has features much like those found in a true magnetosphere surrounding a strongly magnetized planet. This is because the interaction causes the magnetic field of the solar wind to drape around the planet. The draped field stretches out downstream (away from the Sun), forming a magnetotail. The symmetry of the magnetic configuration within such a tail is governed by the orientation of the magnetic field in the incident solar wind, and that orientation changes with time. For example, if the interplanetary magnetic field (IMF) is oriented northward, the east–west direction lies in the symmetry plane of the tail and the northern lobe field (see Figure 7.1 for the definition of lobe) points away from

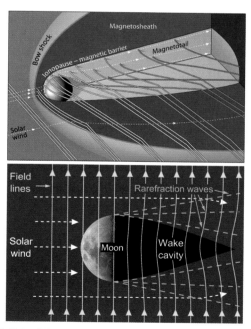

FIGURE 7.4 Schematic illustrations of the interaction regions surrounding, top, a planet like Mars or Venus, which is sufficiently conducting that currents close through the planet or its ionosphere (solar magnetic field lines are shown in yellow to red and are draped around and behind the planet) and, bottom, a body like the Moon, which has no ionosphere and low surface and interior conductivity. *Credit: Steve Bartlett.*

the Sun, while the southern lobe field points toward the Sun. A southward-oriented IMF would reverse these polarities, and other orientations would produce rotations of the symmetry axis.

Much attention has been paid to magnetic structures that form in and around the ionospheres of unmagnetized planets. Magnetic flux tubes of solar wind origin pile up at high altitudes at the dayside ionopause where, depending on the solar wind dynamic pressure, they may either remain for extended times, thus producing a magnetic barrier that diverts the incident solar wind, or penetrate to low altitudes in localized bundles. Such localized bundles of magnetic flux are often highly twisted structures stretched out along the direction of the magnetic field. Such structures, referred to as flux ropes, are illustrated in Figure 7.5.

Although, in the present epoch, Mars has only a small global scale magnetic field and interacts with the solar wind principally through currents that link to the ionosphere, there are portions of the surface over which local magnetic fields block the access of the solar wind to low altitudes. "Mini-magnetospheres" extending up to 1000 km form above the regions of intense crustal magnetization in the southern hemisphere; these mini-magnetospheres protect portions of the atmosphere from direct interaction with the solar wind. As a result, the crustal magnetization may have modified the evolution of the atmosphere and may still contribute to the energetics of the upper atmosphere.

2.3. Interactions of the Solar Wind with Asteroids, Comets, and Pluto

Asteroids are small bodies (<1000 km radius and more often only tens of kilometers) whose signatures in the solar wind were first observed by the *Galileo* spacecraft in the early 1990s. (See Main-Belt Asteroids.)

Asteroid-related disturbances are closely confined to the regions near to and downstream of the magnetic field lines that pass through the body, and thus the interaction region is fan shaped as illustrated in Figure 7.6 rather than bullet shaped like Earth's magnetosphere and there is no shock standing upstream of the disturbance in the solar wind. The signature found by *Galileo* in the vicinity of the asteroid Gaspra suggested that the asteroid is magnetized at a level similar to the magnetization of meteorites. Because the measurement locations were remote from the body, its field was not measured directly, and it is possible that the putative magnetic signature was a fortuitous rotation of the IMF. Data from other asteroids do not establish unambiguously the strength of their magnetic fields. A negligibly small magnetic field was measured by the

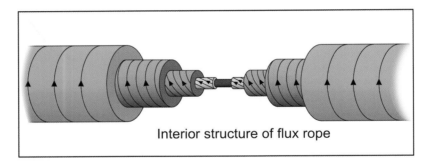

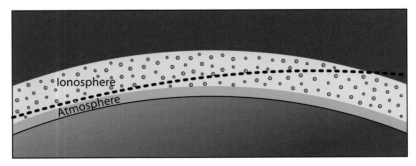

FIGURE 7.5 Schematic illustration of a flux rope, a magnetic structure that has been identified in the ionosphere of Venus (shown as black dots within the ionosphere) and extensively investigated (a low-altitude pass of the *Pioneer Venus Orbiter* is indicated by the dashed curve). The rope (see earlier) has an axis aligned with the direction of the central field. Radially away from the center, the field wraps around the axis, its helicity increasing with radial distance from the axis of the rope. Structures of this sort are also found in the solar corona, near the magnetopause, and in the magnetotails of magnetized planets. *Credit: Steve Bartlett.*

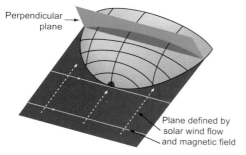

FIGURE 7.6 Schematic of the shape of the interaction between an asteroid and the flowing solar wind. The disturbance spreads out along the direction of the magnetic field downstream of the asteroid. The disturbed region is thus fan shaped, with greatest spread in the plane defined by the solar wind velocity and the solar wind magnetic field. The curves bounding the intersection of that plane with the surface and with a perpendicular plane are shown. *Credit: Steve Bartlett.*

NEAR-Shoemaker mission close to and on the surface of asteroid Eros, possibly because it is formed of magnetized rocks of random orientation. The *DAWN* spacecraft to Ceres and Vesta unfortunately lacks instrumentation to establish the magnetic properties of these two large asteroids. Other missions under discussion would add to our knowledge of asteroid magnetic properties. In the future we expect to have better determinations of asteroidal magnetic fields and to be able to establish how they interact with the solar wind.

Comets are also small bodies. The spectacular appearance of an active comet, which can produce a glow over a large visual field extending millions of kilometers in space on its approach to the Sun, is somewhat misleading because comet nuclei are no more than tens of kilometers in diameter. It is the gas and dust released from these small bodies by solar heating that we see spread out across the sky. Some of the gas released by the comet remains electrically neutral, with its motion governed by purely mechanical laws, but some of the neutral matter becomes ionized either by photoionization or by exchanging charge with ions of the solar wind. The newly ionized cometary material is organized in interesting ways that have been revealed by spacecraft measurements in the near neighborhood of comets Halley, Giacobini-Zinner, Borrelly, and others. Figure 7.7 shows schematically the types of regions that have been identified, illustrating clearly that the different gaseous regions fill volumes of space many orders of magnitude larger than the actual solid comet. The solar wind approaching the comet first encounters the expanding neutral gases blown off the comet. As the neutrals are ionized by solar photons, they extract momentum from the solar wind, and the flow slows a bit. Passing through a shock that further decelerates the flow, the solar wind encounters ever-increasing densities of newly ionized gas of cometary origin, referred to as pickup ions. Energy is extracted from

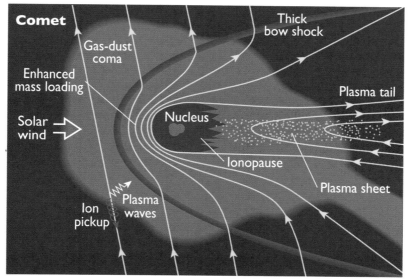

FIGURE 7.7 Schematic illustration of the magnetic field and plasma properties in the neighborhood of a comet. The length scale is logarithmic. The nucleus is surrounded by a region of dense plasma into which the solar wind does not penetrate. This region is bounded by a contact surface. Above that lies an ionopause or cometopause bounding a region in which ions of cometary origin dominate. Above this, there is a transition region in which the solar wind has been modified by the addition of cometary ions. As ions are added, they must be accelerated to become part of the flow. The momentum to accelerate the picked-up ions is extracted from the solar wind; consequently, in the transition region, the density is higher and the flow speed is lower than in the unperturbed solar wind. The newly picked up ions often generate plasma waves. The region filled with cometary material imposes the large-scale size on the visually observable signature of a comet. Similar to Venus-like planets, the solar wind magnetic field folds around the ionopause, producing a magnetic tail that organizes the ionized plasma in the direction radially away from the Sun and produces a distinct comet tail with a visual signature. The orientation of the magnetic field in the tail is governed by the solar wind field incident on the comet, and it changes as the solar wind field changes direction. Dramatic changes in the structure of the magnetic tail are observed when the solar wind field reverses direction. *Credit: Van Allen and Bagenal (1999).*

the solar wind as the pickup ions are swept up, and the flow slows further. Still closer to the comet, in a region referred to as the cometopause, a transition in composition occurs as the pickup ions of cometary origin begin to dominate the plasma composition. Close to the comet, at the **contact surface**, ions flowing away from the comet carry enough momentum to stop the flow of the incident solar wind. Significant asymmetry of the plasma distribution in the vicinity of a comet may arise if strong collimated jets of gas are emitted by the cometary nucleus. Such jets have been observed at Halley's comet and at comet Borrelly.

Pluto is also a small body that was classified as a planet until 2006 and more recently reclassified as a dwarf planet. Pluto's interaction with the solar wind has not yet been observed, but it is worth speculating about what that interaction will be like to test our understanding of comparative planetology. (See Pluto.) The solar wind becomes tenuous and easily perturbed at large distances from the Sun (near 30 AU), and either escaping gases or a weak internal magnetic field could produce an interaction region many times Pluto's size. At some phases of its 248-year orbital period, Pluto moves close enough to the Sun for its surface ice to sublimate, producing an atmosphere and possibly an ionosphere. Models of Pluto's atmosphere suggest that the gases would then escape and flow away from the planet. If the escape flux is high, the solar wind interaction would then appear more like that at a comet than like that at Venus or Mars. Simulations show a very asymmetric shock surrounding the interaction region for a small but possible neutral escape rate. Pluto's moon, Charon, may serve as a plasma source within the magnetosphere, and this could have interesting consequences of the type addressed in Section 6 in relation to the moons of Jupiter and Saturn. As is the case for small asteroids and comets, ions picked up in the solar wind at Pluto have **gyroradii** and ion inertial lengths that are large compared with the size of the obstacle, a situation that adds asymmetry and additional complexity to the interaction. For most of its orbital period, Pluto is so far from the Sun that its atmosphere disappears and its interaction with the solar wind is more likely to resemble that of the Moon, with absorption occurring at the sunward surface and a void developing in its wake. It seems unlikely that a small icy body will have an internal magnetic field large enough to produce a magnetospheric interaction region, but one must recognize that actual observations of the magnetic fields of small bodies have repeatedly challenged our ideas about magnetic field generation.

2.4. Magnetospheres of Magnetized Planets

In a true magnetosphere, the scale size is set by the distance, R_{MP}, along the planet–Sun line at which the sum of the pressure of the planetary magnetic field and the pressure exerted by plasma confined within that field balance the dynamic pressure of the solar wind. (The dynamic pressure is ρu^2 where ρ is the mass density and u is its flow velocity in the rest frame of the planet. The thermal and magnetic pressures of the solar wind are small compared with its dynamic pressure.) Assuming that the planetary magnetic field is dominated by its dipole moment and that the plasma pressure within the magnetosphere is small, one can estimate R_{MP} as $R_{MP} \approx R_P (B_0^2 / 2\mu_0 \rho u^2)^{1/6}$. Here B_0 is the surface equatorial field of the planet and R_P is its radius. Table 7.1 gives the size of the magnetosphere, R_{MP}, calculated from its internal field and observed for the different planets and shows the vast range of scale sizes both in terms of the planetary radii and of absolute distance.

Within a magnetosphere, the magnetic field differs greatly from what it would be if the planet were placed in a vacuum. The field is distorted, as illustrated in Figure 7.1, by currents carried on the magnetopause and in the plasma trapped within the magnetosphere. Properties of the trapped plasma and its sources are discussed in Section 4. An important source of magnetospheric plasma is the solar wind. Figure 7.1 makes it clear that, along most of the boundary, solar wind plasma would have to move across magnetic field lines to enter the magnetosphere. The Lorentz force of the magnetic field opposes such motion. However, in the polar cusp shocked solar wind plasma of the magnetosheath easily penetrates the boundary by moving along the field. Other processes that enable solar wind plasma to penetrate the boundary are discussed in Section 5.

3. PLANETARY MAGNETIC FIELDS

Because the characteristic timescale for **thermal diffusion** is greater than the age of the solar system, the planets tend to have retained their heat of formation. At the same time, the characteristic timescale for diffusive decay of a magnetic field in a planetary interior is much less than the age of the planets. Consequently, primordial fields and permanent magnetism on a planetary scale are small and the only means of providing a substantial planetary magnetic field is an internal dynamo (see Magnetic Field Generation).

For a planet to have a magnetic dynamo, it must have a large region that is fluid, electrically conducting, and undergoing convective motion. The deep interiors of the planets and many larger satellites are expected to contain electrically conducting fluids: terrestrial planets and the larger satellites have differentiated cores of liquid iron alloys; at the high pressures in the interiors of the giant planets Jupiter and Saturn, hydrogen behaves like a liquid metal; for Uranus and Neptune, a water–ammonia–methane mixture forms a deep conducting "ocean". (See Interiors of the Giant Planets.) The fact that some planets and satellites do not have dynamos tells us that their interiors are stably stratified and do not convect or that the interiors have solidified. Models of the thermal evolution of terrestrial planets show that as the object cools, the liquid core ceases to convect, and further heat is lost by conduction alone. In some cases, such as the Earth, convection continues because the nearly pure iron solidifies out of the alloy in the outer core, producing an inner solid core and creating compositional gradients that drive convection in the liquid outer core. The more gradual cooling of the giant planets also allows convective motions to persist.

Of the eight planets, six are known to generate magnetic fields in their interiors. Exploration of Venus has provided an upper limit to the degree of magnetization comparable with the crustal magnetization of the Earth, suggesting that its core is stably stratified and that it does not have an active dynamo. The question of whether Mars does or does not have a weak internal magnetic field was disputed for many years because spacecraft magnetometers had measured the field only far above the planet's surface. The first low-altitude magnetic field measurements were made by *Mars Global Surveyor* in 1997. It is now known that the surface magnetic field of Mars is very small ($|\mathbf{B}| < 10$ nT or 1/3000 of Earth's equatorial surface field) over most of the northern hemisphere but that in the southern hemisphere there are extensive regions of intense crustal magnetization as already noted. Pluto has yet to be explored. Models of Pluto's interior suggest that it is probably differentiated, but its small size makes one doubt that its core is convecting and any magnetization is likely to be remanent. Earth's moon has a negligibly small planet-scale magnetic field, although localized regions of the surface are highly magnetized. Jupiter's large moons are discussed in Section 6.

The characteristics of the six known planetary fields are listed in Table 7.2. Assuming that each planet's magnetic field has the simplest structure, a dipole, we can characterize the magnetic properties by noting the equatorial field strength ($\mathbf{B}_0$) and the tilt of the axis with respect to the planet's spin axis. For all the magnetized planets other than Mercury, the surface fields are on the order of a Gauss $= 10^{-4}$ T, meaning that their dipole moments are of order $R_P^3 10^{-4} T$, where R_P is the planetary radius (i.e. the dipole moments scale with planetary size).

The degree to which the dipole model is an oversimplification of more complex structure is indicated by the ratio of maximum to minimum values of the surface field. This ratio has a value of 2 for a dipole. The larger values, particularly for Uranus and Neptune, are indications of strong nondipolar contributions to the planets' magnetic fields. Similarly, the fact that the magnetic axes of these two planets are strongly tilted (see Figure 7.8) also suggests that the dynamos in the icy giant planets may be significantly different than those of the planets with aligned, dipolar planetary magnetic fields.

The size of a planet's magnetosphere (R_{MP}) depends not only on the planet's radius and magnetic field but also on the ambient solar wind density, which decreases as the inverse square of the distance from the Sun. (The solar wind speed is approximately constant with distance from the Sun.) Thus, it is not only planets with strong magnetic fields that have large magnetospheres but also the planets Uranus and Neptune whose weak magnetic fields create moderately large magnetospheres because of the weak force exerted by the tenuous solar wind far from the Sun. Table 7.1 shows that the measured sizes of planetary magnetospheres generally agree quite well with the theoretical R_{MP} values. Jupiter, for which the plasma pressure inside the magnetosphere is sufficient to further "inflate" the magnetosphere, is the only notable exception. Jupiter's strong internal field combines with the relatively low solar wind

TABLE 7.2 Planetary Magnetic Fields

	Mercury	Earth	Jupiter	Saturn	Uranus	Neptune
Magnetic moment, (M_{Earth})	4×10^{-4}	1[1]	20,000	600	50	25
Surface magnetic field at dipole equator (nT)	195	30,600	430,000	21,400	22,800	14,200
Maximum/minimum[2]	2	2.8	4.5	4.6	12	9
Dipole tilt and sense[3]	<3° S	9.92° S	9.6° N	0.0° N	59° N	47° N
Obliquity[4]	0.2°	23.5°	3.1°	26.7°	97.9°	29.6°
Solar wind angle[5]	90°	67–114°	87–93°	64–117°	8–172°	60–120°

[1] $M_{Earth} = 7.906 \times 10^{25}$ Gauss cm^3 = 7.906×10^{15} T m^3.
[2] Ratio of maximum surface field to minimum (equal to 2 for a centered dipole field).
[3] Angle between the magnetic axis and rotation pole (S or N).
[4] The inclination of the equator to the orbit.
[5] Range of angle between the radial direction from the Sun and the planet's rotation axis over an orbital period.

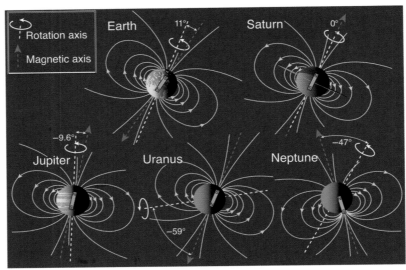

FIGURE 7.8 Orientation of the planets' spin axes (dashed white lines) and their magnetic fields (magnetic field lines shown in yellow, axes in red) with respect to the ecliptic plane (horizontal). The larger the angle between these two axes, the greater the magnetospheric variability over the planet's rotation period. The variation in the angle between the direction of the solar wind (close to radial from the Sun) and a planet's spin axis over an orbital period is an indication of the degree of seasonal variability. *Credit: Steve Bartlett.*

density at 5 AU to make the magnetosphere of Jupiter a huge object—about 1000 times the volume of the Sun, with a tail that extends at least 6 A.U. in the antisunward direction, beyond the orbit of Saturn. If the jovian magnetosphere were visible from Earth, its angular size would be much larger than the size of the Sun, even though it is at least four times farther away. The magnetospheres of the other giant planets are smaller (although large compared with the Earth's magnetosphere), having scales of about 20 times the planetary radius, comparable with the size of the Sun. Mercury's magnetosphere is extremely small because the planet's magnetic field is weak and the solar wind close to the Sun is very dense. Figure 7.9 compares the sizes of several planetary magnetospheres.

Although the size of a planetary magnetosphere depends on the strength of a planet's magnetic field, the configuration and internal dynamics depend on the field orientation (illustrated in Figure 7.8). At a fixed phase of planetary rotation, such as when the dipole tilts toward the Sun, the orientation of a planet's magnetic field is described by two angles (tabulated in Table 7.2): the tilt of the magnetic field with respect to the planet's spin axis and the angle between the planet's spin axis and the solar wind direction, which is generally within a few degrees of being radially outward from the Sun. Because the direction of the spin axis with respect to the solar wind direction varies only over a planetary year (many Earth years for the outer planets), and the planet's magnetic field is assumed to vary only on geological timescales, these two angles are constant for the purposes of describing the magnetospheric configuration at a particular epoch. Earth, Jupiter and Saturn have small dipole tilts and relatively small obliquities. This means that changes of the orientation of the magnetic field with respect to the solar wind over a planetary rotation period and seasonal variations, though detectable, produce only subtle magnetospheric effects. Thus, Mercury, Earth, Jupiter, and Saturn have reasonably symmetric, quasi-stationary magnetospheres, with the first three exhibiting a wobble at the planetary rotation period owing to their ∼10° dipole tilts. In contrast, the large dipole tilt angles of Uranus and Neptune imply that the orientation of their magnetic fields with respect to the interplanetary flow direction varies greatly over a planetary rotation period, resulting in highly asymmetric magnetospheres that vary at the period of planetary rotation. Furthermore, Uranus' large obliquity means that its magnetosphere undergoes strong seasonal changes of its global configuration over its 84-year orbital period.

4. MAGNETOSPHERIC PLASMAS

4.1. Sources of Magnetospheric Plasmas

Magnetospheres contain considerable amounts of plasma whose sources are both internal and external (see Table 7.3). The main source of plasma in the solar system is the Sun. The solar corona, the upper atmosphere of the Sun (which has been heated to temperatures of 1−2 million Kelvin), streams away from the Sun at a more or less steady rate of 10^9 kg/s carrying approximately equal numbers (8×10^{35} s^{-1}) of electrons and ions. The boundary between the solar wind and a planet's magnetosphere, the magnetopause, is not entirely plasma tight. Wherever the IMF

Chapter | 7 Planetary Magnetospheres

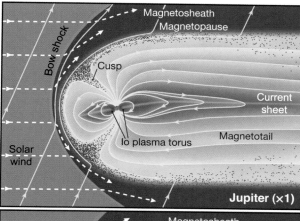
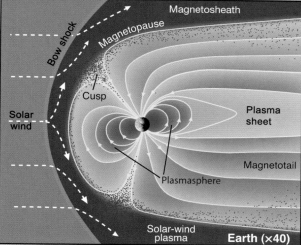
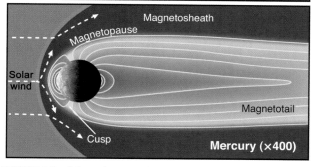

FIGURE 7.9 Schematic comparison of the magnetospheres of Jupiter, Earth, and Mercury. Relative to the Jupiter schematic, the one for Earth is blown up by a factor of 40, and the one for Mercury is blown up by a factor of 400. The planetary radii are given in Table 7.1. *Credit: Steve Bartlett.*

A secondary source of plasma is the ionosphere. Although ionospheric plasma is generally cold and gravitationally bound to the planet, a small fraction of particles can acquire sufficient energy to escape up magnetic field lines and into the magnetosphere. In some cases, field-aligned potential drops accelerate ionospheric ions and increase the escape rate. Ionospheric plasma has a composition that reflects the composition of the planet's atmosphere (e.g., abundant O+ for the Earth and H+ for the outer planets).

The sulfur and oxygen ions that dominate Jupiter's magnetospheric plasma are formed by breakdown and ionization of sulfur and sulfur dioxide whose source is the volcanoes of Io. Energetic ions impacting Io's surface or atmosphere can sputter off ions of lower energy through a direct interaction and can also create an extensive cloud of neutral atoms that are subsequently ionized, possibly far from the satellite. Although the sputtering process, which removes at most a few microns of surface ice per thousand years, is probably insignificant in geological terms, sputtering has important consequences for the optical properties of satellite or ring surfaces.

Water-product ions are ubiquitous in Saturn's magnetosphere. Initially it was thought that these ions were formed from neutrals sputtered from ring particles and from the surfaces of icy satellites, but it was hard to account for the observed densities. The dominant source of the water group ions was established only after measurements and images from the *Cassini* orbiter revealed that plumes or geysers spouting from the surface of the tiny moon Enceladus are the dominant source of the water group ions at Saturn.

In the hot tenuous plasmas of planetary magnetospheres, collisions between particles are very rare. By contrast, in the cold dense plasmas of a planet's ionosphere, collision rates are high enough to allow ionospheric plasmas to conduct currents. Cold, dense, collision-dominated plasmas are expected to be in thermal equilibrium, but such equilibrium was not originally expected for the hot, tenuous collisionless plasmas of the magnetosphere. Surprisingly, even hot tenuous plasmas in space are generally found not far from equilibrium (i.e. their particle distribution functions are observed to be approximately *Maxwellian*, although the ion and electron populations often have different temperatures). This fact is remarkable because some of the source mechanisms tend to produce particles whose initial energies fall in a very narrow range and timescales for equilibration by means of *Coulomb collisions* are usually much longer than transport timescales. A distribution close to Maxwellian is achieved by interaction with waves in the plasma. Space plasmas support many different types of plasma waves that can grow in amplitude when free energy is present in the form of non-Maxwellian energy distributions, unstable spatial

has a component antiparallel to the planetary magnetic field near the magnetopause boundary, magnetic *reconnection* (discussed in Section 5) is likely to occur, and solar wind plasma can enter the magnetosphere across the magnetopause. Solar wind material is identified in the magnetosphere by its energy and characteristic composition of protons (H^+) with $\sim 4\%$ alpha particles (He^{2+}) and trace heavy ions, many of which are highly ionized.

TABLE 7.3 Plasma Characteristics of Planetary Magnetospheres

	Mercury	Earth	Jupiter	Saturn	Uranus	Neptune
Maximum density (cm^{-3})	~1	1–4000	>3000	~100	3	2
Composition	H^+	O^+, H^+	O^{n+}, S^{n+}	O^+, OH^+, H_2O^+, H^+	H^+	N^+, H^+
Dominant source	Solar wind	Ionosphere[1]	Io	Enceladus	Atmosphere	Triton
Strength (ions/s)	~10^{26}	2×10^{26}	$1-4 \times 10^{28}$	$1-3 \times 10^{27}$	10^{25}	10^{25}
(kg/s)	~5	5	260–1400	30–80	0.02	0.2
Lifetime	Minutes	Days[1] Hours[2]	10–100 days	30 days	1–30 days	~1 day
Plasma motion	Solar wind driven	Rotation[1] Solar wind[2]	Rotation	Rotation	Solar wind + rotation	Rotation (+solar wind?)

[1]Inside plasmasphere.
[2]Outside plasmasphere.
Based on Bagenal (2009) and Kivelson (2006)

distributions, or anisotropic velocity–space distributions of newly created ions. Interactions between plasma waves and particle populations not only bring the bulk of the plasma toward thermal equilibrium but also accelerate or scatter suprathermal particles.

Plasma detectors mounted on spacecraft can provide detailed information about the particles' velocity distribution, from which bulk parameters such as density, temperature, and flow velocity are derived, but plasma properties are determined only in the vicinity of the spacecraft. Data from planetary magnetospheres other than Earth's are limited in duration and spatial coverage, so there are considerable gaps in our knowledge of the changing properties of the many different plasmas in the solar system. Some of the most interesting space plasmas, however, can be remotely monitored by observing emissions of electromagnetic radiation. Dense plasmas, such as Jupiter's plasma torus, comet tails, Venus's ionosphere, and the solar corona, can radiate collisionally excited line emissions at optical or UV wavelengths. Radiative processes, particularly at UV wavelengths, can be significant sinks of plasma energy. Figure 7.10 shows an image of optical emission from the plasma that forms a ring or torus deep within Jupiter's magnetosphere near the orbit of its moon, Io (see Section 6). Observations of these emissions give compelling evidence of the temporal and spatial variability of the Io plasma torus. Similarly, when magnetospheric particles bombard the planets' polar atmospheres, various auroral emissions are generated from radio to X-ray wavelengths and these emissions can also be used for remote monitoring of the system. (See Atmospheres of the Giant Planets.) Thus, our knowledge of space plasmas is based on combining the remote sensing of plasma phenomena with available spacecraft measurements that provide "ground truth" details of the particles' velocity distribution and of the local electric and magnetic fields that interact with the plasma.

4.2. Energetic Particles

Significant populations of particles at keV–MeV energies, well above the energy of the thermal population, are found in all magnetospheres. The energetic particles are largely trapped in long-lived radiation belts (summarized in Table 7.4) by the strong planetary magnetic field. Where do these energetic particles come from? Since the interplanetary medium contains energetic particles of solar and galactic origins an obvious possibility is that these energetic particles are "captured" from the external medium. In most cases, the observed high fluxes are hard to explain without identifying additional internal sources. Compositional evidence supports the view that some fraction of the thermal plasma is accelerated to high energies, either by tapping the rotational energy of the planet, in the cases of Jupiter and Saturn, or by acceleration in the distorted and dynamic magnetic field in the magnetotails of Earth, Uranus, and Neptune. If the energy density of the energetic particle populations is comparable with the magnetic field energy density, currents develop that significantly modify the planetary magnetic field. Table 7.4 shows that this occurs at Jupiter and Saturn, where the high particle pressures inflate and stretch out the magnetic field and generate a strong azimuthal current in the magnetodisc. Even though

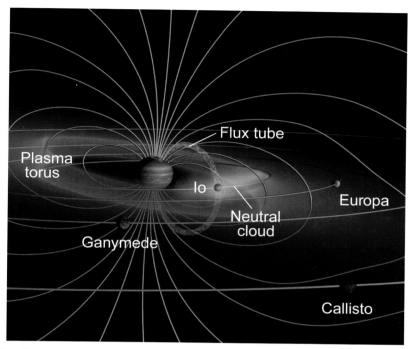

FIGURE 7.10 The ionization of an extended atmosphere of neutral atoms (yellow) around Jupiter's moon Io is a strong source of plasma, which extends around Jupiter in a plasma torus (red). Electrical currents generated in the interaction of Io with the surrounding plasma couple the moon to Jupiter's atmosphere where they stimulate auroral emissions. The main ring of auroral emissions is associated with currents generated as the plasma from the Io torus spreads out into the vast, rotating magnetosphere of Jupiter. *Credit: John Spencer.*

Uranus and Neptune have significant radiation belts, the energy density of particles remains small compared with the magnetic energy density and the azimuthal current is very weak. In Earth's magnetosphere, the azimuthal current, referred to as the *ring current*, is extremely variable, as discussed in Section 5. Relating the magnetic field produced by the azimuthal current to the kinetic energy of the trapped particle population (scaled to the dipole magnetic energy external to the planet), we find that even though the total energy content of magnetospheres varies by many orders of magnitude and the sources are very different, the net particle energy builds up to only 1/1000 of the magnetic field energy in each magnetosphere. Earth, Jupiter, and Saturn all have energetic particle populations close to this limit. The energy in the radiation belts of Uranus and Neptune is much below this limit, perhaps because it is harder to trap particles in nondipolar magnetic fields.

We have commented on ways in which particles gain energy. Where do these energetic particles go? Most appear to diffuse inward toward the planet. Loss processes for energetic particles in the inner magnetospheres of the terrestrial planets arise largely from inward diffusion to low

TABLE 7.4 Energetic Particle Characteristics in Planetary Magnetospheres

	Earth	Jupiter	Saturn	Uranus	Neptune
Phase space density[1]	20,000	200,000	60,000	800	800
Plasma beta[2]	<1	>1	>1	~0.1	~0.2
Ring current, ΔB (nT)[3]	10–200	200	10	<1	<0.1
Auroral power (W)	10^{10}	10^{12}	10^{11}	5×10^9	$2-8 \times 10^7$

[1] The phase space density of energetic particles (in this case 100 MeV/Gauss ions) is measured in units of $(cm^2 \, s \, sr \, MeV)^{-1}$ and is listed near its maximum value.
[2] The ratio of the particle pressure to the magnetic pressure of a plasma, $nkT/(\mathbf{B}^2/2\mu_o)$. These values are typical for the body of the magnetosphere. Higher values are often found in the tail plasma sheet and, in the case of the Earth, at times of enhanced ring current.
[3] The magnetic field produced at the surface of the planet due to the ring current of energetic particles in the planet's magnetosphere.

altitudes or charge exchange with neutral clouds, and scattering by waves so that the particles stream into the upper atmospheres of the planets. At Jupiter and Saturn, energetic particles can be lost through absorption by rings and satellites. Some energetic particles precipitate into the atmosphere at high latitudes where they excite auroral emission and deposit large amounts of energy, at times exceeding the local energy input from the Sun. The presence of high fluxes of energetic ions and electrons of the radiation belts must be taken into account in designing and operating spacecraft. At Earth, relativistic electron fluxes build to extremely high levels during magnetically active times referred to as storm times. High fluxes of relativistic electrons affect sensitive electronic systems and have caused anomalies in the operation of spacecraft. At Earth damaging levels of relativistic electrons occur intermittently but in the inner part of Jupiter's magnetosphere, such high fluxes are always present. Proposed missions to Jupiter's moon Europa must be designed with attention to the fact that the energetic particle radiation near Europa's orbit is punishingly intense.

5. DYNAMICS

Magnetospheres are ever-changing systems. Changes in the solar wind, in plasma source rates, and in energetic cosmic ray fluxes can couple energy, momentum, and additional particle mass into the magnetosphere and thus drive magnetospheric dynamics. Sometimes the magnetospheric response is direct and immediate. For example, an increase of the solar wind dynamic pressure compresses the magnetosphere. Both the energy and the pressure of field and particles then increase even if no particles have entered the system. Sometimes the change in both field and plasma properties is gradual, similar to a spring being slowly stretched. Sometimes, as for a spring stretched beyond its breaking point, the magnetosphere responds in a very nonlinear manner, with both field and plasma experiencing large-scale abrupt changes. These changes can be identified readily in records of magnetometers (a magnetometer is an instrument that measures the magnitude and direction of the magnetic field), in scattering of radio waves by the ionosphere or emissions of such waves from the ionosphere, and in the magnetic field configuration, plasma conditions and flows, and energetic particle fluxes measured by a spacecraft moving through the magnetosphere itself.

Auroral activity is the most dramatic signature of magnetospheric dynamics and it is observed on distant planets as well as on Earth. Accounts of the terrestrial aurora (the lights flickering in the night sky that inspired fear and awe) date to ancient days, but the oldest scientific records of magnetospheric dynamics are measurements of fluctuating magnetic fields at the surface of the Earth. Consequently, the term *geomagnetic activity* is used to refer to magnetospheric dynamics of all sorts. Fluctuating magnetic signatures with timescales from seconds to days are typical. For example, periodic fluctuations at frequencies between ~ 1 mHz and ~ 1 Hz are called magnetic pulsations. In addition, impulsive decreases in the horizontal north–south component of the surface magnetic field (referred to as the H-component) with timescales of tens of minutes occur intermittently at latitudes between 65° and 75° often several times a day. The field returns to its previous value typically in a few hours. These events are referred to as *substorms*. A signature of a substorm at a $\sim 70°$ latitude magnetic observatory is shown in Figure 7.11. The H-component decreases by hundreds to 1000 nT (the Earth's surface field is 31,000 nT near the equator). Weaker signatures can be identified at lower and higher latitudes. Associated with the magnetic signatures and the current systems that produce them are other manifestations of magnetospheric activity including particle precipitation and auroral activation in the polar region and changes within the magnetosphere previously noted.

The auroral activity associated with a substorm can be monitored from above by imagers on spacecraft at high polar altitudes. The dramatic intensification of the brightness of the aurora as well as its changing spatial extent can thereby be accurately determined. Figure 7.12 shows an image of the aurora taken by the Far Ultraviolet Imaging System on the IMAGE spacecraft on July 15, 2000. Note that the intense brightness is localized in a high-latitude band surrounding the polar regions. This region of auroral activity is referred to as the auroral oval. Only during very

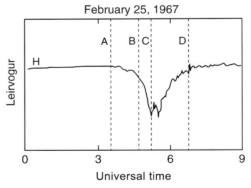

FIGURE 7.11 The variation of the H component of the surface magnetic field of the Earth at an auroral zone station at 70° magnetic latitude plotted versus universal time in hours during a 9-h interval that includes a substorm. Perturbations in H typically range from 50 to 200 nT during geomagnetic storms. Vertical lines mark: (A) The beginning of the growth phase during which the magnetosphere extracts energy from the solar wind, and the electrical currents across the magnetotail grow stronger. (B) The start of the substorm expansion phase during which currents from the magnetosphere are diverted into the auroral zone ionosphere and act to release part of the energy stored during the growth phase. Simultaneously, plasma is ejected down the tail to return to the solar wind. (C) The end of the substorm onset phase and the beginning of the recovery phase during which the magnetosphere returns to a stable configuration. (D) The end of the recovery phase.

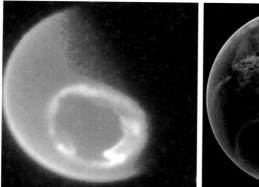
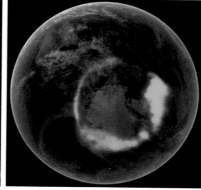

FIGURE 7.12 (Left) The image shows Earth's aurora observed with the Far Ultraviolet Imaging System on the *IMAGE* spacecraft during a major geomagnetic storm that occurred on July 15, 2000. The picture was obtained when the *IMAGE* spacecraft was at a distance of 7.9 Earth radii, and was looking down onto the northern polar region. The Sun is to the left. The auroral emissions are from molecular nitrogen that is excited by precipitating electrons. (Photo credit: S. Mende and H. Frey, University of California, Berkeley.) (Right) An ultraviolet image of aurora overlaid on an NASA visible image of the Earth. The aurora occurred during a strong geomagnetic storm on September 11, 2005.

intense substorms does the auroral region move far enough equatorward to be visible over most of the United States.

The intensity of substorms and other geomagnetic activity is governed to some extent by the speed of the solar wind but of critical importance is the orientation of the magnetic field embedded in the solar wind incident on a magnetized planet. The fundamental role of the magnetic field in the solar wind may seem puzzling. It is the orientation of the IMF that is critical, and at Earth substorm activity normally occurs following an interval during which the interplanetary field has been tilted southward. The issue is subtle. Magnetized plasma flowing through space is frozen to the magnetic field. The high conductivity of the plasma prevents the magnetic field from diffusing through the plasma, and, in turn, the plasma particles are bound to the magnetic field by a "$v \times \mathbf{B}$" Lorentz force that causes the particles to spiral around a field line. How, then, can a plasma ion or electron move from a solar wind magnetic field line to a magnetospheric field line?

The coupling arises through a process called reconnection, which occurs when plasmas bound on flux tubes with oppositely directed fields approach each other sufficiently closely. The weak net field at the interface may be too small to keep the plasma bound on its original flux tube and the field connectivity can change. Newly linked field lines will be bent at the reconnection location. The curvature force at the bend accelerates plasma away from the reconnection site. At the dayside magnetopause, for example, solar wind magnetic flux tubes and magnetospheric flux tubes can reconnect in a way that extracts energy from the solar wind and allows solar wind plasma to penetrate the magnetopause. A diagram first drawn in a French café by J. W. Dungey in 1961 (and reproduced frequently thereafter) provides the framework for understanding the role of magnetic reconnection in magnetospheric dynamics (Figure 7.13). Shown in the diagram on the top are southward-oriented solar wind field lines approaching the dayside magnetopause. Just at the nose of the magnetosphere, the northern ends of the solar wind field lines break their connection with the southern ends, linking instead with magnetospheric fields. Accelerated flows develop near the reconnection site. The reconnected field lines are dragged tailward by their ends within the solar wind, thus forming the tail lobes. When the magnetic field of the solar wind points strongly northward at Jupiter or Saturn, reconnection is also thought to occur at the low-latitude dayside magnetopause, but the full process has not yet been documented by observations, although there is some evidence that auroral displays intensify at Jupiter as at Earth when magnetopause reconnection is occurring. At Earth, if the reconnection is persistent, disturbances intensify. Energetic particle fluxes increase and significant fluxes of energetic particles may appear even at low latitudes and the ring current (see Section 4.3) intensifies. If dayside reconnection occurs at Earth, the solar wind transports magnetic flux from the dayside to the nightside. The path of the foot of the flux tube crosses the center of the polar cap, starting at the polar edge of the dayside auroral zone and moving to the polar edge of the nightside auroral zone as shown schematically in Figure 7.13(a). Ultimately that flux must return, and the process is also shown, both in the magnetotail where reconnection is shown closing a flux tube that had earlier been opened on the dayside and in the polar cap (Figure 7.13(b)) where the path of the foot of the flux tube appears at latitudes below the auroral zone, carrying the flux back to the dayside. In the early stage of a substorm (between A and B in Figure 7.11), the rate at which magnetic flux is transported to the nightside is greater than the rate at

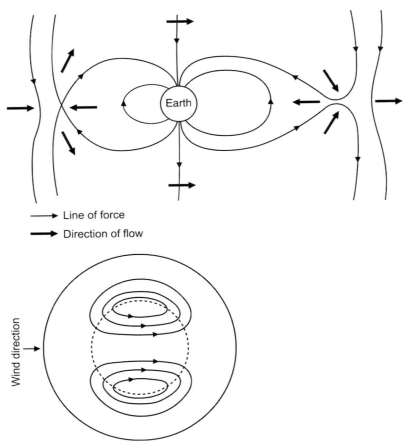

FIGURE 7.13 Adapted from the schematic view of reconnection sketched by J. W. Dungey in 1961. (a) A noon–midnight cut through the magnetosphere showing from left to right, in addition to two dipole-like field lines (rooted at two ends in the Earth): a solar wind field line with plasma flowing earthward; a newly reconnected pair of field lines, one of solar wind origin and one dipole-like field line, with plasma flowing toward the reconnection point from two sides near the midplane and accelerated both north and south away from the reconnection point; two reconnected field lines with one end in the solar wind and one end in the Earth flowing over the polar caps; two field lines about to reconnect in the magnetotail carried by plasma flow toward the midplane of the diagram; and a newly reconnected field line moving further away from the Earth in the solar wind. (b) A view down on the northern polar cap showing flow lines moving from day to night near the center, above the auroral zone, and returning to the dayside at latitudes below the auroral zone.

which it is returned to the dayside. This builds up stress in the tail, reducing the size of the region within the tail where the magnetic configuration is dipole like and compressing the plasma in the plasma sheet (see Figure 7.1). Only after reconnection starts on the nightside (at (B) in Figure 7.11) does flux return to the dayside. Complex magnetic structures form in the tail as plasma jets both earthward and tailward from the reconnection site. In some cases, the magnetic field appears to enclose a bubble of tailward-moving plasma called a *plasmoid*. At other times, the magnetic field appears to twist around the earthward- or tailward-moving plasma in a flux rope (see Figure 7.5). Even on the dayside magnetopause, twisted field configurations seem to develop as a consequence of reconnection, and, because these structures are carrying flux tailward, they are called *flux transfer events or FTEs*.

The diversity of the processes associated with geomagnetic activity, their complexity and the limited data on which studies of the immense volume of the magnetosphere must be based have constrained our ability to understand details of substorm dynamics. However, both new research tools and anticipated practical applications of improved understanding have accelerated progress toward the objective of being able to predict the behavior of the magnetosphere during a substorm. The new tools available in this century include a fleet of spacecraft in orbit around and near the Earth (*ACE, Wind, Polar, Geotail, Cluster, Double Star Themis*, and several associated spacecraft) that make coordinated measurements of the solar wind and of different regions within the magnetosphere, better instruments that make high time resolution measurements of particles and fields, spacecraft imagers covering a broad spectral range, ground radar systems, and networks of magnetometers. The anticipated applications relate to the concept of forecasting *space weather* (See Space Weather) much as we forecast weather on the ground. An ability to anticipate an imminent storm and take precautions to protect spacecraft in orbit, astronauts on space stations, and electrical systems on the surface (which can experience power surges during big storms) has been adopted as an important goal by the space science community, and improvements in our understanding of the dynamics of the magnetosphere will ultimately translate into a successful forecasting capability.

Dynamical changes long studied at Earth are also expected in the magnetospheres of the other planets. In passes through Mercury's magnetosphere, the *Mariner* spacecraft observed substorms that lasted for minutes and at the present epoch the Messenger spacecraft is compiling measurements that further characterize the dynamics of the system. FTEs can occur on Mercury's magnetopause with such frequency that the observations are described as showers of FTEs. The occurrence of FTEs at Jupiter's magnetopause is infrequent, suggesting that the formation process is controlled by the plasma parameters of the solar wind.

Substorms or related processes should also occur at the outer planets, but the timescale for global changes in a system is expected to increase as its size increases. For a magnetosphere as large as Jupiter's, the equivalent of a substorm is not likely to occur more often than every few days or longer, as contrasted with several each day for Earth. Until December 1995 when *Galileo* began to orbit Jupiter, no spacecraft had remained within a planetary magnetosphere long enough to monitor its dynamical changes. Data from Galileo's 8-year orbital reconnaissance of Jupiter's equatorial magnetosphere demonstrate unambiguously that this magnetosphere like that of Earth experiences intermittent injections of energetic particles and, in the magnetotail, unstable flows correlated with magnetic perturbations of the sort that characterize terrestrial substorms. Yet the source of the disturbances is not clear. The large energy density associated with the rotating plasma suggests that centrifugally driven instabilities must themselves contribute to producing these dynamic events. Plasma loaded into the magnetosphere near Io may ultimately be flung out down the magnetotail, and this process may be intermittent, possibly governed both by the strength of internal plasma sources and by the magnitude of the solar wind dynamic pressure that determines the location of the magnetopause. Various models have been developed to describe the pattern of plasma flow in the magnetotail as heavily loaded magnetic flux tubes dump plasma on the nightside, but it remains ambiguous what aspects of the jovian dynamics are internally driven and what aspects are controlled by the solar wind.

Whether or not the solar wind plays a role in the dynamics of the jovian magnetosphere, it is clear that a considerable amount of solar wind plasma enters Jupiter's magnetosphere. One way to evaluate the relative importance of the solar wind and Io as plasma sources is to estimate the rate at which plasma enters the magnetosphere when dayside reconnection is active and compare that estimate with the few $\times 10^{28}$ ions/s whose source is Io. If the solar wind near Jupiter flows at 400 km/s with a density of 0.5 particles/cm^3, it carries $\sim 10^{31}$ particles/s onto the circular cross-section of a magnetosphere with $>50\ R_J$ radius. If reconnection is approximately as efficient as it is at Earth, where a 10% efficiency is often suggested, and if a significant fraction of the solar wind ions on reconnected flux tubes enter the magnetosphere, the solar wind source could be important, and, as at Earth, the solar wind may contribute to the variability of Jupiter's magnetosphere. *Galileo* data are still being analyzed in the expectation that answers to the question of how magnetospheric dynamics are controlled are contained in the archives of the mission and the *Juno* mission that will reach Jupiter in October, 2016, will provide additional useful data.

Cassini has been monitoring Saturn's magnetosphere since mid-2004. Earlier passes through the magnetosphere (*Voyager 1* and *2* and *Pioneer 10*) were too rapid to provide insight into the dynamics of Saturn's magnetosphere or even to identify clearly the dominant sources of plasma. *Cassini*, in orbit at Saturn since 2004, has characterized such dynamical features as boundary oscillations, current sheet reconfigurations and other aspects of magnetotail dynamics (Figure 7.14).

Saturn's magnetic moment is closely aligned with its spin axis, and there is no evident longitudinal magnetic asymmetry. Nonetheless, both the intensity of radio emissions and magnetic perturbations are modulated at approximately the planetary rotation period. The period changes slowly with Saturn season, which makes it likely that the source of periodicity is not linked to the deep interior of the planet, but there is not yet consensus on the source of the periodicity.

The energetic particle detector on *Cassini* is capable of "taking pictures" of particle fluxes over large regions of the magnetosphere. The technique relies on the fact that if an energetic ion exchanges charge with a slow-moving neutral, a fast-moving neutral particle results. The energetic particle detector then acts like a telescope, collecting energetic neutrals instead of light and measuring their intensity as a function of the look direction. The images show that periodic intensifications and substorm-like acceleration are present at Saturn.

It is still uncertain just how particle transport operates at Saturn and how the effects of rotation compare in importance with convective processes imposed by interaction with the solar wind. It is already clear that with a major source of plasma localized close to the planet (see discussion of the plume of Enceladus in Section 6), Saturn's magnetospheric dynamics resemble those of Jupiter more closely than those of Earth. Observations scheduled

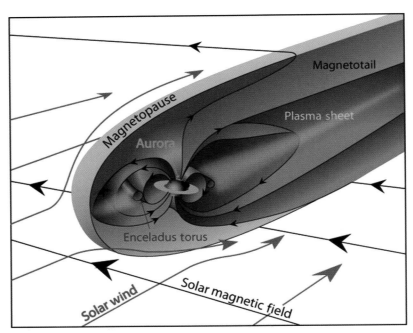

FIGURE 7.14 Saturn's magnetosphere shown in three dimensions. Water vapor from Enceladus' plumes is dissociated and ionized to form a torus of plasma that diffuses out into an equatorial plasma sheet. *Credit: Steve Bartlett.*

6. INTERACTIONS WITH MOONS

Embedded deeply within the magnetosphere of Jupiter, the four Galilean moons (Io, Europa, Ganymede, and Callisto whose properties are summarized in Table 7.5) are immersed in magnetospheric plasma that corotates with Jupiter (i.e. flows once around Jupiter in each planetary spin period). At Saturn, Titan, shrouded by a dense atmosphere, Enceladus, and the other icy moons are also embedded within the flowing plasma of a planetary magnetosphere. (See Titan.) In the vicinity of these moons, interaction regions with characteristics of induced or true magnetospheres develop. The scale of each interaction region is linked to the size of the moon and to its electromagnetic properties. Ganymede, Callisto, and

TABLE 7.5 Properties of Selected Moons of Jupiter and Saturn

Moon	Orbit Distance (R_P)	Rotation Period (Earth days)[1]	Radius (km)	Radius of Core (moon radii)[2]	Mean Density (kg/m³)	Surface B at Dipole Equator (nT)	Approx. Average B_{ext} (nT)[3]
Io	5.9	1.77	1821	0.25–0.5	3550	<200	−1900
Europa	9.4	3.55	1570		2940	0 or small	−420
Ganymede	15	7.15	2631	0.25–0.5	1936	750	−90
Callisto	26	16.7	2400		1850	0 or small	−30
Enceladus	4	1.37	252		1609	0	−325
Titan	20	15.9	2575		1900	0 or small	−5.1

[1] *Jupiter's rotation period is 9 h 55 min, so corotating plasma moves faster than any of the moons.*
[2] *Core densities are assumed in the range from 5150 to 8000 kg/m³. This corresponds to maximum and minimum core radii, respectively.*
[3] *The magnetic field of Jupiter at the orbits of the moons oscillates in both magnitude and direction at Jupiter's rotation period of 9 h 55 min. The average field over a planetary rotation period is southward oriented (i.e. antiparallel to Jupiter's axis of rotation). Neither the orbits nor the spin axes of the moons are significantly inclined to Jupiter's equatorial plane, so we use averages around the moon's orbit from the model of Khurana (1997).*

Titan are similar in size to Mercury; Io and Europa are closer in size to Earth's Moon; Enceladus is tiny (mean radius 252 km) but its interaction region is greatly expanded by a cloud of vapor that it introduces into the local magnetosphere. This means that Enceladus as well as Io is a major source of the plasma in which it is embedded. A plume at Enceladus injects $\sim 10^{28}$ water molecules/s or 0.3 tons/s into Saturn's inner magnetosphere. Dust, too, is ejected in the plume and has been found to modify plasma dynamics in critical ways. Approximately 1 ton per second of ions is introduced into Jupiter's magnetosphere by the source at Io, thus creating the Io plasma torus alluded to in Section 4. The other moons, particularly Europa and Titan, are weaker plasma sources.

The magnetospheric plasma sweeps by the moons in the direction of their orbital motion because the Keplerian orbital speeds are slow compared with the speed of local plasma flow. Plasma interaction regions develop around the moons, with details depending on the properties of the moon. Only Ganymede, which has a significant internal magnetic moment, produces a true moon magnetosphere.

The interaction regions at the moons differ in form from the model planetary magnetosphere illustrated in Figure 7.1. An important difference is that no bow shock forms upstream of the moon. This difference can be understood by recognizing that the speed of plasma flow relative to the moons is smaller than either the sound speed or the Alfvén speed, so that instead of experiencing a sudden decrease of flow speed across a shock surface, the plasma flow can be gradually deflected by distributed pressure perturbations upstream of a moon. The ratio of the thermal pressure to the magnetic pressure is typically small in the surrounding plasma, and this minimizes the changes of field geometry associated with the interaction. Except for Ganymede, the magnitude of the magnetic field changes only very near the moon. Near each of the unmagnetized moons the magnetic field rotates because the plasma tied to the external field slows near the body but continues to flow at its unperturbed speed both above and below, producing a draped magnetic field. The effect is that expected if the field lines were "plucked" by the moon. The regions containing rotated field lines are referred to as Alfvén wings. Within the Alfvén wings, the field connects to the moon and its surrounding ionosphere. Plasma on these flux tubes is greatly affected by the presence of the moon. Energetic particles may be depleted as a result of direct absorption, but low-energy plasma densities may increase locally because the moon's atmosphere serves as a plasma source. In many cases, strong plasma waves, a signature of anisotropic or non-Maxwellian particle distributions, are observed near the moons.

In the immediate vicinity of Io, both the magnetic field and the plasma properties are substantially different from those in the surrounding torus because currents associated with ionization of pickup ions greatly affect the plasma properties in Io's immediate vicinity. When large perturbations were first observed near Io it seemed possible that they were signatures of an internal magnetic field, but multiple passes established that the signatures near Io can be interpreted purely in terms of currents flowing in the plasma.

Near Titan, the presence of an extremely dense atmosphere and ionosphere also results in a particularly strong interaction whose effects on the field and the flow were observed initially by *Voyager 1* and have been extensively explored by the *Cassini* orbiter. Saturn's magnetospheric field drapes around the Titan's ionosphere much as the solar wind field drapes to produce the magnetosphere of Venus, a body that like Titan has an exceptionally dense atmosphere.

It was field draping that provided the first hint of the presence of an ion source at Enceladus, deep within the magnetosphere at 4 R_S. The anomalous bending of the magnetic field alerted investigators to the likelihood that high-density ionized matter was present below the south pole of the moon. *Cassini's trajectory* was modified on some orbits to enable imaging instruments to survey the region. A plume of vapor, largely water, was observed to rise far above the surface. This geyser is a major source of Saturn's magnetospheric plasma and is the reason that Enceladus plays a role at Saturn much like that of Io at Jupiter.

One of the great surprises of the *Galileo* mission was the discovery that Ganymede's internal magnetic field not only exists but is strong enough to stand off the flowing plasma of Jupiter's magnetosphere and to carve out a bubble-like magnetospheric cavity around the moon. A schematic of the cross-section of the magnetosphere in the plane of the background field and the upstream flow is illustrated in Figure 7.15. Near Ganymede, both the magnetic field and the plasma properties depart dramatically from their values in the surroundings. A true magnetosphere forms with a distinct magnetopause separating the flowing jovian plasma from the relatively stagnant plasma tied to the moon. Within the magnetosphere, there are two types of field lines. Those from low latitudes have both ends linked to Ganymede and are called closed field lines. Little plasma from sources external to the magnetosphere is present on those field lines. The field lines in the polar regions are linked at one end to Jupiter. The latter are the equivalent of field lines linked to the solar wind in Earth's magnetosphere and are referred to as open field lines. On the open field lines, the external plasma and energetic charged particles have direct access to the

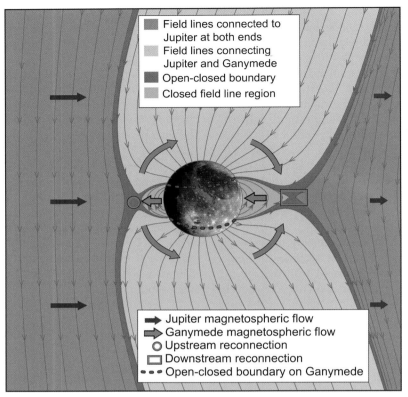

FIGURE 7.15 A schematic view of Ganymede's magnetosphere embedded in Jupiter's magnetospheric field in a plane that is normal to the direction of corotation flow. The thick purple line that bounds the region in which field lines link to Ganymede is the equivalent of the magnetopause and the polar cusp in a planetary magnetosphere. *Credit: Steve Bartlett.*

interior of the magnetosphere. The particle distributions measured in the polar regions are extremely anisotropic because the moon absorbs a large fraction of the flux directed toward its surface. Where the energetic particles hit the surface, they change the reflectance of the ice, so the regions of open field lines can be identified in images of Ganymede's surface and compared with the regions inferred from magnetic field models. The two approaches are in good agreement. As expected, the angular distribution of the reflected particles has also been found to be modified by Ganymede's internal dipole field.

Ganymede's dipole moment is roughly antiparallel to Jupiter's, implying that the field direction reverses across the near equatorial magnetopause. This means that magnetic reconnection is favored. There is some evidence that such reconnection occurs intermittently. Should future missions allow a systematic study of this system, it will be of interest to establish directly how the properties of the upstream plasma control the reconnection process and whether, with steady upstream conditions, it occurs relatively steadily or whether it occurs with some periodic or aperiodic modulation.

7. CONCLUSIONS

We have described interactions between flowing plasmas and diverse bodies of the solar system. The interaction regions all manifest some of the properties of magnetospheres. Among magnetospheres of magnetized planets, one can distinguish (a) the large, symmetric, and rotation-dominated magnetospheres of Jupiter and Saturn; (b) the small magnetosphere of Mercury where the solar wind drives rapid circulation of material through the magnetosphere (see Mercury); and (c) the moderate-sized and highly asymmetric magnetospheres of Uranus and Neptune, whose constantly changing configuration does not allow substantial densities of plasma to build up. The Earth's magnetosphere is an interesting hybrid of the first two types, with a dense corotating plasmasphere close to the planet and tenuous plasma, circulated by the solar wind driven convection, in the outer region. All of these magnetospheres set up bow shocks in the solar wind. The nature of the interaction of the solar wind with nonmagnetized objects depends on the presence of an atmosphere that becomes electrically conducting when ionized. Venus and Mars have tightly bound atmospheres so that the region of interaction with the solar wind is close to the planet on the sunward-facing side, with the IMF draped back behind the planet to form a magnetotail. Bow shocks form in front of both these magnetospheres. The regions on the surface of Mars where strong magnetization is present produce mini-magnetospheres whose properties are being explored. Comets cause the solar wind field to drape much as at Venus and Mars; they produce clouds extended over millions of kilometers. The interaction of the solar wind with the cometary neutrals weakens or eliminates a bow shock. Small bodies like asteroids disturb the solar wind without setting up shocks. Within the magnetospheres of Saturn and Jupiter, the large moons interact with the subsonic magnetospheric flow, producing unique signatures of interaction with fields that resist draping. No shocks have been observed in these cases.

The complex role of plasmas trapped in the magnetosphere of a planetary body must be understood as we attempt to improve our knowledge of the planet's internal structure, and this means that the study of magnetospheres links closely to the study of intrinsic properties of planetary systems. Although our understanding of the dynamo process is still rather limited, the presence of a planetary magnetic field has become a useful indicator of properties of a planet's interior. As dynamo theory advances, extensive data on the magnetic field may provide a powerful tool from which to learn about the interiors of planets and large satellites. For example, physical and chemical models of interiors need to explain why Ganymede has a magnetic field while its neighbor of similar size, Callisto, does not, and why Uranus and Neptune's magnetic fields are highly nondipolar and tilted while Jupiter's and Saturn's fields are nearly dipolar and aligned.

Continued exploration of the plasma and fields in the vicinity of planets and moons is needed to reveal features of the interactions that we do not yet understand. We do not know how effective reconnection is in the presence of the strong planetary fields in which the large moons of Jupiter are embedded. We have not learned all we need to know about moons as sources of new ions in the flow. We need many more passes to define the magnetic fields and plasma distributions of some of the planets and all of the moons because single passes do not provide constraints sufficient to determine more than the lowest order properties of the internal fields. Temporal variability of magnetospheres over a wide range of timescales makes them inherently difficult to measure, especially with a single spacecraft. Spurred by the desire to understand how the solar wind controls geomagnetic activity, space scientists combine data from multiple spacecraft and from ground-based instruments to make simultaneous measurements of different aspects of the Earth's magnetosphere or turn to multiple spacecraft missions like *Cluster* and *Themis* and the much anticipated Magnetospheric Multiscale Mission. As it orbited Jupiter, the *Galileo* spacecraft mapped out different parts of the jovian magnetosphere, monitoring changes and measuring the interactions of magnetospheric plasma with the Galilean satellites. *Cassini* in orbit around Saturn for many years provides ever more complete coverage of the properties of another magnetosphere and its interaction with Enceladus, Titan, and the other moons. The properties of the magnetic and plasma environment of Mars are still being clarified by spacecraft measurements. *Messenger* at Mercury will fully characterize the mysterious magnetic field of this planet and its magnetosphere. And finally, Pluto beckons as the prototype of an important new group of solar system bodies that the New Horizons spacecraft with explore in 2016. As new technologies lead to small lightweight instruments, we look forward to missions that will determine which of the many small bodies of the solar system have magnetic fields and help us understand the complexities of magnetospheres large and small throughout the solar system.

BIBLIOGRAPHY

Bagenal, F. (2009). Comparative planetary environments. In C. J. Schrijver, & G. L. Siscoe (Eds.), *Heliophysics: Plasma physics of the local cosmos*. Cambridge University Press.

Bagenal, F., Dowling, T., & McKinnon, W. (Eds.), (2004). *Jupiter: The planet, satellites and magnetosphere*. Cambridge, U.K: Cambridge Univ. Press.

Brain, D. A., Barabash, S., Bougher, S., Duru, F., Jakosky, B., & Modolo, R. (2004). Solar wind interaction and atmospheric escape. In R. Haberle, T. Clancy, F. Forget, & R. Zurek (Eds.), *Mars atmosphere*.

Dungey, J. W. (1961). Interplanetary magnetic field and the auroral zones. *Physical Review Letters, 6*, 47–48.

Fisk, L. A. (23 September 2005). Journey into the unknown beyond. *Science, 2016*(309). www.sciencemag.org.

Jia, X., Kivelson, M. G., Khurana, K. K., & Walker, R. J. (2010). Magnetic fields of the satellites of Jupiter and Saturn. *Space Science Reviews, 152*, 271–305.

Khurana, K. K. (1997). Euler potential models of Jupiter's magnetospheric field. *Journal of Geophysical Research, 102*, 11295–11306.

Kivelson, M. G. (2006). Planetary magnetospheres. In Y. Kamide, & A. C.-L. Chian (Eds.), *Handbook of solar-terrestrial environment*. New York: Springer Verlag.

Kivelson, M. G., & Russell, C. T. (Eds.), (1995). *Introduction to space physics*. Cambridge, U.K: Cambridge Univ. Press.

Luhmann, J. G. (1986). The solar wind interaction with Venus. *Space Science Reviews, 44*, 241.

Nagy, A. F., Winterhalter, D., Sauer, K., Cravens, T. E., Brecht, S., Mazelle, C., et al. (2004). The plasma environment of Mars. *Space Science Reviews, 111*, 33–114.

Sundberg, T., & Slavin, J. A. (2013). Mercury's magnetotail. In *Magnetotails*. Washington, DC: AGU Chapman Monograph.

Van Allen, J. A., & Bagenal, F. (1999). Planetary magnetospheres and the interplanetary medium. In Beatty, Petersen, & Chaikin (Eds.), *The new solar system* (4th ed.). Cambridge, U.K: Sky Publishing and Cambridge Univ. Press.

Chapter 8

Rotation of Planets

Véronique Dehant
Royal Observatory of Belgium, Brussels, Belgium

Tim Van Hoolst
Instituut voor Sterrenkunde, KU Leuven, Celestijnenlaan 200D, B-3001 Leuven, Belgium; Royal Observatory of Belgium, Brussels, Belgium

Chapter Outline

Introduction	159	6. Nutation	169
1. Observed Rotation State of Planets	159	7. LOD Variations	171
2. Origin and Long-Term Spin Evolution	160	8. Libration	175
2.1. Origin	160	9. Wobbles and the Interiors of Terrestrial Planets	177
2.2. Tidal Dissipation	161	9.1. Polar Motion	178
2.2.1. Tidal Torque	161	9.2. Forced Nutation and Free Core Nutation	181
2.2.2. Long-Term Spin Evolution	161	10. Observation of the Rotation of Terrestrial Planets	181
2.2.3. Effects on the Orbit	163	10.1. Ground-Based Observations	181
3. Long-Term Evolution of the Orientation	164	10.2. Spacecraft Observation	182
4. Rotational Flattening of Planets	165	10.3. The Particular Case of the Earth	183
5. Precession	166	Bibliography	184

INTRODUCTION

In this part of the book we describe the rotation of planets and moons of the solar system. Planetary rotation can be divided into the rotation speed around an axis and the orientation of this axis (or another axis of the planet) in space. Here, we summarize the main observed rotational characteristics of the planets of the solar system and explain how the rotation might have evolved during the history of the solar system. On short timescales, planetary rotation is variable and yields information on the interior structure of planets. Most of us know that the rotation of a boiled egg noticeably differs from that of a raw egg. This simple observation shows that information on the inside of an object, here an egg, can be obtained from its rotation. The same idea applies to the rotation of celestial bodies.

The chapter is organized in 10 sections: first we describe the observed rotation states of the planets (Section 1); next we discuss the origin of the spin and its long-term evolution (see Section 2), as well as the long-term evolution of the orientation (Section 3); and Section 4 explains the flattening of planets induced by the rotation. The next five sections (Sections 5, 6, 7, 8, and 9) are dedicated to a description

and study of the variable components of rotational motion of planets (precession, nutation, length-of-day (LOD) variations, libration, and polar motion) with a particular emphasis on terrestrial planets. These sections also provide insight into how information on the interior and dynamics of a planet can be obtained from observations of the rotation. Methods for observing the rotation of planets are described in the last section (Section 10); the particular case of the Earth is also treated there.

1. OBSERVED ROTATION STATE OF PLANETS

As Kepler (1571–1630) has shown, all planets of the solar system move on approximately elliptical orbits. The dimensions of these orbits are characterized by the **semimajor axis a** and the **semiminor axis b** (see Figure 8.1). The elliptical form of the orbit is uniquely determined by the distance between the center of the ellipse and one of its foci, also referred to as the parameter of the ellipse and denoted by f. Alternatively, it is described by the *eccentricity e*, which is defined as the ratio of the distance between the two foci to the length of the major axis: $e = 2f/2a = f/a$. The

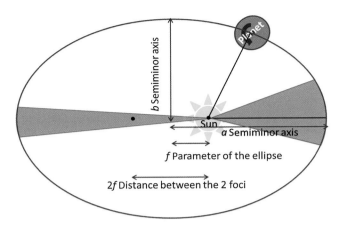

FIGURE 8.1 Kepler's law and planet orbits.

eccentricity of an ellipse is between 0 and 1. When the eccentricity is 0, the foci coincide with the center point and the figure is a circle. The Sun for planets or the central planet for satellites occupies one of the focal points of the ellipse. Since the planets and most satellites of the solar system revolve in an almost circular motion around the Sun or their parent planet, their orbital eccentricity is generally much closer to 0 than to 1. An eccentricity of 1 characterizes a parabola, and $e > 1$, a hyperbola. Some comets are on parabolic and hyperbolic orbits (See Comet Populations and Cometary Dynamics).

Besides the orbital motion, the planets and satellites (or moons; we will use those two words designating the same objects of our solar system interchangeably) also rotate around an internal axis. This *rotation axis* or *spin axis* differs from the perpendicular to the orbital plane by an angle called the **obliquity**. The obliquities, rotation periods, and revolution periods are provided in Table 8.1 for the planets of our solar system. The rotation period given is the sidereal rotation period, which is defined as the period between two passes of a given point at the surface of the planet to the same direction in space. It must be mentioned that the rotation of the surface layers of the giant gaseous planets, which can be obtained from direct visual observations, does not necessarily represent the internal rotation rate. It is usually accepted that the rotation axis orientation can be obtained from the magnetic dipole axis and that the rotation speed can be obtained from the rotation of the magnetic field (magnetically linked to the deep interior) or from natural radio emissions varying periodically with the rotation. For a planet such as Venus with a thick atmosphere (or a moon such as Titan) surface details can hardly be seen in the visible and the rotation is determined from radar observations of the surface and from measurements of the gravity field rotating with the planet.

The planets rotate about their axis in about 1 day or less, except for the two planets closest to the Sun, Mercury and Venus (see Table 8.1). Both these planets and Jupiter rotate about an axis almost perpendicular to their orbital plane, in contrast to most other planets. These observations suggest that the innermost planets have significantly changed their rotation in the about 4.6 billion years since the origin of the solar system, as will be further explained in the next sections (see Section 2 and Section 3). The near alignment of Jupiter's rotation axis with that of the Sun is primordial and a not unexpected result of conservation of angular momentum during the formation of this by far the most massive planet of the solar system. Surprisingly, half of the planets have an obliquity close to 25°, including the Earth. Uranus is peculiar in the sense that its rotation axis is almost in the orbital plane. Most probably, Uranus did not retain its original spin, but has been tilted onto its side by a large impact event. Equally unique is that Venus and Uranus have a retrograde rotation instead of a counter-clockwise rotation as seen from above the Sun's North Pole as for all other planets.

2. ORIGIN AND LONG-TERM SPIN EVOLUTION

2.1. Origin

The solar system planets acquired rotational angular momentum during the formation process. As a consequence of conservation of angular momentum, the planets are expected to form with the same sense of rotation as their orbital motion. Gas and ice giant planets mainly exchange angular momentum with the gas disk, whereas terrestrial planets can achieve fast rotation speeds by collisions with planetesimals. The spin state after formation is expected to be mainly

TABLE 8.1 Orbit, Rotation, and Orientation Characteristics of the Eight Planets of the Solar System at Present

Planets	Orbital Periods (Earth Days or Years)	Rotation Periods (Earth Hours or Days or Years)	Obliquity
Mercury	87.97 days	58.65 days	2.0 arcmin
Venus	224.70 days	−243.02 days (retrograde)	177.36°
Earth	365.256 days or 1 year	23 h 56 m 4 s	23.439°
Mars	686.98 days or ~2 year	24 h 37 min 23 s or 1.026 days	25.19°
Jupiter	11.86 year	9.55 h	3.1°
Saturn	29.46 year	10.32 h	26.7°
Uranus	84.02 year	−17.24 h	97.8°
Neptune	164.79 year	16.11 h	28.3°

determined by the random giant impacts that terrestrial planets encountered at the final stages of planet formation.

Not all current spin characteristics summarized in Table 8.1 correspond to the initial values after formation since both the obliquities and the rotation rates changed on long timescales. As already mentioned, for Mercury and Venus, in particular, even important changes probably occurred. The long-term changes in the rotation of planets and satellites are mainly due to friction associated with tides raised by the central body (the Sun for planets or the parent planet for satellites). Tidal dissipation also explains why most large natural satellites rotate with a period equal to their orbital period.

2.2. Tidal Dissipation

2.2.1. Tidal Torque

Planets are periodically deformed by the gravitational attraction of the Sun and their natural satellites. These tides are never perfectly elastic responses to the tidal forcing and lead to the dissipation of a small amount of energy in the planets. This input of energy to the planets is so small that it can generally be neglected for the thermal evolution and internal dynamics of the planets of the solar system (although it can be important for exoplanets near their mother star). Nevertheless, it can drastically change the orbital and rotational dynamics of planets. For satellites, the effect of tidal dissipation (the tides are mainly raised by the parent planet) is more pronounced and can also be important for internal processes. For example, Io's volcanism is due to dissipation of energy associated with the large tides raised by Jupiter on Io (See Io: The Volcanic Moon).

Deviations from perfectly elastic behavior of the materials inside planets delay the response of the planet to an applied tidal force. The tidal bulge therefore does not form instantaneously and, because of the rotation of the planet relative to the Sun, it becomes slightly misaligned with respect to the direction to the body raising the tides (see

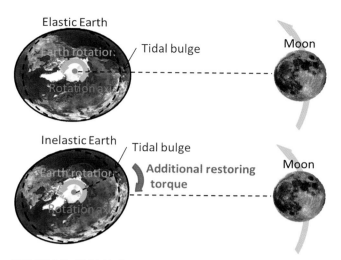

FIGURE 8.3 Tidal friction due to the inelasticity of the planet (here the Earth). First representation for the elastic case and second representation for the inelastic case.

Figure 8.3). As a result of the asymmetrical orientation of the tidal bulge with respect to the Sun, the Sun exerts a gravitational torque on the planet, which tends to alter the rotation of the planet (see Figure 8.3). The magnitude of the torque can be expressed as $\Gamma = \frac{3}{2} k_P \frac{GM_S^2}{R_P} \left(\frac{R_P}{d}\right)^6 \sin 2\varepsilon$, where G is the universal gravitational constant, M_S is the mass of the Sun, R_P is the radius of the planet, d is the distance between the planet and the Sun, ε is the angle between the long symmetry axis of the tidal bulge and the direction to the Sun, and k_P is the tidal Love number of the planet specifying the response of the planet to a unit tidal potential. In general, this torque is small because of the large distances in the solar system. It rapidly decreases with increasing distance d between the planet and the Sun according to d^{-6}. As it is a gravitational torque that is acting on a tidal bulge, which itself is created by tidal forces, the torque is also proportional to the square of the mass of the Sun (see Figure 8.2). A gravitational torque can be exerted by any body on a planet, such as the Moon for the Earth (see Figure 8.3). The torque can be large for small distance d between the planet and the perturbing body and for large mass of the tide-raising body. This mass dependence also explains why Io, which is at about the same distance from Jupiter as the moon is from the Earth, is geologically very active with volcanism on a much larger scale than on Earth, whereas the Earth's moon is geologically quiet (See The Moon).

2.2.2. Long-Term Spin Evolution

The tides slow down the rotation of a planet or a satellite when its rotation period is shorter than the orbital period. In that case, the tidal bulge closest to the perturbing body will be behind the direction to the tide-raising body (e.g. the Sun

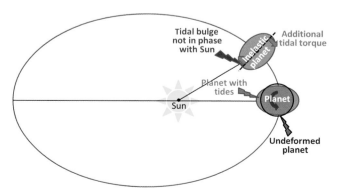

FIGURE 8.2 A misaligned tidal torque leads to a gravitational torque that slows down the rotation.

in the case of Mercury, the Earth in the case of the Moon, and Jupiter in the case of Io) with respect to the direction of orbital motion (see Figures 8.2 and 8.3). The gravitational tidal torque from the central body then acts to slow down the rotation. For a circular orbit ($d = a$), the despinning timescale τ can be obtained directly from the change in the angular momentum of the planet as

$$\tau = -\frac{2}{3}\Omega_P \frac{Q_P}{k_P} \frac{R_P}{GM_S^2} C_P \left(\frac{a}{R_P}\right)^6.$$

Here Ω_P is the planet's or satellites' rotation frequency, C_P is its polar moment of inertia, R_P is its radius, and M_S is the mass of the primary. Q_P, the quality factor of tidal dissipation, is defined as the inverse of the sine of the lag angle. It is of the order of 100 for terrestrial planets but can be as small as 10 if the interior is partially molten, as is likely the case for Io. The quality factor is several orders of magnitude larger for the giant planets since gases are much less dissipative than liquids or solids. On Mercury, the closest planet, the Sun has the largest tidal effect of all planets in the solar system. Tides can despin Mercury from an initially faster spin with a period of about a day to rotation periods commensurate with the orbital period on a timescale of a few 100 million years, much shorter than the age of the solar system (see Figure 8.2). Although the initial spin of Mercury is unknown, this shows that Mercury must necessarily be spinning slowly. Besides Mercury, tidal friction by the Sun could also significantly despin Venus over the age of the solar system. It also affects the rotation of the Earth, but has only a very small influence on the rotation of the more distant Mars and the gas and ice giants.

In contrast to the decelerating effect of the tidal torque for rapidly rotating planets, the rotation is accelerated when the orbital period is shorter than the rotation period. Taken together, this suggests that planets and satellites will ultimately evolve to a situation in which their rotation period is equal to the orbital period.

In contrast to planets, most large and medium-sized satellites are indeed observed to have rotation periods that are almost exactly equal to their revolution periods. This is the most common example of a **spin—orbit resonance** in the solar system. The spin and orbital motion are said to be resonant because the ratio of their periods is a ratio of integers. The resonance is referred to as the 1:1 spin—orbit resonance (see Figure 8.4) and is an end state of tidal friction processes. Once captured in such a state, the satellite always shows approximately the same face to the central planet (see Figure 8.4).

The despinning timescale for most satellites is less than 1 million years and often even much shorter. The main tide-raising body here is the central planet. Although the planets are much less massive than the Sun (its mass is about 1000 times larger than that of the largest planet Jupiter), the

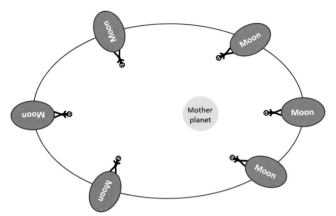

FIGURE 8.4 Rotation and revolution of a moon around a mother planet in a spin—orbit 1:1 resonance.

distance between the planets and their satellites is also much shorter. Typical distances are of the order of 100 thousand kilometers, whereas distances to the Sun are on the order of 100 million kilometers and more. Because of the strong dependence on distance, the planets can have a larger tidal effect on the rotation of their satellites than the Sun has on planets.

In a circular orbit, tidal friction will ultimately drive the rotation of the planet to a state synchronous with the orbital motion (we neglect obliquity here). The final equilibrium rotation rate for a satellite in an eccentric orbit is slightly faster than synchronous due to the larger accelerating effect on rotation near pericenter than the decelerating effect near apocenter. Rotation accelerates near pericenter because the orbital motion there is faster than the rotational motion and the tidal bulge will be ahead of the satellite—planet line. Near apocenter the orbital motion is slower than the rotational motion resulting in a gravitational torque tending to slow down the rotation. The synchronous rotation state is, nevertheless, stable for a slightly eccentric orbit if the gravitational torque of the central planet on a permanent asymmetry of the satellite is larger than the tidal torque at synchronous rotation averaged over an orbital period. For the largest satellites of the solar system for which the ellipsoidal shape has been measured, this condition is indeed fulfilled.

Besides the Sun, the Moon also contributes to the deceleration of the rotation of the Earth by tidal friction. Because the Moon is so much closer to the Earth than is the Sun, its decelerating effect is even about a factor 2 larger than that of the Sun.

Friction at the core—mantle boundary (CMB) associated with differential rotation between a liquid **core** and a solid **mantle** can also change the rotation of a planet on long timescales. The friction is associated with either the viscosity of the liquid core or with the electromagnetic interaction between mantle and core (See Earth: Surface And Interior). In the latter case, the rotational motions of the fluid core relative to the neighboring solid regions (mantle or inner core) at any of the tidal frequencies results in an oscillatory sweep of the magnetic field through the fluid, producing a **Lorentz force** which causes the fluid itself to be partially dragged with the field. Since the dissipated energy must come from the rotation, core—mantle friction torques cause the rotation to slow down.

In addition, a dense atmosphere can have an effect on the long-term evolution of planets. This is especially the case for Venus, for which the atmospheric pressure at the surface is 92 times that on Earth (See Venus Atmosphere). The atmosphere and surface of a planet like the Earth and Venus is heated mostly by solar radiation at the subsolar point. This causes pressure changes and atmosphere motions, or thermal tides, away from the subsolar point. This creates an atmospheric "bulge", which is almost perpendicular to the planet—Sun direction, but not exactly because of the associated dissipation and a finite response time. For a rapidly rotating planet such as the Earth, the torque on the thermal atmospheric tides tends to accelerate the rotation but the effect is more than a factor 10 smaller than the decelerating effect on rotation from gravitational tides. For the slowly rotating planet Venus, however, the atmospheric tides have a large influence on its rotation. The rotation of Venus is thought to be the result of a balance between solid body tidal torques, which drive Venus to synchronous rotation, and atmospheric torques, which drive it away.

Tidal dissipation can occur everywhere in a planet or satellite, but some layers may have a dominant contribution. For Io, a partially molten asthenosphere may be important. On the Earth, dissipation induced by the tides in the oceans and in particular in the shallow seas plays the most important role. The presence of oceans therefore increases the change in the rotation period of a planet. In the Devonian, the LOD on Earth was about 22 h. In 100 years, it will take two more milliseconds for the Earth to complete a full rotation around its axis. At that rate, the day will be an hour longer after more than 100 million years.

Both Pluto and Charon have evolved to a double-synchronous state in which the orbital period and the rotation period of both bodies are equal. Both Pluto and Charon always show the same face to the other body.

2.2.3. Effects on the Orbit

Tidal interactions change not only the rotation of the planet or satellite but also the orbital motion. As long as the system of the planet and the tide-raising body can be considered isolated, the total angular momentum of the system must be conserved. Since the planet loses rotational angular momentum due to friction of tides in the planet, the angular momentum of orbital motion must increase. As a result the semimajor axis increases, making the orbit wider. This effect due to the Sun is very small for planets. Since the angular momentum associated with orbital motion is many orders of magnitude larger than that of rotational motion, tidal braking is much less efficient in changing the orbit of planets than in changing their rotational motion.

Because of the shorter distance between planets and satellites than between the Sun and planets, this process of orbital change is more important for planet—satellite systems. For example, because of dissipation of tidal energy in the Earth associated with tides raised by the Moon, the Moon recedes from the Earth and the Earth slows down its rotation. After one century, the semimajor axis of the Moon increases by 3—4 m (~ 3.5 cm/year).

The Earth also raises tides on the Moon and as a result, tidal friction in the Moon also changes the semimajor axis of the orbit. This change, however, is much smaller than that induced by tides in the Earth mainly because periodic tides in the Moon only exist for an eccentric orbit. For an exactly circular orbit, any material point in a satellite in a 1:1 resonance would exactly feel the same gravitational force from the central planet and there would be no periodic tides. Since the deviation of a circular orbit, measured by the eccentricity, is small, tidal dissipation in the Moon contributes less than 1% to the change in its semimajor axis. For satellites in the outer solar system orbiting a gas or ice giant, the situation can be different since tidal dissipation in a solid body is more efficient than in a gas. For example, tidal dissipation in Io, the closest of the Galilean satellites to Jupiter, contributes about as much to the change in orbital motion as dissipation in Jupiter.

Besides changing the semimajor axis, tidal friction also changes the orbital eccentricity. Since despinning is a faster process than circularization because of the smaller angular momentum involved in rotation than in orbital motion, satellites still have an elliptical orbit upon reaching an equilibrium rotation state. Nevertheless, their eccentricity usually is very small. Mercury, on the other hand, has a large eccentricity of 0.2059. With such a large eccentricity, the equilibrium rotation rate, at which the mean tidal torque averaged over one orbit is equal to zero, is about 1.26 times the mean motion. This, however, is not the rotation state occupied by Mercury. Observations show that Mercury is in a 3:2 spin—orbit resonance, meaning that its orbital period is 1.5 times longer than the rotation period (see Figure 8.5 and also the beginning of Movie 6). As a result, a solar day on Mercury—the time between two consecutive passes of the Sun through the local meridian—is twice as long as a year, a quite remarkable situation compared to our situation on Earth.

The 3:2 **spin—orbit resonance** is dynamically stable for the shape of Mercury and the current eccentricity, and Mercury has probably been captured into this 3:2 resonant state while spinning down from an earlier faster rotation. Because of large oscillations in the eccentricity of Mercury driven by other planets, the resonance may have been

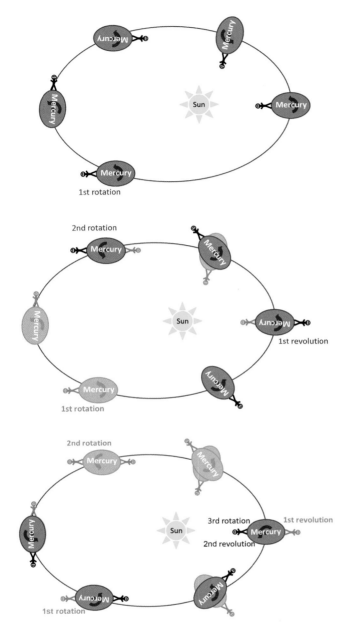

FIGURE 8.5 Rotation and revolution of Mercury around the Sun in a spin–orbit 3:2 resonance.

3. LONG-TERM EVOLUTION OF THE ORIENTATION

Since dissipation associated with tidal deformation delays the appearance of the tidal bulge with respect to a perfectly elastic body that reacts instantaneously, the tidal bulge is carried outside the orbital plane by rotation during the time lag for **obliquities** different from 0° to 180°. The gravitational torque of the central body on the tidal bulge therefore also has a component in the orbital plane. As a result, besides changing the rotation rate of a planet, the tidal torque can also change the obliquity. The effect can be important for those planets and satellites that also changed their rotation rate as a result of tidal dissipation. For those slowly rotating bodies, meaning that their rotation period is not too different form their orbital period, the tidal torque tends to drive the obliquity to zero in the case of zero eccentricity since the obliquity is the cause of torque. Mercury, Venus, and most satellites are indeed observed to have a small obliquity.

For Mercury and satellites in **spin–orbit** resonance, the tidal evolution has also led to a stable equilibrium state for the orientation of the body, called the Cassini state. In this state, the rotation axis and the orbit normal remain coplanar with the normal to the Laplace plane, while the rotation axis and the orbit normal rotate (or precess) about the normal to the Laplace plane with the same period. The Laplace plane is defined as the plane about which the orbit rotates (or precesses) and is equal to the mean orbital plane (see Figure 8.6). The obliquity is then constant but nonzero as tides try to bring the spin axis to the orbit normal which is itself moving on long timescales.

Radar observations have shown that Mercury occupies the Cassini state within the observational error. Mercury's obliquity is equal to 2.04 ± 0.08 arcmin. The observed spin pole position differs by 2.7 arcsec from the predicted Cassini position, within the 5-arcsec uncertainty on the orientation. This observation is very useful for constraining the interior structure of Mercury because it allows the polar moment of inertia of Mercury to be calculated. The latter is a measure of the internal mass distribution in the radial direction (See Mercury). The obliquity in the Cassini state is

crossed many times before final capture. This increases the likelihood of capture in the 3:2 spin–orbit resonance. Capture into the **3:2 spin–orbit resonance** is the most probable capture but Mercury might also have been captured before in another resonance, which subsequently, probably during the period of Late Heavy Bombardment, was destabilized by large impacts. If Mercury initially rotated retrogradely, an initial capture into the 1:1 resonance is even the most likely. The observed asymmetrical distribution of large impact basins may be an indication for such a different primordial resonant state.

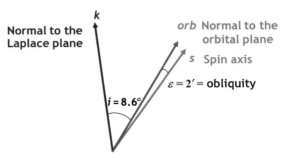

FIGURE 8.6 Spin axis (*s*), normal to the orbital plane (*orb*), both being very close to one another, and normal to the Laplace plane (*k*) with numerical values for Mercury's case.

theoretically connected to the polar moment of inertia by a relation which also involves the degree-two gravitational coefficients J_2 and C_{22}, and some orbital parameters. For the recently determined values from MErcury Surface, Space ENvironment, Geochemistry and Ranging (MESSENGER) radio tracking ($J_2 = (5.03 \pm 0.02) \times 10^{-5}$ and $C_{22} = (0.809 \pm 0.006) \times 10^{-5}$), we have $C = (0.346 \pm 0.014) M_P R_P^2$, where M_P is the mass of the planet Mercury and R_P is its radius. A further rotational constraint and joint inferences on the core of Mercury are described in Section 8.

For most satellites the obliquity has not yet been accurately determined but is generally small. Based on radar images obtained from the Cassini spacecraft, the obliquity of Titan is estimated to be $0.32 \pm 0.02°$. The observations indicate that Titan is close to but does not exactly occupy the Cassini state since the rotation axis deviates by $0.12 \pm 0.02°$ from the plane formed by the orbit normal and the normal to the Laplace plane. The obliquity of Titan is a few times larger than expected for a rigid Titan in the Cassini state, suggesting that either Titan has a subsurface ocean or that the obliquity is excited by other causes like its atmosphere.

Since obliquity is defined as the angle between the rotation axis and the normal to the orbital plane, changes in obliquity can have two different causes. Obliquity can change not only due to a change in the orientation of the planet (or spin axis) in inertial space but also due to variations in the orientation of the orbital plane, called orbital precession. The total motion of the spin axis with respect to the orbit normal depends critically on the ratio of the orbital precession period to the period of the precession of the planet. Precession is the slow change in the orientation of a planet due to gravitational torques exerted by other solar system bodies in which the tip of the rotation axis describes a large circle on the celestial sphere around the normal to the orbital plane. The physics of precession will be explained in Section 5. For precession much faster than the precession of the orbit, the spin axis precesses about the slowly moving orbit normal and tracks that motion effectively such that the obliquity remains nearly constant. On the other hand, if the precession of the spin axis is much slower than changes in the orientation of the orbit, the spin axis will precess about the averaged orbit normal. If the ratio is about one, a resonant situation occurs and the obliquity variations can be large and chaotic. Mars is in such a situation and shows large changes in obliquity over a relatively short period of time. For example, over the past million years the obliquity of Mars fluctuated between about 15° and 35°. The other three terrestrial planets could also have experienced large chaotic variations in obliquity during some period of their history. Therefore, the obliquities of the terrestrial planets cannot be considered as primordial.

The presence of the Moon increases the Earth's precession rate, or decreases the precession period, by about a factor 3 with respect to an Earth without a Moon. As a result the precession period of the Earth is much shorter than the orbital precession. Therefore the obliquity of the Earth nowadays remains nearly constant. It is 23.44° at present and oscillates between 22.1 and 24.5° with a period of 41 000 years. These changes in the orientation of the Earth create changes in the insolation from the Sun on the Earth's surface, producing climate change. Milanković (1879–1958) first put forward the idea that variations in eccentricity, axial tilt, and precession of the Earth's orbit determine climate changes of the Earth through orbital forcing. Such climate changes are particularly important for Mars because of its large obliquity variations. Note that Mars has no large moon (the moons Phobos and Deimos are rocky bodies of only about 20 km and 10–15 km size) to increase the precession rate and Mars' precession is also slower than that of the Earth since it is further from the Sun (see Section 5). Its period is 171,000 years, close to some planetary perturbations on the orbit.

Besides the gravitational torque on the gravitationally forced tides, the orientation of a planet can also evolve as a result of a gravitational torque on the thermal tides in the atmosphere and of core–mantle friction torques. For the slowly rotating Venus, core–mantle friction torque is thought to dominate the obliquity evolution and to drive Venus to an orientation with its polar axis perpendicular to its orbital plane. If Venus initially had a rapid prograde rotation like Mars and the Earth, it may have slowed down so much that it started developing a retrograde rotation. In an alternative scenario, however, the rotation axis of Venus has flipped direction under the influence of strong atmospheric torques during the rapid rotation phase and slowed down later. Due to planetary perturbations, the obliquity never reaches a zero value.

4. ROTATIONAL FLATTENING OF PLANETS

Let us consider a spherical deformable planet rotating with angular velocity Ω. A point P on the surface ($r = a_0$ where a_0 is the mean radius of the planet) or inside the planet at a distance r from the center of the planet and at a **colatitude** θ experiences a centrifugal acceleration of magnitude:

$$a_{cf} = \Omega^2 r \sin \theta.$$

The centrifugal acceleration can be written as the gradient of the centrifugal potential V_{cf} as:

$$\vec{a_{cf}} = -\vec{\nabla} V_{cf} \quad \text{with} \quad V_{cf} = -\frac{1}{2}\Omega^2 r^2 \sin^2 \theta.$$

The centrifugal acceleration is largest at the equator and decreases to zero at the poles. Since the centrifugal acceleration is perpendicular to the rotation axis and oriented outward, rotation will flatten planets and expand them in the equatorial regions into the shape of an oblate spheroid.

The gravitational acceleration resulting from self-gravitation also derives from a potential. The gravitational potential at the point P is proportional to the mass M of the planet and inversely proportional to the distance to the center[1]. As we consider the planet to be of an oblate spheroidal form, the potential also depends on latitude. For positions exterior to the planet we have

$$V = \frac{GM}{r}\left(1 + J_2\left(\frac{r_{eq}}{r}\right)^2 \frac{3\cos^2\theta - 1}{2}\right) + \frac{1}{2}\Omega^2 r^2 \sin^2\theta,$$

where $J_2 = \frac{C-\bar{A}}{Mr_{eq}^2}$ is the form factor of the planet with $\bar{A} = \frac{A+B}{2}$, A and B being the equatorial moments of inertia, C being the polar moment of inertia ($A < B < C$), and r_{eq} being the equatorial radius of the planet; for an oblate ellipsoidal planet we have $\bar{A} = \frac{A+B}{2} = A = B$.

The total gravity at the surface of the planet can be written:

$$\vec{g} = -\vec{\nabla}V.$$

A planet is said to be in hydrostatic equilibrium when gravity is balanced by a differential pressure (p) force at each position inside the planet with density ρ:

$$\vec{\nabla}p = -\rho\vec{g}.$$

The centrifugal potential then forces the constant gravity surfaces or constant potential surfaces to be ellipsoidal. The constant potential surfaces coincide with surfaces of constant pressure and constant density. By using the equation for the total potential and expressing the radial coordinate of the surface as $r = a_0\left(1 - \frac{2}{3}\alpha P_2(\cos\theta)\right)$, the flattening α of the equipotential surface of the planet is seen to be given by $\alpha = \frac{3}{2}J_2 + \frac{q}{2}$ where $q = \frac{\Omega^2 r_{eq}^3}{GM}$ is the ratio of the centrifugal acceleration to the gravity at the equator of radius r_{eq}. Since J_2, too, is nonzero due to the rotation, the flattening is proportional to the square of the rotation frequency, as is the centrifugal acceleration. Therefore, fast rotating planets show a more distinct equatorial bulge than the slower rotating planets Mercury and Venus.

For bodies rotating synchronously with their orbital motion like most of the large natural satellites including the Moon, the tidal forces also have an important static component, which further deforms the satellites in addition to the centrifugal acceleration. The static tidal bulge also takes the form of an oblate ellipsoid but with the long axis in the direction to the central planet at pericenter. As a result, synchronous satellites have the shape of a triaxial ellipsoid with three different principal moments of inertia A, B, and C. These principal moments of inertia are computed from the mass repartition inside the planet:

$$A = \int \rho(\vec{r})(y^2 + z^2)d^3r,$$

$$B = \int \rho(\vec{r})(x^2 + z^2)d^3r,$$

$$C = \int \rho(\vec{r})(x^2 + y^2)d^3r.$$

When the coordinate axes are chosen to coincide with the principal axes of inertia, the external gravitational potential for a triaxial planet can be expressed as

$$\frac{GM}{r}\left(1 - J_2\left(\frac{r_{eq}}{r}\right)^2 \frac{3\cos^2\theta - 1}{2}\right.$$
$$\left. + C_{22}\left(\frac{r_{eq}}{r}\right)^2 (3\sin^2\theta)\cos 2\lambda\right)$$

where θ and λ are the colatitude and longitude and C_{22} is the degree-two sectorial coefficient of the gravitational potential and is related to the equatorial moment of inertia by $C_{22} = \frac{B-A}{4Mr_{eq}^2}$.

5. PRECESSION

For a planet whose rotation axis is tilted with respect to the orbital plane (Table 8.1), the equatorial bulge is out of the equatorial plane during the orbital motion. As a result, the Sun exerts a gravitational torque on the planet tending to twist the equator toward the orbital plane of the planet. All other objects in the solar system in principle also exert a torque on the planet, but apart from torques of large and nearby satellites (such as the Moon for the Earth) these torques are several orders of magnitude smaller. As the planet is rotating, it acquires additional angular momentum in the direction of the torque and reacts as a spinning top. The main effect is **precession**, which is the slow motion of the rotation axis and the planet in space around the perpendicular to the orbital plane (see Figure 8.7).

---

1 Let us consider a small mass element dm at a point P inside a planet; the potential that is felt at a point Q outside the planet due to dm is $G\,dm/r'^2$ where r' is the distance from P to Q. When integrating over all the dm in the volume of the planet, one obtains the gravitational potential at Q due to the whole planet. Considering that the reference frame is tied to the center of mass of the planet and to the principal moments of inertia (the products of inertia are zero in that frame), and that the mass repartition is mainly due to rotation (the equatorial moment of inertia in any direction are identical), one obtains the total gravitational potential at the first order: $\frac{GM}{r}\left(1 - J_2\left(\frac{r_{eq}}{r}\right)^2 \frac{3\cos^2\theta - 1}{2}\right)$.

Chapter | 8 Rotation of Planets

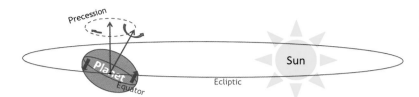

FIGURE 8.7 Representation of the precession motion (in dashed line) of the rotation axis of the planet in orbit around the Sun (straight line).

Movie 1 shows first the motion of a top. It then shows the long-term variation of the orientation of the Earth in space, the precession.

Supplementary video related to this chapter can be found at http://dx.doi.org/10.1016/B978-0-12-415845-0.00008-6.

The following is/are the supplementary data related to this chapter: Movie 1: Precession of the Earth in space. (file EarthPrecession.mpg)

The gravitational torque exerted by the Sun on a planet can be computed from the relative position of the celestial bodies. From the point of view of the center of mass of the planet, the Sun describes an apparent motion in the orbital plane with angular revolution velocity n, commonly referred to as the **mean motion**, around the planet. According to Kepler's third law, we have

$$n^2 = \frac{G(M_{Sun} + M_{Planet})}{a^3} \cong \frac{GM_{Sun}}{a^3}$$

where M_{Sun} and M_{Planet} are the masses of the Sun and the planet, respectively. In order to describe the position of the Sun, we define a reference frame $(\widehat{X}, \widehat{Y}, \widehat{Z})$ with origin in the mass center of the planet and three orthogonal axes. Two axes $(\widehat{X}, \widehat{Y})$ are chosen in the orbital plane, and the third axis $\widehat{Z}$ is perpendicular to the orbit. The position of the Sun can then be expressed by the following three coordinates with respect to that frame (see Figure 8.8):

$$\xi = d \cos(nt)$$
$$\eta = d \sin(nt)$$
$$\zeta = 0$$

Here d is the distance between the centers of mass of the planet and the Sun at a given position in the elliptical orbit.

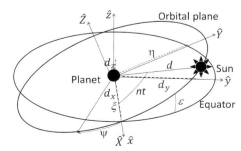

FIGURE 8.8 Representation of the apparent motion of the Sun around a planet for the computation of precession.

The X-axis of the frame $\widehat{X}$ is chosen in the direction of the ascending equinox, the intersection between the equator and the orbital plane where the Sun crosses from below to above the orbital plane. We also consider here that the Sun is at the equinox at time $t = 0$.

The gravitational acceleration of a material element at $\vec{r}$ inside the planet due to the Sun is $-\nabla W^{Sun}(\vec{r}, t)$, where W^{Sun} is the gravitational potential of the Sun. The force on the element occupying a volume dV at that position inside the planet is $-\rho(\vec{r})\nabla W^{Sun}(\vec{r}, t)dV$, where $\rho(\vec{r})$ is the density of matter at $\vec{r}$. In a reference frame at the planet's mass center, this acceleration is the centripetal acceleration of the planet as a whole toward the Sun, which is responsible for the relative orbital motion of the planet around the Sun. We neglect deviations from spherical symmetry in the Sun and then have

$$W^{Sun}(\vec{r}, t) = -\frac{GM_{Sun}}{|\vec{d} - \vec{r}|}$$

$$= -\frac{GM_{Sun}}{\sqrt{d^2 - 2(\vec{d} \cdot \vec{r}) + r^2}} \cong -\frac{GM_{Sun}}{d}$$

$$\times \left(1 + \frac{\vec{d} \cdot \vec{r}}{d^2} - \frac{r^2}{2d^2} + \frac{3(\vec{d} \cdot \vec{r})^2}{2d^4} + \cdots \right)$$

where $\vec{d}$ is the position vector of the Sun with respect to the planet and is time dependent because of the motion of the Sun relative to the planet frame, where $|\vec{d}| = d$ and $|\vec{r}| = r$.

In the reference frame $(\widehat{x}, \widehat{y}, \widehat{z})$ tied to the planet, the torque $\vec{\Gamma}$ exerted on the planet by the Sun may be calculated by expressing the force (and hence the torque) on a mass element in the planet in terms of the gravitational potential due to the Sun at the location of the mass element, making vectorial multiplication with $\vec{r}$, and then integrating over the whole planet. The torque $\vec{\Gamma}$ that it produces on an element of matter at $\vec{r}$ is

$$\vec{\Gamma} = -\int \rho(\vec{r})\vec{r} \times \vec{\nabla} W^{Planet}(\vec{r}, t)dV$$

$$= -\frac{GM_{Sun}}{d} \int \rho(\vec{r})$$

$$\times \left(\frac{\vec{r} \times \vec{d}}{d^2} + \frac{3(\vec{d} \cdot \vec{r})(\vec{r} \times \vec{d})}{d^4}\right)dV$$

where the integration is over the volume of the planet. The three components of the torque integral acting on the rotating planet can then be expressed as

$$\begin{pmatrix} \Gamma_x \\ \Gamma_y \\ \Gamma_z \end{pmatrix} = \frac{3GM_{\text{Sun}}}{d^5} \int \rho(\vec{r})\{d^2 + 3(xd_x + yd_y + zd_z)\}$$

$$\times \begin{pmatrix} (yd_z - zd_y) \\ (zd_x - xd_z) \\ (xd_y - yd_x) \end{pmatrix} dV$$

The components (d_x, d_y, d_z) are the coordinates of $\vec{d}$ and (x, y, z) are the coordinates of $\vec{r}$ in the $(\hat{x}, \hat{y}, \hat{z})$ reference frame. We choose the axes x, y, and z to coincide with the **principal axes of inertia** of the static planet, which we refer to as the principal axis frame ("static" means that variations in the density distribution due to tidal deformations and other causes are not considered). The first term of the above integral drops out due to the fact that the reference frame is tied to the center of mass and the equator of the planet. Since we have chosen a principal axis coordinate system, the matrix of inertia reduces to a diagonal matrix; therefore, $\int \rho(\vec{r})xy dV = \int \rho(\vec{r})yz dV = \int \rho(\vec{r})xz dV = 0$. By using the definitions of the principal moments of inertia, we then have

$$\begin{pmatrix} \Gamma_x \\ \Gamma_y \\ \Gamma_z \end{pmatrix} = \frac{3GM_{\text{Sun}}}{d^5} \begin{pmatrix} (C - B) d_y d_z \\ (A - C) d_x d_z \\ (B - A) d_x d_y \end{pmatrix}$$

For rapidly rotating planets, the difference between the two equatorial moments of inertia is a few orders of magnitude smaller than the difference between the polar moment of inertia and the mean equatorial moment of inertia. Therefore, the planet can be considered as axially symmetric with $A \approx B$. In this case, $\Gamma_z = 0$, and so the torque vector lies in the equatorial plane. It, moreover, is perpendicular to the equatorial projection (d_x, d_y) of $\vec{d}$, so it is perpendicular to the direction to the Sun.

The torque is maximal when the Sun is at the highest points above the equator (at summer and winter solstices). At both points, the torque tends to align the equator with the orbit plane. It is zero at the equinoxes. The torque will change the angular momentum, and thus the rotation of the planet according to extension of Newton's third law to rotational motion is $\frac{d\vec{H}}{dt} = \vec{\Gamma}$.

For a rigid body rotating around an axis of symmetry (say the polar axis $\hat{z}$), the angular momentum can be expressed as the product of the polar moment of inertia, C, and its angular velocity Ω_Z: $\vec{H} = C\Omega_z \hat{z}$, where $\hat{z}$ is the unit vector in the z direction. Since the torque $\vec{\Gamma}$ is in the equatorial plane and the angular momentum is perpendicular to the equatorial plane, we have

$$\vec{H} \cdot \vec{\Gamma} = 0,$$

and hence

$$\vec{H} \cdot \frac{d\vec{H}}{dt} = 0 \quad \text{or} \quad \frac{d(\vec{H} \cdot \vec{H})}{dt} = 0.$$

Therefore, the magnitude of angular momentum does not change with time. As a result, the torque can only change the orientation of the planet but not its rotation speed.

The average torque is oriented in the direction of the line joining the equinoxes ($\hat{X}$). For a circular orbit, it can easily be calculated from the above expression of the torque by using expressions for the position of the Sun with respect to the planet. Instead of considering a reference frame with rotating axes coinciding with principal axes of inertia as we did above, we will now change to nonrotating equatorial x- and y-axes and choose the x-axis to coincide with the X-axis of the reference frame tied to the orbital plane $(\hat{X}, \hat{Y})$ (see Figure 8.8). We can do so because we consider a biaxial planet. Any equatorial axis can then be considered as a principal axis of inertia. The coordinates of the Sun d_x, d_y, and d_z in the $\hat{x}$, $\hat{y}$, and $\hat{z}$ directions of this frame can then easily be seen to be

$$d_x = d \cos(nt)$$
$$d_y = d \cos\varepsilon \sin(nt)$$
$$d_z = d \sin\varepsilon \sin(nt).$$

For the average torque we then have,

$$\overline{\Gamma}_x = \frac{3n^2}{2(C - \overline{A})\sin\varepsilon} \cos\varepsilon, \quad \overline{\Gamma}_y = 0 \text{ and } \overline{\Gamma}_z = 0.$$

The fact that there is only an average torque component in the X-axis can be understood from the geometry of the problem by realizing that the torque is oriented along the line joining the equinoxes when it takes its maximum value at the solstices and diminishes to zero and changes its orientation to the perpendicular direction when the Sun moves to the equinoxes. The average torque is therefore perpendicular to the plane formed by the rotation axis and the orbit normal. As a result, the angular momentum of the planet cannot acquire an additional component in the plane of the orbit normal and rotation axis. The average torque, therefore, cannot change the obliquity of an oblate spheroidal (biaxial) planet and results in a gyroscopic motion in which the rotation axis moves on the surface of a cone with the orbit normal as its axis (precession). In a triaxial case, there will be an additional change in the orientation of the cone with respect to space.

As shown by the above expression, the magnitude of the average solar torque on the planet is determined by the obliquity ε, the mean motion n, and the parameter $(C - \overline{A})$. The torque would vanish if the planet were spherically symmetric or if the obliquity ε were zero, i.e. if the orbit of the Sun were to lie in the equatorial plane itself. It also decreases for smaller mean motion, or increasing distance to the Sun.

The change in the rotation can be calculated from the angular momentum equation $\frac{d\vec{H}}{dt} = \vec{\Gamma}$, which is valid only in an inertial reference frame. We therefore need an expression for the torque in an inertial reference frame. We have shown that the average torque has only an X-component, but the X-axis considered is not an inertial axis as it changes with the precession we are studying here. Nevertheless, during one orbital evolution of the planet it almost does not change as precession is very slow, which we will see below. Therefore, the X-component will be to a very good approximation equal to the x-component of an inertial frame.

Because of precession, the line of the equinoxes and thus the $\widehat{X}$-axis defined above changes its orientation in space (see Figure 8.9).

Therefore, as seen from Figure 8.9, the rate of change in the X-direction due to precession is $\dot{\psi} \sin \varepsilon$. The rate of change in the X-component of the angular momentum vector, $\dot{H}_X$, is thus equal to the rate of change in the celestial longitude angle $\dot{\psi} \sin \varepsilon$ times the angular momentum amplitude $C\Omega$, as can be seen by projecting the rate of change of the unit angular momentum vector $\frac{H_X}{(C\Omega)}$ on the equator. Therefore we have

$$\Gamma_x = \dot{H}_x = C\Omega \dot{\psi} \sin \varepsilon.$$

By using this and using the expression of the X-component of the torque, the precession rate can be expressed as

$$\dot{\psi} = \frac{3n^2}{2\Omega} \left(\frac{C - \overline{A}}{C} \right) \cos \varepsilon.$$

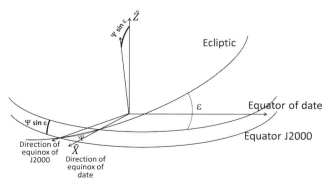

FIGURE 8.9 Representation of precession and changes in the equator. The direction of the equinox at a given time ("the equinox of date") changes with respect to any chosen fixed direction, here the direction of equinox of J2000, due to precession.

Or equivalently by using $J_2 = \frac{C - \overline{A}}{M r_{eq}^2}$ as

$$\dot{\psi} = \frac{3n^2}{2\Omega} \frac{M r_{eq}^2}{C} J_2 \cos \varepsilon.$$

For the Earth, the time needed to perform one cycle around the orbit normal is about 25,600 years for a dynamical flattening around $1/300 \left(\frac{C-\overline{A}}{C} = 3.274 \times 10^{-3} \text{ and } J_2 = 1.082 \times 10^{-3} \right)$. For the other planets, the periods of precession may be very different and increase with increasing dynamical flattening defined in Section 4 and decrease with increasing distance from the Sun. For example for Mars, as seen above, the precession period is about 171,000 years and corresponds to a dynamical flattening of about $1/900 \left(\frac{C-\overline{A}}{C} = 1.074 \times 10^{-3} \text{ and } J_2 = 3.913 \times 10^{-4} \right)$. For Venus, the precession period is thought to be about 29,000 years, for Mercury about 550 years.

Precession is a very useful observable for the internal geophysics of the Earth and Mars because its rate is inversely proportional to the polar moment of inertia C. Since all other quantities influencing the precession rate are well known (the mean rotation rate, the mean orbital motion, the obliquity, and the difference in moments of inertia from the J_2 coefficient), observation of the precession yields an estimate of the polar moment of inertia, which is a measure of the radial mass distribution in the planet. For Mars, precession is one of the main and best-determined constraints on the interior structure. Also for the Earth, precession is used in addition to seismology, to determine the internal structure (See Earth: Surface and Interior and Probing the Interiors of Planets with Geophysical Tools).

6. NUTATION

As explained above, the torque changes with time during the orbital motion. Besides having a component perpendicular to the plane formed by the angular momentum and the orbit normal, the torque also has a component perpendicular to the line joining the equinoxes. The former will induce periodic changes to the precession, while the latter will cause the obliquity to change periodically. In addition, due to interaction with other solar system bodies, the orbit changes. All the orbital parameters of the orbit of the planet around the Sun, such as the eccentricity, change with time. As a consequence, the gravitational torque acting on the planet changes with time, and thus also the orientation in space. The changes are periodic with periods equal to harmonics of the orbital motion around the Sun and to the periods with which the orbital elements of the planet around the Sun change, which are themselves related to orbital periods of the planets perturbing the orbital motion. The situation is similar if one considers moon(s) around a planet. Figure 8.10 shows these changes in the orientation of the planet in space.

The Movie 2 shows the periodic variations of the orientation of the Earth in space, the nutation.

Supplementary video related to this chapter can be found at http://dx.doi.org/10.1016/B978-0-12-415845-0.00008-6.

The following is/are the supplementary data related to this chapter: Movie 2: Nutation of the Earth in space. (file EarthNutation.mpg)

While precession carries the pole of the axis at a uniform rate in an anticlockwise sense along a circle on the surface of the celestial sphere, centered on the normal to the ecliptic plane, nutation consists of small deviations from this uniform motion, both over the precessional path (called nutation in longitude) and perpendicular to it on the celestial sphere (called nutation in obliquity). The resulting path of the pole appears wiggly, as in Figure 8.10. For a hypothetical planet with a circular orbit, the torque variation has a period equal to half the orbital period. Since the planets of the solar system all have orbits close to circular, the main nutation will also have a semiannual period.

The amplitude of the nutation with a period half the orbital period depends on the shape of the planet, the obliquity, the rotation speed, and the orbital speed. For a circular orbit, this can be shown as follows.

As we have done for the constant part of the torque, starting from its above expression (see Section 5) and substituting the expressions for the position of the Sun with respect to the planet, the time-dependent torque as a function of the angular velocity n and the obliquity ε can thus be written as:

$$\Gamma_x = \frac{3n^2}{2}(C - \overline{A}) \sin \varepsilon \cos \varepsilon (1 - \cos 2nt),$$

$$\Gamma_y = -\frac{3n^2}{2}(C - \overline{A}) \sin \varepsilon \sin 2nt, \quad \text{and} \quad \Gamma_z = 0$$

For $F = \frac{3n^2}{2}(C - \overline{A}) \sin \varepsilon$, the torque becomes, $\Gamma_x = F \cos \varepsilon (1 - \cos 2nt)$, $\Gamma_y = -F \sin 2nt$, and $\Gamma_z = 0$.

As shown above, the constant term leads to precession. The time-dependent part causes nutation. The magnitude of the time-dependent part of the solar torque on the planet is determined by the mean motion n, the obliquity ε, and the

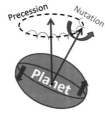

FIGURE 8.10 Nutation of the planet's principal axis in space.

parameter $(C - \overline{A})$. As previously for precession, the torque would vanish if the planet were spherically symmetric or if the obliquity ε were zero, i.e. if the orbit of the Sun were to lie in the equatorial plane itself.

We also see that the torque has a period of half a year (frequency $2n$). The time variation of (Γ_x, Γ_y) describes an elliptical path in the equatorial plane. The periodic part of the torque components are

$$\Gamma_x = -F \cos \varepsilon \cos 2nt, \quad \Gamma_y = -F \sin 2nt.$$

The semiannual nutation, which is the largest of the nutations of solar origin, is the result of this periodic term in the torque. Both the torque and the nutation may be resolved, like any periodic elliptical motion, into two counterrotating circular motions:

$$\begin{pmatrix} \Gamma_x \\ \Gamma_y \end{pmatrix} = -\frac{F}{2} \left[(1 + \cos \varepsilon) \begin{pmatrix} \cos 2nt \\ \sin 2nt \end{pmatrix} - (1 - \cos \varepsilon) \begin{pmatrix} \cos 2nt \\ -\sin 2nt \end{pmatrix} \right].$$

The first of the two column vectors on the right-hand side of this equation represents a vector in prograde (counterclockwise) motion, while the second one is a retrograde motion. The prograde part of the nutation is due to the prograde part of the torque and the retrograde part of the nutation results from the retrograde part of the torque. Prograde means that the angle increases with time counterclockwise like the rotation of the Earth; retrograde means that the angle increases clockwise with time. The superposition of two motions, one clockwise and the other counterclockwise, gives rise to an elliptical motion. The direction of the elliptical motion depends on which amplitude of the prograde and retrograde terms is largest. For the semiannual nutation, the first (prograde) term is larger than the second (retrograde) term $((1 + \cos \varepsilon)$ is larger than $(1 - \cos \varepsilon))$, and the elliptical semiannual nutation is prograde.

A lot of nutation terms other than semiannual arise when one takes into account the deviations from the simple model that we considered here: the orbit of the Sun relative to the planet is not circular but elliptical, so that d is not constant, the simple harmonic time dependence of d_x and d_y is a simplification, and there are other planetary perturbations of the orbit of the Earth around the Sun. The present simplified picture serves to highlight the essential aspects of the nutation–precession phenomenon.

Nutations have up to now only been unambiguously observed for the Earth. Very long baseline interferometry (VLBI, see Section 10.3) observations provide the precession and nutation of the Earth with a precision below the tenth of a milliarcsecond (see Section 10). For this semiannual nutation, the amplitudes of the nutations in

TABLE 8.2 Periods and Amplitudes of the Main Earth Nutations

Periods of Nutations	Amplitudes of Non-Rigid-Earth Nutations in Longitude	Amplitudes of Non-Rigid-Earth Nutations in Obliquity
18.6 years (lunar origin)	17 208 mas	9205 mas
0.5 year (solar origin)	1317 mas	573 mas
9.3 years (lunar origin)	207 mas	90 mas
1 year (solar origin)	148 mas	7 mas
13.7 days (lunar origin)	228 mas	98 mas

longitude and obliquity are about 1300 mas and 570 mas, respectively, which corresponds to a quasicircular motion at the Earth's surface of about 16 m as seen from space (see Table 8.2). The largest nutations are of lunar origin and have a period of about 18.6 years (and 9.3 years to a minor extent), which arises from the precession of the lunar orbit around the ecliptic. The amplitudes of nutation in longitude and obliquity of the 18.6-year nutation are about 17 200 mas and 9200 mas, respectively (for the 9.3-year nutation they are about 200 mas and 100 mas, respectively). The nutations of solar origin at 1 year and half a year have amplitudes provided in Table 8.2.

Nutation for a rigid planet can easily be computed from the torque as performed above for precession, given that the ephemerides (relative positions of the celestial bodies) are known. Those theoretical values, however, do not all correspond to the observed values since the planets are deformable and thus nonrigid and contain a liquid layer such as the outer core of the Earth. Thanks to the high precision of the observations, information on the interior of the Earth can be obtained. In order to do so, models for the precession and nutation of a nonrigid body have been developed. A short description of such a model is given in Section 9 on the wobbles. That section also discusses results of a comparison between observational results and theoretical modeling for the Earth and explains what theory predicts for the nutations of Mars.

7. LOD VARIATIONS

The rotation of a planet is approximately uniform, meaning that the planet rotates with an almost constant rate. Small variations in the rotation rate occur due to various reasons. For the terrestrial planets with an atmosphere, the largest of these changes are due to the atmosphere dynamics. For Mercury the largest rotation variations are most likely due to the gravitational torque exerted by the Sun. This will be further developed in the next section (see Section 8). The gas and ice planets, which are primarily composed of H, He, and hydrides of the light elements O, C, and N (i.e. water H_2O, methane CH_4, and ammonia NH_3) are mainly fluid. Different large atmospheric patterns such as the great Red Spot on Jupiter and large-scale circulations such as banded structures have been observed. Since the global mean rotation rate of the giant planets is not well known (e.g. within a few minutes for Saturn) here we only consider how exchanges of angular momentum within the planet can affect the rotation of the surface of the terrestrial planets.

If we consider an isolated planet or neglect external torques, the angular momentum of the system is conserved:

$$\frac{d\vec{H}}{dt} = 0.$$

Any change in the angular momentum a layer of the planet must necessarily be balanced by a change in angular momentum of another layer. Therefore, the rotation of the solid outer part of the terrestrial planet (*solid*) can change due to angular momentum exchange with other layers. For example, the Earth's rotation rate changes due to interactions and angular momentum exchange of the mantle and crust with the atmosphere, the liquid outer core, and the solid inner core (See Earth: Surface and Interior). As an example, we consider angular exchange with the atmosphere (*atm*) only. Because of conservation of angular momentum, we have

$$\frac{d\vec{H}_{\text{solid}}}{dt} = -\frac{d\vec{H}_{\text{atm}}}{dt}.$$

Changes in the winds and in the mass distribution in the atmosphere both affect the angular momentum of the atmosphere ($\vec{H}_{\text{atm}}$) and consequently change the angular momentum and rotation of the solid planet ($\vec{H}_{\text{solid}}$). For example, stronger winds from the west will increase the angular momentum of the atmosphere and reduce the angular momentum of the solid planet. Likewise, if mass is transferred to the polar regions, the atmospheric angular momentum decreases and the planet rotates faster. This effect is particularly important for Mars, in particular since during winter about one-fourth of the total mass of the CO_2 atmosphere of Mars condenses at the winter pole. If the planet can be considered to be rigid, any change in the angular momentum of the solid planet transfers immediately into changes in the angular velocity with respect to a uniform rotation Ω along the z-axis:

$$-\frac{d\vec{H}_{\text{atm}}}{dt} = \bar{\bar{I}} \frac{d\vec{\Omega}}{dt}$$

FIGURE 8.11 Variations of the LOD in millisecond (ms) over several decades showing in addition to the seasonal variations the decadal changes of Earth rotation related to the core. This series has been computed from observation gathered at the International Earth Rotation and Reference Frame Service (IERS)—Earth Orientation Center at Observatoire de Paris (http://hpiers.obspm.fr/eop-pc/).

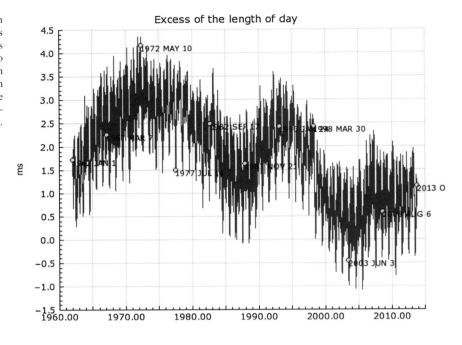

For a rigid body rotating around an axis of symmetry (say the polar axis $\hat{z}$), the angular momentum can be expressed as the product of the polar moment of inertia, C, and its angular velocity Ω_z: $H_Z = C\Omega_z$. We then have

$$\frac{dH_z}{dt} = C\frac{d\Omega_z}{dt}.$$

The change in the rotation of the planet is therefore proportional to the change in the angular momentum of the atmosphere and inversely proportional to the polar moment of inertia, which expresses the rotational inertia. The most important changes in the angular momentum in the atmosphere are due to the seasons. The Earth's LOD changes at seasonal timescales by a few milliseconds (see Figure 8.12 and Figure 8.13). Variations due to weather systems on much shorter timescales can change the rotation of the Earth by up to a few milliseconds.

As an illustration, Movie 3 shows the uniform rotation of the Earth viewed from space and a comparison with the nonuniform case. Movie 4 shows first the

FIGURE 8.12 Variations of the LOD in millisecond over several years showing the seasonal variations of Earth rotation related to the atmosphere and oceans. This series has been computed from observation gathered at the IERS — Earth Orientation Center at Observatoire de Paris (http://hpiers.obspm.fr/eop-pc/).

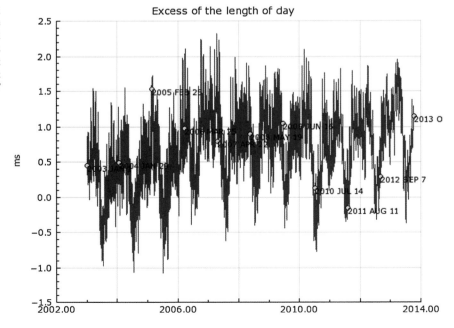

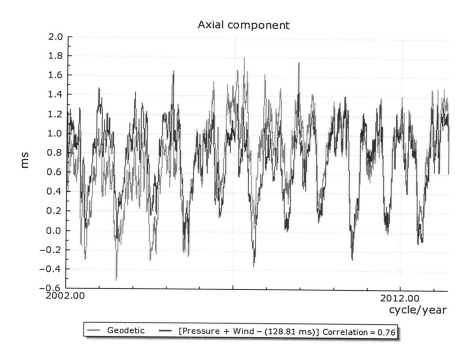

FIGURE 8.13 Variations of the LOD in millisecond (ms) over several years showing the seasonal variations of Earth rotation (in red) and the LOD computed from the atmospheric angular momentum. These series have been computed from observation gathered at the IERS — Earth Orientation Center at Observatoire de Paris (http://hpiers.obspm.fr/eop-pc/).

nonuniform rotation, then the nutation, and next the polar motion (defined in Section 9 as the motion of the rotation axis in a frame tied to the planet). Movie 5 shows first the combination of all motions, then the nonuniform rotation, and separately the nutation, then the polar motion. In these animations we have exaggerated the motions and did not respect the real timing for the Earth. They are only intended to illustrate the physical phenomena.

Supplementary video related to this chapter can be found at http://dx.doi.org/10.1016/B978-0-12-415845-0.00008-6.

The following is/are the supplementary data related to this chapter: Movie 3: Earth uniform rotation. (file EarthUniformRotation.wmv)
Supplementary video related to this chapter can be found at http://dx.doi.org/10.1016/B978-0-12-415845-0.00008-6.

The following is/are the supplementary data related to this chapter: Movie 4: Earth nonuniform rotation nutation. (file.EarthNonUniformRotationNutationPM.wmv)
Supplementary video related to this chapter can be found at http://dx.doi.org/10.1016/B978-0-12-415845-0.00008-6.

The following is/are the supplementary data related to this chapter: Movie 5: Earth nonuniform orientation. (file.EarthNonUniformOrientation.wmv)

Exchange of angular momentum with the liquid outer core is also an important cause of rotation variations for the Earth and is a likely cause of rotation variations for the other terrestrial planets, in particular Mercury. At decadal timescales, flow in the Earth's core related to variations in the **geodynamo** (See Magnetic Field Generation) leads to changes in the angular momentum of the core which are partially transferred to the solid parts of the planet through several coupling mechanisms at the CMB and inner core boundary (ICB). In a first approximation, the motion in the core is often supposed to be geostrophic, such that the flow has constant velocity on cylinders coaxial with the rotation axis. Each cylinder is rotating at its own rotation angular velocity and these rotation speeds are varying with time and linked through the so-called torsional oscillations of which the amplitudes varies with time. From the velocity field in the whole core, one can compute the angular momentum of the core, and hence, the variation of the LOD induced by the time variation of this quantity.

The coupling mechanisms are mainly due to the gravitational torque resulting from the core—mantle gravitational interactions, the electromagnetic torque, and the pressure torque mainly related to the differential motions of the core with respect to the mantle on the bumpy boundary of the core. Figure 8.14 illustrates these coupling mechanisms in the case of the atmosphere. Figure 8.11 shows the LOD variations of the Earth over several decades. Both the decadal variations related to the decadal changes in the core angular momentum and seasonal variations due to interaction with the atmosphere are clearly visible. Figure 8.12 considers Earth rotation variations over 12 years and better shows the seasonal and shorter variations due to interaction with the atmosphere. At that timescale, angular momentum exchange with the oceans contributes about 10% of the rotation variations.

Variations in the angular momentum of a fluid layer can be computed if the velocity field in the fluid is known. For

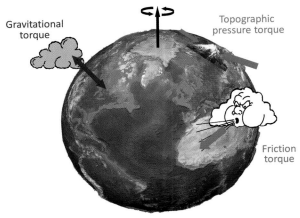

FIGURE 8.14 Coupling mechanism acting on the solid Earth. This results in an angular momentum exchange between the solid planet and its atmosphere.

the atmosphere or the ocean, general circulation models can be used and the angular momentum can be calculated with the help of the following integral over the volume of the fluid layer

$$\vec{H}_{\text{fluid}} = \int_{\text{Fluid volume}} \vec{r} \times \rho \vec{v} \, dV$$

where fluid particles have the velocity $\vec{v}$ at the position $\vec{r}$ in the fluid layer. For an atmosphere or ocean on Earth, the fluid can be considered as a thin layer allowing to neglect the gravity variations with the distance to the center along a fluid column. One considers the hydrostatic equilibrium along the vertical: $dp = -\rho g \, dz$ (where ρ is the fluid density, p is the pressure in the fluid layer, and g is the gravity), so that

$$\vec{H}_{\text{fluid}} = \frac{r_{\text{surface}}^2}{g} \int_0^{2\pi} \int_0^{\pi} \int_0^{P_{\text{surface}}} \vec{r} \times \vec{v} \sin\theta \, dp \, d\theta \, d\lambda$$

where r_{surface} is the mean radius of the planet; θ and λ are the latitude and longitude respectively; and P_{surface} is the surface pressure. Considering that $\vec{v} = \vec{\omega} \times \vec{r} + \vec{u}$ and transforming the volume integral into a surface integral, this yields

$$\vec{H}_{\text{fluid}} = \frac{r_{\text{surface}}^4}{g} \int_0^{2\pi} \int_0^{\pi} \hat{r} \times (\vec{\omega} \times \hat{r}) P_{\text{surface}}(\theta, \lambda) \sin\theta \, d\theta \, d\lambda$$
$$+ \frac{r_{\text{surface}}^3}{g} \int_0^{2\pi} \int_0^{\pi} \int_0^{P_{\text{surface}}} \hat{r} \times \vec{u} \sin\theta \, dp \, d\theta \, d\lambda$$

or more explicitly for the atmosphere

$$\vec{H}_{\text{atmosphere}} = \frac{r_{\text{surface}}^4}{g} \int_0^{2\pi} \int_0^{\pi} P_{\text{surface}}(\theta, \lambda) \sin^2\theta \begin{pmatrix} \cos\theta \cos\lambda \\ \cos\theta \sin\lambda \\ \sin\theta \end{pmatrix} d\theta \, d\lambda + \frac{r_{\text{surface}}^3}{g}$$

$$\times \int_0^{2\pi} \int_0^{\pi} \int_0^{P_{\text{surface}}} \sin\theta \begin{pmatrix} u_\theta \sin\lambda + u_\lambda \cos\theta \cos\lambda \\ -u_\theta \cos\lambda + u_\lambda \cos\theta \sin\lambda \\ -u_\lambda \cos\lambda \end{pmatrix} dp \, d\theta \, d\lambda$$

$$= \vec{H}_{\text{atmosphere}}^{\text{mass}} + \vec{H}_{\text{atmosphere}}^{\text{wind}}$$

The first term depends on the pressure and thus on the total atmospheric mass rotating with the planet. It is the angular momentum associated with a global rotation of the atmosphere with the planet and is called the mass term or inertia term. The second term involves the relative velocity with respect to the solid planet and is called the wind term. The sum of time derivative of these two terms can also be expressed in terms of torques acting on the surface.

The changes in the rotation of the planet can be computed from Global Circulation models (GCMs) providing maps of the hydrostatic pressure on the surface of the planet and maps of winds at the surface of the planet.

These exchanges of angular momentum exist on Mars as well. Mars' atmosphere consists of 95% carbon dioxide (CO_2) and, although it is much more diluted and has a surface pressure hundred times smaller than that on the Earth, it undergoes large seasonal changes related to the CO_2 sublimation and condensation process (See Mars Atmosphere: History and Surface Interactions). About one-fourth of the atmosphere is participating in this process, forming large ice caps in the winter of each hemisphere (the maximum mass of the north and south seasonal caps that is produced is 3.0×10^{15} and 5.5×10^{15} kg, respectively; the total mass of the atmosphere is about 2.5 10^{16} kg). Using GCMs, it is possible to compute the angular momentum in the Martian atmosphere and therewith to estimate the large changes in the LOD. LOD variations have amplitudes at the same level as that on Earth, at the level of several tenths of millisecond with annual and semiannual periods. Wind is found to induce rotation angle variations with an amplitude of 14 mas for the annual period and 76 mas for the semi-annual period. The effect of the ice caps is small because mass at the polar ice caps has only a small contribution to the polar moment of inertia C, whereas atmospheric masses can be a Mars radius further from the polar axis and cause a larger change in the polar moment of inertia. The amplitude of the rotation variations is at the level of 500 mas on the equator over 1 year corresponding to annual changes in the LOD at the level of 0.3 ms. For the semiannual amplitude, we have 170 mas on the equator, corresponding to LOD changes at the level of 0.2 ms. A global dust storm as it might happen on Mars may induce variations in the LOD of several percents of the total seasonal effect.

LOD variations on Venus are expected to be small because Venus does not have seasons as the Earth and Mars (See Venus Atmosphere).

8. LIBRATION

No planet or satellite is exactly spherically symmetric. In a lowest order approximation, planets and satellites take the form of an ellipsoid (see Section 4). For rapidly rotating planets, an oblate **spheroid** (biaxial ellipsoid) flattened at the poles represents well the shape (the equator is a perfect circle but any meridian circle passing through the poles is an ellipsoid flattened at the equator). For a **triaxial ellipsoid** planet or satellite, the equator is also an ellipse. The three principal moments of inertia are then different ($A < B < C$), while for the spheroid $A = B$. Due to the rotational and orbital motion the long axis of the planet or satellite in the equatorial plane, coinciding with the axis of smallest principal moment of inertia A (see Section 4), is generally not in the direction to the central body. Therefore, the central body exerts a gravitational torque on the planet or satellite, which therefore accelerates or decelerates its rotation depending on its orientation with respect to the central body. When the planet's rotation is much faster than the orbital motion as, for example, for the Earth and Mars, the effects on rotation are very small. However, in the case of a **spin–orbit resonance**, the torque of the central body can cause relatively large variations in the rotation rate. These longitudinal librations are important for Mercury and also for satellites in a 1:1 spin–orbit resonance. Figure 8.15 illustrates the geometry

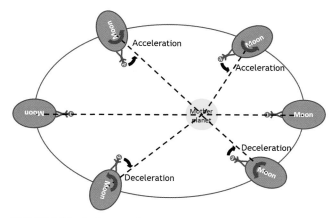

FIGURE 8.15 In the case of a moon in a 1:1 spin–orbit resonance with its mother planet, the gravitational attraction of the mother planet tends to align the equatorial bulge with respect to the Sun's direction indicated with the black dashed line; there is resulting acceleration or deceleration of the moon's rotation.

of the problem for satellites in a 1:1 spin–orbit resonance. Although the rotational period is exactly equal to the rotational period, the satellites do not exactly show the same face to the central planet because the speed of orbital motion varies according to Kepler's second law as a result of the eccentricity of the orbit.

We neglect the very small obliquity of Mercury and the synchronous satellites. The polar component of the gravitational torque exerted by the central body can then be expressed as (see Section 5)

$$\Gamma_3 = \frac{3}{2}(B - A)\frac{GM_{\text{central}}}{r^3}\sin 2\xi$$

where r is the distance between the mass centers of the central body and the planet or satellite, and ξ is the angle between the direction of the long axis and the direction to the central body ($\xi = f - \varphi$, where f is the true **anomaly** and φ is the rotation angle between the long axis of the planet or satellite and the major axis of the orbit) (see Figure. 8.16).

For satellites in a 1:1 spin–orbit resonance, the rotation angle is close to the mean **anomaly** describing the mean angle associated with orbital motion and defined as $M = \frac{2\pi}{n}(t - t_0)$, where t_0 is the time when the satellite passes through pericenter. One can then introduce a small libration angle $\gamma = \varphi - M$.

For Mercury, the rotation angle changes faster than the orbital angle by a factor of about 1.5 and the small libration angle is defined as $\gamma = \varphi - 1.5M$. At the pericenter, the libration angle describes the difference between the direction of the long axis and the direction to the central body. It can be determined by solving the equation for the change in angular momentum due to the above torque.

By expressing the true anomaly and the distance between the mass centers in terms of the semimajor axis, the

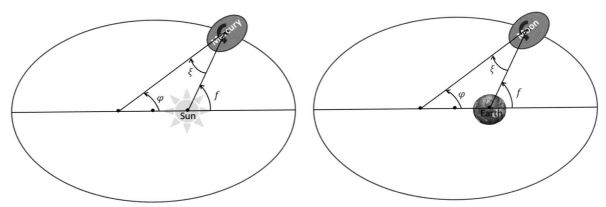

FIGURE 8.16 Definition of the angles used to describe the libration of Mercury around the Sun or of the Moon around the Earth.

eccentricity, and the mean anomaly for a Keplerian orbit, the torque can be seen to consist of a term proportional to the libration angle and periodic terms with period equal to the orbital period and its harmonic frequencies. If we restrict ourselves to the largest term with period equal to the orbital period and assume that Mercury behaves rigidly, the equation for angular momentum of Mercury can be expressed as the pendulum equation

$$\frac{d^2\gamma}{dM^2} + \frac{3}{2}\frac{B-A}{C}e\left(7 - \frac{123}{8}e^2\right)\gamma = -\frac{3}{2}\frac{B-A}{C}[(1 - 11e^2 + \ldots)\sin M].$$

Also, for synchronously rotating satellites, the governing equation for libration can be expressed as a pendulum equation. The amplitude of the libration angle of the main libration at orbital period of 88 days, $\gamma_{88\text{days}}$, can then be approximately expressed as

$$\gamma_{88\text{days}} = \frac{3}{2}\frac{B-A}{C}(1 - 11e^2 + \ldots).$$

For rigid synchronous satellites, the amplitude of libration at the orbital period, $\gamma_{\text{orbital period}}$, can be approximately expressed as

$$\gamma_{\text{orbital period}} = 6\frac{B-A}{C}e.$$

An important difference with respect to the libration amplitude $\gamma_{88\text{days}}$ of Mercury in 3:2 spin–orbit resonance is that the libration amplitude $\gamma_{\text{orbital period}}$ for a 1:1 spin–orbit resonance decreases linearly with the eccentricity. The decrease with eccentricity for a satellite in a 1:1 spin–orbit resonance can be understood by considering the limit case of a circular orbit. In that case, there would not be any forced libration in longitude since the orbital speed is then constant and the satellite always shows exactly its same face to the central planet.

As is typical for a pendulum equation, a free libration is also possible. When the long axis does not point toward the Sun at perihelion, the averaged gravitational torque on Mercury tends to restore the alignment, and the long axis will librate around the direction to the Sun at perihelion. The free libration period of a rigid Mercury is given by

$$P_{\text{free}} = \sqrt{\frac{1}{\frac{3}{2}\frac{B-A}{C_m}e\left(7 - \frac{123}{8}e^2\right)}}\frac{2\pi}{n}$$

Free libration is also due to the gravitational interaction between the Sun and Mercury, as is forced libration, but the distinctive feature of free libration as opposed to forced libration is that the amplitude and phase of free libration cannot be determined without knowing its excitation and dissipation, as is the case for a free harmonic oscillator. Free libration has a much longer period of several years than the main forced libration at 88 days and has not yet been unambiguously observed.

An illustration of the libration of Mercury is shown in Movie 6.

Supplementary video related to this chapter can be found at http://dx.doi.org/10.1016/B978-0-12-415845-0.00008-6.

The following is/are the supplementary data related to this chapter: Movie 6: Libration of Mercury. (file MercuryLibration.wmv)

Like nutation, libration depends on the interior. The main effect, and probably the only effect that is substantially larger than the precision of current observation techniques, is due to the liquid outer core of Mercury. A liquid outer core effectively decouples the libration of the mantle from the core. The total torque on the mantle from the Sun and the core is equal to the total torque on an entirely rigid Mercury but the polar moment of inertia, which represents the resistance to rotational forcing, is smaller than the total planetary polar moment of inertia. Therefore, the libration of Mercury with a liquid outer core is larger than that of an entirely solid Mercury by about a factor C/C_m. The libration period with a liquid core is also shorter than that without by a factor $(C_m/C)^{1/2}$. Based on radar observations using two different large radio telescopes (see Section 10), the amplitude of the 88-day libration of Mercury has been determined as (38.5 ± 1.6) arcsec. This means that a point at the equator of Mercury will be periodically offset from its equilibrium position at constant rotation rate with an amplitude of (455 ± 19) m. By using $B - A = (3.235 \pm 0.024) \times 10^{-5} \, MR^2$, estimated from radio tracking observations of the National Aeronautics and Space Administration (NASA) MESSENGER mission currently in orbit around Mercury, the polar moment of the mantle can then be determined as $C_m = (0.148 \pm 0.006) \, MR^2$, less than half the value for the total planetary moment of inertia determined from the obliquity value. Therefore, Mercury cannot be entirely solid and a global liquid layer must exist. For an entirely solid Mercury, the libration amplitude at the equator would be about 190 m. The comparison of the rotational constraints on C and C_m with theoretical moment of inertia values for models of the interior structure of Mercury shows that the liquid core has a radius of about 2000 km. The core of Mercury is probably larger than the core of the larger planet Mars. It is relatively much larger than the core of the Earth, which has a radius of about 55% of the total radius of the Earth. For Mercury, the relative core radius is about 80%. This clearly shows that Mercury has had a different formation history.

9. WOBBLES AND THE INTERIORS OF TERRESTRIAL PLANETS

Besides precession and nutation which describe the changes in the orientation of a planet in space, the position of the rotation axis can also change with respect to a reference frame tied to the planet. The term wobble is used in a very broad sense for any periodic or quasiperiodic motion of the instantaneous rotation axis of a planet or moon with respect to the figure axis, i.e. conventionally the z-axis of the frame tied to the planet, irrespective of the frequency or the physical origin of the motion. In general, wobble cannot occur without accompanying nutation, and equivalently, nutation is always associated with wobble, as we explain below. Depending on the period, nutation is either much larger or much smaller than wobble for planets, like the Earth, rotating much faster than the orbital motion. More specifically, at periods much longer than a day in the terrestrial frame, the wobble has a larger amplitude than the corresponding motion of the instantaneous rotation axis in space. The frequency of the latter motion is quasidiurnal due to the rotation of the planet. Consider for instance a static or zero-frequency phenomenon in a frame tied to the planet. With respect to inertial space, the signal will have a diurnal frequency due to the rotation of the planet. In general, if the wobble frequency in the terrestrial frame is σ, the corresponding frequency in the celestial frame is $\sigma' = \sigma + \Omega$, the nutation frequency. Since the gravitational torque inducing nutation has long periods with respect to an inertial reference frame, the nutation frequencies σ' are much smaller than Ω and the associated wobble has retrograde quasidiurnal frequencies. At these periods, the nutation amplitude is much larger than that of the wobble.

Traditionally, the two motions, wobble and nutation, are represented by two cones, with the smaller one rotating inside the larger one without gliding (see Figure 8.17).

The study of the wobble of a planet is performed by observing the instantaneous position of the rotation axis (i.e. the direction of the instantaneous angular velocity vector $\vec{\Omega}$) from the direction of the mean angular velocity vector $\Omega\hat{z}$, which is also the direction of the figure axis. The components of $\vec{\Omega}$ in the terrestrial frame are usually denoted by

$$\vec{\Omega} = \Omega \begin{pmatrix} 0 \\ 0 \\ 1 \end{pmatrix} + \Omega \begin{pmatrix} m_1 \\ m_2 \\ m_3 \end{pmatrix}$$

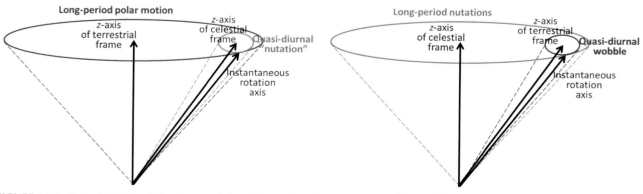

FIGURE 8.17 On the left, Euler–Poinsot representation of the rotation axis motion in space (blue) and in the terrestrial frame (red) for the Earth; on the right, representation of the nutation and diurnal wobble for the Earth.

where m_1, m_2, and m_3 are small quantities; variations of m_3 represent the fractional variations in the spin rate, which manifest themselves in LOD variations as seen in Section 7; m_1 and m_2 express the deviation of the instantaneous rotation axis with respect to the uniform rotation axis in the plane perpendicular to the figure axis $\hat{z}$ and so describes wobble (or polar motion, see Section 9.1) in a reference frame $(\hat{x}, \hat{y}, \hat{z})$ tied to the planet and rotating uniformly as introduced before.

In order to demonstrate the link between nutation and wobble, we introduce the Euler angles relating the inertial $(\hat{X}, \hat{Y}, \hat{Z})$ and body-fixed $(\hat{x}, \hat{y}, \hat{z})$ reference frames (see Figure 8.18).

Three rotations are needed to pass from one frame to the other:

1. A rotation of angle Ψ around $\hat{z}$:

$$R(\Psi) = \begin{pmatrix} \cos \Psi & \sin \Psi & 0 \\ -\sin \Psi & \cos \Psi & 0 \\ 0 & 0 & 1 \end{pmatrix},$$

2. A rotation of angle θ around $\hat{x}$:

$$R(\theta) = \begin{pmatrix} 1 & 0 & 0 \\ 0 & \cos \theta & -\sin \theta \\ 0 & \sin \theta & \cos \theta \end{pmatrix},$$

3. A rotation of angle φ around $\hat{Z}$:

$$R(\varphi) = \begin{pmatrix} \cos \varphi & \sin \varphi & 0 \\ -\sin \varphi & \cos \varphi & 0 \\ 0 & 0 & 1 \end{pmatrix}.$$

The body-fixed reference frame rotates with respect to the inertial reference frame with the instantaneous rotation vector $\vec{\Omega}$. Therefore, the Euler angles change with time. The rotation vector $\vec{\Omega}$ can thus be expressed as a function of the time derivatives of the Euler angles. Over an infinitesimal time dt the change in the relative positions of the two reference frames can be obtained from the infinitesimal changes in the Euler angles by performing the sum of three rotations over (1) the angle $d\Psi \; (= \dot{\Psi} dt)$ around the axis $\hat{Z}$ of the frame $(\hat{X}, \hat{Y}, \hat{Z})$ in space, (2) the angle $d\theta = \dot{\theta} dt$ around the axis of intersection between the plane $(\hat{X}, \hat{Y})$ and the plane $(\hat{x}, \hat{y})$, and (3) the angle $d\varphi = \dot{\varphi} dt$ around the $\hat{z}$ of the frame $(\hat{x}, \hat{y}, \hat{z})$. The instantaneous rotation vector can thus be expressed as

$$\vec{\Omega} = \dot{\Psi} \hat{Z} + \dot{\theta} \hat{x'} + \dot{\varphi} \hat{z}.$$

For the Earth (and Mars), the equatorial terrestrial frame $(\hat{x}, \hat{y}, \hat{z})$ is tied to the planet and the celestial frame $(\hat{X}, \hat{Y}, \hat{Z})$ in space is tied to the ecliptic. The relative orientation of these frames differs from the representation in Figure 8.18 classically used for Euler angles as the $\hat{Y}$-axis in the ecliptic lies in the Northern Hemisphere and the obliquity ε is defined as $\varepsilon = -\theta$. We then have

$$\vec{\Omega} = \dot{\Psi} \hat{Z} + (-\dot{\varepsilon}) \hat{x'} + \dot{\varphi} \hat{z}.$$

In the terrestrial reference frame, the instantaneous rotation vector is given by

$$\vec{\Omega} = \begin{pmatrix} -\dot{\Psi} \sin \varepsilon \sin \varphi - \dot{\varepsilon} \cos \varphi \\ -\dot{\Psi} \sin \varepsilon \cos \varphi + \dot{\varepsilon} \sin \varphi \\ \dot{\Psi} \cos \varepsilon + \dot{\varphi} \end{pmatrix}.$$

Identifying this expression with the above expression for the instantaneous rotation vector involving m_1, m_2, and m_3, we obtain the famous kinematic equation of Euler:

$$\dot{\varepsilon} + i \sin \varepsilon \, \dot{\Psi} = -(\Omega_1 + i\Omega_2) e^{i\varphi}$$
$$= -\Omega (m_1 + i m_2) e^{i\varphi} \approx -\Omega (m_1 + i m_2) e^{i\Omega t}.$$

This equation provides the relation between the nutation angles Ψ and ε, and the wobble components (m_1, m_2) and proves that nutation and wobble are intimately related.

9.1. Polar Motion

The wobble of the instantaneous rotation axis has several components grouped in different frequency bands. One of them is the small-amplitude retrograde diurnal frequency component related to nutation. Larger wobble, commonly referred to as polar motion, occurs at long periods in the body-fixed reference frame. For the Earth, for example, the position where the rotation axis intersects the surface changes seasonally. The separation from the mean position remains below 20 m (see Figure 8.19).

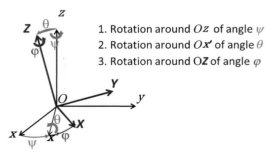

FIGURE 8.18 Euler angles; in blue the body-fixed frame and in black the inertial frame.

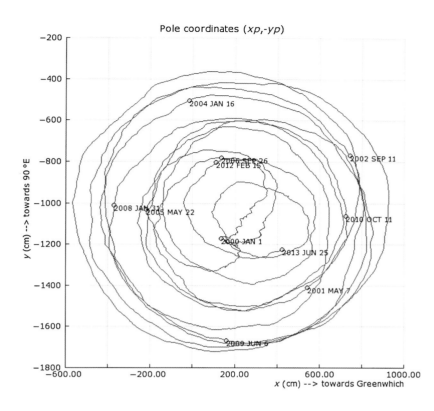

FIGURE 8.19 Polar motion as computed from observation on the IERS Website of the Observatoire de Paris (http://hpiers.obspm.fr/eop-pc/).

Movie 7 shows the polar motion of the Earth viewed from space.

Supplementary video related to this chapter can be found at http://dx.doi.org/10.1016/B978-0-12-415845-0.00008-6.

The following is/are the supplementary data related to this chapter: Movie 7: Polar motion for the Earth. (file EarthPolarMotion.mpg)

As for LOD variations, seasonal polar motion of the Earth is mainly due to the effect of the atmosphere and the other geophysical fluids such as the ocean and the hydrosphere. As the rotation changes, it can be computed from the angular moment exchange between the solid Earth and the fluid layers (conservation of the total angular momentum).

The wobbles of a planet are computed from the Liouville equations describing the angular momentum conservation equation for a nonrigid planet. The angular momentum of a rotating body ($\vec{H}$) is the sum of two terms, one related to the global rotation (as seen previously in Section 0) and defined as the product of the inertia matrix ($\bar{I}$) and the rotation vector ($\vec{\Omega}$) and the other related to the relative moment of inertia $\vec{h}$, due to the relative motion with respect to the corotating frame in the fluid parts of the planet:

$$\vec{H} = \bar{I}\vec{\Omega} + \vec{h}$$

Angular momentum conservation can be expressed in a reference frame corotating with the mantle as

$$\frac{\partial \vec{H}}{\partial t} + \vec{\Omega} \times \vec{H} = \vec{T}$$

where $\vec{T}$ is the external torque, and $\partial/\partial t$ expresses the time derivative in the corotating reference frame. The inertia matrix in a reference frame tied to the moment of inertia can be written as:

$$\bar{I} = \begin{pmatrix} A & 0 & 0 \\ 0 & B & 0 \\ 0 & 0 & C \end{pmatrix} + \bar{c}$$

where $\bar{c}$ is the incremental inertia matrix due to the deformation of the planet and A, B, and C are the equatorial and polar moments of inertia as seen before (see Section 4). For a biaxial ellipsoidal planet (typical for fast rotating planets), $A = B$.

Using the expressions:

$$\vec{h} = \begin{pmatrix} h_1 \\ h_2 \\ h_3 \end{pmatrix} \quad \vec{T} = \begin{pmatrix} T_1 \\ T_2 \\ T_3 \end{pmatrix} \quad \bar{c} = \begin{pmatrix} c_{11} & c_{12} & c_{13} \\ c_{21} & c_{22} & c_{23} \\ c_{31} & c_{32} & c_{33} \end{pmatrix}$$

and limiting ourselves to terms of the first order in the small quantities, one obtains the so-called Liouville equations:

$$\begin{cases} \Omega A \dot{m}_1 + \Omega^2 (C-A) m_2 + \Omega \dot{c}_{13} - \Omega^2 c_{23} + \dot{h}_1 - \Omega h_2 = \Gamma_1 \\ \Omega A \dot{m}_2 - \Omega^2 (C-A) m_1 + \Omega \dot{c}_{23} + \Omega^2 c_{13} + \dot{h}_2 + \Omega h_1 = \Gamma_2 \\ \Omega A \dot{m}_3 + \Omega \dot{c}_{33} + \dot{h}_3 = \Gamma_3 \end{cases}$$

For a rigid planet (meaning that $c_{13} = c_{23} = 0$, and $h_1 = h_2 = 0$ in the above equation), the equations simplify to:

$$\begin{cases} \Omega A \dot{m}_1 + \Omega^2 (C-A) m_2 = \Gamma_1 \\ \Omega A \dot{m}_2 - \Omega^2 (C-A) m_1 = \Gamma_2 \\ \Omega A \dot{m}_3 = \Gamma_3 \end{cases}$$

The Liouville equations can be brought to a simpler form by using complex notation. For one wobble component, we write the time dependence of the wobble as $\exp(i\sigma t)$, where σ is the wobble frequency. The time derivative of the variables can then be expressed as $i\sigma$ times the variable. On taking the complex sum of the first two equations and introducing $m = m_1 + im_2$ and $\Gamma = \Gamma_1 + i\Gamma_2$, one obtains:

$$\Omega A \dot{m} - \Omega^2 (C-A) m = \Gamma$$

Without forcing ($\Gamma = 0$), polar motion (if excited) is possible for a particular frequency, the free Euler frequency

$$\sigma_{\text{Euler}} = \frac{(C-A)}{A}\Omega.$$

The Euler frequency depends only on the dynamical flattening (and thus on the moments of inertia) and the rotation rate of the planet. If the free polar motion is excited, the rotation axis has a circular motion in the $(\hat{x}, \hat{y})$-plane at the Euler period. The amplitude depends on the excitation and damping.

In reality, terrestrial planets contain a liquid core and some also a solid inner core. For those internal layers, additional small deviations from the mean rotation must be considered since they will not, in general, rotate precisely with the mantle. Considering in a first approximation that the planet can still be considered rigid (deformations are neglected) and neglecting a possible solid inner core, two complex equations can be derived for the complex polar motion of the mantle m and the complex polar motion of the core m_f. Those Liouville equations can be expressed as

$$i\sigma\Omega A m + i\sigma\Omega A_f m_f - i\Omega^2 e A m + i\Omega^2 A_f m_f = \Gamma$$
$$= -ie\Omega^2 A \varphi$$

$$i\sigma\Omega A_f m + i\sigma\Omega A_f m_f + i\Omega^2 (1 + e_f) A_f m_f = 0$$

These two equations are two linear equations in m and m_f. They can be expressed in matrix form as

$$\begin{pmatrix} \sigma - e\Omega & (\sigma + \Omega)\left(\frac{A_f}{A}\right) \\ \sigma & \sigma + \Omega(1 + e_f) \end{pmatrix} \begin{pmatrix} m \\ m_f \end{pmatrix} = \begin{pmatrix} -e\Omega\varphi \\ 0 \end{pmatrix}.$$

In the absence of any forcing Γ or φ, polar motion is possible again for a particular frequency, the frequency of the free wobble. The frequencies providing a solution of the above equations for no external forcing are computed from equating to zero the determinant of the matrix. Since polar motion occurs mainly at seasonal periods, the frequency of each polar motion contribution is small with respect to the rotation frequency for rapidly rotating planets. Therefore, the matrix simplifies further and the frequency of the free wobble σ_{free} can be expressed as

$$\sigma_{\text{free}} = \left(\frac{A}{A_m}\right) e\Omega = \frac{(C-A)}{A_m}\Omega.$$

The existence of a liquid core decouples, to first order, the mantle wobble from that of the core. Therefore, the period of the free wobble is proportional to the moment of inertia of the mantle and not that of the whole planet, thereby decreasing the period with respect to an entirely solid planet. The free frequency is, therefore, different from the Euler frequency. Moreover, deformation changes the frequency, which for terrestrial planets is called the Chandler frequency. The associated polar motion is known as the Chandler wobble. For the Earth, the Euler period is 305 days and the Chandler period is 433 days, showing that non-rigid effects are very important. The different non-rigid contributions to the period are shown in Figure 8.20 for the Earth. Besides elasticity, the existence of the ocean and the mantle inelasticity also increase the period.

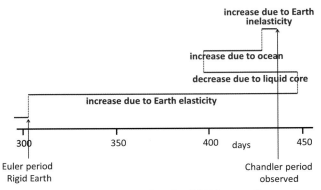

FIGURE 8.20 Period of the Chandler Wobble, contributions from different geophysical sources, starting from the rigid Earth computation (the Euler period).

Besides the free polar motion, the atmosphere and other fluid layers of the Earth also excite polar motion at seasonal periods. Both the annual component and the Chandler Wobble contribution have amplitudes of about a few meters. The seasonal and Chandler components sometimes add up constructively providing a large amplitude, or sometimes subtract destructively providing a small amplitude in polar motion. At the beginning of 2013, polar motion was very small (see Figure 8.19).

In principle, as Mars has an atmosphere and seasons, the polar motion of Mars will also consist of a *CW* contribution and an annual/seasonal contribution. Because the atmosphere is much less dense than that of the Earth, the expected motion is below a meter, instead of the motion in a 15 m × 15 m square that we have for the Earth (see Figure 8.19 for the Earth). It has, therefore, not yet been observed. Mars is also expected to have a Chandler wobble with a period of about 200 days.

9.2. Forced Nutation and Free Core Nutation

In order to obtain the nutation forced by a gravitational torque (involving the gravitational φ) for a nonrigid planet with a possible liquid core, one can also start from the Liouville equations established above, but instead of considering a small frequency as for polar motion, one considers frequencies σ for the wobble in the retrograde diurnal band (near $-\Omega$; with the relation $\sigma = -\Omega + \sigma'$, where σ' is small).

The matrix that corresponds to the Liouville equations in that case

$$\begin{pmatrix} \sigma - e\Omega & (\sigma + \Omega)\left(\frac{A_f}{A}\right) \\ \sigma & \sigma + \Omega(1 + e_f) \end{pmatrix},$$

then simplifies to

$$\cong \begin{pmatrix} -\Omega & \sigma'\left(\frac{A_f}{A}\right) \\ -\Omega & \sigma' + e_f\Omega \end{pmatrix}.$$

The frequency of nonforced ($\varphi = 0$) free solution of the Liouville equations is obtained by setting the determinant of the above matrix equal to zero. The frequency is called the Free Core Nutation Frequency (FCN), which for a nondeformable planet can be expressed as:

$$\sigma_{FCN} = -\Omega\left(1 + \frac{A}{A_m}e_f\right),$$

The FCN is also called the Nearly Diurnal Free Wobble which has nearly diurnal frequency in the terrestrial frame as its name implies. Without liquid core, the FCN does not exist. In the frame of nutation computation, only the retrograde quasidiurnal frequencies are of interest. The forced response to the external forcing can be computed from

$$\begin{pmatrix} -\Omega & \sigma'\left(\frac{A_f}{A}\right) \\ -\Omega & \sigma' + e_f\Omega \end{pmatrix}\begin{pmatrix} m \\ m_f \end{pmatrix} = \begin{pmatrix} -e\Omega\varphi \\ 0 \end{pmatrix}$$

The solutions of this system show that both m and m_f are resonantly amplified by the FCN frequency. As the nutation can be computed from m, as shown above, one immediately sees that the FCN amplifies the planet orientation changes induced by the tidal gravitational torque. Observing the nutation will thus provide information on whether the core is liquid or entirely solid and even, when the FCN can be exactly determined, on the core moment of inertia A_f and the core flattening e_f, and hence on the core radius and density, as the core hydrostatic dynamical flattening and core moment of inertia depend on those quantities.

For the Earth, the FCN has a period $\left(1/\sigma'_{FCN}\right)$ of about -430 days in the celestial frame. For Mars, the FCN period has not yet been observed but theoretical computations based on plausible interior models for Mars provide a range of values between -220 and -280 days. The FCN period could thus be very close to the ter-annual Martian nutation at 229 days and thus largely amplify it, making it better observable and usable for deducing information on the Martian core.

10. OBSERVATION OF THE ROTATION OF TERRESTRIAL PLANETS

10.1. Ground-Based Observations

Rotations of solid planets can be observed from Earth-based telescopes. The principle uses the transmission by Earth-based radar telescopes of a circularly polarized monochromatic X-band radio signal to the planet. Radar echoes from the planet surface are reflected back to the Earth and received at several large telescopes. The signal exhibits spatial irregularities in the wave front caused by the constructive and destructive interference of waves scattered by the irregular surface. Because of the rotation of the planet, the irregularities in the wave front, also called speckles, sweep over the different receiving stations with time (see Figure 8.21). As the radar speckle patterns are tied to the surface rotation, the observations determine the rotation of the top surface layer.

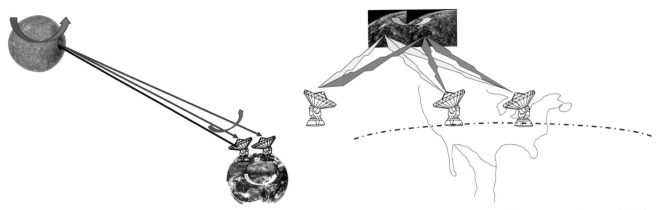

FIGURE 8.21 Representation of the principle of RADAR observations of Mercury's rotation as seen from space (left-hand side) and as seen from the Earth (right-hand side).

10.2. Spacecraft Observation

Rotation observations can be performed by using spacecraft orbiting a planet and even by spacecraft during flybys. Several methods can be used to determine the rotation. For example, as the orbit of a spacecraft is sensitive to the gravity of the planet and as the planet with its mass anomalies rotates and the spacecraft moves, the spacecraft is sensitive to the planet rotation as well. In order to determine the gravity field and the planet rotation, one uses the radio link between the spacecraft and the Earth. A radio signal is sent from Earth to the spacecraft and sent back by a transponder on the spacecraft to Earth after a coherent turnaround. One measures on Earth the Doppler shift on the radio signal in order to determine the relative velocity between the Earth and the spacecraft. Knowing the Earth rotation and orientation almost perfectly in space (at centimeter level from VLBI measurements), one determines the spacecraft velocity, position, and orbit.

Successive images taken from the spacecraft of the surface of the planet can also be used to determine the relative changes between the images with respect to space. This obviously requires good knowledge of the position of the spacecraft, which can be determined from radioscience. The ESA BepiColombo mission to Mercury, in particular, will be able to determine the libration of Mercury in this way. This rotation experiment is represented in Movie 8.

Supplementary video related to this chapter can be found at http://dx.doi.org/10.1016/B978-0-12-415845-0.00008-6

The following is/are the supplementary data related to this chapter: Movie 8: Libration experiment. (file LibrationExperiment.mpg)

A very valuable way of observing the rotation, the rotation variations, and the orientation changes is to use landers (or rovers when they are fixed for a long period). A two-way direct radio link from the Earth to the lander provides the relative velocity of the lander with respect to the ground stations on Earth and therewith, provides, again knowing were the Earth is in space, the orientation and rotation of the moon or planet as a function of time. The geometry coverage of the experiment and the precision that can be reached with such observations are such that this is the best method to get precession and nutation; in particular, it is and will be used for Mars in the future (see Figure 8.22).

The radio links are normally in **S-band, X-band or Ka-band**. The X-band is mostly used and preferred with respect to the S-band as it is less sensitive to perturbations from the ionosphere and plasma. The ground stations on the Earth are quite big: they have a diameter of up to 70 m (see Figure 8.23). NASA has installed DSN antennas (Deep Space Network), and European Space Agency, ESTRACK stations (ESA TRACKing station) to communicate with spacecraft.

For measurements of the rotation of the Moon, one also uses ranging by lasers (Lunar Laser Ranging, LLR). In this method, a laser station on the Earth (see Figure 8.25) sends

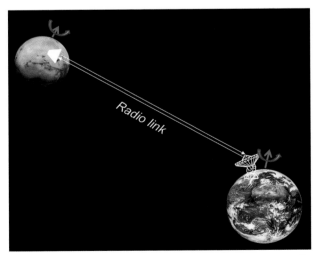

FIGURE 8.22 Direct to Earth link with a lander or a rover at the surface of Mars in order to get the rotation and orientation changes.

Chapter | 8 Rotation of Planets

FIGURE 8.23 DSN Goldstone antennas (left 70-m antenna and right 34-m antenna).

a laser beam to the Moon. The beam is reflected back to the Earth with retroreflectors (corner cubes reflecting the laser pulses back in the same direction as the received direction, see Figure 8.24). The Apollo missions have deposited several retroreflectors (Figure 8.24) on the surface of the Moon (locations in Figure 8.26), which are still used for determining the Moon–Earth distance and the librations of the Moon.

10.3. The Particular Case of the Earth

The Earth is a particular case as we are able to build relatively large equipment such as VLBI antennas, LLR stations, and satellite laser ranging (SLR) instruments. LLR is a technique explained in the previous paragraph used mainly for obtaining the distance between the Earth and the Moon, and SLR is essentially used for computing the gravity field of the Earth. VLBI is a technique based on multiple radio astronomy telescopes on Earth, at which the signals from very distant astronomical radio sources, such as quasars, are collected simultaneously and processed. The distances and changes of distance between the radio telescopes are then computed using the time differences between the arrivals of the radio signal at the different telescopes. This allows determining the rotation of the Earth and the Earth's orientation in space, since the radio sources are essentially fixed in space on account of their large distance (or with a very small and well-determined proper motion). This technique is very precise as it also uses very precise frequency references (hydrogen maser), thereby providing the orientation of the Earth at the sub-centimeter level.

FIGURE 8.24 Retroreflector deposited on the Moon by the Apollo missions.

FIGURE 8.25 LLR station.

FIGURE 8.26 Location of the reflectors on the Moon.

In parallel, there is a well-known geodetic technique used on the Earth: the Global Navigation Satellite System (GNSS). It provides positions of receivers with respect to satellites orbiting around the Earth with a very high precision (below the centimeter for geodesic receivers). As the satellites are orbiting in space and the receivers are on the Earth's surface, this in turn can provide additional information on the Earth's rotation and orientation. And, as polar motion expresses the variation of the rotation axis with respect to the figure axis of the Earth, GNSS also provides information on polar motion. Moreover, we have high spatial coverage of measuring GNSS devices, which ensures a precise determination of polar motion, also at the subcentimeter level.

BIBLIOGRAPHY

Dehant, V., & Mathews, P. M. (2007). Earth rotation variations. In T. Herring, & G. Schudert (Eds.), *Geodesy: Vol. 3. Treatise on Geophysics* (pp. 295–349). Amsterdam: Elsevier.

Dehant, V., & Mathews, P. M. (2014). *Precession, nutation, and wobble of the Earth*. Cambridge University Press, in press.

de Pater, I., & Lissauer, J. J. (2010). *"Planetary sciences"*, textbook (2nd ed.). Cambridge University Press, 663 pp.

Elkins-Tanton, L. T. (2006). *The solar system (The Sun, Mercury, and Venus, The Earth and the Moon, Mars, Asteroids, Meteorites, and Comets, Jupiter and Saturn, Uranus, Neptune, Pluto, and the outer solar system)*. six-book reference series. published by Chelsea House, an imprint of Facts on File, Inc, 1st ed. 2006; 2nd ed. 2010.

Lambeck, K. (2005). *The Earth's variable rotation*. Cambridge University Press, 525 pp.

Lissauer, J. J., & de Pater, I. (2013). *"Fundamental planetary science; physics, chemistry and habitability"*, textbook. Cambridge University Press, 616 pp.

Munk, W. H., & McDonald, G. J. F. (1960). *The rotation of the Earth - A geophysical discussion*. Cambridge University Press (reprinted with corrections 1975).

Van Hoolst, T., The rotation of the terrestrial planets. In: T. Spohn & G. Schudert (Eds.), *Planets and Moons: Vol. 10. Treatise on Geophysics* (pp. 123–164). Amsterdam: Elsevier.

Chapter 9

Evolution of Planetary Interiors

Nicola Tosi, Doris Breuer and Tilman Spohn
Institute for Planetary Research, German Aerospace Center (DLR), Berlin, Germany

Chapter Outline

1. Introduction 185
2. Formation and Early Evolution of Terrestrial Bodies 186
3. Subsolidus Convection 189
4. Rock Rheology and Modes of Convection 193
5. Modeling Interior Dynamics and Evolution 197
6. Constraints on and Models of the Evolution of Planetary Interiors 198
   6.1. Crust Formation and Volcanic History 199
   6.2. Surface Heat Flow and Mantle Heat Budget 201
   6.3. Core–Mantle Boundary Heat Flow and Magnetic Field History 202
   6.4. Lithosphere and Surface Deformation, and Thermal Evolution 204
   6.5. Life, Plate Tectonics and the Evolution of the Interior 205
7. Concluding Remarks and Perspectives 207
Bibliography 208

1. INTRODUCTION

The evolution of the interior of terrestrial bodies is the result of the balance between the heat accumulated during planetary formation and generated by slowly decaying radioactive isotopes present in mantle rocks and the heat that is lost and radiated to space over billions of years. Understanding this balance requires the characterization of the processes that control the way heat is transported from the **core** and **mantle** to the surface.

At the extreme temperature and stress levels of deep interiors, the thermally activated migration of crystalline defects allows rocky materials to flow over geological timescales like extremely viscous liquids. Powered by internal heat sources and heat from the core, terrestrial mantles undergo **convection** in response to buoyancy forces due to temperature variations, which cause thermal expansion and contraction. Convection transports thermal energy very efficiently and ultimately controls planetary cooling.

While the Earth's surface is fragmented into plates whose slow movement away from **midocean ridges** toward convergent margins is a direct expression of the underlying convection, all other terrestrial bodies of the solar system are characterized by a single immobile plate. Convection occurs beneath a thick **stagnant lid**, which isolates the mantle retarding its cooling, and experiences deformation to a much lower extent than the Earth's surface. The thermal history of planetary bodies crucially depends on whether the cold upper layers are mobile and participate in convection, as for the Earth with **plate tectonics**, or stagnant and separated from the interior, as for Mercury, Venus, Mars, and the Moon. The consequences of mantle convection are multiple and profound: volcanism, the generation of magnetic fields via **dynamo action**, or the formation of the tectonic landforms that shape planetary surfaces and **crusts** all involve physical processes intimately related to convective heat transport at depth.

Apart from the notable exception of **seismic tomography** (see Chapter 21), which provides at present a unique window on the structure of the Earth's mantle, the depths of planets and moons remain largely inaccessible to direct inspection. The study of mantle rocks at high-pressure and high-temperature conditions combined with observational constraints on the composition, geology, magnetization, and thermal state of planetary surfaces provide a framework upon which theoretical models of mantle convection can be built. These models are a fundamental tool to characterize the numerous processes that govern the dynamics of planetary interiors and influence their evolution.

Whether or not life has influenced the evolution of the interior of the Earth to an extent even remotely rivaling its effect on the evolution of the atmosphere and oceans (see Chapter 20) is presently a matter of debate. With plate tectonics cycling matter between the surface reservoirs and the interior, there is a venue for cycling biologically altered rock and volatiles with the interior where these may physically and chemically interact with mantle rock. As the concentration of volatiles, water in particular, may have profound effects on the mantle **rheology**, and thus on the vigor of the convective engine, there is a potential venue for life having a profound effect on tectonics.

2. FORMATION AND EARLY EVOLUTION OF TERRESTRIAL BODIES

The terrestrial bodies of the solar system are the end products of a complex process of **accretion** (see Chapter 2). The gravitational collapse that led to the formation of the Sun around 4.5 billion years ago was accompanied by the generation of a disk of cold gas and dust—the **solar nebula**. This so-called protoplanetary disk, which orbited the newly born Sun owing to the conservation of angular momentum, was the source of the materials from which all planetary bodies formed. Although the existence of the solar nebula can only be inferred indirectly from the physical and chemical evidence left in the solar system, protoplanetary disks orbiting young stars are now routinely detected via infrared- and radioastronomy, which strongly substantiate the hypothesis that the solar system itself developed from similar conditions.

The accretion of planetary bodies consists of three main phases. First, the collective gravity of localized swarms of centimeter-sized dust particles leads to the formation of planetesimals—solid bodies ranging in size from about 1 to 100 km. Once initiated, this process is expected to take place rapidly (less than about 10^5 years). However, the occurrence of sizable particle concentrations in the disk can be sporadic both in space and time, with the consequence that planetesimal formation can proceed over a relatively extended period of a few million years. Afterward, within 10^5-10^6 years, planetesimals undergo a **runaway growth** (also called oligarchic growth), with large bodies growing faster than small ones until few tens of planetary **embryos** of lunar to Martian mass are formed. Finally, over a timescale of tens of millions of years, relatively rare collisions between embryos of comparable size, and between embryos and leftover planetesimals, give rise to a small number of terrestrial planets.

Intimately related to the formation of terrestrial bodies, from small asteroids to satellites and planets, is their **differentiation** into a metallic core enriched in iron and a rocky mantle consisting of **silicate** compounds (Figure 9.1). Core—mantle differentiation, which occurred contemporary with or shortly following the accretion, is a simple consequence of the larger density of iron and its alloys with respect to silicates. The most stable

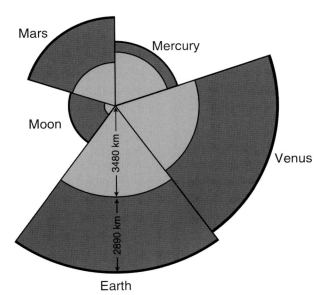

FIGURE 9.1 Basic interior structure of the terrestrial planets and the Moon. In first approximation, all terrestrial bodies possess a layered structure consisting of an innermost metallic core (light gray layer), a silicate rocky mantle (dark gray layer), and a thin crust of volcanic origin, which is chemically distinct from the mantle (black layer not drawn to scale). While planetary radii are known precisely, core radii (light gray layer) are not (see Table 9.2), apart from the case of the Earth's core, which consists of a liquid part surrounding a solid inner core (not drawn) with a radius of 1220 km.

configuration of a rotating mass subject to its own gravity is in fact an oblate spheroid with the material of highest density located at its center. Albeit conceptually simple, this picture is complicated by problems related to the availability of the materials involved in the differentiation process and to the conditions and time at which these were added to the accreting mass.

The separation of solid metal from solid silicate due to gravitational instability is too slow to have played an important role during accretion. Therefore, core—mantle differentiation requires some form of melting. Core materials, having a lower melting temperature than silicates, can coexist as a liquid phase with a solid silicate matrix and sink through it via grain-scale percolation or in the form of large molten diapirs descending through fractures and dykes. Alternatively, metal—silicate separation can take place if both phases are in a liquid state. The feasibility of core formation requires thus an understanding of whether and how, during accretion, it was possible to reach temperatures large enough to cause melting of metallic and silicate materials.

Three main processes have been recognized as capable to provide a sufficient amount of energy to produce widespread melting during the early history of terrestrial bodies. First, the radioactive **isotopes** ^{26}Al and ^{60}Fe, known to be abundant in the solar nebula from geochemical analyses of primitive meteorites, have short half-lives (see Table 9.1). Because of the energy released by the decay of these radionuclides, planetesimals that accreted within the first few million years of the solar system could have experienced the degree of melting required for core—mantle differentiation. As a consequence, at least part of the planetesimals that contributed to the formation of embryos first, and planets later, were likely already differentiated, and their cores built up the larger cores of the bodies to which they accreted. Second, it is important to recognize that the late stages of accretion involving impacts between large bodies resulted in strongly increased temperatures due to the conversion of kinetic and gravitational energy into heat. The upper bound to the total heat of accretion can be calculated based on the gravitational binding energy E, the energy required to pull apart to infinity all the accreted material, or, from the opposite point of view that is relevant here, the amount of energy liberated upon accretion from material pulled together from infinity:

$$E = \frac{3}{5}\frac{GM^2}{R}, \qquad (9.1)$$

where G is the universal gravitational constant and M and R are the total mass and radius of the accreted body, respectively. The temperature change ΔT associated with the energy E is

$$\Delta T = \frac{E}{Mc_p}, \qquad (9.2)$$

where c_p is the heat capacity ($\sim 10^3$ J/kg/K for silicate minerals and ~ 800 J/kg/K for metallic core materials). According to Eqn (9.2), if all accretionary energy were delivered instantaneously, the resulting temperature rise would be $\sim 10^3 - 10^4$ K for bodies ranging in size and mass from Mercury to Earth. From Figure 9.2, which shows the solidus and liquidus curves for the Earth's mantle, it can be quickly realized that a temperature increase of a few thousand degrees would be sufficient to cause large-scale melting, possibly even partial vaporization, of the silicate materials. Indeed, it is now widely accepted that, because of violent impacts with embryos such as the one of Martian size from which the Moon probably originated (so-called giant impacts, Figure 9.3), **terrestrial planets** likely experienced one or multiple episodes of widespread melting that led to the formation of deep **magma oceans**. However, it should be also noted that accretionary heat is neither delivered through a single episode nor homogeneously. Also, depending on the size and velocity of the impactor, while part of its energy can be actually buried at depth and retained by the target body, part of it is also quickly radiated back to space. Nevertheless, even though it only gives a crude maximum heating estimate, the simple analysis above illustrates that the energy involved in the formation of planetary bodies was prodigious.

The third main source of energy that contributed to the early melting of planetary interiors is the process of core—mantle differentiation itself. No matter whether iron sinks toward the center of the planet through a fully or partially molten magma ocean, the associated redistribution of mass results in the release of gravitational potential energy in the form of additional heat. Let us consider, for example, a hypothetical Moon-forming impact involving

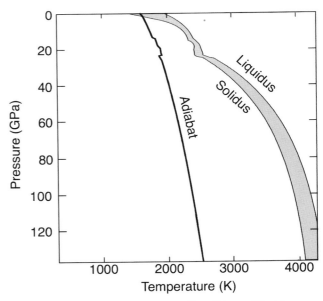

FIGURE 9.2 Adiabatic temperature profile, solidus, and liquidus curves for the Earth's mantle as a function of pressure. The solidus denotes the temperature at which the mineral with the lowest melting temperature among those that form the mantle mixture melts. The liquidus coincides with the melting temperature of the mineral with the highest melting point. Between solidus and liquidus (gray envelope), rocks are partially molten, i.e. they form a mixture in which liquid magma coexists with a solid matrix. *After Stixrude et al. (2009).*

TABLE 9.1 Heat Producing Elements in Terrestrial Bodies

Isotope	Specific Heat Production (W/kg)	Half-life (years)
^{26}Al	0.355	7.2×10^5
^{60}Fe	0.068	2.6×10^6
^{238}U	9.46×10^{-5}	4.47×10^9
^{235}U	5.69×10^{-4}	7.04×10^8
^{232}Th	2.64×10^{-5}	1.40×10^{10}
^{40}K	2.92×10^{-5}	1.25×10^9

Isotopes of Al and Fe have very short half-lives and were only important for internal heating of planetesimals during the early stages of the solar system. Isotopes of U, Th, and K have half-lives comparable with the age of the solar system and are relevant for the long-term evolution of planetary interiors.

FIGURE 9.3 Hydrodynamic simulation of the collision between two bodies of similar masses as an example of the Moon-forming impact. Time is shown in hours, distance in units of 10^3 km, and temperature in Kelvin. The final planet is surrounded by a disk with very similar composition, from which the Moon would later originate. This simulation illustrates the extreme heating caused by energy transfer associated with large-scale impacts. *After Canup (2012).*

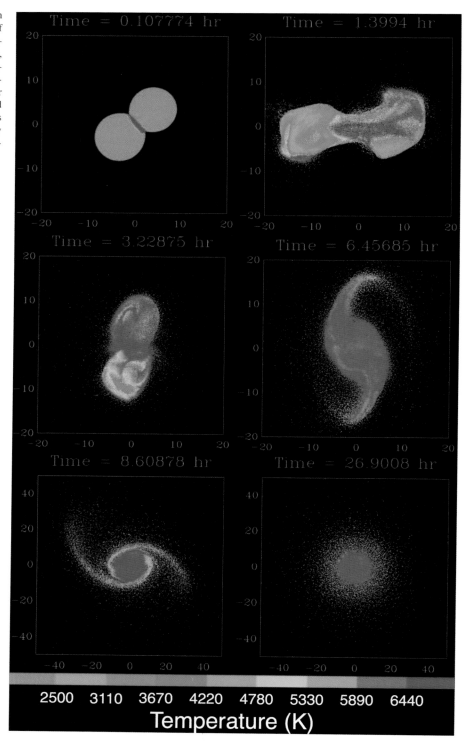

the proto-Earth and an embryo having, respectively, a mass 80–90% and 10–20% of that of the present-day Earth. Assuming that both bodies already possess a core that occupies half of their radius and that the two cores will completely merge, the mean temperature change in the final core can be estimated to be ∼1000–2000 K, sufficient to induce substantial melting of the lower mantle and to generate a strongly superheated core.

In synthesis, thanks to the heat due to the decay of short-lived radioactive isotopes as well as to the conversion of kinetic energy of impactors and of gravitational potential energy associated with core formation, near the end of the accretion period, terrestrial bodies were likely hot and chemically differentiated into a fully or partially molten silicate mantle and a liquid metallic core possibly superheated with respect to the overlying mantle. The time for reaching this stage is relatively short compared with the age of the solar system. It varies from a few million years for small asteroids and planets like Mars, to tens of millions of years for the Earth. This condensed timescale predicted by theoretical models of planetary accretion and differentiation is supported by the study of isotopic systems. In particular, during the past decade, the application of the Hf−W decay system to the available inventory of meteorites has repeatedly proved to be a very effective tool for timing various stages of the early evolution of rocky bodies (see Chapters 2 and 28).

The idea that planets were extremely hot shortly after their formation is not new. The famously wrong calculation of the age of the Earth made by William Thompson (better known as Lord Kelvin) in 1862 was based on the assumption that our planet was initially completely molten and that, after solidifying, reached its present temperature by cooling via conduction. After all, temperature measurements in mines suggested a geothermal gradient of 20−30 K/km, which clearly indicated that the Earth was cooling. Kelvin's estimates that ranged from few ten to few hundred million years, however, were far too short because the continental crust, where the above temperature measurements were taken, is in a nearly steady-state balance as a consequence of internal heat production due to the decay of radiogenic isotopes and mantle heat flow from below (Section 3), both of which were ignored by Kelvin. The fact that the history of planetary bodies started from a hot state, possibly associated with the existence of vast regions of molten silicates, can have important consequences. The process of metal−silicate separation is certainly the most dramatic differentiation event in the history of a planet, but the solidification of a magma ocean can also represent a fundamental step in the development of the interior structure and composition. Although the existence of a terrestrial magma ocean is considered as highly probable, its evidence is not compelling, particularly because plate tectonics and mantle convection have contributed to erase all the traces that this event may have left at the surface. Nevertheless, a basal magma ocean, which crystallized slowly enough to allow localized patches of dense melt to survive until present, has been invoked to explain the existence of a deep primordial reservoir of incompatible species, which could account for the ultralow seismic shear wave velocities detected in some regions of the lowermost Earth's mantle (see Chapter 21). Geological and geochemical arguments suggest that smaller bodies such as Mars or the Moon experienced a high degree of silicate melting early in their history. In this respect, the Moon provides the most notable and widely accepted example. The widespread presence of plagioclase minerals all over the Moon's highlands (see Chapters 23 and 24) is best explained in the framework of a solidifying magma ocean. In fact, assuming its fractional crystallization, i.e. the formation upon cooling of specific rock compositions due to the segregation of certain elements in the liquid magma out of the solid phase, plagioclase, having a relatively low density, is positively buoyant and tends to rise to the surface where it solidifies and forms a primordial crust (see Section 6.1).

Completion of large-scale mantle solidification ideally defines the end of the cataclysmic events associated with the formation and early differentiation of terrestrial bodies. The subsequent evolution of their interiors is characterized by **secular cooling**, the slow and continuous loss of the primordial heat accumulated during accretion and core formation and of the heat generated by the decay of long-lived radiogenic isotopes present in the mantle.

3. SUBSOLIDUS CONVECTION

Shortly after the discovery of natural radioactivity in 1896 by Henri Becquerel, Pierre Curie noticed that a speck of radium continuously and spontaneously emits a vast amount of heat, much higher than that released in any known chemical reaction involving the same quantity of matter. Geologists quickly realized that the decay of radioactive elements could be responsible for heating the interior of the planet and in turn affect Kelvin's estimate of the age of the Earth. Accurate dating techniques based on radioactive decay were rapidly developed and in 1911 the British geologist Arthur Holmes demonstrated that the most ancient rock sample from the large collection he examined was 1.6 billion years old. Holmes laid in this way the foundations of modern geochronology, although it was not until 1955 that the first meteorite was dated, making it finally possible to place the age of the Earth at 4.5 billion years.

Kelvin's idea according to which the Earth was simply undergoing secular cooling was initially replaced by the concept of steady-state conductive balance. The heat flowing out of the Earth was still believed to be lost only by conduction, but also balanced by the heat generated by radioactive sources, which were thought to be concentrated exclusively in the crust overlying a solid interior depleted of heat producing elements (see Section 6.1). Using heat flow measurements from continental regions, this model could actually well predict typical concentrations of heat producing elements in continental rocks. The assumption that the heat sources were concentrated solely in the crust also implied that the heat flow in oceanic regions, where the crust was known to be thinner than under **continents** (Figure 9.4(a)), would be much lower. However, the first measurements carried out in the 1950s in the Pacific and the Atlantic showed that the oceanic heat flow away from midocean ridges is very similar to that measured in continental regions (Figure 9.4(b)). This surprising equivalence was then attributed to the existence of convection currents in the mantle. The high thermal gradients observed near the Earth's surface are the expression of the upper **thermal boundary layer** associated with mantle convection (Figure 9.7). Across this layer, heat is transported by **thermal diffusion** (i.e. by conduction), while transport by convection dominates beneath it, with the bulk temperature, well below its melting value, increasing with depth along a nearly **adiabatic gradient** (see Figure 9.2).

The process of thermal (or natural) convection refers to the transport of heat due to bulk fluid motion. Although the

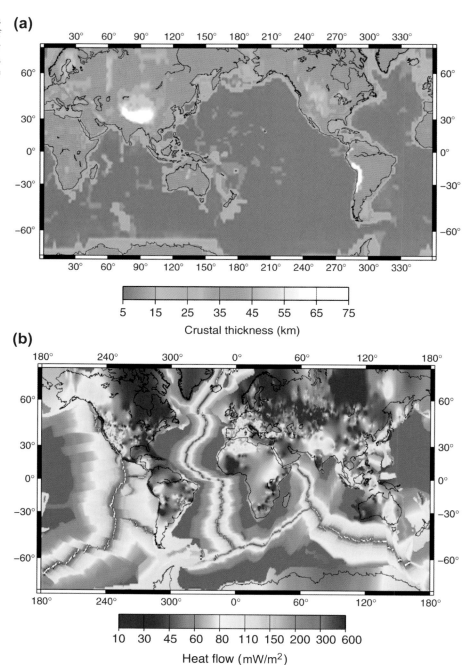

FIGURE 9.4 Global maps of the Earth's crustal thickness from a compilation of seismic data (a) and surface heat flow combining heat flux measurements and a cooling model for the oceanic lithosphere (b). *After Jaupart and Mareschal (2011).*

mantle is by all means solid in that it propagates elastic waves on short timescales, on geological timescales of the order of millions of years it deforms like a highly viscous fluid in a way similar to the more familiar one with which the ice in glaciers flows on timescales of years. One way to study the fluid behavior of the Earth, which led to the first empirical estimates of rock **viscosity** in the 1930s, is to investigate the viscous response of the mantle to the last deglaciation. The vast ice sheets that covered large portions of the northern hemisphere around 20,000 years ago depressed the Earth's surface below sea level. As a consequence of their quite abrupt melting, the mantle has been responding by flowing to compensate the mass deficit caused by the disappearance of the ice load (Figure 9.5(a)).

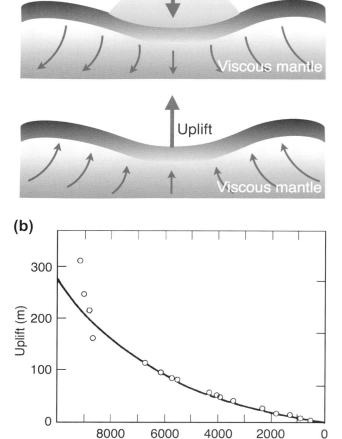

FIGURE 9.5 (a) Surface subsidence due to loading of an ice sheet of characteristic wavelength λ and subsequent viscous uplift following rapid ice melting. (b) Data points for the uplift h of the mouth of the Angerman River, Sweden, as a function of time before present compared with an exponential relaxation model of the kind $h \sim \exp(-t/\tau)$, with τ defined as in Eqn (9.3). *After Turcotte and Schubert (2002).*

The characteristic time τ for the **viscous relaxation** of the surface topography associated with this process can be calculated as

$$\tau = \frac{4\pi\eta}{\rho g \lambda}, \quad (9.3)$$

where η is the viscosity, ρ is the density, g is the gravity acceleration, and λ is a characteristic spatial length of the surface load (i.e. the ice sheet). From observations of elevated beach terraces, it is possible to determine τ and λ. Assuming $\rho = 3300$ kg/m^3, $g = 9.8$ m/s^2, and $\tau = 4400$ year and $\lambda = 3000$ km as based on the data of Figure 9.5(b), we obtain $\eta = 1.1 \times 10^{21}$ Pa s for the viscosity of the upper mantle. This value is enormous when compared with the viscosity of more ordinary materials. For example, at ambient conditions, water has a viscosity of $\sim 10^{-3}$ Pa s, while the typical value for honey is ~ 10 Pa s and that for ice is $\sim 10^{13}$ Pa s.

Mantle rocks, similarly as they deform in response to surface loading, also flow at subsolidus temperatures as a consequence of temperature differences, thereby advecting heat. An increase in the temperature of a fluid parcel produces a reduction in density due to volumetric thermal expansion. If the fluid is subject to a gravitational field, buoyancy forces arise that tend to lift the heated fluid (the opposite argument clearly applies in the presence of a decrease in temperature). Let us consider a homogeneous plane layer of thickness D whose upper and lower surfaces are maintained at constant temperatures T_0 and T_1, respectively, with $T_1 > T_0$. Fluid near the hotter lower boundary will become lighter than the overlying fluid and tend to rise, while fluid near the colder upper boundary will become heavier and tend to sink. By performing an analysis of **linear stability** of this system, it can be established under which conditions small disturbances will decay and lead to a stable configuration, or grow by overcoming the viscous resistance of the fluid and lead instead to a motion of finite amplitude. Only one nondimensional parameter is sufficient to characterize the stability of a fluid layer heated from below and cooled from above. This parameter is the **Rayleigh number**:

$$Ra = \frac{\rho^2 c_\mathrm{p} g (T_1 - T_0) \alpha D^3}{k\eta}, \quad (9.4)$$

where ρ, η, and g have been defined above; c_p is the isobaric heat capacity; α is the coefficient of thermal expansion; and k is the thermal conductivity. If Ra is greater than a certain critical threshold Ra_cr, thermal convection will set in. Figure 9.6 shows this critical Rayleigh number as a function of the nondimensional wave number (i.e. the inverse of the wavelength) of the applied perturbation. At a wave number of $\pi\sqrt{2}/2$, the curve attains its minimum

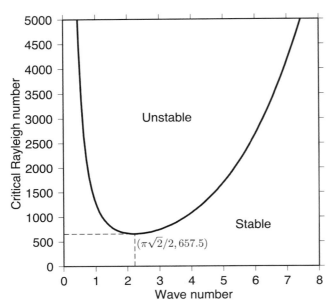

FIGURE 9.6 Critical Rayleigh number as a function of the nondimensional wave number for a horizontal layer heated from below and cooled from above. The curve separates stable regions in which perturbations tend to decay from unstable regions in which initial small disturbances will grow to motion of finite amplitude.

$Ra_{cr} = 657.5$. A similar analysis can be carried out for a layer heated from within and cooled from above. The appropriate Rayleigh number is

$$Ra_H = \frac{\rho^3 c_p g \alpha H D^5}{k^2 \eta}, \quad (9.5)$$

where H is the rate of internal heat production per unit mass. The minimum critical value for the onset of convection in this case is 867.8. Also note that although the above estimates are only valid for a fluid enclosed in a horizontal layer, in the case of spherical shells, the critical Rayleigh number for both bottom and internally heated convection is not much different, being of the order of $\sim 10^3$. Essentially on the grounds of these analyses, Arthur Holmes proposed as early as in 1931, well before the acceptance of the plate tectonics paradigm in the 1960s, that the Earth's interior experiences solid-state convection. Indeed, by setting in Eqn (9.5) $D = 2890$ km (the thickness of the Earth's mantle), $g = 9.8$ m/s^2, $\rho = 4000$ kg/m^3, $c_p = 1200$ J/kg, $\alpha = 3 \times 10^{-5}$/K, $k = 3$ W/m/K, $H = 10^{-11}$ W/kg (see Eqn (9.6) and Table 9.1), and $\eta = 10^{21}$ Pa s as derived from the analysis of postglacial uplift, we obtain $Ra_H \sim 10^7$. This value is much greater than the critical threshold predicted by the theory and thus indicates that the mantle of the Earth is in a highly supercritical regime and undergoes vigorous thermal convection.

A similar argument applies to the other terrestrial bodies of the solar system (Mercury, Mars, Venus, and the Moon) whose mantle is also thought to undergo thermal convection. Internal heating is provided by the decay of the long-lived radioactive isotopes of uranium, thorium, and potassium whose specific heat production rates and half-lives are reported in Table 9.1. Their relative concentration in the mantle can vary significantly among different planets. Nevertheless, an order-of-magnitude estimate can be made considering the relative abundance of heat producing elements as measured for **chondrites**, i.e. stony meteorites that have not been modified by melting and differentiation and that are thought to form the building blocks of terrestrial planets. In one of the

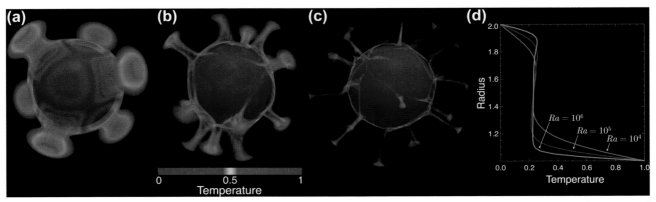

FIGURE 9.7 Planforms of isoviscous thermal convection in a spherical shell heated from below and cooled from above for (a) $Ra = 10^4$, (b) $Ra = 10^5$, and (c) $Ra = 10^6$. Temperature isosurfaces between 0.4 and 0.6 are shown in nondimensional units. (d) Laterally averaged temperature profiles corresponding to the snapshots shown in panels (a–c). The profiles are characterized by two thermal boundary layers across which the temperature changes rapidly and heat is transported by conduction. The interior is instead nearly isothermal (the effects of adiabatic compression were neglected in these simulations). As the Rayleigh number increases, the characteristic wavelengths of convection become smaller and thermal boundary layers thinner. Also note that while the temperature distribution is in steady state for $Ra = 10^4$, for $Ra = 10^5$ and $Ra = 10^6$, it is time dependent. The color scale is strongly saturated to facilitate visualization. *Simulations courtesy of Plesa, A., DLR Berlin.*

most primitive type, the carbonaceous Ivuna (CI), radiogenic isotopes have the following proportions: 7.4×10^{-9} kg/kg uranium, 29×10^{-9} kg/kg thorium, and 550×10^{-6} kg/kg potassium. Radioactive substances decay exponentially according to their time constant and the heat production as a function of time is given by

$$H(t) = \sum_i C_i H_i \exp\left(-\frac{\ln(2)t}{\tau_i}\right), \quad (9.6)$$

where C_i is the concentration of the ith element and H_i and τ_i are the corresponding heat production rate and half-life, respectively. Using Eqn (9.6), we find that the heat production of a CI chondritic body would vary between 2.4×10^{-11} W/kg 4.5 Ga ago and 0.34×10^{-11} W/kg today.

All terrestrial planets, with the possible exception of Mercury, have a mantle characterized by a supercritical Rayleigh number, which is indicative of a convecting interior. In Table 9.2, we report ranges of possible Rayleigh numbers for the terrestrial planets and the Moon. On the one hand, differences in the Rayleigh number among different bodies are primarily due to the thickness of their mantles, which varies from ~400 km for Mercury to 2890 km for the Earth, and scales with the third power for bottom-heated convection (Eqn (9.4)) or with the fifth power for internally heated convection (Eqn (9.5)). On the other hand, for a given body, the ranges reported in the table are the consequence of a rather poor knowledge of the viscosity (see Section 4).

TABLE 9.2 Order-of-Magnitude Estimate of the Reference Rayleigh Number for the Terrestrial Planets and the Moon

Planetary Body	Radius (km)	Outer Core Radius (km)	Rayleigh Number
Mercury	2440	1940–2140	Subcritical–10^5
Venus	6052	3089	10^7–10^9
Earth	6371	3480	10^7–10^9
Moon	1737	150–400	10^5–10^8
Mars	3390	1400–1900	10^6–10^8

Rayleign numbers are calculated from Eqn (9.5) using the following parameters: $\rho \sim 10^3$ kg/m^3, $c_p \sim 10^3$ J/kg/K, $g \sim 1$ m/s^2, $\alpha \sim 10^{-5}$/K, $H \sim 10^{-11}$ W/kg, $k \sim 1$ W/m//K, $\eta \sim 10^{19}$–10^{21} Pa s (see Table 9.3), and mantle thickness D calculated as the difference between the radius of a given body and the radius of its core. Note that, apart from the Earth, the lack of seismic observations and uncertainties in the gravity field measurements from which interior structure models can be derived, the size of the outer core of the other terrestrial bodies is not well constrained. This is particularly relevant for Venus whose core size was simply obtained by rescaling Earth's parameters.
Core radii are from Breuer, et al. (2010). Thermal evolution and magnetic field generation in terrestrial planets and satellites, Space Science Reviews, 152, 449–500.

4. ROCK RHEOLOGY AND MODES OF CONVECTION

The way rocks deform in response to applied **stresses** (i.e. their **rheology**) depends on several factors such as rock composition, temperature, pressure, and level of deviatoric stress (the portion of the stress field responsible for actual distortions in a continuum as opposed to isotropic volume changes caused by the mean stress, i.e. the pressure). In the crust and the shallow mantle, rocks undergo primarily elastic and **brittle deformation**, while at the higher pressures and temperatures characteristic of the deep interior, ductile behavior becomes dominant.

At low stresses, the relation between stress and **strain** is linear and solids behave elastically (the strain is a measure of deformation that accounts for the relative displacement between particles in a stressed body). Upon compressing or placing under tension a crystalline material, interatomic forces tend to resist compression or expansion and to keep atoms in their lattice position. This elastic deformation is completely reversible. As the stress level increases, a critical value known as **yield stress** can be reached. Above this value, deformation becomes irreversible and is accommodated through brittle failure or ductile flow (Figure 9.8). The former manifests itself either through the creation of new cracks and faults, or through frictional sliding along existing fractures. The latter consists instead in a fluidlike motion characterized by an effective viscosity that controls how temporal variations of the strain (strain rates) vary as a function of applied stresses.

Viscous flow is the principal mode of subsolidus deformation in planetary mantles and takes place via two main mechanisms: diffusion and dislocation creep. These mechanisms control the thermally activated migration of crystalline defects. In diffusion creep, which dominates at relatively low stresses, atomic vacancies, i.e. empty sites in the crystal lattice, can move in response to applied stresses within grains (Herring–Nabarro creep), or along grain boundaries (Coble creep). In dislocation creep, which becomes relevant at higher stresses, deformation is due to the motion of one-dimensional linear imperfections of the lattice (so-called dislocations). Both theory and laboratory experiments indicate that the dependence of the strain rate $\dot{\varepsilon}$ on the applied stress τ can be expressed in the following form, valid for both diffusion and dislocation creep:

$$\dot{\varepsilon} = A\tau^n \exp\left(-\frac{E_a + V_a P}{RT}\right), \quad (9.7)$$

where A is a material prefactor, n is the stress exponent, P is the hydrostatic pressure, T is the absolute temperature, R is the universal gas constant, E_a is the activation energy, and V_a is the activation volume. In diffusion creep the relationship between stress and strain rate is linear (Newtonian)

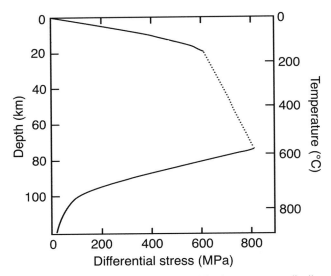

FIGURE 9.8 Strength envelope calculated for the temperature distribution in a 60-Ma-old oceanic lithosphere assuming a dry olivine rheology. Differential stress here indicates the difference between the maximum and minimum stresses necessary to cause failure. The straight line at shallow depths indicates brittle failure occurring in the crust. The dashed segment refers to combined brittle–ductile behavior and is not well constrained. The bottom segment plotted with solid line describes the transition to fully ductile behavior. *After Kohlstedt et al. (1995).*

orders of magnitude. It should be also noted that the uncertainties in the parameters that enter the rheological relation (Eqn (9.8)) are also large. Laboratory experiments in which these parameters can be determined are particularly difficult. In order for deformations to be accurately measured, the strain rates considered in the laboratory are typically orders of magnitude greater than those expected in planetary interiors, with the result that significant extrapolations must be applied.

Diffusion and dislocation creep occur simultaneously. However, at given conditions of temperature, pressure, and strain rate, the effective viscosity of a stressed material is determined by the weakest mechanism, i.e. the one that delivers the smallest viscosity. In Figure 9.9 we show a map of viscosity as a function of temperature for dry olivine obtained using parameters from Table 9.3 and different strain rates, from 10^{-20} to 10^{-12}/s, as appropriate for the slow creeping flow of mantle rocks. Diffusion creep tends to prevail at low strain rates and relatively high temperatures, while, as the strain rate is raised, the effective viscosity associated with dislocation creep dominates over the whole temperature range considered. The diagram also illustrates the large sensitivity of viscosity to temperature changes. For example, for diffusion creep, a decrease in temperature from 1200 to 1000 K causes an increase in viscosity of more than two orders of magnitude.

At the relatively low temperatures characteristic of planetary surfaces and shallow interiors, the exponential dependence in Eqn (9.8) causes the viscosity to become so large that ductile flow is no longer possible. At these conditions, terrestrial bodies develop a so-called stagnant lid, an immobile upper layer that cannot participate in mantle convection. This situation is the most common in the solar

and $n = 1$. Dislocation creep is instead a nonlinear (non-Newtonian) deformation mechanism with $n > 1$ (typically $n = 3.5$). In the exponential term of Eqn (9.7) (Arrhenius law), the activation enthalpy $H_a \equiv E_a + V_a P$ accounts, through the term E_a, for the energy necessary to form vacancies and for the energy barrier that atoms must overcome to migrate into a vacancy site, and, through the term $V_a P$, for the fact that pressure tends to render these two processes more difficult. Introducing the viscosity η as the proportionality factor relating stress and strain rate, i.e. $\tau = 2\eta\dot{\varepsilon}$, we have from Eqn (9.7):

$$\eta = \frac{1}{2} A^{-1/n} \dot{\varepsilon}^{(1-n)/n} \exp\left(\frac{E_a + V_a P}{nRT}\right). \quad (9.8)$$

Although Eqn (9.8) depends explicitly only on strain rate, pressure, and temperature, the prefactor A is itself a function of several parameters that can change the viscosity by several orders of magnitude. The most significant are **grain size** (relevant for diffusion creep only), melt fraction, and concentration of water. Large grains tend to increase the viscosity, while **partial melt** and water content tend to decrease it. Table 9.3 lists numerical values of the parameters that appear in Eqn (9.8) and the corresponding viscosity calculated at a reference temperature of 1600 K and at a pressure of 3 GPa for water-free (dry) and water-saturated (wet) olivine, the most abundant mineral of the Earth's upper mantle. The effect of water is large: its presence can reduce the effective viscosity almost by two

TABLE 9.3 Rheological Parameters for Diffusion and Dislocation Creep in Dry and Wet Olivine

	Diffusion Creep		Dislocation Creep	
	Dry	Wet	Dry	Wet
A (Pa^{-n}/s)	1.9×10^{-11}	1.2×10^{-11}	2.4×10^{-16}	3.9×10^{-15}
n	1	1	3.5	3
E_a (J/mol)	3×10^5	2.4×10^5	5.4×10^5	4.3×10^5
V_a (m^3/mol)	6×10^{-6}	5×10^{-6}	1.5×10^{-5}	1.0×10^{-5}
η (Pa s)	6.3×10^{20}	1.1×10^{19}	2.1×10^{20}	3.2×10^{19}

The prefactor A was calculated assuming a shear modulus of 80 GPa, grain size of 10^{-3} m, Burgers vector's length of 5×10^{-10} m. To calculate the viscosity from Eqn (9.8), we also assumed reference temperature and pressure of 1600 K and 3 GPa, respectively, and a strain rate of 10^{-15}/s.
Parameters after Karato & Wu (1997). Rheology of the upper mantle: A synthesis. Science 260, 771–778.

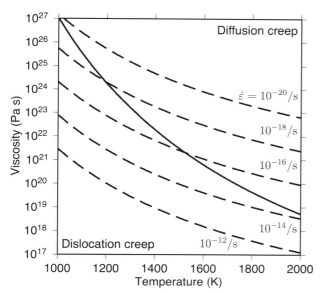

FIGURE 9.9 Viscosity as a function of temperature for dry olivine calculated from the parameters of Table 9.3. The solid line refers to the viscosity of diffusion creep, while the dashed lines to the viscosity of dislocation creep computed for different values of the strain rate $\dot{\varepsilon}$. At a given temperature, the effective viscosity is that of the weakest mechanism. For example, with $\dot{\varepsilon} = 10^{-18}/s$, the material will deform via dislocation creep for $T < 1200$ K and via diffusion creep for $T > 1200$ K.

system. Little doubts exist that Mercury, Mars, and the Moon have been one-plate bodies throughout all or most of their history. Although Venus, because of its young surface (Chapter 15), is generally believed to have experienced an episodic regime in which short events of surface mobilization are interspersed between long phases of quiescence associated with a stagnant lid, at present it also possesses an immobile surface. The Earth is the only planetary body in a mobile lid regime with a surface split in tectonic plates that are an active part of the mantle convection system. The consequences of this difference in tectonic modes on the dynamics and evolution of the interior are profound. A qualitative glimpse of the impact that a stagnant lid regime has on mantle convection can be obtained from Figure 9.10, which shows the temperature field from numerical simulations of an Earth-like planet in stagnant (Figure 9.10(a)) and mobile lid regimes (Figure 9.10(b)) (see also Section 5). In both cases, we assumed the rheology to be only governed by diffusion creep of dry olivine with temperature- and pressure-dependent viscosity. In the first case, the upper layers are immobile because of their high viscosity. Convection is driven primarily by bottom boundary layer instabilities that result in upwelling plumes, and, to a lesser extent, by sinking downwellings originating from the bottom part of the lid, which are only slightly colder than the surrounding mantle. The stagnant lid clearly acts as a thermal insulator that tends to keep the interior of the planet warm.

In order for surface mobilization to be possible, other deformation mechanisms must be taken into account. A convective behavior like the one portrayed in Figure 9.10(b) results from considering, in addition to diffusion and/or dislocation creep, a mechanism of brittle failure. Recall from the discussion at the beginning of this

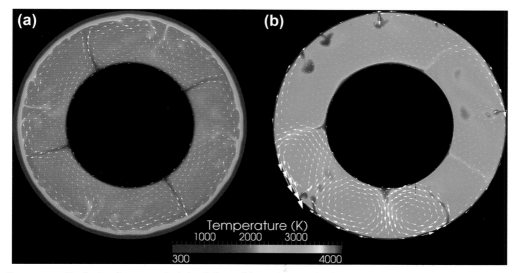

FIGURE 9.10 Temperature distribution from numerical simulations of thermal convection in a two-dimensional cylindrical geometry. (a) Stagnant lid and (b) mobile lid regime. In both, cases the system is cooled from above and heated from both below and within. In the first, the uppermost cold layers are immobile because of their high viscosity and exert a blanketing effect that tends to keep the interior warm. In the second, the introduction of an yielding mechanism allows for surface mobilization and subduction, which in turn causes a strong cooling of the interior. Directions of flow velocity are indicated by white arrows whose length is proportional to the magnitude of the velocity vector. The central region is occupied by the core, which is not modeled. *Simulations courtesy of Hüttig C., DLR Berlin.*

section that deformation becomes irreversible when the yield stress is exceeded. The brittle nature of rocks at relatively low pressures at which fracture or frictional sliding can occur can be described with the Byerlee's law:

$$\sigma_y = C + \mu P, \quad (9.9)$$

where σ_y is the yield stress, C is the cohesive strength, μ is the friction coefficient, and P is the hydrostatic pressure. Remembering now that the effective viscosity η_{eff} is controlled by the weakest mechanism, we can write

$$\eta_{eff} = \left(\frac{1}{\eta_{diff}} + \frac{1}{\eta_{disl}} + \frac{1}{\eta_y} \right)^{-1}, \quad (9.10)$$

where η_{diff} and η_{disl} are diffusion and dislocation creep viscosities as calculated from Eqn (9.8) and $\eta_y \equiv \sigma_y/2\dot{\varepsilon}$ is a plastic viscosity associated with the yielding mechanism (Eqn (9.9)). When stresses generated by mantle convection are large enough, the stagnant lid can locally fail. The effective viscosity (Eqn (9.10)) drops, making it possible to strongly localize deformation in regions that serve as nucleation points for the initiation of **subduction**. Figure 9.10(b) was obtained using a rheological model accounting for diffusion creep and brittle deformation (see also Figure 9.11(a)). Two important features emerge from the resulting mode of mobile lid convection that render it remarkably different from the stagnant lid mode. First, the mobilization and accompanying foundering of the cold thermal boundary layer is a significantly stronger driver of convection than the sublid instabilities arising in the presence of a stagnant lid. Indeed, it is widely recognized that Earth's mantle convection is driven primarily by buoyancy forces associated with the subduction of oceanic plates. Second, the fact that the cold upper thermal boundary layer actively participates in convection yields a continuous cooling of the interior as opposed to the stagnant lid case in which the mantle tends to remain homogeneously hot and to cool largely because of the decay of radioactive heat sources. As we shall see in the following section, the ability of a planet to cool more effectively has important consequences for the evolution of the interior and its geophysical implications.

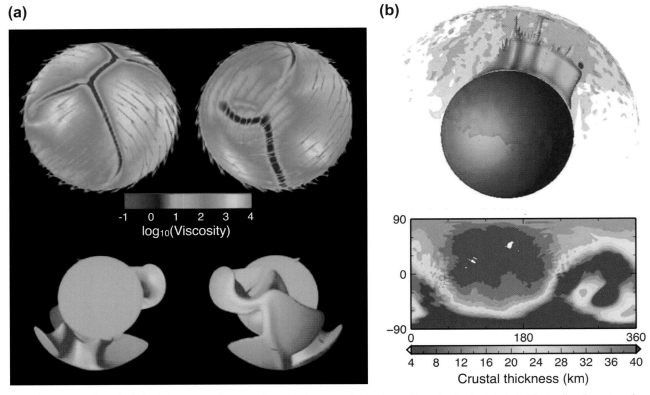

FIGURE 9.11 (a) Numerical simulations of mantle convection with plate tectonics in a three-dimensional spherical shell. The top line shows two views of the viscosity field at the surface. Rigid plates moving with uniform velocity are separated by both divergent and convergent narrow boundaries with low viscosity at which plates form and are recycled into the mantle. The bottom line shows temperature isosurfaces representing the subduction of cold downwellings. *(After van Heck & Tackley (2008)*. (b) Numerical simulation of the formation of Mars' crustal dichotomy by mantle convection. The top panel shows a hot upwelling plume (yellow) rising from the core–mantle boundary (red) able to generate a large region of partial melt (light blue). Migration of the melt toward the surface leads to the formation of a crust (Section 6.1) whose thickness in the southern hemisphere is much greater than in the northern one (bottom panel). *After Šrámek & Zhong (2012)*.

5. MODELING INTERIOR DYNAMICS AND EVOLUTION

Heat transport by conduction and subsolid convection are the main physical processes underlying the dynamics and thermal evolution of terrestrial planetary interiors. Plate tectonics, which has been operating on the Earth for billions of years; magma production and volcanism, which have shaped the surfaces of all terrestrial bodies; and the generation of a magnetic field, currently operating on the Earth and Mercury only but likely active in a remote past also on Mars and the Moon, are all ultimately related to the way heat is transported in the interior.

Since solid mantles exhibit a fluidlike behavior over geological timescales, their dynamics is governed by the same set of fundamental equations employed for describing the motion of more ordinary fluids like, for example, water. A crucial difference is due to the large viscosity of rocks. Because of it, mantle materials respond instantaneously to applied stresses, causing viscous forces, which tend to resist motion, to be constantly in equilibrium with the buoyancy forces that drive it. As a consequence, inertial forces are negligible and convective flows in the mantle are always laminar, as opposed to the turbulent flows that frequently characterize convection in fluids with low viscosity, such as flows in the ocean and atmosphere. Note however, that at sufficiently high Rayleigh numbers (see Figure 9.7 and Table 9.2), mantle flow, albeit laminar, is highly time dependent, so that small variations in initial conditions are able to induce large differences in the flow at subsequent times, which renders mantle convection an intrinsically chaotic process.

In general, modeling mantle convection requires to solve the conservation equations of mass, linear momentum, thermal energy, and chemical composition, appropriate for viscous media with negligible inertia. These build a set of coupled partial differential equations whose analytical solution is only available under a handful of highly simplifying assumptions. Solutions to these equations at conditions relevant for planetary interiors are usually based on numerical models, which represent today the primary tool for investigating mantle dynamics and evolution. The conservation equations must be supplemented by a suitable equation of state relating density to temperature and chemical composition, by a rheological law describing the viscosity field (see Eqns (9.8) and (9.10)), and by a series of material parameters (e.g. coefficients of thermal expansion and conduction, heat capacity, etc.) that in first approximation can be considered constant but, in general, like the viscosity, are themselves functions of pressure, temperature, and composition. Since the atmosphere and the (liquid) outer core do not exert any significant traction on the solid mantle, it is usually assumed that shear stresses vanish at the planetary surface and at the core—mantle boundary. The average surface temperature of a planetary body is controlled by the atmosphere, or the lack thereof, and remains approximately constant throughout the evolution governed by solid-state convection, although Venus may represent an exception in this sense (see Chapter 15). The bottom mantle temperature at the interface with the core, instead, can change considerably over long timescales. In first approximation, in fact, a liquid core underlying the mantle can be treated as an isothermal heat bath of given density and heat capacity from which as much heat can be extracted as required by the heat flowing through the bottom thermal boundary layer associated with the convecting mantle (see Figure 9.7(d)). In addition, heat production decreases with time because of the decay of long-lived radioactive isotopes (Eqn (9.6)), ultimately causing the mantle to cool over billions of years of evolution.

In recent years, numerical simulations have witnessed dramatic progress thanks to the steadily increasing availability of large computational resources. Dynamic approaches based on the solution of the full set of conservation equations of mantle convection allow now for the treatment of complex problems in three-dimensional spherical shell geometry. For example, self-consistent models of Earth's plate tectonics on the sphere have started to appear (Figure 9.11(a)). Or, in the framework of Mars' interior dynamics, theories based on numerical simulations have been put forth supporting the idea that its hemispheric dichotomy, the strong difference in the thickness of the crust beneath the smooth northern lowlands and the rugged southern highlands (see Chapter 17), is the result of an endogenic process associated with a peculiar planform of mantle convection (Figure 9.11(b)).

Nevertheless, despite the relative easiness with which computing resources can be now accessed, the growing complexity of modern three-dimensional simulations presents significant challenges related both to technical difficulties associated with the development of adequate numerical methodologies and to the long calculation times that such simulations typically demand. For these reasons, a simpler approach based on so-called parametrized convection models is often adopted as a valuable alternative to fully dynamic simulations. These are one-dimensional models based on the solution of energy balance equations for the mantle and core. Instead of resolving the flow responsible for convective heat transport, appropriate scaling laws are employed to parameterize it. Being derived from experiments and numerical models, these scaling laws can describe the energetics of stagnant or mobile lid convection. Boundary layer theory is then used to determine the thickness of the thermal boundary layers across which heat is transported by conduction. In this way, the interior evolution can be obtained by constructing a radial thermal profile for the entire planet assuming that the temperature increases linearly in the boundary layers and adiabatically in the mantle and core.

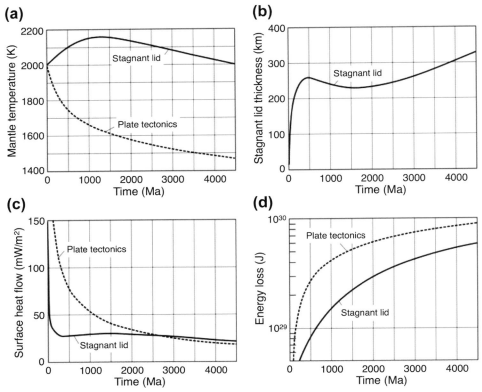

FIGURE 9.12 Thermal evolution model based on parametrized convection and parameters appropriate for Mars assuming scaling laws for heat transport by stagnant lid (solid lines) and plate tectonics (i.e. mobile lid) convection (dashed lines). The four panels show the time evolution of the mantle temperature (a), stagnant lid thickness (b), surface heat flow (c), and energy loss (d) as a function of time. *After Breuer (2009).*

Figure 9.12 illustrates a simple thermal evolution model calculated from parameters typical of Mars' mantle and considering parameterizations of heat transport appropriate for stagnant lid convection, as expected for Mars, and for mobile lid or plate tectonics-like convection, for illustrative purposes. In the first case, the lid grows rapidly (Figure 9.12(b)) and the mantle temperature remains confined within a relatively narrow range (solid line in Figure 9.12(a)). On the contrary, in the plate tectonics case in which the lid can not grow but is continuously recycled into the mantle, cooling of the interior is much more efficient (see also Figure 9.10) and the mantle temperature exhibits a large decrease over time (solid line in Figure 9.12(a)). At the end of the evolution, the two temperatures differ by as much as ~ 600 K, corresponding to a difference of 3×10^{29} J in total energy loss between the two models (Figure 9.12(d)). The absence of a stable insulating layer in the plate tectonics case also leads to an initially higher, but steadily decreasing, surface heat flow with respect to the stagnant lid case. After about 2.5 Ga, however, heat flows become comparable and decrease with similar rates until today (Figure 9.12(c)).

An important characteristic of the evolution of vigorously convecting mantles with strongly temperature-dependent viscosity is that the thermal state and surface heat flow at present are only slightly dependent on the initial temperature distribution (Figure 9.13). In fact, any increase of the mantle temperature is accompanied by a reduction in mantle viscosity (see Eqn (9.8)) that increases convective vigor and leads to a more rapid heat loss. On the contrary, upon mantle cooling, the viscosity tends to increase, reducing convective vigor and rendering heat transfer less efficient. As a result of this feedback mechanism, small temperature changes can produce large variations in the heat flux, with the consequence that, after a sufficiently long time, the temperature is buffered at nearly constant values.

6. CONSTRAINTS ON AND MODELS OF THE EVOLUTION OF PLANETARY INTERIORS

The variety and complexity of physical and chemical processes, as well as the number of poorly known parameters that potentially affect the evolution of the interior over a wide range of spatial and temporal scales is extremely large. In order to understand the processes that govern the dynamics of the mantle and make inferences on the actual history or present-day status of planetary interiors, it is thus

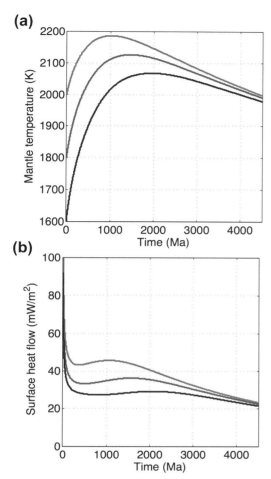

FIGURE 9.13 Mantle temperature (a) and surface heat flow (b) from parametrized models of the thermal evolution of Mars' mantle assuming stagnant lid convection and different initial temperatures of 1600 K (blue), 1800 K (green), and 2000 K (red). Because of the feedback between temperature and viscosity, the large differences in the initial thermal state and heat flow decrease steadily with time and reduce to a maximum of about 20 K and 2 mW/m^2 at the end of the evolution.

temperature rises above the solidus (see Figure 9.2), partial melt forms. At relatively low pressures (less than about 10 GPa), silicate melts are less dense than solids and can buoyantly percolate via **porous flow** or migrate upward through dykes and fractures over significantly shorter timescales (up to about 10^4 years) than those that characterize subsolidus deformation ($\sim 10^6$ years). Subsequent cooling and solidification of the extracted magma, either at depth or following eruption at the surface, lead to the formation of the crust. The composition of the crust is representative of the degree of melting experienced by the parent material, and hence it contains hints on the thermal state of the mantle from which the melt originated. In addition, the thickness of the crust and its time of emplacement are further constraints that can be employed to characterize the evolution of the interior.

The crust is essentially the result of a differentiation process that alters the chemical composition of the rocks involved. Upon melting, so-called **incompatible elements** tend to partition in the liquid phase. The term compatibility refers to the tendency of a trace element (i.e. an element present in concentrations much smaller than 1%) to remain in the solid phase or to move preferentially into the liquid one. This depends on the ability of a given element's size and charge to fit the crystal structure in which it finds itself. Incompatible elements are, for example, volatile substances, such as water and carbon dioxide, which are also greenhouse gases. Volatiles-enriched magmas can eventually feed volcanoes whose eruptions release the volatiles in the atmosphere with important consequences for planetary climates. Heat producing elements are also incompatible, with the consequence that planetary crusts tend to be enriched in these elements, while the residual mantle is depleted in them. Because of partial melting, the depleted residuum can acquire a lower density than the original mantle, as for harzburgite formation upon extraction of basaltic melt, and also a higher viscosity if hydrated rocks are involved from which water is removed by the melt (see Section 4 and Table 9.3). Furthermore, the upward transport of melt represents an additional mechanism that quickly removes heat from the interior. The dynamics and evolution of the mantle are thus affected in several fundamental ways by the processes related to the generation of crustal material. On the one hand, the redistribution of heat sources and the extraction of melt influence the heat budget of the interior and, in turn, secular cooling. On the other hand, the creation of compositional heterogeneity can affect mantle rheology and generate additional density differences, which can enhance or retard thermally driven mantle flow and induce complex effects such as the emergence of convection in separate layers. As a consequence, mantle dynamics cannot be simply treated as a problem of thermal convection, but rather becomes one of thermo-compositional (or thermochemical) convection.

essential to integrate a number of observational constraints into numerical simulations. To this end, geophysical, geological, and geochemical evidence collected remotely or in situ by artificial satellites and lander spacecrafts, and, as far as the Earth is concerned, in the field or in the laboratory, needs to be considered in theoretical evolution models. In the following subsections, we discuss selected examples in which key aspects and processes of the evolution of the Earth and the other terrestrial bodies can be elucidated with the methods described in the previous section combined with different types of constraints.

6.1. Crust Formation and Volcanic History

The crust is the brittle layer that occupies the outermost part of terrestrial bodies (see Figure 9.1). When the mantle

The fractional crystallization of a cooling magma ocean (see Section 2) may lead to the formation of the first primordial crust. For this, however, observational evidence among terrestrial bodies is scarce. As noted before, an exception is represented by the Moon with its ancient anorthositic crust that likely formed because of the flotation of buoyant plagioclase magma over denser iron-rich melts (see Section 2 and Chapters 23 and 24). The bulk of planetary crusts is thought to form later as a consequence of partial melting in the convecting solid mantle. On Earth, the most common way to build this secondary crust is through decompression melting below spreading centers, where hot material melts while rising adiabatically (despite being cooled) because of the steeper slope of the mantle adiabat with respect to the solidus (see Figure 9.2). This mechanism is responsible for the formation of oceanic crust (typically basaltic). Thanks to plate tectonics, the crust is continuously recycled into the mantle at convergent margins via subduction, whose associated melting processes, including remelting of basalt, lead to the formation of continental crust. This last process is absent in one-plate planets, which lack a mechanism of surface recycling. Decompression melting is also more difficult because a thick stagnant lid obstructs the rise of material near the surface where the solidus temperature is lowest. Partial melting is more likely induced by mantle heating, which in turn can occur in two forms. On the one hand, because of the combined effects of inefficient heat transport through the stagnant lid and of internal heat generation, the mantle temperature below the lid can rise above the solidus, thereby creating a global layer of partial melt. Its surface extrusion can be made difficult by the thick stagnant lid through which it has to percolate. Indeed, it is estimated that only a small part of the melt can reach the surface, while the rest leads to intrusive volcanism forming crust at depth. On the other hand, the mantle temperature can increase locally above the solidus because of ascending hot mantle plumes (see Figures 9.7 and 9.10), which are also thought to be responsible for **intraplate volcanism** on Earth.

A direct and accurate determination of the crustal thickness is only possible using seismological methods. With the exception of a limited amount of data collected in the framework of the Apollo program of lunar exploration, seismic measurements are available at present only for the Earth. Alternatively, the thickness of the crust can be obtained from the joint analysis of gravity and topography data, which, however, suffers from problems of nonuniqueness. While the study of the Earth's crust obviously benefits from the easiness of collecting data directly from the field, remote sensing methods need to be employed to investigate the surface of the other terrestrial bodies in order to derive information about their crusts. Image-based analyses of lava flows as well as visible-, infrared-, and nuclear spectroscopy (see Chapters 50 and 54) can all be used to characterize the composition, mineralogy, and chemistry of planetary surfaces and shed light on the magmatic processes that may have led to the observations.

The way theoretical models are employed to gain insight into the thermochemical history of terrestrial planets can be well exemplified by their application to the evolution of the mantle and crust of Mars. Although it is widely accepted that the planet has been volcanically active throughout most of its history—indeed the detection of lava flows as young as a few tens of millions of years is indicative of volcanic activity that lasted until a recent past, it is also believed that the bulk of the Martian crust, including the prominent dichotomy (see Chapter 17 and Figure 9.11(b)), formed approximately within the first 500 million years of evolution. Figure 9.14 shows results from simulations of Mars' thermochemical evolution in which suitable parameterizations have been employed to describe

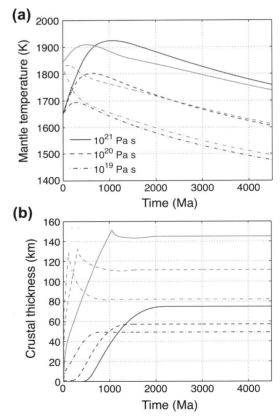

FIGURE 9.14 Thermochemical evolution of Mars mantle from parametrized models. (a) Mantle temperature and (b) crustal thickness for different reference viscosities and initial temperatures of 1650 K (blue lines) and 1850 K (red lines). Models started from a high temperature rapidly produce large amounts of crust, which can grow thicker than the stagnant lid and is subsequently recycled in the mantle. Low initial temperatures lead instead to lower crustal thicknesses and no recycling. The timing of crustal production is mainly affected by the mantle viscosity. *After Grott et al. (2013).*

not only the heat transport by stagnant lid convection but also partial melting and accompanying crust formation. Two families of models distinguished by low (blue) and high (red) initial temperatures are shown for three different reference viscosities. As already discussed above, a lower viscosity not only promotes more efficient convection and cooling but also earlier formation of the crust. An initially hot mantle leads to a rapid generation of large partial melt zones and to the growth of the crustal thickness, which reaches values in excess of 120 km or more depending on the viscosity. The insulating effect of the crust due to its low thermal conductivity and its enrichment in heat producing elements let heat accumulate at the base of the stagnant lid, which tends to thin, causing erosion of the crust from below and recycling in the mantle (this is marked by the abrupt change in the slope of red curves in Figure 9.14(b)). Isotopic characteristics of the Martian meteorites (see Chapters 2 and 28) indicate an early mantle differentiation event about 4.5 Ga ago accompanied by the formation of distinct geochemical reservoirs with little mixing occurring afterward. Therefore, crustal recycling predicted by these models is generally believed to be incompatible with this observation. Crustal erosion is not observed in models started from low temperatures that predict thicknesses between 50 and 75 km, which are within the range of estimates derived from analyses of gravity and topography data. Finally, timing of crust formation is a function of the viscosity. Low values, indicative of a wet rheology and hence of a hydrated mantle, tend to be preferred as they lead to the emplacement of the bulk crust at earlier times.

6.2. Surface Heat Flow and Mantle Heat Budget

Understanding the global energy budget of a planet requires assessing to what extent the loss of primordial heat (secular cooling) is balanced by the heat production due to the decay of long-lived radiogenic elements. In this context, the Earth offers probably the best example for testing models and ideas on planetary dynamics and evolution because of the availability of relatively tight observational constraints. The most recent estimates place the global present-day surface heat flow at 46 TW, corresponding to 90 mW/m^2. This value is the sum of the contribution from ocean basins (32 TW, corresponding to 106 mW/m^2) and continental regions (14 TW, corresponding to 66 mW/m^2). The former is estimated by combining a cooling model of the oceanic lithosphere with the ages of the sea floor deduced from marine magnetic anomalies. The latter is based on actual heat flow measurements from continents and their margins. The total heat production of the mantle and continental crust is placed by geochemical models at about 20 TW, of which continents are responsible for approximately 7 TW.

Continental heat sources, however, are stored in the continental lithosphere, which, being highly buoyant and stable, does not participate in convective heat transfer. Therefore, subtracting the continental heat production from the global heat flow yields $46 - 7 = 39$ TW for the heat flowing out of the mantle, and $20 - 7 = 13$ TW for its heat production. The ratio of mantle heat production to heat loss is termed **Urey ratio** (Ur). For the Earth, we thus have $Ur = 13/39 = 0.33$. However, this is only a preferred value; consideration of various uncertainties in the above estimates, the largest of which are related to the actual mantle composition and hence its heat sources content, yields Urey ratios between about 1/5 and 1/2.

Present-day surface heat flow and interior heat production can be employed to constrain models of the thermochemical history of the Earth's interior. Figure 9.15 shows results from one such model. The authors of this study conducted dynamic calculations in a two-dimensional spherical annulus. In their simulations, they considered two end-member initial mantle temperatures (1600 and 2500 K) representative of a "cold" and "hot" initial state, accounted for a viscoplastic rheology able to simulate a surface with platelike behavior (see Figure 9.10(b)), and modeled the effects associated with magmatism and crustal production along with the accompanying recycling into the mantle via plate tectonics and subduction. Because of the self-regulation effect induced by the strong temperature dependence of the viscosity, after about 1.5 Ga, average mantle temperatures (Figure 9.15(a)) and surface heat flows (Figure 9.15(b)) tend to converge despite the large differences in their initial values (see Figure 9.13). At the end of the evolution, surface heat flows lie in a narrow range close to the present-day estimate of 32 TW for ocean basins, which is the appropriate value in this context as the model does not account for the presence of continents. For the case with an initial mantle temperature of 2500 K, Figure 9.15(c) and (d) illustrate how the heat flow is partitioned between different mechanisms and the resulting Urey ratios, respectively. The total surface heat flow is the result of a complex heat budget that involves several contributors (Figure 9.15(c)): heat escaping the core, radioactive heat production, heat flux due to magma cooling and solidification, and secular cooling. Variations in surface plate motion and in the formation of hot plumes able to generate significant amounts of partial melt are responsible for the largest oscillations in the evolution of the heat flow associated with secular cooling and magmatism. This strong variability clearly illustrates the highly time-dependent behavior of the mantle caused by the chaotic nature of convection. Figure 9.15(d) shows that the total Urey ratio oscillates around an approximately constant value that is compatible with the range predicted on the basis of observations and geochemical models. Without considering the contribution of magma transport (convective heat flow in

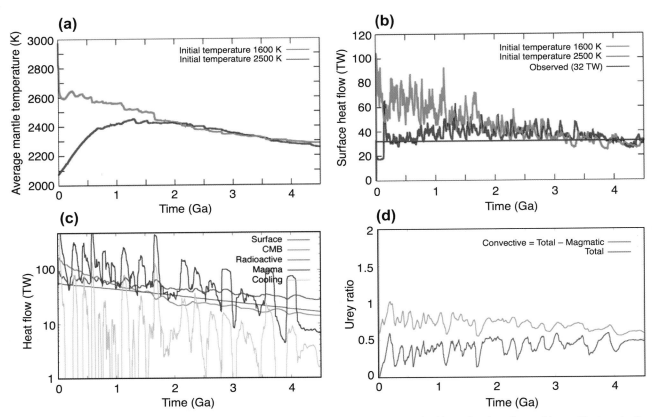

FIGURE 9.15 Thermochemical evolution of the Earth's mantle from numerical models constrained by surface heat flow and internal heat production. (a) Average mantle temperature and (b) total surface heat flow for models started at initial temperatures of 1600 K (green lines) and 2500 K (red lines). The blue line in panel b indicates the expected surface oceanic heat flow. (c) Contributions to the total heat budget and (d) Urey ratio for the model with initial temperature of 2500 K. *After Nakagawa & Tackley (2009).*

Figure 9.15(d)), the Urey ratio lies systematically above the expected range, indicating that, particularly at early time, magmatism is an essential contributor to the global heat budget of a planet and to its cooling.

6.3. Core–Mantle Boundary Heat Flow and Magnetic Field History

The generation of magnetic fields in terrestrial bodies is the result of a so-called dynamo action. A necessary condition for a magnetic dynamo to be feasible is that the electrically conductive, iron-rich fluids, which make up the liquid part of metallic cores, undergo vigorous convection. The existence of a liquid core is certain for the Earth, and Mercury, which, incidentally, are also the only terrestrial bodies possessing at present an active magnetic field (note, however, that Jupiter's icy moon Ganymede also generates a self-sustained field). Nonetheless, several lines of evidence suggest that the cores of Mars, Venus, and the Moon should also be, at least partly, liquid. Although the possibility that the cores of these bodies have completely solidified by now cannot be excluded, the fact that they do not possess a global magnetic field is most likely explained by the absence of an adequate source of buoyancy able to drive intense convection. For a dynamo-generated field to be feasible, it is necessary that the core is cooled by the mantle at a sufficiently high rate. On the one hand, if the heat flowing from the core into the mantle exceeds the heat conducted along the core adiabatic temperature profile, thermal buoyancy can be large enough to overcome ohmic losses (the dissipation of energy into heat caused by electrical resistance). In this case, a thermal dynamo can be realized. On the other hand, a large heat flow at the core–mantle boundary can also cool the liquid core below its liquidus temperature, leading to the onset of the solidification of the inner core. Upon solidification, lighter elements such as sulfur and oxygen, which are usually thought to be alloyed with iron because of their cosmochemical abundance and siderophile nature (i.e. their tendency to dissolve preferentially in iron), tend to be expelled from the newly forming solid phase and be enriched in the remaining liquid. Owing to their lower density with respect to the alloyed composition, light elements provide a source of chemical buoyancy able to drive convection. In this case, we speak of chemical or compositional dynamos. Therefore, the evolution of the mantle and, in particular, of the core—mantle boundary temperature play a central role in the generation of planetary magnetic fields: if the mantle cools efficiently, e.g. via plate tectonics and subduction, a dynamo is possible; if instead heat escape is difficult, e.g. because of the presence of a thick stagnant lid, a dynamo is still possible but likely less efficient.

Despite its similarity with the Earth in terms of size and, probably, of interior structure, it is known from orbital measurements that Venus does not possess at present an Earth-like magnetic field. This fact does not rule out the possibility that a dynamo operated in the past. Indeed, different scenarios have been proposed such as the existence of an early dynamo or of an intermittent one. The latter would be due to episodic events of plate tectonics, spaced out by long periods of stagnant lid convection, which may have cooled the mantle sufficiently to make a dynamo action possible. However, any reconstruction of the magnetic field history of Venus remains rather speculative. In fact, the high temperature of the surface (~740 K), and hence also of the crust, are close to the specific temperature (Curie temperature) above which a permanent magnetization can no longer be recorded from magnetic minerals such as magnetite and hematite. Therefore, information on Venus' magnetic activity from analyses of the magnetization of its surface cannot be expected to be retrieved.

The situation is different for Mars and the Moon. In the former case, a strong remnant magnetization of the crust induced by a global magnetic field has been observed in ancient terrains. The absence of magnetic imprint in younger regions corresponding to large impact basins has been then interpreted as an indication that a Martian dynamo ceased to operate around 4 billion years ago. As far as the Moon is concerned, although it also has no internally generated magnetic field at present, paleomagnetic data combined with radiometric ages of lunar samples retrieved by the Apollo missions indicate that a global field did operate approximately between 4.2 and 3.2 billion years ago, suggesting that an internal field started about 300 million years after the formation of the lunar core. Nevertheless, an even earlier start cannot be ruled out since the very first crust, which might have recorded a magnetic activity, may now be lost.

Timing of magnetic field activity is an important constraint for thermal evolution models as these can be used to predict whether the heat flow from the core into the mantle is compatible with the generation of a (thermally driven) dynamo and its onset and cessation. Figures 9.16 and 9.17 show two models that have been proposed to explain the magnetic field histories of Mars and the Moon, respectively. In the first, which is based on a parametrized description of Mars' thermochemical evolution, a

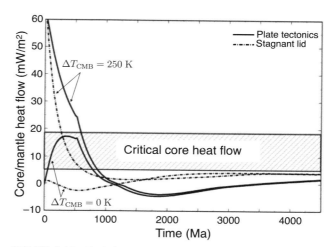

FIGURE 9.16 Evolution of the core—mantle boundary (CMB) heat flow from parametrized models of the evolution of Mars interior. The shaded area indicates the range of critical heat flow for magnetic field generation. Values above or within this critical range are compatible with the onset of a thermal dynamo. Solid lines indicate models assuming that Mars initially had plate tectonics and subsequently switched to a stagnant lid mode of convection marked by the inflection in the solid lines. Dashed-dotted lines refer to models assuming that stagnant lid convection operated over the entire evolution. In both cases, it is assumed that the core either initially has the same temperature of the mantle ($\Delta T_{CMB} = 0$ K) or is superheated with respect to it ($\Delta T_{CMB} = 250$ K). *After Breuer & Spohn (2003).*

supercritical heat flow that exceeds the heat conducted along the core adiabat can be obtained in two different ways: either by invoking a plate tectonics mode of convection for the early evolution of Mars (solid lines) or by considering stagnant lid convection for the entire evolution (dashed-dotted lines), but assuming that the core, right after its formation, is superheated with respect to the mantle (by 250 K in this case). On the one hand, plate tectonics is such an efficient mechanism for cooling the interior that it allows for a supercritical heat flow even when assuming that the mantle and core have the same initial temperature. On the other hand, an initially superheated core is a rather natural consequence of the process of core formation. Given the absence of unambiguous evidence for early plate tectonics on Mars, the last simpler scenario appears thus more plausible also in light of its ability to provide a history of crust formation and evolution more consistent with geological and geophysical observations (not shown in the figure).

The case of the lunar magnetic field presented in Figure 9.17 well illustrates the usefulness of two- and three-dimensional models to study processes in which the dynamics of the interior with its complex time dependence plays a central role. Beside the formation of plagioclase, which gave rise to the anorthositic crust (see Sections 2 and 6.1), models of crystallization of the lunar magma ocean also predict the formation of a dense layer of ilmenite (titanium-rich) cumulates at relatively shallow depths.

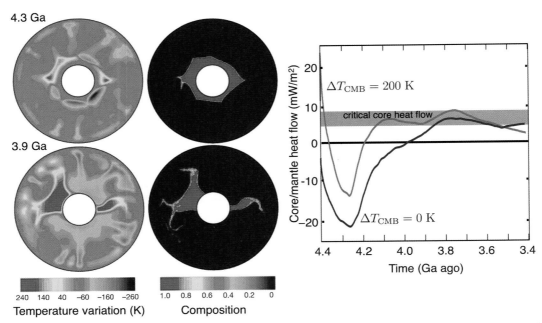

FIGURE 9.17 Three-dimensional model of thermochemical convection in the Moon's mantle. Equatorial slices of the temperature difference with respect to the mean profile (left slices) and of the composition field representing the ilmenite layer enriched in heat producing elements (right slices). The right panel shows the evolution of the heat flow at the core–mantle boundary assuming that the core initially has the same temperature of the mantle ($\Delta T_{CMB} = 0$ K), or is superheated with respect to it ($\Delta T_{CMB} = 200$ K). After an initial phase during which the dense layer acts as a thermal blanket and causes the core–mantle heat flow to rapidly decrease, heating from radioactive sources destabilizes the basal layer (top slices) until this can be fully entrained by mantle convection (bottom slices), thereby causing a rapid loss of heat from the core which can exceed the critical value required for the onset of a thermal dynamo (gray area). *After Stegman, et al. (2003).*

Because of gravitational instability, ilmenites may have sunk to the core–mantle boundary carrying along an overlying layer strongly enriched in heat producing elements. Accumulation of this mixed and dense layer at the bottom of the mantle creates a thermal blanket that initially prevents core cooling and magnetic field generation. Subsequent heating due to radioactive heat sources progressively increases the thermal buoyancy of the layer, which can ultimately become unstable and rise buoyantly, generating a sufficiently high heat flux to overcome the critical value for the onset of a dynamo. Depending on the actual density of the sunken layer, this model can actually predict the correct timing for the onset and disappearance of the lunar magnetic field and also for the eruption of mare basalts early in the lunar history.

6.4. Lithosphere and Surface Deformation, and Thermal Evolution

Theoretical inferences and observations on the state of deformation of the surface and interior of planetary bodies can also serve to constrain their evolution. In particular, the determination of the elastic response of the lithosphere to surface loads due to geological structures (ice fields, volcanoes, rift systems, etc.) can be linked to the thermal state of the mantle, thereby helping to reconstruct the history of the interior. The lithosphere represents the mechanical boundary layer of convective terrestrial bodies whose viscosity is a strong function of temperature. It comprises the crust and the upper layers of the mantle that undergo elastic and plastic–brittle deformation, and lies above the ideal depth below which fully ductile behavior becomes dominant (see Figure 9.8). The thickness of the elastic part of the lithosphere can be calculated by combining flexural models based on given rheological assumptions with gravity and topography data. The base of the elastic lithosphere, which marks a transition in the mechanical behavior of the mantle, can also be identified with a specific isotherm (~ 1050 K assuming a dry olivine rheology). Knowledge of the mantle temperature at a given depth allows for an estimate of the surface heat flow at the loading site. Since, in addition, the elastic lithosphere can support loading stresses over geologically long time intervals ($\sim 10^8$ years), such heat flow is not representative of the current thermal state but rather of the conditions at the time of formation of the load responsible for the deformation, which remained "frozen" in the lithosphere. This information can be employed to constrain the thermal history of a planet.

Figure 9.18 shows results of combined analyses of lithospheric flexure and gravity/topography data for Mars (Figure 9.18(a)) and Venus (Figure 9.18(b)). In the former

Chapter | 9 Evolution of Planetary Interiors

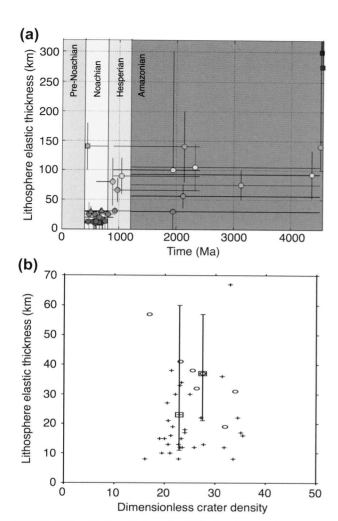

FIGURE 9.18 (a) Compilation of published estimates of the elastic thickness of Mars' lithosphere as a function of the age of the corresponding region for which the studies were made. The two points at ~300 km correspond to loading of the polar regions and are somewhat anomalous as they require lithospheric temperature colder than expected. *After Grott, et al. (2013).* (b) Thickness of the elastic lithosphere at several regions over Venus' surface as a function of the (dimensionless) crater density, which is a measure of surface age. No correlation between age and elastic thickness can be recognized, and an average value of ~25 km yields a satisfactory fit to the gravity and topography data of all analyzed areas. *After Barnett, et al. (2002).*

uniform age (see Chapter 15), no correlation can be recognized between elastic thickness and age of the surface (Figure 9.18(b)), which renders this kind of analyses useful to characterize the recent state of the planet but not its long-term history.

On stagnant lid bodies, whose surface has been preserved for billions of years, the deformation caused by certain tectonic phenomena can also help to understand processes associated with the evolution of the interior. An interesting example in this sense is offered by Mercury, whose surface is characterized by a widespread system of tectonic landforms termed lobate scarps (Chapter 13). These are crust-breaking thrust faults that are interpreted as the result of periods of global planetary contraction. Since the vertical displacement on a fault scales with the length of the fault itself, measurements of the length of lobate scarps from photogeological mapping can be used to retrieve the total amount of contractional strain experienced by the planet and in turn the corresponding decrease in planetary radius. The latest estimates based on the analysis of the images delivered by the camera onboard of the MESSENGER (MErcury Surface Space ENvironment Geochemistry and Ranging) spacecraft indicate that Mercury experienced a global contraction of ~2–4 km throughout its history, depending on whether certain secondary tectonic features are also an expression of global contraction or not. As the interior heats up or cools, the corresponding increase or decrease in volume due to thermal expansion and contraction causes the planet's radius to change accordingly. The observed amount of contraction can be thus used as a constraint in evolution models. Figure 9.19 shows results from one such model. Besides variations due to heating and cooling of the interior, the production of crust also affects the calculation of global radius changes because of the variations in density (and hence in volume) associated with the extraction of partial melt (Section 6.1). During the first 1.5–2 billion years of evolution, mantle heating, accompanied by the production of partial melt and crust formation, causes the planet to expand. Afterward, cooling takes over causing contraction to continue at a nearly constant rate until present. At the end of the evolution, a decrease in planetary radius of ~4 km following the expansion phase is obtained, in agreement with the observations.

case, the elastic thickness correlates well with the age of the surface. While during the Noachian period elastic thicknesses are around 20 km, they quickly increase to values between 50 and 100 km in the Hesperian, and further rise in the Amazonian, with present-day values reaching 300 km. Small elastic thicknesses during the earliest evolution, followed by larger ones at later times are generally consistent with models of Mars' thermal history that predict the heat flow to decrease rapidly at the beginning of the evolution and more slowly later on (see Figure 9.13). On the contrary, for Venus, which possesses a surface of nearly

6.5. Life, Plate Tectonics and the Evolution of the Interior

The continuous cycling of rock and volatiles between the surface reservoirs and the mantle and crust through plate tectonics opens pathways of biologically altered rock to the interior where this may react with the mantle (compare Figure 9.20(a)). Norman Sleep and colleagues (see

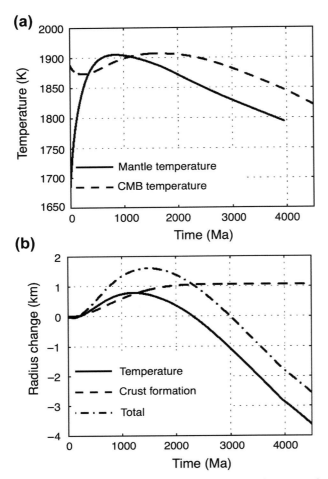

FIGURE 9.19 (a) Evolution of the mantle (solid line) and core−mantle boundary (dashed line) temperatures from a parametrized model of the thermochemical evolution of Mercury. The interruption of the line corresponding to the mantle temperature at ∼4 Ga corresponds to the cessation of heat transport by convection. According to this model, Mercury would be at present in a conductive state. (b) Total changes of planetary radius (dashed-dotted line) as a function of time resulting from the contributions of thermal expansion and contraction (solid line), and mantle differentiation due to crustal production (dashed line). *After Tosi, et al. (2013).*

bibliography) have argued that the mantle should thus have acquired biosignatures dating back some billions of years, in particular to times after the rise of photosynthesis (which has caused the biomass to dramatically increase). The biologically altered mantle would vent its bisosignatures through island arcs and oceanic volcanoes to the surface where they could be read and interpreted—a work that still needs to be done.

Dennis Höning and colleagues (see Figure 9.20) have used the well-established fact that the biosphere enhances the erosion rate on continents to consider the possible effects on the mantle water budget and continental growth rate of a biologically enhanced sedimentation rate. It is well established that the rate of water transfer to the mantle will increase with the thickness of the sedimentary layer on a subducting slab (Figure 9.20(a)). On the one hand, the sedimentary layer by virtue of its own porosity will carry water. But it will also reduce the rate of shallow dewatering of the oceanic crust due to overlying and interbedded sediments of low permeability (e.g. clay-rich deposits) sealing off the oceanic crust. An increased flux of water to depth would increase the rate of melting of the slab and adjacent mantle in the source region of andesitic volcanism (see Figure 9.20(a)) thereby increasing the rate of continental growth. The increased rate of subduction of water would further increase the water content of the mantle. Since the viscosity of the mantle will decrease with increasing water content as we have discussed in Section 4, the convection rate would increase and thus the rate of turnover of the plate tectonic engine and the rate of subduction indicating a self-regulatory system. The process would also factor in the carbon silicate cycle that buffers the surface temperature of the Earth. Taking everything else constant, an increased turnover rate of plate tectonics should result in a cooling of the atmosphere.

Figure 9.20(b) shows how the plate tectonics system will evolve in a simplified model that is assumed to evolve toward stable states. The figure shows which stable states the model can attain in a phase plane defined by the mantle water content and the continental surface coverage. The solid line joins states for which the continental surface area is stable (it will not change with time). The dashed line joins states for which the mantle water content is stable. Where the two lines cross, both continental surface coverage and mantle water content are constant. The two points labeled F_{wet} and F_{dry} are attractors. The arrows show how states away from these points will evolve toward one of these. The third intersection point between F_{wet} and F_{dry} is not an attractor. Rather, the system would evolve away from this point should it ever be reached. For the model parameter combination shown—representative of the present Earth—almost all initial states will evolve toward F_{wet}. The area of attraction for F_{dry} is small and is labeled by the Roman letter I. Thus, for the present Earth, F_{wet} is the stable attractor toward which the Earth has evolved. Reducing the weathering rate will enlarge the zone of attraction of F_{dry} and may even lead to the disappearance altogether of F_{wet}. Thus it has been concluded that an abiotic Earth would likely have only a small coverage of continents and a comparatively dry mantle.

It should be said that the model described above assumes plate tectonics to be operating. The conditions under which plate tectonics is working are not completely understood but it is largely held that it requires a wet and sufficiently vigorously convecting mantle. It can thus be speculated that an abiotic world would not only have a significantly dryer mantle than a biotic one but also may lack plate tectonics altogether. Plate tectonics could thus be a biosignature.

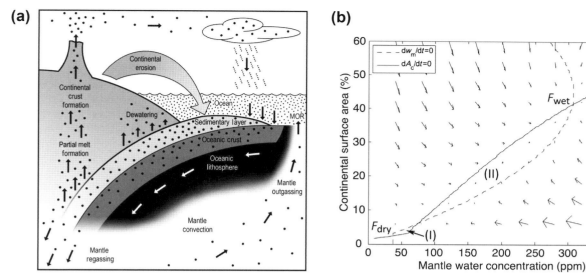

FIGURE 9.20 (a) Schematic cartoon depicting Earth's global plate tectonics and water cycle. Water is represented by large and small dots, its path by black arrows, and movement of the oceanic plate by white arrows. Initial water uptake occurs within the submarine oceanic crust and sediments. Water loss first occurs after the subduction trench through dewatering, followed by the formation of the water-rich partial melt. The partial melt drives arc volcanism and continental crust formation. A fraction of the water contained in the subducting plate is regassed into the mantle. The water leaves the convecting mantle at midoceanic ridges (MOR) as free volatiles closing the cycle or becomes part of the newly formed oceanic crust. (b) Phase plan spanned by mantle water concentration and continental surface area for a model of the present biotic Earth. The dashed and the solid isolines indicate a steady state for mantle water concentration and the continental area, respectively. The arrows indicate state vectors for a trajectory. Points labeled F_{wet} and F_{dry} are attractors toward which the system will evolve given enough time. The parameters of F_{wet} are close to the parameters of the present Earth. The zone of attraction for F_{dry} is small and is indicated by the Roman letter I, while the Roman letter II indicates the zone of attraction of F_{wet} and comprises basically the rest of the phase plane. The size of zone I will increase with decreasing sedimentation rate. For an abiotic Earth with a considerably smaller erosion rate than for the present Earth, the system will likely evolve toward the dry state depending to some extent on initial conditions. It is possible, if not likely, that such a world would lack plate tectonics altogether. *After Höning, et al. (2013).*

7. CONCLUDING REMARKS AND PERSPECTIVES

The combination of theoretical models with a variety of observational constraints has been dramatically improving our understanding of the evolution of terrestrial bodies. Ever more sophisticated numerical simulations have been helping to unravel the basic principles governing the workings of planetary interiors, despite their inaccessibility to direct observation. Nevertheless, the number and complexity of the physical and chemical processes at play in the mantle and core over a wide range of spatial and temporal scales requires continuous advancements in the models as well as in the way these make use of observational data, both in terms of inputs and constraints. Several fundamental questions are still open. Research on the early evolution following planetary accretion and magma ocean solidification is in its infancy, although these processes could significantly affect the modes of interior heat transfer and hence the entire evolution. Despite the possibility to simulate the behavior of the Earth as a plate tectonics planet (Figure 9.11), a consistent physical theory of this phenomenon explaining its initiation and the ultimate reason why it occurs on Earth but not on the other known bodies is still lacking. Coupling between interior and atmospheric processes (see Section 6.1) has received little attention so far, but it could be a key aspect to understanding the evolution of planets, like e.g. Venus, on which variations in surface conditions, of temperature in particular, can be large enough to induce significant rheological changes.

Progress from mineral physics in the analysis of rocks at conditions resembling those of deep interiors is fundamental to restrict the number and range of free parameters that enter evolution models. Despite the somewhat simplistic view according to which terrestrial planets consist of ordered shells with near-constant properties (Figure 9.1), the study and imaging of the Earth's interior demonstrates that important heterogeneities exist at all scales, and there is no reason to believe that this should not also be the case of other terrestrial bodies. The integration of consistent descriptions of mantle mineralogy in simulations of thermochemical convection will be thus an essential aspect of the forthcoming generation of evolution models. The quality and amount of surface observations from spacecraft missions continue to increase and to be a major driver of research on planetary interiors. The level of realism with which numerical models will be able to reconstruct the evolution of terrestrial bodies will crucially depend on our ability to simultaneously account for the largest possible number of observational constraints.

BIBLIOGRAPHY

Barnett, D. N., et al. (2002). Flexure of Venusian lithosphere measured from residual topography and gravity. *Journal of Geophysical Research, 107*, E2. http://dx.doi.org/10.1029/2000JE001398.

Breuer, D. (2009). Dynamics and thermal evolution, in Landolt-Börnstein Astronomy and Astrophysics. *Springer Veralg*, 254–270.

Breuer, D., & Moore, W. B. (2007). Dynamics and thermal history of the terrestrial planets, the Moon, and Io. *Treatise on Geophysics, 10*, 299–348. Elsevier.

Breuer, D., & Spohn, T. (2003). Early plate tectonics versus single-plate tectonics on Mars: Evidence from magnetic field history and crust evolution. *Journal of Geophysical Research, 108*, E7. http://dx.doi.org/10.1029/2002JE001999.

Canup, R. (2012). Forming a Moon with an Earth-like composition via a giant impact. *Science, 338*, 1052–1055.

Grott, M., et al. (2013). Long-term evolution of the martian crust-mantle system. *Space Science Reviews, 174*, 49–111.

Höning, D., et al. (2013). Biotic vs abiotic Earth: A model for mantle hydration and continental coverage. *Planetary and Space Science*. http://dx.doi.org/10.1016/j.pss.2013.10.004.

Ismail-Zadeh, A., & Tackley, P. J. (2010). *Computational methods for geodynamics*. Cambridge University Press.

Jaupart, C., & Mareschal, J.-C. (2011). *Heat generation and transport in the Earth*. Cambridge University Press.

Karato, S. I. (2008). *Deformation of Earth materials: an introduction to the rheology of solid earth*. Cambridge University Press.

Kohlstedt, D. L., et al. (1995). Strength of the lithosphere: Constraints imposed by laboratory experiments. *Journal of Geophysical Research, 100*(B2), 1, 7587–17, 602.

Nakagawa, T., & Tackley, P. J. (2009). Influence of magmatism on mantle cooling, surface heat flow and Urey ratio. *Earth and Planetary Science Letters, 329–330*, 1–10.

Rubie, D. C., Nimmo, F., & Melosh, H. J. (2007). Formation of the Earth's core. *Treatise on Geophysics, 9*, 51–90. Elsevier.

Schubert, G., Turcotte, D. L., & Olson, P. (2001). *Mantle convection in the Earth and planets*. Cambridge University Press.

Sleep, N. H., Bird, D. K., & Pope, E. (2012). Paleontology of Earth's mantle. *Annual Review of Earth and Planetary Science, 40*, 277–300.

Šramek, O., & Zhong, S. (2012). Martian crustal dichotomy and Tharsis formation by partial melting coupled to early plume migration. *Journal of Geophysical Research, 117*. http://dx.doi.org/10.1029/2011JE003867. E01005.

Stegman, D. R., et al. (2003). An early lunar core dynamo driven by thermochemical mantle convection. *Nature, 421*, 143–146.

Stixrude, L., et al. (2009). Thermodynamics of silicate liquids in the deep Earth. *Earth and Planetary Science Letters, 278*, 226–232.

Tosi, N., et al. (2013). Thermochemical evolution of Mercury's interior. *Journal of Geophysical Research, 108*. http://dx.doi.org/10.1002/jgre.20168.

Turcotte, D. L., & Schubert, G. (2002). *Geodynamics*. Cambridge University Press.

van Heck, H., & Tackley, P. J. (2008). Planforms of self-consistently generated plate tectonics in 3-D spherical geometry. *Geophysical Research Letters, 35*. http://dx.doi.org/10.1029/2008GL035190. L19312.

Chapter 10

Astrobiology

Christopher P. McKay
Space Science Division, NASA Ames Research Center, Moffett Field, CA, USA

Wanda L. Davis
Space Science Division, NASA Ames Research Center, Moffett Field, CA, USA

Chapter Outline

1. Introduction 209
2. What is Life? 210
   2.1. The Ecology of Life: Liquid Water 211
   2.2. Generalized Theories for Life 212
3. The History of Life on Earth 214
4. The Origin of Life 217
5. Limits to Life 219
6. Life in the Solar System 219
   6.1. Mercury and the Moon 220
   6.2. Venus 220
   6.3. Mars 220
      6.3.1. The Viking Results 220
      6.3.2. Early Mars 222
      6.3.3. Subsurface Life on Mars 224
      6.3.4. Meteorites from Mars 226
   6.4. The Giant Planets 226
      6.4.1. Europa 227
      6.4.2. Titan 227
      6.4.3. Enceladus 228
   6.5. Asteroids 228
   6.6. Comets 228
7. How to Search for Life on Mars, Europa, or Enceladus 228
8. Life About Other Stars 229
9. Conclusion 230
Bibliography 230

1. INTRODUCTION

Life is widespread on the Earth and appears to have been present on the planet since early in its history. Biochemically all life on Earth is similar and seems to share a common origin. Throughout geological history life has significantly altered the environment of the Earth while at the same time adapting to this environment. It would not be possible to understand the Earth as a planet without the consideration of life. Thus, life is a planetary phenomenon and is arguably the most interesting phenomenon observed on planetary surfaces.

Everything we know about life is based on the example of life on Earth. Generalization to other areas or extended forms of life must proceed with this caveat. Although we remain uncertain of the process or the time for its origin, the advent of life on Earth was established within 1 billion years after the formation of the planet. While life also requires energy and nutrients, liquid water is the defining ecological requirement for life on Earth. Thus, a liquid water environment is currently the best indicator of where to search for extraterrestrial life. Looking out into the Solar System we see evidence for liquid water. Europa appears to have a liquid water ocean underneath a global ice surface—the evidence is indirect but persuasive. Enceladus has geysers erupting from its South Polar area powered by subsurface liquid water. There are several lines of evidence that suggest past liquid water on Mars. Direct images from orbiting spacecraft show fluvial features on the surface of Mars. Orbital infrared spectrometers have found local regions that show minerals formed in liquid water environments. The Mars Exploration Rovers and the Curiosity Rover have also found evidence for past aqueous activity at their landing sites on Mars. Our understanding of life, albeit limited to one example and one planet, would suggest that life is possible on other planets whenever conditions allow for environments like those on Earth—energy, nutrients, and most critically liquid water. This suggests the possibility of early microbial life on Mars and forms the basis for a search for Earth-like planets orbiting other stars. Studies of a second example of life—a second genesis—to which we can compare and contrast terrestrial biochemistry will be the beginning of a more general understanding of life as a process in the universe. This implies a search for more than just fossils but a search for the biochemical remains of organisms, dead or alive.

2. WHAT IS LIFE?

Our understanding of life as a phenomenon is currently based only on the study of life on Earth. One of the profound results of biology is the realization that all life forms on Earth share a common physical and genetic makeup. The impression of vast diversity that we experience in nature is a result of manifold variations on a single fundamental biochemistry. The biochemistry of life is based on 20 **amino acids** and five nucleotide bases. Added to this are the few sugars, from which are made the **polysaccharides**, and the simple alcohols and fatty acids that are the building blocks of lipids. This simple collection of primordial biomolecules (Figure 10.1) represents the set from which the rest of biochemistry derives.

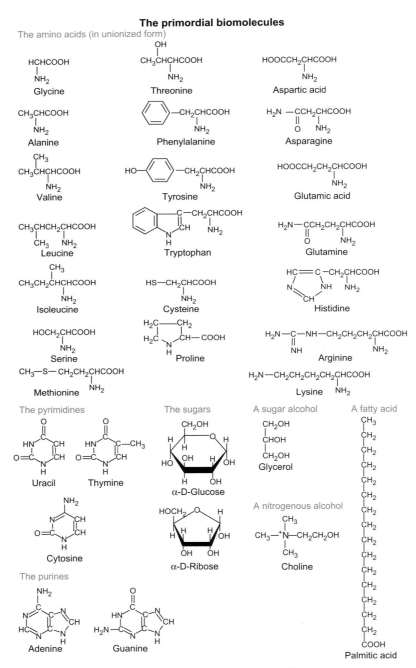

FIGURE 10.1 The basic molecules of life.

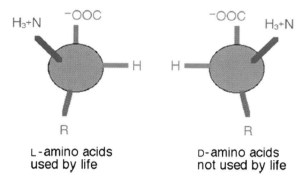

FIGURE 10.2 The L and D forms of the amino acid alanine.

Except for glycine, the **amino acids** in Figure 10.1 can have either left handed (L) or right handed (D) symmetry. Figure 10.2 shows the two versions, known as enantiomers (from the Greek *enantios* meaning opposite), for alanine. Life uses only the L-enantiomer to make proteins, although there are some bacteria that use certain D-forms in their cell walls and many others have **enzymes** that can convert the D-form to the L-form. In addition, L-**amino acids** other than the 20 listed in Figure 10.1 are occasionally used in proteins and are sometimes used directly, for example, as toxins by fungi and plants. We do not yet understand how and why life acquired a preference for the L-amino acids over the D-amino **acid**s; this is one of the key observations that theories for the origins of life seek to explain.

The genetic materials of life—DNA (deoxyribonucleic acid) and RNA (ribonucleic acid)—are both constructed from nucleotide bases that form the alphabet of life's genetic code. In DNA these are adenine (A), thymine (T), cytosine (C), and guanine (G). In RNA thymine is replaced by uracil (U). The nucleic acids each provide a four-letter alphabet in which the codes for the construction of proteins are based. This information recording system is found in all living systems.

The biochemical unity of life, in particular the genetic unity, strongly suggests that all living things on Earth are descendant from a common ancestor. This is the phylogenic unity of life as shown in Figure 10.3. These genetic trees are obtained by comparing the ribosomal RNA within each organism. There are sections within the RNA that are remarkably similar within all life forms. These conserved sections show only random point changes and not evolutionary trends. Thus, the similarity between the genetic sequences of any two organisms is a measure of their evolutionary distance, or more precisely the time elapsed since they shared a common ancestor. When viewed in this way, life on Earth is divided into three main groups: the eucarya, the bacteria, and the archaea. The eucarya include the multicellular life forms encompassing all plants and animals. The bacteria are the familiar bacteria including intestinal bacteria, common soil bacteria, and the pathogens. The archaea are a different class of microorganisms that are found in unusual and often harsh environments such as hypersaline ponds and H_2 rich anaerobic sediments. All methane-producing microbes are archaea. Archaea are also found in soils and grow on and in humans, producing methane in the gut. Archaea are not known to be human pathogens or to produce substances that are toxic to humans. Why some bacteria but no archaea are pathogenic is not yet understood.

2.1. The Ecology of Life: Liquid Water

In addition to describing the building blocks of life it is instructive to consider what life does. In this regard it is

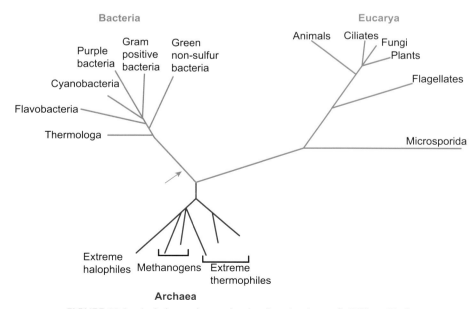

FIGURE 10.3 A phylogenetic tree showing the relatedness of all life on Earth.

TABLE 10.1 Ecological Requirements for Life

Requirement	Occurrence in the Solar System
Energy	Common
Predominately sunlight	Photosynthesis at 100 AU light levels
Chemical energy	e.g. $H_2 + CO_2 \rightarrow CH_4 + H_2O$
Carbon	Common as CO_2 and CH_4
Liquid water	Rare, only on Earth for certain
N, P, S, and other elements	Common

possible to define a set of ecological or functional requirements for life. There are four fundamental requirements for life on Earth: energy, carbon, liquid water, and a few other elements. These are listed in Table 10.1 along with the occurrence of these environmental factors in the Solar System.

Energy is required for life from basic thermodynamic considerations. Typically on the Earth this energy is provided by sunlight, which is a thermodynamically efficient (low **entropy**) energy source. Some limited systems on Earth are capable of deriving their energy from chemical reactions (e.g. methanogenesis, $CO_2 + 4H_2 \rightarrow CH_4 + 2H_2O$) and do not depend on photosynthesis. On Earth these systems are confined to locations where the more typical photosynthetic organisms are not able to grow, and it is not known if an ecosystem that was planetary in scale or survived over billions of years could be based solely on chemical energy. There are no known organisms on Earth that make use of temperature gradients to derive energy. These organisms would be analogous to a Carnot heat engine. Table 10.2 lists some of the most important metabolic reactions by which living systems generate energy. This list includes autotrophs (which derive energy from nonbiological sources) as well as heterotrophs (which derive energy by the consumption of organic material, usually other life forms).

Elemental material is required for life, and on Earth carbon has the dominant role as the backbone molecule of biochemistry. Life almost certainly requires other elements as well. Life on Earth utilizes a vast array of the elements available on the surface. However, this does not prove that these elements are absolute requirements for life. Other than H_2O and C, the elements N, S, and P are probably the leading candidates for the status of required elements. Table 10.3 lists the distribution of elements in the cosmos and on the Earth and compares these with the common elements in life.

As indicated in Table 10.1, sunlight and the elements required for life are common in the Solar System. What appears to be the ecologically limiting factor for life in the Solar System is the stability of liquid water. Liquid water is a necessary requirement for life on Earth. Liquid water is key to biochemistry because it acts as the solvent in which biochemical reactions take place and furthermore it interacts with many biochemicals in ways that influence their properties. For example, water forms hydrogen bonds with some parts of a large molecule, the hydrophilic groups, and repels other parts, the hydrophobic groups, thereby forcing these molecules to curl up with their hydrophobic groups in the interior and the hydrophilic groups on the exterior in contact with the water. Certain organisms, notably lichen and some **algae**, are able to utilize water in the vapor phase if the relative humidity is high enough. Many organisms can continue to metabolize at temperatures well below the freezing point of pure water because their intracellular material contains salts and other solutes that lower the freezing point of the solution. No microorganism currently known is able to obtain water directly from ice. Many organisms, such as the snow **algae** *Chlamydomonas nivalis*, thrive in liquid water associated with ice but in these circumstances the organisms are the beneficiaries of external processes that melt the ice. There is no known occurrence of an organism using metabolic methods to overcome the latent heat of fusion of ice thereby liquefying it.

Because liquid water is universally required for known life and because it appears to be rare in the Solar System the search for life beyond the Earth begins first with the search for liquid water.

2.2. Generalized Theories for Life

There have been many attempts at a definition of life and perhaps such a definition would aid in our investigation for life on other planets and help unravel the origins of life on Earth. However, it is probable that there will never be a simple definition of life and it may not be necessary in a search for life on other worlds. Despite the fundamental unity of biochemistry and the universality of the genetic code, no single definition has proven adequate in describing the single example of life on Earth. Many of the attributes that we would associate with life, for example,

Chapter | 10 Astrobiology

TABLE 10.2 Examples of Metabolic Pathways

Heterotrophy

1. Fermentation $\quad C_6H_{12}O_6 \rightarrow 2CO_2 + 2C_2H_5OH$
2. Anaerobic respiration $\quad C_6H_{12}O_6 + 12NO_3^- \rightarrow 6CO_2 + 6H_2O + 12NO_2^-$
3. Aerobic respiration $\quad C_6H_{12}O_6 + 6O_2 \rightarrow 6CO_2 + 6H_2O$

Photoautotrophy

1. Anoxic photosynthesis $\quad 12CO_2 + 12H_2S + h\nu \rightarrow 2C_6H_{12}O_6 + 9S + 3SO_4$
2. Oxygenic photosynthesis $\quad 6CO_2 + 6H_2O + h\nu \rightarrow C_6H_{12}O_6 + 3O_2$

Chemoautotrophy

Anaerobic

1. Methanogens $\quad CO_2 + 4H_2 \rightarrow CH_4 + 2H_2O$
$\quad CO + 3H_2 \rightarrow CH_4 + H_2O$
$\quad 4CO + 2H_2O \rightarrow CH_4 + 3CO_2$
2. Acetogens $\quad 2CO_2 + 4H_2 \rightarrow CH_3COOH + 2H_2O$
3. Sulfate reducers $\quad H_2SO_4 + 4H_2 \rightarrow H_2S + 4H_2O$
4. Sulfur reducers $\quad S + H_2 \rightarrow H_2S$
5. Thionic denitrifiers $\quad H_2S + 2NO_3^- \rightarrow SO_4^{2-} + H_2O + N_2O$
$\quad 3S + 4NO_3^- + H_2 \rightarrow 3SO_4^{2-} + 2N_2 + 2H^+$
6. Iron reducers $\quad 2Fe^{3+} + H_2 \rightarrow 2Fe^{2+} + 2H^+$

Aerobic

1. Sulfide oxidizers $\quad 2H_2S + 3O_2 \rightarrow 2SO_4S + 2H_2O$
2. Iron oxidizers $\quad 4FeO + O_2 \rightarrow 2Fe_2O_3$

TABLE 10.3 Elemental Abundances by Mass

Rank	Cosmic (%)	Earth's Crust (%)	Humans (%)	Bacteria (%)
1	H 70.7	O 46.6	O 64	O 68
2	He 27.4	Si 29.7	C 19	C 15
3	O 0.958	Al 8.13	H 9	H 10.2
4	C 0.304	Fe 5.00	N 5	N 4.2
5	Ne 0.174	Ca 3.63	Ca 1.5	P 0.83
6	Fe 0.126	Na 2.83	P 0.8	K 0.45
7	N 0.110	K 2.59	S 0.6	Na 0.40
8	Si 0.0706	Mg 2.09	K 0.3	S 0.30
9	Mg 0.0656	Ti 0.44	Na 0.15	Ca 0.25
10	S 0.0414	H 0.14	Cl 0.15	Cl 0.12

self-replication, self-ordering, response to environmental stimuli, can be found in nonliving systems, fire, crystals, bimetallic thermostats, respectively. Furthermore, there are various and peculiar life forms such as viruses and giant cell-less slime molds that defy even a biological definition of life in terms of the cell or the separation of internal and external environments. In attempting a resolution of this problem the most useful definition of life is that it is a system that develops Darwinian evolution: reproduction, mutation, selection (Table 10.4). This is an answer to the question what does life do?

We are able to answer the questions, what does life need and what does life do, even if we do not have a closed form compact definition of life. Thus, the requirements for life listed in Table 10.1 and the functions of life in Table 10.4 are therefore very general and it is probably unwise to apply more restrictive criteria. For example, for evolution to occur some sort of information storage mechanism is required. However, it is not certain that this information mechanism needs to be a DNA/RNA-based system or even that it be expressed in structures dedicated solely for replication. While on the present Earth all life uses dedicated DNA and RNA systems for genetic coding there is evidence that at one time genetic and structural coding were combined into one molecule, RNA. In this so-called RNA world there would have been no distinction between genotype (genetic) and phenotype (structural) molecular replicating systems—both of these processes would have been performed by an RNA replicating molecule. In present biology the phenotype is composed of proteins for the most part. This example illustrates the difficulty in determining which aspects of biochemistry are fundamental and which are the result of the peculiarities of life's history on Earth.

In basing our consideration of life, on the distribution we observe here on Earth as a general phenomenon, we suffer simultaneously from the problem that there is only one kind of life on this planet while the variety of that life is too complex to allow for precise definitions or characterizations. Thus, we can neither extrapolate nor be specific in our theories for life.

Some scientists have suggested that living systems elsewhere in the universe may exhibit vast differences from terrestrial biology, and have proposed a variety of alternative life forms. One postulated alternative life form is based on the substitution of ammonia for water. Certainly ammonia is an excellent solvent—in some respects better than water. The range of temperatures over which ammonia is liquid is prevalent in the universe (melting point: $-78°$, normal boiling point $-33°$, liquid at room temperature when mixed with water) and the elements that compose it are abundant in the cosmos. Other scientists have suggested the possibility that silicon may be used as a substitute for carbon in alien life forms. However, silicon does not form polymeric chains either as readily or as long as carbon does and its bonds with oxygen (SiO_2) are much stronger than carbon bonds (CO_2) rendering its oxide essentially inert.

Although speculations of alien life capable of using silicon in place of carbon or ammonia in place of water are intriguing, no specific experiments directed toward alternate biochemistries have been designed. Thus, we have no strategies for where or how to search for such alternate life or its fossils. More significantly, these speculations have not contributed to our understanding of life. One can only conclude that our unique understanding of terrestrial life is based on Earth systems and wide ranging speculations regarding alternate chemistries are currently too limited to be fruitful. Perhaps some day we will develop general theories for life or, more likely, have many sources of life to compare thereby allowing for complete theories. Basing our theories on Earth-like life should be considered a necessary first approach and not a fundamental limitation.

3. THE HISTORY OF LIFE ON EARTH

There are several sources of information about the origin of life on Earth. These include the physical record, the genetic record, the metabolic record, and laboratory simulations. The physical record includes the collection of sedimentary and fossil evidence of life. This record is augmented by theoretical models of the Earth and the Solar System all of which provide clues to conditions billions of years ago when the origin of life is thought to have occurred. There is also the record stored in the genomes of living systems that comprises the collective gene pool of our planet. Genetic information tells us the path of evolution as shaped by environmental pressures, biological constraints, and random events that connect the earliest genomic organism, through the **last universal common ancestor** to the present tree of life (Figure 10.3). There is also the record of metabolic pathways in the biochemistry of organisms that have evolved in response to changes in the environment while simultaneously causing changes to that environment. All these records are palimpsests in that they have been overwritten—often repeatedly—over time. Laboratory simulations of prebiotic chemistry—the chemistry assumed present before life—can provide clues to the conditions and chemical solutions leading up to the origin of life. Experiments of DNA/RNA replication sequences can provide

TABLE 10.4 Properties of Life

Properties of Life
Mutation
Selection
Reproduction

clues to the selection process that optimizes mutations as well as provide a basic understanding of reproduction. Perhaps one day the process that initiates life will be studied in the laboratory or discovered on another planet.

The major events in the history of life are shown in Figure 10.4. As the Earth was forming about 4.5 Gyr ago its surface would have been inhospitable to life. The gravitational energy released by the formation of the planet would have kept surface temperatures too high for liquid water to exist. Eventually as the heat flow subsided rain would have fallen for the first time and life could be sustained in liquid water. However, it is possible that subsequent impacts could have been large enough to sterilize the Earth by melting, excavating, and vaporizing the planetary surface, removing all liquid water. Thus, life may have been frustrated in its early starts. Following a sufficiently large impact, the entire upper crust of the Earth would be ejected into outer space and any remnant left as a magma ocean. Barring these catastrophic events, however, sterilizing the Earth is a difficult task since it is not sufficient to merely heat the surface to high temperatures. At the present time, microorganisms survive at the bottom of the ocean and even kilometers below the surface of the planet. An Earth-sterilizing impact must not only completely evaporate the oceans but also must then heat the surface and subsurface of the Earth such that the temperature does not fall anywhere below about 200 °C—which is required for heat sterilization of dry, dormant organisms. This is a difficult requirement since the time it takes heat to diffuse down a given distance scales as the square of the distance. Thus, heat must be applied a million times longer to sterilize to a depth of 1 km compared to a depth of 1 m.

It is not known when the last life-threatening impact occurred. As shown schematically in Figure 10.4 the rate of impact, extrapolated from the record on the moon, rises steeply before 3.8 Gyr ago. Thus, it is likely that the Earth was not continuously suitable for life much before 3.8 Gyr ago. There is persuasive evidence that microbial life was present on the Earth as early as 3.4 Gyr ago. This evidence includes microbial fossils and **stromatolites**. Stromatolites are large features—often many meters in size—that can be formed by the lithification of laminated microbial mats—although physical processes can result in similar forms (See Figure 10.5). **Phototactic** microorganisms living on the bottom of a shallow lake or ocean shore may be periodically covered with sediment carried in by spring runoff, for example. To retain access to sunlight the organism must move up through this sediment layer and establish a new microbial zone. After repeated cycles a layered series of mats are formed by lamination of the sediments containing the organic material. One characteristic of these biogenic mats that distinguishes them from nonbiologically caused layering is that the response is phototactic not gravitational. Thus, the layered structure is

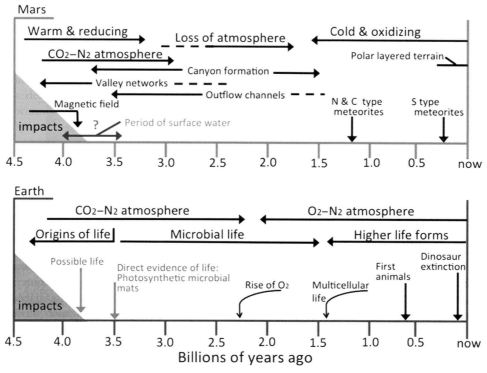

FIGURE 10.4 Major events in the history of the Earth and Mars. The period of moist surface conditions on Mars may have corresponded to the time during which life originated on Earth. The similarities between the two planets at this time raise the possibility of the origin of life on Mars.

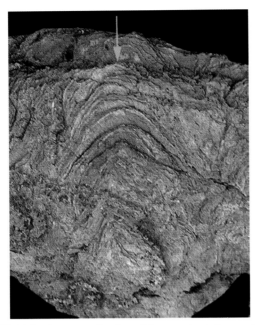

FIGURE 10.5 A stromatolite formed by cyanobacteria over 1 billion years ago from the Crystal Springs formation, Inyo County, California. Stromatolites are an important form of fossil evidence of life because they form macroscopic structures that could be found on Mars. It is therefore possible that a search for stromatolites near the shores of an ancient Martian lake or bay could be conducted in the near future. Expecting microbial communities to have formed stromatolites on Mars is not entirely misplaced geocentricism. The properties of a microbial mat community that results in stromatolite formation need only be those associated with photosynthetic uptake of CO_2. There are broad ecological properties that we expect to hold on Mars even if the details of the biochemistry and community structure of Martian microbial mats were quite alien compared to their terrestrial counterparts. Within stromatolites trace microfossils can sometimes be found.

not usually flat but is more often domed-shaped because covered microorganisms in a lower layer on the periphery of the structure move more toward the side to reach light. In this way stromatolites can sometimes be distinguished from similar but nonbiological laminae. Often stromatolites contain microfossils—further testimony to their biological origin.

Microbial life—possibly capable of photosynthesis and mobility—appears to have originated early in the history of the Earth possibly before the end of the late bombardment 3.8 Gyr ago and almost certainly not later than 3.4 Gyr ago. This suggests that the time required for the onset of life was brief. If the Greenland sediments are taken as evidence for life it suggests that, within the resolution of the geological record, life arose on Earth as soon as a suitable habitat was provided. The microbial mats at 3.4 Gyr ago put an upper limit of 400 million years on the length of time it took for life to arise after clement conditions were present.

It is possible, in principle, to determine which organism on the Earth is the most similar to the last universal **common ancestor**. To do so one must determine which organism has changed the least compared to all other organisms. For example, if some **taxon** of organism contains a certain mutation but many do not one can trace the mutation to an ancestor common to all organisms in that taxon. Within this related group of organisms the most primitive traits can be established based on how widespread they are. Traits that are found in all or most of the major groupings should be primitive, particularly if these traits are found in groups that diverged early. Traits found in only a few recently related groups are probably younger traits. This line of reasoning applied to the entire **phylogenetic** tree would indicate which organism extant today has the most primitive set of traits. This organism would therefore be the most similar to the common ancestor. Studies of this type have indicated that the organisms alive today that are most similar, genetically and hence presumably ecologically, to the common ancestor are the thermophilic hydrogen-metabolizing bacteria and perhaps the sulfur-metabolizing bacteria. The arrow in Figure 10.3 represents the suggested position of the last common ancestor.

It is important to note here that the last universal common ancestor is not necessarily representative of the first organisms on Earth but was merely the last organism (or group of organisms) from which all life forms today are known to have descended. The common ancestor may have existed within a world of multiple lineages none of which are in evidence today. If all life on Earth has indeed descended from a sulfur bacterium living in a hot springs environment this could be the result of at least three possibilities. First, it may be the case that hot sulfurous environments are important in the origin of life and the common ancestor may represent this primal cell. Second, the common ancestor may have been a survivor of a catastrophe that destroyed all other life forms. The survival of the common ancestor may have been the result of its ability to live deep within a hydrothermal system. Third, the nature of the common ancestor may be serendipitous with no implications as to origin or evolution of the biosphere.

For over 2 Gyr after the earliest evidence for life, life on the Earth was composed of microorganisms only. There were certainly bacteria and possibly one-celled eukaryotes as well. There seemed to be a major change in the environment of the Earth with the rise of photosynthetically produced oxygen beginning at about 2.5 Gyr ago, reaching significant levels about 1 Gyr ago and culminating about 600 Myr ago. (Figure 10.4 shows a timeline of Earth's history with these events.) Soon after the development of high levels of oxygen in the atmosphere multicellular life forms appeared. These rapidly radiated into the major phylum known today (as well as many that have no known living representatives). In time organisms adapted to land environments in addition to aqueous environments, and plants and animals appeared.

4. THE ORIGIN OF LIFE

There are numerous and diverse theories for the origin of life currently under serious consideration within the scientific community. A diagram and classification of current theories for the origin of life on Earth are shown in Figure 10.6. At the most fundamental level, theories may be characterized within two broad categories: theories that suggest that life originated on Earth (Terrestrial in Figure 10.6) and those that suggest that the origin took place elsewhere (Extraterrestrial in Figure 10.6). The extraterrestrial or **panspermia** theories suggest that life existed in outer space and was transported by meteorites, asteroids, or comets to a receptive Earth. In this case the origin of life is not related to environments possible on the early Earth. Along similar lines, life may have been ejected by impacts from another planet in the Solar System and jettisoned to Earth—or vice versa. Furthermore, it has been suggested in the scientific literature that life may have been purposely directed to Earth (Directed Panspermia in Figure 10.6) by an intelligent species from another planet.

The terrestrial theories are further subdivided into organic origins (carbon based) and inorganic origins (mineral based). Mineral-based theories suggest that life's first components were mineral substrates that organized and synthesized clay organisms. These organisms have evolved via natural selection into the organic-based life forms visible on Earth today. The majority of theories that do not invoke an extraterrestrial origin require an organic origin for life on Earth. Theories postulating an organic origin suggest that the initial life forms were composed of the same basic building blocks present in biochemistry today—organic material. If life arose in organic form then there must have been a prebiological source of organics. The **Miller–Urey experiments** and their successors have demonstrated how organic material may have been produced naturally in the primordial environment of Earth (endogenous production in Figure 10.6). An alternative to the endogenous production of organics on early Earth is the importation of organic material by celestial impacts and debris—comets, meteorites, interstellar dust particles, and comet dust particles. A comparison of these sources is shown in Table 10.5. Table 10.6 lists the organics found in the Murchison meteorite and compares these with the organics produced in a Miller–Urey abiotic synthesis. Organic origins differ mainly in the type of primal energy sources: photosynthetic, chemosynthetic, or heterotrophic. The phototrophs and chemotrophs (collectively called autotrophs) use energy sources that are inorganic (sunlight and chemical energy, respectively), whereas heterotrophs acquire their energy by consuming organics (Table 10.2).

Hydrothermal vent environments have been suggested for the subsurface origin of chemotrophic life. In the absence of sunlight these organisms must utilize chemical energy (e.g. $CO_2 + 4H_2 \rightarrow CH_4 + 2H_2O +$ energy). Alternatively phototrophic life utilizes solar radiation from the surface for prebiotic synthesis. These organisms with the ability to chemosynthesize and photosynthesize can assimilate their own energy from materials in their

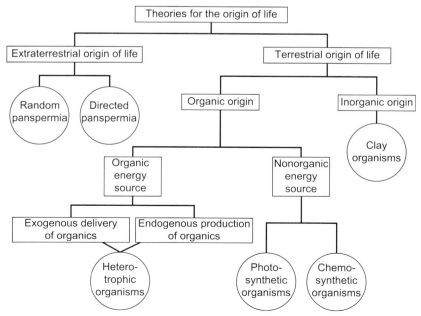

FIGURE 10.6 Diagrammatic representation and classification of current theories for the origin of life.

TABLE 10.5 Sources of Prebiotic Organics on Early Earth

Source	Energy Dissipation (J/year)	Organic Production (in a Reducing Atmosphere (kg/year)
Lightning	1×10^{18}	3×10^{9}
Coronal discharge	5×10^{17}	2×10^{8}
Ultraviolet light ($\lambda < 270$ nm)	1×10^{22}	2×10^{11}
Ultraviolet light ($\lambda < 200$ nm)	6×10^{20}	3×10^{9}
Meteor entry shocks	1×10^{17}	1×10^{9}
Meteor postimpact plumes	1×10^{20}	2×10^{10}
Interplanetary dust	—	6×10^{7}

TABLE 10.6 Comparison of the Amino Acids in Murchison Meteorite and in an Electric Discharge Synthesis, Normalized to Glycine

Amino Acid	Murchison Meteorite	Electric Synthesis
Glycine	100	100
Alanine	>50	>50
α-Amino-n-butyric acid	>50	>50
α-Aminoisobutyric acid	10	>50
Saline	10	1
Norvaline	10	10
Isovaline	1	1
Proline	10	0.1
Pipecolic acid	0.1	<1
Aspartic acid	10	10
Glutamic acid	10	1
β-Alanine	1	1
β-Amino-n-butyric acid	0.1	0.1
δ-Aminoisobutyric acid	0.1	0.1
γ-Aminobutyric acid	0.1	1
Sarcosine	1	10
N-Ethyl glycine	1	10
N-Methyl alanine	1	0

environment. One feature that the various theories for the origin of life have in common is the requirement for liquid water. This is because the chemistry of even the earliest life requires a liquid water medium. This is true if the primal organism appears fully developed (**panspermia**), if it engages in organic chemistry, as well as for the clay inorganic theories.

For many years the standard theory for the origin of life posited a terrestrial organic origin requiring endogenous production of organics leading to the development of heterotrophic organisms. This was generally known as the primordial "soup" theory. Recently, there has been serious consideration for the chemotrophic origin of life and at the present time the scientific community is split between these two views.

5. LIMITS TO LIFE

In considering life beyond the Earth it is useful to quantitatively determine the limits that life has been able to reach on this planet with respect to environmental conditions. Life is not everywhere. There are environments on Earth in which life has not been able to effectively colonize even though these environments could be suitable for life. Perhaps the largest life-free zone on Earth is the polar ice sheets. Here there is abundant energy, carbon, and nutrients (from atmospheric deposition) to support life. However, water is available only in the solid form. No organism on Earth has adapted to use metabolic energy to liberate water from ice even though the energy required per molecule is only ~1% of the energy produced by photosynthesis per molecule. Table 10.7 lists the limits to life as we currently know them. The lower temperature limit clearly ties to the presence of liquid water while the higher temperature limits seem to be determined by the stability of proteins, also in liquid water. Life can survive at extremely low light levels—corresponding to 100 AU, roughly three times the distance between Pluto and the Sun. Salinity and pH also allow for a wide range. Water activity, effectively a measure of the relative humidity of a solution or vapor, can support life only for values above 0.6 for yeasts, lichens, and molds. Bacteria require levels above 0.8. Radiation-resistant organisms such as *Deinococcus radiodurans* can easily survive radiation doses of 1–2 Mrad and higher when in a dehydrated or frozen state.

6. LIFE IN THE SOLAR SYSTEM

Because the knowledge of life is restricted to the unique but varied case found here on Earth, the most practical approach to the search for life on the other planets has been to proceed by way of analogy with life on Earth. The argument for the origin of life on another world is then based on the similarity of other planetary environments with the postulated environments on early Earth. Whatever process led to the establishment of life in one of these environments on Earth could then be logically expected to have led to the origin of life on this comparable world. The more exact the comparison between the early Earth and another planet the more compelling is the argument by analogy. This comparative process should be valid for all the theories for the origin of life listed in Figure 10.6—ranging from panspermia to the standard theory.

Following this line of reasoning further we can conclude that if similar environments existed on two worlds and life arose in both of them then these life forms should be comparable in their broad ecological characteristics. If sunlight was the available energy source, CO_2 the available carbon source, and liquid water the solvent then one could expect phototrophic autotrophs using sunlight to fix carbon dioxide with water as the medium for chemical reactions. Our knowledge of the Solar System suggests such an

TABLE 10.7 Limits to Life

Parameter	Limit	Note
Lower temperature	~ −15 °C	Liquid water
Upper temperature	122 °C	Thermal denaturing of proteins
Low light	~ 10^{-4} S	Algae under ice and deep sea
pH	0–11	
Salinity	Saturated NaCl	Depends on the salt
Water activity	0.6	Yeasts and molds
	0.8	Bacteria
Radiation	1–2 Mrad	May be higher for dry or frozen state

environment could have existed on Mars early in its history as well as Earth early in its history. Thus, while life forms independently originating on these two planets would have different biochemical details they would be recognizably similar in many fundamental attributes. This approach—by analogy to Earth life and the early Earth—provides a specific search strategy for life elsewhere in the Solar System. The key element of that strategy is the search for liquid water habitats.

Spacecrafts have now visited or flown past comets, asteroids, and most of the large worlds in the Solar System except Pluto—and one is en route to Pluto at the time of this writing. Observatory missions have studied all the major objects in the Solar System as well. We can do a preliminary assessment of the occurrence of liquid water habitats—and indirectly life—in the Solar System.

6.1. Mercury and the Moon

Mercury and the Moon appear to have few prospects for liquid water, now or anytime in the past. These virtually airless worlds have negligible amounts of the volatiles (such as water and carbon dioxide) essential for life. There are no geomorphological features that indicate fluid flow. There is evidence that permanently shaded regions of the polar areas on the Moon and Mercury act as traps for water ice. However, there is no indication that the pressure and temperature were ever high enough for liquid water to exist at the surface on either body. (*See* Mercury, Moon.)

6.2. Venus

Venus currently has a surface that is clearly inhospitable to life. There is no liquid water on the surface and the temperature is over 450 °C at an atmospheric pressure of 92 times the Earth's. There is water on Venus but only in the form of vapor and clouds in the atmosphere. The most habitable zone on Venus is at the level in the atmosphere where the pressure is about half of the sea level on Earth. At that location, there are clouds composed of about 25% water and 75% sulfuric acid at a temperature of about 25 °C; these might be reasonable conditions for life. It is possible therefore to speculate that life can be found, or survive if implanted, in the clouds of Venus. What argues against this possibility is the fact that clouds on Earth—at similar pressures and temperatures—do not harbor life. We do not know of any life forms that thrive in cloud environments. Perhaps the essential elements are there but a stable environment is required. (*See* Venus, Atmosphere.)

Theoretical considerations suggest that Venus and Earth may have initially had comparable levels of water. In this case Venus may have had a liquid water surface early in its history when it was cooler—4 billion years ago, due to the reduced brightness of the fainter early sun. Unfortunately, all record of this early epoch has been erased on Venus and the question of the origin of life during such a liquid water period remains untestable. (*See* Venus, Surface and Interior.)

6.3. Mars

Of all the extraterrestrial planets and smaller objects in the Solar System, Mars is the one that has held the most fascination in terms of life. Early telescopic observations revealed Earth-like seasonal patterns on Mars. Large white polar caps that grew in the winter and shrunk in the summer were clearly visible. Regions of the planet's surface near the polar caps appeared to darken beginning at the start of each polar cap's respective spring season and then spreading toward the equator. It was natural that these changes, similar to patterns on the Earth, would be attributed to like causes. Hence, the polar caps were thought to be water ice and the wave of darkening was believed to have been caused by the growth of vegetation. The nineteenth century arguments for the existence of life, and even intelligent life, on Mars culminated in the book *Mars as the Abode of Life* by Percival Lowell in 1908 and the investigations of the celebrated canals. The Mars revealed by spacecraft exploration is decidedly less alive than Lowell's anticipation but its standing as the most interesting object for biology outside Earth still remains.

6.3.1. The Viking Results

In 1976, the Viking landers successfully reached the Martian surface while the two orbiters circled the planet repeatedly photographing and monitoring the surface. The primary objective of the Viking mission was the search for microbial life. Previous reconnaissance of Mars by the Mariner flyby spacecraft and the photographs returned from the Mariner 9 orbiter had already indicated that Mars was a cold dry world with a thin atmosphere. There were intriguing features indicative of past fluvial erosion but there was no evidence for current liquid water on the surface. It was thought that any life to be found on Mars would be microbial. The Viking biology package consisted of three experiments shown schematically in Figure 10.7.

The pyrolytic release (PR) experiment searched for evidence of photosynthesis as a sign of life. The PR was designed to see if Martian microorganisms could incorporate CO_2 under illumination. The experiment could be performed under dry conditions—similar to the Martian surface—or it could be run in a humidified mode. The CO_2 in the chamber was labeled with radioactive carbon, which could then be detected in any organic material synthesized during the experiment. The very first run of the PR experiment produced a significant response. It was well below the typical response observed when biotic soils from Earth had

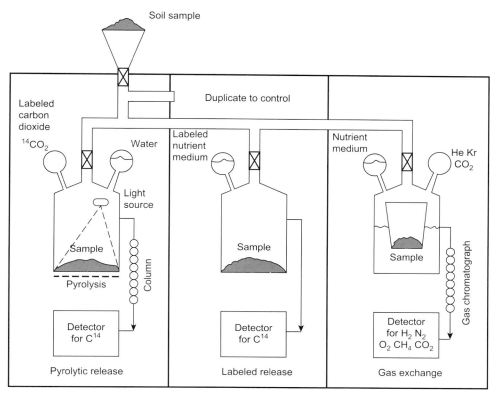

FIGURE 10.7 Schematic diagram of the Viking biology experiments.

been tested in the experiment but it was much larger than the noise level. Subsequent trials did not reproduce this high result and this initial response was attributed to a start-up anomaly, possibly some small prelaunch contamination.

The gas exchange (GEx) experiment searched for heterotrophs, which are microorganisms capable of consuming organic material. The GEx was designed to detect any gases that the organisms released as a by-product of their metabolism—bacterial flatulence. After a sample was placed in the chamber the soil was first equilibrated with water vapor and then combined with a nutrient solution. At prescribed intervals, a sample of the gas above the sample was removed and analyzed by a **gas chromatograph**.

The GEx results were startling. When the Martian soil was merely exposed to water vapor it released oxygen gas at levels of 70–770 nmol per gram soil, much larger than could be explained by the release of ambient atmospheric oxygen that had been absorbed onto the soil grains. The GEx results are summarized in Table 10.8. It was clear that some chemical or biological reaction was responsible for the oxygen release. A biological explanation was deemed unlikely since the reactivity of the soil persisted even after it had been heat sterilized to temperatures of over 160 °C. Furthermore, adding the nutrient solution did not change the result that some chemical in the soil was highly reactive with water.

TABLE 10.8 A Comparison of GEx O_2 and LR ^{14}C Results[1]

Sample	GEx O_2 (nmol/cm³)	Source[2] (Trapped O_2)	LR CO_2 (nmol/cm³)	Oxidant[2] (Hypochlorite[3])
Viking 1 (surface)	770	16 ppm/m	~30	1 ppm/m
Viking 2 (surface)	194	4	~30	1
Viking 2 (subrock)	70	1.5	~30	1

[1]After Klein, 1979.
[2]Assuming a bulk soil density of 1.5 g/cm³.
[3]Based on Quinn et al., 2013.

The labeled release (LR) experiment also searched for evidence of heterotrophic microorganisms. In the LR experiment, a solution of water containing seven organic compounds was added to the soil. The carbon atoms in each organic compound were radioactive. A radiation detector in the headspace detected the presence of radioactive CO_2 released during the experiment. Any carbon metabolism in the soil would be detected as organisms consumed the organics and released radioactive CO_2.

When the LR experiment was performed on Mars there was a steady release of radioactive CO_2; the results are summarized in Table 10.8. When the soil sample was heat sterilized before exposure to the nutrient solution no radioactive CO_2 was detected. The results of the LR experiment were precisely those expected if there were microorganisms in the soil sample. Taken alone the LR results would have been a strong positive indication for life on Mars.

In addition to the three biology experiments there was another instrument that gave information pivotal to the interpretation of the biological results. This was a combination of a **gas chromatograph** and a mass spectrometer (GCMS). This instrument received Martian soil samples from the same sampling arm that provided soil to the biology experiments. The sample was then heated to release any organics. The decomposed organics were carried through the gas chromatograph and identified by the mass spectrometer. No Martian organics were reported and all signals were attributed to cleaning agents used on the spacecraft before launch. The limit on the concentration of organics that would remain undetectable by the GCMS was one part per billion. A part per billion of organic material in a soil sample represents over a million individual bacterium, each the size of a typical *Escherichia coli*. This may not seem to rule out a biological explanation for the LR results. However, all life is composed of organic material and it is constantly exuded and processed in the biosphere. On Earth, it is difficult to imagine life without a concomitant matrix of organic material. This apparent absence of organic material is the main argument against a biological interpretation of the positive LR results.

The initial explanation for the reactivity of the Martian soil and the apparent absence of organics focused on possible atmospheric oxidants such as hydrogen peroxide produced by ultraviolet light in the atmosphere and deposited onto the soil surface. However, the finding by the Phoenix mission to Mars that the dominant form of chlorine on Mars is as perchlorate has provided the key to understanding the Viking results. Perchlorates in the Viking samples when heated to $500\,°C$ would have decomposed into reactive O and Cl oxidizing any organics present and producing the trace chlorinated organic compounds detected. The presence of perchlorates has been confirmed by results from the Curiosity Rover at the Martian equator.

Furthermore, ionizing radiation can decompose perchlorate in the soil on Mars and result in the formation of hypochlorite, other lower oxidation state oxychlorine species, with a concomitant production of O_2 gas that remains trapped in the salt crystal. The hypochlorite provides an explanation for the LR results and the trapped O_2 gas provides an explanation for the GEx results. These reactive forms of chlorine would have broken down any naturally occurring organic material or any material carried in by meteorites on the Martian surface. Table 10.8 also lists the concentration of the sources necessary to explain the Viking results for perchlorate-based models of the chemistry of the LR oxidant.

Amplifying the apparently negative results of the Viking biology experiments, the environment of Mars appears to be inhospitable to life. Although the atmosphere contains many of the elements necessary for life—it is composed of 95% CO_2 with a few percent N_2 and Argon and trace levels of water—the mean surface pressure is less than 1% of sea level pressure on the Earth, and the mean temperature is $-60\,°C$. The mean surface pressure is close to the triple point pressure of water. This is the minimum pressure at which a liquid state of water can exist. The low pressures and low temperatures make it unlikely that water will exist as a liquid on Mars. Due to seasonal transport, the available surface water on Mars is trapped as ice in the polar regions. In the locations at low elevation where the pressures and temperatures are sufficient to support liquid water, the surface is desiccated. Even saturated brine solutions cannot exist in equilibrium with the atmosphere near the equator. The absence of liquid water on the surface of Mars is probably the most serious argument against the presence of life anywhere at the surface of the planet. A second significant hazard to life on the Martian surface is the presence of solar ultraviolet light in the wavelengths between 190 and 300 nm. This radiation, which is largely shielded from Earth's surface by atmospheric oxygen and the ozone layer, is highly effective at destroying terrestrial organisms. Wavelengths below 190 nm are absorbed even by the present thin Martian CO_2 atmosphere. Compounding the effects of UV irradiation, and perhaps caused by it, are possible chemical oxidants that are thought to exist in the Martian soil. Such strong oxidants have been suggested as the causative agent for the chemical reactivity observed at the Viking sites. (*See* Mars: Atmosphere and Volatile History.)

6.3.2. Early Mars

There is considerable evidence that early in its history Mars did have liquid water on its surface. Images from the many orbiters show complex dendritic valley networks that are believed to have been carved by liquid water. These valleys are predominantly found in the heavily cratered, hence ancient, terrains in the southern hemisphere. This would suggest that the period of liquid water on Mars occurred contemporaneously with the end of the last stages of heavy cratering, about 3.8 Gyr ago. This epoch is the same at which life is thought to have originated on Earth (Figure 10.4). (*See* Mars, Surface and Interior.)

Figure 10.8 shows part of Nanedi Vallis on Mars. The canyon snakes back and forth—characteristic of liquid flow. On the floor of the canyon there appears a small channel. Presumably this channel was the flow of the river that carved the canyon. It would have taken considerable flow, although not necessarily continuous flow, for this river to have carved the much larger canyon. This image provides what is perhaps the best evidence from orbit that liquid water that flowed on the surface of Mars is stable flow for long periods of time.

The presence of liquid water habitats on early Mars at approximately the time that life is first evident on Earth suggests that life may have originated on Mars during the same time period. Liquid water is the most critical environmental requirement for life on Earth and the general similarity between Earth and Mars leads us to assume that life on Mars would be similar in this basic environmental requirement. More exotic approaches to life on Mars cannot be ruled out nor are they supported by any available evidence.

It is interesting therefore to consider how evolution may have progressed on Mars by comparison with the Earth. The histories of Earth and Mars are compared in Figure 10.4. In this figure it is seen that the period between 4.0 and 3.5 Gyr ago is the time when life is most likely to have evolved on both planets. On Earth life persists and remains essentially unchanged for several billion years until the cumulative effects of O_2 production induces profound changes on the atmosphere of that planet. On Mars, conditions become unsuitable for life (no liquid water on the surface) in a billion years or less. Thus, it is likely that if there were any life on early Mars it remained microbial.

The evidence of liquid water on early Mars, particularly that provided by the valley networks, suggests that the climate on early Mars may have been quite different than the present. It is generally thought that the surface temperature must have been close to freezing, much warmer than the present $-60\,°C$. These warmer temperatures are thought to have occurred as a result of a greatly enhanced greenhouse due to a thick (1–5 atm) CO_2 atmosphere. However, CO_2 condensation may have limited the efficacy of the CO_2 greenhouse but theoretical models indicate that CO_2 clouds or CH_4 could enhance the greenhouse and maintain warmer temperatures. The detection of clays in ancient sediments on Mars is another indication that liquid water was present.

If Mars did have a thick CO_2 atmosphere this strengthens the comparison to the Earth, which is thought to have also had a thick CO_2 atmosphere early in its history. The duration of a thick atmosphere on Mars and the concomitant warm, wet surface conditions are unknown but simple climate models suggest that significant liquid water habitats could have existed on Mars for ~ 0.5 Gyr after the mean surface temperature reached freezing. This model is based on the presence of deep ice-covered lakes (over 30 m) such as those in the dry valleys of the Antarctic where mean annual temperatures are $-20\,°C$.

If we divide the possible scenario for the history of water on the surface of Mars into four epochs, the first epoch would have warm surface conditions and liquid water—this is the epoch of clay formation. As Mars gradually loses its thick CO_2 atmosphere the second and third epochs would be characterized by low temperatures but still relatively high atmospheric pressures. This is because the temperature would drop rapidly as the pressure decreased. During the second epoch temperatures would rise above freezing during some of the year and liquid water habitats would require a perennial ice cover. However, by epoch

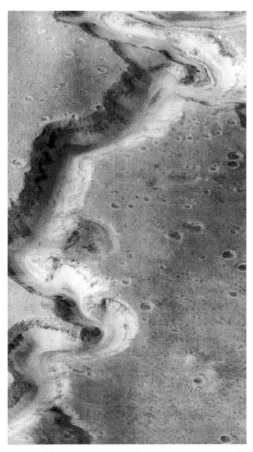

FIGURE 10.8 Liquid water on another world. Mars Global Surveyor image showing Nanedi Vallis in the Xanthe Terra region of Mars. Image covers an area 9.8 km by 18.5 km; the canyon is about 2.5 km wide. This image is the best evidence we have of liquid water anywhere outside the Earth. *Photo from NASA/Malin Space Sciences.*

three the temperature would never rise above freezing; the only liquid water would be found in porous rocks with favorable exposures to sunlight. In epoch four the pressure would fall too low for the presence of liquid water. These epochs might possibly be associated with the production of sulfates as shown in the timeline in Figure 10.4.

A point worth emphasizing here is that the biological requirement is for liquid water per se. Current difficulties in understanding the composition and pressure of the atmosphere need not lessen the biological importance of the direct evidence for the presence of liquid water. In fact, as we observe in the Antarctic dry valleys, ecosystems can exist when the mean temperatures are well below freezing. Mars need not have ever been above freezing for life to persist.

The particular environment on the early Earth in which life originated is not known. However, this does not pose as serious a problem to the question of the origin of life on Mars as might be expected. The reason is that all the environments found on the early Earth would be expected to be found on Mars; these include hydrothermal sites, hot springs, lakes, oceans (that is planetary scale water reservoirs), volcanoes, tidal pools (solar tides only), wetlands, salt flats, and others. Thus, whatever environment or combination of environments that was needed for life to get started on Earth should have been present on Mars as well—and at the same time.

Since the rationale for life on Mars early in its history is based on analogs with fossil evidence for life on the early Earth it is natural to look to the fossil record on Earth as a guide to how relics of early Martian life might be found. The most persuasive evidence for microbial life on the early Earth comes from **stromatolites** as discussed before. The resulting structures can be quite large—they are macroscopic fossils generated by microorganisms.

6.3.3. Subsurface Life on Mars

Although there is currently no direct evidence to support speculations about extant life on Mars, there are several interesting possibilities that cannot be ruled out at this time. Protected subsurface niches associated with hydrothermal activity could have continued to support life even after surface conditions became inhospitable. Liquid water could be provided by the heat of geothermal or volcanic activity melting permafrost or other subsurface water sources. Gases from volcanic activity deep in the planet could provide reducing power (as CH_4, H_2, or H_2S) percolating up from below and enabling the development of a microbial community based upon chemolithoautotrophy (see **chemoautotrophy**). An example is a methanogen (or acetogen) that uses H_2 and CO_2 in the production of CH_4. Such ecosystems have been found deep underground on the Earth consuming H_2 produced by the reaction of water with **basaltic** rock—a plausible reaction for subsurface Mars. However, their existence is neither supported, nor excluded, by current observations of Mars. Tests for such a subsurface system involve locating active geothermal areas associated with ground ice or detecting trace quantities of reduced atmospheric gases that would leak from such a system. It is interesting to consider the recent reports of CH_4 in the atmosphere of Mars at the tens of parts per billion level, and highly variable in space and time. If these reports are confirmed, it may be that this CH_4 may be related to subsurface biological activity. However, nonbiological sources of CH_4 are also possible. The reports are unlikely to be correct as CH_4 is expected to be a long-lived, and hence well-mixed gas in the Martian atmosphere.

While it certainly seems clear that volcanic activity on Mars has diminished over geological time, intriguing evidence for recent (on the geological timescale) volcanic activity comes from the young crystallization ages (all less than 1 Gyr) of the Shergotty meteorite (and other similar meteorites thought to have come from Mars). Volcanic activity by itself does not provide a suitable habitat for life—liquid water that may be derived from the melting of ground ice is also required. Presumably, the volcanic source in the equatorial region would have depleted any initial reservoir of ground ice and there would be no mechanism for renewal—although there are indications of geologically recent volcano/ground ice interactions at equatorial regions. Closer to the poles, ground ice is stable. It is conceivable that a geothermal heat source could result in cycling of water through the frozen ice-rich surface layers. The heat source would be melting and drawing in water from any underlying reservoir of groundwater or ice that might exist. (*See* Meteorites.)

Another line of reasoning also supports the possibility of subsurface liquid water. There are outflow channels on Mars that appear to be the result of the catastrophic discharge of subsurface aquifers of enormous sizes. There is evidence based on craters and stratigraphic relations that these have occurred throughout Martian history. If this is the case then it is possible that intact aquifers remain. This would have profound implications for exobiology (as well as human exploration). Furthermore, it suggests that the debris field and outwash regions associated with the outflow channel may hold direct evidence of life that existed within the subsurface aquifer just prior to its catastrophic release.

The collection of available water on Mars in the polar regions naturally suggests that summer warming at the edges of the permanent water ice cap may be a source of meltwater, even if short lived. In the polar regions of Earth there are complex microbial ecosystems that survive in transient summer meltwater. However, on Mars the temperature and pressures remain too low for liquid water to form. Any energy available is lost due to sublimation of the ice before any liquid is produced. It is unlikely that there are even seasonal habitats at the edge of the polar caps. This situation may be different over longer timescales. Changes in the obliquity axis of Mars can significantly increase the amount of insolation reaching the polar caps in summer. If the obliquity increases to over about 50° then the increased temperatures, atmospheric pressures, and polar insolation that result may cause summer liquid water melt streams and ponds at the edge of the polar cap.

In addition to the discovery of perchlorate discussed above, the Phoenix mission, at 68°N, confirmed the presence of ice-cemented ground at a depth below the surface that appeared consistent with vapor-deposited ice. The depth was 4–6 cm. However, in addition to ice-cemented soil there was relatively pure light-toned ice (Figure 10.9). This ice was unexpected, and it appears consistent with the formation of excess ice by soil ice accretion, such as would occur by vapor deposition during times of thermal expansion and contraction. Figure 10.9 shows the light-toned ice at the Phoenix landing site. The change, due presumably to evaporation over the four-sol period, indicates that the light-toned material is indeed ice and not salt or carbonate.

The ice-cemented soils in the northern plains of Mars, such as the Phoenix site, are possibly the best location on Mars for recent habitability. The presence of ice near the surface provides a source of H_2O. The atmospheric surface pressure over the northern plains is well above the triple point of water, so the liquid phase even of pure water would be stable against boiling. This situation is in contrast with the ice-rich southern polar regions, which are at high elevation. Note that the pressure at the Viking 2 lander site located at 49°N never fell below 750 Pa; the triple point of water is 610 Pa. Thus, all that would be needed to provide liquid water activity capable of supporting life is sufficient energy to melt the subsurface ice. This may have occurred as recently as 5 Myr ago, which is when calculations indicate that Mars had an orbital tilt of 45°, compared to the present value of 25°. The summer insolation in the polar regions of Mars at summer solstice for an obliquity of 45° is about twice that for an obliquity of 25°. When Mars had an obliquity of 45°, the polar regions (especially 68°N) received roughly the same level of summer sunlight as Earth's polar regions do at the present time.

The polar regions may harbor remnants of life in another way. Tens of meters beneath the surface the temperature is well below freezing ($<-70\,°C$) and does not change from summer to winter. It is likely that these permafrost zones have remained frozen—particularly in the southern hemisphere—since the end of the intense crater formation period. In this case there may be microorganisms frozen within the permafrost that date back to the time when liquid water was common on Mars, over 3.5 Gyr ago. On Earth permafrost of such age does not exist, but there are sediments in the polar regions that have been frozen for many millions of years. When these sediments are exhumed and samples extracted using sterile techniques viable bacteria are recovered. The sediments on Mars have been frozen much longer (1000 times) but the temperatures are also much colder. Thus, it may be possible that intact microorganism could be

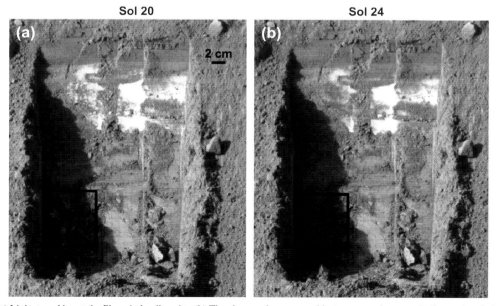

FIGURE 10.9 (a) Light-toned ice at the Phoenix landing site. (b) The change, due presumably to evaporation over the four-sol period, indicates that the light-toned material is indeed ice and not salt or carbonate. *Image credit NASA/University of Arizona.*

recovered from the Martian permafrost. Natural radiation from U, Th, and K in the soil would be expected to have killed any organism but their biochemical remains would be available for study. The southern polar region seems like the best site for searching for evidence of ancient microorganisms since the terrain there can be dated to the earliest period of Martian history as determined by the number of observed craters.

6.3.4. Meteorites from Mars

Of the thousands of meteorites known, there are over 30 that are thought to have come from Mars. It is certain that these meteorites came from a single source because they all have similar ratios of the oxygen isotopes—values distinct from terrestrial, lunar, or asteroidal ratios. These meteorites can be grouped into four classes. Three of these classes contain all but one of the known Mars meteorites and are known by the name of the type specimen; the S (Shergotty), N (Nakhla), and C (Chassigny) class meteorites. The S, N, and C meteorites are relatively young, having crystallized from lava flows between 200 and 1300 million years ago (see Figure 10.4). Gas inclusions in two of the S-type meteorites contain gases similar to the present Martian atmosphere as measured by the Viking Landers—proving that this meteorite, and by inference the others as well, came from Mars. The fourth class of Martian meteorite is represented by the single specimen known as ALH84001. Studies of this meteorite indicate that it formed on Mars about 4.5 Gyr ago in warm, reducing conditions. There are even indications that it contains Martian organic material and appears to have experienced aqueous alteration after formation. This rock formed during the time period when Mars is thought to have had a warm, wet climate capable of supporting life.

It has been suggested that ALH84001 contains evidence for life on Mars based on four observations. (1) Polycyclic aromatic hydrocarbons similar to molecules found in interstellar space are present inside ALH84001. (2) Carbonate globules are found in the meteorite that are enriched in ^{12}C over ^{13}C. The isotopic shift is within the range that on Earth, and indicates organic matter derived from biogenic activity. (3) Magnetite and iron sulfide particles are present that are similar to those produced by microbial activity. (4) Features are seen that could be fossils of microbial life, except that they are much smaller than any bacteria on Earth. As a result of more than a decade of study, most scientists currently prefer a nonbiological explanation for all these results. Only the magnetite result is generally considered relevant, although not conclusive, evidence related to life.

ALH84001 does not provide convincing evidence of past life on Mars, when compared to the multiple lines of evidence for life on Earth 3.4 Gyr ago including fossil evidence. However, the ALH84001 results do provide strong support to the suggestion that conditions suitable for life were present on Mars early in its history. When compared to the **SNC meteorites**, ALH84001 indicates that Mars experienced a transition from a warm reducing environment with organic material present to a cold oxidizing environment in which organic material was unstable.

6.4. The Giant Planets

The "habitable zone" in the inner Solar System provides the temperature conditions that can support liquid water on a planetary surface, but the outer Solar System is richer in the organic material from which life is made. This comparison, which shows the ratio of carbon to heavy elements (all elements other than H and He) for various objects in the Solar System, is shown in Figure 10.10. Earth is in fact depleted in carbon with respect to the average Solar System value by a factor of about 10^4. It may be interesting then to consider life in the organic-rich outer Solar System.

The giant planets Jupiter, Saturn, Uranus, and Neptune, do not have firm surfaces on which water could accumulate and form a reservoir for life. Here the only clement zone would be that region of the clouds in which temperatures were in the range suitable for life. Cloud droplets would provide the only source of liquid water. Such an environment might provide the key elements needed for life as well as an energy source in the form of sunlight. (See Atmospheres of the Giant Planets.)

There have been speculations that life, including advanced multicellular creatures, could exist in such an environment. However, such speculations are not supported by considerations of the biological state of clouds on Earth.

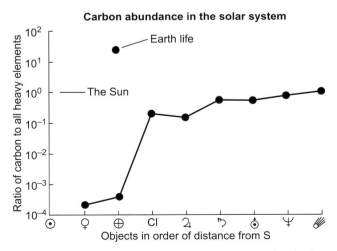

FIGURE 10.10 Ratio of carbon atoms to total heavy atoms (heavier than He) for various solar system objects illustrating the depletion of carbon in the inner solar system. The x-axis is not a true distance scale but the objects are ordered by increasing distance from the sun. Mars is not shown since the size of its carbon reservoir is unknown.

There are no organisms that have adapted themselves to live exclusively in clouds on Earth even in locations where clouds are virtually always present. This niche remains unfilled on Earth and by analogy is probably unfilled elsewhere in the Solar System.

Following this line of thought leads us to search for environments suitable for life on planetary bodies with surfaces. In the outer Solar System this focuses us on the moons of the giant planets. Of particular interest are Europa, Titan, and Enceladus.

6.4.1. Europa

Europa, one of the moons of Jupiter, appears to be an airless ice-shrouded world. However, theoretical calculations suggest that under the ice surface of Europa there may be a layer of liquid water sustained by tidal heating as Europa orbits Jupiter. The Galileo Spacecraft imaging showed features in the ice consistent with a subsurface ocean and the magnetometer indicated the presence of a global layer of slightly salty liquid water. The surface of Europa is crisscrossed by streaks that are slightly darker than the rest of the icy surface. If there is an ocean beneath a relatively thin ice layer then these streaks may represent cracks where the water has come to the surface. (See Small Satellites.)

There are many ecosystems on Earth that thrive and grow in water that is continuously covered by ice. These are found in both the Arctic and Antarctic regions. In addition to the polar oceans where sea ice diatoms perform photosynthesis under the ice cover, there are perennially ice-covered lakes in the Antarctic continent in which microbial mats based on photosynthesis are found in the water beneath a 4 m ice cover. The light penetrating these thick ice covers is minimal—about 1% of the incident light. Using these Earth-based systems as a guide it is possible that sunlight penetrating through the cracks (the observed streaks) in the ice of Europa could support a transient photosynthetic community. Alternatively, if there are hydrothermal sites on the bottom of the Europan ocean it may be possible that chemosynthetic life could survive there—by analogy to life at hydrothermal vent sites at the bottom of the Earth's oceans. The biochemistry of hydrothermal sites on Earth does depend on O_2 produced at the Earth's surface. On Europa, a chemical scheme like that suggested for subsurface life on Mars would be appropriate ($H_2 + CO_2$).

The main problem with life on Europa is the question of its origin. Lacking a complete theory for the origin of life, and lacking any laboratory synthesis of life, we have to base our understanding of the origin of life on other planets on analogy with the Earth. It has been suggested that hydrothermal vents may have been the site for the origin of life on Earth and in this case the prospects for life in a putative ocean on Europa are improved. However, the early Earth contained many environments other than hydrothermal vents, such as surface hot springs, volcanoes, lake and ocean shores, tidal pools, and salt flats. If any of these environments were the locale for the origin of the first life on Earth then the case for an origin on Europa is weakened considerably.

6.4.2. Titan

Titan, the largest moon of the planet Saturn, has a substantial atmosphere composed primarily of N_2 and CH_4 with many other organic molecules present. The temperature at the surface is close to 94 K and the surface pressure is 1.5 times the pressure of Earth at sea level. The surface does not appear to have expansive oceans as once suggested but numerous small lakes have been discovered in the north polar region. However, the ground beneath the Huygens Probe was wet with liquid CH_4, the result of a slight but constant drizzle. (See Titan.)

The spacecraft data from the Voyager and Cassini/Huygens missions, as well as ground-based studies, indicate that there is an optically thick haze in the upper atmosphere. The haze is composed of organic material and the atmosphere contains many organic molecules heavier than CH_4. Photochemical models suggest that these organics are produced from CH_4 and N_2 through chemical reactions driven by solar photons and by magnetospheric electrons. The observed organic species and even heavier organic molecules are predicted to result from these chemical transformations. Laboratory simulations of organic reactions in Titan-like gas mixtures produce solid refractory organic substances (tholin) and similar processes are expected to occur in Titan's atmosphere.

Conditions on Titan are much too cold for liquid water to exist, although the pressure is in an acceptable regime. For this reason it is unlikely that Earth-like life could originate or survive there. However, the organic material in Titan's atmosphere provides a potential source of energy and the liquid methane on the surface could provide a possible liquid medium for an alternate type of life. Life in liquid methane could use active transport and large size to overcome the low solubility of organics in liquid methane and **enzymes** to catalyze reactions at the low temperatures. If carbon-based life in liquid methane existed on Titan it could be widespread. With or without life, Titan remains interesting because it is a naturally occurring **Miller–Urey experiment** in which simple compounds are transformed into more complex organics. A detailed study of this process may yield valuable insight into how such a mechanism might have operated on the early Earth.

There is also some speculation that under unusual conditions Titan may have liquid water on or near the surface. This could have occurred early in its formation when the gravitational energy released by the formation of Titan would have heated it to high temperatures. More recently, impacts could conceivably melt local regions generating warm subsurface temperatures that could last for thousands of years. Whether such brief episodes of liquid water could have led to water-based life remains to be tested.

6.4.3. Enceladus

The Cassini mission has documented geysers erupting from the south polar region of Enceladus. (*See* Small Satellites.) Associated with this outflow of water, CH_4, other organics, and NH_3 have been identified. The source of the water has been shown to be a slightly salty subsurface liquid water reservoir heated and pressurized by subsurface heat flow. Such a subsurface habitat could support the sort of anaerobic chemoautotrophic life that has been found on Earth. These systems are based on methanogens that consume H_2 produced by geochemical reactions or by radioactive decay. The age or lifetime of any subsurface liquid water on Enceladus is not known, which adds uncertainty to speculation about the origin of life. The theories for the origin of life on Earth, shown in Figure 10.6, which would apply to Enceladus are **panspermia** and a chemosynthetic origin of life. The same theories that would apply to Europa.

If there is subsurface life in the liquid water reservoirs on Enceladus then the geysers would be carrying these organisms out into space. Here they would quickly become dormant in the cold vacuum of space and would then be killed by solar ultraviolet radiation. But these dead, frozen microbes would still retain the biochemical and genetic molecules of the living forms. Thus, a Stardust-like mission moving through the plume of Enceladus' geysers might collect life forms for return to Earth. This might provide the easiest way to get a sample of a second genesis of life.

6.5. Asteroids

Asteroids seem like unlikely locations for life to have originated. Certainly they are too small to support an atmosphere sufficient to allow for the presence of liquid water at the present time. However, asteroids, particularly the so-called carbonaceous type, are thought to contain organic material. Thus, they might have played a role in the delivery of organics to the prebiotic Earth. A more intriguing aspect of some asteroids is the presence of hydrothermally altered materials. This seems to indicate that the asteroids were once part of a larger parent body. Furthermore, conditions on this larger parent body were such that liquid water was present—at least in thin films. Containing both organic material and liquid water, the parent bodies of these asteroids are thus interesting targets in the search for extraterrestrial life forms. However, a thorough assessment of this possibility will require a more detailed study of carbonaceous asteroids in the asteroid belt. Meteorites found on the Earth provide only a glimpse of small fragments of these objects and no signs of extraterrestrial life have been found. But the samples are small and the potential for contamination by Earth life is great.

6.6. Comets

Comets are also known to be rich in organic material. However, unlike asteroids, comets also contain a large fraction of water. In their typical state this water is frozen as ice—unsuitable for life processes. As a comet approaches the sun its surface is warmed considerably, but this leads only to the sublimation of the water ice. Liquid does not form because the pressure at the surface of the comet is much too low.

There has been the suggestion that soon after their formation the interior of large comets would have been heated by short-lived radioactive elements (^{26}Al) to such an extent that the core would have melted. In this case there would have been a subsurface liquid water environment similar to that postulated for the present day Europa. Again the question of the origin of life in such an environment rests on the assumption that life can originate in an isolated deep dark underwater setting.

7. HOW TO SEARCH FOR LIFE ON MARS, EUROPA, OR ENCELADUS

If we were to find organic material in the subsurface of Mars, or in the ice of Europa, or entrained in the geysers of Enceladus, how could we determine if it was the product of a system of biology or merely abiotic organic material from meteorites or photochemistry? If the life is related to Earth life it should be easy to detect. We now have very sensitive methods, such as the amplification of DNA, florescent antibody markers, etc., for detecting life from Earth. The case of Earth-like life is the easiest but it is also the least interesting. If the life is not Earth-like then the probes specific to our biology are unlikely to work. We need a general way to determine a biological origin. The question is open and possibly urgent. As we plan missions to Mars and Europa we may have the opportunity to analyze the remains of alien biology.

One practical approach makes use of the distinction between biochemicals and organic matter that is not dependent on a particular organic molecule but results from considering the pattern of the organics in a sample. Abiotic processes will generate a smooth distribution in molecular types without sharp distinctions between similar molecules, isotopes, or chemical chirality. If we consider a generalized phase space of all possible organic molecules then for an abiotic production mechanism the relative concentration of different types will be a smooth function. In contrast to abiotic mechanisms, biological production will not involve a wide range of possible types. Instead, biology will select a few types of molecules and build biochemistry up from this restricted set. Thus, organic molecules that are chemically very similar may have widely different concentrations in a sample of biological organics. An example of this on Earth is the 20 **amino acids** used in proteins and the selection of life for the left-handed version of these **amino acids**. To maximize efficiency life everywhere is likely to evolve this strategy of using a few molecules repeatedly. It may be that other life forms discover the same set of biomolecules that Earth life uses because these are absolutely the most efficient and effective set under any planetary conditions. But it may also be that life elsewhere uses a different set that is optimal given the specific history and conditions of that world. We can search for the repeated use of a set of molecules without knowing in advance what the members of that set will be.

We can apply this approach to the search for biochemistry in the Solar System. Samples of organic material collected from Mars and Europa can be tested for the prevalence of one chirality of **amino acid** over the other. More generally a complete analysis of the relative concentration of different types of organic molecules might reveal a pattern that is biological even if that pattern does not involve any of the biomolecules familiar from Earth life. Interestingly, if a sample of organics from Mars or Europa shows a preponderance of D-amino acids this will suggest the presence of extant or extinct life and at the same time show that this life is distinct from Earth life. This same conclusion would apply to any clearly biological pattern that is distinct from the pattern of Earth life.

The pattern of biological origin in organic material can potentially persist long after the organisms themselves are dead. Eventually, this distinctive pattern will be destroyed as a result of thermal and radiation effects. Below the surface of Mars, both temperature and radiation are low so this degradation should not be significant. On Europa the intense radiation may destroy the biological signature after several million years at depths to about 1 m below the surface ice.

8. LIFE ABOUT OTHER STARS

In the Solar System we find only our own planet with clear signs of life. Mars, Europa, and Enceladus provides some hopes of finding past or present liquid water but not comparable to the richness of water and life on Earth. Our understanding of life as a planetary phenomenon would clearly benefit from finding another Earth-like planet, around another Sun-like star, that harbored life.

One way of formulating the probability of life, and intelligent life, elsewhere in the galaxy is known as the Drake equation after Frank Drake, a pioneer in the search for extraterrestrial intelligence. The equation and the terms that comprise it are listed in Table 10.9. The most accurately determined variable in the Drake equation at this time is R^*, the number of stars forming in the galaxy each year. Since we know that there are about 10^{11} stars in our galaxy and that their average lifetime is about 10^{10} years, then $R^* \sim 10$ stars per year. All the other terms are uncertain and can be only estimated by extrapolating from what has occurred on Earth. Estimates by different authors for N, the number of civilizations in the galaxy capable of communicating by radiowaves, range from one to millions. Perhaps the most uncertain term is L, the length of time that a technologically advanced civilization can survive.

TABLE 10.9 The Probability of Life, and Intelligent Life, Elsewhere in the Galaxy

The Drake Equation $N = R_* \times f_p \times n_e \times f_l \times f_i \times f_c \times L$

N	The number of civilizations in the galaxy
R_*	The number of stars forming each year in the galaxy
f_p	The fraction of stars possessing planetary systems
n_e	The average number of habitable planets in a planetary system
f_l	The fraction of habitable planets in which life originates
f_i	The fraction of life forms that develop intelligence
f_c	The fraction of intelligent life forms that develop advanced technology
L	The length of time, in years, that a civilization survives

The primary criterion for determining whether a planet can support life is the availability of water in the liquid state. This in turn depends on the surface temperature of the planet that is controlled primarily by the distance to a central star. Life appeared so rapidly on Earth after its formation that it is possible that other planets may only have had to sustain liquid water for a short period of time for life to originate. However, it is important to note that the origin of life is not understood and its probability is completely unknown. Planets orbiting a variety of star types could satisfy this criterion at some time in their evolution. The development of advanced life on Earth, and in particular intelligence, took much longer, almost 4 billion years. Earth maintained habitable conditions for the entire period of time. Locations about stars in which temperatures are conducive to liquid water for such a long period of time have been called continuously habitable zones (CHZs). Calculations of the CHZ about main sequence stars indicate that the mass of the star must be less than 1.5 times the mass of our sun for the CHZ to persist for more than 2 billion years.

An interesting result of these calculations is that the current habitable zone for the sun has an inner limit at about 0.8 AU and extends out to between 1.3 and 1.6 AU, depending on the way clouds are modeled. Thus, while Venus is not in the habitable zone, Earth and Mars both are. This calculation would suggest that Mars is currently habitable. But we see no indication of life. This is owing to the fact that the distance from the sun is not the only determinant for the presence of liquid water on a planet's surface. The presence of a thick atmosphere and the resultant greenhouse effect is required as well. On Earth the natural greenhouse effect is responsible for warming the Earth by 30 °C; without the greenhouse effect the temperature would average −15 °C. Mars does not have an appreciable greenhouse effect and hence its temperature averages −60 °C. If Earth were at the same distance from the sun that Mars is it would probably be habitable. The reason is the thermostatic effect of the long-term carbon cycle. This cycle is driven by the burial of carbon in seafloor sediments as organic material and carbonates. The formation of carbonates is due to chemical erosion of the surface rocks. Subduction carries this material to depths where the high temperatures release the sedimentary CO_2 gas. These gases escape to the surface in volcanoes that lie on the boundary arc of the subduction zones. The thermostatic action of this cycle results because the erosion rate is strongly dependent on temperature. If the temperature were to drop, erosion would slow down. Meanwhile, the outgassing of CO_2 would result in a buildup of this greenhouse gas and the temperature would rise. Conversely, higher temperatures would result in higher erosion rates and a lowering of CO_2 again stabilizing the temperature.

Mars became uninhabitable because it lacks plate tectonics and hence has no means of recycling the carbon-containing sediments. As a result, the initial thick atmosphere that kept Mars warm has dissipated, presumably into carbonate rocks located on the floor of ancient lake and ocean basins on Mars. Mars lacks plate tectonics because it is too small, 10 times smaller than the Earth, to maintain the active heat flows that drive tectonic activity. The low gravity of Mars and the absence of a magnetic field also contributed to the loss of its atmosphere. Hence, planetary size and its effect on geological activity also play a role in determining the surface temperature and thereby the presence of liquid water and life.

The concept of a galactic habitable zone (GHZ) extends the CHZ to the Milky Way galaxy. The defining feature of the GHZ is the probability of planet formation rather than planet habitability for the CHZ. Planet formation appears to correlate with the concentration of heavy elements and indicates that many galaxies should have a GHZ.

9. CONCLUSION

Life is a planetary phenomenon. We see its profound influences on the surface of one planet—the Earth. Its origin, history, present reach, and global scale interactions remain a mystery primarily because we have only one datum. Many questions about life await the discovery of another life form with which to compare. Mars early in its history is probably the best prospective target in the search for extraterrestrial life forms, although Europa and Enceladus are also promising candidates due to the likely presence of liquid water beneath a surface ice shell and the possibility of associated hydrothermal vent activity. In any case it is likely that our true understanding of life is to be found in the exploration of other worlds—both those with and without life forms. We have only just begun to search.

BIBLIOGRAPHY

Allwood, A. C., Walter, M. R., Kamber, B. S., Marshall, C. P., & Burch, I. W. (2006). Stromatolite reef from the Early Archaean era of Australia. *Nature*, 714–718.

Davis, W. L., & McKay, C. P. (1996). Origins of life: a comparison of theories and application to Mars. *Origins of Life and Evolution of Biospheres, 26*, 61–73.

Goldsmith, D. (1997). *The hunt for life on Mars*. Penguin.

Klein, H. P. (1979). The Viking mission and the search for life on Mars. *Reviews of Geophysics and Space Physics, 17*, 1655–1662.

Lederberg, J. (1960). Exobiology: approaches to life beyond the Earth. *Science, 132*, 393–400.

Lehninger, A. L. (1975). *Biochemistry*. New York: Worth.

McKay, C. P. (2011). The search for life in our Solar System and the implications for science and society. *Philosophical Transactions of the Royal Society A: Mathematical, Physical and Engineering Sciences, 369*, 594–606.

Miller, S. L. (1992). The prebiotic synthesis of organic compounds as a step toward the origin of life. In J. W. Schopf (Ed.), *Major events in the history of life* (pp. 1–28). Boston, MA: Jones and Bartlett Publishers.

Navarro-González, R., Vargas, E., de La Rosa, J., Raga, A. C., & McKay, C. P. (2010). Reanalysis of the Viking results suggests perchlorate and organics at midlatitudes on Mars. *Journal of Geophysical Research, 115*(E12).

Postberg, F., Schmidt, J., Hillier, J., Kempf, S., & Srama, R. (2011). A salt-water reservoir as the source of a compositionally stratified plume on Enceladus. *Nature, 474*, 620–622.

Quinn, R. C., Martucci, H. F. H., Miller, S. R., Bryson, C. E., Grunthaner, F. J., & Grunthaner, P. J. (2013). Perchlorate radiolysis on Mars and the origin of the martian soil reactivity. *Astrobiology, 13*, 515–520.

Shapiro, R. (1986). *Origins: A skeptics guide to the creation of life on Earth.* Summit Books.

Suthar, F., & McKay, C. P. (2012). The galactic habitable zone in elliptical galaxies. *International Journal of Astrobiology, 11*, 157–161.

Part III

The Sun

Chapter 11

The Sun

Markus J. Aschwanden
Lockheed Martin ATC Solar and Astrophysics Laboratory, Palo Alto, CA, USA

Chapter Outline

1. Introduction   235
2. The Solar Interior   236
   2.1. Standard Models   236
   2.2. Thermonuclear Energy Source   238
   2.3. Neutrinos   238
   2.4. Helioseismology   238
   2.5. Solar Dynamo   239
3. The Photosphere   239
   3.1. Granulation and Convection   239
   3.2. Photospheric Magnetic Field   240
   3.3. Sunspots   240
4. The Chromosphere and Transition Region   241
   4.1. Basic Physical Properties   241
   4.2. Chromospheric Dynamic Phenomena   242
5. The Corona   243
   5.1. Active Regions   243
   5.2. Quiet-Sun Regions   243
   5.3. Coronal Holes   244
   5.4. Hydrostatics of Coronal Loops   244
   5.5. Dynamics of the Solar Corona   244
   5.6. The Coronal Magnetic Field   245
   5.7. MHD Oscillations of Coronal Loops   246
   5.8. MHD Waves in Solar Corona   247
   5.9. Coronal Heating   248
6. Solar Flares and CMEs   248
   6.1. Magnetic Reconnection   249
   6.2. Filaments and Prominences   249
   6.3. Solar Flare Models   249
   6.4. Flare Plasma Dynamics   252
   6.5. Particle Acceleration and Kinematics   253
   6.6. Hard X-Ray Emission   254
   6.7. Gamma-Ray Emission   254
   6.8. Radio Emission   256
   6.9. Coronal Mass Ejections   256
7. Final Comments   258
Bibliography   259

1. INTRODUCTION

The Sun is the central body and energy source of our solar system. The Sun is our nearest star, but otherwise it represents a fairly typical star in our galaxy, classified as G2-V spectral type, with a radius of $r_\odot \approx 700,000$ km, a mass of $m_\odot \approx 2 \times 10^{33}$ g, a luminosity of $L_\odot \approx 3.8 \times 10^{26}$ W, and an age of $t_\odot \approx 4.6 \times 10^9$ years (Table 11.1). The distance from the Sun to the Earth is called an Astronomical Unit (AU) and amounts to $\sim 150 \times 10^6$ km. The Sun lies in a spiral arm of our galaxy, the Milky Way, at a distance of 8.5 kiloparsecs from the galactic center. Our galaxy contains $\sim 10^{12}$ individual stars, many of which are likely to be populated with similar solar systems, according to the rapidly increasing detection of extrasolar planets over the past years; the binary star systems are very unlikely to harbor planets because of their unstable, gravitationally disturbed orbits. The Sun is for us humans of particular significance, first because it provides us with the source of all life and second because it furnishes us with the closest laboratory for astrophysical plasma physics, magnetohydrodynamics (MHD), atomic physics, and particle physics. The Sun still represents the only star from which we can obtain spatial images, in many wavelengths.

The basic structure of the Sun is sketched in Figure 11.1. The Sun and the solar system were formed together from an interstellar cloud of molecular hydrogen some 5 billion years ago. After gravitational contraction and subsequent collapse, the central object became the Sun, with a central temperature hot enough to ignite thermonuclear reactions, the ultimate source of energy for the entire solar system. The chemical composition of the Sun is of 92.1% hydrogen and 7.8% helium by number (or 27.4% He by mass), and 0.1% of heavier elements (or 1.9% by mass,

TABLE 11.1 Basic Physical Properties of the Sun

Physical Parameter	Numerical Value
Solar radius, $R_\odot$	695,500 km
Solar mass, $m_\odot$	1.989×10^{33} g
Mean density, $\rho_\odot$	1.409 g/cm^3
Gravity at solar surface, $g_\odot$	274.0 m/s^2
Escape velocity at solar surface, v	617.7 km/s
Synodic rotation period, P	27.3 days (equator)
Sidereal rotation period, P	25.4 days (equator)
	35.0 days (at latitude $\pm 70°$)
Mean distance from Earth	1 AU = 149,597,870 km
Solar luminosity, $L_\odot$	3.844×10^{26} W (or 10^{33} erg/s)
Solar age, $t_\odot$	4.57×10^9 years
Temperature at the Sun center, T_c	15.7×0^6 K
Temperature at the solar surface, T_{ph}	6400 K

Source: Cox, 2000.

mostly C, N, O, Ne, Mg, Si, S, Fe). The central core, where hydrogen burns into helium, has a temperature of ~15 million K (Figure 11.1). The solar interior further consists of a radiative zone, where energy is transported mainly by radiative diffusion, a process where photons with hard X-ray (kiloelectronvolt) energies get scattered, absorbed, and reemitted. The outer one-third of the solar interior is called the convective zone, where energy is transported mostly by convection. At the solar surface, photons leave the Sun in optical wavelengths, with an energy that is about a factor of 10^5 lower than the original hard X-ray photons generated in the nuclear core, after a random walk of $\sim 10^5 - 10^6$ years.

The irradiance spectrum of the Sun is shown in Figure 11.2, covering all wavelengths from gamma rays, hard X-rays, soft X-rays, extreme ultraviolet (EUV), ultraviolet, white light, infrared, to radio wavelengths. The quiet Sun irradiates most of the energy in visible (white light) wavelengths, to which our human eyes have developed the prime sensitivity during evolution. Emission in EUV is dominant in the solar **corona** because it is produced by ionized plasma in the coronal temperature range of $\sim 1-2$ million K. Emissions in shorter wavelengths require higher plasma temperatures and thus occur during **flares** only. Flares also accelerate particles to nonthermal energies, which cause emission of hard X-rays, gamma rays, and radio wavelengths, but to a highly variable degree.

2. THE SOLAR INTERIOR

The physical structure of the solar interior is mostly based on theoretical models that are constrained (1) by global quantities (age, radius, luminosity, total energy output; see Table 11.1), (2) by the measurement of global oscillations (helioseismology), and (3) by the neutrino flux, which now constrains for the first time elemental abundances in the solar interior, since the neutrino problem has been solved in the year 2001.

2.1. Standard Models

There are two types of models of the solar interior: (1) hydrostatic equilibrium models and (2) time-dependent numerical simulations of the evolution of the Sun,

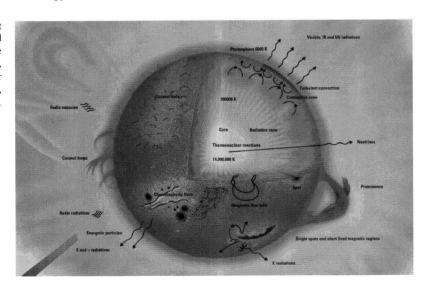

FIGURE 11.1 A cutaway view of the Sun, showing the three internal (thermonuclear, radiative, and convective) zones, the solar surface (photosphere), the lower (chromosphere) and upper atmosphere (corona), and a number of phenomena associated with the solar activity cycle (filaments, prominences, and flares). IR, infrared; UV, ultraviolet. *Courtesy of Calvin J. Hamilton and NASA/ESA.*

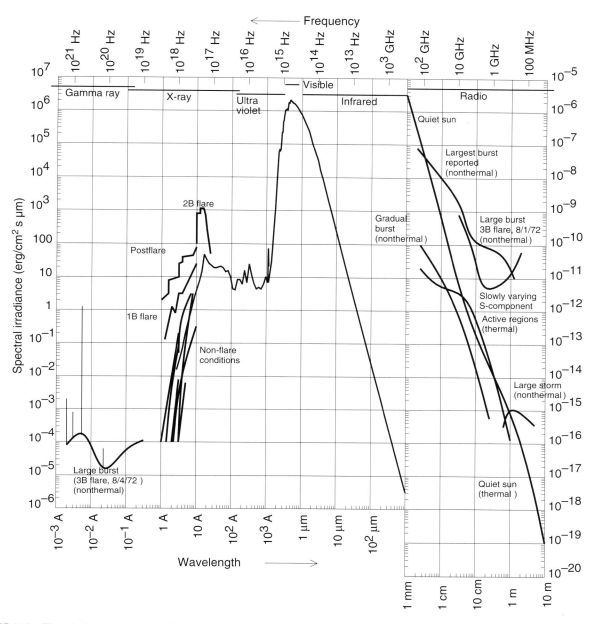

FIGURE 11.2 The solar irradiance spectrum from gamma rays to radio waves. The spectrum is shifted by 12 orders of magnitude in the vertical axis at $\lambda = 1$ mm to accommodate for the large dynamic range in spectral irradiance. *Courtesy of H. Malitson and NASA/NSSDC.*

starting from an initial gas cloud to its present state today, after ∼ 8% of the hydrogen has been burned into helium.

The standard hydrostatic model essentially calculates the radial run of temperature, pressure, and density that fulfill the conservation of mass, momentum, and energy in all internal spherical layers of the Sun, constrained by the boundary conditions of radius, temperature, and radiation output (luminosity) at the solar surface, the total mass, and the chemical composition. Furthermore, the ideal gas law and thermal equilibrium is assumed, and thus the radiation is close to that of an ideal blackbody. The solar radius has been measured by triangulation inside the solar system (e.g. during a Venus transit) and by radar echo measurements. The mass of the Sun has been deduced from the orbital motions of the planets (Kepler's laws) and from precise laboratory measurements of the gravitational constant. The solar luminosity is measured by the heat flux received at the Earth. From these standard models, a central temperature of ∼ 15 million K, a central density of ∼ 150 g/cm^3, and a central pressure of 2.3×10^{17} dyne/cm^2 have been inferred. Fine-tuning of the standard model is obtained by including convective transport and by varying the (inaccurately known) helium abundance.

2.2. Thermonuclear Energy Source

The source of solar energy was understood in the 1920s, when Hans Bethe, George Gamow, and Carl Von Weizsäcker identified the relevant nuclear chain reactions that generate solar energy. The main nuclear reaction is the transformation of hydrogen into helium, where 0.7% of the mass is converted into radiation (according to Einstein's energy equivalence, $E = mc^2$), the so-called p−p chain, which starts with the fusion of two protons into a nucleus of deuterium (^{2}He), and, after chain reactions involving ^{3}He, ^{7}Be, and ^{7}Li produces helium (^{4}He),

$$p + p \rightarrow {}^2He + e^+ + \nu_e$$
$$^2He + p \rightarrow {}^3He + \gamma$$
$$^3He + {}^3He \rightarrow {}^4He + p + p$$

or

$$^3He + {}^4He \rightarrow {}^7Be + \gamma$$
$$^7Be + e^- \rightarrow {}^7Li + \nu_e$$
$$^7Li + p \rightarrow {}^8Be + \gamma \rightarrow {}^4He + {}^4He$$

One can estimate the Sun's lifetime by dividing the available mass energy by the luminosity, where we assumed that only about a fraction of 0.1 of the total solar mass is transformed because only the innermost core of the Sun is sufficiently hot to sustain nuclear reactions.

$$t_\odot \approx 0.1 \times 0.007 m_\odot c^2 / L_\odot \approx 10^{10} \text{years}$$

An alternative nuclear chain reaction occurring in the Sun and stars is the carbon−nitrogen−oxygen (CNO) cycle,

$$^{12}C + p \rightarrow {}^{13}N + \gamma$$
$$^{13}N \rightarrow {}^{13}C + e^+ + \nu_e$$
$$^{13}C + p \rightarrow {}^{14}N + \gamma$$
$$^{14}N + p \rightarrow {}^{15}O + \gamma$$
$$^{15}O \rightarrow {}^{15}N + e^+ + \nu_e$$
$$^{15}N + p \rightarrow {}^{12}C + {}^4He$$

The p−p chain produces 98.5% of the solar energy, and the CNO cycle produces the remainder, but the CNO cycle is faster in stars that are more massive than the Sun.

2.3. Neutrinos

Neutrinos interact very little with matter, unlike photons, and thus most of the electronic neutrinos (ν_e), emitted by the fusion of hydrogen to helium in the central core, escape the Sun without interactions and a very small amount is detected at the Earth. Solar neutrinos have been detected since 1967, pioneered by Raymond Davis, Jr., using a chlorine tank in the Homestake Gold Mine in South Dakota, but the observed count rate was about one-third of the theoretically expected value, causing the puzzling neutrino problem that persisted for the next 35 years. However, Pontecorvo and Gribov predicted already in 1969 that low-energy solar neutrinos undergo a "personality disorder" on their travel to the Earth and oscillate into other atomic flavors of muonic neutrinos (ν_μ) (from a process involving a muon particle) and tauonic neutrinos (ν_t) (from a process involving a tauon particle), which turned out to be the solution of the missing neutrino problem for detectors that are only sensitive to the highest energy (electronic) neutrinos, such as the Homestake chlorine tank and the gallium detectors Gallium Experiment (GALLEX) in Italy and SAGE in Russia. Only the Kamiokande and Super-Kamiokande-I pure-water experiments and the Sudbury Neutrino Observatory (SNO, Ontario, Canada) heavy-water experiments are somewhat sensitive to the muonic and tauonic neutrinos. It was the SNO that measured in 2001 for the first time all three lepton flavors and, in this way, brilliantly confirmed the theory of neutrino (flavor) oscillations. Today, after the successful solution of the neutrino problem, the measured neutrino fluxes are sufficiently accurate to constrain the helium abundance and heavy element abundances in the solar interior.

2.4. Helioseismology

During 1960−1970, global oscillations were discovered on the solar surface in visible light, which became the field of helioseismology. Velocity oscillations were first measured by R. Leighton and then interpreted in 1970 as standing sound waves in the solar convection zone by R. Ulrich, C. Wolfe, and J. Leibacher. These acoustic oscillations, also called p-modes (pressure-driven waves), are detectable from fundamental up to harmonic numbers of ∼1000 and are most conspicuous in dispersion diagrams, $\omega(k)$, where each harmonic shows up as a separate ridge, when the oscillation frequency (ω) is plotted as function of the wavelength λ (i.e. essentially the solar circumference divided by the harmonic number). Frequencies of the p-mode correspond to periods of ∼5 min. An example of a p-mode standing wave is shown in Figure 11.3 (left), which appears like a standing wave on a drum skin. Each mode is characterized by the number of radial, longitudinal, and latitudinal nodes, corresponding to the radial quantum number n, the azimuthal number m, and the degree l of spherical harmonic functions. Since the density and temperature increase monotonically with depth inside the Sun, the sound speed varies as a function of radial

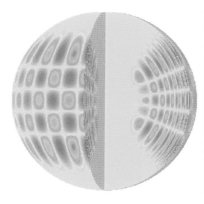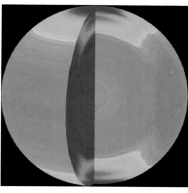

FIGURE 11.3 Left: A global acoustic p-mode wave is visualized: the radial order is $n = 14$, the angular degree is $l = 20$, the angular order is $m = 16$, and the frequency is $\nu = 2935.88 \pm 0.1\,\mu\text{Hz}$ with Solar and Heliospheric Observatory/Michelson Doppler Image (SoHO/MDI). The red and blue zones show displacement amplitudes of opposite sign. Right: The internal rotation rate is shown with a color code, measured with SoHO/MDI during May 1996–April 1997. The red zone shows the fastest rotation rates ($P \approx 25$ days), and dark blue the slowest ($P \approx 35$ days). Note that the rotation rate varies in latitude differently in the radiative and convective zones. *Courtesy of SoHO/MDI and NASA.*

distance from the Sun center. P-mode waves excited at the solar surface propagate downward and are refracted toward the surface. The low harmonics penetrate very deep, whereas high harmonics are confined to the outermost layers of the solar interior. By measuring the frequencies at each harmonic, the sound speed can be inverted as a function of the depth; in this way, the density and temperature profile of the solar interior can be inferred and unknown parameters of theoretical standard models can be constrained, such as the abundance of helium and heavier elements. By exploiting the Doppler effect, frequency shifts of the p-mode oscillations can be used to measure the internal velocity rates as a function of depth and latitude, as shown in Figure 11.3 (right). A layer of rapid change in the internal rotation rate was discovered this way at the bottom of the convection zone, the so-called tachocline (at 0.693 ± 0.002 solar radius, with a thickness of 0.039 ± 0.013 solar radius).

Besides the p-mode waves, gravity waves (g-modes), where buoyancy rather than pressure supplies the restoring force, are suspected in the solar core. These gravity waves are predicted to have long periods (hours) and very small velocity amplitudes, but they have not yet been convincingly detected.

Global helioseismology detects p-modes as a pattern of standing waves that encompass the entire solar surface; however, local deviations of the sound speed can also be detected beneath sunspots and active regions, a diagnostic that is called local helioseismology. Near sunspots, p-modes are found to have oscillation periods in the order of 3 min, compared to 5 min in active region plages and quiet-Sun regions.

2.5. Solar Dynamo

The Sun is governed by a strong magnetic field (much stronger than those on planets), which is generated with a magnetic field strength of $B \approx 10^5$ G in the tachocline, the thin shear layer sandwiched between the radiative and the convective zone. Buoyant magnetic flux tubes rise through the convection zone (due to the convective instability obeying the Schwarzschild criterion) and emerge at the solar surface in active regions, where they form sunspots with magnetic field strengths of $B \approx 10^3$ G and coronal loops with field strengths of $B \approx 10^2$ G at the photospheric footpoints, and $B \approx 10$ G in larger coronal heights. The differential rotation on the solar surface is thought to wind up the surface magnetic field, which then fragments under the magnetic stress, circulates meridionally to the poles, and reorients from the toroidally stressed state (with field lines oriented in the east–west direction) at solar maximum into a poloidal dipole field (connecting the North with the South Pole) in the solar minimum. This process is called the solar dynamo, which flips the magnetic polarity of the Sun every ~ 11 years (the solar cycle), or returns to the same magnetic configuration every ~ 22 years (the Hale cycle). The solar cycle controls the occurrence rate of all solar activity phenomena—from sunspot numbers, active regions, to flares, and **coronal mass ejections (CMEs)**.

3. THE PHOTOSPHERE

The **photosphere** is a thin layer at the solar surface that is observed in white light. The irradiance spectrum in Figure 11.2 shows the maximum at visible wavelengths, which can be fitted with a blackbody spectrum with a temperature of $T \approx 6400$ K at wavelengths of $\lambda \geq 2000$ Å, which is the solar surface temperature. The photosphere is defined as the range of heights from which photons directly escape, which encompasses an optical depth range of $0.1 \leq \tau \leq 3$ and translates into a height range of $h \approx 300$ km for the visible wavelength range.

3.1. Granulation and Convection

The photospheric plasma is only partially ionized; there are fewer than 0.001 electrons per hydrogen atom at the

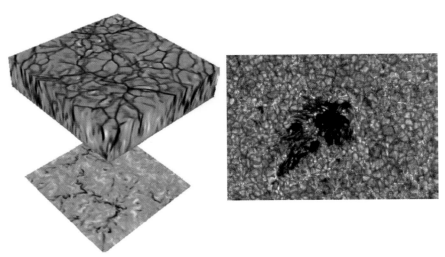

FIGURE 11.4 (Left) Numerical simulation of cellular convection at the solar surface, performed by Fausto Cattaneo and Andrea Malagoli. (Right) High-resolution observation of the granulation pattern in the solar photosphere. A granule has a typical size of 1000 km, representing the surface of an elementary convection cell. The large black area represents a sunspot, where the temperature is cooler than the surroundings. *This image was taken by Tom Berger with the Swedish Solar Observatory.*

photospheric temperature of $T = 6400$ K at $\lambda = 5000$ Å. These few ionized electrons come mostly from less abundant elements with a low ionization potential, such as magnesium, while hydrogen and helium are almost completely atomic. The magnetic field is frozen into the gas under these conditions. However, the temperature is rapidly increasing below the photospheric surface, exceeding the hydrogen ionization temperature of $T = 11,000$ K at a depth of 50 km, where the number of ionized electrons increases to 0.1 electrons per hydrogen atom, and the opacity increases by a similar factor. The high opacity of the partially ionized plasma impedes the heat flow. Moreover, a stratification with a temperature gradient steeper than an adiabatic gradient is unstable to convection (Schwarzschild criterion). Thus the partially ionized photosphere of the Sun, as well as of other low-mass stars (with masses $m < 2m_\odot$ are therefore convective.

The observational manifestation of subphotospheric convection is the granulation pattern (Figure 11.4, right), which contains granules with typical sizes of ~ 1000 km and lifetimes of $\tau \approx 7$ min. The subphotospheric gas flows up in the bright centers of granulation cells, and then cools by radiating away some heat at the optically thin photospheric surface, and, while cooling, becomes denser and flows down in the intergranular lanes. This convection process can now be reproduced with numerical simulations that include hydrodynamics, radiative transfer, and atomic physics of ionization and radiative processes (Figure 11.4, left). The convection process is also organized on larger scales, exhibiting cellular patterns on scales of $\sim 5000-10,000$ km (mesogranulation) and on scales of $\sim 20,000$ km (supergranulation).

3.2. Photospheric Magnetic Field

Most of what we know about the solar magnetic field is inferred from observations of the photospheric field, from the Zeeman effect of spectral lines in visible wavelengths (e.g. Fe 5250 Å). From two-dimensional (2D) maps of the photospheric magnetic field strength, we extrapolate the coronal three-dimensional (3D) magnetic field, or try to trace the subphotospheric origin from emerging magnetic flux elements. The creation of magnetic flux is thought to occur in the tachocline at the bottom of the convection zone, from where it rises upward in the form of buoyant magnetic flux tubes and emerges at the photospheric surface. The strongest fields emerge in sunspots, amounting to several kilogauss field strengths, and fields with strengths of several 100 G also emerge all over in active regions, often in the form of a leading sunspot trailed by following groups of opposite magnetic polarity. Due to the convective motion, small magnetic flux elements that emerge in the center of granulation cells are then swept to the intergranular lanes, where often unresolved small concentrations are found, with sizes of less than a few 100 km. The flow velocities due to photospheric convection are on the order of ~ 1 km/s. In the quiet Sun, away from active regions, the mean photospheric magnetic field amounts to a few Gauss.

3.3. Sunspots

Sunspots are the areas with the strongest magnetic fields, and therefore a good indicator of the solar activity (Figure 11.5, bottom). The butterfly diagram shows that sunspots (or active regions) appear first at higher latitudes

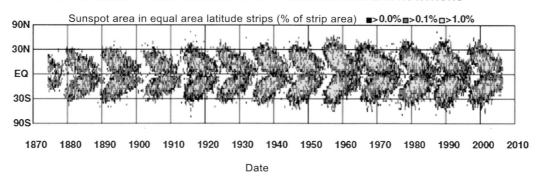
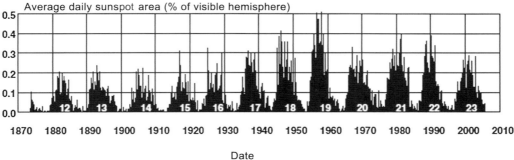

FIGURE 11.5 Top: Butterfly diagram of sunspot appearance, which marks the heliographic latitude of sunspot locations as a function of time, during the solar cycles 12–23 (covering the years 1880–2000). Bottom: Sunspot area as a function of time, which is a similar measure of the solar cycle activity as the sunspot number. *Courtesy of D. Hathaway and NASA/MSFC.*

early in the solar cycle and then drift equatorward toward the end of the solar cycle (Figure 11.5, top). Since all solar activity phenomena are controlled by the magnetic field, they have a similar solar cycle dependence as sunspots, such as the flare rate, active region area, global soft X-ray brightness, and radio emission. The appearance of dark sunspots lowers the total luminosity of the Sun only by about 0.15% at sunspot maximum, and thus the variation of the sunlight has a negligible effect on the Earth's climate. The variation of the EUV emission, which affects the ionization in the Earth's ionosphere, however, has a more decisive impact on the Earth's climate.

An individual sunspot consists of a very dark central umbra, surrounded by a brighter, radially striated penumbra. The darkness of sunspots is attributed to the inhibition of convective transport of heat, emitting only about 20% of the average solar heat flux in the umbra and being significantly cooler (~4500 K) than the surroundings (~6000 K). Their diameters range from 3600 to 50,000 km, and their lifetime ranges from a week to several months. The magnetic field in the umbra is mostly vertically oriented, but it is strongly inclined over the penumbra, nearly horizontally. Current theoretical models explain the interlocking comb structure of the filamentary penumbra with outward submerged field lines that are pumped down by turbulent, compressible convection of strong descending plumes.

Sunspots are used to trace the surface rotation since Galileo in 1611. The average sidereal differential rotation rate is

$$\omega = 14.522 - 2.84 \sin^2 \Phi \ °/\text{day}$$

where Φ is the heliographic latitude. The rotation rate of an individual feature, however, can deviate from this average by a few percent because it depends on the anchor depth to which the feature is rooted, since the solar internal differential rate varies radially (Figure 11.3, right).

4. THE CHROMOSPHERE AND TRANSITION REGION

4.1. Basic Physical Properties

The **chromosphere** (from the Greek word χρωμοσ, color) is the lowest part of the solar atmosphere, extending to an average height of ~2000 km above the photosphere. The first theoretical concepts conceived the chromosphere as a spherical layer around the solar surface (in the 1950s; Figure 11.6, left), while later refinements included the diverging magnetic fields (canopies) with height (in the 1980s; Figure 11.6, middle), and finally ended up with a very inhomogeneous mixture of cool gas and hot plasma, as a result of the extremely dynamic nature of chromospheric

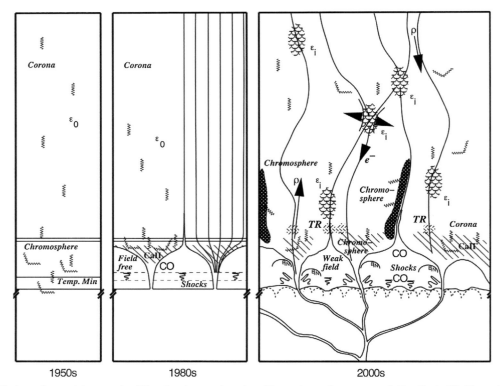

FIGURE 11.6 Cartoon of geometric concepts of the solar chromosphere, transition region, and corona: gravitationally stratified layers in the 1950s (left), vertical flux tubes with chromospheric canopies in the 1980s (middle), and a fully inhomogeneous mixing of photospheric, chromospheric, and coronal zones by dynamic processes such as heated upflows, cooling downflows, intermittent heating (electron), nonthermal electron beams (electron), field line motions and reconnections, emission from hot plasma, absorption and scattering in cool plasma, acoustic waves, and shocks (right). *Courtesy of Carolus J. Schrijver.*

phenomena (in the 2000s; Figure 11.6, right). According to hydrostatic standard models assuming local thermodynamic equilibrium (LTE), the temperature first reaches a minimum of $T = 4300$ K at a height of $h \approx 500$ km above the photosphere and rises then suddenly to $\sim 10,000$ K in the upper chromosphere at $h \approx 2000$ km, but the hydrogen density drops by about a factor of 10^6 over the same chromospheric height range. These hydrostatic models have been criticized because they neglect the magnetic field, horizontal inhomogeneities, dynamic processes, waves, and non-LTE conditions.

Beyond the solar limb (without having the photosphere in the background), the chromospheric spectrum is characterized by emission lines; these lines appear dark on the disk as a result of photospheric absorption. The principal lines of the photospheric spectrum are called the Fraunhofer lines, including, for example, hydrogen lines (H I; with the Balmer series Hα (6563 Å), Hβ (4861 Å), Hγ (4341 Å), Hδ (4102 Å)), calcium lines (Ca II; K 3934 Å, H 3968 Å), and helium lines (He I; D$_3$ 5975 Å).

4.2. Chromospheric Dynamic Phenomena

The appearance and fine structure of the chromosphere varies enormously depending on which spectral line, wavelength, and line position (core, red wing, or blue wing) is used because of their sensitivity to different temperatures (and thus altitudes) and Doppler shifts (and thus velocity ranges). In the H and K lines of Ca II, the chromospheric images show a bright network surrounding supergranulation cells, which coincide with the large-scale subphotospheric convection cells. In the Ca II K2 or in ultraviolet continuum lines (1600 Å), the network and internetwork appear grainier. The so-called bright grains have a high contrast in wavelengths that are sensitive to the temperature minimum (4300 K), with an excessive temperature of 30–360 K and with spatial sizes of ~ 1000 km. The bright points in the network are generally associated with magnetic elements that collide, which then heat the local plasma after **magnetic reconnection**. In the intranetwork, bright grains result from chromospheric oscillations that produce shock waves. There are also very thin spaghetti-like elongated fine structures visible in Hα spectroheliograms (Figure 11.7, left), which are called fibrils around sunspots. More vertically oriented fine structures are called mottles on the disk or spicules above the limb. Mottles appear as irregular threads, localized in groups around and above supergranules, at altitudes of 700–3000 km above the photosphere, with lifetimes of 12–20 min, and are apparently signatures of upward and downward motions of plasmas

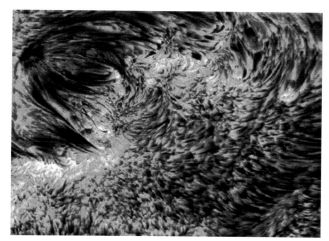

FIGURE 11.7 High-resolution image of Active Region 10380 on June 16, 2003, located near the limb, showing chromospheric spicules in the right half of the image. The image was taken with the Swedish 1-m Solar Telescope on La Palma, Spain, using a tunable filter, tuned to the blue-shifted line wing of the Hα 6563 Å line. The spicules are jets of moving gas, flowing upward in the chromosphere with a speed of ~15 km/s. The scale of the image is 65,000 × 45,000 km. *Courtesy of Bart DePontieu.*

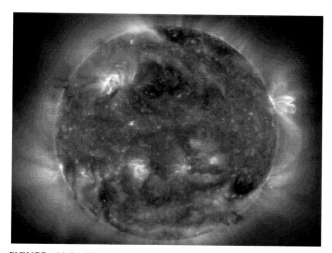

FIGURE 11.8 The multitemperature corona, recorded with the Extreme-Ultraviolet Imaging Telescope instrument on board the Solar and Heliospheric Observatory spacecraft. The representation shown here is a false-color composite of three images all taken in EUV light. Each individual image highlights a different temperature regime in the upper solar atmosphere and was assigned a specific color; red at 2 million, green at 1.5 million, and blue at 1 million K. The combined image shows active regions in white color (according to Newton's law of color addition), because they contain many loops with different temperatures. Also, nested regions above the limb appear in white, because they contain a multitude of loops with different temperatures along a line of sight, while isolated loops on the disk show a specific color according to their intrinsic temperature. *Courtesy of EIT/SoHO and NASA.*

with temperatures of $T = 8000-15,000$ K and velocities of $v \approx 5-10$ km/s. Spicules (Figure 11.7, right) are jetlike structures of plasma with temperatures of $T \approx 10,000$ K that rise to a maximum height of $h \approx 10,000$ km into the lower corona, with velocities of $v \approx 20$ km/s. They carry a maximum flux of 100 times the solar wind into the low corona. Recent numerical simulations by DePontieu and Erdelyi show that global (helioseismic) p-mode oscillations leak sufficient energy from the global resonant cavity into the chromosphere to power shocks that drive upward flows and form spicules. There is also the notion that mottles, fibrils, and spicules could be unified, being different manifestations of the same physical phenomenon at different locations (quiet Sun, active region, above the limb), in analogy to the unification of **filaments** (on the disk) and **prominences** (above the limb).

5. THE CORONA

It is customary to subdivide the solar corona into three zones, which all vary their size during the solar cycle: (1) active regions, (2) quiet-Sun regions, and (3) coronal holes.

5.1. Active Regions

Active regions are located in areas of strong magnetic field concentrations, visible as sunspot groups in optical wavelengths or magnetograms. Sunspot groups typically exhibit a strongly concentrated leading magnetic polarity, followed by a more fragmented trailing group of opposite polarity. Because of this bipolar nature, active regions are mainly made up of closed magnetic field lines. Due to the permanent magnetic activity in terms of magnetic flux emergence, flux cancellation, magnetic reconfigurations, and magnetic reconnection processes, a number of dynamic processes such as plasma heating, flares, and CMEs occur in active regions. Consequences of plasma heating in the chromosphere are upflows into coronal loops, which give active regions the familiar appearance of numerous filled loops, which are hotter and denser than the background corona, producing bright emission in soft X-rays and EUV wavelengths. In the EUV image shown in Figure 11.8, active regions appear in white.

5.2. Quiet-Sun Regions

Historically, the remaining areas outside of active regions were dubbed quiet-Sun regions. Today, however, many dynamic processes have been discovered all over the solar surface, so that the term quiet Sun is considered to be a misnomer, only justified in relative terms. Dynamic processes in the quiet Sun range from small-scale phenomena such as network heating events, nanoflares, explosive events, bright points, and soft X-ray jets, to large-scale structures, such as transequatorial loops or coronal arches. The distinction between active regions and quiet-Sun regions becomes more and more blurred because most of the large-scale structures that overarch quiet-Sun regions are rooted in active regions. A good working definition is that quiet-Sun regions encompass all closed magnetic field regions (excluding active regions), which demarcates the quiet-Sun territory from coronal holes (that encompass open magnetic field regions).

5.3. Coronal Holes

The northern and southern polar zones of the solar globe have generally been found to be darker than the equatorial zones during solar eclipses. Max Waldmeier thus dubbed those zones as coronal holes (i.e. Koronale Löcher in German). Today it is fairly clear that these zones are dominated by open magnetic field lines, which act as efficient conduits for flushing heated plasma from the corona into the solar wind, whenever they are fed by chromospheric upflows at their footpoints. Because of this efficient transport mechanism, coronal holes are empty of plasma most of the time, and thus appear much darker than the quiet Sun, where heated plasma flowing upward from the chromosphere remains trapped, until it cools down and precipitates back to the chromosphere. A coronal hole is visible in Figure 11.8 at the North Pole, where the field structures point radially away from the Sun and show a cooler temperature ($T \leq 1.0$ MK; dark blue in Figure 11.8) than the surrounding quiet-Sun regions.

5.4. Hydrostatics of Coronal Loops

Coronal loops are curvilinear structures aligned with the magnetic field. The cross-section of a loop is essentially defined by the spatial extent of the heating source because the heated plasma distributes along the coronal magnetic field lines without cross-field diffusion, since the thermal pressure is much less than the magnetic pressure in the solar corona. The solar corona consists of many thermally isolated loops, where each one has its own gravitational stratification, depending on its plasma temperature. A useful quantity is the hydrostatic pressure scale height λ_p, which depends only on the electron temperature T_e,

$$\lambda_p(T_e) = \frac{2k_B T_e}{\mu m_H g_\odot} \approx 47{,}000 \frac{T_e}{1 \text{ MK}} \text{ (km)}.$$

where m_H denotes hydrogen mass.

Observing the solar corona in soft X-rays or EUV, which are both optically thin emissions, the line-of-sight integrated brightness intercepts many different scale heights, leading to a hydrostatic weighting bias toward systematically hotter temperatures in larger altitudes above the limb. The observed height dependence of the density needs to be modeled with a statistical ensemble of multi-hydrostatic loops. Measuring a density scale height of a loop requires careful consideration of projection effects, loop plane inclination angles, cross-sectional variations, line-of-sight integration, and the instrumental response functions. Hydrostatic solutions have been computed from the energy balance between the heating rate, the radiative energy loss, and the conductive loss. The major unknown quantity is the spatial heating function, but analysis of loops in high-resolution images indicate that the heating function is concentrated near the footpoints, say at altitudes of $h \leq 20{,}000$ km. Of course, a large number of coronal loops are found to be not in hydrostatic equilibrium, while nearly hydrostatic loops have been found preferentially in the quiet corona and in older dipolar-active regions. An example of an active region (recorded with the Transition Region and Coronal Explorer (TRACE) about 10 h after a flare) is shown in Figure 11.9, which clearly shows super-hydrostatic loops where the coronal plasma is distributed over up to four times larger heights than expected in hydrostatic equilibrium (Figure 11.9, bottom).

5.5. Dynamics of the Solar Corona

Although the Sun appears lifeless and unchanging to our eyes, except for the monotonic rotation that we can trace

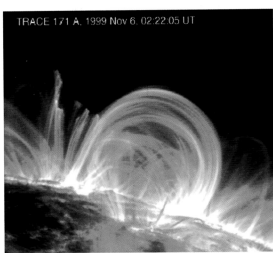

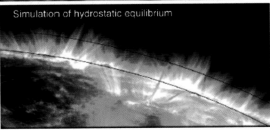

FIGURE 11.9 An active region with many loops that have an extended scale height of $\lambda_p/\lambda_T < 3-4$ (top) has been scaled to the hydrostatic thermal scale height of $T = 1$ MK (bottom). The pressure scale height of the 1 MK plasma is $\lambda_T = 47{,}000$ km, but the observed flux is proportional to the emission measure ($F \rightarrow \text{EM} \rightarrow n_e^2$), which has the half pressure scale height $\lambda_T/2 = 23{,}000$ km.

from the sunspot motions, there are actually numerous vibrant dynamic plasma processes continuously happening in the solar corona, which can be detected mainly in EUV and soft X-rays. There is currently a paradigm shift stating that most of the apparently static structures seen in the corona are probably controlled by plasma flows and intermittent heating. It is, however, not easy to measure and track these flows with our remote sensing methods, like the apparently motionless rivers seen from an airplane. For slow flow speeds, the so-called laminar flows, there is no feature to track, while the turbulent flows may be easier to detect because they produce whirls and vortices that can be tracked. A similar situation happens in the solar corona. Occasionally, a moving plasma blob is detected in a coronal loop; it can be used as a tracer. Most of the flows in coronal loops seem to be subsonic (like laminar flows) and thus featureless. Occasionally, we observe turbulent flows, which clearly reveal motion, especially when cool and hot plasma mixes by turbulence and thus yields contrast by emission and absorption in a particular temperature filter. Motion can also be detected with Doppler shift measurements, but this yields only the flow component along the line of sight. There is increasing evidence that flows are ubiquitous in the solar corona.

There are a number of theoretically expected dynamic processes. For instance, loops at coronal temperatures are thermally unstable when the radiative cooling time is shorter than the conductive cooling time, or when the heating scale height falls below one-third of a loop half-length. Recent observations show ample evidence for the presence of flows in coronal loops, as well as evidence for impulsive heating with subsequent cooling, rather than a stationary hydrostatic equilibrium. High-resolution observations of coronal loops reveal that many loops have a superhydrostatic density scale height, far in excess of hydrostatic equilibrium solutions (Figure 11.9, top). Time-dependent hydrodynamic simulations are still in a very exploratory phase, and hydrodynamic modeling of the **transition region**, coronal holes, and the solar wind remains challenging due to the number of effects that cannot easily be quantified by observations, such as unresolved geometries, inhomogeneities, time-dependent dynamics, and MHD effects.

The coronal plasma is studied with regard to hydrostatic equilibria in terms of fluid mechanics (hydrostatics), with regard to flows in terms of fluid dynamics (hydrodynamics), and including the coronal magnetic field in terms of MHD. The coronal magnetic field has many effects on the hydrodynamics of the plasma. It can play a passive role in the sense that the magnetic geometry does not change (e.g. by channeling particles, plasma flows, heat flows, and waves along its field lines or by maintaining a thermal insulation between the plasmas of neighboring loops or flux tubes). On the other hand, the magnetic field can play an active role (where the magnetic geometry changes), such as exerting a Lorentz force on the plasma, building up and storing nonpotential energy, triggering an instability, changing the topology (by various types of magnetic reconnection), and accelerating plasma structures (filaments, prominences, and CMEs).

5.6. The Coronal Magnetic Field

The solar magnetic field controls the dynamics and topology of all coronal phenomena. Heated plasma flows along magnetic field lines and energetic particles can only propagate along magnetic field lines. Coronal loops are nothing other than conduits filled with heated plasma, shaped by the geometry of the coronal magnetic field, where cross-field diffusion is strongly inhibited. Magnetic field lines take on the same role for coronal phenomena as do highways for street traffic. There are two different magnetic zones in the solar corona that have fundamentally different properties: open-field and closed-field regions. Open-field regions (white zones above the limb in Figure 11.10), which always exist in the polar regions, and sometimes extend toward the equator, connect the solar surface with the interplanetary field and are the source of the fast solar wind (~ 800 km/s). A consequence of the open-field configuration is efficient plasma transport out into the **heliosphere**, whenever chromospheric plasma is heated at the footpoints. Closed-field regions (gray zones in Figure 11.10), in contrast, contain mostly closed-field lines in the corona up to heights of about one solar radius, which open up at higher altitudes and connect eventually to the heliosphere, but produce a slow solar wind component of ~ 400 km/s. It is the closed-field regions that contain all the bright and overdense coronal loops, produced by filling with chromospheric plasma, that stays trapped in these closed-field lines. For loops reaching altitudes higher than about one solar radius, plasma confinement starts to become leaky, because the thermal plasma pressure exceeds the weak magnetic field pressure that decreases with height (plasma-β parameter <1).

The magnetic field on the solar surface is very inhomogeneous. The strongest magnetic field regions are in sunspots, reaching field strengths of $B = 2000-3000$ G. Sunspot groups are dipolar, oriented in an east–west direction (with the leading spot slightly closer to the equator), and with opposite leading polarity in both hemispheres, reversing every 11-year cycle (Hale's law). Active regions and their plages comprise a larger area around sunspots, with average photospheric fields of $B \approx 100-300$ G, containing small-scale pores with typical fields of $B \approx 1000$ G. The background magnetic field in the quiet Sun and in coronal holes has a net field of $B \approx 0.1-0.5$ G, while the absolute field strengths in

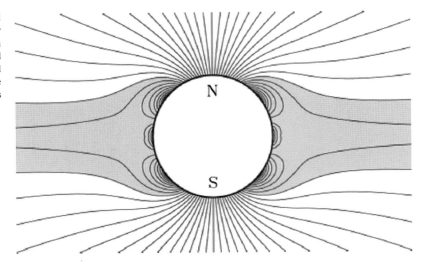

FIGURE 11.10 The global coronal magnetic field can be subdivided into open-field regions (mostly near the polar regions) and closed-field regions (mostly in latitudes of $\Phi \leq 70°$). The analytical magnetic field model shown here, a multipole-current sheet coronal model of Banaszkiewicz, approximately outlines the general trends. The high-speed solar wind originates and leaves the Sun in the unshaded volume.

resolved elements amount to $B = 10-50$ G. Our knowledge of the solar magnetic field is mainly based on measurements of Zeeman splitting in spectral lines, whereas the coronal magnetic field is reconstructed by extrapolation from magnetograms at the lower boundary, using a potential or force-free field model. The extrapolation through the chromosphere and transition region is, however, uncertain due to unknown currents and non–force-free conditions. The fact that coronal loops exhibit generally much less expansion with height than potential field models underscores the inadequacy of potential field extrapolations. Direct measurements of the magnetic field in coronal heights are still in their infancy.

5.7. MHD Oscillations of Coronal Loops

Much like the discovery of helioseismology four decades ago, it was recently discovered that the solar corona also contains an impressively large ensemble of plasma structures that are capable of producing sound waves and harmonic oscillations. Thanks to the high spatial resolution, image contrast, and time cadence capabilities of the Solar and Heliospheric Observatory (SoHO) and TRACE spacecraft, oscillating loops; prominences, or sunspots; and propagating waves have been identified and localized in the corona and transition region and studied in detail since 1999. These new discoveries established a new discipline that became known as coronal seismology. Even though the theory of MHD oscillations was developed several decades earlier, only the new imaging observations provide diagnostics on length scales, periods, damping times, and densities that allow a quantitative application of the theoretical dispersion relations of MHD waves. The theory of MHD oscillations has been developed for homogeneous media, single interfaces, slender slabs, and cylindrical flux tubes. There are four basic speeds in flux tubes: (1) the Alfvén speed $v_A = B_0/\sqrt{4\pi\rho_0}$, (2) the sound speed $c_s = \sqrt{\gamma P_0/\rho_0}$, (3) the cusp or tube speed $c_T = (1/c_s^2 + 1/v_A^2)^{-1/2}$, and (4) the kink or mean Alfvén speed $c_k = [(\rho_0 v_A^2 + \rho_e v_{A_e}^2)/(\rho_0 + \rho_e)]^{1/2}$. For coronal conditions, the dispersion relation reveals a slow-mode branch (with acoustic phase speeds) and a fast-mode branch of solutions (with Alfvén speeds). For the fast-mode branch, a symmetric (sausage) mode and an asymmetric (kink) mode can be distinguished. The fast kink mode produces transverse amplitude oscillations of coronal loops, which have been detected with TRACE (Figure 11.11), having periods in the range of $P = 2-10$ min, and can be used to infer the coronal magnetic field strength, thanks to its nondispersive nature. The fast sausage mode is highly dispersive and is subject to a long-wavelength cutoff, so that standing wave oscillations are only possible for thick and high-density (flare and postflare) loops, with periods in the range of $P \approx 1$ s to 1 min. Fast sausage-mode oscillations with periods of $P \approx 10$ s have recently been imaged for the first time with the Nobeyama radioheliograph, and there are numerous earlier reports on nonimaging detections with periods of $P \approx 0.5-5$ s. Finally, slow-mode acoustic oscillations have been detected in flarelike loops with Solar Ultraviolet Measurements of Emitted Radiation having periods in the range of $P \approx 5-30$ min. All loop oscillations observed in the solar corona have been found to be subject to strong damping, typically with decay times of only one or two periods. The relevant damping mechanisms are resonant absorption for fast-mode oscillations (or alternatively phase mixing, although requiring an extremely low Reynolds number), and thermal conduction for slow-mode acoustic oscillations. Quantitative modeling of

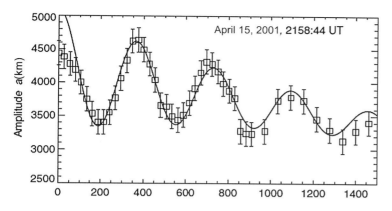

FIGURE 11.11 The transverse amplitude of a kink-mode oscillation measured in one loop of a postflare loop arcade observed with TRACE on April 15, 2001, 21:58:44 UT. The amplitudes are fitted by a damped sine plus a linear function, $a(t) = a_0 + a_1 \sin(2\pi^*(t-t_0)/P)(\exp(-t/\tau_D)) + a_2^* t$, with a period of $P = 365$ s and a damping time of $t_D = 1000$ s. *Courtesy of Ed DeLuca and Joseph Shoer.*

coronal oscillations offers exciting new diagnostics on physical parameters.

5.8. MHD Waves in Solar Corona

In contrast to standing modes (with fixed nodes), propagating MHD waves (with moving nodes) have also been discovered in the solar corona recently. Propagating MHD waves result mainly when disturbances are generated impulsively, on timescales faster than the Alfvénic or acoustic travel time across a structure.

Propagating slow-mode MHD waves (with acoustic speed) have been recently detected in coronal loops with TRACE and SoHO/EIT (Figure 11.12); they are usually being launched with 3-min periods near sunspots, or with 5-min periods in plage regions. These acoustic waves propagate upward from a loop footpoint and are quickly damped; they have never been detected in downward direction at the opposite loop side. Propagating fast-mode MHD waves (with Alfvénic speeds) have recently been discovered in a loop in optical (Solar Eclipse Coronal Imaging System eclipse) data, as well as in (Nobeyama) radio images. Ubiquitous low-amplitude waves with Alfvénic phase speeds have also been detected in the corona with the Coronal Multi-Channel Polarimeter at the National Solar Observatory (NSO).

Besides from coronal loops, slow-mode MHD waves have also been detected in plumes in open-field regions in coronal holes, while fast-mode MHD waves have not yet been detected in open-field structures. However, spectroscopic observations of line broadening in coronal holes provide strong support for the detection of **Alfvén waves**, based on the agreement with the theoretically predicted height-dependent scaling between line broadening and density, $\Delta v(h) \propto n_e(h)^{-1/4}$.

The largest manifestations of propagating MHD waves in the solar corona are global waves that spherically propagate after a flare and/or CME over the entire solar surface. These global waves were discovered earlier in Hα, called Moreton waves, and recently in EUV, called EIT

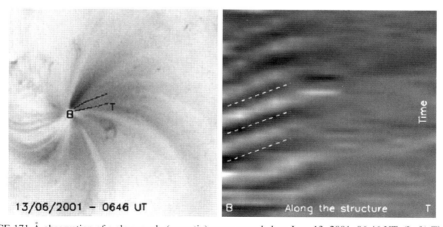

FIGURE 11.12 TRACE 171-Å observation of a slow-mode (acoustic) wave recorded on June 13, 2001, 06:46 UT. (Left) The diverging fanlike loop structures emerge near a sunspot, where the acoustic waves are launched and propagate upward. (Right) A running difference plot is shown for the loop segment marked in the left frame, with time running upward in the plot. Note the diagonal pattern, which indicates propagating disturbances. *Courtesy of Ineke De Moortel.*

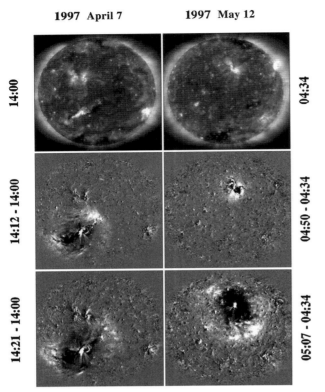

FIGURE 11.13 Two global wave events observed with SoHO/EIT 195 Å, on April 7, 1997 (top row) and May 12, 1997 (bottom row). The intensity images (right) were recorded before the eruption, while the difference images (left and middle) show differences between the subsequent images, enhancing emission measure increases (white areas), and dimming (black areas). *Courtesy of Yi-Ming Wang.*

waves (Figure 11.13), usually accompanied with a coronal dimming behind the wave front, suggesting evacuation of coronal plasma by the CME. The speed of Moreton waves is about three times greater than that of EIT waves, which still challenges dynamic MHD models of CMEs.

5.9. Coronal Heating

When Bengt Edlén and Walter Grotrian identified Fe IX (nine-times ionized iron) and Ca XIV (14-times ionized calcium) lines in the solar spectrum in 1943, a coronal temperature of $T \approx 1$ MK was first inferred from the formation temperature of these highly ionized atoms. A profound consequence of this measurement is the implication that the corona then consists of a fully ionized hydrogen plasma. Comparing this coronal temperature with the photospheric temperature of 6400 K, we are confronted with the puzzle of how the 200 times hotter coronal temperature can be maintained, the so-called coronal heating problem. Of course, there is also a chromospheric heating problem and a solar wind heating problem. If only thermal conduction were at work, the temperature in the corona should steadily drop down from the chromospheric value with increasing distance, according to the second law of thermodynamics. Moreover, since we have radiative losses by EUV emission, the corona would just cool off in a matter of hours to days, if the plasma temperature could not be maintained continuously by some heating source.

The coronal heating problem has been narrowed down by substantial progress in theoretical modeling with MHD codes; new high-resolution imaging with the Yohkoh Soft X-ray Telescope (SXT), EIT, TRACE, Hinode, and the Atmospheric Imaging Assembly onboard the Solar Dynamics Observatory (SDO); and with more sophisticated data analysis using automated pattern recognition codes. The total energy losses in the solar corona range from $F = 3 \times 10^5$ erg/cm^2 s in quiet-Sun regions to $F \approx 10^7$ erg/cm^2 s in active regions. Two main groups of direct current (DC) and alternating current models involve as a primary energy source chromospheric footpoint motion or upward leaking Alfvén waves, which are dissipated in the corona by magnetic reconnection, current cascades, MHD turbulence, Alfvén resonance, resonant absorption, or phase mixing. There is also strong observational evidence for solar wind heating by cyclotron resonance, while velocity filtration seems not to be consistent with EUV data. Progress in theoretical models has mainly been made by abandoning homogeneous flux tubes, but instead including gravitational scale heights and more realistic models of the transition region and taking advantage of numerical simulations with 3D MHD codes (by Boris Gudiksen and Aake Nordlund). From the observational side, we can now unify many coronal small-scale phenomena with flarelike characteristics, subdivided into microflares (in soft X-rays) and nanoflares (in EUV) solely by their energy content. Scaling laws of the physical parameters corroborate their unification. They provide a physical basis to understand the frequency distributions of their parameters and allow estimation of their energy budget for coronal heating. Synthesized data sets of microflares and nanoflares in EUV and soft X-rays have established that these impulsive small-scale phenomena match the radiative loss of the average quiet-Sun corona (Figure 11.14), which points to small-scale magnetic reconnection processes in the transition region and lower corona as primary heating sources.

6. SOLAR FLARES AND CMEs

Rapidly varying processes in the solar corona, which result from a loss of magnetic equilibrium, are called eruptive phenomena, such as flares, CMEs, or eruptive filaments and prominences. The fundamental process that drives all these phenomena is magnetic reconnection.

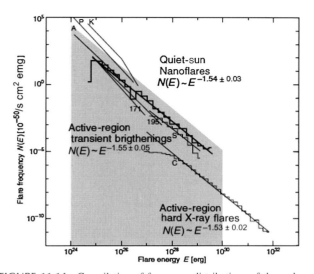

FIGURE 11.14 Compilation of frequency distributions of thermal energies from nanoflare statistics in the quiet Sun, active region transient brightenings, and hard X-ray flares. The overall slope of the synthesized nanoflare distribution, $N(E) \propto E^{-1.54 \pm 0.03}$, is similar to that of transient brightenings and hard X-ray flares. The gray area indicates the coronal heating requirement of $F = 3 \times 10^5$ erg/cm^2 s for quiet-Sun regions. Note that the observed distribution of nanoflare energies, which only includes the radiative losses, accounts for about one-third of the heating rate requirement of the quiet Sun.

6.1. Magnetic Reconnection

The solar corona has dynamic boundary conditions: (1) the solar dynamo in the interior of the Sun constantly generates new magnetic flux from the bottom of the convection zone (i.e. the tachocline) which rises by buoyancy and emerges through the photosphere into the corona, (2) the differential rotation as well as convective motion at the solar surface continuously wrap up the coronal field, and (3) the connectivity to the interplanetary field has to constantly break up to avoid excessive magnetic stress. These three dynamic boundary conditions are the essential reasons why the coronal magnetic field is constantly stressed and has to adjust by restructuring the large-scale magnetic field by topological changes, called magnetic reconnection processes. Of course, such magnetic restructuring processes occur wherever magnetic stresses build up (e.g. in filaments, in twisted sigmoid-shaped loops, and along sheared neutral lines). Topological changes in the form of magnetic reconnection always liberate free nonpotential energy, which is converted into heating of plasma, acceleration of particles, and kinematic motion of coronal plasma. Magnetic reconnection processes can occur in a slowly changing quasi-steady way, which may contribute to coronal heating (Section 5.9), but more often happen as sudden violent processes that are manifested as flares and CMEs.

Theory and numerical simulations of magnetic reconnection processes in the solar corona have been developed for steady 2D reconnection (Figure 11.15, top), bursty 2D reconnection, and 3D reconnection. Only steady 2D reconnection models can be formulated analytically; they provide basic relations for inflow speed, outflow speed, and reconnection rate, but represent oversimplifications for most (if not all) observed flares. A more realistic approach seems to be bursty 2D reconnection models (Figure 11.15, bottom), which involve the tearing mode and coalescence instability and can reproduce the sufficiently fast temporal and small spatial scales required by solar flare observations. The sheared magnetic field configurations and the existence of coronal and chromospheric null points, which are now inferred more commonly in solar flares, require ultimately 3D reconnection models, possibly involving null point coalescence, spine reconnection, fan reconnection, and separator reconnection. Magnetic reconnection operates in two quite distinct physical parameter domains: (1) in the chromosphere during magnetic flux emergence, magnetic flux cancellation, and the so-called explosive events and (2) under coronal conditions during microflares, flares, and CMEs.

6.2. Filaments and Prominences

Key elements in triggering flares and/or CMEs are erupting filaments. A filament is a current system above a magnetic neutral line that builds up gradually over days and erupts during a flare or CME process. The horizontal magnetic field lines overlying a neutral line (i.e. the magnetic polarity inversion line) of an active region are filled with cool gas (of chromospheric temperature), embedded in the much hotter tenuous coronal plasma. On the solar disk, these cool dense features appear dark in Hα or EUV images, in absorption against the bright background, and are called filaments, while the same structures appear bright above the limb, in emission against the dark sky background, where they are called prominences. Thus, filaments and prominences are identical structures physically, while their dual name just reflects a different observed location (inside or outside the disk). A further distinction is made regarding their dynamic nature: Quiescent filaments/prominences are long-lived stable structures that can last for several months, while eruptive filaments/prominences are usually associated with flares and CMEs (see example in Figure 11.16).

6.3. Solar Flare Models

A flare process is associated with a rapid energy release in the solar corona, believed to be driven by stored nonpotential magnetic energy and triggered by an instability in the magnetic configuration. Such an energy release process results in acceleration of nonthermal particles and in

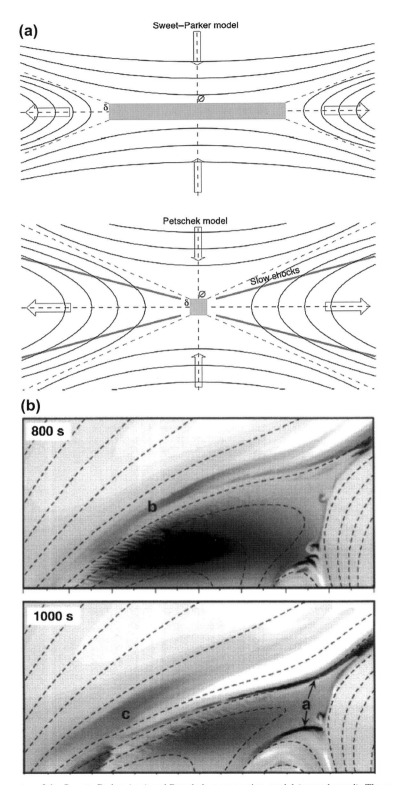

FIGURE 11.15 (a) Top: Geometry of the Sweet–Parker (top) and Petschek reconnection model (second panel). The geometry of the diffusion region (gray box) is a long thin sheet ($\Delta \gg d$) in the Sweet–Parker model, but much more compact ($\Delta \approx d$) in the Petschek model (second panel). The Petschek model also considers slow-mode MHD shocks in the outflow region. (b) Numeric MHD simulation of a magnetic reconnection process in a sheared arcade. The grayscale represents the mass density difference ratio, and the dashed lines show the projected magnetic field lines in the vicinity of the reconnection region, at two particular times of the reconnection process. Location **a** corresponds to a thin compressed region along the slowly rising inner separatrix and location **b** to a narrow downflow stream outside of the left outer separatrix, and **c** indicates a broader upflow that follows along the same field lines. *Courtesy of Judith Karpen.*

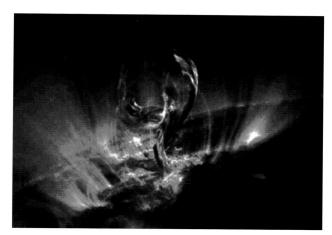

FIGURE 11.16 Erupting filament observed with TRACE at 171 Å on July 19, 2000, 23:30 UT, in Active Region 9077. The dark filament mass has temperatures around 20,000 K, while the hot kernels and threads contain plasma with temperatures of 1.0 MK or more. The erupting structure extends over a height of 75,000 km here. *Courtesy of TRACE and NASA.*

content of 10^{-6} to 10^{-9} of the largest flares fall into the categories of microflares and nanoflares (Figure 11.14), which are observed not only in active regions but also in quiet-Sun regions. Some of the microflares and nanoflares have been localized above the photospheric network and are thus also dubbed network flares or network heating events. There are also a number of small-scale phenomena with rapid time variability for which it is not clear whether they represent miniature flare processes (e.g. active region transients, explosive events, blinkers). It is conceivable that some are related to photospheric or chromospheric magnetic reconnection processes, in contrast to flares that always involve coronal magnetic reconnection processes.

The best known flare/CME models entail magnetic reconnection processes that are driven by a rising filament/prominence, flux emergence, converging flows, or shear motion along the neutral line. Flare scenarios with a driver perpendicular to the neutral line (rising prominence, flux emergence, convergence flows) are formulated as 2D reconnection models, while scenarios that involve shear along the neutral line (tearing-mode instability, quadrupolar flux transfer, the magnetic breakout model, and sheared arcade interactions) require 3D descriptions. A 2D reconnection model involving a magnetic X-point is shown in Figure 11.17 (left); a generalized 3D version involving a

heating of coronal/chromospheric plasma. These processes emit radiation in almost all wavelengths: radio, white light, EUV, soft X-rays, hard X-rays, and even gamma rays during large flares. The energy range of flares extends over many orders of magnitude. Small flares that have an energy

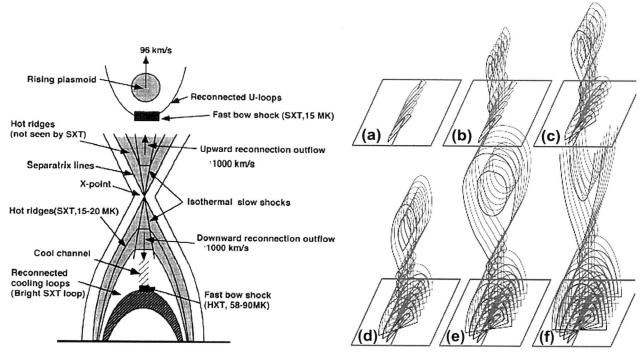

FIGURE 11.17 Left: A version of the standard 2D X-type reconnection model for two-ribbon flares, pioneered by Carmichael, Sturrock, Hirayama, and Kopp–Pneumann (CSHKP), which also includes the slow and fast shocks in the outflow region, the upward-ejected plasmoid, and the locations of the soft X-ray bright flare loops. (*Courtesy of Saku Tsuneta.*) Right: 3D version of the two-ribbon flare model, based on the observed evolution during the Bastille Day (July 14, 2000) flare: (a) low-lying, highly sheared loops above the neutral line first become unstable; (b) after loss of magnetic equilibrium, the filament jumps upward and forms a current sheet according to the model by Forbes and Priest. When the current sheet becomes stretched, magnetic islands form and coalescence of islands occurs at locations of enhanced resistivity, initiating particle acceleration and plasma heating; (c) the lowest lying loops relax after reconnection and become filled due to chromospheric evaporation (loops with thick linestyle); (d) reconnection proceeds upward and involves higher lying, less-sheared loops; (e) the arcade gradually fills up with filled loops; (f) the last reconnecting loops have no shear and are oriented perpendicular to the neutral line. At some point, the filament disconnects completely from the flare arcade and escapes into interplanetary space.

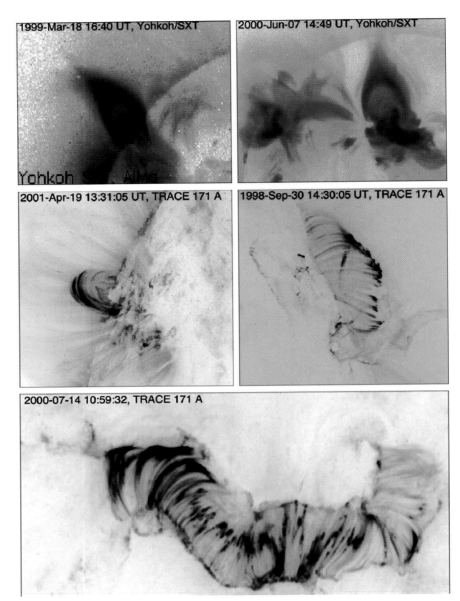

FIGURE 11.18 Soft X-ray and EUV images of flare loops and flare arcades with bipolar structure. Yohkoh/SXT-observed flares (March 18, 1999, 16:40 UT, and June 7, 2000, 14:49 UT) with "candle flame"-like cusp geometry during ongoing reconnection, while TRACE sees postflare loops once they cool down to 1–2 MK, when they already relaxed into a near-dipolar state. Examples are shown for a small flare (the April 19, 2001, 13:31 UT, Geostationary Orbiting Earth Satellite (GOES) class M2 flare) and for two large flares with long arcades, seen at the limb (September 30, 1998, 14:30 UT) and on the disk (the July 14, 2000, 10:59 UT, X5.7 flare). *Courtesy of Yohkoh/ISAS and TRACE/NASA.*

highly sheared neutral line is sketched in Figure 11.17 (right). There are more complex versions like the magnetic breakout model, where a second arcade triggers reconnection above a primary arcade. Observational evidence for magnetic reconnection in flares includes the 3D geometry, reconnection inflows, outflows, detection of shocks, jets, ejected plasmoids, and secondary effects like particle acceleration, conduction fronts, and chromospheric evaporation processes. Flare images in soft X-rays often show the cusp-shaped geometry of reconnecting field lines (Figure 11.18, top), while EUV images invariably display the relaxed postreconnection field lines after the flare loops cooled down to EUV temperatures in the postflare phase (Figure 11.18, middle and bottom).

6.4. Flare Plasma Dynamics

The flare plasma dynamics and associated thermal evolution during a flare consists of a number of sequential processes: plasma heating in coronal reconnection sites, chromospheric flare plasma heating (either by precipitating nonthermal particles or by downward propagating heat conduction

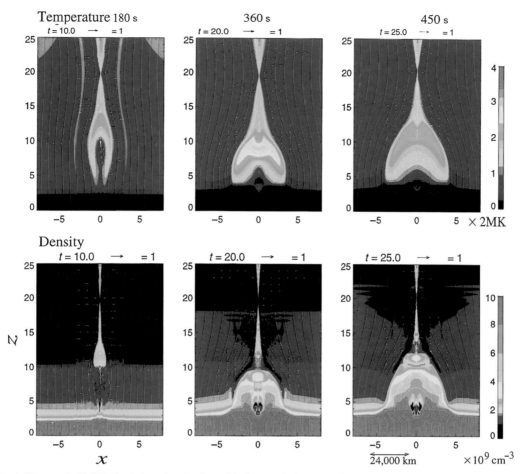

FIGURE 11.19 A 2D numerical MHD simulation of a solar flare with chromospheric evaporation and anisotropic heat conduction in the framework of a 2D magnetic reconnecting geometry. The temporal evolution of the plasma temperature (top row) and density (bottom row) is shown. The temperature and density scale are shown in the bars on the right side. The simulation illustrates the propagation of thermal conduction fronts and the upflows of chromospheric plasma in response. *Courtesy of Takaaki Yokoyama and Kazunari Shibata.*

fronts), chromospheric evaporation in the form of upflowing heated plasma, and cooling of postflare loops. The initial heating of the coronal plasma requires anomalous resistivity because Joule heating with classical resistivity is unable to explain the observed densities, temperatures, and rapid timescales in flare plasmas. Other forms of coronal flare plasma heating, such as slow shocks, electron beams, proton beams, or inductive currents, are difficult to constrain with currently available observables. The second stage of chromospheric heating is more thoroughly explored, based on the theory of the thick-target model, with numeric hydrodynamic simulations, and with particle-incell simulations. Important diagnostics on chromospheric heating are also available from Hα, white light, and UV emission, but quantitative modeling is still quite difficult because of the chromospheric opacities and partial ionization. The third stage of chromospheric evaporation has been extensively explored with hydrodynamic simulations, in particular to explain the observed Doppler shifts in soft X-ray lines, while application of spatial models to imaging data is quite sparse. Also, certain types of slow-drifting radio bursts seem to contain information on the motion of chromospheric evaporation fronts. The fourth stage of postflare loop cooling is now understood to be dominated by thermal conduction initially and by radiative cooling later on. However, spatiotemporal temperature modeling of flare plasmas (Figure 11.19) has not yet been fitted to observations in detail.

6.5. Particle Acceleration and Kinematics

Particle acceleration in solar flares is mostly explored by theoretical models because neither macroscopic nor microscopic electric fields are directly measurable by remote sensing methods. The motion of particles can be described in terms of acceleration by parallel electric fields, drift velocities caused by perpendicular forces (i.e. $E \times B$ drifts), and gyromotion caused by the Lorentz force of the magnetic field. Theoretical models of particle acceleration in solar flares can be broken down into three groups: (1) DC electric field acceleration, (2) stochastic or second-order Fermi acceleration, and (3) shock acceleration. In the models of the first group, there is a paradigm shift from large-scale DC electric fields (of the size of flare loops) to small-scale electric fields (of the size of magnetic islands produced by the tearing mode instability). The acceleration and trajectories of particles is studied more realistically in the inhomogeneous and time-varying electromagnetic fields around magnetic X-points and O-points of magnetic reconnection sites, rather than in static, homogeneous, large-scale Parker-type current sheets. The second group of models entails stochastic acceleration by gyroresonant wave–particle interactions, which can be driven by a variety of electrostatic and electromagnetic waves, supposed that wave turbulence is present at a sufficiently enhanced level and that the MHD turbulence cascading process is at work. The third group of acceleration models includes a rich variety of shock acceleration models, which is extensively explored in magnetospheric physics and could cross-fertilize solar flare models. Two major groups of models are studied in the context of solar flares (i.e. first-order Fermi acceleration or shockdrift acceleration and diffusive shock acceleration). New aspects are that shock acceleration is now applied to the outflow regions of coronal magnetic reconnection sites, where first-order Fermi acceleration at the standing fast shock is a leading candidate. Traditionally, evidence for shock acceleration in solar flares came mainly from radio type II bursts. New trends in this area are the distinction of different acceleration sites that produce type II emission: flare blast waves, the leading edge of CMEs (bow shock), and shocks in internal and lateral parts of CMEs. In summary, we can say that (1) all three basic acceleration mechanisms seem to play a role to a variable degree in some parts of solar flares and CMEs, (2) the distinctions among the three basic models become more blurred in more realistic (stochastic) models, and (3) the relative importance and efficiency of various acceleration models can only be assessed by including a realistic description of the electromagnetic fields, kinetic particle distributions, and MHD evolution of magnetic reconnection regions pertinent to solar flares.

Particle kinematics, the quantitative analysis of particle trajectories, has been systematically explored in solar flares by performing high-precision energy-dependent time delay measurements with the large-area detectors of the Compton Gamma-Ray Observatory (CGRO). There are essentially five different kinematic processes that play a role in the timing of nonthermal particles energized during flares: (1) acceleration, (2) injection, (3) free-streaming propagation, (4) magnetic trapping, and (5) precipitation and energy loss. The time structures of hard X-ray and radio emission from nonthermal particles indicate that the observed energy-dependent timing is dominated either by free-streaming propagation (obeying the expected electron time-of-flight dispersion) or by magnetic trapping in the weak diffusion limit (where the trapping times are controlled by collisional pitch angle scattering). The measurements of the velocity dispersion from energy-dependent hard X-ray delays allows then to localize the acceleration region, which was invariably found in the cusp of postflare loops (Figure 11.20).

6.6. Hard X-Ray Emission

Hard X-ray emission is produced by energized electrons via collisional **bremsstrahlung**, most prominently in the form of thick-target bremsstrahlung when precipitating electrons hit the chromosphere. Thin-target bremsstrahlung may be observable in the corona for footpoint-occulted flares. Thermal bremsstrahlung dominates only at energies of ≤ 15 keV. Hard X-ray spectra can generally be fitted with a thermal spectrum at low energies and with a single or double power law nonthermal spectrum at higher energies. Virtually all flares exhibit fast (subsecond) pulses in hard X-rays, which scale proportionally with flare loop size and are most likely spatiotemporal signatures of bursty magnetic reconnection events. The energy-dependent timing of these fast subsecond pulses exhibit electron time-of-flight delays from the propagation between the coronal acceleration site and the chromospheric thick-target site. The inferred acceleration site is located about 50% higher than the soft X-ray flare loop height, most likely near X-points of magnetic reconnection sites (Figure 11.20). The more gradually varying hard X-ray emission exhibits an energy-dependent time delay with opposite sign, which corresponds to the timing of the collisional deflection of trapped electrons. In many flares, the time evolution of soft X-rays roughly follows the integral of the hard X-ray flux profile, which is called the Neupert effect. Spatial structures of hard X-ray sources include: (1) footpoint sources produced by thick-target bremsstrahlung, (2) thermal hard X-rays from flare loop tops, (3) above-the-loop-top (Masuda-type) sources that result from nonthermal bremsstrahlung from electrons that are either trapped in the acceleration region or interact with reconnection shocks, (4) hard X-ray sources associated with upward soft X-ray ejecta, and (5) hard X-ray halo or albedo sources due to backscattering at the photosphere. In spatially extended flares, the footpoint sources assume ribbonlike morphology if mapped with sufficient sensitivity. The monthly hard X-ray flare rate varies about a factor of 20 during the solar cycle, similar to magnetic flux variations implied by the monthly sunspot number, as expected from the magnetic origin of flare energies.

6.7. Gamma-Ray Emission

The energy spectrum of flares (Figure 11.21) in gamma-ray wavelengths (0.5 MeV–1 GeV) is more structured than in

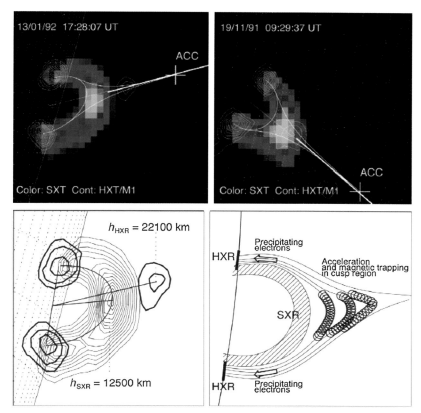

FIGURE 11.20 Top: The geometry of the acceleration region inferred from direct detections of above-the-loop-top hard X-ray sources with Yohkoh/Hard X-Ray Telescope (HXT) (contours) and simultaneous modeling of electron time-of-flight distances based on energy-dependent time delays of 20–200 keV hard X-ray emission measured with Burst and Transient Source Experiment/CGRO (crosses marked with Acceleration source (ACC)). Soft X-rays detected with Yohkoh/SXT or thermal hard X-ray emission from the low-energy channel of Yohkoh/HXT/Lo are shown in colors, outlining the flare loops. Bottom: The observations in the left panel show a Yohkoh/HXT 23- to 33-keV image (thick contours) and Be119 SXT image (thin contours) of the Masuda flare, January 13, 1992, 17:28 UT. The interpretation of the above-the-loop-top source is that temporary trapping occurs in the acceleration region in the cusp region below the reconnection point (bottom right).

hard X-ray wavelengths (20–500 keV) because it exhibits both continuum emission as well as line emission. There are at least six different physical processes that contribute to gamma-ray emission: (1) electron bremsstrahlung continuum emission, (2) nuclear deexcitation line emission, (3) neutron capture line emission at 2.223 MeV, (4) positron annihilation line emission at 511 keV, (5) pion-decay radiation at ≥ 50 MeV, and (6) neutron production. The ratio of continuum to line emission varies from flare to flare, and gamma-ray lines can completely be overwhelmed in electron-rich flares or flare phases. When gamma-ray lines are present, they provide a diagnostic of the elemental abundances, densities, and temperatures of the ambient plasma in the chromosphere, as well as of the directivity and pitch angle distribution of the precipitating protons and ions that have been accelerated in coronal flare sites, presumably in magnetic reconnection regions. Critical issues that have been addressed in studies of gamma-ray data are the maximum energies of coronal acceleration mechanisms, the ion/electron ratios (because selective acceleration of ions indicate gyroresonant interactions), the ion/electron timing (to distinguish between simultaneous or second-step acceleration), differences in ion/electron transport (e.g. neutron sources were recently found to be displaced from electron sources), and the first ionization potential effect of chromospheric abundances (indicating enhanced abundances of certain ions that could be preferentially accelerated by gyroresonant interactions). Although detailed modeling of gamma-ray line profiles provides significant constraints on elemental abundances and physical properties of the ambient chromospheric plasma, as well as on the energy and pitch angle distribution of accelerated particles, little information or constraints could be retrieved about the timescales and geometry of the acceleration mechanisms, using gamma-ray data. Nevertheless, the high spectral and imaging resolution of the Ramaty High-Energy Spectroscopic Solar Imager (RHESSI) spacecraft facilitates promising new data for a deeper understanding of ion acceleration in solar flares.

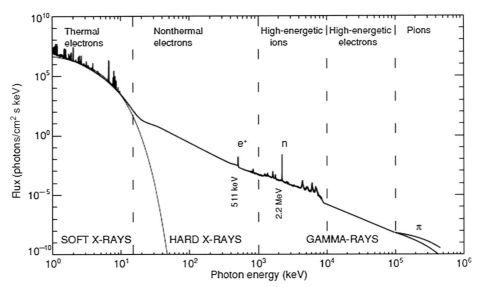

FIGURE 11.21 Composite photon spectrum of a large flare, extending from soft X-rays (1–10 keV), hard X-rays (10 keV–1 MeV), to gamma rays (1 MeV–100 GeV). The energy spectrum is dominated by different processes: by thermal electrons (in soft X-rays), bremsstrahlung from nonthermal electrons (in hard X-rays), nuclear deexcitation lines (in ~0.5- to 8-MeV gamma rays), bremsstrahlung from high-energetic electrons (in ~10- to 100-MeV gamma rays), and pion decay (in ≥100 MeV gamma rays). Note also the prominent electron–positron annihilation line (at 511 keV) and the neutron capture line (at 2.2 MeV).

6.8. Radio Emission

Radio emission in the solar corona is produced by thermal, nonthermal, and high-relativistic electrons, and thus provides useful diagnostics complementary to EUV, soft X-rays, hard X-rays, and gamma rays. Thermal or Maxwellian distribution functions produce in radio wavelengths either free–free emission (bremsstrahlung) for low magnetic field strengths or gyroresonant emission in locations of high magnetic field strengths, such as above sunspots, which are both called incoherent emission mechanisms. Since EUV and soft X-ray emission occurs in the optically thin regime, the emissivity adds up linearly along the line of sight. Free–free radio emission is somewhat more complicated because the optical thickness depends on the frequency, which allows direct measurement of the electron temperature in optically thick coronal layers in metric and decimetric frequencies up to $\nu \leq 1$ GHz. Above ~2 GHz, free–free emission becomes optically thin in the corona, but gyroresonance emission at harmonics of $s \approx 2, 3, 4$ dominates in strong-field regions. In flares, high-relativistic electrons are produced that emit gyrosynchrotron emission, which allows for detailed modeling of precipitating and trapped electron populations in time profiles recorded at different microwave frequencies.

Unstable non-Maxwellian particle velocity distributions, which have a positive gradient in parallel (beams) or perpendicular (losscones) direction to the magnetic field, drive gyroresonant wave–particle interactions that produce coherent wave growth, detectable in the form of coherent radio emission. Two natural processes that provide these conditions are dispersive electron propagation (producing beams) and magnetic trapping (producing losscones). The wave–particle interactions produce growth of Langmuir waves, upper-hybrid waves, and electron–cyclotron maser emission, leading to a variety of radio burst types (type I, II, III, IV, V, decimetric radio bursts (DCIM); Figure 11.22), which have been mainly explored from (nonimaging) dynamic spectra, while imaging observations have been rarely obtained. Although there is much theoretical understanding of the underlying wave–particle interactions, spatiotemporal modeling of imaging observations is still in its infancy. A solar-dedicated, frequency-agile imager with many frequencies (Frequency-Agile Solar Radio (FASR) telescope) is in the planning stage and might provide more comprehensive observations.

6.9. Coronal Mass Ejections

Every star is losing mass, caused by dynamic and eruptive phenomena in its atmosphere, which accelerate plasma or particles beyond the escape speed. Inspecting the Sun, our nearest star, we observe two forms of mass loss: the steady solar wind outflow and the sporadic ejection of large plasma structures, or CMEs. The solar wind outflow amounts to $\sim 2 \times 10^{-10}$ (g/cm² s) in coronal holes, and to $\leq 4 \times 10^{-11}$ (g/cm² s) in active regions. The phenomenon of CME occurs with a frequency of about one event per day, carrying a mass in the range of $m_{\mathrm{CME}} \approx 10^{14} - 10^{16}$ g,

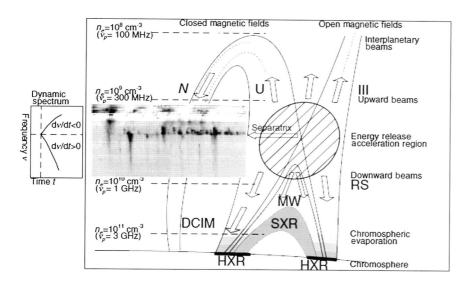

FIGURE 11.22 Radio burst types in the framework of the standard flare scenario: the acceleration region is located in the reconnection region above the soft X-ray-bright flare loop, accelerating electron beams in the upward direction (type III, U, N bursts) and in the downward direction (type reverse drifting radio bursts (RS), DCIM bursts). Downward moving electron beams precipitate to the chromosphere (producing hard X-ray emission and driving chromospheric evaporation), or remain transiently trapped, producing microwave (MW) emission. Soft X-ray loops become subsequently filled up, with increasing footpoint separation as the X-point rises. The insert shows a dynamic radio spectrum (ETH Zurich) of the September 6, 1992, 11:54 UT, flare, showing a separatrix between type III and type RS bursts at ~600 MHz, probably associated with the acceleration region.

which corresponds to an average mass loss rate of $m_{CME}/(\Delta t \cdot 4\pi R_0^2 \approx 2) \times 10^{-14} - 2 \times 10^{-12} (g/cm^2 s)$, which is $\leq 1\%$ of the solar wind mass loss in coronal holes, or $\leq 10\%$ of the solar wind mass in active regions. The transverse size of CMEs can cover a fraction up to more than a solar radius, and the ejection speed is in the range of $v_{CME} \approx 10^2 - 10^3$ (km/s). A CME structure can have the geometric shape of a fluxrope, a semishell, or a bubble (like a light bulb, see Figure 11.24), which is the subject of much debate, because of ambiguities from line-of-sight projection effects and the optical thinness. Recent 3D reconstructions with data from the dual Solar Terrestrial Relationships Observatory (STEREO) spacecraft, however, clarified the 3D geometry of CMEs considerably. There is a general consensus that a CME is associated with a release of magnetic energy in the solar corona, but its relation to the flare phenomenon is controversial. Even big flares (at least GOES M-class) have no associated CMEs in 40% of the cases. A long-standing debate focused on the question of whether a CME is a by-product of the flare process, or vice versa. This question has been settled in the view that flares and CMEs are two aspects of a large-scale magnetic energy release, but the two terms evolved historically from two different observational manifestations (i.e. flares, which mainly denote the emission in hard X-rays, soft X-rays, and radio waves, and CMEs, which refer to the white-light emission of the erupting mass in the outer corona and heliosphere). Recent studies, however, clearly established the coevolution of both processes triggered by a common magnetic instability. A CME is a dynamically evolving plasma structure, propagating outward from the Sun into interplanetary space, carrying a frozen-in magnetic flux and expanding in size. If a CME structure travels toward the Earth, which is mostly the case when launched in the western solar hemisphere, due to the curvature of the Parker spiral interplanetary magnetic field, such an Earth-directed event can engulf the Earth's magnetosphere and generate significant geomagnetic storms. Obviously such geomagnetic storms can cause disruptions of global communication and navigation networks, can cause failures of satellites and commercial power systems, and thus are the subject of high interest.

Theoretical CME models include at least seven categories: (1) thermal blast models, (2) dynamo models, (3) mass loading models, (4) tether release or straining models, (5) quadrupolar breakout models, and (6) kink or torus instability models. Numerical MHD simulations of CMEs are currently produced by combinations of a fine-scale grid that entails the corona and a connected large-scale grid that encompasses propagation into interplanetary space, which can reproduce CME speeds, densities, and the coarse geometry. The trigger that initiates the origin of a CME seems to be related to previous photospheric shear motion and subsequent kink instability of twisted structures (Figure 11.23). The geometry of CMEs is quite complex, exhibiting a variety of topological shapes from spherical semishells to helical fluxropes (Figure 11.24), and the density and temperature structure of CMEs is currently investigated with multiwavelength imagers. The height-time, velocity, and acceleration profiles of CMEs seem to establish two different CME classes: gradual CMEs associated with propagating interplanetary shocks and impulsive CMEs caused by coronal flares. The total energy of CMEs (i.e. the sum of magnetic, kinetic, and gravitational energy) seems to be conserved in some events, and the total energy of CMEs is comparable to the energy range estimated from flare signatures. A phenomenon closely associated with CMEs is coronal dimming (Figure 11.13),

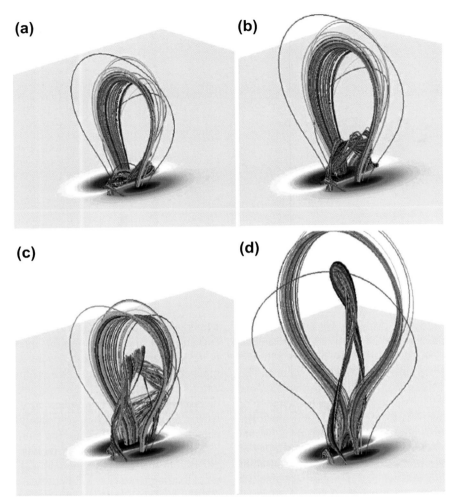

FIGURE 11.23 Numerical MHD simulation of the evolution of a CME, driven by turbulent diffusion. The four panels correspond to the times (a) $t = 850$, (b) $t = 950$, (c) $t = 1050$, and (d) $t = 1150$, where viscous relaxation is started at $t = 850$, triggering a global disruption involving opening, reconnection through the overlying arcade and below, and the formation of a current sheet, associated with a high dissipation of magnetic energy and a strong increase of kinetic energy. *Courtesy of T. Amari.*

which is interpreted in terms of an evacuation of coronal mass during the launch of a CME. The propagation of CMEs in interplanetary space provides diagnostic information on the heliospheric magnetic field, the solar wind, interplanetary shocks, solar energetic particle events, and interplanetary radio bursts.

7. FINAL COMMENTS

The study of the Sun, our nearest Star, is systematically moving from morphological observations (sunspots, active regions, filaments, flares, CMEs) to a more physics-based modeling and theoretical understanding, in terms of nuclear physics, magnetoconvection, MHD, magnetic reconnection, and particle physics processes. The major impact of physics-based modeling came from the multiwavelength observations from solar-dedicated space-based (Hinode, Solar Maximum Mission (SMM), Yohkoh, CGRO, SoHO, TRACE, RHESSI, Hinode, (STEREO, SDO, and ground-based instruments (in radio, Hα, and white-light wavelengths). Major achievements over the past decades are the advancement of new disciplines such as helioseismology and coronal seismology and the solution of the neutrino problem; however, there are still unsolved outstanding problems such as the coronal heating problem and particle acceleration mechanisms. We can optimistically expect substantial progress from future solar-dedicated space missions (Solar Orbiter, Solar Probe) and ground-based instruments (The Advanced Technology Solar Telescope (ATST) and the FASR telescope).

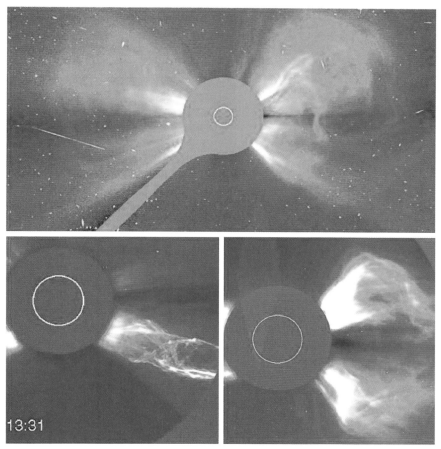

FIGURE 11.24 Large Angle Solar Coronagraph C3 image of a halo CME of May 6, 1998 (top); an erupting prominence of June 2, 1998, 13:31 UT (bottom left); and a large CME of November 6, 1997, 12:36 UT (bottom right). *Courtesy of SoHO/LASCO and NASA.*

BIBLIOGRAPHY

Aschwanden, M. J. (2004). *Physics of the solar corona—an introduction.* Chichester, England: Praxis Publishing Ltd (Springer: New York).

Benz, A. O. (2003). *Plasma astrophysics, kinetic processes in solar and stellar coronae* (2nd ed.). Dordrecht, Netherlands: Kluwer Acad. Publ.

Cox, A. N. (Ed.). (2000). *Allen's astrophysical quantities* (4th ed.). New York: American Institute of Physics Press/Springer.

Dwivedi, B. N. (2003). *The dynamic Sun.* Cambridge, England: Cambridge University Press.

Foukal, P. V. (2003). *Solar astrophysics* (2nd ed.). New York: John Wiley and Sons.

Golub, L., & Pasachoff, J. M. (1997). *The solar corona.* Cambridge, Massachusetts: Cambridge Univ. Press.

Golub, L., & Pasachoff, J. M. (2001). *Nearest Star: The surprising science of our Sun.* Cambridge, Massachusetts: Harvard Univ. Press.

Lang, K. R. (2001). *The Cambridge encyclopedia of the Sun.* Cambridge, England: Cambridge Univ. Press.

Murdin, P. (Ed.). (2000). *Encyclopedia of astronomy and astrophysics.* Institute of Physics Publishing/Grove's.

Schrijver, C. J., & Zwaan, C. (2000). *Solar and Stellar magnetic activity.* Cambridge, England: Cambridge Univ. Press.

Stix, M. (1989−2002). *The Sun* (2nd ed.). New York: Springer.

Zirker, J. B. (2002). *Journey from the center of the Sun.* Princeton, New Jersey: Princeton Univ. Press.

Chapter 12

The Solar Wind

J.T. Gosling
Laboratory for Atmospheric and Space Physics, University of Colorado, Boulder, CO, USA

Chapter Outline

1. Discovery 261
   1.1. Early Indirect Observations 261
   1.2. Parker's Solar Wind Model 262
   1.3. First Direct Observations of the Solar Wind 262
2. Statistical Properties in the Ecliptic Plane at 1 AU 262
3. Nature of the Heliospheric Magnetic Field 263
4. Coronal and Solar Wind Stream Structure 264
5. The Heliospheric Current Sheet and Solar Latitude Effects 265
   5.1. The Sun's Large-Scale Magnetic Field and the Ballerina Skirt Model 265
   5.2. Solar Latitude Effects 265
6. Evolution of Stream Structure with Heliocentric Distance 266
   6.1. Kinematic Stream Steepening and the Dynamic Response 266
   6.2. Shock Formation 266
   6.3. Stream Evolution in Two and Three Dimensions 267
7. Coronal Mass Ejections and Transient Solar Wind Disturbances 268
   7.1. Coronal Mass Ejections 268
   7.2. Origins, Associations with Other Forms of Solar Activity, and Frequency of Occurrence 268
   7.3. Heliospheric Disturbances Driven by Fast CMEs 269
   7.4. Characteristics of ICMEs 270
   7.5. The Magnetic Field Topology of ICMEs and the Problem of Magnetic Flux Balance 270
   7.6. Field Line Draping About Fast ICMEs 271
8. Variation with Distance From the Sun 271
9. Termination of the Solar Wind 272
10. Kinetic Properties of the Plasma 273
    10.1. The Solar Wind as a Marginally Collisional Plasma 273
    10.2. Kinetic Aspects of Solar Wind Ions 274
    10.3. Kinetic Aspects of Solar Wind Electrons 274
11. Heavy Ion Content 276
12. Energetic Particles 276
13. Turbulence and Magnetic Field and Velocity Fluctuations 277
14. Conclusion 278
Bibliography 279

The Solar Wind is a **plasma**, that is an ionized gas, that permeates interplanetary space. It exists as a consequence of the supersonic expansion of the Sun's hot outer atmosphere, the **solar corona**. The solar wind consists primarily of electrons and protons, but **alpha particles** and many other ionic species are also present at low abundance levels. At the orbit of Earth, 1 Astronomical Unit (AU) from the Sun, typical solar wind densities, flow speeds, and temperatures are of the order of 8 protons/cm^3, 440 km/s, and 1.2×10^5 K respectively; however, the solar wind is highly variable in both space and time. A weak magnetic field embedded within the solar wind plasma is effective both in excluding some low-energy cosmic rays from the solar system and in channeling energetic particles from the Sun into the **heliosphere**. The solar wind plays an essential role in shaping and stimulating planetary magnetospheres and the ionic tails of comets. (See Planetary Magnetospheres.)

1. DISCOVERY

1.1. Early Indirect Observations

In 1859 R. Carrington made one of the first white light observations of a **solar flare**. He noted that a major geomagnetic storm began approximately 17 h after the flare and tentatively suggested that a causal relationship might exist between the solar and geomagnetic events. Subsequent observations revealed numerous examples of associations between solar flares and large geomagnetic storms. In the early 1900s F. Lindemann suggested that this could be explained if large geomagnetic storms result from an interaction between the geomagnetic field and plasma clouds ejected into interplanetary space by solar activity. Early studies of geomagnetic activity also noted that some geomagnetic storms tend to recur at the ~27 day rotation period of the Sun as observed from Earth, particularly during declining years of solar activity. This observation led to the suggestion that certain regions on the Sun, commonly called M (for magnetic)-regions, occasionally produce long-lived charged particle streams in interplanetary space. Furthermore, because some form of auroral and geomagnetic activity is almost always present at high geomagnetic latitudes, it was inferred that charged particles from the Sun almost continuously impact and perturb the geomagnetic field.

Observations of modulations in **galactic cosmic rays** in the 1930s also suggested that plasma and magnetic fields are ejected from the Sun during intervals of high solar activity. For example, S. Forbush noted that cosmic ray intensity often decreases suddenly during large geomagnetic storms, and then recovers slowly over a period of several days. Moreover, cosmic ray intensity varies in a cycle of ~11 years, but roughly 180° out of phase with the **solar activity cycle**. One possible explanation of these observations was that magnetic fields embedded in plasma clouds from the Sun sweep cosmic rays away from the vicinity of Earth.

In the early 1950s, L. Biermann concluded that there must be a continuous outflow of charged particles from the Sun to explain the fact that ionic tails of comets always point away from the Sun. He estimated that a continuous particle flux of the order of 10^{10} protons/cm^2 s was needed at 1 AU to explain the comet tail observations. He later revised his estimate downward to a value of $\sim 10^9$ protons/cm^2 s, closer to the average observed solar wind proton flux of $\sim 3.8 \times 10^8$ protons/cm^2 s at 1 AU.

1.2. Parker's Solar Wind Model

Apparently inspired by these diverse observations and interpretations, E. Parker, in 1958, formulated a radically new model of the solar corona in which the solar atmosphere is continually expanding outward. Before Parker's work most theories of the solar atmosphere treated the corona as static and gravitationally bound to the Sun except for sporadic outbursts of material into space at times of high solar activity. S. Chapman had constructed a model of a static solar corona in which heat transport was dominated by electron thermal conduction. For a 10^6 K corona Chapman found that even a static solar corona must extend far out into space. Parker realized, however, that a static model leads to pressures at large distances from the Sun that are seven to eight orders of magnitude larger than estimated pressures in the interstellar plasma. Because of this mismatch at large heliocentric distances, he reasoned that the solar corona could not be in hydrostatic equilibrium and must therefore be expanding. His consideration of the hydrodynamic (i.e. fluid) equations for mass, momentum, and energy conservation for a hot solar corona led him to unique solutions for the coronal expansion that depended on the coronal temperature close to the surface of the Sun. Parker's model produced low flow speeds close to the Sun, supersonic flow speeds far from the Sun, and vanishingly small pressures at large heliocentric distances. In view of the fluid character of the solutions, Parker called this continuous, supersonic coronal expansion the "solar wind". The region of space filled by the solar wind is now known as the "heliosphere".

1.3. First Direct Observations of the Solar Wind

Several Russian and American space probes in the 1959–1961 era penetrated interplanetary space and found tentative evidence for a solar wind. Firm proof of the wind's existence was provided by C. Snyder and M. Neugebauer, who flew a plasma experiment on Mariner 2 during its epic 3-month journey to Venus in late 1962. Their experiment detected a continual outflow of plasma from the Sun that was highly variable, being structured into alternating streams of high- and low-speed flows that lasted for several days each. Several of the high-speed streams recurred at roughly the rotation period of the Sun. Average solar wind proton densities (normalized for a 1 AU heliocentric distance), flow speeds, and temperatures during this 3-month interval were 5.4 cm^{-3}, 504 km/s, and 1.7×10^5 K respectively, in essential agreement with Parker's predictions. The Mariner 2 observations also showed that helium, in the form of alpha particles, is present in the solar wind in variable amounts; the average alpha particle abundance relative to protons of 4.6% being about a factor of 2 lower than estimates of the helium abundance within the Sun. Finally, measurements made by Mariner 2 confirmed that the solar wind carried a magnetic field whose strength and orientation in the ecliptic plane were much as predicted by Parker (see Section 3).

Despite the good agreement of observations with Parker's model, we still do not fully understand the processes that heat the solar corona and accelerate the solar wind. Parker simply assumed that the corona is heated to a very high temperature, but he did not explain how the heating was accomplished. Moreover, it is now known that electron heat conduction is insufficient to power the coronal expansion. Present models for heating the corona and accelerating the solar wind generally fall into two classes: heating and acceleration by waves generated by convective motions below the photosphere; and bulk acceleration and heating associated with transient events in the solar atmosphere such as **magnetic reconnection**. Present observations are incapable of distinguishing between these and other alternatives.

2. STATISTICAL PROPERTIES IN THE ECLIPTIC PLANE AT 1 AU

Table 12.1 summarizes a number of statistical solar wind properties derived from spacecraft measurements in the

TABLE 12.1 Statistical Properties of the Solar Wind at 1 AU

Parameter	Mean	STD	Most Probable	Median	5–95% Range
n (cm^{-3})	8.7	6.6	5.0	6.9	3.0–20.0
V_{sw} (km/s)	468	116	375	442	320–710
B (nT)	6.2	2.9	5.1	5.6	2.2–9.9
A(He)	0.047	0.019	0.048	0.047	0.017–0.078
T_p (×10^5 K)	1.2	0.9	0.5	0.95	0.1–3.0
T_e (×10^5 K)	1.4	0.4	1.2	1.33	0.9–2.0
T_α (×10^5 K)	5.8	5.0	1.2	4.5	0.6–15.5
T_e/T_p	1.9	1.6	0.7	1.5	0.37–5.0
T_α/T_p	4.9	1.8	4.8	4.7	2.3–7.5
nV_{sw} (×10^8/cm^2 s)	3.8	2.4	2.6	3.1	1.5–7.8
C_s (km/s)	63	15	59	61	41–91
C_A (km/s)	50	24	50	46	30–100

ecliptic plane at 1 AU. The table includes mean values, standard deviations about the mean values, most probable values, median values, and the 5–95% range limits for the proton number density (n), the flow speed (V_{sw}), the magnetic field strength (**B**), the alpha particle abundance relative to protons (A(He)), the proton temperature (T_p), the electron temperature (T_e), the alpha particle temperature (T_a), the ratio of the electron and proton temperatures (T_e/T_p), the ratio of alpha particle and proton temperatures (T_a/T_p), the number flux (nV_{sw}), the sound speed (C_s), and the **Alfvén speed** (C_A). All solar wind parameters exhibit considerable variability; moreover, variations in solar wind parameters are often coupled to one another. Proton temperatures are considerably more variable than electron temperatures, and alpha particle temperatures are almost always higher than electron and proton temperatures. Alpha particles and the protons tend to have nearly equal thermal speeds and therefore temperatures that differ by a factor of about 4. The solar wind flow is usually both supersonic and super-Alfvénic. Finally, we note that the Sun yearly loses ~6.8×10^{19} g to the solar wind, a very small fraction of the total solar mass of ~2×10^{33} g.

3. NATURE OF THE HELIOSPHERIC MAGNETIC FIELD

In addition to being a very good thermal conductor, the solar wind plasma is an excellent electrical conductor. The electrical conductivity of the plasma is so high that the solar magnetic field is "frozen" into the solar wind flow as it expands away from the Sun. Because the Sun rotates, magnetic field lines in the equatorial plane of the Sun are bent into Archimedean spirals (Figure 12.1) whose inclinations relative to the radial direction depend on heliocentric distance and the speed of the wind. At 1 AU the average field line in the equatorial plane is inclined ~45° to the radial direction.

In Parker's simple model the magnetic field lines out of the equatorial plane take the form of helices wrapped about cones of constant latitude. These helices are evermore elongated at higher solar latitudes and eventually approach radial lines over the solar poles. The equations describing Parker's model of the magnetic field far from the Sun are

$$\mathbf{B}_r(r,\phi,\theta) = \mathbf{B}(r_o,\phi_o,\theta)(r_o/r)^2,$$

$$\mathbf{B}_\phi(r,\phi,\theta) = -\mathbf{B}(r_o,\phi_o,\theta)(\omega r_o^2/V_{sw}r)\sin\theta,$$

$$\mathbf{B}_\theta = 0$$

Here r, ϕ, and θ are radial distance, longitude, and latitude in a Sun-centered spherical coordinate system, respectively, $\mathbf{B}_r$, $\mathbf{B}_\phi$, and $\mathbf{B}_\theta$ are the magnetic field components, ω is the Sun's angular velocity (2.9×10^{-6} radians/sec), V_{sw} is the flow speed (assumed constant with distance from the Sun), and ϕ_o is an initial longitude at a reference distance r_o from Sun center. This model is in reasonably good agreement with suitable averages of the **heliospheric magnetic field** measured over a wide range of heliocentric distances and latitudes. However, the instantaneous orientation of the

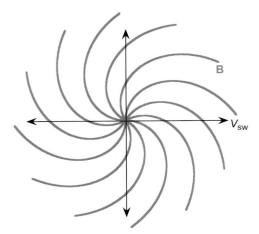

FIGURE 12.1 Configuration of the heliospheric magnetic field in the solar equatorial plane for a uniform radial solar wind flow.

field usually deviates substantially from that of the model field at all distances and latitudes. Moreover, there is evidence that the magnetic field lines wander in latitude as they extend out into the heliosphere. This appears to be a result of field line foot point motions associated with differential solar rotation (the surface of the Sun rotates at different rates at different latitudes) and convective motions in the solar atmosphere.

4. CORONAL AND SOLAR WIND STREAM STRUCTURE

The solar corona is highly nonuniform, being structured by the complex solar magnetic field into arcades, rays, holes (regions relatively devoid of material), and streamers. (See The Sun.) The strength of the Sun's magnetic field falls off sufficiently rapidly with height above the solar surface that it is incapable of containing the coronal expansion at altitudes above ~0.5–1.0 solar radii. The resulting solar wind outflow produces the "combed-out" appearance of coronal structures above those heights in eclipse photographs.

As first seen in observations of the solar wind by Mariner 2 in the ecliptic plane, the solar wind is also highly nonuniform. In the ecliptic plane it tends to be organized into alternating streams of high- and low-speed flows. Figure 12.2 illustrates certain characteristic aspects of this **stream structure**. Five high-speed streams are clearly evident in the figure. The fourth and fifth streams were reencounters with the first and second streams, respectively, on the following solar rotation. Each high-speed stream was asymmetric with the speed rising more rapidly than it fell and each stream was essentially unipolar in the sense that $\mathbf{B}_r$ was either positive or negative throughout the stream. Reversals in field polarity occurred in the low-speed flows between the streams. Those polarity reversals

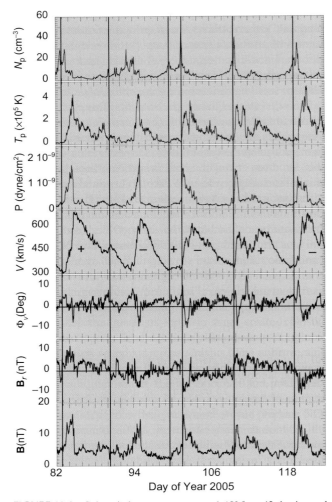

FIGURE 12.2 Solar wind stream structure at 1 AU for a 42-day interval in 2005. From top to bottom: 1-hr averages of solar wind proton density, proton temperature, total (plasma + field) pressure, bulk flow speed, flow azimuth angle (positive in the sense of Earth's motion about the Sun), radial component of the heliospheric magnetic field, and field magnitude. Vertical lines mark crossing of the heliospheric current sheet. Plus and minus signs in the fourth panel indicate magnetic polarities, outward from and inward toward the Sun, respectively. Changes in magnetic field polarity are best determined by reversals in the solar wind suprathermal electron strahl flow polarity (parallel or antiparallel to the magnetic field; see Section 10.3). *Adapted from Gosling (2010).*

correspond to crossings of the **heliospheric current sheet** (discussed in more detail in the following section) that separates solar wind regions of opposite magnetic polarity. The magnetic field strength, proton temperature and density, and total pressure all peaked on the leading edges of the streams and the solar wind flow there was deflected first westward (positive flow azimuth) and then eastward. This pattern of variability is highly repeatable from one stream to the next and is the inevitable consequence of the evolution of the streams as they progress outward from the Sun (see Section 6).

Recurrent high-speed streams originate primarily in coronal holes, which are large nearly unipolar regions in the solar atmosphere having relatively low density. Low-speed flows, on the other hand, tend to originate in the coronal streamer belt that straddles regions of magnetic field polarity reversals in the solar atmosphere. Both coronal and solar wind stream structures evolve considerably from one solar rotation to the next as the solar magnetic field, which controls that structure, continuously evolves. It is now clear that the mysterious M-regions, hypothesized long before the era of satellite X-ray observations of the Sun, are to be identified with coronal holes, and the long-lived particle streams responsible for recurrent geomagnetic activity are to be identified with high-speed solar wind streams. (See Sun—Earth Connection.)

5. THE HELIOSPHERIC CURRENT SHEET AND SOLAR LATITUDE EFFECTS

5.1. The Sun's Large-Scale Magnetic Field and the Ballerina Skirt Model

During the declining phase of the solar activity cycle and near solar activity minimum the Sun's large-scale magnetic field well above the photosphere often appears to be approximately that of a dipole. The solar magnetic dipole is tilted with respect to the Sun's rotation axis; this tilt changes with the advance of the solar cycle. As illustrated in the left portion of Figure 12.3, near solar activity minimum the solar magnetic dipole tends to be aligned nearly with the rotation axis, while during the declining phase of activity it is generally inclined at a considerable angle relative to the rotation axis. Near solar maximum the Sun's large-scale field is probably not well approximated by a dipole.

When the solar magnetic dipole and the solar rotation axis are closely aligned, the heliospheric current sheet, which is effectively the extension of the solar magnetic equator into the solar wind, coincides roughly with the solar equatorial plane. On the other hand, at times when the dipole is tilted substantially, the heliospheric current sheet is warped and resembles a ballerina's twirling skirt, as illustrated in the right portion of Figure 12.3. Successive outward ridges in the current sheet (folds in the skirt) correspond to successive solar rotations and are separated radially by about 4.7 AU when the flow speed at the current sheet is 300 km/s. The maximum solar latitude of the current sheet in this simple picture is equal to the tilt angle of the magnetic dipole axis relative to the solar rotation axis.

5.2. Solar Latitude Effects

During the declining phase of the solar activity cycle and near solar activity minimum stream structure and solar wind variability are largely confined to a relatively narrow latitude band centered on the solar equator. This is illustrated in the upper left and right panels of Figure 12.4, which show solar wind speed as a function of solar latitude measured by Ulysses during the declining phases of the last two solar cycles. (Ulysses was in a ~ 6.2-year solar orbit that took it to solar latitudes of $\pm 80°$.) At this phase of the solar cycle the solar wind is dominated by stream structure at low latitudes, but flows at a nearly constant speed of ~ 850 km/s at high latitudes. This latitude effect is a consequence of the following: (1) solar wind properties change rapidly with distance from the heliospheric current sheet, with flow speed generally being a minimum in the vicinity of the current sheet; and (2) the heliospheric current sheet is commonly tilted relative to the solar equator, but is usually found within about $\pm 30°$ of it during the declining phase of the solar cycle. The width of the band of solar wind variability changes as the solar magnetic dipole tilt changes. The upper middle panel of Figure 12.4 demonstrates that, in contrast, in the years surrounding solar activity maximum the band of solar wind variability extends up to the highest latitudes sampled by Ulysses.

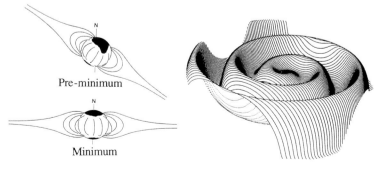

FIGURE 12.3 Right, schematic illustrating the configuration of the heliospheric current sheet when the solar magnetic dipole is tilted substantially relative to the rotation axis of the Sun. The heliospheric current sheet separates magnetic fields of opposite magnetic polarity and is the heliospheric extension of the solar magnetic equator. Left, schematic illustrating the changing tilt of coronal structure and the solar magnetic dipole relative to the rotation axis of the Sun as a function of phase of the solar activity cycle. *Adapted from Jokipii and Thomas (1981) and Hundhausen (1977).*

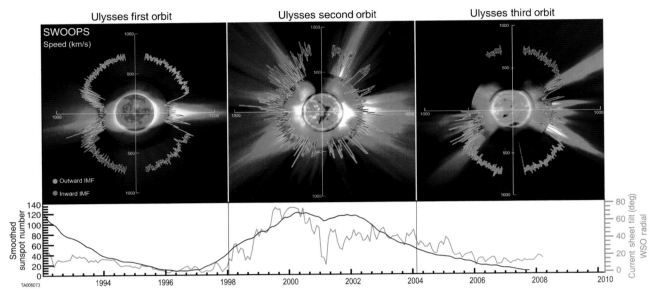

FIGURE 12.4 (Top) Polar plots of solar wind speed as a function of latitude for almost three complete Ulysses orbits about the Sun plotted over solar images characteristic of solar minimum (8/17/1996 on left; 3/28/2006 on right) and characteristic of solar maximum (12/07/2000 middle panel). Color coding of the speed plots indicates the polarity of the magnetic field. Note the reversal of the field polarities in the polar regions between the first and third orbits. In each plot the earliest times are near aphelion on the left (nine o'clock position) and time progresses counterclockwise. (Bottom) Contemporaneous values for the smoothed sunspot number (black) and the tilt of the heliospheric current sheet (red) relative to the solar equator. *From McComas et al. (2008).*

6. EVOLUTION OF STREAM STRUCTURE WITH HELIOCENTRIC DISTANCE

6.1. Kinematic Stream Steepening and the Dynamic Response

Because the coronal expansion is spatially variable, at low latitudes alternately slow and fast plasma is directed outward along any radial line from the Sun as the Sun rotates (with a period of 27 days as seen from Earth). Faster moving plasma overtakes slower moving plasma ahead while outrunning slower moving plasma behind. Because radially aligned parcels of plasma within a stream originate from different locations on the Sun, they are threaded by different magnetic field lines and thus cannot interpenetrate one another except during relatively infrequent magnetic reconnection events. The result is that the leading edges of high-speed streams steepen with increasing distance from the Sun, producing the asymmetric stream profiles obvious in Figure 12.2. As the streams steepen, plasma and field on the leading edge of a stream is compressed, causing an increase in plasma density, temperature, field strength and pressure there, while plasma and field on the trailing edge becomes increasingly rarefied. The buildup of pressure on the leading edge of a stream produces forces that accelerate the low-speed wind ahead and decelerate the high-speed wind within the stream itself. The net result is a transfer of momentum and energy from the fast-moving wind to the slow-moving wind.

6.2. Shock Formation

As long as the amplitude of a high-speed solar wind stream is sufficiently small, it gradually damps with increasing heliocentric distance in the manner just described. However, when the difference in flow speed between the crest of a stream and the trough ahead is greater than about twice the local fast mode speed, C_f (the fast mode speed is the characteristic speed with which small amplitude pressure signals propagate in a plasma: $C_f = (C_s^2 + C_A^2)^{0.5}$), ordinary pressure signals do not propagate sufficiently fast to move the slow wind out of the path of the oncoming high-speed stream. In that case the pressure eventually increases nonlinearly and **shock** waves form on either side of the high-pressure region (see Figure 12.5). The leading shock, known as a forward shock, propagates into the low-speed wind ahead and the trailing shock, known as a reverse shock, propagates back through the stream. Both shocks are, however, convected away from the Sun by the highly supersonic and super-Alfvénic flow of the wind. The major accelerations and decelerations associated with stream evolution then occur discontinuously at the shocks, giving a stream speed profile the appearance of a double sawtooth wave. The stream amplitude decreases and the compression region expands with increasing heliocentric distance as the shocks propagate. Observations indicate that the shocks often do not form until the streams are well beyond 1 AU. Nevertheless, because C_f generally decreases with

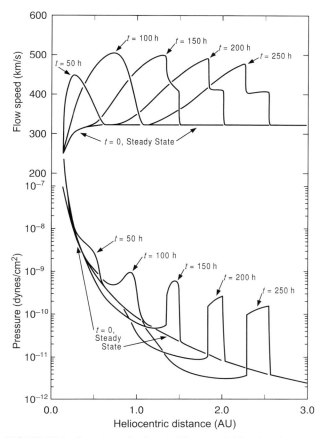

FIGURE 12.5 Snapshots of solar wind flow speed (above) and pressure (below) as functions of heliocentric distance at different times during the evolution of a large-amplitude high-speed solar wind stream as calculated from a simple one-dimensional numerical model. *Adapted from Hundhausen (1973).*

Figure 12.6 develops in the equatorial plane. This entire pattern corotates with the Sun and the compression regions are known as corotating interaction regions, or CIRs; however, only the pattern rotates—each parcel of solar wind plasma moves outward nearly radially as indicated by the black arrows. The region of high pressure associated with a CIR is nearly aligned with the magnetic field line spirals in the equatorial plane and the pressure gradients are thus nearly perpendicular to those spirals. Consequently, at 1 AU the pressure gradients that form on the rising speed portions of high-speed streams have transverse as well as radial components. In particular, not only is the low-speed plasma ahead of a high-speed stream accelerated to a higher speed, but also it is deflected in the direction of solar rotation. In contrast, the high-speed plasma near the crest of the stream is both decelerated and deflected in the direction opposite to solar rotation. These transverse deflections produce the systematic west–east flow direction changes observed near the leading edges of quasi-stationary high-speed streams (see Figure 12.2).

There is an interesting three-dimensional aspect to stream evolution, ultimately associated with the fact that the solar magnetic dipole typically is tilted relative to the solar rotation axis. That tilt causes CIRs in the northern and southern solar hemispheres to have opposed meridional tilts that, particularly beyond about 3 AU, can be discerned in plasma data as systematic north–south deflections of the flow at CIRs. The meridional tilts are such that the forward waves in both hemispheres propagate toward the opposite

increasing heliocentric distance, virtually all large-amplitude solar wind streams steepen into shock wave structures at heliocentric distances beyond ~3 AU. At heliocentric distances immediately beyond the orbit of Jupiter (~5.4 AU) a large fraction of the mass in the solar wind is found within compression regions bounded by shock waves on the rising portions of damped high-speed streams. The basic structure of the solar wind in the solar equatorial plane in the distant heliosphere thus differs considerably from that observed at 1 AU. Stream amplitudes are severely reduced and short wavelength structure is damped out. The dominant structures at low latitudes in the outer heliosphere are expanding compression regions that interact and merge with one another to form what are commonly called global merged interaction regions.

6.3. Stream Evolution in Two and Three Dimensions

When the coronal expansion is spatially variable but time stationary, a steady flow pattern such as sketched in

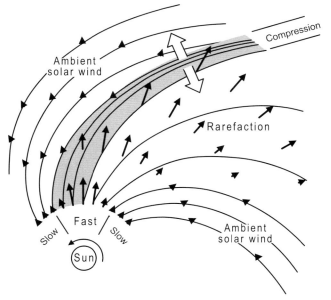

FIGURE 12.6 Schematic illustrating two-dimensional quasi-stationary stream structure in the ecliptic plane in the inner heliosphere. The compression region on the leading edge of a stream is nearly aligned with the spiral magnetic field and the forces associated with the pressure gradients have transverse as well as radial components. *From Pizzo (1978).*

hemispheres, whereas the reverse waves in both hemispheres propagate poleward. As a result, forward shocks in the outer heliosphere near solar minimum are generally confined to the low-latitude band of solar wind variability, while the reverse shocks are commonly observed both within the band of variability and poleward of it. However, the reverse waves seldom reach latitudes more than ~15° above the low-latitude band of variability.

7. CORONAL MASS EJECTIONS AND TRANSIENT SOLAR WIND DISTURBANCES

7.1. Coronal Mass Ejections

The solar corona evolves on a variety of timescales closely connected with the evolution of the coronal magnetic field. (See The Sun.) The most rapid and dramatic evolution in the corona occurs in events known as **coronal mass ejections** or **CMEs** (Figure 12.7(a)). CMEs originate in closed field regions in the corona where the magnetic field normally is sufficiently strong to constrain the coronal plasma from expanding outward. Typically these closed field regions are found in the coronal streamer belt that encircles the Sun and that underlies the heliospheric current sheet. The outer edges of CMEs often have the optical appearance of closed loops such as the event shown in Figure 12.7(a). Few CMEs ever appear to sever completely their magnetic connection with the Sun. During a typical CME, somewhere between 10^{15} and 10^{16} g is ejected into the heliosphere. Ejection speeds near the Sun range from less than 50 km/s in some of the slower events to greater than 2500 km/s in the fastest ones. The average CME speed at ~5 solar radii is close to the median ecliptic solar wind speed of ~440 km/s. Since observed solar wind speeds near 1 AU are never less than ~280 km/s, the slowest CMEs are further accelerated enroute to 1 AU.

7.2. Origins, Associations with Other Forms of Solar Activity, and Frequency of Occurrence

The processes that trigger CMEs and that determine their sizes and outward speeds are only poorly understood; there is presently no consensus on the physical processes responsible for initiating or accelerating these events, although it is clear that stressed magnetic fields are the underlying cause of these events and that CMEs play a fundamental role in the long-term evolution of the structure of the solar corona. They appear to be an essential part of the way the corona responds to the evolution of the solar magnetic field associated with the advance of the solar activity cycle. Indeed, the release of a CME is one way that the solar atmosphere reconfigures itself in response to changes in the solar magnetic field. CMEs are commonly, but not always, observed in association with other forms of solar activity such as eruptive prominences and solar flares.

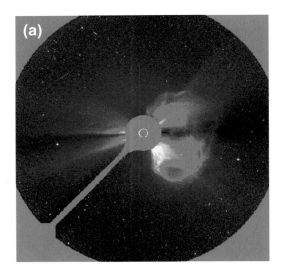
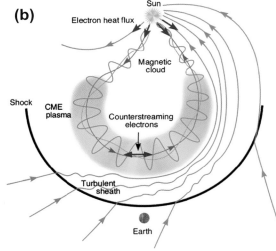

FIGURE 12.7 (a) A CME as imaged by the LASCO/C3 coronagraph on SOHO on 20 April 1998. The Sun, indicated by the white circle, has been occulted within the instrument. The field of view of the image is 30 solar diameters. (The SOHO/LASCO data are produced by a consortium of the Naval Research Laboratory (USA), Max-Planck-Institut fur Sonnensystemforschung (Germany), Laboratoire d'Astrophysique de Marseille (France), and the University of Birmingham (UK). SOHO is a project of international cooperation between ESA and NASA.) (b) A sketch of a solar wind shock disturbance produced by a fast ICME directed toward Earth. Red and magenta arrows indicate the ambient magnetic field and that threading the ICME, respectively. Blue arrows indicate the suprathermal electron strahl flowing away from the Sun along the magnetic field. The ambient magnetic field is compressed by its interaction with the ICME and is forced to drape around the ICME. *Adapted from a sketch originally in Crooker (2000) and from Zurbuchen and Richardson (2006).*

From a historical perspective one might be led to expect that large solar flares are the prime cause of CMEs; however, it is now clear that flares and CMEs are separate, but closely related, phenomena associated with magnetic disturbances on the Sun. Like other forms of solar activity, CMEs occur with a frequency that varies in a cycle of ~11 years. On average, the Sun emits about 3.5 CMEs/day near the peak of the solar activity cycle, but only about 1 CME every 10 days near solar activity minimum.

7.3. Heliospheric Disturbances Driven by Fast CMEs

As illustrated in Figure 12.7(b), fast CMEs produce transient solar wind disturbances that, in turn, often are the cause of large geomagnetic storms and major space weather events in general. (See Sun—Earth Connections.) Figure 12.8 shows calculated radial speed and pressure profiles of a simulated solar wind disturbance driven by a fast CME at the time the disturbance first reaches 1 AU. As indicated by the insert in the top portion of the figure, the disturbance was initiated at the inner boundary of the one-dimensional fluid calculation by abruptly raising the flow speed from 275 to 980 km/s, sustaining it at this level for 6 h, and then returning it to its original value of 275 km/s. The initial disturbance thus mimics a uniformly fast spatially limited CME with an internal pressure equal to that of the surrounding solar wind plasma. A region of high pressure develops on the leading edge of the disturbance as the CME overtakes the slower wind ahead. This region of higher pressure is bounded by a forward shock on its leading edge that propagates into the ambient solar wind ahead and by a reverse shock on its trailing edge that propagates backward into and eventually through the CME. Both shocks are, however, carried away from the Sun by the highly supersonic flow of the solar wind. A rarefaction forms on the trailing edge of the disturbance as the CME outruns slower plasma behind. This rarefaction produces a deceleration of the rear portion of the CME and an acceleration of the trailing wind. The overall interaction with the ambient wind produces an expanding CME whose radial width at ~0.8 AU is greater than its width when introduced into the simulation.

Except for the reverse shock, which observations and more detailed three-dimensional calculations indicate are ordinarily present only near the central portions of the disturbances, the simple calculation shown in Figure 12.8 is consistent with observations of many solar wind disturbances obtained near 1 AU in the ecliptic plane and illustrates to first order the radial and temporal evolution of an interplanetary disturbance driven by a fast CME (now commonly called an interplanetary coronal mass ejection, ICME, when observed in the solar wind). In the example illustrated, the ICME slows from an initial speed of 980 km/s to less than 600 km/s by the time the leading edge

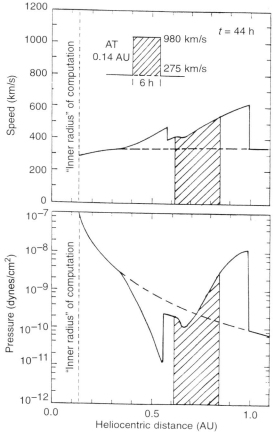

FIGURE 12.8 Solar wind speed and pressure as functions of heliocentric distance for a simple one-dimensional gas-dynamic simulation of a CME-driven disturbance. The dashed line indicates the steady state before introduction of the temporal variation in flow speed imposed at the inner boundary of 0.14 AU and shown at the top of the figure. The hatching identifies material that was introduced with a speed of 980 km/s at the inner boundary, and therefore identifies the CME in the simulation. *Adapted originally from Hundhausen (1985).*

of the disturbance reaches 1 AU. This slowing is a result of momentum transfer to the ambient solar wind ahead and behind and proceeds at an ever-slower rate as the disturbance propagates outward. Figure 12.9 displays selected plasma and magnetic field data from a solar wind disturbance driven by an ICME observed near 1 AU. The shock is distinguished in the data by discontinuous increases in flow speed, density, temperature and field strength. The plasma identified as the ICME had a higher flow speed than the ambient solar wind ahead of the shock. In this case it was also distinguished by counterstreaming suprathermal electrons (indicative of a closed magnetic field topology, see Section 10.3), anomalously low proton temperatures, somewhat elevated helium abundance, and a strong smoothly rotating magnetic field that indicates that the field topology was that of a **magnetic flux rope** (see Figure 12.7(b)).

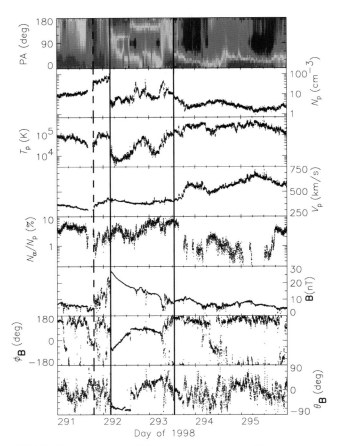

FIGURE 12.9 A solar wind disturbance associated with a moderately fast ICME observed by the Advanced Composition Explorer (ACE) in October 1998. From top to bottom the quantities plotted are color-coded pitch angle distributions of 256–288 eV electrons, proton density, proton temperature, bulk flow speed, alpha-proton density ratio, and magnetic field strength, azimuth, and polar angle in solar ecliptic coordinates. The color scale for $f(v)$ extends from 5×10^{-32} (dark purple) to 2×10^{-29} s^3/cm^6 (dark red). Dashed and solid vertical lines respectively mark the shock and the edges of the ICME. *From Gosling et al. (2002).*

7.4. Characteristics of ICMEs

The identification of ICMEs in solar wind plasma and field data is still something of an art; however, shocks serve as useful fiducials for identifying fast ICMEs. Table 12.2 provides a summary of plasma and field signatures that qualify as unusual compared with the normal solar wind, but that are commonly observed as a number of hours after shock passage. Most of these anomalous signatures are observed elsewhere in the solar wind as well where, presumably, they serve to identify those numerous relatively low-speed ICMEs that do not drive shock disturbances. Few ICMEs at 1 AU exhibit all of these characteristics, and some of these signatures are more commonly observed than are others.

Most ICMEs expand as they propagate outward through the heliosphere. ICME radial thicknesses are variable; at 1 AU the typical ICME has a radial width of ~ 0.2 AU whereas at Jupiter's orbit ICMEs can have radial widths as large as 2.5 AU. Magnetic reconnection occurs relatively rarely at the leading and/or trailing edges of ICMEs, but when it does occur there it erodes away portions of those ICMEs. Approximately one-third of all ICMEs in the ecliptic plane have sufficiently high speeds relative to the ambient solar wind to drive shock disturbances at 1 AU; the slower ICMEs do not drive shock disturbances and simply coast along with the rest of the solar wind. Typically ICMEs cannot be distinguished from the normal solar wind at 1 AU on the basis of either their speed or density (the event in Figure 12.9 is an example). Near solar activity maximum ICMEs account for 15–20% of the solar wind in the ecliptic plane at 1 AU, while near solar activity minimum they account for less than 1%. The Earth intercepts about 72 ICMEs/yr near solar activity maximum and ~ 8 ICMEs/yr near solar activity minimum. ICMEs are much less common at high heliographic latitudes, particularly near activity minimum when ICMEs are confined largely to the low-latitude band of solar wind variability.

7.5. The Magnetic Field Topology of ICMEs and the Problem of Magnetic Flux Balance

The coronal expansion carries a portion of the solar magnetic field outward to form the heliospheric magnetic field. In the quiescent wind these field lines are usually "open" in

TABLE 12.2 Characteristics of ICMEs at 1 AU

Common signatures:
Counterstreaming (along the field) suprathermal electrons (energy > 70 eV)
Counterstreaming (along the field) energetic (energy > 20 keV) protons
Helium abundance enhancement
Anomalously low proton and/or electron temperatures
Strong magnetic field
Low *plasma beta*
Low magnetic field strength variance
Anomalous field rotation (magnetic flux rope)
Anomalous ionic composition (for example, Fe^{16+}, He^+)
Cosmic ray depression
Average radial thickness: 0.2 AU
Range of speeds: 300–2000 km/s
Single point occurrence frequency:
~ 72 events/yr at solar activity maximum
~ 8 events/yr at solar activity minimum
Magnetic field topology: Predominantly closed magnetic loops rooted in Sun
Fraction of events driving shocks: $\sim 1/3$
Fraction of earthward-directed events producing large geomagnetic storms: $\sim 1/6$

the sense that they connect to field lines of the opposite polarity only in the very distant heliosphere. CMEs, on the other hand, originate in closed field regions in the corona not previously participating directly in the solar wind expansion and inject new closed magnetic flux into the heliosphere. In the absence of magnetic reconnection, such injections would lead to a continual buildup of magnetic flux in the heliosphere, which is not observed: measurements reveal that solar rotation averages of the heliospheric magnetic field strength in the ecliptic plane at 1 AU vary by a factor of about 2 roughly in phase with the 11-year solar activity cycle. That variation is determined by the competition between flux ejection into the heliosphere by CMEs and flux removal by reconnection in the magnetic legs of CMEs and at the heliospheric current sheet. Such reconnection is effective in reducing the magnetic flux in the heliosphere only when it occurs sunward of the **Alfvén point** where the solar wind flow becomes super-Alfvénic. It remains to be determined which of these reconnection sites is dominant in balancing the new flux ejected into the heliosphere by CMEs.

Figure 12.10 illustrates that reconnection within the magnetic legs of CMEs is inherently three-dimensional in nature and, when occurring between adjacent magnetic loops, produces helical magnetic field lines that are partially disconnected from the Sun as well as new closed field lines relatively low in the solar atmosphere. Sustained three-dimensional magnetic reconnection in the magnetic legs eventually produces a mixture of closed, open, and disconnected field lines threading an ICME as well as additional closed magnetic loops low in the solar atmosphere. All of the types of reconnection illustrated in Figure 12.10 reduce the amount of magnetic flux that a CME adds to the heliosphere and all of the magnetic topologies produced by such reconnection are apparent in suprathermal electron observations of various ICMEs at 1 AU. Recent sequences of coronagraph and heliospheric images indicate that reconnection also commonly occurs at the heliospheric current sheet inside the Alfven point, producing heliospheric field lines that are disconnected from the Sun.

7.6. Field Line Draping About Fast ICMEs

Because ICME plasma and ambient wind plasma are threaded by different field lines, they cannot, in general, interpenetrate one another. Consequently, the ambient plasma and magnetic field ahead must be deflected away from the path of a fast ICME in much the same manner as the solar wind is deflected around Earth's magnetosphere. Figure 12.7(b) illustrates that such deflections cause the ambient magnetic field to drape about the ICME. The degree of draping and the resulting orientation of the field ahead of an ICME depend upon the relative speed between the ICME and the ambient plasma, the shape of the ICME and the original orientation of the magnetic field in the ambient plasma. Draping plays an important role in reorienting the magnetic field ahead of a fast ICME. On the other hand, conditions and processes back at the Sun largely determine field orientations within ICMEs. As a final point of interest, Figure 12.7(b) also illustrates that, just as the bow wave in front of a boat moving through water is considerably broader in extent than is the boat that produces it, so too is the shock in front of a fast ICME somewhat broader in extent than is the ICME that drives it. As a result, spacecraft often encounter ICME-driven shocks without also encountering the ICMEs that drive them.

8. VARIATION WITH DISTANCE FROM THE SUN

For a structureless solar wind, the speed remains nearly constant beyond the orbit of Earth, the density falls off with heliocentric distance, r, as r^{-2}, and the magnetic field decreases with distance as described by the equations in Section 2. The temperature also decreases with increasing heliocentric distance due to the spherical expansion of the plasma; however, the precise nature of the decrease depends upon particle species and the relative importance of such things as collisions, heat conduction, turbulence dissipation, and plasma instabilities. Protons and electrons evolve differently with increasing heliocentric distance. For an adiabatic expansion of an isotropic plasma the temperature falls off as $r^{-4/3}$; for a plasma dominated by heat conduction the temperature falls as $r^{-2/7}$. Observations reveal that both proton and electron temperatures inside

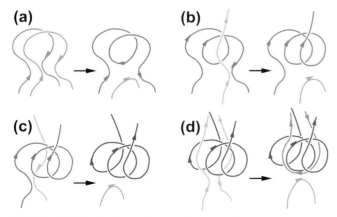

FIGURE 12.10 Sketches of successive steps in three-dimensional magnetic reconnection in the corona beneath a departing CME. The sketches are not to scale and are intended only to illustrate successive changes in CME/ICME magnetic topologies resulting from reconnection. *From Gosling et al. (1995).*

5 AU decrease with distance somewhere between the adiabatic and conduction-dominated extremes.

Of course, the solar wind is not structureless. The continual interaction of high- and low-speed flows with increasing heliocentric distance produces a radial variation of speed that differs considerably from that predicted for a structureless wind. High-speed flows decelerate and low-speed flows accelerate with increasing heliocentric distance as a result of momentum transfer (see Sections 6 and 7). Consequently, at low solar latitudes far from the Sun (beyond ~ 15 AU) the solar wind flows at 400–500 km/s most of the time (Figure 12.11). Only rarely are substantial speed perturbations observed at these distances; these relatively rare events usually are associated with disturbances driven by very large and fast ICMEs that require a greater-than-usual distance to share their momentum with lower speed wind.

9. TERMINATION OF THE SOLAR WIND

Interstellar space is filled with a dilute gas of neutral and ionized particles and is threaded by a weak magnetic field. In the absence of the solar wind, the interstellar plasma would penetrate deep into the solar system in the same fashion as do interstellar neutral particles. However, because of the magnetic fields embedded in both, the interstellar and solar wind plasmas cannot easily interpenetrate one another. The result is that the solar wind creates a cavity in the interstellar plasma; however, magnetic reconnection at the boundary between the two plasmas may allow limited interpenetration of the interstellar and solar wind plasmas in thin layers there.

The details of the solar wind's interaction with the interstellar plasma are still somewhat speculative largely because, until recently, we lacked direct observations of this interaction. Figure 12.12 shows what are believed to be the major elements of the interaction. The Sun and heliosphere move at a speed of ~ 23 km/s relative to the interstellar medium. A bow wave must stand in the interstellar plasma upstream of the heliosphere to initiate the slowing and deflection of the plasma around the heliosphere. Recent remote measurements indicate that this bow wave is probably not a shock, although labeled that way in Figure 12.12. The **heliopause** is the outermost boundary of the heliosphere. Sunward of the heliopause is a **termination shock** where the solar wind flow becomes subsonic so that it ultimately can be turned to flow roughly parallel to the heliopause. Direct observations of the termination shock by the two Voyager spacecraft reveal that the termination shock is unusual in the sense that the bulk solar wind plasma is

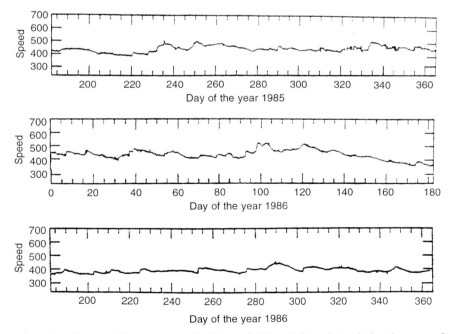

FIGURE 12.11 Solar wind speed as a function of time as measured by Voyager 2 during a 1.5-year interval when the spacecraft was beyond 18 AU from the Sun. Because stream amplitudes are severely damped at large distances from the Sun, the solar wind speed there generally varies within a very narrow range of values. Compare with the speed variations evident in Figure 12.2 that were obtained at 1 AU during a comparable period of the solar cycle. *Adapted from Lazarus and Belcher (1988).*

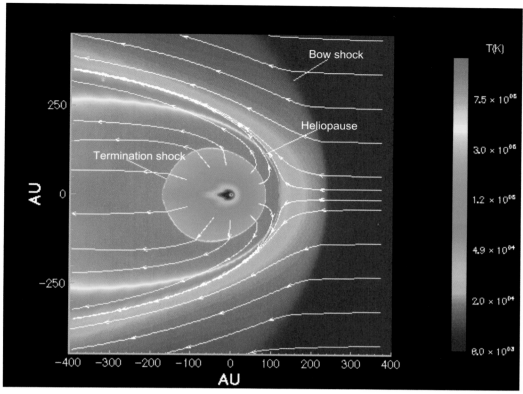

FIGURE 12.12 Simulated structure of the solar wind's interaction with the interstellar plasma. Color coding represents the proton temperature and arrows indicate the direction of the solar wind and interplanetary plasma flows. *Adapted from Zank and Mueller (2003).*

significantly slowed as it crosses the shock but is only slightly heated. A large fraction of the energy associated with the bulk plasma motion lost at the shock is transferred to suprathermal and energetic particles, which dominate the solar wind's internal pressure in the outer heliosphere.

The shape of the heliosphere is asymmetric because of its motion relative to the interstellar gas; it is compressed in the direction of that motion and is greatly elongated in the opposite direction. The shape is also affected by the draping of the interstellar magnetic field about the heliopause. Observations in the outer heliosphere indicate that the termination shock is constantly in motion relative to the Sun, owing at least in part to an ever-changing solar wind momentum flux. The size and shape of the heliosphere depend on the momentum flux carried by the solar wind, the dynamic pressure of the interstellar plasma, the strength and orientation of the interstellar magnetic field, and the motion of the heliosphere relative to the interstellar medium. Voyager 1 crossed the termination shock in December 2004 at a heliocentric distance of about 94 AU; Voyager 2 crossed it in August 2007 at a distance of about 84 AU. Both Voyager trajectories are directed within about 45° of the heliosphere's motion relative to the interstellar medium.

It is currently believed that the nose of the heliopause lies at a heliocentric distance of about 120 AU.

Voyager 1 observed significant and, ultimately, long-lasting changes in the magnetic field and in energetic particle populations beginning on 26 July 2012 at a heliocentric distance of ~121 AU. A recent interpretation of those observations that invokes reconnection between the heliospheric and interstellar magnetic fields suggests that those changes signaled Voyager 1's crossing of the heliopause. That interpretation is consistent with subsequent plasma wave measurements of large increases in plasma density after 26 July 2012.

10. KINETIC PROPERTIES OF THE PLASMA

10.1. The Solar Wind as a Marginally Collisional Plasma

On a large scale the solar wind behaves like a compressible fluid and is capable of supporting relatively thin structures such as shocks. It is perhaps not obvious why the solar wind should exhibit this fluid-like behavior since the wind is a dilute plasma in which collisions are relatively rare. For example, using values given in Table 12.1, we find that the time between collisions for a typical solar wind proton at 1 AU is several days. (These collisions do not result from direct particle impacts such as colliding billiard balls, but rather from the long distance **Coulomb interactions** characteristic of charged particles.) The time between collisions is thus comparable with the time for the solar wind to expand from the vicinity of the Sun to 1 AU; this is the basis for statements that the solar wind is a marginally collisional plasma.

There are several reasons why the solar wind behaves like a fluid even in the absence of particle collisions to effect fluid-like behavior. First, when the temperature is low and the density is high collisions are more frequent than noted above. Second, the presence of the heliospheric magnetic field causes charged particles to gyrate about the field and they thus do not travel in straight lines between collisions. For typical conditions at 1 AU, solar wind electrons and protons have **gyro radii** of ~ 1.4 km and ~ 60 km respectively, which are small compared with the scale size of most structures in the solar wind. Third, the solar wind plasma is subject to a variety of instabilities that are triggered whenever particle distribution functions depart significantly from Maxwellian distributions (see next section). These instabilities produce collective interactions that mimic the effects of particle collisions. Finally, because the magnetic field is frozen into the solar wind flow, parcels of plasma originating from different positions on the Sun cannot interpenetrate one another in the absence of magnetic reconnection.

10.2. Kinetic Aspects of Solar Wind Ions

Collisional gases can usually be described by a single isotropic (i.e. the same in all directions) temperature, T, with the distribution of particle speeds, v, obeying the Maxwellian $f(v) \sim \exp(-m(v-V_0)^2/2\,kT)$, where f is the number of particles per unit volume of velocity space, k is Boltzman's constant (1.38×10^{-23} J/K), m is the particle mass and V_0 is the bulk speed of the gas. In contrast, proton distribution functions in the solar wind are usually anisotropic because of the paucity of collisions and because the magnetic field provides a preferred direction in space. Moreover, solar wind proton and alpha particle distributions often exhibit significant non-Maxwellian features such as the double-peaked distributions illustrated in Figure 12.13(a). The secondary proton and alpha particle peaks are associated with beams streaming relative to the main solar wind component along the heliospheric magnetic field. The relative streaming speed of such beams is usually comparable with or less than the local Alfvén speed, suggesting that the streaming is limited by an ion beam streaming instability. Closer to the Sun, where the Alfvén speed is higher, relative streaming speeds between the beams and the main components can be as large as several 100 km/s. Secondary proton beams are common in the solar wind in both low- and high-speed flows and may play a role in the overall acceleration and heating of the wind; however, their origin in solar and/or heliospheric processes is presently uncertain. Figure 12.13(b) illustrates that solar wind ion distributions in the low-speed wind also commonly have extended nonthermal tails of uncertain origin. Particles in these extended tails are easily accelerated to much higher energy when they encounter shocks (see Section 12).

10.3. Kinetic Aspects of Solar Wind Electrons

Electron distributions in the solar wind consist of a relatively cold and dense thermal "core" population that is electrically bound to the solar wind ion population and a much hotter and freer running suprathermal population that becomes collisionless close to the Sun. At 1 AU the breakpoint between these populations typically occurs at an energy of ~ 70 eV (Figure 12.14(a)). This breakpoint moves steadily to lower energies with increasing heliocentric distance as the core population cools. Typically the core contains about 95% of the electrons, and at 1 AU has a temperature of $\sim 1.3 \times 10^5$ K. The core electrons typically are mildly anisotropic, with the temperature parallel to the field exceeding the temperature perpendicular to the field by a factor of ~ 1.1 on average at 1 AU. However, the temperature anisotropy for core electrons varies systematically with density such that at very low densities (<2 cm^{-3}) the temperature ratio often exceeds 2.0, while at very high densities (>10 cm^{-3}) the temperature ratio is often slightly less than 1.0. Such systematic variations of core electron temperature anisotropy with plasma density reflect the marginally collisional nature of the thermal electrons and their nearly adiabatic expansion in the spiral magnetic field.

The suprathermal electrons consist of a beam of variable width and intensity, known as the "strahl," directed outward from the Sun along the heliospheric magnetic field and a more tenuous and often roughly isotropic "halo" (Figure 12.14(b)). The angular width of the strahl results from a competition between focusing associated with conservation of an electron's magnetic moment in the diverging heliospheric magnetic field and defocusing associated with particle scattering. The strahl carries the solar wind electron heat flux; variations in strahl intensity largely reflect spatial variations in the corona from which it arises. In addition, brief (hours) strahl intensifications

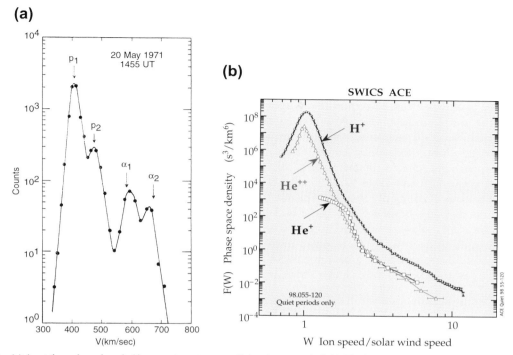

FIGURE 12.13 (a) A cut through a solar wind ion count spectrum parallel to the magnetic field. The first two peaks are protons and the second two peaks are alpha particles. (The velocity scale for the alpha particles has been increased by a factor of 1.4.) Both the proton and alpha particle spectra show clear evidence for a secondary beam of particles streaming along the field relative to the main solar wind beam at about the Alfvén speed. Such secondary beams, not always well resolved, are common in both the low and the high-speed wind (Asbridge et al., 1974). (b) Solar wind speed distributions of H^+, He^{2+} and He^+ observed in the low-speed solar wind at 1 AU, averaged over a 65-day period in 1998 and excluding intervals of shocks and other disturbances. Such extended suprathermal tails appear to be ubiquitous in the low-speed solar wind. The He^+ ions are primarily of interstellar origin. *From Gloeckler et al. (2000).*

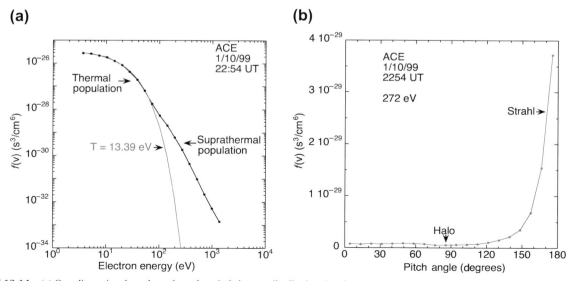

FIGURE 12.14 (a) One-dimensional cut through a solar wind electron distribution showing the thermal and suprathermal populations. (b) Suprathermal electron pitch angle distribution (relative to the magnetic field) showing the field-aligned strahl and a nearly isotropic halo.

commonly occur during solar electron bursts associated with solar activity (see Section 12). The strahl serves as an effective tracer of magnetic field topology in the interplanetary medium since its usual unidirectional nature arises because field lines in the normal solar wind are "open" (see Section 7.5) and are thus effectively connected to the solar corona at only one end. In contrast, field lines threading ICMEs are often attached to the Sun at both ends (see Sections 7.4 and 7.5), and counterstreaming strahls are commonly observed there. Indeed, counterstreaming strahls are one of the more reliable signatures of ICMEs (see Figures 12.7(b) and 12.9 and Table 12.2). Finally, at 1 AU the electron halo results primarily from backscattered strahl electrons from distances beyond 1 AU. Those backscattered electrons are subsequently mirrored (i.e. magnetically reflected) inside 1 AU by the stronger magnetic fields that reside there, producing a halo population that often is roughly isotropic, as in the example shown in Figure 12.14(b).

11. HEAVY ION CONTENT

Although the solar wind consists primarily of protons (hydrogen), electrons, and alpha particles (doubly ionized helium), it also contains traces of ions of a number of heavier elements. Table 12.3 provides estimates of the relative abundances of some of the more common solar wind elements summed over all ionization states. After hydrogen and helium, the most abundant elements are carbon and oxygen. The ionization states of all solar wind ions are "frozen in" close to the Sun because the characteristic times for ionization and recombination are long compared with the solar wind expansion time. Commonly observed ionization states include He^{2+}, C^{5+}, C^{6+}, $O^{6+}-O^{8+}$, $Si^{7+}-Si^{10+}$, and $Fe^{8+}-Fe^{14+}$. Ionization state temperatures in the low-speed wind are typically in the range $1.4-1.6 \times 10^6$ K, while ionization state temperatures in the high-speed wind are typically in the range $1.0-1.2 \times 10^6$ K. Unusual ionization states such as Fe^{+16} and He^{+1}, which are not common in the normal solar wind, are often abundant within ICMEs, reflecting the unusual coronal origins of those events.

The relative abundance values in Table 12.3 are long-term averages; however, abundances vary considerably with time. Such variations have been extensively studied for the He^{2+}/H^+ ratio, A(He). The most probable A(He) value is ~ 0.045, but the A(He) ranges from less than 0.01 to values of 0.35 on occasion. The average A(He) is about twice that commonly attributed to the solar interior, for reasons presently unknown. Much of the variation in A(He) and in the abundance of heavier elements is related to the large-scale structure of the wind. For example, Fe/O and Mg/O ratios are systematically lower in high-speed streams than in low-speed flows. A(He) tends to be relatively constant at ~ 0.045 within the cores of high-speed streams from coronal holes, but tends to be highly variable within low-speed flows. Particularly low (<0.02) abundance values are commonly observed in the vicinity of the heliospheric current sheet. A(He) values greater than about 0.10 are relatively rare and account for less than 1% of all the measurements. At 1 AU enhancements in A(He) above 0.10 occur almost exclusively within ICMEs. The physical causes of these variations are uncertain for the most part, although thermal diffusion, gravitational settling, and Coulomb friction in the chromosphere and corona all probably play roles.

12. ENERGETIC PARTICLES

A proton moving with a speed of 440 km/s has an energy of ~ 1 keV. Thus, by most measures solar wind ions are low-energy particles. The heliosphere is, nevertheless, filled with a number of energetic ion populations of varying intensities with energies ranging upward from ~ 1 keV/nucleon to $\sim 10^8$ keV/nucleon. These populations include galactic cosmic rays, anomalous cosmic rays (see discussion that follows), and energetic particles associated with CIRs, CMEs, solar flares, and the planetary bow shocks. All but the galactic cosmic rays are energized within the heliosphere.

Shocks are particularly effective particle accelerators and all but one of the above populations have shock origins. The physical process by which a collisionless shock accelerates a small fraction of the ions it intercepts to high energy is reasonably well understood, although complex

TABLE 12.3 Average Elemental Abundances in the Solar Wind

Element	Abundance Relative to Oxygen
H	1900 ± 400
He	75 ± 20
C	0.67 ± 0.10
N	0.15 ± 0.06
O	1.00
Ne	0.17 ± 0.02
Mg	0.15 ± 0.02
Si	0.19 ± 0.04
Ar	0.0040 ± 0.0010
Fe	$0.19 + 0.10 - 0.07$

in detail. The effectiveness of the acceleration process depends upon factors such as shock speed and strength, the angle between the magnetic field and the shock normal, time of field line connection to the shock, and the local reservoir of particles available for acceleration. Recent work indicates that shocks in the solar wind most easily accelerate ions that already exceed solar wind thermal energies when they encounter the shocks. These so-called "seed" particles include the suprathermal ion tails always present in the low-speed wind (Figure 12.13(b)), but also "pickup ions" (see discussion that follows), and energetic particles remaining in the heliosphere from previous solar flares and CME-driven disturbances.

Anomalous cosmic rays have energies per nucleon that are lower than that of galactic cosmic rays, are predominantly singly ionized H, He, N, O, and Ne and, like galactic cosmic rays, have an intensity that varies slowly with time. They are associated with a particularly interesting seed population—neutral atoms from the local interstellar cloud that penetrate deep into the heliosphere. As the neutrals approach the Sun some of them are ionized by solar EUV radiation, electron impact, or charge exchange with solar wind protons, are then picked up by the solar wind magnetic field (the pickup process accelerates them to ~ 4 keV/nucleon), and are swept into the outer reaches of the heliosphere by the solar wind flow. It long has been thought that the pickup ions are accelerated to high energies as they encounter the termination shock and then diffuse back into the interior of the heliosphere as anomalous cosmic rays. However, this idea was questioned when the Voyager 1 spacecraft observed that anomalous cosmic ray intensities continued to increase well after the spacecraft crossed the termination shock. One possible explanation for this unexpected result is that the pickup ions are accelerated primarily along the far flanks of the termination shock where the Archimedean spiral magnetic field line connection times are considerably longer than in the region around the nose of the shock where the Voyagers crossed it.

Of the energetic ion populations in the heliosphere that associated directly with solar flares appears to be the only population that is not obviously shock associated, although even in this case shock acceleration cannot be ruled out conclusively. Flare events are usually impulsive and short lived (hours), are overabundant in ^{3}He, appear to originate relatively low in the solar atmosphere, occur at a rate of ~ 1000 events/year near solar activity maximum and generally occur in association with impulsive energetic solar electron bursts. The latter have energies ranging from several hundred eV up to several hundred keV. Recent work suggests that solar electron bursts originate at a variety of altitudes in the solar atmosphere and can be triggered by more than one process.

13. TURBULENCE AND MAGNETIC FIELD AND VELOCITY FLUCTUATIONS

Figure 12.15 illustrates that fluctuations in velocity and magnetic field are observed throughout the solar wind at 1 AU on a variety of spatial and temporal scales; however, fluctuation amplitudes tend to be greatest on the rising speed portions and within the cores of high-speed streams from coronal holes. As illustrated in Figure 12.16(a), velocity and magnetic field fluctuations in the high-speed wind are largely Alfvénic (coupled changes in velocity and magnetic field components). These Alfvénic fluctuations propagate predominantly anti-sunward in the solar wind rest frame, indicating that they are largely remnants of Alfvénic fluctuations present close to the Sun. In contrast, as illustrated in Figure 12.16(b), in the low-speed wind and on the rising speed portions of high-speed streams fluctuations in velocity and magnetic field often are not coupled to one another. Presumably, such non-Alfvénic fluctuations do not propagate in the solar wind rest frame. Particularly sharp changes in field orientation, such as that at 1853 UT in Figure 12.16(b), are where magnetic reconnection commonly occurs in the solar wind. Both Alfvénic and non-Alfvénic fluctuations typically appear to be stochastic in nature, reflecting the turbulent nature of the solar wind flow, and have amplitudes that decrease with increasing heliocentric distance. The turbulent nature of the fluctuations is graphically illustrated in Figure 12.17, which shows a

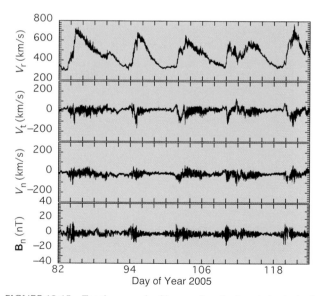

FIGURE 12.15 Top three panels: 64-s samples of solar wind velocity in spacecraft centered **r**, **t**, **n** coordinates, where **r** is the Sun to spacecraft unit vector, **t** is the unit vector in the direction of $\Omega \times \mathbf{r}$, and **n** completes a right-handed system. Here Ω is the spin axis vector. Bottom panel: 64-s averages of the n-component of the magnetic field. The 42-day interval shown is the same as in Figure 12.2. Fluctuations in the other two field components are similar to those in $\mathbf{B}_n$.

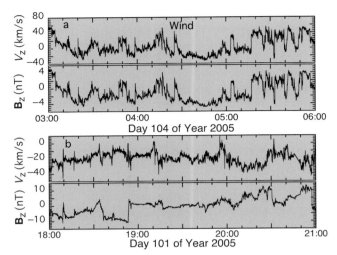

FIGURE 12.16 (a) Three-second samples of the geocentric solar eclipt ic z-components of velocity and magnetic field during a 3-hr interval within a high-speed stream on Day 114 2005. Similar coupled velocity and magnetic field fluctuations occurred in the other two components. (b) Three-second samples of geocentric solar ecliptic z-components of velocity and magnetic field during a 3-hr interval within the low-speed wind on Day 111 2005. Similar uncoupled fluctuations occurred in the other two velocity and magnetic field components.

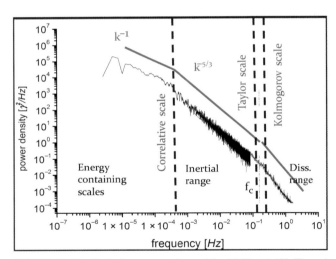

FIGURE 12.17 Typical power spectrum of the HMF at 1 AU. The red line is a schematic representation illustrating the characteristic spectral slopes in the different frequency ranges. The correlative scale is the largest separation distance over which turbulent eddies are correlated. The Taylor scale is the scale size at which dissipation begins to affect the eddies. It marks the transition from the inertial range to the dissipation range. The Kolmogorov scale characterizes the smallest dissipation-scale eddies (See Figure 25 in Bruno and Carbone (2013)).

representative power spectrum of fluctuations in the heliospheric magnetic field obtained near Earth's orbit. The spectral slope varies in a fashion characteristic of a turbulent cascade of energy via eddies from the lowest frequencies (the energy containing scales) through the so-called inertial range, down to the highest frequencies where the turbulent energy is dissipated. In addition to its probable, but poorly understood, role in heating and accelerating the solar wind, turbulence strongly affects energetic particle transport in the heliosphere and is an essential element of most current models of particle acceleration at shocks in the heliosphere.

14. CONCLUSION

The solar wind is a magnificent natural laboratory for studying and obtaining understanding of processes and phenomena that also occur in a variety of other astrophysical contexts. These include kinetic and fluid aspects of plasmas, plasma heating and acceleration, collisionless shock physics, particle acceleration and transport, magnetic reconnection, and turbulence and waves. Proof of the existence of the solar wind was one of the first great triumphs of the space age, and much has been learned about the physical nature of the wind and related processes in intervening years. Nevertheless, our understanding of the solar wind is far from complete. For example, we still do not know what physical processes heat and accelerate the solar wind or what determines its flow speed. We do not yet know if the low-speed wind arises primarily from quasi-stationary processes or from a series of transient reconnection events in the solar atmosphere. Likewise, the physical origins of CMEs are still being debated. We do not yet fully understand how a rough balance is obtained between new magnetic flux injected into the solar wind by CMEs and magnetic flux removed from the heliosphere by reconnection inside the Alfvén point. Nor do we understand how the magnetic topologies of ICMEs evolve with time. In general, our ideas about the structure of the heliospheric magnetic field are still developing and need testing with further observations. The generally weak heliospheric magnetic fields and reduced solar wind dynamic pressures that characterized the recent extended solar activity minimum have provided a new, and as yet not fully understood, perspective on the long-term variation of the solar wind. Ideas about the solar wind's interaction with the interstellar medium, the physical nature of the heliopause, and the role of the termination shock in accelerating anomalous cosmic rays are presently being tested with both in situ and remote observations. The physical origin of variations in elemental abundances in the solar wind is beginning to be understood, as are temporal changes in the charge states of the heavier elements. Origins of double ion beams and suprathermal ion tails in the solar wind are still being debated, as are the mechanisms by which solar wind turbulence evolves and eventually dissipates. Moreover, we do not yet fully understand why different ionic species have different speeds and temperatures in the solar wind. Further analysis of existing data, new types of measurements, numerical simulations, and fresh theoretical insights should lead to understanding in these and other areas of solar wind research in the years ahead.

BIBLIOGRAPHY

Asbridge, J. R., et al. (1974). *Solar Physics, 37*, 451.

Balogh, A., Gosling, J. T., Jokipii, J. R., Kallenbach, R., & Kunow, H. (Eds.), (1999). *Corotating interaction regions*. Dordrecht, Netherlands: Kluwer Academic Publishers.

Bruno, R., & Carbone, V. (2013). *Living Reviews in Solar Physics, 2*, 4 Accessed 20.08.13 http://www.livingreviews.org/lrsp-2013-2.

Crooker, N., Joselyn, J. A., & Feynman, J. (Eds.), (1997). *"Coronal mass ejections", geophysical monograph 99*. Washington D.C: American Geophysical Union.

Crooker, N. U. (2000). *Journal of Atmospheric and Solar-Terrestrial Physics, 62*, 1071.

Fleck, B., & Zurbuchen, T. H. (Eds.), (2005). *Proceedings of solar wind 11///SOHO 16, Connecting sun and heliosphere*, ESA SP-592. Noordwijk, Netherlands: European Space Agency.

Florinski, V., Pogorelov, N. V., & Zank, G. P. (Eds.), (2004). *Physics of the outer heliosphere*. Melville, New York: AIP Conference Proceedings 719.

Gloeckler, G., et al. (2000). In R. A. Mewaldt, J. R. Jokipii, M. A. Lee, E. Mobius, & T. H. Zurbuchen (Eds.), *Acceleration and transport of energetic particles observed in the heliosphere: ACE 2000 Symposium*. (pp. 221–228). Melville, New York: AIP Conference Proceedings, 528.

Gosling, J. T., et al. (1995). *Geophysical Research Letters, 22*, 869.

Gosling, J. T., et al. (2002). *Geophysical Research Letters, 29*(12). http://dx.doi.org/10.1029/2001GL013949.

Gosling, J. T. (2010). In C. J. Schrijver, & G. L. Siscoe (Eds.), *Heliophysics III: Evolving solar activity and the climates of space and Earth* (pp. 217–242). Cambridge, UK: Cambridge University Press.

Hansteen, V. H. (2009). Stellar winds and magnetic fields. In C. J. Schrijver, & G. L. Siscoe (Eds.), *Heliophysics I: Plasma physics of the local cosmos* (pp. 225–255). Cambridge, UK: Cambridge University Press.

Hundhausen, A. J. (1973). *Journal of Geophysical Research, 78*, 1528.

Hundhausen, A. J. (1977). In J. Zirker (Ed.), *Coronal holes and high speed wind streams*. Boulder Colorado: Colorado Associated University Press.

Hundhausen, A. J. (1985). In R. G. Stone, & B. T. Tsurutani (Eds.), *Collisionless shocks in the heliosphere: A tutorial review*. Washington DC: Geophysical Monograph 34, American Geophysical Union.

Jokipii, J. R., & Thomas, B. (1981). *Astrophysics Journal, 243*, 1115.

Jokipii, J. R. (2010). The heliosphere and cosmic rays. In C. J. Schrijver, & G. L. Siscoe (Eds.), *Heliophysics III: Evolving solar activity and the climates of space and Earth* (pp. 243–268). Cambridge, UK: Cambridge University Press.

Lazarus, A. J., & Belcher, J. (1988). In V. J. Pizzo, T. E. Holzer, & D. G. Sime (Eds.), *Proceedings of the sixth international solar wind conference*. Boulder Colorado: National Center for Atmospheric Research.

Marsch, E. (2006). Kinetic physics of the solar corona and solar wind. *Living Reviews in Solar Physics, 8*(1).

Maksimovic, M., Issautier, K., Meyer-Vernet, N., Moncuquet, M., & Pantellini, F. (Eds.), (2010). *Twelfth international solar wind conference*, AIP Conference Proceedings 1216. Melville, New York: American Institute of Physics.

McComas, D. J., et al. (2008). *Geophysical Research Letters, 35*, (L18103). http://dx.doi.org/10.1029/2008GL034896.

Mewaldt, R. A., Jokipii, J. R., Lee, M. A., Mobius, E., & Zurbuchen, T. H. (Eds.), (2000). *Acceleration and transport of energetic particles observed in the heliosphere*. Melville, New York: AIP Conference Proceedings 528.

Parker, E. N. (1963). *Interplanetary dynamical processes*. New York: Interscience Publishers.

Pizzo, V. J. (1978). *Journal Geophysical Research, 83*, 5563.

Smith, C. W. (2009). Turbulence in space plasmas. In C. J. Schrijver, & G. L. Siscoe (Eds.), *Heliophysics I: Plasma physics of the local cosmos* (pp. 163–194). Cambridge, UK: Cambridge University Press.

Zank, G. P., & Mueller, H. R. (2003). *Geophysical Research Letters, 108*(1240). http://dx.doi.org/10.1029/2002JA009689.

Zurbuchen, T. H., & Richardson, I. G. (2006). *Space Science Reviews, 123*(31). http://dx.doi.org/10.1007/s11214-006-9010-4.

Part IV

Earthlike Planets

Chapter 13

Mercury

Scott L. Murchie, Ronald J. Vervack, Jr., and Carolyn M. Ernst
Space Department, Johns Hopkins University Applied Physics Laboratory, Laurel, MD, USA

Robert G. Strom
Department of Planetary Sciences, University of Arizona, Tucson, AZ, USA

Chapter Outline

1. Exploration of Mercury — 283
2. General Planetary Characteristics — 284
3. Motion and Temperature — 284
4. Internal Structure and Magnetic Field — 286
5. Exosphere and Magnetosphere — 288
   5.1. Exosphere — 288
   5.2. Magnetosphere — 290
6. Geologic Features — 292
   6.1. Impact Craters and Basins — 293
   6.2. Volcanic Plains and Vents — 294
   6.3. Tectonics and Topography — 297
   6.4. Surface Composition — 299
7. Recent Surface Features — 301
   7.1. Radar-bright Polar Deposits — 301
   7.2. Hollows — 302
8. History — 303
   8.1. Geologic History — 303
   8.2. Thermal History — 303
   8.3. Origin — 303
   Bibliography — 304

1. EXPLORATION OF MERCURY

Mercury is one of the five planets known to the ancients (along with Venus, Mars, Jupiter, and Saturn), to whom it was the messenger of the gods. The planet's small size and proximity to the Sun have historically made ground-based telescopic observations difficult, so until the 1970s very little was known about it. Since then, three phases of exploration have peeled back Mercury's veil of mystery. The *Mariner 10* spacecraft performed three flybys of Mercury on March 29 and September 21, 1974, and March 16, 1975. Mariner 10 imaged about 45% of the surface at an average resolution of about 1 km per pixel, but less than 1% at resolutions between 100 and 500 m per pixel, which are needed to see landforms diagnostic of a variety of surface processes. This coverage and resolution are comparable to telescopic Earth-based coverage and resolution of the Moon before the advent of space flight. Mariner 10 also discovered that Mercury is the only terrestrial planet besides Earth with an internally generated, global magnetic field, and it first detected Mercury's thin atmosphere, or exosphere. Over the next 30 years, ground-based observations provided all the new measurements of the topographic, radar, and reflectivity characteristics of Mercury's surface; measured its exosphere; and helped to constrain its surface composition. Finally, the *MESSENGER* spacecraft performed three flybys of its own on January 14, 2008, October 6, 2008, and September 29, 2009; on March 18, 2011, it entered Mercury orbit. Mercury thus became the last of the five "classical" planets to be studied by an orbiting spacecraft.

New data from MESSENGER have revolutionized knowledge of Mercury. This mission is one of the National Aeronautics and Space Administration's *Discovery* series of planetary exploration missions. The name MESSENGER, besides reminding us of the ancients' understanding of Mercury, is an acronym for MErcury Surface, Space ENvironment, GEochemistry, and Ranging. The spacecraft was launched on August 3, 2004. Over its 7-year cruise it used gravity assists from multiple flybys of terrestrial planets to slow its "fall" toward Mercury, so that the main rocket engine could finish slowing the spacecraft enough for orbit insertion. The flybys included an Earth flyby in August 2005, two flybys of Venus in October 2006 and June 2007, and the three Mercury flybys. MESSENGER's initial science objectives were (1) to determine the nature of polar deposits including their composition; (2) to determine the properties of Mercury's core including its diameter and whether there is an outer fluid core; (3) to determine the geologic history of Mercury; (4) to determine the nature of the magnetic field and how it is generated; (5) to measure the composition of the surface, to help in understanding the planet's formation and to test competing hypotheses for why it is so dense; and (6) to measure its exosphere and how it interacts with the magnetosphere and surface. As with any first orbital mission to a planet, some of MESSENGER's initial findings were unexpected and drove new, more specific objectives that have been the mission's new focus since the primary 1-Earth-year orbital investigation was completed.

Upon orbit insertion, MESSENGER was in a highly elliptical, 12-h orbit (Figure 13.1) — the main engine could slow the spacecraft enough to be captured into Mercury orbit, but it was not possible to carry enough fuel to make the orbit circular. The orbit is near-polar to provide global viewing and to measure the planet's libration. As MESSENGER transitioned to its extended mission, the orbit apoapsis was lowered in altitude, while the periapsis stayed the same, changing the orbital period to 8 h and enabling more time to be spent at lower altitudes taking high-resolution measurements. There are eight science experiments onboard the spacecraft (Figure 13.1 and Table 13.1): (1) a dual imaging system, (2) a gamma-ray and neutron spectrometer, (3) a magnetometer, (4) a laser altimeter, (5) atmospheric (0.105–0.6 μm) and surface (0.3–1.45 μm) spectrometers, (6) an energetic particle and plasma spectrometer, (7) an X-ray spectrometer, and (8) a radio science experiment that uses the telecommunication system. These instruments address the science objectives discussed previously. More information on the mission and its findings at Mercury are available at http://messenger.jhuapl.edu.

For the next phase of Mercury's exploration, the European Space Agency and the Japan Aerospace Exploration Agency will jointly conduct a Mercury orbital mission called *BepiColombo*, currently planned for a 2016 launch.

2. GENERAL PLANETARY CHARACTERISTICS

Mercury has been compared with the Moon because both are heavily cratered, but Mercury has had more extensive volcanism, and has an internal magnetic field and active surface processes that the Moon lacks. Mercury's diameter is 4880 km, 28% smaller than the next largest planet, Mars, and its mass is 3.301×10^{23} kg. Because of this large mass in relation to its volume, Mercury has an exceptionally high mean density of 5440 kg/m^3, second among planets only to the Earth (5520 kg/m^3). Without the effects of gravitational self-compression, however, Mercury's density would be even greater than that of the Earth's. Mercury's brightness (albedo) is greater than that of iron-rich basaltic plains on the Moon, but less than that of low-iron lunar highlands. Like the Moon, Mercury is covered with a **regolith** consisting of fragmental material formed by impacting meteoroids over billions of years.

Global maps of Mercury's surface from MESSENGER orbital imaging (Figure 13.2) show that the surface is divided into two major **geologic units**, regions that share similar measurable properties: mostly lighter colored **smooth plains** covering about a quarter of the planet, and darker, more heavily cratered **intercrater plains** (the most extensive terrain type) filling most regions between clusters of large craters. A third unit is **ejecta**, debris excavated from depth by impact craters and basins and deposited onto the surface. All these units are cut by tectonic features. Long **lobate scarps** traverse the surface for hundreds of kilometers, and arrays of **graben** occur in the interiors of some of the larger impact basins.

On Mercury, the prime meridian (0°) was chosen to coincide with the subsolar point during the first perihelion passage after January 1, 1950. In most maps constructed prior to MESSENGER, longitude increases to the west; map products generated from MESSENGER data use the convention of longitude increasing to the east, also widely used for modern spacecraft missions to the Moon and Mars. Craters are mostly named after famous authors, artists, and musicians such as Dickens, Michelangelo, and Beethoven, whereas valleys are named after prominent radio telescopes including Arecibo and Goldstone. Prominent ridges and scarps are named after ships associated with exploration and scientific research such as Discovery and Victoria. Plains are named after the name of planet Mercury in various languages, for example, Odin (Scandinavian) and Tir (Germanic). Borealis Planitia (Northern Plains) and Caloris Planitia (Plains of Heat) are exceptions. The largest well-preserved impact basin, named the **Caloris Basin** (Basin of Heat) because it nearly coincides with one of the "hot poles" of Mercury, figures conspicuously in the planet's geology.

No natural satellite of Mercury has been discovered. Any satellite that might exist would likely be smaller than 1.6 km in size, to have escaped detection in surveys prior to MESSENGER.

3. MOTION AND TEMPERATURE

Mercury has the most eccentric (0.205) and inclined (7°) orbit of any major planet, although over periods of a few million years, eccentricity varies from about 0.1 to 0.28 and inclination varies from about 0° to 11°. Its average distance from the Sun is 0.387 AU (5.79×10^7 km), but because of its large eccentricity, at present times, the distance varies from 0.308 AU (4.60×10^7 km) at perihelion to 0.467 AU

Chapter | 13 Mercury

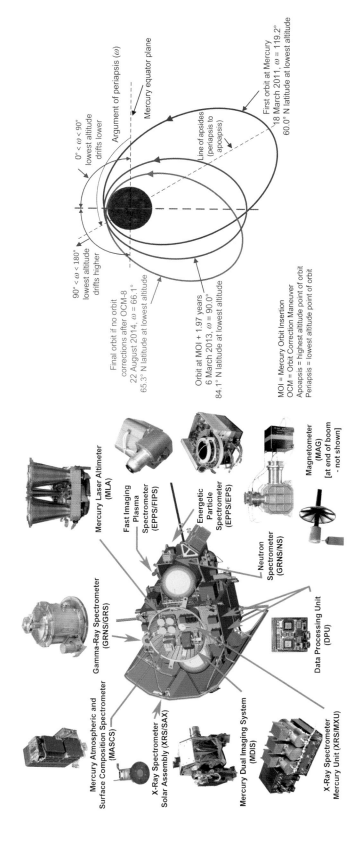

FIGURE 13.1 (Left) Drawing of the MESSENGER spacecraft showing the placement of the science instruments listed in Table 13.1. (Right) Illustration of the orbit around Mercury shows the very elliptical orbit used for the primary mission and the lower orbits used during the extended mission. *These and other images in this article are thanks to NASA/Johns Hopkins University Applied Physics Laboratory/Carnegie Institution of Washington unless otherwise noted.*

TABLE 13.1 MESSENGER Instruments and Primary Measurements

Instrument	Observation
Dual imaging system (1.5° and 10.5° fields of view)	Surface mapping in stereo and in 11 color filters
Gamma-ray and neutron spectrometer	Surface composition (O, Si, Fe, K, Na, U, Th)
X-ray spectrometer (1–10 keV)	Surface composition (Mg, Al, Fe, Si, S, Ca, Ti)
Atmospheric and surface composition spectrometer	Surface and exosphere composition
Magnetometer	Magnetic field
Laser altimeter	Topography, reflectivity of northern hemisphere
Energetic particles and plasma spectrometer	Energetic particles and plasma
Radio science (X-band transponder)	Gravity field and physical libration

(6.98×10^7 km) at aphelion. As a consequence, although Mercury's orbital velocity averages 47.6 km/s, it varies from 56.6 km/s at perihelion to 38.7 km/s at aphelion. At perihelion, the Sun's apparent diameter is over three times larger than its apparent diameter as seen from the Earth.

Mercury's sidereal rotation period (relative to the stars) is 58.65 Earth days, and its orbital period is 87.97 Earth days. It has a unique 3:2 resonance between its rotational and orbital periods, making exactly three rotations on its axis for every two orbits around the Sun. This resonance was probably acquired as a consequence of the dissipative processes of **tidal heating** and the relative motion between a solid mantle and a liquid core. As a consequence of this resonance, a solar day (from sunrise to sunrise) lasts exactly 2 Mercurian years, or 176 Earth days. The **obliquity** of Mercury is close to 0°; therefore, it does not experience seasons like those of the Earth and Mars. Some topographic depressions in the polar regions thus never receive direct rays of sunlight and can be permanently colder than −163 °C (−262 °F).

Another effect of the 3:2 resonance is that the same hemisphere always faces the Sun at alternate perihelion passages. This happens because the hemisphere facing the Sun at one perihelion will rotate one-and-a-half times by the next perihelion, so that it faces away from the Sun; after another orbit, it rotates another one-and-a half times so that it directly faces the Sun again. Because the subsolar points at perihelion are the 0 and 180° longitudes, they are called **hot poles**. The 90° and 270° longitudes are called **warm poles** because they are the subsolar points at aphelion. Yet another consequence of the 3:2 resonance, combined with the large eccentricity and variable orbital velocity, is that the Sun briefly backtracks in the sky during perihelion passage. At the warm poles after sunrise, the Sun reverses direction, sets, and then rises again.

Although Mercury is the planet closest to the Sun, it is not the hottest planet. The surface of Venus is hotter because of its atmospheric greenhouse effect. However, Mercury experiences the greatest range (day to night) in surface temperatures (650 °C = 1170 °F) of any planet or satellite in the solar system, because of its close proximity to the Sun, its long solar day, and its lack of an insulating atmosphere. Near the equator, surface temperature reaches about 467 °C (873 °F) at perihelion, hot enough to melt zinc. At night just before dawn, surface temperature at the same location plunges to about −183 °C (−297 °F).

4. INTERNAL STRUCTURE AND MAGNETIC FIELD

Mercury's large core is unique in the solar system and imposes severe constraints on the origin of the planet. Mercury's mean density of 5440 kg/m^3 is only slightly less than that of the Earth (5520 kg/m^3) and larger than that of Venus (5250 kg/m^3). Without Earth's large internal pressures, however, Earth's uncompressed density is only 4400 kg/m^3 compared to Mercury's uncompressed density of 5300 kg/m^3. This means that Mercury contains a much larger fraction of iron than any other planet or satellite in the solar system. MESSENGER's measurement of Mercury's moment of inertia suggests that the core is about 4060 km in diameter, or about 83% of the planet's diameter, and some 58% of its volume. The silicate mantle and crust together are only about 410 km thick. For comparison, Earth's iron core is only 54% of its diameter and just 16% of its volume. Assuming an appropriate density for the thin silicate mantle, the density of the core must be less than that of pure iron–nickel metal, requiring that it contain lighter elements such as silicon and sulfur.

The thickness of the crust in the northern hemisphere has been estimated using MESSENGER's low-altitude measurements of the gravitational field and topography, by subtracting the component of gravity due to topography and modeling the subsurface contours of a low-density crust that would be necessary to explain remaining variations in the gravitational field. A density contrast between the mantle and crust and an average crustal thickness have to be assumed: 200 kg/m^3 based on geochemical results discussed below and 50 km based on tectonic models for the depth extent of faulting, respectively. Given these assumptions, the resulting crustal thickness is generally greater near the equator (50–80 km) and less toward the north polar region (20–40 km). The thinnest crust is

Chapter | 13 Mercury

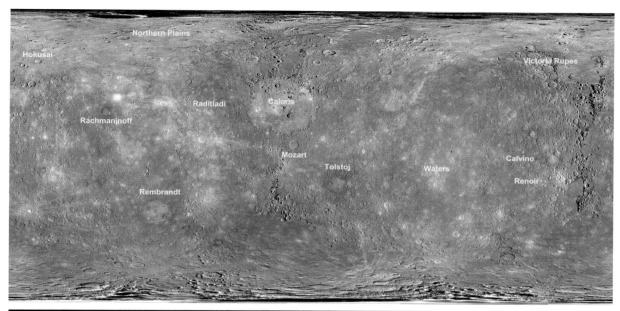

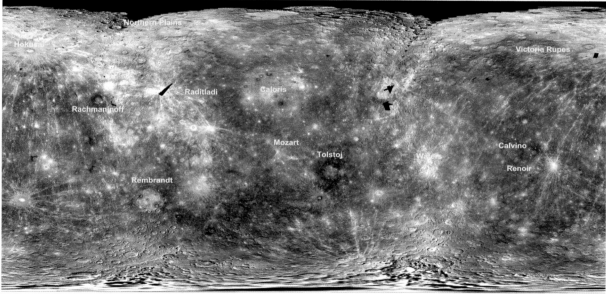

FIGURE 13.2 (Top) MESSENGER morphology photomosaic of Mercury, assembled from images taken with the Sun low in Mercury's sky to accentuate topography, showing the names of major features discussed in this chapter. The data are in simple cylindrical projection centered at 180° E. (Bottom) MESSENGER false-color photomosaic in the same map projection, assembled from images taken with the Sun high in Mercury's sky to accentuate color and albedo variations. The red, green, and blue image planes show images taken at 1.00, 0.75, and 0.43 μm, respectively.

thought to be located beneath the northern lowlands, the Caloris basin, and other northern hemisphere impact basins. Crustal thickness in the southern hemisphere cannot be estimated in this way due to the lack of low-altitude MESSENGER measurements.

Aside from Earth, Mercury is the only terrestrial planet with an internally generated magnetic field. The field was first detected by Mariner 10 during its flybys, and later measured in detail by MESSENGER. The magnetic field can be modeled as a dipole field like a bar magnet, with a strength near Mercury's surface of about 190 nanotesla (nT). The dipole is tilted less than 0.8° from the planet's rotational axis, and its center is offset northward from the center of the planet by about 480 km (nearly 20% of the planet's radius). In comparison, Earth's dipole field is 25,000–65,000 nT at the surface (130–340 times stronger than Mercury's field at the planet's surface), tilted 11° from the axis of rotation, and offset only about 8% of the planetary radius.

The maintenance of a terrestrial planet dipole magnetic field is thought to require an electrically conducting, fluid outer core surrounding a solid inner core. Earth-based radar measurements of the magnitude of Mercury's **librations** indicate that the mantle is detached from the core, confirming that the outer core is fluid. If the core were pure Fe and Ni metal, cooling of Mercury's interior over the planet's history would have allowed the core to solidify and shut down the magnetic field. The presence of the light elements, such as silicon or sulfur, in the core would lower its melting point, allowing the outer core to remain fluid to the present day.

5. EXOSPHERE AND MAGNETOSPHERE

The exosphere and magnetosphere of Mercury represent a highly dynamic, coupled system that is unique among the terrestrial planets.

5.1. Exosphere

The **exosphere** is an atmosphere so thin that its few atoms or molecules are unlikely to collide with one another. In Earth's atmosphere, the exosphere is the highest part of the atmosphere where the density of gas molecules is very low. At Mercury the exosphere is the only atmosphere, so the planet has what is called a surface-bounded exosphere, whose gas molecules collide with the surface (or escape from the planet) rather than colliding with each other.

A primary method by which the exosphere of Mercury is studied is observation of **resonant emission** from atoms, in which solar photons of specific energies or wavelengths are absorbed and then reemitted at the same wavelength. Because the combinations of energies at which such emissions occur vary between elements, observed emission spectra provide unique spectral fingerprints for elements that are present. Mariner 10's ultraviolet spectrometer discovered Mercury's exosphere through observations of emission from both hydrogen (H) and helium (He) atoms. Mariner 10 measurements imply a surface pressure 1 trillion times smaller than that of the Earth's atmosphere. Nearly a decade after Mariner 10's flybys, advances in telescopes and instrumentation led to the discovery of sodium (Na) and potassium (K) in the exosphere; calcium (Ca) was detected in 2000. MESSENGER added magnesium (Mg) to the known elements in the exosphere during its second flyby.

In contrast to the denser atmospheres of Earth, Venus, and Mars, the contents of Mercury's exosphere are transient and must be continually replenished. If the source processes for Mercury's exosphere suddenly stopped, the exosphere would dissipate in just 2–3 days. Also in contrast to other terrestrial planet atmospheres, Mercury's exosphere is composed almost entirely of atoms rather than molecules, a result primarily of the manner in which the exosphere is generated and maintained. Any molecules that are present in the exosphere are quickly photodissociated (i.e. broken apart) by sunlight, which is intense at Mercury owing to its proximity to the Sun and the lack of a thick upper atmosphere to absorb the sunlight.

Mercury's exosphere originates from the planet's surface, partly from material native to Mercury, partly from material implanted in Mercury's surface by the stream of charged particles from the Sun known as the **solar wind**, and partly from the impacts of comets and meteoroids. The generation and maintenance of Mercury's exosphere is summarized in Figure 13.3. The first of three major sources of exospheric atoms is sunlight striking the surface, releasing material in either of two ways. **Photon-stimulated desorption**, or PSD, occurs when solar photons hit the surface and release their energy, breaking the bonds that hold the surface materials together and ejecting atoms from the surface. **Thermal desorption**, or evaporation, occurs when sunlight heats the surface and loosely bound, volatile material is boiled off. Both processes are low-energy processes, so trajectories of the ejected atoms do not carry them very high or very far.

The second major source of exospheric atoms is a process known as **sputtering**, which occurs when ions from the solar wind or Mercury's magnetosphere impact the surface. The energy in these impacts is higher than in the case of PSD or thermal desorption, so that atoms ejected by sputtering have larger velocities and their trajectories carry them higher and farther than atoms released through low-energy thermal processes. Ion sputtering can fracture the surface on atomic scales, liberating volatile species such as Na. This leads to greater release of material through PSD than in the absence of ion sputtering, by a process known as ion-enhanced PSD.

Meteoroid impacts are the third primary source of exospheric material. Although large impacts release much material, they are rare and an influx of small dust particles from interplanetary space that collide with Mercury's surface is more responsible for day-to-day maintenance of the exosphere. The energy of these collisions vaporizes both the dust particles and some of the surface, releasing high-energy atoms to high altitudes.

Atoms released with low velocities follow ballistic trajectories under the influence of gravity. Because they do not go very high, these atoms mostly fall back to the surface where they either bounce or stick. Some atoms undergo multiple bounces before sticking (known as ballistic hops), and in this manner redistribute volatile material across Mercury's surface, gradually transferring it from the hotter, equatorial regions to the colder, polar regions.

Atoms released with high velocities also follow ballistic trajectories; however, the longer time of residence of these atoms in the exosphere allows two other processes to affect them. The first is solar radiation pressure, in which solar photons push the atoms in the antisunward direction. If the solar radiation pressure pushes them far enough, atoms will

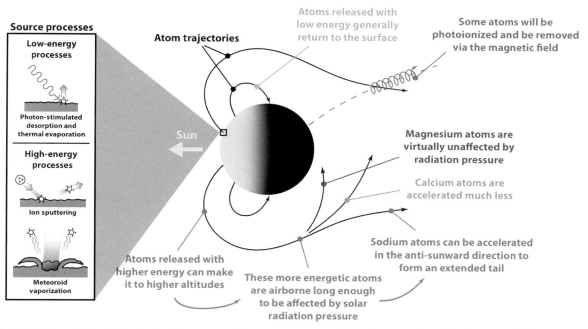

FIGURE 13.3 Schematic illustration showing the source and loss processes responsible for the generation and maintenance of Mercury's exosphere.

not return to the surface but will be pushed "behind" the planet to become part of a neutral, cometlike tail. Atoms pushed into the tail will escape the planet unless influenced by the second process redirecting exospheric atoms, ionization from solar photons striking the atoms (photoionization), and removing electrons. The positively charged atoms will be picked up by Mercury's magnetic field and rapidly accelerated either toward or away from the planet, depending on the orientation of the local magnetic field lines. Those atoms accelerated toward the planet impact the surface and can drive sputtering; those accelerated away from the planet will be lost to interplanetary space. Thus, atoms in the exosphere of Mercury ultimately either return to the surface or are lost to space, explaining why the exosphere would dissipate so quickly if not resupplied.

The processes that release atoms to Mercury's exosphere and that subsequently affect them differ in magnitude depending on the element. Ca and Mg are **refractory elements** (having strong chemical bonds requiring higher energies to release them from the surface), whereas Na is a **volatile element** (with weak bonds broken at lower energies). At the same time, Ca atoms have a lifetime against photoionization a factor of 10 smaller than for Na atoms and 100 smaller than for Mg atoms. Thus, neutral Ca atoms do not survive long in Mercury's exosphere (typically 1 h), whereas Mg atoms last a much longer time (typically 2–3 days). In addition, the effects of radiation pressure vary with each element, Na being affected strongly, Ca more weakly, and Mg hardly at all. Radiation pressure is also proportional to solar flux, and the Sun's spectrum contains deep absorptions, known **Fraunhofer lines**, at the wavelengths of most resonance emissions. In the parts of Mercury's elliptical orbit where the planet is accelerating toward or away from the Sun, the Doppler shift between the Sun and Mercury shifts the exospheric resonance emissions away from the Fraunhofer lines, providing increased radiation pressure. This Doppler shifting creates "seasons" in the exosphere, as Mercury orbits the Sun. MESSENGER saw these seasonal variations clearly during its flybys of Mercury when it viewed the exosphere from a distance (Figure 13.4).

These varying effects on different elements lead to different distributions in Mercury's exosphere. Sodium is seen everywhere in Mercury's exosphere, a consequence of its volatile nature and relatively easy release from the surface. Because solar radiation pressure has a large effect on Na, it is also the primary constituent in Mercury's tail and has been observed as far from Mercury as 2 million miles. At times there is so much Na in the tail that an observer on the nightside of Mercury might see a yellow-orange tinge to the night sky: the intensity of the Na emission, at the same wavelength as Na vapor street lamps, is similar in strength to a moderate aurora on Earth. The Na distribution is mostly symmetric about the Sun—Mercury line, indicating that it has a large PSD source; however, there are often local enhancements due to other processes,

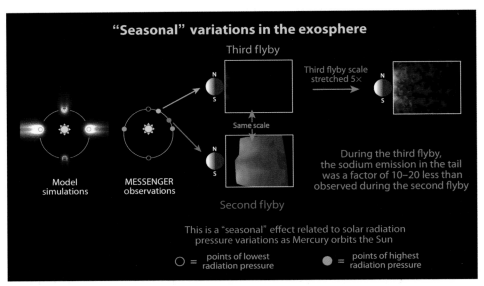

FIGURE 13.4 Illustration of "seasonal" variation in Mercury's neutral sodium tail. At the time of the second MESSENGER flyby the tail was well developed, whereas it was effectively "missing" during the third MESSENGER flyby. (The black spaces in the image from the second flyby are gaps in the data.) Variation in the sodium tail is related to variations in solar radiation pressure along Mercury's orbit, driven by Doppler shifting of the solar spectrum. Model simulations (left) show how the tail is expected to vary over one orbit around the Sun.

and the altitude distribution of Na exhibits a distinct two-component profile, consistent with its release from both low-energy and high-energy processes.

Calcium, on the other hand, has a very different distribution. MESSENGER observations reveal altitude profiles with only a high-energy component; they also show a persistent, strongly asymmetric distribution about the Sun—Mercury line, with peak densities near the equator at dawn. There are several possibilities for the difference between Ca and Na density with time of day. There may be more meteoroid impacts on the dawn side, which is the leading side of Mercury as it plows through the dust in the inner solar system. Alternatively, there may be differences in sputtering and photodissociation with time of day that affect the two elements differently.

Prior to MESSENGER, Mg had been predicted to be a part of Mercury's exosphere, but was not discovered until MESSENGER's second flyby. Mg has a distribution that contrasts with both Na and Ca. The overall Mg distribution is mostly isotropic about Mercury and characteristic of a high-energy release process; however, there is some evidence for localized enhancements. It remains a puzzle why Mg and Ca, both refractory species, are distributed so differently in the exosphere.

Other elements have also been observed in Mercury's exosphere, including hydrogen, helium, potassium, and possibly oxygen and aluminum. They are more difficult to observe because their emissions are weak and/or they are not particularly abundant. What limited information we do have on these elements shows that there could be even more puzzling aspects to Mercury's exosphere, whose understanding requires further observations.

5.2. Magnetosphere

The interaction of Mercury's magnetic field with the solar wind creates a **magnetosphere** qualitatively similar in several ways to Earth's magnetosphere, although roughly 20 times smaller (Figure 13.5). The field is strong enough to stand off the solar wind on the dayside of the planet under normal conditions, with the **magnetopause**—the boundary between the magnetosphere and the solar wind—generally located about 1000 km above the surface near the subsolar point. In the polar regions, funnel-shaped **cusps** form, exposing portions of Mercury's surface to direct bombardment by the solar wind. Because the magnetic field is so asymmetric about the planet's geographic equator, the southern polar region exposed to the solar wind is roughly four times larger than the corresponding region in the north (Figure 13.6). The solar wind sweeps a portion of the magnetic field downstream to form the **magnetotail** of Mercury. **Reconnection** events—releases of energy that occur when Mercury's magnetic field and the interplanetary magnetic field (IMF) coalign and "splice" together—happen regularly. The magnetosphere is also populated with **plasma**, or hot, ionized gas.

The overall structures of Mercury's and Earth's magnetospheres are similar, whereas the time and spatial scales of their dynamic phenomena are very different. Reconnection at Mercury's dayside magnetopause, which occurs

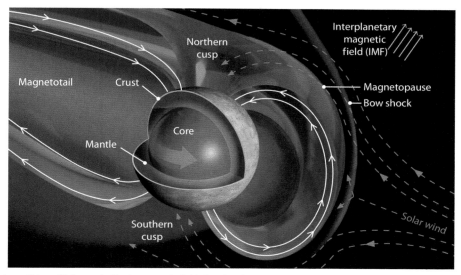

FIGURE 13.5 Artist's rendition of Mercury's dipole magnetic field showing the structure of the magnetosphere and how it interacts with the solar wind. *Image courtesy of James Slavin, now at the University of Michigan.*

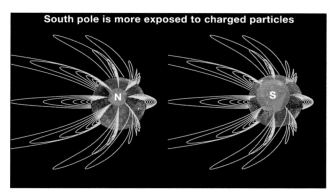

FIGURE 13.6 As a result of the north–south asymmetry in Mercury's internal magnetic field, the geometry of magnetic field lines is different in Mercury's north and south polar regions. In particular, the magnetic "polar cap" where field lines are open to the interplanetary medium is much larger near the south pole.

when there is an IMF component antiparallel to the local planetary magnetic field, occurs at rates ~10 times the typical rate observed at Earth. Such rapid reconnection can significantly shrink the magnetosphere on the dayside and push the magnetopause toward the surface. Under extreme solar wind conditions, the two magnetospheric cusps in the north and south can migrate equatorward, merge, and expose large regions of the dayside directly to the solar wind. Such dayside reconnection events increase the overall energy levels of the magnetosphere, but whereas the fractional increase at Earth for similar events is 10–30%, the increase at Mercury is typically 200–300%!

Reconnection also occurs in the magnetotail, where the magnetic field exhibits variations on time scales of seconds to minutes. Such fast magnetic field reconfiguration leads to vigorous heating of plasma trapped in the magnetic field, acceleration of ions to high energy levels, and other phenomena. One important phenomenon is "**plasmoids**", quasi-loop-like magnetic "islands" that form in the tail during reconnections. Some plasmoids move away from the "X-line" where they form, in a sunward direction (i.e. toward the nightside of the planet), and deposit energetic particles on the planet's surface; others move antisunward, down the tail away from the planet, and remove material from the magnetosphere. The circulation of plasma and energy from the X-line at the dayside magnetopause to the X-line in the magnetotail constitutes the "Dungey cycle" that powers Earth-type magnetospheres. At Earth, the cycle time is on the order of 1 h; at Mercury it is on the order of 1–2 min. Mercury's magnetosphere is extremely dynamic, and in that respect has no equal in the solar system.

MESSENGER has studied Mercury's magnetosphere not only by measuring the magnetic field but also by measuring ions and electrons in the magnetosphere, collectively called magnetospheric plasma. The plasma has two sources: the solar wind and the planet's surface. Solar wind plasma enters the magnetosphere through the cusps or by "leaking" through the magnetopause, mostly in concert with reconnection events. Plasma originating from the planet's surface not only derives from photoionization of neutral atoms in the exosphere but also includes ions produced directly from sputtering or micrometeoroid impacts.

The distribution of ions in Mercury's magnetosphere is controlled by motion of charged particles within the magnetic field. Ions gyrate, or rotate in a helical fashion, around local magnetic field lines and drift in directions driven by the laws of electromagnetics. Motion of the ions within the magnetosphere is called magnetospheric convection. For heavier ions (i.e. those with atomic masses larger than He^+, such as Na^+), the **gyro radius** around the field lines is large relative to the size of the magnetosphere. Many heavier ions thus collide with the planet during magnetospheric convection or cross the magnetopause and are picked up by the solar wind and swept away. The net result is that anisotropies are created in ion distributions—that is, differences in ion density as a function of latitude and local time of day. MESSENGER first mapped out these anisotropies and discovered three persistent features: (1) a large ion population at high northern latitudes on the dayside near the magnetospheric cusp, (2) an ion population near the equator on the nightside, and (3) an increase in ion abundance near the magnetopause, spanning the magnetopause boundary.

Ions observed by MESSENGER fall into five categories based on atomic mass and charge, that are distinguishable by the spacecraft's plasma spectrometer: H^+ (protons), He^+ (ionized helium atoms), He^{2+} (alpha particles), O^+-group ions (ionized atoms or molecules having atomic masses close to oxygen), and Na^+-group ions (ionized atoms or molecules having atomic masses close to Na, including Mg and Al). Protons are the most abundant ions (average density, $10/cm^{-3}$), followed by He^{2+} ions ($3.9 \times 10^{-2}/cm^3$), Na^+-group ions ($5.1 \times 10^{-3}/cm^3$), O^+-group ions ($8.0 \times 10^{-4}/cm^3$), and finally He^+ ($3.4 \times 10^{-4}/cm^3$). When ions are divided into these categories, anisotropies in plasma distribution are highlighted.

A look at Na^+-group ions illustrates the types of anisotropies discovered by MESSENGER. Two enhancements in ion density occur at high northern latitudes (Figure 13.7), with one feature ("feature 1") centered at a local time of ~10.5 h, corresponding to the northern magnetic cusp, and a second ("feature 2") centered at a local time of ~19 h. Enhancements are also evident at equatorial latitudes near the dawn terminator ("feature 3", local time 6 h) and at premidnight local times ("feature 4", centered around ~20 h), at altitudes above ~2000 km. This latter feature continues to high southern latitudes at an altitude of ~6000 km ("feature 5"). Features 2, 4, and 5 appear to be part of a single larger structure that represents an asymmetry between the dawn and dusk hemispheres. O^+-group ions show the same major features as do the Na^+-group, but at lower densities. In contrast, He^+ ions do not show these anisotropies, and are more uniformly distributed about the planet.

In contrast to the populations of positive ions, electrons exhibit no persistent distributions as a function of location or local time. Rather, their striking characteristic is localized energetic events. First observed by Mariner 10, and later studied in detail by MESSENGER, these events are recurring, intense bursts of high-energy electrons that increase in intensity by orders of magnitude above background in times as short as a few seconds. Individual events typically last for a few seconds to several minutes and come in groups spread over as many as several hours. Two regions of electron events are regularly observed: one at high northern latitudes on the nightside and another, less energetic one near the equator at most local times. An explanation for these electron bursts remains elusive.

6. GEOLOGIC FEATURES

Mercury's surface is dominated by plains having various densities of superimposed craters that surround and fill impact basins. Younger **smooth plains** infill some impact basins and also cover vast expanses outside recognized basins. Older, more cratered **intercrater plains** occupy the space between well-preserved impact basins. The largest well-preserved impact basin, Caloris, is infilled with distinctly colored smooth plains (Figure 13.2). The north polar region is the largest topographic low and contains the

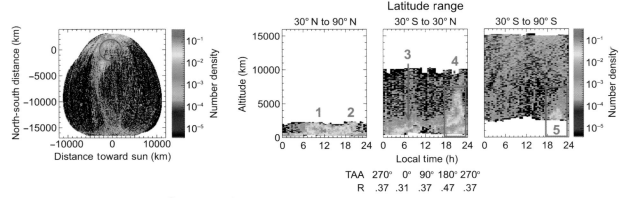

FIGURE 13.7 Average densities (cm^{-3}) for the Na^+-group ions. Left panel: average density projected onto the noon–midnight plane, binned 100 km × 100 km. The Sun is to the right, and the red circle shows the approximate size of the planet in the projection. The shape of the sampled volume is driven by MESSENGER's orbit around Mercury. Right panels: average density as a function of altitude (km) and local time (h) for three different latitude ranges. Mercury's position in its orbit ("true anomaly angle", or TAA) and its heliocentric distance (R, in astronomical units (AU)) at the times of the measurements are shown under the middle panel only, but apply to all three panels. The numbers indicate magnetospheric features discussed in the text. *Figure courtesy of Jim Raines, University of Michigan.*

largest expanse of relatively young smooth plains. Mercury's surface is also traversed by compressional **thrust faults** that form lobate scarps. Locally within impact basins, radial and concentric patters of tectonic ridges and **graben** occur.

Thermal infrared measurements from Mariner 10 showed that the surface is a good insulator and, therefore, consists of fine-grained, fragmental **regolith**, which is formed by fragmentation of surface rocks by eons of impact cratering. The highest resolution images from MESSENGER show small-scale landforms produced where the regolith has gradually moved down topographic slopes by a process called **mass wasting**, in which loose material migrates from topographically higher to lower regions (Figure 13.8). In the absence of a thick atmosphere and flowing water, mass wasting is a major process for gradually eroding Mercury's surface, imparting a smoothed appearance to the surface at small scales.

6.1. Impact Craters and Basins

Mercury's cratered surface records the **late heavy bombardment** by crater-forming projectiles. Heavily cratered regions on the Moon and Mars also record this bombardment, but volcanism and tectonics have removed its record from the surfaces of Venus and Earth. Based on radioactive dating of *Apollo* samples from the surface of the Moon and asteroid dynamical studies, the late heavy bombardment appears to have started about 4.0 billion years ago, peaked 3.9 billion years ago, and then declined rapidly for about 100-200 million years. After this time the impact rate by objects left over from the late heavy

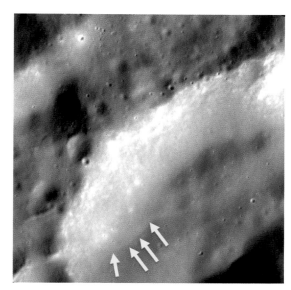

FIGURE 13.8 This very high resolution image shows the interior wall of a crater. The subtle curving line, indicated by the arrows, represents the edge of a lobe of material that has likely formed by movement of loose regolith down the steeper parts of the crater wall. The image is centered at 45.4° N latitude, 298.8° E longitude and is about 13 km across.

bombardment was much less, ending about 2 billion years ago. The same population of impacting bodies is thought to have affected the whole inner solar system, allowing dates from *Apollo* samples to be used to estimate ages of heavily cratered surfaces on Mercury and Mars.

Effects of young impacts are illustrated by the large, fresh crater Hokusai (Figure 13.9, left). Hokusai has an extensive system of **crater rays** extending nearly halfway around Mercury. Rays form where clouds of material

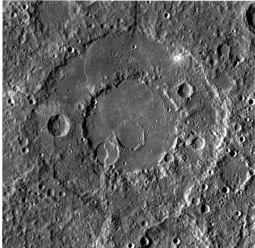

FIGURE 13.9 In the distant view of Mercury at left, the crater Hokusai (58° N, 17° E) is located on the planet's limb in the upper center of the image. Its rays extend nearly halfway around the planet. The ringed basin Renoir at right is located at 18° S, 308° E and is approximately 215 km in diameter. Note the two large, partly infilled craters within the inner ring; their occurrence shows that the smooth plains inside the inner ring formed from volcanism a geologically long time after the basin formed. The north-south-oriented lobate scarps cut across the smooth plains showing that compressional deformation dates from later in Mercury's history.

broken up during ejection from the parent crater create strings of reimpacting ejecta that form small **secondary craters**. The secondary craters in turn excavate brighter regolith from below a very thin surface layer that has been darkened and reddened by a process called **space weathering**. The tiny secondary craters in the rays will lose obvious association with Hokusai over millions of years as space weathering causes the rays to fade. On the Moon, space weathering is caused by iron in silicate minerals being reduced to metallic blebs only nanometers in size that coat the regolith grains. Micrometeoroid impacts and hydrogen from implanted solar wind drive the process. How the process works on Mercury is a subject of ongoing study.

Fresh impact craters on Mercury exhibit morphologies similar to those on the other terrestrial planets. Small craters are bowl shaped, but with increasing size, craters develop central peaks, flat floors, and terraces on their inner walls. The transition diameter from simple (bowl-shaped) craters to complex craters (with central peak and terraces) occurs at about 10 km. Beginning at a diameter of ~150 km and continuing through diameters up to ~350 km, Mercurian craters have an interior concentric ring, or **peak ring**, instead of a central peak. At even larger diameters, they have multiple, concentric rings, forming **multiringed basins**. For a given crater diameter, the radial extent of continuous ejecta outside the crater is uniformly smaller than on the Moon by a factor of about 0.65. The maximum density of secondary impact craters also occurs closer to the crater rim than for similarly sized lunar craters, at about 1.5 crater radii from the rim compared with 2–2.5 crater radii on the Moon. Both differences are due largely to the greater surface gravity of Mercury (3.70 m/s^2) than that of the Moon (1.62 m/s^2).

Mercury has 46 recognized impact basins 300 km or more in size. Per unit surface area, this population yields a lower density of impact basins than on the Moon, with a greater discrepancy between the two bodies at larger diameters: the Moon has an average of one 500-km basin per 2,700,000 km^2, but Mercury only has 1 per 4,300,000 km^2. The disparate population densities highlight two major differences between the superficially similar-looking cratered surfaces. First, it is generally thought that Mercury's lower basin density results from the oldest basins being buried by intercrater plains, which are widespread on Mercury but not the Moon. Second, Mercury has three times as many peak-ring basins per unit area as the Moon, with a total of 110 recognized peak-ring basins, like Renoir (Figure 13.9, right), compared with only 17 on the Moon. Although the reason for this difference is debated, a leading hypothesis is that the higher mean velocity of Sun-orbiting impactors at Mercury versus at the Moon leads to greater melting at the center of impact basins, resulting in a larger central melt cavity. There is widespread evidence for

FIGURE 13.10 Relatively blue-colored impact melt is evident at this rayed, 15-km-diameter. Waters crater is located at 9° S, 255° E. The great extent of the melt from Waters suggests flow of melt with low viscosity. The color inset uses images collected through 1.00, 0.75, and 0.43 μm filters as the red, green, and blue image planes; the black-and-white image is part of the global morphology mosaic.

impact melting being prevalent on Mercury in smaller craters as well. For example, at the young rayed crater Waters (15 km in diameter), a large and distinctly colored "tongue" of impact melt splashed out of the crater (Figure 13.10).

6.2. Volcanic Plains and Vents

Based on Mariner 10 images, there was strong evidence that some plains deposits originated as lavas, but uncertainty as to whether others may have formed as impact melt or impact basin ejecta. MESSENGER's higher resolution images, multicolor imaging, and global coverage revealed that relatively young, smooth plains cover about 27% of Mercury's surface. The smooth plains exhibit two characteristics of having formed from volcanic lavas. First, most large impact basins are at least partially infilled by smooth plains, and those smooth plains commonly partly fill impact craters superimposed on the basins, for example, in the interior of Renoir (Figure 13.9). This relation requires that a geologically long time transpired between the formation of the host basin and the plains, to allow time for the formation of the infilled craters, thus ruling out an impact melt origin. Second, the plains typically have a distinct color that indicates a compositional difference from the underlying basin. The large basins Caloris, Tolstoj, Rembrandt, and Rachmaninoff all have ejecta of dark-colored **low-reflectance material** (LRM) but are filled with smooth **high-reflectance plains** (HRP) (Figure 13.2).

In some places, the colors of materials excavated by craters suggest that different types of plains materials form layers in Mercury's upper crust, at least locally. For example, the crater Calvino (Figure 13.11) is formed in intermediate-colored plains that infilled an unnamed, highly degraded impact basin. That basin formed in a surface layer that contained LRM, which constitutes the basin's ejecta. Small craters within the degraded basin expose more of the infilling intermediate plains. Calvino is the largest crater within the degraded basin and exposes HRP in its rim material; only material forming the central peak of Calvino excavated deeply enough to expose LRM. This suggests a three-layer sequence: LRM that the degraded basin excavated, overlain by HRP that Calvino excavated, overlain by intermediate material that infills the degraded basin.

Smooth plains have morphologic features suggesting that lavas forming them had very low viscosity and were comparable in age to, or older than, dark lava plains on the Moon. They mostly lack obvious flow fronts that sometimes mark the edges of Earth's or the Moon's lava flows. Instead, their emplacement appears to have eroded channels into preexisting rock, forming streamlined islands (Figure 13.12, left). In addition, their burial of craters in the northern plains suggests that at least those plains formed as massive floods up to 1 km or more in thickness, covering many hundreds of thousands of square kilometers. These characteristics indicate extremely fluid lavas, which is consistent with expectations based on the elemental compositions of the lavas (discussed below). Densities of superimposed craters suggest typical ages of 3.7–3.9 billion years, comparable to the oldest lava deposits that constitute the lunar maria. This means that Mercury's youngest large volcanic deposits, forming the smooth plains, occurred only during the early phases of emplacement of the Moon's widespread maria.

FIGURE 13.11 Enhanced-color view of the 68-km diameter crater Calvino, below and left of center, distinguished by its orange-colored rim material in this view. Calvino is centered near 4° S, 304° E. Darker, bluer colors indicate lower albedo and a less red-sloped spectrum. Calvino formed in intermediate-colored plains material infilling an old, unnamed basin that excavated LRM, exposed in ejecta along the right, upper, and left edges of this image. Calvino's rim exposes a layer of high-reflectance material buried beneath the intermediate-colored plains, and its central peak—representing the deepest material excavated by the crater—contains low-reflectance material possibly from the floor of the old basin.

Mercury has at least 49 irregular, scallop-rimmed, steep-walled depressions surrounded by haloes of bright material having color properties like those of HRP. These depressions are interpreted to be **pyroclastic vents**, formed

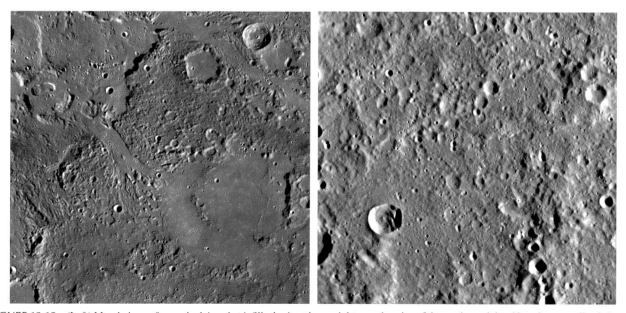

FIGURE 13.12 (Left) Morphology of smooth plains, that infill a basin at lower right near the edge of the northern plains. Note the streamlined channel suggesting flooding by very low viscosity lavas. (Right) Morphology of intercrater plains. Each scene is about 300 km across.

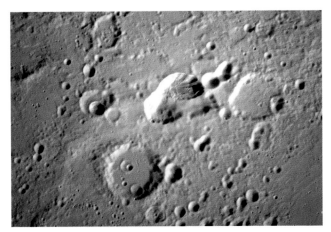

FIGURE 13.13 The 36-km-long pyroclastic vent northeast of Rachmaninoff basin, surrounded by a diffuse, bright halo of material even higher in albedo and redder than HRP. The red, green, and blue image planes show images taken at 1.00, 0.75, and 0.43 µm. The vent is located at 35.8° N latitude, 63.7° E longitude.

where ascending volatile-rich magma degassed and erupted explosively before falling back to mantle the surrounding surface (Figure 13.13). Similar features occur on the Earth and Mars, where water is the dominant volatile, and more rarely on the Moon, where the dominant volatile may instead be carbon monoxide. The driving volatile on Mercury is uncertain, but geochemical evidence discussed below suggests that sulfur or sulfur compounds may be important. The vents occur predominantly along basin rings, thrust faults, and at crater central peaks, suggesting that impact and tectonic fractures are important conduits for the eruptions.

Mercury's youngest volcanic features are not uniformly distributed. Their mapping from MESSENGER images (Figure 13.14) shows that smooth plains are concentrated in the northern hemisphere, in the topographically low northern plains and within and around Caloris. In contrast, pyroclastic vents occur mostly outside the smooth plains.

Older, intercrater plains (Figure 13.12, right) are the most extensive terrain on Mercury. They differ from smooth plains mainly in having a greater density of superimposed small craters. They are probably responsible for burying a substantial number of impact basins, yet the ejecta of other large basins are also superimposed on the intercrater plains. These relations indicate that intercrater plains were emplaced over a range of ages contemporaneous with the late heavy bombardment. Craters in the intercrater plains also excavate layers of differently colored material from depth. Some intercrater plains infill extremely degraded large impact basins that are barely recognizable from peaks in their rings that poke through the plains. For these reasons, many intercrater plains are probably older, more cratered versions of the volcanic smooth plains.

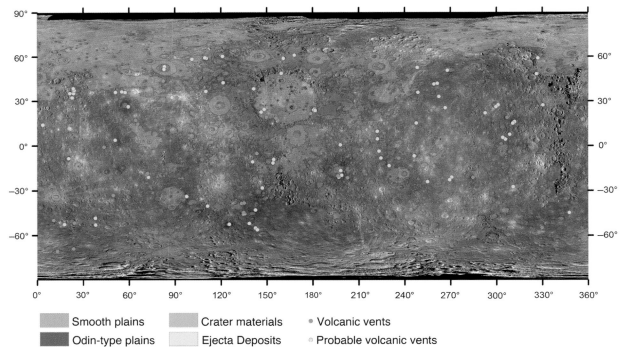

FIGURE 13.14 Simple cylindrical map of Mercury's smooth plains and pyroclastic vents. Crater materials that bury parts of the smooth plains are also shown. "Odin-type plains" are a special geologic unit ringing Caloris that may or may not be volcanic in origin. *Courtesy of Brett Denevi, Applied Physics Laboratory.*

6.3. Tectonics and Topography

No other planet or satellite in the solar system has tectonics so dominated by features formed by compression as does Mercury. On the Earth, for example, although compression occurs in folded mountain belts, extension occurs in rift zones and at the midocean ridges so that our planet's surface area is conserved. Mercury has globally distributed compressional thrust faults called lobate scarps (Figure 13.15, left). Individual scarps vary in length from ~20 to >600 km and have heights from a few hundred meters to about 3 km. They are nearly globally distributed, but tend to occur in broad belts separated by large regions having fewer scarps. In the equatorial region, the scarps have a predominantly north–south orientation. Lobate scarps are conspicuously absent from the vast northern expanse of smooth plains. There, the more common landform is smaller **wrinkle ridges** (Figure 13.15, right). This landform is also found in thick lava flows on Venus, the Earth, the Moon, and Mars and is thought to form by cooling and contraction of the lava. The dominance of compressional features on Mercury is widely thought to result from cooling and **global contraction** of Mercury throughout most of the planet's history. The total length of the lobate scarps is ~42,000 km, and reduction in Mercury's surface area associated with their formation is estimated to be ~0.08–0.12%, corresponding to a decrease in planetary radius of ~1–1.5 km. The dominant orientation of lobate scarps has been speculated to be a consequence of the reuse of ancient fractures formed by tidal stresses during slowing of an earlier more rapid rotation rate, called **tidal despinning**. Mercury's earlier, faster spin would have formed an equatorial bulge that collapsed as rotation slowed; at low latitudes, north–south compressional faults would be expected.

Mercury's topography is also unlike that of the Moon and Mars. All three bodies have large impact basins, but on the Moon and Mars, large basins are also vast depressions that are obvious and unambiguous in topographic maps, despite being partially filled by lavas (Moon and Mars) or

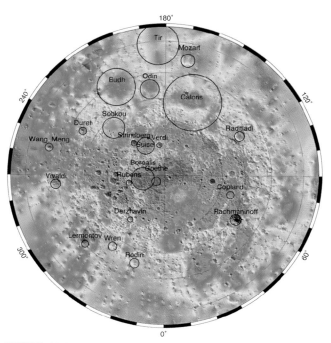

FIGURE 13.16 Map of topography in the northern hemisphere constructed from data collected by MESSENGER's laser altimeter, showing outlines of major impact basins. Caloris is above and to the right of center. The data are smoothed, and shown in a north polar stereographic projection. Red colors indicate higher topography and blue colors lower topography. The difference between the highest and lowest spots is approximately 9.85 km.

sediments (Mars). Only a few of the youngest large impact basins, including Rachmaninoff and Raditladi (Figure 13.2), are clearly recognizable in a topographic map of Mercury's northern hemisphere (Figure 13.16, constructed using data from MESSENGER's laser altimeter). Caloris Basin, the largest well-preserved basin on Mercury, does not even appear as a low area in the topographic map—parts of its floor stand above terrain outside the basin! This unexpected topography is thought to result from a combination of infilling by volcanic plains and subsequent contraction of the planet that warped Mercury's

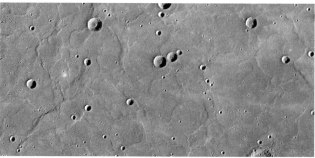

FIGURE 13.15 Victoria Rupes, at left, shows the asymmetric, steeper on one side profile typical of lobate scarps. Wrinkle ridges, right, are widespread in the northern plains. Each scene is 500 km across.

rigid, outer layer—the **lithosphere**—into broad, gentle dips and swells. Mercury also has comparatively low topographic relief; the range of elevations in the northern hemisphere measured by MESSENGER's laser altimeter is 9.85 km, considerably less than the elevation range on the Moon (19.9 km) or Mars (30 km).

Interiors of impact basins are the locations where extensional tectonic features are common on Mercury. In smaller basins, such as Mozart (Figure 13.17, left) there are a few graben, but the larger basins Rembrandt and Caloris (Figure 13.17, right) contain complicated patterns of graben and ridges. Figure 13.18 shows maps of tectonic structures that illustrate the change in tectonic style with basin size. In Figure 13.18(a), an unnamed northern plains crater about 120 km in diameter is completely buried by plains materials, but its rim is traced by wrinkle ridges. In the crater's interior, randomly oriented graben break the surface up into polygonal blocks a few kilometers in size. The larger, 235-km diameter Mozart basin (Figure 13.18(b)) has a few randomly oriented ridges at the center of smooth plains within its inner ring, surrounded by roughly concentric graben. Rachmaninoff (290 km) and Raditladi (257 km) (Figure 13.2) have a similar size and tectonic pattern. The even larger 720-km-diameter Rembrandt basin (Figure 13.18(c)), the second-largest well-preserved basin, has radial graben *and* ridges within its inner ring, concentric graben near the location of the basin ring, and randomly oriented ridges outside that. The largest well-preserved basin, 1550-km-diameter Caloris (Figure 13.18(d)), has breathtakingly complex tectonics. In the central part of the basin, a radial system of graben called Pantheon Fossae occurs together with roughly concentric ridges; they are surrounded by a ring of concentric graben. Outside this, as in Rembrandt, randomly oriented wrinkle ridges occur. The processes responsible for this transition in tectonic patterns with basin size are uncertain, but two plausible ones are cooling and sagging of the smooth plains fill and flow of the soft upper mantle toward the center of the basin early in Mercury's history as early basin topography reached **isostasy**.

Impact basin formation is also thought to have formed tectonic features directly. Opposite the Caloris basin on the other side of Mercury (around the **antipode** of Caloris) is hilly and lineated terrain, also called weird terrain, that disrupts preexisting landforms including crater rims (Figure 13.19). The hills are 4–10 km wide and typically 0.1–1.8 km high. Linear depressions that are probably graben form a roughly orthogonal pattern. Geologic relationships suggest that the age of this terrain is comparable to that of Caloris. Similar terrains occur at the antipodes of the Imbrium and Orientale impact basins on the Moon. The hilly and lineated terrain is thought to be the result of shock waves generated by the Caloris impact and focused at the antipodal region. Computer simulations of shock wave propagation indicate that focused shock waves from an impact of this size can cause vertical ground motions of about 1 km or more and fracturing of crustal rock to depths of tens of kilometers below the antipode. The antipodal regions of the Rembrandt and Tolstoj basins lack hilly and

 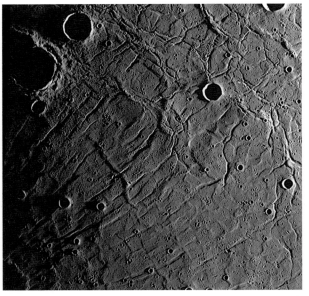

FIGURE 13.17 The interior smooth plains of Mozart basin (left) are deformed by a relatively few, roughly concentric extensional graben, seen in this image covering an area approximately 150 km across. In contrast, the inner parts of plains infilling the much larger Caloris basin (right) are densely fractured by radial graben and roughly concentric ridges, seen in this image of an area approximately 280 km across.

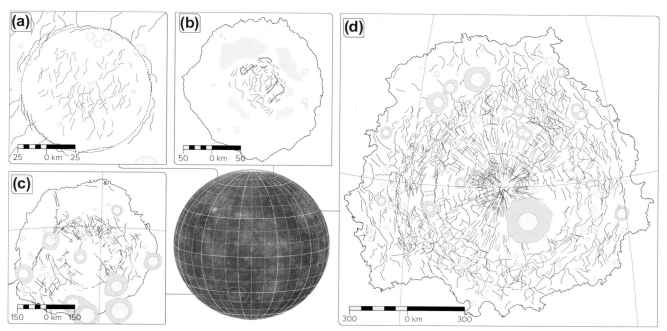

FIGURE 13.18 Tectonic maps of the interiors of selected impact craters and basins, whose locations are shown in the inset. Graben are shown in red and ridges in blue. Black outlines show the outer boundaries of basin-filling smooth plains; gray shading indicates superimposed craters, and yellow shading basin-ring massifs. (a) Unnamed buried crater in the northern plains. (b) Mozart basin. (c) Rembrandt basin. (d) Caloris basin. *Image courtesy of Paul Byrne, Carnegie Institute of Washington. Both the maps and the inset use orthographic map projections.*

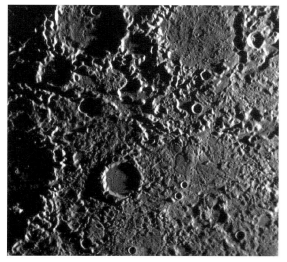

FIGURE 13.19 Hilly and lineated terrain antipodal to Caloris basin is densely fractured, as seen in this image of an area approximately 230 km across.

lineated terrain, suggesting that formation of these smaller basins did not affect their antipodal regions as strongly as Caloris' formation did.

6.4. Surface Composition

During exploration of most planetary bodies, the first information on surface composition typically comes from the manner in which the surface reflects the Sun's visible and near-infrared light, measured using a technique called **reflectance spectroscopy**. Transition metal cations, especially ferrous iron in the common silicates olivine and pyroxene and traces of it in feldspar, cause absorptions—preferential absorption of light at specific energies—at wavelengths that are diagnostic of different minerals. Space weathering tends to mask these absorptions, but on the Moon, even a very low iron content of a few percent in feldspar-rich highland rocks creates an easily detected absorption. Mercury's spectrum, in contrast, is utterly smooth at wavelengths where iron absorptions occur on other planetary bodies, even in crater rays where space weathering has had the least effect. The absence of iron absorptions shows that Mercury's silicates probably contain no more than several tenths of a percent iron. A complementary technique called **thermal emission spectroscopy** uses reduced heat emission at specific wavelengths as the fingerprint of different minerals, with the wavelengths governed by vibrations of mineral lattices. This technique does not depend on the presence of transition metal cations. Thermal emission spectra of Mercury taken from the Earth suggest abundant pyroxene, but with a low iron abundance. The very low iron content in silicates from spectroscopic measurements was the first key evidence that Mercury's surface is chemically highly reduced, with iron in chemical forms other than silicates.

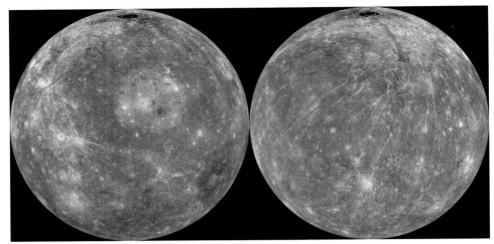

FIGURE 13.20 Enhanced-color mosaics of images acquired during MESSENGER's primary orbital mission, which mapped color variations in eight wavelengths at about 1 km per pixel on an average. Red and blue image planes show regions of redder and less red spectral slope between 0.4 and 1.0 μm. The green image plane shows bright, less red materials including crater rays less affected by space weathering. Opposite hemispheres (centered left and right at 150 and 330° longitude, 20° N latitude) are shown in orthographic projection. In this composite, HRP is light orange, intermediate-colored materials are dull brown, LRM is blue, and material least affected by space weathering (exposed in crater rays) appears light blue-green.

Color imaging of Mercury, first by Mariner 10 and then by MESSENGER, does show that there are compositional variations among surface materials (Figure 13.20). However, those variations manifest themselves as differences in albedo and **red slope**—the steepness of the spectrum of reflected sunlight—leaving differences in mineralogy obscure. Many smooth plains consist of higher albedo, more red-sloped HRP material, whereas older plains contain interbedded HRP and intermediate-colored plains material. Low-albedo, less red LRM is concentrated in ejecta of impact basins, particularly Rembrandt and Tolstoj, suggesting that LRM is more prevalent at depth. The nature of the mineral phase(s) creating color variations is uncertain: Mercury's low-iron silicates should be relatively bright, even after space weathering, whereas most of Mercury's surface is low in albedo compared to low-iron lunar highlands.

The first quantitative information on surface composition comes from MESSENGER's orbital measurements of elemental composition using **X-ray** and **gamma-ray spectroscopy**. These techniques measure interactions of high-energy radiation with different chemical elements and are not confounded by the reduced state of the surface. One surprising result, for a hot planet close to the Sun, is that the surface is not depleted and is even relatively rich in volatile elements. For example, on average, volatile potassium is more abundant relative to nonvolatile thorium than on the surface of Venus, Earth, and the volatile-depleted Moon and is comparable in abundance to that on volatile-rich Mars (Figure 13.21). Total contents of potassium and sodium are estimated to be up to 0.2 wt-% and 1–5% wt-% depending on location. Mercury's surface also contains

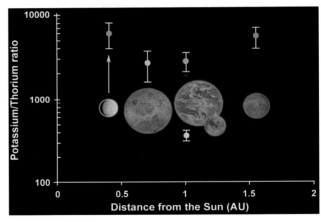

FIGURE 13.21 Relative abundance of potassium and thorium, a proxy for volatile content of planetary materials, as a function of planet and distance from the Sun.

up to ~4 wt-% sulfur. Total iron content of ~1% is more than can be contained in silicates, with most probably occurring as sulfide or free metal; total iron content is very low compared to ~14% on Mars.

Spatial variations in major element abundances support the idea that Mercury's plains were formed by different lava compositions. Figure 13.22 shows measured abundances of the major, nonvolatile elements magnesium, silicon, and aluminum measured from orbit in the upper centimeters of the surface. HRP plots close to a basaltic composition, although it is low in iron compared to terrestrial basalts. Less red units dominating the older, intercrater plains are higher in magnesium and lower in aluminum, suggesting

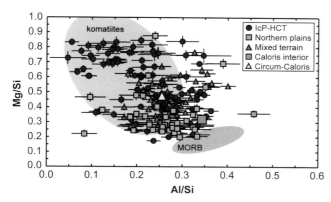

FIGURE 13.22 Major element composition in different plains units on Mercury. Lower albedo, less red plains include intercrater plains and heavily cratered terrain (IcP-HCT), mixed terrain, and circum-Caloris plains. Higher albedo, redder HRP include the northern plains and Caloris interior. "MORB" shows the field occupied by midocean ridge (seafloor) basalts on the Earth.

more pyroxene and/or olivine and less feldspar compared to basalt, more consistent with a type of volcanic rock called komatiite, which is rich in olivine and forms on Earth by melting of a larger fraction of its source region deep in the mantle than does basalt. These differences suggest that Mercury's volcanism evolved over time, from komatiitic or a related composition to more basaltic. Mercury also exhibits spatial variations in abundance of the relatively volatile elements sodium and potassium in the upper centimeters of the surface. Both elements are lower in abundance closer to the equator, suggesting gradual evaporation of them over billions of years.

7. RECENT SURFACE FEATURES

Like the Earth and Mars, Mercury has ongoing surface processes related to migration of volatile materials, beyond the "dry", gravitationally driven processes of mass wasting.

7.1. Radar-bright Polar Deposits

High-resolution radar images of Mercury from both the Arecibo and Goldstone radar facilities discovered patches with high radar reflectivity clustered around the poles. The reflectivity characteristics of the deposits are similar to those of outer-planet icy satellites and the residual polar water ice caps of Mars, so Mercury's polar radar-bright patches have long been thought to be water ice, which in many places may be buried by a few centimeters of regolith cover. Water ice is stable in Mercury's polar region inside some topographic depressions, especially craters, that remain in permanent shadow due to Mercury's low obliquity; these areas are illuminated only indirectly by sunlight reflecting off crater walls. How much reflected sunlight reaches a crater floor depends on the shape of the crater interior, which varies with crater size. Larger craters have shallower depth-to-diameter ratios, so they have a lesser solid angle of illuminated crater wall radiating onto their floors. Thermal models suggest that large craters (>40 km in size) poleward of 82° latitude have shadowed regions so cold (colder than −163 °C, or −262 °F) that water ice can remain stable at the surface over the age of the solar system. In contrast, 10-km craters can preserve water ice at the surface only poleward of 88°, assuming typical fresh crater shapes. Ice can persist further equatorward under a thin cover of only a few centimeters of regolith, to insulate the ice from the warmest daytime temperatures.

MESSENGER images of Mercury's polar regions show that craters with permanently shadowed portions enclose most of the radar-bright material. Radar-bright material around the south pole is largely confined within the 170-km crater Chao Meng-Fu (Figure 13.23). In the north polar region, the deposits reside within multiple craters (Figure 13.24). Close to both poles, nearly all permanently shadowed regions contain radar-bright material. However, radar-bright materials also occur in permanently shadowed regions equatorward of 80° latitude; those more equatorward deposits show a strong preference for longitudes near 90° E and 270° E, Mercury's "warm poles", where local noon occurs at aphelion, instead of at the "hot poles" at 0 E and 180° E, where local noon occurs at perihelion.

Differences in characteristics of the north polar radar-bright material have been revealed by MESSENGER orbital neutron spectrometer and laser altimeter measurements; presumably similar variations occur in south polar material but spacecraft altitude over high southern latitudes

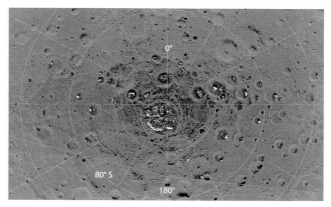

FIGURE 13.23 Map of Mercury's south polar region, showing a radar image taken from Earth, overlain on the percent of time the surface is illuminated, derived from MESSENGER images acquired over the course of a Mercury solar day. The color scale indicates the fraction of the solar day during which the surface is illuminated by the Sun; black areas are in permanent shadow, and redder areas are illuminated approximately 50% of each solar day. The blue line shows the limit of radar imaging from Earth. Note the preferential occurrence of outlying radar-bright spots in craters along the greenish-colored "warm pole" longitudes of 90° E and 270° E.

FIGURE 13.24 MESSENGER morphology mosaic of the north polar region, in orthographic map projection, with a yellow overlay showing radar-bright materials detected by Earth-based radar. The radar-bright materials are located in permanent shadows cast by crater rims and other topographic obstacles.

is too high for either instrument to resolve those deposits. The neutron spectrometer, which is highly sensitive to hydrogen, reveals concentrations of hydrogen centered on the pole. The strength of the neutron signal suggests that the deposits are nearly pure water ice. The laser altimeter measures reflectivity of the surface at 1064 nm, as well as spacecraft distance from the surface. Radar-bright materials in craters very close to the north pole have a high albedo, suggesting that water ice is present at the surface. Radar-bright materials farther from the pole are darker than surrounding terrain. This difference in reflectivity may indicate a darker, possibly organic component covering or mixed with the ice, at latitudes where a thin cover is required to preserve the ice. The total mass of ice preserved at the poles is estimated to be 20 to 2000 trillion kg—to within a factor of 10— comparable to the amount of water in Lake Tahoe.

The radar-bright deposits of frozen volatiles may have originated from impacting comets or water-rich asteroids, which released water and organic compounds that became cold-trapped in permanently shadowed craters. Comets and asteroids also impact the Moon. The neutron and gamma-ray spectrometers on the *Lunar Prospector* spacecraft discovered enhanced hydrogen signals at very high lunar latitudes that correspond with locations of permanently shadowed craters. These high-hydrogen regions have been interpreted as water ice with a concentration of only $1.5 \pm 0.8\%$ weight fraction, a much lower ice fraction than may exist in high-albedo, probably nearly pure ice very close to Mercury's north pole. The *Lunar Crater Observation and Sensing Satellite* impacted into a permanently shadowed region in the crater Cabeus, creating a plume of cold-trapped volatiles and dust ejected into sunlight that was observed from the Earth and from the impactor's shepherding spacecraft. This experiment confirmed the presence of water ice and suggested a variety of other volatiles. In the lunar south polar region, the crater Shackleton has a nearly permanently shadowed interior; measurements by the *Lunar Reconnaissance Orbiter*'s laser altimeter of the crater interior did not detect the same magnitude of brightening within it as within Mercury's polar craters. Thus, Mercury's polar regions resemble those of the Moon in being enriched with volatiles including water ice, but unlike at the Moon, there are large deposits of nearly pure ice and large exposures of it at the surface.

7.2. Hollows

An unexpected landform discovered in the highest resolution MESSENGER orbital images is "hollows", flat-floored depressions hundreds of meters wide and tens of meters deep, which commonly coalesce into interconnected groups kilometers to tens of kilometers across. Many hollows are surrounded by bright, less red material. This surrounding bright material had been detected in Mariner 10 and MESSENGER flyby images taken at lower resolution and had been called "bright crater floor material". Hollows occur where LRM has been exposed from depth, usually within craters and impact basins such as in the peak-ring and floor of Raditladi (Figure 13.25). They occur even in geologically young craters, and fresh-appearing hollows have few if any superimposed craters. These relations suggest that hollows may be actively forming even today. Hollows occur preferentially on equator- or hot-pole-facing slopes, and where impact melt or HRP-forming lava has contacted LRM, consistent with formation due to the loss of volatile material by heating; the surrounding bright material may be a residue of devolatilization. In the shallow

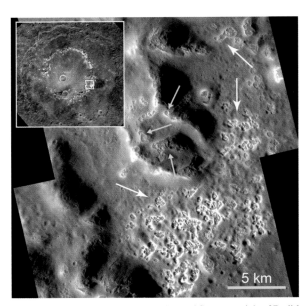

FIGURE 13.25 Hollows in the peak ring and floor materials of Raditladi basin.

subsurface under centimeters of regolith, temperature is buffered to a daily average value of 150 °C (303 °F) or less; on sunward facing slopes the surface can reach 467 °C (872 °F), and lavas and impact melts would be hundreds of degrees hotter. Although the volatile composition is uncertain, sulfur-containing phases are the leading candidates because of sulfur's abundance and the temperature regime in which the hollows form.

8. HISTORY

8.1. Geologic History

The earliest well-preserved surface features on Mercury are large basins and intercrater plains that formed concurrently during the late heavy bombardment ≥3.9 billion years ago. Older basins exist, but are recognizable only from basin-ring massifs that protrude through the intercrater plains, from remnants of ejecta with sculptured textures radial to their basins, and from regions of thinned crust. Near the end of late heavy bombardment, the youngest large impact basin, Caloris, was formed. From 3.7 to 3.9 billion years ago further eruption of lava occurred within and surrounding Caloris and other smaller basins and in the northern lowlands to form smooth plains. The global system of thrust faults formed after the intercrater plains, but how long after is unclear; there are no recognized examples of smooth plains partially burying lobate scarps that would constrain the beginning of the scarps' formation. However, the scarps do cross-cut relatively fresh craters; from estimates of the age of such craters, lobate scarps probably were still forming ∼1.5 billion years ago. Rayed craters are estimated to be about 100–200 million years old, and there are no examples of lobate scarps disrupting crater rays, suggesting that lobate scarp formation has either stopped or greatly decreased. Smaller peak-ring basins continued to form relatively late into Mercury's history, including Raditladi, which may have formed as recently as 1 billion years ago. Present-day surface processes include impact cratering, mass wasting, cold trapping of volatiles in permanently shadowed craters, space weathering, and formation of hollows.

8.2. Thermal History

Thermal history models of planetary interiors depend on compositional assumptions and are only as good as those assumptions. Mercury thermal models were generated during the time between Mariner 10 and MESSENGER and examined (among other factors) how the amount of sulfur in the core would affect the amount of global contraction that occurred as Mercury cooled from an initially molten state and the inner core solidified. The models have yet to be updated with new interpretations of other light elements occurring in the core, principally Si. With this important caveat in mind, the models predicted a total amount of planetary radius decrease due to cooling between 6 and 10 km: ∼6 km due to cooling of the mantle, and up to 4 km due to cooling of the core, with greater amounts corresponding to lower contents of sulfur in the core. The 6 km of contraction due to mantle cooling is expected to have occurred mostly before the late heavy bombardment and should not be evident in Mercury's present geology; contraction due to core cooling is thought to have occurred subsequently, and thus could be preserved. To the extent that the models remain valid, the ∼1 km radius decrease inferred from lobate scarps may be consistent with a core sulfur abundance of ∼5%.

Thermal models also suggest that Mercury has been contracting throughout its history, creating compression in the lithosphere that would tend to close fractures and make it difficult for magmas to ascend to the surface. However, large impacts would be expected to strongly fracture the lithosphere, providing egress for lavas to reach the surface.

8.3. Origin

A major question about the origin of Mercury is how it acquired such a large fraction of core-forming metal compared to the other terrestrial planets. Prior to MESSENGER's orbital measurements, four recent hypotheses had been put forward to explain Mercury's enrichment in iron. One (selective accretion) invoked an enrichment of iron due to mechanical and dynamical processes in the innermost part of the solar nebula during Mercury's accretion. Two more (postaccretion vaporization and giant impact) invoked removal of a large fraction of the silicate mantle from a once larger proto-Mercury. A final one (accretion from carbon-rich dust) invoked carbon-bearing dust desiccated of water by high inner nebular temperatures modifying the chemistry of accreting Mercury. In the selective accretion model, the different responses of iron and silicates to impact fragmentation, and aerodynamic sorting of fragments in the nebula, led to iron enrichment owing to the higher gas density and shorter dynamical timescales in the innermost part of the solar nebula. In this model, the removal process for silicates from Mercury's present position is more effective than for iron, leading to iron enrichment. The postaccretion vaporization hypothesis proposed that intense bombardment by solar electromagnetic and particle radiation in the earliest phases of the Sun's evolution vaporized and drove off much of the silicate fraction of Mercury leaving the core intact. In the giant impact hypothesis, a planet-sized object impacted Mercury and blasted away much of the planet's silicate mantle leaving the core largely intact. In the carbon-rich dust hypothesis, as the solar nebula cooled, abundant carbon bonded with silicon and kept it in a gaseous phase until iron had condensed and accreted to form Mercury's core. As a result, once silicates condensed and formed the mantle, they were nearly iron free; some silicon and oxygen were lost to more distant parts of the nebula, enriching Mercury in iron. The very reducing environment caused sulfur to be concentrated in early forming calcium and magnesium sulfide, enriching Mercury in sulfur compared to Earth.

These models predict somewhat different chemical compositions for the silicate part of Mercury, due to the different processes experienced by the early planet. For the selective accretion model, there is no reason for Mercury's silicate portion to be highly depleted in either alkali oxides (Na and K) or FeO. In contrast, postaccretion vaporization should lead to depletion of more volatile elements like sulfur and alkali oxides. For the giant impact model, the prediction depends on the composition of crust stripped away late in accretion: if there were a primordial crust analogous to the feldspar-rich lunar highlands, then alkali oxides may be depleted. However, there is no reason to expect a depletion of FeO. The carbon-rich dust hypothesis predicts extremely low iron in silicates, and unique among the four hypotheses, specifically predicts a high content of sulfur, around 4%.

MESSENGER's findings provide tests for these hypotheses. Mercury's high contents of alkali oxides and sulfur are inconsistent with postaccretion vaporization. The high alkali oxides are also inconsistent with impact stripping of a lunarlike primordial crust, but are consistent with the selective accretion model. However, this hypothesis offers no explanation for the high sulfur content or low iron in silicate. The carbon-rich dust hypothesis explains both the low iron content of silicates and the high content of sulfur, and at present, seems best able to explain Mercury's composition. The origins of Mercury's unique and surprising composition will be better understood as analysis of MESSENGER results continues and, eventually, new results are obtained by *BepiColumbo*.

BIBLIOGRAPHY

Anderson, B. J., Johnson, C. L., Korth, H., Purucker, M. E., Winslow, R. M., Slavin, J. A., et al. (2011). *Science, 333*, 1859–1862.

Blewett, D. T., Chabot, N. L., Denevi, B. W., Ernst, C. M., Head, J. W., Izenberg, N. R., et al. (2011). *Science, 333*, 1856–1859.

Chabot, N. L., Ernst, C. M., Denevi, B. W., Harmon, J. K., Murchie, S. L., Blewett, D. T., et al. (2012). *Geophysical Research Letters, 39*, L09204. http://dx.doi.org/10.1029/2012GL051526.

Denevi, B. W., Robinson, M. S., Solomon, S. C., Murchie, S. L., Blewett, D. T., Domingue, D. L., et al. (2009). *Science, 324*, 613–618.

Evans, L. G., Peplowski, P. N., Rhodes, E. A., Lawrence, D. J., McCoy, T. J., Nittler, L. R., et al. (2012). *Journal of Geophysical Research, 117*, E00L07. http://dx.doi.org/10.1029/2012JE004178, 2012.

Lawrence, D. J., Feldman, W. C., Goldsten, J. O., Maurice, S., Peplowski, P. N., Anderson, B. J., et al. (2013). *Science, 339*, 292–296.

Neumann, G. A., Cavanaugh, J. F., Sun, X., Mazarico, E., Smith, D. E., Zuber, M. T., et al. (2013). *Science, 339*, 296–300.

Nittler, L. R., Starr, R. D., Weider, S. Z., McCoy, T. J., Boynton, W. V., Ebel, D. S., et al. (2011). *Science, 333*, 1847–1850.

McClintock, W. E., Vervack, R. J., Bradley, E. T., Killen, R. M., Mouawad, N., Sprague, A. L., et al. (2009). *Science, 324*, 610–613.

Slavin, J. A., Anderson, B. J., Baker, D. N., Benna, M., Boardsen, S. A., Gold, R. E., et al. (2012). *Journal of Geophysical Research, 117*, A01215. http://dx.doi.org/10.1029/2011JA016900.

Smith, D. E., Zuber, M. T., Phillips, R. J., Solomon, S. C., Hauck, S. A., II, Lemoine, F. G., et al. (2012). *Science, 336*, 214–217.

Solomon, S. C., McNutt, R. L., Jr., Gold, R. E., & Domingue, D. L. (2007). *Space Science Review, 131*, 3–39.

Vervack, R. J., McClintock, W. E., Killen, R. M., Sprague, A. L., Anderson, B. J., Burger, M. H., et al. (2010). *Science, 329*, 672–675.

Zuber, M. T., Smith, D. E., Phillips, R. J., Solomon, S. C., Neumann, G. A., Hauck, S. A., II, et al. (2012). *Science, 336*, 217–220.

Chapter 14

Venus: Atmosphere

Fredric W. Taylor
University of Oxford, Oxford, UK

Donald M. Hunten*
University of Arizona, USA

Chapter Outline

1. **Introduction and Observations** — 306
   1.1. Earth-Based Observations — 306
   1.2. Space Missions — 307
2. **Atmospheric Temperatures** — 308
   2.1. Surface — 308
   2.2. Lower Atmosphere — 309
   2.3. Middle Atmosphere — 310
   2.4. Upper Atmosphere — 310
3. **Composition** — 311
   3.1. Carbon Dioxide and Nitrogen — 311
   3.2. Oxygen and Ozone — 311
   3.3. Water Vapor — 312
   3.4. Deuterium to Hydrogen (D/H) Ratio — 312
   3.5. Sulfur Dioxide — 312
   3.6. Carbon Monoxide — 313
   3.7. Noble Gases and Isotopes — 313
   3.8. Ionosphere — 313
4. **Clouds and Hazes** — 314
   4.1. Appearance and Motions — 314
   4.2. Vertical Layering — 314
   4.3. Global Variability — 315
   4.4. Cloud Chemistry — 316
   4.5. Lightning — 316
5. **General Circulation and Dynamics** — 316
   5.1. Surface and Lower Atmosphere Wind Profiles — 316
   5.2. Zonal Winds and Superrotation — 317
   5.3. Meridional Wind Field — 317
   5.4. Meteorology — 318
   5.5. Polar Vortex — 318
   5.6. Tides — 319
   5.7. Upper Atmosphere — 319
6. **Evolution of the Atmosphere and Climate** — 319
   6.1. Sources of Atmospheric Gases: Volcanism — 319
   6.2. Surface Pressure and the CO_2 Budget — 320
   6.3. Escape Processes and Loss of Water — 320
   6.4. Evolutionary Climate Models — 321
Bibliography — 322

Venus possesses a dense, hot atmosphere, composed primarily of carbon dioxide. A surface pressure of nearly 100 bars sustains the mean surface temperature of 740 K, which is essentially globally uniform except for topographic effects. The surface is totally hidden at visible wavelengths by multiple cloud decks extending from about 48 km altitude to about 65 km above the surface, above which the particle concentration falls off gradually with a scale height of about 3 km. The clouds are approximately bounded by the evaporation temperature of H_2SO_4 below and the top of the convectively mixed **troposphere** above. Their composition is primarily liquid droplets of concentrated sulfuric acid, with an additional ultraviolet (UV) absorber in the upper layers and large, possibly solid, particles near the base level, both of unknown composition. The middle atmosphere (**stratosphere and mesosphere**) extends from 65 to about 95 km and the upper atmosphere (**thermosphere and exosphere**) from 95 km up. Although the rotation period of the solid planet is 243 Earth days (**sidereal**), tracking of pronounced markings visible in the clouds at UV wavelengths (Figure 14.1) shows that the atmosphere in the cloud region rotates in about 4 days in the same **retrograde** direction. Evidence for complex and very active dynamics and meteorology is seen globally in variations in these markings and in the temperature, composition, and cloud density distributions in the upper troposphere up to the thermosphere. Giant polar vortices with complex, variable

---

* Dr Hunten died on December 14, 2010 and is much missed by his former colleagues. This chapter reuses some of the material from his original version in the previous edition of the Encyclopaedia, and is dedicated to his memory.

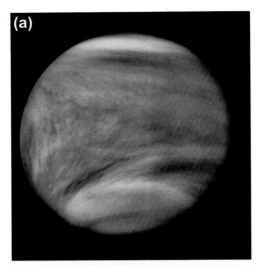

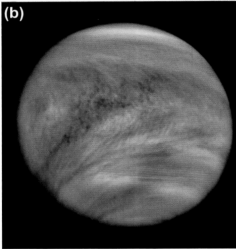

FIGURE 14.1 UV images of Venus' clouds as seen from *Pioneer Venus* Orbiter on February 5, 1979 (a) and on February 26, 1979 (b). The image has been processed to bring out the subtle structure seen in the blue and near UV and flattened to remove the limb darkening.

morphologies (most commonly dominated by wavenumber 2) are found at high latitudes (>60°) in both hemispheres.

1. INTRODUCTION AND OBSERVATIONS

1.1. Earth-Based Observations

The study of Venus through telescopes was frustrating for centuries due to the complete cloud cover, until Earth-based radars began to penetrate the cloud to map the surface in the 1960s. The presence of CO_2 in the atmosphere had been established in 1932, as soon as spectrometers equipped with infrared (IR)-sensitive photographic plates could be applied to the problem. Determining the abundance proved difficult because the radiation observed is scattered among the cloud particles and also because it was assumed that nitrogen would be abundant, as it is on Earth. This gas cannot be detected in the spectral range available from the ground, so its proportion remained speculative for many years.

Following the detection of CO_2, relatively little was achieved until the mid-1960s, when improved techniques including the development of Fourier spectroscopy led to the discovery of small proportions of water vapor and carbon monoxide in Venus' atmosphere in and above the clouds. Traces of hydrogen chloride and hydrogen fluoride were also found, and a tight upper limit was set on the amount of O_2. Similar studies of the atmosphere below the clouds became possible from 1983 onward, following the discovery of several narrow spectral "windows" in the near-IR region. These are parts of the spectrum that fall between the strong absorption bands of CO_2 and H_2O, so that the radiation from deep layers can be detected from above. Their existence had not been predicted theoretically because it was assumed that absorption in the clouds, and the far wings of strong lines, would obscure the windows. In fact, the line wings are weaker than simple theory predicts, and the sulfuric acid clouds are conservatively scattering with little absorption at short IR wavelengths (below ~2.3 μm).

In the shortest wavelength windows, as at microwave radio wavelengths, radiation from the planet's surface can escape to space. At longer IR wavelengths (greater than ~5 μm), the emission from the night side is characteristic of the temperature of the cloud tops, about 240 K. In the windows, the brightness, and therefore the temperature of the emitting region, is characteristic of the lower atmosphere and surface and therefore considerably higher. Images taken in a window reveal horizontally banded structures formed of silhouettes of the lowest part of the cloud (around 50 km) against the hotter atmosphere below (Figure 14.2).

The two most prominent near-IR windows are at 1.74 and 2.3 μm (Figure 14.3), and others are at 1.10, 1.18, 1.27, and 1.31 μm. From spectra of the absorption lines and bands in these windows, like the example in Figure 14.3, inferences about the composition at various levels all the way to the surface are possible. Some of the composition data in Table 14.1 were obtained by this technique, and the rest mainly by mass spectrometer measurements from entry probes. Earlier, millimeter-wave spectroscopy had provided estimates of the water and carbon monoxide abundances in the deep atmosphere, initially using radiation from the whole disk. In the early 1990s, modest spatial resolution became available for ground-based radio work by use of interferometric techniques in which the signals from several antennas are combined. The breakthrough in the near IR produced improved spatial resolution on the subcloud atmosphere, although only at equatorial- and midlatitudes since the polar regions are virtually inaccessible due to the small obliquity of Venus' spin axis.

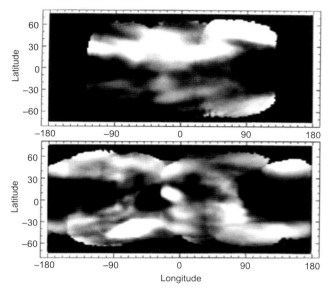

FIGURE 14.2 Near-IR images (2.36 μm) of the night side combined into maps for (above) December 31 to January 7, 1991, and (below) February 7–15, 1991. Bright areas are thinner parts of the cloud through which thermal radiation from deeper layers can shine. *From Crisp et al. (1991).*

TABLE 14.1 Composition of the Venusian Atmosphere	
Carbon Dioxide	96.5%
Nitrogen	3.5%
Sulfur dioxide	180 ppm
Water vapor	30 ppm
Carbon monoxide	17 ppm
Helium	12 ppm
Neon	7 ppm
Argon	70 ppm
Krypton	52 ppb
Xenon	2 ppb
Hydrogen chloride	0.5 ppm
Hydrogen fluoride	5 ppb

Considerable uncertainty applies to many of these numbers, especially the variable species like SO_2 and HCl, and the noble gases.

Cloud patterns detected in blue and near-UV images were used from the 1930s onward to establish the presence of the 4-day rotation at the cloud tops. Measurements of the Doppler shifting of spectral lines confirmed that this is indeed due to bulk motions, corresponding to winds of the order of 100 m/s at pressure levels similar to those near the Earth's surface. Later, IR imaging disclosed the 6-day period of a deeper region. The same patterns, seen in more detail from spacecraft, have revealed wave motions and other meteorological activity in the cloudy layers on a variety of scales.

The composition of the clouds was another important question that was answered first from analysis of ground-based observations of the polarization of light reflected from the planet. Although such measurements were first made in the 1930s, the computers and programs to carry out the analysis did not exist until the mid-1970s. The results pinned down the refractive index and showed that the upper cloud particles are spherical; these two properties eventually led to the identification of supercooled droplets of concentrated sulfuric acid (H_2SO_4). The nature of the compositional and microstructure variations in the clouds that gives rise to the UV and near-IR patterns remains obscure, however.

Radio astronomers, observing Venus's emission at the microwave wavelength of 3.15 cm, discovered in 1958 that it appears to be much hotter than expected, and this was confirmed by later results at other wavelengths. The radiation at this wavelength is expected to originate at the surface, which should be warmed, as Earth is, by the **greenhouse effect**, but the warming required to explain the new microwave data was so extreme compared to expectations that other hypotheses, such as auroral emissions, were debated. Spacecraft measurements, including radiometry from close flybys and then direct measurements from the early landers, finally settled the issue in favor of the greenhouse effect and showed that the pressure at the mean surface is 93 bars.

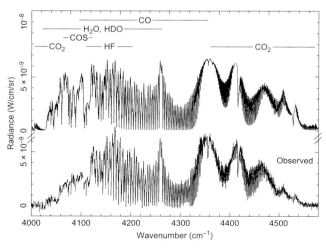

FIGURE 14.3 Near-IR spectrum in the 2.3-μm window (bottom); the spectrum above it was calculated by making use of laboratory data for the six different molecules shown. *From Bézard, de Bergh, Crisp, and Maillard (1990).*

1.2. Space Missions

A large number of experiments on 25 spacecraft have been devoted to studies of the atmosphere. United States missions, starting in 1962, were the flybys *Mariner* 2, 5, and 10 (which went on to Mercury); *Pioneer Venus Multiprobe* and *Orbiter* in 1978; the radar mapper *Magellan*; and *Galileo* en route to Jupiter, followed by the Saturn-bound *Cassini*. The Soviet Union had success with *Venera 4–14*, which included entry and descent probes as well as flybys or orbiters; *Venera* 15 and 16, which were radar mappers; and *Vega* 1 and 2, which dropped both probes and balloon-borne payloads during Venus encounters en route to Halley's comet. Early missions were devoted to reconnaissance, in particular to confirmation of the high surface pressure and temperature inferred from the microwave radio measurements, and basic composition measurements. The *Pioneer Venus* probes confirmed the cloud composition and gave unprecedented detail on the densities, sizes, and layering of the particles. The European Space Agency sent its first mission to Venus in 2006, a polar orbiter based on the design of its successful *Mars Express*, which had reached the red planet 3 years earlier. The payload for *Venus Express* focused on high-resolution imagery and spectroscopy of the atmosphere and surface, measuring winds, detecting lightning, and mapping the escape of water from the planet's exosphere and magnetosphere.

Many of the same techniques used from the Earth, principally spectroscopy, radiometry, and imaging, have been applied from flyby and orbiting spacecraft. An important addition is the radio occultation experiment, which tracks the effect of the atmosphere on the telemetry carrier as the spacecraft disappears behind the atmosphere or reappears from behind it. On Venus, the neutral atmosphere can be observed in this way from about 34 to 90 km and the ionosphere from 100 to 400 km. At greater depths, the **refraction** of the waves by the atmosphere is so great that the beam strikes the surface and never reappears. In addition to using radio occultation and carrying several instruments for remote sensing, *Pioneer Venus Orbiter* actually physically penetrated the upper atmosphere once per orbit down to as low as 135 km above the surface, carrying a suite of instruments to make measurements in situ. Two mass spectrometers measured individual gases and positive ions; a **Langmuir probe** and a retarding potential analyzer measured electron and ion densities, temperatures, and velocities; and a fifth instrument measured plasma waves. Higher energy ions and electrons, both near the planet and in the solar wind, were measured by a plasma analyzer, and important auxiliary information was provided by a magnetometer. In addition, the atmospheric drag on the spacecraft gave an excellent measure of the density as a function of height. This drag experiment was repeated by *Venus Express* in 2014, and stellar and solar occultation techniques, which have allowed spectroscopic observations up to 160 km altitude, added to new radio occultation measurements.

A number of probes have descended part or all the way through the atmosphere, and the *Vega* balloons carried out measurements in the middle of the cloud region. All of them have carried an atmospheric structure package measuring pressure, temperature, and acceleration as a function of height. The height was determined on the early *Venera* probes by radar and on all probes by integration of the hydrostatic equation. Gas analyzers have increased in sophistication from the simple chemical cells on *Venera 4* to mass spectrometers and gas chromatographs on later Soviet and US missions. In some cases, however, there are suspicions that the composition was significantly altered in passage through the sampling inlets, especially below 40 km, where the temperature is high.

A variety of instruments measured the clouds and their optical properties, most recently the visible-infrared thermal imaging spectrometer (VIRTIS) on the *Venus Express* orbiter. Previously, radiometers on entry probes had observed the attenuation of solar energy during their descent through the atmosphere, and others measured the thermal IR fluxes as well. Winds were obtained by tracking the horizontal drifts of the probes as they descended and the balloons as they floated, following direct measurement of wind, pressure, and temperature at the surface by the *Venera* landers. *Venera 11–14* carried radio receivers to seek evidence of lightning activity and *Venus Express* detected "whistler"-mode emissions attributed to lightning bursts.

Venera, *Pioneer Venus*, and *Venus Express* orbiters obtained a great deal of information on the dynamics of the atmosphere with UV and near-IR imaging and thermal IR mapping. *Pioneer Venus* discovered the north polar vortex with its double "eye", the polar warming trend in the middle atmosphere, and the principal wave modes and tides in the atmosphere at cloud-top level. Spectacular images and movies were obtained of the corresponding vortex over the South Pole by *Venus Express*, revealing its detailed structure and its dynamical variations. Still operating at the time of writing, the European mission built on the earlier explorations to paint a picture of an Earthlike planet that apparently lost its oceans of water and gained a dense, volcano-fueled atmosphere as a result.

2. ATMOSPHERIC TEMPERATURES

A representative mean temperature profile is illustrated in Figure 14.4, with an indication of the names given to the different regions and the levels occupied by the cloud layers.

2.1. Surface

The air temperature at the surface of 740 K and the apparent absence of any significant diurnal or latitudinal variation, are consistent with energy balance calculations

Chapter | 14 Venus: Atmosphere

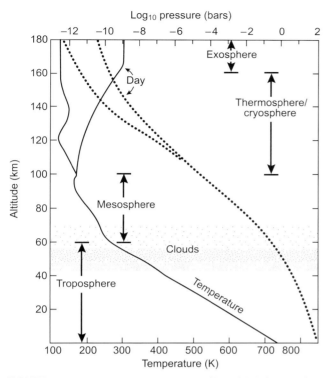

FIGURE 14.4 Temperature and pressure profiles, with their day–night variations, from the surface to 180 km altitude. The approximate locations of the main layers of cloud are also shown.

particular, is strongly absorbing at long IR wavelengths, in contrast to its behavior in the near IR where scattering dominates over absorption, allowing around 2% of the solar energy incident on the planet to reach the surface.

2.2. Lower Atmosphere

The lower atmosphere or troposphere extends from the surface up to the level of the visible cloud tops at about 65 km. The latter marks the tropopause, defined as the upper boundary of the region where vertical heat transport by convection dominates over radiative exchange with space. The lower atmospheric temperature profile has been measured in detail by numerous descent probes, with most results in close agreement, and above 35 km also by radio occultation. The gradient of temperature versus height is basically consistent with simple thermodynamic theory, being close to the dry adiabatic lapse rate of ~10 K/km from the surface to a few kilometers below the tropopause, where it tends toward the constant value with height that characterizes the overlying stratosphere (Figure 14.4). Small deviations from this lapse rate exist and have significance for the atmospheric dynamics, as discussed below.

Venus and Earth actually have rather similar vertical atmospheric temperature profiles if the comparison is restricted to the range of pressures common to both (Figure 14.5). The reason for Venus being so much hotter at the surface is the existence of the high-pressure region below 1 bar. Considering the profile of temperature versus height from space downward, once Venus' atmosphere becomes opaque, at a pressure of around 50–100 mbar, the profile is constrained by the laws of thermodynamics and hydrostatics to follow an adiabat with a gradient of approximately 10 K/km. The windows that are so useful for spectroscopic studies obviously correspond to leakage of radiative energy from the "greenhouse" in the near IR, but quantitative calculations show that the effect on vertical heat transport is essentially negligible and that the observed

for Venus' distance from the Sun and an overlying atmosphere with the observed density and composition. The high surface temperature is produced by the small percentage of solar energy that reaches the surface, trapped by the opacity of the overlying atmosphere in an extreme version of the phenomenon we know on Earth as the greenhouse effect. Radiation from the lower atmosphere in the thermal IR is ineffective for cooling to space because of the opacity of the atmospheric gases at long IR wavelengths. The species principally responsible are CO_2, SO_2, H_2O, and the solids and liquids in the clouds; H_2SO_4, in

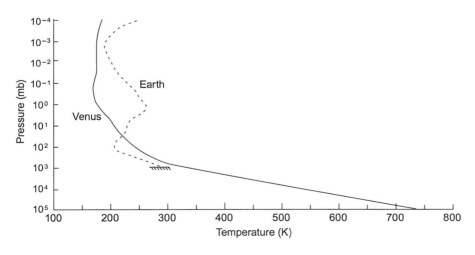

FIGURE 14.5 Temperature profiles for Venus and Earth, on a common pressure scale (From Taylor and Grinspoon (2009)). The main difference, where they overlap, is the stratospheric heating on Earth due to the ozone layer.

solar and IR net fluxes can be closely reproduced without considering them.

The uniformity with both latitude and longitude of the surface temperature, for a given elevation, is consistent with the high time constant of around 100 years for the dense near-surface atmosphere and very slow (<1 m/s) near-surface winds, again as observed. In this regard, the deep Venusian atmosphere is somewhat analogous to the oceans on Earth. The vertical relief of the surface is about 15 km and, taking into account the vertical temperature gradient, is responsible for surface temperature differences of around 150 K across the planet. Again, this has an effect on the general circulation.

2.3. Middle Atmosphere

The middle atmosphere (stratosphere and mesosphere) extends from the tropopause near 65 km to the temperature minimum or mesopause at about 95 km (see Figure 14.4). Here, as expected for a region in radiative balance between the hot tropopause below and cold space above, the temperature is approximately constant with height. The distinction between the stratosphere and mesosphere is not as useful as it is on Earth, where the middle atmosphere temperature profile is strongly affected by the ozone layer, and most authors use "middle atmosphere" or mesosphere (which of course means the same thing) for the combined 30-km-deep layer.

Unlike the lower atmosphere, there is considerable temperature structure and variability in the middle atmosphere (Figure 14.6). The latitudinal gradient above the clouds is such that the polar region is warmer by 15–20 K than the equator, that is, in the opposite sense to what would be expected from radiative balance. This is in contrast to the upper troposphere, where a large equator-to-pole temperature contrast of ~30 K in the other sense is found at the 1000 hPa level. Model studies indicate that the polar warming is a feature of the rapid global superrotation of the atmosphere, being associated with the pressure gradients required to balance the flow.

In the circumequatorial (zonal) direction, the most remarkable feature of the temperature field is a prominent solar-fixed wavenumber-2 structure with an eastward, i.e. upwind, tilt with increasing altitude at a mean rate of 6 K/km. At the cloud tops, temperature maxima are found just before local noon and midnight, and these have moved eastward by more than 180° in longitude by the 100-km level. This clearly is an upward-propagating solar tide, which is to be expected, although the dominance of wavenumber-2 is less obvious as a response to predominantly wavenumber-1 forcing. In fact, wave 1 does dominate inside the clouds where most of the solar energy is deposited, and the cloud tops themselves rise and fall by about 2 km during the solar day, but tidal models show that this component propagates vertically much less efficiently than wave-2.

2.4. Upper Atmosphere

The upper atmosphere (thermosphere and exosphere) lies above the mesopause. Here, temperatures can no longer be measured directly, but are inferred from the

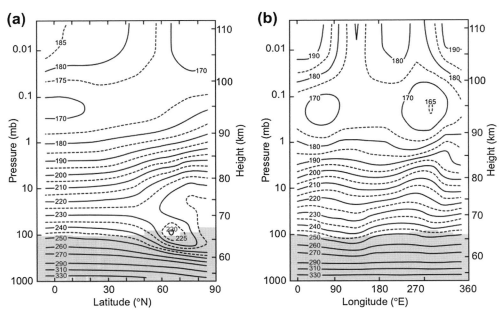

FIGURE 14.6 Time-averaged temperature fields in the middle atmosphere of Venus (Schofield & Taylor, 1982). (a) The zonal mean field and (b) the variations around a latitude belt from 0 to 30° N, both plotted against pressure and approximate height. The horizontal stepped line represents the retrieved mean cloud-top height.

scale heights of various gases with use of the **hydrostatic equation**. On Earth, this region is called the thermosphere because temperatures as high as 1000 K are reached in the outermost layer, or exosphere. The exospheric temperature is much more modest on Venus, no more than 350 K on the day side and not far above 100 K on the night side, a result of strong cooling by the IR bands of carbon dioxide (Figure 14.4). The large temperature difference translates into a pressure difference that drives strong winds from the day to the night side at all levels above 100 km.

On Earth, the exospheric temperature changes markedly with solar activity, being around 700 K at sunspot minimum and 1400 K at maximum. The corresponding change at Venus is much more modest, perhaps 50 K. These differences are again traceable to the fact that CO_2, the principal radiator of heat, is just a trace constituent of Earth's atmosphere but is the major constituent for Venus (and also Mars). Venus's slow rotation also contributes to the very cold temperatures on the night side, although the atmosphere does rotate substantially faster than the solid planet.

3. COMPOSITION

A summary of the bulk composition of Venus' atmosphere from all measurements is given in Table 14.1 and model vertical profiles are given in Figure 14.7.

3.1. Carbon Dioxide and Nitrogen

Confirmation that carbon dioxide is indeed the major gas came from a simple chemical analyzer on the *Venera 4* entry probe. The mole fraction was found to be about 97%, in reasonable agreement with the currently accepted value. The next most abundant gas is nitrogen; although it is only 3.5% of the total, the absolute quantity is about three times that in the Earth's atmosphere. Many of the differences between Venus and Earth can be traced to a relative scarcity of water in the atmosphere of Venus and the total absence of liquid water on the surface. On Earth, carbon dioxide and sulfuric, hydrochloric, and hydrofluoric acids are all carried down by precipitation, a process that is absent in the hot, dry lower atmosphere of Venus. They all then react and are incorporated in geological deposits; estimates of the total amount of carbonate rocks in the Earth give a quantity of CO_2 almost equal to that seen in the atmosphere of Venus.

3.2. Oxygen and Ozone

Free oxygen is undetectable at the Venus cloud tops where one molecule in 10 million could have been seen. At higher levels, the oxygen in carbon dioxide is liberated by dissociation by sunlight and is readily detected (along with CO) by spacecraft instruments orbiting through the upper atmosphere and in airglow measurements. The free oxygen is removed before it can diffuse down to the cloud level by the action of a strong mechanism in the middle atmosphere that converts O_2 and CO back to CO_2. The nature of this mechanism is revealed by observations of around 1 ppm of hydrogen chloride in the region. The HCl molecules are also subject to dissociation by solar radiation, yielding a chlorine concentration that is nearly 1000 times greater than that on Earth. This participates in a **catalytic cycle** that can completely eliminate free oxygen from the middle atmosphere and prevent the formation of ozone in significant quantities.

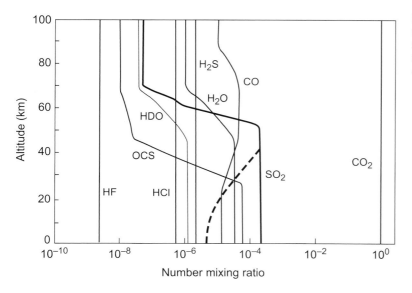

FIGURE 14.7 Model profiles of atmospheric composition based on various measurements. In the case of SO_2, the results near the surface are controversial with *Vega* 1 and 2 measurements (Bertaux, Widemann, Hauchecorne, Moroz, & Ekonomov, 1996) shown as a dashed line.

On Venus the chlorine-based chemistry is closely coupled to the sulfur cycle that maintains the clouds, as further discussed below.

3.3. Water Vapor

Measurements of water abundance above, within, and below the clouds show a great deal of variance, some attributable to difficulties with the measurement technique but some probably real. Overall, it is clear that Venus has between ten and one hundred thousand times less water in its atmosphere than exists in the oceans and atmosphere of the Earth.

The fact that water is depleted while, at the same time, deuterium is more than 100 times more abundant on Venus than Earth (see next section) suggests that Venus had much more water initially but that most of it has been lost by dissociation to form hydrogen and oxygen which then escape from the planet. Models suggest that Venus could have lost an ocean of present-day terrestrial proportions in only a few hundred million years in this way. The ratio of heavy to light hydrogen (D/H) on Venus would be much larger if all of the deuterium in a primordial Venusian ocean had been retained; however, deuterium as well as hydrogen can escape from the atmosphere, only more slowly due to its greater mass which affects the mixing ratio of vapor in the upper atmosphere falls and leads to fractionation of the two isotopes.

Studies of the water vapor abundance in the middle atmosphere of Venus, both within and above the main cloud deck, have yielded very diverse results. Earth-based observations, made from high-altitude sites and with high-resolution airborne spectrometers, resulted in mixing ratios between 0 and 40 ppm and evidence of higher values in localized and temporary wet spots. The infrared radiometer experiment on board the *Pioneer Venus* Orbiter discovered the presence of a wet area (~ 100 ppm) in the afternoon equatorial region, presumed to be produced by the heating maximum at the subsolar point, with very small amounts (a few ppm) of water vapor at other times of the day and at higher latitudes.

Such large variations do not seem to occur in the lower atmosphere. Although in situ measurements by the *Venera* and *Pioneer Venus* entry probes and the two *Vega* balloons reported mixing ratios ranging from 30 to 5000 ppm, direct sampling measurements of water are notoriously difficult and the higher value probably can be discounted. Since the discovery of the near-IR windows, high-resolution Earth-based spectroscopic observations of the night side have found a near-constant global abundance of 40 ± 20 ppm over a range of subcloud altitudes. The near-IR mapping spectrometer operating during the *Galileo* flyby of Venus confirmed this with a deep atmosphere water vapor mixing ratio of 30 ± 15 ppm, subsequently endorsed by *Venus Express*.

3.4. Deuterium to Hydrogen (D/H) Ratio

The D/H ratio was first measured on ions in the ionosphere then confirmed in the middle and lower atmosphere by mass spectrometer data and by analysis of spectra taken from Earth in the near-IR windows. For the direct sampling instruments, deuterium provided a valuable signature for distinguishing Venus water vapor in the mass spectrometer from any contaminants carried along from Earth. Values for the D/H isotopic ratio from the accumulated measurements averaged around 0.019, 120 times larger than Earth.

The very high spectral resolution obtained by the Spectroscopy for Investigation of Characteristics of the Atmosphere of Venus (SPICAV)/Solar Occultation at Infrared (SOIR) spectrometer on *Venus Express* allows the measurement of simultaneous vertical profiles of H_2O and HDO above the clouds and D/H fractionation. The averaged HDO/H_2O ratio equals a factor of 240 ± 25 times the ratio in Earth's ocean, or around two times the bulk atmospheric value measured in the lower atmosphere. This could be due to preferential destruction of H_2 relative to HD, preferential escape of H relative to D, or possibly selective condensation, a process that has recently been found to be important for fractionating D and H on Mars and Earth.

3.5. Sulfur Dioxide

The high sulfur content of the atmosphere, including the H_2SO_4 clouds, is a powerful indicator of recent volcanic activity, since gases like sulfur dioxide have a short lifetime in the atmosphere before they are removed by interaction with the surface. The measured abundance of SO_2 in the deep atmosphere is about 180 ppm, which is more than 100 times too high to be at equilibrium with the surface. The time constant for the decline of the sulfur abundance in the atmosphere if the source were removed is a few million years, indicating that the atmospheric sulfur must be of recent origin. *Pioneer Venus* UV spectra showed a decline by more than a factor of 10 in sulfur dioxide abundance at the cloud tops over a 5-year period, and more recently, *Venus Express* has also detected very large, long- and short-term variations in SO_2 at all altitudes from the clouds to the thermosphere.

The high level of SO_2 in the atmosphere is the source for the concentrated sulfuric acid that is the dominant component of the clouds (see Section 4.4 below). Although less well understood, it is probably the nonuniform distribution of SO_2 and the formation of trace amounts of elemental sulfur and possibly other sulfur compounds that gives rise to the UV markings in the clouds that the visible face of Venus. Apart from forming the highly reflective clouds that tend to cool the planet, sulfur dioxide is a greenhouse gas contributing to the warming of the surface (Section 6).

3.6. Carbon Monoxide

The principal source of carbon monoxide in the atmospheres of both Venus and Mars is the dissociation of CO_2 by solar UV radiation, which occurs primarily at high levels on Venus since the energetic photons required do not penetrate far into the atmosphere. During its encounter with Venus in February 1990, the near-infrared mapping spectrometer NIMS on the Galileo spacecraft used observations of the CO 2-0 vibration-rotation band in the 2.3-μm spectral region to obtain the concentration of CO in the deep atmosphere, around 30 km above the surface. These data indicated relatively high concentrations of CO at high northern latitudes, a mean increase of ∼50% from 23 ± 2 ppm to 32 ± 2 ppm, behavior which was shown to occur in a symmetric manner in the other hemisphere by observations from *Venus Express* in 2006. The findings raise the question of what is the source of the lower atmospheric CO and why its proportion in the well-mixed lower atmosphere increases systematically from equator to pole in both hemispheres. Vertical diffusion from the upper atmosphere is too slow, since the chemical lifetime is probably only a few days, and current understanding of atmospheric chemistry on Venus suggests *depletion* at all levels in the lower atmosphere. Probably the answer lies with the global circulation, which produces enhanced downward transport from the upper atmosphere at high latitudes, as discussed below.

3.7. Noble Gases and Isotopes

Most of the advanced studies and proposals for future missions to Venus emphasize the potential value of obtaining more accurate and comprehensive measurements of the isotopic ratios of the chemical elements within the gases in the atmosphere. Some current values are listed in Table 14.2. These can give important clues about the early history of the Solar System, the formation of the planets, and the evolution of their climates. Particularly interesting cases, in addition to the deuterium to hydrogen ratio ($^2H/^1H$ or D/H) discussed above, are the isotopic ratios of helium ($^2He/^3He$), nitrogen ($^{15}N/^{14}N$), carbon ($^{12}C/^{13}C/^{14}C$), and oxygen ($^{16}O/^{18}O$). The noble gases argon ($^{36}A/^{38}A/^{40}A$), krypton, neon, and xenon are all both heavy (hence less likely to escape) and inert (hence not prone to removal by chemical reactions with the crust) and so the ratios of their abundances on different planets should provide particularly good tests of evolutionary models. In this way we would hope, for example, to distinguish between initial capture from the solar nebula or later from the solar wind, radioactive decay in the interior, followed by release into the atmosphere, and the contribution from the collision of volatile-rich bodies such as comets with the planet.

TABLE 14.2 Some Isotopic Abundance Ratios for Venus, with Comparisons with Earth and Mars

Isotopic Ratio	Venus	Earth	Mars
$^2H/^1H$	0.02	0.00015	0.0008
$^{18}O/^{16}O$	0.002	0.002	0.002
$^{13}C/^{12}C$	0.01	0.01	0.01
$^{15}N/^{14}N$	0.003	0.003	0.006
$^{22}Ne/^{20}Ne$	0.08	0.1	0.1
$^{38}Ar/^{36}Ar$	0.18	0.19	0.18

Interpretations rely on quite complex models, the integrity of which is hampered by the inadequacy of present data. For example, xenon, the heaviest of the relatively abundant noble gases, has not been measured dependably on Venus at all. Still, some important inferences are possible. Primordial argon and neon are more than an order of magnitude less abundant on Earth than on Venus and less common on Mars than Earth by a similar large factor. This observation is interpreted to mean that the present atmospheres accreted along with the planetary bodies and were later outgassed, with the differences in argon abundance explained in terms of differences in the density of the solar nebula at the time of formation. At the same time, the ratios between the noble gas abundances on Venus are more similar to the ratios found in the Sun than those of Earth and Mars. This suggests less modification and fractionation than the highly processed atmospheres of Earth and Mars.

The isotopic ratios of carbon and oxygen are approximately the same on Venus, Mars, and Earth, but $^{15}N/^{14}N$ in the Martian atmosphere is about 1.7 times that on Earth. If the difference on Mars is a result of the preferential loss of the lighter isotope, models require that nitrogen was initially of the order of 100 times more abundant on Mars than it is now. Venus has roughly the same (actually about three times as much) total amount of atmospheric nitrogen as Earth and a similar $^{15}N/^{14}N$ ratio; this tends to suggest that both planets may have retained most of their original inventory.

3.8. Ionosphere

The principal heat source for the thermosphere is the production of ions and electrons by far-UV solar radiation. The most abundant positive ions are O_2^+, O^+, and CO_2^+. As part of these processes, CO_2 is dissociated into CO and O and N_2 into N atoms. All of these ions, molecules, and atoms

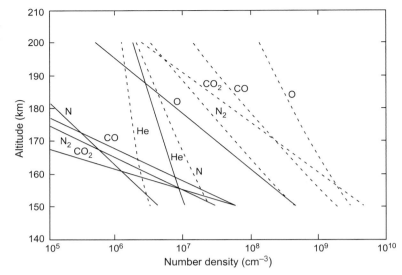

FIGURE 14.8 Daytime (dashed) and nighttime (solid) number densities of the major gases in the thermosphere obtained by fitting a large number of measurements by the mass spectrometer on *Pioneer Venus Orbiter*. After Hunten, Colin, Donahue, and Moroz (1984).

have been observed or directly inferred (Figure 14.8). Some of the O^+ ions (with an equal number of electrons) flow around to the night side and help to maintain a weak ionosphere there. Venus lacks any detectable magnetic field, and the dayside ionosphere is therefore impacted by the solar wind, a tenuous medium of ions (mostly H^+), and electrons flowing from the Sun at about 400 km/s. Electrical currents are induced in the ionosphere, and they divert the solar wind flow around the planet. The boundary between the two media, called the ionopause, is typically at an altitude of a few hundred kilometers near the subsolar point, flaring out to perhaps 1000 km above the terminators and forming a long, tail-like cavity behind the planet (*See The Solar Wind*).

4. CLOUDS AND HAZES

4.1. Appearance and Motions

The view of the planet from Earth or from an orbiting spacecraft shows Venus completely and permanently shrouded in cloud. The surface apparently formed by the cloud is not a discrete upper boundary, but extends upward as a gradually thinning haze to least 80 km. The "cloud top" is usually defined as the level at which the **optical depth** reaches unity at some specified wavelength, and for visible light occurs near 65 km altitude. At this level, the range of visibility (the horizontal distance within which objects are visible) is still several kilometers.

Studies of sequences of images like those in Figure 14.1 reveal that the cloud-top region is rotating with a period of about 4 days, corresponding to an equatorial east–west wind speed of about 100 m/s. This varies with latitude, usually reaching around 140 m/s in a localized high-speed jet near 60° in each hemisphere, although secular variations have also been observed and sometimes the jet is absent, leaving near solid-body rotation speeds. Near-IR images like Figure 14.3 show a longer period consistent with the idea that the dark features are silhouettes of the lower cloud, where entry probes have measured wind speeds of 70–80 m/s.

4.2. Vertical Layering

Several entry probes have made measurements of cloud scattering as they descended, and spectroscopic and other optical measurements add information about what is obviously a very complex and variable regime. Three regions (upper, middle, and lower) can be distinguished in the main cloud, and there is also a thin haze extending down to 30 km. Descent probe and spectroscopic and polarimetric data suggest that the particle size distributions form four distinct "modes", which have different mean particle radii and different distributions with height, and possibly different constituents. Figure 14.9 shows a simplified interpretative model that shows a mean vertical structure and inferred composition for the clouds.

The upper cloud, the one that can be most readily studied from the Earth or from orbit, is in a region of high convective stability and intense solar UV irradiation, indicating a photochemical production regime. Here, most of the opacity is due to "Mode 2" particles with radii around 1 μm. The same particles extend throughout the clouds, but become somewhat larger in the middle and lower clouds (Mode 2′). In the deepest layer, which the vertical stability data from temperature profile measurements suggest is in a tropospheric convective regime, a third population of large Mode 3 (6–35 μm in diameter) particles is also found. The existence of multiple, distinct modes is still neither

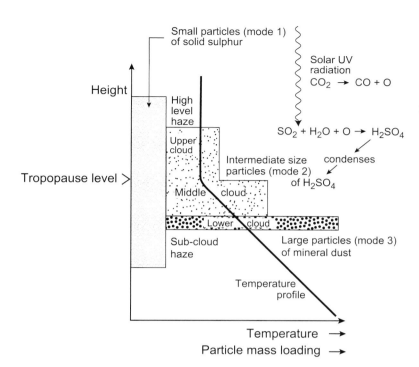

FIGURE 14.9 A model for the vertical structure and possible composition of the cloud layers on Venus, typical of most of the global cloud cover but excluding the polar regions. Most of the opacity, and most of the global variability, is in the lower layer, which is physically thin but contains the largest radii and the highest concentration of particles.

understood nor well delineated and depends to a considerable extent on the interpretation of the single profile obtained by the *Pioneer Venus* Large Probe. However, the dominance in the upper cloud of a "monodispersion" of spherical sulfuric acid droplets with a tight size distribution around a radius of ∼1 μm is well established from polarimetry measurements. There is some evidence, also from *Pioneer*, that the Mode 3 particles in the lower cloud are nonspherical and this, and their large size, suggests that they might contain solid crystals.

4.3. Global Variability

Ground-based observations first revealed enormous variations in the optical thickness of the lower cloud deck, which were later studied in detail from spacecraft. The clustered appearance of the deep clouds is consistent with tropospheric cumulus, in contrast to the muted variability and small contrasts seen in the upper clouds. The spatial variation of sulfuric acid concentration in the cloud particles has been estimated from *Venus Express* VIRTIS spectral maps and found to be higher in regions of optically thick cloud. The retrieved cloud base altitude varies with latitude, reaching a maximum height near −50° before falling by several kilometers toward the pole, along with a similar fall in the cloud-top height (Figure 14.10). The cloud particles in the polar region have different scattering properties from elsewhere on the planet and this has been interpreted as being due to an increase in average particle size, and possibly a difference in composition, near the pole.

The latitudinally variable CO abundance at 35–40 km altitude originally found by *Galileo* has been confirmed by *Venus Express*. An increase in CO collocated with a decrease in tropospheric H_2O abundance is observed at high latitudes, which is probably due to strong downwelling between ±60° and ±75° latitude marking the poleward extent of the Hadley cell circulation. In addition, tentative evidence for long-term secular change, over a period of 2

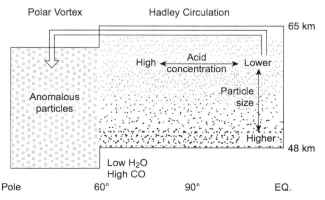

FIGURE 14.10 A schematic diagram of the global variability in cloud structure and related gaseous species, plus major circulation features, inferred from long-term spectral mapping by *Venus Express*. After Barstow et al. (2012).

Earth years, was observed in the acid concentration and the CO and H_2O abundances.

4.4. Cloud Chemistry

The high abundance of sulfur dioxide in the atmosphere leads to the formation of the concentrated sulfuric acid cloud layers via a chemical system involving the photolytic destruction of carbon dioxide by solar UV radiation, summarized by

$$CO_2 \rightarrow CO + O$$

followed by reactions equivalent to

$$SO_2 + H_2O + O = H_2SO_4.$$

This sequence forms the acid near the visible cloud tops, where it combines with other H_2O molecules to produce the hydrated acid droplets that are the main constituent of the clouds. The degree of hydration varies between perhaps 10% and 25%, with 20% ($4H_2SO_4 \cdot H_2O$) typical.

A cloud particle of the observed mean radius (~ 1 μm) has a sedimentation velocity of 7.5 m/day at 60 km; this velocity varies as the square of the size. Although small, these velocities, aided by coagulation, eventually carry the particles out of the cloud to lower altitudes and higher temperatures, where they will evaporate and, at still lower heights, decompose back into water and sulfur dioxide. Atmospheric mixing carries these gases back upward where they can again contribute to the formation of H_2SO_4. An important intermediate is the reactive free radical SO, and probably some elemental sulfur is produced. UV spectra (pertaining to the region above the clouds) reveal the presence of small amounts of SO_2 shown in Table 14.1, but much less than the amounts that have been measured below the clouds.

Sulfuric acid is perfectly colorless in the blue and near-UV regions, and the yellow coloration that provides the contrasts of Figure 14.1 must be caused by something else. The most likely thing is elemental sulfur, but yellow compounds are abundant in nature, and the identification remains tentative. The photochemical models do predict production of some sulfur, but it is a minor by-product, and the amount produced is uncertain. It is also unclear what constitutes the large Mode 3 particles in the lower cloud. Optical data suggests solid, irregular particles coated with sulfuric acid; the most likely candidate for the solid material is volcanic ash.

4.5. Lightning

Electromagnetic pulses, attributed to lighting bursts, have been observed by several entry probes and orbiters. Most recently, *Venus Express* confirmed the existence of "whistler" mode waves with burst durations of about 100 ms and properties similar to signals generated by atmospheric lightning in the terrestrial ionosphere. The frequency of occurrence, mapped over more than 4 years, suggests a level of lightning activity on Venus that is also similar to Earth's. However, searches for the corresponding optical flashes have been negative, except for one ambiguous inference from *Venera 9* and a few optical events reported from Earth-based observations. A close flyby by the *Cassini* spacecraft saw no evidence of any impulses with a sensitive instrument that, in a later Earth encounter, found them in abundance. Theoreticians have opined that conditions on Venus do not seem propitious for large-scale charge separation, noting that on Earth, lightning is seen during intense precipitation and in volcanic explosions. In thunderstorms, large drops are efficient at carrying charge of one sign away from the region where it is produced, and the gravitational force is large enough to resist the strong electric fields, but this is not the case for small particles. The evidence for large, precipitating particles on Venus remains incomplete, as indeed are direct observations of volcanoes, although both are difficult to detect and either or both may turn out to be the source of the lightning behind the radio bursts.

5. GENERAL CIRCULATION AND DYNAMICS

5.1. Surface and Lower Atmosphere Wind Profiles

The Soviet landers *Venera* 9 and 10 used a simple cup anemometer, i.e. a rotating vane device similar to those seen on most earthbound meteorological stations, and found velocities of ≤ 1 m/s. As noted above, such slow winds are in line with expectations based on the atmospheric density and small thermal contrasts at these levels. It is also possible to track the drift of descent probes as they pass through the atmosphere, and so to obtain vertical coverage and directional data, as well as wind strength. The *Venera* landers were tracked by the measurement of the Doppler shift in the radio signal from the spacecraft, while the *Pioneer Venus* probes used an interferometric technique involving more than one receiving station.

The results (Figure 14.11) confirm the low surface winds but show a dramatic increase with height to speeds near 100 m/s near the cloud tops, more than 50 times faster than the rotation rate of the surface below. This zonal "superrotation" also manifests itself in the observed cloud structure, which moves rapidly around the planet in a direction parallel to the equator. The cloud markings, which appear with high contrast through an UV filter, have their origin at heights near 60 km above the surface (where the pressure is of the order of 100 mb). Motions in the deeper cloud layers can be observed by near-IR imaging on the

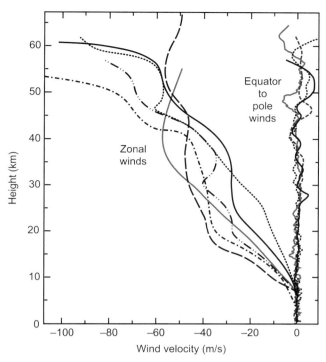

FIGURE 14.11 Profiles of the zonal and meridional winds on Venus as measured by tracking the *Pioneer Venus* entry probes during their descent at different latitudes and local solar times (Counselman, Gourevich, King, Loriot, & Ginsberg, 1980).

parameterization of viscosity, particularly that due to eddies, and assumptions of cyclostrophic balance that break down near the equator, has shown that the temperature field is consistent with the observed winds.

Attempts have been made to explain the high zonal wind speeds in the relatively dense atmospheric regions on Venus by mechanisms that convert the slow motion of the Sun, relative to a fixed point on Venus, into a much more rapid motion of the atmosphere. Currently prevailing opinion favors a mechanism in which momentum from the solid planet is transported by eddies whose interaction with the main flow is complex and in which the mean meridional circulation plays an important role. Experiments with general circulation models based on this principle suggest that global superrotation is always a characteristic of optically thick atmospheres on slowly rotating planets, with the predicted wind speed depending critically on the detailed energy deposition profile of the atmosphere. Such models can also explain the observed superrotation of Titan's atmosphere. Their validation requires more information about wind speed versus latitude and height, cloud variability and wave modes in the atmosphere below the visible cloud tops, and the role of the surface topography in maintaining or opposing the superrotation, some of which is currently being provided by UV and near-IR mapping from *Venus Express*.

night side of the planet in the "windows" at wavelengths from 1 to 3.5 μm. These originate in the main cloud deck, illuminated from behind by the hot lower atmosphere. The typical velocities inferred near the equator were about half as fast as those from UV markings, consistent with the vertical profiles of wind and cloud opacity measured by the Pioneer and *Venera* probes. Because the density increases by a large factor over this height range, the angular momentum is a maximum at 20 km.

5.2. Zonal Winds and Superrotation

Superrotation is observed in many atmospheres, usually at the low pressures found in thermospheres, and superposed on a rapid planetary rotation. Venus exhibits the most dramatic example, with zonal winds in excess of 100 m/s seen in the circulation of UV cloud markings at relatively high pressure levels of around 100 mb. The time-averaged global maps of the Venusian middle atmosphere temperature field obtained by remote sensing (Figure 14.6) show several features clearly related to the general circulation, including the temperature increase from pole to equator over a broad altitude range, against the trend in radiative heating, and the variation with local time of day of the air temperature. A model-dependent analysis, requiring the

5.3. Meridional Wind Field

The cloud motions which trace the zonal winds also reveal the pattern of the meridional circulation on Venus, although with much larger proportional errors, since the poleward component is much slower than that parallel to the equator (Figure 14.11). Despite the uncertainties, the data fairly consistently show mean poleward motions of up to 10 m/s at most latitudes, tending to confirm the theoretical expectation that Hadley cells exist in each hemisphere, i.e. global-scale circulation cells characterized by rising motion at low latitudes and descending motion nearer to the poles. It has been suggested that the *Pioneer Venus* probe tracking data, showing alternations in the direction, as well as the magnitude, of the meridional wind, marks the passage of the probe through the different components of a stack of Hadley cells, each extending from the equator to high latitudes. This notion is supported by the fact that the Hadley cell seen at the cloud tops appears to be thermally indirect, that is to say, carries heat from the equator to the pole against the observed temperature gradient. It may be driven by a stronger, direct cell underneath, and the layered eddy sources and sinks which could drive the zonal superrotation may be related to the cell interfaces, although the current data is inadequate to establish this with any certainty.

5.4. Meteorology

The tracking of meteorological features on Venus was, for many decades, limited to the transient and quasi-permanent features seen in the UV images of the cloud-top region, where they revealed structures identified with Rossby and gravity wave activity, and the measurement of temperature anomalies by remote sensing. The spectral imaging instruments on *Venus Express* have exploited the fact that the near-IR windows permit imaging of the deep cloud structure to investigate the meteorological activity that is clearly present in the deep atmosphere of Venus. Along with high-resolution UV images, these reveal chaotic convective and wave activity near the equator where most of the solar energy is deposited in the clouds, with an abrupt switch to a more laminar flow at midlatitudes, and then finally a further transition to the polar vortex complex near the poles.

5.5. Polar Vortex

Vortex behavior occurs in the polar region of any terrestrial planet, due to the subsidence of cold, dense air and the propagation of zonal angular momentum in the meridional flow. On Venus, the small obliquity and the equatorial superrotation lead to an extreme version of this effect, manifest by a sharp transition in the circulation regimes in both hemispheres at a latitude of about 65°. There, a complex instability develops, resulting in dramatic long-lived wave structures. The *circumpolar collar* takes the form of a belt of very cold air that surrounds the pole at a radial distance of about 2500 km and has a predominantly wavenumber-1 structure locked to the Sun. The vertical extent of the collar must be much less than its 5000 km diameter, and the indications from *Pioneer Venus* and *Venus Express* data are that it may be only about 10 km deep, with a complex vertical structure. The cloud-top temperatures that characterize the collar are about 30–40 °C colder than at the same altitude outside, so the feature generates pressure differences that would cause it to dissipate rapidly were it not continually forced.

Poleward of the collar, the air at the center of the vortex must descend rapidly to conserve mass, and we expect to find a relatively cloud-free region at the pole, analogous to the eye of a terrestrial hurricane but much larger and more permanent. Interestingly, however, the "eye" of the Venus polar vortex is not circular but elongated, and with typically two brightness maxima (possibly corresponding to maxima in the downward flow) at either end of a quasi-linear feature connecting the two (Figure 14.12). This wave-2 characteristic gives the polar atmosphere a "dumbbell" appearance in IR images that use the thermal emission from the planet as a source and has led to the name *polar dipole* for the feature. A dipole was first seen at the North Pole by *Pioneer Venus*, and a similar feature has been discovered and extensively studied at the South Pole as well by *Venus Express*. In particular, it has become clear that the "dipole" description is too simplistic: more complicated shapes, as well as monopoles and tripoles, also occur, with remarkably rapid (for such a large feature) morphing between them, although wave-2 does seem to dominate as some theories predict.

The northern dipole was observed in successive images obtained in 1979–1980 to be rotating about the pole with a period whose dominant component, among several, was 2.7 Earth days, i.e. with about twice the angular velocity of the equatorial cloud markings. If angular momentum were being conserved by a parcel of air as it migrated from equator to pole, the dipole might be expected to rotate five or six times faster. In fact, the UV markings are observed to keep a roughly constant zonal velocity (solid-body rotation) from the equator to at least 60° latitude, and must be accelerating poleward of this if the rotation of the dipole represents the actual speed of mass motions around the pole

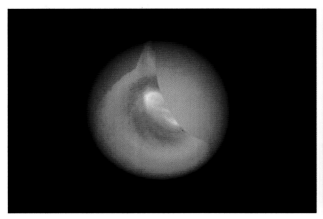 

FIGURE 14.12 The Venus polar "dipole", left the North Pole, by *Pioneer Venus*, and right, the South Pole, by *Venus Express*. The bright feature is approximately 2000 km across in both cases.

and not simply the phase speed of a wavelike disturbance superimposed on the polar vortex. Southern dipole observations by *Venus Express* found a rotation rate that varied from 2.2 to 2.5 days, and confirmed that, in addition to a variable rotation rate, the position of the apparent center of the vortex can wander several degrees away from the rotation pole of the planet.

5.6. Tides

A particularly important form of wave motion is the *solar tide*, the temperature and density cycle induced by the apparent motion of the Sun overhead. This contains a whole spectrum of Fourier components, because the forcing is nonsinusoidal; the actual atmospheric response depends on the mean wind and the interference between the various components. The Earth's atmosphere has a small wavenumber-2 component superposed on the familiar early afternoon maximum to postmidnight minimum cycle, but this component dominates on Venus. In fact, the dynamical theory of atmospheric tides, as developed for Earth, shows when applied to Venus that the observed state of affairs can be explained as primarily a consequence of the long solar day on Venus, provided that a realistic representation of the zonal wind is incorporated.

5.7. Upper Atmosphere

There are no direct wind measurements above the cloud tops, but deductions from temperature measurements and the measurement of Doppler-shifted emission lines from atmospheric gases suggest a slowing of the 100 m/s flow up to perhaps the 100-km level. At still greater heights, the dominant circulation switches to a rapid day–night flow, first suggested on theoretical grounds and confirmed by the large observed temperature difference. But the flow is not quite symmetrical; maxima in the hydrogen and helium concentrations, and in several airglow phenomena, are systematically displaced from the expected midnight location toward morning. The thermospheric winds carry the photochemical products O, CO, and N from the day side to the night side, where they descend into the middle atmosphere in a region ~ 2000 km in diameter and generally centered near the equator at 2 a.m. local time. This region can be observed by the airglow emitted during the recombination of N and O atoms into NO molecules, which then radiate in the UV, and O_2 molecules, which radiate in the near IR. The light gases hydrogen and helium are also carried along and accumulate over the convergent point of the flow; for these gases, the peak density is observed at about 4 a.m. These offsets are the principal evidence that this part of the atmosphere rotates with a 6-day period, a rotation that is superposed on the rapid day-to-night flow.

6. EVOLUTION OF THE ATMOSPHERE AND CLIMATE

Earth and Venus have a common origin if, as is generally believed, the Sun, the planets, and their atmospheres condensed, about 4.6 billion years ago, from the same solar nebula (*See* The Origin of the Solar System). An intermediate stage was the formation of planetesimals, typically Moon-sized objects in noncircular orbits that collided and merged, so that those in the inner solar system might end up as part of any of the terrestrial planets. One would therefore expect them to begin with similar atmospheric compositions, and indeed those of Venus and Earth have many similarities and the same gases are present in each, although in different proportions.

The key question is whether the large divergences in the surface climate we now see have arisen from different evolutionary paths, starting from quite similar early atmospheric conditions, and to what extent the two planets may always have been very different. In either case, the amounts of the various gases in the atmosphere are expected to vary in time, including those that contribute to heating the surface, especially carbon dioxide. Water vapor and sulfur dioxide are also "greenhouse" gases, and are the main precursors of the clouds, which reflect some of the solar energy that would otherwise heat the planet. These and other gases like carbon monoxide and argon are released from the interior through volcanic activity and outgassing, while water in particular may also be brought in by icy debris from space.

At the same time as gases are being added, they are lost by various processes. Carbon dioxide and sulfur dioxide have well-studied reactions with rocky minerals that are likely to be common on the surface of Venus. Water and carbon dioxide are dissociated by solar UV radiation, but the CO_2 tends to recombine, while the light H atoms from H_2O escape relatively easily into space from the top of the atmosphere. *Venus Express* recently discovered that oxygen atoms also escape in large quantities, assisted by the solar wind flux impinging directly onto the planet with no interceding magnetic field. Current knowledge of the various key processes and their effect on the climate of Venus over time are reviewed by Taylor and Grinspoon (2009) and summarized below.

6.1. Sources of Atmospheric Gases: Volcanism

The existence of the atmosphere on Venus, as on Earth, depends primarily on the venting of gases from the interior, which has continued at some level up to the present, accompanied by an unknown contribution from the infall of icy material as cometary (i.e. volatile rich) and meteoritic dust and larger fragments. The main evidence for recent volcanism on Venus is the abundance of volcanic structures seen on the surface, many of them of pristine appearance with apparently fresh lava flows, and the high level of reactive, and therefore relatively short-lived, volcanic gases in the atmosphere, especially SO_2. Searches for "hot spots" on the surface, and plumes of volcanic effluvia, which would confirm current active volcanism and help establish its level compared to Earth, have been attempted with tantalizingly suggestive, but so far inconclusive, results.

If Venus has the same total heat flux as Earth and, in the absence of tectonics, it is all accounted for by volcanoes, then we would expect 1000 times as much gas, in particular sulfur dioxide, to be released. In fact, there is approximately 100,000 times as much SO_2 in Venus' atmosphere compared to that of the Earth's. The difference of a factor 100 could conceivably be explained by the fact that this is approximately the ratio of the lifetimes of the gas on the two planets when the efficient rainout mechanism that applies near the surface on Earth is taken into account. Such crude attempts at balancing the budget must admit additional unknowns; for instance, volcanism on Venus may release a mix of gases different from that on the Earth, where each volcano is at least slightly different from every other in any case.

The most that can be said with reasonable certainty at present is that the existence of SO_2 and H_2SO_4 in the amounts observed must imply substantial outgassing in recent geological time. The present flux of volcanic gases into the atmosphere remains unknown and could conceivably be as low as zero, although the latter seems very unlikely. The time constant for the decline of the sulfur abundance in the atmosphere if the source was removed is of the order of 1 Myr, much shorter than the period since the hypothetical global resurfacing event 750–500 Myr ago (see Venus Surface), indicating that the atmospheric sulfur must be of recent origin. The very high variability in sulfur dioxide abundance measured at and above the cloud tops has been extensively interpreted as indicating current volcanism. While it is not possible at present to associate these with specific eruptions on the surface, and transport effects due to local meteorology need also to be considered, large SO_2 variations are seen at comparable pressure levels in the terrestrial upper atmosphere following large eruptions. Short-lived but very bright plumes in the stratosphere, seen in UV images of reflected sunlight, and dark plumes in deep tropospheric clouds in near-IR images are also possible evidence for active volcanoes.

6.2. Surface Pressure and the CO_2 Budget

A plausible first-order explanation for the apparent superabundance of CO_2 on Venus relative to Earth is not particularly difficult to find. It has been estimated that the carbonate rocks on the Earth, of which the White Cliffs of Dover are a noted example, hold in total the equivalent of roughly 90 bars of CO_2 which would in the distant past have been in the atmosphere, although not necessarily all at once. The fact that the conversion of atmospheric to crustal carbonate occurs much more efficiently in the presence of liquid water to dissolve the CO_2 first suggests that the relatively water-depleted state of Venus may be responsible for so much of the gas remaining in the atmosphere there but not on Earth.

The evidence for such processes in Earth's history, presumably accompanied by large changes in surface pressure (although this is hard to verify), raises the crucial question of whether the current surface pressure on Venus is stable. The possible effects of even quite small variations in the CO_2 abundance on the climate on Earth have received a great deal of recent attention and clearly similar or very much larger changes can be contemplated on past or future Venus, resulting from changes in the volcanic CO_2 output, for example. If the climate on Venus is in fact stable in the long term, then it is likely that some mechanism provides a buffer that stabilizes the atmospheric carbon dioxide content. It was suggested by Urey as long ago as 1953 that the exchange between atmospheric CO_2 and common minerals in the surface, viz.

$$CaCO_3(\text{calcite}) + SiO_2(\text{quartz}) \leftrightarrow CaSiO_3(\text{wollastonite}) + CO_2,$$

might perform this function. Intriguingly, it has since been shown that this reaction reaches equilibrium at precisely the temperature and pressure found on the surface of Venus. However, the equilibrium is unstable, so additional factors and a more complex model of surface–atmosphere interactions are required to make a model of the climate on Venus that moves to, and stays at, the status quo. It is worth noting that we have no such model for Earth either.

6.3. Escape Processes and Loss of Water

Venus is generally taken to have essentially the same internal structure as the Earth, with sufficient differences to account for the apparent absence of dynamo action, as evidenced by the lack of an intrinsic magnetic field. The history of the field, if there was one in the past, is crucial for determining the rate at which atmospheric gases including water vapor have escaped due to solar wind erosion.

The loss processes involve dissociation to form hydrogen and oxygen ions (Figure 14.7) which then escape from the planet. Hydrogen, but not oxygen, can escape by purely thermal processes (**Jeans escape**) but several other mechanisms are active for both. For the heavier atoms like oxygen, the dominant processes are probably sputtering and electric fields produced by charge exchange processes that accelerate the exospheric ions. Each of these has a different solar cycle average loss rate and depends ultimately on the abundance of water in the middle atmosphere. The Earth is protected to some extent from water loss by the very low temperatures near the tropopause, which form a cold trap. On Venus, the enhanced solar heating of its presumed primordial ocean would have evaporated additional water into the atmosphere, increasing greenhouse warming and raising the humidity still higher. This feedback may have continued until the oceans were gone and the atmosphere contained up to several hundred bars of steam. Whether it was ever cool enough for this to form a liquid water ocean is a matter of ongoing debate.

Either way, water vapor was probably for a long time the major atmospheric constituent, extending to high altitudes where it would be efficiently dissociated into hydrogen and oxygen by UV sunlight. Rapid escape of hydrogen would ensue, accompanied by the slower escape of the heavier deuterium, helium, and oxygen. This could explain why Venus now has 100,000 times less water, but 150 times more deuterium relative to hydrogen, in its atmosphere than exists in the oceans and atmosphere of the Earth. The difference in D/H is attributable to differences between the escape rates of H and D under given conditions, plus the fact that the mixing ratio of vapor in the upper atmosphere is different for the two water isotopes.

If total water on Venus is currently in a steady state between source (volcanic outgassing and cometary infall) and loss (dissociation and escape) processes, then the escape flux measures the time-averaged sum of these sources. All three components of the equation can, in principle, be measured, although recent progress is mainly limited to measuring the escape rates. While sampling the hydrogen (and helium) fluxes, the Analyser of Space Plasmas and EneRgetic Atoms (ASPERA) experiment on *Venus Express* found surprising evidence that a large flux of oxygen ions is also escaping from the upper atmosphere of Venus. Provisional estimates suggest a planetary average column escape flux in roughly the ratio of 2:1 for hydrogen and oxygen, seemingly indicating water as the parent. If confirmed, this would solve an earlier puzzle about the fate of the large quantities of oxygen that were thought to have been left behind following massive hydrogen escape.

The rates of atmospheric water loss that the ASPERA data imply are two orders of magnitude too small to balance the water budget of the planet even if the level of current volcanic activity on Venus is no more than that on Earth. As we saw above, estimates based on imaging data, SO_2 abundances, and heat fluxes would place it several orders of magnitude higher. The explanation may be that the volcanic effluvia are much drier on Venus than they are here. Terrestrial ocean waters seep continuously into the hot crust, to be subsequently expelled as vapor, a process that is not available on Venus. Sulfur dioxide may be the dominant volatile driving explosive volcanism on Venus, rather than water as on Earth.

6.4. Evolutionary Climate Models

Attempts have been made to trace the origins and evolution of Venus' atmosphere using simplified one-dimensional evolutionary climate models. These incorporate the large-scale processes and their interrelations in a globally averaged sense, neglecting or simplifying dynamics so that they can model the complex set of time-dependent feedbacks that control the planetary climate. Bullock and Grinspoon (2001) used a radiative transfer code to calculate the radiative–convective equilibrium temperature structure as a function of atmospheric composition and coupled this to models of the chemistry and microphysics of Venus' clouds, volcanic outgassing, heterogeneous reactions of atmospheric gases with surface minerals, and the escape of hydrogen from the exosphere.

A key feature of such models is how they simulate the clouds, since these have two important effects on climate. They alter the visual albedo of the planet, changing the magnitude and profile of solar heating, and they alter the thermal IR opacity of the atmosphere, affecting the cooling at and above the surface. The formation of clouds depends on the abundance of sulfur dioxide, which in turn depends on the volcanic source and the loss processes at the surface. The current measured mixing ratio of SO_2 is 180 ppm, which is more than two orders more abundant than that required for equilibrium with calcite, but it is close to equilibrium with pyrite and magnetite. The actual reaction rate will depend on the abundances of these and other potential reactants in the surface, on the relevant chemical kinetics, and on the ability of the gas to diffuse to new reaction sites on buried grains once the easily available surface has reacted. The choice of diffusion coefficient requires assumptions about soil porosity and the effectiveness with which forming sulfate rinds will reduce pore space.

A case study that modeled the rapid supply of gases from the putative massive volcanic event responsible for depositing the extensive lava plains found that this could raise the surface temperature by as much as 100 °C for half a billion years, before relaxing to conditions similar to present ones. A second study that ran the model into the future predicted that the present situation is unstable and that after a billion years the clouds may disappear altogether. The surface temperature is predicted to fall by about 50 K since the reduced opacity more than offsets the increased absorption of incoming solar energy. A large number of other scenarios can be imagined, depending on the rate and timing of the events that might supply extra gases and the ratio of water to sulfur dioxide in each event.

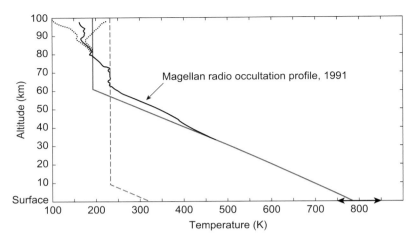

FIGURE 14.13 Two simple radiative–convective models, one for the current atmosphere and (dashed line) the other for a hypothetical future "Earth-like" Venus, compared to a measured temperature profile for the middle atmosphere of Venus from the *Magellan* radio occultation experiment.

For example, the impact of a large comet would supply mostly water vapor, with relatively little sulfur dioxide.

Figure 14.13 shows two simple radiative–convective models and a measured temperature profile for the middle atmosphere of Venus from the *Magellan* radio occultation experiment. The simple models have a stratosphere in radiative equilibrium with the Sun, overlying a deep atmosphere in which the profile follows a dry adiabat. The solid line is such a model calculated assuming present-day conditions; the dashed line is an imaginary scenario in which the surface pressure on Venus falls to 1 bar and the planetary albedo falls to 0.52, that is, to a less cloud-reflective state, as pictured by Arrhenius a century ago. This may yet occur, perhaps as the result of continued exospheric loss and chemical erosion of the atmosphere following a cessation of the volcanic source at some distant point in the future. If so, a surface temperature in the region of 320 K is predicted, the Earthlike "tropical" scenario envisaged by Arrhenius and other early visionaries long ago.

BIBLIOGRAPHY

Barstow, J. K., Tsang, C. C. C., Wilson, C. F., Irwin, P. G. J., Taylor, F. W., McGouldrick, K., et al. (2012). Models of the global cloud structure on Venus derived from Venus Express observations. *Icarus, 217*(2), 542–560.

Bertaux, J.-L., Widemann, T., Hauchecorne, A., Moroz, V. I., & Ekonomov, A. P. (1996). *Vega*-1 and *Vega*-2 entry probes: an investigation of local UV absorption (220–400 nm) in the atmosphere of Venus (SO_2 aerosols, cloud structure). *Journal of Geophysical Research, 101*, 12,709–12,745. http://dx.doi.org/10.1029/96JE00466.

Bézard, B., de Bergh, C., Crisp, D., & Maillard, J.-P. (1990). The deep atmosphere of Venus revealed by high-resolution nightside spectra. *Nature, 345*, 508–511.

Bougher, S. W., Phillips, R. J., & Hunten, D. M. (Eds.). (1997). *Venus II*. Tucson: Univ. Arizona Press.

Bullock, M. A., & Grinspoon, D. H. (2001). The Recent Evolution of Climate on Venus. *Icarus, 150*, 19–37.

Counselman, C. C., III, Gourevich, S. A., King, R. W., Loriot, G. B., & Ginsberg, E. S. (1980). Zonal and meridional circulation of the lower atmosphere of Venus determined by radio interferometry. *Journal of Geophysical Research, 85*, 8026–8031.

Crisp, D., McMuldroch, S., Stephens, S. K., Sinton, W. M., Ragent, B., Hodapp, K.-W., et al. (1991). Ground-based near-infrared imaging observations of Venus during the Galileo encounter. *Science, 253*, 1538–1541.

Esposito, L. W., Stofan, E. R., & Cravens, T. E. (Eds.). (2007). *Exploring Venus as a terrestrial planet* (pp. 157–170). American Geophysical Union. Geophysical Monograph No. 176.

Hunten, D. M., Colin, L., Donahue, T. M., & Moroz, V. I. (Eds.). (1984). *Venus*. Tucson: Univ. Arizona Press.

Krasnopolsky, V. I. (1986). *Photochemistry of the atmospheres of Mars and Venus*. New York: Springer-Verlag.

Lee, Y. J., Titov, D. V., Tellmann, S., Piccially, A., Ignatiev, N., Paetzold, M., et al. (2012). Vertical structure of the Venus cloud top from the VeRa and VIRTIS observations onboard Venus express. *Icarus, 217*, 599–609.

Russell, C. T. (Ed.). (1991), *Space science review: Vol. 55. Venus Aeronomy* (pp. 1–489).

Schofield, J. T., & Taylor, F. W. (1982). Net global thermal emission from the Venus atmosphere. *Icarus, 52*, 245–262.

Svedhem, H., Titov, D. V., Taylor, F. W., & Witasse, O. (2007). Venus as a more Earth-like planet. *Nature, 450*(7170), 629–633.

Taylor, F. W., Crisp, D., & Bézard, B. (1997). Near-infrared sounding of the lower atmosphere of Venus. In S. W. Bougher, D. M. Hunten, & R. J. Phillips (Eds.), *Venus 2* (pp. 325–351). Tucson, AZ: University of Arizona Press.

Taylor, F. W., & Grinspoon, D. (2009). Climate evolution of Venus. *Journal of Geophysical Research, 114*, E00B40. http://dx.doi.org/10.1029/2008JE003316.

Tellmann, S., Pätzold, M., Häusler, B., Bird, M. K., & Tyler, G. L. (2008). The structure of the Venus neutral atmosphere as observed by radio science experiment VeRa on Venus Express. *Journal of Geophysical Research, 114*. http://dx.doi.org/10.1029/2008JE003204.E00B36.

Chapter 15

Venus: Surface and Interior

Suzanne E. Smrekar
Earth and Space Sciences, Jet Propulsion Laboratory/Caltech, Pasadena, CA, USA

Ellen R. Stofan
National Aeronautics and Space Administration, Headquarters, Washington, DC, USA

Nils Mueller
German Aerospace Center (DLR), Institute of Planetary Research, Berlin, Germany

Chapter Outline

1. Introduction	323	7. Volcanism	335
2. History of Venus Exploration	324	8. Tectonics	337
3. General Characteristics	325	8.1. Tessera and Crustal Plateaus	338
3.1. Orbital Rotations and Motions	325	8.2. Chasmata and Fracture Belts	339
3.2. Radius, Topography, and Physiography	325	8.3. Coronae	339
3.3. Surface Conditions	326	8.4. Ridge Belts and Wrinkle Ridges	340
3.4. Views of the Surface	326	8.5. Plains Fractures, Grids, and Polygons	340
4. Impact Craters and Resurfacing History	326	9. Summary	341
5. Interior Processes	329	Bibliography	341
6. Composition	332	Books	341
6.1. Global Implications	332	Journal Articles Special Issues	341
6.2. Surface Weathering	334	Web Sites	341

1. INTRODUCTION

Venus plays a pivotal role in understanding the evolution of the terrestrial planets, the four rocky bodies closest to the sun. Venus is the planet most similar to Earth in terms of radius and density, implying a very similar bulk composition. Since the terrestrial or inner planets have all formed via the same process, condensing out of the solar nebula, the primary factor that distinguishes them is their size, and to a lesser extent, distance from the sun. The energy available to drive geologic evolution comes from the heat of accretion and from decay of radiogenic isotopes. The majority of geologic activity on the smaller bodies of the inner solar system, Mars, Mercury, and the Moon, occurred during the first 1–2 billion years. Larger planets like Earth and Venus have a greater volume of radiogenic elements and can be expected to be geologically active longer. Earth has abundant geologic activity today. We are uncertain about the present-day level of activity on Venus, but it has clearly been extremely active within the last billion years since current data reveal no evidence of impact craters that must have accumulated previously. However, Venus does not currently have a system of **plate tectonics** that governs the pattern of geologic activity on Earth (see Earth as a Planet: Surface and Interior). Clearly, size is important in determining the duration and extent of geologic activity, but other factors must affect the overall style of geologic evolution. The atmospheric conditions on Venus are also wildly different from those on Earth. Venus' atmosphere contains a runaway greenhouse effect, in which abundant carbon dioxide causes the atmosphere to heat up, trapping more water vapor. The water vapor increases the infrared opacity, which further heats up the atmosphere (see Venus: Atmosphere). The resulting thick, dense atmosphere gives

Venus a surface temperature of about 468 °C (874 °F), and a pressure 90 times greater than that of Earth's. For this reason, Venus has been called Earth's "evil twin".

Volatiles, such as water and carbon dioxide, are essentially the link between the atmosphere, the surface, and the interior, as well an essential element in the habitability of a planet. A planet's atmosphere forms primarily through the outgassing of volatiles from the interior. Outgassing results from the eruption and degassing of lava onto the surface. Volcanic and tectonic resurfacing on a planet is driven by heat loss from the interior, which is primarily fueled by decay of radioactive elements. The interiors of the larger terrestrial planets are hot enough to convect, allowing hot material to rise and cold material to sink on timescales of millions of years. On Earth, convection is linked to surface processes via the process of plate tectonics. The presence of water in the interior of Earth acts to reduce the strength of the rock, which in turn allows the exterior shell of the Earth to be broken up into plates. As plates are pushed back into the interior, water is recycled back into the interior. Volatiles on Earth are also strongly affected by both the hydrosphere and the biosphere, both lacking on Venus.

Although plate tectonics has controlled the evolution of Earth for at least 3 billion years, the surface of Venus shows no clear evidence of such a process. Venus lacks the interconnected network of crustal spreading zones, where new **crust** is continually created as at the midocean ridges of Earth, and on the other side **subduction** troughs, where crust is pushed back into the mantle. Although it lacks the clear plate boundaries evident in Earth's topography, there are major rifts and even possible subduction sites. Most explanations of why Venus does not have plate tectonics point to the very low amounts of water currently present on Venus. The water in the atmosphere is equivalent to a surface layer less than 10 cm thick. The abundance of heavy hydrogen, or deuterium, in the atmosphere relative to the normal hydrogen population indicates that a huge amount of water was lost from Venus atmosphere early in its history. The dry atmosphere could imply a dry interior for Venus, which is believed to make the outer shell on Venus too stiff to break into the plates observed on Earth.

Although plate tectonics does not operate on Venus, it is clearly an active planet with a relatively young surface and a wealth of volcanic and tectonic features. The majority of the planet is covered with volcanic features such as shield volcanoes and lava plains directly analogous to Earth's volcanic features. Many of the highland areas appear to form over **mantle plumes**, where hot material from the interior rises to the surface creating "hotspots" on the surface similar to Hawaii. In contrast, many of the tectonic features are unique to Venus. Examples include **coronae**, which are believed to result from small-scale plumes deforming the surface, and **tessera**, which are intensely deformed regions with multiple intersecting fracture sets.

2. HISTORY OF VENUS EXPLORATION

Venus has been long observed as one of the brightest objects in the evening or morning sky. Transits of Venus had been used to determine a value for the astronomical unit, and thus for the absolute scale of the solar system. During the transit of 1761, Lomonosov discovered that Venus had an atmosphere. But it was not until the 1960s that the modern exploration of the surface of Venus began, with observation by Earth-based radio telescopes. Radio telescopes at Arecibo in Puerto Rico and at Goldstone in California were used to accurately measure the rotation period and diameter of Venus. They also produced images of the surface that showed large, continent-size regions. However, Earth-based radio telescopes were hindered by only being able to image the same side of Venus that faced Earth at inferior conjunction (see Solar System at Radio Wavelengths).

Spacecraft observation of Venus began in 1962 with a flyby by the Mariner 2 spacecraft. It observed Venus from 34,833 km, determining a 468 °C (874 °F) surface temperature and that Venus lacked a magnetic field. In 1967, Mariner 5 flew by Venus at an altitude of 4023 km returning data on atmospheric composition and surface temperature. Also in 1967, the first probe entered the Venus atmosphere, when the Soviet Union's Venera 4 returned data for 93 min. The Venera 5 and 6 probes followed in 1969, sending back more atmospheric measurements. Two more Venera probes followed in 1970 and 1972 making soft landings on the surface, with Venera 8 in 1972 transmitting data on surface temperature, pressure, and composition. The Venera 8 measurements were initially thought to be consistent with a granitic composition (see Section 5 for more discussion).

The next US mission to observe Venus was Mariner 10 in 1973, which was on its way to Mercury. Mariner 10 provided observations of the atmospheric circulation of Venus with both visible and ultraviolet wavelengths. In 1975, the Soviet Union landed two more probes on the surface of Venus, Veneras 9 and 10, sending back panoramas of the surface for the first time (see Section 2), and making detailed geochemical measurements. These landers measured surface compositions similar to terrestrial **basalts**.

The US Pioneer Venus mission in 1978 consisted of an orbiter plus four atmospheric probes. The probes returned data on atmospheric circulation, composition, pressure, and temperature. The orbiter provided radar images of the surface, as well as a detailed global topographic map with a resolution of about 150 km. Major topographic regions such as Aphrodite Terra and Bell Regio were mapped, as well as the 11-km-high Maxwell Montes. The spacecraft was also used to map the gravity field of Venus.

The Soviets followed with four more soft landers between 1978 and 1981, with three of the landers returning surface panoramas and surface compositional information. The last two soft landers (Veneras 13 and 14) returned color panoramas (see Section 2), and drilled into the surface for samples. The next two Soviet missions were orbiters, Veneras 15 and 16, and returned synthetic aperture radar (SAR) images of the northern hemisphere of Venus in 1983, with resolutions of about 5–10 km. This rich data set revealed new types of features on the surface of Venus, including tessera terrain and coronae (discussed below). Vega 1 and Vega 2 in 1984 carried balloon probes into the atmosphere, and were the Soviet Union's last missions to Venus.

The National Aeronautics and Space Administration (NASA)'s Magellan mission to Venus was launched in 1989 from the space shuttle and arrived at Venus in August of 1990. It obtained SAR images and altimetry of the surface between 1990 and 1994, mapping over 98% of the surface. The spacecraft also obtained high-resolution gravity field measurements, especially after the orbit was lowered and circularized in 1993. The 120 to 250 m-resolution SAR images and 10 to 27-km-resolution altimetry data completely unveiled the surface of Venus and provided a global data set that could be used to test models of the interior and surface evolution of the planet.

In 2005, the European Space Agency launched Venus Express. The primary mission objectives are the composition and circulation of the atmosphere of Venus. In addition, this mission has provided a valuable new data set for the surface of Venus. Ground-based observations and infrared images acquired during the Venus flyby of the NASA mission Galileo in 1991 showed that some spectral windows permit thermal emission of the surface to escape to space. The visible and infrared thermal imaging spectrometer (VIRTIS) was able to image most of the southern hemisphere in a window near 1 μ. The derived near-infrared surface emissivity provides information on differences in surface composition. Additionally, local maps of surface emissivity in more equatorial regions were derived from the images of the Venus Monitoring Camera.

The Japanese Space Agency, JAXA, launched Akatsuki to study the atmosphere and surface of Venus in 2010. However, it did not achieve orbit as planned. Another attempt at orbit insertion is planned for 2016.

3. GENERAL CHARACTERISTICS

3.1. Orbital Rotations and Motions

Venus orbits the Sun in a nearly circular path once every 224.7 Earth days. It is the second planet from the sun, located between Mercury and Earth. The plane of Venus' orbit is inclined to that of the Earth by 3.4°. Analysis of the obliquity of Venus reveals that it has a liquid core, similar to that of Earth. One day on Venus lasts 116.7 Earth days. The rotation of Venus on its axis is not only extremely slow, but occurs in the opposite direction from all the other planets (retrograde rotation), so that the sun rises in the west.

The location of topographic features based on their surface brightness in Venus Express VIRTIS data has indicated that the length of day on Venus averaged over 16 years is ∼0.002 days longer than when it was estimated over 20 years ago by Magellan. This change in the length of day could indicate angular momentum exchange between the Venus' dense, superrotating atmosphere and the solid body.

When visible, Venus is the brightest planet in the night sky due to its size, albedo, and proximity to both the Sun and the Earth. Its easy visibility and the unusual pattern it makes in the night sky have given Venus a special place in **astrology** and the mythology of ancient civilizations, as well as made it an easy target for stargazers. Its proximity to the Sun means that it never rises very high in the sky, but can often be seen as either the "evening star" in the west, or as the "morning star" in the east.

3.2. Radius, Topography, and Physiography

The radius of Venus is 6052 km, only 5% less than the equatorial radius of the Earth, 6378 km. The average density of Venus is 5230 kg/m^3, somewhat higher than Earth's density. Thus, the acceleration of gravity at the surface is 8.87 m/s^2, 90% of Earth's. The radius of the Earth measured at the poles is approximately 21 km less than the radius at the equator. This difference is called the "rotational bulge". The Earth's spin accelerates the equator more than the pole, causing the pole to be flattened and the equator to bulge out. The very slow rotation of Venus means that no such flattening occurs, making it, on average, nearly spherical.

The topography on Venus is dominated by plains, which cover at least 80% of the planet. There are also major highlands, including plateaus and topographic rises, as well as rifts and ridge belts that stand out from the background plains (see Figure 15.1). Venus and Earth both have a large topographic range due to the intense geologic activity that the two planets have experienced. However, the distribution of elevations on the two planets is very different (see Figure 15.2). Earth's topography is bimodal, while Venus' topography is unimodal. The two peaks on Earth reflect the division between oceans and continents. Venus has no ocean, and as we will discuss below, arguably no continents. The topography on Venus differs from that on Earth in other significant ways. Most importantly, Venus lacks the

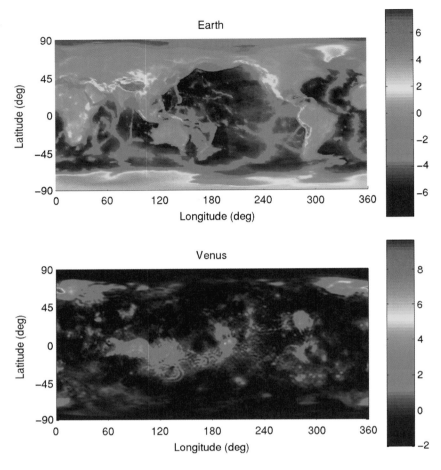

FIGURE 15.1 Topography of the Earth and Venus in a sinusoidal projection at a resolution of 1 pixel/degree. Note the long ridges that dissect many of Earth's ocean basins and the long mountain belts that are the signature of plate tectonics. Venus has numerous large highland regions, but the only long, quasi-linear mountain belts occur in the northernmost highland region, Isthar Terra.

interconnected system of narrow midocean ridges and long linear mountain belts that are the hallmark of plate tectonics on Earth (see Figure 15.1). The absence of these features on Venus reflects fundamental differences in evolution between the two planets, and will be discussed in greater detail below.

3.3. Surface Conditions

The surface conditions on Venus can best be described as hellish. The surface temperature at the mean planetary elevation is 437 °C (867 °F), with most of this high temperature due to the Venus greenhouse, not proximity to the sun. The surface temperature at the highest elevations is approximately 10 °C less. The surface pressure is 95 bars, equivalent to the pressure under almost 1 km of water. The surface temperature varies by only about 1 °C over the course of a day (243 Earth days) due to the dense, insulating atmosphere and the very low obliquity. Note that a year on Venus (225 Earth days) is actually shorter than 1 day. The atmosphere is 96.5% carbon dioxide, with lesser amounts of nitrogen, sulfur dioxide, argon, carbon monoxide, and water. The clouds are largely sulfuric acid (see chapter on Venus Atmosphere).

3.4. Views of the Surface

Four Soviet landers have returned views of the surface of Venus, Veneras 9, 10, 13, and 14. These panoramas showed relatively similar sites: rocky surfaces with varying amounts of sediment (Figure 15.3). Rocks at each site tend to be relatively angular, suggesting minimal erosion and possible ejection from an impact crater. All the sites are consistent with a volcanic origin, showing platy lava flows that have been covered to varying extents by sediments. The sediments may be of impact origin, produced by aeolian erosion or by chemical weathering.

4. IMPACT CRATERS AND RESURFACING HISTORY

There are approximately 940 identified impact craters on the surface of Venus. They range in diameter from approximately 1.5–268.7 km. The dense atmosphere on

Chapter | 15 Venus: Surface and Interior

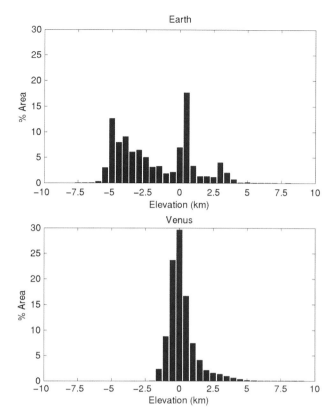

FIGURE 15.2 Histogram of the elevation in 0.5 km bins for Earth and Venus, normalized by area.

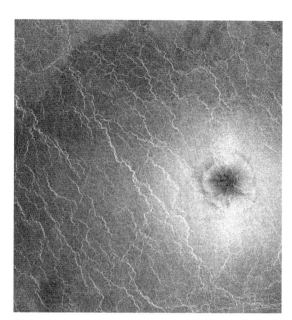

FIGURE 15.4 This radar image (approximately 125 by 140 km in size) shows an impact splotch with a dark center and a bright halo. The splotch is superimposed on a set of predominantly NW-trending wrinkle ridges. The spacing between major ridges is roughly 10–20 km. These wrinkle ridges are part of the set of ridges that wraps around Western Eistla Regio.

Venus causes impactors ≤1 km in diameter to break up before impacting the ground, reducing the number of craters ≤30 km in diameter. The shock waves that travel through these small objects can cause them to explode in a manner analogous to the Tunguska event on Earth (see Planetary Impacts). Atmospheric breakup and explosion, or other dynamic effects in the atmosphere, is believed to produce both radar-bright and radar-dark splotches on the surface (Figure 15.4). The brightness of a radar image is primarily a function of how rough the surface is at the scale of the radar wavelength (for the Magellan radar, 12.6 cm).

FIGURE 15.3 These photographs of the surface of Venus were obtained by the Soviet Venera 13 spacecraft. Venera 13 was the first of the Venera lander missions to include a color camera. The Venera 13 lander touched down on March 3, 1982, near 305° E 5° S, in the plains east of Phoebe Regio. The arm on the surface in the top image is a soil mechanics experiment. A color bar for calibration is visible in each image, as well as other spacecraft parts.

The darker the image is, the smoother the surface. Very rough areas appear very bright. Rough areas reflect the signal back to the spacecraft while smooth areas allow the radar waves to bounce off in a direction away from the spacecraft. Approximately 400 of these "splotch" regions have been identified. These regions are believed to be either areas where fine-grained material has been scoured away (radar-bright areas), or regions where relatively fine-grain material has settled out of the atmosphere (smooth, radar-dark areas). Additionally, most impact craters have dark parabolas associated with them, which are also part of fine-grained ejecta that are deposited out of the atmosphere.

In the absence of samples returned from planetary bodies, the only means of dating the surface is the analysis of the impact crater population. A great deal of work has been done on assessing the population of comets and asteroids available to impact the larger planetary bodies (see chapters on Comets, Asteroids, and Planetary Impacts). Dating of samples returned from the Moon has been used to tie the record of lunar craters to an absolute age. The estimated flux of impactors on the Moon must be extrapolated to other bodies in the solar system, which have different dynamical environments and thus different expected rates for impacts. This introduces a major uncertainty into the estimated age of a surface based on impact crater counts. Another major factor is the history of the surface itself. Modification of a surface by erosion, deposition, volcanic resurfacing, or tectonism can decrease the number of identifiable craters. Erosion can also remove deposits that had covered a crater. Additionally, secondary craters can form when large blocks of material are ejected during an impact. Impactors can break up during entry to the atmosphere, producing multiple smaller impacts rather than a single large impact. Despite these issues and the resulting uncertainties, estimated surface age is a very important clue in deciphering the geologic history of a planet.

The estimated age of resurfacing on Venus is ~750 My. Given all the possible uncertainties in this age, estimates between 300 My. and 1 billion years are permissible. This age is in contrast to ages of 3—4 Gy on average for Mars and the Moon. On Earth, new crust is continually forming along spreading centers in the oceanic crust. Continental crust can be old as 4 Gy, but craters are erased by water and wind erosion much more rapidly than on the other terrestrial planets.

There are two highly intriguing characteristics of the Venus crater population. The first is that the distribution of craters cannot be distinguished from a random population. The second is that apparently very few of the craters are modified by either volcanism or tectonism. Only ~17% of the total population is either volcanically embayed and/or tectonized. An example of a crater that is both embayed and tectonized is Baranamtarra (Figure 15.5). The low number

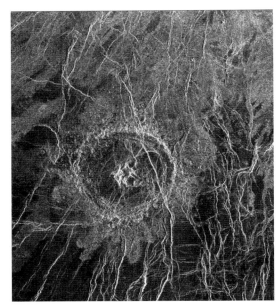

FIGURE 15.5 Crater Baranamtarra is both heavily embayed by volcanic flows and fractured. It is 25.5 km in diameter, and centered at 17.94° N, 267.80° E.

of modified craters on the surface of Venus means that there is little record of the process or processes that reset the surface age to be less than 1 billion years. If volcanic flows had covered the surface of Venus at a uniform rate, there would be more partially buried craters. This observation of the crater population initially led to the hypothesis of global, catastrophic resurfacing. Subsequent detailed modeling of resurfacing showed that the population is consistent with a wide range of resurfacing models, allowing for different size areas to resurface at different rates. Analysis of both crater density and modification of craters provides additional information that is most consistent with resurfacing occurring as small diameter regions, on the order hundreds of kilometers in diameter rather than on global scale. Another potential clue to the resurfacing history is whether or not craters with dark floors may have been volcanically flooded. Ultimately, more data are needed to provide constraints on the nature of processes that have resurfaced the planet.

We can estimate the rate of volcanic resurfacing if we assume that craters have been removed by burial under volcanic flows. Crudely, if one takes the characteristic resurfacing age to be 750 million years and the average crater rim height to be 0.5 km, then the rate of volcanic production is ~0.3 km^3/yr. Alternatively, if we consider the hypothesis that Venus resurfaced more quickly, in perhaps 50 Ma, the production rate is ~4.6 km^3/yr. The relative volume of lava extruded on the surface is believed to be small compared to the volume intruded into the subsurface, perhaps 10% of the total. For comparison, the estimated rate of volcanism for combined intrusive and extrusive volcanism on Earth is 20 km^3/yr.

On planets with large numbers of craters, such as Mars, the surface age of local regions can be estimated from the crater populations. On Venus, some attempts have been made to determine the relative ages of either populations of specific types of geologic features or large areas on Venus. However, statistical analysis of this approach indicates that the very small number of craters on Venus makes attempts at dating particular landforms or even large areas not reliable. Although traditional crater-counting methods are not very useful, both the distribution of modified craters and the distribution of dark crater parabolas suggest some variation in surface age. In particular, the region with the highest density of volcanoes, coronae, and rifts appears to have a lower density of haloes and more modified craters, suggesting a younger age.

Overall, the crater population on Venus indicates it is a comparatively active planet, completely resurfaced within the last 1 billion years, possibly with resurfacing ongoing today. Volcanic resurfacing rates are likely on the same order of magnitude as those on Earth, but are a function of the poorly constrained rate of resurfacing, which could be either constant or variable. The distribution and modification of the craters imply that there are limited differences in the ages of large regions on Venus, unlike the dichotomy between the age of oceanic and continental crust on the Earth. The small number of modified impact craters leaves few clues as to the process(es) that obliterated the earliest surface of Venus. Below we discuss the implications of resurfacing for the overall geologic evolution of Venus.

5. INTERIOR PROCESSES

One of the greatest curiosities about Venus is that global-scale geologic processes are totally unlike that of Earth. The system that shapes the Earth's large-scale physiography and the majority of geologic features is plate tectonics. The surface of the Earth is broken into dozens of plates that move over the surface of the Earth at rates of up to a few centimeters per year. The plates are tens to hundreds of kilometers thick. Mountain belts form where plates meet, such as where they collide, slide at an angle past each other, or where one plate is pushed into the **mantle** beneath another at subduction zones. Hot material wells up from the mantle below along narrow ridges in the ocean crust, creating new oceanic crust. These characteristic features are easily seen in the topography of the Earth, even at the relatively low resolution available for Venus (Figure 15.1). Venus clearly does not have plate tectonics. There is no evidence for this type of geologic process in the topography or in the radar images (see Earth Surface and Interior).

The energy that drives plate tectonics and other geologic processes is dominantly generated by the decay of radioactive elements. For the terrestrial planets, the primary contributors to radioactive decay are uranium (U), thorium (Th), and potassium (K). Based on estimates of the abundance of these elements on Earth and in chondrites (see Meteorites), radioactive decay cannot account for the total amount of energy. In addition, a significant amount, perhaps 25%, of the heat lost from the interior results from cooling of the planet over time, with some additional contribution from the heat of initial planetary accretion. The heat in the interior of the planet is dominantly transmitted to the surface via convection in the interior. Convection in the mantle brings hot, low-density material from the interior to the surface, or near the surface, allowing it to cool.

Generally speaking, the larger a planet, the longer it will continue to lose energy and be geologically active. However, the details of the thermal evolution are complex. Numerous factors affect thermal evolution, including accretion, differentiation, composition, convective style, and amounts of volcanism. Venus and Earth provide perhaps the quintessential example of variations in evolution. Most explanations of how Venus and Earth ended up on different geologic paths have to do with the history of volatiles. Volatiles, mainly in the form of water, play a key role in enabling plate tectonics on Earth. The presence of even a small amount of water in rock has a major effect on its strength and on the temperature at which it will melt. The water in the **lithosphere** is believed to be essential to making it weak enough to break into plates and subduct in response to the motions of convection in the interior. The **asthenosphere** is the upper part of the mantle, directly below the lithosphere, which has a lower viscosity than the rest of the mantle and acts to lubricate the motion of the plates at the surface of the Earth. The low viscosity of the asthenosphere may be a result of small amounts of melt. Melt would not be expected in the asthenosphere unless at least a small percentage of water is present. Thus, water appears to be an essential ingredient in the development of plate tectonics.

Measurements made to date indicate that the atmosphere of Venus has very little water, on the order of tens of parts per million. The upper crust is inferred to be dry as well, although Ar isotopes in the atmosphere indicate that only about 25% of interior volatiles have been lost. In terms of the strength of the crust, the extremely high surface temperatures might be expected to offset the lack of water, making the crust extremely weak. However, laboratory studies of rock strength at Venus temperatures have shown that dry basalt (see Section 5 below) is stronger than wet basalt at Earth temperatures. This extreme strength of the crust on Venus likely contributes to the apparent lack of lithosphere scale breaks that are required to form plates. As we discuss below, there is also evidence suggesting that Venus has no asthenosphere.

Convection studies have proposed that Venus exists in a "stagnant lid" mode rather than the "active lid" mode predicted for Earth. When convective stresses exceed the lithospheric strength, an active lid such as the terrestrial system of plate tectonics is predicted. On Earth, conditions such as weak, narrow fault zones, or the presence of a low-viscosity asthenosphere, allow the convective stresses to exceed the lithospheric strength. On Venus, the present-day lithospheric strength is apparently too high to allow plates to develop. This model is consistent with the loss of volatiles as key to differences on Venus and Earth.

Given the similarity in heat-producing elements and size between Earth and Venus and the absence of plate tectonics on Venus, how does Venus lose its heat? Venus must be convecting in its interior. Although there is no evidence for present-day plate tectonics, some numerical simulations suggest that Venus may have had plate tectonics in the past, or may have experienced multiple transitions active and stagnant lid regimes. The greater heat loss during plate tectonics causes sufficient cooling of the mantle and lithosphere to initiate a stagnant lid regime. The insulating lid then causes the mantle to heat up and revert to plate tectonics. The lack of plate tectonics today suggests that Venus may be heating up today.

In addition to the deformation of surface plates, hot plumes from the interior can affect surface geology. On Earth, hot blobs of material form within the overall convecting pattern in the interior. These plumes form "hotspots", such as the Hawaiian Island chain. The hot mantle material pushes up on the lithosphere, creating a broad topographic swell. The heat causes the lithosphere and crust to melt locally, thickening the crust and forming surface volcanoes. On Earth, the majority of the heat is lost where the upwelling mantle creates new crust at midocean rises and cold lithosphere is pushed back into the mantle at subduction zones. Hotspots account for <10% of Earth's heat loss. Venus appears to have a similar number of hotspots as Earth, providing evidence of current convection and contributing to heat loss.

There are approximately 10 such hotspot features on Venus. These rises are Atla, Bell, Beta, Dione, W. Eistla, C. Eistla, E. Eistla, Imdr, Themis, and Laufey Regiones (Figure 15.6). Those features believed to be active today, such as Atla, Beta, and Bell Regiones, have broad topographic swells, abundant volcanism, and strong, positive gravity signatures. New evidence from surface emissivity gives further evidence of their present-day activity (see Composition). Several rises also have rifts, such as Guor Linea at W. Eistla Regio (Figure 15.7). These features are characteristic of hotspots above a mantle plume. However, there are too few hotspot features on Venus (~10 on Venus vs 10–30 on Earth) to account for a major portion of Venus' heat budget. In addition to the large-scale (1000–2000 km diameter) hotspots on Venus, there are also smaller scale (mean diameter of ~250 km) features called coronae (see Section 7). There are ~515 of these features, which are unique to Venus. There is considerable evidence that many of these features form above small-scale plumes. Some may be a result of cold lithosphere becoming gravitationally unstable and sinking back into the interior, which also acts to cool the planet. However, all these heat loss mechanisms together would not be able to account for more than about one-quarter of the interior heat loss on Venus.

The relationship between the gravity and topography provides evidence that Venus does not have a low-viscosity asthenosphere. On Earth, a mantle plume must pass through the asthenosphere before reaching the lithosphere. (Note that there is not an asthenospheric layer beneath the very thick continental lithosphere on Earth.) The plume tends to spread out in the relatively weak asthenospheric layer, resulting in a reduced amount of topographic uplift for a given plume size. Comparing the observed amount of uplift to the estimated size and depth of the low-density plume provides evidence for this behavior on Earth but not on Venus. On Venus, plumes strike the lithosphere directly, thus causing more uplift for a given plume size.

The relationship between the gravity and the topography provide some insight into interior structure and convection. The magnitude of variations in the gravity field as compared to a given topographic feature is an indication of the interior structure that supports a given topographic feature. The strength of the lithosphere can support topography. Variations in density in the interior can also support topography. A mountain can be supported by a thick "root" of low-density crust, analogous to an iceberg floating in denser water. Variations in the mantle temperature associated with convection can also support topography. The gravity field of Venus has been carefully studied to estimate the thickness of the strong, or elastic, part of the lithosphere, the thickness of the crust, and the location of low-density, relatively hot regions in the mantle. Clearly, some highlands, such as tessera plateaus (see Section 7), are compensated by crustal roots. Many other highlands appear to be compensated by mantle plumes.

In addition to plumes, conduction through the lithosphere must contribute to the heat loss on Venus. The thinner the lithosphere, the more rapidly the planet loses heat. Estimates of the thickness of the lithosphere on Venus, derived from gravity and topography, are typically 100 to 200+ km. This is comparable to the lithospheric thickness on Earth, and is too large to account for the majority of Venus' heat loss. There is growing evidence that the recycling of the lower lithosphere back into the mantle may help cool Venus, just as subduction helps cool the Earth. New models for corona formation show that at least some coronae may form above sites where the thickening, cold lithosphere becomes too dense and breaks off into the

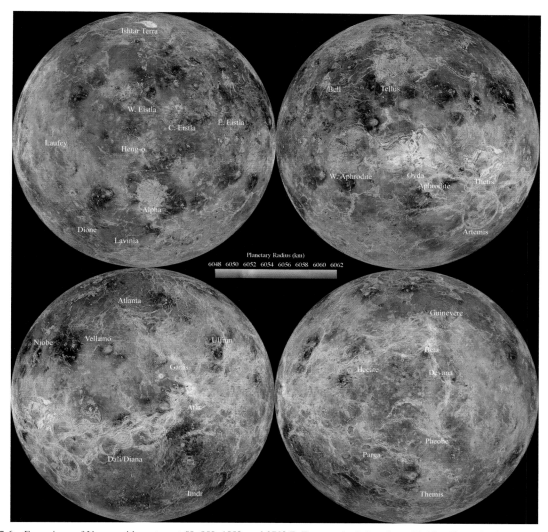

FIGURE 15.6 Four views of Venus, with centers at 0°, 90°, 180°, and 270° E. Topography is in color, with Magellan radar images overlain on top.

mantle. Estimates of lithospheric thickness variations also suggest that the lower lithosphere may thicken and become unstable locally. Although possibly important, such a process is not going to be nearly as efficient a cooling mechanism as subduction.

Volcanism, resulting from melting of the mantle and/or lithosphere and the rise of hot magma, can contribute to heat loss. As discussed above, Venus was completely resurfaced, most likely by volcanism within the last billion years. Present-day rates of volcanism are not constrained, but are unlikely to be sufficient to be a major contributor to heat loss.

Another constraint on interior processes is the absence (or very low level) of a magnetic field. The Mariner flyby missions measured no magnetic field, indicating that, if present, the field must be <500 nT at the surface. More recent data from Venus Express suggest a limit of 10 nT on the strength of the magnetic field at the surface. Most models of interior dynamos indicate that a planet must be losing large amounts of heat from the planet's metal core to provide enough energy for a dynamo. Some models have suggested that relatively rapid heat loss through plate tectonics is a good method of driving a dynamo. Thus, one possible scenario is that Venus had early plate tectonics and an active dynamo, but eventually lost much of its water from the crust through volcanism to the atmosphere, where it was subsequently lost to space. This decrease in water increased the strength of the lithosphere to the point that tectonics ceased and the dynamo shut down. Heat is then lost primarily by conduction through the lithosphere, causing the mantle to heat up and increase the rate of volcanism, causing the planet to resurface. This idea is

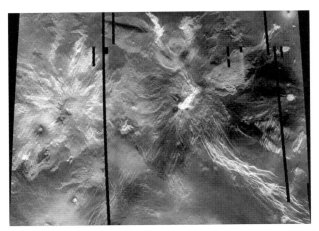

FIGURE 15.7 Radar image of W. Eistla Regio centered at 22° N, 354.5° E with dimensions of approximately 1725 by 1260 km. The western volcano is Sif Mons, 350 km in diameter, and the eastern volcano is Gula Mons, 450 km in diameter. Radial radar-bright and dark flows surround both volcanoes; radar-bright linear fractures of Guor Linea are seen in the southeastern corner. Black areas are data gaps.

speculative, as there is no direct evidence for an early plate tectonic period. The possibility of a low-level magnetic field provides a new challenge for understanding dynamo processes and could indicate that the core of Venus is losing heat but at a lower rate than that of Earth.

The unusual cratering record on Venus indicates that the first 3.5 billion years of geologic history has been somehow erased, possibly with a lower rate of resurfacing occurring subsequently. In contrast, Mars, Mercury, and the Moon have surfaces that preserve the large impact basins from early bombardment and reflect a gradual loss of heat and decline in geologic activity. Some models have proposed that resurfacing on Venus occurs episodically. In one scenario, the lithosphere thickens and becomes denser due to both cooling and chemical phase transitions. The lithosphere is predicted to founder, or get mixed into the mantle, when it becomes gravitationally unstable. However, how the lithosphere actually breaks and initiates this process is unclear. In another scenario, the stagnant lid heat insulates the mantle, causing it to heat up to the point that widespread melting occurs, eventually erupting on the surface. Other models show that volcanism that is globally distributed and resurfaces small regions in each event can produce the observed distribution. High mantle temperatures could facilitate this kind of widespread volcanism.

6. COMPOSITION

6.1. Global Implications

The similarity between Venus and Earth in terms of size and location in the solar system indicates that their bulk compositions should be comparable. The exact composition of the crust is related to the composition and temperature in the interior of the planet when the rock melts, as well how much of the original rock is melted. The typical rock type that forms on Earth when the interior melts and erupts is basalt. Thus, it is not surprising that geochemical measurements on the surface of Venus have a gross composition similar to terrestrial basalts, with some variation. On Earth, basalts make up the majority of the oceanic crust and are found in volcanic regions of continents. When processes such as subduction remelt basalts the resulting rocks are enriched in silica (SiO_2). Continental rocks are a result of billions of years of remelting of a basaltic crust driven by convective and plate tectonic processes. They are of lower density than basalt due to the enrichment of silica relative to iron and magnesium. The presence of at least small amounts of water may be essential to the formation of such silica-rich rocks. Continents stand higher than the oceanic crust due to both their lower density and the greater thickness of continental crust. As we will discuss, there is possible evidence for silica-rich rock on Venus.

The abundances of primary mineral-forming and radiogenic elements were measured by spectrometers on Venera landers. Venera landers 8, 9, and 10 and Vega landers 1 and 2 measured the amounts of uranium (U), thorium (Th), and potassium (K) using a gamma-ray spectrometer (Table 15.1). The Venera landers 13 and 14 measured these elements as well as the major-element forming minerals (see Table 15.2). Due to the orbital dynamics of delivering probes to the surface of Venus in any given time period, Venera landers 8–14 are all located in a relatively small region on Venus within 270°–330° E and 15° S to 30° N. This area includes the eastern flank of Beta Regio, a major hotspot, and the plains to the east of Beta and Pheobe Regiones. The Vega 1 and 2 landers, sent at an earlier time, are located near 170° E, 10° N, and 180° E, 10° S, to the west of Atla Regio.

The silica content and the relative abundances of iron and magnesium for rocks at the Venera lander sites

TABLE 15.1 The Abundances for Each Element Represent an Interpretation of the Most Likely Minerals on the Surface of Venus, as is Standard Practice in Geochemical Analysis

Lander	U (ppm)	Th (ppm)	K (wt%)
Venera 8	2.2 ± 0.7	6.5 ± 0.2	4.0 ± 1.2
Venera 9	0.6 ± 0.2	3.6 ± 0.4	0.5 ± 0.1
Venera 10	0.5 ± 0.3	0.7 ± 0.3	0.3 ± 0.2
Vega 1	0.68 ± 0.47	1.5 ± 1.2	0.5 ± 0.3
Vega 2	0.68 ± 0.38	2.0 ± 1.0	0.4 ± 0.2

TABLE 15.2 The Raw Data are Converted from Measurements of Elemental Abundance into Likely Chemical Combinations

Constituent	Venera 13 (wt%)	Venera 14 (wt%)	Vega 2 (wt%)
SiO_2	45.1 ± 3.0	48.7 ± 3.6	45.6 ± 3.2
TiO_2	1.59 ± 0.45	1.25 ± 0.41	0.2 ± 10.1
Al_2O_3	15.8 ± 3.0	17.9 ± 2.6	16.0 ± 1.8
FeO	9.3 ± 2.2	8.8 ± 1.8	7.74 ± 1.1
MnO	0.2 ± 0.1	0.16 ± 0.08	0.14 ± 0.12
MgO	11.4 ± 6.2	8.1 ± 3.3	11.5 ± 3.7
CaO	7.1 ± 0.96	10.3 ± 1.2	7/5 ± 0.7
K_2O	4.0 ± 0.63	0.2 ± 0.07	0.1 ± 0.08
S	0.65 ± 0.4	0.35 ± 0.31	1.9 ± 0.6
Cl	<0.3	<0.4	<0.3

(Table 15.1) are characteristic of basalt. Although some variations in composition do exist, when the overall abundance of elements is considered in the context of minerals that occur stably together, all the rock compositions are consistent with a basaltic composition. Early analysis of the relatively high value of U, Th, and K at Venera 8 and 13 sites suggested that these locations were composed of a more silica-rich rock, possibly even **granite**. However, subsequent analyses have discounted this idea and concluded that Venera 8 and 13 sites are most likely basaltic, although more alkaline. Variations in elemental abundance do suggest that some real differences exist. The bulk composition of Venus can be extrapolated from these measures. Within the uncertainties, the composition is similar to that of Earth. Similarly available data on Fe/Mg and Fe/Mn suggest that the core composition is similar to Earth's. Some variation may occur after the rock forms. For example, the amount of Al, Ti, Ca, or Si may change through chemical weathering or metamorphism when the rock experiences changes in pressure and/or temperature.

The initial chemical measurements of the surface have provided invaluable constraints on the surface composition. However, the overall number and geographic diversity of sites remains limited. The precision of the measurements that were possible with instrumentation built in the 1970s is very low compared with measurements possible today. The uncertainties in the measurements mean that numerous questions such as the size of the core (which is constrained by the ratios of Fe/Mn/Mg) and the amount of crustal recycling cannot be addressed. In fact, the uncertainties in the Venusian measurements are so large that they encompass the entire range of composition for basalts on Earth, Mars, the Moon, and meteorites. In contrast, basalts from the Moon and Mars (as represented in meteorites) have a distinct chemical signature from those on Earth (see Meteorites). These variations represent key differences in the formation and evolution of these bodies, such as the formation of a magma ocean on the Moon.

The emissivity of the surface at 1.02 μm, obtained by the VIRTIS instrument on Venus Express (Figure 15.8), is primarily a function of variation in the iron mineralogy, as well as any grain size differences. Both surface temperature variations and the effects of scattering and absorption must be removed from the radiance measured by VIRTIS to estimate surface emissivity. Temperature is a very strong function of topography on Venus because the insulating atmosphere precludes temperature variations due to either annual or diurnal changes. Thus, surface emissivity anomalies could, in principle be due to local temperature variations such as active lava flows. Making the assumption that lava flows on Venus have a similar size and effusion rate as typical terrestrial flows, a thermal signature that could be resolved at a scale of 100 km is likely to last only hours to days. VIRTIS data show no definitive evidence of such transient anomalies.

Thermal emissivity anomalies that are significantly lower and higher than the average are found to correlate with tessera terrains (see Figure 15.8). Low emissivity is consistent with a high silica composition, such as granite. There are two significant uncertainties in this interpretation. One is the uncertainty in the Magellan altimetry measurement in rugged areas, like the tesserae (see below). Another is the nonuniqueness of the interpretation of a single spectral band. However, if this interpretation is correct, it is extremely significant. On Earth, granite forms when basalt melts in the presence of water, such as when subducted plates melt at high temperatures. As discussed further below, individual tessera plateaus are up to ~2000 km across, indicating significant water would have been present during their formation if they are granitic in composition.

The highest emissivity anomalies are correlated with volcanic flows at hotspots, where gravity data indicate a plume at depth. These anomalies are interpreted to indicate relatively recent flows. The contrast between the average emissivity of most of the southern hemisphere and the high emissivity flows is consistent with the signatures of minerals expected to form when basalt is exposed to Venus surface conditions versus fresh basalt (see Weathering below). Although rates of weathering are not well constrained, the lack of weathering indicates that the flows formed in geologically recent times (<2 My) and may even be currently active. Although a difference in the iron content of the initial magma eruption could also produce an emissivity variation, higher initial iron contents would also

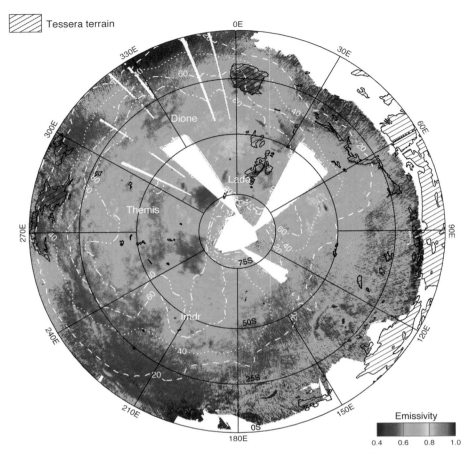

FIGURE 15.8 Surface emissivity at 1.02 μm derived from VIRTIS surface brightness. This map is derived in the same manner as that in Mueller et al. (2008) but using an improved topographic corrections. An average surface emissivity of 0.62 is used to provide a physical range of values. Contours show the number of VIRTIS images acquired in each area. The data are reduced using first-order corrections for stray light, clouds, and topography based on data statistics. A greater number of looks therefore mean not only a reduction of instrumental noise but also a lower likelihood of deviations due to variable observing effects. White areas are data gaps.

weather to minerals with a lower emissivity, indicating that the flows are likely young in that case as well.

In addition to direct measurements of the composition, morphology can be used as a very crude indication of composition. For example, lavas with a basaltic composition tend to be very fluid, forming long, narrow flows, and broad, low volcanoes. As the silica content increases, the viscosity of the lava increases. The thickness of flows increases, their length decreases, and the slopes of volcanoes formed increases. Terrestrial examples are Mauna Loa in Hawaii (basaltic) and Mt. St. Helens in Washington (more silica rich). On Venus, the morphology of flows is generally consistent with low-viscosity basaltic compositions. There are some features that appear to represent much thicker, shorter flows (see description of "pancakes" in Section 6 below). However, these morphologies cannot be considered diagnostic of composition as factors such as the volume and rate of material erupting, the atmospheric pressure during eruption, and the amount of gas in the lava also shape the morphology of the flow.

6.2. Surface Weathering

Although there is little evidence of weathering of the surface by wind consistent with Venus' dense atmosphere, the environment for chemical weathering is extremely harsh. In addition to the searing temperature and high pressure, the atmosphere contains highly corrosive and chemically active gases such as SO_2 (sulfur dioxide), CO, OCS, HCl (hydrochloric acid), and CO_2. A variety of minerals form in laboratory experiments that simulate Venus conditions, such as calcite, dolomite, anhydrite, and hematite, but no landers have measured actual minerals. Measurement of the specific minerals present and their abundances is highly desirable as they provide insight into the nature of the chemical interaction between the surface and the atmosphere. This information is a critical piece of understanding the larger problem of how Venus arrived at the hellish climate that now exists.

One of the key questions is how much CO_2 is trapped in minerals on the surface of Venus. Most of the CO_2 found on Earth is trapped as carbonates via biological processes, specifically the formation and accumulation of seashells (although carbonates would precipitate out of sea water even without biological processes). This process is an important element of the overall balance that makes Earth habitable. Available information from surface composition and laboratory experiments suggests that significant amounts of carbonates could be present on the surface of Venus, perhaps up to 10%. If so, this would mean that CO_2 in surface rocks is an important part of determining the atmospheric pressure and composition. Another key question is how atmospheric SO_2 interacts with the surface. On Earth, most of the SO_2 is dissolved in the oceans. Rates of chemical reactions involving SO_2 are known for the conditions in the atmosphere of Venus and predict that the SO_2 in the present-day sulfuric acid clouds on Venus should disappear over time. Note this reaction requires abundant surface calcite, which may or may not be present. If this reaction is occurring, atmospheric SO_2 is not in equilibrium and must be resupplied. The fact that sulfuric acid clouds are present today implies that new sulfur gases have been added to the atmosphere. Both the Pioneer Venus Orbiter and Venus Express have observed significant changes in the concentration of SO_2 in the atmosphere over the timescale of years. One possible explanation is the release of SO_2 due to recent volcanism. Atmospheric dynamics could also be responsible. Thus, determination of surface mineralogy is key to understanding the surface—atmosphere interaction and any associated contribution to atmospheric stability and climate change.

7. VOLCANISM

With the exception of Jupiter's moon Io, Venus is the most volcanic world in the solar system. Volcanic features of a broad range in morphology cover the surface, from sheetlike expanses of lava flows to volcanoes shaped like pancakes and ticks, as illustrated below. The high surface temperature and pressure on Venus make explosive volcanism less likely, although some possible deposits produced by explosive volcanism have been mapped. Magellan data illustrated that volcanic features do not occur in chains or specific patterns, indicating the lack of plate tectonics on Venus.

The plains or low-lying regions on Venus are covered by sheet and digitate deposits that are interpreted to be volcanic in origin (Figure 15.9). These extensive deposits are likely to be flood basalts, formed in similar ways to the Columbia River Basalts or the Deccan Traps on Earth. In some plains regions, the surface is clearly built up of

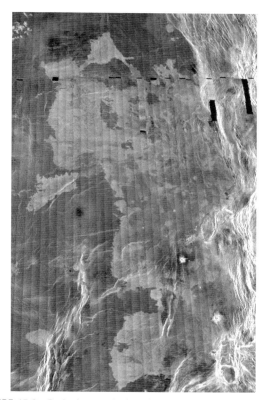

FIGURE 15.9 Radar image of a lava flow field at 60° N, 183° E in the plains of Venus. The flow field is approximately 540 km by 900 km. The name of the flow field is Mamapacha Fluctus, and it is made up of lava flows of moderate radar brightness or moderate roughness.

multiple, superposed lava flow deposits, while other regions are more featureless. Lava flows have varying brightness in the Magellan SAR images. Most lava flows are of intermediate brightness. Comparisons to radar images of lava flows in Hawaii indicate that the Venusian flows have similar roughness, although some flows on Venus are unusually smooth.

The plains are also covered with abundant small (<5 km across) shield and cone-shaped volcanoes (e.g. Figure 15.10). Thousands of these volcanoes have been mapped, and they may contribute as much as 15% of the plains volcanic deposits. Other flows in the plains may have originated at fissures, which were then obscured by later eruptions. Timing of the plains flows is a subject of debate, with some advocating that the plains formed relatively synchronously across Venus in a single resurfacing event. Others argue that the data support a slower, nonsynchronous formation for the plains. Unfortunately, the impact crater population can be interpreted to support either hypothesis, and it will take future mission data to constrain plains formation on Venus.

Large volcanoes on Venus (those with diameters >100 km) are found at topographic rises, along rift zones,

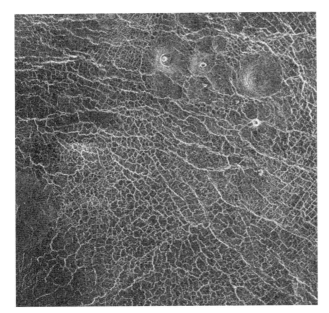

FIGURE 15.10 Radar image of small shield volcanoes and polygonal terrain, ~30 by 30 km, centered at 28.8° N, 142.2° E. Polygons range in size from the limit of resolution to several kilometers in diameter. The volcanoes at the north overly the polygons. Polygons are superimposed on the volcanoes in center right of the image, where calderas indicate the top of the volcanoes. On the western side of the image, various volcanic flows bury polygons. Thus, the formation of the polygons appears synchronous with the volcanism in this region.

FIGURE 15.11 Lava flows extend for hundreds of kilometers across the fractured plains shown in the foreground, to the base of Maat Mons, which is located at about 0.9° N latitude, 194.5° longitude. Magellan data was combined with radar altimetry to develop a three-dimensional map of the surface. The vertical scale in this perspective has been exaggerated 22.5 times. The simulated red color is based on images recorded by the Soviet Venera 13 and 14 spacecraft that indicate the atmosphere on Venus would make the surface appear red to our eyes. *The image was produced at the JPL Multimission Image Processing Laboratory.*

and concentrated in the region bounded by Beta Regio, Atla Regio, and Themis Regiones. Over 100 large volcanoes have been identified. Large volcanoes have average heights of about 1.5 km and aprons of lava flows that extend hundreds of kilometers from their summits. Maat Mons, the largest volcano on Venus is about 8.5 km high and 400 km across (Figure 15.11). In comparison, Mauna Loa, the largest volcano on Earth, is about 9 km high and 100 km across. Detailed studies of individual large volcanoes have revealed their complex histories. Many volcanoes show evidence of multiple eruptions from their summits as well as sites on their flanks. Some large volcanoes have calderas at their summits similar to volcanoes on Earth and Mars, formed by collapse of the underlying magma chamber. Others have radially fractured summits, with the radial fractures interpreted as the surface expression of subsurface dikes. These dike sets provide evidence that many large volcanoes have undergone multiple episodes of intrusion and extrusion.

At the smaller end of the scale, volcanoes 5–50 km across are also abundant on the surface of Venus (Figure 15.12). Many of the volcanoes resemble their terrestrial counterparts, with summit calderas and radiating digitate flows. Venus also has several types of volcanic features that differ from those on Earth and other planets. Steep-sided or pancake domes are flat-topped, steep-sided

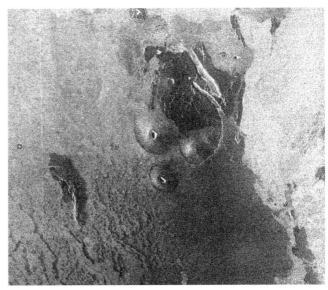

FIGURE 15.12 Radar image of small volcanoes on the flank of Maat Mons. The image is centered at about 3.2° N, 194.9° E, and is 90 km wide and 80 km long.

features (Figure 15.13), similar to flat-topped domes like the Inyo domes in California that are formed by silicic lavas. The Venus domes may have a different composition, however, as they are much larger, and have smooth rather

FIGURE 15.13 This image shows two steep-sided and one scalloped-margin domes in the plains of Venus. At the center of the image is a 50 km dome that overlaps another feature to the southwest that is about 45 km in diameter. This volcano is cut by many fractures. The southeastern volcano (25 km diameter) has scalloped edges that give this feature a bottle cap- or ticklike appearance. The scalloped edges are interpreted to form when material slides off the volcano margin.

than blocky surfaces in comparison to the terrestrial domes. Other unusual volcanoes on Venus resemble ticks, or bottle caps. These small domes have scalloped margins, and are interpreted to be steep-sided domes whose margins have collapsed.

The Magellan radar also imaged channels, a few kilometers wide and hundreds of kilometers long. The channels are found in many places within the plains, tend to be very sinuous, and in places show evidence of levees and flow breakouts. The channels have formed by lava of some unusual composition, so fluid that it behaved like water and is able to flow long distances without cooling. A number of compositions have been proposed, including carbonate- or sulfur-rich lavas and ultramafic silicate melts. Others have suggested that the channels were formed by erosion of the surface by lava, similar to lunar rilles on the Moon (See The Moon). Some of the channels extend for long distances allowing them to be used as a time marker, as it can be assumed that the channel formed over a relatively short period of time. For example, the channel may superpose one feature, but be overlain or cut by another. Also, a few channels now trend uphill, indicating that the surface deformed after they formed.

Surface thermal emissivity data create a new way to monitor volcanic activity. Gravity studies indicate that plumes exist under ∼10 hotspot regions, suggesting that the associated volcanoes could be active. However, a thermal signature can take on the order of 100 My to conduct through the lithosphere. Interpreting the emissivity data for four hotspot regions as indicating unweathered basalts suggests that these volcanoes were active recently, perhaps up to the present day. Venus Express mapped only the southern hemisphere, but all the hotspots observed showed high emissivity anomalies associated with flows. The implication is that hotspots in the northern hemisphere are also active given the similarity in their gravity signatures.

The resolution of the thermal emissivity data is on the order of 100 km. Thus, it is entirely possible that smaller, unresolved volcanic features are also active. However, if volcanism today is confined to hotspots, and plains volcanism is inactive, this could have very important implications for interior dynamics. For example, it is possible that prolonged melting of the upper mantle due to high mantle temperatures induced by an insulating stagnant lid could have dried out the upper portion of the mantle. Thus, melting might occur more easily in hotspot settings where new material is brought up from depth in the mantle, localizing new melting.

Unfortunately, currently available data preclude a strong constraint on the rate of volcanism. Thus, it is not possible to resolve the resurfacing debate. However, the size of the observed areas of recent volcanism is consistent with ongoing resurfacing.

8. TECTONICS

For the larger terrestrial planets, Venus, Earth, and Mars, mantle convection is the primary driving force for tectonic processes. On Mars, most tectonic structures are associated with either the gigantic Tharsis rise or the global dichotomy (See Mars Surface and Interior). The global dichotomy divides the smoother northern lowlands from the heavily cratered southern highlands. On Earth, plate tectonics is clearly dominant. Tectonic features on Venus are highly variable and enigmatic. Tessera terrains are unique to Venus, and are defined as having multiple intersecting deformational structures with different directions. One possible factor in creating these highly deformed regions is that Venus experiences very little surface erosion. In contrast, most continental regions have experienced multiple episodes of deformation but surface structures are often eroded between events, leaving evidence of only the most recent occurrence. Many of the tectonic features on Venus are continuous for thousands of kilometers, and likely reflect underlying mantle processes including upwelling, downwelling, and horizontal flow. Tectonic deformation takes many forms and is distributed across the surface of Venus, in contrast to the concentration of deformation along plate boundaries on Earth. Below we describe the characteristics and likely origins of the key types of tectonic features on Venus.

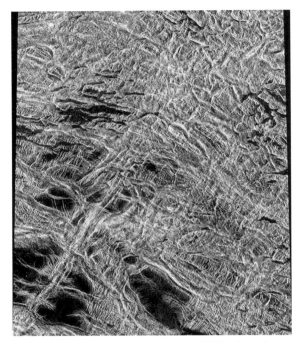

FIGURE 15.14 This radar image of a portion of Tellus Regio is centered at 36° N, 79.4° E and is approximately 340 by 420 km. The area is deformed by an NE and an NW set of lineations. Locally, each set contains both narrow, linear fractures resembling extensional graben and areas where the fractures coalesce into ridges, and appear to be compressional ridges. In the northern section of the image a third set of very narrow NNW-trending fractures crosscuts the other sets. The dark regions are volcanically flooded valleys, with two small vents visible in the SW corner of the image. Black areas are data gaps.

8.1. Tessera and Crustal Plateaus

Tessera terrains are highly deformed and thus stand out as very bright in radar images (Figure 15.14). They are made up of both extensional and compressional deformational features. In some cases the sequence of events can be determined, but more often it is ambiguous. Tesserae occur both as isolated fragments embayed by later plains material and in major plateaus. There are six major crustal plateaus: Alpha, Ovda, Pheobe, Thetis, and Tellus Regiones plus Ishtar Terra. Figure 15.15 shows Alpha Regio, one of the smaller highland plateaus. Western Ovda Regio may be a relaxed crustal plateau. These plateaus are 1000–3000 km in diameter and 0.5–4 km higher than the surrounding plains. Their gravity signature indicates that they are supported by crustal roots rather than active mantle processes, thus the name "crustal" plateaus.

Ishtar Terra is unique among the highland plateaus. It is the largest of the crustal plateaus, and is surrounded by significant mountain belts on three sides, with large areas of tesserae occurring on their exterior flanks. They are Venus' only real mountain belts. Lakshmi Planum makes up the interior of Ishtar Terra. This smooth plateau is

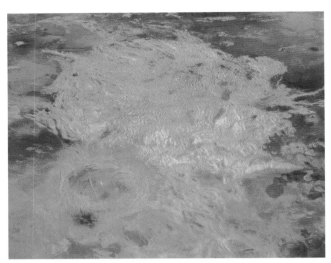

FIGURE 15.15 This false color, perspective radar image of Alpha Regio is approximately 2000 km across and is centered at 25° S, 5° E. The blank strip is a data gap. The texture of the deformed regions is similar to that of Tellus Regio (Figure 15.13). A corona is located at the SW edge of Alpha. To the west are several small pancake domes. An impact crater is seen on the western margin of Alpha.

elevated 3–4 km above the surrounding plains, and is covered by volcanic flows that emanated primarily from two large calderas. The Maxwell Montes to the east of Lakshmi Planum contain the highest point on Venus, at approximately 11 km above the mean planetary radius (see Figures 15.1 and 15.6). Although other crustal plateaus tend to have relatively flat interiors and rims of higher topography, no other crustal plateau is as extreme as Ishtar Terra in terms of its diameter, elevation, undeformed interior volcanic plains, and circumferential deformation features.

Crustal plateaus have been proposed to form over mantle upwellings and over mantle downwellings. In the mantle upwelling scenario, a plume creates a crustal plateau through decompression melting above the plume head, analogously to plateaus formed on the terrestrial seafloor. Deformation occurs as the topography viscously relaxes. The alternative model forms the plateaus above a cold, sinking mantle downwelling. On Earth, both subduction zones and local sites of downwelling form below cold mountain roots. Venusian crustal plateaus are proposed to form as a downwelling causes sinking of the lower lithosphere and accumulation and compression of the crust at the surface. The mechanism for forming small, local regions of tessera is not clear. In many cases these regions are embayed and thus appear to be old and possibly inactive. There are few clues as to original processes that cause deformation. One possibility is that these areas represent sections of tessera plateaus that were once elevated but have topographically relaxed. If plateaus formed in an earlier, hotter time period, relaxation may have proceeded more rapidly, allowing for complete relaxation of plateaus.

Chapter | 15 Venus: Surface and Interior

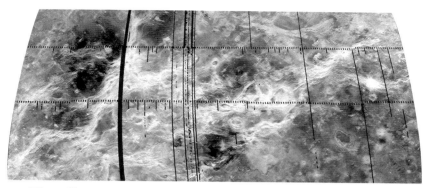

FIGURE 15.16 Radar image of Hecate Chasma, approximately 7000 by 3000 km, centered at 16° N, 240° E. Hecate Chasma is a huge tectonic feature, stretching from Atla Regio to Beta Regio. The rift is very bright in radar, and has a wispy appearance. The rift comprises numerous branches at a range of orientations, with coronae present throughout the region, both on and off the rift. The black stripes are data gaps.

The semicircular rim of Western Ovda Terra could be the remnant of a relaxed plateau. Alternatively, small tessera terrains may be a result of an entirely different type of tectonic event, such as ridge belt formation.

A key question for tesserae is their composition. Some researchers have suggested that crustal plateaus may be analogous to terrestrial continents due to their size, their compensation by crustal roots, their highly deformed surfaces, and their local embayment by the surrounding plains. Tesserae are relatively old, and could be made of more silica-rich crust. As discussed above, the low thermal emissivity anomalies observed by both Venus Express and the Galileo Orbiter flyby support this interpretation. If tesserae are silica-rich, perhaps granitic rocks like Earth's continents, it implies that they are remnants of an earlier, wetter time period on Venus.

8.2. Chasmata and Fracture Belts

Chasmata (chasma means canyon) are regions of extensional deformation, as indicated by their locally low topography and **graben** or grabenlike morphology. There are five major chasmata on Venus that extend for thousands of kilometers and are several kilometers deep: Parga, Hecate (see Figure 15.16), Dali/Diana, Devana, and Ganis Chasmata. The fracture zones in these regions are typically ∼200 km wide, with topographic troughs that are generally narrower, with widths of ∼50–80 km. There are seven smaller chasmata, with lengths of hundreds of kilometers and proportionately narrower fracture belts and troughs. Several of the chasmata occur on the flanks of hotspot rises and may be a result of topographic uplift above a plume. The majority of other chasmata form synchronously with coronae, as discussed above. Although chasmata are not required for coronae to form, or vise-versa, it is clear that the presence of one increases the likelihood of the other. Both extension and upwelling plumes can thin the lithosphere, which may focus additional extension and upwelling in an area. Chasmata are analogous to terrestrial continental rift zones, which have relatively small amounts of crustal extension, in the range of several percent.

Fracture belts appear similar to minor chasmata, but are less intensely fractured, implying less extension. A curious feature of fracture belts is that they are topographically broad swells rather than topographic lows. The positive relief suggests that they went through a compressional stage, and that the fractures may be due to topographic uplift rather than regional extension.

8.3. Coronae

Coronae are large (>100 km across) circular features surrounded by concentric ridges and fractures (Figure 15.17).

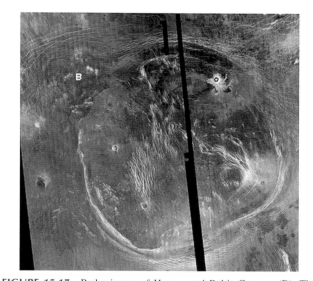

FIGURE 15.17 Radar image of Heng-o and Beltis Coronae (B). The topographic rim of Heng-o, which corresponds approximately to the fracture annulus deforms the local regional plains. To the west lie extensive flow fields interpreted to originate from Beltis Corona (to the NW of Heng-o) and the western annulus of Heng-o. Three volcanic centers of different ages lie within the annulus of Heng-o. Curved black edges result from the sinusoidal projection. Black areas are data gaps.

Over 500 coronae have been identified on Venus; the largest one is Artemis Corona at 2500 km across. Coronae often have volcanoes in their interiors and many are surrounded by extensive lava flows. Coronae tend to be raised at least 1 km above the surrounding plains, but others are depressions, rimmed depressions, or rimmed plateaus. Most coronae are located along rift or chasmata systems, although some are at topographic rises and others occur in the plains away from other features. Coronae are thought to form over thermal plumes or rising hot blobs, smaller in scale and probably rising from shallower depths than the plumes that form topographic rises. The wide range in corona topographic shapes indicates that coronae evolution also involves delamination or sinking of lithospheric material in its later stages. Studies of the gravity signatures of some large coronae indicate that many coronae are likely to be isostatically compensated, and thus probably inactive. The fact that we do not see coronae on Earth may be due to the lack of an asthenosphere on Venus.

8.4. Ridge Belts and Wrinkle Ridges

Ridge belts occur in a variety of morphologies and are distributed around the planet. Based on the morphology of individual fractures and the long, narrow topographic highs that comprise individual ridges, ridge belts are interpreted to be a result of compressional stresses (see Figure 15.18). Individual ridges are typically less than 0.5 km in height, 10–20 km wide, and 100–200 km long with a spacing of ~25 km. The two largest concentrations of ridge belts occur in Atalanta/Vinmara Planitiae and Lavinia Planitia. The belts in Atalanta/Vinmara Planitiae are roughly an order of magnitude larger than those elsewhere. Belts in Lavinia are unusual in that they have extensional fractures roughly parallel to compressional features within the same belt, possibly due to topographic uplift along the ridge. Larger belts are believed to result from mantle downwelling, similar to the proposed downwelling origin for crustal plateaus, but with lower strain. Smaller belts may be associated with more local-scale tectonics.

Wrinkle ridges are extremely common features on Venus and are also interpreted as simple compressional folds and/or faults but are much narrower (~1 km or less in width) than ridges. They have positive relief, based on the fact that lava flows can be seen to pond against some wrinkle ridges, but that relief is too small to be seen in Magellan altimetry. Most ridges occur in evenly spaced set, 20–40 km apart. These sets of wrinkle ridges can be local in nature, associated with a corona, for example, but more commonly cover thousands of kilometers. These larger sets are likely to be gravitational spreading of high topography into lower regions and can be seen to form rings around some large topographic features (see Figure 15.4). Other sets cannot be clearly associated with topographic highs.

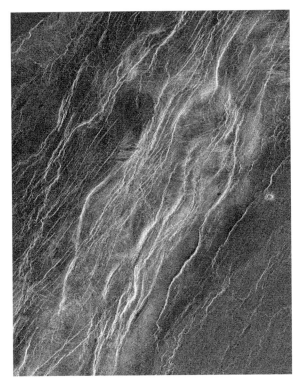

FIGURE 15.18 Radar image of a ridge belt in Atalanta/Vinmara Planitia, approximately 85 by 110 km in dimension and centered at 41° N, 196° E. The belt comprises a series of NE-trending ridges. There are both very narrow ridges, down to the resolution of the data, and ridges several kilometers wide.

One hypothesis is that these features result from thermal contraction due to climate change-driven atmospheric temperature changes. In some regions there are two sets of wrinkle ridges, although one set is usually better developed.

8.5. Plains Fractures, Grids, and Polygons

A wide range of long, narrow, approximately straight fractures occur in the plains. Some fractures are wide enough to be resolved as graben, but most are too narrow (less than 0.5 km) to be resolved as more than fractures. Most are interpreted as extensional fractures because they parallel resolvable graben and because of their shape. Some are clearly associated with local features such as volcanoes or corona, and are probably due to extension above dikes. In some locations there are either single sets or intersecting grids of fractures that cover hundreds of kilometers (Figure 15.19). They are very regularly spaced, with separations of 1–2.5 km. The narrow spacing suggests that a thin layer is involved in the deformation. It is not obvious how a uniform stress can be transmitted to such a thin layer over such a broad region. Shear deformation is required to produce grids of intersecting lineations.

Another type of extensional feature observed on Venus is polygons, which are found in over 200 locations on Venus.

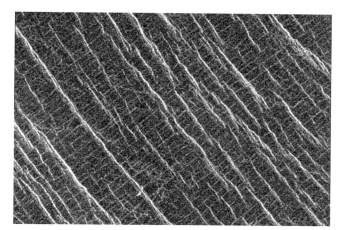

FIGURE 15.19 Radar image of a section of lineated plains approximately 35 km across, centered at 30° N, 333.3° E. These gridded plains are located in Guinevere Planitia and are incredibly uniform in orientation, size, and space over nearly 1000 km.

These features are analogous to mud cracks in that they form in a uniform, extensional stress field. However, they form not as water is lost but instead when rock cools and contracts. The typical diameter is ~2 km, but some are up to 25 km across. Some areas have multiple scales of deformation. Again, some of these features can be associated with local events such as volcanoes, but others cover very broad regions and do not have an obvious origin. Polygons are most commonly associated with small volcanic edifices, and frequently appear to form synchronously (Figure 15.10). Some may form by actual cooling of lava flows. Such basaltic columns are common on Earth, but the scales of the features found on Venus are orders of magnitude larger, implying that the flow thickness on Venus would probably be too large to be plausible. Another mechanism, as proposed for wrinkle ridges, is the possible heating and cooling of the upper crust due to climate change.

9. SUMMARY

Venus provides a unique window into the evolution of terrestrial planets. It is essentially identical to Earth in size and bulk composition, yet its geologic history is entirely different. Venus' level of geologic activity over the last billion years is comparable to that of Earth and exhibits many of the same geologic processes. The convecting interior drives geologic activity at the surface, creating a dozen major highlands. These highlands include "hotspots", which form above mantle plumes, and the more enigmatic and intensely deformed highland plateaus. Venus' hotspots appear to be sources of recent or even active volcanism today, providing clues about the interior, surface, and atmospheric processes. The majority of the surface is composed of vast volcanic plains along with nearly ubiquitous tectonic features. There are tens of thousands of volcanic features from small-scale (hundreds of meters) flows, vents, and shields, to hundreds of large-scale (>100 km) shield volcanoes that blanket the surface. The pervasive volcanism may have buried the earliest, heavily cratered surfaces, or they may have been destroyed through tectonic processes. Tectonic features range in scale from pervasive linear fractures and polygons at the limit of resolution to highland plateaus composed of tessera terrain 1000—2000 km in diameter.

Despite the similarities between Venus and Earth, Earth is the only body in our solar system that developed the system of plate tectonics that has so shaped the geologic and environmental evolution of our planet. The atmosphere of Venus lost nearly all its water early in its evolution. The loss appears to have affected the interior as well, causing the lithosphere to be too strong to break into the plates observed on Earth, and the asthenosphere to be too strong to facilitate rapid horizontal plate motion. This same loss of water has contributed to the dominance of CO_2 in the atmosphere and the resulting greenhouse effect that created the scorching surface conditions. Why Venus lost its water is not understood, but as with Mars, the absence of a strong magnetic field exposes the atmosphere to erosion by solar wind. In turn a planet must be losing heat rapidly enough to drive the formation of a magnetic dynamo. The interior volatile content affects the processes through which planets lose heat, and appears to be the key to whether or not plate tectonics develops. Was Venus originally on the same evolutionary path as Earth? What was the pivotal event or process that sent Venus down an alternate path to the hellish, uninhabitable planet we observe today? We can begin to address these questions, thus better understanding the evolution of our own planet, through future missions to understand the coupled evolution of the atmosphere, surface, and interior.

BIBLIOGRAPHY

Mueller, N., Helbert, J., Hashimoto, G. L., Tsang, C. C. C., Erard, S., Piccioni, G., et al. (2008). Venus surface thermal emission at 1 μm in VIRTIS imaging observations: evidence for variation of crust and mantle differentiation conditions. *Journal of Geophysical Research, 113.* http://dx.doi.org/10.1029/2008JE003118.

Books
Brougher, S. W., Hunten, D. M., & Phillips, R. J. (Eds.). (1997). *Venus II.* Tucson: Univ. of Arizona Press.

Journal Articles Special Issues
Magellan at Venus, parts 1 and 2. (1992). *Journal of Geophysical Research, 97*(E8 and E10).
Advances in Venus science. (2012). *Icarus, 217*(2).

Web Sites
Venus data from US missions are available through the Planetary Data System at pds.nasa.gov and from ESA missions at www.sciops.esa.int.

Chapter 16

Mars Atmosphere: History and Surface Interactions

David C. Catling
University of Washington, Dept. of Earth and Space Sciences/Astrobiology Program, Seattle, WA, USA

Chapter Outline

1. Introduction — 343
2. Volatile Inventories and Their History — 344
   2.1. Volatile Abundances — 344
   2.2. Sources and Losses of Volatiles — 346
3. Present and Past Climates — 348
   3.1. Present Climate — 348
   3.2. Past Climates — 349
   3.3. Mechanisms for Producing Past Wetter Environments — 352
      3.3.1. Carbon Dioxide Greenhouse — 352
      3.3.2. Impact Heating — 353
      3.3.3. Sulfur Dioxide Greenhouse — 353
      3.3.4. Methane-Aided Greenhouse — 353
      3.3.5. Hydrogen-Aided Greenhouse — 353
      3.3.6. Mechanisms for Producing Fluvial Features in Cold Climates — 354
   3.4. Milankovitch Cycles — 354
   3.5. Wind Modification of the Surface — 355
4. Concluding Remarks — 356
Bibliography — 357

A fundamental question about the surface of Mars is whether it was ever conducive to life in the past, which is related to the broader questions of how the planet's atmosphere evolved over time and whether past climates supported widespread liquid water. Taken together, geochemical data and models support the view that much of the original atmospheric inventory was lost to space prior to about 3.7 billion years ago. Before and around this time, the erosion of valley networks by liquid water suggests a past climate that was warmer. But exactly how the early atmosphere produced warmer conditions and the extent to which it did so remain open questions. Suggestions include an ancient greenhouse effect enhanced by various gases, impacts that created may temporary wet climates by turning ice to vapor and rainfall, and periodic melting of ice under moderately thicker atmospheres as Mars' orbit and axial tilt changed. For the last 3.7 billion years, it is likely that Mars has been predominantly cold and dry so that outflow channels that appeared later were probably formed by fluid release mechanisms that did not depend on a warm climate. Very recent gullies and narrow, summertime dark lineae that form on steep slopes are features that form in the current cold climate. In addition, wind erosion, dust transport, and dust deposition have been modulated by changes in Mars' orbital elements over time, which complicates the interpretation of climate and volatile history. In the past, surface modification by winds in a denser atmosphere may have been significant.

1. INTRODUCTION

The most interesting and controversial questions about Mars revolve around the history of liquid water. Because temperatures are low, the current, thin Martian atmosphere only contains trace amounts of water as vapor or ice clouds. In larger quantities, water is present as ice and hydrated minerals near the surface. Some geological structures resemble dust-covered glaciers or rock glaciers, while others strongly suggest the flow of liquid water relatively recently as well as in the distant past. But the present climate does not favor liquid water near the surface. Surface temperatures range from about 140 to 310 K. Temperatures above freezing occur only under highly desiccating conditions in a thin layer at the interface between the soil and atmosphere. Also, the surface air pressure over much of the planet is below the triple point of water (611 Pa or 6.11 mbar); under these conditions, at temperatures above freezing, liquid water would boil away. If liquid water is present near the surface of Mars today, it is confined to thin adsorbed layers on soil particles or highly saline solutions. No standing or flowing liquid water, saline or otherwise, has been unambiguously proven.

Conditions appear to have been more favorable for liquid water in the ancient past. The landscape has a number of fluvial (stream-related) features, of which the most important for climate are the valley networks, which are dried-up riverlike depressions fed by treelike branches of tributaries. Deltas exist at the end of a small fraction of valleys. Fluvial features that occurred later than the valleys are giant outflow channels. The valley networks indicate wetter past climates, while the outflow channels are commonly interpreted as massive release of liquid water from subsurface aquifers or the melting of underground ice (although a minority opinion argues in favor of runny lavas as the primary erosive agent). In addition to fluvial features, the soil and sedimentary rocks incorporate hydrous (water-containing) minerals that are interpreted to have formed in the presence of liquid water. The extent and timing of the presence of liquid water are central to the question of whether microbial life ever arose and evolved on Mars.

Atmospheric volatiles are substances that tend to form gases or vapors at the temperature of a planet's surface and so could have influenced the past climate and the occurrence of liquid water. Here we review the current understanding of volatile reservoirs, the sources and sinks of volatiles, the current climate, and evidence for different climates in the past. We consider the hypothesis that there have been one or more extended warm and wet climate regimes in the past, the problems with that hypothesis, and the alternative possibility that Mars has had a cold, dry climate similar to the present climate over nearly all of its history, while still allowing for some fluid flow features to occur on the surface. The possible relevance of very large orbital variations (Milankovitch cycles) for Mars' climate history is also examined.

Whether or not extended periods of warm, wet climates have occurred in the past, wind is certainly an active agent of surface modification at present and has probably been even more important in the past. Consequently, we also discuss evidence for how the surface has been changed by wind erosion, burial, and exhumation, and the resulting complications for interpreting Mars' surface history. We conclude with a brief overview of open questions.

2. VOLATILE INVENTORIES AND THEIR HISTORY

2.1. Volatile Abundances

Mars' thin atmosphere is dominated by carbon dioxide (CO_2), and in addition to the major gaseous components listed in Table 16.1, the atmosphere contains a variable amount of water vapor (H_2O) up to 0.1%, minor concentrations of photochemical products of carbon dioxide and

TABLE 16.1 Basic Properties of the Present Atmosphere

Average surface pressure	~6.1 mbar, varying seasonally by ~30%
Surface temperature	Average 215–218 K, range: 140–310 K
Major gases	Viking Landers: CO_2 95.3%, $^{14}N_2$ 2.6%, ^{40}Ar 1.6% Mars Science Lab: CO_2 96%, $^{14}N_2$ 1.9%, ^{40}Ar 1.9%
Significant atmospheric isotopic ratios relative to the terrestrial values	D/H in water ≈ 5 $^{15}N/^{14}N$ = 1.7 $^{38}Ar/^{36}Ar$ = 1.3 $^{13}C/^{12}C$ in CO_2 = 1.05 $^{18}O/^{16}O$ in CO_2 = 1.05

water vapor (e.g. CO, O_2, H_2O_2, and O_3), and trace amounts of the noble gases neon (Ne), argon (Ar), krypton (Kr), and xenon (Xe). Methane (CH_4), averaging about 10 parts per billion by volume (ppbv), has been reported based on spectra from ground-based telescopes (which are complicated by having to remove the effect of viewing Mars through the Earth's atmosphere) and relatively low resolution spectra obtained by European Space Agency's (ESA's) *Mars Express* orbiter. However, in situ measurements by National Aeronautics and Space Administration's (NASA's) *Mars Science Lab* rover have found no methane with an upper limit of about 1 ppbv, which must be taken as more definitive.

Volatiles that can play important roles in climate are also stored in the **regolith** and near-surface sediments. The regolith is a geologic unit that includes fine dust, sand, and rocky fragments comprising the Martian soil together with loose rocks, but excluding bedrock. Approximate estimates of the inventories of water, carbon dioxide, and sulfur are given in Table 16.2.

Water is stored as ice in the permanent north polar cap and its surrounding layered terrains, in layered terrains around the South Pole, and as ice, hydrated minerals, or adsorbed water in the regolith. The 5-km-deep residual northern polar cap consists of a mixture of ≥95% water ice and fine soil or dust, while layered south polar terrains contain water ice and about 15% dust. Taking account of their volumes and ice fractions, each cap and associated layered terrain contains water ice equivalent to a global ocean about 10 m deep.

Measurements of the energy of neutrons emanating from Mars into space by NASA's *Mars Odyssey* orbiter has also provided evidence for abundant water ice, adsorbed water, and/or hydrated minerals in the upper 1–2 m of regolith at high latitudes and in some low-latitude regions (Figure 16.1). Cosmic rays enter the surface of Mars and cause neutrons to be ejected with a variety of energies depending on the elements in the subsurface and their

TABLE 16.2 Volatile Reservoirs

Water (H_2O) Reservoir	Equivalent Global Ocean Depth:
Atmosphere	10^{-5} m
Polar caps and layered terrains	20 m
Ice, adsorbed water, and/or hydrated salts stored in the regolith	<100 m
Alteration minerals in 10-km crust assuming 1–3 wt% hydration	150–900 m
Deep aquifers	None found by radar
Carbon Dioxide (CO_2) Reservoir	**Equivalent Surface Pressure:**
Atmosphere	6 mbar
Carbonate in weathered dust	~200 mbar/100 m global average layer of weathered dust
Adsorbed in regolith	<40 mbar
Carbonate sedimentary rock	~3 mbar (from known outcrops)
Crustal subsurface carbonates	<250 mbar per km depth
Sulfur Dioxide (SO_2) Reservoir	**Global Mass:**
Atmosphere	0
Sulfate in weathered dust	$<0.9 \times 10^{16}$ kg SO_3 per m of global average soil (assuming <8 wt% sulfur as SO_3)
Sulfate sedimentary rock reservoirs	$\sim 10^{17}$ kg SO_3 (assuming 20 vol % SO_3 in observed volumes of sulfate deposits)

distribution. Abundant hydrogen serves as a proxy for water and/or hydrated minerals. In 2008, the robotic arm and thrusters on NASA's *Phoenix* Lander exposed ice at 68°N some 5–10 cm below the surface, verifying the inferences from *Mars Odyssey*'s neutron measurements. However, water ice probably extends to no more than ~20–30 m depth in the mid- to high-latitude regolith based on radar and the morphology of small craters. Consequently, the total water inventory appears to be dominated by hydrated minerals rather than ice and has a depth of 200–1000 m of a global equivalent ocean.

The inventory of CO_2 mainly depends on how much is locked up in carbonates hidden in the subsurface. Weathering of dust has occurred over billions of years even in the prevailing cold dry climate, and as a consequence some CO_2 appears to have been irreversibly transferred from the atmosphere to carbonate minerals in dust particles. The total amount depends on the global average depth of dust. Some CO_2 is likely to be adsorbed in the soil also, but the quantity is limited by competition for **adsorption** sites with water. From orbital spectra, some carbonate sedimentary rock outcrops have been identified but with an area of only 10^5 km^2 assuming a subsurface extent underneath associated geologic units. Taking an average thickness of 100 m, the CO_2 inventory in these outcrops is only 3 mbar. However, the carbonate inventory in the subsurface remains unknown. Also, carbonate outcrops that are smaller than orbital resolution can detect are likely present. Indeed, instruments on NASA's *Spirit Rover* identified a small carbonate outcrop in the Columbia Hills region of Gusev Crater.

Table 16.2 also lists sulfates. Although there are presently no detectable sulfur-containing gases in the atmosphere, sulfur gases should have existed in the atmosphere in the past when Mars was volcanically active. Measurements by NASA's landers and rovers show that sulfur is a substantial component in soil dust (~7–8% by mass) and surface rocks. Hydrated sulfate salt deposits have been identified in numerous layered deposits from near-infrared spectral data collected by *Mars Express* and NASA's *Mars Reconnaissance Orbiter* (*MRO*). About two-thirds of these deposits are within 10° latitude of the equator. Several sulfate minerals have been detected and the total sulfur abundance can be estimated. Notable sulfate minerals include kieserite ($MgSO_4 \cdot H_2O$) and gypsum ($CaSO_4 \cdot 2H_2O$). Jarosite ($XFe_3(SO_4)_2(OH)_6$, where "X" is a singly charged species such as Na^+, K^+, or hydronium (H_3O^+)) has been identified by the *Opportunity* rover in Meridiani Planum. Additional sulfate as gypsum is present in northern circumpolar dunes. Anhydrous sulfates, such as anhydrite ($CaSO_4$), are probably present but would give no signature in near-infrared spectra.

The total sulfur in visible deposits on Mars is around 10^{17} kg SO_3, which is within an order of magnitude of Earth's oceanic sulfate of 3.2×10^{18} kg SO_3 but well below Earth's sulfur inventory of 2.4×10^{19} kg SO_3 that includes sedimentary sulfur in the form of pyrite and sulfates. Thus, even accounting for its lower surface area, Mars' surface apparently has a smaller sulfur inventory than the Earth, which is presumably because of less extensive volcanic outgassing.

Evidence of volatile abundances also comes from Martian meteorites [*See* Meteorites]. These meteorites are known to be from Mars because of their igneous composition, unique oxygen isotope ratios, spread of ages, and gaseous inclusions whose elemental and isotopic compositions closely match the present Martian atmosphere. Ages of crystallization of these basaltic rocks, i.e. the times when the rocks solidified from melts, range from 4.4 to 0.15 billion years, which implies a parent body with active volcanism during this entire interval. Many of the Martian meteorites contain salt minerals, up to 1% by volume, which can include halite (NaCl), gypsum, anhydrite, and carbonates of magnesium, calcium and iron. The bulk meteorite compositions are generally dry, 0.05–0.3 wt% water, compared to terrestrial H_2O contents from 0.1 wt% in midocean ridge basalts to 2 wt% in basaltic magmas from subduction zones. There is debate about the extent to which Martian magmas may have degassed on eruption and lost their water. Consequently, estimates for the preeruptive volatile contents of Martian magmas vary from nearly anhydrous to about 2 wt% H_2O, which is a range from lunarlike to Earth-like. On the other hand, the Martian mantle is generally inferred to be sulfur rich, with 0.06–0.09 wt% S compared to 0.025 in Earth's mantle.

One important Martian meteorite, ALH84001, is a sample of 4.1 billion-year-old crust and contains about 1% by volume of distributed, 3.9-billion-year-old carbonate. ALH84001 has been heavily studied because of a controversial investigation in which four features associated with the carbonates were considered of possible biological origin: the carbonates themselves, traces of organic compounds, 0.1 μm-scale structures identified as microfossils, and crystals of the mineral magnetite (Fe_3O_4) (McKay et al., 1996). However, the biological nature of all these features has been strongly disputed and alternative abiotic origins have been proposed.

2.2. Sources and Losses of Volatiles

Volatile acquisition began during the formation of Mars. Planetary formation models indicate that impacting bodies that condensed from the evolving solar nebula near Mars' orbit were highly depleted relative to solar composition in the atmospheric volatiles: carbon, nitrogen, hydrogen, and noble gases. Nonetheless, formation of Jupiter and the outer planets would have gravitationally deflected volatile-rich asteroids from the outer asteroid belt and **Kuiper Belt** comets into the inner solar system. Analyses of the compositions of the Martian meteorites indicate that Mars acquired a rich supply of the relatively volatile elements during its formation. However, carbon, nitrogen, and noble gases are severely depleted in Mars' atmosphere and surface compared with Earth and Venus, apparently because loss processes efficiently removed these elements from Mars, as they did for hydrogen.

Two processes, **hydrodynamic escape** and impact erosion, must have removed much of any early Martian atmosphere. Hydrodynamic escape is pressure-driven escape that occurs when a planet's upper atmosphere is sufficiently warm to expand, accelerate through the speed of sound, and attain escape velocity *en masse*. Because this process is easiest for hydrogen-rich atmospheres, the general conception of hydrodynamic escape on Mars is of an early hydrogen-rich atmosphere flowing outward in a planetary wind (analogous to the "solar wind") that entrains and removes other gases. Heavy atoms are carried upward by collisions with hydrogen faster than they diffuse down under gravity, and the downward diffusion gives only weak selectivity to atomic mass. Nonetheless, the high $^{38}Ar/^{36}Ar$ ratio (Table 16.2) could be a sign of early hydrodynamic escape, although later escape processes can also drive this ratio high (see below).

Intense solar ultraviolet radiation and soft X-rays provide the energy needed to drive hydrodynamic escape. These fluxes would have been at least two orders of magnitude larger than at present during the first $\sim 10^7$ years after the solar system formed as the evolving sun moved toward the

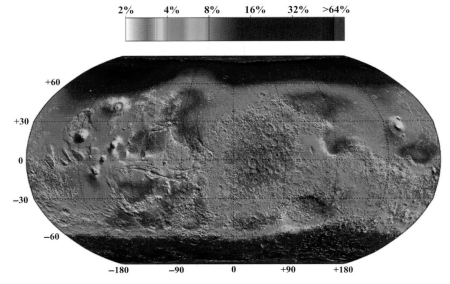

FIGURE 16.1 Water-equivalent hydrogen content of subsurface water-bearing soils derived from the Mars Odyssey Neutron Spectrometer. *From Feldman et al. (2004).*

main sequence. Although the early Sun was 25–30% less luminous overall, studies of early stars suggest that the early Sun was rotating more than 10 times faster than present, which would have caused more magnetic activity, associated with over a 100 times more emission in the extreme ultraviolet portion of the spectrum than today. Consequently, hydrodynamic escape would have been a very efficient atmospheric removal mechanism if hydrogen had been a major atmospheric constituent during this period.

The amount of hydrogen in the early atmosphere of a terrestrial planet depends on the chemical reaction of iron and water during accretion and the segregation of the core and mantle. If water brought in by impactors could mix with free iron during this period, it would oxidize free iron, releasing large amounts of hydrogen to the atmosphere and fostering hydrodynamic escape. Interior modeling constrained by Mars' gravitational field and surface composition together with analyses of the composition of the Martian meteorites indicates that Mars' mantle is rich in iron oxides relative to the Earth, consistent with the hypothesis that a thick hydrogen-rich atmosphere formed at this early stage. It has been suggested that hydrodynamic escape removed the equivalent of an ocean at least 1 km deep together with most other atmospheric volatiles from Mars, although this estimate is based on extrapolation from the current value of the deuterium–hydrogen ratio (D/H), which is uncertain because D/H may reflect geologically recent volatile exchange rather than preferential loss of hydrogen compared to deuterium over the full history of Mars. Comets arriving after the completion of hydrodynamic escape may have brought in much of the atmospheric volatiles in the current inventory.

Early Mars was also potentially vulnerable to impact erosion—the process where atmospheric gases are expelled as a result of the large-body impacts. Big impacts release enough energy to accelerate atmospheric molecules surrounding the impact site to speeds above the escape velocity. A large fraction of these fast molecules escape. Since escape is easier with a smaller gravitational acceleration of the planet, impact erosion would have been far more efficient on Mars than on Earth. The early history of the inner solar system is characterized by a massive flux of large asteroids and comets, and most models suggest net atmospheric erosion for early Mars rather than accumulation of volatiles. Based on dating of lunar rocks and impact features, bombardment by massive objects is known to have declined rapidly with time after planet formation. The interval from 4.1 to 3.7 billion years ago is the Noachian eon in Martian geologic time, so that massive bombardment effectively ceased around the end of the Noachian. The base of the Noachian is defined by the time of formation of the Hellas impact basin around 4.1 billion years ago, before which is the Pre-Noachian. Interestingly, the lunar record suggests a spike in the impact flux 4.0–3.8 billion years ago, called the Late Heavy Bombardment, when most impact melt rocks formed. Thus, the late Noachian was probably a time of particularly intense impacts.

Bombardment by massive bodies during the Noachian has left an imprint in the form of large impact craters that are obvious features of the southern hemisphere (Figure 16.2). More subtle "ghost" craters and basins that have been largely erased by erosion and/or filling in the relatively smooth northern plains provide further evidence of Noachian impact bombardment. Calculations suggest that impact erosion should have removed all but ~1% of an early CO_2-rich atmosphere (e.g. see Carr, 1996, p.141). Water in ice and carbon in carbonates would have been relatively protected, however, and the exact efficiency of the removal of volatiles by impact erosion is unknown.

What was the size of Mars' volatile reservoirs after Late Heavy Bombardment, some 3.7 billion years ago? The isotopic ratios $^{13}C/^{12}C$, $^{18}O/^{16}O$, $^{38}Ar/^{36}Ar$, and $^{15}N/^{14}N$ are heavy compared with the terrestrial ratios (see Table 16.1). This has been interpreted to indicate that 50–90% of the initial reservoirs of CO_2, N_2, and cosmogenic argon have been lost over the past 3.7 billion years by mass-selective **nonthermal escape** from the upper atmosphere (mainly **sputtering** produced by the impact of the solar wind on the upper atmosphere). Considering the possible current reservoirs of CO_2 in Table 16.2, the resulting CO_2 available 3.7 billion years ago could have been as much as ~1 bar or as little as a few tens of millibars.

Another approach to estimating the CO_2 abundance at the end of the Noachian is based on the abundance of ^{85}Kr in the present atmosphere. Since this gas is chemically inert and too heavy to escape after the end of massive impact bombardment, its current abundance probably corresponds closely to the abundance at the end of the Noachian. Impact erosion would have effectively removed all gases independent of atomic mass, so the ratio of ^{85}Kr abundance to C in plausible impactors (Kuiper Belt comets or outer solar system asteroids) can then yield estimates of the total available CO_2 reservoir at the end of the Noachian. The corresponding atmospheric pressure, if all CO_2 were in the atmosphere, would be only ~0.1 bar, in the lower range of estimates from the isotopic and escape flux analysis. This low estimate is consistent with the low modern nitrogen abundance after allowing for mass-selective escape as indicated by the high $^{15}N/^{14}N$ ratio (Table 16.1).

If ~0.1 bar was left at the end of the Noachian, besides nonthermal escape, slow carbonate weathering of atmospheric dust could also have removed CO_2 from the atmosphere (as mentioned previously). This irreversible mechanism may account for the fate of a large fraction of the CO_2 that was available in the late Noachian, along with adsorbed CO_2 in the porous regolith (Table 16.2). It has long been speculated that much of the CO_2 that was in the

FIGURE 16.2 Elevation map of Mars derived from the Mars Orbiter Laser Altimeter (MOLA) on NASA's Mars Global Surveyor, with some major features labeled. *NASA/MOLA Science Team.*

Noachian atmosphere got tied up as carbonate sedimentary deposits beneath ancient water bodies. However, so far only small outcrops and no significant quantities of carbonate sedimentary rocks have been found (see further discussion below).

Escape of water in the form of its dissociation products H and O takes place now, and must have removed significant amounts of water since the Noachian. Isotopic ratios of D/H and $^{18}O/^{16}O$ in the atmosphere and in Martian meteorites and escape flux calculations provide rather weak constraints on the amount that has escaped over that period. Upper bounds on the estimates of water loss are 30–50 m of equivalent global ocean. These amounts are roughly comparable to estimates of the water currently stored as ice in the polar caps and regolith (Table 16.2).

Sulfur is not stable in the current Martian atmosphere as either sulfur dioxide (SO_2) or hydrogen sulfide (H_2S) because both gases oxidize and ultimately produce sulfate aerosols that fall to the surface; however, significant sulfur gases must have been introduced into the atmosphere by volcanism. Estimated ages of volcanic surfaces on Mars indicate that rates of volcanism declined and became more intermittent after about 3.5 billion years ago. Formation of the Tharsis ridge volcanic structure, believed to have occurred in the late Noachian eon, must have corresponded with outgassing of large amounts of sulfur as well as water from the mantle and crust. The probable quantity of sulfur released is consistent with the relatively high mass fraction of sulfur in the soil and the presence of large deposits of sedimentary sulfates. Martian meteorites are ~5 times as rich in sulfur as in water and it is likely that the regolith contains more sulfur than water. The volatile elements chlorine and bromine are also abundant in rocks and soils, but more than an order of magnitude less so than sulfur.

An important observation in the sulfates found in Martian meteorites is that sulfur and oxygen isotopes are found in relative concentrations that are mass-independently fractionated. Most kinetic processes fractionate isotopes in a mass-dependent way. For example, the mass difference between ^{34}S and ^{32}S means that twice as much fractionation between these isotopes is produced as between ^{33}S and ^{32}S in a mass-dependent isotopic discrimination process such as diffusive separation. Mass-independent fractionation (MIF) is a deviation from such proportionality. MIF is found to arise when ultraviolet radiation interacts with certain atmospheric gases in photochemistry. On Earth, the MIF of oxygen in sulfates in the extraordinarily dry Atacama Desert is taken to prove that these sulfates were deposited by photochemical conversion of atmospheric SO_2 to submicron particles and subsequent dry deposition. The MIF signature in sulfates in Martian meteorites suggests that a similar process produced these sulfates on Mars, and implies that the sulfur cycled through the atmosphere at some ancient time.

3. PRESENT AND PAST CLIMATES

3.1. Present Climate

The thin, predominantly carbon dioxide atmosphere produces a small greenhouse effect, raising the average surface temperature of Mars only 5–8 K above the 210 K temperature that would occur in the absence of an atmosphere. Carbon dioxide condenses out during winter in the polar caps, causing a seasonal range in the surface pressure of about 30%. There is a small residual CO_2 polar cap at the South Pole, which persists all year round in the current epoch; it represents a potential increase in the CO_2 pressure of 4–5 mbar if it were entirely sublimated into the atmosphere. The atmospheric concentration of water vapor is controlled by saturation and condensation, and so varies seasonally and daily. Water vapor exchanges with the polar caps over the course of the Martian year, especially with the North Polar cap. After **sublimation** of the winter CO_2 polar cap, the summertime central portion of the cap surface is water ice. Water vapor sublimates from this surface in northern spring to early summer, and is transported southward, but most of it is precipitated or adsorbed at the surface before it reaches southern high latitudes.

In addition to gases, the atmosphere contains a variable amount of dust as well as icy particles that form clouds. Dust loading can become quite substantial, especially during northern winter. Transport of dust from regions where the surface is being eroded by wind to regions of dust deposition occurs in the present climate. Acting over billions of years, wind erosion, dust transport, and dust deposition strongly modify the surface (see Section 3.5). Visible optical depths can reach ∼5 in global average and even more in local dust storms. A visible **optical depth** of 5 means that direct visible sunlight is attenuated by a factor of $1/e^5$, which is roughly 1/150. Much of the sunlight that is directly attenuated by dust reaches the surface as scattered diffuse sunlight. Median dust particle diameters are ∼1 μm, so this optical depth corresponds to a column dust mass ∼3 mg/m^2. Water ice clouds occur in a "polar hood" around the winter polar caps and over low latitudes during northern summer, especially over uplands. Convective carbon dioxide ice clouds occur at times over the polar caps, and they occur rarely as high-altitude cirrus clouds of CO_2 ice particles.

Orbital parameters cause the cold, dry climate of Mars to vary seasonally in somewhat the same way as intensely continental climates on Earth. The present tilt of Mars' axis (25.2°) is similar to that of the Earth (23.5°), and a Martian year is 687 Earth days long or about 1.9 Earth years. Consequently, seasonality bears some similarity to that of the Earth but Martian seasons last about twice as long on average. However, the **eccentricity** of Mars' orbit is much larger than that of the Earth's (0.09 compared with 0.015) and perihelion (the closest approach to the sun) currently occurs near northern winter solstice. As a consequence, asymmetries between northern and southern seasons are much more pronounced than on the Earth. Mars' rotation rate is similar to that of the Earth's, and like the Earth, the atmosphere is largely transparent to sunlight so that heat is transferred upward from the solid surface into the atmosphere. These are the major factors that control the forces and motions in the atmosphere, i.e. atmospheric dynamics, which is similar on Mars and Earth. Both planetary atmospheres are dominated by a single meandering mid-latitude jet stream, strongest during winter, and a **Hadley circulation** in lower latitudes. The Hadley circulation is strongest near the solstices, especially northern winter solstice, which is near perihelion, when strong rising motion takes place in the summer (southern) hemisphere and strong sinking motion occurs in the winter (northern) hemisphere.

Mars lacks an ozone layer, and the thin, dry atmosphere allows very short wavelength ultraviolet radiation to penetrate to the surface. In particular, solar ultraviolet radiation in the range 190–300 nm, which is shielded on Earth by the ozone layer and oxygen, can reach the lower atmosphere and surface on Mars. This allows water vapor dissociation close to the Martian surface (H_2O + ultraviolet photon → H + OH). As a consequence of photochemical reactions, oxidizing free radicals (highly reactive species with at least one unpaired electron, such as OH or HO_2) are produced in near-surface air. In turn, any organic material near the surface rapidly decomposes and the soil near the surface is oxidizing. These conditions as well as the lack of liquid water probably preclude life at the very surface on present-day Mars.

Although liquid water may not be completely absent from the surface, even in the present climate, it is surely very rare. This is primarily because of the low temperatures. Even though temperatures of the immediate surface rise above freezing at low latitudes near midday, above-freezing temperatures occur only within a few centimeters or millimeters on either side of the surface in locales where the relatively high temperatures would be desiccating. A second factor is the relatively low pressure. Over large regions of Mars, the pressure is below the triple point for which exposed liquid water would rapidly boil away.

Because the present atmosphere and climate of Mars appear unsuitable for the development and survival of life, at least near the surface, there is great interest in the possibility that Mars had a thicker, warmer, and wetter atmosphere in the past.

3.2. Past Climates

Several types of features suggest that fluids have shaped the surface during all eons—the Noachian (4.1–3.7 billion years ago), Hesperian (3.7 to 3.5–3.0 billion years ago), and Amazonian (from 3.5 to 3.0 billion years ago to the present). In terrains whose ages are estimated on the basis of crater distributions and morphology to be Noachian to early Hesperian, "valley network" features are abundant (Figure 16.3). The morphology of valley networks is very diverse, but most consist of dendritic networks of small valleys, often with V-shaped profiles in their upper reaches becoming more U-shaped downstream. Their origin is

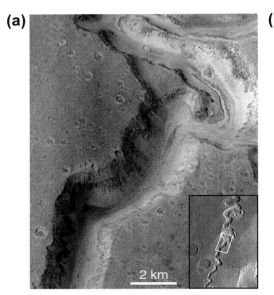

FIGURE 16.3 (a) An image of Nanedi Vallis (5.5° N, 48.4° W) from the Mars Orbiter Camera (MOC) on NASA's Mars Global Surveyor (MGS) spacecraft. The sinuous path of this valley at the top of the image is suggestive of meanders. In the upper third of the image, a central channel is observed and large benches indicate earlier floor levels. These features suggest that the valley was incised by fluid flow (the inset shows a lower resolution Viking Orbiter image for context). (*From image MOC-8704, NASA/Malin Space Science Systems.*) (b) Valley networks that illustrate dissected Noachian terrain (14° S, 61° E). Part of the rim of Huygens Crater is shown on the left of the image. Note that the valley networks incise a Noachian landscape with large craters that were that were degraded by some process prior to the formation of the valleys. *HRSC image from orbit 532, ESA/DLR/FUB.*

attributed to surface water flows or groundwater sapping. The latter is when underground water causes erosion and collapse of overlying ground. Although often much less well developed than valley network systems produced by fluvial erosion on Earth, Martian valley networks are suggestive of widespread precipitation and/or groundwater sapping that would have required a much warmer climate, mainly but not entirely, contemporaneous with termination of massive impact events at the end of the Noachian (~3.7 billion years ago). In Figure 16.3, we show two examples of valley network features. Figure 16.3(a) is a high-resolution image that shows a valley without tributaries in this portion of its reach (although some tributary channels are found farther upstream), but its morphology strongly suggests repeated flow events. Figure 16.3(b) shows fairly typical valley networks that are incised on Noachian terrain. A relatively small number of valleys on Mars terminate in deltas, which provides strong evidence of liquid water. However, the deltas are undissected, which implies an abrupt end to the era of valley formation. Some deltaic deposits also contain clay minerals such as the one in Jezero crater (18.4° N, 282.4° W) (Figure 16.4), which is fed by a valley network northwest of Isidis.

Valley networks are incised on top of a Noachian landscape of craters with heavily degraded rims and infilling or erosion (Figure 16.3(b)). Such crater morphologies indicate relatively high erosion rates. Some models suggest that erosion and deposition was caused by fluvial activity, at least in part. However, the interpretation is

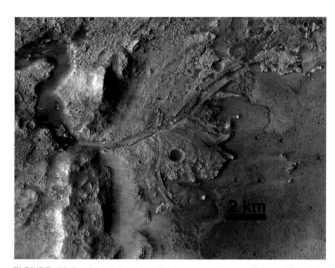

FIGURE 16.4 A deltaic deposit in western Jezero Crater. The delta feature is in positive relief, which shows that the material in the deposit was more resistant to erosion than surroundings. Yellow and blue colors indicate basaltic minerals, while clay minerals are green. Purple-brown surfaces have no distinctive spectral features. *NASA/JPL/JHUAPL/MSSS/Brown University.*

complex because the image data suggests that craters were also degraded or obscured by impacts, eolian transport, mass wasting and, in some places, airfall deposits such as volcanic ash or impact ejecta.

Besides valley networks, a second class of geomorphic features suggesting liquid flow is a system of immense

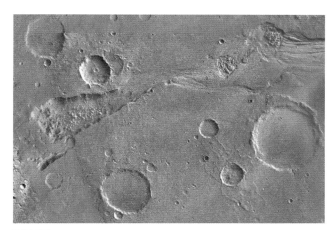

FIGURE 16.5 The head of the channel Ravi Vallis, about 300 km long. An area of chaotic terrain on the left of the image is the apparent source region for Ravi Vallis, which feeds into a system of channels that flow into Chryse Basin in the northern lowlands of Mars. Two further such regions of chaotic collapsed material are seen in this image, connected by a channel. The flow in this channel was from west to east (left to right). This false color mosaic was constructed from the Viking Mars Digital Image Map. *From NASA/Lunar and Planetary Institute Contribution No. 1130.*

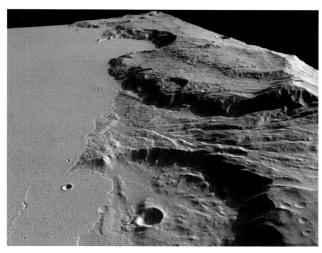

FIGURE 16.6 A perspective view looking toward the south of the edge of the outflow channel Mangala Valles (17° S, 213° E), which is located southwest of the Tharsis volcanic region. The two eroded craters in the foreground and background each have diameters of about 30 km. The smooth floor of the channel is covered in a lava flow with distinct edges and the surrounding cratered terrain is old compared to both the channel and the lava. *HRSC image from orbit 4117, ESA/DLR/FUB.*

channels that formed during the late Hesperian (Figure 16.5). These features, referred to as "**outflow channels**" or "catastrophic outflow channels", are sometimes more than 100 km in width, up to ∼1000 km in length, and as much as several kilometers deep. They occur mainly in low latitudes (between 20° north and south) around the periphery of major volcanic provinces such as Tharsis and Elysium, where they debauch northward toward the low-lying northern plains. The geomorphology of these channels has been compared with the scablands produced by outwash floods in Eastern Washington State from ice age Lake Missoula, but if formed by flowing water, flow volumes must have been larger by an order of magnitude or more. It has been estimated that the amount of water required to produce them is equivalent to a global ocean at least a few hundred meters deep. Many of these channels originate in large canyons or jumbled chaotic terrain that was evidently produced by collapse of portions of the plateau surrounding Tharsis. The origin of outflow channels is unknown, but the dominant hypothesis is that they were generated by catastrophic release of water from subsurface aquifers or rapidly melting subsurface ice. Alternatively, volcanic or impact heating caused catastrophic dehydration of massive hydrated sulfate deposits and resulting high-volume flows of liquid water. If water was released by these flows, its fate is unknown, although a number of researchers have proposed that water pooled in the northern plains and may still exist as ice beneath a dust-covered surface. A minority view is that the outflow channels were not carved by water but runny lavas. Some outflow channels (e.g. Mangala Valles, Athabasca Valles) have lava flows on their floors (Figure 16.6) and the source of some outflow channels are also sources of lava, e.g. Cerberus Fossae for Athabasca Valles or Memnonia Fossae for Mangala Valles. However, water and lava are not mutually exclusive. Magmatic intrusions could have melted subsurface ice, triggering floods of water in association with lava flows.

Unlike the valleys and channels, gullies are very recent features that have been interpreted to imply fluvial flow, but evidence from *MRO* images suggests that gullies can form today in association with carbon dioxide ice rather than liquid water. Gullies are incisions of tens to hundreds of meters length commonly found on poleward facing sloping walls of craters, plateaus, and canyons, mainly at southern midlatitudes (∼30°−55°S) (Figure 16.7). The gullies typically have well-defined alcoves above straight or meandering channels that terminate in debris aprons. Their setting and morphology has led to comparisons with debris flows in terrestrial alpine regions that are produced by rapid release of meltwater and consist typically of ∼75% rock and silt carried by ∼25% water.

High-resolution images from *MRO* show that gullies are forming on Mars today when carbon dioxide frost turns to vapor at the end of winter (Figure 16.7). Sublimation of CO_2 frost presumably causes a fluidlike, dry flow of rock and soil with the CO_2 gas providing lubrication. This mechanism explains the predominant location of gullies in the southern hemisphere where there is a greater seasonal accumulation and distribution of carbon dioxide frost than in the north because of prolonged southern winters. Consequently, gullies do not require a warm climate or low-latitude reservoirs of subsurface water or ice. In fact, gullies

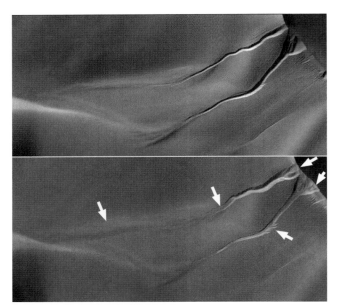

FIGURE 16.7 Changes in gullies on a sand dune inside Matara Crater (49.4° S, 34.7° E) that run leftward downhill from a dune crest in the upper right corner. Each image is 1.2 km across. The top image was taken in 2008, during midautumn in the southern hemisphere, while the bottom one was taken in 2009, during the beginning of the following summer. Over the Martian winter, the alcoves at the crest of the dune (arrows) and the channel beds (arrows) widened as material moved downward. The apron at the bottom of the gullies (arrow) also lengthened. Illumination is from the upper left. *HiRISE images PSP_007650_1300 and ESP_013834_1300, NASA/JPL-Caltech/Univ. of Arizona.*

FIGURE 16.8 Summer season dark streaks called RSL on a steep wall in Newton Crater (41.6° S, 202.3° E). The RSL are 0.5—5 m wide. False color. *A reprojection of HiRISE image ESP_022689_1380, NASA/JPL-Caltech/Univ. of Arizona.*

induced by CO_2 sublimation illustrate how features that look very similar to fluvially eroded analogs on Earth can actually have other causes.

Other curious features that develop on Mars today are Recurring Slope Lineae (RSL), which may be associated with seasonal melting of water ice. RSL are dark, narrow (0.5—5 m) lineaments that develop during warm seasons on steep rocky slopes that are equator facing in equatorial or mid latitude regions (Figure 16.8). The RSL fade and disappear during the cold season. Their formation is not understood, but an association with peak surface temperatures ranging from 250 to 300 K supports the idea that melting of water ice is probably involved. Salts dissolved in the water would also allow flow below 273 K.

Minerals on very ancient surfaces provide evidence for a wetter early Mars that bolsters the inference of liquid water from valley networks and outflow channels (if water eroded). Much of the surface on Mars is **basalt**, a dark-colored igneous rock rich in iron and magnesium silicate minerals. When basalt reacts with water, "alteration minerals" are produced, such as clays. So alteration minerals can be diagnostic of the past presence of liquid water and sometimes they suggest its pH. For example, clay minerals tend to be produced when alkaline water reacts with basaltic minerals. Notably, clay minerals have been detected in mudstone by NASA's *Curiosity Rover* in an area interpreted as an ancient dried-up lake bed in Gale Crater.

Alteration minerals on Mars have a broad trend in their distribution in geologic time. Orbital spectroscopy suggests that hydrous alteration minerals cover about 3% of Noachian surfaces in the form of clays and some carbonates. Sulfate minerals tend to be found on late Noachian or Hesperian surfaces, while younger Amazonian surfaces have reddish, dry iron oxides. To some, this broad pattern suggests three environmental epochs, starting in Noachian when alkaline or neutral pH waters made clay minerals from basalt. In the second epoch, sulfates were presumably derived from sulfur-rich volcanic gases. The third epoch is the cold, dry environment with rust-colored surfaces that continues today.

3.3. Mechanisms for Producing Past Wetter Environments

Despite extensive investigation, the causes of early warm climates in the Noachian or Hesperian remain to be identified. Here we review several possibilities.

3.3.1. Carbon Dioxide Greenhouse

A suggestion put forward after the *Mariner 9* orbiter mission in 1972 is that the early atmosphere contained much more CO_2 than it does now. The idea is that substantial CO_2 caused an enhanced greenhouse effect through its direct infrared radiative effect and the additional greenhouse effect of increased water vapor that the atmosphere would have held at higher temperatures. Applied to the late Noachian period of valley network formation, this theory runs into difficulty because of the lower solar output at 3.7 billion years ago (about 75% of the present flux), and consequent large amount of CO_2 required to produce an adequate CO_2-H_2O greenhouse effect. At least several bars of CO_2 would have been required to produce widespread surface temperatures above freezing. However, such thick atmospheres are not physically possible because CO_2 condenses into clouds at ~ 1 bar and also into permanent ice caps at pressures exceeding ~ 3 bar. CO_2 ice clouds can have some greenhouse effect because they scatter infrared radiation, but three-dimensional climate models show that the clouds are never opaque enough in the infrared to warm the surface to a mean temperature above freezing anywhere on the planet.

Apart from the difficulties with climate simulations, if a massive CO_2 atmosphere ever existed at the same time as abundant liquid water, it would have eventually collapsed due to removal of the CO_2 by dissolving in the water and subsequently forming carbonate sediments. However, despite extensive efforts, few outcrops of carbonate sediments have been found and their equivalent global pressure of CO_2 is insubstantial (Table 16.2). Carbonates are absent even in areas in which water is interpreted to have flowed, such as valley networks, and in areas of extensive erosion where we would expect exposures of carbonate sediments buried beneath regolith. In contrast, sulfate sedimentary deposits are widespread at low latitudes, some in terrains that have been exhumed by wind erosion. In retrospect, it is not surprising that carbonate reservoirs have not been found. In the presence of abundant sulfuric acid, carbonate would be quickly converted to sulfate with release of CO_2 to the atmosphere, where it would be subject to various loss processes discussed earlier.

Although a future discovery of a large carbonate sediment reservoir cannot be ruled out, it seems doubtful, and the amount of CO_2 available seems inadequate to have produced a warm enough climate to account by itself for the valley networks by surface in the late Noachian.

3.3.2. Impact Heating

The largest asteroid or comet impacts would vaporize large quantities of rock. Vaporized rock would immediately spread around the planet, condense, and, upon reentry into the atmosphere, would flash heat the surface to very high temperature. This would quickly release water from surface ice into the atmosphere. Upon precipitation, this water could produce flooding and rapid runoff over large areas. Water would be recycled into the atmosphere as long as the surface remained hot, anywhere from a few weeks to thousands of years depending on impact size. It has been proposed that this is an adequate mechanism for producing most of the observed valley networks and that the drop off in impact flux explains why valley formation declines after the early Hesperian. Although a very extended period of warm climate would not be produced this way, repeated short-term warm climate events could have occurred during the late Noachian to early Hesperian. Detailed questions of timing of large impact events and formation of the valley network features needed to test this hypothesis remain to be resolved. The effect of water clouds on the postimpact climate also needs further study. However, impact heating of ice must have released water to the atmosphere and caused subsequent precipitation at some times during the Noachian.

3.3.3. Sulfur Dioxide Greenhouse

The high abundance of sulfur in surface rocks and dust as well as in the Martian meteorites suggests that Martian volcanism may have been very sulfur rich. In contrast to Earth, Martian volcanoes may have released sulfur in amounts equal to or exceeding water vapor. In the atmosphere in the presence of water vapor, reduced sulfur would rapidly oxidize to SO_2 and then form aerosols of sulfate and, in more reducing atmospheres, elemental sulfur also. Sulfur dioxide is a powerful greenhouse gas, but in the presence of liquid water, it dissolves and is removed from the atmosphere by precipitation very rapidly. More importantly, the sulfate and sulfur aerosols that form reflect sunlight so that the net effect of sulfur gases would be to cool Mars. Such cooling has been measured as an effect of volcanic eruptions on Earth that inject SO_2 into the stratosphere where the resulting sulfate aerosols increase the albedo of the Earth.

3.3.4. Methane-Aided Greenhouse

Methane is also a greenhouse gas, but because of its long-term instability in the atmosphere resulting from photolysis and oxidation, it is not an attractive option for contributing to an early warm climate. To have a role of any significance, early Mars would require a global methane flux similar to that produced by the present-day biosphere on Earth. Even so, calculations suggest that the net warming would be limited and inadequate to solve the early climate problem.

3.3.5. Hydrogen-Aided Greenhouse

In an atmosphere that is thick and hydrogen-rich, hydrogen behaves as a greenhouse gas and this might have relevance for early Mars. Although hydrogen (H_2) is a nonpolar molecule, collisions with other molecules can cause it to acquire a temporary dipole or absorb infrared photons in transitions that are normally forbidden; the effect is "collision-induced absorption" (CIA). Indeed, hydrogen is an important greenhouse gas in all the giant planet atmospheres because of CIA. If a thick atmosphere of early Mars had five to tens of percent H_2, the greenhouse warming might be significant. However, hydrogen easily escapes from Mars into space and so very large volcanic fluxes of H_2 would be required. Furthermore, such a solution to the early Mars climate problem, if true, would disfavor life on the early planet because hydrogen is a food for primitive microbes and so its high abundance in the early air would imply an absence of biological consumption.

3.3.6. Mechanisms for Producing Fluvial Features in Cold Climates

Although some precipitation must have occurred due to impacts and short-lived greenhouse warming is plausible, other factors may have been conducive to widespread fluvial features. One proposal concerns seasonal snowmelt under atmospheres that have surface pressures of a few hundred millibars. A second idea is that many of the fluids were low-temperature brines that can exist at temperatures far below 0 °C.

On Mars today, at a particular latitude with the same solar heating, there is a little variation of temperature with altitude, but when the pressure exceeds a few hundred millibars, vertical convection maintains a temperature gradient in the atmosphere that influences the temperature of topography, so that the tops of mountains become much colder than low altitudes. As a result, the southern highlands can become ice or snow covered and periodic melting from changes in orbital characteristics (see below) might be sufficient to erode valleys. If salty fluids rather than pure water were prevalent on early Mars, melting could have occurred at temperatures many tens of degrees below 0 °C. For example, the eutectic temperature (at which a salty solution freezes into ice and solid salt) is −50 °C for calcium chloride ($CaCl_2$), −75 °C for calcium perchlorate ($Ca(ClO_4)_2$), and −57 °C for magnesium perchlorate ($Mg(ClO_4)_2$). In addition, laboratory experiments show that brines of some salts (such as the aforementioned perchlorates) can supercool tens of degrees Celsius below the eutectic temperature for periods exceeding a Martian day.

The outflow channels remain enigmatic because of the large flows of water that are required relatively late in geologic time, in the late Hesperian, when the atmosphere was probably thin. The idea that very fluid lava flows might have been responsible for channel erosion is the only concept that has no need of a warmer climate to allow flows to persist for hundreds of kilometers. Extensive outflow channels, some of which strongly resemble Martian outflow channel features, are found on Venus and are explained by low-viscosity lava flows. The spatial relationship between the Martian outflow channels and the major volcanic constructs is suggestive of the idea that very fluid lavas may have played some role in the formation of outflow channels.

3.4. Milankovitch Cycles

As on Earth, Mars' orbital elements (**obliquity, eccentricity, argument of perihelion**) exhibit oscillations known as Milankovitch cycles at periods varying from 50,000 to several million years. The obliquity and eccentricity oscillations are much larger in amplitude on Mars than on Earth (Figure 16.9). Milankovitch cycles cause climate variations in two ways. First, they control the distribution of incoming solar radiation (insolation) on both an annual average and seasonal basis as functions of latitude. Second, because Milankovitch-driven changes of insolation force variations of annual average surface temperature, they can cause exchanges of volatiles between various surface reservoirs and the atmosphere. Water vapor can migrate between polar cap ice deposits, and ice and adsorbed water in the regolith. Carbon dioxide can move between the atmosphere, seasonal residual polar caps, and the surface adsorption reservoir. Milankovitch variations are believed to be responsible for complex layered structure in both the north polar water ice cap and terrains surrounding the south polar residual carbon dioxide ice cap. Also, rhythmic variations in the layering of some sedimentary deposits in low latitudes have been interpreted as a probable sign of influence from orbital cycles, perhaps because such cycles shifted ice to the tropic in times of high obliquity that were followed by periods of melting.

In general, annual average polar cap temperatures increase relative to equatorial temperatures as obliquity increases. At very low obliquity (<10°−20° depending on the precise values of polar cap **albedo** and thermal **emissivity**), the carbon dioxide atmosphere collapses onto permanent carbon dioxide ice polar caps. Orbital calculations indicate that this collapse could occur ∼1−2% of the time. At high obliquity, atmospheric pressure may increase due to warming and release of adsorbed carbon dioxide from high-latitude regolith. Calculations indicate, however, that the maximum possible pressure increase is likely to be small, only a few millibars, so Milankovitch cycles are unlikely to have been responsible for significant climate warming.

Evidence for the active influence of Milankovitch-type cycles includes a thin, patchy mantle of material, apparently consisting of cemented dust, that has been observed within a 30°−60° latitude band in each hemisphere, corresponding to places where near-surface ice has been stable in the last few million years due to orbital changes. The material is interpreted to be an atmospherically deposited ice−dust mixture from which the ice has sublimated.

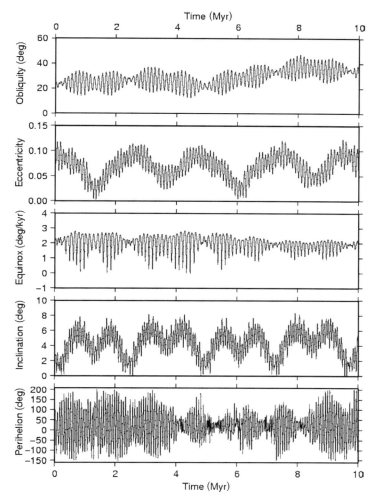

FIGURE 16.9 (a) Orbital elements. Mars, like other planets, moves in an elliptical orbit with a semimajor axis, a. The eccentricity e, defines how much the ellipse is elongated. The plane of the orbit is inclined by angle i, to the ecliptic, which is the geometrical plane that contains the orbit of the Earth. The ascending node is the point where the planet moves up across the ecliptic plane and the descending node is where the planet moves below it. The vernal equinox, marked γ, represents a reference direction that defines the longitude of the ascending node, Ω. Angle ω is the argument of perihelion. (b) Calculated variations in Martian orbital parameters over the past 10 million years. *Reprinted from Armstrong et al. (2004).*

3.5. Wind Modification of the Surface

Orbital and landed images of the surface show ubiquitous evidence of active wind modification of the surface, which complicates the interpretation of climate and volatile history. The action of wind erosion, dust transport, and dust deposition is modulated by Milankovitch cycles and must have strongly changed the surface over the last few billions of years and during the Noachian, as we discuss below.

Today, dunes, ripples, and other bedforms are widespread. Wind-modified objects, known as ventifacts, are very evident in the grooves, facets, and hollows produced by the wind in rocks at the surface. Yardangs are also common, which are positive relief features in coherent materials sculpted by wind on scales from tens of meters to kilometers. Strong winds exert stress on the surface that can initiate saltation (hopping motion) of fine sand grains (diameter $\sim 100-1000$ μm) and creep of larger particles. Saltating grains can dislodge and suspend finer dust particles (diameters $\sim 1-10$ μm) in the atmosphere, thereby initiating dust storms. Minimum wind speeds required to initiate saltation are typically ~ 30 m/s at the level 2 m above the surface, but this saltation threshold wind speed decreases with increasing surface pressure.

Strong winds needed for saltation are rare. Wind observations at the *Viking* Lander sites and computer simulations of the atmospheric circulation suggest that they occur at most sites <0.01% of the time. Nevertheless, over the planet as a whole, dust storms initiated by saltation are common; they tend to occur with greater frequency in low-elevation regions than uplands because the relatively high surface pressure in lowlands lowers the saltation threshold wind speed. They are favored by topographic variations, including large and small-scale slopes and are common over ice-free surfaces near the edges of the seasonally varying polar caps and in "storm track" regions where the equator-to-pole gradient of atmospheric temperature is strong. Dust storms generated by strong winds and saltation are common in some tropical lowland regions, especially close to the season of perihelion passage when the Hadley circulation is strong (near the southern summer solstice at the current phase of the Milankovitch cycle). During some years, these perihelion season storms expand and combine to such an extent that high dust opacity spreads across almost the entire planet. These planetwide dust events are fostered by positive feedbacks between dust-induced heating of the atmosphere, which contributes to driving wind systems, and the action of the wind in picking up dust.

Dust can also be raised at much lower wind speeds in dust devils, which are small-scale quasivertical convective vortices called dust devils. Because the atmosphere is so thin, convective heating per unit mass of atmosphere is much greater on Mars than anywhere on Earth, and Martian dust devils correspondingly tend to be much larger sizes (diameters up to several hundred meters and depths up to several kilometers). Since the winds required to raise dust in dust devils are lower than saltation threshold winds, dust devils are common in some regions of Mars during the early afternoon and summer when convective heating is strongest. They are often associated with irregular dark tracks produced by the removal of a fine dust layer from an underlying darker stratum. The relative importance of large saltation-induced dust storms and dust devils to the overall dust balance is unclear, but modeling studies suggest that the former are substantially more important.

Over the past 4 billion years, there must have been substantial systematic wind transport of fine soil particles from regions in which erosion is consistently favored to regions of net deposition. Models of Martian atmospheric circulation and the saltation process suggest that net erosion must have taken place in lowland regions, particularly in the northern lowlands, the Hellas basin, and some tropical lowlands (e.g. Isidis Planitia and Chryse Planitia), with net deposition in upland regions and in some moderate elevation regions where the regional slope is small and westward facing, such as portions of Arabia Terra and southern portions of Amazonis Planitia. The distribution of surface **thermal inertia** inferred from the measured surface diurnal temperature variation supports these inferences. Regions of high thermal inertia, corresponding to consolidated or coarse-grained soils, exposed surface rocks, and bedrock patches, are found where the circulation-saltation models predict net erosion over Milankovitch cycles, and regions of very low thermal inertia corresponding to fine dust are found where net deposition is predicted by the models.

There are no terrestrial analogs of surfaces modified by wind erosion and deposition over 4 billion years, so it is difficult to fully comprehend the modifying effect of Martian winds over such a long time. However, it is clear from the surface imagery that in some areas repeated burial and exhumation events have taken place. Based on the heights of erosionally resistant mesas, the Meridiani Planum site of the *Opportunity* rover activities appears to have been exhumed from beneath at least several hundred meters and perhaps as much as several kilometers of soil. Many of the sulfate layer deposits described above appear to be undergoing exhumation. Since surface features can be repeatedly buried, exposed, and reburied over time, inferences of event sequences and surface ages from crater size distributions are rendered complex.

Because the saltation process operates on the extreme high velocity tail of the wind speed distribution, it is very sensitive to surface density or pressure changes. Model results indicate that an increase in surface pressure up to only 40 mbar would increase potential surface erosion rates by up to two orders of magnitude. If, as is likely, Mars had a surface pressure $\sim$100 mbar or higher during the late Noachian, rates of surface modification by wind should have been orders of magnitude greater than today. Indeed, it has long been observed than late Noachian surfaces were undergoing much more rapid modification than during later periods. This has generally been attributed to precipitation and runoff under a warmer climate regime, as discussed earlier. But surface modification by winds under a denser atmosphere should also have contributed to the observed rapid modification of late Noachian age surfaces.

4. CONCLUDING REMARKS

Although ice is now known to be widespread near the surface and there is considerable evidence that liquid water once flowed across the surface in dendritic valley networks, a major outstanding problem is that we still do not know the exact conditions responsible for releasing liquid water on early Mars or what controlled the early ancient climate (see Figure 16.10 for a timeline). As the sophistication of climate models for early Mars have grown, it has perhaps become increasingly difficult (rather than easier) to explain how Mars could have had a sustained, warm, and wet climate during the Noachian or Hesperian. The basic problem is that the early Martian surface needs about 80 °C of greenhouse warming to raise its mean global temperature above freezing, which is more than double the greenhouse warming of 33 °C of the modern Earth. The discovery of only a few outcrops of sedimentary carbonates provides little support for the idea that a large CO_2 reservoir exists on Mars today and was derived from an earlier thick atmosphere; in any case, three-dimensional climate models are unable to raise global mean surface temperatures above freezing for any CO_2-H_2O atmosphere. More generally, no widely accepted solution to the climate of early Mars has been found despite numerous modeling permutations of an enhanced greenhouse effect with various gases on early Mars.

In view of the new data and theoretical constraints, other candidate mechanisms for the release of fluids at the surface to form valley networks and outflow channels need to be considered. During the Noachian, large impacts

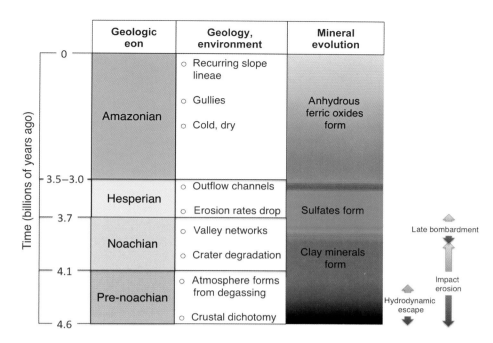

FIGURE 16.10 An overview of the Martian geologic timescale and its relationship to geologic features, predominant minerals, and events affecting the atmosphere.

would have provided sufficient heat to vaporize subsurface volatiles, such as water and CO_2 ice. Consequently, impacts may have generated many temporary warm, wet climates, which would be accompanied by erosion from rainfall or the recharge of aquifers sufficient to allow groundwater flow and sapping. Such a scenario might explain why valley networks appear to peak in occurrence around the era of the Late Heavy Bombardment (the late Noachian, early Hesperian) and why there is subsequently an apparently large drop in erosion rates. The periodic release of meltwater from high-altitude ice and snow that occurs under moderately thicker atmospheres is another mechanism that seems likely to have produced fluvial erosion on early Mars. Also, the potential of thicker Noachian atmospheres to cause wind erosion requires further consideration.

Geochemical data and models suggest that most of Mars' original volatile inventory was lost early by hydrodynamic escape and impact erosion (Figure 16.10). However, we still do not know the degree to which some volatiles were sequestered into the subsurface as minerals and protected. Future landed and orbital missions can refine our understanding of the distribution and properties of hydrated minerals and subsurface ices. However, determining the amount of sulfate and carbonate that has been sequestered into the subsurface will require drilling into the subsurface. Further study of the geology of Mars from orbit and the surface will also help establish the amount of fluvial erosion, its duration, or episodicity. Finally, resolving questions of timing ultimately requires absolute (radiometric) dating of Martian surfaces.

BIBLIOGRAPHY

Armstrong, J. C., Leovy, C. B., & Quinn, T. (2004). A 1 Gyr climate model for Mars: New orbital statistics and the importance of seasonally resolved polar processes. *Icarus, 171*, 255—271.

Bell, J. (Ed.). (2008). *The Martian surface: composition, mineralogy and physical properties*. Cambridge University Press.

Barlow, N. (2008). *Mars: an introduction to its interior, surface and atmosphere*. Cambridge University Press.

Carr, M. H. (1996). *Water on Mars*. Oxford University Press.

Carr, M. H. (2007). *The surface of Mars*. Cambridge University Press.

Catling, D. C. (2013). *Astrobiology: a very short introduction*. Oxford University Press.

Catling, D. C., & Kasting, J. F. (2015). *Atmospheric evolution on inhabited and lifeless worlds*. Cambridge University Press.

Feldman, W. C., Prettyman, T. H., Maurice, S., Plaut, J. J., Bish, D. L., Vaniman, D. T., et al. (2004). *Journal of Geophysical Research, 109*, E09006. http://dx.doi.org/10.1029/2003JE002160.

Haberle, R. M., Clancy, R. T., Forget, F., Smith, M. D., & Zurek, R. W. (Eds.). (2014). *The atmosphere and climate of Mars*. Cambridge University Press.

Hartmann, W. K. (2003). *A traveler's guide to Mars*. University of Arizona Press.

Jakosky, B. M., & Phillips, R. J. (2001). Mars volatile and climate history. *Nature, 412*, 237—244.

Kieffer, H. H., Jakosky, B. M., Snyder, C. W., & Matthews, M. S. (Eds.). (1992). *Mars*. Tucson, AZ: University of Arizona Press.

Leovy, C. B. (2001). Weather and climate on Mars. *Nature, 412*, 245—249.

McKay, D. S., Gibson, E. K., Thomas-Keprta, K. L., Romanek, C. S., Clemmett, S. J., Chillier, X. D. F., et al. (1996). Search for past life on Mars: possible relic biogenic activity in Martian meteorite ALH84001. *Science, 273*, 924—930.

Chapter 17

Mars: Surface and Interior

Michael H. Carr
U. S. Geological Survey, Menlo Park, CA, USA

James F. Bell, III
School of Earth and Space Exploration, Arizona State University, Tempe, AZ, USA

Chapter Outline

1. Mars Exploration 359
2. General Characteristics 361
   2.1. Orbital and Rotational Constants 361
   2.2. Surface Conditions 362
   2.3. Planet Formation and Global Structure 362
   2.4. Global Topography and Physiography 363
3. Impact Cratering 364
   3.1. Cratering Rates and the Martian Timescale 364
   3.2. Crater Morphology 365
4. Volcanism 365
5. Tectonics 367
6. Canyons 367
7. Water 368
   7.1. Erosion and Weathering 368
   7.2. Valley Networks, Lakes, Deltas 368
   7.3. Outflow Channels 370
   7.4. Gullies 371
   7.5. Dark Streaks 371
8. Ice 372
9. Wind 372
10. Poles 373
11. The View from the Surface 373
12. Summary 377
Bibliography 377

Mars, the outermost of the four terrestrial planets—Mercury, Venus, Earth, and Mars—is intermediate in size between the Earth and the Moon. The terrestrial planets all have solid surfaces, each of which preserves a partial record of how each planet has evolved. Successive events, such as volcanic eruptions or meteorite impacts, both create a new record and partly destroy the old. The task of the geologist is to reconstruct the history of the planet from what is preserved at the surface. Both Mercury and the Earth's moon appear to have become geologically inactive early in their history, so most of the preserved record dates from very early in the history of the solar system, 3.5 Gyr ago. The geologic record on Venus is relatively young, most of the surface apparently having formed in the past 1 Gyr. The record on Earth is also mostly young, although ancient records are preserved on some continents. On Mars we have a record that spans almost the entire history of the solar system. While much of the Martian surface dates back to the first billion years, volcanism, tectonism, fluvial activity, glaciation, and impact processes appear to have continued at a low rate until the recent geologic past, allowing us to follow the evolution of the planet for almost its entire history.

Our knowledge of the geologic evolution of the Earth has been largely derived from the study of the lithology, chemistry, mineralogy, and distribution of rocks at the surface. Geomorphology has played a relatively minor role. On Mars, however, much of what we know about the geology is derived from the morphology of the surface. While geomorphologic data are being increasingly supplemented by orbital mineralogy measurements as well as data from Martian meteorites and landers/rovers, our global perspective is still largely based on the appearance of the surface from orbit, which is the main subject of this chapter.

1. MARS EXPLORATION

The modern era of Mars exploration began on July 14, 1965, when the *Mariner 4* spacecraft flew by the planet and transmitted to the Earth 22 close-up pictures of the surface with resolutions of several kilometers per pixel

(Table 17.1). Prior to this, we depended on telescopic observations, with resolution at best of 100–200 km. The observations revealed no topography, only surface markings. Nonetheless, prior to the space age we knew that Mars has a thin CO_2 atmosphere, polar caps that advance and recede with the seasons, and surface markings that undergo annual and secular changes. Geologic studies of the planet, however, could realistically begin only when we acquired spacecraft data.

The *Mariner 4* pictures, covering only a small fraction of the southern hemisphere, revealed only an ancient surface that resembled the lunar highlands. These results were disappointing because it had been speculated that Mars, having an atmosphere and being larger than the Moon, might be more Earth-like than Moon-like. *Mariner 4* was followed by two more Mariner spacecraft in 1969 (Table 17.1), which seemed to confirm Mars' lunar-like characteristics. However, our perception of Mars changed dramatically in 1972 when systematic mapping by the *Mariner 9* orbiter spacecraft revealed the planet that we know today. As mapping progressed, huge volcanoes, deep canyons, enormous dry riverbeds, and extensive dune fields came into view and a complex geologic history became apparent. Exploration of Mars continued in the 1970s as both the USSR and the United States sent landers to the surface and other vehicles to the planet. Exploration in the 1970s culminated with the *Viking* mission, which successfully placed two landers on the surface and two other spacecraft into orbit. By the end of the *Viking* mission, almost all the surface had been photographed from orbit at a resolution of about 250 m/pixel and small fractions with resolutions as good as 10 m/pixel. In addition, the *Viking* landers had carried out a variety of experiments directed mostly toward detecting life and understanding the chemistry of the soil.

In the early 1980s, our understanding of Mars was further enhanced when it became clear that we had samples of Mars in our meteorite collections here on the Earth. A group of meteorites, called **SNCs** (pronounced snicks and standing for "Shergottites, Nahklites, and Chassigny") were initially suspected to be of Martian origin because they were **basaltic** and had ages of ∼1.3 billion years. These meteorites could not have come from the Earth because their oxygen isotope ratios are distinctly different from terrestrial ratios. The only plausible body that could have been volcanically active so recently and supplied the meteorites was Mars. Martian origin was later confirmed by

TABLE 17.1 Successful Mars Missions (from http://nssdc.gsfc.nasa.gov/planetary/chronology_mars.html)

Mission	Nation	Launch Date	Fate
Mariner 4	US	November 18, 1964	Flew by July 15, 1965; first close-up images
Mariner 6	US	February 24, 1969	Flew by July 31, 1969; imaging and other data
Mariner 7	US	March 27, 1969	Flew by August 5, 1969; imaging and other data
Mariner 9	US	May 30, 1971	Into orbit November 3, 1971; mapped planet
Mars 5	USSR	July 25, 1973	Into orbit February 12, 1974; imaged surface
Viking 1	US	August 20, 1975	Landed on surface July 20, 1976; orbiter mapping
Viking 2	US	September 9, 1975	Landed on surface September 3, 1976; orbiter mapping
Phobos2	USSR	July 12, 1988	Mars and Phobos remote sensing
Pathfinder	US	December 4, 1996	Landed July 4, 1997; lander and rover data
Global Surveyor	US	November 7, 1996	Into orbit September 11, 1997; imaging and other data
Mars Odyssey	US	April 7, 2001	In orbit October 24, 2001; imaging, remote sensing
Spirit Rover	US	June 10, 2003	Landed in Gusev January 3, 2004
Opportunity Rover	US	July 7, 2003	Landed in Meridiani January 24, 2004
Mars Express	Europe	June 2, 2003	In orbit December 25, 2003; imaging, remote sensing
Mars Reconnaissance Orbiter	US	August 12, 2005	In orbit March 10, 2006; imaging, remote sensing
Phoenix Lander	US	August 4, 2007	Landed in northern plains, May 25, 2008
Curiosity Rover	US	November 26, 2011	Landed in Gale Crater August 6, 2012
MAVEN	US	November 18, 2013	Mars orbiter, en route

finding gasses trapped within the meteorites that are identical in composition to gasses in the Martian atmosphere as measured by the Viking landers. The meteorites are believed to have been ejected from Mars by large impacts and subsequently captured by the Earth after spending several million years in space. We have since added to the collection, and there are now about 50 known Martian meteorites.

All these meteorites are basaltic and all but one are aged significantly less than the presumed ~4.6 Gyr age of the planet. The one exception known so far, ALH84001, has a crystallization age of ~4 Gyr. In 1996, it was tentatively suggested that carbonate globules within this meteorite, together with some disequilibrium mineral assemblages, polycyclic aromatic hydrocarbons and a number of different types of very small segmented rods that resemble some terrestrial nanofossils, might all be the result of biologic activity. This suggestion has, however, received little support from subsequent investigations by the general science community.

More recently, the exploration of Mars has continued with a series of landers (Pathfinder, Phoenix), rovers (Spirit, Opportunity, Curiosity) and orbiters (Mars Global Surveyor, Mars Odyssey, Mars Express, and Mars Reconnaissance Orbiter), as listed in Table 17.1. These missions have documented the mineralogical diversity of the planet; mapped the global chemistry and topography, the distribution of water ice, and the strength of the remanent magnet field; as well as provided higher resolution imaging. The scientific analyses of these data sets has confirmed the active role that water has played in the planet's evolution and has documented numerous ongoing changes on the surface, some possibly related to water action.

2. GENERAL CHARACTERISTICS

2.1. Orbital and Rotational Constants

The Martian day is almost the same as the Earth's day, but the year is almost twice as long (Table 17.2). Because its rotational axis is inclined to the orbit plane, Mars, like the Earth, has seasons. But the Martian orbit has significant eccentricity, causing the pole that is tilted toward the Sun at perihelion to have warmer summers than the other pole. At present, the south has the warmer summers, but, because of a slow change in the direction of tilt of the rotational axis and a slow change in the orientation of perihelion, the hot and cold poles change on a 51,000-year cycle. The eccentricity also causes the seasons to have significantly different lengths (see

TABLE 17.2 Earth and Mars: General Characteristics Compared

	Earth	Mars
Mean equatorial radius (km)	6378	3396
Mass ($\times 10^{24}$ km)	5.98	0.624
Mean distance from the Sun (10^6 km)	150	228
Orbit eccentricity	0.017	0.093
Obliquity	23.5°	25.2°
Length of day	24 h	24 h, 39 m, 35 s
Length of year (Earth days)	365.3	686.9
Seasons (Earth days)		
Northern spring	92.9	199
Northern summer	93.6	183
Northern fall	89.7	147
Northern winter	89.1	158
Atmosphere	78% N_2, 21% O_2, 1% Ar	95% CO_2, 3% N_2, 2% Ar
Surface pressure (mbar)	1013	7
Mean surface temperature (K)	288	215
Surface gravitational acceleration (cm/s^{-2})	981	371
Moons	1	2

Table 17.2). At present, the Martian **obliquity** is similar to that of the Earth. Yet while the Earth experiences only minor changes in its obliquity, the obliquity of Mars changes significantly, and relatively rapidly, over long timescales. For example, during the past 10 Myr it has been as low as 15° and as high as 45°. It has been estimated that there is a 63% probability that the obliquity reached 60° in the past 1 Gyr. At low obliquities the atmosphere thins because most of the CO_2 in the atmosphere condenses out onto the poles. At high obliquities the water ice polar caps dissipate in summer and ice condenses at lower latitudes.

2.2. Surface Conditions

Mars has a thin atmosphere that provides almost no thermal blanketing. As a result, temperatures at the surface have a wide diurnal range, controlled largely by latitude, the reflectivity of the surface, and the thermal properties of the surface materials. Typically, surface temperatures in summer at latitudes ±60° range from 180 K at night to 290 K at midday, but can range more widely if the surface consists of unusually low-density, fine-grained material. However, these temperatures are somewhat deceiving because at depths of just a few centimeters below the surface, temperatures at the equator are at the diurnal mean of 210–220 K. At the poles in winter, temperatures drop to 150 K, at which point CO_2 condenses out of the atmosphere to form a decimeter- to meter-thick dry ice seasonal cap. The atmospheric pressure at the surface ranges from about 14 mbars at the bottom of the Hellas basin to about 3 mbars at the top of the tallest volcanoes, and it changes seasonally as a result of the formation and sublimation of the polar caps. Winds typically have a velocity of a few meters per second but there may be gusts up to 50 m/s. Dust devils and local dust storms are common, and almost every year, regional- or global-scale dust storms can occur.

The stability of water is of profound importance for understanding Martian geology. Under the conditions just described, the planet has a thick permafrost layer that extends a few kilometers deep at the equator and several kilometers deep at the poles. Any unbound water present will exist as ice in this zone. There may be liquid water beneath the permafrost, depending on the magnitude of the planet's largely unknown geothermal heat flux and other subsurface thermophysical properties. Water ice caps roughly 3 km thick are present at both poles, although at the South Pole the water ice cap is largely masked by a remnant summer CO_2 cap. At latitudes between about 40° and the edge of the water ice cap, abundant ice has been detected just below a dehydrated zone a few tens of centimeters thick. At latitudes less than about 40°, ice is unstable at all depths (i.e. a block of pure water ice placed in the ground at these latitudes will slowly sublime into the atmosphere). The small amounts of water that have been detected at low latitudes may be water bound in minerals or water inherited from an earlier era of higher obliquity when water ice was stable at these latitudes. Under present conditions, liquid water, although unstable near the surface, can exist transiently, particularly if very salty.

2.3. Planet Formation and Global Structure

Like the other planets, Mars formed from materials that condensed out of the early solar nebula, a disk of gas and dust that surrounded the early Sun. A class of meteorites called carbonaceous chondrites, which are almost identical in composition to the photosphere of the Sun, are believed to closely resemble the early nebula in composition. Radioisotopes date the formation of the nebula at 4.567 billion years ago. The planets formed as the dust and gas accumulated into discrete bodies, and gravitational attraction favored growth of larger bodies over smaller bodies. Mars appears to have formed remarkably quickly based on evidence from short-lived radioisotopes. The high rate of accretion resulted in global melting, which enabled settling of heavy iron-rich melts to the center of the planet to form a core separated from the silicate-rich mantle. During this differentiation process, siderophile elements (those which dissolve preferentially in iron-rich melts over coexisting silicate-rich melts) became depleted in the mantle and enriched in the core. This enables formation of the core to be dated because the daughter products of some short-lived, strongly siderophile elements are present in the mantle, as indicated by the composition of Martian meteorites. For example, ^{182}Hf decays to ^{182}W with a half-life of 9 Myr. W is highly siderophilic so should mostly enter the core, yet there is an excess of ^{182}W in the mantle, implying that not all the Hf had decayed before the core formed. This and other isotopic evidence indicates that the core formed within 20 Myr of the formation of the elements that comprise the solar system. Isotopic evidence indicates that some crust formed very early but that new crust continued to form through Mars' history, as indicated by volcanoes and extensive volcanic plains.

The Earth's core is inferred to be iron-rich from (1) the core's density as deduced from the core's size and the planet's **moment of inertia**, (2) computer models of the bulk composition of the Earth and compared with the composition of the chondritic meteorites from which the Earth formed, and (3) depletion of siderophile elements in mantle-derived rocks as compared with chondritic meteorites. Similar reasoning can be used for Mars except that, while the size of the Earth's core is accurately known from seismic data, the size of Mars' core must be inferred indirectly. The best estimate is that the core radius is between 1300 and

1500 km. In addition, the Martian core may be more sulfur rich than the Earth's core because Mars' mantle is more depleted of chalcophile elements (those that preferentially dissolve in sulfur-rich melts) than is the Earth's.

One of the more surprising results of the Mars Global Surveyor mission was the discovery of large magnetic anomalies in the crust despite the absence of a global planetary magnetic field today. Their presence indicates that Mars had a magnetic field in the past, but that it switched off at some time. The size of the anomalies suggests that they must result from sources in the outer few tens to several tens of kilometers of the crust and that their magnetizations are higher by an order of magnitude than magnetizations typically encountered in terrestrial rocks. The anomalies probably formed when rocks, containing magnetically susceptible Fe-bearing minerals, crystallized in the presence of a strong magnetic field. Most of the anomalies, and all of the largest ones, are in the southern highlands. They are particularly prominent on either side of the 180° longitude where there are several broad, east–west stripes. One interpretation of the linear anomalies is that they result from injection of dikes or dike swarms several tens of kilometers wide and hundreds of kilometers long in the presence of a strong magnetic field. Anomalies are mostly absent around the youngest large-impact basins (Utopia, Hellas, Isidis, and Argyre). The simplest explanation is that there was no longer a magnetic field when these basins formed, formation of the basins melted and reset/destroyed any preexisting anomalies, and no new ones formed when the affected materials cooled after the basin-forming events. The ages of the basins are not known, but, by analogy with the Moon, they are likely to have formed toward the end of the so-called late heavy bombardment period in solar system formation history, around 3.8–4 Gyr ago. Thus, the magnetic field may have turned off by around 4 Gyr ago.

The Earths magnetic field is generated by convection within its core. Mars' early **dynamo** probably had a similar cause. Possible causes for cessation of the dynamo are loss of core heat, solidification of most of the core, and/or changes in the mantle convection regime. Magnetization of minerals within 3.9- to 4.1-Gyr-old carbonates in the Martian meteorite ALH84001 suggests that there was still a magnetic field at this time. If true, it implies that Mars had a magnetic field for the first 500 My of its history and that the field turned off around 4 Gyr ago, just before formation of the youngest impact basins.

Like the Earth's mantle, Mars' mantle is chondritic in composition except for the depletion of siderophile and chalcophile elements as noted above and depletion of volatile elements which would have been largely lost from the interior during the early global melting phase. It consists mainly of iron–magnesium silicates. One difference between the mantles of the two planets that is suggested by the compositions of the Martian meteorites is that the Fe/Mg ratio is higher in the Martian mantle.

The crust of Mars appears to be essentially a melt extract from the mantle and is mostly basaltic in composition. The thickness of the crust varies considerably, ranging from 5 to 100 km, as estimated from the relations between the global gravity field and the global topography. The thickest crust is under the high-standing cratered terrain in the southern hemisphere, whereas the thinnest is under the large impact basins of Isidis and Hellas.

2.4. Global Topography and Physiography

The topography and physiography of Mars have a marked north–south asymmetry, which is referred to as the global dichotomy (See Figure 17.1). The dichotomy is expressed in three ways: as a change in elevation, a change in crustal thickness, and a change in crater density. The southern uplands have an average elevation 5.5 km higher than the northern plains, the crust is roughly 25 km thicker in the uplands, and most of the upland terrain is heavily cratered, dating back to the period of the late heavy bombardment. The northern plains are mostly younger, but remnants of buried craters poke through the surface indicating that there is an older surface at some depth beneath the younger plains. The low-lying plains constitute roughly one-third of the planet and are mostly in the north. The cause of the dichotomy is not known. Suggestions include a very large impact, soon after the planet formed, or internal convection sweeping most of the light, crustal material into one half of the planet.

Superimposed on the global dichotomy is the Tharsis bulge, over 5000 km across and 10 km high, centered on the equator at 260 °E. Most of the planet's volcanic activity has been centered on the bulge, which has the five largest volcanoes (Montes Olympus, Alba, Arsia, Ascreus, and Pavonis) on its northwest flank. Tharsis is also at the center of a vast array of radial faults and circumferential ridges that affect over half the planet's surface. The bulge, an accumulation of 3×10^8 km^3 of volcanic rocks, deformed the lithosphere, thereby affecting slopes around the bulge. Ancient valley networks incised into these slopes indicate that most of the bulge formed very early in the planet's history. To the east of the center of the bulge are a series of vast canyons thousands of kilometers long and up to 10 km deep. They are roughly radial to the bulge and appear to have formed largely by faulting, although they also have been extensively modified by fluvial and mass-wasting processes. At the east end of the canyons, extensive areas of terrain have seemingly collapsed to form **chaotic terrain** from which emerge large dry river beds that extend for thousands of kilometers downslope into the northern plains. A much smaller bulge centered in Elysium at 25 °N, 213 °W has also been a center of volcanic, tectonic, and

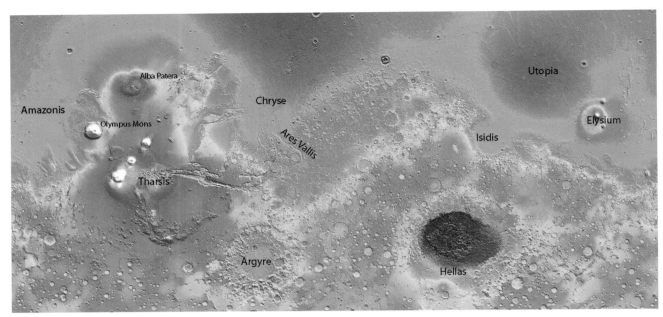

FIGURE 17.1 Topographic map of Mars between latitudes 65 °S and 65 °N. The highest elevations (whites and grays) are in Tharsis. The lowest elevations (blues) are in the northern plains and Hellas. The dominant feature of the planet is the global dichotomy between the low-lying northern plains and the cratered southern uplands. The main positive features are the volcanic provinces of Tharsis and Elysium. The main negative features are the large-impact basins Hellas, Argyre, and Isidis and the buried basin Utopia. The canyons extending eastward from Tharsis and large outflow channels such as Ares Vallis are visible even at this global scale (Mars Orbiter Laser Altimeter (MOLA)).

fluvial activity. Other prominent topographic features are large impact basins; the largest are Hellas (2600 km diameter), Isidis (1600 km), and Argyre (1500 km).

The physiography of the poles is distinctively different from that of the rest of the planet. At each pole, extending out to the 80° latitude circle, is a stack of finely layered deposits a few kilometers thick. In the north they rest on plains; in the south they rest on cratered uplands. The small number of superimposed impact craters suggests that these sediments are only a few tens of millions of years old.

3. IMPACT CRATERING

3.1. Cratering Rates and the Martian Timescale

All solid bodies in the solar system are subject to impact by asteroidal and cometary debris. Current cratering rates are low. On Earth, in an area the size of the United States, a crater larger than 10 km across is expected to form every 10–20 million years and one larger than 100 km across every billion years. The rates on the other terrestrial planets are likely to be within a factor of two or three of these rates. On the Moon, surfaces are either densely covered by large craters (lunar highlands) or sparsely affected by large craters (maria) with no surfaces of intermediate crater densities. This contrast arises because of the Moon's cratering history.

Very early, cratering rates were much higher, but around 3.7 billion years ago they declined rapidly to roughly the present rate. Accordingly, surfaces that formed more than 3.7 billion years ago are heavily cratered, and those that formed afterward are much less cratered. Mars has had a similar cratering history, hence the contrast between the heavily cratered uplands and the sparsely cratered plains.

Craters provide a means of estimating the ages of surfaces. The solar system's most densely cratered surfaces formed more than 3.7 billion years ago, and the cratering rate has been roughly constant since that time. Consequently, a 3 billion-year-old surface will have roughly three times more craters on it than a 1 billion-year-old surface. However, there is considerable uncertainty in estimating absolute ages in this way because we do not know exactly what the cratering rate on Mars has been for the past few billion years. Nevertheless, by counting craters over key geologic areas, we can put surfaces in a time-ordered relative age sequence and perhaps even make rough estimates of their absolute ages. Craters have thus been used to divide the history of Mars into different epochs. The Noachian refers to the period of heavy bombardment that ended around 3.7 billion years ago. The rest of the planet's history is divided into the Hesperian, roughly 3.7 to 3.0 billion years ago, and the Amazonian, roughly 3.0 billion years ago to the present. The Noachian period is characterized by high cratering rates, formation of valley networks, and the presence of hydrated minerals such as phyllosilicates. The Hesperian period is characterized by large outflow channel floods and extensive lava plains and the presence of abundant sulfate deposits. During the Amazonian, most of the processes that occurred earlier continued, but at much lower rates, enabling less-energetic processes such as wind erosion to exert a strong influence on the preserved landforms.

3.2. Crater Morphology

Impact craters have similar morphologies on different planets. Small craters are simply bowl-shaped depressions with constant depth-to-diameter ratios. With increasing size, the craters become more complex as central peaks appear, terraces form on the walls, and the depth-to-diameter ratio decreases. At very large diameters, the craters become multiringed, and it is not clear which ring is the equivalent of the crater rim of smaller craters. On Mars, the transition from simple to complex takes place at 6—7 km, and the transition from complex craters to multiringed basins takes place at 130—150 km diameter.

Although impact craters on Mars resemble those on the Moon, the patterns of ejecta are quite different. Lunar craters generally have continuous hummocky ejecta near the rim crest, outside of which is a zone of radial or concentric ridges, which merge outward into string or loops of **secondary craters**, formed by material thrown out of the main crater. In contrast, the ejecta around most fresh-appearing Martian craters, especially those in the 5—100 km size range, occur in discrete, clearly outlined lobes (See Figure 17.2). Various patterns are observed. The ejecta around craters smaller than 15 km in diameter are enclosed in a single, continuous lobate ridge or rampart, situated about one crater diameter from the rim. Around larger craters, there may be many lobes, some superimposed on others, but all surrounded by a rampart. The distinctive Martian ejecta patterns have been attributed to two possible causes. The first suggestion, based on experimental craters formed under low atmospheric pressures, is that the patterns are formed by interaction of the ejecta with the atmosphere. The second is that the ejecta contained water and had a mudlike consistency and so continued to flow along the ground after ejection from the crater and ballistic deposition. This view is supported by the resemblance of Martian craters to those produced by impacts into mud.

The previous discussion refers to fresh-appearing craters. Erosion rates at low latitudes for most of Martian history are very low—typically 0.1—10 nm/year, although rates may be higher locally. However, early in the planet's history erosion rates were much higher. As a consequence, in the cratered uplands, craters range in morphology from fresh-appearing craters to barely discernible, rimless depressions. In contrast, on volcanic plains in equatorial

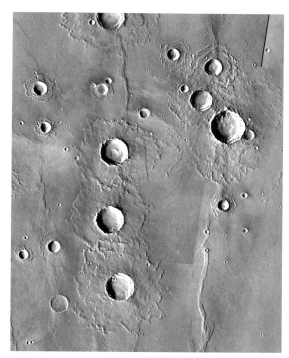

FIGURE 17.2 Impact craters in Lunae Planum. The ejecta are distributed around the craters in lobes, each surrounded by a low ridge or rampart. The largest crater is 35 km across (Thermal Emission Imaging System (THEMIS)).

regions, almost all the craters are fresh-appearing, even though they may be billions of years old. Obliteration rates have been higher at high latitudes. This has been attributed to ice-abetted creep of the near-surface materials, but other factors such as repeated burial and removal of possibly ice-rich materials by sublimation and the wind, may have contributed to modification of the craters. Such processes have been invoked to explain the so-called pedestal craters that are particularly common at high latitudes. These craters are inset into a platform or pedestal that has the same or a slightly larger aerial extent as the ejecta. The simplest explanation is that the region in which these craters are found was formerly covered with a layer of loose material or ice that has since been removed except around craters where the surface was protected by the ejecta.

4. VOLCANISM

Mars has had a long and varied volcanic history. Crystallization ages of Martian meteorites as young as 150 million years, and the scarcity of impact craters on some volcanic surfaces, suggest that the planet is still volcanically active, although the rates must be very low compared with the Earth. The tectonic framework within which Martian volcanism occurs is very different from that in which most volcanism occurs on the Earth. Most terrestrial volcanism takes place at plate boundaries, but these have no Martian equivalents, there being no plate tectonics on Mars. Perhaps the closest terrestrial analogs to Martian volcanoes are those, such as the Hawaiian volcanoes, that occur within plates rather than on the boundaries. Most Martian volcanism is **basaltic**, but basaltic volcanism expresses itself somewhat differently on Mars because of the lower heat flow, gravity, and atmospheric pressure. Eruptions are expected to be larger and less frequent and more likely to produce ash, and ash clouds are more likely to collapse and produce ash-rich surface flows.

The large **shield volcanoes** of Tharsis and Elysium present the most spectacular evidence of volcanism (See Figure 17.3). Shield volcanoes, such as those in Hawaii, are broad domes with shallow sloping flanks that form mainly by eruption of fluid basaltic lava. Each has a summit depression formed by collapse following eruptions on the volcano flanks or at the summit. In contrast, stratovolcanoes such as Mt Fujiyama, tend to be much smaller and have steeper flanks and a summit depression that is a true volcanic vent. Explosive, ash-rich eruptions tend to be more common in the building of a stratovolcano and the lava tends to be more volatile rich, more siliceous, and more viscous than that which forms shields. In Tharsis, three large shield volcanoes form a northeast–southwest trending line and 1500 km to the northwest of the line stands the largest shield of all, Olympus Mons, 550 km across and reaching a height of 21 km above the Mars datum. The three aligned Tharsis Montes shields are only slightly smaller. Olympus Mons has a summit caldera 80 km across and the flanks have a fine striated pattern caused by long linear flows, some with central channels. The main edifice is surrounded by a cliff, in places 8 km high. Outside the main edifice is the aureole that consists of several huge lobes with a distinctively ridged texture.

The lobes are thought to have formed as a result of successive collapses of the periphery of a previously much larger Olympus Mons. The collapses, which could have been catastrophic or gradual, left a cliff around the main edifice. The largest lobe has roughly the same area as France. The edifice is thought to have been built slowly over billions of years by large eruptions, widely spaced in time, and fed from a large magma chamber within the edifice that was itself fed by a magma source deep within the mantle. Although huge, Olympus Mons is not the largest volcano in aerial extent. Alba **Patera**, at the north end of Tharsis is 2000 × 3000 km across, almost the size of the United States. The large size of the Martian shields results partly from the lack of plate tectonics. The largest shield volcanoes on the Earth, those in Hawaii, are relatively short lived. They sit on the Pacific plate, and the source of the lava is below the rigid plate. As a Hawaiian volcano grows, movement of the Pacific plate carries it away from the lava source, so it becomes extinct within a few 100,000 years. A trail of extinct volcanoes across the Pacific attests to the long-term supply of magma from the mantle source presently below Hawaii. On Mars, a volcano remains stationary and will continue to grow as long as magma continues to be supplied, so the volcanoes are correspondingly larger.

The Elysium province is much smaller than Tharsis, having only three sizeable volcanoes. One unique attribute of the Elysium province is the array of large channels that start in **graben** around the volcanoes and extend thousands of kilometers to the northwest. They may have been formed by dikes injected into ice-rich frozen ground. Other volcanoes occur near Hellas and in the cratered uplands. Not all the volcanoes are formed by fluid lava. Some appear to be surrounded by extensive ash deposits and some have

FIGURE 17.3 View looking southeast across Tharsis. Olympus Mons, in the foreground, is 550 km across; 21.2 km high; and surrounded by a cliff 8 km high. Lobes of the aureole can be seen extending from the base of the cliff; 10× vertical exaggeration (MOLA).

densely dissected flanks as though they were composed of easily erodible materials such as ash.

Lava plains may constitute the bulk of the planet's volcanic products. There are several kinds of volcanic plains. On some plains, found mostly between the volcanoes in Tharsis and Elysium, volcanic flows are clearly visible. On others, mostly found around the periphery of Tharsis and in isolated patches in the cratered uplands, ridges are common but flows are rare. Others with numerous low cones may have formed when lava flowed over water-rich sediments. Finally, some young, level plains, such as those in Cerberus, estimated to be only a few million years old, appear to consist of thin plates that have been pulled apart, for they can be reconstructed like a jigsaw puzzle. The plates may indicate rafting of pieces of crust on a lava lake. Geomorphic evidence of volcanic activity that occurred prior to the end of the late heavy bombardment has been largely destroyed by the effects of impacts.

5. TECTONICS

Most of the deformation of the Earth's surface results from the movement of the large lithospheric plates with respect to one another. Linear mountain chains, transcurrent fault zones, rift systems, and oceanic trenches all result directly from plate tectonics. There are no plate tectonics on Mars, so most of the deformational features familiar to us here on the Earth are absent. The tectonics of Mars are instead dominated by the Tharsis bulge. The enormous pile of volcanics that constitute the Tharsis bulge has stressed the **lithosphere** and caused it to flex under the load. Modeling suggests that around the bulge tensional stresses should be circumferential and compressional stresses should be radial. This is entirely consistent with what is observed. The bulge is surrounded by arrays of radial, tensional fractures and circumferential compressional ridges. Some of the tensional fractures, particularly those to the southwest of the bulge, extend for several thousand kilometers. Development of some of the fractures may have been accompanied by emplacement of dikes. The fractures clearly started to form very early in the planet's history, since many of the young lava plains are only sparsely fractured, whereas the underlying plains, visible in windows through the younger plains, may be heavily fractured.

Not all the deformational features result from the Tharsis load, however. Ridges, suggestive of compression, are common on ridged plains, such as Hesperia Planum and Syrtis Major, which are far removed from Tharsis. Some arcuate faults around Isidis and Hellas clearly result from the presence of the large basins. Circular fractures around large volcanoes, such as Elysium Mons and Ascreus Mons, have formed as a result of bending of the lithosphere under the volcano's load. Finally, large areas of the northern plains are cut by fractures that form polygonal patterns at a variety of scales. Polygonal fracture patterns are common in the terrestrial arctic where they form as a result of seasonal contraction and expansion of ice-rich permafrost. Some of the polygonal patterns on Mars, those with polygons up to a few tens of meters across, may have also formed in this way. However, polygons that are several kilometers across are unlikely to have formed in this way and may be the result of regional warping of the surface. Despite these examples, the variety of deformation features is rather sparse compared with the Earth because of the lack of plate tectonics. In particular, folded rocks, although present, are rare.

6. CANYONS

On the eastern flanks of the Tharsis bulge is a vast system of interconnected canyons (See Figure 17.4). They extend just south of the equator from Noctis Labyrinthus at the crest of the Tharsis bulge eastward for about 4000 km until they merge with some large channels and chaotic terrain. The characteristics of the canyons change from east to west. Noctis Labyrinthus at the western end consists of numerous intersecting closed, linear depressions. The depressions are generally aligned with faults in the surrounding plateau. Further east the depressions become deeper, wider, and more continuous to form roughly east–west trending canyons. Still further east the canyons become shallower; fluvial features become more common; both on the canyon floor and on the surrounding plateau; and finally, the canyons end as the canyon walls merge into walls containing chaotic terrain and evidence of floods. The canyons almost certainly formed largely by faulting and not by fluvial erosion, as is the case with the Grand Canyon in Arizona.

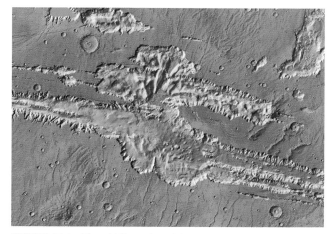

FIGURE 17.4 The middle section of the canyons. In the upper left is the completely enclosed Hebes Chasma, within which a mound of layered, sulfate-bearing sediments has been detected. The main part of the canyon consists of three parallel canyons each 200 km across, also partly filled with mounds of sulfate-bearing sediments. Some of the sediments may have been deposited in lakes which drained catastrophically to the east (MOLA and THEMIS).

Faulting is indicated by the partial merger of numerous closed depressions in the western end of the canyons and by straight walls in the east. While faulting created the initial relief, the canyons have been subsequently enlarged by failure of the walls in huge landslides and by fluvial action. The faulting was on such an enormous scale that it probably involved the entire lithosphere.

Thick sequences of layered (and unlayered) deposits are present in many places throughout the canyons, including some closed canyons completely isolated from the main depression. One possibility is that the canyons formerly contained lakes and that the layered sediments were deposited in these lakes. The lakes drained to the east, hence the continuity eastward from the canyons into several large flood channels. Orbital detection of sulfates within the canyons and some of the outlying depressions supports the lake hypothesis. If climatic conditions were similar to the present, such lakes would have frozen over, although the lake beneath the ice could have been sustained for extended periods if fed by groundwater. While the lake hypothesis is plausible, there are many unanswered questions, such as: Were the lakes fed mostly by surface runoff or groundwater? Where did the sediment in the layered deposits come from? What caused the layering?

7. WATER

Water-formed features present some of the most puzzling problems of Martian geology. Valley networks likely formed when the climate was significantly warmer than at present, yet how the climate might have changed is unclear. Huge floods have episodically moved across the surface, yet there is little trace left of the vast amounts of water that must have been involved, and gullies are forming on steep slopes during the present epoch despite the cold conditions. Perhaps most puzzling of all is the question of whether there were ever oceans present, and if so how big they were; when did they form; and where did all the water go?

7.1. Erosion and Weathering

All Noachian terrain has undergone extensive erosion, whereas younger terrains are only barely eroded for the most part. It thus appears that there were continuously high or episodically high rates of erosion during the Noachian, that the rates fell precipitously at the end of the period, around 3.7 Gyr ago, and that then the rates have remained low for the rest of the planet's history. Average erosion rates during the Noachian are estimated to be around $10^{-6}-10^{-5}$ m/yr, at the low end of continental denudation rates on Earth.

Weathering rates may also have been higher during the Noachian. Hydrated secondary silicate minerals occur throughout the Noachian terrains but are generally not found in younger terrains. A strong possibility, consistent with the presence of valley networks, is that warm and wet conditions at the surface resulted in weathering of primary igneous rocks to produce the hydrated minerals (mainly clays) that we see. An alternative explanation is that the hydrous minerals were formed at depths below the surface and were brought to the surface as a result of excavation by impacts.

7.2. Valley Networks, Lakes, Deltas

Much of the ancient cratered uplands are dissected by branching valley networks that in plan resemble terrestrial river valleys (See Figure 17.5). However, although all the uplands are highly eroded, not all are densely dissected. The valleys mostly have rectangular to U-shaped cross-sections, are a few kilometer across, 50—300 m deep, and tens to hundreds of kilometers long, although a few extend for thousands of kilometers (See Figures 17.6 and 17.7). Drainage densities vary considerably by location but many areas have such high densities that precipitation and surface runoff is implied. Precipitation could have been rain or snow; in either case, climatic conditions significantly warmer than the present are needed. The distribution of the valleys, their excellent preservation, and the poor development of drainage basins suggest that there was a period of enhanced incision at the end of the Noachian. The extent to which valley formation contributed to the universal degradation of the earlier Noachian terrain remains unclear.

Lakes were likely common throughout the poorly graded Noachian terrain while it was undergoing fluvial erosion. Most valleys terminate in closed depressions such as craters or low areas between craters, where at least transient, closed lakes almost certainly formed. Many of these areas are underlain by seemingly fine-grained, horizontally layered, easily erodible sediments. Possible chlorine-bearing (salt) deposits and sulfates in local depressions within the Noachian uplands may be the result of evaporation from such lakes. Many depressions in the highlands have both inlet and outlet valleys indicating the former presence of a lake that overflowed, as might be expected from a flood event. Other lakes have no outlets indicating that infiltration and evaporation kept pace with runoff supply. Comparisons of lake volumes with drainage basin areas suggest that most lakes formed by modest-sized fluvial events spread over extended periods.

Most valleys terminate at grade in local depressions with little or no deposits at their mouths, which suggests that in most cases the materials eroded to form the valleys were either too fine grained to form a delta or were distributed across the depression to form alluvial fans. If a lake was present, its level may have fluctuated. However, a few tens of deltas have been recognized (See Figure 17.8). The deltas typically are fan shaped, outlined by an outward facing scarp, and have the remains of distributary channels on their upper surface. Lack of incision on the delta surface,

Chapter | 17 Mars: Surface and Interior

FIGURE 17.5 Global distribution of the larger valley networks. Most valleys are in the cratered uplands. Not all the cratered uplands are dissected. Extensive areas between Argyre and Hellas are poorly dissected, as is western Arabia.

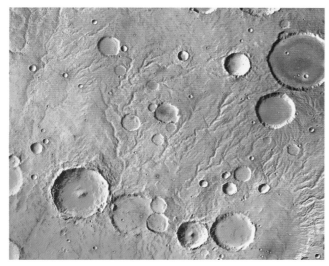

FIGURE 17.6 Typical valley networks in the ancient cratered terrain northeast of Hellas. The regional slope is to the southwest down into Hellas. The dense branching patterns indicate that the valleys formed by surface runoff following precipitation rather than seepage of groundwater. The image is 170 km across (MOLA and THEMIS).

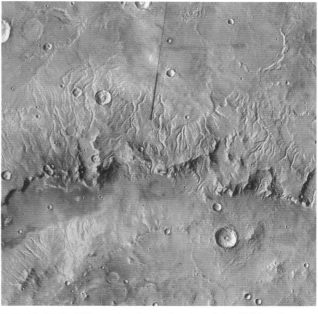

FIGURE 17.7 Valley networks on the rim of the 320-km-diameter, multiring basin Huygens (MOLA and THEMIS).

which would have resulted when the lake level dropped, indicates that fluvial activity terminated abruptly. The valley and delta dimensions, coupled with lake volumes and drainage basin areas suggest that the fluvial episode at the end of the Noachian involved peak discharges comparable to terrestrial rivers. Failure to overflow most closed basins implies that the valley was not formed by a few deluge events but by intermittent modest-sized events spread over an extended period.

No satisfactory explanation has been proposed for how early Mars could have been warmed to allow precipitation and stream flow. The output of the Sun is thought to have been perhaps as much as 20–30% less than at present during this early era. **Greenhouse** models suggest that even a very thick CO_2–H_2O atmosphere could not warm the surface enough. The possible role of other greenhouse gases is being explored. One possibility is that large

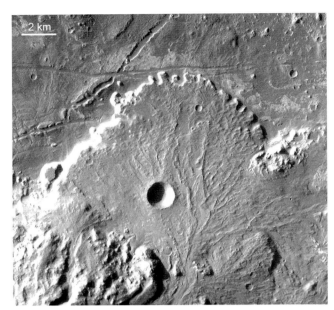

FIGURE 17.8 A delta on the floor of a large-impact crater, just north of where a valley breaches the crater rim. Differential erosion has left remnants of distributary channels as positive features on the delta surface (THEMIS).

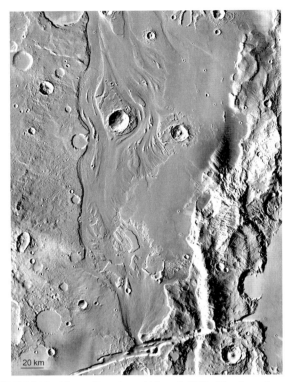

FIGURE 17.9 The outflow channel Mangala Vallis. The channel starts at a 7-km-wide gap in a graben wall, widens northwards, and can be traced for over 1000 km into the northern plains.

impacts injected massive amounts of water into the atmosphere, which then precipitated out as hot acid rain. The idea is attractive in that it might explain why the valley networks formed mainly in the old terrains when impact rates were high but it does not appear to be consistent with the pattern of precipitation implied by the geometry of the fluvial features. Other possibilities include massive injection of greenhouse gases during volcanic eruptions, or melting of snow deposited at low latitudes during high obliquity periods.

A major uncertainty is whether large bodies of water were present while the valley networks were forming. If rainfall was involved in their formation, then large bodies of water must have been present as a source for the precipitation. But if the valleys formed by melting of snow during periods of high obliquities, then ice at the poles could have been the source. Observational evidence for ocean-sized bodies of water remains ambiguous, particularly for this early era.

7.3. Outflow Channels

Outflow channels are very different from valley networks. They are tens of kilometers wide, thousands of kilometers long, have streamlined walls and scoured floors, and contain teardrop-shaped islands (See Figures 17.9 and 17.10). Most start full size and have few if any tributaries. They closely resemble large terrestrial flood features and have almost universally been accepted to be the result of massive floods. Most start around the Chryse basin, emerging either from the canyons or from closed rubble-filled depressions, and extending northward for thousands of kilometers until all traces are lost in the northern plains. While the largest flood features are in the Chryse region, others occur in Elysium, Hellas, and elsewhere, commonly starting at faults. As already indicated above, the channels that merge with the canyons may have formed by catastrophic drainage of lakes within the canyons. Other outflow channels appear to have formed by massive eruptions of groundwater that may have been stored under pressure beneath kilometers-thick permafrost. Release occurred when the permafrost seal was broken such as by impact, volcanic activity, or faulting. Most of the outflow channels formed in the Hesperian period (middle Mars history), well after the time of formation of most of the valley networks, but some formed much more recently. Cold surface conditions and thick permafrost were probably required for their formation.

Major issues are how much water was involved and where did it all go. The size of the channels suggests that the discharges were enormous: 1000 to 10,000 times the discharge of the Mississippi. But we do not know how long the floods lasted and so do not know the total volume of each flood. Nevertheless, large bodies of water, or seas, must have been left in low-lying areas when the floods were over. Efforts to find evidence for these seas has led to mixed

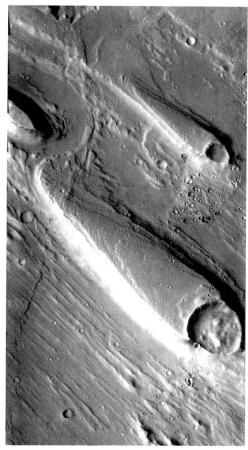

FIGURE 17.10 Teardrop-shaped islands and scour in Ares Vallis. The islands formed where flow was diverted around preexisting craters. Flow is from lower right to upper left. The image is 19 km across (Mars Orbiter Camera (MOC)).

FIGURE 17.11 Water was probably involved in the formation of these deeply incised gullies on the inner wall of a crater, but precisely how they formed remains unclear. The scene is 2 km across (High Resolution Imaging Science Experiment (HiRISE)).

results. Some researchers claim that Mars must have had oceans as extensive as those on the Earth; others claim that seas larger than the Mediterranean were unlikely. Under the present conditions, such seas would have frozen and the ice would have slowly sublimed, thereby adding to the ice at the poles. However, estimates of the amount of water currently in the polar ice caps falls far short of even the lowest estimates of the amounts of water involved in the floods, so a mystery remains as to where the water went. Most likely, it is in the subsurface as either ground ice or groundwater.

7.4. Gullies

Gullies are by far the most common fluvial feature that have formed in the past few billion years of Martian history (See Figure 17.11). They are common on steep slopes in the 30°–60° latitude belts with a preference for poleward-facing slopes. They typically consist of an upper theater-shaped alcove that tapers downslope to converge on a channel that extends further downslope to terminate in a debris fan. The channels are mostly several meters wide and hundreds of meters long. They appear to be forming today. Their origin is controversial. Although initially attributed to groundwater seeps, this origin now seems unlikely given the probable thick cryosphere during the second half of Mars' history and the common presence of gullies at locations where groundwater is unlikely, as at crater rim crests and on slopes around mesas and central peaks. Dry mass-wasting may contribute to their formation but this also seems to be an unlikely cause since many of the gullies cut through bedrock ledges. All the morphologic attributes are consistent with water erosion, and the broad consensus is that this is their cause. One possibility, consistent with their preference for pole-facing slopes, is that they formed mainly by summer melting of snow during periods of high obliquity. But that does not explain the changes that are occurring today. Another possibility may be slope failure aided by the volatilization of interstitial H_2O or CO_2 ice when heated by the summer Sun.

7.5. Dark Streaks

Dark streaks occur on many slopes at low latitudes in dust-covered areas (See Figure 17.12). They are forming today, are dark when first formed, brighten with age, and ultimately disappear over decadal timescales. Although they likely

FIGURE 17.12 Dark streaks on a valley wall. The darker the streak, the more recently it formed. They are most likely caused by dust avalanches, perhaps trigged by rock falls or volatilization of interstitial ice (HIRISE).

result from slope failure, precisely how they form is uncertain. Hypotheses proposed for their formation include subsurface liquid water seepage and flow, disturbance and flow of surface materials as a result of devolatilization of ground ice by solar illumination, and dry granular flow of dust- or sand-sized grains initiated by rockfalls resulting from nearby small impacts or minor internal tectonic activity.

8. ICE

As discussed in more detail below, at each pole is a 3-km-thick water ice cap that extends out to about the 80° latitude circle. Water ice is also present within 2 m of the surface at latitudes down to about 60°, but how deep the ice-rich zone extends is uncertain. Most of the terrain in the 30°−55° latitude belt is covered with a 10 m-thick veneer of ice-rich material that appears to be undergoing removal. The veneer is thought to have been deposited within the past few millions years during a period of high obliquity. Other ice-rich deposits are found at latitudes as low as 30°, despite the present stability relations. Between latitudes 30° and 55°, flows called lobate debris aprons occur at the base of many steep slopes (See Figure 17.13). They have long been suspected to contain significant fractions of ice, but recently, ground-penetrating radar has shown that they consist almost entirely of ice. They have been interpreted as remanents of a more extensive ice cover that formed at midlatitudes during high-obliquity periods. Glacier-like flows on volcanoes and elsewhere may have formed similarly. A wide range of other observations, particularly in the low-lying northern

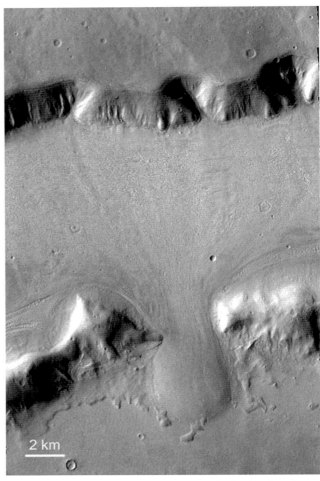

FIGURE 17.13 Flow of ice in the fretted terrain. Ice has flowed away from the cliff and converged on a gap in the hills to the south. At the latitude of this image (40 °N), similar flows occur at the base of almost all cliffs and hills. Orbital radar measurements show that the flows consist of ice (THEMIS).

plains, have been interpreted to be the result of ground ice or glaciers. These include polygonally fractured ground (analogous to arctic-patterned ground?); closely spaced, curvilinear, parallel ridges (moraines?); local hollows (left by removal of ice?); branching ridges (sites of former subglacial streams?); and striated ground (glacial scour?).

9. WIND

We know that the wind redistributes material across the Martian surface. We have observed dust storms from the orbits of the Earth and Mars and the changing patterns of surface markings that they cause. The 2004 Mars Exploration Rovers has made movies of dust devils, and tracks made by dust devils are visible on many high-resolution images taken from the orbit. Dust can be seen draped over rocks in many lander and rover images. Wind streaks

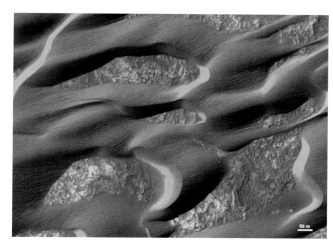

FIGURE 17.14 Dunes. Dunes are common in low areas. Here we see crescentic barchanoid dunes merging to form longitudinal dunes. The wind is from right to left (HIRISE).

caused by eolian deposition or erosion in the lee of craters and other topographic features are common. Dunes are visible in almost all orbiter images with resolutions of a few meters per pixel or better (See Figure 17.14), and in some areas such as around the North Pole, dunes cover vast areas. Given all this evidence, it is somewhat surprising that wind erosion is not more widespread. Fine details of lava flows and impact ejecta, even though billions of years old, are generally well preserved. The wind appears to mostly move loose material around the surface. Additions to the inventory of loose material by erosion of consolidated primary igneous rocks or impact ejecta appears be proceeding only slowly.

Although the average large-scale effects of wind erosion in most places are trivial, locally the effects may be substantial. This is particularly true where friable deposits are at the surface. In southern Amazonis and south of Elysium Planitia, thick, easily erodible deposits cover the plains/upland boundary. Eroded into these deposits are arrays of curvilinear, parallel grooves that resemble terrestrial wind-cut grooves called yardangs. Wherever such wind erosion is observed, other evidence indicates that a deposit that blankets the local bedrock is being eroded.

10. POLES

During fall and winter, CO_2 condenses onto the polar regions to form a seasonal cap that can extend as far equatorward as $40°$ latitude. In summer the CO_2 cap sublimates. The seasonal north polar cap sublimates completely to expose a residual water ice cap, at which time the temperature at the pole rises from the frost point of CO_2 (150 K) to the frost point of water (200 K), and the amount of water vapor over the pole rises dramatically. In the south, the CO_2 cap does not dissipate completely, but water ice has still been detected under the seasonal cap.

At both poles, kilometers-thick layered deposits extend down to roughly the $80°$ latitude. Individual layers are best seen in the walls of valleys cut into the sediments, where layering is observed at a range of scales down to the resolution limit of our best pictures. The frequency of impact craters on the upper surface of the deposits suggests that the sediments are young, of the order of 10^8 years or less. The poles act as a cold trap for water. Any water entering the atmosphere as a result of geologic processes such as volcanic eruptions or floods will ultimately freeze out at the poles. The poles may also be a trap for dust, in that dust can be scavenged out of the atmosphere as CO_2 freezes onto the poles each fall and winter. The layered deposits are, therefore, probably mixtures of dust and ice, with ice predominating. The layering is thought to be caused in some way by periodic changes in the thermal regimes at the poles, induced by variations in the planet's orbital and rotational motions (see Section 2.1). These cyclical motions affect temperatures at the poles, the stability of CO_2 and H_2O, the pressure and circulation of the atmosphere, the incidence of dust storms, and so forth, hence the belief that they are responsible in some way for the observed layering.

11. THE VIEW FROM THE SURFACE

At the time of this writing, we had successfully landed at seven locations on the Martian surface: two Viking lander spacecraft in 1976, Mars Pathfinder in 1997, the Mars Exploration Rovers Spirit and Opportunity in 2004, the Phoenix lander in 2008, and the Mars Science Laboratory Curiosity rover in 2012. Viking 1 landed on a rolling, rock-strewn plain partly covered with dunes in the Chryse basin. Viking 2 landed on a level, rocky plain in Utopia. The main goal of the Viking landers was life detection. They carried a complex array of experiments designed to detect metabolism in different ways and to determine what organics there might be in the soil. Neither metabolism nor organics was detected. The lack of organics was somewhat surprising since organics should have been there from meteorite infall. However, the soil turned out to be oxidizing and the solar ultraviolet radiation environment on the surface harsh, which probably caused decomposition of any organics that might at one time have been present. Mars Pathfinder also landed on a rock-strewn plain in Chryse and deployed a small microrover named Sojourner. The site is at the mouth of one of the large outflow channels. It was hoped that evidence of floods might be observed there. However, the only sign of floods were some rocks stacked on edge and terraces on nearby hills that could have been shorelines. Phoenix landed at $68°N$ on a plain where ground ice was expected to be found, and indeed water ice was found just below the surface.

The two rovers Spirit and Opportunity, launched in 2003, provided the first solid evidence from the surface for pooling of water and aqueous alteration. Spirit landed on the flat floor of the 160-km-diameter crater Gusev. The site was chosen because the southern wall of Gusev is breached by a large channel called Ma'adim Vallis. Water from the channel must at one time have pooled in Gusev, and it was hoped that the rover would be able to sample sediments from the postulated Gusev lake. The floor of Gusev turned out to be another rock-strewn plain. The rocks are basalts but they have alteration rinds with varying amounts of water-soluble components such as S, Cl, and Br. The alteration is minor and has been attributed to the action of acid fogs. Erosion rates estimated from craters superimposed on the plains indicate that the rates have been several orders of magnitude less than typical terrestrial rates. These somewhat disappointing results spurred a move to drive to and explore some nearby hills, where it was hoped different materials would be found, and indeed they were. Most of the rocks on the Columbia Hills are very different from those on the plains (See Figure 17.15). Many different classes of rocks have been identified, ranging from almost unaltered olivine basalts like those on the plains to almost completely altered, soft rocks enriched throughout with mobile elements such as S, Cl, and Br. In these altered rocks, primary **basalt** minerals are almost absent, having been replaced by secondary minerals such iron oxides and oxyhydroxides that have high Fe^{3+}/Fe^{2+} ratios compared with the unaltered rocks. Hydrated sulfates, opaline (hydrated) silica, or carbonates may also be present. A sulfate cement in some rocks suggests evaporation of sulfate-bearing (low pH) waters. Some rocks have soft interiors that have been hollowed out by the wind to leave only the hard outer shell. Layered rocks are common, and a coarse stratification appears to follow the contours of the hills. The origin of the Columbia Hills rocks and minerals is still being debated. Some may have formed by aqueous alteration of newly deposited impact or volcanic debris. Some may have been hydrothermally altered either during or perhaps even long after deposition. For others, waters from the postulated Gusev lake may have been implicated. Whatever the cause, aqueous processes were involved. Regolith deposits of almost pure silica are also indicative of hydrothermal activity.

Opportunity landed in Meridiani Planum on a thick stack of layered rocks that had been observed from orbit (See Figures 17.16 and 17.17). The site was chosen because a particular coarse-grained form of the iron oxide mineral hematite, which often forms in aqueous environments, had been detected there. The number of impact craters superimposed on the layered rocks suggests that they formed at the end of the heavy bombardment period around 3.8 billion years ago. The rover data demonstrated unequivocally that the local rocks are reworked evaporitic sandstones with roughly equal proportions of basaltic debris and evaporitic minerals such as Mg, Ca, Fe, Na sulfates, and chlorides. Although most of the rocks were initially deposited by the wind, there had to be a nearby source for the evaporites, which form by evaporation of bodies of water. The source had to be substantial because the layered sequence on which the rover landed extends for several hundred kilometers. A small fraction of the rocks have depositional textures that indicate that they were deposited in standing water. The environment in which the Meridiani sequence accumulated is thus thought to be one in which there were wind-blown dunes with interdune ponds.

Both Spirit and Opportunity, although landing on very different geologic materials, are telling a somewhat similar story. The oldest rocks, those that formed during or near the end of the late heavy bombardment, have abundant evidence for aqueous processes, but the evidence for

FIGURE 17.15 View from the Spirit rover in the Columbia Hills. The level plains of Gusev are in the background. The hills in the distance are part of delta-like deposits at the mouth of a large channel that enters the crater from the south. The rocks in the foreground have been aqueously altered to varying degrees. The origin of the hills is unknown, but they may have been uplifted by an impact event that postdated the formation of Gusev itself (Mars Exploration Rover (MER)/Pancam).

Chapter | 17 Mars: Surface and Interior

FIGURE 17.16 View of Endurance crater from the Opportunity Rover in Meridiani. The impact crater formed in a sequence of horizontally layered rocks, which are exposed in the foreground and in the walls of the crater. The horizon in the background gives and indication of how level the rock sequence is. The rover entered the crater and made measurements down section, almost to the center of the crater. Burns Cliff, seen in the next figure, is on the far wall (MER/Pancam).

FIGURE 17.17 View of Burns Cliff from the Opportunity rover in Meridiani. The rocks consist of a mixture of evaporites, such as sulfates and chlorides, and basaltic debris. The bedding patterns indicate that they were mostly deposited by the wind. However, the evaporites must originally have been derived by evaporation of a nearby lake or sea (MER/Pancam).

FIGURE 17.18 Curiosity rover telephoto view of the layered rocks within the central mound of Gale crater, named Mount Sharp by the rover team. The rover is expected to traverse up into these layers to investigate their geology, chemistry, and mineralogy. The triangular mound near the center of the image is about 300 m wide and 100 m tall (Mars Science Laboratory (MSL)/Mastcam).

such processes well after the end of the late heavy bombardment is sparse or absent. Designed to last just 90 Martian days, Spirit's mission lasted more than 2200 days and Opportunity's mission continues as of this writing, more than 3100 days after landing. The longevity and mobility of these remote-controlled vehicles has enabled significantly more diverse scientific discoveries than originally envisioned. Indeed, Opportunity is now extensively exploring phyllosilicate-bearing terrains within the rim of an ancient Noachian crater, pushing our understanding of Martian geology and geochemistry back even further into the past.

Most recently, the Curiosity rover landed in August 2012 and began its mission within the 150-km-diameter crater named Gale, along the boundary between the southern highlands and northern lowlands. Gale was chosen because it is a deep and closed depression that may once have hosted a crater lake (like Gusev) and which also contains an enormous (4 km tall) central mound of finely layered sedimentary rocks that appear to span a large fraction of early Martian history (See Figure 17.18). The floor of the crater and the lower layers of the mound contain evidence for fluvial transport as well as hydrated phyllosilicate and sulfate minerals. It is hoped that over the course of the mission Curiosity will be able to traverse up the mound, going from older to younger deposits, looking for evidence of changing environmental conditions. Already, as of this writing, solid evidence of stream-deposited sediments has been found (See Figure 17.19), confirming the former presence of liquid water on the crater floor.

FIGURE 17.19 Part of a fractured, gravelly outcrop rock investigated by the Curiosity rover near the initial landing site on the floor of Gale crater. The rounded sand- to gravel-sized clasts occur within a finer-grained whiter matrix of cementing material, characteristic of a conglomeritic sandstone deposit typical of those found in lithified ancient streambed environments. The clasts here are too large to have been transported or rounded by the wind, providing evidence for vigorous liquid water flow in Gale crater early in Martian history (MSL/Mastcam).

12. SUMMARY

Mars is a geologically heterogeneous planet on which have operated many of the geologic processes familiar to us here on the Earth. It has been volcanically active throughout its history; the crust has experienced extensive tectonic deformation, largely as a result of massive surface loads; and the surface has been eroded by wind, water, and ice. Despite these similarities, the evolution of Mars and Earth has been very different. The lack of plate tectonics on Mars has prevented the formation of linear mountain chains and cycling of crustal material through the mantle, and climatic conditions that hindered the flow of water across the surface have limited erosion and deposition to almost negligible levels for most of the planet's history. As a consequence, a geologic record is preserved on the surface that spans almost the entire history of the planet. For the late heavy bombardment period we have compelling chemical and mineralogic evidence for aqueous alteration and compelling geomorphologic evidence of widespread fluvial erosion and transient lakes. The following Hesperian period was characterized by large floods, widespread volcanism, and accumulation of sulfate minerals indicative of lower pH weathering conditions. In the second half of Mars' history, geologic activity declined significantly, although there were still occasional floods and other fluvial and glacial events. The climatic implications of the geologic observations remain uncertain. While early Mars must have had at least episodic warm, wet climatic episodes, any warm episode after the end of the late heavy bombardment must have been very short, because the cumulative amounts of erosion and weathering are so small. Significant mysteries remain. What caused the early warm conditions? What processes were responsible for the early massive amounts of erosion and deposition indicated by the geologic record? Where did all of the water go? How much of the Martian surface and subsurface was (or perhaps still is) habitable, by terrestrial standards? Most important of all, did some form of life ever evolve on the planet, and if so is it extant today?

BIBLIOGRAPHY

Bell, J. F., III (2008). *The Martian surface: Composition, mineralogy, and physical properties*. Cambridge: Cambridge University Press.

Cabrol, N. A., & Grin, E. A. (2010). *Lakes on Mars*. Amsterdam: Elsevier.

Carr, M. H. (2006). *The surface of Mars*. Cambridge: Cambridge University Press.

Carr, M. H., & Head, J. W. (2009). Geologic history of mars. Earth planet. *Science Letters, 294*, 185–203. http://dx.doi.org/10.1016/j.epsl.2009.06.042.

Hartmann, W. K., & Neukum, G. (2001). Cratering chronology and the evolution of Mars. *Space Science Review, 96*, 165–194. http://dx.doi.org/10.1023/A:10111945222010.

Malin, M., & Edgett, K. (2001). Mars global surveyor mars orbiter camera: interplanetary cruise through primary mission. *Journal Geophysics Research, 106*, 23429–23570.

McSween, H. Y., Jr. (1994). What we have learned about Mars from SNC meteorites. *Meteoritics, 29*, 757–779.

Murchie, S. L., Mustard, J. F., Ehlmann, B. L., Milliken, R. E., Bishop, J. L., McKeown, N. K., et al. (2009). A synthesis of martian aqueous mineralogy after 1 Mars year of observations from the Mars reconnaissance orbiter. *Journal Geophysics Research, 114*. http://dx.doi.org/10.1029/2009JE003342. E00D06.

Chapter 18

Interior Structure and Evolution of Mars

Tim Van Hoolst
Royal Observatory of Belgium, Brussels, Belgium,
Instituut voor Sterrenkunde, KU Leuven, Celestijnenlaan 200D, B-3001 Leuven, Belgium

Attilio Rivoldini
Royal Observatory of Belgium, Brussels, Belgium

Chapter Outline

1. Introduction	379
2. Formation and Differentiation of Mars	381
3. Core	382
4. Mantle	383
5. Crust	384
6. Principles of Global Interior Structure and Evolution	385
6.1. Basic Equilibrium Equations	385
6.2. Heat Sources	385
6.3. Heat Transport by Conduction and Convection	387
6.3.1. Conduction	387
6.3.2. Convection	388
7. Global Interior Structure of Mars	390
7.1. Introduction	390
7.2. Global Geodesy Data	390
7.3. Model Results	392
8. Evolution of Mars	393
8.1. Thermal Evolution	393
8.2. Early Dynamo	395
8.3. Chemical Evolution	395
8.3.1. Crust Formation	395
8.3.2. Extraction of Water	396
Bibliography	**396**

1. INTRODUCTION

The planet Mars is situated further from the Sun than the other three terrestrial planets of the solar system and has the lowest average surface temperature. Because its surface layers are, nevertheless, much warmer than the interplanetary medium, Mars, like the other terrestrial planets, loses heat and, as its decreasing internal heat sources cannot balance the loss, slowly cools. The rate at which heat is transferred from the deep interior to space is currently unknown but is much smaller than the energy emitted by a blackbody in thermal equilibrium with the surface temperature of Mars. The surface temperature is mainly determined from the balance between the absorbed solar energy and the reemitted blackbody radiation. Although the energy lost by Mars is small, the thermal evolution of the planet and the slow transfer of heat from the deep interior to the surface is the major driver for the general evolution and dynamics of the planet. The internal dynamics of Mars is affected by the internal temperature and internal heat transfer because heat is often transported by macroscopic mass motion in terrestrial planets. For example, convection is the main mechanism for redistributing heat in the mantle of Mars and is ultimately responsible for a wide range of phenomena including volcanism. Convective motions inside the core have even caused a **dynamo** to operate in the early history of Mars and explain the existence of large regions of magnetized rock on the surface of Mars.

All four terrestrial planets are thought to be principally composed of only four elements: iron, oxygen, silicon, and magnesium. For the Earth, the first two elements constitute about 30% each of the total planetary mass, whereas the latter two contribute about 15% each. Information on the bulk composition of Mars is much more limited than for the Earth, with main data obtained from analyses of meteorites from Mars, in situ data from landers, remote sensing data from orbiters, and **cosmochemistry**. In terms of bulk chemical composition it is thought that Mars is rather similar to the Earth. This may, however, not be taken as an indication that all terrestrial planets would have such a bulk composition. In the case of Mercury, more than half of its mass is in the form of iron, suggesting a different formation history (see the chapter on Mercury).

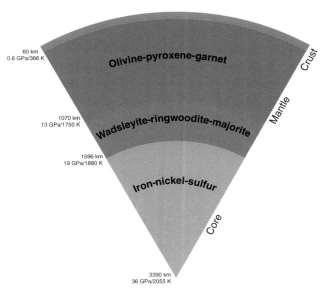

FIGURE 18.1 Section through the interior structure of Mars depicting the 3 principal reservoirs: crust, mantle, and core. At each interface, depth, pressure, and temperature are given.

The chemical elements are distributed unevenly in three main reservoirs of the terrestrial planets: the core consists mainly of iron and the mantle and crust consist of silicate rock, which consists primarily of the four elements mentioned above (see Figure 18.1). In the Earth, the core is subdivided into a solid inner core and a liquid outer core. The physical state and structure of the core of the other terrestrial planets is not known with certainty. For Mars, recent geodesy data suggest that the core is entirely liquid.

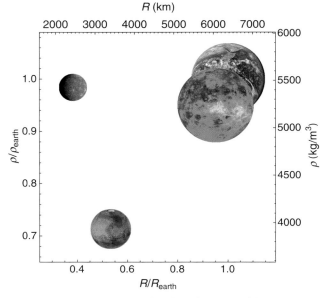

FIGURE 18.2 Radius and density of the four terrestrial planets (Mercury, Venus, Earth, and Mars (relative size to scale)) of the solar system and relative values compared to the Earth.

Compared to the Earth and Venus with a radius over 6000 km, Mars is much smaller with a mean radius of 3389.5 ± 0.2 km (Figure 18.2). In terms of mass, Mars is almost 10 times less massive than its larger sister planets (9.3 times for the Earth and 7.6 times for Venus, see Table 18.1 for an overview of general characteristics of Mars). However, compared to Mercury, Mars is somewhat larger and almost twice as massive. Because Mars is smaller than the Earth, it is expected to form at a lower temperature. One might also expect that smaller planets like Mars cool faster than the Earth because the heat content of a planet is proportional to the planetary volume (or to the third power of the radius) and the heat loss is proportional to the surface (or to the square of the radius) so that the heat source to heat loss ratio is proportional to the radius. However, heat transport in planets is a complex phenomenon and can be organized in different ways in different planets, causing, for example, Mars to cool somewhat slower than the Earth.

The mean density of Mars is 3933 kg/m^3, which is about 28% less than that of the Earth (5515 kg/m^3, see Figure 18.2). This smaller value, however, does not mean that Mars' bulk chemical composition is depleted in heavier elements and enriched in lighter chemical elements with respect to the Earth, instead that the lower density results from the planet's smaller size. The pressure in smaller planets, like Mars, is on average lower than that in the larger planets Venus and the Earth. Therefore, planetary material in small planets is on average less compressed than in large planets, and small planets are expected to be less dense than large terrestrial planets if they were to have the same bulk composition. Since this compression effect is important in terrestrial planets, the interior of terrestrial planets cannot be properly understood without good knowledge of how much planetary materials can be compressed at typical planetary pressure conditions. For Mars, internal pressures increase with depth up to about 40 GPa. At those pressures iron is compressed by about 20%. Moreover, planetary materials expand with increasing temperatures, which in Mars increases with depth to about 2000 K, but the effect on the material density is smaller than the compression effect of pressure. For example, at a pressure of 40 GPa and a temperature of 2000 K, iron is about 12% denser than at standard conditions of 298 K and 10^5 Pa. Material properties at the high internal pressures and temperatures of Mars can currently be measured in the laboratory. Moreover, theoretical calculations based on the basic principles of quantum mechanics (*ab initio* methods) are currently being used to study how planetary material behaves at high pressures and temperatures.

This chapter is organized as follows. We present a brief overview of the formation process and internal differentiation of Mars into a core, mantle, and crust in Section 2. General properties of these three main reservoirs of Mars are provided in the next three sections. Section 6

TABLE 18.1 General Characteristics of Mars

Quantity	Symbol and Unit	Value
Mass	M (10^{23} kg)	6.4186 ± 0.0008
Radius	R (km)	3389.5 ± 0.2
Mean density	ρ (kg/m^3)	3935.0 ± 0.8
Mean moment of inertia factor	$I/(MR^2)$	0.3645 ± 0.0005
Mean surface gravitational acceleration	g (m/s^2)	3.7379 ± 0.0007
Semimajor axis	a (AU)	1.5237
Orbital period	P_{orb} (days)	686.98
Rotation period	P_{rot} (hours)	24 h 37 min 22.662993 s $\pm$ 0.000003
Obliquity	ε (degrees)	$25.189379242 \pm 0.00001°$
Eccentricity	e	0.0934

explains the general physical principles governing the interior structure and evolution of terrestrial planets. Recent results about the global interior structure of Mars and the thermochemical evolution are described in Sections 7 and 8.

2. FORMATION AND DIFFERENTIATION OF MARS

Terrestrial planets like Mars form when kilometer-sized planetesimals, which originate from the accumulation of dust grains in less than about 10,000 years, collide under the influence of the gravitational attraction between them and gas drag, a process called accretion. In less than a million years, tens to hundreds of planetary embryos are formed with masses between those of the Moon and Mars. By bringing planetesimal material with a Mars mass M from a far distance to a small region of space with Mars radius R, the gravitational potential energy decreases enormously by about $0.6GM^2/R$, where G is the universal gravitational constant. This change in energy is mostly converted into heat. The associated temperature rise is about $0.6GM/(RC) \approx 6000$ K, where $C \approx 1200$ J/K/kg is the specific heat of Mars, suggesting that Mars formed hot.

However, part of the massive amount of energy liberated by accretion will be radiated to space. For slow accretion over about 1 My of small planetesimals, even most of the gravitational energy could be lost by radiation and planets would not form hot but would be cold with temperatures of a few hundred Kelvin close to the temperature of the protoplanetary disk. The key to the hot origin of planets lies in how deep the energy can be deposited in a planet. If energy can be brought to the deep interior, a very efficient internal heat transport mechanism would be required for a cold formation to occur. The general consensus is that in particular the large impacts at the end of formation strongly heat the deep interior of a planet in a fraction of geological time and can melt at least part of the planet interior producing a magma ocean.

The final formation of Mars took less than about 10 My. This is faster than for the Earth, for which the last large impact that formed the Moon is thought to have occurred 30–50 My after the formation of the solar system. The much smaller mass of Mars and fast formation compared to the Earth suggest that Mars could be a remnant embryo. This embryo status of Mars could be explained in the accretion scenario if the outer edge of the planetesimal disk of the initial solar system was at about 1 AU (an Astronomical Unit is the distance between the Sun and the Earth). Such a small inner disk possibly formed by inward migration of Jupiter to the Sun in the early phases of the solar system to a distance of only about 1.5 AU.

Because of its fast formation, Mars probably suffered less violent impacts with respect to the Earth resulting in a more limited heating. Nevertheless, the decay of the short-lived radioactive isotope ^{26}Al with a **half-life** of 0.72 My could produce up to half the energy of accretion, more than sufficient for mantle melting and the formation of a magma ocean (half-life is the length of time needed for half of the parent atoms to decay into daughter atoms). Iron droplets in the magma ocean could then descend to form the core and at the same time mantle material crystallized at the cold surface layers to form the primordial crust. In the core formation process, additional gravitational energy is released and converted into heat, further increasing the internal temperature of Mars and facilitating core formation. Mars, therefore, quickly differentiated into a core, a mantle, and a crust, a process that already started during the formation. After solidification of the magma ocean, at least part of the mantle is thought to be gravitationally unstable because the lightest material most rich in magnesium solidifies first at the bottom of the magma ocean and the more iron-rich and denser silicates solidify later. As a result, the mantle overturns: the denser materials sink and the lighter materials rise producing a gravitationally stable stratification, a process that could have latest up to 100 My after core formation.

The chronology of the planetary formation has been possible to unravel thanks to the study of radioactive parent–daughter systems with half-lives of the order of 10 My. In particular, the hafnium–tungsten system is widely used to constrain the accretion timescale and to date core formation of planetary bodies. ^{182}Hf is a short-lived isotope that decays to ^{182}W with a half-life of 9 My. Hafnium is a lithophile, or "rock-loving", element meaning that it will stay in the mantle when the iron core forms. Tungsten on the other hand is a siderophile ("iron-loving") element and will follow the iron to the core on core formation. Therefore, part of the ^{182}W produced from ^{182}Hf will be in the core if the core formation time is comparable to the half-life of hafnium. By comparing the Hf/W ratio from the Martian mantle (estimated from Martian meteorites) with the initial ratio (derived from chondrite meteorites), the age of core formation of less than 10 My can be deduced.

3. CORE

Seismic data have shown that the core of the Earth has an outer liquid part and a solid inner part and that both parts have densities that are only several percent lower than iron at core pressure and temperature. Iron is the most likely constituent element of planetary cores since it is the only element abundant enough in the universe with a density and elastic properties closely agreeing with seismic data. Besides iron, analyses of iron meteorites—thought to be representative for cores of planetary bodies—suggest that the Earth's core also contains a few percent of the somewhat denser element nickel. Since the core is less dense than pure iron, it also contains light elements. The identity of those elements is still debated, but the most prominent candidates are sulfur, silicon, and oxygen. The light elements not only lower the density and change the elastic properties of pure iron but also decrease the melting temperature of the core material significantly compared to the melting temperature of pure iron. About 1220 km of the 3480 km of the core radius have solidified due to cooling of the Earth over the past 4.5 By. On cooling, iron-rich material solidifies at the boundary of the solid inner core and liquid outer core. Seismic data have revealed that the solidified inner core material is denser than solid material of outer core composition at inner core pressures and temperatures. This implies that some of the light elements do not partition in equal amounts into the solid and liquid parts of the core. As a result, material immediately above the inner core–outer core boundary becomes enriched in light elements. The buoyancy force acting on this less dense material forming at the bottom of the outer core generates convective motions in the whole liquid outer core and provides the major part of the energy necessary to drive the dynamo responsible for the Earth's global magnetic field.

Similarly as for the Earth, the core of Mars is thought to be principally made of iron with unknown but small fractions of nickel and light elements. Direct information on the density, elastic properties, and size of the core is lacking because of absence of seismic data. Based on the chemical affinity of light elements to iron–nickel mixtures, the main candidate light elements for the Martian core are sulfur, silicon, oxygen, carbon, and hydrogen, in binary, ternary or more complicated systems with iron and nickel. Analyses of rock samples representative of the bulk composition of the mantle of Mars, such as Martian meteorites, allow further constraining the core composition, although the results depend on core formation scenarios. They indicate that iron, nickel, and the light element sulfur are the principal constituents of the core. Additional evidence for the presence of sulfur comes from the fact that it has been found in many nickel–iron meteorites. Moreover, of all the light elements, sulfur and silicon are most easily incorporated into planetary cores of terrestrial planets as their solubility in molten iron is high over an extended pressure range. In contrast, for example, oxygen solubility in liquid iron is below 1wt% at ambient pressure and increases with increasing pressure and sulfur concentration. At the pressure and temperature conditions at the bottom of the Martian mantle, at most a few weight percent of oxygen could partition in the liquid core. Silicon and carbon can only have been incorporated in the core if Mars formed under reducing conditions which is, however, rather unlikely given the highly oxidized state of the Martian surface. Finally, hydrogen might also enter the core if the magma ocean contained water and the pressure at the bottom of the magma ocean was higher than about 5 GPa.

Given the presumed formation conditions and the above data for liquid iron systems, it is unlikely that Mars' core contains significant amounts of light elements other than sulfur. Moreover, sulfur has the important property that it strongly reduces the melting temperature of the iron system with respect to pure iron. For example, at 21 GPa, the melting temperature of Fe decreases by about 60 K for each additional weight percent of S as long as the concentration of S is below the eutectic concentration, at which the melting temperature is the lowest for a Fe–FeS system. Low melting temperature of the core might be important since the outer part of the core is liquid since the solid tides of Mars, estimated by observing orbital variations of spacecraft motion, are several times larger than for an entirely solid Mars. Whether or not the inner part of the core is solid like in the Earth depends on the composition and temperature of the core. If the melting temperature of the core is below the core temperature, as is possible for a large sulfur concentration in the core, the core will be entirely liquid. A fully liquid core is consistent with the observation that Mars does currently not possess a global magnetic field as the Earth but only localized magnetized rocks. Dynamo action in the core, which is responsible for the global magnetic field of the Earth, is not possible at the present day without a growing inner core. In Section 7, it will be shown that tidal observations can be used to put further constraints on the size, density, physical state, and composition of the core.

4. MANTLE

The silicate part of the Earth consists mainly of rocks (~21wt% Si and ~44wt% O) that are rich in magnesium (~23wt%) and poor in iron (~7wt%). Together with the elements Al (~2.3wt%) and Ca (2.5wt%) these four elements are responsible for more than 98% of the total mass of the silicate Earth. The upper mantle rocks are principally made of the minerals olivine and pyroxenes. With increasing depth, or increasing pressure and temperature, these minerals experience phase transitions to denser polymorphs or dissociate to other minerals that are stable at those pressure and temperature conditions. The principal lower mantle minerals are magnesium perovskite and magnesiowüstite. At the bottom of the mantle close to the core, the still denser mineral post-perovskite dominates. The distribution of the minerals as a function of depth in the mantle is not accurately known but can be constrained from the known seismic velocities and density in the Earth's mantle and from measurements of the electrical conductivity of the mantle.

Unlike for the Earth, seismic data are not yet available for Mars. As Mars is also a terrestrial planet it is assumed that the principal minerals in the Martian mantle are the same as those in the terrestrial mantle. Since the pressures in the mantle of the smaller planet Mars are much smaller than those in the Earth (maximum pressure of about 20 GPa in the Martian mantle compared to about 135 GPa at the bottom of the mantle of the Earth), the densest mineral phases in the Earth's mantle are not present (post-perovskite) or only marginally (perovskite and magnesiowüstite) present. Therefore, the minerals in the mantle of Mars are likely the same as those in the upper mantle of the Earth: olivine and its high-pressure polymorphs, pyroxenes and garnet (Figure 18.3). The chemical composition of the silicate part of Mars has been constrained by analyzing Martian meteorites, performing surface monitoring from orbiting spacecraft, and analyzing surface rocks in situ by means of robotic rovers. One of the most striking differences compared to the Earth is the significantly higher concentration of iron (>20wt%) in the minerals. This difference could be explained by a difference in the chemical composition of the materials that formed the planets and by the significantly lower pressure at the bottom of the magma ocean on Mars compared to the Earth, which strongly reduces the dissolution of oxidized iron in the iron melt forming the core.

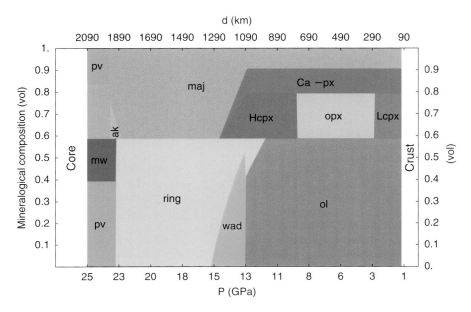

FIGURE 18.3 Phase diagram for the mantle of Mars derived from the chemical analysis of Martian meteorites (Dreibus and Wänke, 1985) assuming a cold end-member mantle temperature profile. The upper mantle is rich in olivine (ol) and pyroxenes (clinopyroxene LP (Lcpx), clinopyroxene HP (Hcpx), orthopyroxene (opx), and Ca-pyroxene (Ca-px)). With increasing pressure the pyroxenes dissociate to majorite (maj) and olivine transforms to its higher pressure polymorphs wadsleyite (wad) and ringwoodite (ring). At the highest possible pressure, close to the core–mantle boundary, ringwoodite starts to dissociates to (Mg,Fe)-perovskite (pv) and (Mg,Fe)-wüstite (mw). At those high pressures, majorite first takes the akimotoite (ak) structure, and subsequently transforms into perovskite.

In terms of oxides, the mantle of Mars is thought to consist of 45% SiO_2, 30% MgO, 17% FeO, 3% Al_2O_3, 2% CaO, and small contributions from other elements. For the Earth, we have of 47% SiO_2, 37% MgO, 8% FeO, 4% Al_2O_3, and 3.5% CaO.

5. CRUST

The crust is the thin upper layer of a terrestrial planet. It consists of silicate rocks like the mantle but is chemically different from the mantle. The crust is more silica rich (SiO_2 contributes 50% or more in mass) and has a lower density than the underlying mantle as a result of its formation from the mantle by melting and crystallization. Studies of radioactive parent–daughter isotopes in Martian meteorites that separate differently between the liquid and the solid silicate phases show that the bulk of the crust was created within 100 My after formation of the planet. This primordial crust presumably formed from crystallization of the cooling magma ocean. Additional crust is formed later in Mars' evolution when ascending hot material from the mantle partially melts and rises to the crust and surface by volcanism (see Section 8). The crust is **basaltic** in composition with the older crust being more silica rich. However, rocks with the highest silica content on the Earth, felsic rocks such as granite, are almost absent on Mars. On the Earth, plate tectonics, which is absent on Mars, continuously changes the crust. A division in continental crust and oceanic crust as for the Earth can, therefore, not be made for Mars.

Mars' crust exhibits a very notable crustal dichotomy: the northern hemisphere is almost flat and covered with volcanic rocks, whereas the southern hemisphere is a few kilometers higher and cratered by ancient impacts (Figure 18.4). The crust is about 25 km thicker in the southern highlands than in the northern lowlands. The thickness of the crust has been estimated from the topography data measured by the laser altimeter onboard the Mars Global Surveyor (operational between 1997 and 2006) and from the gravity field determined by radio tracking orbiting spacecraft. The average crustal thickness is estimated to be between 38 and 62 km, and the average crust density is about 2700–3100 kg/m^3. The dichotomy was formed shortly after formation of Mars within the first half billion year and maybe even within the first 50 My. The dichotomy could have been created by large impactors but could also be due to internal processes such as the crystallization of the magma ocean, an early phase of plate tectonics, or convective motions in the mantle characterized by a large upwelling plume beneath the southern hemisphere.

One of the most prominent topographic features on the surface of Mars besides the dichotomy is the volcanic plateau of Tharsis which is located near the equator in the western hemisphere. The plateau or bulge is about 5000 km across and up to 7 km high and harbors the largest and highest volcanoes of the solar system. It is thought that Tharsis is created by a volcanic plume, quite similar to the one found beneath the island of Hawaii. The idea is that a hot column of mantle material rose from the core–mantle boundary through the mantle and delivered substantial volumes of basaltic lava to the surface. Because of the absence of plate tectonics, high volcanoes could develop over long timescales, with the highest of all, Olympus Mons, rising 22 km above the reference surface. The lower gravity on Mars (see Table 18.1) explains why higher mountains can exist on Mars than on the Earth.

Analyses of the mineralogy of surface rocks and water erosion features such as outflow channels on Mars' surface indicate that Mars had large amounts of liquid water on its

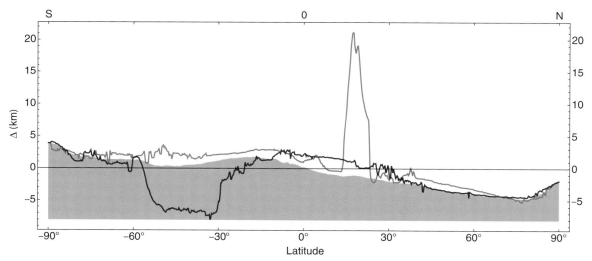

FIGURE 18.4 Longitudinally averaged topography of Mars showing the elevation difference between the southern highlands and northern lowlands. The red and the blue curves show the topography at the latitudes with the highest (Olympus Mons) and lowest features (Hellas basin) on the surface of Mars.

surface in the very distant past during intermittent periods. Nowadays, water on the surface is mostly found in the form of solid water ice below the CO_2 ice-covered polar ice caps. A review of the geology and water history of Mars can be found in Chapter 19 of this encyclopedia.

6. PRINCIPLES OF GLOBAL INTERIOR STRUCTURE AND EVOLUTION

6.1. Basic Equilibrium Equations

Although a terrestrial planet such as Mars is not exactly spherically symmetric, its global interior structure in the radial direction can very well be described by assuming that the physical quantities do not depend on the angular variables but only on the radial distance to the planet's mass center. A major mathematical advantage of this approximation is that the problem of the description of the interior structure is reduced from a three-dimensional problem to a one-dimensional (1D) problem.

Although the cold and stiff upper layers of Mars can support shear stresses on very long timescales, the bulk of the interior cannot and behaves as a viscous fluid. The nearly spherical shape of Mars is a clear indication of the fluidlike behavior of Mars on very long timescales. Even on timescales of the order of 10,000 years mantle material of terrestrial planets is known to behave as a viscous fluid as convincingly follows from studies of postglacial rebound on Earth. A basic physical model that exhibits elastic behavior on short timescales and viscous behavior on long timescales is the Maxwell model, which can be described in terms of a spring and a dashpot in series. When an outward force is applied at both sides of such a system, the spring immediately responds elastically and stretches. When the system is subsequently kept at the same length, the dashpot gradually pulls apart, and the spring decreases in length and relieves stress. After a long time, the spring returns to its original length without stress.

Because of the fluidlike behavior on geological timescales terrestrial planets are close to hydrostatic equilibrium. Hydrostatic equilibrium expresses that the downward gravitational force is balanced by the upward differential pressure force at any point in the planet at a radial distance r to the planet's mass center:

$$\frac{dP(r)}{dr} = -\rho(r)g(r), \quad (18.1)$$

where $P(r)$ is the pressure, $\rho(r)$ is the mass density, and $g(r)$ is the gravity.

Newton's theory of gravitation shows that anywhere in the planet, gravity and density must also satisfy Poisson's equation:

$$\frac{dg(r)}{dr} + \frac{2}{r}g(r) = 4\pi G\rho(r). \quad (18.2)$$

Besides pressure and gravity, Eqns (18.1) and (18.2) depend on the density. A third equation is therefore needed to solve for these three variables. This is given by an equation of state (EoS) specifying the dependence of the density on pressure, temperature, and composition. For solids and liquids, EoSs are more complicated than the ideal gas law. An often used EoS for solids is the **Birch–Murnaghan** equation.

For a given temperature profile and composition, the radial profiles of density, pressure, and gravity can then be calculated from the center to the surface (Figure 18.5). The temperature profile depends on the heat sources and on the method of heat transport, which are discussed below. Although Mars loses heat at its surface to interplanetary space and globally cools, the heat loss of Mars is small and the average mantle temperature of Mars only decreases by 30–50 K per billion year. It is then justified to assume that, on short timescales, Mars is in thermal equilibrium expressing a balance between the energy lost by outward energy flux and the internal energy generation. This energy balance can be expressed as

$$\frac{dq(r)}{dr} + \frac{2}{r}q(r) = \rho(r)\varepsilon(r), \quad (18.3)$$

where $\varepsilon(r)$ is the specific heat production rate per unit mass and unit time and $q(r)$ is the outward heat flux per unit area and unit time. Expressions for $q(r)$ are given below.

6.2. Heat Sources

Several radioactive isotopes occur in nature but only a few of them play an important role in the global energy budget of terrestrial planets on long timescales. The most important are the uranium isotopes ^{235}U and ^{238}U, thorium ^{232}Th, and the potassium isotope ^{40}K. All these isotopes have a long half-life, of 7.04×10^8 years, 4.47×10^9 years, 1.40×10^{10} years, and 1.25×10^9 years, respectively, so that they still contribute to the heat budget of Mars up to the present day. Currently on the Earth, natural uranium consists of 99.28 wt% ^{238}U and 0.71 wt% ^{235}U and natural potassium contains only 0.0119 wt% ^{40}K. As the energy liberated by a unit of mass differs between all these isotopes by at most one order of magnitude, ^{238}U and ^{232}Th are the most important radiogenic heat sources of Mars in the current era, but both ^{235}U and especially ^{40}K were more important in the early phases of Mars (see Figure 18.6).

Radioactive isotopes can be found in mantle and crust rocks. As they are incompatible elements, meaning that on partial melting of rocks they will be concentrated in the melt phase, radiogenic elements are expected to be more abundant in crustal rocks than in mantle rocks. On Earth, the concentration of radioactive elements in the crust is

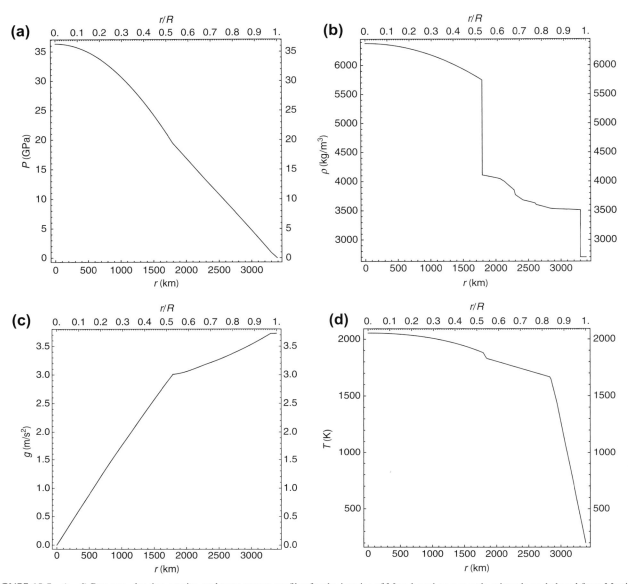

FIGURE 18.5 (a–d) Pressure, density, gravity, and temperature profiles for the interior of Mars based on a mantle mineralogy deduced from Martian meteorites (Dreibus and Wänke, 1994). At the center of the planet the pressure is larger than 35 GPa. The location of the discontinuity in the slope of the pressure profile marks the core–mantle boundary. With the exception of the two boundary layers and the stagnant lid, the temperature profile is adiabatic and decreases only slightly with increasing radius. The heat transport through the stagnant lid is the least efficient and the temperature drops by more than 1000 K from the adiabatic mantle to the surface. The jumps in the density profile are due to changes in composition or phase transitions. The largest density difference is caused by the transition from the liquid iron core to the silicate mantle. The density there drops by about 70%.

more than 10 times higher than in the mantle. Little information is available to constrain the amount of heat-producing elements in Mars. Based on analyses of meteorites from Mars, it is usually assumed that their concentration is close to that in chondritic meteorites, the most primitive type of meteorites. Data on the occurrence of radioactive elements in the crust are compatible with this assumption. For an essentially chondritic concentration the radioactive elements have up to now produced an energy of about 7×10^{29} J in Mars.

Although radiogenic isotopes keep on producing energy, the dominant source of energy for Mars is the release of gravitational energy on formation (see Section 2). When a total mass M is accreted to form Mars, a gravitational energy of about 5×10^{30} J is released and mostly converted into heat. In addition to the two major energy sources, several other phenomena contribute to Mars' heat balance. In Mars' early history, core formation released about 5% of the gravitational energy production of formation and short-lived radioactive isotopes such as ^{26}Al also

Chapter | 18 Interior Structure and Evolution of Mars

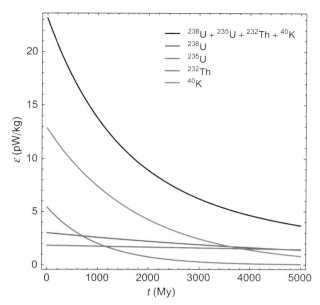

FIGURE 18.6 Time evolution of the energy produced per kilogram by the principal heat-producing radioactive elements (^{232}Th, ^{238}U, ^{40}K, ^{235}U) in the mantle of Mars and the sum of the contributions. A current concentration of 305 ppmnbsp;K, 56 ppb Th, and 16 ppb U is assumed.

delivered an important amount of energy. In the later evolution, contraction on cooling and further differentiation of the planet, in particular the formation of an inner core, release additional gravitational energy.

6.3. Heat Transport by Conduction and Convection

Heat is transported in Mars by two different mechanisms. Like in any medium with a spatial variation in temperature, heat is conducted from hot to colder regions. Conduction is a diffusive process wherein molecules transmit their kinetic energy to other molecules by colliding with them. Besides conduction, macroscopic motions in the medium can also transfer heat when flows carry material of a certain temperature into a region with a different temperature.

6.3.1. Conduction

According to Fourier's law, the conductive heat flux $q(r)$ (the flow of heat through a unit area per unit of time) is linearly proportional to the temperature gradient:

$$q(r) = -k(r)\frac{dT(r)}{dr}, \qquad (18.4)$$

where k is the coefficient of thermal conductivity. The minus sign indicates that heat flows in the direction of decreasing temperature. The temperature distribution in the crust and the upper part of the mantle of terrestrial planets like Mars is controlled by conduction. Equation (18.4) shows that the heat flux at the surface can be determined from the surface temperature gradient. Accurate measurements of the surface temperature gradient require deep drill holes. On the Earth, such measurements combined with measurements of the coefficient of thermal conductivity of crustal rocks have shown that the mean continental heat flow is 65 ± 1.6 mW/m^2. The heat flow out of Mars has up to now not been measured, but will be by the *National Aeronautics and Space Administration* (NASA) InSight mission planned for launch to Mars in 2016. From theoretical models of the thermal evolution of Mars (see Section 8) the heat flow is estimated to be about 20 mW/m^2 and the temperature increase with depth in the upper crust to be about 7 K/km.

Lord Kelvin (1862) assumed that conduction was the only heat transport mechanism in the Earth and tried to calculate the age of the Earth by considering a simple model. He approximated heat transfer in the Earth by that in an infinite half-space and assumed that surface heat flow results from the cooling of an initially hot Earth. The temperature in such a system obeys the diffusion equation

$$\frac{\partial T(x,t)}{\partial t} = \kappa \frac{\partial^2 T(x,t)}{\partial x^2}, \qquad (18.5)$$

where $\kappa = k/(\rho C)$ is the thermal diffusivity and x is the distance from the surface. A large thermal diffusivity indicates a fast adjustment of temperature to the surrounding temperature. Assuming that the Earth was initially at a hot homogeneous temperature T_i and the surface at a cold temperature T_s, a solution of Eqn (18.5) for the temperature can be expressed as

$$T(x,t) - T_i = (T_s - T_i)\,\mathrm{erfc}\left(\frac{x}{2\sqrt{\kappa t}}\right), \qquad (18.6)$$

where erfc is the complementary error function. Since erfc(z) decreases rapidly with increasing argument z—its value decreases from 1 at the origin to 0.1 for an argument of about 1.16, and is nearly equal to 0 for larger arguments—only the surface layer cools significantly (Figure 18.7). The physical reason for this behavior is that the characteristic time for conductive propagation of a temperature change over the radius of the planet, which can be expressed as R^2/κ, is much larger than the age of the solar system. For Mars, the characteristic time for cooling over its entire radius is about 100 times longer than the age of Mars. The thermal boundary layer over which Mars can efficiently cool over its lifetime is less than 1000 km. As a result, Mars, like the Earth, cannot efficiently change its deep interior temperature by conduction. A purely conductive planet can, therefore, not maintain a large heat flux over the age of the solar system. From the high observed surface heat flux, Lord Kelvin estimated the age

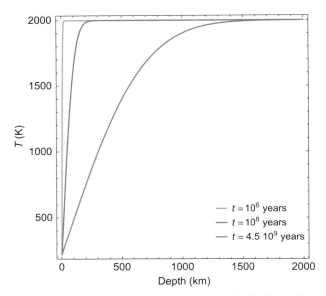

FIGURE 18.7 Temperature profiles in a purely conductive Mars without internal heat source as a function of time, starting from an initial uniform temperature of 2000 K.

of the Earth to be of the order of 100 My, much smaller than the age of the solar system as we now know. Addition of heat-producing radioactive isotopes can increase the age estimate, but the main reason why the Earth can maintain a high geothermal gradient is that the principal means of heat transfer in the deeper Earth layers is not by conduction but by convection. For Mars too, convection is the dominant heat transport mechanism in the mantle.

6.3.2. Convection

Convection in the mantle is driven by radiogenic heat sources and the cooling of Mars. A fluid that is heated from below and within and cooled from above is denser (cooler) at the top than at depth. A gravitationally unstable situation then develops and the cool fluid tends to sink and the hot fluid tends to rise due to buoyancy. As explained above, the mantle behaves as a viscous fluid on long timescales and approximately as an elastic solid on short timescales. On microscopic level, the fluidlike behavior of the mantle of terrestrial planets is due to solid-state creep processes. The dominant creep process at low stress levels is the diffusion of ions and vacancies through the crystal lattice. For a material with a Maxwell rheology, the characteristic timescale that separates predominantly elastic from predominantly viscous behavior is given by the Maxwell timescale $\tau = 2\eta/E$, where η is the dynamic viscosity and E is the Young's elastic modulus. For a viscosity of 10^{20} Pa s and $E = 70$ GPa, typical estimates for Earth mantle rocks expected in Mars, the Maxwell timescale is about 900 years. Mantle convection occurs on much longer timescales so that a viscous fluid behavior is a very good description for the mantle rocks. As a result viscosity will tend to resist the convective motion driven by buoyancy. Typical convection velocities are on the order of a few centimeters per year only, so it takes on the order of 10 My or more for a cold mass element to descend to the core–mantle boundary.

Mantle convection can be described by the laws of fluid mechanics: the equation of motion, the continuity equation expressing conservation of mass, and an energy equation expressing the change in internal energy due to heat sources, fluid flow, and conduction. These equations cannot be solved analytically for planets and are therefore solved numerically. Nevertheless, analytical solutions have been derived for some simplified cases. These studies, as well as laboratory measurements and numerical models, show that an isoviscous layer will be convecting if a certain quantity called the Rayleigh number Ra exceeds a critical value, the critical Rayleigh number Ra_{cr}. The Rayleigh number is defined for a plane-parallel layer without internal heat sources and heated from below and cooled from above as

$$\text{Ra} = \frac{\rho g \alpha \Delta T b^3}{\eta \kappa}, \quad (18.7)$$

where g is gravity and α is the coefficient of thermal expansion, describing the relative volume change with changing temperature of a material. The thickness of the layer is denoted by b and ΔT is the nonadiabatic temperature difference over the layer. The Rayleigh number can be thought of as a ratio between two timescales: a cooling timescale due to conduction and a timescale of motion due to buoyancy and viscosity. If the cooling timescale is large with respect to the timescale of motion, convection can develop ($\text{Ra} > \text{Ra}_{cr}$). If, on the other hand, conductive cooling is fast enough, the mantle can cool before convection can set in ($\text{Ra} < \text{Ra}_{cr}$). The critical Rayleigh number Ra_{cr} depends on the wavelength λ in the horizontal direction of the velocity and temperature perturbations resulting from the convective motions:

$$\text{Ra}_{cr} = \frac{\left(\pi^2 + \frac{4\pi^2 b^2}{\lambda^2}\right)^3}{\frac{4\pi^2 b^2}{\lambda^2}}. \quad (18.8)$$

The minimum Ra_{cr} for this situation with free surfaces is about 658 and is obtained at a wavelength of $2^{3/2} b$. The most unstable horizontal wavelength is therefore a few times the thickness of the layer. When convection first sets in at a Rayleigh number equal to the critical Rayleigh number, convection cells of that size will therefore develop. For larger Rayleigh number convection cells with both larger and smaller wavelengths than the most unstable will appear. For a mantle thickness of about 1500 km for Mars, a temperature increase of about 1500 K over the mantle, and the values $\eta = 10^{20}$ Pa s,

$k = 4$ W/(mK), $\kappa = 10^{-6}$ m²/s, $\alpha = 3 \times 10^{-5}$ K^{-1}, $\rho = 3500$ kg/m³, we have Ra $\approx 2 \times 10^7 \gg$ Ra$_{cr}$, strongly suggesting that the mantle of Mars is convecting.

In the convective mantle, the temperature increases approximately adiabatically with depth. An adiabatic temperature change describes how a material element adapts its temperature to the surrounding temperature when rapidly rising or sinking in the layer without exchanging energy with the surrounding fluid. For convection to set in, the temperature gradient in the layer must be larger than the adiabatic temperature gradient, as can be understood by considering the dynamics of a fluid parcel that is locally heated. The parcel expands and starts rising due to buoyancy. If it does not exchange heat with the surrounding, the fluid parcel changes its temperature and density adiabatically due to the different pressure it encounters. If the temperature gradient is larger than the adiabatic temperature gradient, the adiabatic density change of the fluid parcel will be smaller than the change in density of the environment and the parcel will have the tendency to keep on rising. Viscosity and conduction have a stabilizing effect on convective motions, as expressed by the Rayleigh number. However, due to the efficient energy transport of convection only a very small temperature difference of a few Kelvin on top of the adiabatic temperature gradient is needed for convection. As a result, the temperature gradient of convection is nearly adiabatic for a layer in convection. In the convecting part of the Martian mantle, we therefore have approximately

$$\frac{dT}{dr} = -\frac{\alpha g T}{C}. \quad (18.9)$$

This adiabatic temperature increases only slightly with depth by about 1.2 K per 10 km, much slower than in the crust where convection is absent (Figure 18.8). Over the entire range of about 1200 km in the convective mantle, the temperature increases by only about 140 K.

Heat flow by convection is much more efficient than by conduction. The heat flow for a convective planet is customarily described in terms of the Nusselt number, which is defined as the ratio of the total heat flux and the heat flux that would be transported by conduction only. It is generally expressed in terms of the Rayleigh number as

$$\text{Nu} = c \left(\frac{\text{Ra}}{\text{Ra}_{cr}}\right)^{\beta}, \quad (18.10)$$

where c and β are constants. Experimental and advanced numerical studies show that typically c is smaller than 1 and that β is approximately 1/3. Since the conducted heat flux is linearly proportional to the temperature difference over the convecting layer (Fourier's law), the total heat flux at the surface of Mars is approximately proportional to that temperature difference to the power 4/3. The larger the

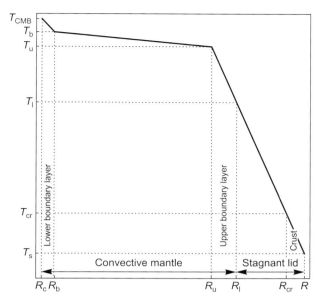

FIGURE 18.8 Sketch of the temperature in the mantle and crust of Mars. Here, R_c, R_b, R_u, R_l, R_{cr}, and R denote the radial coordinate of the core–mantle boundary, of the top of the boundary layer at the bottom of the mantle, of the bottom of the upper boundary layer, of the bottom of the stagnant lid, of the crust–mantle boundary, and of the outer surface, respectively. Temperatures at the core–mantle boundary, at the top of the boundary layer at the bottom of the mantle, at the bottom of the upper boundary layer, at the bottom of the stagnant lid, at the crust–mantle boundary, and at the outer surface are indicated by T_{CMB}, T_b, T_u, T_l, T_{cr}, and T_s, respectively.

temperature difference is over the convective system, the larger the heat flux.

Since the surface temperature is much lower than the temperatures in the convecting mantle and the temperature change in the convecting region is very small, a transition layer, or boundary layer, will develop at the top of the mantle in which temperature increases much faster than the adiabatic temperature and heat is transported by conduction. At the bottom of the mantle, a second boundary layer exists above the core (Figure 18.8). These boundaries have thicknesses of the order of several 10 km in Mars. The change in temperature in the mantle occurs mainly over the boundary layers. Since convection is only driven by a temperature difference on top of the difference in adiabatic temperature, the temperature difference considered in the Rayleigh number is approximately the sum of the temperature differences over the boundary layers. As heat in the boundary layers diffuses upward by conduction, the heat flux out of the mantle can be expressed by Fourier's law as $q = k_m \Delta T / \delta$, where k_m is the coefficient of thermal conductivity of the mantle and ΔT is the temperature difference over the boundary layer with thickness δ.

In Mars, in contrast to the Earth, the outer layers of the planet do not participate in the convection and are essentially rigid. This type of mantle convection is usually referred to as stagnant lid convection, as the upper lid is immobile or stagnates, and heat transport in the outer lid is through conduction only. In an almost isoviscous fluid with limited viscosity contrast, the whole fluid takes part in the convection. The stagnant lid convection regime is typical for planets with a large viscosity ratio between the cold surface layers and the hot mantle interior of the order of at least 10^4 between the surface layers and the deep mantle. Viscosity decreases with depth in Mars because of the increase in temperature. The decrease can be large from a cold surface to a hot interior since the dependence on temperature is exponential (viscosity can be expressed as $\eta = \eta_0 \exp((A/R)(1/T - 1/T_r))$, where T_r is a reference temperature, η_0 the viscosity at the reference temperature, R is the gas constant, and A is the activation energy of the rocks; the small dependence of viscosity on pressure for Mars is not made explicit here). For example, the viscosity in the mantle of Mars at a temperature of 1500 K is already a factor 40 smaller than at a temperature of 1300 K. Therefore, a stagnant lid will develop above the convective layer. Most of the radial viscosity variation occurs in the stagnant lid, whereas the convective layer below is approximately isoviscous as a result of its small temperature gradient. Since the stagnant lid does not belong to the convection system and the temperature increases significantly over it, the temperature difference to be considered in the Rayleigh number for the convective layer below the stagnant lid is much smaller than for the larger convective system including the upper layers if the lid would be mobile (Figure 18.8). As a result, the Rayleigh number and the Nusselt number are smaller for stagnant lid convection than for mobile lid convection and the heat flux of the mantle is also smaller for a given temperature profile. Stagnant lid convection therefore cools a planet less effectively than mobile lid convection. For most of its history, Mars did not have plate tectonics, as the old crust convincingly shows.

Both numerical and laboratory experiments have shown that the temperature difference over the thermal boundary layer beneath the stagnant lid is of the order of 100 K. For a mantle temperature of about 1500 K, this means that the stagnant lid extends to a depth at which the temperature is about 1400 K. The thickness of the stagnant lid can then be estimated from the temperature gradient to be about a few hundred kilometers thick. The stagnant lid thus consists of the crust and an upper part of the mantle.

7. GLOBAL INTERIOR STRUCTURE OF MARS

7.1. Introduction

Most information on the interior structure of the Earth is provided by seismic data. Since the propagation of seismic waves depends on material properties like density and elastic parameters, observation of seismic wave arrival times and **seismic normal-mode** frequencies informs on the Earth's interior. Since normal modes can involve motion of the whole Earth and seismic waves can penetrate down to the Earth's center, the whole interior can be sampled. For Mars, no seismic data are yet available, but NASA plans to send a lander called InSight equipped with a seismometer to Mars in 2016. In the absence of these seismic data, geodesy data provide the strongest constraints on the deep interior structure of Mars.

7.2. Global Geodesy Data

Besides surface data and data from Mars meteorites, which are related to the interior but on itself cannot determine the interior structure, the best current data available to constrain Mars' deep interior are global geodesy data derived from the gravity field, rotation, and tides of Mars. Four main data constrain the interior structure of Mars: the radius, the mass, the mean **moment of inertia**, and the tidal Love number k_2, which describes the reaction of Mars to the tidal forcing of the Sun (see below).

The mean radius of Mars $R = 3389.5 \pm 0.2$ km, defined as the radius of a sphere defining the same volume as Mars, has been determined by measuring the time needed for laser pulses to travel from the Mars Global Surveyor to the surface of Mars and back. The distance to the surface can then be calculated and by subtracting it from the spacecraft's location with respect to the center of mass of Mars, the distance of the surface from the mass center can be determined. A spherical volume with the mean Martian radius is used to describe global 1D interior structure models of Mars, although Mars itself is only nearly spherical. A better approximation for the shape of Mars is a biaxial ellipsoid flattened at the poles due to rotation, as is the case for the Earth. The mean polar radius of Mars is 3376.2 ± 0.1 km, whereas the mean equatorial radius is 3396.2 ± 0.1 km.

The gravitational field of Mars can nowadays be determined very accurately by tracking orbiting spacecraft. By precisely measuring the Doppler shift on radio links between the spacecraft and the Earth the orbital motion of the spacecraft can be modeled. From the knowledge of the orbit the gravitational field of Mars can be determined since the spacecraft's orbit depends essentially on the gravitational attraction from Mars. The most accurate representation of the external gravitational field of Mars reaches an average spatial resolution of 120 km at the surface and yields an estimate of the product GM of Mars with 10 significant digits. Since the universal gravitational constant is only known with limited precision, the mass of Mars can be determined with five significant digits only.

For a value of $G = (6.67259 \pm 0.00085)10^{-11}$ m^3/s^2/kg, the mass $M = (6.4186 \pm 0.0008)\ 10^{23}$ kg.

The moment of inertia of Mars can be determined by combining data from the gravitational field of Mars and the rotation. It can in general be expressed by a 3 × 3 matrix I_{ij} as

$$I_{ij} = \int_V \rho(\vec{r})(r^2 \delta_{ij} - x_i x_j) dV, \quad (18.11)$$

where $\vec{r}$ is the position vector of a point P in Mars with respect to the origin of the Cartesian coordinate system, chosen here to be at the mass center of Mars, r is the distance to the mass center, x_i and x_j are the i and j coordinates, and δ_{ij} is the delta Kronecker, which is equal to 1 when $i = j$ and else equal to 0. Three principal moments of inertia exist which express the moment of inertia of Mars with respect to the three principal axes of inertia. When the three coordinate axes are chosen along these three axes of inertia, the moment of inertia matrix I_{ij} can be expressed entirely in terms of the three principal moments of inertia A, B, and C, describing the moment of inertia with respect to the three axes. We have

$$A = \int_V \rho(\vec{r})(y^2 + z^2) dV,$$

$$B = \int_V \rho(\vec{r})(x^2 + z^2) dV, \quad C = \int_V \rho(\vec{r})(x^2 + y^2) dV.$$

$$(18.12)$$

The mean moment of inertia $I = (A + B + C)/3$ is the third quantity used to constrain the interior structure.

The gravitational field can be related to differences in moments of inertia if it is described with respect to the inertia axes defined above. If we restrict ourselves to a model that can describe a triaxial ellipsoidal shape, the external gravitational field Φ can be expressed as

$$\Phi(\vec{r}) = -\frac{GM}{r}\left[1 + \left(\frac{a}{r}\right)^2 \left(\frac{(A+B)/2 - C}{Ma^2} \cos(m\phi)\right.\right.$$
$$\left.\left. \times P_{20}(\cos\theta) + \frac{B-A}{4Ma^2}\sin(m\phi)P_{22}(\cos\theta)\right)\right],$$
$$(18.13)$$

where a is the length of the longest equatorial principal axis, θ is colatitude, ϕ is longitude, and P_{20} and P_{22} are two associated Legendre functions. Since the gravitational field is accurately known, the moment of inertia differences in Eqn (18.13) can be determined precisely. In order to be able to determine the three principal moments of inertia and the mean moment of inertia, a third equation is needed and can be obtained from the rotation of Mars.

Like the Earth, Mars performs a **precessional** motion in space on a long timescale. Its orientation in space slowly changes, like a spinning top, and the tip of its polar axis would describe a large cone in space every 171,000 years if the orbital plane of Mars were to be fixed in space. During this motion, the obliquity, or the angle between the polar axis and the normal to the orbital plane, remains constant. Precession is due to the gravitational torque of the Sun on Mars, which differs from zero because Mars is flattened at its poles. The precession rate is related to the three moments of inertia in the following way. It is proportional to the difference of the polar and mean equatorial moment of inertia since this difference is a measure of the polar flattening and it is inversely proportional to the polar moment of inertia because that quantity describes the rotational inertia for Mars rotating around its polar axis.

Since a change in the orientation of Mars also changes its external gravitational field, precession can be measured by tracking the orbital motion of a spacecraft around Mars. More directly it can be measured by recording the Doppler shifts on radio signals between a lander on Mars and the Earth. The best current value for the precession rate is $(7594 \pm 10) \times 10^{-3}$ arcsec/year. This estimate has been determined from the Doppler and range data to the Mars Pathfinder lander and the Viking landers and from radio tracking Mars Global Surveyor, Mars Odyssey, and the Mars Reconnaissance Orbiter. The polar moment of inertia factor $C/(MR^2)$ of Mars derived from the precession rate is then $C/(MR^2) = 0.3644 \pm 0.0005$. Since the polar moment of inertia is much less precisely known than the moment of inertia differences from the gravitational coefficients, the mean moment of inertia factor is known with similar precision as the polar moment of inertia factor: $I/(MR^2) = 0.3645 \pm 0.0005$.

The Sun raises tides on Mars because the acceleration induced on each mass element due to the Sun's gravitational force is not equal to the acceleration of the mass center of Mars responsible for the orbital motion on account of the different distance to the Sun for different positions in Mars. As Mars is not a rigid body, it deforms in response to this differential gravitational force. The deformations or solid body tides are smaller than for the Earth because Mars is further away from the Sun and is smaller than the Earth. Moreover, the Moon raises larger tides on the Earth than the Sun. The radial tidal displacement on the surface of Mars is of the order of 1 cm only compared to about 40 cm for the Earth. Currently, this is too small to be measured directly by either a laser altimeter on board an orbiter or by radio tracking of a lander. Nevertheless, tides have been observed indirectly by their effect on the orbital motion of spacecraft. As tides slightly change the mass distribution in Mars, the external gravitational potential changes and so does the spacecraft orbit. The Love number k_2 describes the reaction of Mars to the tidal forcing and is defined by

$$\delta\Phi(\vec{r},t) = \left[1 + k_2\left(\frac{R}{r}\right)^3\right]\Phi_t(r,t). \quad (18.14)$$

Here, t is time, $\delta\Phi$ is the change in the external gravitational potential due to the tides, and we have restricted the tidal potential Φ_t to its degree-two part, which is a factor d/R, where d is the distance between the Sun and Mars, larger than the degree-three part. By measuring small secular changes in the orbital inclination of Mars Global Surveyor, Mars Odyssey, and the Mars Reconnaissance Orbiter, the Love number has been determined to be $k_2 = 0.164 \pm 0.009$.

7.3. Model Results

Viable models of the interior structure of Mars must be consistent with the four observed quantities mass, radius, mean moment of inertia, and Love number k_2. Obviously, four quantities are not sufficient to determine all details of the interior structure and many properties cannot be constrained by them. The latter property can be used to our advantage by allowing to simplify the model. For example, the detailed structure of the crust has only a small effect on the moment of inertia and Love number. As a result, in modeling the global interior structure of Mars, the crust can be described by two parameters only: its mean thickness and mean density.

In the lowest order approach useful to understand the Martian interior, it is even possible to neglect the crust and to consider that Mars consists of only two layers with a homogeneous density: the mantle and the core. Since the radius is well known, such a two-layer model is described by three parameters: the radius R_c and density ρ_c of the core and the density ρ_m of the mantle. The mean density ρ and moment of inertia I of such a Mars model can then be expressed as

$$\rho = \rho_m + (\rho_c - \rho_m)\left(\frac{R_c}{R}\right)^3 \quad (18.15)$$

$$\frac{I}{MR^2} = \frac{\rho_m}{\rho} + \frac{\rho_m - \rho_m}{\rho}\left(\frac{R_c}{R}\right)^5 \quad (18.16)$$

The Love number k_2 is not uniquely determined for a basic two-layer model because it also requires knowledge of the rigidity of the layers. The rigidity for silicate rocks in the Martian mantle is thought to be in the range 70–80 GPa. Since both the moment of inertia and the Love number k_2 are determined with a certain error and the mantle rigidity is also not exactly known, the geodesy data cannot uniquely determine the interior structure of Mars, not even for a basic two-layer model. Nevertheless, the data show that the

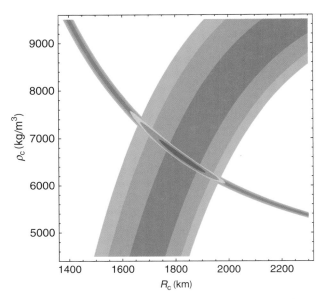

FIGURE 18.9 Average core density as a function of core radius for the two-layer model for the interior structure of Mars. The thin region that decreases with increasing core radius represents the constraint of the moment of inertia on the core density and radius. The thicker region that increases with increasing core radius is the constraint provided by the Love number k_2, which is less accurately known than the moment of inertia. The colored areas represent the confidence areas of the solutions (decreasing confidence from darker to lighter shades). The areas delimited by the blue shaded colors represent the joint solution of both constraints on the core radius and core average density.

radius of the core of Mars is 1840 ± 65 km and that the density of the core is 6715 ± 270 kg/m³ (at 1σ, see Figure 18.9). The Love number puts the strongest constraint on the core because tides are highly sensitive to the distance of the liquid core from the surface. The results show that the core of Mars is about equally large as that of the Earth relative to the total size of the planet.

The core density is much smaller than the density of pure iron at core pressure and temperature. At a pressure of 30 GPa and a temperature of 2000 K, the density of pure iron is 8670 kg/m³, which is 29% higher than the mean estimated density, suggesting a large amount of light elements in the core. More realistic and sophisticated models confirm the above findings for a basic two-layer model. They indicate that the core radius is 1794 ± 65 km (at 1σ) and that the core density is 6265 ± 200 kg/m³. The smaller core radius and density than for the two-layer model is mainly due to the neglect of the crust and the assumption of homogeneous layers in the simplified model. If it is assumed that sulfur is the only light element in the core, the sulfur concentration is 16 ± 2 wt%, which is a larger light element concentration than that for the Earth. These high sulfur concentrations confirm results obtained from compositional studies of Mars, although much smaller concentrations have also been proposed for Mars' core. As a result of the large sulfur concentration in the core, the melting temperature of the core is strongly reduced with respect to a pure iron composition and is below the expected temperatures at the core–mantle boundary for current models of the thermal evolution of Mars (see next section). As a consequence, the core of Mars is not expected to be differentiated into a solid inner core and a liquid outer core as for the Earth, but to be entirely liquid. A lower temperature in Mars than expected or a different chemical composition of the core, with a large amount of other light elements than sulfur, which has the strongest effect on the reduction of the melting temperature of an iron alloy, might, however, be consistent with an inner core. Nevertheless, the absence of an inner core is in agreement with the absence of a global magnetic field, which is thought to require a growing inner core.

The mantle density and composition cannot be determined from the global data, although the estimate of the core radius suggests that the mantle of Mars does most likely not contain minerals occurring in the lower mantle of the Earth. Mars' core seems to be too large to have a lower mantle like the Earth, although a thin lower mantle cannot be excluded.

8. EVOLUTION OF MARS

8.1. Thermal Evolution

Mars formed hot and differentiated into a mantle, a core, and a crust (see Section 2). The subsequent evolution is much less dramatic: during the longest part of its evolution Mars slowly cools altering only slightly and gradually its core, mantle, and outer layers. Obviously, Mars can only cool since it loses more energy than it produces in its interior, implying that its internal energy transport is effective in carrying heat from the deep interior to the surface. If we only consider the major energy sources, the energy lost from the solid surface of Mars per second, or total surface heat flow L, is equal to the difference of the total energy production per second and the sum of the internal energy U, consisting of the heat content and the strain energy, and the gravitational energy E_G:

$$L = \int_0^M \varepsilon(m) dm - \frac{d}{dt}(U + E_G) \qquad (18.17)$$

Here, t is time and dm an infinitesimal mass element. Compression of Mars as a result of cooling increases the internal energy and decreases the gravitational energy, but it can be shown that these contributions almost cancel. Therefore, the total surface heat flow is approximately equal to the sum of the energy production per second by heat-producing radioactive isotopes and the loss of thermal energy.

If radiogenic heating were the only heat source of the planet and were to supply energy at a constant rate, a balance between the internal energy source and the loss of energy through the surface could exist and Mars would not need to cool. However, the radioactive isotopes decay so that the amount of heat-producing elements decreases with time. Since the internal heat produced in Mars declines with time, the surface heat flux also declines with time. The heat transported by convection thus diminishes, and therefore the temperature decreases with time, as follows from the relation between the Nusselt number and temperature. Mars therefore necessarily cools. About 30% of the heat flux of Mars is estimated to be due to cooling. A value of 25–30% is characteristic for planets with stagnant lid convection and is smaller than for the Earth, which has plate tectonics and cools more efficiently.

The cooling of Mars can be described by considering the changes in energy in the core and the mantle. Since the specific heat describes the amount of energy needed to raise the temperature of 1 kg of a material by 1 K, the thermal energy of the core is equal to the product of the mass M_c, the mean specific heat C_c, and the mean temperature T_c of the core. The change in this energy with respect to time is due to heat loss to the mantle, which can be expressed as

$$M_c C_c \frac{dT_c}{dt} = 4\pi R_c^2 q_c, \qquad (18.18)$$

where q_c is the heat flux out of the core. If Mars was to have an inner core, this equation would have to be extended in order to account for the latent heat and gravitational energy release of inner core formation.

For the mantle several additional energy sources and sinks have to be taken into account in the energy equation. The mantle gains energy from the decay of radioactive isotopes and the flow of heat from the core into the mantle. When mantle rock melts and solidifies or undergoes mineral phase changes, latent heat is also consumed and released. Moreover, the mantle loses heat through its upper surface. The energy equation for the mantle expresses that the change in internal energy is equal to the sum of the energy gains and the energy losses. If we neglect the contribution of the latent heat, the energy equation for the mantle can be expressed as

$$M_m C_m \frac{dT_m}{dt} = M_m \varepsilon + 4\pi R_c^2 q_c - 4\pi R_m^2 q_m, \qquad (18.19)$$

where T_m is the mean mantle temperature, C_m is the mean specific heat of the mantle, and M_m is the mass of the mantle. The two energy equations for the core and the mantle are the basic equations for the study of the thermal evolution of Mars. For a given initial temperature in the core and the mantle, they describe the evolution of the temperature with time.

The current heat flux out of Mars has not yet been observed and can therefore not be used to constrain the thermal evolution. However, this situation will improve in

2016 when the NASA InSight mission is foreseen to be launched to Mars. This mission carries a heat probe that will measure the heat flow in the upper few meters of Mars at the landing location. For the Earth, the current mean total surface heat flow is known to be about 45 TW. From this value, an upper limit to the decrease in temperature over recent times can easily be estimated by neglecting the contribution of the radioactive isotopes. The total change in internal energy of the Earth over the last billion year, $MC\Delta T$, where ΔT is the change in temperature, must be equal to the product of 1 billion year and the total surface heat flow. We here assume that the heat flow has not changed much, as thermal evolution models indicate. It then follows that the internal temperature of the Earth has decreased approximately by about 200 K over the last billion years. The real temperature decrease is somewhat less than half that value since about half or somewhat less of the total surface heat flow comes from the radioactive elements. For Mars, the heat flux from the surface is not yet known, but the temperature decrease can be estimated from the energy produced by the radioactive elements. For an essentially chondritic concentration, the radioactive elements currently produce about 2 TJ per second. Since about 30% of the heat flow, or about 0.86 TW, is due to cooling, the temperature reduction over the past billion years is about 35 K/Gyr, which corresponds to estimates of cooling rates from geochemical observations of the melting history in the mantle.

Since the initial temperature of Mars was probably high (see Section 2), both the mantle and the core were vigorously convecting shortly after core formation. Even if the mantle was not fully molten, the viscosity of the mantle rocks was low because of the high temperature. The Rayleigh number is then high and convection can support a high heat flux (high Nusselt number). As a result, the planet initially cools rapidly (Figure 18.10). In case of a mantle overturn after magma ocean solidification, the onset of subsolidus mantle convection may be delayed as a result of the gravitationally stable configuration of the mantle produced by the overturn. The phase of fast cooling lasted only a few hundred million years until the lower temperature led to a larger viscosity resulting in a lower Rayleigh number and less vigorous convection. Analyses of meteorites from Mars have shown that chemical heterogeneities in the silicate part are still present indicating that mantle convection has not been sufficiently vigorous to homogenize the mantle.

In the phase of slow cooling, which is still ongoing, the mantle is in a state in which temperature and viscosity take values that facilitate the removal of the heat produced by radioactivity plus part of the primordial heat. If less heat would be transported than is created internally, the mantle temperature would rise, viscosity would decrease, and more heat would be transported until at least as much

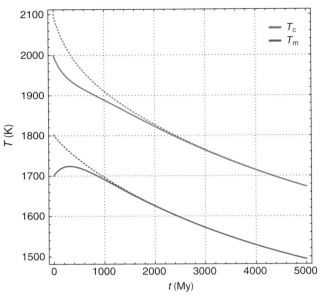

FIGURE 18.10 Temporal evolution of the average mantle (T_m) and core temperature (T_c) of Mars for cold (continuous curves) and hot (dotted curves) initial temperatures. The reference viscosity of the hot initial temperature model is 20% larger than of the cold model.

heat would be transported as is produced. It is often argued that the exact initial temperature is not very important for the present-day evolution of Mars since, for any hot initial temperature, Mars will have cooled rapidly and reached the slow cooling regime in a short timescale. The higher the initial temperature is and the lower the initial viscosity, the faster will Mars cool in its early phase. Mars therefore has little memory of its initial thermal state. Nevertheless, this principle does not always apply, and temperature evolutions starting from different initial temperatures do not always converge and lead to different present-day temperatures, for example, for low Rayleigh numbers or low initial temperatures.

Mantle viscosity is very important for the thermal evolution of Mars. However, it is not well known and the reference mantle viscosity η_0 may vary over several orders of magnitude. In particular, viscosity increases with increasing grain size of the rocks and water reduces significantly the viscosity of mantle rocks. The grain size of the upper mantle of the Earth is about 1 mm but is not well constrained for Mars. For dry olivine, an increase in grain size from 1 mm to 1 cm increases the reference viscosity by a factor of about 300. The viscosity of rock with a water content of the Earth's mantle is about two orders of magnitude smaller than the viscosity of dry rock. Studies of Martian meteorites show water contents ranging from as low as 1–1000 ppm of water per silicon atom. Higher reference viscosities, for example, due to less water or larger grain size, imply more resistance to convection and therefore less vigorous convection and a lower heat flux (smaller Rayleigh number and smaller Nusselt number). In order for Mars to be able to transfer the heat generated by the radioactive elements to the surface, a higher mantle temperature is needed. For an increase in the reference viscosity of one order of magnitude, the expected current mantle temperature is about 100 K higher, independent of the initial temperature. The cooling rate in the slow cooling phase is, however, almost independent of the reference viscosity because the same energy produced by the radioactive isotopes must be transferred and the relative contribution of cooling to the heat flux has a characteristic value of about 30%.

If the core cools sufficiently, the local temperature somewhere in the core will reach the freezing temperature of the core material. At that depth in the core material will start to solidify, initiating the formation of a solid iron core. Models of the thermal evolution of Mars indicate that the present-day temperature at the core—mantle boundary could be as low as 1500 K, but higher values up to over 2000 K are also not excluded. For a core composed of iron and sulfur and assuming that sulfur makes up about 16% of the mass of the core as indicated by the geodesy measurements, the core—mantle boundary of Mars still has to cool at least by about 200 K before an inner core will form. Since the cooling rate is only at most 50 K per billion year, the formation of an inner core is expected to start only in at least about 4 Gyr from now. However, if the core composition is different and other light elements are present or even contribute more to the core mass than sulfur, the melting temperature of the core can be several hundred Kelvin higher. In that case, the core could already have formed an inner core, although that might be difficult to reconcile with the observations that Mars does not contain a global magnetic field since such a field is supposed to develop once an inner core starts growing because the production of less dense liquid core material enriched in light elements at the inner core boundary is a driver for convection in the outer core. Nevertheless, a sufficiently large increase in light element concentration of the liquid at the inner core boundary is required for the maintenance of a dynamo, and this depends on how light elements partition between the solid and the liquid. At the Mars core pressures and temperatures, almost all sulfur remains in the liquid on solidification for sulfur concentrations below the eutectic concentration, but this is not the case for all other light elements.

When Mars cools, the stagnant lid thickens because less heat flows into the stagnant lid. The layers below the base of the stagnant lid will then cool, thereby thickening the lid. The upper mantle layers that become part of the stagnant lid during the evolution of Mars cool much more than the deeper mantle layers, which stay relatively hot. During the past 4 Gyr, the stagnant lid of Mars has steadily grown and may now be several hundred kilometers thick.

8.2. Early Dynamo

Based on first observations of magnetic patterns in the Southern highlands with the magnetometer onboard the Mars Global Surveyor, it has been suggested that plate tectonics might have existed on early Mars. A consequence of early plate tectonics is that it could have driven a core dynamo. Early plate tectonics is, however, difficult to reconcile with the age of the crust. During the transition from plate tectonics to stagnant lid convection the mantle would heat, leading to increased partial melting and substantial crust formation. This peak in crust production would happen at a later time than the bulk of the crust is thought to have formed, rendering the hypothesis of early plate tectonics unlikely.

Without plate tectonics, the deep interior of Mars does not cool sufficiently fast to sustain a purely thermally driven dynamo. An early magnetic field on Mars must, however, have been present since otherwise the observation of remanent magnetization of Martian surface rocks cannot be explained. The magnetization was acquired in the first 500 My of Mars' history when rocks crystallized in the presence of a magnetic field. Younger rocks do not show any magnetization as the dynamo had already shut down. A more likely possibility is that the release of gravitational energy associated with core formation heated the core to temperatures a few hundred Kelvin higher than the temperature at the bottom of the mantle. In that case, the core could cool rapidly enough to be convecting, providing a sufficiently large source of kinetic energy to drive the dynamo. Alternatively, a large impactor penetrating to the core might have superheated the core and triggered core convection and a dynamo. Such dynamos could only operate for a short time, at most a few hundred million years because of the fast energy transport and cooling. Once the decreasing heat flux out of core reaches the value that can be conducted through the iron core at the temperature gradient corresponding to an adiabatic profile, the dynamo shuts down.

8.3. Chemical Evolution

8.3.1. Crust Formation

During its cooling history, the mantle of Mars chemically evolves because of partial melting. When mantle material ascends beneath the stagnant lid as a result of mantle convection, both the temperature and the melting temperature experienced by the upwelling material decrease because of the lower pressure exerted on the rocks. Since the temperature in the rising material element decreases slower than the melting temperature, the rocks can reach a region where their temperature is higher than the melting temperature and partially melt. Only part of a rock melts because rock is composed of several minerals, each with a different melting temperature. Through volcanism the molten rock can rise above the base of the upper thermal boundary layer and will recrystallize in the crust or at the surface, thereby contributing to the growth of the crust.

For rocks to partially melt, the temperature in Mars must be higher than the solidus temperature. As a result of cooling the occurrence of partial melting therefore decreases with time. Moreover, the thickening of the stagnant lid also moves the melt zone to greater depths. As the melting temperature increases faster with depth than the mantle temperature, melting is less likely in deeper mantle layers. Because most of the crust was formed in the first 100 My, partial melting must have been strongly reduced from then on. Initial mantle temperatures well above 2000 K and high Rayleigh numbers are thought to be implausible as they would lead to a thicker crust than observed and might even destroy the crustal dichotomy. The state of the crust therefore sheds light on the initial conditions of Mars.

Locally, volcanism has, nevertheless, continued to create new crust, in particular in the Tharsis region, with evidence of lava flows not older than a few million years. Volcanism requires melting and penetration of melt into the stagnant lid, which becomes increasingly unlikely with increasing stagnant lid thickness but local melting due to hot upwelling mantle plumes seems not excluded. Local magmatism observed in the Tharsis region might also be a consequence of the locally thick crust. If the crust is sufficiently enriched in radioactive elements and has a significantly lower thermal conductivity than the mantle, the crust locally better insulates the mantle leading to higher local temperatures in the upper mantle and partial melting.

8.3.2. Extraction of Water

Besides crustal formation, partial melting also leads to the extraction of incompatible elements, such as the radioactive elements and also water. About 40–80% of the water of the mantle is thought to be removed in the first 500 million years and probably outgassed through extrusive volcanism. The loss of water from the mantle increases the mantle viscosity, resulting in less vigorous convection. The chemical evolution of the mantle, which is driven by the cooling, therefore also affects the thermal evolution.

Although the initial water concentration is not known, estimates of the volume of water outgassed from the mantle correspond to a global water layer of tens to hundreds of meters on the surface of Mars. This water extracted in the first half billion year may have been important for the early surface conditions and atmosphere of Mars, and therefore for the habitability of early Mars.

BIBLIOGRAPHY

Breuer, D., & Moore, W. B. (2007). Dynamics and thermal history of the terrestrial planets, the Moon, and Io. In G. Schubert, & T. Spohn (Eds.), *Planets and Moons: Vol. 10. Treatise on geophysics* (pp. 299–348).

Davies, G. F. (2010). *Mantle convection for geologists*. Cambridge, UK: Cambridge University Press.

Dreibus, & Wänke. (1985). Mars, a volatile-rich planet. *Meteoritics, 20*, 367–381.

Grott, M., Baratoux, D., Hauber, E., Sautter, V., Mustard, J., Gasnault, O., et al. (2013). Long-term evolution of the crust-mantle system of Mars. *Space Science Reviews, 174*, 49–111.

Jaupart, C., Labrosse, S., & Mareschal, J.-C. (2007). Temperatures, heat and energy in the mantle of the Earth. In G. Schubert, & D. Bercovici (Eds.), *Mantle dynamics: Vol. 7. Treatise on geophysics* (pp. 253–303).

Konopliv, A. S., Asmar, S. W., Foiles, S. M., Karatekin, Ö., Nunes, D. C., Smrekar, S. E., et al. (2011). Mars high resolution gravity fields from MRO, Mars seasonal gravity, and other dynamical parameters. *Icarus, 211*(1), 401–428.

Morbidelli, A., Lunine, J. I., O'Brien, D. P., Raymond, S. N., & Walsh, K. J. (2012). Building terrestrial planets. *Annual Review of Earth and Planetary Sciences, 40*, 251–275.

Nimmo, F., & Tanaka, K. (2005). Early crustal evolution of Mars. *Annual Review of Earth and Planetary Sciences, 33*, 133–161.

Rivoldini, A., Van Hoolst, T., Verhoeven, O., Mocquet, A., & Dehant, V. (2011). Geodesy constraints on the interior structure and composition of Mars. *Icarus, 213*, 451–472.

Schubert, G., Turcotte, D. L., & Olson, P. (2001). *Mantle convection in the Earth and planets*. Cambridge, UK: Cambridge University Press.

Sohl, F., & Schubert, G., Interior structure, composition, and mineralogy of the terrestrial planets. In G. Schubert, T. Spohn (Eds.), *Treatise on geophysics, Vol. 10. Planets and Moons*, (pp. 27–68).

Stacey, F. D., & Davis, P. M. (2008). *Physics of the Earth*. Cambridge, UK: Cambridge University Press.

Verhoeven, O., Rivoldini, A., Vacher, P., Mocquet, A., Choblet, G., Menvielle, M., et al. (2005). Interior structure of terrestrial planets. I. Modeling Mars' mantle and its electromagnetic, geodetic and seismic properties. *Journal of Geophysical Research (Planets), 110*(E4), E04009.

Wieczorek, M. A. (2007). Gravity and topography of the terrestrial planets. In G. Schubert, & T. Spohn (Eds.), *Planets and Moons: Vol. 10. Treatise on geophysics* (pp. 165–206).

Wänke, H., & Dreibus, G. (1994). Chemistry and accretion history of Mars. *Royal Society of London Philosophical Transactions Series A, 349*, 285–293.

Chapter 19

Mars: Landing Site Geology, Mineralogy, and Geochemistry

Matthew P. Golombek
Jet Propulsion Laboratory, California Institute of Technology, Pasadena, CA, USA

Harry Y. McSween
Department of Earth & Planetary Sciences, University of Tennessee, Knoxville, TN, USA

Chapter Outline

1. Introduction to Mars Exploratiion — 397
2. Landing Sites on Mars — 400
3. Mars Landing Sites in Remotely Sensed Data — 404
   3.1. Surface Physical Properties — 404
   3.2. Global Compositional Units — 407
4. Landing Site Geology — 409
   4.1. Introduction — 409
   4.2. Rocks — 409
   4.3. Outcrops — 410
   4.4. Soils — 410
   4.5. Eolian Deposits — 411
   4.6. Craters — 412
5. Landing Site Mineralogy and Geochemistry — 412
   5.1. Rocks — 412
   5.2. Soils — 416
6. Implications for the Evolution of Mars — 417
   6.1. Origin of Igneous Rocks — 417
   6.2. Chemical Evolution and Surface Water — 417
   6.3. Weathering on Mars — 417
   6.4. Eolian Processes — 418
   6.5. Geologic Evolution of the Landing Sites and Climate — 418
   6.6. Implications for a Habitable World — 419
Bibliography — 419

1. INTRODUCTION TO MARS EXPLORATIION

Most of our detailed information about the materials that make up the Martian surface comes from in situ investigations accomplished by seven successfully landed spacecraft (Table 19.1). The focus of these spacecraft and the era in which they explored Mars have varied, but all have been preceded by orbiters that acquired remote sensing data that helped frame the questions they addressed and the locations where they landed. The first successful landings were the Viking spacecraft in 1976, part of two orbiter/lander pairs that were launched in 1975. These landers were preceded by a number of failed Soviet and United States spacecraft, several successful "flyby" missions, and the Mariner 9 orbiter that provided basic imaging and spectral information that gave an early view of the surface and atmosphere of Mars. The overriding impetus for the Viking landers was to determine if life existed on Mars. Both immobile, legged landers carried sophisticated life detection experiments as well as imagers, seismometers, atmospheric science packages, and magnetic and physical properties experiments. The Viking mission was done in the post-Apollo era (after 1972) and involved a massive mobilization of engineering and scientific talent (as well as a budget befitting a major mission). Both landers carried arms with scoops that collected soil and fed them into the life detection experiments. No unequivocal evidence for life was found in the soil (although gases released from the soil suggested a significant oxidizing component), but the spacecraft imaged the landing sites, determined the chemistry of the soils, and provided a long record of surface meteorology.

The successful landings and operations of the orbiters (that lasted years) set the stage for the systematic study of Mars that fueled our modern view of the Red Planet. The spacecraft also left a legacy for landing using aeroshields and supersonic parachutes that have been

TABLE 19.1 Landing Sites on Mars

Site	Latitude (Degree +N)	Longitude (Degree +E)	Elevation (km, MOLA)	Region
Viking Lander 1 (VL1)	22.27	311.81	−3.6	Chryse Planitia
Viking Lander 2 (VL2)	47.67	134.04	−4.5	Utopia Planitia
Mars Pathfinder (MPF)	19.09	326.51	−3.7	Ares Vallis
MER Spirit (SPI)	−14.57	175.47	−1.9	Gusev Crater
MER Opportunity (OPP)	−1.95	354.47	−1.4	Meridiani Planum
Phoenix (PHX)	68.22	234.25	−4.1	Vastitas Borealis (high northern plains)
MSL Curiosity (MSL)	−4.59	137.44	−4.5	Gale Crater

MOLA, Mars Orbiter Laser Altimeter on Mars Global Surveyor.

employed by all subsequent landers. Orbiter and lander data defined the atmosphere and basic geology of Mars. By mapping morphologic units and crater density, three main eras of Martian geologic history were defined: Noachian (>3.6 billion years ago), Hesperian (3.6–3.0 billion years ago), and Amazonian (since 3.0 billion years ago). The mapping showed that Mars was very geologically and tectonically active during the Noachian, with decreasing activity into the Hesperian and Amazonian. The Viking orbiters returned images of valley networks and eroded ancient craters in Noachian terrain that suggested an earlier wetter and possibly warmer environment and the onset of freezing conditions in the Hesperian, leading to the present climate in the Amazonian that is generally too cold and thin (and dry) to support liquid water (current atmospheric pressure and temperature are so low that water is typically stable only in solid and vapor states).

The Mars Pathfinder (MPF) mission, launched 20 years later in 1996, was an engineering demonstration of a low-cost lander and small mobile rover and on landing on July 4, 1997, ushered in our modern era of Mars exploration. The spacecraft was a small free flyer that used a Viking-derived aeroshell and parachute, but employed newly developed robust airbags surrounding a tetrahedral lander, rather than retrorockets and legged landers as did Viking. The lander carried a stereoscopic color imager (Imager for Mars Pathfinder (IMP)), which included a magnetic properties experiment and wind sock and an atmospheric structure and meteorology experiment. The 10-kg, microwave-size rover (Sojourner) carried engineering cameras, 10 technology experiments, and an Alpha Proton X-ray Spectrometer (APXS) for measuring the chemical composition of surface materials, and conducted 10 technology experiments. The MPF lander and rover operated on the surface for about 3 months (well beyond their design lifetime) and the rover traversed about 100 m around the lander, exploring the landing site and characterizing surface materials in a couple of hundred square meter area. Rocks analyzed by the APXS appeared relatively high in silica, similar to andesites; tracking of the lander fixed the spin pole and polar **moment of inertia** that indicates a central metallic **core** and a **differentiated** planet, and the atmosphere was observed to be quite dynamic with water ice clouds, abruptly changing near-surface morning temperatures, and the first measurement of small wind vortices or dust devils. The mission captured the imagination of the public, garnered front-page headlines during the first week of operations, and became one of NASA's most popular missions as the largest Internet event in history at the time. Much of the flight system, lander, and rover design were used for the next two successful landings.

Launching before MPF, but arriving later, was the Mars Global Surveyor (MGS) orbiter, which was a partial reflight of instruments on the Mars Observer orbiter that was lost when attempting to enter into orbit around Mars in 1993. This spacecraft defined the global topography and magnetic field and identified different rock types and minerals that make up the surface. It also identified layered sedimentary rocks in high-resolution images suggesting deposition in standing bodies of water and fresh gullies suggesting recent flow of liquid water. Mars Odyssey (2001) followed MGS and the failed Mars Climate Orbiter and Mars Polar Lander launched in 1999. Instruments on Odyssey identified ground ice at high latitudes, produced the highest resolution global image mosaic (100 m/pixel) to date, and with MGS improved our knowledge of the atmosphere and global physical and mineralogical properties of surface materials by measuring their thermal properties and infrared spectral characteristics.

TABLE 19.2 Instruments Used to Process and Analyze Rocks and Soils at Spacecraft Landing Sites

Alpha Particle X-ray Spectrometer (APXS) on Mars Exploration Rovers and MSL: measures rock elemental chemistry using interactions of alpha particles with the target

Alpha Proton X-ray Spectrometer (APXS) on MPF: measured rock elementary chemistry, using interactions of alpha particles and protons with the target

ChemCam on MSL: fires a laser and analyzes the elemental abundances of vaporized areas on rocks and soils

ChemMin on MSL: a powder X-ray diffraction instrument used to identify minerals

Gas Chromatograph/Mass Spectrometer (GCMS) on Viking: instruments that analyzed chemical compounds in soils

IMP: a lander-mounted digital imaging system for stereo, color images and visible near-infrared reflectance spectra of minerals

Mars Hand Lens Imager (MAHLI) on MSL: a camera that provides close-up views of the textures of rocks and soil

Mast Camera (MASTCAM) on MSL: a digital imaging system for stereo color images and visible near-infrared reflectance spectra of minerals

MI on Mars Exploration Rovers: a high-resolution camera used to image textures of rocks and soil

Microscopy, Electrochemistry, and Conductivity Analyzer (MECA) on Phoenix: includes a wet chemistry laboratory, optical and atomic force microscopes, and a thermal and electrical conductivity probe

Mini-TES on Mars Exploration Rovers: identifies minerals via thermal infrared spectral characteristics produced by crystal lattice vibrations

MB on Mars Exploration Rovers: identifies iron-bearing minerals and distribution of iron oxidation states by measuring scattered gamma rays

Pancam on Mars Exploration Rovers: digital imaging system for stereo color images and visible near-infrared reflectance spectra of minerals

RAT on Mars Exploration Rovers: brushes or grinds rock surfaces to reveal fresh interiors

SAM on MSL: suite of three instruments (mass spectrometer, gas chromatograph, tunable laser spectrometer) used to identify carbon compounds and to analyze hydrogen, oxygen, and nitrogen

Sampling System (SA/SPaH) on MSL: includes a drill, brush, soil scoop, and sample processing device

Surface Stereo Imager (SSI) on Phoenix: digital imaging system for stereo color images and visible near-infrared reflectance spectra of minerals

Thermal and Evolved Gas Analyzer (TEGA) on Phoenix: furnace and mass spectrometer to analyze ice and soil

X-ray Fluorescence Spectrometer (XRFS) on Viking: instrument that analyzed elemental composition of soils

The Mars Exploration Rover (MER) mission landed twin golf cart-sized rovers in early 2004 that have explored over 40 km of the surface at two locations. Each rover carried a payload that contains multiple imaging systems including the color, stereo Panoramic Camera (Pancam) and Miniature Thermal Emission Spectrometer (Mini-TES) for determining mineralogy. The rovers also carried an arm that can brush and grind away the outer layer of rocks (the Rock Abrasion Tool (RAT)) and can place an APXS, Mössbauer Spectrometer (MB), and Microscopic Imager (MI) against rock and soil targets (Table 19.2). The rover and payload partially mimics a field geologist, being able to identify interesting targets using the remote sensing instruments (a field geologist's eyes), rove to those targets (legs), remove the outer weathering rind of a rock (equivalent to a rock hammer), and identify the rock type (equivalent to a geologist's hand lens and analysis in the laboratory) using the chemical composition (APXS), iron mineralogy (MB), and rock texture (MI). These rovers have lasted years (well beyond their 3-month design lifetime) and returned a treasure trove of basic field observations along their traverses as well as sophisticated measurements of the chemistry, mineralogy, and physical properties of the rocks and soils encountered. They have returned compelling information that indicates an early wet and likely warm environment on Mars.

Mars Express, the first European Space Agency mission, also carried the British Beagle 2 exobiology lander to Mars, arriving in late 2003. Although the lander was not successful, the orbiter has observed Mars for almost 10 years. Mars Express carries imagers, imaging spectrometers, radar sounders, and atmosphere and exosphere sensors. Stereo color images have refined the geologic history of Mars and the first visible to near-infrared imaging system discovered clay minerals that formed by alteration of primary volcanic minerals in neutral waters in the ancient terrains in agreement with an early warmer and wetter Mars.

The Mars Reconnaissance Orbiter (MRO) was launched in 2005 and carries imagers capable of resolving meter-size features on the surface (25 cm/pixel), images that cover broad regions at 6 m/pixel, and a higher resolution (18 m/pixel) visible and near-infrared spectral imager. It has confirmed widespread deposits of clay minerals in the ancient highlands, refining our understanding of water activity on Mars. It has also sounded the atmosphere to provide a much better understanding of its temperature, pressure, and density variations with altitude, which has dramatically improved our knowledge of the atmosphere that is important in landing spacecraft.

The Phoenix lander was a low-cost refly of a lander originally developed to be launched in 2001 that landed in the high northern plains in 2008. It carried a variety of imagers and meteorology instruments, but its main goal was to measure the chemistry of the soil and shallow ground ice believed to be in equilibrium with the present-day climate. It did find ice several centimeters beneath the surface and found a surface that is heavily modified by the ice. The instruments discovered low levels of calcium carbonate and perchlorate salts in the soils, both arguing for aqueous processes in the past.

The Mars Science Laboratory (MSL) rover is a major mission designed to determine if Mars was habitable in the past. MSL is a mobile laboratory with remote sensing instruments and in situ instruments that can be placed against rocks and surface materials. MSL carries a drill designed to feed material to sophisticated laboratory instruments that measure the mineralogy and geochemistry of surface materials and, for the first time since Viking, organic molecules. It landed on Mars in 2012 in Gale crater and is designed to last several years and traverse tens of kilometers. It is the first spacecraft that used aeromaneuvering and entry guidance during flight on the aeroshell to dramatically reduce the size of the landing ellipse (25 km compared to >100 km for all previous landers). The small landing ellipse (the uncertainty from entry, descent, and landing to a targeted location) and long roving capability make this mission the first to consider "go to" landing sites in which landing occurs in smooth, flat terrain next to areas of prime scientific interest (that are too hazardous to land). As of this writing, the Curiosity rover is in the middle of its surface exploration, but has already discovered conglomerates that formed in surface running water, sandstones and mudstones deposited in streams and lakes, and clays, indicative of a habitable environment.

Two missions are presently under development that will continue the exploration of Mars. The low-cost MAVEN orbiter, launched in 2013, will study the upper Martian atmosphere to determine atmospheric escape rates as a clue to how the atmosphere evolved from a possibly warmer and wetter (thicker) state early on to its current cold and dry (thin) state. Finally, the low-cost Interior Exploration Using Seismic Investigations, Geodesy, and Heat Transport (InSight) mission will land a **seismometer**, **heat flow** probe, and precision tracking station in 2016 to measure the overall structure of the interior to better understand the accretion and differentiation of the rocky planets.

2. LANDING SITES ON MARS

The seven landing sites (Table 19.1) that constitute the "ground truth" for orbital remote sensing data on Mars were all selected primarily on the basis of science and safety considerations. Because a safe landing is required for a successful mission, the surface characteristics must meet the engineering constraints based on the designed entry, descent, and landing system. The most important factor controlling the selection of the seven landing sites is elevation, as all landers used an aeroshell and parachute to slow them down and sufficient atmospheric density and time are required to carry out entry and descent. This favored landing at low elevations is shown in Figure 19.1, which illustrates the locations of the landing sites on a topographic map of Mars.

The map shows that the southern hemisphere is dominated by ancient heavily cratered terrain estimated to be more than 3.6 billion years old (Noachian). The northern hemisphere is dominated by younger (Hesperian and Amazonian), smoother, less-cratered terrain that is on an average 5 km lower in elevation. Astride the hemispheric dichotomy is the enormous Tharsis volcanic province, which rises to an elevation of 10 km above the datum, covers one quarter of the planet, is surrounded by tectonic features that cover the entire western hemisphere, and is topped by five giant volcanoes and extensive volcanic plains (active during the Hesperian and Amazonian). The elevated Tharsis province and the cratered highlands have been too high for landing of existing spacecraft. The Viking landers landed in the northern lowlands, as did MPF and Phoenix; the Mars Exploration Rovers and MSL landed at relatively low elevations in the transition between the highlands and lowlands. The next most important factor in landing site selection is latitude, with low latitudes ($\pm 30°$) favored for greater solar power (Pathfinder, Spirit, and Opportunity) and thermal management (Curiosity).

Landing site selection for the seven landers included intensive periods of data analysis of preexisting and incoming information. The Viking lander/orbiter pairs were captured into Mars orbit and the orbiter cameras started a concentrated campaign to image prospective landing sites (at tens to hundreds of meters per pixel) selected on the basis of previous Mariner 9 images. A large site selection science group assembled mosaics (using paper cutouts pasted together by hand) in real time and, after waiving off several landing sites on the basis of rough terrain and radar scattering results (and missing the intended July 4th landing), Viking 1 landed on ridged plains in Chryse Planitia. The site is downstream from Maja and Kasei Valles, giant catastrophic **outflow channels** that originate north of Valles Marineris, the huge extensional rift or canyon that radiates from Tharsis (Figure 19.1). The site's low elevation and proximity to the channels suggested that water and near-surface ice might have accumulated there, possibly leading to organic molecules and life. Viking 2 was sent to the middle northern latitudes where larger amounts of atmospheric water vapor were detected, thereby ostensibly improving the chance for life. Landing was deferred for Viking 2 as well, as the site

Chapter | 19 Mars: Landing Site Geology, Mineralogy, and Geochemistry

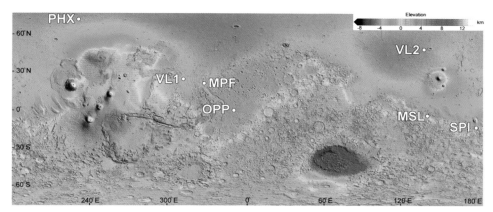

FIGURE 19.1 Mars Orbiter Laser Altimeter on Mars Global Surveyor (MOLA) topographic map of Mars showing the seven successful landing sites. Elevations are reported with respect to the geoid (or geopotential surface) derived from the average equatorial radius extrapolated to the rest of the planet via a high-order and high-degree gravity field. The resulting topography faithfully records downhill as the direction that liquid water would flow. Longitudes are measured positive to the east according to the most recent convention. The locations of the landers, their elevations, and their three-letter acronyms are reported in Table 19.1. Prior to MOLA, which provided excellent global topography and an accurate cartographic grid, elevations and locations were poorly known for landing spacecraft on Mars. The map shows three fundamental terrains of Mars: the southern highlands, northern lowlands, and Tharsis, an enormous elevated region of the planet (located southwest of VL1 on the map). Tharsis is surrounded by a system of generally radial extensional tectonic features (including the huge Valles Marineris canyon that extends to the east of Tharsis) and generally concentric compressional tectonic features that both imprint the entire western hemisphere of the planet. Located at the edges of Tharsis and the highland–lowland boundary are the catastrophic outflow channels that funneled huge volumes of water into the northern plains (including Chryse Planitia where the VL1 and MPF landing sites are located) intermediate in Mars history (during the Hesperian). Note that all of the landing sites are at low elevation and many are near the equator.

selection team analyzed images and thermal observations before landing in the midnorthern plains, just west of the crater Mie (Figure 19.1). Although predictions of the surfaces and materials present at the Viking landing sites were incorrect (likely due to the newness of the data and the coarse resolution of the orbital images), the atmosphere was within specifications and both landed successfully.

The MPF site selection effort involved little new data since the Viking mission 20 years earlier, but there was a much better understanding of how the two Viking landing sites related to the remote sensing data acquired by the Viking orbiters. The site selection effort took place over a two-and-a-half-year period prior to launch and included extensive analysis of all existing data as well as the acquisition of Earth-based radar data. An Earth analog in the Ephrata fan near the mouth of a catastrophic outflow channel in the Channeled Scabland of western and central Washington State was identified as an analog and studied as an aid to understanding the surface characteristics of the selected site on Mars. Important engineering constraints, in addition to the required low elevation, were the narrow latitudinal band $15° N \pm 5°$ for solar power and the large landing ellipse (300 km by 100 km), which required a relatively smooth flat surface over a large area. This and the requirement to have the landing area covered by high-resolution Viking Orbiter images (<50 m/pixel) severely limited the number of possible sites to consider (∼approximately seven). The landing site selected for MPF was near the mouth of a catastrophic outflow channel, Ares Vallis, that drained into the Chryse Planitia lowlands from the highlands to the southeast (Figure 19.2). Ares Vallis formed during the Hesperian (after the early warm and wet period) and involved outpourings of huge volumes of water (roughly comparable to the water in the Great Lakes) in a relatively short period of time (a few weeks). The surface appeared acceptably safe, and the site offered the prospect of analyzing a variety of rock types from the ancient cratered terrain and intermediate aged ridged plains. Surface and atmospheric predictions were correct and Pathfinder landed safely.

Landing site selection for the Mars Exploration Rovers took place over a two-and-a-half-year period involving an unprecedented profusion of new information from the MGS (launched in 1996) and Mars Odyssey (launched in 2001) orbiters. These orbiters supplied targeted data of the prospective sites that made them the best-imaged, best-studied locations up to that time in Mars exploration history. For comparison, most of the ellipses were covered by ∼3 m/pixel Mars Orbiter Camera (MOC) images, whereas the MPF ellipse was covered by ∼40 m/pixel Viking images.

All major engineering constraints were addressed by data and scientific analyses that indicated that the selected sites were safe. Important engineering requirements for landing sites for these rovers included relatively low elevation, a latitude band of $10° N$ to $15° S$ for solar power, and landing ellipse sizes that were ultimately less than 100 km long and 15 km wide. Because of the smaller

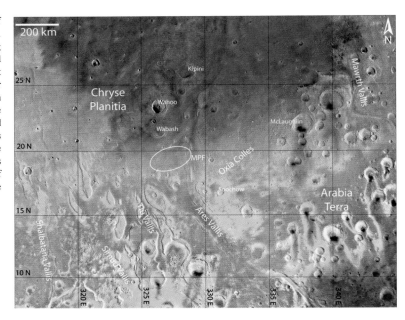

FIGURE 19.2 Regional enhanced color mosaic of Chryse Planitia, Ares Vallis, and the MPF landing ellipse. Viking mosaic shows catastrophic outflow channels cutting the heavily cratered (ancient) terrain to the south and flowing to the lower northern plains. Ares Vallis is about 100 km wide and 2 km deep and by analogy with similar features on Earth formed in about a 2-week period when roughly the volume of water in the Great Lakes carved the valley in about 2 weeks. Note streamlined islands produced during the flooding. The MPF landing ellipse shown is 200 km by 100 km and lies about 100 km north of the mouth of the channel where it exits the highlands and thus was interpreted to be a depositional plain composed of materials dropped by the flood. Characterization of the surface after landing supports this interpretation.

ellipse size compared to Pathfinder, ~150 sites were initially possible from which high-science-priority sites were selected for further investigation. Both sites selected showed strong evidence for surface processes involving water to determine the aqueous, climatic, and geologic history of sites where conditions may have been favorable to the preservation of prebiotic or biotic processes. The site selected for the Spirit rover was within Gusev crater, an ancient 160-km-diameter impact crater at the edge of the cratered highlands in the eastern hemisphere. The southern rim of Gusev is breached by Ma'adim Vallis, an 800-km-long branching valley network that drains the ancient cratered highlands to the south (Figure 19.3). The smooth flat floor of Gusev was interpreted as sediments deposited in a crater lake, so that the rover could analyze fluvial sediments deposited in a lacustrine environment (Figure 19.4).

The site selected for the Opportunity rover is in Meridiani Planum in which thermal infrared spectra from orbiting Thermal Emission Spectrometer (TES) instrument indicated an abundance (somewhat unique) of dark, gray coarse-grained hematite, a mineral that typically forms in the presence of liquid water. Layers associated with the hematite deposit in Meridiani Planum suggested a sequence of sedimentary rocks that could be interrogated by the rover. Meridiani Planum is a unique portion of the ancient heavily cratered terrain in western Arabia Terra that was downwarped in response to the formation of Tharsis and heavily eroded early in Mars history and thus stands at a lower elevation than the adjacent southern highlands (Figure 19.5). The atmospheric and surface characteristics inferred from the extensive remote sensing data were correct for both, and Spirit and Opportunity landed safely.

The Phoenix lander was designed to land at the northern polar region (65°–72° N) where ground ice overlain by a few centimeters of soil had been detected by instruments on the Mars Odyssey orbiter. In addition to the low elevation of the northern plains, landing sites were initially identified (before MRO was operational) that had low rock abundance, low slopes, and a calm atmosphere. Once MRO became operational, about 1 year before Phoenix launched, the High-Resolution Imaging Science Experiment (HiRISE) began imaging these areas at about 25 cm/pixel. To everyone's surprise and dismay, most of the high northern plains are covered by areas with dense boulder fields that could not be avoided with the large landing ellipse of Phoenix (~100 km long) and were far too rocky to safely land. Because of the low Sun angle of the HiRISE images, large rocks cast long shadows, which could be easily measured. An anxious search for rock-free areas using HiRISE ultimately identified a suitable landing site just before launch. Phoenix landed safely in an ellipse completely covered by HiRISE images and the site matched expectations with few rocks, a smooth flat plain, and a few centimeters of soil over ground ice. The mission lasted several months until the darkness and cold of polar winter encased the lander in ice.

The selection of Gale crater as the MSL landing site took over 5 years with prospective landing sites heavily targeted by MRO instruments. Engineering constraints included low latitude for thermal management of the rover and instruments, low elevation and relief, and low rock abundance. Science criteria important for the selection included the ability to assess past habitable environments, which include diversity, context, and potential biosignature

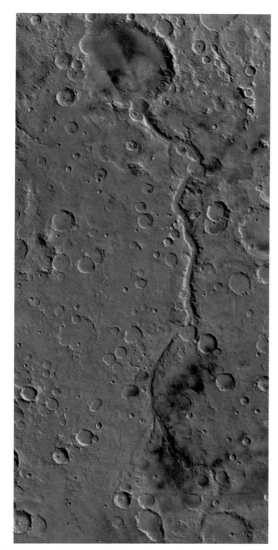

FIGURE 19.3 Viking regional color mosaic of Ma'adim Valles and Gusev crater. The 800-km-long Ma'adim Valles, one of the largest branching valley networks on Mars, drains the heavily cratered terrain to the south and breaches the southern rim of Gusev crater. Gusev crater, which formed much earlier, is 160 km in diameter and the smooth flat floor strongly suggests that it was a crater lake that filled with water and sediments. Spirit did not identify any sediment associated with Ma'adim Valles. The cratered plains are composed of **basalt** flows modified by impact and eolian processes and so represent a late volcanic cover. Rocks in the Columbia Hills have been altered by water, but cannot be related to deposition in a lake associated with Ma'adim discharge.

(including organic molecules) preservation. Over 50 prospective landing sites were studied and downselected to four finalists (three of which were "go to" sites), all of which have layered sedimentary rocks with spectral evidence for clays. All four sites were covered with unprecedented imaging data, dominantly from MRO, including spectral and stereo images and derived high-resolution HiRISE-derived topographic and rock maps that were

FIGURE 19.4 Mosaic of Gusev crater showing the landing ellipse, landing location for the Spirit rover, and the extensive data sets that were obtained to evaluate the Mars Exploration Rover landing sites. Ma'adim Valles breaches the southern rim and hills immediately downstream have been interpreted as delta deposits. The blue ellipse is the final targeted ellipse and the red X is the landing location. Background of mosaic is Viking 230 m/pixel mosaic, overlain by MOLA elevations in color. Thin image strips mostly oriented to the north−northwest are MOC high-resolution images typically at 3 m/pixel. Wider image strips mostly oriented to the north−northeast are Mars Odyssey THEMIS-visible images at 18 m/pixel. Mosaic includes 13° S−16° S latitude and 174° E−177° E longitude; solid black lines are 0.5° ($\sim$30 km) and dashed black grid is 0.1° ($\sim$6 km).

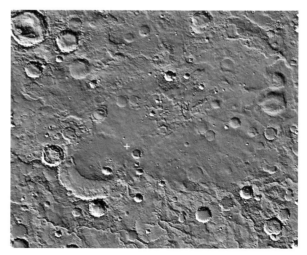

FIGURE 19.5 Regional setting of Meridiani Planum in Mars Orbiter Laser Altimeter shaded relief map ($\sim$850 km wide). Note that smooth lightly cratered plains on which Opportunity landed (cross), which bury the underlying heavily cratered (ancient) terrain with valley networks to the south. Note that large degraded craters in the smooth plains indicate the sulfate rocks below the basaltic sand surface are very old (>3.6 billion years). In contrast, the lightly cratered basaltic sand surface that Opportunity has traversed is young. Opportunity has traversed 35 km to the large crater, and Endeavour to the southeast of the landing location.

FIGURE 19.6 Regional setting of Gale crater, the MSL landing ellipse, and the landing location. Gale crater is 150 km in diameter with a 5 km high mound of material (Mount Sharp) in its interior. The landing ellipse is on smooth cratered plains to the northwest. Final landing ellipse (black) is 20 km by 7 km and the rover landed at the yellow X. Dark material in the southeastern part of the ellipse are active basaltic sand dunes and the layered rocks of clays and sulfates at the base of the mound are due south of the landing ellipse. As a result, this is a "go to" site in which the landing ellipse is on the smooth flat terrain nearby and the rover must traverse to the material of greatest interest by leaving the ellipse.

used to run detailed landing simulations that indicated all four sites were safe. In addition, the traversability of the landing sites and target areas outside of the ellipse were evaluated, indicating that all are trafficable and that "go to" sites could be accessed within the lifetime of the mission. The Gale crater site (Figure 19.6) has a 5-km-high mound adjacent to the landing site that has layered strata that contains clays and sulfates at its base, which will be studied by the Curiosity rover after traversing out of the landing ellipse. The landing site explored so far is consistent with expectations from remote sensing data.

3. MARS LANDING SITES IN REMOTELY SENSED DATA

3.1. Surface Physical Properties

Understanding the relationship between orbital remote sensing data and the surface is essential for safely landing spacecraft and for correctly interpreting the surfaces and kinds of materials globally present on Mars. Safely landing spacecraft on the surface of Mars is obviously critically important for future landing missions. Understanding the surfaces and kinds of materials globally present on Mars is also fundamentally important to deciphering the erosional, weathering, and depositional processes that create and affect the Martian surface layer. This surface layer or regolith, composed of rocks and soils, although likely relatively thin (of order meters thick), represents the key record of geologic processes that have shaped it, including the interaction of the surface and atmosphere through time via various alteration (weathering) and eolian (wind-driven) processes.

Remote sensing data available for selecting landing sites have varied for each of the landed missions, but most used visible images of the surface as well as thermal inertia and **albedo**. Thermal inertia is a measure of the resistance of surface materials to a change in temperature and can be related to particle size, thermal conductivity, bulk density, and cohesion. Albedo is a measure of the solar reflectance of a surface in which the viewing geometry has been taken into account. A surface composed mostly of rocks will change temperature more slowly, remaining warmer in the evening and night, than a surface composed of fine-grained loose material that will change temperature rapidly, thereby achieving higher and lower surface temperatures during the warmest part of the day and the coldest part of the night, respectively. As a result, surfaces with high thermal inertia will be composed of more rocks or cohesive, cemented material than surfaces with low thermal inertia. Thermal inertia can be determined by measuring the surface temperature using a spectrometer that measures the thermal infrared radiance at several times during the day or by fitting a diurnal thermal model to a single radiance-derived temperature measurement. Thermal observations of Mars have been made by many orbiters, including the Mariners, Viking, MGS, and Mars Odyssey, with increasingly higher spatial resolution. Thermal inertia data have been used to map areas of the surface covered by high-inertia materials or rocks from areas covered by lower inertia materials or soil.

Global thermal inertia and albedo data combine in ways that reveal several dominant surface types. One has high albedo and very low thermal inertia and is likely dominated by substantial thicknesses (centimeters to a meter or more) of high albedo, reddish dust that is neither load bearing nor trafficable. These areas have very few rocks and have been eliminated for landing solar-powered or surface missions interested in investigating rocks or outcrop. Regions with moderate to high thermal inertia and low albedo are likely relatively dust free and composed of dark eolian sand and/or rock. Regions with moderate to high thermal inertia and intermediate to moderately high albedo are likely dominated by cemented crusty, cloddy, and blocky soil units that have been referred to as duricrust with some dust and various abundances of rocks. Coarse-resolution global abundance of rocks on Mars, derived by thermal differencing techniques that remove the high-inertia (rocky) component, shows that the high-albedo, low-inertia type of surface has almost no rocks and the other two types of surfaces have rock abundances that vary from about 5% (the global mode of rock abundance of Mars) to a maximum of about 50% of the surface covered by rocks.

The seven landing sites sample the latter two types of surfaces in thermal inertia and albedo combinations that cover most of Mars. Along with variations in their rock abundance, they sample the majority of likely safe surfaces that exist and are available for landing spacecraft on Mars. The Viking landing sites both have intermediate to relatively high albedo, high rock abundance ($\sim 17\%$), and intermediate thermal inertia. On the surface, these sites are consistent with these characteristics, with both being rocky and somewhat dusty plains with a variety of soils, some of which are cohesive and cemented (Figures 19.7 and 19.8).

Prior to landing, the MPF site was expected to be a rocky plain composed of materials deposited by the Ares Vallis catastrophic flood that was safe for landing and roving and was less dusty than the Viking landing sites based on the intermediate to high thermal inertia, high rock

(a)

(b)

FIGURE 19.7 The Viking 1 landing site. (a) Mosaic of the Viking Lander 1 landing site showing bright drifts and dark rocks. Large rock to the left is Big Joe and is subrounded. Smaller angular dark rocks are sitting on soil and have been interpreted as impact ejecta blocks. Bright drift in the center of the image shows layers and some particles that may be large enough to require deposition by running water rather than the wind. (b) Color mosaic of the Viking 1 landing site showing dusty reddish surface, darker pitted rocks nearby, and rim of crater on the left horizon. Jointed slightly lighter toned low rock mass in the middle distance appears to be outcrop. The location of the site on ridged plains suggests that the outcrop is basalt, with angular rocks as ejecta and drift materials deposited by either the wind or floodwaters from Maja or Kasei Valles.

FIGURE 19.8 Color mosaic of the Viking 2 landing site showing a flat, rocky, and dusty plain. Pitted rocks in foreground suggest that they are volcanic basalts, and angular homogeneous rock field suggests that they are distal ejecta from the fresh crater Mie to the east of the landing site. The lighter toned trough in the middle of the image, in front of the large rocks, has been interpreted to result from the thermal contraction of subsurface ground ice.

abundance (18%), slightly lower albedo, and relation to an analogous catastrophic outflow depositional plain in the Channel Scabland. All these predictions were confirmed by data gathered by the MPF lander and rover (Figure 19.9).

The Spirit landing site in Gusev crater has comparable thermal inertia and fine-component thermal inertia and albedo to the two Viking sites and so was expected to be similar to these locations, but with fewer rocks (8%). Dark dust devil tracks in orbital images suggested that some of the surfaces would be lower albedo, where the dust has been preferentially removed (Figure 19.10). Spirit landed and traversed across both dusty (Figure 19.11) and dust devil track surfaces. It found that the average rock abundance is similar to expectations. In darker dust devil tracks,

FIGURE 19.9 Color mosaic of the MPF landing site showing an undulating, ridge-trough, moderately dusty, and rocky plain. Large rocks in the middle left of the image appear stacked or imbricated on a ridge with a trough behind it that trends toward the northeast. Streamlined hills on the horizon, the ridge-trough topography, and angular to subrounded boulders are consistent with depositional plains deposited by catastrophic floods as expected from the setting of the site downstream from the mouth of Ares Vallis outflow channel. Note dust coating the tops of rocks.

FIGURE 19.10 Color mosaic of the Spirit landing site on the cratered plains of Gusev crater. Note the soil-filled hollows that are impact craters filled in by sediment. Dark angular blocks are consistent with ejecta, and the pebble-rich surface is similar to a desert pavement in which the sand-sized particles have been moved by the wind leaving a lag deposit. The landing site is in a dust devil track explaining its lower albedo and less dusty surface. The plain is relatively flat with Grissom Hill on the horizon. Note dark wind tails behind rocks in the lower middle foreground.

FIGURE 19.12 False-color mosaic of the Opportunity landing site showing dark, basaltic sand plain and the rim of Eagle crater in the foreground (brighter). Note the light-toned pavement outcrop near the rim, which is brighter than the plains. Parachute and 1-m-high backshell that Opportunity used to land are 450 m away and demonstrate the exceptionally smooth flat surface as expected from orbital data. The relatively dust-free surface of the plains is in agreement with their very low albedo from orbital data. Even though dust has rapidly fallen on the solar panels, the basalt surface is relatively dust free, indicating that the dust is being swept off the surface at a rate that roughly equals its deposition rate.

Spirit found that the albedo is low and the surface is relatively dust free (at the landing site) compared to areas outside of dust devil tracks, where the albedo is higher and the surface is more heavily coated with bright atmospheric dust that has fallen from the sky (Figure 19.11).

The Meridiani Planum site has moderate thermal inertia, very low albedo, and few rocks. This site was expected to look very different from the three landing sites with a dark surface, little high albedo dust, and few rocks. Opportunity has traversed across a dark, **basaltic** sand surface with very few rocks and almost no dust (Figure 19.12). The Phoenix landing site has moderate thermal inertia and intermediate albedo and was expected to be slightly dusty with low rock abundance (from HiRISE images), all of which were confirmed at the surface (Figure 19.13). The MSL landing site (Figure 19.14) has intermediate albedo and relatively high thermal inertia (comparable to Pathfinder). The landing site is as expected, slightly dusty with low rock abundance and cemented surface materials.

FIGURE 19.11 Color mosaic of the eastern part of Bonneville crater showing the dusty and rocky surface of this part of the Gusev cratered plains. Note that the wall of crater is composed of dark rubble suggesting that it formed in a regolith of basalt ejecta. This location is not in a dust devil track and so is much dustier with much higher albedo, consistent with inferences made from orbital images. Hills in the background are the Columbia Hills, which are 90 m high and composed of older rocks. Spirit traversed the cratered plains and climbed to the top of the Columbia Hills (highest peak shown is Husband Hill).

FIGURE 19.13 Color mosaic of the Phoenix landing site showing trenches dug by the robotic arm exposing ground ice (white) that appears in equilibrium with the present climate. Note smooth, flat, somewhat dusty surface with few rocks and polygonal troughs as expected from remote sensing data. Thermal contraction of the ground ice has produced the polygons, sorted the rocks, and destroyed most small craters. Note the pitted surface and dust-covered footpad and leg caused by rocket exhaust moving fines beneath the lander (also exposing ground ice). Brownish yellow color of the atmosphere is caused by the suspension of fine-grained (micron size) dust that is omnipresent on Mars.

FIGURE 19.14 Surface mosaic of the MSL landing site with the 5-km-high Mount Sharp in the background. Note the relatively smooth, flat, relatively rock-free plain as expected from remote sensing data. The pebble-rich surface and lack of fines suggests that it is a lag deposit in which the sand-sized particles have been removed by the wind similar to a desert pavement. Darker terrain in the distance is due to the dark, basaltic sand dunes seen in Figure 19.6. The clay and sulfate layers are at the base of Mount Sharp. Lighter toned surfaces higher up in the mound are dominated by potentially thick deposits of non–load-bearing dust and have untraversably steep slopes, so the rover will not be able to climb it.

The slopes and relief at various length scales that are important to landing safely were also estimated at the seven landing sites using a variety of altimetric, stereo, shape-from-shading, and radar backscatter remote sensing methods. Results estimated from these data are in accord with what was found at the surface. Of the seven landing sites, Meridiani Planum was judged to be the smoothest, flattest location ever investigated at 1 km, 100 m, and several meter length scales, which is in agreement with the incredibly smooth flat plain traversed by Opportunity (Figure 19.12). On the other extreme, the MPF landing site (Figure 19.9) was expected at the time of landing to be the roughest at all three of these length scales, which agrees with the undulating ridge and rough terrain and the more distant streamlined islands visible from the lander. The other five landing sites are in between these extremes at the three length scales, with Viking 2 (Figure 19.8) and portions of Gusev (Figure 19.10) fairly smooth at the 100-m and 1-km scale, Viking 1 slightly rougher at all three length scales, Viking 2 and portions of Gusev in between in roughness, and the Columbia Hills in Gusev roughest at the several meter length scale. Phoenix (Figure 19.13) is comparable to Viking 2 at all three scales and Gale crater (Figure 19.14) is the roughest site at the two longer scales and similar to Pathfinder in roughness at the several meter scale. All these observations are consistent with the relief observed at the surface.

The close correspondence between surface characteristics inferred from orbital remote sensing data and that found at the landing sites argues that future efforts to select safe landing sites will be successful. Linking the seven landing sites to their remote sensing signatures indicates that surface types with moderate to high thermal inertia and moderate to low albedo are both suitable for landing on Mars. Such surfaces constitute almost 60% of the planet, suggesting that to first order most of Mars is likely safe for suitably engineered landers. These results show that basic engineering parameters important for safely landing spacecraft such as elevation, atmospheric profile, bulk density, rock distribution and slope can be well constrained using available and targeted remote sensing data.

3.2. Global Compositional Units

The compositions of surface materials on Mars can be determined from infrared measurements of the planet's surface. The TESs on the MGS and THEMIS (Thermal Emission Imaging System) on Mars Odyssey orbiting spacecraft revealed two broad spectral classes representing different compositional units. Based on spectral similarity to rocks measured in the laboratory on Earth, "Surface Type 1" material is interpreted as basaltic rock and/or sand derived from basalt (Figure 19.15). Basalt consists mostly of silicate **minerals—pyroxene**, **feldspar** (**plagioclase**) and **olivine**—and forms by partial melting of the upper **mantle** producing a **mafic** (magnesium- and iron-rich) magma that erupts on the surface as a dark lava flow (or shallow intrusion). Basalt is the most abundant type of rock on Earth, comprising the floors of the oceans and significant flooded

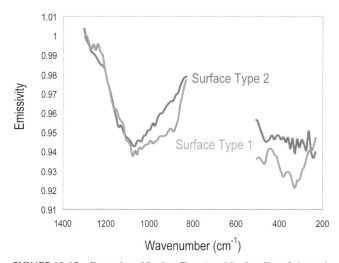

FIGURE 19.15 Examples of Surface Type 1 and Surface Type 2 thermal emission spectra, from the MGS spacecraft. Surface Type 1 spectra match laboratory spectra of basalt. Surface Type 2 spectra could be either andesite, a more silica-rich volcanic rock, or slightly weathered basalt.

areas of the continents, and it is no surprise that it is common on Mars as well. The giant shield volcanoes of Olympus Mons and the Tharsis Montes are likely composed of basalts based on their similar morphology to shield volcanoes as well as many plains that resemble basalt plains on Earth. "Surface Type 2" material is variously interpreted as either andesite or partly weathered basalt—the spectrum is consistent with either possibility (Figure 19.15). Andesite is another common lava type on the Earth, occurring primarily at subduction zones. Andesite contains pyroxene (or amphibole) and feldspar. Andesite can form when mafic crystals form in cooling basaltic magma and are extracted from the liquid, leaving a more silica-rich andesitic liquid behind. The spectra of Surface Type 2 can also be explained as a mixture of basaltic minerals plus clays or other weathering products, which commonly form when basalt is weathered by interaction with water. This latter interpretation has been widely adopted.

The TES data have fairly large footprints (about 3 by 6 km/pixel), so they cover big regions. Mars surface spectra (Figure 19.15) represent mixtures of spectra for the individual minerals that comprise the rocks and soil within each pixel. The TES spectra can be unmixed ("deconvolved") into the spectra of constituent minerals, allowing not only their identification but also an estimate of their proportions. Because we know the chemical compositions of the minerals in the spectral library and the proportions needed to produce the measured spectra, it is possible to calculate the approximate chemical composition of the mixture. This is important because volcanic rocks are usually classified based on their chemistry rather than their mineralogy (minerals in volcanic rocks are small and hard to identify, and quickly solidified magmas often form glass rather than crystalline minerals). The commonly used chemical classification for volcanic rocks, based on the measured abundances of the alkali elements (sodium and potassium, expressed as oxides) versus silica (silicon dioxide), is shown in Figure 19.16. The estimated chemical compositions of Surface Type 1 and Surface Type 2 are illustrated in this figure.

In addition to these major units, a few areas on Mars show the distinctive thermal infrared spectra of hematite, an iron oxide usually formed by interaction with water. The Meridiani Planum region has the highest concentration of hematite measured from orbit, which as discussed earlier led to its selection as a landing site for the Opportunity rover.

Visible and near-infrared (as opposed to thermal infrared) spectrometers on orbiting spacecraft provide further information on the composition of the Martian surface. Certain minerals, including iron-bearing silicates (like pyroxene or clays), sulfates, carbonates, and silica, can be readily identified by their characteristic spectra. Most of these are secondary minerals, formed by aqueous

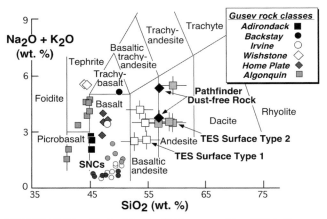

FIGURE 19.16 Alkalis ($Na_2O + K_2O$) versus silica (SiO_2) diagram, commonly used to classify volcanic rocks. Several estimates of the compositions of Surface Type 1 and Surface Type 2 materials are shown, along with the measured compositions of Martian **meteorites** and APXS analyses of rocks from the MPF and Spirit landing sites (Gusev). See text for discussion of rock types and classes.

alteration of primary igneous rocks in neutral pH conditions (pH refers to hydrogen ion concentration). Orbital surveys reveal that these alteration minerals are geographically widespread but concentrated in ancient, mostly Noachian terrains of the southern highlands. Some occur in layered sequences, often showing mineralogical changes within succeeding strata (like Gale crater). It has been suggested that the Martian sedimentary record consists of distinct mineralogical epochs reflecting changes in aqueous conditions from wet, neutral pH conditions in the Noachian to highly acidic (or low pH) conditions later, to cold and dry in the Amazonian. In this scenario, the neutral aqueous conditions would lead to the production of clay minerals, and the highly acidic conditions would lead to the deposition of sulfates via evaporation, with cold and dry conditions in the past ~3 billion years. However, some of the clays and secondary minerals apparently formed under conditions suggesting subsurface hydrothermal alteration rather than surface weathering, so they would not necessarily be indicators of climate change.

The global distribution of Type 1 and 2 spectrally identified units on Mars is distinctive (Figure 19.17). The heavily cratered, ancient southern hemisphere of Mars is mapped mostly as Surface Type 1. In contrast, the younger northern lowlands are mapped mostly as Surface Type 2 materials. The distribution of global geochemical units is illustrated in Figure 19.17. About half of the surface of Mars is covered with a thin layer of dust, which precludes the infrared spectrometers from mapping the compositions of the rocks that underlie the dust. Some of the spacecraft landing sites on Mars are located in dusty regions. Consequently, it is difficult to compare interpretations of orbital spectra with rocks actually on the ground. The two MER

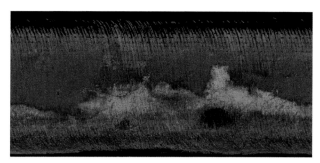

FIGURE 19.17 Global map showing the distributions of Surface Type 1 (green) and Surface Type 2 (red) materials, based on thermal emission spectroscopy from MGS. Dust-covered areas where this technique cannot distinguish rock units are shown in blue. The preponderance of Surface Type 2 in the northern lowlands is consistent with these materials being slightly weathered basalts.

landing sites are exceptions—Spirit landed in a region mapped as Surface Type 1 and the Opportunity site in Meridiani was selected because of its hematite spectral signature. The Curiosity landing site was chosen for its alteration minerals (dominantly clays and sulfates), identified from visible to near-infrared spectra, and the Phoenix landing site in the high northern plains in Type 2 materials was selected for its ground ice.

4. LANDING SITE GEOLOGY

4.1. Introduction

The geology of the seven landing sites has been investigated from color, stereo, panoramic imaging that provides information on the morphology of the landing sites; on the lithology, texture, distribution, and shape of rocks and eolian soil deposits; and on other local geologic features and landforms that are present. Landing sites on Mars are composed of rocks, outcrops, eolian bedforms, and soils, many of which are cemented. Craters and eroded crater forms are also observed at almost all the landing sites and other hills have been observed at some of the landing sites.

Our knowledge of how the surfaces at the different landing sites developed and the important geological processes that have acted on them is directly related to the mobility of the lander (arm) or rover and the ability of the lander or rover to make basic field geologic observations. The lack of mobility of the two Viking landers and their inability to analyze rocks at these sites hampered our ability to constrain their geologic evolution. Because the Phoenix lander was able to dig down into the ground ice and observe variations (Figure 19.13), it was better able to characterize the materials and important geological processes that shaped the surface. The mobility of the MPF Sojourner rover and its ability to make basic field observations over a couple of hundred square meter area resulted in a much better understanding of the geology and the events that shaped the Ares Vallis surface. The two Mars Exploration Rovers that collectively traversed over 40 km have amassed a robust suite of geologic observations over a wide area that have resulted in much better knowledge of the geologic evolution of the rocks and surfaces investigated (and the same is true for MSL). This section will review the basic geological materials found at the seven landing sites and discuss the landforms present.

4.2. Rocks

Rocks are common at all the landing sites (except Meridiani). At most sites, they are distinct dark, angular to subrounded clasts that range in size from several meters in diameter down to small pebbles that are a centimeter or less in diameter. Most appear as float or individual rocks not associated with a continuous outcrop or a body of rock. Many appear dust covered and there is evidence at Gusev and Meridiani for some surface chemical alteration as is common on the Earth (see next section), where rocks exposed to the atmosphere develop an outer rind of weathered material. Although the composition of rocks could not be measured at the Viking 1 and 2 and Phoenix landing sites (Figures 19.7, 19.8 and 19.13), their dark angular and occasionally pitted appearance is consistent with a common igneous rock known as basalt. Rocks making up the cratered plains on which Spirit landed and traversed (Figures 19.10 and 19.11) for the first few kilometers are clearly basalts (see next section). The distribution and shape of many of the rocks at the Viking 1 and 2 landing sites and the Gusev cratered plains are all consistent with a surface that has experienced impact cratering with the rocks constituting the ejected fragments. Many subrounded rocks at the MPF (Figure 19.9) and Viking 1 landing site have been attributed to deposition by catastrophic floods in which motion in the water partially rounded the clasts. Some rocks at the Pathfinder site had textures that looked like layers (perhaps sedimentary or volcanic), one resembled a pillow basalt in which hot lava cools rapidly in the presence of water, and several rocks resembled conglomerates, in which rounded pebbles and cobbles were embedded in a rock. The cobbles were rounded by running water and later cemented in a finer grained matrix. Curiosity has also observed conglomerates near the surface of its landing site (Figure 19.18). Some rocks at most of the landing sites appear polished, fluted, and grooved. These are interpreted as ventifacts in which sand-sized grains, entrained by the wind, have impacted and eroded the rocks. Rocks at the Phoenix landing site are size sorted within polygonal troughs formed by thermal contraction processes in the ground ice (Figure 19.13).

FIGURE 19.18 Conglomerate rock imaged by the Curiosity rover near its landing site. The rock is composed of rounded pebbles of various sizes that have been cemented together. The pebbles were rounded by collisions in running water indicating that water flowed across the surface at some time in the past. These rocks are associated with an alluvial fan that was fed by a channel that cuts through the rim of Gale crater to the northwest of the landing site (Figure 19.6).

4.3. Outcrops

Continuous expanses of rocks typically referred to as outcrop (or bedrock) have been observed at three of the landing sites. An area of continuous jointed rocks has been observed at the Viking 1 landing site, but little else is known about it (Figure 19.7). Outcrop has been discovered in the Columbia Hills by Spirit where there appear to be coherent stratigraphic layers in and near the Cumberland Ridge on the flank of Husband Hill (Figure 19.19). These rocks, described in the next section, appear to be layers of ejecta or explosive volcanics deposited early in Mars history. In some places the rocks are finely layered and in other places they appear massive. At Meridiani Planum light-toned outcrops are exposed in crater walls and areas

FIGURE 19.19 Color mosaic of the northwest flank of Husband Hill showing layered strata of the outcrop Methuselah dipping to the northwest. These rocks are clastic rocks, consistent with impact ejecta that have been highly altered by liquid water. Hills on the horizon are the rim of the 20-km-diameter Thira crater near the eastern end of the landing ellipse shown in Figure 19.4.

FIGURE 19.20 False-color mosaic of a promontory, called Cape St Vincent, on the north wall of Victoria crater exposing outcrop. Lower strata are finely bedded sandstones formed by evaporation of acid-rich waters and later deposited by the wind. Note the lighter toned unit above and the cross-beds (curved layers) below. The topmost **breccia** unit is composed of widely different sizes of angular rocks all jumbled together. This breccia is ejecta deposited by the impact that formed Victoria crater.

where the covering dark, basaltic sand sheet is thin (Figure 19.20). These outcrops appear to be thinly laminated evaporites that formed via evaporation of subaerial salt water (see next section) early in Mars history. The layers are composed of sand-sized grains of fairly uniform composition that appear to have been reworked by the wind in sand dunes before being diagenetically altered by acid groundwater of differing compositions (see next section). Finally, continuous layers or strata of sedimentary rocks have been found and investigated by the Curiosity rover.

4.4. Soils

All the landing sites have soils composed of generally small fragments of granules, sand, and finer materials. Except where they have been sorted into bedforms by the wind, they have a variety of grain sizes and cohesion, even though their composition appears remarkably similar at all the landing sites (dominantly basaltic). Crusty to cloddy and blocky soils are also present at most of the landing sites and are distinguished as more cohesive and cemented materials. These materials appear to be the duricrust inferred to be present over much of Mars based on higher thermal inertia, but generally low rock abundance. Strong cemented light-toned duricrust was uncovered at the MPF site by Sojourner and may contribute to the higher thermal inertia at this site than the others. Some bright soil deposits outside the reach of the arm at the Viking 1 landing site (Figure 19.7(a)) show layers and hints of coarse particles that could be fluvial materials deposited by the Maja or Kasei Valles floods. Most fine particles (roughly sand sized) appear rounded to subrounded suggesting that they have been entrained in the wind and rounded when they impact the surface (see next section). Ground ice has been observed beneath the soil at the Phoenix landing site (Figure 19.13).

4.5. Eolian Deposits

Most of the landing sites have examples of eolian bedforms, or materials that have been transported and sorted by the wind. Sand-sized particles that average several hundred microns in diameter can be moved by saltation in which they are picked up by the wind and hop in parabolic arcs across the surface. Because these particles can be preferentially moved by the wind, they are effectively sorted into bedforms. Sand dunes form when sand-sized particles are sorted into a large enough pile to move across the surface. Sand dunes take a variety of forms such as barchan or crescent shaped (horns pointing downwind), star shaped from reversing winds, transverse to the wind, and longitudinal or parallel to the wind, all of which are generally diagnostic enough to be identified from orbit. Sand dunes have been identified at the MPF landing site where a small barchan dune was discovered in a trough by the rover and at Meridiani Planum where star dunes were found at the bottom of Endurance (Figure 19.21) and Victoria craters.

Ripples are eolian bedforms formed by saltation-induced creep of granules, which are millimeter-sized particles. They typically have a coarse fraction of granules at the crest and poorly sorted interiors indicating a lag of coarser grains after the sand-sized particles have been removed (Figure 19.22). Ripples have been found at the MPF, Spirit, Opportunity, and Curiosity sites. Drifts of eolian material have also been identified at many of the landing sites behind rocks as wind tails and in other configurations.

Finally, the reddish dust on Mars is only several microns in diameter and is carried in suspension in the atmosphere, giving rise to the omnipresent reddish color. Although it takes high winds to entrain dust-sized particles in the atmosphere, once it is in the atmosphere it takes a long time to settle out. Dust has been identified on the surface at all the landing sites (in addition to being in the atmosphere) giving everything a reddish color and has fallen steadily on the solar panels decreasing solar power at a similar rate.

FIGURE 19.21 False-color mosaic of star sand dunes in the bottom of Endurance crater. Dark bluish surface is basalt with a surface lag of hematite spherules. Lighter sides of dunes are likely covered by dust that has settled from the atmosphere. Light-toned outcrop in the foreground.

FIGURE 19.22 Large ripple called Serpent that was studied by Spirit on the cratered plains. (A) A hazard camera image showing the rover front wheels and the tracks produced by a wheel wiggle maneuver to section the drift. (B) Color image of the dusty (reddish) surface and darker more poorly sorted interior. (C) MI image of the brighter (dust cover) granule-rich surface (millimeter-sized particles) and poorly sorted, but generally finer grained basaltic sand interior. The dusty, granule-rich surface indicates that the eolian feature is an inactive (dust covered) ripple formed by the saltation-induced creep of granules, which are left as a lag.

FIGURE 19.23 False-color mosaic of the rim of the 20-km-diameter Endeavour crater at Meridiani Planum. Opportunity drove about 30 km to get to this ancient Noachian highly eroded impact crater rim, which found clays where indicated by orbital data. The sulfates investigated by Opportunity that formed in acid-rich surface waters are younger than the clays exposed in the crater rim.

FIGURE 19.24 Color mosaic of Sojourner with APXS instrument measuring the chemical composition of the rock Yogi. Note dusty surface darkened by the rover wheels. Lighter toned soil in the wheel tracks is cemented soil or duricrust. Note tabular rock on the left horizon, called Couch, and other tabular and partially rounded boulders as expected if deposited by catastrophic floods.

Dust devils, or wind vortices, have been observed at the MPF and Spirit sites and appear to be an important mechanism for lifting dust into the atmosphere.

4.6. Craters

Impact craters are ubiquitous on Mars, so it is no surprise that craters have been imaged at most of the landing sites. At Viking 1 (Figure 19.7) and the MPF landing sites, the uplifted rims of craters have been imaged from the side. At Gusev (Figure 19.11) and Meridiani (Figure 19.23), the rovers have investigated a number of craters of various sizes during their traverses, including the interiors of some. Because impact craters resemble nuclear explosion craters and because many fresh craters have been characterized on the Moon, much is known about the physics of impact cratering and the resulting shape and characteristics of fresh craters (see chapter on Planetary Impacts). Fresh primary impact craters less than 1 km in diameter have well-understood, bowl-shaped interiors whose depth is about 0.2 times their diameter; they also have uplifted rims and ejecta deposits (Figures 19.11 and 19.20) that get less rocky and thin with distance from the crater. As a result, imaging impact craters provides clues to the geomorphologic changes that have occurred at the site such as the amount of erosion and/or deposition.

5. LANDING SITE MINERALOGY AND GEOCHEMISTRY

5.1. Rocks

Based on their appearance, rocks at the Viking and Phoenix landing sites (Figures 19.7, 19.8, and 19.13) were inferred to be basalts, but the Viking lander arms could not reach and collect rocks small enough to analyze and Phoenix could not analyze rocks, so little is known about their composition. Rocks at the MPF, Spirit, Opportunity, and Curiosity landing sites have been analyzed by a variety of rover-mounted instruments, as described in Table 19.2.

Pathfinder rock chemical compositions were analyzed by the APXS (Figure 19.24), and partial mineralogy was inferred from IMP spectra on the lander. The APXS analyzes only the outer surface (generally just a few tens of microns) of rocks. IMP images showed that the rocks were variably coated with dust. Plots of different elements versus sulfur yield straight lines, with soils plotting at the sulfur-rich end, best interpreted as mixing lines between the compositions of rocks and soil. The composition of the dust-free rock interior was inferred by extrapolating the rock composition trends to zero sulfur. The dust-free rocks have concentrations of alkalis and silica that would classify them as andesite (two different calibrations of the APXS instrument data are shown in Figure 19.16), and it was inferred from the rocks' appearance that these were volcanic rocks. However, because the APXS analyzes only the rock surface, it is also possible that this andesitic composition represents a silica-enriched weathering rind beneath the dust rather than the composition of the rock interiors. The IMP spectra indicate the presence of iron oxides, but a more comprehensive spectral interpretation was hampered by the dust coatings.

Rocks at the Spirit landing site in Gusev crater were analyzed using a greater variety of analytical instruments (Table 19.2), aided by the RAT that can brush or grind the outer rock surface. Rocks on the plains in the vicinity of the Spirit lander are clearly basalts, in agreement with the location of Gusev crater within an area mapped by TES as Surface Type 1. Some of these rocks are vesicular—pocked with holes that were once gas bubbles exsolved from magma, and most rocks are coated with dust (Figure 19.19). Spectra from Pancam, Mini-TES, and MB of relatively dust-free or RAT-abraded rocks provide a consistent picture

Gusev Plains Rocks

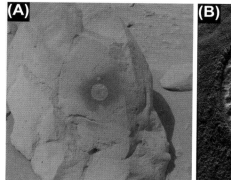
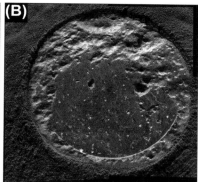

FIGURE 19.25 The Gusev cratered plains rock Humphrey studied by the Spirit rover. (A) Pancam color image of rock after RAT grinding showing darker interior and thus the presence of a dusty and slightly weathered surface. (B) Microscopic image of Humphrey RAT hole, illustrating dark grains thought to be olivine crystals and holes likely to be vesicles, consistent with the basaltic chemistry and mineralogy determined by the APXS, MB, and Mini-TES.

of the minerals that comprise these basalts—olivine, pyroxene, and iron oxides. All the spectra from these instruments are dominated by minerals containing iron and magnesium. Chemical compositions of basalts of plains measured by APXS not only support the presence of olivine, pyroxene, and oxides but also suggest abundant feldspar (plagioclase) and phosphate, which cannot be seen by other spectra. The APXS analyses confirm that the rocks on the plains of Gusev (Figure 19.16) are basalts (called the Adirondack class), especially rich in olivine (and hence lower in silica). Abundant dark crystals interpreted to be olivine can be seen in MI images of RAT holes in the rocks (Figure 19.25). Surface alteration rinds and veins cutting through the interiors of these rocks can also be clearly seen in some MI images, suggesting limited interactions of the rocks with water.

After analyzing rocks near the landing site, the Spirit rover traversed about 3 km across the plains and climbed Husband Hill, a promontory within Gusev crater (one of the Columbia Hills in Figure 19.11). The Hills outcrops are distinct from the basalts of the plains. Some are massive, others are laminated, and most are altered and deeply weathered (Figure 19.26). Pancam, Mini-TES, and MB spectra suggest highly varying mineralogy. Some rocks contain combinations of olivine, pyroxene, feldspar, and iron oxides (as on the plains), whereas others contain large amounts of glass, sulfate, ilmenite, and phosphate. APXS analyses have been used to divide the rocks into several different classes according to their chemistry, but the mineralogy can vary considerably even within a class. Some rocks appear to be relatively unaltered, but most show very high contents of sulfur, phosphorus, and chlorine, suggesting a high degree of alteration. The chemical compositions of these rocks are not illustrated in Figure 19.16, because this classification is only applicable to unaltered igneous rocks. The textures of Hills rocks, as revealed by the MI, are also highly variable but commonly indicate alteration of rocks composed of angular particles and clasts (Figure 19.26). RAT grinding indicates that these rocks are much softer than the basalts of the plains. They have been interpreted as mixtures of materials formed by impacts or explosive volcanic eruptions, and subsequently altered by fluids. Two classes of rocks on the northwest flank of Husband Hill have what appear to be roughly concordant dips to the northwest suggesting a **stratigraphy** (Figure 19.19). The lower rock has layered materials and angular to rounded clasts in a matrix that compares favorably to impact ejecta that has been altered by water to various extents. The upper rock is a finely layered sedimentary rock that has been cemented by sulfate, but the aqueous alteration did not affect the basaltic character of the sediment. A few distinctive rock types found as loose stones (geologists call these "float") in the Hills include Backstay, Irvine, and Wishstone, which are dark, fine-grained basaltic rocks with compositions distinct from the basalts of the plains (Figure 19.16) and only limited signs of alteration by water. These rocks appear to have formed by removal of crystals from magmas similar in composition to basalts of the plains.

Once Spirit gained the crest of Husband Hill, it traveled down the south face, encountering olivine-rich rocks of the Algonquin class (Figure 19.16). Although not recognized until later, carbonate-bearing rocks were also discovered on Husband Hill. Upon reaching the bottom, Spirit traversed an area containing highly vesicular rocks (scoria) to Home Plate, tentatively interpreted as a small volcanic edifice formed of ash. Outcrops of silica-rich rocks at Home Plate are thought to have formed by precipitation under hydrothermal conditions. Similar environments on Earth are habitable and the deposition of silica provides a ready mechanism for preservation of fossils. The compositions of all the relatively unaltered igneous rocks in Gusev crater are rich in alkalis and low in silica (Figure 19.16), allowing their classification as alkaline rocks. These are the first alkaline rocks recognized on Mars.

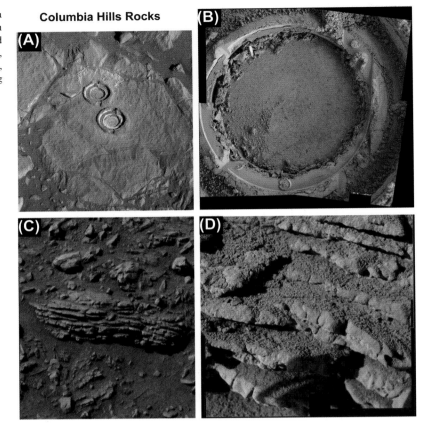

FIGURE 19.26 Images of rocks from the Columbia Hills in Gusev crater. (a) Pancam image of Wooly Patch after several RAT grinds showing darker interior and natural dusty surface. (b) MI of Wooly Patch RAT hole, showing clastic texture. (c) Pancam image of Tetl, which exhibits fine layering. (d) MI of Tetl, illustrating coherent layers separated by finer grained material.

Rocks at the Opportunity landing site are mostly exposed in the walls of impact craters and where the sand is thin. Outcrops in Eagle crater were studied extensively after landing (Figure 19.12), and thicker stratigraphic sections in Endurance and Victoria craters were analyzed later in the mission (Figure 19.20). Pancam and MI images (Figure 19.27) show that the rocks are finely laminated and sometimes exhibit cross-bedding (Figure 19.28), and RAT grinds indicate that they are very soft. At the microscopic scale, they consist of sand grains bound together by fine-grained cement. Small gray spherules, called "blueberries" (Figures 19.27 and 19.28), are embedded within the rock (the spherules are actually gray, but appear bluish in many false-color images). Some parts of the outcrop also exhibit tabular voids (Figure 19.27). APXS analyses of these rocks indicate very high concentrations of sulfur, chlorine, and bromine (highly water-soluble elements), demonstrating that the cement and sand (partially) consists of sulfate and halide salts. MB spectra reveal the presence of iron sulfate, and Mini-TES spectra suggest that magnesium and calcium sulfates also occur. The spherules are at least half hematite, the mineral seen from orbital TES spectra of the Meridiani region. The rocks are interpreted as sandstones composed of dirty evaporites of basaltic and sulfate composition formed by the evaporation of brines. Their textures suggest repeated cycles of flooding, exposure, and desiccation. Exposure and desiccation allowed some of the sediments to be mobilized into sand dunes (Figure 19.29). After deposition, the rocks underwent a number of different phases of diagenesis by groundwater of varying composition that circulated through the rocks, mobilizing and reprecipitating iron in the form of hematite spherules (concretions) and dissolving highly soluble minerals to leave the voids. Subsequently studied rocks in Victoria crater, 6 km from Endurance crater, are very similar in composition, indicating that the same stratigraphic sequence occurs over a wide area. Certain iron-bearing minerals present in the sulfates indicate that the water involved in their formation was highly acidic.

Several unusual rocks discovered by Opportunity deserve special mention. Bounce Rock, so named because the lander bounced on it as it rolled to a stop, was discovered on the Meridiani plains as the rover exited Eagle crater. Its chemical composition, as measured by APXS, is remarkably like the compositions of a group of Martian basaltic **meteorites** called shergottites (Figure 19.16). Its mineralogy is dominated by

Chapter | 19 Mars: Landing Site Geology, Mineralogy, and Geochemistry   415

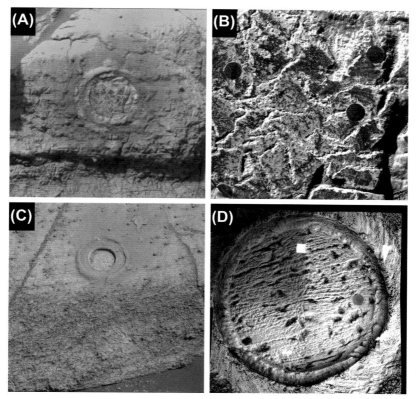

FIGURE 19.27 Images of Meridiani outcrops acquired by the Opportunity rover. (A) Pancam image of Guadalupe in Eagle crater, after RAT grinding. Notice slightly redder, dustier surface around the circular RAT hole and small hematite spherules protruding from the outcrop. (B) Microscopic image of Guadalupe RAT hole, showing blueberries (dark circles) and tabular voids produced by dissolution of soluble minerals. (C) Pancam image of Ontario in Endurance crater, after RAT grinding (circular smooth area). (D) Mosaic of microscopic images of Ontario, showing fine laminations, tabular voids, and a few blueberries (dark circles).

FIGURE 19.28 MI image mosaic of the Upper Dells in Eagle crater showing fine sand-sized particles making up the laminations, blueberries, and cuspate or curved cross-laminations that indicate that the sand-sized particles were deposited by running water.

FIGURE 19.29 Color image mosaic of evaporite outcrop of Burns Cliff at the rim of Endurance crater. The lower unit exposed in the lower left shows steeply dipping layers that are truncated by a middle layered unit with shallow dipping beds. The uppermost unit is lighter toned. The lowermost unit has been interpreted as eolian cross-beds that are truncated by the flatter beds of a sand sheet. The uppermost layer is interpreted as the unit deposited in running water of an ephemeral playa or saltwater lake.

pyroxenes and plagioclase, as are shergottites. This rock is obviously not in place and was probably lofted in as ejecta from a large impact crater to the south. Marquette Island is an unusual mafic rock that also was probably ejected from elsewhere on Mars. Heat Shield Rock, named for its proximity to the heat shield discarded during descent of the Opportunity lander, is likewise an interloper in this terrain. The Opportunity instruments revealed that it is an iron meteorite, composed of iron—nickel alloys, similar to some iron meteorites that fall to the Earth (see Meteorites chapter). Several other iron meteorites, some large in size and displaying ablation features formed as they came through the atmosphere as well as subsequent weathering features, were also found by Opportunity along its extended traverse.

Rocks in the rim of Endeavour crater studied by Opportunity (Figure 19.23) are impact **breccias**, similar to suevites in terrestrial impact craters. Localized enrichments in zinc suggest that some breccia materials were affected by hydrothermal alteration, and veins of gypsum arise from low-temperature aqueous fluids in these deposits. Light-toned clay bearing rocks were found beneath the breccias, pointing to low-pH aqueous conditions in the Noachian.

The Curiosity landing site is littered with small rocks that appear similar to a desert pavement or lag, similar to the Spirit landing site, left by winnowing away smaller sand-sized particles. Rover images show subsurface layers of strong conglomerates composed of subrounded particles cemented together that are related to the alluvial fan observed in orbital images (Figure 19.18). An angular, dark rock analyzed with APXS has a composition similar to an alkali-rich volcanic rock (mugearite). Clay minerals, composed of at least 20% of a mudstone outcrop in a network of stream channels, have been identified definitively by the ChemMin instrument. Curiosity's Sample Analysis at Mars (SAM) instrument also found a mixture of chemicals representing different oxidation states in the rock implying a habitable environment. Orbital spectra indicate that more clay- and sulfate-bearing sedimentary rocks will be found at the base of the Gale crater mound.

5.2. Soils

In addition to numerous soil analyses by the MPF, Spirit, Opportunity, and Curiosity rovers, soils were collected by scoops and analyzed at the two Viking landing sites and the Phoenix landing site. As defined by soil scientists on the Earth, "soil" usually contains a component of organic matter formed by decayed organisms. Soils on Mars do not contain measurable organic materials, but the term "soil" is nonetheless commonly used in planetary science ("regolith" is also used for the surface layer formed by the destruction of rocks).

Soil and dust on Mars are distinguishable based on particle size and spectral and thermal properties, although these materials are often comingled. Soil, normally dark, represents deposited materials, commonly of sand-sized grains (Figure 19.22). Bright reddish dust is much finer grained (several microns in size), and can either be suspended in the atmosphere or deposited on the ground. The top surface of soil is usually a thin layer of reddish dust, as seen by the color change when it is disturbed in rover tracks (Figure 19.22) or airbag bounces. Most measurements of soil mineralogy or chemistry represent a mixture of soil and dust, sometimes with an admixture of small particles of the local rocks.

At all these sites, the soils have broadly similar compositions, consisting of basaltic sands mixed with fine-grained dust and salts. Pancam and MB spectra of bright dust are dominated by nanophase ferric oxides, especially hematite, while Mini-TES spectra show evidence for plagioclase, minor carbonate, and an unidentified hydrous phase. MB spectra of dark soils indicate abundant olivine, pyroxene, and magnetite at the MER landing sites. The degree of alteration appears to be limited, but fractionation of chlorine and bromine in some soils suggest some mobilization by water. APXS chemical analyses show that plagioclase is also an important component of soils, and that their compositions resemble basalts with extra sulfur, chlorine, and bromine. At the Pathfinder site, local andesitic rock fragments are present in varying amounts, and at the Opportunity site, hematite spherules occur abundantly at the surface as a lag of granules. Trenches dug by the Spirit and Opportunity rovers reveal clods, suggesting greater proportions of salts that precipitated in the subsurface have sand bound into weakly cohesive near-surface layers or clumps, and APXS analyses of some subsurface soils show high concentrations of magnesium sulfate salt. Soils also contain significant amounts of nickel, which may reflect admixture of meteorite material into the regolith. Dust appears similar in composition to the soil (basaltic). Analysis of dust adhered to magnets on the rovers indicates that it contains olivine, magnetite, and a nanophase iron oxide (likely hematite) that suggests the dust is an oxidation or alteration product of fine-grained basalt. The presence of olivine in the dust suggests that liquid water was not heavily involved in its formation as it would have readily changed to other minerals (especially serpentine) in the presence of water.

An unusual silica- and titanium-rich soil, likely the result of abrasion of a silica-rich outcrop by the Spirit rover's stuck wheel, was discovered at Home Plate in Gusev crater. Soil at the Phoenix polar landing site is also distinct from low-latitude surface sediments. Although Phoenix did not have the capability to analyze bulk soil chemistry, it did measure unusual ratios of water-soluble elements, the presence of calcium carbonate, and the surprising occurrence of perchlorate ions, possibly formed by photochemical reactions in the atmosphere.

6. IMPLICATIONS FOR THE EVOLUTION OF MARS

6.1. Origin of Igneous Rocks

Igneous rocks form by partial melting of the planet's deep interior. The significance of the olivine-rich basaltic compositions found by Spirit on the Gusev cratered plains is that they appear to represent "primitive" magmas formed by melting in the mantle. Most magmas partly crystallize as they ascend toward the surface, losing the crystals in the process, so that the liquid progressively changes composition. Primitive magmas retain their original compositions and thus reveal the nature of their mantle source regions.

It is unlikely that rocks with andesitic composition at the MPF landing site formed by partial melting of the mantle, unless the mantle contains large quantities of water-bearing minerals. More likely, andesite lavas would form by partial melting of previously formed basaltic crustal rocks (the crust forms an outer layer above the mantle). A more likely alternative, previously mentioned, is that these rocks are not really andesites at all, but instead are basalts with silica-rich weathering rinds. The latter idea seems especially plausible considering that Surface Type 2 (andesitic) rocks are found primarily in places (like the northern lowlands) where surface waters would have collected and the sediments they carried would have been deposited. If this is correct, the orbital data and the samples of rocks at the various landing sites strongly argue that Mars is a basalt-covered world. Basalts, sediments derived from basalts, and dust derived from mildly weathered basalts are confirmed or suspected of the landing sites. Adding the thermal emission spectra of Type 1 and Type 2 materials as basalt and weathered basalt would suggest that most of Mars is made of this primitive volcanic rock.

6.2. Chemical Evolution and Surface Water

The iron-bearing sulfate jarosite is one of the minerals that formed in outcrops of evaporites at the Opportunity landing site. This mineral could only have precipitated from highly acidic water. Any sea at Meridiani was more like battery acid than drinking water. Given the abundance of basaltic lavas on the Martian surface, it is surprising that these waters would be so acidic. Reactions between water and basalt on the Earth tend to produce neutral to basic solutions. On Mars, huge volumes of sulfur and chlorine emitted from volcanoes must have combined with water to make sulfuric and hydrochloric acids. Only a few locations on the Earth mimic this kind of fluid—mostly areas devastated by acidic waters released by weathering of sulfides that drained from mines. Acidic water dissolves and precipitates different minerals than the neutral waters we are more familiar with on the Earth. Carbonates are not precipitated, and iron sulfates are more common in acidic solutions.

The presence of significant amounts of sulfate and chloride in soils from all the landing sites further suggests that acidic waters may have been common at one time in many places on Mars. Either evaporites like those at Meridiani Planum were abundant and have been redistributed as small particles throughout the planet's regolith or they occur as cements formed by groundwater leaching all over Mars. Results from the visible to near-infrared spectrometers on Mars orbiters support the finding of abundant sulfates elsewhere on Mars.

The occurrence of clay minerals in ancient highlands rocks is inconsistent with acidic fluids and instead suggests neutral to slightly basic water. This could signify a change in environmental conditions. Because the clay minerals generally appear in terrains older than the sulfates, this change in chemistry could be associated with decreasing amounts of water that became more acid at the tail end of a possibly wet period, prior to the dry modern era. The presence of carbonates in the soil, carbonate discovered by Spirit in the Columbia Hills, and alteration carbonates in the ancient, heavily cratered terrain suggest that acidic fluids were not present and/or pervasive everywhere (or the carbonates would have been destroyed).

6.3. Weathering on Mars

There is considerable controversy about the degree to which Mars rocks are weathered. Weathering by acidic water preferentially attacks olivine, and the surface layers and rinds of rocks at the MER sites appear to be depleted in olivine. However, remote sensing indicates that olivine is a common mineral in many places on Mars, and olivine appears to be a ubiquitous constituent of Martian soils and dust. Perhaps weathering was more common in the distant past, when acidic waters were abundant and produced outcrops like those found by Opportunity. Then the acid waters disappeared, and since that time, the lavas that were erupted and the soils that formed have only experienced limited weathering.

Visible and near-infrared spectral data indicate that clay minerals occur in some localities in the ancient terrains of Mars. Clays have been suggested to be present in some rocks on Husband Hill (Gusev crater) and at Endeavour crater (Meridiani Planum), based on aspects of their chemistry. Their occurrence was measured directly by ChemMin in Gale crater. Clay minerals can form by weathering, but they may also form by subsurface hydrothermal activity. Weathering processes clearly occurred at the Phoenix landing site. The soils there are clearly chemically altered, and soil particles are bound together with subsurface ice.

6.4. Eolian Processes

The remarkable uniformity of soil compositions at all the landing sites, some separated by thousands of kilometers, suggests an efficient homogenization process, although soils nearly everywhere on Mars were probably made from similar (basaltic) rocks. Eolian transport of rock particles along the surface by the wind has apparently mixed these materials very efficiently, so that the soil everywhere on Mars represents a rather homogeneous stratigraphic layer.

In contrast to sand, tiny dust particles can be suspended in the atmosphere and circulated globally, which has also created a homogeneous material that is distributed globally. A dust cycle can be inferred from the omnipresent dusty atmosphere being supplied by dust devils and other processes that occasionally lead to globe-encircling storms. Dust deposition has been observed on most of the landed spacecraft at a rate that is so high that it must be removed at a similar rate (or the surfaces would be quickly buried by thick accumulations of dust). Dust currently may be or previously has been deposited at a higher rate overall in broad areas of the planet that have very low **thermal inertia** and very high albedo.

Sand-sized particles created by impact and other processes have been harnessed by the wind to form sand dunes and other eolian bedforms. The consistent basaltic composition of the soil and dust all over Mars further argues that Mars is dominated by basaltic rocks and that the soil and dust form by physical weathering and minor oxidation without large quantities of water. This further argues that these weathering products have formed and been mobilized by the wind in the current dry and desiccating environment.

6.5. Geologic Evolution of the Landing Sites and Climate

Study of the geology, geomorphology, and geochemistry of the seven landing sites in context with their regional geologic setting allows some constraints to be placed on the environmental and climatic conditions on Mars through time. The Viking 1 landing site shows sedimentary drift and soil deposits over angular, dark, presumably volcanic rocks with local outcrops (Figure 19.7). The location of this site on the ridged plains terrain downstream from the mouth of Maja and Kasei Valles suggests that the site is on layered basalts (the preferred interpretation of the ridged plains) with rocks, soils, and drifts derived from impact ejecta, flood, and eolian processes. The rocks at the Viking 2 landing site (Figure 19.8), in the midnorthern plains, are angular and pitted consistent with being volcanic rocks as part of the distal ejecta from Mie crater. High-resolution orbiter images show that the surface has a small-scale hummocky character and lander images show small polygonal sediment-filled troughs, both suggesting that the surface has been partially shaped by the presence of ground ice. The density of craters observed from orbit at both sites places them as Hesperian in age and constraints on the geomorphologic development of the sites suggest very little erosion or change of the surfaces since they formed.

Many characteristics of the MPF landing site (Figures 19.9 and 19.24) are consistent with its being a plain composed of materials deposited by catastrophic floods as suggested by its location near the mouth of the Ares Vallis catastrophic outflow channel. Some of the rocks potentially identified (conglomerate, pillow basalt) are suggestive of a wetter past. However, given that the surface still appears similar to that expected for a fresh depositional fan, any erosional and/or depositional process appears to have been minimal since it formed around 3 billion years ago.

The cratered surface of Gusev that Spirit has traversed (exclusive of the Columbia Hills) is generally low-relief, moderately rocky plains dominated by hollows, which appear to be small craters filled with soil (Figure 19.10). The plain formed by basaltic volcanism with impacts producing an unconsolidated regolith greater than 10 m thick (Figure 19.11). The observed gradation and deflation of ejected fines and deposition in craters to form hollows thus provides a measure of the rate of erosion or redistribution of mobile sediment since the plains formed about 3.5 billion years ago. These rates of erosion are so slow that they provide a broad indicator of a climate that has been cold and dry. Taken together, the slow rates of change inferred from the Viking-, Pathfinder-, and Gusev cratered plains landing sites argues for a dry and desiccating climate similar to today's for the past ~3.6 billion years.

Rocks in the Columbia Hills (Figure 19.19) sampled by the Spirit rover reveal an earlier period in which liquid water was present. The Columbia Hills appear to be older materials that were either uplifted or eroded before deposition of the basalts responsible for the cratered plains. The basalts of the cratered plains are Hesperian in age and so the Columbia Hills rocks are likely older (Noachian in age). These rocks record impact and explosive volcanic processes, but many have been heavily altered or deeply weathered by water. In contrast, soils in the Columbia Hills are similar to basaltic soils elsewhere, suggesting that these formed and were deposited later in the cold and dry Martian climate.

The geology and geomorphology of the Meridiani Planum landing site explored by the Opportunity rover shows clear evidence for an earlier warmer and wetter environment followed by a drier period dominated by eolian activity. The layered rocks examined by Opportunity are older than 3.6 billion years based on the density of highly eroded large craters observed in orbital images (Figure 19.5). These rocks are dirty evaporites composed of materials that have precipitated from salty water and been mobilized and moved by the wind (Figures 19.20 and 19.29) before being deposited and altered by groundwater. On Earth, this sequence of events and resulting rocks is common in hot and dry saltwater playa or sabkha environments such as the Persian Gulf, the Gulf of California, and some inland enclosed basins. By analogy, the environment on Mars was warmer and wetter when these rocks were deposited more than 3.6 billion years ago. Because the evaporites are part of a sedimentary sequence that outcrops throughout the broad Meridiani region, these climatic conditions were operative over an area that was at least 1000 km wide, arguing that the environment was both warmer and wetter and the atmosphere was thicker. Later in Mars' history, the environment changed and Meridiani Planum was dominated by eolian activity that eroded and filled in impact craters and concentrated the hematite spherules as a lag on the top of the layer of basaltic sand. The presence of olivine in the basaltic sand suggests that these materials were not weathered by liquid water and that the saltation of the sand appears to have efficiently eroded the weak sulfates.

The Phoenix landing site clearly reveals that the polar environment is distinct from other regions. The presence of perchlorate in soils also reveals atmospheric processes that are not so common elsewhere. The surface has pervasive polygonal troughs and very few impact craters suggesting a near surface that is constantly being modified by thermal contraction processes in the ground ice. The polar ice caps are even more dynamic, with annual cycles of sublimation and condensation of ices, and probably represent a much more geologically active environment than the rest of present-day Mars.

6.6. Implications for a Habitable World

The Meridiani Planum evaporites and Columbia Hills rocks in Gusev crater indicate a warmer and wetter environment before about 3.6 billion years ago. This is consistent with a variety of coeval geomorphic indicators such as valley networks, degraded and filled ancient craters, highly eroded terrain, and layered sedimentary rocks that point to an early warm and wet climate. The presence of clays and other alteration products that formed in neutral aqueous conditions are consistent with this early wet period. The highly acidic conditions that the sulfates formed in are consistent with a drying environment with less water. A warmer and wetter environment would also imply a thicker atmosphere capable of supporting liquid water. In contrast, the surficial geology of the landing sites younger than about 3.6 billion years all indicate a dry and desiccating environment in which liquid water was not stable and eolian and impact processes dominated. This further indicates that a major climatic change occurred around 3.6 billion years ago.

A warmer and wetter environment before 3.6 billion years ago suggests that Mars was possibly habitable at a time when life started on the Earth. The highly acidic nature of water at some Mars landing sites may not have been conducive to the appearance of early organisms, but clays and carbonates, which formed in earlier neutral aqueous conditions, may indicate more habitable conditions. In any case, the earliest chemical evidence for life on Earth is about 3.6 billion years old and the most important ingredient for life on Earth is liquid water. If liquid water was stable on Mars when life began on Earth, could a second genesis on Mars have occurred? Is it possible that life actually started on Mars earlier when it was more clement than Earth, which was subject to early giant possibly sterilizing impacts, and was later transported to the Earth via meteorites ejected off the Martian surface? Will life form at any place where liquid water is stable or is it a rare occurrence? These are the compelling questions that can be addressed by missions in our ongoing exploration of Mars.

BIBLIOGRAPHY

Bell, J. (Ed.). (2008). *The Martian surface: Composition, mineralogy, and physical properties.* Cambridge University Press, 636 pages. This book presents a wealth of information on what has been learned about Mars from orbiters and landers through the MER missions.

Golombek, M. P., Cook, R. A., Economou, T., Folkner, W. M., Haldemann, A. F. C., Kallemeyn, et al. (1997). Overview of the Mars

Pathfinder mission and assessment of landing site predictions. *Science*, *278*, 1743−1748, and the next 5 papers in Science (Vol. 278, pp. 1734−1774, December 2, 1997) in which the scientific results of the Mars Pathfinder mission were first reported.

Golombek, M., Arvidson, R. E., Bell, J. F., III., Christensen, P. R., Crisp, J. A., & Crumpler, L. S. (7 July 2005). Assessment of Mars Exploration Rover landing site predictions. *Nature*, *436*, 44−48. http://dx.doi.org/10.1038/nature03600. The next 5 papers in Nature, Vol. 436, Issue 7047, pp. 42−70, in which further results from the Mars Exploration Rovers were reported.

Golombek, M., Grant, J., Kipp, D., Vasavada, A., Kirk, R., et al. (2012). Selection of the Mars science laboratory landing site. *Space Science Reviews*, *170*, 641−737. http://dx.doi.org/10.1007/s11214-012-9916-y. which describes the selection of the landing site. See also other papers in this volume (pp. 1−860) that describe the pre-landing mission, instruments and various aspects of the landing site.

Golombek, M. P., Anderson, R. C., Barnes, J. R., Bell, J. F., III., Bridges, N. T., et al. (1999). Overview of the Mars Pathfinder mission: launch through landing, surface operations, data sets, and science results. *Journal of Geophysical Research*, 104, 8523−8553. Special issues of the Journal of Geophysical Research, Planets (Vol. 104, pp. 8521−9096, April 25, 1999; and Vol. 105, pp. 1719−1865, January 25, 2000) also featured the scientific results of the mission.

Grotzinger, J. P., Sumner, D. Y., Kah, L. C., Stack, K., Gupta, S., et al. (2014). A Habitable Fluvio-Lacustrine Environment at Yellowknife Bay, Gale Crater, Mars. *Science*, 343(6169) [DOI:10.1126/science.1242777], in which geology of Gale crater is described along with 5 other papers in the same volume.

Kieffer, H. H., Jakosky, B. M., Snyder, C. W., & Matthews, M. S. (Eds.). (1992). *Mars*. University of Arizona Press (This volume summarizes our knowledge of Mars through the Viking era of exploration).

Smith, P. H., et al. (2009). H_2O at the phoenix landing site. *Science., 325*, 58−61. and the next three papers (pp. 61−70) in which the initial results from the phoenix landing were reported.

Squyres, S. W., Arvidson, R., Bell J. F., III., Brückner, J., Cabrol, N. A., et al. (2004). The spirit rover's Athena science investigation at Gusev crater, Mars. *Science*, *305*(5685), 794−799, DOI: 10.1126/science.1100194, and the next 10 papers in Science: Vol. *305*(5685), pp. 793−845, in which the first results of the Spirit rover were reported.

Squyres, S. W., Arvidson, R., Bell J. F., III., Brückner, J., Cabrol, N. A., et al. (2004), The Opportunity rover's Athena science investigation at Meridiani Planum, Mars. *Science*, *306*(5702), 1698−1703, DOI: 10.1126/science.1106171 and the next 10 papers in Science, Vol. 306, Issue 5702, pp. 1697−1756, in which the first results of the Opportunity rover were reported.

Part V

Earth and Moon as Planets

Chapter 20

Earth as a Planet: Atmosphere and Oceans

Adam P. Showman
Department of Planetary Sciences, Lunar and Planetary Laboratory, University of Arizona, Tucson, Arizona

Timothy E. Dowling
Department of Physics and Astronomy, University of Louisville, Louisville, Kentucky

Chapter Outline

1. Overview of Planetary Characteristics — 424
  1.1. Length of Day — 424
2. Vertical Structure of the Atmosphere — 425
  2.1. Troposphere — 426
  2.2. Stratosphere — 426
  2.3. Mesosphere — 427
  2.4. Thermosphere — 427
  2.5. Exosphere and Ionosphere — 427
3. Atmospheric Circulation — 428
  3.1. Processes Driving the Circulation — 428
  3.2. Influence of Rotation — 428
  3.3. Observed Global-Scale Circulation — 429
  3.4. Insights from Other Atmospheres — 432
4. Oceans — 433
  4.1. Oceanic Structure — 433
  4.2. Ocean Circulation — 434
  4.3. Salinity — 435
  4.4. Atmosphere–Ocean Interactions — 436
  4.5. Oceans on Other Worlds — 436
5. Climate — 436
  5.1. Basic Processes: Greenhouse Effect — 437
  5.2. Basic Processes: Feedbacks — 437
  5.3. Recent Times — 438
  5.4. Ice Ages — 439
  5.5. Volatile Inventories of Terrestrial Planets — 441
6. Life in the Atmosphere–Ocean System — 441
  6.1. Interplanetary Spacecraft Evidence for Life — 441
    6.1.1. Radio Emissions — 442
    6.1.2. Surface Features — 443
    6.1.3. Oxygen and Methane — 443
7. Conclusions — 444
Bibliography — 444

Earth is the only planet that orbits the Sun in the distance range within which water occurs in all three of its phases at the surface (as solid ice caps, liquid oceans, and atmospheric water vapor), which results in several unusual characteristics. Earth is unique in the solar system in exhibiting a global ocean at the surface, which covers almost three-quarters of the planet's area (such that the total amount of dry land is about equal to the surface area of Mars). The ocean exerts a strong control over the planet's climate by transporting heat from equator to pole, interacting with the atmosphere chemically and mechanically, and, on geological timescales, influencing the exchange of **volatiles** between the planet's atmosphere and interior. The Earth's atmosphere follows the general pattern of a troposphere at the bottom, a stratosphere in the middle, and a thermosphere at the top. There is the usual east–west organization of winds, but with large north–south and temporal fluctuations. Many of the atmospheric weather patterns (jet streams, Hadley cells, vortices, thunderstorms) occur on other planets too, but their manifestation on the Earth is distinct and unique. The Earth's climate has varied wildly over time, with atmospheric CO_2 and surface temperature fluctuating in response to ocean chemistry, planetary orbital variations, feedbacks between the atmosphere and interior, and a 30% increase in solar luminosity over the past 4.6 billion years (Ga). Despite these variations, the Earth's climate has remained temperate, with at least partially liquid oceans, over the entire recorded ~3.8-Ga geological record of the planet. Life has had a major influence on the ocean–atmosphere system, and as a result it is possible to discern the presence of life from remote spacecraft data. Global biological activity is indicated by the presence of atmospheric gases such as oxygen and methane that are in extreme thermodynamic disequilibrium, and by the widespread presence of a red-absorbing pigment (chlorophyll) that does not match the spectral signatures of any known rocks or minerals. The presence of intelligent life on Earth can be discerned from stable radio-wavelength signals emanating from the planet that do not match naturally occurring signals but do contain regular pulsed modulations that are the signature of information exchange.

1. OVERVIEW OF PLANETARY CHARACTERISTICS

Atmospheres are found on the Sun, eight planets, and seven of the 60-odd satellites, for a total count of 16—in addition to the atmospheres that exist around the ~1000 known gas giant planets orbiting other stars. Each has its own brand of weather and its own unique chemistry. They can be divided into two major classes: the terrestrial planet atmospheres, which have solid surfaces or oceans as their lower boundary condition, and the gas giant atmospheres, which are essentially bottomless. Venus and Titan form one terrestrial subgroup that is characterized by a slowly rotating planet, and interestingly, both exhibit a rapidly rotating atmosphere. Mars, Io, Triton, and Pluto form a second terrestrial subgroup that is characterized by a thin atmosphere, which in large measure is driven by vapor—pressure equilibrium with the atmosphere's solid phase on the surface. Both Io and Triton have active volcanic plumes. Earth, along with Mars and the giant planets, is in the rapidly rotating regime where the Coriolis force plays a dominant role. And although regional lakes of methane—ethane mixtures exist near the poles of Titan, Earth is also the only planet with a global (planet-encircling) ocean at the surface (see Venus: Atmosphere; Io: The Volcanic Moon; Triton; and Pluto).

Earth has many planetary attributes that are important to the study of its atmosphere and oceans, and conversely, there are several ways in which its physically and chemically active fluid envelope directly affects the solid planet. Earth orbits the Sun at a distance of only 108 times the diameter of the Sun. The warmth from the Sun that the Earth receives at this distance, together with a 30 K increase in surface temperature resulting from the atmospheric greenhouse effect, leads to temperatures allowing H_2O to appear in all three of its phases. This property of the semimajor axis of Earth's orbit is the most important physical characteristic of the planet that supports life.

Orbiting the Sun at just over 100 Sun diameters is not as close as it may sound; a good analogy is to view a basketball placed just past first base while standing at home plate on a baseball diamond. For sunlight, the Sun-to-Earth trip takes 499 s or 8.32 min. Earth's semimajor axis, $a_3 = 1.4960 \times 10^{11}$ m = 1 AU (astronomical unit), and orbital period, $\tau_3 = 365.26$ days = 1 year, where the subscript 3 denotes the third planet out from the Sun, are used as convenient measures of distance and time. When the orbital period of a body encircling the Sun, τ, is expressed in years, and its semimajor axis, a, is expressed in astronomical units, then Kepler's third law is simply $\tau = a^{3/2}$, with a proportionality constant of unity (see Solar System Dynamics: Regular and Chaotic Motion).

1.1. Length of Day

The Earth's rotation (see Solar System Dynamics: Rotation of the Planets) has an enormous effect on the motions of its fluid envelope that accounts for the circular patterns of large storms like hurricanes, the formation of **western boundary currents** like the Gulf Stream, the intensity of jet streams, the extent of the Hadley cell, and the nature of fluid instabilities. All these processes are discussed in Sections 2—5. Interestingly, the reverse is also true: The Earth's atmosphere and oceans have a measurable effect on the planet's rotation rate. For all applications but the most demanding, the time the Earth takes to turn once on its axis, the length of its day, is adequately represented by a constant value equal to 24 h or 1440 min or 86,400 s. The standard second is the Système International (SI) second, which is precisely 9,192,631,770 periods of the radiation corresponding to the transition between two hyperfine levels of the ground state of the ^{133}Cs atom. When the length of day is measured with high precision, it is found that Earth's rotation is not constant. The same is likely to hold for any dynamically active planet. Information can be obtained about the interior of a planet, and how its atmosphere couples with its surface, from precise length-of-day measurements. Earth is the only planet to date for which we have achieved such accuracy, although we also have high-precision measurements of the rotation rate of pulsars, the spinning neutron stars often seen at the center of supernova explosions.

The most stable pulsars lose only a few seconds every million years and are the best-known timekeepers, even better than atomic clocks. In contrast, the rotating Earth is not an accurate clock. Seen from the ground, the positions as a function of time of all objects in the sky are affected by Earth's variable rotation. Because the Moon moves across the sky relatively rapidly and its position can be determined with precision, the fact that Earth's rotation is variable was first realized when a series of theories that should have predicted the motion of the Moon failed to achieve their expected accuracy. In the 1920 and the 1930s, it was established that errors in the position of the Moon were similar to errors in the positions of the inner planets, and by 1939, clocks were accurate enough to reveal that Earth's rotation rate has both irregular and seasonal variations.

The quantity of interest is the planet's three-dimensional angular velocity vector as a function of time, $\Omega(t)$. Since the 1970s, time series of all three components of $\Omega(t)$ have been generated by using very long baseline interferometry for purposes ranging from accurately determining the positions of quasars and laser ranging to accurately determining the positions of man-made satellites and the Moon, the latter with corner reflectors placed on the Moon by the *Apollo* astronauts (*see* Planetary Exploration Missions; The Moon).

The theory of Earth's variable rotation combines ideas from geophysics, meteorology, oceanography, and astronomy. The physical causes fall into two categories: those that change the planet's moment of inertia (like a spinning skater pulling in her arms) and those that torque the planet by applying stresses (like dragging a finger on a spinning globe). Earth's moment of inertia is changed periodically by tides raised by the Moon and the Sun, which distort the solid planet's shape. Nonperiodic changes in the solid planet's shape occur because of fluctuating loads from the fluid components of the planet, namely, the atmosphere, the oceans, and, deep inside the planet, the liquid iron-rich core. In addition, shifts of mass from earthquakes and melting ice cause nonperiodic changes. Over long timescales, plate tectonics and mantle convection significantly alter the moment of inertia and hence the length of day.

An important and persistent torque that acts on the Earth is the gravitational pull of the Moon and the Sun on the solid planet's tidal bulge, which, because of friction, does not line up exactly with the combined instantaneous tidal stresses. This torque results in a steady lengthening of the day at the rate of about 0.0014 s/century and a steady outward drift of the Moon at the rate of 3.7 ± 0.2 cm/year, as confirmed by lunar laser ranging. On the top of this steady torque, it has been suggested that observed 0.005-s variations that have timescales of decades are caused by stronger, irregular torques from motions in Earth's liquid core. Calculations suggest that viscous coupling between the liquid core and the solid mantle is weak, but that electromagnetic and topographic coupling can explain the observations. Mountains on the core–mantle boundary with heights around 0.5 km are sufficient to produce the coupling and are consistent with seismic tomography studies, but not much is known about the detailed topography of the core–mantle boundary. Detailed model calculations take into account the time variation of Earth's external magnetic field, which is extrapolated downward to the core–mantle boundary. New improvements to the determination of the magnetic field at the surface are enhancing the accuracy of the downward extrapolations (*see* Earth as a Planet: Surface and Interior).

Earth's atmosphere causes the strongest torques of all. The global atmosphere rotates faster than the solid planet by about 10 m/s on an average. Changes in the global circulation cause changes in the pressure forces that act on mountain ranges and changes in the frictional forces between the wind and the surface. Fluctuations on the order of 0.001 s in the length of day, and movements of the pole by several meters, are caused by these meteorological effects, which occur over seasonal and interannual timescales. General circulation models (GCMs) of the atmosphere routinely calculate the global atmospheric angular momentum, which allows the meteorological and non-meteorological components of the length of day to be separated. All the variations in the length of day over weekly and daily timescales can be attributed to exchanges of angular momentum between Earth's atmosphere and the solid planet, and this is likely to hold for timescales of several months as well. Episodic reconfigurations of the coupled atmosphere–ocean system, such as the **El Niño-Southern Oscillation**, cause detectable variations in the length of day, as do changes in the stratospheric jet streams.

2. VERTICAL STRUCTURE OF THE ATMOSPHERE

The Earth may differ in many ways from the other planets, but not in the basic structure of its atmosphere (Figure 20.1). Planetary exploration has revealed that essentially every atmosphere starts at the bottom with a **troposphere**, where temperature decreases with height at a nearly constant rate up to a level called the tropopause, and then has a **stratosphere**, where temperature usually increases with height or, in the case of Venus and Mars, decreases much less quickly than in the troposphere. It is interesting to note that atmospheres are warm both at their bottoms and their tops, but do not get arbitrarily cold in their interiors. For example, on Jupiter and Saturn there is significant methane gas throughout their atmospheres, but nowhere does it get cold enough for methane clouds to form, whereas in the much colder atmospheres of Uranus and Neptune, methane clouds do form. Details vary in the middle atmosphere regions from one planet to another, where photochemistry is important, but each atmosphere is topped off by a high-temperature, low-density thermosphere that is sensitive to solar activity and an exobase, the official top of an atmosphere, where molecules float off into space when they achieve escape velocity (*see* Venus: Atmosphere; Mars Atmosphere: History and Surface Interactions; Atmospheres of the Giant Planets).

Interestingly, the top of the troposphere occurs at about the same pressure, about 0.1–0.3 bar, on most planets (Figure 20.1). This similarity is not coincidental but instead results from the pressure dependence of the atmospheric opacity on solar and especially infrared radiation. In the high-pressure regime of tropospheres, the gas is relatively opaque at infrared wavelengths, which inhibits heat loss by radiation from the deep levels and hence promotes a profile where temperature decreases strongly with altitude. In the low-pressure regime of stratospheres, the gas becomes relatively transparent at infrared wavelengths, which allows

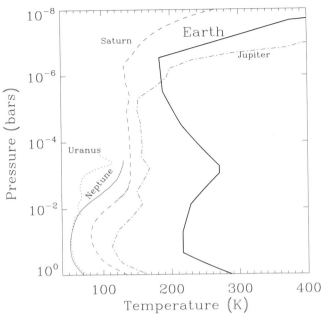

FIGURE 20.1 Representative temperature structure for the Earth (thick solid line) as compared with those of several other planets, including Jupiter (dash-dot), Saturn (dashed), Uranus (dotted), and Neptune (solid). For Earth, the altitude scale runs from the surface to about 130 km altitude. Atmospheres have high pressure at the bottom and low pressure at the top, so pressure is a proxy for altitude. Starting at the bottom of Earth's atmosphere and moving up, the troposphere, stratosphere, mesosphere, and thermosphere correspond to layers where temperature decreases, increases, decreases, and then increases with height, respectively. The top of Earth's troposphere, stratosphere, and mesosphere are at altitudes of about 10–15, 50, and 100 km, respectively. Note that other planets also generally have tropospheres and thermospheres, although the details of the intermediate layers (the stratosphere and mesosphere) differ from planet to planet.

the temperature to become more constant—or in some cases even increase—with altitude. This transition from opaque to transparent tends to occur at pressures of 0.1–0.3 bar for the compositions of most planetary atmospheres in our solar system.

In the first 0.1 km of a terrestrial atmosphere, the effects of daily surface heating and cooling, surface friction, and topography produce a turbulent region called the planetary boundary layer. Right at the surface, molecular viscosity forces the "no slip" boundary condition and the wind reduces to zero, such that even a weak breeze results in a strong vertical wind shear that can become turbulent near the surface. However, only a few millimeters above the surface, molecular viscosity ceases to play a direct role in the dynamics, except as a sink for the smallest eddies.

Up to altitudes of about 80 km, Earth's atmosphere is composed of 78% N_2, 21% O_2, 0.9% Ar, and 0.002% Ne by volume, with trace amounts of CO_2, CH_4, and numerous other compounds. Water exists in abundances up to $\sim 1\%$ at the surface in the tropics, less at the poles, and dropping to a few parts per million in the stratosphere. Diffusion, chemistry, and other effects substantially alter the composition at altitudes above ~ 90 km.

2.1. Troposphere

The troposphere is the lowest layer of the atmosphere, characterized by a temperature that decreases with altitude (Figure 20.1). The top of the troposphere is called the tropopause, which occurs at an altitude of 18 km at the equator but only 8 km at the poles (the cruising altitude of commercial airliners is typically 10 km). Gravity, combined with the compressibility of air, causes the density of an atmosphere to fall off exponentially with height, such that Earth's troposphere contains 80% of the mass and most of the water vapor in the atmosphere, and consequently most of the clouds and stormy weather. Vertical mixing is an important process in the troposphere. Temperature falls off with height at a predictable rate because the air near the surface is heated and becomes light and the air higher up cools to space and becomes heavy, leading to an unstable configuration and convection. The process of convection relaxes the temperature profile toward the neutrally stable configuration, called the adiabatic temperature lapse rate, for which the decrease of temperature with decreasing pressure (and hence increasing height) matches the drop-off of temperature that would occur inside a balloon that conserves its heat as it moves, that is, moves adiabatically. In reality, latent heating due to water vapor—and horizontal heat transports—causes the temperature profile to decrease slightly less with height than such an adiabat. As a result, the troposphere is slightly stable to convection. Nevertheless, the adiabat provides a reasonable reference for the troposphere.

In the troposphere, water vapor, which accounts for up to $\sim 1\%$ of air, varies spatially and decreases rapidly with altitude. The water vapor mixing ratio in the stratosphere and above is almost four orders of magnitude smaller than that in the tropical lower troposphere.

2.2. Stratosphere

The nearly adiabatic falloff of temperature with height in the Earth's troposphere gives way above the tropopause to an increase of temperature with height. This results in a rarified, stable layer called the stratosphere. Observations of persistent, thin layers of aerosol and of long residence times for radioactive trace elements from nuclear explosions are direct evidence of the lack of mixing in the stratosphere. The temperature continues to rise with altitude in the Earth's stratosphere until one reaches the stratopause at about 50 km. The source of heating in the Earth's stratosphere is absorption of solar ultraviolet (UV) light by ozone. Ozone itself results from photochemistry, and exhibits abundances that peak at about 25 km. The Sun's UV radiation causes stratospheres to form in other atmospheres, but instead of the absorber being ozone, which is plentiful on the Earth because of the high concentrations of O_2 maintained by the biosphere, other gases absorb the UV radiation. On the giant planets, methane, hazes, and aerosols do the job.

The chemistry of Earth's stratosphere is complicated. Ozone is produced mostly over the equator, but its largest concentrations are found over the poles, meaning that dynamics is as important as chemistry to the ozone budget. Some of the most important chemical reactions in Earth's stratosphere are those that involve only oxygen. Photodissociation by solar UV radiation involves the reactions $O_2 + h\nu \to O + O$ and $O_3 + h\nu \to O + O_2$, where $h\nu$ indicates the UV radiation. Three-body collisions, where a third molecule, M, is required to satisfy conservation of momentum and energy, include $O + O + M \to O_2 + M$ and $O + O_2 + M \to O_3 + M$, but the former reaction proceeds slowly and may be neglected in the stratosphere. Reactions that either destroy or create "odd" oxygen, O or O_3, proceed at much slower rates than reactions that convert between odd oxygen. The equilibrium between O and O_3 is controlled by fast reactions that have rates and concentrations that are altitude dependent. Other reactions that are important to the creation and destruction of ozone involve minor constituents such as NO, NO_2, H, OH, HO_2, and Cl. An important destruction mechanism is the catalytic cycle $X + O_3 \to XO + O_2$ followed by $XO + O \to X + O_2$, which results in the net effect $O + O_3 \to 2O_2$. On the Earth, human activity has led to sharp increases in the catalysts $X = Cl$ and NO and subsequent sharp decreases in stratospheric ozone, particularly over the polar regions. The Montreal Protocol is an international treaty signed in 1987 that is designed to stop and eventually reverse the damage to the stratospheric ozone layer; regular meetings of the parties, involving some 175 countries, continually update the protocol.

2.3. Mesosphere

Above Earth's stratopause, temperature again falls off with height, although at a slower rate than in the troposphere. This region is called the mesosphere. Earth's stratosphere and mesosphere are often referred to collectively as the middle atmosphere. Temperatures fall off in the mesosphere because there is less heating by ozone and emission to space by carbon dioxide is an efficient cooling mechanism. The mesopause occurs at an altitude of about 80 km, marking the location of a temperature minimum of about 130 K.

2.4. Thermosphere

As is the case for ozone in Earth's stratosphere, above the mesopause, atomic and molecular oxygen strongly absorb solar UV radiation and heat the atmosphere. This region is called the thermosphere, and temperatures rise with altitude to a peak that varies between about 500 and 2000 K depending on solar activity. Just as in the stratosphere, the thermosphere is stable to vertical mixing. At about 120 km, molecular diffusion becomes more important than turbulent mixing, and this altitude is called the homopause (or turbopause). Rocket trails clearly mark the homopause—they are turbulently mixed below this altitude but mixed primarily by molecular diffusion above it, causing the rocket trails to appear differently above and below the interface. Molecular diffusion is mass-dependent and each species falls off exponentially with its own scale height, leading to elemental fractionation that enriches the abundance of the lighter species at the top of the atmosphere.

For comparison with Earth, the structure of the thermospheres of the giant planets has been determined from *Voyager* spacecraft observations, and the principal absorbers of UV light are H_2, CH_4, C_2H_2, and C_2H_6. The thermospheric temperatures of Jupiter, Saturn, and Uranus are about 1000, 420, and 800 K, respectively. The high temperature and low gravity on Uranus allow its upper atmosphere to extend out appreciably to its rings (*see* Atmospheres of the Giant Planets).

2.5. Exosphere and Ionosphere

At an altitude of about 500 km on the Earth, the mean free path between molecules grows to be comparable to the density scale height (the distance over which density falls off by a factor of $e \approx 2.7128$). This defines the exobase and the start of the exosphere. At these high altitudes, sunlight can remove electrons from atmospheric constituents and form a supply of ions. These ions interact with a planet's magnetic field and with the solar wind to form an ionosphere. On Earth, most of the ions come from molecular oxygen and nitrogen, whereas on Mars and Venus most of the ions come from carbon dioxide. Because of the chemistry, however, ionized oxygen atoms and molecules are the most abundant ions for all three atmospheres. Mercury and the Moon have exospheres right down to the planetary surface, with ions supplied from the surface crust and the solar wind.

Mechanisms of atmospheric escape fall into two categories, thermal and nonthermal. Both processes provide the kinetic energy necessary for molecules to attain escape velocity. When escape velocity is achieved at or above the exobase, such that further collisions are unlikely, molecules escape the planet. In the thermal escape process, some fraction of the high-velocity wing of the Maxwellian distribution of velocities for a given temperature always has escape velocity; the number increases with increasing temperature. An important nonthermal escape process is dissociation, both chemical and photochemical. The energy for chemical dissociation is the excess energy of reaction, and for photochemical dissociation, it is the excess energy of the bombarding photon or electron, which is converted into kinetic energy in the dissociated atoms. A common effect of electrical discharges of a kilovolt or more is "sputtering", where several atoms can be ejected from the spark region at high velocities. If an ion is formed very high in the atmosphere, it can be swept out of a planet's atmosphere by the solar wind. Similarly, at Io, ions are swept away by Jupiter's magnetic field. Other nonthermal escape mechanisms involve charged particles. Charged particles get trapped by magnetic fields and therefore do not readily escape. However, a fast proton can collide with a slow hydrogen atom and take the electron from the hydrogen atom. This charge exchange process changes the fast proton into a fast, hydrogen atom that is electrically neutral and hence can escape.

Nonthermal processes account for most of the present-day escape flux from Earth, and the same is likely to be true for Venus. They are also invoked to explain the $62 \pm 16\%$ enrichment of the $^{15}N/^{14}N$ ratio in the Martian atmosphere. If the current total escape flux from thermal and nonthermal processes is applied over the age of the solar system, the loss of hydrogen from the Earth is equivalent to only a few meters of liquid water, which means that Earth's sea level has not been affected much by this process. However, the flux could have been much higher in the past, since it is sensitive to the structure of the atmosphere (*see* Mars Atmosphere: History and Surface Interaction).

3. ATMOSPHERIC CIRCULATION

3.1. Processes Driving the Circulation

The atmospheric circulation on Earth, as on any planet, involves a wealth of phenomena ranging from global weather patterns to turbulent eddies only centimeters across and varies over periods of seconds to millions of years. All this activity is driven by absorbed sunlight and loss of infrared (heat) energy to space. Of the sunlight absorbed by the Earth, most ($\sim 70\%$) penetrates through the atmosphere and is absorbed at the surface; in contrast, the radiative cooling to space occurs not primarily from the surface but from the upper troposphere at an altitude of 5–10 km. This mismatch in the altitudes of heating and cooling means that, in the absence of air motions, the surface temperature would be much hotter than temperatures in the upper troposphere. However, such a trend produces an unstable density stratification, forcing the troposphere to overturn. The hot air rises, the cold air sinks, and thermal energy is thus transferred from the surface to the upper troposphere. This energy transfer by air motions leads to surface temperatures cooler than they would be in radiative equilibrium (while still being significantly hotter than the upper troposphere). This vertical mixing process is fundamentally responsible for near-surface convection, turbulence, cumulus clouds, thunderstorms, hurricanes, dust devils, and a range of other small-scale weather phenomena.

At global scales, much of Earth's weather results not simply from vertical mixing but from the atmosphere's response to horizontal temperature differences. Earth absorbs most of the sunlight at low latitudes, yet it loses heat to space everywhere over the surface. Hot equatorial air and cold polar air results. This configuration is gravitationally unstable—the hot equatorial air has low density and the cold polar air has high density. Just as the cold air from an open refrigerator slides across your feet, the cold polar air slides under the hot equatorial air, lifting the hotter air upward and poleward while pushing the colder air downward and equatorward. This overturning process transfers energy between the equator and the poles and leads to a much milder equator-to-pole temperature difference (about 30 K at the surface) than would exist in the absence of such motions. On average, the equatorial regions gain more energy from sunlight than they lose as radiated heat, while the reverse holds for the poles; the difference is transported between equator and pole by the air and ocean. The resulting atmospheric overturning causes many of Earth's global-scale weather patterns, such as the 1000-km-long fronts that cause much midlatitude weather and the organization of thunderstorms into clusters and bands. Horizontal temperature and density contrasts can drive weather at regional scales too; examples include air–sea breezes and monsoons.

3.2. Influence of Rotation

The horizontal pressure differences associated with horizontal temperature differences cause a force (the "pressure gradient force") that drives most air motion at large scales. However, how an atmosphere responds to this force depends strongly on whether the planet is rotating. On a nonrotating planet, the air tends to directly flow from high to low pressure, following the "nature abhors a vacuum" dictum. If the primary temperature difference occurs between equator and pole, this would lead to a simple overturning circulation between the equator and pole. On the other hand, planetary rotation (when described in a noninertial reference frame rotating with the solid planet) introduces new forces into the equations of motion: the centrifugal force and the Coriolis force. The centrifugal force naturally combines with the gravitational force and the resultant force is usually referred to as simply the gravity. For rapidly rotating planets, the Coriolis force is the dominant term that balances the horizontal pressure gradient force in large-scale circulations (a balance called **geostrophy**). Because the Coriolis force acts perpendicular to the air motion, this leads to a fascinating effect—the horizontal airflow is perpendicular to the horizontal pressure gradient. A north–south pressure gradient (resulting from a hot equator and a cold pole, for example) leads primarily not to north–south air motions but to east–west air motions! This is one reason why east–west winds dominate the circulation on most planets, including the Earth. For an Earth-sized planet with Earth-like wind speeds, rotation dominates the large-scale dynamics as long as the planet rotates at least once every 10 days.

Two other important effects of rapid rotation are the suppression of motions in the direction parallel to the rotation axis, called the Taylor–Proudman effect, and the coupling of horizontal temperature gradients with vertical wind shear, a three-dimensional relationship described by the thermal wind equation.

3.3. Observed Global-Scale Circulation

As described earlier, the atmospheric circulation organizes primarily into a pattern of east–west winds, and perhaps the most notable feature is the eastward-blowing jet streams in the midlatitudes of each hemisphere (Figure 20.2). In a longitudinal and seasonal average, the winter hemisphere wind maximum reaches 40 m/s at 30° latitude, and the summer hemisphere wind maximum reaches 20–30 m/s at 40°–50° latitude. In between these eastward wind maxima, from latitude 20 °N to 20 °S, the tropospheric winds blow weakly westward. The jet streams are broadly distributed in height, with peak speeds at about 12 km altitude. Although the longitudinally and seasonally averaged winds exhibit only a single tropospheric eastward wind maximum in each hemisphere, instantaneous three-dimensional snapshots of the atmosphere illustrate that there often exist two distinct jet streams, the subtropical jet at ∼30° latitude and the so-called eddy-driven jet at ∼50° latitude. These jets are relatively narrow—a few 100 km in latitudinal extent—and can reach speeds up to 100 m/s. However, the intense jet cores are usually less than a few thousand kilometers in longitudinal extent (often residing over continental areas such as eastern Asia and eastern North America), and the jets typically exhibit wide, time-variable wavelike fluctuations in position. When averaged over longitude and time, these variations in the individual jet streams smear into the single eastward maximum evident in each hemisphere in Figure 20.2.

Although the east–west winds dominate the time-averaged circulation, vertical and latitudinal motions are nevertheless required to transport energy from the equator to the poles. Broadly speaking, this transport occurs in two distinct modes. In the tropics exists a direct thermal overturning circulation called the *Hadley cell*, where, on average, air rises near the equator, moves poleward, and descends. This is an extremely efficient means of transporting heat and contributes to the horizontally homogenized temperatures that exist in the tropics. However, planetary rotation prevents the Hadley cell from extending all the way to the poles (to conserve angular momentum about the rotation axis, equatorial air would accelerate eastward to extreme speeds as it approached the pole, a phenomenon that is dynamically inhibited). On Earth, the Hadley cell extends to latitudes of ∼30°. The subtropical jet lies at the poleward edge of the Hadley cell at ∼30°. Poleward of ∼30°, the surface temperatures decrease rapidly toward the pole. Although planetary rotation inhibits the Hadley cell in this region, north–south motions still occur via a complex three-dimensional process called **baroclinic instability**. Meanders on the jet stream grow, pushing cold high-latitude air under warm low-latitude air in confined regions ∼1000–5000 km across. These instabilities grow, mature, and decay over ∼5-day periods; new ones form as old ones disappear. These structures evolve to form regions with a sharp thermal gradient called *fronts*, as well as 1000–5000-km-long arc-shaped clouds and precipitation that dominate much of the winter weather in the United States, Europe, and other midlatitude regions.

Water vapor in Earth's troposphere greatly accentuates convective activity because latent heat is liberated when moist air is raised above its lifting condensation level, and this further increases the buoyancy of the rising air, leading to moist convection. Towering thunderstorms get their energy from this process, and hurricanes are the most dramatic and best-organized examples of moist convection. Hurricanes occur only on the Earth because only the Earth provides the necessary combination of high humidity and surface friction. Surface friction is required to cause air to spiral into the center of the hurricane, where it is then forced upward past its lifting condensation level.

The Hadley cell exerts a strong control over weather in the tropics. The upward transport in the ascending branch of the Hadley circulation occurs almost entirely in localized thunderstorms and cumulus clouds whose convective towers cover only a small fraction (perhaps ∼1%) of the total horizontal area of the tropics. Because this ascending branch resides near the equator, equatorial regions receive abundant rainfall, allowing the development of tropical rainforests in Southeast Asia/Indonesia, Brazil, and central Africa.

On the other hand, this condensation and rainout of water dehydrates the air, so the descending branch of the Hadley cell, which occurs in the subtropics at ∼20°–30° latitude, is relatively dry. Because of the descending motion and dry conditions, little precipitation falls in these regions, which explains the abundance of arid biomes at 20°–30° latitude, including the deserts of the African Sahara, southern Africa, Australia, central Asia, and the southwestern United States. However, the simple Hadley cell is to some degree a theoretical idealization, and many regional three-dimensional

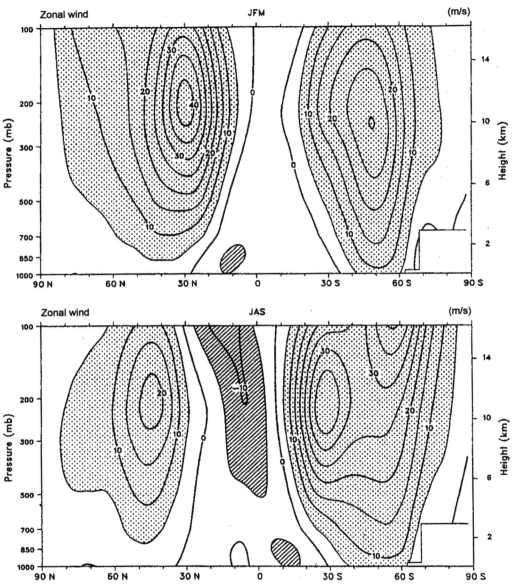

FIGURE 20.2 Longitudinally averaged zonal (i.e. east–west) winds in Earth's troposphere, showing the midlatitude maxima associated with the jet streams. *From Hurrell, van Loon, and Shea (1998).*

time-variable phenomena—including monsoons, equatorial waves, El Niño, and longitudinal overturning circulations associated with continent–ocean and sea-surface-temperature contrasts—affect the locations of tropical thunderstorm formation and hence the climatic rainfall patterns.

Satellite images (Figure 20.3) dramatically illustrate the signature of the Hadley cell and midlatitude baroclinic instabilities as manifested in clouds. In Figure 20.3, the east–west band of clouds stretching across the disk of the Earth just north of the equator corresponds to the rising branch of the Hadley cell (this cloud band is often called the intertropical convergence zone). These clouds are primarily the tops of thunderstorm anvils. In the midlatitude regions of both hemispheres (30°–70° latitude), several arc-shaped clouds up to 3000–5000 km long can be seen. These are associated with baroclinic instabilities. These clouds, which can often dominate midlatitude winter precipitation, form when large regions of warm air are forced upward over colder air masses during growing baroclinic instabilities. In many cases, the forced ascent associated with these instabilities produces predominantly sheetlike stratus clouds and steady rainfall lasting for several days, although sometimes the forced ascent can trigger local convection events (e.g. thunderstorms).

FIGURE 20.3 Visible-wavelength image of the Earth from the *Geostationary Operational Environmental Satellite (GOES)* geostationary weather satellite, illustrating the clouds associated with the Hadley cell, baroclinic instabilities, and other weather systems. North America can be seen at the upper right and South America (mostly obscured by clouds) is at the lower right.

What causes the jet stream? This is a subtle question. At the crudest level, poleward-moving equatorial air deflects eastward due to the **Coriolis acceleration**, so the formation of eastward winds in the midlatitudes is a natural response to poleward-moving air in the upper troposphere. Because the low latitudes are warm and the high latitudes are cold, the horizontal pressure gradients at the top of the troposphere point on average from the equator to pole, and in steady state, are balanced by a Coriolis force (associated with atmospheric winds) that points toward the equator. Such an equatorward-pointing Coriolis force can only occur if the upper tropospheric winds in midlatitudes flow to the east. However, these processes alone would tend to produce a relatively broad zone of eastward flow rather than a narrow jet. Nonlinear turbulent motions, in part associated with baroclinic instabilities, pump momentum upgradient into this eastward-flowing zone and help to produce the narrow jet streams.

Identifying the particular mechanisms that cause the jet streams is aided by examining the force balance in the *longitudinal* direction. For example, in the Hadley cell, air moves toward the poles in the upper troposphere, and as it does so, the Coriolis force acts on the air to accelerate it in the eastward direction—causing the **subtropical jet**. On the other hand, forces due to waves and turbulence cause a westward acceleration at this latitude. Thus, in the subtropical jet, the Coriolis forces accelerate the jet, and turbulent forces act as a drag that tries to slow it down. The balance between these two opposing forces leads to a jet stream whose speeds remain relatively steady, on average, over time.

On the other hand, at the latitudes of the baroclinic instabilities, the situation is the exact opposite. Baroclinic instabilities lead to complex three-dimensional wave structures that transport momentum from their surroundings into the latitudes of the baroclinic instabilities. As this wave-transported momentum builds up at the instability latitude, it leads to an eastward acceleration—causing a jet stream called the **eddy-driven jet**. This is the latitude of the **Ferrel cells**, where air in the upper troposphere flows equatorward, causing a westward Coriolis force. Thus, in the eddy-driven jet, waves and turbulence act to accelerate the jet, and Coriolis forces act as a drag that slows it down. This is the exact opposite force balance as occurs in the subtropical jet. Again, the balance between these two opposing forces leads to a jet stream whose speeds remain relatively steady, on average, over time.

Although the Earth's equator is hotter than the poles at the surface, it is noteworthy that, in the upper troposphere and lower stratosphere (~ 18 km altitude), the reverse is true. This seems odd because sunlight heats the equator much more strongly than the poles. In reality, the cold equatorial upper troposphere results from a dynamical effect: large-scale ascent in the tropics causes air to expand and cool (a result of decreased pressure as the air rises), leading to the low temperatures despite the abundant sunlight. Descent at higher latitudes causes compression and heating, leading to warmer temperatures. Interestingly, this means that, in the lower stratosphere, the ascending air is actually denser than the descending air. Such a circulation, called a thermally indirect circulation, is driven by the absorption of atmospheric waves that are generated in the troposphere and propagate upward into the stratosphere. There is a strong planetary connection because all four giant planets—Jupiter, Saturn, Uranus, and Neptune—are also thought to have thermally indirect circulations in their stratospheres driven by analogous processes.

3.4. Insights from Other Atmospheres

Planetary exploration has revealed that atmospheric circulations come in many varieties. The goal of planetary meteorology is to understand what shapes and maintains these diverse circulations. The *Voyager* spacecraft provided the first close-up images of the atmospheres of Jupiter, Saturn, Uranus, and Neptune and detailed information on the three satellites that have atmospheres thick enough to sport weather—Io, Titan, and Triton. *Galileo*, *Cassini*, and *New Horizons* have visited Jupiter, and *Cassini* has obtained a wealth of information about Saturn and Titan. The atmospheres of Venus and Mars have been sampled by entry probes, landers, orbiting spacecraft, and telescopic studies. Basic questions like why Venus' atmosphere rotates up to 60 times faster than does the planet, or why Jupiter and Saturn have superrotating equatorial jets, do not have completely satisfactory explanations. However, by comparing and contrasting each planet's weather, a general picture has begun to emerge.

Theoretical studies and comparative planetology show that planetary rotation rate and size exert a major control over the type of global atmospheric circulations that occur. When the rotation rate is small, Hadley cells are unconfined and stretch from the equator to the pole. Venus, with a rotation period of 243 days, seems to reside in such a state. Titan rotates in 16 days and, according to circulation models, its Hadley cell extends to at least $\sim 60°$ latitude, a transitional regime between Venus and Earth. On the other hand, fast rotation confines the Hadley cell to a narrow range of latitudes ($0°-30°$ on Earth) and forces baroclinic instabilities to take over much of the heat transport between low latitudes and the poles. Increasing the rotation rate still further—or making the planet larger—causes the midlatitudes to break into a series of narrow latitudinal bands, each with their own east–west jet streams and baroclinic instabilities. The faster the rotation rate, the straighter and narrower are the bands and jets. This process helps explain the fact that Jupiter and Saturn, which are large and rapidly rotating, have ~ 30 and 20 jet streams, respectively (as compared to only a few jet streams for the Earth). Fast rotation also contributes to smaller structures because it inhibits free movement of air toward or away from pressure lows and highs, instead causing the organization of vortices around such structures. Thus, a planet identical to the Earth but with a faster or slower rotation rate would exhibit different circulations, equatorial and polar temperatures, rainfall patterns, and cloud patterns, and hence would exhibit a different distribution of deserts, rainforests, and other biomes.

The giant planets Jupiter and Saturn exhibit numerous oval-shaped windstorms that superficially resemble terrestrial hurricanes. However, hurricanes can generate abundant rainfall because friction allows near-surface air to spiral inward toward the low-pressure center, providing a source of moist air that then ascends inside thunderstorms; in turn, these thunderstorms release energy that maintains the hurricane's strength against the frictional energy losses. In contrast, vortices like Jupiter's Great Red Spot and the hundreds of smaller ovals seen on Jupiter, as well as the dozens seen on Saturn and the couple seen on Neptune, do not directly require moist convection to drive them and hence are not hurricanes. Instead, they are simpler systems that are closely related to three types of long-lasting, high-pressure "storms", or coherent vortices, seen on the Earth: blocking highs in the atmosphere and Gulf Stream rings and Mediterranean salt lenses ("meddies") in the ocean. Blocking highs are high-pressure centers that stubbornly settle over continents, particularly in the United States and Russia, thereby diverting rain from its usual path for months at a time. For example, the serious 1988 drought in the US Midwest was exacerbated by a blocking high. Gulf Stream rings are compact circulations in the Atlantic that break off from the meandering Gulf Stream, which is a "river" inside the Atlantic Ocean that runs northward along the eastern coast of the United States and separates from the coast at North Carolina, where it then jets into the Atlantic in an unsteady manner. Seen in three dimensions, the Gulf Stream has the appearance of a writhing snake. Similar western boundary currents occur in other ocean basins, for example, the Kuroshio Current off the coast of Japan and the Agulhas Current off the coast of South Africa. Jet streams in the atmosphere are a related phenomenon. When Gulf Stream rings form, they trap phytoplankton and zooplankton inside them, which are carried large distances. Over the course of a few months, the rings dissipate at sea, are reabsorbed into the Gulf Stream, or run into the coast, depending on which side of the Gulf Stream they formed. The ocean plays host to another class of long-lived vortices,

Mediterranean salt lenses, which are organized high-pressure circulations that float under the surface of the Atlantic. They form when the extrasalty water that slips into the Atlantic from the shallow Mediterranean Sea breaks off into vortices. After a few years, these meddies eventually wear down as they slowly mix with the surrounding water. The mathematical description of these long-lasting vortices on the Earth is the same as that used to describe the ovals seen on Jupiter, Saturn, and Neptune (*see* Atmospheres of the Giant Planets).

Given that we know that atmospheric motions are fundamentally driven by sunlight, and we know that the problem is governed by Newton's laws of motion, why then are atmospheric circulations difficult to understand? Several factors contribute to the complexity of observed weather patterns. In the first place, fluids move in an intrinsically nonlinear fashion that makes paper-and-pencil analysis formidable and often intractable. Laboratory experiments and numerical experiments performed on high-speed computers are often the only means for making progress on problems in geophysical fluid dynamics. Second, meteorology involves the intricacies of moist thermodynamics and precipitation, and we are only beginning to understand and accurately model the microphysics of these processes. And for the terrestrial planets, a third complexity arises from the complicated boundary conditions that the solid surface presents to the problem, especially when mountain ranges block the natural tendency for winds to organize into steady east–west jet streams. For oceanographers, even more restrictive boundary conditions apply, namely, the ocean basins, which strongly affect how currents behave. The giant planets are free of this boundary problem because they are completely fluid down to their small rocky cores. However, the scarcity of data for the giant planets, especially with respect to their vertical structure beneath the cloud tops, provides its own set of difficulties (*see* Interiors of the Giant Planets).

4. OCEANS

Earth is the only planet in the solar system with a global ocean at the surface. The oceans have an average thickness of 3.7 km and cover 71% of Earth's surface area; the greatest thickness is 10.9 km, which occurs at the Marianas Trench. The total oceanic mass—1.4×10^{21} kg—exceeds the atmospheric mass of 5×10^{18} kg by nearly a factor of 300, implying that the oceans dominate Earth's surface inventory of volatiles (one way of visualizing this fact is to realize that, if Earth's entire atmosphere condensed as ices on the surface, it would form a layer only ∼10 m thick). The Earth therefore sports a greater abundance of fluid volatiles at its surface than any other solid body in the solar system. Even Venus' 90 bar CO_2 atmosphere contains only one-third the mass of Earth's oceans. On the other hand, Earth's oceans constitute only 0.02% of Earth's total mass; the mean oceanic thickness of 3.7 km pales in comparison to Earth's 6400 km radius, implying that the oceans span only 0.06% of Earth's width. The Earth is thus a relatively dry planet, and the oceans truly are only skin deep.

It is possible that Earth's solid mantle contains a mass of dissolved water (stored as individual water molecules inside and between the rock grains) equivalent to several oceans' worth of water. Taken together, however, the total water in Earth probably constitutes less than 1% of Earth's mass. In comparison, most icy satellites and comets in the outer solar system contain ∼40–60% H_2O by mass, mostly in solid form. This lack of water on Earth in comparison to outer solar system bodies reflects the relatively dry conditions in the inner solar system when the terrestrial planets formed; indeed, the plethora of water on Earth compared to Venus and Mars has raised the question of whether even the paltry amount of water on Earth must have been delivered from an outer solar system source such as impact of comets onto the forming Earth.

The modern oceans can be subdivided into the Pacific, Atlantic, Indian, and Arctic Oceans, but these four oceans are all connected, and this contiguous body of water is often simply referred to as the global ocean.

4.1. Oceanic Structure

The top meter of ocean water absorbs more than half of the sunlight entering the oceans; even in the sediment-free open ocean, only 20% of the sunlight reaches a depth of 10 m and only ∼1% penetrates to a depth of 100 m (depending on the angle of the Sun from vertical). Photosynthetic single-celled organisms, which are extremely abundant near the surface, can thus only survive above depths of ∼100 m; this layer is called the photic zone. The much thicker aphotic zone, which has too little light for photosynthetic production to exceed respiration, extends from ∼100 m to the bottom of the ocean. Despite the impracticality of photosynthesis at these depths, the deep oceans, nevertheless, exhibit a wide variety of life fueled in part by dead organic matter that slowly sediments down from the photic zone (*see* Astrobiology).

From a dynamical point of view, the ocean can be subdivided into several layers. Turbulence caused by wind and waves homogenizes the top 20–200 m of the ocean (depending on weather conditions), leading to profiles of density, temperature, salinity, and composition that vary little across this layer, which is therefore called the mixed layer. Below the mixed layer lies the thermocline, where the temperature generally decreases with depth down to ∼0.5–1 km. The salinity also often varies with depth between ∼100 and 1000 m, a layer called the halocline. For example, regions of abundant precipitation but lesser evaporation, such as the North Pacific, have relatively fresh surface waters, so the salinity increases with depth below the mixed layer in those regions. The variation of temperature and salinity between ~100 and 1000 m implies that density varies with depth across this layer too; this is referred to as the pycnocline. Below the thermocline, halocline, and pycnocline lies the deep ocean, where temperatures are usually relatively constant with depth at a chilly 0–4 °C.

The temperature at the ocean surface varies strongly with latitude, with only secondary variations in longitude. Surface temperatures reach 25 °C–30 °C near the equator, where abundant sunlight falls, but plummet to 0 °C near the poles. In contrast, the deep oceans (>1 km) are generally more homogeneous and have temperatures between 0 °C and 4 °C all over the world (when enjoying the bathtub-temperature water and coral reefs during a summer vacation to a tropical island, it is sobering to think that if one could only scuba dive deep enough, the temperature would approach freezing). This latitude-dependent upper ocean structure implies that the thermocline and pycnocline depths decrease with latitude: They are about ~1 km near the equator and reach zero near the poles.

Because warmer water is less dense than colder water, the existence of a thermocline over most of the ocean implies that the top ~1 km of the ocean is less dense than the underlying deep ocean. The implication is that, except for localized regions near the poles, the ocean is stable to vertical convective overturning.

4.2. Ocean Circulation

Ocean circulation differs in important ways from atmospheric circulation, despite the fact that the two are governed by the same dynamical laws. First, the confinement of oceans to discrete basins separated by continents prevents the oceanic circulation from assuming the common east–west flow patterns adopted by most atmospheres (topography can cause substantial north–south deflections in an atmospheric flow, which may help explain why Earth's atmospheric circulation involves more latitudinal excursions than that of the topography-free giant planets; nevertheless, air's ability to flow over topography means that atmospheres, unlike oceans, are still fundamentally unbounded in the east–west direction.) The only oceanic region unhindered in the east–west direction is the Southern Ocean surrounding Antarctica, and, as might be expected, a strong east–west current, which encircles Antarctica, has formed in this region.

Second, the atmosphere is heated from below, but the ocean is heated from above. Because air is relatively transparent to sunlight, sunlight penetrates through the atmosphere and is absorbed primarily at the surface, where it heats the near-surface air at the bottom of the atmosphere. In contrast, liquid water absorbs sunlight extremely well, so that 99% of the sunlight is absorbed in the top 3% of the ocean. This means, for example, that atmospheric convection—thunderstorms—predominates at low latitudes (where abundant sunlight falls) but is rare near the poles; in contrast, convection in the oceans is totally inhibited at low latitudes and instead can occur only near the poles.

Third, much of the large-scale ocean circulation is driven not by horizontal density contrasts, as in the atmosphere (although these do play a role in the ocean), but by the frictional force of wind blowing over the ocean surface. In fact, the first simple models of ocean circulation developed by Sverdrup, Stommel, and Munk in the 1940s and the 1950s, which were based solely on forcing caused by wind stress, did a reasonably good job of capturing the large-scale horizontal circulations in the ocean basins.

As in the atmosphere, the Earth's rotation dominates the large-scale dynamics of the ocean. Horizontal Coriolis forces nearly balance pressure gradient forces, leading to **geostrophy.** As in the atmosphere, this means that ocean currents flow perpendicular to horizontal pressure gradients. Rotation also means that wind stress induces currents in a rather unintuitive fashion. Because of the existence of the Coriolis force, currents do not simply form in the direction of the wind stress; instead, the three-way balance between Coriolis, pressure gradient, and friction forces can induce currents that flow in directions distinct from the wind direction.

Averaged over time, the surface waters in most mid-latitude ocean basins exhibit a circulation consisting of a basin-filling gyre that rotates clockwise in the Northern Hemisphere and counterclockwise in the Southern Hemisphere. This circulation direction implies that the water in the western portion of the basin flows from the equator toward the pole, while the water in the eastern portion of the basin flows from the pole toward the equator. However, the flow is extremely asymmetric: The equatorward flow comprises a broad, slow motion that fills the eastern 90% of the ocean basin; in contrast, the poleward flow becomes concentrated into a narrow current (called a western boundary current) along the western edge of the ocean basin. The northward-flowing Gulf Stream off the US eastern seaboard and the Kuroshio Current off Japan are two examples; these currents reach speeds up to ~1 m/s in a narrow zone 50–100 km wide. This extraordinary asymmetry in the ocean circulation results from the increasing strength of the Coriolis force with latitude; theoretical models show that in a hypothetical ocean where Coriolis forces are independent of latitude, the gyre circulations do not exhibit western intensification. These gyres play an important role in Earth's climate by transporting heat from the equator toward the poles. Their clockwise (counterclockwise) rotation in the northern (southern) hemisphere helps explain why the water temperatures tend to be colder along continental west coasts than continental east coasts.

In addition to the gyres, which transport water primarily horizontally, the ocean also experiences vertical overturning. Only near the poles does the water temperature become cold enough for the surface density to exceed the deeper density. Formation of sea ice helps this process, because sea ice contains relatively little salt, so when it forms, the remaining surface water is saltier (hence denser) than average. Thus, vertical convection between the surface and deep ocean occurs only in polar regions, in particular in the Labrador Sea and near parts of Antarctica. On average, very gradual ascending motion must occur elsewhere in the ocean for mass balance to be achieved. This overturning circulation, which transports water from the surface to the deep ocean and back over ~1000 year timescales, is called the thermohaline circulation.

The thermohaline circulation helps explain why deep ocean waters have near-freezing temperatures worldwide: all deep ocean water, even that in the equatorial oceans, originated at the poles and thus retains the signature of polar temperatures. Given the solar warming of low-latitude surface waters, the existence of a thermocline is thus naturally explained. However, the detailed dynamics that control the horizontal structure and depth of the thermocline are subtle and have led to major research efforts in physical oceanography over the past four decades.

Despite the importance of the basin-filling gyre and thermohaline circulations, much of the ocean's kinetic energy resides in small eddy structures only 10–100 km across. The predominance of this kinetic energy at small scales results largely from the natural interaction of buoyancy forces and rotation. Fluid flows away from pressure highs toward pressure lows, but Coriolis forces short circuit this process by deflecting the motion so that fluid flows perpendicular to the horizontal pressure gradient. The stronger the influence of rotation relative to buoyancy, the better this process is short circuited, and hence the smaller are the resulting eddy structures. In the atmosphere, this natural length scale (called the deformation radius) is 1000–2000 km, but in the oceans, it is only 10–100 km. The rings and meddies described earlier provide striking examples of oceanic eddies in this size range.

4.3. Salinity

When one swims in the ocean, the leading impression is of saltiness. The ocean's global mean salinity is 3.5% by mass but varies between 3.3% and 3.8% in the open oceans and can reach 4% in the Red Sea and Persian Gulf; values lower than 3.3% can occur on continental shelves near river deltas. The ocean's salt would form a global layer 150 m thick if precipitated into solid form. Sea salt is composed of 55% chlorine, 30% sodium, 8% sulfate, 4% magnesium, and 1% calcium by mass. The ~15% variability in the salinity of open ocean waters occurs because evaporation and precipitation add or remove freshwater, which dilutes or concentrates the local salt abundance. However, this process cannot influence the relative proportions of elements in sea salt, which therefore remain almost constant everywhere in the oceans.

In contrast to seawater, most river and lake water is relatively fresh; for example, the salinity of Lake Michigan is ~200 times less than that of seawater. However, freshwater lakes always have both inlets and outlets. In contrast, lakes that lack outlets—the Great Salt Lake, the Dead Sea, and the Caspian Sea—are always salty. This provides a clue about processes determining saltiness.

Why is the ocean salty? When rain falls on continents, enters rivers, and flows into the oceans, many elements leach into the water from the continental rock. These elements have an extremely low abundance in the continental water, but because the ocean has no outlet (unlike a freshwater lake), these dissolved trace components can build up over time in the ocean. Ocean–seafloor chemical interactions (especially after volcanic eruptions) can also introduce dissolved ions into the oceans. However, the composition of typical river water differs drastically from that of sea salt—typical river salt contains ~9% chlorine, 7% sodium, 12% sulfate, 5% magnesium, and 17% calcium by mass. Although sodium and chlorine comprise ~85% of sea salt, they make up only ~16% of typical river salt. The ratio of chlorine to calcium is 0.5 in river salt but 46 in sea salt. Furthermore, the abundance of sulfate and silica is much greater in river salt than in sea salt. These differences result largely from the fact that processes act to remove salt ions from ocean water, but the efficiency of these processes depends on the ion. For example, many forms of sea life construct shells of calcium carbonate or silica, so these biological processes remove calcium and silica from ocean water. Much magnesium and sulfate seems to be removed in ocean water–seafloor interactions. The relative inefficiency of such removal processes for sodium and chlorine apparently leads to the dominance of these ions in sea salt despite their lower proportion in river salt.

It is often suggested that ocean salinity has been stable over the past billion years. If so, this would imply that the ocean is near a quasisteady state where salt removal balances salt addition via rivers and seafloor–ocean interactions. Nevertheless, evidence from fluid inclusions in marine halites, among other sources, suggests significant changes of oceanic chemistry (including salinity) over time. Although the salinity in early Earth history is not well known, indirect evidence suggests that it may have exceeded the current ocean salinity by up to a factor of two. These temporal fluctuations in salinity result from imbalances in salinity input and removal processes to the ocean over time. Salt removal processes include biological sequestration in shell material, abiological seafloor–ocean water chemical interactions, and physical processes such as formation of evaporate deposits when shallow seas dry up, which has the net effect of returning the water to the world ocean while leaving salt behind on land.

4.4. Atmosphere–Ocean Interactions

Many weather and climate phenomena result from a coupled interaction between the atmosphere and ocean and would not occur if either component were removed. Two major examples are hurricanes and El Niño.

Hurricanes are strong vortices, 100–1000 km across, with warm cores and winds often up to ~70 m/s; the temperature difference between the vortex and the surrounding air produces the pressure differences that allow strong vortex winds to form. In turn, the strong winds lead to increased evaporation off the ocean surface, which provides an enhanced supply of water vapor to fuel the thunderstorms that maintain the warm core. This enhanced evaporation from the ocean must continue throughout the hurricane's lifetime because the thermal effects of condensation in thunderstorms inside the hurricane provide the energy that maintains the vortex against frictional losses. Thus, both the ocean and atmosphere play crucial roles. When the ocean component is removed—say, when the hurricane moves over land—the hurricane rapidly decays.

El Niño corresponds to an enhancement of ocean temperatures in the eastern equatorial Pacific at the expense of those in the western equatorial Pacific; increased rainfall in western North and South America result, and drought conditions often overtake Southeast Asia. El Niño events occur every few years and have global effects. At the crudest level, "normal" (non-El Niño) conditions correspond to westward-blowing equatorial winds that cause a thickening of the thermocline (hence producing warmer sea surface temperatures) in the western equatorial Pacific; these warm temperatures promote evaporation, thunderstorms, and upwelling there, drawing near-surface air in from the east and thus helping to maintain the circulation. On the other hand, during El Niño, the westward-blowing trade winds break down, allowing the thicker thermocline to relax eastward toward South America, hence helping to move the warmer water eastward. Thunderstorm activity thus becomes enhanced in the eastern Pacific and reduced in the western Pacific compared to non-El Niño conditions, again helping to maintain the winds that allow those sea surface temperatures. Although El Niño differs from a hurricane in being a hemispheric-scale long-period fluctuation rather than a local vortex, El Niño shares with hurricanes the fact that it could not exist where either the atmosphere or the ocean component prevented from interacting with the other. To successfully capture these phenomena, climate models need accurate representations of the ocean and the atmosphere and their interaction, which continues to be a challenge.

4.5. Oceans on Other Worlds

The *Galileo* spacecraft provided evidence that subsurface liquid-water oceans exist inside the icy moons Callisto, Europa, and possibly Ganymede (*see* Ganymede and Callisto; Europa; Titan; Planetary Satellites). The recent detection of a jet of water molecules and ice grains from the south pole of Enceladus raises the question of whether that moon has a subsurface reservoir of liquid water. Theoretical models suggest that internal oceans could exist on a wide range of other bodies, including Titan, the smaller moons of Saturn and Uranus, Pluto, and possibly even some larger Kuiper Belt objects. These oceans of course differ from Earth's ocean in that they are ice covered; another difference is that they must transport the geothermal heat flux of those bodies and hence are probably convective throughout. Barring exotic chemical or fluid dynamical effects, then, one expects that such oceans lack thermoclines. In many cases, these oceans may be substantially thicker than Earth's oceans; estimates suggest that Europa's ocean thickness lies between 50 and 150 km.

The abundant life that occurs near deep-sea vents ("black smokers") in Earth's oceans has led to suggestions that similar volcanic vents may help power life in Europa's ocean. (In contrast to Europa, any ocean in Callisto and Ganymede would be underlain by high-pressure polymorphs of ice rather than silicate rock, so such silicate–water interactions would be weaker.) However, much of the biological richness of terrestrial deep-sea vents results from the fact that Earth's oceans are relatively oxygenated; when this oxidant-rich water meets the reducing water discharged from black smokers, sharp chemical gradients result, and the resulting disequilibrium provides a rich energy source for life. Thus, despite the lack of sunlight at Earth's ocean floor, the biological productivity of deep-sea vents results in large part from the fact that the oceans are communicating with an oxygen-rich atmosphere. If Europa's ocean is more reducing than Earth's ocean, then the energy source available from chemical disequilibrium may be smaller. Nevertheless, a range of possible disequilibrium reactions exist that could provide energy to drive a modest microbial biosphere on Europa (*see* Astrobiology).

5. CLIMATE

Earth's climate results from a wealth of interacting physical, chemical, and biological effects, and an understanding of current and ancient climates has required a multidecadal research effort by atmospheric physicists, atmospheric chemists, oceanographers, glaciologists, astronomers, geologists, and biologists. The complexity of the climate system and the interdisciplinary nature of the problem have made progress difficult, and even today many aspects remain poorly understood. "Climate" can be defined as the mean conditions of the atmosphere/ocean system—temperature, pressure, winds/currents, cloudiness, atmospheric humidity, oceanic salinity, and atmosphere/ocean chemistry in three dimensions—when time-averaged over intervals longer than those of typical weather patterns. It also refers to the distribution of sea ice, glaciers, continental lakes and streams, as well as coastlines, and the spatial distribution of ecosystems that result.

5.1. Basic Processes: Greenhouse Effect

Earth as a whole radiates with an effective temperature of 255 K, and therefore, its flux peaks in the thermal infrared part of the spectrum. This effective temperature is 30 K colder than the average temperature on the surface, and quite chilly by human standards.

What ensures a warm surface is the wavelength-dependent optical properties of the troposphere. In particular, infrared light does not pass through the troposphere as readily as visible light. The Sun radiates with an effective temperature of 5800 K and therefore, its peak flux is in the visible part of the spectrum (or stated more correctly in reverse, we have evolved such that the part of the spectrum that is visible to us is centered on the peak flux from the Sun). The atmosphere reflects about 31% of this sunlight directly back to space, and the rest is absorbed or transmitted to the ground. The sunlight that reaches the ground is absorbed and then reradiated at infrared wavelengths. Water vapor (H_2O) and carbon dioxide (CO_2), the two primary greenhouse gases, absorb some of this upward infrared radiation and then emit it in both the upward and downward directions, leading to an increase in the surface temperature to achieve balance. This is the **greenhouse effect**. Contrary to popular claims, the elevation of surface temperature by the greenhouse effect is not a situation where "the heat cannot get out". Instead, the heat must get out, and to do so in the presence of the blanketing effect of greenhouse gases requires an elevation of surface temperatures.

The greenhouse effect plays an enormous role in the climate system. A planet without a greenhouse effect, but otherwise identical to Earth, would have a global mean surface temperature 17 °C below freezing. The oceans would be mostly or completely frozen, and it is doubtful whether life would exist on Earth. We owe thanks to the greenhouse effect for Earth's temperate climate, liquid oceans, and abundant life.

Water vapor accounts for between one-third and two-thirds of the greenhouse effect on Earth (depending on how the accounting is performed), with the balance resulting from CO_2, methane, and other trace gases. Steady increases in carbon dioxide due to human activity seem to be causing the well-documented increase in global surface temperature over the past ~100 years. On Mars, the primary atmospheric constituent is CO_2, which together with atmospheric dust causes a modest 5 K greenhouse effect. Venus has a much denser CO_2 atmosphere, which, along with atmospheric sulfuric acid and sulfur dioxide, absorbs essentially all the infrared radiation emitted by the surface, causing an impressive 500 K rise in the surface temperature. Interestingly, if all the carbon held in Earth's carbonate rocks were liberated into the atmosphere, Earth's atmospheric CO_2 abundance and greenhouse effect would approach that on Venus (*see* Mars Atmosphere: History and Surface Interaction; Venus: Atmosphere).

5.2. Basic Processes: Feedbacks

The Earth's climate evolves in response to volcanic eruptions, solar variability, oscillations in Earth's orbit, and changes in internal conditions such as the concentration of greenhouse gases. The Earth's response to these perturbations is highly nonlinear and is determined by feedbacks in the climate system. Positive feedbacks amplify a perturbation and, under some circumstances, can induce a runaway process where the climate shifts abruptly to a completely different state. In contrast, negative feedbacks reduce the effect of a perturbation and thereby help maintain the climate in its current state. Some of the more important feedbacks are as follows.

Thermal feedback: Increases in the upper tropospheric temperature lead to enhanced radiation to space, tending to cool the Earth. Decreases in the upper tropospheric temperature cause decreased radiation to space, causing warming. This is a negative feedback.

Ice-albedo feedback: Ice caps and glaciers reflect visible light easily, so the Earth's brightness (**albedo**) increases with an increasing distribution of ice and snow. Thus, a more ice-rich Earth absorbs less sunlight, promoting colder conditions and growth of even more ice. Conversely, melting of glaciers causes Earth to absorb more sunlight, promoting warmer conditions and even less ice. This is a positive feedback.

Water vapor feedback: Warmer surface temperatures allow increased evaporation of water vapor from the ocean surface, increasing the atmosphere's absolute humidity. Because water vapor is a greenhouse gas, it promotes an increase in the strength of the greenhouse effect and hence even warmer conditions. Cooler conditions inhibit evaporation, lessen the greenhouse effect, and cause additional cooling. This is a positive feedback.

Cloud feedback: Changes in climate can cause changes in the spatial distribution, heights, and properties of clouds. Greater cloud coverage means a brighter Earth (higher albedo), leading to less sunlight absorption. Higher altitude clouds have colder tops that radiate heat to space less well, promoting a warmer Earth. For a given mass of condensed water in a cloud, clouds with smaller particles reflect light better, promoting a cooler Earth. Unfortunately, for a specified climate perturbation (e.g. increasing the CO_2 concentration), the extent to which the coverage, heights, and properties of clouds will change remains unclear. In the current Earth climate, clouds cause a net cooling effect (relative to an otherwise similar atmosphere with no clouds). Sophisticated GCMs suggest that the cloud feedback for the modern Earth climate is positive, although significant uncertainties remain.

The sum of these and other feedbacks determine how Earth's climate evolved during past epochs and how Earth will respond to current human activities such as emissions of CO_2. Much of the uncertainty in current climate projections results from uncertainty in these feedbacks. A related concept is that of thresholds, where the climate undergoes an abrupt shift in response to a gradual change. For example, Europe enjoys temperate conditions despite its high latitude in part because of heat transported poleward by the Gulf Stream. Some climate models have suggested that increases in CO_2 due to human activities could suddenly shift the ocean circulation in the North Atlantic into a regime that transports heat less efficiently, which could cause widespread cooling in Europe (although this might be overwhelmed by the expected global warming that will occur over the next century). The rapidity with which ice ages ended also suggests that major re-organizations of the ocean/atmosphere circulation occurred during those times. Although thresholds play a crucial role in past and possibly future climate change, they are notoriously difficult to predict because they involve subtle nonlinear interactions.

5.3. Recent Times

A wide range of evidence demonstrates that Earth's global mean surface temperature rose by about 0.6 °C between 1900 and 2000 (see Figure 20.4). Over the past 50 years, the global mean rate of temperature increase has been ~0.13 °C per decade (with a greater rate of warming over land than ocean). As of 2006, 20 of the hottest years measured since good instrumental records started in ~1860 have occurred within the past 25 years, and the past 25 years has been the warmest 25-year period of the past 1000 years.

There is widespread consensus among climate experts that the observed warming since ~1950 has been caused primarily by the release of CO_2 due to human activities, primarily the burning of oil, coal, natural gas, and forests:

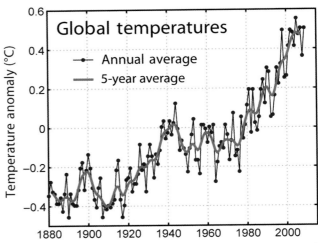

FIGURE 20.4 Observed global mean near-surface temperatures of Earth over the past 130 years. Blue points show annual averages, whereas red line shows the 5-year average. Although short-term climate fluctuations cause the temperature to vary from year to year, the overall trend has been inexorably upward over the past 50 years.

the greater CO_2 concentration has increased the strength of the greenhouse effect, modified by the feedbacks discussed in Section 5.2. Before the Industrial Revolution, the CO_2 concentration was ~280 ppm (i.e. a mole fraction of 2.8×10^{-4}), as shown in Figure 20.5. Starting approximately in 1800, however, the atmospheric CO_2 abundance began rising rapidly, and in 2012, the CO_2 concentration was 391 ppm—a 40% increase over pre-Industrial Revolution values. Evidence indicates that the sharp rise of CO_2 since 1800—starkly visible in Figure 20.5—is not a natural climate cycle but the result of human activity. Interestingly, only half of the CO_2 released by human activities each year remains in the atmosphere; the remainder is currently absorbed by the biosphere and especially the oceans.

Superposed on top of the mean rise of temperature with time since ~1950 are numerous short-term fluctuations associated with weather and short-period regional or global climate cycles such as El Niño, temporary shifts in the latitudes and strengths of the jet streams, and other effects. These short-term, year-to-year fluctuations are visible as the jittery year-to-year variation of the blue points in Figure 20.4. This necessarily means that, in some years, the mean climate is warmer than the previous year, while in other years, it is cooler than the previous year. This fact is often quoted by climate skeptics in the popular press as being evidence against global warming. Figure 20.4, however, shows that this argument is specious. Despite the year-to-year fluctuations, the overall long-term trend is clearly toward a warmer climate.

This increase in mean surface temperature has been accompanied by numerous other climate changes, including retreat of glaciers worldwide, thawing of polar

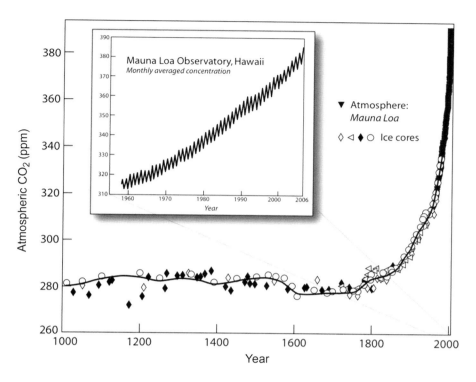

FIGURE 20.5 Observed abundance of carbon dioxide (CO_2) over the past 1000 years, from ice core records and direct atmospheric measurements. Inset shows CO_2 abundance since 1958 as measured in Hawaii; the periodic oscillation is the seasonal cycle, superposed on a near-linear increase due to human activities. (Doney & Schimel, 2007).

permafrost, early arrivals of spring, late arrivals of autumn, changes in the Arctic sea ice thickness, approximately 0.1–0.2 m of sea level increase since 1900, and various effects on natural ecosystems. These changes are expected to accelerate in the twenty-first century.

5.4. Ice Ages

The repeated occurrence of ice ages, separated by warmer interglacial periods, dominates Earth's climatic record of the past 2 million years. During an ice age, multikilometer-thick ice sheets grow to cover much of the high-latitude land area, particularly in North America and Europe; most or all of these ice sheets melt during the interglacial periods (however, ice sheets on Antarctica and Greenland have resisted melting during most interglacials, and these two ice sheets still exist today). The sea level varies by up to 120 m between glacial and interglacial periods, causing migration of coastlines by hundreds of kilometers in some regions. The time history of temperature, ice volume, and other variables can be studied using stable isotopes of carbon, hydrogen, and oxygen as recorded in glacial ice, deep-sea sediments, and land-based records such as cave calcite and organic material. This record shows that glacial/interglacial cycles over the past 800,000 years have a predominant period of ~100,000 years (Figure 20.6). During this cycle, glaciers gradually increase in volume (and air temperature gradually decreases) over most of the 100,000-year period; the glaciers then melt, and the temperature increases over a relatively short ~5000-year interval. The cycle is thus extremely asymmetric and resembles a sawtooth curve rather than a sinusoid. The last ice age peaked 18,000 years ago and ended by 10,000 years ago; the modern climate corresponds to an interglacial period. Analysis of ancient air trapped in air bubbles inside the Antarctic and Greenland ice sheets shows that the atmospheric CO_2 concentration is low during ice ages—typically about 200 ppm—and rises to ~280 ppm during the intervening interglacial periods (Figure 20.6).

Ice ages seem to result from changes in the strength of sunlight caused by periodic variations in Earth's orbit, magnified by several of the feedbacks discussed in Section 5.2. A power spectrum of the time series in Figure 20.6 shows that temperature, ice volume, and CO_2 vary predominantly on periods of 100,000, 41,000, 23,000, and 19,1000 years (ka; the summation of sinusoids at each of these periods leads to the sawtooth patterns in Figure 20.6). Interestingly, these periods match the periods over which Northern Hemisphere sunlight varies due to orbital oscillations. The Earth's orbital eccentricity oscillates on periods of 100 ka, the orbital obliquity (the tilt of Earth's rotation axis) oscillates on a period of 41 ka, and the Earth's rotation axis precesses on periods of 19 and 23 ka. These variables affect the difference in sunlight received at Earth between winter and summer and between the equator and pole. In turn, these sunlight variations determine the extent

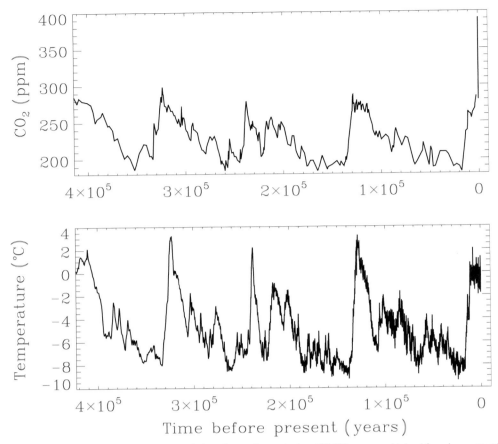

FIGURE 20.6 CO_2 concentrations (top) and temperature variations (bottom) over the last 420,000 years as obtained from ice cores at Vostok, Antarctica (data from Petit et al., 1999). The approximate 100,000-year period of the ice ages is evident, although many shorter period fluctuations are superimposed within the record. Prominent ice age terminations occurred at ~410, 320, 240–220, 130, and 15 ka in the past. Also note the correlation between temperature and CO_2 concentration during these cycles, which shows the influence of changes in the greenhouse effect on ice ages. The vertical line at the right side of the top plot shows the increase in CO_2 caused by humans between ~1800 and 2012.

to which snowpack accumulates in high northern latitudes during winter, and the extent to which this snowpack resists melting during summer; glaciers build up when snow that falls during winter cannot melt the following summer. The idea that these orbital variations cause ice ages has become known as the Milankovitch theory of ice ages.

By themselves, however, orbital variations are only part of the story. Sunlight variations due to the 100-ka eccentricity variations are much weaker than sunlight variations due to the 41-, 23-, and 19-ka obliquity and precession variations. Thus, if the orbit-induced sunlight variations translated directly into temperature and ice variations, ice ages would be dominated by the 41-, 23-, and 19-ka periods, but instead, the 100-ka period dominates (as can be seen in Figure 20.6). This means that some nonlinearity in the climate system amplifies the climatic response at 100 ka much better than at the shorter periods. Furthermore, the observed oscillations in CO_2 between glacial and interglacial periods (Figure 20.6) indicates that ice ages are able to occur partly because the greenhouse effect is weak during ice ages but strong during interglacial periods. Most likely, atmospheric CO_2 becomes dissolved in ocean water during ice ages, allowing the atmospheric CO_2 levels to decrease; the ocean then rejects this CO_2 at the end of the ice age, increasing its atmospheric concentration. Recent analyses of Antarctic ice cores show that, at the end of an ice age, temperature rise precedes CO_2 rise in Antarctica by about 800 years, indicating that CO_2 variation is an amplifier rather than a trigger of ice age termination. Interestingly, however, both of these events precede the initiation of deglaciation in the Northern Hemisphere. These observations suggest that the end of an ice age is first triggered by a warming event in the Antarctic region; this initiates the process of CO_2 rejection from the oceans to the atmosphere, and the resulting increase in the greenhouse effect, which is global, then allows deglaciation to commence across the rest of the planet. The ice-albedo and water vapor feedbacks (Section 5.2) help amplify the transition. However, many details, including the exact mechanism that allows CO_2 to oscillate between the ocean and atmosphere, remain to be worked out.

Figure 20.6 shows how the increase in CO_2 caused by human activities compares to the natural variability in the past. The saw-toothed variations in CO_2 between 200 and 280 ppm over 100,000-year-periods indicate the ice age/interglacial cycles, and the vertical spike in CO_2 at the far right of Figure 20.6 (from 280 to 391 ppm, also visible in the last ~ 200 years of the time series in Figure 20.5) shows the human-induced increase. The current CO_2 concentration far exceeds that at any previous time over the past 420,000 years, and is probably the greatest CO_2 level the Earth has seen since 20 million years ago. The fact that CO_2 rises by 30–40% at the end of an ice age indicates that very large magnitude climate changes can accompany modest CO_2 variations; it is noteworthy that human activities have so far increased CO_2 by an additional 36% beyond preindustrial values. The relationship between CO_2 and global temperature during ice ages may differ from the relationship these quantities will take over the next century of global warming; however, it is virtually certain that additional CO_2 will cause global temperature increases and widespread climate changes. Current economic and climate projections indicate that, because of continued fossil fuel burning, the atmospheric CO_2 will reach 500–1000 ppm by the year 2100 unless drastic measures are adopted to reduce fossil fuel use.

5.5. Volatile Inventories of Terrestrial Planets

Venus, Earth, and Mars have present-day atmospheres that are intriguingly different. The atmospheres of Venus and Mars are both primarily CO_2, but they represent two extreme fates in atmospheric evolution: Venus has a dense and hot atmosphere, whereas Mars has a thin and cold atmosphere. It is reasonable to ask whether Earth is ultimately headed toward one or the other of these fates, and whether these three atmospheres have always been so different.

The history of volatiles on the terrestrial planets includes their origin, their interactions with refractory (nonvolatile) material, and their rates of escape into space. During the initial accretion and formation of the terrestrial planets, it is thought that most or all of the original water reacted strongly with the iron to form iron oxides and hydrogen gas, with the hydrogen gas subsequently escaping to space. Until the iron cores in the planets were completely formed and this mechanism was shut down, the outflow of hydrogen probably took much of the other solar-abundance volatile material with it. Thus, one likely possibility is that the present-day atmospheres of Venus, Earth, and Mars are not primordial, but have been formed by outgassing and by cometary impacts that have taken place since the end of core formation.

The initial inventory of water that each terrestrial planet had at its formation is a debated question. One school of thought is that Venus formed in an unusually dry state compared with Earth and Mars; another is that each terrestrial planet must have started out with about the same amount of water per unit mass. The argument for an initially dry Venus is that water-bearing minerals would not condense in the high-temperature regions of the protoplanetary nebula inside of about 1 AU. Proponents of the second school of thought argue that gravitational scattering caused the terrestrial planets to form out of materials that originated over the whole range of terrestrial planet orbits, and therefore that the original water inventories for Venus, Earth, and Mars should be similar.

An important observable that bears on the question of original water is the enrichment of deuterium (D) relative to hydrogen. A measurement of the D/H ratio yields a constraint on the amount of hydrogen that has escaped from a planet. For the D/H ratio to be useful, one needs to estimate the relative importance of the different hydrogen escape mechanisms and the original D/H ratio for the planet. In addition, one needs an idea of the hydrogen sources available to a planet after its formation, such as cometary impacts. The initial value of D/H for a planet is not an easy quantity to determine. A value of 0.2×10^{-4} has been put forward for the protoplanetary nebula, which is within a factor of 2 or so for the present-day values of D/H inferred for Jupiter and Saturn. However, the D/H ratio in Standard Mean Ocean Water (a standard reference for isotopic analysis) on Earth is 1.6×10^{-4}, which is also about the D/H ratio in hydrated minerals in meteorites, and is larger by a factor of 8 over the previously mentioned value. At the extreme end, some organic molecules in carbonaceous chondrites have shown D/H ratios as high as 20×10^{-4}. The enrichment found in terrestrial planets and most meteorites over the protoplanetary nebula value could be the result of exotic high-D/H material deposited on the terrestrial planets, or it could be the result of massive hydrogen escape from the planets early in their lifetimes through the hydrodynamic blowoff mechanism (which is the same mechanism that currently drives the solar wind off the Sun).

6. LIFE IN THE ATMOSPHERE–OCEAN SYSTEM

6.1. Interplanetary Spacecraft Evidence for Life

An ambitious but ever-present goal in astronomy is to detect or rule out life in other solar systems, and in planetary science that goal is to detect or rule out life in our own solar system apart from the Earth. Water in its liquid phase is one of the few requirements shared by all life on Earth, and so the hunt for life is focused on the search for liquid water. We know that Mars had running water on its surface at some point in its history because we can see fluvial channels in high-resolution images, and because the Mars rovers *Spirit* and *Opportunity* have discovered aqueous geochemistry on the ground; there is even some evidence suggesting present-day seepage in recent orbiter images. Farther out in the solar system, we know that Europa, a satellite of Jupiter, has a smooth icy surface with cracks and flow features that resemble Earth's polar ice fields and suggest a liquid water interior, while its larger sibling, Ganymede, exhibits a conductive reaction to Jupiter's magnetic field that is most easily explained by a salty liquid water interior (*see* Mars: Surface and Interior; Meteorites; Planetary Satellites).

However, to date we have no direct evidence for extraterrestrial life. This includes data from landers on Venus, Mars, and the Moon, and flyby encounters with eight planets, a handful of asteroids, a comet (Halley in 1986), and over 60 moons. Are the interplanetary spacecraft we have sent out capable of fulfilling the goal of detecting life? This question has been tested by analyzing data from the *Galileo* spacecraft's two flyby encounters with the Earth, which, along with a flyby encounter with Venus, were used by the spacecraft's navigation team to provide gravity assists to send *Galileo* to Jupiter. The idea was to compare ground-truth information to what we can learn solely from *Galileo* (*see* Atmospheres of the Giant Planets; Io: the Volcanic Moon; Planetary Satellites; Planetary Exploration Missions).

Galileo's first Earth encounter occurred on December 8, 1990, with closest approach 960 km above the Caribbean Sea; its second Earth encounter occurred on December 8, 1992, with closest approach 302 km above the South Atlantic. A total of almost 6000 images were taken of Earth by *Galileo*'s camera system. Figure 20.7 shows the Earth—Moon system as seen by *Galileo*. Notice that the Moon is significantly darker than the Earth. The spacecraft's instruments were designed and optimized for Jupiter; nevertheless, they made several important observations that point to life on Earth. These strengthen the null results encountered elsewhere in the solar system. The evidence for life on Earth includes complex radio emissions, nonmineral surface pigmentation, disequilibrium atmospheric chemistry, and large oceans.

6.1.1. Radio Emissions

The only clear evidence obtained by *Galileo* for intelligent life on Earth was unusual radio emissions. Several natural radio emissions were detected, none of which were unusual, including solar radio bursts, auroral kilometric radiation, and narrowband electrostatic oscillations excited by thermal fluctuations in Earth's ionospheric plasma. The first unusual radio emissions were detected at 1800 UT and extended through 2025 UT, just before closest approach. These were detected by the plasma wave spectrometer on

FIGURE 20.7 The Earth—Moon system as observed by the *Galileo* spacecraft.

the nightside, in-bound pass, but not on the day side, outbound pass. The signal strength increased rapidly as Earth was approached, implying that Earth itself was the source of the emissions. The fact that the signals died off on the dayside suggests that they were cut off by the day side ionosphere, which means we can place the source below the ionosphere.

The unusual signals were narrowband emissions that occurred in only a few distinct channels and had average frequencies that remained stable for hours. Naturally occurring radio emissions nearly always drift in frequency, but these emissions were steady. The individual components had complicated modulations in their amplitude that have never been detected in naturally occurring emissions. The simplest explanation is that these signals were transmitting information, which implies that there is advanced technological life on Earth. In fact, the radio, radar, and television transmissions that have been emanating from Earth over the past century result in a nonthermal radio emission spectrum that broadcasts our presence out to interstellar distances (see The Solar System at Radio Wavelengths).

6.1.2. Surface Features

During its first encounter with the Earth, the highest resolution mapping of the surface by *Galileo's* Solid-State Imaging System (SSI) covered Australia and Antarctica with 1–2 km resolution. No usable images were obtained from Earth's nightside on the first encounter. The second encounter netted the highest resolution images overall of Earth by *Galileo*, 0.3–0.5 km/pixel, covering parts of Chile, Peru, and Bolivia. The map of Australia from the first encounter includes 2.3% of Earth's total surface area, but shows no geometric patterns that might indicate an advanced civilization. In the second encounter, both the cities of Melbourne and Adelaide were photographed, and yet no geometric evidence is visible because the image resolution is only 2 km. The map of Antarctica, 4% of Earth's surface, reveals nearly complete ice cover and no signs of life. Only one image, taken of Southeastern Australia during the second encounter, shows east–west and north–south markings that would raise suspicions of intelligent activity. The markings in fact were caused by boundaries between wilderness areas, grazing lands, and the border between South Australia and Victoria. Studies have shown that it takes nearly complete mapping of the surface at 0.1-km resolution to obtain convincing photographic evidence of an advanced civilization on Earth, such as roads, buildings, and evidence of agriculture.

On the other hand, many features are visible in the *Galileo* images that have not been seen on any other body in the solar system. The SSI camera took images in six different wavelength channels. A natural-color view of Earth was constructed using the red, green, and violet filters, which correspond to wavelengths of 0.670, 0.558, and 0.407 μm, respectively. The images reveal that Earth's surface is covered by enormous blue expanses that specularly reflect sunlight, and end in distinct coastlines, which are both easiest to explain if the surface is liquid. This implies that much of the planet is covered with oceans. The land surfaces show strong color contrasts that range from light brown to dark green.

The SSI camera has particular narrowband infrared filters that have never been used to photograph Earth before, and so they yielded new information for geological, biological, and meteorological investigations. The infrared filters allow the discrimination of H_2O in its solid, liquid, and gaseous forms; for example, clouds and surface snow can be distinguished spectroscopically with the 1-μm filter. False-color images made by combining the 1-μm channel with the red and green channels reveal that Antarctica strongly absorbs 1-μm light, establishing that it is covered by water ice. In contrast, large regions of land strongly reflect 1 μm without strongly reflecting visible colors, which conflicts with our experience from other planetary surfaces and is not typical of igneous or sedimentary rocks or soil. Spectra made with the 0.73- and 0.76-μm channels reveal several land areas that strongly absorb red light, which again is not consistent with rocks or soil. The simplest explanation is that some nonmineral pigment that efficiently absorbs red light has proliferated over the planet's surface. It is hard to say for certain if an interstellar explorer would realize that this is a biological mechanism for gathering energy from sunlight, probably so, but certainly we would recognize it on another planet as the signature of plant life. We know from ground truth that these unusual observations are caused by the green pigments chlorophyll a ($C_{55}H_{72}MgN_4O_5$) and chlorophyll b ($C_{55}H_{70}MgN_4O_6$), which are used by plants for photosynthesis. No other body in the solar system has the green and blue colorations seen on Earth (see The Solar System at Ultraviolet Wavelengths; Infrared Views of the Solar System From Space).

6.1.3. Oxygen and Methane

Galileo's Near-Infrared Mapping Spectrometer (NIMS) detected the presence of molecular oxygen (O_2) in Earth's atmosphere with a volume mixing ratio of 0.19 ± 0.05. Therefore, we know that the atmosphere is strongly oxidizing. (It is interesting to note that Earth is the only planet in the solar system where one can light a fire.) In light of this, it is significant that NIMS also detected methane (CH_4) with a volume mixing ratio of $3 \pm 1.5 \times 10^{-6}$. Because CH_4 oxidizes rapidly into H_2O and CO_2, if thermodynamical equilibrium holds, then there should be no detectable CH_4 in Earth's atmosphere.

The discrepancy between observations and the thermodynamic equilibrium hypothesis, which works well on other planets (e.g. Venus), is an extreme 140 orders of magnitude. This fact provides evidence that Earth has biological activity and that it is based on organic chemistry. We know from ground truth that Earth's atmospheric methane is biological in origin, with about half of it coming from nonhuman activity like methane bacteria and the other half coming from human activity like growing rice, burning fossil fuels, and keeping livestock. NIMS also detected a large excess of nitrous oxide (N_2O) that is most easily explained by biological activity, which we know from ground truth comes from nitrogen-fixing bacteria and algae.

The conclusion is that the interplanetary spacecraft we have sent out to explore our solar system are capable of detecting life on planets or satellites, both the intelligent and primitive varieties, if it exists in abundance on the surface. On the other hand, if there is life on a planet or satellite that does not have a strong signature on the surface, as would probably be the case if Europa or Ganymede harbor life, then a flyby mission may not be adequate to decide the question. With regard to abundant surface life, we have a positive result for Earth and a negative result for every other body in the solar system.

7. CONCLUSIONS

Viewing Earth as a planet is the most important change of consciousness that has emerged from the space age. Detailed exploration of the solar system has revealed its beauty, but it has also shown that the home planet has no special immunity to the powerful forces that continue to shape the solar system. The ability to remotely sense Earth's dynamic atmosphere, oceans, biosphere, and geology has grown up alongside our ever-expanding ability to explore distant planetary bodies. Everything we have learned about other planets influences how we view Earth. Comparative planetology has proved in practice to be a powerful tool for studying Earth's atmosphere and oceans. The lion's share of understanding still awaits us, and in its quest we continue to be pulled outward.

BIBLIOGRAPHY

Doney, S. C., & Schimel, D. S. (2007). Carbon and climate system coupling on timescales from the Precambrian to the Anthropocene. *Annual Review of Environment and Resources, 32*, 31–66.

Dowling, T. E. (2001). Oceans. In *Encyclopedia of astronomy and astrophysics* (pp. 1919–1928). Bristol: IOP Publishing Ltd and Nature Publishing Group.

Geissler, P., Thompson, W. R., Greenberg, R., Moersch, J., McEwen, A., & Sagan, C. (1995). Galileo multispectral imaging of Earth. *Journal of Geophysical Research, 100*(16), 16895, 906.

Hide, R., & Dickey, J. O. (1991). Earth's variable rotation. *Nature, 253*, 629–637.

Holton, J. R., Pyle, J., & Curry, J. A. (Eds.). (2002). *Encyclopedia of atmospheric sciences*. Academic Press.

Hurrell, J. W., van Loon, H., & Shea, D. J. (1998). The mean state of the troposphere. In D. J. Karoly, & D. G. Vincent (Eds.), *Meteorological Monograph: 27(49). Meteorology of the southern hemisphere* (pp. 1–46). Boston: American Meteorological Society.

Petit, J. R., Jouzel, J., Raynaud, D., Barkov, N. I., Barnola, J.-M., Basile, I., et al. (1999). Climate and atmospheric history of the past 420,000 years from the Vostok ice core, Antarctica. *Nature, 399*, 429–436.

Showman, A. P., & Malhotra, R. (1999). The Galilean satellites. *Science, 286*, 77–84.

Chapter 21

Earth as a Planet: Surface and Interior

David C. Pieri
Jet Propulsion Laboratory, California Institute of Technology, Pasadena, CA, USA

Adam M. Dziewonski
Department of Earth and Planetary Sciences, Harvard University, Cambridge, MA, USA

Chapter Outline

1. Introduction: The Earth as a Guide to Other Planets 445
2. Physiographic Provinces of Earth 447
    2.1. Basic Divisions 447
    2.2. Landform Types 450
        2.2.1. Submarine Landforms 450
        2.2.2. Subaerial Landforms 453
    2.3. Summary: Terrestrial vs Planetary Landscapes 456
3. Earth Surface Processes 456
    3.1. Constructive Processes in the Landscape 456
    3.2. Destructive Geomorphic Processes 457
4. Tools for Studying Earth's Deep Interior 462
5. Seismic Sources 466
6. Earth's Radial Structure 469
    6.1. Crust 469
    6.2. Upper Mantle: Lithosphere and Asthenosphere (25–400 km Depth) 470
    6.3. Transition Zone (400–660 km Depth) 471
    6.4. Lower Mantle (660–2890 km) 471
    6.5. Outer Core (2981–5151 km) 472
    6.6. Inner Core (5251–6371 km) 473
7. Earth in Three Dimensions 473
8. Earth as a Rosetta Stone 477
Bibliography 478

1. INTRODUCTION: THE EARTH AS A GUIDE TO OTHER PLANETS

The surface of the Earth is perhaps the most geochemically diverse and dynamic among the planetary surfaces of our solar system. Uniquely, it is the only one with liquid water oceans under a stable atmosphere and—as far as we now know—it is the only surface in our solar system that has given rise to life. The Earth's surface is a dynamic union of its solid crust, its atmosphere, its hydrosphere, and its biosphere, all having acted in concert to produce a constantly renewing and changing symphony of form (Figure 21.1).

The unifying theme of the Earth's surficial system is water—in liquid, vapor, and solid phases—which transfers and dissipates solar, mechanical, chemical, and biological energy throughout global land and submarine landscapes (e.g. Ritter, Kochel, & Miller, 2011). The surface is a window to the interior processes of the Earth, as well as the putty that atmospheric processes continually shape. It is also the Earth's interface with extraterrestrial processes and, as such, has regularly borne the scars of impacts by meteors, comets, and asteroids, and will continue to do so. The February 15, 2013, impact of an approximately 17- to 20-m-diameter ($12-13 \times 10^3$ kg) body over Chelyabinsk, Russia, is a potent reminder that impacts are still modifying the earth and other bodies in the solar system. Much larger impacting bodies (e.g. >1 km diameter), although now rare, are still possible, and pose a serious threat to society. (See Planetary Impacts.)

Our solar system has a variety of terrestrial planets and satellites in various hydrologic states with radically differing hydrologic histories. Some appear nearly totally desiccated, such as the Moon, Mercury, and Venus. Even in those places, water may yet prove to be more prominent. For instance, recently, evidence of probable magmatic water in the form of hydroxyl molecules found in spacecraft observations of equatorial lunar impact craters, combined with reanalyses of Apollo lunar samples, may argue for a source of magmatic water within the moon. Isotopic data from these samples are also somewhat problematic. Such findings may cause revision of theories of the origin and early evolution of the earth–moon system.

FIGURE 21.1 Blue Marble view of the Earth from Apollo 17. Earth as seen from the outbound Apollo 17, showing Mediterranean Sea to the north and Antarctica to the south. The Arabian Peninsula and the northeastern edge of Africa can also be seen. Asia is on the northeast (upper right) horizon. Most striking is the prevalence of liquid water (thus evidence of an average surface temperature >273 K), not now present in the arid landscapes of the other solid bodies within our solar system. *Courtesy of NASA.*

In some places water is very abundant now at the surface, such as on the Earth, on the Jovian Galilean satellite Europa (solid at the surface and possibly liquid underneath, with strong indications of surface water eruption plumes at its south pole), on the Saturnian satellite Enceladus (erupting water vapor into space through an icy surface), and on Titan, Saturn's largest moon (where a 94K surface temperature makes water ice at least as hard as **granite**). In other places, such as Mars and Ganymede, it appears that water may have been very abundant in liquid form on the surface in the distant past. Also, in the case of Mars, water may yet be abundant in solid and/or liquid form in the subsurface today. Thus, for understanding geological (and, where applicable, biological) processes and environmental histories of terrestrial planets and satellites within our solar system, it is crucial to explore the geomorphology of surface and submarine landforms and the nature and history of the land—water interface where it existed. Such an approach and "lessons learned" from this solar system will also be key in future reconnaissance of extrasolar planets. (See Mars: Surface and Interior.)

For the planets, remote sensing techniques, especially as implemented at optical (e.g. visible, near-infrared, and thermal infrared) and at radio frequencies (e.g. radars), have been a primary exploration tool. Starting with telescopic observations from earth-based observatory, through the phase of flyby and orbital observations, and now by landers and rovers, the planets and their satellites have become very real places to interested observers on this planet. Over the last 20 years, a number of planetary data archives have been created, and are now conveniently online, accessible over the Internet. It is thus particularly ironic, that effective, accessible, systematically organized digital data archives containing earth remote sensing image (and other) data, have only come to the fore within about the last 10 years. A discussion of how such data are archived, accessed, and standardized, both within the planetary and terrestrial contexts, is an important one, but beyond the scope of this treatment. Nevertheless, it is significant to point out that for many reasons, some historical, some technical, and some psychological, because of our intimate familiarity with the Earth, practitioner's of Earth Science, have tended to separate the various realms of scientific investigations of our planet, as the realms themselves seem separate to us humans from the vantage point of our existence and history on the Earth's land surface. The lack of emphasis on studying the Earth as a system—land, ocean, atmosphere, and near space—has hampered our full understanding of the complex interplay, history, and evolution of processes within these milieus.

Now, however, and especially within the last decade, new systematically organized archives of earth data, acquired using remote sensing techniques and predominately time-series image data, including topographic and seismic data, have emerged. This is, in large part, because of a conscious attempt of national scientific and space agencies to address the Earth as a system. Such has been particularly true in the United States within the decades-long orbitally focused National Aeronautics and Space Administration (NASA) Earth Observing System (EOS) project. Currently the EOS data are warehoused within the EOS Data Information System with a variety of subarchives representing general discipline areas, such as the Land Surface Data Acquisition and Analysis Center (LPDAAC; Sioux Falls, South Dakota, USA, operated under an agreement between NASA and the US Geological Survey). The LPDAAC provides online access to data from a variety of US earth orbital missions, acquired over many years.

For workers in a specific discipline, data access can be challenging within such a generalized large volume "Big Data" archive as the LPDAAC, so smaller more focused specialty archives have developed. One specific archive example that draws on data from the United States—Japan Advanced Spaceborne Thermal Emission and Reflection (ASTER) radiometer mission (1999—present) is the ASTER Volcano Archive at NASA's Jet Propulsion Laboratory in Pasadena, California, USA (http://ava.jpl.nasa.gov). It puts data that reflect ASTER's unique capabilities to see volcanic activity worldwide (14 optical channels between 0.4 and 12 μm; Pieri & Abrams, 2004) for over

1500 volcanoes in the Smithsonian Global Volcanism catalog for the period 2000—present, at the fingertips of online professional and general users. Archive holdings include basic and higher level data products (e.g. topography and thermal emission analyses) in convenient formats that can be displayed in Google Earth™ and Google Maps™ clients. Also included in such archives is instrument-specific in situ data (e.g. atmospheric profiles and ground-based and low-altitude surface temperature and emissivity measurements), acquired by both manned aircraft and newly employed unmanned aerial vehicles (e.g. Pieri et al., 2013).

The ready accessibility of such data, worldwide, tends to militate against geoscientific parochialism, and allows a more unified planetary perspective from which to assess earth science data, in this case volcanological data, more consistent with a comparative planetology approach advocated by researchers in solar system studies—examples of which are rife throughout this volume. Sister Federal agencies in the United States, and in other space-faring countries, significantly within Europe, Japan, and India, have also similarly pursued this approach. The net effect is an increasing emphasis on interdisciplinary investigations that necessarily cross traditional subject boundaries and promote a more general understanding of how the Earth system has responded in the past to various external stimuli (e.g. the solar Milankovic cycles) and how it will respond in the future (e.g. increased anthropogenic CO_2 inputs). It is becoming more and more clear that such understanding will be a requirement as mankind moves forward through the current century, with profound forecast impacts on terrestrial land surfaces, on our atmosphere, and on our oceans, and thus on the habitability of our global environment. (See Earth as a Planet: Oceans and Atmospheres.)

2. PHYSIOGRAPHIC PROVINCES OF EARTH

2.1. Basic Divisions

From a geographic and geomorphologic point of view, especially when seen from space, the surface of the Earth is dominated by its oceans of liquid water; approximately 75% of the Earth's surface is covered by liquid or solid water. The remaining 25% of nonmarine subaerial land, the subject of nearly all historical geological and geomorphological study, lies mainly in its Northern Hemisphere, where most of the world's population lives. The Southern Hemisphere is dominated by oceans, some subaerial continental and archipelago land masses (mainly parts of Africa, South America, southeast Asia, and Australia), and the large, currently subglacial, island continent of Antarctica (Figure 21.2(a)).

Remarkably, despite the fact that geological and geographical sciences have been practiced on the Earth for about 200 years, it has been mainly since the Second World War that scientists have begun detailed mapping and geophysical explorations of the submarine land surface. Subsea remote-sensing technology has provided one of the most profound discoveries in the history of geological science: the paradigm of "plate tectonics". The extent, morphology, and dynamics of the Earth's massive tectonic plates were only realized after careful topographic and geomagnetic mapping of the intensely volcanic mid-oceanic ridges and their associated parallel-paired geomagnetic domains.

Similar topographic mapping of the corresponding submarine trenches along continental or island-arc margins was equally revealing. The mid-oceanic ridges were found to be sites of accretion of new volcanically generated plate material, and the trenches the sites of deep **subduction**, where oceanic crust is consumed beneath other overriding crustal plates. Tectonic plates represent the most fundamental and largest geomorphic provinces on Earth.

The Earth's crustal plates come in two varieties: oceanic and continental (Figure 21.3(a)).

Oceanic plates comprise nearly all of the Earth's ocean floors, and thus most of the Earth's crustal area. They are composed almost exclusively of iron- and magnesium-rich rocks derived from volcanic processes (called "**basalts**"). Oceanic plates are created by volcanic eruptions along the apices of the Earth's mid-oceanic ridges, 1000-km-long sinuous ridges that rise from the flat ocean floor (called "abyssal plains") in the middle of oceans. Oceanic plates are typically less than 10 km thick. Here, nearly continuous volcanic activity from countless submarine volcanic centers (far more than the 1000 or so active subaerial volcanoes) provides a steady supply of new basalt, which is accreted and incorporated into the interior part of the plate.

At plate edges, roughly the reverse occurs, where the outer, oldest plate margins are forced below overriding adjacent plate edges. Usually, when two oceanic plates collide, the resulting subduction zone forms an island arc along the trace of the collision. The islands, in this case, are the result of the eruption of lighter, more silica-rich magmas generated as part of the subduction process. The subducted plate margin is consumed along the axis of the resulting trench. Because the more silicic island arcs tend to be less dense and thus more resistant to subduction, they can be accreted onto plate margins and can thus increase the areal extent at the edges of oceanic plates or can enlarge the margins of existing continental plates.

Continental plates tend to consist of much more silicic material, and are thus lighter, as compared with oceanic plates. Because of their lower density and the fact that they are isostatically compensated, they are much thicker than oceanic plates (30—40 km thick) and tend to "float" over the denser, more mafic (ferromagnesian—consisting of mostly of the metals iron and magnesium) subjacent

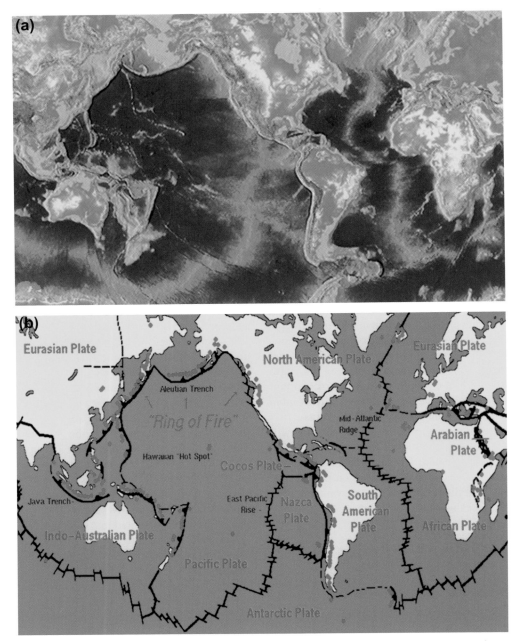

FIGURE 21.2 (a) Physiographic map of the Earth. This image was generated from digital databases of land and seafloor elevations on a 2 min latitude/longitude grid. Assumed illumination is from the west, and the projection is Mercator. Spatial resolution of the gridded data varies from true 2 min for the Atlantic, Pacific, and Indian Ocean floors and all land masses to 5 min for the Arctic Ocean floor. *(Courtesy of NOAA.)* (b) Volcanoes and the Crustal Plates Global map of the major tectonic plate boundaries and locations of the world's volcanoes. *Courtesy of the US Geological Survey.*

material in the Earth's upper mantle. When continental plates collide with oceanic plates, deep subduction trenches, such as the Peru—Chile trench along the west coast of South America, occur, as the oceanic plate is forced under the much thicker and less dense continental plate. Usually, the landward side of the affair is marked by so-called Cordillieran belts of mountains, including andesitic-type volcanoes, which parallel the coastline. The Andes Mountains are an example of this type of tectonic arrangement.

When continental plates collide, a very different tectonic and geomorphic regime ensues. Here, equally buoyant and thick continental plates crush against each other, resulting in the formation of massive fold belts and towering mountains, as long as the tectonic zone is active (e.g. the Himalayan Range in Asia). When aggregate stresses are

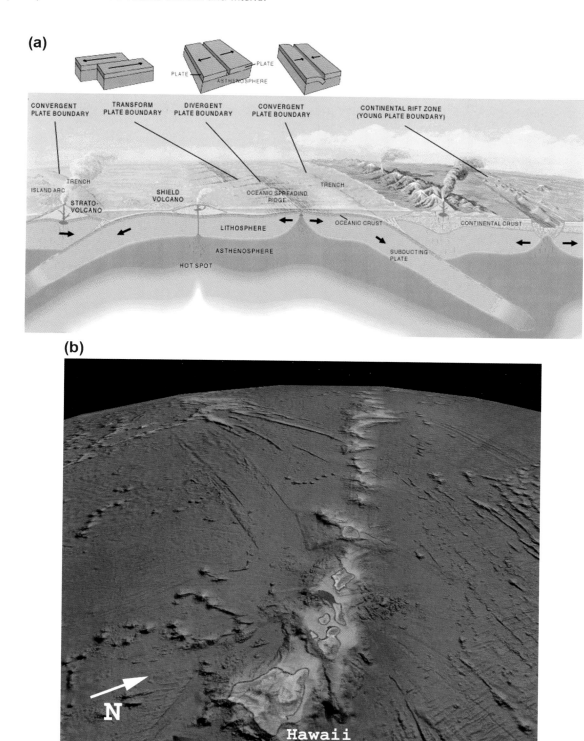

FIGURE 21.3 (a) Tectonic plate interactions. Tectonic plate interactions and the three fundamental kinds of plate boundaries. (Left) A convergent boundary caused by the subduction of oceanic material as it is overridden by another oceanic plate. (Center left) A subplate hot spot capped by a shield volcano (e.g. Hawaiian Islands). (Center right) A divergent plate boundary, in particular, a mid-oceanic spreading ridge. (Right) Another kind of convergent plate boundary, where the oceanic crust is being subducted by overriding continental crust, producing a chain of volcanic mountains (e.g. Andes Mountains). (Far right) A continental rift zone, another kind of divergent plate boundary (e.g. East African Rift). Finally, a transform plate boundary is shown at the upper middle of the scene, where two plates are sliding past each other without subduction. The three relationships are shown as block diagrams at the top of the figure. *(Courtesy of the US Geological Survey.)* (b) Emperor seamount chain spans Pacific plate. Perspective view of the Emperor seamount chain that spans the central and northwest sector of the Pacific Basin. The southeastern end of the hot spot track terminates in the Hawaiian Islands, and the predominate trend of plate motion has been to the northwest over time. Deeply rooted, persistent hot spots are probably the result of persistent hot upwelling plumes of lower density material from the upper mantle. "Petit spot" subsea volcanoes may form as hot spots in oceanic plates above or near the intersection of flexure cracks. *Courtesy of Google Earth™.*

tensional rather than compressive, extensional mountain ranges can form, as tectonic blocks founder and rotate. The western US Basin and Range Province is a good example of that type of mountain terrane. Another large subaerial extensional tectonic landform is the axial rift valley and associated inward-facing fault scarps, which form when aggregate tensional stresses tend to pull a continental plate apart (e.g. the East African Rift Valley). Such rift valleys are often characterized by ubiquitous mafic volcanism (e.g. Afar Triangle).

The geomorphic provinces just discussed generally tend to be very dynamic, with lifetimes that are intrinsically short (100–200 million years) relative to the age of the Earth (4.56 billion years). Some of the stable interior areas of continental plates, or cratons, however, do possess landforms and associated lithologic regions with ages comparable within a factor of two or three to the age of the Earth (2–3 billion years). The interior of the Canadian Shield and the Australian continent are two such special areas. Despite having been scoured repeatedly by continental ice sheets, the granitic craton of the Canadian Shield possesses a record of giant asteroidal and cometary impacts that are about 2 billion years old. (See Planetary Impacts.) These interior cratonic areas, in contrast to most of the rest of the Earth, which is mobile and active, provide a chance to view a part of the long sweep of the Earth's surface history. They are thus important, particularly in trying to understand how the environmental history of the Earth compares to that of the other terrestrial planets.

The distribution of the earth's landscape altitudes, relative to the mean geoid, is bimodal—continental and seafloor (Figure 21.4(a)). Although limited in percentage of surface area coverage, the interface between the two modes is a relatively high-energy place called the littoral or tidal zone. Ocean tides in this zone generate frequent (twice daily) environmental stresses on its residents that profoundly encourage evolution and natural selection, and may have been a key influence on the origin and early evolution of life here. It is interesting that Mars is another planet with a global bimodal highland/lowland dichotomy and may have had early oceans, although the absence of large lunar tides may be significant in this context. (See Planets and the Origin of Life.)

2.2. Landform Types

2.2.1. Submarine Landforms

Geomorphically, submarine oceanic basins comprise the areally dominant landform of the Earth, but ironically, they are probably less well explored than the well-imaged surfaces of Mars, Venus, and the satellites of the outer planets. Dominant features of oceanic basins are the oceanic ridge and rise systems, which have a total length of about 60,000 km ($\sim$1.5 times the equatorial circumference of the Earth), rise to 1–3 km above the average depth of the ocean, and can be locally rugged. In the Atlantic Ocean, oceanic rises exhibit a central rift valley that is at the center of the rise, whereas in the Pacific Ocean this is not always present (Figure 21.2(a)).

Older crust within oceanic basins can have gently rolling abyssal hills, which are generally smoother than the ridge and rise systems. These may have been much more rugged originally, but are now buried beneath accumulated sediment cover. Perhaps the most areally dominant feature of ocean basins (with the largest ones occurring in the Atlantic Ocean) is the predominantly flat abyssal plains that stretch for thousands of kilometers, usually also covered with accumulated marine sediments. Generally characterized by little topographic relief, in places they are punctuated by seamounts (Figure 21.4(b)), which are conical topographic rises sometimes topped by coral lagoons, or which sometimes do not reach the oceans' surface. These features are subsea volcanoes associated with island arcs or with midplate hot spots, such as the famous Emperor seamount chain, the southeastern end of which terminates in the Hawaiian Islands (Figure 21.3(b)). Such large hot spots are probably the result of persistent hot upwelling **plumes** from the upper mantle. Smaller "petit spot" subsea volcanoes may form above flexure cracks in oceanic plates. (See Planetary Volcanism.)

Oceanic margins represent another important, although more areally restricted, submarine landform province (Figure 21.4(b) and (c)). Because nearly half of the world's people live within 100 km of them and because seafood is a major food source for most of the world's population, they comprise a suite of landforms especially critical to the health and well-being of humanity.

"Atlantic style" continental margins tend to exhibit substantial ancient sediment accumulations and a shelf–slope–rise overall morphology, which probably represents submerged subaerial landscapes remnant from the last Ice Age, when the sea level was lower (about 135 m below current sea level, worldwide). Nevertheless, many such margins, and those of related basins (e.g. Hudson's Bay) appear to us now as "emergent shorelines" (e.g. Figure 21.4(d)), as they undergo postglacial rebound (PGR).

Ice ages were manifested by expansion, then contraction, of the Earth's ice sheets and mountain glaciers in most high-latitude and high-altitude zones. The most recent global deglaciation event was essentially complete by 6000 years ago, but relative sea levels have continued to change. This continuing change is generally thought to be the result of the earth's latent viscoelastic response to deglaciation (PGR), as its surface mass was redistributed. Regions that were most heavily glaciated (e.g. Canada and Northwestern Europe) show relative sea level falling at a rate controlled by postglacial crustal isostatic upward

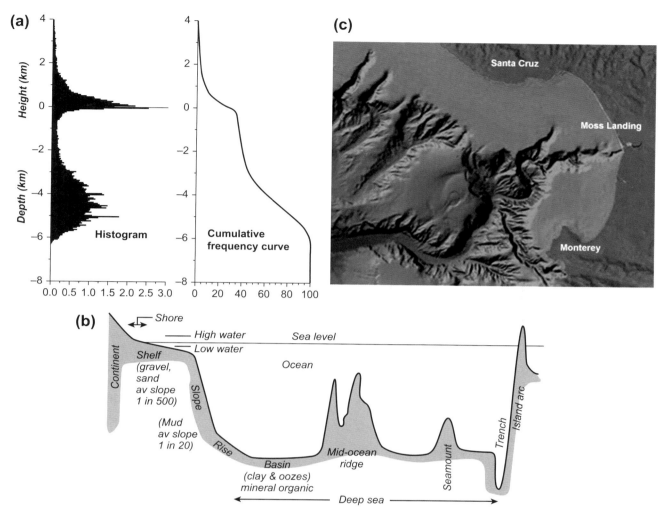

FIGURE 21.4 (a) Global altitude diagrams. At left are histograms of land altitudes and seafloor depth as a percentage of the Earth's surface area (50 m intervals), illustrating the classic continent—seafloor dichotomy. The interface between the two, subject to tidal and climatic fluctuation stress, is thought to have provided, in part, stimuli for biological evolutionary adaptations. At right is the global hypsometric curve, showing cumulative frequency of global topographic heights. (b) Ocean basin schematic. Principal features of the ocean floor shown in schematic form—height is greatly exaggerated. (c) Topography of the submarine Monterey Canyon, California, USA. The continental shelf offshore of Monterey California showing the Monterey and other canyons. Such canyons are common on shelves on both Atlantic and Pacific margins, often cutting through the shelf and down the continental slope to deep water. *(Figures used with permission of the Monteray Bay Aquarium Research Institute (MBARI).)* (d) Rebounding Canadian beach. A systematically striated sand beach near Nunavut, Canada (68° 05′ 50.74″ N, 108° 16′ 54.97″ W) seen on July 01, 2013. Each striation marks an episode of isostatic uplift, illustrating how the Arctic Ocean coastline has continued to rebound after the last glacial period. Tides in this area are weak enough such that the strands are preserved. *(With permission of P.D. Tillman.)* (e) Map of Gravity Recovery and Climate Experiment (GRACE)-derived global postglacial isostatic rebound. Shown is the distribution of global postglacial isostatic rebound as derived from GRACE data, expressed as changes in the surface mass distribution that would cause the changes in gravity if the mass were concentrated at the surface. It is expressed here in millimeters per year of equivalent water thickness. The mass estimates are provided on a $1 \times 1°$ grid and have an estimated $\pm 20\%$ accuracy. With permission of NASA and the GRACE Team. *(From Gerou et al., (2013); http://grace.jpl.nasa.gov.)* (f) Postglacial sea rise plotted as a function of time. Rise in sea level since the most recent global glaciations. *(From Fleming et al. (1998), Fleming (2000), and Milne et al. (2005).)* The existence of significant short-term fluctuations versus smooth and gradual change is disputed, although rapid deglaciation, "meltwater pulse 1A", by consensus is indicated on the plot. Lowest sea level occurred at about the last glacial maximum. Before this, waxing ice sheets resulted in almost continuously decreasing sea level during an approximately 100,000 year interval. *(With permission of Robert A. Rohde.)* (g) Most recent global sea level rise plotted as a function of time—1870 to present day. Sea level increases illustrated here indicate an average of approximately 0.15 cm/year during the term 1879—2008; however, since 2008 the rate has increased to about 0.30 cm/year. Regional and local trends may be variable depending on postglacial land movement and coastal current variations. *(With permission of the U.S. Environmental Protection Agency.)*

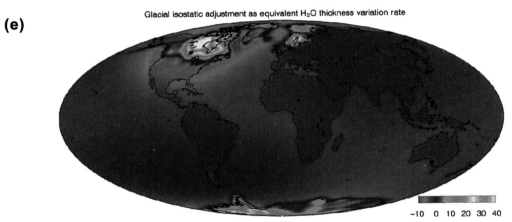

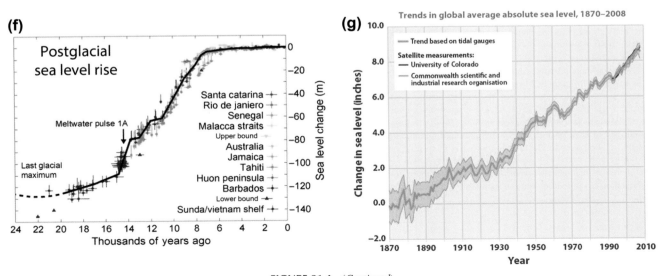

FIGURE 21.4 (*Continued*).

adjustment, greater than 1 cm/year in some places (e.g. Hudson Bay). Even in zones distant from where past glaciations occurred, rates of relative sea level adjustment are substantial (Peltier, 1999). The current distribution of areas affected by isostatic adjustment are strikingly displayed in the data from the Gravity Recovery and Climate Experiment twin satellite orbital gravimeter mission, launched in 2002 and operated jointly by NASA and the German Aerospace Center (e.g. Figure 21.4(e); Hanna et al., 2013). Along shorelines where PGR is observed, it somewhat offsets the eustatic (i.e. ocean volume is increasing as land ice melts) sea level rise now being observed as a result of global climate change (Figure 21.4(f); Fleming, 2000; Fleming et al., 1998; Ivins et al., 2013; Milne, Long, & Bassett, 2005). Recently absolute ocean levels have generally risen dramatically and systematically, but observed trends can be variable, with local relative sea levels not uniformly increasing due to idiosyncratic changes in land isostatic adjustment, as well as changes in coastal circulation patterns over extended periods due to long-term changes in weather patterns, and possibly climate.

Such costal changes, as related to a systematic rise in sea level (e.g. Figure 21.4(g)), are some of the biggest environmental challenges facing mankind in the twenty-first century. In countries that are not landlocked, populations tend to be most concentrated near shorelines, due to benefits derived from harbors and ocean transport, harvesting from fisheries, and recreational uses and tourism. Current global estimates of 40% population within 100 km of the world's shorelines, suggest that with additionally increased population densities and economic activity will come increasingly severe pressures on coastal human infrastructure and natural habitats. Changes in indigenous land cover, introduction of exotic species, and general increase in pollution will lead to narrower biodiversity, destruction of coral reefs, as well as a range of microbiota changes that will result in negative impacts on human health. Increases in pathogens, particularly cholera and hepatitis A, are associated with the expected decreased potability of drinking water sources.

Continental shelves are usually less than about 100 km in width and have very shallow ($\sim 0.1°$) topographic slopes. They typically end in a slope break that merges into the continental slope ($\sim 4°$ slope, about 50 km wide), which in turn merges into a gentle continental rise ($\sim 0.2°$ slope, about 50 km wide), which then typically transitions into an abyssal plain. Submarine canyons (also probably remnant from the last Ice Age, e.g. Hudson Canyon of the coast of New York) can deeply cut the continental shelf and slope and terminate in broad submarine sediment fan deposits at the seaward canyon outlet. "Pacific style" oceanic margins can be even narrower. Along the margins of continents of the Pacific Rim, a short shelf and slope can terminate into deep submarine trenches, manifested by subduction zones (e.g. South America and Kamchatka), up to 10 km depth. Similar fore-arc submarine morphology is observed along the margins of Pacific island arcs (e.g. Aleutians and Kurile Is). Much shallower "back-arc" basins occur behind the arcs, on the overriding plate (e.g. Sea of Okhotsk). (See Earth as a Planet: Atmosphere and Oceans.)

2.2.2. Subaerial Landforms

The subject of classic geomorphological investigations, and historically far more well studied because they are where people on Earth live, are the "subaerial" landscapes—the quarter of the earth's surface that is not submerged. These terranes exist almost exclusively on continents; however, some important subaerial landscapes (particularly volcanic ones, e.g. Hawaii and Galapagos Islands) exist on oceanic islands. Most continental landscapes are predominantly Cenozoic to late Cenozoic in age, because over that timescale (65 million years or so), the combined action of plate tectonics, constructive landscape processes (e.g. volcanism and sedimentary deposition), and destructive landscape processes (e.g. erosion and weathering) have tended to rearrange, bury, or destroy preexisting continental landscapes at all spatial scales. Thus, while often retaining the imprint of preexisting forms, subaerial landscapes on the Earth are constantly being reinvented.

Because the Earth's crust is so dynamic, one must realize from the planetary perspective that any geomorphic survey of the Earth's surface may be representative only of the current continental plate arrangement, and currently associated climatic and atmospheric circulation regimes. Plate tectonics is a powerful force in setting scenarios for continental geomorphology. For instance, during early Cenozoic times the global continental geography was characterized by the warm circumglobal Tethys Sea and higher sea levels than now (possibly linked to higher rates of mid-oceanic spreading), which strongly biased the overall terrestrial climate toward the tropical range (Figure 21.5).

The rearrangement of continental landmasses in the later Cenozoic closed the Tethys Sea, produced a circum-Antarctic ocean, and set up predominantly north–south circulation regimes within the Atlantic and Pacific Oceans. This global plate geography, combined with greater ocean basin volume (linked to lower ridge spreading rates) and the onset of continental glaciation, lowered sea levels, exposing large marine continental self-environments to subaerial erosion. Our current global surface environment reflects a kind of "oceanic recovery" after the last Ice Age, with somewhat higher sea levels. Thus, our current perception of the Earth's subaerial geomorphic landform inventory is strongly biased by our temporal observational niche in its

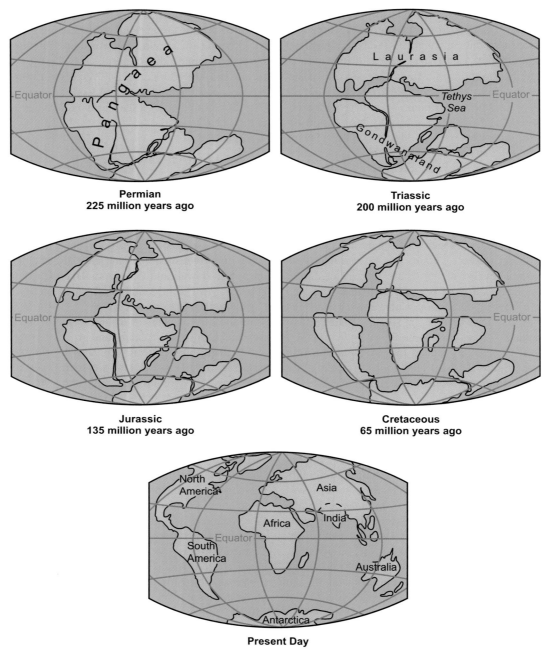

FIGURE 21.5 Continental geography through time. Modern plate tectonic theory is consistent with, and provides the scientific framework for observations of continental drift. Geologic evidence records the breakup of the supercontinent Pangaea about 225–200 million years ago, eventually fragmenting over time to create our familiar continental geography. The Tethys Sea referred to in the text is labeled. *Courtesy of the US Geological Survey.*

environmental history. Hypothetical interstellar visitors who arrived here 50 million years ago or may arrive 50 million years in the future would likely have a much different perception because of this distinctive dynamic character.

Terrestrial subaerial landform suites are the classic landscapes studied in geomorphology. These are listed in Table 21.1 (modified from Baker (1986), and Bloom (1998)). Currently, on Earth, globally dominant subaerial geomorphic regimes are related to the surface transport of liquid water and sediment due to the action of rainfall. Thus drainage basins dominate terrestrial landscapes at nearly all scales, from the continental scale to sub-100 m scales. These include currently active drainage basins in humid and semiarid climatic zones, to only occasionally active or relict drainages in arid zones. Drainage basin topographies

TABLE 21.1 Classification of Terrestrial Geomorphological Features by Scale

Order	Approximate Spatial Scale (km²)	Characteristic Units (with Examples)	Approximate Timescale of Persistence (years)
1	10^7	Continents, ocean basins	10^8-10^9
2	10^5-10^6	Physiographic provinces, shields, depositional plains, continental-scale river drainage basins (e.g. Amazon, Mississippi rivers, Danube, and Rio Grande)	10^8
3	10^4	Medium-scale tectonic units (sedimentary basins, mountain massifs, domal uplifts)	10^7-10^8
4	10^2	Smaller tectonic units (fault blocks, volcanoes, troughs, sedimentary subbasins, individual mountain zones)	10^7
5	$10-10^2$	Large-scale erosional/depositional units (deltas, major valleys, and piedmonts)	10^6
6	$10^{-1}-10$	Medium-scale erosional/depositional units or landforms (floodplains, alluvial fans, moraines, smaller valleys, and canyons)	10^5-10^6
7	10^{-2}	Small-scale erosional/depositional units or landforms (ridges, terraces, and dunes)	10^4-10^5
8	10^{-4}	Larger geomorphic process units (hillslopes and sections of stream channels)	10^3
9	10^{-6}	Medium-scale geomorphic process units (pools and riffles, river bars, solution pits)	10^2
10	10^{-8}	Microscale geomorphic process units (fluvial and eolian ripples and glacial striations)	$10^{-1}-10^4$

Modified from Baker (1986).

and network topologies, however, are strongly influenced by the interplay of the orogenic aspects of plate tectonics (i.e. mountain building) and prevailing climatic regimes, including the biogenic aspects of climate (e.g. vegetative ground cover). Clearly, areas of rapid uplift (e.g. San Gabriel Mountains, California) have characteristically steep bedrock drainages, where gravitational energies are high enough to scour stream valleys, generally have parallel or digitate (handlike) drainage patterns, have high local flood potentials, and respond strongly to local weather (e.g. spatial scales 10–100 km in characteristic dimension). At the other spatial extreme, major continental drainages (e.g. Amazon River, Mississippi River, and Ob River in Siberia—Table 21.1), with highly dendritic (tree-like) overall pattern organization, are low average gradient systems that integrate the effects of a variety of climatic regimes at different spatial scales and tend to respond to mesoscale and larger climatic and weather events (e.g. 100–1000 km scale).

Subaerial volcanic processes produce characteristic landforms in all terrestrial climate zones (Figure 21.2(b)). They tend to occur in belts, mainly at plate boundaries, with a few notable oceanic (e.g. Hawaiian Islands) and continental (e.g. the San Francisco volcanic field in Northern Arizona, the Columbia and Snake River volcanic plains in the US Pacific Northwest, and the Deccan Traps in India) exceptions that occur within plate interiors. Although not as massive or as topographically high as their planetary counterparts (e.g. Martian volcanoes such as Olympus Mons), they provide some of the most spectacular and graceful landforms on the Earth's surface (e.g. Mt Fujiyama, Japan, and Mt Kilamanjaro, Kenya). Our planet's central vent volcanic landforms range from the majestic stratocone volcanic structures just mentioned to large collapse and resurgent caldrons or caldera features (e.g. Valles Caldera, New Mexico; Yellowstone Caldera, Wyoming; Campi Flegrei, Italy; and Krakatau, Indonesia). More areally extensive and lower subaerial shield volcanoes, formed by more fluid lavas (and thus with topographic slopes generally less than 5°) exist in the Hawaiian Islands, at Piton de la Fournaise (Reunion Island), in Sicily at Mt Etna (compound shield with somewhat higher average slopes, up to ∼20°), and the Galapagos Islands (Equador), for example. Often their areal extent corresponds strongly to the rate of their effusion. Subaerial and submarine volcanoes occur on the Earth at nearly all latitudes. Indeed some of the world's most active volcanoes occur along the Kurile-Kamchatka-Aleutian arc, in subarctic to arctic environments, often with significant volcano—ice interaction. High-altitude volcanoes that occur at more humid, lower latitudes (e.g. Andean volcanoes like Nevado del Ruiz in Columbia) can also have significant magma or lava—ice interactions. Volcanoes also occur in Antarctica, Mt Erebus being the most active, with a perennial lava pond. (See Planetary Volcanism.)

2.3. Summary: Terrestrial vs Planetary Landscapes

Overall, the Earth's geomorphic or physiographic provinces, as compared to those of the other planets in our solar system, are distinguished by their variety, their relative youth, and their extreme dynamism. Many of the other terrestrial-style bodies, such as the Moon, Mars, and Mercury, are relatively static, with landscapes more or less unchanging for billions of years. Although this may not have been the case early in their histories, as far as we can tell from spacecraft exploration, this is the case now. Other landscapes, such as those on Venus and Europa and a few of the other outer planets' satellites, appear younger and appear to be the result of very dynamic planetwide processes, and possibly for Venus, a planetwide volcanic "event". Currently most of these bodies appear relatively static, although this point may be credibly debated. For instance, the Jovian satellite Io has vigorous on-going volcanic activity as was first discovered in Voyager spacecraft imaging, and the Saturnian satellite Enceladus appears to be erupting water from relatively warm spots in its southern hemisphere, as seen in recent Cassini spacecraft data. Nevertheless, it seems that the crusts of all these bodies are currently somewhat less variegated than that of the Earth. Be aware, however, that this last statement may turn out to be just another example of "Earth chauvinism", and will be proved wrong once we eventually know the lithologies and detailed environmental histories of these bodies as well as we know the Earth's. (See Venus: Surface and Interior.)

3. EARTH SURFACE PROCESSES

The expenditure of energy in the landscape is what sculpts a planetary surface. Such energy is either "interior" (endogenic) or "exterior" (exogenic) in origin. The combined gravitational and radiogenic thermal energy of the Earth (endogenic processes) powers the construction of terrestrial landscapes. Thus, the Earth's main constructional landscape processes, plate tectonics and resulting volcanism, are endogenic processes.

Destructional processes, such as rainfall-driven runoff and streamflow, are essentially exogenic processes. That is, the energy that drives the evaporation of water that eventually results in precipitation, and the winds that transport water vapor, comes from an exterior source—the Sun (with the possible exception of very local, but often hazardous, weather effects near explosive volcanic eruptions, and endogenic energy source). In familiar ways, such destructional geomorphic processes work to reduce the "gravitational disequilibria" that constructive landscapes represent. For instance, the relatively low and ancient Appalachian Mountains, pushed up during one of the collisions between the North American and European continental landmasses, were probably once as tall as the current Himalayan chain. Their formerly steep slopes and high altitudes represented a great deal of gravitational disequilibria, and thus a great deal of potential energy that was subsequently expended as kinetic energy by erosive downhill transport processes (e.g. rainfall runoff and streamflow). Once the processes of continental collision ebbed and tectonic uplift ceased, continuing erosion and surface transport processes (such as rainfall, associated runoff, snowfall, and glaciation) over only a few tens of millions of years reduced the proto-Appalachian Mountains to their present gently sloping and relatively low-relief state.

Volcanic landforms provide myriad illustrations of the competition between destructive and constructive processes in the landscape. For example, Mt Fuji, the most sacred of Japanese mountains, is actually an active volcano that erupts on the order of every 100—150 years. Its perfectly symmetrical conical shape is the result of volcanic eruptions that deposit material faster than it can be transported away, on average. If Fuji stopped erupting, it would become deeply incised by stream erosion and it would lose its classic profile over a geologically short time interval (Figure 21.6).

3.1. Constructive Processes in the Landscape

Over the geologic history of the Earth, volcanism has been one of the most ubiquitous processes shaping its surface. Molten rock (lava) erupts at the Earth's surface as a result of the upward movement of slightly less dense magma. Its melting and upward migration are triggered by convective instabilities within the upper mantle. Volcanic processes very likely dominated the earliest terrestrial landscapes and competed with meteorite impacts as the dominant surface process during the first billion years of Earth history. With the advent of plate tectonics, multiphase melting of ultramafic rocks tended to distill more silicic lavas. Because silicate-rich rocks tend to be less dense than more mafic varieties, they tend to "float" and resist **subduction**, thus continental cores (cratons) were generally created and enlarged by island-arc accretion.

FIGURE 21.6 Mt Fuji, Japan. Mt Fuji at sunrise from Lake Kawaguchi. Perhaps the world's quintessential volcano, the perfect conical shape of Mt Fuji has inspired Japanese landscape artists for centuries. It is considered a sacred mountain in Japanese tradition and thousands of people hike to its summit every year. Volcanologically, Mt Fuji is termed a "stratovolcano" and rises to an altitude of 3776 m above sea level. It erupts approximately every 150 years, on average (public domain).

Most volcanism tends to occur on plate boundaries. Subaerial plate boundary volcanism tends to produce island arcs (e.g. Aleutian Islands and Indonesian archipelago) when oceanic plates override one another or subaerial volcanic mountain chains (e.g. Andes) underride more buoyant continental plates. Such volcanism tends to be relatively silica-rich (e.g. andesites), producing lavas with higher viscosities, thus tending to produce steeper slopes. Rough lava flows on these volcanoes tend to be classified as aa or blocky lavas. High interior gas pressures contained by higher **viscosity** magmas can produce very explosive eruptions, some of which can send substantial amounts of dust, volcanic gas, and water vapor into the stratosphere.

Another kind of volcanic activity tends to occur within continental plates. As is thought to have been widespread on the Moon, Mars, and Venus and to a lesser degree within impact basins on Mercury, continental flood eruptions have erupted thousands of cubic kilometers of layered **basalts** (e.g. Deccan and Siberian Traps in India and Russia; Columbia River Basalt Group in the United States). These are among the largest single subcontinental landforms on the Earth. Such lavas were mafic, of relatively low viscosity, and are thought to have erupted from extended fissure vents at very high eruption rates over relatively short periods (1–10 years). Recent work on the 100-km-long Carrizozo flow field in New Mexico, however, suggests that such massive deposits may have formed at much lower volume effusion rates over much longer periods than previously thought (10–100 years or more). The same may be true for lava flows of similar appearance on other planets. (See Planetary Volcanism.)

Perhaps the most familiar kind of subaerial volcanism is the well-behaved, generally nonexplosive, Hawaiian-style low-viscosity eruptions of tholeiitic basalts that form shield volcanoes, erupting in long sinuous flows. Typically such flows are either very rough ("aa") (Figure 21.7(a)) with well-defined central channels and levees or very smooth, almost glassy ("pahoehoe") (Figure 21.7(b)).

These lavas are thought to be comparable to lavas observed in remote sensing images of Martian central vent volcanoes (e.g. Alba Patera and Olympus Mons). Shield volcanoes on both planets tend to exhibit very low slopes (i.e. ∼5°). Active submarine basaltic volcanoes tend to occur along mid-oceanic ridges. Often the hot sulfide-rich waters circulating at erupting submarine venting sites provide habitats for a wide variety of exotic chalcophile (sulfur-loving) biota found nowhere else on Earth and proposed as a model for submarine life on Europa.

The transport of water across the land surface also has a hand in forming constructional landforms. Sediment erosion, transportation, and deposition can set the stage for a variety of landscapes, especially in concert with continental-scale tectonic ("epirogenic") uplift. The Colorado Plateau in the southwestern United States is perhaps the best example of this type of landscape. The Grand Canyon of the Colorado River slices through the heart of the Colorado Plateau and exposes over 5000 vertical feet of sedimentary layers, the oldest of which date to the beginning of the Cambrian era (Figure 21.8(a)).

Water itself can form constructive landforms on the Earth. In its solid form, water can be thought of as another solid component of the Earth's crust, essentially as just another rock. Under the present climatic regime, the Earth's great ice sheets—Antarctica and Greenland—along with numerous valley glaciers scattered in mountain ranges across the world in all climatic zones, compose a distinct suite of landforms. Massive (up to kilometers thick) deposits of perennial ice form smooth, crevassed, plastically deforming layers of glacial ice. Continental ice sheets depress the upper crust upon which they reside and can scour the subjacent rocky terrains to bedrock, as during the Wisconsin Era glaciation in Canada (i.e. last Ice Age in North America). Valley glaciers, mainly by mechanical and chemical erosion in concert, tend to carve out large hollows (cirques) in their source areas and have large outflows of meltwater at their termini (Figure 21.8(b)).

3.2. Destructive Geomorphic Processes

Friction probably represents the largest expenditure of energy as geologic materials move through the landscape: friction of water (liquid or solid) on rock, friction of the

FIGURE 21.7 (a) Aa flow from Mauna Loa Volcano, Hawaii, USA. Advancing flow of incandescent aa lava. Generally, aa flows are very rough and meters to tens of meters thick. They form broad toes and lobes and can advance kilometers per day, as often happens during eruptions of large aa flows on Mauna Loa volcano in Hawaii (e.g. Mauna Loa 1984 eruption). *(Courtesy of the US Geological Survey.)* (b) Pahoehoe from Kilauea Volcano, Hawaii, USA, cascading over scarp. Incandescent ($\sim$1400 K) fluid pahoehoe flows near the coast south of Kilauea Volcano, showing a lava breakout from an upstream lava tube cascading into two main branches. The cliff is approximately 15 m high. Fields of pahoehoe lava tend to form in a very complex intertwined fashion, and old cooled flows are often smooth enough to walk on in bare feet. *Courtesy of the US Geological Survey.*

wind, friction of rock on rock, or friction of rock on soil. All these processes are driven by the relentless force of gravity and generally express themselves as transport of material from a higher place to a lower one. Erosion (removal and transport of geologic materials) is the cumulative result, over time reducing the average altitude of the landscape and often resculpting or eliminating preexisting landforms of positive relief (e.g. mountains) and incising landforms of negative relief (e.g. river valleys or canyons). Overall, the source of potential energy for these processes (e.g. the height of mountain ranges) is provided by the tectonic activity of plates as they collide or subduct.

Subaerial landscapes on the Earth are most generally dominated by erosive processes, and subaqueous landscapes are generally dominated by depositional processes. Thus, from a planetary perspective, it is the ubiquitous availability and easy transport of water, mostly in liquid form, that makes it the predominant agent of sculpting terrestrial landscapes on Earth. Based on the geologic record of ancient landscapes, it appears that this has been the case for eons on the Earth. Such widespread and constant erosion does not appear to have happened for such a long time on any other planet in the solar system, although it appears that Mars may have had a period of time when aqueous erosion was important and even prevalent.

Fluvial erosion and transport systems (river and stream networks) dominate the subaerial landscapes of the Earth, including most desert areas. Even in deserts where aeolian (wind-driven, e.g. sand dunes) deposits dominate the current landscape, the bedrock signature of ancient river

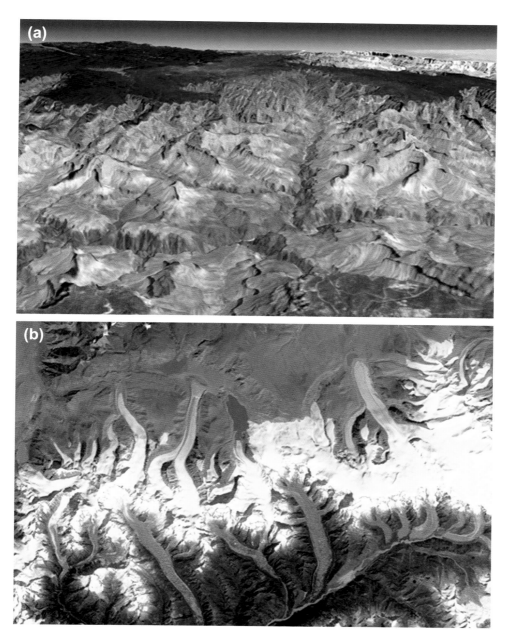

FIGURE 21.8 (a) Classic view of the Grand Canyon, Colorado, USA. Classic view of the Grand Canyon of the Colorado River in Arizona. The massive layering records the local geologic history for at least the last 500 million years. Comparable layering has also been observed recently in canyons on Mars. This simulated true color perspective view over the Grand Canyon was created from Advanced Spaceborne Thermal Emission and Reflection (ASTER) radiometer data acquired on May 12, 2000. The Grand Canyon Village is in the lower foreground; the Bright Angel Trail crosses the Tonto Platform, before dropping down to the Colorado River and then to the Phantom Ranch (green area across the river). Bright Angel Canyon and the North Rim dominate the view. At the top center of the image the dark blue area with light blue haze is an active forest fire. *(Courtesy NASA/GSFC/METI/ERSDAC/ JAROS, and U.S./Japan ASTER Science Team.)* (b) Bhutan Glaciers, Himalayan Mountains, Asia. Classic Himalayan valley glaciers in Bhutan, showing theaterlike "cirque" source areas, long debris-covered ice streams, and terminal meltwater lakes. ASTER data have revealed significant spatial variability in glacier flow, with velocities from 10 to 200 m/year. Meltwater volumes have been increasing in recent years and threaten to breach terminal moraine deposits with consequent dangerous downstream flooding. This ASTER scene acquired on November 20, 2001, is centered near 28.3° N latitude, 90.1° E longitude, and covers an area of 32.3 × 46.7 km. *Courtesy NASA/GSFC/METI/ERSDAC/JAROS, and U.S./Japan ASTER Science Team.*

systems, relict from more humid past climatic epochs, can be detected in optical and radar images taken from orbiting satellites. Surface runoff, usually due to the direct action of rainfall occurs in nearly all climatic zones, except the very coldest.

On the Earth, such network forms resulting from this process tend to be scale-independent and take on a nearly fractal character. That is, network patterns tend to be replicated at nearly all scales, with regular geometric relationships that tend to be similar, no matter what the physical size of the network. In contrast to the situation on the Earth, the most visible and well-expressed Martian valley networks tend to be highly irregular in their network geometries, probably reflecting very restricted source areas of seepage or melt-driven runoff, rather than rainfall, and strong directional control by fractures and faults that was not overcome easily by river erosional processes. In addition, they are distributed very sparsely and are primitive in their branching, very much like the canyon networks arid areas of the world like Northern Africa (Figure 21.9) and the desert Southwest of the United States (Figure 21.8(a)). Thus, in contrast to Mars, for most of its discernable history, the Earth's landscapes have been distinguished, overall, by well-integrated and complexly branched fluvial drainage networks driven primarily by rainfall.

Uniquely on the Earth (within this solar system at least) it is the competition between constant fluvial erosion and constant tectonic uplift (and in some land areas, frequent volcanic eruptions) that is the predominant determinator of the landscape's appearance. For instance, the present terrestrial landscape is not dominated by impact scars. Plate tectonic processes are, in part, responsible; however, fluvial erosion is probably the dominant factor for subaerial landscapes in this regard. Also, without constant tectonic reinforcement, rainfall would probably reduce a Himalayan-style, or Alpine range to Appalachian-style mountains within 10 million years or so. On the Earth, when tectonic forces subside, constant fluvial erosion wins out and hilly landscapes are flattened.

Other erosive processes, independently or in concert with fluvial activity, also clearly play a role on the Earth, including seepage-induced collapse (called "groundwater sapping"), which can result in networks of steep-walled gulleys and canyons. In addition, the chemical action of groundwater can form landscapes of caves and sinkholes in limestone areas (called "karsts"). Groundwater sapping and karst formation on the Earth may be relatively less important than fluvial erosion, whereas the opposite case may be true for Mars. Another process regime that dominates arid and polar deserts on the Earth, and apparently is

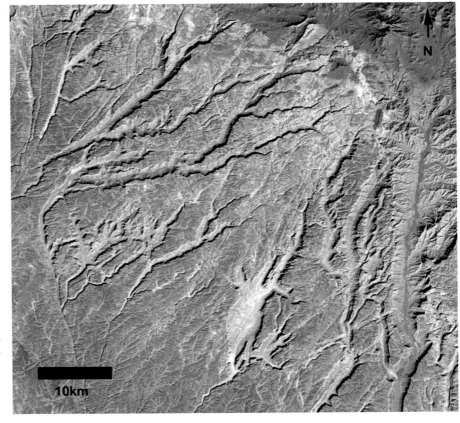

FIGURE 21.9 Desert drainage networks in Chad, North Africa. Shown here are deeply incised canyons on the southwest slope of the Tarso Voon Volcano located in the west-central part of the Tibesti Volcanic Range, in northern Chad (20.5° N latitude, 17° W longitude, approximately 3400 ft above sea level). Characteristic steep-walled theater-headed canyons form as overlying relatively soft Tarso Voon ignimbrites are stripped back over more resistant basement schists, through the action of groundwater seepage and surface runoff during infrequent storms, or during previous eras of wetter climate. Such differential erodability very likely also played a part in the formation of ancient complex ramified canyon networks on Mars of similar scale and appearance, and may reflect the former presence of more abundant supplies of near-surface water during warmer periods on Mars in its distant past. The ASTER image was acquired on January 12, 2003. Spatial resolution is 15 m/pixel and the image as shown is an RGB composite of three visible bands (1N, 2N, 3N—0.52—0.82 μm). *Courtesy NASA/GSFC/METI/ERSDAC/JAROS, and U.S./Japan ASTER Science Team.*

FIGURE 21.10 (a) Sand dunes in Namibia. Namib-Naukluft National Park is an ecological preserve in Namibia's vast Namib Desert, and is the largest game park in Africa. Coastal winds create the tallest sand dunes in the world here, with some dunes reaching 300 m in height. This ASTER perspective view was created by draping an ASTER color image over an ASTER-derived digital elevation model. The image was acquired on October 14, 2002. In the great deserts of the world, sand sheets are the dominant morphology and are wind driven. In open desert areas (e.g. Sahara or Arabian Peninsula), dune trains may stretch for tens or hundreds of miles. *(Courtesy NASA/GSFC/METI/ERSDAC/JAROS, and U.S./Japan ASTER Science Team.)* (b) Deadly landslide in La Conchita, California. Large 1995 landslide and more recent 2005 debris flow that initiated from the slide above the town of La Conchita, California. It destroyed or seriously damaged 36 houses and killed 10 people. Loss of coherence in water-saturated marine sediments was triggered by heavy rain. Landslides observed on Mars are typically one to as much as three orders of magnitude larger and may indicate the past presence of water. Alternatively substantial atmospheric lubrication is possible as is thought to have occurred during the ancient gigantic Blackhawk Slide on the slopes of the San Bernardino Mountains in California. *Courtesy of the US Geological Survey.*

highly active, even today, on Mars, is that of wind-driven erosion and transport of fine dust and sand (called "aeolian", after the Roman god of the winds). On the earth, aeolian processes are dominant only in certain restricted areas, such as the desert sand seas of Africa and Asia (Figure 21.10(a)). On Mars, however, fine dust and sand dune and drift morphologies appear everywhere and can reveal important information on current wind regimes and on the constitution of the fine material based on observations and models of terrestrial dune morphologies.

Another important terrestrial geomorphic process is weathering—the breakdown of consolidated material into constituent grains. Rock can be broken down in several ways. Chemical weathering can occur when natural acids act on carbonates in susceptible rocks, such as limestone or sandstones, releasing the residual silicate grains. Mechanical weathering of rock can occur when the hydrostatic pressures of ice in freeze–thaw cycles overcome rock brittle strength thresholds at microscopic and macroscopic scales. The formation of salt crystals also exerts mechanical energy to break up rocks and can chemically weather rocks. Oxidation of minerals, particularly iron-containing minerals, is another form of chemical weathering. Biological weathering occurs through chemical weathering caused by biogenic acids, particularly in tropical areas. It can also occur mechanically, by bioturbation of soils and sediments, as well as by the physical pressure of root and stem turgor in cracks and fissures within solid rock. It is of significance that on the Earth, all three major forms of weathering are enhanced or enabled by the ubiquitous presence of water.

Perhaps some of the most dramatic forms of nonvolcanic landscape alteration that we see on the Earth today fall into the category that geomorphologists call mass wasting. Generally, the term mass wasting is applied to processes such as landslides, creep, snow and debris avalanches, submarine slides and slumps, volcanotectonic sector collapses, and scour related to the action of glaciers. Mass-wasting processes tend to affect a relatively minor proportion of the Earth's surface at any given time, however, such as volcanic eruptions (with which they are often associated); when they occur near population areas, their effects can be devastating (Figure 21.10(b)). On Mars, massive landslides, similar in morphology and scale to the largest terrestrial submarine landslides, are commonly seen within Vallis Marineris and its tributary canyons.

4. TOOLS FOR STUDYING EARTH'S DEEP INTERIOR

In comparison with other planets, the interior of the Earth can be studied in unprecedented detail. This is because of the existence of sources of energy, such as earthquakes or magnetic and electric disturbances. Seismic waves, for example, can penetrate deep inside the Earth, and the time they travel between the source (earthquake or an explosion) and the receiver (seismographic station) depends on the physical properties of the Earth. The same is true with respect to electromagnetic induction, although observations are different in this case.

Observation and interpretation of seismic waves provide the principal source of information on the structure of the

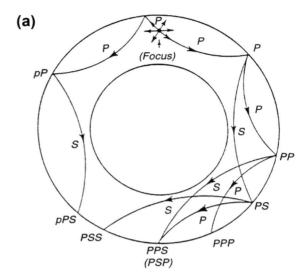

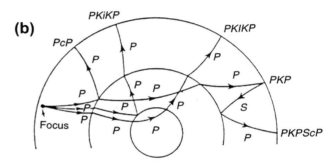

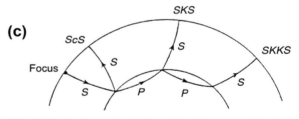

FIGURE 21.11 Ray paths. (a) Ray paths of the compressional waves (P) in the mantle, including their conversion to shear waves (S). (b) Ray paths of the P-waves interacting with the outer and inner core. (c) Ray paths of the S-waves interacting with the core; the S-waves are converted into P-waves in the outer core.

deep interior of the Earth. Both compressional (P-waves) and shear (S-waves) can propagate in a solid, only P-waves in a liquid. Compressional waves propagate faster than shear waves by, roughly, a ratio of $\sqrt{3}$. Velocities, generally, increase with depth because of the increasing pressure, hence the curved ray paths (Figure 21.11).

At the discontinuities (which include the Earth's surface) waves may be converted from one type to another. Figure 21.11(a) shows P-waves emanating from the source ("Focus"). The P-waves can propagate downward (right

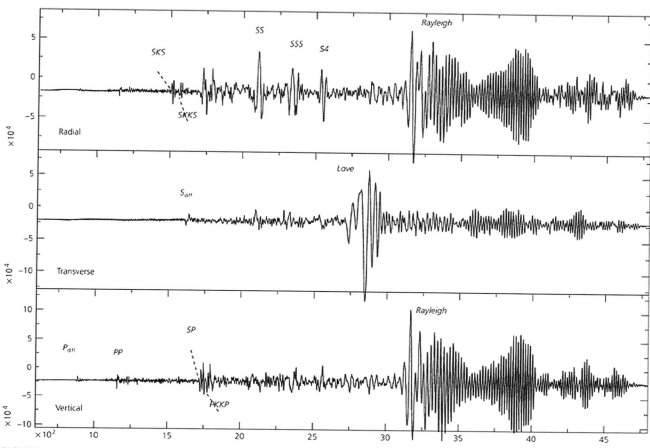

FIGURE 21.12 Global Seismic Network (GSN) Recording. Three-component recording at a GSN digital, high dynamic range station. Note identification of various phases. The dispersed Rayleigh waves are seen on the radial and vertical components and Love waves are seen on the transverse component.

part of the figure) and are observed as PP, PS, PPP, PPS, for example. They can also propagate upward, be reflected from the surface, and then observed as so-called depth phases: pP, pPS. Depth phases are very helpful for a precise determination of the depth of focus.

Figure 21.11(a) shows rays in the mantle; there are also the outer core and inner core. The outer core is liquid and has distinctly different composition; the P-wave speed is some 40% lower than at the bottom of the mantle; also, there are no S-waves. The inner core is solid, with a composition similar to that of the outer core. Figure 21.11(b) shows the rays (mostly P-waves) that are reflected from the core–mantle boundary (CMB; a letter c is inserted, e.g. PcP) or that are transmitted through the outer core (letter K: PKP) or also through the inner core (letter I: PKIKP). Figure 21.11(c) shows S-wave rays interacting with the CMB, reflected (ScS) or converted at the CMB into a P-wave and then again reconverted into an S-wave: SKS and SKKS. The latter indicates one internal reflection from the underside of the CMB.

Figure 21.12 shows an example of an earthquake recorded on a three-component seismograph system and then rotated such that the "radial" component shows horizontal motion along the great circle from the earthquake to the station; "transverse" component is also horizontal motion but in the direction perpendicular to the ray path, and "vertical" component shows up-and-down motion.

Figure 21.13 compares observed travel times, reported by the International Seismological Centre with those predicted by an Earth model. The scatter around the predicted values is caused by the effects of lateral heterogeneity and measurement errors.

Measurements of the travel times of the waves such as shown in Figures 21.11 and 21.13 have led to the derivation of models of the seismic wave speed as a function of depth. These, in turn, were used to improve the location of earthquakes and further refine the models. The first models were constructed early in the twentieth century; the models published by Beno Gutenberg and Sir Harold Jeffreys in the 1930s are very similar in most depth ranges to current ones. The model of Jeffreys is compared with a recent model (iasp91) in Figure 21.14. The upper mantle (the topmost 700 km) with its discontinuities and the inner core are exceptions.

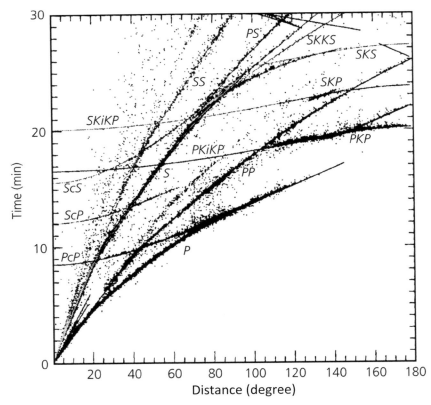

FIGURE 21.13 Observed travel times vs predictions. Observed travel times from a Bulletin of International Seismological Centre are compared with predictions for model IASP91. There are additional observed branches, such as PPP and SSS, for which travel times have not been computed.

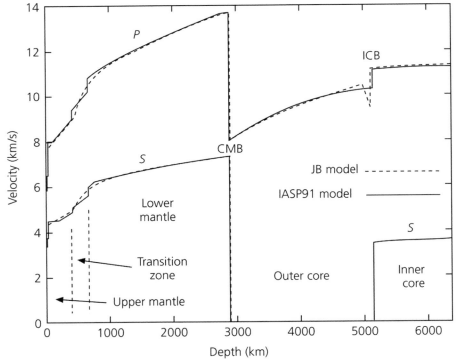

FIGURE 21.14 Model Comparisons. Comparison of a velocity model by Jeffreys (c. 1937) with model IASP91. Notice that for the most part changes have been minor, except for the discontinuities in the transition zone, solidity of the inner core, and structure just above inner core boundary.

In addition to the body waves, which propagate through the volume of the Earth, there are also surface waves, whose amplitude is the largest at the surface and decreases exponentially with depth. Surface waves are important in studying the crust and upper mantle and, in particular, their lateral variations, as the Earth is most inhomogeneous near the surface. There are Rayleigh waves with the particle motion in the vertical plane (perpendicular to the surface; second and third trace in Figure 21.12) and Love waves whose particle motion is in the horizontal plane (parallel to the surface). Surface waves are dispersed in the Earth because of the variation of the physical parameters with depth; notice that the longer period surface waves in Figure 21.12 arrive before shorter period waves.

Very long-period surface waves (>100 s) are sometimes called "mantle waves", have horizontal wavelengths in excess of 1000 km, and maintain substantial amplitudes (and, therefore, sensitivity to the physical properties) down to depths as large as 600–700 km. Because of their long periods, mantle waves are attenuated relatively slowly and can be observed at the same station as they travel around the world several times along the same great circle (both in the minor and major arc direction). Figure 21.15 shows a three-component recording of mantle waves (note the timescale); the observed seismograms are shown at the top of each pair of traces; the bottom trace is a synthetic seismogram computed for a three-dimensional (3-D) Earth model.

Superposition of free oscillations of the Earth (known also as the normal modes) in the time domain will yield mantle waves. First spectra of the vibrations of the Earth were obtained following the Chilean earthquake of 1960, the largest seismic event ever recorded on seismographs. The measurements of the frequencies of free oscillations lead to the renewed interest in the Earth's structure. In particular, they, unlike body waves, are sensitive to the density distribution and thus provide additional constraints on the mass distribution other than the average density and moment of inertia. Figure 21.16 shows an example of a spectrum of a vertical component recording of a very large deep earthquake under Bolivia; the lowest frequency mode shown has a period of about 40 min.

Sometime in the 1970s it became clear that further refinements in one-dimensional Earth models cannot be achieved, and perhaps do not make much sense, without considering the three dimensionality of the Earth's structure.

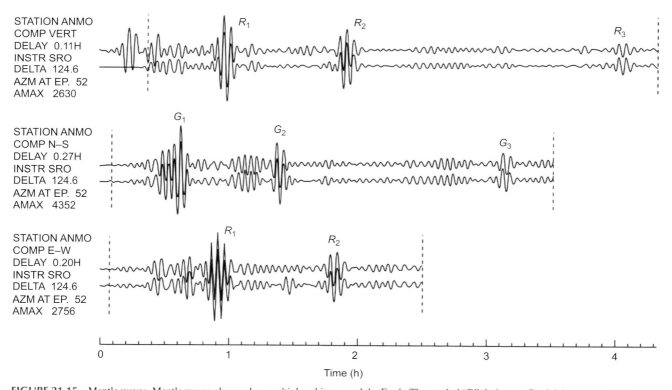

FIGURE 21.15 Mantle waves. Mantle waves observed on multiple orbits around the Earth. The symbol "R" designates Rayleigh waves and "G" Love waves. Odd-numbered (1,3) arrivals correspond to minor arc arrivals plus an integer number complete number of paths around the Earth. Even-numbered wavegroups correspond to initial propagation in the major arc direction. The signal between arrivals of the fundamental mode wavegroups represents contribution of overtones. Top traces are observed seismograms; bottom traces are synthetic seismogram computed for 3-D model of upper mantle M84C; if one-dimensional model (PREM) was used, there would be significant differences between observed and computed traces.

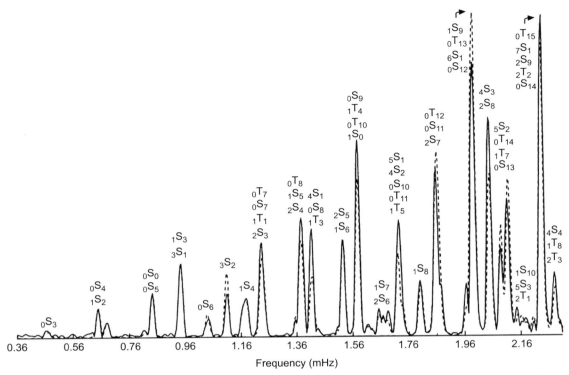

FIGURE 21.16 Amplitude spectrum. Amplitude spectrum of a vertical component seismogram of the great deep Bolivia earthquake of 1994. The peaks in the spectrum correspond to periods of free oscillations (vibrations) of the Earth. The symbols designate the specific normal modes. Some of them appear in groups, which indicate a possibility of coupling between modes close in frequency. Usually the fundamental modes (presubscript "0") are excited most strongly.

All three types of data described earlier are sensitive to the lateral heterogeneity. Travel times will be perturbed by slight variations of the structure along a particular ray path, compared to the prediction by a one-dimensional model. All we need is many observations of travel times along crisscrossing paths. Many millions of such data are available from the routine process of earthquake location; they are assembled from some 6000 stations around the world by the National Earthquake Information Center in Golden, Colorado, and by the International Seismological Centre in England (see Figure 21.13). Surface waves, mantle waves, and periods of free oscillations in a 3-D Earth also depend on the location of the source and the receiver. Progress during the last decade in global seismographic instrumentation, in terms of the quality and distribution of the observatories and exchange and accessibility of the data, makes the required observations much more readily available.

5. SEISMIC SOURCES

Even though the field of seismology can be divided into studies of seismic sources (earthquakes, explosions) and of the Earth's structure, they are not fully separable. To obtain information on an earthquake, we must know what happened to the waves along the path between the source and receiver, and this requires the knowledge of the elastic and anelastic Earth structure. The reverse is also true; in studying the Earth structure, we need information about the earthquake, at least its location in space and time, but sometimes also the model of forces acting at the epicenter.

Most of the earthquakes can be described as a process of release of shear stress on a fault plane. Sometimes the stress release can take place on a curved surface or involve multiple fault planes; the radiation of seismic waves is more complex in these cases. Also, explosions, such as those associated with nuclear tests, have a distinctly different mechanism and generate P- and S-waves in different proportions, which is the basis for distinguishing them from earthquakes.

Figure 21.17 shows three principal types of stress release, sometimes also called the earthquake mechanism. The top part of Figure 21.17(a) is a view in the horizontal plane of two blocks sliding with respect to each other in the direction shown by the arrows. Such a mechanism is called strike slip, and the sense of motion is left-lateral; there is also an auxiliary plane, indicated by a dashed line; a ground motion generated by a slip on the auxiliary plane (right lateral) cannot be distinguished from that on the principal plane. The bottom part of Figure 21.17(a) is a stereographic

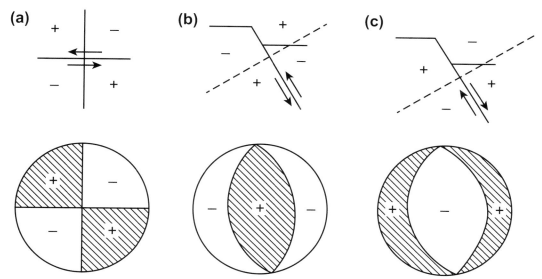

FIGURE 21.17 Classical types of earthquakes. Three classical types of earthquakes (top) and the distribution of the signs of the P-wave arrivals: (a) strike slip, (b) thrust or reverse fault, and (c) normal fault. The beach balls represent the equal area projection of the signs of first motion of the P-waves. The motion would be positive within the shaded areas. The lines separating shaded areas with the unfilled ones are called "nodal planes".

projection of the sign of P-wave motion observed on the lower hemisphere of the focal sphere (a mathematical abstraction in which we encapsulate the point source in a small uniform sphere). The plus sign corresponds to compressive arrivals and minus sign to dilatational zones; quadrants with compressive arrivals are shaded.

The top part of Figure 21.17(b) is a section in the vertical plane. In this case, the block on the right moves upward on a plane that dips at a 45° angle with respect to the block on the left; this mechanism is called thrust and is associated with compression in the horizontal plane and tension in the vertical plane and corresponds to the convergence of the material on both sides of the fault. Such processes are responsible for mountain building. The shaded central region in the bottom part of Figure 21.17(b), with the dilatational arrivals on the sides, is characteristic of the thrust—or reverse faulting—events. Figure 21.17(c) illustrates the opposite mechanism, in which tension is horizontal and compression vertical; this is called normal faulting and is associated with extension, which can lead to the development of troughs or basins.

The "beach-ball" diagrams are commonly used as a graphic code to represent the tectonic forces. Some earthquakes are a combination of two different types of motion, e.g. thrust and strike slip; in this case the point at which the two planes intersect would be moved away from either the rim or the center of the beach-ball diagram.

The size of the earthquake is measured by magnitude. There are several different magnitude scales depending on the type of a wave whose amplitude is being measured. In general, magnitude is a linear function of the logarithm of the amplitude; thus a unit magnitude increase corresponds to a 10-fold increase in amplitude. Most commonly used magnitudes are the body wave magnitude, m_b, and surface wave magnitude, M_S. The frequency of occurrence of earthquakes, i.e. a number of earthquakes per unit time (year) above a certain magnitude M, satisfies the Gutenberg–Richter law: $\log_{10} N = aM + b$. The value of a is close to -1, which means that there are, on average, 10 times more earthquakes above magnitude five than above magnitude six. A new magnitude, M_W, based on the estimates of the released seismic moment (shear modulus × fault area × offset (slip) on the fault) is becoming increasingly popular; it is more informative for very large earthquakes, for which M_S may become saturated.

Figure 21.18 is a map of the principal tectonic plates, as defined in plate tectonic theory. The direction of the arrows shows the relative motion of the plates; their length corresponds to the rate of motion. At a plate boundary where the blue arrows converge, we expect compression and, therefore, thrust faulting; one of the plates is subducted, hence the term "subduction zones". At a plate boundary where the red arrows diverge, there is normal faulting and creation of a new crust, mid-ocean ridges. For boundaries that slip past each other in the horizontal plane (green arrows), also called the transform faults, there is strike-slip faulting.

Figure 21.19 shows the source mechanism of approximately 4000 shallow earthquakes from 1993 through 1997 determined at Harvard University using the centroid-moment tensor (CMT) method; the center of each beach ball is at the epicenter—many earthquakes have been plotted on top of each other. It is easy to see that thrust faulting is dominant at the converging boundaries

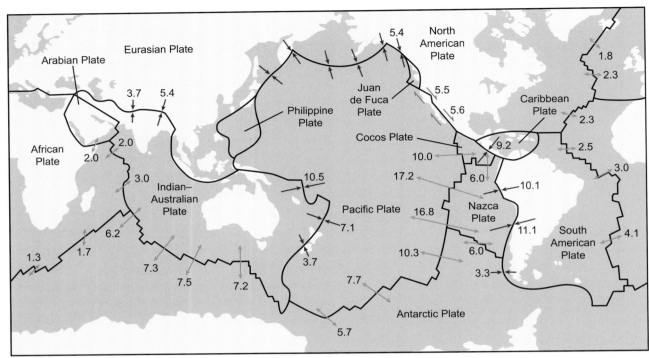

FIGURE 21.18 Motion of principal tectonic plates. Principal tectonic plates and relative plate motion rates. Red arrows signify spreading, blue arrows, convergence, and green arrows, strike-slip motion.

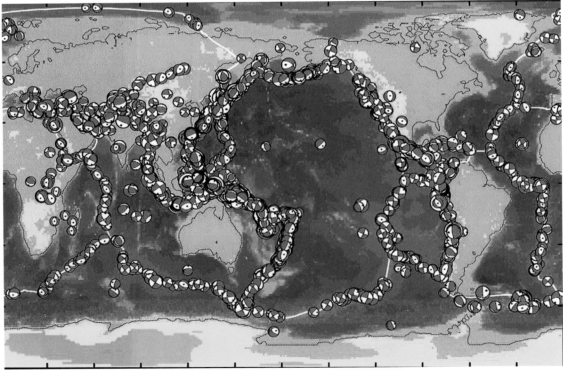

FIGURE 21.19 Earthquake source mechanisms. Source mechanisms of approximately 4000 earthquakes from 1993 to 1997 obtained through the CMT analysis. The center of a beach ball is plotted at the epicenter. Only a small fraction of earthquakes are visible. Note the preponderance of earthquakes occurring on plate boundaries (Figure 21.18) and their mechanism corresponding closely to the type of the boundary (convergent, thrust faulting; divergent, normal faulting; transform, strike-slip faulting). Some earthquakes occur away from plate boundaries. They are particularly numerous in Asia and Africa along the east African rift system, but there are some in eastern North America and the center of the Pacific.

(subduction zones), there are exceptions related to bending of the plates, plate motion oblique to the boundary, and other causes. The CMT project moved in 2006 from Harvard to Columbia; the catalog, available online (globalcmt.org) contains now over 40,000 earthquakes that occurred between 1976 and 2013.

At mid-ocean ridges, we see predominantly normal faulting, the faults where a mid-ocean ridge is offset, show strike-slip faulting, in accordance with the plate tectonic theory. The exception is where the fault is complex. Along the San Andreas Fault, the most famous transform fault, we see many complexities that led to earthquakes other than the pure strike slip. For example, the Northridge earthquake of January 1994 was a thrust, and the Loma Prieta earthquake of October 1989 was half-thrust, half-strike slip. There are also earthquakes away from the plate boundaries. These are called intraplate earthquakes and their existence demonstrates the limits of the validity of the plate tectonic theory, as there should be no deformation within the plates. A very wide zone of deformation is observed in Asia; the rare large earthquakes in eastern North America are sometimes associated with isostatic adjustment following the last glaciation. If we compare the distribution of earthquakes along a mid-ocean ridge, including its transform faults, with that of the Alpide belt, we notice that for the oceanic plates the region in which earthquakes occur is very narrow, while in Eurasia it may be 3000 km wide. A part of the reason that the theory of plate tectonics has been put forward is because of observations (bathymetry, magnetic stripes, and seismicity) in the oceans.

There are also deep earthquakes, with the deepest ones just above 700 km depth; earthquakes with a focal depth from 50 to 300 km are said to be of an intermediate depth and are called "deep" when the focal depth is greater than 300 km. Intermediate and deep earthquakes are explained as occurring in the subducted **lithosphere** and are used to map the position of the subducted slab at depth. Not all subduction zones have very deep earthquakes; for example, in Aleutians, Alaska, and Middle America the deepest earthquakes are above 300 km depth. The variability of the maximum depth and the mechanism of deep earthquakes have been attributed in the late 1960s to the variation in the resistance that the subducted plate encounters; more recent studies indicate more complex causes, often invoking the phase transformations (change in the crystal structure) that the slab material subjected to the relatively rapidly changing temperature and pressure may undergo.

6. EARTH'S RADIAL STRUCTURE

A spherically symmetric Earth model (SSEM) approximates the real Earth quite well; the relative size of the 3-D part with respect to SSEM varies from several percent in the upper mantle to a fraction of a percent in the middle mantle and increases again above the CMB.

A concept of an SSEM, often referred to as an "average" Earth model, is a necessary tool in seismology. Such models are used to compute functionals of the Earth's structure (such as travel times), and their differential kernels are needed to locate earthquakes and to determine their mechanism. Knowledge of the internal properties of the Earth is needed in geodesy and astronomy. Important inferences with respect to the chemical composition and physical conditions within the deep interior of the Earth are made using information on radial variations of the elastic and anelastic parameters and density.

An SSEM is a useful mathematical representation that is not necessarily completely representative of the real Earth. This is most obvious at the Earth's surface, where one must face the dilemma of how to reconcile the occurrence at the same depth, or elevation, of water and rocks; the systems of equations governing the wave propagation in liquid and in solid are different. The commonly adopted solution is to introduce a layer of water whose thickness is such that the total volume of water in all the oceans and that calculated for the SSEM are equal. It is a reasonable decision, but it will be necessary to introduce corrective measures even when constructing the model, as practically all seismographs that record ground motion are located on land.

This chapter uses the preliminary reference Earth model (PREM) published in 1981 by Dziewonski and Anderson as an example. It has been derived using a large assembly of body wave travel time data, surface wave dispersion, and periods of free oscillations, collected through the end of 1970s. An effort to revise it is now under way; a large body of very accurate data has been assembled in the nearly 20 years since the publication of PREM. However, with the exception of the upper mantle, no substantial differences are expected. A reference model designed to fit the travel times of body waves (ak135) has been developed by Kennett and Engdahl in 1995.

Figure 21.20(a) shows the density, compressional velocity, and shear velocity in the model PREM. To illustrate the complexities in the uppermost 800 km of the model, its expansion is shown in Figure 21.20(b). In what follows, we shall give a brief summary of our knowledge and significance of the individual shells in the Earth's structure.

6.1. Crust

This is the most variable part of the Earth's structure, both in terms of its physical properties as well as history. Large areas of the Earth's surface are covered by soils, water, and sediments. These provide support for life and economic activity. However, the vast proportion of what is called "the crust of the Earth" consists of crystalline rocks, mostly of igneous origin.

The primary division is between the continental and the oceanic crust. The former can be very old, with a significant

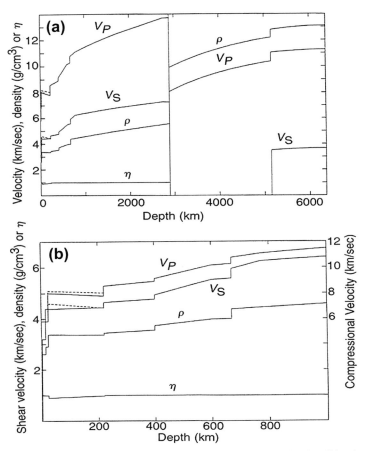

FIGURE 21.20 The Preliminary Earth Reference Model (PREM) of Dziewonski and Anderson (1981) describing the compressional velocity (v_s), shear velocity (v_S), and density (ρ). (a) Model for the entire Earth, and (b) an expansion of the uppermost 1000 km. From Moho to 220 km depth the model is characterized by transverse anisotropy, in which the waves propagating in the vertical (solid line) and horizontal (dashed lines) planes have different velocities. Parameter η, characterizing the propagation of P-waves at intermediate angles, is unity in an isotropic medium and is about 0.95, just under the Moho. Below 220 km depth the model is isotropic.

fraction being older than 1.5 Ga. It is light, with an abundance of calcium, potassium, sodium, and aluminum. Its average thickness is 40 km, but varies substantially, from about 25 km in the areas of continental thinning due to extension (the Basin and Range province in the Western United States, for example) to 70 km under Tibet, in the area of continent—continent collision.

The oceanic crust is thin (7 km, on average, covered by some 4.5 km of the ocean), young (from 0 to 200 Ma), and somewhat more dense, with a greater abundance of elements such as magnesium and iron. It is created at the mid-ocean ridges and is consumed in subduction zones, with trenches being their surficial manifestation. The difference between oceanic and continental crusts is called by some the most important fact in Earth sciences, as it is related intimately to plate tectonics. The thinner, denser oceanic crust provides conditions more favorable for initiation of the subduction process.

Overall, crustal thickness follows the Airy's hypothesis of isostasy closely, thick roots under mountains and a thin crust under "depressed" areas—oceans. The seismic velocities in the crust increase with depth. It is a subject of debate whether this increase is gradual or the crust is layered; recently, the latter view has begun to prevail. The knowledge of global crustal structure is important in studying Earth's structure at greater depths. The most recent result provides the crustal structure for each $1 \times 1°$ grid (Laske et al., 2013) for a total of 648,000 entries. Each of the grid point entries contains information for up to eight crustal layers and specifies layer thickness, seismic velocities, and density.

6.2. Upper Mantle: Lithosphere and Asthenosphere (25–400 km Depth)

The boundary between the crust and the upper mantle was discovered in 1909 by a Yugoslavian geophysicist Andreiji Mohorovicic. It represents a 30% increase in seismic velocities and some 15% increase in density. It is a chemical boundary with the mantle material primarily composed of minerals olivine and pyroxene, being much richer in heavier elements, such as magnesium and iron.

The terms **lithosphere** and **asthenosphere** refer to the rheological properties of the material. The lithosphere, strong and brittle, is characterized by very high **viscosity**. It is often modeled as an elastic layer. It includes the crust and some 30–100 km of the upper mantle. The asthenosphere is hotter (>1573 K by convention), its viscosity much lower, and in modeling is represented by yielding. Under loads, such as glacial caps, the lithosphere bends elastically, whereas the asthenosphere flows. The difference of rheological properties is explained by differences in temperature; the viscosity is an exponential function of temperature. The lithosphere is relatively cool; the transport of heat is mostly through conduction. The asthenosphere is hotter, and the convective processes are believed to become important. Low viscosity of the asthenosphere is used to explain the mechanical decoupling between the plates (in the plate tectonic theory) and the underlying mantle. The depth of this decoupling varies with position; it is shallow near mid-ocean ridges and increases as the plate cools with time and its lithosphere grows in thickness.

The continents, with their very old and cold shield regions, may be significantly different. If the hypothesis of the "tectosphere" is correct, it may have roots that are 400 km deep and move as coherent units over long periods of the Earth's history. The depth of roots is still subject to a debate (most recent results would indicate their depth extent as 200–250 km). As the seismic velocities decrease with increasing temperature, the vertical gradient of seismic velocities in the transition between the lithosphere and the asthenosphere may become negative. This is called the "low-velocity zone"; its presence creates a shadow zone in seismic wave propagation, making interpretation of data complex and nonunique.

Measurements of attenuation of seismic waves led to the determination of models of quality factor (Q) for the shear and compressional energy. Anelastic dissipation of shear energy, due to grain boundary friction, is most important. Attenuation in the range of depths corresponding to the low-velocity zone is several times stronger than in the lithosphere.

Somewhere below 200 km depth the velocities and Q (attenuation parameter) begin to increase slowly; the effect of increasing pressure begins to dominate over the increase in temperature. The so-called Lehmann discontinuity is elusive and does not appear to be a global feature; this is one of the elements of PREM, where it shows a sudden increase in both P and S velocities, that will be changed in the next version of the reference Earth model.

6.3. Transition Zone (400–660 km Depth)

Knowledge of the composition of the transition zone is essential to the understanding of the composition, evolution, and dynamics of the Earth. In seismic models, this depth range has been known for a long time to have a strong velocity gradient; much too steep for an increase under pressure of the elastic moduli and density of a homogeneous material. It was first postulated in the 1930s that this steep gradient may be due to phase transformations: changes in the crystal lattice that for a given material take place at certain temperatures and pressures.

In the 1960s, when major improvement in seismic instrumentation took place, two discontinuities were discovered: one at 400 km and the other at 670 km (the current best estimate of the global average of their depth is 410 and 660 km, respectively). Their existence has been well documented by nearly routine observations of reflected and converted waves. There is still some uncertainty of how abrupt the velocity changes are; the 410 km discontinuity is believed to be spread over some 5–10 km, whereas the 660 km discontinuity appears to be abrupt. The estimates of the velocity and density contrasts are still being studied by measuring the amplitudes of the reflected and converted waves; the values of these contrasts are important for understanding the mineralogical composition of the transition zone.

In general terms, the seismological models are consistent with the hypothesis that olivine is the main (up to 60%) constituent of the upper mantle. Laboratory experiments under pressures corresponding to depths up to 750 km show that olivine undergoes phase transformations to denser phases with higher seismic wave speeds. At pressures roughly corresponding to 400 km depth, the α-olivine transforms into β-spinel. The latter will transform to γ-spinel at about 500 km depth, with only a minor change in seismic velocities. Indeed, a seismic discontinuity at 520 km has been reported, although some studies indicate that in some parts of the world it may not be substantial enough to be detected. At 660 km γ-spinel transforms into perovskite and magnesio-wüstite.

Although olivine may be the dominant constituent, it is not the only one. The presence of other minerals complicates the issue. Also, there are other hypotheses of the bulk composition of the upper mantle, "piclogyte model", for example.

6.4. Lower Mantle (660–2890 km)

The uncertainties in the mineralogy of the upper mantle and the bulk composition of the Earth have created one of the most stubborn controversies in the Earth sciences: are the upper mantle and lower mantle chemically distinct? A "yes" answer means that there has not been an effective mixing between these two regions throughout the Earth's history, implying that the convection in the Earth is layered. The abrupt cessation of seismic activity at about 660 km depth, coinciding with the phase transformation described earlier, and geochemical arguments—mostly with respect to differences in isotopic composition of the mid-ocean ridge **basalts** and ocean island basalts—are used as strong arguments in favor of the layered convection. New evidence, gathered within the tomographic studies to be discussed later, gives support to a significant impedance to the flow between the upper mantle and lower mantle.

The whole mantle convection is favored by geodynamicists who develop kinematic and dynamic models of the mantle flow. For example, the geometry and motions of the known motions of the plates are much easier to explain assuming whole mantle circulation. Evidence has been presented for penetration of slabs into the lower mantle, based on the presence of fast velocity anomalies in the regions of the past and current **subduction**. At the same time, there is evidence for stagnation and "ponding" in the transition zone of some of the subducted slabs. The recent results from seismic tomography seem to support the concept of at least partial separation of the upper and lower mantle flow.

In the early 1990s a model of mantle avalanches was developed; the subducted material is temporarily accumulated in the transition zone as the result of an endothermic phase transformation at the 660 km discontinuity. Once enough material with the negative buoyancy collects, however, a penetration can occur in a "flushing event", where most of the accumulated material sinks into the lower mantle. The calculations, originally performed in two-dimensional geometry, indicated the possibility of such events causing major upheavals in the Earth's history. However, when calculations were extended to 3-D spherical geometry (Tackley et al., 1993), their distribution in space and time turned out to be rather uniform.

The computer models of the mantle convection are still tentative. There are many parameters that control the process. Some, such as the generation of the plates and plate boundaries at the surface, are difficult to model. Others, such as the variation of the thermal expansion coefficient with pressure- or temperature-dependent viscosity, are poorly known; even one-dimensional viscosity variation with depth is subject to major controversies.

The lower mantle appears mineralogically uniform, with the possible exception of the uppermost and lowermost 100–150 km. There is a region of a steeper velocity gradient in the depth range of 660–800 km, which may be an expression of the residual phase transformations. Also, at the bottom of the mantle, there is a region of a nearly flat, possibly slightly negative gradient. This region, just above the CMB, known as D'', is the subject of intense research. Its strongly varying properties, both radially and horizontally, are being invoked in modeling mantle convection, chemical interaction with the core, possible chemical heterogeneity (enrichment in iron), and as evidence for partial melting. In 2004, the existence of a new phase "post-perovskite" has been proposed; its existence may affect the complexities in the D'' region. The seismic velocities and density throughout the bulk of the lower mantle appear to satisfy the Adams–Williamson law, describing the properties of the homogeneous material under an **adiabatic** increase in pressure.

6.5. Outer Core (2981–5151 km)

The outer core is liquid; it does not transmit shear waves. Consideration of the average density and the **moment of inertia** pointed to a structure with a core that would be considerably heavier, possibly made of iron, judging from cosmic abundances. We now know that the core is mostly made of iron, with some 10% admixture of lighter elements, needed to lower its density. It has formed relatively early in the Earth's history (first 50 Ma) in a melting event in which droplets of iron gravitationally moved toward the center. Although difficult to estimate, some current models place its temperature in the range of 3000–5000 K.

The presence of a liquid with a very high electrical conductivity creates conditions favorable to self-excitation of a magnetic **dynamo**. It is important to know that the magnetic field we observe at the surface is only a small fraction of the fields present in the core. Actually we see only one class of the field, the poloidal, whereas the toroidal field, possibly much stronger, is confined to the core.

Numerical models of the dynamo predicted several key phenomena observed at the surface: the primary dipolar structure with the alignment of the dipole axis close to the axis or rotation of the Earth, the westward drift of secular variations, and reversals of the polarity of the magnetic field. The later phenomenon is the cause of the magnetic anomalies on the ocean floor, which allowed estimating the rate of ocean spreading. Numerical simulations of geodynamo has begun in mid-1990s some being able to obtain the polarity reversal. Yet, the realistic parameter space has not yet been adequately sampled because of the computational challenges.

Seismological data are consistent with the model of the core as that of a homogeneous fluid under adiabatic temperature conditions. As often near major discontinuities, there is difficulty with pinning down the values near the end of the interval, just below the CMB and just above the inner core boundary (ICB). Some seismic models of the compressional velocity just under the CMB have gradient too steep for a homogeneous material. This implies chemical heterogeneity, with the intrinsically denser material at the top, which is unlikely. Also, some models contain a 100-km-thick layer just above the ICB, which has a nearly zero gradient. Using the velocity-density systematics, this would imply enrichment in iron and indicate that mixing of the material in the outer core is not as complete as the low viscosity of the fluid outer core predicts.

6.6. Inner Core (5251–6371 km)

An additional seismic discontinuity deep inside the core, which came to be called the inner core, was discovered by Inge Lehmann in 1936. The fact that it is solid was postulated soon afterward, but satisfactory proof required 35 years, when observations and modeling of the free oscillations of the Earth showed that it indeed must have finite rigidity (Dziewonski & Gilbert, 1971). However, mineral physicists have difficulty with explanation of the very high Poisson ratio of 0.44.

Estimates of the energy required to maintain geodynamo require that the inner core formed a long time after accretion of the Earth, perhaps some 2 Ga ago or even more recently. As the Earth was cooling, the temperatures at the Earth's center dropped below the melting point of iron (at the pressure of 330 GPa) and the inner core began to grow. The release of the gravitational energy associated with the precipitation of solid iron is believed to be an important source of the energy driving the dynamo. Again, estimates are difficult, but models yield a current temperature range of 5000–7500 K.

The inner core might have been considered quite uninteresting, with a very small variation of the physical parameters across the region. This all changed in the mid-1980s when it was discovered that this region is anisotropic, with the symmetry axis roughly parallel to the rotation axis. A deviation from that ray's symmetry and an observation of temporal variation of travel times through the inner core brought forward an interpretation that the inner core rotates at a slightly (1°/year) higher rate than the mantle. This could be explained by the electromagnetic coupling with the dynamo field of the other core. However, the inference of the rate of the differential rotation soon becomes very controversial. Several studies now indicate that this differential rotation must be much less. Another controversial issue is the "hemispheric asymmetry", with the western hemisphere having stronger anisotropy than the eastern one. In 2002, it was proposed that there exists an "innermost inner core", the central region with some 300 km radius in which the anisotropy is distinctly different than in the bulk of the inner core. Since then, the anomalous properties of this region have been confirmed by other studies.

It is likely that some of these controversies originate because of the insufficient, or uneven, data coverage. The inner core occupies less than 1% of the Earth's volume. Introduction of a new parameter such as the cylindrical anisotropy imposes additional requirements. Unlike in modeling an isotropic structure when travel time depends only on the bottoming depth of a ray, modeling the anisotropy requires also data coverage as a function of the location of the ray's bottoming point as well as the azimuth of the ray's path. The data requirement may be significantly reduced if simplifying assumptions are made, for example, the transverse isotropy with the symmetry axis aligns with the rotation axis. But if such an assumption turns out to be incorrect, the results may be very misleading.

7. EARTH IN THREE DIMENSIONS

Figure 21.21 is an example of results obtained using global seismic tomography (GST). It shows a triangular cut into an Earth model of the shear velocity anomalies in the Earth's mantle and shows only deviations from the average; if the Earth were radially symmetric, this picture would be entirely featureless. The surface is the top of the mantle (Mohorovicic discontinuity, or Moho) and the bottom is the CMB. Seismic wave speeds higher than average are shown with blue colors, whereas slower than average are shown as yellow and red colors. Seismic velocities decrease with increasing temperature; the inference is that the light areas are hotter than average and dark are colder. Seismic wave speeds also vary with chemical composition, but there are strong indications that the thermal effect is dominant.

Density is also a function of temperature. Material hotter than average is lighter and, in a viscous Earth, will tend to float to the surface, whereas colder material is denser and will tend to sink. Thus our picture can be thought to represent a snapshot of the temperature pattern in the convecting Earth's mantle. In particular, the picture implies a downwelling under the Indian Ocean and an upwelling originating at the CMB under Africa; sections passing through this anomaly indicate that this upwelling may continue to the surface. This "window into the Earth" shows the outer core (blue), inner core (pink), and the innermost inner core (red); the latter represents only 0.01% of the Earth's volume.

The GST is limited by the distribution of globally detected earthquakes and by the locations of seismographic stations. There is not much that we can do about the distribution of seismicity, except that now and then an earthquake occurs in an unexpected place, so the coverage is expected to improve with time. Generally, the earthquake distribution is more even in the Northern Hemisphere. Much has been done in the last decade to improve the distribution and the quality of the seismographic stations, and recent results show considerably better resolution of the details in the top 200 km, for example. However, even using the available oceanic islands (which are very noisy, because of the wave action), there are oceanic areas with dimensions of several 1000 kilometers where no land exists. A series of experiments by Japanese, French, and

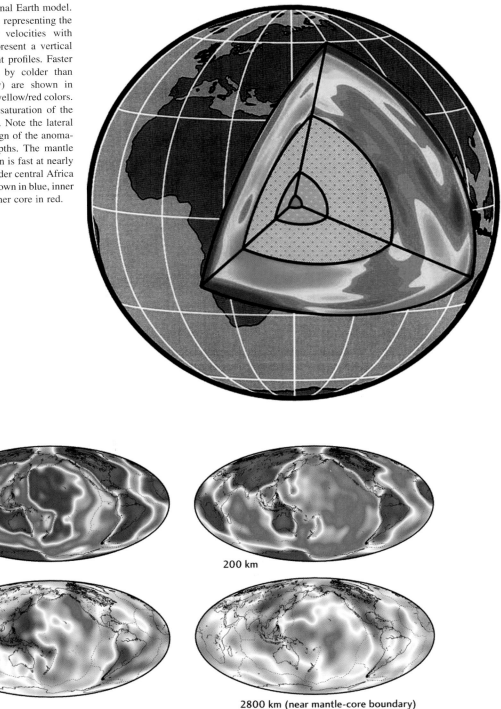

FIGURE 21.21 Three-dimensional Earth model. A 3-D model S362D1 of Gu et al. representing the lateral deviations of the shear velocities with respect to PREM. The sides represent a vertical cross-section along three different profiles. Faster than average velocities (caused by colder than normal temperature, presumably) are shown in green/blue and slower (hotter) in yellow/red colors. The scale is ±1.5%; significant saturation of the scale occurs in the upper mantle. Note the lateral and vertical consistency of the sign of the anomalies over large distances and depths. The mantle underneath Asia and Indian Ocean is fast at nearly all depths, whereas the mantle under central Africa is slow. The liquid outer core is shown in blue, inner core in red, and the innermost inner core in red.

FIGURE 21.22 Lateral variations of the Earth's shear wave velocity. Maps of lateral variations of S velocities at four depths in a shear velocity model of Ekström and Dziewonski. The yellow/red colors indicate slower than average velocities and blue, the faster. The range of variations is about 7% at 70 km and 3% at near the core—mantle boundary.

American seismologists have demonstrated that the establishment of a permanent or semipermanent network of ocean bottom high-quality seismographic stations is now a real, even though expensive, possibility.

Figure 21.22 is a collection of maps of the shear velocity anomalies from a recent model of the mantle by Ekström and Dziewonski published in 1998, built using a wide range of types of data (travel times, surface wave

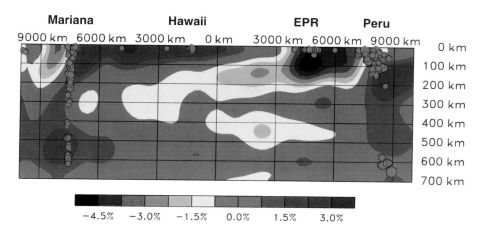

FIGURE 21.23 Cross-section of upper mantle model. Cross-section through the upper mantle of model of Ekström and Dziewonski. Note that velocities change as a function of distance from the East Pacific Rise (proportional to age of the plate) to depths greater than 200 km. Red dots indicate earthquakes. There is vertical exaggeration by a factor of about 20.

dispersion, and waveforms). The nominal resolution of this model is about 100 km in depth and 1500 km horizontally near the surface.

At 70 km depth, the model agrees with the predictions of the plate tectonics and the thermal history of the continents. The stable continental areas (old and cool) are very fast (up to +7%), whereas material under the mid-ocean ridges is much slower than normal (up to −7%). This negative anomaly decreases with the increasing age of the oceanic plate to become faster than average for ages greater than 100 Ma. The depth to which the anomalies associated with the mid-ocean ridges persists in the tomographic maps (>200 km) puzzles geodynamicists who think that mid-ocean ridges are passive features and that conditions below about 100 km depth are isothermal.

Figure 21.23 shows a cross-section through the upper mantle of the Pacific from the model of Ekström and Dziewonski; the direction of the cross-section follows the direction of motion of the Pacific plate. Going from East to West, we see higher seismic velocities associated with **subduction** under South America; very slow **lithosphere** at the East Pacific rise; increase in velocities with the distance from the ridge; and subduction under the Mariana trench; the red dots are earthquakes. It is clear that the velocities change with age to depths below 200 km. The map in

Figure 21.22 at 200 km depth shows diminished variability of velocities under ocean but still very strong anomalies under the continents, the old cratons, in particular. Thus, the tomographic model is consistent with the tectosphere hypothesis, but the amplitude of the velocity anomalies decreases rapidly below 200–250 km depth.

The map in Figure 21.22 at 500 km depth represents average shear velocity anomalies in the transition zone. The most characteristic features are the fast anomalies in the western Pacific and Eastern Asia and in the east, under South America and the Atlantic, reaching to western Africa. In the western Pacific they can be associated with subduction zones, although they are much wider than an anomaly associated with a 100-km-thick slab. Studies of the topography of the 660 km discontinuity show that the areas of high seismic velocity are correlated with a depressed boundary, yielding credence to an interpretation that these anomalies are indicative of an accumulation (temporary, perhaps; see earlier discussion on the models of flow in the mantle) of the subducted material in the transition zone. Figure 21.24 shows comparison of lateral variation in velocities obtained in a model named S362D1. The two maps one just above and the other just below the 660 km discontinuity are very different; the map representing the transition zone shows features similar to that at

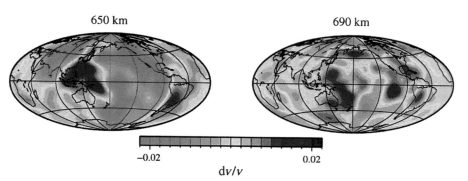

FIGURE 21.24 Shear velocity anomalies. Shear velocity anomalies just above and below the boundary between the upper mantle and lower mantle in model S262D1 of Gu et al. The differences indicate a drastic change in the pattern of the anomalies, most likely associated with a serious impedance to flow.

500 km map in Figure 21.22; the lower mantle map is quite different and has distinctly different spectral content; it is dominated by relatively short wavelength features. This result, and similar maps obtained by other modeling groups, supports the concept of a separation—perhaps not absolute—between the upper mantle and lower mantle.

In the middle mantle the anomalies are not well organized. This observation contrasts with the results of "high-resolution" tomography, which in this depth range shows two narrow high-velocity features: one stretching from the Hudson Bay to Bolivia and the other from Indonesia to the Mediterranean. Even though elements of these two structures are present in our model, they are not equally well defined. Also, there are many other features of comparable amplitude. This is also true with respect to models published by scientists at the University of California Berkeley and at Caltech/Oxford, who used parameterization similar to that in Figure 21.22. Intensive efforts are made to understand the differences between the results of two different approaches to tomography.

The map at 2800 km depth shows the velocity anomalies as the CMB is approached. The ring of high velocities circumscribing the Pacific basin is already visible at 2000 km; it strengthens considerably over the next 500 km and increases even further toward the CMB. In the wave number domain of spherical harmonics, the spectrum of lateral heterogeneities is very red, being dominated by degree 2 and 3. These two harmonics account for 70% of the variance of the heterogeneity near the CMB. The location of the ring of fast velocities corresponds to the location of subduction zones during the past 200 Ma. The large red (slow) regions are sometimes called the African and the Pacific "superplumes"; some seismologists prefer Large Low Shear Velocity Provinces. Their origin is unknown; they, most likely, represent both thermal and chemical heterogeneity. There is a good correlation between the location of the two superplumes and distribution of hot spots at the Earth's surface, indicating a degree of connection between the tectonics at the surface and conditions near the core—mantle boundary.

Figure 21.25 gives two views of low-pass filtered anomalies in the lower mantle in a model by Ritsema, van Heijst, Woodhouse (1999), plotted in Cartesian coordinates; the red is a 0.6% isosurface and blue is +0.6%. We see the circum-Pacific ring of fast anomalies and the two low-velocity anomalies: one very concentrated under the Pacific and a more diffuse one under the Atlantic and Africa. Their radial continuity throughout the lower mantle indicates that they cannot be explained by processes at the core—mantle boundary alone. The origin of this large-amplitude, very large-wavelength signal has not yet been explained by geodynamic modeling, although an assumption that the velocity and gravity anomalies are correlated leads to a good prediction of the geoid at the gravest harmonics.

Top view

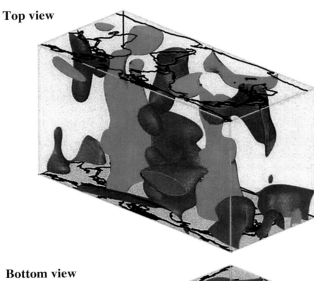

Bottom view

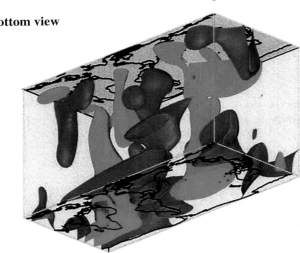

FIGURE 21.25 Low-pass filtered S-velocity model. Low-pass filtered S-velocity model of Ritsema et al. in a 3-D projection; the top 800 km of the structure is removed.

Dziewonski, Lekic, and Romanowicz (2010) hypothesized that the degrees 2 and 3 velocity anomalies in the lowermost mantle are indicative of a thermal and/or compositional heterogeneity that imposes a long-term control on the global tectonics. Close examination of the long-wavelength shear velocity signal in the lowermost mantle in the wavenumber domain ties several geophysical observations together and leads to fundamental inferences.

Figure 21.26 compares degree 2 shear velocity anomalies at 2800 km depth in three relatively recent 3-D shear velocity models; the maps for that harmonic are nearly indistinguishable. The power in this degree is more than one half of the total power contained in harmonics from 1 to 18. Even though no constraints were imposed a priori, the pattern—for all practical purposes—is described by a single spherical harmonic coefficient, the "recumbent" Y20 spherical harmonic, a Y20 with its axis of symmetry rotated

FIGURE 21.26 Comparison of shear velocity anomaly components. Comparison of degree 2 component of shear velocity anomalies at 2800 km depth for three relatively recent 3-D models published by research groups at (from the left) Harvard, Berkeley, and Caltech/Oxford. Subduction during the last 200 Ma occurred within the blue area, hot spots, over red (slow anomalies). The TPW positions of Besse and Courtillot (2002) during the last 200 Ma are indicated by small orange circles.

to the equatorial plane. There is less than 1/1000 probability that such a result could be obtained by chance. This degree 2 signal, which continues with decreasing amplitude throughout the lower mantle, is characterized by two antipodal regions of low velocities, separated by a circumpolar torus of velocities higher than average. If the slow regions are associated with net excess mass, then any axis of rotation located in the plane of the polar great circle will be the maximum moment of inertia axis; this includes, of course, the current axis of rotation. The polar great circle would be also the preferred direction of true polar wander (TPW); the open circles show the locations of the TPW axes during the most recent 200 Ma determined by Besse and Courtillot (2002). It is suggested that the recumbent Y20 is a very stable feature; once established, it is difficult to erase, and only relatively small departures from this equilibrium configuration are possible. This anomaly correlates strongly with the degree 2 terms of the residual geoid expansion, distribution of the hot spots above the slow regions, high attenuation in the transition zone, and position of subduction zones above the fast band during the last 200 Ma. Also, the preferred paths of the virtual geomagnetic pole and TPW locations for the last 200 Ma lie within the fast band. Since the nonhydrostatic perturbation of the moment of inertia tensor depends only on degree 2 anomalies in the density distribution and deformation of discontinuities, it is natural to infer that rotational dynamics of the Earth have influenced the distribution of heterogeneities in the Earth's deep interior. It is proposed that the large-scale heterogeneity at the base of the mantle, named "mantle anchor structure" may have formed early in the history of the convecting mantle, remained locked in place with respect to the Earth's rotation axis ever since, and is currently imposing the planform of flow in the mantle and—on a shorter timescale—plate tectonics at the surface.

It was believed since 1977, the time of publication of the first large-scale GST study, that 3-D images of lateral heterogeneity in the mantle will be an essential tool in addressing some of the fundamental problems in earth sciences. The results accumulated since then confirm that statement, even though much progress is still to be made. Cooperation among the different fields of Earth sciences (geodynamics, mineral physics, geochemistry, seismology, and geomagnetism) is the requisite condition to fulfill this goal.

8. EARTH AS A ROSETTA STONE

The Earth is unique among its partners in our solar system in that it has had liquid water oceans for most of its history, has a highly mobile crust, and a dynamically convecting interior. This combination means that the surface is and has been constantly driven by the movement of the interior, such that the oldest terrestrial subaerial landscapes are at most $\sim 10\%$ of the age of the planet, and the oldest submarine landscapes are only a little more than 10% of that. Thus, the Earth not only has one of the most globally dynamic surfaces in the solar system, but its interior is also one of the most dynamic. Only the tidally wracked and volcanically incessant surface of Io, Jupiter's innermost satellite, may be younger and more active. Driven by internal forces, the periodic conglomeration and separation of continental landmasses, causing opening and closing of oceans, and construction and destruction of mountain ranges profoundly impact the global climate. The environmental stresses caused by such reshuffling of the surface may themselves have influenced the progress of evolution on the planet—evolution that was possibly reset every 100 million years or so by devastating asteroidal impacts. In the final analysis, the Earth is the only planetary body with which the human species has had intimate experience—for millennia. Thus, beyond being our home, the Earth is for us a crucial yardstick—a Rosetta stone—by which we will measure and interpret the processes, internal structure, and overall histories of other planets in this solar system and, someday, of other planets around other stars.

BIBLIOGRAPHY

Baker, V. R. (1986). *Introduction: Regional landforms analysis.*

Besse, J., & Courtillot, V. (2002). Apparent and true polar wander and the geometry of the geomagnetic field over the last 200 Myr. *Journal of Geophysical Research, 107*, 2300. http://dx.doi.org/10.1029/2000JB000050.

Bloom, A. L. (1998). *Geomorphology: A systematic analysis of late cenozoic landforms* (3rd ed). Upper Saddle River, NJ: Prentice Hall.

Dziewonski, A. M., Hager, B. H., & O'Connell, R. J. (1977). Large scale heterogeneities in the lower mantle. *Journal of Geophysical Research, 82*, 239–255.

Dziewonski, A. M., & Anderson, D. L. (1984). Seismic tomography of the Earth's interior. *American Scientist, 72*(5), 483–494.

Dziewonski, A. M., Lekic, V., & Romanowicz, B. (2010). Mantle anchor structure: an argument for bottom up tectonics. *Earth and Planetary Science Letters, 299*, 69–79.

Fleming, K., Johnston, P., Zwartz, D., Yokoyama, Y., Lambeck, K., & Chappell, J. (1998). Refining the eustatic sea-level curve since the Last Glacial Maximum using far- and intermediate-field sites. *Earth and Planetary Science Letters, 163*(1–4), 327–342. http://dx.doi.org/10.1016/S0012-821X(98)00198-8.

Fleming. K. M. (2000). *Glacial rebound and sea-level change constraints on the Greenland ice sheet* (Ph.D. thesis). Australian National University.

Francis, P. W., & Oppenheimer, C. (2004). *Volcanoes.* Oxford, UK: Oxford University Press.

Geruo, A., Wahr, J., & Zhong, S. J. (2013). Computations of the viscoelastic response of a 3-D compressible Earth to surface loading: an application to Glacial Isostatic Adjustment in Antarctica and Canada. *Geophys. J. Int., 192*, 557–572. http://dx.doi.org/10.1093/gji/ggs030.

Hanna, E., Navarro, F., Pattyn, F., Domingues, C. M., Fettweis, X., Ivins, E. R., et al. (2013). Ice sheet mass balance and climate change. *Nature, 498*, 51–56.

Heezen, B., & Tharp, M. (1997). *Panoramic maps of the ocean floor.*

Ivins, E. R., James, T. S., Wahr, J., Schrama, E. J. O., Landerer, F. W., & Simon, K. M. (2013). Antarctic contribution to sea-level rise observed by GRACE with improved GIA correction. *Journal of Geophysical Research - B: Solid Earth and Planets*, 3126–3141. http://dx.doi.org/10.1002/jgrb.50208.

King, L. C. (1967). *Morphology of the Earth* (2nd ed). Edinburgh: Oliver and Boyd Ltd.

Milne, G. A., Long, A. J., & Bassett, S. E. (2005). Modelling Holocene relative sea-level observations from the Caribbean and South America. *Quaternary Science Reviews, 24*(10–11), 1183–1202. http://dx.doi.org/10.1016/j.quascirev.2004.10.005.

Peltier, W. R. (1999). Global sea level rise and glacial isostatic adjustment. *Global and Planetary Change, 20*, 93–123.

Pieri, D., & Abrams, M. (2004). ASTER watches the world's volcanoes: a new paradigm for volcanological observations from orbit. *Journal of Volcanology and Geothermal Research, 135*(1–2), 13–28.

Pieri, D., Diaz, J. A., Bland, G., Fladeland, M., Madrigal, Y., Corrales, E., et al. (2013). In situ observations and sampling of volcanic emissions with NASA and UCR unmanned aircraft, including a case study at Turrialba Volcano, Costa Rica. Geological Society. In D. M. Pyle, T. A. Mather, & J. Biggs (Eds.), *Remote sensing of volcanoes and volcanic processes: Integrating observation and modelling* (p. 380). London: Special Publications. http://dx.doi.org/10.1144/SP380.13.

Ritsema, J., van Heijst, H. H., & Woodhouse, J. H. (1999). Complex shear wave velocity structure imaged beneath Africa and Iceland. *Science, 286*, 1925–1928.

Ritsema, J., Deuss, A., van Heijst, H. J., & Woodhouse, J. H. (2011). S40RTS: a degree-40 shear-velocity model for the mantle from new Rayleigh wave dispersion, teleseismic traveltime and normal-mode splitting function measurements. *Geophysical Journal International, 184*, 1223–1236.

Ritter, D. F., Kochel, R. C., & Miller, J. R. (2011). *Process Geomorphology* (5th ed.). Waveland Press, 652 pp.

Schumm, S. A. (2005). *River variability and complexity.* New York: Cambridge University Press, 234pp.

Short, N. M., & Blair, R. W., Jr. (Eds.). (1968). *Geomorphology from space: A global overview of regional landforms.* Washington, DC: NASA Scientific and Technical Information Branch.

Snead, R. E. (1980). *World Atlas of geomorphic features.* New York: Robert E. Krieger Co., Huntington, NY, and Van Nostrand Reinhold, 301pp.

Stein, S., & Wysession, M. (2003). *An introduction to seismology, earthquakes, and earth structure.* Oxford, UK: Blackwell Publishing, 498pp.

Ward, P., & Brownlee, D. (2002). *The life and death of planet Earth: How the new science of astrobiology charts the ultimate fate of our world.* New York: Henry Holt and Company, 256pp.

Woodhouse, J. H., & Dziewonski, A. M. (1984). Mapping the upper mantle: three dimensional modeling of earth structure by inversion of seismic waveforms. *Journal of Geophysical Research, 89*, 5953–5986.

Chapter 22

Space Weather

J.G. Luhmann
Space Sciences Laboratory, University of California, Berkeley, CA, USA

S.C. Solomon
High Altitude Observatory, National Center for Atmospheric Research, Boulder, CO, USA

Chapter Outline

1. The Solar and Heliospheric Roles in Space Weather 481
2. The Geospace Role in Space Weather 484
3. Atmospheric Effects of Space Weather 487
4. Practical Aspects of Space Weather 489
5. Implications for Planetary Astronomy and Astrophysics 491
6. Epilogue 491
Bibliography 492

The Sun has profound effects on the Earth through its primarily visible and infrared photon emissions. This radiated energy, generated as a by-product of the nuclear reactions in the Sun's core (see The Sun), is absorbed or reflected at different wavelengths by the sea and land surfaces and the atmosphere. The result is the atmospheric circulation system that generates tropospheric weather through the diurnal and seasonal cycles caused by Earth's rotation and axis tilt. (See Earth: Atmosphere and Oceans). The climate of the Earth is the result of the long-term interaction of solar radiation, weather, surface, oceans, and human activity.

These influences are not the only ways the Sun affects the Earth. Ultraviolet (UV) and X-ray light from the Sun are much less intense, but more energetic and variable than the visible emissions. The UV radiation is absorbed in the stratosphere where it affects the production of the ozone layer and other atmospheric chemistry, while the extreme UV (EUV) photons and X-rays are absorbed in the thermosphere (above $\sim$90 km), creating the ionized component of the upper atmosphere known as the **ionosphere**. Even more variable is the emission of charged particles and magnetic fields by the Sun. One form of this output is the magnetized solar wind **plasma** and its gusty counterpart, the **Coronal Mass Ejection** or CME. CMEs interact with the Earth to create major **geomagnetic storms**. These and other forms of matter, energy and momentum transfer couple the physical domains of the connected Sun–Earth system, which is illustrated in Figure 22.1. A brief summary of the subject of this chapter, whose focus is this system, follows here.

Space weather begins in the solar interior where dynamo activity (see The Sun) generates the solar magnetic field. The solar magnetic field, coupled with the mechanical and radiative energy outputs from core fusion reactions, ultimately determines both the variability of the Sun's energetic (EUV, X-ray) photon outputs, and the interplanetary conditions at the orbit of Earth. The latter include the solar wind plasma properties, the interplanetary magnetic field magnitude and orientation, and the energetic particle radiation environment. Both the energetic photon outputs and interplanetary conditions vary with the $\sim$11-year solar cycle, which is characterized by changing frequencies of solar flares and CMEs, the two primary forms of solar activity. These in turn determine conditions in near-Earth space or **geospace**, the region composed of the **magnetosphere**, the upper atmosphere and the ionosphere. Only in the 1960s was it appreciated that the interplanetary magnetic field orientation relative to Earth's own dipolar field plays a major role in solar wind–magnetosphere couplings as described in more detail in the main text below.

The magnetosphere, the region of near-Earth space dominated by the magnetic field of the Earth and shaped by its interaction with the solar wind (see Figure 22.1), organizes geospace. Various particle populations in the magnetosphere, including the plasmas originating in the

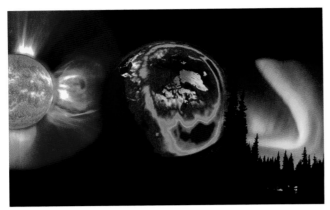

FIGURE 22.1 Triptych illustrating the coupled Sun–Earth system, showing from left to right an image of the erupting **solar corona** from the SOHO spacecraft, and images of the Earth's auroral emissions from space (center) and from the ground (right). See http://sohowww.nascom.nasa.gov/hotshots/2003_03_14/.

solar wind and Earth's ionosphere, and the more energetic particles trapped in the **radiation belts**, are constantly modified by changing interplanetary conditions. The ionosphere acts as a conducting inner boundary affecting the magnetosphere's response to those conditions, but is also a source of ions and electrons for the magnetosphere. Under the disturbed local interplanetary conditions following an Earth-directed CME, a collection of major magnetospheric modifications called a geomagnetic storm occurs. The population of trapped energetic particles in the radiation, or **Van Allen, belts** surrounding the Earth undergoes enhancements, losses, and redistribution. Current systems and particle exchanges couple the magnetosphere and ionosphere to a greater than normal degree. The result is enhanced solar wind energy transfer into geospace, causing **auroral** emissions and related changes in the high-latitude dynamics of the ionosphere, and the density and composition in the thermosphere. Evidence of atmospheric influences of geomagnetic storms and other solar effects down to the stratosphere has been reported, although it remains controversial. On the other hand, induced currents in conductors on the ground from storm-associated magnetic field changes are unarguable proof of the depth of influence of extreme space weather.

Studies of space weather investigate the physics that makes the Solar Wind, Magnetosphere, and Upper Atmosphere/Ionosphere, a highly coupled system. Figure 22.2 shows an attempt to diagram its various components and their relationships. There are also practical aspects to understanding the connections shown. Specifications of radiation tolerances for spacecraft electronics components, designs of protective astronaut suits and on-orbit shielding, and definitions of the surge limits for power grids on the ground can be made with a better understanding of space weather effects. Forecast models can help predict the changes in the magnetosphere that alter the radiation belts and the changes in the ionosphere that disrupt radio communications and GPS navigation. Observational and theoretical research in space weather processes also increases our understanding of other areas of Astronomy and Astrophysics such as planet–solar wind interactions, extra-solar planetary systems, stellar activity, and the acceleration of particles in the universe.

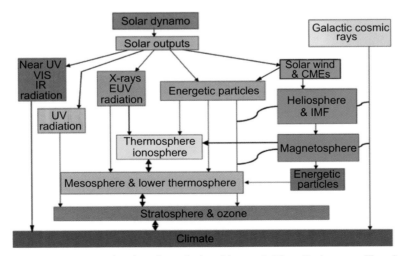

FIGURE 22.2 Flow diagram illustrating the connections in and complexity of the coupled Sun–Earth system. The solar **dynamo** (top) generates the solar magnetic field, which modulates the solar outputs of EUV and X-ray emissions, as well as the solar wind plasma. The solar wind and its gusty counterpart, CMEs (green box, upper right), directly determine the state of the local heliosphere, which controls the state of the magnetosphere (including its energetic particle or radiation belt populations). In the meantime, both solar photons (upper right boxes) and solar energetic particles directly affect the state of the upper atmosphere. The possible connection to climate, suggested at the bottom, is currently a matter of speculation. *Adapted from http://lws-trt.gsfc.nasa.gov/lika_radtg.ppt.*

1. THE SOLAR AND HELIOSPHERIC ROLES IN SPACE WEATHER

Solar radiation in the UV, EUV and X-ray wavelengths are the primary sources of ionization in the Earth's atmosphere. Of these solar EUV fluxes are the most important source of the ionosphere. Figure 22.3 illustrates the relatively large variability of this part of the solar spectrum, compared with the visible and infrared wavelengths that dominate the "solar constant". As mentioned before, this variability is a result of the control of these emissions by the solar magnetic field, which undergoes significant evolution during the course of the ~11-year solar activity cycle (see The Sun). The EUV emissions come largely from bright plage areas seen on the photosphere and from the chromospheric network, while the X-rays come mainly from hot plasma-containing coronal loops structured by the coronal magnetic field. The plages and X-ray bright loops are related to active regions, areas with the strongest photospheric magnetic fields, that are nonuniformly distributed over the solar surface. An important result of having magnetic field observations of the Sun was the appreciation that not all active regions have fields strong enough to produce sunspots on the visible disk. The changing numbers of active regions, and their areas, determine the solar activity cycle. Thus the solar EUV flux experienced at Earth undergoes variations on both the 27-day timescale of solar rotation (due to the nonuniform distribution of active regions on the surface) and the near-decadal timescale of solar activity (see The Sun). The transient brightenings in active regions called solar flares occasionally produce solar EUV and X-ray emission enhancements of up to several orders of magnitude at photon energies extending into the gamma ray range. These outbursts affect Earth's atmosphere and ionosphere at depths depending on their wavelengths as indicated in Figure 22.4. The magnetosphere responds to changes in the ionosphere and upper atmosphere, but its primary solar controller is the magnetized solar wind plasma.

The solar wind is the outflowing ionized gas or plasma of the solar upper atmosphere (see The Solar Wind). This outermost extension of the corona fills a space up to about ~100 AU in radial extent, defining the region surrounding the Sun, the **heliosphere**. The mainly hydrogen solar wind flows primarily from places in the corona that are magnetically "open" to interplanetary space. These open field regions are often called coronal holes because of their dark appearance in soft X-ray and EUV images (see The Sun). The solar wind also carries with it the stretched out coronal magnetic field that takes on an average outward or inward orientation depending on the magnetic field direction or polarity at its photospheric base. There is also a component of the quiet solar wind that comes from the edges of coronal closed magnetic field regions, producing the equivalent of a boundary layer between outflows from different open field regions. On the average the solar wind speed is slowest in these boundary layers and fastest where it flows from the center of large

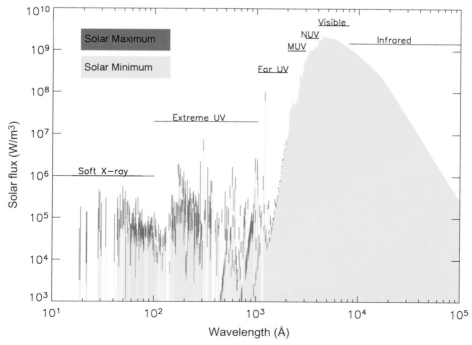

FIGURE 22.3 Illustration of the solar spectrum, showing the intensities of various wavelength emissions and their variation from active (red) to quiet times. Notice that order of magnitude variations from solar minimum to maximum occur at the short (<1000 Å) wavelengths. (1 Å is equivalent to 0.1 nm.)

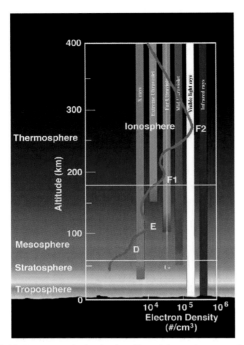

FIGURE 22.4 Atmospheric depths affected by various wavelengths in the solar spectrum. (Courtesy of Windows to the Universe, http://www.windows.ucar.edu.) The longer wavelength radiation affects mainly heating, while the shorter wavelengths can produce ionization and associated chemistry changes. The letters F1, F2, E, and D are used to designate different ionospheric layers.

open field regions. Typical solar wind speeds range from ~300 km/s to ~800 km/s and are roughly constant with radial distance. Undisturbed solar wind magnetic field strengths at 1 AU range from ~5 to 10 nT and densities from ~5 particles/cm^3 to 15 particles/cm^3. Heliospheric images that capture the faintest coronal structures indicate that coronal hole boundary outflows are at least partly in the form of small transients or "blobs", the result of the constant evolution of solar surface magnetic fields at the base of the corona. Because the open and closed field regions change with the distribution of active regions on the Sun, the solar wind stream structure and field polarity pattern evolve with the solar activity cycle. They are simplest at the quietest times of the cycle, during which the corona usually exhibits two main solar wind sources near the Sun's polar regions, one with positive (outward) and one with negative (inward) magnetic polarity. As more active regions appear on the solar surface, the coronal magnetic structure becomes more complicated and with it the distribution of open fields. In particular, low and mid-latitude coronal holes appear that can produce most of the solar wind the Earth experiences.

In fact, a critical aspect of the solar wind stream structure for the impact on geospace is Earth's location near the solar rotational equator. This region is often dominated by the presence of the slow wind and the related heliospheric current sheet that separates the solar wind from open coronal field sources with outward and inward magnetic field polarities. This circumstance, together with the rotation of the Sun, produces local interplanetary conditions that at low solar activity exhibit repeating or "corotating" 27-day variations in solar wind speed and density, and interplanetary magnetic field polarity. E. Parker, who first proposed the existence of the solar wind, also recognized that the Sun's rotation would wind up the interplanetary magnetic field into a spiral shape. Figure 22.5 illustrates this "Parker Spiral" orientation of the near-ecliptic field in the heliosphere, and its typical 45° (from radial) orientation at 1 AU (see The Solar Wind). At solar minimum, adjacent streams of different speeds from different coronal source regions may interact, producing spiral density and field ridges at their boundaries. When these ridges, which are called stream interaction regions or corotating interaction regions, rotate past the Earth, they can cause modest geomagnetic activity. At solar activity maximum the more complicated coronal field mentioned above makes the corresponding solar wind conditions more variable and structured, and less organized by solar latitude. The active phase solar wind also includes many transient disturbances caused by rapid changes in coronal structure, including CMEs, whose effects are further described below. Figure 22.6 illustrates the general morphology of solar wind characteristics from periods around solar minimum and solar maximum. These solar cycle-dependent interplanetary conditions shape the

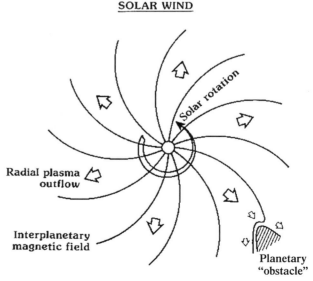

FIGURE 22.5 Illustration of the Parker Spiral interplanetary magnetic field carried in the outflowing solar wind, and wound up by solar rotation. At 1 AU the typical angle the field makes with respect to the Sun—Earth line is ~45°. *See The Solar Wind for further information.*

Chapter | 22 Space Weather

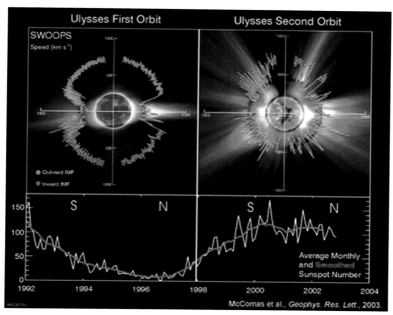

FIGURE 22.6 Solar wind velocities at solar minimum and maximum as measured on the Ulysses spacecraft which passes over the Sun's poles. The speed is shown in a polar coordinate system with zero speed at the center of the Sun. The blue and red indicate interplanetary magnetic field polarity. *McComas et al. (2003).*

Earth's magnetosphere, and regulate its internal particles, energization, and stresses.

When the Sun is active, it produces greater numbers of both the flares mentioned earlier and CMEs. CMEs have the greatest effects on geospace, and so we focus on them here. The details of the CME initiation process, as well as CME structure, are subjects of intensive current research. As seen in white light images from coronagraphs like the SOHO LASCO instrument (an example of which is shown in Figure 22.1), CMEs appear to be eruptions of large magnetic bubbles or twisted "flux ropes" of coronal magnetic fields. These structures, referred to as drivers or ejecta, travel outward at speeds ranging from tens of km/s to several thousands of km/s. As they travel they interact with the surrounding solar wind, compressing it ahead if they are moving faster. If they move fast enough relative to their surroundings they create an interplanetary shock wave. Figure 22.7 illustrates the effect of a CME on the solar wind and interplanetary magnetic field at 1 AU. These propagating disturbances are experienced by the magnetosphere as sudden increases of solar wind density, velocity, and magnetic field at the shock passage, followed by several hours of enhanced solar wind parameters, and then the ejecta passage characterized by a period with normal densities but high magnetic field strengths and, often, high inclinations. The entire structure may take hours to days to pass Earth depending on its speed. Enhanced solar wind pressure associated with a CME is usually from the sheath portion between the shock front and the ejecta. The ejecta

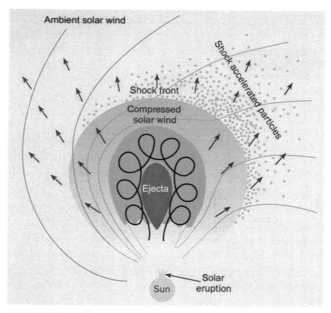

FIGURE 22.7 Illustration of the interplanetary effects of a CME. The CME produces an ejection of coronal material (ejecta) that may include a helical magnetic field structure or flux rope (illustrated by the black line). This structure plows into the ambient solar wind and may produce a shock in the solar wind plasma ahead of the ejecta. The region of compressed solar wind between the shock and ejecta is referred to as the sheath. Some solar wind particles are accelerated at the shock and speed out ahead of it along interplanetary field lines (red, green). *Luhmann (2000).*

fields can occasionally be modeled as a passing magnetic flux rope configuration as suggested in Figure 22.7. Around the minimum of solar activity the local interplanetary medium is disturbed by one of these interplanetary CMEs or ICMEs once every few months, but at solar maximum they can occur about once a week. The most extreme (largest, fastest) ICMEs usually follow CMEs associated with large, complex active regions on the Sun that also produce solar flares.

In addition to the magnetized solar wind plasma, the heliosphere also contains a population of energetic (tens of kiloelectron Volts (keV) to hundreds of Megaelectron Volts (MeV)) charged particles that varies with time. Ions and electrons are accelerated at both flare sites on the Sun and by the shock waves formed in the corona and interplanetary space by the fast-moving CME ejecta or by interacting high- and low-speed solar wind streams. CME shocks produce the most intense and long-lived (several day) episodes. The particles race ahead of their shock source along the spiral interplanetary field lines, surrounding the magnetosphere with a sea of potentially hazardous radiation within tens of minutes of the events at the Sun. Sometimes the fluxes of these particles increase by several orders of magnitude when the CME shock itself arrives several days after the event in the corona. The contributions of **solar energetic particles** to local interplanetary conditions are related to the level of flare and CME activity on the Sun. In an interesting opposite effect of solar activity, other more permanently present energetic charged particles called cosmic rays, which arrive at Earth from the heliospheric boundary and beyond, show locally decreased fluxes when solar activity is high. These decreases are likely due to the sweeping action of the highly structured solar wind around the time of solar activity maximum, when interplanetary field disturbances carried outward present effective barriers to incoming charged particles. In fact especially notable episodes called "Forbush Decreases" can occur during the passages of CME ejecta. Under certain conditions solar energetic particles can enter the magnetosphere where they contribute to the radiation belts and produce layers of enhanced ionization deep in the Earth's polar atmosphere.

2. THE GEOSPACE ROLE IN SPACE WEATHER

The Earth's space environment is determined by its nearly dipolar internal magnetic field that forms an obstacle to the solar wind, creating the magnetosphere (see Planetary Magnetospheres). Spreiter, Summers, and Alksne (1966) and Axford and Hines (1961) were among the pioneering researchers to recognize the fluid-like aspects of the solar wind interaction with Earth's compressible field, describing it in terms of a blunt body in a hypersonic flow. The size and shape of this blunt body, the magnetopause, can be calculated from the assumption of pressure balance between the Earth's internal magnetic field pressure and the incident dynamic pressure of the solar wind. (Dynamic Pressure = mass density × velocity squared.) It typically occurs at ~10 Earth radii along the line connecting the centers of the Earth and the Sun and at ~15 Earth radii in the terminator plane. In contrast to the compressed, solar wind pressure-confined dayside, the nightside magnetosphere stretches out into an elongated structure called the magnetotail. These features, confirmed by decades of observations, are illustrated in Figure 22.8. The magnetopause separates geospace and the solar wind plasma-dominated regions outside. As seen in Figure 22.8, the outermost features associated with

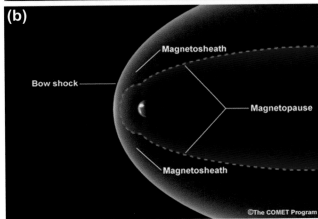

FIGURE 22.8 (a) Illustration of the blunt body shape of the magnetosphere, showing some gossamer "shells" of magnetic field surfaces along which charged particles drift, and the magnetopause (Rice University). (b) Magnetospheric boundaries described in the text. The magnetopause nominally separates solar wind and magnetospheric domains. *The source of this material is the Cooperative Program for Operational Meteorology, Education, and Training (COMET) Web site at http://meted.ucar.edu/ of the University Corporation for Atmospheric Research (UCAR) pursuant to a Cooperative Agreement with National Oceanic and Atmospheric Administration. 1997–2004 University Corporation for Atmospheric Research. © All Rights Reserved.*

the Earth's magnetic obstacle are actually the bow shock that forms in the solar wind ~5 Earth radii upstream of the dayside magnetopause, and the magnetosheath. The magnetosheath is the slowed, deflected solar wind between the bow shock and the magnetopause. Thus when the Earth's field interacts with the solar wind, it does so through the altered solar wind in the magnetosheath.

Dungey (1961) first recognized that the magnetopause is not a complete barrier to the solar wind, and that magnetospheric field topology is also controlled by its interconnection, or **reconnection**, with the interplanetary field. This leap of understanding revolutionized the study of solar-wind magnetosphere coupling and geomagnetic activity. Figure 22.9 reproduces Dungey's original cartoon suggesting the different appearances of the magnetospheric field topology for the extreme cases of steady Northward and Southward interplanetary fields. Similar pictures can be obtained by adding background uniform fields of both directions to a dipole field. The Northward interplanetary field, which is parallel to the Earth's dipole field at the equator, produces a magnetically closed magnetosphere. The Southward field, which is antiparallel to the Earth's dipole field at the equator, produces a magnetically open configuration with the polar region fields of the Earth connected to the interplanetary field. These differences

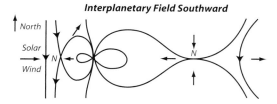

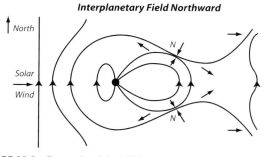

FIGURE 22.9 Dungey's original 1961 cartoon suggesting the reason for the association of greater geomagnetic activity with southward interplanetary magnetic fields Southward interplanetary fields can reconnect or merge with the Earth's dipole field at the dayside magnetopause. Another reconnection in the magnetotail returns the Earth's opened fields to their original "closed" state, so that all of the Earth's field is not permanently opened by dayside reconnection. The process of reconnection drives magnetospheric circulation (see Figure 22.10) and is thus a means by which solar wind energy is transferred to the magnetosphere. *Reprinted figure with permission from Dungey (1961). Copyright 2005 by the American Physical Society.*

greatly affect the transfer of both energy and particles from the solar wind into geospace. For the Northward case the solar wind interaction resembles a viscous boundary interaction at the magnetopause and there is minimal exchange of energy and particles. For the Southward case the charged particles in interplanetary space have access to the polar regions along interplanetary field lines. An electric field associated with solar wind convection $E = -V \times \mathbf{B}$, where V is the solar wind velocity and $\mathbf{B}$ its magnetic field, maps along open field lines into polar regions where it drives vigorous magnetosphere and ionosphere circulation as in Figure 22.10. The two-celled vortical convection pattern has been observed in the ionosphere by high-latitude radars and can be inferred from magnetometer measurements. Solar wind–magnetosphere coupling is thus greatly enhanced at times when the interplanetary magnetic field is Southward.

The physics of the reconnection or magnetic field merging process that results in this configuration change for Southward interplanetary fields is still a subject of intensive research. Because space is not a vacuum, simple superposition of the external (interplanetary) and internal (Earth dipole) fields is not a physically correct explanation. Somehow the solar wind plasma that carries the interplanetary field "frozen" into its flow must allow the field to merge with the magnetospheric field at the magnetopause when the two have antiparallel components. The interested reader is referred to the review by Drake and Shay (2005) for further details on current theories of magnetic field reconnection in space plasmas. When CME effects reach the Earth, the solar wind dynamic pressure incident on the magnetosphere can increase by an order of magnitude or more, primarily due to the compression of the ambient solar wind plasma by the driver or ejecta from the CME. The onset of this increase may be sudden if a leading shock is present. The solar wind magnetic field is also compressed with the plasma, and can become significantly inclined with respect to the Earth's equatorial plane. The ejecta fields are also highly inclined, and often strong and steady, or slowly rotating over intervals of about a day. Thus, larger than normal Northward and/or Southward interplanetary field components result from both passing segments of the disturbance. The magnetosphere's response to these disturbed heliospheric conditions includes increased compression of the dayside magnetosphere, sometimes to within a few Earth radii of the surface, and increased reconnection between the Earth's magnetic field and interplanetary field during the passage of the southward-oriented portions. The time-dependent nature of these boundary conditions can introduce additional complexity into the solar wind–magnetosphere coupling.

Resulting geospace consequences of CMEs are numerous and varied, as illustrated by Figure 22.11. The associated magnetospheric compression is accompanied by enhancements of the energetic radiation belt particles

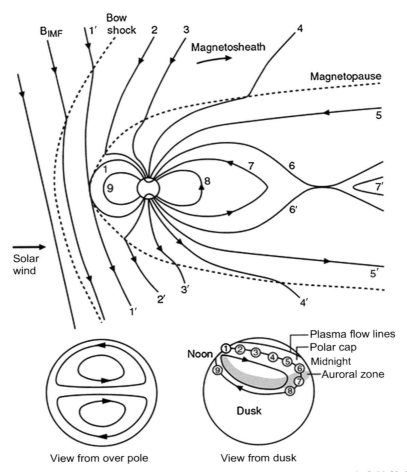

FIGURE 22.10 Illustration of magnetospheric circulation during periods of southward interplanetary magnetic field. Various key features of the solar wind interaction are shown, including magnetospheric field line connections to the interplanetary field, and their mapping to the high-altitude atmosphere. The numbers indicate a time sequence. The driven circulation occurs all the way down to the polar ionosphere as shown by the inset, which shows the dusk half of the double-celled ionosphere convection pattern. The aurora occurs mainly in the regions of convection reversals. *Kivelson & Russel (1995)*.

trapped in the Earth's dipole field, due to a combination of inward diffusion and energization of the existing particle populations (see Planetary Magnetospheres). Solar energetic particles accelerated at the CME-driven shock or in associated solar flares can also leak into the magnetosphere along newly reconnected field lines at the magnetopause or along open field lines into the polar regions, as these particles tend to stream along field lines. Magnetic reconnection between the stretched out antiparallel fields in the magnetotail causes currents to flow through the high-latitude ionosphere. As magnetospheric charged particles move toward Earth with the field lines, they are accelerated in some cases by electric fields parallel to the magnetic field. These energized particles include electrons, protons, and other heavier ions. When they reach the upper atmosphere, they collide with neutral gases at altitudes of ~ 100 to ~ 200 km, causing ionization and excitation of atoms and molecules. The ionization enhances the flow of magnetosphere–ionosphere currents, and when the excited atoms and molecules decay back to their ground states they emit the light known as the aurora. (Further information about the aurora can be found in Section 3.)

Magnetotail reconnection also triggers injection of particles toward Earth at low latitudes that form a **ring current** at ~ 4 to ~ 7 Earth radii. In the polar regions, protons and ions, including ionospheric oxygen ions, O^+, are driven upward from the base of open field lines and flow into the magnetosphere, changing the composition of the magnetosphere and ring current ion populations. The ring current noticeably changes the magnetic field in the magnetosphere and at Earth's surface. Altogether these phenomena characterize a geomagnetic storm, whose magnitude is characterized by the ring current-related reduction of the field at the ground, defined by the index Dst. (Another index, Kp, is also used, but is more a measure of the auroral current systems.) Eventually, the magnetosphere and ionosphere return to their pre-storm states. Most effects are gone after a few days, but some trapped particle

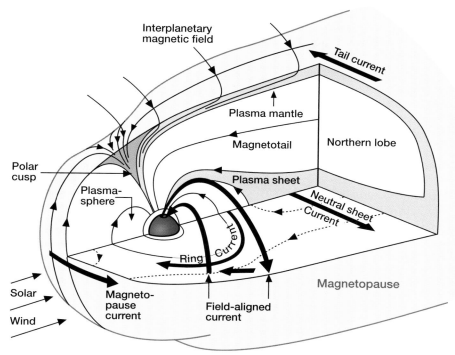

FIGURE 22.11 A more detailed illustration of the magnetosphere, showing further features mentioned in the text, such as the ring current and magnetotail current sheet. The plasmasphere is a region of denser, corotating magnetospheric plasma of ionospheric origin. The plasma sheet is a denser region of magnetotail plasma that participates in the physics of magnetotail reconnection and ring current formation (Rice University).

populations may last much longer. This complex geospace response to a CME has recently been simulated by several research groups using numerical models of geospace, with solar wind measurements defining the time-dependent boundary conditions on the magnetosphere. Some results from one of these models are illustrated in Figure 22.12.

There are also weaker, more frequent geomagnetic disturbances known as substorms. Substorms may occur during storms, as periodic enhancements of the storm–time geospace responses, or as stand-alone disturbances when the quiet interplanetary magnetic field has a southward component. In some cases they appear to follow a sudden change in the interplanetary north–south field component or a dynamic pressure pulse in the solar wind. Current ideas on the reasons for substorms, which have been debated for decades, include internal instabilities of the magnetosphere that occur in response to a variety of triggers. However, geomagnetic storms involve the largest episodic energy transfers from the solar wind, and are thus responsible for the strongest space weather effects.

3. ATMOSPHERIC EFFECTS OF SPACE WEATHER

The atmospheric responses to solar activity and its magnetospheric consequences are the closest counterparts of space weather to traditional weather. They are therefore of special interest to space weather research. Direct effects are largely confined to the thermosphere and ionosphere, above the mesopause at ~ 90 km. They fall into two main categories: the effects of particles entering or "precipitating" into the atmosphere, and the effects of high-latitude ionospheric convection from magnetosphere–ionosphere coupling.

There are several types of precipitating particles: $\sim 1-20$ keV auroral electrons, $\sim 10-100$ keV ring current ions (protons and some oxygen ions), $\sim 1-10$ MeV radiation belt electrons, and $\sim 1-100$ MeV solar energetic particles (primarily protons). The more energetic the particles, the deeper they penetrate; the altitude ranges to which these various particles penetrate to deposit their energy are illustrated in Figure 22.13. As mentioned earlier, when these particles encounter atmospheric atoms and molecules, they cause impact ionization, and dissociate molecules into their atomic elements. They also excite bound electrons to unstable states, which then radiatively decay to produce photons with specific energies and thus wavelengths that give the aurora its colors. Chemical reactions caused by the interactions of ions with the dissociated and excited atomic products also excite particular emission features. The characteristic green and red auroral emissions at 557.7 and 630.0 nm are produced by the excitation of the upper atmosphere oxygen atoms; other auroral emission features in the blue and near-UV spectral regions are formed from excitation of molecular nitrogen and its ion.

Most auroral emissions occur in an oval-shaped band just equatorward of the open field lines at high latitude, giving the auroral oval (shown in Figures 22.1 and 22.10) its name. Ring current ion precipitation can produce high-altitude red aurora at lower latitudes. In contrast, radiation belt electrons and solar energetic protons leave mainly chemical signatures. Along with other chemical by-products, all particle precipitation produces nitric oxide (NO) from molecular nitrogen dissociation. Very energetic particles can produce NO in the mesosphere (50–90 km) and even the stratosphere (15–50 km). Increases in NO affect ozone levels, because chemical reactions involving NO, ozone, nitrogen dioxide, and atomic oxygen form a catalytic cycle that reduces ozone. For some major solar proton events ∼30% depletions of ozone in the mesosphere and upper stratosphere have been detected. While the reduction of mesospheric ozone does not have the biological impact of reductions in the denser lower stratospheric ozone, it can modify the temperature and thus the dynamics of the mesosphere. These alterations can in turn modify the transmission of energy from the stratosphere and troposphere to the upper atmosphere. Whether these effects have significant consequences for the lower atmosphere, especially over the long term, is unknown.

As noted above, the solar wind electric field within the open magnetospheric field regions at high latitudes typically stirs the polar ionosphere in a twin vortex pattern, illustrated in Figure 22.10. Ionospheric ions and electrons are dragged antisunward over the polar caps, and then forced into a return flow at lower latitudes. The differential motion of the ions and electrons caused by the interplay between ion–neutral collisions and the electric and magnetic fields that control their motion leads to an ionospheric current. This auroral electrojet current has a strength dependent on the combination of the solar wind electric field and the level of ionization in the auroral ionosphere. Collisional dissipation or friction within the volume occupied by the electrojet heats the auroral zone atmosphere. This resistive "joule" heating results in large density perturbations called auroral gravity waves in the upper atmosphere. These travel equatorward, in some cases depositing significant energy and modifying upper atmosphere circulation globally. Traveling ionospheric disturbances are one manifestation of the passage of these waves. Magnetic field perturbations associated with the time-varying electrojet current and the ring current mentioned earlier are detected on the ground, giving the geomagnetic storm its name. A particular geomagnetic index called AE is a widely used measure of the level of ground magnetic field modifications by the auroral electrojet currents.

Solar and magnetospheric events are not the only source of space weather in the thermosphere–ionosphere system. Wave motions and atmospheric tides propagating upward from the lower and middle atmosphere have important effects on the ionosphere, and contribute to its variability

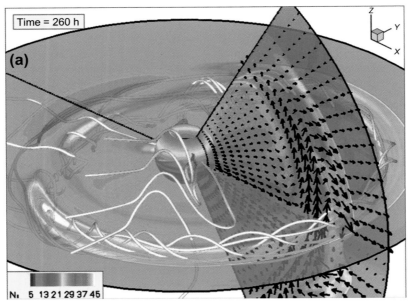

FIGURE 22.12 (a) Results from a numerical simulation of a CME, showing the distortion of the interplanetary plasma density (contours) and magnetic field (white lines) as it travels toward the Earth. The vectors indicate directions of the velocity in a selected meridional slice. Note the flux rope ejecta that drives the leading interplanetary shock (sharp red contour outer boundary) (courtesy of George Mason University). *(Courtesy: Charles Goodrich, Boston University) Luhmann et al. (2004).*

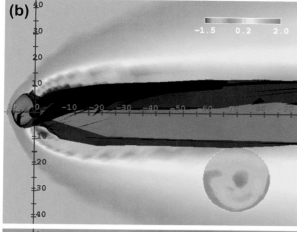
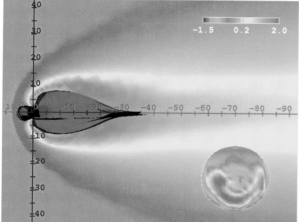
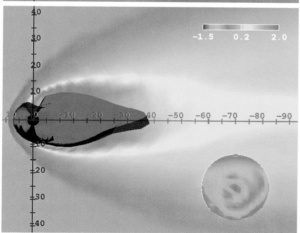

FIGURE 22.12 (continued) (b) Geospace response to the CME in (a), showing density contours (log scale), in the local solar wind, the surface of outermost closed magnetospheric field lines, and a view of the resulting energy input into the earth's high-latitude atmosphere at three different times. (Courtesy: Charles Goodrich, Boston University) Luhmann et al. (2004).

even when the external space environment is very quiet. An important mechanism of atmospheric influence on the ionosphere is change in the global electric field, especially at low latitude, where atmospheric tides drive high-altitude

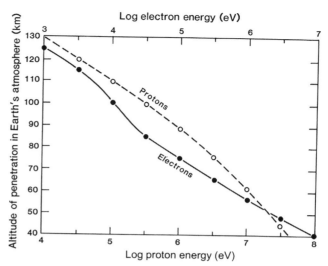

FIGURE 22.13 Plot showing the approximate depth of penetration into the atmosphere of energetic electrons and protons. Kvelson & Russell (1995).

winds that in turn generate currents in the ionosphere. Meteorological events in the troposphere—conventional weather—can disrupt these winds and currents. The turbulence above storms and frontal systems is also transmitted to high altitude, where it can alter the composition of the thermosphere and subsequently the structure of the ionosphere. These effects from the coupled atmospheric system are now recognized as an important component of space weather effects near the Earth.

4. PRACTICAL ASPECTS OF SPACE WEATHER

Space weather is a complex and fascinating physical system that also has many practical consequences. Society is increasingly dependent on space-based telecommunications and satellite systems that monitor tropospheric weather, global resources, and military activity. The satellite environment is part of the design of these spacecraft, which can suffer radiation damage to electronics if the extremes they may encounter are not taken into account. Satellite orbits are affected by drag as they pass through the upper thermosphere, where changes in density caused by EUV radiation and geomagnetic storms affect their tracking and lifetime. This is particularly true of large, relatively low-altitude vehicles such as the Hubble Space Telescope and the International Space Station. In addition, satellite orientation controls often rely on Earth's magnetic field, which can be highly variable during geomagnetic storms. Changes in the ionosphere disrupt radio communications by changing ionospheric transmission or reflection characteristics. GPS navigational signals, which pass through and are altered by the ionosphere as they are transmitted from very high-altitude satellites to ground receivers, can degrade during disturbed conditions, giving inaccurate locations. On the ground, currents induced in power system transformers and in oil pipelines by storm-related magnetic field perturbations respectively lead to overload and corrosion.

In the era of human space flight, there is also great concern over space radiation hazards from energetic particles. Manned vehicle orbits at low latitude and low altitude are largely protected from the radiation belts and the less common but potentially dangerous major solar energetic particle events. However, the orbit of the International Space Station is sufficiently inclined with respect to Earth's equator that it is occasionally exposed to solar energetic particles at high latitudes, when the magnetosphere is disturbed as it often is during solar particle events. Astronaut radiation exposure is carefully monitored and is limited by NASA. To minimize it, plans for extravehicular activities take into account conditions on the Sun and the likelihood of a major solar event that might affect the Earth. Even commercial and military aircraft on polar routes monitor major solar events as a precaution. For future space travel outside of the effective but imperfect magnetospheric shield, protection from solar particle radiation is a major problem to be solved.

In response to both international civilian interests and military needs, the National Oceanic and Atmospheric Administration (NOAA) runs a Space Environment Center (SEC) that collects, analyzes, and distributes information on the Earth's space environment and solar activity. Space weather reports are regularly issued via the Internet (swpc.noaa.gov), where one can also find access to the archives of solar, heliospheric, magnetospheric, and upper atmosphere/ionosphere data that are used. Customers of these services seek information on subjects ranging from interference to radio transmissions by solar radio bursts or ionospheric scintillations, to satellite orbit decay rates based on solar EUV emission intensities. Alerts are posted when a forecaster interprets behavior in the relevant data to mean a solar energetic particle event, geomagnetic storm, or ionospheric disturbance will occur within the next minutes to days. One of the most useful geomagnetic storm forecasting methods takes advantage of the SOHO spacecraft, which allows the forecaster to identify CMEs and the location of active regions on the solar disk. When a CME is headed toward Earth, it sometimes appears in the SOHO coronagraph images as a ring around the Sun called a halo CME. These events are known to have an increased probability of causing a geomagnetic storm. However storm forecasts are still extremely difficult, with false alarms a major problem. For example some halo CMEs are actually heading away from the Earth, having originated on the far side of the Sun, while others have ejecta that are deflected away from Earth enroute or include mainly northward magnetic fields that are not particularly "geoeffective".

Geomagnetic indices, calculated from ground-based measurements of magnetic field perturbations, are routinely used as a measure of the level of space weather disturbance. Different indices emphasize particular Earth responses depending on how and from what stations they are calculated. Several of these were mentioned above. The auroral electrojet index AE is primarily a measure of auroral zone ionospheric currents obtained from high and mid-latitude monitors, while the ring current index Dst is mainly a measure of the ring current obtained at lower latitudes. The planetary index Kp uses ground stations in both regions. These indices were developed and have been recorded since before the space age. They are used to both parameterize empirical models (e.g. of the auroral zone ionosphere), and maintain a continuous long-term historical record of space weather in concise form.

A major goal of space weather research today is a physics-based model of the coupled heliosphere—magnetosphere—upper atmosphere/ionosphere system, including CMEs. Such a model could provide both a forecast tool for space weather events based on solar observations and a numerical experiment framework to gain greater insight into the physics of the coupled system and its extremes. For example, severe space weather events are occasionally observed, but even larger events have been inferred from records of the Earth's historical cosmic ray exposure present in ice cores. What is the worst that could happen to our planet and space assets after one or more of these greatest of solar activity episodes? It is an intriguing question. As a practical matter, the plan to again send humans into deep space, to the Moon and to Mars, also renews the concerns of space radiation hazard issues faced to a lesser degree on the International Space Station. The unpredictability of CMEs, and the fact that historically the strongest space weather events have not been at solar maximum, helps motivate applied space weather research.

For many decades there has also been research on and discussion about the connection of solar activity and Earth's climate. It is possible that the very small changes in total solar irradiance, the so-called solar constant, or changes in UV radiation can have measurable climatic effects. It is also possible that some Sun Earth connection-related phenomena can reach the troposphere. One highly cited example is the coincidence of the Maunder Minimum in solar activity with the Little Ice Age in Europe during the fifteenth to sixteenth centuries. Ideas on how nonradiative effects might play a role include mechanisms such as cloud—cover alteration by low-altitude ionization effects of energetic cosmic rays reaching Earth (Tinsley, 2000). The Maunder Minimum climate change has also inspired speculation about the climatic and other consequences of geomagnetic field reversals. Analyses of ice age records and geomagnetic field reversals have as yet produced no definitive results, although it may be significant that the Earth's field still maintains an important higher order harmonic component (e.g. a quadrupole moment) during reversals.

5. IMPLICATIONS FOR PLANETARY ASTRONOMY AND ASTROPHYSICS

Space weather, and the physics it encompasses, is often invoked in the planetary sciences in connection with solar wind interaction issues. Our knowledge of the solar wind coupling to the Earth's magnetosphere and upper atmosphere is far greater than our comparable knowledge for any other planet due to both the wealth of available observations and the efforts that have been put into their interpretation. Planetary spacecraft found that there are magnetospheres around Mercury, Jupiter, Saturn, Uranus, and Neptune. Mars does not have a strong dipolar internal field, but it has patchy crustal magnetism that makes a rather unique obstacle to the solar wind. Venus has essentially no planetary field and thus represents another extreme contrast in solar wind interaction styles. One of the main goals of the Messenger mission to Mercury is to better observe the magnetosphere there in terms of its response to solar wind conditions, and its particle content and dynamics. Mercury has no substantial atmosphere or ionosphere; so it represents an interesting contrast to Earth that may tell us more about the atmosphere's role in space weather. In addition, the relative strengths of its internal field close to its surface and the local interplanetary field are more comparable than Earth's, making the coupling of these two fields particularly close. Mercury may represent an example of a strongly driven planet—solar wind interaction. Jupiter's giant magnetosphere was found by the Galileo spacecraft to be dominated by the internal mass content contributed by the volcanic satellite Io. Similarly, Saturn's magnetosphere, currently under scrutiny by the Cassini Orbiter, also seems to have an important internal mass source in its satellite Enceladus. The auroras at both these planets thus have different reasons than Earth's for occurring where and when they do. Torques provided by the internally produced heavy ions seem to exert considerable control over the magnetospheres' dynamics compared with the external conditions in these cases. The comparative analysis of planet—solar wind interactions and the related atmospheric effects is extremely valuable for achieving maximum understanding from necessarily limited planetary data.

In the world of astrophysics, extrasolar planetary system research strives to infer from poorly resolved observations the details of individual planets. One possibility for remote sensing is provided by the stellar wind interaction with the planets, which may produce detectable emissions from the planetary atmospheres. To be useful, these emissions must be interpretable in terms of familiar examples in our own solar system. Of particular interest is the detection of Earth-like planets. There are a range of possible remote signatures for applications to these "origins" investigations. Similarly, the identification of the effects of stellar winds around other stars is enabled in part by our own heliospheric experience. The interaction of the stellar wind and the surrounding interstellar medium produces a feature like the magnetosheath that is remotely detectable in Lyman-alpha emission. Some stellar outbursts suggest the occurrence of CMEs, and the associated space weather around remote worlds.

Finally, fundamental astrophysical processes are involved in energetic particle acceleration as well as in magnetic reconnection. Much of what has been learned about particle acceleration at shocks in plasmas has come from the analysis of the observations from the region around the Earth's bow shock. Similarly, reconnection processes at the magnetopause and in Earth's magnetotail have been examined using spacecraft data from both single spacecraft and small constellations. These difficult observations of a dynamical and nonuniform space plasma system with many scales are slowly yielding information about the process that suggests that it occurs when and where electrons are no longer controlled by the magnetic field. In addition, numerical simulations have been carried out using fluid, kinetic (particle), and hybrid (mixed particle and fluid) codes to shed light on the microphysics of how oppositely directed magnetic fields in space plasmas undergo major topological changes. Laboratory work has also contributed to these investigations, all under the umbrella of space weather research.

6. EPILOGUE

The phrase "space weather" is used to describe the physically rich and dynamic system by which processes at the Sun affect near-Earth space via other than the solar constant radiative emissions. It is necessary to use a combination of theory, measurements and modeling to study the system components: the heliosphere, the magnetosphere, and the upper atmosphere and ionosphere, to learn how they are coupled. As described above, the couplings are numerous and diverse. They are sometimes subtle like cosmic ray effects on clouds, and sometimes overt like CMEs and the related topological changes reconnection imposes on the magnetosphere in response to their associated large southward oriented interplanetary magnetic fields. The consequences of these couplings are only partly understood. Practical applications of space environment knowledge are in the meantime growing in popularity and demand. Other fields are beneficiaries of the planetary and astrophysical "laboratory". And in an era of new human

Exploration Initiatives, and a society increasingly dependent on space-based technologies, space weather may one day become part of the weather report on your local news.

BIBLIOGRAPHY

Axford, W. I., & Hines, C. O. (1961). A unifying theory of high-latitude geophysical phenomena and geomagnetic storms. *Canadian Journal of Physics, 39*, 1433–1464.

Drake, J. F., & Shay, M. A. (2005). The fundamentals of collisionless reconnection. In J. Birn, & E. R. Priest (Eds.), *Reconnection of magnetic fields: Magnetohydrodynamics and colliosionless theory and observations*. Cambridge Univ. Press Cambridge.

Dungey, J. W. (1961). Interplanetary magnetic field and the auroral zones. *Physical Review Letters, 6*, 47–48.

Kivelson, M. G., & Russell, C. T. (1995). *Introduction to space physics*. New York: Cambridge University Press.

Luhmann. (July 2000). *Physics World*.

Luhmann, J. G., Solomon, S. C., Linker, J. A., Lyon, J. G., Mikic, Z., Odstrcil, D., Wang, W., & Wiltberger, M. (2004). Coupled model simulation of a Sun-to-Earth space weather event. *Journal of Atmospheric and Solar-Terrestrial Physics, 66*(15–16), 1243–1256.

McComas, D. J., Elliott, H. A., Schwadron, N. A., Gosling, J. T., Skoug, R. M., & Goldstein, B. E. (2003). The three-dimensional solar wind around solar maximum. *Geophysical Research Letters, 30*. http://dx.doi.org/10.1029/2003GL017136.

Parker, E. N. (1963). *Interplanetary dynamical processes*. New York: Interscience Publishers.

Spreiter, J. R., Summers, A. L., & Alksne, A. Y. (1966). Hydromagnetic flow around the magnetosphere. *Planetary and Space Science, 14*, 223–253.

Tinsley, B. A. (2000). Influence of solar wind on the global electric circuit and inferred effects on cloud microphysics, temperature and dynamics in the troposphere. *Space Science Reviews, 94*, 231–258.

Chapter 23

The Moon

Harald Hiesinger
Institut für Planetologie, Westfälische Wilhelms-Universität, Münster, Germany

Ralf Jaumann
Deutsches Zentrum für Luft- und Raumfahrt (DLR), Berlin, Germany, Freie Universität Berlin, Institut für Geologische Wissenschaften, Berlin, Germany

Chapter Outline

1. Introduction — 493
2. The Orbit of the Moon — 495
   2.1. General Characteristics — 495
   2.2. Lunar Phases and Eclipses — 495
   2.3. Librations and Tidal Forces — 496
3. Physical Properties — 496
4. Origin of the Moon — 496
5. The Magma Ocean — 497
6. The Interior of the Moon — 499
   6.1. Seismicity — 500
   6.2. Gravity — 500
   6.3. Magnetic Field — 501
   6.4. Thermal Evolution — 502
7. The Lunar Crust and Lunar Terranes — 503
8. Lunar Rocks — 505
   8.1. Highland Rocks — 506
   8.2. The Magnesian Suite — 507
   8.3. Mare Basalts — 507
   8.4. Impact Breccias — 507
9. Surface of the Moon — 508
   9.1. Impact Processes — 508
   9.2. Volcanic Processes — 509
   9.3. Tectonic Processes — 511
   9.4. Space Weathering, Erosion, and Degradation — 512
10. Lunar Stratigraphy and Surface Ages — 513
    10.1. Lunar Stratigraphy — 513
    10.2. Late Heavy Bombardment — 513
    10.3. Age Determination — 515
    10.4. Age Record of Lunar Maria — 515
11. The Regolith — 517
    11.1. Regolith Thickness — 517
    11.2. Regolith Properties and Sources — 518
12. The Apollo and Luna Landing Sites — 518
    12.1. Overview — 518
    12.2. Apollo 11 (July 1969) — 519
    12.3. Apollo 12 (November 1969) — 520
    12.4. Apollo 14 (January–February 1971) — 522
    12.5. Apollo 15 (July–August 1971) — 525
    12.6. Apollo 16 (May 1972) — 525
    12.7. Apollo 17 (December 1972) — 526
    12.8. Luna 16 (September 1970) — 527
    12.9. Luna 20 (February 1972) — 529
    12.10. Luna 24 (August 1976) — 529
    12.11. Luna 17/Lunokhod 1 (November 1970–October 1971) — 531
    12.12. Luna21/Lunokhod 2 (January–June 1973) — 532
13. Significance of Landing Sites for the Interpretation of Global Data Sets — 533
14. Lunar Volatiles — 536
15. Lunar Atmosphere and Environment — 537
Acknowledgement — 538
Bibliography — 538

1. INTRODUCTION

The Moon is the single most important planetary body for understanding not only Earth but also all **terrestrial planets** in our solar system in terms of planetary processes. Building on earlier telescopic observations, our knowledge about the Moon was drastically expanded by the wealth of information provided by Apollo and other missions of the 1960 and 1970s, as well as several recent space missions, including Smart-1, Chandrayaan, Chang'e 1-3, the Selenological and Engineering Explorer (SELENE (Kaguya)), the Lunar Reconnaissance Orbiter (LRO), the Lunar Crater Observation and Sensing Satellite (LCROSS), the Gravity Recovery and Interior Laboratory (GRAIL), and the Lunar Atmosphere and Dust Environment Explorer (LADEE).

Telescopic observations from the Earth, observations by spacecraft from lunar **orbit**, measurements on the lunar surface by manned and unmanned landed missions, the analyses of lunar samples including those returned by the Apollo and Luna missions, and lunar **meteorites** found on the Earth are important sources of information that were used to develop and test various hypotheses and models that were subsequently applied to other terrestrial bodies. Remote sensing using passive sensors in the optical wavelength range, as well as active laser and radar measurements supported by in situ field work, allow the entire lunar history to be studied with respect to its impact-related, volcanic, tectonic, and space weathering evolution. Spectral measurements in all wavelength ranges from high-energy **gamma rays** to mid-infrared provide the overall surface composition and major mineralogical content of **rocks**, whereas geochemical details and the origin, **differentiation**, and evolution of the lunar rocks have been deduced from the returned samples and geophysical data. The vast amount of knowledge gained from samples brought to the Earth by the Apollo and Luna missions and the lunar meteorites and the in situ geophysical measurements made by the Apollo Lunar Surface Experiment Packages demonstrate how valuable the Moon is for understanding our solar system. Today, the Moon is unique in that we have a rich data set for geology, geochemistry, mineralogy, petrology, chronology, and internal structure that is unequaled for any planetary body other than the Earth. These data are crucial for understanding planetary surface processes and the geologic evolution of a planet. They are also essential for linking these processes with internal and thermal evolution. Because the Moon's surface has not been affected by plate recycling, an atmosphere, water, or living organisms, the Moon recorded and preserved evidence for geologic processes such as impact cratering, magmatism, and tectonism that were active over the past 4.5 Ga and offers us the unique opportunity to look back into geologic times for which evidence on Earth has already been erased. Thus, the Moon is the Rosetta stone for understanding the fundamental processes that drive planetary formation and evolution.

With respect to impact processes, the Moon allows us to study, for example, the depths of excavation, the role of oblique impact, modification stages, composition and production of **impact melt**, **ejecta** emplacement dynamics, and the role of **volatile** element addition. The lunar samples returned from known geological units provide the calibration of crater size—frequency distribution chronologies for the entire solar system. These data are important for further understanding the importance of impact cratering in shaping planetary **crusts**, particularly early in solar system history. For example, crater counts indicate that the impactor flux was much higher in the early history of the Moon, the period of the "heavy bombardment", which lasted until ∼3.8 Ga ago.

The Moon also allows us to study planetary magmatic activity in an unobscured form, that is, without the influence of plate tectonics, an atmosphere and associated erosion, and biological activity. We have detailed knowledge of many aspects of lunar volcanism (extrusion) and to a lesser amount of plutonism (**intrusion**) and can assess the role of magmatism as a major crust building and resurfacing process throughout the Moon's geologic history. For example, the ages, distribution, and volumes of volcanic materials indicate the distribution of **mantle** melting processes in space and time. In addition, the detailed magmatic record coupled with the samples permit an assessment of the processes in a manner that can be used to infer similar processes on other planets. The sample-based and remote sensing data reveal the role of magmatic activity during the heavy bombardment (intrusion, extrusion) and more recently in lunar history, the **mare** stratigraphic record, the distribution of **basalt** types, and the implied spatial and temporal distribution of mantle melting. Stratigraphic information and crater ages also provide an emerging picture of volcanic volumes and fluxes. In addition, the Moon allows us to assess a wide range of eruption styles, including **pyroclastics**, and their petrogenetic significance, owing to the existence and preservation of a diverse suite of volcanic deposits.

The Moon is also a type locality for tectonic activity on a one-plate planet. Tectonic processes and tectonic activity can be understood in the context of the complete lunar data set, including the internal structure and thermal evolution. **Graben** illustrate deformation associated with **mascon** loading, and **wrinkle ridges** appear to document the change in the net state of **stress** in the **lithosphere** from initially extensional to contractional in early lunar history. Finally small-scale scarps formed by thrust faulting probably indicate a global change in diameter and can be linked to the lunar thermal evolution, i.e., the cooling of the planetary body.

In summary, the Moon is a complex differentiated planetary object and much remains to be explored and discovered, especially regarding the origin of the Moon, the history of the Earth—Moon system, and processes that have operated in the inner solar system over the past 4.5 Ga. The Moon remains an extremely interesting target scientifically and technologically because although the current data have helped to address some of our questions about the Earth—Moon system, many questions remain unanswered. Returning to the Moon is therefore the critical next stepping-stone to further exploration and understanding of our planetary neighborhood.

In this chapter, we first discuss the astronomical aspects of the Moon, including its orbit, eclipses, and **librations** (Section 2). We briefly review the Moon's physical properties (Section 3), followed by the description of the latest understanding of the origin of the Moon (Section 4), and the formation of the magma ocean and its related processes (Section 5). In the succeeding sections, we start with the interior and move outward to the exterior (i.e., the surface), ending with a description of the lunar atmosphere/exosphere. For example, Section 6 deals with the interior of the Moon, specifically seismicity, gravity, magnetic field, and the thermal evolution, while Section 7 describes the lunar crust and its terranes. In Section 8, we provide a review of lunar rock types, including highland rocks, the magnesian-suite rocks, mare basalts, and impact **breccias**. In Section 9 we describe the lunar surface, which is characterized by impacts, volcanism, tectonism, space weathering, erosion, and degradation. Section 10 provides information on the stratigraphy and surface ages. Here, we introduce the reader to the lunar stratigraphy, the **late heavy bombardment (LHB)**, the methods of age determination, and the age record of lunar maria. The thickness of the **regolith**, as well as its properties and sources, are discussed in Section 11. Section 12 provides detailed information on the geology of the Apollo and Luna landing sites and Section 13 discusses the significance of the landing sites for the interpretation of global data sets. The recent identification of lunar volatiles in samples and remote sensing data merits a separate section (Section 14). Finally, the lunar atmosphere and environment are presented in Section 15. In the bibliography section, we point the reader to seminal papers on various aspects of lunar research.

2. THE ORBIT OF THE MOON

2.1. General Characteristics

As seen from a north polar orientation, the Moon revolves counterclockwise around the Earth, as the Earth and the other planets do around the Sun. Moving at a velocity of 1.03 km/s along a slightly elliptical orbit ($e = 0.0549$), it takes the Moon on average 27.32166 days to revolve around the Earth (sidereal period). Because the Earth also moves along its own orbit, the synodical period of the Moon, i.e., the time between successive new moons, is longer (29.5306 days). The lunar rotation slowed down early in its history owing to tidal friction associated with deformations and became locked into a synchronous state, orienting the Moon such that the same hemisphere faces the Earth at all times. Besides locking the Moon into a **synchronous orbit**, tidal effects cause the Moon to recede from Earth at 3.79 cm/year. With respect to the ecliptic, the orbit of the Moon is inclined by 5.15°; however, the orientation of its rotation axis is tilted only 1.53° with respect to the ecliptic pole (Figure 23.1). The **inclination** of the lunar orbit to the equatorial plane of the Earth varies between 18.4° and 28.6°. On average, the Moon is 384,400 km from the Earth, which corresponds to about 60 Earth radii. However, due to the **eccentricity** of its orbit, this distance varies from 363,000 km at **perigee** to 406,000 km at **apogee**.

2.2. Lunar Phases and Eclipses

Over the course of a month, the appearance of the Moon undergoes regular changes, the so-called lunar phases. These phases are caused by the amount of illuminated lunar surface visible from Earth and thus depend on the relative orientation of the Sun, the Earth, and the Moon. Starting from new Moon when the Moon is not visible in the sky, the Moon appears to grow (wax) every night and is visible as a crescent. One week after new Moon, half of the lunar disk is visible, known as quarter Moon. During the next week, the Moon goes through the gibbous waxing phase. Two weeks after new Moon, the Moon is completely illuminated (full Moon) because the Sun and the Moon are in opposite directions from the Earth. During the next 2 weeks, the Moon appears to shrink (wane), passing through the gibbous, quarter, and crescent waning phases until it eventually becomes new again.

In rare occasions, the Sun, the Earth, and the Moon line up precisely to cause eclipses. If the Moon and the Sun are exactly in opposite directions, the Earth's shadow darkens the Moon to cause a lunar eclipse. Such lunar eclipses usually last for less than 100 min, occur typically twice a year, and can either cover the entire Moon or parts of it. If the Moon and the Sun are exactly in the same direction, the Moon casts a shadow onto the Earth, causing a solar

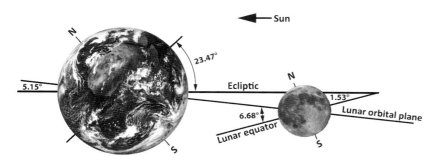

FIGURE 23.1 The orbit of the Moon. *From Hiesinger and Jaumann (2014).*

eclipse. Such solar eclipses are possible because the angular diameter of the Moon (29.3–34.1′) is close to the angular diameter of the Sun (31.6–32.7′) as seen from Earth. Consequently, the Moon can block our view of the Sun either completely or partially, resulting in total, partial, or annular eclipses.

2.3. Librations and Tidal Forces

The Moon is in **synchronous rotation** with respect to the Earth, i.e., it is in a 1:1 spin–orbit coupling. However, because of the orbital eccentricity of $e = 0.0549$ and the resulting variation of the Moon's orbital angular velocity along its elliptic trajectory around the Earth, exact synchronism is given only at two points along the orbit. Near perigee, the rotational angular velocity lags behind the orbital angular velocity, whereas near apogee, the opposite occurs. This situation causes geometric (sometimes called optical) librations in longitude of about $\pm 8.6°$. The inclination of the Moon's orbit with respect to the ecliptic plane causes latitudinal librations of about $6.9°$. These geometric librations mean that we can actually see about 60% of the lunar surface from the Earth. Geometrical librations cause librational tides that are essential for the dissipation of orbital energy within the Moon. Radial tides also occur owing to the variation of the Earth–Moon distance because of the elliptical orbit.

Constraints on the rate of dissipation by tidal friction in lunar rocks are essential for calculating the long-term evolution of the lunar orbit. Tidal dissipation in the Moon tends to decrease the orbital eccentricity and the **semimajor axis**, whereas tidal torques exerted by the Earth tend to increase the Moon's orbital eccentricity and semimajor axis. Although lunar tides only contribute about 1% to the Moon's orbital acceleration, they do significantly contribute to changes in eccentricity. Better constraints on the current dissipation of tidal energy in the lunar interior are needed to fully understand the present orbital state of the Moon and the long-term orbital evolution of the Earth–Moon system.

The dissipation rate and thus the orbital evolution are closely connected to the thermal state of the lunar interior. The presence of partial melt in the deep lunar interior significantly enhances the dissipation rate and is consistent with the orbital characteristics of the Moon. Important clues on the constitution of the lunar interior can be gained by monitoring tidally induced surface deformation from orbiting and landed spacecraft, and Earth-based, highly precise lunar laser ranging observations. In addition to the Apollo lunar seismic data record, these observations provide useful constraints on mantle elasticity and size, as well as composition and physical state of a lunar **core**. The Moon is subject to physical librations because of the tidal contribution to its permanent equatorial bulge. These tidal effects are periodic variations of the Moon's rotation rate caused by torques exerted by not only the Earth but also other planetary bodies and the Sun. In addition to the free librations that depend only on the ratio of the Moon's moments of inertia, there are various forced librations that do depend on the frequency of the external forcing. Examples of forced periods are the 1-year period caused by the Sun, a 3-year period that is in resonance with the free libration period of 2.89 years, a monthly period caused by the Earth, and a 273-year term that arises from perturbations by Venus. The amplitudes of the physical librations are on the order of several tens of **arc seconds** and are thus significantly smaller than variations in longitude caused by geometrical librations. However, physical librations have a significant effect on the dissipation of energy inside the Moon.

3. PHYSICAL PROPERTIES

Within the solar system, Earth and Moon form a unique celestial system with a common center of mass. Although the large moons of Jupiter and Saturn have similar masses, the satellite/parent ratio is the largest for the Earth-Moon system—with respect to its own mass no other planet has such a large moon in orbit. With a radius of 1738 ± 0.1 km, the Moon is the fifth largest satellite in the solar system and has a surface gravity of 1.622 m/s^2, about 17% of that on Earth. Its surface area is less than one-tenth that of the Earth's surface; its volume is about 2%, and its mass is only 1.2% that of the Earth. The lunar bulk density is 3.344 ± 0.003 g/cm^3 and is considerably smaller than the bulk density of the Earth (5.52 g/cm^3). The moment of inertia is 0.3932 ± 0.0002 and indicates a small increase in density with depth. The spin **angular momentum** of the Earth–Moon system is high compared to that of Mars and Venus, which has been interpreted as evidence for some process or event such as a large impact that spun up the system.

4. ORIGIN OF THE MOON

The origin and formation of the Moon is still debated and is one of the key scientific questions because the associated conditions and processes have profound implications for both the initial thermal state of the Earth and the Moon, and for their subsequent thermal and magmatic evolution. There are also major questions about the specific implications for the Earth–Moon system with respect to the uniqueness of Earth, such as why is the Earth the only terrestrial planet with a large moon and what are the gravitational influences on the origin and evolution of life and on the stability of Earth's climate? Prior to the Apollo program, models attempted to explain the origin of the Moon by **accretion** or condensation along with the Earth as a double planetary system, by fission from a rapidly rotating Earth, or by capture by the Earth of a body that formed elsewhere in the solar system. However, all these models fail to explain modern observations. For example, the coaccretion model is inconsistent with the angular momentum of the two bodies and their contrasting densities. Similarly, the fission model cannot account for the orbital dynamics of the Earth—Moon system and would require unrealistically large tidal forces to extract the Moon. Finally, the capture model proposed that elsewhere in the solar system a cool and relatively undifferentiated Moon was formed with a composition similar to chondritic meteorites and was captured by the Earth's gravity. However, the samples of the Apollo and Luna missions demonstrated that the Moon is not the source of **chondrites** because it is not a primitive body; it is differentiated. Furthermore, dynamical considerations argue against the capture model.

Important constraints for the formation of the Moon are its bulk composition, particularly its depletion in metallic iron (i.e., small core), the low abundances of volatiles, and the high angular momentum of the Earth—Moon system. Such constraints are best explained by a giant collision of the proto-Earth with a large object in early solar system history. This Mars-sized object, like the proto-Earth, must have already differentiated, with the iron and siderophile elements concentrated in the core. Following the collision, most of the dense material was incorporated in the Earth's mantle and core, while the outer portions of the impactor and the crust and upper mantle of the Earth were ejected into Earth orbit where the material reaccreted to form the Moon (Figure 23.2). Much work is currently aimed at understanding the chemical and isotopic **fractionations** that would accompany the giant impact and subsequent accretion of the Moon from a hot disk. Comparison of model results to measured chemical and isotopic signatures of lunar materials is a key area of ongoing work, especially in light of recent determinations of lunar indigenous H contents and stable isotope fractionation (or lack thereof) for volatile elements such as K and Zn.

Although the giant impact hypothesis is widely accepted in the scientific community, it is not perfect. For example, numerical simulations indicate that lunar accretion might be relatively inefficient in that only 10—55% of the ejected material would be available for the formation of the Moon. These simulations imply that much more than one lunar mass must be ejected into orbit by the collision. Furthermore, simulations of lunar accretion predict that the initial inclination of the lunar orbit relative to the Earth's equator had to be about 1°, whereas orbital reconstructions indicate that the initial inclination was at least 10°. In addition to uncertainties concerning the distribution of ejecta in the circumterrestrial disk from the giant impact, there are also uncertainties about the amount of time required for the disk to evolve between the giant impact and the onset of accretion.

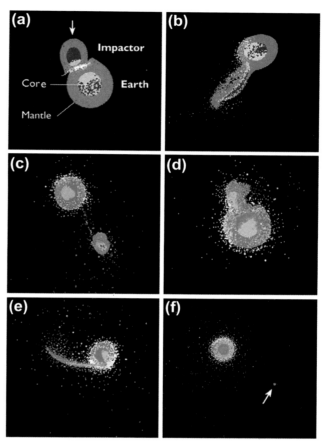

FIGURE 23.2 Mathematical simulation of the giant impact hypothesis. The model demonstrates the feasibility of the formation of the Moon by the oblique collision of a Mars-sized object with proto-Earth (a). The giant impact ejected crustal and mantle material (b—e), which later accreted to the Moon (f, white arrow). © *Alastair Cameron, Harvard-Smithsonian Center for Astrophysics and Sky Publishing.*

In summary, whether the giant impact hypothesis is capable of accurately explaining the origin of the Moon is an area of continuing investigation. However, increasingly sophisticated numerical models demonstrate that a giant impact can eject sufficient mass into Earth orbit, which accretes very quickly to form the Moon, consistent with observed orbital and compositional properties. Hence, these new numerical models strongly support the giant impact hypothesis.

5. THE MAGMA OCEAN

Geochemical evidence suggests that large parts of the Moon were initially molten to form a magma ocean (Figure 23.3) (see Chapter 8). Energy release associated with the giant impact and accretion that formed the Moon produced large-scale melting, accompanied by density segregation of crystals from the melt and the formation of a low-density, plagioclase-rich crust. While the details of the

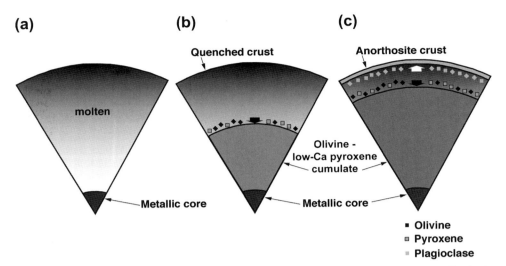

FIGURE 23.3 Schematic cross-section through the magma ocean at three different times. As the molten Moon (a) crystallized, minerals denser than the magma sank to the bottom (b). Eventually, the lighter minerals floated to the top of the magma ocean to form the primary anorthositic crust of the Moon (c). © http://www.psrd.hawaii.edu/Aug11/LMO-crystallization.html.

crystallization of such a magma ocean are still under debate, a likely scenario is that crystallization of olivine and orthopyroxene resulted in **cumulates** in the deeper parts of the magma ocean because they were denser than coexisting melt. As early Fe- and Mg-rich **minerals** crystallized, the melt became richer in Al and Ca, resulting in crystallization of plagioclase after about 75–80% solidification. Plagioclase had a lower density than the melt and thus accumulated upward to form the anorthositic highland crust (Figure 23.3). Rare earth element (REE) patterns of crustal rocks show the effect of crystallization of plagioclase-rich cumulates in their strongly positive Eu anomalies and low trivalent REE concentrations, and corresponding enrichment of REE^{3+} relative to Eu in the late-stage magma ocean melt (Figure 23.4). As the crystallization of the magma ocean continued, KREEP (rocks rich in K, P, and REEs) was eventually produced. The distribution of this KREEP was globally asymmetric as shown by the concentration of thorium on the lunar nearside, as determined from orbital remote sensing. The lunar samples suggest that the crystallization of the main phases of the magma ocean was probably completed by about 4.4 Ga ago and that the KREEP residue was solid at about 4.36 Ga ago (Figure 23.5). In summary, evidence for the existence of a lunar magma ocean and its fractional crystallization comes from the anorthositic flotation crust, the ages of crustal rocks, and trace elements in several suites of rocks, including KREEP formed by high degrees of fractional solidification.

Crystallization models predict that the cumulate mantle that results from solidification of a magma ocean is gravitationally unstable, with later-formed, Fe-rich and therefore denser cumulates overlying less dense, early formed

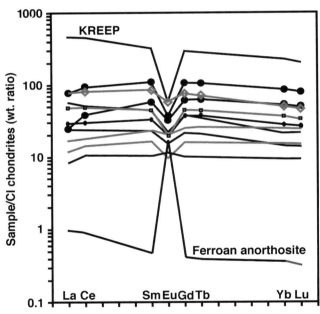

FIGURE 23.4 Chondrite-normalized REE concentrations in mare basalts, KREEP, and a representative ferroan anorthosite. *From Warren. (2003).*

Mg-rich cumulates. Geophysical models predict that the development of such instability and the overturn of the cumulate mantle might deliver cold, dense, incompatible-element-rich material to the core—mantle boundary. In these models, this situation may cause a core flux that was sufficient to drive an early magnetic **dynamo** and may have produced a long-lived dense layer at the core—mantle boundary that presumably did not (and does not) participate

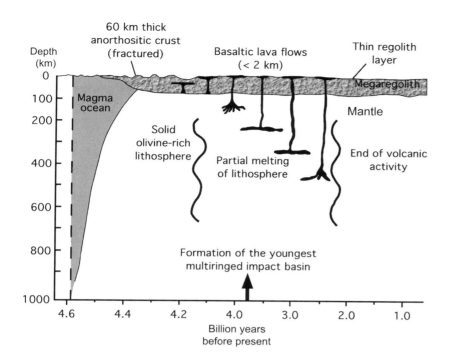

FIGURE 23.5 Evolution of lunar magma ocean, crust, and mantle as a function of time. The simplified geological cross-section of the outer 1000 km of the Moon illustrates the early formation of an olivine-rich residual mantle and plagioclase-rich flotation crust by fractional crystallization of the magma ocean and the simultaneous impact brecciation of the crust to form a megaregolith and a regolith. Later, partial melting of the mantle results in the formation of chemically distinct basaltic lavas due to mineralogical zoning of the mantle established about 4.3 Ga. Finally, volcanic activity ceases while regolith formation due to impacts continues. Timescale not linear. *From Hiesinger, H. & Head, J.W., III. (2006).*

in thermally driven **convection**. Simultaneously, hot rising **diapirs** may have melted **adiabatically** to produce the first basaltic resurfacing crust of the Moon.

6. THE INTERIOR OF THE MOON

The lunar interior is key to better understanding the formation, differentiation, and evolution of the early solar system, in general, and the Earth–Moon system, in particular. Seismic and remotely sensed data, as well as the lunar samples, suggest that the Moon differentiated into a crust, mantle, and possibly a small core (Figure 23.6). The formation of a global low-density crust coupled with a lack of prolonged and significant heating in the interior led to the absence of plate tectonics on the Moon, and a predominantly conductive cooling through the low-density crustal layer. Accordingly, the Moon has a globally continuous lithosphere (crust plus rigid part of mantle), i.e., it is a one-plate planet.

The low mean lunar density of 3344 kg/m^3 indicates that the Moon has only a small iron core compared to that of Earth or any other terrestrial planet. Geophysical data suggest that the central portion of the Moon is molten and might have a solid inner core. The core has a high electrical conductivity and is denser than the overlying mantle. Although the composition of the core is not well constrained, the moment of inertia, the k_2 Love number, the quality factor Q, and lunar rotation and magnetic induction studies are all consistent with a core of 375 km radius that

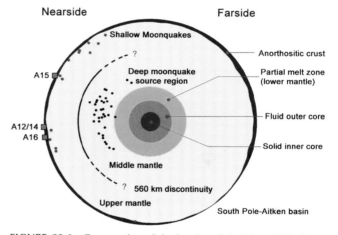

FIGURE 23.6 Cross-section of the interior of the Moon. This figure summarizes information gained by geophysical measurements performed at the landing sites and from orbit. It is believed that the lunar interior is characterized by a solid inner core, a fluid outer core, and a partially molten mantle, superposed by a rigid crust. This crust is substantially thicker on the farside compared to the nearside. Detected moonquake hypocenters (nests) are projected along the 0° meridian. *From Wieczorek, M.A., et al. (2006).*

is composed of Fe–FeS–C metallic alloy or a slightly larger core of molten silicates. Thermal models indicate that at least some portions of the core, regardless of its composition, should have crystallized over the past 4.5 Ga, thus supporting the concept of a solid inner core. Further support for a small lunar core also comes from geochemical considerations, i.e., modeling of lunar siderophile element abundances, which suggests an iron-nickel-rich core of 260–293 km radius. Lunar laser ranging data show that the true spin axis of the Moon is displaced from the Cassini alignment (mean direction of the spin axis) by 0.26 arcsec. A possible interpretation is that internal dissipation in the presence of partially molten rock in the deep mantle and a fluid core causes this displacement. Recent reprocessing of the seismic data supports the presence of a partially molten layer at the base of the mantle and shows the presence of not only a 330-km liquid core but also a small solid inner core.

Global models of lunar topography improved considerably in terms of spatial resolution and accuracy since the availability of maps based on accurate laser altimetry and improved coverage with geometric stereo imaging. Doppler tracking of the Lunar Prospector and SELENE (Kaguya) spacecraft led to global maps of the Moon's gravitational field, as well as information on the mean moment of inertia factor and tidal potential Love number. These data were further improved by several orders of magnitude by the GRAIL spacecraft, which used two orbiters with accurate tracking and knowledge of their motions relative to each other to obtain very high resolution global gravity data.

6.1. Seismicity

To date, the Moon is the only body besides the Earth on which extensive seismological experiments have been done. The Apollo 12, 14, 15, and 16 astronauts installed a network of four three-component, long-period seismometers on the Moon, which detected more than 13,000 seismic events. Although the Apollo seismometers were highly sensitive state-of-the-art instruments and recorded lunar seismic events for almost 8 years, the relatively small network aperture and their restriction to only the nearside limited both the detection capability and the accuracy with which seismic sources could be located. Unfortunately, such location uncertainties are propagated into uncertainties of the interior structure. Nevertheless, the Apollo data revealed that lunar seismicity is different from that of the Earth due to the absence of tectonic plate movement. Sources of seismic activity on the Moon are related to deep moonquakes, shallow moonquakes, and meteoroid impacts.

Deep moonquakes occur in so-called "nests", which repeatedly release seismic energy in specific cycles correlated with lunar tides (Figure 23.6). More than 7000 individual deep moonquakes were recorded, originating from more than 250 of these nests, which are situated at depths between 700 and 1100 km. From the limited data available, neither the spatial nor the temporal patterns of deep lunar quakes are fully understood. However, it is apparent that the activity of these nests correlates with the tidal cycles caused by the Earth and the Sun. The shallow lunar quakes are even more enigmatic. Compared to the deep moonquakes, they are relatively rare; during almost 8 years of continuous data collection, only 28 such events were recorded and their epicenters were not correlated with known geologic or geographic features. One interpretation is that these moonquakes were formed by the release of thermoelastic stresses in the lunar crust or the upper mantle, possibly connected to large impact basins.

While early seismic studies favored crustal thicknesses of about 60 km beneath the Apollo network, more recent studies favor thinner crusts of only 30–45 km. The mean thickness of the lunar crust is estimated at 49 ± 15 km if an Airy-type compensation mechanism applies. These crustal thicknesses are in agreement with estimates of 34–43 km on the basis of GRAIL gravity data and assumed mantle densities of 3220 kg/m^3 and 3150 kg/m^3.

For the upper mantle, the Apollo seismic data indicate almost constant seismic velocities, whereas they are more variable in the underlying middle mantle. At about 560 km below the surface of the Moon, the seismic data suggest a discontinuity in the mantle, which might be related to a compositional change to more aluminous and MgO-rich **mafic** silicates. If this discontinuity exists globally, it may represent the initial depth of melting during the magma ocean phase of the Moon. If it only exists on the nearside, it can be interpreted as the maximum depth of melting of the mare basalt source region beneath the Procellarum KREEP Terrane (PKT). In the lower mantle below a depth of about 1150 km, seismic waves are strongly attenuated, indicating a partially molten zone extending to the core–mantle boundary.

With few exceptions, almost all recorded seismic events within the Moon occurred on the nearside, a **dichotomy** that might be either observational bias or related to geology. Because of the depth distribution of the deep moonquakes, P and S waves do not sample depths below 1200 km and thus the core region remains poorly understood. Among the ~1700 meteoroid impacts detected by Apollo seismometers, only one was at a distance great enough to allow for investigations of the deepest mantle. This observation constrains the lunar core to have a radius between 170 and 360 km.

6.2. Gravity

The gravitational field of the Moon provides important information on the spatial distribution of mass at the surface and within the lunar interior. Gravitational field investigations are commonly based on measurements of the Doppler shift in the radio tracking signal caused by accelerations acting on orbiting spacecraft and are augmented by earthbound lunar laser ranging observations. Early Lunar Orbiter gravitational field models revealed major mass concentrations (mascons) on the lunar nearside, and Apollo 15 and 16 measurements were consistent with a thicker crust on the farside hemisphere as compared to the nearside. The first reliable global characterizations of lunar surface elevations and gravity were based on the Clementine topographic and gravity maps as well as radio tracking of the Lunar Prospector spacecraft. The first reliable gravity field determination for the lunar farside was accomplished by SELENE (Kaguya), which also considerably improved the spatial resolution of the lunar gravitational field data. Recently, the GRAIL spacecraft obtained the highest resolution global gravity data (Figure 23.7).

Gravity data reveal a 1.9-km offset of the center of mass relative to the center of figure of the Moon in the direction toward the Earth. This offset is attributed to a pronounced dichotomy between nearside and farside crustal thicknesses possibly established during the magma ocean stage. Analyses of global gravity and topography data indicate that the thickness of the farside crust exceeds that of the nearside by about 15 km on average. However, the gravity-based results are model dependent; laterally varying crustal compositions are also part of the dichotomy and may contribute to the offset.

Undulations of the current lunar geoid (or selenoid) are larger than 500 m and thus are much rougher than the geoid of the Earth (200 m). The lunar highlands appear to be nearly isostatically compensated, yet basin structures show a wide range of compensation states that are independent of their sizes or ages. Circular maria in many cases display large positive gravity anomalies (mascons) of a few 100 mGal, indicating mass surpluses, whereas some basins are surrounded by concentric gravity lows. Mascons are best explained by superisostatic uplift of the crust—mantle boundary beneath basins and the emplacement of dense mare basalts within basins.

6.3. Magnetic Field

Currently, the Moon does not have an internally generated magnetic field. However, its crust exhibits regions of weak **remanent magnetization** (Figure 23.8). The origin of this remanent magnetization might be related to an ancient internally generated magnetic field, which requires the presence of a highly conductive fluid in the lunar interior and motion within this fluid. Such motion within the core could have been driven by either thermal or chemical convection caused by core freezing. Because these processes depend on the amount of heat extracted from the lunar interior, knowledge of the rate of heat loss places important constraints on models of lunar magnetic field generation. Some thermal history calculations demonstrate that a thermal dynamo could have generated an early magnetic field for about 1 Ga. The generation of a dynamo could also be powered by the **precession** of the lunar spin axis, or by changes in spin rate of the mantle following large impact events. The recent detection of a strongly magnetized mare basalt sample is consistent with a relatively long-lived lunar core dynamo, operating roughly

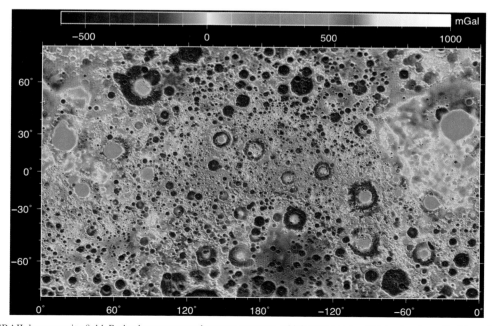

FIGURE 23.7 GRAIL lunar gravity field. Red colors correspond to mass excesses, which create areas of higher local gravity; blue colors represent mass deficits, which result in areas of lower local gravity. © *PIA 16587, NASA/JPL-Caltech/GSFC/MIT.*

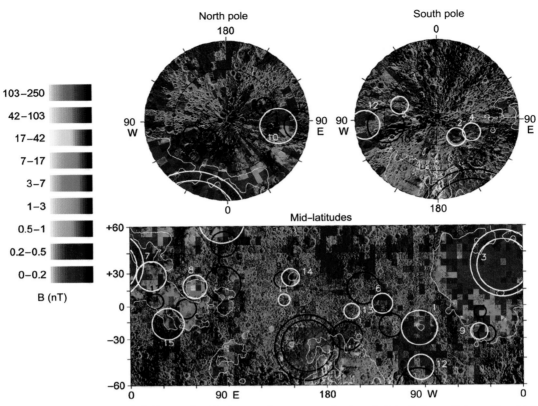

FIGURE 23.8 Lunar Prospector magnetic field strength of the lunar surface superposed onto a shaded relief map. Data are binned into 5 × 5 degree elements at the equator. Thin white line is an optical albedo contour that separates mare from highland regions. White circles mark major lunar impact basins, black circles are antipodal to these basins. Red colors indicate strong magnetization, while purple colors show weakly magnetized regions. *From Mitchell et al., (2008).*

between 4.2 and 3.7 Ga. However, the recent detection of an inner core presents a problem for models of the magnetic history of the Moon. In particular, if the inner core had been growing over an extended period until recently, chemical buoyancy in the core should have extended the lifetime of the dynamo much beyond the lifetime of the thermal dynamo.

Alternatively, an impact-related origin appears plausible. Because the strongest magnetic anomalies are often found at the antipodes of large impact structures, it has been argued that either shock-remanent magnetization or an expanding impact-generated **plasma** cloud could amplify ambient magnetic fields near the antipodes or magnetize the rocks in those locations. The patchiness of the remanent crustal magnetization indicates the presence of strong, localized magnetic fields on a regional scale, which is consistent with transient magnetic fields generated by impact events. The induced lunar magnetic moment as the Moon passes through the geomagnetic tail lobes of the Earth is consistent with a lunar core radius of 340 ± 90 km.

The depth variation of electrical conductivity from which mantle temperature and composition can be inferred hints at a partially molten mantle zone at 800–1500 km depth. This finding is in agreement with the observed attenuation of seismic waves (Figure 23.6).

6.4. Thermal Evolution

Magmatic evolution and lunar volcanism are closely related to the thermal evolution of the Moon, which is linked to the heat released by the decay of radioactive elements in the lunar interior and **secular** cooling, i.e., the loss of **primordial** heat acquired during accretion. A fundamental parameter characterizing the thermal state of the Moon is the amount of heat loss given by the surface **heat flow**. Thermal models predict that the lunar surface cooled mostly by thickening of a lithosphere, while the deep interior cooled by only a few tens of Kelvin. Seismic data from deep moonquakes may indicate the presence of a partially molten layer in the deep mantle (Figure 23.6). Although temperatures in the lunar mantle have also been calculated from magnetic studies, lunar surface heat flow measurements at the Apollo landing sites are still the most direct indicator of the thermal state of the Moon.

From the Earth it is known that heat flow varies widely owing to variations in crustal composition and thickness, tectonic activity, terrane age, climatic effects, and hydrothermal circulation. On the Moon, gamma-ray **spectrometer** data indicate that the distribution of heat sources on the lunar surface is highly non-uniform. In particular, very high K, Th, and U abundances have been identified in a region known as the Procellarum KREEP Terrane (PKT) and a few other isolated areas.

Although the average surface heat flow of the Moon is important for understanding the thermal evolution of the Moon (Figure 23.5), only two heat flow measurements were done successfully during the Apollo missions (15 and 17). Unfortunately, several factors complicate the interpretation of these heat flow measurements. First, both the Apollo 15 and 17 sites are located at the boundary between the feldspathic highlands and the PKT, thus the measured heat flow values are likely not representative of the entire lunar surface. Second, the experimental setup in which a high-conductivity bore stem was inserted into a predrilled hole resulted in considerable shunting of heat and a change of the regolith thermal conductivity due to compaction by the rotary-concussion drill system. Accordingly, the Apollo 15 and 17 heat flow measurements vary widely between 14 and 21 mW/m^2, and this variation might be related to lateral and possibly vertical variations in the concentrations of heat-producing elements. Because of this dependency, the average surface heat flow allows one to estimate the abundances of the heat-producing elements K, Th, and U. On the basis of such calculations, bulk U contents of 20–46 ppb have been estimated. The concentration of 20 ppb is close to the estimated U content of the bulk silicates on the Earth and is thus consistent with the hypothesis that the Moon formed largely from material similar to Earth's mantle excavated during a giant impact into the proto-Earth.

The canonical view of a simplified lunar thermal evolution can be summarized as: (1) a relatively rapid accretion from vaporized and pulverized materials originating both from a Mars-sized impactor and from the proto-Earth itself; (2) development of a magma ocean whose fractional crystallization resulted in the sinking of mafic minerals to form an ultramafic cumulate mantle, and upward accumulation of plagioclase after about 75–80% crystallization to form an anorthositic rigid crust, followed by the late concentration of KREEP materials in the last zones of crystallization, presumably at the base of the crust; (3) overturn of the early formed cumulate pile due to density instability, bringing hot, magnesian cumulates to the top of the mantle; (4) extrusion of melts from this heterogeneous interior facilitated by crustal weakening and excavation due to massive impacts; and (5) generation of impact melts of various compositions dependent mostly on the make-up of target materials (Figure 23.5). Cumulate overturn (3) may have occurred prior to complete solidification of the magma ocean (2), and might have produced some of the hybrid rock suites such as KREEP-bearing, plagioclase-rich magnesian-suite materials.

7. THE LUNAR CRUST AND LUNAR TERRANES

To a first order, **albedo** differences on the Moon reflect major differences in the composition and origin of surface materials. The traditional and most likely too simple view of the lunar surface suggests that the more highly reflective lunar highlands consist mainly of anorthositic rocks and plagioclase-rich magnesian-suite rocks, including norites, gabbronorites, and troctolites. Mg-suite dunites and other magnesian rocks such as recently discovered Mg-spinel **anorthosite** are less abundant. Even rarer in the sample collection are alkali-suite rocks and KREEP basalts, which are enriched in potassium (K), REEs, and phosphorus (P). The dark mare **terrains** consist of younger basaltic rocks, including **lava flows** and pyroclastic deposits (Figure 23.9).

However, orbital gamma-ray spectrometer data allow division of the lunar crust and the underlying mantle into distinct "terranes" that possess unique geochemical, geophysical, and geological characteristics (Figure 23.10). Among those, the Procellarum KREEP Terrane (PKT) and the Feldspathic Highlands Terrane (FHT) are the two most extensive. The FHT is characterized by Th values of less than 3 ppm and covers approximately 60% of the lunar surface. This terrane consists mostly of ancient anorthositic rocks, presumably mainly of the ferroan anorthositic suite. Basaltic flows in the FHT terrane (Figure 23.10) erupted exclusively during the Imbrian period, are volumetrically small, and generally exhibit low to moderate concentrations of TiO$_2$ (<6 wt%). The PKT (Figure 23.10) in the Oceanus Procellarum and Mare Imbrium regions is characterized by high Th concentrations of 3–12 ppm (at the spatial resolution of orbital mapping) and extensive mare basaltic volcanism that erupted from at least 4.2 Ga to about 1 Ga. It has been argued that the high concentrations of heat-producing elements within the PKT are responsible for the longevity of extensive magmatic activity within this region. Global Th mass balance calculations indicate that as much as ~30% of the Moon's heat-producing elements may be located within the PKT, increasing the heat flow and enhancing the rates of viscous relaxation of basins within this terrane. A third terrane has been identified in association with the South Pole–Aitken (SPA) basin (Figure 23.10). The SPA terrane is characterized by FeO abundances of ~6–12 wt% and Th abundances derived from orbital data that are slightly elevated compared to the surrounding FHT (~2–3 ppm versus ~1 ppm). Spectral investigations indicate that the SPA impact mostly sampled the lower noritic crust and that mantle rocks, if excavated at all, are not abundant. Recent petrologic modeling coupled with results from the GRAIL mission suggest the possibility that the exposed interior of SPA basin reflects the upper part of a thick, differentiated impact melt body produced by the SPA impact.

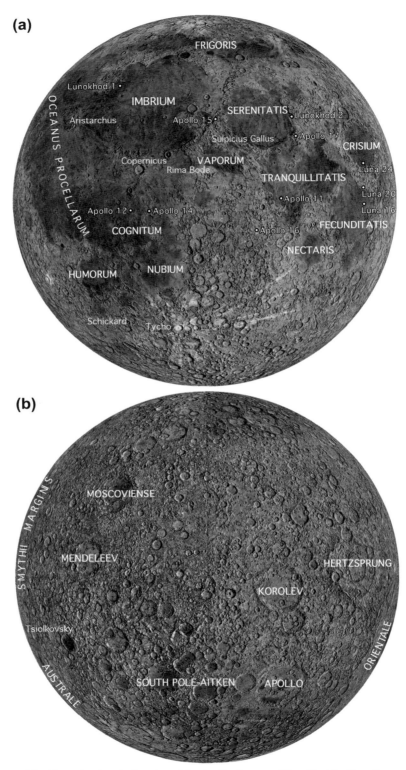

FIGURE 23.9 Global views of the Moon showing the location of the Apollo, Luna, and Lunokhod landing sites, major impact basins, and other prominent morphologic features superposed on a U.S. Geological Survey (USGS) shaded relief map: (a) nearside and (b) farside. *Adapted from Hiesinger, H. & Head, J.W., III. (2006).*

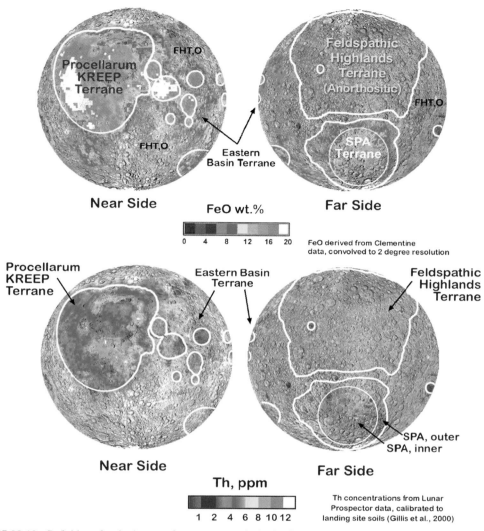
FIGURE 23.10 Definition of major lunar surface terranes on the basis of FeO and Th abundances. *From Jolliff et al. (2001).*

8. LUNAR ROCKS

Information about the mineralogy and chemical composition of the lunar crust has been obtained from the Apollo and Luna samples, and lunar meteorites, as well as telescopic observations and orbital remote sensing. The returned samples provide first-order ground truth because they have enabled detailed studies of the age, mineralogy, chemistry, and petrology of lunar rocks, as well as their physical properties. These samples have allowed remotely sensed data to be properly calibrated, interpreted, and extrapolated to areas from which no samples have been returned.

On the basis of their textures and compositions, lunar samples can be classified into four distinct groups: (1) pristine highland rocks that are primordial igneous rocks; uncontaminated by impact mixing; (2) pristine basaltic volcanic rocks, including lava flows and pyroclastic deposits; (3) **polymict** clastic breccias, impact melt rocks, and thermally metamorphosed granulitic breccias; and (4) lunar regolith and regolith breccias.

As a consequence of the depletion of key volatiles, in particular H_2O and OH^-, the lunar surface mineralogy is relatively simple compared the mineralogy of other planetary bodies such as the Earth, Mars, and presumably Venus. In general, lunar surface mineralogy and petrology are characterized by four main mineral phases: plagioclase, pyroxene(s), olivine, and ilmenite, with other relatively rare silicates (SiO_2 **polymorphs**, K-**feldspar**, etc.), oxides, phosphates, and sulfides as accessory minerals. The relative fractions of these minerals permit classification of lunar rocks into distinct petrological groups, albeit within a continuous compositional **spectrum**. In particular, rocks containing large fractions of plagioclase are classified according to the type and mass fraction of their mafic component, e.g., "troctolite" contains olivine, "norite" contains orthopyroxene, and "gabbro" contains clinopyroxene and orthopyroxene.

8.1. Highland Rocks

Pristine highland rocks can be divided into two chemical groups by their molar $Ca/(Ca + Na + K)$ abundance versus the molar $Mg/(Mg + Fe)$ abundance of their bulk rock compositions (Figure 23.11). Radiometric age determinations indicate that the ferroan anorthosites are extremely old (4.56–4.29 Ga). However, recent results based on more accurate geochronologic analyses call some of the older dates into question. Magnesian-suite rocks (high Mg/Fe) also have very old ages (4.46–4.18 Ga), overlapping with the ferroan anorthosites, but extending to somewhat younger ages. Ages determined on alkali suite rocks are younger still (4.37–3.80 Ga) and include rocks such as alkali anorthosite, norite, gabbronorite, granite, and quartz monzodiorite or monzogabbro. From these age relationships, it appears that the earliest Mg-suite magmatism occurred contemporaneously with the formation of at least some ferroan anorthosites. However, such a contemporaneous crystallization of the ferroan anorthosites and Mg-suite rocks is not consistent with the idealized magma ocean model. In such a model, ferroan anorthosites crystallized first to form the oldest crust, which was later intruded by younger Mg-suite plutonic rocks. Consequently, the complexity and overlap of the age relationship of the two lithologies have been interpreted to support an alternative crust building mechanism to the global magma ocean hypothesis, i.e., serial magmatism. However, massive, highly pure anorthosites in many regions of the lunar surface where fresh rocks from depth are exposed are indicative of a plagioclase accumulation from a magma ocean rather than from serial magmatism. Also, given the violent early impact bombardment and other events such as localized overturn of the mantle and possible concentration of Mg-suite intrusions in the PKT, or at the base of thick crust elsewhere, it appears likely that the crystallization of the magma ocean and distribution of primary crustal materials resulted in a wide variety of rocks now exposed at the lunar surface. A third alternative for explaining the overlapping ages is that some sort of reprocessing, e.g., heating owing to large impacts has modified some of the oldest ages.

Since the early inspection of Apollo samples it has been known that anorthosite is common in the lunar highlands with a plagioclase composition of typically An_{96}. In comparison to plagioclase in terrestrial rocks (e.g., An_{35} to An_{65}), such anorthitic contents are much higher and ultimately reflect the depletion of the Moon in volatile elements such as sodium. The Mg# ($=MgO/MgO + FeO$) of pyroxene and olivine in lunar anorthosite is more ferroan than in terrestrial rocks of such high Ca/Na ratio and any other non-mare lunar rocks, such as impact melt breccias and troctolites (En_{44-76}). For this reason, lunar anorthosite with plutonic or relic plutonic textures has been called ferroan anorthosite. The ferroan anorthositic suite consists of ferroan anorthosites (>90% plagioclase), as well as their more mafic but less common variations, ferroan noritic anorthosite and ferroan anorthositic norite. Although some ferroan anorthosites contain olivine, pyroxene is more abundant. However, geochemically, ferroan anorthosites contain little FeO and incompatible trace elements such as REE, Zr, U, Th, Ni, and Co, compared to other lunar rocks. Despite the fact that the ferroan anorthosites are usually heavily brecciated, a number of samples indicate that they are coarse-grained intrusive igneous rocks, formed during slow cooling at depth below the surface. Because the ferroan anorthosites contain high concentrations of plagioclase feldspar (Figure 23.11), they were interpreted as cumulate rocks produced by the separation and accumulation of crystals from the remaining melt. The Apollo missions returned several large nearly monomineralic plagioclase rocks and outcrops of "pure" anorthosite have been identified in Earth-based and spacecraft observations. However, the perceived importance of ferroan anorthosite to lunar crustal formation might have been overestimated. Although analyses of Apollo samples indicated a key role for ferroan anorthosite in the feldspathic lunar highlands, recent analyses of lunar meteorites suggest that more magnesian anorthosite, which is extremely rare in Apollo samples, might play a more significant role. The petrologic

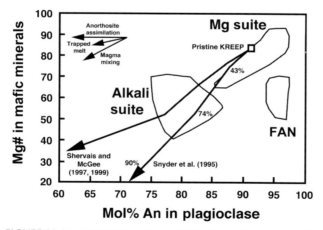

FIGURE 23.11 Composition of lunar highland plutonic rocks: anorthite content (An) in plagioclase versus Mg# of coexisting mafic silicates. *From Shearer, C.K., et al. (2006).*

significance of abundant magnesian anorthosite is not yet understood.

8.2. The Magnesian Suite

Most lunar highland rocks were presumably formed during the differentiation of a magma ocean when buoyant plagioclase accumulated in the upper crust to produce a thick anorthositic crust. Magnesian-suite rocks, once thought to make up a significant proportion of the crust, may instead be more localized products of intrusive activity within the PKT or localized to regions where overturn of mantle cumulates brought magnesian rocks to the upper mantle where melting, mixing with, and assimilation of crustal rocks produced hybrid, magnesian magmas. Some investigations indicate that magnesian-suite rocks may also (or instead) have formed as part of thick differentiated basin-formed impact melt bodies.

8.3. Mare Basalts

Geochemically, mare basalts are rich in FeO and some in TiO_2, depleted in Al_2O_3, and have higher CaO/Al_2O_3 ratios than highland rocks. Mineralogically, they contain more olivine and/or pyroxene, especially clinopyroxene, and less plagioclase than crustal rocks. Mare basalts are interpreted as products of partial melting of mantle cumulates produced during the early lunar differentiation. KREEP basalts differ from mare basalts and are thought to have formed by remelting or assimilation of mantle melts of late-stage magma ocean residua, the so-called urKREEP. Commonly, mare basalts are categorized on the basis of their petrography, mineralogy, and chemistry. Particularly useful is the subdivision using their TiO_2 contents, which distinguishes three mare basalt types: very low Ti basalts (<1.5 wt% TiO_2), low-Ti basalts (1.5–6 wt% TiO_2), and high-Ti basalts (>6 wt% TiO_2) (Figure 23.12). The TiO_2 abundances of the lunar samples have been studied extensively and the application of this knowledge to remote sensing data resulted in the derivation of global maps of the major mineralogy/chemistry of the lunar surface.

An early interpretation of the Apollo and Luna data suggested that Ti-rich basalts were generally older than Ti-poor basalts. Thus, models were proposed in which lunar mare volcanism began with high-TiO_2 content but decreased with time, and that this change was linked to the depth of melting. However, remote sensing data show that young basalts exist with high TiO_2 concentrations. Moreover, there are also some old (mostly >3 Ga) lunar basaltic meteorites that have very low TiO_2 contents. Combining iron and titanium maps determined from orbital data with crater size–frequency distribution ages across the nearside did not reveal a distinct correlation between mare ages and composition. Instead, FeO and TiO_2 concentrations vary

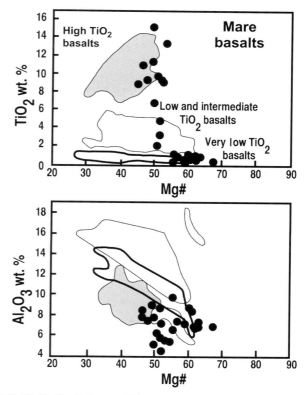

FIGURE 23.12 Mg# versus TiO_2 and Al_2O_3 for crystalline mare basalts and picritic glasses. The fields for high-Ti, low-Ti, and very low Ti basalts were derived from more than 500 analyses. The filled circles represent picritic glasses, thought to represent melt compositions. *Adapted from Papike (1998).*

independently with time, and TiO_2 (FeO)-rich and TiO_2 (FeO)-poor basalts have erupted contemporaneously. Hence, today it appears that the early interpretations were simply biased by the fact that two of the six Apollo landing sites happened to be in areas of high-Ti basalts.

8.4. Impact Breccias

Lunar impact breccias are mixtures of materials from different locations and different rock types and are produced by single or multiple impacts. Breccias contain various amounts of clastic rock fragments and impact-melted materials and show a wide variety of textures, grain sizes, and chemical compositions. On the basis of these characteristics, fragmental, glassy melt, crystalline melt, clast-poor impact melt, granulitic, dimict, and regolith breccias have been discriminated. Fragmental breccias consist of angular clasts in a porous, mostly friable clastic matrix of rocks, minerals, and rare glass debris. Glassy melt breccias are characterized by a coherent glassy or devitrified glass matrix with or without clasts. Crystalline melt breccias exhibit rock or mineral clasts or both in an igneous (extrusive)-textured matrix. Those breccias might be fine or coarse grained, and clast-poor or clast-rich. Melt-poor impact melt breccias are igneous (extrusive)-textured rocks that contain meteoritic siderophile-element contaminations. Granulitic breccias have rock and/or mineral clasts in an equilibrated granoblastic to poikiloblastic matrix. Textures of these breccias are similar to plutonic rocks and compositions reflect siderophile-element contamination. Dimict breccias have a veined texture of intrusive, dark, fine-grained crystalline melt breccias with coarser-grained light-colored breccias consisting of plutonic or metamorphic fragments or both. Finally, regolith breccias can be described as lithified regolith. They contain regolith fragments, including impact glass, volcanic glass, and volcanic debris with glassy matrix.

9. SURFACE OF THE MOON

On a macroscopic scale there are three major geological processes that influence the lunar surface: impacts, volcanism, and tectonics.

9.1. Impact Processes

The Moon has been bombarded by **asteroids** and comets that range over 35 orders of magnitude in mass with impact velocities that in some cases exceed 30 km/s, resulting in impact craters with diameters from micrometers to ~2500 km. The final **morphology** of a crater depends on several factors, including the size of a specific crater, the rheologic properties of the surface, and erosional and degradational processes. Figure 23.13 shows size-dependent morphologies of typical lunar craters from microcraters found in lunar rock samples to large impact basins. On the Moon, fresh simple craters occur at diameters smaller than approximately 18 km and are characterized by (1) a paraboloid (or bowl-shaped) crater interior, (2) sharp rims, and (3) a well-developed ejecta blanket (Figure 23.13). With time, erosion and degradation processes, mostly related to impacts, as well as the deposition of mare lavas result in **mass wasting** and the obliteration of crater rims and ejecta.

With increasing impact energy and crater size, crater morphology becomes more complex. At crater sizes larger than the lunar simple-to-complex transition diameter at ~18 km (with a broader transition from 15 to 25 km), craters are characterized by (1) flat floors and a lower depth-to-diameter ratio compared to simple craters, (2) wall terraces, (3) central peaks, (4) well-developed continuous and discontinuous ejecta, and, (5) solidified impact melt pooled inside the crater and ejected from it (Figure 23.13). The relaxation of the compressed crust produces flat floors and central peaks. Like simple craters, resurfacing processes, which result in partial degradation of rims, central peaks, or ejecta, and partial or even complete filling by later basaltic lavas also influence the final

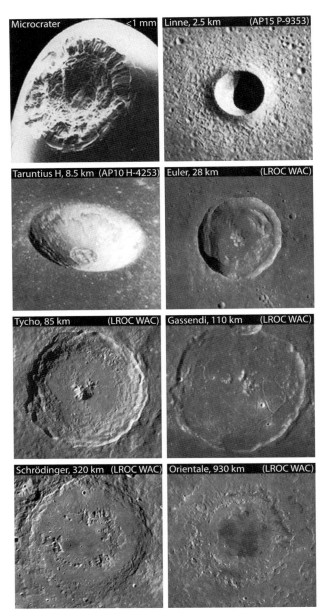

FIGURE 23.13 Size-dependent morphology of lunar impact craters. Microcrater on a lunar glass spherule. Craters Linne and Taruntius H are simple, bowl-shaped craters while craters Euler, Tycho, and Gassendi represent complex craters. Schrödinger and Orientale are multiring basins. For Orientale, the outermost (Cordillera) ring is about 930 km in diameter. All images are LROC Wide Angle Camera (WAC) images with the exception of the microcrater, Linne and Taruntius H. The latter two were imaged by the cameras on the Apollo 15 and 10 command modules. *From Jaumann et al. (2012).*

complex crater morphologies (Figure 23.13). Craters younger than about 1 Ga often show bright, radial ejecta rays, which can extend up to several hundreds of kilometers across the lunar surface. These bright rays are either caused by differences in maturity between the ray and surrounding material or reflect compositional differences.

At diameters larger than 100 km, craters show a transition from central peaks to peak-rings. Two or more concentric ridges or scarps facing inward toward the crater center are characteristic of very large craters, i.e., basins that are larger than approximately 300 km (Figure 23.13). Peak-ring and **multiring basins** were formed by massive impact events during the first ~800 Ma of lunar history. While an early interpretation of Apollo radiometric ages suggested that many of these basins might have formed during a spike in the cratering rate around 3.9–4.0 Ga ago (i.e., the LHB or cataclysm), there is mounting evidence that this interpretation was influenced by a sampling bias (see Chapter 10.2).

9.2. Volcanic Processes

Because the Moon lacks plate tectonics it has preserved evidence of its long volcanic history. Lunar basalts occur preferentially on the nearside and in the interiors of many low-lying impact basins and cover about 17% or 7×10^6 km^2 of the total lunar surface amounting to ~1% of the lunar crustal volume (Figure 23.14). High volumes of low-**viscosity**, high-temperature basaltic lava flooded the topographically low-lying impact basin interiors and reshaped approximately 30% of the lunar nearside hemisphere. The radioactive decay of mainly K, U, and Th resulted in partial melts of the ultramafic mantle at depths between ~60 and 500 km, which led to the generation of basaltic magmas. Because of the enlarged concentration of heat-producing elements in the PKT of the nearside hemisphere, volcanism was active longer on the nearside than on the farside. In addition, geophysical models suggest that massive basin-forming impacts stripped away low-density upper crustal materials on the nearside, to allow dense basaltic magmas to erupt onto the surface. Individual volcanic features in areas resurfaced by mare basalts tend to be obscured by (1) the flatness of the deposits of the basalts due to the low viscosity of the lavas and (2) degradation from several billion years of impact cratering. However, within the lunar maria areas, several geomorphologic features of volcanic origin can be observed, including (1) lava flows, (2) sinuous rilles, (3) volcanic domes and cones, and (4) pyroclastic deposits (Figure 23.15). Volcanic vents in the mare regions are rarely observed. However, with the LRO imaging and topography data sets, many more are being found. Lava flow fronts several tens of kilometers long are visible, especially in low-sun and multispectral imaging, but at high resolution, flow fronts are heavily degraded by the impact cratering process (Figure 23.15). Very common in the mare areas are wrinkle ridges of presumably volcanic/tectonic origin. Meandering lava channels (i.e., sinuous rilles) are between a few tens of meters and ~3 km wide, a few kilometers to up to 300 km long, and on average ~100 m deep (Figure 23.15). One such sinuous rille (Hadley Rille) was examined in detail by the Apollo 15 mission. Geophysical modeling of sinuous rilles suggest that they were formed by high-effusion, high-temperature, low-viscosity lavas that thermally eroded into the substrate.

While already recognized, for example, in Apollo 15 images of the wall of Hadley Rille, new high-resolution imaging data reveal extensive layering within the mare basalt deposits. For example, in at least three mare regions, 100- to 150-m large holes or pits in the mare surface, presumably formed by collapse into voids created by subsurface movement of volcanic melt or by tectonism (Figure 23.16), were identified. Similar features were also observed in impact melt ponds of some craters. Within the pit walls,

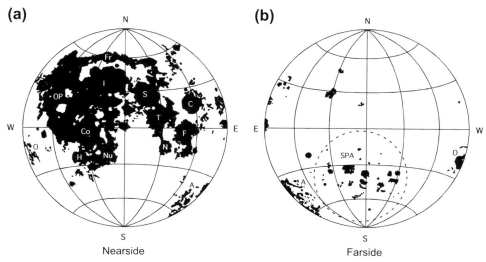

FIGURE 23.14 Map of the highly asymmetric distribution of mare basalts on the lunar nearside (a) and the farside (b) South Pole-Aitken (SPA), Australe (A), Crisium (C), Cognitum (Co), Fecunditatis (F), Frigoris (Fr), Humorum (H), Imbrium (I), Nectaris (N), Nubium (Nu), Orientale (O), Oceanus Procellarum (OP), Serenitatis (S), Tranquillitatis (T). *Adapted from Head (1976).*

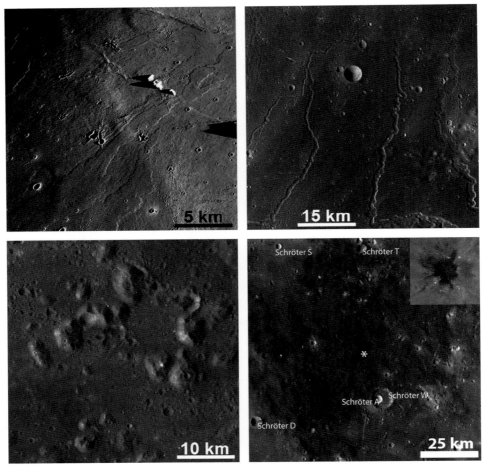

FIGURE 23.15 Volcanic landforms. Basaltic lava flow fronts in Mare Imbrium (top left; AS15-M-1556); sinuous lava channels (Rimae Prinz) in Mare Imbrium (top right, LROC WAC); basaltic domes in Mare Imbrium (bottom left, LROC WAC); and pyroclastic deposits (dark mantle deposits) south of Sinus Aestum; asterisk indicates location of insert, which shows a 170-m-diameter crater that excavated fresh pyroclastic material (bottom right, LROC Wide Angle Camera (WAC) and Narrow Angle Camera (NAC)). *From Jaumann et al. (2012).*

multiple basalt layers of several meters and up to ~15 m thickness were observed. These observations indicate that the basalts in the mare areas were emplaced by multiple thin flow events with individual flows ~10 m thick, rather than by a small number of flows of tens or hundreds of meters thickness. This interpretation is consistent with the Apollo 15 observations of the wall of Hadley Rille and observations of the interior rim walls of several craters.

Dark mantle deposits on the Moon were produced by pyroclastic eruptions, which may cover large areas on the order of >2500 km². Examples of such large pyroclastic deposits include the Rima Bode, Sulpicius Gallus, and Aristarchus Plateau deposits. However, most are concentrated in smaller areas, some only several kilometers in diameter; such deposits comprise the smallest currently known pyroclastic deposits on the Moon.

Mare domes, cones, and shields, which measure 3–17 km across and up to several hundred meters high, are mostly basaltic in composition and occur mostly within the extensive mare areas (Figure 23.15). For example, the Marius Hills complex in Oceanus Procellarum contains more than 100 domes and shields and exhibits a wide range of volcanic materials such as olivine-rich basalts and distinct lava flows. Changes in slope at the flanks of the domes are best explained by changes in viscosity due to effusion rate, temperature, and/or degree of crystallization. Blocks and boulders observed on the domes and cones, and along lava flow margins, are consistent with changes in flow rate or temperature. In addition, evidence in the form of layering in some of the cones suggests a mixture of lava and pyroclastic materials. Spatially much less abundant than landforms associated with basaltic volcanism are domes, which are characterized by steeper slopes, higher albedo, and strong absorption in the ultraviolet (UV) region (Figure 23.15). Known as red spots because of their specific spectral behavior in the UV, they were formed by much more viscous lava of dacitic or rhyolitic composition. In general, volcanic edifices having shallow flank slopes (<10°; e.g., in the

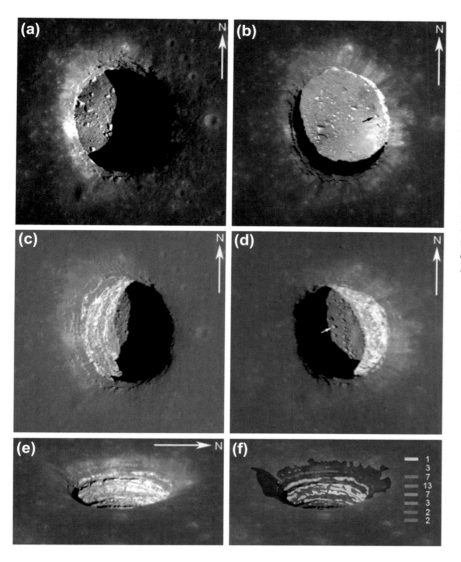

FIGURE 23.16 Pit interpreted as skylight into an underlying lava tube in Mare Tranquillitatis. The topmost images are two LROC NAC near-nadir images with opposite Sun azimuth angles: (a) M126710873R and (b) M155016845R. Together these two images show more than 90% of the pit floor; both images are approximately 175 m wide. (c) Layering in the west wall; portions of the pit floor are beneath the overhanging mare (29° emission angle; M175057326R); (d) a significant portion of the illuminated area is beneath the eastern overhanging mare (26° emission angle; M152662021R); white arrow indicates the same boulder marked with black arrow in panel b. Detailed layering is shown in (e) and (f) (M155023632R and M144395745L, respectively). Outcropping bedrock layer thickness estimates (in meters, ±1 m) are presented in (f). *Robinson et al. (2012).*

Rümker and Marius Hills) were preferentially produced by low-viscosity eruptions, while steeper slopes (>20°) indicate eruptions of more silica-rich, higher viscosity lava (e.g., the Gruithuisen domes) (Figures 23.15, 23.38). Detailed morphologic and spectral studies of non-mare domes, such as the Gruithuisen domes and Hansteen Alpha, and domes in the Compton–Belkovich volcanic complex on the lunar farside, corroborated highly viscous, silicic non-mare volcanism in these areas.

9.3. Tectonic Processes

On the Moon, tectonism is associated with both impacts and volcanism and is manifested by extensional and compressional features, including faults, graben, dikes, and wrinkle ridges (Figure 23.17). Compared to the Earth, landforms created by tectonic stress are less abundant on the Moon. The large surface/volume ratio resulted in a very efficient cooling of the Moon, which rapidly produced a thickening lithosphere that today is about 800–1000 km thick. Early in its history, the Moon formed a stagnant lid from the solidification of the global magma ocean, creating a crust of low density. Hence, the Moon is a so-called one-plate or stagnant-lid planetary body with sluggish convection in its interior and has lost most of its heat through conduction as opposed to volcanism and plate tectonics.

Lunar tectonic features are mainly related to (1) impact-induced stress, (2) stress induced by load of basaltic materials within impact basins, (3) tidal forces, and (4) thermal effects. Impacts induce rock failure decreasing with distance from the impact site, creating radial and/or concentric extensional troughs or graben. At the edges of lava-filled impact basins, extensional stresses develop due to loading with basaltic deposits in the basin interior, creating arcuate

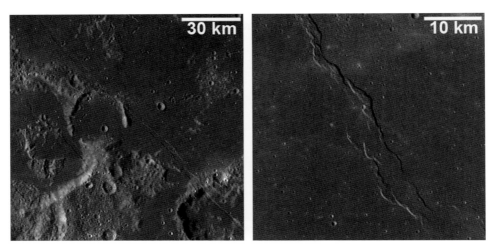

FIGURE 23.17 Tectonic landforms. Extensional graben (Rimae Goclenius) in Mare Fecunditatis (left, LROC WAC) and mare ridges (Dorsa Whiston) in Oceanus Procellarum (right, LROC WAC). © *Jaumann et al. (2012)*.

troughs or rilles. Subradial and concentric ridges toward basin interiors are likely formed by compressional stress associated with the loading and inward movement of basalts. Such wrinkle ridges are common landforms in the mare areas and range from several kilometers up to 10 km in width, tens to hundreds of kilometers in length, and average heights on the order of 100 m (Figure 23.17). The origin of these wrinkle ridges is either related to lava extruding along fissures or to thrust faulting by compression in the interior of a lava-filled basin.

Craters with fractured floors have been explained by isostatic uplift of the floor or the emplacement of sills resulting in an extensional stress field, the latter being the interpretation preferred by most researchers.

On the basis of recent LRO images, a range of tectonic features not readily visible in older images has been observed, including narrow extensional troughs or graben, splay faults, and multiple low-relief terraces. **Lobate scarps** in mare as well as in the highlands, for example, have now been observed globally, including at high latitudes, suggesting a global, late-stage contraction of the Moon. These lobate scarps are among the youngest landforms on the Moon and are interpreted as compressional, low-angle thrust faults.

Earth—Moon gravitational interaction causes tides that are the main source of present-day, deeply seated moonquakes. Deformation by tidal stress was likely more intense early in lunar history when the Moon was closer to the Earth. However, today's spatial distribution of lineaments mapped on the lunar surface is similar on the nearside and farside and does not support an origin by tidal forces, i.e., the collapse of a tidal bulge. Rather, the thermal history, i.e., the shrinking diameter, has influenced the development of compressional stresses and style of tectonic deformation with time. Extensional structures (graben) on the order of 1–2 km long and 500 m wide and superimposing or crosscutting craters only 10 m in diameter are presumably only a few millions of years up to 50 Ma old. The existence of such recent extensional structures is inconsistent with thermal history models that assume a totally molten early Moon because the models predict late-stage compressional stress that would prevent graben formation.

9.4. Space Weathering, Erosion, and Degradation

At small scales, mainly two processes, space weathering and impact "gardening", have reshaped the upper surface of the Moon. The Moon does not possess a global magnetic field and has only a negligible atmosphere, i.e., exosphere, thus the surface regolith is constantly bombarded with (1) micrometeorites, (2) energetic particles of the solar wind, (3) photons of solar radiation, and (4) cosmic rays. This particle bombardment results in physical and chemical processes that are called "space weathering". Space weathering is an important process on airless planetary bodies that is responsible for an increasing maturation of the regolith material by (1) lithification, (2) mechanical comminution, (3) melting and sublimation, (4) formation of agglutinates (partially melted soil **aggregates**), (5) implantation of ions, (6) **sputtering**, and (7) incorporation of meteoritic material. In some lunar regions with strong local magnetic anomalies, solar wind and cosmic rays may have been deflected, thus decreasing the rates of space weathering. For example, large swirl-like albedo features, such as Reiner Gamma, were likely formed by the partial standoff of the solar wind. The amount of trapped solar wind particles not only is an important maturity parameter of the lunar regolith but also records the history and evolution of the Sun over extended periods of time. In particular, areas where younger lava flows or impact ejecta have protected older regolith material ("paleoregoliths") from space weathering, it may be possible to reconstruct the evolution of the solar wind and thus that of the Sun, and the galactic environment of the solar system, throughout much of solar system history.

On the Moon, micrometeorite impacts produce glasses that weld together small particles to form larger aggregates, termed "agglutinates". These micrometeoroid impacts, in combination with impacts of solar wind particles (primarily protons), also result in an autoreduction of FeO to metallic Fe, which increases the abundance of nanophase iron in a "mature" regolith. Maturity is a quality that is roughly proportional to the time the soil has been exposed to micrometeorite bombardment and the agglutinate abundance and varies widely across the lunar surface. The products of these continuous space weathering processes result in spectral reddening (increased reflectance toward longer wavelengths), a general darkening, and an attenuation of diagnostic absorption features. The longer any given surface material is exposed to space weathering, the more its spectral characteristics will be influenced. With the exception of a few small-scale exposures of bedrock, the lunar regolith covers more or less the entire lunar surface. Thus, for optical remote sensing techniques such as the mapping of TiO_2 and FeO concentrations it is crucially important to understand the optical properties of the lunar regolith and how they might be affected by, for example, grain size and maturity.

Impact gardening results in the abrasion of regolith material by small projectiles (micrometeorites) and the fragmentation of rock-sized material on the surface by centimeter-sized projectiles. As a result, fine-grained regolith consists mostly of poorly sorted silt to sand-sized material with variable amounts of rocky fragments and with average grain sizes of about 60–80 μm. Lunar erosion rates are generally low and depend on the size of the regolith material constituents. For example, a boulder with a mass of ∼1 kg has an average survival time against impact destruction of about 10 Ma.

10. LUNAR STRATIGRAPHY AND SURFACE AGES

Cratered landscapes and impact basins are evidence of the extensive bombardment from space over time by projectiles ranging from micrometeorites to asteroids and/or comets. The lunar surface is ideally suited to study craters and the populations of solar system objects forming them. Using knowledge of the rate by which craters form over time, it is possible to date unsampled geological units by the number and size of craters that accumulated on them with time.

10.1. Lunar Stratigraphy

The lunar history is divided into five chronostratigraphic time periods: Pre-Nectarian, Nectarian, Imbrian, Eratosthenian, and Copernican. The Imbrian Period is further subdivided into the Early and Late Imbrian Epochs by the formation of the Orientale basin at the end of the Early Imbrian Epoch. However, because there are no rock samples, that unambiguously date the Orientale event, this subdivision is based on **stratigraphy** (e.g., superposition of **geologic units**) and on impact crater frequencies. Traditionally, the absolute boundaries of the lunar time periods and epochs differ among various authors because there is no unique solution to assigning a certain age of a given Apollo rock sample to a specific impact event. Commonly, more than one age may be extracted from rocks of a given landing site, and ages may vary widely even within a specific sample. Also, different radiometric systems have different closure characteristics and so date different types of events, e.g., crystallization resetting versus shock resetting. In addition, for the assignment of an absolute age to a specific basin impact, either the youngest age represented in a rock sample or the "peak age" (i.e., the age with the highest occurrence) has been chosen to represent the event. Consequently, estimates of the absolute ages of the lunar basins such as Nectaris or Imbrium and the time boundaries, which are defined by them, vary. Thus, according to the different chronostratigraphic models (Figure 23.18), the pre-Nectarian comprises the time before 3.92 or 4.1 Ga, the Nectarian Period spans the time from 3.92/4.1 to 3.85 or 3.91 Ga, the Imbrian Period lasts from 3.85/3.91–3.2 Ga, the Eratosthenian Period lasts from 3.2 to 1.0 or 0.8 Ga, and the Copernican Period comprises the time since 1.0/0.8 Ga. New cratering model ages of the young lunar craters Copernicus and Tycho derived from crater counts on LROC images are in excellent agreement with exposure ages of material collected at the Apollo 12 and 17 landing sites and are consistent with a beginning of the Copernican Period about 800 Ma ago.

Recent improvements in our understanding of the lunar sample collection, the applied dating methods, as well as, for example, a new accurate Ar decay constant have resulted in significantly better constraints of absolute ages assigned to the chronostratigraphic time periods. For example, one key recent result is the determination of 3.91 Ga formation ages of rocks that almost certainly formed during the Imbrium impact. These ages, coupled with required modifications to all Ar–Ar ages derived with the old decay constant, are critical for an accurate absolute chronostratigraphy of the Moon.

10.2. Late Heavy Bombardment

Apollo rock samples show an apparent clustering of radiometric ages at about 3.8–3.9 Ga. This clustering was interpreted as indicating a spike in the impact rate of large bolides around that time, known as the terminal lunar cataclysm or the Late Heavy Bombardment (LHB). Alternatively an exponential decay in the impact rate since the earliest times until about 3.3–3.5 Ga ago has been

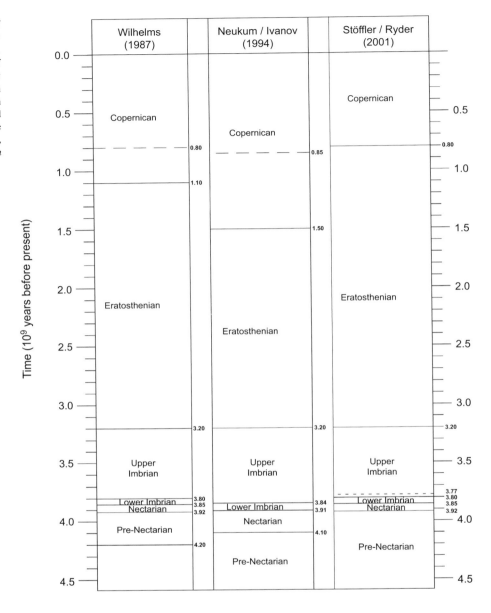

FIGURE 23.18 The stratigraphies of Wilhelms (1987), Neukum and Ivanov (1994), and Stöffler and Ryder (2001). Dashed lines in the stratigraphies of Wilhelms (1987) and Neukum and Ivanov (1994) indicate radiometric ages, which these authors attributed to the formation of the crater Copernicus. In Stöffler and Ryder (2001), two formation ages for the Imbrium basin were discussed, i.e., 3.85 Ga and 3.77 Ga (dashed line). *From Hiesinger et al. (2003).*

proposed. The so-called Nice model was initially developed by researchers at the Observatoire de la Côte d'Azur to test the plausibility of the LHB hypothesis. In order to explain an intense bombardment of the inner solar system around 3.9 Ga, the Nice model required dramatic changes in the orbits of the giant planets to cause a disruption of the primordial asteroid belt, which resulted in the ejection of many of those bodies, some out of the solar system and some into the inner solar system. Although the Nice model posits that an LHB is feasible under certain assumptions and also provides a plausible explanation of the evolution of the asteroid belt, it neither proves nor disproves this hypothesis. Hence, establishing the age of formation of the oldest and largest lunar basin, i.e., the SPA basin and other stratigraphically old basins remains key to testing the cataclysm hypothesis. In addition, careful analyses of ancient clast components of the lunar meteorites are critical to further test the LHB hypothesis. Lunar meteorites are considered to be random samples of the lunar surface and show a broad distribution of ages, thus not necessarily supporting the hypothesis of an LHB. Rather, a majority of the samples collected at the Apollo landing sites might have been influenced by the Imbrium impact, hence showing the observed spike in ages. In addition, attempts to date the SPA basin using impact crater size—frequency distributions have yielded absolute model ages of at least 4.2 Ga, and these ages are inconsistent with some models of the LHB. However, the scientific question whether an LHB existed or not is still under debate and probably requires the return of new samples from areas unaffected by the Imbrium impact, such as from the SPA basin.

10.3. Age Determination

From the Moon we only have a small number of samples that can help to decipher its geologic history and evolution. For example, accurate radiometric ages for mare basalts are available only for the spatially restricted Apollo and Luna landing sites. Owing to the lack of samples and their radiometric ages, most mare basalts remain undated even after the Apollo and Luna missions. However, it is possible to derive relative and absolute model ages for these regions from remote sensing data. In particular, inspection and interpretation of superposition of geologic units onto one another, embayment, and cross-cutting relationships in high-resolution images can be used to obtain relative ages of lunar surface units. Furthermore, crater degradation stages and crater size-frequency distribution measurements (CSFDs) are useful for deriving relative and absolute model ages. Such studies also allow testing dynamical models of impactor populations and thus of the evolution of the solar system.

Age determinations with CSFDs are possible because for a given surface unit, the number of its impact craters correlates with the time this unit was exposed to the bombardment of asteroids and comets: the higher the crater frequency, the older the age of the unit. Thus, the number of craters superposed on a specific surface unit at a given diameter, or range of diameters, is a direct measure of the relative age of the unit. Determining the absolute age of a geologic unit with CSFD measurements is possible because the crater frequencies of key geological units could be calibrated with radiometric and exposure ages of returned Apollo and Luna samples (Figure 23.19). However, there are some uncertainties associated with this method, including the lack of calibration points for the impact flux prior to 3.9 Ga and radiometrically dated surfaces with ages between one and 3 Ga. Furthermore, while the method assumes a homogeneous distribution of impact craters across the lunar surface, there is evidence that the synchronous rotation of the Moon and the preference for asteroidal orbits to lie close to the ecliptic plane, result in spatial variations in the distribution of craters. The absolute chronology of geologic events on the Moon is based on the combination of radiometric and exposure ages of the lunar samples, regional stratigraphic relationships, and crater degradation and size–frequency distribution data. Accurate age estimates, for example, of mare basalts are crucial to constrain the duration and the flux of lunar volcanism, as well as the petrogenesis of mare basalts and its relationship to the thermal evolution of the Moon.

10.4. Age Record of Lunar Maria

Prior to and following the US and Russian lunar missions, the lunar stratigraphy was extensively investigated. This early work indicated that the lunar highlands are generally older than the mare regions; that mare volcanism did not occur within a short time interval, but shows a wide range

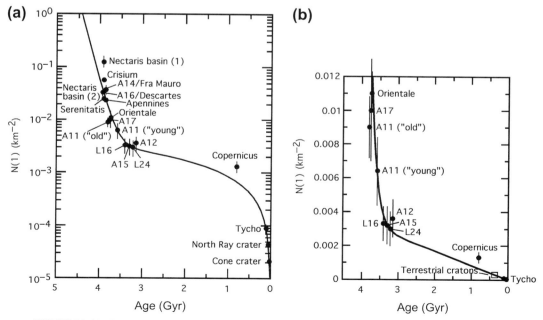

FIGURE 23.19 Lunar cratering chronology in (a) log form and (b) linear form. *From Neukum et al. (2001).*

in ages; and that there is significant variation in the mineralogy and composition of basalts of different ages.

Dating of mare basalts in extensive areas across the lunar nearside using CFSDs has shown that (1) in the investigated basins volcanism was active for almost 3 Ga, starting at about 3.9–4.0 Ga ago and ceasing at ~1.2 Ga ago; (2) most basalts erupted during the Late Imbrian Period at about 3.6–3.8 Ga ago; (3) significantly fewer basalts were emplaced during the Eratosthenian Period (3.2–1.2 Ga); and (4) basalts of possible Copernican age (~1.2 Ga) were only found in limited areas in Oceanus Procellarum (Figure 23.20). These results confirm and extend the general distribution of ages of mare basalt volcanism seen in the samples and show the preferred occurrence of older mare basalts on the eastern and southern nearside and in patches peripheral to the larger mare. In contrast, the younger basalt ages are preferentially located on the western nearside, i.e., in Oceanus Procellarum. Mare volcanism on the central lunar farside exhibits a range of absolute model ages between 3.5 and 2.7 Ga, which is well within the range of ages found for the nearside mare basalts. However, farside mare volcanism ceased earlier than on the nearside, which might be related to either a thicker crust or reduced abundances of radioactive elements in the farside mantle.

CSFDs indicate that the Gruithuisen domes in the northern Oceanus Procellarum region erupted from at least about 3.85–3.7 Ga ago. Therefore, the Gruithuisen domes postdate the Imbrium impact but predate the mare materials in the vicinity, which show ages of 3.55–2.4 Ga. Volcanic features such as the red spots in southern Oceanus Procellarum and Mare Humorum have a wider range in ages compared to the Gruithuisen domes. For example, in the region north of Mare Humorum, red spot light plains associated with a feature named "The Helmet" range in age from 3.94 Ga (Darney χ) to 2.08 Ga. The red spot dome "Hansteen Alpha" has an absolute model age range of 3.74–3.56 Ga, slightly less than the Gruithuisen domes, and postdates craters Billy (3.88 Ga) and Hansteen (3.87 Ga) but predates younger mare materials (3.51 Ga). The ages of the Gruithuisen domes and Hansteen Alpha suggest that at least in these two regions high-silica, viscous, non-mare volcanism was active only in the Late Imbrian Epoch, hence a much shorter interval than mare volcanism.

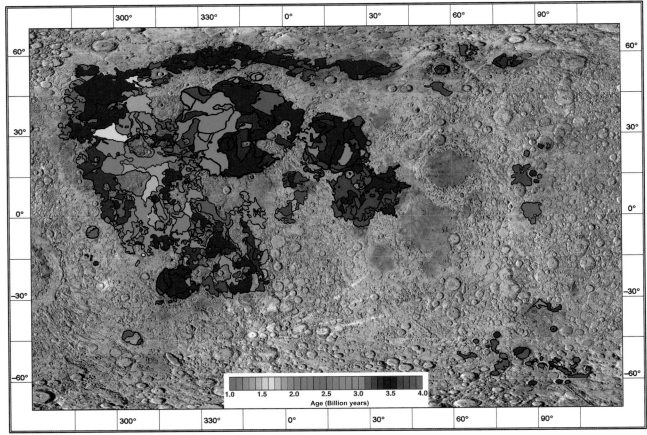

FIGURE 23.20 Ages of mare basalts derived from CSFDs on spectrally and morphologically defined mare units. *From Hiesinger et al. (2012).*

11. THE REGOLITH

The heavy lunar bombardment resulted in fracturing and fragmentation of the lunar crust down to several kilometers and produced a global layer of chaotically mixed impact debris termed the "megaregolith" (Figure 23.21). The fine-grained fraction (mostly <1 cm) of unconsolidated surface material is called "lunar regolith" or "lunar soil".

At the end of the heavy bombardment at ∼3.8–3.9 Ga, the size distribution of impacting projectiles changed significantly and smaller projectiles became increasingly important for the evolution of the lunar surface. These impacts generated a layer of unconsolidated, fine-grained, and poorly sorted material, the lunar regolith. While small-sized impacts homogenized the regolith by continuously turning over the surface layer, larger impacts excavated fresh material from the subsurface resulting in layering of the regolith with numerous discontinuities. Despite the fact that the regolith is mostly produced from the lithologies of the underlying substrate, remotely sensed data of highland/mare boundary regions indicate that lateral transport of material owing to 0.1–10 m large meteorite impacts occurs as well.

The composition of the lunar regolith generally ranges from basaltic to anorthositic (with a small meteoritic component of usually <2%). The regolith consists of fragments of igneous intrusive and extrusive rocks, crystalline impact melt rocks, various types of crystalline and glassy breccias, mineral fragments, glassy and crystallized spherules of volcanic origin, impact glasses, meteorites, meteoritic metal, and agglutinates (see Section 9.4) with an average grain size of ∼60–80 μm.

11.1. Regolith Thickness

On the basis of seismic data and models of the effects of large-scale impacts during the heavy bombardment, the cumulative ejecta thickness on the lunar highlands has been estimated to be on the order of 2.5–10 km, possibly up to tens of kilometers in places. While the average thickness of the mechanically disturbed lunar crust by impacts is not well known, conservative estimates imply a thickness of at least 2–3 km (megaregolith), a structural disturbance to depths of more than 10 km, and fracturing of the in situ crust down to about 25 km. Smaller impacts since the end of the heavy bombardment produced a fine-grained powdery regolith layer on the lunar surface above the megaregolith. This uppermost part is continuously "gardened" or modified by small impacts. Rare large craters play an important role because they can create a mixed regolith layer consisting of preexisting substrate plus fresh ejecta

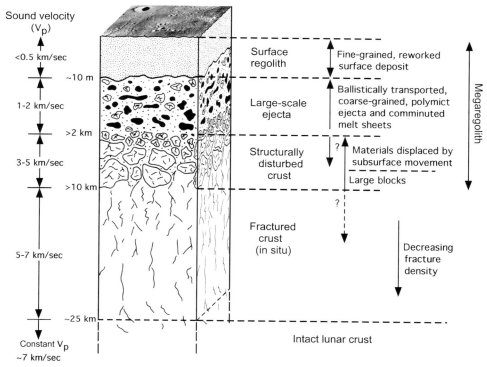

FIGURE 23.21 Idealized cross-section of the upper crust illustrating the internal structure of the megaregolith. The depth scale of this figure is debatable because regional variations are expected depending on the degree to which a region has been influenced by basin-sized impacts. *From Hiesinger, H. & Head, J.W., III. (2006) and Hörz et al. (1991).*

despite the tendency of the smaller craters to homogenize the upper parts of the regolith (Figure 23.21).

Crater morphologies, spectral information, and radar data can be used to estimate the regolith thickness. These studies show that in mare areas the regolith is typically only a few meters thick, whereas in the highlands it is ∼10 m thick (Figure 23.21). Regolith thickness is roughly correlated with surface age and evidence indicates that a higher regolith production rate existed ∼3.5 Ga ago. With time, the production rate of regolith decreased because the increasing regolith thickness protects underlying bedrock from being incorporated into the regolith. In areas with a thick regolith layer, only rare large craters can penetrate the regolith to incorporate "fresh" bedrock, while smaller impacts only redistribute the existing regolith.

11.2. Regolith Properties and Sources

Space weathering processes (Section 9.4) act directly on the lunar surface because the Moon does not have a magnetic field or a thick atmosphere to shield the surface from incoming charged particles and micrometeorites. Thus, the exposure of the surface to intense space weathering results in strong physical and chemical alterations or maturation of the regolith, which involves lithification, mechanical comminution, melting, vaporization, formation of agglutinates, ion implantation, and sputtering. Several products of this maturation are scientifically important and could be economically significant, including trapped solar wind particles to decipher the history of the Sun, cosmogenic nuclei resulting from galactic cosmic rays (and thus the galactic environment of the solar system), ^{3}He as a possible future energy resource, metallic iron, and hydrogen/water.

From a scientific point of view, understanding the space weathering process and its products is essential because space weathering changes the optical properties of the lunar regolith, which must be considered when analyzing remote sensing data (see Chapter 9.4). While the spectral characteristics of the main mineralogical species are well understood from detailed laboratory measurements, remote sensing spectra of the lunar surface in comparison to the laboratory spectra are influenced by several parameters, including topography (e.g., shadowing), illumination angles, grain size, thermal environment, and space weathering. In addition, lunar soils may not be entirely representative of the underlying lithologies because they are produced by impacts and mixing, which have modified their spatial and mineralogical characteristics. Finally, returned lunar samples that are used to calibrate remote sensing instruments offer a rather biased representation of surface materials because they were all collected within just 35° latitude and 95° longitude on the nearside.

12. THE APOLLO AND LUNA LANDING SITES

12.1. Overview

Between 1969 and 1976, six manned American Apollo missions (1969–1972) and three automated Soviet Luna missions (1970–1976) returned samples from the Moon (Figure 23.9). These landing sites are key for understanding

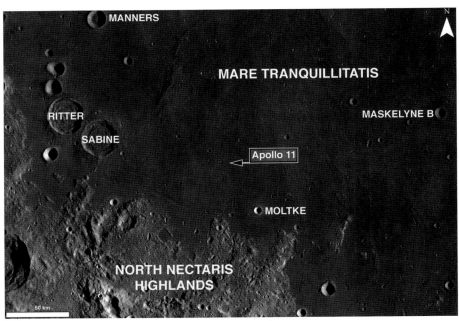

FIGURE 23.22 The geologic setting of the Apollo 11 landing site (LRO WAC mosaic). *From Hiesinger and Jaumann (2014).*

and interpreting the global orbital data sets and provide ground truth for geologic investigations and experiments, data calibration, and new techniques of data reduction.

12.2. Apollo 11 (July 1969)

Apollo 11 was the first lunar mission to return samples of basalt from Mare Tranquillitatis, thereby confirming the hypothesis that extruded lavas, formed by partial melting of the lunar mantle, flooded the lunar impact basins. Compared to terrestrial standards, the returned lunar basalts are characterized by very high TiO_2 concentrations, mostly hosted in the mineral ilmenite. These high TiO_2 abundances are consistent with the spectrally "blue" character of the basalts, due to the interaction of solar irradiation with mineralogical compounds. The ages of the basalts range from 3.57 to 3.88 Ga and provided a first calibration point for the lunar chronology. In addition, they showed that lunar volcanism was ancient by terrestrial standards. In the Apollo 11 sample collection, there are also pieces of non-volcanic material, including fragments of plagioclase-rich anorthosites, and these were interpreted to be from the adjacent highlands. These pieces of highland rocks were particularly important because they indicated the general feldspathic character of the lunar crust and led to the first correct concepts of lunar differentiation. The landing site of Apollo 11 within Mare Tranquillitatis, located about 40–50 km from the nearest mare/highland boundary (0.674158° N, 23.473146° E), was chosen primarily for safety reasons (Figures 23.9, 23.22).

During their stay on the lunar surface, astronauts Neil Armstrong and Edwin "Buzz" Aldrin performed a 2.5-h-long extravehicular activity (EVA), during which they collected 21.6 kg of lunar samples about 400 m west of the West crater, a sharp-rimmed, rayed crater ~180 m in diameter and ~30 m deep (Figure 23.23). The thickness of the regolith at the landing site is about 3–6 m. Detailed geochemical analyses in the laboratory revealed that five

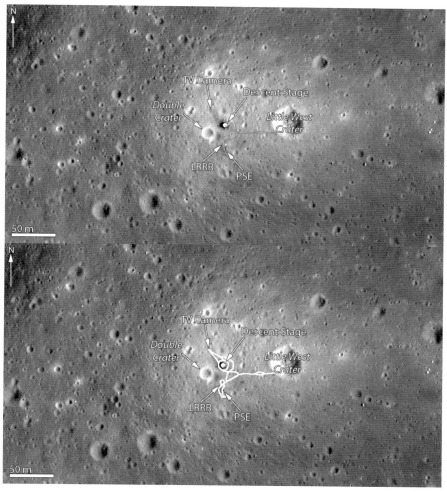

FIGURE 23.23 Detailed view of the Apollo 11 landing site and the location of experiments and traverses (top); detailed view of the EVA traverses (bottom). © *Images from the LROC webpage.*

distinctive groups of mare basalts were sampled, together with regolith samples consisting of feldspathic lithic and mineral fragments that were derived from highland regions. Situated in an extensive mare region, the Apollo 11 regolith nevertheless contains up to ~28% of non-mare material that is similar in composition to the regolith of the Cayley Plains at the Apollo 16 landing site (Section 12.6). It has been suggested that this non-mare material is most likely derived from Imbrium ejecta deposits excavated from beneath the basalt flows by subsequent impact craters.

12.3. Apollo 12 (November 1969)

Apollo 12 with astronauts Charles Conrad and Alan Bean was sent to the Surveyor 3 site, a relatively flat mare site with only a few large boulders to demonstrate the capability of pinpoint landings. This objective was successfully accomplished because Apollo 12 touched down within 200 m of the Surveyor 3 landing site in southeastern Oceanus Procellarum (3.012499° S, 336.578158° E) (Figures 23.9, 23.24). This landing site is less cratered, hence younger than the Apollo 11 site, and the sampled mare basalts differ in their composition and spectral characteristics from the Apollo 11 basalts. The differences in age and chemistry between Apollo 11 and Apollo 12 basalts were interpreted to show variability in mantle sources and basalt production processes on the Moon. Exposures of non-mare materials near the landing site, mostly part of the Fra Mauro Formation, indicate that the basalts are relatively thin, forming a complex topography. Crater counts and nominal crater degradation values or D_L values (defined as the diameter of craters with wall slopes of 1°) demonstrate that the landing site basalts are older than basalts 1 km to the east and west. Although the Apollo 12 landing site is dominated by the ejecta of several >100 m large craters (Figure 23.25), its regolith is only half as thick as the Apollo 11 regolith and craters only 3 m deep penetrate into basaltic bedrock. A prominent ray from crater Copernicus crosses the Apollo 12 location, delivering abundant non-mare materials to the site. There are probably multiple sources of the non-mare materials at the Apollo 12 site, including material mixed into the surface from impact melt formations that underlie the basalts. However, analyses of the non-mare materials show a major disturbance at about 800–900 Ma ago. This disturbance has been interpreted as being related to the emplacement of Copernicus ray material. Thus, this age is inferred to be the age of Copernicus, although no specific sample has been unambiguously related to this crater.

During two EVAs, the astronauts collected 34.3 kg of samples, mostly basalts. The radiometric ages of these basalts range between 3.29 and 3.08 Ga and at least three chemically distinctive groups of mare basalts were identified. The mare basalt flows (20–21% FeO) cover the older Fra Mauro Formation, which contains about 10% FeO. Apollo 12 was the first mission to collect KREEP (rocks and glasses enriched in potassium, rare earth

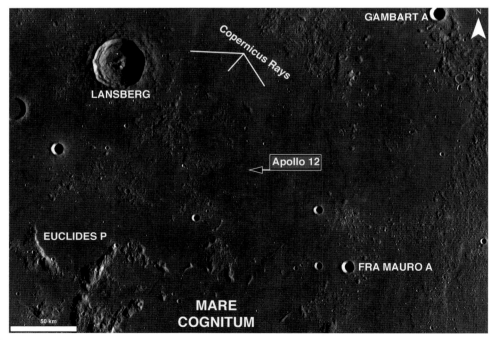

FIGURE 23.24 The geologic setting of the Apollo 12 landing site (LRO WAC mosaic). *From Hiesinger and Jaumann (2014).*

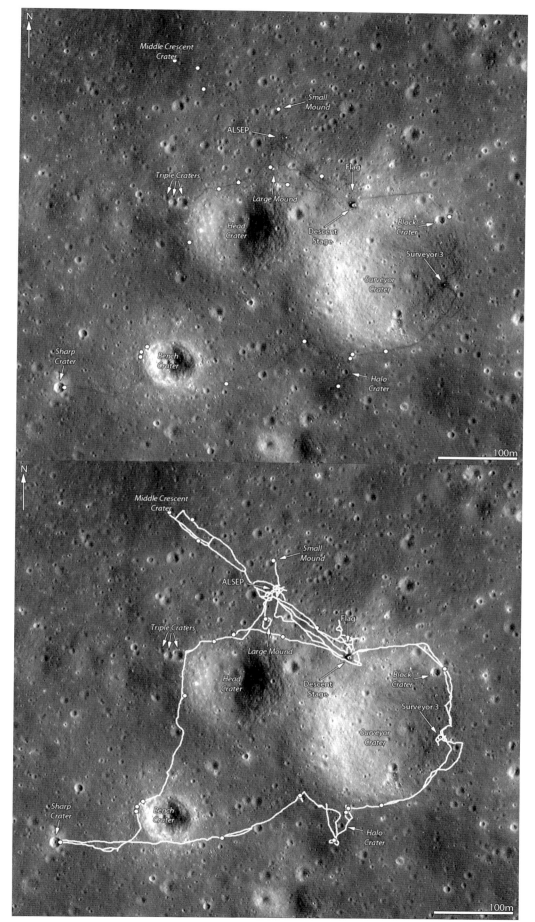

FIGURE 23.25 Detailed view of the Apollo 12 landing site and the location of experiments and traverses (top); detailed view of the EVA traverses (bottom). © *Images from the LROC webpage.*

elements, phosphorus, and other incompatible elements). The very distinct and unique geochemical signature of KREEP requires substantial magmatic differentiation, consistent with a strongly differentiated Moon. KREEP material occurs as dark, ropy glasses similar in composition to Apollo 14 soils and a subset of Apollo 14 impact glasses, and several crystalline rock and lithic fragments. These glasses appear to represent impacted surface material similar to that exposed in the Apollo 14 Fra Mauro Formation. The KREEP material is derived mostly from the PKT and was probably incorporated into the regolith of the landing site by a combination of lateral transport and vertical mixing.

12.4. Apollo 14 (January–February 1971)

By landing on a hummocky terrain north of the Fra Mauro crater (3.645890° S, 342.528050° E) ~550 km south of the Imbrium basin, Apollo 14 became the first mission to sample the lunar "highlands" (Figures 23.9, 23.26). The landing site was selected primarily to sample the Imbrium basin ejecta, which presumably came from deep crustal levels. It was also chosen to date the Imbrium event, which is a major stratigraphic marker in lunar history. During two EVAs, astronauts Alan Shepard and Edgar Mitchell collected 42.3 kg of samples, including complex fragmental breccias, impact melt breccias, and clast-poor impact melts with generally basaltic and KREEP-rich compositions. Radiometric dating revealed that these breccias were formed about 3.9–3.8 Ga ago. However, it is still under debate whether they date the Imbrium event itself, or later, smaller impacts into Imbrium ejecta. Although the Fra Mauro Formation has been interpreted to consist primarily of Imbrium ejecta, some studies suggest that it is mostly dominated by local material with only 15–20% Imbrium ejecta mixed in.

The Apollo 14 landing site is commonly considered a "highland" landing site. However, its elevation is some 3200 m below the lunar mean and, from a geochemical point of view it is neither feldspathic highlands nor mare. Rather, unknown at the time of the site selection, the Apollo 14 site is located within the PKT, which contains exceptionally high thorium concentrations. At the Apollo 14 landing site relatively aluminous basalts have been identified, some of them also enriched in K. These aluminous and high-K basalts are not found at other Apollo mare landing sites. However, one of the Soviet Luna missions, Luna 16, also returned aluminous basalts (Section 12.8). A "model" mixture of the regolith at the Apollo 14 site contains ~60% impact melt breccias, ~20–30% noritic lithologies, ~5–10% each of mare basalts and troctolitic anorthosites, and minor amounts of meteoritic components. Regolith from different stations at Apollo 14 shows little variation in composition, both vertically (tens of centimeters) and laterally (kilometers around the landing site). The landing site is situated ~1100 m west of the ~340-m-diameter and 75-m-deep Cone crater, which ejected blocks of up to ~15 m in size (Figure 23.27). In addition, the landing site

FIGURE 23.26 The geologic setting of the Apollo 14 landing site (LRO WAC mosaic). *From Hiesinger and Jaumann (2014).*

Chapter | 23 The Moon

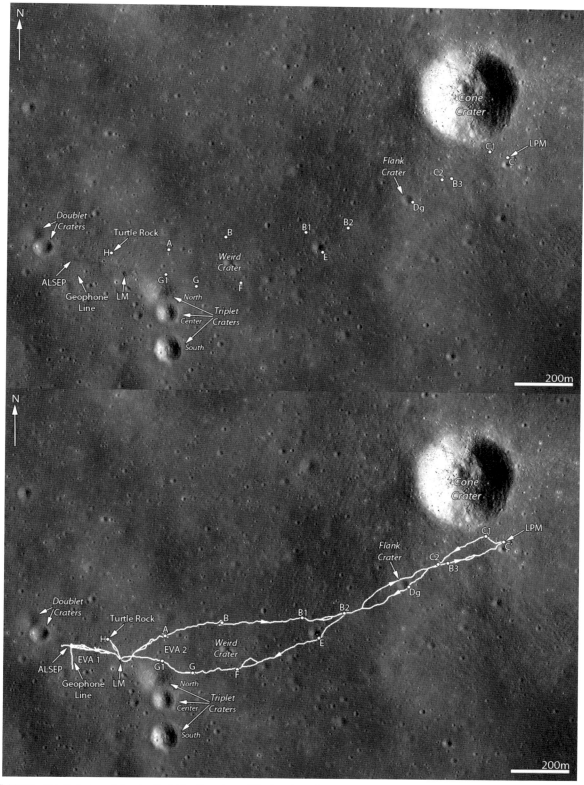

FIGURE 23.27 Detailed view of the Apollo 14 landing site and the location of experiments and traverses (top); detailed view of the EVA traverses (bottom). © *Images from the LROC webpage.*

FIGURE 23.28 The geologic setting of the Apollo 15 landing site (LRO WAC mosaic). *From Hiesinger and Jaumann (2014).*

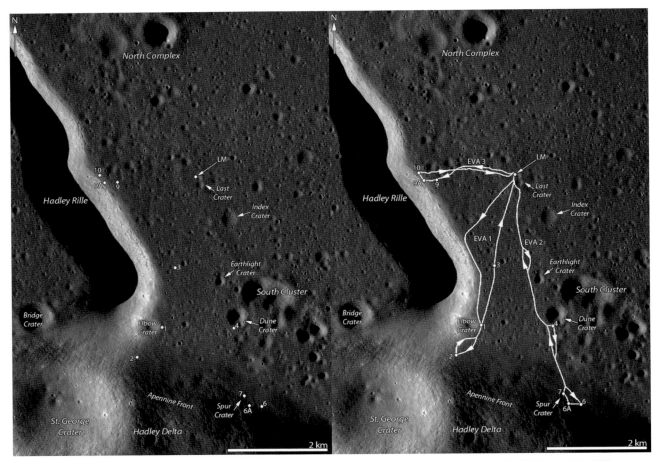

FIGURE 23.29 Detailed view of the Apollo 15 landing site and the location of experiments and traverses (top); detailed view of the EVA traverses (bottom). © *Images from the LROC webpage.*

shows numerous subdued craters up to several hundreds of meters across and the regolith is estimated to be 10–20 m thick.

12.5. Apollo 15 (July–August 1971)

The Apollo 15 mission was the first advanced ("J") mission that carried the Lunar Roving Vehicle (LRV). The complex multiobjective landing site in the Hadley-Apennine region (26.132379° N, 3.633303° E) (Figures 23.9, 23.28) was chosen to sample and study the massifs and highlands of the Imbrium rim, as well as the mare lavas and landforms of Palus Putredinis (e.g., Hadley Rille). The main Imbrium ring rises ~3.5 km above the plains at Hadley Delta, only 4 km south of the Apollo 15 landing site, and extensive lava plains are exposed west of the landing site. The so-called Apennine Bench Formation, a light plains unit that probably underlies the Late Imbrian basalts at the Apollo 15 landing site is exposed within a few kilometers of the landing site and is located inside the Imbrium ring. **Crater rays** from Autolycus and Aristillus cross the landing site. The regolith thickness, which varies widely depending on the local terrain is only ~5 m thick at the landing site, and the wall of Hadley Rille exposes a layered basalt flow sequence. The Apollo 15 astronauts Dave Scott and James Irwin performed three EVAs with the LRV collecting a total of 77.3 kg of mare and non-mare rocks (Figure 23.29). In particular, they collected two types of basalts (quartz normative and olivine normative), which have ages of ~3.3 Ga. Non-volcanic rocks include anorthosites, magnesian-suite plutonic rocks, impact melts, and granulites, many of which occur as individual clasts in regolith breccias. In addition, the Apollo 15 sample collection contains KREEP-rich non-mare basalts and green ultramafic volcanic glasses. Together with Apollo 14 samples, rocks from the Apollo 15 site have been interpreted to indicate that the Imbrium basin is 3.91 Ga old.

12.6. Apollo 16 (May 1972)

The objective of the Apollo 16 mission was to explore the smooth Cayley Formation and the hilly and furrowed Descartes Formation, both of which were initially thought to be volcanic in origin (Figures 23.9, 23.30). Further objectives were to sample highland material at great distance from any mare region and to study two nearby, fresh 1–2 km diameter craters. Apollo 16 landed at 8.973476° S latitude and 15.501019° E longitude near Descartes crater. Three long EVAs were performed with the LRV, and astronauts John Young and Charles Duke returned 95.7 kg of samples (Figure 23.31). The Apollo 16 landing site is the most highland-like landing site of the Apollo program, but surface morphology, vicinity to the Imbrium basin and thus possible presence of PKT material, composition of the regolith, and lithologic components of the regolith compared to those of the feldspathic regolith breccia lunar meteorites argue against Apollo 16 having landed on "typical" highlands.

FIGURE 23.30 The geologic setting of the Apollo 16 landing site (LRO WAC mosaic). *From Hiesinger and Jaumann (2014).*

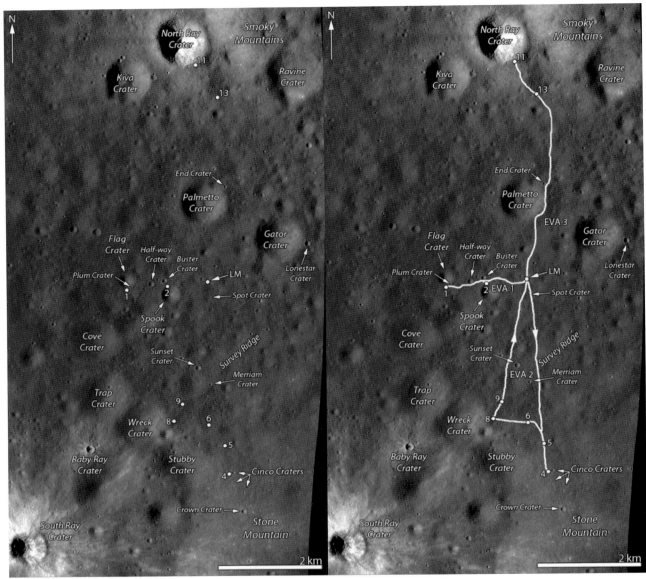

FIGURE 23.31 Detailed view of the Apollo 16 landing site and the location of experiments and traverses (left); detailed view of the EVA traverses (right). © *Images from the LROC webpage.*

The landing site includes numerous overlapping subdued craters ~500 m size; two young fresh craters, North Ray (1 km wide, 230 m deep) and South Ray (680 m wide, 135 m deep); as well as Stone Mountain (Figure 23.31). The regolith thickness on both the Cayley and the Descartes Formations varies from 3 to 15 m and averages about 6–10 m. In contrast to pre-mission interpretations, the Apollo 16 samples are mostly rocks of impact origin. Many are impact melts or fragmental breccias, but there are also some anorthositic rocks. From the returned samples it is apparent that neither the Cayley Formation nor the Descartes Formation is volcanic in origin, but instead both are related to the deposition of ejecta from the Imbrium, Serenitatis, and Nectaris basins. The relative amounts of ejecta contributed to the Apollo 16 site by each basin, however, are still under debate. Some Apollo 16 impact melts have been interpreted as products of the Nectaris event, indicating a formation of the Nectaris basin 3.92 or 4.21 Ga ago.

12.7. Apollo 17 (December 1972)

Apollo 17 was the last mission of the Apollo program and was sent to the highland/mare boundary near the southeastern rim of the Serenitatis basin, the Taurus-Littrow valley (20.190,850° N, 30.772,247° E) (Figures 23.9, 23.32). The goals of Apollo 17 were to examine

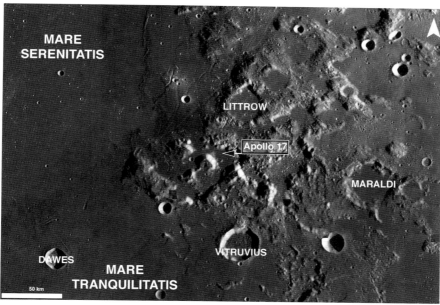

FIGURE 23.32 The geologic setting of the Apollo 17 landing site (LRO WAC mosaic). *From Hiesinger and Jaumann (2014).*

two highland massifs where rocks from deep crustal levels could be collected, to investigate the presumably basaltic valley floor, and to study the low-albedo deposit that mantles both highlands and mare at the site. The Apollo 17 and 15 sites were the most geologically complex landing sites. Using the LRV, astronauts Eugene Cernan and Harrison Schmitt performed three EVAs and collected 110.5 kg of samples, bringing the total amount of samples collected by Apollo missions to 381.7 kg (Figure 23.33). Samples from this site show that the valley floor is covered with TiO_2-rich basalts. Specifically, several chemical subgroups of basalts were identified, ranging in age from 3.8 to 3.7 Ga.

It has been argued that the Apollo 17 impact melts are likely of Serenitatis origin because they were collected from thick ejecta deposits on the Serenitatis basin rim. However, some of the impact melt breccias have trace element signatures similar to some groups from Apollo 15 and Apollo 14, suggesting a minor component from the later Imbrium basin. Deep-seated magnesian-suite rocks and granulitic breccias were also identified in the highland massif regolith.

At the Apollo 17 landing site, the regolith is on the order of 15 m thick. The highlands along this part of the Serenitatis rim are relatively FeO-rich because of abundant mafic impact melt breccias and addition of basaltic material as impact debris and pyroclastic deposits. The compositions of the highland massifs are consistent with mixtures of noritic impact melt and feldspathic granulitic material, plus variable amounts of high-Ti basalts on the flanks at low elevations and pyroclastic deposits at high elevations. In summary, the highlands at the Apollo 17 site consist of complex impact melt breccias and plutonic rocks of the Mg-suite. Impact melt breccias are 3.87 Ga old and have been interpreted to indicate the age of the Serenitatis basin. Orange and black volcanic glass beads of high-Ti basaltic composition are incorporated in the dark mantling deposits, which are approximately 3.64 Ga old. A light-colored mantle unit presumably results from an avalanche at the slope of the South Massif triggered by secondary impacts from the 2200-km-distant Tycho crater (Figure 23.33). According to cosmic ray exposure ages, this landslide is ~100 Ma old and has been related to the age of the Tycho event.

12.8. Luna 16 (September 1970)

The first successful automated Soviet sample return mission, Luna 16, landed in northern Mare Fecunditatis at 0.513° S and 56.364° E (Figures 23.9, 23.34, 23.37). Luna 16 landed on a series of relatively thin basalt flows ~300 m in thickness. The landing site is geologically complex as it incorporates ejecta and ray material from Eratosthenian crater Langrenus (132 km diameter) 220 km to the southeast, Copernican crater Tauruntius (56 km diameter), and craters Theophilus and Tycho. Luna 16 samples were taken from a shallow regolith drill core, which reached a depth of 35 cm and provided 101 g of dark gray regolith with preserved stratigraphy. While no visible layering was recognized in the core, five zones of increasing grain sizes with depth were identified. Luna 16 samples consist of moderately high Ti, high-Al basalt fragments, approximately

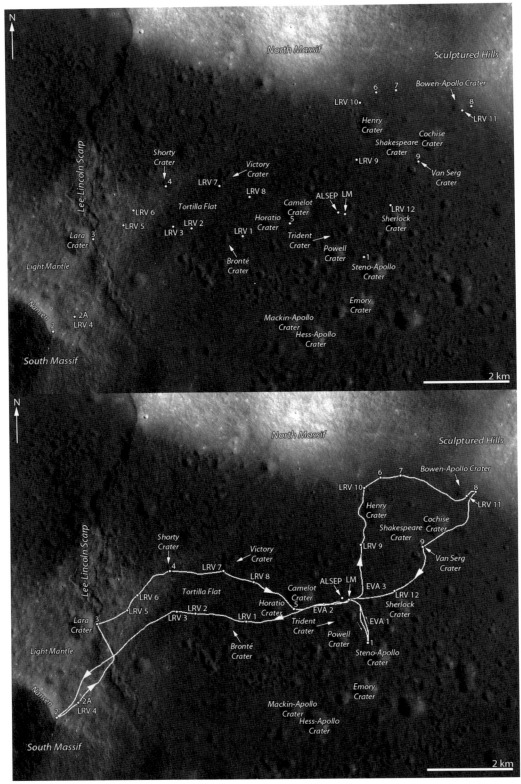

FIGURE 23.33 Detailed view of the Apollo 17 landing site and the location of experiments and traverses (top); detailed view of the EVA traverses (bottom). © *Images from the LROC webpage.*

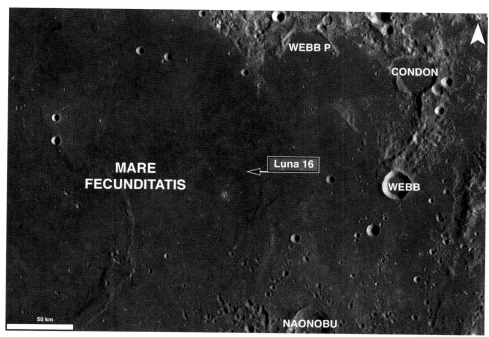

FIGURE 23.34 The geologic setting of the Luna 16 landing site (LRO WAC mosaic). *From Hiesinger and Jaumann (2014).*

3.41 Ga old. While mare basalts from the Luna 16 landing site are among the most Fe- and Mg-poor lunar basalt samples, they are also the most Al-rich basalts ($\sim$13.5% Al_2O_3) yet sampled.

12.9. Luna 20 (February 1972)

Luna 20 landed in the highlands south of Mare Crisium (3.787° N, 56.625° E) and returned samples of the rim of the Nectarian-aged Crisium basin (Figures 23.9, 23.35, 23.37). The Luna 20 landing site is topographically $\sim$1 km above the basaltic surface of Mare Fecunditatis and is influenced by Apollonius C, a 10-Ka fresh Copernican crater located only a few kilometers to the east. Morphologically, the landing site is characterized by smooth rounded hills and shallow linear valleys that give the region a hummocky appearance. Luna 20 returned a drill core that contained $\sim$50 g of fine-grained light gray regolith. While no stratification has been observed within the core, mixing during transport to Earth cannot be excluded. The mission sampled the anorthositic highland regolith (Figures 23.9, 23.35) that contains lithic fragments of granulites, anorthosites, impact melts, and polymict breccias. Most of the lithic regolith fragments are breccias of anorthositic–noritic–troctolitic composition and impact melt rocks of noritic–basaltic composition. The Luna 20 regolith contains less K and P than the regolith at the Apollo 16 landing site, anorthosite fragments are rare, and MgO-rich spinel troctolites are common. Of particular importance are the higher concentrations of MgO because they may reflect the addition of an unknown mafic highland rock material to the Luna 20 site that is absent at the Apollo 16 landing site. Owing to the low K and P concentrations, it has been suggested that the Luna 20 samples have not been contaminated with KREEP-rich Imbrium ejecta. Instead, they may represent middle to lower crustal material ejected by the Crisium impact.

12.10. Luna 24 (August 1976)

Luna 24 was the last and most successful Luna mission. Luna 24 landed in southern Mare Crisium (12.714° N, 62.213° E) (Figures 23.9, 23.36, 23.37), about 40 km north of the main basin ring, which is $\sim$3.5–4.0 km higher than the landing site. The landing site is characterized by wrinkle ridges, and several distinctive basalt types of Mare Crisium were identified by remote sensing techniques. The thickness of these basalts was estimated to be $\sim$1–2 km. Several bright patches and rays imply that non-mare material has been dispersed across the basin by impacts such as Giordano Bruno and Proclus. Luna 24 returned a 1.6-m-long drill core that contained 170 g of mostly fine-grained mare regolith. In contrast to the drill cores of Luna 16 and Luna 20, stratification was well preserved in the Luna 24 core sample. On the basis of color and grain size differences, four layers were

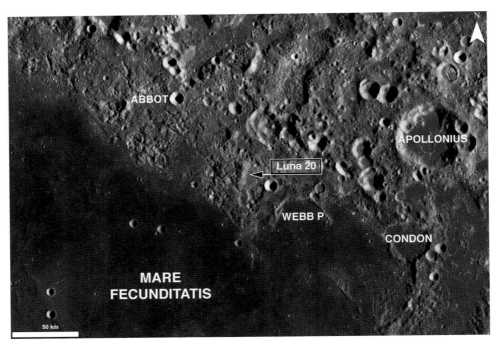

FIGURE 23.35 The geologic setting of the Luna 20 landing site (LRO WAC mosaic). *From Hiesinger and Jaumann (2014).*

FIGURE 23.36 The geologic setting of the Luna 24 landing site (LRO WAC mosaic). *From Hiesinger and Jaumann (2014).*

identified. The basaltic fragments from this core are very low in TiO_2, low in MgO, and high in Al_2O_3 and FeO and are 3.6–3.4 Ga old. Mare basalts and soils of Luna 24 contain similar concentrations of FeO and this observation indicates that only minor amounts of non-mare material occur at the landing site. In the Luna 24 drill core, there is also evidence for about 0.1 wt% of water at a depth of 143 cm. Although the abundance of water appeared to increase with depth, a possible contamination of the sample on Earth could not be excluded.

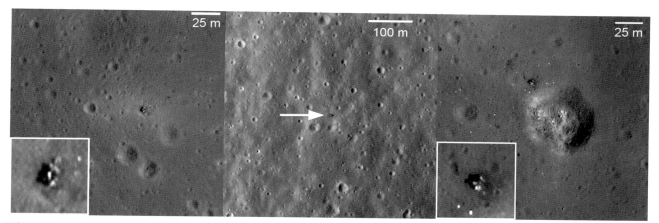

FIGURE 23.37 Detailed views of the Luna 16, 20, and 24 landing sites; (left) Luna 16, insert shows the spacecraft; (center) Luna 20; (right) Luna 24, insert shows the spacecraft. *From Robinson et al. (2012).*

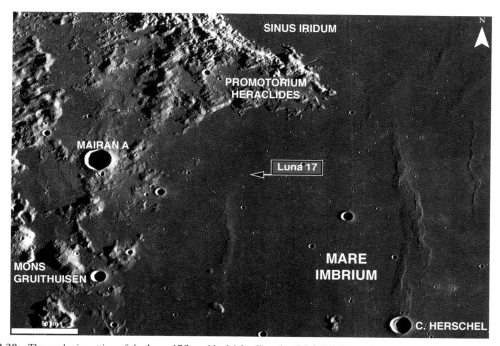

FIGURE 23.38 The geologic setting of the Luna 17/Lunokhod 1 landing site (LRO WAC mosaic). *From Hiesinger and Jaumann (2014).*

12.11. Luna 17/Lunokhod 1 (November 1970–October 1971)

Lunokhod 1 was the first of two unmanned Soviet rovers. Equipped with four television cameras, an X-ray spectrometer, an X-ray telescope, cosmic ray detectors, a laser retroreflector, and a device to determine the density and mechanical characteristics of the lunar soil, Lunokhod 1 was transported to the lunar surface by the Luna 17 spacecraft. Lunokhod 1 landed in northern Mare Imbrium (Sinus Iridum, 38.238° N, 325.003° E) and operated for about 11 months, 8 months longer than originally planned.

While roving about 10.5 km (Figures 23.9, 23.38, 23.39), Lunokhod 1 took about 20,000 images and 200 television panoramas and performed roughly 500 soil investigations. After losing contact with the rover in 1971, Lunokhod 1 was recently rediscovered in LRO images. Having a laser reflector on board, the position of Lunokhod 1 is now known to about 1 cm accuracy. Due to its high northern latitude and its location close to the limb, the Lunokhod 1 landing site is far from other landing sites with laser reflectors. Thus, Lunokhod 1 is particularly useful in determining the lunar librations and has become a significant new addition to lunar and gravity science.

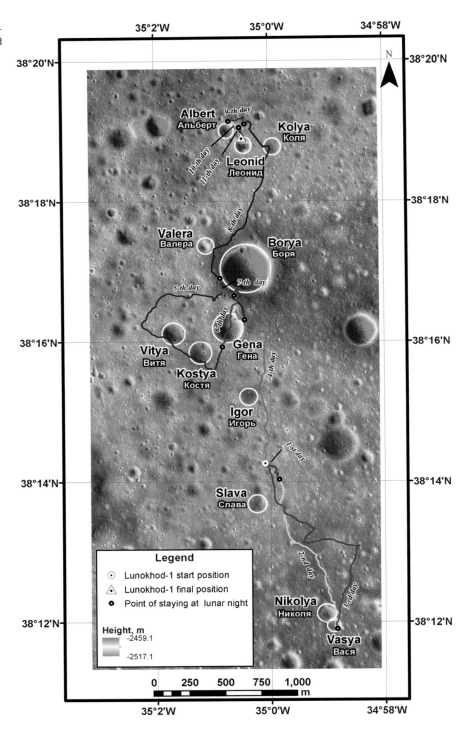

FIGURE 23.39 Detailed view of the Lunokhod 1 rover traverses with superposed topography. *Courtesy of I. Karachevtseva.*

12.12. Luna21/Lunokhod 2 (January–June 1973)

In 1973, the Soviet spacecraft Luna 21 delivered the second lunar rover, Lunokhod 2, to the surface of the Moon. Lunokhod 2 landed in crater Le Monnier (26.003° N, 30.408° E) at the northeastern edge of Mare Serenitatis. During 4 months of operation, Lunokhod 2 roved more than 42 km across the lunar surface (Figures 23.9, 23.40, 23.41). The main objective of Lunokhod 2 was to study the transitional zone between mare and highland areas and to investigate the southern part of a linear furrow located on

FIGURE 23.40 The geologic setting of the Luna 21/Lunokhod 2 landing site (LRO WAC mosaic). *From Hiesinger and Jaumann (2014).*

the mare surface of crater Le Monnier. The landing site consists of gently undulating terrain with numerous bowl-shaped impact craters. Around craters larger than 10 m in diameter, boulders are common. Lunokhod 2 discovered that close to the craters the wheels would sink deeper (20 cm) into the regolith compared to areas between the craters (5 cm), indicating local differences in the physical properties of the regolith. Despite the fact that the landing site is about 6 km from the highlands, even on the mare surface the X-ray spectrometer discovered a highland-like composition. Reported Al_2O_3 abundances are $\sim 9 \pm 1\%$ and FeO abundances are $\sim 6 \pm 0.6\%$. Because the composition became more highland-like the closer Lunokhod 2 approached the highlands, it was concluded that significant lateral transport of material occurs on the Moon.

13. SIGNIFICANCE OF LANDING SITES FOR THE INTERPRETATION OF GLOBAL DATA SETS

Our understanding of the geology, geophysics, history, and evolution of the Moon depends heavily on the information derived from the lunar landing sites. Landing site exploration and sample return are particularly beneficial because of the strength and synergy of having samples with known geologic context. Samples returned from the landing sites can be analyzed in detail in sophisticated laboratories with state-of-the-art techniques. In addition, portions of samples reserved for future analysis with even more sophisticated methods offer numerous advantages over in situ robotic measurements that are restricted to techniques available at the time of the mission. Numerous geophysical experiments deployed by astronauts are equally important and improved our knowledge of the physical properties and the internal structure of the Moon. In fact, laser reflectors at the landing sites are still used to accurately measure the Earth–Moon distance and to study lunar librations. Generally, the Apollo era data are still used in new innovative investigations. As a result, these data allow new models to be developed and tested long after the end of the Apollo missions. And of course, the landing sites provide important ground truth for all remote sensing experiments. In fact, the calibration of remote sensing data to the landing sites and the lunar samples allows derivation of information about unsampled areas.

The Apollo landing sites were carefully selected on the basis of safety considerations and scientific questions. In particular, the selection of each individual landing site can be viewed in the context of the scientific goal to obtain comprehensive understanding of impact basin structures and materials. This is crucial because the impact basin process is largely responsible for shaping the first-order features of the lunar surface. The schematic cross-section in Figure 23.42 illustrates the geologic setting of the Apollo and Luna landing sites relative to a multiringed impact structure, specifically in relation to morphologic features of the youngest lunar impact basin, Orientale. From this figure it is apparent that each landing site sampled very specific terrain types associated with large

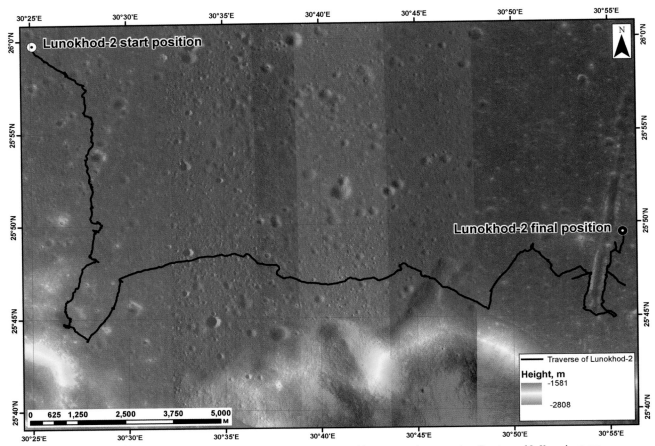

FIGURE 23.41 Detailed view of the Lunokhod 2 rover traverses with superposed topography. *Courtesy of I. Karachevtseva.*

lunar impact basins. In particular, Apollo 16 sampled old pre-basin highland terrain, Apollo 14 landed on radially textured basin ejecta deposits, and Apollo 15 samples contain polymict breccias, which were formed by a post-basin crater. Besides other investigations, Apollo 17 examined dark mantle deposits in detail, and Apollo 11 and 12 investigated different types of basalts. With the exception of Apollo 16 and Luna 20, all Apollo and Luna missions returned basaltic samples, and even these two missions returned small fragments of basaltic rock and glass in their regolith samples. As no mantle xenoliths have been identified in the returned samples, these basalts and the volcanic glasses are our best window into the lunar mantle.

In the late 1990s, Lunar Prospector identified a large area with elevated thorium concentrations on the lunar nearside. Several models for the origin of this thorium anomaly have been proposed, including accumulation of KREEP-rich magma beneath a region of thinned crust corresponding to the region of Oceanus Procellarum or accumulation above a degree-1 downwelling of dense ilmenite cumulates. While the precise carrier of the Th-rich compositional signature in broad areas of the Procellarum region cannot be identified from remote sensing alone, this question is crucial for understanding the magmatic and thermal evolution of the Moon due to radiogenic heat production. Fortunately, the landing sites of Apollo 12, 14, and 15 are located within the Procellarum KREEP Terrane (PKT) and thus allow us to study the KREEP signature from returned samples. For example, the Apollo 12 and 14 samples indicate that impact melt breccias and derivative breccias, such as regolith breccias and agglutinates, are the main carriers of KREEP. At the Apollo 15 landing site, however, basalt is an important carrier of KREEP. It has been proposed that the Imbrium basin impacted into the PKT and distributed KREEP-rich material globally across the lunar surface. This hypothesis is supported by the observation that abundant KREEP-rich material occurs in impact melt breccias of landing sites located outside the PKT, including Apollo 11, 16, and 17. In addition, smaller amounts of KREEP-rich material were found in soils from the Luna 16, 20, and 24 landing sites, which are located at greater distances from the PKT.

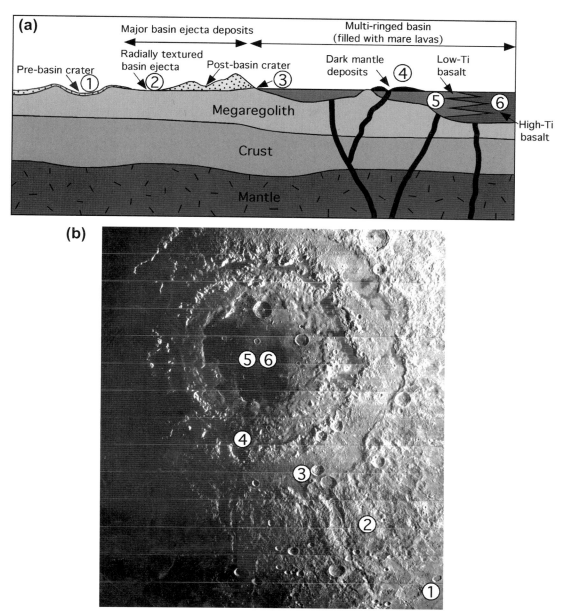

FIGURE 23.42 Schematic cross-section showing the Apollo landing sites in the context of a multiringed impact basin (a). Plan view of the Orientale basin with superposed locations similar to the Apollo landing sites (b); (1) similar to Apollo 16, (2) similar to Apollo 14, (3) similar to Apollo 15, (4) similar to Apollo 17, (5) similar to Apollo 12, (6) similar to Apollo 11. *From Hiesinger, H. & Head, J.W., III, (2006).*

In 1994, Clementine was the first spacecraft to provide high-resolution multispectral remote sensing images. These images permitted, for the first time, researchers to locate individual sampling stations at the Apollo 15, 16, and 17 landing sites. A comparison of spectra from these sampling stations with their average soil composition, revealed a strong correlation between spectral variations and compositions. On the basis of this correlation, algorithms were developed that allow calibrated global estimates of the FeO and TiO_2 concentrations on the lunar surface. Knowledge of the variability of these two compositional parameters in lunar basalts and their spatial distribution is important for understanding the petrogenesis of mare basalts and the history and evolution of magmatic processes on the Moon.

The lunar landing sites and the returned samples are essential for understanding the absolute chronology to anchor the Moon's geologic record. While the returned samples were dated radiometrically (e.g., U–Pb, Rb–Sr, Sm–Nd, and ^{40}Ar–^{39}Ar), with high accuracy, they only represent a few small areas on the Moon (i.e., the landing sites). Consequently, remote sensing techniques, i.e., CSFDs, are required to derive absolute model ages for the vast unsampled lunar regions. In order to derive absolute model ages from crater counts for any unsampled lunar region, the radiometric ages from returned samples have to be correlated with the crater counts from the landing sites to establish the lunar cratering chronology. For this purpose, crater counts for the Apollo 11, 12, 14, 15, 16, and 17 and the Luna 16 and 24 landing sites were performed and correlated with the corresponding radiometric ages of these sites. There are several slightly different empirically derived lunar impact chronologies based on a variety of interpretations as to which radiometric sample ages are most representative for a given landing site.

Finally, the lunar landing sites are important to understand the physical properties of the lunar surface. In fact, the Moon is the only extraterrestrial body for which seismic activity, heat flow, and thermal and electric conductivity have been measured. In addition, the Apollo laser reflectors allow us even today, more than 40 years after the Apollo program, to accurately measure the distance between the Earth and the Moon and its librations. Both sets of measurements are crucial for our understanding of the lunar internal structure. The rediscovery of the Lunokhod 2 retroreflector adds to this data set.

Although it is apparent that samples and surface exploration are important as ground truth for remote sensing, information derived from orbital data provide the necessary context for the landing sites, thus feeding back into our understanding of the entire planetary body. With increasing spatial and spectral capabilities of modern remote sensing instruments, the gap between in situ exploration and orbital observations continues to close. For example, with images of the Lunar Reconnaissance Orbiter (LRO) that have a **pixel** scale on the order of 0.5 m, one can study the lunar surface on the scale approaching that of field observations on the Earth.

Some measurements are best done globally from orbit. High-resolution global gravity, topography, and magnetic field measurements all provide important input for studies of the lunar interior. From global compositional data it is apparent that the vast area of the Feldspathic Highland Terrane is not well represented by the returned samples and that the South Pole-Aitken (SPA) Terrane is not at all represented in the sample collection. The compositions of the increasing number of lunar meteorites better represent the compositional variety of the Moon than is reflected in the Apollo and Luna samples. These statistically significant and more or less random meteorite samples of the Moon are dominated by feldspathic breccias and by low- to intermediate-Ti basalts. The feldspathic breccia meteorites appear to represent the typical FHT better than Apollo 16 samples. The distribution of basalt compositions in the meteorite collection also appears to be more like what is anticipated from remote sensing, whereas the Apollo basalts disproportionately sampled high-Ti regions. In summary, an optimal understanding of the Moon comes from integration of all available data sets.

14. LUNAR VOLATILES

Compared to the Earth, the bulk Moon is significantly depleted in volatile elements, including water, and this has been interpreted to be a direct consequence of the violent impact origin of the Moon. While small amounts of water were reported in lunar soils returned from each of the Apollo missions, these were generally attributed to terrestrial contamination. Consequently, in the post-Apollo era, the consensus was that the Moon was primarily an anhydrous planetary body that contained <1 ppb H_2O. However, modern data returned by instruments onboard several *National Aeronautics and Space Administration* (NASA) missions indicate the occurrence of water ice, particularly in the permanently shadowed polar regions of the Moon (Figure 23.43). For the south polar crater Cabeus, the LCROSS mission confirmed the existence of H_2O and OH^-. On the basis of the estimated total excavated mass of regolith that was observable, the concentration of water ice in the regolith at the LCROSS impact site was calculated to be $5.6 \pm 2.9\%$ by mass. In addition, several other volatile species were detected in the impact vapor plume, including H_2S, NH_3, SO_2, C_2H_4, CO_2, CH_3OH, and CH_4. Remote sensing data also indicate that nearly the entire lunar surface, at least in the uppermost few micrometers to millimeters depth, is hydrated at least during some portions of the lunar day. This water may be present either as molecular H_2O adsorbed onto the lunar soil grains or as structurally bound OH^-. In addition, remote sensing data identified stronger absorptions at 2.8–3.0 μm at cooler high latitudes and associated with fresh feldspathic craters. These absorptions are typically attributed to OH^-- and/or H_2O-bearing materials and are thought to be the result of hydration processes at the lunar surface or else created continuously by solar wind protons interacting with oxygen-rich surfaces during the formation of lunar soil particles.

Complementing the discovery of water on the Moon by orbiting and impacting spacecraft has been the detection and quantification of water contents in lunar samples returned by Apollo missions and lunar meteorites. In addition, studies of the drill core of the Luna 24 mission revealed at least 0.1 wt% of water at depths below 143 cm. The H content and the D/H isotopic composition were measured in lunar volcanic glasses, derived from deep in the mantle, as well as apatite, the main OH^--bearing mineral in lunar samples. These estimates suggest that the lunar mantle might contain as much H as the terrestrial mantle. Some of the measured D/H ratios are high, consistent with cometary sources for at least some of the

FIGURE 23.43 Moon Mineralogy Mapper (M^3) daytime near-infrared (NIR) color composite of the lunar nearside illustrating the spatial extent of diagnostic absorptions. Shown in blue is the 3-μm absorption associated with OH/H_2O; green indicates the reflected solar radiation (brightness) at 2.4 μm, and red represents the absorption at 2 μm due to the presence of iron-bearing pyroxenes. Small amounts of surficial OH/H_2O appear to be a function of the surface thermal and radiation environment and perhaps composition. During daytime, volatiles were most prominently detected at the higher, cooler latitudes. However, it cannot be excluded that a 3-μm feature at lower latitudes is masked by a minor thermal emission component. *From Pieters et al. (2009).*

water present in the lunar interior, and/or terrestrial-like original values with fractionation due to outgassing leading to high D/H values. In contrast, on the basis of chlorine isotope data from lunar samples it has been concluded that the lunar interior is dry.

Up to seven sources have been proposed for lunar volatiles including the Sun (solar wind reduction of lunar regolith), Earth (from the Moon-forming giant impact), Moon (internal degassing), comets, asteroids, interplanetary dust, and giant interstellar molecular clouds. The dominant sources of water are likely to be primordial acquisition during lunar accretion, "water ice" delivered to the Moon by cometary and asteroidal impacts after the completion of the main phase of lunar magma ocean crystallization, water produced by the interaction of solar wind hydrogen with oxygen-bearing minerals of the lunar soil, or a combination of these processes. For example, while the polar ice deposits on the Moon are most likely primarily cometary in origin, migration of solar wind-derived water from lower latitudes to polar cold traps during the **diurnal** cycle of the sunlit side of the Moon is also possible.

In the context of future lunar exploration, water is of particular interest and importance because water plays a fundamental role in planetary accretion and subsequent evolution. Finally, the discoveries of water on the Moon—in polar regions, in surface regolith, and in lunar rocks—allow investigating key astrobiological questions and enable future lunar exploration by allowing water production with in situ resource utilization techniques.

15. LUNAR ATMOSPHERE AND ENVIRONMENT

The atmosphere of the Moon is extremely tenuous. Important constituents of the lunar environment are neutral gases, plasma, and ejected dust particles that are generated by the interaction of the lunar surface with the space environment. These components and their characterization are not only of scientific interest but also have gained importance from an engineering perspective because of their effects on the proposed human and robotic activity on the lunar surface.

The atmospheric conditions on the Moon are similar to those of interplanetary space. While 10^5 to 10^7 particles/cm^3 or the total of 10^4 kg of gaseous species—14 orders of magnitude less than the atmosphere of Earth—are contained in a neutral lunar "atmosphere", interactions between particles due to collisions are negligible. Thus, the Moon has an exosphere. Known constituents are He, Ar, Na, K, and Rn that are captured from the solar wind or originate from the Moon by radioactive decay, sputtering, or micrometeorite impacts. A variety of energetic particles (particles associated with solar wind and solar **flares**, as well as galactic cosmic rays mostly consisting of protons, electrons and some heavier nuclei) also characterize the Moon's environment. The absence of a significant atmosphere and the absence of a global magnetic field cause this radiation to directly interact with the surface, yielding measurable alteration of the uppermost material as well as secondary radiation from the surface.

Recent measurements by Kaguya (SELENE), Chandrayaan 1, and LRO (i.e., the Cosmic Ray Telescope for the Effects of Radiation (CRaTER)) have shown that not all of the protons impinging on the lunar surface are absorbed or implanted. Instead, about 20% of the protons are backscattered or reflected, most of them as neutral hydrogen and only <1% as charged ions. Furthermore, reductions in the backscattered neutral hydrogen flux are observed in association with lunar magnetic anomalies, which may act as "minimagnetospheres" and reduce the impacting proton flux. Embedded in the plasma environment, the lunar surface becomes charged to an electrostatic potential, which is either positive or slightly negative on the dayside and significantly negative on the nightside. Thus, lateral changes in the electrostatic potential especially close to the terminator may bear on the still open question of dust levitation.

Dust generated by hypervelocity impacts of fast projectiles (10−72 km/s) striking the lunar surface may also be present in the lunar environment. A total of 10^6 kg of micrometeoroids impact the lunar surface every year and each projectile excavates about 1000 times its own mass. Most of the ejecta return to the lunar surface, but a fraction could reach escape velocities forming a dust envelope around the Moon, similar to the Galilean satellites.

ACKNOWLEDGEMENT

This chapter is based on the articles "Geology, Geochemistry, and Geophysics of the Moon: Status of Current Understanding" published in the *Planetary Space Science* special issue "*Scientific Preparations For Lunar Exploration*", and "New Views of Lunar Geoscience: An Introduction and Overview", "Understanding the Lunar Surface and Space-Moon Interactions", and "Thermal and Magmatic Evolution of the Moon" published in *Reviews in Mineralogy and Geochemistry* that summarized the work of numerous lunar researchers. It is not possible to give credit to all lunar researchers whose work was included in this encyclopedia article, but we would like to refer the reader to these articles to find extensive references. The selected bibliography and the references therein further allow the reader to find the relevant primary literature, which has been used for this chapter. Particularly, we greatly appreciate the work of M. Anand, I. Antonenko, L. E. Borg, B. A. Campbell, R. M. Canup, J. Carpenter, I.A. Crawford, L. T. Elkins-Tanton, R. Elphic, B. Feldman, O. Gasnault, J. J. Gillis, M. Grott, T. L. Grove, B. H. Hager, A. N. Halliday, S. Hempel, P. C. Hess, H. Hiesinger, J. W. Head III, H. Hoffmann, L. L. Hood, D. Hunten, H. Hussmann, R. Jaumann, B. L. Jolliff, M. Knapmeyer, U. Köhler, R. L. Korotev, K. Krohn, D. Lawrence, S. Lawson, D.-C. Lee, J. Longhi, P. Lucey, S. Maurice, M. Mendillo, C. R. Neal, S. Noble, J. Oberst, J. J. Papike, E. M. Parmentier, T. Prettyman, M. E. Pritchard, R. C. Reedy, M. S. Robinson, N. Schmitz, F. Scholten, C. K. Shearer, F. Sohl, T. Spohn, L. A. Taylor, R. Wagner, U. Wiechert, and M. Wieczorek, as well as A. T. Basilevsky, C. P. Florensky, I. Karachevtseva, E. Speyerer, G. J. Taylor, and P. Warren. Special thanks to C. H. van der Bogert and J. H. Pasckert.

Last but not least, we are very thankful for Brad Jolliff's very thorough and extremely constructive review, which helped tremendously in improving the manuscript.

BIBLIOGRAPHY

Ambrose, W. A., & Williams, D. A. (Eds.). (2011). *Recent Advances and current Research Issues in Lunar Stratigraphy. Special paper 477*, Boulder, CO: Geol. Soc. Am.

Basaltic Volcanism Study Project (1981). *Basaltic volcanism on the terrestrial planets*. New York: Pergamon.

Canup, R. M., & Righter, K. (Eds.). (2000). *Origin of the Earth and the Moon*. Tucson, AZ: The University of Arizona Press.

Head, J. W. (1976). Lunar volcanism in space and time. *Reviews of Geophysics and Space Physics, 14*, 265−300.

Heiken, G. H., Vaniman, D. T., & French, B. M. (Eds.). (1991). *The lunar sourcebook*. Cambridge, England: Cambridge Univ. Press.

Hiesinger, et al. (2012). *GSA. Special Paper 477*.

Hiesinger, H., & Head, J. W., III (2006). New Views of Lunar Geoscience: an introduction and overview. *Reviews in Mineralogy and Geochemistry, 60*, 1−81.

Hiesinger, H., & Jaumann, R. (2014). The Moon. In T. Spohn, D. Breuer, & T. Johnson (Eds.), *Encyclopedia of the Solar System* (pp. 493−538).

Jolliff, et al. (2001). *JGR*.

Jolliff, B. L., Wieczorek, M. A., Shearer, C. K., & Neal, C. R. (Eds.). (2006), *Reviews in Mineralogy & Geochemistry: 60. New views of the moon*. Min. Soc. Am.

Mitchell, et al. (2008). *Icarus, 194*, 401−409.

Neukum, G., & Ivanov, B. A. (1994). Crater size distributions and impact probabilities on Earth from lunar, terrestrial-planet, and asteroid cratering data. In T. Gehrels (Ed.), *Hazard Due to Comets and Asteroids* (pp. 359−416). Tucson: Univ. of Ariz. Press.

Neukum, G., Ivanov, B. A., & Hartmann, W. K. (2001). Cratering records in the inner solar system in relation to the lunar reference system. *Space Science Reviews, 96*, 55−86.

Neukum, et al. (2001). *ISSI Buch, Chronology of Mars*.

Papike (1998). *Planetary materials, Reviews in Mineralogy and Geochemistry*.

Papike, J. J. (Ed.). (1999), *Rev. Miner: 36. Planetary materials*. Miner. Soc. Amer.

Pieters, et al. (2009). *Science, 23*, 568−572. supplemental material.

Robinson, et al. (2012). *Planetary Space Science, 69*, 76−88.

Shearer, C. K., Hess, P. C., Wieczorek, M. A., Pritchard, M. E., Parmentier, E. M., Borg, L. E., et al. (2006). Thermal and Magmatic Evolution of the Moon. *Reviews in Mineralogy and Geochemistry, 60*, 265−518.

Stöffler, D., & Ryder, G. (2001). Stratigraphy and isotope ages of lunar geologic units: Chronological standard for the inner solar system. In R. Kallenbach, J. Geiss, & W. K. Hartmann (Eds.), *Space Science Reviews: Vol. 96. Chronology and Evolution of Mars* (pp. 9−54). Dordrecht: Space Science Series of ISSI, Kluwer Academic Publishers.

Taylor, S. R. (1982). *Planetary science: A lunar perspective*. Houston, TX: Lunar and Planetary Institute.

Taylor, S. R. (2007). The moon. In L.-A. McFadden, P. R. Weissman, & T. V. Johnson (Eds.), *Encyclopedia of the solar system*. Elsevier.

Warren (2003). *Treatise in Geochemistry, 1*, 559−599.

Wieczorek, M. A., Jolliff, B. L., Khan, A., Pritchard, M. E., Weiss, B. P., Williams, J. G., et al. (2006). The Constitution and Structure of the Lunar Interior. *Reviews on Mineralogy and Geochemistry, 60*, 325.

Wilhelms, D. E. (1987). *The geologic history of the moon*. U.S. Geol. Surv. Prof. Paper No. 1348. Washington, D.C.: U.S. Geological Survey.

Further information on lunar science can be found in the journal *Elements, 5*, pp. 11−46, including articles by J. W. Delano, G. J. Taylor, M. D. Norman, T. L. Grove, M. J. Krawcynski, M. A. Wieczorek, and P. G. Lucey.

Results of recent lunar missions are summarized in *Science, 323* (2009, SELENE), *Science, 326* (2009, Chandrayaan), *Science, 329* (2010, LRO), *Science, 330* (2010, LCROSS), and *Science, 339* (2013, GRAIL).

Chapter 24

Interior of the Moon

Renee C. Weber
NASA Marshall Space Flight Center, Huntsville, AL, USA

Chapter Outline

1. Introduction 539
2. Bulk Lunar Properties 539
3. Methods Used to Probe the Lunar Interior 541
   - 3.1. Apollo Core Samples 542
   - 3.2. Gravity Measurements 542
   - 3.3. Laser Ranging 543
   - 3.4. Magnetic Techniques 544
   - 3.5. Heat Flow 545
   - 3.6. Compositional Studies 546
   - 3.7. Seismology 546
4. Lunar Internal Structure 549
   - 4.1. Regolith 549
   - 4.2. Crust 550
   - 4.3. Mantle 551
   - 4.4. Core 551
5. Implications for Lunar formation and Evolution 552
Bibliography 554

1. INTRODUCTION

Understanding the internal structure of the Moon is of key importance to deciphering its early history. The current consensus is that the Moon formed following the collision of a Mars-sized body with the Earth about 4.5 billion years ago. The rocky mantle of the impactor spun out to form the Moon, while the core of the impactor fell into the growing Earth. This model explains the high angular momentum of the Earth–Moon system, the low density of the Moon relative to the Earth, and the Moon's depletion of lighter **volatile** elements, which were likely lost during heating associated with the impact event. The model also provides a source of energy to melt the early Moon.

The geochemical and petrological evidence strongly supports the theory that the molten Moon floated an anorthositic crust about 4.45 billion years ago. This forms the present high-**albedo** highland crust. As the initially hot, molten Moon cooled, the mantle likely crystallized into a sequence of mineral zones by about 4.4 billion years ago. Heavier elements sank to form a small metallic core. Following the formation of the crust, major impacts on the surface produced many craters and multiring basins. The oldest basin observed is the South Pole–Aitken Basin and the youngest is the Orientale Basin, which formed 3.85 billion years ago. Some evidence suggests that a spike or "cataclysm" created many of these basins around 3.9–4.0 billion years ago.

Beginning about 4.3 billion years ago, and peaking between 3.8 and 3.2 billion years ago, partial melting occurred in the lunar interior, and basaltic lavas flooded the low-lying basins on the surface. This occurred mostly on the nearside, where the crust is thinner, resulting in the low-albedo lunar mare. Major volcanic activity ceased around 3.0 billion years ago, although minor activity may have continued until 1.0–1.3 billion years ago. The Moon has suffered only a few major impacts since that time (forming, for example, the young **rayed craters** such as Copernicus and Tycho).

An overview of the commonly accepted model of the Moon's present day internal structure, illustrating the crust, mantle, and core layers, is shown in Figure 24.1, and includes the probable depths of the interfaces between the core layers.

2. BULK LUNAR PROPERTIES

The mass of the Moon, determined from the orbital periods of various spacecrafts using **Kepler's third law**, is 7.35×10^{22} kg, which is 1/81 of the mass of the Earth. Although the Galilean satellites of Jupiter (Io, Europa, Ganymede, and Callisto), and Saturn's moon Titan are comparable in mass, the Moon/Earth ratio is the largest satellite-to-parent mass ratio in the solar system (Table 24.1). The lunar radius is 1738 ± 0.1 km, or 27% of the Earth's radius. This radius is intermediate between that of

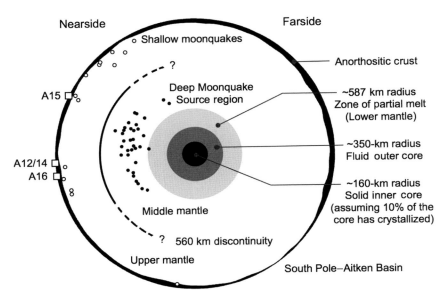

FIGURE 24.1 Schematic cross-section of the Moon showing the approximate depths of layering inferred using various geophysical methods. The nearside is to the left. The thickness of the crust is shown for a pole-to-pole profile at 0 and 180° longitude; the nearside crust is known to be thinner than the farside crust. The source regions of natural moonquakes (dots) detected by the Apollo seismometers (squares) have been projected onto the nearside hemisphere as a function of depth and latitude (see Section 3.7). *Figure reproduced from Wieczorek et al. (2006).*

TABLE 24.1 Radius, Density, and Satellite-to-Parent Mass Ratio of the Largest five Satellites in the Solar System, Sorted by Radius

Moons	Radius (km)	Density (kg/m³)	Satellite-to-Parent Mass Ratio
Europa	1561	3014	1/39544
Moon	1738	3344	1/81
Io	1818	3529	1/21256
Callisto	2410	1834	1/17575
Titan	2576	1880	1/4225
Ganymede	2634	1942	1/12825
Materials			
Wood (pine)	—	0500	—
Water	—	1000	—
Stone (granite)	—	2700	—
Metallic iron	—	7874	—

The densities of several common earth materials are given for comparison.

Europa (radius = 1561 km) and Io (radius = 1818 km). The Moon is much smaller than Ganymede (radius = 2634 km), which is the largest satellite in the solar system. The lunar mean density is 3344 ± 0.003 kg/m³, while the Earth has a much higher mean density of 5515 kg/m³. The lunar density is also intermediate between that of Europa (density = 3014 kg/m³) and Io (density = 3529 kg/m³). Most of the other known satellites in the solar system are ice–rock mixtures and so are much less dense.

The Moon's moment of inertia (MOI) factor is related to its second-degree gravitational harmonics, which have been measured to high precision by orbiting spacecraft. Current estimates indicate that the Moon's mean MOI is 0.3931 ± 0.0002, very close to that of a uniform density sphere, which has an MOI of 0.4. This value is met with a Moon that has a slight density increase toward its center, in addition to the presence of a low-density crust. In comparison, the mean MOI value for the Earth, with its dense metallic core that constitutes 32.5% of the Earth's total mass, is 0.3315.

The mass of the Moon is distributed in a nonsymmetrical manner, with the center of mass (CM) lying 1.9 km closer to the Earth than the geometrical center of figure (CF) (Figure 24.2). This offset is due to the presence of a thicker farside crust. This is a contributing factor in placing the Moon into synchronous orbit with the Earth, such that the Moon always presents the same face to the Earth. The gravitational influence of the Earth (and to a lesser extent, the Sun) on the Moon's asymmetric mass distribution resulted in torques that slowed down the rotation of the early Moon. These torques acted in combination with torques on the Moon's tidal bulges to slow the Moon's rotation until it became tidally locked, always presenting the same face to the Earth. However, the lunar longitudinal and latitudinal **librations** in combination allow a total of 57% of the Moon's surface to be visible at different times in the orbital cycle.

Various explanations have been advanced to account for the offset of the Moon's CM from its CF. Dense mare basalts erupted from the lunar interior cover about 17% of the lunar surface, mostly on the nearside, but they are generally less than 1 or 2 km thick and constitute only about 1% of the total volume of the crust—insufficient by about an order of magnitude to account for the effect. It has also been suggested that the offset could arise if the lunar core is displaced from the CM. However, such a displacement would generate shear stresses that could not be supported by the hot, likely molten (or partially molten) deep interior. Another suggestion is that some form of density asymmetry developed in the mantle during crystallization of the magma ocean, with a greater thickness of lower density materials being concentrated within the farside mantle. However, it is unlikely that such density irregularities would survive stress relaxation in the hot interior, unless actively maintained by convection (for which there is no present-day evidence).

The conventional explanation for the CM/CF offset is that the farside highland low-density crust is thicker, probably a consequence of an asymmetry developed during crystallization of the magma ocean. This explanation is supported by crustal thickness estimates derived from gravity mapping (see Section 3.2). The crust is massive enough and sufficiently irregular in thickness to account for the CM/CF offset. An equipotential surface is closer to the actual surface on the nearside. Magmas that originate at equal depths below the surface will thus have greater difficulty in reaching the surface on the farside, where the crust is thicker. This explains the scarcity of observed mare basalts on the farside. Lavas rise owing to the relative low density of the melt and do not possess sufficient hydrostatic head to reach the surface on the farside, except in craters in some very deep basins.

3. METHODS USED TO PROBE THE LUNAR INTERIOR

The Apollo Lunar Surface Experiments Package (ALSEP), deployed across the lunar surface by the astronauts on

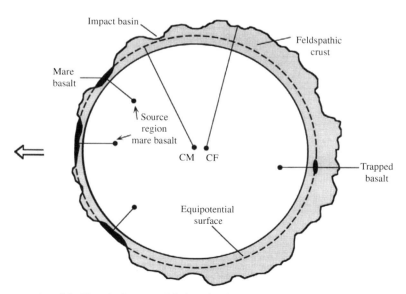

FIGURE 24.2 Schematic cross-section of the Moon in the equatorial plane showing the displacement of the Moon's center of mass toward Earth (figure left), due to the presence of a thicker farside crust (figure right). The crustal thickness is exaggerated for clarity. An equipotential surface is indicated with a dashed line. *Figure reproduced from "The Moon" chapter in the previous edition of the Encyclopedia of the Solar System.*

Apollo missions 12, 14, 15, 16, and 17, gathered several years' worth of data relevant to the lunar interior. Each ALSEP installation consisted of a set of geophysical instruments connected to a central base station. The base station acted as the command center for the entire package. It received commands and transmitted data to and from Earth, and distributed power to each experiment. The astronauts also gathered a wide collection of samples from the lunar surface, which were returned to Earth for analysis.

Instruments both onboard spacecraft in lunar orbit and Earth-bound also gather measurements that are useful for deciphering the Moon's internal structure. These include gravity and magnetic field data measured from orbit and laser ranges originating from Earth.

This section will review results of the active and passive seismic experiments and the heat flow experiment from ALSEP, analyses performed on samples gathered from the surface and shallow subsurface, and a variety of orbital and Earth-based measurements, and discuss interpretations of the lunar interior made from these data.

3.1. Apollo Core Samples

The near-surface structure of the Moon was revealed by core samples taken by the Apollo astronauts (Figure 24.3). Core tubes were either 2 or 4 cm in diameter and were pounded into the surface with a hammer. The deepest core was nearly 3 m at the Apollo 17 landing site, and a total of 24 cores were collected over all six Apollo surface sites. These cores revealed that the shallowest lunar layer, known as the regolith, is a complex array of overlapping ejecta blankets resulting from meteor bombardment on the lunar surface throughout the Moon's history. This process is known as impact gardening, and results in a shallow layer of particles of varied size and texture (see Section 4.1).

3.2. Gravity Measurements

The Moon's internal structure can be inferred through analyses of the lunar gravity field as measured from orbit (Figure 24.4). Variations in surface gravity across the Moon are caused by density heterogeneity in the subsurface, and these variations affect the position of the orbiting spacecraft. First noticed during analysis of tracking data from National Aeronautics and Space Administration (NASA)'s Lunar Orbiter program in the 1960s, the Moon's gravity field has been mapped in successively higher resolution by missions such as NASA's Lunar Prospector in the 1990s, the Japanese space agency's SELENE orbiter (Selenological and Engineering Explorer) in the 2000s, and NASA's GRAIL mission (Gravity Recovery and Interior Laboratory) in the 2010s. The GRAIL mission mapped the Moon's gravity in unprecedented detail, resulting in the highest resolution gravity map of any body in the solar system, including Earth.

The biggest features resolved in the lunar gravity field are known as **mascons**, or mass concentrations. They are associated with giant impact basins and are caused by the uplift of a central plug of dense mantle material during impact, followed by the much later addition of dense mare basalt. Smaller shallow features are also resolved in the gravity data, including tectonic structures, volcanic landforms, basin rings, complex crater central peaks, and simple bowl-shaped craters. Young ray craters have negative gravity anomalies because of the mass deficit associated with excavation of the crater, combined with the low density of the fallback rubble. Craters less than

 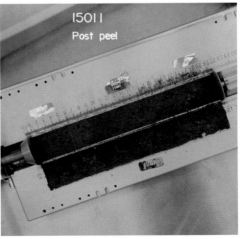

FIGURE 24.3 (Left) Core sample being taken by Apollo 12 astronaut Alan Bean at the Bench Crater site (photo AS12-49-7243 from the Apollo Image Archive). (Right) Apollo 15 core sample during analysis at the Lunar Sample Laboratory Facility, NASA Johnson Space Center (photo S79-37,062 from Apollo Image Archive).

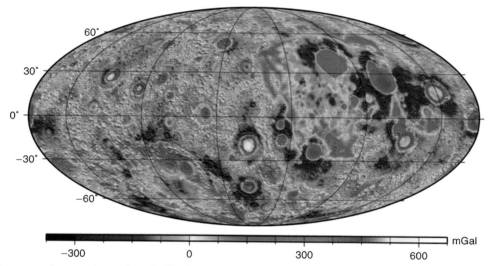

FIGURE 24.4 Bouguer gravity anomaly map from the GRAIL lunar gravity model GL0420A. The map is a Molleweide projection centered on 270° E longitude, with the nearside on the right and farside on the left. *Figure reproduced from Zuber et al. (2013).*

200 km in diameter have negative gravity anomalies for the same reason (e.g. Sinus Iridum has a negative anomaly of −90 mGal). Volcanic domes such as the Marius hills have positive anomalies (+65 mGal), indicating support by a rigid lithosphere. The gravity signature of young, large, ringed basins, such as Mare Orientale, shows a "bull's-eye" pattern with a central positive anomaly (+200 mGal) surrounded by a ring of negative anomalies (−100 mGal) with an outer positive anomaly collar (+30 to +50 mGal).

In combination with topography data, the gravity field can also be used to infer the depth of the crust−mantle interface (known as the **Moho**). The Moho deflects in response to surface loads, and the resulting flexural signature contributes to the observed gravitational field. Crustal thickness largely correlates with topography, with the exception of the lunar mare regions. These areas of low elevation were resurfaced by high-density basaltic lava flows, resulting in more complex flexural signals. The average density of the highland crust calculated from GRAIL-derived crustal thickness estimates is 2250 kg/m^3.

The lunar highland crust is strong. High mountains such as the Apennines (7 km high), formed during the Imbrium collision 3.85 billion years ago, are uncompensated and are supported by a strong cool interior. The gravity data are consistent with an initially molten Moon that cooled quickly and became rigid enough to support loads such as the circular mountainous rings around the large, younger, ringed basins as well as the mascons. Even if some farside lunar basins do not show mascons, this may merely be a consequence of the greater thickness of the farside crust. The South Pole−Aitken Basin (the largest and oldest basin, age at least 4.3 billion years) is particularly significant in this respect.

3.3. Laser Ranging

Additional information about the interior of the Moon can also be inferred from data gathered by the ongoing Lunar Laser Ranging (LLR) experiment. This experiment consists of Earth-based laser ranges to an array of retroreflectors emplaced on the lunar surface 30 years ago by both US and Russian missions.

A laser pulse is fired from the Earth to the Moon, where it bounces off a retroreflector and returns back to Earth (Figure 24.5). The round-trip travel time can be used to measure the Moon's shape and position with accuracy better than 2 cm. The analysis of LLR data provides a wealth of information concerning the dynamics and internal structure of the Moon.

The distances between the retroreflectors and the Earth change in part because of lunar rotation (physical librations) and tides. Values of the gravitational harmonics, the moments of inertia, the lunar Love number k_2 (which measures the tidal change in the Moon's moments of inertia and gravity), and variations in the lunar physical librations are related to the Moon's composition, mass distribution, and internal dynamics.

A range of internal structure models is compatible with the MOI values constrained by LLR. For example, a 60-km-thick lunar crust with density of 2750 kg/m^3, a constant-density lunar upper mantle, a lower mantle with a similar change in density relative to the upper mantle, and a variable-radius iron core with density of 7000 kg/m^3 produces an appropriate MOI. In this case the maximum

FIGURE 24.5 (Left) The Lunar Laser Ranging Station at the University of Texas McDonald Observatory. Photo by Randall L. Ricklefs. (Right) Laser retroreflector installed on the lunar surface by the Apollo 11 astronauts. *Photo AS11-40-5952 from the NASA Apollo Archive.*

core size is in the range of 220–350 km, and an increase in crustal density to 2959 kg/m^3 raises the maximum core size to 400 km, consistent with other estimates (see Section 4.4). All layers can be adjusted in thickness and density to produce a suite of plausible lunar structure models.

For a perfectly rigid Moon, the mean direction of the lunar spin axis would be expected to precess with the Earth–Moon orbit plane. The LLR data show, however, that the true spin axis of the Moon is displaced from the expected direction. This is the result of ongoing active dissipation in the lunar interior, which has been proposed to be due in part to friction at the interface between the solid lower mantle and a fluid core.

3.4. Magnetic Techniques

At present, the Moon does not possess an internally generated magnetic field. However, some samples returned by the Apollo astronauts from the lunar surface retain natural remanent magnetism. In addition, orbital estimates of surface magnetic field strength reveal regions of increased magnetic intensity (Figure 24.6), albeit with field strengths of only about 1/100th of the terrestrial field. Based on the age dating of Apollo samples, this magnetic signature suggests that between about 3.6 and 3.9 billion years ago, there was a planetary-wide magnetic field that has now vanished. The field appears to have been much weaker both before and after this period.

Although taken from the lunar surface, these observed present-day remanent magnetic anomalies are relevant to the lunar internal structure since one interpretation is that the Moon once possessed a lunar dipole field of internal origin. The favored mechanism is that the field was produced by dynamo action in a liquid iron core, similar to the way Earth's magnetic field is generated. A core about 400 km in diameter could produce a field at the lunar surface with strength comparable to the observations. In early lunar history, this magnetization would be impressed into the cooling lunar crust.

An alternative interpretation suggests that the magnetic signature may not be internally generated but rather results from shock magnetization in transient fields generated following the basin-forming impacts in lunar history. This theory is supported by the observation that the largest crustal magnetizations appear to be located at or near the antipodes of the largest impact basins. In addition, some localized strong magnetic anomalies are associated with patterns of swirls—high-albedo features that impart no observable topography. These swirls have likewise been suggested to form by some focusing effect of the seismic waves that resulted from the large basin-forming impacts. More work is clearly needed to substantiate this hypothesis and to understand the association of swirls and magnetic fields.

The internal structure of the Moon can also be inferred by measuring the lunar-induced magnetic dipole moment. This is the residual response of the lunar interior to the

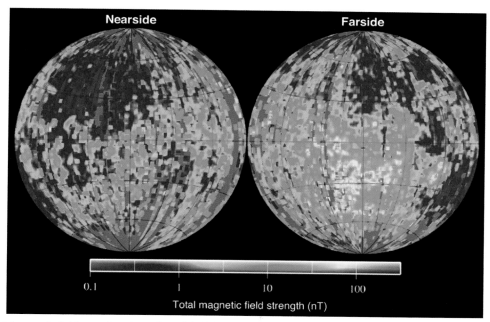

FIGURE 24.6 Total magnetic field strength at the surface of the Moon as derived from the Lunar Prospector electron reflectometer experiment, which was in orbit around the Moon during the years 1998–1999. *Figure courtesy of M. Wieczorek.*

sudden exposure of the Moon to a uniform magnetic field in a near-vacuum environment, which happens every month as the Moon passes through the Earth's geomagnetic tail. The external field is perturbed by an induced magnetic field caused by currents at the surface of a highly electrically conducting (iron) core, and these perturbations can be measured by orbiting spacecraft. Data gathered by the Lunar Prospector magnetometer were analyzed to conclude that the Moon likely does possess an iron-rich core, with a preferred radius of 340 ± 90 km.

3.5. Heat Flow

The rate at which a planetary body loses heat to space is an important indicator of the level of tectonism and volcanic activity on that planet. Two measurements of the lunar heat flow are available, as measured by the ALSEP's Heat Flow Experiment during the Apollo 15 and Apollo 17 mission's surface operations. The Heat Flow Experiment involved drilling a hole into the lunar regolith and inserting a probe that measured temperature at several depths within the hole. The rate at which temperature increases with depth provides a measure of the total heat flowing from the Moon's interior: 21 mW/m^2 at the Apollo 15 site and 16 mW/m^2 at the Apollo 17 site, respectively. These surface heat flow measurements are close to Earth-based estimates from microwave observations.

Unlike the Earth, which dissipates most of its heat by convective volcanism at the midocean ridges, the Moon transports its heat to the surface by conduction. A lack of observed present-day active volcanism or tectonism on the Moon indicates that most of its original internal heat has been lost, so any observed heat flow must be instead predominantly due to the radioactive decay of heat-producing elements, with a small percent of the total heat flow consisting of the loss of residual heat from lunar formation.

If the Apollo heat flow measurements are considered to represent the average heat loss characteristic of the entire Moon, they can be used to provide constraints on the bulk lunar abundances of elements that release heat through radioactive decay. The heat-producing elements K, U, and Th were concentrated just below the crust by differentiation during lunar formation. However, the constraints on these abundances are only mild, as the distribution of heat-producing elements is not symmetric across the lunar surface.

The heat flow measurements made by Apollo could indicate bulk lunar uranium values as high as 45 ppb, over twice the terrestrial abundances. A more likely scenario is that uranium and other heat-producing elements are concentrated beneath the lunar crust. This is a consequence of magma crystallization. Potassium (K), rare earth elements (REE), and phosphorus (P) (KREEP), along with thorium and uranium, are among the last trace elements to crystallize from a melt. As the early molten Moon cooled, various minerals crystallized from the melt. Heavy olivines sank to the bottom, while lighter anorthosites floated to the top. The remaining incompatible trace elements probably remained molten for a much longer period of time and were eventually exposed to the surface through impact processes. It is unknown whether KREEP is globally distributed in the Moon. It is only exposed on the nearside, but large impact basins did not excavate below the thick farside crust to the same extent. It is possible that KREEP is a global layer; it is also possible it was somehow concentrated on the nearside, which may help explain the asymmetric distribution of lunar mare.

3.6. Compositional Studies

The Moon is depleted of volatile elements compared to the Earth. It also lacks ferric iron, as determined by both orbital measurements and sample analyses. This lack of iron is reflected in the low lunar density. The Earth contains about 25% metallic Fe; the Moon, less than about 2–3%. However, the bulk Moon contains between 12 and 13% FeO, or 50% more than current estimates of 8% FeO in the terrestrial mantle. Along with its depletion in iron, the Moon also has a low abundance of **siderophile** or "metal-seeking" elements. These elements are extracted into metallic phases according to their metal/silicate partition coefficients during accretion. The lunar depletion of these elements has been used to argue that they have likely been segregated into a metallic core.

The other major element abundances are mostly model-dependent. Si/Mg ratios are commonly assumed to be **chondritic** (CI), although the Earth and many meteorite classes differ from this value. The lunar Mg value is generally estimated to be about 0.80, lower than that of the terrestrial mantle value of 0.89.

The Moon is probably enriched in refractory elements such as Ti, U, Al, and Ca, a conclusion consistent with geophysical studies of the lunar interior. This conclusion is reinforced by the data from the Galileo, Clementine, and Lunar Prospector missions, which indicate that the highland crust is dominated by anorthositic rocks. This requires that the bulk lunar composition contains about 5–6% Al_2O_3, compared with a value of about 3.6% for the terrestrial mantle and so is probably enriched in refractory elements (e.g. Ca, Al, Ti, and U) by a factor of about 1.5 compared to the Earth. Both the Cr and O isotopic compositions are identical in the Earth and Moon, probably indicating an origin in the same part of the nebula, consistent with the single-impact hypothesis that derives most of the Moon from the silicate mantle of the impactor.

The Moon has a composition that is unlikely to have been made by any single-stage process from the material of the primordial solar nebula. The compositional differences from the primitive solar nebula, from the Earth, from Phobos and Deimos (almost certainly of carbonaceous chondritic composition), and from the satellites of the outer planets (rock/ice mixtures, with the exception of Io) thus call for a distinctive mode of origin (see Section 5).

3.7. Seismology

The Apollo astronauts deployed four seismometers on the lunar surface between 1969 and 1972 (Figure 24.7). These

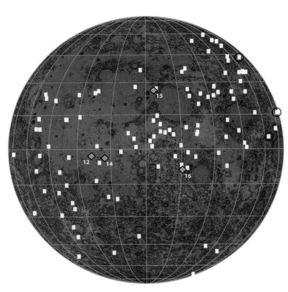
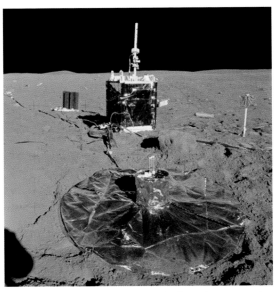

FIGURE 24.7 (Left) Map of the lunar nearside (15° increments of latitude and longitude) showing the locations of the four seismometers that comprised the Apollo Passive Seismic Experiment (red diamonds) and the nominal epicenter locations of the deep moonquake source regions (white rectangles). Farside source regions (outer black circles) are projected on the nearside. The deep moonquakes are roughly constrained to a wide swath trending northeast–southwest across the entire nearside of the Moon. (Right) Photograph of the Apollo 16 seismometer as installed on the lunar surface. It is covered with a mylar shield that was intended to thermally protect the instrument. Other instruments from the Apollo Lunar Surface Experiments Package can be seen in the background, as well as the central station that telemetered data continuously to Earth. *Photo AS16-113-18,347 from the NASA Apollo Archive.*

instruments gathered data continuously until 1977, making the Moon the only extraterrestrial body for which extensive seismic data has been gathered.

The Moon is much less seismically active than the Earth, due to its lack of oceans and plate tectonics. Still, the Apollo network recorded several types of both naturally occurring and artificial seismic events, resulting in a total number of approximately 13,000 catalogued events over the 8-year span of the experiment. Because the Moon has no atmosphere to burn off potential impactors, there were a significant number of meteoroid impacts on the surface that registered on the seismic array. The booster rockets and lunar modules from the Apollo spacecraft were also purposely impacted onto the surface, in part to test and calibrate the seismic array. Observed naturally occurring moonquakes include the relatively large but rare shallow moonquakes of unknown origin (similar to intraplate earthquakes), and the relatively small but frequent deep moonquakes, (triggered with monthly periodicity by the lunar tides). Observed deep moonquakes generally had body wave equivalent magnitudes less than three, with most less than magnitude one; shallow moonquakes were larger, with the larger recorded events having moment magnitudes ranging between 2.4 and 4.1. In addition, the network detected many noiselike thermal events that were associated with the large temperature fluctuations between lunar day and night.

Deep moonquakes are the most numerous type of seismic event, comprising approximately half of the event catalog. They are known to originate from distinct source regions located in a wide swath across the nearside, at depths between approximately 700 and 1200 km (Figure 24.7). Events from a single source are periodic at monthly (tidal) periods, and exhibit high degrees of waveform similarity, likely representing repeated failure on existing fault structures at depth.

Compared to terrestrial seismograms, lunar seismic signals exhibit characteristics typical of a large degree of wave scattering and very low attenuation, due in part to the fractured nature of the upper few hundred meters of lunar regolith. During events, seismic energy reverberates within the Moon, resulting in recorded signals of extremely long duration, sometimes an hour or more, with P- and S-wave codas that mask secondary arrivals. The small number of seismic stations and lack of high-quality seismic events limited the types of analyses that could be performed. Despite these limitations, data from the Apollo passive seismic network have been extensively analyzed to reveal details on the Moon's internal structure, confirming the presence of separate crust, mantle, and core layers (Figure 24.8).

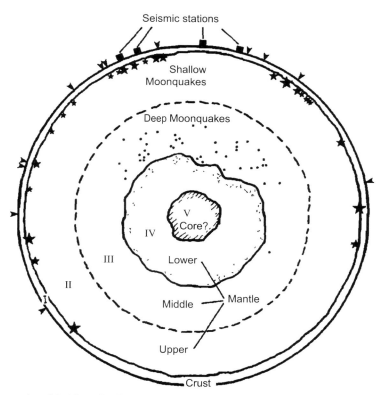

FIGURE 24.8 Schematic cross-section of the Moon showing the crust, mantle, and core layers inferred through analyses of the Apollo seismic data. The Apollo station locations and seismically active regions are indicated; arrowheads show a sample of meteoroid impact locations. *Figure reproduced from Nakamura, Latham, and Dorman (1982).*

The detailed structure of the upper kilometer of the lunar crust was determined by two additional seismic experiments: the Active Seismic Experiment on Apollo 14 and 16, and the Lunar Seismic Profiling Experiment on Apollo 17. In both experiments, the astronauts detonated a series of small explosives on the lunar surface (Figure 24.9). A network of geophones then recorded the ground motions generated by these explosions. On Apollo 14 and 16, up to 19 explosions were detonated by an astronaut using a "thumper" device along a 90-m-long geophone line. Additionally, on Apollo 16, three mortar shells were used to launch explosive charges to distances of up to 900 m from the ALSEP. On Apollo 17, the astronauts were able to position eight explosive charges at distances of up to 3.5 km from the Lunar Module, with the assistance of the Lunar Roving Vehicle. Both the Apollo 16 mortar shells and the Apollo 17 explosives were detonated by radio control after the astronauts left the lunar surface.

These experiments showed that the seismic P-wave velocity is between 0.1 and 0.3 km/s in the upper few hundred meters of the crust at all three landing sites. These velocities are much lower than observed for intact rock on Earth, but are consistent with a highly fractured material produced by the prolonged meteoritic bombardment of the Moon. At the Apollo 17 landing site, the surface basalt layer was determined to have a thickness of 1.4 km.

The lower crust and mantle seismic velocities have been estimated using the classical nonlinear inversion of compressional and shear wave arrival time readings made from the Apollo seismograms (Table 24.2). Shallow structure is constrained largely using surface and near-surface events (impacts and shallow moonquakes), while deep structure is constrained using midmantle and deeper

TABLE 24.2 Lunar Seismic P- and S-wave Velocity and Density Structure with Depth

Depth (km)	v_p (km/s)	v_s (km/s)	ρ (kg/m³)
0.0	1.0	0.5	2600
1.0	1.0	0.5	2600
1.0	3.2	1.8	2700
15.0	3.2	1.8	2700
15.0	5.5	3.2	2800
40.0	5.5	3.2	2800
40.0	7.7	4.4	3300
238.0	7.7	4.4	3300
238.0	7.8	4.4	3400
488.0	7.8	4.4	3400
488.0	7.6	4.4	3400
738.0	7.6	4.4	3400
738.0	8.5	4.5	3400
1257.1	8.5	4.5	3400
1257.1	7.5	3.2	3400
1407.1	7.5	3.2	3400
1407.1	4.1	0.0	5100
1497.1	4.1	0.0	5100
1497.1	4.3	2.3	8000
1737.1	4.3	2.3	8000

FIGURE 24.9 (Left) The Apollo 16 mortar package mounted on its base. The cable running off to the left connects the experiment to the ALSEP Central Station. The red flag at the top of the mast provided a visual warning for the crew to steer clear when driving the Lunar Rover (photo AS16-113-18,378 from the NASA Apollo Archive). (Right) Apollo Lunar Module Pilot Edgar Mitchell walks along the geophone line during the Apollo 14 mission, operating the "thumper". *Photo AS14-67-9374 from the NASA Apollo Archive.*

events (deep moonquakes). Seismic velocities increase steadily down to approximately 20 km. At that depth, there is a change in velocities within the crust that probably represents the depth to which extensive fracturing, due to massive impacts, has occurred (see Figure 24.11 and Section 4.1). At an earlier stage of seismic data analysis, this velocity change was thought to represent the base of the mare basalts, but these are now known to be much thinner. The main section of unbroken crust has rather uniform velocities that correspond to the velocities expected from the average anorthositic composition of the lunar samples.

Very few seismic rays detected by Apollo traverse the region below the deep moonquake zone. Evidence for a highly attenuating region in the deep interior such as a layer of partial melt or a fluid lunar core is implied in part by a lack of observation of seismic signals originating from the farside of the Moon (Figure 24.10). Since deep moonquakes are generally small, their energy cannot penetrate the attenuating region to reach the nearside Apollo array. An additional interpretation of the lack of farside signals is that the farside is aseismic, which, given the other global nearside/farside asymmetries (e.g. crustal thickness and mare distribution), is not outside the realm of possibility. Further seismic exploration is needed to resolve this issue.

4. LUNAR INTERNAL STRUCTURE

Decades of research following the Apollo era has led to a model of the lunar interior that consists of a silicate crust and mantle, and a small iron core (Figure 24.1). Overlying the crust is a very thin layer of extremely pulverized material known as the regolith, resulting from the initial heavy and continued bombardment of meteorites on the lunar surface. As discussed previously, the crust is globally asymmetric in thickness, with the nearside on average thinner than the farside. The mantle is considered to be largely homogeneous, with increases in seismic velocity and density of only a few percent from the base of the crust down to the partial melt boundary layer between the mantle and core. The core itself likely consists of a fluid outer layer and a solid inner layer. This section will provide details on the structure of the lunar interior.

4.1. Regolith

The surface of the Moon is covered with a debris blanket, called the regolith, produced by the impacts of meteorites (Figure 24.11). It ranges in scale from fine dust to blocks several meters across. Although there is much local variation, the average regolith thickness on the maria is 4–5 m, whereas the highland regolith is about 10 m thick.

Seismic velocities are only about 100 m/s at the surface, but increase to 4.7 km/s at a depth of 1.4 km at the Apollo 17 site. The density is about 1.5 g/cm^3 at the surface, increasing with compaction to about 1.7 g/cm^3 at a depth of 60 cm. The porosity at the surface is about 50% but is strongly compacted at depth. The individual crater ejecta blankets that comprise the regolith typically range in thickness from a few millimeters up to about 10 cm, derived from the multitude of meteorite impacts at all scales. These have little lateral continuity even on scales of a few meters. Most of the regolith is of local origin: lateral mixing occurs only on a local scale so that the mare–highland contacts are relatively sharp over a kilometer or so. The rate of growth of the regolith is very slow, averaging about 1.5 mm/million years or 15 Å/year, but it was more rapid between 3.5 and 4 billion years ago during the late heavy bombardment.

The regolith consists of fragments of minerals, crystalline rocks, and **breccias**, as well as impact glasses and

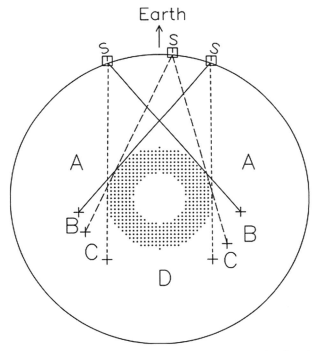

FIGURE 24.10 Schematic cross-section of the Moon showing the bounding seismic rays for shear wave shadows from the deep attenuating region. The direction to Earth is indicated. Small squares denote the Apollo seismic station locations (Apollo 12 and 14 are approximately colocated, so only one square is shown for clarity). The crosses mark the locations of hypothetical deep moonquake hypocenters for which the seismic rays to a station at a corner of the triangular station network graze the lower mantle, in which shear waves are severely attenuated. For events in Zone A, clear shear wave arrivals may be observed at all three corners of the network, i.e. none of the seismic stations are in the S-wave shadow. For events in Zone B, clear shear wave arrivals may be observable at two corners of the network, i.e. one corner of the array is in the S-wave shadow. For events in Zone C, clear shear wave arrivals may be observable only at one corner of the network, i.e. two corners of the array are in the S-wave shadow. For events in Zone D, no clear S-wave arrivals are observable at all three corners of the network, i.e. all corners of the array are in the S-wave shadow. *Figure reproduced from Nakamura (2005).*

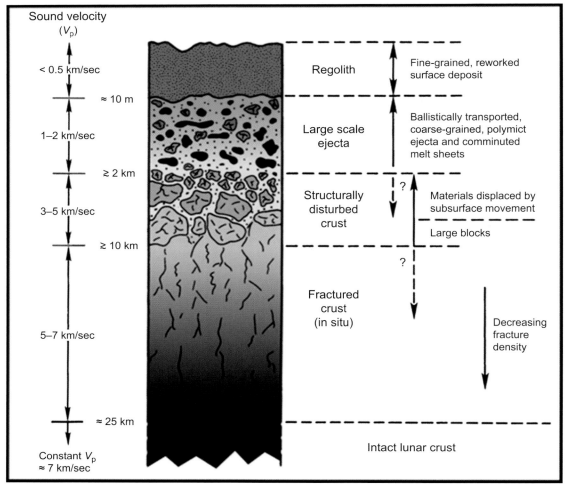

FIGURE 24.11 Schematic cross-section of the upper 25 km of the lunar surface, illustrating the effects of large-scale cratering on the structure of the lunar crust. *Figure reproduced from the Lunar Source Book.*

agglutinates. The latter are aggregates of smaller particles welded together by glasses. They may compose 25–30% of a typical regolith sample and tend to an equilibrium size of about 60 μm. Their abundance in a sample is a measure of its maturity, or length of exposure to meteoritic bombardment. Most lunar regolith samples reached a steady state in particle size and thickness. Agglutinates contain metallic iron droplets (typically 30–100 Å) referred to as "nanophase" iron, produced by surface interaction with the solar wind during melting of the regolith by meteorite impact.

A "megaregolith" of uncertain thickness covers the heavily cratered lunar highlands. This term refers to the debris sheets from the craters and particularly those from the large impact basins that have saturated the highland crust. The aggregate volume of ejecta from the presently observable lunar craters amounts to a layer about 2.5 km thick. Earlier bombardment may well have produced megaregolith thicknesses in excess of 10 km. Related to this question is the degree of fracturing and brecciation of the deeper crust due to the large basin collisions (Figure 24.11). Some estimates equate this fracturing with the leveling off in seismic compression-wave velocities (v_p) to an approximately constant 7 km/s at 20–25 km depth. In contrast to the highlands, bedrock is present at relatively shallow depths (tens of meters) in the lightly cratered maria.

4.2. Crust

Recent analyses of both seismic and gravimetric data from the Moon indicate that the average crustal thickness lies between 34 and 43 km. The farside crust averages about 15 km thicker than that of the nearside. The crust thus constitutes about 9% of lunar volume. The maximum relief on the lunar surface is over 16 km. The deepest basin (South Pole–Aitken) has a 12 km relief.

The mare basalts cover 17% of the lunar surface, mostly on the nearside (Figure 24.12). Although prominent

Chapter | 24 Interior of the Moon

FIGURE 24.12 A composite full-Moon photograph that shows the contrast between the heavily cratered highlands and the smooth, dark basaltic plains of the maria. Mare Imbrium is prominent in the northwest quadrant. The dark, irregular, basalt-flooded area on the west is Oceanus Procellarum. Mare Crisium is the dark circular basalt patch on the eastern edge. *Figure courtesy of UCO/Lick Observatory, photograph L9.*

visually, they are usually less than 1 or 2 km thick, except near the centers of the basins. These basalts constitute only about 1% of the volume of the crust and make up less than 0.1% of the volume of the Moon.

Seismic velocities increase steadily down to around 25 km. At that depth, there is a change in velocities within the crust, which probably represents the depth to which extensive fracturing due to massive impacts has occurred (Figure 24.11). At an earlier stage, this velocity change was thought to represent the base of the mare basalts, but these are now known to be much thinner. The main section of the crust from approximately 20–40 km has rather uniform shear (S) and compression (P) wave velocities, corresponding to the velocities expected from the average anorthositic composition of the lunar samples.

4.3. Mantle

The structure of the mantle has been difficult to evaluate on account of the complexity of interpreting the lunar seismograms. From MOI considerations alone we know it is largely homogeneous at least in density. The average P-wave velocity is 7.7 km/s and the average S-wave velocity is 4.45 km/s down to about 1200 km. Most models postulate a pyroxene-rich upper mantle that is distinct from an olivine-rich lower mantle beneath about a depth of 500–600 km. Seismic data are ambiguous regarding the nature of the lunar mantle below 500 km. They may be interpreted as representing magnesium-rich olivines or indicate the presence of garnet. If the latter is present, this implies that the Moon has a larger bulk aluminum content than has been predicted previously, which has important implications for reconstructing the early evolution of the Moon. However, this distinction cannot be made on the basis of the Apollo seismic data alone.

The seismically active deep moonquake zone lies deep within the lower mantle at about 800–1000 km depth (Figure 24.8). The very low seismic attenuation observed in the outer 800 km of the mantle is indicative of a volatile-free rigid lithosphere. Solid-state mantle convection is thus extremely unlikely in the Moon.

Below about 800 km, P- and S-waves become attenuated ($v_S = 2.5$ km/s). P-waves are transmitted through the center of the Moon, but S-waves are missing, possibly suggesting the presence of a melt phase (Figure 24.10). It is unclear, however, whether the S-waves were not transmitted or were so highly attenuated that they were not recorded by the nearside Apollo array.

4.4. Core

The evidence for a metallic core is suggestive but inconclusive. As discussed previously, current (indirect) constraints on core properties arise from MOI considerations, the LLR experiment, magnetic field strength studies, and analyses of elemental abundances in mare basalts. These estimates are varied, and the presence of a lunar core (and its properties, if existent) is a topic of debate among the planetary science community.

Electromagnetic sounding can also constrain deep lunar structure by estimating the electrical conductivity as a function of depth. This approach applied to data from the Apollo orbital magnetometers results in an upper limit of 400–500 km radius for a highly conducting (e.g. metallic) core. The MOI value of 0.3931 ± 0.0002 is low enough to require a small density increase in the deep interior, in addition to the low-density crust. Although a metallic core with a radius of about 400 km (4% of lunar volume) is consistent with the available data, denser silicate phases might be present. The resolution of these problems requires improved seismic data.

A direct seismic constraint on the size and state of the lunar core (through observation of reflected and/or converted core phases on Apollo seismograms) has not been achieved, due in part to the strong scattering of seismic energy in the lunar regolith and the limited sensitivity of the instruments. Many deep moonquake signals occurred at or just slightly above the Apollo instrument detection threshold (approximately 5.4×10^{-9} cm of ground motion at 1 Hz), and if any seismic phases were observed at all, these were typically the main P- and S-wave arrivals. Since

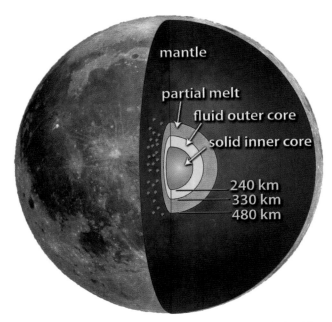

FIGURE 24.13 Schematic meridional cross-section of the Moon showing the approximate distribution of deep moonquakes (red circles) and the radii of physical layers in the deepest lunar interior, as resolved from recent reanalyses of the Apollo seismic data. *Figure reproduced from Weber, et al. (2011).*

the predicted amplitudes of the lunar core phases are many times smaller, the extended coda of the primary arrivals obscured their identification on individual Apollo seismograms. However, recent application of a terrestrial seismology technique known as "array stacking" revealed the presence of several core-reflected phases, arguing strongly for a solid inner core, a fluid outer core, and a partial melt boundary layer that likely accounts for the lack of observed farside deep moonquake signals (Figure 24.13).

5. IMPLICATIONS FOR LUNAR FORMATION AND EVOLUTION

A practical model for lunar formation must result in a Moon with present-day internal structure that is consistent with the observations presented in this chapter. In addition, it must be able to explain the high value for the angular momentum of the Earth–Moon system, the lunar orbit inclined at 5.09° to the plane of the ecliptic, the high mass relative to that of its primary planet, and the low bulk density of the Moon, much less than that of the Earth or the other inner planets. The chemical age and isotopic data revealed by the returned lunar samples added additional complexities to these classic problems because the lunar composition is unusual by either cosmic or terrestrial standards. It is perhaps not surprising that previous theories for the origin of the Moon failed to account for this diverse set of properties and that only recently has something approaching a consensus been reached.

Hypotheses for lunar origin can be separated into five categories:

1. Capture of an intact Moon from an independent orbit
2. Simultaneous Earth–Moon formation as a double planet (or "coaccretion")
3. Fission of the Moon from a rapidly rotating Earth
4. Disintegration of incoming planetesimals
5. Earth impact by a Mars-sized planetesimal and capture of the resulting debris into Earth orbit

These are not all mutually exclusive, and elements of some hypotheses occur in others. For example,

1. Capture of an already formed Moon from an independent orbit has been shown to be highly unlikely on dynamic grounds. The hypothesis provides no explanation for the peculiar composition of our satellite. In addition, it could be expected that the Moon might be an example of a common and primitive early solar system object, similar to the captured rock-ice satellites of the outer planets, particularly since the Moon's density is similar to that of primitive carbonaceous chondrites. It would, however, be an extraordinary coincidence if the Earth had captured an object with a unique composition, in contrast to the many examples of icy satellites captured by the giant planets.
2. Formation of the Earth and the Moon in association as a double-planet system immediately encounters the problems of differing density and composition of the two bodies. Various attempts to overcome the density problem led to coaccretion scenarios in which disruption of incoming differentiated planetesimals formed from a ring of low-density silicate debris. Popular models to provide this ring involved the breakup of differentiated planetesimals as they come within a **Roche limit** (about three Earth radii). The denser and tougher metallic cores of the planetesimals survived and accreted to the Earth, while their mantles formed a circumterrestrial ring of broken-up silicate debris from which the Moon could accumulate. This attractive scenario has been shown to be flawed because the proposed breakup of planetesimals close to the Earth is unlikely to occur. It is also difficult to achieve the required high value for the angular momentum in this model. Such a process might be expected to have been common during the formation of the terrestrial planets, and Venus, in particular, could be expected to have a satellite.
3. In 1879, George Darwin proposed that the Moon was derived from the terrestrial mantle by rotational fission (see cartoon illustration in Figure 24.14). Such fission hypotheses have been popular since they produced a

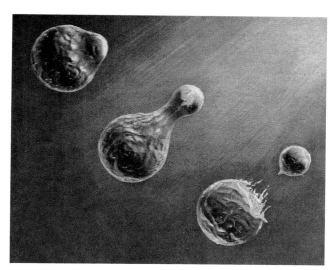

FIGURE 24.14 Artist's rendition of the fission theory of lunar formation, in which the Moon was formed from the mantle of a hot, early Earth. In this theory, the molten Earth rotates rapidly, causing a blob of material to spin out, which later rounded into the Moon. *Figure reproduced from Binder (1961).*

low-density, metal-poor Moon. The lunar sample return provided an opportunity to test these hypotheses because they predicted that the bulk composition of the Moon should provide some identifiable signature of the terrestrial mantle. The O and Cr isotopic compositions are similar, and this is sometimes used to argue for a lunar origin from the Earth's mantle. However, other significant chemical differences remain. The Moon contains, for example, 50% more FeO and has distinctly different trace siderophile element signatures. It also contains higher concentrations of refractory elements (e.g. Al and U) and lower amounts of volatile elements (e.g. Bi and Pb). The similarity in V, Cr, and Mn abundances in the Moon and the Earth is nonunique since CM, CO, and CV chondrites show the same pattern. These differences between the chemical compositions of the Earth's mantle and the Moon are fatal to theories that wish to derive the Moon from the Earth. But perhaps more importantly, the angular momentum of the Earth-Moon system, although large, is insufficient by a factor of about four to allow for rotational fission. If the Earth had been spinning fast enough for fission to occur, there is no available mechanism for removing the excess angular momentum following lunar formation.

4. One proposed modification of the fission hypothesis uses multiple small impacts to place terrestrial mantle material into orbit that sequentially accretes into a moon. However, it is exceedingly difficult to obtain the required high angular momentum by such processes because multiple impacts are random in both direction and energy, and the angular momentum they impart to the Earth over time should average out.

Most of these Moon-forming hypotheses should be general features of planetary and satellite formation and should produce Moon-like satellites around the other terrestrial planets. They either fail to account for the high angular momentum (relative to the other terrestrial planets) of the Earth–Moon system and the volatile-depleted composition of the Moon, or they do not account for the differences between the lunar composition and that of the terrestrial mantle.

The single-impact hypothesis was developed to solve the angular momentum problem, but, in the manner of successful hypotheses, it has accounted for other parameters as well and has become a consensus in the lunar science community. The theory proposes that, during the final stages of accretion of the terrestrial planets, a body about the size of Mars collided with the Earth and spun out a disk of material from which the Moon formed. This giant impact theory resolves many of the problems associated with the origin of the Moon and its orbit. The following scenario is one of several possible, although restricted, variations on the theme.

In the closing stages of the accretion of the terrestrial planets from the protoplanetary disk surrounding our Sun, the Earth suffered a grazing impact with an object of about 0.10 Earth mass. This body is assumed to have been differentiated into a silicate mantle and a metallic core. Because the oxygen and chromium signatures of the Earth and Moon are identical and the impact velocities are required to be low in the formation models, the impactor likely came from the same general region of the initial planet-forming nebula as the Earth.

The impactor was disrupted by the collision and the resulting debris mostly went into orbit about the Earth. Gravitational torques, due to the asymmetrical shape of the Earth following the impact, assisted in accelerating material into orbit. Expanding gases from the vaporized part of the impactor also promoted material into orbit. Following the impact, the mantle material from the impactor was accelerated, but its metallic core remained as a coherent mass and was decelerated relative to the Earth, so that it fell into the Earth within about 4 h. A metal-poor mass of silicate material, mostly from the mantle of the impactor, remained in orbit.

In some variants of this hypothesis, the orbiting material immediately coalesced to form a totally molten Moon. In others, it broke up into several "moonlets" that subsequently accreted to form a partly molten Moon. This highly energetic event accounts for the geochemical evidence that indicates that at least half the Moon was molten shortly after accretion. Figure 24.15 illustrates several stages of a computer simulation of the formation of the Moon

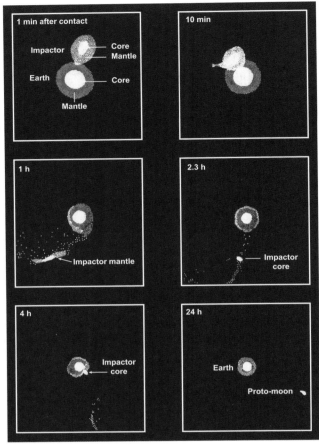

FIGURE 24.15 A computer simulation of the origin of the Moon by a glancing impact of a body approximately the same size of Mars with the early Earth. This event occurred about 4.5 billion years ago during the final stages of accretion of the terrestrial planets. By that time, both the impactor and the Earth had differentiated into a metallic core and rocky silicate mantle. Following the collision, the mantle of the impactor is ejected into orbit. In this simulation, the metallic core of the impactor clumps together and falls into the Earth within about 4 h. Most terrestrial mantle material ejected by the impact follows a ballistic trajectory and is eventually reaccreted by the Earth. The metal-poor, low-density Moon is thus derived mainly from the silicate mantle of the impactor. *Figure courtesy of A. G. W. Cameron.*

according to one version of the single giant impact hypothesis.

Unique events are notoriously difficult to accommodate in most scientific disciplines. An obvious requirement in this model is that a suitable population of impactors existed in the early solar system. Evidence in support of the previous existence of large objects in the early solar system comes from the ubiquitous presence of heavily cratered ancient planetary surfaces, from the large number of impact basins with diameters up to 2000 km or so, and from the obliquities or tilts of the planets, all of which demand collisions with large objects in the final stages of accretion. The extreme example is that an encounter between Uranus and an Earth-sized body is required to tip that planet on its side. Thus, the possibility of many large collisions in the early solar system is well established, one of which had the right parameters to form the Moon. The single-impact scenario is thus consistent with the planetesimal hypothesis for the formation of the planets from a hierarchical sequence of smaller bodies.

BIBLIOGRAPHY

Binder, O. O. (1961). *The Moon, our neighboring world. Series: The Golden Library of Knowledge.* Golden Press.

Dickey, J. O., Bender, P. L., Faller, J. E., Newhall, X. X., Ricklefs, R. L., Ries, J. G., et al. (1994). Lunar laser ranging: a continuing legacy of the Apollo program. *Science, 265*(5171), 482–490.

Heiken, G. H., Vaiman, D. T., & French, B. M. (1991). *The Lunar Source Book.* Cambridge University Press. An updated electronic version is available from http://www.lpi.usra.edu/publications/books/lunar_sourcebook/.

McKay, D. S., Heiken, G., Basu, A., Blanford, G., Simon, S., Reedy, R., et al. (1991). The lunar regolith. In G. H. Heiken, D. T. Vaniman, & B. M. French (Eds.), *The lunar source book.* Cambridge University Press.

Nakamura, Y. (2005). Farside deep moonquakes and deep interior of the Moon. *Journal of Geophysical Research, 110.* http://dx.doi.org/10.1029/2004JE002332.

Nakamura, Y., Latham, G. V., & Dorman, H. J. (1982). Apollo lunar seismic experiment – final summary. In *Proceedings of the 13th lunar and planetary science conference* (pp. 117–123).

Warren, P. H., & Rasmussen, K. L. (1987). Megaregolith insulation, internal temperatures, and bulk uranium content of the Moon. *Journal of Geophysical Research, 92*(B5), 3453–3465.

Weber, R. C., Lin, P., Garnero, E. J., Williams, Q., & Lognonné, P. (2011). Seismic detection of the lunar core. *Science, 331,* 309–312.

Wieczorek, M. A., Jolliff, B. L., Khan, A., Pritchard, M. E., Weiss, B. P., Williams, J. G., et al. (2006). The Constitution and structure of the lunar interior. *Reviews in Mineralogy and Geochemistry, 60,* 221–364.

Zuber, M. T., Smith, D. E., Watkins, M. M., Asmar, S. W., Konopliv, A. S., Lemoine, F. G., et al. (2013). Gravity field of the moon from the gravity Recovery and interior Laboratory (GRAIL) mission. *Science, 339,* 668–671.

Chapter 25

Lunar Exploration

Ian A. Crawford
Department of Earth and Planetary Sciences, Birkbeck College, University of London, UK

Katherine H. Joy
School of Earth, Atmospheric and Environmental Sciences, University of Manchester, UK

Mahesh Anand
Department of Physical Sciences, The Open University, UK

Chapter Outline

1. Introduction — 555
2. Telescopic Exploration of the Moon — 556
3. The Early Space Age — 557
4. The Apollo Program — 559
   4.1. Apollo Samples and Their Analysis — 559
   4.2. Apollo Surface Experiments — 560
   4.3. Apollo Orbital Remote Sensing — 562
   4.4. Apollo and the Benefits of Human Exploration — 565
5. Post-Apollo Exploration — 566
   5.1. Clementine — 566
   5.2. Lunar Prospector — 566
   5.3. Small Missions for Advanced Research in Technology-1 — 567
   5.4. Kaguya — 568
   5.5. Chandrayaan-1 — 568
   5.6. Chang'e 1 and 2 — 569
   5.7. Lunar Reconnaissance Orbiter — 569
   5.8. LCROSS — 571
   5.9. Gravity Recovery and Interior Laboratory — 571
   5.10. Lunar Atmosphere and Dust Environment Explorer — 572
   5.11. Chang'e 3 Lunar Lander and Rover — 572
6. Letting the Moon Come to Us: The Importance of Lunar Meteorites for Lunar Exploration — 572
7. Future Lunar Exploration Objectives — 573
   7.1. Science of the Moon — 573
   7.2. Science on the Moon — 575
   7.3. Science from the Moon: Lunar-Based Astronomy — 576
   7.4. Lunar Resources — 576
   7.5. The Moon as a Stepping Stone for the Exploration of Mars and other Solar System Destinations — 578
8. Conclusion — 578
Acknowledgments — 578
Bibliography — 579
   Websites — 579

1. INTRODUCTION

Owing to the Moon's relative proximity to the Earth, lunar exploration has played a pivotal role in the exploration of the solar system. This was true of Galileo's first telescopic observations, through to the first spacecraft to orbit, land on, and return samples from another celestial body. And, of course, the Moon was the first, and is to date the only, celestial body that human beings have explored in person—during the Apollo program between 1969 and 1972. During the Apollo missions, samples were collected, measurements were made, and instruments were deployed that have revolutionized lunar and planetary science, and which continue to have a major scientific impact today. And although Apollo was followed by a 20-year hiatus in lunar missions, the past two decades have seen a renaissance in lunar exploration conducted from orbit. This has resulted in a wealth of new data which have in many ways revolutionized our knowledge of the Moon and has also taught us much about the early history of the solar system, the geological evolution of rocky planets, and the near-Earth cosmic environment throughout solar system history. It has also become clear that the Moon still has much more to tell us about all these areas of planetary science and that strong arguments exist for a renewed program of lunar exploration in the twenty-first century.

2. TELESCOPIC EXPLORATION OF THE MOON

The scientific exploration of the Moon essentially began with Galileo's first telescopic observations in 1609, as reported in his book *Sidereus Nuncius*. For the first time these observations demonstrated that the Moon is an essentially Earth-like planetary body, with mountains and valleys, rather than being a perfect sphere made of special celestial material (Figure 25.1). Galileo was also able to estimate the heights of lunar features by measuring the lengths of shadows and thus initiated a program of the quantitative measurement of lunar topography which continues to this day. It was also from this early period that the smooth, low-lying, dark areas of the lunar surface became identified with seas ("maria"), and the rougher, higher, brighter areas with land ("terrae"), introducing the top-level lunar geographical nomenclature that is still in use.

In the decades following Galileo's initial telescopic observations, European astronomers began mapping the nearside of the Moon in earnest. Essentially all the major maria, and most of the prominent craters, had been mapped and assigned their current names by the middle of the seventeenth century, when Giovanni Riccioli published his influential lunar map. Telescopic cartography of the lunar nearside continued for more than 300 years, employing increasingly sophisticated techniques, and indeed played an important part in planning the Apollo missions during the 1960s. Especially during the latter part of this period, telescopic observations were used to make geological maps of the nearside and to address key questions in lunar geology. These included studies of the origin of craters and the nature of lunar volcanism, the identification and interpretation of tectonic features such as faults and graben, and, perhaps most importantly, the definition of the lunar stratigraphic column which still underpins much of lunar geology.

More recently, the recognition that different minerals have different reflectance spectra provided another tool for telescopic lunar geology. This technique made it possible to identify dominant surface rock types spectroscopically and to explore subsurface compositional variations by observing the ejecta from large craters. However, the relatively low spatial resolution achievable with ground-based telescopes, and the fact that only the nearside is accessible from the Earth, means that spectral studies of this kind have now largely been superseded by remote sensing from orbiting spacecraft. Indeed, despite the advances made by telescopic observations over the centuries, most of our knowledge about the Moon has been obtained

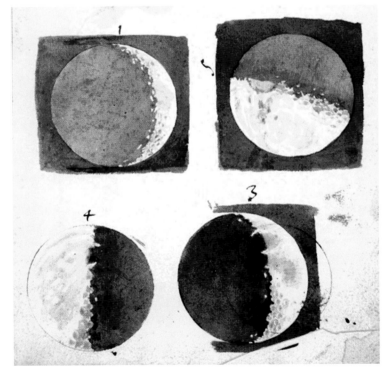

FIGURE 25.1 A set of Galileo's Moon drawings based on his early telescopic observations. Note the correct rendition of the heavily cratered lunar surface, and the "islands" of illuminated mountain peaks beyond the terminator (most easily seen in 3), from which Galileo calculated the heights of lunar mountains. *Biblioteca Nazionale Centrale, Florence.*

Chapter | 25 Lunar Exploration

since space technology made it possible to conduct observations from lunar orbit, make in situ measurements on the surface, and collect rock samples for analysis in terrestrial laboratories.

3. THE EARLY SPACE AGE

A summary of the most important spacecraft to have visited the Moon is given in Table 25.1. The first spacecraft to reach the Moon was the Soviet Luna 2, which impacted the lunar surface on September 13, 1959. Luna 2 also made the first attempt to measure the lunar magnetic field (which it failed to detect), and the charged particle and micrometeoroid environments around the Moon. Of greater significance for lunar geology was the flight of Luna 3, in October that same year, which completed the first flyby of the Moon and obtained the first ever images of the farside (Figure 25.2). Even these first observations were sufficient to show that the surface of the farside is different from that of the nearside, with

TABLE 25.1 Highlights of Lunar Exploration by Spacecraft

Spacecraft/Program Name	Nationality	Launch Year	Mission Description
Luna 2	USSR	1959	First lunar impact
Luna 3	USSR	1959	Flyby: first farside images
Ranger probes (3/7)*	USA	1962–1965	Impact probes: near-surface imagery
Luna 9	USSR	1966	Soft lander: first surface images
Luna 10	USSR	1966	First lunar orbiter
Surveyor landers (5/7)	USA	1966–1968	Soft landers: surface properties
Lunar orbiters (5/5)	USA	1966–1967	Orbiters: orbital photography
Apollo 8, 10	USA	1968	Manned lunar orbiters: orbital photography
Apollo 11, 12, 14–17	USA	1969–1972	Manned landings: surface and interior properties, sample return, orbital remote sensing
Lunokhod 1, 2 (Luna 17, 21)	USSR	1970, 1973	Robotic rovers: surface properties
Luna 16, 20, 24	USSR	1970, 1972, 1976	Robotic sample return
Hiten (MUSES-A)	Japan	1990	Lunar orbiter: dust detection
Clementine	USA	1994	Orbital remote sensing
Lunar Prospector	USA	1998	Orbital remote sensing
SMART-1	Europe	2003	Orbital remote sensing
Kaguya	Japan	2007	Orbital remote sensing
Chang'e 1	China	2007	Orbital remote sensing
Chandrayaan-1	India	2008	Orbital remote sensing
Lunar Reconnaissance Orbiter	USA	2009	Orbital remote sensing
Lunar Crater Observation and Sensing Satellite (LCROSS)	USA	2009	Impact probe, polar volatile detection
Chang'e 2	China	2010	Orbital remote sensing
Gravity Recovery and Interior Laboratory (GRAIL)	USA	2011	Orbital gravity mapping
Lunar Atmosphere and Dust Environment Explorer (LADEE)	USA	2013	Orbital exospheric gas and dust characterization
Chang'e 3	China	2013	Soft lander with rover: surface properties

*For programs consisting of several spacecraft, the numbers in parentheses denote the fraction of successful missions.

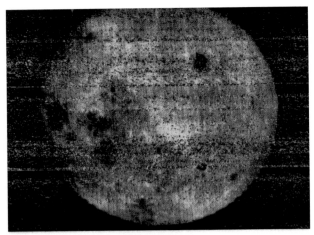

Фотография 1

FIGURE 25.2 The first image ever taken of the lunar farside by Luna 3 on October 7, 1959, from a distance of 63,500 km. It showed that the farside is different from the nearside in lacking extensive areas of dark maria. Although noisy, many features can be recognized: the dark spot at upper right is Mare Moscoviense, the two dark areas in the lower left are Mare Marginis and Mare Smythii, and the small dark feature at lower right with the white dot in the center is the crater Tsiolkovsky and its central peak. *NASA NSSDC Data Set ID 59-008A-01A.*

few of the large, dark expanses of basaltic lava that dominate the latter.

There was a gap of over 6 years until the next two major milestones in lunar exploration, both in 1966 and both again by Soviet space probes: on February 3, 1966, Luna 9 successfully soft landed and obtained the first surface images (Figure 25.3), and on April 3, Luna 10 became the first spacecraft to enter orbit about the Moon. During this apparent hiatus, however, the US lunar exploration program started ramping up, initially with the Ranger series of "hard landers", designed to take ever increasing resolution images of the surface before crashing into it. Although the Ranger project had limited success (with only three of the seven probes launched functioning as intended), it paved the way for the much more successful Surveyor program which soft landed five robotic landers (of the seven attempts) on the lunar surface between 1966 and 1968 (Figure 25.4). These vehicles used a variety of instruments to measure regolith properties and composition.

In parallel, between 1966 and 1967 the United States flew a highly successful series of Lunar Orbiter spacecraft that were designed to obtain high-resolution images of the lunar surface. With surface resolutions of several tens of meters (occasionally as high as 2 m) for some areas of the Moon, these images long remained unsurpassed as a resource for lunar geology (although they are rapidly becoming superseded by images obtained by the Lunar Reconnaissance Orbiter (LRO) Camera discussed below). To a large extent, the Lunar Orbiter missions were designed to identify potential landing sites for the manned Apollo missions then under development, just as the Surveyors were designed to provide knowledge of the surface environment with the manned landings in mind.

The Apollo program is discussed in detail in Section 4 owing to its pivotal importance in lunar exploration. However, before leaving the discussion of this early phase of lunar exploration, it is necessary to highlight two important Soviet robotic programs that overlapped with it. These were the two "Lunokhod" rovers (Luna 17 and 21) that landed on the Moon in November 1970 and January

FIGURE 25.3 Mosaic of images of the lunar surface obtained by Luna 9 in February 1966 from its landing site in Oceanus Procellarum (at approximately 7.08° N, 64.37° W). The horizon is approximately 1.4 km away. *'First Panoramas of the Lunar Surface According to the Material from the Automatic Station Luna-9', NASA NSSDC/Akademiia nauk SSSR.*

FIGURE 25.4 Charles "Pete" Conrad, Commander of Apollo 12, stands next to Surveyor III in November 1969. Surveyor III had landed two and a half years earlier in April 1967. Conrad's right arm is pointing to the Surveyor III camera system, and the soil sampling scoop passes in front of his knees; the Apollo 12 Lunar Module is in the background. *NASA image AS12-48-7134.*

1973, respectively. These were the first teleoperated robotic rovers to operate on another planetary body. Lunokhod 1 operated for 322 days and traversed a total distance of 10.5 km; the corresponding numbers for Lunokhod 2 were 115 days and 37 km. During their traverses, the Lunokhods made measurements of the regolith properties and composition, as well as the surface radiation environment. They also carried laser reflector panels which, similar to those deployed by the Apollo 11, 14, and 15 missions, have been used to measure the Earth–Moon distance and the Moon's physical librations (from which information about its internal structure can be inferred).

The other significant Soviet missions during this period were the three successful robotic sample return missions Luna 16, 20, and 24. These collected, and returned to the Earth, a total of approximately 320 g from three sites close to the eastern limb of the nearside. Although the quantity of material collected was small compared to that returned by *Apollo* (see below), their geographical separation from the Apollo landing sites (Figure 25.5) makes them important for our understanding of lunar geological diversity and the calibration of remote sensing measurements.

4. THE APOLLO PROGRAM

On May 25, 1961, the US President John F. Kennedy announced before a joint session of Congress his proposal to land a man on the Moon by the end of the 1960s. This initiated the Apollo program, which, at the peak of its development phase in 1965 consumed almost 4% of the US federal budget and which succeeded in landing Neil Armstrong and Edwin "Buzz" Aldrin on the surface of Mare Tranquillitatis with Apollo 11 on July 20, 1969. In the three and a half years between Armstrong's "one small step" in 1969 and the departure of the Apollo 17 astronauts, Gene Cernan and Harrison "Jack" Schmitt, from the Moon in December 1972, a total of 12 astronauts explored the lunar surface in the immediate vicinity of six Apollo landing sites (Figure 25.5). The total cumulative time spent on the lunar surface was 12.5 days, with just 3.4 days spent performing extravehicular activities (EVAs) outside the lunar modules. Yet during this short time samples were collected, measurements made, and instruments deployed which have revolutionized lunar and planetary science and which continue to have a major scientific impact today.

In their cumulative 12.5 days (25 man-days) on the lunar surface, the 12 Apollo moonwalkers traversed a total distance of 95.5 km from their landing sites (heavily weighted to the last three missions that were equipped with the Lunar Roving Vehicle; Figure 25.6), collected and returned to Earth 382 kg of rock and soil samples (from over 2000 discrete sampling localities), drilled three geological sample cores to depths greater than 2 m, obtained over 6000 surface images, and deployed over 2100 kg of scientific equipment. These surface experiments were supplemented by wide-ranging remote sensing observations conducted from the orbiting Command and Service Modules (CSM; Figure 25.7), which are equally part of the Apollo scientific legacy. Table 25.2 lists the Apollo surface and orbital science experiments.

4.1. Apollo Samples and Their Analysis

The most important scientific legacy of Apollo has resulted from analysis of the 382 kg of rock (Figure 25.8(a)–(e)) and

FIGURE 25.5 Nearside lunar mosaic constructed from Clementine 750-nm albedo data, with the landing sites of the six Apollo and three Luna sample return missions indicated. *USGS/K.H. Joy*

FIGURE 25.6 Lunar Module Pilot Jim Irwin next to the Apollo 15 Lunar Roving Vehicle (LRV) with the 4.6-km-high Mt Hadley in the background; note the sample bags attached to the rear of the LRV. *NASA image AS15-86-11,603.*

FIGURE 25.7 The Scientific Instrumentation Module (SIM) bay of the Apollo 15 Command and Service Module (CSM). On Apollo 15, the SIM included mapping cameras, a laser altimeter, a mass spectrometer for exosphere composition measurements, X-ray and gamma ray spectrometers for surface composition studies, and a subsatellite for measuring charged particles and magnetic fields (see Table 25.2). *NASA image AS15-88-11971.*

soil (Figure 25.9) samples returned to Earth. These samples have been analyzed in laboratories around the world to address many questions about the Moon's formation, differentiation, geological evolution, resource potential, and the records that they contain of wider solar system processes. Analytical techniques that have been employed to study lunar materials are summarized in Table 25.3.

Probably the most important result based on the returned samples has been the calibration of the lunar cratering rate. Not only has this facilitated the dating of lunar surfaces from which samples have yet to be obtained, but it is used, with assumptions, to estimate the ages of cratered surfaces throughout the solar system from Mercury to the moons of the outer planets.

Perhaps the next most important result of Apollo sample analysis from a planetary science point of view has been the evidence provided for the origin of the Moon. In particular, the discovery that lunar materials have compositions broadly similar to those of the Earth's mantle (including nearly identical isotope ratios for oxygen and several other elements), but that the Moon is highly depleted of volatiles compared to the Earth and has only a small iron core, led to the current paradigm that the Moon formed from debris resulting from a giant impact of a Mars-sized (or possibly larger) planetesimal with the early Earth. Constraining theories of lunar origins are of wide significance for planetary science because they inform our understanding of the general processes of planet formation through the merger of planetesimals in the early solar system. It is very doubtful that we would have sufficient geochemical evidence usefully to constrain theories of lunar origins without the quantity and diversity of samples provided by Apollo.

Beyond this, the Apollo samples have been vital to our understanding of the Moon's geological history and evolution, and have, therefore, helped elucidate important geological processes (including core formation, magma ocean evolution, and primary and secondary crust formation through magmatic and volcanic activity) that are of general relevance to our understanding of terrestrial planet evolution. In addition, Apollo samples of the lunar regolith have demonstrated the importance of the lunar surface layers as an archive of material that has impacted the Moon throughout its history. These include records of solar wind particles, the cosmogenic products of cosmic ray impacts, and meteoritic debris (Figure 25.9(f)). Extracting meteoritic records from lunar regolith samples is especially important for planetary science as it potentially provides a means of determining how the flux and composition of asteroidal material in the inner solar system has evolved through time.

4.2. Apollo Surface Experiments

Although the study of the Apollo samples has been, and continues to be, important for lunar and planetary science, many other areas of scientific investigation were also performed by the Apollo missions. Many of these were conducted as part of the Apollo Lunar Surface Experiment Packages (ALSEPs), which were deployed at all the Apollo landing sites apart from Apollo 11 (which deployed a cut down version). The ALSEPs consisted of a central communications hub and radioisotope power source (so as to permit operation during the lunar night) connected to a variety of instruments. The instruments varied between missions, but typically included some combination of seismometers, magnetometers, heat flow probes and atmospheric and ionospheric detectors. These instruments were supplemented by a variety of non-ALSEP instruments and experiments (Table 25.2).

Probably the most influential set of Apollo surface experiments were those related to various geophysical investigations, including both passive (Figure 25.10) and active (Figure 25.11) seismology studies, surface gravimetry and magnetometry, heat flow measurements (Figure 25.12), and the deployment of laser reflectors to measure the changing Earth—Moon distance and the

TABLE 25.2 Apollo Science Experiments

Mission Phase		Experiment/Activity	Apollo 11	Apollo 12	Apollo 14	Apollo 15	Apollo 16	Apollo 17
Lunar orbit	Command and Service Module (CSM)	Alpha particle spectrometer				•	•	
		Bistatic radar			•	•	•	
		Gamma ray spectrometer				•	•	
		Infrared scanning radiometer						•
		Laser altimeter				•	•	•
		Mass spectrometer (atmosphere studies)				•	•	
		Photography: Hasselblad (frames)	760	795	758	2350	1060	1170
		Photography: mapping (frames)				3375	2514	2350
		Photography: multispectral			•			
		Photography: panoramic (frames)				1570	1415	1580
		Radar sounder						•
		S-band transponder (gravity measurements)		•	•	•	•	•
		UV photography of the Earth and Moon				•	•	•
		UV spectrometer (exosphere studies)						•
		X-ray fluorescence spectrometer				•	•	
	Sub-satellites	Magnetometer				•	•	
		Plasmas and energetic particles				•	•	
		S-band transponder (gravity studies)				•	•	
Lunar surface	ALSEP	Charged particle lunar environment experiment (CPLEE)			•			
		Cold cathode gage experiment (atmosphere studies; CCG)			•	•	•	
		Dust detector experiment (DTREM)	•	•	•	•		
		Gravimeter (fixed; LSG)						•
		Heat flow experiment (HFE)				•	•	•
		Lunar atmosphere composition experiment (LACE)						•
		Lunar ejecta and meteorites experiment (LEAM)						•
		Magnetometer (fixed; LSM)		•		•	•	
		Seismic experiment (Active; ASE)			•		•	
		Seismic experiment (passive; PSE)	•	•	•	•	•	
		Solar wind spectrometer (SWS)		•		•		
		Suprathermal ion detector experiment (ionospheric studies; SIDE)		•	•	•		

(Continued)

TABLE 25.2 Apollo Science Experiments—cont'd

Mission Phase	Experiment/Activity	Apollo 11	Apollo 12	Apollo 14	Apollo 15	Apollo 16	Apollo 17
Non-ALSEP	Cosmic ray detector experiment (CRD/LSCRE)					•	•
	Far-UV camera/spectrograph (astronomy; UVC)					•	
	Gravimeter experiment (traverse; TGE)						•
	Laser ranging retroreflector	•		•	•		
	Magnetometer (portable; LPM)			•		•	
	Neutron probe experiment (LNPE)						•
	Seismic profiling experiment (LSPE)						•
	Soil Mechanics experiment (ASP/SRP)	•	•	•	•	•	•
	Solar wind composition experiment (SWC)	•	•	•	•	•	
	Surface electrical properties (SEP)						•
	Surveyor 3 inspection and partial retrieval		•				
	Thermal degradation sample experiment (TDS)				•		
Field geology	Number of EVAs	1	2	2	3	3	3
	Total EVA duration (hours)	2.4	7.5	9.4	18.6	20.2	22.1
	Lunar Roving Vehicle				•	•	•
	Total distance traversed (km)	0.3	2.0	3.3	27.9	27.0	35.0
	Hasselblad photography (frames)	325	583	417	1150	1774	2200
	Close-up regolith photography (stereopairs)	17	15	17.5			
	Returned sample mass (kg)	21.6	34.3	42.3	77.3	95.7	110.5

Notes: Acronyms are the official NASA designations for these experiments. As is well known, Apollo 13 did not land on the Moon, but did acquire 112 frames of orbital Hasselblad imagery. The Apollo 16 heat flow experiment (HFE) was rendered inactive before any data could be collected when its cable became detached from the central ALSEP station; the Apollo 17 surface gravimeter (LSG) also did not work as intended, owing to a design flaw, but did return useful data acting as a pseudoseismometer.
Sources: Harland (1999); Wilhelms (1993); NASA Catalog of Apollo Experiment Operations: http://ares.jsc.nasa.gov/HumanExplore/Exploration/EXlibrary/docs/ApolloCat/apollo.htm

Moon's physical librations (Figure 25.13). With the exception of an ineffective seismic experiment sent to Mars on the Viking landers in 1976, the Moon remains the only planetary body apart from the Earth on which these geophysical techniques have been applied in situ at the surface.

Key results of the Apollo geophysics experiments include the discovery of natural moonquakes and their exploitation to probe the structure of the nearside crust and mantle, geophysical constraints on the existence and physical state of the lunar core (from both seismic data and laser reflection studies of lunar rotation), the use of active seismic profiling to determine the near-surface structure (Figure 25.11), and measurements of the lunar heat flow at the Apollo 15 and 17 localities. It is important to recognize that although these data are for the most part over 30 years old (the ALSEPs were switched off on September 30, 1977), advances in interpretation, and especially in computing power and numerical computational techniques, means that they continue to give new insights into the interior structure of the Moon. For example, in 2011, an apparently definitive seismic detection of the Moon's core, and strong evidence that, like the Earth's, it consists of solid inner and liquid outer layers, was made by a reexamination of Apollo seismic data.

4.3. Apollo Orbital Remote Sensing

Supplemental to the Apollo surface scientific payloads (see above) were a series of remote sensing experiments mounted onto the Scientific Instrumentation Module (SIM)

Chapter | 25 Lunar Exploration

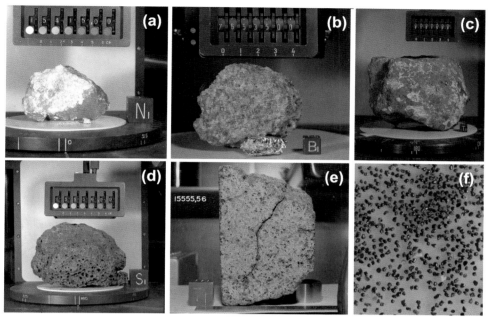

FIGURE 25.8 NASA curation photographs of Apollo main rock types. (a) Pristine lunar anorthosite sample 15415, which is also known as the "Genesis Rock" (NASA photo S-71−42,951). (b) Igneous intrusive sample 76535, which is a troctolite and part of the lunar High Magnesian Suite (NASA photo S73-19,458). (c) Impact melt breccia 61015, which is dimict (two toned) in nature with both dark impact melt and included anorthositic fragments (NASA photo S72-40585). (d) Mare basalt 15556, which was erupted in a lava flow at ∼3.4 billion years ago and is highly vesicular in nature (NASA photo S71-43,325). (e) Sliced view of mare basalt 15555 showing pyroxene and plagioclase crystals (NASA photo 5−33,419). Scale bars in (a)−(e) are 1 cm. (f) Orange- and black-colored volcanic glass beads (pyroclastic material) collected from Shorty Crater (NASA photo S73-15,085; individual beads are 90−150 μm in diameter.).

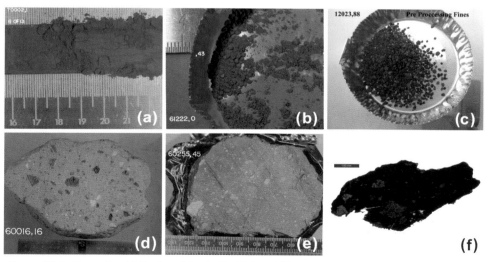

FIGURE 25.9 Apollo regolith (soil) samples. (a) Section of the Apollo 17 deep drill core 70002 (NASA photo S77-20158). (b) Apollo 16 soil 61222 prior to sieving (NASA photo S80-40445). (c) Sieved fine fraction (1−2 mm) of Apollo 12 soil 12023 (Image NASA). (d) Regolith breccia sample 60016, which is believed to have been fused from a soil to a rock ∼3.7 billion years ago (NASA photo S78-34417). (e) Regolith breccia sample 60255, which is believed to have been fused from a soil to a rock ∼1.7 billion years ago (NASA photo S79-34528). Scale bar ticks in (a)−(e) are 1 cm. (f) Optical light image of carbonaceous chondritic meteorite fragment "Bench Crater" collected in Apollo 12 soil 12032 (*Image courtesy: K.H. Joy*).

on the orbiting CSM of the later J-class missions (Figure 25.7). These instruments varied between the Apollo 15, 16 and 17 missions but typically included some combination of panoramic and mapping cameras, laser altimeters, gamma ray and X-ray spectrometers, and, for Apollo 17, a ground-penetrating radar sounder. Apollo 15 and 16 also carried small subsatellites to permit more accurate measurements of charged particles and magnetic

TABLE 25.3 Examples of Some Common Analytical Techniques Used in Laboratories to Investigate Lunar Samples

Investigation	Method/Laboratory Technique	Information Obtained	Type of Sample Studied or Requirement for Measurement
Mineralogy and bulk-sample chemistry	X-ray fluorescence (XRF)	Bulk-sample chemistry	Powdered samples often dissolved in strong acids and analyzed in solution.
	Instrumental neutron activation analysis (INAA)	Bulk-sample chemistry	
	Inductively coupled plasma mass spectrometry (ICP-MS)	Bulk-sample trace element chemistry	
	Inductively coupled atomic emission spectrometry (ICP-AES)	Bulk-sample major and trace element chemistry	
	Accelerator mass spectrometry	Radionuclides (exposure information)	
	X-ray diffraction (XRD)	Mineral crystal structure	
	Gas chromatography mass spectrometry (GC–MS)	Concentration of light elements (e.g. H, C, N), organic compounds and their isotopic composition	
	Thermal ionization mass spectrometry (TIMS), and multicollector ICP-MS (MC-ICP-MS)	Mainly used for acquiring high-precision isotopic composition	
Mineral/phase chemistry	Electron probe microanalysis (EPMA) and secondary electron microscopy (SEM)	X-ray elemental maps, concentrations of major elements, some minor, and trace elements	Typically performed on polished thin section (30 μm thick) or polished thick sections (>30 μm thick) or polished blocks
	Secondary ion mass spectrometry (SIMS) including nanoSIMS and time of flight (TOF)–SIMS	Trace element concentrations and isotopic composition at high spatial resolutions	
	Laser ablation ICP-MS (LA-ICP-MS)	Trace element concentrations and isotopic composition	
	Raman spectroscopy	Mineral structure and composition	
	Transmission electron microscope (TEM)	Nanoscale detail of minerals	
	Fourier transform infrared (FTIR) microscopy	H and C abundances and their speciation in minerals	
Age dating (radioisotopes)	Potassium–argon (K–Ar)	Isotopic information used to infer formation or reset ages of a sample	Requires knowledge of K content of sample
	Argon–argon (Ar–Ar)		Requires irradiation in a reactor prior to analysis
	Samarium–neodymium (Sm–Nd)		Ideally requires analysis on bulk sample as well as its mineral constituents
	Rubidium–strontium (Rb–Sr)		
	Uranium–lead (U–Pb)		Can be a bulk measurement or performed using in situ (e.g. by SIMS or LA-ICP-MS techniques)
	Lead–lead (Pb–Pb)		
	Hafnium–tungsten (Hf–W)	Used to infer timing of core–mantle separation	Requires analysis on bulk samples
Magnetic properties	Magnetometer	Detection of remanent magnetization and paleointensity of magnetic field	Chip–collection orientation ideally known

Chapter | 25 Lunar Exploration 565

FIGURE 25.10 Apollo 14 seismometer deployed on the lunar surface; the silvery skirt provided thermal stability. These instruments, also deployed at the Apollo 12, 15, and 16 landing sites, constituted the Apollo passive seismic network which remained active until 1978. *NASA image AS14-67-9363.*

FIGURE 25.12 Commander David Scott deploys one of the Apollo 15 heat flow probes *NASA image AS15-92-12,407.*

FIGURE 25.11 One of eight explosive packages deployed by the Apollo 17 astronauts to provide data for the lunar seismic profiling experiment, which measured the thickness of regolith and the underlying lava in the Taurus—Littrow Valley. The Apollo 17 Lunar Roving Vehicle is in the foreground and the lunar module, where a geophone array was deployed to collect the signals, in the middle distance about 300 m away. *NASA image AS17-145-22,184.*

FIGURE 25.13 Apollo 14 laser reflector deployed on the lunar surface. *NASA image AS14-67-9385.*

fields. These orbital instruments are summarized in Table 25.2.

4.4. Apollo and the Benefits of Human Exploration

The Apollo program was the first, and to date the only, planetary exploration program involving astronauts, and its success has some lessons for future exploration policy. In particular, the collection of a large quantity of diverse geological samples by Apollo, as well as its deployment of an extensive range of massive and bulky geophysical instrumentation, was a direct beneficiary of the relatively generous mass budgets that are an inherent feature of human space missions compared to robotic ones. It seems most unlikely that our geological and geophysical knowledge of the Moon would be as developed as it is had the Apollo missions not taken place. Looking forward, the efficiency demonstrated by the Apollo astronauts augurs well for the scientific returns which may be anticipated from future human expeditions to the Moon, as discussed in Section 7.

5. POST-APOLLO EXPLORATION

After the Apollo program came to an end in 1972, and the last Soviet mission to the Moon (Luna 24 in 1976), there was a long gap in lunar exploration. This was only broken in the 1990s, when the Clementine and Lunar Prospector missions flew to the Moon and heralded a renewed era of lunar exploration.

5.1. Clementine

The Clementine mission was a small (140 kg) satellite launched on January 25, 1994. Clementine was designed to enter lunar orbit, map the Moon's surface, and then travel onto an asteroid. The spacecraft successfully completed two months of mapping at the Moon, but upon leaving the lunar orbit on May 3, 1994, a software failure resulted in the firing of the altitude-control thrusters, causing the spacecraft to spin uncontrollably. The mission was finally abandoned and Clementine did not rendezvous with its asteroid target.

Clementine carried a suite of instruments designed to investigate the mineralogy and topography of the lunar surface. These included:

- Multispectral cameras operating in the ultraviolet–visible (UV–VIS) and near-infrared (IR) spectral ranges. Near-complete global multispectral mapping was completed at an average resolution of 100 m/pixel (UV–VIS) and 200 m/pixel (near IR). The Clementine spectral bands were sensitive to the dominant mineralogy within each pixel (for example, among mafic minerals they may discriminate between low- and high-calcium pyroxenes and olivine), and to the iron and titanium oxide (FeO and TiO_2) concentrations in the surface. The resulting global maps (e.g. Figure 25.14) reveal the compositional diversity of the lunar crust and have been particularly important in determining the crustal composition outside sampled areas (especially high latitudes and the farside).

- The Clementine laser altimeter successfully mapped lunar topography at a resolution of 1–2 km along track with 40 m height resolution. This was the first accurate lunar topographic map to be obtained and, among other things, it revealed the true extent of many lunar craters and basins, including the giant ($\sim 2400 \times 2000$ km diameter and ~ 13 km deep) farside South Pole - Aitken basin.

- The Clementine bistatic radar experiment aimed radio signals at the lunar surface, which were reflected and collected back on the Earth by ground-based radio telescopes. Scattered reflections from the lunar poles suggested the presence of icy materials within some permanently shadowed craters.

5.2. Lunar Prospector

The NASA Lunar Prospector mission was launched on January 6, 1998, and orbited the Moon for a year at an

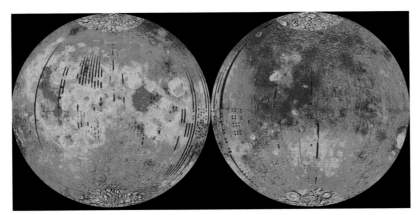

FIGURE 25.14 Distribution of rock types on the lunar nearside (left) and the farside (right) based on Clementine multispectral imaging data. Blue: anorthositic highlands; yellow: low-Ti basalts; red: high-Ti basalts. The large yellow/greenish area in the southern hemisphere of the farside is the South Pole-Aitken Basin, where the colors mostly reflect the more Fe-rich nature of the lower crust exposed by the basin rather than basaltic material. *Image courtesy of Dr Paul Spudis/LPI; LPSC 33, 1104, 2002.*

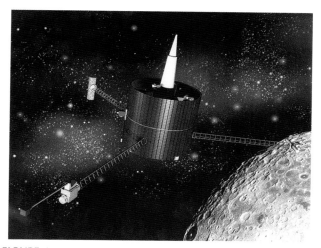

FIGURE 25.15 Artist's drawing of the Lunar Prospector spacecraft, showing instrument booms.

altitude of 100 km before being dropped into two lower orbits of 30 km and 10 km. The satellite was crashed into a south polar crater in a controlled experiment on July 31, 1999, in an attempt to see if ejected material was ice rich, although the results were inconclusive.

Lunar Prospector carried five experiments on three radial science booms (Figure 25.15), including a gamma ray spectrometer, a neutron spectrometer, a magnetometer, an electron reflectometer, and an alpha particle spectrometer (damaged at launch); spacecraft tracking also provided gravity data. The key results from the mission were:

- Gamma ray spectra were used to determine the abundance of major elements (O, Mg, Al, Si, Ca, Ti, and Fe) and radioactive elements (K, Th, and U) globally across the lunar surface. In particular, the Th distribution clearly showed that the northwest nearside of the Moon around the Imbrium Basin is compositionally unusual, with an enhanced abundance of Th (and, thus, other incompatible trace elements). This region has been dubbed the Procellarum KREEP Terrane (PKT). The term KREEP is used to describe lithologies that are enriched in potassium (K), rare earth elements (REE), and potassium (P). As these lithologies are also the dominant carriers of radioactive elements, this suggests that a large portion of the Moon's heat-producing elements may be concentrated in this region of the Moon.
- The neutron spectrometer provided independent evidence for ice at the lunar poles. Neutrons are produced by cosmic ray interaction with atoms about a meter below the regolith surface. If, on their route back through the regolith toward the surface, these neutrons encounter hydrogen nuclei an attenuated signal will be recorded by an orbiting neutron spectrometer. Such a signal was observed by Lunar Prospector, apparently supporting the Clementine bistatic radar results that ice may be present at the lunar poles. However, the data lacked sufficient spatial resolution to confirm that the areas of attenuated neutron flux correspond to permanently shadowed craters, and not all workers agree that the results prove that water ice is present in these areas (an alternative possibility could be that enhanced concentrations of solar wind-implanted hydrogen preferentially retained in the cold polar regolith).
- The Lunar Prospector magnetometer made detailed investigations of lunar crustal remanent magnetization, confirming the presence of isolated magnetic anomalies discovered by earlier missions. Combined with electron reflectometer measurements, these observations indicate that lunar crustal magnetic anomalies create mini-magnetospheres that give some protection to the underlying surface from the solar wind.

5.3. Small Missions for Advanced Research in Technology-1

Following Lunar Prospector, there was another 5-year gap in lunar exploration, until the European Space Agency's (ESA's) Small Missions for Advanced Research in Technology-1 (SMART-1) mission initiated a burst of activity in the early years of the twenty-first century. SMART-1 was ESA's first mission to the Moon. It was launched on September 27, 2003, and traveled to the Moon over a 14-month cruise period while testing an innovative solar electric propulsion (ion) drive, arriving in lunar orbit on November 15, 2004. The spacecraft carried seven technology demonstrator miniaturized instruments, designed for scientific investigations of the Moon and the surrounding space environment. SMART-1 spent almost 2 years mapping the surface of the Moon before being crashed onto its surface on September 3, 2006, when its impact into Lacus Excellentiae (34° 24′ S, 46° 12′ W) was observed from Earth.

The Advanced Moon Micro-Imager Experiment camera obtained intermediate-resolution (about 250 m/pixel globally, but better than 50 m/pixel at perilune) images of the lunar surface. The highest resolution images were of the south polar region and proved to be especially valuable in mapping the changing illumination conditions around the pole over the course of almost 2 years. These illumination maps will be useful in landing site selection for future missions designed to land at the lunar South Pole. SMART-1 also carried two instruments designed to investigate the chemical and mineralogical composition of the lunar surface. These were the SMART-1 infrared spectrometer (SIR) and demonstration of a compact X-ray spectrometer (D-CIXS). Intended primarily as technology demonstrators, these pioneering instruments resulted in greatly improved instruments flown on the Chandrayaan-1 mission (see below).

5.4. Kaguya

The Kaguya mission, known as Selene prior to launch on September 14, 2007, was the Japanese Aerospace Exploration Agency's (JAXA's) second mission to the Moon (Japan had previously sent the small (197.4 kg) Hiten-Hagoromo probe into lunar orbit in 1990 to measure the ambient dust density). Kaguya was a much larger (nearly 3 t) spacecraft and included two subsatellites designed for radio science and Doppler tracking experiments. In all, 13 instruments were carried by Kaguya, including a high-resolution television camera for public engagement. The latter has resulted in some truly spectacular images (e.g. Figure 25.16). Scientific packages included a terrain mapping camera, multiband imager and spectral profiler for mineral identification and mapping, X-ray and gamma ray spectrometers for compositional mapping, a laser altimeter for measuring lunar topography, and plasma analyzers and a charged particle detector for exosphere measurements. The two relay radio satellites helped to map lunar gravity variations, notably improving our understanding of farside gravity anomalies and crustal thickness. The Kaguya mission ended on June 10, 2009, when the main spacecraft impacted the lunar nearside surface at $\sim 65°\ 5'$ S, $80°\ 4'$ E.

Key results from Kaguya included:

- A greatly improved topographic map based on the laser altimeter measurements (Figure 25.17).
- The detection of a diverse set of as yet unsampled lunar lithologies, including outcrops of pure anorthosite (which may represent pristine magma ocean flotation cumulates) and olivine-rich outcrops (which may sample mantle material).
- Measurements of the farside gravity field. This cannot be determined by tracking spacecraft directly from the Earth, but was possible with Kaguya by using the subsatellites as relays.

FIGURE 25.16 The Earth over the South Pole of the Moon, imaged by the high-definition television camera onboard Japan's Kaguya spacecraft on November 7, 2007. The 21-km-diameter crater Shackleton, almost entirely filled with shadow, lies in the right foreground *JAXA/NHK*

5.5. Chandrayaan-1

India's first mission to the Moon, Chandrayaan-1, was launched on October 22, 2008. The mission was lost prematurely on August 29, 2009, when Chandrayaan-1 ceased communication with the Earth. The spacecraft included 11 instrument packages, five of which were provided by Indian institutes and six by overseas space agencies, including two from NASA, three from ESA, and one from Bulgaria. The Indian contributions included a small Moon Impact Probe that separated from the main satellite and hard-landed near to the lunar South Pole on November 14, 2008. This probe carried a mass spectrometer (known as CHACE), which

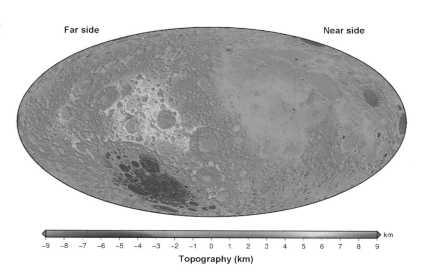

FIGURE 25.17 Lunar global topographic map obtained from Kaguya laser altimetry. *Araki et al. (2009); AAAS©.*

measured the composition of the lunar exosphere during descent, in the process detecting the presence of small quantities of atmospheric water molecules.

Other instruments on board included the Moon Mineralogy Mapper (M^3) imaging spectrometer, the Miniature Synthetic Aperture Radar (Mini-SAR), the Chandrayaan-1 X-ray Spectrometer (C1XS), and an IR spectrometer (SIR-2). Key results include:

- The discovery by M^3 of apparently hydrated minerals in high-latitude regoliths (Figure 25.18). It is hypothesized that this OH/H_2O, which cannot exist as ice in these diurnally sunlit areas, is produced by the reduction of iron oxides in the regolith by solar wind-implanted hydrogen, with OH/H_2O being retained in the relatively cold high-latitude regolith. It is, however, possible that, over time, this high-latitude OH/H_2O may migrate to permanently shadowed polar craters and contribute to ice deposits there.
- The M^3 instrument also discovered the presence of spinel-bearing lunar lithologies, in addition to those rich in olivine and pyroxene.
- The possible discovery of ice deposits in (some) permanently shadowed lunar craters by Mini-SAR. However, this interpretation, while broadly consistent with the Clementine bistatic radar and Lunar Prospector neutron spectrometer results, is inconsistent with some Earth-based radar observations and requires confirmation by future missions.
- C1XS, while benefiting from experience gained on D-CIXS on SMART-1, suffered from a low level of solar activity during the mission (solar flares are required to produce the X-rays that excite surface fluorescence). Nevertheless, the instrument was able to make compositional measurements over several Chandrayaan-1 ground tracks with a surface resolution of about 50 km.

5.6. Chang'e 1 and 2

The Chinese Lunar Exploration Program (also known as the Chang'e series) was initiated on October 24, 2007, with the launch of the Chang'e 1 spacecraft. A follow-up mission, Chang'e 2, followed on October 1, 2010. The two spacecraft served as the first phase of an ambitious Chinese lunar exploration program to move from an orbiter series, to soft landers to robotic sample return, with the long-term goal of sending people to the Moon. Chang'e 1 orbited in a 200-km orbit, conducting scientific experiments to image and map the composition of the lunar surface and space environment, including a microwave radiometer to map global temperatures at different depths down to several meters. Chang'e 2 operated in a lower 100-km orbit and included similar instrument suite as Chang'e 1, but with improved spatial resolution to perform detailed analysis of future landing sites for the planned future landers and investigations of lunar resources. Key results included:

- The derivation of global major element maps.
- Publication of the first global microwave maps of the lunar surface.

The Chang'e 1 mission concluded on March 1, 2009, with a controlled crash on the lunar nearside in Mare Fecunditatis (1° 30′ S, 52° 22′ E). The Chang'e 2 spacecraft at the end of its mission was transferred successfully from lunar orbit to the Earth—Sun L2 Lagrangian point, the first satellite successfully to make this maneuver, and was subsequently sent on an extended mission to rendezvous with a near-Earth asteroid.

5.7. Lunar Reconnaissance Orbiter

LRO was launched on the June 18, 2009, as a joint launch with the Lunar Crater Observation and Sensing Satellite (LCROSS) mission (see Section 5.8). LRO was designed initially to be an exploration mission to map and study the Moon in preparation for future human exploration efforts, but after 1 year it entered an extended scientific mapping

FIGURE 25.18 False color image of the lunar nearside based on data obtained by the M^3 instrument on India's Chandrayaan-1 mission. Blue represents areas where a 2.8- to 3.0-μm absorption band attributed to bound H_2O/OH was detected; red represents areas where the 1-μm band of the mineral pyroxene is strong, and picks out the basaltic maria; and green represents albedo. Note that evidence for hydration is restricted to high latitudes, but is much more extensive than areas of permanent shadow at the poles. *ISRO/NASA/JPL-Caltech/Brown Univ./USGS*

phase. The payload included seven scientific instruments: (1) the Lunar Reconnaissance Orbiter Camera (LROC), composed of three separate elements: the wide-angle camera fitted with seven color bands with a spatial resolution of 100 m/pixel and two narrow-angle cameras (NACs) that provide panchromatic images down to a spatial resolution of ~0.5 m/pixel; (2) the Lunar Orbital Laser Altimeter (LOLA), designed to produce the most accurate lunar topographic map to date; (3) the Diviner radiometer, designed to measure surface temperatures; (4) the Lunar Exploration Neutron Detector (LEND), a neutron spectrometer designed to search for possible near-surface water ice deposits at the lunar poles; (5) the Lyman-Alpha Mapping Project (LAMP), designed to image the interiors of permanently shadowed regions in reflected ultraviolet (UV) starlight; (6) the Cosmic Ray Telescope for the Effects of Radiation (CRaTER) instrument designed to characterize the lunar radiation environment; and (7) the Miniature Radio Frequency (Mini-RF) technology demonstration of an advanced synthetic aperture radar capable of detecting ice deposits at the lunar poles.

Key results include:

- LROC NACs have yielded many insightful images of the lunar surface. These include images of fine-scale layering within lava flows (Figure 25.19), which will be exciting targets for future exploration because in situ sampling will enable studies of the temporal evolution of lunar magmatism and the search for buried paleo-regolith deposits (see below). The LROC NACs have also imaged the Apollo (Figure 25.20), Luna, and Lunokhod landing sites. The location of the Lunokhod 1 laser reflector has made it possible to expand the Lunar Laser Ranging (LLR) network. Moreover, the identification of the precise landing sites of Luna 16, 20, and 24 has provided local geological context for the samples returned by these missions that was previously poorly constrained.
- LOLA has produced extremely detailed and high-resolution topographic data with a horizontal resolution of 10 m and a vertical resolution of about 2 m, along the ground tracks. Owing to the polar orbit of LRO, the poles now have essentially complete

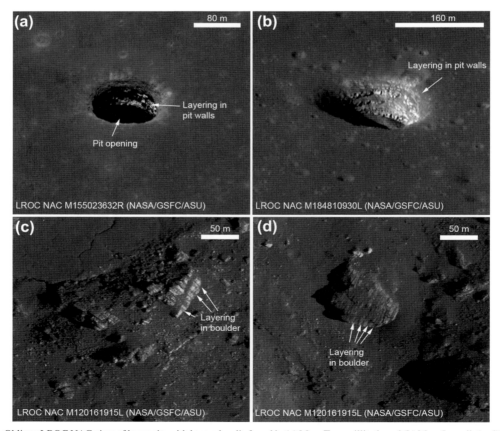

FIGURE 25.19 Oblique LROC NAC view of lunar pits with layered walls found in (a) Mare Tranquillitatis and (b) Mare Ingenii. (c, d) Layered boulders found on the lunar surface within Aristarchus crater. Scale bars in all cases have been estimated from pixel resolution of the NAC images. *NASA/GSFC/Arizona State University, modified by K.H. Joy*

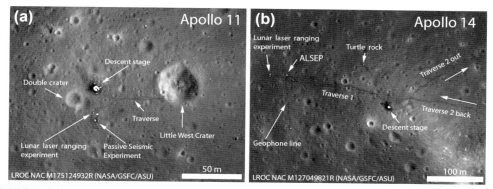

FIGURE 25.20 LROC NAC high-resolution images of the (a) Apollo 11 (NAC frame: M175124932R, 0.48 pixels/m) and (b) Apollo 14 landing sites (NAC frame: M127049821R, 0.4 pixels/m). Features are annotated based on the USGS published traverse maps. *Images: NASA/GSFC/Arizona State University, modified by K.H. Joy*

topographic coverage at this resolution, with coverage of lower latitudes constantly improving as the mission progresses.
- Diviner has measured global surface temperatures and has demonstrated that permanently shadowed regions at the poles have temperatures as low as 40 K (with some areas possibly as low as 29 K, which would be the coldest surfaces yet measured in the solar system). Water ice is stable at these low temperatures.
- The LEND instrument has confirmed the attenuation of epithermal neutrons at high latitudes first discovered by Lunar Prospector, but not the one-to-one correlation with permanently shadowed regions which would be expected if subsurface ice is responsible.
- Mini-RF, the next-generation model of the Mini-SAR radar that flew on Chandrayaan-1, has returned data about the structure of the lunar regolith and possibly detected ice-bearing deposits in permanently shadowed polar craters (but, as for the Mini-SAR results, there are alternative interpretations and the presence of polar ice deposits cannot yet be considered to have been unambiguously proved).

Many of the instruments on board LRO were used to help in mission planning for the impact of the LCROSS mission (see below) and were also involved in collecting data from the impact site.

5.8. LCROSS

The LCROSS mission was co-launched with LRO. It was designed to detect volatiles in the ejecta plume caused by the impact of the launch vehicle's spent Centaur upper stage into a permanently shadowed near-polar crater. The target selected was a region of Cabeus crater (84° 42′ S, 48° 42′ W), and impact occurred on the October 9, 2009. The Centaur rocket stage had a nominal impact mass of about 2300 kg and an impact velocity of about 2800 m/s, releasing the kinetic energy equivalent of detonating approximately 2 t of TNT (9 GJ). The LCROSS Shepherding Spacecraft passed through the resulting ejecta plume and collected and relayed data before itself impacting the surface 6 min later.

The Shepherding Spacecraft instrument payload consisted of a total of nine instruments: one visible, two near-IR, and two mid-IR cameras; one visible and two near-IR spectrometers; as well as a photometer. Near-IR spectral features attributed to water vapor and ice, and UV emissions attributed to hydroxyl radicals, were observed, and supported the presence of water in the near subsurface of Cabeus. The maximum total water vapor and water ice within the instrument field of view was estimated to be 155 ± 12 kg. Given the estimated mass of the excavated regolith, the concentration of water ice at the LCROSS impact site was estimated to be $5.6 \pm 2.9\%$ by mass. In addition to H_2O and OH, spectral bands of a number of other volatile compounds were observed, including NH_3, CO_2, sulfur-bearing compounds (H_2S, SO_2), and light hydrocarbons (CH_4, C_2H_4, and CH_3OH), although the column densities of the latter have large uncertainties associated with them.

5.9. Gravity Recovery and Interior Laboratory

Gravity Recovery and Interior Laboratory (GRAIL) was a NASA mission that was launched on September 10, 2011. The mission consisted of two small satellites GRAIL-A (known as Ebb) and GRAIL-B (known as Flow) to map the gravity field of the Moon using high-precision range-rate measurements of the distance change between the two spacecraft. Ebb and Flow were placed in circular polar orbits, with a mean altitude of 55 km above the lunar surface, and very precise tracking of these spacecraft has produced the most accurate map of the lunar gravity field yet obtained. Key results included:

- Measurement of the mean density of the lunar crust to be only 2550 kg/m^3, which indicates significant porosity (~12%) to depths of several kilometers as a result of intense bombardment.
- Determination of the thickness of the lunar crust, which with an average thickness of 34—43 km was found to be substantially thinner than previously thought. This in turn implies that the bulk abundance of aluminum in Moon is nearly the same as that in the Earth, consistent with the hypothesis that the Moon is largely constituted of material derived from the Earth (as predicted by some versions of the giant impact hypothesis of lunar origin).
- Discovery of large (i.e. hundreds of kilometers long, tens of kilometers wide) magmatic intrusions (dikes) deep within the lunar crust, and possibly associated with an early expansion of the Moon as predicted by some thermal models.

The GRAIL mission ended on December 17, 2012, when the Ebb and Flow spacecraft crashed onto an unnamed mountain at 75.62° N, 26.63° W.

5.10. Lunar Atmosphere and Dust Environment Explorer

Lunar Atmosphere and Dust Environment Explorer (LADEE) is a NASA mission launched on 7 September 2013. The mission is designed to measure the composition of the lunar exosphere and dust environment and carries an UV ultraviolet spectrometer, a neutral mass spectrometer, and the Lunar Dust Experiment.

5.11. Chang'e 3 Lunar Lander and Rover

The Chinese Chang'e 3 mission was launched on 1 December 2013 and successfully landed in northern Mare Imbrium (approximate coordinates: 44.12 degrees north; 19.51 degrees west) on 14 December 2013. This marked the first controlled soft landing of a spacecraft on the lunar surface since Luna 24 in 1976. Chang'e 3 carried a rover which made in situ measurements of regolith structure and composition in the vicinity of the landing site. Instrumentation included a ground-penetrating radar, allowing for the first direct measurement of the structure of the regolith down to a depth of about 30 m, and an alpha particle X-ray spectrometer for compositional measurements.

6. LETTING THE MOON COME TO US: THE IMPORTANCE OF LUNAR METEORITES FOR LUNAR EXPLORATION

The Apollo and Luna collection has been supplemented by the discovery of lunar meteorites (Figure 25.21). These are pieces of lunar rocks that have been launched off the Moon at >2.4 km/s (lunar escape velocity) during the impact of an asteroid or comet projectile and that have been transported to the Earth, surviving entry through the atmosphere and found as meteorites in hot and cold (i.e. Antarctica) deserts on Earth. The first lunar meteorite, collected in November 1979, was Yamato 791197 but its lunar origin was not recognized at the time. It was not until Allan Hills (ALHA) 81005 was collected in January 1982, and classified as being lunar in origin, that meteorites became an important addition to the lunar sample collection.

Radionuclide studies indicate that the majority of known lunar meteorites have been launched from the Moon

FIGURE 25.21 Examples of lunar meteorites. (a) Antarctic feldspathic regolith breccia ALHA 81005 showing clast-rich interior and brown fusion crust (Photo: NASA). (b) Antarctic feldspathic regolith breccia 791197 (Photo: NIPR). (c) Polished slab of hot desert (Libya) feldspathic regolith breccia Dar al Gani 400 (Photo: K.H. Joy). (d) Antarctic low-Ti mare basalt meteorite LaPaz (LAP) 02224 showing crystalline interior and dark black fusion crust (Photo: NASA), (e) Scanned thin section of Antarctic low-Ti mare basalt meteorite Miller Range (MIL) 05035 (Photo: K.H. Joy). (f) Polished face of hot desert KREEP-rich regolith breccia North West Africa (NWA) 4472 (Photo: K.H. Joy). Cube scale is 1 cm and other scale bars are denoted on images.

within the past 10 million years and that all have been launched within the past 20 million years. As no large lunar craters are believed to have formed in this time period, it has been theorized that lunar meteorites are launched from small craters of only a few kilometers or less in diameter. The meteorites potentially have been launched from anywhere on the lunar surface and so arguably represent a global sampling of lunar rock types. There are currently (as of January 2014) 177 named stones, which, when paired or grouped together, represent ∼85 separate meteorite arrivals on the Earth with a collective mass of ∼70 kg.

Mineralogical, chemical, and isotopic studies of lunar meteorites, using the same techniques applied to study other lunar samples (Table 25.3), have helped to advance our knowledge of lunar geology. In particular, the random sampling of the lunar surface by the meteorites implies that many of them will be derived from previously unsampled regions, including the farside. They therefore give us a better understanding of the *diversity* of lunar crustal rocks than do the Apollo and Luna samples, even though generally we do not know the source region of any given lunar meteorite. Important discoveries include evidence for Mg-rich anorthosites in the farside crust, which may challenge the prevailing global flotation cumulate model for the origin of the lunar crust, and the presence of very-low-titanium basalt clasts in highland breccias. Discoveries of such basalt samples may imply that some areas of the lunar limbs and farside are underlain by mare material (as the so-called cryptomaria) that were erupted or intruded into the crust early in lunar history.

Careful comparisons between lunar meteorite compositions and crustal compositions determined by orbital remote sensing measurements are now making it possible for the source regions for some lunar meteorites to be geographically constrained. This is important because, once we know where a given meteorite comes from, it's detailed mineralogical, geochemical, and isotopic composition, which can only be measured in laboratories on Earth, can be related to a specific geological context on the Moon.

7. FUTURE LUNAR EXPLORATION OBJECTIVES

Future lunar exploration objectives can logically be divided into three categories: (1) science of the Moon (i.e. studies of the Moon itself), (2) science on the Moon (i.e. studies using the lunar surface as a platform for scientific investigations not directly related to the Moon itself), and (3) science from the Moon (i.e. studies utilizing the lunar surface as a platform for astronomical observations). These scientific objectives, with detailed recommendations for how to achieve them, are described in more detail in the Lunar Exploration Roadmap maintained by the Lunar Exploration Analysis Group (LEAG), an advisory body chartered by NASA to assist in planning for future lunar exploration.

7.1. Science of the Moon

In 2007, a US National Research Council Report on "The Scientific Context for Exploration of the Moon" identified, and prioritized, eight top-level scientific "concepts" (each of which can be broken down into multiple individual science goals) and identified the capabilities that would be required of space missions designed to address them. We summarize these here, and then describe how future surface exploration (human and/or robotic) will aid in meeting these scientific objectives.

- The Bombardment History of the Inner Solar System

The vast majority of lunar terrains have never been directly sampled, and their ages are based on the observed density of impact craters calibrated against the ages of Apollo and Luna samples. However, the current calibration of the cratering rate, used to covert crater densities to absolute ages, is not as complete as it is often made out to be. For example, there are no calibration points that are older than about 3.85 Ga, and crater ages younger than about 3 Ga are also uncertain. Improving the calibration of the lunar cratering rate would be of great value for planetary science for the following three reasons: (1) it would provide better estimates for the ages of unsampled regions of the lunar surface; (2) it would provide us with a more reliable estimate of the impact history of the inner solar system, including that of the Earth; and (3) the lunar impact rate is used, with various assumptions, to date the surfaces of other planets for which samples have not been obtained. Obtaining an improved cratering chronology is straightforward in principle: it requires the sampling, and radiometric dating, of surfaces having a wide range of crater densities, supplemented where possible by dating of impact melt deposits from individual craters and basins. However, in practice, this is likely to require the implementation of multiple missions to many different sites. These might be robotic missions, ideally involving rover-facilitated mobility and either in situ radiometric dating or, preferably, a sample return capability.

- The structure and composition of the lunar interior

The structure of the lunar interior provides fundamental information on the evolution of differentiated planetary bodies. Despite data from the Apollo geophysical instruments, key aspects of the Moon's interior structure, composition, and evolution are left unresolved. The two heat flow measurements made during Apollo were obtained near the edge of the atypical PKT, so they may not be representative of the Moon as a whole. Resolving these questions will require making further geophysical measurements. While a few such measurements can be made from orbit (such as the measurement of the Moon's gravity and magnetic fields), most require geophysical instruments to be placed on, or below, the lunar surface. These should be configured to give global coverage and should be located at least 100 km from major terrain boundaries. Key instruments in this respect are seismometers to probe the structure of the deep interior, heat flow probes to measure the heat loss from the lunar interior, and magnetometers to measure remanent magnetizations and internal electrical conductivity variations.

- The diversity of lunar crustal rocks

Quantifying and understanding the diversity of the lunar crust will require detailed chemical and mineralogical analysis of rocks and soils from as yet unsampled regions of the lunar surface. In particular, no sample has yet been returned from the polar regions or the farside, greatly limiting our knowledge of lunar geological processes. Although, statistically, many of the lunar meteorites discussed in Section 6 presumably originate from these areas, the value of these materials is limited by lack of knowledge of their source regions and thus geological context. The diversity of lunar crustal materials has been demonstrated most recently by the orbital remote sensing instruments (see Section 5), and it is important to confirm the interpretation of these remote sensing observations. It is also important to obtain measurements of minor and trace elements in these materials, which cannot be detected by orbital remote sensing but which would help discriminate between different suggested origins and formation mechanisms. Sample return missions to currently unsampled regions would be the preferred means of furthering our knowledge of lunar geological diversity, although an alternative would be to make robotic in situ geochemical measurements.

- Volatiles at the lunar poles

The lunar poles potentially bear witness to the flux of volatiles present in the inner solar system throughout much of solar system history. As discussed in Section 5, a range of remote sensing observations suggest the presence of water ice in the floors of permanently shadowed polar craters, an interpretation supported by the LCROSS impact experiment. It seems likely that this water is ultimately derived from the impacts of comets and hydrated asteroids with the lunar surface, although solar wind implantation and endogenic sources might also contribute. However, the inferred quantity of water is sensitive to a number of assumptions and ideally needs to be confirmed by future in situ measurements. The confirmation of significant quantities of water and other volatiles at the lunar poles would help facilitate future human exploration of the Moon and of the inner solar system more generally (see Section 7.4 below).

- Lunar volcanism

The characterization of lunar volcanism is a high lunar science priority because of the window it provides into the thermal and compositional evolution of the lunar mantle. Recent remote sensing observations have demonstrated two very important points: (1) that the samples of volcanic products returned by the Apollo and Luna missions are not fully representative of all the types of volcanism present on the Moon and (2) that the presence of nonmare silicic volcanism on both the near- and farsides of the Moon shows that lunar magmatic processes are able to produce high-silica magmas in addition to the more common mare basalts. There is no doubt that these as yet unsampled areas would benefit from in situ field investigations. However, from a lunar exploration perspective, this is just one aspect of the wider requirement to sample a diverse set of lunar rocks (described above) and the implications for exploration capabilities are essentially the same.

- Impact processes

Impact cratering is a fundamental planetary process, an understanding of which is essential for our knowledge of planetary evolution. Yet our knowledge of impact processes is based on a combination of theoretical modeling, small-scale laboratory hypervelocity impact experiments, and field geological studies of generally poorly preserved terrestrial impact craters. The Moon provides a unique record of essentially pristine impact craters of all sizes (from micron-sized pits up to >300 km impact basins). Field studies, combining sample collection (including drill cores) and in situ geophysical studies of the ejecta blankets and subfloor structures of pristine lunar craters of a range of sizes would greatly aid in our understanding of the impact cratering process.

- Regolith processes

The lunar surface is a natural laboratory for understanding regolith processes and space weathering on airless bodies throughout the solar system. The nature of cold high-latitude regoliths, which have never been sampled or studied in situ, and which may contain a volatile component, are of particular interest. Another important aspect of the lunar regolith is the record it contains of the early solar system history. Studies of Apollo samples have revealed that solar wind particles are efficiently implanted in the lunar regolith and therefore contain a record of the composition and evolution of the Sun throughout the solar system history. Recently, it has been suggested that samples of Earth's early crust may also be preserved in the lunar regolith in the form of terrestrial meteorites. Meteorites derived from elsewhere in the solar system have already been found on the Moon, preserving a record of the dynamical evolution of small bodies throughout solar system history. Last but not least, the lunar regolith may contain a record of galactic events, by preserving the signatures of ancient cosmic ray fluxes and the possible accumulation of interstellar dust particles during passages of the Sun through interstellar clouds. From the point of view of accessing the ancient solar system history it will be especially desirable to find layers of ancient regoliths (*paleoregoliths*) that were formed and buried billions of years ago and thus protected from more recent geological processes (Figure 25.22). Locating and sampling such deposits will likely be an important objective of future lunar exploration activities.

- Atmospheric and dust environment

The extent to which transient releases of gasses into the lunar exosphere may occur is of interest because this may correlate with ongoing low-level geological activity. The surface dust environment, and especially the extent to which dust grains may become electrostatically charged and transported, is another important research topic, one which will require surface instruments to be deployed. The processes involved are likely to be common on other airless bodies, and quantifying them on the relatively accessible lunar surface will give us better insight into regolith/exosphere interactions throughout the solar system. It is, however, important to note that landed missions have the potential to significantly disturb the tenuous lunar atmospheric environment, and it is therefore important that the lunar atmosphere/exosphere be properly characterized before renewed surface operations are initiated; the orbital LADEE mission (Section 5.10) is designed with this objective in mind.

7.2. Science on the Moon

The lunar surface is a potential platform for a number of scientific investigations unrelated to the Moon itself (see, for example, the LEAG Roadmap referred to above). These include investigations in the life sciences, human physiology, astrobiology, and fundamental physics.

- Life sciences and astrobiology

Although the Moon has, almost certainly, never supported any life of its own, lunar exploration will nevertheless inform our searches for life elsewhere. Organic molecules delivered to the inner solar system by comets and asteroids, and protected either in the lunar subsurface or in permanently shadowed craters, could provide insights into the inventory of organic material in the inner solar system. Insofar as these organics may have provided an exogenous source of prebiotic organics necessary to initiate life on the

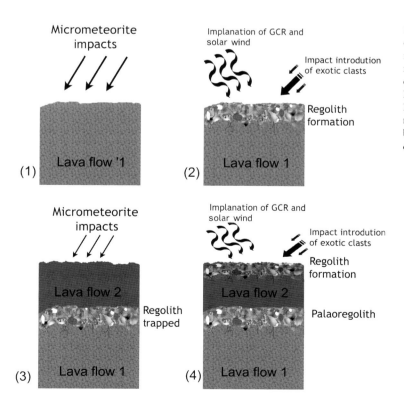

FIGURE 25.22 Schematic representation of the formation of a paleoregolith layer: (1) a new lava flow is emplaced, and meteorite impacts immediately begin to develop a surficial regolith; (2) solar wind particles, galactic cosmic ray particles, and "exotic" material derived from elsewhere on the Moon (and perhaps elsewhere) are implanted; (3) the regolith layer, with its embedded historical record, is buried by a more recent lava flow, forming a paleoregolith; (4) the process begins again on the upper surface. *Royal Astronomical Society /K.H. Joy, reproduced with permission*©.

Earth, investigating organics on the Moon has important contributions to make to understanding the origin of life on the Earth. Indigenous organic processing on the Moon may also yield insights into the origin and fate of organics in interplanetary space. Moreover, the lunar surface now contains the crashed remains of a large number unsterilized spacecraft and the study of microorganisms, or microbial spores, found within these spacecraft would answer many questions about the survivability of microorganisms in the space environment.

- Human physiology

There is medical interest in the long-term effects of reduced gravity and enhanced radiation environments on the human body. This research is needed partly to enhance our understanding of fundamental biological processes, with potential feedback into the design of medical therapies for use on Earth, and partly to support future human space operations. With regard to potential medical applications, human adaptation to prolonged exposure to partial gravity, such as exists on the Moon, may offer significant insights into vestibular disorders, cardiovascular disease, and a range of processes associated with aging. Physiological research in lunar gravity would thus supplement the knowledge accrued by microgravity exposure of astronauts in low Earth orbit. With regard to paving the way for future space exploration, for example, sending astronauts to Mars (see below) or near-Earth asteroids, research into the long-term health of a human crew operating under reduced gravity, and after a long period in microgravity, will be required. A lunar base, perhaps in combination with microgravity research on the International Space Station, is probably the only location where such research could be safely conducted.

- Fundamental physics

Although not a major driver for lunar exploration, it is recognized that a number of research fields in the area of fundamental physics may also benefit from the ability to place scientific instruments on the lunar surface. These include tests of general relativity through improved LLR measurements, tests of quantum entanglement over large baselines, and searches for exotic subatomic particles (including dark matter candidates).

7.3. Science from the Moon: Lunar-Based Astronomy

The Moon is a potentially valuable platform for astronomical observation. The lunar farside, in particular, is probably the best site in the inner solar system for radio astronomy, as it is continuously shielded from the Earth. The Moon is especially attractive as a site for low-frequency radio astronomy, because at frequencies below 30–100 MHz radio waves are seriously distorted by the Earth's ionosphere. As a consequence, the low-frequency universe is the last uncharted part of the electromagnetic spectrum. Of particular importance would be the investigation of the so-called cosmic dark ages, dating from several hundred million years after the Big Bang but before the formation of the first stars, when the cosmos was mainly filled with dark matter and neutral hydrogen. This epoch, which will contain important information on the early universe, can only be studied through radio emission from atomic hydrogen red shifted to a frequency of several tens of megahertz and would greatly benefit from radio telescopes on the Moon.

The lunar surface also lends itself to cosmic ray astronomy (as it lacks an atmosphere and lies outside the Earth's magnetosphere); other astronomies requiring large, bulky detectors (e.g. gamma ray and neutrino astronomy); and of course to observations of the Earth itself and its magnetosphere. For optical and IR astronomy, there is an argument that free-flying spacecraft offer better observing platforms than the lunar surface. Nevertheless, in the context of future human exploration of the Moon, it is important to note that the maintenance and upgrading of astronomical instruments is likely to benefit from the proximity to a human infrastructure. Thus, when and if a lunar base is established, the Moon may become a *more* attractive astronomical location than free-flying locations precisely because a human-tended infrastructure will exist to transport, service, and upgrade the instruments.

7.4. Lunar Resources

Exploration of space is an expensive undertaking in purely economic terms, with return on investment mostly measured in terms of scientific output. Identification of extraterrestrial resources, and the development of techniques to use them, could reduce our dependence on Earth-based resources and aid in the establishment of financially sustainable space exploration programs. In situ resource utilization (ISRU) is the term used to refer to the generation of consumables for autonomous or human activities from raw materials found in situ on the Moon or other planetary bodies.

Currently, the most important ISRU processes for future lunar exploration are perceived to be the production of O_2 and H_2O for life support, and/or H_2 and O_2 for fuel and propellant. At least 20 different processes have been put forward to produce oxygen from indigenous lunar resources. The most widely studied oxygen production process is ilmenite ($FeTiO_3$) reduction in the presence of hydrogen. Moreover, if ilmenite is exposed at the lunar surface (e.g. in high-Ti mare basalts) over time it accumulates solar wind-implanted species, especially hydrogen nuclei (ilmenite appears to be especially conducive to retaining such volatiles). Lunar regolith exposed at the surface for a long time is said to be "mature" and this can be seen in the reflected spectra observed from orbit, which shifts to the longer wavelengths (i.e. it becomes "redder") the more mature the regolith is. Correlation of Ti concentration with "maturity" maps will therefore help guide future resource exploration. Moreover, as our knowledge and understanding of lunar resources improves, other elements and compounds on the Moon may become equally attractive for ISRU activities, enabling sustainable lunar exploration in the longer term.

- Regolith as a resource

The lunar regolith is the most likely primary feedstock in processes aimed at extraction of in situ resources on the Moon. The regolith particles, composed of a variety of mineral assemblages and glass, have been processed by millions to billions of years of (micro-) meteoroid impacts, with subsequent space weathering by thermal cycling, solar wind erosion, and implantation, and impacts leading to their comminution and agglutination. In general, the basaltic terrains of lunar mare tend to be richer in the minerals ilmenite, olivine, and pyroxene, while the lunar highlands are dominated by the mineral plagioclase. The actual composition of lunar soils is, however, observed to be very localized, even on the scale of a few kilometers, indicating a limited extent of lateral mixing. Within this complex mixture of materials and minerals lie several possible resources, including metal oxides, solar wind-implanted volatiles and, at high latitudes, possible water ice and/or hydrated minerals.

- Water ice in permanently shadowed craters

The possible presence of large quantities of water in the permanently shadowed regions near the lunar poles has major implications for ISRU as a potential source of water and oxygen for life support and hydrogen for fuel. As a first step, the extent, quantity, distribution, and nature of this potential ice resource must be better constrained and this will require in situ surface exploration. It is recognized that accessing permanently dark craters and identifying the presence or otherwise of in situ ice as a potential resource poses major technical challenges. Nevertheless, confirmation of exploitable quantities of water at the lunar poles would undoubtedly be a major boost for lunar ISRU, especially in the context of developing a space-faring infrastructure.

- Water on the lunar surface away from the polar regions and in the lunar interior

The recent discoveries of OH/H_2O on and near the lunar surface by infrared remote sensing measurements (Figure 25.18) have been complemented by a number of new ground-based findings of OH/H_2O in lunar samples that originated at depth within the Moon, although the sources of these species are almost certainly quite different in the two cases. The extent to which these deposits have practical utility remains to be determined, and dedicated lander/rover or sample return missions will be required to better constrain this issue.

- Solar wind-implanted volatiles

Solar wind-implanted volatiles are another potential resource, with the advantage that their extraction from the regolith can be achieved by heating alone. These volatiles have a number of potential applications, including their use as a feed product for the reduction of metal oxides such as FeO for the production of O and H_2O. The main solar wind-implanted volatiles include the elements H, N, C, and He. These elements can be important for various aspects of lunar exploration and the maintenance of a lunar outpost, and volatile extraction from the regolith may reduce the requirements for replenishment of these elements from terrestrial sources. Volatile extraction is probably the simplest example of potential ISRU, but because the abundance of volatile elements in the regolith is in general fairly low (e.g. of the order of 50 ppm by mass in the case of H), their practical application to future exploration requires further investigation.

Much previous speculation has centered on the possible use of ^{3}He in the lunar regolith as a fuel for future nuclear fusion reactors. However, the concentration of ^{3}He in the regolith samples returned by Apollo is very low (in the range 2—10 parts per billion by mass) and it is far from clear whether significant exploitation could ever be economic in terms of mining it on the Moon and exporting it back to the Earth.

- Future outlook

Lunar oxygen, hydrogen, and water may be the most crucial resources in the near-term human exploration of the Moon, but other chemical and mineral resources such as metals and rare gases could also become valuable resources (especially if they are produced as by-products of other metallurgical processes). Continued innovation in mining, processing, and manufacturing techniques could further increase the resource potential of the Moon. The existing lunar sample collection provides unprecedented access to lunar materials for improving our understanding of the chemical/mineralogical makeup of the Moon and for devising experiments to develop and test new technologies for extracting lunar resources. High-resolution global (ideally meter-scale) mapping of lunar surface elemental and mineralogical compositions, complemented by in situ measurements by landed spacecraft (including rover-enabled mobility) and sample return missions, is needed to identify the optimal locations for lunar ISRU. In the current phase of expanding global exploration initiatives, ISRU is likely to benefit from a synergistic approach between governmental and commercial sectors.

7.5. The Moon as a Stepping Stone for the Exploration of Mars and other Solar System Destinations

Future human operations on the Moon may pave the way for the eventual, and far more challenging, human exploration of Mars, near-Earth asteroids, and other inner solar system destinations. In the case of Mars, while there remains much essential reconnaissance work to be performed robotically, in the longer term, Mars exploration would benefit from a human presence for the same reasons discussed here in the context of lunar exploration. This is likely to be especially true if evidence for past or present life is found on Mars, because the subsequent demand for follow-up investigations is likely to outstrip the limited capabilities of robotic exploration.

That said, there is still much to learn about human physiological and psychological responses to long-term immersion in the space environment before astronauts could safely be sent to Mars, and much of the required knowledge could be built up in the course of a human lunar exploration program. In addition, many of the operational techniques that will be required for exploring Mars (such as field geology using space-suited astronauts, the use of pressurized rovers, and the development of planetary drilling technologies) could also be perfected on the Moon, where they will be required to address the scientific issues discussed above. For these reasons, lunar exploration can be seen as a stepping stone for the scientific exploration of Mars and other locations in the inner solar system.

8. CONCLUSION

The Moon's proximity to the Earth has historically placed lunar exploration at the forefront of solar system exploration. This was true of Galileo's first telescopic observations (when the Moon was the only "planet" large enough to be properly resolved by the instrumentation available at the time), right through to the first spacecraft to orbit, land on, and return samples from another celestial body. Moreover, the Moon was the first, and is to date the only, celestial body that human beings have explored in person. The past two decades have seen a renaissance in lunar exploration conducted from orbit, and continued studies of the Apollo and Luna samples, supplemented by studies of lunar meteorites. This has resulted in a wealth of new data which has in many ways revolutionized our

FIGURE 25.23 Artist's concept of astronauts supervising a drill. Such a capability would permit access to the subsurface, for example, to extract buried paleoregolith samples containing ancient solar wind and galactic cosmic ray records, and is an example of how science will benefit from returning humans to the Moon.

knowledge of Earth's natural satellite and has also taught us much about the early history of the solar system (including the origin and evolution of the Earth—Moon system), the geological evolution of rocky planets more generally, and the near-Earth cosmic environment throughout solar system history.

However, it is also clear that the Moon still has much more to tell us about all these areas of planetary science. In addition, the Moon also offers outstanding opportunities for research in astronomy, astrobiology, fundamental physics, life sciences, and human physiology and medicine. Addressing these objectives will require an end to the 40-year hiatus of lunar surface exploration with the placing of scientific instruments on, and the return of samples from, the surface of the Moon. Some of these scientific objectives can be addressed, at least partially, by a new generation of robotic spacecraft dispatched to the lunar surface. However, for many lunar exploration objectives, the requirements for mobility, deployment of complex instrumentation, subsurface drilling (Figure 25.23), and sample return capacity are likely to outstrip the capabilities of robotic or telerobotic exploration. Insofar as these aspects of lunar exploration would benefit from renewed human operations on the Moon, lunar exploration may once again fulfill its historical role by helping to lay the foundations for the future human exploration of the entire solar system.

ACKNOWLEDGMENTS

We thank the referee, Dr Clive Neal, for his detailed comments which have improved the quality of this chapter. Section 4 is based in part on material previously published in the Royal Astronomical Society's journal, *Astronomy and Geophysics* (Vol. 53, pp. 6.24—6.28, 2012), and is reproduced here with permission. Section 7 is based on material previously published by two of the authors (IAC and MA) in papers published in *Planetary and Space Science* (74, 3—14, and 42—48, 2012) and is reproduced here with permission.

BIBLIOGRAPHY

Anand, M., Crawford, I. A., Balat-Pichelin, M., Abanades, S., van Westrenen, W., Péraudeau, G., et al. (2012). A brief review of chemical and mineralogical resources on the Moon and likely initial in situ resource utilization (ISRU) applications. *Planetary and Space Science, 74*, 42–48.

Araki, H., Tazawa, S., Noda, H., Ishihara, Y., Goossens, S., Sasaki, S., et al. (2009). Lunar Global Shape and Polar Topography Derived from Kaguya-LALT Laser Altimetry. *Science, 323*, 897–900.

Beattie, D. A. (2001). *Taking science to the Moon: Lunar experiments and the Apollo program*. Baltimore, USA: Johns Hopkins University Press.

Crawford, I. A. (2012). The scientific legacy of Apollo. *Astronomy and Geophysics, 53*, 6.24-6.28.

Crawford, I. A., Anand, M., Cockell, C. S., Falcke, H., Green, D. A., Jaumann, R., et al. (2012). Back to the moon: the scientific rationale for resuming lunar surface exploration. *Planetary and Space Science, 74*, 3–14.

Ehrenfreund, P., McKay, C., Rummel, J. D., Foing, B. H., Neal, C. R., Masson-Zwaan, T., et al. (2012). Toward a global space exploration program: A stepping stone approach. *Advances in Space Research, 49*, 2–48.

Galilei, G. (1610). *Sidereus Nuncius*. Venice. (English translation by Van Helden, A., 1989, University of Chicago Press, Chicago, USA).

Harland, D. M. (1999). *Exploring the Moon: The Apollo expeditions*. Chichester, UK: Praxis Publishing.

Heiken, G. H., Vaniman, D., & French, B. M. (Eds.). (1991). *The Lunar sourcebook: A user's guide to the Moon*. UK: Cambridge University Press.

Jaumann, R., Hiesinger, H., Anand, M., Crawford, I. A., Wagner, R., Sohl, F., et al. (2012). Geology, geochemistry, and geophysics of the moon: status of current understanding. *Planetary and Space Science, 74*, 15–41.

Jolliff, B. L., Wieczorek, M. A., Shearer, C. K., & Neal, C. R. (Eds.). (2006), New views of the moon. *Reviews of Mineralogy and Geochemistry, 60*; (pp. 1–721).

Korotev, R. L. (2005). Lunar geochemistry as told by lunar meteorites. *Chemie der Erde, 65*, 297–346.

Lewis, J. S., Matthews, M. S., & Guerrieri, M. L. (Eds.). (1993). *Resources of near-Earth space*. Tucson, Arizona, USA: University of Arizona Press.

Neal, C. R. (2009). The Moon 35 years after Apollo: what's left to learn? *Chemie der Erde, 69*, 3–43.

NRC. (2007). *The scientific context for exploration of the moon*. Washington DC: National Research Council, National Academies Press.

Orloff, R. W., & Harland, D. M. (2006). *Apollo: The definitive sourcebook*. Chichester, UK: Springer-Praxis.

Spudis, P. D. (1996). *The Once and Future Moon*. Washington D.C.: Smithsonian Institution Press.

Whitaker, E. A. (1999). *Mapping and naming the moon*. Cambridge, UK: Cambridge University Press.

Wilhelms, D. E. (1993). *To a rocky moon: A geologist's history of lunar exploration*. Tucson, USA: University of Arizona Press.

Websites

The Apollo Lunar Surface Journal, maintained by E.M. Jones and K. Glover, provides full details of the Apollo missions, including all the Apollo imagery and communications transcripts, and can be found at: http://www.hq.nasa.gov/alsj/.

A complete list of Apollo science experiments is given in NASA's "Catalog of Apollo Experiment Operations" (NASA RP-1317) at: http://ares.jsc.nasa.gov/HumanExplore/Exploration/EXlibrary/docs/ApolloCat/apollo.htm.

Full details of the Apollo and Luna rock and soil samples can be found at: http://curator.jsc.nasa.gov/lunar/lsc/index.cfm.

A continuously updated list of lunar meteorites can be found at: http://curator.jsc.nasa.gov/antmet/lmc/.

The LEAG Lunar Exploration Roadmap can be obtained at: http://www.lpi.usra.edu/leag/ler_draft.shtml.

The Global Exploration Roadmap, which outlines an international approach to the future exploration of the inner Solar System, including the Moon, can be found at: http://www.globalspaceexploration.org/web/isecg/news/2013-08-20

Part VI

Asteroids, Dust and Comets

Chapter 26

Main-Belt Asteroids

Daniel T. Britt
University of Central Florida, Orlando, FL, USA

Guy Consolmagno, S. J.
Specola Vaticana, Vatican City State

Larry Lebofsky
Planetary Science Institute, Tucson, AZ, USA

Chapter Outline

1. Introduction to Asteroids	583
1.1. What are Asteroids?	583
1.2. Discoveries, Numbers, and Names	585
1.3. Sizes and Shapes	585
1.4. Asteroid Density, Porosity, and Rotation Rates	586
2. Locations and Orbits	587
2.1. Zones, Orbits, and Distributions	587
2.2. Special Orbital Classes	590
2.3. The Evolution of Orbits: Yarkovsky and YORP	591
2.4. Asteroid Families	591
2.5. Asteroids and Meteorites	592
3. Physical Characteristics and Composition	593
3.1. The Surfaces of Asteroids	593
3.2. Asteroid Satellites	594
3.3. Telescopic Observations of Composition	595
3.4. Composition, Taxonomy, and the Distribution of Classes	595
4. Puzzles and Promise	599
4.1. Asteroids and Earth	599
4.2. Origins of Asteroids	599
4.3. Spacecraft Missions to Asteroids	600
Bibliography	601

1. INTRODUCTION TO ASTEROIDS

1.1. What are Asteroids?

Asteroids (or more properly, minor planets) are small, naturally formed solid bodies that orbit the Sun, are airless, and show no detectable outflow of gas or dust. Shown in Figure 26.1 are eight asteroids that have been imaged in detail by spacecraft. The difference between asteroids and the other naturally formed Sun-orbiting bodies — planets, dwarf planets, and comets — is largely historical and to some extent arbitrary. To the ancient Greeks and other peoples, there were three kinds of bright objects populating the heavens. The first and most important group was the stars, or *astron* in Greek, which are fixed relative to each other. The English word star is an Old English and Germanic derivation of the Indo-European base word *stêr*, which provided the source of the Greek *astron* and the Latin *astrum*. The terms for the study of stars were based on the Greek root, i.e. astronomy or astrophysics. The second group of objects is planets, or Greek *planetos*, meaning "wanderer" since the planets were not fixed but moved relative to the background of the stars. For the ancients *planetos* included the Sun, Moon, Mercury, Venus, Mars, Jupiter, and Saturn. The final group is comets or *kometes* meaning "long haired" because of their long tails or comas and their unpredictable paths and appearances.

Asteroids were not known to the ancients and the first asteroid, 1 Ceres, was discovered in 1801 by the Sicilian astronomer Giuseppe Piazzi. While observing stellar positions he noticed a slowly moving object, which he originally reported as a comet but suggested might be "more than a comet". Its relatively slow apparent orbital motion suggested that it orbited in the gap between Mars and Jupiter, where theorists at the time had speculated would be the location of a "missing planet". The asteroid 1 Ceres was thought initially to be this new planet. However, other astronomers disputed this designation because of Ceres' apparently small size. Soon after William Olbers discovered the second such object, Pallas, in 1802, Sir William Herschel (who had discovered Uranus 20 years earlier)

FIGURE 26.1 The family portrait of asteroids. These are the asteroids that have been imaged by spacecraft shown to scale. *Image courtesy of NASA/JPL-Caltech/JAXA/ESA.*

proposed that, because these new objects were planetlike in their sun-centered orbits, but starlike in that they were unresolvable points of light in a telescope, the disused Greek root for a single star, "aster" should be used to describe this new addition to the celestial population. However, this term was not universally adopted at that time. By the mid-1800s, after several dozen of these bodies had been discovered, the French and Germans referred to them as "small" (*petit* or *kleine*) planets, while the British Royal Astronomical Society officially called them "minor planets". Until modern times, the term "asteroid" was only used by astronomers in America.

In 2006, the International Astronomical Union (IAU) added additional terms to the mix by defining a group of "dwarf planets". The IAU was attempting to precisely define a planet given the increasing evidence that Pluto was just one of the larger members of the **Kuiper Belt** (See also Kuiper Belt: Dynamics, Kuiper Belt: Physical Studies, and Pluto) and substantially different from the terrestrial or gas giant planets. A dwarf planet orbits the Sun, is not a satellite of another body, has sufficient mass to assume a hydrostatic equilibrium (nearly spheroidal) shape, but does not have sufficient mass to have "cleared its neighborhood" of small bodies. Under this definition 1 Ceres joins Pluto as a dwarf planet. However, for the purposes of this chapter, both Ceres and Pluto can also be considered large asteroids.

Although asteroids share many of the characteristics of planets (Sun-centered orbits, seemingly solid bodies), the primary distinction is that they are simply much smaller than the known planets or dwarf planets. Similarly, the distinction between asteroids and comets is also based on their observational qualities rather than any inherent difference in physical properties or composition. Comets are characterized by their coma, or cloud of sublimating gas and expelled dust. This gives them their characteristic diffuse "fuzzy" halo and long streaming tail (see chapter Comets). Compared to the fuzzy look of comets, an asteroid is a "starlike" sharp point of light. But comets only become "cometary" when they enter the inner solar system and are heated sufficiently by the Sun to vaporize their volatile materials. The point at which frozen volatiles begin to sublimate can vary depending on composition, but for most comets this is approximately at 4 **astronomical units (AU)**. A number of outer solar system objects that could be called asteroids may be composed of the same collection of volatile ices, dust, metal, and carbonaceous organics as comets. Since their orbits are less eccentric than currently active comets, they never travel close enough to the Sun to warm their surfaces, cause their ices to flash into gas, and thus appear cometary. These objects are "solid" bodies because their surfaces stay cold enough to keep their gases frozen.

Fundamentally both asteroids and comets are "small bodies", whose stable lifetimes depend on their location in the solar system.

In the final analysis, asteroids are defined by what they are not. They move against the celestial background so they are not stars. They are not large enough to be planets, dwarf, or otherwise. They are not actively shedding gas and dust so are not comets.

1.2. Discoveries, Numbers, and Names

Because asteroids appear as relatively small and dim points of light moving slowly against the stellar background, finding and identifying an object as an asteroid is fundamentally a question of observation coupled with precise "bookkeeping". The field of view seen through a telescope at any one moment may be filled with hundreds of points of light and rarely will one be an asteroid. The asteroid may move a small amount relative to the stars during the course of a night's observations, but the trick is to know the relative positions of all the viewed stars precisely enough to know when one of the points of light is out of place. Today the viewing through the telescope is done by extremely sensitive charge-coupled devices (CCD) that feed their digital data directly to computers to do the "bookkeeping" of the stars and known asteroids. In the days when Giuseppe Piazzi discovered 1 Ceres, all the observations were done with an eye to the telescope, and the bookkeeping was done by hand drawings of the star fields. Discoveries were made by visually comparing each point of light in the telescope field with a chart that was drawn on a previous observation (there was a premium on being able to draw accurately in the dark when cold). With these methods it is not surprising that only four more asteroids were found in the 45 years after Piazzi found 1 Ceres.

The application of photography to astronomy revolutionized the search for asteroids in the last half of the nineteenth century and the early part of the twentieth century. A photographic plate is essentially an instant and precise local star chart that is far more light sensitive than the human eye, far more accurate than what could be drawn by hand, and able to take advantage of long exposures that compensate for the Earth's rotation. As a result stars appear as fixed as bright dots, and asteroids become streaks that stand out since they move relative to the stars. Modern searches have replaced photographic plates with highly sensitive electronic imaging and computers.

A newly discovered asteroid is given a temporary "name" based on the date of discovery. The first four characters are the year of discovery, followed by a letter indicating which half month of the year the discovery took place. The final character is a letter assigned sequentially to the asteroids discovered in the half month in question. Thus, asteroid 2006 CE would be the fifth asteroid discovered in the first two weeks of February in 2006. If a half month has more than 25 discoveries (the letter "I" is not used), then the letter sequence starts over with additional numerical characters, usually added as a subscript. The 26th object discovered in the first half of February would be 2006 CA_1.

However, discovery is just the first step. Unless an asteroid is tracked and its orbit reliability determined, it will be "lost". This tracking process takes weeks and sometimes months of additional observations. Once an object has an accurate orbit, it is given a permanent number. The numbers are not assigned in order of discovery, but sequentially by order of orbit determination. As of the beginning of 2013 there were 353,926 numbered asteroids and another 251,384 awaiting an accurate-enough orbit to merit a number. With the assignment of a number, the asteroid's discoverer has the right to suggest a name for the object. There are rules, of course, set by the IAU and asteroids are properly referred to by their number and the name (if unnamed, by the temporary date-based name). Asteroids are unique in that they can be named after persons or things living (for example, the authors of this chapter: 4395 Danbritt, 4597 Consolmagno, and 3439 Lebofsky) or dead (2272 Montezuma), real (4457 van Gogh, 12838 Adamsmith) or imaginary (2598 Merlin), mythological characters (5731 Zeus) or characters in operas (558 Carmen), and in several cases, pets (although this is now discouraged); however, political and military leaders must have been dead 100 years before an asteroid can bear their names, and asteroids cannot be named to advertise commercial products. As of the beginning of 2013 there were 17,698 named asteroids.

1.3. Sizes and Shapes

Shown in Figure 26.2 are the 20 largest main-belt asteroids. Asteroid sizes drop rapidly, with the largest asteroid 1 Ceres being almost twice as large as the next largest. There are only four asteroids with diameters greater than 400 km and only three with diameters between 400 and 300 km. The asteroid population becomes relatively abundant only below 300 km diameter.

The number of asteroids increases exponentially as the size decreases in a "power-law" size distribution. This is consistent with an initial population of strong, solid bodies that have been ground down by repeated impacts over the age of the solar system. Today most asteroids are fragments of larger parent bodies that have been collisionally shattered into much smaller pieces. This power law is seen not only in the sizes of asteroids but also in the sizes of the craters on the Moon, Mars, and the asteroids themselves, as well as the moons of the Jupiter and Saturn, reflecting the population of the asteroids whose impacts made those craters.

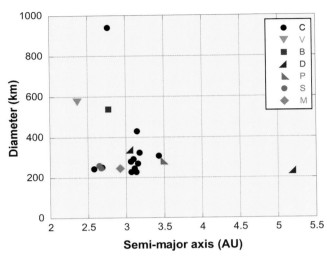

FIGURE 26.2 The 20 largest asteroids by location in the solar system and compositional type. Note that the sizes of asteroids drop rapidly with only four asteroids larger than 400 km diameter.

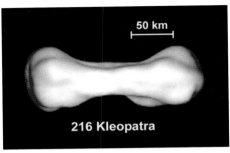

FIGURE 26.3 Radar image of the asteroid 216 Kleopatra. This irregularly shaped object resembles a 200-km-long dog bone. *Image courtesy of NASA/JPL-Caltech.*

Given the conditions in the asteroid belt today, only the largest asteroids are large enough to have survived from the beginning of the solar system. The power law predicts, and observations confirm, that by far the most common asteroids are the smallest. Asteroid search programs using powerful telescopes, extremely sensitive CCD sensors, and state-of-the-art software regularly find asteroids in near-Earth space with diameters as small as only 5–10 m. The primary limitation on our ability to find asteroids is their apparent brightness. Smaller (and more distant) objects reflect less light and, after a point, a small object is not observable because the light it reflects drops below the limiting sensitivity of the telescopic system trying to detect it. The good news is that we have probably discovered and tracked all asteroids in the main asteroid belt larger than 20 km and all those in near-Earth space larger than 4 km. The bad news is that there are thousands of smaller asteroids in Earth-crossing orbits, a few as large as several kilometers in diameter, that remain undiscovered and potential threats to Earth (see chapter Near-Earth Objects).

Since most asteroids are presumably collisionally produced fragments of larger asteroids it should not be a surprise that they are not perfect spheres. Many asteroids that have been directly imaged optically or by radar tend to show very irregular shapes (Figure 26.1). The exception is the largest asteroids (or dwarf planets), 1 Ceres and 4 Vesta, which are (or once were) large enough for hydrodynamic forces to maintain a spheroidal shape. The asteroid 4 Vesta, the only large asteroid known to have been formed of molten lava, is an interesting boundary case. When it was molten it could form itself into a sphere. But after its lava was completely frozen it became rigid and today it appears as a flattened spheroid due to a large crater near its south pole.

Other large asteroids are far from spherical. For example, shown in Figure 26.3 is a radar image of the asteroid 216 Kleopatra, which has a strong resemblance to a 200-km long bone. Most asteroid shapes can be approximated as triaxial ellipsoids, which are objects that have different dimensions on each of their principal axes. In the case of Kleopatra the long dimension is over four times greater than the short dimension.

Star/asteroid **occultation** provides a direct measurement of an asteroid's shape and an opportunity for amateur astronomers to become involved in significant scientific research. The principle is simple: when an asteroid passes through (or "occults") the light from a star, the asteroid creates a "shadow" in the starlight projected on the Earth. Observers in different locations time the disappearance of the occulted star and trace out the shape of this shadow by reconstructing their "chords" or time-tagged observations of the star disappearing behind the asteroid and reappearing on the other side. This is illustrated in Figure 26.4. When done skillfully with modern equipment such as CCD detectors, computer-driven imaging systems, precise time, and the global positioning system, these measurements can be taken with very high accuracy and provide an excellent "snapshot" of the projected two-dimensional shape of the asteroid at the moment of occultation.

1.4. Asteroid Density, Porosity, and Rotation Rates

A fundamental physical property of an asteroid is its density. To first order, asteroid density is related to its composition and should be similar to the densities of meteorites thought to be derived from those asteroids (see chapter Meteorites). However, as is often the case, such expectations can be frustrated by unexpected results from actual measurements. Asteroids in general appear to be significantly underdense relative to their meteorite analogs.

The primary complication is porosity. Asteroids appear to have significant porosity; some may be as much as 50% empty space while their meteorite analogs have only small

Chapter | 26 Main-Belt Asteroids

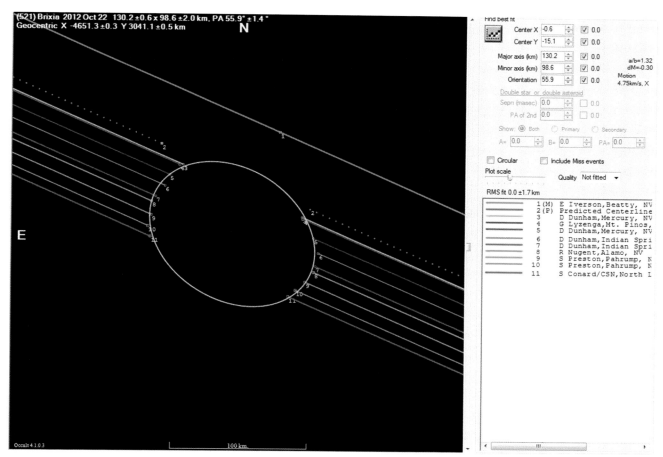

FIGURE 26.4 Occultation observation of asteroid 521 Brixia. The asteroid blocks out the star as viewed by observers at different locations on the Earth. The sum of the observations yields the shape of the asteroid. *Image courtesy of the International Occultation Timing Association.*

to moderate porosities. The observed power law of asteroid sizes, and studies of the collisional dynamics of the asteroid belt, has suggested a history of intense collisional evolution that resulted in only a few of the largest asteroids retaining their primordial masses and surfaces. Asteroids below 300 km in diameter have been shattered by energetic collisions. Some objects reaccrete to form gravitationally bound rubble piles, while the rest are broken into smaller fragments to be further shattered or fragmented. Thus, most asteroids appear to be shattered heaps of loosely bound rubble with significant porosity in the form of large fractures, vast internal voids, and loose-fitting joints between major fragments. Thus, it is not surprising that the average asteroid would have a very large porosity.

Another line of evidence supporting the rubble pile model for asteroids are the images of 253 Mathilde. This object, whose density is only half the density of typical meteorite material, has six identified impact craters that are larger than the size necessary to shatter the asteroid. The only way that Mathilde could have survived these repeated huge impacts is if it were already a shattered rubble pile that dissipates much of the energy of large impacts in the friction of the pieces of rubble grinding against each other.

On the opposite extreme are small rapidly rotating asteroids that are also rubble piles. Shown in Figure 26.5 is asteroid (66391) 1999 KW_4, which is about 1.5 km in diameter and spins with a period of 2.8 h. The bulk porosity of this object is approximately 50% and the rapid spin forces the loose material to assume a shape reminiscent of a spinning top that narrows at the ends and bulges at the equator. The rapid spin forces material from the poles and deposits it at the equator resulting in this remarkable shape.

2. LOCATIONS AND ORBITS

2.1. Zones, Orbits, and Distributions

Minor planets can be found in almost any region of the solar system, but as shown in Figure 26.6(a), one of the largest concentrations of asteroids is located in the "belt" between 1.8 and 4.0 AU. Figure 26.6(b), showing the outer solar system, gives the distribution of known short-period

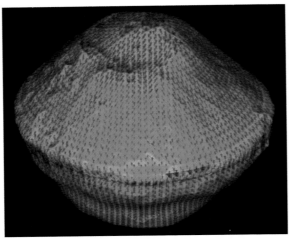

FIGURE 26.5 Asteroid (66391) 1999 KW$_4$. The rapid spin forces material from the poles to the equator, producing a pronounced bulge as material accumulates. *Image courtesy of NASA/JPL-Caltech.*

comets, Centaurs, and Kuiper Belt objects as well as the Trojan asteroids (discussed below).

A more detailed analysis of the average distances of asteroids from the Sun (the asteroids' **semimajor axes**) as shown in Figure 26.7 reveals a subtle structure to the asteroid belt. First, there appears to be a sharp inner boundary to the asteroid belt at about 2.2 AU. But note that this boundary curves to higher AU for asteroids with higher orbital **inclinations**. Second, there is a sharp gap in the number of asteroids whose average distance from the Sun (semimajor axis) is 3.28 AU. Asteroids orbiting here would have exactly half the orbital period of Jupiter and are said to be in a 1:2 mean-motion resonance with Jupiter. Similar gaps can be seen elsewhere in the asteroid population as well, most notably at the locations of the 1:3 and 2:5 mean-motion resonances. The gaps in the distribution of asteroid semimajor axes are called **Kirkwood gaps** for Daniel Kirkwood who first pointed them out in 1886. Unlike the gaps in Saturn's rings, however, these gaps are not directly visible within the asteroid belt in Figure 26.6(a) because asteroid orbits have a wide range of eccentricities and are constantly crossing through the region of these gaps. Third, there is a dearth of asteroids in orbits with semimajor axes beyond 3.5 AU, with two exceptions; there are clusters of asteroids at 3.97 AU, corresponding to the 2:3 mean-motion resonance with Jupiter, and at 5.2 AU, where asteroids share the same orbit as Jupiter (especially visible in Figure 26.6(b)).

These boundaries and gaps are formed by the steady influence of the gravitational attraction of the planets on the orbits of the asteroids. These interactions are strongest when a planet is closest to an asteroid, which normally occurs at random time intervals and at random locations of the asteroid's orbit; on average they cancel out without causing a significant change in the asteroid's orbit. However, an asteroid whose orbital period is a simple fraction of Jupiter's 11.86-year period will be in resonance with Jupiter and have a close approach in the same place in its orbit over and over again. For an asteroid with a 6-year period (in a 1:2 resonance), this closest approach will occur at the same place every other asteroid orbit. (Similarly, asteroids in the 1:3 resonance encounter Jupiter at the same place in their orbits, every third orbit, and so forth with higher resonances.) Jupiter's pull at this point, imparting some energy to the asteroid's orbit, will then compound itself rather than cancel out. The largest effect of

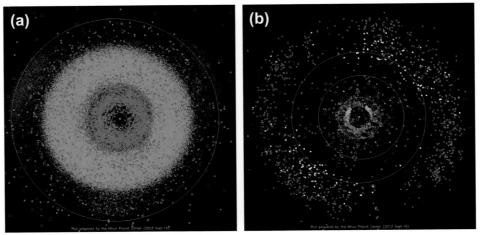

FIGURE 26.6 (a) The location of asteroids in the inner solar system. The outer circle is the orbit of Jupiter. The thick main asteroid belt (green) is readily visible just outside the orbit of Mars; the "swarms" before and after Jupiter are the Trojans. (b) The location of asteroids in the outer solar system (at or beyond Jupiter). The outer circle is the orbit of Neptune; the well-populated region outside of Neptune is the Kuiper Belt. *Image courtesy of Gareth Williams/Minor Planet Center.*

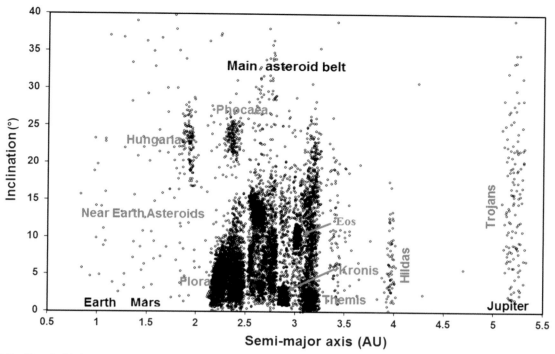

FIGURE 26.7 Plot of orbital parameters of numbered asteroids in semimajor axis vs. inclination space. The major asteroid families and the planets are identified in the plot.

this sort of perturbation is to increase the **eccentricity** of the asteroid's orbit. This does not change its average distance from the Sun, but it makes the perihelion move closer to the Sun, and the aphelion move farther out. Once its eccentricity reaches a value of about 0.3, a main-belt asteroid's orbit begins to approach or even cross the orbit of Mars. Close encounters with Mars can further alter its orbit, leading to interactions with the other inner planets or with Jupiter, which eventually results in a collision with either a planet or the Sun, or ejection from the solar system. For asteroids, orbital life in the Kirkwood gaps is (relatively) short, but exciting.

This kind of resonance explains the Kirkwood gaps. But it does not explain the inner boundary and its dependence on the inclination of the asteroid orbit, the lack of asteroids with semimajor axes outside 3.5 AU, or the concentration of asteroids at the outer resonances. More indirect effects give rise to these patterns. The shape of the inner boundary is the result of a subtle but surprisingly powerful effect. Every asteroid has an orbit that is at least slightly eccentric, and the orientation of its perihelion slowly drifts with time. This precession of the perihelion is caused by the perturbations of the other planets. Likewise, the orientations of the major planets' orbits, which are not perfectly circular, also drift with time. A subtle interaction arises when the precession of Saturn's orbit is in resonance with the precession of an asteroid's orbit. This secular resonance (so called because it builds up in the same direction over time, regardless of where the asteroid and Saturn are in their orbits) is called the ν_6 resonance; ν is the Greek letter that represents the precession rate, and the 6 represents Saturn, the sixth planet from the Sun. Its effect is to increase an asteroid orbit's eccentricity, as with the Jupiter mean-motion resonances. The position of this resonance depends on both the location and the inclination of the asteroid orbit. For asteroids orbiting in the plane of the planets, it occurs at around 2.2 AU; as the inclination of the asteroid orbit increases, the location of this resonance moves further from the Sun. This resonance sculpts the inner edge of the asteroid belt.

The lack of asteroids in the outer regions of the asteroid belt may be the result of the migration of Jupiter's orbit early in the history of the solar system. Models and observations of disks around nearby young stars suggest that the planets were formed from a relatively smooth cloud of gas and dust called the **solar nebula**. (See chapter Origin of the Solar System.) Jupiter, the innermost gas giant, may have been the first planet formed, and as we have seen, planetesimals in resonance with Jupiter would have been ejected from the inner solar system by Jupiter's gravity. The motion of these objects outward would have caused Jupiter to move inward, sweeping the location of its resonances (especially the strongest resonance, 2:1) across that region of the nebula and ejecting yet more material, effectively depleting material outside the present location of the 2:1 resonance. (Some of that material may today be residing in the far-distant **Oort Cloud** (see Physics and Chemistry of Comets).) Asteroids in the 2:3 resonance would be stabilized against ejection and thus carried along with Jupiter as it moves.

It is possible, however, that the motion of Jupiter may have been sufficiently great that the entire region of the asteroid belt (and much of the material that would have gone into Mars) was completely removed (as described below, in Section 4.2). In this case, the material in the asteroid belt today would have migrated into this region from elsewhere in the solar system after this sweeping was complete. In this case, the resonances noted above would mark the boundaries of stable orbits for this migrating material.

2.2. Special Orbital Classes

While most asteroids are found in the main asteroid belt between Jupiter and Mars, there are a number of other asteroid groups. The "asteroids" beyond the orbit of Jupiter are probably volatile-rich and would become cometary if they were moved to the inner solar system, but for the purposes of this discussion we will list these groups of small bodies as asteroids here. The asteroids that circle the Sun at the same orbital distance as Jupiter are called Trojan asteroids. They reside in dynamically stable zones 60° ahead and behind Jupiter. These positions are the last two of the five "Lagrangian points", named by the nineteenth-century mathematician J. L. Lagrange. He first described the orbital behavior of small bodies affected by the gravitation pull of two large objects such as the Sun and any planet. He found that along with three unstable equilibrium points (L_1 through L_3), a small body like an asteroid could share a planet's orbit so long as both formed an equilateral triangle with the Sun. There are two such points; the L_4 point lies ahead of the planet, while L_5 trails behind it. In theory, any planet can have Trojan asteroids, but in practice the powerful gravitational pull of Jupiter means that it absolutely dominates the Trojan population.

The Trojans derive their name from the first such asteroid discovered in Jupiter's orbit, named Achilles after the hero of the Trojan War. For Jupiter Trojans, the L_4 region asteroids are named after Greek heroes of the Iliad, while Trojan heroes populate the L_5 region. (The exceptions, named before this rule was adopted, include two of the largest Trojans: 617 Patroclus, named for the Greek hero, orbits among the Trojans at L_5, while 624 Hektor, the largest Trojan and a hero of Troy, orbits at L_4 with the Greeks.) A total of 5928 Jupiter Trojans have been discovered to date; oddly, the L_4 region is nearly twice as populated as the L_5 region. Remember that any planet can, in theory have Trojans; as of the beginning of 2013, the census includes one suspected Earth Trojan, three Mars Trojans, and nine Neptune Trojans (and again the Neptune L_4 population is twice as large as the L_5).

Another major group of minor planets is the Centaurs. Named as a class after the discovery of Chiron, a small body orbiting between Saturn and Uranus, the term has eventually grown to include any noncometary body beyond Saturn whose orbit crosses the orbit of a major planet, and even the noncomet part must be relaxed, as Chiron itself has been seen on occasion to have a cometlike coma. These "asteroids" are most likely large, volatile-rich objects (i.e. comets) perturbed inward from the Kuiper Belt. But since the Centaurs orbit deep in the outer solar system, they cannot warm sufficiently to allow volatiles to sublimate off and show cometary activity, so they are considered asteroids until proven otherwise. In terms of their orbits, this group includes the classical Centaurs (some two dozen objects known to orbit like Chiron between Saturn and Uranus), roughly 50 objects whose orbits cross Uranus' or Neptune's orbit, and the 448 objects (discovered as of 2013) that lie in highly eccentric orbits ranging out beyond the Kuiper Belt. All are considered "scattered disk objects", which have been dynamically scattered by Neptune's gravity out of the disk of the Kuiper Belt. (See chapter Kuiper Belt: Dynamics.)

The Kuiper Belt itself is the outermost set of minor bodies. It is made up of objects populating space beyond the orbit of Neptune but inside about 1000 AU. The first object was discovered in 1992 (1992 QB_1) with a semimajor axis of 44 AU and an estimated diameter of several 100 kilometers. Besides the scattered disk objects noted above, other dynamical classes of Kuiper Belt objects include others like 1992 QB_1 in low-inclination, low-eccentricity orbits (sometimes called "cubewanos" after their first example) and others orbiting like Pluto (and so-called plutinos) in a 2:3 resonance with Neptune. Again, all these objects are probably cometary. In fact the existence of a belt of material like this was first suggested in 1949 as a source area for short-period comets. Given the nearly 1300 Kuiper Belt "asteroids" discovered as of 2013, there are probably hundreds of thousands of objects larger than a kilometer populating this belt (see chapter Kuiper Belt: Dynamics).

Inward from the main asteroid belt are the asteroids that cross the orbits of the inner planets: the Amor, Apollo, and Aten asteroids. Amor asteroids are asteroids whose eccentric orbits dip in from the asteroid belt to cross the orbit of Mars, but without reaching the orbit of the Earth. Apollos are those that do cross Earth's orbit, but whose semimajor axis is always ≥ 1 AU. This differentiates them from Atens, which also cross the Earth's orbit but have semimajor axes inside of Earth's orbit. The Apollo, Aten, and Amor objects are collectively called **Near-Earth Objects** or NEOs (See Near Earth Objects). They are relatively small objects; the largest known NEO is the Amor object 1036 Ganymed, with a diameter of 38.5 km. NEOs are also subject to a power-law distribution, so as their sizes drop the population increases rapidly. As of 2013, there are roughly 850 NEOs with diameters greater than 1 km out of a population of approximately 9600 known NEOs. It is estimated that there are approximately 920 total NEOs that are larger than 1 km. These are the objects that can and (in the course of geologic time) do frequently collide with Earth. Indeed, computer calculations indicate that most NEOs could only survive in their present orbits for roughly 10 million years before falling into the Sun, colliding with a planet, or being ejected. Thus, the NEO population must be continually replenished from the asteroid belt. Compositional data indicates that NEOs are drawn from every zone of the asteroid belt and have been perturbed into the inner solar system by a variety of mechanisms including the Yarkovsky effect described below. (See chapter Near-Earth Objects.)

2.3. The Evolution of Orbits: Yarkovsky and YORP

The gravitational perturbations of the planets are not the only forces acting on the asteroids. Although the Kirkwood gaps show that resonances are the most effective way to clear material from the asteroid belt, the generally low population of asteroids throughout the belt (even in its most heavily-populated regions), the replenishment of asteroids into short-lived NEO orbits, and the constant delivery of meteorites from the asteroid belt to the Earth (see below) all indicate that some other forces must be moving material from the main belt to the resonance regions.

One early hypothesis was that collisions between asteroids could impart enough momentum to scatter the collision products into a wide variety of new orbits, some of which would lie in resonance with Jupiter and thus be delivered out of the asteroid belt. However, detailed computer modeling of both the collisions and the ensuing orbits of the collisional products conclusively shows that this process alone fails by many orders of magnitude to move nearly enough material from the asteroid belt to match the observed population of NEOs or the meteorite flux. Some other force or forces must be involved.

Another early suggestion first proposed by the Russian theorist I. O. Yarkovsky in the late nineteenth century is that sunlight itself could provide a surprisingly effective way of changing the orbits of asteroids. The general idea is simple enough. Since light carries momentum, as sunlight is absorbed or reflected by an asteroid there is a small momentum transfer from the light to the asteroid. However, since sunlight comes from the same direction as the force of the Sun's gravity (and, like gravity, varies as $1/r^2$) this effect by itself will merely change the effective pull of the Sun, without changing the energy (or semimajor axis) of an asteroid's orbit. (There is a small relativistic effect called "Poynting-Robertson" drag, but it is ineffectual for anything larger than small grains of dust.) However, when an asteroid absorbs sunlight, the energy of that light heats the asteroid and that heat must eventually be reradiated to space as infrared photons. When each infrared photon is emitted, it exerts a tiny amount of recoil momentum to the asteroid itself. And, unlike the direct reflection of sunlight, this recoil is not necessarily in the same direction as the pull of the Sun's gravity, since there is always a small time lag between the absorption and the reradiation of the energy.

For example, the afternoon side of a spinning body will always be slightly warmer than the morning side. This means that more infrared energy is radiated from the afternoon side; that side of the asteroid experiences greater recoil from those photons' emissions than the morning side does. The way the spin axis is tilted or the differences in heating between perihelion and aphelion are other examples of situations that will lead to the asymmetric radiation of infrared photons. Depending on how the asteroid spins, this difference can add or subtract energy from the asteroid's orbit and thus continually change its semimajor axis. It can also change the way the asteroid itself spins. An elaborate theory based on the work of Yarkovsky, as further elaborated by O'Keefe, Radzievskii, and Paddack, dubbed the "YORP" effect, suggests a number of ways in which the momentum of emitted radiation can alter both the speed and the direction of an asteroid's spin. More than just a mathematical curiosity, the predictions of this work have been confirmed in a number of cases, including asteroids whose spin rates have been observed to change or be aligned in a way predicted by this theory.

2.4. Asteroid Families

As discoveries of asteroids accumulated in the early part of the twentieth century astronomers noted that it was common for several asteroids to have very similar orbital elements and that asteroids tended to cluster together in semimajor axis, eccentricity, and inclination space. In 1918, K. Hirayama suggested that these clusters were "families" of asteroids. Hirayama suggested five families, and this number has been greatly increased by the work of generations of orbital dynamicists.

These families are probably the result of the collisional breakup of a large parent asteroid into a cloud of smaller fragments sometime in the distant past. Time and the gravitational influence of other solar system objects have gradually dispersed the orbits of these fragments, but not enough to erase the characteristic clustering of families.

It has been suggested that families could provide a glimpse at geologic units that are usually deeply hidden in the interiors of planets. If a differentiated asteroid were broken into family members, for example, that family should have members that represent the metallic core, others coming from the metal—rock transition zone called the core—mantle boundary, yet others made of the dense, iron-rich units in the mantle, and others originating from the crust of the former planetesimal. In fact, however, no such elaborate collection of different asteroid types has been seen in a family.

Families may, in fact, be relatively short lived. The Yarkovsky effect described above has proved to be very effective in moving family members out of their original orbits. Understanding and defining the dynamics of asteroid families remains an active and rapidly changing field of study.

2.5. Asteroids and Meteorites

There are a number of lines of evidence that show the ultimate source region for meteorites is the asteroid belt. (See chapter Meteorites.) As will be discussed in more detail later (Section 3.3), meteorite **reflectance spectra** can be matched with several classes of asteroids. In addition, camera networks, or many well-separated video images of fireballs, have recorded the falls of about a dozen recovered meteorites with sufficient detail that their orbits before hitting the atmosphere can be calculated; each of these meteorites have eccentric orbits that reach back into the main asteroid belt.

The physical state of the meteorites themselves also indicates an origin in the asteroid belt. Cosmic ray exposure ages of meteorites show that they have spent most of their existence shielded from cosmic rays, deep in asteroid-sized parent meteoroids. A brecciated meteorite of one given type often contains fragments of other meteorite types, called xenoliths, which require that the source region have the mineralogical diversity found in the asteroid belt. And solar-wind implanted gases found in **regolith** meteorites indicate that implantation took place at a distance from the Sun consistent with the location of the asteroid belt.

Meteorites do not automatically provide the location and taxonomic class of particular parent bodies, however. The very fact that a meteorite is "in our hands" suggests the occurrence of some violent event that may have fragmented and perhaps destroyed the parent body. Usually, the best that can be done is to link individual asteroid spectral classes with meteorite compositional groups.

One exception is the relatively rare basaltic meteorites of the interrelated howardite, eucrite, diogenite (HED) classes. (All have similar ages and isotope abundances, and howardites are clearly a mixture of eucrites and diogenites.) Their spectra are very well matched by the spectra of asteroid 4 Vesta, and the composition of the surface of Vesta as determined by the Dawn spacecraft confirms this match between the asteroid and the HEDs. Since these meteorites are igneous rocks, one can use standard geochemical models to compute that they were made from lavas in equilibrium with a large mantle of dunite. However, while hundreds of HEDs are known, no corresponding dunite meteorites have been found, suggesting that these meteorites are sampling the surface of an otherwise intact asteroid. Thus, we infer that Vesta (perhaps via smaller asteroids chipped from the surface of Vesta) can be identified as the source of the HED meteorites.

The task of correlating other asteroid spectra with meteorite composition is less certain. Not only are there many possible candidate asteroids for any given meteorite type, but there is also no certainty that the spectra we see of asteroid surfaces is necessarily correlated with that of meteorites. Most meteorites were originally buried beneath the surface of an asteroid, the actual asteroid surface conditions are unknown, and the effects of space weathering on asteroids are poorly understood. Thus, spectral matches between asteroids and meteorites, including the ones detailed here, should be viewed with healthy skepticism.

Still, the return in 2010 of a few thousand tiny fragments (less than 100 microns diameter) from asteroid 25143 Itokawa by the Japanese Hyabusa mission, as shown in Figure 26.8, has provided the first confirmation that such identification may well be accurate. Spectra of Itokawa had been interpreted as indicating its composition was matched by LL-class ordinary chondritic material. The iron oxide, trace element, and isotope abundances of the returned samples were also well matched by LL chondrite material. In addition, some of these small particles had evidence of nanometer metal and sulfide particles, just as models of "space weathering" had predicted. The fact that LL chondrites are the least common of the ordinary chondrite classes suggests the match is more than just coincidence, and gives us confidence that other spectral identifications may also be reliable.

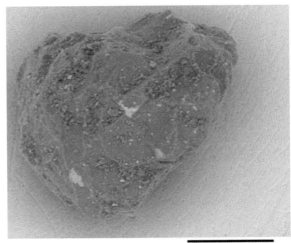

FIGURE 26.8 A fragment of the Itokawa asteroid returned to Earth by the Hayabusa spacecraft. Its composition is similar to LL chondrites and it shows evidence of space weathering. *From Noguchi et al. (2011).*

An event in 2008 showed both the connection between meteorites and asteroids, and the complications that simple spectra identification may mask. A near-Earth asteroid, designated 2008 TC_3, was discovered and observed just hours before it hit the Earth, scattering meteorites across the desert in northern Sudan. Spectra at that time suggested it had a dark and relatively featureless spectrum consistent with an F or B type; no meteorite type had been definitively identified with such asteroids, but they were thought to represent primitive material. The first meteorites collected from the strewn field near Almahata Sitta were identified as belonging to the rare ureilite class, which is unusual for having an igneous texture while being rich in carbon. As more samples were recovered from the desert, however, it was found that a number of samples actually represented other classes of meteorites, including a variety of different ordinary, carbonaceous, and enstatite chondrite meteorites, and samples that are chemically unique among meteorites. By testing samples for radioactive elements produced by cosmic rays, it has been shown convincingly that samples of these many different types nonetheless landed on the Earth at the same time, in the fall of this asteroid, and may have been incorporated together in the same asteroid for at least the past 3.8 billion years. Thus, while confirming that meteorites come from asteroids, it is not at all clear that a given asteroid can necessarily be identified with only one particular meteorite type.

In addition to the complication of multiple meteorite types coming from a given asteroid, there are also several factors that bias the population of meteorites arriving on Earth and therefore limit our sample of the asteroid belt. First, the dynamical processes that deliver meteorites from the asteroid belt to Earth are probably strongly biased toward sampling relatively narrow zones in the asteroid belt. Calculations demonstrate that the vast majority of meteorites and planet-crossing asteroids originate from just two resonances in the belt, the 1:3 Kirkwood gap and ν_6 resonance. Both these zones are in the inner asteroid belt where the asteroid population is dominated by S-type asteroids. However, the Yarkovsky effect significantly increases the chances of fragments from anywhere in the asteroid belt working their way into Earth-crossing orbits. A second factor is the relative strength of the meteorites. To survive the stress of impact, acceleration, and then deceleration when hitting the Earth's atmosphere, without being crushed into dust, the meteorite must have substantial cohesive strength. Large iron meteorites are more likely to survive until they hit the surface of the Earth; they may form a crater (like Meteor Crater in Arizona) when they hit, but in that process most of the iron is vaporized and lost. The Earth's atmosphere is probably the most potent filter for meteorites. The relatively weak, volatile-rich meteorites from the outer asteroid belt stand little chance of surviving the stress and heating of atmospheric entry. It is very likely that the meteorites available to us represent only a small fraction of the asteroids, and it is possible that most asteroids either cannot or only rarely contribute to the meteorite collections.

3. PHYSICAL CHARACTERISTICS AND COMPOSITION

3.1. The Surfaces of Asteroids

As shown in Figures 26.1, 26.9, and 26.10 the surfaces of asteroids appear cratered, lined with fractures, and covered

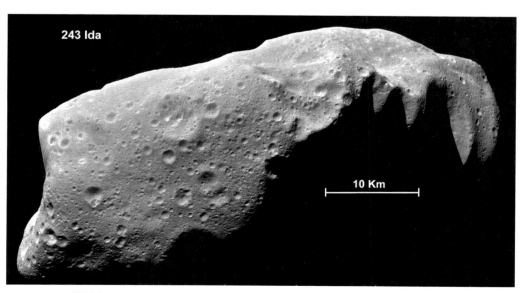

FIGURE 26.9 The surface of asteroid 243 Ida. *Image courtesy of NASA/JPL-Caltech.*

FIGURE 26.10 Asteroid 25143 Itokawa. The asteroid is approximately 700 m in it longest dimension. The smooth areas in the center and on the lower left center are examples of "ponding" of fine regolith. *Image courtesy of ISAS/JAXA.*

with regolith. These surfaces are dominated by impact processes. As discussed in earlier sections, asteroids are strongly affected by collisional disruption and have a complex history of impact fracturing and fragmentation. Objects in the size range shown in the figures are probably formed as disrupted fragments from larger objects and some are likely rubble piles themselves. Since asteroids are far too small to retain an atmosphere that could offer some protection from the exposure to space, the surfaces of asteroids are exposed to an extremely harsh environment. There is a range of processes associated with exposure to the space environment; high levels of hard radiation, high-energy cosmic rays, ions and charged particles from the solar wind, impacts by micrometeorites, impacts by crater-forming objects, and finally impacts by other asteroids large enough to destroy the parent asteroid. The overall result of these processes is threefold. First, large impacts shatter the parent asteroid creating substantial internal fracturing, porosity, and an extremely rough and irregular surface. Second, small impacts and micrometeorites create a regolith that blankets the asteroid in a fine soil of debris from the bedrock. Finally, micrometeorites, radiation, and solar wind produce chemical and spectral alterations in the regolith soil and exposed bedrock that "weather" the surface of the asteroids.

All the small asteroids viewed by spacecraft show significant regoliths, and the power of radar waves reflected by asteroids large and small (especially those passing near the Earth, and so more easily observed by radar) also shows that their surfaces are comparable to dry soil or sand (see chapter Planetary Radar). On several of these asteroids the regoliths appear to have been altered by space weathering processes, although just how this alteration affects asteroidal material is still not completely understood. In the asteroid population, there are general spectral trends that appear to be associated with the age of an asteroid's surface, with younger less altered surfaces tending to be less red. This effect is seen in the meteorite population.

Meteorites that have evidence of residing on the surfaces of asteroids have strong spectral differences from meteorites that were not exposed on asteroid surfaces.

Another major surface effect is the development of small "ponds" of regolith material as shown in Figure 26.10. These ponds have been seen on Eros and Itokawa and consist of ground-up material that has been somehow mobilized on the surface and accumulated in local "depressions" or gravitational lows. The actual magnitude and direction of gravity on a body as small and irregularly shaped as an asteroid is not at all intuitive, but the effect is still strong enough to drive surface processes. Ponds appear to develop over time and appear to bury the boulders and cobbles within them.

Another process that affects the surfaces of asteroids is the reaccretion of ejecta debris. Impacts of other small asteroids produce the abundant craters seen on all these objects. While much of the impact debris escapes the low gravity of an asteroid, a large amount is reaccreted by the asteroid. The abundance and location of boulders on objects such as Eros and Itokawa (Figure 26.9) have been explained by the low-velocity ejecta debris slowly "falling" back onto the rotating asteroid.

3.2. Asteroid Satellites

While it had been long suspected that some asteroids had satellites, this was spectacularly confirmed when the Galileo spacecraft flew by asteroid 243 Ida and discovered its moon Dactyl. As of 2013, a total of 232 asteroid satellites have been announced in 209 systems including at least three triple systems and one sextuple. Figure 26.11 is an image of asteroid 22 Kalliope and its satellite Linus.

NEOs tend to have small separation distances from their satellites, which are probably the result of formation by "fission". Many NEOs have rotation rates close to the fission limit and additional collisions or the YORP effect can enhance asteroid spin enough to cause fission. Once fission occurs the new satellite carries away some of the primary's angular momentum, thus dropping the rotation

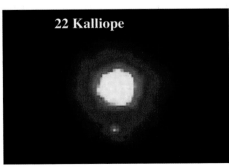

FIGURE 26.11 Asteroid 22 Kalliope and its satellite Linus. *Image courtesy of SWRI/W. Merline.*

back below the fission limit. The lack of distant NEO companions may be the result of gravitational encounters with planets. Distant satellites would be much more likely to be stripped from their primaries during close planetary encounters common with NEOs.

Although only about 209 have been discovered, asteroid satellites are thought to be fairly common with a small percentage of all asteroids having satellites. With over 350,000 numbered asteroids, a large number of satellites remain to be discovered. This is another area where amateur astronomers can make a significant contribution to science. While some satellites have been discovered by direct imaging, either from spacecraft or **adaptive optics** (i.e. Figure 26.11), most main-belt asteroid satellites are discovered by analysis of asteroid **lightcurves**. The principle is that the satellite in its orbit will periodically add or subtract its illumination from the brightness of the asteroid. By precisely tracking the change in brightness, it is possible to identify the satellite and determine its orbit and period. With CCD imagers available commercially and modest-sized telescopes, a skilled amateur can successfully compete in discovering asteroid satellites.

3.3. Telescopic Observations of Composition

Our understanding of the composition of asteroids rests on two pillars: the detailed study of meteorite mineralogy and geochemistry, and the use of remote sensing techniques to analyze asteroids. The meteorites provide, as discussed in a previous section, an invaluable but limited sample of asteroidal mineralogy. To extend this sample to what are effectively unreachable objects, remote sensing uses a variety of techniques to determine asteroid composition, size, shape, rotation, and surface properties. The best available technique for the remote study of asteroid composition is visible and near-infrared **reflectance spectroscopy** using ground-based and Earth-orbiting telescopes.

Reflectance spectroscopy is fundamentally the analysis of the "color" of asteroids over the wavelength range $0.2-3.6$ μm. An experienced rock hound limited to the three colors of the human eye can identify a surprisingly wide variety of minerals. For example, the silicate olivine is green, and important copper minerals such as azurite (blue) and malachite (green) are vividly colored. These colors are a fundamental diagnostic property of the mineralogy because the atoms of a mineral's crystal lattice interact with light and absorb specific wavelengths depending on its structural, ionic, and molecular makeup, producing a unique reflectance spectrum. The reflectance spectrum is essentially a set of colors, but instead of three colors our remote sensing instruments "see" very precisely in 8, 52, or even several thousand colors. What can be seen are the details of the major rock-forming minerals olivine, pyroxene, spinel, the presence of phyllosilicates, organic compounds, hydrated minerals, and the abundance of free iron and opaque minerals.

In addition to a spectroscopic inventory of minerals, telescopic measurements yield several other critical pieces of information. The **albedo** or reflectivity of the asteroid can be determined by measurements of the visible reflected light and the thermal emission radiated at longer wavelengths. A dark asteroid will absorb much more sunlight than it reflects, but will heat up and radiate that extra absorbed energy at thermal wavelengths. Finding the ratio of the reflected and emitted flux at critical wavelengths provides an estimate of just how dark that asteroid is, its albedo. Measuring how the reflectance properties of the surface change as we observe sunlight reflecting off it at different angles (the "solar phase angle") can be used to model the photometric properties of the surface material and estimate physical properties like the surface roughness, surface soil compaction, and the light-scattering properties of the asteroidal material. Measurements of polarization as a function of solar phase angle can be used to infer albedo and also provide insight into the texture and mineralogy of the surface.

3.4. Composition, Taxonomy, and the Distribution of Classes

Our basic knowledge of asteroids is primarily limited to ground-based telescopic data, usually broadband colors in the visible and near-infrared wavelengths and albedo that is indicative of composition; this forms the basis of asteroid taxonomy. Asteroids that have similar color and albedo characteristics are grouped together in a class denoted by a letter or group of letters. Asteroids in particularly large classes tend to be broken into subgroups with the first letter denoting the dominant group and the succeeding letters denoting less prominent spectral affinities or subgroups.

Asteroid taxonomy has developed in tandem with the increase in the range and detail of asteroid observational data sets. Early observations were often limited in scope to the larger and brighter asteroids, and in wavelength range, to filter sets originally designed to be used for stellar astronomy. As observations widened in scope and more specialized filter sets and observational techniques were applied to asteroids, our appreciation of the variety and complexity of asteroid spectra has also increased. The asteroid classification system has evolved to reflect this complexity, and the number of spectral classes has steadily increased. Shown in Table 26.1 is a listing of the expanded "Tholen" asteroid classes and the current mineralogical interpretation of their reflectance spectra. While the Tholen classification is widely used it is not by any means the only asteroid classification system. Other widely accepted

TABLE 26.1 Meteorite Parent Bodies

Asteroid Class	Inferred Major Surface Minerals	Meteorite Analogs
D	Organics + anhydrous silicates? (+ice??)	None (cosmic dust?)
P	Anhydrous silicates + organics? (+ice??)	None (cosmic dust?)
C (dry)	Olivine, pyroxene, carbon (+ice??)	"CM3" chondrites, gas-rich/black chondrites?
K	Olivine, orthopyroxene, opaques	CV3, CO3 chondrites
Q	Olivine, pyroxene, metal	H, L, LL chondrites
C (wet)	Clays, carbon, organics	CI1, CM2 chondrites
B	Clays, carbon, organics	None (highly altered CI1, CM2??)
G	Clays, carbon, organics	None (highly altered CI1, CM2??)
F	Clays, opaques, organics	None (altered CI1, CM2??)
W	Clays, salts????	None (opaque-poor CI1, CM2??)
V	Pyroxene, feldspar	Basaltic achondrites
R	Olivine, pyroxene	None (olivine-rich achondrites?)
A	Olivine	Brachinites, Pallasites
M	Metal, enstatite	Irons (+EH, EL chondrites?)
T	Troilite?	Troilite-rich irons (Mundrabilla)?
E	Mg-pyroxene	Enstatite achondrites
S	Olivine, pyroxene, metal	Stony irons, irons, lodranites, winonites, siderophyres, ureilites, H, L, LL chondrites

classifications include the Barucci system, the Bus-DeMeo system, and the Howell system.

To explain the compositional meaning of asteroid reflectance spectra and color data, we can treat the asteroid belt as a series of zoned geologic units, starting at the outer zones of the main belt and working inward toward the Sun. This is illustrated in the distribution of asteroid classes shown in Figure 26.12. The outer asteroid belt is dominated by the low-albedo P and D classes. The analogs most commonly cited are cosmic dust or carbonaceous chondrites that are enriched in organics like CI chondrites or the Tagish Lake meteorite. However, the spectral characteristics of these asteroids are difficult to duplicate with material that is delivered to the inner solar system. Probably P and D asteroids are composed of primitive materials that have experienced a somewhat different geochemical evolution than cosmic dust or CI chondrites. Their spectra indicate increasing amounts of complex organic molecules with increasing distance from the Sun. These objects are also probably very rich in volatiles including water ice.

Dark inner asteroid belt asteroids include the B, C, F, and G classes whose meteorite analogs are the dark CI and CM carbonaceous chondrite meteorites. The spectral differences between these classes are thought to represent varying histories of aqueous alteration or thermal metamorphism. The CI carbonaceous chondrites are rich in water, clay minerals, volatiles, and carbon; they represent primitive material that has been mildly heated and altered by the action of water (for more details on all the meteorites see chapter Meteorites).

Sunward of 3 AU, bright rocky asteroids become much more common. This zone was strongly affected by the early solar system heating event and contains those classes most likely to represent differentiated and metamorphosed meteorites. The best asteroid/meteorite spectral match is that of the V-class asteroids with the basaltic achondrite meteorites. V types are interpreted to be a differentiated assemblage of primarily orthopyroxene with varying amounts of plagioclase, which makes them very close analogs to the basaltic HED association of meteorites. These meteorites are basaltic partial melts, essentially surface lava flows and near-surface intrusions originating on an asteroid (like Vesta) that underwent extensive heating, melting, and differentiation.

While the V-class asteroids represent the surface and near-surface lava flows of a differentiated asteroid, the A-class asteroids are thought to represent the next zone deeper. These rare asteroids are interpreted to be nearly

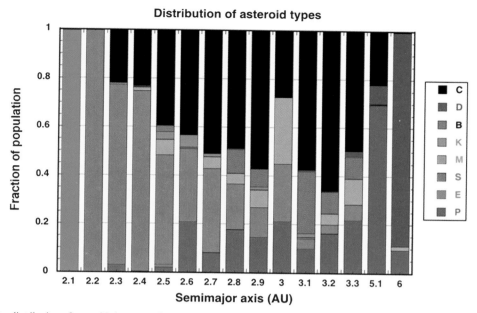

FIGURE 26.12 The distribution of asteroids by spectral class; each color represents the volume fraction of asteroids of a given class orbiting (within 0.1 AU) at the same average distance from the Sun. (The largest four asteroids, Ceres, Vesta, Pallas, and Hebe, are not included as their volumes would dominate the regions where they orbit.) Stony classes (red and green) dominate the inner belt; the darker and more volatile-rich classes (blue, purple, and black) dominate the outer belt.

pure olivine and may be derived from the mantle of extensively differentiated parent bodies. The Earth's mantle is dominated by olivine and theoretical studies show that differentiation of asteroids with a bulk composition similar to ordinary chondrite meteorites should produce olivine-rich mantles. Another possible mantle-derived asteroid is the R class, which is a single-member class made up of the asteroid 349 Dembowska. Analysis of its reflectance spectra suggests a mineralogy that contains both olivine and pyroxene and may be a partial melt residue of incomplete differentiation.

A more common asteroid class is the M class, which has the spectral characteristics of almost pure iron–nickel metal; several of them show high radar reflections consistent with metal. These objects are thought to be direct analogs to the metallic meteorites and may represent the cores of differentiated asteroids. Isotopic and chemical studies indicate that iron meteorites could come from as many as 60 different parent bodies indicating a wide variety of differentiated bodies in the asteroid belt. However, some M-class asteroids have been shown to have hydrated minerals on their surfaces and others have low radar reflections that are inconsistent with metallic compositions. The spectral characteristics of M asteroids can also be consistent with some clay-rich silicates and this raises the possibility that the "wet" M asteroids are assemblages of clays, like the CI carbonaceous chondrites, but without the carbon-rich opaques that darken the CIs. The W (or "wet") class—think of the W as an upside-down M—was coined to classify these unusual objects.

The E-class asteroids are another example of the perils of extrapolation from limited information to a convenient meteorite analog. Looking at the spectrum of the "type" asteroid for the E class, 44 Nysa, it was easy to assume that these asteroids were excellent analogs for the enstatite achondrites. The only problem was that enstatite meteorites are entirely anhydrous, and 44 Nysa was later observed to be strongly hydrated. Although some E-class asteroids are probably composed of the same differentiated enstatite assemblages as the enstatite achondrites, about half of the observed E asteroids are hydrated and cannot be composed of anhydrous enstatite. The "wet" E asteroids like Nysa may be related to the W asteroids and have surfaces rich in hydrated silicate clays. Making matters worse, the E, M, and P classes have essentially the same reflectance spectrum. What makes it possible to tell them apart is their albedo: E's are very bright, P's very dark, and M's, in the middle. But, as discussed above, these asteroids are thought to represent radically different mineralogies with E's being the high-temperature silicate enstatite, M's being metal, and P's being rich primitive organics and clays. This does show the limitations of remote sensing.

Perhaps the most complex class of asteroids is the very large S class. S-class spectra, on average, indicate varying amounts of olivine and pyroxene with a substantial metallic component but the mineralogy of these asteroids varies from almost pure olivine to almost pure pyroxene, to a variety of mixtures of these two end members. With this wide range of mineralogies comes a wide range of meteorite analogs and possible formation scenarios. The S class could represent a range of asteroid material from the core—mantle boundary, the mantle, and the lower crust of differentiated asteroids—material not represented in any meteorite class—as well as including undifferentiated but metamorphosed asteroids that are the parent bodies of ordinary chondrite meteorites.

Ordinary chondrites are by far the largest meteorite type, accounting for approximately 80% of observed meteorite falls, but so far only a few small asteroids have been identified as Q class, direct analogs for ordinary chondrites. A number of S-class asteroids have spectral absorption bands roughly similar to those of ordinary chondrites, but S asteroids typically have a moderate spectral red slope that is not seen in ordinary chondrites. However, it has been shown in laboratory experiments that ordinary chondrite material can redden in response to "space weathering" by micrometeorite bombardment. The small ordinary chondrite parent bodies are probably relatively young fragments that have not had enough time to redden their surfaces. The larger ordinary chondrite parents have older, reddened surfaces and are members of the S class. This explanation has been strengthened by the return of ordinary chondrite (LL class) material sampled from the S-class asteroid Itokawa, as described above.

In general, the differentiated asteroids of the V, A, R, S, and M classes may represent examples of a geologic transect from the crust to the core of differentiated asteroids and can tell us a great deal about the geochemical evolution of a differentiated body. In this scenario, the V-class asteroids would be the surface and crustal material. The A asteroids would be from a completely differentiated mantle, while the R asteroids would represent a mantle that experienced only partial differentiation. Some S asteroids, particularly the olivine-rich members, would be either material from some region in the mantle or the core—mantle boundary. And finally, M-class materials represent samples of the metallic cores of these asteroids. However, as noted above, this attractive interpretation is weakened by the fact that no asteroid family, presumably made of fragments from the breakup of a larger body, shows anything like this range of spectral types among its members.

From the preceding discussion it is clear that the asteroid classes were not uniformly distributed throughout the asteroid belt. The S class dominates the inner asteroid belt while the C class is far more abundant in the outer asteroid belt. The most populous taxonomic classes (the E, S, C, P, and D classes) peak in abundance at different heliocentric distances. If we assume that the spectral and albedo differences between the asteroid classes reflect real differences in mineralogy, then we are seeing rough compositional zones in the asteroid belt. According to models of solar system condensation the high-to-moderate-temperature silicate minerals would tend to dominate the inner solar system, while lower temperature carbonaceous minerals would be common in the cooler, outer regions of the solar system. The transition between moderate and low-temperature nebular condensates is apparently what we are seeing in the taxonomic zonation of the asteroid belt. The innermost major group of asteroids, peaking at 2 AU, is the E class, which is rich in iron-free silicate enstatite, indicating formation under high-temperature, relatively reducing conditions. The next group out is the S class, thought to be rich in the moderate temperature silicates olivine and pyroxene, which contain substantial amounts of iron oxide, which indicate more oxidizing conditions; however, they still also have significant amounts of free iron—nickel. The C class, which peaks in abundance at 3 AU, shows a major transition in asteroid mineralogy to less free metal, more oxidized silicates, important low-temperature carbon minerals, and significant amounts of volatiles such as water. The P asteroids peak at about 4 AU, and the D asteroids at 5.2 AU; they are probably richer in low-temperature materials such as carbon compounds, complex organics, clays, water, and volatiles and represent the transition between the rocky asteroids of the main belt and the volatile-rich comets in the Kuiper Belt and the Oort Cloud.

Several processes have blurred the taxonomic imprint from the original condensation. Apparently, a thermal event heated much of the asteroid belt soon after accretion. Evidence from meteorites shows that some parent asteroids were completely melted (basaltic achondrites, irons, and stony irons), some asteroids were strongly metamorphosed (ordinary chondrites), and some were heated only enough to boil off volatiles and produce aqueous alteration (CI and CM carbonaceous chondrites). This event seems to have been much more intense in the inner asteroid belt and strongly affected the E- S-, A-, R-, V-, and M-class asteroids. The dynamical interaction of asteroids with each other and the planets, particularly Jupiter, has altered and blurred the original orbital distribution of the asteroids and cleared whole sections of the belt. The net result probably has been to expand the original compositional zones and produce orbital overlaps of zones that once may have been distinct from each other.

The recent flyby of 21 Lutetia by the Rosetta spacecraft points up the difficulty of characterizing asteroid mineralogy and the complexity of asteroid evolution. Lutetia has a spectrum that puts it firmly in the M class, but has a low radar reflectivity suggesting a metal-poor surface. The rich collection of flyby data included a confirmation of the flat, featureless spectrum, a measurement of the asteroid's bulk density that is higher than the densities of the possible meteorite analogs, and a structure that shows intense collisional evolution. One interpretation of these data is that Lutetia had a complex, multistage history that began with the melting and differentiation of the inner part of the asteroid, the formation of an iron core, the preservation of a primitive low-iron silicate mantle and crust made of enstatite chondritelike material, and then the "stripping" of some portion of the crust and mantle from a major impact, leaving the higher density core intact. This points up that all but the largest asteroids are most likely fragments of originally larger bodies. Many of these bodies (and perhaps most) have survived as gravitationally bound loose piles of rubble. Lutetia is a stark reminder of how little we yet know about this dazzling array of thousands of small planets we call asteroids.

4. PUZZLES AND PROMISE

4.1. Asteroids and Earth

It is a rare but exciting event in science when a single idea by a small group of scientists ignites an entirely new field of study and redefines the scientific debate. That is exactly what happened to such diverse fields as impact physics, asteroid observations, and paleontology after Alvarez and their colleagues hypothesized that the iridium anomaly found in Cretaceous-Tertiary (K/T) boundary sediments was the mark of an impact event that destroyed the dinosaurs (see chapter Planetary Impacts).

Asteroid impacts are a consistent and steady state fact in the solar system. One just has to look at the extensively cratered surface of any solid body to realize that impacts happen. To some extent, the fact that the Earth has active geological processes that erase the scars of impact craters rapidly and a thick atmosphere that filters out the smaller impactors, has lulled us into a false sense of security.

The real question is not whether asteroids hit the Earth, but rather how often it happens. Before they hit, these impactors are comets and asteroids with the same power-law distribution of sizes that we see in the asteroid belt, so small impacts will be more frequent and large "species-killing" impacts will be much rarer. However, as those who live near dormant volcanoes should realize, rare events on human timescales can be common and frequent events on geologic timescales.

There is plenty of evidence in the geologic and fossil record for repeated major impacts, some of which are associated with mass extinctions. For instance, there were 5 mass extinctions during the last 600 million years, about what would be predicted by a purely impact-driven extinction model. The bottom line is that asteroid impacts should be treated as one of the steady state processes that result from a dynamic solar system. Although the chances of a cratering event like the one that dug the almost 1-mile-diameter Meteor Crater in Arizona happening on any random day are small, the probability is 100% that it will happen sometime. The only question is, when? When faced with predictable dangers, it is sensible to take precautions.

In the same way that people who live on the Gulf coast of North America track hurricanes and people who live in tornado-prone Oklahoma build houses with cellars, it seems a reasonable precaution to identify, track, and study the asteroids in near-Earth space. The telescope search programs that are engaged in this activity are detailed in the Near-Earth Objects chapter.

4.2. Origins of Asteroids

As pieces of material left over from the accretion of the planets, the asteroids represent important chemical and physical clues about the origin of the planets. But these clues can only be interpreted by having a reliable theory for how the asteroids themselves were formed. What processes shaped both the structure of the individual asteroids and the characteristics of the asteroid belt as a whole? Where did the material in the asteroid belt today come from originally?

We do have a reasonably complete census of asteroids in the main belt, down to a size of a few kilometers, and from that we can infer (see Section 2.5) how the perturbing gravity of Jupiter and Saturn has shaped the distribution of asteroids today. And we know that asteroids come in distinct spectral classes, and that there is a tendency for S-type asteroids to be found in the inner belt and C types to be found in the outer belt. But while we recognize that those inferences are very uncertain, and that there could well be material in the asteroid belt that is not sampled in our meteorite collections, we can sketch out a testable scenario for the formation and evolution of the asteroid belt. This is not a final answer but rather a best guess, which we will continue to test and refine as we learn more about the asteroids.

The solar system formed in a solar nebula of gas and dust that smoothly varied in density and temperature from the hot, dense center where the Sun was forming to the thin, cold outer edges where the nebula bordered interstellar space. (See chapter The Origin of the Solar System.) It is possible to calculate the rate at which dust in this cloud would encounter and stick to other bits of dust. These calculations indicate that it is possible in the early solar nebula for very loose balls of dust (more than 90% empty space) as large as a kilometer across to be formed. Relatively low-speed collisions between such dust balls would lead to further compression and accretion into objects big enough to not be carried away with the gas when the last of the solar nebula was pulled into the Sun or ejected in a massive early solar wind. But it seems probable that these planetesimals were very different from the asteroids we see today.

If one takes the present-day masses of the planets, adding a solar proportion of hydrogen and helium to the rocky planets' compositions, and then imagines spreading this material in a disk around the Sun to simulate the smallest possible nebula capable of making planets, one can see that the amount of material in such a nebula varies smoothly from the Sun to the outer reaches of the solar system, with boundaries to the nebula inside Mercury and outside Neptune. But in the region of Mars and the asteroid belt, there should have been a significant amount of mass that appears to be missing today.

We saw in Section 2 how Jupiter and Saturn perturb asteroids out of the asteroid belt. But modeling the early solar nebula allows us to estimate just how much material was so perturbed. It suggests that Mars is made up of less than 10% of the material originally available in its region of the solar nebula, while the mass of the asteroid belt is less than 0.1% of the inferred original material present. The perturbations of asteroidal material by Jupiter and Saturn must have been extremely efficient, at least in the earliest stages of the solar system's history.

Indeed, as material in resonance with Jupiter was ejected to the outer solar system, the effect on Jupiter would be to drag it into the inner part, eventually sweeping out much if not all of the material in this region. Recent calculations based on this concept, the Nice Model, suggest that interactions between Jupiter and Saturn at this time may have resulted in the wide-scale reshuffling of the gas giant planets. These radical changes would have scattered material into the Kuiper Belt, and removed nearly all the original material between Mars and Jupiter.

Thus, it has been suggested that the asteroids today could represent material scattered into this region from elsewhere. It is possible that the high-temperature, differentiated asteroids like 4 Vesta or the metallic M-class asteroids were part of a group of asteroids that formed in the region of the terrestrial planets (near Earth) and were perturbed into the asteroid belt, ending up near objects with much different mineralogies and histories. Conversely, the darker, low-temperature, carbonaceous asteroids may have come originally from the outer solar system and were likewise perturbed into the belt, creating the wildly varying collection we see today. Other variants on this theory suggest that as Jupiter and the other giant planets migrate, material originally formed between Mars and Jupiter was first scattered out, but then scattered back into the area we now call the asteroid belt.

One inevitable result of having the mass of the asteroid belt excited into such orbits is that there must have been a very high collision rate among asteroids in the early solar system. These collisions would break larger asteroids into smaller pieces and destroy the smaller pieces entirely. For the largest asteroids—many tens of kilometers in radius—impacts energetic enough to shatter them may not have enough energy to disperse the pieces completely. Instead, the fragments were likely to reaccrete into piles of rubble, consistent with the structure that asteroids are inferred to have today.

As the population of the asteroid belt dissipated, the rate of collision likewise would have dropped. Given the present-day population, collisions that are capable of breaking pieces of an asteroid into Earth-crossing orbits or creating families of asteroids where one asteroid once orbited still do occur. We do see young families of asteroids today. Likewise, by measuring short-lived radioactive isotopes formed in meteorites by cosmic rays, we can see peaks in the ages of certain meteorite classes that imply they were broken off a parent body at a specific moment some tens to hundreds of millions of years ago. But these events must be many, many times less frequent today than when the asteroid belt was much more heavily populated.

One result of this scattering of asteroids by Jupiter and Saturn may have been that a few rare bodies originally from the asteroid belt may have been captured into orbits around other planets. Among the moons suspected of being captured asteroids are the Martian moons Phobos and Deimos, the irregular moons of the gas giant planets, and even Neptune's large moon Triton.

4.3. Spacecraft Missions to Asteroids

Although telescopic studies are by far the most prolific source of data on asteroids, critical science questions on asteroid composition, structure, and surface processes can only be addressed by spacecraft missions getting close to these objects. The range of spacecraft encounters includes flybys, rendezvous, and sample return missions, which provide information of ever-increasing detail and reliability. As shown in Table 26.2 we have now seen the results of 10 missions (color coded green), one is currently on the way (colored red), and two more are awaiting launch (colored orange).

The Near-Earth Asteroid Rendezvous spacecraft, the first dedicated asteroid mission, flew past asteroid 253 Mathilde and arrived in orbit around 433 Eros in 2001. After orbiting Eros for 1 year and mapping its morphology, elemental abundances, and mineralogy with an X-ray/gamma ray spectrometer, imaging camera, near-infrared reflectance spectrometer, a laser range finder and a magnetometer, the spacecraft ended its mission by landing on the surface of Eros (see chapter Near-Earth Objects).

In late 2005, the ambitious Hayabusa asteroid sample return mission of the Japanese Aerospace Exploration Agency (JAXA) rendezvoused with asteroid 25143 Itokawa. This NEO turned out to have an extremely rough surface, as shown in Figure 26.10. After several months of mapping and analysis, the spacecraft collected samples by shooting a small projectile into the surface and collecting some of the fragments that splashed off, and successfully returned its load of tiny fragments to Earth (Figure 26.8), parachuting its sample pod onto the Australian desert in 2010.

TABLE 26.2 Spacecraft Missions to Asteroids. The 10 missions already flown are color coded green, one currently on the way is colored red, and two more awaiting launch are colored orange

Asteroid	Mission	Encounter
101955 Bennu	OSIRIS-REx	Sample return
162173 1999 JU_3	Hayabusa 2	Sample return
1 Ceres	DAWN	Orbit
4 Vesta	DAWN	Orbit
21 Lutetia	Rosetta	Flyby
2867 Steins	Rosetta	Flyby
25143 Itokawa	Hayabusa	Sample return
5535 Annefrank	Stardust	Flyby
433 Eros	NEAR	Orbit
9969 Braille	Deep space 1	Flyby
253 Mathilde	NEAR	Flyby
243 Ida	Galileo	Flyby
951 Gaspra	Galileo	Flyby

There are three exciting missions "on the way" or awaiting launch to asteroid encounters. The DAWN spacecraft has spent 15 months in orbit mapping 4 Vesta during 2011–2012 and is currently on its way to orbit 1 Ceres starting in early 2015. JAXA is planning to launch another sample return, Hayabusa 2 to asteroid 162173 (1999 JU_3), in July 2014. National Aeronautics and Space Administration is in the final stages of development in the OSIRIS-REx sample return mission. This mission is planned for a 2016 launch to encounter asteroid 101955 Bennu (1999 RQ_{36}).

While we have made great strides in exploring asteroids they are still largely unexplored; indeed, in the case of the smaller NEOs, still largely undiscovered. They have great potential for science, for destruction, as resources in space, and for exploration. We are only just starting to understand these numerous objects that share our solar system.

BIBLIOGRAPHY

Bottke, W. F., Cellino, A., Paolicchi, P., & Binzel, R. P. (Eds.). (2002). *Asteroids III* (p. 785). University of Arizona Press.

Gehrels, T. (Ed.). (1994). *Hazards due to comets and asteroids*. University of Arizona Press.

Chapter 27

Near-Earth Objects

Alan W. Harris and Line Drube
German Aerospace Center (DLR), Institute of Planetary Research, 12489 Berlin, Germany

Lucy A. McFadden
Planetary Systems Laboratory, NASA Goddard Space Flight Center, Greenbelt, MD 20771 USA

Richard P. Binzel
Department of Earth, Atmospheric and Planetary Sciences, Massachusetts Institute of Technology, Cambridge, MA 02139, USA

Chapter Outline

1. Introduction — 603
2. Significance — 606
   2.1. Remnants of the Early Solar System — 606
   2.2. Hazard Assessment — 606
   2.3. Exploration Destinations and Resource Potential — 606
3. Origins — 607
   3.1. Relationship to Main-Belt Asteroids — 607
   3.2. Dynamical History — 607
   3.3. Relationship to Meteorites — 608
   3.4. Meteor Shower Associations — 608
   3.5. Relationship to Comets — 609
      3.5.1. Tisserand Parameter — 609
      3.5.2. Dynamical and Physical Evidence For Extinct Comets — 610
4. Population — 610
   4.1. Search Programs and Techniques — 610
   4.2. How Many? — 611
5. Physical Properties — 612
   5.1. Brightness — 613
   5.2. Shape — 613
   5.3. Rotation Rate — 614
   5.4. Size and Albedo — 614
   5.5. Density — 616
   5.6. Color and Taxonomy — 616
   5.7. Mineralogy — 617
6. In Situ Studies — 618
   6.1. NEAR Shoemaker — 618
   6.2. Hayabusa — 618
7. Impact Hazards — 619
   7.1. Collision Magnitude — 620
   7.2. Collision Frequency — 621
   7.3. Mitigation Measures — 621
Appendix — 622
   Asteroid Numbering and Naming Conventions — 622
Bibliography — 623

1. INTRODUCTION

A **near-Earth object (NEO)** is an asteroid or comet orbiting the Sun with a perihelion distance of less than 1.3 AU (1 AU, an "astronomical unit", is the mean distance between the Earth and the Sun, around 150 million km). If the orbit of an NEO can bring it to within 0.05 AU of the Earth's orbit, and it is larger than about 120 m, it is termed a potentially hazardous object (PHO); an object of this size is likely to survive passage through the atmosphere and cause extensive damage on impact. (The acronyms NEA and PHA are used when referring specifically to asteroids.)

The recognition that a giant asteroid or comet perhaps 5–10 km across most likely caused, or at least contributed to, the extinction of the dinosaurs in a geological episode known as the Cretaceous–Tertiary Event has highlighted the hazard to our civilization presented by NEOs. The energy involved in collisions of NEOs with the Earth can be much larger than that released in the detonation of nuclear weapons or naturally occurring phenomena on Earth (e.g. volcanoes, earthquakes, or tsunamis).

Scientists cannot accurately predict what effects a major NEO impact would have on today's technically sophisticated and highly networked world. Computer simulations of impacts provide some insight but natural phenomena

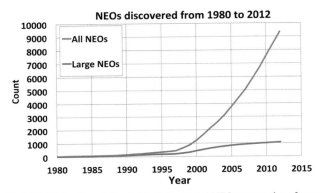

FIGURE 27.1 Cumulative total of discovered NEOs versus time. Large NEOs are defined as those with a diameter of at least 1 km.

elsewhere in the solar system provide real proof of the destructive potential of collisions between planets and small bodies. The collision of comet Shoemaker-Levy 9 with Jupiter, observed worldwide through telescopes in 1994, created scars in Jupiter's atmosphere larger than the Earth.

Even relatively small impactors can cause considerable damage on Earth. The object that exploded over the Russian city of Chelyabinsk in February 2013 had a diameter of only 17–20 m, yet it produced a blast wave that damaged buildings and injured some 1500 people. The potentially devastating effects on Earth of a collision with a large asteroid or comet are now well recognized by scientists and policy makers.

One of the pioneering programs in the search for NEOs, Spacewatch, founded by Tom Gehrels and Robert McMillan at the University of Arizona, started to detect NEOs in earnest in 1990, with the first automatic detection of a NEO. More programs have since joined in, using sensitive **charge-coupled devices (CCDs)** and sophisticated software to automate the process of identifying moving objects. These programs, mainly US-based, have increased the number of known NEOs to around 10,000 at the time of this writing, compared to 56 known in 1980 (Figure 27.1). Currently, there are some 1400 known objects qualifying as PHOs, although the total population of such objects is estimated to be around 4700.

It is now certain that most of the NEOs originated in the Main Asteroid Belt, located between the orbits of Mars and Jupiter. The range of composition and physical characteristics of NEOs spans those found among asteroids in the main belt. However, some NEOs probably evolved into their current orbits from the reservoir of short-period comets extending beyond Jupiter and into the outer solar system and may be extinct or dormant cometary nuclei.

NEOs are assigned to one of four subgroups according to their orbital types (Figures 27.2 and 27.3):

- Amors whose orbits are completely outside the Earth's orbit.
- Apollos whose orbits cross the Earth's orbit but are mostly outside it.
- Atens, whose orbits also cross the Earth's orbit but are mostly inside it.
- Atiras (also termed inner Earth objects or Apohele asteroids), which have orbits entirely inside the Earth's orbit.

Only Apollos and Atens can collide with the Earth at the present time, although the orbits of all NEOs evolve as a result of planetary perturbations and may intersect the Earth's orbit at some time in the future.

The Amor asteroid (433) Eros was the first NEO to be discovered. Gustav Witt of Berlin, Germany, recorded its position on a photographic plate in 1898. It is also one of the largest NEOs, having a longest dimension of 33 km and a cross-section of 10.2 × 10.2 km. Since then some 4000 Amors have been discovered. The group's namesake, (1221) Amor, was discovered in 1932. In the same year, the first Earth-crossing asteroid, (1862) Apollo, was discovered. Around 4700 Apollos have been discovered since. It was not until 44 years later that (2062) Aten was discovered

FIGURE 27.2 Apollo orbits cross that of the Earth but most of the orbit is external to the Earth's orbit. Atens are also Earth crossers but with orbits largely inside that of the Earth. Amor orbits are external to that of the Earth. Atira asteroids have orbits entirely inside the Earth's orbit. The perihelion and aphelion are the points in an orbit that have the shortest and longest distances from the Sun, respectively.

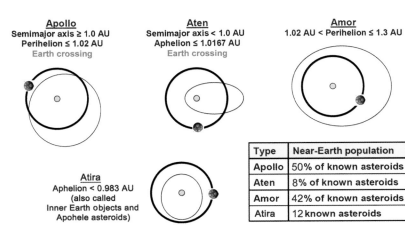

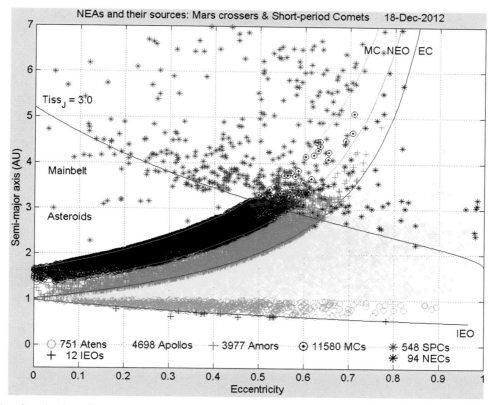

FIGURE 27.3 Plot of **semimajor axis** versus **eccentricity** of the near-Earth asteroid population, together with populations of related objects: Mars crossers (MC), short-period comets (SPC), and near-Earth comets (NEC). The Earth's orbit has a **semimajor axis** of 1 AU. The Atens are just below the 1 AU line, the Apollos just above, and the Amors fall above the Apollos as they do not cross the orbit of the Earth at all. The curves labeled MC, NEO, and EC mark the upper boundaries of the populations of Mars crossers, NEOs, and Earth crossers, respectively. The Tisserand parameter, $Tiss_j$, is explained in Section 3.5.1. *Credit G. Hahn/EARN, DLR Institute of Planetary Research.*

by Eleanor Helin, still using photographic plates for the search. Over 750 Atens have been discovered to date. In 1998, the first Atira or inner Earth object, 1998 DK_{36}, was discovered. Atiras are difficult to detect because they orbit inside the Earth's orbit and are therefore not normally observable in the night sky. For this reason, only about 12 Atiras have been observed to date. Several space telescope missions for NEO search and discovery have been proposed that can observe the daytime sky, unlike Earth-bound observers. Such telescopes would be better placed to detect the elusive Atiras.

Figure 27.3 is a plot of semimajor axis against eccentricity, illustrating the orbital distributions of the different dynamic classes of NEO, together with the orbital characteristics of related objects.

In 2005, the asteroid (1862) Apollo was found to be a binary system: two asteroids orbiting around a common center of mass, in this case with diameters of 1.5 km and 75 m. Of the known NEOs, only about 40 have been found to be binary objects, but it is expected that many more await discovery. According to recent estimates, some 15% of all NEOs with a diameter larger than 300 m are binary or multiple asteroid systems. Binary asteroids are scientifically fascinating objects, providing clues to some puzzling aspects of NEO physical properties.

To date, there have been two **rendezvous** missions to NEOs. National Aeronautics and Space Administration's (NASA's) Near-Earth Asteroid Rendezvous-Shoemaker (NEAR Shoemaker) mission was the first designed to orbit an asteroid. The NEAR Shoemaker spacecraft orbited its target, (433) Eros, for 1 year from February 2000. The mission ended with a descent of the spacecraft to the asteroid's surface. The Japanese Hayabusa spacecraft was launched on May 9, 2003, on a mission to investigate asteroid (25143) Itokawa and to demonstrate the technology necessary to return samples to Earth. The spacecraft arrived at Itokawa in September 2005 and performed remote sensing measurements for 3 months. In November 2005, there were two scheduled touchdowns in which some grains of surface material were collected. After a journey back to Earth lasting more than 3 years, the return capsule was released from the main spacecraft a couple of hours before both entered the Earth's atmosphere. The Hayabusa spacecraft disintegrated on reentry in a fireball, while the thermally shielded capsule containing the grains from Itokawa was recovered in the south Australian outback on June 13, 2010.

2. SIGNIFICANCE

2.1. Remnants of the Early Solar System

From a scientific point of view, NEOs are studied for the same reason as comets and main-belt asteroids: they are remnants of the early solar system (Figure 27.4). As such, they contain information that has been lost in the planets through large-scale, planetary processes such as accretion, tectonism, volcanism, and metamorphism. Asteroids and comets have undergone less processing since their formation in the early solar nebula; therefore, knowledge of them derived from astronomical observations and studies of direct samples in the form of meteorites or collected by space missions is critical to piecing together a scenario for the formation of the solar system. (See The Origin of the Solar System.)

Most NEOs are asteroid-like in their nature, being derived from the Main Asteroid Belt. The Asteroid Belt forms an important boundary in the solar system; the planets that formed closer to the Sun, the terrestrial planets, are dominated by rocky, lithophile materials. Beyond the Asteroid Belt, the planets are composed predominately of nebula gases. Some NEOs almost certainly originated elsewhere in the solar system, such as in the cometary reservoirs lying at great distances from the Sun, beyond the gaseous planets. Knowledge of the materials in these reservoirs provides insight into the chemical and physical processes that were active in the outer regions of the solar system, both in the near and distant past. A major objective of the scientific study of NEOs is to determine the proportions of the population that derive from various regions of the asteroidal and cometary reservoirs.

2.2. Hazard Assessment

The phenomenon of collisions in the history of our solar system is very fundamental, having played the major role in forming the planets we observe today. Asteroids may have contributed to the delivery of water and organic materials to the early Earth necessary for the development of life, but later impacts of asteroids probably played a role in mass extinctions, and they currently pose a small but significant threat to the future of our civilization. Collisions of asteroids with the Earth have taken place frequently over geological history, and it is an undeniable fact that major collisions of asteroids and comets with the Earth will continue to occur at irregular, unpredictable intervals in the future. The risk of a comet impact is thought to be much lower than that of an NEA impact, although given the potentially high relative velocities, the effects in the case of a comet impact could be much more devastating.

As a result of modern observing techniques and directed efforts, thousands of NEOs have been discovered over the past 20 years, and the reality of the impact hazard has been laid bare. Future observation programs and space missions will be crucial for a better understanding of the orbits, composition, and physical nature of asteroids, and the techniques that would be most effective in preventing a collision of an asteroid with the Earth. (See Planetary Impacts; Cometary Dynamics.)

2.3. Exploration Destinations and Resource Potential

NEOs come closer to Earth than any other planetary bodies, and those with low orbital inclinations are very accessible targets for spacecraft and are considered attractive as training venues for missions to Mars. As civilization moves beyond the Earth, knowledge of materials in space is critical to their efficient use in situ. It could eventually become more

FIGURE 27.4 Artist's impression of planets forming in a young planetary system. According to theory, collisions between dust grains and small bodies at the dawn of our solar system led to the formation of "planetesimals" with diameters of up to several hundred kilometers. The significant gravitational fields of the planetesimals caused more material to be accreted and led in some cases to the growth of planets. On the other hand, dramatic collisions between bodies at the time of planet formation led to the release of countless fragments, many of which are still present in the solar system as asteroids and comets. *Credit: NASA/JPL-Caltech/T. Pyle, SSC.*

economical to use space resources than transporting material from the Earth. A number of groups have been formed over the years, including commercial companies, with the aim of eventually extracting valuable materials, such as platinum and other precious metals, from the Moon, Mars, and NEOs. For future human space exploration, water is also a precious resource that may be retrievable from certain types of NEO, both for human consumption and for the production of hydrogen and oxygen as rocket propellant.

3. ORIGINS

In the widely accepted scenario of the formation of the solar system 4.6 billion years ago, a cloud of gas and dust collapsed into a disk-shaped nebula from which planetesimals and eventually planets formed (Figure 27.4). The process of planet formation started with molecules and dust grains colliding and combining with each other. **Aggregates** and clumps grew in size by sweeping up smaller dust and ice grains. Further growth continued via collisions and accretion of material forming objects with diameters of up to a few hundred kilometers, called planetesimals. Both electromagnetic and gravitational forces came into play to oppose the destructive forces of erosion from collisions. Some planetesimals continued to increase in mass, attracting ever more material via their gravitational fields. It was primarily the strong gravitational field of the forming Jupiter that disrupted neighboring planetesimals so much that a planet never formed between the orbits of Mars and Jupiter, in the region of the so-called Main Asteroid Belt. Some large main-belt asteroids may be original planetesimals that were prevented from growing to the size of the major planets by the effects of collisions. Smaller asteroids are most likely fragments produced in collisions. The largest object that formed in the main belt, Ceres, has a diameter of 950 km and a volume of only one-fiftieth that of the Moon; Ceres constitutes one-third of the mass in this region. Over time, collisions between asteroids together with various dynamical processes have led to many objects being ejected from the Main Asteroid Belt.

3.1. Relationship to Main-Belt Asteroids

In the late 1970s it became clear from telescope observations that asteroid composition changes as a function of the distance from the Sun, and hence temperature. Therefore, the exact nature of asteroidal material holds clues to the temperature and location where the material formed, valuable information for scientists attempting to understand the processes that led to the formation of our solar system and are possibly involved in the formation of extrasolar planetary systems.

Studies of the composition of NEOs led to the conclusion that NEO composition spans the range found in the Main Asteroid Belt, thus establishing that most NEOs are derived from the main belt. Therefore physical information derived from NEOs can be reasonably considered to apply to main-belt asteroids.

Statistical analysis of the evolution of many asteroid orbits over the age of the solar system indicates that the lifetime of an Earth-crossing body against gravitational perturbations is relatively short, on the order of 10 million years or less. Within this time frame, the bodies will collide with a planet, fall into the Sun, or be dynamically ejected from the solar system. This time interval applies to the average of the entire population and does not refer to the exact lifetime of any particular asteroid. It turns out that the orbital evolution of a specific **NEO** cannot actually be determined very far into the future or the past (i.e. more than a few hundred years) owing to the difficulty of knowing the exact starting conditions and accurately predicting frequent close approaches between the NEO and the planets. (See Solar System Dynamics: Regular and Chaotic motion.)

3.2. Dynamical History

Dynamicists have simulated the pathways that objects might take from unstable regions of the Asteroid Belt using computations of dynamical forces acting in the solar system. In some cases, fragments from asteroid collisions may be violently cast into these regions of instability. However, a softer touch may play an even bigger role. Constant warming by the Sun causes asteroids of all sizes to reradiate their heat back into space. Since all asteroids rotate, the reradiation does not occur in the same direction as the incoming sunlight, resulting in a small force acting on the asteroid. This force acts as a very gentle push on the asteroid, which over many millions of years can cause the asteroid to slowly drift inward or outward from its original main-belt location. The phenomenon is called Yarkovsky drift and is especially effective on small objects; it may be particularly important for supplying meteoroids to Earth. An object undergoing Yarkovsky drift may eventually enter regions where resonances with Jupiter's orbit are particularly strong, such as the Kirkwood gaps, located where an asteroid's orbital period is shorter than Jupiter's by the ratio of two small integers, such as 3:1, 5:2, or 2:1 (see Figure 27.6 below). Any asteroid or debris that migrates into such a region finds Jupiter to be especially effective in increasing its orbital **eccentricity**. As the orbit becomes increasingly elongated, it can intersect the orbit of the Earth. There are regions of chaotic motion associated with resonances with both Jupiter and Saturn. The two gas giant planets are believed to play a significant role in directing meteoroids to Earth, and presumably also many of the NEOs.

Other NEOs evolve from Jupiter-family comets or Halley-type short-period comets. Life as a Jupiter-family comet is not long, as Jupiter imparts changes to the orbits on timescales of 10^4-10^6 years. Leaving Jupiter's gravitational sphere of influence, the soon-to-be NEOs may sometimes be perturbed by Mars and other terrestrial planets and also affected by the influences of nongravitational forces, such as volatile outgassing or splitting of the cometary nucleus. Nongravitational forces also contribute to orbital changes that result in planet-crossing orbits.

3.3. Relationship to Meteorites

Small rocky fragments, hand-sized chunks, and large boulders continually fall to Earth from space; most fall unnoticed in the oceans or in remote areas, but occasionally one is found lying on the ground as a meteorite. Exploring the relationship between NEOs and meteorites is motivated by the possibility of making insightful connections between the geochemical, isotopic, and structural information on meteorites available from laboratory studies and the NEOs.

In January 2000, an exceptionally bright **bolide** was seen by eyewitnesses in the Yukon, Northern British Columbia, parts of Alaska, and the Canadian Northwest Territories. Nearly 10 kg of precious samples were recovered from the surface of frozen Tagish Lake. On the basis of eyewitness reports and the **bolide's** detection by military satellites, the orbit of the impacting body was traced back to the Main Asteroid Belt. Prior to striking the Earth, the body is estimated to have been about 4 m across with a mass of 56 metric tons. The determination of meteorite orbits also serves as a constraint on the mechanisms that result in meteoroid delivery to Earth.

October 2008 marked the first time that a NEO was discovered by a search program and found to be on collision course with the Earth. The NEO, 2008 TC$_3$, was discovered at 6:39 UT on the morning of October 6, 2008, by Richard Kowalski of the Catalina Sky Survey, using the Mt Lemmon 1.5-m-aperture telescope near Tucson, Arizona. The International Astronomical Union's Minor Planet Center (MPC) houses the world's repository of asteroid observations and uses them to automatically compute orbits for each potential discovery. Upon receiving the discovery data for 2008 TC$_3$, the MPC calculated a preliminary orbit that showed the object to be heading for a collision with the Earth the very next day! Dynamics experts at the NASA/JPL Near-Earth Object Program Office determined that 2008 TC$_3$ would enter the Earth's atmosphere above northern Sudan around 02:46 UT on October 7. Luckily, although its speed as it entered the Earth's atmosphere was calculated to be 12.4 km/s, the estimated diameter of the object was only around 4 m, too small to leave any mark on the ground, save for a scattering of meteorites. Nevertheless, the event produced a spectacular fireball. Predictions of the location of entry into the atmosphere were quite accurate, as witnessed by US government satellites, images from the Meteosat 8 weather satellite, and a sighting by an airline pilot flying over Chad. Some 280 small chunks of 2008 TC$_3$ have since been retrieved from the Nubian Desert in Sudan (Figure 27.5). Scientific investigations of the meteorites, named "Almahata Sitta", will continue for many years, aided by the fact that some very alert observers were able to acquire spectroscopic and light curve observations of 2008 TC$_3$ while it was still in orbit, in the few hours between discovery and entry into the atmosphere.

FIGURE 27.5 Peter Jenniskens, meteor expert and member of the Almahata Sitta meteorite recovery team, finds a piece of the ex-asteroid 2008 TC$_3$, during a search on February 28, 2009. *Credit: Peter Jenniskens, SETI Institute.*

At the time of writing the most recent case of an asteroid being discovered and subsequently impacting the Earth is that of 2014 AA, the first asteroid to be discovered in 2014. The car-sized object entered the Earth's atmosphere over the Atlantic Ocean at about 03:00 UT on January 2, 2014, less than a day after being discovered by the Catalina Sky Survey. The object presumably disintegrated harmlessly in the atmosphere.

The fireball over Chelyabinsk on February 15, 2013, was caught on many security cameras, allowing the circumstances of the event to be investigated in unprecedented detail. Information on the trajectory and brightness of the **bolide** and the orbit of the original 18-m-diameter asteroid, combined with details of the composition and internal structure of the resulting meteorites, will allow a very complete picture to be constructed of this event, providing unique insight into the nature and origin of the object that caused the worst damage on the ground since the Tunguska event of 1908 (see Section 7). (See Meteorites.)

3.4. Meteor Shower Associations

Streams of material can develop in the orbits of comets and asteroids that eject particles as they orbit the Sun. If such a stream of particles crosses the Earth's orbit, it can give rise to a meteor shower at the same time each year, as particles in the stream are swept into the Earth's atmosphere. An icy comet nucleus warmed by the Sun ejects material in the form of gas molecules and dust grains, which feed its coma and tail in a process called outgassing. Comet nuclei are generally fragile bodies that can spontaneously lose large fragments or even disintegrate altogether. Asteroids can also lose material as a result of collisions or an increase in rotation rate caused, for example, by a close encounter with a planet. Many asteroids may be loosely bound **agglomerates** of collisional fragments that formed by gravitational reaccumulation of debris after a dramatic collision between two larger objects. Such objects are termed "rubble piles" (see Sections 5.3 and 6.2); it is easy to understand how such an object can lose material if it spins rapidly. Other mechanisms potentially causing the loss of particles from an asteroid include surface cracking and dehydration under thermal stress, which may occur if the object approaches very close to the Sun.

In 1983, Fred Whipple recognized the orbital elements of an asteroid found by an Earth-orbiting infrared telescope to be essentially the same as the Geminid meteor shower, which occurs in mid-December. (See Infrared Views of the Solar System from space.) There is little doubt that this asteroid, (3200) Phaethon, is the parent body of the Geminid meteors. While Phaethon has never exhibited a coma or tail, it has orbital, and some physical, characteristics that resemble those expected of a dormant or extinct comet nucleus. The perihelion distance of Phaethon's orbit is very small (0.14 AU), implying that surface temperatures rise to around 1000 °C at its closest approach to the Sun. Some researchers suspect that the ejection of particles via thermal fracturing and the dehydration of certain minerals may be the mechanisms by which Phaethon feeds the Geminid meteor stream. There are several other NEOs with orbital elements implying associations with the paths of existing meteor showers.

3.5. Relationship to Comets

Comets are icy and dusty objects that come from the outer reaches of the solar system. In contrast to asteroids, which are predominantly mixtures of rock and metals, a comet nucleus is a rocky body that contains frozen volatile materials, such as water, carbon monoxide, carbon dioxide, and methane, which give rise to a halo and coma when the object approaches the Sun. The orbital periods of comets are long, their orbital eccentricities are high, and they may have large or small orbital inclinations. What is their relationship to NEOs? In the 1950s, Ernst Öpik concluded that comets must be a partial source of NEOs because he could not produce the number of observed meteorites from the Asteroid Belt alone via his calculations. Building on Öpik's work, George Wetherill predicted that 20% of the NEO population consists of extinct cometary nuclei. Some now find evidence that the fraction of comets is smaller, closer to 5%. (See Cometary Dynamics; Physics and Chemistry of Comets.)

Are there hints that any particular NEO that looks like an asteroid was once a comet? If an object sometimes has a tail like a comet and sometimes looks just like an asteroid (no coma or tail), what is it: an asteroid or a comet? There is both dynamical and physical evidence that addresses this question:

3.5.1. Tisserand Parameter

A strong hint that an asteroid-like object may be a comet in disguise comes from its orbit. The orbital elements of asteroids and comets tend to occupy different regions in a plot of **semimajor axis** against orbital **eccentricity** (Figure 27.6). Another way to characterize an orbit is to calculate its Tisserand parameter from the equation:

$$T = \frac{a_J}{a} + 2\sqrt{\frac{a}{a_J}(1 - e^2)} \cos i$$

In this equation, a and a_J refer to the **semimajor axis** values for the object and Jupiter, respectively. The parameters i and e are the **inclination** and **eccentricity** of the object's orbit. The Tisserand parameter is useful because it is a constant even if the comet's orbit is perturbed by Jupiter. Also, it helps to determine whether an object is in an orbit

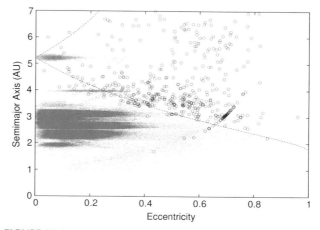

FIGURE 27.6 Tisserand parameter. The dashed line represents the Tisserand parameter with a value of 3. To the left of the line, T is greater than 3 and to the right, T is less than 3. Red dots and blue circles represent known asteroids and comets, respectively. The curved string of circles centered near **eccentricity** = 0.7, **semimajor axis** = 3, represents fragments of comet 73P/Schwassmann–Wachmann, which began to disintegrate in 1995. The Kirkwood gaps (see Section 3.2) are clearly evident in the distribution of asteroid **semimajor axes** between 2 and 3.5 AU. *Credit: Michael Mommert, DLR Institute of Planetary Research.*

that is strongly controlled by Jupiter or not. Most objects that display the characteristics of comets have a value $T < 3$, while most objects that are asteroid-like have $T > 3$. The value of $T = 3$ is represented by the dashed line in Figure 27.6. Due to diverse influences on the orbits of asteroids and comets, the value $T = 3$ does not represent a hard boundary, but "asteroids" with $T < 3$ are good candidates for being comets in disguise; the reason they do not currently display any tell-tale coma or tail is their dormancy or depletion of volatile materials.

3.5.2. Dynamical and Physical Evidence For Extinct Comets

A powerful way to investigate the mystery of how many extinct comets reside in the NEO population is to explore both dynamical factors and physical measurements to identify possible candidates. For example, numerical simulations of the orbits of short-period comets can reveal how likely it is that gravitational interactions with Jupiter and the other planets can send them into the NEO population. In these simulations, many thousands of hypothetical comets, each with slightly different initial orbits, can be tracked for millions of years to see how they are tossed around chaotically by the gravitational tugs and pulls of the planets. In the same way, thousands of different starting places for main-belt asteroid orbits can be modeled to reveal the effectiveness of resonances for sending asteroids into near-Earth space. Alessandro Morbidelli, William Bottke, and coworkers have done extensive computer calculations to assess the relative effectiveness of these dynamical processes. Their calculations suggest that, when considering NEOs with diameters of 250 m or larger, about 6% of them have a cometary origin.

Spacecraft and telescopic measurements of known comets reveal what characteristics to look for when trying to determine if a given asteroid-like NEO is an inactive comet nucleus. For example, spacecraft data show that surface regions of comets Halley, Borrelly, Wild 2, and Tempel 1 that are not actively outgassing are very dark (low albedo) and have gray color. Some other comets go through periods of very low activity, allowing astronomers to measure the albedos and colors of the nucleus unobscured by the gas and dust of a coma. All these measurements consistently show low albedos (the surface reflects back only about 4% or less of the incoming light) and gray or reddish colors. Spectroscopic analysis of reflected sunlight reveals no major absorption bands, indicating an absence of the minerals olivine or pyroxene, commonly observed in asteroid spectra, on their surfaces.

Recent estimates that combine dynamical information with albedo and color measurements indicate that perhaps a few to 10% of NEOs are in fact inactive comet nuclei.

4. POPULATION

4.1. Search Programs and Techniques

Organized, telescopic search programs for NEOs have led to a steady increase in discoveries over the past 15 years (Figure 27.7). The search programs supported by NASA include the Lincoln Near-Earth Asteroid Research program, the Panoramic Survey Telescope and Rapid Response System, the Near-Earth Asteroid Tracking system, Lowell

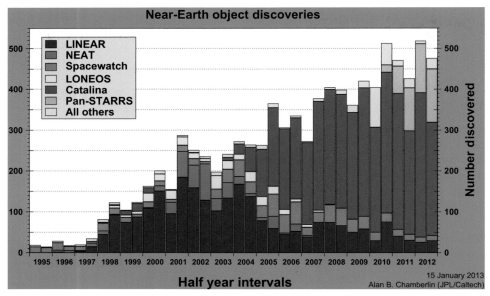

FIGURE 27.7 Numbers of NEO discoveries made by the various search programs. Improvements in detector technology and computing capabilities have led to a dramatic increase in the discovery rate since 1997. *Credit: Alan Chamberlin, JPL.*

Observatory's Near-Earth Object Search, the Wide-field Infrared Survey Explorer's (WISE) NEO search project, and the University of Arizona's Catalina Sky Survey and Spacewatch programs. Although the objectives of these programs are all similar, to inventory the objects in the vicinity of Earth, each has its own design and approach. Modern ground-based search programs employ digital imaging devices known as **CCDs** that cover large areas of the sky in a single exposure. Typically, a given area of sky is imaged and re-imaged three to five times at intervals of 10 min–1 h. With digital images, fast computers can compare the images and identify and subtract all the "uninteresting" objects that remain fixed, leaving behind the tracks of moving asteroids or comets. By rapidly repeating this process for many patches of sky throughout a night, nearly the whole sky can be scanned in the course of about 1 or 2 weeks. More sensitive and rapid search systems are under study that could significantly expand the capability to discover small NEOs, even down to diameters of 100 m or less.

When an NEO is first discovered, astronomers initially trace only a short piece of its orbit as measured over a few hours or even over a few weeks. With each new NEO discovery, astronomers wish to assess whether the object poses any immediate or future impact threat. Orbit calculations for most objects can be made reliably for many decades into the future, but of course if only a tiny part of the orbit has been observed, the extrapolation into the future is very uncertain. Sometimes this extrapolation shows that the Earth itself resides within the overall uncertainty region for a NEO's future position. If the cross-section of the Earth occupies 1/10,000th of this space, then there is a 1 in 10,000 chance of an impact with the Earth. Even though headlines may proclaim the end of the world, statistically speaking, the odds are actually 10,000 to 1 in our favor that continued observations refining the orbit will show a collision is ultimately ruled out.

There are currently two automatic systems operating that independently update the orbital parameters of NEOs and calculate future close Earth approaches and impact probabilities. Sentry is run by NASA's Jet Propulsion Laboratory in Pasadena, California, and NEODyS by the University of Pisa in Italy. Although independent, the two systems are designed to be complementary, and there is constant communication between the two sites to enable cross-checking of their results.

Two scales have been developed to facilitate assessment and comparison of impact probabilities of NEOs: The Palermo Technical Impact Hazard Scale and the Torino Scale. The Palermo Scale was developed to assess whether any NEO discovery merits concern or response. The Palermo Scale value (PS) is a measure of how much more likely an impact is than the general "background" risk of an impact from all such objects, or larger, over the period preceding the potential impact. For each potential impact of an NEO, the PS is simply the impact probability (P_i) divided by the background risk (P_b) on a logarithmic scale: $PS = \log_{10}(P_i/P_b)$. P_b is given by $0.03 \times T \times E^{-0.8}$, where T is the time in years until the potential impact, and E is the energy release in megatons of **trinitrotoluene (TNT)** associated with the potential impact, which depends on the object's size and its velocity at impact. For example, PS of $+2$ indicates that an impact is 100 times more likely than the general background risk, which would merit serious concern. On the other hand, $PS = -2$ indicates that the likelihood of an impact is only 1% of the background risk. $PS = 0$ implies that the risk associated with the potential impact is identical to the background risk.

The less technical Torino Scale is a 10-point scale designed for use by the media and for communicating impact risks to the public. On the Torino Scale, 0 indicates no likelihood of impact, or the impact presents no hazard (e.g. a small object that will burn up in the atmosphere), 1 indicates that an impact is extremely unlikely and there is no cause for concern, 2 indicates that the object merits attention by astronomers, and 10 implies that a collision is certain and will result in a global catastrophe.

At the time of this writing, the highest values on the Palermo and Torino Scales are -1.5 and 1, respectively. For more information on the Palermo and Torino Scales and to check the latest list of possible impactors see http://neo.jpl.nasa.gov/risk/.

4.2. How Many?

A frequently asked question is: how many NEOs are there in total? Being products of collisions, small NEOs are much more numerous than large ones. As only a small fraction of the total population has been discovered to date by the various survey programs, the observed samples must be extrapolated to determine the total number of NEOs. The largest NEO has a diameter of about 40 km. There are some 30 NEOs with diameters above 5 km, about 1000 with diameters above 1 km, and estimates run to several tens of thousands for objects with diameters above 100 m. But the size distribution does not stop there: there are billions of NEOs with diameters around 1 m (Figure 27.8). Small objects of around 1 m diameter enter the Earth's atmosphere every week and disintegrate harmlessly as they shoot Earthwards. Search programs are constantly adding to the inventory of NEOs and have found about 10,000 NEOs of all sizes to date, some 1400 of which are PHOs. The biggest objects appear brightest and are most easily found. So far the inventory is complete down to about 2 km, but only about 25% of the total population of NEOs with diameters of 100 m or more have been discovered. For sizes comparable to the Chelyabinsk impactor (17–20 m), it is estimated that less than 1% of the population has been discovered to date.

The frequency of new discoveries compared to the frequency at which already discovered objects turn up again in

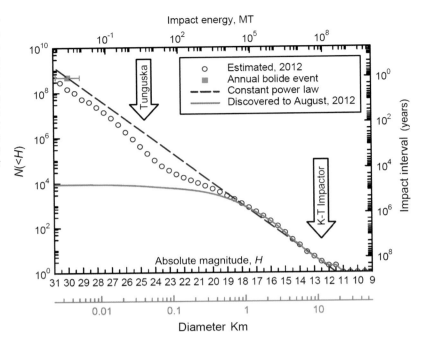

FIGURE 27.8 Estimated number of NEOs larger than a given diameter/brighter than a given absolute magnitude. (Absolute magnitude, H, is a log-scale measure of absolute brightness, on which H decreases as the brightness increases; see Section 5.1.) The red curve represents discoveries to August 2012; the blue circles represent the estimated total number of objects in the NEO population. The estimated mean interval between impacts of objects larger than a given diameter can be read off on the right-hand scale using the curve traced by the blue circles. *Credit: Alan W. Harris the elder, no relation to the significantly younger first author.*

the searches provides an estimate of how near the search is to completion, to the sensitivity limit of the telescope system. From knowledge of the sky-coverage characteristics of the search programs and how sensitive the searches are, it is possible to estimate how many objects are left to find.

When considering impact hazards on Earth, most scientists consider 1 km as the size large enough for an impact to present a global threat to human survival. Thus, the most immediate goal of search efforts was to find all objects larger than 1 km, a task that is almost complete (see Section 5.4). In the process, many smaller objects are found, and these contribute to the long-term goal of completeness to all sizes of NEOs that could cause significant damage on impact. Searchers have a long way to go to complete the survey of all the estimated hundreds of thousands of objects with diameters down to about 50 m. Even NEOs with diameters of only 30–50 m are capable of destroying a city or an urban region. The task of completing the surveys down to these sizes would benefit from new, large, specialized telescopes with huge **CCD** arrays, capable of scanning the skies more frequently and with greater sensitivity. Another possibility would be to conduct the search using telescopes in space, such as more capable versions of the Wide-field Infrared Survey Explorer (WISE) and Spitzer space telescopes discussed below in Section 5.4.

5. PHYSICAL PROPERTIES

The most accurate data on the physical nature of individual asteroids are provided by **rendezvous** missions. The NEAR mission studied the physical and chemical properties of asteroid (433) Eros from orbit and at the spacecraft's landing site. From its shape and surface morphology, astronomers deduced information about its global structure; an X-ray and gamma ray spectrometer provided information about its surface chemistry. The Japanese mission Hayabusa, which touched down briefly on the surface of the NEO (25143) Itokawa, provided valuable insight into that object's mineralogy, structure, and history. Hayabusa successfully returned a small quantity of material from the surface of Itokawa for investigation on Earth, which was the first time that material from a solar system body other than the Moon has been brought to Earth by a space mission. See Section 6 for details.

Space missions to asteroids are not only scientifically extremely productive but also extremely costly. An alternative means of providing information on the physical characteristics of NEOs is astronomical observations with telescopes (either on the ground or in orbit). While telescopic observations cannot match the results of **rendezvous** missions in terms of the wealth and accuracy of information on any particular object, they enable a significant fraction of the NEO population in all its diversity to be investigated. The first physical measurement an astronomer might make, after the position of an NEO has been established, is its brightness measured on the astronomical **magnitude** scale. The changing cross-section of an object as viewed from the Earth affects its brightness and with time reflects the shape and rotation rate of the object. Analysis of this changing brightness in the form of a so-called light curve

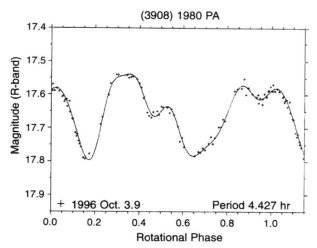

FIGURE 27.9 Light curve for Amor asteroid (3908) Nyx indicating its irregular shape. *Courtesy of Petr Pravec, Astronomical Institute, Academy of Sciences of the Czech Republic.*

(see Figure 27.9), accounting for the observational geometry, results in constraints on its shape and the determination of its rotation rate and orientation of the spin axis in space. From analyses of sunlight reflected off asteroid surfaces at different wavelengths, NEO colors are classified into different taxonomic types. (See Main-Belt Asteroids.) Further analysis can determine surface mineralogy, and, from that, constraints can be placed on the temperatures and locations in the solar system at which these objects formed.

5.1. Brightness

An asteroid's spectrum consists of reflected sunlight as well as heat that it emits as a result of being warmed by the Sun. The standard asteroid photometric magnitude system compensates for the distance and **phase angle** at which the object is observed at visible wavelengths (the solar phase angle is the angle between the observer and the Sun as seen from the asteroid). The magnitude scales by the inverse square law: as the distance from both the Sun and the observer increases, the brightness decreases by a factor equal to the inverse square of those distances. Scattering properties of the surface are expressed in the so-called phase function, which is compensated for by extrapolating the magnitude to 0° phase angle. For comparison purposes, a magnitude measurement is converted to an absolute scale, H, which is defined as the brightness of an object at a distance of 1.0 AU from both the Earth and Sun, and viewed at 0° phase angle (note that the scale is logarithmic and defined such that smaller H values correspond to brighter objects). The relation between an object's brightness and the solar phase angle at which it is observed has been measured for some of the brighter NEOs. Large changes in magnitude with changing phase angle indicate a very rough surface with significant effects due to shadowing. A relatively small dependence of brightness on phase angle indicates either a very dark surface, against which the impact of shadows is not significant, or that the surface is relatively smooth with minimal shadowing. When observations are made over a range of phase angles, fits to theoretical models with multiple variables can be made. Combined with other observational techniques (e.g. radar, polarimetry), constraints on the physical characteristics of the surface regolith can be made. A further parameter that contributes to brightness, the reflectivity of the surface expressed in terms of albedo, is discussed in Section 5.4.

5.2. Shape

Light curves are measurements of brightness as a function of time (Figure 27.9). If the object is perfectly spherical, such that its cross-section does not change with time, there is no variation and the light curve is flat. There are no such objects known, although there are light curves with very small amplitudes (not commonly found among NEOs, however). Light curves of NEOs often show two or more maxima and minima, often with inflections embedded within them. The shape of an NEO in terms of a **triaxial ellipsoid** can be modeled on the basis of observational data. Inflections in the light curves represent changes in the object's cross-section that reflect either the large-scale shape or albedo variations across the surface, or both.

Radar measurements are also analyzed to produce images that reveal the shape of asteroids. Coded wave packets transmitted from Earth to an asteroid reflect back and are received as a radar echo. The bandwidth of the echo power spectrum is proportional to the cross-section of the asteroid presented to the Earth and normal to the line of sight at the time of interaction with the surface, convolved with Doppler shifts in the returned signals caused by the object's rotation. The signal can be built up as the asteroid rotates, producing an image that represents its shape. For objects that have approached the Earth at close enough range to employ this technique, such as (4769) Castalia, (4179) Toutatis, (1627) Ivar, (1620) Geographos, and (433) Eros, the results show shapes varying from slightly nonspherical to very irregular (Figure 27.10). (See Planetary Radar.)

Knowledge of the objects' shapes provides clues to the collisional history of this population. The fact that many NEOs are irregularly shaped implies that they are products of collisions that have knocked off significant chunks of material from a larger body. Images of (433) Eros (Figure 27.11) show it as an ellipsoid measuring $33 \times 10.2 \times 10.2$ km. Its shape is irregular and dominated by large impact craters.

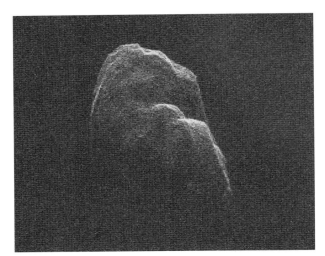

FIGURE 27.10 Radar image of NEO (4179) Toutatis from observations with the 70-m Goldstone Solar System Radar in December 2012 when Toutatis was 6.9 million km from the Earth. The elongated asteroid is about 5 km in length. *Credit: NASA/JPL-Caltech.*

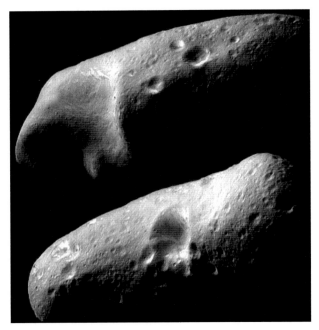

FIGURE 27.11 Asteroid (433) Eros's eastern and western hemispheres. Two mosaics created from six images when the NEAR spacecraft was orbiting 355 km above the surface. Smallest detail is 35 m across. The large depression on the top image is Himeros (10 km across). In the bottom image, the 5.3-km crater Psyche is prominent. Bright exposures can be seen on interior walls of craters. *Credit: NASA/JHU/APL.*

5.3. Rotation Rate

As a result of their proximity to Earth, NEOs are the smallest objects in space for which we can measure rotational properties. In some cases, the rotation rates for NEOs smaller than about 150 m are 100 revolutions per day or faster (i.e. they have rotation periods of just a few minutes). These objects are likely to be relatively strong and intact rock fragments. Larger objects that spin substantially slower may be less strong "rubble piles" composed of individual fragments or fractured rock held together only by gravity. A rubble pile without any cohesion must spin at a rate slower than once every 2.2 h (the so-called "spin barrier"), or else it will fly apart (Figure 27.12). Recent results indicate that dust particles between the components of a rubble pile may provide some cohesion, in which case some rubble piles may spin faster than the 2.2-h limit. Interestingly, the primary bodies of binary NEOs tend to have spin rates close to the spin barrier, consistent with the idea that a binary NEO forms when the spin rate of a rubble pile increases to the point of partial breakup (see Section 5.5). Thus, studies of the rotation rates of NEOs provide insight into the likely range of internal structures occurring within small bodies in our solar system.

5.4. Size and Albedo

Size and albedo are two of the most fundamental physical properties of asteroids, especially for considerations of the impact hazard. The albedo of a surface is a measure of its reflectivity. The so-called visual geometric albedo, normally denoted p, with a subscript for the photometric band (e.g. p_V for visual geometric albedo), is the ratio of the body's brightness at zero solar phase angle (i.e. as seen from the direction of the Sun) to the brightness of a perfectly diffusing disk with the same apparent size and at the same position as the body. The albedos of planetary surfaces depend on their compositions and physical properties. A crude method of estimating the sizes of asteroids uses their measured brightness and an assumed albedo. This method is referred to as a photometric diameter. It is used when only visual magnitudes are available. The diameter, D, is given by the equation

$$D(\text{km}) = 1329 \times 10^{-H/5}/\sqrt{p_V}$$

where p_V, the visual geometric albedo, is assumed, and H is the magnitude defined by the International Astronomical Union magnitude system for asteroids in the V, or visual bandpass. Unfortunately, the range of asteroid albedos is large, from only a few percent up to 50% or more, producing considerable uncertainty in diameters thus derived. Observations in the thermal infrared ("heat radiation") part of the spectrum allow more accurate determinations of size and albedo to be made. For an object illuminated by the Sun alone, the sum of the reflected and emitted (thermal) radiation from the object (assuming no internal energy sources or sinks) is equal to the total solar radiation on its surface. If the object's distance from the Sun is known, the amount of incident energy on the object's surface can be calculated

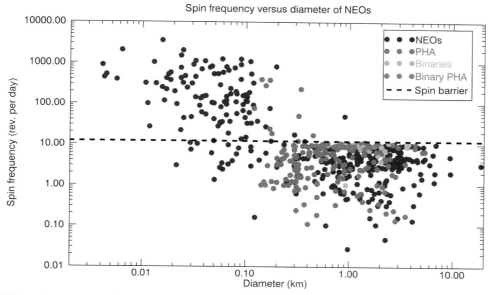

FIGURE 27.12 NEO rotation rates versus size. Very few objects with diameters larger than 200 m spin faster than one rotation per 2.2 h or 10.9 rev. per day, suggesting that the majority have insufficient cohesion to prevent break up at faster spin rates and may be rubble piles or monolithic objects with low tensile strength.

from knowledge of the Sun's radiation output. By measuring the reflected (visible) and reemitted (thermal infrared) components of radiation, and with some assumptions on the body's shape and physical characteristics of its surface, one can derive its diameter and estimate its albedo (Figure 27.13). The two parameters, diameter and albedo, are derived in tandem, with the requirement that the sum of the reflected and emitted components is equal to the incident **solar flux**. The albedo is an important indicator of the mineralogical composition of an asteroid.

For main-belt asteroids, two models incorporating the above procedure have proved to be useful. The asteroid standard thermal model (STM) assumes that the asteroid is spherical and slowly rotating (and/or has practically no thermal inertia, as would be the case for a surface of fine dust). The so-called fast rotating model (FRM) caters to the other extreme, in which the object rotates rapidly (and/or has a large thermal inertia, as would be the case for solid rock). For most large main-belt asteroids, the STM appears to be applicable and gives reasonably accurate diameters and albedos. However, due to their proximity to the Earth, observed NEOs tend to be smaller than most observed main-belt asteroids and many appear to have surfaces dominated by rubble rather than dust (e.g. see the image of Itokawa in Figure 27.17 below).

Furthermore, in contrast to main-belt asteroids, NEOs are often observed at large solar phase angles, which cause much of the thermal emission to be directed away from the observer. It was found that the simple approach to size and albedo measurements based on the STM and FRM often gave large errors when applied to NEOs. A more sophisticated approach is offered by the near-Earth asteroid thermal model (NEATM), developed by one of us (AWH), which more accurately accounts for the observing geometry and allows an additional adjustment to the model surface temperature resulting in better agreement with the observed infrared thermal emission.

Surveys of the sizes and albedos of hundreds of NEOs using the NEATM have been carried out by the NASA

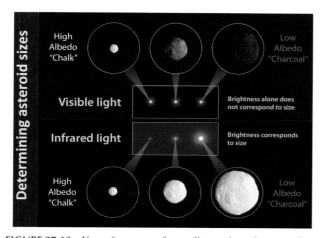

FIGURE 27.13 Upper frame: seen from a distance in a telescope, a large dark asteroid appears as a point source of light with the same brightness as a small, light one. The brightness of an asteroid viewed in visible light is the result of both its albedo and size. Lower frame: in infrared light, the brightness of the object is mainly determined by its size and is not strongly affected by its albedo, i.e. how light or dark its surface is. When visible and infrared measurements are combined, the sizes and albedos of asteroids can be more accurately determined. *Credit: NASA/JPL-Caltech.*

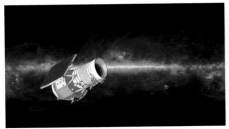 

FIGURE 27.14 Artist's impressions of the WISE (left) and Spitzer Space Telescopes (right). *Credit: NASA/JPL-Caltech/R. Hurt, SSC.*

WISE and Spitzer (Figure 27.14) infrared space telescopes. WISE was launched to Earth orbit in December 2009 carrying a 40-cm-diameter telescope and infrared detectors. WISE surveyed the sky for 12 months and observed a total of at least 584 NEOs, of which more than 130 were new discoveries. Spitzer, carrying an 85-cm-diameter telescope, was launched in August 2003 and inserted into an Earth-trailing orbit (i.e. it orbits the Sun just outside the Earth's orbit and thus trails behind the Earth in space). Spitzer has been used to observe some 600 previously detected NEOs to study the size and albedo distributions of the NEO population and identify NEOs that may have a cometary origin (see Section 3.3.2) or that could serve as accessible and scientifically interesting targets for future robotic or manned space missions.

Extrapolation of the WISE data implies that there are a total of 981 NEOs with a diameter of 1 km or more in the NEO population, of which nearly all have been found. The number of objects rises dramatically with decreasing size (see Figure 27.8): the number of objects with diameters larger than 100 m is estimated to be 20,500, of which approximately 25% have been found.

5.5. Density

Knowledge of the bulk density of an asteroid can provide insight into its internal structure. Many asteroids are known to have bulk densities much lower than the densities of the minerals of which they are composed, as evidenced by the relatively high densities of meteorites. The commonly accepted reason is that there are voids and cavities between the individual fragments making up rubble pile asteroids and/or the highly porous nature of some types of asteroid.

There are various methods of determining asteroid densities, which in effect all require measurement of the object's mass and size, or their ratio. Asteroid masses can be determined by means of, for example, a spacecraft, observations of the perturbations of the orbits of other asteroids or Mars (applicable to large asteroids only), or observations of a satellite or companion asteroid in a binary system by means of precision optical or radar observations. Asteroid sizes can be determined from, for example, spacecraft, thermal-infrared measurements (Section 5.4), radar observations, and **occultation** observations. In the special case of a binary asteroid, the density, ρ, can be estimated from telescope observations using Kepler's third law. For example, if we assume for simplicity that the moon's mass is much smaller than that of the primary asteroid, and the latter is spherical, $\rho = 24\pi a^3/D^3 GP^2$ where P is the period of revolution, G is the universal gravitational constant, and a/D is the ratio of the **semimajor axis** of the moon's orbit to the diameter of the primary. Both the period, P, and the ratio a/D can be derived from observations of the light curve of an eclipsing binary system. Using this method, the densities of a number of asteroids in binary systems have been estimated, including that of the NEO 1996 FG$_3$, which at 1.4 g/cm^3 is very low compared to values of 3–5 g/cm^3 typical of most meteorites. Similar density results for other binary systems suggest that rubble pile structures may be a characteristic of binary systems and, indeed, a necessary prerequisite for their formation: a rubble pile that is spun up by the gravitational effects of a close approach to a planet, for example, may partially break up with the result that a large fragment ends up in orbit around the primary mass.

5.6. Color and Taxonomy

Since the early part of the twentieth century, astronomers have recognized that small bodies come in different colors. As observational techniques evolved and the ability to investigate them improved, the number of observable characteristics increased. Sorting objects into meaningful groups is the process of classification or taxonomy. Asteroid taxonomy developed in response to advances in observing techniques and new technology in the field of stellar photometric astronomy. The intention of a classification scheme is to reflect the compositional variations and thus provide clues on a body's origin and evolution. Astronomers are constantly attempting to test and refine the asteroid taxonomy by employing new statistical methods and extending the number of meaningful parameters that are included in the classification process, while eliminating meaningless or redundant parameters. Today, the alphabet soup of asteroid taxonomy extends to about 13 main types denoted by letters and about the same number of subtypes denoted by subscripted letters. The taxonomy too has evolved, and it is important to be aware of which system is being referred to and what its exact definitions are. (See Asteroids.)

NEOs have representatives from practically all taxonomic types, indicating that many locations in the Asteroid Belt feed the near-Earth population. About 90% of known NEOs fall in the S-, Q-, C-, and X-complexes (a complex is a grouping of taxa from different instrument types and different taxonomies combined into a general category that can encompass all available observations). Some 65% of observed NEOs are bright and members of the S- (40%) or Q- (25%) complexes. When considering the observed ratio of dark objects to bright, there are around twice as many bright objects observed compared to dark ones in the NEO population. However, darker objects are more difficult for visible light telescopes to discover and measure. Accounting for discovery biases is important when studying NEOs using an incomplete sample.

5.7. Mineralogy

By measuring the relative amount of reflected sunlight from the surface of an object as a function of wavelength of the light (a reflectance spectrum), it is possible to constrain its surface mineralogy. Astronomers find that the spectra of some 80% of NEOs studied spectroscopically contain two strong absorption bands, one in the ultraviolet (UV) with a band centered below 0.35 μm and the other in the near infrared near 1 μm. Sometimes a second near-infrared band is observed at a wavelength of 2 μm. Other objects do not display prominent absorption bands: they are found to be featureless and either flat or sloped. Most often an object with a featureless spectrum also has a low albedo. Figure 27.15 shows spectral reflectance measurements of some NEOs. Three spectra have prominent UV and near-infrared absorption bands that are common in silicate minerals. The broad band at 1 μm of asteroid (5641) McCleese is diagnostic of a mineral called olivine. Subtle differences in the position of the center of the band constrain the chemistry of the olivine, which can accommodate a range of magnesium and iron in its mineralogical structure. The presence of a second absorption near 2 μm indicates that a second silicate, pyroxene, is present.

The spectrum of (433) Eros contains both olivine and two types of pyroxene. Detailed spectral analysis and modeling suggest the presence of an additional component that may be a glassy material, or possibly vapor-deposited coatings of nanometer-sized iron grains. They are inferred because the brightness of the spectrum is lower than that of mixtures of only crystalline silicates. These mineral

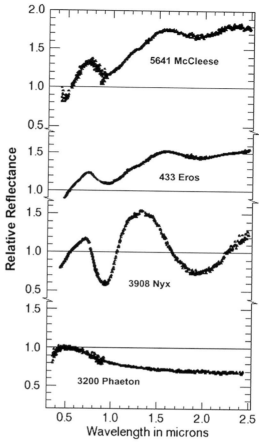

FIGURE 27.15 Spectral reflectance measurements of four NEOs. The range of spectra reflects the range of surface characteristics, including mineralogy and particle sizes, of the surface material.

constituents are present in ordinary chondrite meteorites (See Meteorites.); the deviation from ordinary chondritic composition and the processes controlling it have been studied and ascribed to **space weathering** and/or partial melting.

The spectrum of asteroid (3908) Nyx in Figure 27.15 is dominated by pyroxene and has the same spectral characteristics as the basaltic achondrite meteorites. This asteroid may be a fragment of the large main-belt asteroid, (4) Vesta. (See Main-Belt Asteroids.)

The lower spectrum in Figure 27.15 is characteristic of a subgroup of C types, labeled B. There is no UV absorption and not much of an infrared absorption. Interpretation of this spectrum is uncertain. This asteroid, (3200) Phaethon, is a candidate for an extinct comet, although its albedo (about 10%) is higher than most comets observed to date (about 4%). The range of variations in the mineral composition of NEOs reflects that seen in the Main Asteroid Belt, indicating that NEOs are mostly derived from the main belt.

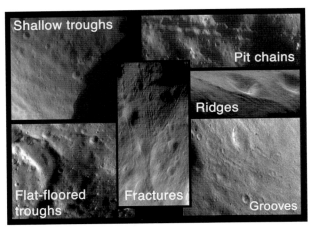

FIGURE 27.16 A montage of images from the NEAR Shoemaker mission showing structural features on (433) Eros. Shallow troughs are partially filled-in linear structures; pit chains are crater chains. A ridge winding almost around the entire asteroid is called Rahe Dorsum; fractures are at the end of Rahe Dorsum, and grooves are evenly spaced and may have raised rims. *Credit: NASA/JHU/APL.*

6. IN SITU STUDIES

6.1. NEAR Shoemaker

The NEAR Shoemaker spacecraft, named in honor of planetary scientist Gene Shoemaker, was launched from Cape Canaveral, Florida, in February 1996, on a 3-year journey to asteroid (433) Eros. NEAR orbited Eros for 1 year in 2000–2001, training its six scientific instruments on the asteroid's surface. It provided the first detailed characterization of an NEO's chemical and physical properties. The objective was to study Eros' relationship to meteorites, the nature of its surface and collisional history, as well as aspects of its interior state and structure.

Orbital imaging of Eros revealed an irregularly shaped body dominated at the global scale by both convex and concave forms, including a 10-km-diameter depression named Himeros, and a 5.3-km bowl-shaped crater named Psyche. At scales of 1 km–100 m, (Figure 27.16) there are grooves and ridge patterns superimposed on a heavily cratered surface, mostly covered by overlapping craters. At the <100-m scale, the surface is dominated by boulders, evidence of downslope movement, and ponding of material in crater bottoms, all of which indicate regolith accumulation and transport. There are not many craters with diameters less than 100 m. Evidence of structural strength on the asteroid includes chains of craters, sinuous and linear depressions, ridges and scarps, and rectilinear craters.

Eros is not a gravitationally bound rubble pile. Rather, it is a fragment of a once larger body, possessing cohesive strength throughout. Both gravitational forces and mechanical strength play a role in the formation and evolution of Eros. Its density of 2670 ± 30 kg/m^3 is low compared to ordinary chondrite meteorites of 3400 kg/m^3 that have approximately the same composition as Eros. By considering the porosity of each meteorite and its relation to a much larger asteroid, the macroporosity of Eros is determined to be 20%, most likely due to collisional fragmentation throughout Eros' interior.

Figure 27.17 shows an image of Eros from the NEAR mission alongside the much smaller NEO (25143) Itokawa. Itokawa was the target of the Japanese **rendezvous** mission Hayabusa.

6.2. Hayabusa

The Japanese Hayabusa spacecraft was launched in 2003 and started its 3-month **rendezvous** with the NEO Itokawa in 2005. It observed Itokawa with four instruments: an imaging camera with eight filters, a near-infrared spectrometer, a laser ranging instrument, and an X-ray fluorescence spectrometer. At the end of this period, the spacecraft collected a sample from the surface of Itokawa and returned it back to the Earth, where the sample capsule containing ~1500 grains (mostly in the 10–50 μm size range) landed in 2010. Since then these grains have undergone analysis in laboratories around the world.

Itokawa is a stony-type (=S-type) asteroid with a 12.1-h rotation period. The size of the asteroid is 535 × 294 × 209 m. The science team remarked that it was shaped like a

FIGURE 27.17 Near-Earth asteroid (433) Eros (left) was the target of the NASA NEAR Shoemaker spacecraft in 2000–2001. Eros is 34 km in length. The large number of impact craters indicates an age of 1–2 billion years. (Credit: JHUAPL, NASA.) Near-Earth asteroid (25143) Itokawa (right) was the target of the Japanese Hayabusa spacecraft in 2005–2006. Itokawa is 535 m in length. Itokawa, which lacks craters, is much younger than Eros and may be an **aggregate** of components weakly bound by gravity ("rubble pile"). *Credit: JAXA.*

sea otter with a head and a body (Figure 27.17). About 80% of the surface is considered to be rough terrain, with many boulders ranging in size from a few meters to 50 m; the largest, named Yoshinodai, is one-tenth the size of the whole asteroid. It has been proposed that Itokawa is the result of a gravitational reaccumulation of some of the pieces of a larger parent body that underwent a collisional breakup, i.e. a rubble pile. Scientists count boulders, rather than craters, on Itokawa to understand its impact history.

The rough terrains of Itokawa have experienced less processing, e.g. less breaking up, sorting, and transporting of material over its lifetime, than other small bodies such as Eros and the Martian moon Phobos. The smooth terrain (called seas) covers 20% of the body, is featureless, and has few boulders. At closest range, the spacecraft resolved centimeter- to millimeter-sized grains, the size of pebbles, in the smooth areas. Itokawa has few craters, although the remnants of some are just visible, almost erased by debris from both impacts and global shaking, which has filled in craters with fine material over time. Evidence of grain migration can be seen in the shapes of the particles returned to the Earth by Hayabusa: some of them have rounded edges, in contrast to the shapes of lunar grains. The composition of the grains is 64% olivine, 22% pyroxene, 11% plagioclase, and 3% other minerals.

One of the main mission goals was to test the theory that the most common meteorite type found on the Earth, the ordinary chondrites (See Meteorites), comes from S-type asteroids, and that the differences in the spectra of ordinary chondrite meteorites and S-type asteroids are due to "**space weathering**", i.e. the action of solar radiation and micrometeorite bombardment on the optical characteristics of the surface material. Analyses of the grains from Itokawa in the laboratory are in agreement with the theory. The mineral chemistry of the grains is very similar to that of thermally processed ordinary chondrites, and half the grains tested showed the surface modification expected from **space weathering**.

With a mass of 3.510×10^{10} kg and a volume of 1.840×10^7 m^3, Itokawa's density is 1900 kg/m^3, which is lower than that of other S-type asteroids. Based on the measured bulk density of ordinary chondrites, the macroporosity of Itokawa is estimated to be 39%, twice as high as that of Eros and, in fact, soils on the Earth. The absence of linear features extending the length of the body, the presence of local flat areas tens of meters long, and the high porosity are consistent with the assessment that Itokawa is a rubble pile asteroid.

7. IMPACT HAZARDS

While this chapter was in preparation, an asteroid with a diameter of about 17–20 m entered the Earth's atmosphere

FIGURE 27.18 The trail left by the Chelyabinsk **bolide**. The left part of the image shows the condensation trail of at least two pieces of the original **bolide**. One piece apparently disintegrated near the center of the image. *Credit: Wikimedia Commons.*

over Russia at a speed of 18 km/s producing a brilliant fireball and a blast wave that injured some 1500 people in the town of Chelyabinsk and caused considerable damage to buildings (Figure 27.18). The event, which occurred on February 15, 2013, was estimated to have an explosive energy equivalent to 500 kt of **TNT** and was the most damaging of its kind since June 1908, when a somewhat larger object exploded in the atmosphere, destroying 80 million trees in the Tunguska region of Siberia. It is a rather unsettling coincidence that the Chelyabinsk event should happen on the very same day that a 30-m-diameter asteroid, 2012 DA$_{14}$, passed exceptionally close to the Earth at a distance from the surface of only 28,000 km! It has since been established that there is no connection between the Chelyabinsk asteroid and 2012 DA$_{14}$, which were on completely different orbits around the Sun and of different taxonomic types. The Chelyabinsk asteroid was not detected prior to its appearance as a fireball due to the fact that it approached from a direction too close to the Sun (Figure 27.19). In contrast, the nature of the orbit of 2012 DA$_{14}$, and its larger size, enabled it to be discovered 12 months before its close approach. The events of February 15, 2013, were a dramatic reminder that space near the Earth is not empty, and that NEOs can cause devastation on our planet.

Much larger objects frequently pass uncomfortably close to our planet: (99942) Apophis, a potentially hazardous asteroid with a diameter of about 300 m discovered in June, 2004, is scheduled to miss the Earth's surface by only about 31,000 km in April 2029. Apophis will pass by at a lower altitude than that of **geostationary** communications and television satellites, as did 2012 DA$_{14}$. Furthermore, the presence of large impact craters on the Earth's surface (around 180 have been confirmed to date) serve to warn us that occasionally very large objects strike the Earth. There are two aspects of the collision hazard to be considered: the magnitude of the collisions and their frequency.

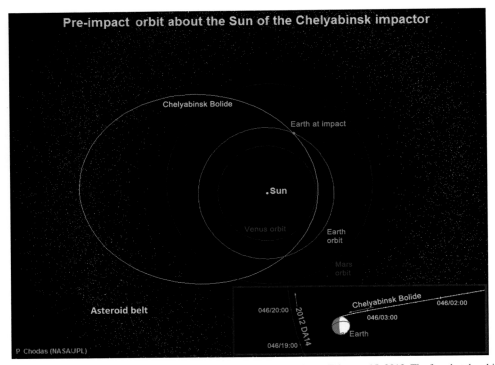

FIGURE 27.19 Heliocentric orbit of the asteroid that impacted near Chelyabinsk, Russia, on February 15, 2013. The fact that the object approached the Earth from the daylight sky prevented it from being discovered prior to its entry into the atmosphere. The inset, bottom right, shows the tracks of the Chelyabinsk bolide and 2012 DA$_{14}$ near the Earth on February 15, 2013 (day 46). The tick marks on the tracks give the positions of the objects at various Universal Times. The inset demonstrates that the two objects had completely different orbits, implying that they were unrelated. *Credit: Paul Chodas, Courtesy NASA/JPL-Caltech; inset adapted from a diagram by Dan Adamo. Both figures are taken from the NASA/JPL Near-Earth Object Program website: http://neo.jpl.nasa.gov/news/fireball_130301.html.*

7.1. Collision Magnitude

The primary physical parameter of concern is the energy of the collision and particularly the kinetic energy transferred to the Earth. The controlling parameters are mass and velocity, according to the relation

$$E = 1/2 \, mv^2$$

The energy of an impact is proportional to the mass of the impactor. Objects that are larger than a kilometer in diameter can cause significant damage on a global scale by triggering changes in the climate that would affect vital human activities such as agriculture. Objects smaller than a kilometer, down to 30–50 m in diameter, still pose a significant regional threat, having the potential to lay waste to large cities or urban areas.

Energy is also proportional to the square of the velocity, so a high-velocity object will have considerably higher impact energy than relatively slow-moving objects. Most NEOs have orbital velocities similar to that of the Earth when nearby, about 30 km/s, but because their orbits are often inclined or more eccentric than the Earth's orbit, there is still a significant relative velocity. Objects in highly eccentric and/or inclined orbits, such as comets, can have tremendous impact energy.

Any object approaching the Earth is accelerated by the Earth's gravity. The minimum velocity of any object entering the Earth's atmosphere is equal to the Earth's escape velocity, 11.8 km/s. So even NEOs with an orbital velocity differing little from that of the Earth's will have quite significant kinetic energy when they hit the ground. Assessment of the damage that a particular impact will impart to the Earth is based on how much energy any particular location can absorb and whether or not that location can recover from an impact. Small objects, in the size range of a meter to a few tens of meters, normally break up in the Earth's atmosphere, leaving perhaps only scattered fragments.

Craters are produced by impacts with energies greater than about 4.2×10^{16} J, or 10 Mt of **TNT**. (See Planetary Impacts; Meteorites.) Impacts with energies of an order of magnitude greater, corresponding roughly to an impactor diameter of 100 m (and the energy released in the largest H bomb detonation), can impart regional damage, potentially destroying areas within a 25 km radius. Of further concern is that an impact into the ocean might induce tsunamis that would destroy coastal areas. The Cretaceous–Tertiary Event 65 million years ago has been estimated at $>4.2 \times 10^{23}$ J, or 100,000,000 Mt!

Such large impacts occur very infrequently, but they have occurred continuously over the lifetime of the Earth and will continue to occur in the future. Most of the energy from the Tunguska and Chelyabinsk impactors was dissipated in the form of an air burst; it was the shock waves from both objects being torn apart high in the Earth's atmosphere that produced the bulk of the ground damage.

7.2. Collision Frequency

A complete assessment of the situation requires knowledge of the frequency of collisions by objects of different sizes. Objects in the range of hundreds of kilometers in diameter were swept up and incorporated into the planets or dynamically ejected as the solar system formed during the period called the Late Heavy Bombardment. The lunar basins formed during this time, which ended ∼3.8 billion years ago. No terrestrial collisions are expected from such large objects today. (See The Moon.)

The impact in the Tunguska region of Siberia in 1908 was most likely a NEO with a diameter of 30−50 m. Interestingly, the Tunguska object did not make a crater because it probably was a weak (heavily cracked) rocky body that broke apart in the atmosphere. Only the shock wave of air from the ∼12-Mt explosion reached the ground, felling thousands of square kilometers of remote forest. The frequency of impacts increases exponentially with decreasing size and energy (see Figure 27.8). Small objects a few meters in diameter, such as 2008 TC$_3$ or 2014 AA (see Section 3.3), enter the atmosphere once or twice a year. Conversely, for larger objects the frequency decreases. On the basis of present, albeit incomplete, knowledge of NEO numbers, we should reckon with an impact that could cause serious damage on the ground (comparable to the Tunguska event) every several hundred years. Present data on the NEO population suggest that the average time interval between impacts of NEOs with diameters of 1 km or more (a catastrophic impact with global consequences) is around 600,000 years. A civilization-threatening impact, such as the one to which the extinction of the dinosaurs 65 million years ago is attributed (diameter of some 10 km, see K-T impactor arrow in Fig. 27.8), would be expected only about every 100 million years. It should be remembered that these estimates are statistical and based on incomplete knowledge of the NEO population and are continually being revised. To assess the overall impact risk, it is imperative to have accurate knowledge of the numbers, sizes, and orbits of NEOs. An object's orbit bears directly on its velocity relative to the Earth in its motion around the Sun. PHOs are monitored closely, worldwide.

There is no need for individuals (except perhaps astronomers) to lose sleep over the impact risk, but preventing occasional serious disruption to our very complex and highly networked society will depend on us developing techniques to deal with the impact hazard. While major impacts in the past have probably altered the evolutionary course of life on the Earth, and paved the way for the dominance of mankind, we would now rather not remain at the mercy of this natural process. Can we protect our civilization from the next serious impact?

Our surveys for NEOs at present remain incomplete, and once an object is discovered, it must be observed over a sufficiently long period for its orbit to be characterized with enough precision to allow it to be tracked into the future. While more than 90% of 1 km and larger near-Earth asteroids have been discovered at present, less than 1% of Chelyabinsk-sized objects have been found, so more work remains to discover, track, and characterize the majority of NEOs. Research into developing improved survey and follow-up systems is ongoing.

7.3. Mitigation Measures

At present, there is no general agreement on the most effective strategy to adopt in the case of a predicted impact. In the case of an object with a diameter below 50 m, the best course of action may be to simply evacuate the region around the predicted impact point, assuming there would be sufficient advance warning (only a small fraction of the asteroids in this size category have been discovered to date). For objects larger than 100 m, serious scientific and technological research is in progress to investigate methods of mitigation. A technique that has been the subject of several studies is the so-called kinetic impactor (Figure 27.20), which involves accurately guiding a massive spacecraft to the target at a high relative velocity, causing a transfer of momentum to it. Such an impactor would thereby slightly change the velocity and orbit of the NEO so that it misses the Earth. The change of momentum of the target asteroid would depend on its structure (rubble pile?) and the amount of ejecta produced in the impact, predictions of which would require some prior knowledge of the asteroid's physical properties. A second spacecraft could be used for reconnaissance purposes, to study the mitigation-relevant physical properties of the asteroid prior to the collision and to monitor the course of the asteroid afterward.

Alternative approaches include the "gravity tractor". The gravity tractor relies on the force of gravity between the target asteroid and a spacecraft hovering under power in close proximity to gradually modify the asteroid's orbit. A significant advantage of the gravity tractor is that no contact with the target is required, and the deflection process is largely independent of the structural properties of the NEO.

FIGURE 27.20 Artist's impression of a kinetic impactor spacecraft deployed to modify the orbit of a near-Earth asteroid. The second spacecraft at the bottom of the picture is orbiting the asteroid or "hovering" near it to observe the effects of the impact and monitor the altered trajectory of the asteroid. The scenario depicted derives from the Don Quijote mission studies commissioned by ESA in 2006 (ESA-AOES Medialab).

It should be noted that the frequency at which a political decision will be taken to launch a space mission to modify the orbit of a NEO may far exceed the estimated frequency of impacts on the Earth. The distribution of NEOs in near-Earth space is random: For every direct hit there will inevitably be many uncomfortable near misses. Predicting the exact course of a NEO years or decades ahead is subject to considerable uncertainty, therefore decision makers may wish to play safe and deflect a NEO before a direct hit becomes a certainty.

A unique research program funded by the European Commission and involving 13 partner organizations in six countries, including Russia and the United States, was established in 2012 to investigate mitigation techniques in detail. The project, called NEOShield (Figure 27.21), aims to provide solutions to scientific and technical issues that will enable the feasibility of promising mitigation options to be demonstrated in the future via test missions. Research into the mitigation-relevant physical properties of NEOs, including laboratory experiments and associated modeling, is aimed at supporting technological development work leading to the detailed design of demonstration missions.

FIGURE 27.21 The logo of the European Commission-funded NEO-Shield project. *Credit: NEOShield.*

APPENDIX

Asteroid Numbering and Naming Conventions

Once at least two nights of observations of a newly discovered object are available, the Minor Planet Center (MPC) assigns a provisional designation consisting of the year of discovery followed by a two-letter code (often with a numerical appendage). The first letter of the code indicates the half-month period (Universal Time) in which the object was discovered (A = January 1−15, B = January 16−31, C = February 1−15, etc.; the letters I and Z are not used in this sequence); the second letter is a sequential counter incremented for each discovery made in the same half-month period. Due to the productivity of modern search programs, the number of objects discovered in a half-month period can run into hundreds. To enable provisional designations to be assigned to the 26th, 27th, etc. discoveries in the same half-month period, the sequential part of the designation may contain a numerical appendix. If more than 50 discoveries are made in a half-month period, the second letter is recycled again with the number 2 appended, etc. The appended numbers should, wherever possible, be written as subscript characters, e.g. 1998 SF_{36}. Once an object has been redetected during several subsequent oppositions, it may be assigned a permanent designation, i.e. a number, which is normally enclosed in parentheses and precedes the provisional designation, e.g. (25143) 1998 SF_{36}. As of the end of 2012, there were about 350,000 numbered asteroids. The discoverer of a numbered asteroid may suggest a name and an accompanying citation, briefly describing the significance of the name. The name and citation have to be approved by the Committee for Small Body Nomenclature of the International Astronomical Union, established in 1994. The assigned name replaces the provisional designation, e.g. (25143) Itokawa (the parentheses are often dropped).

BIBLIOGRAPHY

Bell, J. F., & Minton, J. (Eds.). (2002). *Asteroid rendezvous: NEAR Shoemaker's adventures at Eros*. New York: Cambridge Univ. Press.

Belton, M. J. S., Morgan, T. H., Samarasinha, N., & Yeomans, D. K. (Eds.). (2004). *Mitigation of hazardous comets and asteroids*. Cambridge: Cambridge Univ. Press.

Bobrowsky, P., & Rickman, H. (Eds.). (2007). *Comet/asteroid impacts and human society*. Berlin Heidelberg: Springer-Verlag.

Bottke, W. F., Cellino, A., Paolicchi, P., & Binzel, R. P. (Eds.). (2002). *Asteroids III*. Tucson: Univ. Arizona Press.

Chandler, D., Harris, A. W., & Steel, D. (2008). News feature: cosmic impacts. *Nature, 453*. June http://www.nature.com/nature/journal/v453/n7199/full/4531143a.html.

Lazzaro, D., Ferraz-Mello, S., & Fernández, J. A. (Eds.). (2006). *Asteroids, comets, meteors. Proceedings of IAU Symposium 229*. Cambridge: Cambridge Univ. Press.

Shapiro, I. I., Vilas, F., A'Hearn, M., Cheng, A. F., Culbertson, F., Jewitt, D. C., et al. (2010). *Defending planet Earth: Near-earth object surveys and hazard mitigation strategies final report*. Washington, D. C: National Research Council of the National Academies, National Academies Press. http://www.nap.edu/catalog.php?record_id=12842.

Chapter 28

Meteorites

Michael E. Lipschutz
Purdue University, West Lafayette, IN, USA (Professor Emeritus)

Ludolf Schultz
Max-Planck-Institut für Chemie, Mainz, Germany (retired)

Chapter Outline

1. Introduction — 625
   1.1. General — 625
   1.2. From Parent Body to Earth — 626
2. Meteorite Classification — 628
   2.1. General — 628
   2.2. Characteristics of Specific Classes — 632
   2.3. Oxygen Isotopes and Their Interpretation — 633
   2.4. Chondrites — 635
      2.4.1. Petrographic Properties — 635
      2.4.2. Breccias — 638
      2.4.3. Carbonaceous Chondrites — 638
      2.4.4. Shock — 639
      2.4.5. Weathering — 641
3. Meteorites of Asteroidal Origin — 641
   3.1. The Meteorite–Asteroid Connection — 641
   3.2. Have Meteorite Populations Changed through Time? — 642
4. Meteorites from Larger Bodies — 643
   4.1. Historical Remarks — 643
   4.2. Lunar Meteorites — 644
   4.3. Martian Meteorites — 644
5. Chemical and Isotopic Signatures — 644
   5.1. Chemical Elements — 644
      5.1.1. Cosmochemical Fractionations — 644
      5.1.2. Meteorites — 645
   5.2. Noble Gases — 646
6. Components of Chondrites — 647
   6.1. Chondrules — 647
   6.2. Ca-Al-Rich Inclusions — 648
   6.3. Interstellar Grains — 648
7. Meteorite Chronometry — 649
   7.1. Terrestrial Ages — 650
   7.2. Cosmic-Ray Exposure Ages — 650
   7.3. Gas Retention Ages — 652
   7.4. Crystallization Ages — 652
   7.5. Extinct Radionuclides: Chronology of the Early Solar System — 654
8. Epilogue — 655
Bibliography — 655

1. INTRODUCTION

1.1. General

In the Western world, 1492 marked the discovery of the New World by the Old, the Spanish Expulsion, and, the oldest documented, preserved, and scientifically studied meteorite fall, a 127-kg stone, which fell in 1492 at Ensisheim in Alsace. A meteorite is named for the nearest post office or geographic feature. The oldest preserved meteorite fall might be Nogata (Japan), an L6, which allegedly fell in 861 (but all associated documentation is more recent) and is in a Shinto shrine there. Recovered meteorites, whose fall was unobserved, are finds, some having been discovered (occasionally artificially reworked) in archaeological excavations in such Old World locations as Ur, Egypt, and Poland, and in New World burial sites. Obviously, prehistoric and early historic man recognized meteorites as unusual, even venerable, objects.

Despite this history, and direct evidence for meteorite falls, scientists began to generally accept them as genuine samples of other planetary bodies only at the beginning of the nineteenth century, initiated by a book published by E. F. F. Chladni in 1794. Earlier, acceptance of meteorites as being extraterrestrial, and thus of great scientific interest, was spotty. One might laboriously assemble a meteorite collection only to have someone later dispose of this invaluable material. This occurred, for example, when the noted mineralogist, Ignaz Edler von Born, discarded the imperial collection in Vienna as "useless rubbish" in the latter part of the eighteenth century. With the recognition that meteorites sample extraterrestrial planetary bodies, collections of them proved particularly important. In 1943, with the imminent invasion of Germany, the Russian Government planned for "trophy brigades" to accompany their armies and collect artistic, scientific, and production materials as restitution for Russian property seized or destroyed by Nazi armies during their occupation of parts of Russia. Meteorites that fell in Russia, fragments of which were acquired by and housed in German collections, were explicitly identified as material to be seized.

Apart from its recovery and preservation, Ensisheim is a typical fall. For finds, some peculiarity must promote recognition, hence, the high proportion of high-density iron meteorites outside of Antarctica (Table 28.1). Observed falls are taken to best approximate the contemporary population of near-Earth meteoroids.

A meteoroid's minimum initial entry velocity into the Earth's atmosphere is 11.2 km/s. While its maximum velocity could reach 70 km/s, the observed average value is 17 km/s. During atmospheric passage, the meteoroid's surface melts and ablates by frictional heating. Heat generation and **ablation** rates are rapid and nearly equivalent, so detectable heat effects only affect a few millimeters below the surface. The meteorite's interior is preserved in its cool, preterrestrial state. Ablation and fragmentation—causing substantial mass loss and deceleration, often to terminal velocity—leave a dark brown-to-black, sculpted fusion crust as the surface, diagnostic of a meteorite on Earth (Figure 28.1(a)). If it is appropriately shaped perhaps by ablation, a meteoroid may assume a quasi-stable orientation late in its atmospheric traversal. In this case, material ablated from the front can redeposit as delicate droplets or streamlets on its sides and rear (Figure 28.1(b)). The delicate droplets on Lafayette's fusion crust would have been erased in a few days' weathering: it must have been recovered almost immediately after it fell.

The majority of meteorites derive from **asteroids** and, less commonly, from larger parent bodies. In 2012, about 160 individual samples representing about 76 separate falls are known to come from Earth's Moon and about 60 others (120 individual specimens) from planet Mars. Some interplanetary dust particles may also come from these sources, and/or **comets**. Meteorites are rocks and therefore polymineralic (Table 28.2), with each of the 100 or so known meteoritic minerals generally having some chemical compositional range, reflecting its formation and/or subsequent alteration processes. Important episodes during meteorite genesis are shown in Figure 28.2.

1.2. From Parent Body to Earth

If a meteoroid is small enough to be decelerated significantly during atmospheric passage, it may land as an individual or as a shower. A recovered individual can have a mass of a gram or less (as the 1965 fall of the Revelstoke stone (CI) in British Columbia), or up to 60 metric tons (e.g. the Namibian Hoba iron meteorite found in 1920). A meteorite shower results from a meteoroid fragmenting high in the atmosphere, usually leaving a particle trail down to dust size. Shower fragments striking the earth define an ellipse whose long axis—perhaps extending for tens of kilometers—is a projection of the original trajectory. Typically, the most massive fragments travel farthest and fall at the farthest end of the ellipse.

Some falls are signaled by both light and sound displays, others, like the Peekskill meteorite (Figure 28.3(a)),

TABLE 28.1 Number of Falls and Finds of Major Meteorite Groups

	Nondesert Meteorites		Desert Meteorites	
	Falls	Finds	Antarctica*	"Hot" Deserts[§]
Chondrites				
Carbonaceous	43	83	947	399
H-group	353	1274	11,834	3373
L-group	393	1013	10,652	3254
LL-group	94	162	4507	866
Enstatite	17	26	378	112
Rumuruti	1	3	34	80
Achondrites				
Diogenites	11	16	139	88
Eucrites	34	50	263	294
Howardites	16	20	111	91
Primitive[¶]	18	37	219	266
Martian	5	9	24	70
Lunar	0	4	33	117
Irons	49	763	143	148
Stony-irons				
Pallasites	4	48	31	13
Mesosiderites	7	26	51	96
Others**	55	81	7	15
Total	1100	3534	29,373	9282

*Numbers are fragments recovered, not corrected for pairings.
[§]Meteorites from main desert find locations (Northwest Africa, "Sahara", Algeria, Australia, Libya, Oman). Not corrected for pairing. Finds from other "dry" locations are not included.
[¶]Acapulcoites, aubrites, lodranites, brachinites, ureilites, winonaites.
**Includes members of smaller groups and unclassified specimens.
Source: Database of the Meteoritical Bulletin (www.lpi.ursa.edu/meteor), April 1, 2012.

Chapter | 28 Meteorites

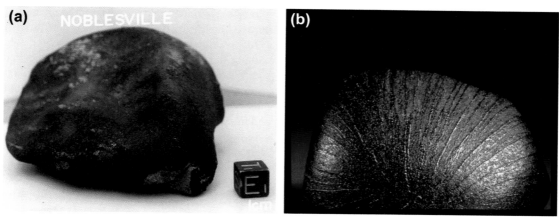

FIGURE 28.1 Fusion crusts: (a) Noblesville H chondrite; (b) Lafayette Martian meteorite. Noblesville, which fell on 31 August, 1991, has nearly complete fusion crust but exposed surface at lower right next to the 1-cm cube shows a genomict (H6 in H4) breccia. (Photo courtesy of NASA Johnson Space Center.) Lafayette exhibits very delicate, redeposited droplets on its sides, indicating an orientation with its top pointing Earthward late in atmospheric traversal. *Photo courtesy of the Smithsonian Institution.*

exhibit a spectacular fireball trail observed over many states. Small falls, like Noblesville (Figure 28.1(a)) fall silently and unspectacularly, and, when recovered immediately after fall, have cold to slightly warm surfaces. The temperature of the interior of a meteorite is not affected by the passage through the atmosphere.

Large meteoroids—tens of meters or larger—are not decelerated much by atmospheric transit and, with an appropriate trajectory, may ricochet off the Earth's atmosphere (Figure 28.3(b)) or strike it at full velocity, which is >11.2 km/s. Such explosive, crater-forming impacts can do considerable damage. The 1.2-km-diameter Meteor Crater (Figure 28.3(c)) in northern Arizona formed about 50,000 years ago by the impact of a 30- to 40-m-diameter meteoroid, yielded fragments now surviving as Canyon Diablo iron meteorites. At least 180 terrestrial craters exhibit

TABLE 28.2 Common Meteoritic Minerals

Mineral	Formula	Mineral	Formula	Mineral	Formula
Anorthite	$CaAl_2Si_2O_8$	Graphite	C	Orthopyroxene sol. soln.	$(Mg,Fe)SiO_3$
Apatite	$Ca_3(PO_4)_3(F,Cl,OH)$	Hibonite	$CaAl_{12}O_{19}$	Enstatite (En)	$MgSiO_3$
Chromite	$FeCr_2O_4$	Ilmenite	$FeTiO_3$	Ferrosilite (Fs)	$FeSiO_3$
Clinopyroxene sol. soln.	$(Ca,Mg,Fe)SiO_3$	Kamacite	$\alpha\text{-}(Fe,Ni)$	Wollastonite (Wo)	$CaSiO_3$
Augite	$Mg(Ca,Fe,Al)_2(Si,Al)_2O_6$	Lonsdalite	Diamond (h)	Pentlandite	$(Fe,Ni)_9S_8$
Diopside	$CaMgSi_2O_6$	Magnetite	Fe_3O_4	Perovskite	$CaTiO_3$
Hedenbergite	$CaFeSi_2O_6$	Melilite sol. soln.		Schreibersite	$(Fe,Ni)_3P$
Cohenite	$(Fe,Ni)_3C$	Åkermanite (Ak)	$Ca_2MgSi_2O_7$	Serpentine (chlorite)	$(Mg,Fe)_6Si_4O_{10}(OH)_8$
Cristobalite	SiO_2	Gehlenite (Ge)	$Ca_2Al_2SiO_7$	Spinel	$MgAl_2O_4$
Diamond	C	Oldhamite	CaS	Taenite	$\gamma\text{-}(Fe,Ni)$
Feldspar solid soln.	$(Ca,K,Na,Al)Si_3O_8$	Olivine sol. soln.	$(Mg,Fe)_2SiO_4$	Tridymite	SiO_2
Orthoclase (Or)	$KAlSi_3O_8$	Fayalite (Fa)	Fe_2SiO_4	Troilite	FeS
Plagioclase		Forsterite (Fo)	Mg_2SiO_4	Whitlockite	$Ca_3(PO_4)_2$
Albite (Ab)	$NaAlSi_3O_8$				
Anorthite (An)	$CaAl_2Si_2O_8$				

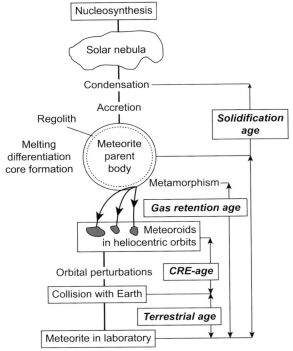

FIGURE 28.2 From nebula to meteorite: genetic processes and the corresponding age determinable for each process. Nuclides of nearly all elements were formed by nuclear reactions in interiors of large stars, which then ejected them in very energetic supernova events. Ejected nebular gas and dust subsequently nucleated, condensed, and accreted into primitive bodies. Source bodies for most meteorites were heated, causing solid state metamorphism or, at higher temperatures, differentiation involving separation of solids, liquids, and gases. As a body evolved, it suffered numerous impacts and, if atmosphere-free, its surface was irradiated by solar and galactic particles that embedded in the skins of small grains and/or caused nuclear reactions. Larger impacts ejected fragments that orbited the Sun. Subsequently, orbital changes caused by large-body gravitational attraction placed meteoroids into Earth-crossing orbits allowing their landing and immediate recovery (as a fall) or later (as a find). Each process can alter elemental and/or isotopic contents: which affected a given meteorite and the time elapsed since it occurred are definable.

features believed to be produced only by intensive explosive impact of a large meteoroid or perhaps even a comet nucleus. The 180-km-diameter Chicxulub feature in Yucatan, Mexico, is the impact site of a 10-km meteorite. By consensus, this impact generated the climatic consequences responsible for the extinction of ~60% of then-known species of biota—including dinosaurs—ending the Cretaceous period and beginning the Tertiary, 65 Ma ago. Other, less well-established events are suggested as having caused extinctions at other times.

Some meteorites have struck man-made objects. The Peekskill stone meteorite with a recovered mass of 12.4 kg ended its journey in a car trunk (Figure 28.3(d)). Its descent in 1992 was videotaped over a five-state area of the eastern United States by many at Friday evening high school football games (Figure 28.3(d)), yielding a well-determined orbit (see Figure 28.13). Two authenticated reports of humans hit by meteorite falls exist. The first involved a 3.9-kg stone (the larger of two fragments), which, after passing through her roof in Sylacauga, Alabama, in 1954 struck a recumbent woman's thigh, badly bruising her. The second involved a 3.6-g piece of the Mbale (Uganda) meteorite shower in 1992, which bounced off a banana tree and hit a boy on the head. An undocumented report tells of a dog being killed by a piece of the 40-kg Nakhla meteorite shower in 1911 near Alexandria, Egypt. This, incidentally, is one of the martian meteorites.

Meteorites may impact anywhere on Earth and, as of April 2012, the numbers of known falls and isolated, non-desert-cluster finds are 1100 and 3534, respectively (cf. Table 28.1). For these, it is readily established whether meteorite fragments found nearby are from the same meteoroid. Such linkages are difficult for the very numerous meteorite pieces found clustered in hot or cold (Antarctic) deserts since 1969. So far, starting in 1969, but mainly since 1976, Antarctica yielded about 30,000 fragments, hot desert clusters in Australia, North Africa (mainly Algeria, Libya, and Morocco), and Oman about 10,000 more. These discoveries are possible in these areas because dark meteorites can be readily distinguished from the local, light-colored terrestrial rocks, "meteorwrongs". The 14 million km^2 ancient Antarctic ice sheet is a meteorite trove because of the continent's unique topography and its effect on ice motion, which promotes the meteorites' collection, preservation, transportation, and concentration (Figure 28.4). Assuming four fragments per meteoroid, Antarctic meteorites recovered thus far correspond to more than 8000 different impact events.

After **chondrites** fall on Earth, their survival times as recognizable objects are limited by oxidation, mainly of small metal grains in them. Antarctic survival times are longer than those of other desert finds because most of their terrestrial residence is spent frozen in the ice sheet, where chemical reaction rates are much slower. The high number of desert finds is also the result of slower weathering rates.

2. METEORITE CLASSIFICATION

2.1. General

Meteorites, like all solar system matter, ultimately derive from primitive materials that condensed and accreted from the gas- and dust-containing presolar disk. Most primitive materials were altered by postaccretionary processes—as in lunar, terrestrial, and martian samples—but some survived essentially intact, as specific chondrites or inclusions in them. Some primitive materials are recognizable unambiguously (albeit with considerable effort), usually from isotopic abundance peculiarities; others are conjectured as unaltered primary materials. Postaccretionary

FIGURE 28.3 Large meteoroids: (a) From the videotape record of the Peekskill meteoroid during its atmospheric traverse on 9 October, 1992. During fragmentation episodes such as this one (over Washington D.C.), large amounts of material fell but none were recovered. (b) Meteor of the Sutter's Mill carbonaceous chondrite that fell on May 3, 2012. (c) The 1-km diameter Meteor Crater in Arizona formed by the explosive impact of the Canyon Diablo IA octahedrite meteoroid about 50 ka ago (photo by Allan E. Morton). (d) Landing site of Peekskill chondrite in the right rear of an automobile. *Photo by Peter Brown, University of Western Ontario.*

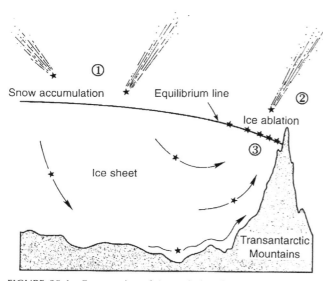

FIGURE 28.4 Cross-section of Antarctic ice sheet and sub-ice topography: meteorites fall (1), are collected by the ice sheet and buried (i.e. preserved), transported, and concentrated near a barrier to the ice sheet (2), and exposed by strong South Polar winds that ablates the stagnant ice (3). *Reprinted from Workshop on Antarctic Glaciology and Meteorites, C. Bull and M. E. Lipschutz (Eds.) LPI Tech. Rept. 82-03, 57 pp. Copyright 1982 with kind permission from the LPI, Houston, TX.*

processes produced obvious characteristics that permit classification of the thousands of known meteorites into a much smaller number of types. Many classification criteria contain genetic implications, which we now summarize.

At the coarsest level, we class meteorites as irons, stones, or stony-irons from their predominant constituent (Figure 28.5(a)); each can then be classified by a scheme with genetic implications (Figure 28.5(b)). Stones include the very numerous, more-or-less primitive chondrites (Figures. 28.6(a) and (b)) and the **achondrites** (Figure 28.6(d)), of igneous origin. Irons (Figure 28.6(e)), stony-irons (Figure 28.6(c)), and achondrites are differentiated meteorites, formed from melted chondritic precursors by secondary processes in parent bodies. During melting, physical (and chemical) separation occurred, with high-density iron sinking to form pools or a core below the lower density achondritic parent magma. Ultimately, these liquids crystallized as parents of the differentiated meteorites, the irons forming parent body cores or, perhaps, dispersed "raisins" within their parent.

Stony-iron meteorites are taken to represent metal-silicate interface regions. Pallasites (Figure 28.6(c)), having large (centimeter-sized) rounded olivines embedded in well-crystallized metal, resemble an "equilibrium" assemblage that may have solidified within a few years but that

FIGURE 28.5 Meteorite classifications: (a) the most common classes and some chemical-petrologic classification criteria; (b) genetic associations involving meteorites.

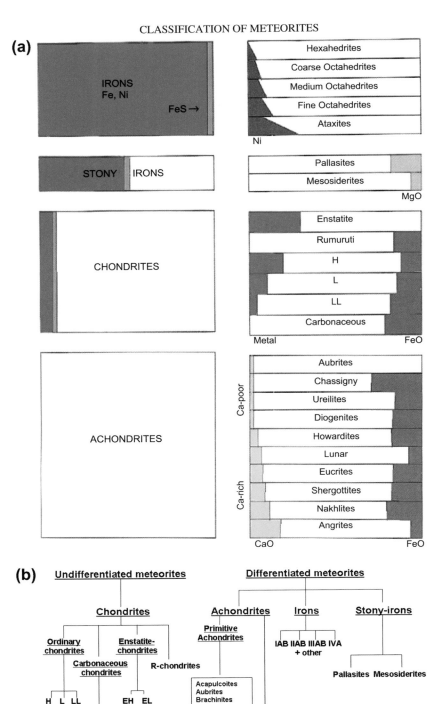

FIGURE 28.6 Common meteorite types (approximate longest dimension, in centimeters): (a) Whitman, H5 (6 cm); (b) Allende; C3V (8 cm)—note 1 cm chondrule in center, (c) Springwater pallasite (18 cm); (d) Sioux Co. eucrite (8 cm); (e) Sanderson IIIB medium octahedrite (13 cm)—note large FeS inclusions.

cooled slowly at iron meteorite formation rates, a few degrees per million years. Mesosiderite structures suggest more rapid and violent metal and silicate mixing, possibly by impacts.

During **differentiation**, siderophilic elements follow metallic iron geochemically and combine with iron into metal alloys. Such elements (e.g. Ga, Ge, Ni, or Ir) are thus depleted in silicates and enriched in metal to concentrations well above those in precursor chondrites. Conversely, magmas become enriched in **lithophilic** elements—like rare earth elements (REE), Ca, Cr, Al, or Mg — above chondritic levels: concentrations of such elements approach zero in metallic iron. During substantial heating, noble gases and other atmophile elements—like carbon and nitrogen—are vaporized and lost from metallic or siliceous regions. Chalcophilic elements that geochemically form sulfides, like troilite, include Se, Te, Tl, or Bi. Chalcophiles and a few siderophiles and lithophiles are also often quite easily mobilized (i.e. vaporized from condensed states of matter) so that they may be enriched in sulfides in the parent body or lost from it. Concentrations of these elements in specific meteorites then depend in part on the fractionation histories of their parents and are markers of heating.

2.2. Characteristics of Specific Classes

It is obvious, even to the naked eye, that most iron meteorites consist of large metallic iron and nickel crystals, which are usually single-crystal α-Fe (kamacite) lamellae 0.2–50 mm thick with decimeter to meter lengths (Figure 28.6(e)). These relatively wide Ni-poor lamellae are bounded by thin, Ni-rich γ-Fe (taenite). The solid-state nucleation and diffusive growth process by which kamacite grew at slow cooling rates from taenite, previously nucleated from melt, is well understood. The 1-atm Fe-Ni phase diagram and measurement of Ni-partitioning between kamacite and taenite permits cooling rate estimation between 900 and 400 °C. These typically are a few degrees or so per million years, depending on iron meteorite group, consistent with formation in objects of asteroidal size. The Ni concentration in the melt determines the temperature of incipient crystallization and this, in turn, establishes kamacite orientation in the final meteorite. These orientations are revealed in iron meteorites by brief etching (with nitric acid in alcohol or ferric chloride) of highly polished cut surfaces: Baron Alois von Widmanstätten discovered this in the eighteenth century and the etched structure is called the "Widmanstätten pattern". Quite independently, an Englishman in Italy, William or Guglielmo Thomsen, simultaneously discovered this but his contribution was unrecognized.

Meteorites containing <6% Ni are called hexahedrites because they yield a hexahedral etch pattern of large, single-crystal (centimeter-thick) kamacite. Iron meteorites containing 6–16% Ni crystallize in an octahedral pattern and are octahedrites. Lower Ni meteorites have the thickest kamacite lamellae (>3.3 mm) and yield the very coarsest Widmanstätten pattern, while those highest in Ni are composed of very thin (<0.2 mm) kamacite lamellae and are called very fine octahedrites. Iron meteorites containing >16% Ni nucleate kamacite at such low temperatures that large single crystals could not form over the 4.55 billion years of solar system history; they lack a Widmanstätten pattern and are called Ni-rich ataxites (i.e. without structure). The Ni-poor ataxites are hexahedrites or octahedrites reheated in massive impacts, or artificially after they fell on Earth.

When primitive parent bodies differentiated, siderophilic elements were extracted into molten metal. During melt crystallization, fractionation or separation of siderophiles could occur. About 60 years ago, Ga and Ge contents of iron meteorites were found to be quantized, not continuous: they could then be used to classify irons into groups denoted as I to IV. Originally, these Ga-Ge groups, which correlate well with Ni content and Widmanstätten pattern, were thought to sample core materials from a very few parent bodies. Subsequent studies of many additional meteorites and some additional elements, especially Ni and Ir, modified this view. At present, the chemical groups (Figure 28.7) suggest that iron meteorites sample perhaps 120 parent bodies, although many, if not most, irons derive from but five parents represented by the IAB, IIAB, IIIABCD, IVA, and IVB irons. (The earlier Roman numeral notation for Ga-Ge groups was retained to semi-quantitatively indicate the meteorite's Ga or Ge content. However, a letter suffix was added to indicate whether siderophiles fractionated from each other.) In addition to the major minerals (kamacite, taenite, and mixtures of them) minor amounts of other minerals like troilite and graphite may be present. Also, silicates or other oxygen-containing inclusions exist in some iron meteorites.

In most cases, chondrites contain spherical millimeter-sized chondrules or their fragments. These chondrules were silicates that melted rapidly at temperatures near 1600 °C and rapidly cooled, early in the solar system's history, some ~1000 °C/h, others more slowly at 10–100 °C/h. Rapid heating and cooling are relatively easy to do in the laboratory, but are difficult on a much larger solar system scale. Yet, large volumes of chondrules must have existed in the solar system because chondrites are numerous. Chondrites (and many achondrites) date back to the solar system's formation—indeed provide chronometers for it (see Sections 7.4 and 7.5)—and represent

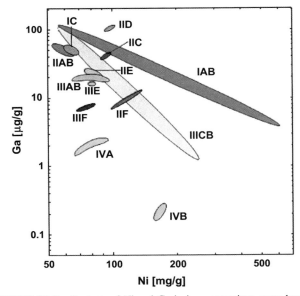

FIGURE 28.7 Contents of Ni and Ga in iron meteorites; some larger chemical groups are indicated by Roman numerals and letters.

accumulated primary nebular condensate and accretionary products. A portion of this condensate formed from the hot nebula as millimeter-sized Ca-Al-rich inclusions (CAI), mineral aggregates predicted as vapor-deposition products by thermodynamic calculations. These CAI, found mainly in chondrites rich in carbonaceous material, exhibit many isotopic anomalies and contain atoms with distinct nucleosynthetic histories. Other inclusions (like SiC and extremely fine diamond) represent relict presolar material. Other condensates formed at much lower temperatures. Some—perhaps even many—CAI may be refractory residues, not condensates.

Although most chondrites contain the same minerals, the proportions of these and their compositions differ in the principal chondritic chemical groups. The primary bases for chondrite classification involve proportions of iron as metal and silicate (in which oxidized iron—expressed as FeO—may be present), and total iron (from Fe, FeO, and FeS) content. The last (Figure 28.8) defines meteorites with high and low total iron (H and L, respectively) or low total iron and low metal (LL). Numbers of H, L, and LL chondrites are so large (see Table 28.1) that these are called the ordinary chondrites. Obviously, chondrite compositions (typically, as in Table 28.3, with elements apportioned by chemical form) are not continuous but, rather, quantized. Table 28.3 lists also major element ratios diagnostic of specific chondritic groups. The total iron in some enstatite (E) chondrites exceeds that in the H group of ordinary chondrites, denoting them as EH chondrites, the EL chondrite designation is self-evident.

Achondrites, formed at high temperatures, contain essentially no metal or sulfide and are enriched in refractory lithophiles, which, with their constituent minerals, allow classification into specific groups. Most groups are named for a specific prototypical meteorite, others—howardites, eucrites, and diogenites (HED meteorites)—were named nonsystematically. Some achondrite groups were associated in the same parent body but derive from different regions: the HED and the Shergottites-Nakhlites-Chassigny (SNC) associations. The HED meteorites are thought to come from 4 Vesta, and/or other V class asteroids produced from it. SNC meteorites and the orthopyroxenite ALHA 84001 come from Mars and are often called martian meteorites.

2.3. Oxygen Isotopes and Their Interpretation

Meteorites "map" the solar system by isotopic composition of oxygen (Figure 28.9), a major element in all but the irons. Since its high chemical reactivity causes oxygen to form numerous compounds, it exists in many meteoritic minerals, even in silicate inclusions in iron meteorites. In standard references, such as the Chart of the Nuclides, the terrestrial composition of its three stable (i.e. nonradioactive) isotopes is given as 99.762% ^{16}O, 0.038% ^{17}O, and 0.200% ^{18}O. In fact, any physical or chemical reaction alters its isotopic composition slightly by mass fractionation. Since the mass difference between ^{16}O and ^{18}O is about twice that existing between ^{16}O and ^{17}O, a mass-dependent reaction (e.g. physical changes and most chemical reactions) increases or decreases the $^{18}O/^{16}O$ ratio by some amount, and will alter the $^{17}O/^{16}O$ ratio in the same direction, but by about half as much. Accordingly, in a plot of $^{17}O/^{16}O$ vs $^{18}O/^{16}O$ or units derived from these ratios (i.e. $\delta^{17}O$ and $\delta^{18}O$; cf. Figure 28.9 caption), all mass-fractionated samples derived by chemical or physical processes from an oxygen reservoir with a fixed initial isotopic composition will lie along a line of slope $\sim 1/2$.

Data from terrestrial samples define the terrestrial fractionation line (TFL) in Figure 28.9, whose axes are like those described above, but normalized to a terrestrial reference material, standard mean ocean water. Not only do all terrestrial data lie along the TFL line, but so too do the oxygen isotopic compositions of lunar samples, which occupy a small part of it. The single Earth–Moon line (defined by data covering the solid line's full length) suggests that both bodies sampled a common oxygen isotopic reservoir, thus supporting the idea that the Moon's matter spun off during the massive impact of a Mars-sized projectile with a proto-Earth.

One important feature of Figure 28.9 is that many chondrite and achondrite groups defined by major element composition and mineralogy occupy their own regions in oxygen isotope space. These data suggest that the major chondritic groups (H, L, LL, R, CH, CI, CM, CR, and E) and

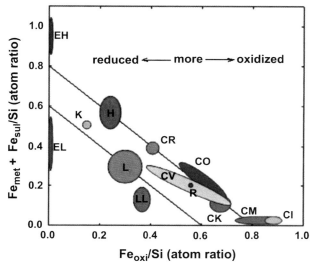

FIGURE 28.8 Silicon-normalized contents of Fe as metal and in FeS (ordinate) vs Fe in ferromagnesian silicates (abscissa) in various chondritic groups; each diagonal defines constant total iron content.

TABLE 28.3 Average Chemical Compositions and Elemental Ratios of Carbonaceous and Ordinary Chondrites and Eucrites

Species*	C1	C2M	C3V	H	L	LL	EUC	Species*	C1	C2M	C3V	H	L	LL	EUC
SiO_2	22.69	28.97	34.00	36.60	39.72	40.60	48.56	NiO	1.33	1.71					
TiO_2	0.07	0.13	0.16	0.12	0.12	0.13	0.74	CoO	0.08	0.08					
Al_2O_3	1.70	2.17	3.22	2.14	2.25	2.24	12.45	NiS			1.72				
Cr_2O_3	0.32	0.43	0.50	0.52	0.53	0.54	0.36								
Fe_2O_3	13.55							CoS			0.08				
FeO	4.63	22.14	26.83	10.30	14.46	17.39	19.07	SO_3	5.63	1.59					
MnO	0.21	0.25	0.19	0.31	0.34	0.35	0.45	CO_2	1.50	0.78					
MgO	15.87	19.88	24.58	23.26	24.73	25.22	7.12								
CaO	1.36	1.89	2.62	1.74	1.85	1.92	10.33	Total	98.86	99.82	99.84	99.99	99.99	99.92	100.07
Na_2O	0.76	0.43	0.49	0.86	0.95	0.95	0.29	ΣFe	18.85	21.64	23.60	27.45	21.93	19.63	15.04
K_2O	0.06	0.06	0.05	0.09	0.11	0.10	0.03								
P_2O_5	0.22	0.24	0.25	0.27	0.22	0.22	0.05	Ca/Al	1.08	1.18	1.10	1.11	1.12	1.16	1.12
H_2O^+	10.80	8.73	0.15	0.32	0.37	0.51	0.30	Mg/Si	0.90	0.89	0.93	0.82	0.80	0.80	0.19
H_2O^-	6.10	1.67	0.10	0.12	0.09	0.20	0.08	Al/Si	0.085	0.085	0.107	0.066	0.064	0.062	0.290
FeE		0.14	0.16	15.98	7.03	2.44	0.13	Ca/Si	0.092	0.100	0.118	0.073	0.071	0.072	0.325
Ni		0.29	1.74	1.24	1.07	0.01		CaTi/Si	0.004	0.006	0.006	0.004	0.004	0.004	0.0019
Co		0.01	0.08	0.06	0.05	0.00		$\Sigma Fe/Si$	1.78	1.60	1.48	1.60	1.18	1.03	0.66
FeS	9.08	5.76	4.05	5.43	5.76	5.79	0.14	$\Sigma Fe/Ni$	18.12	16.15	16.85	15.84	17.73	18.64	
C	2.80	1.82	0.43	0.11	0.12	0.22	0.00	FeE/Ni			9.21	5.67	2.29		
S (elem)	0.10							FeE/ΣFe			0.58	0.32	0.12		

*ΣFe includes all iron in the meteorite whether existing in metal (FeE), FeS, or in silicates as Fe^{2+} (FeO) or Fe^{3+} (Fe_2O_3). The symbol H_2O^- indicates loosely bound (adsorbed?) water removable by heating to $110°C$: H_2O^+ indicates chemically bound water that can be lost only above $110°C$.
Source: Data courtesy of Dr E. Jarosewich, Smithsonian Institution.

acapulcoites, brachinites, the two achondrite associations (SNC and HED), ureilites (U), and the silicate inclusions in group IAB iron meteorites derive from different "batches" of nebular material. The HED region also includes data for most pallasites and many mesosiderites suggesting derivation from a common parent body. Extension of the HED region by a line with slope 1/2 passes through the isotopic region of the oxygen-containing silicate inclusions from IIIAB irons, suggesting that they, too, may be related to the HED association. Perhaps these irons come from deeper in the HED parent body but this would imply complete disruption than V-class asteroids, like 4 Vesta exhibit (see also Section 3). While oxygen isotopic compositions of the rare angrites and brachinites resemble those of the HED association, differences in other properties weaken the connection. Other possible links indicating common nebular reservoirs (based upon limited oxygen isotopic data) are silicate inclusions in IIE irons with H chondrites; silicates in IVA irons with L or LL chondrites; aubrites with E chondrites; winonaites (primitive meteorites modified at high-temperatures) with silicates from IAB and IIICD irons; and the very rare, highly metamorphosed—even melted—primitive acapulcoites and lodranites.

One interpretation of Figure 28.9 is that the solar system was isotopically inhomogeneous, since each "batch" of nebular matter seems to have its characteristic oxygen isotopic composition. Isotopic homogenization of gases is more facile than is chemical homogenization, so that the isotopic inhomogeneity demonstrated by Figure 28.9 implies that the solar system condensed and accreted from a chemically inhomogeneous presolar nebula.

The other important feature to be noted from Figure 28.9 is the "carbonaceous chondrite anhydrous minerals line", with slope near 1. A feature distinguishing

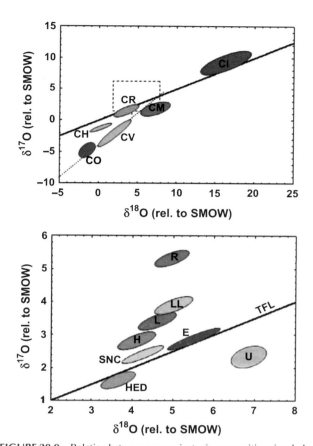

FIGURE 28.9 Relation between oxygen isotopic compositions in whole-rock and separated mineral samples from the Earth, Moon, and various meteorite classes. Units, $\delta^{17}O$ (‰) and $\delta^{18}O$ (‰), are those used by mass spectrometrists and are, in effect, $^{17}O/^{16}O$ and $^{18}O/^{16}O$ ratios, respectively. The $\delta^{17}O$ (‰) is referenced to Standard Mean Ocean Water (SMOW), as $\delta^{18}O$ is defined analogously. Oxygen isotopic compositions for carbonaceous chondrites are much more variable than for other meteorite classes (dashed box in the upper part expanded in the lower one).

graphite—diamond mixtures) in amounts intermediate to those of CV or CO chondrites and CM. Ureilite data do not indicate formation by differentiation of material with uniform oxygen isotopic composition. Rather, ureilite formation may reflect carbonaceous chondritelike components mixed in various proportions.

As originally interpreted, the anhydrous minerals line represented a mixture of nebular material containing pure ^{16}O with others higher in ^{17}O and ^{18}O. If so, the former reflected a unique nucleosynthetic history, perhaps material condensed from an expanding, He- and C-burning supernova shell. Subsequently, photochemical reactions of molecular oxygen with a given isotopic composition were shown to yield oxygen molecules with isotopic composition defining a slope 1 line as in Figure 28.9.

Which process—nebular or photochemical—produced the trends in Figure 28.9 is unknown. Even so, Figure 28.9 still serves to link meteorites or groups of them produced from one batch of solar system matter. Moreover, the position of any sample(s) could reflect some combination of the mass-fractionated and mixing (slope 1) lines. For example, primary matter that ultimately yielded L chondrites (or any ordinary chondrite group) and HED meteorites could have had a single initial composition, subsequently mass fractionated and/or mixed or reacted photochemically to produce meteorite groups with very different oxygen isotopic compositions. However, suitable meteorites with intermediate oxygen isotopic compositions are unknown.

2.4. Chondrites

The available data suggest that heat sources for melting primitive bodies (presumably compositionally chondritic) that formed differentiated meteorites were within rather than external to parent bodies. Important sources include radioactive heating from radionuclides—both extant (^{40}K, ^{232}Th, ^{235}U, and ^{238}U) and extinct (e.g. ^{26}Al)—that were more abundant in the early solar system, and impact heating. Calculations show that ^{26}Al was important in heating small (a few kilometers) primitive parents; other heat sources were effective in differentiating larger ones.

2.4.1. Petrographic Properties

Major element and/or oxygen isotope data demonstrate that differences between parent materials of chondrites of the various chemical groups (e.g. H, CM, or EH) are of primary nebular—preaccretionary origin. Parent body differentiation, on the other hand, is secondary (postaccretionary). Such heating does not necessarily melt the entire parent body and it is thus reasonable to expect an intermediate region between the primitive surface, and the molten differentiated interior. Properties of many chondrites

C1 and C2 chondrites from all others is evidence for preterrestrial aqueous alteration or hydrolysis of some phases in them. Evidence for hydrous alteration of minerals is also observed in some unequilibrated ordinary chondrites (UOC). Anhydrous minerals (including CAI) in carbonaceous chondrites were seemingly never exposed to water, so that these chondrites are regarded as a mixture of materials with different histories. As seen from Figure 28.9, oxygen isotopic compositions of anhydrous minerals in CM, CV, and CO chondrites are consistent with a line defined by CAI whose slope cannot reflect the mass-fractionation process indicated by a slope 1/2 line like TFL. Instead, the anhydrous minerals line seems to represent a mixture of two end-members (batches of nebular material), which, at the ^{16}O-rich (i.e. low ^{17}O, ^{18}O) end lie at or beyond the CO region. Ureilite oxygen isotopic compositions lie on an anhydrous minerals line near CM, suggesting a link. These achondrites contain carbon (as support this expectation and suggest that solid-state alteration of primary chondritic parent material (similar to type 3 chondrites) occurred during secondary heating. Eight characteristics observed during petrographic study of optically thin sections (Figure 28.10) serve to estimate the degree of thermal metamorphism experienced by a chondrite and to categorize it into the major three to seven types (Table 28.4). The absence of chondrules and the presence of abnormally large (≥ 100 μm) **feldspar** characterize rare type 7. These pigeon holes approximate a chondritic thermal metamorphic continuum. Petrographic properties (with bulk carbon and water contents) suggest increasing aqueous alteration of type 3 material into types 2 and 1.

Two of these characteristics are illustrated in Figure 28.10: the opaque matrix and distinct chondrules of the type 3 chondrite Sharps (Figure 28.10(a)) should be contrasted with the recrystallized matrix and poorly defined chondrules of extensively metamorphosed (type 6) Kernouve (Figure 28.10b). Chemically, Fe^{2+} contents of the ferromagnesian silicates—olivine and pyroxene—are almost completely random in a chondrite like Sharps and quite uniform in one like Kernouve. Chondrites of higher numerical types could acquire their petrographic characteristics by extended thermal metamorphism of a more primitive, i.e. lower type, chondrite of the same chemical group.

The petrography of achondrites, like the martian meteorite Nakhla, clearly indicate igneous processes in parent bodies at temperatures $>1000\,°C$. The resultant melting and differentiation erased all textural characteristics of the presumed chondritic precursor (Figure 28.10c) so its nature can only be inferred.

Chemical changes involving loss of a constituent, like carbon or water in chondrites, require an open system; other changes in Table 28.4 could occur in open or closed systems. We emphasize that thermal metamorphism can only affect secondary (parent body) characteristics—those listed horizontally in Table 28.4—not primary ones. Post-accretionary processes by which H chondritelike material can form from L or vice versa are unknown.

Because properties of a given chondrite reflect both its primary and subsequent histories, a chondritic classification scheme reflecting both is used: chondrites already mentioned are Ensisheim, LL6; Nogata, L6; Sharps, H3

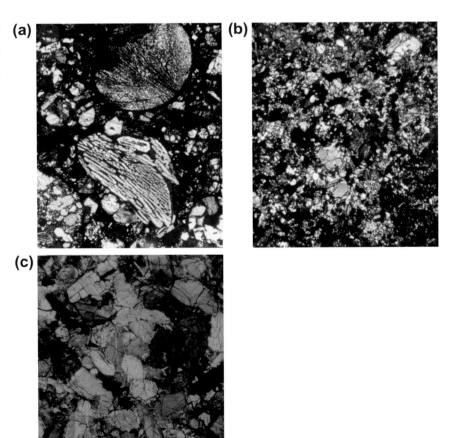

FIGURE 28.10 Petrographic (2.5 mm wide) thin sections in polarized transmitted light. Partial large chondrules are obvious in the H3 chondrite Sharps (a) but barely recognizable in the H6 chondrite Kernouve. (b) (H6); Nakhla, (c) is of Martian origin. *Photos courtesy of R. Hutchison, Natural History Museum, London.*

Chapter | 28 Meteorites

TABLE 28.4 Definitions of Chondrite Petrographic Types*

	1	2	3	4	5	6	7
Homogeneity of olivine composition	—	>5% mean deviations		>5% mean deviations to uniform	Uniform		
Structural state of low-Ca pyroxene	—	Predominantly monoclinic		Monoclinic >20%	<20%	Orthorhombic	
Feldspar	—	Minor primary grains		Secondary <2 μm	Secondary <50 μm	Secondary >50 μm	
Chondrule glass	—	Mostly altered	Clear isotropic	Devitrified	Absent		
Metal: Maximum Ni content (wt%)	—	<20 (taenite absent or very minor)	>20 kamacite and taenite present				
Sulfide minerals: Average Ni content (wt%)	—	>0.5%	<0.5%				
Chondrule-matrix integration	No chondrules	Very sharply defined chondrules		Well-defined chondrules	Chondrules readily delineated	Poorly defined chondrules	Primary textures destroyed
Matrix	Fine grained, opaque	Mostly fine grained, opaque	Opaque to transparent	Transparent, recrystallized			
Carbon (wt%)	3–5	0.8–2.6	0.2–1.0	<0.2			
Water (wt%)	18–22	2–16	0.3–3	<1.5			

*According to Weisberg et al. (2006) in Meteorites and the early solar system (eds. Lauretta D.S. & McSween Jr. H.Y.)

(Figure 28.10(a)); Sylacauga, H4; and Kernouve and Peekskill, H6 (Figure 28.10(b)). No ordinary (or enstatite) type 1 or 2 chondrite is known. Type 3 ordinary chondrites, the UOC, vary the most among themselves and from chondrites of other petrographic types. Within UOC, a variety of properties—for example, the chemical heterogeneity of ferromagnesian silicates, highly volatile trace element (mainly Bi, Tl, In, and noble gases) contents, and thermoluminescence sensitivity—subdivide UOC into subtypes 3.0 to 3.9. Sharps is a very primitive H chondrite, being an H3.0 or H3.4, depending on the classification criteria used. A similar subclassification of CO chondrites also exists.

Many properties of ordinary chondrites demonstrate that each group has its special history, even in something as simple as the numbers of each chemical-petrographic type (Figure 28.11), e.g. the plurality of H falls are H5, while type 6 dominates L and LL chondrites. Stony-iron and, especially, iron finds are very numerous because they are obviously "strange", hence more likely to be brought to someone knowledgeable enough to identify them as meteoritic. Achondrites grossly resemble terrestrial igneous rocks and are less likely to be picked up. Only their fusion crust permits ready recognition of their origin.

2.4.2. Breccias

Even though most chondrites are readily pigeonholed, a few consist of two or more meteorite types, each readily identifiable in the lithified breccia. Noblesville, for example, consists of light H6 clasts embedded in dark H4 matrix. Such an assemblage—two petrographic types of the same chondritic chemical group—is a genomict breccia. A polymict breccia contains two or more chemically distinct meteorite types, implying the mixing of materials from two (or more) parent bodies, each with its own history.

Of the other sorts of breccias, perhaps the most important is the regolith breccia. Noblesville is such a meteorite, and its typically dark and fine-grained matrix contains large quantities of light noble gases—He and Ne—of solar origin (cf. Section 5.2). In addition to these gases, radiation damage is present as **solar-flare** tracks (linear solid-state dislocations) in a 10-nm-thick rim on the myriad matrix crystals. However, solar gases and flare tracks are absent in the larger, lighter colored clasts of regolith breccias. Clearly, dark matrix is lithified fine dust originally dispersed on the very surface of the meter-thick regolith or fragmental rocky debris layer produced by repeated impacts on bodies with no protective atmosphere. The lunar regolith is both thicker, ~ 1 km, and more mature and gardened, or better mixed by impacts than are asteroidal regoliths. This dust acquired its gas- and track-component from particles with kiloelectron volt/nucleon energies streaming outward as **solar wind** or solar flares with megaelectron volt energies, so that the dust sampled the solar photospheric composition. The irradiated dust, often quite rich in volatile trace elements from another source, was mixed with coarser, unirradiated pebblelike material and formed into a breccia by mild impacts that did not heat or degas the breccia to any great extent. Regolith breccias occur in many meteoritic types but are especially encountered as R and H chondrites, aubrites, and howardites.

2.4.3. Carbonaceous Chondrites

2.4.3.1. Composition

The only type 1 or 2 chondrites are carbonaceous chondrites, nearly all non-Antarctic ones being observed falls. A dominant process recorded in them involves hydrolysis, the action of liquid water (in the nebula or on parent bodies) that altered preexisting grains, producing various hydrated, claylike minerals. The chondrites' petrography and the decidedly nonterrestrial $^2H/^1H$ ratios in water from them show that this hydrolysis was preterrestrial. As noted earlier, oxygen isotopic compositions of hydrated minerals demonstrate that the two groups derive from different batches of nebular matter: thus, C1 (or CI) could not form C2 (or CM) by thermal metamorphism nor could C1 have formed by hydrolysis of C2 parent material. For this reason some specialists prefer the CM designation: others prefer a hybrid classification like C2M or CM2 because other C2-like chondrites exist. Tagish Lake, although very primitive, is unique.

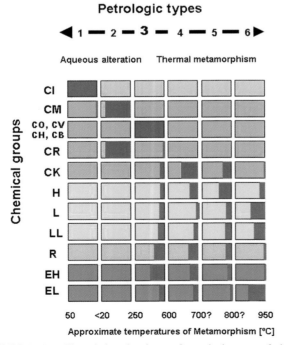

FIGURE 28.11 The relative abundance of petrologic types of chondrites; estimated temperatures of metamorphism or alteration are shown. *Adapted from McSween and Huss (2010).*

C1 chondrites contain no chondrules, but their obvious compositional and mineralogic similarities to chondrule-containing meteorites prompt this classification. Compositionally, C1 chondrites closely resemble the solar photosphere, but a few differences exist (Figure 28.12). Elements depleted in C1 chondrites relative to the Sun's surface (e.g. hydrogen, noble gases, or carbon) are gaseous or easily form volatile compounds that largely remained as vapor in the nebular region where C1 chondrite parent material condensed and accreted. Other elements (e.g. lithium, beryllium, and boron) are easily destroyed by low-temperature nuclear reactions during pre-main-sequence stellar evolution; consequently, they are depleted in the solar photosphere relative to C1 chondrites.

Because chemical analysis of C1 chondrites (or any planetary material) on Earth is more precise and accurate than is spectral analysis of the solar photosphere, "cosmic abundance" tables of chemical and isotopic data for most elements mainly derive from C1 chondrite analyses. Generally these data are used to estimate our solar system's composition. Only where such processes, as incomplete nebular condensation, are suspected, do such compilations adopt solar photosphere values. Recall, however, that earlier we inferred that chemical heterogeneity of the presolar nebula existed, so cosmic abundances may not have been the same in all nebular regions.

2.4.3.2. Organic Constituents

Although chondrites are depleted in carbon, hydrogen, and nitrogen relative to the solar photosphere, C1 and, to a lesser extent, C2 chondrites contain large amounts of organic matter (Table 28.3). They are visible in situ (as globules) only in the unique carbonaceous chondrite, Tagish Lake. Over 400 different extraterrestrial organic compounds have been identified in C1 and C2 chondrites, but their concentrations are very low. Molecular characteristics demonstrate that many are preterrestrial, but the problem of terrestrial contamination is ever-present.

Polyaromatic hydrocarbons (PAH) were found inside two martian meteorites, but not near their surfaces, suggesting that the PAH are not terrestrial contaminants but, rather, originated on Mars. Particles identified as microfossils were reported in at least one martian meteorite and, decades earlier, in CI and CM chondrites. Some advocate biogenic formation of these, but their arguments fail to alter the consensus view that meteoritic organics formed abiogenically.

Since many organic compounds in meteorites can be altered or destroyed by even brief exposure to temperatures of 200–300 °C, their presence in meteorites constitute a thermometer for postaccretionary heating during metamorphism, shock, or atmospheric transit.

2.4.4. Shock

A meteorite parent body cannot be disrupted by internal processes, but only by collision with another similarly sized object. Accordingly, many meteorites evidence exposure to significant shock. A few decades ago, chondrites were qualitatively classed "shocked" if the hand-specimen interior exhibited blackening, veining, or brecciation. Now, petrographic and mineralogic characteristics provide a semiquantitative estimate of the shock-exposure level. Such characteristics reflect changes induced directly, by the peak pressure wave, or indirectly, by the shock-associated, high residual temperature. Specific shock-pressure indicators ("shock barometers") have been calibrated against characteristics produced by laboratory shock-loading experiments. Using these criteria, the degree of shock-loading is known for almost all ordinary chondrites (Table 28.5).

The current scheme to estimate shock histories of equilibrated ordinary chondrites and enstatite chondrites involves the addition of S1, S2,...S6 to its chemical-petrographic classification. It relies upon shock effects observed in olivine and plagioclase. The shock pressures to produce these criteria are S1: <5 GPa; S2: 5–10 GPa; S3: 10–15 GPa; S4: 30–35 GPa; S5: 45–60 GPa; S6: 75–90 GPa. Shock pressures higher than S6 will melt the meteorite. Thus, the Noblesville H4 regolith breccia is S1 as a whole with some H6 clasts being S2. Other chondritic compositional data (radiogenic gases and thermally mobile trace elements, i.e. easily volatilized and lost in an open system) also give information on shock histories. Equilibrated L chondrites exhibit the highest proportion of heavily shocked chondrites, almost half having been shocked above 20 GPa.

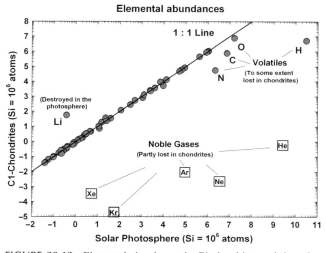

FIGURE 28.12 Elemental abundances in C1-chondrites and the solar photosphere both normalized to 10^6 atoms of silicon; the 1:1 line describes equal abundances. Most elements agree within 10%; exceptions are the volatile noble gases and other volatiles like H, C, N, and O, which are not completely retained in C1-chondrites. Li is destroyed in the Sun by nuclear processes.

TABLE 28.5 Degrees of Shock Loading in Chondrites as Function of Chemical-Petrological Type*

Type (Total)	S1	S2	S3	S4	S5	S6
H3 (290)	25.9	51.4	14.8	6.6	1.0	0.3
H4 (659)	19.6	52.2	21.2	5.8	0.8	0.5
H5 (1141)	11.0	45.1	35.2	7.8	0.4	0.4
H6 (605)	17.5	33.6	33.9	12.4	2.1	0.5
L3 (137)	10.2	40.1	31.4	15.3	2.9	0.0
L4 (284)	12.0	36.6	30.6	16.2	2.1	2.5
L5 (519)	7.1	25.0	36.2	24.9	4.4	2.3
L6 (1062)	3.4	14.2	33.5	33.6	8.8	6.5
L7 (7)	14.3	14.3	14.3	28.6	28.6	0.0
LL3 (85)	14.1	55.3	15.3	14.1	1.2	0.0
LL4 (78)	10.3	56.4	26.9	3.8	2.6	0.0
LL5 (112)	8.0	43.8	27.7	16.1	4.5	0.0
LL6 (175)	3.4	26.9	38.9	24.6	4.6	1.7
LL7 (5)	20.0	20.0	20.0	20.0	20.0	0.0
EH (34)	0.0	20.6	55.9	20.6	2.9	0.0
EL (44)	4.5	70.5	18.2	2.3	4.5	0.0
R (34)	5.9	61.8	26.5	5.9	0.0	0.0

*Data from METBASE 7.3 in percent of total number of chondrites listed with each type. Desert meteorites not corrected for pairing. Dual classifications like S3/4 or H4/5 are doubly counted, e.g. in each respective member.

Lesser but significant proportions of H and LL chondrites show substantial degrees of shock loading (Table 28.5). The mineralogy (really, metallography) of iron meteorites is relatively simple. For shock studies, kamacite and troilite phases give the most information. Etched kamacite usually show shock features called Neumann bands, which indicate relatively low levels of shock (about 1 GPa), however, about half of all iron meteorites were shocked at ≥ 13 GPa, nearly all during collisions that disrupted their parent bodies. Only large meteoroids that formed terrestrial explosion craters can generate pressures as high as 13 GPa when they hit Earth.

The best preserved, perhaps the only, case of strong shock loading during terrestrial impact involves the Canyon Diablo meteoroid that produced Meteor Crater, Arizona. Some Canyon Diablo fragments contain millimeter- to centimeter-sized graphite–diamond aggregates, indicating partial transformation of graphite to diamond. Highly unequilibrated chondrites contain very tiny (~ 0.002 μm) vapor-deposited diamond grains that do not have a high-pressure origin; see Section 6.3. These aggregates contain lonsdaleite, a hexagonal diamond **polymorph** produced, so far as known, only by shock transformation of graphite, which also is hexagonal. Diamond-containing Canyon Diablo specimens always show metallographic evidence for exposure to shock ≥ 13 GPa. They are mainly found on the crater rim, not in the surrounding plain, and contain low levels of cosmogenic stable and radionuclides indicating derivation from the interior, but near the front of the impacting meteoroid where the greater explosive shock existed. The mutual correlations between degree of shock loading, depth in the impacting meteoroid, and geographic locations around Meteor Crater, argue that strongly shocked Canyon Diablo specimens experienced this during terrestrial impact.

The percentages of strongly shocked (i.e. ≥ 13 GPa) members of iron meteorite chemical groups differ widely. The IIIAB irons have virtually all been shocked preterrestrially in the 13–75 GPa range. Nearly 60% of IVA irons, the next largest group, show such shock. A similar proportion of IIB iron meteorites have been shocked at ≥ 13 GPa but this group is small. No other chemical group of iron meteorites shows an especially high proportion of shocked members. Shock-loading experiments show that pressures of 13–75 GPa acting on metallic Fe impart a free-surface velocity of 1–3 km/s. This shock impulse was important, maybe essential, in bringing large numbers of strongly shocked meteorites to Earth.

Semiquantitative petrographic shock indicators in **basaltic** achondrites, i.e. mainly the HED association and shergottites, suggest also a six-stage shock scale. The full range is seen in HED samples—primarily in clasts in howardites. These and other data, mainly compositional, suggest that howardites are a shock-produced, near-surface mixture of two deeper eucritic and diogenitic igneous layers in the HED parent body.

Shergottites have been heavily shocked, in keeping with their accepted derivation from a massive object, like Mars with its 5 km/s **escape velocity**. Nakhlites are less shocked. Lunar meteorites are breccias in most cases. Otherwise, they show no unusual evidence for shock greater than that evident in rocks returned by the Apollo programs.

Nearly all ureilites show petrographic evidence for very substantial shock. Most also contain large graphite–diamond aggregates generally believed to have formed during preterrestrial impacts.

Shock indices of olivine and kamacite in pallasites correspond to moderate shock intensities, but their parent body or bodies suffered at least one strong shock, which produced these stony-iron meteoroids. Most mesosiderites have not been shocked to more than 10 GPa, but some mineral fragments show shock stages S3 to S6. Apparently, the mesosiderites formed by intrusion of shock-loaded silicates into or onto preexisting, generally unshocked metal, possibly after excavation from parent body interiors.

2.4.5. Weathering

Meteoritic material survived in space for billions of years in near vacuum and without contact with water or oxygen. The terrestrial environment has therefore severely altered the chemistry of stone meteorite finds by oxidation or leaching of mobile elements. Also, fresh falls can experience alterations during long exposure under "normal" conditions. For Antarctic meteorites, a simple three-class-system is used based on the degree of rustiness, from minor to severe rustiness. A more detailed weathering scale for chondrites was developed in 1993 and is now generally applied to new finds. This scale designates seven progressive weathering stages from W0 (no visible oxidation) to W6 (extensive replacement of mafic minerals by iron oxides and clay minerals). This classification scheme is, however, qualitative and somewhat subjective. Different observers may come to different classifications, and within one meteorite different weathering stages may occur.

Today, the complete classification of a chondrite will contain class and type, as well as shock and weathering grade, e.g. the Canadian fall, Grimsby, in 2006 is classified as H5, S2, W0-W1; the Saharan find Dar al Gani 001, is LL6, S3, W4.

3. METEORITES OF ASTEROIDAL ORIGIN

3.1. The Meteorite–Asteroid Connection

Several arguments suggest or imply an asteroidal origin for most meteorites: (1) Instrumentally determined orbits (Figure 28.13) for most meteorites—chondrites, and achondrites—pass into the asteroidal belt, indicating that their parent bodies are located in this region of the solar system. (2) Mineralogic evidence indicates the origin of meteorites in asteroidal-sized objects. This includes iron meteorite cooling rates (implying formation depths of asteroidal dimensions), the presence of minerals (e.g. tridymite) and phase relations (e.g. the Widmanstätten pattern), and the absence of any mineral indicating high lithostatic (generated by the rocky overburden) — rather than shock pressures.

Another property linking meteorites and certain asteroids is spectral reflectance. The reflectivity (**albedo**) wavelength variation for an asteroid, involving white solar incident light, can characterize its mineralogy and mineral chemistry. To uncover possible links, asteroidal spectral reflectance are compared with possible meteoritic candidates, both as-recovered or treated in the laboratory to simulate effects of extraterrestrial processes. According to reflection spectra, asteroids are classified into different types, which are not randomly distributed in the asteroidal belt. For example, the E-type asteroids with the highest albedo are mainly found in the inner belt, the most abundant C-type asteroids with the lowest albedo occupy the middle belt to the outer edge. Most S-type asteroids, with medium albedo values, are found in the inner belt at about 2.4 AU.

The best matches (Figure 28.14) exist between the HED association and rare V-type asteroids (4 Vesta and its smaller progeny), iron meteorites, and M-type asteroids;

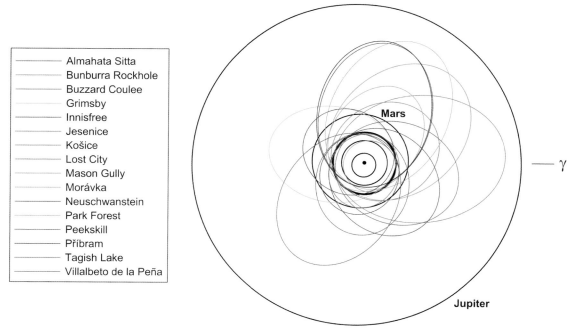

FIGURE 28.13 Instrumentally determined orbits for 16 recovered meteorite falls, as a projection into the ecliptic plane (orbits of the terrestrial planets and Jupiter are included with γ, the vernal equinox). *Figure courtesy of J. Borovicka, Ostrava, The Czech Republic.*

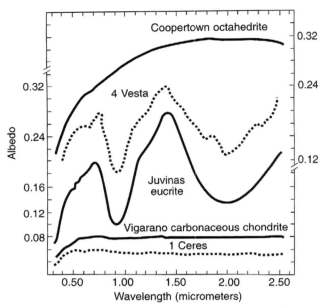

FIGURE 28.14 Spectral reflectances of the Coopertown IIIE coarse octahedrite, Juvinas eucrite and V-class asteroid 4 Vesta, and Vigarano C3V chondrite and G-class asteroid 1 Ceres. The albedo scale for all but Coopertown is on the left: Coopertown's is at right. Solid and dashed lines delineate meteorite and asteroid spectra, respectively. *Courtesy of Dr Lucy-Ann McFadden, University of Maryland.*

CI and CM chondrites thermally metamorphosed at temperatures up to 900 °C with the numerous C-class and related (B-, F-, and G-) asteroids. Aubrites match with the somewhat unusual E-class asteroids and pallasites with a few of the very abundant and diverse S class.

Unfortunately, relative frequencies with which meteorites of a given type and asteroids of a supposedly similar type are encountered do not agree. Specifically, there is the ordinary chondrite-S-asteroid paradox: why are there so few asteroidal candidates for the very numerous ordinary chondrites and so few olivine-dominated stony-irons from the very numerous S asteroids? One possible answer is that "space weathering" by energetic dust impingement on an asteroid's surface—causing metal reduction and dispersion—as well as the action of solar wind ions on mineral surfaces, change their surface reflectivity. This effect could mask ordinary chondrite-like interiors. Another is that Earth collects a biased meteorite sampling compared with the asteroid population, in either near-Earth space or in the asteroid belt. Alternatively, most meteoroids from some organic-rich and friable asteroids might not survive atmospheric passage. For example, Tagish Lake and Sutter's Mill, two unusual carbonaceous chondrites, are extremely friable.

A direct approach to match meteoritic samples with asteroidal material was carried out by the Japanese Spacecraft Hayabusa ("Peregrine Falcon"). Launched in 2003, it rendezvoused with the near-Earth asteroid 25143 Itokawa in 2005 and a lander collected about 2000 tiny dust particles, which were returned to Earth in 2010. Itokawa is an S-type asteroid and the first measurements (oxygen isotopic composition and model mineral abundances) show that the surface matter of this asteroid is consistent with LL-chondrite material.

The NASA spacecraft Dawn was launched in 2007 and reached 4 Vesta in July 2011: orbits of this asteroid strengthen its proposed connection to the HED meteorites. First results reveal a giant impact basin that is sufficiently large to have produced the smaller asteroids of the Vesta family (Vestoids). Furthermore, it was confirmed that Vesta is a differentiated asteroid. Both support the suggestion that the HED meteorites originated from Vesta and/or from the Vestoids.

3.2. Have Meteorite Populations Changed through Time?

The mass of extraterrestrial material falling on Earth is estimated at 40,000–60,000 t/year, most arriving as small dust grains. The number and size distribution of meteorites landing on Earth can be determined from fireball observations, and many attempts have been made to determine the flux from finds in cold and hot desert accumulation zones. The latter, however, requires the determination of terrestrial ages, weathering rates, and pairing in the respective area. In particular, data from the Meteorite Observation and Recovery Project suggest a flux of about nine meteorite falls greater than 1 kg per year and million km^2. The desert find estimates, covering the last 50,000 years, agree within a factor of two to three with this value.

The contemporary flux of meteorites, however, may be biased and unrepresentative of the meteoroid population in near-Earth space. It has been argued that compositional and mineralogical differences exist between recent meteorite falls and finds and older finds in Antarctica. Chemical differences are reported between Antarctic H4-6 and L4-6 chondrites with terrestrial ages >50,000 years and those that are younger. Other studies of Antarctic meteorites reveal more preterrestrial genetic differences between them and falls, but detailed interpretations of these differences remain controversial.

Evidence from fossil meteorites, however, shows clearly that the flux of meteoroids has changed during the geological time of the Earth. About 470 million years ago, during the Ordovician period, a shallow sea covered the area what is now southern Sweden. Limestone beds formed, which have been quarried for many centuries because it is a decorative building stone with beautifully preserved fossils. Among these fossils, ~80 meteorites were also found, as well as some remnants in the form of chromite minerals. Their chemical composition and structure show that all these fossil meteorites and grains derive from L-chondrites. Precise noble gas analyses allow calculation of their exposure ages (see 7.2), which are only a few 100,000 years—much smaller than those of other meteorites. Based on the abundance and distribution of the fossil meteorites and the geological setting, the meteorite flux 470 million years ago could be estimated. Compared to today, this flux in the mass range 10–1000 g is enhanced by a factor of at least 100—and all these meteorites were L-chondrites. Today only about 40% of all stony meteorite falls belong to this group (Table 28.1).

It is suggested that the fossil meteorites are the result of a massive collision in the asteroid belt 470 million years ago producing a huge amount of debris of different sizes. Part of the smaller bodies arrived on Earth as an early surge and was cached in the sea sediments. Modern falls of L-chondrites, excavated later from larger parent bodies, also show the result of the 470-million-year collision in their radiogenic noble gases (see Figure 28.25).

4. METEORITES FROM LARGER BODIES

4.1. Historical Remarks

Through the first half of the twentieth century, it was assumed that the lunar craters were of volcanic origin. The only terrestrial circular structures familiar to geologists were volcanoes. Before Chladni's pioneering paper in 1794, it was suggested that meteorites might come from lunar volcanic eruptions. In that case, large lunar "volcanoes" might have been able to eject stones out of the lunar gravitational field. Some 60–70 years ago it became clear that the craters on the Moon were impact craters and ejecta from their formation could have been accelerated to the Moon's escape velocity of 2.38 km/s and beyond.

The Moon is the only astronomical body on which humans have landed and returned samples that are extensively investigated on Earth. When the first lunar rocks were returned in 1969, no meteorite was known that match their chemistry and mineralogy. Nine years after the last Apollo mission to the moon, during the 1981–82 field season of the US Antarctic Search for Meteorites, a 31-g stone was collected, known as ALHA (Allan Hills) 81005 (Figure 28.15(a)). Its identification as a lunar meteorite was uncontroversial. In the 1979/80 season, three lunar meteorites had been recovered in Antarctica by Japanese teams (Yamato 791197, 793169, and 793274), but their lunar origin was not recognized then. Today, more than 160 lunar meteorites are known, mainly found in hot deserts and Antarctica. No falls are observed.

There are two kinds of martian meteorites. (1) In 2005, the Mars Rover, Opportunity, discovered an IAB iron meteorite on the surface of Mars and (2) martian meteorites have reached the Earth after impacts ejected rocks from the surface layers of that planet. This latter group of achondrites was named SNC meteorites after the three falls Shergotty, Nakhla, and Chassigny. They have similar oxygen isotopic compositions, and—with one exception—have unusually low crystallization ages as igneous rocks (0.6–1.3 billion years), uncommon for chondrites and other achondrites. The young ages were suggestive of a large parent body, since only a planetary body with relatively low surface/volume could retain interior temperatures sufficient to maintain igneous melts that recently. Asteroidal-sized objects could have been differentiated early but would have cooled rapidly. However, without lunar meteorites it seemed unlikely that a rock could have been ejected by an impact on Mars (escape velocity: 5.03 km/s).

But the SNC meteorites are linked to Mars specifically by gases (Ne, Ar, Ar, Kr, Xe, N_2, and CO_2) trapped in shock-formed glass in the Antarctic meteorite Elephant Moraine (EET) A79001 (Figure 28.15(b)), the only meteorite showing a contact between two igneous regions. Contents of these gases in EET A79001 match those in the

FIGURE 28.15 a) The first recognized lunar meteorite Allan Hills (ALH) A 81005. It is a regolith breccia, the white inclusions are anorthositic rock fragments. The cube is 1 cm on each side. (b) A cut face of the martian meteorite Elephant Moraine (EET) A 79001. The dark melt pockets are glass in which gases of the atmosphere of the planet are trapped. These gases are compositionally identical to the atmospheric gases measured by the Viking spacecrafts. The meteorite is 15 cm across.

martian atmosphere measured in 1976 by the Viking landers. These data and the first lunar meteorite find were the reasons that the martian origin of these achondrites became widely accepted.

4.2. Lunar Meteorites

The minerals, textures, chemical compositions, and isotope ratios of these about 160 individuals, belonging to about 75 falls, are similar to those of samples brought to Earth by the Apollo and Luna missions and unlike those of terrestrial rocks or martian and other meteorites. All were ejected from unknown locations on the Moon by meteoroid impacts, while the Apollo and Luna samples were all collected from globally unrepresentative areas. The lunar meteorites, however, more closely match the geochemical composition of the Moon's nearside and farside. Indeed, when compared with lunar orbital geochemical analyses from the Lunar Prospector or Clementine spacecrafts (e.g. distribution of FeO contents), lunar meteorites themselves parallel the overall lunar geochemical character.

Most lunar meteorites are polymict breccias, the majority being regolith and fragmental breccias, consisting of lithified surface material from the upper few meters. As a consequence, regolith breccias contain the products of solar particle radiation (e.g. solar wind gases) and of micrometeorite impacts (e.g. glass-welded aggregates, called agglutinates). Other breccias are dominated by feldspathic material, the major minerals of the lunar highlands. Several lunar meteorites are not brecciated, but are basalts. By terrestrial standards their crystallization ages are great and range from 2.8–3.9 billion years.

Transfer times from the Moon to Earth ("exposure age") are rather short for lunar meteorites compared with other meteorites. Most lunar meteorites were ejected less than 0.5 million years ago. Extreme values are a few 100 years (Kalahari 008 and 009) and about 8 million years (the paired Yamato 82192/3 and 86032). Some lunar meteorites seem to come from the same impact but arrive on Earth at different times (crater- or launch-paired).

The returned Apollo and lunar samples amount to about 383 kg, the total mass of all recovered lunar meteorites is about 62 kg. The largest lunar meteorites are Kalahari 009 (13.5 kg) and Northwest Africa 5000 (11.5 kg); the smallest is Groves Nunatak 06157 at 0.79 g. Many lunar meteorites from hot deserts have been subdivided into small pieces and sold at prices that are much higher per gram than the price of gold.

4.3. Martian Meteorites

According to mineral composition and texture, the 108 martian meteorites represent ~ 60 igneous rocks of basaltic and ultramafic origin. They are classified as shergottites (subdivided as basaltic [22], lherzolitic [13], and olivine-phyric shergottites [15]), nakhlites [8], chassignites [2], and orthopyroxenites [1] (in brackets are the number of falls and finds known). The main minerals in basaltic shergottites are pyroxenes and plagioclase (converted to maskelynite by shock), olivine-phyric shergottites are similar but they also contain olivine. Lherzolites are similar to gabbros, composed mainly of olivine and pyroxene. Nakhlites are augite-rich basaltic rocks, chassignites are olivine cumulates, and the unique orthopyroxenite ALHA 84001 consists mostly of orthopyroxene crystals.

The ejection from the planet's surface must have been by large-scale impacts, which produced craters with diameters >3 km. A special spallation mechanism provides escape velocities >5 km/s. Therefore, all martian meteorites are metamorphosed by moderate to strong shock pressures (15–45 GPa), which also produced local melts.

The cosmic-ray exposure (CRE) ages (see Section 7.2) show that the martian meteorites were ejected in discrete events. There is a good correlation between exposure age and petrographic class: nakhlites and chassignites have exposure ages of 10.5 Ma, all lherzolites lie between 3.8 and 4.7 Ma, basaltic shergottites between 2.4 and 3.0 Ma and five of six olivine-phyric shergottites cluster around 1.2 Ma. Only three meteorites have other ages: EETA 79001 (0.7 Ma), ALHA 84001 (15 Ma), and Dhofar 019 (20 Ma).

In 1996, ALHA 84001 became the object of hot debate, which continues today. This meteorite contains carbonates, which are associated with very small, submicron size magnetite and sulfide formations. The carbonates are complex in their chemistry and shape and their origin is still argued. A team of scientists reported tiny structures that resembling fossilized bacteria in shape, together with organic compounds. The organics have proved to be terrestrial contamination, but these reports have triggered a new research area—astrobiology—which proved very influential for the Mars Exploration Program during the past decade. The question of possible biological activity in the history of Mars is still debated. However, it is not likely that live martian microbes (if they exist) will be found in any sample on Earth since they would be destroyed by the high temperatures produced by the high shock pressures that launched the rocks earthward.

5. CHEMICAL AND ISOTOPIC SIGNATURES

5.1. Chemical Elements

5.1.1. Cosmochemical Fractionations

Earlier, we summarized meteorite compositions and genetic processes necessary to understand general meteoritic properties. Here, we focus upon these topics in greater detail.

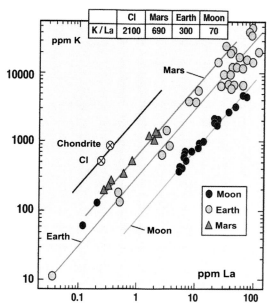

FIGURE 28.16 Lithophile element concentrations (K vs La) in ordinary (chondrites) and Cl chondrites and samples of evolved bodies: Lunar samples from various Apollo missions, and lunar meteorites, terrestrial rocks, and martian meteorites. Data for HED achondrites parallel and lie between the Earth and Moon lines.

The solar system formed from a cloud of gas and dust. The bulk chemical composition is most likely given by the solar abundances (Figure 28.12). The processes that produced the individual chemical composition of planets, asteroids, and meteorites fractionated or separated at least some of the original elements. Such chemical and isotopic fractionations are studied in detail on Earth but conditions for forming meteoritic material may well have been different from terrestrial analogs.

Many of the chemical fractionations observed in meteorites took place in the early solar system. The condensation of each element from vapor to liquid or solid phases takes place over a limited temperature range and is the reason for major chemical fractionation. The volatility of an element plays an essential role here. Elements that condense at high temperatures are called refractory elements; volatile elements condense at much lower temperatures. Other processes that fractionate elements are crystallization of minerals from melts, physical sorting of solids during the building of larger bodies, or planetary differentiation in a mantle and core. Meteorites are the result of all these fractionations and some elemental ratios are characteristic for individual classes (Figure 28.16). Several cosmochemical models try to understand these processes, which converted the originally solar material into that of individual meteorite classes.

5.1.2. Meteorites

Most elements in the periodic table are present in a meteorite at very low levels—microgram per gram (parts per million), nanogram per gram (parts per billion), or picogram per gram (parts per trillion) concentrations. Such low concentrations exist because nucleosynthesis produced stable isotopes of trace elements in only small amounts, and because their geochemical and/or physical properties prevent enrichment during genetic episodes or may even cause significant depletions. Their geochemistry (cosmochemistry) may cause some trace elements to be sited in specific hosts of particular meteorites—e.g. siderophiles like Ir, Ga, or Ge are enriched relative to Cl levels in iron meteorites—while others are dispersed among a variety of minerals. The same element may be dispersed in one meteorite class, but sited in a particular host in another. For example, REE are found in phosphates in achondrites, but some are dispersed elements in chondrites. They concentrate in whitlockite grains in eucrites, and are even more enriched in CAI (see 6.2). Trace elements convey important information because a small absolute concentration change induced by a genetic process will result in a large relative effect.

An example is the element iridium in the Cretaceous-Tertiary (K-T) boundary layer around the Earth. The fall of a massive iron meteorite enriched in siderophiles like Ir, will distribute this element around the globe and redeposit it in a thin layer. Enrichments in the K-T boundary layer suggests that dinosaurs (and many other biota) died off from sudden severe environmental changes created by a large meteoroid impact 65 Ma ago. The resulting "bullet hole" is identified as the 180-km Chicxulub crater in the Gulf of Mexico. While initially controversial, this idea is now generally accepted. In other instances, enrichments of siderophile elements in impact-breccias at an explosion crater on Earth or Moon provide a "fingerprint" of the meteoritic type that created the crater.

Volatile elements condensable at very low temperatures (like the noble gases) may not have similar contents in C1 chondrites and the solar photosphere (Figure 28.12). Meteorite compositions are referenced to readily condensable material by normalization to a refractory lithophile element—most commonly Si (as in Figure 28.8), sometimes Mg or Al—rather than hydrogen as in the astronomical scale for the solar photosphere. For meteorites, normalized ratios can be on a weight or atom basis: in the latter, trace element contents are usually referred to as atomic abundances and are often normalized to C1 contents. In some meteorites, particular C1-normalized abundances can approach or exceed C1 levels. On this basis, we say that moderately to highly refractory siderophiles are enriched in iron meteorites, or that refractory lithophiles are enriched in achondrites. Contents of the more refractory trace elements are characteristic of, hence can define, achondrite associations.

A priori identification of a trace element as refractory or volatile is impossible since its chemical form in a meteorite is usually unknown. For example, indium metal or gaseous oxygen are each quite volatile as elements, but more refractory when chemically bonded in InO. Because In exists only at parts per billion levels in even the most volatile-rich meteorites, neither InO nor any other In compound is identifiable. Several approaches have been used to obtain at least a qualitative elemental volatility order. Criteria used include calculation of theoretical condensation temperatures in a nebular gas of solar composition at pressures of $10^{-3}-10^{-6}$ atm, determination of C1-normalized atomic abundances in equilibrated (petrographic type 5 and 6) ordinary chondrites, and laboratory studies of elemental mobility (ease of vaporization and loss) during long heating of primitive chondrites under conditions simulating parent body metamorphism (400–1000 °C, 10^{-4} atm H_2). By these criteria, elements considered as moderately volatile include (in increasing order) Ni, Co, Au, Mn, As, P, Rb, Cu, K, Na, Ga, and Sb, while strongly volatile ones include Ag, Se, Cs, Te, Zn, Cd, Bi, Tl, and In.

More volatile elements exhibit great variability in stony meteorites. Concentrations of the three to four most volatile elements are several orders of magnitude higher in UOC than in their equilibrated analogues and decrease by one to two orders of magnitude with increasing homogenization of ferromagnesian silicates in UOC. Contents of most strongly volatile elements in L chondrites of petrographic types 4–6 are highly variable and do not correlate with petrographic type. However, in H chondrites, concentrations of many moderately volatile elements vary as H4 > H5 > H6, consistent with loss at progressively higher metamorphic temperatures in stratified parent(s).

Such a model cannot be established for the L chondrites because late shock affected thermometric characteristics, thus obscuring earlier histories. In addition to the petrographic evidence, strongly shocked L4-6 chondrites exhibit loss of some noble gases, highly mobile elements and siderophiles, and lithophile enrichments.

The H chondrite regolith breccias, like Noblesville, differ from "normal" H chondrites in that the dark, gas-rich portions of the breccias are also quite rich in volatile trace elements, sometimes exceeding C1 levels. These volatiles, distributed very heterogeneously in the dark matrix, were apparently not implanted by the solar wind but rather occur in black clasts. During exposure on the asteroidal surface, these dark clasts and light ones (containing "normal" levels of volatiles) were apparently gardened by repeated impacts, ultimately forming the regolith breccia.

In contrast to ordinary chondrites, volatile trace elements in carbonaceous chondrites are very homogeneously distributed. These elements are unfractionated from each other in almost all carbonaceous chondrites, implying that their parent material incorporated greater or lesser amounts of C1-like matter during **accretion**. The proportions define a continuum from 100% C1 down to about 20% in C5 or C6. As in enstatite chondrites, volatile-rich samples have higher proportions of more siderophile trace elements. These trends accord with oxygen isotope data implying a continuum of formation conditions for parent materials of carbonaceous chondrites. Contents of mobile trace elements and noble gases, and the petrography of some C1 to C3 chondrites provide evidence for open-system thermal metamorphism in their parent bodies.

For meteorites of less common types, meteorites from hot deserts and, particularly Antarctica, provide a broader sampling of extraterrestrial materials than do contemporary falls. Systematic and reproducible differences involving moderately to highly volatile elements suggest that differences between falls and Antarctic meteorites may exist and may reflect variations in the near-Earth meteoroid flux with time. However, these suggestions remain controversial (cf. Section 3.2).

5.2. Noble Gases

The chemical inertness of noble gases allows their ready separation from all other chemical elements. Thus, gas mass spectrometers can determine very small noble gas concentrations in a meteorite and, in addition, measure the isotopic composition. Most analyses are carried out on meteorite samples of <100 mg but with effort, samples as small as 5 μg provide essential data.

Noble (rare) gases in meteorites have different origins and each component has a specific isotopic or elemental composition. Some components like the *radiogenic* gases were produced in situ in meteorites, e.g. radiogenic ^{40}Ar is produced by spontaneous, radioactive decay of long-lived, naturally occurring ^{40}K, while ^{4}He is produced mainly from Th and U decay. Fission Kr and Xe components derive from spontaneous or induced fission of heavy nuclei, e.g. stable U isotopes—each with a characteristic fission-fragment distribution. In addition, decay products of extinct radionuclides (e.g. ^{129}I or ^{244}Pu: $t_{1/2}$ = 15.7 and 81 Ma, respectively) exist in meteorites.

Other in situ produced gases are cosmogenic nuclides formed by nuclear reactions of high energy galactic or solar particles with meteoroids. Cosmogenic nuclides are limited to the surface (<1 m depth) of larger bodies and to meter-sized objects in space. Inert gases found in iron meteorites are mainly cosmogenic, but stony meteorites contain a mix of many components.

Trapped gases include a whole family of noble gas components, which were not produced in situ but incorporated in the meteoroid when it formed. Trapped gases are of three main varieties, solar, planetary, and "exotic". Elemental solar gas ratios are similar to those observed in the Sun and are introduced into meteoritic mineral grains by direct implantation of solar-wind ions in the regolith of asteroids or the Moon. The planetary noble gas pattern shows a systematic fractionation in which the lightest noble gases—He and Ne—are depleted relative to Ar, Kr, and Xe.

As an example, let us consider the element Ne. Because a meteorite can sample Ne from any or all of above-mentioned sources, its isotopic composition represents a weighted average of the isotopic compositions of its component sources. These can be recognized on a three-isotope plot (Figure 28.17). A sample consisting of essentially one component is represented by one point in such a diagram, while a neon mix of two components will lie on a line connecting the isotopic compositions of these components. Included in Figure 28.17, as an example, is the Ne isotopic composition of samples of the meteoritic breccia ALH 85151, which contains solar and cosmogenic gases. Lunar soils also contain solar Ne with variable isotopic composition. The proposed SEP component (Solar Energetic Particles) are interpreted now as implantation-fractionated solar wind.

Addition of Ne from other sources, like Ne-E from the decay of extinct ^{22}Na, can complicate this picture. A mixture of three Ne components will fall within a triangle whose apexes each have the Ne isotopic composition of a pure component. In addition, many chondrites contain one or more trapped Ne components, examples of which are in Figure 28.17 inset.

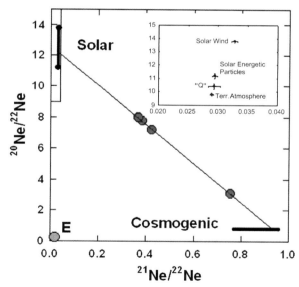

FIGURE 28.17 Three-isotope plot for stable Ne isotopes in ALH 85151 chondrite. Data for grain size separates lie on lines connecting solar Ne (composed of kiloelectron volts/nucleon solar wind particles) with cosmogenic Ne produced by gigaelectron volt galactic cosmic rays. The box in the upper left is expanded to show isotopic compositions of individual Ne components in meteorites. The inset depicts isotopic compositions of Ne from solar wind, solar energetic particles (recently interpreted as implantation-fractionated solar wind), the terrestrial atmosphere, and a trapped (planetary) component [Q]. Pure ^{22}Ne (so-called Ne-E) is formed by radioactive decay of very short-lived ^{22}Na in the protoplanetary nebula or presolar grains.

Krypton and Xe systematics are more complicated for several reasons. The Kr and Xe isotopes derive from several nucleosynthetic sources, two of which are especially important. One is now-extinct ^{129}I, which decayed to produce ^{129}Xe and gives chronometric information (cf. Section 7.5). The second involved fission of now-extinct ^{244}Pu, which produced a Xe component with a characteristic fission-yield curve. In addition to induced and spontaneous U fission products, different trapped components exist. Kr and Xe in presolar grains provide almost pure gas from individual nucleosynthetic events.

Each solar system body has its particular formation history and, thus, its own noble gas isotopic "fingerprint". Gases on the Earth, Moon, Venus, and Mars are distinguishable from each other and from those in chondrites. As discussed in Section 4.1, glass in the EET A79001 shergottite contains Martian atmospheric gas, providing incontrovertible proof that it (and the other martian meteorites) formed there.

Our brief Ne discussion outlined, in principle, how to disentangle several noble gas components from an average meteoritic datum. Actually, the situation is more complicated because each "component" may, in fact, be resolvable into constituents from specific sources, each with reproducible isotopic patterns involving more than one noble gas. Ingenious laboratory treatments can yield a phase enriched in one true gaseous constituent from others. These include investigation of individual grains, selective acid dissolution of specific minerals, enrichment by mineral density using heavy liquids, stepwise heating and mass analysis of gases evolved in some temperature interval or some combination of these steps.

6. COMPONENTS OF CHONDRITES

Chondrites are a mixture in varying proportions of chondrules, metal, sulfides, refractory inclusions, and matrix material. Because of postaccretionary alteration by thermal metamorphism, aqueous alteration, or shock, the original texture has changed. Additionally, in many cases, new minerals have been formed. Chondrules and refractory inclusions are of special interest because of their origin and their memory of processes in the early solar system. A very minor component of some meteorite classes turns out to be presolar grains that have formed in and around stars.

6.1. Chondrules

Chondrules exhibit a large variety of compositions and textures. Most chondrules, 0.1- to 1-mm-sized spherules (Figures 28.6(b) and 28.10(a)), consist primarily of pyroxene and olivine but other minerals may also be present.

Chondrules are often surrounded by small grains that accreted to them before the meteoritic rock formed.

Since their first description about 200 years ago, the formation of chondrules has been fiercely debated. There is agreement that they were once molten or partly molten droplets that rapidly cooled "like drops of fiery rain". Decay products of extinct radionuclides (see Section 7.5) suggest that chondrules are formed early in the solar system from melts initially at temperatures around 1800 °C. Cooling rates were between 100 and 2000 °C/hour. However, no consensus has been reached as to the cause that produced the liquids necessary for chondrule formation, or the necessary energy source.

Generally, two types of models have been developed: (1) Direct condensation from the material of the hot solar nebula; (2) Manufacture from preexisting solids. Processes suggested include collisions in space between asteroids or planetesimals, lightning in the solar nebula, shock waves and formation in the vicinity of the young Sun. After 200 years of research and numerous special conferences on this topic, no generally accepted theory has resulted.

6.2. Ca-Al-Rich Inclusions

In addition to low-temperature materials, like the matrix of C1 chondrites, refractory grains like CAI record early solar history. The CAI are millimeter- to centimeter-sized refractory inclusions especially recognizable in C2 and C3 chondrites, but also identifiable in some UOC and in R and E3 chondrites. They were first described in the Allende chondrite, which fell in 1969.

Typically, CAI consist of refractory silicate and oxide mineral assemblages, mainly hibonite, perovskite, melilite, spinel, diopside, and anorthite. The refractory mineralogy of CAI makes them unique among cosmochemical samples. Several isotope systems indicate that CAI are the oldest objects to have formed in the solar system (see Section 7.5).

Major-element compositions of CAI agree with calculations by equilibrium vapor-deposition evaporation models to represent the first 5% of condensable nebular matter solidifying at ≥ 1400 K from a gas of solar composition at a pressure of 10^{-3} atm, or at 0.3 atm, if the dust/gas ratio is 40-fold enriched. Most individual CAI contain tiny particles (usually $<50\,\mu m$) very rich in refractory siderophiles (Re, W, Mo, Pt, Pd, Os, Ir, and Rh) and, occasionally, refractory lithophiles like Zr and Sc. Sometimes, even smaller (micrometer-sized) refractory metal nuggets are found, consisting of single-phase pure noble metals or their alloys.

The textural and mineralogic complexities of CAI indicate a variety of formation and alteration processes in their history. Undoubtedly, CAI formed at high temperatures; properties of some suggest vapor condensation as crystalline solids, while others seemingly reflect liquid or **amorphous** intermediates. Volatilization, melting, solid-state metamorphism, and/or alteration in the nebula or after accretion may also have affected some/many CAI. Clearly, CIA had complicated histories that obscured their primary textural properties but left their chemical and isotopic properties relatively unaltered.

Volumetrically, fine-grained CAI are encountered more often than coarse-grained ones but the latter are more easily studied. Coarse-grained CAI are grouped into different types, defined mainly by mineralogy, formed at progressively lower temperatures, from a melilite-dominated type that apparently condensed as solid from vapor, to other types formed from partly molten mixtures or melt droplets. Compositionally, CAI reflect a high-temperature origin, refractory lithophiles like REE are generally enriched 20 times or more relative to C1 compositions.

Many to all of these CAI exhibit isotopic anomalies (both in positive and/or negative directions) for O, Ca, Ti, and Cr. A few CAI, mineralogically and texturally indistinguishable from others, are called FUN inclusions because they exhibit *F*ractionated and *U*nidentified *N*uclear isotopic effects involving not only Kr and Xe but also elements like Mg, Si, Sr, Ba, Nd, and Sm. Six FUN inclusions contain mass-fractionated oxygen (i.e. follow slope 1/2 lines in Figure 28.9); two of these six exhibit isotopic anomalies for every element thus far studied.

Another type of refractory-element-rich inclusions are amoeboid olivine aggregates (AOA). Like CAI, they are depleted in volatile and moderately volatile elements. AOA are interpreted as aggregates of grains that condensed from the solar nebula gas.

6.3. Interstellar Grains

Until about 1970, the solar system was considered "isotopically homogeneous", objects in it having formed from a well-mixed and chemically and isotopically homogenized primordial nebula. The later discovery of oxygen isotopic variations, e.g. Figure 28.9, disproved this. However, even then, rare samples extracted from meteorites exhibited anomalous isotopic compositions of, for example, Ne or Xe isotopes. These anomalies cannot be explained by well-established processes like decay of naturally occurring radionuclides, cosmic ray interaction with matter, or mass-dependent physical or chemical fractionation.

These isotopic anomalies, usually orders of magnitude larger than in other solar system materials, are associated with very minor mineral phases of primitive chondrites distributed irregularly in unequilibrated meteorites. These minerals include diamond, graphite, silicon carbide, and aluminum oxide, with typical grain sizes being $0.1-10\,\mu m$, with diamond being much smaller ($\sim 0.002\,\mu m$). All these minerals are rare in meteorites: e.g. SiC is about $10-15$ ppm

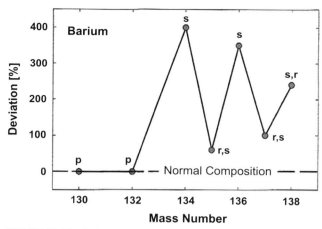

FIGURE 28.18 Stable Ba isotopes in a SiC separate from the Murchison CM chondrite normalized to those in normal terrestrial Ba. Letters indicate nucleosynthetic processes by which individual isotopes are produced. It is presolar neutron-capture isotopes (on slow s, and rapid, r, timescales that formed in presupernova and supernova stages, respectively), which are anomalously high, by up to a factor of four.

by mass, diamond about 1500 ppm. Figure 28.18 depicts such an anomaly, the Ba isotopic composition in presolar SiC separated from the carbonaceous chondrite Murchison. The data are normalized to terrestrial values of ^{130}Ba and ^{132}Ba, the anomalous s- and r-process isotopes (see below) lying far above the horizontal line.

Since the isotopic composition of these grains differs wildly from those of ordinary solar system matter, they must derive from outside our solar system. These grains were incorporated into the solar nebula with intact memories of their individual nucleosynthetic sources, then accreted into meteoritic matter and obviously survived all later episodes in their parent bodies' histories. The isotopic anomalies identified thus far point at specific genetic processes. Most SiC grains probably formed in stars on the Asymptotic Giant Branch (i.e. AGB stars) in the Hertzsprung—Russell diagram. This is the source of isotopes produced by neutron capture on a slow timescale (or so-called s-process) nuclides, with rapid neutron capture (r-process) nuclides forming immediately prior to the supernova stage. Supernovae also seem required to explain the isotopic anomalies in tiny diamonds.

Isotopic anomalies of many trace elements in these presolar grains provide a wealth of unique information regarding the evolution of stars and nucleosynthetic processes. This information is only obtainable by exhaustive, detailed, highly sensitive and highly accurate analyses of rare interstellar grains from primitive samples in terrestrial laboratories. Undoubtedly, isotopic anomalies in these rare meteoritic constituents tell us many details about stellar formation and evolution, and the formation and early history of the solar system.

7. METEORITE CHRONOMETRY

How old are meteorites? What is the "age" of the solar system? Modern cosmochronological methods give answers using naturally occurring radioactive nuclides. An age is a time interval between two events marked by specific chronometers. The "clock" starts by an event beginning the time interval and its end must be clearly and sharply recorded. Chronometers used in modern geo- and cosmochronology usually involve radioactive isotopes such as the U-isotopes, ^{87}Rb or ^{40}K. Radioactive decay allows calculation of an age if the concentrations of both parent and daughter nuclide are known, the time interval beginning is defined and—an important condition—the system is not disturbed (i.e. it is a "closed system") during the time interval. Some meteorite ages involve production of particular stable or radioactive nuclides, or decay of the latter. Typically, the chronometer's half-life should be comparable with the time interval being measured.

Meteoritic matter yield a variety of ages, each reflecting a specific episode in its history. Some of these are shown in Figure 28.2: the first formation of solids in the solar system, melt crystallization in parent bodies, excavation of meteoroids from these bodies, and the meteorite's fall to Earth. Other events, like volcanism or metamorphism on parent objects can be established, as can dating of processes in the early solar system (based on extinct radionuclides). CRE ages date the exposure of a meteoroid as a small body (about 1 m radius) in interplanetary space, as well as the meteorite's terrestrial age, the time elapsed since it landed on the Earth's surface (Figure 28.19). In the following

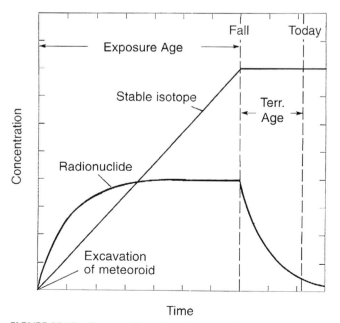

FIGURE 28.19 Concentrations of cosmic ray produced radioactive and stable nuclides during cosmic ray exposure and after the meteorite's fall on Earth.

sections we discuss some of these different ages, starting with the most recent one.

7.1. Terrestrial Ages

Terrestrial ages are determined from amounts of cosmogenic radionuclides found in meteorite falls and finds, ^{14}C ($t_{1/2} = 5.73$ ka), ^{81}Kr ($t_{1/2} = 200$ ka), ^{36}Cl ($t_{1/2} = 301$ ka), and ^{26}Al ($t_{1/2} = 730$ ka) being the nuclides most frequently employed for this purpose. In Section 3.2, we mentioned possible changes of meteorite populations with time. Terrestrial ages are important for this question. A meteorite's survival time during terrestrial residence is determined by the weathering conditions where the meteorite resides. Survival times (terrestrial ages) for meteorites are much lower for warm and/or wet areas than for the arid ones.

Stony meteorites in Antarctica have terrestrial ages in excess of 2 Ma (Figure 28.20) and age distributions depend on their locations, presumably reflecting ice sheet dynamic differences. Terrestrial ages of meteorites from the Allan Hills average ~ 300 ka, while those from Frontier Mountain have younger ages (<100 ka). For Antarctic iron meteorites, however, much longer terrestrial ages have been reported (Derrick Peak: ~ 1 Ma; Lazarev: ~ 2.4 Ma).

Meteorites from hot deserts generally have terrestrial ages up to 50 ka (Figure 28.21) but the terrestrial age distributions depend on the hot desert recovery site. A few stony meteorites, especially those without metal grains, can survive longer on the Earth's surface: e.g. the lunar meteorite Dhofar 025 from Oman has a terrestrial age of about 500 ka.

7.2. Cosmic-Ray Exposure Ages

The transit of a meteoroid from the asteroidal belt to the Earth takes place in several steps, dynamical aspects and

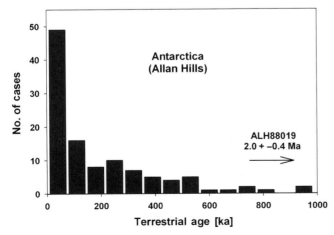

FIGURE 28.20 Terrestrial age distributions of meteorites from the Allan Hills region of Victoria Land (Antarctica), which, on average, are older than meteorites from any other part of Antarctica.

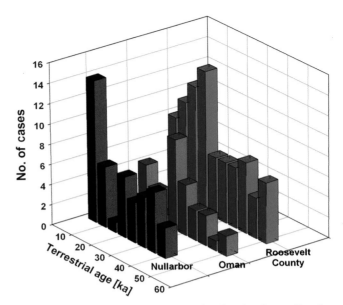

FIGURE 28.21 Terrestrial ages of chondrites from hot deserts. Data for the Saharan samples are from Welten et al. (2004); the others are from Jull (pers. commun.). Age distributions tend to vary in the four sites and, while most maximum ages are 40–50 ka, those for lunar and martian meteorites tend to be somewhat older (~ 150 ka).

orbital evolution considerations are required for its understanding. The asymmetrical heating of the meteoroid by solar radiation (**Yarkovsky effect**) may change its orbit into a resonance with that of Jupiter. This process is rather slow and may take tens of millions of years. Due to their optical properties, iron meteorites are affected less by this effect. Calculations show that once a meteoroid's orbit is in resonance, it quickly evolves into an Earth-crossing orbit.

In principle, we need to determine both a cosmogenic radionuclide and a stable nuclide to establish a CRE age, assuming constant cosmic-ray fluxes. In practice, however, production rates of stable cosmogenic noble gas nuclides in stony meteorites (e.g. ^{3}He, ^{21}Ne or ^{38}Ar) are well known as a function of their chemical composition. It usually suffices to measure just their concentrations with appropriate corrections. An especially useful and precise method is based on the measurement of cosmogenic krypton. ^{81}Kr/Kr CRE ages are almost independent of the chemical composition and location of the measured sample within the meteoroid. However, the concentrations of ^{81}Kr are small and Kr analyses are a challenge in noble gas spectrometry.

Absent contrary evidence, we generally assume that irradiation by cosmic rays of solar and galactic origin is simple, that is, the meteoroid was completely shielded in a parent body until an impact ejected it as a meter-sized object, remaining essentially undisturbed until collision with Earth. Some stones (e.g. the chondrite Jilin) and irons (e.g. Canyon Diablo) exhibit complex irradiation histories involving pre-irradiation on the parent body surface. In such cases, different samples of a meteorite exhibit different CRE ages.

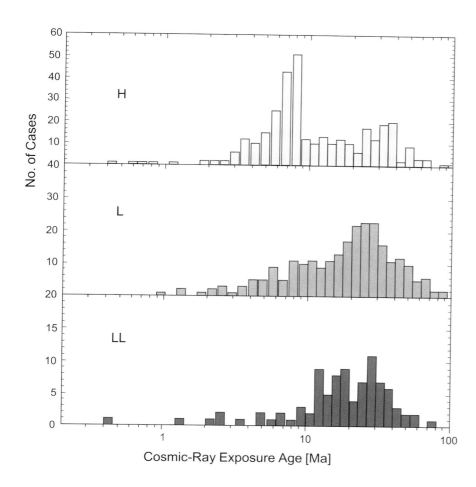

FIGURE 28.22 Cosmic ray exposure ages for ordinary H, L, and LL chondrites. Peaks in the histograms indicate major collisional events on parent bodies that generated many meter-sized fragments.

The data for ordinary chondrites in Figure 28.22 show that all groups have CRE ages ranging up to 100 Ma, but the distributions differ markedly. For H-chondrites, there is one major peak, at ~7 Ma and a smaller one at ~33 Ma. For L chondrites, major peaks are not obvious, but clusters occur at 20–30 Ma and 40 Ma. Contrary to the H-chondrite situation, nearly 2/3 of the L chondrites have CRE ages >10 Ma. For LL chondrites, the major peak at 15 Ma includes ~30% of all measured samples, and another is at 30 Ma. Major peaks correspond to major collisional breakups on/of chondrite parent bodies. The distributions of these CRE ages imply that meteoroids are not continuously produced from their parent bodies but reflect a limited number of impacts.

Most CRE ages for CI and CM chondrites tend to be short (<5 Ma). These ages are mainly based on data from cosmogenic radioisotopes alone. Either their parent bodies are close to a resonance or already in an Earth-crossing orbit. Also, most lunar meteorites have very short exposure times of <1 Ma but a few highland breccias have longer CRE-ages. Although lunar meteorites fell at different locations at different times, they may originate from the same impact on the Moon (source-crater pairing). Clustering of exposure ages is also observed for other meteorite groups, like E or R chondrites. HED meteorites display clustering for diogenites at 22 and 39 Ma, coincident with those of eucrites, and howardites. Most specimens of the acapulcoites/lodranites cluster around 5 Ma.

The martian meteorites were ejected from the planet in a number of discrete events (Figure 28.23). All nakhlites have ejection ages (exposure age + terrestrial age) of 10.5 ± 1.1 Ma. This cluster also includes the two chassignites. An ejection age of about 2.7 ± 0.4 Ma is observed for the eight basaltic shergottites measured: possibly all lherzolithic shergottites were ejected about 4 ± 0.7 Ma ago. Nine olivine-phyric shergottites measured have ages close to 1.1 ± 0.2 Ma; there are, however, three "outliers" with different ejection ages. The only orthopyroxenite has an ejection age unlike that of any other martian meteorite. At least eight individual impacts delivered all these meteorites to Earth.

The first reliable CRE age method for iron meteorites was a difficult, tedious, and no longer practiced technique involving ^{40}K ($t_{1/2} = 1.28$ Ga) and stable ^{39}K and ^{41}K.

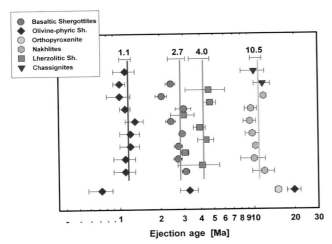

FIGURE 28.23 Ejection ages (CRE age + terrestrial age) for martian meteorites. Most meteorite types cluster at specific values, indicating distinct ejections by impacts on Mars. Only the olivine-pyric shergottites have "outliers". This group of SNC meteorites was ejected in at least four impacts.

About 70 iron meteorites were dated by the $^{40}K/^{41}K$ method and the resulting ages range from 100 Ma to 1.2 Ga (Figure 28.24). These data disagree with those determined using a stable isotope and radioactive nuclides with different half-lives (e.g. $^{36}Cl/^{36}Ar$). These discrepancies have been interpreted as time variations in the galactic cosmic ray flux. However, also complex irradiation histories as well as destructive collisions (so-called space erosion) must also be taken into account.

CRE exposure age peaks are evident for a few chemical groups, e.g. for group IIIAB a cluster around 650 Ma, for IVA around 450 Ma, both suggesting major collisional events involving the parents of the chemical groups III and IV irons. CRE ages of stony-iron meteorites are somewhat longer than those of chondrites but show no obvious clustering.

7.3. Gas Retention Ages

From measurements of U, Th, and radiogenic ^{4}He or of ^{40}K and radiogenic ^{40}Ar, one can calculate a gas retention age or the time elapsed since a meteorite sample cooled sufficiently low to retain these noble gases, if the system was closed during this period. This radiogenic age records primary formation of the meteorite's parent material, but, in most cases, subsequent episodes (metamorphic and/or shock) accompanied by substantial heating, partially or completely degassed the primary material. A variant of the K/Ar age, the $^{40}Ar-^{39}Ar$ method, involves conversion of some stable ^{39}K to ^{39}Ar by fast-neutron bombardment, i.e. $^{39}K(n,p)^{39}Ar$, in a nuclear reactor followed by stepwise heating and mass spectrometric analysis. From the $^{39}Ar/^{40}Ar$ ratios in different temperature steps, it is possible to correct for gas loss. This method has several advantages over the conventional $^{40}K/^{40}Ar$ method: it eliminates the need for a separate measurement of potassium and the effect of argon loss can be detected.

Gas retention ages of chondrites, achondrites, and silicate inclusions in iron meteorites range up to about 4.5 Ga. Many meteorites, particularly L chondrites, have young gas retention ages, ~500 Ma, while H chondrites cluster at higher ages (Figure 28.25). The discovery of many fossil meteorites in limestone beds in Swedish quarries (see Section 3.2) imply that the meteorite flux on Earth ~470 Ma ago was much higher than the contemporary flux. A major collisional breakup at that time and the subsequent loss of gases in parts of the parent body may explain the many L-chondrites showing deficits of radiogenic helium and argon. Furthermore, meteorites with young gas retention ages generally exhibit petrographic evidence for strong shock loading, implying diffusive gas loss from material that experienced quite high residual temperatures generated in major destructive collisions. Almost always, meteorites having young K/Ar ages have lower U,Th/He ages. This occurs because He is more easily lost from most minerals than is Ar.

7.4. Crystallization Ages

Crystallization or solidification ages establish the time elapsed since the last homogenization of parent and daughter nuclides, normally by forming a mineral or rock. Nuclides used to establish crystallization ages are long-lived radioisotopes and their decay products that might have not be affected by later geological events. Some

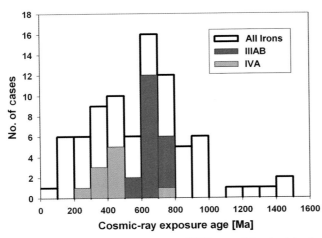

FIGURE 28.24 Cosmic-ray exposure (CRE) ages determined by the $^{40}K-^{41}K$ method for iron meteorites. The group IIIAB irons nearly all apparently produced by a single, massive collision on their parent body 650 ± 60 Ma ago. CRE ages for IV A irons are more diverse but average 400 ± 60 Ma.

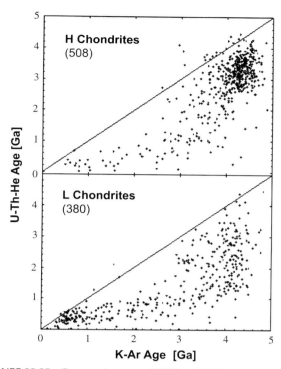

FIGURE 28.25 Gas retention ages of 508 H and 380 L chondrites. Data from the U, Th-He, and K-Ar methods are plotted against each other. The 45° line represents concordant ages. The two major chondrite types exhibit strong thermal history differences. The dominant concordant long ages of H chondrites suggest that their parent bodies generally remained thermally unaltered since forming 4–4.5 Ga ago. The concentration of data defining concordant short ages of L chondrites suggests strong shock heating in a major collision(s) about 0.6 Ga ago. Nearly all discordant meteorites lie below the 45° lines because radiogenic ^{4}He is lost far more easily than is radiogenic ^{40}Ar.

TABLE 28.6 Long-Lived Radionuclides Used Mostly in Cosmochronology

Radionuclide	Half-life (years)	Daughter
^{40}K	1.27×10^9	^{40}Ar/^{40}Ca
^{87}Rb	4.88×10^{10}	^{87}Sr
^{147}Sm	1.06×10^{11}	^{143}Nd
^{176}Lu	3.75×10^{10}	^{176}Hf
^{187}Re	4.12×10^{10}	^{187}Os
^{238}U	4.47×10^9	^{206}Pb
^{235}U	7.04×10^8	^{207}Pb
^{232}Th	1.40×10^{10}	^{208}Pb

techniques, as the Pb/Pb system, which involves the ultimate decay products of U and Th isotopes, is based upon relatively mobile Pb that is more easily redistributed than nuclides of other dating systems. However, these effects are detectable since a comparison with other dating systems (e.g. Sm–Nd) can yield somewhat different ages, depending upon the sample's postformation thermal history.

Common systems found to yield useful solidification ages include a number of nuclide pairs (Table 28.6). In most cases, it is difficult to determine the age of a rock just from measurements of the concentrations of a radioactive nuclide and its daughter isotope alone, because a nonradiogenic contribution can be present. However, by measuring mother and daughter isotopes in various minerals of a rock it is possible to establish a crystallization age and determine the initial isotopic ratio of the daughter element. Figure 28.26 shows as an example the evolution diagram (also known as isochron diagram) of Sm-Nd measurements in different separates of the martian meteorite, Governador Valaderes. The slope of the isochron establishes the age of the meteorite. Such isochrons can also be constructed from whole rock samples of different meteorites, which formed from the same source material. The Rb-Sr system, for example, defines a line for basaltic achondrites that gives an age of about 4.3 Ga and provides an initial $(^{87}Sr/^{86}Sr)_0$ isotope ratio at the time their parent body differentiated. The most "primitive" value of $(^{87}Sr/^{86}Sr)_0$ was found in the chondrite Allende; those of Angra dos Reis, an angrite, and the basaltic achondrites are somewhat larger. This indicates a time difference in the formation of their parent bodies. The Re-Os system provides good isochrons for iron meteorites of 4.537 ± 0.008 Ga.

The U,Th–Pb system has proved to be a powerful tool to determine absolute ages. To obtain absolute ages, several methods have been developed for geochronology. In cosmochemistry, the Pb–Pb method provides the most accurate dates. It is based on the measurement of radiogenic lead generated from the two uranium isotopes and an assumed initial lead composition, measured in troilite of iron meteorites. This method works well for undisturbed early solar system samples with very low abundances of common lead, like CAIs. The current best measurement of a CAI yields a value of 4568.2 ± 0.5 Ma. This age is corroborated by other methods described in the next section. However, some additional uncertainty of the absolute age exists because of uncertainties in the decay rates of the uranium isotopes. CAIs are the oldest material so far dated and their age is taken as the age of the solar system.

Solidification ages for most meteoritic samples are "old", i.e. close to 4.56–4.57 Ga. A well studied example is the Shallowater aubrite with a crystallization age of 4563.3 ± 0.4 Ma. This age is also used as an "anchor" for precise dating of events in the early solar system by short-lived—now extinct—radionuclides.

While most meteorites have solidification ages around 4.56 Ga, there is clear evidence that all martian meteorites but one have "young" crystallization ages of ≤ 1.4 Ga

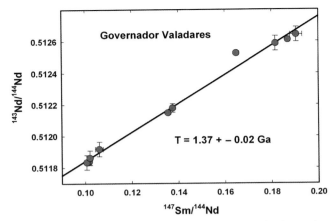

FIGURE 28.26 Samarium-neodymium isochron diagram for the martian meteorite Governador Valaderes. The regression line through the data of whole rock samples, mineral separates, and leachates defines a crystallization age of 1.37 ± 0.02 Ga. *Data from Shih et al. (1999)*.

TABLE 28.7 Extinct Radionuclides Important as Early Solar System Chronometers

RadioNuclide	Half-Life (Ma)	Daughter	Initial Abundance
^{26}Al	0.73	^{26}Mg	$5 \times 10^{-5} \times {}^{27}$Al
^{10}Be	1.5	^{9}Be	$7 \times 10^{-4} \times {}^{9}$Be
^{53}Mn	3.74	^{53}Cr	$2\text{-}4 \times 10^{-5} \times {}^{55}$Mn
^{107}Pd	6.5	^{107}Ag	$5 \times 10^{-5} \times {}^{108}$Pd
^{182}Hf	9	^{182}W	$10^{-4} \times {}^{182}$Hf
^{129}I	15.7	^{129}Xe	$1.1 \times 10^{-4} \times {}^{127}$I
^{146}Sm	103	^{142}Nd	$8 \times 10^{-3} \times {}^{144}$Sm

(Figure 28.26). Most shergottites have crystallization ages between 165 and 475 Ma, while those of Nakhlites and Chassignites cluster around 1.3–1.4 Ga. The exception is ALH88001 (about 4.5 Ga), which may represent a sample from the ancient crust of Mars.

It is likely that the source craters of lunar meteorites are randomly distributed on the surface of the Moon. Some lunar meteorites are probably samples from the lunar farside, where no samples were collected by landing missions. Crystallization ages of lunar meteorites are similar to those of samples collected by Apollo or Luna missions. Ages of mare basalts cluster between 3.8 and 3.9 Ga, but the later impact bombardment of the Moon has reset the radiometric clocks to varying degrees. This crystallization-age cluster of lunar meteorites does not resolve the debate about whether the Moon experienced an increase in impact rates around 3.9 Ga ("lunar cataclysm") or the impact formation of the large Imbrium basin.

7.5. Extinct Radionuclides: Chronology of the Early Solar System

Short-lived, extinct radionuclides originally in the early solar system have been now completely decayed away. The decay products of these extinct nuclides, however, provide valuable information allowing determination of the order in which objects formed within the early solar system. This is based on the amount of radionuclides in objects that were present during their formation and are now detectable in their decay products. Short-lived radioisotopes used in cosmochronology are given in Table 28.7.

Measurements of decay products of an extinct radionuclide do not provide absolute dates in the sense discussed in the previous section, but do permit relative chronologies on timescales comparable with the half-life of the radionuclide. Often, relative ages are calculated from three-isotope plots involving decay products of the extinct radionuclide. For example, the "canonical" initial ratio in CAIs of the extinct nuclide ^{26}Al to the stable ^{27}Al is 5×10^{-5}. This value was determined from an excess of ^{26}Mg in several CAI samples, the oldest solar system material. Any material with lower $(^{26}\text{Al}/^{27}\text{Al})_0$ must have been formed later than the CAIs.

This relative chronologic information can be combined with absolute age data to establish small absolute time differences in the early solar system. The "anchor" is an object for which reliable measurements exist both for the Pb–Pb system and for the short-lived nuclide system. Such objects are CAIs, and individual meteorites like the Shallowater aubrite. Several isotope systems consistently indicate that the CAIs are the oldest objects formed in the early solar system over a time span of less than 1 Ma. Chondrules, the most abundant constituents of chondrites, followed a few million years later: for ordinary chondrites values are about 2 Ma while CO3 and CV3 chondrules formed 2–3 Ma after CAIs. Chondrules of metal-rich chondrites (CH, CB) formed significantly later, 5–6 Ma after CAIs.

Until recently, it was believed that differentiated meteorites formed from chondritic precursors and thus later than chondrules. Several isotope systems now indicate, however, that for the HED parent body, differentiation and crust formation began about 2.5–3 Ma after CAIs formed. The formation of the HED's parent body was probably very rapid, within about 1 Ma after formation of the solar system. The crystallization ages of individual eucrites are not very well known because of later shock overprinting. However, magmatic activity on Vesta, the proposed parent asteroid of these meteorites, seems to have continued over a longer time.

FIGURE 28.27 Representations of a French Southern and Antarctic Territories stamp illustrating a micrometeorite or cosmic dust particle (left) collected by melting Antarctic ice cores, the coring drill being at the meteor trails (of cometary origin) and a fireball are at the top.

For iron meteorites results from the ^{182}Hf–^{182}W system indicates that the parent bodies of magmatic iron meteorites differentiated within about 1 Ma after CAIs. For non-magmatic irons, however, the Hf–W systematics was modified by later impacts. The timing of these events is poorly constrained.

The thermal metamorphism and dates of aqueous alteration of chondrites have been studied by several isotopic systems. These processes continued for more than 30 Ma after CAI formation.

8. EPILOGUE

The most accessible source of extraterrestrial material are meteorites. The meteoritic record is read best in an interdisciplinary light. Results of one type of study give insight to another. Early experience gained from meteorite studies provided guidance for proper handling, preservation, and analysis of Apollo Lunar samples. Studies of these samples, in turn, led to the development of extremely sensitive techniques now being used to analyze meteorites and microgram-sized interplanetary dust particles collected by high-flying aircraft or found in Antarctic ice samples (Figure 28.27). The Stardust mission returned samples from a comet, the Genesis spacecraft collected particles of the solar wind, and Hayabusa brought to Earth samples from an asteroid. These samples have been analyzed with techniques developed for and tested with meteorite materials. Undoubtedly, this experience will prove invaluable as samples from other planets, their satellites, and small solar system bodies are brought to Earth for study.

Previous studies of meteorites have provided an enormous amount of knowledge about the solar system, and there is no indication that the scientific growth curve in this area is beginning to level off. Predictions about future developments are very hazardous but we can expect future surprises. New investigators, new ideas, and new techniques will contribute to the exploration of our solar system and—via presolar grains—study also the history of stars. Organic compounds present in carbonaceous chondrites, comets, and interplanetary dust particles may have contributed to the first prebiotic building blocks of life on Earth. Any new meteorite sample may reveal an unknown chapter in the history of the complete solar system.

BIBLIOGRAPHY

Bevan, A. W. R., & de Laeter, J. R. (2002). *Meteorites, a journey through space and time*. Washington, D.C: Smithsonian Institution Press.

Burke, J. G. (1986). *Cosmic debris — Meteorites in history*. Berkeley, CA: Univ. of California Press.

Hutchison, R. (2004). *Meteorites, a petrologic, chemical and isotopic synthesis*. Cambridge, U.K: Cambridge Univ. Press.

Lauretta, D. S., & McSween Jr., H. Y. (Eds.). (2006). *Meteorites and the early solar system II*. Tucson, AZ: Univ. of Arizona Press.

Lodders, K., & Fegley, B., Jr. (2011). *Chemistry of the solar system*. Cambridge, U.K: RSC Publishing.

McSween, H. Y., Jr., & Huss, G. R. (2010). *Cosmochemistry*. Cambridge, U.K: Cambridge Univ. Press.

Norton, O. R. (2002). *The Cambridge encyclopedia of meteorites*. Cambridge, U.K: Cambridge Univ. Press.

Norton, O. R., & Chitwood, L. A. (2008). *Field guide to meteors and meteorites*. London, U.K: Springer.

Papike, J. J. (Ed.). (1998). *Planetary materials*. Washington, D.C., U.S.A: Mineralogical Soc. of America.

Porcelli, D. P., Ballentine, C. J., & Wieler, R. (2002). *Noble gases in geochemistry and cosmochemistry*. Washington, D.C., U.S.A: Mineralogical Soc. of America and Geochemical Soc.

Shih, et al.. (1999). *Meteoritics & Planetary Science, 24*, 647–655.

Smith, C., Russel, A., & Benedix, G. (2009). *Meteorites*. London, U.K: Natural History Museum.

Welten, et al.. (2004). *Meteoritics & Planetary Science, 39*, 481–498.

Chapter 29

Dust in the Solar System

Harald Krüger
Max-Planck-Institut für Sonnensystemforschung, Göttingen, Germany

Eberhard Grün
Max-Planck-Institut für Kernphysik, Heidelberg, Germany and LASP, University of Colorado, Boulder, CO, USA

Chapter Outline

1. Introduction	657
2. Manifestations of Cosmic Dust	659
2.1. Meteors	660
2.2. Interplanetary Dust Particles Collected in the Stratosphere	660
2.3. Zodiacal Light	663
2.4. Lunar Microcraters and the Near-Earth Dust Environment	664
2.5. Comet Dust	665
2.6. Spacecraft Measurements	667
2.6.1. Interplanetary Dust	667
2.6.2. Planetary Dust Streams	669
2.6.3. Tenuous Dusty Planetary Rings	670
2.6.4. Interstellar Dust in the Heliosphere	673
2.7. Extrasolar Debris Disks	674
3. Dynamics and Evolution	675
3.1. Gravity and Keplerian Orbits	675
3.2. Radiation Pressure and the Poynting–Robertson Effect	676
3.3. Collisions	677
3.4. Charging of Dust and Interaction with the Interplanetary Magnetic Field	677
3.5. Evolution of Dust in Interplanetary Space	678
4. Future Studies	678
Bibliography	682

Solar system dust is finely divided particulate matter that exists between the planets. These cosmic dust particles are also often called **micrometeoroids** and range in size from assemblages of a few molecules to tenth-millimeter-sized grains, above which size they are called **meteoroids**. Sources of this dust are larger **meteoroids**, comets, asteroids, the planets and their moons and rings; there is interstellar dust sweeping through the solar system. Because of their small sizes, forces additional to solar and planetary gravity affect their trajectories. Radiation pressure and the interactions with ubiquitous magnetic fields disperse dust particles in space away from their sources. In this way, **micrometeoroids** become messengers of their parent bodies in distant regions of the solar system. A tablespoon of finely dispersed micrometer-sized dust grains scatters about 10 million times more light than a single **meteoroid** of the same mass. Therefore, a tiny amount of dust becomes recognizable, while the parent body from which it derived may remain undetected.

1. INTRODUCTION

One of the earliest known phenomena caused by solar system dust is the **zodiacal light**. Zodiacal light is visible to the human eye in the morning and evening sky in non-polluted areas (Figure 29.1). Already in 1683, Giovanni Domenico Cassini presented the correct explanation for this phenomenon: It is sunlight scattered by dust particles orbiting the Sun. The relation to other "dusty" interplanetary phenomena, like comets, was soon suspected. Comets shed large amounts of dust, visible as dust tails, during their passage through the inner solar system. The genetic relation between **meteors** and comets was already known in the nineteenth century. It turned out that **meteoroids** are the link between interplanetary dust and the larger objects: **meteorites**, asteroids, and comets.

Cosmic dust can have different appearances in different regions of the solar system. It consists not only of **refractory** rocky or metallic material as in stony and iron meteorites, but also of **carbonaceous** material; dust in the outer solar system can even be ice particles.

Individual dust particles in interplanetary space have much shorter lifetimes than the age of the solar system. Several dynamic effects disperse the material in space and size (generally going from bigger to smaller particles). Therefore, interplanetary dust must have contemporary sources, namely, bigger objects like meteoroids, comets,

FIGURE 29.1 A wedge of interplanetary dust. The dusk twilight sky (pink) toward the northwest horizon shows zodiacal light (blue), framed by the Pleiades (upper left), Comet Hale–Bopp (upper right), and Mercury in Aries (left of center above horizon). *Courtesy M. Fulle.*

and asteroids in interplanetary space and also planetary moons and rings. In addition, there are dust particles immersed in the local interstellar cloud through which the solar system currently passes. Such interstellar dust particles penetrate the planetary system.

Dust is often a synonym for dirt, which is annoying and difficult to quantify. This is also true for interplanetary dust. Astronomers who want to observe extrasolar system objects have to struggle removing the foreground scattered light from the zodiacal light. Theoreticians who want to model interplanetary dust have the difficulty of representing these particles by simplified models, for example, a spherical particle of uniform composition and optical properties of a pure material. True interplanetary dust particles (IDPs) can be very different from these simple models (Figure 29.2).

Another practical aspect of dust is its danger to technical systems. A serious concern of the first spaceflights was the hazard from meteoroid impacts. Among the first instruments flown in space were simple dust detectors, many of which were unreliable devices that responded not only to impacts but also to mechanical, thermal, or

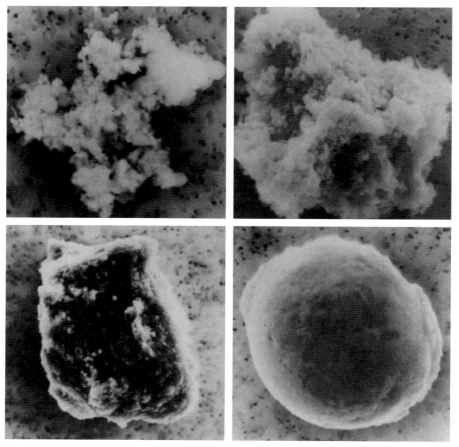

FIGURE 29.2 Interplanetary dust particles collected in the stratosphere by NASA's cosmic dust program. Three grains are of chondritic composition and of various degrees of compactness and there is one Fe–S–Ni sphere. The width of the photographs is 15 μm (upper left, lower right) and 30 μm (upper right, lower left). *NASA photo.*

electrical interference. A dust belt around the Earth was initially suggested, which was dismissed only years later when instruments had developed enough to suppress this noise by several orders of magnitude. Modern dust detectors can reliably measure dust impact rates from a single impact per month up to a 1000 impacts per second.

In the early days of spaceflight, measures were taken to protect spacecraft against the suspected heavy bombardment by meteoroids. The bumper shield concept found its ultimate verification in the European Space Agency's *Giotto* mission to comet Halley. This spacecraft was designed to survive impacts of particles of up to 1 g mass at an impact speed of 70 km/s. These grains carry energies comparable to cannon balls that are 1000 times more massive. Heavy metal armor was not possible because spacecraft are required to be lightweight. The *Giotto* bumper shield combined a 1-mm-thick aluminum sheet positioned 23 cm in front of a 7-cm-thick lightweight composite rear shield. A dust particle that struck the thin front sheet was vaporized or disrupted. The cloud made of vapor and particle fragments then expanded into the empty space between the two sheets and penetrated the rear shield, where its energy was absorbed by being distributed over a large area. In this way, 2.7 m^2 front surface of the spacecraft was effectively protected by armor that weighed only 50 kg.

Only recently has the dust hazard become important again, because of man-made **space debris** in Earth orbit. Each piece of equipment carried into space becomes, after disruption by an explosion due to malfunctioning batteries or fuel systems or by an impact, the source of small projectiles, which endanger other satellites. Some estimates indicate that, in 50 years, the continuous increase in man-made space activity will lead to a runaway effect that will make the near-Earth space environment unhabitable to humans and equipment.

However, we are not concerned with this aspect of interplanetary dust; rather, the topic of this chapter is interplanetary dust as an exciting object of astrophysical research: through its wide distribution over the solar system, cosmic dust can tell stories about its parent bodies (comets, asteroids, moons, even interstellar matter) that are otherwise not easily accessible. This view, however, requires that dust particles be traced back to their origins. To do this, we must understand their dynamics. Dust particles not only follow the gravitational pull of the Sun and the planets but also feel the interplanetary magnetic field and the electromagnetic radiation that fills the solar system. In addition, they interact with the solar wind and with other dust particles that they encounter in space, generally at high speeds. These collisions lead to erosion or to disruption of both particles, thus generating many smaller particles. The dynamics of interplanetary dust cannot be described solely in terms of position and velocity; their size or mass must also be considered.

2. MANIFESTATIONS OF COSMIC DUST

Different methods are available to study cosmic dust (Figure 29.3). They are distinguished by the size or mass range of particles that can be studied. The earliest methods were ground-based zodiacal light and *meteor* observations. About 60 years ago, radar observations of meteor trails became available. With the advent of spaceflight, in situ detection by space instrumentation provided new information on small IDPs. Among the first reliable instruments were simple penetration detectors; modern impact ionization detectors allow not only the detection but also the chemical analysis of **micrometeoroids**. Deep space probes equipped with dust detectors identified micrometeoroids in interplanetary space from 0.3 AU out to 25 AU from the Sun. Natural (e.g. lunar samples) and artificial surfaces exposed to micrometeoroid impacts have been returned from space and analyzed. High-flying aircraft collected dust in the stratosphere that was identified as extraterrestrial material and was analyzed in the laboratory with the most advanced microanalytic tools. Modern space-based infrared observatories now allow the observation of the thermal emission from interplanetary dust in the outer solar system. Beyond our solar system, hundreds of so-called **debris disks** have been identified around other stars by their emission at infrared and longer wavelengths. They are naturally produced dust disks, forming the extrasolar counterparts of our own zodiacal dust cloud. Brightness asymmetries and other structures

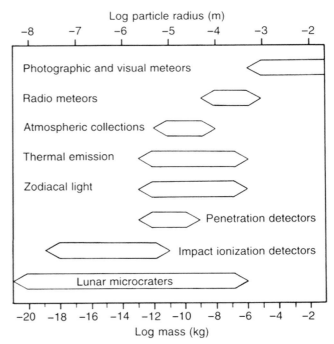

FIGURE 29.3 Comparison of meteoroid sizes and masses covered by different observational methods.

sometimes observed in these disks trace the presence of one or more planets orbiting their central star.

2.1. Meteors

Looking up at the clear night sky, one can record about 10 faint meteors (or shooting stars in colloquial language) per hour. Once in a while, a brighter streak or trail of light or "fireball" will appear. Around the year 1800, the extraterrestrial nature of meteors was established when triangulation was used to deduce their altitude and speed. This technique is still used in modern meteor research by employing specifically equipped cameras and telescopes. About 60 years ago, radar techniques were developed to observe faint meteor trails even during daylight.

Visible meteors result when centimeter-sized meteoroids enter the Earth's atmosphere at a speed greater than 10 km/s. At this speed, the energy of motion, which is converted to heat, is sufficient to totally vaporize the meteoroid. During its deceleration in the atmosphere at about 100 km altitude, the meteoroid heats up and atoms from its surface are ablated until it is completely evaporated. A luminous trail several kilometers in length follows the meteoroid. It is this ionized and luminous atmospheric gas and material from the meteoroid that is visible and that scatters radar signals. From triangulation of the meteor trail by ground stations (several cameras or a radar station), the interplanetary meteoroid orbit can be obtained with high accuracy.

During the atmospheric entry of objects larger than several tens of kilograms or about 10 cm in diameter, a surface layer several centimeters in thickness burns away, and the object gets decelerated. The surviving object which reaches Earth's surface is called a meteorite. Meteorites of 1 kg to several tons are sufficiently decelerated and fall on the Earth with their interior remaining mostly unaltered by atmospheric entry. These meteorites are the source of our earliest knowledge about extraterrestrial material (*see* Meteorites; Near-Earth Objects; Planetary Impacts).

Much of the material ablated from a meteoroid condenses into small droplets, which cool and form cosmic spherules that subsequently rain down to the Earth. These cosmic spherules can be found and identified in abundance in deep-sea sediments and on the large ice masses of Greenland, the Arctic, and Antarctica. An average of 40 tons of extraterrestrial material per day falls onto the surface of Earth in the form of fine dust.

At certain times, meteor showers can be observed at a rate that is a 100 (and more) times higher than the average sporadic meteor rate (Table 29.1). Figure 29.4 shows several meteors in a photograph of the night sky. The visible rate was about one meteor per second. Because all these meteoroids travel on parallel trajectories in space, to an observer on the Earth, they seem to arrive from a common point in the sky (the radiant), which in this case lies in the constellation Leo. Therefore, this meteor shower is called the Leonid shower.

The explanation for the yearly occurrence of meteor showers is that all meteoroids in one stream closely follow the elliptical orbit of their parent comet around the Sun but are spread out all around the orbit. Each year when the Earth crosses this orbit on the same day, some meteoroids of the stream hit the atmosphere and cause the shower.

Many meteor streams have orbits similar to those of known comets (cf. Table 29.1). It is generally accepted that meteor streams are derived from comets. Millimeter- to centimeter-sized particles that are emitted from comets at low speeds (meters per second) are not visible in the normal comet tail but form so-called **comet trails** along a short segment of the comet's orbit. Their different speeds slowly spread the particles out over the full orbit. Infrared observations by the *Infrared Astronomical Satellite* (*IRAS*) identified many such trails connected to short-period comets (Figure 29.5).

Gravitational interactions with planets and collisions with other cosmic dust particles scatter meteoroids out of their stream, and they eventually become part of the sporadic background cloud of meteoroids. The fact that some meteor showers display strong variations of their intensities indicates that they are young streams still concentrated in a small segment of their parent's orbit. The meteor streams of the Leonids and its parent, the periodic comet Tempel–Tuttle, have the same periodicity of 33.3 years. The parent object of one of the strongest yearly meteor showers, the Geminids, is 3200 Phaethon. It was classified as an asteroid because it shows no **cometary** activity. However, its association with a meteor stream indicates that it is an inactive, dead comet that at some time in the past emitted large quantities of meteoroids (*see* Physics and Chemistry of the Comets; Comet Populations and Cometary Dynamics; Near-Earth Objects).

Less than 1 of 10,000 radar meteors has been identified to be caused by interstellar meteoroids that pass through the solar system on hyperbolic orbits. Their heliocentric speed is significantly higher than the solar system escape speed, confirming that they are of truly interstellar origin. The radius of these interstellar meteoroids is about 20 μm. These particles generally arrive from southern ecliptic latitudes with enhanced fluxes from discrete sources.

2.2. Interplanetary Dust Particles Collected in the Stratosphere

There is another "window" through which extraterrestrial material reaches the Earth's surface in a more or less undisturbed state. Small **IDPs** of a few to 50 μm diameter are

TABLE 29.1 Major Meteor Showers, Date of Shower Maximum, Radiant in Celestial Coordinates (Right Ascension, RA, and Declination, DEC, in Degrees), Geocentric Speed (km/s), Maximum Hourly Rate, and Parent Objects (if Known, Short-Period Comets are Indicated by P/)

Name	Date	Radiant RA	DEC	Speed	Rate	Parent Object
Quadrantids	January 3	230	+49	42	140	
April Lyrids	April 22	271	+34	48	10	Comet 1861 I Thatcher
Eta Aquarids	May 3	336	−2	66	30	1P/Halley
June Lyrids	June 16	278	+35	31	10	
S. Delta Aquarids	July 29	333	−17	41	30	
Alpha Capricornids	July 30	307	−10	23	30	45P/Honda–Mrkos–Pajdusakova
S. Iota Aquarids	August 5	333	−15	34	15	
N. Delta Aquarids	August 12	339	−5	42	20	
Perseids	August 12	46	+57	59	400 (1993)	109P/Swift–Tuttle
N. Iota Aquarids	August 20	327	−6	31	15	
Aurigids	September 1	84	+42	66	30	Comet 1911 II Kiess
Giacobinids	October 9	262	+54	20	10	21P/Giacobini–Zinner
Orionids	October 21	95	+16	66	30	1P/Halley
Taurids	November 3	51	+14	27	10	2P/Encke
Taurids	November 13	58	+22	29	10	2P/Encke
Leonids	November 17	152	+22	71	3000 (1966)	55P/Tempel–Tuttle
Geminids	December 14	112	+33	34	70	Phaethon
Ursids	December 22	217	+76	33	20	8P/Tuttle

Source: After Cook, A. F. (1973). In C. L. Hemenway, P. M. Millman, & A. F. Cook (Eds.), *Evolutionary and physical properties of meteoroids* (pp. 183–191). NASA SP-319.

decelerated in the tenuous atmosphere above 100 km. At this altitude, the deceleration is so gentle that the grains do not reach the temperature of substantial evaporation ($T \sim 800\,°C$). The high surface area to mass ratio of these small particles enables them to effectively radiate away excess heat. These dust particles subsequently sediment through the atmosphere and become accessible to collection and scientific examination. The abbreviation IDP (or "Brownlee particle" after Don Brownlee, who first reliably identified their extraterrestrial nature) is often used for such extraterrestrial particles that are collected in Earth's atmosphere.

Early attempts to collect IDPs by rockets above about 60 km were not successful because of the very low influx of micrometeoroids into the atmosphere and the short residence times of IDPs at these altitudes. Airplane collections were more successful in the stratosphere at or above 20 km. At this altitude, the concentration of 10-μm-diameter particles is about 10^6 times higher than in space and terrestrial contamination is still low. Only micrometer- and submicrometer-sized terrestrial particles (e.g. from volcanic eruptions) can reach these altitudes in significant amounts. Another type of interference is caused by man-made contamination: About 90% of all collected particles in the 3 to 8 μm size range are aluminum oxide spheres, which are products of solid rocket fuel exhaust. Because of this overwhelming contamination problem for small particles, the lower size limit of IDPs collected by airplanes is a few micrometers in diameter.

Since 1981, IDP collection by airplanes has been routinely performed by National Aeronautics and Space Administration (NASA) using high-flying aircraft which can cruise at 20 km altitude for many hours. On their wings they carry dust collectors that sweep huge amounts of air. Dust particles stick to the collector surface that is coated with silicone oil. After several hours of exposure, the collector is retracted into a sealed storage container and returned to the laboratory. There, all particles are removed from the collector plate and the silicone oil is washed off. A wide variety of microanalytic tools are used to examine and analyze the IDPs. For example, scanning electron microscopes (SEMs) can image atomic lattice layer structures. Focused ion beams in combination with an SEM are used for sample preparation and secondary ion mass spectrometers can measure the distribution of individual elements and isotopes at submicrometer resolution, deriving important information on the composition of the samples.

According to their elemental composition, IDPs come in three major types: chondritic, 60% (cf. Table 29.2);

FIGURE 29.4 An unusually strong meteor shower (Leonids) was observed on November 17, 1966. The meteor trails seem to radiate from the constellation Leo.

FIGURE 29.5 Celestial structure mapped by IRAS. This false-color map plots the 12-μm fluxes as *blue* and 100-μm fluxes as *red*. The comet Temple 2 dust trail streaks across the bottom half of the image, while the diagonal broad *light blue bar* is the very bright midinfrared emission from the zodiacal dust along the ecliptic plane; the pronounced edges to this feature highlight the inner zodiacal dust bands. The more complex structure is the emission from interstellar dust, which is cool and, therefore, dominantly red away from the ecliptic plane.

TABLE 29.2 Average Elemental Composition (All Major and Selected Minor and Trace Elements) of Several Chondritic Interplanetary Dust Particles (IDPs) is Compared with CI Chondrite Composition

Element	CI	IDP	Variation	T_c
Mg	1,191,000	0.9	0.6–1.1	1067
Si	1,112,000	1.2	0.8–1.7	1311
Fe	1,000,000	1	1	1336
S	572,700	0.8	0.6–1.1	648
Al	94,400	1.4	0.8–2.3	1650
Ca	67,900	0.4	0.3–0.6	1518
Ni	54,800	1.3	1.0–1.7	1354
Cr	15,000	1.1	0.9–1.4	1277
Mn	10,620	1.1	0.8–1.6	1190
Cl	5830	3.6	2.8–4.6	863
K	4200	2.2	2.0–2.5	1000
Ti	2650	1.5	1.3–1.7	1549
Co	2500	1.9	1.2–2.9	1351
Zn	1400	1.4	1.1–1.8	660
Cu	580	2.8	1.9–4.2	1037
Ge	132	2.3	1.6–3.4	825
Se	69	2.2	1.6–3.0	684
Ga	42	2.9	2.1–3.9	918
Br	13	34	23–50	690

The CI abundance is normalized to Fe = 1,000,000; the IDP abundances are normalized to iron (Fe) and to CI. Condensation temperatures T_c (°C).
Source: After Jessberger, E. K., et al. (1992). *Earth planet. Science Letters*, 112, 91.

iron–sulfur–nickel, 30%; and mafic silicates (iron-magnesium-rich silicates, i.e. olivine and pyroxene), 10%. Most chondritic IDPs are porous aggregates, but some smooth chondritic particles are found as well. Chondritic aggregates may contain varying amounts of carbonaceous material of unspecified composition. Table 29.2 shows a significant enrichment of volatile (low-condensation-temperature) elements in IDPs when compared to CI chondrites. This observation is used to support the argument that these particles consist of some very primitive solar system material that had never seen temperatures above about 500 °C, as is the case for some cometary material. This and the compositional similarity with comets argue for a genetic relation between comets and chondritic IDPs.

A remarkable feature of IDPs is their large variability in isotopic composition. Extreme isotopic anomalies have been found in some IDPs, whereas under typical solar system conditions, isotopic variations of only fractions of percent can occur. These huge isotopic variations indicate that some grains were not homogenized with other solar system material but have preserved much of their presolar character. Submicrometer-sized grains known as GEMS (glass with embedded metal and sulfides) are major constituents of the chondritic porous class of IDPs. Several GEMS with nonsolar oxygen isotopic compositions were identified, confirming that at least some are indeed presolar grains. These **amorphous** interstellar silicates are considered one of the fundamental building blocks of the solar system.

2.3. Zodiacal Light

The wedge-shaped appearance of the zodiacal light (see Figure 29.1) demonstrates its concentration in the **ecliptic plane**. For an observer on the Earth, the zodiacal light extends along the ecliptic plane all the way around the sky to the antisolar direction, although at strongly reduced intensities. In the direction opposite to the Sun, this light forms a hazy area of a few degrees in dimension known as the gegenschein, or counterglow. If seen from the outside, the zodiacal dust cloud would have a flattened, lenticular shape that extends along the ecliptic plane about seven times farther from the Sun than perpendicular to the ecliptic plane.

The brightness of the zodiacal light is the result of light scattered by a huge number of particles in the direction of observation. The observed zodiacal brightness is a mean value, averaged over all sizes, compositions, and structures of particles along the line of sight. Zodiacal light brightness can be traced clearly into the solar corona. However, most of this dust is foreground dust close to the observer because of a favorable scattering function. Nevertheless, the vicinity of the Sun is of considerable interest for zodiacal light measurements because it is expected that close to the Sun the temperature of the dust rises, and the dust particles start to sublimate, first the more volatile components and closer to the Sun, even the refractory ones. Inside about four solar radii distance, dust should completely sublimate. Some observers indeed found a sharp edge of a dust-free zone at four solar radii, while others did not see such a sharp edge. Perhaps the inner edge of the zodiacal cloud changes with time.

The large-scale distribution of the zodiacal dust cloud is obtained from zodiacal light measurements on board interplanetary spacecraft spanning a distance ranging from 0.3 to approximately 3 AU from the Sun. Even though the intensity decreases over this distance by a factor of 150, the spatial density of dust needs to decrease by only a factor of 15. The radial dependence of the number density is slightly steeper than an inverse distance dependence. A slight **inclination** of about 3° of the symmetry plane of zodiacal light with respect to the ecliptic plane has been determined from zodiacal light measurements.

At visible wavelengths, the spectrum of the zodiacal light closely follows the spectrum of the Sun. A slight reddening (i.e. the ratio of red to blue intensities is larger for zodiacal light than for the Sun) indicates that the majority of particles are larger than the mean visible wavelength of 0.54 μm. In fact, most of the zodiacal light is scattered by 10- to 100-μm-sized particles. Therefore, the dust seen as zodiacal light is only a subset of the interplanetary dust cloud. Submicrometer- and micrometer-sized particles, as well as millimeter-sized and bigger particles, do not contribute much to the zodiacal light at optical wavelengths but they exist.

Above about 1 μm in wavelength, the intensities in the solar spectrum rapidly decrease. The zodiacal light spectrum follows this decrease up to about 5 μm, while at longer wavelengths the thermal emission of the dust particles prevails. Because of the low albedo (fraction of incident sunlight reflected and scattered in all directions is smaller than 10%) of IDPs, most visible radiation (>90%) is absorbed and emitted at infrared wavelengths. The maximum of the thermal infrared emission from the zodiacal dust cloud lies between 10 and 20 μm. From the thermal emission observed by the IRAS and Cosmic Background Explorer (*COBE*) satellites, an average dust temperature at 1 AU distance from the Sun between 0 and 20 °C has been derived. Some spatial structure has been observed at thermal infrared wavelengths. Asteroid bands mark several **asteroid families** as significant sources of solar system dust just as comet trails identify dust emitted from individual comets.

Optical and infrared observations of other extraterrestrial dusty phenomena have also provided important insights into the zodiacal dust complex. Cometary and **asteroidal** dust is considered to be an important source of the zodiacal cloud. The study of circumplanetary dust and rings has stimulated much research on the dynamics of dust clouds. Interstellar dust is believed to be the ultimate source of all refractory material in the solar system. Circumstellar dust clouds like the one around β Pictoris are "zodiacal clouds" of their own right, the study of which may eventually give information on extrasolar planetary systems (*see* Infrared Views of the Solar System from Space; Planetary Rings; Extra-Solar Planets).

2.4. Lunar Microcraters and the Near-Earth Dust Environment

The size distribution of IDPs is represented by the lunar microcrater record. Microcraters on lunar rocks were found ranging from 0.02 μm to millimeters in diameter (Figure 29.6). Laboratory simulations of high-velocity impacts on lunarlike materials were performed to calibrate crater sizes with projectile sizes and impact speeds. Submicrometer- to centimeter-sized projectiles were used with speeds above several kilometers per second. The typical impact speed of interplanetary meteoroids on the Moon is about 20 km/s. For the low-mass particles, electrostatic dust accelerators were used that reach projectile speeds of up to 100 km/s. The high-mass projectiles were accelerated with light-gas guns, which reached speeds up to about 10 km/s. For the intermediate mass range, plasma drag accelerators reached impact speeds of 20 km/s. The crater diameter to projectile diameter ratio varies from 2 for the smallest microcraters to about 10 for centimeter-sized projectiles.

The difficulty in deriving the impact rate from a crater count on the Moon is that the degree to which rocks shield other rocks and thus the exposure time of any surface is generally unknown. Therefore, the crater size or meteoroid distribution has to be normalized with the help of an impact rate or meteoroid flux measurement obtained by other means. In situ detectors or recent analyses of impact plates that were exposed to the meteoroid flux for several years on NASA's *Long Duration Exposure Facility* provided this flux calibration (Figure 29.7). The smallest particles dominate the flux, and the mass flux of meteoroids peaks at 10^{-5} g. The total mass density of interplanetary dust at 1 AU is 10^{-16} g/m^3 and the total mass of the zodiacal cloud inside Earth's orbit is between 10^{16} and 10^{17} kg, which corresponds to the mass of a single object (comet or asteroid) of about 20 km in diameter.

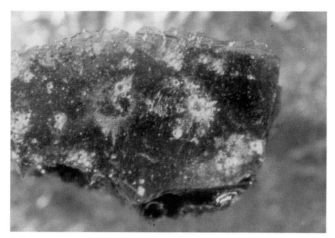

FIGURE 29.6 Microcraters on the glassy surface of a lunar sample. Bright spallation zones surround circular central pits.

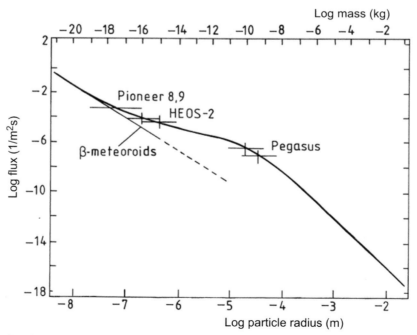

FIGURE 29.7 Cumulative flux of interplanetary meteoroids on a spinning flat plate at 1 AU distance from the Sun. The solid line has been derived from lunar microcrater statistics and it is compared with satellite and spaceprobe measurements.

In low Earth orbit the meteoroid flux is about a factor of two higher than in deep space because of the Earth's gravitational concentration. However, man-made space debris outnumbers micrometer-sized natural meteoroids (by a factor of three). Craters produced by space debris particle impacts are identified by chemical analyses of residues in the craters. Residues have been found from space materials and signs of human activities in space, such as paint flakes, plastics, aluminum, titanium, and human excretion.

2.5. Comet Dust

The inner **coma** of a comet (*see* Physics and Chemistry of Comets) is one of the most dust-rich environments in the solar system. Almost everything we see from a comet with the naked eye is dust. Both the coma and the tail are seen as sunlight scattered by micron-sized dust (cf. Figure 29.1). Particles ejected from a comet form different dust populations. Submicron-sized grains have high ejection velocity (~ 1 km/s), and on ejection from the comet they may rapidly assume hyperbolic orbits and escape from the solar system. Grains of size 1–100 µm are ejected at a speed of several hundred meters per second and under the action of solar radiation pressure they form the comet tail which disperses them far from the comet orbit within a short time period. Even bigger particles stay in a trail close to the comet's orbit. If the Earth crosses such a trail, it is observed as a meteor stream (Table 29.1).

One of the problems in characterizing the dust environment of a comet is that information on the nucleus, its dust, and gas release is very limited. Before 1986, observations of cometary dust were the domain of astronomers. High-resolution images of cometary comae revealed jets and other structures in the inner parts. Some of these structures formed spirals which rotated like water from a lawn sprinkler, indicating discrete dust emissions from localized active parts of the nucleus surface. A consequence of observing in visible light is that the results are biased by particle sizes in the range of 1–10 µm, because much smaller and much larger particles do not contribute significantly to the scattered light in the visible range. With the extension of the observable spectral range to infrared wavelengths using space-based telescopes, the thermal emission of dust also became accessible to astronomers. It revealed information on the abundance of larger grains and on the **mineralogical** composition of the dust.

A breakthrough in understanding cometary constituents came with space missions to several comets: Giotto and two VeGa spacecraft to comet 1P/Halley in 1986, Deep Space 1 to comet 19P/Borelly in 1999, Stardust to comet 81P/Wild 2 in 2004, Deep Impact to comets 9P/Tempel 1 in 2005 and 103P/Hartley 2 in 2010, and most recently Stardust to 9P/Tempel 1 in 2011. Water and CO were identified as the main species in the gas, and dust particles made of carbonaceous and silicate materials ranging from nanometer to millimeter sizes were detected. Active areas on resolved images of cometary nuclei and corresponding dust jets were identified for some of the visited comets.

The Stardust mission was the first space mission designed to return extraterrestrial material from interplanetary space. The primary goal of Stardust was to collect dust samples during its flyby of comet Wild 2. The spacecraft flew through the coma at a speed of 6.1 km/s within 236 km from the nucleus and comet particles were collected in **aerogel** of 50 and 20 kg/m^3 density. The collector consisted of 0.1 m^2 aerogel and of 0.015 m^2 aluminum foil. Micrometer-sized dust particles were decelerated gradually by the aerogel, forming several millimeter long tracks, and minimizing the damage to the dust grains. The collector was stored in a Sample Return Capsule which was released from the spacecraft just before reentry into Earth's atmosphere, for a landing on a parachute.

In January 2006, the Stardust sample capsule returned safely to the Earth with thousands of particles from comet 81P/Wild 2 for laboratory study. Impact tracks in aerogel created by particles ranging from dense **mineral** grains to loosely bound, polymineralic aggregates ranging from 0.01 to 100 µm in size displayed diverse impact features. Residues in impact craters on the structure supporting the aerogel were also analyzed.

The collected particles are chemically heterogeneous; however, the mean elemental composition of comet Wild 2 particles is consistent with CI meteorite composition (cf. chapter meteorites). The particles are weakly constructed mixtures of nanometer-scale grains with occasionally much larger Fe–Mg silicates, Fe–Ni sulfides, and Fe–Ni metal phases. A very wide range of olivine and low-Ca pyroxene compositions was also found.

The collected samples are dominated by high-temperature materials that closely resemble meteoritic components. These materials include chondrule and **calcium-aluminum-rich inclusion (CAI)**-like fragments (Figure 29.8). The abundance of high-temperature minerals such as forsterite and enstatite appears to have formed in the hot inner regions of the solar nebula. From there they were transported beyond the **orbit** of Neptune where they accreted together with ice and organic components to form comet Wild 2.

Hydrogen, carbon, nitrogen, and oxygen isotopic compositions are heterogeneous among particle fragments; however, extreme isotopic anomalies are rare, indicating that this comet is not a pristine aggregate of presolar materials. An extreme oxygen ratio $^{17}O/^{16}O = 10^{-3}$ was found which is a factor 2.6 higher than the Solar System value and is similar to that of some presolar grains found in meteorites. Only five presolar grains have been discovered in the Wild 2 samples so far. The presolar grain content appears to be lower than in chondrites and in most IDPs.

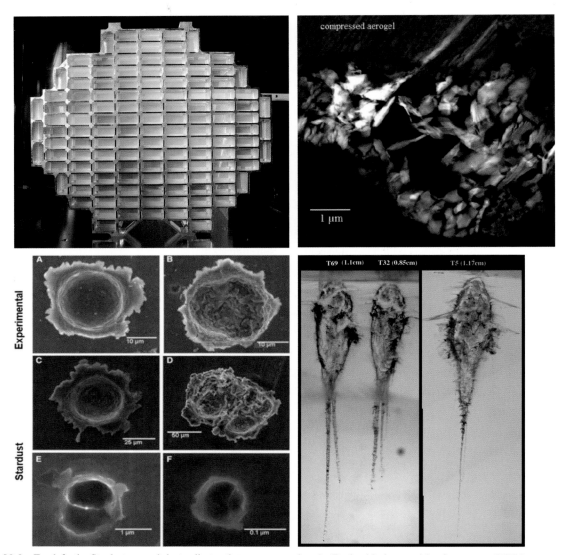

FIGURE 29.8 Top left: the Stardust aerogel dust collector that was returned to the Earth with dust particles from comet Wild 2 on one side of the collector and with interstellar dust particles on the other. Top right, a CAI particle found in the Stardust collection. Bottom left, comparison of experimental and Stardust impacts into the aluminum structure of the Stardust collectors. Bottom right, tracks of cometary particles in aerogel. The particles approached from the top. *Courtesy NASA, Science.*

The organics are rich in oxygen and nitrogen compared with **meteoritic** organics. Aromatic compounds are present but less abundant than in meteorites and IDPs. The organics found in comet Wild 2 shows a heterogeneous and unequilibrated distribution in abundance and composition. Even glycine, a fundamental building block of life, was found in samples of comet Wild 2 which supports the idea that these precursors of life are prevalent in space (*see* Physics and Chemistry of Comets; Meteorites).

Small bodies in the inner solar system were traditionally classified as either asteroids or comets. Comets are active objects, with activity showing up as a prominent coma and tail(s) when they approach the Sun. Cometary activity is largely driven by the sublimation of near-surface water ice. On the contrary, asteroids lack such signatures of activity. Recently, a small number of intermediate-type objects were discovered which are called "main belt comets" or "active asteroids". Dynamically their orbits resemble those of asteroids, while they show clear signs of activity similar to comets. The activity is usually evidenced by the light scattering off dust particles emitted from the object, forming a comalike structure and sometimes a tail. Their mass loss, however, is likely driven by a surprising diversity of mechanisms. Besides sublimation of water ice, impact ejection, rotational instability due to spin up, dehydration stresses and thermal fracture, as well as electrostatic repulsion and solar radiation pressure sweeping of dust particles from the surfaces of these objects have been suggested. Even though no single mechanism can explain the varied examples of activity observed, coma and tail morphology as evidenced by the emitted dust particles can give valuable information about the dominant processes. Coma and tail structures may constrain the duration of activity, e.g. whether the emission event was impulsive due to an impact or long lasting or even repetitive. Astronomers may also be able to constrain particle sizes and ejection velocities from the coma and tail morphology (*see* Main-Belt Asteroids; Comet Populations and Cometary Dynamics).

2.6. Spacecraft Measurements

In situ measurements by dust impact detectors on board interplanetary spacecraft are complementary to ground-based and astronomical dust observations. In situ measurements have been performed in interplanetary space between 0.3 and 25 AU heliocentric distance (Table 29.3).

Two types of impact detectors were mainly used for interplanetary dust measurements: penetration detectors and impact ionization detectors. Penetration detectors record the mechanical destruction from a dust particle's impact, e.g. the penetration of a 25- or 50-μm-thick steel film has a detection threshold of 10^{-9} or 10^{-8} g (approximately 10 or 20 μm particle radius) at a typical impact speed of 20 km/s. At lower impact speeds the minimum detectable particle mass is bigger, and vice versa. A more sensitive penetration detector is the polyvinylidene fluoride (PVDF) film. PVDF is a polarized material (i.e. all dipolar molecules in the material are aligned so that they point in the same direction). When a dust particle impacts the film, it excavates some polarized material. This depolarization generates an electric signal which is then detected. The pulse height of the signal is a function of the mass and the speed of the dust particle. A typical measurement range is from 10^{-13} to 10^{-9} g (approximately 1–10 μm radius).

The most sensitive dust detectors are impact ionization detectors. Figure 29.9 shows a photograph of the dust detector flown on the *Cassini* spacecraft. The detector has an aperture of 0.1 m^2 and is based on the impact ionization effect: a dust particle that enters the detector and hits the hemispherical target in the back at speeds above 1 km/s produces an impact crater and part or all of the projectile's material is vaporized. Because of the high temperature at the impact site, some electrons are stripped off atoms and molecules and generate a vapor that is partially ionized. These ions and electrons are separated in an electric field within the detector and are collected by electrodes. Coincident electric pulses on these electrodes signal the impact of a high-velocity dust particle. The strength and the waveform of the signal are measures of the mass and speed of the impacting particle. The small central part of the Cassini detector is a time-of-flight mass spectrometer: a strong electric field between the target and a grid 3 mm in front of the target accelerates the ions to high speeds. During the flight between the grid and the ion collector, ions of different masses separate and arrive at different times at the multiplier. The lightest ions arrive first and heavier ones appear later. In this way, a mass spectrum is measured which represents the elemental composition of the dust grain. Entrance grids in front of the target pick up any electric charge of dust particles. Measurements of the electric charge carried by IDPs have been accomplished for the first time by the Cassini detector. Dust detectors incorporating a mass spectrometer were flown on the *Helios* spacecraft, the Giotto and VeGa missions to Comet Halley, the Stardust mission to Comet Wild 2, and the Cassini mission to Saturn. Electrostatic dust accelerators are used to calibrate these detectors with micrometer- and submicrometer-sized projectiles at impact speeds up to 100 km/s.

2.6.1. Interplanetary Dust

The radial profile of the dust flux in the inner solar system between 1 and 0.3 AU from the Sun was determined by the *Helios 1* and 2 space probes. Three dynamically different interplanetary dust populations were identified in the inner solar system. First, particles on low-eccentricity orbits about the Sun had already been detected by the *Pioneer 8* and *9* and *Highly Eccentric Orbit Satellite (HEOS) 2* dust experiments. They relate to particles originating in the asteroid belt that spiral toward the Sun under the **Poynting–Robertson** effect. Second, there are particles on highly eccentric orbits that have, in addition, large semimajor axes and that derive from short-period comets. Third, the *Pioneer 8* and *9* dust experiments detected a significant flux of small particles from approximately the solar direction which were called beta meteoroids. The existence of these particles was recently confirmed by measurements with the Japanese *Hiten* satellite.

Recently, the *Galileo* and *Ulysses* spacecraft carried dust detectors through interplanetary space between the orbits of Venus and Jupiter and above the ecliptic plane. In a swingby of Jupiter, the Ulysses spacecraft was brought into a heliocentric orbit almost perpendicular to the ecliptic plane that carried it under the South Pole, through the ecliptic plane, and over the North Pole of the Sun. The Galileo spacecraft always remained close to the ecliptic plane and became the first man-made satellite of Jupiter.

Interplanetary dust measurements were obtained by the Galileo spacecraft in the ecliptic plane between Venus orbit and the asteroid belt. The dust impact rate was generally higher closer to the Sun than farther away. After flybys at Earth and Venus, the spacecraft moved away from the Sun. At these times, the impact rate was more than an order of magnitude higher than before the flyby when the spacecraft moved toward the Sun. This observation is explained by the

TABLE 29.3 In situ Dust Detectors and Dust Analyzers Flown in Interplanetary Space (A) and on Cometary Missions (B): Distance of Operation, Mass Sensitivity, Sensitive Area, and Compositional Resolution

(A)

Mission	Launch Year	Distances (AU)	Mass Threshold (g)	Area (m^2)
Pioneer 8	1967	0.97–1.09	2×10^{-13}	0.0094
Pioneer 9	1968	0.75–0.99	2×10^{-13}	0.0074
HEOS 2	1972	1	2×10^{-16}	0.01
Pioneer 10	1972	1–18	2×10^{-9}	0.26
Pioneer 11	1973	1–10	10^{-8}	0.26
Helios 1/2	1974/76	0.3–1	10^{-14}	0.012
Galileo	1989	0.7–5.3	10^{-15}	0.1
Hiten	1990	1	10^{-15}	0.01
Ulysses	1990	1–5.4	10^{-15}	0.1
Cassini CDA	1997	0.7–10	2×10^{-16}	0.1
Cassini HRD	1997	0.7–10	3×10^{-13}	0.0006
Nozomi	1998	1–1.5	10^{-15}	0.01
New Horizons	2006	2.6–25 (February 2013)	2×10^{-12}	0.11
Interplanetary Kite-craft Accelerated by Radiation Of the Sun (IKAROS)	2010	0.72–1.1	10^{-9}	0.54
Lunar Atmosphere and Dust Environment Explorer (LADEE)	2013	1	2×10^{-16}	0.01

(B)

Mission Instrument	Encounter Year	Target Comet	Mass Threshold (g)	Area (m^2)	Compositional Resolution $M/\Delta M$
Giotto Dust Impact Detection System (DIDSY)	1986 1992	Halley, Grigg–Skjellerup	4×10^{-9}, 10^{-6}	2, 0.7	
Giotto Particle Impact Analyzer (PIA)	1986	Halley	2×10^{-15}	0.0005	100
VeGa 1/2 Dust Counter and Mass Analyzer (DUCMA)	1986	Halley	10^{-11}	0.0075	
VeGa 1/2 Particle Impact Mass Analyzer (PUMA) (translated from Russian)	1986	Halley	2×10^{-15}	0.0005	100
VeGa 1/2 Solid Particle Experiment (SP)-1	1986	Halley	2×10^{-15}	0.0081	
VeGa 1/2 SP-2	1986	Halley	10^{-11}	0.05	
Stardust aerogel collector	2004	Wild 2	$\sim 10^{-12}$	0.1	>1000 (in laboratory)

TABLE 29.3 In situ Dust Detectors and Dust Analyzers Flown in Interplanetary Space (A) and on Cometary Missions (B): Distance of Operation, Mass Sensitivity, Sensitive Area, and Compositional Resolution—cont'd

(B)

Mission Instrument	Encounter Year	Target Comet	Mass Threshold (g)	Area (m^2)	Compositional Resolution $M/\Delta M$
Stardust Cometary and Interstellar Dust Analyzer (CIDA)	2004 2011	Wild 2 Tempel 1	2×10^{-15}	0.009	200
Stardust Dust Flux Monitor Instrument (DFMI)	2004 2011	Wild 2 Tempel 1	$10^{-5}, 10^{-12}$ $10^{-8}, 3 \times 10^{-12}$	0.7, 0.002	
Rosetta Grain Impact Analyzer and Dust Accumulator (GIADA)	2014	Churyumov−Gerasimenko	10^{-7}	0.01	
Rosetta Cometary Secondary Ion Mass Analyzer (COSIMA)	2014	Churyumov−Gerasimenko	10^{-7}	0.0003	2000
Rosetta Micro-Imaging Dust Analysis System (MIDAS)	2014	Churyumov−Gerasimenko	10^{-16}	10^{-5}	
Rosetta/Philae Surface Electric Sounding and Acoustic Monitoring Experiment-Dust Impact Monitor (SESAME-DIM)	2014	Churyumov−Gerasimenko	10^{-4}	0.006	

fact that interplanetary dust inside the asteroid belt orbits the Sun on low-inclination (<30°) and in low-eccentricity **bound orbits**. Thus, the detector that looked away from the Sun all the time detected more dust impacts when the spacecraft moved in the same direction (outward) than in the opposite case when the spacecraft moved inward. The spatial dust density follows roughly an inverse radial distance dependence. Close passages of the asteroids Gaspra and Ida did not exhibit increased dust impact rates.

In the outer solar system, the dust detectors on board *Pioneers 10* and *11*, Galileo, Ulysses, Cassini, and *New Horizons* measured the flux of IDPs. The flux of micrometer-sized particles decreased from 1 AU going outward. No sign of a flux enhancement was detected in the asteroid belt. Outside Jupiter's orbit, *Pioneer 10* recorded a flat flux profile (Figure 29.10), which indicates a constant spatial density of micrometer-sized dust in the outer solar system. This observation has been interpreted to be due to the combined input of dust from the **Kuiper belt** and comets like Halley and Schwassmann−Wachmann 1.

2.6.2. Planetary Dust Streams

Inside a distance of about 3 AU from Jupiter, both the Ulysses and the Galileo spacecraft detected unexpected swarms of submicrometer-sized dust particles arriving from the direction of Jupiter. Figure 29.11 shows the strongly time-variable dust flux observed by Ulysses during its first flyby of Jupiter in 1992. About 1 month after closest approach to Jupiter, Ulysses encountered the most intense dust burst at a distance of about 40 million km from Jupiter. For about 10 h, the impact rate of submicrometer-sized particles increased by a factor of 1000 above the background rate. The measurements indicated that the particles in the burst were moving in collimated streams at speeds of several 100 km/s. Even stronger and longer lasting dust streams were observed in 1995 by the Galileo dust detector during its approach to Jupiter. Later, dust measurements inside the Jovian magnetosphere showed a modulation of the small particle impact rate with a period of 10 h, which is the rotation period of Jupiter and its magnetic field. Positively charged dust particles in the 10-nm-size range are coupled to the magnetic field and are thrown out of Jupiter's magnetosphere in the form of a warped dust sheet. Sources of these dust particles are the volcanoes on Jupiter's moon Io and, to a smaller extent, Jupiter's ring. During Cassini's flyby of Jupiter, this phenomenon was also observed, and mass spectra of the particles were obtained (Figure 29.12(a)). Both sodium chloride and sulfurous components were identified in the mass spectra, which is consistent with spectral measurements of Io's volcano-induced environment. In 2004 during its second flyby at Jupiter, Ulysses measured the dust streams even to a distance of 4 AU from Jupiter.

At Saturn, Cassini observed dust streams emanating from this system as well. Here, Saturn's dense A ring and the extended E ring were identified as sources. The ejection mechanism is very similar to that acting at Jupiter. Freshly generated nanometer-sized dust grains get charged and—if the charge is positive—thrown out by Saturn's magnetic

FIGURE 29.9 The Cassini Cosmic Dust Analyzer consists of two types of dust detectors, the High Rate Dust Detector, (HRD), and the Dust Analyzer, (DA). The cylindrical DA (upper center) has a diameter of 43 cm. The bottom of the sensor contains the hemispherical impact target; in the center are charge-collecting electrodes and the multiplier for the measurement of the impact mass spectrum. Two entrance grids sense the electric charge of incoming dust grains. The detector records impacts of sub-micron- and micron-sized dust particles above 1 km/s impact speed. HRD consist of two circular film detectors which record impacts of micron-sized dust particles at a rate of 10,000 per second. The detectors are carried by the electronics box which is mounted on top of a turntable bolted to the spacecraft.

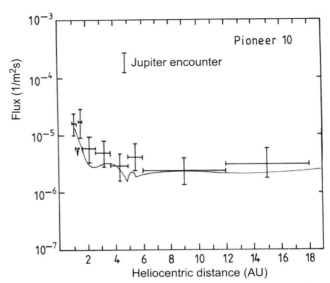

FIGURE 29.10 Flux of meteoroids with masses above 8×10^{-10} kg (about 10 μm in size) in the outer solar system measured by the Pioneer 10 penetration detector. At 18 AU distance from the Sun, the instrument quit operation. The measurements are in agreement with a model of constant spatial dust density in the outer planetary system. *From Humes (1980).*

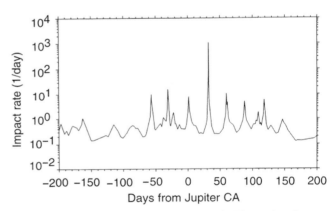

FIGURE 29.11 Dust impact rate observed by the Ulysses dust detector during 400 days around Jupiter closest approach (CA, February 8, 1992). At the beginning and end of the period shown, Ulysses was 240 million km (1.6 AU) from Jupiter, while at CA, the distance was only 450,000 km. Except for the flux peak at CA, when bigger particles were detected, the peaks at other times consisted of submicron-sized dust particles.

field. In some parts of the magnetosphere, dust particles become negatively charged; these particles remain bound to the magnetic field and stay in the vicinity of Saturn. The Saturnian stream particles primarily consist of silicate materials (Figure 29.12(b)) implying that the particles are the contaminations of icy ring material rather than the ice particles themselves (*see* Planetary Rings).

2.6.3. Tenuous Dusty Planetary Rings

All four giants planets have their own ring systems. Of these, Saturn's ring is by far the most prominent and most well known. In addition to its dense main ring system that is described in another chapter of this book (*see* Planetary Rings), Saturn also hosts a set of diffuse faint rings in and around its main rings. Similarly Jupiter, the largest of the giant planets, also has its own gossamer ring system. The dusty rings of both planets are composed primarily of particles smaller than 100 μm in radius. Interparticle collisions are rare in these tenuous rings, and the small sizes of the particles make them sensitive to nongravitational forces. Jupiter's and Saturn's dusty rings were studied in situ by dedicated dust detectors on board the Galileo and Cassini spacecraft and with Earth- and space-based imaging. This offered unique opportunities to combine in situ and remote imaging observations for the first time. The rings of Uranus and Neptune have not yet been studied with in situ dust detectors and are not covered in this chapter.

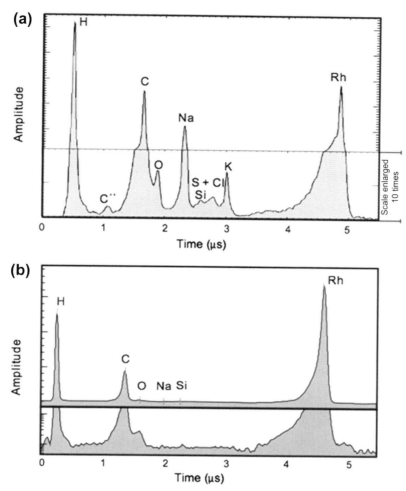

FIGURE 29.12 Cassini CDA mass spectra of (a) Jovian and (b) Saturnian stream particles. Due to the high impact velocity and small impactor mass, stream particle spectra are usually dominated by target and contaminant ions. (a) The mass spectrum is co-added from 30 Jovian stream particle spectra recorded upon Cassini's approach to Jupiter in 2000. The Na^+, K^+, S^+, Cl^+, and partly Si^+ and O^+ lines are particle constituents. (b) Saturnian stream particle mass spectrum recorded in 2004. The lower panel is on a logarithmic scale to show the weak lines. The major peaks in the spectrum are target (Rh^+) and target contamination ions (H^+, C^+). *From Postberg et al. (2009).*

2.6.3.1. Jupiter's Gossamer Rings

Jupiter's ring system consists of at least five distinct components: the main ring projecting inward from the orbits of the two moonlets Metis and Adrastea, the vertically extended halo interior to the main ring, two gossamer rings associated with the small moons Amalthea and Thebe, and a faint outward protrusion called the Thebe extension (Figure 29.13). The last of these is visible in edge-on ring images out to at least 270,000 km from the giant planet. During two passages of the ring system, the in situ dust detector on board the Galileo spacecraft measured dust impacts further away out to at least 360,000 km. The measured ring structure implies that the ring is maintained by dust particles released from the small moons embedded in the ring via impacts of micrometeoroids onto the surfaces of these moons and that evolve inward under Poynting–Robertson drag. The grain size distribution measured in situ ranges from 0.2 to 5 μm, with a steep increase of the particle number density toward smaller grains. In the ring region inside Amalthea's orbit, the grain size distribution was derived from in situ measurements and from imaging, and both agreed very well.

Observed ring structures, in particular gaps in the particle spatial density, large grain **orbit inclinations**, and the radial outward extension of the ring can be explained by electromagnetic forces having a strong influence on particle dynamics. Particles orbiting the giant planet attain variable electric charges on the day- and nightside of the planet. On the dayside of Jupiter, photoelectric charging by solar radiation dominates over charging from the ambient plasma in Jupiter's magnetosphere. When a dust grain enters Jupiter's shadow, photoelectric charging switches off, and the grain's

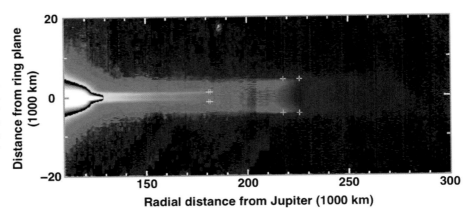

FIGURE 29.13 Jupiter's diffuse ring system seen edge-on. The main ring and the halo interior to the main ring are shown in white, the Amalthea gossamer ring is shown in yellow, and the Thebe ring in red. Crosses mark the extremes of the radial and vertical motions of the two small moons Amalthea and Thebe caused by their eccentric and inclined orbits. The Thebe extension is the material located outside Thebe's orbit (shown in blue). *False color; adapted from Burns et al. (1999).*

electric potential decreases. This leads to an oscillating particle charge due to the switch on and off of photoelectric charging on the day- and nightside of the planet (shadow resonance). It changes the electromagnetic force acting on the particle and results in coupled oscillations of the orbital eccentricity and semimajor axis of the particle's orbit. The oscillations cause the rings to extend significantly outward, but only slightly inward, of their source moons while preserving their vertical thicknesses. Electromagnetic forces seem to be crucial for determining the structure and dust transport in Jupiter's tenuous gossamer rings.

2.6.3.2. Saturn's Dusty Rings

Saturn's dusty rings are located in and around the planet's dense main rings. There are diffuse rings inside the main ring, dusty ringlets in various gaps in the main ring, B ring spokes, the narrow G ring, dusty ring arcs, and finally the very extended E ring. The last of these is maintained by particles ejected from Saturn's volcanically active moon Enceladus (Figure 29.14). Much farther from the planet, a dusty ring was recently discovered stretching along the orbit of the moon Phoebe that orbits Saturn on an eccentric orbit between 4 and 13 million km distance from Saturn. Only the B ring spokes and the E ring are discussed here.

2.6.3.2.1. B Ring Spokes Spokes are intermittent, approximately radial markings in Saturn's B ring. These perplexing features were first recognized in images taken by *Voyager 1* and *2*. Later, Hubble Space Telescope (HST) observations showed that spoke formation can be suspended for extended periods of time, implying that they are a seasonal phenomenon. Remote sensing observations from the early Cassini mission (2004–2008) showed spokes that were much fainter and less frequent than during the *Voyager* era, while by late 2008, spoke activity had recovered to values observed by the *Voyagers*.

Several spoke formation theories have been proposed, although there is still no generally accepted model for how they are triggered. The most widely accepted are those of meteoroid impacts onto the rings and field-aligned electron beams originating from the auroral regions of Saturn: both could transiently increase the plasma density above a critical threshold and trigger the formation of spokes. Seasonal variations in spoke activity may be a consequence of the variable plasma density near the ring as a function of the solar elevation angle above the ring plane. Other spoke formation scenarios include dusty plasma waves and impact-induced avalanches of small charged dust particles. The electrostatic charging of grains may also be linked to electrical storms in the atmosphere of Saturn itself.

2.6.3.2.2. E Ring The E ring stretches from about 180,000 km out to at least 700,000 km from Saturn. Already in the late 1970s it was recognized that in edge-on images the maximum brightness occurs near Enceladus' orbit, implying that this icy moon must be the dominant source of the ring particles. Since then the ring was extensively studied with ground- and space-based imaging and more

FIGURE 29.14 An arc of Saturn's E ring with the moon Enceladus (dark spot) ejecting dust particles into the ring. In this image the faint ring appears especially bright because the small particles that mostly populate the ring scatter light very efficiently in this observing geometry.

recently by the remote sensing instruments on board the Cassini spacecraft. In situ spacecraft measurements provide a complementary view by measuring dust impacts during passages through the ring. The first to identify dust impacts in the E ring were the radio and plasma wave instruments on board *Voyager*, as the spacecraft traversed the E ring in 1980. The Cosmic Dust Analyzer (CDA) on board Cassini was the first dedicated dust instrument to investigate the local properties of the E ring in situ. CDA measured the spatial and size distributions of the dust particles, their charge state, as well as their chemical composition.

The Cassini measurements revealed that Enceladus is a geologically active moon (*see* Enceladus). Plume particles launched from the vents at Enceladus' south pole are the likely primary source for most of the E ring (Figure 29.14). Plume particles condense from water vapor expanding inside fractures in the moon's surface, while wall collisions within the vents likely play a role in establishing the speed and size distribution of the escaping grains. The overall ring structure was derived from CDA measurements and can be described by a set of power laws. The vertical profiles are in good agreement with those derived from plasma wave measurements and edge-on ring images. Dust densities in the ring change by about a factor of two from one ring traversal to the next, reflecting time variability of the plumes, the plasma properties, and the magnetosphere configuration. In HST images, the ring blends with the background at about 450,000 km from the planet, while CDA measurements revealed that the ring extends much further away, reaching even Titan's orbit at 1.2 million km from the planet. CDA measured particle size distributions in the E ring for grains larger than 0.9 μm, in agreement with those inferred in the Enceladus plumes from remote imaging.

For the first time, the time-of-flight mass spectrometer of CDA allowed the composition of the ring particles to be determined in-situ (Figure 29.15). It turned out that water ice dominates the particle composition in the entire E ring. Besides a minor grain population which consist of pure minerals three major types of icy ring particles were identified. Type I spectra contain almost pure water ice and a tiny amount of sodium; Type II spectra exhibit impurities of organic compounds and/or silicate minerals within the ice particles. Type III particles contain a substantial amount (about 1–2%) of sodium salts, primarily sodium chloride (NaCl), sodium bicarbonate ($NaHCO_3$), sodium carbonate (Na_2CO_3) and a minor amount of potassium chloride (KCl). These are exactly the four most abundant species predicted for an oceanic composition on Enceladus. Therefore, salt-rich Type III grains suggest evaporation of salty liquid water from a reservoir below Enceladus' ice crust in contact with this moon's rocky core as the most plausible plume-producing process. All types of icy E ring grains almost entirely originate from Enceladus' ice geysers. The production of icy dust from these plumes seems to be sufficient to maintain the E ring and to fill the entire Saturnian magnetosphere with grains far beyond the optically visible boundary of the E ring.

2.6.4. Interstellar Dust in the Heliosphere

The solar system currently passes through a region of low-density weakly ionized interstellar medium of our galaxy. This material shows a larger abundance of heavy refractory elements in the gas phase such as iron, magnesium, and silicon than is found in cold dense interstellar clouds. Interstellar dust is part of the interstellar medium, although it was not directly observed by astronomical means in the tenuous local interstellar cloud. Interstellar dust is formed as stardust in the cool atmospheres of giant stars and in nova and supernova explosions.

In the 1990s, interstellar dust was positively identified inside the planetary system. At the distance of Jupiter, the dust detector on board the Ulysses spacecraft detected impacts predominantly from a direction that was opposite to the expected impact direction of interplanetary dust grains. The impact velocities exceeded the local solar system escape velocity, even if radiation pressure effects were considered. The motion of interstellar grains through the solar system turned out to be parallel to the flow of neutral interstellar hydrogen and helium gas, both traveling at a speed of 26 km/s with respect to the Sun. The interstellar dust flow persisted at higher latitudes above the ecliptic plane, even over the poles of the Sun, whereas interplanetary dust is strongly concentrated toward the ecliptic plane (Figure 29.16).

Since that time, Ulysses monitored the stream of interstellar dust grains through the solar system at higher latitudes. It was found that the flux of small interstellar grains varied with the period of the solar cycle, which indicates a coupling of the grains to the solar wind magnetic field. Interstellar dust was initially identified outside 3 AU out to Jupiter's distance. However, refined analyses showed that both Cassini and Galileo recorded interstellar grains in the region between 0.7 and 3 AU from the Sun as well. Even in the Helios dust data, interstellar grains were identified down to 0.3 AU distance from the Sun.

The radii of clearly identified interstellar grains range from 0.1 to above 1 μm with a maximum at about 0.3 μm. The deficiency of small grain masses (<0.3 μm) compared to astronomically observed interstellar dust indicates a depletion of small interstellar grains in the heliosphere. Interstellar particles even bigger than 1 μm were reliably identified by their hyperbolic speeds in radar meteor observations. The flow direction of these bigger particles varies over a much wider angular range than that of small (submicrometer-sized) grains observed by spacecraft.

There are significant differences in the particle sizes that were recorded at different heliocentric distances.

FIGURE 29.15 (a) Mass spectrum of a sodium-poor ice particle (Type I), dominated by a sequence of water cluster ions of the form $H(H_2O)_n^+$, ($n = 2-8$). A weak Na^+ mass line is also present. Within the water matrix, some of the Na ions immediately form hydrates $Na(H_2O)_n^+$. Since the sensitivity of the detector to Na is at least several hundred times higher than to water, the respective ice grains only contain traces of sodium. (b) Coadded mass spectra of sodium-rich water ice particles (Type III). These spectra typically show very few pure water and Na-hydrate clusters, if any. They are characterized by an abundant Na^+ mass line followed by a peak sequence of hydroxy-cluster ions $Na(NaOH)_n^+$. Ions with masses of $Na(NaCl)_n^+$, ($n = 1-3$), and $Na(Na_2CO_3)^+$ give evidence for NaCl followed by $NaHCO_3$ and/or Na_2CO_3 as the main Na-bearing compounds. *From Postberg et al. (2009).*

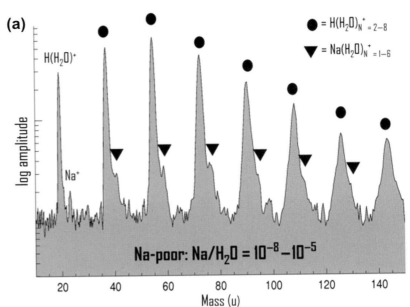

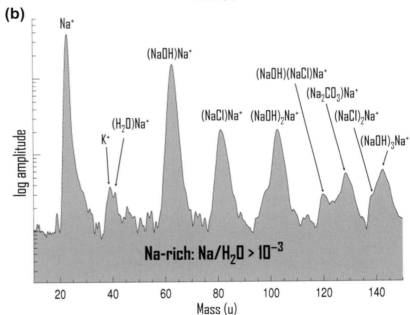

Measurements of the interstellar particle mass distribution revealed a lack of small grains inside 3 AU heliocentric distance. Measurements by Cassini and Galileo in the distance range between 0.7 and 3 AU showed that interstellar particles were bigger than 0.5 μm with average particle masses increasing closer to the Sun. The flux of these bigger particles did not exhibit temporal variations due to the solar wind magnetic field like the flux of smaller particles observed by Ulysses. The trend of increasing masses of particles continues as demonstrated by Helios measurements, which recorded particles of about 1 μm radius down to 0.3 AU. These facts support the idea that the interstellar dust stream is filtered by both radiation pressure and electromagnetic forces. It is concluded that interstellar particles with optical properties of grains consisting of astronomical silicates or organic refractory materials are consistent with the observed radiation pressure effect.

2.7. Extrasolar Debris Disks

Circumstellar disks play a fundamental role in the formation of stars and planets and the evolution of young planetary

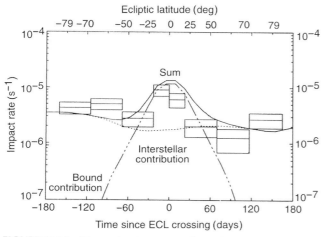

FIGURE 29.16 Ulysses dust impact rate observed around the time of its ecliptic plane crossing (ECL). ECL occurred on March 12, 1995, at a distance of 1.3 AU from the Sun. The boxes indicate the mean impact rates and their uncertainties. The top scale gives the spacecraft latitude. Model calculations of the impact rate during Ulysses' south–north traverse through the ecliptic plane are shown by the lines. Contributions from interplanetary dust on bound orbits and interstellar dust on hyperbolic trajectories and the sum of both are displayed. From these measurements, it is concluded that interstellar dust is not depleted to a distance of 1.3 AU from the Sun.

systems. Primordial disks form from the contraction and conservation of **angular momentum** of the densest regions of molecular clouds. A new star is born in the central condensation of the disk, and planet formation takes place by the accretion of gas and dust particles onto larger and larger bodies, resulting in a few massive planetary cores and a swarm of small objects, the so-called planetesimals. The formation of planets continues for approximately 10–100 million years until the thick prestellar disk is cleared and blown away by the strong wind of the newly formed young star.

Observations with the *Spitzer Space Telescope* show that at least 15% of solarlike stars show infrared emission in excess of that expected from the stellar photosphere. This excess is thought to arise from a circumstellar dust disk. However, given that typical lifetimes of dust grains are much shorter (0.01–1 million years) than the age of the star (10 million years–10 billion years), these disks cannot be leftovers from the primordial disks out of which the systems formed. Instead they must be of secondary origin, formed later after planet formation had ceased long ago. These disks are thought to result from ongoing dust production by planetesimals, like asteroids, comets, or the Kuiper Belt Objects in our solar system. This is why these dust disks are often called **debris disks**. This term must not be confused with man-made space debris described above!

The study of debris disks can give us information about the planetesimals' characteristics and about the planet populations in these disks. For example, the disks surrounding the stars β Pictoris, ε Eridani, or Fomalhaut show asymmetries and brightness enhancements that are thought to arise from one or more planets orbiting their central star within the disk. The structure of debris disks is sensitive to planets with a wide range of masses and semimajor axes. The study of debris disks and, hence, their source planetesimals can give us a more complete picture of the diversity of planetary systems, shedding light on their formation and dynamical histories. The study of debris disk structure can serve as a planet detection method, covering a parameter space complementary to that of the "classical" planet detection methods, and therefore may help us place our solar system in context with other planetary systems (*see* Extra-Solar Planets).

3. DYNAMICS AND EVOLUTION

3.1. Gravity and Keplerian Orbits

In the planetary system, solar gravity determines the orbits of all bodies larger than dust particles for which other forces become important. But even for dust, gravity is an important factor. Near planets, planetary gravity takes over. However, the basic orbital characteristics remain the same. Two types of orbits are possible: bound and unbound orbits around the central body. Circular and elliptical orbits are bound to the Sun; the planets exert only small disturbances to these orbits. Planets, asteroids, and comets move on such orbits. Objects on unbound orbits will eventually leave the solar system. Typically, interstellar dust particles move on unbound, hyperbolic orbits through the solar system. Similarly, interplanetary particles are unbound to any planetary system and traverse it on hyperbolic orbits with respect to the planet (*see* Solar System Dynamics).

A Keplerian orbit is a conic section that is characterized by its semimajor axis a, eccentricity e, and inclination i. The Sun (or a planet) is in one focus. The **perihelion** distance (closest to the Sun) is given by $q = a(1 - e)$. Circular orbits have eccentricity $e = 0$, elliptical orbits have $0 < e < 1$, and hyperbolic orbits have $e > 1$ and a is taken negative. The **aphelion distances** (furthest from the Sun) are finite only for circular and elliptical orbits. The inclination is the angle between the orbit plane and the ecliptic (i.e. the orbit plane of Earth).

Dust particles in interplanetary space move on very different orbits, and several classes of orbits have been identified. One class of meteoroids moves on orbits that are similar to those of asteroids, which peak in the asteroid belt. Another class of orbits that represents the majority of zodiacal light particles has a strong concentration toward the Sun. Both orbit populations have low to intermediate eccentricities ($0 < e < 0.6$) and low inclinations ($i < 40°$). These asteroidal and zodiacal core populations satisfactorily describe meteors, the lunar crater size distribution, and a major portion of zodiacal light observations. Also, spacecraft measurements inside 2 AU are well represented by the core population. (see Main-Belt Asteroids).

3.2. Radiation Pressure and the Poynting–Robertson Effect

Electromagnetic radiation from the Sun (most intensity is in the visible wavelength range at $\lambda_{max} \approx 0.5$ μm) is absorbed, scattered, or diffracted by any particulate, exerting pressure on this particle. Because solar radiation is directed outward from the Sun, radiation pressure also points away from the Sun. Thus, the radiation pressure force reduces the gravitational attraction. Both radiation pressure and gravity have an inverse square dependence on the distance from the Sun. Radiation pressure depends on the cross-section of the particle and gravity on the mass; therefore, for the same particle, the ratio β of radiation pressure, F_R, over gravitational force, F_G, is constant everywhere in the solar system and depends only on particle properties: $\beta = F_R/F_G \sim Q_{pr}/s\rho$, where Q_{pr} is the efficiency factor for radiation pressure, s is the particle radius, and ρ is its density.

Figure 29.17 shows the dependence of β on the particle size for different shapes. For big particles ($s \gg \lambda_{max}$), the radiation pressure force is proportional to the geometric cross-section, giving rise to the $1/s$ dependence of β. For particle sizes comparable to the wavelength of sunlight ($s \sim \lambda_{max}$), β values peak, and they decline for smaller particles as their interaction with light decreases.

A consequence of the radiation pressure force is that particles with $\beta > 1$ are not attracted by the Sun but rather are repelled by it. If such particles are generated in planetary space either by a collision or by release from a comet, they are expelled from the solar system on hyperbolic orbits. But even particles with β values smaller than 1 eventually leave the solar system on hyperbolic orbits if their speed at formation is high enough so that the reduced solar attraction can no longer keep the particle on a bound orbit. A particle with $\beta > \frac{1}{2}$ that is released from a parent body moving on a circular orbit will leave on a hyperbolic orbit. These particles are termed beta meteoroids.

Because of the finite speed of light ($c \sim 300,000$ km/s) radiation pressure does not act perfectly radial but has an aberration in the direction of motion of the particle around the Sun. Thus, a small component (approximately proportional to v/c, where v is the speed of the particle) of the radiation pressure force always acts against the orbital motion, reducing its orbital energy. This effect is called Poynting–Robertson effect. As a consequence of this drag force, the particle is decelerated. This deceleration is largest at its perihelion distance where both the light pressure and the velocity peak. Consequently, the eccentricity (aphelion distance), is reduced, and the orbit is circularized. Subsequently, the particle spirals toward the Sun, where it finally sublimates.

The lifetime τ_{pr} of a particle on a circular orbit that spirals slowly to the Sun is given by $\tau_{pr} = 7 \times 10^5 \, \rho s \, r^2/Q_{pr}$, where τ_{pr} is in years, r is given in astronomical units, and all other quantities are in SI units. Even a centimeter-sized ($s = 0.01$ m), stony ($\rho \sim 3000$ kg/m^3, $Q_{pr} \sim 1$) particle requires only 21 million years to spiral to the Sun if it is not destroyed earlier by a collision. This example shows that all interplanetary dust has to be recently generated; no dust particles remain from the times of the formation of the solar system. The dust we find today must originate from bigger objects (asteroids and comets) which have sufficient lifetimes.

The effect of solar wind impingement on particulates is similar to radiation pressure and Poynting–Robertson effect. Although direct particle pressure can be neglected with respect to radiation pressure, solar wind drag is about 30% of Poynting–Robertson drag. Particle orbits that evolve under Poynting–Robertson drag eventually cross the orbits of the inner planets and, thereby, are affected by planetary gravity. During the orbit evolution of particles, resonances with planetary orbits may occur even if the orbit periods of the particle and the planet are not the same but form a simple integer ratio. This effect is largest for big particles, the orbits of which evolve slower and which spend more time near the resonance position. Density enhancements of interplanetary dust have been found (i.e. the Earth-resonant ring was identified in IRAS data and later confirmed by COBE).

Dust near other stars also evolves under Poynting–Robertson drag and forms dust disks around these stars. Such disks were found around many stars (e.g. β Pictoris, ε Eridani, Fomalhaut). Brightness enhancements in these disks are indicative of resonances due to one or more planets in orbit around these stars (see Extra-Solar Planets).

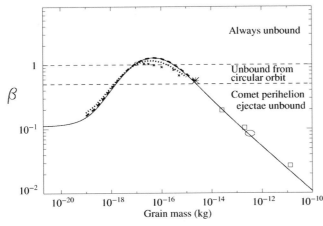

FIGURE 29.17 Ratio® of the radiation pressure force over solar gravity as a function of particle radius. Values are given for particles made of astronomical silicates *(from Gustafson et al., 2001)* with various shapes: sphere (solid curve), long cylinders (dashes), and flat plates (dots).

3.3. Collisions

Mutual high-speed ($v > 1$ km/s) collisions among dust particles lead to grain destruction and generation of fragments. By these effects, dust grains are modified or destroyed, and many new fragment particles are generated in interplanetary space. From impact studies in stony material, we know that, at a typical collision speed of 10 km/s an impact crater is formed on the surface of the target particle if it is more than 50,000 times more massive than the projectile. This mass ratio is strongly speed and material dependent. A typical impact crater in brittle stony material (Figure 29.6) consists of a central hemispherical pit surrounded by a shallow spallation zone. The largest ejecta particle (from the spallation zone) can be many times bigger than the projectile; however, it is emitted at a very low speed on the order of meters per second. The total mass ejected from an impact crater at an impact speed of 10 km/s is about 500 times the projectile mass.

However, if the target particle is smaller than the stated limit, the target is catastrophically destroyed, and the material of both colliding particles is transformed into a huge number of fragment particles (Figure 29.18). Thus, catastrophic collisions are a very effective process for generating small particles in interplanetary space. Interplanetary particles bigger than about 0.1 mm in diameter are destroyed by a catastrophic collision rather than transported to the Sun by Poynting–Robertson drag.

3.4. Charging of Dust and Interaction with the Interplanetary Magnetic Field

Any meteoroid in interplanetary space carries an electric charge, and several competing charging processes determine the actual charge of a meteoroid (Figure 29.19). Irradiation by solar ultraviolet light releases photoelectrons which leave the grain. The grains collect electrons and ions from the ambient solar wind plasma. Energetic ions and electrons then cause the emission of secondary electrons. Whether electrons or ions can reach or leave the grain depends on their energy and on the polarity and electrical potential of the grain. Because of the predominance of the photoelectric effect in interplanetary space, meteoroids are mostly charged positively to a potential of a few volts. Only at times of very high solar wind densities does the electron flux to the particle dominate and the particle gets negatively charged. The final charging state is reached when all currents to and from the meteoroid cancel. The timescale for charging is seconds to hours depending on the size of the particle; small particles charge slower. Electric charges on dust particles in interplanetary space were measured by the Cassini CDA. These measurements indicate a dust potential of $+5$ V. In the dense plasma of the inner Saturnian magnetosphere dust particles were found with a potential of -2 V.

The outward-streaming (away from the Sun) solar wind carries a magnetic field away from the Sun. Due to the rotation of the Sun (with a period of 25.7 days), magnetic field lines are drawn in a spiral, like water from a lawn sprinkler. The polarity of the magnetic field can be positive or negative, depending on the polarity at the base of the field line in the solar corona, which varies spatially and temporally. For an observer or a meteoroid in interplanetary space, the magnetic field sweeps outward with the speed of the solar wind (400–600 km/s) (*see* The Sun). In the magnetic reference frame, the meteoroid moves inward at about the same speed because its orbital speed is comparatively small. The *Lorentz* force on a charged dust particle

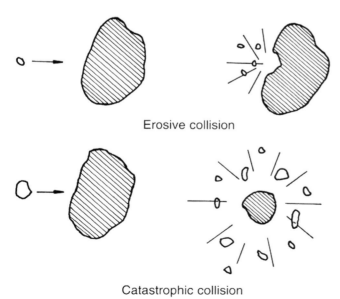

FIGURE 29.18 Schematics of meteoroid collisions in space. If the projectile is very small compared to the target particle, only a crater is formed in the bigger one. If the projectile exceeds a certain size limit, the bigger particle is shattered into many fragments. The transition from one type to the other is abrupt.

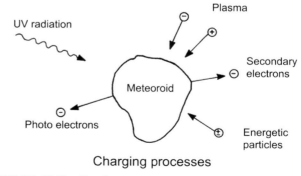

FIGURE 29.19 Charging processes of meteoroids in interplanetary space. UV radiation releases photoelectrons, electrons and ions are collected from the solar wind plasma; the impact of energetic particle radiation releases secondary electrons.

near the ecliptic plane points mostly either upward or downward, depending on the polarity of the magnetic field. Near the ecliptic plane, the polarity of the magnetic field changes with periods (days to weeks) that are much faster than the orbital period of an IDP, and the net effect of the Lorentz force on micrometer-sized particles is small. Only secular effects on the bigger zodiacal particles are expected to occur, which could have an effect on the symmetry plane of the zodiacal cloud close to the Sun. For nanometer-sized particles, like the ones that have been found in the dust streams, the Lorentz force dominates all other forces, and as a result the particles gyrate about the magnetic field lines and are eventually convected out of the solar system with the solar wind.

The overall polarity of the solar magnetic field changes with the solar cycle of 22 years. For one solar cycle, positive magnetic polarity prevails away from the ecliptic plane in the northern hemisphere and negative polarity in the southern hemisphere. Submicrometer-sized interstellar particles that enter the solar system are deflected either toward the ecliptic plane or away from it depending on the overall polarity of the magnetic field. Interstellar particles entering the heliosphere from one direction with a speed of 26 km/s need about 20 years (one solar cycle) to get close to the Sun. Therefore, trajectories of small interstellar grains (0.1 μm in radius) are strongly diverted: in some regions of space, their density is strongly increased; in others, they are depleted. At the time of the initial Ulysses and Galileo measurements (1992–1996), the overall solar magnetic field was in an unfavorable configuration; therefore, only big (micrometer-sized) interstellar particles reached the positions of Ulysses and Galileo. By 2003, the magnetic field had changed to the focusing configuration and the interstellar dust flux had recovered (*see* The Solar Wind).

3.5. Evolution of Dust in Interplanetary Space

Forces acting on interplanetary particles are compared in Table 29.4. The force from solar gravity depends on the mass of the particle; therefore, it depends on the size as $F_G \sim s^3$. Radiation pressure depends on the cross-section of the particle, hence, $F_R \sim s^2$. The electric charge on a dust grain depends on the size directly, as does the Lorentz force $F_L \sim s$. Therefore, these latter forces become more dominating at smaller dust sizes. At a size comparable to the wavelength of visible light ($s \sim 0.5$ μm), radiation pressure dominates over gravity, and below that size, the Lorentz force dominates the particles' dynamics. While gravity is attractive to the Sun, radiation pressure is repulsive. The net effect of solar wind interactions on small particles is that they are convected out of the solar system.

Besides energy-conserving forces, there are also dissipative forces: the Poynting–Robertson effect and the ion drag from the solar wind. They cause a loss of orbital energy and force particles to slowly spiral to the Sun, where they eventually evaporate. These atoms and molecules become ionized and are flushed out of the solar system by the solar wind.

Figure 29.20 shows the flow of meteoritic matter through the solar system as a function of the meteoroid size. There is a constant input of mass from comets and asteroids. From the intensity enhancement of zodiacal light toward the Sun it was deduced that, inside 1 AU, significant amounts of mass have to be injected into the zodiacal cloud by short-period comets. While comets shed their debris over a large range of heliocentric distances but preferentially close to the Sun, asteroid debris is mostly generated in the asteroid belt, between 2 and 4 AU from the Sun. Collisions dominate the fate of big particles and are a constant source of smaller fragments. Meteoroids in the range of 1–100 μm are dragged by the Poynting–Robertson drag to the Sun. Smaller fragments are driven out of the solar system by radiation pressure and Lorentz force.

Estimates of the mass loss from the zodiacal cloud inside 1 AU give the following numbers. About 10 tons per second are lost by collisions from the big (meteor-sized) particle population. A similar amount (on average) has to be replenished by cometary and asteroidal debris. About 9 tons per second of the collisional fragments are lost as small particles to interstellar space, and the remainder of 1 ton per second is carried by the Poynting–Robertson effect toward the Sun, evaporates, and eventually becomes part of the solar wind. Interstellar dust transiting the solar system becomes increasingly important farther away from the Sun. Already at 3 AU from the Sun, the interstellar dust flux seems to dominate the flux of submicrometer- and micrometer-sized interplanetary meteoroids. A summary of the various manifestations of cosmic dust presented in this chapter is given in Table 29.5.

4. FUTURE STUDIES

New techniques will generate new insights. These techniques will include innovative observational methods, new space missions to unexplored territory, and new experimental and theoretical methods to study the processes affecting solar system dust. Questions to be addressed are: the composition (elemental, molecular, and isotopic) and spatial distribution of interplanetary dust, the quantitative understanding of processes affecting dust in interplanetary space, and the quantitative determination of the contributions from different sources (asteroids, comets, planetary environments, and interstellar dust).

Analyses of brightness measurements at infrared wavelengths up to 200 μm by the COBE satellite result in refined models of the distribution of dust mostly outside

TABLE 29.4 Comparison of Various Forces (Dominating Forces in Bold) Acting on Dust Particles of Size s under Typical Interplanetary Conditions at 1 AU Distance from the Sun

s (μm)	F_G (N)	F_R (N)	F_L (N)	F_{PR} (N)	F_{ID} (N)
0.01	9×10^{-23}	1.4×10^{-21}	**1.5×10^{-20}**	1.4×10^{-25}	4×10^{-26}
0.1	9×10^{-20}	**1.4×10^{-19}**	**1.5×10^{-19}**	**1.4×10^{-23}**	4×10^{-24}
1	**9×10^{-17}**	1.4×10^{-17}	1.5×10^{-18}	1.4×10^{-21}	4×10^{-22}
10	**9×10^{-14}**	1.4×10^{-15}	1.5×10^{-17}	1.4×10^{-19}	4×10^{-20}
100	**9×10^{-11}**	1.4×10^{-13}	1.5×10^{-16}	1.4×10^{-17}	4×10^{-18}

Subscripts G, R, L, PR, and ID refer to gravity, radiation pressure, Lorentz force, Poynting–Robertson drag, and ion drag, respectively.

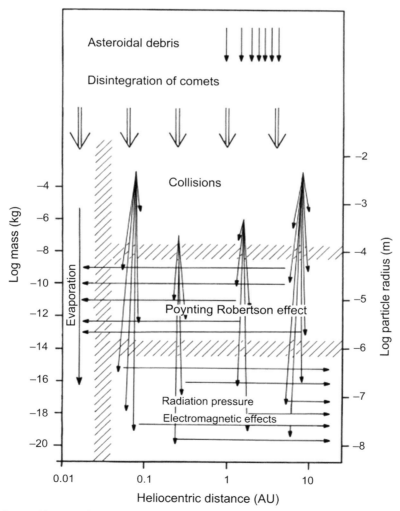

FIGURE 29.20 Mass flow of meteoritic matter through the solar system. Most of the interplanetary dust is produced by collisions of larger meteoroids, which represent a reservoir continually being replenished by disintegration of comets or asteroids. Most of it is blown out of the solar system as submicrometer-sized grains. The remainder is lost by evaporation after being driven close to the Sun by the Poynting–Robertson effect. In addition to the flow of interplanetary matter shown, there is a flow of interstellar grains through the planetary system.

TABLE 29.5 Manifestations of Cosmic Dust in Our Solar System Discussed in This Chapter

Dust Types	Details	Observations	Known Sources	Characteristics
Zodiacal dust	—	Visible and thermal IR observations, SC	Comets, asteroids	Dust mostly within Jupiter's orbit
Comet dust	Coma	Visible observations, SC	Comets	Within ~30,000 km from nucleus controlled by emission process
	Tail	Visible observations	Comets	Micron-sized particles. Several million kilometers; controlled by solar radiation pressure and gravity
	Trail	IR observations	Comets	Millimeter and bigger particles all along orbit
Asteroid dust	Bands	IR observations	Asteroids	Millimeter-sized grains created by asteroid collisions several million years ago
	Active asteroids	Visible observations	Asteroids	Currently ongoing emissions of micron-sized dust in contrast to old dust seen in dust bands
Meteoroids	—	SC, astronomical observations	Comets, asteroids	Any solid object in interplanetary space below the size of comets and asteroids
Meteors	Background	Visible, radar	Comets, asteroids	Meteoroids entering a planetary atmosphere and generating a visible or ionization trail at rates ~1/h
	Streams/storms	Visible	Comets	Meteors originating from a single object (comet or asteroid) at rates ≫ 1/h
Meteorites	—	Terrestrial collection	Comets, asteroids, Earth Moon, Mars	Residue of an extraterrestrial object found on the surface of the Earth
	IDPs, micrometeorites	Atmospheric collections, ice collections, deep-sea sediments	Comets, asteroids	Micrometeoroid of 5–50 μm size collected in the Earth stratosphere
Beta meteoroids	—	SC	Collisions of bigger meteoroids	Flux of submicron-sized dust arriving from the solar direction interacting with solar radiation and interplanetary magnetic field

Nanodust	—	SC	Unknown	High flux of potentially nanometer-sized dust observed by plasma wave instruments
Planetary rings	Dense rings	SC, astronomical observations	Planetary moons	Centimeter- to meter-sized objects orbiting gas planets within Roche zone
	Diffuse rings	SC, astronomical observations	Planetary moons	Micron-sized dust generated by planetary moons and interacting with the magnetosphere
Planetary moons	Volcanism: Io, Enceladus, Triton	SC	Planetary moons	Submicron-sized dust emitted by volcanic action energized by tidal effects
	Ejecta clouds	SC	Planetary moons	Micron-sized dust ejected by meteoroid impacts onto moons
Lunar dust	Electrostatically levitated dust	SC, manned missions	Lunar regolith	Electrostatic forces are believed to both mobilize lunar dust and to cause strong clinging to other surfaces
	Ejecta	Lunar samples, SC	Lunar regolith	Lunar soil consists of ejecta material generated in impacts
Kuiper belt dust	—	SC	Collisions of Kuiper-Belt (KB) objects	Dust generated by collisions in the Kuiper belt populates the outer planetary system
Interstellar dust	—	SC, astronomical observations	Stars, molecular clouds	The solar system currently passes through a tenuous cloud of interstellar dust and gas
Space debris	Aluminum oxide, slack	SC	Solid rocket motors	The dominant source of micron-sized space debris are reaction products of solid rocket motors
	Defunct and broken up satellites	SC, radar, optical	Man-made satellites, space vehicles	Material brought into space feeds a collisional cascade in which fragments are generated that become projectiles for more collisions

Abbreviations: IR, infrared; SC, spacecraft measurements.

1 AU. Spectrally resolved observations of asteroids, comets, and zodiacal dust by infrared space observatories (*Infrared Space Observatory (ISO), Spitzer*, and *Herschel*) show the genetic relation between these larger bodies and interplanetary dust. Improved observations of the inner zodiacal light and the edge of the dust-free zone around the Sun will provide some clues to the composition of zodiacal dust. Improved optical and infrared observations of extra-solar systems will bring new insights to zodiacal clouds around other stars.

The Cassini spacecraft has become the first man-made satellite of Saturn and will continue investigating the dust environment of the gas giant including its dusty ring system. The detailed study of cometary and interstellar dust was the goal of NASA's Stardust mission, which returned samples of dust from comet Wild 2. Similarly, the Japanese *Hyabusa* mission returned dust from asteroid Itokawa. The analyses of these cometary and asteroidal dust samples are ongoing. The European Space Agency's *Rosetta* mission will follow comet 67P/Churyumov−Gerasimenko through its perihelion and investigate its dust release to interplanetary space. The Rosetta lander Philae will make the first attempt ever to land on a cometary nucleus. New Horizons will pass by Pluto and its icy moons and head on to the Kuiper Belt. It is the first spacecraft investigating the dust environment in the outer solar system with a dedicated dust detector since *Pioneer 10* and *11* launched in the 1970s.

Dust particles, like photons, are born at remote sites in space and time, and carry from there information that may not be accessible to direct investigation. From knowledge of the dust particles' birthplace and the particles' bulk properties, we can learn about the remote environment out of which the particles were formed. This approach is called dust astronomy and is carried out by means of dust telescopes on dust observatories in space. Targets for dust telescopes are dust from the local interstellar medium, cometary and asteroidal dust, and space debris. Dust particles' trajectories are determined by the measurements of the electric charge signals that are induced when the charged grains fly through charge-sensitive grid systems. Modern in situ dust detectors are capable of providing mass, speed, and physical and chemical information of dust grains in space. A dust telescope can, therefore, be considered as a combination of detectors for dust particle trajectories along with detectors for physical and chemical analysis of dust particles. Both dust trajectory sensors and large-area dust analyzers have been developed recently and await their use in space.

In near-Earth space, ambitious new techniques will be applied to collect meteoritic material which is not accessible by other methods. High-speed meteoroid catchers which permit the determination of the trajectory as well as the recovery of material for analysis in ground laboratories have been developed and await deployment in space.

Laboratory studies are instrumental in improving our understanding of planetary and interplanetary processes in which dust plays a major role. The study of dust−plasma interactions is a new and expanding field that is attracting considerable attention. New phenomena are expected to occur when plasma is loaded with large amounts of dust. Processes of this type are suspected to play a significant role in cometary environments, in planetary rings, and in protoplanetary disks.

BIBLIOGRAPHY

Burns, J. A., Showalter, M. R., Hamilton, D. P., Nicholson, P. D., de Pater, I., Ockert-Bell, M. E., & Thomas, P. C. (1999). The formation of Jupiter's faint rings. *Science, 284,* 1146−1150.

Conference series. In Gustafson, B. A. S., & Hanner, M. S. (Eds.). (1996), *Physics, chemistry, and dynamics of interplanetary dust* (Vol. 104). San Francisco: Astronomical Society of the Pacific.

Giese, R. H., & Lamy, P. (Eds.). (1985). *Properties and interactions of interplanetary dust.* Dordrecht/Boston: Reidel.

Green, S. F., Williams, I. P., McDonnell, J. A. M., & McBride, N. (Eds.). (2002). *Dust in the solar system and other planetary systems.* Amsterdam: Pergamon.

Grün, E., Gustafson, B. A. S., Dermott, S., & Fechtig, H. (Eds.). (2001). *Interplanetary dust.* Heidelberg: Springer.

Gustafson, B. A. S., Mayo Greenberg, J., Kolokolova, Ludmilla, Xu, Yu-lin, & Stognienko, Ralf. (2001). "Interactions with Electromagnetic Radiation: Theory and Laboratory Simulations". In E. Grün, B. A. S. Gustafson, S. Dermott, & H. Fechtig (Eds.), *"Interplanetary Dust"*, (pp. 509−567). Heidelberg: Springer.

Horányi, M., Burns, J. A., Hedman, M. M., Jones, G. H., & Kempf, S. (2009). Diffuse rings. In M. K. Dougherty, et al. (Eds.), *Saturn from Cassini-Huygens*. http://dx.doi.org/10.1007/978-1-4020-9217-6_16.

Humes, D. (1980). *Results of Pioneer 10 and 11 meteoroid experiments − Inter-planetary and near-Saturn. J. Geophys. R., 85* (A/II), 5841.

Krüger, H., & Graps, A. (Eds.). (2007). *Dust in planetary systems.* ESA SP-463.

Mann, I. (2010). Interstellar dust in the solar system. *Annual Reviews of Astronomy & Astrophysics* (Vol. 48), 173−203.

McDonnell, J. A. M. (Ed.). (1978). *Cosmic dust.* Chichester, England: John Wiley & Sons.

Moro-Martin, A. (2012). Dusty planetary systems. In P. Kalas & L. French (Eds.), *Solar and planetary systems:* Vol. 3 of the series *Planets, stars and stellar systems* (T.D. Oswalt, Editor-in-chief). Springer.

Leinert, C., & Grün, E. (1990). In R. Schwenn, & E. Marsch (Eds.), *Physics of the inner heliosphere I* (pp. 207−275). Berlin: Springer-Verlag.

Levasseur-Regourd, A. C., & Hasegawa, H. (Eds.). (1991). *Origin and evolution of interplanetary dust.* Dordrecht: Kluwer.

Postberg, F., Kempf, S., Schmidt, J., Brillantov, N., Beinsen, A., Abel, B., Buck, U., & Srama, R. (2009). Sodium Salts in E Ring Ice Grains from an Ocean below the Surface of Enceladus, *Nature, 459*(7250), 1098−1101.

Chapter 30

Physics and Chemistry of Comets

John C. Brandt
Astronomy Department, University of Washington, Seattle, Washington, USA

Chapter Outline

1. Space Missions to Comets — 683
2. A Brief History of Comet Studies — 685
3. Physics of the Nucleus — 689
4. Coma and Hydrogen Cloud — 693
5. Tails — 695
6. Comet Chemistry — 698
7. Formation and Ultimate Fate of Comets — 700
8. Summary — 702
Bibliography — 703

The spectacular sight of a bright comet with a tail stretching across the sky (Figure 30.1) prompts questions about the nature of the object and the physical processes at work. The current era is one of major comet research, with several space missions to comets producing pioneering results. The images and data that are becoming available often prompt new questions and challenge old ideas. A quarter century has passed since the first space missions to comets, and comet science has reached a level of maturity that was unimaginable not long ago.

1. SPACE MISSIONS TO COMETS

Many lines of evidence indicate that the source of all cometary phenomena is a rather small central body called the **nucleus**. Typical dimensions are in the range 1–10 km. Comets are typically observed when they are near the Earth. Thus, viewing an object, say, 3 km across from a distance of 0.2 AU (or 3×10^7 km) means that the object subtends an angle 1/50th of an arc second. Typical resolution from ground-based observatories is about 1.0 arc second and, for large telescopes, is determined by the effects of the Earth's atmosphere. Mountaintop observatories in good locations can do better, and the *Hubble Space Telescope* (HST) has a resolution of about 0.1 arc seconds. From Earth, except in extraordinary circumstances, the nucleus cannot be resolved, and no detail on the surface can be seen. The solution is to send spacecraft with imaging systems close to the cometary nuclei. In situ measurements of gas, dust, plasma, magnetic fields, and energetic particles can be obtained while the spacecraft is near the comet. The imaging and in situ data provide a major source of information on comets.

Table 30.1 summarizes completed missions to comets. Of course, analysis often continues for years. In this section, only the missions with imaging are discussed. The missions to comet Halley in 1986 were collectively called the Halley Armada, and three of them had imaging. Two *VEGA* spacecraft were sent by the Soviet Union. These spacecraft first went to Venus, and the name is a contraction of the Russian language words for Venus(VEnera) and Halley(GAllei). They passed within 8890 km (*VEGA 1*) and 8030 km (*VEGA 2*) of the comet. The images from the *VEGA*s are valuable, but they were somewhat noisy and were taken from larger distances than those taken by *Giotto*.

The European Space Agency (ESA) sent the *Giotto* spacecraft to pass the nucleus of Halley's comet within 596 km. The spacecraft carried the Halley Multicolor Camera (HMC), which obtained images of the nucleus until approximately the time of closest approach (CA) when it was damaged by the impacts of dust particles. Figure 30.2(a) is an overall view of the nucleus composed of 68 individual images. The nucleus was not spherical but was a potato-shaped object with a long axis of approximately 15 km and short axes of approximately 7 km. The nucleus showed features that appeared to be valleys, hills, and craters. The average albedo or reflectivity of the surface was only 0.04; the surface was very dark. The **jets** containing the dust and gas emission from the nucleus came from approximately 10% of the entire surface and were active when their location was in

FIGURE 30.1 Comet Hale–Bopp on April 8, 1997, showing the whitish dust tail and the blue plasma tail. *Courtesy of H. Mikuz, Crni Vrh Observatory, Slovenia and the Ulysses Comet Watch.*

sunlight. The direction of emission was generally sunward.

The National Aeronautics and Space Administration (NASA) sent the *Deep Space 1* spacecraft to within about 2171 km of the nucleus of comet Borrelly on September 22, 2001, (Figure 30.2(b) shows a close-up view of the nucleus). The long axis of the nucleus is approximately 8 km and the short axes are about 3.2 km. The surface showed features and was also very dark. The albedo varied between 0.01 and 0.03 over the surface. The jets with dust and gas emission, which were clearly seen, occupied 10% or less of the surface area.

NASA's *Stardust* spacecraft passed within approximately 236 km of comet Wild 2 on January 2, 2004. Excellent images were obtained, and an example, taken just after CA, is shown in Figure 30.2(c). The nucleus is roughly a rounded body with a diameter of 4 km. Jets of dust and gas emission were seen, and the albedo determined was 0.03 ± 0.015. Features with steep slopes have been identified, providing clues to the history of the surface. The main goal of the *Stardust* mission, which is to return to the Earth dust samples collected in the comet's **coma**, has been achieved with the return of samples that parachuted to the Utah desert on January 15, 2006 (see the discussion in Section 4).

The *Deep Impact* impactor spacecraft collided with comet Tempel 1 on July 4, 2005. The impactor spacecraft separated from the flyby spacecraft 24 h before impact, and the flyby spacecraft passed the nucleus at a distance of 500 km. An image of the nucleus taken from the impactor is shown in Figure 30.2(d). The average diameter of the nucleus is close to 6.0 km, the longest dimension is 7.6 km, and the shortest dimension is 4.9 km. The surface shows both smooth and rough terrain, scarps (a line of cliffs usually produced by faulting), and impact craters. The surface is generally homogeneous in color and albedo, which varied from 0.02 to 0.06, and the temperature of the surface indicates an equilibrium with sunlight. Observations on approach detected numerous short outbursts that can be associated with specific regions on the surface.

The impactor spacecraft delivered 19 GJ of kinetic energy to comet Tempel 1. The spectacular impact is shown in Figure 30.3, and a view of the ejecta plume containing $\sim 10^6$ kg of material is shown in Figure 30.4. In addition to observations from *Deep Impact*, the event was extensively observed by ground-based and space-based observatories. By 9 July the comet had returned to its preimpact state and the impact crater was not seen until the revisit in 2011 discussed below. The ejecta consisted of fine particles (1–100 μm) and individual species, including water, **water ice**, carbon, carbon dioxide, hydrocarbons, and crystalline silicates. The spectra of the ejecta are a good match to the spectra of material ejected from comet Hale–Bopp and to the dusty disk spectrum of a young stellar object.

The flyby spacecraft from the *Deep Impact* mission was retargeted to comet Hartley 2 as part of Extrasolar Planet Observation and Deep Impact Extended Investigation (EPOXI). Close approach to comet Hartley 2 took place on November 4, 2010, at a distance of 694 km. An image of this remarkable comet is shown in Figure 30.2(e). The nucleus is bilobed and small, with a length of 2.3 km. The average albedo is approximately 0.04. The surface has two principal types of terrain: (1) areas containing knobby features with typical dimensions in the range 50–100 m and (2) smooth areas on part of the larger lobe and the waist area. Comet Hartley is considered hyperactive in the sense that the gas production is much higher than expected from simple sublimation from a comet its size. Volatiles, primarily CO_2, drag chunks of ice out of the nucleus and these sublimate to enhance the gas production rate. Individual chunks were seen in many images during *EPOXI*'s CA.

The *Stardust* spacecraft was retargeted for a revisit of comet Tempel 1 as part of the *Stardust-NExT* mission. The

TABLE 30.1 Missions to Comets

Spacecraft	Comet	Encounter Date	Imaging
International Cometary Explorer (ICE)	Giacobini–Zinner	September 11, 1985	No
VEGA 1	Halley	March 6, 1986	Yes
Suisei	Halley	March 8, 1986	No
VEGA 2	Halley	March 9, 1986	Yes
Sakigake	Halley	March 11, 1986	No
Giotto	Halley	March 14, 1986	Yes
ICE	Halley	March 25, 1986	No
Giotto Extended Mission (GEM)	Grigg–Skjellerup	July 10, 1992	No
Deep Space 1	Borrelly	September 22, 2001	Yes
Stardust	Wild 2	January 2, 2004	Yes
Deep Impact	Tempel 1	July 4, 2005	Yes
EPOXI	Hartley 2	November 4, 2010	Yes
Stardust-NExT	Tempel 1	February 14, 2011	Yes

name NExT is derived from New Exploration of Tempel 1. Close approach was on February 14, 2011, at a distance of 178 km. The revisit provided an opportunity to complete the *Deep Impact* experiment by reexamining the impact site and looking for changes in comet Tempel 1's complex surface after an interval of about 5.5 years. The *Stardust-NExT* imagery increased the coverage of the surface to 60%. With the new coverage the so-called smooth flow areas were found to cover some 30% of the surface. These appear to form from eruptions of ice-dust material that flows to areas of gravitational lows. The surface also shows thicker layers probably caused by impact events.

Comparison of images shows that most of the comet's surface was unchanged in 5.5 years. But, there was a major exception. A scarp, the edge of a smooth flow area (called S2), was 10–15 m high and receded by as much as 50 m as shown in Figure 30.5. This result is supported by the result that many jets appear to come from the edges of eroding scarps. Pitted terrain covers more than 50% of the surface. These are not produced by impacts, but are probably the result of CO-driven minioutbursts.

The *Deep Impact* impact site was imaged by *Stardust-NExT*; the before and after images are shown in Figure 30.6. The crater is about 50 m in diameter and is unremarkable. Possible changes during the 5.5 years since the impact, including filling in by ejecta falling back, are unknown. But, the crater size is consistent with an impact into lightly packed snow.

All these images confirm the basic view of the nucleus as a single, sublimating (direct phase transition from the solid to the gas state) body as proposed by F. L. Whipple. As the solid body approaches the Sun, energy supplied by solar radiation raises the temperature of the near-surface layers, sublimation of ices (mostly water ice) takes place, and the emission of gas and entrained dust produces the large features seen in the sky. The five comets with surfaces imaged in detail represent a currently unknown combination of intrinsic diversity and postformation evolution.

Before leaving space missions, it is important to note that ESA's *Rosetta* mission to comet Churyumov–Gerasimenko was launched on March 2, 2004, to begin its 10-year journey to the comet. The plan is for the main spacecraft to rendezvous with the comet in May 2014, to spend approximately 2 years in the vicinity of the comet, and to place the lander *Philae* on the surface in November 2014.

2. A BRIEF HISTORY OF COMET STUDIES

The realization that the nucleus of a comet was a single, sublimating body prior to the confirmation by direct imaging was the result of several lines of reasoning. In the seventeenth century, it was known that the part of a comet's orbit near the Sun could often be accurately represented by a parabola with the Sun at the focus. This idea was used by Isaac Newton to determine a parabolic orbit for the comet of 1680. Edmond Halley refined the calculation and showed

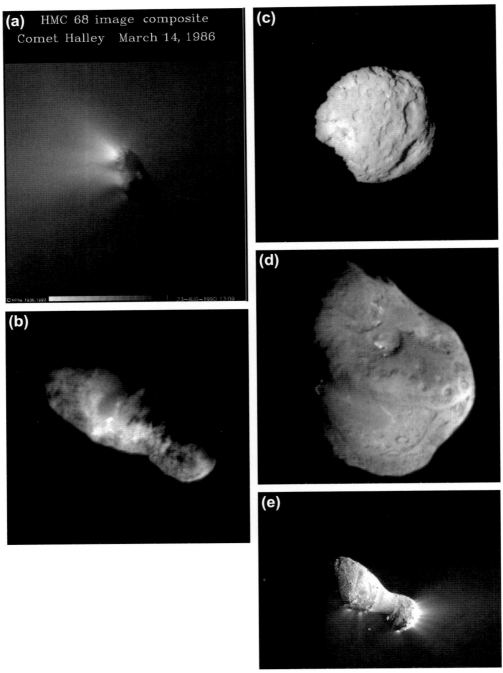

FIGURE 30.2 Images of comet nuclei. (a) Comet Halley nucleus composite. (*Courtesy of H. U. Keller, Max-Planck-Institut für Aeronomie, Katlenburg-Lindau, Germany © MPAE.*) (b) Comet Borrelly. (c) Comet Wild 2. (*Courtesy of NASA and the Stardust Mission Team.*) (d) Comet Tempel 1. (*Courtesy of NASA/JPL-Caltech/UMD.*) (e) Comet Hartley 2. *(Courtesy of NASA/JPL-Caltech/UMD).*

that an ellipse of high eccentricity very accurately represented the comet's orbit. Comet orbits generally are ellipses with high eccentricities. Halley continued to determine the orbits of comets and found that the orbits of comets observed in 1531, 1607, and 1682 were quite similar and had periods of approximately 75–76 years. This was the basis of his famous prediction that the comet that now bears his name would return in 1758.

A complication in the detailed orbit calculations was that Jupiter and Saturn would perturb the orbit through their gravitational attraction. Halley's comet passed perihelion in early 1759. The successful prediction of the return of

FIGURE 30.3 Spectacular image of comet Tempel 1 taken from *Deep Impact's* flyby spacecraft 67 s after the impactor spacecraft's impact. The linear spokes of light radiate away from the impact site. Light from the collision site saturated the camera's detector. Compare with Figure 30.2(d). *Courtesy of NASA/JPL-Caltech/UMD.*

FIGURE 30.4 Image of comet Tempel 1 taken from *Deep Impact's* flyby spacecraft 50 min after impact showing the plume of ejected material. The comet's nucleus is mostly in shadow with the sunlit portion visible on the left-hand side. *Courtesy of NASA/JPL-Caltech/UMD.*

Halley's comet began the development of celestial mechanics and the positional astronomy of comets that flourished in the eighteenth and nineteenth centuries. But the orbit of comet Encke presented another problem. The comet had a very short period of 3.3 years. Many orbits were observed, and it would typically arrive at perihelion about 0.1 day early. The only explanation for this behavior was some sort of nongravitational force, and the only version that has stood the test of time is a "rocket effect" produced by the ejection of material in a preferential direction. Such an effect was suggested by F. W. Bessel based on his observations of a sunward plume of material in Halley's comet in 1835. But how would such a plume of material be produced?

Another problem was the persistence of comets after many passes through the inner solar system. Comets are rich in water ice (discussed later), and small icy clumps or a surface layer of ice on dust grains would not persist.

The Whipple model solves these problems by postulating that the nucleus is a single, rotating, icy body. Ices are poor conductors of heat, and only a relatively thin layer is lost during a perihelion passage. The rocket effect is produced by the reaction force on the nucleus due to the sublimating ices. Historically, the mass loss due to sublimation of ices was assumed to come preferentially from the afternoon side. Just as on the Earth, the warmer temperatures would occur in the afternoon, and the sublimation rate is higher. This type of mass loss would accelerate or retard the comet in its orbit. This basic type of nongravitational force model was used for decades and was successful in producing accurate ephemeris predictions. Nevertheless, the basic model is not realistic when complications are considered, such as the mass loss occurring in jets and precession of the rotation axis. Physically sound models require detailed models of the outgassing surface features and the nucleus rotation. The sublimation of the ices produces the gas molecules that form the gas coma and subsequently the **plasma tail**. When the ices sublimate, the embedded dust particles are released to form the dust coma and the **dust tail**. The dust particles that are not carried away or that fall back onto the nucleus form an insulating crust on the surface.

The bright coma and tails of comets are the features that distinguish them from other solar system objects. Their study was greatly facilitated during the twentieth century by the development of photography. Images and spectra of comets could be accurately recorded and analyzed. The gas and dust comas could extend to approximately $10^5 - 10^6$ km. The nucleus and the coma surrounding it form the comet's head. Dust tails could achieve lengths of roughly 10^7 km, and plasma tails often could achieve lengths of tenths of astronomical units (or several times 1.5×10^7 km). In exceptional cases, plasma tails can exceed 1 AU (or 1.5×10^8 km) in length. Figure 30.7 shows a summary of comet features. Subsequent sections

FIGURE 30.5 Before and after images of comet Tempel 1 showing recession of the smooth flow area (S2), shown at the upper right. The arrows mark the receeding scarp. *Courtesy of NASA; JPL-Caltech; University of Maryland; Cornell University.*

FIGURE 30.6 Before and after images of the *Deep Impact* collision site, marked by arrows in the right-hand image. *Courtesy of NASA; JPL-Caltech; University of Maryland; Cornell University.*

present the physical processes that produce features with these large dimensions, all originating from the small icy bodies shown in Section 1.

Traditionally, comet orbits were classified as short period or long period with the dividing line at periods (P) of 200 years. This evolved to a system with three groups. The Jupiter family contains comets with periods $P \leq 20$ years. These orbits are direct (in the same sense as the Earth's revolution around the Sun) and generally have low inclinations with respect to the plane of the ecliptic. Halley-type comets (HTCs) have periods $20 < P \leq 200$ years. The long-period comets (LPCs) have $P > 200$ years, and their orbital inclinations to the plane of the ecliptic are approximately isotropic. A newer classification system, based on the **Tisserand** *parameter*, T, will be used (Section 7) in the discussion of cometary formation scenarios. The older

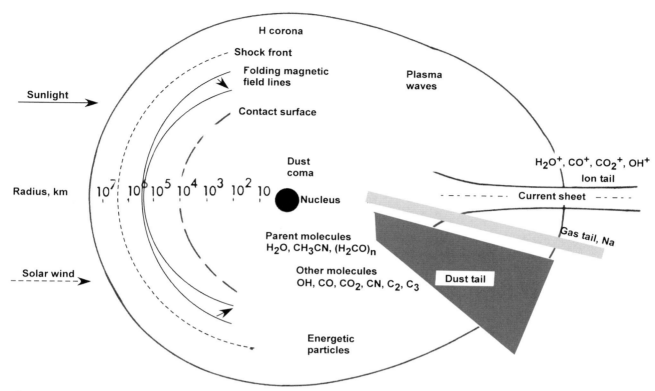

FIGURE 30.7 Summary schematic on a logarithmic scale of cometary features and phenomena. *Reprinted with permission from John C. Brandt and Robert D. Chapman. Copyright © Cambridge University Press, 2004.*

classification schemes were not based on physical parameters, but were mostly for convenience. The newer scheme of H. Levison uses a dividing line of $T = 2$ to neatly divide comets into two classes: ecliptic and nearly isotropic. For details, see Comet Populations and Cometary Dynamics. Basically, the ecliptic comets ($T > 2$) are mostly the Jupiter-family comets (JFCs). A small fraction have $T > 3$ and these cannot cross the orbit of Jupiter. In this case, Encke-type comets have orbits interior to Jupiter's and Chiron-type comets have orbits exterior to Jupiter's. The nearly isotropic comets ($T < 2$) include the HTCs and the LPCs. The latter are nearly isotropic. The HTCs have semimajor axes small enough to fall into a mean motion resonance with a major planet. Their orbital inclinations are not isotropic.

3. PHYSICS OF THE NUCLEUS

The basic physical process—the one that ultimately produces the cometary features (e.g. the tails)—is sublimation of ices. Sublimation is the phase transition that goes directly from the solid to the gaseous state without passing through the liquid state. The evidence for the ice composition of the nucleus—80–90% H_2O (water) ice, roughly 10% CO (carbon monoxide) ice, and small amounts of other ices—is presented in Section 6. The ice in a cometary interior is almost surely amorphous ice. This comes about because ices formed by condensation on a surface at low temperatures do not have energy available to change into the crystalline forms that minimize energy.

When the water ice or snow sublimates, water vapor is produced, and embedded dust particles are released. The energy sources for the sublimation are solar radiation, ice phase transitions, and radioactive decay. Solar radiation deposits energy on the surface or in the near-surface layers. This energy affects the deeper layers by producing a heat wave that moves inward. The transition from amorphous ice to crystalline ice releases energy. Amorphous ice undergoes a transition to cubic ice at approximately 137 K, and cubic ice undergoes a transition to hexagonal ice at approximately 160 K. Model calculations usually treat both transitions as a single energy release event. Radioactive decay is primarily from short-lived isotopes, such as ^{26}Al with a half-life of $\approx 7 \times 10^5$ years. These sources are most important in the deep interior and during approximately the first million years. Diffusion of volatiles could result.

In simple terms, the energy balance of the surface layers is as follows. When a comet is far from the Sun, the energy balance is achieved by the solar radiant energy being reradiated by blackbody (infrared) radiation. The temperature of the surface layers is not high enough to produce significant sublimation. At intermediate distances, the surface temperature is high enough for sublimation, and the solar radiant energy input is balanced both by blackbody reradiation and by sublimation. At closer distances to the Sun, the surface temperature increases further, and essentially the entire solar radiation input is balanced by sublimation. Of course, blackbody reradiation takes place, but it is small in terms of the energy balance. For water ice, sublimation becomes important around 3 AU, and it dominates the energy balance near 1 AU. The sublimation rate is roughly proportional to an exponential containing the temperature, T. Thus, it is extremely sensitive to T. This copious production of material drives cometary activity and produces cometary features as described later.

Naturally, there are complications to this simple picture. When the surface layer ices are sublimated, not all the dust is liberated, and a porous dust mantle is formed. The mantle insulates the ices beneath the surface. This idea has been confirmed observationally. Infrared observations of the surface layers indicate temperatures reasonably close to values expected for a nonsublimating, low-albedo object bathed by sunlight. These temperatures are much higher than the temperatures for sublimating water ice. The ice sublimation probably takes place a few centimeters below the surface. Also, there is no reason to believe that sublimation takes place uniformly over the surface. Regions of enhanced sublimation are expected, a view consistent with the images of comet nuclei that show dust and gas emission predominantly in jets. These jets can produce some of the surface features on the nucleus, and, along with impact craters, they can produce an irregular shape for the nucleus.

Figure 30.8 shows how the surface layers of a comet can become stratified and illustrates the potential complexity of accurate modeling. These layers include many intermediate stages—from the pristine composition of the deep interior to the ejected gas and dust—and these must be modeled accurately. The details of the gas flow through the porous dust layers are important. In recent years, the trend has been to think of the nucleus as a fairly porous body. The porosity is defined as the fraction of the volume occupied by the pores, and values of roughly 0.5 are often discussed. At present, such values can apply to some, but probably not all, comet nuclei.

The rotation of comet nuclei provides an example of how complex some situations can become. Given the extensive ground-based observations of comet Halley and the close-up images taken by *VEGA 1*, *VEGA 2*, and *Giotto*, the determination of the rotation was expected to be straightforward. An initial complication was the reports of different periods of brightness variation. Sorting things out was a major effort. In short, the rotation was complex, and a model with five jets was needed to reproduce the observations. Figure 30.9 shows views of the rotating nucleus through an entire period. The solution was consistent with a constant internal density.

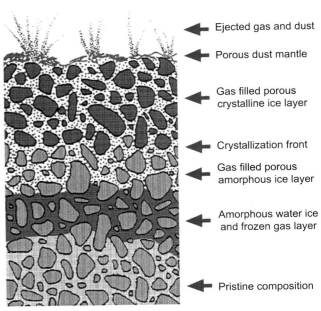

FIGURE 30.8 Schematic showing the layered structure of a cometary nucleus from the pristine composition up to the porous dust mantle. The vertical scale is arbitrary. *Courtesy of D. Prialnik 1997–1999, Modeling gas and dust release from comet Hale-Bopp. Earth, Moon, and Planets, 77:223–230, Figure 1. Copyright © 1999, with kind permission of Springer Science and Business Media.*

The rotation state determined for comet Halley is interesting because it is not in the lowest rotational energy state for a given angular momentum. This would be rotation only around the short axis. The excited rotational state is probably not primordial because estimates of the relaxation time due to frictional dissipation in the comet's interior (caused by flexing as the angular momentum vector moves through the interior) are in the range 10^6-10^8 years. It is probably due to jet activity or splitting of the nucleus.

The splitting of comet nuclei has been observed many times. A recent example is the case of comet LINEAR in early August 2000 (see Figure 30.10). LINEAR is the acronym for Lincoln Near-Earth Asteroid Research. The Lincoln Laboratory is run by the Massachusetts Institute of Technology. Large pieces and fragments of the nucleus are visible in the images. Most of the fragments have an estimated size of less than 500 m. This is an example of "spontaneous" splitting (i.e. there is no apparent correlation with orbital parameters or time in the orbit relative to perihelion). This type of splitting occurs for roughly 10% of dynamically new comets on the first perihelion passage. Splitting can also occur when the nucleus passes close enough to the Sun or a planet and is tidally disrupted. Comet Shoemaker–Levy 9 passed close to Jupiter in July 1992. The disruption produced about 20 fragments (see Figure 30.11). These crashed into Jupiter over several days in July 1994. The tide-induced splittings

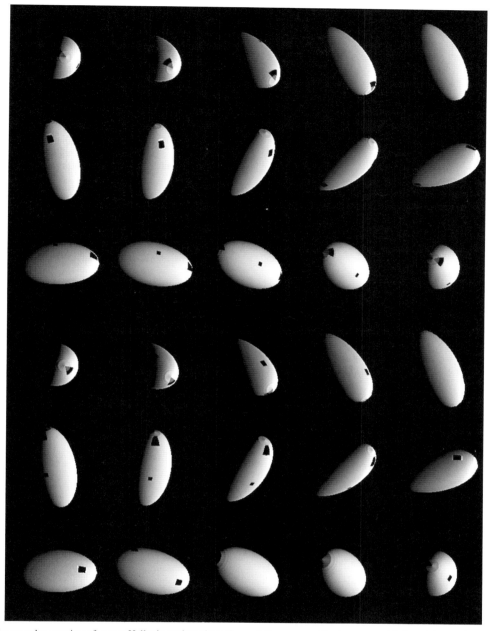

FIGURE 30.9 The complex rotation of comet Halley's nucleus through one full sequence. The images read left to right starting at top left. The time between images is 0.25 days and the sequence repeats after approximately 7.25 days. The five active areas (jets) are marked as low-albedo features. *Courtesy of M. J. S. Belton, Belton Space Initiatives.*

have been used to estimate the tensile strength of the nuclei, and very low values were found. The units of tensile strength are force per unit area (N/m^2) or the Pascal (Pa). The inferred values from splittings are in the range 10^2–10^4 Pa. For comparison, rocks have values $\sim 4 \times 10^6$ Pa, and the value for steel is $\sim 4 \times 10^8$ Pa.

The splittings are consistent with the view of the cometary interior as being porous, having a weak structure, and consisting of agglomerated building blocks called cometesimals. Available evidence indicates that the interior consists of volatile ices (mostly H$_2$O ices, probably amorphous) and dust. The interior does not appear to be differentiated, the compositions are surprisingly uniform, and the ratio of ice to dust does not vary with depth.

Some cometary outbursts may be related to splittings. In a major outburst, the brightness of a comet increases by a factor typically of 6–100, and the outburst lasts for weeks. The observational evidence indicates that the increase in

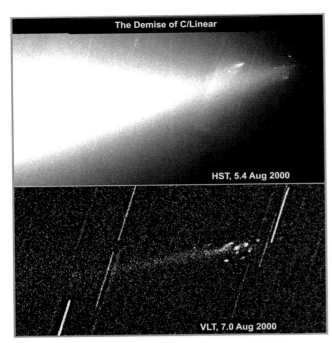

FIGURE 30.10 Splitting of comet LINEAR. (Top) The *HST* image on August 5.4, 2000, showing the dust tail (extending to the left) and several large remnants near the tip of the tail; July 22, 2000, is the estimated date of disintegration of the nucleus. (Bottom) The *Very Large Telescope (VLT)* image on August 7.0, 2000, showing fragments. Image processing was used to suppress light from the diffuse tail. The streaks are star trails. *Courtesy of H. Weaver, Johns Hopkins University; C. Delahodde, O. Hainaut, R. Hook, European Southern Observatory; Z. Levay, Space Telescope Science Institute; and the HST/VLT observing team; NASA/ESA, ESO.*

brightness is due to an increase in the number of dust particles that scatter sunlight. Comet Halley displayed an extraordinary outburst on February 12, 1991, when it was 14.3 AU from the Sun.

Splitting exposes fresh ice surfaces and hence produces enhanced loss of material. An impact from an interplanetary boulder would have much the same effect. A plausible mechanism not involving splitting or impacts uses the crystallization of amorphous ices as the energy source. On this picture, a heat wave propagates inward, triggering the energy release from the amorphous ice and producing pockets of gas that break through to the surface to produce the outburst. This mechanism is plausible for the outburst in comet Halley (mentioned earlier) and in comets that have repeated outbursts, such as comet Schwassmann–Wachmann 1.

Our knowledge of interiors is insecure, but there is general agreement that it is a loosely bound agglomeration of smaller, icy cometesimals with a bulk density less than 1 g/cm^3. Some hints about nucleus structure have come from the *Stardust* images of comet Wild 2. The nucleus shows a highly structured surface that can be described as pockmarked. Some of the features, possibly impact craters, have steep slopes, and the surface must have some cohesive strength. How did these sharp features persist if layers were peeled off by sublimation during every perihelion passage? For comet Wild 2, at least part of the answer lies in its orbital history. Comet Wild 2 was captured into its current (Jupiter-family) orbit by a close encounter with Jupiter only 30 years ago. With an orbital period of 6.4 years, this comet has probably made only a handful of passes through the inner solar system. By comparison, comet Halley has probably made hundreds or thousands of inner solar system passes and thus has a surface smoothed by many sublimation episodes. The surface of comet Wild 2 appears young in terms of sublimation exposure. The steep slopes, which imply some cohesive strength, mean that the surface does not resemble a pile of material held together by gravity. The results from *Deep Impact* have raised new questions and begun the process of understanding the

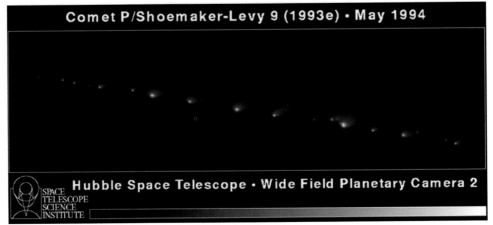

FIGURE 30.11 Comet Shoemaker–Levy 9 on May 17, 1994, as imaged by the *HST*. The fragments extended for over 1.1 million km. *Courtesy of H. A. Weaver and T. E. Smith, Space Telescope Science Institute/NASA.*

nucleus. The surface geology of comet Tempel 1 shows clearly distinct layers that seem to be discrete blocks like geologic strata. The surfaces of the three JFCs (Borrelly, Wild 2, and Tempel 1) are quite different, and the existence of the vastly different surface of the hyperactive come Hartley 2, challenge the notion of a typical comet. Analysis of the ejecta and its evolution yields the following results. The fine particles seen in the ejecta must be from a surface layer at least tens of meters deep. The tensile strength was estimated at 65 Pa or less. This is comparable to the strength of talcum powder or lightly packed snow. The density of the nucleus is about 0.6 g/cm^3 meaning that the interior must be porous with some 50–70% of the volume consisting of empty space.

4. COMA AND HYDROGEN CLOUD

The gas and dust liberated by the sublimation of the ice is the origin of comet features with large dimensions. The coma is the essentially spherical cloud around the nucleus of neutral molecules and dust particles. It is visible in images of comets with low gas production rates (Figure 30.12) or in short-exposure images of comets with high production rates. The size of comas can range up to 10^5–10^6 km. The dusty gas expands at speeds of ≈ 1 km/s, and the flow is transonic in that the flow begins subsonic and becomes supersonic. This is similar to the flow of the solar wind (see The Solar Wind). Because the gas is dragging the dust along, the gas flows faster than the dust. Images and in situ measurements show that the material emission from the nucleus is structured into jets in the near-nucleus region. On the mechanism proposed by M. J. S. Belton, pockets of volatile ices such as CO and CO_2 can be sublimated when the solar energy wave from the surface reaches them. The gases punch through the dust layer to form jets. Collimation

FIGURE 30.12 Comet Giacobini–Zinner on October 26, 1959, showing the coma and a plasma tail extending some 450,000 km. *Photograph by E. Roemer, University of Arizona: official U.S. Navy photograph.*

is provided by the ambient coma and visibility is enhanced by the turbulent flow picking up additional dust particles from the edges of the jet on the surface. The image of comet Hartley 2 (Figure 30.2(e)) shows excellent examples of the jets. Well away from the nucleus, this structure is not usually important.

Most of our observations and measurements of composition in comets refer to the coma region. For some species, the variation with radial distance from the nucleus can be modeled by including creation and destruction mechanisms for parent and daughter molecules. For a molecular gas expanding radially at constant speed, the density falls off as r^{-2} (r is the distance from the nucleus), and the surface brightness (proportional to an integral along a line of sight through the coma) falls off as ρ^{-1} (ρ is the projected distance from the nucleus). The slope on a $\log B$ (brightness) vs $\log \rho$ plot would be -1. Shallower slopes indicate a creation process, and steeper slopes indicate a destruction process. This behavior is observed in molecules such as C_2. These results and results from more detailed modeling lead to an important conclusion.

The molecules measured and observed in the coma are not necessarily the molecules coming directly from the nucleus, but they are part of a chain of creation and destruction of species, presumably from complex molecules in the nucleus to progressively simpler molecules with increasing distance from the nucleus. Thus, the molecules observed are simply the ones that are caught at some specific distance from the nucleus or with the method of observation.

Calculations that include the various changes in composition with the goal of understanding the composition of the original material from the nucleus are very complex and must include gas-phase reactions and photolytic (involving photons) reactions as well as possible interactions between the gas and dust. While progress has been made, final resolution of this problem may require measurements obtained at a cometary surface. As discussed in Section 6, the knowledge of the bulk composition of comets seems secure and is consistent with condensation from a cloud that initially had solar abundances.

Table 30.2 lists chemical species observed spectroscopically and measured by mass spectrometry in comets. This table shows the variety of species in comets and the similarity to interstellar material. This relationship is discussed in Section 4.

The **hydrogen cloud** around comets is much larger than the coma but was not observed until the 1970s. Its existence was predicted in 1968 by L. Biermann. Observations above the Earth's atmosphere were needed because the hydrogen cloud is best seen in Lyman-α (121.6 nm), the resonance line of hydrogen. Figure 30.13 shows the hydrogen cloud of comet Hale–Bopp along with a visible light image. The huge size of the cloud is shown by the yellow disk at the

TABLE 30.2 Measured and Observed Species in Comets

Atoms + Molecules	Ions
H, C, O, S, Na, Fe, Ni, CO, CS, NH, OH, C_2, $^{12}C^{13}C$, CH, CN, ^{13}CN, S_2, SO, H_2, CO_2, HDO, CHO, HCN, DCN, $H^{13}CN$, OCS, SO_2, C_3, NH_2, H_2O, H_2S, HCO, H_2CS, C_2H_2, HNCO, H_2CO, CH_4, HC_3N, CH_3OH, CH_3CN, NH_2CHO, C_2H_6	C^+, N^+, O^+, Na^+, CO^+, CH^+, CN^+, OH^+, NH^+, H_2O^+, HCO^+, CO^+, C^+, CH^+, H_2S^+, NH^+, HCN^+, DCN^+, CH^+, H_3O^+, H_3S^+, NH^+, C_3H^+, CH^+, H_3CO^+, CH^+, C_3H^+

right. This disk is the angular size of the Sun at the comet's distance. The hydrogen cloud has the largest size; however, smaller clouds of oxygen and carbon are also seen.

Modeling the outflow of hydrogen (the lifetime of the H atoms is determined primarily by the proton flux in the solar wind) to produce the observed cloud size shows that the required outflow speed is 8 km/s. This is much larger than the outflow speed in the coma, ≈1 km/s. An additional energy source is needed. If H_2O were photodissociated, a speed of 19 km/s would result, and this value is too high. The likely scenario is that OH is produced by photodissociation and then is further dissociated into H outside the thermalization region. These H atoms and the thermalized H atoms from H_2O photodissociation combine to give the deduced outflow speed of 8 km/s.

The outflow rate of hydrogen, Q_H, provides a good surrogate for the total gas production rate from a comet. For large comets, this rate can approach 10^{31} atoms/s, and the general range is $10^{27}-10^{30}$ atoms/s. The heliocentric variation is roughly $r_h^{-1.3}$ (r_h = the heliocentric distance). This expression follows the practice of basing variations on the value at 1 AU (where comets are most easily observed) and using a power law to give the heliocentric variation.

Early dust measurements were made in the coma of comet Halley by dust detectors on the *VEGA* spacecraft and on *Giotto*. Three basic types of dust composition were found. The CHON particles have only the light elements **C**arbon, **H**ydrogen, **O**xygen, and **N**itrogen. The silicate particles are rich in **S**ilicon, **M**agnesium, and **I**ron. The third type is essentially a mixture of the CHON and silicate types. The differential size distribution can be represented by a power law in size, r^a, with $a \sim -3.5$, for grain sizes

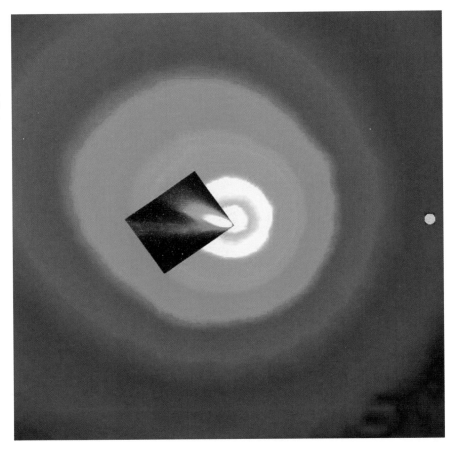

FIGURE 30.13 Hydrogen Lyman-α image taken on April 1, 1997, showing the hydrogen cloud of comet Hale–Bopp (contours in shades of blue) along with a visible image showing the plasma and dust tails. The image is approximately 40° on a side. The small yellow disk shows the angular size of the Sun and the solar direction. *Courtesy of M. Combi, University of Michigan; visual photograph by Dennis di Cicco and Sky and Telescope.*

greater than 20 μm. This implies that most of the dust mass is emitted in large grains. There was also evidence for large numbers of small dust grains down to sizes of 0.01 μm. The results are compatible with the sizes needed in models of the dust tail.

Interest in the dust particles from the coma has increased with the return to the Earth of the dust collected by *Stardust*. Having particles collected in the coma and available for analysis in the laboratory opens a whole new era. The sample return portion of the *Stardust* mission to comet Wild 2 was accomplished by catching the particles in an ultralow density glasslike material called aerogel. The collection exceeded expectations with thousands of particles embedded in the aerogel. The mineral structure has been preserved for many of the grains. Some first results indicate the presence of high-temperature minerals such as olivine, one of the most common minerals in the universe. It certainly did not form inside the comet's cold body. It probably formed near the Sun or from hot regions around other stars. In any event, the discovery that cometary material contains substances formed in hot and cold environments adds a new constraint to formation scenarios.

5. TAILS

The dust and gas in the coma are the raw materials for the comet's tails. The prominent dust and gas (plasma) tails are the traditional identifying characteristic of comets. Dust tails are flat, curved structures and, compared to plasma tails, are relatively featureless. They can reach lengths $\sim 10^7$ km.

Dust particles, once they are decoupled from the coma gas, are in independent orbits around the Sun (see Solar System Dust). But the solar gravitational attraction is not the full value because the dust particles generally stream away from the Sun. An extra force, solar radiation pressure, is acting on the particles. Because both solar gravity and radiation pressure vary as r_h^{-2}, the orbit is determined by initial conditions and an effective gravity. The parameter μ is the ratio of the net force on the tail particle to the gravitational force. Or, the parameter $(1 - \mu)$ gives the normalized nongravitational force.

For a constant emission rate of dust particles with a single size or a small range of sizes, the *syndyne* (or same force) from the Bessel–Bredichin theory is a good description. The tails are tangent to the radius vector (the prolonged Sun–comet line) at the head, and the curvature of the tail increases with decreasing $(1 - \mu)$. An important concept is the fact that the shape of a particle's orbit is not the observed shape of the tail. The observed tail shape is the location of dust particles at a specific time of previously emitted particles.

Another case from the Bessel–Bredichin theory is the *synchrone* (or same time). It is produced by particles with many sizes [or values of $(1 - \mu)$] being emitted at the same time. These features are rectilinear, and the angle with the radius vector increases with time. This type of feature is occasionally observed as synchronic bands.

In practice, comets emit dust particles with a range of sizes and at a rate that varies with time. Several computational approaches that accurately model observed dust tails with reasonable assumptions are available. The size distribution generally peaks at a diameter around 1 μm.

Besides the synchronic bands (mentioned earlier), fine structure in the form of *striae* occasionally appear in dust tails. They are a system of parallel, narrow bands found at large distances from the head. So far, striae appear at heliocentric distances greater than 1 AU and always after perihelion. Figure 30.14 shows a spectacular example in comet Hale–Bopp. Currently, there is no satisfactory explanation. Organization by the solar wind's magnetic field acting on electrically charged dust particles or dust particle fragmentation has been proposed.

Two other dust features are sometimes observed. Antitails or sunward spikes are produced by large dust particles in the plane of the comet's orbit. These particles do not experience the relatively large force that sends the smaller dust particles into the dust tail. They remain near the comet and, when seen in projection, appear to point in the sunward direction. If the Earth is close to the plane of the comet's orbit, a sunward spike is observed. If the Earth is away from the orbital plane but reasonably close, a sunward fan is observed.

The most famous sunward spike of the twentieth century was observed in comet Arend–Roland during April 1957 (Figure 30.15). Comets Kohoutek (December 1973/January 1974) and Halley (February 1986) also showed sunward spikes. Some of these are produced by large ejection speeds in the sunward direction, but most only appear to be sunward in projection.

The neck-line structure is a long, narrow dust feature observed when the comet is past perihelion and the Earth is close to the comet's orbital plane. Dust particles emitted from the comet at low speeds are, in fact, in orbit around the Sun. These orbits return to the orbital plane to produce a dust concentration. The neck-line structure has been observed in comets Bennett, Halley, and Hale–Bopp (Figure 30.16). The neck-line structure in comet Halley was stable and was a major feature for over a month in May and June 1986.

Although the existence of sodium gas tails was implied by spectroscopic observations of earlier comets, the definitive detection was comet Hale–Bopp's dramatic example. Figure 30.17 shows the long, narrow sodium tail. There is also a wide sodium tail superimposed on the dust tail. The source for the narrow tail is probably sodium-bearing

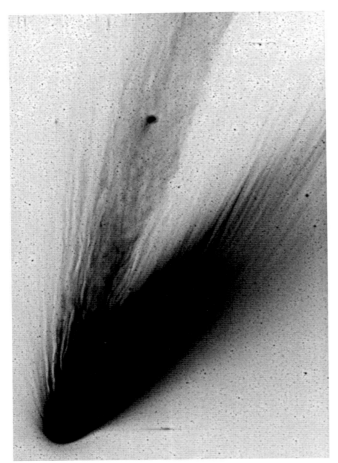

FIGURE 30.14 Comet Hale–Bopp on March 17, 1997, showing well-defined striae in the dust tail at right. The plasma tail is at left. *Courtesy of Kurt Birkle, Max-Planck-Institut für Astronomie, Heidelberg, Germany.*

FIGURE 30.15 Comet Arend–Roland on April 25, 1957, showing the sunward spike. *Photo © UC Regents/Lick Observatory.*

FIGURE 30.16 Comet Hale–Bopp on June 6, 1997, showing the neckline structure, the narrow feature extending to the left from the head. *Image taken by G. Pizarro, European Southern Observatory.*

molecules in the inner coma that are dissociated. The source for the wide tail is probably the dust tail itself.

Sodium tails may well be a common feature of comets. Comet Hale–Bopp's nucleus was very large, with a diameter 60 ± 20 km. Estimates for the total gas production rate near perihelion are as high as 10^{31} molecules/s. Visibility of the sodium tail was enhanced by the sodium atom's high oscillator strength (one of the highest in nature), but the exceptional brightness of comet Hale–Bopp greatly increased the likelihood of observing the sodium tails.

Comet McNaught in January and February of 2007 was the brightest comet seen from the Earth in 40 years and the major visible structure was dust. Figure 30.18 shows an extraordinary image of the comet taken from an observatory in Chile overlooking the Pacific Ocean. Normally, the dust structures (the long streaks) point toward the nucleus. When they do not, the cause is larger particles breaking up and the smaller fragments moving on new orbits under the influence of solar radiation pressure. The size of the comet's nucleus must be quite large to produce this extraordinary dust tail. This view is supported by the large, downstream solar wind disturbance from this comet (discussed below). A curiosity is that the usual, blue plasma tail is not clearly visible in Figure 30.18.

The plasma tails of comets are long and generally straight and show a great deal of fine structure that constantly changes. They are typically 10^5-10^6 km wide, and the lengths recorded optically are routinely several tenths of astronomical units (or several times 1.5×10^7 km). The structure of the plasma tail may extend much farther. Measurements of magnetic fields and ions made on board

Chapter | 30 Physics and Chemistry of Comets

FIGURE 30.17 Images of comet Hale–Bopp in April 1997. The left-hand image records the fluorescence emission from sodium atoms and clearly shows the thin, straight sodium tail. Compare to the right-hand image, which shows the traditional plasma and dust tails. *Courtesy of Gabriele Cremonese, INAF-Astronomical Observatory Padova, and the Isaac Newton Team.*

the *Ulysses* spacecraft (when the Sun, the comet, and the spacecraft were radially aligned) have detected the signatures of comet Hyakutake's plasma tail 550 million km (or 3.7 AU) from the head and comet McNaught's plasma tail 225 million km (or 1.5 AU) from the head.

These tails are composed of electron-molecular ion plasmas. As the neutral molecules in the coma flow outward, they are ionized. Photoionization is the traditional process and easiest to include in models. Impact ionization by solar wind and cometary electrons and ionization by charge exchange also need to be considered. The result is to produce the molecular ions H_2O^+, OH^+, CO^+, CO_2^+, CH^+, and N_2^+. Images of plasma tails, particularly those taken with photographic emulsion, usually show the plasma tail to be bright blue because of strong bands of CO^+ (e.g. see Figure 30.1).

These molecular ions cannot continue their simple outward flow because they encounter the solar wind magnetic field. The **Larmor radius** gives the radius for an ion spiraling around the magnetic field lines, and a typical value is ~ 100 km. Thus, the solar wind and the cometary ions are joined together. The magnetic field lines are said to be loaded with the addition of the pickup ions and their motion slows down. This effect is strong near the comet and weak well away from the comet. The effect causes the field lines loaded with ions to wrap around the comet like a folding umbrella. This behavior is observed. These bundles of field lines loaded with molecular ions form the plasma tail. The central, dense part of the plasma tail contains a current sheet separating the field lines of opposite magnetic polarity. Because the tail is formed by an interaction with the solar wind flow, the tail points approximately antisunward but makes an angle of a few degrees with the prolonged radius vector opposite to the comet's orbital motion. The flow direction is given by the **aberration angle** produced by the solar wind speed and the comet's motion perpendicular to the radius vector. This aberration effect was used by L. Biermann to discover the solar wind in 1951 and to estimate its speed. H. Alfvén introduced the magnetic field into the interaction and gave the basic view of plasma tails presented here. Spacecraft measurements have verified this view. Note that plasma tails usually should be considered as attached to the head of the comet. This contrasts with dust tails where the tail emanates from the head region but the dust particles are on independent orbits. Additional complications from the interaction with the solar wind are a bow shock and plasma waves, which are present over very large volumes of space (also see Planetary Magnetospheres).

The interaction between the solar wind and a comet is clearly shown in Figure 30.19, which is a plot of results from the ion analyzer on the *Deep Space 1* mission. The undisturbed solar wind flow is shown at approximately

FIGURE 30.18 Comet McNaught as imaged from the Paranal Observatory in Chile on January 21, 2007. The Pacific Ocean forms the horizon and the bright object at lower right is Venus. *Courtesy of Sebastian Deiries, European Southern Observatory.*

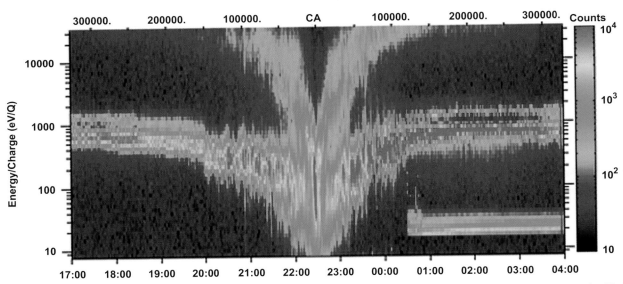

FIGURE 30.19 Plasma results from comet Borrelly measured by the ion analyzer on the *Deep Space 1* mission. The times refer to September 22–23, 2001. The bar at lower right was produced by xenon ions from the spacecraft thruster. See text for discussion. *Courtesy of Los Alamos National Laboratory.*

700 eV/Q, and it steadily decreases toward CA as the solar wind flow is loaded by the addition of cometary ions. The situation reverses as the spacecraft passes through the comet. The higher energy ions are the pickup molecules from the comet.

The exception to the picture of plasma tails usually being attached to the comet's head is when disconnection events (DEs) occur. Here, the entire plasma tail disconnects from the head and drifts away. The comet forms a new plasma tail. Many DEs have been observed over the past century, and Figure 30.20 shows a spectacular example in comet Hyakutake. DEs occur when a comet crosses the heliospheric current sheet (HCS). The HCS is an important feature in the solar wind. It separates "hemispheres" of opposite magnetic polarity and is, in essence, the magnetic equator of the heliosphere. When a comet crosses the HCS, the field lines being captured by the comet (as described earlier) are of opposite polarity. Thus, field lines of opposite polarity are pressed together in the comet causing the field lines to be severed by the process of magnetic reconnection. The old plasma is no longer attached to the head and moves away. Meanwhile, the comet develops a new plasma tail. The sequence is a regular process and repeats at each HCS crossing.

The HCS separates the heliosphere into regions of opposite magnetic polarity and defines the latitudinal structure of the solar wind. Well away from solar maximum, the solar wind is organized into a dense, gusty, slow equatorial region and a less dense, steady, fast polar region. These solar wind properties are clearly reflected in plasma tails. In the polar region, plasma tails have a smooth appearance, show aberration angles corresponding to a fast solar wind, and do not exhibit DEs. In the equatorial region, plasma tails have a disturbed appearance, show aberration angles corresponding to a slow solar wind, and exhibit DEs.

Although cometary X-rays properly belong in the coma discussion, they are included here because they are produced by a solar wind interaction. X-rays in the energy range 0.09–2.0 keV were unexpectedly discovered in comet Hyakutake; see Figure 30.21 for a false-color X-ray image of comet LINEAR. When databases were searched, several more comets were seen as X-ray sources. X-ray emission is an expected phenomenon of all comets.

The principal mechanism is charge exchange between heavy minor species in the solar wind and neutral molecules in the coma. The heavy species in the solar wind are multiply ionized. For example, six-time ionized oxygen can charge exchange with a neutral molecule to produce an ionized molecule and a five-time ionized oxygen in an excited state. X-ray lines are produced when the excited ions spontaneously decay. Spectroscopic X-ray observations have confirmed this mechanism. Some contribution to the total flux may come from electron-neutral thermal bremsstrahlung.

6. COMET CHEMISTRY

The overall chemical composition of comets seems to be rather uniform. Exceptions to this general statement are discussed later. Ultraviolet spectra of comets (see

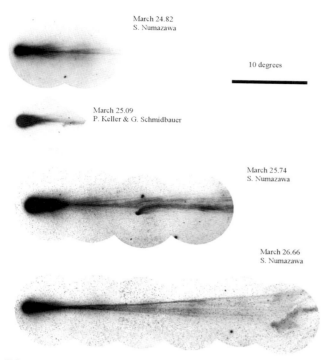

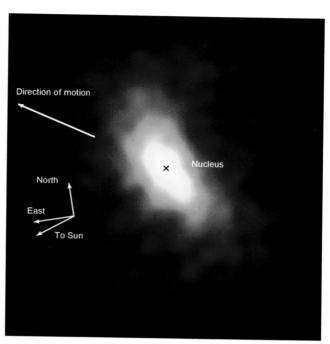

FIGURE 30.20 The spectacular 1996 disconnection event in comet Hyakutake. The 24.82 March, 25.74 March, and 26.66 March images appeared on the July 1996 cover of *Sky and Telescope* and are courtesy of *Sky and Telescope* and S. Numazawa, Japan. The 25.09 March image is courtesy of P. Keller and G. Schmidbauer, *Ulysses Comet Watch*. Image sequence courtesy of the Ulysses Comet Watch.

FIGURE 30.21 False-color rendering of an X-ray image of comet LINEAR obtained on July 14, 2000, by the Chandra X-ray Observatory. *Courtesy of C. M. Lisse, Applied Physics Laboratory, Johns Hopkins University; D. J. Christian, Queens University, Belfast, United Kingdom; K. Dennerl, Max-Planck-Institut für Extraterrestrische Physik, Garching, Germany; and S. J. Wolk, Chandra X-ray Center, Harvard-Smithsonian Center for Astrophysics.*

Figure 30.22) are dominated by the hydrogen (H) Lyman-α line at 121.6 nm and by the hydroxyl (OH) bands at 309.0 nm. This is certainly compatible with the conclusion that the nucleus is composed of roughly 80–90% water ice, 10% carbon monoxide (CO), and many minor constituents.

Table 30.2 lists species in comets that have been observed spectroscopically or measured in situ by mass spectrometers on spacecraft. The list is not exhaustive.

Providing a detailed explanation of the abundances of these species is a formidable task and is subject to many processes in the coma. But, as argued by W. F. Huebner, the situation is comprehensible if we assume a condensation process in the primordial solar nebula at a temperature of 30 K and solar abundances except for H and N. The abundance of hydrogen is determined by the capability to chemically bind to other species. Much is lost from the solar system. Some nitrogen is also lost; for example, when N_2 is formed, the nitrogen is in a form that is not chemically active. A gas mixture consisting of C, O, Mg, Si, S, and Fe in solar abundances with reduced amounts of H and N can condense into molecules at 30 K. The silicates Fe_2SiO_4 and Mg_2SiO_4 are formed from Fe, Mg, Si, and O. Then, the remainder of O goes into H_2O and into HCO and CO compounds. Finally, the remainder of the C, N, and S goes into HCNS compounds.

The result of this fairly straightforward condensation sequence is a material that, when formed into a substantial solid body, resembles comets. By mass, the relative abundance of H_2O:silicates:carbonaceous molecules plus hydrocarbons is approximately 1:1:1. Also, by mass, the abundances of ices:dust is about 1:1.

The temperature of 30 K used in the previous discussion is not only the approriate temperature for the condensation sequences but also consistent with direct determinations of the interior temperatures of cometary nuclei using the *ortho*- to *para*-hydrogen ratio (OPR). Hydrogen in water (and some other compounds) can have the spin of their nuclei in the same direction (*ortho*-water) or in the opposite direction (*para*-water). The OPR depends on the temperature of the water molecules at the time of formation, and the OPR can only be changed by chemical reactions. Thus, the ice can be sublimated in a comet's subsurface layers and flow through the crust into the coma while retaining its original OPR.

Infrared measurements of the OPR for comets Halley, Hale–Bopp, and LINEAR are all consistent with an interior temperature near 30 K. These results are important in discussing formation scenarios. The existence of S_2 in comets may require a formation temperature as low as

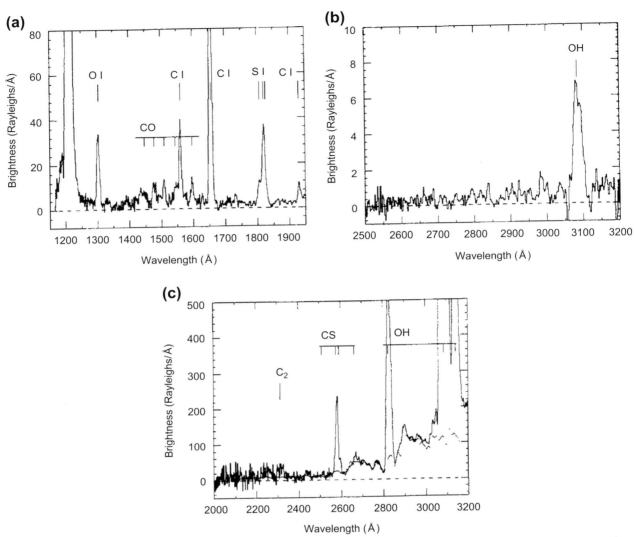

FIGURE 30.22 *International Ultraviolet Explorer* spectra of comet Halley. (a) Spectrum on March 9, 1986: the very strong line close to 1200 Å is the Lyman-α line of neutral hydrogen. (b) Spectrum on September 12, 1985. (c) Spectrum on March 11, 1986. *Courtesy of P. D. Feldman, Johns Hopkins University.*

15 K. While there is some uncertainty in the exact temperature, cold temperatures are required.

A monumental study using narrow-band photometry with major results for the chemical compositions of comets was led by astronomer M. F. A'Hearn. Standardized techniques were used to characterize 85 comets with filters that covered emission bands from CN, C_2, C_3, OH, and NH as well as selected continuum regions. As with the ultraviolet results described previously, the compositions are surprisingly uniform. Barring some unusual event, a comet's production of gases and dust from orbit to orbit (and position in the orbit) is essentially the same. This implies a basically homogeneous interior. When the sample of comets was divided into old and new comets based on their orbital properties, no compositional differences were found.

The preponderance of evidence implies chemical homogeneity for comets. Still, there were some exceptions to the similarity in compositions. A class of comets shows depletions in the carbon chain molecules C_2 and C_3 relative to CN. Comet Giacobini–Zinner is the prototype for this class. Almost all the members of this class are JFCs, but not all JFCs are members of the class. And, there may be some differences in the deuterium-to-hydrogen ratios determined for comets.

7. FORMATION AND ULTIMATE FATE OF COMETS

The icy bodies of the solar system formed as part of the process that produced the Sun, the terrestrial planets, and the giant planets. The icy bodies include some of the asteroids (including the Centaurs, which are bodies with eccentric orbits generally between Saturn and Neptune), comets, and Kuiper Belt Objects (KBOs) (see Kuiper Belt Objects: Physical Studies).

The solar system is thought to have formed from the collapse of an interstellar gas cloud. The collapse process produced a newly formed star with a circumstellar disk of gas and dust, the solar nebula (see The Origin of the Solar System). As discussed in Section 6, cometary material can condense at temperatures of roughly 30 K. Models of the early solar nebula have temperatures of approximately 30 K in the (current) Uranus—Neptune region, and it is reasonable to conclude that comets formed near there, meaning that the material condensed and agglomerated into comet-sized (most with radii in the range 1—10 km) bodies. Note, however, that the uncertainty in the temperatures for models of the presolar nebula may be as much as a factor of 2.

But the story does not end there because most comets are not in the Uranus—Neptune region today. The classical view is that dynamical processes dispersed the icy bodies. Gravitational perturbations by the giant planets sent some of the comets to large distances from the Sun and some into the inner solar system. The latter comets faded long ago. Many of the comets sent to large distances escaped from the solar system, but the ones that are barely bound form a roughly spherical cloud with dimensions of 10^4—10^5 AU. This is the cloud of comets, the Oort cloud, postulated by J. Oort many years ago. It is the source of the LPCs with $P > 200$ years. They are perturbed and sent into the inner solar system by passing stars, passing giant molecular clouds, and the tidal gravitational field of the Milky Way galaxy (see Cometary Dynamics). The HTCs were captured from the population of LPCs. Further study indicates that the Oort cloud probably has two components: the spherical outer cloud discussed previously and a more flattened inner cloud. The boundary between the inner and outer Oort cloud is at approximately 20,000 AU.

The JFCs with $P \leq 20$ years cannot come from the Oort cloud. Their origin requires a close-in, flattened source. The logical source is the Kuiper Belt. Most observed KBOs are much larger than observed JFCs, but this is almost surely due to observational selection. It is reasonable to assume that the size distribution of objects in the Kuiper Belt includes comets. Note that most KBOs are currently found with semimajor axes between 35 and 50 AU. They were not always there but were moved outward along with the outward migration of Uranus and Neptune early in the history of the solar system. The sharp outer boundary for the region of the KBOs was thought to originally be at about 30 AU; it is now at 50 AU.

A different view is evolving from the viewpoint of chemical composition advocated by M. F. A'Hearn and from the viewpoint of orbital dynamics advocated by H. Levison (recall the discussion of comet classification schemes in Section 2). While there are some differences in comet chemistry as discussed in Section 6, the overall differences are small. It is likely that all comets were formed in a region of limited extent with the same physical properties, such as the Uranus—Neptune region of the solar nebula. If the view of essentially all comets being chemically homogeneous is accepted, mechanisms are required to deliver the same comet bodies as ecliptic comets and as nearly isotropic comets. The required dynamical evolution can be done if comets formed in the present Uranus—Neptune region and were moved to the scattered disk population of the Kuiper Belt by Neptune's outward migration. The scattered disk population has perihelia around the orbit of Neptune (perihelion < 40 AU) and their eccentricities are high (e ~ 0.5). These objects are not in stable orbits and the scattered disk is said to be dynamically active. Objects scattered inward come from a flattened source area (as required to match the orbital inclinations) and produce the ecliptic comets. Objects scattered outward evolve with increasing semimajor axes until they are far enough out that galactic perturbations become important. Then, the comet joins the Oort cloud. Thus, scattering near perihelion by Neptune ejects objects from the scattered disk inward to produce ecliptic comets and outward to produce the source of the LPCs or the nearly isotropic comets, the Oort cloud. On this picture, the HTCs are considered part of the nearly isotropic comet population. This scenario is not yet established, but is very attractive. Final acceptance can occur when the dynamical histories and observed chemical compositions of comets converge with models of the solar nebula.

Several possible KBOs with semimajor axes greater than of 200 AU are known. The trans-Neptunian object Sedna has a semimajor axis of about 530 AU. If it is a KBO, it could indicate additional objects at large distances (see Kuiper Belt: Dynamics). And these objects could indicate complications in the basic scheme just outlined.

The dynamical processes that involve comets eject many of them from the solar system. Some estimates suggest that the number lost can be as high as 30—100 for every comet in the Oort cloud. There are many stars similar to the Sun in the solar neighborhood and throughout the galaxy, and if the formation of comets is an integral part of star and planetary system formation, there should be many interstellar comets. Some of these should pass through the solar system. They would reveal themselves by having clearly hyperbolic orbits. A quantitative calculation yields the result that six or more comets should have traveled through the solar system at distances within the orbit of Mars during the past 150 years. None has been observed so far. The newer view of populating the Oort cloud via the scattered disk may help to resolve this apparent discrepancy. J. A. Fernández has noted that the orbital evolution

FIGURE 30.23 *Infrared Astronomical Satellite* false-color image constructed from infrared scans showing the long dust trail of comet Tempel 2. The trail appears as the thin blue line stretching from the comet's head at upper left to lower right. *Courtesy of Mark Sykes, Planetary Science Institute.*

from the scattered disk to the Oort cloud is quite smooth and the process produces a ratio of Oort cloud comets to ejected comets close to 2:1. In other words, most comets moving outward are trapped in the Oort cloud before they could be ejected.

Active comets have a limited life because the volatile materials sublimated away are not replenished. Eventually, the volatiles are gone and the body is inactive. Such objects would be classified as asteroids, and some "asteroids" are clearly dead comets because examples of the transition from comet to asteroid have been documented (see Near-Earth Objects; Main-Belt Asteroids).

Remnants of comets in the solar system include the dust particles on bound Keplerian orbits that, along with an asteroidal contribution, constitute the cloud that produces the zodiacal light from scattered sunlight. The remnants also include the meteoroid streams that produce meteor showers. These streams have long been known to be closely associated with the orbits of comets. Perturbations distribute the rocky or dusty pieces of the comet along its orbit. When the Earth encounters the stream, the pieces enter our upper atmosphere and are observed as meteor showers.

Infrared observations of comets show many long trails of dust, and several were associated with known comets. Figure 30.23 shows the long dust trail of comet Tempel 2. The false-color image from the *Infrared Astronomical Satellite* was constructed from 12-, 60-, and 100-μm scans. The dust trail is the thin blue line stretching from the comet's head at upper left to lower right. The particle sizes are estimated to be in the range 1 mm–1 cm. These dust trails appear to be meteoroid streams in the making (see Solar System Dust).

Comets can also be destroyed by collisions with the Sun, moon, planets, and satellites. The collision of the train of fragments from comet Shoemaker–Levy 9 (see Figure 30.11) with Jupiter in July 1994 is a spectacular example. Collisions of comets with the Earth have been invoked as a source of terrestrial water and possibly a source of complex organic molecules that could be important for the origin of life. At present, there is no consensus on these ideas.

8. SUMMARY

Comets are a diverse population of icy, sublimating bodies that display large-scale phenomena. The central body, the nucleus, has typical dimensions of 1–10 km. The bulk composition is mostly H_2O ice and dust, and the composition of comets is remarkable uniform. The physical processes involved—sublimation of ices in the interior, the flow of gases away from the nucleus, the dissociation and ionization of molecules, and the interaction with the solar wind—continue to provide challenges for scientists. Comets are important to our understanding of other solar system phenomena such as meteors and the zodiacal light. Many problems in comet physics can be solved only by sending spacecraft to the immediate vicinity for close-up imaging and in situ measurements. The past few years have seen several space missions to comets and an extraordinary increase in our knowledge of comets and their diversity. The interiors of comets are not well understood, but results from the *Deep Impact* and *EPOXI* missions provide an important first step. These missions showed that comet Tempel 1's nucleus is porous and that at least the outer layers have very low tensile strength. Ultimately, samples of cometary material must be returned to the Earth for analysis in the laboratory. This has begun with the return of dust particle samples from the *Stardust* mission in 2006. Although the *Rosetta* mission to comet Churyumov–Gerasimenko is expected to greatly expand our knowledge of comets, with the main spacecraft spending an extended time period near the comet and the lander spacecraft landing on and anchoring itself to the nucleus, the return of icy materials to the Earth for analysis is probably well in the future. Comet Tempel 1 has been suggested as a good candidate for a sample return mission.

BIBLIOGRAPHY

A'Hearn, M. F., Belton, M. J. S., Delamare, W. A., Feaga, L. M., Hampton, D., Kissel, J., et al. (June 17, 2011). EP0XI at comet Hartley. *Science, 332*, 1396−1400.

Brandt, J. C. (1999). Comets. In J. K. Beatty, C. C. Petersen, & A. Chaikin (Eds.), *The new solar system*. Cambridge, Massachusetts: Sky Publishing.

Brandt, J. C., & Chapman, R. D. (2004). *Introduction to comets* (2nd ed.). Cambridge, United Kingdom: Cambridge Univ. Press.

Deep impact at comet Tempel 1. Science, 310, (October 14, 2005), 257−283. Special Section.

Fernández, J. A. (2005). *Comets: Nature, dynamics, origin, and their cosmogonical relevance*. Dordrecht: Springer.

Festou, M., Keller, H. U., & Weaver, H. A. (Eds.). (2005). *Comets II*. Tucson: Univ. Arizona Press.

Huebner, W. F. (Ed.). (1990). *Physics and chemistry of comets*. Berlin: Springer-Verlag.

ICARUS. (2013). *Special issue on the Stardust-NExT and EPOXI missions (comets Tempel 1 and Hartley 2)*.

Prialnik, D. (1997−1999). Modeling gas and dust release from comet Hale-Bopp. *Earth, Moon, and Planets, 77*, 223−230.

Stardust at comet Wild 2. Science, 304, (June 18, 2004), 1760−1780. Special Section.

Yeomans, D. K. (1991). *Comets: A chronological history of observations, science, myth, and folklore*. New York: Wiley.

Chapter 31

Comet Populations and Cometary Dynamics

Harold F. Levison and Luke Dones
Southwest Research Institute, Boulder, CO, USA

Chapter Outline

1. Basic Orbital Dynamics of Comets — 706
2. Taxonomy of Cometary Orbits — 709
   2.1. Nearly Isotropic Comets — 711
   2.2. Ecliptic Comets — 712
   2.3. Orbital Distribution of Comets — 712
3. Comet Reservoirs — 713
   3.1. The Oort Cloud — 713
   3.2. The Scattered Disk — 714
   3.3. Formation of the Oort Cloud and Scattered Disk — 717
4. Conclusions — 719
Bibliography — 719

The Solar System formed from a collapsing cloud of dust and gas. Most of this material fell into the Sun. However, since the primordial cloud had a little bit of angular momentum or spin, a flattened disk also formed around the Sun. This disk contained a small amount of mass, as compared with the Sun, but most of the cloud's original angular momentum. This disk, known as the **protoplanetary nebula**, contained the material from which the planets, satellites, asteroids, and comets formed.

The first step in the planet formation process was that the dust, which contained ice in the cooler, distant regions of the nebula, settled into a thin central layer within the nebula. Although the next step has not been fully explained (see *The Origin of the Solar System*), as the dust packed itself into an ever-decreasing volume of space, larger bodies started to form. First came the objects called **planetesimals** (meaning small planets), which probably ranged in size from roughly a kilometer across to tens of kilometers across. As these objects orbited the Sun, they would occasionally collide with one another and stick together. Thus, larger objects would slowly grow. This process continued until the planets or the cores of the gas giant planets formed. (See *Interiors of the Giant Planets*.)

Fortunately for us, planet formation was a messy process and was not 100% efficient. There are a large number of remnants floating around the Solar System. Today we call these small bodies comets and asteroids. These pieces of refuse of planet formation are interesting because they can tell us a lot about how the planets formed. For example, because comets and asteroids are the least chemically processed objects in the Solar System (there is a lot of chemistry that happens on planets), studying their composition tells us about the composition of the protoplanetary nebula.

From our perspective, however, comets and asteroids are most interesting because their orbits can tell us the story of how the planets came together. Just as blood spatters on the wall of a murder scene can tell as much, or more, about the event than the body itself, the orbits of asteroids and comets play a pivotal role in unraveling the planetary system's sordid past.

In this chapter we present the story of where comets originated, where they have spent most of their lives, and how they occasionally evolve through the planetary system and move close enough to the Sun to become the spectacular objects we sometimes see in the night sky.

However, to tell this story, we must work backward because the majority of observational information we have about these objects comes from the short phase when they are close to the Sun. The rest of the story is gleaned by combining this information with computer-generated dynamical models of the Solar System. Thus, in Section 1 we start with a discussion of the behavior of the orbits of comets. In Section 2 we present a classification scheme for comets.

This step is necessary because, as we will show, there are really two stories here. Comets can follow either one of them, but we must discuss each of them separately. In Section 3, we describe the cometary reservoirs that are believed to exist in the Solar System today. In addition, we discuss our current understanding of how these reservoirs came to be. We conclude in Section 4.

1. BASIC ORBITAL DYNAMICS OF COMETS

For the most part, comets follow the basic laws of orbital mechanics first set down by Johannes Kepler and Isaac Newton. These are the same laws that govern the orbits of the planets. In this section, we present a brief overview of the orbits of small bodies in the Solar System. (For a more detailed discussion, see *Solar System Dynamics: Regular and Chaotic Motion*.)

In the Solar System there are eight major planets, many smaller dwarf planets, and vast numbers of smaller bodies, each acting to perturb gravitationally the orbits of the others. The major planets in the Solar System follow nearly circular orbits. They also all lie in nearly the same plane, and so it has been long assumed that the planets formed in a disk. The planets never get close to each other. So, the first-order gravitational effect of the planets on one another is that each applies a torque on the other's orbit, as if the planets were replaced by rings of material smoothly distributed along their orbits. These torques cause both the longitude of perihelion, ϖ, and longitude of the ascending node, Ω, to precess. In particular, $\dot{\varpi} > 0$ and $\dot{\overline{\Omega}} < 0$. The periods associated with these frequencies range from 47,000 to 2,000,000 years in the outer planetary system. Because the masses of the planets are much smaller than the Sun's mass, this is much longer than the orbital periods of the major planets, which are all less than 170 years.

There are four main differences between the orbits of the comets that we see and those of the planets. First, unlike planets, visible comets usually are on eccentric orbits, and so they tend to cross the orbits of the planets. So, they can suffer close encounters with the planets. While these encounters sometimes lead to direct collisions, like the impact of the comet D/Shoemaker-Levy 9 on Jupiter in 1994, more frequently the planet acts as a gravitational slingshot, scattering the comet from one orbit to another. The solid curve in Figure 31.1 shows the temporal evolution of comet 95P/Chiron's semimajor axis according to a numerical integration of the comet's orbit (black curve). This comet currently has $a = 14$ AU, which means it is between Saturn and Uranus, $e = 0.4$, and $i = 7°$. All the changes seen in the figure are because of gravitational encounters with the giant planets. Individual distant encounters lead to small changes, while close encounters lead to large changes. According to this integration, the comet will be ejected

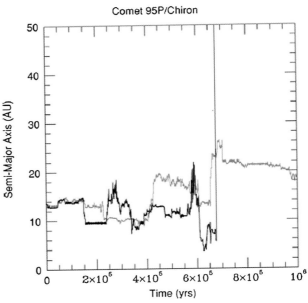

FIGURE 31.1 The long-term evolution of the semimajor axis of comet 95P/Chiron (black curve) and a *clone* of this comet (red curve). These trajectories were determined by numerically integrating the equations of motion of these comets, the Sun, and the four giant planets. The clone was an object with almost the exact same initial conditions as 95P/Chiron, but the position was offset by 1 cm. The fact that the two trajectories diverge shows that the orbit is chaotic.

from the Solar System by a close encounter with Jupiter in 675,000 years.

This calculation illustrates that the orbits of objects on planet-crossing orbits, and thus the comets that we see, are generally unstable. This means that, on timescales very short compared with the age of the Solar System, most of these objects will be ejected from the Solar System by a gravitational encounter with a planet, or hit the Sun or a planet. (Some comets appear to disintegrate spontaneously, for reasons that are not well understood.) So, the comets that we see could not have formed on the orbits that we see them on, because if they had, they would no longer be there. They must have formed, or at least been stored, for long periods of time in a reservoir or reservoirs where their orbits are long lived and they remain cold enough so that their volatiles are, for the most part, preserved. These reservoirs are mainly hidden from us because they are far from the Sun. We discuss cometary reservoirs in more detail in Section 3.

Figure 31.1 also shows that cometary orbits are formally **chaotic**. If the Solar System consisted of only the Sun and one planet, interacting through Newton's law of gravity, the planet's orbit would remain a Keplerian ellipse for all time. The distance between the planet and the Sun would vary periodically, akin to a pendulum. This is an example of **regular** motion. For regular motion, if there were two planetary systems that were exactly the same, except that the position of the planet was slightly offset in one versus the other, this offset would increase linearly with time. However, if three or more bodies are present in the system, **chaos** is possible, meaning that any offset between two nearly identical systems would increase exponentially. In certain cases, such as if the orbit of a comet or asteroid crosses that of a planet, chaos leads to gross unpredictability. That is, in these cases it is impossible to foretell, even qualitatively, the orbit of a comet or asteroid very far into their future or past.

For example, in Figure 31.1, the black curve shows the predicted evolution of 95P/Chiron's semimajor axis, using its nominal orbit. The red curve shows the evolution of an object (the clone) that initially had exactly the same velocity as 95P/Chiron, and an initial position that differed by 1 cm! In less than a million years, a tiny fraction of the age of the Solar System, the orbits are totally different. One clone has been ejected from the Solar System, while the other continues to orbit within the planetary region. This **sensitivity to initial conditions** means that we can never predict where any object in the Solar System will be over long periods of time. By "long periods" we mean at most tens of millions of years for the planets, but for many comets less than a few hundred years. On timescales longer than this, we can only make statistical statements about the ultimate fate of small bodies on chaotic orbits.

The chaotic nature of cometary orbits has important implications for our study of cometary reservoirs. Once we determine the current orbit of a comet, it would be ideal if we could calculate how the orbit has changed with time and trace it backward to its source region. Thus, by studying the physical characteristics of these comets, we could determine what the cometary reservoirs are like. Unfortunately, the unpredictability of chaotic orbits affects orbital integrations that go backward in time as well as those that go forward in time. Thus, it is impossible to follow a particular comet backward to its source region. To illustrate this point, consider the analogy of an initially evacuated room with rough walls and a large open window into which molecules are injected through a narrow hose. Once the system has reached a steady state (i.e. the number of molecules entering through the hose is equal to the number leaving through the window), suppose that the position and velocity of all the particles in the room were recorded, but with less than perfect accuracy. If an attempt were made to integrate the system backward, the small errors in our initial positions and velocities would be amplified every time a molecule bounced off a wall. Eventually, the particles would have "forgotten" their initial state, and thus, in our backward simulation of the gas, more particles would leave through the window than through the hose, simply because the window is bigger. In our case, injection through the hose corresponds to a comet's leaving its reservoir, and leaving through the window corresponds to the many more avenues of escape available to a comet.

So, it is not possible to directly determine which comet comes from which reservoir. Therefore, the only way to use visible comets to study reservoirs is to dynamically model the behavior of comets after they leave the reservoir, and follow these hypothetical comets through the Solar System, keeping track of where they go and what kind of comets they become. By comparing the resulting orbital element distribution of the hypothetical comets to real comet types, we can determine, at least statistically, which type of comets come from which reservoir.

A second major difference between cometary and planetary orbits is that many comets are active. That is, since they are mainly made of dust (or rock) and water ice, and water ice only sublimates within ~ 4 AU of the Sun, comets that get close to the Sun spew out large amounts of gas and dust. This activity is what makes comets so noticeable and beautiful in the night sky. However, outgassing also acts like a rocket engine that can push the comet around and change its orbit. The most obvious effect of these so-called **nongravitational forces** is to change the orbital period of the comet. For example, nongravitational forces increase the orbital period (ΔP) of comet 1P/Halley by roughly 4 days every orbit.

The magnitude, direction, and variation with time of nongravitational forces are functions of the details of an individual comet's activity. Most of the outflow is in the sunward direction; however, the thermal inertia of the spinning nucleus delays the maximum outgassing toward the afternoon hemisphere. Thus, there is a nonradial component of the force. This delay is a function of the angle between the equator of the cometary nucleus and its orbital plane and will vary with time because of seasonal effects. Also, localized jetting can also produce a nonradial force on the comet and will also change the spin state and orientation of the nucleus.

As a result, there is a huge variation of nongravitational forces from comet to comet. For example, for many comets there is no measurable nongravitational force because they are large and/or relatively inactive. Some active comets, like Halley, have nongravitational forces that behave similarly from orbit to orbit. For yet other comets, the magnitude of these forces has been observed to change over long periods of time. A good example of this type of behavior is comet 2P/Encke, which had $\Delta P = -0.13$ days in the early nineteenth century, but now has ΔP of -0.008 days.

In general it is possible to describe the nongravitational accelerations $\vec{a}_{ng}$ that a comet experiences by:

$$\vec{a}_{ng} = g(r)\,[A_1\hat{r} + A_2\hat{t} + A_3\hat{n}],$$

where the A's are constants fit to each comet's behavior, r is the instantaneous heliocentric distance, and $\hat{r}$, $\hat{n}$, and $\hat{t}$ are unit vectors in the radial direction, the direction normal to the orbit of the comet, and the transverse direction, respectively. The value $g(r)$ is related to the gas production rate as a function of heliocentric distance and is usually given as

$$g(r) = 0.111262 \left(\frac{r}{r_0}\right)^{-2.15} \left[1 + \left(\frac{r}{r_0}\right)^{5.093}\right]^{-4.6142}$$

where the parameter $r_0 = 2.808$ AU is the heliocentric distance at which most of the solar radiation goes into sublimating water ice.

A third difference between a planetary orbit and a cometary orbit arises because visible comets tend to be on eccentric (sometimes very eccentric) orbits and on orbits that are inclined with respect to the ecliptic (sometimes even **retrograde** orbits with inclinations greater than 90°). The rates at which the **apse** and **node** of a comet (ϖ and Ω) precess depend upon the comet's eccentricity and inclination. Thus, although cometary orbits precess, like the orbits of the planets, their behavior can be very different from the subtle behavior of the planets. Of particular interest, if the inclination of a comet is large, it can find itself in a situation in which, on average ($\dot{\varpi} = \dot{\Omega}$), that is ϖ and Ω are said to be in resonance with one another. Since these two frequencies are linked to changes in eccentricity and inclination, this resonance allows eccentricity and inclination to become coupled, and allows each to undergo huge changes at the expense of the other. And, since a comet's semimajor axes is preserved in this resonance, changes in inclination also lead to changes in perihelion distance.

An example of this so-called **Kozai resonance** can be seen in the behavior of comet 96P/Machholz 1 (Figure 31.2). 96P/Machholz 1 currently has an eccentricity of 0.96 and an inclination of 60°. Its perihelion distance, q, is currently 0.12 AU, well within the orbit of the planet Mercury. Figure 31.2 shows the evolution of the orbit of 96P/Machholz 1 over the next few thousand years. The Kozai resonance is responsible for the slow systematic oscillations in both inclination and eccentricity (or q, which equals $a \times (1 - e)$). These oscillations are quite large; the inclination varies between roughly 10° and 80°, while the perihelion distance gets as large as 1 AU. According to these calculations, the Kozai resonance will drive this comet into the Sun ($e = 1$) in less than 12,000 years! Similarly, the Kozai resonance was important in driving comet D/Shoemaker-Levy 9 to collide with Jupiter. However, in that case, the comet had been captured into orbit around Jupiter, and the oscillations in i and e were with respect to the planet, not the Sun.

The final gravitational effect that we want to discuss in this section is the effect that the galactic environment has on cometary orbits. Up to this point, our discussion has assumed that the Solar System was isolated from the rest of

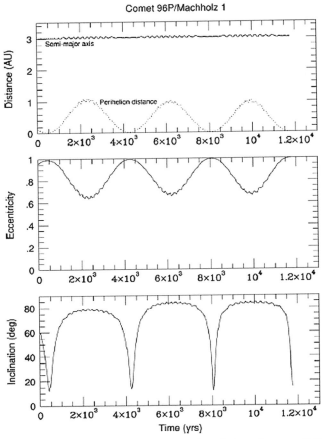

FIGURE 31.2 The long-term dynamical evolution of comet 96P/Machholz 1, which is currently in a Kozai resonance. Three panels are shown. The top presents the evolution of the comet's semimajor axis (solid curve) and perihelion distance (dotted curve). The middle and bottom panels show the eccentricity and inclination, respectively. Because of the Kozai resonance, the eccentricity and inclination oscillate with the same frequency, but are out of phase (i.e. eccentricity is large when inclination is small and vice versa). According to this calculation, this comet will hit the Sun in less than 12,000 years.

the Universe. This, of course, is not the case. The Sun, along with its planets, asteroids, and comets, is in orbit within the Milky Way Galaxy, which contains hundreds of billions of stars. Each of these stars is gravitationally interacting with the members of the Solar System. Luckily for the planets, the strength of the Galactic perturbations varies as a^2, so the effects of the Galaxy are not very important for objects that orbit close to the Sun. However, if a comet has a semimajor axis larger than a few thousand AU, as some do (see Section 2), the Galactic perturbations can have a major effect on its orbit.

For example, Figure 31.3 shows a computer simulation of the evolution through time of the orbit of a hypothetical comet with an initial semimajor axis of 20,000 AU, roughly 10% of the distance to the nearest star. (For scale remember that Neptune is at 30 AU.) For the sake of discussion, it is useful to

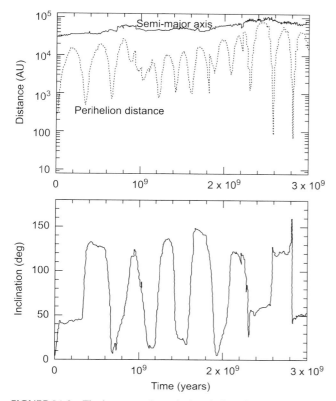

FIGURE 31.3 The long-term dynamical evolution of a fictitious object initially at 20,000 AU from the Sun under the gravitational perturbations of the Galaxy. Two panels are shown. The top presents the evolution of the comet's semimajor axis (solid curve) and perihelion distance (dotted curve; recall that $e = 1 - q/a$). The bottom panel shows the inclination.

divide the evolution into two superimposed parts: (1) a slow secular change in perihelion distance (i.e. eccentricity) and inclination, and (2) a large number of small, but distinct jumps leading to a **random walk** in the orbit.

The secular changes are because of the smooth background gravitational potential of the Galaxy as a whole. If we define a rectangular coordinate system $(\tilde{x}, \tilde{y}, \tilde{z})$ centered on the Sun, such that $\tilde{x}$ points away from the galactic center, $\tilde{y}$ points in the direction of the galactic rotation, and $\tilde{z}$ points toward the south, it can be shown that the acceleration of a comet with respect to the Sun is where $\Omega_0 = 27.2 \pm 0.9$ km/s/kpc is the Sun's angular speed about the Galactic center, $\delta \equiv -\frac{A+B}{A-B}$ and $A = 14.5 \pm 1.5$ km/s/kpc and $B = -12 \pm 3$ km/s/kpc are Oort's constants of Galactic rotation, $\rho_0 = 0.1 \, M_\odot/\text{pc}^3$ is the density of the galactic disk in the solar neighborhood, and G is the gravitational constant. The value of δ is usually assumed to be zero.

Because of the nature of the above acceleration, it acts as a torque on the comet. As a result, the smooth part of the Galactic perturbations can change a comet's eccentricity and inclination, but not its semimajor axis. In addition, the eccentricity and inclination oscillate in a predictable way.

In this example, in Figure 31.3 the oscillation period is approximately 300 million years (Myr). However, this period scales as $a^{-3/2}$, and thus the oscillations are faster for large semimajor axes. The small jumps are because of the effects of individual stars passing close to the Sun. Since these stars can come in from any direction, the kick that the comet feels can affect all the orbital elements, including the semimajor axis. The apparent random walk of the comet's semimajor axis seen in the figure is because of this effect.

2. TAXONOMY OF COMETARY ORBITS

The first step toward understanding a population is to construct a classification scheme that allows one to place like objects with like objects. This helps us begin to construct order from the chaos. However, before we talk about comet classification, we need to make the distinction between what we see and what is really out there. As we describe in much more detail below, most of the comets that we see are on orbits that cross the orbits of the planets. For example, the most famous comet, 1P/Halley (the "1P" stands for the first known *periodic* comet, see below), has $q = 0.6$ AU and an **aphelion distance** (farthest distance from the Sun) of 35 AU. Thus, it crosses the orbits of all the planets except Mercury. But planet-crossing comets represent only a very small fraction of the comets in the Solar System, because we can only easily see those comets that get close to the Sun.

Comets are very small compared with the planets. As a result, we cannot see comets very far away. For example, 1P/Halley, a relatively large comet, is a roughly (American) football-shaped object roughly 16 km long and 8 km wide. The farther away an object is, the fainter it is. The brightness (b) of a light-bulb decreases as the square of the distance d from the observer ($b \propto 1/d^2$). However, this is not true for objects in the Solar System that shine by reflected sunlight. To first approximation, the brightness of a solid sphere seen from the Earth is proportional to $1/(d_\odot^2 d_\oplus^2)$, where $d_\odot$ and $d_\oplus$ are the distance between the object and the Sun and Earth, respectively. As objects get farther from the Sun, they get less light from the Sun and so reflect less (that is the $1/d_\odot^2$ term). Also, the further they get from us, the fainter they appear (that is the $1/d_\oplus^2$ term). In the outer Solar System, $d_\oplus$ and $d_\odot$ are nearly equal and thus $b \sim 1/d^4$.

It is even worse for a comet since it is not simply a solid sphere. As described above, as a comet approaches the Sun, its ice begins to sublimate. The resulting gas entrains dust from the comet's surface, forming a halo known as the **coma**. Because the dust is made of small objects with a lot of surface area, it can reflect a lot of sunlight. So, this cometary activity makes the comet much brighter. Observational studies show that as a comet approaches the Sun, its brightness typically increases as $1/(d_\odot^4 d_\oplus^2)$. The result of all this activity is that it can make an object that would normally be very difficult to see, even through a telescope, into a body visible with the naked eye. Thus, we know of only a very small fraction of comets in the Solar System and this sample is **biased** because it represents only those objects that get close to the Sun. However, before we can try to understand the population as a whole, we need to first try to understand the part that we see.

The practice of developing a classification scheme or taxonomy is widespread in astronomy, where it has been applied to everything from Solar System dust particles to clusters of galaxies. Classification schemes allow us to put the objects of study into a structure in which we can look for correlations between various physical parameters and begin to develop evolutionary models. In this way, classification schemes have played a crucial role in advancing our understanding of the universe. However, we must be careful not to confuse these schemes with reality. In many cases, we are forcing a classification scheme on a continuum of objects. Then we argue over where to draw the boundaries. The fact that we astronomers find cubbyholing objects convenient does not imply that the universe will necessarily cooperate. With this caveat in mind, in the remainder of this section we present a scheme for the classification of cometary orbits.

Historically, comets have been divided into two groups: long-period comets (with periods greater than 200 years) and short-period comets (with $P < 200$ years). This division was developed to help observers determine whether a newly discovered comet had been seen before. Since orbit determinations have been reliable for only about 200 years, it may be possible to link any comet with a period less than this length of time with previous apparitions. Conversely, it is very unlikely to be possible to do so for a comet with a period greater than 200 years, because even if it had been seen before, its orbit determination would not have been accurate enough to prove the linkage. Thus this division has no physical justification and is now of historical interest only. Unfortunately, there does not yet exist a physically meaningful classification scheme for comets that is universally accepted. Nonetheless, such schemes exist. Here we present a scheme developed by one of the authors roughly 10 years ago. A flowchart of this scheme is shown in Figure 31.4.

The first step is to divide the population of comets into two groups. Astronomers have found that the most physically reasonable way of doing this is to employ the so-called **Tisserand parameter**, which is defined as

$$T \equiv a_J/a + 2\sqrt{(1-e^2)a/a_J} \; \cos i,$$

where a_J is Jupiter's semimajor axis. This parameter is an approximation to the **Jacobi constant**, which is an **integral of the motion** in the **circular restricted three-body problem**. The circular restricted three-body problem, in turn, is a well-understood dynamical problem consisting of two massive objects (mainly the Sun and Jupiter in this context) in circular orbits about one another, with a third,

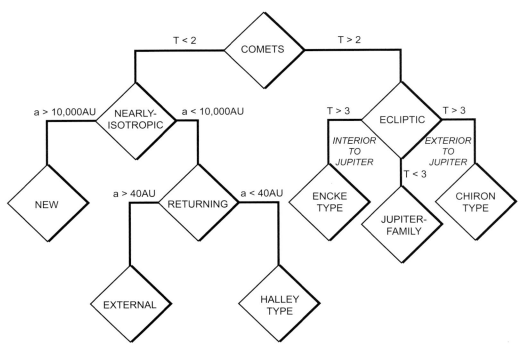

FIGURE 31.4 A flowchart showing the cometary classification scheme used in this chapter.

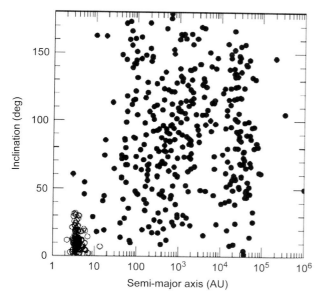

FIGURE 31.5 The inclination—semimajor axis distribution of all comets in the 2003 version of Marsden and Williams' *Catalogue of Cometary Orbits*. Comets with $T > 2$ are marked by the open circles, while comets with $T < 2$ are indicated by the filled circles.

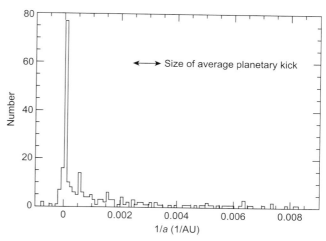

FIGURE 31.6 The distribution of inverse semimajor axis a, which measures the strength with which comets are gravitationally bound to the Solar System, for the known nearly-isotropic comets.

very small, body in orbit about the massive pair. If, to zeroth order, a comet's orbit is approximately a perturbed Kepler orbit about the Sun, then, to first order, it is better approximated as the small object in the circular restricted three-body problem with the Sun and Jupiter as the massive bodies. This means that as comets gravitationally scatter off Jupiter or evolve because of processes like the Kozai resonance, T is approximately conserved. The Tisserand parameter is also a measure of the relative velocity between a comet and Jupiter during close encounters, $v_{rel} \sim v_J \sqrt{3 - T}$, where v_J is Jupiter's orbital speed around the Sun. Objects with $T > 3$ cannot cross Jupiter's orbit in the circular restricted case, being confined to orbits either totally interior or totally exterior to Jupiter's orbit.

Figure 31.5 shows a plot of inclination versus semimajor axis for known comets. Astronomers put the first division in our classification scheme at $T = 2$. Objects with $T > 2$ are shown as open circles in the figure, while those with $T < 2$ are the filled circles. The bodies with $T > 2$ are confined to low inclinations. Thus, we call these objects **ecliptic comets**. We call the $T < 2$ objects **nearly isotropic** comets to reflect their broad inclination distribution. We now discuss each of these in turn.

2.1. Nearly Isotropic Comets

Nearly isotropic comets (hereafter NICs) are divided into two groups: dynamically "new" comets and "returning" comets. This division is one that has its roots in the dynamics of these objects and is based on the distribution of their semimajor axes, a. Figure 31.6 shows a histogram of $1/a$, which is proportional to orbital binding energy $\overline{E} = -\frac{GM_\odot}{2a}$. These values of semimajor axes were determined by numerically integrating the observed trajectory of each comet backward in time to a point before it entered the planetary system. Taken at face value, a comet with $1/a < 0$ is unbound from the Sun, that is it follows a hyperbolic orbit. However, all of the negative values of $1/a$ are because of errors in orbit determination either due to poor astrometry or due to uncertainties in the estimates of the nongravitational forces. Thus, we have yet to discover a comet from interstellar space. The fraction of comets that suffer from this problem is small and we will ignore them for the remainder of this chapter.

The most striking feature of this plot is the peak at about $1/a \sim 0.00005/\text{AU}$, that is $a \sim 20{,}000$ AU. In 1950, this feature led Jan Oort to conclude that the Solar System is surrounded by a spherically symmetric cloud of comets, which we now call the Oort cloud. The peak in the $1/a$ distribution of NICs is fairly narrow. And yet, the typical kick that a comet receives when it passes through the planetary system is approximately $\pm 0.0005/\text{AU}$, that is a factor of 10 larger than the energy of a comet initially in the peak (Figure 31.6). Thus it is unlikely that a comet that is in the peak when it first passes through the Solar System will remain there during successive passes. We conclude from this argument that comets in the peak are dynamically "new" in the sense that this is the first time that they have passed through the planetary system.

Comets not in the peak ($a \lesssim 10{,}000$ AU) are most likely objects that have been through the planetary system before. Comets with $a \ll 20{,}000$ AU that are penetrating the planetary system for the first time cannot make it into the inner Solar System where we see them as active comets without first encountering a planet (see Section 3.1 for a more complete discussion). Therefore, we should expect to see few comets directly from the Oort cloud with semimajor axes smaller than this value. We can conclude that an NIC not in the peak is a comet that was initially in it but has evolved to smaller a during previous passes through the planetary system. These comets are called "returning" comets. The boundary between new and returning comets is usually placed at $a = 10{,}000$ AU.

Returning comets are, in turn, divided into two groups based on their dynamics. Long-term numerical integrations of the orbits of returning comets show that a significant fraction of those with semimajor axes less than about 40 AU are temporarily trapped in what are called **mean motion resonances** with one of the giant planets during a significant fraction of the time they spend in this region of the Solar System. Such a resonance is said to occur if the ratio of the orbital period of the comet to that of the planet is near the ratio of two small integers. For example, on average Pluto orbits the Sun twice every time Neptune orbits three times. So, Pluto is said to be in the 2:3 mean motion resonance with Neptune. Comet 109P/Swift–Tuttle, with a semimajor axis of 26 AU, is currently trapped in a 1:11 mean motion resonance with Jupiter. Mean motion resonances can have a large effect on the orbital evolution of comets because they can change eccentricities and inclinations, as well as protecting the comet from close encounters with the planet it is resonating with. This is true even if the comet is only temporarily trapped. In our classification scheme, comets that have a small enough semimajor axis to be able to be trapped in a mean motion resonance with a giant planet are designated as **Halley-type** comets, named for its most famous member comet 1P/Halley. Returning comets that have semimajor axes larger than this are known as **external** comets. Although it is not really clear exactly where the boundary between these two type of comets should be, we place the boundary at $a = 40$ AU.

2.2. Ecliptic Comets

Recall that ecliptic comets are those comets with $T > 2$. These comets are further divided into three groups. Comets with $2 < T < 3$ are generally on Jupiter-crossing orbits and are dynamically dominated by that planet. Thus, we call these **Jupiter-family** comets. This class contains most of the known ecliptic comets. As described above, comets with $T > 3$ cannot cross the orbit of Jupiter and thus should not be considered members of the Jupiter family. A comet that has $T > 3$ and whose orbit is interior to that of Jupiter is designated a *Encke type*. This class is named after its best-known member, 2P/Encke. 2P/Encke is a bright active comet that is decoupled from Jupiter. Its aphelion distance is only 4.2 AU.

A comet that has $T > 3$ and has a semimajor axis larger than that of Jupiter is known as a **Chiron type**, again named after its best-known member, 95P/Chiron. As we discussed in Section 2.2, Chiron has a semimajor axis of 14 AU and a perihelion distance of 8 AU, putting it well beyond the grasp of Jupiter. Indeed, 95P/Chiron is currently dynamically controlled by Saturn. Although 95P/Chiron has a weak coma and is designated as a comet by the International Astronomical Union (IAU), it is also considered to be part of a population of asteroids known as **Centaurs**, which are found on orbits beyond Jupiter and that cross the orbits of the giant planets. The IAU distinguishes between a comet and an asteroid based on whether an object is active or not. This distinction is therefore not dependent on an object's dynamical history or where it came from. Thus, Chiron is simply a member of the Centaurs, of which there are currently a few dozen known members. For the remainder of this chapter, we will not distinguish between the **Chiron-type** comets and the Centaur asteroids, and will call both Centaurs.

2.3. Orbital Distribution of Comets

Figure 31.7 shows the location of the comet classes described above as a function of their Tisserand parameter and semimajor axis. Also shown is the location of all comets in the 2003 version of Marsden and Williams' *Catalogue of Cometary Orbits*. The major classes of ecliptic and NICs are defined by T and are independent of a. The ranges of these two classes are thus shown with arrows only. The extent of the subclasses is shown by different

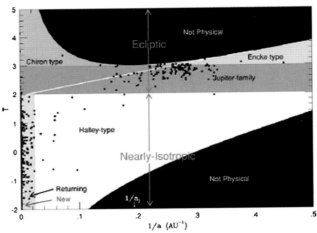

FIGURE 31.7 The location of the classes in our adopted comet taxonomy as a function of the Tisserand parameter (T) and semimajor axis (a). The major classes of ecliptic comets and NICs are defined by their values of T. The ranges of these two classes are thus shown with arrows only. The extent of each subclass is shown by different shadings. Also shown is the location of all the comets with $1/a > 0$ in the 2003 version of Marsden and Williams' *Catalogue of Cometary Orbits*. The white curve shows the relationship of T versus a for a comet with $q = 2.5$ AU and $i = 0$. Comets above and to the left of this line have $q > 2.5$ AU and thus are difficult to detect.

shadings. Also shown is the location of all the comets with $1/a > 0$ in the catalog. The white curve shows the relationship of T versus a for a comet with $q = 2.5$ AU and $i = 0$. Comets above and to the left of this line have $q > 2.5$ AU and thus are difficult to detect. By far, most comets in the plot are new or returning NICs. The second largest group consists of the Jupiter-family comets.

We end this section with a short discussion of the robustness of this classification scheme. Long-term orbital integrations show that comets rarely change their primary class (*ecliptic* versus *nearly isotropic*), but do frequently change their subclass (i.e. *new* versus *returning* or *Jupiter family* versus *Chiron type*). This result suggests that ecliptic comets and NICs come from different source reservoirs. In particular, as we will now describe, the NICs come from the Oort cloud, while the ecliptic comets are thought to originate in a structure that we call the **scattered disk**.

3. COMET RESERVOIRS

As we discussed above, the active comets that we see are on unstable short-lived orbits because they cross the orbits of the planets. For example, the median dynamical lifetime of a Jupiter-family comet (defined as the span of time measured from when a comet first evolves onto Jupiter-family comet-type orbit until it is ejected from the Solar System, usually by Jupiter) is only about 300,000 years. So, these comets must have been stored in one or more reservoirs, presumably outside the planetary region, for billions of years before being injected into the inner Solar System where they can be observed. These reservoirs are far from the Sun (and they would have to be to store an ice ball for 4 billion years), and thus much of what we know about them has been learned by studying the visible comets and linking them to their reservoirs through a theoretical investigation of the orbital evolution of comets. As we currently understand things, there are two main cometary reservoirs: the Oort cloud and the scattered disk. We discuss each of these separately.

3.1. The Oort Cloud

NICs originate in the Oort cloud, which is a nearly spherical distribution of comets (at least in the outer regions of the cloud), centered on the Sun. The position of its outer edge is defined by the Solar System's tidal truncation radius at about 100,000–200,000 AU from the Sun. At these distances, the gravitational effect of stars and other material in the Galaxy can strip a comet away from the Solar System. This edge can be seen in the distribution of NICs shown in Figure 31.6. For reasons described below, we have no direct information about the location of the Oort cloud's inner edge, but models of Oort cloud formation (see Section 3.3) predict that it should be between 2000 and 5000 AU.

The orbits of comets stored in the Oort cloud evolve because of the forces from the Galaxy. As shown in Figure 31.3, the primary role of the Galaxy is to change the angular momentum of the comet's orbit, causing large changes in the inclination and, more importantly, the perihelion distance of the comet. Occasionally, a comet will evolve so that its perihelion distance falls to within a few AU of the Sun, thus making it visible as a new NIC. As we discussed above, the new comets that we see have semimajor axes larger than 20,000 AU, as illustrated by the spike in Figure 31.6. This led Jan Oort to suggest that the inner edge of the Oort cloud was at this location. However, this turns out not to be the case. In order for us to see a new comet from the Oort cloud, it has to get close to the Sun, which generally means that its perihelion distance, q, must be less than 2 or 3 AU.[1] However, during the perihelion passage before the one on which we see a comet for the first time, its perihelion distance must have been outside the realm of the gas giants ($q > 15$ AU), because if the comet had q near either Jupiter or Saturn when it was near perihelion, it would have received a kick from the planets that would have knocked it out of the spike. Thus, new comets can only come from the region in the Oort cloud in which the Galactic tides are strong enough that the change in perihelion in one orbit (Δq) is greater than ~ 10 AU. It can be shown that the timescale on which a comet's perihelion changes is

$$\tau_q = 6.6 \times 10^{14} \text{ year } a^{-2} \Delta q / \sqrt{q},$$

in the current galactic environment where a, Δq, and q are measured in AU. Thus, only those objects for which τ_q is larger than the orbital period can become a visible new comet. For $\Delta q = 10$ AU and $q = 15$ AU, this occurs when $a > 20,000$ AU.

The above result does not imply that Oort comets far inside of 20,000 AU do not contribute to the population of NICs. In fact, they do. It is simply that these objects do not become active comets until their orbits have been significantly modified by the giant planets.

Figure 31.8 shows the cumulative inclination distribution for a combination of new and external comets (solid curve) and Halley-type comets (dotted curve). The solid curve is what would be expected from an isotropic Oort cloud. The curve follows a roughly $\sin(i)$ distribution, which has a median inclination of 90° and thus has equal numbers of prograde and retrograde orbits. It is these data

---

1. Comets are sometimes discovered at larger perihelion distances because the comet is unusually active because of the sublimation of ices, such as carbon monoxide, that are more volatile than water ice. The current record holder, the new comet C/2003 A2 Gleason, had $q = 11$ AU. Some become returning comets. Indeed, from modeling the inclination distribution of the Halley-type comets, we think that some objects from the inner regions of the Oort cloud eventually become NICs.

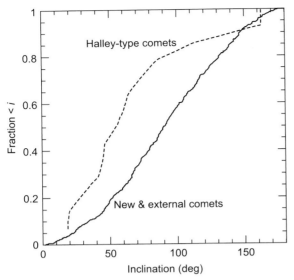

FIGURE 31.8 The cumulative inclination distribution of the NICs in Marsden and Williams' catalog. We divide the population into two groups: Halley types ($a < 40$ AU) and a combination of new and external comets.

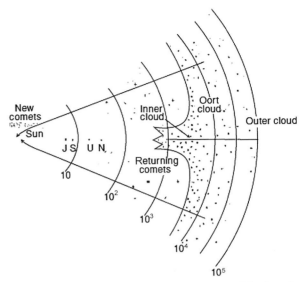

FIGURE 31.9 An artist's conception of the structure of the Oort cloud. In particular, the locations of the inner and outer edges of the Oort cloud, and where the cloud is flattened, are shown with respect to the location of the giant planets. Note that the radial distance from the Sun is spaced logarithmically. The location of the returning comets and the source for the new comets are also illustrated.

that astronomers use to argue that the outer Oort cloud is basically spherical.

The inclination distribution of the Halley-type comets is quite different from that of the rest of the NICs. Almost 80% of Halley-type comets are on prograde orbits ($i < 90°$); the median inclination is only 55°. Numerical simulations of the evolution of comets from the Oort cloud to Halley-type orbits show that the inclination distribution of the comets is approximately conserved during the capture process. This means that the source region for these comets should have the same inclinations, on average, as the dotted curve in Figure 31.8. The only way to reconcile this with the roughly spherical shape of the outer Oort cloud is if the inner regions of the Oort cloud are flattened into a disk-like structure. Indeed, simulations suggest that the inner Oort cloud must have a median inclination of between 10° and 50° for it to match the observed inclination distribution of Halley-type comets. Figure 31.9 shows an artist's conception of what the Oort cloud may look like in cross-section.

3.2. The Scattered Disk

To start the discussion of the scattered disk, we turn our attention back to Figure 31.5, which shows the semimajor axis–inclination distribution of the known comets. There is a clear concentration of comets on low-inclination orbits near $a \sim 4$ AU. Indeed, 27% of all the comets in the catalog lie within this concentration. As we described above, we call these objects ecliptic comets, and most are Jupiter-family comets.

Until the 1980s, the origin of these objects was a mystery. Even at that time it was recognized that the inclination distribution of comets does not change significantly as they evolve from long-period orbits inward. This is a problem for a model in which these comets originate in the Oort cloud, as most astronomers believed, because the median inclination of the Jupiter family is only 11°. So, dynamicists argued that Jupiter-family comets could not come from the Oort cloud, but must have originated in a flattened structure. Indeed, it was suggested that these objects originated in a disk of comets that extends outward from the orbit of Neptune. Spurred on by this argument, observers discovered the first trans-Neptunian object in 1992. Although this object is about a million times more massive then the typical ecliptic comet (it needs to be much larger than a typical comet, or we would not have seen it that far away), it was soon recognized that it was part of a population of objects both large and small—mainly small.

Since 1992, the trans-Neptunian region has been the focus of intense research, and over a 1000 objects are now known to reside there. The diversity (both physical and dynamical) of its objects make it one of the most puzzling and fascinating places in the Solar System. As such, a complete discussion is beyond the scope of this chapter and, indeed, chapters on the Kuiper Belt are dedicated to this topic (See *Kuiper Belt: Dynamics; Kuiper Belt Objects: Physical Studies*). For our purposes, it suffices to say that the trans-Neptunian region is inhabited by at least two populations of objects that roughly lie in the same region of physical space, but have very different dynamical properties.

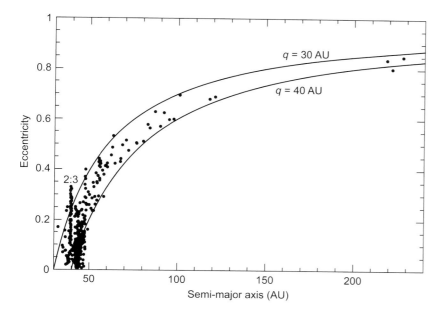

FIGURE 31.10 The eccentricity—semimajor axis distribution for the known trans-Neptunian objects with good orbits as of November 2005. We truncated the plot at 250 AU to resolve the inner regions better. Two curves of constant perihelion distance (q) are shown. In addition, the location of Neptune's 2:3 mean motion resonance is marked.

These are illustrated in Figure 31.10, which shows the semimajor axis and eccentricity of all known trans-Neptunian objects with good orbits as of November 2005.

The first population of interest consists of those objects which are on orbits that are stable for the age of the Solar System. These objects mostly have perihelion distances (q) larger than 40 AU, or are in mean motion resonances with Neptune. Of particular note are the bodies in Neptune's 2:3 mean motion resonance, which are marked in the figure. Pluto is a member of this group. Even though some objects in the resonances are on orbits that cross the orbit of Neptune, they are stable because the resonance protects them from close encounters with that planet. All in all, we call this population the **Kuiper Belt**.[2]

The second population is mainly made up of objects with small enough perihelion distances that Neptune can push them around as they go through perihelion. Because of this characteristic, we call this population the **scattered disk**. These are mainly nonresonant objects with $q < 40$ AU. (See *Kuiper Belt: Dynamics* for a more detailed definition.) Although most of the trans-Neptunian objects thus far discovered are members of the Kuiper Belt as defined here, it turns out that this is because of observational bias, and the Kuiper Belt and scattered disk contain roughly the same amount of material. In particular, the scattered disk contains about a billion objects that are comet sized (roughly kilometer sized) or larger.

Since the scattered disk is a dynamically active region, objects are slowly leaking out of it with time. Indeed, models of the evolution of scattered disk objects show that the scattered disk contained about 100 times more objects when it was formed roughly 4 billion years ago than it does today (see below). Objects can leave the scattered disk in two ways. First, they can slowly evolve outward in semimajor axis until they get far enough from the Sun that Galactic tides become important. These objects then become part of the Oort cloud. However, most of the objects evolve inward onto Neptune-crossing orbits. Close encounters with Neptune can then knock an object out of the scattered disk. Roughly one comet in three that becomes Neptune crossing, in turn, evolves through the outer planetary system to become a Jupiter-family comet for a small fraction of its lifetime.

Figure 31.11 shows what we believe to be the evolution of a typical scattered disk object as it follows its trek from the scattered disk to the Jupiter family and out again. The figure shows this evolution in the perihelion distance (q)—aphelion distance (Q) plane. The positions are joined by blue lines until the object first became "visible" (which we take to be $q < 2.5$ AU) and are linked in red thereafter. Initially, the object spent considerable time in the scattered disk, that is with perihelion near the orbit of Neptune (30 AU) and aphelion well beyond the planetary system. However, once an object evolves inward, it tends to be under the dynamical control of just one planet. That planet will scatter it inward and outward in a random walk, typically handing it off to the planet directly interior or exterior

---

2. There are two meanings of the phrase "Kuiper Belt" in the literature. There is the one employed above. In addition, some researchers use the phrase to describe the entire trans-Neptunian region. In this case the term "classical Kuiper Belt" is used to distinguish the stable regions. We prefer the former definition.

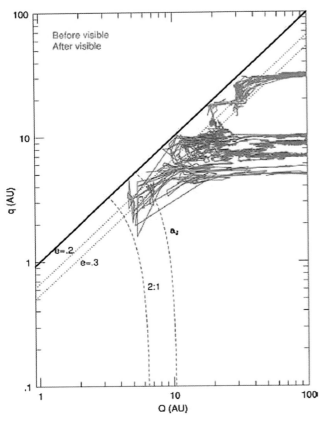

FIGURE 31.11 The orbital evolution of a representative object originating in the scattered disk. In particular, the locations of the object's orbit in the $q-Q$ (perihelion–aphelion) plane are joined by blue lines until the object became "visible" ($q < 2.5$ AU) and are linked in red thereafter. The sampling interval was every 10,000 years in the previsibility phase and every 1000 years thereafter. Also shown in the figure are three lines of constant eccentricity at $e = 0$, 0.2, and 0.3. In addition, we plot two dashed curves of constant semimajor axis, one at Jupiter's orbit and one at its 2:1 mean motion resonance. Note that it is impossible for an object to have $q > Q$, so objects cannot move into the region above and to the right of the solid diagonal line.

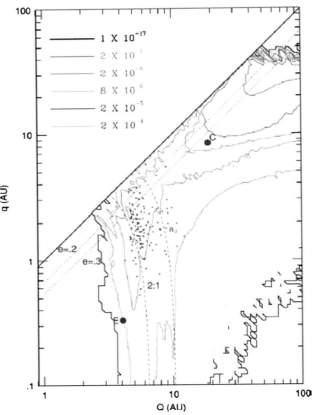

FIGURE 31.12 A contour plot of the relative distribution of ecliptic comets in the solar system as a function of aphelion (Q) and perihelion (q). The units are the fraction of comets per square AU in $q-Q$ space. Also shown in the figure are three lines of constant eccentricity at $e = 0$ (solid), 0.2, and 0.3 (both dotted). In addition, we plot two dashed curves of constant semimajor axis, one at Jupiter's orbit and one at its 2:1 mean motion resonance. They gray dots labeled "E" and "C" show the locations of comets 2P/Encke and 95P/Chiron. The small gray dots show the orbits of the Jupiter-family comets.

to it. Because of the roughly geometric spacing of the giant planets, comets tend to have eccentricities of about 25% between "handoffs" and spend a considerable amount of time with perihelion or aphelion near the semimajor axis of Saturn, Uranus, or Neptune.

However, once comets have been scattered into the inner Solar System by Jupiter, they can have much larger eccentricities as they evolve back outward. The post-visibility phase of the object in Figure 31.11 is reasonably typical of Jupiter-family comets, with much larger eccentricities than the previsibility comets and perihelion distances near Jupiter or Saturn. This object was eventually ejected from the Solar System by a close encounter with Saturn.

Numerical models, like the one used to create Figure 31.11, show that most of the ecliptic comets and Centaurs most likely originated in the scattered disk. Figure 31.12 shows the distribution of the ecliptic comets derived from these simulations. The figure is a contour plot of the relative number of comets per square AU in perihelion–aphelion ($q-Q$) space. Also shown are the locations of 95P/Chiron and 2P/Encke (big dots marked "C" and "E", respectively), and the known Jupiter-family comets (small gray dots).

There are two well-defined regions in Figure 31.12. Beyond approximately $Q = 7$ AU, there is a ridge of high density extending diagonally from the upper right to the center of the plot, near $e \approx 0.25$. The peak density in this ridge drops by almost a factor of 100 as it moves inward, having a minimum where the semimajor axes of the comets are the same as Jupiter's (shown by a dotted curve and marked with a_J). This region of the plot is inhabited mainly by the Centaurs. Inside of $Q \approx 7$ AU, the character of the distribution is quite different. Here there is a ridge of high density extending vertically in the figure at $Q \sim 5-6$ AU that extends over a wide range of perihelion distances. Objects in this region are the Jupiter-family comets. This characteristic of a very narrow distribution in Q is seen in the real Jupiter-family comets and is a result of the narrow range in T which, in turn, comes from the low to moderate inclinations and eccentricities of bodies in the scattered disk.

Figure 31.12 shows the relationship between the Centaurs and the Jupiter-family comets and illustrates the distribution of objects throughout the outer Solar System. The simulations predict that the inclinations of this population should be small everywhere, which is consistent with observations.

3.3. Formation of the Oort Cloud and Scattered Disk

Let us take stock of where we have come thus far. Active comets can be divided into two groups based on the value of the Tisserand parameter, T. The NICs have $T < 2$ and originate in the Oort cloud. The ecliptic comets have $T > 2$ and originate in the scattered disk. The Oort cloud is a population of comets that lie very far from the Sun, with semimajor axes extending from tens of thousands of AU down to thousands of AU. It also is roughly spherical in shape. The scattered disk, on the other hand, lies mainly interior to ~ 1000 AU and is flattened. It may be surprising, therefore, that modern theories suggest that both of these structures formed as a result of the same process and therefore the objects in them formed in the same region of the Solar System.

First, we must address why we think that these structures did not form where they are. The answer has to do with the comets' eccentricities and inclinations. Although comets are much smaller than planets, they probably formed in a similar way. The Solar System formed from a huge cloud of gas and dust that initially collapsed to a protostar surrounded by a disk. The comets, asteroids, and planets formed in this disk. However, initially the disk only contained very small solid objects, similar in size to particles of smoke, and much smaller than comets. Although it is not clear how these objects grew to become comet sized, all the processes thus far suggested require that the relative velocity between the dust particles was small. This, in turn, requires the dust particles to be on nearly circular coplanar orbits. So, the eccentric and inclined orbits of bodies in the cometary reservoirs must have arisen because they were dynamically processed from the orbits in which they were formed to the orbits in which they are found today.

Astronomers generally agree that comets originally formed in the region of the Solar System now inhabited by the giant planets. Although comets formed in nearly circular orbits, their orbits were perturbed by the giant planets as the planets grew and/or the planets' orbits evolved. Figure 31.13 shows the behavior of a typical comet as it evolves into the Oort cloud. At first, the comet is handed off from planet to planet, remaining in a nearly circular orbit (Region 1 in the figure). However, eventually Neptune scatters the body outward. It then goes through a period of time when its semimajor axis is changing because of encounters with Neptune (Region 2). During this time its perihelion distance is near the orbit of Neptune, but its semimajor axis can become quite large. (If this reminds you of the scattered disk, it should.) When the object gets into the region beyond 10,000 AU, galactic perturbations lift its perihelion out of the planetary system, and it is then stored in the Oort cloud for billions of years (Region 3).

Figure 31.14 shows the result of a numerical model of the formation of the Oort cloud and scattered disk. The simulation followed the orbital evolution of a large number of comets initially placed on nearly circular low-inclination orbits between the giant planets, under the gravitational influence of the Sun, the four giant planets, and the Galaxy. The major steps of Oort cloud formation can be seen in this figure. Initially the giant planets start scattering objects to large semimajor axes. By 600,000 years, a massive scattered disk has formed, but only a few objects have evolved far enough outward that Galactic perturbations are important.

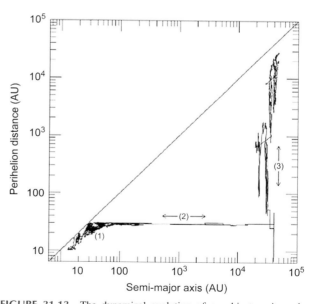

FIGURE 31.13 The dynamical evolution of an object as it evolves into the Oort cloud. The object was initially in a nearly circular orbit between the giant planets. Its evolution follows three distinct phases. During Phase 1 the object remains in a relatively low eccentricity orbit between the giant planets. Neptune eventually scatters it outward, after which the object undergoes a random walk in semimajor axis (Phase 2). When it reaches a large enough semimajor axis, galactic perturbations lift its perihelion distance to large values (Phase 3).

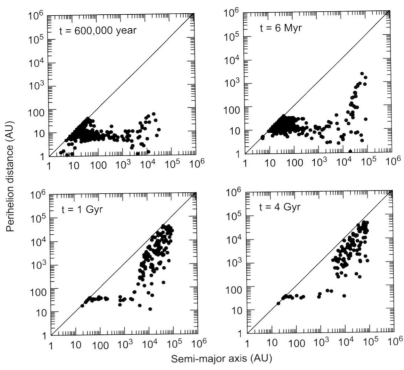

FIGURE 31.14 Four snapshots of comets in a simulation of the formation of the scattered disk and the Oort cloud.

At $t = 6$ Myr the Oort cloud is beginning to form. The Galactic perturbations have started to raise the perihelion distances of the most distant comets, but a complete cycle in q has yet to occur (see Figure 31.3). Note that the scattered disk is still massive. By 1 billion years, the Oort cloud beyond 10,000 AU is inhabited by objects on moderate-eccentricity orbits (i.e. where $a \sim q$). Note also that a scattered disk still exists. There is also a transition region between ~ 2000 and ~ 5000 AU, where objects are beginning to have their perihelia lifted by the Galaxy, but have not yet undergone a complete cycle in perihelion distance. By 4 billion years, the Oort cloud is fully formed and extends from 3000 to 100,000 AU. The scattered disk can easily be seen extending from Neptune's orbit outward. If our current understanding of comet reservoirs is correct, these are the two source reservoirs of all the known visible comets.

The above calculations assume that the Sun has always occupied its current Galactic environment, that is it is isolated and not a member of a star cluster. However, almost all stars form in dense clusters. The gravitational effects of such a star cluster on a growing Oort cloud is similar to that of the Galaxy except that the torques are much stronger. This would lead to an Oort cloud that is much more compact if the Sun had been in such an environment at the time that the cloud was forming. However, models of the dynamical evolution of star clusters show that the average star spends less than 5 Myr in such an environment and the giant planets might take that long to form. Additionally, even if the planets formed very quickly, Figure 31.14 shows that the Oort cloud is only partially formed after a few million years. In particular, only those objects that originated in the Jupiter–Saturn region have evolved much in semimajor axis. Therefore, the Oort cloud probably formed in two stages. Before ~ 5 Myr a dense *first generation* Oort cloud formed from Jupiter–Saturn planetesimals at roughly $a \sim 1000$ AU because of the effects of the star cluster. After the Sun left the cluster, a normal Oort cloud formed at $a \sim 10{,}000$ AU from objects that originated beyond Saturn. Figure 31.15 shows an example of such an Oort cloud as determined from numerical experiments. There is some observational evidence that the Solar System contains a first-generation Oort cloud. In 2004, the object known as Sedna was discovered. Sedna has $a = 468$ AU and $q = 76$ AU, placing it well beyond the planetary region. Numerical experiments have shown that the most likely way to get objects with perihelion distances as large as Sedna is through external torques (as in Figure 31.15). And, since the current Galactic environment is too weak to place Sedna on its current orbit, Sedna's orbit probably formed when the Sun was in its birth star cluster. If true, Sedna's orbit represents the first observational constraint we have concerning the nature of this star cluster. If such a structure really exists, it does not contribute to the

Chapter | 31 Comet Populations and Cometary Dynamics

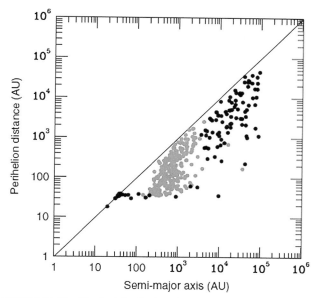

FIGURE 31.15 The final distribution of comets in the scattered disk and the Oort cloud according to a numerical experiment where the Sun spent 3 Myr in a star cluster. The gray and black dots refer to objects that formed interior to or exterior to 14 AU.

population of observed comets because it is in a part of the Solar System which is currently stable: objects in this region do not get close to the planets and the Galactic tides are too weak.

4. CONCLUSIONS

Comets are only active when they get close to the Sun. However, they must come from more distant regions of the Solar System where it is cold enough for them to survive the age of the Solar System without sublimating away. Dynamical simulations of cometary orbits argue that there are two main source regions in the Solar System. One, known as the Oort cloud, is a roughly spherical structure located at heliocentric distances of thousands to tens of thousands of AU. The NICs come from this reservoir. The scattered disk is the other important cometary reservoir. It is a disk-shaped structure that extends outward from the orbit of Neptune. The ecliptic comets come from the scattered disk.

However, there are substantial reasons to believe that these two cometary reservoirs are not primordial structures and that their constituent members formed elsewhere and were dynamically transported to their current locations.

Indeed, current models suggest that objects in both the Oort cloud and scattered disk formed in the region between the giant planets and were delivered to their current locations by the action of the giant planets as these planets formed and evolved. Comets, therefore, represent the leftovers of planet formation and contain vital clues to the origin of the Solar System.

BIBLIOGRAPHY

Brandt, J. C., & Chapman, R. D. (2004). *Introduction to comets* (2nd ed.). Cambridge University Press, 450 pp.

British Astronomical Association. (2014). *BAA comet section*. Updated December 2013. http://www.ast.cam.ac.uk/~jds/

Dones, L., Weissman, P. R., Levison, H. F., & Duncan, M. J. (2004). Oort cloud formation and dynamics. In M. C. Festou, H. U. Keller, & H. A. Weaver (Eds.), *Comets II* (pp. 153–174).

Duncan, M., Levison, H., & Dones, L. (2004). Dynamical evolution of ecliptic comets. In M. C. Festou, H. U. Keller, & H. A. Weaver (Eds.), *Comets II* (pp. 193–204).

Fern'ández, J. A. (2005). *Comets—nature, dynamics, origin, and their cosmogonical relevance*. Springer, 383 pp.

Fernández, Y. (2014). *List of Jupiter-family and Halley-family comets*. Updated April 8, 2014. http://www.physics.ucf.edu/~yfernandez/cometlist.html

Festou, M. C., Keller, H. U., & Weaver, H. A. (Eds.). (2004). *Comets II*. Univ. Arizona Press, 745 pp.

Jet Propulsion Laboratory. (2014). *JPL solar system dynamics*. http://ssd.jpl.nasa.gov/.

Kinoshita, K. (2014). *Comet orbit home page*. Updated March 27th, 2014. http://jcometobs.web.fc2.com

Kresák, L. (1982). Comet discoveries, statistics, and observational selection. In L. L. Wilkening (Ed.), *Comets* (pp. 56–82). Tucson: Univ. Arizona Press.

Kronk, G. W. (1999). *Cometography: A catalog of comets*. In *Ancient to 1799* (Vol. 1). New York: Cambridge Univ. Press, 563 pp.

Kronk, G. W. (2003). *Cometography: A catalog of comets* (Vol. 2: 1800–1899). New York: Cambridge Univ. Press, 852 pp.

Kronk, G. W. (2014). *Cometography*. http://cometography.com/.

Marsden, B. G., Sekanina, Z., & Yeomans, D. K. (1973). Comets and nongravitational forces. V. *The Astronomical Journal, 78*, 211–225.

Marsden, B. G., & Williams, G. V. (2003). *Catalogue of cometary orbits* (15th ed.). Cambridge, MA: Smithsonian Astrophysical Observatory, 169 pp.

Minor Planet Center. (2014). *IAU: Minor planet center*. http://www.cfa.harvard.edu/iau/mpc.html.

Oort, J. H. (1950). The structure of the cloud of comets surrounding the solar system and a hypothesis concerning its origin. *Bulletin of the Astronomical Institutes of the Netherlands, 11*, 91–110.

Part VII

Giant Planets and their Satellites

Chapter 32

Atmospheres of the Giant Planets

Robert A. West

Jet Propulsion Laboratory, California Institute of Technology, Pasadena, California, USA

Chapter Outline

1. Introduction 723
2. Chemical Composition 724
3. Clouds and Aerosols 728
4. Dynamical Meteorology of the Troposphere and Stratosphere 733
5. Energetic Processes in the High Atmosphere 738
6. A Word about Extrasolar Planets 741
Acknowledgment 742
Bibliography 742

1. INTRODUCTION

To be an astronaut explorer in Jupiter's atmosphere would be strange and disorienting. There is no solid ground to stand on. The temperature would be comfortable at an altitude where the pressure is eight times that of the Earth's surface, but it would be perpetually hazy overhead, with variable conditions (dry or wet, cloudy or not) to the east, west, north, and south. One would need to carry oxygen as there is no free oxygen and wear special clothing to protect the skin against exposure to ammonia and hydrogen sulfide gases, which form clouds and haze layers higher in the atmosphere. A trip to high latitudes would offer an opportunity to watch the most powerful, vibrant, and continuous auroral displays in the solar system. On the way, one might pass through individual storm systems the size of Earth or larger and be buffeted by strong winds alternately from the east and west. One might be sucked into a dry downwelling sinkhole like the environment explored by the *Galileo* probe. The probe fell to depths where the temperature is hot enough to vaporize metal and rock. It is now a part of Jupiter's atmosphere.

Although the atmospheres of the giant planets share many common attributes, they are at the same time very diverse. The roots of this diversity can be traced to a set of basic properties, and ultimately to the origins of the planets. The most important properties that influence atmospheric behavior are listed in Table 32.1. The distance from the Sun determines how much sunlight is available to heat the upper atmosphere. The minimum temperature for all these atmospheres occurs near the 100-mbar level and ranges from 110 K at Jupiter to 50 K at Neptune. The distance from the Sun and the total mass of the planet are the primary influences on the bulk composition. All the giant planets are enriched in heavy elements, relative to their solar abundances, by factors ranging from about 3 for Jupiter to 1000 for Uranus and Neptune. The latter two planets are sometimes called the ice giants because they have a large fraction of elements (O, C, N, and S) that were the primary constituents of ices in the early solar nebula.

The orbital period, axial tilt, and distance from the Sun determine the magnitude of seasonal temperature variations in the high atmosphere. Jupiter has weak seasonal variations; those of Saturn are much stronger. Uranus is tipped such that its poles are nearly in the orbital plane, leading to more solar heating at the poles than at the equator when averaged over an orbit. The ratio of radiated thermal energy to absorbed solar energy is diagnostic of how rapidly convection is bringing internal heat to the surface, which in turn influences the abundance of trace constituents and the morphology of eddies in the upper atmosphere. Vigorous convection from the deeper interior is responsible for the unexpectedly high abundances of several trace species on Jupiter, Saturn, and Neptune, but convection on Uranus is sluggish. All these subjects are treated in more detail in the sections that follow.

TABLE 32.1 Physical Properties of the Giant Planets

Property	Jupiter	Saturn	Uranus	Neptune
Distance from the Sun (Earth distance = 1[1])	5.2	9.6	19.2	30.1
Equatorial radius (Earth radius = 1[2])	11.3	9.4	4.1	3.9
Planet total mass (Earth mass = 1[3])	318.1	95.1	14.6	17.2
Mass of gas component (Earth mass = 1)	254–292	72–79	1.3–3.6	0.7–3.2
Orbital period (years)	11.9	29.6	84.0	164.8
Length of day (hours, for a point rotating with the interior)	9.9	10.7[4]	17.4	16.2
Axial inclination (degrees from normal to orbit plane)	3.1	26.7	97.9	28.8
Surface gravity (equator–pole, m/s^2)	22.5–26.3	8.4–11.6	8.2–8.8	10.8–11.0
Ratio of emitted thermal energy to absorbed solar energy	1.7	1.8	~1	2.6
Temperature at the 100-mbar level (K)	110	82	54	50

[1] Earth distance = 1.5×10^8 km.
[2] Earth radius = 6378 km.
[3] Earth mass = 6×10^{24} kg.
[4] Saturn's internal rotation rate is not accurately known. The value indicated is approximate.

2. CHEMICAL COMPOSITION

This section is concerned with chemical abundances in the observable part of the atmosphere, a relatively thin layer of gas near the top (where pressures are between about 5 bar and a fraction of a microbar). To place the subject in context, some mention will be made of the chemical makeup of the interior (*see* Interiors of the Giant Planets).

The chemical constituents of the interior of a planet cannot be directly observed, but must be inferred from information on its mean density, its gravity field, and the abundances of constituents that are observed in the outer layers. The more massive planets were better able to retain the light elements during their formation, and so the chemical makeup of Jupiter resembles that of the Sun. When the giant planets formed, they incorporated relatively more rock and ice fractions than a pure solar elemental mix would allow, and the fractional amounts of rocky and icy materials increase from Jupiter through Neptune (*see* The Origin of the Solar System). Most of the mass of the heavy elements is sequestered in the deep interior. The principal effects of this layered structure on the observable outer layers can be summarized as follows.

On Jupiter, the gas layer (a fluid molecular envelope) extends down to about 40% of the planet's radius, where a phase transition to liquid metallic hydrogen occurs. Fluid motions that produce the alternating jets and vertically mix gas parcels may fill the molecular envelope but probably do not extend into the metallic region. Thus, the radius of the phase transition provides a natural boundary that may be manifest in the latitudinal extent of the zonal jets (see Section 4), whereas vertical mixing may extend to levels where the temperature is quite high. These same characteristics are found on Saturn, with the additional possibility that a separation of helium from hydrogen is occurring in the metallic hydrogen region, leading to enrichment of helium in the deep interior and depletion of helium in the upper atmosphere.

Uranus and Neptune contain much larger fractions of ice and rock-forming constituents than do Jupiter and Saturn. A large water ocean may be present in the interiors of these planets. Aqueous chemistry in the ocean can have a profound influence on the abundances of trace species observed in the high atmosphere.

In the observable upper layers, the main constituents are molecular hydrogen and atomic helium, which are well mixed, up to the **homopause** level, where the mean free path for collisions becomes large enough that the lighter constituents are able to diffuse upward more readily than heavier ones. Other constituents are significantly less abundant than hydrogen and helium, and many of them condense in the coldest regions of the atmosphere. Figure 32.1 shows how temperature varies with altitude and pressure, and the locations of the methane, ammonia, and water cloud layers.

The giant planets have retained much of the heat generated by their initial collapse from the solar nebula. They cool by emitting thermal infrared radiation to space. Thermal radiation is emitted near the top of the atmosphere, where the opacity is low enough to allow infrared photons to escape to space. In the deeper atmosphere, heat is

Chapter | 32 Atmospheres of the Giant Planets

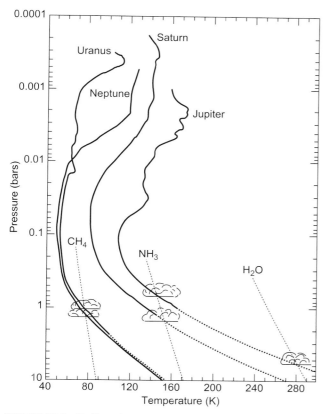

FIGURE 32.1 Profiles of temperature as a function of pressure in the outer planet atmospheres derived from measurements by the Voyager Radio Sciences experiment (solid curves). The dashed parts of the temperature profiles are extrapolations using the adiabatic lapse rate. At high altitudes (not shown), temperatures rise to about 1200 K for Jupiter, 800 K for Saturn and Uranus, and 300 K for Neptune. The dotted lines show vapor pressure curves divided by observed mixing ratios for water, ammonia, and methane. Condensate clouds are located where the solid and dotted curves cross. *From Gierasch, & Conrath, (1993). Copyright American Geophysical Union.*

Hydrogen is the main constituent in the observable part of the giant planet atmospheres, but not until recently was it recognized as especially important for thermodynamics. The hydrogen molecule has two ground-state configurations for its two electrons. The electrons can have their spins either parallel or antiparallel, depending on whether the spins of the nuclei are parallel or antiparallel. These states are called the ortho and para states. Transitions between ortho and para states are slow because, unlike most molecules, the nuclear spin must change when the electron spin changes. At high temperature (about 270 K or higher), the ortho:para relative abundance is 3:1. At lower temperature, a larger fraction is converted to the para state. Heat release from conversion of *ortho-* to *para-*hydrogen can act in the same way as **latent heat** release from condensation. The relative fractions of *ortho-* and *para-*hydrogen are observed to be close to thermal equilibrium values in the giant planet atmospheres, leading to the question of how equilibrium is achieved. Catalytic reactions on the surfaces of **aerosol** particles are thought to be important in equilibrating the ortho and para states.

The competition between convective cooling and solar heating produces a temperature minimum near the 100 mbar level (the tropopause). At pressures between about 100 and 0.1 mbar, the temperature is determined primarily by equilibrium between thermal radiative cooling and solar heating. At even lower pressures, other processes, including auroral heating, dump energy into the atmosphere and produce higher temperatures. More will be said about this in Section 5.

The current inventory of observed gaseous species is listed in Table 32.2. Molecular hydrogen and helium are the most abundant. Helium is in its ground state in the **troposphere** and **stratosphere** and therefore does not produce spectral lines from which its abundance can be determined. The mixing ratio for Saturn, Uranus, and Neptune is inferred from its influence on the broad collision-induced hydrogen lines near the 45 μm wavelength and from a combined analysis of the infrared spectrum and refractivity profiles retrieved from spacecraft radio occultation measurements. Helium on Jupiter is accurately known from measurements made by the *Galileo* probe, which descended through the atmosphere. It is a little smaller than the mixing ratio inferred for the primitive solar nebula from which the planets formed. Helium is depleted in Saturn's upper atmosphere, consistent with the idea that helium is precipitating out in the metallic hydrogen region. For Uranus and Neptune, the helium mixing ratio is close to the mixing ratio (0.16) in the primitive solar nebula. There is still some uncertainty in the helium mixing ratio for Uranus, Neptune, and Saturn because additional factors, such as aerosol opacity and molecular nitrogen abundance, affect the shapes of the collision-induced spectral features, and we do not have a completely consistent set of values for all these parameters.

transported by convective fluid motions from the deep, hot interior to the colder outer layers. In this region, upwelling gas parcels expand and subsiding parcels contract adiabatically (e.g. with negligible transport of heat through their boundaries by radiation or conduction). Therefore temperature depends on altitude according to the adiabatic law $T = T_0 + C(z - z_0)$, where T_0 is the temperature at some reference altitude z_0, C is a constant (the adiabatic lapse rate) that depends on the gas mixture, and z is the altitude. The adiabatic lapse rate for dry hydrogen and helium on Jupiter is -2.2 K/km. On Uranus it is -0.8 K/km. The adiabatic lapse rate is different in regions where a gas is condensing or where heat is released as *ortho*-hydrogen and is converted to *para*-hydrogen. Both of these processes are important in the giant planet atmospheres at pressures between about 30 and 0.1 bar.

TABLE 32.2 Abundances of Observed Species in the Atmospheres of the Giant Planets

Constituent	Peak Mixing Ratio (by Number) or Upper Limit			
	Jupiter	Saturn	Uranus	Neptune
Species with Constant Mixing Ratio below the Homopause				
H_2	0.86	0.90	0.82	0.79
HD	4×10^{-5}	4×10^{-5}		
He	0.14	0.10	0.15	0.18
CH_4	2×10^{-3}	2×10^{-3}		
CH_3D	3.5×10^{-7}	2×10^{-7}		
^{20}Ne	2×10^{-5}			
^{36}Ar	1×10^{-5}			
Condensable Species (Estimated or Measured below the Condensation Region)				
NH_3	2.5×10^{-4}	2×10^{-4}		
H_2S	7×10^{-5}			
H_2O	6×10^{-4}			
CH_4			0.04	0.04
CH_3D			2×10^{-5}	2×10^{-5}
Disequilibrium Species in the Troposphere				
PH_3	5×10^{-7}	2×10^{-6}		
GeH_4	7×10^{-10}	4×10^{-10}		
AsH_3	2.4×10^{-9}	3×10^{-9}		
CO	2×10^{-9}	$1-25 \times 10^{-9}$	$<1 \times 10^{-8}$	1×10^{-6}
HCN			$<1 \times 10^{-10}$	1×10^{-9}
Photochemical Species (Peak Values)				
C_2H_2	1×10^{-7}	3×10^{-7}	1×10^{-8}	6×10^{-8}
C_2H_4	7×10^{-9}			
C_2H_6	7×10^{-6}	7×10^{-6}	$<1 \times 10^{-8}$	2×10^{-6}
C_3H_4	2.5×10^{-9}			
C_6H_6	2×10^{-9}			

Mixing ratios of **deuterated** hydrogen and methane (HD and CH_3D) also provide information on the formation of the planets. **Deuterium**, which once existed in the Sun, has been destroyed in the solar atmosphere, and the best information on its abundance in the primitive solar nebula comes from measurements of the giant planet atmospheres. On Jupiter, the deuterium mixing ratio is thought to be close to that of the primitive solar nebula. On Uranus and Neptune, it is enhanced because these planets incorporated relatively more condensed material on which deuterium preferentially accumulated through isotopic fractionation. Isotopic fractionation (the enhancement of the heavier isotope over the lighter isotope during condensation) occurs because the heavier isotope has a lower energy than the lighter isotope in the condensed phase.

The elements oxygen, carbon, nitrogen, and sulfur are the most abundant molecule-forming elements in the Sun (after hydrogen), and all are observed in the atmospheres of the giant planets, mostly as H_2O, CH_4, NH_3, and (for Jupiter) H_2S. Water condenses even in Jupiter's atmosphere, at levels that are difficult to probe with infrared radiation (6 bars or deeper). A straightforward interpretation of Jupiter's spectrum indicated its abundance to be about a 100 times less than what is expected from a planet formed from the same material as the sun. The *Galileo* probe measurements indicated that water was depleted relative to solar abundance by roughly a factor of two at the deepest level measured (near 20 bars of pressure) and even more depleted at higher altitude. However, the probe descended in a relatively dry region of the atmosphere, analogous to a desert on the Earth, and the bulk water abundance on Jupiter may well be close to the solar abundance. Water is not observed on any of the other giant planets because of the optically thick overlying clouds and haze layers. It is thought to form a massive global ocean on Uranus and Neptune based on the densities and gravity fields of those planets, coupled with theories of their formation.

Methane is well mixed, up to the homopause level, in the atmospheres of Jupiter and Saturn, but it condenses as ice in the atmospheres of Uranus and Neptune. Its mixing ratio below the condensation level is enhanced over that expected for a solar-type atmosphere by factors of 2.6, 5.1, 35, and 40 for Jupiter, Saturn, Uranus, and Neptune, respectively. These enhancements are consistent with ideas about the amounts of icy materials that were incorporated into the planets as they formed. The stratospheres of Uranus and Neptune form a cold trap, where methane ice condenses into ice crystals that fall out, making it difficult for methane to mix to higher levels. Nevertheless, the methane abundance in Neptune's stratosphere appears to be significantly higher than its vapor pressure at the temperature that the tropopause would allow (and also higher than the abundance in the stratosphere of Uranus), suggesting some mechanism such as convective penetration of the cold trap by rapidly rising parcels of gas. This mechanism does not appear to be operating on Uranus, and this difference between Uranus and Neptune is symptomatic of the underlying difference in internal heat that is available to drive convection on Neptune but not on Uranus.

Ammonia is observed on Jupiter and Saturn, but not on Uranus or Neptune. Ammonia condenses as an ammonia ice cloud near 0.6 bar on Jupiter and at higher pressures on the colder outer planets. Ammonia and H_2S in solar abundance would combine to form a cloud of NH_4SH (ammonium hydrosulfide) near the 2-bar level in Jupiter's atmosphere and at deeper levels in the colder atmospheres of the other giant planets. Hydrogen sulfide was observed in Jupiter's atmosphere by the mass spectrometer instrument on the *Galileo* probe. Another instrument (the nephelometer) on the probe detected cloud particles in the vicinity of the 1.6-bar pressure level, which would be consistent with the predicted ammonium hydrosulfide cloud. Evidence from thermal emission at radio wavelengths has been used to infer that H_2S is abundant on Uranus and Neptune. Ammonia condenses at relatively deep levels in the atmospheres of Uranus and Neptune and has not been spectroscopically detected. A dense cloud is evident at the level expected for ammonia condensation (2–3 bar) in near-infrared spectroscopic observations, but the microwave spectra of those planets are more consistent with a strong depletion of ammonia at those levels. An enhancement of H_2S relative to NH_3 could act to deplete ammonia by the formation of ammonium hydrosulfide in the deeper atmosphere. In that case, H_2S ice is the most likely candidate for the cloud near 3 bars.

Water, methane, and ammonia are in thermochemical equilibrium in the upper troposphere. Their abundances at altitudes higher than (and temperatures colder than) their condensation level are determined by temperature (according to the vapor pressure law) and by meteorology, as is water in the Earth's atmosphere. Some species (PH_3, GeH_4, and CO) are not in thermochemical equilibrium in the upper troposphere. At temperatures less than 1000 K, PH_3 would react with H_2O to form P_4O_6 if allowed to proceed to thermochemical equilibrium. Apparently, the timescale for this reaction (about 3 years) is longer than the time to convect material from the 1000 K level to the tropopause. A similar process explains the detection of GeH_4. Yet another phenomenon (impact of a comet within the past 200 years) probably accounts for the detection of CO in the stratosphere.

Ammonia and phosphine are present in the stratospheres of Jupiter and Saturn, and methane is present in the stratospheres of all the giant planets. These species are destroyed at high altitudes by ultraviolet (UV) sunlight and by charged particles in auroras, producing N, P, and C, which can react to form other compounds. Ammonia photochemistry leads to formation of hydrazine (N_2H_4), and phosphine photochemistry leads to diphosphine (P_2H_4). These constituents condense in the cold tropospheres of Jupiter and Saturn and may be responsible for much of the UV-absorbing haze seen at low latitudes. Nitrogen gas and solid P are other by-products of ammonia and phosphine chemistry. Solid phosphorus is sometimes red and has been proposed as the constituent responsible for the red color of Jupiter's Great Red Spot (GRS). This suggestion (one of several) has not been confirmed, and neither N_2H_4 nor P_2H_4 has been observed spectroscopically.

Organic compounds derived from the dissociation of methane are present in the stratospheres of all the giant planets. The photochemical cycle leading to stable C_2H_2 (acetylene), C_2H_4 (ethylene), C_2H_6 (ethane), and C_4H_2 (diacetylene) is shown schematically in Figure 32.2. The chain may progress further to produce polyacetylenes ($C_{2n}H_2$). These species form condensate haze layers in the cold stratospheres of Uranus and Neptune. More complex

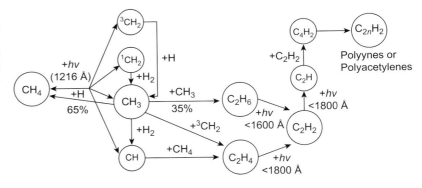

FIGURE 32.2 Summary of CH_4 (methane) photochemical processes in the stratospheres of the giant planets. Photodissociation by ultraviolet light is indicated by $+hv$ at the indicated wavelength. Methane photodissociation is the starting point in the production of a host of other hydrocarbons. *Revised by Atreya, S. K. from Figure 5.3 from Pollack, J. B., & Atreya, S. K. (1992).*

hydrocarbon species (C_3H_8, C_3H_4) are observed in Jupiter's atmosphere primarily in close proximity to high-latitude regions, where auroral heating is significant. The abundant polar aerosols in the atmospheres of Jupiter and Saturn may owe their existence to the ions created by auroras in the upper atmosphere.

As instruments become more sensitive, new species are detected. These include C_2H_4, C_3H_4, and C_6H_6 in the atmospheres of Jupiter and Saturn, and C_3H_8 for Saturn. The methyl radical CH_3 (an unstable transition molecule in the reaction chain) has been detected on Jupiter, Saturn, and Neptune.

Hydrogen cyanide (HCN) is observed in the stratospheres of Jupiter and Neptune. On Jupiter, HCN was emplaced high in the stratosphere as a result of the 1994 impacts of comet Shoemaker–Levy 9. During the 3 years after the impacts, it was observed to spread north of the impact latitude (near $45°S$), eventually to be globally distributed. It is expected to dissipate over the span of a decade or so. Impact by cometary and interplanetary dust particles may also be responsible for HCN in Neptune's stratosphere.

Quantitative thermochemical and photochemical models are available for many of the observed constituents and provide predictions for many others that are not yet observed. These models solve a set of coupled equations that describe the balance between the abundances of species that interact and include important physical processes such as UV **photolysis**, condensation/sublimation, and vertical transport. Current models heuristically lump all the transport processes into an effective eddy mixing coefficient, and the value of that coefficient is derived as part of the solution of the set of equations. As we gain more detailed observations and more comprehensive laboratory measurements of reaction rates, we will be able to develop more sophisticated models. Some models are beginning to incorporate transport by vertical and horizontal winds. Figures 32.3 and 32.4 show vertical profiles calculated from models for a number of photochemically produced species.

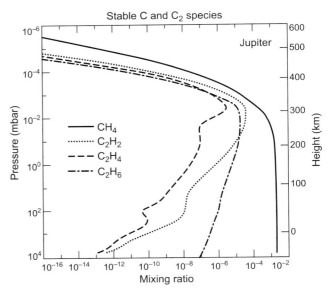

FIGURE 32.3 Vertical profiles of some photochemical species in Jupiter's stratosphere. The mixing ratios (horizontal axis) are plotted as a function of pressure. *From Gladstone, G. R., et al. (1996). Copyright Academic Press.*

3. CLOUDS AND AEROSOLS

The appearance of the giant planets is determined by the distribution and optical properties of cloud and aerosol haze particles in the upper troposphere and stratosphere. Cameras on the *Voyager* spacecraft provided detailed views of all the giant planets, whose general appearances can be compared in Figure 32.5. Their atmospheres show a banded structure (which is difficult to see on Uranus) of color and shading parallel to latitude lines. These were historically named belts and zones on Jupiter and Saturn, with belts being relatively dark and zones being relatively bright. Specific belts and zones were named in accordance with their approximate latitudinal location (Equatorial Belt, North and South Tropical Zones near latitudes $\pm20°$, North

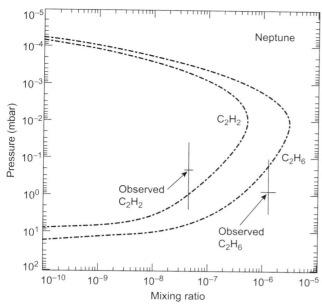

FIGURE 32.4 Vertical profiles of photochemical species in the Neptunian stratosphere. *From Romani, P., et al. (1993). Copyright by Academic Press.*

dark. On Jupiter, there is a correlation between visible albedo and temperature, such that bright zones are usually cool regions and dark belts are usually warm near the tropopause. Cool temperatures are associated with cooling of upwelling gas, and the correlation of cool temperatures with bright clouds points to enhanced condensation of ice particles as condensable gases flow upward and cool. This correlation does not hold completely on Jupiter and almost not at all on the other giant planets. The mechanisms responsible for producing reflectivity contrasts and color remain largely mysterious, although a number of proposals have been advanced. These will be discussed in more detail.

Our understanding of aerosols and clouds is rooted in thermochemical equilibrium models that predict the temperature (and hence pressure and altitude) of the bases of condensate clouds. The cloud base occurs where the vapor pressure of a condensable gas equals its partial pressure. Model predictions for the four giant planets are shown in Figure 32.6. The deepest cloud to form is a solution of water and ammonia on Jupiter and Saturn, with dissolved H_2S as well on Uranus and Neptune. At higher altitudes, an ammonium hydrosulfide cloud forms, and its mass depends on both the amounts of H_2S and NH_3 available and the ratio of S to N. At still higher altitudes, an ammonia or hydrogen sulfide cloud can form if the S/N ratio is less than or greater than 1, respectively. If the ratio is greater than 1, all the N will be taken up as NH_4SH, with the remaining sulfur available to condense at higher altitudes. This seems to be the situation on Uranus and Neptune, but the reverse is true for Jupiter and Saturn. Only the atmospheres of Uranus and Neptune are cold enough to condense methane, which occurs at 1.3 bar in Uranus and at about 2 bar in Neptune. It is predominantly the uppermost clouds that we see at visible wavelengths.

Observational evidence to support the cloud stratigraphy shown in Figure 32.6 is mixed. The *Galileo* probe detected cloud particles near 1.6-bar pressure and sensed cloud opacity at higher altitudes corresponding to the ammonia cloud. With data only from remote sensing experiments, it is difficult to probe to levels below the top cloud, and the evidence we have for deeper clouds comes from careful analyses of radio occultations and of gaseous absorption lines in the visible and near infrared, and from thermal emission at 5, 8.5, and 45 μm. Contrary to expectation, spectra of the planets show features due to ice in only a small fraction of the cloudy area. The *Voyager* radio occultation data showed strong refractivity gradients at locations predicted for methane ice clouds on Uranus and Neptune, essentially confirming their existence and providing accurate information on the altitude of the cloud base. Ammonia gas is observed spectroscopically in Jupiter's upper troposphere, and its abundance decreases with altitude above its cloud base in accordance with expectation. There is no doubt that ammonia ice is the major

and South Temperate Zones and Belts near ±35°, and polar regions).

The nomenclature should not be construed to mean that low latitudes are relatively warmer than high latitudes, as they are on Earth and Mars. Nor is it true that the reflectivities of these features remain constant with time. Some features on Jupiter, such as the North and South Tropical Zones, are persistently bright, whereas others, like the South Equatorial Belt, are sometimes bright and sometimes

FIGURE 32.5 *Voyager* images of Jupiter, Saturn, Uranus, and Neptune, scaled to their relative sizes. Earth and Venus are also shown scaled to their relative sizes.

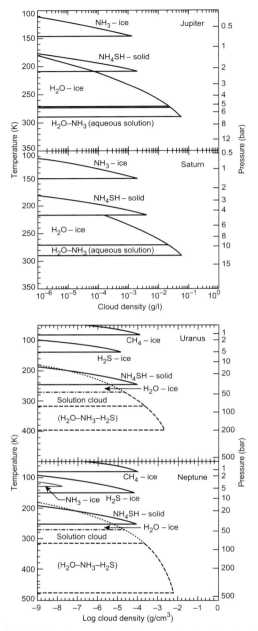

FIGURE 32.6 The diagrams in the four panels show the locations of condensate cloud layers on Jupiter, Saturn, Uranus, and Neptune. These figures indicate how much cloud material would condense at various temperatures (corresponding to altitude) if there were no advective motions in the atmosphere to move vapor and clouds. They are based on simple thermochemical equilibrium calculations, which assume, for Jupiter and Saturn, that the condensable species have mixing ratios equal to those for the sun. *Figures for Jupiter and Saturn were constructed from models by Atreya, S. K., & Wong, M. based on Atreya, S. K., & Romani, P. N. (1985). Those for Uranus and Neptune were first published by de Pater, I., et al. (1991). Copyright by Academic Press.*

component of the visible clouds on Jupiter and Saturn, but it cannot be the only component and is not responsible for the colors (pure ammonia ice is white). In fact, all the ices shown in Figure 32.6 are white at visible wavelengths. The colored material must be produced by some disequilibrium process like photochemistry or bombardment by energetic particles from the magnetosphere.

Colors on Jupiter are close to white in the brightest zones, gray yellow to light brown in the belts, and orange or red in some of the spots. The colors in Figure 32.5 are slightly and unintentionally exaggerated owing to the difficulty of achieving accurate color reproduction on the printed page. Colors on Saturn are more subdued. Uranus and Neptune are gray-green. Neptune has a number of dark spots and white patchy clouds. Part of the green tint on Uranus and Neptune is caused by strong methane gas absorption at red wavelengths, and part is due to aerosols that also absorb preferentially at wavelengths longer than 0.6 μm.

Candidate materials for the **chromophore** material in outer planet atmospheres are summarized in Table 32.3. All candidate materials are thought to form by some

TABLE 32.3 Candidate Chromophore Materials in the Atmospheres of the Giant Planets

Material	Formation Mechanism
Sulfur	Photochemical products of H_2S and NH_4SH. Red allotropes are unstable.
H_2S_x, $(NH_4)_2S_x$, $N_2H_4S_x$	Photochemical products of H_2S and NH_4SH.
N_2H_4	Hydrazine, a photochemical product of ammonia, a candidate for Jupiter's stratospheric haze.
Phosphorus (P_4)	Photochemical product of PH_3.
P_2H_4	Diphosphine, a photochemical product of phosphine, a candidate for Saturn's stratospheric haze.
Products of photodecomposition or charged particle decomposition of CH_4	Includes acetylene photopolymers (C_xH_2), proton-irradiated methane, and organics with some nitrogen and/or sulfur. Confined to stratospheric levels where ultraviolet photons and auroral protons or ions penetrate.

nonequilibrium process such as photolysis or decomposition by protons or ions in auroras, which acts on methane, ammonia, or ammonium hydrosulfide. Methane is present in the stratospheres of all the giant planets. Ammonia is present in the stratosphere of Jupiter. Ammonium hydrosulfide is thought to reside near the 2-bar level and deeper in Jupiter's atmosphere, which is too deep for UV photons to penetrate.

There are two major problems in understanding which, if any, of the proposed candidate chromophores are responsible for the observed colors. First, no features have been identified in spectra of the planets that uniquely identify a single candidate material. Spectra show broad slopes, with more absorption at blue wavelengths on Jupiter and Saturn and at red wavelengths on Uranus and Neptune. All the candidates listed in Table 32.3 produce broad blue absorption. None of them can account for the red and near-infrared absorption in the spectra of Uranus and Neptune. Second, our understanding of the detailed processes that lead to the formation of chromophores is inadequate. Gas-phase photochemical theory cannot account for the abundance of chromophore material. It is likely that UV photons or charged particle bombardment of solid, initially colorless particles like acetylene and ethane ice in the stratospheres of Uranus and Neptune or ammonium hydrosulfide in Jupiter's atmosphere breaks chemical bonds in the solid state, paving a path to the formation of more complex hydrocarbons or inorganic materials that seem to be required. Additional laboratory studies are needed to address these questions (*see* The Solar System at Ultraviolet Wavelengths).

Haze particles are present in the stratospheres of all the giant planets, but their chemical and physical properties and spatial distributions are quite different. Jupiter and Saturn have UV-absorbing aerosols abundant at high latitudes and high altitudes (corresponding to pressures ranging from a fraction of a millibar to a few tens of millibars). The stratospheric aerosols on Uranus and Neptune do not absorb much in the UV and are not concentrated at high latitude. The polar concentration of UV-absorbing aerosols on Jupiter and Saturn suggests that their formation may be due to chemistry in auroral regions, where protons and/or ions from the magnetosphere penetrate the upper atmosphere and deposit energy. Association with auroral processes may help explain why UV absorbers are abundant poleward of about 70° latitude on Saturn, extend to somewhat lower latitudes on Jupiter, and show a hemispheric asymmetry in Jupiter's atmosphere. Saturn's magnetic dipole is nearly centered and parallel to Saturn's spin axis, but Jupiter's magnetic dipole is both significantly offset and tilted with respect to its spin axis, producing asymmetric auroras at lower latitudes than on Saturn. Other processes, such as the **meridional circulation**, also influence the latitudinal distribution of aerosols, so more work needs to be done to establish the role of auroras in aerosol formation.

Photochemistry is responsible for the formation of diacetylene, acetylene, and ethane hazes in the stratospheres of Uranus and Neptune. The main steps in the life cycle of stratospheric aerosols are shown in Figure 32.7. Methane gas mixes upward to the high stratosphere, where it is photolyzed by UV light. Diacetylene, acetylene, and ethane form from gas-phase photochemistry and diffuse downward. Temperature decreases downward in the stratosphere, so ice particles form when the vapor pressure equals the partial pressure of the gas. On Uranus, diacetylene ice forms at 0.1 mbar, acetylene at 2.5 mbar, and ethane at 14 mbar. The ice particles sediment to deeper levels on a timescale of years and evaporate in the upper troposphere at 600 mbar and deeper. Polymers that form from solid-state photochemistry in the ice particles are probably responsible for the little UV absorption that does occur. They are less volatile than the pure ices and probably mix down to the methane cloud and below.

Photochemical models predict formation of hydrazine in Jupiter's stratosphere and diphosphine in Saturn's atmosphere. If these are the only stratospheric haze constituents, it is not apparent why the UV absorbers are concentrated at high latitude. As discussed earlier, auroral bombardment of methane provides an attractive candidate process for the abundant high-latitude aerosols on Jupiter and Saturn. However, we do not know enough to formulate a detailed chemical model of this process.

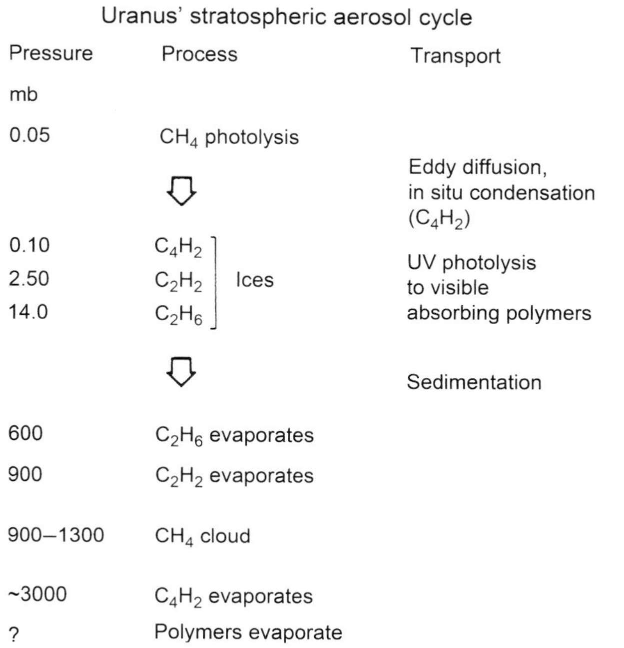

Uranus' stratospheric aerosol cycle

Pressure mb	Process	Transport
0.05	CH_4 photolysis	
	⇩	Eddy diffusion, in situ condensation (C_4H_2)
0.10	C_4H_2 ⎤	
2.50	C_2H_2 ⎬ Ices	UV photolysis to visible absorbing polymers
14.0	C_2H_6 ⎦	
	⇩	Sedimentation
600	C_2H_6 evaporates	
900	C_2H_2 evaporates	
900–1300	CH_4 cloud	
~3000	C_4H_2 evaporates	
?	Polymers evaporate	

FIGURE 32.7 Life cycle for stratospheric aerosols on Uranus. *From Pollack, J., et al. (1987). Copyright American Geophysical Union.*

Thermochemical equilibrium theory serves as a guide to the location of the bases of tropospheric clouds, but meteorology and cloud microphysical processes determine the vertical and horizontal distribution of cloud material. These processes are too complex to let us predict to what altitudes clouds should extend, and so we must rely on observations. Several diagnostics are available to measure cloud and haze vertical locations. At short wavelengths, gas molecules limit the depth to which we can see. In the visible and near infrared are methane and hydrogen absorption bands, which can be used to probe a variety of depths depending on the absorption coefficient of the gas. There are a few window regions in the thermal infrared where cloud opacity determines the outgoing radiance. The deepest probing wavelength is 5 μm. At that wavelength, thermal emission from the water cloud region near the 5-bar pressure level provides sounding for all the main clouds in Jupiter's atmosphere (see Infrared Views of the Solar System from Space).

The results of cloud stratigraphy studies for Jupiter's atmosphere are summarized in Figure 32.8. There is spectroscopic evidence for the two highest tropospheric layers in Jupiter's atmosphere. There is also considerable controversy surrounding the existence of the water-ammonia cloud on Jupiter. The *Galileo* probe descended into a dry region of the atmosphere and did not find a water cloud, but water clouds may be present in moister regions of the atmosphere that are obscured by overlying clouds. There is evidence for a large range of particle sizes. Small particles (less than about 1 μm radius) provide most of the cloud opacity in the visible. They cover belts and zones, although their optical thickness in belts is sometimes less than in zones. Most of the contrast between belts and zones in the visible comes from enhanced abundance or greater visibility of chromophore material, which seems to be vertically, but not horizontally, well mixed in the ammonia cloud. The top of this small-particle layer extends up to about 200 mbar, depending on latitude. Jupiter's GRS is a location of relatively high-altitude aerosols, consistent with the idea that it is a region of upwelling gas.

Larger particles (mean radius near 6 μm) are also present, mostly in zones. This large-particle component appears to respond to rapid changes in the meteorology. It is highly variable in space and time and is responsible, together with the deeper clouds, for the richly textured appearance of the planet at 5 μm wavelength (Figure 32.9). Some of the brightest regions seen in Figure 32.9 are called 5-μm hot spots, not because they are warmer than their surroundings but because thermal radiation from the 5-bar region emerges with little attenuation from higher clouds. The *Galileo* probe sampled one of these regions. The dark regions in the image are caused by optically thick clouds in the NH_4SH and NH_3 cloud regions. The thickest clouds are generally associated with upwelling, bright (at visible wavelengths) zones, but many exceptions to this rule are observed. Until we understand the chemistry and physics of

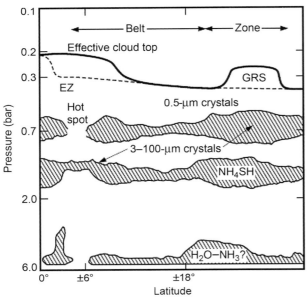

FIGURE 32.8 Observations of Jupiter at wavelengths that sense clouds lead to a picture of the Jovian cloud stratigraphy shown here. There has been no direct evidence for a water-ammonia cloud near the 6-bar pressure level, but it is likely that such a cloud exists from indirect evidence. The hot spots are named from their visual appearance at a wavelength of 5 μm. They are not physically much warmer than their surroundings, but they are deficient in cloudy material (see Figure 32.9). *From West, R., et al. (2004). Cambridge Univ. Press.*

FIGURE 32.9 At a wavelength of 5 μm, most of the light from Jupiter is thermal radiation emitted near the 6-bar pressure level below the visible cloud. Places where the clouds are thin permit the deep radiation to escape to space, making these regions appear bright. Thicker clouds block the radiation and these appear dark. Jupiter's Great Red Spot is the dark oval just below the center. This image was taken with the NASA Infrared Telescope Facility. *Courtesy of J. Spencer.*

chromophores, we should not expect to understand why or how well albedo is correlated with other meteorological parameters.

Most of Jupiter's spots are at nearly the same altitude. Some notable exceptions are the GRS, the three white ovals just south of the GRS, and some smaller ovals at other latitudes. These anticyclonic features extend to higher altitudes, probably up to the 200 mbar level, compared to a pressure level of about 300 mbar for the surrounding clouds. Some of the anticyclonic spots have remarkably long lifetimes compared to the terrestrial norm. The GRS was recorded in drawings in 1879, and reports of red spots extend back to the seventeenth century. The three white ovals in a latitude band south of the GRS formed from a bright cloud band that split into three segments in 1939. The segments shrunk in longitude over the course of a year, until the region (the South Temperate Belt) was mostly dark except for three high-albedo spots that remain to the present. Anticyclonic ovals tend to be stable and long-lived, whereas cyclonic regions constantly change.

Similar features are observed in Saturn's atmosphere, although the color is much subdued compared to Jupiter, and Saturn has nothing that is as large or as long lived as the GRS. The reduced contrast may be related to Saturn's colder tropopause temperature. The distance between the base of the ammonia cloud and the top of the troposphere (where the atmosphere becomes stable against convection) is greater on Saturn than on Jupiter. The ammonia ice cloud on Saturn is both physically and optically thicker than it is on Jupiter. Occasionally (about two or three times each century), a large, bright cloud forms near Saturn's equator. One well-observed event occurred in 1990, but its cause is unknown. It appears to be a parcel of gas that erupts from deeper levels, bringing fresh condensate material to near the top of the troposphere. It becomes sheared out in the wind shear and dissipates over the course of a year.

Uranus as seen by *Voyager* was even more bland than Saturn, but recent images from the *Hubble Space Telescope* and from the ground show a much richer population of small clouds (see Figure 32.10). Midlatitude regions on Uranus and Neptune are cool near the tropopause, indicating upwelling. But cloud optical thickness may be lower there than at other latitudes. The relation between cloud optical thickness and vertical motion is more complicated than the simple condensation model would predict.

Neptune's clouds are unique among the outer planet atmospheres. *Voyager* observed four large cloud features that persisted for the duration of the *Voyager* observations (months). The largest of these is the Great Dark Spot (GDS) and its white companion. Because of its size and shape, the GDS might be similar to Jupiter's GRS, but the GDS had a short life compared to the GRS.

There is no explanation yet of what makes the dark spot dark. The deepest cloud (near the 3-bar level) is probably

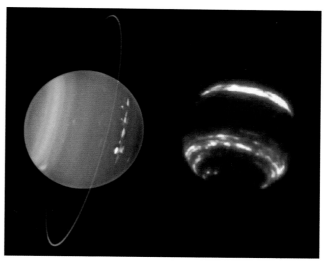

FIGURE 32.10 Images of Uranus (left) and Neptune (right) taken in 2004 and 2000, respectively. Both were obtained at the Keck telescope with filters in the near infrared. Many cloud features that were not seen during the *Voyager* flyby can be seen. The Uranus ring can also be seen (a red ellipse in this false-color representation). *The Uranus image appeared on the cover of* Icarus *(December 15, 2005, issue) and was provided by L. Sromovsky. The Neptune image is from de Pater, I., et al. (2005). Copyright Academic Press).*

H_2S ice, since ammonia is apparently depleted and NH_4SH would be sequestered at a deeper level. At higher altitudes there is an optically thin methane haze (near 2 bar) and stratospheric hazes of ethane, acetylene, and diacetylene. At high spatial resolution, the wispy white clouds associated with the companion to the GDS and found elsewhere on the planet form and dissipate in a matter of hours. It was difficult to estimate winds from these features because of their transitory nature. Individual wisps moved at a different speed than the GDS and its companion, suggesting that these features form and then evaporate high above the GDS as they pass through a local pressure anomaly, perhaps a standing wave caused by flow around the GDS. Cloud shadows were seen in some places, a surprise after none was seen on the other giant planets. The clouds casting the shadows are about 100 km higher than the lower cloud deck, suggesting that the lower cloud is near 3 bar and the shadowing clouds are near 1 bar, in the methane condensation region. More recent Hubble and ground-based images show clouds not seen in *Voyager* images (Figure 32.10).

4. DYNAMICAL METEOROLOGY OF THE TROPOSPHERE AND STRATOSPHERE

Our understanding of giant planet meteorology comes mostly from *Voyager* observations, with observations from *Galileo*, *Cassini*, the *Hubble Space Telescope*, and ground-based data adding to the picture. Although we have theories and models for many of the dynamical features, the fundamental nature of the dynamical meteorology on the giant planets remains puzzling chiefly because of our inability to probe to depths greater than a few bars in atmospheres that go to kilobar pressures and because of limitations in spatial and time sampling, which may improve with future missions to the planets.

Thermodynamic properties of atmospheres are at the heart of a variety of meteorological phenomena. In the terrestrial atmosphere, condensation, evaporation, and transport of water redistribute energy in the form of latent heat. The same is true for the outer planet atmospheres, where condensation of water, ammonia, ammonium hydrosulfide, hydrogen sulfide, and methane takes place. Condensables also influence the dynamics through their effects on density gradients. In the terrestrial atmosphere, moist air is less dense than dry air at the same temperature because the molecular weight of water vapor is less than that of the dry air. Because of this fact, and also because moist air condenses and releases latent heat as it rises, there can be a growing instability leading to the formation of convective plumes, thunderstorms, and anvil clouds at high altitudes. On the giant planets, water vapor is significantly heavier than the dry atmosphere and so the same type of instability will not occur unless a strongly upwelling parcel is already present. Some researchers proposed that the Equatorial Plumes on Jupiter and the elongated clouds on Uranus are the outer planet analogs of terrestrial anvil clouds.

Terrestrial lightning occurs most frequently over tropical oceans and over a fraction of the land surface. Its distribution in latitude, longitude, and season is indicative of certain properties of the atmosphere, especially the availability of liquid water. Lightning has been observed on the giant planets as well, either from imaging on the night side (Jupiter) or from signals recorded by plasma wave instruments. A somewhat mysterious radio emission from Saturn (the so-called Saturn Electrostatic Discharge events) has been interpreted as a lightning signature. Combined imaging and plasma wave observations from *Cassini* in 2004 revealed a large cloud complex associated with this source. The intensity and size of the lightning spots in the images imply that they are much more energetic than the average lightning bolt in the terrestrial atmosphere and that they occur in the water-ammonia cloud region as expected. The *Galileo* probe did not detect lightning in Jupiter's atmosphere within a range of about 10,000 km from its location at latitude $6.5°N$ (*see* The Solar System at Radio Wavelengths).

The heat capacity of hydrogen, and therefore the dry lapse rate of the convective part of the atmosphere, depends on the degree to which the ortho/para states equilibrate. The lapse rate is steepest when equilibration is operative. The observed lapse rate for Uranus, as measured by the *Voyager* radio occultation experiment, is close to the "frozen" lapse rate—the rate when the relative fractions of ortho- and para-hydrogen are fixed. How can the observed relative fractions be near equilibrium when the lapse rate points to nonequilibrium? One suggestion is that the atmosphere is layered. Each layer is separated from the next by an interface that is stable and that is thin compared to the layer thickness. The air within each layer mixes rapidly compared to the time for equilibration, but the exchange rate between layers is slow or comparable to the timescale for conversion of ortho to para and back.

How can layers be maintained in a convective atmosphere? In the terrestrial ocean, two factors influence buoyancy: temperature and salinity. If the water is warmer at depth, or if the convective amplitude is large, the different timescales for diffusion of heat and salinity lead to layering. In the atmospheres of the outer planets, the higher molecular weight of condensables acts much as salinity in ocean water. Layering can be established even without molecular weight gradients. Layering in the terrestrial stratosphere and mesosphere has been observed. Layers of rapidly convecting gas occur where gravity waves break or where other types of wave instabilities dump energy. Between layers of rapid stirring are stably stratified layers with transport by diffusion rather than convection.

Some of the variety of the giant planet meteorology, as well as our difficulty to understand it, is nicely illustrated by observations of the wind field at the cloud tops. Wind vectors of all the giant planet atmospheres are predominantly in the east—west (zonal) direction (Figure 32.11). These are determined by tracking visible cloud features over hours, days, and months. Jupiter has an abundance of small features and the zonal winds are well mapped. Saturn has fewer features, and they are of less contrast than those on Jupiter, but there is still a large enough number to provide detail in the wind field. Only a few features were seen in *Voyager* images of the Uranus atmosphere, and all but one of these were between latitudes $20°S$ and $40°S$. More recent images from the *Hubble Space Telescope* show new features at many other latitudes. The *Voyager 2* radio occultation provided an additional estimate for wind speed at the equator. Neptune has more visible features than Uranus, but most of them are transitory and difficult to follow long enough to gauge wind speed.

Figure 32.11 reveals a great diversity in the zonal flow among the giant planet atmospheres. Wind speed is relative to the rotation rate of the deep interior as revealed by the magnetic field and radio emissions. Jupiter has a series of jets that oscillate with latitude and are greatest in the prograde direction at latitude $23°N$, and near $\pm10°$. The pattern of east—west winds is approximately symmetric about the equator except at high latitude. Saturn has a very strong prograde jet at low latitudes (within the region

Chapter | 32 Atmospheres of the Giant Planets

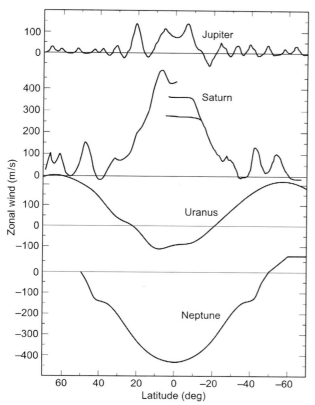

FIGURE 32.11 Zonal (east–west) wind velocity for the giant planets as a function of latitude. *For Jupiter, the data are from Porco et al. (2003). Copyright American Association for the Advancement of Science). For Saturn's northern hemisphere, the data are from Gierasch, P., & Conrath, B. (1993). Copyright American Geophysical Union). For Saturn's southern hemisphere, data are from Porco et al. (2005). Copyright American Association for the Advancement of Science). Two branches are shown for the southern low latitudes. Both are from Cassini observations, with similar values from Hubble Space Telescope images. The higher wind speeds were observed for deepest clouds, while the lower winds were observed for higher clouds. Both branches are moving more slowly than clouds at similar latitudes in the north observed by Voyager. This apparent change in the wind speed must have involved a large energy exchange. Data for Uranus and Neptune are mostly from analyses of Hubble and Keck data (Sromovsky, L., & Fry, P. (2005). Copyright Academic Press. Sromovsky, L., et al. (2001). Copyright Academic Press.).*

despite the many small-scale features that evolve with much shorter lifetimes. An interesting exception to this rule occurred at equatorial latitudes on Saturn between the time of the *Voyager* observations (around 1981) and observations in the 1990s and later by the *Hubble Space Telescope* and beginning in 2004 by the *Cassini* cameras. Current equatorial jet speeds are significantly less than those measured on Saturn by *Voyager*. It is difficult to understand how such a large change of momentum could occur, and another explanation has been sought. Possibly the equatorial atmosphere was clearer (less haze) during the *Voyager* epoch, permitting observations to deeper levels where the wind speed is higher. Detailed analyses of haze altitudes show that the haze is thicker and higher in more recent times than it was in 1981, but probably not enough to account for the difference in wind speed from this effect alone.

Some of the key observations that any dynamical theory must address include: (1) the magnitude, direction, and latitudinal scale of the jets; (2) the stability of the jets, at least for Jupiter and Saturn, where observations over long periods show little or no change except for Saturn's equatorial jet, which was mentioned earlier; (3) the magnitude and latitudinal gradients of heat flux; and (4) the interactions of the mean zonal flow with small spots and eddies. One of the controversies during the past two decades concerns how deep the flow extends into the atmosphere. It is possible to construct shallow-atmosphere models that have approximately correct jet scales and magnitudes. A shallow-atmosphere model is one in which the jets extend to relatively shallow levels (100 bar or less), and the deeper interior rotates as a solid body, or at least as one whose latitudinal wind shear is not correlated with the wind shear of the jets. The facts that the jets and some spots on Jupiter are very stable, that there is approximate hemispheric symmetry in the zonal wind pattern between latitudes ±60°, and that the Jovian interior has no density discontinuities down to kilobar levels suggested to some investigators that the jets extend deep into the atmosphere. A natural architecture for the flow in a rotating sphere with no density discontinuities is one in which the flow is organized on rotating cylinders (Figure 32.12).

Apart from the stability and symmetry noted here, there is little evidence to suggest that the zonal wind pattern really does extend to the deep interior. The conductivity of Jupiter's atmosphere at depth is probably too high to allow the type of structure depicted in Figure 32.12 to exist. The strength of the zonal jet at the location where the *Galileo* probe entered (6.5°N) increased with depth, consistent with the idea of a deeply rooted zonal wind field on Jupiter. One way to test that hypothesis is to make highly precise measurements of the gravity field close to the planet. There are density gradients associated with the winds, and these produce features in the gravity field close to the planet. The

±15°). It also has alternating but mostly prograde jets at higher latitudes, with the scale of latitudinal variation being about 10°. Uranus appears to have a single prograde maximum near 60°S, and the equatorial region is retrograde. Neptune has an enormous differential rotation, mostly retrograde except at high latitude. Various theories have been advanced to explain the pattern of zonal jets. None of them can account for the great variety among the four planets.

The zonal jets are stable over long time periods (observations span many decades for Jupiter and Saturn),

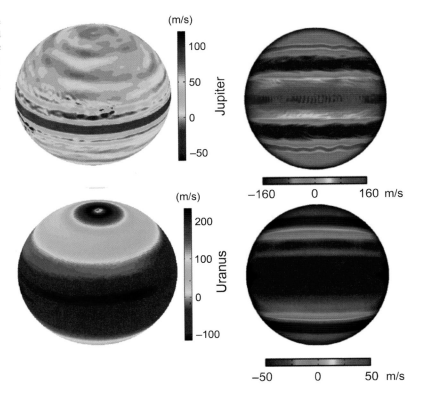

FIGURE 32.12 Models for the zonal wind fields of the giant planets attempt to explain the observed zonal wind profiles shown in Figure 32.11. *The left panels are from Lian, & Showman, (2010). (Copyright by Elsevier). Panels on the right are from Liu, & Schneider, (2010). (Copyright by the American Meteorological Society). Color bars indicate wind speeds.*

largest signature is produced by Neptune's remarkable differential rotation. The *Voyager 2* spacecraft flew just above Neptune's atmosphere and provided the first evidence that the differential rotation cannot extend deep into the atmosphere. Gravity-field tests of the deep-wind hypothesis for the other giant planets are more difficult because the differential rotation is much weaker. No spacecraft have come close enough to make the measurements but one (the Juno mission) will do so for Jupiter and another (the Cassini mission) will do so for Saturn.

What process maintains the zonal wind pattern? *Voyager* measurements shed some light on this question, but provided some puzzles as well. The ultimate energy source for maintaining atmospheric motions is the combination of internal thermal and solar energy absorbed by the atmosphere. Jupiter, Saturn, and Neptune all have significant internal energy sources, whereas Uranus has little or none. A measure of the amount of energy available for driving winds is the escaping radiative energy per square meter of surface area. Almost 20 times as much energy per unit area is radiated from Jupiter's atmosphere as from Neptune's, yet the wind speeds (measured relative to the interior as determined from the magnetic field rotation rate) on Neptune are about three times higher than those on Jupiter. Rather than driving zonal winds, the excess internal energy may go into driving smaller scale eddies, which are most abundant on Jupiter.

The widths and speeds of the zonal jets are very different for the different planets (see Figure 32.11). Is there a way to understand these differences in terms of fundamental atmospheric physical parameters, to achieve a unifying framework? Three hypotheses have emerged in recent years. One postulates that the important parameter is the ratio of convective heat flow to radiative heating (Figure 32.12 right panels). Another postulates that latent heat transfer associated with water condensation is key (Figure 32.12 left panels). A third proposes that the interaction of the Coriolis force with convective motions determines how the zonal bands are distributed and whether the equatorial flow is eastward or westward. It is possible that all these processes are important.

What role do eddies have in maintaining the flow? Measurements of the small spots on Jupiter and Saturn have provided an estimate of the energy flow between the mean zonal wind and the eddy motions. For Jupiter, the eddies at the cloud top appear to be pumping energy into the mean zonal flow. Recent measurements from cameras on the Cassini spacecraft show convincingly that eddies do pump energy into the zonal jets of Saturn. Ongoing studies try to understand why the jets are so stable when there is apparently enough energy in eddy motions to significantly modify the wind field. At the same time, other observations imply dissipation and decay of zonal winds at altitudes above the cloud tops.

Chapter | 32 Atmospheres of the Giant Planets

The relationship known to atmospheric physicists as the thermal wind equation provides a means of estimating the rate of change of zonal wind with height (which is usually impossible to measure remotely) from observations of the latitudinal gradient of temperature (which is usually easy to measure). One of the common features of all the outer planet atmospheres is a decay of zonal wind with height in the stratosphere, tending toward solid-body rotation at high altitudes. The decay of wind velocity with height could be driven by eddy motions or by gravity wave breaking, which effectively acts as friction on the zonal flow.

Thermal contrasts on Jupiter are correlated with the horizontal shear and with cloud opacity as indicated by 5-μm images (see Figure 32.9). Cool temperatures at the tropopause level (near 100 mbar) are associated with upwelling and anticyclonic motion, and warmer temperatures are associated with subsidence. Jupiter's GRS is an anticyclonic oval with cool tropospheric temperatures, upwelling flow, and aerosols extending to relatively high altitudes. Enhanced cloud opacity and ammonia abundance in cooler anticyclonic latitudes (mostly the high-albedo zones on Jupiter) are predicted in upwelling regions. The correlation is best with cloud opacity in the 5-μm region. At shorter wavelengths (in the visible and near infrared), there is a weaker correlation between cloud opacity and **vorticity**. Perhaps the small aerosols near the top of the troposphere, sensed by the shorter wavelengths but not at 5 μm, are transported horizontally from zone to belt on a timescale that is short compared to their rainout time (several months).

The transport of heat may well be more complicated than the previous paragraph implies. There may be at least two regions, an upper troposphere where heat transport is determined by slow, large-scale motions as previously depicted, and a lower troposphere at pressures between 2 and 10 bar, where heat is transported upward mostly in the belts, by small convective storms which are seen in the belts. There is evidence from the *Galileo* and *Cassini* observations that this is the case.

The upwelling/subsidence pattern at the jet scale in the upper troposphere penetrates into the lower stratosphere. We have relatively little information on the stratospheric circulation for the giant planets. Most of it is based on the observed thermal contrasts and the idea that friction is a dominant driver for stratospheric dynamics. We are beginning to appreciate the role of forcing by gravity or other dissipative waves. A model for the Uranus stratospheric circulation is based on the frictional damping and the observed thermal contrast as a function of latitude. The coldest temperatures in the lower stratosphere are at mid-latitudes, indicating upwelling there and subsidence at the equator and poles. A different pattern is expected if the deposition of solar energy controlled the circulation. Momentum forcing by vertically propagating waves from the deeper atmosphere is apparently more important than solar energy deposition.

The mean meridional circulation in Jupiter's stratosphere differs from that predicted by the frictional damping model at pressure levels less than about 80 mbar. The zonal pattern of upwelling/sinking extends to about 100 mbar, giving way at higher altitude to a two-cell structure with cross-equatorial flow. There is also a hemispheric asymmetry. The high latitudes (poleward of 60°S and 40°N) are regions of sinking motion at the tropopause. Images from the *Hubble Space Telescope* indicate that the optical depth of the ammonia cloud decreases rapidly with latitude poleward of 60°S and 40°N and is well correlated with the estimated downward velocity. The descending dry air inhibits cloud formation. To produce that circulation, there must be momentum forcing in the latitude range 40°S to 80°S and 30°N to 80°N at pressures between 2 and 8 mbar. Dissipation of gravity waves propagating from the deep interior is the most likely source of momentum forcing.

Superimposed on the long-term mean are much faster processes such as horizontal eddy mixing, which can transport material in the north–south direction in days or weeks. The impacts of comet Shoemaker–Levy 9 on Jupiter in 1994 provided a rare opportunity to see the effects of eddy transport on small dust particles and trace chemical constituents deposited in the stratosphere immediately after impact. Particles spread rapidly from the impact latitude (45°S) to latitude 20°S, but there has been almost no transport farther toward the equator. Trace constituents at higher altitude such as HCN were observed to move across the equator into the northern hemisphere (*see* Physics and Chemistry of Comets).

Long-term monitoring of the Jovian stratosphere has yielded some interesting observations of an oscillating temperature cycle at low latitudes. At pressures between 10 and 20 mbar, the equator and latitudes ±20° cool and warm alternately on timescales of 2–4 years. The equator was relatively (1–2 K above the average 147 K) warm and latitudes ±20° were relatively (1–2 K below average) cool in 1984 and 1990. The reverse was true in 1986 and 1987. Changes in temperature must be accompanied by changes in the wind field, and these must be generated by stresses induced by wave forcing or convection. The similarities of the Jovian temperature oscillations to low-latitude temperature oscillations in the terrestrial atmosphere led some researchers to propose that the responsible mechanism is similar to that driving the quasi-biennial oscillation on Earth: forcing by vertically propagating waves. The period of the oscillation is about 4 (Earth) years and so the phenomenon has been called the quasi-quadrennial oscillation. Recent measurements from the ground and from the Cassini spacecraft have identified the same kind of wave structure on Saturn.

The *Voyager* cameras and more recently *Hubble* and ground-based images provided much information about the shapes, motions, colors, and lifetimes of small features in the atmospheres of the giant planets. In terms of the number of features and their contrast, a progression is evident from Jupiter, with thousands of visible spots, to Uranus, with only a few. Neptune has a few large spots that were seen for weeks and an abundance of small ephemeral white patches at a few latitudes. We do not have a good explanation for the contrasts and color because the thermochemical equilibrium ices that form these clouds (NH_3, NH_4SH, H_2O, CH_4, and H_2S) are colorless. We need to know more about the chemical nature, origin, and location of the colored material before we can understand how the contrasts are produced.

Fortunately, it is not necessary to understand how the contrasts are produced to study the meteorology of these features. One of the striking attributes of some of the clouds is their longevity. Jupiter's GRS has been observed since 1879 and may have existed much earlier. A little to the south of the GRS are three white ovals, each about one-third the diameter of the GRS. These formed in 1939–1940, beginning as three very elongated clouds (extending 90° in longitude) and rapidly shrinking in longitude. They survived as three distinct ovals until 1998 when two of them merged. In 2000, the remaining two merged, leaving one. There are many smaller, stable ovals at some other Jovian latitudes. All these ovals are anticyclonic and reside in anticyclonic shear zones. Because they are anticyclonic features, there is upwelling and associated high and thick clouds, and cool temperatures at the tropopause. Sinking motion takes place in a thin boundary region at the periphery of the clouds. The boundary regions are bright at 5 μm wavelength, consistent with relatively cloud-free regions of sinking. The GRS as revealed by *Galileo* instruments is actually much more complex, with cyclonic flow and small regions of enhanced 5-μm emission (indicating reduced cloudiness) in its interior.

Another attribute of many of the ovals is the oscillatory nature of their positions and sometimes shape. The most striking example is Neptune's GDS, whose aspect ratio (ratio of shortest to longest dimension) varied by more than 20% with a period of about 200 h, with a corresponding oscillation in orientation angle. Neptune's Dark Spot 2 drifted in latitude and longitude, following a sinusoidal law with amplitude 5° in latitude (between 50°S and 55°S) and 90° (peak to peak) in longitude. Other spots on Neptune and Jupiter, including the GRS, show sinusoidal oscillations in position. The Jovian spots largely remain at a fixed mean latitude, but the mean latitude of the GDS on Neptune drifted from 26°S to 17°S during the 5000 h of observations by the *Voyager 2* camera. Ground-based observations in 1993 did not show a bright region at methane absorption wavelengths in the southern hemisphere, unlike the period during the *Voyager* encounter when the high-altitude white companion clouds were visible from the Earth. The GDS may have drifted to the northern hemisphere and/or may have disappeared. *Hubble Space Telescope* images and ground-based images since the *Voyager* encounter show new spots at new latitudes.

Jupiter's GRS is often and incorrectly said to be the Jovian analog of a terrestrial hurricane. Hurricanes are cyclonic vortices. The GRS and other stable ovals are anticyclones. Hurricanes owe their (relatively brief) stability to energy generated from latent heating (condensation) over a warm ocean surface, where water vapor is abundant. Upwelling occurs in a broad circular region, and subsidence is confined to a narrow core (the eye). The opposite is true for anticyclonic spots in the giant planet atmospheres, where subsidence takes pace in a narrow ring on the perimeter of the oval. The key to their stability is the long-lived, deep-seated background latitudinal shear of the jets. The stable shear in the jets provides an environment that is able to support the local vortices. Latent heat, so important for a terrestrial hurricane, seems to play no role. However, the ephemeral bright small clouds seen in some locations may be places where strong upwelling is reinforced by release of latent heat analogous to a terrestrial thunderstorm.

5. ENERGETIC PROCESSES IN THE HIGH ATMOSPHERE

At low pressure (less than about 50 μbar), the mean free path for collisions becomes sufficiently large that lighter molecules diffusively separate from heavier ones. The level where this occurs is called the homopause. The outer planet atmospheres are predominantly composed of H_2 and He, with molecular hydrogen dissociating to atomic hydrogen, which becomes the dominant constituent at the exobase (the level where the hottest atoms can escape to space). This is also the region where solar extreme ultraviolet (EUV) radiation can dissociate molecules and ionize molecules and atoms. Ion chemistry becomes increasingly important at high altitudes. Some reactions can proceed at a rapid rate compared to neutral chemistry. Ion chemistry may be responsible for the abundant UV-absorbing haze particles (probably hydrocarbons) in the polar stratospheres of Jupiter and Saturn.

The high atmospheres of the giant planets are hot (400–800 K for Jupiter to 300 K for Uranus and Neptune), much hotter than predicted on the assumption that EUV radiation is the primary energy source. Estimates prior to the *Voyager* observations predicted high-altitude temperatures closer to 250 K or less. One of the challenges of the post-*Voyager* era is to account for the energy balance of the high atmosphere. Possible sources of energy in addition to EUV radiation include (1) Joule heating, (2) currents induced by a planetary dynamo mechanism, (3) electron precipitation from the magnetosphere (and also proton and S and O ion precipitation in the Jovian auroral region), and (4) breaking inertia-gravity waves.

Joule heating requires electric currents in the ionosphere that accelerate electrons and protons. It is a major source of heating in the terrestrial thermosphere. We do not have enough information on the magnetosphere to know how important this process or the others mentioned are for the giant planet atmospheres. The planetary dynamo current theory postulates that currents are established when electrons and ions embedded in the neutral atmosphere move through the magnetic field, forced by the neutral wind tied to the deeper atmosphere. Electric fields aligned with the magnetic field are generated by this motion and accelerate high-energy photoelectrons that collide with neutrals or induce plasma instabilities and dissipate energy. Similar mechanisms are believed to be important in the terrestrial atmosphere.

Electron precipitation in the high atmosphere was one of the first mechanisms proposed to account for bright molecular hydrogen UV emissions. There is recent evidence for supersonic pole-to-equator winds in the very high atmosphere on Jupiter driven by auroral energy. These winds collide at low latitudes, producing supersonic turbulence and heating. Electron and ion precipitation outside the auroral regions undoubtedly contributes to the heating, but the details remain unclear. The possible contribution from breaking planetary waves is difficult to estimate, but *Galileo* probe measurements, details of the radio occultation profiles, and less direct lines of evidence point to a significant energy density in the form of inertia-gravity waves in the stratosphere and higher. How much of that is dissipated at pressures less than 50 μbar is unknown but could be significant to the energy budget of the high atmosphere.

The giant planets have extensive ionospheres. Like the neutral high atmospheres, they are hotter than predicted prior to the *Voyager* encounters. As for Earth, the ionospheres are highly structured, having a number of high-density layers. Layering in the terrestrial ionosphere is partially due to the deposition of metals from meteor ablation. The same mechanism is thought to be operative in the giant planet ionospheres. The Jupiter and Saturn ionospheres are dominated by the H_3^+ ion, whereas those of Uranus and Neptune are dominated by H^+.

Auroras are present on all the giant planets. Auroras on the Earth (the only other planet in the solar system known to have auroras) are caused by energetic charged particles streaming down the high-latitude magnetic field lines. The most intense auroras on Earth occur when a solar flare disturbs the solar wind, producing a transient in the flow that acts on Earth's magnetosphere through ram pressure. As the magnetosphere responds to the solar wind forcing, plasma instabilities in the tail region accelerate particles along the high-latitude field lines.

The configuration of the magnetic field is one of the key parameters that determines the location of auroras. Jupiter's magnetosphere is enormous compared to the Earth's. If its magnetosphere could be seen by the naked eye from the Earth, it would appear to be the size of the Moon (about 30 arc minutes), whereas Jupiter's diameter is less than 1 arc minute. To a first approximation, the magnetic fields of the Earth and the giant planets can be described as tilted dipoles, offset from the planet center. Table 32.4 lists the strength, tilt, and radial offsets for each of these planets. Earth and Jupiter have relatively modest tilts and offsets, Saturn has virtually no tilt and almost no offset, whereas Uranus and Neptune have very large tilts and offsets. Such diversity presents a challenge to planetary dynamo modelers (*see* Planetary Magnetospheres).

The mapping of the magnetic fields onto the upper atmosphere determines where auroral particles intercept the atmosphere. Maps for Jupiter, Uranus, and Neptune are shown in Figure 32.13, along with locations of field lines connected to the orbits of some satellites that may be important for auroral formation. The configuration for Saturn is not shown because contours of constant magnetic field magnitude are concentric with latitude circles owing to the field symmetry. Because of the large tilts and offsets for Uranus and Neptune, auroras on these planets occur far from the poles.

The Jovian aurora is the most intense and has received the most scrutiny. The remainder of this section will focus on what is known about it. It has been observed over a remarkable range of wavelengths, from X-rays to the infrared, and possibly in the radio spectrum as well. Energetic electrons from the magnetosphere dominate the energy input, but protons and S and O ions contribute as well. Sulfur and oxygen k-shell emission seems to be the most plausible explanation for the X-rays. Models of energetic electrons impacting on molecular hydrogen provide a good fit to the observed molecular hydrogen emission spectra. Secondary electrons as well as UV photons are emitted when the primary impacting electrons dissociate the molecules, and these secondaries also contribute to the

TABLE 32.4 Magnetic Field Parameters (Offset Tilted Dipole Approximation)

	Earth	Jupiter	Saturn	Uranus	Neptune
Tilt (degrees)	11.2	9.4	0.0	58.6	46.9
Offset (planetary radius)	0.076	0.119	0.038	0.352	0.485

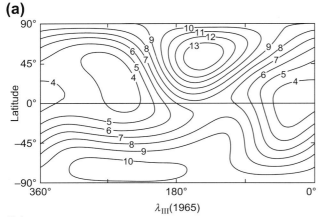
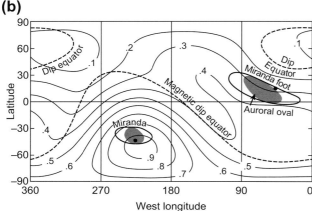
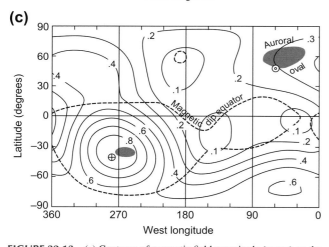

FIGURE 32.13 (a) Contours of magnetic field magnitude (gauss) on the surface of Jupiter (using the Goddard Spaceflight Center (GSFC) Model D_4). (b) Contours of constant magnetic field on the upper atmosphere of Uranus, along with the location of the auroral oval and the lines connected to the orbit of the satellite Miranda (Model Q_3). The magnetic dip equator is the location where the field lines are tangent to the surface. (c) Contours of constant magnetic field magnitude and pole locations (circled cross and dot) for Neptune (Model O_8). *(From Connerney, J. (1993). Copyright American Geophysical Union.)*

UV emissions. Some of the UV-emitted radiation is reabsorbed by other hydrogen molecules, and some is absorbed by methane molecules near the top of the homopause. From the detailed shape of the spectrum, it is possible to infer the depth of penetration of electrons into the upper atmosphere. In the near infrared (2–4 μm), emissions from the H_3^+ ion are prominent. Attempts to account for all the observations call for more than one type of precipitating particle and more than one type of aurora.

UV auroras from atomic and molecular hydrogen emissions are brightest within an oval that is approximately bounded by the closed field lines connected to the middle magnetosphere (corresponding to a region some 10–30 Jupiter radii from the planet) rather than the orbit of Io or open field lines connected to the tail. Weaker, diffuse, and highly variable UV emissions appear closer to the pole. They are produced by precipitation of energetic particles originating from more distant regions in the magnetosphere. There is also an auroral hot spot at the location where magnetic field lines passing through Io enter the atmosphere (the Io flux tube footprint). All these features are evident in Figure 32.14.

Io is a significant source of sulfur and oxygen, which come off its surface. The satellite and magnetosphere produce hot and cold plasma regions near the Io orbit, which may stimulate plasma instabilities. High-spatial-resolution, near-infrared H_3^+ images show emission from a region that maps to the last closed field lines far out in the magnetosphere (Figure 32.15). This and evidence for auroral response to fluctuations in solar wind ram pressure indicate that at least some of the emission is caused by processes that are familiar to modelers of the terrestrial aurora (*see* Io: The Volcanic Moon).

Auroral emission is strongest over a small range of longitudes. In the north, longitudes near 180°, System III coordinates (which rotate with the magnetic field) show enhanced emission in the UV and also in the thermal infrared. The spectrum of the aurora in the UV resembles electron impact on molecular hydrogen, except that the shortest wavelengths are deficient. This deficit can be accounted for if the emission is occurring at some depth in the atmosphere (near 10 μbar) below the region where methane and acetylene absorb UV photons. By contrast, the Uranian high atmosphere is depleted in hydrocarbons and does not produce an emission deficit.

Energy deposition at depth is also required to explain the warm stratospheric temperatures seen in the 7.8-μm methane band. At 10 μbar of pressure, the hot spot region near longitude 180° appears to be 60–140 K warmer than the surrounding region, which is near 160 K. Undoubtedly such temperature contrasts drive the circulation of the high atmosphere. Auroral energy also contributes to anomalous chemistry. An enhancement is seen in acetylene emission in the hot spot region, whereas ethane emission decreases

Chapter | 32 Atmospheres of the Giant Planets

FIGURE 32.14 (Top) Image of Jupiter at ultraviolet wavelengths taken with the Wide Field and Planetary Camera 2 on the *Hubble Space Telescope*. Bright auroral ovals can be seen against the dark UV-absorbing haze and in the polar regions. Jupiter's north magnetic pole is tilted toward the Earth, making it easier to see the northern auroral oval as well as some diffuse emission inside the oval. Small bright spots just outside the oval in both hemispheres are at the location of the magnetic field lines connecting to Io, depicted by a blue curve. Io is dark at UV wavelengths. (Bottom) Image taken a few minutes after the one above in a filter that samples the violet part of the spectrum just within the range that the human eye can detect. The Great Red Spot appears dark at this wavelength and can just be seen in the top image as well. Io's small disk appears here along the blue curve, which traces the magnetic field lines in which it is embedded. *Courtesy of J. Trauger and J. Clarke.*

there. A significant part of the acetylene enhancement could be due simply to the higher emission from a warmer stratosphere, but a decrease in ethane requires a smaller ethane mole fraction.

Future work on the auroras of Jupiter and the other giant planets will focus on which types of particles are responsible for the emissions, the regions of the magnetosphere or

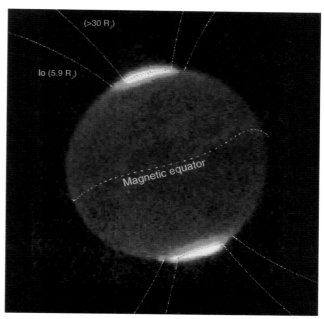

FIGURE 32.15 Auroral regions are bright in this image at wavelength 3.4 μm, where the H_3^+ ion emits light. Magnetic field lines connecting to Io and to the 30-Jupiter-radius equator crossing are shown. The brightest emissions are poleward of the $30R_J$ field line, which means the precipitating particles responsible for this emission come from more distant regions on the magnetosphere. *Reprinted with permission from Connerney, J., et al. (1993). Copyright 1993 American Association for the Advancement of Science.*

torus from which they originate, the acceleration mechanisms, and how the deposited energy drives circulation and chemistry in the high atmosphere.

6. A WORD ABOUT EXTRASOLAR PLANETS

The discovery and study of planets in orbit around other stars (extrasolar planets) is an exploding field, with a rapidly increasing inventory of new planets, with new information from observations and theory. This field is likely to continue to blossom in the near future. So far most of the discoveries are in the category of the giant planets, and Neptune-size planets seem to be abundant. How can knowledge of the giant planets in our solar system inform about the properties of newly discovered giant planets?

What we know about the giant planets of our solar system provides a framework from which to extend our understanding of the atmospheres of planets around other stars. We have learned that two very different processes (rapid convection from the deep interior and photochemistry in the high atmosphere) produce molecular species unexpected from local chemical equilibrium. We have learned that the colors of clouds and hazes are influenced by similar disequilibrium processes. We have learned that large vortices can be stable over long time periods and that the wind pattern is dominated by east–west jets. We have learned that energetic processes in the high atmosphere produce temperatures higher than expected from the deposition of solar energy.

In short, we have learned that the atmospheres of the giant planets are more complicated than we once thought. Dealing with these issues has forced us out of our comfort zone with respect to our understanding of how atmospheres work built on familiarity with the Earth's atmosphere.

Complexity will surely become even more evident for the atmospheres of giant planets orbiting other stars. A large fraction of those discovered so far are orbiting very close to their central star. These planets have hotter atmospheres, so the clouds in the upper atmospheres would be composed of rocky minerals instead of ices of water, ammonia, hydrogen sulfide, and methane. Photochemistry for the close planets will proceed at a much more vigorous rate than for the planets in our solar system. These planets will have some peculiar flow patterns due to tidal forcing and intense local heating at the substellar point. Some systems are far richer in carbon than ours is. Chemical composition and the appearance of planets in such systems will be strikingly different. This is all good because it forces us even farther out of our comfort zone and will stimulate yet a deeper understanding of how atmospheres work.

ACKNOWLEDGMENT

The writing for this article was carried out at the Jet Propulsion Laboratory, California Institute of Technology, under a contract with the National Aeronautics and Space Administration. Copyright Statement: Copyright 2012 California Institute of Technology. Government sponsorship acknowledged.

BIBLIOGRAPHY

Atreya, S. K., & Romani, P. N. (1985). In G. E. Hunt (Ed.), *Planetary meteorology (pp. 17–68)*. Cambridge, United Kingdom: Cambridge Univ. Press.

Atreya, S. K., Pollack, J. B., & Matthews, M. S. (Eds.). (1989). *Origin and evolution of planetary and satellite atmospheres*. Tucson: Univ. Arizona Press.

Bagenal, F., Dowling, T., & McKinnon, W. (Eds.). (2004). *Jupiter: The planet, satellites and magnetosphere*. Cambridge, UK: Cambridge Univ. Press.

Beatty, J. K., & Chaikin, A. (Eds.). (1990). *The new solar system* (3rd ed.). Cambridge, Massachusetts: Sky Publishing.

Beebe, R. (1994). *Jupiter: The giant planet*. Washington, DC: Smithsonian Institution Press.

Bergstralh, J. T., Miner, E. D., & Matthews, M. S. (Eds.). (1991). *Uranus*. Tucson: Univ. Arizona Press.

Chamberlain, J. W., & Hunten, D. M. (1987). *Theory of planetary atmospheres: An introduction to their physics and chemistry* (2nd ed.). Orlando, Florida/San Diego: Academic Press.

Connerney, J. (1993). *Journal of Geophysical Research, 98*, 18,659–18,679.

Connerney, J., et al. (1993). *Science, 262*, 1035–1038.

Cruikshank, D. P. (Ed.). (1995). *Neptune*. Tucson: Univ. Arizona Press.

de Pater, I., et al. (1991). *Icarus, 91*, 220–233.

de Pater, I., et al. (2005). *Icarus, 174*, 263–373.

Dougherty, M. K., Esposito, L. W., & Krimigis, S. M. (Eds.). (2009). *Saturn from Cassini–Huygens*. New York: Springer.

Gierasch, P., & Conrath, B. (1993). *Journal of Geophysical Research, 98*, 5459–5469.

Gladstone, G. R., et al. (1996). *Icarus, 119*, 1–52.

Gombosi, T. I., & Ingersoll, A. P. (2010). Saturn: atmosphere, ionosphere, and magnetosphere. *Science, 327*, 1476–1479.

Lian, & Showman. (2010). *Icarus, 207*, 373–393.

Liu, & Schneider. (2010). *Journal of the Atmospheric Science, 67*, 3652–3672.

Pollack, J. B., & Atreya, S. K. (1992). In G. Carle, et al. (Eds.), *Exobiology in solar system exploration (pp. 82–101)*. NASA-SP 512.

Pollack, J., et al. (1987). *Journal of Geophysical Research, 92*, 15,037–15,066.

Porco, et al. (2003). *Science, 299*, 1541–1547.

Porco, et al. (2005). *Science, 307*, 1243–1247.

Rogers, J. H. (1995). *The planet Jupiter*. Cambridge, United Kingdom: Cambridge Univ. Press.

Romani, P., et al. (1993). *Icarus, 106*, 442–462.

Sromovsky, L., & Fry, P. (2005). *Icarus, 179*, 459–484.

Sromovsky, L., et al. (2001). *Icarus, 150*, 244–260.

West, R., et al. (2004). *Jupiter: the planet, satellites and magnetosphere (pp. 79–104)*. In F. Bagenal, T. Dowling, & W. McKinnon (Eds.). Cambridge, United Kingdom: Cambridge Univ. Press.

Chapter 33

Interiors of the Giant Planets

Mark S. Marley
Space Science Division, NASA Ames Research Center, Moffett Field, CA, USA

Jonathan J. Fortney
Department of Astronomy and Astrophysics, University of California, Santa Cruz, CA, USA

Chapter Outline

1. General Overview 743
2. Constraints on Planetary Interiors 745
   2.1. Gravitational Field 745
   2.2. Atmosphere 746
   2.3. Magnetic Field 747
3. Equations of State 747
   3.1. Overview 747
   3.2. Hydrogen 748
   3.3. Helium 749
   3.4. Ices 749
   3.5. Rock 749
   3.6. Mixtures 749
4. Planetary Interior Modeling 750
5. Planetary Interior Models 751
   5.1. General Overview 751
   5.2. Jupiter 752
   5.3. Saturn 754
   5.4. Uranus and Neptune 754
   5.5. Extrasolar Giant Planets 755
6. Jovian Planet Evolution 756
7. Future Directions 757
Bibliography 758

The giant, or Jovian planets—Jupiter, Saturn, Uranus, and Neptune—account for 99.5% of all the planetary mass in the solar system. The internal composition and structure of all these planets thus provide important clues about the conditions in the solar nebula during the time of planet formation. But such information does not come easily. The familiar faces of these planets, such as the cloud-streaked disk of Jupiter, tell relatively little about what lies beneath. Knowledge of these planetary interiors must instead be gained from analysis of the mass, radius, shape, and gravitational fields of the planets. For the majority of giant planets around other stars, at best only the mass and radius can be discerned. Thus for both solar and **extrasolar planets**, theoretical models of the planetary interiors must be compared to the available data in order to infer what lies within the planets. The study of the behavior of planetary materials at high densities and pressures provides additional constraints for connecting mass, radius, and internal composition. Once constructed, interior models provide a window into the internal structure of these planets and shed light on processes that led to planet formation in our solar system and others.

1. GENERAL OVERVIEW

Several lines of observational evidence provide information on the composition and structure of the giant planets. The first and most easily obtained quantities are mass (known from the orbits of natural satellites), radius (polar and equatorial radii), and rotation period (obtained originally from telescopic observations, now derived from remote and in situ observations of planetary magnetic fields). By the 1940s, these fundamental observations, coupled with the advances in understanding the high-pressure behavior of matter in the 1920s and 1930s, constrained the composition of Jupiter and Saturn to be predominantly hydrogen. Direct measurement of the planets' high-order gravity fields, interior rotation states, and heat flow, along with spacecraft and ground-based spectroscopic detection of atmospheric elemental composition, has since allowed the construction of progressively more refined models of the interior structure of these planets.

These models divide the giant planets into two broad categories. Jupiter and Saturn are predominantly hydrogen—helium gas giants with a somewhat enhanced abundance of

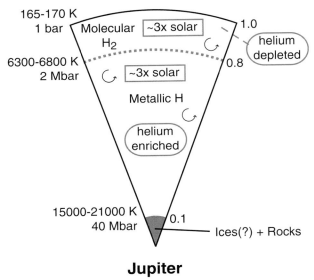

Jupiter

FIGURE 33.1 Highly schematic, idealized cross section of the interior structure of Jupiter. The numbers to the right refer to the relative radius (r/R) of the core and the molecular-to-metallic hydrogen phase transition. On the left are listed the approximate temperatures and pressures at which these interfaces occur. Arrows indicate convection. Boxes denote approximate enhancement of elements other than H and He over the abundance found in the Sun. The core mass is uncertain, but likely has a mass between 0 and 10 Earth masses. The real Jupiter is undoubtedly more complex. It is likely that interfaces are gradual and the composition of the various regions is inhomogeneous.

heavier elements and dense **cores**. Uranus and Neptune are **ice** giants with hydrogen–helium envelopes and dense cores. The following description of Jupiter's interior, as illustrated schematically in Figure 33.1, is qualitatively valid for Saturn and serves as a point of departure for understanding the interiors of Uranus, Neptune, and the extrasolar giant planets. Individual planetary interior structures are discussed in Section 5.

The interior begins at the base of the outermost atmospheric envelope that we can see directly. The Jovian atmospheres consist of a gaseous mixture of molecular hydrogen, helium, methane, ammonia, and water. At 1 **bar** pressure (the pressure at sea level on Earth), the temperature in Jupiter's atmosphere is 165 K. Near this level, the ammonia condenses into clouds; the water condensation level is even deeper. In the colder atmospheres of Uranus and Neptune, methane also condenses into clouds. Deeper into the planet the pressure of the overlying atmosphere compresses the gas, increasing its temperature and density. This process, **adiabatic compression**, is the same one responsible for the increase in temperature with decreasing altitude on Earth. About 100 km beneath the cloud tops, the temperature has reached 350 K.

As pressures and temperatures increase, the gas begins to take on the characteristics of a liquid. Since the **critical point** of the dominant constituent, molecular hydrogen, lies at 13 bars and 33 K, there is not a distinct gas–liquid phase boundary. By several hundred thousand bars the envelope closely resembles a hot liquid. This characteristic of the giant planets—they exist in the supercritical regime of their primary constituent—leads to their most fundamental property: these planets have essentially bottomless atmospheres.

Deeper into the planet the temperature and pressure continue to steadily increase. By 15,000 km beneath the cloud tops the temperature reaches 6,500 K and the pressure is 2 Mbar. Experiments and theory suggest that by this point hydrogen, previously present as molecules of H_2, has begun to undergo a phase transition to a liquid, metallic state. Most of the mass of Jupiter consists of this **metallic hydrogen**: protons embedded in a sea of electrons. Helium and other constituents exist as impurities in the hydrogen soup. For the remaining 50,000 km to Jupiter's core, the pressure and temperature continue to rise, reaching 40 Mbar and nearly 20,000 K in the deep interior. Near the center of the planet the composition changes, perhaps gradually, from a predominantly hydrogen/helium mixture to a combination of **rock** and **ice**. The density of this rock and ice core is $10{,}000-20{,}000$ kg/m^3, higher than the metallic hydrogen density of about 1000 kg/m^3 (uncompressed water, like the one that comes out of a tap, also has a density of 1000 kg/m^3).

The interiors of giant planets are hot, thousands to tens of thousands of degrees Kelvin, while their observable atmospheres are cold, 100 K or so and cooler. To understand the evolution of a giant planet through time, one has to understand how the energy from the hot interior is transported out to the much cooler atmosphere. The three possible choices would be by radiation, conduction, and **convection**. Throughout most of the interior, the transport of energy by radiation is severely hampered by the high **opacity** of dense hydrogen. Other constituents such as methane and water effectively block energy transport by radiation in those regions of the spectrum where the hydrogen is a less powerful absorber. Since conduction of heat by the thermal motion of molecules is also inefficient, convection is the dominant energy transport mechanism throughout the interior. The rising and sinking convective cells in the interior move slowly, at velocities of just centimeters per second or less.

Convective transport of heat may be responsible for the banded appearance of the planets. Because of the rapid rotation of these planets, it is easier for convective cells to rise parallel to the rotation axes, from the center toward the poles, than in the perpendicular direction, out toward the equator. This tendency combined with the continuous nature of the atmosphere connects the atmosphere to the deep interior. The wind patterns seen in the belts and zones of Jupiter and Saturn may thus have roots that reach into the deep, convective interior of the planet. Indeed, the winds measured by Galileo spacecraft's atmosphere probe continued to blow steadily at the deepest levels reached by the probe, about 20 bars.

The interior of Saturn may be similar to that of Jupiter, and that of Uranus similar to that of Neptune, although in both cases the different heat flows would allow for dissimilar interiors. Saturn's lower mass and consequently lower pressures produce a smaller metallic hydrogen region than Jupiter's. It is likely that Uranus and Neptune lack a metallic hydrogen region; instead, at about 80% of their radius the abundance of methane, ammonia, and water increases markedly. In this region, temperatures of over 2500–10,000 K produce an ocean of electrically charged fluid water, ammonia, methane molecules, and molecular fragments, along with more complex compounds. Most of the mass of Uranus and Neptune exists in such a state. Deep in their interiors, all the planets likely have cores of primarily rocky material.

This picture of the interiors of the Jovian planets has been painstakingly pieced together since the 1930s. This chapter discusses the components of observation, experiment, and theory that are combined to reach these conclusions.

2. CONSTRAINTS ON PLANETARY INTERIORS

2.1. Gravitational Field

A variety of observations yield information about the makeup and interior structure of the Jovian planets. The mass of each of the four Jovian planets (Table 33.1) has been known with some precision since the discovery of their natural satellites. The masses range from 318 times the Earth's mass ($M_\oplus$) for Jupiter to 14.5 $M_\oplus$ for Uranus. A second fundamental observable property is the radius of each planet measured at a specified pressure, typically the 1-bar pressure level. Radii are most accurately measured by the occultation technique, in which the attenuation of the radio signal from a spacecraft is measured as the spacecraft passes behind the planet. Jovian planet radii range from 11 times the Earth's radius ($R_\oplus$) for Jupiter to 3.9 $R_\oplus$ for Neptune. The combination of mass and radius allows calculation of mean planetary density, ρ. Although a surprising amount can be learned about the bulk composition of a planet from just ρ (as we will later see for extrasolar giant planets), more subtle observations are required to probe the detailed variation of composition and density with radius.

If the Jovian planets did not rotate, they would assume a spherical shape and their external gravitational field would be the same as that of a point of the same mass. No information about the variation in density with radius could be extracted. Fortunately, the planets do rotate and their response to their own rotation provides a great deal more information. This response is observed in their external gravitational field.

For a uniformly rotating body in hydrostatic equilibrium, the external gravitational potential, Φ, is

$$\Phi = -\frac{GM}{r}\left(1 - \sum_{n=1}^{\infty}\left(\frac{a}{r}\right)^{2n} J_{2n} P_{2n}(\cos\theta)\right)$$

Here G is the gravitational constant, M is the planetary mass, a is the planet's equatorial radius, θ is the colatitude (the angle between the rotation axis and the radial vector $\mathbf{r}$), P_{2n} are the Legendre polynomials, and the dimensionless numbers J_{2n} are known as the gravitational moments. The assumption of hydrostatic equilibrium means that the

TABLE 33.1 Observed Properties of Jovian Planets

Quantity	Jupiter	Saturn	Uranus	Neptune
M (kg)	1.8981×10^{27}	5.6833×10^{26}	8.683×10^{25}	1.024×10^{26}
a (km)	$71,492 \pm 4$	$60,268 \pm 4$	$25,559 \pm 4$	$24,766 \pm 15$
P_s (hours)	9.92492	10.78	17.24	16.11
$J_2 \times 10^6$	$14,697 \pm 1$	$16,290.7 \pm 0.3$	$3,516 \pm 3$	$3,539 \pm 10$
$J_4 \times 10^6$	-584 ± 5	-936 ± 3	-35 ± 4	-28 ± 22
$J_6 \times 10^6$	31 ± 20	86 ± 10	–	–
q	0.0892	0.151	0.0295	0.026
Λ_2	0.1647	0.108	0.1191	0.136
$\bar{\rho}$ (g cm^{-3})	1.328	0.687	1.27	1.64
Y	0.238 ± 0.007	0.18 – 0.25	0.26 ± 0.05	0.26 ± 0.05
T_1 (K)	165	135	76	74

planet is in a fluid state, responding only to its rotation, and there are no permanent, nonaxisymmetric lumps (e.g. mountains) in the interior. This assumption is believed to be quite good for the Jovian planets.

The gravitational harmonics are found from observations of the orbital motion of natural satellites, **precession** rates of elliptical rings, and perturbations to the trajectories of spacecraft. As a spacecraft flies by a planet, it samples the gravitational field at a variety of radii. Careful tracking of the spacecraft's radio signal reveals the Doppler shift due to its acceleration in the gravitational field of the planet. Inversion of these data yields an accurate determination of the planet's mass and gravitational harmonics (see Table 33.1). In practice, it is difficult to measure terms of order higher than J_4, and the value of J_6 is generally quite uncertain. Progressively higher order gravitational harmonics reflect the distribution of mass in layers progressively closer to the surface of the planet. Thus, even if they could be measured accurately, terms such as J_8 would not contribute greatly to understanding the deep interior.

A planet's response to its own rotation is characterized by how much a surface of constant total potential (including the effects of both gravity and rotation) is distorted. The amount of distortion on such a surface of constant potential, known as a level surface, depends on the distribution of mass inside the planet, the mean radius of the level surface, and the rotation rate. The distortion, or oblateness, of the outermost level surface is measured from direct observations of the planet and is given by $\varepsilon = (a - b)/a$, where b is the polar radius. The equatorial and polar radii can be found from direct telescopic measurement or, more accurately, from observations of spacecraft or stellar occultations. Distortion of level surfaces cannot be described simply by ellipses. Instead, the distortion is more complex and must be described by a power series of shapes, as illustrated in Figure 33.2. The most obvious distortion of a spherical planet ($n = 0$) is elliptical ($n = 2$). More subtle distortions are described by harmonic coefficients of ever increasing degree, as illustrated up to $n = 10$ in Figures 33.2.

A nonrotating, fluid planet would have no J_{2n} terms in its gravitational potential. Thus the gravitational harmonics provide information on how the shape of a planet responds to rotating-frame forces arising from its own spin. Since the gravitational harmonics depend on the distribution of mass of a particular planet, they cannot be easily compared between planets. Instead, a dimensionless linear response coefficient, Λ_2, is used to compare the response of each Jovian planet to rotation. To lowest order in the square of the angular planetary rotation rate, ω^2, $\Lambda_2 \approx J_2/q$, where $q = \omega^2 a^3/GM$. Table 33.1 lists the Λ_2 calculated for each planet. The Jovian planets rotate rapidly enough that the nonlinear response of the planet to rotation is also important and must be considered by computer models.

Since the gravitational harmonics provide information about the planet's response to rotation, interpretation of the harmonics requires accurate knowledge of the rotation rate of the planet. Before the space age, observations of atmospheric features as they rotated around the planet provided rotation periods. This method, however, is subject to errors introduced by winds and weather patterns in the planet's atmosphere. Instead, rotation rates are now found from the rotation rate of the magnetic field of each planet, generally as measured by the Voyager spacecraft (radio emissions arising from charged particles in Jupiter's **magnetosphere** can be detected by radio telescopes on Earth). This approach assumes that the magnetic field is generated by convective motions deep in the electrically conducting interior of the planet and that the field's rotation consequently follows the rotation of the bulk of the interior. This assumption is not certain and the measurement of the magnetic rotation period itself can also be challenging. As a result, in recent years there have been several suggestions that traditional rotation periods are in error by as much as several tens of minutes. This controversy has not yet been fully resolved and we quote here in Table 33.1 the generally accepted rotation periods.

2.2. Atmosphere

The observable atmospheres of the Jovian planets provide further constraints on planetary interiors. First, the atmospheric temperature at 1 bar pressure, or T_1, constrains the

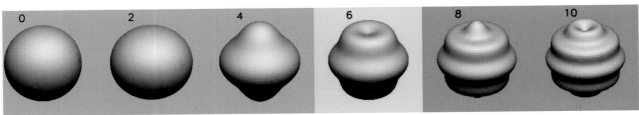

FIGURE 33.2 Illustration of the ways that a planet changes shape owing to its own rotation. A nonrotating planet is purely spherical (leftmost). The distortions due to second through tenth order in the gravitational harmonics J_n are also shown (for $n = 2$ through 10). The first three gravitational harmonics (mass, J_2 and J_4, cyan background) are well constrained. J_6 (yellow) is known with a larger uncertainty, while the values of J_8 and J_{10} are currently unknown but will be measured by the *Juno* spacecraft. Figure courtesy William Hubbard (Univ. Ariz.).

temperature of the deep interior. The interior temperature distribution of the Jovian planets is believed to follow a specified pressure–temperature path known as an **adiabat**. For an adiabat, knowledge of the temperature and pressure at a single point uniquely specifies the temperature as a function of pressure at all other points along the adiabat. Thus T_1 gives information about the temperature structure throughout the convective interior of the planet. Both the amount of sunlight that the atmosphere absorbs and the amount of heat carried by convection up from the interior of the planet to the atmosphere control T_1. For each planet, save Uranus, T_1 is higher than expected if the atmosphere were simply in equilibrium with sunlight. Thus each Jovian planet, save Uranus, exhibits a significant internal heat flow, ranging from 0.3 W/m^2 at Neptune, to 2.0 W/m^2 at Saturn, to 5.4 W/m^2 at Jupiter. For Uranus, only an upper limit to the heat flow, 0.08 W/m^2, has been measured.

Second, the composition of the observable atmosphere also holds clues to the internal composition. This is because of the supercritical nature of the Jovian atmospheres. The principal component of the Jovian atmosphere, hydrogen, does not undergo a vapor–liquid phase change above 33 K. Since the planets are everywhere warmer than this temperature, the observed atmosphere is directly connected to the deep interior. Knowledge of the composition of the top of the atmospheres therefore provides some insight into the composition at depth. (See Atmospheres of the Giant Planets.)

The Galileo spacecraft atmosphere probe returned direct measurements of the composition of Jupiter's atmosphere. The composition of the remaining planetary atmospheres is inferred from spectroscopy. In planetary science, compositions are often stated relative to "solar" abundances. Solar abundances are the relative quantities of elements present in the solar nebular at the time of planetary formation. The solar abundances of hydrogen and helium are about 70% and 28% by mass, respectively. Oxygen, carbon, nitrogen, and the other elements make up the remainder. These elements are collectively called the "heavy elements" to distinguish them from hydrogen and helium. Measurements of the rate at which the atmospheric pressure decreases with height in these atmospheres require that hydrogen and helium must be the dominant components of the atmospheres of all four Jovian planets. Spectroscopy supports this conclusion and gives the relative abundance of hydrogen and helium. The helium mass fraction of each atmosphere, Y, is listed in Table 33.1. The heavier elements are generally enriched in the Jovian atmospheres over their solar abundances, which must be explained by any formation scenario for these planets.

2.3. Magnetic Field

All four Jovian planets possess a magnetic field. Jupiter's is large and complex; Saturn's is less complex and weaker. The magnetic fields of both Uranus and Neptune are very complex: they deviate substantially from a dipole and their field axes are tilted strongly with respect to their rotation axes. The only known mechanism for producing global planetary magnetic fields, the hydromagnetic **dynamo** process, requires nonuniform motion of a large electrically conductive region. Convection in the highly conductive interior of the Jovian planets is presumed to be responsible for the formation of their fields. In Jupiter and Saturn, this is the liquid metallic hydrogen region. In Uranus and Neptune, this is the ionized fluid predominantly "icy" region. The level of complexity of each field plausibly relates to the depth of the electrically conducting region. Magnetic fields formed by relatively small, deep sources may be simpler and smaller than fields formed by large, shallow dynamos, albeit simple fields may require complex explanations. (See Planetary Magnetospheres.)

3. EQUATIONS OF STATE

3.1. Overview

Beyond observations of the planets themselves, a second major ingredient in interior models is an **equation of state** (EOS). An EOS is a group of equations, derived from laboratory observations and theory, that relate the pressure (P) of a mixture of materials to its temperature (T), composition (x), and density (ρ). Any attempt to model the interior structure of a giant planet must rely on an EOS. The construction of accurate EOSs is a primary activity in planetary interior modeling.

For an ideal gas, the well-known EOS is $P = nkT$. Here k is the Boltzmann's constant and n is the number density of the gas. The composition of an ideal gas does not affect the pressure; only the number of molecules and atoms in a given volume, n, enters the equation. Under the conditions of high temperature and pressure found in the interiors of the giant planets, atoms and molecules interact strongly with one another, thus violating the conditions under which the ideal gas EOS holds. Additionally, the typical pressures reached in the interiors of the giant planets (tens to hundreds of megabars) are also amply sufficient to modify the electronic structure of individual atoms and molecules. This further adds to the challenge of understanding the EOS. In short, the properties of planetary materials at high pressures will differ substantially from those encountered in their low-pressure, and more familiar, forms. In practice, the behavior of planetary materials must be understood from both experiments and theory.

For pressures less than about 3 Mbar, depending on the material, shock wave experiments provide guidance in the construction of EOSs. Gas guns, powerful lasers, explosives, or magnetic pressure can produce shock waves that produce both high temperatures and high pressures in the

FIGURE 33.3 The Z Machine at the Sandia National Laboratory in Albuquerque, Mexico. The facility uses intense magnetic pressure to launch flyer plates at ~20 km/s into samples of hydrogen, deuterium, water, and other materials. The shocked samples are monitored during the collision so that pressure, density, and sometimes temperature can be determined. *Photo courtesy: Sandia National Laboratory.*

sample. High-speed measuring devices record the temperatures, pressures, and densities achieved during the brief experiments. A photograph of the "Z Machine" at Sandia National Laboratory, which employs magnetic pressure and has measured the EOS of many planetary materials under extreme pressure and temperature conditions, is shown in Figure 33.3.

The temperatures and pressures reached in these experiments are the closest that terrestrial laboratories can come to reliably duplicating the conditions in the interiors of the Jovian planets. Hydrogen experiments model conditions about 80–90% of the way out from the Jupiter's center and 60–70% for Saturn. An important recent advance has been the exploration of the EOS of water up to 7 Mbar at Sandia, which is similar to the central pressure of Uranus and Neptune.

Diamond anvils are used in another type of experiment to squeeze microscopically small samples of planetary materials to very high pressure. Although these experiments can reach higher pressures than the shock experiments, they have historically only been conducted at room temperature, making them less applicable to Jovian planets. Very recently, however, teams have begun dynamically shocking samples that have been precompressed by anvils, to allow a much wider range of final pressure and temperature states to be reached for shocked samples.

The EOSs that are used to model giant planets are becoming ever more sophisticated in the era of abundant computing power. Theoretical physicists are now able to directly simulate the quantum mechanical interactions of hydrogen atoms, helium atoms, their mixtures, and water and rock. When these direct computer simulations can be compared to laboratory work, excellent agreement has been found. This lends some confidence when these calculated EOSs are accurate when applied to deep planetary interiors, beyond the realm of current experiment. As we will see later, our understanding of the structure of Jupiter, in particular, is only as good as our understanding of the hydrogen EOS.

3.2. Hydrogen

For pressures less than about 1 Mbar, the behavior of molecular hydrogen, H_2, is understood fairly well from theory and the shock experiments. At higher pressures such as those encountered deeper in the interiors of Jupiter and Saturn, the hydrogen molecules are squeezed so closely together that they begin to lose their individual identities, as adjacent hydrogen molecules become as closely spaced as the two atoms within a given molecule. The electrons become free to move between atoms and are unbound. This "pressure-ionized" phase of hydrogen is termed metallic hydrogen. In giant planets, this metallic hydrogen is always fluid, not solid. A shock wave experiment suggests that this transition occurs gradually near 1.4 Mbar at 3000 K, however, more work is needed to fully understand this phase transition. All recent theoretical calculations show that the transition is continuous under planetary interior conditions and may not be complete until a pressure of 10 Mbar. Liquid metallic hydrogen in Jupiter and Saturn consists of a dense mixture of ionized protons and electrons at temperatures over about 6000 K. The EOS of liquid metallic hydrogen is understood well theoretically for pressures above about 10 Mbar, but the EOS is not well constrained from 0.5 to 10 Mbar, the transition region. A hydrogen phase diagram and temperature/pressure profiles for each giant planet are shown in Figure 33.4. The detailed

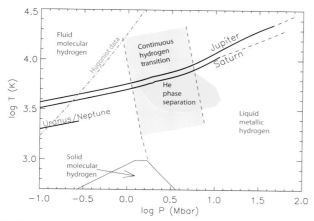

FIGURE 33.4 Phase diagram for hydrogen, the main constituent of Jupiter and Saturn. The approximate domains of liquid metallic hydrogen and molecular hydrogen are shown along with approximate interior temperature profiles for the Jovian planets. The light blue shaded region shows the approximate molecular to metallic transition region for hydrogen. The yellow shaded area indicates the approximate region in the interior of Saturn and, possibly, Jupiter where helium begins to separate out from liquid metallic hydrogen. The locations of phase boundaries are highly uncertain except for the liquid to solid transition of H_2. "Hugoniot data" shows the pressure–temperature curve where single shock experiments reach. High-pressure regions at lower temperatures can be probed by multiply shocking a sample. Most of the interior of Jupiter and Saturn exists at pressures greater than can currently be probed in laboratory experiments.

behavior of hydrogen near the phase transition is not fully known. Thus in the past various simplifying assumptions had to be made when considering these regions of giant planets. It is only recently that detailed computer calculations can simulate this important region in giant planets, although more data is nevertheless sorely needed.

3.3. Helium

Helium has not been as well studied as hydrogen, but shock wave data do provide information to around 1 Mbar. Above this pressure, theory must guide models of the behavior of this element. Helium is thought to be entirely neutral in Saturn but may become ionized in the deeper interior, just above the core, in Jupiter. Although the EOSs of hydrogen and helium individually are reasonably understood, the behavior of mixtures of these two constituents is less well constrained. This is a serious theoretical void because the hydrogen–helium mixture composes most of the mass of Jupiter and Saturn and is an important component in Uranus and Neptune.

Current calculations of the behavior of hydrogen and helium mixtures show that helium is not soluble in liquid metallic hydrogen at all mass fractions and temperatures. At the temperatures predicted in the interior of Saturn (and perhaps also in Jupiter), hydrogen and helium do not mix. According to this model, droplets of helium-rich material are constantly forming in the molecular to metallic transition region of the planet. Because they are denser than their surroundings, the drops fall to deeper, warmer levels of the envelope, where temperatures may be high enough to again allow mixing, and the droplets dissolve. Thus at certain depths in Saturn's interior, it is always raining helium. This remarkable conclusion is discussed in Section 5.3 in the context of the Saturn interior models.

3.4. Ices

The term "ices" is applied to mixtures of volatile elements in the form of water (H_2O), methane (CH_4), and ammonia (NH_3) in solar proportions, not necessarily present as intact molecules. Ices are the primary constituent of Uranus and Neptune, but are less abundant in Jupiter and Saturn. As the planetary interior temperatures are over several thousand degrees Kelvin, they are present as liquids.

EOS data are more sparse for these compounds, although shock wave experiments conducted on a mixture of water, isopropanol, and ammonia (dubbed "synthetic Uranus") have helped establish the EOS of such a mixture at pressures less than about 2 Mbar and temperatures less than about 4000 K, and water alone has been probed to 7 Mbar. These experiments helped confirm that ices are the primary constituent of Uranus and Neptune. The shock wave data on this mixture show that at pressures exceeding ~ 200 kbar, the planetary ice constituents ionize to form an electrically conductive fluid. At pressures ≥ 1 Mbar, the ice constituents dissociate and the EOS becomes quite "stiff", meaning the density is not particularly sensitive to the pressure. Modern computational work on icy mixtures is focusing on high-pressure chemistry, meaning that they investigate how mixtures of the ices behave differently than the pure phases. The rainout of pure carbon from the icy mixture in Uranus and Neptune has been suggested.

3.5. Rock

The remaining planetary constituents are lumped into the category "rock". Rock is presumed to consist of a solar mixture of silicon, magnesium, and iron, with uncertain additions of oxygen and the remaining elements. Although the rock EOS is not well known, it is also expected to be quite "stiff". The lack of a detailed rock EOS is not a serious limitation for planetary interior models, as the rock component is not believed to be a major fraction of the mass, at least for Jupiter and Saturn.

3.6. Mixtures

Since all the planetary components, including gas, ice, and rock, are likely mixed throughout the interiors, EOSs of such mixtures are required for interior modeling.

Hydrogen—helium mixtures, considered earlier, may not exist at all temperatures, pressures, and concentrations. The solubility of other mixtures, for example, rock or oxygen in metallic hydrogen, is less well known. From the limited data, it appears that the planetary constituents other than hydrogen and helium do mix well under the temperature and pressure conditions typically found in planetary interiors. This is because delocalization of electrons at high pressure diminishes the well-defined intermolecular bonds present at lower pressures. Thus the separation of planetary materials into distinct layers of "pure" rock or ice is highly unlikely. If correct, such considerations also have important cosmogonic implications. For example, the dense cores of the planets likely did not "settle" from an initially well-mixed planet, but instead the gaseous components likely collapsed onto a preexisting rocky nucleus that formed in the protosolar nebula. The water/rock cores of the planets may also slowly become dredged up and dissolve into the overlying H/He mixture, although the timescale for this process is not yet well constrained.

Since the EOS of all possible mixtures has not been studied, either experimentally or theoretically, approximations must be employed. One approximation, the additive volume law, weights the volumes of individual components in a mixture by their mass fraction. An implication of such approximations is that the computed densities of mixtures of rock, ice, and gas can be similar to that of pure ice. Thus it is not currently possible to differentiate between models of Uranus and Neptune with mantles of pure ice and models with mantles of a mixture of rock, ice, and gas.

4. PLANETARY INTERIOR MODELING

In addition to an EOS for the material in the interior of a planet, two more components are required to produce an interior model. The temperature and composition in the interior as a function of pressure, $T(P)$ and $x(P)$, respectively, must also be known. (These quantities are described as functions of pressure since the pressure increases monotonically toward the center of the planet.) The first of these ingredients, $T(P)$, is not difficult to find. If the Jovian planets are fully convective in their interiors, transporting internal heat to the surface by means of convection, the relation between temperature and pressure in their interiors is known as an adiabat. An adiabat has the property that knowledge of a single temperature and pressure at any point allows specification of T as a function of P at any other point (assuming the material's EOS is known). Since the temperature and pressure in the convecting region of each Jovian atmosphere have been measured, a unique $T(P)$ relation for each planet can be found. If there is indeed a large nonconvective region in the giant planets (see below) significant corrections may be required.

More difficult to specify is the variation in composition through each planet, $x(P)$. The composition of each planet's atmosphere is known, but there is no guarantee that this composition is constant throughout the planet. Earth's core, for example, has a very different composition from the crust. For the Jovian planets, an $x(P)$ relation is typically guessed at, an interior model computed, and the results compared to the observational constraints. With multiple iterations, a variation in composition with pressure that is compatible with the observations is eventually found. Modern theoretical work is also now addressing how composition gradients (due to He settling or core erosion) may lead to interior temperature profiles that significantly deviate from an adiabat.

The combination of these three ingredients, an EOS $P = P(T, x, \rho)$; a temperature—pressure relation, $T = T(P)$; and a composition—pressure relation, $x = x(P)$; completely specifies pressure as only a function of density, $P = P(\rho)$. Since the Jovian planets are believed to be fluid to their centers, the pressure and density are also related by the equation of hydrostatic equilibrium for a spherical rotating planet:

$$\frac{\partial P}{\partial r} = -\rho(r)g(r) + \frac{2}{3}r\omega^2\rho(r)$$

where g is the gravitational acceleration at radius r and ω is the angular rotation rate. This relation simply says that, at equilibrium, the pressure gradient force at each point inside the planet must support the weight of the material at that location. Combining the equation of hydrostatic equilibrium with the $P(\rho)$ relation finally allows determinations of the variation of density with radius in a given planetary model, $\rho = \rho(r)$.

The computed model must then satisfy all the observational constraints discussed in Section 2. Total mass and radius of the model are easily tested. The response of the model planet to rotation and the resulting gravitational harmonics must be calculated and compared with observations. Figure 33.5, showing the relative contribution versus the depth from the center of the planet, illustrates the regions of a Saturn model that contribute to the calculation of the gravitational harmonics J_2, J_4, and J_6. Higher degree modes provide information about layers of the planet progressively closer to the surface.

The construction of computer models that meet all the observational constraints and use realistic EOSs requires several iterations, but the calculation does not strain modern computers. The current state of the art is to calculate dozens of interior models, while varying the many parameters within theoretically or experimentally determined boundaries. The EOSs of hydrogen and helium both have uncertainties due to differences in experimental data and theory. In addition, the size and composition of the heavy element core, as well as the heavy element enrichment in

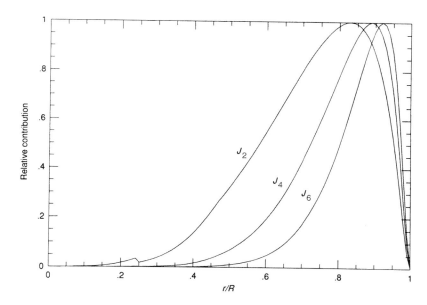

FIGURE 33.5 Gravitational harmonics are computed from integrals over density and powers of radius of a rotating planet. The curves illustrate the integrands for the harmonics J_2, J_4, and J_6 of a Saturn interior model. Higher-degree terms are proportional to the interior structure in regions progressively closer to the surface. All curves have been normalized to unity at their maximum value. The bump in the J_2 curve near 0.2 is due to the presence of the core.

the envelope, is varied with different EOSs for ices and rocks. A subset of these possible models fits all available planetary constraints. By necessity, each modeler begins with an ad hoc set of assumptions that limit the range of models that can be calculated. This inherent limitation of models should always be borne in mind when considering their results. The consensus for the structure of Jovian planet interior models is presented in the next section.

5. PLANETARY INTERIOR MODELS

5.1. General Overview

Even early "cosmographers" recognized that the giant planets of the solar system were distinct from the inner **terrestrial planets**. The terrestrial planets have mean densities of 4000–5000 kg/m³, intermediate between the density of rocks and iron, whereas the giant planets have mean densities closer to that of water (1000 kg/m³), between 700 and 1700 kg/m³. From this single piece of information, it is clear that the bulk composition of the giant planets must be substantially different from that of the terrestrial planets.

It has been known since the 1940s that if the interiors of Jupiter and Saturn are "cold", the primary component of these planets must be hydrogen. In this context, "cold" means that the densities throughout the interior must not deviate significantly from the values they would assume at the same pressures if the temperature was 0 K. The approximation is relevant because the behavior of substances at 0 K and high pressure can be calculated analytically. Hydrogen is then a likely dominant constituent because at the high pressures prevalent in the interiors of Jupiter and Saturn it would be a metallic fluid with a density of about 1000 kg/m³, not the more familiar molecular gas. Since the density of "cold" metallic hydrogen is close to the bulk densities of Jupiter and Saturn, it was recognized as a plausible major constituent of these planets.

Mass–radius calculations provide a more compelling demonstration of the dominance of hydrogen in the interiors of Jupiter and Saturn. For a given composition, there is a unique relation between the radius of a spherical body in hydrostatic equilibrium and its mass. These relations can be calculated analytically for all elements at high pressure and zero temperature. Although the interiors of Jovian planets are not at zero temperature, they are cool when measured on an atomic temperature scale. This is adequate for a qualitative calculation, but zero-temperature EOSs are insufficiently accurate for the calculation of detailed interior models.

Mass–radius curves for several likely planetary constituents are shown in Figure 33.6. For low masses, the interior pressures are small compared to intermolecular forces and the volume of an object is just proportional to its mass, thus $R \propto M^{1/3}$. This is a realm with which we are familiar in daily life. At larger masses, the greater interior pressures ionize the material, liberating many electrons. In this regime, $R \propto M^{-1/3}$: when mass is added to an object, it shrinks. For intermediate masses where the curves meet, there is a region where the radius is not highly sensitive to the mass. At sufficiently high masses, the hydrogen in the core of the object will undergo fusion, the temperature will rise, and the zero-temperature relations shown in Figure 33.6 are no longer applicable. However, for planets and white dwarf stars, Figure 33.6 is applicable. An important consequence of these considerations is that for

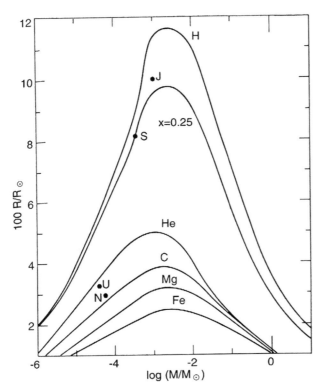

FIGURE 33.6 Mass-radius curves for objects of various compositions at zero temperature. Curve labeled $x = 0.25$ is for an approximately solar mixture of hydrogen and helium. Points J, S, U, and N represent Jupiter, Saturn, Uranus, and Neptune, respectively. Radius is in units of hundredths of a solar radius and mass is in units of solar masses ($1R_\odot = 6.96 \times 10^5$ km and $1\ M_\odot = 1.99 \times 10^{33}$ g). Jupiter and Saturn are clearly composed predominantly of hydrogen and helium; Uranus and Neptune must have a large complement of heavier elements. Figure from Zapolsky and Salpeter (1969).

any given composition, there is a maximum radius that a planet lacking a significant internal heat source can have. For solar composition, the maximum radius is about 80,000 km for a planet with about four times Jupiter's mass. Very young giant planets, which are still hot from their formation, can be larger. As we discuss below, some extrasolar planets also have unexpectedly large radii.

The total mass and radius of each Jovian planet are plotted on Figure 33.6 as well. This figure immediately proves that Jupiter must be composed primarily of hydrogen and helium. The maximum radii of planets composed of heavier, cosmically abundant elements are all much smaller. For example, only if Jupiter were very hot and very thermally expanded could carbon be a dominant constituent. But Jupiter's observed heat flux rules out a very hot ($>10^7$ K) internal state. Thus, Jupiter must primarily consist of a mixture of hydrogen and helium. Saturn's position on the graph implies a greater abundance of elements heavier than hydrogen, but still a primarily hydrogen bulk composition. Uranus and Neptune lie well below the mass–radius curve for hydrogen, thus revealing an appreciable component of heavier elements in their interiors. In Section 5, we will discuss giant planets around other stars, for which the only properties we can determine are radii and masses. For these planets, we can get a good estimate of the percentage of their mass that is made of hydrogen and helium, compared to the **heavy elements**.

Although the mass–radius relations clearly reveal the bulk composition of Jupiter and Saturn, they do not reveal information about the distribution of material inside the planet. It is here that the shape and gravitational harmonics enter the calculation. The response coefficient Λ_2 measures the response of the planet to its own rotation. For a uniform, hydrogen-rich material, $\Lambda_2 = 0.17$. Values smaller than 0.17 indicate a reduced gravitational response to rotation compared with that of the uniform composition hydrogen-rich planet. Such a reduced response results when more mass is concentrated toward the center of the planet. Thus, the smaller the Λ_2, the greater the degree of central condensation.

Λ_2 Varies (see Table 33.1) from 0.16 for Jupiter to 0.11 for Saturn. The mass–radius relations show that the Jovian planets are not pure hydrogen and their Λ_2 values suggest that they are centrally condensed. Hence the heavier constituents are not uniformly distributed in radius but are concentrated toward the center of each planet. Jupiter exhibits the least central condensation; Saturn and Uranus are most centrally condensed. Thus we begin to construct an elementary interior model.

Finally, the gravitational harmonics, J_2, J_4, and J_6, probe the detailed variation of the various planetary constituents. To simplify the interpretation of these harmonics, early interior models for Jupiter and Saturn tended to employ three distinct compositional zones: an inner rocky core, an icy core surrounding the rock one, and a hydrogen/helium envelope. More modern models allow the composition of various zones to vary gradually between layers and allow the outer envelopes to be enriched over solar abundance. The primary unknowns to be found from interior modeling are the mass of the rocky/icy core and the abundance of helium and other heavy elements in the envelope.

5.2. Jupiter

Jupiter contains more mass than all the other planets combined. Since its gravitational harmonics are also well measured, Jupiter serves as a test bed for theoretical understanding of Jovian interiors. The observed physical characteristics of Jupiter are listed in Table 33.1. From *Galileo Entry Probe* data, abundance of methane in Jupiter's atmosphere is about 3.5 times the solar abundance and the abundance of ammonia is about five times the solar abundance. Water does not show such enrichment, but it has been argued that the *Galileo Entry Probe* fell into an anomalously dry region of Jupiter's atmosphere.

The general structure of Jupiter's interior was briefly described in Section 1. Modern interior models attempt to specifically determine the degree of enrichment of heavy elements in the hydrogen/helium envelope of the planet. The atmospheric enrichment of methane and ammonia provides some indication that heavy element enrichment in the deeper interior may be expected. Jupiter's Λ_2 implies that Jupiter is not homogeneous but is slightly centrally condensed. Indeed, detailed modeling has shown that Jupiter's current core is less than 15 Earth masses and there may not be a core at all. The size and composition of Jovian planet cores and the amount of heavy element enrichment in the envelopes have a bearing on the scenarios by which they are supposed to have formed.

The variation of density with radius for two typical Jupiter models is shown in Figure 33.7. It should be emphasized that these are two Jupiter models that are consistent with all available constraints. Other equally valid interior models exist. Figure 33.8 shows the mass of heavy elements in the core vs heavy elements in the hydrogen—helium envelope for a large number of Jupiter (and Saturn) models. Any model within the hatched region is a valid interior model for each planet, given the current model uncertainties. For Saturn, models with a homogeneous H/He envelope and distinct core are shown, which match the planet's gravity field. For both Jupiter and Saturn, modelers have recently applied state-of-the-art first principles H/He EOS. A feature of Jupiter modeling over the past 30 years is that as we have gained better knowledge of the EOS of hydrogen, the calculated mass of the core has shrunk.

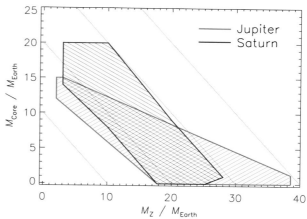

FIGURE 33.8 Earth masses of heavy elements in the hydrogen—helium envelope vs Earth masses of heavy elements in the core for Jupiter and Saturn. Models for Saturn include an H/He envelope that is uniform in its heavy element enrichment. Models for Jupiter include similar uniform models (models with large cores) and those where the H/He envelope is broken up into two layers, with an inner layer more enriched in heavy elements. These models have smaller cores and better fit the gravity field.

Surrounding the core is an envelope of hydrogen and helium. The temperature and pressure at the bottom of the hydrogen—helium envelope are near 15,000—20,000 K and 40 Mbar, respectively, for typical models. The gravitational harmonics require the envelope to be denser at each pressure level than a model that has only a solar mixture of elements. Thus the envelope must be enriched in heavy elements compared to a purely solar composition. The total mass of heavy elements is constrained between 10 and 40 Earth masses. If Jupiter had only a solar abundance of heavy elements, this value would be six Earth masses. This means that, averaged throughout the planet, Jupiter is enriched in heavy elements over solar abundances by a factor of 1.5—6.

Jupiter's atmospheric mass fraction of helium, $Y = 0.238 \pm 0.007$, is less than the solar abundance of about 0.28. This depletion is likely an indication that the process of helium differentiation, described more fully in Section 6, may have recently begun on Jupiter. The interior models do not provide a sufficiently clear view into the interior structure to determine if this is the case. The inferred interior structure is, however, compatible with limited helium differentiation.

Hydrogen and helium compose about 90% of Jupiter's mass. Most of the hydrogen exists in the form of metallic hydrogen. Jupiter is the largest reservoir of this material in the solar system. Convection in the metallic hydrogen interior is likely responsible for the generation of Jupiter's magnetic field. The transition from molecular to metallic hydrogen takes place about 10,000 km beneath the cloud tops, compared to about 30,000 km at Saturn. The exceptionally large volume of metallic hydrogen is likely responsible for the great strength of Jupiter's magnetic field.

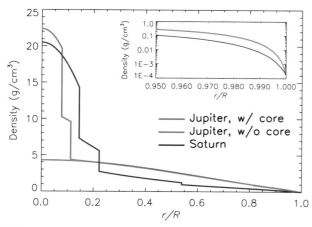

FIGURE 33.7 Density as a function of normalized radius for Jupiter and Saturn models. A helium deficit in molecular hydrogen regions and corresponding helium enrichment in metallic regions is responsible for the small density change near 0.55 *r/R* is Saturn. For Jupiter, a model with and without a core is shown. For the models with a core, the core is assumed to be ices overlying rock.

5.3. Saturn

The observational constraints for Saturn are listed in Table 33.1. Although Saturn has less than one-third of Jupiter's mass, it has almost the same radius. This is a consequence of the relative insensitivity of radius to mass for hydrogen planets in Jupiter and Saturn's mass range (see Figure 33.6). Saturn's atmosphere, like Jupiter's, is enriched in methane and ammonia. The atmosphere's carbon enrichment (in the form of methane) was recently determined to be seven times the solar abundance. There are some indications that Saturn's atmosphere has less helium than Jupiter's but the error bars are large, since there has never been a Saturn entry probe. Since there is no known process by which Saturn could have accreted less helium than Jupiter, another process must be at work.

Saturn's interior is grossly similar to Jupiter's. The biggest difference is that Saturn's core may be as large as 20 Earth masses. A sample Saturn model is shown in Figure 33.7. Temperatures inside Saturn are also cooler. In the model shown in Figure 33.7, the temperature and pressure at the base of the metallic hydrogen envelope are 9000 K and 10 Mbar. Like Jupiter, there is strong evidence that Saturn's envelope is enriched in heavy elements over solar abundance. The mass of the core and the heavy elements in the hydrogen–helium envelope are also shown in Figure 33.8. The total mass fraction of heavy elements in Saturn is about 2½ times greater than in Jupiter. On the whole, Saturn is enhanced in heavy elements by a factor of 6–14, relative to the Sun. This may be an indication that more condensed icy material was available to be incorporated into Saturn at its location in the solar nebula. Nevertheless, as at Jupiter, hydrogen and helium are the dominant component of Saturn's mass ($\sim$75%).

Saturn's somewhat low atmospheric helium abundance implies that the process of helium differentiation (see Section 6) has begun inside the planet. This process results in removal of helium from the outer molecular hydrogen envelope of the planet and enhancement of helium in the deep interior. Thus the helium fraction should increase with depth in Saturn's interior. The inferred density structure is consistent with this widely accepted explanation for Saturn's low atmospheric helium abundance. If helium is presumed to be uniformly depleted from the outer molecular envelope of the planet, it can be self-consistently accounted for in the deeper interior. The unmixed helium may have actually been removed from molecular *and* metallic regions of the planet and settled down on top of the core. The models lack the sensitivity to confirm that this is definitely the case, however.

The inferred interior structure of Saturn is most consistent with the giant planet formation scenario known as nucleated collapse. In this scenario, a nucleus of rock and ice first forms in the solar nebula. When the nucleus has grown to about ten Earth masses, the gas of the nebula collapses down on the core, thus forming a massive hydrogen/helium envelope surrounding a rock/ice core. Planetesimals that accrete later in time cannot pass through the thick atmosphere surrounding the core. Instead, they break up and dissolve into the hydrogen/helium envelope. This scenario accounts for both the core of the planet and the enrichment of heavy elements in the envelope. It is possible that Jupiter formed via a different mechanism, such as the direct gravitational collapse of nebular gas. It is perhaps more likely that both Jupiter and Saturn had larger cores in the past, but have been dredged up by convective plumes over the past 4.5 Gyr. This mechanism could plausibly be more efficient in the hotter interior of Jupiter, where convection is more vigorous.

5.4. Uranus and Neptune

Before the *Voyager* encounters, Uranus and Neptune were assumed to have similar interior structures. This assumption was well justified given their similar radii, masses, atmospheric compositions, and location in the outer solar system. Uranus and Neptune were modeled as having three distinct layers: an inner rocky core, a large icy mantle, and a methane-rich, hydrogen–helium atmosphere. Little more could be said with precision since their atmospheric oblateness and interior rotation rates were not accurately known.

Upon its arrival at Uranus in 1986 and Neptune in 1989, *Voyager 2* provided the measurements needed to constrain interior models and provide individual identities for each planet. *Voyager* observed the structure of the magnetic field of both planets and measured their rotation rates. In both cases, the fields were off-center, tilted dipoles of similar strengths. *Voyager* also measured the higher order components of the gravitational fields of both planets. The abundance of carbon in both atmospheres is about 50 times the solar value. Although Uranus and Neptune have similar radii and masses, the differences are such that the mean density of Neptune is 24% higher than that of Uranus.

Voyager data revealed that, although similar, the interior structures of the two planets are not identical. As with Jupiter and Saturn, Λ_2 provides information on the distribution of mass inside each planet. If Uranus and Neptune had a similar distribution of mass in their interiors, their Λ_2 parameters would be similar. As Table 33.1 shows, for Uranus $\Lambda_2 = 0.119$, whereas for Neptune $\Lambda_2 = 0.136$. The larger value of Λ_2 for Neptune implies that Neptune is less centrally condensed than Uranus. Models show that this difference can be understood in terms of equal relative amounts of ice, rock, and gas that are simply distributed differently within the two planets. The two planets also

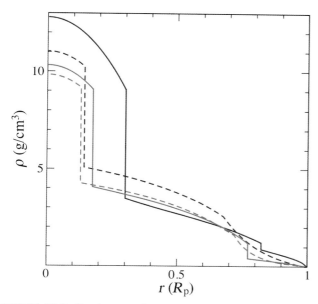

FIGURE 33.9 Density as a function of normalized radius for two Neptune models (blue) and two Uranus models (red). Solid curves use three distinct homogeneous layers, including a rocky core, a water-rich middle layer (but which includes some H/He), and an H/He-rich outer layer (but which includes some water). Dashed curves are four-layer models where the additional layer includes a gradient in the water abundance. A much wider variety of models are possible than can be shown here, including those with small or nonexistent rocky cores. Figure courtesy Nadine Nettelmann.

follow virtually the same pressure–density law, another indication that they have similar composition.

Models (Figure 33.9) of Uranus and Neptune's interior begin with a hydrogen-rich atmosphere that extends from the observable cloud tops to about 85% of Neptune's radius and 80% of Uranus' radius. The composition in this region does not vary significantly from the hydrogen-rich atmospheric composition. Near 0.3 Mbar and 3000 K ($0.85 R_{Neptune}$ and $0.80 R_{Uranus}$), the density rises rapidly to over 1000 kg/m³. The density then increases steadily into the deep interior of both planets, where the pressure reaches 6 Mbar at 7000 K. The variation of density with pressure in this region is very similar to that found in the laboratory shock wave experiments on the artificial "icy" mixture known as "synthetic Uranus". The composition of this region is likely predominantly "icy", although models less dense then 100% icy composition are preferred. This may indicate that H/He may be mixed into the middle icy layer, or that the notion of a simple three-layer model is too simple. Since the density of rock/ice/gas mixtures can mimic the density of pure ice, the exact composition cannot be known with precision. Any hydrogen present in the deep interior would be in the metallic phase.

Interestingly, Uranus and Neptune models can be constructed that do not have rock cores. Other models with cores as large as 1 $M_\oplus$ are also consistent with the available data.

The total mass of hydrogen and helium in Uranus and Neptune is about 2 $M_\oplus$, compared to about 300 $M_\oplus$ at Jupiter. Given the relatively small amounts of gas compared to ices in Uranus and Neptune, these planets are aptly termed "ice giants", whereas Jupiter and Saturn are indeed "gas giants".

Shock compression measurements show that the fluids of the hot, ice-rich region of Uranus and Neptune are expected to be substantially ionized and dissociated. The large electrical conductivities of such fluids, coupled with the modest convective velocities predicted for the interiors of Uranus and Neptune, can generate and sustain the observed magnetic fields of the planets. One possible explanation for the complexity of their magnetic fields is that the electrically conductive region of these planets is comparatively close (within about 4000 km) to the cloud tops, a consequence of the ionization behavior of water, ammonia, and methane. This is consistent with the trend in field complexity seen at Jupiter and Saturn.

Uranus and Neptune likely represent failed gas giant planets. The time to accrete solid objects onto the growing ice and rock planetary cores was much longer in the outer solar nebula than at the orbital distances of Jupiter and Saturn. Thus Uranus and Neptune took longer to grow. By the time the nebular gas was swept away, these planets had not yet grown massive enough to capture substantial amounts of hydrogen and helium gas from the nebula. Perhaps if the nebular gas had persisted for a longer time, Uranus and Neptune would have grown large enough to complete the capture of a hydrogen/helium envelope. In that case, these planets might now more closely resemble the current Jupiter and Saturn.

5.5. Extrasolar Giant Planets

With current technology, we can learn very little of the physical state of most of the thousands of planets of planets that have been detected around other stars. The minimum masses of planets are obtained by observing the wobble induced on the parent star's orbit by the gravitational tug of the planet. But just knowing the minimum mass (since orbital inclinations are unknown) does not tell us much about the structure of a planet. However, there are hundreds of planets in orbit around other stars for which we have derived accurate masses *and* radii. The radii can be measured if the extrasolar planetary system has a favorable alignment, and the planet passes in front of its parent star (a transit), blocking a small fraction of the star's light. The planet's orbital inclination is then constrained to be essentially edge-on, so then the mass is also known. As was shown in Figure 33.6, with a determination of only the mass and radius of a planet, we can get to a reasonable understanding of its interior composition.

Since planets that orbit close to their parent stars are more likely to transit, as seen from the Earth, most discovered transiting planets are highly irradiated. New theoretical procedures have had to be developed to understand the structure of giant planets that are 100 times closer to their stars than Jupiter is to the Sun and hence receive intense stellar irradiation. This irradiation slows the cooling of the planet and hence slows the contraction of a giant planet with time. Our understanding of giant planets at small orbital distances is progressing, but the results are very surprising. Figure 33.10 shows the radius and mass of nearly 200 transiting planets, as compared to our solar system's planets. The density of these planets can be read from the dotted curves.

It is clear that there is *tremendous* diversity among the sample of detected planets. Two planets with the same mass can have radii that differ by a factor of two. There are two immediate implications. Many of the hottest transiting gas giants have been affected by some mechanism that has dramatically slowed, or even halted, planetary contraction. In addition, since some planets are smaller than Jupiter and Saturn, some of these planets are even more strongly enhanced in heavy elements than our Jovian planets.

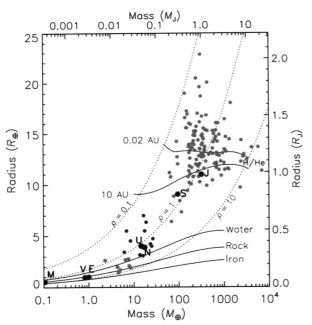

FIGURE 33.10 Mass vs radius for nearly 200 transiting planets. Jupiter, Saturn, Uranus, Neptune, Earth, Venus, and Mars are shown by black capital letters. Gas giants are shown in red, planets that may have thinner H/He envelopes (like Uranus and Neptune) are shown in blue, and low-mass, relatively dense planets are shown in green. Models for pure iron, rock, and water planets are shown as black curves. Models for 4.5-Gyr-old pure H/He gas giants at 10 AU and 0.02 AU from the Sun are shown as black curves as well. Dotted curves of constant density (in g/cm^3) are overplotted.

The "radius anomaly" question has not yet been solved, but it is now reasonably clear that the most strongly irradiated planets have the largest radii, and that the relatively cool planets below 1000 K do not appear inflated. The most promising postulated radius inflation mechanisms invoke ways to transport a small fraction of the large amount of absorbed stellar flux into the deep interior. These transport mechanisms generally involve the absorbed stellar flux driving fast winds in the atmosphere, with a small fraction of this wind kinetic energy dissipated in the deep atmosphere or the top of the convective interior. Studies of cooler, noninflated planets show that Jupiter- and Saturn-like enrichments of heavy elements, over parent star abundances, are extremely common. At the extreme enrichment end, the relatively small radii of some planets that are several Jupiter masses imply several hundred Earth masses of heavy elements within their interiors.

6. JOVIAN PLANET EVOLUTION

The amount of energy radiated by each of the Jovian planets, except possibly for Uranus, is greater than the amount of energy that they receive from the Sun (see Table 33.1). This internal heat source is far too large to be explained by decay of radioactive elements in the rock cores of the planets. Temperatures, even in the deep interior, are far below the 10,000,000 K required for thermonuclear fusion. The source of the excess energy is gravitational potential energy that was converted to heat during the planets' formation and stored in their interiors. (See The Origin of the Solar System.)

The potential energy of gas and solids in the solar nebula was converted to thermal energy when they were accreted onto the forming planet. Over time, the planets radiated energy into space and cooled, slowly losing their primordial energy content. Thus all four Jovian planets were initially warmer than they are now. During the early evolutionary stages the planets contracted as they cooled, thereby releasing even more gravitational potential energy. Today, the planets all cool at essentially constant radius because the internal pressures depend only slightly on temperature. The coupled contraction and cooling is known as Kelvin–Helmholtz cooling.

Evolutionary models test whether Kelvin–Helmholtz cooling can account for the current observed heat flows of the Jovian planets. In these calculations, a series of sequentially cooler planetary interior models is created with the last model representing the present-day planet. The time elapsed between each static model is calculated and thus the evolutionary age of the planet is found.

Models predict that Jupiter should have cooled from an initially hot state (accompanied by an atmospheric temperature greater than about 700 K) to its current temperature in about 4.5 billion years. Since this is about the age of the solar system, the Kelvin–Helmholtz model is judged a success for Jupiter. For Saturn, however, the model is less successful. The models suggest that Saturn, with its current heat flow, should be about 2 billion years old. Since there is no reason to believe that Saturn formed 2.5 billion years later than Jupiter, another heat source must be adding to Saturn's Kelvin–Helmholtz luminosity. This leads to the hypothesis that differentiation of helium in the interior provides additional thermal energy to the planet.

The helium depletion hypothesis holds that as Saturn has cooled from an initially warmer state with the solar abundance of helium throughout, its interior reached the point (near 2 Mbar and 8000 K) at which hydrogen and helium no longer mix in all proportions. Like oil and vinegar in salad dressing, the hydrogen and helium are separating into different phases.

As the helium-rich drops form in Saturn's envelope and fall to deeper, warmer layers of its interior, the helium eventually again mixes with hydrogen. Over time, this rainfall is depleting the supply of helium in the outer envelope and visible atmosphere and enriching the helium content deeper in the interior, close to the core. The overall planetary inventory of helium remains constant. This model is compatible with the observed depletion at Saturn. Jupiter, with a warmer interior and smaller helium depletion, has apparently only recently begun this process.

This process of helium differentiation liberates gravitational potential energy as the drops fall. The helium abundance in a giant planet's atmosphere is difficult to determine without an entry probe. If the value for Saturn's atmosphere is correct, then the helium droplets in Saturn must be raining down very far into the planet, possibly all the way down the planet's core. No other process can simultaneously explain Saturn's anomalously high heat flow and the observed atmospheric depletion of helium. Observations from the *Cassini* spacecraft will allow for a better determination of the helium abundance (and the abundances of many other compounds) in Saturn's atmosphere.

The problem for Uranus and Neptune is somewhat different. The Kelvin–Helmholtz hypothesis predicts a correct age for Neptune, but an age that is too large for Uranus. In other words, the model predicts that Uranus should have a higher heat flow at the current time than it is observed to have. The problem is quite severe, since Uranus has no detectable heat flow. There are several possible resolutions to this contradiction.

One possibility is that gradients in the composition of Uranus with radius have served to impede convection in the deep interior. Composition gradients, for example, a gradual increase in the rock abundance with depth, can severely limit heat flow from the planet. In such a case, only the outermost layers could transport energy by convection to the atmosphere and cool effectively to space, thus producing a lower than expected heat flow. More of Neptune's interior than that of Uranus might be convective, thus explaining its higher current heat flow. Of course if this hypothesis were correct, then the existing interior models of these planets would have to be revised, since an initial assumption that the planets are fully convective would have been violated. Inhibition of convection in the deep interior by this mechanism has been proposed as one explanation for the strong nondipole component of both planets' magnetic fields.

The strongly irradiated extrasolar gas giants, in the absence of an additional energy source, are thought to evolve much like Jupiter, except that their interiors cool, and the planets contract, more slowly. The incident stellar flux leads to a radiative zone with a shallow temperature gradient to pressure of up to ∼1 kbar, 1000 times deeper than in Jupiter. This limits how quickly the interior heat flux can escape from the planet. However, this simple picture should limit giant planet radii to values 20% larger than Jupiter, whereas planets twice the radius of Jupiter have been found. Clearly there is missing physics. Finding more transiting planets, with a variety of radii, masses, and orbital separations will allow for a better understanding of how stellar irradiation effects the cooling of giant planets.

7. FUTURE DIRECTIONS

Models of Jovian planetary interiors have constrained the mass of each planet's core and the approximate composition of their envelopes. These results have provided important constraints on the processes by which these planets form. In turn, formation models place limits on the mass, composition, and evolution of the solar nebula. Further progress, however, requires even tighter limits on the interior structure of these planets. Sufficiently detailed interior models may even provide constraints on the EOS of hydrogen. Since Jupiter is the largest reservoir of metallic hydrogen in the solar system, it may potentially resolve issues such as the exact pressure of the transition between molecular and metallic hydrogen.

One might expect that in the future, more accurate measurements of each planet's gravitational harmonics would help to address questions such as these. The higher order moments, however, are most sensitive to the density distribution in the outer 10 or 20% of the planetary radius. Thus, little additional information about the deep interior is likely to be forthcoming from such observational improvements. The higher order harmonics do, however, provide some information about the state of rotation of the outer layers and may help address questions regarding the degree of differential rotation in the Jovian planets. For example, it is unknown if Jupiter rotates completely as a solid body, or if different cylindrical regions of its interior rotate at different rates.

The NASA spacecraft Juno, now heading toward a 2016 arrival at Jupiter, will answer this and other questions. Juno will be placed into a low polar orbit such that the spacecraft will readily be able to measure additional higher order harmonics up to J_{12}, which will allow for a determination of planet's interior rotation. Juno may measure Jupiter's **moment of inertia** which would help to constrain the core size. In addition, the spacecraft will observe microwave emission from below the "weather layer" of the planet's atmosphere (100 bars) to determine the deep abundance of water and ammonia. The planet's magnetic field will also be mapped in unprecedented detail. Together, these new measurements should shed new light on the interior structure of the planet.

Further improvements in delineating the EOSs of Jovian planetary components will help to clarify their interior structures. Better knowledge of the behavior of planetary constituents and their mixtures at high pressure will enable more accurate interior models to be constructed. Yet any possible dramatic changes in understanding are unlikely to result from such improvements. Only significantly new and different sources of information offer the potential of providing fundamentally new insights into the interior structure of these planets.

Jovian seismology is one particularly promising new avenue of research into these planetary interiors. Much of our knowledge of the interior structure of the Earth arises from study of seismic waves that propagate through the interior of the planet. The speed and trajectory of these waves carry information about the composition and structure of the Earth's interior. During the collisions of the fragments of **comet** Shoemaker-Levy 9 with Jupiter, several experiments attempted to detect seismic waves launched by the impacts. If these waves had been detected, they would have provided a direct probe into the interior structure of Jupiter.

Another avenue for Jovian seismology is to detect resonant acoustic modes trapped inside Jupiter. The frequency of a given Jovian oscillation mode depends on the interior structure of the planet within the region in which the mode propagates. Thus measurement of the frequencies of a variety of modes would provide information on the overall interior structure of the planet. Indeed, the study of such modes on the Sun, a science known as helioseismology, has revolutionized our knowledge of the solar interior. A number of attempts have been made to detect the Jovian oscillations with various techniques. Such observations and data analysis are difficult and interpretation of the available data has been limited by the restricted number of observing nights on large telescopes. Nevertheless, an analysis of the available data strongly suggests that Jupiter's oscillations have indeed been detected. There is also emerging evidence that certain wavelike features in Saturn's C-ring are created by periodic perturbations to the gravitational field of the planet induced by seismic oscillations of Saturn. The exact locations of these ring features depend upon the seismic mode frequencies and thus the internal structure of Saturn.

Definitive detection of oscillations of any Jovian planet would first serve to accurately determine the core size and rotation profile of the planet. Since such determinations would remove two important sources of uncertainty surrounding the interior structure, more information could then be gleaned from the traditional interior model constraints. Seismology might also help to constrain more accurately the location of the transition from molecular to metallic hydrogen in Jupiter's interior. If so, seismology may ultimately provide the tightest constraints on the hydrogen EOS and interior structure of Jovian planets.

Together with a refined understanding of the interiors of our solar system's giant planets, additional understanding of giant planet interiors will come from extrasolar planets. Determinations of the radii of transiting extrasolar planets will allow us to build up a statistical sample to learn how the radius of planets change as a function of mass, age, the amount of heavy elements available in the system (which can be estimated from spectra of the parent star), and the amount of irradiation the planet receives from its parent star. The interiors of giant planets will likely yield many additional surprises as more extrasolar planets are found. (See Extra-Solar Planets.)

BIBLIOGRAPHY

Fortney, J. J., & Nettelmann, N. (2010). *Space Science Reviews, 152*, 423−447.

Baraffe, I., Chabrier, G., Barman, T., 2010. Reports in Progress in Physics, 73, 016901.

Guillot, T., Stevenson, D. J., Hubbard, W. B., & Saumon, D. (2004). In Fran Bagenal, Timothy E. Dowling, & William B. McKinnon (Eds.), *Cambridge planetary science: Vol. 1. Jupiter. The planet, satellites and magnetosphere* (pp. 35−57). Cambridge, UK: Cambridge University.

Guillot, T. (2005). *Annual Review of Earth and Planetary Sciences, 33*, 493−530.

Hubbard, W. B., Burrows, A., & Lunine, J. I. (2002). *Annual Review of Astronomy and Astrophysics, 40*, 103−136.

Stevenson, D. J. (1982). *Annual Review of Earth and Planetary Sciences, 10*, 257−295.

Chapter 34

Planetary Satellites

Bonnie J. Buratti
Jet Propulsion Laboratory, California Institute of Technology, Pasadena, CA, USA

Peter C. Thomas
Department of Astronomy, Center for Radiophysics & Space Research, Cornell University, Ithaca, NY, USA

Chapter Outline

1. Summary of Characteristics — 759
   1.1. Discovery — 759
   1.2. Physical and Dynamical Properties — 764
2. Formation of Satellites — 765
   2.1. Theoretical Models — 765
   2.2. Evolution — 766
3. Observations of Satellites — 767
   3.1. Telescopic Observations — 767
      3.1.1. Spectroscopy — 767
      3.1.2. Photometry — 767
      3.1.3. Radiometry — 767
      3.1.4. Polarimetry — 768
      3.1.5. Radar — 768
   3.2. Spacecraft Exploration — 768
4. Individual Satellites — 769
   4.1. The Satellites of Mars: Deimos and Phobos — 769
   4.2. The Small Satellites of Jupiter — 770
   4.3. The Saturnian System — 770
      4.3.1. The Medium-Sized Icy Satellites of Saturn (Other than Enceladus): Rhea, Dione, Tethys, Mimas, and Iapetus — 770
      4.3.2. The Small Satellites of Saturn — 773
   4.4. The Satellites of Uranus — 774
      4.4.1. The Medium-Sized Satellites of Uranus: Miranda, Ariel, Umbriel, Titania, and Oberon — 774
      4.4.2. The Small Satellites of Uranus — 775
   4.5. The Satellites of Neptune — 776
      4.5.1. The Inner Moons — 776
      4.5.2. The Outer Irregular Moons — 777
Acknowledgment — 777
Bibliography — 777

A planetary satellite is any one of the celestial bodies in orbit around a planet, which is known as the **primary body**. They range from large, planetlike, geologically active worlds with significant atmospheres, such as Neptune's satellite Triton and Saturn's satellite Titan, to tiny irregular objects tens of kilometers in diameter. The satellites in the inner solar system—the two moons of Mars and the Earth's Moon—are composed primarily of rocky material. The satellites of the outer solar system, with the exception of Io, all have as major components some type of frozen volatile, primarily water ice, and also methane, ammonia, nitrogen, carbon monoxide, carbon dioxide, or sulfur dioxide existing alone or in combination with other volatiles. As of the end of 2013, the planets have among them a total of 173 known satellites (compare the list of satellites in the appendix). There undoubtedly exist many more undiscovered small satellites in the outer solar system. The relative sizes of the main satellites are illustrated in Figure 34.1. Table 34.1 is a summary of their characteristics. This chapter covers the satellites of Mars, Jupiter, Saturn, Uranus, and Neptune, except the Galilean satellites (the four largest moons of Jupiter), Triton, Titan, and Enceladus (*see* The volcanic Moon; Galilean Satellites; Titan; Enceladus; Triton; Pluto and Charon).

1. SUMMARY OF CHARACTERISTICS

1.1. Discovery

None of the satellites of the outer planets was known before the invention of the telescope. When Galileo turned his telescope to Jupiter in 1610, he discovered the four large satellites in the Jovian system. His observations of their orbital motion around Jupiter in a manner analogous to the motion of the planets around the Sun provided important evidence for the acceptance of the heliocentric (Sun-centered) model of the solar system. These four moons—Io, Europa, Ganymede, and Callisto—are sometimes called the Galilean satellites.

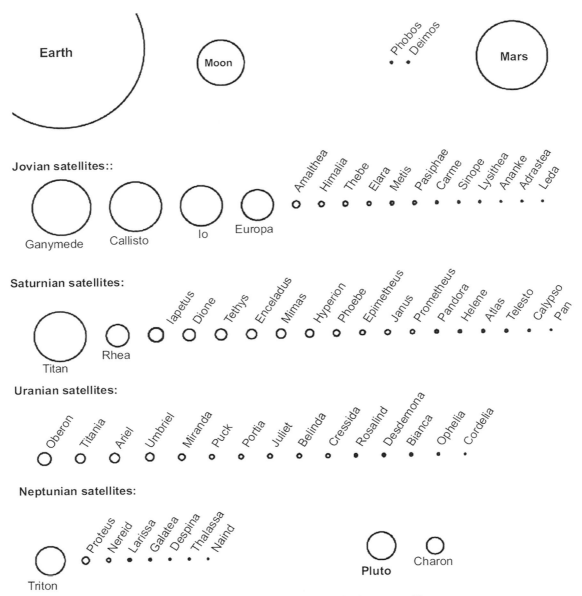

FIGURE 34.1 The relative sizes of the main planetary satellites.

In 1655, Christian Huygens discovered Titan, the giant satellite of Saturn. Later in the seventeenth century, Giovanni Cassini discovered the four next largest satellites of Saturn. More than 100 years would pass before the next satellite discoveries were made: the Uranian satellites Titania and Oberon and two smaller moons of Saturn. As telescopes acquired more resolving power in the nineteenth century, the family of satellites grew (see Table 34.1 and Figure 34.1). The smallest satellites of Jupiter and Saturn, and all the small satellites of Uranus and Neptune (except Nereid), were discovered during flybys of the *Pioneer* and *Voyager* spacecraft (see Table 34.2). Many small satellites have been recently discovered by sensitive charge-coupled device (CCD) cameras attached to large ground-based telescopes, the *Galileo*, *Voyager*, and Cassini spacecraft, and the Hubble Space Telescope.

The natural planetary satellites are generally named after figures in world mythologies that were associated with the namesakes of their primaries. They are also designated by the first letter of their **primary** and an Arabic numeral assigned in order of discovery: Io is J1, Europa is J2, and so on. When satellites are first discovered but not yet confirmed or officially named, they are known by the year in which they were discovered, the initial of the primary, and a number assigned consecutively for all solar system discoveries, for example,

Chapter | 34 Planetary Satellites

TABLE 34.1 Properties of the Main Planetary Satellites (Compare the list of satellites and their properties in the appendix)

Satellite	Distance from Primary (10^3 km)	Revolution Period (days) R = Retrograde	Orbital Eccentricity	Orbital Inclination (Degrees)	Radius (km)	Density (g/cm^3)	Visual Geometric Albedo	Discoverer	Year Discovered
Mars (2)									
M1 Phobos	9.38	0.32	0.018	1.1	13 × 11 × 9	1.87	0.05	Hall	1877
M2 Deimos	23.5	1.26	0.0002	1.8	8 × 6 × 5	1.4	0.05	Hall	1877
Jupiter (67)									
J15 Adrastea	129	0.30	0.002	0.054	10		<0.1	Voyager	1979
J16 Metis	128	0.30	0.001	0.019	20		<0.1	Voyager	1979
J5 Amalthea	181	0.50	0.003	0.4	131 × 73 × 67	1.0	0.05	Barnard	1892
J14 Thebe	222	0.67	0.018	1.08	50		<0.1	Voyager	1979
J1 Io	422	1.77	0.004	0.04	1818	3.53	0.6	Galileo	1610
J2 Europa	671	3.55	0.009	0.5	1560	2.99	0.6	Galileo	1610
J3 Ganymede	1070	7.15	0.001	0.2	2634	1.94	0.4	Galileo	1610
J4 Callisto	1883	16.69	0.007	0.2	2409	1.85	0.2	Galileo	1610
J13 Leda	11,165	241	0.1636	267.5	10	2.6	0.04	Kowal	1974
J6 Himalia	11,480	251	0.163	27.6	85		0.03	Perrine	1904
J10 Lysithea	11,720	259	0.107	29.0	12			Nicholson	1938
J7 Elara	11,737	260	0.207	24.8	40		0.03	Perrine	1904
J12 Ananke	21,200	631R	0.17	147	10			Nicholson	1951
J11 Carme	22,600	692R	0.21	163	15			Nicholson	1938
J8 Pasiphae	23,500	735R	0.38	145	18			Melotte	1908
J9 Sinope	23,700	758R	0.28	153	14			Nicholson	1914
Additional outer satellites (51)	7507–28332	130–979; 48R	0.11–0.48	43–165	0.5–4.3			Various	1975–2011
Saturn (62)									
S/2009 S1	117	0.47	0.0	0.0	0.3			Cassini	2009
S18 Pan	134	0.57	0.0	0.0	17 × 15 × 10	0.43	0.5	Showalter	1990
S35 Daphnis	137	0.59	0	0.003	5 × 5 × 3	0.34		Cassini	2005
S15 Atlas	138	0.60	0.000	0.0	21 × 18 × 9	0.46	0.4	Voyager	1980
S16 Prometheus	139	0.61	0.002	0.0	68 × 42 × 28	0.48	0.6	Voyager	1980
S17 Pandora	142	0.63	0.004	0.05	52 × 41 × 32	0.49	0.5	Voyager	1980

(Continued)

TABLE 34.1 Properties of the Main Planetary Satellites (Compare the list of satellites and their properties in the appendix)—cont'd

Satellite	Distance from Primary (10³ km)	Revolution Period (days) R = Retrograde	Orbital Eccentricity	Orbital Inclination (Degrees)	Radius (km)	Density (g/cm³)	Visual Geometric Albedo	Discoverer	Year Discovered
S10 Janus	151	0.69	0.007	0.14	102 × 93 × 76	0.65	0.6	Dollfus	1966
S11 Epimetheus	151	0.69	0.009	0.34	65 × 57 × 53	0.63	0.73	Fountain and Larson	1978
S53 Aegaeon	168	0.81	0.0002	0.001	0.3		0.5	Cassini	2008
S1 Mimas	186	0.94	0.020	1.5	199	1.15	0.8	Herschel	1789
S32 Methone	194	1.01	0.00	0.013	1.5			Cassini	2004
S49 Anthe	198	1.036	0.0011	0.015	0.3			Cassini	2007
S33 Pallene	212	1.154	0.004	0.0	2.2			Cassini	2004
S2 Enceladus	238	1.37	0.004	0.0	252	1.61	1.4	Herschel	1789
S3 Tethys	295	1.89	0.000	1.1	536	0.96	0.8	Cassini	1684
S14 Calypso	295	1.89	0.0	1.1	15 × 9 × 6	1.0	0.6	Pascu et al.	1980
S13 Telesto	295	1.89	0.0	1.0	16 × 12 × 10	1.0	0.9	Smith et al.	1980
S4 Dione	377	2.74	0.002	0.02	563	1.47	0.55	Cassini	1684
S12 Helene	377	2.74	0.005	0.15	23 × 20 × 13	1.5	0.5	Laques and Lecacheux	1980
S34 Polydeuces	377	2.74	0.0191	0.175	1.3			Cassini	2004
S5 Rhea	527	4.52	0.001	0.35	734	1.23	0.65	Cassini	1672
S6 Titan	1222	15.94	0.029	0.33	2575	1.88	0.2	Huygens	1655
S7 Hyperion	1481	21.28	0.104	0.4	180 × 133 × 103	0.54	0.3	Bond and Lassell	1848
S8 Iapetus	3561	79.33	0.028	14.7	718	1.09	0.4–0.08	Cassini	1671
S9 Phoebe	12,952	550.4R	0.163	150	107	1.63	0.06	Pickering	1898
Additional outer satellites (37)	11,110–25,110		0.11–0.53	34–180	2–20			Various	2000–2009
Uranus (27)									
U6 Cordelia	49.7	0.33	0.0005	0.14	13			Voyager 2	1986
U7 Ophelia	53.8	0.38	0.010	0.09	15			Voyager 2	1986
U8 Bianca	59.2	0.43	0.001	0.16	21			Voyager 2	1986
U9 Cressida	61.8	0.46	0.0002	0.04	31		~0.04	Voyager 2	1986
U10 Desdemona	62.7	0.47	0.0002	0.16	27		~0.04	Voyager 2	1986

Chapter | 34 Planetary Satellites

Satellite	a	e	i	radius	albedo	Discoverer	Year
U11 Juliet	64.4	0.49	0.0006	0.06	42	~0.06	Voyager 2	1986
U12 Portia	66.1	0.51	0.0002	0.09	54	~0.09	Voyager 2	1986
U13 Rosalind	69.9	0.56	0.00,009	0.28	27	~0.04	Voyager 2	1986
U27 Cupid	74.4	0.61	0.0013	0.099	5		Showalter and Lissauer	2003
U14 Belinda	75.3	0.62	0.0001	0.03	33		Voyager 2	1986
U25 Perdita	76.4	0.64	0.0116	0.47	10		Karkoschka	1999
U15 Puck	86.0	0.76	0.00,005	0.31	77	0.07	Voyager 2	1985
U26 Mab	97.7	0.92	0.0025	0.134	5		Showalter and Lissauer	2003
U5 Miranda	130	1.41	0.003	3.4	236	1.2	0.35	Kuiper	1948
U1 Ariel	191	2.52	0.003	0.0	579	1.6	0.36	Lassell	1851
U2 Umbriel	266	4.14	0.005	0.0	585	1.5	0.20	Lassell	1851
U3 Titania	436	8.71	0.002	0.0	789	1.7	0.30	Herschel	1787
U4 Oberon	583	13.46	0.001	0.0	761	1.6	0.22	Herschel	1787
U16 Caliban	7775	654R	0.2	146	36		Gladman et al.	1997
U17 Sycorax	8846	795R	0.34	154	75		Gladman et al.	1,997
Additional outer satellites (7)	4276–20900	267–2790; 6R	0.13–0.68	57–170			Various	1999–2003
Neptune (13)								
N8 Naiad	48.2	0.29	0.000	0.0	29		Voyager 2	1989
N7 Thalassa	50.1	0.31	0.0002	4.5	40		Voyager 2	1989
N5 Despina	52.5	0.33	0.0001	0.0	74	0.05	Voyager 2	1989
N6 Galatea	62.0	0.43	0.0001	0.0	79		Voyager 2	1989
N4 Larissa	73.6	0.55	0.000	0.0	104 × 89	0.06	Voyager 2	1989
N14 S/2004 N 1	105.3	0.94	0.0	0.0	~9		Showalter et al.	2013
N3 Proteus	117.6	1.12	0.0004	0.0	208	0.06	Voyager 2	1989
N1 Triton	354.8	5.87R	0.000,015	157	1353	2.08	0.73	Lassell	1846
N2 Nereid	5513	360.1	0.751	29	170	0.16	Kuiper	1949
Additional outer satellites (5)	15,730–48,390	1880–9374; 3R	0.293 –0.571	35–137			Various	2002–2003

See web-based sources in the bibliography for additional information.

TABLE 34.2 Major Flyby Missions to Planetary Satellites

Mission	Objects	Encounter Dates
Mariner 9	Martian satellites	1971
Viking 1 and 2	Martian satellites	1976
Pioneer 10	Jovian satellites	1979
Pioneer 11	Jovian satellites	1979
	Saturnian satellites	1979
Voyager 1	Jovian satellites	1979
	Saturnian satellites	1980
Voyager 2	Jovian satellites	1979
	Saturnian satellites	1981
	Uranian satellites	1986
	Neptunian satellites	1989
Phobos 2 (Russia)	Martian satellites	1989
Galileo	Jovian satellites	1996–1998
Mars Global Surveyor	Martian satellites	1998-present
Mars Express	Phobos and Deimos from Mars orbit	2003–present
Cassini-Huygens	Saturnian satellites	2004–2017 (planned)
Mars Reconnaissance Orbiter	Phobos and Deimos from Mars orbit	2006–present

1980 J27. Official names for all satellites are assigned by the International Astronomical Union (IAU).

After planetary scientists were able to map geologic formations of the satellites from spacecraft images, they named many of the features after characters or locations from Western and Eastern mythologies. These names are also approved by the IAU.

1.2. Physical and Dynamical Properties

The motion of a satellite around its center of mass and its primary defines an ellipse with the primary at one of the foci. The orbit is defined by three primary orbital elements: (1) the semimajor axis, (2) the eccentricity, and (3) the angle made by the intersection of the plane of the orbit and the plane of the primary's spin equator (the angle of inclination). The orbits are said to be regular if they are in the same sense of direction (the prograde sense) as that determined by the rotation of the primary, and if their eccentricities and inclinations are low. The orbit of a satellite is irregular if its motion is in the opposite (or retrograde) sense of motion, if its orbit is highly eccentric (greater than $\sim 5\%$ of circular), or if it has a high angle of inclination (greater than $\sim 5°$). The majority of the outer planets' main satellites move in regular, prograde orbits. Many of the smaller satellites that move in irregular orbits are thought to be captured objects.

Except for the small, outer irregular moons, most of the planetary satellites present the same hemisphere toward their primaries, a state that is the result of tidal evolution. When two celestial bodies orbit each other, the gravitational force exerted on the nearside is greater than that exerted on the far side. The result is an elongation of each body to form tidal bulges, which can consist of solid, liquid, or gaseous (atmospheric) material. The primary tugs on the satellite's tidal bulge to lock its longest axis onto the primary–satellite line. The satellite, which is said to be in a state of **synchronous rotation**, keeps the same face toward the primary. Since this despun state occurs rapidly (usually within a few million years) and more easily at shorter distances from the primary, most of the inner natural satellites are in synchronous rotation. Satellites that are far away from the primary often maintain their original rotational period. One satellite of Saturn, Hyperion, rotates chaotically. Satellites that are in synchronous rotation generally exhibit an oscillating motion with respect to their primary to reveal more than half of their surfaces (for example, 59% of the Moon is visible from the Earth at some point). This motion, known as libration, is caused by differences between the satellite's average spin and the changing orbital velocity in elliptical orbits. Tidal forces can also induce pendulumlike swings away from an average rotation (forced libration).

The satellites of the outer solar system are unique worlds, each representing a vast panorama of physical processes. The two satellites of Mars and the small satellites of Jupiter and Saturn are irregular chunks of rock, ice, or mixtures of the two. They are perhaps captured asteroids or even objects from the Kuiper Belt that have been subjected to intensive meteoritic bombardment. Several of the satellites, including the Saturnian satellite Phoebe and areas of the Uranian satellites, are covered with **C-type** material, the dark, unprocessed, carbon-rich material found on the C **(carbonaceous)** class of asteroids. The surfaces of other satellites such as Hyperion and the dark side of Iapetus contain primitive matter that is spectrally red and is thought to be rich in organic compounds. Because these materials, which are common in the outer solar system, represent the material from which the solar system formed, understanding their occurrence and origin will yield clues on the state and early evolution of the solar system. In addition, the transport of organic matter from the outer solar system to the inner solar system, perhaps by comets, is generally concluded to be an essential step in the formation of life. Iapetus historically presented a particular enigma: one hemisphere is 10 times more reflective than the other (*see* Asteroids; Kuiper Belt).

Before the advent of spacecraft exploration, planetary scientists expected satellites to be geologically dead worlds. They assumed that heat sources were not sufficient to have melted their mantles to provide a source of liquid or semi-liquid ice or ice—silicate slurries. Reconnaissance of the icy satellite systems of the four outer giant planets by the two *Voyager* spacecraft uncovered a wide range of geologic processes, including currently active volcanism on Io and Triton. *Cassini* discovered active tectonic processes on Enceladus, a small satellite of Saturn. At least one additional satellite (Europa) may have current activity. The medium-sized satellites of Saturn and Uranus are large enough to have undergone internal melting with subsequent **differentiation** and resurfacing. Among the Galilean satellites, only Callisto lacks evidence for periods of such activity after formation.

Recent work on the importance of tidal interactions and subsequent heating has provided the theoretical foundation to explain the existence of widespread activity in the outer solar system. Another factor is the presence of nonice components, such as ammonia hydrate or methanol, which lower the melting point of near-surface materials. Partial melts of water ice and various contaminants—each with its own melting point and viscosity—provide material for a wide range of geologic activity. The realization that such partial melts are important to understanding the geologic history of the satellites has spawned interest in the rheology (viscous properties and resulting flow behavior) of various ice mixtures and exotic phases of ices that exist at extreme temperatures or pressures. Conversely, the types of features observed on the surfaces provide clues to the likely composition of the satellites' interiors. One key observable regarding the bulk composition of satellites is the density: lower densities imply higher ice content and lower rock content.

Because the surfaces of so many outer planet satellites exhibit evidence of geologic activity, planetary scientists have begun to think in terms of unified geologic processes that function throughout the solar system. For example, partial melts of water ice with various contaminants could provide flows of liquid or partially molten slurries that in many ways mimic terrestrial or lunar lava flows formed by the partial melting of mixtures of silicate rocks. The ridged and grooved terrains on satellites such as Ganymede, Enceladus, Tethys, and Miranda may all have resulted from similar tectonic activities. Finally, explosive volcanic eruptions occurring on Io, Triton, Earth, and possibly Enceladus may all result from the escape of volatiles released as the pressure in upward-moving liquids decreases (*see* Planetary Volcanism.).

2. FORMATION OF SATELLITES
2.1. Theoretical Models

Because the planets and their associated moons condensed from the same cloud of gas and dust at about the same time, the formation of the natural planetary satellites must be addressed within the context of the formation of the planets. The solar system formed 4.6 Ga. This age is derived primarily from radiometric dating of meteorites, which are thought to consist of primordial, unaltered matter (*see* The Origin of the Solar System).

The Sun and planets formed from a disk-shaped rotating cloud of gas and dust known as the protosolar nebula. When the temperature in the nebula cooled sufficiently, small grains began to condense. The difference in solidification temperatures of the constituents of the protosolar nebular accounts for the major compositional differences of the satellites. Since there was a temperature gradient as a function of distance from the center of the nebula, only those materials with high melting temperatures (e.g. silicates, iron, aluminum, titanium, and calcium) solidified in the central (hotter) portion of the nebula. Earth's Moon consists primarily of these materials. Beyond the orbit of Mars, carbon, in combination with silicates and organic molecules, condensed to form the **carbonaceous** material found on C-type asteroids. Similar carbonaceous material is found on the surfaces of the Martian moon Phobos, several of the Jovian and Saturnian satellites, regions of the Uranian satellites, and possibly Triton. In the outer regions of the asteroid belt, formation temperatures were sufficiently cold to allow water ice to condense and remain stable. Thus, the Jovian satellites are primarily ice—silicate admixtures (except for Io, which has apparently outgassed all its water). On Saturn and Uranus, these materials are predicted to be joined by methane and ammonia, and their hydrated forms. For the satellites of Neptune and Pluto, formation temperatures were low enough for other volatiles, such as nitrogen, carbon monoxide, and carbon dioxide, to exist in solid form. In general, the satellites that formed in the inner regions of the solar system are denser than the outer planets' satellites, because they retained a lower fraction of volatile materials.

After small grains of material condensed from the protosolar nebula, electrostatic forces caused them to stick together. Collisions between these larger aggregates caused meter-sized particles, or planetesimals, to be accreted. Finally, gravitational collapse occurred to form larger, kilometer-sized planetesimals. The largest of these bodies swept up much of the remaining material to create the protoplanets and their companion satellite systems. One important concept of planetary satellite formation is that a satellite cannot accrete within the **Roche limit**, the distance at which the tidal forces of the primary become greater than the self-gravity of the satellite.

The formation of the regular satellite systems of Jupiter, Saturn, and Uranus is sometimes thought to be a smaller scale version of the formation of the solar system. A density gradient as a function of distance from the primary does exist for the regular system of small, inner Neptunian satellites and for the Galilean satellites (see Table 34.1). This implies that more volatiles (primarily ice) are included in the bulk composition as the distance increases. However, this simple scenario cannot be applied to Saturn or Uranus because their regular satellites do not follow this pattern.

The retrograde satellites are probably captured asteroids, Kuiper Belt Objects, or large planetesimals left over from the major episode of planetary formation. Except for Titan and Triton, the satellites are too small to possess gravitational fields sufficiently strong to retain an appreciable atmosphere against thermal escape. At least one satellite (Ganymede) has a magnetic field.

2.2. Evolution

Soon after the satellites accreted, they began to heat up from the release of gravitational potential energy. An additional heat source was provided by the release of mechanical energy during the heavy bombardment of their surfaces by remaining debris. The satellites Phobos, Mimas, and Tethys all have impact craters caused by bodies that were nearly large enough to break them apart; probably such catastrophes did occur. The decay of radioactive elements found in silicate materials provided another major source of heat. The heat produced in the larger satellites was sufficient to cause melting and chemical fractionation; the dense material, such as silicates and iron, went to the center of the satellite to form a core, while ice and other volatiles remained in the crust. A fourth source of heat is provided by tidal interactions. When a satellite is being tidally despun, the resulting frictional energy is dissipated as heat. Because this process happens very quickly for most satellites (~10 Million years), another mechanism involving orbital resonances among satellites is probably required to produce the heat for more recent resurfacing events. Gravitational interactions tend to evolve the orbital periods of the satellites within a system into multiples of each other. If two neighboring moons have orbital periods that are close to a ratio of small integers, a periodic extragravitational tug exerted by the moons when they are close will cause the periods of the orbits to evolve to exactly that ratio, a dynamical state known as a mean-motion resonance. In the Galilean system, for example, Io and Europa complete four and two orbits, respectively, for each orbit completed by Ganymede. The result is that the satellites meet each other at the same point in their orbits. The resulting flexing of the tidal bulge induced on these bodies by their mutual gravitational attraction causes significant heat production in some cases; this heating is related to an orbital eccentricity that is maintained by the resonance (*see* Planetary Impacts; Solar System Dynamics).

Some satellites, such as the Earth's Moon, Ganymede, and several of the Saturnian satellites, underwent periods of melting and active geology within a billion years of their formation and then became quiescent. Others, such as Io, Triton, Enceladus, and possibly Europa, are currently geologically active. For nearly a billion years after their formation, the satellites all underwent intense bombardment and cratering. The bombardment tapered off to a slower rate and presently continues. By counting the number of craters on a satellite's surface and making certain assumptions about the flux of impacting material, geologists are able to estimate when a specific portion of a satellite's surface was formed. Continual bombardment of satellites causes the pulverization of both rocky and icy surfaces to form a covering of fine material known as a **regolith**.

Many scientists expected that most of the craters formed on the outer planets' satellites would have disappeared owing to viscous relaxation. The two *Voyager* spacecraft revealed surfaces covered with craters that in many cases had morphological similarities to those found in the inner solar system, including central peaks, large ejecta blankets, and well-formed outer walls. Recent research has shown that the elastic properties of very cold ice provide enough strength to offset viscous relaxation. Silicate mineral contaminants or other impurities in the ice may also provide extra strength to sustain surface topography.

Planetary scientists classify the erosional processes affecting satellites into two major categories: endogenic, which includes all internally produced geologic activity, and exogenic, which encompasses the changes brought by outside agents. The latter category includes the following processes: (1) meteoritic bombardment and resulting gardening and impact volatilization; (2) magnetospheric interactions, including sputtering and implantation of energetic particles; (3) alteration by high-energy ultraviolet (UV) photons; and (4) accretion of particles from sources such as planetary rings.

Meteoritic bombardment acts in two major ways to alter the optical characteristics of the surface. First, the impacts excavate and expose fresh material (cf. the bright ray craters on the Moon, Ganymede, and the Uranian satellites). Second, impact volatilization and subsequent escape of volatiles result in a lag deposit enriched in opaque, dark materials. The relative importance of the two processes depends on the flux, size distribution, and composition of the impacting particles, and on the composition, surface temperature, and mass of the satellite. For the Galilean satellites, older geologic regions tend to be darker and redder, but both the Galilean and Saturnian satellites tend to be brighter on the hemispheres that lead in the direction of orbital motion (the so-called "leading" side, as opposed to the "trailing" side); this effect is thought to be due to preferential micrometeoritic gardening on the leading side due to systematic differences in impact velocities caused by the orbital motion of the satellite. The Uranian satellites show no similar dichotomy in albedo, but their leading sides do tend to be redder, possibly due to more accretion of reddish meteoritic material on that hemisphere.

For satellites that are embedded in planetary **magnetospheres**, their surfaces are affected by magnetospheric interactions in three ways: (1) chemical alterations; (2) selective erosion, or sputtering; and (3) deposition of magnetospheric ions. In general, volatile components are more susceptible to sputter erosion than refractory ones. The overall effect of magnetospheric erosion is thus to enrich surfaces in darker, redder opaque materials. A similar effect is probably caused by the bombardment of UV photons, although much fundamental laboratory work remains to be done to determine the quantitative effects of this process (*see* Planetary Magnetospheres).

Sublimation and redeposition of frosts has also altered the surfaces of satellites. Triton, which has a thin atmosphere, undergoes seasonal transport of nitrogen on its surface. Thermal segregation, which is the tendency of frosts to collect in brighter, colder regions of the surface and accentuate existing albedo differences, is a significant effect for Iapetus and perhaps other moons.

3. OBSERVATIONS OF SATELLITES

3.1. Telescopic Observations

3.1.1. Spectroscopy

Before the development of interplanetary spacecraft, all observations from the Earth of objects in the solar system were obtained by naked eye observers and telescopes. One particularly useful tool of planetary astronomy is spectroscopy, or the acquisition of spectra from a celestial body.

Each component of the surface or atmosphere of a satellite has a characteristic pattern of absorption and emission bands. Comparison of the astronomical spectrum with laboratory spectra of materials that are possible components of the surface yields information on the composition of the satellite. For example, water ice has a series of absorption features between 1 and 4 μm. The detection of these bands on three of the Galilean satellites and several satellites of Saturn and Uranus demonstrated that water ice is a major constituent of their surfaces. Other examples are the detections of SO_2 frost on the surface of Io, methane in the atmosphere of Titan, nitrogen and carbon dioxide on Triton, and carbon monoxide on Pluto.

3.1.2. Photometry

Photometry of planetary satellites is the accurate measurement of radiation from their surfaces or atmospheres. These measurements can be compared to light-scattering models that depend on physical parameters, such as the porosity of the optically active upper surface layer, the albedo of the material, and the degree of topographic roughness. These models predict brightness variations as a function of solar **phase angle** (the angle between the observer, the Sun, and the satellite). Like the Earth's Moon, the planetary satellites present changing phases to an observer on the Earth. As the face of the satellite becomes fully illuminated to the observer, the integrated brightness exhibits a nonlinear surge in brightness that most likely results from the disappearance of mutual shadowing among surface particles. The magnitude of this surge, known as the "**opposition effect**", is greater for a more porous surface.

One measure of how much radiation a satellite reflects is the **geometric albedo**, p, which is the disk-integrated brightness at "full moon" (or a phase angle of 0°) compared to a perfectly reflecting, diffuse disk of the same size. The **phase integral**, q, defines the angular distribution of radiation over the sky:

$$q = 2 \int_0^\pi \Phi(\alpha) \sin \alpha \, d\alpha$$

where $\Phi(\alpha)$ is the disk-integrated brightness and α is the phase angle.

The **Bond albedo**, which is given by $A = p \times q$, is the ratio of the integrated flux reflected by the satellite to the integrated flux received. The geometric albedo and phase integral are wavelength dependent, whereas a true (or bolometric) Bond albedo is the total power obtained by integrating energy at all wavelengths. Thus, it is the fraction of the incident power reflected, and $(1 - A)$ is the fraction absorbed.

Another ground-based photometric measurement that has yielded important information on satellites' surfaces is the integrated brightness of a satellite as a function of orbital angle. For a satellite in synchronous rotation with its primary, the subobserver geographical longitude on the satellite is equal to the longitude of the satellite in its orbit. Observations showing significant albedo and color variegations for Io, Europa, Rhea, Dione, and especially Iapetus suggest that diverse geologic terrains coexist on these satellites. This view was confirmed by images obtained by the *Voyager* spacecraft.

Another important photometric technique is the measurement of radiation as one celestial body occults, or blocks, another body. Time-resolved observations of occultations yield the flux emitted from successive regions of the eclipsed body. This technique has been used to map albedo variations on Pluto and its satellite Charon and to map the distribution of infrared emission—and thus volcanic activity—on Io.

3.1.3. Radiometry

Satellite radiometry is the measurement of radiation that is absorbed and reemitted at thermal wavelengths. The distance of each satellite from the Sun determines the mean temperature for the equilibrium condition that the power of the absorbed radiation is equal to the power of the emitted radiation:

$$\pi R^2 (F/r^2)(1 - A) = 4\pi R^2 \varepsilon \sigma T^4$$

$$T = \left(\frac{(1-A)F}{4\sigma \varepsilon r^2}\right)^{1/4}$$

where R is the radius of the satellite, r is the Sun–satellite distance, ε is the emissivity, σ is Stefan–Boltzmann's constant, A is the Bond albedo, and F is the incident solar flux (this equation assumes rotation sufficiently rapid to produce a constant surface temperature; for more slowly rotating bodies it is assumed that only the surface facing the sun is heated and thus radiates over $2\pi R^2$). Typical mean temperatures in Kelvin for the satellites are as follows: the Earth's Moon, 280; Europa, 103; Iapetus, 89; the Uranian satellites, 60; and the Neptunian satellites, 45. For thermal equilibrium, measurements as a function of wavelength yield a blackbody curve characteristic of T: in general, the temperatures of the satellites closely follow the blackbody emission values. Some discrepancies are caused by a weak **greenhouse** effect (in the case of Titan), or the existence of volcanic activity (in the case of Io).

Another possible use of radiometric techniques, when combined with photometric measurements of the reflected portion of the radiation, is the estimate of the diameter of a satellite. A more accurate method of measuring the diameter of a satellite from Earth involves measuring the light from a star as it is occulted by the satellite. The time the starlight is dimmed is proportional to the satellite's diameter.

A third radiometric technique is the measurement of the thermal response of a satellite's surface as it is being eclipsed by its **primary**. The rapid loss of heat from a satellite's surface indicates a thermal conductivity consistent with a porous surface. Eclipse radiometry of Phobos, Callisto, and Ganymede suggests that these objects all lose heat rapidly. The measurement of the thermal conductivity of the Saturnian satellites, particularly Mimas and Tethys, by the *Cassini* spacecraft suggests that parts of their surfaces may have been annealed by high-energy electrons in Saturn's magnetosphere.

3.1.4. Polarimetry

Polarimetry is the measurement of the degree of polarization of radiation reflected from a satellite's surface. The polarization characteristics depend on the shape, size, and optical properties of the surface particles. Generally, the radiation is linearly polarized and is said to be negatively polarized if it lies in the scattering plane and positively polarized if it is perpendicular to the scattering plane. Polarization measurements as a function of solar phase angle for atmosphereless bodies are negative at small phase angles; comparisons with laboratory measurements indicate that this is characteristic of complex, porous surfaces consisting of multisized particles. In 1970, ground-based polarimetry of Titan that showed it lacked a region of negative polarization led to the correct conclusion that it has a thick atmosphere.

3.1.5. Radar

Planetary radar is a set of techniques that involve the transmission of radio waves to a remote surface and the analysis of the echoed signal. Among the outer planets' satellites, the Galilean satellites, Titan, and Iapetus have been observed with radar (*see* Planetary Radar).

3.2. Spacecraft Exploration

Interplanetary missions to the planets and their moons have enabled scientists to increase their understanding of the solar system more in the past 35 years than in all of previous scientific history. Analysis of data returned from spacecraft has led to the development of whole new fields of scientific endeavor, such as planetary geology. From the earliest successes of planetary imaging, which included the flight of a Soviet *Luna* spacecraft to the far side of the Earth's Moon to reveal a surface unlike that of the visible side, devoid of smooth lunar plains, and the crash landing of a United States *Ranger* spacecraft, which sent back pictures showing that the Earth's Moon was cratered down to meter scales, it was evident that interplanetary imaging experiments had immense capabilities. Table 34.2 summarizes the successful spacecraft missions to the planetary satellites, not including the Moon.

The return of images from space is very similar to the transmission of television images. A camera records the level of intensity of radiation incident on its focal plane, which holds an array of detectors. A computer onboard the spacecraft records these numbers and sends them by means of a radio transmitter to the Earth, where another computer reconstructs the image.

Although images are the most spectacular data returned by spacecraft, a whole array of equally valuable experiments are included in each scientific mission. For example, a gamma-ray spectrometer aboard the lunar orbiters was able to map the abundance of iron and titanium on the Moon's surface. The *Voyager* spacecraft included an infrared spectrometer capable of mapping temperatures; an UV spectrometer; a photopolarimeter, which simultaneously measured the color, intensity, and polarization of light; and a radio science experiment that was able to measure the pressure of Titan's atmosphere by observing how radio waves passing through it were attenuated. The *Cassini* spacecraft includes sensitive spectrometers covering wavelengths from the UV to the far infrared.

The *Pioneer* spacecraft, which were launched in 1972 and 1973 toward an encounter with Jupiter and Saturn, returned the first disk-resolved images of the Galilean satellites. But even greater scientific advancements were made by the *Voyager* spacecraft, which returned thousands of images of the satellite systems of all four outer planets, some of which are shown in Section 4. Color information for the objects was obtained by means of six broadband filters attached to the camera. The return of large numbers of images with resolution down to a kilometer has enabled geologists to construct geologic maps, to make detailed crater counts, and to develop realistic scenarios for the structure and evolution of the satellites.

Further advances were made by the *Galileo* spacecraft, which was launched in 1990 and began obtaining data at Jupiter in 1996. The mission consisted of a probe that explored the Jovian atmosphere and an orbiter designed to make several close flybys of the Galilean satellites. The orbiter contained both visual and infrared imaging devices, a UV spectrometer, and a photopolarimeter. The visual camera was capable of obtaining images with better than 20-m resolution. The spacecraft was intentionally crashed into Jupiter in September 2003 to avoid possible contamination with Europa in the future. (Europa is thought to have a subsurface ocean that may be an appropriate habitat for primitive life.) The *Cassini/Huygens* mission to Saturn was launched in 1997 and entered into orbit around Saturn in 2004 for at a 4-year in-depth study of the planet, its rings, satellites and magnetosphere. Its instruments include a camera, an imaging spectrometer, infrared and UV spectrometers, a radar system, and a suite of fields and particles experiments. In January 2005, the spacecraft jettisoned its *Huygens* probe into Titan's atmosphere where it parachuted to the surface; valuable data on the **ionosphere**, atmosphere, and surface of this unique world were obtained. The prime mission was followed by a 2-year extended mission, and then by a 7-year solstice mission, during which scientists planned to continue a study of seasonal changes in the system as well as changes in activity on Enceladus (*see* Planetary Exploration Missions).

4. INDIVIDUAL SATELLITES

4.1. The Satellites of Mars: Deimos and Phobos

Mars has two small satellites, Phobos and Deimos (fear and terror), which were discovered by the American astronomer Asaph Hall in 1877. They were named after the attendants of Mars in Greek mythology. The two bodies are barely visible in the scattered light from Mars in Earth-based telescopes. Most of what is known about Phobos and Deimos was obtained from the *Mariner 9, Viking 1* and *2, and Mars Express* missions (see Table 34.2). Their physical and orbital properties are listed in Table 34.1. Both satellites are shaped approximately like ellipsoids, and they are in **synchronous rotation**. Phobos, and possibly Deimos, has a **regolith** of dark material similar to that found on D or CM-type (**carbonaceous**) asteroids common in the outer asteroid belt. This similarity poses problems for theories of the satellites' origin, as their regular orbits suggest a formation around Mars rather than capture from asteroidal orbits. Late-stage addition of more primitive materials from an impact of one or more objects originating beyond the orbit of Mars, for example, from the asteroid belt or even the Kuiper Belt, might alleviate this problem.

Both satellites are heavily cratered, which indicates that their surfaces are at least 3 Billion years old (Figure 34.2). Deimos is distinctively smooth, possibly because of ejecta from its large, southern crater, possibly in combination with different regolith mechanical properties from Phobos. The surface of Phobos is extensively scored by linear grooves that define several planes parallel to the intermediate axis of Phobos (the leading—trailing direction) and also cut the huge impact crater Stickney (named after the Asaph Hall's wife, Angeline Stickney Hall, who collaborated with him in many of his astronomical observations). The groove origins

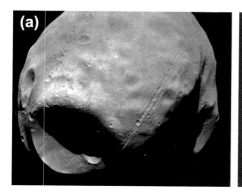
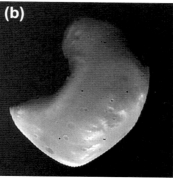

FIGURE 34.2 The two moons of Mars. (a) Phobos and (b) Deimos. Both pictures were obtained by the *Viking* Spacecraft.

remain controversial, possibly involving fractures or effects of secondary debris on Phobos' regolith. Tidal action is shrinking Phobos' orbit; the satellite will either disintegrate (perhaps to form a ring) or crash into Mars in about 100 Million years.

4.2. The Small Satellites of Jupiter

As late as 1979, Jupiter had only a dozen known small satellites (in addition to the Galilean satellites), including three discovered by the *Voyager* mission. As of November 2012, the planet had 67 known satellites; most of the new small satellites were discovered by sensitive CCD cameras and large telescopes. The small satellites are irregular in shape, and many may be captured objects.

Within the orbit of Io are at least four satellites: Thebe, Amalthea, Adrastea, and Metis (see Figure 34.3). Amalthea is a dark, reddish, heavily cratered object reflecting less than 5% of the visible radiation it receives; the red color is probably due to contamination by sulfur particles from Io. Little else is known about its composition except that the dark material may be carbonaceous. In addition to two large craters, Pan (100 km wide) and Gaea (80 km wide), Amalthea has mountains that are about 20 km high. In 2002, the *Galileo* spacecraft swooped to within 150 km of the moon's surface to find that its density is anomalously low, about that of water ice (1.0 g/cm^2). This result suggests that Amalthea has much internal void space and may be a "rubble pile" composed of an agglomeration of debris reaccreted from a collision long ago.

Adrastea and Metis, both discovered by *Voyager*, are the closest known satellites to Jupiter and move in nearly identical orbits just outside the outer edge of the thin Jovian ring, for which they may be a source of material. Between Amalthea and Io lies the orbit of Thebe, also discovered by *Voyager*. Little is known about the composition of these satellites, but they are most likely primarily rock—ice mixtures. The three inner satellites sweep out particles in the Jovian **magnetosphere** to form voids at their orbital positions.

Exterior to the Galilean satellites, there is a class of four satellites moving in highly inclined orbits (Lysithea, Elara, Himalia, and Leda). They are dark objects, reflecting only 2% or 3% of incident light, and may be similar to C- and **D-type** asteroids. Another family of objects is the outermost four satellites, which also have highly inclined orbits, except that they move in the retrograde direction around Jupiter. They are Sinope, Pasiphae, Carme, and Ananke, and they may be captured asteroids.

The 47 additional satellites discovered since 2000 are all small outer satellites less than about 10 km in diameter, the majority of which have retrograde orbits, indicating they are probably captured objects. These objects suggest a large reservoir of additional undiscovered satellites.

4.3. The Saturnian System

4.3.1. The Medium-Sized Icy Satellites of Saturn (Other than Enceladus): Rhea, Dione, Tethys, Mimas, and Iapetus

The Saturnian system contains 62 known satellites. Excluding the giant Titan, the six largest satellites of Saturn are smaller than the Galilean satellites but still sizable—as such they represent a unique class of icy satellite. Earth-based telescopic measurements showed the spectral signature of ice for all six satellites. The satellites' low densities and high albedos (see Table 34.1) imply that their bulk composition is largely water ice, possibly combined with ammonia or other volatiles. They have smaller amounts of rocky silicates than the Galilean satellites. Resurfacing has occurred on several of the satellites. Most of what is presently known of the Saturnian system was obtained from the *Voyager* flybys in 1980 and 1981 and the *Cassini—Huygens* exploration of Saturn's satellites. *Cassini* images of the main satellites of Saturn are shown in Figure 34.4.

The innermost medium-sized satellite Mimas is covered with craters, including one (named Herschel) with a size nearly one-third of the satellite's diameter, or about 130 km, and 10 km deep (see upper left of Figure 34.5). The craters on Mimas tend to be high-rimmed, bowl-shaped pits; apparently, surface gravity is not sufficient to have caused slumping. No components other than water ice have been detected on Mimas. Through orbital resonances, Mimas is responsible for many of the gaps in Saturn's ring, including the Cassini Division. Mimas is also in a mean-

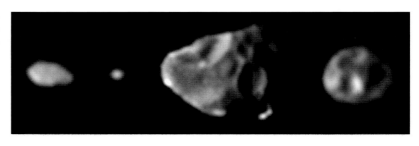

FIGURE 34.3 Four small satellites of Jupiter: Metis, Adrastea, Amalthea, and Thebe, shown to their correct relative sizes. All four satellites orbit between Jupiter's ring and the Galilean satellites.

FIGURE 34.4 The major satellites of Saturn as imaged by the *Cassini* spacecraft. They are, from the upper left corner clockwise around Saturn: Enceladus, Hyperion, Titan, Tethys, Rhea, Mimas, Dione, and at the lower left corner, Iapetus.

motion resonance of 3:2 with Pandora, and a 2:1 resonance with Tethys. The infrared instrument on *Cassini* discovered a global, arcuate-shaped thermal anomaly that is thought to be due to increased thermal inertia, possibly from annealing by high-energy electrons.

Tethys is covered with impact craters, including Odysseus, which is shown in the upper left of Figure 34.6. The craters tend to be flatter than those on Mimas or the Moon, probably because of viscous relaxation and flow over the eons under the stronger gravitational field of Tethys. Evidence for resurfacing episodes is seen in regions that have fewer craters and higher albedos and in the huge trench formation, the Ithaca Chasma. This chasm is about 2000 km in length, 100 km in width, and 3 km deep. It may have formed during a global freezing and subsequent expansion of Tethys. The density of Tethys is lower than that of the other main Saturnian moons—only 0.98 g/cc—which implies a very small fraction of nonice material and possibly some internal void space. It is not known whether Tethys is differentiated, and no surface components other than water ice have been identified. A thermal anomaly similar to the one on Mimas appears on its surface, but it is less distinct.

Dione, which is about the same size as Tethys, exhibits a wide diversity of surface morphology. Most of the surface is heavily cratered (Figure 34.7), but gradations in crater density indicate that several periods of resurfacing occurred during the first billion years of its existence. The leading side of the satellite is about 25% brighter than the trailing side, possibly due to more intensive micrometeoritic bombardment on this hemisphere, in addition to the accretion of E-ring particles. Bright wispy streaks seen by the *Voyager* spacecraft were revealed by *Cassini* to be deep and extensive tectonic faults, or chasmata. Dione modulates the radio emission from Saturn, but the mechanism for this phenomenon is unknown. There is some evidence that Dione has been recently active, including geologic features that appear to be cryovolcanic and that are surrounded by smooth plains. The detection of an anomaly in Saturn's magnetic field can be explained by an outflow of material from Dione equal to 0.6% of the atmospheric loading by material from Enceladus.

Rhea is superficially very similar to Dione. Bright wispy streaks cover one hemisphere: as on Dione, *Cassini*

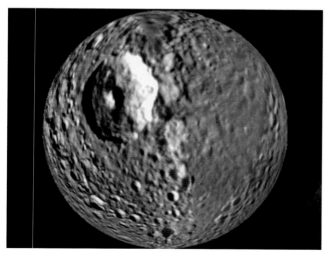

FIGURE 34.5 *Cassini* photomosaic of Mimas, showing its heavily cratered surface and the Herschel crater, which is 139 km in diameter. The bluish tinge occurs in the region of unusual thermal properties, which may be due to the impact of high-energy electrons on the moon's surface.

FIGURE 34.6 The best Cassini spacecraft view of the Odysseus impact crater on Tethys, which is about 450 km in diameter.

FIGURE 34.7 A *Cassini* image of Dione, showing both heavily cratered terrain and the bright wispy streaks, which are raised tectonic features up to several hundred meters in height.

FIGURE 34.8 *Cassini* image of Iapetus, showing both bright and dark terrains. The image was obtained at a distance of 173,000 km, with a resolution of about 2 km.

showed they are tectonic features formed by extensional forces. There is a dichotomy between crater sizes on Rhea's surface, with some regions lacking craters larger than 40 km in diameter, while other regions have craters only larger than that. The larger craters may be due to a population of larger debris more prevalent during an earlier episode of collisions. A subsequent resurfacing may have covered up these features on a portion of the moon.

The most recent evidence from the *Cassini* spacecraft is that Rhea is an almost undifferentiated body in hydrostatic equilibrium. The spacecraft also detected a thin atmosphere with a surface pressure only a few trillionths that of the Earth and composed of oxygen and carbon dioxide. The discovery in 2008 of possible rings around Rhea has not been confirmed by later observations.

When G. D. Cassini discovered Iapetus in 1672, he noticed that at one point in its orbit around Saturn it was very bright, but on the opposite side of the orbit it nearly disappeared from view. He correctly deduced that one hemisphere is composed of highly reflective material and that the other side is much darker. *Voyager* images show that the bright side, which reflects about 60% of the incident visible sunlight from its brightest regions, is fairly typical of a heavily cratered icy satellite. The other side, which is centered on the direction of motion, is coated with a material with a reflectivity of about 3–4% (Figure 34.8). Other aspects of Iapetus are notable. It is the only large Saturnian satellite in a highly inclined orbit, and its density is only slighter higher than that of water ice, implying that very small amounts of rock exist in its interior.

Historically, two models were proposed for the origin of the dark material: that it was derived from an exogenic (external) source or that it was endogenically (internally) created. The discovery of a large dust ring around Saturn, consisting of material ejected from Phoebe due to meteoritic impacts, provided a plausible source for exogenous particles. The redder color of Iapetus implies that some nonvolatile thermal native lag deposit and perhaps material from other sources is mixed with the low-albedo material from the Phoebe ring. The pattern of the low-albedo side is consistent with material being swept up in the orbit of Iapetus and subsequently altered by thermal migration, The *Cassini* visual and infrared spectrometer detected carbon dioxide and organics in the dark material.

The surface of Iapetus is heavily cratered, implying an age >4 Ga, with at least five large-impact basins between 300 and ~800 km in diameter—the largest in the Saturnian system—and associated with landslides. The shape is less spherical than would be expected from its current rotation period of 79 days: it is more in agreement with an original rotation period of 10 h, which may have been its spin period as the crust cooled. The moon also has a unique equatorial ridge up to 20 km in height that extends nearly the entire circumference of the body; in the bright regions, the ridge breaks up into scattered mountains (see Figure 34.8). No completely satisfactory theory for the origin of this ridge exists.

4.3.2. The Small Satellites of Saturn

The Saturnian system has many unique small satellites. Telescopic observations showed that the surface of Hyperion, which lies between the orbits of Iapetus and Titan, is covered with ice. Because Hyperion has a visual **geometric albedo** of 0.30, this ice must be mixed with a significant amount of darker material rich in rocky or carbonaceous substances. Hyperion's highly irregular shape (see Table 34.1) suggests, along with the satellite's battered appearance, that it has been subjected to intense bombardment and fragmentation. Hyperion is the only satellite known to be in chaotic rotation—perhaps a collision within the last few million years knocked it out of a tidally locked orbit (*see* Chaotic Motion in the Solar System).

Saturn's outermost satellite, Phoebe, a dark object (see Table 34.1) with a surface composition similar to that of **C-type** asteroids (but apparently with more organic material), moves in a highly inclined, retrograde orbit, strongly suggesting that it is a captured object. The spectral signature of water ice was detected by ground-based telescopes. Although it is smaller than Hyperion, Phoebe has a more nearly spherical shape that could indicate very early heating and relaxation. Figure 34.9 shows an image of the satellite obtained by *Cassini–Huygens* in July 2004. The heavily battered surface has a number of unusual conical craters, and the largest crater reveals ice cliffs on its rims. Carbon dioxide and organic material was detected by *Cassini*, suggesting that the satellite formed in the outer solar system, perhaps as far out as the Kuiper Belt.

Saturnian small satellites occur in several dynamic niches: the shepherding satellites, the co-orbitals, the **Lagrangians**, and satellites that orbit in ring gaps. All these objects are irregularly shaped (see Figure 34.10) and probably consist primarily of ice. Those whose masses have been measured are less dense than water ice, suggesting substantial porosity. The three shepherds, Atlas, Pandora, and Prometheus, help define the edges of Saturn's A- and F-rings, but recent work shows that resonances among satellites also define the sharp edges of the rings. The orbit of Atlas lies several hundred kilometers from the outer edge of the A-ring. The other two shepherds, which orbit on either side of the F-ring, not only constrain the width of this narrow ring but also may cause its kinky appearance. *Cassini* discovered two new small satellites that are associated with the F-ring.

The co-orbital satellites Janus and Epimetheus, which were discovered in 1966 and 1978, exist in an unusual dynamical situation. They move in almost identical orbits at about 2.5 Saturn radii. Every 4 years the inner satellite (which orbits slightly faster than the outer one) overtakes its companion. Instead of colliding, the satellites exchange orbits. The 4-year cycle then begins over again. Perhaps

FIGURE 34.9 *Cassini* image of Phoebe, obtained at a distance of 32,500 km, with a corresponding resolution of about 300 m. The heavily cratered surface shows no hint of geological resurfacing. Icy cliffs are evident in the large crater at the top of the image, and the large crater near the center shows evidence for layering near its rim.

these two satellites were once part of a larger body that disintegrated after a major collision.

Four other small satellites of Saturn orbit in the **Lagrangian points** of larger satellites: two are associated with Dione (including Polydeuces, which was discovered by *Cassini* in October 2004) and two with Tethys. The Lagrangian points are locations along an object's orbit in which a less massive body can move in an identical, stable orbit. They lie about 60° in front of and at the back of the larger body. Although no other known satellites in the solar system are Lagrangians, the Trojan asteroids orbit in two of the Lagrangian points of Jupiter.

The final class of unusual Saturnian satellite is the one that dwells in ring gaps and sweeps and clears particles from the gaps. Pan, which was discovered in 1990 from *Voyager* images, sits in the Encke Gap. The *Cassini* spacecraft discovered Daphnis, a small satellite in the Keeler Gap, in May 2005.

Like Jupiter, Saturn has a large family of outer irregular satellites, most of which have been recently discovered with large telescopes. The 25 known outer small satellites move in eccentric inclined orbits, and most of their orbits

FIGURE 34.10 Some of the small satellites of Saturn. From the left they are: Pan, Atlas, Prometheus (top), Pandora (bottom), Epimetheus (top), Janus (bottom), Pallene, Methone, Calypso (top), Telesto (bottom), and Helene.

are retrograde, implying that they are captured objects. The farthest satellites orbit more than 20 million km from Saturn.

4.4. The Satellites of Uranus

4.4.1. The Medium-Sized Satellites of Uranus: Miranda, Ariel, Umbriel, Titania, and Oberon

Uranus has a total of 27 known satellites that can be grouped into three types: small inner moons, small outer irregular moons, and a main satellite system composed of medium-sized bodies that orbit the planet at distances from 130,000 to 583,000 km. The orbits of Ariel, Umbriel, Titania, and Oberon are regular, whereas Miranda's orbit is slightly inclined. Figure 34.11 is a telescopic image of the satellites typical of the quality attainable before the advent of spacecraft missions.

Theoretical models suggest that the satellites are composed of water ice, possibly in the form of methane clathrates or ammonia hydrates, and silicate rock. Water ice has been detected spectroscopically on all five satellites. Carbon dioxide has been detected on Ariel, Umbriel, and Titania. The relatively dark visual albedos of the satellites, ranging from 0.13 for Umbriel to 0.33 for Ariel (see Table 34.1), and gray spectra indicate that their surfaces contain a dark component such as graphite or carbonaceous material. Another darkening mechanism that may be important is bombardment of the surface by UV radiation. The higher density of Umbriel implies that it has a larger fraction of rocky material than the other four satellites. Heating and **differentiation** have occurred on Miranda and Ariel and possibly on some of the other satellites. Models indicate that tidal interactions may provide an important heat source in the case of Ariel.

Miranda, Ariel, Oberon, and Titania all exhibit large opposition surges, indicating surfaces composed of very porous material, perhaps resulting from eons of micrometeoritic "gardening". Umbriel lacks a significant surge, which

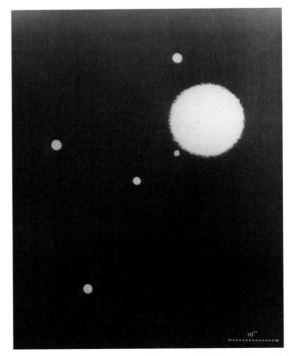

FIGURE 34.11 Telescopic view of Uranus and its five satellites obtained by Ch. Veillet on the 154-cm Danish-ESO telescope. Outward from Uranus they are: Miranda, Ariel, Ubriel, Titania, and Oberon. *Photograph courtesy of Ch. Veillet.*

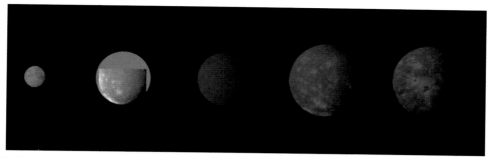

FIGURE 34.12 The five major satellites of Uranus, shown to relative size based on *Voyager 2* images. They are, from the left, Miranda, Ariel, Umbriel, Titania, and Oberon.

suggests that its surface properties are in some way unusual. Perhaps its **regolith** is very compacted, or it is covered by a fine dust that scatters optical radiation in the forward direction.

The *Voyager 2* spacecraft encountered Uranus in January 1986 to reveal satellites that have undergone melting and resurfacing (Figure 34.12). Three features on Miranda, known as "coronae", consist of a series of ridges and valleys ranging from 0.5 to 5 km in height (Figure 34.13). Their origin is uncertain: some geologists favor a compressional folding interpretation, whereas others invoke a volcanic origin or a faulting origin. Both Ariel, which is the geologically youngest of the five satellites, and Titania are covered with cratered terrain transected by grabens, which are fault-bounded valleys. Umbriel is heavily cratered and it is the darkest of the satellites: both these observations suggest that its surface is very old, although the moderate-resolution images obtained by *Voyager 2* cannot rule out geologic activity on a smaller scale. Some scientists have in fact interpreted small albedo variegations on its surface as evidence for melting events early in its history. Oberon is similarly covered with craters, some of which have very dark deposits on their floors. On its surface are situated faults or rifts, suggesting resurfacing events (*Voyager 2* provided ambiguous, medium-resolution views of the satellite). In general, the Uranian satellites appear to have exhibited more geologic activity than the Saturnian satellites (except Enceladus) and Callisto, possibly because of the presence of methane, ammonia, nitrogen, or additional volatiles, as well as more rocky material that provided heat from radioactive decay.

There is some evidence that Umbriel and Oberon, as well as certain regions of the other satellites contain the organic-rich primordial constituents that seem to be ubiquitous in the outer solar system. This material is found on primitive asteroids, and may contain some of the same molecules that are seen on the dark side of Iapetus, on Hyperion, and on specific areas of the larger satellites.

4.4.2. The Small Satellites of Uranus

Voyager 2 discovered 10 new small satellites of Uranus, including two that act as shepherding satellites for the outer (epsilon) ring of Uranus (see Table 34.1). All these satellites lie inside the orbit of Miranda. Images of two satellites, Puck and Cordelia, provided sufficient resolution to directly determine their radii (see Table 34.1). The sizes of the other bodies were estimated by assuming that their values of surface brightness are equal to those of the other inner satellites and calculating the projected area required to yield their observed integral brightness. Puck is slightly nonspherical in shape. It is likely that the other small satellites are irregularly shaped. The satellites' visual **geometric albedos** range from 0.04 to 0.09, which is slightly higher than that of Uranus's dark ring system. No reliable color information was obtained by *Voyager 2* for any of the small satellites,

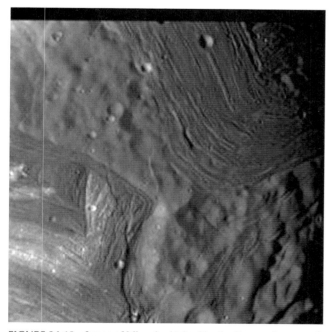

FIGURE 34.13 Image of Miranda obtained by the *Voyager 2* spacecraft at 30,000–40,000 km from the Moon. Resolution is 560–740 m. Older, cratered terrain is transected by ridges and valleys, indicating more recent geologic activity.

although their low albedo suggests that they are **C-type** objects. Ground-based observers have discovered another 12 outer irregular satellites to bring the total of known Uranian satellites to 27.

4.5. The Satellites of Neptune

4.5.1. The Inner Moons

Neptune has 14 known satellites: one is the large moon Triton and the remaining 13 are small, irregularly shaped bodies (see Table 34.1). The small satellites can be divided into two categories: the seven inner bodies, which move in highly regular, circular orbits close to Neptune (<5 planetary radii), and the irregular outer satellites. Triton has an appreciable atmosphere, seasons, and currently active geologic processes (*see* Triton).

Six of the seven small inner satellites were all discovered within a few days during the *Voyager 2* encounter with Neptune in August 1989. They were given names of mythical nautical figures by the IAU. For four of these satellites (Proteus, Larissa, Galatea, and Despina), as well as Nereid, *Voyager 2* images provided sufficient resolution to determine their dimensions (see Table 34.1). All five bodies are irregularly shaped. The sizes of Thalassa and Naiad were derived by assuming that their albedos are equal to those of the other inner satellites. The size of the six satellites discovered by *Voyager 2* increases with the distance from Neptune. Proteus is the largest known satellite with an irregular shape. The seventh inner moon, S/2004 N 1, was discovered on archived *Hubble Space Telescope* Images in 2013.

Spacecraft tracking of the six inner satellites observed by *Voyager 2*, ground-based observations of Nereid, and *Hubble* observations of the seventh inner moon provided accurate orbit determinations, which are listed in Table 34.1. All the small inner satellites except Proteus orbit inside the so-called synchronous distance, which is the distance from Neptune at which the rotational spin period equals the Keplerian orbital period. However, Proteus has been tidally despun so that its rotational period equals its orbital period.

The masses of the satellites were not measured directly by *Voyager 2*. Limits may be obtained by assuming reasonable values for their bulk densities. These values range from 0.7 g/cm^3, corresponding to water ice with a bulk porosity of about 30%, to 2 g/cm^3, corresponding to water ice with a significant fraction of rocky material. If the satellites were formed from captured material, the higher density is more reasonable. In any case, the small satellites have less than 1% of the mass of Triton. The ring system of Neptune contains only a very small amount of mass, possibly one-millionth of the small satellites' combined masses.

Figure 34.14 depicts the best *Voyager 2* images obtained for Proteus, with a resolution of 1.3 km/pixel. The large feature—possibly an impact basin—has a diameter of about 250 km. Close scrutiny of this image reveals a concentric structure within the impact basin. Possible ridgelike features appear to divide the surface. The regions of Proteus outside of the impact basin show signs of being heavily cratered. The best image of Larissa was obtained at a resolution of 4.2 km/pixel, insufficient to depict surface features.

Analysis of calibrated, integral *Voyager 2* measurements of the four inner satellites reveals that their geometric albedos are about 0.06, in the *Voyager 2* clear filter with an effective wavelength of about 480 nm. The limited spectral data obtained by *Voyager 2* suggest that Proteus and Larissa are gray objects. The dark albedos and spectrally neutral character of the inner satellites suggest that they are carbonaceous objects, similar to the primitive **C-type** asteroids, possibly the Uranian satellite Puck, the satellites of Mars, and several other small satellites. The evolution of the inner satellites was likely punctuated by the capture of Triton. Initially, the inclinations and eccentricities of the satellites would have been increased by the capture, and subsequent collisions would have occurred. The resulting debris would then have reaccreted to form the present satellites. Models of the collisional history of the satellites suggest that the reaccreted satellites are much younger than the age of the solar system, with the exception of Proteus. The heavily cratered surface that appears in the one resolvable *Voyager 2* image of these bodies (see

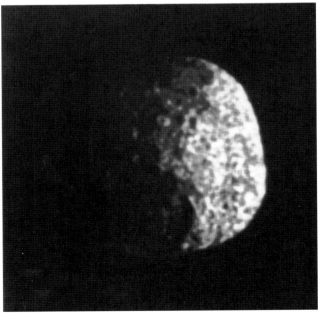

FIGURE 34.14 The best Voyager image of Proteus, obtained at 1.3 km/pixel.

Figure 34.14) does suggest that they have undergone vigorous bombardment.

The only satellite that has been shown to have a dynamical relationship with the rings of Neptune is Galatea, which confines the ring arcs. The orbits of the satellites have probably evolved under the influence of tidal evolution and resonances. For example, the inclination of Naiad is possibly due to its escape from an inclination resonance state with Despina. Larissa was probably detected in 1981 when it blocked a star that astronomers were measuring for possible occultations by planetary rings. Because it was not subsequently observed or tracked, it was not classified as a satellite until its existence was confirmed by *Voyager 2* observations.

4.5.2. The Outer Irregular Moons

The small outer moon Nereid was discovered in 1949 by Gerard P. Kuiper at the McDonald Observatory in Texas. In keeping with the theme of water and oceans for the Neptunian system, the satellite was named after the sea nymphs known in Greek mythology as Nereids. At the time of the *Voyager 2* encounter in 1989, Nereid, which moves in an eccentric orbit bringing it from 57 to 385 planetary radii from Neptune, was the only known small outer satellite of Neptune. Reliable ground-based observations of Nereid were limited to estimates of its visual magnitude. The integral brightness of Nereid is almost three times that of Proteus, which is slightly smaller; its geometric albedo is therefore ~0.20. This markedly higher albedo, coupled with its gray color, implies a surface of water frost contaminated by a dark, spectrally neutral material. It is more similar to the differentiated satellites of Uranus than to the dark C-type objects. It is also similar in albedo, size, and color to Phoebe, the outer satellite of Saturn, which moves in an inclined, retrograde orbit, suggesting that it is a captured object.

Voyager 2 images of Nereid attained 43 km/pixel, not sufficient to clearly depict surface features. In 1988, 1989, and 1991, several ground-based observers reported large light curve amplitudes (up to factors of four), which they interpreted as significant albedo variegations on Nereid. Although there is some uncertainty (a factor of two at most) in comparing *Voyager 2* to ground-based observations because the orientation of Nereid's spin axis is unknown, a light curve for Nereid from *Voyager 2* images over a 12-day period shows an amplitude of less than 15%. No rotation period has yet been derived for this moon, although the *Voyager 2* observations suggest that it is not in **synchronous rotation**. Nereid has the largest eccentricity of any known satellite (0.75).

In 2003, five more moons were discovered, including one with a period of over 25 years, which corresponds to a distance of 48 million km from Neptune. Three of the five outer satellites discovered in 2003 have retrograde orbits, and may thus be captured objects. Jupiter, Saturn, and Neptune all appear to have families of outer, captured satellites.

ACKNOWLEDGMENT

Portions of this work were performed at the Jet Propulsion Laboratory, California Institute of Technology, under contract with the National Aeronautics and Space Administration. P. Smith provided technical assistance.

BIBLIOGRAPHY

Beatty, J. K., Petersen, C. C., & Chaikin, A. (Eds.). (2000). *The new solar system* (4th ed.). Cambridge, Mass: Sky Publishing.

Belton, M. J. S., & the Galileo Science Teams. (1996). *Science, 274,* 377−413.

Bergstralh, J., & Miner, E. (Eds.). (1991). *Uranus*. Tucson: Univ. Arizona Press.

Burns, J., & Matthews, M. (Eds.). (1986). *Satellites*. Tucson: Univ. Arizona Press.

de Pater, I., & Lissauer, J. (2010). *Planetary sciences* (2nd ed.). Cambridge, UK: Cambridge Univ. Press.

Dougherty, M., Esposito, L., & Krimigis, S. (2009). *Saturn from Cassini-Huygens*. New York: Springer.

Fraeman, A. A., & 10 colleagues. (2012). *Journal of Geophysical Research (Planets), 117,* E00J15.

Gehrels, T. (Ed.). (1984). *Saturn*. Tucson: Univ. Arizona Press (Up-to-date listing of satellite discoveries and physical and dynamical properties) http://nssdc.gsfc.nasa.gov/planetary/planetfact.html.

Hartmann, W. K. (2004). *Moons and planets* (4th ed.). Belmont, Calif: Wadsworth.

Lang, K. R. (2011). *The Cambridge guide to the solar system*. Cambridge: Cambridge University Press.

Stone, E., & the Voyager Science Teams. (1989). *Science, 246,* 1417−1501.

Willner, K., et al. (2010). *Earth and Planetary Science Letters, 294,* 541−546.

Chapter 35

Io: The Volcanic Moon

Rosaly M.C. Lopes
Jet Propulsion Laboratory, California Institute of Technology, Pasadena, CA, USA

Chapter Outline

1. Introduction 779
2. Io Exploration 780
3. Io's Surface 783
4. Io's Volcanic Eruptions 787
5. Heat Flow and Interior 788
6. Atmosphere, Torus, and the Jupiter Environment 790
7. Outstanding Questions and Future Exploration 791
Bibliography 792

Io, the innermost of Jupiter's four Galilean satellites, is the only body outside the Earth so far known to have large-scale active volcanism. Io's heat flow is much higher than that of the Earth's and its interior is thought to contain a magma ocean. Io's lavas may be hotter than any that erupt on the Earth's surface today.

iron in Io's interior than in the Moon's. Io's mantle composition is thought to be predominantly **silicates**. However, sulfur compounds are abundant on the surface. After the discovery of volcanism in 1979, a major question was the composition of the erupting material: silicates or sulfur? Ground-based observations and, later, Galileo's

1. INTRODUCTION

Io (Figure 35.1) was discovered by the Italian scientist and astronomer Galileo Galilei on January 8, 1610, and was named after one of the ancient Roman god Jupiter's illicit lovers. The discovery of active volcanism was made by the Voyager 1 spacecraft, which flew close to Io in 1979. Images showed volcanic plumes up to 300 km in height and a vividly colored surface dominated by large **caldera**-like and flowlike features. The study of Io's remarkable volcanism has continued since then using observations by telescopes on the Earth, by the Hubble Space Telescope, and the Galileo, Cassini, and New Horizons spacecraft.

Io's unusual spectroscopic characteristics, due to its volcanic activity and widespread covering of sulfur dioxide, were recognized in the 1970s. In 1979, just prior to the two Voyager flybys, Io's 4:2:1 orbital resonance with Europa and Ganymede was predicted to induce severe tidal heating and subsequent active volcanism on Io. The two Voyager spacecraft confirmed the prediction that Io is volcanically active.

Io's size (Table 35.1) is similar to that of the Earth's Moon but its density is higher, indicating that there is more

FIGURE 35.1 Io imaged by Galileo's Solid State Imaging System on September 7, 1996, at a range of about 487,000 km. The image is centered on the side of Io that always faces away from Jupiter. The black and bright red materials correspond to the most recent volcanic deposits. The near-infrared filter makes Jupiter's atmosphere (in the background) look blue. The active volcano Prometheus is seen as a dark sinuous feature near the right-center of the disk.

TABLE 35.1 Io's Basic Orbital and Physical Properties
Mean radius: 1821.6 ± 0.5 km
Bulk density: 3528 ± 3 kg/m^3
Orbital period: 1.769 days
Orbital eccentricity: 0.0041
Orbital inclination: 0.037
Orbital distance a: 421,800 km
Rotational period: synchronous with orbit
Maximum moment of inertia: 0.3769 ± 0.0004
Potential love number k^2: 1.292 ± 0.003
Mass: $(8.9320 \pm 0.0013) \times 10^{22}$ kg
Surface gravity: 1.80 m/s^2
Global average heat flow: >2.5 W/m^2
Radius of core: 656 km (if pure iron) and 947 km (iron and iron sulfide mixture)
Surface equatorial magnetic field strength: <50 nT
Geometric albedo: 0.62
Local topographic relief: up to ~17 km
Active volcanic centers: at least 166
Typical surface temperature (away from hot spots): 85 K (night) to 140 K (day)
Atmospheric pressure: <10^{-9} bar, higher at locations of plumes
Escape velocity: 2.56 km/s
Crustal thickness (estimated): 30–50 km
Atmospheric composition: SO_2 (main), SO, S_2, Na

results showed that eruption temperatures were too high to be sulfur and were consistent with silicates, but some sulfur flows may exist.

Sulfur dioxide is ubiquitous on Io's surface, and sulfur and sulfur dioxide are known to be present in Io's volcanic plumes.

Io's **heat flow** (Table 35.1) is very large compared with that of the Earth and other planets. Io's heat flow is about 200 times what could be expected from heating due to the decay of radioactive elements, illustrating how crucial tidal heating is to driving Io's active volcanism. The effect of Io's volcanic eruptions extends well beyond the surface, and there is considerable interaction of Io with the Jovian magnetic field. Io has both a patchy, very low density atmosphere and an ionosphere. Sulfur dioxide is the main constituent of the atmosphere, and it is thought to be supplied largely by volcanic plumes, with a lesser amount coming from sublimation of frost deposits on the surface.

An important discovery made during the Galileo mission was Io's aurora, caused by collisions between Io's atmospheric gases and energetic charged particles trapped in Jupiter's magnetic field.

Materials escaping from Io form a cloud of neutrals along Io's orbital path. Escaping materials also populate the Io torus, a donut-shaped region along Io's path, held by Jupiter's powerful magnetic field. Materials escaping from Io include sulfur, sulfur dioxide, sodium, potassium, and dust particles. Io is therefore a wonderful natural laboratory for the study of geological and geophysical processes, and its location within Jupiter's magnetic field makes it a rich source for studies of the fields and particle environment in space.

2. Io EXPLORATION

Since its discovery in 1610, Io has been important to our understanding of the solar system, along with the other Galilean satellites, Europa, Ganymede, and Callisto. After a few observations, Galileo Galilei concluded that the four objects were not stars as he originally thought, but satellites in orbit around Jupiter. Galileo's studies of the motion of the newly discovered satellites had a profound effect on human history becoming, along with Galileo's discovery of the phases of Venus, key evidence in favor of the Copernican theory of the universe. Another major step for science came in 1675, when Danish astronomer Olaus Roemer noted that the times of the eclipses and occultations of the four moons by Jupiter showed a phase shift with a periodicity of about 6.5 months. He concluded that, when Jupiter is at opposition (when Jupiter and Earth are closest, on the same side as the Sun), light from the Jovian system must travel a distance of approximately 4 astronomical units (AU) to reach the Earth. However, when Jupiter is at conjunction (when the Earth and Jupiter are farthest apart, on opposite sides of the Sun), light traveled about 6 AU on its journey to the Earth. Roemer concluded that this phase shift in the arrival time of jovicentric events meant that light has a finite velocity, and he used the motions of the Galilean satellites to determine the speed of light.

In 1805, Laplace demonstrated that the Galilean satellites have an orbital configuration (known as the Laplace resonance), which suggested a special dynamical relationship among Io, Europa, and Ganymede. For each time that Ganymede orbits Jupiter, Europa orbits almost exactly twice and Io orbits four times. Later studies would reveal that this resonance plays a key role in the existence of active volcanism on Io.

Even before the first close-up images of Io were returned by Voyager in 1979, there were indications that Io was a remarkably different world from our Moon and other moons in the solar system. Telescopic observations showed that Io's brightness varied according to its position in its orbit, suggesting that the moon always keeps one face toward Jupiter (now referred to as the sub-Jovian hemisphere). During the mid-twentieth century, photometric and color data showed that Io is the reddest object in the solar system and has a marked color variation with orbital phase angle. These observations also showed Io to be very different from the other Galilean satellites (and most other satellites in the outer solar system) because of the absence of water bands in its spectra.

The peculiar nature of Io's surface became more evident in 1964, when astronomers A. P. Binder and D. P. Cruikshank reported an anomalous brightening of Io's surface as it emerged from eclipse. This first report of "posteclipse brightening" and the suggestion of a possible atmosphere spurred more telescopic observations but, even though the presence of an atmosphere was confirmed, posteclipse brightening has remained controversial and has not been confirmed to this day.

The first evidence of an electromagnetic link between Io and the Jovian magnetosphere was put forward in 1964 by E. K. Bigg, who found that bursts of decametric radio emission by Jupiter were apparently controlled by Io's orbital position. Models of electrodynamic interaction between Jupiter and Io addressed the coupling mechanism between Io and Jupiter's inner magnetosphere.

The first spacecraft to fly by the Jupiter system was Pioneer 10 in 1973. These observations revealed that Io has an ionosphere and thin atmosphere. Pioneer measurements also showed a cloud of neutrals along Io's orbital path. Ground-based measurements in the mid-1970s revealed ionized sulfur emission in the inner Jovian magnetosphere, but on the opposite side of Jupiter from the position of Io at the time. Subsequent studies revealed this to be a plasma torus. The Io torus is a donut-shaped trail along Io's orbital path, made up almost exclusively of various charged states of sulfur and oxygen, thought to be derived from the break up of volcanic sulfurous compounds (SO_2 and S_2). The ionized particles are held within the torus by Jupiter's magnetic field, in a similar way to the mechanism that holds charged particles in the Van Allen radiation belts around the Earth.

The first clues to Io's bulk composition came from measurements of Io's mean radius using a stellar occultation and from mass derived from the Pioneer flyby in 1973. The bulk composition of 3.54 g/cm^3 indicated that silicates were dominant on Io, but the surface's high albedo and cold temperatures indicated frosts. Telescopic observations using improved spectral reflectance techniques were used to attempt to determine Io's surface composition, and polysulfides were suggested as a possible coloring agent for the surface. The idea of sulfur on Io was strongly supported by laboratory experiments by W. Wamsteker, which showed that sulfur and its compounds matched the strong UV absorption and reflectance spectrum of Io, suggesting that these compounds might be abundant on Io's surface. However, the discovery of a strong absorption band near 4 μm could not be explained by sulfur. It was later found to be due to sulfur dioxide (SO_2), which is now known to be the dominant compound covering Io's surface. Other key discoveries during the 1970s were those of the Io sodium cloud in 1973 by R. Brown and in 1975 of a potassium cloud by L. Trafton.

The first indications of volcanic activity were given by infrared photometry and radiometry that showed higher brightness temperatures at 10 μm than at 20 μm, but the thinking at the time was that Io was a cold and dead world, and these observations remained puzzling. However, shortly before Voyager 1 arrived at the Jupiter system in March 1979, several scientists published results that, in retrospect, suggested active volcanism. In 1978, R. Nelson and B. Hapke reported a spectral edge at 0.33 μm and proposed that sulfur was the major contributor to this spectral feature. They suggested that the presence of allotropes of sulfur could explain this and several other spectral features and that these allotropes could be produced by melting yellow sulfur and subsequently quenching it, possibly "in the vicinity of a volcanic fumarole or hot spring". Astronomers F. Witteborn and colleagues reported a telescopic observation of an intense temporary brightening of Io in the infrared wavelengths from 2 to 5 μm. They explained it, although with some skepticism, as thermal emission caused by part of Io's surface being at a temperature of about 600 K, much hotter than the average expected daytime temperature of about 130 K. A few days before the Voyager 1 flyby of Io, a seminal theoretical paper by Stan Peale and colleagues was published. They had studied the tidal stresses generated within Io as a result of the gravitational "tugs" from Jupiter and Europa. Their calculations showed that the possible heat generated by tidal stresses was on the order of 10^{13} W, much greater than heat that could be released from normal radioactive decay. Their prediction—that Io might have "widespread and recurrent volcanism"—was spectacularly confirmed by Voyager 1.

Io's tidal heat comes from the orbital energy of the Io—Jupiter system (resulting in orbital acceleration), whereas dissipation of energy in Jupiter causes Io's orbital motion to decelerate. If Europa and Ganymede were not interacting with Io, its orbit would be circular and tidal deformation would not occur. Due to Laplace resonance between the satellites, Io's orbit is elliptical and Jupiter's differential gravitational pull causes Io to continuously and repeatedly deform as it orbits Jupiter. It is estimated that Io's surface moves up and down by as much as 100 m over one orbit around Jupiter. If Laplace resonance is ever broken, as some studies by V. Lainey and colleagues indicate, then Io will, in the distant future, become dormant.

Active volcanoes were not immediately obvious in the first images returned by Voyager 1. The most striking aspect of Io shown in the first images was its colorful surface, with yellows, oranges, reds, and blacks. Scientists on the imaging team nicknamed Io the "pizza moon" and suggested the colors were likely due to large quantities of sulfur on the surface. Another surprising aspect was the absence of impact craters. The obvious conclusion was that Io's surface was very young and the craters must have been obliterated—but how? The answer came soon after, when a navigation engineer at the Jet Propulsion Laboratory, Linda Morabito, noticed a peculiar umbrella-shaped feature emanating from Io's limb in one of the images that was taken to aid navigation of the spacecraft (Figure 35.2). The pattern turned out to be an eruption plume rising about 260 km above the surface. A second plume was found on the same image, and more plumes were seen on close examination of various other images. Additional evidence for active volcanism came from another of Voyager's instruments, the infrared interferometer spectrometer (IRIS), which detected enhanced thermal emission from parts of Io's surface—some areas had temperatures of about 400 K, much higher than the rest of the surface, which has noontime equatorial temperatures of about 107–124 K. When one of the hot areas was found to coincide with one of the plumes, there was no doubt that active volcanism was taking place.

About 18 weeks after Voyager 1's dramatic discovery, the companion spacecraft, Voyager 2, flew close to Io.

FIGURE 35.3 Voyager 1 image showing the Loki plume on the limb and the heart-shaped Pele plume deposit in the lower part of the image. When Voyager 2 arrived 18 weeks later, the "heart" had become an oval, as material from the plume had filled out the area.

Intense activity was still taking place, but significant changes had occurred between the two flybys, including the cessation of the largest plume, Pele, and the altered shape of the deposits associated with this plume (Figure 35.3). An area of about 10,000 km^2 had been filled in, presumably by fresh material falling down from the plume. It became evident that dramatic changes of Io's surface could occur over short timescales.

Initial analysis of the Voyager observations showed nine plumes and nine hot spots (active volcanoes), although not all plumes coincided with hot spots, and vice versa. "Hot spot" is a term used by Io researchers to define a region of enhanced thermal emission, a sign of active volcanism. The Voyager IRIS experiment did not observe the whole surface, so it was suspected that other hot spots existed. The surface showed many features with morphologies similar to volcanic landforms on other planets, such as calderas (volcanic craters) and flows.

After the two Voyager spacecraft left the Jupiter system on their way to Saturn and beyond, the study of Io's volcanism was continued from the Earth by astronomers using infrared detectors mounted on telescopes. These observations showed that brightenings and fadings of hot spots occur, indicating variations in the level of volcanic activity. Telescopic observations were also used to analyze the reflected light from Io's surface to determine surface composition, confirming that it was dominated by sulfur

FIGURE 35.2 Io's active volcanoes were discovered from this image, taken by Voyager 1 on March 8, 1979, looking back 4.5 million km. The Pele plume is seen on the lower right, rising nearly 300 km above the surface. The bright spot near the terminator (shadow between day and night) is the top of the Loki plume, illuminated by the Sun.

dioxide (SO_2). Io was also observed by the International Ultraviolet Explorer satellite and by the Hubble Space Telescope.

The first spacecraft to orbit Jupiter was Galileo, which was able to image Io and monitor its volcanic activity from 1996 through early 2002. Galileo was designed to orbit the planet Jupiter for 2 years (1996–1997) and collect data on the planet's atmosphere, its moons, rings, and magnetic field. However, the failure of Galileo's high-gain antenna to deploy (a problem discovered while the spacecraft was on its way to Jupiter) drastically reduced the quantity of images and data that could be returned to the Earth. The mission still accomplished its objectives, thanks to the successful reconfiguration of the spacecraft's software to utilize a lower gain antenna and perform data compression on board. These measures, along with changes made in the Deep Space Network, maximized the amount of data that could be returned. Galileo observations were so successful and spectacular that two mission extensions to gather additional data were approved. The Galileo Europa Mission lasted from 1998 to 1999 and the Galileo Millennium Mission from 2000 to 2002. These extensions were particularly important for Io because all the high-resolution remote sensing observations obtained by Galileo were collected during these mission extensions. The close Io flybys during which high-spatial-resolution remote sensing observations were collected happened in October and November 1999, February 2000, and August and October 2001. The main remote sensing instruments observing Io were the solid-state imaging system (SSI), the near-infrared mapping spectrometer (NIMS), and the photopolarimeter radiometer (PPR).

Io was observed by the Cassini spacecraft on its way to Saturn in 2000 and in 2007 by the New Horizons spacecraft on its way to Pluto. Both spacecraft detected large plumes and New Horizons detected a previously unknown hot spot. Continued monitoring from ground-based telescopes continued to find variations in the level of activity, including those of Io's most powerful hot spot, Loki, which exhibits brightenings that switch on in 1 month or less and last several months before fading.

3. Io'S SURFACE

Io's phenomenal volcanic activity makes it the most geologically active object in the solar system. Remote sensing observations from Galileo revealed its surface in unprecedented detail and substantially changed our understanding of Io's geology and geophysics. Galileo's remote sensing instruments (visible and infrared) were used to study the surface features and volcanic activity, while the tracking of the spacecraft itself provided new constraints on the interior. Gravity measurements from tracking indicated that Io has a large iron/iron sulfide core and a silicate mantle. Galileo's close flybys of Io failed to reveal an intrinsic magnetic field, suggesting that little core convection is taking place.

The surface of Io contains three primary types of features (Figure 35.4): (1) broad, flat, layered plains, which are partially covered with visible, diffuse **pyroclastic materials**; (2) volcanic structures including **paterae** (caldera-like depressions), **flucti** (lava flow fields), and **tholi** (shield volcanoes and other positive-relief structures); and (3) mountains of volcano–tectonic origin. The complementary imaging coverage of Galileo and Voyager has allowed these features to be mapped in a global scale, thus giving us a window into not only local but also global processes.

Between the paterae, mountains, and other major geologic features, Io's surface appears smooth except for scarps that cut across the plains. Some scarps are linear and occur in parallel groups, which suggests a tectonic origin. Other scarps, however, are irregular and appear to be erosional, sometimes forming a series of mesas or large plateaus. The presence of these features on Io is somewhat puzzling because of the lack of a significant atmosphere or flow of liquid water. Sulfur and sulfur dioxide, possibly escaping explosively from a subterranean "aquifer", have been suggested as the main eroding agent on Io's surface.

Volcanic features dominate Io's surface, and the volcanoes cover a wide range of sizes and present varying characteristics such as power output, persistency of activity, and association with plumes. Interestingly, most of Io's volcanoes manifest themselves as caldera-like depressions, referred to as paterae. Unlike terrestrial volcanoes, those on Io rarely build large topographic structures such as tall shields (like Mauna Loa) or stratovolcanoes (like Mount St Helens). There are only a few tholi scattered across Io.

Io's surface shows a few remarkably large flucti. The lava flow field from the Amirani volcano is ∼300 km long, the largest active flow field known in the solar system. Io's large lava flows are possibly analogs of the continental flood basalt lavas on the Earth, such as the Columbia River Basalts in the United States. These ancient terrestrial flows were never directly observed, but they are suspected of producing major climatic effects.

A major question about Ionian volcanism after Voyager was the nature of volcanism—whether sulfur or silicates were predominant. Although temperature measurements from Galileo clearly showed that many hot spots have temperatures far too high for sulfur, the possibility that some sulfur flows occur on the surface cannot be ruled out. At the time of the Voyager flybys, Carl Sagan argued that the colorful flows around Io's Ra Patera volcano were sulfur (Figure 35.5). Unfortunately, the flows could not be studied by Galileo as the area had been covered over by new eruptions before Galileo's first observations in 1996. However, other locations may have sulfur flows. Most Ionian flows appear dark, but a few locations show pale

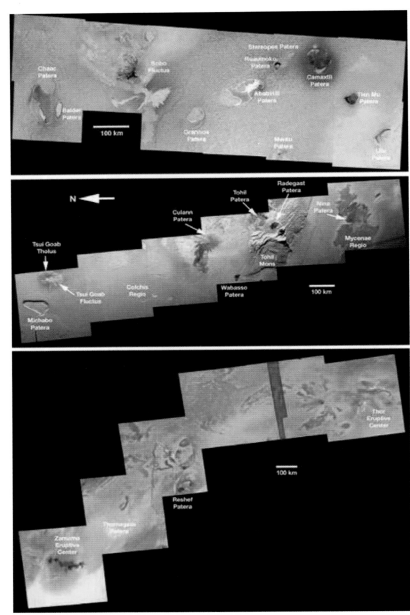

FIGURE 35.4 Mosaics of images acquired by Galileo's camera of three regions that accentuate the different types of geologic materials and terrains on Io. (Top) The Chaac-Camaxtli region shows paterae in various sizes, shapes, and colors, indicating varying volcanic and tectonic influences on their formation *(from Williams et al., 2002)*. (Middle) The Culann–Tohil region, which contains paterae, fluctii (lava flow fields), a mountain (Tohil Mons), and a volcanic construct (Tsui Goab Tholus) *(from Williams et al., 2004)*. (Bottom) The Zamama–Thor region, dominated by two eruptive centers, Zamama and Thor. Zamama has a long lava flow field and Thor was the site of the tallest eruptive plume seen on Io. *From Williams et al., 2005 (Figure courtesy of David Williams, Alfred McEwen, and Moses Milazzo).*

yellow or white flows that may well be sulfur. D. Williams and colleagues proposed that flows radiating from Emakong Patera may be sulfur and that low-temperature liquid sulfur (~450 K) could explain many of the morphological features seen around Emakong Patera, such as a meandering channel 105 km in length that appears to feed a gray–white flow some 270 km in length. Infrared measurements using Galileo NIMS indicated temperatures less than 400 K inside Emakong caldera, and much cooler conditions (below the instrument's detection capabilities) over the flows. However, Galileo's instruments could not distinguish between sulfur flows and cooled silicates coated by bright sulfurous materials after erupting. One possibility, suggested by R. Greeley and colleagues, based on studies of a sulfur flow at Mauna Loa in 1984, is that rising silicate magma may melt sulfur-rich country rock as it

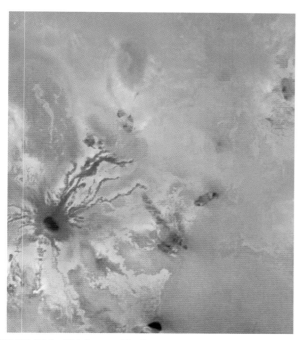

FIGURE 35.5 This image of Ra Patera volcano was taken by Voyager 1 at a range of 128,500 km (77,100 miles). The width of the picture is about 1000 km. Ra Patera is the dark spot with the irregular radiating pattern of flows, which were interpreted as being sulfur.

nears the surface, producing "secondary" sulfur flows (as opposed to "primary" flows that originate from molten magmas at depth). Sulfur dioxide is ubiquitous on Io, and the colorful surface is thought to be the result of **sulfur allotrope** deposits, making the possibility of secondary sulfur flows likely. Galileo data of the volcanoes Balder and Tohil Paterae suggest that sulfur dioxide could be mobilized as "flows" in very cold regions. However, the presence of sulfur and SO_2 flows on Io has not been confirmed and whether these flows exist on Io remains an open question.

The most common type of volcanic feature on Io is the patera. Although the origin of paterae is still somewhat uncertain, they are thought to be similar to terrestrial volcanic calderas, formed by collapse over shallow magma chambers following partial removal of magma. Some paterae show angular shapes that suggest some structural control, indicating that they may be structural depressions that were later used by magma to travel to the surface. At least 400 Ionian paterae have been mapped. Their average diameter is ~40 km, but Loki, the largest patera known in the solar system, is over 200 km in diameter. In contrast, the largest caldera on Earth, Yellowstone, is ~80 km by 50 km in size. The larger sizes of the Ionian features probably reflect the much larger sizes of magma chambers.

Mountains are the major structural landforms on Io and tower over the surrounding plains. Ionian mountains are defined as steep-sided landforms rising more than ~1 km over the plains. At least 115 mountains have now been identified and mapped. Io's mountains rise, on an average, about 6 km high, with the highest rising 17 km above the surrounding plains. Galileo images revealed that many mountains are partly or completely surrounded by debris aprons, plateaus, and layered plains. Mountains appear to be unstable and are thought to be relatively short-lived features. They are not active volcanoes, but their origin is still uncertain. Various models have been proposed to explain the origin of the mountains. Their asymmetrical shapes suggest the uplift and rotation of crustal blocks, implying that compressional uplift is probably the dominant mechanism.

Neither the volcanic features nor the mountains appear to follow a distinct global pattern such as seen on the Earth, suggesting that, on Io, surface expressions of internal dynamics are subtle. However, the distribution of mountains and paterae is not totally random; both features are concentrated toward lower latitudes and follow a bimodal distribution with longitude. According to P. Schenk and colleagues, the greatest frequency of mountains occurs in two large antipodal regions near the equator at about 65° and 265°. In contrast, J. Radebaugh and colleagues studied the distribution of Ionian patera, and although they found the paterae to follow a similar bimodal distribution, the highest concentrations are 90° out of phase with that of the mountains. When only the hot spots known to be currently or recently active are studied, their distribution appears random, although no active (or inactive) volcanic center has been detected at latitudes greater than 78°. The distribution pattern for volcanic centers is consistent with the pattern of heat flow from tidal heating in Io's asthenosphere predicted from simulations. The anticorrelation in the distribution of mountains and volcanic centers is further evidence that the two are not related, but the reasons for the anticorrelation are still unknown.

Because of the dynamic nature of Io's volcanism, its surface appearance can change in dramatic ways over time. Detectable changes occurred in the years between the Voyager and Galileo observations (1979–1995); however, many surface changes at the timescale of months have also been detected. One example is the change in the Pele plume deposit between the two Voyager flybys, which were spaced about 4 months apart. Surface changes are mostly due to new volcanic eruptions, particularly sulfur and sulfur dioxide from volcanic plumes and pyroclastic (ash and **tephra**) deposits. Other changes include new lava flows, increases in the area of flows, and changes in surface color. Volcanic materials have been observed to fade or disappear due to burial, alteration, radiation exposure, or erosion. Most surface changes have been detected at visible wavelengths; however, within individual volcanic centers, changes in temperature and sulfur dioxide coverage have been detected at infrared wavelengths. Most surface changes are localized and take place inside dark volcanic paterae that cover only 1.4% of Io's surface, or are ephemeral volcanic plume deposits that fade or change color on timescales of a few months to years. One surprise from the first Galileo observations was that Io's surface appearance remained largely the same since the last Voyager flyby. Based on the changes observed between the two Voyager flybys (4 months apart), major changes were expected in the years between Voyager and Galileo. Instead, more than 90% of Io's surface remained unchanged between Voyager (1979) and the end of the prime Galileo mission (1999). New Horizons imaged almost all of Io in 2007, at conditions suitable for comparison with previous surface maps, and detected only one-fourth of the number of surface changes detected by Galileo. Although this is partly due to the limited spatial resolution of the images, it is consistent with previous results indicating that Io's surface remains largely unchanged.

Localized changes from major eruptions, however, can be dramatic. Some of these were particularly useful in the study of surface changes from Galileo. The eruption of the Pillan volcanic center in 1997 left a conspicuous "black eye" on Io's surface (Figure 35.6), covering an area of about 200,000 km^2 and reaching distances up to 260 km from the source (Figure 35.6). Later observations from Galileo's SSI showed a spectral absorption at 0.9 μm in these and other dark materials on Io, suggesting silicate composition (possibly magnesium-rich orthopyroxene). The dark deposit at Pillan slowly faded between 1997 and 1999 as it was covered by red sulfurous deposits from nearby Pele.

Surface colors are the most easily observed manifestations of surface change. Galileo results brought new insights into the intriguing question of what causes the vivid colors of Io's surface. The global distribution of the different color deposits gives some clues to their origin and Galileo's repeated flybys allowed observations at different illumination angles, which affect how colors appear in images. Io's surface has four primary color units: most of the surface is yellow (about 40%), white—gray (about 27%), or red—orange (about 30%), while black deposits are localized around volcanic centers. Red and orange materials are interpreted as deposits of short-chain sulfur molecules (S_3, S_4). These are concentrated at latitudes higher than 30° north and south and, where they are thought to result from the breakdown of sulfur (cyclo-S_8), by charged particle irradiation. These red deposits at high latitudes appear to last longer than those at equatorial regions. At lower latitudes, patches of red materials are associated with hot spots and plumes and are thought to be formed by condensation from sulfur-rich plumes. These red plume deposits are ephemeral, lasting perhaps a few years if the deposit is not replenished.

The yellow materials that cover a lot of the surface are interpreted to be sulfur (cyclo-S_8), with or without a covering of sulfur dioxide (SO_2) frosts deposited by plumes, or alternatively polysulfur oxide and sulfur dioxide without large quantities of elemental sulfur. White—gray materials are interpreted to be composed of coarse- to moderate-grained sulfur dioxide that condensed from plumes and later recrystallized. Black areas (<2% of

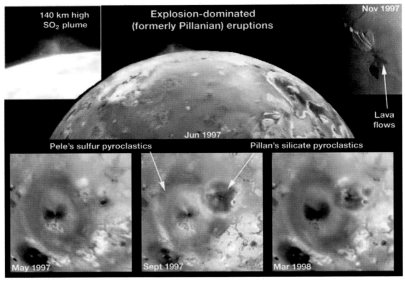

FIGURE 35.6 Explosion-dominated or Pillanian eruptions on Io occur in relatively brief (few months or less), intense outbursts that produce very high (possibly ultramafic) temperatures, plumes (top left), and rapidly emplaced lava flows (top right). Plumes can reach great heights (several hundred kilometers), and their deposits have produced black (Pillan), red (Pele, Tvashtar), and white (Thor) rings. These compositions are thought to be associated with silicate, sulfur, and SO_2 pyroclastic materials, respectively. *Figure courtesy of David Williams.*

surface) mostly correlate with active hot spots and occur as patera floors, as lava flow fields, or as dark diffuse materials near or surrounding active vents. These materials are consistent with magnesium-rich orthopyroxene, indicative of silicate lava flows or lava lakes (within paterae) or diffuse silicate pyroclastic deposits near paterae. Perhaps the most intriguing materials on Io's surface are the small greenish yellow deposits seen in a few isolated patches in or near active vents, which are thought to be composed of either sulfur compounds contaminated by iron or silicates such as olivine or pyroxene with or without sulfur-bearing contaminants.

Detection of other substances on Io's surface has been difficult because sulfur dioxide condensed from volcanic plumes blankets most of the surface and hinders detection of other species. Galileo NIMS detected a broad absorption at about 1 µm, which had been seen from telescopic observations. However, it is still not known what substance this spectral absorption is due to, although NIMS observations showed that it is anticorrelated with recently emplaced lavas. NIMS also detected local patches of almost pure SO_2, in one case, in Balder Patera, topographically confined, raising the possibility that it was emplaced as a fluid.

4. Io'S VOLCANIC ERUPTIONS

Shortly after the Voyager mission, the major controversy about Io's volcanic activity concerned the nature of the volcanism: sulfur or silicates? Io's surface colors were interpreted as sulfur deposits and this, among other factors, made the sulfur volcanism hypothesis attractive. One way to distinguish between sulfur and silicate volcanism is to measure the temperature of the molten material because sulfur has a lower melting temperature than silicate lavas. Sulfur volcanism would not produce temperatures exceeding ~700 K (427 °C), whereas basaltic lavas on the Earth range from 1300 to 1450 K (1027—1177 °C). The temperatures of the hot spots measured by the Voyager IRIS instrument were relatively low (below ~650 K) and could be consistent with either molten sulfur or silicates. However, Voyager instruments lacked the sensitivity and wavelength coverage needed to detect small areas at higher temperatures; hence, Voyager was "seeing" only the cooler areas, perhaps cooling silicate lava flows. Between the Voyager observations in 1979 and the Galileo observations that started in 1996, several of Io's hot spots were detected by ground-based telescopes. Temperature measurements using infrared detectors mounted on telescopes showed higher temperatures than had been measured by IRIS—such as 900 K reported by T. Johnson and colleagues in 1988, and 1225 and 1500 K reported by G. Veeder and colleagues in 1991. These measurements are consistent with silicate magmas but not with sulfur volcanism.

Galileo included much more sensitive instruments than Voyager, such as the SSI sensitive from ~400 to 1000 nm wavelengths and the NIMS sensitive from 700 to 5200 nm, but both had limitations. SSI was able to detect only spots hotter than ~700 K and only when Io was in eclipse (in Jupiter's shadow) to eliminate reflected and scattered light. NIMS had the ideal spectral coverage for detecting both the temperatures and spectral reflectances expected from silicate lavas, but it had limited spatial resolution (120 km or more) except during the close Io flybys. However, Galileo's instruments soon showed the hot spot temperatures to be indeed consistent with silicate rather than sulfur volcanism. The greatest surprise was the detection of very high temperature volcanism on Io when Galileo's NIMS and SSI instruments observed a vigorous eruption at Pillan in 1997 (Figure 35.6). The results, reported by A. McEwen and colleagues in 1998, provided evidence of temperatures exceeding 1500 K at several hot spots and, in the case of the Pillan eruption, temperatures of about 1800 K. The Pillan eruption temperatures are higher than any seen on lavas erupting on Earth now or in recent times and suggested compositions similar to those of ancient lavas on Earth. More recent analysis of the same data for the Pillan eruption suggests lower temperatures, about 1600 K, which could be due to unusually hot (superheated) basalts. Magma can be superheated by rapid ascent from a deep, high-pressure source. Melting temperatures of dry silicate rocks increase with pressure; therefore, the erupted lava can be significantly hotter than its melting temperature at surface pressure. Rapid ascent of basaltic magmas resulting in ~100 K of superheating should be possible. However, no record of such an eruption is known on the Earth.

Io eruption temperatures measured from observations of other hot spots by Cassini and New Horizons have been in the temperature range of basalts. The composition of Io's lavas remains questionable, as the temperatures measured could be consistent either with unusually hot basalts or with ultramafic (komatiite-like) lavas that had cooled by a couple of hundred degrees Celsius at the time the measurements were made. Komatiites and komatiitic basalts are ultramafic volcanic rocks on Earth that are rich in magnesium and dominated by olivine or pyroxene. These lavas have very rarely erupted on the Earth since the Proterozoic, about 1.8 billion years ago. Therefore, if Io's lavas are indeed ultramafic, studying Io's current volcanism may lead to a better understanding of the emplacement of lavas on the ancient Earth.

It is important to note that at present there are no direct measurements of the composition of Io's lavas. The most critical question about Io's volcanism—the composition of the erupting magma and crust—remains open.

Not all Ionian eruptions are vigorous like Pillan. On Earth, volcanic eruptions are often classified depending on their character—effusive, explosive, very explosive—and

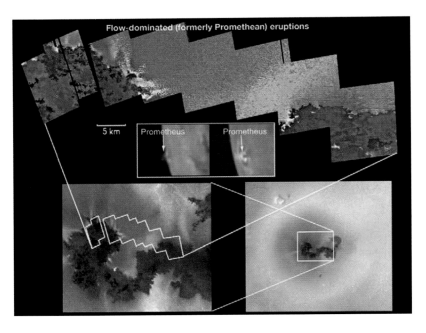

FIGURE 35.7 Flow-dominated or Promethean eruptions are relatively long lived (months to years) and are associated with long-lived plumes and flow fields. In this montage, Galileo SSI views of the Prometheus plume (center) are surrounded by increasingly higher resolution views of the Prometheus flow field. Note also the bright plume deposit forming an annulus in the lower right image.

these eruptions are often named after volcanoes or locations where they have been studied (e.g. Hawaiian, Icelandic, Strombolian, Pelean). High-resolution observations and temporal data on Io's hot spots showed that some broad generalizations can also be made for Ionian eruptions. The majority of Ionian eruptions can be placed in three classes initially designated "Promethean", "Pillanian", and "Lokian", although a single hot spot can exhibit more than one eruption style over time. Explosion-dominated ("Pillanian") eruptions (Figure 35.6) have an intense, short-lived phase that may correspond to the outbursts detected from the Earth. These eruptions originate from either paterae or fissures and produce extensive dark lava flow fields and dark pyroclastic deposits through short-lived, high-effusion-rate, vigorous activity. These events may or may not include eruption episodes with large (>200 km high) explosive plumes, which can produce large plume deposits such as that around the Pillan itself.

Less intense but more persistent flow-dominated eruptions ("Promethean") are named after the Prometheus hot spot (Figure 35.7), which has a persistent plume about 100 km high, active during both Voyager encounters in 1979, throughout the Galileo mission, and spacecraft observations since then. Surprisingly, distant images obtained by Galileo in 1996 showed that the Prometheus plume site had moved about 80 km west since 1979 but that its size and appearance had not changed. A new lava flow linked the old and new plume sites. Images and infrared observations obtained in 1999 showed that the main vent of this volcano was near the Voyager plume site and that the plume, not the volcano, had moved west. The plume's movement was modeled in terms of the interaction between the advancing hot lava and the underlying sulfur dioxide snowfield by Susan Kieffer and colleagues. The movement of lava flows on Earth over marshy ground can give rise to small, short-lived explosive activity, but nothing on the scale of the Prometheus plume has ever been observed. This type of eruption may be common on Io and, once the flow stops moving, the plume eventually shuts off, as has been observed at the Amirani volcano. The lava flows associated with these eruptions can be quite extensive and are thought to be emplaced through repeated small breakouts of lava, similar to the slowly emplaced flow fields at Kilauea in Hawaii.

Intrapatera ("Lokian") eruptions (Figure 35.8) are confined within the caldera-like paterae. These eruptions are thought to be lava lakes, some of which are possibly overturning. Observations from Galileo flybys showed that lava lakes are abundant on Io, and they may be a significant mechanism for heat loss from the interior. Io's most powerful hot spot, Loki, is thought to be a giant lava lake that perhaps undergoes periodic overturning, leading to brightenings that have been observed from the Earth for decades. Many other hot spots on Io appear to be persistent lava lakes.

5. HEAT FLOW AND INTERIOR

Observations of Io have also provided knowledge about the satellite's interior, where the tidal heat is being dissipated, driving the volcanic eruptions. Observations from the Earth and spacecraft have shown that the heat that comes

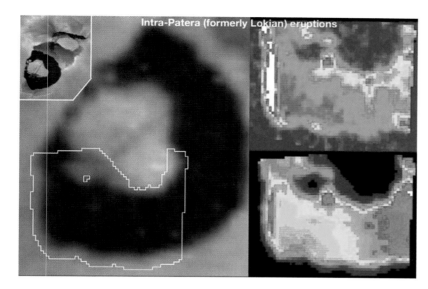

FIGURE 35.8 Intrapatera or Lokian eruptions are confined to paterae (caldera-like depressions) and are thought to represent the resurfacing of paterae floors by lava flows or overturning lava lakes. The style of eruption is exemplified by Loki (inset upper left, Voyager image). In this montage, Galileo NIMS images track the temperature changes across the floor of Loki. The top right image is at 2 μm, showing the hottest areas, while the bottom image, at 4 μm, shows the distribution of cooler areas. Note that the center of Loki (which appears white in the SSI image) is cold at infrared wavelengths.

out of Io and is radiated into space, called the heat flow, is very large compared with that of the Earth and other planets. The heat flow is measured at infrared wavelengths and the portion due to reflected sunlight is calculated and subtracted, taking into account Io's surface albedo. The difference is the heat flow due to volcanism and originates in Io's interior.

The first estimate of Io's heat flow was done by D. Matson and colleagues in 1981, who reported a value of 2 ± 1 W/m^2. In 1991, G. Veeder and colleagues reported measurements of Io's heat flow compiled from 10 years of ground-based photometric observations in the range 5–20 μm and estimated a minimum average heat flow of 2.5 W/m^2. The latest estimates of Io's heat flow are from J. Spencer and colleagues in 2002, using Galileo PPR measurements. They reported 2.2 ± 0.9 W/m^2, which is in close agreement with the first estimate by Matson. This range of values is very high even compared to geothermal and volcanic areas on the Earth such as Yellowstone. Io's heat flow is about 200 times what could be expected from heating due to the decay of radioactive elements and illustrates how crucial tidal heating is to drive Io's active volcanism. The heat flow is not uniform over the surface; in fact, it is dominated by the Loki hot spot.

An unresolved problem for Io is that there is a discrepancy between observed values of heat flow and theoretical estimates expected from steady-state tidal heating models over the course of Io's history. The current estimates of heat flow from observations are about twice the predicted value. If the theoretical estimates are correct, then Io's heat flow must have varied over time due to its orbital evolution. Other studies also suggest that Io's current heat flow and tidal heating rate are higher than the long-term equilibrium value. One suggestion is that Io is spiraling slowly inward, losing more energy from internal dissipation than it gains from Jupiter's tidal torque. The resolution of the apparent discrepancy between the observed and theoretical heat flow will have important implications for understanding not only the evolution of Io but also that of Europa and Ganymede.

What is happening deep within Io? Studies of Io's interior, including the lithosphere and mantle, are still in their early stages, but the Galileo spacecraft made some significant contributions. The properties of Io's interior determine how tidal forces deform the body. Therefore, one can use measurements of the deformation (variations in shape) of Io to get information on internal structure. Data obtained from images and radio tracking of the spacecraft as it came close to Io during several flybys provided this information. Variations in the spacecraft's motion revealed distortions in Io's gravitational field, which provided evidence that Io is differentiated.

Io is about the same size as the Earth's Moon, but it has a higher density. On the basis of density alone, it can be inferred that Io has a large metallic core. The size of this core can be inferred from the density and spacecraft measurements of Io's shape, assuming Io is in hydrostatic equilibrium. Work by M. Segatz and colleagues using Voyager observations revealed the basic structure of the interior, Galileo measurements have been used to refine this knowledge. Io is thought to be a two-layer body, consisting of a large metallic core of iron and sulfur and a silicate mantle. Galileo's magnetometer instrument measurements were complex to analyze because of the influence of Jupiter's strong magnetic field and the question of whether

Io had a magnetic field. However, more recent reexamination of the data by K Khurana and colleagues showed that Io has an induced magnetic field. Induced magnetic fields are created when a time-variable magnetic field sweeps through an electrically conducting material, which Khurana and colleagues suggest is a global magma ocean about 50 km beneath Io's surface. A model for Io's interior of a completely molten core and a crystal-rich ("mushy") magma ocean had been suggested by L. Kezthelyi and colleagues a few years earlier in 2004.

Other key measurements made by Galileo, including the discovery of widespread, high-temperature, silicate volcanism and tall mountains have contributed to a model of Io's interior. If the mountains are formed as thrust blocks, then Io's lithosphere must be at least as thick as the tallest mountains (~ 15 km).

The possibility that ultramafic volcanism exists on Io has strong implications for the interior. The idea that Io's crust is ultramafic (magnesium-rich) seems inconsistent with the well-understood process of magmatic **differentiation**. Heat flow on Io is sufficiently high that Io was expected to have undergone partial melting and differentiation hundreds of times, producing a low-density crust, depleted of heavy elements like magnesium (as mantle rocks begin to melt, the first component to melt has a lower density and segregates and rises toward the surface, while the heavier components sink). The "mushy" magma ocean model of Kezthelyi and colleagues suggested that widespread ultramafic volcanism would be present on Io, as the upper mantle would consist of orthopyroxene-rich magma with about the same density as the overlying crust. As lavas are deposited, the crustal layers sink and are eventually mixed back into the magma ocean, so a low-density crust cannot form. However, this model may not allow for sufficient tidal heat generation to occur and thus may be inconsistent with the heat flow observed. Another possibility, suggested by W. Moore and colleagues in 2005, is that local processes such as tidal forcing through cracks may account for the very high temperatures (>1400 K) observed at some hot spots on Io.

6. ATMOSPHERE, TORUS, AND THE JUPITER ENVIRONMENT

Io orbits Jupiter at a distance of about 421,800 km, which is deep within the Jovian magnetopause. Io has both a patchy but relatively large atmosphere and an ionosphere, and there is considerable interaction of Io with the Jovian magnetic field. Io's atmospheric density is low (about 10^{-9} bar), equivalent to good laboratory vacuums on the Earth, but the density is greater at the locations of active volcanic plumes. The main constituent of the atmosphere is SO_2, which is supplied largely by volcanic plumes, with a lesser amount coming from evaporation of the SO_2 frost deposits on the surface. Io's low gravity allows some of the atmosphere to escape, but it is continuously replenished by volcanic outgassing.

Since the time of the Voyager flybys, Io has been known to produce volcanic plumes hundreds of kilometers high, which serve as an efficient delivery mechanism for gas and dust particles into the magnetosphere and the space surrounding Io, although only a relatively minor amount of atmospheric gas is lost to space. The dynamics of Io's plumes are very complex, particularly because models of plume emplacement have to take into account the very low atmospheric pressure on Io.

Large plumes were detected by Galileo, Cassini, and New Horizons. The largest plume known on Io (500 km high) was detected from images obtained in August 2001, shortly after the Galileo spacecraft had flown through it. Observations by the plasma science experiment indicated the presence of SO_2 molecules in the plume. This in situ measurement is consistent with others that show the presence of SO_2 in plumes and SO_2 frost in plume deposits. Sulfur (S_3 and S_4), in addition to SO_2, was detected in the Pele plume from measurements made from the Hubble Space Telescope by J. Spencer and colleagues.

The temperatures of the frost deposits on Io's surface are sufficiently low that cold trapping of SO_2 by condensation is a very important process. Some material does escape Io, forming a corona and neutral clouds, and the Io torus. The corona refers to the region within Io's gravitational pull, where bound and escaping atoms and molecules populate a low-density shell. The neutral clouds of sodium, oxygen, and sulfur extend from the corona to distances of many times the radius of Jupiter.

An important discovery made during the Galileo mission was Io's aurora. The aurora (Figure 35.9) was detected through color eclipse imaging with the camera while Io was in Jupiter's shadow. The vivid colors detected (red, green, and blue) are caused by collisions between Io's atmospheric gases and energetic charged particles trapped in Jupiter's magnetic field. The green and red emissions are probably produced by mechanisms similar to those in the Earth's polar regions that produce terrestrial aurorae. The green (actually yellow) glow comes from emission from sodium ions, whereas the red glow is associated with oxygen ions. The bright blue glows mark the sites of dense plumes of volcanic vapor and may represent the locations where Io is electrically connected to Jupiter via a flux tube. Subsequent observations of Io's aurora by the New Horizons spacecraft as Io went in and out of eclipse by Jupiter showed variations in auroral brightness and morphology.

Observations from the Pioneer spacecraft were the first to reveal a cloud of neutrals along Io's orbital path. The most easily observed of these neutral clouds around Io is

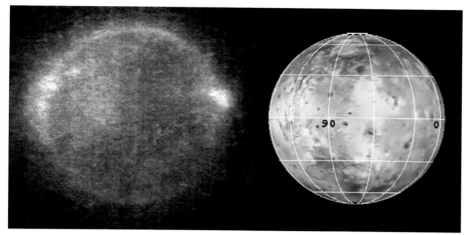

FIGURE 35.9 Io aurora. Bright blue glows represent sulfur dioxide excited by electrical currents flowing between Io and Jupiter in the flux tube. The blue glow on the right is over the Acala hot spot, which is thought to be the site of a "stealth plume" (composed of mostly gas and therefore hard to detect from images). Red and green (actually yellow) glows represent atomic oxygen and sodium, respectively.

the sodium cloud. The cloud is populated by sodium atoms escaping Io at about 2.6 km/s. It appears as a diffuse yellowish emission produced by scattered light from volcanic plumes and Io's lit crescent. This emission comes from neutral sodium atoms within Io's extensive material halo that scatter sunlight at the yellow wavelength of about 589 nm. Although neutral sodium atoms are most easily detectable by spectroscopy, it has been determined through extensive Earth-based telescopic studies that sodium is a minor component of the neutral material escaping from Io. The primary neutral elements in the cloud are oxygen and sulfur, which are thought to have dissociated from sulfur dioxide gas (SO_2) expelled from many of Io's active volcanoes at a rate of ~ 1 t/s. So far, sodium has not been detected on Io's surface or plumes, but its existence is inferred because of its detection in the cloud.

The Io torus is a donut-shaped trail about 143,000 km wide along Io's orbital path. The torus is made up almost exclusively of various charged states of sulfur and oxygen, thought to be derived from the breakup of volcanic SO_2 and S_2. The ionized particles are held within the torus by Jupiter's magnetic field, in much the same way that charged particles are held in the Van Allen radiation belts around the Earth. Measurements made by the Galileo spacecraft during its close flybys showed that the plasma in the torus is slowed by Io's ionosphere, redirected around Io, and then reaccelerated in Io's wake. Other Galileo measurements showed that Io strongly perturbs Jupiter's magnetic field. These perturbations vary with time, suggesting that Io's variable volcanic activity influences the density of the plasma torus and the strength of its interactions with the Jovian magnetic field. (See Planetary Magnetospheres.)

7. OUTSTANDING QUESTIONS AND FUTURE EXPLORATION

The Galileo mission significantly advanced our knowledge of Io, but the failure of the high-gain antenna to open (and subsequent low data rates) prevented all but a very small part of Io's surface to be imaged at high resolution. The last spacecraft to date to fly by Io was New Horizons, which provided a valuable data set but limited in temporal coverage. Future exploration by spacecraft is needed to reveal Io's surface in detail at a variety of different wavelengths and to answer many outstanding questions. The geometry of Galileo's orbit around Jupiter resulted in lack of coverage at high resolution of Io's Jupiter-facing side, which should be a priority for future missions to observe. However, even parts of the surface previously imaged by Galileo and New Horizons are likely to change because of the dynamic nature of Io, so new missions will always reveal new features.

One of the most significant questions raised by Galileo which is still outstanding concerns the nature of Io's high-temperature volcanism. If ultramafic compositions are involved, it is difficult to explain how the magma composition would have stayed ultramafic throughout Io's history because differentiation would have been expected, leading to evolved types of magmas such as those that we find in present-day Earth. It is possible that the current style of volcanic activity is a geologically recent phenomenon (i.e. Io has only recently attained its resonant orbit with resulting tidal heating) or that the response of Io's lithosphere mantle to tidal heating has prevented extreme differentiation. Perhaps the magmas are not ultramafic, but are basaltic, possibly superheated during ascent. Compositional measurements of Io's fresh magmas would be invaluable for future missions to obtain. Another intriguing question considers what the **volatiles** in the magma are dissolved in. The presence of explosive volcanism on Io is evident from the plumes and dark deposits that are thought to be ash and magma fragments. On Earth, the most common volatile is water; on Io, sulfur and sulfur dioxide have been detected in the plumes. But are there other compounds?

Other aspects of Io's geology are also intriguing. How are the mountains formed? Nearly half of Io's mountains are located adjacent to volcanic centers (paterae), but they do not appear to be part of the volcanic system. What can that relationship tell us about Io's crust? Questions also abound about Io's atmosphere and the interaction between Io and the Jovian magnetosphere, particularly the interaction of the flux tube that allows the transfer of charged particles between the two bodies. What are the sources of the atoms, neutrals, and ions that are released into the plasma torus and magnetosphere, and what physical processes allow them to escape? These are just some of the key questions that have developed about Io after Galileo and New Horizons and that require further analysis of existing data sets, and likely new data obtained by future missions. At the time of this writing, there are no missions to Io planned, but the European Jupiter Icy Moons Explorer is expected to observe Io and make significant contributions to outstanding questions. Advances are possible from ground-based and space telescope-based programs. Ground-based observations using Adaptive Optics have been a major step forward in the study of Io because the spatial resolution of these observations can now rival some of those obtained from Galileo. Io has been successfully observed by the Hubble Space Telescope and can potentially be observed from future orbiting telescopes.

BIBLIOGRAPHY

Bagenal, F., McKinnon, W., & Dowling, T. (Eds.). (2004). *Jupiter: Planet, satellites and magnetosphere.* Cambridge, UK: Cambridge Univ. Press.

Geissler, P. E. (2003). Volcanic activity on Io during the Galileo era. *Annual Review of Earth and Planetary Sciences, 31*, 175–211.

Kargel, J. S., et al. (2003). Extreme volcanism on Io: latest insights at the end of the Galileo era. *EOS, 84*(33).

Khurana, K. K., et al. (2011). Evidence of a global magma ocean in Io's interior. *Science, 332*, 1186. doi:1126/science.1201425.

Lainey, V., et al. (2009). Strong tidal dissipation in Jupiter from astrometric observations. *Nature, 459*, 957–959.

Lopes, R., & Williams, D. (2005). Io after Galileo. *Rep. Prog. Phys., 68*, 303–340.

Lopes, R., & Spencer, J. R. (Eds.). (2007). *Io after Galileo.* Praxis Publishing Company (Springer-Verlag).

McEwen, A. S., Lopes-Gautier, R., Keszthelyi, L., & Kieffer, S. W. (2000). Extreme volcanism on Jupiter's moon Io. In J. Zimbelman, & T. Gregg (Eds.), *Environmental effects on volcanic eruptions: From deep oceans to deep space* (pp. 179–204). New York: Plenum.

Morrison, D. (Ed.). (1982). *Satellites of Jupiter.* Univ. Arizona Press.

Nelson, R. M. (1997). Io. In J. H. Shirley, & R. W. Fairbridge (Eds.), *Encyclopedia of planetary sciences* (pp. 345–351). Chapman and Hall.

Retherford, K. D., et al. (2007). Io's atmospheric response to eclipse: UV aurorae observations. *Science, 318*, 237–240.

Spencer, J. R., & Schneider, N. M. (1996). Io on the eve of the Galileo mission. *Annual Review of Earth and Planetary Sciences, 24*, 125–190.

Spencer, J. R., et al. (2007). Io volcanism seen by new horizons: a major eruption of the Tvashtar volcano. *Science, 318*, 240–243.

Williams, et al. (2002). Geological Mapping of the Chaac-Camaxtli Region of Io from Galileo Imaging Data. *J. Geophys. Res., 107*, 5068.

Williams, et al. (2004). Geologic Mapping of the Culann-Tohil Region of Io from Galileo Imaging Data. *Icarus, 169*, 80–97.

Williams, et al. (2005). The Zamama-Thor Region of Io: Insights from a Synthesis of Mapping, Topography, & Galileo Spacecraft Data. *Icarus, 177*, 69–88.

Chapter 36

Europa

Louise M. Prockter
Johns Hopkins University Applied Physics Laboratory, Laurel, MD, USA

Robert T. Pappalardo
Jet Propulsion Laboratory, California Institute of Technology, Pasadena, CA, USA

Chapter Outline

1. Introduction and Exploration History — 793
2. Formational and Compositional Models — 794
3. Internal Structure, Tides, and Global Tectonics — 795
   - 3.1. Europa's Internal Structure — 795
   - 3.2. Tidal Evolution — 795
   - 3.3. Tidal Heating — 796
   - 3.4. Diurnal Stressing — 796
   - 3.5. Nonsynchronous Rotation — 797
4. Landforms on Europa — 797
   - 4.1. Ridges, Troughs, and Bands — 797
     - 4.1.1. Troughs — 798
     - 4.1.2. Double Ridges and Ridge Complexes — 798
     - 4.1.3. Cycloidal Structures — 798
     - 4.1.4. Triple Bands — 799
     - 4.1.5. Pull-Apart Bands — 799
     - 4.1.6. Folds — 800
   - 4.2. Lenticulae and Chaos — 801
     - 4.2.1. Lenticulae — 801
     - 4.2.2. Chaos — 803
     - 4.2.3. Impact Structures — 805
5. Surface Composition and Thermal State — 806
6. Surface Physical Processes — 807
7. Surface Age and Evolution — 808
   - 7.1. Surface Age — 808
   - 7.2. Surface History and Geological Evolution — 809
8. Astrobiological Potential — 810
9. Future Exploration — 810
Bibliography — 811

1. INTRODUCTION AND EXPLORATION HISTORY

Europa and her sibling satellites (See also chapters on Io, Ganymede and Callisto, and Planetary Satellites) were famously discovered by Galileo in 1610, and more controversially by Simon Marius at essentially the same time, but it took almost four centuries before any detailed views of their surfaces were seen and the grandeur of the **Galilean satellites** was revealed. In the 1960s, ground-based telescopic observations determined that Europa's surface composition is dominated by water ice, as are most other solid bodies in the far reaches of the solar system.

The Pioneer 10 and 11 spacecraft flew by Jupiter in the 1970s, but the first spacecraft to image the surfaces of Jupiter's moons in detail were the Voyager twins. Voyager 1's closest approach to Jupiter occurred in March 1979, and Voyager 2's, in July of the same year.

Both Voyagers passed farther from Europa than from any of the other Galilean satellites, with the best imaging resolution limited to 2 km per **pixel**. These images revealed a surface brighter than that of most other moons, crossed with numerous bands, ridges, cracks, and a surprising lack of large impact craters or high-standing topography (see Galileo spacecraft image in Figure 36.1). Despite the distance from which the images were acquired, they were of sufficiently high resolution that researchers noted some of the dark bands had opposite sides that matched each other extremely well, like pieces of a jigsaw puzzle. These cracks appeared to have separated, and **ductile** icy material

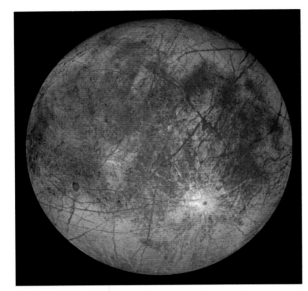

FIGURE 36.1 Global view of Europa's trailing hemisphere, acquired early in the Galileo spacecraft's tour of the Jovian system. The colors in this image have been enhanced to emphasize detail. This image shows the extent of the reddish-brown material that mottles Europa's surface, along with numerous linear features, many over 1000 km long. Two large impact features are also visible: Pwyll crater, surrounded by bright rays, is thought to be the youngest large crater on Europa, and the Callanish multiringed structure is the large circular feature toward the west (left).

TABLE 36.1 Properties of Europa

Discovered	1610
Discoverers	Galileo Galilei, Simon Marius
Mean distance from Jupiter	671,100 km
Radius	1560.8 km
Mass	4.79955×10^{22} kg
Density	3.013 g/cm^3
Orbital period (sidereal)	85.224 h (3.551 Earth days)
Rotational period	85.224 h (3.551 Earth days)
Synodic period	11.23 h (0.4649 Earth days)
Orbital eccentricity	0.0094
Orbital inclination	0.466°
Visual geometric albedo	0.67
Escape velocity	2.026 km/s
Spacecraft visitors	Voyager 1 (March 1979)
	Voyager 2 (July 1979)
	Galileo (July 1996 to 2000)
Predicted average surface temperature	~50 K (poles) to ~110 K (equator)

appeared to have flowed into the opened gaps. This suggested that the surface could have once been mobile. The relative youth of Europa's surface is demonstrated by the lack of large impact structures—Voyager images suggested only a handful. These are expected to build up over time as a planetary surface is constantly bombarded by asteroidal and cometary debris over billions of years, until the surface is covered in craters. A lack of craters implies that something has erased them—such as volcanic (or in Europa's case, **cryovolcanic**) flows or tectonic features, or perhaps **viscous relaxation** of the icy crust. Researchers studying the Voyager data also noted that the patterns of some of the longest linear features on the surface did not fit with predicted simple models of global stresses that might arise from tidal interactions with Jupiter. However, if the shell was rotated backward a few tens of degrees, the patterns fit well to a model of **nonsynchronous rotation**, by which the icy surface had slowly migrated eastward with respect to the satellite's tidal axes. This mechanism requires a ductile or liquid layer between the surface ice and the deeper interior. Combined with the observations of dark bands, these were tantalizing hints that Europa had a warm interior at some time in the past, and perhaps still today. Increasingly sophisticated theoretical models of **tidal heating** of Europa, discussed later in more detail, suggested that a global subsurface ocean might yet persist within Europa.

Such intriguing findings meant there was much anticipation for the Galileo mission, which launched from the space shuttle Atlantis in 1989 and entered orbit around Jupiter in 1995. The primary mission included dropping a probe into Jupiter's atmosphere, as well as observing all its large satellites, its atmosphere, and the local environment. Despite severe data rate limitations during the mission (because the spacecraft's main antenna failed to open), the data from Galileo were so intriguing that the mission was extended in 1997 for a further 3 years, to enable eight further close flybys of Europa, and to study its volcanically active neighbor, Io. Data from the extended Galileo Europa Mission afforded many more high-resolution images of Europa, as well as magnetic data that strongly imply the presence of a briny water layer beneath the surface today. After running low on fuel, the Galileo spacecraft was intentionally commanded to crash into Jupiter in 2003, ending its successful mission. The physical and orbital properties of Europa are summarized in Table 36.1.

2. FORMATIONAL AND COMPOSITIONAL MODELS

During the formation of our solar system, the growing gas giant planet Jupiter pulled material from the solar nebula. It is now understood that, in contrast to early models of satellite **accretion**, the solids of the Jovian circumplanetary disk were probably grabbed from the **solar nebula** in nearly **primordial** form, and that the circumplanetary disk may have been gas-poor. Thus, the material incorporated into the Galilean satellites was probably similar in composition to the asteroids of the outer asteroid belt, containing ice, **anhydrous silicates, carbonaceous** material, and nickel–iron metal alloy.

The Galilean satellites formed by aggregation of these solids, with the proportion of ice varying with distance from the warm protoplanet Jupiter. Io formed relatively close to Jupiter, so it was not able to accrete and retain significant amounts of water ice. As the next moon outward, Europa formed as a mostly rocky satellite, able to accrete sufficient volatiles to form a ~100 km thick outer layer of H_2O. In the colder reaches of the Jovian circumplanetary disk, Ganymede and Callisto formed with near-equal amounts of rock and ice. If the Jovian subnebula were cold enough, some lower temperature condensates such as CO_2 could have been incorporated as Europa and the other Galilean satellites formed.

Europa's early heat of **accretion** combined with heat from radioactive decay in the silicate component would have warmed the satellite's interior and could have formed a primordial ocean, which was likely reduced and sulfidic. Thermal and geochemical evolution would have caused some oxidation of the ocean through time, forming sulfates. Refined models of Europa's accretion and chemical evolution are bringing improved understanding of the satellite's initial conditions.

3. INTERNAL STRUCTURE, TIDES, AND GLOBAL TECTONICS

3.1. Europa's Internal Structure

Measurements of Europa's gravity field from the Galileo spacecraft constrain Europa's axial **moment of inertia** to 0.346 ± 0.005 (a value 0.4 represents a constant density body), indicating significant concentration of mass toward its center. Further assumptions of likely composition and density of its internal layers suggest that Europa probably has a ~100 km thick layer of H_2O overlying a rocky **mantle**, which surrounds an iron-rich **core** (Figure 36.2). The gravity data do not constrain the nature of the water-rich layer, i.e. whether it is solid or liquid.

The most definitive evidence that there is a liquid water ocean within Europa at the present time comes from the Galileo spacecraft's magnetometer, combined with theoretical analyses. Because Jupiter's powerful magnetic field is tilted by 10° relative to the planet's equatorial plane in which the satellites revolve, the satellites experience Jupiter's magnetic field as time-varying. For Europa, each 5.6 h (half the synodic period) the satellite finds itself

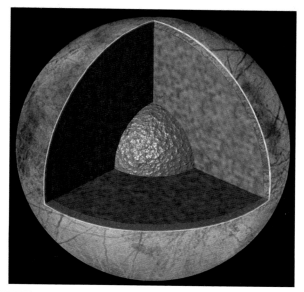

FIGURE 36.2 Interior structure of Europa. Rocky mantle (brown) and iron-rich core (gray) are synchronously locked in position with respect to Jupiter, but the ice shell (white) may rotate nonsynchronously—slightly faster than the interior—if decoupled by the water layer (blue). Layer thicknesses are approximately to scale.

alternately above and below the magnetic equator of Jupiter. Surprisingly, the Galileo magnetometer measured a magnetic signature in the vicinity of Europa that alternately flips to oppose the external Jovian magnetic field. This implies that Europa is behaving as a conductor, generating an induced magnetic field in response to the external Jovian field. Modeling of the Galileo observations suggests that there is a conductive layer—probably a briny ocean—within the outer portion of Europa.

The thickness of the ice shell overlying the ocean is significant for models of Europa's thermal evolution, geological processes, and astrobiology. With future missions we will want to sample material from the ocean to understand Europa's potential for life, but accomplishing this task is dependent upon the ease of accessing material from the ocean, and the ways this material may have been processed. To estimate Europa's ice shell thickness, we must search for clues in the geophysical history and geological record preserved in the icy surface.

3.2. Tidal Evolution

The principal energy source that heats Europa's interior and drives its tectonics today comes from its orbital interaction with Jupiter. An **orbital resonance** can occur when two satellites have orbital periods that are related by integer relationships, allowing them to exert a gravitational influence over each other and affecting the eccentricity of their orbits. The three Galilean satellites are locked in the

FIGURE 36.3 Mutual conjunctions resulting from the Laplace resonance of Io, Europa, and Ganymede. Io completes approximately four orbits to every two of Europa's, while Ganymede orbits approximately once during the same time. (Left) Mutual conjunction of the Europa–Ganymede pair. (Right) Io and Europa experience a mutual conjunction one Io day later, while Europa has moved along half an orbit, and Ganymede has progressed through one-quarter of its orbit. The resonance forces Europa's eccentricity to be nonzero, causing tidal heating and geological consequences.

Laplace resonance, in which the orbital periods of Ganymede:Europa:Io are in a near 1:2:4 ratio (Figure 36.3), but more important, the mutual **conjunctions** of the Io–Europa pair and of the Europa–Ganymede pair **precess** around Jupiter at precisely the same rate. Like a child on a swing pushed at the optimal moment, the recurring mutual conjunctions force and maintain eccentricities in their orbits (Figure 36.3). Although it is not known exactly when or how the moons came to form the precise clock that is the Laplace resonance, one model suggests this resonance was progressively achieved after Io moved outward into a near 2:1 resonance with Europa, and then the Io–Europa pair moved outward until Ganymede was captured into its own near 2:1 resonance with Europa. An alternative model suggests early (primordial) formation of the resonance closely connected to formation scenarios of the Galilean satellite system. Europa's forced eccentricity is key to its youthful and complex surface, as will be described in the following sections.

3.3. Tidal Heating

Europa's resonance with siblings Io and Ganymede causes it to have a slightly **eccentric** orbit ($e = 0.0094$). This eccentricity causes Europa to move closer to and farther from Jupiter (at **perijove** and **apojove**, respectively) as it moves along its 85 h orbit, causing the satellite to undergo increasing and decreasing gravitational pull from Jupiter. At the same time, Europa undergoes **libration**, its tidal bulge rocking from side to side as Europa orbits Jupiter. If an ocean is present decoupling the shell (as described below), Europa could deform by up to ~30 m over each orbital period, and the dissipation of **strain** energy resulting from this deformation would cause the interior to warm. Dissipation of tidal energy can occur in several ways, such as by friction along **faults**, turbulence at liquid/solid boundaries, and **viscoelastic heating** at the scale of individual ice or mineral grains. It is likely that a great degree of tidal energy is currently dissipated near the base of Europa's icy shell, just above the interface between the ice and the underlying liquid ocean, where the ice is warmest and most deformable on the timescale of the satellite's orbit. Combined with radiogenic heating from the deep interior, this regular input of energy is believed to be sufficient to keep Europa's ocean liquid; additional heating might result if there is also tidal dissipation in the rocky mantle.

3.4. Diurnal Stressing

In addition to heating, the tides induced by Europa's eccentric orbit are believed to be responsible for many of its tectonic processes, including formation of cracks on its surface. As Europa orbits Jupiter, its radial and librational deformation results in **diurnal stresses** (so named because Europa's day is equal to its orbital period). These stresses are relatively small (<0.1 MPa), but they are apparently sufficient to crack Europa's ice shell, producing regions of extensional and compressional stresses that migrate across the surface, changing in direction and magnitude as Europa moves through its eccentric orbit. The magnitude of distortion, and thus stress, due to tidal flexing depends on a satellite's interior structure. If Europa's ice shell were frozen to the rocky mantle, there would be very little tidal

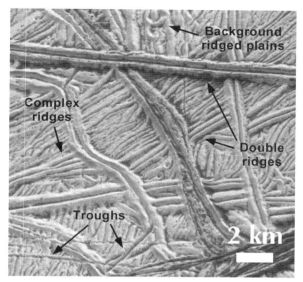

FIGURE 36.4 Galileo image of a typical portion of Europa's ridged plains, showing ridges and troughs crisscrossing and overprinting each other. Double ridges are most common, but ridge complexes (also known as "complex ridges") with more than two crests also occur, along with simple troughs (cracks). The background terrain is so heavily overprinted that it is difficult to distinguish individual features.

distortion overall, with a tidal amplitude of about 1 m; however, if there is a liquid water ocean beneath the ice shell, the surface is predicted to distort by up to 30 m during an orbit. (In comparison, the Earth's rocky moon, which has a relatively cold interior and high eccentricity of 0.055, deforms by just ~10 cm over each orbit due to tides raised by the Earth.)

Diurnal stresses have been invoked to explain some of Europa's unusual surface features, including cracks and ridges (Figure 36.4), and likely contribute to the relatively youthful surface age. Most dramatically, diurnal stresses have been invoked to explain the unusual **cycloidal** shapes of some ridges and other structures on Europa, due to the changing direction and magnitude of stresses, as described further in Section 4.1.3. Moreover, diurnal stresses tend to rotate anticlockwise in the northern hemisphere, and clockwise in the southern. These stress rotations may be responsible for the observed preponderance of left-lateral **strike-slip** faults observed in Europa's northern hemisphere and right-lateral strike-slip faults in the southern.

3.5. Nonsynchronous Rotation

If a satellite is in a perfectly circular orbit around its primary, it keeps the same face toward its parent planet, and this tidal bulge remains fixed relative to the planet. Most of the solar system's large satellites, including the Earth's moon, rotate synchronously, keeping the same hemisphere always facing the parent planet. Because of Europa's combination of eccentricity and a decoupling global ocean, a net torque results due to eccentricity, causing Europa's decoupled ice shell to rotate slightly faster than synchronously. Europa's rocky interior is expected to maintain a permanent mass asymmetry to counter this effect, so beneath the icy shell, Europa is probably synchronously locked (as is the Earth's moon). The rate of this predicted "nonsynchronous rotation" is not known, and it might not be constant through time, but a lower limit time for one complete rotation of the shell is ~10,000 years, based on nondetection from comparisons of Voyager and Galileo images. Nonsynchronous stresses are expected to be large—perhaps an order of magnitude larger than the diurnal stresses and potentially sufficient to open deep (kilometer-scale) cracks in the ice shell. The orientations of Europa's major lineaments do not correspond to the current patterns expected from tidal stresses alone, but if the shell is backrotated by moving it "back in time" by ~30° westward in longitude, there is an overall good fit of lineaments to the predicted stresses. Mapping of crosscutting relationships among lineaments in some areas of Europa suggest that the ice shell may have completed one or many rotations. Some geological features, including several arcuate troughs, suggest that the icy shell probably has also undergone some degree of **polar wander**, i.e. a tilt of the ice shell relative to the spin axis.

4. LANDFORMS ON EUROPA

Europa exhibits two primary types of terrain: (1) the bright-ridged plains, which are crisscrossed by bright and dark linear features, and (2) mottled terrain, which shows evidence for **endogenic** disruption and modification of the surface. Each type of terrain and the **morphologies** of their constituent **landforms** are discussed in this section.

4.1. Ridges, Troughs, and Bands

Europa's linear features are ubiquitous, covering most of the satellite's surface. These landforms exist at a variety of sizes and scales, and exhibit several morphologies, some of which are unique to Europa. Many linear features have overprinted and offset one another, sometimes by several kilometers, making it difficult to piece together the history of the surface. Some ridges have shallow topographic depressions and/or fine-scale fractures alongside them, suggestive of loading of the **lithosphere** by the weight of the ridge material from above and/or from withdrawal of material from below. Understanding how ridged plains form and evolve is important to the question of where liquid water exists within Europa, whether and how water is involved in the formation of surface landforms, and possible niches for life.

4.1.1. Troughs

The simplest of Europa's landforms, troughs (commonly called "cracks"), may be several hundred kilometers long and less than a few hundred meters wide (Figure 36.4). They can have subtle rims or none, and are generally V shaped, suggesting an origin as **tension fractures**. Some have undergone **mass wasting** along their sides, and some troughs have elevated flanks and appear to be transitional forms between troughs and double ridges. As discussed previously, Europa's troughs probably originate by tensile cracking due to diurnal and/or nonsynchronous rotation stresses.

4.1.2. Double Ridges and Ridge Complexes

The most ubiquitous landform on Europa, ridges are most commonly found in a "double ridge" form, with two parallel ridges separated by an axial valley (Figure 36.4). Double ridges can be from ~ 0.5 to ~ 2 km wide, and some span thousands of kilometers in length. Slopes of these ridges tend to be near the **angle of repose** ($\sim 30°$), and at some, preexisting topography can be traced up the flank, suggesting that they may have formed by upwarping of the surface. Most ridges tend to be linear or only gently curved, and mass wasting is prevalent along ridge flanks. The cycloidal ridges discussed below (Section 4.1.3) have notably arcuate shapes, but are otherwise morphologically identical to other double ridges on Europa. "Ridge complexes" have three or more subparallel ridge crests (Figure 36.4). Some ridge complexes appear to be sets of several double ridges, running parallel to, or in some cases intertwined with, each other, while others seem to be composed of bundles of ridge crests separated by intervening troughs.

Ridge formation on Europa is not yet fully understood, and several models have been suggested. Europa's troughs are probably modified by additional processes to form double ridges with distinct and uniform crests. In one model, ridges form through the buildup of cryovolcanic material erupted from fissures. A major drawback with this model is that it is hard to explain the remarkable uniformity of ridge crests, and the distinct V-shaped trough along their axes. An alternative model suggests that double ridges form in response to cracking and subsequent rise of warm or compositionally buoyant ice. Possibly aided by tidal heating, the buoyant ice intrudes and lifts the surface to form ridges. However, this model does not explain how multiple ridges might form within ridge complexes.

Another model proposes that the ridges form in a manner similar to pressure ridges in Arctic sea ice. In this model, cracks created by diurnal tidal stresses allow water to seep up from the ocean below, filling the crack and partially freezing into a slurry. It is envisioned that diurnally varying tidal stresses would then push the crack margins back together, and this partially frozen ice is easily smashed up, forming a jumbled pile of ice that squeezes out of the crack. Although pressure ridge formation is well understood in the Earth's sea ice, where the ice is thin and ocean currents cause movement of the ice, it is unknown whether Europa's ice is sufficiently thin to pull apart atop the liquid layer. Even if so, it is not clear that this model can explain the morphology of Europa's ridges, including distinct V-shaped troughs along double ridges, their uniform parallel ridge crests, and the apparently upwarped features along some ridge flanks.

Alternative models have suggested that water-filled cracks instead penetrate upward from the ocean into the ice shell and upwarp the surface. This model might explain the general morphology of ridges, but it has difficulty explaining the uniformity of the crests and the morphologies of ridge complexes.

Another model is one in which a fracture forms along the surface and then undergoes **shear stresses** and strike-slip motion as a result of Europa's diurnal tides. This strike-slip motion along the crack produces frictional heating as the walls of the crack rub past each other, warming the subsurface ice (Figure 36.5). This shear heating may trigger warm ice to well up beneath the crack, forming ridges, and perhaps inducing partial melting beneath the ridge axis at the same time. This model predicts that a ridge a few hundred meters high could be built by upwelling warm ice in only a decade or so, and because both sides of the crack are subject to the heating, the ridges would be expected to be of uniform width and height, as is observed. If shear heating were sufficient to induce partial melting below the ridge, the melt would tend to drain downward, perhaps forming the V-shaped axial trough above as material is removed. This model predicts softened ice along the ridge, perhaps enabling contractional deformation to occur in response to compressional stress. A model in which contraction occurs across some ridges is viable based on reconstruction of preexisting features and kinematic arguments. If ridges do hide contraction along initially extensional structures, then ridges could help to balance the abundant extension on Europa that is represented by its pull-apart bands (see Section 4.1.5). Ultimately it may be found that ridge formation is a combination of several processes, but currently these features remain an enigma.

4.1.3. Cycloidal Structures

While most double ridges are linear in overall planform, cycloidal ridges are shaped like a chain of distinct arcs (Figure 36.6). Cycloidal fractures, ridges, and some other structures on Europa's surface are likely explained by the action of diurnal stresses. If a fracture propagates slowly enough—at about walking speed—the rotation of **tensile**

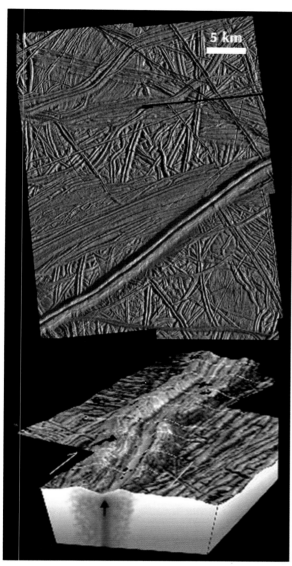

FIGURE 36.5 Double ridges such as Androgeous Linea (top) are ubiquitous on Europa, but their origin is not well understood. In the shear heating model (bottom), strike-slip, or shear, motion along a fracture results from diurnal tidal stresses. Friction between the fracture walls warms the ice, softening or partially melting it. Warm ice close to the fracture rises buoyantly, upwarping the ridge crests. Downward drainage of melt may aid formation of the axial depression. *Image: NASA/JPL. Diagram: Topography courtesy B. Giese, DLR.*

stresses over a Europan day occurs on a timescale such that the propagating fracture can be affected by these changing stresses, tracing out an arc instead of a straight path. As Europa moves in its orbit, the tensile diurnal stresses will drop below a critical value needed for fracture propagation, until the next orbit, when tensile diurnal stresses again increase above the critical value for fracture propagation, generating the next cycloidal arc. In this model, the diurnal stresses that crack the ice to create cycloidal fractures are relatively small, just a few tens of kilopascals. Cycloidal ridges would evolve from cycloidal fractures in a manner similar to the formation of other ridges. Some pull-apart bands (Section 4.1.5) with arcuate and scalloped margins may have opened along preexisting cycloidal ridges. Tides imparted by Jupiter's gravitational pull would be insufficient to crack the surface into cycloidal patterns if Europa had no ocean, so the presence of cycloid structures is strong argument for the existence of an underlying ocean at the time these structures formed.

4.1.4. Triple Bands

One specific type of lineament consists of a bright central ridge, flanked by patchy, diffuse, low-albedo margins, hence the term "triple band" (Figure 36.7). These are most commonly larger, more prominent ridges. It has been suggested that the dark flanks were created by the eruption of icy cryovolcanic material (similar to some explosive volcanoes on Earth), which either seeped out along the ridge flanks or rained dark **pyroclastic** material onto the surface alongside the ridge. Another possibility is that **intrusions** of ice that is warmer than its surroundings might result in local **sublimation** of icy surface materials, leaving a layer of more **refractory** dark deposits.

4.1.5. Pull-Apart Bands

Polygonal dark and gray bands on Europa's surface have margins that can be closed together almost perfectly, reconstructing structures that were apparently laterally displaced when the bands formed along fractures. Many such bands are bounded along each of their margins by an individual ridge, suggesting that a double ridge was split along its axis during band formation (Figure 36.8). These structures have been termed "pull-apart" bands, or dilational bands, and they are a clear indication of movement of a brittle surface layer atop a mobile subsurface. Where the bands pulled apart, dark, probably low-**viscosity** subsurface material moved up to fill the gap. Limited topographic data across bands suggest that many stand somewhat higher than the surrounding terrain, consistent with formation atop buoyant ice, rather than liquid water. Bands have been shown to have brightened over time, possibly because of frost deposition or radiation damage. This has led to a wide range of band brightnesses, ranging from relatively dark, through gray, to as bright as the brightest background plains on the surface.

Many bands exhibit bilateral symmetry, with V-shaped central troughs and hummocky textures, and some have zones of ridges and troughs parallel to the central axis, which may include normal faults (Figure 36.9). Morphological comparisons between bands and terrestrial mid-ocean ridges suggest that band formation may have been analogous to seafloor spreading centers on the Earth, where

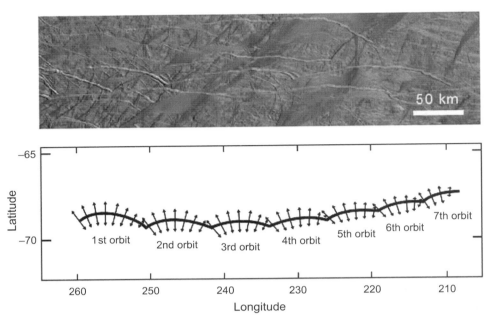

FIGURE 36.6 (Top) Cycloidal ridges on Europa. (Bottom) Model for cycloidal ridge formation, in which each arc forms during one orbital cycle. A crack initiates when stresses reach a critical value and propagates slowly enough that the changing diurnal stress field affects its orientation, causing it to follow a curved path. When the stress drops below a critical level, the crack ceases propagation until the stresses are once again sufficiently large to reinitiate cracking. When the crack reinitiates, Europa has moved along in its orbit, and the stresses are now in a different orientation, leading to a sharp cusp as the next arc begins to propagate. *Top: NASA/JPL. Bottom: After Hoppa et al. (1999).*

plates are pulled apart and new volcanic material erupts along the spreading axis. Features on both planets exhibit central troughs and subparallel ridges where volcanic (in Europa's case, cryovolcanic) material has apparently erupted intermittently throughout the spreading process. Newly formed terrain at terrestrial midocean ridges undergoes normal faulting as it cools and moves away from the ridge crest. On Europa, normal faults parallel to the central axis may have similarly formed as new band material cooled sufficiently for faulting to take place (Figure 36.9).

A major difference between terrestrial plate tectonics and Europan band formation is the lack of subduction zones on Europa. Thus, because band formation has clearly resulted in a large amount of extension (many tens of percent in some areas), there must be some mechanism for balancing this extension elsewhere on Europa's surface. Some fraction of Europa's extension could be related to net global expansion, as would be the case if the ocean were freezing and the ice shell were thickening with time.

4.1.6. Folds

Analysis of high-resolution images of Europa has identified regional-scale contractual folds in a handful of regions on Europa. The most apparent are in the band Astypalaea Linea, where they appear as subtle hills and valleys; several have warped the band at a wavelength of ~25 km, with fine-scale fractures along the crests of the hills and small contractional ridges within the valleys (Figure 36.10). It is unclear whether such folds can represent the primary mechanism by which the icy satellite's considerable surface

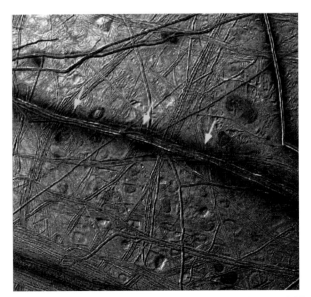

FIGURE 36.7 An example of a triple band, consisting of a central ridge about 5 km wide that is flanked on each side by diffuse, dark material (arrows). This flanking material can be patchy and discontinuous, and may be related to cryovolcanic eruptions during formation of the band, although the exact mechanism is not yet understood.

Chapter | 36 Europa

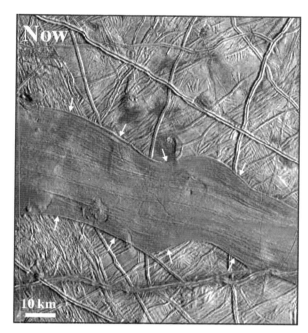

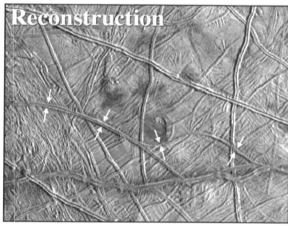

FIGURE 36.8 Points along a prominent gray band (arrows, top) can be reconstructed if the band is removed (arrows, bottom), with the preexisting terrain matching up perfectly along the margins. The band appears to have exploited two existing double ridges during its formation. Reconstructions like this show that a completely new surface has been created by band formation, suggesting that bands represent a considerable amount of extension of Europa's surface. *After Prockter et al. (2002).*

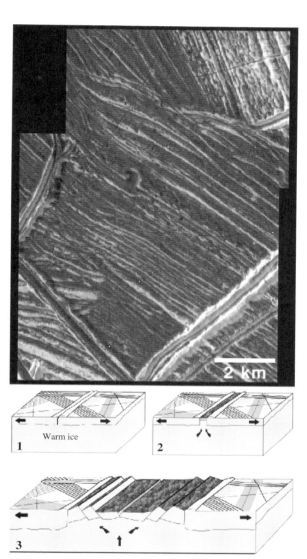

FIGURE 36.9 Model for band formation suggests they are analogous to midocean ridges on Earth. (Top) Distinct morphological zones are mirrored on either side of the central axis of this band. Closest to a central trough, the terrain is hummocky and relatively fine-textured. Further away from the axis, the terrain breaks into normal faults. These types of terrains are also found at spreading centers on midocean ridges on Earth, leading to the suggestion that Europa's bands form in a similar way (cracking followed by extension), allowing new, warmer ice to well up to fill the gap. As this new material cools and moves away from the central axis, it thickens sufficiently that it can form normal faults. This process is analogous to the way new seafloor crust forms on Earth. *After Prockter et al. (2002).*

extension has been accommodated. Other features (such as ridges), along with net global expansion (from freezing of its ice shell), may play roles in explaining how Europa's surface extension is balanced.

4.2. Lenticulae and Chaos

Much of Europa's surface is covered with dark terrain with a mottled appearance, termed "mottled terrain" from Voyager images. High-resolution Galileo images show that in these areas the surface has been endogenically disrupted at small and large scales.

4.2.1. Lenticulae

Portions of Europa's surface are disrupted by subcircular to elliptical pits, spots, and domes, and microchaos regions (collectively termed "lenticulae"), with size a distribution

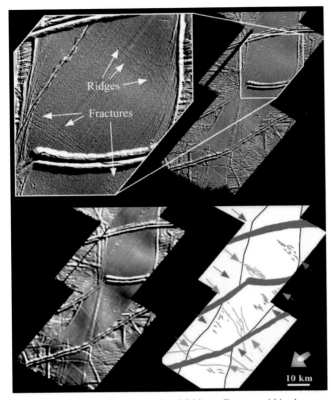

FIGURE 36.10 The best example of folds on Europa, within the gray band Astypalaea Linea. (Top) Close examination of the band reveals fine-scale ridges and fractures. (Bottom left) If a low-pass filter is applied to the image, the 25 km wavelength folds can be distinguished. (Bottom right) Map showing sets of small-scale ridges (blue arrows) within the fold valleys, and fractures (green arrows) that mark the fold crests. *Prockter and Pappalardo, 2000.*

that peaks at ~7 km in diameter, and with a variety of specific morphologies (Figure 36.11). Domes can be convex with upwarped and faulted or unbroken margins against the plains. Pits are topographically low areas where the surface has downwarped while preserving the preexisting terrain. Many pits are associated with dark plains material that **embays** surrounding valleys in the ridged terrain, so it was probably relatively fluid when emplaced. Spots were apparently flooded with dark plains material. Lenticulae known as "microchaos" typically consist of a fine-scale hummocky material, including embedded small plates of preexisting material, commonly with some associated dark plains material. These microchaos regions resemble the larger chaos terrains described in Section 4.2.2.

Although a range of dome, pit, and spot sizes exists, there is a preferred diameter of ~7 km. This consistency in size and the range in their morphologies suggests that they are genetically related; the size and range are consistent with an origin from convective upwelling of buoyant ice **diapirs** within Europa's icy shell (Figure 36.12). Convection is predicted within a tidally heated ice shell greater than about 20 km thick overlying a liquid water ocean. The ice may be thermally buoyant (commonly referred to by the counterintuitive term "warm" ice) and/or compositionally buoyant, where the rising diapiric ice is "clean" relative to its surroundings. Compositional buoyancy of diapirs is possible if they are cleaned out of low melting temperature substances, specifically salts (see Section 5), allowing the clean ice to be more buoyant than the surroundings. In this model, domes form by buoyant diapirs that would reach and break through the surface, and pits may form when a

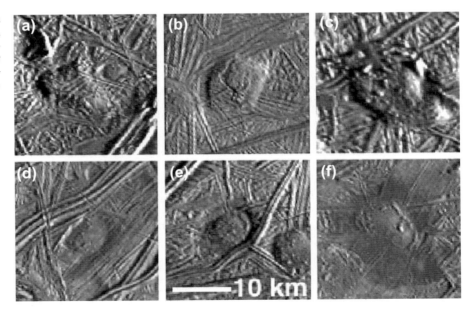

FIGURE 36.11 Lenticulae are found in a range of morphologies, including domes (a, b), microchaos (c), pits (d), and combinations of these morphologies (e), which may or may not have dark plains material associated with them (f). *After Pappalardo et al. (1998).*

4.2.2. Chaos

Chaos regions are areas in which kilometer-scale blocks of existing ridged plains material have translated and rotated with respect to one another within a matrix of hummocky material (Figure 36.13). The matrix material can be low-lying or high-standing relative to the surrounding plains. In one chaos area, Conamara Chaos, at least 60% of the preexisting terrain has been replaced with or converted into matrix material, and the matrix in part stands above the surroundings. Some of the broken plates have been rotated and/or moved horizontally by as much as several kilometers. These plates can be reassembled like a jigsaw puzzle, reconstructing portions of preexisting ridges and troughs (Figure 36.14), although much of the original surface has been destroyed.

Chaos regions have been interpreted as places where Europa's heat flow has been enhanced. One model suggests that local melt-through of ocean water to the surface may have occurred, in which case the blocks are analogous to icebergs floating buoyantly on top of the watery matrix. This model requires that Europa has a very thin shell, less than ~6 km; otherwise, the warm base of the ice shell would flow to maintain its thickness faster than the ice shell could melt from below. Moreover, this model requires that the ocean is only weakly **stratified** in temperature and salinity because strong stratification would prevent heat from being transferred from the ocean floor to the base of the ice shell. In addition, a large, concentrated mantle heat source would need to be stable for hundreds of years to melt the overlying ice.

A proposed alternative model for chaos formation is analogous to that for lenticulae, where ice diapirs have risen buoyantly through the ice crust, breaking or otherwise interacting with the surface (Figure 36.15). This mechanism would explain why some chaos areas stand several hundred meters above the surrounding plains, something that is hard to explain if they formed atop liquid water, but

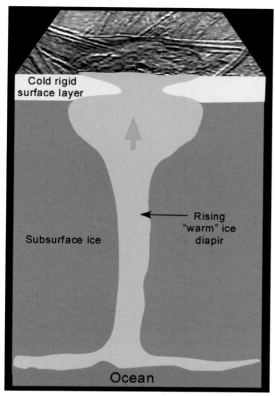

FIGURE 36.12 Model for the formation of lenticulae through diapiric upwelling of buoyant warm ice.

diapir does not quite make it to the surface but melts out impurities from the ice above it, allowing the surface to sag downward.

The range of morphologies and levels of degradation observed in microchaos regions supports the suggestion that upwelling diapirs may partially melt pockets of briny ice as they rise to the surface, causing disaggregation into matrix material and local flooding by dark low-viscosity melt.

 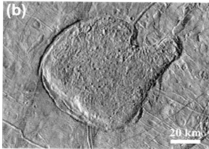 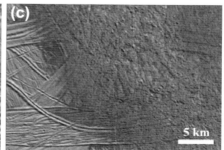

FIGURE 36.13 Examples of chaos on Europa. (a) Conamara Chaos exhibits distinct plates of preexisting terrain (see also Figure 36.14). (b) Murias Chaos, a region of fine-textured chaos that has apparently overflowed its margins on one side, depressing the surrounding plains. (c) The edge of Thrace Macula, showing fine-textured material and a hint of preexisting ridged terrain within, suggesting that the preexisting plains material has disaggregated in place. Dark plains material from Thrace has embayed the surrounding ridged plains, indicating that it was relatively fluid when emplaced. *After Prockter et al. (2004).*

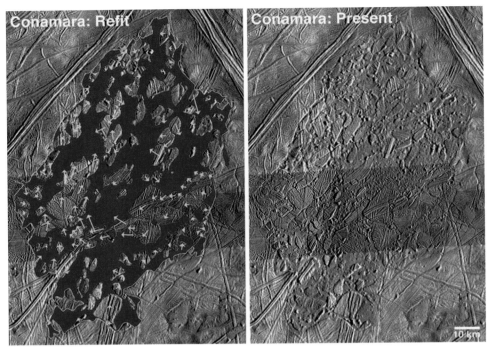

FIGURE 36.14 Broken plates of material within the hummocky matrix of Conamara Chaos (right) can be reconstructed into their original positions like a jigsaw puzzle (left), by matching up older lineaments. This exercise shows that the plates may have moved by several kilometers (arrows show approximate amount of displacement), indicating that the matrix material was originally mobile. Most of the original terrain is missing, however, and may have been subsumed or disaggregated during formation of the chaos. *After Spaun et al. (1998).*

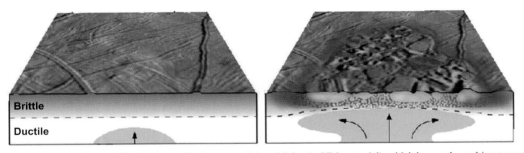

FIGURE 36.15 Model showing how a rising diapir may impinge upon brine-rich ice (reddish material), which lowers the melting temperature, thereby thermally disaggregating the surface (right). *See Collins, Head, Pappalardo, and Spaun (2000).*

feasible if buoyant diapirs rose to their level of neutral buoyancy. Given the morphological similarities between lenticulae and chaos, it seems entirely plausible that they have similar origins through diapiric upwelling. In such a model, chaos terrains may form from a number of separate lenticulae that link together by fractures, forming distinct plates that then separate and mobilize. The diapiric model for chaos formation, however, has difficulty explaining partial melting of the matrix because initially warm ice diapirs would be expected to cool significantly as they approach the surface, before being able to rotate and translate surface crustal blocks.

A recent hybrid model suggests that chaos formation occurs above lakes that have melted out within Europa's icy shell, potentially due to diapirism in briny ice. Disaggregation and collapse of ice above such a lake is envisioned as analogous to the rapid collapse observed of some terrestrial ice shelves, initiated by hydrofracture due to water-filled cracks. This lake model predicts that chaos topography should be sunken relative to the surroundings while a subsurface water lens exists; later freezing would cause the topography to rise upward above the surroundings. Thus, low-lying Thera Macula (Figure 36.16) may signal the presence of a lake the size of Lake Superior within Europa's ice shell today. How a lake could melt out within Europa's ice shell remains uncertain. It is possible that tidal heating would concentrate in the warm ice of a rising diapir, potentially countering its cooling, and melting salty ice.

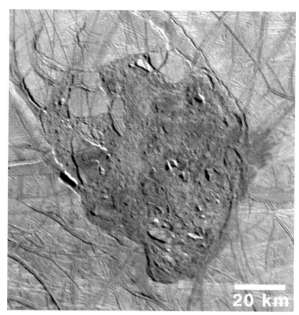

FIGURE 36.16 Thera Macula is a chaos region that lies below the level of the surrounding terrain, suggesting a huge lake lies beneath it within Europa's icy shell today. *ASU/Univ. Arizona/JPL.*

If so, chaos would represent yet another manifestation of the tidal effects imposed by Jupiter and the Laplace resonance.

4.2.3. Impact Structures

Although formed in ice, rather than silicate rock, Europa's craters have a similar range of morphological features as craters on other planetary bodies, including bowl shapes, central peaks, bright ray systems, and **secondary crater** fields. Europa's craters are shallower than those formed on silicate bodies, however, probably because of viscous behavior of the ice in which they form. Another difference is the size at which the transition from simple, bowl-shaped craters to more complex craters with central peaks occurs. On the Moon, a rocky body with similar gravity, this transition occurs at ∼15–20 km in diameter, while on Europa it occurs at only 5–6 km, presumably because the ice crust is relatively weak compared to rock. Simple or bowl-shaped craters are too small to undergo gravity-driven collapse or other significant modifications during formation.

The 24-km-diameter crater Pwyll (Figure 36.17) is thought to be the youngest large impact crater on Europa because it exhibits a bright ray system that extends for over 1000 km and can be seen in global views of the satellite. These rays overlie everything in their path, and their brightness suggests they are so young that they have not been darkened or significantly eroded by charged particle irradiation or micrometeorite bombardment. Pwyll's relaxed topography and the presence of a central peak imply that it formed in warm but solid ice.

One of Europa's largest impact structures is Tyre (Figure 36.18), with a diameter of ∼44 km. Tyre and one other known feature named Callanish (Figure 36.1) are multiringed structures, somewhat analogous to impact basins on the terrestrial worlds. While Tyre's rim crest is difficult to identify, the structure exhibits a complex interior with a smooth, bright central patch interpreted to be impact

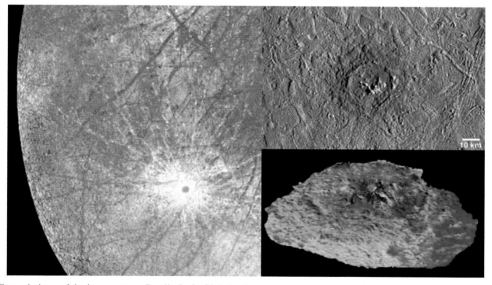

FIGURE 36.17 Several views of the impact crater Pwyll. (Left) Global color-enhanced view showing bright ejecta and rays from material thrown over 1000 km by the impact, the white color indicating exposure of fresh, icy material. (Top right) Higher resolution image of Pwyll, showing distinct rim and central peak, along with a blanket of proximal ejecta around the impact. (Bottom right) Topographic model of Pwyll, created from stereo imaging. This shows the fresh but shallow topography of the crater. *NASA/JPL/DLR.*

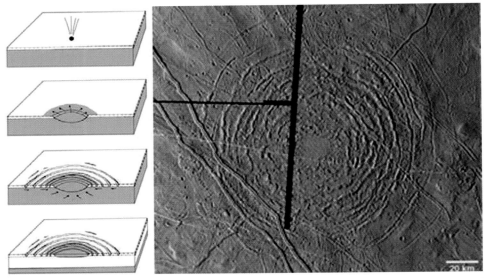

FIGURE 36.18 (Right) The Tyre impact structure has numerous concentric rings around a smoother central region. Outside the rings are many small craters, which are secondary craters caused by ejecta from the impact. This false-color image highlights the reddish material associated with the Tyre rings. (Left) Model for formation of multiring impact structures on the icy Galilean satellites. This model may be applicable to structures such as Tyre, which are inferred to have formed in a relatively thin, brittle layer (white) over a fluid (potentially liquid) layer (gray). *Image: NASA/JPL; Sketch: after McKinnon and Melosh (1980).*

melt or frozen remnants of fluid material that may have been emplaced from below during the impact event. Tyre's most striking characteristic is its concentric troughs, tectonic features resulting from the impact process. These structures are thought to originate when an impact occurs through a brittle layer into relatively fluid material, allowing for rapid collapse and infill of the **transient crater**, and dragging the cold and brittle overlying crust inward to break along concentric faults (Figure 36.18). Galileo near-infrared mapping spectrometer (NIMS) observations of Tyre show that dark material associated with the troughs is similar to the reddish material observed elsewhere on the surface.

Europa's simple and complex craters have morphologies consistent with impact into a solid (although warm and weak) ice target. In contrast, the larger (~ 40 km) impacts inferred to have formed Tyre and Callanish, with their distinctive rough topography and concentric ring systems, imply penetration of the transient crater to a fluid layer at a depth of ~ 20 km. These observations are consistent with Europa's solid ice shell being ~ 20 km thick, overlying Europa's liquid water ocean.

5. SURFACE COMPOSITION AND THERMAL STATE

It has long been known from Earth-based telescopic observations that Europa's surface is predominantly water ice, as amply confirmed by Galileo's NIMS instrument. However, Europa's darker regions show a distinctly different composition, associated with many landforms such as ridges and chaos (Figure 36.19). This material is thought to contain impurities including **hydrated** salts, along with a reddish component.

Spectra from the NIMS instrument show highly distorted water bands in the dark regions, indicative of one or more hydrated minerals (Figure 36.20). Some thermal evolution models of Europa predict large quantities of

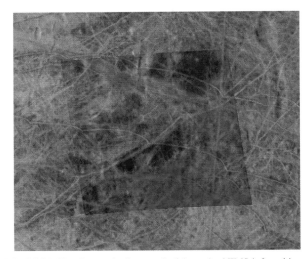

FIGURE 36.19 Composite image of a false-color NIMS infrared image overlain on a monochrome camera image. Blue areas represent relatively clean, icy surfaces, while redder areas have high concentrations of dark, nonice materials, which may originate in a subsurface ocean. The infrared image is about 400 km across.

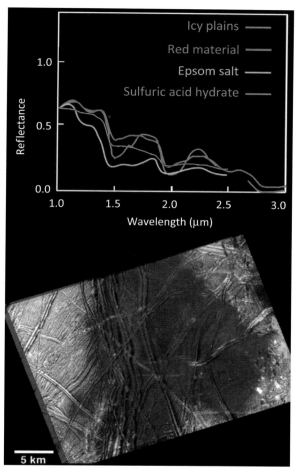

FIGURE 36.20 Infrared spectrum (top after Pappalardo, Head, & Greeley, 1999; Scientific American) of material such as found in Castalia Macula (bottom), one of the reddest, darkest spots on Europa. The spectrum of the icy plains material is distinct from the red material of the spot, which is more similar to something like Epsom salt or hydrated sulfuric acid.

magnesium sulfate hydrate (such as Epsom salt) within the ocean, and mixtures of this and sodium hydrate are predicted at Europa. Another candidate for the surface material is hydrated sulfuric acid (H_2SO_4), more commonly known as battery acid. Although these candidate compounds are colorless, irradiation of the surface by charged particles may be why Europa's dark areas appear reddish. Because of its proximity to Jupiter, Europa's surface is constantly irradiated by high-energy ions and electrons. This irradiation is sufficient to rip apart molecules of water ice and other compounds, allowing them to recombine in a process known as **radiolysis**. This could allow sulfur ions (at least some of which likely originate at Io) and sulfur-containing compounds such as sulfuric acid to synthesize long molecular chains that are ochre in color. These sulfur chains may be responsible for the reddish color of materials that have been emplaced on the surface relatively recently.

Sulfuric acid itself could result from the breakdown and recombination of ice and sulfur dioxide frost, which has also been detected on Europa.

Generally the stratigraphically youngest features on Europa are the darkest, implying that the darkening and reddening process is rapid relative to the age of observable surface features. Also, the fact that Europa's older features are relatively bright implies that some other process brightens features over time, as discussed in Section 6. (The exception to this is young craters, such as Pwyll, which have bright crater rays, probably as a result of depositing fresh materials onto and stirring the preexisting surface upon secondary impact.) Whatever their specific origin, the close association of these hydrated minerals with areas of presumed surface disruption suggests they are related to endogenic processes and may have originated, at least in part, in the subsurface ocean.

Strong absorptions in the infrared region of the spectrum by Europa's H_2O-bearing minerals easily mask the signatures of minor constituents; however, hydrogen peroxide (H_2O_2) is observed and is probably a radiolysis product of water ice. An ultraviolet absorber identified on the trailing side of Europa has been attributed to sulfur from Io, delivered to Europa's surface via the Jovian magnetosphere.

The Galileo spacecraft carried a photopolarimeter radiometer (PPR) instrument that showed that temperatures at low latitudes are in the range 86−132 K, with higher temperatures where the surface is dark, and colder temperatures where it is bright. This correlation between brightness and temperature holds on a global scale, but significant local temperature variations are inferred beyond the spatial resolution of the PPR instrument. These may be due to local-scale variations in surface physical properties. A distinct anomaly around the crater Pwyll may imply a relatively warm (blocky) ejecta blanket. Other thermal variations such as lower than expected temperatures along the equator at dusk are harder to explain. These may be due to variations in grain sizes and structures of ice, but endogenic heat fluxes indicative of interior activity cannot be ruled out.

6. SURFACE PHYSICAL PROCESSES

Processes affecting Europa's surface materials are dominated by thermal processing and radiation bombardment, with meteorite bombardment playing a lesser role.

Jupiter's magnetosphere sweeps up and traps particles including electrons, protons, and heavy ions such as S and O. Because of its close proximity to Jupiter, these particles result in a high-energy ($<$10 MeV) radiation flux at the surface of Europa. The heavy ions in particular are responsible for **sputtering**, where molecules are physically blasted from the surface, creating an **exosphere** of sputtered products, including sodium and low-energy electrons.

There is much still to be learned about the effects of irradiation of ices and the stability of hydrated salt minerals at Europa's surface temperatures.

Europa's water ice exhibits a variety of grain sizes and is particularly abundant and fine grained (<100 μm diameter) between ±60° latitude on the **leading hemisphere**, but it is less abundant with coarser grains (>400 μm) on the trailing hemisphere. The polar regions have a mixture of particles with a range of grain sizes. Bright regions of Europa's surface are topped with a 1 μm layer of **amorphous** ice, which is probably the result of radiolytic disruption of the regular crystalline structure. This disruption can be counteracted by thermal annealing and recrystallization, but these processes are impeded by Europa's cold surface temperatures.

Sputtering and **thermal desorption** act to remove water from water ice and hydrated minerals. These water molecules may either escape to space, but more typically, recondense elsewhere on the satellite as frost. This deposition will vary depending on temperature and surface albedo, so the frost will be more likely to be deposited at high latitudes and on bright surfaces than at the warmer equatorial latitudes on darker materials. Frost deposition may be responsible for the brightening and whitening of Europa's dark reddish surface features over time. In addition, radiolysis itself may cause chemical changes that brighten the surface.

Mass wasting, which is movement of material downslope under the influence of gravity, is less significant on Europa than on the other Galilean satellites because the surface is so young. Mass-wasted material is commonly dark and is likely the nonice debris that remains after the surrounding ice has been removed by sputtering and sublimation. This lag material may be salts and impactor contaminants.

Sublimation (Figure 36.21) has played a significant role in shaping and muting the topography of Callisto and, to a lesser extent, of Ganymede, and has occurred only in darker warmer regions on Europa, potentially including where warm material has been in close proximity to the surface. Sublimation lags have been suggested to result as water molecules are driven off by the intrusion of warm water or ice along ridges and chaos regions. This process has been suggested as the origin of low-albedo spots along triple bands and of dark material along the flanks of ridges. Some craters have dark material in their floors, which is consistent with a thermal lag produced during the impact cratering process, when very hot material from the impact would have rapidly sublimated any water ice off the surface, and by the downslope movement of dark lag material onto the crater floor.

7. SURFACE AGE AND EVOLUTION

7.1. Surface Age

Europa's surface age can be estimated from the number of recent large impact craters on its surface, if accurate estimates of the impactor flux can be made. Modeling of the dynamics of small solar system bodies suggests that the impactor population at Jupiter's orbit is dominated by **Jupiter-family comets**. From the paucity of large (>10 km diameter) craters on Europa, this model implies a surface age of ~60 Ma (million years), with uncertainties of about a factor of 3.

Another way to estimate Europa's age is to use estimates of ice sputtering, which occurs when high-energy particles swept along with Jupiter's magnetic field impact Europa's surface, causing ice particles to be dislodged, most of which then escape to space. This process has a number of uncertainties, but measurements from Galileo's energetic particle detector have lead to estimates that a couple of centimeters to over half a meter of ice may be removed every million years. High-resolution imaging has

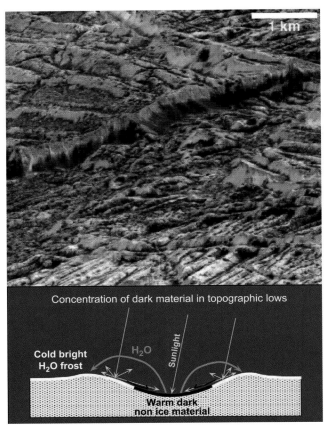

FIGURE 36.21 Sublimation of water molecules by sunlight results in a dark lag deposit, which can move down slopes to collect at their bases. This process appears to have occurred at the cliff in the center of this high-resolution Galileo image. The water molecules may be "cold-trapped" on brighter, icier surfaces, forming frost deposits.

shown numerous examples of topography on vertical scales of tens of meters; this observation is consistent with an age similar to that predicted by the comet impactor model.

The dearth of impact craters on Europa makes it an excellent place to study the ratio of primary to secondary craters, something that is very difficult to accomplish on heavily cratered bodies like the Moon or Mercury. Although high-resolution imaging of Europa's surface is limited, studies suggest that most of the small (<1 km diameter) craters on Europa are secondaries.

Comparisons of images from Voyager and Galileo, acquired 20 years apart, show no definitive evidence for current activity on Europa's surface, although such comparisons are hampered by a lack of overlapping high-resolution image data taken at similar lighting geometries. Similarly, searches for plumes such as those observed on Io and Enceladus have so far proved unsuccessful. Nevertheless, if the surface is only ~60 Ma old, it seems very likely that Europa is still geologically active today.

7.2. Surface History and Geological Evolution

Mapping of Europa's landforms and their interactions with each other yields a rough time history, or **stratigraphy**, of surface evolution and can reveal whether the **resurfacing** style has changed over time. Several areas across Europa's surface have been mapped at a variety of scales, and most researchers infer that there has been a change in geological activity and style through the decipherable time history of the surface. The oldest type of terrain is the "ridged plains", a mélange of ridges and linear structures that are uniformly bright overall, where it is difficult to distinguish distinct feature types. Bands are intermediate in the stratigraphic column, while chaos and lenticulae are among the youngest surface features, commonly disrupting bands and ridged plains. Troughs and double ridges have formed throughout Europa's surface history and crosscut bands and are younger than some lenticulae and chaos. There appears to have been more activity in the earlier surface record, with a waning in the number and width of features in the later stratigraphy.

Stratigraphic mapping therefore suggests that Europa's geological style has generally changed over time, from ridged plains formation, to band formation, to chaos and lenticulae formation, with the activity level simultaneously waning. The mechanism for this change is uncertain, but one plausible model that fits the observations is one in which Europa and its ocean have been slowly cooling, such that the ice shell has thickened as the ice shell reached a critical thickness, **solid-state convection** may have initiated, allowing ice diapirs to rise toward the surface. Thus, a thickening ice shell could be related to a waning intensity

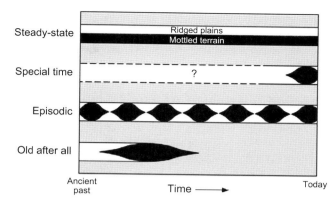

FIGURE 36.22 Possible schematic evolutionary models for Europa's surface. White represents epochs dominated by ridged plains formation, and black represents mottled terrain formation. Current analyses suggest that Europa is either at a special time in its history or, more likely, that it undergoes episodic resurfacing. *After Pappalardo, Belton, et al., 1999.*

of geological activity, and with a transition from a predominance of ridged plains to formation of chaos and lenticulae.

Because Europa's surface is probably relatively young, such a fundamental change in style might seem unlikely over the last ~1% of the satellite's history, and we must speculate on Europa's activity over the rest of its ~4.5 billion year existence. Four possible scenarios have been proposed (Figure 36.22): (1) Europa resurfaces itself in a steady state and relatively constant, but patchy style; (2) Europa is at a unique time in its history, having undergone a recent major resurfacing event; (3) global resurfacing is episodic or sporadic; or (4) the satellite's surface is actually much older than our cratering models suggest.

From the standpoint of the dynamical evolution of the Galilean satellite system, there is good reason to believe that Europa's surface evolution could be cyclical (i.e. scenario (3) above). As participants in the Laplace resonance, the orbital characteristics of Io, Europa, and Ganymede are inherently linked to one another, and also to their interior thermal characteristics. Io experiences the greatest amount of tidal heating and largely drives the predicted cycling. The eccentric orbit of Io can cause a great amount of tidal heating in that satellite, which tends to drive its orbit toward circularity, and in turn decreases its tidal heating. The decreased tidal heating causes Io to cool, but it also allows its eccentricity and internal heating to increase again, thereby completing the cycle. This cyclical evolution of Io's tidal heating and orbital characteristics pulls Europa (and Ganymede) along for the ride via the Laplace resonance. In this way, Europa can experience cyclical variations in its orbital characteristics and tidal heating on timescales of perhaps 100 Ma. Thus, Europa may undergo changes in its activity level and perhaps its resurfacing style on approximately these timescales. In this scenario, Io and

Europa are currently in a diminishing phase of activity. The observed surface characteristics of Europa may represent the latest, waning stage of a long cyclical thermal and geological history.

8. ASTROBIOLOGICAL POTENTIAL

Based on our terrestrial view, the primary ingredients for life are water (See also chapter on Astrobiology), the elements to build organic compounds, and energy. Europa may have all three: water of the ocean, organic compounds that have been delivered to the satellite, and chemical energy from the products of radiolysis at the surface and possibly hydrothermal activity at the ocean floor. The evidence for liquid water within Europa is strong, as discussed in the sections above, and Europa's subice ocean may have a volume twice that of all Earth's surface water. Cometary and asteroidal impactors have rained onto the surfaces of the Galilean satellites throughout solar system history. Just as Ganymede and Callisto have been darkened by impactor material through time, similar material must have been delivered to Europa, where its young and bright surface implies that much of this material is now incorporated into the ice shell and ocean. Moreover, the original accretion of Europa may have delivered carbon in the form of CO and CO_2.

Metabolic reactions within living cells depend upon chemical reactions between **oxidants** and **reductants**. For animals, this depends on taking in oxygen, which is combined with sugars to produce CO_2 and water. For plants, CO_2 is combined with water to form sugars and oxygen. In extreme environments on Earth, and possibly within Europa, more exotic materials such as hydrogen sulfide (H_2S), formaldehyde (HCOH), methane (CH_4), or even sulfuric acid (H_2SO_4) can be key molecules of metabolism. If chemical disequilibrium exists, then organisms may be able to exploit chemical reactions to create and store the energy needed for metabolism.

Whether Europa has sufficient chemical energy to support life is the most significant unknown in understanding Europa's potential for life. Irradiation of surface ice can create free oxygen and hydrogen, with most of the hydrogen floating off but much of the oxygen and other oxidants remaining behind, like a condensed atmosphere frozen into the uppermost centimeters of ice. If these oxidants can be delivered to the ice shell and ocean, they may be able to power the chemical reactions necessary for life. Geological processes such as chaos formation may be able to deliver near-surface materials to the deep ice shell and ultimately the ocean, but the means of surface–ocean communication remain poorly understood.

At Europa's ocean floor, reaction between rock and water should result in serpentinization, a weathering reaction that produces hydrogen as a by-product. Furthermore, if Europa's rocky mantle is tidally heated, then hydrothermal systems could exist on Europa's ocean floor. On Earth, hot chemical-laden water pours into the oceans, delivering organic materials and reductants into the water. If reductants are produced at the bottom of Europa's ocean, and if oxidants are delivered from the ice shell above, then the necessary chemical disequilibrium to support life could exist.

Another important consideration is whether Europa's interior environment is stable enough through time, such that if life ever developed it would still exist today. Europa's ocean may have persisted for eons thanks to internal radioactive heating and the tidal dissipation resulting from Jupiter's gravitational tug. However, the internal heating induced by the Laplace resonance is not necessarily ancient, and (as discussed in Section 7.2) the intensity of tidal heating may have varied (perhaps cyclically) through time. It is an open question whether chemical energy sources for life exist within Europa and have been sufficiently stable to permit life to persist through time.

9. FUTURE EXPLORATION

The harsh radiation environment near Jupiter makes exploration of the satellite technically and financially challenging. Nevertheless, the possibilities for life on this icy moon are sufficiently intriguing and significant that such a mission has a very high priority. Key scientific issues remain to be addressed, including confirmation that there is indeed a liquid water ocean, the characteristics and composition of the ocean, the nature of surface—ice ocean material exchange, and whether Europa has the chemical energy to support life.

Possible means to explore Europa are with a spacecraft in Europa orbit, or from a spacecraft in Jupiter orbit that makes many close flybys of Europa. A spacecraft in orbit around Europa could make continuous gravimetric and topographic measurements of the fluctuating tidal bulge and magnetic measurements of the conductive layer below the ice shell. Either from orbit or from flybys, an ice-sounding radar would be able to sense shallow water deposits and brines and may be able to probe to the bottom of the ice if the ice is relatively cold and thin.

The composition of Europa's surface is not well known, so high-resolution spectral and spatial compositional measurements are also needed to understand Europa's evolution and surface processes. Using a mass spectrometer, it is also possible to determine surface composition by "sniffing" the particles of Europa's very tenuous atmosphere, which are sputtered off the surface by radiation.

Experiments designed to determine Europa's potential for life are ultimately best made with a lander on the surface, which could directly sample and analyze material from beneath the upper ~10 cm of the surface, which is most strongly affected by irradiation. Moreover, the only way to acquire an unambiguous measurement of the thickness of Europa's ice shell is to make seismic measurements, by landing a seismometer on the surface. Spacecraft data has yielded tantalizing insights into Europa's history and evolution, but there is still much we do not know. Further exploration is the only way by which we will learn Europa's deepest and most intriguing secrets.

BIBLIOGRAPHY

Bierhaus, E., Chapman, C., Merline, W., Brooks, S., & Asphaug, E. (2001). Pwyll secondaries and other small craters on Europa. *Icarus, 153*, 264–276.

Carlson, R. W., Johnson, R. E., & Anderson, M. S. (1999). Sulfuric acid on Europa and the radiolytic sulfur cycle. *Science, 286*, 97–99.

Chyba, C. F., & Phillips, C. B. (2001). Possible ecosystems and the search for life on Europa. *Proceedings of the National Academy of Sciences of the United States of America, 98*, 801–804.

Collins, G. C., Head, J. W., Pappalardo, R. T., & Spaun, N. A. (2000). Evaluation of models for the formation of chaotic terrain on Europa. *Journal of Geophysics Research, 105*, 1709–1716.

Greeley, R., Chyba, C. F., Head, J. W., III, McCord, T. B., McKinnon, W. B., Pappalardo, R. T., et al. (2004). Geology of Europa. In F. Bagenal (Ed.), *Jupiter: The planet, satellites, and magnetosphere* (pp. 329–363). Cambridge, United Kingdom: Cambridge Univ. Press.

Greenberg, R., Geissler, P. E., Hoppa, G., Tufts, B. R., Durda, D. D., Pappalardo, R., et al. (1998). Tectonic processes on Europa: tidal stresses, mechanical response, and visible features. *Icarus, 135*, 64–78.

Head, J. W., & Pappalardo, R. T. (1999). Brine mobilization during lithospheric heating on Europa: implications for formation of chaos terrain. *Journal of Geophysics Research, 104*, 27143–27156.

Hoppa, G. V., Tufts, B. R., Greenberg, R., & Geissler, P. E. (1999). Formation of cycloidal features on Europa. *Science, 285*, 1899–1902.

McKinnon, W. B., & Melosh, H. J. (1980). Evolution of planetary lithospheres: evidence from multiringed basins on Ganymede and Callisto. *Icarus, 44*, 454–471.

Moore, J. M., Asphaug, E., Belton, M. J. S., Bierhaus, B., Breneman, H. H., Brooks, S. M., et al. (2001). Impact features on Europa: results from the *Galileo* Europa mission. *Icarus, 151*, 93–111.

Pappalardo, R. T., Belton, M. J. S., Breneman, H. H., Carr, M. H., Chapman, C. R., Collins, G. C., et al. (1999). Does Europa have a subsurface ocean? Evaluation of the geological evidence. *Journal of Geophysics Research, 104*, 24015–24055.

Pappalardo, R. T., Head, J. W., & Greeley, R. (1999). The hidden ocean of Europa. *Scientific American, 281*(4), 54–63.

Pappalardo, R. T., Head, J. W., Greeley, R., Sullivan, R. J., Pilcher, C., Schubert, G., et al. (1998). Geological evidence for solid-state convection in Europa's ice shell. *Nature, 391*, 365–368.

Prockter, L. M. (2004). Chapter 10: Ice Volcanism on Jupiter's moons and beyond, In R. Lopes, & T. Gregg (Eds.), *Volcanic Worlds* (p. 256). Springer Praxis. ISBM: 3-540-00431-9.

Prockter, L. M., Head, J. W., Pappalardo, R. T., Sullivan, R. J., Clifton, A. E., Giese, B., et al. (2002). Morphology of Europan bands at high resolution: a mid-ocean ridge-type rift mechanism. *Journal of Geophysics Research, 107*. http://dx.doi.org/10.1029/2000JE001458.

Prockter, L. M., & Pappalardo, R. T. (2000). Folds on Europa: Implications for crustal cycling and accommodation of extension. *Science, 289*, 941–943.

Schenk, P. M., Chapman, C. R., Zahnle, K., & Moore, J. M. (2004). Ages and interiors: the cratering record of the Galilean satellites. In F. Bagenal (Ed.), *Jupiter: The planet, satellites, and magnetosphere* (pp. 427–457). Cambridge, United Kingdom: Cambridge Univ. Press.

Schenk, P. M., & McKinnon, W. B. (1989). Fault offsets and lateral crustal on Europa—evidence for a mobile ice shell. *Icarus, 79*, 75–100.

Schenk, P., & Pappalardo, R. (2004). Topographic variations in chaos on Europa: implications for diapiric formation. *Geophysical Research Letters, 31*. http://dx.doi.org/10.1029/2004GL019978.

Schmidt, B. E., Blankenship, D. D., Patterson, G. W., & Schenk, P. M. (2011). Active formation of 'chaos terrain' over shallow subsurface water on Europa. *Nature, 479*, 502–505.

Spaun, N. A., Head, J. W., Collins, G. C., Prockter, L. M., & Pappalardo, R. T. (1998). Conamara Chaos region, Europa: Reconstruction of mobile polygonal ice blocks. *Geophys. Res. Lett., 25*, 4277–4280.

Weiss, J. W. (2004). Planetary parameters. In F. Bagenal, et al. (Eds.), *Jupiter: The planet, satellites & magnetosphere* (pp. 699–706). United Kingdom: Cambridge Univ. Press Cambridge.

Zahnle, K., Schenk, P., Levison, H., & Dones, L. (2003). Cratering rates in the outer solar system. *Icarus, 163*, 263–289.

Chapter 37

Ganymede and Callisto

Geoffrey Collins
Physics and Astronomy Dept., Wheaton College, Norton, Massachuse, USA

Torrence V. Johnson
Jet Propulsion Laboratory, California Institute of Technology, Pasadena, California, USA

Chapter Outline

1. Introduction — 813
2. Exploration — 813
   2.1. Discovery — 813
   2.2. Astronomical Observations — 813
      2.2.1. Masses and Densities — 815
   2.3. Spacecraft Exploration — 815
3. Interiors — 816
   3.1. Interior Structures — 816
   3.2. Internal Oceans — 816
   3.3. Magnetic Fields — 818
      3.3.1. Intrinsic Fields — 818
      3.3.2. Induction Fields and Oceans — 819
   3.4. Formation and Evolution — 819
4. Surface Materials — 820
   4.1. Composition of Surfaces — 820
   4.2. Surface and Atmosphere Interactions with Local Environment — 821
   4.3. Regolith — 822
5. Impact Craters — 823
   5.1. Crater Structures — 823
   5.2. Distribution of Craters and Surface Ages — 824
6. Tectonism and Volcanism — 825
   6.1. Bright Terrain — 825
   6.2. Dark Terrain — 827
   6.3. Callisto — 828
7. Unanswered Questions and Future Exploration — 828
Bibliography — 829

1. INTRODUCTION

Ganymede and Callisto (Figure 37.1) are the largest and outermost of Jupiter's four Galilean satellites. Similar in size to Mercury, and with surfaces dominated by dirty water ice, they are prime examples of planet-sized icy bodies. Although Ganymede and Callisto are neighbors and they share many bulk characteristics such as size and density, they have followed divergent evolutionary paths. The striking differences in their interior structure and surface geology provide insight into which processes are common and which processes are unique in the development of a large icy world.

2. EXPLORATION

2.1. Discovery

Ganymede and Callisto were discovered by Galileo Galilei in 1610, when he first trained his telescope on Jupiter and shortly thereafter published his results in the *Siderius Nuncius*. Along with Io and Europa, they became the first natural satellites, other than the Moon, known to science. Galileo immediately recognized the significance of the "new stars" traveling with Jupiter and changing their positions every night. The orbits of what are now known as the Galilean satellites were rapidly calculated and found to be essentially circular and in the same plane as Jupiter's equator. Because Galileo made these observations centuries ago, his records of satellite eclipses provide a long time line to compare with modern measurements, and they are still used to constrain calculations of the dynamical evolution of Jupiter's satellite system under the influences of tidal dissipation and the satellites' mutual gravitational interactions.

2.2. Astronomical Observations

The Galilean satellites are large enough to exhibit distinct disks (on the order of ∼1 arcsec in angular diameter) when

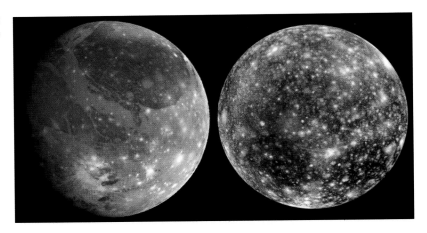

FIGURE 37.1 Global views of Ganymede (left) and Callisto (right), obtained by the camera on the Galileo spacecraft.

viewed through even moderate power telescopes, and it was thus known from simple geometry that they must be bodies comparable in size to the Moon. Precise measurements of their sizes proved difficult with conventional astronomical techniques, with published estimates from different observers disagreeing significantly. Even these relatively uncertain size estimates were sufficient, when combined with the satellites' brightness, to indicate that their surfaces are highly reflective compared with the Moon.

In the two decades leading up to the first spacecraft exploration of the Jupiter system, astronomical techniques advanced rapidly, particularly in the area of sensors in the visible and near-infrared spectral range (~0.3−2.5 μm). The pioneering planetary astronomer Gerard Kuiper used early infrared detectors to show that Ganymede's reflectance at 2 μm was much lower than that in the visible range and suggested that water ice might be responsible. Vassily Moroz, a planetary scientist working at the Crimea Observatory in the Soviet Union, made even more detailed infrared color measurements and concluded that water ice was the best explanation for Ganymede's spectrum.

At the spectral resolution and signal-to-noise ratio of these pioneering measurements, however, a conclusive identification of the surface composition could not be made, since several other candidate materials, including ices of carbon dioxide and ammonia, were known to have **spectral absorptions** in the same part of the infrared spectrum. The issue was settled conclusively for Ganymede and Europa by a team led by Carl Pilcher, then at MIT, who published the first high-resolution infrared reflection spectra for these satellites in 1972 and compared them in detail with laboratory spectra of ices at low temperature. All the significant absorption features in the 1- to 2.5-μm region matched spectra of water ice and ruled out any major contribution from other ices. Callisto's spectrum also displays water ice and hydrated silicate features, although the water signature is subdued compared with Ganymede's strong water ice spectral absorptions, due to the larger amount of dark material mixed with the ice on Callisto's surface.

Figure 37.2 shows a compilation of the best telescopic spectra of Ganymede and Callisto compared with Io and Europa. The dominant features in all the spectra except Io's are the deep absorptions at wavelengths longer than 1 μm due to the presence of hydrated minerals and water ice. Laboratory studies of water ice reflectance and theoretical simulations of spectra from mixtures of material have demonstrated that the observed spectra can be explained by water ice/frost, mixed with varying amounts of a spectrally neutral darker component with a reddish color in the visible portion of the spectrum (that is, increasingly absorptive at shorter wavelengths). The nature of the non−water ice component in the satellites' surfaces is still under investigation, but spectra of different regions on both satellites

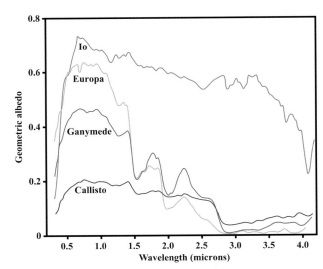

FIGURE 37.2 Compilation of telescopic spectra of Ganymede (green) and Callisto (blue) compared with Io (red) and Europa (yellow). All the satellites except for Io show absorptions on infrared light due to water ice and hydrated minerals, notably near 1.5 and 2 μm. *Modified from R. Clark & T. McCord, (1980).*

taken by the near-infrared mapping spectrometer (NIMS) instrument on the Galileo mission are providing clues to the identification of this material (see Section 3).

2.2.1. Masses and Densities

The mass of a distant planetary object is normally impossible to determine from remote astronomical observations alone, unless it happens to have a companion whose orbit can be determined, as is the case for the giant planets with their satellite systems, Mars, numerous asteroids, and several trans-Neptunian objects including Pluto. The Galilean satellites represent a more difficult case. They do not themselves have satellites, but the mutual gravitational attraction among these large satellites produces significant and measurable changes in their orbits about Jupiter. The mathematician Pierre Laplace studied these interactions in the late nineteenth century, and subsequent developments in this new branch of dynamical astronomy permitted reasonable estimates of the satellites' masses to be made in the early twentieth century. When combined with the still uncertain size estimates, the best estimates of masses prior to 1970 suggested that the inner satellites, Io and Europa, had rocklike densities, similar to the Moon's, while Ganymede and Callisto appeared to be less dense, suggesting the possible presence of large amounts of ice in their constituent materials.

In 1972, observations from two stations on the Earth of a stellar occultation by Ganymede provided the first high-precision measurement of its diameter. This was closely followed by the first spacecraft exploration of Jupiter by the Pioneer 10 and 11 missions in 1973 and 1974, which greatly improved the mass estimates of the satellites from tracking the gravitational perturbations in the spacecraft trajectories caused by the satellites. This led to the first accurate determination of Ganymede's density of about 1900 kg/m^3, adding more evidence to the hypothesis that its bulk composition is a mixture of rock and ice. The Voyager 1 and 2 Jupiter encounters in 1979 provided even more data on satellites' masses and accurate determination of their shapes and volumes. These data showed that Callisto is very similar to Ganymede in its bulk properties, with a density of about 1800 kg/m^3. Interior structure models estimate that the average bulk composition for both satellites is 50—60% (by mass) dry silicate rock, mixed with 40—50% water, which exists as normal water ice at the surface, but transforms into higher density phases at high pressures deeper in the interior (see Section 2).

2.3. Spacecraft Exploration

Eight spacecraft have visited the Jupiter system to date: Pioneer 10 and 11, Voyager 1 and 2, Galileo, Ulysses, Cassini, and New Horizons. Ulysses, a joint European Space Agency/National Aeronautics and Space Administration (NASA) mission to study the sun's environment at high latitudes, made measurements of Jupiter's magnetic field, radiation belts, and dust environment, but did not study the satellites directly. Cassini, on its way to its rendezvous with Saturn, flew by Jupiter in 2000 and returned spectacular observations of its atmosphere and **magnetosphere**, but its trajectory was too far from the Galilean satellite system to provide high-resolution views of the satellites. New Horizons also flew through the Jupiter system on its way to Pluto, but Ganymede and Callisto were on the far side of Jupiter during the flyby, so only very low resolution images were obtained, as well as infrared spectra of a portion of Ganymede's Jupiter-facing hemisphere.

The first Jupiter-focused missions, Pioneer 10 and 11, were designed to provide an initial reconnaissance of the system and to establish the intensity of the radiation belts. The Pioneer program's major contribution to knowledge of Ganymede and Callisto was improving the mass estimates of the satellites (as mentioned earlier), leading to the first precision bulk density measurements.

In 1979, Voyager 1 and 2, with powerful remote sensing payloads and close targeted flybys of each Galilean satellite, provided the first in-depth reconnaissance of the satellites and set the stage for the geological and geophysical exploration of these worlds. Voyager's cameras showed that Ganymede and Callisto, alike in many large-scale properties, have divergent geological histories. Callisto's surface is heavily cratered at all scales, from large-impact scars over 1000 km in diameter down to craters a few kilometers in diameter, the smallest scale resolvable on Callisto by the Voyager cameras. This battered, uniform surface stands in stark contrast to Ganymede's varied landscape. Ganymede's surface can be divided into two distinct types of terrain, based on a sharp **albedo** contrast. The darker areas (named "dark terrain") are heavily cratered and exhibit **palimpsests**, much like the surface of Callisto. The brighter parts of Ganymede's surface (named "bright terrain") form wide lanes through the dark terrain and are less heavily cratered, implying a younger surface. Voyager images showed the bright terrain to have some areas that appeared to be smooth, while other areas exhibit sets of parallel ridges and troughs.

One of the major objectives of the Galileo mission was to perform detailed observations of the Galilean satellites. The mission design allowed multiple close flybys at ranges 100—1000 times closer than the Voyager encounters, enabling high-resolution studies of their surfaces and detailed measurements of their gravity fields and interactions with Jupiter's magnetospheric environment. High-resolution images of the different terrains first identified by Voyager have illuminated their origins, described in detail in subsequent sections of this chapter. The close flybys also enabled more detailed spectroscopic observations, which identified some of the non—water ice components on the satellite surfaces, including carbon dioxide embedded in the surface and evidence for carbon compounds.

Repeated close flybys of Ganymede and Callisto enabled Galileo to make precise gravity and magnetic measurements, resulting in several major discoveries. First, Ganymede has a strongly layered internal structure, with heavier rock and metal concentrated in the center, while Callisto has a more homogenous structure. Second, Ganymede was found to have a relatively strong internal magnetic field, creating its own "mini-magnetosphere" embedded within Jupiter's vast magnetosphere. Finally, the interactions of Ganymede and Callisto with Jupiter's rotating, tilted magnetic field show that both satellites exhibit an induced magnetic field interpreted as evidence for an electrically conducting liquid water ocean beneath their icy crusts.

3. INTERIORS

3.1. Interior Structures

The ice/rock bulk composition inferred for Ganymede and Callisto from their densities naturally led to the suggestion that even modest heating from **accretion** and the decay of radioactive elements in the rocks would melt the ice in the interior and allow the rocks to sink, leading to differentiated interiors—that is, a layered structure with the denser rock and metal constituents concentrated closer to the center of the satellite and the ice in the outer layers. Most analyses following the Voyager mission operated on the assumption that Ganymede and Callisto had similar differentiated interior structures, but the data to test this assumption would not come until the Galileo mission.

Determining the interior structure of a planetary object is intrinsically difficult, particularly from remote observations alone. Most of the information about the interior of our Earth, for instance, comes from over a century of study of seismic data, where waves created by earthquakes travel deep through the Earth and provide clues to the density and composition throughout the interior [see *Earth Surface and Interior*]. So far the only other world for which we have seismic data is the Moon, acquired with seismometers left by the Apollo astronauts (see *The Moon*).

An extremely important quantity that can be used to assess the distribution of mass inside an object is a dimensionless number known as the moment of inertia; a sphere with uniform density throughout has a moment of inertia of 0.4, with lower values indicating increasing degrees of mass concentration near the center. The moments of inertia for Ganymede and Callisto were measured indirectly by the radio experiment on Galileo, which measured the perturbations of the spacecraft's trajectory as it flew by the satellites on multiple occasions at low altitudes and different latitudes. Although perfect spheres with different moments of inertia have identical external gravity fields, the key to this experiment is that the distribution of mass in the interior of a satellite does affect the way its shape is perturbed from a perfect sphere by rotation and tides. The rotation rates of Ganymede and Callisto, although slow by terrestrial standards (a little over a week for Ganymede, and over 2 weeks for Callisto), are still sufficient to cause a slight equatorial bulge and polar flattening, while Jupiter's strong gravity raises tidal bulges on the sub- and anti-Jupiter hemispheres. The combination of these two effects leads to distinctly nonspherical components to the external gravity field (in mathematical terms, the description of the satellites' gravity in a spherical harmonic expansion contains significant J_2 and C_{22} terms). The magnitude of these nonspherical terms depends on the degree of internal mass **differentiation**, and they are related directly to the moment of inertia as long as the object responds as a fluid to spin and tidal distortion (i.e. hydrostatically).

The surprising results of the Galileo tracking experiment showed that Ganymede and Callisto have distinctly different interiors. The derived moments of inertia for both satellites were lower than they would be for bodies of uniform density, as expected. Ganymede's measured value of 0.31 is so low that it implies essentially complete separation of its water ice from the heavier rock and metal. However, Callisto has a significantly larger moment of inertia, 0.35. This is small enough to imply some differentiation, but too large to be compatible with full separation of light and heavy components. Callisto probably has some significant portion of its interior composed of a rock—ice mixture.

The measured moments of inertia can be combined with the values for the total mass, size, and properties of ice and rock under pressure to construct models of the satellite's interiors that match all the known quantities. As there are several unknown parameters regarding interior chemistry, the exact compositions and thicknesses of layers in such models are inherently uncertain, but reasonable assumptions can be made to construct a general model. Figure 37.3 shows the best current estimates of their internal structures. Ganymede is shown with a three-layer structure: a metallic core, a rock mantle, and a deep water-rich upper layer; Callisto is shown with a two-layer structure: a large rock-ice core, with the fraction of dense material increasing toward the center, and an upper water-rich layer. Note that the water-rich layers on both satellites are shown with liquid water oceans trapped below thick layers of surface ice, as discussed in the next section.

3.2. Internal Oceans

A major question regarding these icy worlds is whether they possess subsurface oceans of liquid water. This intriguing possibility was first raised in the early 1970s by planetary geochemist John Lewis, who pointed out that radioactive

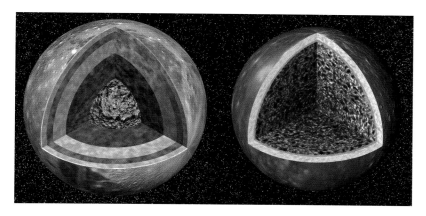

FIGURE 37.3 Cutaway diagrams showing current models for the interior structures of Ganymede and Callisto based on Galileo gravity data. Ganymede (left) is highly differentiated, with a molten iron core surrounded by a rocky mantle, in turn surrounded by a thick outer layer of ice. An interior ocean of liquid water may exist sandwiched between the surface layer of Ice-I and the higher pressure phases of ice below. Callisto (right) has an interior composed of a mixture or rock and ice, slowly increasing in density toward the center. The outermost layer is relatively clean water ice, with a liquid water ocean at its base. *Zareh Gorjian and Eric De Jong, NASA/Jet Propulsion Laboratory.*

heating of satellites' interiors might result in their internal temperatures reaching the water ice melting point at some depth below their surfaces. With the satellites' densities known, a relatively simple calculation of internal temperature from the heating produced by the decay of radioactive nuclides in the rock fraction (primarily U, Th, and K) shows that indeed the ice melting point should be reached about 75–100 km below the surface. More detailed calculations are complicated by several additional factors.

The behavior of water as a function of temperature and pressure is complex. At the surface of the Earth, only normal low-density ice, known as Ice-I, exists. It floats in liquid water and melts at 273 K (0 °C, 32 °F). Increased pressure decreases the melting temperature, but under terrestrial conditions the solid form remains as low-density Ice-I. Laboratory studies show that at the high pressures reached deep in the interiors of icy satellites the size of Ganymede and Callisto, ice transforms to various high-density forms over a wide range of temperatures (Figure 37.4). These phases of ice, as they are known, are denser than liquid water and would sink in a liquid ocean. Calculations of the temperature and pressure as a function of depth within the satellites show that if temperatures reach the required melting point the resulting subsurface oceans are strange indeed—a liquid layer sandwiched between low-density Ice-I on the top and higher density ice phases on the bottom (or possibly a mixture of high-density ice and rock in the case of Callisto).

The other major complication is whether the interior will ever actually warm up to the ice melting point. Simple calculations that reach the melting temperature are based on the heat produced by radioactive decay, escaping the interior by thermal conduction through the ice crust. However, as the temperature of ice approaches the melting point within the satellite, another heat transfer process comes into play, convection. Ice near its melting point is not stiff and brittle, but can flow and deform under pressure, particularly over long periods of time. In geophysical

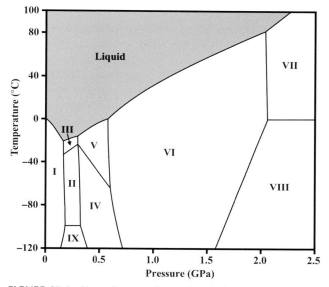

FIGURE 37.4 Phase diagram of water ice. At low pressures near the surface, Ice-I is less dense than liquid water. At higher pressures, ice converts to denser phases, with higher melting points.

terms, the solid ice may be considered as a low-viscosity fluid. Low-viscosity ice under some conditions can begin to convect, with warmer lower density ice rising toward the surface, exchanging with cooler higher density ice sinking into the interior. This glacially slow solid ice circulation is similar to what occurs in the Earth's rock mantle. The important point is that convection is much more efficient at transporting heat than conduction alone. In simple terms, as the ice heats up from the radioactive energy from below, it will begin to convect, taking heat to the surface, but never allowing the temperature within the ice to rise above the melting point. Under these conditions, even if an ocean formed early in the satellite's history, the convection process in the ice crust could rapidly freeze it solid.

On Ganymede, the low-viscosity ice may also play an alternative role as a source of thermal energy due to tidal interactions with Europa and Io (see Section 2.4). Tidal heating is currently a minor source of energy on Ganymede compared to the large role it plays on Europa and Io, but it may have been more significant in the past. This additional heating process does not affect Callisto.

A final complication is the issue of ammonia, NH_3. In many formation models, ammonia is a possible minor constituent of the icy satellites. If present, it has a major effect on the melting point of a water–ammonia mixture, depressing the temperature of the solid–liquid transition to about 173 K, 100° below the point at which pure water melts. Although ammonia has not been detected on the surfaces of the satellites, even small amounts can affect the results of theoretical thermal and convection calculations, and most discussions of the satellites' interiors include both ammonia and nonammonia cases.

Whether convective cooling "wins" over heating determines whether a liquid ocean can exist at the present time. The calculations for interior models including convection are quite complex and depend on some properties of ice that are poorly known. Current models for Ganymede and Callisto show that liquid layers are possible under some conditions, but these models cannot definitively demonstrate their existence. Measurements pointing strongly to the presence of liquid oceans in both satellites came from an unexpected source—magnetic field measurements made by the Galileo mission, which is the subject of the next section.

3.3. Magnetic Fields

The Galileo magnetometer experiment had two major objectives for studying Ganymede and Callisto: (1) to determine whether they possess intrinsic magnetic fields of their own and (2) to study the interactions of the satellites with Jupiter's huge and powerful magnetosphere. These two objectives are closely coupled because the satellites orbit deep within the region of space controlled by Jupiter's magnetic field and its associated trapped radiation and **plasma**. Measurements of magnetic fields in the vicinity of the large satellites thus have to take into account the large background field from Jupiter, which is continually changing due to Jupiter's fast rotation sweeping the field past the satellites and magnetic perturbations from large-scale electrical currents flowing within the magnetosphere. Once these effects are measured and understood, the experimenters can search for the smaller perturbations in the local magnetic field produced by any intrinsic field and from local currents set up by the interactions of the satellites and their tenuous atmospheres with the magnetosphere.

3.3.1. Intrinsic Fields

On the very first Galileo close encounter with Ganymede, the space physics instruments detected strong evidence for both an intrinsic field and complex interactions with Jupiter's environment. As the spacecraft flew by the satellite, the magnetometer recorded a marked change in both the magnitude and direction of the magnetic field. At the same time the plasma wave spectrometer (which receives natural radio "noise" produced by the interactions of charged particles and magnetic fields) showed sharp changes in the nature of the radio signals they received, coinciding closely in time with the observed magnetic deflections. To the investigators on these experiments, these observations were familiar, a "fingerprint" indicating that the spacecraft had passed though a planetary magnetosphere. Due to the complexities discussed earlier, it took observations on subsequent flybys to confirm the discovery, but it soon became clear that Ganymede possesses a relatively strong intrinsic magnetic field, oriented in the opposite sense to Jupiter's field, which produces a "mini-magnetosphere" embedded within Jupiter's magnetosphere. This nested magnetosphere arrangement is a unique situation in our solar system.

An intrinsic field at Ganymede was not totally unexpected. UCLA space physicist Margaret Kivelson, the head of the Galileo magnetometer team, had suggested prior to the Galileo mission that the big satellites might be able to generate their own internal fields. Nevertheless, the discovery of an intrinsic field at Ganymede raises a number of issues for our understanding of planetary magnetic field generation (see *Planetary Magnetospheres*).

How planetary fields, including the Earth's, are generated and maintained is an active area of research. It is believed that some form of what is called a "geodynamo" is responsible for producing a magnetic field within a planetary core. The exact requirements for generating a field by this dynamo process in a given planet are the subject of debate. Ganymede's internal field is consistent with its high degree of differentiation and favors a three-layer model with a metallic iron/iron sulfide core. However, merely having a metallic iron core is not sufficient to produce a planetary magnetic field. Although the Earth's field and other planetary fields are frequently described in textbooks as "bar magnet" fields, this only describes the field's mathematical description (having a dipolar—N and S—configuration with field lines connecting the poles). The solid iron bar magnet analogy is misleading in terms of the source of the field, since it has long been known that iron will lose its magnetization at the temperatures typical of planetary cores (temperatures above the Curie point, at which a magnetic material loses its magnetism).

Current theories of planetary dynamos suggest that the basic requirement for generating a field is continual convective motion of an electrically conducting fluid. Theoretical models of Ganymede's thermal evolution suggest that it could have a fluid, electrically conducting, iron or iron sulfide core at the present time. However, the same models show that although there could have been convective motion in the fluid core early in Ganymede's history, at present the core should be stable against thermal convection, thus explaining the source of Ganymede's magnetic field requires additional factors. Possibilities discussed in the literature include some event, such as tidal heating, stirring up the core in recent geological history, or that Ganymede has heated over a longer timescale than earlier models suggest, reaching the required conditions for core convection and planetary dynamo formation only recently. Another possible explanation is due to a change in the phase behavior of iron sulfide at the lower pressures experienced in Ganymede's core, solid iron snow may form at the top of the liquid outer core and stir up convective motions as it falls through the liquid.

3.3.2. Induction Fields and Oceans

Callisto shows no evidence for an intrinsic dipole field like the one observed at Ganymede. When Galileo flew close to Callisto, the magnetometer recorded perturbations to the background field, but comparisons of data from several encounters showed there was no pattern consistent with single dipole field. However, when the investigators correlated the data with Callisto's position with respect to Jupiter's field, they found another intriguing pattern. Since Jupiter's dipole field is tilted about 10° to its rotation axis, the background field seen by a satellite orbiting in the equatorial plane exhibits a periodic rocking motion as Jupiter rotates. The observed magnetic perturbations correlated with times when this tilt was at different angles.

The key to understanding this type of perturbation lies in the basic theory of electromagnetism—moving magnetic fields can produce electrical currents and electrical currents can produce magnetic fields (electromagnets and electric motors are among the practical applications of this principle). A classic laboratory physics experiment demonstrates that placing an electrically conducting sphere (such as a copper ball) in an oscillating magnetic field will set up electrical currents in the surface of the sphere that produce an induced magnetic field, countering the imposed field. The magnetometer investigators found that the Callisto perturbations closely matched those expected for an induced field in response to the changing Jupiter field. In other words, Callisto was acting as if it had an electrically conducting layer at or under its surface.

What is the conducting layer on Callisto? The electrical conductivity required to produce the observed perturbations is much larger than the known conductivities of ice or rock, the major surface constituents. Going back to the theoretical possibility of a subsurface ocean, the investigators found a possible explanation for Callisto's behavior. The electrical conductivity of salty ocean water is in the right range to produce the required induction field.

Although the argument is indirect, these magnetic results are the best evidence to date that the hypothesized ocean exists under Callisto's icy crust. After this discovery, investigators looked closely at the Ganymede magnetic data and found that there are small deviations from the best-fit intrinsic dipole model, which indicate the presence of an induced field from a conducting ocean layer on Ganymede as well. This same type of induced magnetic field evidence was used to infer a liquid water ocean under the ice on Europa, but ironically, the evidence for a conducting ocean layer on Callisto is stronger than on Europa, because the background field is smaller at Callisto.

3.4. Formation and Evolution

The Galilean satellites have been viewed as a sort of "miniature solar system" since the time of Galileo. Their coplanar, nearly circular orbits strongly suggest that they formed as part of the same process that formed Jupiter. The water-rich, low-density composition of the outer satellites, Ganymede and Callisto, compared with the rock-rich, high-density inner satellites, Io and Europa, suggest that there was a gradient in the conditions within the circumplanetary gas and dust nebula from which the satellites formed, much like the gradient in the solar system as a whole that produced rocky inner planets and volatile-rich outer planets.

The mixed rock/ice composition of Ganymede and Callisto is very similar to that expected from condensation from a nebula with solar composition. Early models for the formation of the satellites envisioned a circumplanetary nebula of gas and dust, which was heated by the growing Jupiter at the center. In this scenario, Ganymede and Callisto formed in the cooler outer portion of the system, under conditions similar to the surrounding solar nebula, which permitted the condensation of water ice. Io and Europa, on the other hand, formed further inside the nebula, under warmer conditions with little to no condensation of water.

This relatively simple Jovian subnebular theory explains the major characteristics of the system in the context of the formation of Jupiter itself. However, there are problems with the details of the model when the evolution of the forming satellites is considered. One problem is that as the satellites form they are subjected to drag from remaining gas and dust in the subnebula. This drag can quickly cause a protosatellite to spiral inward and be swallowed up by the growing proto-Jupiter. Current calculations show that this is a serious problem with early forms of the subnebular models since the timescale for the accretion of the satellites is much shorter than the times for dissipation of the nebula and decay of the satellite orbits. Another issue is the differences between Ganymede and Callisto. In the simple subnebula accretion models, they should have similar histories. However, Voyager and

Galileo observations show major differences in their interior structures and geologic histories. The most difficult point to reconcile is that Callisto's incompletely differentiated interior implies a longer accretion time to prevent melting and differentiation triggered by accretional heating.

The latest formation models attempt to address these issues in a number of ways. Current models for Jupiter's formation suggest that the Jovian subnebula interacts strongly with the surrounding solar nebula as the growing giant planet opens a "gap" in the solar nebula and material is continually fed from the solar nebula surroundings to the outer parts of the subnebula. This class of models can account for longer satellite accretion times and allows them to form without being dragged into the proto-Jupiter. Another type of formation model proposes that the inner satellites formed in a hot dense subnebula, but avoided destruction by opening gaps themselves in the subnebula, slowing their orbital decay. In this type of model, Callisto forms more slowly in a thinner outer nebula environment, accounting for some of its differences.

A final factor which may have affected the apparently different histories of Ganymede and Callisto is the existence of what is known as the **Laplace resonance**. This is a dynamical relationship between the orbital periods of the inner three satellites, first studied by the French mathematician Laplace in the nineteenth century. Io, Europa, and Ganymede currently exhibit a simple numerical relationship (1:2:4) in their orbital periods, causing them to always encounter each other at the same points in their orbits. The gravitational tugs from this repeating encounter pattern continually perturb their orbits, resulting in significantly noncircular orbits. The noncircular orbits cause tidal heating in each of these satellites, resulting most notably in the violent volcanic activity on Io, which has the largest dose of tidal heating due to its proximity to Jupiter (*see* Io). Callisto does not participate in this celestial dance and apparently has never experienced tidal heating.

Despite its participation in the Laplace resonance, Ganymede does not currently experience significant tidal heating because of its distance from Jupiter and the relatively small degree of noncircularity of its orbit. However, calculations of the dynamical evolution of the satellite system suggest that Ganymede's orbit may have been more **eccentric** during its capture into the resonance, possibly resulting in a pulse of tidal heating, which could have triggered differentiation and/or stirred up the core and started magnetic field generation. While the question of why Ganymede and Callisto have experienced such different interior and geological evolution has not been conclusively solved, it seems likely that the key to the solution lies in some combination of differences in their formation conditions and subsequent orbital evolution.

4. SURFACE MATERIALS
4.1. Composition of Surfaces

As noted in the discussion of astronomical discoveries (Section 1.2), water ice was identified as a primary surface constituent on the surface of Ganymede and Callisto (and Europa as well) in the 1970s by obtaining infrared spectra of these bodies. Seen with the eye, the surfaces are darker and redder than pure water ice, so there must be some other material mixed with the ice, but the composition of this material has been difficult to determine. Based on analogy to meteorite and asteroid spectra as well as cosmochemical arguments, most researchers have assumed that the nonwater component of the surface is similar to the material found in primitive, carbon-rich meteorites—a mixture of hydrated silicates (clays) and dark, complex organic compounds (dubbed "tholins" by the astronomer Carl Sagan, who studied the production of organic material in laboratory simulations of planetary environments). Laboratory studies of ice and mineral mixtures show that even small amounts of dark material will disproportionately lower the reflectance (albedo) of the mixture and weaken the spectral signature of water ice, producing reflectances consistent with the observed spectra of the satellites. Unfortunately, the more subtle spectral signatures of the dark minerals are themselves obscured in the mixed spectra by the much stronger water features, making identification of the dark constituents difficult.

The **Near Infrared Mapping Spectrometer (NIMS)** on Galileo provided new insights into the composition of the nonwater constituents. This instrument not only covered the spectral range accessible to Earth-based telescopes but also returned spectra in the 3–5 μm spectral region. This part of the infrared spectrum is inaccessible from the surface of the Earth due to strong absorptions in the Earth's atmosphere by water vapor and carbon dioxide. It is also a key part of the spectrum for studying non–water ice components mixed into the satellite surfaces, since water ice is essentially black at these wavelengths and whatever signal is seen arises primarily from the non–water ice component of the surface mixture.

NIMS spectra of Ganymede and Callisto indeed proved their value in the 3–5 μm range, exhibiting a number of detectable absorption features (see Figure 37.5). The strongest feature is a relatively sharp absorption of infrared light centered at about 4.25 μm wavelength, with weaker, but still easily detectable, absorptions at 3.88, 4.05, and 4.57 μm. There is also a weak absorption seen centered near 3.4 μm. These absorptions are seen in the spectra from both satellites, but are most easily seen in the Callisto spectra, where there is more of the dark material exposed on the surface.

The 4.25-μm feature has been identified as being caused by the presence of CO_2 on the surface. The location of the

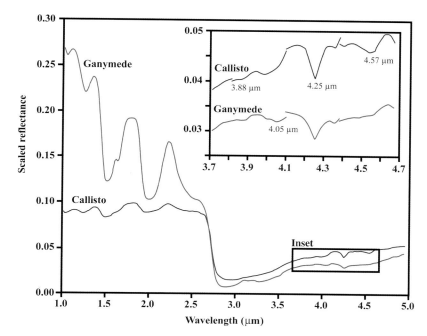

FIGURE 37.5 Galileo NIMS spectra of Ganymede and Callisto, showing absorption of infrared light by various surface materials. The lower reflectance of Callisto at wavelengths shorter than 2.7 μm and the higher reflectance of Callisto at longer wavelengths are both indicative of the lower fraction of water ice exposed on Callisto's surface as compared to Ganymede. Several absorptions at other wavelengths indicative of minor chemical species are highlighted in the inset, including CO_2, SO_2, C–H bonds, and C≡N bonds. See text for details. *Modified from T. McCord et al., (1998).*

center of the absorption indicates that the CO_2 is not in the form of either solid ice or liquid, but rather occurs in microdeposits, bonded to some other material in the soil. The 4.57-μm absorption is believed to be due to a carbon–nitrogen compound based on its frequency, which corresponds to that expected for C≡N (a triple bond of carbon and nitrogen). The weaker features near 3.4 μm are also believed to be due to carbon bonds with hydrogen (C–H hydrocarbons). These features have also been identified in spectra of interstellar ice grains obtained by the European Space Agency's Infrared Space Observatory (ISO) mission. Observations of the satellites of Saturn by a similar instrument on the Cassini spacecraft show nearly identical CO_2 absorption and similar C≡N and C–H features (*see* Planetary Satellites). The other two absorption features (3.88 and 4.05 μm) seen on Ganymede and Callisto appear to be unique to the Jupiter system and are thought to be due to S–H bonds and sulfur dioxide, respectively.

4.2. Surface and Atmosphere Interactions with Local Environment

The environment around Jupiter is awash in radiation from charged particles trapped in Jupiter's intense magnetic field. Since they are trapped in the field, which rotates rapidly with Jupiter's spin, the particles sweep past the satellites in their comparatively slow orbits. Thus, the side of a satellite facing away from the direction of orbital motion (the trailing hemisphere) is exposed to a much higher dose of radiation than the leading hemisphere. When the charged particles strike the surfaces of the satellites, they can send surface molecules flying (a process called sputtering), and they can break molecular bonds in the surface material, causing new chemical reactions to occur and creating new compounds.

Laboratory studies suggest that the CO_2, C≡N, and C–H features seen in the spectra of Ganymede, Callisto, and other icy satellites in the solar system may have a common origin due to charged particle irradiation of minerals containing potassium cyanide and possibly other cyanogens (carbon-nitrogen-bearing compounds). Irradiation of these compounds by energetic particles in the presence of water ice is believed to be an important source of the CO_2 found embedded in the mineral/ice matrix on the surfaces of Ganymede and Callisto. On Callisto, the distribution of CO_2 mapped by NIMS shows a marked concentration on the trailing hemisphere of the satellite, as would be expected from radiation-induced CO_2 production. Ganymede shows no such enhancement of CO_2 on the trailing hemisphere, but instead shows an enhancement in the ancient regions of dark terrain.

Sulfur is an important ion in the Jupiter system, continually supplied to space by the escape of sulfur and sulfur dioxide gases from volcanic Io. The sulfur becomes ionized and joins the low-energy plasma streaming through Jupiter's magnetosphere, which then washes up on the other satellite surfaces. Implantation of sulfur into the surfaces of the icy Galilean satellites has been suggested in the past as the likely reason for the low reflectance of the satellites in the ultraviolet part of the spectrum. The sulfur-induced infrared absorption features are also plausible results of bombarding the icy surfaces with sulfur-rich plasma. However, the distribution of SO_2 on the surfaces of Ganymede and Callisto do not show a strong leading–trailing hemisphere asymmetry as one would expect from an external source trapped in Jupiter's magnetosphere, indicating that perhaps there is also sulfur in the ice bedrock.

Both Ganymede and Callisto exhibit other effects from their continual bombardment by charged particles in Jupiter's magnetosphere. Both molecular oxygen, O_2, and ozone, O_3, have been detected in telescopic spectra of their surfaces. These oxygen compounds appear to exist as microscopic bubbles trapped in the matrix of the icy surface material and have also been attributed to irradiation by charged particles.

In addition to the frozen and trapped gases in their surfaces, Ganymede and Callisto have very tenuous atmospheres. The ices on the surfaces of Ganymede and Callisto are weakly warmed by the sun, and exposed to near-vacuum conditions. Even at the cold temperatures in the Jupiter system, ice will slowly sublime and escape as a gas. On Ganymede, Hubble Space Telescope spectra have identified molecular oxygen, and the Galileo ultraviolet instrument detected a thin veil of hydrogen in the surrounding space. Callisto's atmosphere is similarly of very low density, and the only detectable gas so far has been CO_2 identified in NIMS spectra.

A combination of thermal segregation and sputtering has also been invoked to explain the distribution of water frost on Ganymede, where the high-latitude regions appear to be coated with polar caps of frost which may have migrated there from the equatorial areas and been retained there due to lower temperatures. The edges of Ganymede's polar caps (the northern cap is visible in Figure 37.1) closely follow the region where Ganymede's magnetosphere becomes connected with the external Jupiter magnetosphere, indicating that charged particles play an important role in creating the caps. The magnetosphere also plays a role in creating aurora displays on Ganymede, as charged particles funneled into Ganymede's polar regions from Jupiter's magnetosphere impact atomic oxygen in Ganymede's tenuous atmosphere.

Another interaction with the external environment of the satellites is micrometeoroid bombardment. All airless bodies are exposed to the flux of tiny grains from interplanetary space striking their surfaces. In the case of Ganymede and Callisto, the interplanetary particle fluxes are enhanced due to Jupiter's gravity. Galileo carried a sensitive detector that measured the surrounding dust environment as the spacecraft orbited Jupiter, and sampled the population of dust particles near the satellites. The dust investigators found that both satellites have a population of small (micron-sized) particles loosely bound by gravity in the space surrounding the satellites. These measurements are consistent with icy dust grains that have been blasted off the satellite surfaces by the impact of interplanetary micrometeorites.

The **sublimation** of surface ices into the thin atmospheres of Ganymede and Callisto and escape into space or redeposition as frost elsewhere appears to play a crucial role in shaping and redistributing the material seen at their surfaces. At typical temperatures in the Jupiter system, water ice could sublimate at a rate of meters per million years if directly exposed to a vacuum, but on Ganymede and Callisto, sublimation of dirty ice is soon choked off by a blanket of nonice dust left behind as the ice sublimates. A few centimeters of dust cover can slow the process by over a million times and preserve the ice beneath. Sublimation will occur millions of times faster for SO_2 ice than for water ice, and CO_2 ice will sublimate thousands of times faster than SO_2, so incorporation of these compounds into the ice bedrock will drive the sublimation erosion process much faster.

4.3. Regolith

Bright ice crystals and dark nonice dust both exist on the surfaces of Ganymede and Callisto, but they are largely segregated from each other. If one were to pick up a sample of the loose surface material (the regolith), it would probably be composed of mostly ice or mostly dust, and not a mixture of the two. High-resolution images show very high albedo contrasts over small spatial scales, with relatively pure icy material outcropping in patches surrounded by blankets of dark nonice material. This effect is most pronounced on Callisto (Figure 37.6). It appears that the ice

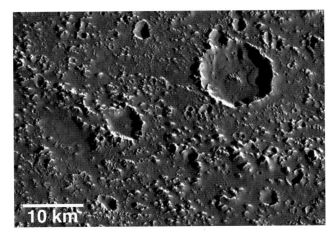

FIGURE 37.6 Callisto's surface is characterized by bright icy hills and impact crater rims surrounded by blankets of dark dust. This surface is thought to result from sublimation of an ice/dust mixture, leaving a lag deposit of loose dust in the low areas and depositing bright frost on steep slopes and hilltops. Note the prominent crater in the upper right section of the image displaying the raised tongue of a landslide deposit consisting of loose dark material originating from the shadowed crater wall. This area is located within the Asgard impact basin.

bedrock is composed of a mixture of ice and nonice dust. When a fresh outcrop of bedrock is exposed at the surface, the ice will begin to sublimate into the near-vacuum atmosphere. Enhanced solar heating of dark material drives faster sublimation of ice from that material, further darkening the material, while any reflective surface will serve as a cold trap, building up a layer of frost and further brightening the surface. These positive feedbacks lead to the effective segregation of ice and nonice materials in the surface regolith.

This process also operates on Ganymede, but it has not modified the surface to the same extent seen on Callisto. Perhaps there is more dark material mixed in with the surface ice on Callisto, or perhaps the ice on Callisto includes a higher proportion of volatile SO_2 and CO_2 ices that enhance the rate of sublimation. Patches of frost are often seen on steep slopes facing away from the sun on many parts of Ganymede's surface. Bright terrain on Ganymede has a higher **thermal inertia** than the dark terrain, indicating that much of the bright ice exposed in this terrain must be more solid or compacted than the loose dust that covers dark terrain.

On both Ganymede and Callisto, the dark material is found filling the topographic lows, while bright material covers the slopes and hilltops. Part of the reason for this is that reflection and emission of light from surrounding terrain tends to make topographic lows slightly warmer, but much of this effect appears to be due to the loose, dusty nature of the dark material. Buildup of loose dark material on steep slopes due to sublimation leads to avalanches of the material into topographic lows. The large crater in the northeastern corner of Figure 37.6 exhibits a thick tongue of dark material that has flowed over the floor; this appears to be a deposit of regolith that has slid from the steep eastern wall of the crater, and its shape indicates that it slid downhill as a dry avalanche of loose debris. Images of dark terrain on Ganymede also show chutes on steep slopes where material has slid downhill, with dark material piled along the bottoms of the slopes. Bright terrain on Ganymede shows the same effect, with dark material filling in the valleys of the grooved terrain, between bright steep icy slopes. The dark dust often appears to form a thick, smooth blanket on Callisto and Ganymede dark terrain, but there are many small craters that penetrate through the dark material, indicating that the layer of loose dark dust may only be meters deep before a solid layer of ice/dust mixture is reached.

5. IMPACT CRATERS

5.1. Crater Structures

Ganymede and Callisto exhibit a wide variety of impact features, including some types unique to these large icy satellites. The smallest craters imaged on the two moons have a classic bowl-shaped morphology, as is the case for small craters on any planet. At a diameter of 2–3 km, central peaks begin to appear (Figure 37.7(a)), again following the normal morphological progression for most planets. However, as crater diameter increases beyond 35 km, instead of the transition to larger central peaks or peak rings seen on the inner planets, large craters on Ganymede and Callisto exhibit central pits (Figure 37.7(b)). Young craters undergo another transition at about 60 km diameter, where the central pits begin to exhibit round domes of material in their centers (Figure 37.7(c)). These central domes have fractured surfaces reminiscent of lava domes, and they may be formed by rapid extrusion of warm, viscous ice into the center of the crater just after its formation. Most large craters on Ganymede and Callisto are very shallow, especially older craters, indicating that warm subsurface ice has flowed in toward the crater depressions and bowed their floors back up to the topographic level of their surroundings (a process known as viscous relaxation). Some central dome craters have been so flattened that they do not exhibit any obvious rim structure; the central dome and surrounding pit wall are the only obvious structures remaining (Figure 37.7(d)). A few large craters, known as penepalimpsests, exhibit only subdued topographic rings, with a smooth patch in the middle (Figure 37.7(e)). Where these occur on Callisto and the dark terrain of Ganymede, they show up as a distinct circular patch of bright material against the dark background. Still other bright circular patches are found within Ganymede's dark terrain and on Callisto that are almost completely flat, except for a subtle outward-facing scarp around the outside and a depressed smooth area in the center (Figure 37.7(f)). These features are called palimpsests, a word for an ancient piece of parchment where the writing has been erased. In a similar way, these large ancient craters have almost been erased by the process of viscous relaxation.

On the Moon, the largest craters form multiring basins (see The Moon). On Ganymede, one large basin called Gilgamesh shares similar characteristics with the lunar basins: a smooth central region surrounded by large irregular massifs, in turn surrounded by a few large concentric mountain ranges. However, most large impact basins on Ganymede and Callisto exhibit a distinctly different morphology, with a large palimpsest in the middle surrounded by many evenly spaced concentric rings. The best example of such an impact basin is Valhalla on Callisto, which is about 1000 km across and exhibits about 20 concentric rings around its central bright palimpsest (Figure 37.8). Most basin rings on Callisto are troughs that appear to have formed by extension of the surface material. These multiring structures are thought to form as subsurface material rapidly flows in from the sides to fill the

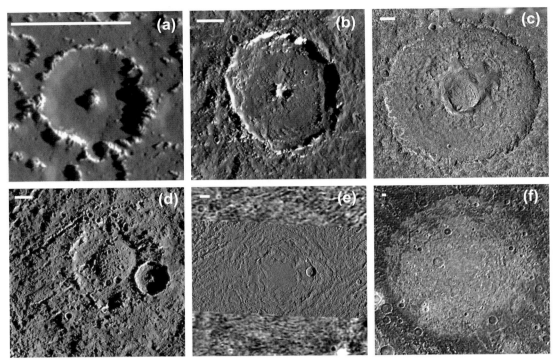

FIGURE 37.7 Diversity of impact crater morphologies on Ganymede and Callisto. All scale bars are 10 km long, and illumination is from the right. (a) Central peak crater on Callisto, (b) central pit crater on Callisto, (c) central dome crater Melkart on Ganymede, (d) anomalous dome crater Har on Callisto, (e) penepalimpsest Buto Facula on Ganymede, and (f) palimpsest Memphis Facula on Ganymede.

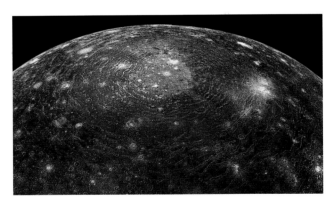

FIGURE 37.8 View from the Voyager 1 spacecraft of the Valhalla multiring basin on Callisto. The extensive system of concentric troughs surrounding the impact site is over 3000 km across.

center of the impact basin, pulling a thin brittle veneer of surface material inward.

The viscous relaxation and modification of craters can inform us about the nature of the subsurface ice during and after crater formation. Older craters are distinctly shallower, showing the action of viscous relaxation through time. However, there is not a continuum of viscously relaxed craters as one might expect if this process was ongoing at a constant rate. Instead, it appears that early craters relaxed quickly, while more recent craters are being preserved in a stiff material. This implies that heat flow was higher in the past, allowing warm ice to flow just below the surface early in solar system history, while more recently the subsurface ice has become colder and stiffer. There is overlap in size between central dome craters, penepalimpsests, and palimpsests, implying that impacts of similar energy formed all these morphologies at different times. Palimpsests are found only in the most ancient terrains, while central dome craters appear to be relatively young. Again, it appears that palimpsests formed early when the subsurface was warm and flowed easily, penepalimpsests record a time when the ice was cooling, and dome craters have formed more recently in a thicker layer of cold stiff ice. On Ganymede, the formation of the bright terrain appears to mark an important transition in crater morphology, with no palimpsests or Valhalla-type multiring basins formed after the formation of bright terrain. Thus, it appears that heat flow was higher on Ganymede in the past, and Ganymede's subsurface became colder and stiffer after the formation of bright terrain.

5.2. Distribution of Craters and Surface Ages

Variations in the aerial density of impact craters are observed on Ganymede and Callisto, giving us information about the population of impactors and the relative ages of different surfaces. In general, the highest crater densities are found on the dark terrain of Ganymede and the plains of Callisto. Bright terrain on Ganymede has a distinctly lower density of craters than the dark terrain, supporting the view that it formed substantially later. The only areas on Callisto with lower crater densities are the interiors of impact craters and large multiring basins, where the surface age has been reset by the impact event.

Translating the spatial density of impact craters into absolute ages of different surfaces on Ganymede and Callisto is a tricky proposition. On the Moon, this has been accomplished by correlating areas of varying crater density on the lunar surface with physical samples of those surface materials that have been returned to the Earth to be precisely dated in the laboratory using radioisotope techniques. Since we have no surface samples from the Galilean satellites, we cannot directly date them. In addition, we cannot be sure that the same population of debris that impacted the Moon also impacted the Galilean satellites, so it is dangerous to directly compare crater densities between these two different parts of the solar system. In general, it is agreed that the surface of Callisto and the dark terrain on Ganymede represent primordial surfaces, formed shortly after the formation of the planets, 4.5 Bya. Bright terrain on Ganymede could have formed shortly after that, or it could have formed only a billion years ago. The current best guess from crater statistics is that bright terrain most likely formed at some time during the middle half of solar system history, but obtaining an exact age is likely to remain elusive for a long time.

Since Ganymede and Callisto are tidally locked and always have the same side facing Jupiter, it is expected that they should gather more of the debris coming from outside the Jupiter system on the sides facing forward in their orbital motion (the bug on the windshield effect), and thus there should be more craters on their leading hemispheres than on their trailing hemispheres. Callisto does exhibit such an asymmetry in crater density, but the asymmetry on Ganymede is much weaker. One hypothesis to explain this is that Ganymede's outer ice shell has rotated with respect to Jupiter in the past and has become locked to Jupiter more recently, while Callisto's surface has always been locked with respect to Jupiter. Another piece of evidence to support this view comes from the study of split comets. In 1994, the comet Shoemaker–Levy 9 impacted into Jupiter, after having been tidally disrupted into a string of fragments by its previous close encounter with Jupiter. If such a string of comet fragments hits one of the satellites on its way out of the Jupiter system, it would form a line of closely spaced, simultaneously formed impact craters called a catena. These catenae are in fact observed on the surfaces on Ganymede and Callisto, recording previous tidally disrupted objects like Shoemaker–Levy 9. On Callisto, all the catenae are found on the Jupiter-facing hemisphere, as one would expect from the impact of a comet on its way out of the system after a close brush with Jupiter. On Ganymede, one-third of the catenae are found on the hemisphere facing away from Jupiter, which would be impossible unless Ganymede's ice shell has rotated in the past.

6. TECTONISM AND VOLCANISM

Like rocky planets, icy worlds also respond to intense surface stresses by cracking and faulting, forming rift valleys and ridges through tectonics. Ice is weaker than rock, but at the cold temperatures in the Jupiter system it is stronger and stiffer than the comparatively warm glacial ice found on the Earth. If liquid water is able to make its way to the cold surface of an icy world, a type of volcanism known as **cryovolcanism** may form lava flows of ice across the surface.

The observed surface record of past tectonic and volcanic activity is the most obvious difference between Ganymede and Callisto. Most of Ganymede's surface has been reworked by some combination of these processes, while Callisto's surface appears to be untouched by internally driven geological activity. Below, we separately consider the roles of tectonism and volcanism in the extensively resurfaced bright terrain of Ganymede, the marginally resurfaced dark terrain of Ganymede, and the relatively pristine surface of Callisto.

6.1. Bright Terrain

Bright terrain covers two-thirds of Ganymede's surface and is composed of a dense network of intersecting and overlapping areas of parallel ridges and troughs, termed "grooved terrain", and other areas with more subdued topography, termed "smooth terrain".

Smooth terrain may occur either as patches bounded by grooved terrain on all sides or as lanes of smooth material tens of kilometers wide cutting across bright and dark terrain. In either case, the terrain appears to be smooth in kilometer-resolution regional images (Figure 37.9(a)), leading to the hypothesis that it formed by low-viscosity cryovolcanic flows flooding the underlying terrain. At higher resolution, it becomes apparent that the smooth terrain is not so smooth after all. In some areas, it appears to be a flat plain crossed by ridges or sets of aligned hills (Figure 37.9(b)). In other areas, especially where the smooth terrain occurs as narrow lanes, it appears to be a flat or gently undulating surface crossed by parallel dark lineations, which may be narrow valleys formed by tensile fracturing of the ice (Figure 37.9(c)). The presence of parallel sets of ridges and valleys in smooth terrain suggests that tectonism plays an important role in shaping this terrain, in addition to possible cryovolcanism.

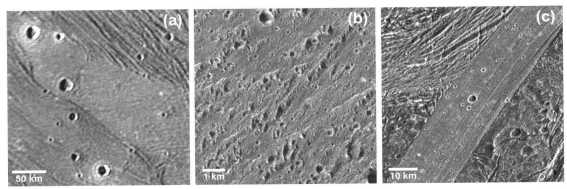

FIGURE 37.9 Views of smooth bright terrain on Ganymede. (a) Voyager 1 image of smooth bright terrain in Harpagia Sulcus; (b) Galileo high-resolution image of smooth terrain from the center of (a), showing ridges and hills not visible in regional-scale views; and (c) Galileo image of Arbela Sulcus, a narrow lane of smooth terrain cutting through the dark terrain of Nicholson Regio.

Although cryovolcanism is an attractive explanation for the smooth, flat areas found within smooth terrain, its role had not been conclusively demonstrated. No obvious volcanic constructs or flows have been observed, although it is unclear if we know what an ice volcano is really supposed to look like. A few features that may possibly be volcanic calderas have been observed (Figure 37.10), but most areas of smooth terrain exhibit no such features. While the smooth regions shown in Figure 37.10 are topographic lows, as one would expect if they were troughs filled by low-viscosity volcanic flows, the smooth regions shown in Figure 37.9 have been found to lie locally higher than parts of their immediate surroundings. Another possible interpretation for the linear bands of smooth terrain is that they formed through separation and spreading of the crust in a manner analogous to Europa's gray bands (see Europa). In either case, the formation of smooth terrain appears to involve the extrusion of liquid water or warm ice from Ganymede's subsurface.

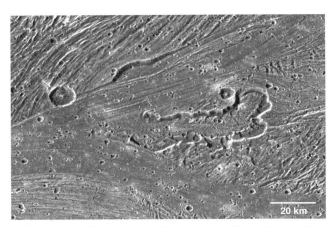

FIGURE 37.10 This irregular pit is one of several found along the edges of smooth terrain in Sippar Sulcus on Ganymede. Faint curved ridges within the pit suggest flow folding in the surface of a cryovolcanic flow emanating from the closed end of the pit and flowing out the open end, into the surrounding smooth terrain.

Tectonism plays a more obvious role in the formation of grooved terrain. In kilometer-resolution images, grooved terrain is characterized by parallel valleys and ridges spaced about 5–10 km apart (Figure 37.11(a)). At higher resolution, each ridge and valley is itself composed of many smaller ridges and valleys (Figure 37.11(b)). Each of these smaller ridges is thought to be a fault block, a piece of the icy crust which has been separated from its surroundings by faults and then moved and tilted as it slid along those faults. The shapes and intersections of the faults are indicative of a style of faulting known as tilt-block normal faulting, in which many parallel faults slice the upper portion of the crust into roughly rectangular blocks which then tilt over and slide against each other as the crust extends, much like books sliding over on a bookshelf when a bookend is removed. This style of faulting creates parallel ridges with a sawtooth topographic profile, matching the triangular ridges on Ganymede with their sharp crests, frosty upper slopes, and dark V-shaped valleys between them.

In a few places on Ganymede, large impact craters have been cut by these networks of faults (Figure 37.12). Since almost all craters are formed in a roughly circular shape, these cut craters offer an opportunity to directly measure how the crust has deformed as their shape becomes progressively distorted by motion along the faults. Measurements of these craters confirm that the development of grooved terrain on Ganymede is dominated by extensional tectonics. At the extreme end of the spectrum, some small parts of the crust appear to have been pulled apart to more than twice their original width, but in most cases, the extension appears to be more moderate. Circumstantial evidence exists for contractional deformation in a few areas, but it is not widespread.

These observations force us to ask how the crust of Ganymede could have undergone a large amount of extension with very little evidence for contraction to balance it out. There are a few possible solutions to this conundrum. One solution is that Ganymede actually expanded during the

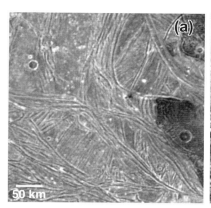
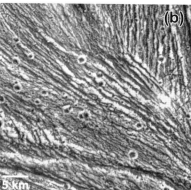

FIGURE 37.11 Views of bright grooved terrain on Ganymede. (a) Voyager 2 imaged this region of grooved terrain in Uruk Sulcus and (b) Galileo image of grooves in the central part of (a), shown at 10 times the scale.

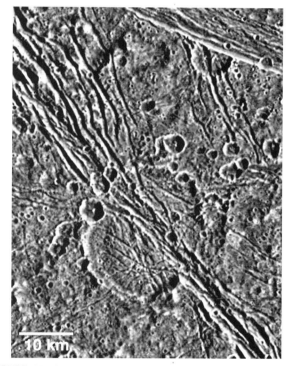

FIGURE 37.12 This crater in the dark terrain of Marius Regio on Ganymede has been cut by faults extending from a nearby region of grooved terrain. Measurement of the postdeformation shape of the crater demonstrates that the faults have extended the terrain and also horizontally translated the eastern part northward relative to the western part.

In places where narrow lanes of grooved terrain cut across dark terrain, it is clear that bright grooved terrain can form simply by extension of the dark terrain, without the cryovolcanism that may have taken place in smooth terrain. Faulting may serve to erase the impact craters on the dark terrain by slicing them up and making their rims unrecognizable. Brightening of the terrain can occur by breaking through the dark regolith layer and exposing bright subsurface ice along the fault scarps.

In larger regions of grooved terrain, tectonism and cryovolcanism may have acted together to shape Ganymede's surface. Small deviations in the trajectory of the Galileo spacecraft as it flew past Ganymede appear to be due to gravity anomalies, which are areas where the local density of the subsurface material is higher or lower than the average for the rest of the satellite. These gravity anomalies can be interpreted to show that bright terrain has anomalously low density, while dark terrain has anomalously high density. This could mean that the icy brightness of bright terrain and the rocky darkness of dark terrain are more than skin-deep.

6.2. Dark Terrain

Dark terrain on Ganymede is dominated by impact craters. Aside from the swaths of bright grooved terrain that cut across the dark terrain, and small peripheral fractures adjacent to the grooved terrain, tectonic features in dark terrain are primarily systems of arcuate to linear features known as furrows. Furrows are usually composed of two bright ridges spaced 10–20 km apart, with a dark trough in between them. Most furrows are arranged in concentric sets of arcs (Figure 37.13), indicating that they are probably ancient multiring basins that originally resembled Valhalla on Callisto, but are now sliced up into fragments by the formation of bright terrain. This interpretation is supported by some small furrow systems that appear to have an impact basin in the center. Some sparse systems of linear

formation of grooved terrain. Differentiation of Ganymede's interior or melting of high-pressure ices can serve to increase the volume of the satellite, leading to an increase in surface area and thus stretching of the crust. Alternatively, we may have missed seeing the contractional features on Ganymede either because the crust shortened mostly in a ductile fashion, leaving few obvious surface features, or because we simply do not recognize the morphology of contractional features formed in ice.

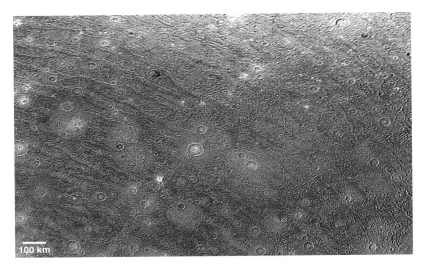

FIGURE 37.13 Furrows in Galileo Regio on Ganymede arc gently from northwest to southeast. Furrows are thought to be the Ganymede equivalent of the concentric rings found around impact basins such as Valhalla on Callisto (see Figure 37.8).

furrows appear to radiate out from a point rather than being concentric arcs. The origin of these radial systems is unclear.

There was speculation based on Voyager images that some areas of dark terrain had a splotchy appearance due to patches of dark cryovolcanic material oozing onto the surface. At higher resolution, however, these splotchy areas were revealed to be plains of dark regolith interrupted by networks of bright fractures.

6.3. Callisto

The story of tectonism and volcanism on Callisto is reminiscent of Ganymede's dark terrain. All the obvious tectonic features are arranged in concentric rings and surround large impact basins. The rings are high scarps or deep troughs with sharp boundaries (Figure 37.14). The scarps and troughs are formed by faults that have extended the crust by a small amount during the formation of the impact basin (see Section 4). Multiring basins on Callisto provide the type examples for the ancient impact basins that formed the furrow systems on Ganymede before they were broken apart. There is also a system of troughs near Callisto's north pole that appears to radiate out from a central point. Unfortunately, the center of this system on Callisto was never imaged at high resolution, so the origin of these features remains mysterious, as does the origin of similar radial furrow systems on Ganymede. Early speculation that smooth dark patches on Callisto might be cryovolcanic in origin has been largely dispelled by evidence that Callisto has a loose regolith that smoothes over the underlying terrain like a thick dark blanket.

7. UNANSWERED QUESTIONS AND FUTURE EXPLORATION

Several interesting unanswered questions remain about Ganymede and Callisto. Since the general properties of these satellites appear to be so similar, understanding the processes and events that have driven their interior

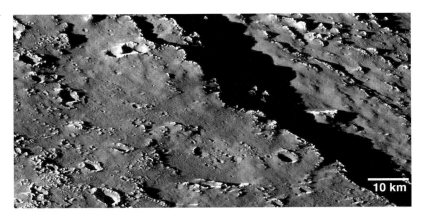

FIGURE 37.14 Oblique view over the surface of Callisto, looking over the edge of one of the concentric ring scarps of the Valhalla impact basin.

evolution and geological records to different states is an important problem in comparative planetology. In the interiors of the satellites, making Ganymede hot enough to generate a **magnetic dynamo** and keeping Callisto cold enough to remain undifferentiated are both challenging problems for our understanding of planetary geophysics. The oceans of liquid water that exist within these bodies, sandwiched between different phases of ice, are interesting phenomena in themselves. On the surfaces of Ganymede and Callisto, we still do not have a clear idea of the composition of some of the materials that are mixed in with the water ice, and which of those materials come from the interiors of the satellites, which ones come from their external environments, and which ones are the products of chemical reactions and radiation processing at the surface.

The unfortunate failure of the main antenna on the Galileo spacecraft left us without a complete global reconnaissance of these bodies at a level of detail sufficient to resolve features at the scale of a kilometer or less. The small target areas that Galileo imaged at high resolution revolutionized our understanding of these bodies, but much of their surfaces will remain relatively unknown for the near future. The Juno mission will arrive to study Jupiter in 2016, but it will not be making significant observations of the satellites.

Our knowledge of Ganymede and Callisto will be revolutionized in the early 2030s when the European Space Agency's Jupiter Icy Moons Explorer (JUICE) spacecraft is scheduled to arrive at the Jupiter system, making several close flybys of Callisto and parking in orbit around Ganymede. The JUICE payload will include instruments for understanding the magnetosphere and plasma environments around the moons, their thin atmospheres, and the detailed spectra of their surface materials. In addition, several instruments will be able to map the surfaces of these bodies in much greater detail, measure the global topography of Ganymede's surface, and use radar to probe the subsurface structure of the regolith and underlying ice. By the end of the JUICE mission, we will have a much more complete survey of the geology and geophysics of Ganymede as a prototypical icy world, and we will have a clearer understanding of its subsurface ocean and its prospects for habitability. In addition to JUICE, a mission to explore Europa is still ranked as a high priority for NASA, and any future mission to Europa would use close flybys of Ganymede and Callisto for gravitational assistance into Europa's orbit. Such a mission would also provide several opportunities to gather more information about these mysterious twin moons of Jupiter.

BIBLIOGRAPHY

Clark, R., & McCord, T. (1980). The Galilean satellites: New near-infrared spectral reflectance measurements (0.65–2.5 μm) and a 0.325–5 μm summary. *Icarus*.

Collins, G. C., Patterson, G. W., Head, J. W., Pappalardo, R. T., Prockter, L. M., Lucchitta, B. K., & Kay, J. P. (2014). *Global geological map of Ganymede, United States Geological Survey Science Investigations Map Series #3237*.

Grasset, O., Dougherty, M. K., Coustenis, A., Bunce, E. J., Erd, C., Titov, D., et al. (2013). JUpiter ICy moons Explorer (JUICE): an ESA mission to orbit Ganymede and to characterize the Jupiter system. *Planetary and Space Science*. http://dx.doi.org/10.1016/j.pss.2012.12.002.

Johnson, T. V. (2004). A look at the Galilean satellites after the Galileo Mission. *Physics Today, 57*, 77–83.

McCord, T., et al. (1998). Non-water-ice constituents in the surface material of the icy Galilean satellites from the Galileo near-infrared mapping spectrometer investigation. *J. Geophys. Res.*.

Moore, J. M., Chapman, C. R., Bierhaus, E. B., Greeley, R., Chuang, F. C., Klemaszewski, J., et al. (2004). Callisto. In F. Bagenal, et al. (Eds.), *Jupiter: The planet, satellites, and magnetosphere* (pp. 397–426). Cambridge, UK: Cambridge University Press.

Pappalardo, R. T., Collins, G. C., Head, J. W., Helfenstein, P., McCord, T., Moore, J. M., et al. (2004). Geology of Ganymede. In F. Bagenal, et al. (Eds.), *Jupiter: The planet, satellites & magnetosphere* (pp. 363–396). Cambridge, UK: Cambridge University Press.

Chapter 38

Titan

Athena Coustenis

LESIA - Observatoire de Paris, CNRS, UPMC Univ. Paris 06, Univ. Paris-Diderot – Meudon, France

Chapter Outline

1. Introduction — 831
  1.1. Titan's Discovery, First Observations, and Models — 831
  1.2. Titan's Exploration — 832
2. The Atmosphere of Titan — 834
  2.1. Thermal Structure — 834
  2.2. Chemical Composition — 835
  2.3. Dynamical Processes — 837
    2.3.1. Zonal Circulation — 837
    2.3.2. Latitudinal and Temporal Variations in the Atmosphere of Titan as Evidence of Meridional Circulation — 837
    2.3.3. A Three-Dimensional View and Waves — 838
  2.4. Haze and Clouds on Titan — 839
    2.4.1. Titan's Haze — 839
    2.4.2. Titan's Clouds — 840
  2.5. Origin and Evolution — 841
3. The Surface of Titan — 841
  3.1. Pre-Cassini Glimpses of an Exotic Ground — 842
  3.2. The View from the Orbiter — 843
  3.3. In Situ Data: the Ground Truth — 845
4. The Interior and Exchange Processes — 847
5. Looking Ahead — 848
Acknowledgments — 849
Bibliography — 849

1. INTRODUCTION

1.1. Titan's Discovery, First Observations, and Models

Titan, Saturn's biggest satellite (second in size among the satellites in our solar system), has attracted the eye of astronomers preferentially ever since its discovery by Dutch astronomer Christiaan Huygens on March 25, 1655. Titan orbits around Saturn at a distance of 1,222,000 km (759,478 mi) in a synchronous rotation, taking 15.9 days to complete. As Titan follows Saturn on its trek around the Sun, one Titanian year equals about 30 Earth years. The sunlight that reaches such distances is only 1/100th of that received by the Earth. Titan is therefore a cold and dark place, but a fascinating one.

It has been known for a long time that Titan possesses a substantial atmosphere: Catalan astronomer Jose Comas i Solá claimed in 1908 to have observed **limb darkening** on Titan. Due to its thick atmosphere, Titan subtends 0.8 **arcsec** in the sky, and it was thought to be the largest of the satellites in the solar system. This explains the name it was given (following a proposition by Herschel, who suggested names of gods associated with Saturn for naming its satellites), until the advent of the **Voyager** missions that showed Ganymede to be a few kilometers larger. Today, we know that this massive atmosphere is the one most similar to the Earth's among the other objects of our solar system, as N_2 is its major constituent and it hosts a complex organic chemistry.

In 1925, Sir James Jeans showed that Titan could have indeed kept an atmosphere, in spite of its small size and weak gravity, because some of the constituents which could have been present in the protosolar nebula (ammonia, argon, neon, molecular nitrogen and methane) would not escape. It was realized later that although ammonia (NH_3) is in solid phase at the current Titan temperatures and could not in principle contribute to its present atmosphere, it could have evaporated in the early atmosphere and been converted into N_2 at the end of the accretion period when the environment was warmer.

On the other hand, methane (CH_4), the second most abundant constituent on Titan, is gaseous at the present Titan's atmospheric temperature range and, unlike molecular nitrogen, exhibits strong absorption bands in the infrared. These bands were first detected in 1944 by Gerard Kuiper of Chicago University. Ethane (C_2H_6), monodeuterated methane (CH$_3$D), ethylene (C$_2$H$_4$), and acetylene (C$_2$H$_2$) were also discovered later.

Prior to spacecraft observations, two models were popular: (1) a "thin methane" atmosphere model, which favored methane as the main component (about 90%) and predicted surface conditions of $T = 86$ K for 20 mbar as well as a temperature inversion in the higher atmospheric levels, illustrated by the presence of emission features of hydrocarbon gases in the infrared spectrum of Titan, and (2) a "thick nitrogen" atmosphere model, which was based on the assumption that ammonia dissociation should produce molecular nitrogen (transparent in the visible and infrared spectrum) in large quantities and held that the surface temperature and pressure could be quite high (200 K for 20 bars). The second model was found to be closer to the truth for the atmospheric composition. Independent of these two models, an explanation of the high observed ground temperatures was advanced: a pronounced greenhouse effect, thought to result from H$_2$–H$_2$ pressure-induced **opacity** at **wavelengths** higher than 15 μm. Later, climate modeling showed that nitrogen and methane also contributed significantly to this greenhouse effect, while the haze caused an antigreenhouse effect (see Section 2.2).

1.2. Titan's Exploration

Titan has since then been extensively studied from the ground and from space. In the latter case, Titan was "blessed" by several space mission encounters in the course of the planetary exploration in our solar system. (See also Planetary Exploration Missions.)

The Pioneer 11 spacecraft was the first to take a close look at the giant planets Jupiter and Saturn, but it flew by Titan at a considerable distance of 363,000 km on September 2, 1979. The Voyager missions that followed were also dedicated to an extended study of the outer solar system. The Voyager 1 (V1) spacecraft (launched in 1977) arrived in the Saturnian system and made its closest approach of Titan on November 12, 1980, at a distance of only 6969 km (4394 miles) to the satellite's center. Voyager 2 flew by Titan 9 months later but at a distance a 100 times greater (663,385 km).

Titan's visible appearance at the time was unexciting—an orange ball, completely covered by thick haze, which allowed no visibility of the surface (Figure 38.1(a)). The most obvious feature seen by Voyager was a difference in the brightness of the two hemispheres. This difference is of the order of 25% at blue wavelengths and falls to a few percent in the ultraviolet (UV) and at red wavelengths. This so-called north–south asymmetry (NSA) is probably related to circulation in the atmosphere pushing haze from one hemisphere to the other. The altitude of unity vertical optical depth is of the order of 100 km. Also noticeable was a dark ring above the north

FIGURE 38.1 Titan observed in 1980 with the cameras of Voyager 1 in the visible region (left) and in 2004 with the Cassini Imaging Science Subsystem (ISS) camera at 0.94 μm (right). In the first case, the bland appearance of the satellite belies a complex world. The only features apparent in the images taken by Voyager were the detached haze layers, the dark polar hood, and generally a difference in brightness between the two hemispheres. In contrast, the recent images by Cassini show Titan's surface features.

(winter) pole. This feature, called polar hood, extending from 70° to 90° N, was most prominent at blue and violet wavelengths, and has since then been associated with the lack of illumination in the polar regions during the winter (since the subsolar latitude goes up to 26.7°) and/or subsidence in global circulation.

Besides the images, the Voyager **radio-occultation** experiment obtained by the Radio Science Subsystem (RSS) provided the basic parameters for Titan (Table 38.1) and the Infrared Radiometer Spectrometer (IRIS) determined the chemical composition and temperature structure in parts of the atmosphere (Table 38.2). Titan's surface radius was found to be 2575 ± 2 km, with a surface temperature of 94 ± 2 K and a pressure of about 1.44 bar,

TABLE 38.1 Titan's Orbital and Body Parameters and Atmospheric Properties

Surface radius	2575 km
Mass	1.35×10^{23} kg (=0.022 × Earth)
Mean density	1.88 g/cm^3
Distance from Saturn	1.23×10^9 m (=20 Saturn radii)
Distance from Sun	9.546 AU
Orbital period	15.95 days
Revolution around Sun	29.5 years
Obliquity	26.7°
Surface temperature	93.6 K
Surface pressure	1.467 bar

Chapter | 38 Titan

TABLE 38.2 Chemical Composition of Titan's Atmosphere from Cassini–Huygens Results (Unless Otherwise Indicated)

Constituent	Mole Fraction (Atmosphere Altitude Level)
Major	
Molecular nitrogen, N_2	0.98
Methane, CH_4	4.9×10^{-2} (surface)
	$1.4–1.6 \times 10^{-2}$ (stratosphere)
Monodeuterated methane, CH_3D	6×10^{-6} (in CH_3D, in stratosphere.)
Argon, ^{36}Ar	2.8×10^{-7}
^{40}Ar	4.3×10^{-5}
Minor	
Hydrogen, H_2	~ 0.0011
Ethane, C_2H_6	1.5×10^{-5} (around 130 km)
Propane, C_3H_8	5×10^{-7} (around 125 km)
Acetylene, C_2H_2	4×10^{-6} (around 140 km)
Ethylene, C_2H_4	1.5×10^{-7} (around 130 km)
Methylacetylene, CH_3C_2H	6.5×10^{-9} (around 110 km)*
Diacetylene, C_4H_2	1.3×10^{-9} (around 110 km)*
Cyanogen, C_2N_2	5.5×10^{-9} (around 120 km)*
Hydrogen cyanide, HCN	1.0×10^{-7} (around 120 km)*
	5×10^{-7} (around 200 km)§
	5×10^{-6} (around 500 km)§
Cyanoacetylene, HC_3N	1×10^{-9} (around 120 km)*
	$1 \times 10+$ (around 500 km)§
Acetonitrile, CH_3CN	1×10^{-8} (around 200 km)¶
	1×10^{-7} (around 500 km)
Water, H_2O	8×10^{-9} (at 400 km)** and 10^{-10} (at 200 km)§§
Carbon monoxide, CO	4×10^{-5} (uniform profile)¶¶
Carbon dioxide, CO_2	1.5×10^{-8} (around 120 km)

*Increasing in the North.
§From ground-based heterodyne microwave observations.
¶Only observed from the ground.
**From ISO observations.
§§From Cassini/CIRS.
¶¶From Cassini and ground-based data.

ground-based and Earth-bound observatories (like the Canada France Hawaii Telescope (CFHT), the Keck, the Very Large Telescope (VLT) or the Hubble Space Telescope (HST), and the Infrared Space Observatory (ISO)) were used to extract complementary information on Titan's inhomogeneous surface and exciting organic chemistry. Cassini–Huygens is a very ambitious mission and an extremely successful collaboration between the European Space Agency (ESA) and National Aeronautics and Space Administration (NASA) (with contributions from 17 countries). It is composed of an orbiter (Cassini) and a probe (Huygens). Although the mission's objectives span the entire Saturnian system, Titan is a privileged target (as for Voyager before it), and the mission is designed to address our principal questions about this satellite and more during its 13-year duration from 2004 onward. The spacecraft is equipped with 18 science instruments (12 on the orbiter and six carried by the probe), gathering both remote sensing and in situ data. It communicates through one high-gain and two low-gain antennas. Power is provided through three radioisotope thermoelectric generators.

The 5650-kg (6 ton) Cassini–Huygens spacecraft was launched successfully on October 15, 1997, from the Kennedy Space Center at Cape Canaveral at 4:43 a.m. EDT. Because of its massive weight, Cassini could not be sent directly to Saturn but used the "gravity assist" technique to gain the energy required by looping twice around the Sun. It also performed flybys of Venus (April 26, 1998, and June 24, 1999), Earth (August 18, 1999), and Jupiter (December 30, 2000). Cassini–Huygens reached Saturn in July 2004 and achieved a flawless Saturn Orbit Insertion, becoming trapped in orbit, like one of Saturn's moons. This situation will last at least until 2017 when it is foreseen to "crash" Cassini into the giant planet.

The Cassini instruments have since then returned a great amount of data concerning the Saturnian system. During its mission so far, the Cassini orbiter has made more than 100 flybys of Titan (up to 2014), some as close as 1000 km (Voyager 1 flew by at 4400 km) from the surface. Cassini performs direct measurements with the visible, infrared, and radar instruments designed to perform remote and in situ (on-site) studies of elements of Saturn, its atmosphere, moons, rings, and magnetosphere. One set of instruments studies the temperatures in various locations, the plasma levels, the neutral and charged particles, the surface composition, the atmospheres and rings, the solar wind, and even the dust grains in the Saturn system, while another performs spectral mapping for high-quality images of the ringed planet, its moons, and its rings.

Additionally, the mission saw the deployment of the European-built Huygens probe. After release from the Cassini orbiter, on December 25, 2004, this 300-kg probe plunged into Titan's atmosphere on January 14, 2005, at

values that were later confirmed by the in situ measurements.

After Voyager, scientists had to wait for about 25 years before getting another close look at Titan. In the meantime,

FIGURE 38.2 Artistic view of the descent of the Huygens probe in Titan's atmosphere and its landing (the real landing site is different from that depicted here). The three parachutes that helped brake the descent and reduce the speed to about 5 m/s on the surface are shown. The total descent lasted 2 h 28 min. The probe spent 1 h 12 min on the surface. The signal from Huygens received on Earth also via radio-telescopes was for a total of 5 h 42 min including 3 h 14 min on the surface.

11:04 UTC and through it descended to the ground. The heat shield decelerated the probe from supersonic speeds down to about 0.5 km/s, followed by several parachute brakes (Figure 38.2), which further slowed the probe to 0.005 km/s at impact. The five batteries onboard the probe lasted much longer than expected, allowing Huygens to collect atmospheric data for 2 h 27 min and surface data for 1 h 12 min. During its descent, Huygens' camera returned more than 750 images, while the probe's other instruments sampled Titan's atmosphere to help determine its composition and structure. The telemetry data from Huygens was stored onboard Cassini's Solid State Recorders at the rate of 8 kbits/s, while the spacecraft was at an altitude of 60,000 km from Titan. Although some measurements (in particular for the wind and the telemetry) from Huygens were lost during its transmission to Cassini, in the end the wind profile was recovered thanks to the fact that Titan's weak signal was captured by Earth-based radio telescopes!

Apart from measuring the atmosphere and surface properties, the probe took samples of the haze and gases. These in situ measurements complement the remote sensing data recorded from the orbiter.

The Cassini–Huygens mission has already provided a wealth of data. Although there are only a few years left until its foreseen termination in 2017, the Cassini orbiter promises to unveil yet more of Titan's secrets in the years to come. In this it is complemented by observations from the large telescopes on the ground and from Earth-orbiting satellites. What follows is an attempt to provide the reader with a brief account of our current understanding of Titan's environment from all such available means of investigation.

2. THE ATMOSPHERE OF TITAN

One of the most interesting features of Titan is its unique atmosphere, a close analog to the Earth's, but more extended and dense and located almost 10 times further away from the Sun.

2.1. Thermal Structure

The first definitive measurement of the atmospheric temperature structure was made by Voyager. The V1 radio-occultation experiment provided density and temperature profiles in Titan's atmosphere from refractivity measurements. Titan's temperature profile was also measured in situ on January, 14, 2005, by the Huygens Atmospheric Structure Instrument (HASI) at the probe's landing site (10° S, 192° W) from 1400 km in altitude down to the surface, where 93.65 ± 0.25 K was measured for a surface pressure of 1467 ± 1 mbar. As Voyager had shown before, Cassini/HASI also found Titan's atmosphere to exhibit the features that characterize the Earth's thermal structure: the atmospheric layers include a thermosphere, a mesosphere, a stratosphere and a troposphere, with two major temperature inversions at 40 and 250 km, corresponding to the tropopause and stratopause, associated with temperatures of 70.43 K (min) and 186 K (max), respectively (Figure 38.3). In addition, HASI recorded several temperature fluctuations due to dynamical (gravity and tidal) phenomena at higher levels of the atmosphere. Indeed, gravity waves signatures of 10–20 K in amplitude were recorded above 500 km around an average temperature of 170 K. Moreover, HASI found a lower ionospheric layer between 140

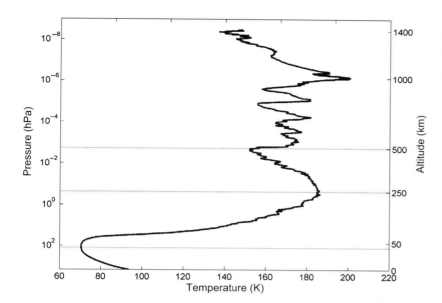

FIGURE 38.3 Titan's temperature profile as derived from Huygens/HASI measurements. The several large inversion layers in the upper atmosphere correspond to gravity waves. The inflection at around 40 km marks the tropopause, whereas the one at around 250 km is due to the stratopause. *Adapted from Fulchignoni et al. (2005).*

and 40 km, with electrical conductivity peaking near 60 km. At the same time, the Composite Infrared Radiometer Spectrometer (CIRS) on the orbiter took spectra that confirmed the presence of a stratopause around 310 km of altitude for a maximum temperature of 186 K. Another inversion region, less contrasted than the previous ones and corresponding to the mesopause, was found at 490 km (for 152 K) in the early years but has gradually disappeared in the recent years.

Besides the Huygens measurements, Titan's higher atmosphere was also explored by the V1 Ultraviolet Spectrometer (UVS) experiment which recorded a temperature of 186 ± 20 K at 1265 km during a solar occulation for a methane mixing ratio of $8 \pm 3\%$ toward 1125 km, placing the homopause level at around 925 ± 70 km. A value of 183 ± 11 K near 450 km was derived from the July 3, 1989, stellar occultation of Titan. The occultation of star 28 Sgr by Titan was observed from places as widely dispersed as Israel, the Vatican, and Paris. This rare event provided information in the 250–500 km altitude range. A mean scale height of 48 km at 450 km altitude (~ 5 µbar level) was inferred. This allowed the mean temperature to be constrained to between 149 and 178 K at that level.

From 1980 V1 infrared disk-resolved measurements, latitudinal variations in temperature were already demonstrated to exist in Titan's stratosphere. At that time, a maximal temperature decrease of 17 K at the 0.4-mbar level (225 km in altitude) was observed between 5° S (the warmest region in the Voyager data) and 70° N, whereas the temperature dropped by only 3 K from 5° S to 53° S. The coldest temperatures, found at high northern latitudes, were associated with enhanced gas concentration and haze opacity (as this may be caused by more efficient cooling) or/and **dynamical inertia**. CIRS has been mapping stratospheric temperatures over much of Titan's disk from the later half of 2004, when it was early southern summer on Titan (solstice was in October 2002) to well after the northern spring equinox (in August 2009). In mid-2010, the epoch on Titan was the same as during the V1 encounter, almost 30 years before (1 Titan year). Temperatures are found to have returned to their 1980 values, with the coldest ones in the north (by 10–20 K with respect to the equatorial ones below 300 km), where the season has been winter moving into spring after mid-2009.

2.2. Chemical Composition

Molecular nitrogen (N_2, detected in the UV) is by far the major component of the atmosphere (average of $\sim 95\%$). The presence of methane (the next most abundant molecule with a mixing ratio of about 1.5% in the stratosphere and 5% at the surface) and of traces of hydrogen give rise to a host of organic gases whose presence in the stratosphere had been established since the twentieth century.

About 90% of the energy at the surface of Titan is held in by a greenhouse effect due to nitrogen, methane, and hydrogen, symmetrical molecules which normally do not have a greenhouse effect on the Earth but have the effect on Titan, due to the dense atmosphere. This opacity blocks the thermal emission reflected by the surface, thus heating up the lower part of the atmosphere, as found on the Earth, where water is the major player. The increase in temperature from the tropopause to the surface of Titan due to this greenhouse effect is of 21 K. Even more interestingly, Titan is the only world in the solar system that has an antigreenhouse effect that lets light in and stops infrared, caused by the haze layers in the atmosphere. The antigreenhouse effect on Titan is half as strong as the greenhouse effect. The troposphere emission temperature (near the tropopause) is determined by the antigreenhouse effect and is 9 K cooler than the effective temperature.

A big leap in understanding Titan's chemical structure came from Cassini and measurements by its Ion and Neutral Mass Spectrometer (INMS) in the ionosphere. Titan was found to have a quite extended ionosphere, due to the lack of a strong intrinsic global magnetic field. High-energy photons (Extreme Ultra Violet and X-rays) and energetic particles from Saturn's magnetosphere are the main energy sources in Titan's upper atmosphere that create an extended ionosphere between 700 and 2700 km. At lower altitudes, galactic cosmic rays are responsible for the production of another ion layer in the atmosphere (between 40 and 140 km), while the neutral atmospheric photochemistry is mainly driven by Far Ultraviolet solar photons.

As Cassini flew through the upper parts of the ionosphere it discovered a multitude of complex positive and negative ions with masses reaching about a few hundreds of atomic mass units for positive and thousands of atomic mass units for negative ions. These molecules are increasing size hydrocarbons and nitrogen-containing species, which diffuse downward. These species are continuously quantified and monitored with precision since 2004 in INMS spectra (with mass detections ranging up to 100 amu and a resolution of 1 amu).

In parallel, the stratospheric chemical composition is revealed from the thermal emission bands of the different molecules observed in the infrared spectrometer CIRS spectra, which cover the 200–1500 cm^{-1} spectral region with a spectral resolution of up to 0.5 cm^{-1} (Table 38.2). In particular, following up on disk-averaged determinations of the chemical abundances on Titan in 1997 by the ISO Short Wavelength Spectrometer, Cassini has recently provided accurate spatial (latitudinal and vertical) distributions of the trace gases (Figure 38.4). The vertical distributions generally increase with altitude, confirming the prediction of photochemical models that these species form in the upper atmosphere and then diffuse downward in the stratosphere. Below the condensation level of each gas, the distributions are assumed to decrease following the respective vapor saturation law. In particular, the ISO spectra had provided the first detection of water vapor in Titan's atmosphere from two emission lines around 40 μm, for an associated

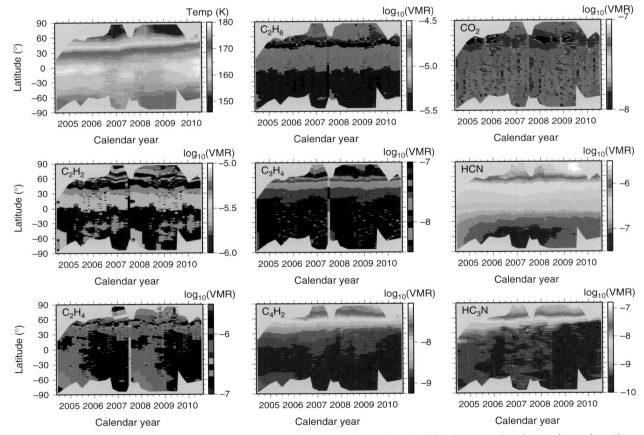

FIGURE 38.4 Titan's chemical composition and variations during the Cassini mission. Note the high values around northern spring equinox (August, 2009) at Titan's northern latitudes. VMR stands for "vertical mixing ratio". *Adapted from Teanby et al. (2010).*

mole fraction derived at 400 km of altitude of about 10^{-8}. ISO had also hinted at the presence of benzene (C_6H_6) at 674 cm^{-1} for a mole fraction on the order of a few 10^{-10}. Since then, benzene detection has been confirmed by Cassini/CIRS and INMS. The water vapor abundance was more precisely inferred by CIRS at stratospheric altitudes (a few 10^{-10} in the 100–200 km range), with strong implications on the whole oxygen chemistry on Titan, which remains yet to be accurately reproduced by photochemical models.

Ground-based high-resolution heterodyne millimeter observations of Titan offered the opportunity to determine vertical profiles and partial mapping in some cases of HCN, CO, HC_3N, and CH_3CN, which showed that the nitrile abundances increase with altitude. Subsidence causes the abundance of these species to decrease in the lower atmosphere.

Closer to the ground, Cassini–Huygens finally allowed for firm determinations of the major components: consistent with CIRS measurements in the stratosphere, the Huygens Gas Chromatograph Mass Spectrometer (GCMS) found a methane mole fraction of $(1.48 \pm 0.1) \times 10^{-2}$ in the stratosphere, increasing below the tropopause and reaching 5.65×10^{-2} near the surface in agreement with the CIRS stratospheric values and the surface estimate given by the Huygens Descent Imager Spectral Radiometer (DISR) spectra (also roughly 5%). The GCMS also saw a rapid increase of the methane signal after landing, which suggests that liquid methane exists on the surface, together with other trace organic species, including cyanogen, benzene, ethane, and carbon dioxide.

2.3. Dynamical Processes

2.3.1. Zonal Circulation

At the time of the Voyager encounter, Titan's northern hemisphere was coming out of winter. During the Cassini observations, so far, Titan's northern hemisphere has moved from winter to spring.

The general faintly banded appearance of Titan's haze suggests rapid zonal motions (i.e. winds parallel to the equator). This impression is reinforced by the infrared temperature maps of the stratosphere, which show very small contrasts in the longitudinal direction and rather large ones (of around 20 K) between the equator and the winter pole. The mean zonal winds inferred from this temperature field are weakest at high southern latitudes and increase toward the north, with maximum values at midnorthern latitudes (20–40° N) of about 160 m/s. On Titan, pressure gradients are in **cyclostrophic balance** with centrifugal forces. The stratospheric zonal winds and temperatures in both hemispheres of Titan are symmetric about a pole that is offset from the surface pole by $\sim 4°$.

Stellar occultations are another indirect means to obtain the zonal winds. The atmospheric oblateness due to the zonal winds can be constrained from the analysis of the central flash, the increase of the signal at the center of the shadow (when the star is behind Titan) due to the focusing of the atmospheric rays at the limb. On July 3, 1989, Titan occulted the bright K-type star 28 Sgr, and fast zonal winds were derived close to 180 m/s at high southern latitudes and close to 100 m/s at low latitudes. Other occultations occurred on December, 20, 2001, and November, 14, 2003. They seem to suggest a seasonal variation with respect to 1989. In 2001, a strong 220 m/s jet was located at 60° N, with lower winds extending between 20° S and 60° S and a much slower motion at midlatitudes. The CIRS data suggest that the strongest northern winds have migrated closer to the equator with respect to previous measurements, while the southern winds have weakened.

Space and occultation wind measurements could not provide the wind direction, a crucial factor for the Huygens probe mission, so different teams of ground-based observers tried to measure the zonal winds directly using alternative methods. The first measurement of prograde winds (in the sense of the rotation of the surface) was performed using infrared heterodyne spectroscopy of Doppler-shifted ethane emission lines. The measured winds were on the order of 210 ± 150 m/s between 7 and 0.1 mbar, a result that has since been refined. Other Doppler studies probing somewhat different stratospheric levels also found prograde winds, using millimeter-wavelength interferometry of nitrile lines or high-resolution spectroscopy of Fraunhofer solar absorption lines in the visible. The recent advances in **adaptive optics** also allowed for the first detections of tropospheric clouds from the ground, mainly at circumpolar southern latitudes, but so far, Titan winds remain poorly constrained due to the sparse data set of cloud positions. Better spatially resolved Cassini ISS observations only indicate slow eastward motions, which, extrapolated to the equator under the assumption of solid-body rotation, yield 19 ± 15 m/s at around 25 km altitude. Finally, in 2005, the Huygens probe provided ground truth measurements of the wind magnitude and direction in the lower stratosphere and troposphere. The Doppler wind experiment shows a marked decrease of winds with decreasing altitude, from 100 m/s at 140 km down to about nil at 80 km, then an increase up to 40 m/s at 60 km before decreasing again to null zonal velocity at the surface, noting also a reversal in the wind direction.

2.3.2. Latitudinal and Temporal Variations in the Atmosphere of Titan as Evidence of Meridional Circulation

Periodic change of Titan's disk-integrated brightness has been monitored from Earth-based observations since the

1970s. Spatially resolved observations, starting with Voyager, have provided an interpretation of the periodic changes of the disk-integrated brightness as the combined action of the high inclination of the rotation axis and the seasonally varying NSA. The NSA that Voyager 1 observed in 1980, with a darker northern hemisphere in visible light, has since been observed to reverse, as Titan's season shifted from northern spring to northern winter. When the HST first observed Titan in 1994, a little over a quarter of a Titan year after the Voyager encounters, the northern hemisphere was found to be brighter than the southern hemisphere. The turnover was later also found to occur gradually, starting at higher altitudes in the atmosphere.

A change in the vertical structure of the photochemical haze has been reported with a detached layer dropping in altitude from 500 km to 350 km from 2007 to 2010. Modeling with a two-dimensional general circulation model provided a qualitative description of the seasonal variations of the haze, where both the gradual inversion of the asymmetry and the detached haze layer can be explained by a seasonally varying **Hadley circulation**. The meridional wind in the upper branch of the Hadley cell is stronger close to the production zone (at 450 km) than below, and particles there are more rapidly transported toward the pole, where they sink. The asymmetry thus reverses first at higher altitudes. But this is not the only effect. As the season changes, shortly after equinox, the circulation reverses and an ascending motion sets in where the particles were previously descending. At the time of the transition, the polar haze, which was previously descending, is then redistributed about a scale height below the production zone, becoming physically separated from the freshly created particles aloft.

Meridional variations were also established for the gases in Titan's stratosphere, and these are also tightly coupled with the circulation. The molecular abundances found by Cassini at this era (1 Titan year after the V1 encounter) indicate an enhancement for most species in the northern stratosphere at high latitudes, with values similar to those observed 30 years ago, with a few exceptions (Figure 38.4). Since 2012, a clear indication of a reversal of the enhancement and the appearance of several gases at southern latitudes is observed.

Such latitudinal contrasts observed in the chemical trace species may be explained by invoking photochemical and dynamical reasons. The UV radiation from the Sun acts on methane and nitrogen to form radicals that combine into nitriles and the higher hydrocarbons. This production occurs in the mesosphere at high altitudes (above 300 km or 0.1 mbar). Eddy mixing transports these molecules into the lower stratosphere and troposphere where most of them condense. Photodissociation by UV radiation occurs on timescales ranging from days to thousands of years. The combination of these processes leads to a vertical variation in the mixing ratio, which usually increases with height toward the production zone. Three-dimensional computation of **actinic fluxes** suggests that this mechanism alone cannot explain the latitudinal contrasts and that circulation must intervene. Simulations coupling photochemistry and atmospheric dynamics provide a consistent view: competition between rapid sinking of air from the upper stratosphere in the winter polar vortex and latitudinal mixing controls the vertical distribution profiles of most species. The magnitude of the polar enrichment is controlled by downwelling over the winter pole, which brings enriched air from the production zone to the stratosphere, and by the level of condensation. Short-lived species are more sensitive to the downwelling due to steeper vertical composition gradients and exhibit higher contrasts.

In the stratosphere, the calculated radiative relaxation time is longer than the Titan season, so the temperature contrasts should be symmetric about the equator. That they are not indicates that the Hadley circulation must be connected with the lower atmosphere, where the time constant is much longer. This is consistent with the small thermal contrasts of 2–3 K in the troposphere, which suggest an efficient heat redistribution. Since Titan's slow rotation and small radius rule out nonaxisymmetric processes, such as **baroclinic** eddies, as a preferred mechanism for heat transport, considerable meridional motions must be inferred. Latitudinal contrasts would be much larger if heat were not being transported poleward by Hadley advection.

Another phenomenon was first reported in 2001 from adaptive optics data taken in 1998 and since then confirmed in Keck observations, among others. A diurnal change was found, manifested in an east–west asymmetry, with a brighter morning limb observed on Titan on several occasions. This dawn haze enhancement could be due to an accumulation of condensates during the Titan night (8 Earth days, although the superrotation of Titan's atmosphere would lead to shorter nights for stratospheric clouds).

2.3.3. A Three-Dimensional View and Waves

Meridional contrasts are apparent in Titan's atmospheric distributions of composition, haze, and temperature, and their seasonal variability is proof for a strong coupling with an underlying meridional circulation that has never been directly detected.

The superrotation observed in the stratosphere, a dynamical state in which the averaged angular momentum is much greater than that corresponding to corotation with the surface, is difficult to explain and has defied our understanding in the much better documented Venus case, the paradigm of a slowly rotating body with an atmosphere in rapid rotation. In recent studies, such a process has been identified under the form of planetary waves, forced by instabilities in the equatorward flank of the high-latitude jet. Two factors play a key role in facilitating the acceleration process. On the one hand, high-altitude absorption processes decouple upper atmosphere dynamics from dissipation occurring at the surface layer, while on the other hand the slow rotation allows the Hadley cell to reach high latitudes by reducing the Coriolis force in the poleward branch. A strong seasonal cycle due to Titan's obliquity of 26.7° was also established: During most of the Titan year, the meridional motion is dominated by a large Hadley cell extending from the winter to the summer pole, with the symmetric two-cell configuration typical of equinoxes occurring only in a limited transition period. In models, the jet is located close to 60° in the winter hemisphere, while the summer zonal circulation is close to solid-body rotation.

The radiative time constant is long in the troposphere, but the surface has a smaller thermal inertia, so the surface temperature does respond to seasonal forcing, albeit by only a few Kelvin. This surface temperature variation is sufficient to reverse the circulation pattern of the Hadley circulation after the equinox when the Sun moves to the opposite hemisphere. Also, the development of convective methane clouds is partly ascribed to seasonal surface heating. The reversal of the Hadley circulation may play an important role in the methane "hydrological" cycle because the vertical and horizontal transport of methane would vary seasonally.

Direct evidence for wave processes in Titan's atmosphere remains scarce, despite their importance in the maintenance of superrotation. Because baroclinic processes are excluded, waves essentially **barotropic** in nature should be expected as the principal carrier of momentum from high to low latitudes. Modeling predicts **wavenumber-2** waves with amplitude of the zonal component about 10% of the mean wind speed, and in principle, they can be inferred from horizontal maps of temperature and trace species exhibiting strong latitudinal contrasts. The first Cassini/CIRS temperature maps at 1.8 mbar do show spatial inhomogeneity, but long time series and better spatial coverage are needed to constrain spatial and temporal variations.

Another relevant nonaxisymmetric phenomenon in Titan's troposphere is the gravitational tide exerted by Saturn. The eccentric orbit of Titan around Saturn gives rise to a tidal force, resulting in periodical oscillation in the atmospheric pressure and wind with a period of a Titan day (16 days), among which the most notable effect is the periodical reversal of the north–south component of the wind. In the lower atmosphere, the effect of this tide is modest, with a maximum temperature amplitude about 0.3 K and winds of 2 m/s.

Temperature inversions have been detected in both the HASI measurements and in stellar occultation data. Inversion layers were present close to 510 km altitude in HASI and 2003 occultation data and at 425 and 455 km in 1989 occultation light curves. Vertical wavelengths were on the order of 100 km.

2.4. Haze and Clouds on Titan

It was recognized quite early on that another important aspect of Titan's atmosphere was the presence of aerosols. Pre-Cassini models treated the dissociation of methane molecules by solar actinic radiation, followed by chemical combination to heavier hydrocarbons that condense into particles. The cloud physics models with sedimentation and coagulation predicted a strong increase in haze opacity with decreasing altitude.

2.4.1. Titan's Haze

The analysis of high-phase Voyager images indicated aerosol radii between 0.2 and 0.5 μm. These "smog" particles form a layer that enshrouds the entire globe of Titan and stretches from the surface to an altitude of about 200 km. A detached haze layer at 340–360 km altitude with large, compact, irregular dark particles was also found. The small haze particles required by Voyager measurements (radii less than or equal to 0.1 μm) produce a strong increase in optical depth with decreasing wavelength shortward of 1 μm. To fit the observations in the methane bands, it was necessary to remove the haze permitted by the cloud physics calculations at altitudes below about 70–90 km (called cutoff altitude) by invoking condensation of organic gases produced at high altitudes as they diffused down to colder levels. The condensation of many organic gases produced by photochemistry at high altitudes on Titan seemed consistent with this view. The next step in the development of Titan haze models included the use of fractal aggregate particles composed of several tens of small (0.06 μm in radius) monomers to produce strong linear polarization. Monomers composed of 45 aggregates with an effective radius of about 0.35 μm matched the Voyager observations.

Starting from the upper atmosphere, the complex chemical composition observed in the thermosphere along with the detection of large mass ions hinted at the formation of aerosols in the ionosphere. This argument was further supported by observations from the Ultraviolet Imaging Spectrometer that detected aerosols up to ~900 km of altitude, as well as by the detection of large-mass negative ions in the ionosphere by the Cassini Plasma Spectrometer.

In the mesosphere, the Cassini ISS camera showed a faint thin haze layer that encircles the denser stratospheric haze (Figure 38.1(b)) and could be the equivalent of the "detached haze layer" observed by Voyager 25 years ago, except for the difference in altitudes: The thin current haze layer is indeed located 150–200 km higher than the one seen by Voyager. In subsequent observations, right after the vernal equinox, this layer appeared to drop in altitude, reaching the altitude levels of the detached layer observed by Voyager. Current models are still unable to render the complexity of seasonal phenomena or circulation patterns on Titan, which could be responsible for such variability in the aerosol distribution.

Cassini images also show a multilayer structure in the north polar hood region and, in some cases, at lower latitudes. These features could be due to gravity waves that have been detected on Titan at lower latitudes. Some of these layers may be related to the two global inversion layers observed in stellar occultations of Titan above 400 km in altitude.

The nature of the haze aerosols measured by Huygens/DISR during the descent through Titan's lower atmosphere came as a surprise to scientists recalling the results from Pioneer and Voyager, as well as predictions by cloud physics models with sedimentation and coagulation. The new observations estimate the monomer radius to be 0.05 μm, in good agreement with previous values. However, contrary to previous assumptions, the DISR data seem to show that the size of the aggregate particles is several times as large as previously supposed.

In addition, measurements by the DISR violet photometer extend the optical measurements of the haze to wavelengths as short as the band from 350 to 480 nm, also helping to constrain the size of the haze particles. The number density of the haze particles does not increase with depth nearly as dramatically as predicted by the older cloud physics models. In fact, the number density increases by only a small factor over the altitude range from 150 km to the surface. This implies that vertical mixing is much less than had been assumed in the older models where the particles were distributed approximately as the gas is with altitude. In any event, the clear space at low altitudes, which was suggested earlier, was not observed.

The methane mole fraction of 1.5–1.6% measured in the stratosphere by the CIRS and the GCMS, is consistent with the DISR spectral measurements. At very low altitudes (20 m), DISR and the GCMS measured $5 \pm 1\%$ for the methane mole fraction.

2.4.2. Titan's Clouds

Cassini–Huygens has provided new information on the role of methane and the methane cycle in Titan's atmosphere. The relative humidity of methane (about 50%) at the surface found by DISR and the evaporation witnessed by the GCMS show that fluid flows have existed and will probably again exist on the surface, implying precipitation of methane through the atmosphere.

Although some discussion took place as to whether Titan's lower atmosphere could support convection and as to whether methane was supersaturated, there is clear evidence today that clouds exist in Titan's troposphere. Methane clouds in Titan's troposphere were first suspected from variability in the methane spectrum observed from the ground. Direct imaging of clouds on Titan has been achieved from Earth-based observatories since the turn of the century.

Most of the clouds detected in the past years are located in Titan's southern hemisphere, as expected given the season on Titan that has been essentially probed (summer in the south), which means that solar heating is concentrated there as are rising motions. In recent years, they have been found to be building up now in the north as the season on Titan changes. Other than the large, bright South Pole system observed for the past 5 years or so, transient, discrete clouds have been detected at midlatitudes (Cassini Visual and Infrared Mapping Spectrometer (VIMS) observations tend to indicate that they rise quickly to the upper troposphere and dissipate through rain within an hour). Keck and Gemini data indicate that they tend to cluster near 350° W and 40° S. They may be related to some surface–atmosphere exchange (such as geysering or **cryovolcanism**) because they do not seem to be easily explained by a shift in global circulation. A dozen or so large-scale zonal streaks have also been observed by Cassini preferentially at low southern latitudes and mostly between 50 and 200° W.

The large south polar system has been visible consistently essentially in the near infrared (at 2.12 μm for instance) since 1999 from ground-based observations, while no previous indication of it was ever reported. It was extremely bright in 2001–2002, and recent Cassini images have shown that it is disappearing (indeed it was visible only during the few first Titan flybys and not afterward, see Figure 38.5). Its shape is irregular and changing with time, recently more resembling a cluster of smaller-scale clouds than a large compact field. Should it prove that this system's life was indeed on the order of 5–6 years (fairly close to a Titan season), stringent constraints can be retrieved on

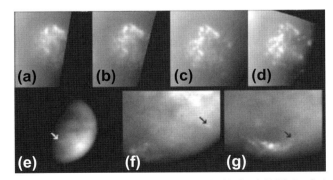

FIGURE 38.5 Titan's meteorology observed with Cassini ISS. (a–d) A sequence of four methane continuum (IRP0-IR3, 928 nm) images showing the temporal evolution over the period 05:05–09:38 of the Titan south polar cloud field on July 2, 2004. (e–g) Three examples of discrete midlatitude clouds (arrows) for which motions have been tracked in CB3 images. (e) 38° S, 81° W (29 May 2004); this image was also viewed through an infrared polarizing filter. (f) 43° S, 67° W (23 October 2004). (g) 65° S, 110° W (October 25, 2004). *From Porco et al. (2005).*

seasonal and circulation patterns on Titan. The cloud made a strong reappearance in 2006, but has been fading again since, although Titan's southern hemisphere still shows a very active meteorology with more than 200 clouds observed in three-and-a-half years.

Note that DISR reported no definite detection of clouds during its descent through Titan's atmosphere. However, the data are compatible with a thin haze layer at an altitude of 21 km, which could be due to methane condensation.

2.5. Origin and Evolution

The only noble gas detected on Titan to date is argon, seen by GCMS in the form of primordial ^{36}Ar (2.1×10^{-7}) and its radiogenic isotope ^{40}Ar (3.4×10^{-5}). The low abundance of primordial noble gases on Titan implies that nitrogen was originally captured as NH_3 rather than N_2. Subsequent photolysis may have created the N_2 atmosphere we see today. The absence of Kr and Xe also has implications on the evolution of the atmosphere, as discussed hereafter. In addition, the detection of neon (^{22}Ne) in Titan's atmosphere requires further investigation.

Isotopic ratios for Titan were determined from Cassini and Huygens instruments: ^{12}C/^{13}C (roughly 90, close to the solar and terrestrial value), ^{14}N/^{15}N (about 170, which is roughly 1.5 times less than on Earth possibly implying large nitrogen escape during Titan's evolution), and D/H (measured in situ in H_2, $1.35 \pm 0.3 \times 10^{-4}$, from the GCMS, and 1.2×10^{-4} in CH_4, from remote sensing of infrared spectra recorded aboard the Cassini orbiter with CIRS). The D/H ratio is very important for Titan's cosmogonical models and the value on Titan tends to suggest a deuterium enrichment in the atmosphere with respect to the protosolar value as well as in the giant planets (D/H $\sim 2-3.4 \times 10^{-5}$). The interpretation of this enrichment is related both to the origin of CH_4 (is this species older than Titan itself or was it produced in the satellite's interior?) and to its evolution in the satellite's atmosphere. In particular, CH_4 is continuously photodissociated so that, in the absence of a substantial reservoir, it would entirely vanish from the atmosphere in 10–50 Myr. This implies that some delivery mechanism needs to be invoked in order to maintain the methane abundance against photolytic depletion throughout Titan's history. A possibility is to have a continuous supply from the interior or by buffering by a surface or near-surface reservoir. Imaging, infrared, and visible observations from the orbiter rule out the presence of a global ocean containing a large amount of CH_4 on the surface of Titan, but lakes and seas (of which the largest are named Kraken Mare, Ligeia Mare, Punga Mare, etc.) have been detected on Titan's surface. However, the composition of these lakes is dominated by ethane and heavier hydrocarbons and the fraction of dissolved methane is too small to account for its atmospheric abundance.

Alternatively, some models suggest that methane can outgas from time to time from the interior of the satellite. Two scenarios for the origin of the internal CH_4 have been proposed. One scenario advocates that CH_4 was produced from hydrothermal reactions in the interior of Titan but the D/H value predicted via this mechanism does not match the measured one. The formation of methane from carbon monoxide in an initially warm Saturn's subnebula is also ruled out because the produced CH_4 would harbor a much lower D/H ratio than the one observed in Titan. A more plausible scenario argues that CH_4 originates from the solar nebula. In this scenario, CH_4 would have come from the interstellar medium and was initially highly enriched in deuterium by ion–molecule reactions in the presolar cloud. Once vaporized in the solar nebula, methane would have continuously exchanged deuterium with protosolar hydrogen as long as it remained in the gas phase. After the solar nebula cooled down to about 50–60 K, methane became trapped in the form of clathrates in the feeding zone of Saturn and these ices were incorporated in the forming of planetesimals. The D/H value measured in Titan's atmosphere would then reflect the one acquired by methane in the gas phase of the solar nebula at the time of its trapping in clathrates. Ice incorporated in planetesimals is composed of pure condensates that formed at extremely low temperatures (about 20 K) and clathrates that resulted from the trapping of some volatiles in crystalline water-based solids at slightly higher temperatures (about 50–60 K) in the nebula. Once embedded within the Saturnian subnebula, planetesimals accreted by proto-Titan would have suffered from partial devolatilization during the processes of migration and accretion that led to the satellite's formation. This scenario is consistent with the fact that Titan's atmosphere is strongly impoverished in carbon monoxide and heavy noble gases but harbors primordial methane and nitrogen derived from primordial ammonia.

3. THE SURFACE OF TITAN

To the eyes of the public and many scientists, the most exciting features revealed by the Cassini–Huygens mission were those found on Titan's surface, finally observed in close-up mode by the Orbiter in 2004 and then in in situ conditions by the Huygens probe instruments on January 14, 2005. The spaceship has offered detailed views of Titan's surface in the visible and the near-infrared regions with its camera, the mapping spectrometer, and the radar. Descending through the atmosphere, the Huygens probe returned extraordinary images of a first-seen domain, the farthest location a human-made vessel has ever landed upon. Although we still have not exactly determined the nature or the composition of the surface, a combined analysis of all the continuously arriving data should eventually force Titan to uncover the nature of its mysterious soil. Undoubtedly, the signs of dried lakes, volcanoes, dunes and channels on Titan's surface, in addition to signs of a possibly active interior, were unexpected. They offer an even more amazing view of a land much fantasized on.

3.1. Pre-Cassini Glimpses of an Exotic Ground

To the Voyager optical cameras, the surface of Titan was obscured by the dense haze in the atmosphere. Glimpses of what lay below were revealed afterward by ground-based radar and infrared images from HST and ground-based observatories.

Theory argued that unless methane supersaturation conditions prevailed on Titan, the organics present in the atmosphere should condense at some level in the lower stratosphere and precipitate out, ending up on Titan's surface and coating the ground in large proportions. Based on the surface conditions believed to prevail on Titan, liquid methane—and its principal by-product, ethane—is expected to exist and could even form an ocean, and in the troposphere, methane clouds (formed by saturation of methane gas) might cause rains. The degree of saturation in the lower atmosphere, however, was unknown, so the methane abundance was difficult to determine.

On the other hand, much of the outer part of the solid body of the satellite must, to be consistent with the observed mean density, consist of a thick layer of ice. The ethane ocean model, developed in 1983, was aesthetically appealing and compatible with all the Voyager-era data. It has since then long been abandoned in view of the spectroscopic and imaging evidence for a heterogeneous surface and the radar echoes indicating the presence of solid material.

Indeed, a shallow, global ocean was shown to be inconsistent with the constraints imposed by Titan's orbital characteristics. The tidal action on an ocean less than 100 m deep would have dissipated Titan's **orbital eccentricity** of 0.03 (where 0 is circular and 1 is parabolic) long ago. Furthermore, the first remote sensing technique to be used for sounding Titan's surface, radar, indicated that the surface should be nonuniform but mostly solid with at most small lakes. Indeed, the radar echoes obtained in 1990 using the National Radio Astronomy Observatory's Very Large Array in New Mexico combined as a receiver of the signal transmitted to Titan by the NASA Goldstone Apple Valley Radio Telescope in California were among the first evidence against the global ocean model of the surface. Radar measurements from Arecibo Observatory in Puerto Rico in 2003, however, revealed a specular component at 75% (12 of 16) of the regions observed (globally distributed in longitude at about 26° S), which was interpreted as indicative of the existence of dark, liquid hydrocarbon somewhere on Titan's surface. The idea of a widespread surface liquid was challenged from more observations from the ground, in particular from spectroscopic data in the near-infrared region (0.8–5 μm). This part of Titan's spectrum, like that of the giant planets, is dominated by the methane absorption bands. At short (blue) wavelengths, light is strongly absorbed by the reddish haze particles. At red wavelengths, light is scattered by the haze, although the column optical depth is still high. In the near-infrared region, the haze becomes increasingly more transparent (since the haze particles are smaller than the wavelength), although absorption by methane in a number of bands is very strong. Where the methane absorption is weak, clear regions or "**windows**", situated near 4.8, 2.9, 2.0, 1.6, 1.28, 1.07, 0.94 and 0.83 μm, permit the sounding of the deep atmosphere and of the surface (Figure 38.6). In between these windows, contrary to the giant planets, solar flux is not totally absorbed but scattered back through the atmosphere by stratospheric aerosols, especially at short wavelengths. The near-infrared spectrum is thus potentially extremely rich in information on the atmosphere and surface of Titan.

The observations all agreed: the **geometric albedo** of Titan, measured over one orbit (16 days), showed significant variations indicative of a brighter leading hemisphere and a darker trailing one and inconsistent with a global ocean. The leading side corresponds to Titan's Greatest Eastern Elongation at about 90° Longitude of the Central Meridian (LCM—as opposed to geographical longitude, which is about 210°), when Titan rotates synchronously with Saturn; the trailing side is near 270° LCM or Greatest Western Elongation. As a consequence, Titan's surface had then to be heterogeneous and rather "dry" and it was proposed that a hydrocarbon reservoir could be stored in the porous, uppermost few kilometers of **methane clathrate** or water ice "bed rock."

The heterogeneity of Titan's surface, indicated in the near-infrared region and with radar light curves (Figure 38.6), was further graphically revealed by images of Titan's surface using the HST and adaptive optics techniques at the CFHT on top of Mauna Kea and its twin at the VLT in Chile. From such images, maps were produced of the surface in several near-infrared wavelengths, showing in some detail the bright leading and dark trailing sides, with notably a large (2500 × 4000 km) bright region, at 114° E and 10° S (nowadays known as Xanadu), as well as at a number of distinct darker regions known today for hosting hydrocarbon liquids or organic deposits.

Although it was thus already essentially demonstrated that Titan's surface was much more complex than initially thought, the arrival of the Cassini–Huygens mission revolutionized our perception of Titan with a particular focus on its mysterious surface, as described hereafter.

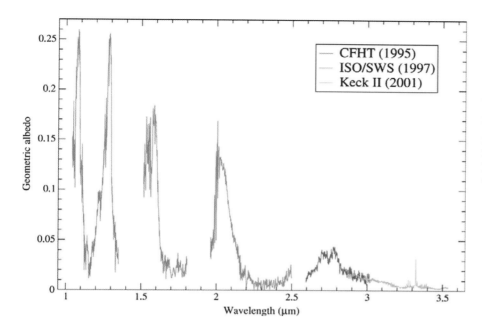

FIGURE 38.6 Titan's albedo observed from ground-based observatories such as the VLT in Chile and the Keck Telescope in Hawaii, as well as with the satellite ISO (in the 2.75-μm window, where the terrestrial turbulence does not allow us to observe Titan from the ground). The spectrum exhibits several strong methane absorption bands, but also "windows" where the methane absorption is weak enough to allow for the lower atmosphere and surface to be probed. *Adapted from Negrao et al. (2006) and Coustenis et al. (2006).*

3.2. The View from the Orbiter

The Cassini ISS and the VIMS cameras first pierced Titan's thick atmosphere and confirmed essentially the results from the ground-based observations showing that the borders of the bright and dark regions were linear but not smooth and that dramatic changes in surface albedo could be noticed. The best resolution achieved by ISS was of a few kilometers on Titan's surface. The large bright area around the equator first observed by the HST and the adaptive optics in 1994 was resolved and finely observed by Cassini instruments. It is centered at 10° S and 100° W and officially named Xanadu Regio (Figure 38.7). The midlatitude regions around the equator on Titan were found to be rather uniformly dark with some extensive bright regions (mostly elevated structures like mountains and ridges), while the poles are relatively bright and also filled with large dark areas now recognized as hydrocarbon "lakes" (mostly in the North Pole) formed possibly by precipitation through the atmosphere. In terms of hypsometry, Titan is considered flat with a range up to 2 km when compared to other planetary bodies such as the Earth (10–30 km). Furthermore, the polar regions were found to be substantially lower in topography than the equator (Figure 38.7).

What exactly is causing the albedo variations is still uncertain but they are more readily attributed to the presence on the surface of constituents with different albedos and possibly a combination with topography. The reason is that the Cassini cameras observing at short wavelengths cannot see shadows and provide information regarding the altitudes, the contours, and in general, the shape of the surface and that Titan's icy bulk does not plead for high topographic structures on the surface (mountains should not exceed 2 km or so).

For the brighter regions, the task of interpreting the data is more difficult. It has been hypothesized that they could be associated with some topography and more exposed ice content, and this tends to be in agreement with findings of the Huygens DISR instrument whose stereoscopic imaging revealed that the brighter terrain was also more elevated than the darker, smoother, and lower ice regions. Alternative interpretation seems to be needed for a number of other areas such as Xanadu, which is low but particularly bright. The exact ice constituent that can satisfy the constraints imposed by all the observations is not easy to determine, but water ice, tholin, CO_2 ice, and hydrocarbon ices have been suggested by various investigators.

Tortola Facula (8.5° N, 143° W), a bright circular structure (about 30 km in diameter) found in the VIMS hyperspectral images is interpreted as a cryovolcanic dome in an area dominated by extension. The VIMS team hypothesized that the dry channels observed on Titan are likely controlled by tectonic processes and related to upwelling "hot ice" and contaminated by hydrocarbons that vaporize as they get close to the surface (to account for the methane gas in the atmosphere), which are similar to the mechanisms operating for silicate volcanism on Earth (using tidal heating as an energy source) and which may lead to flows of non-H_2O ices on Titan's surface. Following such eruptions and the formation of a tectonic pattern on the surface, methane rain could form the dendritic dark structures seen by Cassini–Huygens. Such features could also form solely from heavy localized rainfall, as has been

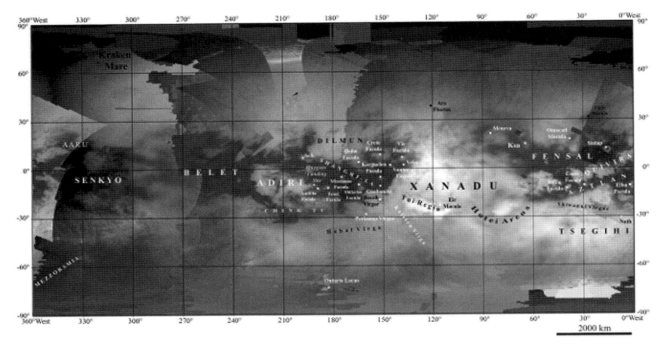

FIGURE 38.7 Global map of Titan's surface at a wavelength of 2 μm produced from Cassini VIMS observations including Titan's recent nomenclature. The map is displayed in a simple cylindrical projection centered at 0° N and 180° W. *From Stephan et al. (2009).*

observed directly through darkening and subsequent rebrightening reported from ISS images of some of Titan's equatorial terrain in the wake of a large low-latitude cloud system early in Titan's northern spring. The changes are interpreted as seasonal methane rainfall that can explain the channels observed at Titan's low latitudes. If these structures are indeed channels, they could have dried out due to climatological phenomena such as evaporation during a dry season at these latitudes. Studying volcanism on Titan (if Cassini definitely yields evidence for it) is important to understand not only the thermal history of Titan (which must surely have evolved differently because it differs in its incorporation of volatiles from the Galilean satellites) but also how volatiles—in particular, methane—were delivered to the surface.

The Cassini instruments have found no obvious evidence for a heavy craterization on the bright or the dark areas of Titan so far (Figure 38.7). A few features interpreted as impact craters have been announced to date: Cassini's RADAR and VIMS saw a 440-km-diameter impact crater on Titan during two separate flybys in early 2005. The coloring of the feature indicates that its terrain is rough, with different material for the crater floor and the ejecta, and tilted toward the radar during the observations. The multiringed impact basin was named Circus Maximus by the science team and later received its official name, "Menrva". A smaller crater of about 40 km was also observed, exhibiting a parabola-shaped ejecta blanket. In spite of the detection of more craterlike features, such formations, identified by the RADAR, VIMS, or the ISS are rare. This may mean that the surface of Titan is young (less than a billion years) or highly eroded/modified.

Other features observed by the Cassini orbiter include areas covered with analogs to terrestrial dunes in a set of linear dark features visible across a large part of the RADAR swath to the west of the large crater. These formations are aligned west to east covering hundreds of kilometers and rising to about 100 m. They are expected to have formed by a process similar to that on the Earth, but the nature of this "sand" is quite different, consisting of fine grains of ice and/or organic material, rather than of silicates (Figure 38.8). These structures and their orientation (west-east) were originally attributed to the influence of Saturn, through tidal forces 400 times greater than on the Earth which could have moved the Titanian "sand" in this world of low gravity, but it is thought now that more likely winds higher than 1 m/s on the surface are needed which points to a more important role of the Hadley circulation for the dune formation.

Additionally, and quite importantly for the hypothesized missing liquid methane or ethane surface reservoir, the RADAR instrument onboard Cassini has discovered lakes sprinkled over the high northern altitudes of Titan (Figure 38.9). In the data recorded since then by several instruments, a variety of dark patches is observed, some of which extended outward (or inward) by means of channels, seemingly carved by liquid. In 2010, specular reflections of the Sun off Titan's northern lakes ("sunglints") were

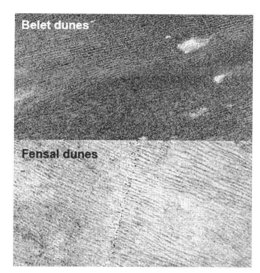

FIGURE 38.8 Dune fields on Titan as captured from Cassini RADAR in the Belet area (upper) and Fensal region (lower). In these images, the dunes are the dark streaks that are 1–2 km wide and the areas between dunes (bright streaks) are 1–4 km wide. Two different types of dune fields are shown: Fensal is at a higher latitude and elevation than Belet and clearly shows thinner dunes with brighter and wider areas in between, suggesting less dune material in this region. *Image Credit: NASA/JPL-Caltech/ASI/ESA and USGS/ESA.*

FIGURE 38.9 Cassini RADAR image of Titan's surface in synthetic aperture mode taken in 2006 and showing the highly contrasted terrain with a variety of geological features like the dark areas which are most probably hydrocarbon lakes. Bolsena Lacus is at the lower right, with Sotonera Lacus just above and to its left. Koitere Lacus and Neagh Lacus are in the middle distance, left of the center and on the right margin, respectively. Mackay Lacus is at the upper left. *Image Credit: Cassini Radar Mapper, JPL, ESA, NASA.*

reported, while Cassini looked back toward a crescent Titan at high phase angle. The data were used to set some constraints on possible surface waves on a north polar lake (Jingpo Lacus), and since then the observed specular reflections have been quite useful, along with spectroscopic detections of liquid ethane in Ontario Lacus by the VIMS instrument, in corroborating the presence of large extents of liquid hydrocarbons on Titan's surface. The major seas on Titan, which are very extended in size and discovered in 2007, are Kraken Mare (surface area of at least 400,000 km^2 centered at 68° N, 310° W), Ligeia Mare (79° N, 248° W), and Punga Mare (85° N, 339° W). Although some of these "seas", or "lakes", or even "oases", are found at midlatitudes, most are located in the northern hemisphere at Titan's current season, but they are expected to build up in the South, following the atmospheric evolution, as the season changes.

3.3. In Situ Data: the Ground Truth

On January 14, 2005, the Huygens probe manufactured by ESA landed at 10° S and 192° W on Titan (Figure 38.7), providing the "ground truth" for the orbital measurements in terms of composition, structure, and geomorphology. The probe flew over an icy surface and then floated down and drifted eastward for about 160 km. Several of the instruments on board contributed to our knowledge of Titan's surface conditions.

At the Huygens landing site, the fact that the surface is solid but unconsolidated was verified by all the data. The first part of the probe to touch the surface was the Surface Science Package (SSP) penetrometer whose data are now interpreted as indicative of the probe first hitting one of the icy pebbles littering the landing area before sinking into the softer, darker ground material. The SSP detected the ground from 88 m in altitude by acoustic sounding, revealing a relatively smooth, but not flat, surface for which our best current hypothesis is gravel, wet sand, wet clay, or lightly packed fine icy particles. With a landing speed of about 5 m/s, the front of the probe followed and penetrated the surface and then slid slightly before settling to allow the DISR camera to take several pictures of a Mars-like landscape, complete with a dark riverbed and brighter pebbles.

No evidence for standing liquid was found at the Huygens landing site, but the surface is expected to be very humid because methane evaporation (a 40% increase of the abundance) was measured by the GCMS after landing. In addition to the GCMS thermal data, the SSP penetrometer, the GCMS composition measurements, and the detection of a likely dewdrop by DISR support the suggestion of the presence of humidity at the place where and at the time when Huygens landed. Thus, either the methane liquid reservoir may not be so far below the surface, but located instead in niches close to the exposed ground, or perhaps Huygens landed on Titan in a "dry" season when the rivers and lakes that may exist near the equator were empty but that could be flowing with hydrocarbons at a different era. Also, the presence of hydrocarbon lakes close to the North Pole may also imply that there are seasonal phenomena that distribute the liquid on the ground. Nevertheless, Huygens landed on an organic-rich surface, with trace organic species such as cyanogens and ethane detected on the ground.

The DISR imager and spectrometer gathered a precious set of data. Starting from the first surface image at 49 km, down to the unprecedented-quality snapshots of the Huygens landing site, and through the lamp-on data recorded below 700 m in altitude, this instrument played a decisive part in untangling the enigma of Titan's surface morphology and lower atmospheric content. Panoramic mosaics constructed from a set of images taken at different altitudes show brighter regions separated by lanes or lineaments of darker material, interpreted as channels, which come in short stubby features or more complex ones with many branches (Figure 38.10). This latter dendritic network can be caused by rainfall creating drainage channels, implying a liquid source somewhere or at some time on Titan's surface. The former stubby channels are wider and rectilinear. They often start or end in dark circular areas suggesting dried lakes or pits.

Stereoscopic analysis was performed on the DISR images indicating that the bright area cut with the dendritic systems is 50–200 m higher than the large darker plane to the south. If the latter feature is a dried lakebed, it seems too large by Earth standards to have been created by the creeks and channels seen on the images and could be due to larger rivers or a catastrophic event in the past. The dark channels visible in Figure 38.10 could be due to liquid methane irrigating the bright elevated terrains before being carried through the channels to the region offshore in southeasterly flows. This migration toward the lower regions probably leads to water ice being exposed along the upstream faces of the ridges. The slopes are generally on the order of 30°. Some of the bright linear streaks seen on the images could be due to icy flows from the interior of Titan emerging through fissures.

The images taken after the probe had landed on Titan's surface show a dark riverbed strewn with brighter round rocks. These "stones", which are 15 cm in diameter at most, could possibly be hydrocarbon-coated water ice pebbles (Figure 38.11).

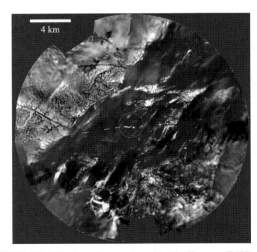

FIGURE 38.10 Titan's surface as viewed by the Huygens/DISR cameras from an altitude of 8 km. *Tomasko et al. (2005). Image Credit: ESA/JPL/ University of Arizona.*

FIGURE 38.11 Titan's surface after the landing of the Huygens probe. The icy pebbles are at most 15 cm in diameter, and the darker riverbed is thought to be methane-wet sand. *Tomasko et al. (2005). Image Credit: ESA/JPL/University of Arizona.*

The spectra acquired during the descent gave information on the atmospheric (Table 38.2) and surface properties. Indeed, it was shown from spectral reflectance data of the region seen from the probe that the differences in albedo were related to differences in topography, which in turn can be connected to the spectral behavior of the ground constituents. Thus, the higher brighter regions were also found to be redder than the lowland lakebeds. The regions near the mouths of the rivers are also redder than the lake regions. The spectra taken by DISR are compatible with the presence of water ice on Titan's surface, something that had already been suggested from ground-based observations. The most intriguing feature found in the spectra was, however, the featureless quasi-linear unidentified blue slope observed between 830 and 1420 nm. No combination of any ice and organic material from laboratory measurements has been adequate in reproducing this characteristic. The jury is still out on the constituent(s) that create(s) this signature.

Although many questions still remain about the sequence of flooding and the formation of all the complex structures observed by DISR, these data tend to clear the picture we have of Titan today and at the same time enhance the impression that by studying this satellite of Saturn we are looking at an environment resembling the Earth more closely than any other place in our solar system.

The reality pictured by the Cassini–Huygens instruments went beyond anything that has been speculated about Titan's environment. The diversity of the terrain includes impact craters, dark plains with some brighter flows, mysterious linear black features possibly related to winds, sand dunes, and a host of possible actors (solids, winds, liquids, ices, volcanism, tectonism, etc.). Titan has been proved to be a much more complex world than originally thought and much tougher to unveil.

4. THE INTERIOR AND EXCHANGE PROCESSES

Titan possesses a complex and dynamic geology as witnessed by its varied surface morphology resulting from aeolian, fluvial, and possibly tectonic and endogenous cryovolcanic processes. Linear features and possible cryovolcanic spots are located close to the equator. In general, even with limited data, Titan's topography seems to be similar to that of Ganymede and Europa. The equatorial regions are considered to be higher in terms of hypsometry than the polar regions.

In particular, elevated as well as fractured crustal features are observed, and the fact that these features are locally regrouped (equator) indicates a morphotectonic pattern. Their shapes, sizes, and morphologies suggest that they are tectonic in origin, although it may be a different form of tectonism than the terrestrial one, originating from internal compressional and/or extensional activities. The triggering mechanism that leads to such dynamic movements is possibly Saturn's tide, whose effects concentrate around the equator.

In addition, the presence of an internal liquid water ocean for Titan is supported by models, radar and gravity Cassini measurements, and the HASI experiment. The extremely low-frequency electric signal recorded by HASI was interpreted as Schumann resonance between the ionosphere and a modestly conducting ocean roughly 50 km below the surface. Current thermal evolution models suggest that Titan may have an icy crust 50–150 km thick, lying atop a liquid water ocean a couple of hundred kilometers deep, with some amount (perhaps ∼10%) of ammonia dissolved in it, acting as an antifreeze. A layer of high-pressure ice could exist underneath (Figure 38.12). The presence of ammonia, from which Titan's nitrogen atmosphere was presumably derived, distinguishes Titan's thermal evolution from that of Ganymede and Callisto. Cassini's measurement of a small but significant asynchronicity in Titan's rotation is most straightforwardly interpreted as a result of decoupling of the crust from the deeper interior by a liquid layer. The recent detection of periodic tidal stresses on Titan requires that its interior be deformable over timescales of the orbital period in a way that is consistent with a global ocean at depth.

Cryovolcanism has also been suggested on Titan, although it has not been proven to this day. This activity can be described as ice-rich volcanism, due to the satellite's conditions (internal and external) which are quite different from the terrestrial ones, while the cryovolcanic ejecta are referred to as cryomagma. The cryomagma could appear in the form of icy cold liquid and, in some cases, as partially crystallized slurry, but still with unknown precise composition. There are currently some cryovolcanic candidate regions on Titan, all of which are located close to the equator, and include Tui Regio, Hotei Regio, and Sotra Patera for instance (Figure 38.7). Indeed, radiative transfer modeling of Cassini/VIMS data taken from these regions from several flybys indicates possible albedo variations with time, suggesting possible fluctuating deposition of material with endogenic origin (i.e. cryovolcanism). On the other hand, an alternative hypothesis, based on spectral similarities between Tui Regio and dry lakebeds at the poles, is that the features at Tui Regio are evaporitic deposits with no interior connection. Other studies focusing on Tui Regio and Hotei Regio describe the areas as paleolake clusters and fluvial or lacustrine deposits indicating an exogenic origin. In particular, some investigators have pointed out the resemblance between Hotei's radar-bright materials and channels with Earth's sediment-filled valleys, thus arguing in favor of the fluvial origin of the observed features.

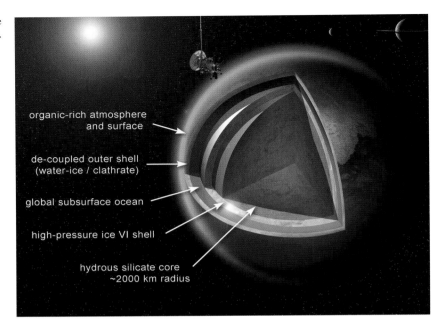

FIGURE 38.12 Artist's concept of a possible Titan internal structure according to Cassini data. *From Fortes, A. D (2012).*

Cryovolcanism may be considered as one of the processes that replenishes methane in the atmosphere and reshapes and changes the surface. However, alternative interpretations of the regions of interest and theories for methane ejection in the atmosphere exist. Among others, one can cite mountain formation by thermal contraction of the interior, isostasy and impacts. In the occasion of such "violent" events, clathrate hydrate in the high-pressure ice of Titan's interior can be destabilized and methane could be released in the atmosphere. Other theories, suggesting the lakes as a source of methane replenishment, are now less in favor since estimates show that only 1% of methane can be provided by that source.

5. LOOKING AHEAD

On Titan we are observing an active hydrologic cycle subjected to seasonal and longer term changes, as on the Earth, and by investigating such a close Earth analog in terms of processes, but with sufficient differences in working fluids and timescales, we can rigorously test the depth of our understanding of planetary climate.

The Cassini data since 2004 show that seasonal variations are quite prominent in Titan's atmosphere with a buildup of trace gases over the south pole approximately 2 years after observing the 2009 postequinox circulation reversal, indicating that middle-atmosphere circulation must extend up to at least 600 km. The primary drivers of this circulation are found to be summer-hemisphere heating of haze by absorption of solar radiation and winter-hemisphere cooling through infrared emission by haze and trace gases.

Furthermore, comparisons made with earlier ground-based and Voyager observations indicate that, although the thermal and chemical structure of Titan's atmosphere is stable after 1 Titanian year, consistent with the solar input being the major energy source on Titan as on the Earth, some interannual variations may exist for a few of the most complex molecules.

The lakes seen by Cassini are signposts of climate change at the Titan poles—with a period of tens of thousands of years— and the cycling of methane and ethane between Titan's poles leads to an active geologic surface modification in the polar latitudes. Seasonal transport of methane from pole to pole may replenish the polar lakes during the wintertime in the North and deplete them in the summer, but more permanent deposits of ethane appear to remain behind.

Indeed, it has been shown that Titan's pole precesses over tens of thousands of years and this (with the eccentric orbit of Saturn around the Sun) leads to an asymmetry in the seasons that biases hydrocarbon lakes toward the north, where we see them. The effect is modulated by, and reverses with, dynamical variations in the orbit of Saturn around the Sun. Much akin to the famous Croll—Milankovitch cycles on the Earth (which are likely also responsible for the Mars polar layered deposits), the resulting cycling of methane and ethane between Titan's poles—with a period of tens of thousands of years—leads to active geologic surface modification in the polar latitudes, but with two interacting fluids, ethane and methane, rather than one as on the Earth. This phenomenon needs to be studied further on longer timescales and with in situ access.

Methane is an essential constituent on Titan, causing a greenhouse effect, much like on Earth, which has important consequences on the surface temperature. As discussed in Section 2.2, methane is normally photolyzed in Titan's atmosphere, and unless it can be replenished by a large reservoir on or beneath the surface, it is bound to disappear in a few million years. What would then be the fate of Titan's atmosphere? In such a case, the surface temperature would drop below the condensation point for nitrogen, and Titan's atmosphere would collapse. Should the absorptivity of the surface increase subsequently (e.g. due to the accumulation of organics), the surface temperature might once again rise and cause the reevaporation of methane and nitrogen, thus rebuilding the atmosphere. Such cycles have been hypothesized to occur on Titan.

Furthermore, should the methane supply become abundant in the future, a small perturbation in the solar flux received on Titan (such as is expected when the Sun becomes a red giant and then a dwarf) would produce a dramatic warming of the climate, raising the temperature on the surface and the pressure to values as high as 180 K (twice what we have today) for several bars. It is not inconceivable to imagine that some day in the distant future, conditions on Titan one day may very closely resemble those found on our own planet today.

But Titan can also be viewed as an analog, albeit with different working materials, of a future state of the Earth when surface conditions, due to the increasing solar luminosity, shall preclude stable equatorial/mid-latitude oceans and the liquid water will be limited to the poles. Thus by studying the evolution of Titan and how the various geologic and meteorological processes interact, we have a glimpse into the Earth's past and future "simulated" on a planetary scale by different materials.

These aspects of Titan, as the fragility and changeability of the climate of our own home world, and the habitable potential of the moon, need to be studied more thoroughly.

In the meantime, the Cassini mission has demonstrated the complexity of this world and our need to further investigate it in order to better comprehend our solar system. Beyond the extended Cassini mission (2017), discussions of future missions to Titan are already underway.

Such concepts are currently in study (even if Cassini is still on the spot, we need in the future a dedicated orbiter, more in situ exploration, and more adapted instrumentation, among others). The use of aerial explorers such as an airplane or balloons and dune and lake landers has been suggested and studied by space agencies (Figure 38.13).

FIGURE 38.13 Artist's concept of a future Titan mission including several concepts studied by space agencies (such as a dedicated orbiter, a lander or a balloon) *ESA/NASA TSSM mission study, 2009. Image credit: C. Waste, NASA/JPL.*

Such missions would allow us to thoroughly explore the unique and complex Titan environment which may have many more surprises in store for us.

ACKNOWLEDGMENTS

The author wishes to thank D. Gautier, M. Hirtzig, F. Ferri, T. Krimigis, P. Lavvas, D. Luz, O. Mousis, A. Solomonidou, T. Tokano, and R. Lorenz for very useful inputs and discussions.

BIBLIOGRAPHY

Brown, R. H., Lebreton, J.-P., & Waite, H. (Eds.). (2009). *Titan from Cassini-Huygens.* New York: Springer-Verlag.
Coustenis, A., & Encrenaz, Th. (2013). *Life beyond Earth: the search for habitable worlds in the Universe.* Cambridge Univ. Press.
Coustenis, A., & Taylor, F. (2008). *Titan: Exploring an Earth-like world.* Singapore: World Scientific Publishers.
Coustenis, et al.. (2006). *Icarus, 180,* 176–185.
ESA/NASA TSSM mission study, 2009.
Fortes, A. D. (2012). *Planet Space Science, 60,* 10–17.
Fulchignoni, et al.. (2005). *Nature 438, 8 December 2005,* 785–791.
Lorenz, R., & Mitton, J. (2008). *Titan unveiled: Saturn's mysterious moon explored.* Princeton University Press.
Major Web sites: http://saturn.jpl.nasa.gov/home/index.cfm http://www.esa.int/SPECIALS/Cassini–Huygens/index.html.
Negrao, et al.. (2006). *Planet Space Science, 54,* 1225–1246.
Porco, et al.. (2005). *Nature, 434,* 159–168.
Several articles in Science and Nature issues of 2004 and 2005 describe in detail the Cassini–Huygens mission first findings. In particular: Cassini arrives at Saturn; Science, 307: February 25, 2005, and Imaging of Titan from the Cassini spacecraft. Nature, 434: March 10, 2005, and Nature, 438: December 8, 2005. Nature, 10 March 2005, 434; Nature, 8 December 2005.
Stephan, et al. (2009). In R. H. Brown, J-P. Lebreton, & H. Waite (Eds.), *Titan from Cassini-Huygens.* Springer.
Teanby, et al. (2010). *Astrophys. J., 724,* L84–L89.
Tomasko, et al.. (8 December 2005). *Nature, 438,* 465–778.
Tomasko, et al.. (8 December 2005). *Nature,* 765–778.

Chapter 39

Enceladus

Francis Nimmo
Dept. Earth and Planetary Sciences, University of California Santa Cruz, CA, USA

Carolyn Porco
CICLOPS, Space Science Institute, Boulder, CO, USA

Chapter Outline

1. Introduction and History — 851
2. Shape, Gravity, Topography — 852
3. Surface Composition — 852
4. Surface Geology and Tectonics — 852
5. The South Polar Region — 854
   5.1. Regional Geology — 854
   5.2. Jets — 854
   5.3. Vapor Component of the Plume — 856
   5.4. Thermal Signature — 856
6. Present-day Structure — 856
7. Evolution — 858
8. Conclusions — 859
9. The Future — 859
Bibliography — 859

1. INTRODUCTION AND HISTORY

Enceladus, a small icy satellite of Saturn (Figure 39.1), is one of the very few solar system bodies that is geologically active at the present day. The discovery, in early 2005, of erupting jets of water vapor, organic compounds, and ice grains, as well as an anomalously large heat flux, from the moon's south polar region represents arguably the most important finding of the Cassini mission.

Voyager, the last spacecraft to encounter Saturn before Cassini, revealed a surface characterized by extensive smooth, crater-free terrains, a clear sign of past internally driven geologic activity; it did not image the south polar region. The persistence of the tenuous E-ring centered on Enceladus had been tentatively ascribed to resupply by geysers, but this idea (now known to be correct) was not generally accepted prior to 2005.

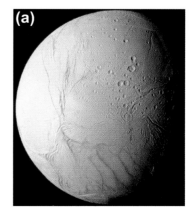

FIGURE 39.1 Cassini ISS images. (a) False-color regional view of Enceladus, including the south polar terrain (SPT) and its famous tiger stripes. Blue indicates larger grain sizes. The diameter of Enceladus is 504 km. (b) Near-terminator high-resolution image. Image scale is ∼75 m/pixel.

TABLE 39.1 Basic Enceladus Parameters

Quantity	Value	Quantity	Value
Mass M	1.08×10^{20} kg	Semimajor axis	237,984 km
Mean radius R	252.1 km	Period	1.370 days
Bulk density	1609 kg/m^3	Eccentricity	0.0047
a	256.6 (+/− 0.3 km)	Inclination	0.019°
b	251.4 (+/− 0.2 km)	Escape velocity	0.239 km/s
c	248.3 (+/− 0.2 km)	Equatorial gravity	0.114 m/s^2

Source: Shape (a, b, c) and density data are from Thomas (2010). *Icarus*, 208, 395−401.

Enceladus is in a 2:1 mean motion eccentricity-type **resonance** with its larger sibling moon, Dione, and thus has a forced nonzero eccentricity (Table 39.1). Tidal heating and flexing arising from the eccentric orbit is the most plausible explanation for the observed activity, but the details and history are unclear (see below).

2. SHAPE, GRAVITY, TOPOGRAPHY

Enceladus' bulk density of 1609 kg/m^3 (the second largest in the Saturnian satellite system after Titan) suggests a compositional mix of ice and silicates. The bright surface material is pure crystalline ice, so Enceladus has generally been presumed to be a differentiated object containing a silicate core beneath an H$_2$O layer.

As with all synchronously rotating bodies—where the rotation period is equal to the orbital period and hence the moon always presents the same face to the host planet—Enceladus is distorted by tides and centrifugal forces into a triaxial ellipsoid. However, its shape is not that expected for a strengthless (hydrostatic) body. A slowly rotating hydrostatic body has a ratio $(b - c)/(a - c)$ of 0.25, where a, b, and c are the ellipsoidal axis lengths. For Enceladus, the value of this quantity is 0.37 ± 0.04; its shape is clearly not hydrostatic. The equivalent gravity coefficients are also not in the hydrostatic ratio, although in this case the deviation is smaller. The nonhydrostatic shape and gravity make it harder to determine the polar moment of inertia C and internal structure of Enceladus. Although interpretations are nonunique, preliminary analyses yield a dimensionless moment of inertia factor of $C/MR^2 = 0.33-0.34$. Assuming a differentiated, two-layer structure with a surface (H$_2$O) layer density of 900−950 kg/m^3, the implied H$_2$O layer thickness is 47−61 km and the silicate core density is 2225−2472 kg/m^3. The latter density range is more consistent with porous or hydrated silicates or silicates containing ~ 10% ice, than with anhydrous, pore-free silicates.

The moment of inertia determination is incapable of distinguishing between liquid and solid H$_2$O, although gravity and topography data have been used together to infer the presence of a subsurface liquid sea beneath the SPT or a global ocean that is thicker under the SPT than elsewhere. A regional sea, at the minimum, can be inferred from other lines of evidence (Section 5.2). It is generally assumed that the majority of the H$_2$O layer is solid ice rather than liquid water, but direct constraints on the ice shell thickness are currently lacking. A global ocean could potentially be detected from its effect on Enceladus' obliquity or the amplitude of the moon's physical **librations**, but neither has yet been measured.

Limb profiles and stereo imaging have quantified Enceladus' regional topography (Figure 39.2(b)). This is dominated by a south polar depression of about 0.5 km in amplitude. This observation strongly suggests (together with the gravity results and other data) the existence of a regional sea beneath the south polar terrain (SPT), in essence making the depression a sink hole. Shallower, 100-km-scale depressions are present elsewhere on the moon, but show no obvious relationship to the surface geology (Figure 39.2(b)) and are not associated with anomalous heat (see Section 5.4). They are therefore not likely to be associated with subsurface liquid water at present but instead may be due merely to shell thickness variations or lateral variations in shell density (e.g. due to differential compaction of porosity).

Despite the fact that parts of Enceladus are heavily cratered, at wavelengths shorter than 100 km the topography is much smoother than all other airless satellites except Europa. This is presumably due to resurfacing and/or relaxation caused by past internal heating. Locally high heat fluxes sometime in the past have been deduced based on the **flexural response** of individual features, and the shallow (relaxed) nature of some large impact craters; the gravity observations are also consistent with a highly compensated (weak) shell, again suggesting past high heat fluxes.

3. SURFACE COMPOSITION

Enceladus' surface is dominated by crystalline water ice with surface grain sizes ~ 10−100 µm. Local variations in inferred grain size correlate with geology; for instance, the tiger stripes exhibit larger grain sizes than the surrounding terrain, perhaps as a result of sintering or the fallout of larger grains. Minor amounts of CO$_2$ ice have also been detected.

4. SURFACE GEOLOGY AND TECTONICS

Enceladus, like Gaul, is divided into three parts (Figure 39.2). Heavily cratered terrain forms a north−south girdle around the moon, encompassing the Saturn-facing,

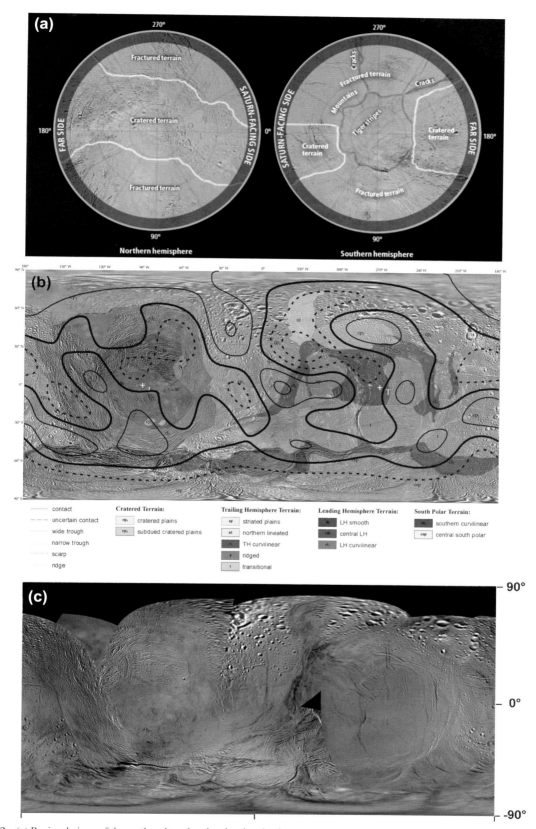

FIGURE 39.2 (a) Regional views of the north and south poles showing the three-part nature of the surface. *(From Porco (2008). Scientific American.)* (b) Global geological map, from Crow-Willard and Pappalardo (submitted). Black contours are spherical harmonic topography from $l = 3$ to $l = 8$. *(From Nimmo et al. (2011))*; bold line is zero contour, solid and dashed lines are positive and negative, and contour interval is 0.5 km (c) Global color (IR-green-UV) map, simple cylindrical projection. *From Schenk et al. (2011).*

north polar, and anti-Saturnian regions, but interrupted by the SPT. The SPT region is tectonically deformed, geologically active, and is described in more detail in Section 5. The leading and trailing hemispheres, in contrast, both consist of tectonically resurfaced, relatively crater-free terrains, although in detail they differ. All the latter regions—leading and trailing hemispheres and the SPT—show a similar overall architecture, consisting of a central tectonized region wrapped by a circumferential belt of mountains. It is currently unclear whether this similarity implies a commonality in origin, but the pronounced symmetry of the terrains about the present-day tidal axis suggests that tidal processes have been important.

The cratered terrain has a nominal age range of 0.6–4.6 Gyr, depending on which model of impact flux is used. The corresponding age ranges from the trailing hemisphere and SPT are 0.04–2 Gyr and 1–100 Myr, respectively. Almost all the large impact craters show signs of viscous relaxation, indicative of ancient elevated heat fluxes in regions other than the SPT.

The surface color of Enceladus shows pronounced spatial variations (Figure 39.2(c)) quite different from the hemispheric asymmetries seen on other Saturnian moons. Two antipodal swaths of blue-pink material protrude northward from the SPT, becoming less pronounced as they do so. This behavior is very similar to modeled patterns of fallout from the south polar jets, and suggests that SPT activity controls the global color distribution. Models of the jet eruptions also show that most of the solid material falls back down to the surface, with some very small percentage—~9% or less—escaping the moon to supply the E-ring (Section 5.2). The extreme brightness overall of Enceladus (**Bond albedo** 0.8) is almost certainly due to this process. Plume fallout has also been inferred at a local scale from the apparent mantling or softening of topography along the tiger stripe fractures seen in high-resolution images (Section 5.2).

5. THE SOUTH POLAR REGION

Because of its importance, we treat the south polar region separately. In discussing this region, it is important to distinguish between "jets" or "geysers"—individual eruption features (Figure 39.1(b))—and the "plume"—the combined, regional envelope of vapor and ice grains.

5.1. Regional Geology

The outermost region of the SPT, at about 55°S, is bounded by arcuate mountain belts, perhaps reflective of compressional relative motion there (Figure 39.3). At several locations along this arcuate boundary, prominent fractures extend northward for hundreds of kilometers. These fractures cross-cut all other features and are thus

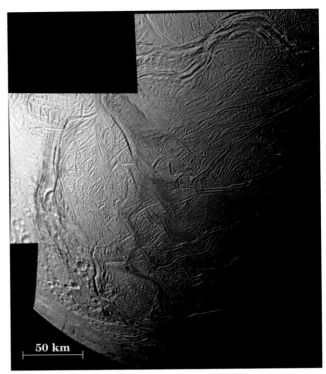

FIGURE 39.3 A false color Cassini mosaic of the SPT taken in August 2008, a year before southern autumnal equinox. The four major tiger stripe fractures, and the belt of mountains circumscribing the entire SPT, are clearly visible.

young. In the center of the SPT are the so-called tiger stripes, prominent linear depressions ~0.3 km deep, 135 km long, and about 2 km wide and separated by ~35 km. These fractures are the source of all the observed jets and the hottest thermal anomalies (see below). In-between the tiger stripes are a dense array of cross-cutting fractures, some of which may represent the remnants of earlier tiger stripe-like features.

5.2. Jets

Cassini images show ~100 distinct, prominent jets emanating from the south pole (Figure 39.1(b); Figure 39.4). The jet locations (dots in Figure 39.4(b)) and tilts (Figure 39.4(a)) can be determined by triangulation. In all cases, jets lie along the main trunks of the four major tiger stripes or their split ends and erupt roughly in a plane defined by the perpendicular to the surface, with some exceptions. Individual jets in some cases turn on and off, but there appears to be no systematic relationship to the orbital phase, as would be expected if tidal stresses alone controlled the individual jets.

The jets consist of vapor (Section 5.3) as well as ice grains having a range of sizes; the grains that escape and supply the E-ring are ~1 μm in diameter but larger grains

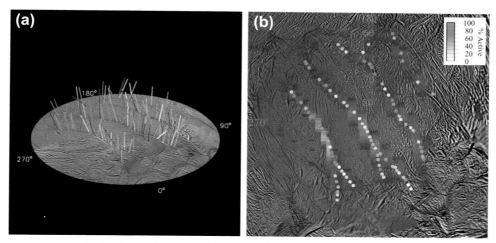

FIGURE 39.4 Tiger stripe region showing the location (b) and (a) tilts of all identified jets, or geysers, and the correlation of low resolution thermal emission with geysering activity (Porco et al. 2013). In part (b), the background is a Cassini ISS mosaic. Colors indicate CIRS temperature map. (*From Howett et al. (2011)*) Dots are individual jet locations derived from triangulation of images, and green color scale indicates the frequency with which each jet is observed to be active. At much higher resolution, individual geysers have been found to be coincident with small-scale thermal hot spots. *Porco et al. (2013)*.

are present and fall back to the surface. The characteristic velocity of the grains exiting the vent is a few tens of meters per second, slow compared to the escaping vapor. Their mass flux is ~30–70 kg/s, comparable to or slightly smaller than the total vapor mass flux: the ratio of solids to vapor is ~0.4–0.7. This value is too large for the solids to be explained by the adiabatic expansion and condensation of vapor alone. Instead, the most plausible explanation is the flash freezing of liquid droplets. Only a small fraction of this mass is required to resupply grains in the E-ring, which have a lifetime against sputtering measured in decades. The velocity of the escaping grains is obviously much higher than typical and greater than the escape velocity from Enceladus.

Moreover, a small fraction of the ice grains are rich in NaCl, with a salinity comparable to the Earth's oceans. The generally accepted explanation for both observations is that the grains form by direct, flash freezing of salty water droplets as they reach the vacuum of space, presumably from a subsurface liquid reservoir that has been in contact with a silicate core. The Na-rich grains and the comparatively large solids/vapor ratio thus represent strong direct evidence for subsurface liquid water. The obvious implication is that the jets are in fact geysers erupting from a liquid water reservoir.

Although individual geysers do not appear to respond in a systematic way to the orbital position of Enceladus, the spatially integrated plume brightness (a reasonable proxy for total mass in solids) does show a correlation with orbital position in both Cassini Visual and Infrared Mapping Spectrometer (VIMS) and Imaging Science Subsystem (ISS) data. The brightness peaks near when Enceladus is at **apoapse**, which is broadly consistent with models in which the tiger stripes open and close in response to tidal stresses.

Geysering activity across the SPT is spatially correlated with the hottest locations in the region as seen at low resolution (Figure 39.4). Individual geysers have also been shown to be spatially coincident with the few, very small-scale (approximately tens of metres) hot spots that have so far been reported. In general, where there are

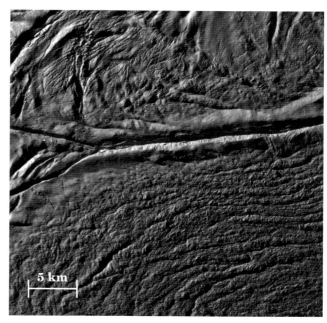

FIGURE 39.5 The region on the tiger stripe fractures that is hottest as seen in low resolution CIRS data (Figure 39.4), and the site of the most geysering activity, is the portion of Damascus fracture captured in this ISS image, with a spatial scale of 32 m/pixel. The ramparts of the fracture are smooth in appearance but eventually give way, at some distance from the fracture, to typical deformed interstripe terrain.

geysers, there is heat and vice versa. The ramparts of jet-active regions on the SPT are smooth in appearance (Figure 39.5), likely caused by local plume fallout.

5.3. Vapor Component of the Plume

Detailed compositional measurements of the plume were obtained by the Cassini Ion and Neutral Spectrometer instrument during multiple fly-throughs of the plume. Preliminary analysis of the results revealed a host of short-chain hydrocarbons, among them acetylene, propane, and benzene. Whether more sophisticated analyses will confirm that these light hydrocarbons are the result of the breakup of larger organic molecules is not yet clear. What is clear is the presence of water vapor at a volume mixing ratio of $\sim 90\%$, and the presence of carbon dioxide, ammonia, and methane. The detection of NH_3 is important because of its potential role as an antifreeze in a subsurface ocean, its effect on ice **rheology**, and as a carrier of **biologically available nitrogen**. The measured D/H ratio is similar to the cometary value and twice the terrestrial value.

The presence of organic materials and biologically available nitrogen in the vapor composing the geysers, and the inference of a water reservoir in contact with a silicate core, make for the very exciting suggestion that the geysers of Enceladus are directly connected to a subsurface habitable zone. (See Section 9.)

Occultation data constrain the upward velocity of the molecules in the vapor component of the plume to be about 300–500 m/s, greater than the escape speed of Enceladus (239 m/s). The vapor mass production rate appears to have held steady to within a factor of two over the course of the Cassini observations.

5.4. Thermal Signature

Because of its high albedo, the surface temperature of Enceladus is about 80 K at the equator, and lower toward the poles. Measurements made by the Cassini Composite Infrared Spectrometer (CIRS) instrument indicate that the tiger stripes are much hotter than their surroundings (Figure 39.4), with the excess power totaling about 5 GW ($1 \text{ GW} = 1 \times 10^9$ W). It is possible that a comparable amount of heat, in total, arises from the entire area of the SPT but with sufficiently low flux to escape detection by CIRS. High-resolution infra-red measurements suggest that the hottest heat flow anomalies along the fractures are confined to very small regions, tens of meters in size. The heat flux emerging from these local hot spots is many times greater than the average heat flux arising from the Yellowstone geothermal area in Wyoming. These hot spots coincide with individual geysers, making the association of thermal emission with geysering activity hold at high as well as low resolution.

There are two main possible explanations for these relationships. One is that the heat is being produced today by tidal dissipation; the other is that we are seeing stored heat, generated over some period in the past and now escaping. An example of the former category is tidally driven frictional heating between opposing walls of the tiger stripe fractures, while an example of the latter is the ascent of warm water from a subsurface ocean. As discussed below, these different hypotheses have different consequences not only for the mechanism that produces the geysers and the source of the water in them, but for the orbital evolution of Enceladus.

Three lines of evidence favor the ascent of heat and water from depth. First, as mentioned earlier, the salty nature of some of the ice grains and the large fraction of solids in the eruptions suggest that the ultimate source is a subsurface ocean, and not sublimation of near-surface ice or the freezing of vapor upon expansion as it exits the surface vents. Second, the narrowness of the imaged thermal anomalies is inconsistent with tidal heating, which would be distributed over kilometers (in the case of frictional heating) or the whole SPT (in the case of distributed tidal heating). Third, the observed correlation between plume strength and orbital phase (see above) very likely implies that cracks are opening and closing on a tidal timescale, allowing subsurface water to erupt as ice and vapor. In this hypothesis, the observed heat flux is mainly due to recondensation of vapor in the cold near-surface ice. Consistent with this explanation is the recognition that liquid water in the cracks would probably inhibit any frictional heating within them. The implied rate of recondensation (~ 2000 kg/s) means that only about $\sim 10\%$ of the vapor produced is actually escaping to form the plume. Individual fractures probably reseal themselves on timescales of a couple of years.

6. PRESENT-DAY STRUCTURE

Here we attempt to reconcile the observations described above into a coherent picture of Enceladus' interior (Figure 39.6).

The presence of liquid water beneath (at least) the SPT is strongly suggested by the detection of salty ice grains and the large solids/vapor ratio (Section 5.2). The depression of the SPT by ~ 0.5 km suggests a regional sea, although it is not yet clear whether this sea forms part of a thin global ocean (see below). The phasing of the time variability of the plume directly implicates the role of cracks opening and closing in response to tides. Such cracks, if water-filled, can readily propagate upward on Enceladus through the majority of the ice shell, despite the relatively low (~ 1 bar) tidal stresses. The ascent through the remainder of the ice shell to the surface could easily be accomplished by the violent exsolution of dissolved gases, such as CO_2, from the

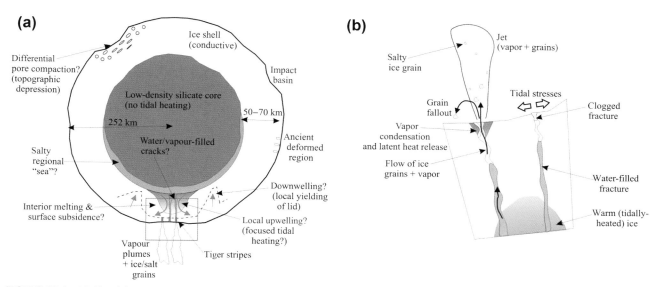

FIGURE 39.6 (a) Sketch interpretation of Enceladus' interior structure (see text). (b) Sketch interpretation of the near-surface structure at the SPT (see text).

liquid that would force water and vapor to the surface. When a crack opens, exposure of the water and vapor to vacuum will result in simultaneous boiling and freezing; a geyser forms. A spectrum of ice grain sizes is produced, with the larger grains readily falling out and the smaller grains in some cases escaping to replenish the E-ring. Much of the vapor produced recondenses on the near-surface walls of the conduit, releasing its latent heat there and generating the majority of the thermal anomaly observed and ultimately clogging the crack, causing the individual geyser to fail. This geysering model can explain how the integrated plume flux might vary with the changing tidal stresses, although individual geysers do not appear to; opening and closing of cracks during the tidal cycle modulates the strengths of individual geysers, but they remain active throughout the cycle. Geysers eventually fail and turn off, not on a daily timescale but when an individual vent becomes clogged, probably on a timescale of a year or two.

In terms of its bulk structure, the relatively low density inferred for the silicate core (Section 2) represents a surprise—most models previously assumed an anhydrous density similar to that of Io (roughly 3500 kg/m^3). However, the pressure at 100 km depth on Enceladus is equivalent to only 300 m depth on the Earth, so it should not be surprising if the silicate core were porous. Alternatively, a hydrothermally altered core is a possibility (especially if Enceladus accreted while radioactive ^{26}Al were still live) and might help to explain the composition of the plume.

Two major unanswered questions concerning Enceladus' interior structure are whether the ice shell is convecting, and whether the subsurface liquid is global in extent.

Although, as noted above, the source of the tiger stripe heat appears to be primarily advection of warm water from depth, a contribution from present-day tidal heating across the entire SPT region is not precluded. Certainly, the extreme tectonic deformation and young age of the surface is suggestive of yielding and lateral motion of material, perhaps in response to deeper seated convection. Convection in tidally heated bodies is subject to strong feedbacks, because warmer ice (e.g. in upwellings) is more easily tidally flexed and therefore more strongly heated. Furthermore, tidal heating is generally greater at the poles than at the equator. Thus, a local, tidally heated convective upwelling at the south pole seems quite likely. Such an upwelling has the additional property of insulating the underlying material, and calculations have shown that a local southern sea can consequently persist indefinitely under such conditions.

Whether convection is currently taking place elsewhere on Enceladus is much more open to debate. If regions other than the south pole lack a subsurface liquid layer, they will be subject to much weaker tidal heating and thus less likely to convect. The long-wavelength topography and ancient deformation features observed, however, may well be an indication of past activity and convection, perhaps suggesting that Enceladus has experienced multiple deformation episodes (see below). The symmetrical distribution of the deformed terrains (Figure 39.2) may perhaps be related to the tendency of convecting regions to undergo true polar wander, modulated by the tidally distorted shape of Enceladus. On the other hand, a globally convective Enceladus would be sufficiently dissipative that it would be hard to match the currently observed orbital configuration. The most likely solution is that convection is localized (very likely in space—the SPT—and perhaps also in time).

Although a regional sea can persist indefinitely under some conditions (see above), a global subsurface ocean encounters severe thermal problems. The silicate core, even if hydrated, is very unlikely to be a significant source of tidal heating, and any initial heat (e.g. from ^{26}Al decay) will have long since diffused away. Thus, the only plausible heat source from the silicates is the decay of long-lived nuclides, and their contribution is simply too small to offset the heat loss through the ice shell, even in the absence of shell convection. A global ocean is thus expected to freeze in a few tens of million years (Myr) Once frozen, subsequent tidal heating is greatly diminished because the ice shell is then locked to the much more rigid silicates beneath. A regional sea, on the other hand, is likely to have none of these thermal problems.

7. EVOLUTION

The long-term evolution of Enceladus' interior is closely coupled to its orbital evolution, because the main heat source is energy extracted from its orbit (and ultimately from Saturn's rotation). Enceladus' current resonance with Dione causes its **eccentricity** to increase, while any tidal dissipation in Enceladus will cause the eccentricity to decrease. There is thus in principle an equilibrium in which the eccentricity remains constant. In this special case heat production in Enceladus does not depend on its interior structure but only on the rate at which energy is being dissipated within Saturn—which also determines the torque on Enceladus's orbit. Taking the dissipation factor of Saturn to be the generally accepted value ($Q > 18,000$) the equilibrium heat production rate within Enceladus has an upper bound of 1.1 GW. This is a factor of five or more less than the measured heat output. There are two possible scenarios to explain this discrepancy.

First, Enceladus is producing heat today at a much greater rate than the equilibrium value, and its eccentricity is consequently rapidly diminishing. Alternatively, Enceladus could be producing heat at the equilibrium rate, and the present-day large thermal output is coming from stored heat that has accumulated over a long period. These two scenarios have quite different consequences for the predicted eccentricity evolution, and can thus (in theory) be distinguished via **astrometry**. Both, however, imply some kind of nonmonotonic behavior, as discussed further below.

It has long been understood that the coupling between thermal and orbital evolution can lead to cyclic behavior, in which the state of the moon oscillates between brief periods of high heat production and rapid (approximately tens of Myr) eccentricity damping, and longer periods of low dissipation and slow (approximately hundreds of Myr) eccentricity growth. Such cyclic behavior could plausibly explain the different surface ages observed (Section 4). The total power output from the tiger stripes and the surrounding SPT regions is still quite uncertain; nor is it clear what fraction (if any) of the measured power output is being produced by present-day tidal heating. If Enceladus is currently in a highly dissipative state with decreasing eccentricity, the tidal stresses will also decrease over time and thus reduce the widths of the cracks that open and close. In this scenario we are seeing Enceladus today in the "active" part of its duty cycle, which probably represents only a few percentage of the total. The perhaps uncomfortable implication is that we are seeing Enceladus today at a very special time, a circumstance that has a rather low probability.

Alternatively, Enceladus may be producing heat at the constant (equilibrium) rate, but only releasing it episodically in bursts—one of which we are seeing now. There are plausible ways in which this might happen. For instance, convection beneath a weak lid can result in episodic, regional outbursts of heat, which share some characteristics with those inferred for the SPT of Enceladus. Similar scenarios can probably be constructed for situations in which the majority of this heat is advected along individual fractures (Figure 39.6(b)). For instance, progressive shell thickening leads to pressurization of the ocean and presumably more vigorous eruptions, which will depressurize the ocean and reset the process. If the surface heat flux is really being supplied by a flux of ~ 2000 kg/s from the ocean, then a 10-km-thick ocean (representing 5% of Enceladus' mass) would be exhausted in 100 Myr. The energy required to remelt that mass (at the equilibrium rate of 1.1 GW) would take roughly 0.5 Gyr to accumulate. These kinds of timescales are roughly consistent with the inferred variations in surface ages. Whichever scenario is correct, it seems necessary to ensure that any body of liquid within the moon never freezes completely, because once fully frozen, subsequent tidal heating is reduced and remelting becomes much more difficult if not impossible. Localized tidal heating is one mechanism for ensuring a persistent regional sea of liquid (see Section 6); tidal heating within a shallow ocean itself may also be a possibility.

The lack of any equivalent to the SPT in the northern polar regions remains a puzzle, because tidal heating is expected to be symmetrical about the equator. Of course, once initiated, a runaway thermal process can take place locally, with tidal heating leading to a weaker rheology and thus more focused tidal heating. Nonetheless, something has to initiate the runaway to begin with. Some possibilities exist. A giant impact, which on other Saturnian moons (such as Iapetus) created large basins, might have deposited sufficient heat to soften the ice and initiate local tidal heating and convection. Or during the moon's formation, stochastic accretion of relatively large **planetesimals** could have formed an irregular silicate core, with liquid water surviving in pockets at the silicate/water interface. Through polar wander and reorientation, in both cases, the region containing liquid water underlying a thinner ice shell would have ended up at one of the poles.

A similar problem exists concerning Mimas; this body is closer to Saturn and more eccentric, yet it shows no signs of activity, either now or in the past. Again, one can appeal to a runaway tidal heating process at Enceladus, perhaps initiated by the fact that Enceladus has a slightly higher rock to ice ratio than Mimas, and thus more radiogenic heating, to explain why the moons are so different. But in all cases, one is left with having to appeal to poorly understood initial conditions to explain the observations. These examples illustrate our incomplete understanding of how Enceladus evolved.

8. CONCLUSIONS

Although our understanding of Enceladus is still evolving rapidly, Figure 39.6 provides a reasonable snapshot of the current state of knowledge. Enceladus has probably gone through several cycles of activity and dormancy, and we are very fortunate in seeing it in one of the former right now. The activity derives from tidally driven opening and closing of cracks, allowing warm water-filled fractures to periodically become exposed to vacuum, thus generating plumes of water vapor and salty ice grains. Enceladus possesses a regional sea beneath the SPT, but probably not a global ocean. Its shell may be undergoing convection locally beneath the SPT, but not globally. At earlier times, other parts of Enceladus' surface were probably undergoing processes similar to what we are seeing at the SPT today.

9. THE FUTURE

Future analysis of Cassini data may help to fill in some of the gaps in our understanding. For instance, astrometry of Enceladus' orbit could reveal whether or not Enceladus' eccentricity, and hence the physical state of its interior, are truly in equilibrium. Similarly, a detection of longitudinal librations by a thorough analysis of Cassini high-resolution images of its surface would help to determine whether the subsurface ocean is global or regional in extent.

Ultimately, future spaceflight missions will be required to make significant advances in understanding the nature and history of the moon's interior and its observed south polar activity. An Enceladus orbiter could more accurately map the moon's gravity field and surface topography, revealing its internal distribution of mass, confirming the thickness of the ice shell and the presence of any subsurface liquid layer and/or concentration of liquid under the SPT. A small lander, equipped with a seismometer, placed within a tiger stripe and close to a locale of high geysering activity could detect the motion of underground liquids. Both mission types, as well as simpler missions that merely fly-through the near-surface regions of the geysers, could be equipped to detect unambiguously, in a way that Cassini cannot, the presence of large organic compounds of biological interest. Such missions could also include a sample-return capsule to collect plume particles and return them to Earth.

The capability of detecting large organic compounds is of critical importance in assessing the astrobiological potential of Enceladus. Organic molecules, **biologically available nitrogen**, an energy source, and a long-lived reservoir of liquid water—four of the key attributes of a habitable environment—are all demonstrably present. Possible anaerobic ecologies that could thrive in a deep, sunlightless environment are present in subsurface environments on the Earth, metabolizing in the absence of sunlight by consuming molecular hydrogen and either carbon dioxide or sulfate. The possibility of oxidants produced at the surface by irradiation by high-energy particles and delivered by convection within the ice shell to the subsurface reservoir only increases its habitability. Since the geysers are likely directly connected to a long-lived water reservoir, itself in contact with silicates, they essentially provide "free samples" of that habitable environment below. This particular set of circumstances makes the habitable zone on Enceladus one of the most accessible in the solar system. If life has taken hold on Enceladus, it is there for the taking.

BIBLIOGRAPHY

Howett, C. J. A., Spencer, J. R., Pearl, J., & Segura, M. (2011). High heat flow from Enceladus' south polar region measured using 10-600 cm^{-1} Cassini/CIRS data. *JGR, 116*. E03003.

Iess, L., et al. (2014). Geophysical implications of the long-wavelength topography of the Saturnian satellites. *The Gravity Field and Interior Structure of Enceladus Science, 344*, 78.

McKay, C. P., Porco, C. C., Altheide, T., Davis, W. L., & Kral, T. A. (2008). The possible origin and persistence of life on Enceladus and detection of biomarkers in the plume. *Astrobiology, 8*, 909–922.

Nimmo, F., Bills, B. G., & Thomas, P. C. (2011). Geophysical implications of the long-wavelength topography of the Saturnian satellites. *JGR, 116*. E11001.

Porco, C. (December 2008). The restless moon of Enceladus. *Scientific American, 299*, 52–63.

Porco, C., et al. (2013). How the jets, heat and tidal stresses across the south polar terrain of Enceladus are related. *Lunar and Planetary Institute Conference Abstracts, 44*, 1775.

Schenk, P., Hamilton, D. P., Johnson, R. E., McKinnon, W. B., Paranicas, C., Schmidt, J., & Showalter, M. R. (2011). Plasma, plumes and rings: Saturn system dynamics as recorded in global color patterns on its midsize icy satellites. *Icarus, 211*, 740–757.

Spencer, J. R., & Nimmo, F. (2013). Enceladus: an active ice world in the Saturn system. *Annual Review of Earth and Planetary Sciences, 41*, 693–717.

Chapter 40

Triton

William B. McKinnon
Department of Earth and Planetary Sciences and McDonnell Center for the Space Sciences, Washington University, Saint Louis, MO, USA

Randolph L. Kirk
U.S. Geological Survey, Flagstaff, AZ, USA

Chapter Outline

1. Introduction — 861
2. Discovery and Orbit — 862
3. Pre-*Voyager* Astronomy — 863
   3.1. Radius, Mass, and Spectra — 863
   3.2. Seas of Liquid Nitrogen? — 863
   3.3. Similarities with Pluto — 864
4. *Voyager* 2 Encounter — 865
5. General Characteristics — 866
6. Geology — 868
   6.1. Undulating, High Plains — 868
   6.2. Walled and Terraced Plains — 869
   6.3. Smooth Plains and Zoned Maculae — 870
   6.4. Cantaloupe Terrain, Ridges, and Fissures — 871
   6.5. Bright Polar Terrains — 871
   6.6. Geological History — 872
7. Atmosphere and Surface — 872
   7.1. Atmosphere — 872
   7.2. Plume Models — 874
      7.2.1. Plumes as Jets — 875
      7.2.2. Eruption Velocity and Temperature — 876
      7.2.3. Subsurface Energy Transport — 876
   7.3. Polar Cap and Climate — 877
8. Origin and Evolution — 878
Bibliography — 881

1. INTRODUCTION

Triton is the major moon of the planet Neptune. It is also one of the most remarkable bodies in the solar system (Figure 40.1). Its orbit is unusual, circular, and close to Neptune, but highly inclined to the planet's equator (by 157°). Furthermore, Triton's sense of motion is retrograde, meaning it moves in the opposite direction to Neptune's spin (Figure 40.2). Triton's history, therefore, must have been quite different from those of "regular" satellites, such as the moons of Jupiter, which orbit in a prograde sense in their primary's equatorial plane. The modern consensus is that Triton originally formed in solar orbit and was subsequently captured by Neptune's gravity.

Like nearly all solar system satellites, Triton has had its spin period increased by tides to be coincident with its orbital period and these effects have also shifted its spin axis to be nearly perpendicular to its orbital plane. Consequently, one hemisphere of Triton permanently faces Neptune. The combination of Neptune's axial tilt (29.6°) and Triton's inclined orbit gives Triton a complicated and extreme seasonal cycle. In the distant geological past, tides associated with Triton's capture may have strongly heated and transformed its interior.

Although Triton was discovered soon after Neptune, little was learned about it until the modern telescopic era, and even so, most of the information we have was acquired during the *Voyager 2* encounter with Neptune in 1989. Triton is a relatively large moon (1352 km in radius), larger than all the midsize satellites of Saturn and Uranus (200–800 km in radius), but not quite as large as the biggest icy satellites—the Galilean satellites and Titan (1570–2630 km in radius). It is a relatively dense world (close to 2 g/cm^3), rock rich, but with a substantial proportion of water and other ices. Ices comprise its reddish visible surface (Figure 40.1), and the freshness of the ices cause Triton to be one of the most reflective bodies in the solar system (its total or Bond albedo is ≈0.85). This, combined with the satellite's distance from the Sun (30 AU), makes Triton's surface a very cold place (≈38 K). Yet, despite these frigid surface conditions, *Voyager 2* discovered a thin atmosphere of nitrogen

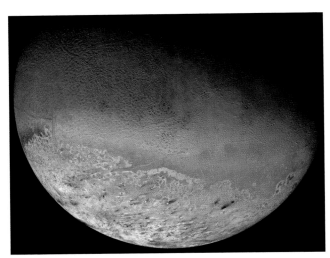

FIGURE 40.1 Digital photomosaic of Triton, centered on the Neptune-facing hemisphere at 15°N, 15°E. The latitude of the subsolar point at the time of the *Voyager* encounter was −45°, so the north polar region was in darkness. Triton's surface is covered with deposits of solid nitrogen with small admixtures of radiation-reddened and radiation-darkened methane; the bluish tinge is characteristic of fresh frosts. Because Triton's spin is tidally locked to Neptune, the eastern hemisphere (to the right) is also the leading hemisphere in its orbit. *Courtesy of the NASA Planetary Data System Photojournal.*

surrounding the satellite (14 μbar surface pressure, where 1 bar is the approximate surface pressure of the Earth's atmosphere). Triton's atmosphere is dense enough to support clouds and hazes and to transport particles across Triton's surface. It is also changing; since 1989 it has been warming and increasing in total mass and pressure.

As with all solid planets and satellites, Triton's history is written into the geological record of its surface. Triton, however, is a geologically young body. Most of the approximately 40% of the satellite's surface that was imaged by *Voyager* at sufficient resolution tells us that it is sparsely cratered. No heavily cratered terrains survive from early solar system times, an absence that may reflect an epoch of severe tidal heating. The geologic terrains that do survive are unique in the solar system. Particularly notable are the enigmatic "cantaloupe" terrain and a hemispheric-scale polar deposit or cap. The polar cap is thought to be predominantly solid nitrogen. Other ices that have been identified on Triton are, in approximate order of abundance, H_2O, CO_2, CH_4, CO, HCN, and possibly C_2H_6. The cap is a site of present-day geological activity, in particular the eruption of plumes or geysers of gas and fine particles.

In the following sections, Triton will be described in greater detail with emphasis on its geology, the interaction of its icy surface and atmosphere (including the plumes), and its probable origin and violent early evolution.

2. DISCOVERY AND ORBIT

Acting on the mathematical prediction of Urbain Le Verrier, astronomers at the Berlin Observatory first identified the planet Neptune on September 23, 1846. It was announced in England on October 1. On that day, Sir John

FIGURE 40.2 Orbits of Neptune's family of satellites, except distant irregulars. Shown is a perspective view along a line of sight inclined 18° to Neptune's equatorial plane. The innermost satellites are all relatively small and were not discovered until *Voyager 2* passed through the Neptune system. N3/N4/N5/N6 are Despina/Galatea/Thalassa/Naiad. They orbit in Neptune's equatorial plane, while the much more massive Triton circles outside them in an inclined, retrograde orbit. All the satellites have virtually circular orbits except for Nereid. The apparent crossing of Nereid's and Triton's orbits is an artifact of the projection. A new, small regular satellite was discovered in 2004 between Larissa and Proteus. *From Kargel, J. S. (1997).*

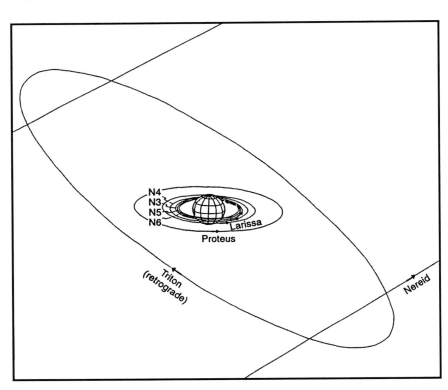

Herschel, son of the discoverer of Uranus, wrote to William Lassell, asking him to look for any satellites of the new planet "with all possible expedition", using his own 24-in reflector. Lassell was a brewer by profession, but he made his own telescopes and was a keen visual observer. Herschel was no doubt seeking to ease some of the sting of Neptune's being found by continental astronomers, given that he was aware of the independent prediction of Neptune's position by John Couch Adams and the unsuccessful search for the planet from English soil. Lassell wasted no time, making his first observations on October 2, and on October 10, 1846, he discovered Triton.

By 1930, it was established that Triton was a most unusual moon. Orbiting at 14.3 Neptune radii, or R_N (using the modern value of 24,760 km for Neptune's equatorial radius), the orbit was circular inasmuch as this could be measured, but distinctly retrograde compared with Neptune's prograde spin or sense of orbital motion. It was also *alone*. No new satellites would be found for over 100 years after Triton. The early contrast with the regular satellite systems of Jupiter, Saturn, and Uranus could hardly have been greater (*see* Neptune).

The year 1930 also marked the discovery of the dwarf planet Pluto. It was soon determined that Pluto actually crosses inside the orbit of Neptune for about 20 years of its 248-year orbital revolution. Although Pluto's orbit is also substantially inclined so that it does not actually intersect Neptune's, British astronomer R.A. Lyttleton argued that different precessions of the orbits could cause them to intersect, either in the future or in the past. In 1936, he published a paper that theoretically explored the possibility that such an orbital configuration once did exist and that Pluto was in reality an escaped satellite of Neptune. Although intriguing, this early theory is now rejected by planetary scientists. The modern view of Triton's origin, and that of Pluto and other bodies in the deep outer solar system, is discussed later in this chapter (*see* Pluto).

3. PRE-*VOYAGER* ASTRONOMY

3.1. Radius, Mass, and Spectra

Through the telescope, Triton is a faint, 14th magnitude object, never more than 17 s of arc from Neptune. Consequently, physical studies of the satellite from the ground have historically been very difficult. Because Triton shows no visible disk, only crude limits could be put on its size, or mass, for many years. But mid-twentieth century estimates implied it was one of the largest moons in the solar system and massive, possibly *the* most massive moon in the solar system. Triton was clearly a moon of mystery.

The first real breakthrough occurred in 1978, when infrared detector technology had improved to the point that a methane (CH_4) band was detected in Triton's infrared spectrum. Soon more bands were found. The relative depths of the new bands, plus their variability as Triton orbited Neptune, indicated that much (if not all) of the methane detected was in solid form, that is, an ice on the surface of Triton. Ices on the surface implied that Triton might be a relatively bright, smaller world, rather than a darker, larger body of the same visual magnitude (*see* Titan).

Methane ice on the surface also offered a potential explanation for Triton's reddish visual color. Experiments had shown that when solid methane is irradiated by solar ultraviolet (UV) rays, or bombarded by charged particles, it turns pink or red as hydrogen is driven off and the remaining carbon and hydrogen form various carbonaceous compounds. However, continued radiation or charged particle bombardment ultimately turns methane into a blackish carbon-rich residue, so Triton's persistent redness also implied that the satellite's methane ice is refreshed on a relatively short timescale.

3.2. Seas of Liquid Nitrogen?

An even more amazing discovery was made in the early 1980s. A single infrared spectral feature was found at ≈ 2.15 μm, a feature that could not be attributed to any of the usual spectral suspects (CH_4, H_2O, silicates, etc.). Nitrogen (N_2) does have an absorption at this wavelength, and because *Voyager 1* had recently determined the dominant atmospheric gas on Titan to be N_2 (not CH_4), finding nitrogen on Triton was not far fetched. The amount of nitrogen gas required to account for the absorption was quite large, however, as nitrogen, a homonuclear diatomic molecule, is a very poor absorber of infrared light. The astronomers concluded that in order to get the necessary path length for the absorption, the nitrogen had to be in condensed form, either solid or liquid. Liquid nitrogen was the favored interpretation, and a fantastic vista emerged—a satellite covered with a global or near-global sea of liquid nitrogen, along with methane-ice-coated islands or even floating methane "icebergs"!

The "problem" with liquid nitrogen is that it freezes at zero pressure at about 63 K. For Triton to have a global ocean at that temperature requires that (1) Triton should absorb most of the sunlight striking it (have a low albedo) and (2) Triton's surface should radiate infrared heat very inefficiently (have a low emissivity). For this and other reasons, planetary chemists offered a competing concept for Triton's surface, one in which both the nitrogen and methane were solid and distributed nonuniformly (Figure 40.3). Because of nitrogen's great volatility, it was argued that crystals of up to centimeter size could grow on Triton's surface over a decades-long season and so provide the path length for the 2.15-μm absorption. Methane, in contrast, would be only a minor surface component; it

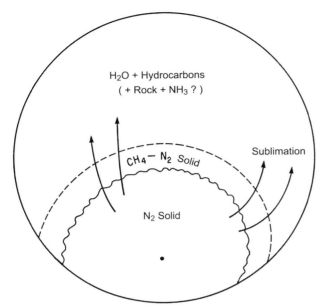

FIGURE 40.3 A pre-*Voyager* prediction for the state of Triton's surface. A subliming N_2 ice cap is centered on the illuminated south pole (dot). From Lunine, J. I., & Stevenson, D. J. (1985).

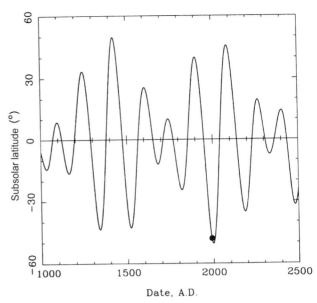

FIGURE 40.4 Seasonal excursion of the subsolar latitude on Triton. Dot shows the subsolar latitude at the time of the *Voyager* encounter. From Kirk, R. L., et al. (1995).

dominates Triton's near-infrared spectrum by virtue of the relative strength of its absorptions.

The difference between the two models for Triton's surface had important implications for the atmosphere and for Triton's seasons. Triton's seasonal cycle is complicated. Because of the precession of its orbit, its seasons vary in intensity and length, and in the decades before the *Voyager* encounter, Triton was moving toward the peak of maximal southern summer (Figure 40.4). Correspondingly, Triton's northern hemisphere was (and is) enduring prolonged darkness. The possibility of long-term cold traps at both poles, with strong seasonal atmospheric flows from pole to pole, was recognized. The illustration in Figure 40.4 was in fact based in part on an analogy with Mars, with N_2 replacing CO_2 as the dominant, and condensable, atmospheric constituent and CH_4 replacing H_2O as the secondary, less volatile component (an analogy that has strengthened, as will be discussed later). Specifically, a large cap of solid nitrogen was predicted for the south pole, sublimating slowly in the feeble summer sun.

3.3. Similarities with Pluto

As Triton was coming into clearer astronomical focus in the 1980s, parallel developments were occurring for other outer solar system bodies, especially Pluto. Methane ice had been discovered on Pluto prior to Triton, and overall, Pluto's visible and near-infrared spectrum bore a strong resemblance to that of Triton, although Pluto's methane absorptions were deeper. Their common bond was reinforced by their similar visual magnitudes (Pluto and its moon together are only ~ 0.3 magnitudes fainter than Triton when referenced to a common distance and solar phase angle).

Pluto's fundamental properties (mass and radius) were by the time of the *Voyager 2* encounter with Neptune (and Triton) relatively well constrained. Pluto's relatively large satellite, Charon, had been discovered in 1978, which allowed determination of the mass of the Pluto–Charon system by means of Kepler's third law. Careful monitoring of the Pluto–Charon system's light curve, plus observations of the occultation of a star by Pluto in 1988, established that Pluto's radius lay between 1150 and 1200 km. Pluto turned out to be a smallish, bright, more-or-less ice-covered world, and a relatively dense one as ice-rock bodies, go close to 2 g/cm^3 (*see* Pluto).

Pluto's density corresponds to a rock/ice ratio of about 70/30, and is, curiously, close to what is predicted for a body accreted in the deep cold reaches of the outer solar system. According to current thinking, the solar nebula at that distance from the Sun, when the Sun and planets were forming, was relatively cold and mostly unprocessed. The outer nebula thus retained many of the chemical signatures of the interstellar gas and dust (molecular cloud) that was the ultimate source of the solar nebula. Specifically, carbon would be in the form of organic matter and carbon monoxide (CO) gas. CO is very volatile, and the solar nebula was unlikely to have ever been cold enough for it to condense in bulk (although small amounts could be adsorbed on or trapped in water ice). The key point is that volatile CO ties up oxygen that would otherwise be available to form water ice. Therefore, bodies formed in the outer solar system, but not near a giant planet, should have relatively *high* rock/water–ice ratios (at least initially). In contrast, in the high-pressure environment near a giant planet, CO combines with H_2 to make H_2O and CH_4, which can both condense. The resulting satellites are predicted to be much icier, with rock/ice ratios of 50/50 or less (*see* The Origin of the Solar System).

That Pluto was so rock rich was one line of reasoning that pointed to Pluto being an original solar-orbiting body and not an escaped satellite of Neptune. Dynamical evidence against Pluto being an escaped satellite also accumulated. By the 1980s it was being argued that Triton and Pluto should be considered as two independent solar system bodies, with independent histories. The link between the two, in terms of brightness (and presumably size) and composition, was that they formed in the same region—the outer solar nebula near or beyond Neptune. Essentially, they are surviving examples of large outer solar system protoplanets. Pluto became locked in a dynamical resonance with Neptune, which preserved its peculiar orbital geometry, whereas Triton was captured by Neptune's gravity (*see* Pluto).

If the analogy with Pluto is correct, then Triton should also be rock rich. If Triton had a relatively bright, icy surface like Pluto, Triton's visible magnitude implied it would probably be somewhat larger than Pluto, but its density would be similar to that of Pluto–Charon. Of course, the surface state and thus the size of Triton could not be pinned down before the *Voyager* encounter, but the consequences for Triton of being captured (as has been alluded to) were potentially spectacular. These include intense tidal heating and wholesale melting of the satellite. These ideas were appreciated by the planetary community on the eve of the *Voyager 2* encounter. So with the observational and theoretical backdrop just described, and with the promise of resolution of fundamental questions and the revelation of novelty, anticipation was high.

4. *VOYAGER* 2 ENCOUNTER

Future history will no doubt record the *Voyager* project as one of humankind's great journeys of discovery. Originally conceived as a "grand tour" of all the giant planets and Pluto, the *Mariner-class* spacecraft that were eventually launched in 1977 (and renamed *Voyager*) were only designed to encounter Jupiter and Saturn. If they worked, although, a highly capable complement of remote sensing instruments for the planets and satellites and in situ detectors for the magnetospheres and plasmaspheres would be carried into the outer solar system for the first time.

At Saturn, *Voyager 1* was targeted to pass close to Titan, a trajectory that sent it out of the ecliptic plane afterward.

The trajectory of *Voyager 2* was carefully chosen to preserve the grand tour option, whereby each successive encounter would boost the spacecraft to a higher velocity and in just the right direction to reach the next giant planet, which were fortuitously arranged in the 1980s. There was no guarantee *Voyager 2* would survive the complete 12-year trip from the Earth to Neptune, many years past its design life. Problems did develop. One radio receiver went out and its backup was failing, and the articulated scan platform, on which the remote sensing instruments were mounted, could no longer move as easily as before. Nevertheless, after successful encounters at Jupiter, Saturn, and Uranus, the stage was set.

Each new *Voyager* encounter increased scientific and public awareness of the richness of the solar system. The *Voyager 2* flyby of Neptune and Triton in late August 1989 was going to be the last, and proved to be perhaps the most exciting of all. But there was one last hurdle. To get to Triton, *Voyager 2* would have to pass very close to Neptune's north pole in order for Neptune's gravity to bend its trajectory southward (Figure 40.5). This would be dangerously close (only 5000 km from the cloud tops) and in an unknown and potentially dangerous environment. To everyone's relief, *Voyager 2* made it past Neptune without incident just after midnight on August 25 (Pacific time), counting the more than 4 h it took for *Voyager's* radio signals to reach the Earth. After 5 h, it passed within 40,000 km of Triton, sending back a sequence of beautiful, mind-boggling images. These images form much of the basis for understanding, to the extent we do, Triton's geology and surface–atmosphere interactions.

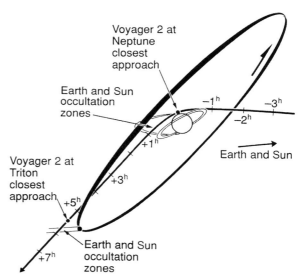

FIGURE 40.5 The trajectory of *Voyager 2* through the Neptune system. *From Chapman, C. R., & Cruikshank, D. P. (1995).*

5. GENERAL CHARACTERISTICS

Voyager 2 determined Triton to be even smaller, brighter, and hence colder than anticipated (Table 40.1). Its average geometric albedo of ≈ 0.7 is extreme even for an icy satellite. Triton's global appearance was revealed during the approach sequence (Figure 40.6). The view, mainly of the southern hemisphere, showed extensive bright polar materials, a bright equatorial fringe with streamers extending to the northeast, and darker low northern latitudes. Radio tracking of *Voyager* yielded a very precise mass for Triton, which when combined with the size, gave a very precise density of ≈ 2.065 g/cm^3. This density is essentially identical to that of the Pluto–Charon system.

With the size and mass known, internal structural models can be created based on a set of plausible chemical components; for bodies formed in the outer solar system, these would be rock, metal, ices, and carbonaceous matter. Such models provide context and to some extent guide interpretations of geological history. A calculation for Triton is illustrated in Figure 40.7. Given

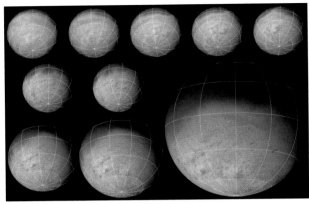

FIGURE 40.6 Triton approach sequence, overlaid with a latitude–longitude grid. Details on Triton's surface unfold dramatically as the resolution changes from about 60 km/pixel at a distance of 5 million km for the image in the upper left to about 5 km/pixel at a distance of 0.5 million km for the image in the lower right. Mainly looking at the southern hemisphere, Triton rotates retrograde (counterclockwise) over an observational period of 4.3 days. *Courtesy of Alfred McEwen, University of Arizona.*

TABLE 40.1 Properties of Triton

Radius, R	1353.5 km
Mass, M	2.140×10^{22} kg
Surface gravity, g	0.78 m/s^2
Mean density, ρ	2060 kg/m^3
Percent rock + metal by mass	65–70%
Distance from Neptune	354.8×10^3 km = 14.33 R_N
Distance from Sun	30.058 AU
Orbit period	5.877 days
Orbit period around Sun	164.8 years
Eccentricity	0.0000(16)
Inclination (present)	156.8°
Obliquity (estimate)	0.4°
Geometric albedo (average)	0.70
Bond albedo (average)	0.85
Surface temperature	38 K (1989), 39 K (2009)
Surface composition	N_2, H_2O, CO_2, CH_4, CO, HCN, C_2H_6 ices
Surface atmospheric pressure	14 µbar (1989), 40 µbar (2009)
Atmospheric composition	N_2, minor CH_4, CO
Tropopause height	8 km

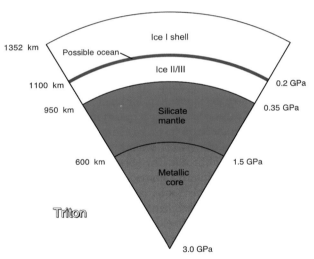

FIGURE 40.7 Internal structure model for present-day Triton. Depending on ocean composition and the amount of current tidal heating, the ocean may be much thicker, potentially reaching to the top of the rocky mantle.

that little direct information exists on the internal makeup of Triton, the model shown simply matches Triton's density and assumes that the interior is hydrostatic (follows the fluid pressure–depth relation) and differentiated (the major chemical components are separated according to density). These last two assumptions are empirically consistent with Triton's surface appearance, which indicates a prolonged history of melting and separation of icy phases. In the model, ice, structurally represented by the most abundant solar system ice (H_2O), forms a deep mantle around a rock + metal core. A metallic (Fe, Ni, and probably S) inner core is also shown. The proportions of rock and metal in the core are fixed to solar composition (carbonaceous chondrite) values, because relatively involatile rock and metal should have been completely condensed in the outer solar system. Melting and separation of metal from rock are justified by theoretical arguments for intense tidal heating in Triton's past, and by the example of Ganymede, where the *Galileo* orbiter's discovery of a dipole magnetic field demands that such an inner metallic core exists.

Whether Triton is also a magnetized body depends on when its tidal heating ended, but *Voyager 2* passed too far away to tell. Triton is, however, a sufficiently rock-rich body that solid-state convection in its icy mantle should be occurring today, powered by the heat released by the decay of U, Th, and ^{40}K in its rocky core and tidal heating due to Triton's small but finite orbital obliquity. Its icy mantle should also be warm enough to mobilize lower melting point ices such as ammonia and methanol, which are among the minor ices a body formed in solar orbit might have accreted. And if Triton formed in solar orbit, it should have also accreted a large carbonaceous component, upward of 10% by mass if comets such as Halley are a guide. But with or without these additional components, the heat flow from Triton today is sufficient to maintain an internal water layer or "ocean". Similar oceans have been discovered within the large Jovian moons Europa, Ganymede, and Callisto by *Galileo* and inferred for Saturn's Titan and especially so for Enceladus by *Cassini*, so there is no fundamental reason why Triton would not possess one as well.

Voyager 2 confirmed the presence of nitrogen ice on Triton's surface. Specifically, a thin nitrogen atmosphere was detected with a surface pressure and temperature consistent with N_2 gas in vapor pressure equilibrium with N_2 ice (see Section 7). All of Triton's surface appears to be icy; even the darker northern hemisphere shown in Figures 40.1 and 40.6 has a geometric albedo of ~0.55. Nitrogen is obviously very volatile, and theoretical models show that nitrogen ice grains on Triton's surface can rapidly (over many decades) anneal and densify into a transparent glaze or sheet. It is thought that such a nitrogen glaze covers much (but not necessarily all) of Triton.

As mentioned earlier, the overall redness of Triton's ices (Figure 40.1) is thought to be due to UV and charged particle processing of CH_4 (along with N_2), which can yield darker, redder chromophores—heavier hydrocarbons, nitriles, and other polymers. CH_4 exists as an atmospheric gas as well as surface ice. *Voyager's* ultraviolet spectrometer (UVS) solar occultation experiment determined the CH_4 mole fraction at the base of the atmosphere to be $\sim 2-6 \times 10^{-4}$, near or at saturation for 38 K. Dark streaks and patches on the polar cap and elsewhere may be methane rich; if they are depleted of N_2 ice, they should be warmer than the global mean surface temperature, which is buffered by the latent heat of nitrogen condensation and sublimation.

The nature and chemistry of Triton's surface ices have been determined by advanced ground-based spectroscopy (Figure 40.8). In 1991, astronomers detected the spectral absorptions of CO and CO_2 ice, along with CH_4 and N_2 ice. Later work confirmed the presence of water ice, and

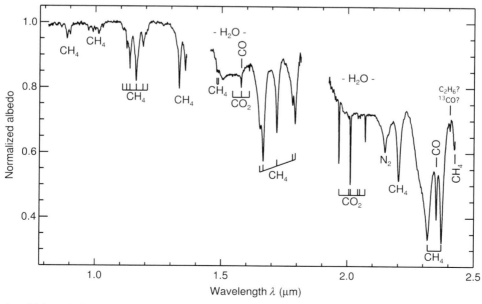

FIGURE 40.8 Modern, high-resolution, near-infrared telescopic reflectance spectra of Triton. Absorptions due to individual species are indicated. The spectral resolution ($\lambda/\Delta\lambda$) is a remarkable 1600–1700. *Modified from Grundy, W. M., et al. (2010).*

more recently, hydrogen cyanide and possibly ethane ice have been detected as well. The shapes of the absorption bands are so well determined that the abundances, grain sizes, and degree of mixing of various components can be modeled. It turns out that CH_4 and CO ice are dissolved in solid solution with the far more abundant N_2 ice, which covers about 55% of Triton's surface. The CH_4 abundance relative to N_2 could be as great as 5%, with the CO abundance less than that. CO is an important tracer of outer solar nebula or cometary chemistry (as discussed in Section 3.3), but is not expected to survive in giant planet satellite-forming nebulae. The detection of CO thus directly supports a capture origin for Triton. Some discrete CH_4 patches also exist, and the ethane ice is one of the "heavier hydrocarbons" predicted to form from methane. CO_2 and H_2O are distributed as discrete units covering the complementary 45% of the rest of the surface. Within these units, CO_2 ice particles represent about 10–20% of the material present. Water ice and CO_2 ice thus represent the composition of Triton's involatile "bedrock".

The geology revealed by the *Voyager* encounter is as remarkable as it was unprecedented. The surface is almost wholly **endogenic** in nature. Intrusive and extrusive volcanism (calderas, flows, diapirs, etc.) dominates the landscape outside the polar terrain, with tectonic structures (mainly ridges) being decidedly subsidiary. Impact cratering is an even more minor process. Triton's surface is geologically young and has apparently been active up until recent times. Triton's topography can be rugged, but does not exceed a kilometer or so in vertical scale (and usually no more than a few 100 m), due to the inherent mechanical weakness of most of the ices that comprise its surface. Polar ices appear to bury much of this topography, and so may in this sense constitute a true polar cap. Details of Triton's geology are pursued in the following section.

Triton's atmosphere is unique as well. It is too thin and cold for radiative processes to play a dominant role. Heat is transported by conduction throughout most of its vertical extent, which is by definition a **thermosphere**, up to an exobase of ~950 km, above which the mean free path of N_2 molecules exceeds the pressure/density scale height. The lowermost atmosphere is characterized by an interhemispheric, seasonal condensation flow and zonal winds. Turbulence near the ground forces the temperature profile to follow a convective, nitrogen-saturated lapse rate of approximately −0.1 K/km up to an altitude of ~8 km (as determined by observations of clouds, hazes, and plume heights), forming a **troposphere** or "weather layer". Unlike in the atmosphere of the Earth and other planets, there is no intervening radiatively controlled **stratosphere** between Triton's troposphere and thermosphere. Triton's atmosphere is further considered in Section 7.

6. GEOLOGY

Triton's surface, at least the 40% seen by *Voyager* at resolutions useful for geological analysis, can be roughly separated into three distinct regions or terrains: smooth, walled, and terraced plains; cantaloupe terrain; and bright polar materials. Each terrain is characterized by unique landforms and geological structures. Substantial variations within each terrain do occur, and the boundaries between each are in many locations gradational, but in general the classification of Triton's surface at any point is unambiguous. Certain geological structures are common to nearly all terrains, specifically, the tectonic ridges and fissures, and impact craters, naturally, can form anywhere.

Although Triton's surface is composed almost entirely of ices, many of the individual geological structures can be readily interpreted as variations of structures terrestrial planet geologists would find familiar, such as volcanic vents, lava flows, and fissures. The volcanic features in particular have inspired a designation **cryovolcanic** in order to distinguish them from those formed by more familiar silicate magmatic processes. The physics and physical chemistry are fundamentally the same, however, whether one deals with silicate or icy volcanism. There are in addition geological structures and features on Triton that are unusual and *not* readily interpretable in terms of terrestrial analogs. Some defy explanation altogether.

6.1. Undulating, High Plains

Plains units are found on Triton's eastern or leading hemisphere (referring to the sense of orbital motion, to the right in Figure 40.1) and to the north of the polar terrain boundary. Figure 40.9 shows a regional close-up of various plains near the terminator in the center of Figure 40.1. To the bottom and right of the image are flat-to-undulating smooth plains centered around circular depressions or linear arrangements of rimless pits. These plains are relatively high standing and bury preexisting topography, with edges that may be well defined or diffuse. There is little doubt that these high plains are the result of icy volcanism and that the various pits and circular structures are the vents from which this material emanated. In general, eruptions along deep-seated fissures or rifts often manifest as a series of vents, and the irregular, ~85-km wide circular depression toward the lower left resembles a terrestrial volcanic caldera complex. This feature, Leviathan Patera (all the features on Triton have been given names drawn from the world's aquatic mythologies), sits at the vertex of two linear eruption trends. Toward the terminator (northeast), one of these trends is anchored by another caldera-like depression of similar scale.

Volcanic activity on the Earth often occurs in cycles, whereby magma formed by partial melting in the mantle

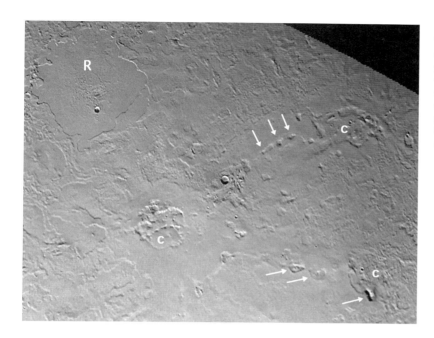

FIGURE 40.9 Young volcanic region on Triton. Toward the bottom and right, smooth undulating flows apparently emanate from complex caldera-like depressions (c) and linear alignments of volcanic pits and vents (arrows at examples), burying preexisting topography. At the upper left, terraced plains surround an exceptionally level plain, Ruach Planitia (R). This region, 675 km across in the image, is very sparsely cratered. North is up. *Courtesy of NASA/Paul Schenk, Lunar and Planetary Institute.*

rises due to buoyancy; accumulates at intermediate, crustal levels to form a magma chamber; and subsequently erupts; things are then quiescent until the magma is replenished and the cycle begins anew. The loss of magma volume often leads to collapse of the vent region over the magma chamber, forming a caldera. Cycles of eruption and collapse can create some complex forms, but calderas are generally composed of quasicircular elements. The two paterae (from the Latin for saucer) in Figure 40.9 are clearly of the caldera type in which renewed volcanism has occurred, because both are partially buried by younger icy lavas or cryoclastic "ash".

The compositions of the icy lavas are, strictly speaking, unknown. *Voyager 2* carried no remote sensing instruments designed to determine compositions. However, the icy plains-forming lavas shown in Figure 40.9 were clearly viscous enough to form thick-enough deposits to bury preexisting topography of a few hundred meters elevation, and the favored composition for viscous lavas on icy satellites has long been ammonia water. As outlined by pioneering planetary chemist J.S. Lewis, ammonia (NH_3) is the chemically stable form of nitrogen in a low-temperature gas of solar composition, and when condensed forms various hydrates with water ice, all of which have low melting points. Triton would not have accreted much ammonia if it formed in solar orbit, because N_2 would have been the dominant original form of nitrogen in the outer solar nebula for the same reasons CO and organic material were favored over CH_4 (see Section 3.3), but it still would have acquired some NH_3 based on cometary compositions (up to 1% or 2% compared with water). The predicted NH_3-H_2O ice mixture would, on heating in Triton's mantle to a rather mild 177 K, yield a lowest melting point (or eutectic) melt. This melt (or cryolava) would be ammonia rich (about 32 mol%) and have a viscosity similar to some types of basaltic magma.

Comets also contain a host of other exotic, presumably interstellar, ices, some of which may have been important in Triton's geological history. For example, methanol (CH_3OH) pushes the minimum melting temperature of ammonia–water ice down to ~ 152 K, and the resulting lava is even more viscous, equivalent to certain types of silicic lavas on the Earth. The range of viscosities available to liquids in the $H_2O-NH_3-CH_3OH$ system is compatible with the appearance of the undulating smooth plains seen in Figure 40.9 (*see* Physics and the Chemistry of Comets).

The abundances of original ices may also have been altered, and new ices created altogether, during Triton's tidal heating epoch (see Section 8). For example, *copious* NH_3 and CO_2 may have been chemically produced within Triton provided there was a sufficient supply of nitrogen and carbon. Despite these exciting possibilities, neither NH_3 nor CH_3OH ice have yet been discovered on Triton by ground-based spectroscopy (Figure 40.8), although they have been identified on other Kuiper belt objects.

6.2. Walled and Terraced Plains

Shown in the northwest corner of Figure 40.9 is a ~ 175-km-wide, remarkably flat plain, Ruach Planitia, that is bounded on all sides by a rougher plains unit that rises in one or more topographic steps (scarps) from the plain floor. It is one of the four so-called walled plains identified on Triton; these are generally quasicircular in outline, with typical relief across the bounding steps or scarps of ~200 m. Ruach Planitia and the other walled plains are the flattest places seen on Triton, which implies infill by a very fluid lava or other liquid. Clusters of irregular, coalesced pits toward the centers of these plains have been likened to eruptive vents or drainage pits.

The planitia themselves have been likened to calderas, but they are generally much larger than the nearby paterae and do not resemble them structurally. Specifically, there is no evidence for collapse at the periphery of any of the walled plains. Rather, the outline of the inward-facing scarps is indented and crenulate, with islands of the bounding plains occurring in the interior. If anything, the outlines of walled plains resemble eroded shorelines. How erosion occurred and under what environmental conditions on Triton is unclear. If the fluid that filled the planitia was responsible for the erosion, it does not explain the similar outline of the plains that overlap the eastern edge of Ruach Planitia (Figure 40.9), which gives this area a terraced appearance and indicates that the rougher plains were laid down in layers. A distinct possibility is that the layers are composed at least in part of a more friable or volatile material, and that over time (or with higher heat flows) the layers disintegrated and the scarps formed by retreat. Similar processes of mass wasting, removal, and scarp retreat are believed to be responsible for the etched plains of the Martian south polar region and similar terrains on Jupiter's Io.

6.3. Smooth Plains and Zoned Maculae

Other plains units can be seen in Figure 40.10, as well as the transition to the bright polar materials. At the top left is a hummocky terrain, composed of a maze of depressions and bulbous mounds. Stratigraphically, it is older than the volcanic plains to the north that overlap it, and appears older (more degraded) as well. The hummocky terrain gives way to a much smoother plains unit to the south. At the available resolution it is unclear whether this smoothness is due to volcanic flooding, volcanic or condensation mantling, or some other form of degradation. These hummocky and smoother units are the most heavily cratered regions on Triton, but by solar system standards are not heavily cratered at all.

Among Triton's most perplexing geological features are the large zoned maculae (spots) close to the eastern limb are shown in Figure 40.10. Each such macula consists of a smooth, relatively dark patch or patches surrounded by a brighter annulus or aureole. The width of any given annulus tends to be relatively constant (20—30 km

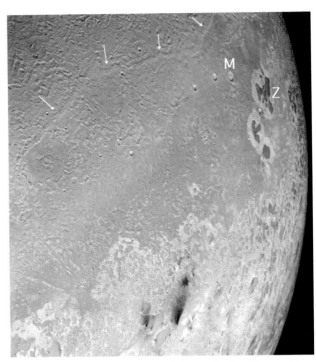

FIGURE 40.10 Southeastern limb of Triton, showing (from top) hummocky terrain, smooth terrain, and bright polar terrain. A prominent bulbous ridge zigzags across the top (arrows), and distinct bright-ringed dark features of uncertain origin, termed maculae (spots), are seen at the right, and more faintly, along the limb and in the bright terrain. The largest crater on Triton, the 25-km-diameter, central-peaked Mozamba (M), is to the left of the largest prominent macula, Zin (Z). *Courtesy of NASA/Paul Schenk, Lunar and Planetary Institute.*

for the three major maculae shown in Figure 40.10). The maculae betray almost no topographic expression, and so must vary in height across their extents by no more than a few tens of meters. The darkness and redness (Figure 40.1) of the central patches implies the presence of carbonaceous material, which probably means some methane ice is or was present. The brightness of the annuli is similar to that of the bright terrain, so they may consist of similar ices (mostly N_2).

The extreme eastern limb shown in Figure 40.10 is composed of a mosaic of maculae, and much of the bright terrain in the rest of the image contains similar, although generally less distinct, features (see also Figure 40.1). Perhaps the maculae are outliers of the southern polar cap, which should have been retreating at the season observed (late southern spring). Furthermore, another walled plain can be seen along the middle left edge of the frame. If this planitia were filled with bright ice, it would passably resemble, in plan and in albedo, the bright terrains to the south, especially those near the boundary with the smoother plains. Perhaps the maculae are planitia underneath.

6.4. Cantaloupe Terrain, Ridges, and Fissures

The entire western half of Triton's nonpolar surface shown in Figure 40.1 is termed cantaloupe terrain, as it appears covered by large dimples and crisscrossed by prominent quasilinear ridges. Much of the terrain displays a well-ordered structural pattern: at high resolution, the dimples become a network of interfering, closely spaced, elliptical and kidney-shaped depressions, termed cavi (Figure 40.11). Unlike impact craters, the cavi are of roughly uniform size, ~25–35 km in diameter, and do not overlap or cross-cut. They are clearly internal in origin, but the leading explanation is not volcanism, but *diapirism*.

Diapirism is triggered by a gravitational instability involving a less dense material rising through overlying denser material. The required buoyancy may be thermal or compositional. Probably the best known terrestrial examples of **diapirs** are salt domes, in which a layer of salt rises as a series of individual blobs, or diapirs, through overlying denser sedimentary strata. In one region of extreme dryness, the Great Kavir in central Iran, the salt diapirs breach the surface, rotating and pushing the overlying strata to the side. The shapes, close spacing, and interference relations of the diapirs of the Great Kavir in fact bear a significant resemblance to the cavi.

The implications of a diapiric origin for cantaloupe terrain are that Triton possesses distinct crustal layering, and based on the spacing of the cavi, that the overlying denser layer or layers are ~20 km thick. This crustal layer could simply be a weaker ice (possibly ammonia rich) that responded to heating from below, or it may be an ice denser than the ammonia–water ices presumably below (such as CO_2 ice).

Triton's surface is cross-cut by a system of ridges and fissures, which are best expressed in the cantaloupe terrain (Figure 40.1). The ridges occur in a variety of forms: pairs of low, parallel ridges bounding a central trough, ~6–8 km across crest to crest and a few hundred meters high; similar but wider ridge-bounded troughs with one or more medial ridges (one, Slidr Sulcus, can be seen in Figure 40.11); and single, broad, bulbous ridges (e.g., Figure 40.10). The fissures, which are less numerous, appear to be simple, long, narrow valleys only 2–3 km wide. All these fundamentally tectonic features appear to result from extension and/or strike-slip faulting of Triton's surface. The medial ridges may be due to dike-like intrusions of icy material, and the bulbous appearance of some may be due to overflow of such injected ice, which could also be a source for smooth plains deposits. Ridges on Triton bear more than a passing resemblance to those on Europa, and a similar mechanical origin has been proposed (*see* Europa).

6.5. Bright Polar Terrains

Most of Triton seen by *Voyager* is actually bright terrain of one type or another, but the imagery is generally not of sufficient quality for geological analysis. Interpretations are further confused by the numerous dark streaks, plumes, and clouds. Nevertheless, the bright terrains represent substantial, not superficial, deposits. The view shown in Figure 40.11 looks across the edge of the cantaloupe terrain, into a band of subdued or mantled cantaloupe-like topography, and then into brighter materials beyond. Cantaloupe-like topographic elements and sections of a linear ridge appear engulfed by bright ice, probably up to a few hundred meters in thickness. The important questions are whether the bright ice thickness increases into the interior of the bright materials in the distance and does it become sufficiently deep to qualify as a true polar cap.

Low-resolution imagery shows that quasicircular elements can be made out at many locations well within the bright materials. Ridges also cross into the bright terrains, and one bright lineament is seen close to the south pole. The implication is that much of the polar topography is

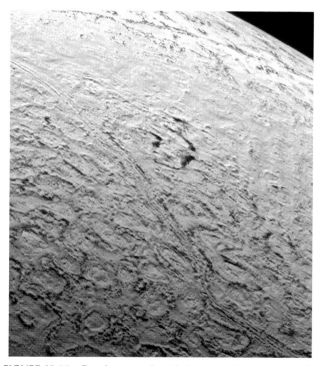

FIGURE 40.11 Cantaloupe terrain at the bottom and polar terrain at the top, in this high-resolution *Voyager* image taken from a distance of 40,000 km. Each cantaloupe "dimple" is about 25–35 km across. A tectonic ridge and fissure set runs through the cantaloupe terrain, probably formed (at least in part) by the extension of Triton's icy crust. Toward the south (upper right), smooth materials, and beyond them, brighter ice, appear to mostly bury cantaloupe and fissure topography.

incompletely buried. On the other hand, there are extensive bright, featureless regions as well (up to several 100 km across), which indicate either complete burial at these locations or obscuration by clouds. Overall thickness of the bright polar ice is therefore probably less than 1 km, but even if not organized as a uniform ice cap or sheet, a thick deposit of a volatile ice such as N_2 could be warm and deformable enough at its base to flow laterally. Although not literally a polar cap, much of the bright polar terrains may behave as if glaciated.

6.6. Geological History

The volcanic province shown in Figure 40.9 is one of two similar ones, with the second occurring to the southeast and together stretching across 1000 km of Triton's surface. The alignments of volcanic vents in both provinces suggest extension and rifting of Triton's relatively strong icy outer shell, or lithosphere. The volcanic plains shown in Figure 40.9 are also very sparsely cratered (the largest crater visible is 16 km across), much less cratered than, say, the lunar maria. Estimates of the rate at which comets bombard Triton suggest that these provinces are no more than 300 million years old, and possibly less than 50 million years old. A broad region of Triton's sublithospheric mantle was thus hot and partially molten very late in solar system history, and probably remains so. Such internal warmth is also consistent with a subsurface ocean (Figure 40.7).

The high volcanic plains postdate most of the other terrains on Triton. They stratigraphically overlie the terraced plains to the west and the hummocky plains to the east. The terraced plains grade into and appear to superpose the cantaloupe terrain. The relative age of the cantaloupe terrain cannot be determined by traditional crater counting methods, because the rugged topography there prevents reliable crater identification in *Voyager* images. Stratigraphically, however, cantaloupe terrain appears to be the oldest unit on Triton. The linear ridges obviously postdate the cantaloupe terrain, yet some ridges fade into the terraced plains to the east and another is discontinuous as it crosses the hummocky and smooth plains near the equator to the east (Figures 40.1 and 40.10); no ridges cut the high volcanic plains.

The eastern hummocky and smooth plains comprise the most heavily cratered region on Triton, and when due account is taken of the concentration of cometary impacts on Triton's leading hemisphere (Figure 40.12), appears to be somewhat older than the high volcanic plains to the north and northwest. The cantaloupe terrain, then, must be even older. The hummocky terrain may be a degraded version of cantaloupe terrain. Indeed, cantaloupe terrain has been suggested to underlie much of Triton's surface. (For

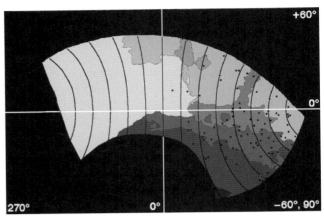

FIGURE 40.12 Distribution of craters (greater than 5 km across) on Triton. Craters are indicated by black dots on a simple cylindrical map, but only those regions with viewing geometries suitable for crater counting are shown. Circles represent concentric zones centered on the apex of motion at 10° intervals. Shadings indicate different geological units, with the major unit at the left being the cantaloupe terrain. *Modified from Schenk, P. M., & Zahnle, K. (2007).*

example, cantaloupe-like topography extends well south into the bright region of the trailing hemisphere.)

The youngest surfaces on Triton, naturally, involve the mobile materials of the bright terrains. These probably include the zoned maculae of the eastern hemisphere. The geological substrate on which the bright materials reside may of course be older. The walled plains themselves are locally the youngest stratigraphic units. Ruach Planitia and a larger planitia immediately to the west are less cratered than the high volcanic plains, albeit with a large statistical uncertainty. The filling of these walled plains may thus represent the most recent volcanic activity on the hemisphere of Triton seen by *Voyager*.

The concentration of craters on Triton's leading hemisphere is pronounced (Figure 40.12), so much so that it has been proposed that the dominant impactors are not comets, but rather, small bodies intrinsic to the Neptune system. Although no plausible source of such "planetocentric" impactors has been identified, if true, Triton's smooth and volcanic plains could be as young as 10 million years old, which would make Triton the most internally active icy satellite after Enceladus. What geologic richness awaits the future spacecraft tasked to more fully survey Triton's surface!

7. ATMOSPHERE AND SURFACE

7.1. Atmosphere

Triton is one of only seven solid bodies in the solar system with an appreciable atmosphere, and one of only four in which the major component of the atmosphere also condenses onto the surface. Triton's atmosphere is composed primarily of nitrogen. The complicated oscillation of the subsolar latitude with time drives an exchange of N_2 and trace species between the atmosphere and surface frost deposits in the two hemispheres that is equally complicated and as yet not fully understood. Internal heating (which is comparatively important because of Triton's extreme distance from the Sun and large proportion of rocky materials containing radioactive elements) and even glacier-like creep of solid nitrogen caps may also play important roles in the interaction of atmosphere and surface (*see* Io; Mars: Atmosphere: History and Surface Interaction).

As described in Section 3, spectroscopic evidence prior to the *Voyager 2* encounter indicated that nitrogen existed on Triton in condensed form. *Voyager* showed Triton to be much smaller, brighter, and colder than had been guessed. Surface temperatures could be inferred from the visible reflectivity as well as measured directly by the Infrared Interferometer Spectrometer. Occultations (passage of the spacecraft or a star behind Triton) observed by the UVS and Radio Science Subsystem probed different parts of the atmosphere, revealing its temperature and density, from which pressure and composition could be deduced. These investigations revealed a consistent picture of a surface and lowermost atmosphere at about 38 K. The pressure at the surface was only 14 μbar, indicating that the gas was in equilibrium with solid nitrogen at the same temperature. The thermal structure of the lower atmosphere is not well constrained, but the temperature probably reaches a minimum at about 8 km height, above which it increases to about 100 K in the upper atmosphere because of heat deposited from space and conducted downward. In meteorological parlance, Triton's thermosphere directly overlays its troposphere.

The *Voyager* images and occultation data revealed a variety of condensates in the lower atmosphere. Most of the atmosphere contains a diffuse haze that can be seen against the background of space at Triton's limbs, and which probably consists of hydrocarbons and nitriles produced by the action of sunlight on trace gases such as methane. Discrete clouds were also seen at the limbs and beyond the terminator, where they formed east−west trending "crescent streaks" roughly 10 km wide, a few hundred kilometers long, and 1−3 km above the unlit surface. At the limbs, clouds could be distinguished from haze by being optically thicker and localized both in height (10 km or less) and in horizontal extent (patchy, and mainly concentrated at mid to high southern latitudes, where they cover one-third of the limb). The sharp upper boundary of the clouds suggests that they consist of condensed nitrogen rather than involatile solids like the haze.

The crescent streaks provide clues to atmospheric motion by their east−west orientation and the apparent eastward motion of the largest, highest cloud seen. Further clues come from markings on the surface. Over 100 dark "streaks" were seen in the southern hemisphere, mainly between latitudes of 15° and 45°S. The streaks range from 4 to over 100 km in length, and many are fan shaped. The vast majority extend to the northeast from their narrow end (presumably the origin); a smaller number are directed westward (Figure 40.13). These streaks resemble the "wind tails" that are common on Mars and are seen on the Earth and Venus as well—and which are created by deposition (or sometimes erosion) of loose material by localized eddies downwind of topographic features. It was initially difficult to understand how wind tails could form on Triton, because even the slightest "stickiness" would prevent dust grains from being lifted by the thin atmosphere.

The interpretation of the streaks as wind created was, nevertheless, strengthened by the discovery, shortly after closest encounter, that some of these features were actually atmospheric phenomena. Stereoscopic viewing of images obtained from varying angles as *Voyager 2* passed by Triton (Figure 40.14) revealed at least two with an altitude of roughly 8 km, the others being on the surface or too low (<1 km) to measure. These features were subsequently named Mahilani Plume (48°S 2°E, with a very narrow, straight cloud 90−150 km long) and Hili Plume (57°S 28°E, actually a cluster of several plumes with broadly tapering clouds up to 100 km long). Thus, it is clear that winds on Triton *do* transport suspended material, but the question is *how* the material becomes suspended.

The plumes were entirely unexpected, and explaining their vigorous activity became a major focus of research. What is clearest is that they complete a coherent picture of winds on Triton at the time of the encounter. Unlike most surface streaks, both plume clouds extend westward from their apparent sources (the plumes proper—narrow, possibly unresolved vertical columns linking the horizontal plume clouds with the surface). Images of Mahilani appear to show kilometer-sized "clumps" within the cloud moving westward at 10−20 m/s and elongation of the cloud from 90 to 150 km at a similar speed. Thus, putting all the descriptions above together (crescent streak clouds, dark surface streaks, and plume tails), the wind is northeast nearest the surface, eastward at intermediate altitudes, and westward at 8 km, the top of the troposphere.

This is precisely the circulation pattern predicted at the time of encounter, the height of summer in the southern hemisphere. Heating by sunlight is presently causing solid nitrogen in the south to sublimate (evaporate); meanwhile in the colder north, the atmosphere is condensing and precipitating. Because of the rotation of Triton once every 5.877 days, however, the wind cannot blow directly from south to north to make up the difference. Instead, gas is transported only in a thin "Ekman layer" near the surface in which the flow is northeastward. The atmosphere above this 1-km-thick layer circulates from west to east. The westward flow at the altitude of the plumes can be explained if Triton's atmosphere is slightly warmer over the equator

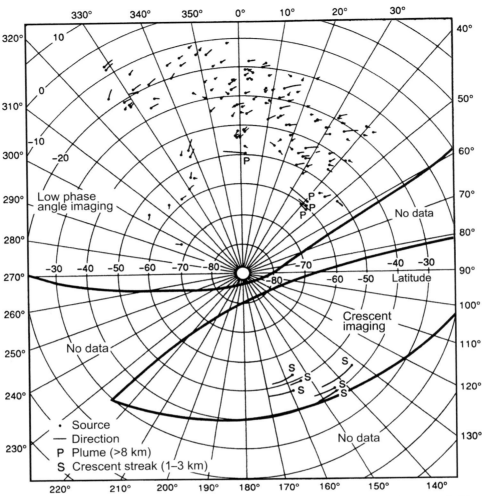

FIGURE 40.13 The geographic distribution and orientation of wind streaks, crescent streaks, and plumes on Triton as seen by *Voyager*. The latitude and longitude of each feature source is plotted as a dot; tails indicate streak or plume length and direction. *Modified from Hansen, C. J., et al. (1990).*

than at the south pole (perhaps because the equator is darker), in which case the temperature gradient will drive a thermal wind that causes the eastward flow to weaken and reverse with increasing altitude.

Basic properties of the plumes can be inferred from the images. The plume clouds do not settle out visibly (at the ~1-km resolution of the best images) over their length, so the suspended particles must be smaller than about 5 μm. From this particle size and the width and contrast of the clouds—about 5% darker when seen against Triton—one can further infer the amount of solids: about 10 kg/s must be discharged if the material is dark or twice as much if it is bright. (Bright material in a cloud would appear relatively dark against Triton's very bright surface, although not as dark as intrinsically dark material. However, bright particles deposited from such a cloud would not show up as a dark streak on the surface.) The cloud moves horizontally at the wind speed, 10–20 m/s, but the vertical velocity in the plume must be faster or the plumes would be blown visibly askew by the wind. The columns may be just barely resolved in the best images, so their diameters are 2 km or less. The source area must have similar (or smaller) dimensions. Little or no structure is visible in the columns, although a "sheath" of descending material around the plume has been described by some authors. The active lifetime of the plumes can be estimated at a few Earth years: shorter, and *Voyager* would have been unlikely to see any plumes active; longer and active plumes should have been more numerous compared with surface streaks.

7.2. Plume Models

Numerous attempts have been made to model the plumes in order to answer the questions of where the particulates, the

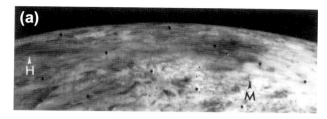
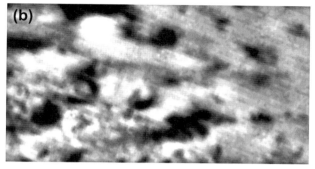
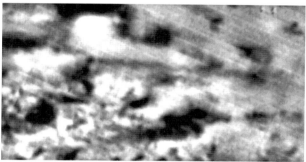

FIGURE 40.14 (a) *Voyager 2* image of the southern polar region of Triton in which geyser-like eruptions were discovered. Here plumes are viewed obliquely with Hili (H) and Mahilani (M) plumes marked. (b) Highly magnified images of Mahilani plume on Triton, taken from increasingly oblique angles and at increasing resolution (top to bottom). The images have been projected onto a spherical surface with a viewing geometry similar to that at the top. The increasing parallax from top to bottom makes the plume "stem" appear to grow taller. *Courtesy of NASA/USGS.*

gas suspending them, and the energy to drive the gas flow originate. Most models have taken their cue from the presence of the active plumes (and surface streaks) at mid to high southern latitudes at a season when the sun was almost directly overhead (Figures 40.4 and 40.13) and assumed that the plumes are somehow solar powered. It is also possible, however, that Triton's internal heat drives the plumes and that their location is determined not by the sun but by a local enhancement of this heat source (i.e., by cryovolcanic activity) or by the thickness of the nitrogen "cap", the equivalent area of the northern hemisphere being hidden in darkness during the encounter.

It is conceivable that the plumes are purely an atmospheric phenomenon. One early suggestion was that the plumes are dust devils, localized regions of spinning and ascending hot atmosphere formed above patches on the surface that are bare of N_2 frost and that can therefore be heated by the sun to higher temperatures than their frosty surroundings. Tritonian dust devils would have difficulty picking up dust from the surface and becoming visible, however, because their winds are not strong enough. If the source areas contained methane frost, though, they would give off clouds of methane gas. Being lighter than nitrogen, the methane would ascend, and might partially condense in the atmosphere, making the rising plume visible. Falling back onto the ground, the methane frost would over time darken from exposure to radiation, explaining the surface wind streaks. This model ingeniously solves the problem of how such a gently rising plume picks up or generates enough solids to become visible, but there is still the fundamental objection that such a plume would be blown sideways (as dust devils on the Earth and Mars are). A final variation on these types of plume model suggests that it is nitrogen rather than methane that is ascending and condensing, with the condensation releasing heat for buoyancy. But again, the nitrogen must somehow start off fast enough to pick up dust and to avoid being blown sideways by winds near the base of the plume.

How could a plume of nitrogen gas get started? One possibility is that they are geysers. Like geysers on the Earth, which consist of water and water vapor, those on Triton would be eruptions of volatile material that has been heated underground. The water in terrestrial geysers starts as a liquid and partially boils, whereas Tritonian geysers would start as hot gas that would partially condense as it expanded to the ambient pressure. This expansion could drive a gas flow powerful enough to pick up dust and form the observed plumes. Solar-powered nitrogen geysers have been studied in some detail. The pieces of the model are as follows.

7.2.1. Plumes as Jets

The energy needed to drive the plumes is determined by how much gas is involved and how fast it has to be erupted. The worst-case assumption is that the nitrogen does not condense as it rises. Instead of becoming buoyant and accelerating, it is denser than its surroundings because of any inert dust entrained in it and the small amount of N_2 (several percent by mass) that crystallizes immediately on eruption. The plume is therefore slowed both by gravity and by interaction with the atmosphere around it. How high it will rise depends on both the size of the eruption and its speed and can be calculated based on laboratory simulations. As an example, a jet with a diameter of 20 m, a velocity of 230 m/s, and 5% solids by mass will reach the observed altitude of 8 km on Triton. The plumes might be this small, but they could be as big as 1—2 km in diameter, in which case they could be somewhat slower. As discussed, plumes erupting more slowly could also reach 8 km if condensation continues after eruption, but in either case the plume will stop at about 8 km because of the increasing atmospheric temperature (buoyancy) above this altitude.

7.2.2. Eruption Velocity and Temperature

Both the initial velocity of the gas and the amount of solid nitrogen that will condense can be calculated from the initial and final temperatures and the thermodynamic properties of nitrogen. The example given above (5% solids, 230 m/s) is attained for nitrogen expanding freely (no change in entropy) and cooling from 42 to 38 K. Thus, the subsurface gas must be heated about 4 K to power the geyser to the right altitude. We also learn from this calculation that the 10 to 20 kg/s of solids estimated to be feeding the plumes is accompanied by as much as 400 kg/s of gas. Given the latent heat of sublimation of nitrogen, about 100 MW of power is needed to convert solid to gas at this rate, which may be supplied by a "solid-state greenhouse."

The greenhouse effect usually describes heating of the Earth's atmosphere (or that of another planet) when sunlight at visible wavelengths penetrates the atmosphere before being absorbed, but longer wavelength thermal radiation is absorbed by the atmosphere and cannot escape to space as easily. A similar effect can take place in a transparent solid, for example, nitrogen ice on Triton, which is such a good insulator that heating by 4 K can be achieved with a greenhouse layer only 1—2 m thick.

7.2.3. Subsurface Energy Transport

What happens after sunlight is absorbed below Triton's surface and before hot gas is erupted? As just estimated, 100 MW are needed to heat the gas in a typical plume. This is the amount of power deposited by sunlight on a region of Triton about 10 km in diameter, much bigger than the 1- to 2-km size of the plume sources. We can therefore conclude that gas (or energy to produce gas by sublimation) is stored over time and then released quickly, or is transported horizontally from the larger area to the geyser, or both. Somewhat counterintuitively, gas is not mainly "stored" in voids in the nitrogen ice, but is produced on demand from hot ice, while heat is mainly transported by flowing gas rather than ordinary thermal conduction. Nitrogen ice can give off more than 100,000 times its own volume of gas as it cools just 4 K. If there are voids in the solid nitrogen, this gas will flow to colder areas and recondense, warming them by releasing its latent heat. This energy flow can be hundreds of times more efficient than conduction if the pores are large enough. Flow between meter-sized blocks of solid could readily supply a geyser, but when a path to the surface was first opened eruption would be vigorous at first and decline over a period of about a year, roughly the estimated lifetime of the plumes. Energy transport by production of gas, its flow through pores, and recondensation at colder points is known on the Earth: "heat pipes" containing a condensable gas (with a wick to return the liquid to the hot end) conduct heat better than metal and are used for baking potatoes from the inside out and for controlling the temperature of spacecraft, including *Voyager*! How a suitably fractured layer of nitrogen ice, overlain by a clear, gas-tight greenhouse layer, might form on Triton is discussed in the next section.

The idea of solar-powered geysers thus seems extremely promising, although much work remains to take the separate pieces that have been modeled so far and make sure that they fit together. Internally powered geysers (more similar to their terrestrial counterparts) have not been studied nearly as thoroughly, but other possibilities exist. As discussed below, the nitrogen "polar caps" on Triton may be so thick near their center (over a kilometer) that they begin to melt at the base. Liquid N_2 finding its way to the surface could erupt as a boiling geyser, with more than enough energy to power the plumes. Gases other than nitrogen could also be erupted from deeper in Triton's water ice mantle, driven by internal heating.

Additional light was unexpectedly shed on the physics of Triton's geysers by the discovery in the early 2000s that some transient, dark spots on the Martian seasonal polar caps are not simply patches of bare ground. Some such patches are fan shaped and typically form clusters with similar orientation corresponding to the prevailing winds predicted by atmospheric models (Figure 40.15). These features form shortly after the first light of spring, only on regions known on the basis of spectral, thermal, and albedo data to be covered by a translucent layer of carbon dioxide frost on the order of a meter thick, and they fade or vanish when the CO_2 layer sublimates away. These circumstances strongly suggest that the Martian fan deposits are formed by venting of dusty gas created by solar heating at the base of the frost layer. The analogy to models of Triton's geysers described above is strong and was noted early on.

The greatest difference between the Martian features and those on Triton is one of scale: typical Martian fans are

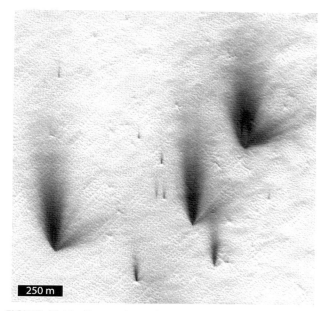

FIGURE 40.15 Fans on the south seasonal CO_2 cap of Mars. Located close to the south pole, these fans are aligned with the prevailing wind direction toward the top of the image, although secondary alignments are also visible. Similar but larger features on Triton (see Figure 40.10) may form from an analogous sunlight-driven process. *From Piqueux, S., & Christensen, P. R. (2008).*

~100 m (give or take a factor of a few) in length, and the elevation of the active plumes above the surface has not been distinguished in stereo images even at 0.3–3 m/pixel. Why should the Tritonian plumes be so much larger? Apart from the quantitative differences in temperature, volatility of the working fluid, insolation, length of season, and of course surface gravity, the greenhouse layer on Triton is believed to overlie a thick, possibly fractured permanent cap of the same (nitrogen) material. This may increase the available gas supply compared with Mars, where the frost layer sits on an inert surface of silicate soil. Radial patterns of channels eroded into this involatile layer by the flowing gas are visible around many of the Martian fan sources and are referred to as araneiforms or "spiders". Unfortunately, Voyager images are of too low resolution to show whether Triton spiders exist, or even if each Triton plume corresponds to a single, large vent or perhaps to a kilometer-scale field of smaller vents resembling those on Mars. If the latter, then the source geometry could be similar and the greater height of the geysers could be explained either by a larger eruption velocity or by buoyancy mechanisms already discussed.

7.3. Polar Cap and Climate

We turn now from the plumes to a consideration of how Triton's surface frosts and atmosphere change over time.

Here, too, the *Voyager* images yielded a surprise: at the height of southern hemisphere summer (Figure 40.4), most of the southern hemisphere was covered with a bright deposit (a polar cap), but the visible portion of the northern, winter hemisphere was darker. Models of the redistribution of N_2 frost with the seasons can be constructed with varying degrees of complexity, but a fundamental expectation is that the summer hemisphere should have less of a polar cap than the winter one!

The basic physics of seasonal frost distribution models is as follows.

1. The whole atmosphere and all frosted areas are at very nearly the same temperature. If a frosted area were colder, more nitrogen would condense there and release of latent heat would raise the temperature. Conversely, a warm frost area would be cooled by sublimation. Winds would quickly even out the atmospheric pressure and temperature.
2. At this fixed temperature, sublimation occurs where frosts are exposed to the sun and condensation where the average input of solar energy is less. Sublimation/condensation rates can be calculated from the amount of sunlight absorbed at each point on Triton.
3. Bare (unfrosted) areas can be warmer than the atmosphere and frosts (if they are dark and/or well exposed to the sun), but they cannot be colder, or frost would immediately condense on them.

Using the albedo of the surface as measured by *Voyager*, models indicate that frost in most of the southern hemisphere is currently subliming, thinning the surface deposits. Nitrogen is presumably being deposited in the northern hemisphere and in a few of the brightest areas of the south where little sunlight is absorbed. Stellar occultations and infrared telescopic observations since *Voyager* have shown that Triton's surface pressure has measurably increased, to around 40 μbar. The surface temperature of the nitrogen ice, which controls the atmospheric pressure, must also have increased by 1–2 K. But what about the long run? By assuming that frost has some given albedo and that the surface underneath has some other albedo, one can model the redistribution of nitrogen over long periods. A layer of nitrogen frost about a meter thick is moved back and forth as the sun shines on one hemisphere and the other, and the pressure and temperature of the atmosphere change as well. Notably, such models predict that all nitrogen deposited in the southern hemisphere the last time it was winter there would have resublimated before *Voyager* arrived. Correspondingly, the northern hemisphere should be extensively frosted.

How can these predictions be reconciled with observation? The frost might deposit mainly in shadows and on north-facing slopes where *Voyager* could not see it, or it could be glassy and transparent, hence invisible. There is some evidence for the last possibility, from laboratory observations of condensing nitrogen; calculations of the rate at which loose frost grains would merge or anneal into a dense, transparent layer; and even from observations of the light-scattering properties of Triton's equator. Such suggestions would each explain the dark, apparently frost-free northern hemisphere, but the bright "cap" in the south must be explained as well. Perhaps it is a much thicker deposit of nitrogen that never completely sublimes away (this is certainly the impression one gets geologically). Although nitrogen frost may be very transparent when first annealed, changing temperatures will make the residual cap expand and contract, fracturing it and making it appear bright. Thus, we are led to the idea of a clear, uncracked (i.e., gastight) seasonal frost layer over a thick, fractured permanent cap: precisely the kind of layering hypothesized above to explain the plumes as solar-powered geysers.

What controls the size of the residual cap, and why is one not seen in the north? A good candidate is solid-state creep, or flow, of the thick nitrogen deposit, similar to the flow of glaciers and spreading of polar caps on the Earth and Mars. Models based on terrestrial polar caps, combined with estimates of the rate at which solid nitrogen would flow, suggest that the permanent cap is about a kilometer thick at the center. Cap spreading also prevents the eventual disappearance of the seasonal frosts predicted by the models discussed above. Because the pole always receives less sunlight than the edges of the seasonal frost deposits, more frost will be deposited at the pole than at the edges, maintaining the cap. There may be a northern as well as a southern permanent cap. If this northern cap extends less than 45° from the pole, it would lie in the dark portion of Triton unseen by *Voyager*. The southern permanent cap might be larger because of hemispheric differences in the heat released from Triton's interior, or it might also extend only 45°, in which case the bright deposits extending almost to the equator have still to be explained. Some of this bright material may be nitrogen "snow" that condenses in the atmosphere into grains that are too big to anneal on a seasonal timescale into a transparent layer. It should be apparent from this discussion that, as with the plumes, we seem to have many pieces of the puzzle of the polar caps (and perhaps a few spurious pieces of unrelated puzzles), but they have yet to be assembled into a final picture of Triton's surface—atmosphere interaction.

Additional clues to the behavior of volatiles on Triton are presently being gathered from Earth-based spectroscopic measurements, and by the occultation of stars by Triton. As noted above, Triton's atmosphere is changing and becoming thicker and slightly warmer. Strong winds aloft are also indicated. Surface—atmosphere interactions on the polar caps of Mars are also being studied in great detail, and the lessons learned are being applied to Triton.

And of course, continued monitoring and study of Pluto provides a valuable second case against which to test theoretical models.

8. ORIGIN AND EVOLUTION

Triton and Pluto turn out to be remarkably similar in size and density and in surface and atmospheric compositions as well. There is little doubt that they share a common heritage. Moreover, they are not isolated in the outer solar system. An entirely new reservoir of minor planets has been found orbiting near and beyond Neptune—the Kuiper belt. The first Kuiper belt object was found in 1992, and as of this writing, over 1400 have been discovered, if one includes the so-called scattered disk objects. A number are nearly as large as Pluto or Triton. One—Eris—is as big as Pluto, denser than Triton and also has methane and nitrogen ice on its surface (*see* Kuiper Belt Objects: Physical Studies).

The link between Triton, Pluto, and the Kuiper belt is strengthened by what is known of the orbital dynamics of this region. For example, a number of Kuiper Belt objects share the same dynamical resonance with Neptune that Pluto occupies (this orbital resonance prevents encounters between Neptune and Pluto and is one of the strong arguments against the Pluto-as-escaped-satellite hypothesis). In this sense, Pluto and its companion "Plutinos" are more like the Trojan or Hilda groups of asteroids (which are locked in orbital resonances with Jupiter), only that Pluto—Charon is the clearly dominant member of its group.

Dynamical calculations show that Pluto and its companions were probably swept into this orbital resonance as Neptune's orbit expanded early in solar system history. During this time the flux close to Neptune of bodies orbiting near and beyond Neptune would have been quite high, and even today Neptune continues to deplete the inner Kuiper belt population. It is perhaps not surprising then that Neptune should have had a catastrophic encounter with at least one escapee from the Kuiper belt: Triton (*see* Kuiper Belt: Dynamics).

Satellite capture does not occur easily. Generally, objects passing near a planet leave with the same speed that they came in with. Even complicated trajectories called temporary gravitational captures (enjoyed by Comet Shoemaker—Levy 9) are just that — temporary. To be permanently captured, a single cosmic body must lose energy (velocity) by running into or through something. In Triton's case, it could have collided with another stray body just passing by Neptune, but the probability of this having happened is quite low. Because Triton orbits close to Neptune, in the region usually occupied by regular satellites, it is much more likely that it ran into a regular satellite or its precursor protosatellite disk.

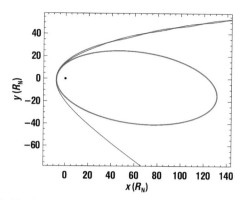

FIGURE 40.16 A possible capture mechanism for Triton. In this example "exchange capture" calculation, an equal mass Triton binary approaches from the upper right and is disrupted by tides from Neptune. One member of the binary (in red) is captured into an elliptical orbit with a semimajor axis of $\approx 70\ R_N$, while the other escapes (R_N is Neptune's radius). *Modified from Agnor, C. B., & Hamilton, D. P. (2006).*

An alternative model proposes that Triton was once part of a binary, and when it passed too close to Neptune, strong tides from the planet split the binary in two. One member of the binary escaped back into solar orbit, while the other stayed behind in Neptune orbit (Figure 40.16). In this case, the captured member of the binary loses orbital energy to the escaping member. While this may at first glance seem far fetched, we now know that a good fraction of Kuiper belt objects are binaries (>10%) and tidal stripping close to a much more massive planet such as Neptune simply requires a close passage. Although permanent capture of one of the original binary members is not assured, the probability is much greater than, say, being captured by colliding with an original Neptune satellite.

The inclination of Triton's postcapture orbit depends on the initial encounter geometry. Triton could have ended up either prograde or retrograde, but there is a preference for retrograde orbits because they are more stable at greater distances from Neptune. After capture, Triton's orbital evolution would be strongly influenced by tides. Every time Triton reapproached Neptune, Neptune's gravity would raise a tidal bulge on Triton. The periodic rise and fall of the bulge would dissipate energy as heat, which would be extracted from the energy of Triton's orbit. Because the tidal couple between Triton's bulge and Neptune would be (on average) radial, no change in Triton's orbital angular momentum would occur. Based on these constraints, and ignoring for the moment any encounters with original satellites, Triton's orbital configuration after capture could evolve as depicted in Figure 40.17. Triton may have begun with a semimajor axis of 1000 R_N or greater, or it may have begun closer in (as in Figure 40.16). The important point is that early on Triton's **periapse** (the closest point to Neptune in its orbit) would lie as low as half its present semimajor axis. Triton's tidal evolution probably took 100 million years or longer, so there would have been sufficient time for Triton's orbit to evolve through and interact with any pre-existing satellites.

This point is emphasized in Figure 40.17(b), which includes the periodic effects of solar tides on Triton's evolving orbit. If Triton's initial capture orbit was very large and eccentric, its periapse would have fluctuated, and may have periodically been as low as 5 R_N! Triton would have had ample opportunity for collisions with Neptune's original satellites (if they were like Uranus' today), possibly accreting them in the process. It may also have scattered original satellites into distant orbits, caused them to crash into Neptune, or perhaps even ejected them from Neptune altogether (all the while speeding up its own orbital evolution).

There is now nothing left of Neptune's original system (if it indeed existed) other than the inner satellites and possibly Nereid. The inner satellites all lie within 5 R_N, however, which is perfectly consistent with this capture scenario. Nereid may also be a survivor of this orbital

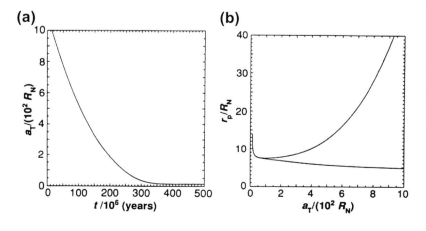

FIGURE 40.17 (a) Example evolution of Triton's semimajor axis, a_T, as a function of time, t, due to tidal dissipation within Triton. (b) Evolution of Triton's minimum and maximum periapse distance, r_p (the closest point to Neptune in its orbit), as a function of semimajor axis due to the combined influence of semiannual solar perturbations and tidal dissipation. The periapse distance oscillates between the two curves shown. *Adapted from Goldreich, P., et al. (1989).*

mayhem. Little is known about this distant moon, save its size (~340 km in diameter), reflectivity (~20%), and presence of surface water ice, but these facts make Nereid more akin to a regular satellite than a dark captured asteroid or comet. Recent dynamical studies have cast doubt on this scenario, however, at least as far as Nereid's stability is concerned. A favored solution is an earlier capture and orbital evolution for Triton, followed by a later capture of Nereid and other distant irregular satellites of Neptune.

The end state of Triton's orbital evolution is an extremely circular orbit. As such, the orbital energy potentially dissipated by tides within Triton represents an absolutely enormous reservoir, about 10^4 kJ/kg. It is sufficient to completely melt all the ice, rock, and metal within Triton 10 times over. The magnitude of Triton's temperature change, however, depends on the heating rate, and somewhat on the size of the initial capture orbit, and how much interaction there is with preexisting regular satellites or nebular gas. Two such models are illustrated in Figure 40.18. Tidal heating after capture in either model is at first modest, as the satellite spends most of its time far from Neptune. As its semimajor axis shrinks and its orbital period decreases, the average heating rate begins to rise. The epoch of greatest heating occurs when the relative change in semimajor axis is the greatest (because orbital energy is inversely proportional to semimajor axis), roughly when the semimajor axis drops below 100 R_N. Because the orbit can only evolve as fast as the tides can convert orbital energy to heat, the response of Triton to tidal flexing is crucial. If Triton responds as a dissipative elastic sphere, then the semimajor axis drops continuously (Figure 40.17(a)) and the tidal heating rises and then falls smoothly as the orbit becomes more circular (Figure 40.18, elastic sphere model). Note that the calculations in Figures 40.17(a) and 40.18 are for two different elastic sphere models, illustrating a range of possible timescales.

A dissipative elastic sphere is clearly an idealized and oversimplified model for Triton. Triton is in reality a complex rock, metal, organic matter, and ice body. More volatile ices especially should be melted and mobilized within Triton early in its history (e.g., ammonia), with or without tidal heating. A partially molten body is a particularly dissipative body, so when capture occurs and tidal heating begins, heat concentrates in the partially liquid regions. This causes more melting, which makes the body more dissipative, which results in greater tidal heating. Thus, within a few hundred million years after capture (perhaps well within), Triton in all probability went through an episode of runaway melting. This is schematically illustrated in Figure 40.18, where in the model labeled thin shells Triton melts spontaneously when enough energy has been accumulated to do so (in reality the runaway occurs much earlier). Thereafter, Triton is a nearly totally molten, but still dissipative body. Its tidal heating curve rises and falls sharply over the course of ~100 million years.

During this epoch of extreme tidal heating Triton's heat flow is an amazing ~2–4 W/m^2, equal to or greater than that measured today from Io. Its surface temperature is governed by this flux and corresponds to a blackbody temperature of 80–90 K. During and after this epoch there would likely have been large chemical exchanges between the global oceanic mantle with its dissolved volatiles and the hot rock core below. Much of Triton's volatiles may have been driven into a massive atmosphere. Atmospheric components plausibly include CO, CH_4, CO_2, and NH_3, or

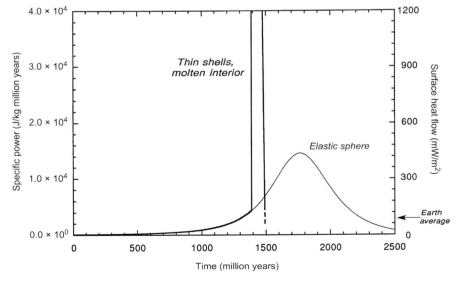

FIGURE 40.18 Power dissipated per unit mass and surface heat flow for Triton as its postcapture orbit shrinks and circularizes due to tides raised by Neptune. Two models are shown. One assumes Triton remains a uniform, undifferentiated sphere, while the second allows for melting. In both cases, the timescales are longer than in the calculations in Figure 40.17(a), due to updated parameters for Triton, but the periapse variations as a function of semimajor axis in Figure 40.17(b) are unchanged. The thin shells model is more realistic than the elastic sphere, but even here the meltdown of Triton has been artificially delayed; in reality a thermal runaway probably occurs much earlier. *From McKinnon, W. B., et al. (1995).*

even H_2 (from photolysis of methane or ammonia or as a minor component in Triton's original ice). Conservative assumptions yield an atmospheric greenhouse with surface temperatures well above 100 K; more extreme possibilities allow for surface temperatures greater than 200 K.

A most intriguing aspect of raising a massive greenhouse atmosphere by tidal heating is that it may persist well after the tidal heating input has tapered off and Triton's interior has begun to freeze. The atmosphere may only collapse after enough of it has been lost to space due to solar-UV-heating-driven hydrodynamic escape, which could have taken in excess of 1 billion years. While the atmosphere existed it would have kept Triton's surface warmer, and enhanced the geological mobility of the satellite's surface layers. Unfortunately, there are as yet no definitive indicators of the atmosphere's former presence (e.g. ancient eolian or fluvial features, peculiar crater shapes, etc.). If a thick atmosphere existed, Triton's continued geological activity has obscured the evidence.

Regardless, once massive tidal heating ended, Triton's interior should have begun to freeze. It would probably have taken a few 100 million years to do so, but even today such freezing would not be complete. Triton's ice mantle is probably warm enough, due to radiogenic heating from the core and continuing dissipative tidal heating from daily obliquity tides, that any ammonia- and methanol-rich fluids are stable (likely in an internal ocean), and Triton's inner core of alloyed iron, nickel, and sulfur should likewise be warm enough (more than ≈ 1250 K) to allow for a eutectic liquid mixture of those elements.

The possible persistence of cryomagmas in Triton's mantle due solely to radiogenic (and some tidal) heating has raised the question as to whether any of the geological observations in Section 6 actually *demand* that Triton was massively tidally heated. Certainly, solar-powered plume models do not require Triton to be internally active at all. Triton's surface, on the other hand, is peculiar, even unique. Furthermore, the extent and intensity of the geological activity recorded there is only seen on satellites that are undergoing active and substantial tidal heating (Io, Europa, and Enceladus). While no ironclad argument can be made, Triton's geology and chemistry in all likelihood indicate that it did indeed experience massive tidal heating, resulting in a geological wonderland.

The proof of Triton's history and provenance requires further exploration of this extraordinary body. For example, determination of the compositions of Triton's icy lavas, and terrains in general, would be key constraints. Detailed exploration of the Neptune system by spacecraft is also a technically feasible proposition, given recent and projected technological advances. Instruments and electronics are being increasingly miniaturized, thereby requiring smaller launch vehicles. Missions to Triton can also take advantage of innovative flight strategies, such as using aerobraking in the Neptune atmosphere to go into initial Neptune orbit. Thereafter, a complement of advanced instruments can be trained on Triton during repeated encounters, filling out our picture of this amazing satellite. For the present, the *New Horizons* encounter with the Pluto system in 2015 promises to reveal much on how Triton-class worlds operate and evolve.

Regarding Triton's ultimate future, as a retrograde satellite its orbit is actually decaying due to tides it raises on Neptune. In the 1960s, it was estimated that Triton would closely approach Neptune and be torn apart by tides in a geologically short time. Present estimates imply less peril: Triton's orbit will perhaps shrink by no more than 15% over the next 5 billion years, giving Triton plenty of time for further geological and atmospheric adventures.

BIBLIOGRAPHY

Agnor, C. B., & Hamilton, D. P. (2006). *Nature, 441*, 192−194.
Barucci, M. A., et al. (Eds.). (2008). *The solar system beyond Neptune*. Tucson: University of Arizona Press.
Beatty, J. K., et al. (Eds.). (1999). *The new solar system* (4th ed.). Cambridge, MA: Sky Publishing.
Chapman, C. R., & Cruikshank, D. P. (1995). In D. P. Cruikshank (Ed.), *Neptune and Triton*. Tucson: University of Arizona Press.
Cruikshank, D. P. (Ed.). (1995). *Neptune and Triton*. Tucson: University of Arizona Press.
Goldreich, P., et al. (1989). *Science,245*, pp. 500−504.
Greeley, R., & Batson, R. (2001). *The compact NASA atlas of the solar system*. Cambridge, UK: Cambridge University Press.
Grundy, W. M., et al. (2010). *Icarus, 205*, 594−604.
Hansen, C. J., et al. (1990). *Science, 250*, 421−424.
Kargel, J. S. (1997). In J. H. Shirley, & R. W. Fairbridge (Eds.), *Encyclopedia of Planetary Sciences*.. London: Chapman & Hall.
Kirk, R. L., et al. (1995). In D. P. Cruikshank (Ed.), *Neptune and Triton*. Tucson: University of Arizona Press.
Littmann, M. (2004). *Planets beyond: Discovering the outer solar system*. Mineola, NY: Dover Publications.
Lunine, J. I., & Stevenson, D. J. (1985). *Nature, 317*, 238−240.
McKinnon, W. B., et al. (1995). In D. P. Cruikshank (Ed.), *Neptune and Triton*.. Tucson: University of Arizona Press.
NASA Planetary Photojournal. http://photojournal.jpl.nasa.gov/index.html.
Piqueux, S., & Christensen, P. R. (2008). *Journal of Geophysical Research, 113*. E06005.
Rothery, D. A. (1999). *"Satellites of the outer planets: Worlds in their own right* (2nd ed.). NY: Oxford Univ. Press.
Schenk, P. M., & Zahnle, K. (2007). *Icarus, 192*, 135−149.
Schmitt, B., et al. (Eds.). (1998). *Solar system ices*. Dordrecht, The Netherlands: Kluwer Academic Publishers.
Smith, B. A., &, the *Voyager* Imaging Team. (1989). Voyager 2 at Neptune: imaging science results. *Science, 246*, 1422−1449.

Chapter 41

Planetary Rings

Matthew S. Tiscareno and Matthew M. Hedman
Center for Radiophysics and Space Research, Cornell University, Ithaca, New York, USA

Chapter Outline

1. Introduction — 883
2. How We Learn about Rings — 884
3. Rings by Location — 885
   3.1. The Rings of Saturn — 885
   3.2. The Rings of Uranus — 887
   3.3. The Rings of Neptune — 887
   3.4. The Rings of Jupiter — 888
   3.5. The Rings of Mars and Pluto — 888
   3.6. The Rings of Rhea and Iapetus — 889
   3.7. The Rings of Exoplanets — 890
4. Dense Broad Disks — 890
   4.1. Structures Formed Spontaneously by the Disk — 891
      4.1.1. Self-Gravity Wakes — 891
      4.1.2. Overstabilities — 892
   4.2. Structures Formed by Nearby or Embedded Moons — 892
      4.2.1. Gap Edges and Moonlet Wakes — 892
      4.2.2. Propellers — 892
   4.3. Structures Formed by Distant Moons — 895
      4.3.1. Spiral Waves — 895
      4.3.2. Confinement of Edges and Ringlets — 896
5. Dense Narrow Rings — 896
6. Dusty Rings — 897
   6.1. Dust Production and Destruction — 897
   6.2. Nongravitational Forces — 898
   6.3. Resonantly Confined Material — 898
   6.4. Azimuthal Clumps and Kinks — 900
7. Using Rings to Probe the Solar System — 901
   7.1. Probing Interplanetary Meteoroids — 902
   7.2. Probing the Planet's Gravity — 903
   7.3. Probing the Planet's Magnetosphere — 903
8. Age and Origins of Ring Systems — 904
Bibliography — 905

1. INTRODUCTION

Planetary rings are swarms of objects orbiting a central planet with vertical motions that are small compared to their motions within a common plane. This characteristic arises because their planets rotate fast enough that they bulge at their equators, thus defining a preferred orbital plane. This is a major contrast between rings and other **astrophysical disks**, such as spiral galaxies and young solar systems, which do not derive their shapes from an external gravity field but from the average **angular momentum** of the disk itself (in both cases, once a preferred plane is established, collisions among particles damp out the motions perpendicular to it). Planetary rings are also distinguished by a large planet/ring mass ratio, which greatly enhances the flatness of rings (their **aspect ratios** are as small as 10^{-7}). However, rings do have a number of similarities with other astrophysical disks, which add to the motivation for studying them. Unlike other known disk systems that are either many light-years away or (like the early stages of our solar system) far back in time, planetary rings can be studied up close and in real time.

By necessity, a dense ring resides inside its planet's **Roche limit**, the distance from a planet within which tides can pull a moon apart.[1] If, contrariwise, a dense ring were outside the Roche limit, it would most likely accrete into one or more moons. The Roche limit is not a sharp boundary; materials that are less dense or more porous can remain dispersed as a ring at the same location where denser material will accrete. Also, dense rings near the transition develop a microstructure as they try to accrete and are frustrated by tides (Section 4.1). Furthermore, this limit operates only for rings dense enough that the particles frequently collide with each other; tenuous rings (Section 6) can resist accretion regardless of their location because their particles do not interact much.

In this chapter, we will first discuss how we learn about rings in Section 2. Then we will give an overview of the

---

1. The Roche limit calculation involves only gravity. Small objects may hold themselves together by material strength despite the pull of tides, and thus may exist inward of the Roche limit. Individual ring particles are an example.

known ring systems (Figure 41.1), as well as systems where rings are unconfirmed but plausible, in Section 3. More detailed descriptions of various ring structures organized by type, with a focus on finding commonalities among rings in different locations that share certain qualities, can be found in Sections 4–6. In Section 7 we will discuss ways by which rings tell us about their surroundings, and in Section 8 we will discuss the age and origins of ring systems.

2. HOW WE LEARN ABOUT RINGS

The most straightforward way to observe rings, now in use for over 400 years, is by images (or, equivalently, with the human eye). When light from the Sun falls on a ring, varying amounts of light are reflected backward (or to the side, depending on the orientation of the surface it encounters), diffracted forward, or absorbed. **Diffraction** is strongest when the particle size is comparable to the

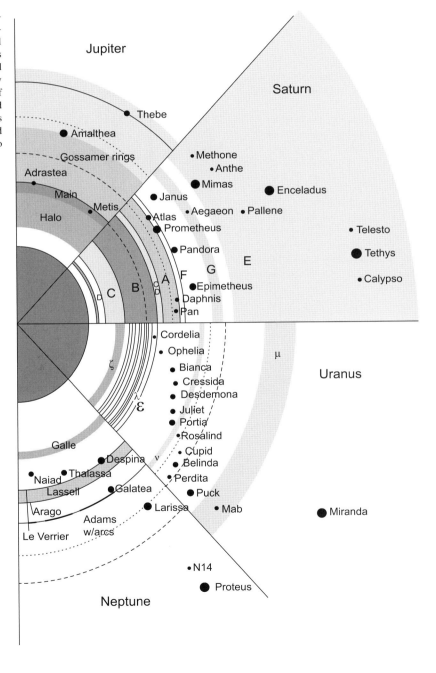

FIGURE 41.1 A graphic schematic of the ring-moon systems of the giant planets scaled to a common planetary radius. The planet is the solid central circle, ring regions are shaded, and nearby satellites are plotted at the correct relative distances. Dotted lines indicate the Roche radius for a satellite of density 0.9 g/cm^3, and dashed lines show the position of synchronous orbit where an object's orbital period matches the planet's rotation period. The Roche radius is outside the synchronous distances for Jupiter and Saturn but inside it for Uranus and Neptune due to the more rapid spins of the larger planets.

wavelength of the light, while larger particles primarily reflect the light that falls on them. Thus, rings with a high fraction of dust[2] tend to be bright at high **phase angles**,[3] when the camera is best positioned to detect the forward-scattered (diffracted) light. By the same token, rings dominated by pebble-sized or larger particles are brightest at low phase angles, when the camera sees light reflected more or less back toward the source.

Because the Earth is much closer to the Sun than is any ring-bearing planet, only low phase angles are accessible from the ground, and one advantage of a spacecraft is the ability to sample a wide variety of phase angles. Because particles of different sizes have different **phase functions** (i.e. the profile of how they scatter light at various phase angles), the ring's **particle-size distribution** can be inferred in this way.

Because rings are generally semitransparent, they can also be studied by means of **occultations**, in which an observer watches a light source pass behind the ring and measures how much of the light the ring blocks. The two primary forms of this are **stellar occultations**, in which a camera on Earth or onboard a spacecraft watches a star pass behind the ring, and **radio occultations**, in which a spacecraft beams a radio signal that passes through the ring and is received on Earth. Occultations often allow more finely detailed observations of ring structure, but only in a single dimension along the projected path of the source. Occultations directly measure the **optical depth** (a measure of the ring's transparency, or equivalently of the fraction of light that it blocks), while imaging observations must infer the optical depth indirectly from the measured brightness.

Although most images of rings are dominated by scattered sunlight, other factors such as **planetshine** can be important, especially when sunlight is reduced (e.g. in the planet's shadow). Rings can be viewed not only on the side illuminated by the Sun (the lit face) but also on the opposite side (the unlit face). On the lit face, a region with higher optical depth appears brighter simply because there is more material to scatter light toward the observer. This is also true on the unlit face for moderately dense rings, but highly dense rings can exhibit a **contrast reversal**, in which higher optical depth leads to a darker appearance on the unlit side as the ring becomes more opaque.

As with any material, much can be learned about a ring from studying its **spectrum**, or the profile of its brightness at different wavelengths of light. These wavelengths can extend beyond **visible light** to include **infrared** or **ultraviolet** and more. A ring's spectrum is akin to a fingerprint, and comparing it to spectra taken of controlled samples in a laboratory can yield important information about its chemical composition, surface texture, temperature, and other characteristics. Different parts of a ring can have different spectra, and these can be investigated using a **spectrometer** with a sufficiently narrow **field of view** and pointing it at different parts of the ring at different times, or by using an **imaging spectrometer**, which builds up an image in which each pixel is a spectrum with anywhere from dozens to thousands of color measurements.

A spacecraft flying near or through a ring can also study it by sensing it directly. A dust detector can measure the velocity, size, and composition of dust particles that strike the spacecraft. Detectors of magnetic fields and charged particles can analyze the interaction between a ring and its planet's **magnetosphere** (e.g. *see* "Planetary Magnetospheres" and Section 3.6).

Rings can also be studied through computer simulations. Advances in simulation techniques, alongside advances in computing hardware and software, have allowed simulations to become an increasingly important part of the study of orbital dynamics in the solar system (including rings) since the early 1990s. For simple systems, it is sufficient to enter a suite of particles with their initial positions and velocities and to see where gravity and other forces will take them. For dense rings (Section 4), however, the number of mutually interacting particles in the entire ring is far too large for a computer to handle, so a line of simulations has been developed that follows a relatively small "patch" of the ring surrounded by "mirror patches" (Figure 41.2). Computer simulations have become a vital tool for understanding not only ring microstructure such as self-gravity wakes (SGWs, see Section 4.1), but also the mechanics of sharp edges (see Section 4.2) and propellers (see Section 4.2) and other ring properties including the rotational states, thermal properties, and particle-size distributions of ring particles.

3. RINGS BY LOCATION

3.1. The Rings of Saturn

Saturn possesses by far the most massive and the most diverse of the known ring systems (Figure 41.3). The only ring system known before recent decades, Saturn's rings have been the focus of much productive study by astronomers over the past four centuries. Saturn's appendages were among the first objects observed through a telescope (by Galileo Galilei in 1610), first understood to be a disk by Christiaan Huygens in 1655, and proved to consist of individual particles on independent orbits by James Clerk Maxwell in 1859. The main part of the rings comprises the solar system's only known broad and dense disk (Section 4),

---

2. To describe a material as "dust" is to describe the size of its particles, regardless of composition (rock, ice, etc.). Particles less than a few tens of microns (μm) across are usually considered dust.
3. The phase angle is formed by the light path from Sun to object to observer.

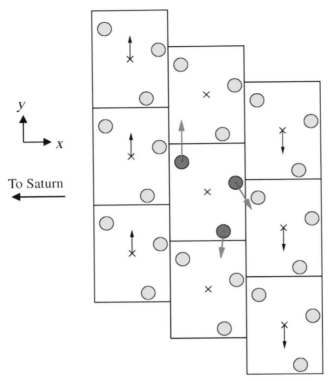

FIGURE 41.2 Schematic representation of a ring-patch simulation with sliding boundary conditions. The simulation cell (center, with green particles) is surrounded on all sides by replicant cells (with yellow particles) for the purpose of calculating the particles' gravitational effects on each other. In this representation, increasing radius is to the right and Keplerian orbital motion is up. When a particle leaves the simulation cell in the azimuthal direction (uppermost and lowermost green particles), it simply reappears at the same location on the other side (equivalently, it enters the neighboring replicant cell). When a particle leaves the simulation cell in the radial direction (right-hand green particle), its appearance on the other side is governed by how the replicant cells slide past the simulation cell according to Kepler's third law (black X's and arrows).

which G. D. Cassini found in 1675 was divided into two parts—now called the A and B rings, separated by what is now called the Cassini Division. The latter is now known to be not an empty gap but simply a region of the disk with more moderate surface density, similar in character to the C ring, which lies inward of the B ring and was discovered in 1850. Subsequent observations from ground-based telescopes and spacecraft have revealed that these rings consist of water-ice-rich particles millimeters to meters wide.

There are a small number of truly empty radial gaps in the dense disk of the main rings, most of them in the C ring and Cassini Division but two in the outer A ring, all of them sharp-edged. The two A ring gaps are held open by moons at their centers, and the C ring's Colombo Gap is held open by a resonance with Titan, but most of the gaps remain unexplained. A diverse array of narrow rings and ringlets resides within ring gaps, some of them dense and sharp-edged and others diffuse and/or dusty. These structures, which can be compared with narrow rings around other planets, are discussed in Section 5.

The Saturn system also contains the most diverse retinue of tenuous dusty ring structures known in the solar system, discussed in Section 6. The main components are the D ring situated innermost between the main rings and Saturn's atmosphere, the dense F ring just off the edge of the A ring, and the G and E rings farther out.

Saturn also possesses the largest known ring in the solar system, the Phoebe ring, which was discovered in 2009 by the Spitzer Space Telescope. The Phoebe ring is some 30 million km across but is exceedingly tenuous, containing only about 10 dust particles per cubic kilometer. It is also the only known ring to be tilted from its planet's equatorial plane (it lies in the plane of Saturn's orbit, as the Sun's perturbations are much more important than Saturn's equatorial bulge at its distance) and is likely the

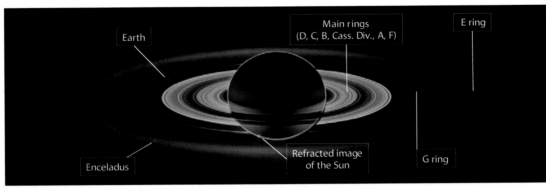

FIGURE 41.3 This Cassini image mosaic shows Saturn's tenuous dusty rings (D, E, and G) with comparable brightness to the main disk, which occurs because the viewing geometry is at high phase angle (in fact, in eclipse) and also views the unlit face of the main disk. The darkest part of the main disk is actually the densest and most opaque, namely, the mid- to outer B ring. The Sun, which is actually behind Saturn, can be seen refracted through the planet's atmosphere at 7 o'clock. Enceladus (actually, only its geyser plume is bright in this geometry) can be seen embedded in the E ring at 8 o'clock. The Earth can be seen as a pinpoint of light between the F and G rings at 10 o'clock.

only known ring whose particles orbit in the retrograde direction (following its presumed source-moon, Phoebe). As particles in the Phoebe ring spiral inward due to electromagnetic effects, they preferentially impact one side of the moon Iapetus, leading to a strong divide between bright and dark regions on the surface of that moon.

3.2. The Rings of Uranus

In contrast to the broad disk of Saturn, all of Uranus' main rings are narrow, and many are eccentric and/or inclined. But in contrast to the dusty rings of Jupiter and Neptune, many of Uranus' rings are dense and sharp-edged. Thus, the Uranian ring system is unique, one that truly deserves the label of "rings" in the plural. The main set of 10 narrow rings is compact and lies inward of Uranus' 13 small inner moons, except for the ε ring, which lies just outside the orbit of Uranus' innermost moon Cordelia (Figure 41.4). A panoply of unnamed dusty rings was seen interspersed with the main rings in the single high-resolution high-phase image taken by Voyager 2 (Figure 41.5).

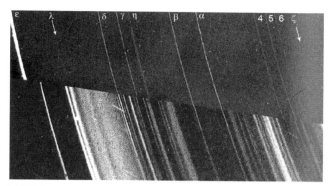

FIGURE 41.5 This composite image of Uranus' main rings in back-scattered (upper) and forward-scattered (lower) light shows that a network of dust structures is interleaved with the planet's dense main rings. The disjoint in the ε ring is due to its eccentricity. Note that the λ ring, the ζ ring, and many other dusty structures are only visible in the lower panel. As the lower panel is the only high-phase image ever successfully taken of Uranus' rings (by the postencounter Voyager 2), the detailed workings of the dust structures remains largely unknown.

The orbits of Uranus' inner moons are also tightly packed, so close that simulations indicate they can persist for only some millions of years (a small fraction of the age of Uranus) without some kind of collision. The dusty ν ring, which lies between two of the moons in this group, may well be the detritus of a recent significant collision. If we could see the Uranian system some 50 million years in the future, we might find that the ν ring has reaccreted into a new moon, while one or more of the present moons may have collided and formed a new dusty ring. In this way, these inner moons of Uranus may be best thought of as a collective system, something like a "ring" except that they are outside the Roche limit—that is, far enough from their planet to be constantly accreting into discrete moons (Section 1).

The composition of the Uranian rings is almost entirely unknown, as Voyager 2 did not carry an infrared spectrometer capable of detecting the rings. However, it is clear from their low albedo that at least the surfaces of the ring particles cannot be primarily water ice. Color imaging indicates that the Uranian rings are dark at all visible wavelengths, indicating a spectrum similar to that of carbon.

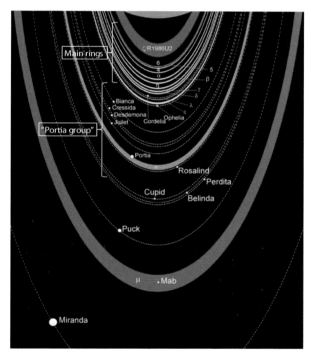

FIGURE 41.4 Uranus' main rings are situated immediately inward of a retinue of small moons. If one were to spread the mass of Uranus' "Portia group" of moons evenly over the annulus they occupy, the surface density would be similar to that of Saturn's A ring. This moon system may be thought of as an accretion-dominated ring, similar in many respects to the known ring systems except that its natural density is larger than the Roche critical density (i.e. it is beyond the "Roche limit"; see Section 1) so that any moon that gets disrupted by a collision (which ought to have happened many times over the age of the solar system) will simply reaccrete.

3.3. The Rings of Neptune

Neptune's ring system, like that of Uranus, consists primarily of a few narrow rings, although Neptune's are generally less dense, higher in dust, less sharp in their edges, and farther from their planet than those of Uranus (Figure 41.6). The most substantial of these is the Adams ring, which is best known for its series of arcs, the first ever discovered (Section 6.3).

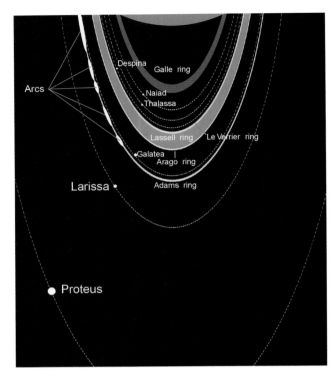

FIGURE 41.6 In Neptune's ring system, uniquely, narrow rings and diffuse rings and moons are all interspersed together.

The composition of the Neptunian rings, like that of the Uranian rings, is unknown due to Voyager's inability to detect them in the infrared. However, again like the Uranian rings, the low albedo of Neptunian ring particles makes it clear that at least their surfaces cannot be primarily water ice.

3.4. The Rings of Jupiter

Jupiter is adorned by the simplest of the known ring systems. All its rings are tenuous and composed of dust-sized particles. As the only confirmed ring system without any dense component, and by far the least massive, Jupiter's is the only ring system to have been discovered by spacecraft without having previously been seen from Earth either by direct imaging (as Saturn's) or by stellar occultations (as Uranus' and Neptune's).

Each component of Jupiter's ring system is closely related to one or two source moons. Dust particles enter the ring as **ejecta** from micrometeoroid impacts onto the moon, then their orbits evolve inward due to electromagnetic effects. As a result of this, the two dusty Gossamer rings are shaped like two nested "tuna cans" enveloping the Main ring (Figures 41.7 and 41.8). The Main ring additionally has a core of centimeter size and larger particles occupying the ~1000 km annulus between the orbits of Adrastea and Metis. This core is the only component of the ring system not composed of dust and the only one that appears bright at low phase angles (Figure 41.9).

Furthermore, the New Horizons spacecraft spotted several azimuthal clumps in one of the core's ringlets (see Section 6.3).

When the dust moving inward from the Main ring reaches a radius of 122,800 km from Jupiter's center, its vertical motions become excited by a 3:2 **resonance** between its orbital period and the rotation period of Jupiter's magnetic field. Inward of this location, the vertical extent of the ring increases dramatically, forming the toroidal Halo ring. Material in the Halo ranges tens of thousands of kilometers above Jupiter's ring plane, although most of its material is concentrated within just a few hundred kilometers.

3.5. The Rings of Mars and Pluto

The four giant planets are the only bodies known to have rings, as just described. But rings may exist in other places as well, although yet unobserved.

Mars is predicted to have a tenuous ring system comprised of dust grains ejected from its moons Phobos and Deimos by meteoroid impactors. The Sun's perturbations should cause

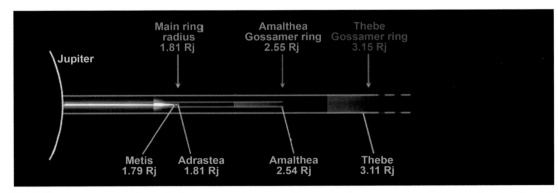

FIGURE 41.7 Galileo image mosaic of Jupiter's rings, seen nearly edge-on at very high phase angle, annotated to show the primary components of the ring system. Image sensitivity increases from left to right, in order to show the increasingly faint structure.

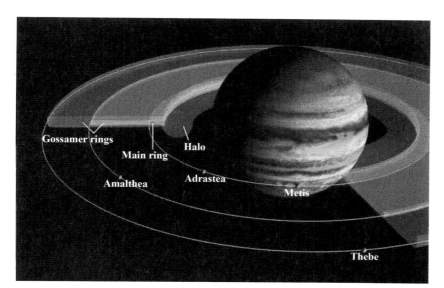

FIGURE 41.8 A schematic of Jupiter, its innermost four moons, and its ring components (shown in different colors). The vertical thickness of the Halo ring is caused by an electromagnetic effect operating on dusty grains, while the vertical thicknesses of the Main ring and the two Gossamer rings are due to the orbital inclinations of the moons that generate those rings. The four well-known Galilean moons are too far from Jupiter to be shown here.

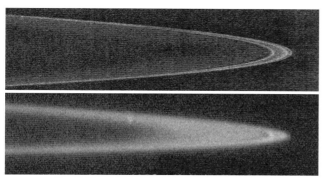

FIGURE 41.9 New Horizons images of Jupiter's Main ring at low phase (upper panel) and at high phase (lower panel), showing respectively the structures composed of macroscopic particles and the dusty envelope.

this ring to be offset away from the Sun and tilted out of Mars' equatorial plane. However, the failure to observe any Martian rings to date, despite sophisticated attempts, may make it necessary to alter the original theory. Perhaps dust is produced at lower rates than expected, or perhaps the destruction of dust as solar radiation pressure pushes it onto orbits that impact the planet proceeds faster than expected.

Compared to the giant planets, Mars (see "Mars: Surface and Interior") is a more challenging place to look for faint rings because it lacks a dense atmosphere. For the giant planets, atmospheric gases like methane absorb light strongly at certain infrared wavelengths; faint rings can more easily be detected at those wavelengths because glare from the planet is dim. Planets also furnish a dark background at high phase angles (when, by definition, we are viewing the night side), but this method is also less effective for detecting rings at Mars because forward-scattering dust particles should be depleted from the Martian rings due to radiation pressure.

Pluto, like Mars, could harbor a tenuous ring system of dust derived from its small moons Styx, Nix, Kerberos, and Hydra. This possibility raises concerns about the safety of the New Horizons spacecraft, set to fly past Pluto (see "Pluto") in 2015, and more sensitive observations are being attempted. However, both the handicaps discussed above for Mars, the lack of a dense atmosphere and the loss of smaller dust grains due to radiation pressure, are also likely to hamper the detection of any Plutonian rings.

3.6. The Rings of Rhea and Iapetus

Rhea, the largest of Saturn's airless moons, is in the opposite situation from Mars and Pluto, with no clear theoretical prediction but a claim that rings have been observed. Cassini scientists reported a symmetric pattern of dropouts in detected electrons (which can be likened to the way a person driving in a blinding rain can perceive having driven under a bridge by the sudden cessation of raindrops hitting the windshield) during Rhea flybys, which were announced as the first evidence for rings around any moon. The following year, Cassini color imaging of Rhea's surface revealed a narrow chain of discrete bluish splotches aligned nearly along Rhea's equator, which was interpreted as the signature of ring material falling out onto the surface.

However, after an extensive search using Cassini images, the present existence of rings around Rhea was practically ruled out. It seems most likely that the electron-absorption signatures are due to some magnetospheric phenomenon, and not to rings of solid material around Rhea. It has been suggested that the chain of bluish splotches is evidence that Rhea (see "Planetary Satellites") had rings in the past, even if it has them no longer, but the best that can be said of this idea is that it has not been disproven.

Saturn's moon Iapetus is another object that could have had a ring in the past. Among its many remarkable features, the best known of which is that one hemisphere is 10× brighter than the other, is one of the highest mountain ranges in the solar system, aligned directly along Iapetus' equator. Here, as for Rhea, the fallout of a past ring system has been invoked by some scientists to explain this equatorial feature; such a theory may even explain some of Iapetus' other oddities as well.

3.7. The Rings of Exoplanets

While hundreds of planets have now been detected in orbit about other stars, no exoplanet (*see* "Exoplanets") yet has a confirmed ring system. The best observed exoplanet systems are mostly "hot Jupiters" that orbit very close to their stars, but several factors combine to make it difficult for rings around such planets both to survive and to be observed. However, the situation should be more agreeable for planets not much closer to their stars than Mercury is to the Sun (which might be called "warm Saturns"). If such a ring system were observed to **transit** its star (e.g. by the Kepler spacecraft), the pattern in which the rings block the starlight may yield information about the planet's spin and its interior.

The only known exoplanet for which a ring system has been specifically proposed is Fomalhaut b, which in 2008 was one of the first exoplanets to be directly seen (rather than perceived indirectly, e.g. by the way its gravity tugs on its star or by the way it blocks its star's light). In order to account for its observed brightness, Fomalhaut b would need to reflect more starlight than a reasonably sized planet can reflect, but this difficulty can be resolved if Fomalhaut b is adorned by a ring system that is so large that it extends far beyond the planet's Roche radius. But as we discussed in Section 1, such a large disk would not be stable against accretion and thus would perhaps be more of a dynamically evolving protosatellite disk than a stable ring system.

4. DENSE BROAD DISKS

Saturn	C ring
	B ring
	Cassini Division
	A ring

Although the idea that Saturn's main ring system is composed of "countless tiny ringlets" continues to appear even in the professional literature, it is inaccurate. It is much better to say that the main ring system is a broad disk; its density does vary in the radial direction, but there are only a few true gaps that would separate one "ring" from another. Furthermore, waves constantly travel radially through the disk (Section 4.3), and ring material can move radially as well, although more slowly.

The components of Saturn's main ring system are quite different from each other in structure, as well as in location

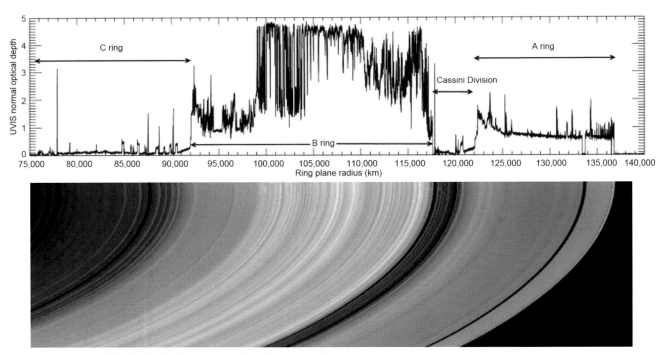

FIGURE 41.10 An optical depth profile (top) and true-color image (bottom) of Saturn's main ring system.

and density. The B ring is by far the most dense, and its particles interact with each other the most vigorously. The A ring is intermediate in both respects, while the C ring and Cassini Division are much more tenuous (Figure 41.10). The B ring's position astride the **synchronous orbit distance**, at which the orbital period of particles equals the rotation period of the planet (and thus of its magnetic field) is thought to influence at least some of its properties (see the discussion of spokes in Section 7.3), while the A ring's outermost position causes it to be more susceptible to both accretion-related processes and perturbations from exterior moons.

Saturn's rings are primarily composed of pebble-sized to house-sized chunks (cf. Section 6). The characteristics of these individual particles, whether they are rotating quickly or slowly and how deeply they are covered by a blanket of fractured material, can be inferred by using infrared observations to "take their temperature".

Broad, dense rings possess a diverse array of structures. Many of these are still not well understood, such as a series of sharp-edged bands called "plateaux" in the C ring (Figure 41.11) and nearly all the multifaceted structure in the B ring. However, other features in the rings have been explained as the result of particle interactions and perturbations from nearby and distant moons.

4.1. Structures Formed Spontaneously by the Disk

4.1.1. Self-Gravity Wakes

The boundary between the two regimes discussed in Section 1—accretion-dominated regions (in which a disk would tend to coalesce into a moon) and disruption-dominated regions (in which disks are stable)—is not a sharp one. In the outer parts of the region of stability for disks, the mutual self-gravity of ring particles can drive temporary accretion that is quickly disrupted again by tides, forming a disk microstructure known as self-gravity wakes. SGWs generally have a webbed structure of elongated particle aggregates extending in a characteristic direction (Figure 41.12, Movie M1), usually a few tens of degrees from azimuthal.

Because of their nonaxisymmetric structure and their finite vertical thickness, the brightness of a disk pervaded by SGWs depends on the observer's longitude. An observer looking along the direction of the elongated wake structures will see more of the gaps between the wakes than an observer looking across the wakes (Figure 41.13), especially at low **elevation angles**. Even before the Voyager encounters, the effects of SGWs were observed as an azimuthal asymmetry in the brightness of the A ring, and considerable light has been shed on their properties in recent years due to a combination of

FIGURE 41.11 The broad bright sharp-edged bands seen in this image are the "plateaux" (singular: "plateau"), named for their high flat appearance in a one-dimensional plot of the ring's brightness with radius. They can be 100 km or more in breadth, and are sometimes called "embedded ringlets" because they are similar in character to dense narrow rings (Section 4), except that they are surrounded not by empty gaps but by more tenuous regions of the ring. It is not known why the plateaux exist, nor why they only occur in the outer C ring.

FIGURE 41.12 Simulated self-gravity wakes (SGWs). This microstructure develops within dense rings as particles clump together under their mutual self-gravity but are ripped apart again by Saturn's tides. *This figure is a still from Movie M1.*

FIGURE 41.13 A ring with simulated self-gravity wakes (SGWs) is viewed (left) along the prevailing direction of the SGWs and (right) across that direction. It is evident that the transparency is higher in the first case. This leads to an azimuthal variation in the reflectivity or transparency of rings containing SGWs, depending on the viewing angle.

sophisticated numerical simulations and repeated occultation observations by Cassini with varying geometries.

SGWs have a dramatic effect on how the ring's density might be inferred from its brightness or optical depth. Numerical simulations have found that adding more mass to a ring merely adds mass to the already-opaque wakes and only weakly increases the overall optical depth, so that the mass of the B ring may be higher than Voyager-era estimates by a factor of 10 or more, approaching twice the mass of Mimas. See Section 8 for a discussion of the impact of this finding on the age and origin of Saturn's rings.

4.1.2. Overstabilities

Regions of axisymmetric microstructure have been found by occultations in the inner A ring and in the B ring. These regions contain patterns with wavelengths of a few hundred meters with such regularity that they can act as a **diffraction grating** for the Cassini radio occultation signal. This structure has been explained in terms of **viscous overstability**, a phenomenon so named because the restoring force that would usually equalize any random variations in density becomes overactive, overshooting the equilibrium and thus driving oscillations instead of damping them (Movie M2). Work is ongoing to characterize the appearance of "overstability waves" in Cassini data, as well as to simulate their behavior in response to various environmental factors.

4.2. Structures Formed by Nearby or Embedded Moons

4.2.1. Gap Edges and Moonlet Wakes

There are 14 named gaps within Saturn's main rings. Most of them have no clear origin, though the Encke and Keeler gaps in the A ring contain known moons (Pan and Daphnis, respectively) that hold the gaps open through their gravity's effect on nearby ring material, and some gaps in the C ring seem to be held open by resonances. Furthermore, even the two gaps that do contain known moons exhibit a surprising amount of unexpected behavior at their edges.

The way a moon holds open a gap is ostensibly straightforward. When a ring particle passes a nearby moon (which is orbiting at a different rate due to **Kepler's third law**), its orbit acquires an eccentricity as well as being very slightly pushed away[4] from the moon's orbit. On its now-eccentric orbit, the ring particle oscillates in and out once per orbit, while Kepler's third law requires it to move laterally over the same period of time in the moon's frame of reference by a distance $3\pi\Delta a$ (where Δa is the radial distance between the two orbits). The result is that the gap edge develops a wavy structure with a wavelength $3\pi\Delta a$ (Figures 41.14 and 41.15). Furthermore, as one moves into the disk from the gap edge, streamlines still have wavelength $3\pi\Delta a$, but that wavelength increases as the distance from the moon Δa increases. The result is a spiraling *moiré* pattern called "moonlet wakes"[5] that rotates with the gap-moon (Figures 41.14 and 41.15).

Wavy gap edges and moonlet wakes are often easier to see than the moons that cause them. In fact, before Pan or any other gap-moon had been discovered, it was the wavy edges of the Encke Gap and the nearby moonlet wakes that were first tracked. Careful analysis of these features led to a prediction of where the responsible moon should be orbiting, which led to the discovery of Pan in decade-old archival Voyager images.

Some edges also exhibit vertical structure. The moon Daphnis, at the center of the Keeler Gap, has an inclined orbit that ventures ~9 km above and below the ring plane. The resulting vertical corrugation of nearby portions of the gap edge were highlighted by their shadows cast during the 2009 **equinox** (Figure 41.16). A region of vertical structure, possibly due to embedded moonlets on inclined orbits, has also been detected and tracked in the outer edge of the B ring (Figure 41.17).

4.2.2. Propellers

A disk-embedded moon that is too small to open a full circumferential gap may still create a local disturbance in the disk. Because of Kepler's third law, the radially inward portion of the disturbance is carried forward and leads the

---

4. This may seem counterintuitive, since gravity is a force of attraction. The moon does indeed attract the ring particle, but there is a paradoxical effect in orbits by which (for example) an object that speeds up is taken farther from its planet, which (by Kepler's third law) puts it in the end on a slower orbit.

5. Moonlet wakes have little in common with the "self-gravity wakes" discussed in Section 4.1, despite an unfortunate similarity in terminology.

Chapter | 41 Planetary Rings

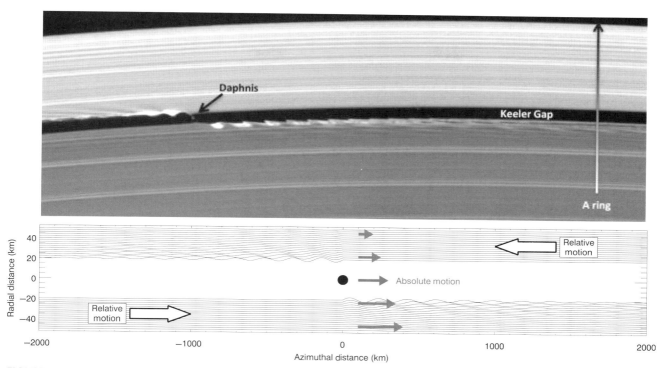

FIGURE 41.14 (Top) Wavy edges and moonlet wakes are seen at the edges of the Keeler Gap, surrounding the position of the gap-moon Daphnis. (Bottom) Streamline diagram showing the $3\pi\Delta a$ wavelength setup as ring material passes a gap-moon. Different wavelengths at different radial distances Δa set up the "moonlet wakes" pattern. By Kepler's third law, inner material is moving faster than outer material (red arrows); by the same token, in the moon's frame of reference, inner material moves forward and outer material moves backward (white arrows).

FIGURE 41.15 In this detailed image of the Encke Gap, Saturn is far to the left and orbital motion is upward. The moon Pan is several frame-widths below, and the $3\pi\Delta a$ wavy edges and the moonlet wakes can be seen on the inward (left-hand) edge. Several faint narrow rings are seen in the gap; the brightest central one is centered on Pan's orbit. A few spiral density waves can also be seen at the far left and far right.

FIGURE 41.16 In this image of the Keeler Gap, taken when the Sun shone nearly edge-on to the rings during the 2009 equinox, not only Daphnis but the vertically corrugated gap edges cast shadows on the nearby A ring.

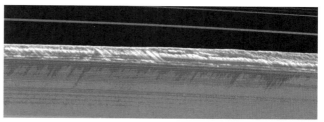

FIGURE 41.17 In this image of the outer edge of the B ring, taken when the Sun shone nearly edge-on to the rings during the 2009 equinox, part of the edge ramps up into a "wall" several kilometers high that casts a shadow.

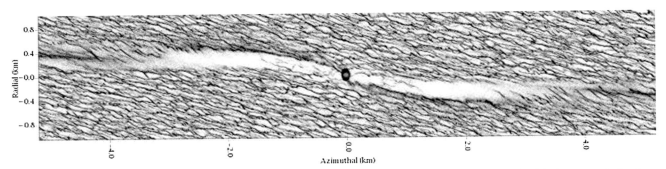

FIGURE 41.18 Simulated "propeller" disturbance in the A ring due to an embedded moonlet. Note the texture of (SGWs) in the unperturbed regions, and the near-horizontal regions of depletion (white) and enhancement (black) in density due to the moonlet.

moon while the radially outward portion trails the moonlet (Figures 41.18 and 41.19). Due to this characteristic shape, such moonlet-caused disturbances have been named "propellers". A propeller-shaped disturbance can be thought of as a moon's unsuccessful attempt to form a full circumferential gap, which is frustrated as collisions among ring particles allow particles to diffuse back into the disturbed region and fill the gap back in before it can fully extend azimuthally. Propellers in a range of sizes have been observed in Cassini images.

In all cases, only the propeller-shaped disturbance is directly seen, while the responsible moon at the center is inferred. The physical size of the moon is a few times smaller than the tell-tale radial offset between the two azimuthally aligned lobes (labeled as Δr in Figure 41.19(b)).

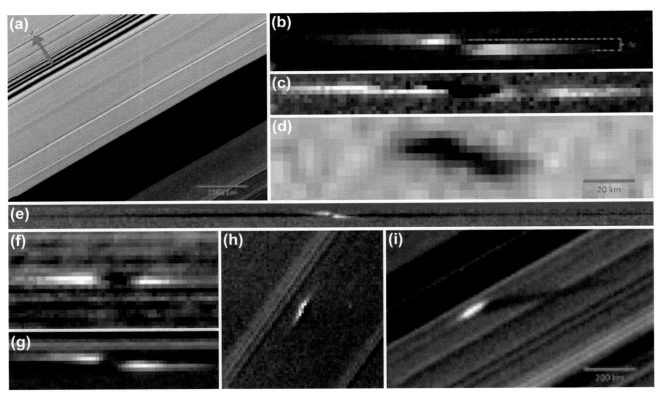

FIGURE 41.19 Propellers as seen in selected Cassini images. Panel (a) shows a propeller in context of the Encke Gap and several density waves. Panel (b) illustrates the radial offset Δr between the two azimuthally aligned lobes, proportional to the size of the unseen central moonlet. Panels (b), (c), and (d) show three views of the same propeller at the same scale, demonstrating how its appearance changes with viewing geometry. Nonequinox views are on the lit (b, e, g) or unlit (a, c, d, f) face of the rings, while panels (h) and (i) show propellers casting shadows near the Saturnian equinox. The scale bar in panel (d) also applies to panels (b), (c), (f), and (g). The scale bar in panel (i) also applies to panels (e) and (h).

Several "Propeller Belts" in the mid-A ring contain swarms of small propellers, due to moons about a few hundred meters in diameter. A separate class of "giant propellers," due to moons up to a few kilometers in diameter, has been found in the outermost A ring, outward of the Encke Gap.

In the Propeller Belts, the propellers are so numerous that, even if the same object were seen on multiple occasions, it would be very difficult to have much certainty that it was the same object due to the many nearby similar objects.[6] By contrast, the giant propellers are larger and more prominent and also much less numerous. Taken together, these factors allow giant propellers to be studied individually and tracked over a period of years. The moons within giant propellers are thus the only objects ever to have had their orbits tracked while embedded in a disk, rather than while orbiting in free space. The interactions between disks and embedded masses, such as has long been thought to occur in other disks such as young planetary systems, are being directly observed for the first time.

Orbital tracking of giant propellers has revealed that their orbits are indeed evolving. The largest and best observed propeller, nicknamed "Blériot",[7] has experienced at least three clear changes in its orbit and may additionally be subject to constant small changes. The leading theories agree that Blériot's orbit evolution is likely driven by random "kicks" that it receives from smaller ring particles and/or SGW clumps (Section 4.1), either through physical collisions or through gravitational interactions during close encounters, while one theory posits that radial density structure in the disk (Section 4.3) also plays an important role. Cassini is expected to continue tracking propeller orbits through 2017, and should further illuminate our understanding of how embedded masses interact with their disks.

The 2009 equinox event revealed that the propeller-shaped disturbance around an embedded moon extends vertically as well as in the ring plane. A giant propeller in the A ring, nicknamed "Earhart", was seen to cast a shadow (Figure 41.19(i)), and the shadow clearly is cast by the entire propeller disturbance and not by the central moon (which remains unseen). A bright feature designated S/2009 S 1, which was seen casting a shadow in the outermost part of the B ring (Figure 41.20), is also most likely an unresolved propeller.

4.3. Structures Formed by Distant Moons

4.3.1. Spiral Waves

Spiral density waves and spiral bending waves are the most widespread well-understood phenomena in dense rings. They were first proposed for the case of galaxies, and are also important for the evolution of protoplanetary disks, but they have been observed in Saturn's rings in far more detail than anywhere else.

Each spiral wave is generated by a **resonance**, which is the "beating in time" of ring-particle orbits with the orbit of a perturbing moon. A common example is a playground swing, which will swing higher if the pusher times the pushes to coincide with the frequency at which the swing is swinging. The two frequencies may be identical (a 1:1 resonance), but more often are related by a whole number ratio (for example, if a ring particle completes 5 orbits around Saturn in the time it takes a moon to complete 3 orbits, that would be a 5:3 resonance). Furthermore, a broad ring contains particle orbits that cover a large array of frequencies (orbits closer to Saturn are faster, those farther from Saturn are slower, in accordance with Kepler's third law); this can be likened to an array of transistor radios, each tuned to a different frequency and together covering the entire FM spectrum so that every radio station is being received on one boom box or another. Similarly, a large array of resonances find expression somewhere or another in the ring, but the lion's share are in the outermost A ring, simply because resonances become more numerous as one moves closer to the moons that cause them (or, continuing the analogy, the radio stations are more common at one end of the spectrum).

---

6. This criterion may be useful for drawing a distinction, should one wish to do so, between a "moon" and a "moonlet". Basing such a distinction upon size is arbitrary and lacks any wide agreement, as there is no major physical threshold crossed by objects in the kilometer-size range. We suggest that any object be called a "moon" if it can be singled out for long-term study, while a "moonlet" is a member of a population or swarm that prevents individual tracking. This may still not be a bright line, but at least it's based on a physical property. The question of whether propeller-causing objects deserve to be considered as full-fledged moons is further complicated by the fact that, although their positions have been tracked over long periods, they are not directly seen but hidden within the surrounding propeller-shaped disturbance.

7. For ease of reference, propellers that are being tracked have been nicknamed after pioneers of aviation.

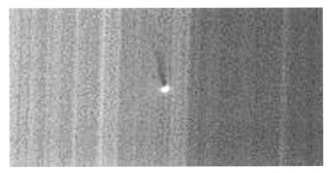

FIGURE 41.20 The feature known as S/2009 S 1, identified by the shadow it cast during equinox, was seen only in this image.

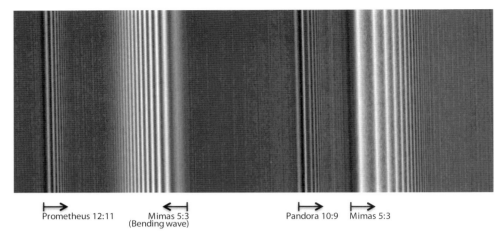

FIGURE 41.21 Spiral density waves and bending waves are seen in this close-up image of Saturn's A ring. The radial width of the area shown is 1000 km, and the center of the image is ∼132,000 km from Saturn's center (which would be far to the left). Three density waves—due to the moons Prometheus, Pandora, and Mimas—are labeled along the bottom with rightward arrows, since density waves propagate outward from the location of the resonance. As an example, the resonance at left occurs where ring particles orbit Saturn 12 times in the same interval that it takes Prometheus to orbit 11 times. Mimas also generates a bending wave, which is labeled with a leftward arrow, since it propagates inward from the resonance. A close look at the image background reveals dozens of fainter waves, mostly due to the smaller moons Pan and Atlas.

The most common type of resonance excites radial motions in the ring particle's orbit (i.e. it makes the ring particle's orbit more **eccentric**). The resulting **spiral density wave** is a compression wave, alternating bands of higher and lower densities that propagate radially outward from the resonant location. Other resonances excite vertical motions in the ring particle's orbit (making the orbit more **inclined**). The resulting **spiral bending wave** is a transverse wave, a corrugation in the ring plane that propagates radially inward. Figure 41.21 shows examples of each kind of wave.

Spiral waves are useful phenomena that can be thought of as scientific instruments that have been placed in the rings for our benefit. The ring's local density and **viscosity** can be inferred by carefully measuring how the wave's **wavelength** and amplitude evolve as it propagates. This has been used to build detailed profiles of these characteristics, especially for the A ring, where spiral waves are most plentiful.

4.3.2. Confinement of Edges and Ringlets

Some resonances are so powerful that the disk cannot coherently respond with a wave, resulting in a gap whose edges are sculpted by the resonance (Section 4.2). One example is the Colombo Gap in the C ring, located at a resonance between ring particle **precession** (the slow rotation of the noncircular orbit) and the orbit period of Titan; in the middle of this gap is an eccentric narrow ringlet (Section 5) that keeps its long axis always pointed toward Titan. The Keeler Gap, in which Daphnis orbits, has a 32-lobed inner edge sculpted by a 32:31 resonance with Prometheus.

Resonances can also provide a barrier to the outward diffusion of ring material, resulting in a sharp outer edge to a ring. The outer edge of the B ring (which is also the inner edge of the Cassini Division) is maintained by a 2:1 resonance with Mimas. In a similar fashion, the outer edge of the A ring is governed by a 7:6 resonance with the co-orbital moons Janus and Epimetheus.

5. DENSE NARROW RINGS

Dense narrow rings (see the upper part of Figure 41.5) are a unique assemblage of matter, behaving as a coherent self-contained object on planetary length scales yet ephemerally thin; they are only 1−100 km in radial width, compared to ∼100,000 km

Saturn	Titan ringlet
	Maxwell ringlet
	Bond ringlet
	Huygens ringlet
	"Strange" ringlet
	Herschel ringlet
	Jeffreys ringlet
	Laplace ringlet
Uranus	6 ring
	5 ring
	4 ring
	α ring
	β ring
	η ring
	γ ring
	δ ring
	ε ring

in diameter. How and why these highly organized

dynamical systems form is almost entirely unknown, and even their present dynamics are only partly understood.

Nearly all known dense narrow rings are either noncircular or inclined to the main ring plane, or both. Furthermore, nearly all dense narrow rings have edges that are quite sharp; as with dense disks, the existence and stability of such sharp edges requires some confinement mechanism to counteract the natural process of radial viscous spreading. Confinement may be due to an external moon or to the ring's own self-gravity, and in some cases to processes yet to be understood.

The "shepherding" mechanism by which a moon opens a gap (Section 4.2) or maintains an edge in a disk (Section 4.3) can also occur with two shepherd moons on either side of a narrow ringlet. However, the only example of a dense narrow ring with a known shepherd on either side is Uranus' ε ring, although Saturn's F ring (see Section 6.1) is a variation on that idea. Indeed, as observations seeking for them continue to pile up, the scarcity of shepherd moons at both Saturn and Uranus is becoming more and more conspicuous, lending indirect support to theories that narrow rings are instead maintained by resonances or by self-sustaining mechanisms.

A number of narrow ringlets are eccentric and appear to precess about their planet as a rigid body. At first blush, this appears to contradict Kepler's third law, by which the inner edge would be expected to precess faster than the outer edge, with the ring eventually smearing out to a circular shape after only a few years. However, the ring's own self-gravity and/or interparticle collisions can counteract this effect, altering the precession rate by different amounts in different parts of the ring so that the whole ring precesses as if it were a rigid body. In its simplest form, this effect will occur only if the ring has a positive "eccentricity gradient", which is to say that the eccentricity increases from the ring's inner edge to its outer edge, and indeed several narrow rings appear to have outer edges that are more eccentric than their inner edges.

6. DUSTY RINGS

All the giant planets possess rings composed primarily of particles less than 100 microns across. The distribution, structure and dynamics of these dusty rings are quite different from those of the dense rings described above.

Narrow Dusty Rings	
Saturn	"Charming" ringlet
	Encke Gap ringlets
	F ring
Uranus	λ ring
Neptune	Le Verrier ring
	Arago ring
	Adams ring

For example, dusty rings are typically much more tenuous than dense rings, so collisions between particles are rare, and the material in these rings cannot efficiently aggregate into larger bodies. Hence these dusty rings can exist outside the planet's Roche limit. Furthermore, dust-sized particles can be destroyed by collisions with micrometeoroids or ions on

Diffuse Dusty Rings	
Jupiter	Halo ring
	Main ring
	Amalthea Gossamer ring
	Thebe Gossamer ring
Saturn	D ring
	Roche Division
	Janus/Epimetheus ring
	G ring
	Methone ring
	Pallene ring
	Anthe ring
	E ring
	Phoebe ring
Uranus	ζ ring
	ν ring
	μ ring
Neptune	Galle ring
	Lassell ring

timescales of years to decades, so these rings must be constantly resupplied with material from larger source bodies. These rings' appearance can therefore respond rapidly to processes that affect the rates of particle production, destruction, or transport (see Section 6.1). The high surface area-to-volume ratios of these small particles also make them extremely sensitive to various nongravitational perturbations (see Section 6.2). These rings therefore are highly dynamic systems.

6.1. Dust Production and Destruction

The primary mechanism for producing orbiting dust particles is micrometeoroid bombardment of larger orbiting bodies. A notable exception is Saturn's E ring, which is supplied by the cryovolcanic plume emerging from the south pole of the 500 km moon Enceladus. The distinctive origin of the E ring manifests itself in its unusual particle-size distribution. Because they respond more readily to the escaping fast-moving gas, smaller particles are launched at higher average speeds, and only particles less than a few microns across escape the moon's gravity to enter the E ring. The dearth of larger particles gives the E ring a distinctive blue color when viewed at low phase angles.

Besides Enceladus, the moons associated with detectable dusty rings are typically rather small. At Saturn, dusty rings are found along the orbits of Pan (with a diameter of 28 km), the co-orbitals Janus and Epimetheus (180 km/120 km), Aegaeon (0.5 km), Methone (3 km), Anthe (2 km), Pallene (5 km), and Phoebe (210 km). Jupiter's main rings are associated with the moons Metis (60 km) and Adrastea (20 km), while its Gossamer rings appear to be derived from Amalthea (170 km) and Thebe (100 km). Finally, Uranus' μ ring is found around the orbit of Mab (25 km). Small moons are probably more efficient sources of dusty rings because their surface gravities are low, which allows impact-generated debris to escape from the moon and enter orbit around the planet. This debris typically contains a broader distribution of particle sizes, giving these rings a neutral or red color (with the still-puzzling exception of the µ ring).

The boulder-sized particles in the dense rings are another potential source of small grains that can form dusty rings. Indeed, such debris is likely the source of a few dusty rings found in gaps within and adjacent to Saturn's and Uranus' main rings. Similarly, the dusty material in Saturn's F ring is probably derived from a population of larger particles embedded within the ring. Neptune's dusty rings may have a similar origin, but the available data for these systems are relatively sparse.

Assuming these rings are more or less in a steady state, then the production of fine grains must be balanced by various loss processes. This includes erosion due to magnetospheric ions and micrometeoroids, as well as reaccretion onto larger particles. The latter is particularly important for the dusty rings found in close association with dense rings. Indeed, aside from the transient spokes (discussed in Section 7.3 below), Saturn's main rings are remarkably dust-free, with dusty ringlets only occupying nearly empty gaps in the A ring, Cassini Division, and C ring.

6.2. Nongravitational Forces

Between their creation and destruction, the orbital properties of these small grains evolve under the influence of various nongravitational forces. For example, several narrow dusty ringlets occupying gaps within Saturn's main rings have the unusual property that the geometric center of the ringlet appears to be consistently displaced away from Saturn's center toward the Sun. This "heliotropic" behavior likely arises because solar **radiation pressure** can significantly perturb the orbits of these tiny particles. There are also structures in the innermost dusty rings around Jupiter and Saturn that can be attributed to resonances with the rotation period of the planet's magnetosphere (see Section 7.3). These structures indicate that the relevant dust grains can be charged enough to make them susceptible to electromagnetic forces.

In other dusty rings, we have evidence that material is being transported radially toward or away from the planet. Jupiter's Gossamer rings and Saturn's enormous Phoebe ring all extend inward from their putative source moons, while Saturn's G ring extends outward from its most likely source. This directional transport of material can be attributed to drag forces. For the Gossamer rings and the Phoebe ring, the drag force can be attributed to the asymmetric scattering of solar photons, also known as **Poynting–Robertson drag**. By contrast, the outward drift of material in the G ring is due to collisions between the dusty ring particles and magnetospheric ions. These ions move around the planet much faster than the ring particles as they are pulled along by the planet's rotating magnetic field (the G ring is well beyond synchronous orbit). Hence the collisions between the ions and the ring particles tend to accelerate the particles and cause them to spiral outward.

6.3. Resonantly Confined Material

Given the number of nongravitational forces affecting dusty rings, and how quickly the small grains can be transported or destroyed, one might naturally expect that all dusty rings would be rather broad and diffuse. However, in reality, many dusty rings are highly structured, and several exhibit significant azimuthal brightness variations.

Ring Arcs	
Saturn	G ring (Aegaeon ring arc)
	Methone ring arc
	Anthe ring arc
Neptune	Galatea ring arc
	Adams ring (?)

Azimuthal Clumps	
Jupiter	Main ring
Saturn	Encke Gap ringlets
	F ring
Neptune	Adams ring (?)

Left to itself, any clump of material orbiting a planet should spread out into a ring fairly quickly. This is because, by Kepler's third law, material at the inner edge of the clump is moving faster than material at the outer edge, and the one will soon "lap" the other by an entire orbit. Narrow clumps will spread into a ring more slowly, but even a compact initial clump ~1 km wide will spread into a ring within a few decades. However, certain azimuthal brightness variations, known as ring-arcs, appear to persist for years or decades without any obvious spreading.

The first-discovered and best known set of these ring arcs are found in Neptune's Adams ring. These are the densest components of Neptune's ring system and were oftentimes the only component detectable in pre-Voyager Earth-based occultations, leading to much confusion as to whether the detected signatures were rings at all until Voyager 2 settled the question. There are five arcs ranging from a few degrees to ~10° in length extending over a region of ~40° in longitude (Figures 41.6, 41.22 and 41.23).

A closer look at these arcs indicated that they could be maintained and confined by a corotation resonance. Unlike the resonances that produce spiral density waves, in which the perturbing moon pumps up a ring particle's eccentricity (Section 4.3), corotation resonances tend to azimuthally confine particles into an orbit commensurate with that of the perturbing moon. Initial analysis of the Voyager 2 images indicated that the Adams ring arcs might be governed by the nearby 43:42 corotation resonance with the

FIGURE 41.22 The brightest two Neptunian rings, Le Verrier (inner curve) and Adams (outer curve) are revealed in this Voyager image. Neptune is overexposed to lower left, indicating the difficulties in searching for faint features near planets. A short-exposure crescent-shaped Neptune has been overlayed to indicate the planet's true size and phase. Three of the famous ring arcs are visible in the outer Adams ring, while the Le Verrier ring has no such features.

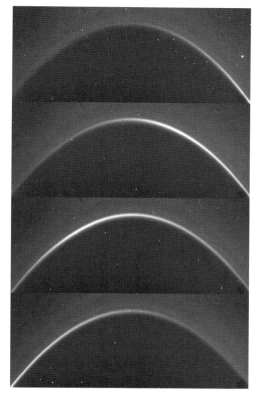

FIGURE 41.24 The G ring arc, passes through the field of view.

moon Galatea, although there was some difficulty in that the azimuthal lengths of the arcs do not easily match the 4° periodicity expected from the resonance. The 1999 reacquisition of the arcs with Earth-based imaging ruled out the simplest version of this model, but a more complex version was suggested and remains plausible. Further Earth-based observations also revealed significant changes in the arcs' structure over the 20 years in which they have been observed in detail (Figures 41.22 and 41.23).

More recently, Cassini has found several examples of ring-arcs in Saturn's dusty rings. These rings are much more tenuous than Neptune's Adams ring, and thus provide somewhat simpler examples of resonantly confined debris.

The most prominent of these arcs is found near the inner edge of Saturn's G ring (Figure 41.24). This arc, along with the tiny moon Aegaeon embedded within it, appears to occupy the 7:6 corotation resonance with Saturn's moon Mimas. The arc orbits the planet seven times in roughly the same time it takes Mimas to orbit the planet six times. Furthermore, the arc's location and azimuthal extent matches the predicted distribution of material trapped by this resonance.

This arc probably represents the source of the G ring, which is much brighter than most other rings generated by small moons. Measurements of charged particles in the vicinity of the arc indicate that in addition to the small moon Aegaeon, this region contains a population of boulder-sized objects that are trapped in the resonance and can serve as source bodies for fine debris. These dust grains can escape from the arc under the influence of nongravitational forces, migrating outward (Section 6.2) to populate the rest of the G ring.

Two other arcs, even more tenuous than the G ring's, surround the small moons Anthe and Methone

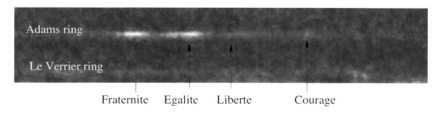

FIGURE 41.23 Reprojected ground-based image of the Adams ring (with arcs) and the Le Verrier ring acquired in October 2003. The orbital direction is to the right and the radial direction (away from Neptune) is up. The leading arc Liberté was originally seen by Voyager 2 (Figure 41.22) to be as bright as the other two main arcs, but here is much dimmer.

FIGURE 41.25 The moons Anthe (top left) and Methone (bottom right) share their orbits with arcs of dust.

(Figure 41.25), and are compatible with the 10:11 and 14:15 corotation resonances with Mimas, respectively.

6.4. Azimuthal Clumps and Kinks

In addition to the relatively stable and persistent arcs that are likely confined by resonances, dusty rings can also exhibit more transient and variable azimuthal structure.

The F ring is the granddaddy of narrow dusty ringlets, and by far the best studied (Figure 41.26, Movie M3). Located a few thousand kilometers off the outer edge of Saturn's main rings, its relatively large vertical thickness effectively frustrates any attempt to see the rest of the main

FIGURE 41.26 The F ring is a multifaceted ribbon lying off the outer edge of Saturn's main disk (see also Movie M3). The moon Prometheus, lying just inward, carves out "streamer channels" as it dips into the F ring (see also Figure 41.27 and Movies M4–M6). Also visible at center-left is the moon Pan, orbiting within the Encke Gap, with the three dusty ringlets that share the gap and the "moonlet wakes" trailing behind it on the outward side of the gap.

rings in edge-on viewing geometries. Its high fraction of forward-scattering dust makes it by far the brightest component of Saturn's ring system when viewed at high-phase geometries.

In general, the F ring's core is a complex region, possessing transient bright "clumps", as well as "kinks" in its apparent radial position. These structures form, merge, split, and dissipate on a wide range of timescales, and only the largest features can be tracked over time with any amount of reliability. Rather than being produced by resonances, these structures instead probably reflect localized sculpting of the dust population by nearby moons and embedded boulders.

As mentioned in Section 5, the F ring is one of only two narrow rings to have known "shepherd" moons orbiting on either side of it. However, it has not been conclusively shown how (or even that) the moons Prometheus and Pandora actually constrain the F ring in its place, and in fact they appear to stir the ring up at least as much as to maintain it. Prometheus, in particular, has an eccentric orbit that allows it to occasionally pass through the core of the F ring once per orbit. These close encounters generate "streamer channel" features in nearby dust strands that subsequently moves downstream and shears under Keplerian motion (Figure 41.27, Movies M4–M6).

Other structures in this ring may be generated by kilometer-sized moonlets hiding within the ring. The moonlets are known to be there because of their absorption of charged particles, from direct detection by occultations, from the characteristic "fan" structures they create in surrounding dust (Figure 41.27, Movie M7), and from shadows cast during the 2009 Saturnian equinox (Figure 41.27). Some of the brightest clumps in the ring, as well as the dusty lanes or "strands" lying on either side of the core may represent debris produced by collisions into or among these objects.

All these interwoven phenomena make the F ring the solar system's foremost natural laboratory for studying accretion and disruption processes. However, even with all this time variability, the F ring core nevertheless maintains over decadal timescales the shape of a freely precessing eccentric inclined ellipse; the orbital solution formulated to account for Voyager and other pre-Cassini data has, somewhat surprisingly, remained a good predictor of the core's position as seen by Cassini.

Besides the F ring, the clumpiest dusty rings are the three narrow ringlets occupying the Encke Gap in Saturn's outer A ring (Figure 41.28). These ringlets all possess "clumps" (azimuthal brightness variations) and "kinks" (radial offsets). However, in this case the clumps move slowly enough relative to each other that they can be tracked over several years. These clumps do not follow the expected trajectories of test particles in the combined gravitational fields of Saturn and Pan, and there is not

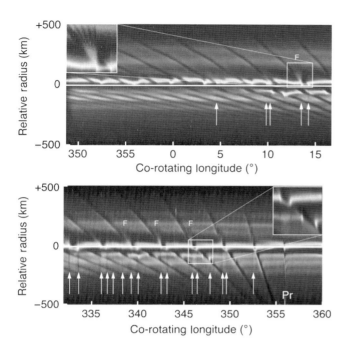

FIGURE 41.27 In each of these two mosaics, ∼8% of the azimuthal extent of the F ring is projected onto a radius-longitude grid. The bright core of the F ring is at zero radius, and parallel strands can be seen on either side of it. The white arrows mark inward-pointing shadows cast by moonlets within the bright F ring core. The letter "F" marks "fan" structures, also seen in the insets and illustrated in Movie M7, which are the dust sheet's response to an embedded moonlet on an eccentric orbit. The narrow dark stripes trending from upper left to lower right are sheared "streamer channels" created as the moons Prometheus and Pandora dip into the ring (see also Movies M4–M6).

FIGURE 41.28 Azimuthal clumps and kinks can be seen both in the central and inner Encke Gap ringlets, here leading Pan in its orbit.

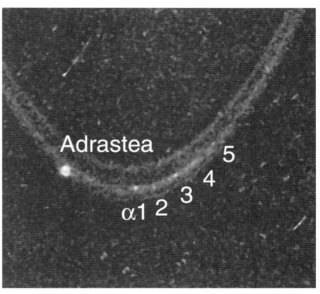

FIGURE 41.29 Azimuthal clumps in the Main ring of Jupiter lie slightly inward of the moon Adrastea's orbit.

yet any clear evidence for larger objects embedded in these ringlets. Hence these clumps could instead reflect collective phenomena or instabilities within the dusty material.

Images from the 2007 New Horizons flyby of Jupiter found several azimuthal clumps in the core of the Main ring (Figure 41.29) that fall close to the 115:116 and 114:115 corotation resonances with Metis, the same kind of resonance invoked to explain Neptune's Adams arcs. Furthermore, continuing the analogy with Neptune's Adams arcs, the azimuthal spacing of the clumps does not correspond to the periodicity expected from the resonances. Thus these features could also represent transient features like the clumps found in Saturn's F ring and Encke Gap ringlets.

It is remarkable that all the obviously clumpy dusty rings (Jupiter's Main ring, Saturn's F ring, and Encke Gap ringlets, and even perhaps Neptune's Adams ring), are located close to the planet's Roche limit. By contrast, narrow dusty rings found closer to the planet, like Uranus' λ ring, Neptune's Le Verrier or Arago rings, appear to be much more sedate.

7. USING RINGS TO PROBE THE SOLAR SYSTEM

As delicate dynamical systems covering vast areas, rings sometimes function as useful detectors of their

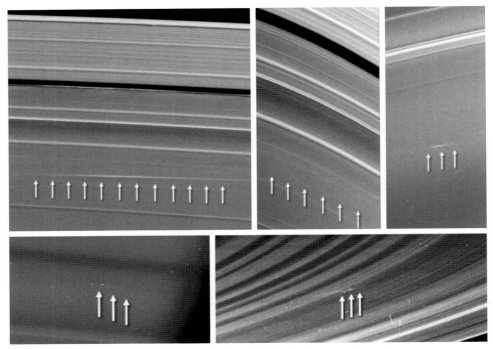

FIGURE 41.30 Ejecta clouds in Saturn's rings are canted from the azimuthal direction by an angle of a few degrees, due to Kepler's third law.

surrounding environment. The structure of the planet's gravity and magnetism, changes in the orbits of its moons, and the population of meteoroids in the outer solar system are all illuminated by phenomena observed in the rings.

7.1. Probing Interplanetary Meteoroids

We are used to the idea that, when a **meteoroid** strikes a planetary surface, **ejecta** flies in all directions and the result is a crater structure such as Meteor Crater in Arizona or the countless craters that punctuate the surface of the Moon. However, the result is very different when the "target" is not a solid surface but rather a ring composed of countless particles that are constantly in motion. Not only does the impact result in no discernible crater (instead, one or more pieces of the target ring are simply pulverized[8]), but also the expanding ejecta cloud soon becomes hijacked by Kepler's third law and assumes a linear shape that is canted from the azimuthal direction by an angle of a few degrees (Figure 41.30).

Several ejecta clouds have been observed by Cassini, yielding the first direct knowledge of particles in the outer solar system that are larger than dust but smaller than moons. These impacts occur continually, but the ejecta clouds are usually too faint to be visible. Cassini saw them best during the 2009 equinox, when the Sun's edge-on illumination yielded a dark ring background while ejecta rising above the ring plane caught full sunlight.

The time elapsed since the cloud was formed can be directly calculated from the cant angle, using Kepler's third law. In one case (the first two panels of Figure 41.30), the same cloud was seen 24 h apart, and the ages derived from the cant angle were also 24 h apart, confirming that we are properly interpreting these observations.

In contrast to the constant influx of meteoroids of meter-size and smaller, larger impacts are rare and can leave reverberations that persist for decades or more. The momentum of the impacting material can give the rings a slight tilt, which then winds up into a spiral corrugation because inclined orbits in the inner part of the disk precess faster than those in the outer part (a principle related to Kepler's third law). The wavelength of the spiral and the rate at which it is winding up give two independent measures of the time elapsed since the impact. One such corrugation appears to have been created in Jupiter's Main ring by debris from comet Shoemaker-Levy 9, which crashed into Jupiter in 1994. Another corrugation, permeating Saturn's C and D ring (Figure 41.31), appears to have been caused by a similar unseen event that occurred around 1983.

---

8. If the meteoroid is much smaller than typical ring particles (e.g. it is a speck of dust), then a target ring particle may indeed survive and sport a small crater. However, in this section we are discussing centimeter-size to meter-size impactors.

7.2. Probing the Planet's Gravity

When a planet's interior structure is not perfectly symmetric, it causes orbits around that planet to precess (that is, the orbit path itself slowly rotates or revolves, similar to the midplane of a spinning top). The rate of precession directly yields information about the planet's internal structure, which is expressed through its gravitational field. Spacecraft can use this method to probe the internal structure of any object they fly past, by precisely tracking changes in its own trajectory, but precision increases dramatically when the orbit is closer to the object being probed. Thus, the most precise gravity measurements for both Saturn and Uranus come from spacecraft observing the precession of their dense narrow rings (Section 5).

Precession is driven by planetary asymmetries in the latitude direction; asymmetries in the longitude direction can drive spiral waves in much the same way as orbiting moons (Section 4.3). Voyager observations revealed a small group of mysterious spiral waves in the C ring, Saturn's closest dense ring, which did not have any connection with the orbit of a known moon that might drive the waves. Suspicion fell on **normal modes**, a hypothetical sloshing back-and-forth of portions of Saturn's interior, but no more could be said given the quality of the Voyager data.

This mystery is starting to yield not only to the greater frequency (measuring the waves at many different points within their oscillations) but also to the greater sensitivity (seeing the waves at each measurement in more detail) of Cassini's observations. The first use of a ring system as a seismometer is underway and is likely to pinpoint specific modes in which Saturn's interior operates and evolves.

7.3. Probing the Planet's Magnetosphere

"Spokes" are near-radial markings on Saturn's B ring, likely composed of dust levitating above the ring plane due to electromagnetic forces (Figures 41.32 and 41.33 and Movie M8). Spokes appear in the central B ring at radial locations near to (and often astride) **synchronous orbit**, where the orbital period of a ring particle matches the rotation period of the planet, and thus also of the magnetic field. Observations by the Hubble Space Telescope and by Cassini have revealed that spokes are a seasonal phenomenon, more abundant near **equinox** and absent near **solstice**.

Spokes are likely triggered by **meteoroid** impacts onto the ring, but their growth is governed by the interaction between impact-ejected dust and Saturn's magnetic field, which causes the dust to levitate above the ring. Strong periodicities in the occurrence of spokes appear to be correlated with periodicities in Saturn's magnetic field.

Canted patterns in the D ring and the Roche Division, the tenuous dusty sheets on either side of Saturn's main rings, are driven by resonances (see Section 4.3) with the rotation period of the planet's magnetic field. Tenuous dusty sheets are populated by tiny grains that are easily charged and thus subject to electromagnetic forces, making them especially sensitive detectors of the planet's magnetism. The complex observed structures pinpoint multiple pattern speeds, furnishing a major source of information for the complex rotation of Saturn's enigmatic magnetic field.

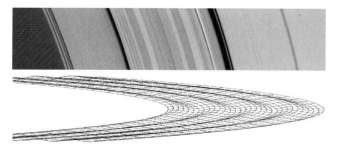

FIGURE 41.31 (Top) Cassini image mosaic of the D (far left) and C rings, taken during the 2009 equinox, showing a pervasive undulation. (Bottom) model demonstrating how such an undulation can be caused by a corrugation in the ring plane.

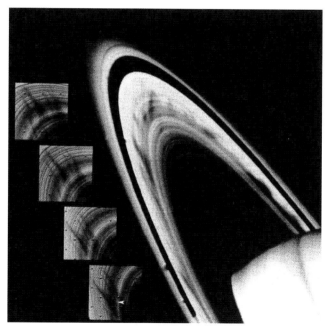

FIGURE 41.32 A Voyager image of dark spokes seen against Saturn's sunlit B ring. Small dust particles appear dark under this lighting condition, hinting and the still poorly understood physical processes behind spoke creation. The inset panels show the change of a given feature with time.

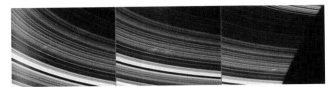

FIGURE 41.33 A three-panel Cassini image of bright spokes seen against the dark side of Saturn's B ring. Small dust particles appear bright under this lighting condition. The motion of the spokes can be seen clearly by comparing the three panels.

8. AGE AND ORIGINS OF RING SYSTEMS

The question of the age of a ring system is actually two separate questions, one being the age of the material and the other the age of the structure currently in place. This distinction is particularly important for diffuse dusty rings (Section 6.1). For example, if Amalthea, Thebe, Adrastea, and Metis were to suddenly disappear, or to suddenly cease emitting dust, Jupiter's rings would dissipate within 100,000 years at the most. Thus, the age of individual particles in a diffuse ring is on the order of the mean residence time, while the age of the overall structure is related to how long the sources have been in place.

The distinction between material age and overall structural age also turns out to be useful for Saturn's main rings. Several aspects of the rings are difficult to reconcile with a ring age comparable to that of Saturn. For example, the exceptionally pure water-ice composition of the A and B rings is difficult to reconcile with the constant pollution of the rings by infall of interplanetary micrometeoroids. Furthermore, the same infall should significantly erode ring particles, especially in more tenuous regions. Finally, the orbits of nearby moons are migrating outward, pushed by the density waves they raise in the rings (Section 4.3), and this process can only be "rewound" for about 10–20 million years. On the other hand, the "young rings" scenario is difficult to imagine because there are few plausible objects in the modern solar system that are large enough to yield Saturn's massive ring system if pulled apart.

Most solid material in the outer solar system is relatively dark because carbon-rich compounds are mixed with the ice. By contrast, Saturn's rings are highly reflective at both visible and radio wavelengths, which indicates that dark organic materials like those seen on comets are relatively rare there, likely composing less than 1% of the material on the ring particles' surfaces. This is surprising because cometary debris is constantly raining down on the rings, which should rapidly cause the rings to darken. Indeed, there is evidence that this darkening is occurring to some limited extent. The particles in the Cassini Division and the C ring are significantly darker than the particles in the A and B rings, which makes sense because the C ring and Cassini Division have less material overall and thus would become contaminated more quickly.

One potentially viable theory suggests that the B ring core is ancient, dating from the early history of the solar system, when collisions with **planetesimals** were much more frequent (*see* "Origin of the Solar System"), but that much of the specific organization of material in Saturn's rings is only about 10–20 million years old. Recent indications that the B ring's mass has previously been underestimated (see Section 4.1) provide an increased buffer against interplanetary pollution, as well as making the ring precursor body even larger and thus a recent ring origin even more unlikely. Erosion of ring particles might also have been slower if ring particles have spent most of their history more densely packed (thus, exposing less surface to erosion) than we now find them. The current ring-moons may also have formed within the rings and emerged only recently.

It is easier to imagine that the rings of Uranus and Neptune, with their lower masses and much darker surfaces, have remained little changed over the age of the solar system. The Uranian rings, as well as the nearby group of moons, may be part of a system that has oscillated between accretion and disruption for billions of years (Section 3.2). But why are the rings where they are? Do their positions simply reflect where source material became available, or did dynamical processes sculpt them into their current configurations? On the planetary level, why does Saturn have its glorious broad dense disk while Uranus and Neptune have much more modest systems and Jupiter has only moon-generated dust? None of these questions has a clear answer as yet.

Supplementary Video

The following supplementary video related to this chapter can be found at http://dx.doi.org/10.1016/B978-0-12-415845-0.00041-4:

Movie M1: Simulated self-gravity wakes (SGWs). This microstructure develops within dense rings as particles clump together under their mutual self-gravity but are ripped apart again by Saturn's tides. Figure 41.12 is a still from this movie.

Movie M2: Simulated viscous overstability.

Movie M3: Bright knots, caused by embedded moonlets, pass continually through the core of the F ring.

Movie M4: As Cassini's camera stares at a single point on the F ring, Prometheus and Pandora march through the field of view, attended by a suite of streamer channels generated by Prometheus.

Movie M5: In this simulation, Prometheus moves in and out as it goes around on its eccentric orbit, dipping into the F ring and generating streamer channels that "breathe" in and out. Compare to the Cassini observations in Movie M6.

Movie M6: Cassini's camera stares at Prometheus as it goes around on its eccentric orbit, dipping into the F ring and generating a streamer channel. Compare to the simulation in Movie M5.

Movie M7: In this simulation, an embedded moonlet on an eccentric orbit generates a "fan". Compare to the Cassini images in Figure 41.27.

Movie M8: As Voyager's camera stares at a single point on the B ring, spokes pass through the field of view. See the related Voyager images in Figure 41.32 and the Cassini images in Figure 41.33.

BIBLIOGRAPHY

The Planetary Rings Node, administered by NASA's Planetary Data System, has a wealth of information, images, and movies at http://pds-rings.seti.org/.

The Cassini Mission web site at http://saturn.jpl.nasa.gov is the source for images, videos, press releases, and other products from all Cassini instruments.

A more detailed review of planetary ring systems, including elementary plots and equations and many references to previous reviews and to primary research articles:

Tiscareno, M. S. (2013). Planetary rings. In T. D. Oswalt, L. French, & P. Kalas (Eds.), *Solar and planetary systems: Vol. 3. Planets, stars, and stellar systems* (pp. 309–370). Springer. Available from http://arxiv.org/abs/1112.3305.

A review of elementary orbital mechanics as it applies to ring systems:

Hedman, M. M. (2013). Planetary ring dynamics. In A. Celletti (Ed.), *Celestial mechanics: Vol. 6. Encyclopedia of life support systems* (pp. 119–155). UNESCO/EOLSS. Available from http://www.eolss.net.

The most recent review of Saturn's rings as seen by Cassini:

Cuzzi, J. N., Burns, J. A., Charnoz, S., Clark, R. N., Colwell, J. E., Dones, L., et al. (2010). An evolving view of Saturn's dynamic rings. *Science, 327*, 1470–1475.

A light textbook intended for the layperson:

Miner, E. D., Wessen, R. R., & Cuzzi, J. N. (2007). *Planetary ring systems*. Springer.

A textbook intended for undergraduates:

Esposito, L. W. (2006). *Planetary rings*. Cambridge University Press.

A comprehensive book on the state of research on the Saturn system (excluding Titan) after the first phase of Cassini's mission, intended for specialists; chapters 13–17 focus on Saturn's rings:

Dougherty, M., Esposito, L., & Krimigis, T. (Eds.). (2009). *Saturn from Cassini-Huygens*. Springer.

A classic book, intended for specialists, whose thorough discussion of fundamental concepts is still useful in many respects:

Greenberg, R., & Brahic, A. (1984). *Planetary rings*. University of Arizona Press.

Detailed reviews of the Jupiter, Uranus, and Neptune ring systems, intended for specialists, as seen by the Galileo (in the case of Jupiter) and Voyager missions:

Burns, J. A., Simonelli, D. P., Showalter, M. R., Hamilton, D. P., Esposito, L. W., Porco, C. C., et al. (2003). Jupiter's ring-moon system. In F. Bagenal, T. Dowling, & W. B. McKinnon (Eds.), *Jupiter: The planet, satellites, and magnetosphere* (pp. 241–262). University of Cambridge Press.

Esposito, L. W., Brahic, A., Burns, J. A., & Marouf, E. A. (1991). Particle properties and processes in Uranus' rings. In J. T. Bergstralh, E. D. Miner, & M. S. Matthews (Eds.), *Uranus* (pp. 410–465). University of Arizona Press.

French, R. G., Nicholson, P. D., Porco, C. C., & Marouf, E. A. (1991). Dynamics and structure of the Uranian rings. In J. T. Bergstralh, E. D. Miner, & M. S. Matthews (Eds.), *Uranus* (pp. 327–409). University of Arizona Press.

Porco, C. C., Nicholson, P. D., Cuzzi, J. N., Lissauer, J. J., & Esposito, L. W. (1995). Neptune's ring system. In D. P. Cruikshank (Ed.), *Neptune and Triton* (pp. 703–804). University of Arizona Press.

A detailed review of dusty ring systems, intended for specialists:

Burns, J. A., Hamilton, D. P., & Showalter, M. R. (2001). Dusty rings and circumplanetary dust: observations and simple physics. In E. Grün, B.Å. S. Gustafson, S. Dermott, & H. Fechtig (Eds.), *Interplanetary dust* (pp. 641–725). Springer.

Beyond the Planets

Chapter 42

Pluto

S. Alan Stern

Space Science and Engineering Division, Southwest Research Institute, Boulder, CO, USA

Chapter Outline

1. **Historical Background** 909
   1.1. Overview 909
   1.2. The Discovery of Pluto's Satellites 910
2. **Pluto's Orbit and Spin** 911
   2.1. Pluto's Heliocentric Orbit 911
   2.2. Pluto's Lightcurve, Rotation Period, and Pole Direction 911
   2.3. Charon's Orbit and the System Mass 912
3. **The Mutual Events** 913
   3.1. Background 913
   3.2. Radii and Average Density of Pluto and Charon 913
4. **Pluto's Surface Properties and Appearance** 914
   4.1. Albedo and Color 914
   4.2. Solar Phase Curve 914
   4.3. Surface Composition 914
   4.4. Surface Temperature 915
   4.5. Surface Appearance and Markings 916
5. **Pluto's Interior and Bulk Composition** 917
   5.1. Density 917
   5.2. Bulk Composition and Internal Structure 917
6. **Pluto's Atmosphere** 918
   6.1. Atmospheric Composition 918
   6.2. Atmospheric Structure 919
   6.3. Atmospheric Escape 920
7. **Charon** 920
8. **The Origin of Pluto's Satellite System** 922
   8.1. The Formation of Pluto's Satellite System 922
   8.2. The Origin of Pluto Itself 922
   8.3. The Context of Pluto in the Outer Solar System 923
   Bibliography 924

Pluto is the prototype of the dwarf planets common in the Kuiper Belt and beyond. It is in an elliptical, 248 year orbit that ranges from 29.6 to 48.8 **astronomical units** (AU) from the Sun. Its largest satellite, Charon, is close enough to Pluto in size that the pair is widely considered to be a double planet. Pluto's four other known satellites, Styx, Nix Kerberos and Hydra, orbit beyond Charon but in Charon's orbital plane, and are relatively small. Almost nothing is known about these moons save their orbits (see Figure 42.1), approximate sizes, and their neutral, Charon-like colors. Both Pluto and Charon are rich in ices, but their surface compositions, **albedos**, and colors are very different. Unlike Charon, Pluto is known to possess distinct surface markings, polar caps, and an atmosphere. Major questions under study about the Pluto system include the fate of Pluto's atmosphere, the degrees of internal activity Pluto and Charon exhibit, and the origin of the system.

1. HISTORICAL BACKGROUND

1.1. Overview

Pluto was discovered in February 1930, at Lowell Observatory in Flagstaff, Arizona. This discovery was made by Clyde Tombaugh (1906–1997), an observatory staff assistant working on a search for a long-suspected perturber of the orbits of Uranus and Neptune. That search, which was first begun in 1905 by the observatory's founder, Percival Lowell, never located the large object originally being searched for because the positional discrepancies of Uranus and Neptune, which prompted that search, were fictitious. Still, the search for Lowell's "Planet X" resulted in the discovery of the tiny planet Pluto, which itself heralded the discovery of the Kuiper Belt some 70+ years later.

Within a year of Pluto's discovery, its orbit was well determined. That orbit is both eccentric and highly inclined

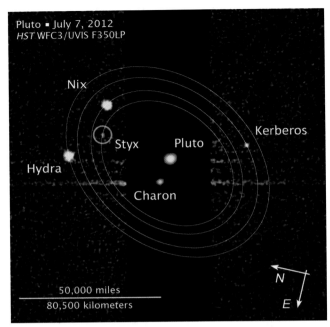

FIGURE 42.1 The Pluto System. This Hubble Space Telescope image shows Pluto and the orbits of it five known moons as of early-2014. *Credit: NASA/STScI.*

to the plane of the **ecliptic**, compared to the orbits of the other planets (see Table 42.1). However, no important discoveries about Pluto's physical properties were made until the early 1950s. This lack of information was largely due to the difficulty of observing Pluto with the scientific instruments available in the 1930s and 1940s. Between 1953 and 1976, however, technological advances in photoelectric photometry made possible several important findings. Among these were the discovery of Pluto's ~6.387 day rotation period, the discovery of Pluto's reddish surface color, and the discovery of Pluto's high axial tilt, or **obliquity**.

Between 1976 and 1989, the pace of discoveries increased more dramatically. In rapid succession, there was the discovery of methane (CH_4) on Pluto's surface; the detection of Pluto's largest satellite Charon; the prediction, detection, and then study of a set of once-every-124-year mutual eclipse events between Pluto and Charon; and the occultation by Pluto of a bright star, confirming the presence of an atmosphere. In addition, the 1989 Voyager 2 encounter with the Neptune system gave us detailed insights into the object believed to be Pluto's closest analog in the solar system, Triton, thereby showing how complex and scientifically interesting Pluto would be under close scrutiny by spacecraft. In the 1990s, it was discovered that Pluto's surface consists of a complex mixture of low-temperature volatile ices, that this surface displays large-scale bright and dark units, and that Pluto's atmosphere consists primarily of nitrogen gas, with trace amounts of carbon monoxide and only a trace of methane. Additionally, Pluto's small moons were discovered, and Pluto's context in the solar system became understood only after the discoveries of many smaller objects in the region of the solar system beyond Neptune called the Kuiper Belt.

1.2. The Discovery of Pluto's Satellites

Charon (pronounced correctly as "Kharon", but more colloquially pronounced as "Sharon") was discovered by J. W. Christy and R. S. Harrington on a series of photographic plates made in 1978 at the US Naval Observatory's Flagstaff Station in Arizona. Interestingly, these images were taken less than 4 miles from Lowell Observatory, where Pluto had been discovered 48 years before. Charon was apparent on the 1978 Naval Observatory images as a bump or elongation in Pluto's apparent shape. This elongation of Pluto had occasionally been seen on photographic plates made in the 1960s, but it had not been recognized to be a satellite. This was because the elongation of Pluto's image by Charon was attributed to turbulence in the Earth's atmosphere causing a distortion of Pluto's pointlike image (the two are <1 **arc second** (arcsec) apart, and blended together by atmospheric seeing). What Christy and Harrington recognized in 1978 was that although Pluto was distorted, none of the stars in the photographs were. This led them to look for a periodicity in the elongations. The recognition that the bump was in fact a close-in satellite was made when it was determined that this bump regularly cycled around Pluto in a 6.387 day period, which matched Pluto's

TABLE 42.1 Pluto's Heliocentric Orbit

Orbital Element	Value
Semimajor axis, a	39.48 AU
Orbital period, P	247.688 years
Eccentricity, e	0.2488
Inclination, i	17.16°
Longitudinal ascending node, ω	110.30°
Longitudinal perihelion, ω	224.07°
Perihelion epoch, T	05.1 September 1989 Universal Time

Note: Osculating elements on JD 2449000.5, referred to the mean ecliptic and equinox of J2000.0.

rotation period, implying that the elongation was due to an object that circled Pluto.

In the first few months after Charon's discovery, Christy and Harrington determined that Charon's orbit is synchronous with Pluto's rotation and also in Pluto's equatorial plane, and therefore highly inclined to the plane of the ecliptic. During that same year, 1978, Leif Andersson recognized that Pluto's orbital motion would cause Charon's orbital plane to sweep through the line of sight to the Earth for a period of several years every half Pluto orbit, or 124 terrestrial years. Mutual eclipses (also called mutual events) would then begin occurring every 3.2 days (half Charon's orbit period). These eclipses were predicted to progress over a period of 5–6 years, from shallow, partial events to central events lasting up to 5 h, then to recede again to shallow grazing events. It was widely recognized that such a series of mutual eclipses and occultations would be scientifically valuable events. Fortuitously, these mutual events began occurring in 1985 and ended in 1990. These events (described in Section 3) yielded a wealth of data on both Pluto and Charon. Searches for other satellites of Pluto were made in the early and mid-1990s by ground-based observatories and Hubble Space Telescope (HST) images obtained for other Pluto studies, but no moons were detected.

In May 2005, however, a much more sensitive, dedicated satellite search by an HST observing team led by H. A. Weaver and S. A. Stern yielded the detections of two small satellites. These bodies, which were subsequently named Nix and Hydra, orbit in circular orbits in Pluto's equatorial plane, as Charon does. Nix and Hydra orbit Pluto somewhat further out than Charon, with Nix being near 48,700 km from Pluto's center and Hydra near 64,800 km. These orbits are close to or in resonance with Charon, with the Charon:Nix:Hydra periods being very close to or at 1:4:6. Figure 42.1 depicts the satellite orbits of Nix and Hydra in relation to Charon. Based on their observed magnitudes, we can make reasonable assumptions about their albedos yield size estimates of approximately 40–160 km diameters for both. Initial color measurements made with HST indicate both satellites are neutrally reflecting, much like Charon. No compositional results are available.

In 2011 and 2012, two even smaller satellites, later named Styx and Kerberos, were also found by HST.

2. PLUTO'S ORBIT AND SPIN

2.1. Pluto's Heliocentric Orbit

Relative to the eight previously discovered planets, Pluto's orbit is unusually eccentric (eccentricity $e \approx 0.25$), highly inclined (inclination $i \approx 17°$), and large (semimajor axis $a \approx 39.4$ AU). Pluto's orbit period is 248 years, during which the planet ranges from inside Neptune's orbit (Pluto's perihelion is near 29.7 AU) to nearly 48.8 AU. The Pluto–Charon **barycenter** passed its once-every-248-year perihelion at 05.1 ± 0.1 September 1989 UT; this will not occur again until AD 2236.

Current orbit integrations using osculating elements are able to predict Pluto's position to 0.5 arcsec accuracy over timescales of a decade. Pluto's perihelion is closer to the Sun than Neptune's. The large change in Pluto's heliocentric distance as it moves around the Sun causes the surface **insolation** on Pluto and Charon to vary by factors of three, which has important implications for Pluto's atmosphere (see Section 6). Pluto's perihelion lies slightly inside Neptune's orbit.

In the mid-1960s, it was discovered through computer simulations that Pluto's orbit librates in a 2:3 resonance with Neptune, which prevents mutual close approaches between the objects. This discovery has been verified by a series of increasingly longer and more accurate simulations of the outer solar system now exceeding 4×10^9 years. It is likely that Pluto was caught in this resonance and had its orbital eccentricity and inclination amplified to current values as Neptune migrated outward during the clearing of the outer solar system by the giant planets.

Pluto and Neptune can never closely approach one another, owing to this resonance, and the fact that the argument of Pluto's perihelion (i.e. the angle between the perihelion position and the position of its ascending node) librates (i.e. oscillates) about 90° with an amplitude of approximately 23°. This ensures that Pluto is never near perihelion when it is in conjunction with Neptune. Thus, Pluto is "protected" because Neptune passes Pluto's longitude only near Pluto's aphelion, never allowing Neptune and Pluto to come closer than ≈ 17 AU. Indeed, Pluto approaches Uranus more closely than Neptune, with a minimum separation of ≈ 11 AU, but still too far to significantly perturb its orbit.

In the late 1980s, it was discovered that Pluto's orbit exhibits a high degree of sensitivity to initial conditions. This is called "orbital chaos" by modern dynamicists. This discovery of a formal kind of chaos in Pluto's orbit does not imply Pluto undergoes frequent, dramatic changes. However, it does mean that Pluto's position is unpredictable on very long timescales. The timescale for this dynamical unpredictability has been established to be 2×10^7 years by Jack Wisdom, Gerald Sussman, and their coworkers.

2.2. Pluto's Lightcurve, Rotation Period, and Pole Direction

As previously indicated, since the mid-1950s Pluto's photometric brightness has been known to vary regularly with a period of about 6.387 days; more precisely, this period is 6.387223 days. Despite Pluto's faintness as seen from Earth, its period was easily determined using photoelectric techniques because the planet displays a large lightcurve amplitude, 0.35 magnitudes at visible wavelengths, which is equivalent to 38%.

From at least 1955 until about 1990, it has also been known that Pluto's lightcurve increases in its amplitude with time but that trend has since reversed. Although the 6.387223 day period is identical to Charon's orbit period, Charon's photometric contribution is too small to account for the lightcurve's amplitude. This in turn implies that the lightcurve's structure is caused by surface features on Pluto. Figure 42.2 shows the shape of the combined Pluto–Charon lightcurve and its evolution over the past few decades.

The first study of Pluto's polar obliquity (or tilt relative to its orbit plane) was reported in 1973. By assuming that the variation of the lightcurve amplitude from the 1950s to the early 1970s was caused by a change in the aspect angle from which we see Pluto's spin vector from Earth, it was then determined that Pluto has a high obliquity (i.e. $90° \pm 40°$). In 1983, additional observations allowed the obliquity to be refined to $118.5° \pm 4°$. Even more recently, the results of the Pluto–Charon mutual events (or eclipses, see following discussion) have given a very accurate value of $122° \pm 1°$.

It is important to note, however, that torques on the Pluto–Charon pair cause Pluto's obliquity to oscillate between $\sim 105°$ and $\sim 130°$ with an $\sim 3.7 \times 10^6$ year period. Thus, although Pluto presently reaches perihelion with its pole vector nearly normal to the Sun and roughly coincident with the orbit velocity vector, this configuration is only coincidental. The pole position executes a $360°$ circulation with a 3.7×10^6 year precession period.

2.3. Charon's Orbit and the System Mass

The discovery that Charon orbits Pluto with a period equal to Pluto's rotation period immediately implied the pair has reached spin-orbit synchronicity. This is an unprecedented situation among the planets in the solar system.

Table 42.2 gives a solution to Charon's orbital elements obtained from various data. This fit relies on a semimajor axis determination of $a = 19,636 \pm 8$ km derived from ground-based and HST data; it is statistically indistinguishable from ground-based results obtained in the mid-1980s of $a = 19,640 \pm 320$ and of $a = 19,558 \pm 153$ km.

Based on Charon's known orbital period and the 19,636 km semimajor axis, the system's (i.e. combined Pluto + Charon) mass is $1.46 \pm 0.003 \times 10^{25}$ g; this is very small, just 2.4×10^{-3} times Earth's mass.

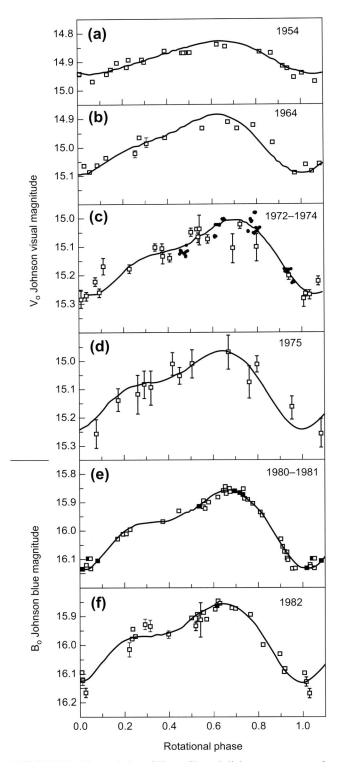

FIGURE 42.2 The evolution of Pluto–Charon's lightcurve over several decades. *Adapted from Marcialis, R. L. (1988).*

TABLE 42.2 Charon's Orbit[1]

Orbital Element	Value
Semimajor axis, a	19636 ± 8 km
Orbital period, P	6.387223 ± 0.00002 days
Eccentricity, e	0.0076 ± 0.003
Inclination, i	$96.2° \pm 0.3°$
Longitudinal perihelion, ω	$222.99° \pm 0.5°$
Mean anomaly, M	$34.84° \pm 0.35°$

[1]*These elements are referred to the epoch 17 January 13.0 UT 1993. See Tholen, D. J. & Buie, M. W. (1997). In S. A. Stern & D. J. Tholen (Eds.), Pluto & Charon. Tucson: Univ. Arizona Press.*

3. THE MUTUAL EVENTS

3.1. Background

After Charon's discovery, the realization that mutual eclipses between Pluto and Charon would soon occur opened up the possibility of studying the Pluto–Charon system with the powerful data analysis techniques developed for eclipses between binary stars. Initial predictions by Leif Andersson indicated that the events could begin as early as 1979. As Charon's orbit pole position was refined, however, the predicted onset date moved to 1983–1986 (this was fortuitous because knowledge of the pole could have changed to indicate that the events had already just ended in the mid-1970s). After a multiyear effort by several groups to detect the onset of these events, the first definitive eclipse detections were made on February 17, 1985, by Richard Binzel at McDonald Observatory and were confirmed during an event 3.2 days later on February 20, 1985, by David Tholen at Mauna Kea. These first, shallow events ($\sim 0.01-0.02\%$ in depth) revealed Pluto and Charon grazing across one another as seen from Earth.

The very existence of these eclipses proved the hypothesis (by 1985 widely accepted) that Charon was in fact a satellite, rather than some incredible topographic high on Pluto. The mutual eclipses persisted until October 1990, and dozens of events were observed. Important results from the 1985–1990 mutual events included reconstructed surface "maps" of Pluto and Charon; individual albedos, colors, and spectra for each object; and improvements in Charon's orbit. First, however, was the opportunity to use event timing to accurately determine the radii of Pluto and Charon.

3.2. Radii and Average Density of Pluto and Charon

Prior to the mutual events, the radii of Pluto and Charon were highly uncertain. Because Pluto and Charon remained unresolved in terrestrial telescopes (their apparent diameters are each <0.1 arcsec), direct measurements of their diameters were not available. A well-observed, near-miss occultation of Pluto in 1965 had constrained Pluto's radius to be <3400 km, but no better observations were available until the mutual events. However, circumstantial evidence that Pluto was smaller than 3400 km was inferred from the combination of Pluto's V astronomical ≈ 14 magnitude and the 1976 discovery of CH_4 frost (see Section 4.3), which exhibits an intrinsically high albedo. The small system mass determined after the discovery of Charon in 1978 strengthened this inference, but Pluto's radius was still uncertain within the bounds 900–2200 km.

The first concrete data to remedy the situation came when a fortuitous **stellar occultation** by Charon was observed on April 07, 1980. The 50 s length of the star's disappearance, observed by a 1-m telescope at Sutherland, South Africa, gave a value for Charon's radius of 605 ± 20 km. Improved results from subsequent stellar occultations yield 603.5 ± 3 km.

As noted earlier, accurate radius measurements for Pluto resulted from both stellar occultations and from fits of mutual event lightcurves, yielding solutions between 1150 and 1200 km. The range of uncertainty in Pluto's radius, which is significantly larger than in Charon's, results primarily from uncertainties in Pluto's atmospheric depth.

The two striking implications of the small radii and comparable masses of Pluto and Charon are (1) that Pluto is a very small planet—even smaller than the seven largest planetary satellites (the Moon, Io, Europa, Ganymede, Callisto, Titan, and Triton), and (2) that Pluto and Charon form the only known example of a binary planet (with the system barycenter outside of Pluto). Based on the radii and the total mass of the pair, it is possible to derive an average density of 2.03 ± 0.06 g/cm^3, where the error bar is dominated by the uncertainty in the radius of Pluto. Any density of 1.8 g/cm^3 or higher implies that the system is compositionally dominated by rocky material, probably hydrated chondrites, as opposed to ices. This result and its implications will be discussed in more detail in Section 5.

4. PLUTO'S SURFACE PROPERTIES AND APPEARANCE

Pluto's surface properties have been studied since the 1950s. Photometric, spectroscopic, and polarimetric techniques have been applied, and the explorable wavelength regime has expanded from the ground-based window to the reflected infrared (IR) and the **space ultraviolet** (UV). Thermal-IR and millimeter-wave measurements have also been made.

4.1. Albedo and Color

Two of the most basic photometric parameters one desires to know for any solid body are its albedo and color. Accurate knowledge of Pluto's albedo was obtained only after the onset of the mutual events because until then Pluto's radius was unknown, and there was no definitive way of removing Charon's contribution.

The very first report of eclipse detections revealed a factor of 2 difference in depth between partial eclipses of Charon and Pluto, indicating Pluto's geometric albedo is substantially higher than Charon's. Once the eclipse season was complete, a more complete data set became available for analysis. Comprehensive models for the analysis of mutual event lightcurve data simultaneously solve for the individual radii of Pluto and Charon, the individual albedos, and Charon's orbital elements. The modeling of these parameters is complicated by solar phase angle effects, the presence of shadows during eclipse events, and instrumental and timing uncertainties. To derive the albedo lightcurve for Pluto alone, Pluto's albedo at the longitude of the total superior eclipses (in which Charon was completely hidden) must first be determined. Albedos at other rotational epochs are then derived from this anchor point, assuming Charon's **rotational lightcurve** contributes only a small constant to the combined Pluto + Charon lightcurve. The assumption that a constant Charon contribution can be removed is not unreasonable, because (1) its geometric cross-section is small (one-fourth of Pluto's), and (2) its eclipsed hemisphere has a geometric albedo only about 50–60% of Pluto's. However, HST observations have shown that Charon does vary somewhat in brightness ($\approx 8\%$) as it rotates on its axis. Analysis of a large set of mutual event data in the way just described has found that Pluto's maximum, disk-integrated, B-bandpass (~ 4360 Å) geometric albedo is 0.61. Rotational variations cause this albedo to range from values as low as 0.44 to values as high as 0.61 as Pluto rotates.

Information on Pluto's color comes from both photometry and the mutual events. As described in Section 1, Pluto's visible-bandpass color slope has been known to be red since the 1950s. Analysis of premutual event photometry yields B-V and U-B color differences of 0.84 and 0.31, respectively, for Pluto + Charon. Eclipse data have revealed that the B-V color of Pluto itself is very close to 0.85 astronomical magnitudes. By comparison, this color is much less red than the **refractory** surfaces of Mars (B-V = 1.36) and Io (B-V = 1.17), and slightly redder than its closest analog in the solar system, Triton (B-V = 0.72).

4.2. Solar Phase Curve

The photometric behavior of a planet or satellite as it changes in brightness on approach to opposition can be used to derive surface scattering properties, and therefore its microphysical properties. Knowledge of the complete solar phase curve is also required to transform geometric albedos into bolometric Bond albedos. HST observations in the 1990s gave linear phase coefficients for Pluto and Charon of 0.029 ± 0.001 magnitudes/deg and 0.866 ± 0.008 magnitudes/deg, respectively, though non-linear effects have now been detected.

Pluto's maximum solar phase angle ($\phi_{\max}$) as seen from Earth is just $\approx 1.9°$. Therefore, no measurements of the large-angle scattering behavior have been possible until recently with New Horizons. Without measurements at large phase angles, no definitive determination of Pluto's phase integral q or Bond albedo A can be made. However, what is really needed are flyby spacecraft measurements of Pluto at high phase angles. For the present, the best available phase integral to use for Pluto is probably Triton's (Pluto and Triton also have similar linear phase coefficients). Triton's q has been measured by Voyager, giving $q = 1.2$ (at green wavelengths) to 1.5 (at violet wavelengths). If Pluto is similar, then its surface may have Bond albedos ranging from 0.3 to 0.7.

4.3. Surface Composition

Progress in understanding Pluto's surface composition required the development of sensitive detectors capable of making moderate spectral resolution measurements in the IR, where most surface ices show diagnostic spectral absorptions. Although this technology began to be widely exploited as early as the 1950s in planetary science, Pluto's faintness (e.g. 700 times fainter than the Jovian Galilean satellites) delayed compositional discoveries about it until the mid-1970s.

The first identification of a surface constituent on Pluto was the discovery by Dale Cruikshank, Carl Pilcher, and David Morrison in 1976 of CH_4 ice absorptions between 1 and 2 µm (a wavelength of 1 µm = 10,000 Å). Cruikshank et al. made this discovery using IR photometers equipped with customized, compositionally diagnostic filters. In their report, Cruikshank et al. also presented evidence against the presence of strong H_2O and NH_3 absorptions in Pluto's spectrum. Confirmation of the methane detection came in 1978 and 1979 when both additional CH_4 absorption bands and true IR spectra of Pluto became available.

In mid-1992, another breakthrough occurred when Toby Owen, Dale Cruikshank, and other colleagues made observations using a new, state-of-the-art IR spectrometer at the UK Infrared Telescope on Mauna Kea. These data revealed the presence of both N_2 and CO ices on Pluto. These molecules are much harder to detect than methane because they produce much weaker spectral features. Their presence on Pluto indicates the surface is chemically more heterogeneous, and more interesting than had previously been thought. Because N_2 and CO are orders of magnitude more **volatile** (i.e. have higher vapor pressures) than CH_4, their presence also implies they play a highly important role in Pluto's annual atmospheric cycle. Abundance inversions of Pluto reflectance spectra make clear that N_2 dominates the composition of much of Pluto's surface, with CO and CH_2 being trace constituents.

In 2006, ethane (C_2H_6) was detected on Pluto's surface. This and other hydrocarbons and nitriles had long been predicted to reside on Pluto as a result of photochemical and radiological processing of Pluto's surface ices and atmosphere.

Rotationally resolved spectra of Pluto's CH_4 absorption bands have been reported by a number of groups. Their studies showed that Pluto's methane is present at all rotational epochs, but the band depths are correlated with the lightcurve so that the minimum absorption occurs at minimum light. Mutual event spectroscopy has now demonstrated that Charon is not the cause of this variation, since Charon's surface is devoid of detectable CH_4 absorptions (see Section 6). This important discovery suggests that Pluto's dark regions could contain reaction products resulting from the photochemical or radiological conversion of methane and N_2 to complex nitriles and higher hydrocarbons.

We thus have the following basic picture of Pluto's surface composition: CH_4 appears rotationally ubiquitous, but with its surface coverage more widespread in regions of high albedo. In many areas, the methane is dissolved in a matrix of other ices, but in some locations the CH_4 is seen as pure ice. CO and N_2 have also been detected. In the bright areas of the planet where these ices are thought to mainly be located, N_2 dominates the surface abundance, and the CO is more abundant than the previously known (but more spectroscopically detectable) CH_4. Ethane, a by-product of CH_4 chemistry, was detected in 2006. Pluto's strong lightcurve and red color demonstrate that other widespread, probably nonvolatile surface constituents exist. This may be due to either rocky material, or hydrocarbons resulting from radiation processing of the CH_4 due to long-term exposure to UV sunlight, or both. Whether the volatile frost we are seeing is a surface veneer or the major component of Pluto's crust has not yet been established.

4.4. Surface Temperature

Results from the Infrared Astronomical Satellite (IRAS) indicated that Pluto's perihelion-epoch surface temperature was in the range of 55–60 K, close to that expected in radiative equilibrium with solar insolation. However, it has subsequently become appreciated that the situation on Pluto's surface is more complicated.

One line of evidence for this conclusion comes from millimeter-wave measurements of Pluto's Rayleigh–Jeans blackbody spectrum. Such measurements, reported first by Wilhelm Altenhoff and collaborators, and then later by Alan Stern, Michel Festou, and David Weintraub, and independently confirmed by David Jewitt, indicate that a significant fraction of Pluto's surface is significantly colder than 60 K, most likely in the range 35–42 K. A second line of evidence came in 1994 from high-resolution spectroscopy of the temperature-sensitive 2.15 µm N_2 ice absorption band, which Kimberley Tryka and her coworkers found indicates a surface temperature of about 40 K for the widespread nitrogen ices on Pluto. As described by Stern et al. in 1993, although the surface pressure of N_2 is not well known, it must be less than ≈ 60 µbar. This is consistent with an N_2 ice temperature of ≈ 40 K, assuming vapor pressure equilibrium between the N_2 ice and the atmosphere. This, combined with the IRAS measurements, led to the conclusion that Pluto's surface must exhibit both warm and cold regions. This was subsequently confirmed by rotationally resolved studies of Pluto's thermal emission spectrum by the Infrared Space Observatory and the Spitzer Infrared Space Telescope Facility. These space telescopes also revealed that Pluto's coldest regions are correlated with bright surface units, and that the warmer regions are correlated with darker surface units with lower abundances of sublimating ices.

It is now well established that Pluto's surface temperature varies from place to place on the surface, with ≈ 40 K regions where N_2 ice is sublimating and $\approx 55-60$ K regions where N_2 ice is not present in great quantities. The strong temperature contrasts across Pluto's surface imply

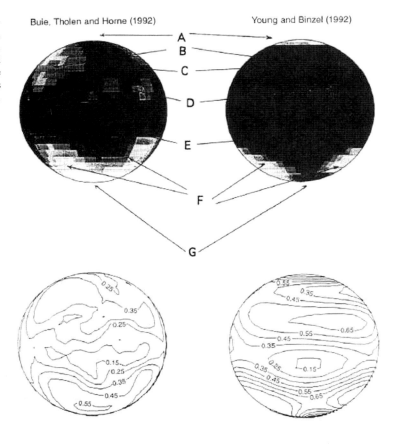

FIGURE 42.3 Two maps of Pluto's Charon-facing hemisphere. The map on the left was derived by M. Buie, K. Horne, and D. Tholen using both mutual event and lightcurve data. The map on the right was derived by E. Young and R. Binzel from their mutual event data. Although the fine details of these maps differ, their gross similarities are striking. Here north is up for the right-hand rule convention. *See Binzel, R. P. et al. (1997).*

strong wind speeds and significant lateral transport of material across the surface.

4.5. Surface Appearance and Markings

Because Pluto is less than 0.1 arcsec across as seen from Earth, its disk could not be resolved until the advent of the HST. However, evidence for surface markings has been available since the mid-1950s, when lightcurve modulation was first detected. Because Pluto is large enough to be essentially spherical (and indeed, mutual event and stellar occultation data show it actually is), the distinct variation in this lightcurve must be related to large-scale albedo features.

From the lightcurve in Figure 42.2, it can be seen that Pluto's surface must contain at least three major longitudinal provinces. Information on the latitudinal distribution of albedo can be gained by observing the evolution of this lightcurve as Pluto moves around its orbit while the pole position remains inertially fixed, assuming, of course, that the surface albedo distribution is time invariant.

The most complete 1990s mapping products obtained from photometric data inversions (variously using rotational lightcurves and mutual event lightcurves) were obtained by two teams. The first team, led by Marc Buie of Lowell Observatory, has used both mutual event lightcurves and rotational lightcurve data compiled from 1954 to 1986 to compute a complete map of Pluto. The second group, consisting of Richard Binzel and Eliot Young, of Massachusetts Institute of Technology (MIT) and Southwest Research Institute (SwRI), numerically fit a spherical harmonic series to each element of a finite element grid using the Charon transit mutual event lightcurve data as the model input. Because Young and Binzel used only mutual event data, their map is limited to the hemisphere of Pluto that Charon eclipses. Because the two groups used different data sets and different numerical techniques, their results are complementary and serve to check one another on the Charon-facing hemisphere they share in common.

These two maps are shown in Figure 42.3. There are differences between the two maps, but it must be remembered that each map has intrinsic noise. The common features of these maps are (1) a very bright south polar cap, (2) a dark band over midsouthern latitudes, (3) a bright band over midnorthern latitudes, (4) a dark band at high northern latitudes, and (5) a northern polar region that is as bright as the southern cap. Later results, including some color information, were subsequently obtained by Eliot Young and colleagues.

In 1990, HST imaged Pluto, but owing to its then-severe optical aberrations, these images, obtained by R. Albrecht

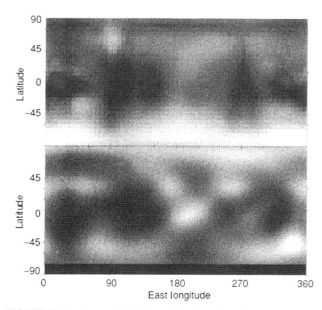

FIGURE 42.4 A map derived from direct imaging of Pluto using HST images made in 1994. Again, north is up. *Adapted from Stern et al. (1997).*

and a team of collaborators, cleanly separated Pluto and Charon, but did not reveal significant details about the surface of Pluto. After HST was repaired by an astronaut crew in late 1993, its optics were good enough to resolve crude details on Pluto's surface. And in mid-1994, it obtained the first actual images of Pluto that revealed significant details about Pluto's surface. These images were made by Alan Stern, Marc Buie, and Laurence Trafton using the Faint Object Camera of the HST. The 20-image HST data set is longitudinally complete and rotationally resolved and obtained at both blue and UV wavelengths. The various images that HST obtained were combined to make blue and UV maps of the planet, such as the one shown in Figure 42.4. The HST images and derived maps reveal that Pluto has (1) a highly variegated surface, (2) extensive, bright, asymmetric polar regions, (3) large midlatitude and equatorial spots, and (4) possible linear features hundreds of kilometers in extent. The dynamic range of albedo features across the planet detected at the Faint Object Camera's resolution in both the 410 and 278 nm bandpasses exceeds 5:1. New HST images were obtained in 2002 by a team led by Marc Buie using HST's Advanced Camera for Surveys, but they do not have better resolution.

5. PLUTO'S INTERIOR AND BULK COMPOSITION

5.1. Density

To determine the separate densities of Pluto and Charon, one must either obtain precise astrometric measurements that detect the barycentric wobble between Pluto and Charon or use orbit solutions for Pluto's small satellites. Since 1992, both HST and ground-based measurements have been gathered to address the mass ratio, and therefore the relative masses and densities of Pluto and Charon. These are very difficult measurements. A good density determination for Pluto due to Buie and coworkers, who analyzed the orbits of Nix and Hydra in a 2006 publication gave 2.03 ± 0.06 g/cm^3 for a reference radius of 1153 km.

5.2. Bulk Composition and Internal Structure

The 1980s discovery that the Pluto–Charon system's average density is near 2 g/cm^3 was a major surprise resulting from the mutual events. Many scientific papers had previously predicted values closer to the density of water ice (~ 1 g/cm^3), or even lower. Thus, contrary to earlier thinking, the Pluto–Charon pair is known to be mass dominated by rocky material. Based on this information, a three-component model for Pluto's bulk composition and internal structure can be derived. In such a model, Pluto's bulk density is assumed to consist of three of the most common condensates in the outer solar system: water ice ($\rho = 1.00$ g/cm^3), "rock" ($2.8 < \rho < 3.5$ g/cm^3, depending upon its degree of hydration), and methane ice ($\rho = 0.53$ g/cm^3).

From three-component models, it is believed that Pluto's rock fraction is in the range of 60–80%, with preferred values close to 70%. By comparison, the large (e.g. $R > 500$ km) icy satellites of Jupiter, Saturn, and Uranus have typical rock fractions in the range 50–60% by mass. Only Io, Europa, and Triton rival Pluto in terms of their computed rock content. Pluto's high rock (i.e. nonvolatile) mass fraction is in contrast to the $\approx 1:1$ rock:ice ratio predicted for objects formed from solar nebula material according to many nebular chemistry models and our present-day understanding of the nebular C/O ratio. This high rock fraction indicates that the nebular material from which Pluto formed was CO rich rather than CH$_4$ rich. As such, roughly half of the available nebular oxygen should have gone into CO, rather than H$_2$O formation, which in turn would lead to a high rock:ice ratio.

There are two possible ways out of the apparent nebular chemistry dilemma imposed by Pluto's high rock fraction. One is that Pluto's minimum estimated radius of 1150 km may be too small; a value near 1200 km, as suggested by some stellar occultation models, would solve the problem. Alternatively, William McKinnon and the late Damon Simonelli independently suggested that a giant impact may have induced volatile loss from an already differentiated Pluto, which may have raised Pluto's rock fraction somewhat (perhaps 20%) to reach its present value. As we

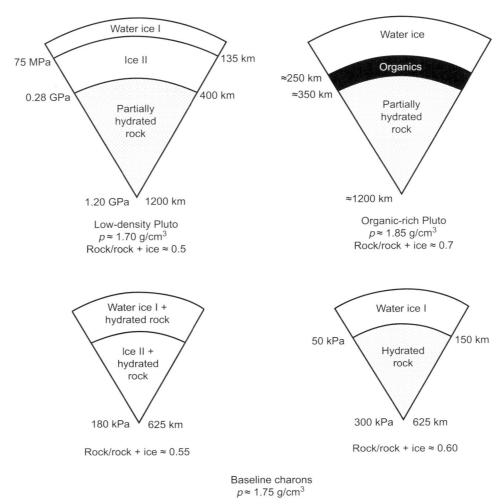

FIGURE 42.5 Typical interior structural models for Pluto and Charon. *Adapted from. McKinnon, W. B. et al. (1997).*

discuss in Section 8, such an impact is thought to be responsible for the formation of Pluto's satellite system.

The gross internal thermal structure of Pluto depends on several factors, virtually all of which are uncertain. These include material viscosities in the interior, the internal convection state, the actual rock fraction and radioisotope content, and the internal density distribution (i.e. most fundamentally, the differentiation state). It would appear likely that Pluto's deep interior reaches temperatures of at least 100–200 K, but not much higher. Whether or not Pluto is warm enough to exhibit convection in its ice mantle depends on both the internal thermal structure and the radial location of water ice in its interior.

Based on the results just given and laboratory equations of state, Pluto's central pressure can be estimated to lie between 0.6 and 0.9 GPa (gigapascals) if the planet is undifferentiated, or 1.1–1.4 GPa if differentiation has occurred. As such, the high-pressure water ice phase Ice VI is expected in the deep interior if the planet has not differentiated. If differentiation has occurred, as is likely, then a higher pressure form of water ice called Ice II may be present, but only near the base of the convection layer. If Pluto did differentiate, then its gross internal structure may be represented by a model like that shown in Figure 42.5.

6. PLUTO'S ATMOSPHERE

6.1. Atmospheric Composition

The existence of an atmosphere on Pluto was strongly suspected after the discovery of methane on its surface in 1976, largely because at the predicted surface temperatures ($\sim$40–60 K), sufficient methane vapor pressure should be obtained to constitute a significant atmosphere. This circumstantial argument was supported by the high reflectivity of Pluto's surface, which suggested some kind of resurfacing—most plausibly due to volatile laundering through an orbitally cyclic atmosphere. Still, however, there was no definitive evidence for an atmosphere until the late 1980s.

The formal proof of Pluto's atmosphere came from the occultation of a 12th magnitude star by Pluto in 1988, by providing the first direct observational evidence for an atmosphere. The best measurements of the occultation were obtained by Robert Millis and James Elliot, and their various MIT, Lowell Observatory, and Australian collaborators. These teams used both National Aeronautics and Space Administration's mobile, Kuiper Airborne Observatory (which contained a 36-inch-diameter telescope), and ground-based telescopes to observe the occultation event. They discovered that light from the star was diminished far more gradually than it would be from an airless body. The apparent extinction of starlight observed during the occultation was caused by atmospheric refraction (i.e. the degree of bending of the starlight by the atmosphere), which varies with height. The rate at which the refractivity of the atmosphere varies with altitude depends on the ratio of atmospheric temperature (T) to atmospheric mean molecular weight (m). The 1989 Pluto occultation data implied $T/m = 3.7 \pm 0.7$ K/g at and above an altitude of 1215 km. If the atmosphere were composed entirely of methane ($m = 16$ g/mole), the implied atmospheric temperature would be 60 K, whereas an N_2 or CO atmosphere ($m = 28$ g/mole) would be at a temperature near 106 K.

From the stellar occultation data alone it was impossible to separately determine the mean atmospheric molecular weight and temperature of Pluto's atmosphere. However, theoretical calculations of the atmospheric temperature made by Roger Yelle and Jonathan Lunine of the University of Arizona indicated a value of 106 K in the upper atmosphere, under a variety of assumed compositions. This is relatively high compared with the surface temperature ($\sim$40–55 K) because the efficiency at which the atmosphere radiates and cools is very small. An upper atmospheric temperature near 106 K implies that the atmospheric mean molecular weight is close to 28 g/mole. This is consistent with an atmosphere dominated by N_2 or CO gas, with trace amounts of other species.

The detection of N_2 ice absorption features on Pluto's surface (see Section 4), coupled with the discovery by Voyager 2 that Triton's atmosphere also consists predominantly of N_2 and only a trace of CO, suggests that Pluto's atmosphere is likely to be N_2 dominated. Nevertheless, if the high-temperature (106 K) atmospheric model is correct, then at least a few percent methane is thought to be required because methane (which is efficient at atmospheric heating) is thought to be responsible for the elevated atmospheric temperatures.

A nitrogen-dominated atmosphere with only a minor amount of methane was significantly strengthened in 1994 when Leslie Young and colleagues at MIT detected CH_4 gas in Pluto's atmosphere for the first time. This discovery, which was made possible by sensitive, high-resolution IR spectroscopy of the 2.3 μm CH_4 band system, indicated a total methane mixing ratio of <1%, and perhaps as little as 0.1% in the atmosphere. Subsequent high-resolution observations by Young and colleagues revealed that the CO abundance in Pluto's atmosphere must also be very low.

6.2. Atmospheric Structure

The 1988 occultation data exhibited interesting behavior at altitudes below 1215 km, as shown in Figure 42.6. The starlight, which was decreasing gradually at higher altitudes, dropped suddenly to a value close to zero below this level; this is called lightcurve steepening. The drop is still not as sudden as would be expected from the setting of a star behind the limb of an airless planet, however. Two possible explanations have been proposed for this change.

In one model, the steepening was caused by the presence of aerosol hazes in the lower atmosphere. (Condensation clouds can be ruled out as an explanation for the aerosol layer because of the temperature structure of the atmosphere.) Because reproducible albedo features

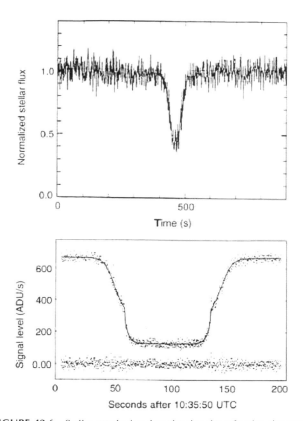

FIGURE 42.6 Stellar occultation data showing the refractive signature of Pluto's atmosphere and the steepening of the lightcurve around the half-light level that is discussed in the article. The upper panel is a ground-based data product; the lower panel was obtained from the Kuiper Airborne Observatory. *Adapted from Elliot et al. (1989).*

have been seen on Pluto's surface, any such aerosol layer must be transparent when viewed from above, but relatively opaque when viewed horizontally. The aerosols must also extend around most of the planet since the steepening of the occultation lightcurve was seen in both immersion and emersion. It has been suggested that the aerosols could be "photochemical smog" similar to the aerosols discovered on Titan and Triton (and a distant cousin to the air pollution in the industrial basins on Earth, such as Los Angeles).

In a second model, the sudden drop in the brightness of starlight below 1215 km was caused by a change in the vertical thermal structure of the atmosphere near the half-light level. Such a gradient is not unexpected from theoretical modeling (see earlier discussion) because atmospheric temperatures are expected to be higher than surface temperatures. Changes in atmospheric temperature cause a variation of refractivity with height in the atmosphere that could be manifested as the accelerated diminution of starlight seen in the occultation.

The haze layer and temperature gradient explanations imply differences in the way that the color of starlight changes during an occultation. Future occultations may help decide between the two explanations if simultaneous observations can be made at two or more well-separated wavelengths. If the temperature gradient explanation is correct, Pluto's surface radius is likely near 1206 ± 11 km. If the haze layer explanation is correct, Pluto's surface radius is more difficult to determine, but it is probably closer to 1180 km.

In either case, the occultation implies a radius that is a few percentage larger than the mutual event solution (1151 km); in the case of the haze model, the radius cannot be much less than 1180 km or else the haze would be so thick as to completely obscure the surface. Clearly, there is a discrepancy between the radii determined from the occultation and those derived from the mutual events, which future research will have to resolve.

The subsequent well-observed occultation of a star by Pluto occurred on July 20, 2002. Both large fixed telescopes and small portable instruments observed the event. Fortuitously, yet another event occurred on August 21, 2002, which was successfully observed from large telescopes on Mauna Kea. From these events, it was determined that the "kink" or "knee" seen in the 1988 data is largely absent from the 2002 data, implying that large changes in Pluto's atmospheric thermal structure, or its haze profile, or both, occurred during the intervening interval.

Further analysis of the data reveals that the pressure in Pluto's atmosphere more than doubled between 1988 and 2002. This is likely due to pressure fluctuations associated with seasonal change, and may even be related to instabilities in the atmosphere prior to complete atmospheric collapse. Further observations will be required to sort this out.

Another occultation was observed on June 12, 2006. This event showed that the lightcurve kink near the half-light level remained less distinct than in 1988 and that the turbulence level in Pluto's lower atmosphere had increased. Fortunately, Pluto is still moving through the dense star fields of Sagittarius, and additional events have now been observed annually through 2014.

6.3. Atmospheric Escape

A particularly interesting feature of Pluto's atmosphere is the very rapid rate at which it escapes to space. Because of Pluto's low mass and consequently weak gravitational binding energy, combined with the 100 K gas temperature in Pluto's upper atmosphere, sufficiently energetic molecules at the top of the atmosphere are able to escape the gravitational pull entirely. This can result in a condition called **hydrodynamic escape**, in which the high-altitude atmosphere achieves an internal thermal energy greater than the planetary gravitational potential energy acting on the atmosphere.

The time-averaged rate of escape from Pluto's atmosphere is likely to be of order $1-10 \times 10^{27}$ molecules/second. This corresponds to a total loss of up to several kilometers of material from the surface over the age of the solar system.

Escape rate estimates also indicate that the present escape rate may be so high that Pluto's tenuous atmosphere may be lost to the escape process (thus requiring replenishment from sublimating surface ices) on timescales possibly as short as a few hundred years. Relatively speaking, the atmosphere of Pluto is escaping at a rate far greater than any other planetary atmosphere in the solar system.

Another interesting feature of Pluto's atmosphere is its strong orbital variability. This is driven by the fact that the strength of solar heating varies by a factor of almost 4 around Pluto's orbit, which in turn causes the vapor pressures of N_2, CO, and CH_4 to vary by factors of hundreds to thousands. Therefore, unlike any other planet, Pluto's atmosphere is thought to be essentially seasonal, with the perihelion pressure being many times the aphelion pressure. Indeed, some models predict that Pluto's atmosphere will largely condense onto the surface, a condition called atmospheric collapse: other models disagree.

7. CHARON

As previously described, Pluto's largest satellite, Charon, was discovered in 1978. Charon's radius of ≈ 604 km is about half of Pluto's, giving a Charon:Pluto mass ratio near 0.12. By comparison, typical satellite:planet mass ratios are

1000:1 or greater, and even the mass ratio of the Moon to the Earth is only 81:1.

The Pluto–Charon mutual event observations resulted in several key discoveries. These included the fact (1) that Charon's average visible surface albedo is 30–35%, much lower than Pluto's, and (2) that Charon's visible surface color is quite neutral, unlike Pluto's clearly reddish tint. Another major set of advances that resulted from the eclipse events was the first set of constraints on Charon's basic surface composition. These came from the subtraction of spectra made just prior to eclipse events from those made when Charon was completely hidden behind Pluto. The resulting "net" spectrum thus contains the Charon-only signal. As shown in Figure 42.7, this technique has been applied both in the visible (0.55–1.0 μm) and IR (1–2.5 μm) bandpasses. The visible light data show that Charon's surface does not display the prominent CH_4 absorption bands that Pluto does, indicating that Charon's surface has little or (more likely) no substantial methane on it.

Additionally, there is no evidence for strong absorptions due to a number of other possible surface frosts, including CO, CO_2, H_2S, N_2, or NH_4HS, on Charon. The IR spectra of Charon show that Charon does, however, display clear evidence of water ice absorptions, which Pluto does not. It is tempting to speculate (as some authors have) that Charon may have lost its volatiles through the escape of a primordial atmosphere or by heating resulting from its formation in a giant impact.

Since the launch of HST, it has been possible to routinely separate Charon's light from Pluto's, and to learn Charon's phase coefficient, UV albedo, and rotational lightcurve. Most notably among these, Marc Buie and Dave Tholen have determined that Charon displays a small but significant lightcurve variation near 8% as it rotates on its axis.

Because the major identified surface constituent of Charon is water ice, which is not volatile at the expected 50–60 K surface radiative equilibrium temperature at perihelion, one does not expect Charon to have an atmosphere. The fact that CH_4 is not present on the surface supports this expectation. However, absence of evidence is not the same as evidence of absence. One published interpretation of the 1980 Charon stellar occultation claims there is some evidence for a weak atmospheric refraction signal. To definitively resolve the issue of Charon's atmosphere, either a better observed stellar occultation event or a spacecraft flyby is required.

In 2001, groups led by Mike Brown of Caltech and Will Grundy of Lowell Observatory used IR ground-based telescopes to find spectroscopic evidence for both crystalline water ice and ammonia (NH_3) or ammonium hydrates on Charon's surface. If these identifications are correct, they imply the possibility of recent geologic activity on Charon.

Table 42.3 compares some basic facts about Pluto and Charon.

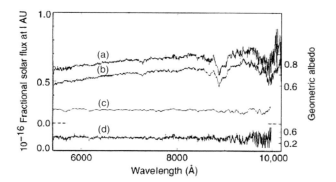
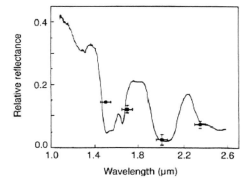

FIGURE 42.7 Pluto and Charon spectra. Top panel shows spectra of (a) Pluto + Charon made prior to eclipse; (b) Pluto-only after second contact with Charon hidden; (c) Charon-only smoothed to 80 Å resolution resulting from the subtraction of (b) from (a); and (d) the raw Charon-only spectrum resulting from the subtraction of (b) from (a). Notice that the strong methane absorption bands present in Pluto's spectrum are not detected in the Charon-only spectrum. (*Adapted from Fink and DiSanti, 1988.*) Bottom panel shows Marcialis et al.'s (1987) detection of water ice in Charon's reflectance spectrum (data points) against a laboratory spectrum of water ice at 55 K. *Adapted from Marcialis et al. (1987).*

TABLE 42.3 Pluto and Charon Comparison

Parameter	Pluto	Charon
Rotation period	6.387223 days	6.387223 days
Radius	1150–1200 km	602–606 km
Density	≈2.1 gm/cm³	≈1.3 gm/cm³
Perihelion, V_0	13.6 magnitude	15.5 magnitude
Mean B geometric albedo	0.55	0.38
Rotational lightcurve	38%	8%
B-V color	0.85 magnitude	0.70 magnitude
V-I color	0.84 magnitude	0.70 magnitude
Known surface ices	CH_4, N_2, CO	H_2O, NH_3OH
Atmosphere	Confirmed	None detected

8. THE ORIGIN OF PLUTO'S SATELLITE SYSTEM

In the past few years, much progress has been made in understanding Pluto's likely origin and its context in the outer solar system. This work began with theoretical consideration in the late 1980s and early 1990s, and was advanced considerably by the discovery of numerous 100- to 1600-km-diameter objects in the Kuiper Belt, where Pluto also resides.

Any scenario for the origin of Pluto must of course provide a self-consistent explanation for the major attributes of the Pluto–Charon system. These include (1) the existence of the exceptionally low, ~0.12 satellite-to-planet mass ratio of Pluto:Charon; (2) the synchronicity of Pluto's rotation period with Charon's orbit period; (3) Pluto's inclined, elliptical, Neptune-resonant orbit; (4) the high axial obliquity of Pluto's spin axis and Charon's apparent alignment to it; (5) Pluto's small mass ($\sim 10^{-4}$ of Uranus's and Neptune's); (6) Pluto's high rock content—the highest among all the outer planets and their major satellites; and (7) the dichotomous surface compositions of Pluto and Charon. This formidable list of constraints on origin scenarios is very clearly dominated by Charon's presence, the unique dynamical state of the binary, and the low mass of Pluto/Charon compared to other planets.

8.1. The Formation of Pluto's Satellite System

Several scenarios have been examined for the origin of the Pluto–Charon system. These include coaccretion in the solar nebula, mutual capture via an impact between proto-Pluto and proto-Charon, and rotational fission. Gravitational capture of Pluto by Charon without physical contact is not dynamically viable. The formation of Pluto and Charon together in a subnebular collapse is not considered realistic because of their small size; standard planetary formation theory suggests bodies in the Pluto and Charon size class formed via solid-body accretion of planetesimals. Similarly, the rotational fission hypothesis is unlikely to be correct because the Pluto–Charon system has too much angular momentum per unit mass to have once been a single body.

The more likely explanation for the origin of the Pluto system is an inelastic collision between two bodies, Pluto and proto-Charon, which were on intersecting heliocentric orbits. A similar scenario has been proposed for the origin of the Earth–Moon binary, based in part on its relatively high mass ratio (81:1) and high specific angular momentum. In the collision theory, Pluto and the Charon-impactor formed independently by the accumulation of small planetesimals and then suffered a chance collision that dissipated enough energy to permit binary formation.

An important qualitative difference between the Pluto–Charon and Earth–Moon giant impacts is that the relative collision velocities, and hence the impact energies of the Pluto–Charon event, were much smaller. This enormously reduced the thermal consequences of the collision. Thus, the Earth may have been left molten by the Mars-sized impactor necessary to have created the Moon, whereas the proto-Charon impactor would probably have raised Pluto's global mean temperature by no more than 50–75 K. This would have been insufficient to melt either body, but may have been sufficient to induce the internal differentiation of either. It would have also produced a substantial short-lived, hot, volatile atmosphere with intrinsically high escape rates. Such an escaping atmosphere could have interacted with the Charon-forming orbital debris, and also perhaps affected Pluto's present-day volatile content.

Until recently, only scaling calculations showing the plausibility of the giant impact hypothesis has been performed, and it was accepted largely because it is the only scenario that remains viable given the various constraints—most particularly the high specific angular momentum of the binary. In 2005, however, Robin Canup of the Southwest Research Institute published the first detailed giant impact simulations demonstrating the viability of Pluto–Charon formation owing to the collision of Pluto with another large body. Canup's work further demonstrated that the most promising candidate impacts involved an oblique collision by an impactor with 30–100% of Pluto's mass, approaching at a relative speed up to 1 km/s.

The discovery of Pluto's small satellites yielded additional support for the giant impact hypothesis. This support comes in two forms: the fact that all three satellites orbit in a single orbital plane and the near or perfect orbital period resonance of the three. The orbital coplanarity would be unlikely for other satellite formation mechanisms like capture, but naturally result from the giant impact scenario. The orbital resonance line of evidence naturally suggests the three bodies were together caught up in the outward tidal migration of Charon following its formation closer in.

8.2. The Origin of Pluto Itself

The presence of volatile ices, including methane, nitrogen, and carbon monoxide on Pluto, and water and other ices on Charon, argues strongly for their formation in the outer solar system. The average density and consequent high rock content of these two bodies also argues for formation from the outer solar nebula, rather than from planetary subnebula material. As described earlier, it is thought that the two objects (or more precisely Pluto and a Charon progenitor) formed independently and subsequently collided, thus forming the binary either through direct, inelastic capture or through the accretion of Charon from debris put in orbit around Pluto by the impact.

The first widely discussed theory for Pluto's origin was R.A. Lyttleton's (1936) suggestion, which was based on the fact that Pluto's orbit is Neptune-crossing. In Lyttleton's well-remembered scenario, Pluto was formerly a satellite of Neptune, ejected via a close encounter between itself and the satellite Triton. According to Lyttleton, this encounter also reversed the orbit of Triton. Variants on the "origin as a former satellite of Neptune" hypothesis were later proposed. However, all these scenarios were dealt a serious blow by the discovery of Charon, which severely complicates the Pluto-ejection problem by requiring either (1) Charon to also be ejected from the Neptune system in such a way that it enters orbit around Pluto, or (2) Charon to be formed far beyond Neptune where Pluto currently orbits and then captured into orbit around Pluto (presumably by a collision).

Other strong objections to scenarios like Lyttleton's also exist. First among these is the fact that any object ejected from orbit around Neptune would be Neptune-crossing and therefore subject to either accretion or rapid dynamical demise. It is implausible that such an object would be transferred to the observed 2:3 Neptune:Pluto resonance, because stable 2:3 libration orbits are dynamically disconnected in orbital phase space from orbits intersecting Neptune. Further, because Pluto is less massive than Triton by about a factor of 2, it is impossible for Pluto to reverse Triton's orbit to a retrograde one, as is observed. Further still, Pluto's rock content is so high that it is unlikely that Pluto formed in a planetary subnebula. Of course, none of these facts were known until decades after Lyttleton made his original (and then quite logical) suggestion that Pluto might be a former satellite of Neptune.

As described in Section 2, it is likely that Pluto was caught in the 2:3 resonance and had its orbital eccentricity and inclination amplified to current values as Neptune migrated outward during the clearing of the outer solar system by the giant planets.

The heliocentric formation/giant collision scenario described earlier for the origin of the satellites can account for most of the major attributes of the system, including the elliptical, Neptune-crossing orbit, the high axial obliquities, and the $\approx 8.5:1$ mass ratio. Further, the present tidal equilibrium state would naturally be reached by Pluto and Charon in 10^8-10^9 years—a small fraction of the age of the solar system.

Still, such a scenario begs two questions. First, why is Pluto so small? And, second, how could Pluto and the Charon progenitor, alone in over 10^3 AU^3 of space, "find" each other in order to execute a mutual collision? That is, the giant impact hypothesis still fails to explain (1) the existence of Pluto and Charon themselves; (2) the very small masses of Pluto and Charon compared to the gas giants in general, and Neptune and Uranus in particular; (3) the fact that the collision producing the impact was highly unlikely; and (4) the system's position in the Neptune resonance.

In 1991, Alan Stern of the Southwest Research Institute suggested that the solution to (1)-(3) lies in the possibility that Pluto and Charon were members of a large population (300-3000) of small ($\sim 10^{-5}$ g) **ice dwarf** planets present during the accretion of Uranus and Neptune in the 20-30 AU zone. Such a population would make likely the Pluto-Charon collision, as well as three otherwise highly unlikely occurrences in the 20-30 AU region: the capture of Triton into retrograde orbit and the tilting of Uranus and Neptune. Similar conclusions based on different considerations were reached by William McKinnon of Washington University in the late 1980s. According to this work, the vast majority of the ice dwarfs were either scattered (with the comets) to the Oort cloud or ejected from the solar system altogether by perturbations from Neptune and Uranus. Only Pluto-Charon and Triton remain in the 20-30 AU zone today, specifically because they are trapped in unique dynamical niches that protect them against loss to such strong perturbations.

If this is correct, it implies that Pluto, Charon, and Triton are important "relics" of a very large population of small planets, dubbed ice dwarfs first by Stern (1991), which by number (but not mass) dominate the planetary population of the solar system. As such, these three bodies would no longer appear as isolated anomalies in the outer solar system and would be genetic relations from an ancient, ice dwarf ensemble, and therefore worthy of intense study as a new and valuable class of planetary body unto themselves.

8.3. The Context of Pluto in the Outer Solar System

When the existence of the ice dwarf population was first suggested, the solar system beyond Neptune appeared to only be inhabited by Pluto and the numerous comets scattered out of the planetary region during the accretion of the giant planets.

Since late 1992, however, our concept of the outer solar system has evolved considerably, owing to a rapid set of discoveries of faint largish bodies orbiting between 30 and 50 AU in what is known as the Edgeworth-Kuiper Belt. The first such objects were detected by David Jewitt and Jane Luu using the University of Hawaii's 2.2 m telescope on Mauna Kea.

As of this writing at the end of 2014, over 1600 worlds with diameters of 100-2000 km have been discovered in the Kuiper Belt, including some objects that have clearly been scattered out of the giant planets' region. Many of these are apparently in the 2:3 mean-motion resonance with Neptune that Pluto also occupies. The largest discovered Kuiper Belt body is almost Pluto's size. Some have satellites. Beyond the Kuiper Belt, in the so-called scattered belt, lie other large bodies, including Eris, almost as large as Pluto.

Because the Kuiper Belt census obtained to date has covered only a tiny fraction of the ecliptic sky, it is estimated that many times the discovered population exists. Current models of the population of the region between 30 and 50 AU from the Sun now indicate that some 100,000 or more objects with diameters larger than approximately 100 km and perhaps several billion comets 1–20 km in diameter reside there. The total mass of bodies currently in the 30–50 AU zone may amount to as much as 0.01 $M_\oplus$, exceeding the mass of the asteroid belt by more than an order of magnitude.

Both the discovery of the rapidly expanding cohort of objects found in the 30–50 AU zone, and the circumstantial evidence that this region of the solar system was much more heavily populated when the solar system was young, finally provide a context for Pluto (and the putative Charon progenitor as well). We now see that Pluto did not form in isolation and does not exist so today. Instead, Pluto is simply one of a large number of significant miniplanets that grew in the region beyond Neptune when the solar system was young. Pluto's presence there today is in large measure due to its location in the stable 2:3 resonance with Neptune.

The question now has moved from why a small planet like Pluto formed in isolation, to why a large population of objects hundreds and thousands of kilometers in diameter formed in the 30–50 AU zone without progressing to the formation of a larger planet there.

BIBLIOGRAPHY

Binzel, R. P. (1990). Pluto. *Scientific American, 252*(6), 50–58.

Binzel, R. P., et al. (1997). In S. A. Stern, & D. J. Tholen (Eds.), *Pluto & Charon*. Tucson: Univ: Arizona Press.

Elliot, et al. (1989). *Icarus,* 77–148.

Marcialis, et al. (1987). *Science,* 237–1349.

Marcialis, R. L. (1988). *Astronomical Journal, 95,* 941.

McKinnon, W. B., et al. (1997). In S. A. Stern, & D. J. Tholen (Eds.), *Pluto & Charon*. Tucson: Univ. Arizona Press.

Stern, S. A. (1992). The Pluto-Charon system. *Annual Review of Astronomy and Astrophysics, 30,* 185–233.

Stern, et al. (1997). *Astronomical Journal,* 113–827.

Stern, S. A., & Mitton, J. (2005). *Pluto and Charon: Ice-dwarfs on the ragged edge of the solar system* (2nd ed.). New York: John Wiley & Sons.

Stern, S. A., & Tholen, D. J. (Eds.). (1997). *Pluto and Charon*. Tucson: Univ. Arizona Press.

Tombaugh, C. W., & Moore, P. (1980). *Out of the darkness: The planet Pluto*. Harrisburg, Pennsylvania: Stackpole Books.

Whyte, A. J. (1980). *The planet Pluto*. Toronto, Canada: Pergamon Press Ltd.

Chapter 43

Kuiper Belt: Dynamics

Alessandro Morbidelli
Observatoire de la Côte d'Azur, Nice, France

Harold F. Levison
Southwest Research Institute, Boulder, CO, USA

Chapter Outline

1. Historical Perspective 925
2. Orbital and Dynamical Structure of the Trans-Neptunian Population 926
3. Correlations between Physical and Orbital Properties 930
4. Size Distribution of the Trans-Neptunian Population and Total Mass 931
5. Ecliptic Comets 933
6. The Primordial Sculpting of the Trans-Neptunian Population 934
7. Concluding Remarks 938
Bibliography 939

The name Kuiper Belt generically refers to a population of small bodies, the orbits of which have a **semimajor axis**—and hence orbital period—larger than those of Neptune. The Kuiper Belt objects (KBOs)—having formed at large distances from the Sun—are rich in water ice and other volatile chemical compounds and have physical properties similar to those of comets. Indeed, the existence of the Kuiper Belt was first deduced from observations of the **Jupiter-family comets**, a population with short orbital periods and small to moderate orbital inclinations, of which the Kuiper Belt is the source.

In 20 years since the discovery of the first object, about 1500 KBOs have been detected. Of these, ~1000 objects have been observed for at least 2 years, a necessary condition to compute their orbital parameters with significant precision. The results of this detailed observational exploration of the Kuiper Belt structure have provided several surprises. Indeed, it was expected that the Kuiper Belt preserved the pristine conditions of the protoplanetary disk. But it is now evident that this picture is not correct: The Kuiper Belt has been affected by a number of processes that have altered its original structure.

The Kuiper Belt may thus provide us with a large number of clues to understand what happened in the outer solar system during the primordial ages. Potentially, the Kuiper Belt might teach us more about the formation of the giant planets than the planets themselves. And, as in a domino game, a better knowledge of the formation of giant planets would inevitably boost our understanding of the subsequent formation of the solar system as a whole. Consequently, Kuiper Belt research is now considered a priority of modern planetary science.

1. HISTORICAL PERSPECTIVE

Since its discovery in 1930, Pluto has traditionally been viewed as the last vestige of the planetary system—a lonely outpost at the edge of the solar system, orbiting beyond Neptune with a 248-year period.

Pluto itself has always appeared to be an oddity among the planets. Traditionally, the planets are divided into two main groups. The first group, the terrestrial planets, formed in the inner regions of the solar system where the material from which the planets were made was too warm for water and other volatile gases to be condensed as ices. These planets, which include the Earth, are small and rocky. Farther out from the Sun, the cores of the planets grew from a combination of rock and condensed ices and captured significant amounts of nebula gas. These are the Jovian planets, the giants of the solar system; they most likely do not have solid surfaces. But, then there is Pluto, unique, small (its radius is only ~1180 km, only two-thirds that of the Earth's Moon) and made of a mix of rock and frozen ices.

The planets formed in a disk of material that originally surrounded the Sun (see "Origin of the Solar System"). The formation of the disk occurred because the molecular cloud that gave origin to the Sun had necessarily a slight global spin. Thus, as it collapsed, the spin rate had to increase in order to conserve **angular momentum**. The cloud could not form a single star with the amount of angular momentum it possessed, so it shed a disk of material (gas and dust) that contained very little mass (as compared with the mass of the Sun), but most of the angular momentum of the system. As such, the planets formed in a narrow disk structure; the plane of that disk is known as the invariable plane. But then there is Pluto, unique, having an orbital inclination of 15.6° with respect to the invariable plane.

The orbits of the planets are approximately ellipses with the Sun at one focus. As the planets formed in the original circumsolar disk, they tended to evolve onto orbits that were well separated from one another. This was required so that their mutual gravitational attraction would not disrupt the whole system. (Or to put it another way, if our system had not formed that way, we would not be here to talk about it!) But, then there is Pluto, unique, having an orbit that crosses the orbit of its nearest neighbor, Neptune.

So, the historical view was that Pluto was an oddity in the solar system, unique for its physical makeup and size as well as its dynamical niche. But, this view changed in September 1992 with the announcement of the discovery of the first of a population of small (compared to planetary bodies) objects orbiting beyond the orbit of Neptune, in the same region as Pluto. Since that time, roughly 1500 objects with radii between a few tens and ~1000 km have been discovered. One object, 136199 Eris, even turned out to be about the same size as Pluto. Moreover, a modeling of the detection efficiency of the performed surveys suggests that there are approximately 100,000 objects larger than a few tens of kilometers occupying this region of space, approximately between 30 and 50 astronomical units (AU) from the Sun. There are inevitably many more smaller ones. As discussed in more detail in the following sections, these objects likely have a similar physical makeup to that of Pluto, and many have similar orbital characteristics. Thus, in the past 20 years, Pluto has been transformed from an oddity, to the founding member of what is perhaps the most populous class of objects in the planetary system.

The discovery of the Kuiper Belt, as it has come to be known, represents a revolution in our thinking about the solar system. First predicted on theoretical grounds and later confirmed by observations, the Kuiper Belt is the first totally new class of bodies to be discovered in the solar system since the first asteroid was found on New Year's day, 1801. Its discovery has radically changed our view of the outer solar system.

Speculation on the existence of a trans-Neptunian disk of icy objects dates back over 100 years. In the early 1900, Campbell, Aitken, and Leuschner considered the possibility of trans-Neptunian planets and speculated on the orbital distribution of small bodies in the outer planetary system.

In the 1940s and early 1950s, a more comprehensive approach to the problem was made independently by Kenneth Edgeworth and Gerard Kuiper. They noticed that if one were to grind up the giant planets and spread out their masses to form a disk, then this disk would have a very smooth distribution, with a density that slowly decreases as the distance from the Sun increases. That holds until Neptune, at which point there is an apparent edge beyond which there was thought to be nothing except tiny Pluto. Edgeworth and Kuiper suggested that perhaps this edge was not real. Perhaps the disk of planetesimals (i.e. small bodies, potentially precursors of planet formation) that formed the planets extended past Neptune, but the density was too low or the formation times too long to form large planets. If so, they argued, these planetesimals should still be there in nearly circular orbits beyond Neptune. The idea of a trans-Neptunian disk received little attention for many years. The objects in the hypothetical disk were too faint to be seen with the telescopes of the time, so there was no way to prove or disprove their existence. Comet dynamicists showed that the lack of detectable perturbations on the orbit of Halley's comet limited the mass of such a disk to no more than 1.3 Earth masses ($M_\oplus$) if it was at 50 AU from the Sun.

However, the idea was resurrected in 1980 when Julio Fernandez proposed that a cometary disk beyond Neptune could be a possible source reservoir for the short-period comets (those with orbital periods <200 years). Subsequent dynamical simulations showed that a comet belt beyond Neptune is the most plausible source for the low-inclination subgroup of the short-period comets, named the Jupiter-family comets. This work led observers to search for KBOs. With the discovery of the first object, 1992 QB$_1$, by D. Jewitt and J. Luu, the Kuiper Belt ceased to be a speculation and became a concrete entity of the solar system.

2. ORBITAL AND DYNAMICAL STRUCTURE OF THE TRANS-NEPTUNIAN POPULATION

Figure 43.1 shows the distribution of the objects with semimajor axis larger than 30 AU whose orbits have been determined from observations spanning over at least 3 years.

A glance at the figure reveals that the orbits of the trans-Neptunian objects can be very diverse. The majority of the discovered objects are clustered in the 36–48 AU range, but several others form a "tail" structure extending beyond 50 AU. Their semimajor axes range up to several 100 AU, but probably go much farther. Their perihelion distances are generally between 30 and 38 AU (dotted curves in the bottom right panel of Figure 43.1), so that on average the orbital eccentricities increase with semimajor axes. Many

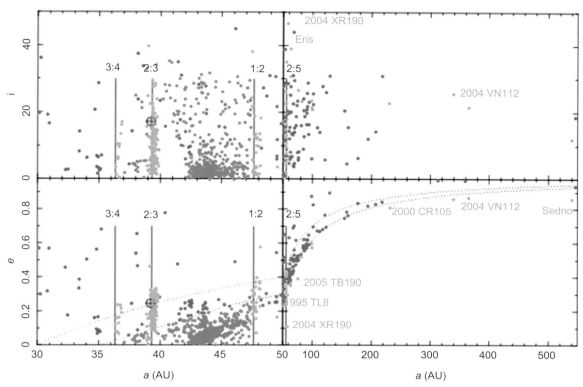

FIGURE 43.1 The distribution of the objects with well-determined orbits, as of October 29, 2013. The upper and lower panels show, respectively, the inclination and the eccentricity vs semimajor axis. Two different semimajor axis scales are used to illustrate the Kuiper Belt (left panels) and the scattered disk (right panels) distributions. Red dots correspond to the scattered disk, orange dots to the extended scattered disk, blue dots to the classical Kuiper Belt, and green dots to the resonant populations. The big, crossed circle denotes the orbit of Pluto. The vertical lines labeled 3:4, 2:3, and 1:2 mark the location of the corresponding mean-motion resonances with Neptune. The two dotted curves on the lower panels correspond to perihelion distances $q = 30$ AU and $q = 35$ AU on the left, and $q = 30$ AU and $q = 38$ AU on the right. These curves approximately bound the scattered disk orbital distribution.

of these objects are dynamically unstable because they suffer sufficiently close encounters with Neptune. At each encounter, they receive an impulselike acceleration, which changes the semimajor axis of their orbits. The perihelion distance remains roughly constant during an encounter, so that the **eccentricity** changes together with the semimajor axis. Thus, under the scattering gravitational action of Neptune (located at 30 AU), these objects move in a sort of random walk in the region confined by the dotted curves. For this reason, the population of unstable objects is now called the **scattered disk**. The name "disk" is justified, because the orbital inclinations, although large, are significantly smaller than 90°, giving this population a disklike structure.

Up to now, we have used "Kuiper Belt" to denote generically the population of objects with $a > 30$ AU. However, the existence of the scattered disk suggests that we should reserve the name "Kuiper Belt" for the population of objects that do not suffer encounters with Neptune and therefore have orbits that either do not significantly change with time or do so very slowly. Adopting this definition, the objects of the Kuiper Belt are plotted with blue and green dots in Figure 43.1, while the scattered disk objects are plotted in red. As one sees, scattered disk objects can also have $a < 50$ AU, provided that they have a small perihelion distance and, for reasons discussed below, are not in one of the more prominent **mean-motion resonances** with Neptune (indicated by the vertical lines labeled 3:4, 2:3, 1:2 and 2:5 in Figure 43.1). All solar system bodies should have accreted on quasicircular and coplanar orbits. This is a necessary condition for small planetesimals being able to stick together and form larger objects. Indeed, if the eccentricities or the **inclinations** are large, the relative encounter velocities are such that, upon collisions, planetesimals do not grow, but fragment into smaller pieces. This consideration suggests that the scattered disk objects formed much closer to Neptune, on quasicircular orbits, and have been transported outward by the scattering action of that planet. The fact that the scattering action is still continuing implies that the origin of the scattered disk does not necessarily require that the primordial solar system was different from the current one.

However, recent observations have revealed that, in addition to the Kuiper Belt and the scattered disk, there is a third category of objects, represented with orange dots in Figure 43.1. Their orbital distribution mimics that of the scattered disk objects, but their perihelion distance is somewhat larger, so that they avoid the scattering action of Neptune. Their orbits do not significantly change over the age of the solar system. Among the objects with these orbital properties are 1995 TL_8 ($a \sim 52$ AU, $q \sim 40$ AU), 2000 CR_{105} ($a \sim 225$ AU, $q \sim 44$ AU), 90377 Sedna ($a \sim 500$ AU, $q \sim 76$ AU), 136199 Eris ($a \sim 67.5$ AU, $q = 38$ AU, the second Pluto-sized trans-Neptunian object known so far), and 2004 XR_{190} ($a \sim 57.4$ AU, $q \sim 51$ AU—exceptional for its inclination of about $45°$). For the previously listed reasons, these bodies also should have formed closer to Neptune on much more circular orbits, and presumably they have been transported outward through close encounters with the planet. However, given that they do not undergo close encounters now, their existence suggests that the solar system was different in the past (either the planetary orbits were different or the environment was different—massive objects crossing the Kuiper belt passing stars, etc.), so that the scattered disk extended further out in perihelion distance during the primordial times. For this reason we call this population the **extended scattered disk**.

In Figure 43.1, the red and orange dots seem to be outnumbered by the blue and green dots. However, the scattered and extended scattered disk objects are more difficult to discover, given that most of them have very elongated orbits and spend most of the time very far from the Sun. Accounting for this difficulty, astronomers have estimated that the scattered disk and the Kuiper Belt should constitute roughly equal populations.

If we look at Figure 43.1 more in detail (left panels), the Kuiper Belt can also be subdivided in a natural way into subpopulations. Several objects (green dots) are located in mean-motion resonances with Neptune. Mean-motion resonances provide a protection mechanism, so that resonant objects can avoid close encounters with Neptune even if their perihelion distance is smaller than 30 AU. Let us take Pluto as an example. Pluto is locked in a mean-motion resonance where it goes around the Sun twice every time Neptune goes around three times. So, every time Pluto crosses the trajectory of Neptune, the giant planet is always in one of three specific locations on its orbit, all very far away from Pluto's position. Thus Pluto and Neptune avoid mutual encounters. Given this resonant protection mechanism, resonant KBOs can be on much more elliptic orbits than the nonresonant ones, the eccentricities of the former ranging up to ~ 0.4. The objects in the 2:3 mean-motion resonance with Neptune are usually called the plutinos (because they share the same resonance as Pluto), while those in the 1:2 resonance are sometimes called twotinos. In Figure 43.1, the resonant population seems to constitute a substantial fraction of the Kuiper Belt population. However, resonant objects are easier to discover because at perihelion they come closer to the Sun than the nonresonant ones. When accounting for this fact, astronomers estimate that, all together, the objects in mean-motion resonances constitute between 10% and 40% of the total Kuiper Belt population.

The nonresonant KBOs (blue dots) are usually referred to as classical. This adjective is attributed because their orbital distribution is the most similar to what the astronomers were expecting, before the discoveries of trans-Neptunian objects began: that of a disk of objects on stable, low-eccentricity, nonresonant orbits. However, even the classical population has unexpected properties. Their eccentricities are moderate—a necessary condition to avoid encounters with Neptune, given that they are not protected by any resonant mechanism. Nevertheless, the eccentricities are definitely larger than those of the protoplanetary disk in which the objects had to form. Some mechanism must have excited the eccentricities, making them grow from almost zero to the current values.

The same is true, and even more striking, for the inclinations (top panel of Figure 43.1). The inclinations are related to the relative encounter velocities among the objects, so that the Kuiper Belt bodies had to grow in a razor-thin disk. Despite this, the current inclinations range up to $30°-40°$. Figure 43.1 gives the impression that large-inclination bodies are a modest fraction among the classical objects. However, one should take into account that the discovery surveys have been concentrated near the ecliptic plane, so that large-inclination bodies have a lower probability of being discovered than low-inclination ones. Accounting for this selection effect, astronomers have computed that the real inclination distribution of the classical objects is bimodal (Figure 43.2). There is a cluster of objects with inclination smaller than $4°$ and a second group of objects with a very distended inclination distribution. The former constitute what is now usually called the cold population and the latter constitute the hot population. The adjectives "hot" and "cold" do not refer to physical temperature (it is always very cold out there) but to the encounter velocities inside each population, in an analogy with gas kinetic theory. The cold and the hot populations should contain roughly the same number of objects.

The last striking property of the Kuiper Belt is its outer edge (Figure 43.1). For very low-eccentricity and low-inclination objects the edge is near 45 AU, although the classical belt as a whole ends at the location of the 1:2 mean-motion resonance with Neptune. For several years, the astronomers suspected that this edge is only apparent, due to the fact that more distant objects are more difficult to discover. However, with an increasing statistical sample, it turned out that this is not true. It has been shown that more distant objects should have been discovered by now, unless

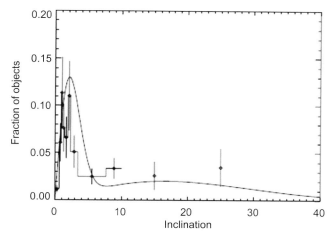

FIGURE 43.2 The inclination distribution (in degrees) of the classical Kuiper Belt after observational biases have been subtracted, according to the classical work of M. Brown (2001). The points with error bars show the model-independent estimate constructed from a limited subset of confirmed classical belt bodies, while the smooth line shows a best fit bimodal population model. In this model, ∼60% of the objects have $i > 4°$.

either the Kuiper Belt population steeply decays in number beyond 48–50 AU or the maximal size of the objects beyond this limit is much smaller than that in the observed Kuiper Belt. For various reasons, astronomers tend to favor the former hypothesis: the existence of a physical outer edge of the Kuiper Belt.

An important issue is to understand which of the orbital properties discussed earlier is due to the dynamical processes that are still occurring in the Kuiper Belt. For instance, do the eccentricities and the inclinations slowly grow due to some dynamical phenomenon? Are the low-eccentricity objects beyond 48 AU unstable? If these are the cases, then the existence of large eccentricities and inclinations, as well as the outer edge of the Kuiper Belt could be simply explained. In the opposite case, these properties—like the existence of the extended scattered disk—reveal that the solar system was different in the past.

Dynamical astronomers have studied in great detail the dynamics beyond Neptune, using numerical simulations and semianalytic models. Figure 43.3 shows a map of the dynamical lifetime of trans-Neptunian bodies on a wide range of initial semimajor axes, eccentricities, and inclinations. This map has been computed numerically, by simulating the evolution of thousands of massless particles

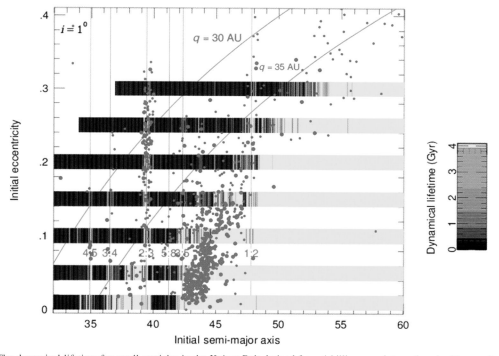

FIGURE 43.3 The dynamical lifetime for small particles in the Kuiper Belt derived from 4-billion-year integrations by Duncan, Levison, and Budd (1995). Each particle is represented by a narrow vertical strip of color, the center of which is located at the particle's initial eccentricity and semimajor axis (initial orbital inclination for all objects was 1°). The color of each strip represents the dynamical lifetime of the particle. Strips colored yellow represent objects that survive for the length of the integration, 4×10^9 years. Dark regions are particularly unstable on these timescales. For reference, the locations of the important Neptune mean-motion resonances are shown in blue and two curves of constant perihelion distance, q, are shown in red. The orbital distribution of the real objects is also plotted. Big dots correspond to objects with $i < 4°$, and small dots to objects with larger inclination. Remember that the dynamical lifetime map has been computed assuming $i = 1°$.

under the gravitational perturbations of the giant planets. The latter have been assumed to be initially on their current orbits. Each particle was followed until it suffered a close encounter with Neptune. Objects encountering Neptune would then evolve in the scattered disk for a time of order $\sim 10^8$ years until they are transported by planetary encounters into the inner planets region or are ejected to the **Oort cloud** or to interstellar space. In Figure 43.3, the colored strips indicate the length of time required for a particle to encounter Neptune as a function of its initial semimajor axis and eccentricity. The initial inclination of the particles was set equal to $1°$. Strips that are colored yellow represent objects that survive for the length of the simulation, 4×10^9 years, the approximate age of the solar system. For comparison, the observed objects with good orbital determination are overplotted with green dots. Big dots refer to bodies with $I < 4°$, consistent with the low inclination at which the stability map has been computed. Small dots refer to objects with larger inclination and are plotted only for completeness.

As can be seen in the figure, the Kuiper Belt can be expected to have a complex structure, although the general trends are readily explained. Objects with perihelion distances less than ~ 35 AU (shown as a red curve) are unstable, unless they are near, and presumably librating about, a mean-motion resonance with Neptune. Indeed, the results in Figure 43.3 show that many of the Neptunian mean-motion resonances (shown in blue) are stable for the age of the solar system. Objects at low inclination with semimajor axes between 40 and 42 AU are unstable. This is presumably due to the presence of three overlapping **secular resonances**. Due to the perturbations exerted by the planets, the orbits of the planets themselves as well as those of small bodies precess in space. A secular resonance arises when the precession rate of the orbit of a body is equal to the precession rate of the orbit of one of the planets. In this case, the 40- to 42-AU region is affected by secular resonances with both Neptune and Uranus. These resonances tend to excite the eccentricities of the KBOs, which then can encounter Neptune and are ultimately removed from the Kuiper belt. Notice from Figure 43.3 that there are some objects in the 40–42 AU semimajor axis range, but they all have high inclination orbits (small dots). In fact, the destabilizing secular resonances are present only at low inclination in this region.

The conclusion to be drawn from a comparison between the distribution of the real low-inclination KBOs and the stability map in Figure 43.3 is that most observed objects (with the exception of scattered disk bodies) are associated with stable zones. Their orbits do not significantly change over the age of the solar system. Thus, their current excited eccentricities and inclination cannot be obtained from primordial circular and coplanar orbits in the framework of the current planetary system orbital configuration. Likewise, the region beyond the 1:2 mean-motion resonance with Neptune is totally stable. Thus, the absence of bodies beyond 48 AU cannot be explained by current dynamical instabilities.

Therefore, it is evident that the orbital structure of the Kuiper Belt has been sculpted by mechanisms that are no longer at work, but presumably were active when the solar system formed. The main goal of dynamical astronomers interested in the Kuiper Belt is to uncover these mechanisms and from them deduce, as far as possible, how the solar system formed and early evolved.

3. CORRELATIONS BETWEEN PHYSICAL AND ORBITAL PROPERTIES

The existence of two distinct classical Kuiper Belt populations, called the hot ($i > 4°$) and cold ($i < 4°$) classical populations, could be caused in one of two general manners. Either a subset of an initially dynamically cold population was excited, leading to the creation of the hot classical population, or the populations are truly distinct and formed separately.

One manner in which we can attempt to determine which of these scenarios is more likely is to examine the physical properties of the two classical populations. If the objects in the hot and cold populations are physically different, it is less likely that they were initially part of the same population.

The first suggestion of a physical difference between the hot and the cold classical objects came from the observation that the intrinsically brightest classical belt objects (those with lowest **absolute magnitudes**) are preferentially found with high inclination. Figure 43.4 shows the distribution of the classical objects in an inclination vs absolute magnitude diagram. As one sees, for an absolute magnitude $H > 5.5$, there is a given proportion between the number of objects discovered in the cold and the hot populations. This ratio is completely different for $H < 5.5$, where cold population objects are almost absent. All the biggest classical objects, such as 50000 Quaoar, 20000 Varuna, 19521 Chaos, 28978 Ixion, 136472 Makemake, and 136108 Haumea, have inclinations larger than $5°$. Their median inclination is $20°$. It has been shown that this result is not an artifact produced by observational biases.

The second possible physical difference between hot and cold classical KBOs is their colors. With the name "color" astronomers generically refer to the slope of the spectrum of the light reflected by a trans-Neptunian object at visible wavelengths, relative to that of the light emitted by the Sun. "Red" objects reflect more at long than at short wavelengths, while "gray" objects have a more or less uniform reflectance. Colors relate in a poorly understood manner to objects' surface composition. It has been shown

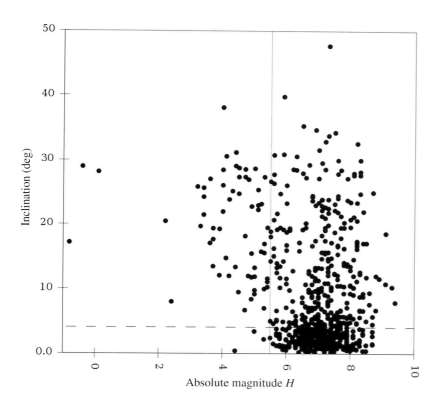

FIGURE 43.4 The inclination of the classical Kuiper Belt objects as a function of their absolute magnitude. The horizontal dashed line at $i = 4°$ separates the cold from the hot population. The vertical dotted line is plotted at $H = 5.5$. The distribution on the left side of the dotted line is clearly different from that on the right-hand side. The largest classical objects are all in the hot population.

and repeatedly confirmed that, for the classical belt, the inclination, and possibly the perihelion distance, is correlated with color. In essence, the low-inclination classical objects tend to be redder than higher inclination objects. More interestingly, colors naturally divide into distinct red and gray populations at precisely the location of the divide between the inclinations of the hot and cold classical objects. These populations differ at a 99.9% confidence level. Interestingly, the cold classical population also differs in color from the plutinos and the scattered objects at the 99.8 and 99.9% confidence level, respectively, while the hot classical population appears identical in color to these other populations.

It has been suggested that the colors of the objects, rather than being markers of different populations, are actually *caused* by the different inclinations. For example, the higher average impact velocities of the high-inclination objects could cause large-scale resurfacing by fresh water ice and carbonaceous materials, which could be gray in color. However, a similar color inclination trend should be observed also among the plutinos and the scattered disk objects, which is not the case. A careful analysis shows that there is no clear correlation between average impact velocity and color.

The idea that the cold population represents a distinct physical class of objects is reinforced by a third observation. Many KBOs are binaries, i.e. consisting of two objects in orbit around one another. Some of these binary objects are composed of a system of small satellites orbiting a major primary, like in the case of 136108 Haumea and 136199 Eris. These binaries are probably of collisional origin. However, the vast majority of known binaries are made of two components of roughly equal size and large separation. These binaries most likely are primordial, in the sense that their members became bound to each other at the time of KBO accretion. Interestingly, the fraction of the KBO population which is "primordial" binary is very different in the cold and hot populations. In the former, about 30% of the objects are binaries. Instead the fraction is only 5% in the hot population. This, again, argues that the cold population has had a distinct evolution.

In summary, the significant differences (color, size, binary fraction) between the hot and cold classical objects imply that these two populations are physically different in addition to being dynamically distinct.

4. SIZE DISTRIBUTION OF THE TRANS-NEPTUNIAN POPULATION AND TOTAL MASS

As briefly described in Section 1, the disk out of which the planetary system accreted was created as a result of the Sun shedding angular momentum as it formed (see also "Origin of the Solar System"). As the Sun condensed from a molecular cloud, it left behind a disk of material (mostly gas with a little bit of dust) that contained a small fraction of the total mass but most of the angular momentum of the system. It is believed that the dust grains sedimented toward the midplane of the disk and started to stick to each other through electrostatic forces, forming pebble-sized objects, of the order of centimeters in diameter. These objects then formed larger objects through a still not fully understood process of accretion, leading to the origin of asteroids, comets, and planets (or the cores of the giant planets, which then accreted gas directly from the solar nebula). Understanding this process is one of the main goals of planetary science today.

There are few clues in our planetary system about this process. We know that the planets formed, and we know how big they are. Unfortunately, the planets have been so altered by internal and external processes that they preserve almost no record of their formation process. Luckily, we also have the Asteroid Belt, the Kuiper Belt, and the scattered disk. These structures contain the best clues to the planet formation process because they are regions where the process started, but for some reason, did not run to completion (i.e. a large planet). Thus, the size distribution of objects in these regions may show us how the processes progressed with time and (hopefully) what stopped them. The Kuiper Belt and the scattered disk are perhaps the best places to learn about the accretion process.

Because the size of the object is not a quantity that can be easily measured and the absolute magnitude is readily obtained from the observations, astronomers generally prefer the absolute magnitude distribution, instead of the size distribution. The absolute magnitude is related to size via the **albedo**, where large H corresponds to small sizes and vice versa.

The magnitude distribution is usually given in the form $\log N(<H) \propto H^{\alpha}$, $\alpha > 0$ where N is the cumulative number of objects brighter than absolute magnitude H. The slope of this distribution, α, contains important clues about the physical strengths, masses, and orbits of the objects involved in the accretion process.

Bernstein et al. (2004, see Bibliography) were the first to argue that the H-distribution of the Kuiper belt population shows a change in α near $H \sim 7-9$. For bodies fainter than this threshold α is small (later it was determined to have $\alpha \sim 0.4-0.5$), whereas for brighter bodies α is much bigger, making the H-distribution steeper in a plot like that of Figure 43.5

This change in α, which is now well established by modern observations, can be interpreted in two ways. One possibility is that it is the result of collisional evolution. There are two extremes to the accretion process. If two large, strong objects collide at low velocities, then the amount of kinetic energy in the collision is small compared to the amount of energy holding the objects together. In this case, the objects merge to form a larger object. If two small,

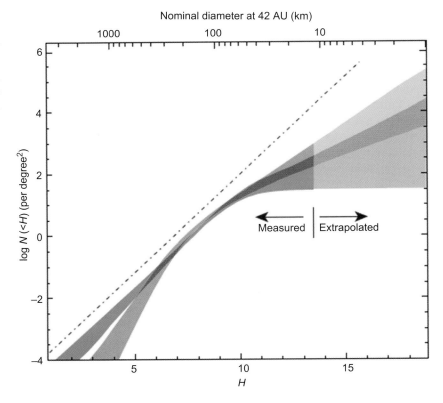

FIGURE 43.5 The cumulative magnitude distribution of the cold population (red) and hot population and scattered disk (green) according to a recent analysis by Bernstein et al. (2004). A turnover of the magnitude distribution is detected around $H \sim 7-9$. The slope of the magnitude distribution was very uncertain beyond this limit at the time this plot was made but is now well determined ($\alpha = 0.4-0.5$).

weak objects collide at high velocities, then the energy in the collision overpowers the gravitational and material binding energies. In this case, the objects break apart, forming a large number of much smaller objects. In this view, originally the H-distribution was steep at all magnitudes and the observed break in the current H-distribution marks the transition from accretional to disruptive collisions. The alternative interpretation is that the observed H-distribution is a primordial signature of the formation process and not the result of collisional erosion over the eons. In this view, the break represents the characteristic size (or H) at which most KBOs formed (i.e. 100–200 km in diameter). If true, this would be an important diagnostic of the overall formation process of KBOs.

Interestingly, the hot and the cold classical population seems to have two different H-distributions in the steep part. Originally, it was thought that the hot population and the scattered disk have a value of α smaller than that of the cold population (Figure 43.5). More recent dedicated observations, however, show that the H-distribution of the hot population is complex, looking like the cold population for $H \gtrsim 5$, but having a very small value of α for brighter objects. This is consistent with the fact that the largest bodies are all in the hot population, and yet the hot and cold populations and the scattered disk contain roughly the same number of bodies bigger than 100 km.

It is possible to integrate under the magnitude distribution shown in Figure 43.5 in order to estimate the total mass in the Kuiper Belt between 30 and 50 AU. Such an integration with limits between $R = 1$ and 1200 km (the approximate radius of Pluto) and assuming a density of 1 g/cm^3 (the measured densities of KBOs fall in the range 0.3–3 g/cm^3) shows that the total mass is about one-hundredth of an Earth mass. Given the uncertainties, it is possible that the mass is of order of 0.1 $M_\oplus$, but not significantly larger.

As with many scientific endeavors, the discovery of new information tends to raise more questions than it answers. Such is the case with the preceding mass estimate. Edgeworth's and Kuiper's original arguments for the existence of the Kuiper Belt were based on the idea that it seemed unlikely that the disk of planetesimals that formed the planets would have abruptly ended at the current location of the outermost known planet. An extrapolation into the Kuiper Belt (between 30 and 50 AU) of the current surface density of nonvolatile material in the outer planets region predicts that there should originally have been about 30 $M_\oplus$ of material there. However, as stated previously, our best estimate is over 200 times less than that figure!

Edgeworth's and Kuiper's argument is not the only indication that the mass of the primordial Kuiper Belt had to be significantly larger in the past. In fact, even if the primordial orbits had been much less eccentric and inclined than they are now (a necessary condition for accretion), the current abundance of solid material in the Kuiper Belt would not allow the growth of $\sim$100-km bodies (the characteristic size of most KBOs) over the age of the solar system. A massive and dynamically cold primordial Kuiper Belt is required.

Therefore, understanding the ultimate fate of the 99% (or 99.9%) of the initial Kuiper Belt mass that appears to be no longer in the Kuiper Belt is a crucial step in reconstructing the history of the outer solar system.

5. ECLIPTIC COMETS

As described in Section 1, the current renaissance in Kuiper Belt research was prompted by the suggestion that the Jupiter-family comets originated there. We now know that there are mainly two populations of small bodies beyond Neptune: the Kuiper Belt and the scattered disk. Which one is the dominant source of these comets?

To answer this question, we need to examine a few considerations on the origin of the scattered disk. We have seen in Section 2 that the bodies in the scattered disk have intrinsically unstable orbits. The close encounters with Neptune move them in semimajor axis, until they either evolve into the region with $a < 30$ AU or reach the Oort cloud at the frontier of the solar system. In both these cases, the bodies are removed from the scattered disk. Despite this possibility of dynamical removal, we still observe scattered disk bodies today. How can this be?

There are a priori two possibilities. The first one is that the scattered disk population is sustained in a sort of steady state by the bodies escaping from the Kuiper Belt. This means that on a timescale comparable to that for the dynamical removal of scattered disk bodies, new bodies enter the scattered disk from the Kuiper Belt. For example, a similar situation occurs for the population of **near-Earth asteroids** (NEAs, see Near Earth Asteroids). NEAs' dynamical lifetimes are only of a few million years because they intersect the orbits of the terrestrial planets. Nevertheless, the population remains roughly constant because new asteroids enter the NEA population from the Main Asteroid Belt at the same rate at which old NEAs are eliminated.

The second possibility is that the scattered disk that we see today is only what remains of a much more numerous population that has been decaying in number since planetary formation. Numerical simulations show that roughly 0.1–1% of the scattered disk bodies can survive in the scattered disk for the age of the solar system. Thus, the primordial scattered disk population should have been about 100–1000 times more numerous.

Which of these possibilities is true? In the first case, we would expect that the Kuiper Belt is much more populated than the scattered disk. For instance, the Asteroid Belt contains about 1000 times more objects than the NEA population, at comparable sizes. However, observations indicate that the population of the scattered disk is a substantial fraction of that of the Kuiper Belt. Thus, the second possibility is more likely. Scattered disk objects most likely formed in the vicinity of the current positions of Uranus and Neptune. When these planets grew massive, they scattered them away from their neighborhoods. In this way, a massive scattered disk of about 10 $M_\oplus$ formed. What we see today is just the last vestige of that primordial population, which is still decaying in number.

The fact that the scattered disk is not sustained in steady state by the Kuiper Belt, but it is still decaying, implies that the scattered disk provides more objects to the giant planet region ($a < 30$ AU) than it receives from the Kuiper Belt. Thus, the outflow from the scattered disk is more important than the outflow from the Kuiper Belt. This implies that the scattered disk, not the Kuiper Belt, is the dominant source of Jupiter-family comets.

A significant amount of research has gone into understanding the dynamical behavior of objects that penetrate into the $a < 30$ AU region from the scattered disk. These studies show that the encounters with the planets spread them throughout the planetary system. These objects are usually called ecliptic comets, even if at large distances from the Sun they typically do not show any cometary activity. The distribution of these objects as predicted by numerical integrations is shown in Figure 43.6.

The ecliptic comets that get close to the Sun become active. When their semimajor axis is smaller than that of Jupiter, they are called Jupiter-family comets. It is somewhat surprising that up to about one-third of the objects leaving the scattered disk in the simulations spend at least some of their time as Jupiter-family comets. The Jupiter-family comets that we see today are, in majority, small, $R \lesssim 10$ km. However, if our understanding of the size distribution of these objects is correct (see Section 5), we should expect to see a 100-km-sized Jupiter-family comet about 0.4% of the time. What a show that would be in the night sky!

Those ecliptic comets between Jupiter and Neptune are called the Centaurs (only the largest of which are observable). The simulations predict that there are $\sim 10^6$ ecliptic comets larger than about 1 km in radius currently in orbits between the giant planets.

6. THE PRIMORDIAL SCULPTING OF THE TRANS-NEPTUNIAN POPULATION

In the previous sections, we have seen that many properties of the Kuiper Belt cannot be explained in the framework of the current solar system:

1. The existence of the resonant populations.
2. The excitation of the eccentricities in the classical belt.
3. The coexistence of a cold and a hot population with different physical properties.
4. The presence of an apparent outer edge at the location of the 1:2 mean-motion resonance with Neptune.
5. The mass deficit of the Kuiper Belt.
6. The existence of the extended scattered disk population.

These puzzling aspects of the trans-Neptunian population reveal that it has been sculpted when the solar system was different, due to mechanisms that are no longer at work. Like detectives on the scene of a crime, trying to reconstruct what happened from the available clues, the astronomers try to reconstruct how the solar system formed and evolved from the traces left in the structure of the Kuiper Belt.

Planet migration has been the first aspect of the primordial evolution of which the astronomers found a signature in the Kuiper Belt. It is widely accepted that the giant planets fully formed during a short period (less than 10 My) when most of the protoplanetary disk was gas. Once the planets formed and the gas disappeared, the planetesimals that failed to be incorporated in the planets' cores had to be removed from the planets' vicinity by the gravitational scattering action of the planets themselves. If a planet scatters a planetesimal outward, the latter gains

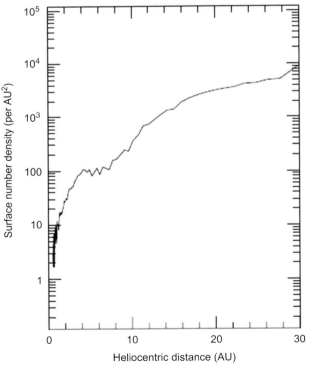

FIGURE 43.6 The surface number density (on the plane of the ecliptic) of ecliptic comets as determined from numerical integrations by Duncan and Levison (1997). There are approximately 10^6 comets larger than about 1 km in radius in this population.

energy. Because of energy conservation, the planet has to lose energy, moving slightly inward. The opposite happens if the planet scatters the planetesimal toward the inner solar system. A planet is much more massive than a planetesimal, thus the displacement of the planet is infinitesimal. However, if the number of planetesimals is large, and their total mass is comparable to that of the planet, the final effect on the planet is not negligible. This is a general process. We now come to what should have happened in our solar system.

Numerical simulations suggest that only a small fraction of the planetesimals originally in the vicinity of Neptune was scattered outward: less than 1% ended up in the scattered disk, and 5%, in the Oort cloud. The remaining 94% of the planetesimals eventually were scattered inward toward Jupiter, which, given its large mass, ejected from the solar system almost everything that came to cross its orbit. Thus, the net effect was that Neptune took energy away from the planetesimals and moved outward, while Jupiter gave energy to them and moved inward. Numerical simulations show that Saturn and Uranus also moved outward. Following Neptune's migration, the mean-motion resonances with Neptune also migrated outward, sweeping the primordial Kuiper Belt until they reached their present position. During this process, some of the KBOs swept by a mean-motion resonance could be captured into resonance. Once captured, these bodies had to follow the resonance in its migration, while their eccentricity had to steadily grow. Thus, the planetesimals that were captured first ended up on very eccentric resonant orbits, while those captured last could preserve a small eccentricity inside the resonance. Numerical simulations show that this process produces an important population of resonant bodies inside all the main mean-motion resonances with Neptune. To reproduce the observed range of eccentricities of resonant bodies, Neptune had to migrate more than 7 AU, thus starting not further than 23 AU (see Figure 43.7). The existence of resonant bodies in the Kuiper Belt thus provides a strong indication that planet migration really happened.

However, as Figure 43.7 also shows, several important properties of the Kuiper Belt cannot be explained by this simple model invoking resonance sweeping through a dynamically cold, radially extended disk. The eccentricity of the classical belt is only moderately excited, and the inclination remains very cold. The planetesimals are only relocated from the classical belt to the resonances, and only a minority of them are lost, which cannot explain the mass depletion of the belt. Finally, the region beyond the 1:2 mean-motion resonance is unaffected by planet migration, and therefore the existence of an outer edge requires a different explanation.

Four plausible models have been proposed so far to explain the formation of an outer edge: (1) the outer part of the disk was destroyed by the passage of a star; (2) it was photoevaporated by the radiation emitted by massive stars originally in the neighborhood of the Sun; (3) planetesimals beyond some threshold distance could not grow because of the enhanced turbulence in the outer disk which prevented the accumulation of solid material; and (4) distant dust particles and/or planetesimals migrated to smaller heliocentric distance during their growth, as a consequence of gas drag, thus forming sizeable objects only within some threshold distance from the Sun. The first two scenarios require that the Sun formed in a dense stellar environment, consistent with recent observations showing that stars tend to form in clusters which typically disperse in about 100 My. The entire protoplanetary disk—both the gas and the planetesimal components—would be truncated by these mechanisms. The third and the fourth scenarios, conversely, form a truncated planetesimal disk out of an extended gaseous disk.

Whatever mechanism formed the edge, it is intriguing that said edge is now at the location of a resonance with Neptune, despite the fact that Neptune did not play any role in the edge formation. Is this a coincidence? Probably not. It may suggest that originally the outer edge of the planetesimal disk was well inside 48 AU, and that the migration of Neptune pushed somehow a small fraction of the disk planetesimals beyond the disk's original boundary. These pushed-out planetesimals are now identified with the current members of the Kuiper Belt. The fact that the Kuiper Belt is mass deficient all over its radial extent (36–48 AU), in addition suggests that the original edge was inside 36 AU. In fact, if the original edge had been somewhere in the 36–48 AU range, we would see a discontinuity in the current radial mass distribution of the Kuiper Belt, which is not the case. An edge of the planetesimal disk close to 30 AU also helps to explain why Neptune stopped there and did not continue its outer migration beyond this limit.

Several mechanisms have been identified to push beyond the original disk edge a small fraction (of order 0.1%) of the disk's planetesimals, and to implant them on stable Kuiper Belt orbits. They are described next. More mechanisms might be identified in the future.

As Neptune moved through the disk on a quasicircular orbit, it scattered the planetesimals with which it had close encounters. Through multiple encounters, some planetesimals were transported outward on eccentric, inclined orbits. A small fraction of these objects still exist today and constitute the scattered disk. Occasionally, some scattered disk objects entered a resonance with Neptune. Resonances can modify the eccentricity of the orbits. If decreased, the perihelion distance is lifted away from the planet; the sequence of encounters stops, and the body becomes "decoupled" from Neptune like a KBO. If Neptune had not been migrating, the eccentricity would have eventually increased back to Neptune-crossing values—the dynamics

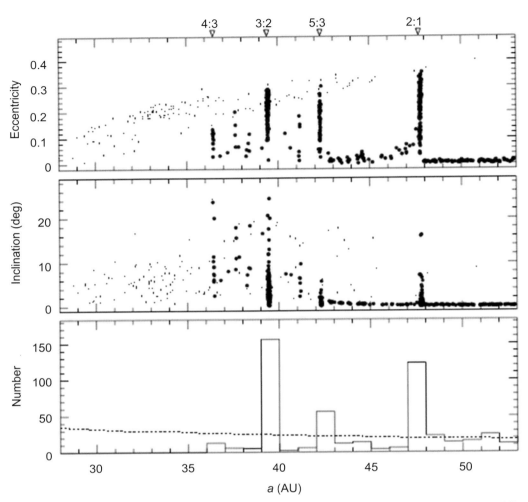

FIGURE 43.7 Final distribution of the Kuiper Belt bodies according to a simulation of the sweeping resonances scenario by Malhotra (1995). The simulation is done by numerical integrating, over a 200-My time span, the evolution of 800 test particles on initial quasicircular and coplanar orbits. The planets are forced to migrate (Jupiter, −0.2 AU; Saturn, 0.8 AU; Uranus, 3 AU; Neptune, 7 AU) and reach their current orbits on an exponential timescale of 4 My. Large solid dots represent "surviving" particles (i.e. those that have not suffered any planetary close encounters during the integration time); small dots represent the "removed" particles at the time of their close encounter with a planet. In the lowest panel, the solid line is the histogram of semimajor axis of the surviving particles, and the dotted line is the initial distribution. Most of the initial mass of the Kuiper Belt is simply relocated from the classical belt to the resonant populations. The mass lost is only a small fraction of the total mass.

being reversible—and the sequence of encounters would have restarted again. Neptune's migration broke the reversibility so that some of the decoupled bodies managed to escape from the resonances and remained permanently trapped in the Kuiper Belt. These bodies preserved the large inclinations acquired during the Neptune-encountering phase, and they can now be identified with the "hot" component of the Kuiper Belt population.

At the same time while Neptune was migrating through the disk, its 1:2 and 2:3 resonances swept through the disk, capturing a fraction of the disk planetesimals as explained earlier. When the 1:2 resonance passed beyond the edge of the disk, it kept carrying its load of objects. Because the migration of Neptune was presumably not a perfectly smooth process, the resonance was gradually dropping objects during its outward motion. Therefore, like a farmer seeding as he advances through a field, the resonance disseminated its previously trapped bodies all along its way up to its final position at about 48 AU. This explains the current location of the outer edge of the Kuiper Belt. Because the 1:2 resonance does not significantly enhance the orbital inclinations, the bodies transported by the resonance preserved their initially small inclination and can now be identified with the cold component of the Kuiper Belt.

This scenario, reproduced in numerical simulations, explains qualitatively the orbital properties of the trans-Neptunian population, but it has difficulties explaining why the hot and the cold classical populations have different physical properties. Indeed, the members of these two populations should have formed more or less in the same region of the disk, although they followed two different dynamical pathways toward the Kuiper Belt. These different pathways can explain the differences in binary fractions because only the hot population suffered close encounters with Neptune, which could dynamically strip the binaries, but definitely they cannot explain the different colors or size distributions.

An alternative possibility is that the hot population formed as explained earlier, but the cold population formed in situ, where it is now observed. Thus, the formation places being well separated, the corresponding physical properties could be different. However, this model has difficulties explaining how the cold population lost most of its primordial mass. It has been proposed that the objects ground down to dust in a collisional cascade process, but the latter is effective and consistent with the available observational constraints only if most of the mass was in subkilometer objects.

A further possibility is offered by a model of the evolution of the outer solar system, that has been developed in order to explain the origin of the so-called **Late Heavy Bombardment** (LHB; see "Solar System Dynamics: Regular and Chaotic Motions and Planetary Impacts") of the terrestrial planets. LHB is a cataclysmic period characterized by huge impact rates on all planets that occurred between 4.1 and 3.8 Bya, namely, about 400–700 My after planet formation. To explain the LHB, this model assumes that when the gas was removed from the solar system, the giant planets were on quasicircular and coplanar orbits, with orbital separations significantly smaller than the current ones. In particular, this model postulates that the planets were all in mean-motion resonance with one another, which is expected as a result of their previous evolution in the disk of gas. The model also assumes that the planets were surrounded by a remaining disk of planetesimals, containing 30–50 $M_\oplus$ out to ~ 35 AU.

Due to the gravitational interactions with this disk of planetesimals, the planets were eventually (i.e. at the time of the LHB) extracted from their multiple resonance. As we said in Section 2, resonances have a strong stabilizing effect against close encounters. Once the planets were extracted from their mutual resonances, this stabilizing effect ended. The planets rapidly became unstable, because they were too close to each other. The planetary orbits became chaotic and started to approach each other. Both Uranus and Neptune were scattered outward, onto large eccentricity orbits ($e \sim 0.3-0.4$) that penetrated deeply into the planetesimal disk. This destabilized the full disk and triggered the LHB. The interactions with the planetesimals damped the planetary eccentricities, stabilizing the planetary system once again and forcing a residual short radial migration of the planets, which eventually reached final orbits when most of the disk was eliminated. Simulations show that this model is consistent with the current orbital architecture of the giant planets of the solar system and several other observational constraints.

In this model, objects can be implanted into the current Kuiper Belt during the large eccentricity phase of Neptune. In fact, the full Kuiper Belt is unstable at that time, so that it can be visited by objects that leave the original planetesimal disk when the latter is destabilized. When Neptune's eccentricity is damped, the Kuiper belt becomes stable so that the objects which, by chance, are in the Kuiper Belt region at that time, become trapped forever. Because the large eccentricity phase of Neptune is short, the inclinations of these objects remain predominantly small, consistent with the cold population of the current Kuiper Belt. The objects with the largest inclinations, conversely, are captured later, during the final bit of Neptune's migration, as explained before. As Figure 43.8 shows, this model reproduces the structure of the Kuiper Belt reasonably well. It is also consistent with its low mass because in the simulations the probability of capture in the Kuiper Belt is roughly of 1/1000 (which predicts a final mass of 0.03 $M_\oplus$). Moreover, the KBOs with final low inclinations and those with final large inclinations are found to come predominantly from different portions of the original planetesimal disk (respectively, outside and inside 29 AU), which can explain, at least at a qualitative level, the correlations with physical properties, with one exception.

Indeed, the main problem with this model is that the objects that end up in the cold population should have suffered a phase of close encounters with Neptune. It has been shown that such encounters would have disrupted the binaries with the widest observed separations.

Consequently, we can conclude that none of the models proposed to date are fully satisfactory. Reconstructing the primordial sculpting of the Kuiper belt is an active field of research.

We finally come to the issue of the origin of the extended scattered disk. Simulations show that, in the same process described earlier, bodies are also delivered to orbits with moderate semimajor axis and perihelion distance, like that of the extended scattered disk object 1995 TL$_8$ and its companions. The origin of Sedna is probably different. The key issue is that bodies with comparably large perihelion distances (~ 80 AU) but smaller semimajor axis ($a < 500$ AU) have never been discovered despite the more favorable observational conditions. Therefore, they probably do not exist. If this is true, and the population of extended scattered disk bodies with large perihelion distance starts only beyond several hundreds of astronomical units, then an "external" perturbation is required. The best candidate is a stellar passage at about 1000 AU from the Sun, lifting the perihelion distance of the distant members of the primordial, massive scattered disk. Such a stellar encounter is very unlikely in the framework of the current galactic environment

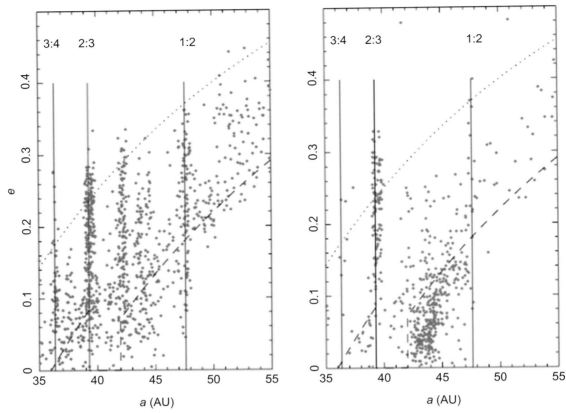

FIGURE 43.8 The distribution of semimajor axes and eccentricities in the Kuiper Belt. Left panel: result of a simulation based on the recent model on the origin of the LHB. Right panel: the observed distribution. The model reproduces fairly well the outer edge of the Kuiper Belt at the 1:2 resonance with Neptune, the characteristic shape of the (a, e) distribution of the classical belt, the scattered and the extended scattered disks, and the resonant populations. The vertical solid lines mark the main resonance with Neptune. The dotted curve denotes perihelion distance equal to 30 AU, and the dashed curve delimits the region above which only high-inclination objects or resonant objects can be stable over the age of the solar system. The overabundance of objects above this curve in the simulation is therefore an artificial consequence of the fact that the final orbits of the giant planets are not exactly the same as the real ones. *From Levison et al. (2008).*

of the Sun, but it would have been probable if the Sun formed in a moderately dense cluster, as mentioned earlier.

7. CONCLUDING REMARKS

At the time of the first edition of this book, 60 objects had been discovered in the Kuiper Belt. Now, we know 25 times more objects, and our view of the Kuiper Belt has become much more precise. It is now clear that the trans-Neptunian population has been sculpted in the primordial phases of the history of the solar system, by processes that are no longer at work.

It has been argued that the explanation of the most important observed properties of the trans-Neptunian population require a "cocktail" with three ingredients: (1) a truncated planetesimal disk; (2) a dense galactic environment, favoring stellar passages at about 1000 AU from the Sun; and (3) the outward migration of Neptune with, presumably, a phase of large eccentricity of the planetary orbits. Some problems still remain open, and the details of some mechanisms have still to be understood, but the basic composition of the cocktail is getting defined. This is a big step forward with respect to our understanding of solar system formation, before the discovery of the Kuiper Belt.

What is next? The upcoming generation of telescopic surveys will probably increase by another order of magnitude the number of discovered trans-Neptunian objects with good orbits. Thus, in the fourth edition of the book, we might have a different story to tell. It is unlikely that our view will totally change with the new discoveries (or at least we hope so!), but certainly there will be some surprises. We are anxious to know more precisely the absolute magnitude distributions of the various subpopulations of the Kuiper Belt, their color properties, the real nature of the outer edge and its exact location, and the orbital distribution of the extended scattered disk. This information will allow us to refine the scenarios outlined earlier, possibly to reject some and design new ones, in an attempt to read with less uncertainty the history of our solar system that is written out there.

BIBLIOGRAPHY

Bernstein, G. M., Trilling, D. E., Allen, R. L., Brown, M. E., Holman, M., & Malhotra, R. (2004). The size distribution of trans-Neptunian bodies. *The Astronomical Journal, 128*, 1364–1390.

Brasser, R., Duncan, M. J., Levison, H. F., Schwamb, M. E., & Brown, M. E. (2012). Reassessing the formation of the inner Oort cloud in an embedded star cluster. *Icarus, 217*, 1–19.

Brown, M. (2001). The inclination distribution of the Kuiper Belt. *The Astronomical Journal, 121*, 2804–2814.

Duncan, M. J., & Levison, H. F. (1997). A scattered comet disk and the origin of Jupiter family comets. *Science, 276*, 1670–1672.

Duncan, M. J., Levison, H. F., & Budd, S. M. (1995). The dynamical structure of the Kuiper Belt. *The Astronomical Journal, 110*, 3073.

Gomes, R. S. (2003). The origin of the Kuiper Belt high inclination population. *Icarus, 161*, 404–418.

Levison, H. F., & Morbidelli, A. (2003). The formation of the Kuiper Belt by the outward transport of bodies during Neptune's migration. *Nature, 426*, 419–421.

Levison, H. F., Morbidelli, A., Tsiganis, K., Nesvorny, D., & Gomes, R. (2011). Late orbital instabilities in the outer planets induced by interaction with a self-gravitating planetesimal disk. *The Astronomical Journal, 142*, 152.

Levison, H. F., Morbidelli, A., Van Laerhoven, C., Gomes, R., & Tsiganis, K. (2008). Origin of the structure of the Kuiper belt during a dynamical instability in the orbits of Uranus and Neptune. *Icarus, 196*, 258–273.

Levison, H. F., & Stern, S. A. (2001). On the size dependence of the inclination distribution of the main Kuiper Belt. *The Astronomical Journal, 121*, 1730–1735.

Malhotra, R. (1995). The origin of Pluto's orbit: Implications for the solar system beyond Neptune. *The Astronomical Journal, 110*, 420–432.

Morbidelli, A., Tsiganis, K., Crida, A., Levison, H. F., & Gomes, R. (2007). Dynamics of the giant planets of the solar system in the gaseous protoplanetary disk and their relationship to the current orbital architecture. *The Astronomical Journal, 134*, 1790–1798.

Noll, K. S., Grundy, W. M., Stephens, D. C., Levison, H. F., & Kern, S. D. (2008). Evidence for two populations of classical transneptunian objects: the strong inclination dependence of classical binaries. *Icarus, 194*, 758–768.

Parker, A. H., & Kavelaars, J. J. (2010). Destruction of binary minor planets during Neptune scattering. *The Astrophysical Journal, 722*, L204–L208.

Trujillo, C. A., & Brown, M. E. (2001). The radial distribution of the Kuiper Belt. *The Astrophysical Journal, 554*, 95–98.

Trujillo, C. A., & Brown, M. E. (2002). A correlation between inclination and color in the classical Kuiper Belt. *The Astrophysical Journal, 566*, 125–128.

Chapter 44

Kuiper Belt Objects: Physical Studies

Stephen C. Tegler
Northern Arizona University, Flagstaff, Arizona, USA

Chapter Outline

1. Discovering Kuiper Belt and Centaur Objects 942
2. Naming Objects 943
3. Databases of Known Objects 943
4. Dynamical Classes 943
5. Brightness 944
   5.1. Apparent Magnitude 944
   5.2. Luminosity Function 945
   5.3. Absolute Magnitude 945
6. Diameter 945
7. Albedo 946
8. Brightness Variation 946
   8.1. Period of Rotation 946
   8.2. Amplitude 947
   8.3. Shape 947
   8.4. Density 948
   8.5. Porosity 948
9. Composition 949
   9.1. Surface Color 949
      9.1.1. Centaur Objects 949
      9.1.2. Classical KBOs 949
      9.1.3. Scattered Disk Objects 949
      9.1.4. Reasons for Color Patterns 950
   9.2. Spectroscopy 951
10. KBO Binaries 952
    10.1. System Mass 952
    10.2. Mutual Events 953
    10.3. Origin of KBO Binaries 954
11. Mass of the Kuiper Belt 954
12. New Horizons 954
13. Future Work 955
Bibliography 955

Our solar system began as a slowly spinning cloud of gas and dust about 4.5 billion years ago. As gravity caused the cloud to shrink in size, conservation of angular momentum required it to spin faster and evolve into a thin disk of gas, ice, and dust surrounding the young Sun. In the outer region of the disk, cold material accreted to first form boulder-sized objects, then mountain-sized objects, and then comet nucleus-sized (1–10 km) objects. Eventually, a small number of objects reached the size of planetary cores. Two cores eventually grew in size to become Uranus and Neptune. As Uranus and Neptune grew in size, their gravitational influence stopped the numerous remaining smaller objects from forming an additional large planet.

The first hint of a debris disk of icy material in the outer solar system came in 1930 with the discovery of Pluto by Clyde Tombaugh of Lowell Observatory in Flagstaff, Arizona. It soon became clear that Pluto was much smaller than any other planet in the solar system. Pluto's small size did not follow the pattern of planetary properties—four small, rocky terrestrial planets (Mercury, Venus, Earth, and Mars) close to the Sun followed by four giant, hydrogen-rich Jupiter-like planets (Jupiter, Saturn, Uranus, and Neptune) farther from the Sun. Why was Pluto not a giant like the other Jupiter-like planets? In 1978, J. W. Christy and R. S. Harrington added to the inventory of small bodies beyond Neptune by discovering Pluto's satellite, Charon (pronounced either "Kharon" or "Sharon"), on images taken at the US Naval Observatory's Flagstaff station. Figure 44.1 illustrates the small sizes of Pluto and Charon by comparing them to the dimensions of the United States.

Perhaps the most important clue to solving the mystery of Pluto's small size came in 1988, when Martin Duncan, Thomas Quinn, and Scott Tremaine presented an extensive series of numerical simulations of the evolution of comet orbits due to the gravitational perturbations of the giant planets. Their simulations provided a dynamical proof that a belt in the outer solar system is a far more likely source of Jupiter-family comets than the Oort cloud. The calculations set David Jewitt and Jane Luu of the University of Hawaii looking for the belt. In 1992, they discovered an object much smaller and fainter than Pluto and Charon orbiting beyond Neptune. At the present time,

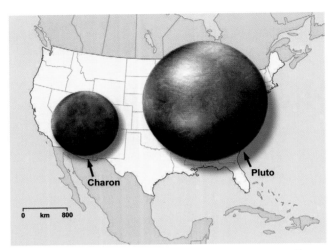

FIGURE 44.1 The diameter of Pluto (2302 km) and its moon Charon (1186 km) in comparison to the size of the United States. The diameters of both Pluto and Charon are smaller than that of Earth's Moon, 3476 km. *Courtesy of Dan Boone and NAU Bilby Research Center.*

~1000 objects ranging in size from a large comet nucleus to Pluto are known. It is now clear that Pluto, Charon, and the numerous smaller objects are what remain of the ancient disk of icy debris that did not accrete into a giant, Jupiter-like planet beyond the orbit of Neptune (Figure 44.2). The discovery of an object slightly larger than Pluto by Michael Brown of the California Institute of Technology in 2003 triggered the International Astronomical Union (IAU) to downgrade Pluto from its status as a planet in 2006. The icy debris disk beyond Neptune is commonly called the **Kuiper Belt** in honor of the Dutch–American astronomer Gerard P. Kuiper, who postulated its existence in 1951.

There are several dozen icy bodies that make up a class of objects closely related to Kuiper Belt objects (KBOs). These **Centaur objects** are recent escapees from the Kuiper Belt. They are on elliptical orbits about the Sun that cross the near-circular orbits of Saturn, Uranus, and Neptune. Within a few tens of millions of years after a Centaur object escapes from the Kuiper Belt, the giant planets scatter it out of the solar system, into the Sun or a planet, or cause it to migrate into the region of the terrestrial planets where it becomes a Jupiter-family comet. Centaurs are quite important because they come closer to the Sun and Earth than KBOs. By virtue of their "close" approach, many of them become bright enough for certain physical studies that are not possible on fainter KBOs. However, it is important to remember that Centaur objects experience a warmer environment than KBOs, and the warmth may alter their physical and chemical properties away from their initial properties at the time of their formation in the Kuiper Belt.

By studying KBOs and Centaurs, we are studying the preserved building blocks of a planet, and we can therefore shed some light on the process of planet building in our solar system as well as extrasolar planetary systems. After more than a decade of study, fundamental physical properties of KBOs and Centaurs—diameter, **albedo**, period of rotation, shape, mass, and surface composition—are being measured with accuracy.

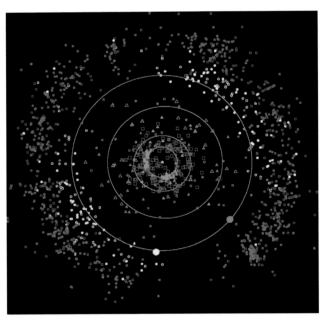

FIGURE 44.2 Positions of known bodies in the outer solar system on September 24, 2006. The orbits of the outer planets are shown in light blue. The location of Neptune is marked by a large blue circle on the outermost orbit. Pluto is marked by a large white circle. Kuiper Belt objects (KBOs) are marked as red (classical KBOs), white (Plutinos), and magenta (SDOs) circles. Centaur objects are marked as orange triangles. Comets are marked as blue squares. *Courtesy of Minor Planet Center.*

1. DISCOVERING KUIPER BELT AND CENTAUR OBJECTS

Centaur objects and KBOs orbit the Sun every 30–330 years. In addition to their own intrinsic motion about the Sun, the Earth's motion about the Sun imparts an apparent (parallactic) motion on these objects as well. These two motions distinguish KBOs and Centaurs from the multitude of background stars and galaxies and thereby make it possible to discover them.

Figure 44.3 illustrates the motion of the Centaur 1994 TA against the "fixed" pattern of much more distant stars and galaxies. Each panel of Figure 44.3 is a 300-s exposure of 1994 TA taken with the Keck II 10-m telescope and a charge-coupled device (CCD) camera in October 1998. The image on the right was taken about an hour after the image on the left.

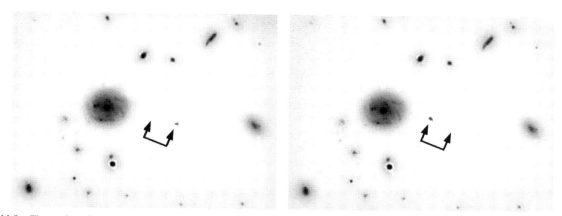

FIGURE 44.3 The motion of a solar system object relative to the background stars and galaxies. Two 300-s images of the Centaur 1994 TA taken with the Keck 10-m telescope and a CCD camera in October 1998. The image on the right was taken about an hour after the image on the left. Because of the Earth's and 1994 TA's revolution about the Sun, 1994 TA moves relative to the fixed pattern of background stars and galaxies. Such motion is how we discover KBOs and Centaur objects; however, the motion complicates physical studies of a known KBO or Centaur when the image of the KBO or Centaur comes close to an image of a background star or galaxy.

2. NAMING OBJECTS

After a Centaur object or KBO is discovered, it needs a name. The IAU is responsible for naming celestial objects. The IAU names KBOs and Centaurs the same way it names asteroids. Upon discovery, an object is given a preliminary designation consisting of a four-digit number indicating the year of discovery, a letter to indicate the half month of discovery, another letter to indicate the order of discovery within the half month, and another number to indicate the number of times the second letter was repeated within the half month period. For example, the provisional name of the KBO 2002 LM_{60} tells us the object was discovered between June 1 and 15 of 2002. After the orbit of a KBO about the Sun becomes well enough known that it is not likely to be lost, the KBO is given a number. It can take observations over several years to establish a good orbit for a KBO. The number of 2002 LM_{60} is 50,000. No other solar system object has the number 50,000. After an object receives a number, it receives a name. For example, 2002 LM_{60} is known as Quaoar. In this case, the same KBO has three names. After an object has a number and name, it is rarely called by its provisional name. If a KBO or Centaur object has a number or name, we know that its orbit about the Sun is well established and that there is very little chance of losing it.

3. DATABASES OF KNOWN OBJECTS

The IAU maintains an Internet listing of known KBOs and Centaurs as well as elements that describe their orbits about the Sun. In addition, the IAU, Lowell Observatory, and National Aeronautics and Space Administration's (NASA's) Jet Propulsion Laboratory in Pasadena, California, provide Internet tools that enable observers to figure out where to point telescopes to see a specific KBO or Centaur object on a specific night. Links to these tools are given in Table 44.1.

4. DYNAMICAL CLASSES

It is possible to divide KBOs into dynamical classes. This section provides a brief discussion of the classes and likely interconnections between the classes. A thorough discussion of KBO dynamics can be found in chapter 43 by Allesandro Morbidelli and Harold Levison in this book. (See Kuiper Belt: Dynamics.)

Classical KBOs are on orbits with perihelion distances, q, larger than 40 AU; semimajor axes, a, between 42 and 45 AU; eccentricities, e, less than 0.1; and inclination angles, i, less than $10°$. It appears that classical KBOs did not experience strong perturbations by Neptune and hence they probably formed at or near their present location. Resonant KBOs are a subset of classical KBOs that became trapped in mean-motion resonances during the primordial migration of the planets. The process of resonance trapping tends

TABLE 44.1 KBO and Centaur Internet Tools

Institution	Web Addresses
Lowell Observatory	http://asteroid.lowell.edu/cgi-bin/asteph
NASA/JPL	http://ssd.jpl.nasa.gov/?tools
IAU	http://mpc.cfa.harvard.edu//mpc/summary

to increase the **eccentricity** and inclination of trapped objects. Plutinos are objects trapped in the 2:3 mean-motion resonance of Neptune at $a = 39.6$ AU, just like Pluto. Scattered disk objects (SDOs) are thought to have originated in the primordial inner belt, $a < 40$ AU, and the primordial Uranus–Neptune region. They were subsequently scattered by Neptune onto orbits with large inclination angles, $i > 15°$; large eccentricities, $e > 0.3$; and large semimajor axes, $a > 45$ AU.

Centaur objects are on outer planet crossing orbits with $q > 5.2$ AU and $a < 30.1$ AU. Relatively recent gravitational interactions between SDOs and Neptune, and to a lesser extent between classical KBOs and Neptune, result in Centaur objects. Because Centaur objects cross the orbits of the outer planets, they are dynamically unstable and have mean lifetimes of $\sim 10^6$ years. As mentioned above, some Centaurs evolve into Jupiter-family comets, others are ejected from the solar system, and yet others impact the giant planets. In addition, some Jupiter-family comets evolve back into Centaurs.

5. BRIGHTNESS

5.1. Apparent Magnitude

The first physical property measured for a KBO is typically its brightness. A KBO is brightest in visible light (4000–8000 Å) by virtue of the sunlight it reflects toward the Earth. It is possible to isolate the brightness of a KBO in a particular bandpass by placing a colored glass filter in front of a CCD camera at the focal plane of a telescope. For example, a blue, green, or red filter in front of a CCD camera makes it possible to measure the brightness of blue, green, or red light from a KBO, i.e. its B ($\lambda_{center} = 4500$ Å), V ($\lambda_{center} = 5500$ Å), or R ($\lambda_{center} = 6500$ Å) magnitudes. Table 44.2 lists V

TABLE 44.2 KBO Magnitudes, Albedos, and Diameters[1]

Name	Number	Prov Des	V[2]	H_v[3]	p_v[4]	D[5]
Triton			13.5	−1.2	75	2707
Eris	136199	2003 UB$_{313}$	18.7	−1.1	>70	<2600
Pluto	134340		14.0	−0.7	61	2290
	136472	2005 FY$_9$	17.0	0.1	70–90	1250–1650
Charon			15.9		37	1242
	136108	2003 EL$_{61}$	17.5	0.4	55–75	1000–1600
Sedna	90377	2003 VB$_{12}$	21.1	1.20	>8.5	<1800
Orcus	90482	2004 DW	19.3	2.3	27	1000
Quaoar	50000	2002 LM$_{60}$	19.2	2.7	12	1300
	55637	2002 UX$_{25}$	19.9	3.6	10	900
	55565	2002 AW$_{197}$	20.2	3.6	12	734
	90568	2004 GV$_9$	19.8	3.7	15	700
Varuna	20000	2000 WR$_{106}$	20.1	3.9	14	586
Ixion	28978	2001 KX$_{76}$	19.9	4.0	19	480
Huya	38628	2000 EB$_{173}$	19.5	5.1	6.6	500
	47171	1999 TC$_{36}$	19.6	5.4	7.9	405
	15874	1996 TL$_{66}$	20.9	5.5	>1.8	<958
	15789	1993 SC	22.4	7.3	3.5	398
	15875	1996 TP$_{66}$	21.1	7.4	1.1	406
	29981	1999 TD$_{10}$	21.1	9.1	5.3	88

[1] *Courtesy John Stansberry.*
[2] *V-band magnitude.*
[3] *Absolute magnitude in V-band.*
[4] *Visual albedo in units of percentage from Spitzer Space Telescope and ISO observations.*
[5] *Diameter in kilometers from Spitzer Space Telescope and ISO observations.*

magnitudes of the brightest KBOs. At the other extreme of brightness, Gary Bernstein used the Hubble Space Telescope (HST) to discover and measure the brightness of the faintest known KBO, $V \sim 28$. The Centaur in Figure 44.3, 1994 TA, has $V = 24.31 \pm 0.05$. For comparison, the Sun has $V = -26.74$ and the faintest star visible in the sky with the unaided eye has $V \sim 6$.

5.2. Luminosity Function

There are many more faint KBOs than bright KBOs. Figure 44.4 comes from KBO discoveries made by a number of surveys, and shows the number of KBOs per unit magnitude per square degree on the sky near the **ecliptic** plane as a function of brightness (R-band magnitude), a luminosity function. Surveys find ~ 100 KBOs with $27 < R < 28$, ~ 2 KBOs with $23 < R < 24$, and only ~ 0.001 KBO with $19 < R < 20$, all per square degree of sky. For reference, the full Moon occupies almost one quarter of a square degree of sky and the Sun has $R = -27.10$.

5.3. Absolute Magnitude

The apparent magnitude of a KBO or Centaur depends on its heliocentric distance, r, and geocentric distance, Δ, in astronomical units. For example, a KBO receding from the Sun and Earth will become fainter and its apparent magnitude will become larger in value. The absolute magnitude, H, of a KBO is a way to compare the intrinsic brightness of one KBO with another KBO and it does not depend on distance. The absolute magnitude of the same KBO receding from the Sun and Earth will not change. The absolute magnitude of a KBO is the brightness it would have if it were located at a distance of 1 AU from the Sun and 1 AU from the Earth, and had a Sun—KBO—Earth (phase) angle, α, of $0°$. The relation between absolute magnitude, H_v, and apparent magnitude, V, is given by

$$H_v = V - 5 \log(r\Delta) + 2.5 \log[(1-G)\Phi_1(\alpha)G\Phi_2(\alpha)],$$

where the last term of the equation is an empirical phase function that describes how H_v of an object varies with phase angle. $G = 0.15$ and Φ_1 and Φ_2 are given by

$$\Phi_i(\alpha) = \exp\left[-A_i\left(\tan\frac{1}{2}\alpha\right)^{B_i}\right]$$

where $i = 1$ and 2, $A_1 = 3.33$, $B_1 = 0.63$, $A_2 = 1.87$, and $B_2 = 1.22$ seem most appropriate for KBOs. H_v values for discovered KBOs and Centaurs range from about -1 to 15. Table 44.2 lists KBOs with the brightest H_v values.

6. DIAMETER

Size is among the most fundamental physical properties of an astronomical object, yet we are only beginning to get accurate diameter measurements for KBOs and Centaurs. The most direct way to measure the diameter of a KBO, D (in kilometer), is to measure its angular diameter, θ (in arcsec), and geocentric distance, Δ (in astronomical units). Geometry gives

$$D = 727\Delta\theta.$$

Unfortunately, KBOs and Centaurs have sufficiently small values for D and sufficiently large values for Δ that the resulting values for θ are too small for measurement even by the HST. Michael Brown pushed the HST to its limits and measured $\theta = 0.0343 \pm 0.0014$ arcsec for the KBO Eris, which was at a geocentric distance of 96.4 AU at the time of their observations. They found a diameter of 2400 ± 100 km for Eris, making it slightly larger than Pluto, $D = 2302$ km. For KBOs and Centaurs with θ too small for measurement, it is possible to estimate their diameters from their brightness,

$$p\Phi D^2 = 9x10^{16}r^2\Delta^2 10^{0.4(m-V)},$$

where, as before, r and Δ are the heliocentric and geocentric distance, respectively, in astronomical units; m is the V-band brightness of the Sun (-26.74); V is the brightness of the KBO; p is the albedo of the object, and

$$\Phi = [(1-G)\Phi_1(\alpha) + G\Phi_2(\alpha)].$$

Since Jupiter-family comets come from Centaurs and the Kuiper Belt, most KBO diameter estimates assume an albedo similar to albedo measurements for a handful of Jupiter-family comets, i.e. $p = 0.04$. Diameter estimates from V magnitudes for about 100 objects range between $D = 25$ km for 2003 BH91 and $D = 2400$ km for Eris. KBO

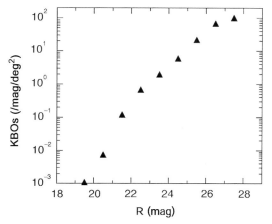

FIGURE 44.4 Number of KBOs per unit magnitude interval per square degree of sky vs R-band magnitude. There are many more faint KBOs than bright KBOs. This is typical of small body populations in the solar system that have been collisionally processed. *Courtesy of Gary Bernstein.*

and Centaur object diameters on the scale of Figure 44.1 range from the tiniest specks to Pluto.

The assumption of a cometlike albedo, although reasonable, is dangerous because Jupiter-family comets come much closer to the Sun than KBOs and Centaur objects. The frequent close proximity of short-period comets to the Sun results in the sublimation of H_2O ice and produces surfaces largely covered by a dark, refractory-rich, lag deposit. The surfaces of Jupiter-family comets may have chemical and physical properties quite different from those of the surfaces of Centaurs and KBOs. Charon has a relatively large albedo of 0.37. If we assume that a KBO has $p = 0.04$, but it actually has $p = 0.4$, we will estimate a diameter that is more than three times too large. Measurements of albedos are essential for accurate measurements of KBO and Centaur object diameters.

7. ALBEDO

By measuring the brightness of sunlight *reflected* from a KBO at visible wavelengths and the brightness of heat emitted by the same KBO at thermal infrared wavelengths, it is possible to disentangle albedo from diameter, and thereby measure separate values for both quantities. The Spitzer Space Telescope, an infrared telescope in orbit about the Sun, is enabling John Stansberry of the University of Arizona, Dale Cruikshank of NASA's Ames Research Center, William Grundy of Lowell Observatory, and John Spencer of Southwest Research Institute to observe much fainter levels of heat from KBOs and Centaurs than is possible with telescopes on the Earth. As a result of their work, we have accurate diameters and albedos for more than a dozen KBOs (Table 44.2).

8. BRIGHTNESS VARIATION

KBOs and Centaurs may have weak internal constitutions (i.e. rubble pile-type interiors) due to fracturing by past impacts between objects. In other words, it is possible that KBOs and Centaurs are nearly strengthless bodies, held together primarily by their own self-gravity. If so, then some objects may deform from spheres into triaxial ellipsoids with axes $a > b > c$ as a result of their rotation. The rotation of an ellipsoid can result in periodic variation of its projected area on the sky and hence a periodic variation of the sunlight it reflects and its brightness (Figure 44.5). Monitoring such a brightness variation can result in a wealth of physical data about the object (e.g. its period of rotation, shape, and perhaps even its density and porosity).

8.1. Period of Rotation

If we can determine the form of the periodic brightness variation (light curve) for a KBO or Centaur, its period of

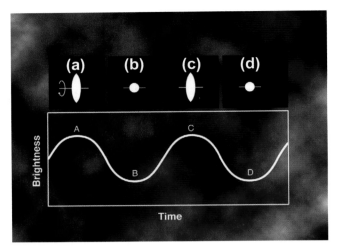

FIGURE 44.5 (a) The rotation of a nonspherical KBO or Centaur object results in a periodic variation of the object's projected area on the plane of the sky and hence a periodic variation in its brightness. (b) Brightness vs time (light curve) for the rotation of a nonspherical object. During one rotation of the object, it goes through two maxima (points A and C) and two minima (points B and D) in brightness. *Courtesy of Ron Redsteer and NAU Bilby Research Center.*

rotation can be determined. Figure 44.6 shows a plot of V magnitude vs time in hours for the Centaur Pholus. At the time of observation in 2003, Pholus was ~18 AU from the Sun, nearly the same distance as Uranus. We see that the two maxima are of nearly equal brightness, but one minimum (at ~5 h) is ~0.03 magnitude (3%) fainter than the other minimum (at 0 h). The pattern of two maxima and two minima repeats every 9.980 ± 0.002 h, Pholus' period of rotation on its axis.

Determining the period of rotation for a KBO or Centaur takes a significant amount of telescope time. In the

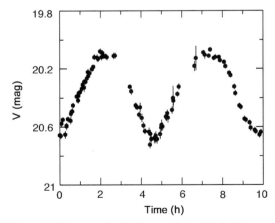

FIGURE 44.6 Light curve for the Centaur Pholus. The brightness pattern of two nearly equal brightness maxima and two brightness minima (at ~0 and ~5 h) that differ by 0.03 magnitude (3%) repeats every 9.980 h. The maximum peak-to-trough brightness variation of Pholus is 0.60 magnitude. From the light curve, we know that Pholus rotates once about its axis every 9.980 h.

case of Pholus, each of the 99 points in Figure 44.6 represents a brightness measurement from a 300-s CCD image. Because of their faintness, measurements of KBO and Centaur **light curves** require telescopes with moderately large apertures, typically with diameters ≥ 2 m. A large amount of time on moderate-sized telescopes is difficult to obtain, so periods of rotations are available for only a handful of objects (Table 44.3). Groups led by William Romanishin of the University of Oklahoma and Scott Sheppard of the University of Hawaii are responsible for many light curve measurements. They find periods of rotation between 4 and 14 h (Table 44.3).

8.2. Amplitude

In Figure 44.6, Pholus has a maximum brightness of $V_{max} = 20.09$ and a minimum brightness of $V_{min} = 20.69$, i.e. each time through its repeating pattern it has a maximum brightness variation or light curve amplitude of $\Delta m = V_{min} - V_m = 0.60$ magnitude. Since

$$\Delta m = 2.5 \log \frac{F_{max}}{F_{min}},$$

the ratio of maximum to minimum brightness is $F_{max}/F_{min} = 1.74$. From Table 44.3, we see that KBOs and Centaurs exhibit $0.1 \leq \Delta m \leq 1.1$ magnitude.

8.3. Shape

If a KBO or Centaur light curve is due to the rotation of a triaxial ellipsoid about its shortest axis, c, we can in principle determine its shape (i.e. the ratio of its axes a/b and c/b). How? As a KBO or Centaur orbits the Sun, we observe it at different aspect angles. Aspect angle is the angle between lines originating at the center of the body and toward the Earth and the north rotational pole of the body. Figure 44.7 illustrates how a change in aspect angle results in a change in light curve amplitude. At point A, we are looking at the object equator-on (aspect angle of 90°), and we see a light curve with an amplitude as large as it gets for the object. A quarter of a revolution about the Sun later, at point B, we are looking down the rotation axis of the body (aspect angle of 0°), and we will not see any brightness variation. If we observe the amplitude change of a body's light curve over a significant portion of the object's revolution about the Sun, it is possible to use a computer to search through all possible combinations of shape and orientation of the rotation axis to find a shape that best simulates the observed amplitude changes.

Figure 44.8 shows three light curves of Pholus from 1992, 2000, and 2003. The x-axis is labeled with the rotational phase of Pholus. The rotational phase interval of 0–1 is equal to a time interval of 9.980 h, the time it takes Pholus to complete one rotation about its axis. The amplitude of the light curves grew from 0.15 to 0.39 to 0.60

TABLE 44.3 Rotation Periods and Light curve Amplitudes

Name	Number	Prov Des	Class[1]	Period[2]	Δm[3]
Varuna	136108	2003 EL$_{61}$	kbo	3.9154	0.28
	15820	1994 TB	kbo	6.0, 7.0	0.30
	20000	2000 WR$_{106}$	kbo	6.34	0.42
	26308	1998 SM$_{165}$	kbo	7.1	0.45
	32929	1995 QY$_9$	kbo	7.3	0.60
	19255	1994 VK$_8$	kbo	7.8, 8.6, 9.4, 10.4	0.42
	19308	1996 TO$_{66}$	kbo	7.9	0.25
	47932	2000 GN$_{171}$	kbo	8.329	0.61
	33128	1998 BU$_{48}$	kbo	9.8, 12.6	0.68
Pholus	5145	1992 AD	cen	9.980	0.60
	40314	1999 KR$_{16}$	kbo	11.858, 11.680	0.18
		2001 QG$_{298}$	kbo	13.7744	1.14

[1] Dynamical class: Kuiper Belt object (kbo) or Centaur object (cen).
[2] Period of rotation in hours. Multiple entries indicate possible periods.
[3] Peak-to-trough amplitude in magnitudes.

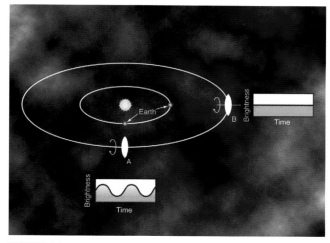

FIGURE 44.7 Changes in aspect angle result in changes in light curve amplitude. When the object is at point A, the angle between two lines originating at the center of the body and toward the Earth and the north rotational pole of the body, the aspect angle, is 90°. We see the object with an equator-on aspect and the light curve of the object has its maximum amplitude. When the body moves to point B, the aspect angle is 0° and we are looking down on the rotational pole. At point B, the object does not exhibit any brightness variation; its light curve amplitude is zero. By monitoring changes in the light curve amplitude of a body as it orbits the Sun, it is possible to calculate the shape of the body. *Courtesy of Ron Redsteer and NAU Bilby Research Center.*

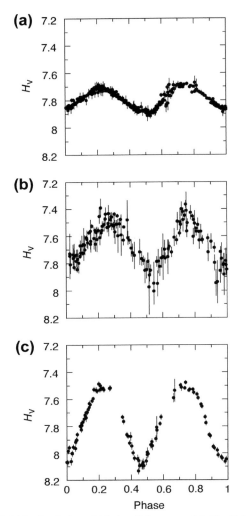

FIGURE 44.8 Evolution of Pholus light curve. (a) The light curve observed by Marc Buie and Bobby Bus in 1992 has an amplitude of 0.15 magnitude. (b) The light curve observed by Tony Farnham in 2000 has an amplitude of 0.39 magnitude. (c) The light curve observed by Bill Romanishin and Guy Consolmagno in 2003 has an amplitude of 0.60 magnitude. The period in 1992, 2000, and 2003 remained constant at 9.980 h; however, the increasing amplitude indicates that we were seeing Pholus more equator-on with each passing year between 1992 and 2003.

magnitude. A computer search of orientations of the rotation axis and shapes for Pholus yields four possible orientations for the rotational axis, all with the same shape of $a/b = 1.9$ and $c/b = 0.9$. Pholus appears to have a significantly elongated shape.

The amplitude measurements of Pholus span little more than 10% of its 92-year period of revolution about the Sun. Confirmation of the shape for Pholus will require additional amplitude measurements two or three decades into the future.

KBO shape measurements require amplitude measurements over more than a century. Yet, we can still say something about the shapes without waiting a century. For example, the KBO Varuna has $\Delta m = 0.42 \pm 0.02$ magnitude. If we assume we are seeing Varuna with an "equator-on" aspect (i.e. an aspect angle of 90°), which corresponds to the largest possible light curve amplitude, we can relate the amplitude of the light curve to an axial ratio,

$$\Delta m = 2.5 \log \frac{a}{b}.$$

Such an assumption gives $a/b = 1.5$ for Varuna. Since we do not know if they are viewing Varuna "equator-on", Δm and a/b are lower limits. At the present time, all KBO axial ratios are lower limits.

8.4. Density

Besides periods of rotation and shapes, light curves allow us to estimate densities for KBOs and Centaurs. If nonspherical shapes are the result of rotational deformation, then we can use the formalism developed by Chandrasekhar that relates the period of rotation and shape to the density of a strengthless ellipsoid. Application of Chandrasekhar's formalism to Pholus, the object with the best shape estimation ($a/b = 1.9$ and $c/b = 0.9$) and period of rotation ($P_{\rm rot} = 9.980$ h), gives an average density of $\rho_{avg} = 0.5$ g/cm^3. It is interesting to note that the similar-sized Saturnian satellites Janus, Epimetheus, Prometheus, and Pandora have average densities of 0.61, 0.64, 0.42, and 0.52 g/cm^3. In the case of KBO Varuna, its assumed shape ($a/b = 1.5$ and $c/b = 0.7$) and period of rotation ($P_{\rm rot} = 6.3442$ h) yield $\rho_{avg} = 1.0$ g/cm^3. Average densities ≤ 1 g/cm^3 suggest that KBOs and Centaurs likely have ice-rich and porous interiors.

8.5. Porosity

Porosity is the fraction of void space in a KBO or Centaur. If a KBO is some mixture of ice, refractory material (dust), and empty space, the average density of a KBO is given by

$$\rho_{\rm avg} = f_i \rho_i + f_r \rho_r,$$

where f_i and f_r are the fractional volumes occupied by icy and refractory material and ρ_i and ρ_r are the densities of icy and refractory material. In addition, the sum of the parts must equal the whole, so

$$f_i + f_r + f_v = 1,$$

where f_v is the fraction of void space or the porosity. The fraction of total mass locked up in refractories, ψ, is given by

$$\psi = \frac{\rho_r f_r}{\rho_r f_r - \rho_i f_i}.$$

By combining the above three equations, it is possible to obtain an algebraic expression for the porosity of a KBO or Centaur,

$$f_v = 1 - \frac{\rho_{avg}}{\rho_i}\left[1 + \psi\left(\frac{\rho_i}{\rho_r} - 1\right)\right].$$

Assuming reasonable values of $\rho_{avg} = 1$ g/cm^3, $\rho_i = 1$ g/cm^3, $\rho_r = 2$ g/cm^3, and $\psi = 0.5$ for Varuna, Jewitt and Sheppard estimate that Varuna has a porosity ~ 0.25. For comparison, beach sand has $f_v \sim 0.4$ and basaltic lunar regolith has $0.4 < f_v < 0.7$.

9. COMPOSITION

9.1. Surface Color

An early expectation was that all KBOs should exhibit a similar red surface color. Why? Initially, KBOs were thought to form over a small range of heliocentric distances where the temperature in the young solar nebula was the same. The similar temperature suggested that KBOs formed out of the solar nebula with the same mixture of molecular ices and the same ratio of dust to icy material. In addition, their similar formation distance from the Sun suggested that KBOs should experience a similar evolution. Specifically, the irradiation of surface CH_4 ice by solar ultraviolet light and solar wind particles should have converted some surface CH_4 ice into red, complex, organic molecules. By their nature, the complex organic molecules were expected to absorb more incident blue sunlight than red sunlight. Therefore, the light reflected from the surfaces of KBOs was expected to consist of a larger ratio of red to blue light than the incident sunlight. It was a surprise to find that KBOs exhibit a range of surface colors rather than just red colors. At one extreme, some KBOs reflect sunlight equally at all wavelengths (i.e. exhibit neutral or gray surface colors). On the Johnson–Kron–Cousins photometric system such KBOs have B-R = 1.0. At the other extreme, some KBOs have extraordinary red colors, i.e. B-R = 2.0.

Because it is a painstaking process to measure the color of a KBO or Centaur, taking as much as 3 h of telescope time to obtain an accurate color for a single object, the first color surveys consisted of only 10 to 20 objects. These small samples lumped KBOs and Centaurs together and resulted in a controversy. Some groups found their samples to exhibit a uniform distribution of colors from gray (B-R = 1.0) to extraordinarily red (B-R = 2.0). These groups suggested that KBOs and Centaurs experienced a steady reddening of their surfaces by solar radiation, and occasional impacts by smaller objects punctured the red surfaces and excavated gray, interior material. Such a radiation-reddening and impact-graying mechanism would explain the uniform distribution of colors. Another group found that their sample of KBOs and Centaurs divided into two distinct color groups—gray objects with $1.0 < $ B-R $ < 1.4$ and red objects with $1.5 < $ B-R $ < 2.0$. They found almost no objects with $1.4 < $ B-R $ < 1.5$. They did not have a physical explanation for their surprising result. Everyone agreed that KBOs and Centaurs did not exhibit only red surface colors.

As the groups pressed hard at telescopes to measure more surface colors and test their initial findings, sample sizes grew from 10 to more than 100 objects. Once the sample sizes became large enough, it became apparent that different dynamical classes of KBOs had different color signatures.

9.1.1. Centaur Objects

Figure 44.9(a) shows a histogram of the number of objects vs B-R color for a sample of 22 Centaur objects. Fourteen objects have B-R < 1.3 and eight objects have B-R > 1.7. Notice that there are no objects with $1.3 < $ B-R $ < 1.7$. Is it possible that Centaurs actually exhibit a uniform distribution of B-R colors and either insufficient sampling or chance is responsible for the apparent split into two B-R color groups? Application of statistical tests like the "dip test" tell us that the probability of making observations in Figure 44.9(a) for an actual uniform distribution of B-R colors is about 1 in 100. The split into two B-R color groups appears to be real. Unlike the earlier controversy, two groups, one led by Nuno Peixinho and the other by this author, find the same highly unusual split. What makes the split so unusual is that there does not seem to be any other physical property that correlates with the color of a Centaur object. For example, if Centaurs that came closest to the Sun were all gray, we might suspect that the warmth of the Sun was chemically or physically altering the surfaces and graying them. But there is no statistically significant correlation between color and **perihelion distance** or any other orbital element.

9.1.2. Classical KBOs

Figure 44.9(b) shows a histogram of the number of objects vs B-R color for a sample of 21 classical KBOs. All 21 classical objects have B-R > 1.5, i.e. there are no gray objects at all in the sample. Classical KBOs exhibit the color signature originally expected for all KBOs.

9.1.3. Scattered Disk Objects

Figure 44.9(c) shows a histogram of the number of objects vs B-R color for a sample of 20 SDOs. Seventeen of the 20 objects exhibit B-R < 1.5. There appears to be a deficit of red objects among this group.

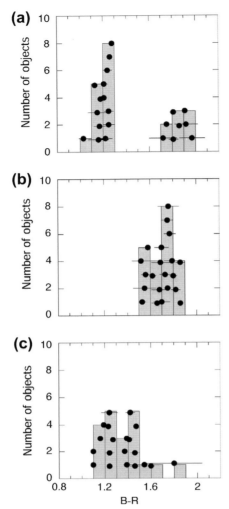

FIGURE 44.9 Correlations between colors and orbital properties of KBOs and Centaurs. (a) A sample of 22 Centaurs neatly divide into two color groups; 14 objects exhibit B-R < 1.3 and eight objects exhibit B-R > 1.7. Surprisingly, there are no Centaurs with 1.3 < B-R < 1.7. (b) All 21 objects of a sample of classical KBOs with $q > 40$ AU, $e < 0.1$, and $e < 10°$ are red (B-R > 1.5). (c) A sample of 20 SDOs are mostly gray (B-R < 1.5). The mechanisms responsible for these correlations between color and orbital properties are not well understood yet.

9.1.4. Reasons for Color Patterns

What could cause these color signatures? One possibility is the radiation-reddening and impact-graying mechanism discussed earlier in Section 1. However, such a mechanism should result in a uniform distribution of B-R colors for Centaurs and not two clusters of B-R colors. In addition, gray impact craters and their ejecta blankets would be randomly distributed on the surface so that one hemisphere might have more than another, resulting in measurable color changes as the object rotates. However, repeated and random measurements of individual rotating KBOs and Centaurs give the same B-R color. Also, extensive observations of Pholus suggest that it has a highly homogeneous surface color. Figures 44.10(a) and 44.10(b) show the R-band brightness and B-band brightness of Pholus as a function of a single rotation phase taking 9.980 h Figure 44.10(c) is the difference of panel a from panel b, yielding the B-R color across the entire surface of Pholus as it makes one rotation about its axis. The solid horizontal line is the average of the points. The dashed lines are ± 1 standard deviation, $\sigma = 0.04$. Any variation in the B-R surface color of Pholus must be smaller than 0.04 magnitude (4%). Again, there is no evidence of gray impact craters on a radiation-reddened surface.

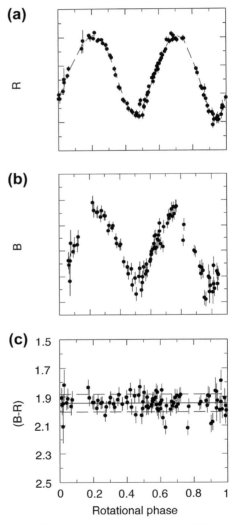

FIGURE 44.10 Homogeneous B-R surface color of Pholus. (a) R-band magnitude vs rotation phase. The x-axis spans a time interval of 9.980 h (b) B-band magnitude vs rotational phase. (c) Difference between the above two panels yields B-R color vs rotational phase. The solid line is the average of the 94 points. The dashed lines are plus or minus one standard deviation, σ, of 0.04 magnitude. Any variation in the surface color of Pholus as it completes one rotation on its axis must be less than 0.04 magnitude (4%). Pholus exhibits a homogeneous surface color.

Another possibility is that the colors of KBOs are the remaining signature of a temperature-induced, primordial composition gradient. The small, rocky terrestrial planets (Mercury, Venus, Earth, and Mars) close to the Sun and the giant, hydrogen-rich gas giant planets (Jupiter, Saturn, Uranus, and Neptune) farther away from the Sun are the result of such a gradient. In the inner solar system, temperatures were so high that only metal- and rock-forming elements could condense from the nebular gas to form small, rocky, and metal-rich solids. At and beyond the orbit of Jupiter, the hydrogen-dominated nebular gas was cold enough for the H_2O to condense out. We may be seeing a similar effect on the colors of KBOs and Centaurs. We now suspect that KBOs did not all form at about the same distance from the Sun. Perhaps the red classical KBOs formed farther out in the nebula where it was cold enough to hang on to their CH_4 ice reddening agent. Perhaps the gray KBOs formed closer to the Sun and were not able to hang on to their CH_4 ice reddening agent.

Additional work is necessary to figure out whether the radiation-reddening and collisional-graying mechanism, the temperature-gradient mechanism, or some other mechanism is responsible for the colors of KBOs and Centaurs.

9.2. Spectroscopy

There are only a handful of KBOs and Centaurs that are known to exhibit ice absorption bands in their spectra. H_2O-ice bands are seen in the spectra of Charon, 19308 (1996 TO_{66}), Varuna, Quaoar, Orcus, Pholus, and Chariklo. CH_4-ice bands are seen in the spectra of Pluto; Neptune's satellite Triton, which may be a captured KBO; Eris; and 136472 (2005 FY_9).

Perhaps one of the most intriguing spectroscopic results comes from David Jewitt and Jane Luu's observations of Quaoar. Specifically, they find not only the H_2O-ice bands at 1.5 and 2.0 μm, but also another H_2O band at 1.65 μm (Figure 44.11). The latter band suggests the surprising result that the H_2O ice has a crystalline rather than an amorphous structure. The H_2O molecules of crystalline ice have a periodic structure, whereas the H_2O molecules of amorphous ice do not. Crystalline H_2O on Quaoar is a surprise because Quaoar's maximum surface temperature is only ~50 K. At such a low temperature, it is difficult for the H_2O molecules to arrange themselves into a coordinated structure of a crystal lattice; somewhere around 100 K, amorphous ice arranges itself into an ordered crystalline lattice. In other words, the 1.65-μm band suggests that the H_2O ice on Quaoar was somehow heated to temperatures above 100 K.

An intriguing possibility for the source of the "warm" H_2O on Quaoar is NH_3-H_2O volcanism. Long ago, long-lived radioactive elements heated the interior of Quaoar,

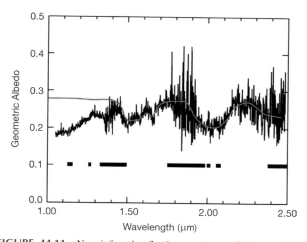

FIGURE 44.11 Near-infrared reflection spectrum of Quaoar (black) compared to a spectrum of H_2O ice (red). The broad absorption bands near 1.5 and 2.0 μm reveal the presence of H_2O ice on the surface of Quaoar. The narrow absorption band near 1.65 μm indicates the presence of crystalline H_2O ice and is not present in amorphous ice. *Courtesy of David Jewitt and Jane Luu.*

and that heat may still be propagating through its interior. The heat may have been sufficient to create a melt of H_2O and NH_3. The lower density melt may have percolated upward, perhaps forming fluid-filled cracks all the way or nearly all the way to the surface in the surrounding, higher density icy–rock mixture. Eventually, the cooling "lava" containing crystalline H_2O ice and crystalline ammonium hydrate might become exposed by occasional impacts on Quaoar's surface. What makes this mechanism even more intriguing is that Jewitt and Luu claim that there is evidence for an ammonia hydrate band in their spectra of Quaoar. Ammonia–water volcanism as the source of the crystalline H_2O ice is highly speculative. Some other mechanism, not requiring a warm interior and volcanoes, may explain the presence of the crystalline H_2O ice on Quaoar.

Figure 44.12 illustrates that Quaoar has a spectrum similar to Charon, but quite different from Pluto. Quaoar and Charon exhibit the 1.5- and 2.0-μm H_2O-ice bands as well as the 1.65-μm crystalline band, but none of the strong CH_4-ice bands seen on Pluto. Note that Quaoar has the 1.65-μm band despite having a larger semimajor axis, $a = 43.6$ AU, than Pluto, $a = 39.8$ AU.

Another intriguing spectroscopic result comes from Javier Licandro's observations of 136472 (2005 FY_9). He finds that CH_4-ice bands in the spectra of 136472 (2005 FY_9) are much deeper than the CH_4-ice bands in the spectra of Pluto, implying that the abundance of CH_4 on the surface of 136472 (2005 FY_9) could be higher than on the surface of Pluto. This author finds that the CH_4-ice bands in his spectrum and Javier Licandro's spectrum of 136472 (2005 FY_9) are blue shifted by

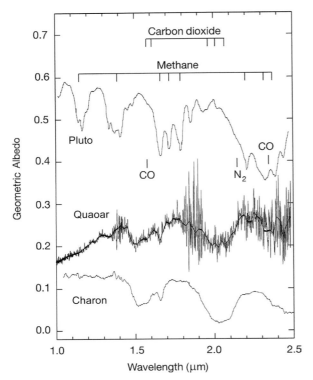

FIGURE 44.12 Near-infrared spectrum of Quaoar compared to near-infrared spectra of Pluto and Charon. The spectra of Quaoar and Charon are similar in that they exhibit three strong H_2O-ice absorption bands at 1.5, 1.65, and 2.0 μm, but no CH_4-ice bands. The spectrum of Pluto exhibits strong CH_4-ice bands. *Courtesy of David Jewitt and Jane Luu.*

3.25 Å relative to the positions of pure CH_4-ice bands (Figure 44.13). Such a shift suggests the presence of another ice component on the surface of 136472 (2005 FY_9), possibly N_2 ice, CO ice, or Ar. In addition,

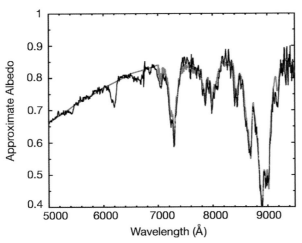

FIGURE 44.13 Optical spectrum of 136472 (2005 EY_9) (black line) and a Hapke model of pure CH_4 ice (red line). The CH_4 absorption bands of 136472 (2005 FY_9) are blue shifted by 3.25 ± 2.25 Å relative to the pure CH_4 model indicating the presence of another molecular ice, possibly N_2, CO, or Ar.

Licandro finds CH_4-ice bands blue shifted in a spectrum of Eris.

It is odd that some KBOs exhibit strong CH_4 bands and others exhibit strong H_2O bands. Pluto and Charon are part of the same system, yet they exhibit very different spectra. Perhaps the difference is due to Pluto's size, it may have experienced some form of methane ice volcanism. In the end, we may find that only the largest KBOs exhibit CH_4-ice bands. Eris, Pluto, and possibly 2005 FY_9 are the three largest KBOs and they all exhibit CH_4-ice bands.

10. KBO BINARIES

In 2001, Christian Veillet announced the discovery of two components to the KBO 1998 WW_{31}. Over the next few years, Keith Noll used the superior imaging resolution of HST to observe 122 KBOs for additional binaries. His survey was sensitive to binaries with separations ≥0.15 arcsec and a magnitude difference between components ≤1 magnitude. Noll discovered six more binaries. Currently, 22 KBO binaries are known (Table 44.4).

10.1. System Mass

Two KBOs of a binary pair revolve about their common center of mass. However, it is far more convenient to observe the position of the fainter of the two components as it makes a complete revolution about the brighter component on the plane of the sky, i.e. to observe the apparent relative orbit. Figure 44.14 illustrates the apparent relative orbit of 1998 WW_{31}. The true orbit of the KBO binary system will not happen to lie exactly in the plane of the sky. Hence, the apparent relative orbit is merely a projection of the true relative orbit onto the plane of the sky. Techniques exist to determine the inclination of the true orbit relative to the plane of the sky. Once the period of revolution, P, and the semimajor axis, a, of the true relative orbit are known, it is possible to use Kepler's Third Law to calculate the combined mass of the binary system,

$$m_1 + m_2 = \frac{4\pi^2 a^3}{GP^2}.$$

From the HST observations in Figure 44.14, Veillet and Noll found that 1998 WW_{31} has a true relative orbit with a semimajor axis of 22,300 km, an eccentricity of 0.8, and a period of revolution of the fainter component about the brighter component of 574 days. The 1998 WW_{31} system has a combined mass of 2.7×10^{18} kg, much smaller than the Pluto–Charon system combined mass of 1.46×10^{22} kg. Table 44.4 lists the true relative orbital properties and combined masses for the better studied binary systems.

TABLE 44.4 Binary KBOs[1]

Name	Number	Prov Des	a[2]	e[3]	Period[4]	Mass[5]
Resonant						
Pluto/Charon	134340		19,636	0.0076	6.38722	14,710
	47171	1999 TC$_{36}$	7640		50.4	13.9
	26308	1998 SM$_{165}$	11,310		130	6.78
Classical						
		2005 EO$_{304}$				
		2003 UN$_{284}$				
		2003 QY$_{90}$				
		2001 QW$_{322}$				
		2000 CQ$_{114}$				
		2000 CF$_{105}$				
		1999 OJ$_{4}$				
		1998 WW$_{31}$	22,300	0.82	574	2.7
	134860	2000 OJ$_{67}$				
	88611	2001 QT$_{297}$	27,300	0.240	825	2.3
	80806	2000 CM$_{105}$				
	79360	1997 CS$_{29}$				
	66652	1999 RZ$_{253}$	4660	0.46	46.263	3.7
	58534	1997 CQ$_{29}$	8010	0.45	312	0.42
Scattered						
		2001 QC$_{298}$	3690		19.2	10.8
Eris						
	136199	2003 UB$_{313}$				
	136108	2003 EL$_{61}$	49,500	0.05	49.12	4200
	82075	2000 YW$_{134}$				
	48639	1995 TL$_{8}$				

[1] Courtesy Keith Noll.
[2] Semimajor axis in kilometers.
[3] Eccentricity.
[4] Period in days.
[5] Mass in units of 10^{18} kg.

10.2. Mutual Events

Between 1985 and 1990, Pluto's orbital motion about the Sun caused the Pluto–Charon orbital plane to sweep through the line of sight to the Earth. As a result, mutual eclipses (also known as mutual events) occurred every 3.2 days (half of Charon's orbital period). Because of the mutual events, observers were able to accurately measure diameters of 2302 ± 12 km and 1186 ± 26 km for Pluto and Charon, and with the total mass of the binary, they were able to derive an average density for the system of 1.95 ± 0.10 g/cm^3.

A key objective of current binary KBO work is to discover as many binaries as possible and to determine their orbits sufficiently well to predict when the onset of mutual events will occur. By observing KBO mutual events, we will obtain radii and density measurements that only a spacecraft encounter could improve upon. At present, no

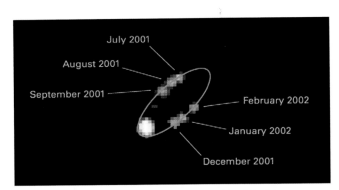

FIGURE 44.14 Binary KBO. The apparent orbit of the fainter component of 1998 WW$_{31}$ relative to the brighter component on the plane of the sky. *Courtesy of Christian Veillet, Keith Noll, and NASA.*

KBO binary orbit (other than Pluto and Charon) is known well enough to predict the onset of a mutual event with confidence.

10.3. Origin of KBO Binaries

Two of the most unusual features of KBO binaries, compared to main belt asteroid and near-Earth asteroid binaries, are the wide separation and similar diameter of each pair of components. These unusual features make it unlikely that collisions between two KBOs created each binary system, as in the case of the Earth and the Moon. Similarly, it is not likely that one KBO gravitationally captured another KBO to form a binary system. A mechanism put forth by Stuart Weidenschilling suggests that it is possible to create a loosely bound KBO binary by collision and capture in the presence of a third body. His mechanism requires many more KBOs than are seen today; perhaps such a mechanism operated long ago in a more densely populated Kuiper Belt (see the next section). Peter Goldreich put forth a mechanism wherein capture takes place during a close encounter as a result of the dynamical friction with the many surrounding small bodies. Each of these mechanisms produces its signature on the population of binaries we see today. For example, Weidenschilling's mechanism favors the production of wide binary pairs, and Goldreich's mechanism favors the production of closer pairs. Only the discovery of many more binaries will allow us to determine whether either of these mechanisms or some other mechanism is responsible for the formation of KBO binaries.

11. MASS OF THE KUIPER BELT

What is the mass of the entire Kuiper Belt? Gary Bernstein combined his HST survey for the faintest KBOs with ground-based telescope surveys for brighter KBOs, and assumed that KBOs have an albedo of 0.04 and a density of 1 g/cm^3, to estimate a Kuiper Belt mass of ~3% of the Earth's mass, or about 14 times the mass of Pluto. A major source of uncertainty in his mass estimate is the uncertainty in the albedos and densities of KBOs.

It appears that the Kuiper Belt did not always have a mass of ~3% of the Earth's mass. Specifically, the present number of KBOs per AU^3 is too small to grow KBOs larger than ~100 km in diameter by accretion in less time than the age of the solar system. Since 1000-km sized KBOs exist, it is likely that the Kuiper Belt initially had many more KBOs per AU^3 than today. Calculations by Alan Stern suggest that the initial Kuiper Belt probably had a mass 10 times the mass of the Earth, and as Neptune grew to a fraction of its present size, it stirred KBOs from their initial circular orbits to more eccentric orbits, resulting in frequent disruptive, rather than accretive collisions especially between KBOs smaller than 40–60 km in diameter. These collisions probably eroded the Kuiper belt mass down to its current value.

12. NEW HORIZONS

Because astronomers can discover and then measure the physical properties of many KBOs, their work is important as it gives us a global view of the Kuiper Belt and context for in situ spacecraft measurements. In January 2006, NASA's New Horizons spacecraft departed the Earth on a journey that will culminate in the first flyby of the Pluto–Charon system in 2015, and hopefully the first flyby of a KBO sometime before 2020. The $500 million spacecraft weighs only 416 kg (917 lb) and has four instrument packages: (1) a CCD camera; (2) an ultraviolet, optical, and near-infrared imaging spectrometer; (3) a charged particle detector; and (4) a radio telescope.

These instrument packages will provide in-depth observations impossible with telescopes on and near the Earth. For example, if New Horizons comes within a few thousand kilometers to a KBO, it could image the surface of the KBO with a resolution of 25 m/pixel. For comparison, HST can only image a KBO at 42 AU with a resolution of about 1200 km/pixel.

What kind of surface might the spacecraft image? If New Horizons visits a small KBO, perhaps it will image a surface with numerous craters, suggestive of an ancient surface bombarded by other small bodies (KBOs and comets) over the age of the solar system? On the other hand, if New Horizons visits a large KBO, perhaps it will see few craters on the surface, suggestive of some process erasing older craters. Perhaps the images of a large KBO will show long linear features in an icy crust, and some roughly round basins that appear flooded by liquids from the interior, much like the Voyager spacecraft images of Triton. Perhaps the spacecraft will catch a geyser erupting, and shooting a plume of gas and ice above the surface.

There are some problems concerning New Horizon's flyby of a KBO. The spacecraft trajectory is fixed since first it will fly by Pluto. In addition, the spacecraft has a limited fuel supply for adjusting its trajectory after the Pluto encounter. At present, none of the almost 1000 currently known KBOs are close to the spacecraft's trajectory. A flyby of a KBO by New Horizons depends on discovering a candidate close to the spacecraft's trajectory. Perhaps New Horizons will have enough fuel to visit one of the smaller (50 km diameter) and more common KBOs. The chances for the spacecraft visiting one of the larger (1000 km diameter) and rarer KBOs appear slim at the moment.

13. FUTURE WORK

It is likely that future work on the physical properties of KBOs and Centaurs will be driven by future state-of-the-art observatories. The 6-m James Webb Space Telescope near the L2 point will be able to obtain images and spectra of very large numbers of KBOs and Centaurs from 0.6 to 27 μm. It should be possible to measure diameters, albedos, surface colors, and optical and infrared spectra for many more objects than possible today. A large increase in the number of objects with physical property measurements will make it possible to look for statistically significant correlations between many more physical properties than possible with today's telescopes, and thereby better constrain the important formation and evolution mechanisms in the outer solar system.

Large ground-based telescopes of the future will likely play a big role in the field too. For example, the Giant Magellan Telescope, a configuration of six off-axis 8.4-m mirror segments around a central on-axis segment that is equivalent to a filled aperture 21.4 m in diameter, and the Thirty Meter Telescope, a configuration of more than 700 hexagonal-shaped mirror segments that is equivalent to a filled aperture 30 m in diameter, will make it possible to obtain higher signal precision optical and infrared spectra than possible with current 10-m telescopes. Better spectra and models will make it possible to map surface concentration of ices (e.g. the CH_4/N_2 concentration) as a function of depth and as a function of rotational phase over the surfaces of numerous objects. Such measurements will provide a wealth of data for constraining cosmochemistry models of the outer solar system. Finally, the Atacama Large Millimeter Array (ALMA), a configuration of about sixty-four 12-m antennas located at an elevation of 16,400 ft in Chile, may reveal extrasolar Kuiper Belts for comparison with our Kuiper Belt. ALMA may provide density and temperature profiles as well as chemical measurements through the detection of spectral lines in the belts. ALMA may initiate a new field of study, comparative Edgeworth–Kuiper Belt object ology, i.e. comparative EKO-logy.

BIBLIOGRAPHY

Davies, J. (2001). *Beyond Pluto: Exploring the outer limits of the solar system*. Cambridge University Press.

Jewitt, D. (2014). *Kuiper Belt*. http://www2.ess.ucla.edu/jewitt/kb.html.

Chapter 45

Extrasolar Planets

Michael Endl
McDonald Observatory, University of Texas at Austin, Austin, TX, USA

Chapter Outline

1. Introduction 957
2. Detection Techniques 958
   2.1. Astrometry 958
   2.2. The Radial Velocity Method 959
   2.3. Transit Photometry 960
   2.4. Microlensing 961
   2.5. Timing Method 961
   2.6. Direct Imaging 962
3. Observational Results of Extrasolar Planets 963
   3.1. The Pulsar Planets 963
   3.2. Planets Around Sun-like Stars: The Success of the Radial Velocity Technique 964
      3.2.1. Hot Neptunes & Super-Earths 966
   3.3. Results from Transit Searches 967
      3.3.1. Results from Microlensing Surveys 969
      3.3.2. The First Direct Images of Extrasolar Planets 970
   3.4. The Atmospheres of Extrasolar Planets 972
4. The *Kepler* Mission 973
5. Summary and Outlook 976
Bibliography 977

1. INTRODUCTION

Extrasolar planets—planets outside the solar system—were a mystery for astronomers for a long time. Are planets also orbiting stars other than the Sun? Is our solar system unique, or is planet formation a natural by-product of star formation and is our galaxy thus teeming with planets? The answers to these questions eluded astronomers for many centuries. It was only over the past two decades that we finally obtained unambiguous evidence for the existence of extrasolar planets. The reason why it took so long to find these objects is the fact that planets are dark objects very close to an extremely bright source, their host star. In visual light, a planet is more than a billion times fainter than a star. But the main problem is not the planet's faintness—today's best telescopes and instruments are sensitive enough to detect them in isolation—but that the light of the close-by star overwhelms the feeble light coming from the companion. Planets are also smaller and much less massive than stars. Astronomers usually rely on indirect methods to discover and characterize extrasolar planets, but a few planets have also been directly detected over the past few years.

The history of research into extrasolar planets consists of three major stages or eras: the first era belongs to the so-called **radial velocity technique**, where tiny variations in the line-of-sight velocity of a star are used to infer the presence of unseen companions. Over the past 15 years, radial velocity surveys have detected hundreds of planetary companions to stars in our galaxy. The majority of these objects are presumably gas giant planets, but some are also planets with lower masses. The structures of most known extrasolar planetary systems are very different from those in our solar system, with giant planets often very close to the star and with a wide range of orbital eccentricities. The second era in extrasolar planet research belongs to planets discovered by the **transit photometry technique**. With this technique, astronomers find the planet when it crosses in front of the star and blocks part of its luminous surface, making the star appear slightly dimmer while the planet moves across. This is the extrasolar equivalent of a Mercury or Venus transit with the obvious difference that the stellar surface is unresolved. To detect a transit of an extrasolar planet, one thus needs very precise measurements of the luminosity of a given star over time. These data can be used to estimate the radius of the transiting planet. In combination with an estimate for the mass, which can often be obtained using radial velocity data, we can derive a mean bulk density. It is precisely this capability, to place planets on a mass–radius diagram, that makes the transit technique so powerful. By doing this, we can compare the measured radii, masses, and densities of planets to models for their

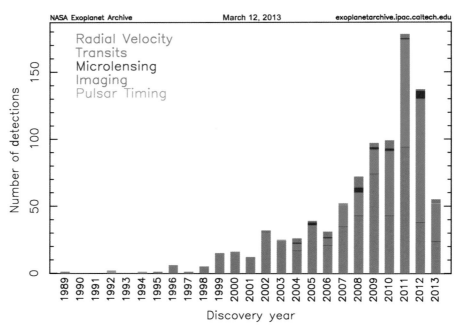

FIGURE 45.1 Detection statistics of the nearly 1000 known exoplanets. The different colored bars show the numbers of planets discovered with the respective astronomical techniques. *Image courtesy of NASA Exoplanet Archive: http://exoplanetarchive.ipac.caltech.edu/.*

internal composition. We can therefore distinguish between gas giant planets, icy giants that contain significant amounts of volatiles, and terrestrial, rocky planets. We can also probe the atmospheres of these planets and measure their temperatures. All this additional information, that we can obtain only for transiting planets, makes this technique so important for our understanding of planetary bodies.

We are currently in the third era of extrasolar planet research, an era that began with the launch of National Aeronautics and Space Administration's **(NASA's) Kepler mission** *in March 2009*. The *Kepler* spacecraft is also using the transit technique but its unprecedented measurement precision, by avoiding Earth's atmosphere, combined with the nearly continuous monitoring of approximately 160,000 stars has revealed a vast population of small planets that have not been seen before. The primary science goal of *Kepler* is to derive the first estimate of the frequency of Earth-size planets in the habitable zone of Sun-like stars. This parameters is called η_{Earth}. The nominal mission duration was 3.5 years and in 2012, the mission was extended for four more years. Unfortunately, in April 2013, a mechanical malfunction ended the extended mission early. Up to this point, *Kepler* collected 4 years of very high-quality photometric transit data and the detailed analysis of all these data will take many years. But this means that we are close to a truly historical moment, not only in astronomy, but in all sciences: we will know in a few years if planets similar to the Earth are common or rare in our galaxy. This measurement of η_{Earth} will have obvious astrobiological implications for how common biospheres on planetary surfaces are, and therefore how likely it is that complex life-forms similar to us have emerged somewhere else in the universe.

So far, the combined efforts of the diverse astronomical methods to detect extrasolar planets have found about 1000 confirmed planets around other stars (Figure 45.1), and *Kepler* alone has produced nearly 4000 candidates for extrasolar planets. All these results have revealed not only a stunning diversity of planetary systems but also a completely new type of planet, the so-called "**super-Earths**". Super-Earths have masses ranging from 1 to 10 Earth masses and radii between 1 and 3—4 Earth radii. Such a planet does not exist in the solar system, but they appear to be quite common around other stars.

The next years will be extremely interesting in the field of extrasolar planets. We move steadily from mere detection to a more detailed characterization of extrasolar planets. But even in detection there are still large unexplored white areas in our galactic map of planetary systems. We will explore these areas and fill in the necessary information to obtain a galactic overview of planetary systems that will allow us to put our own solar system firmly into a cosmic context.

2. DETECTION TECHNIQUES

2.1. Astrometry

Astrometry is the science of positional astronomy, which measures the location of a celestial object and its movement within the plane of the sky. This was one of the first techniques used to search for planets around other stars. As in other indirect methods, astronomers seek to detect the orbit of the central star around the **barycenter** of the star/planet system. The orbit is measured as the change of the position of the star on the plane of the sky, usually compared to a number of more distant background stars, which define an astrometric reference frame.

The amplitude S of an astrometric signal is given by

$$S = \frac{m}{M}\frac{r}{d}$$

where m is the mass of the unseen companion, M is the mass of the central star, r is the semimajor axis of the companion's orbit, and d is the distance to the star. For r in astronomical units and d in parsec (pc), S is given in seconds of arc (1 arcsec is 1/60 of an arcmin, which itself is 1/60 of a degree). For the Sun/Jupiter system with $m/M = 0.001$ and $r = 5.2$ AU, the amplitude of the signal would be 0.001 arcsec (1 mas) seen from a distance of 5 pc and 0.5 mas from 10 pc (1 pc = 3.26 light years).

From the preceding equation, it is obvious that astrometry is most sensitive to companions with large mass ratios (massive planets around less massive stars) at large orbital separations r and nearby stars (d is small). Because of the r-dependence, this technique is a complementary method to the radial velocity technique (which will be discussed next).

As is the case for most of the detection methods, the largest hurdle to overcome in detecting extrasolar planets by astrometry is the need for very precise measurements and the extreme care required to avoid systematic errors, such as instrumental effects. This is necessary to prevent the introduction of spurious signals, which may be misinterpreted as real planets over a long time baseline. The astrometric signals for most extrasolar planets are typically less than 1 mas and are beyond the scope of most current state-of-the-art instruments. The European Space Agency's (ESA's) satellite *Hipparcos* was a space mission entirely dedicated to stellar astrometry. Although its precision was not sufficient to detect planetary companions, the *Hipparcos* data placed very useful upper limits on the masses of some companions detected by other methods.

The highest astrometric precision can be achieved by using interferometry. By letting the light that arrives from the same source at two different locations (two or more telescopes positioned on a well-defined baseline) interfere, one can measure the small difference in the arrival time at these points and thus determine the angle between the source and the baseline very precisely. The Fine Guidance Sensors onboard the *Hubble Space Telescope* (*HST*) can actually be used as an interferometer and they yield currently the best astrometric precision.

The ESA's *Global Astrometric Interferometer for Astrophysics* (*GAIA*)* mission is designed to achieve an astrometric precision of a few microarcseconds. Over its mission lifetime of 5 years, *GAIA* is planned to observe nearly a billion stars and will detect a large number of extrasolar planetary systems with the astrometric technique.

2.2. The Radial Velocity Method

Astronomers, using the radial velocity technique, measure the line-of-sight component of the space velocity vector of a star (hence the term "radial", i.e. the velocity component along the radius between observer and target). The radial velocity of a star can be determined in absolute values, or differentially if only changes of the velocity are of interest.

In order to measure stellar radial velocities, we rely on the well-known Doppler effect. Depending on whether the star moves toward us or away from us, its light will be blue- or redshifted, as compared to a nonmoving source. Such a shift reveals itself as a change in the wavelength position of the absorption lines in the spectrum of the star. Therefore, astronomers use high-resolution spectrometers to perform radial velocity studies. The incoming light of the star is split up into its individual wavelengths, and the spectrum is recorded on a detector. As in astrometry, this method tries to detect the reflex motion of the primary object around the common center of mass with an unseen companion. However, this time this motion reveals itself as a change in the velocity rather than a change in position of the star.

The radial velocity method is traditionally used in stellar astronomy for the discovery and characterization of binary stars. In a binary system, the barycenter of the system is located somewhere between the two stars (the exact location is defined by the mass ratio), and the observed velocity amplitudes are of the order of several kilometers per second. The same method is applied to the search for extrasolar planets, which induce a much smaller reflex orbit on their host star and produce much smaller velocity amplitudes.

For a system of two gravitationally bound objects m_1 and m_2 in a circular orbit, the radial velocity semiamplitude K_1 of m_1 can be calculated by using:

$$K_1 = \frac{(m_2 \sin i)}{(m_1 + m_2)}\sqrt{G\frac{(m_1 + m_2)}{a}}$$

where m_1 is the more massive object and m_2 is the less massive secondary companion, i denotes the angle between

---

* *GAIA* originally stood for Global Astrometric Interferometer for Astrophysics. As the project evolved, the single interferometer concept was replaced by a new payload design. The mission name remained, however, even though it no longer reflects the methods used to perform the science operations.

the orbital plane and the plane of the sky, G is the gravitational constant, and a is the semimajor axis of the orbit. It is immediately clear that for face-on systems ($\sin i = 0$), K_1 is zero.

Using Kepler's famous third law, which relates orbital separation to orbital period P, we can recast this:

$$K_1 = \left(\frac{2\pi G}{P}\right)^{1/3} \frac{(m_2 \sin i)}{(m_1 + m_2)^{2/3}}$$

We are interested in the case of a planet orbiting the star, where $m_2 \ll m_1$ (and thus $m_1 + m_2 \approx m_1$), which simplifies the equation to

$$K_1 = \left(\frac{2\pi G}{P}\right)^{1/3} \frac{(M_2 \sin i)}{m_1^{2/3}}$$

Now we have an expression that relates $m_2 \sin i$ to the observables K_1 (or simply K if only the spectrum of m_1 is detectable) and P. Using units of years for P, m/s for K, and solar mass for m_1, $m_2 \sin i$ is given in Jupiter masses by the following expression:

$$m_2 \sin i = K \frac{(P m_1^2)^{1/3}}{28.4}$$

With a good estimate for m_1, we thus calculate $m_2 \sin i$ of an unseen companion. The $m_2 \sin i$ value represents a lower limit to the true mass of m_2. The $\sin i$ ambiguity is one of the limitations of the radial velocity technique. However, the $m_2 \sin i$ value is probably close to the real value of m_2. Just by assuming a random distribution of orbital planes, we have a 90% statistical probability that m_2 is within a factor of 2.3 of the observed $m_2 \sin i$.

Jupiter induces a K of 12.5 m/s in the Sun when observed in the plane of its orbit ($\sin i = 1$) and Saturn gives a K of only 2.8 m/s. Detection of planets analogous to the two gas giants in our solar system thus calls for measurement uncertainties of a few meters per second or better over many years to decades. More massive planets and also planets at smaller orbital separations produce larger K amplitudes, but the desired velocity precision is still of the order of several meters per second. It is also desirable and often necessary to obtain a sufficiently large number of observations over time to study the noise characteristics of the data in detail.

Obtaining a radial velocity precision of a few meters per second or better is a complex task that requires highly precise and sensitive data reduction techniques and instruments. Over the past years (even decades), two approaches have been successfully used to attain such a high level of precision: (1) the gas absorption cell method and (2) the simultaneous Thorium–Argon (Th–Ar) method in combination with stabilized spectrometers. In the first method, the star light is passed through a small glass cell that is filled with a suitable gas (in most cases iodine vapor), which superimposes its own dense absorption spectrum onto the stellar spectrum. This reference spectrum not only yields a simultaneous wavelength calibration but also can be used to keep track of the imaging properties of the spectrograph. This allows preventing small changes in the image of the stellar absorption lines, which are caused by fluctuations in the light path from the telescope to the detector, from being misinterpreted as Doppler shifts. In the second approach, the emission spectrum of a Th–Ar lamp is imaged parallel to the stellar lines on the charge-coupled device (CCD) frame. Again, this allows a simultaneous wavelength calibration. In recent years, the standard Th–Ar comparison spectrum is sometimes replaced by a more stable reference spectrum generated by a laser, a so-called "laser comb". This method only works in combination with highly stabilized spectrometers. To minimize any instrumental effects, these spectrographs have no movable parts and are placed in pressure- and temperature-stabilized environments. To assure the highest level of precision, some instruments are enclosed in a vacuum tank. This eliminates variations in the light path due to changing refractive index values of air caused by pressure changes. Also, by using optical fibers, the light path from the telescope to the instrument is kept as constant as possible. Both approaches have been demonstrated to reach a radial velocity precision of a few meters per second, and in the best cases even 1–0.5 m/s.

2.3. Transit Photometry

In the special case that the orbital plane of an extrasolar planet is close to perpendicular to the plane of the sky, the planet will appear to move across the disk of the host star. In our own solar system, this phenomenon can be observed from the Earth for the two inner planets, Mercury and Venus. Because we cannot spatially resolve the disk of another star, a transiting extrasolar planet can only be observed as a reduction of the light output coming from the star (i.e. by means of precise photometric measurements).

Assuming random orientations of planetary orbits, the probability P_{trans} of the visibility of a transit event is a function of both the radius of the star and the planet and its orbital separation a:

$$P_{\text{trans}} = \frac{(R_{\text{star}} + R_{\text{planet}})}{a}$$

For a random location in our galaxy, P_{trans} is less than 1% to observe transits of the inner terrestrial planets in the solar system and for the outer planets it decreases from 0.1% (Jupiter) to 0.01% (Pluto). But for giant planets orbiting at very small separations ($a \sim 0.04$ AU), P_{trans} is around

10%. A transit of such a planet produces an ~1% dip in the so-called light curve (i.e. the time series of brightness measurements) of a star. This effect can be detected from the ground with state-of-the-art photometric instruments, which allow a precision of ~0.1%.

Currently, numerous ground-based photometric transit surveys are searching for short-periodic giant planets. These surveys usually use small-aperture telescopes with a wide field of view to survey a large amount of stars, typically hundreds or thousands per CCD image.

Detection of smaller planets requires higher photometric precision than ground-based photometry can achieve because of the limitations imposed by our atmosphere. Space-borne telescopes, on the other hand, should be capable of detecting even the miniscule (~80 ppm) photometric transit of an Earth-like planet orbiting a solartype star at 1 AU.

If photometric data of a transit event can be combined with radial velocity measurements, then the $\sin i$ ambiguity in the planetary mass is removed. Furthermore, the transit depth allows an estimate of the radius of the companion and thus an estimate of the mean density. Comparisons of high-resolution spectroscopic observations during and outside a planetary transit reveal spectral signatures of the planetary atmosphere. Clearly, we can gain a tremendous amount of information from planetary transit observations.

2.4. Microlensing

According to Einstein's theory of general relativity, photons are affected by the presence of a gravitational field. Because gravity can be viewed as the changing curvature of the space–time continuum, the path of a photon follows this shape and is "bent" when it passes close to a massive object. In certain geometric cases, this can lead to a focusing of the light from a distant source by a foreground object. This gravitational bending of light has been measured directly during total solar eclipses when the effect of the Sun's gravitational field can be observed as positional changes of stars close to the Sun's disk.

In astronomy, this effect is also seen on a much larger scale: entire galaxies or even clusters of galaxies act as massive gravitational lenses for the light of more distant objects in the background. However, as already demonstrated by our Sun, every object with mass can be a gravitational lens: a star, a brown dwarf, or even a planet.

As in the transit method, microlensing is caused by a geometric alignment: when a foreground object (the "lens") moves in front of a more distant background object (the "source"), the light of the source passing close to the lens is bent toward the observer. The observer can see several images of the background object separated by milliseconds of arc, which merge into a full ring called the Einstein ring at the moment the lens is directly in front of the source.

Because the gravitational lensing of the source magnifies the image, the total amount of detected light is increased and the apparent brightness of the distant source is enhanced. The magnification factor depends on the exact geometric situation, and the maximum occurs when the lens is at its smallest projected distance from the source. For microlensing events in our galaxy and for stars acting as both sources and lenses, the images cannot be spatially resolved, and only the change in brightness is observed. However, the magnification can be large and theoretically even infinite for point sources. The position where infinite magnification occurs is called a caustic. Because stars are not perfect point sources, the magnification will not be infinite, but can still be very large.

If the lens is not a single object but a binary, then the caustic is no longer a single point but an extended geometric figure, symmetric around the binary axis. Thus, the microlensing technique represents an elegant method to search for planetary companions to stars in our galaxy. Binary lenses reveal themselves by a characteristic shape of their light curve (the time series of the brightness measurements during the lensing event). The light curve of a binary lens contains sharp peaks of even larger magnification due to the crossing of the caustics. The mass ratio q of the two lenses can be derived from modeling the light curve. If the resulting mass ratio is very small (typically $q > 0.001$), the second object in this binary might be a planet.

Like transit surveys, microlensing planet searches have to observe a large number of targets because the probability of observing a single event for a given target is negligible. For microlensing, the situation is even more complex because of the need of a sufficiently large reservoir of lenses moving in front of a high-density sample of background sources. Moreover, microlensing events for a particular lens do not repeat. Each microlensing event is a single isolated transient in the light curve. There is only one chance to observe it. Hence microlensing surveys monitor regions like the bulge of our galaxy (the central cluster of stars that surround the galactic center) where microlensing events can be observed more frequently.

2.5. Timing Method

The timing method is exceptional with respect to the other techniques because it actually is the method that led to the very first detection of planets outside the solar system. As in the astrometric and radial velocity techniques, the fact that a host star has to orbit the common center of mass with an orbiting planet is utilized to detect the unseen companion. But this time the reflex orbit is observed by the change of the arrival time of signals coming from the star. The change is caused by the difference in the distance the signal has to travel from the source to the observer.

If the star is at the location in its orbit where it is the farthest away from the Earth, then the signal needs the longest time to arrive here, and vice versa for the smallest separation. Because reflex motions due to planets are small compared to the speed of light the changes in arrival time are very small.

The timing method can only be applied to cases where (1) a very short duration signal is emitted by a source with a constant periodicity and (2) the observers are able to measure the arrival time of the signal with very high precision. One astrophysical case where these conditions are met is the so-called pulsar. Pulsars are rapidly rotating neutron stars, which are the remnants of a supernova explosion which ended the life of a massive star with 15–30 times the mass of the Sun. A neutron star has a diameter of only 10–20 km, but has a mass of about 1.4 times the mass of the Sun. Hence it is a very dense object which also rotates very fast. Rotation periods of neutron stars can be as short as milliseconds. Strong magnetic fields produce bipolar jets of radio waves and high-energy radiation such as X-rays and gamma rays. Because the magnetic field axis is misaligned with the rotational axis, these stars act like cosmic lighthouses from which we see a pulse every time the jet sweeps over the Earth. Pulsars were first discovered by radio telescopes in 1967, and the timing method can be applied to the fastest rotators (such as the millisecond pulsars) to detect orbiting companions.

A second case where the timing method is applicable is stably pulsating white dwarf stars. White dwarfs are the end stage of the life of stars that are not massive enough to form a neutron star (like our Sun). They are also small (about the size of the Earth) and very dense objects. These stars undergo nonradial pulsations for certain temperature ranges that can be detected by precise photometric observations. The periods of these pulsations are of the order of a few minutes. Some of the white dwarfs exhibit the same pulsation modes over decades and are thus suitable targets for the timing method.

2.6. Direct Imaging

Obtaining a direct image of an extrasolar planet is the type of observation the public expects. Besides the obvious advantage of discovering planets with only a few observations, the images might also allow us to characterize the planets in new depth. From the colors and albedos, we might obtain thermal and chemical information. After the direct detection, follow-up observations can be carried out to collect spectra or other data of the planet.

In many ways, direct imaging of a planet around a nearby star represents the largest challenge in the development of telescope/instrument systems. Surprisingly, it is not the faintness of an irradiated extrasolar planet that is the hurdle to overcome (the HST is sensitive enough to detect

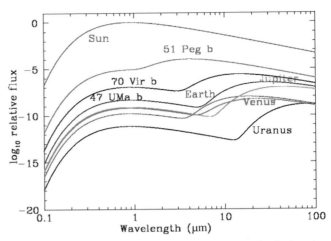

FIGURE 45.2 The relative flux of planets compared to the Sun's emission as a function of wavelength. Four planets of our solar system and three extrasolar planets (51 Peg b, 70 Vir b, and 47 UMa b) are shown. The difference in flux ranges from 10^{-6} to 10^{-12} in the optical (<1 μm) and generally improves toward the infrared (>1 μm), where the planet's thermal emission dominates.

these faint objects) but rather their proximity to a much brighter source of photons: the planet's own host star.

The distances to even the nearest stars are so large that, due to the perspective, any image of a companion orbiting at separations comparable to our solar system would be located in the side wings of the image of the central object. In the optical, the flux difference between a solartype star and a giant planetary companion is of the order of a billion. In the infrared, the difference is more of the order of a million (see Figure 45.2). But the light coming from the planet is completely overwhelmed by the large amount of scattered light from the star.

There are several techniques to minimize the scattered light from the host star. From the ground, the observations are also affected by atmospheric turbulence, called "*seeing*". Seeing usually prevents telescopes from obtaining images at their theoretical resolving power even at the best observing sites in the world. In the near infrared, atmospheric turbulence can be compensated by an adaptive optics (AO) system. AO systems use wave front sensors to measure the wave front errors caused by turbulence in the atmosphere above the telescope and then to adjust the optical path to compensate for these errors using deformable mirrors. This helps to attain images at a spatial resolving power close to the limit set by the diffraction of light. AO imaging improves the situation for high-contrast imaging, but the real goal is to remove the image of the central star entirely from the observations.

The most commonly used instrument to perform this task is the coronagraph. The coronagraph was invented by the French astronomer Bernard Lyot in the 1930s to study the outer parts of the solar atmosphere (the corona) without being totally overwhelmed by the intense glare of the Sun's disk. He managed to remove the light of the Sun's photosphere by introducing an opaque mask (of the same size as the image of the Sun) into the telescope's light path in such a way that it blocked the photons coming from the disk but not from the surrounding environment. This makes a coronagraph the ideal instrument for direct imaging of extrasolar planets. For ground-based searches the highest image quality is achieved by combining a coronagraph with an AO system.

However, no optical system is perfect and even coronagraphic images contain residual scattered light from the central star close to the edge of the opaque mask and other image artifacts produced by diffraction on telescope parts. This makes the detection very close to the central object still very difficult. And even for the nearest stars, the expected angular separations for planetary companions are small compared to the size of typical coronagraphic masks. At a distance of 5.2 pc (=17 light years), an analogous planet to Jupiter would appear 1 arcsec away from the star at maximum projected separation. At 10 pc, the maximum separation is only about 0.5 arcsec. These angles are comparable to the typical dimensions for coronagraphic masks of current state-of-the art instruments (e.g. the HST instrument Near Infrared Camera and Multi-Object Spectrometer (NICMOS) has a coronagraphic mask with a diameter of 0.8 arcsec).

The image area around the central obscuration contaminated by scattered light is called the halo. In very short exposures, this halo is resolved into smaller bright and dark spots called "speckles". Speckles are interference phenomena produced by atmospheric seeing and by the superposition of light coming from all parts of a telescope mirror with imperfect smoothness. At the location of a dark speckle, the light of the star is canceled out by destructive interference. The image of a faint companion can be recovered if it is located at the position of a dark speckle, where the light from the star is severely reduced. By taking a great number of short exposures, the companion can be detected by a proper data analysis algorithm simply by the fact that in every image the speckle pattern is different and that a dark speckle never appears at the location of the companion. This method is called dark speckle coronagraphy. In combination with large aperture ground-based telescopes or the next-generation space telescope, this method should have the sensitivity to detect extrasolar planets around the nearest stars.

Another technique to achieve high-contrast images is nulling interferometry. In theory, it is possible to combine the wave fronts arriving at two or more telescopes in such a way that a wave maximum coming from one telescope is canceled out by a wave minimum from another telescope. In this way, it produces a null image of the central object while it leaves the light from the circumstellar environment unaffected. First trial runs using ground-based telescopes have already been successfully performed. A nulling interferometer is currently built for the Large Binocular Telescope, which consists of two 8-m class telescopes mounted side by side on the same support structure. One of the most ambitious ideas is the plan for a space-based nulling interferometer operating in the infrared that potentially could directly image and characterize Earth-like planets around nearby stars.

3. OBSERVATIONAL RESULTS OF EXTRASOLAR PLANETS

3.1. The Pulsar Planets

The first definitive discovery of planets outside our solar system was announced in 1992. This discovery would set the tone for all subsequent discoveries of extrasolar planets in terms of their strangeness. The planets were found orbiting the millisecond pulsar PSR B1257+12 using the Arecibo radio telescope and the timing method described in Section 2.

The first surprise was that the planets orbit a "dead" star, which had undergone a previous supernova explosion. It is unlikely that these planets existed before the star went supernova. The more plausible scenario is that these planets somehow formed after the explosion. Millisecond pulsars are believed to achieve their high rotation rates due to spin up by infalling material accreted from a companion star. The planets might have formed during this process in the accretion disk around the pulsar.

The planetary companions to PSR B1257+12 are also remarkable in a different way: they have very small masses. Due to the extreme sensitivity of the timing method in the case of millisecond pulsars (where the arrival time of a pulse can be measured with microsecond precision), even companions with the mass of our Moon or less can be detected.

Table 45.1 summarizes the properties of the PSR B1257+12 planetary system. Soon after the system was discovered, the mutual gravitational perturbations, that were predicted by a dynamical study of this system of the

TABLE 45.1 The PSR B1257+12 Planetary System

Planet	$M \sin i$ (Earth mass)	Orbital Period, P (days)	a (AU)
A	0.015	25.34	0.19
B	3.400	66.54	0.36
C	2.8	98.22	0.47

planets on each other, were indeed confirmed by subsequent timing data. This demonstrated that the residuals in the pulse times are indeed due to orbiting planets and not due to a previously unknown effect, intrinsic to the pulsar.

These companions, with masses between ~3 Earth masses and 1 Moon mass, represent some of the lowest mass objects known to orbit a star other than the Sun. However, they must be barren and dead worlds because of the constant bombardment by high-energy radiation coming from the pulsar.

More recently, evidence was presented for a fourth object with a mass of only 15% the mass of Pluto orbiting the pulsar at a distance of 2.7 AU. This new object, however, qualifies as an asteroid or comet rather than a planet.

3.2. Planets Around Sun-like Stars: The Success of the Radial Velocity Technique

In fall 1995, the astronomical community, as well as the public, was stunned by the announcement of the discovery of the first extrasolar planet around a Sun-like star. Precise radial velocity measurements of the Sun-like star 51 Pegasi revealed a periodic variation of 4.2 days and an amplitude consistent with an $m \sin i = 0.5$ Jupiter mass companion (Figure 45.3). The minimum mass of the object firmly places this companion into the gas giant planet mass range. The detection was achieved by precise data from a stabilized spectrograph at the Haute-Provence Observatory in France.

The extremely short orbital period and small orbital separation of 0.05 AU of the proposed planet, named 51 Peg b, were surprising in many ways, and alternative explanations for the radial velocity modulation were put forward. Stars more evolved than the Sun show similar variability, which is caused by pulsations rather than by Keplerian motion. But 51 Peg passed every test for this type of variability, and soon the claim of having found the very first planet orbiting a "normal" star was generally accepted.

The planet 51 Peg b represents the prototype of a new class of planets that soon emerged from the results of the radial velocity surveys, the **hot Jupiters**. Because of their close proximity to the host star, these gas giant planets have estimated upper atmosphere temperatures of more than 1000 K.

In the 2 years following the discovery of 51 Peg, astronomers from the United States announced the discovery of seven more extrasolar planets orbiting Sun-like stars. All these detections were based on years of precise radial velocity measurements of these stars using telescopes and spectrographs at Lick, McDonald, and Whipple Observatories. Table 45.2 lists the first eight extrasolar planets discovered by the radial velocity technique along with their orbital characteristics. In 1999, the teams of Lick and Whipple Observatory Doppler surveys announced the discovery of the first extrasolar multiplanetary system around a Sun-like star. The radial velocities of υ Andromedae deviated progressively from the originally derived, single-planet velocity curve, and with the additional years of data it became apparent that a triple Keplerian model is required to describe the complex reflex motion of this star. In addition to the previously found hot Jupiter, this system contains two more giant planets with $m \sin i = 1.89$ and 3.75 Jupiter masses at separations of 0.8 and 2.53 AU. Also their orbits have significantly nonzero eccentricities (0.28 and 0.27), making this system again quite different from our solar system.

Over the following decade, the radial velocity technique demonstrated its effectiveness in detecting numerous giant planets and nearly a 100 multiplanet systems around nearby stars. At the time of this writing, more than 500 planetary companions in nearly 400 systems were found by the cumulative effort of several radial velocity surveys operating in both hemispheres. The most successful programs were using the 10-m Keck telescope and the High Resolution Echelle Spectrometer (HIRES) spectrograph in Hawaii and the CORALIE and High Accuracy Radial velocity Planet Searcher (HARPS) instruments at La Silla Observatory in Chile. Many characteristics of extrasolar planetary systems detected by radial velocity surveys differ significantly from the planets in our solar system.

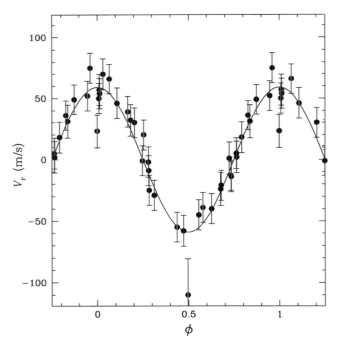

FIGURE 45.3 The radial velocity measurements (dots with error bars) of the solartype star 51 Pegasi phased to the orbital period of its planetary companion. The sinusoidal variation is caused by a companion with $m \sin i = 0.5$ Jupiter masses in a circular orbit with $a = 0.05$ AU and an orbital period of 4.2 days. *Reproduced with permission from Nature.*

TABLE 45.2 The First Eight Radial Velocity Planets

Star	$m \sin i$ (Jupiter mass)	Orbital Period, P (days)	a (AU)	Eccentricity, e
51 Peg	0.5	4.2	0.05	0
70 Vir	7.4	117	0.48	0.4
47 UMa	2.5	1089	2.09	0.06
ρ^1 Cancri	0.84	14.7	0.12	0.02
τ Boo	4.13	3.3	0.05	0.01
υ And	0.69	4.62	0.06	0.012
16 Cyg B	1.69	799	1.67	0.67
ρ Crb	1.11	39.9	0.23	0.13

One of major difference with the giant planets in the solar system is the very small separation where the radial velocity method detected extrasolar planets. These small orbital separations (~ 0.04 AU for a hot Jupiter) are difficult to reconcile with current models of the formation of gas giant planets. In the classical "core accretion" model of planet formation, gas giants form most easily near (and beyond) the so-called "ice line" in the protoplanetary nebula. The ice line is the distance from the star where the temperatures in the nebula drop low enough so that ices can condense out of the gas. This leads to an increase in the surface density of condensed material and allows the formation of massive cores (mixed with rocky material and dust grains) onto which nebula gas can accrete. It was therefore commonly expected to find gas giants only at large orbital separations similar to our Jupiter at 5 AU and more. It appears that, in most cases of the observed extrasolar giant planets, moderate to massive orbital migration has occurred, which moved the planets from the place where they formed to their current location close to the star.

Another remarkable difference is the abundance of moderate to high orbital eccentricities (e) of radial velocity planets. Most of them have more elongated orbits than the planets in our solar system. The eccentricities are distributed quite uniformly and are practically indistinguishable from the eccentricity distribution of stellar binaries, especially for planets at larger separations. The hot Jupiters all have $e = 0$ (or close to 0) because tidal forces between the star and the planet at these small distances tend to circularize the orbit on much shorter timescales than the typical lifetime of the star. The origin of the nonzero eccentricities is not well understood; possible explanations are a more dynamic formation history than in the case of the solar system, in which mutual dynamical interaction between planet embryos pumped up their eccentricities. Also, planet/disk interactions and gravitational perturbation by stellar companions and planet–planet scattering are possible causes of high eccentricity.

About a quarter of the planetary systems detected by the radial velocity method are multiple systems. Some of the orbits in these multiple systems show clear evidence for mean-motion resonances. A mean-motion resonance exists if the ratio of the orbital periods of the two objects are commensurable (equal to the ratios of two integers, e.g. 2:1 or 5:3). This could be explained by the early dynamical evolution of these systems where migration captured the planets in resonances, similar to the resonance belt in the Kuiper belt. A certain number of multiple systems were found to have orbits just near mean-motion resonances but exactly in a resonant configuration.

The overall mass function of radial velocity planets showed a steep rise toward masses of 1 Jupiter mass or less. Thus, although less massive planets are much harder (or impossible) to detect by the radial velocity technique, we expected them to be quite frequent. On the other hand, very massive objects with masses of 10–20 Jupiter masses are very easy to detect with the radial velocity method and therefore seem to be quite rare.

About 10% of the stars surveyed by long-term radial velocity programs have detectable giant planets. The majority of these planets orbit stars of similar spectral type (i.e. effective temperature and mass) as the Sun. They usually have orbital separations less than 5 AU; in fact, about half of them reside within 1 AU from their host star. But these results also reflect strong observational biases.

As already mentioned, the radial velocity technique is more sensitive to close-in planets, and it takes a monitoring timescale of over a decade to discover planets beyond 5 AU. Also, stars hotter and more massive than the Sun are not suitable for the radial velocity technique because they tend to have higher rotation rates and much fewer spectral features, which can be used to measure the velocity. Stars less massive than the Sun are intrinsically fainter, and large-aperture telescopes are needed to collect enough photons to ensure a sufficient data quality. Therefore, combined with the prior knowledge that at least one solartype star, our Sun, produced a planetary system, early radial velocity surveys traditionally focused on Sun-like stars. As radial velocity programs extend their time baselines and expanded their target samples to fainter and lower as higher mass stars, these initial observational biases became somewhat less important.

Main sequence stars with the spectral classification M ("M dwarfs") are smaller and fainter than the Sun. They have masses ranging from roughly 55% to about 0.8% solar masses. Despite the fact that M dwarfs comprise the majority of stars in our galaxy, they formed only subsets in the large target samples of radial velocity surveys, due to their faintness. Based on the data accumulated for a few hundred M dwarfs (compared to >1500 solartype stars) it seems that they host fewer giant planets with small orbital separations than Sun-like stars. This could indicate that the formation efficiency of giant planets is really a function of disk mass, as the protoplanetary disks around M dwarfs are presumably less massive than for bigger stars. But a few systems with M dwarf giant planets have been discovered. The star Gliese 876 was found to have a planetary system of two Jupiter-type companions in a 2:1 mean-motion resonance with periods of 30.12 and 61.02 days. More recently, a third lower mass planet with a 120-day orbit was detected in this system. With this additional third planet (which is also in resonance) the Gliese 876 system now resembles the resonance configuration of the Galilean satellites, Io, Europa, and Ganymede. Studies have also shown that this system is dynamically full, which means that no other planet can exist in this system (the same is true for the solar system). Because of its close proximity to the Sun (15 light years), Gliese 876 is an ideal target for astrometric followup. In 2002, highly precise measurements obtained with the Fine Guidance Sensors on board the HST successfully revealed the astrometric signature of the 60-day planet. By combining the ground-based radial velocity data with the space-based astrometric data, a true mass of 1.9 Jupiter masses was determined for this planet. Another interesting multiplanet system containing several low-mass planets was discovered orbiting the nearby M dwarf Gliese 581 (this system will be discussed in Section 3.2.1)

The radial velocity technique was also used to detect many planets orbiting giant stars. Giants stars are more evolved than solartype stars, and their cooler atmospheres have a spectral signature rich in absorption lines. These stars are thus suitable targets for the radial velocity technique. The progenitor stars (i.e. before they evolve into their current giant status) of most giant stars are more massive than the Sun and the successful detection of planetary companions around them is evidence that planet formation can also occur around more massive stars. This is not a big surprise because several dense dust disks have already been observed around progenitor stars. We shall also see later that the first massive planets were imaged around stars more massive than the Sun (Section 3.3.2).

Another interesting correlation for solartype stars (with spectral types of F, G and K) emerged from the radial velocity results: their detectability is a strong function of the metallicity of the host star. Astronomers call every element heavier than helium a metal. **Stellar metallicity** thus means the abundance of all chemical elements in a star besides hydrogen and helium. In general, the element used for the metallicity determination is iron. By measuring the stellar metal content, we can probe the primordial chemical composition of the gas and dust cloud, out of which the star (and presumably its companions) has formed. It was found that the mean value of the metallicity distribution of planet host stars is offset with respect to the mean metallicity of stars in the solar neighborhood. On average, giant planets are more frequently detected around host stars that have a higher iron to hydrogen abundance ratio ([Fe/H]) than the solar neighborhood mean [Fe/H] value. This can be seen as evidence for the core accretion model for the formation of gas giants. The efficiency of this model is sensitive to the abundance of heavier elements in the protoplanetary disk (more heavy elements → more efficient core formation → more gas giants). Alternatively, this might also be regarded as evidence that orbital migration is a function of the metal content of the planet-forming disk, because close-in planets are easier and faster to detect by radial velocity surveys.

3.2.1. Hot Neptunes & Super-Earths

In 2004, the first radial velocity planets with masses below the gas giant range were discovered. These planets have $m \sin i$ values comparable to the masses of the icy giants of our solar system, Uranus and Neptune (14 and 17 Earth masses). Their very short orbital periods give detectable radial velocity signals despite their low mass. Thus, they have been dubbed hot Neptunes. One of the very first hot Neptunes that were found orbits the M dwarf star Gliese 436. Although the planet has only a minimum mass of 21 Earth masses, it induces an Radio Velocity (RV) perturbation in its host star because the star is much less massive than a solartype star.

With further improvement of the precision of radial velocity measurements to the submeter per second level, it was just a matter of time that the first planets with minimum masses approaching the mass of Earth were found. Planets that have a minimum mass below 10 Earth masses are called "super-Earths". Very precise RV programs have detected several of these low-mass planets in the solar neighborhood, most of them with short orbital periods. One of the most interesting systems that was found orbits another M dwarf: Gliese 581. This system consists of one hot Neptune, and at least two, if not three more, super-Earths. The existence of an additional fifth planet in this system is still controversial. In 2012, the detection of a planet with a minimum mass of only 1.1 Earth masses in a 3.2-day orbit around our close neighbor alpha Centauri B was announced. The detection is based on 4 years of radial velocity measurements of this star using the HARPS spectrograph. After a complex procedure to account and compensate for the signals introduced by the star itself, a weak 3.2 residual signal was found that the discovery team attribute to a very low mass planet orbiting our neighboring star. This, of course, would be the nearest exoplanet detected so far, and also one with the lowest mass. This important result clearly needs confirmation by other groups. Indeed, an independent analysis that used the same data set calls this planet into question.

3.3. Results from Transit Searches

With the discoveries of more and more hot Jupiters, it was just a matter of time until one of them would have a near edge-on orbit so that the planet transits in front of the star. The hot Jupiter companion to HD 209458 was the first transiting extrasolar planet. The planet itself was initially discovered by the radial velocity method, but in this case the photometric follow-up observations revealed—for the first time—the characteristically shaped light curve of a transiting planet (Figure 45.4). With the viewing angle known ($i = 86°$) the $m \sin i$ value transformed into a true mass for the planet of 0.69 Jupiter masses. The depth of the dip in the light curve yielded a radius for the companion of 1.4 Jupiter radii. With a known mass and a known radius, a mean density of 0.31 g/cm^3 was derived. This is an even lower mean density than Saturn, the planet with the lowest mean density in the solar system. The discovery of the HD 209458 transiting planet represented a major milestone in the field of extrasolar planet detection: It demonstrated that the companions discovered by the radial velocity surveys were indeed gas giant planetary companions and not more massive (even stellar) companions seen at a very unfortunate viewing angle i.

The very first planet discovered by the transit method was Optical Gravitational Lensing Experiment (OGLE)-TR-56 (HD 209458 b, the first transiting planet, had been detected by radial velocities before the photometric transit was observed). The OGLE is a precise photometric survey of millions of stars to search for gravitational lensing and planetary transit events. OGLE is a project from astronomers from Warsaw and Princeton and operates a 1.3-m telescope at Las Campanas Observatory in Chile. The companion to OGLE-TR-56 has a mass of 1.4 Jupiter masses and a radius of 1.2 Jupiter radii. The planet has an extremely short orbital period of only 1.2 days or 29 h, and the orbital separation is only 0.023 AU. This is even closer to the host star than the hot Jupiters found by radial velocity surveys.

In order to survey a sufficiently large number of stars, ground-based photometric transit search programs commonly use small-aperture telescopes with a wide field of view. Most of the stars included in these searches are far more distant and thus fainter than the stars included in

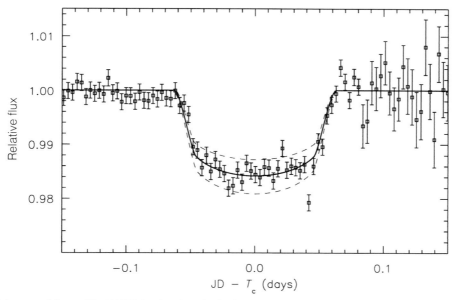

FIGURE 45.4 The light curve of the star HD 209458 showing the reduction in stellar flux by 1.5% due to the transit of its hot Jupiter. From the depth of the transit light curve, a radius of 1.4 Jupiter radii was derived, and combined with the planetary mass—determined from radial velocities—a low mean density of 0.31 g/cm^3 was found for this planet. JD = Julian Date, T_c = Time of center of transit.

typical radial velocity surveys. This makes the spectroscopic follow-up observations more time consuming. Sometimes, the largest telescopes, like the 10-m Keck telescope in Hawaii or the 8-m Very Large Telescope in Chile are necessary for this task. There are many astrophysical phenomena that can mimic the photometric signal of a planetary transit. Thus, in order to determine whether a transiting object is indeed a planet, it is necessary to obtain spectroscopic follow-up observations to rule out false positives, such as eclipsing binary stars in the background, or a giant star with a smaller secondary star or simply binary systems undergoing grazing eclipses. Once a false positive is ruled out, further follow-up observations are done to characterize the host star and, if possible, perform radial velocity measurements to derive a mass for the planetary companion.

In recent years, ground-based transit search programs have started to use entire networks of telescopes distributed over the globe. These networks, like the Trans-Atlantic Exoplanet Survey (TrES) or the Wide Angle Search for Planets (WASP and its successor SuperWASP) have uncovered over a hundred transiting hot Jupiters. Many of these planets share the characteristic of an inflated radius with the prototype case, HD 209458b. The energy source for this large radius is still under debate.

Another advantage of transiting planets is our ability to measure the angle between the stellar spin axis and the orbital plane of the planet. For this purpose astronomers use observations of the so-called Rossiter–McLaughlin (RM) effect, an apparent deviation in the radial velocity of the star from the Keplerian RV orbit due to the planet. This happens when the opaque disk of the planet blocks the blue- and redshifted sections of the rotating stellar surface during the transit. By examining the shape of the RM effect, one can determine if the planet orbits in the same direction as the star spins and if its orbital plane is aligned with the stellar equator (which both is approximately true for the solar system planets). This, of course, would be expected for planets that formed inside the rotating circumstellar disk and then subsequently have migrated inward by exchange of angular momentum with the disk. But again, the results for exoplanets were surprising. While roughly half of the transiting planets are well aligned in prograde orbits, the rest appears to have strongly inclined orbits and some of them were even found with retrograde orbits. This is usually seen as evidence that, at least for these systems, the inward migration was dominating and/or followed by gravitational scattering due to interactions with other planets rather than by mere disk-driven migration.

In December 2006, the French space agency Centre national d'études spatiales (CNES), together with the ESA and other international partners, launched the Convection, Rotation and planetary Transits (CoRoT) spacecraft. CoRoT consists of a telescope with 27-cm aperture and 4 CCD detectors to perform precise photometry. A major part of the science mission of CoRoT was the detection of planetary transits from space. By avoiding the atmospheric noise effects and by being able to monitor a star field for nearly half a year (when the telescope needs to rotate to avoid the Sun), CoRoT had the capability to find much shallower transit signals, and hence smaller planets, than traditional ground-based telescopes could. This was successfully demonstrated with the discovery of CoRoT-7b, the first transiting super-Earth. CoRoT measured a radius of 1.7 Earth radii, while the mass of about 8 Earth masses was determined using the HARPS spectrograph. The orbital period of this super-Earth is just 0.85 days, so it is very close to the parent star and has a high surface temperature. The mass determination of CoRoT-7b was complicated by the fact that the star itself is moderately active and that the RV data revealed the signals of two more low-mass (but not transiting) planets. Another interesting discovery from CoRoT is a 21 Jupiter mass companion with a radius of 1 Jupiter radii. This object is either a supermassive giant planet (as it is located in a typical hot Jupiter orbit with a period of 4.3 days) or a brown dwarf (an object that formed similar to a star but with insufficient mass to start hydrogen fusion in its core). CoRoT-9b was the first transiting giant planet with an orbital period longer than a few days. This planet orbits its host star every 95 days. The superior precision of the photometry from CoRoT allowed very precise radius determinations for these transiting objects and clearly showed the advantages of planetary transit photometry from space.

Ground-based programs also started to find smaller planets, in particular, around smaller stars. The transit signal of a given planet increases if the stellar radius decreases (the transit depth depends only on the surface area ratio of the two bodies involved). The first small transiting planet that was found by ground-based photometry was the radial velocity-detected hot Neptune around the M dwarf Gliese 436. The measured radius turned out to be very similar to the radius of Neptune. Figure 45.5 shows one of the most valuable scientific diagrams that we can only produce for transiting exoplanets: the mass–radius diagram. By placing a planet with measured radius and mass into this diagram, we can compare its location to theoretical models for the internal composition of the planet. Figure 45.5 clearly demonstrates that we can clearly discern between exoplanets that are gas giants, such as Jupiter, and smaller icy giants, such as Neptune, or even rocky planets similar to the Earth.

Another very interesting transiting planet around an M dwarf is GJ 1214b with a radius of 2.7 Earth radii and 7 Earth masses. This super-Earth was discovered by the

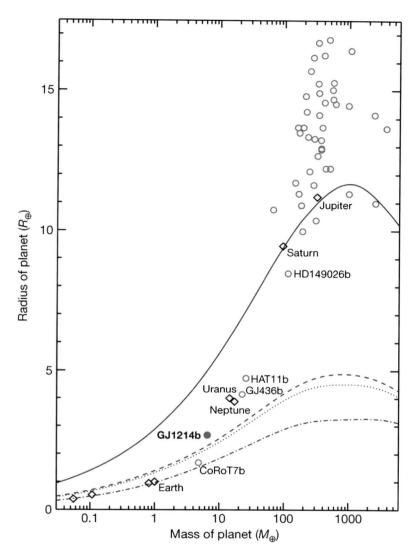

FIGURE 45.5 The mass-radius diagram for some transiting extrasolar planets (red circles). The Solar System planets are shown as black diamonds. Four models for the composition are also displayed: pure H/He (solid blue line), pure H_2O (dashed line), a water-dominated planet (75% H_2O, 22% Si and 3% Fe core; dotted line) and Earth-like (67.5% Si mantle and a 32.5% Fe core; dot-dashed line). This diagram allows us to distinguish between Jovian gas giants, icy giants like Uranus and Neptune and planets with a more Earth-like composition. *From Charbonneau et al. (2009).*

MEarth project that specifically targets small M dwarf stars to exploit the increased transit depth of small planets around this type of star. The host star itself is a cool M dwarf with a radius of only 0.2 solar radii and an effective temperature of 3000 K. Although the orbital period of GJ 1214b is only 1.6 days, the planet could be relatively cool, with an estimated equilibrium temperature of 400–550 K.

By using the Canadian *Microvariability and Oscillations of STars* (*MOST*) satellite and NASA's *Spitzer* space telescope, two independent groups discovered that the super-Earth planet in the ρ^1 Cancri (55 Cnc) system, planet e with an orbital period of only 0.74 days, also transits its star. ρ^1 Cnc e has a radius of 2.2 Earth radii, a mass of 8 Earth masses, and a mean bulk density of 4.5 g/cm^3. In the mass–radius diagram, this object is located outside the models for pure solid, rocky compositions and requires a significant amount of volatiles, e.g. water. This is the only small transiting planet around a star that is visible to the naked eye. Clearly, this planet will allow a much more detailed characterization in the near future.

3.3.1. Results from Microlensing Surveys

In the summer of 2003, the two microlensing surveys, OGLE and Microlensing Observations in Astrophysics (MOA), independently detected a microlensing event toward the galactic bulge. During the close monitoring of this event, a strong 1-week-long deviation from a single-lens light curve was discovered (see Figure 45.6). Careful modeling of the combined photometric data sets showed that the light curve of the OGLE 2003-BLG-235/MOA 2003-BLG-53 event is best described by a

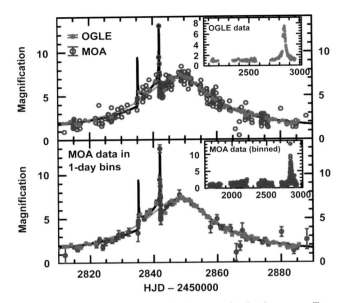

FIGURE 45.6 The first observed planetary microlensing event. Two photometric data sets (MOA survey in blue and OGLE survey in red) were combined to produce this light curve. The best-fit binary lens model is shown as solid black line. The planetary companion causes the strong double-peaked and 1-week-long deviation from the broader microlensing light curve due to its host star. HJD = Heliocentric Julian Date.

binary lens model with an extremely small mass ratio of 0.004. In the probable case that the primary lens is a normal star, the secondary lens would be a ~1.5-Jupiter-mass planet orbiting at ~3 AU. Thus, this event is regarded as the first demonstration of the discovery of an extrasolar planet by microlensing. A better characterization of the star/planet system that caused this lensing event has to await next-generation ground-based or space telescopes, which will be powerful enough to resolve the lens.

Microlensing is a very elegant technique to find exoplanets and it also has the added advantage that the Einstein radius (the lensing zone around a central mass) projects to roughly 1 AU for stars between us and the center of the galaxy. This means that this method is in particular sensitive to planets in or near the habitable zone for solartype stars. The drawback of this method is the complicated shape of the signal that could be fit by several different models with nearly the same goodness of fit. Moreover, the parameters for the planet rely often on statistical arguments on the properties of the lensing star.

About 15 planets have been announced by different groups using the microlensing technique. These surveys derived global statistics from the frequency of particular planetary lensing events compared to the total numbers of lensing events. From these studies, we know that giant planets around M dwarfs are rare and that cold Neptunes (at larger separations) could be relatively common.

3.3.2. The First Direct Images of Extrasolar Planets

Imaging searches for extrasolar planets have yielded the first candidates for successful detections. In these cases, faint companions were found in deep, high-contrast optical or infrared images taken either with powerful AO system, combined with a coronagraph, at large-aperture telescopes on the ground, or with the *HST*. Fainter background sources, mimicking companions, can be ruled out by the fact that these objects are comoving with the central star and are thus likely to be gravitationally bound to it.

One of the first imaging candidates was found near a young "failed" star, a brown dwarf, which itself is not massive enough to start thermonuclear reactions in its core. The central object (called 2MASSWJ 1207334−393254) has only ~25 times the mass of Jupiter and is located at a distance of approximately 230 light years. Its location also gives away its age: the brown dwarf lies within a young star-forming region, the so-called TW Hydrae association. This region contains young stars that are estimated to be only 8 million years old. The fainter (and thus presumably much less massive) companion was detected 0.8 arcsec away from the brown dwarf. At the distance to the brown dwarf, this transforms to a projected separation of 55 AU. First low-resolution spectra of the companion were also obtained. But is it a planet? This is the tricky part to decide. The observed brightness, colors, and spectral information can be compared to theoretical models for young planets. Such a comparison yields a mass of ~5 Jupiter masses for the fainter companion, a mass value placing the object firmly within the range of planets. However, some caution remains because the theoretical models for young planets are not calibrated and their uncertainties are difficult to assess. Moreover, the large separation of (at least) 55 AU raises the question that how massive the protoplanetary disk around a 25 Jupiter mass object must have been to form such a massive planetary companion so far away from the center.

Another candidate for a directly imaged extrasolar planet is the companion to the star GQ Lupi. This time the central object is really a star similar to the Sun, albeit a lot younger. The age of GQ Lupi is estimated to be between 100,000 and 2 million years. These very young ages for the imaging candidates are due to an observational selection effect: imaging searches specifically target young stars because young planets are much brighter at these early evolutionary stages. The companion appears 0.7 arcsec to the west of GQ Lupi, which translates into a projected separation of ~100 AU. A careful comparison of the discovery images with archived images revealed that the companion is indeed a co-moving object (they share the same motion in the plane of the sky, which rules out a background object). Therefore, it can be assumed that it has formed at the same time as the star GQ Lupi has. In this case, comparison of the available photometric and spectral data with models for young planets yielded a mass between 1 and 3 Jupiter masses for the companion. But again, the remaining uncertainties in the values derived from models are large, and it is difficult to distinguish young planets from low-mass brown dwarfs. If GQ Lupi b has indeed formed like a gas giant planet, then it was probably transported from the denser interior of the protoplanetary disk, where sufficient planet-forming material can be found, to its present location of about 100 AU away from the star.

The year 2008 was a milestone in the field of exoplanet imaging. Basically at the same time, two teams announced to have directly imaged planets around two nearby stars. And this time there was additional evidence that supported the planetary status of the candidates in these images. One group presented images from the HST Advanced Camera for Surveys of the star Fomalhaut. These images were taken years apart and show a planet candidate apparently moving according to the expected orbital motion interior to the dense dust ring around this star (see Figure 45.7). In the case of Fomalhaut b, one can calculate an upper mass limit of ~ 3 Jupiter masses, as the dusty disk appears unperturbed by the companion. The planet orbits the star at a very large orbital separation of 119 AU.

The other directly imaged exoplanets were found using ground-based telescopes. High-contrast images obtained with the Keck and Gemini telescope revealed even three companions to the star HR 8799. It is very interesting that a multiplanet system was found; this supports the notion that these companions formed in the circumstellar disk of gas and dust that surrounded the star when it formed. The orbital separations of the three planets around HR 8799 are 24, 38, and 68 AU. It appears to be a scaled-up version of the outer solar system. The two sets of images that were taken 3 years apart even show that these planets all orbit in the same direction, as expected for a planetary system. The masses of these planet are estimated to be between 8 and 13 Jupiter masses. In 2010, the discovery team of the HR 8799 multiplanet system added another planet to the ensemble: HR 8799 e orbits interior to the three already known planets, with a separation of 14.5 AU.

The similarities between the HR 8799 and the Fomalhaut planets are striking: both stars are more massive and younger than the Sun, both are surrounded by a dense dust disk, and both have massive planetary companions at large orbital separations. It was thus not entirely surprising that a planet was also detected by direct imaging, orbiting the prototype of star with a dense debris disk, beta Pictoris, a year following the HR 8799 and Fomalhaut detections.

The basic properties of these first directly imaged exoplanets are quite different to the planets found by the radial velocity or transit technique: they are located at very large orbital separations and are very massive (of course, there is a strong selection effect: these are the planets that we *can* take images of). With the standard giant planet formation model, the core accretion model, it is difficult to explain how such high-mass planets could have formed so far out in the disk. This makes the imaged extrasolar planets good candidates for alternative formation models such as the disk instability model scenario in which giant planets form rapidly from gravitationally unstable and collapsing clumps in a massive disk.

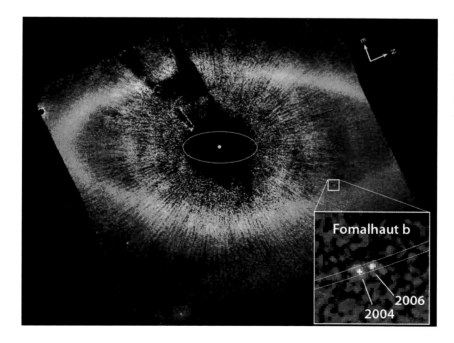

FIGURE 45.7 The directly imaged planet around the star Fomalhaut. Images taken with the Advanced Camera for Surveys camera onboard the Hubble Space Telescope reveal a planetary companion just interior to the massive dust belt around this star. The inset displays the apparent motion of the planet as seen in images taken 2 years apart. *Figure from Kalas et al. (2008).*

3.4. The Atmospheres of Extrasolar Planets

Transiting planets give us the unique opportunity to probe the atmosphere of exoplanets. During a transit one can observe the starlight that passes through the atmosphere of the planet and measure any absorption or scattering that is caused by the planet's atmosphere. With high-resolution spectroscopy we can discern certain chemical elements or molecules that are present. One can also compare the total flux that is received in the infrared in and out of transit to determine the temperature of the exoplanet's atmosphere.

Transmission spectroscopy was first performed with the HST to detect the atmosphere of HD 209548b, the first transiting planet. The *HST* spectra taken during the planetary transit showed a stronger sodium absorption line than the spectra observed without the planet in front of the star. This additional absorption is caused by the sodium in the planet's atmosphere. However, the amount of atmospheric sodium was less than expected from theoretical models, urging the astronomers involved in this study to speculate that a thick cloud cover prevents us from seeing deeper into the planet's atmosphere. Later, sodium absorption was also detected in other hot Jupiters using ground-based facilities.

In another *HST* observation of the HD 209458 planet, it was possible to measure hydrogen escaping from the heated upper layers of the planet's atmosphere. The escaping hydrogen gas forms a kind of cometary coma and tail around the planet and is blown away by the radiation and particle wind of the close-by star.

With the launch of NASA's infrared space telescope Spitzer, a new and very interesting spectral window for observations of extrasolar planets became available in the far infrared, where thermal emission dominates the radiation coming from a planet (as shown in Figure 45.2). Although Spitzer lacks the spatial resolving power to detect planets, its high sensitivity in the infrared can be used to discern between radiation from a star and its planet.

Two independent teams planned basically the same Spitzer observations to observe a transiting extrasolar planet during (and out of) secondary eclipse (the time when the planet is directly behind the star and hidden from view). If the amount of infrared radiation measured by Spitzer during the eclipse is less than that outside the eclipse, then this difference is the radiation coming from the planet itself. This effect was indeed successfully measured for the two transiting planets HD 209458 b and TrES-1 b. The amount of planetary infrared radiation was used to estimate a "surface" temperature for these two planets. The visible upper atmosphere of HD 209458 b has a temperature of 1130 ± 150 K, and for TrES-1 b the respective value is 1060 ± 50 K. Both values are in good agreement with the expected temperature of a giant planet's upper atmosphere heated by the intense irradiation of the nearby host star. The exact timing of the secondary eclipse of HD 209458b also demonstrated that its orbit is indeed circular and that tidal heating cannot be the explanation for its abnormally large radius. Interestingly (and quite ironically, because the information is obtained by the *lack* of photons during eclipse), these observations also represented the first unambiguous detections of photons emitted by extrasolar planets.

Transmission spectra for several transiting extrasolar planets have been obtained using space- as well as ground-based telescopes. Figure 45.8 shows an example of the transmission data for the hot Jupiter XO-1b that were obtained with the NICMOS instrument on the HST. The spectrum is rich on spectral information, showing deep absorption features due to molecules in the atmosphere of the planet. The best-fit atmospheric model contains H_2O, CH_4, and CO_2.

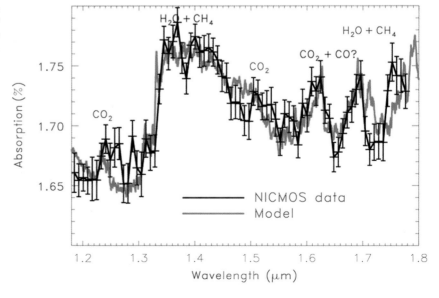

FIGURE 45.8 Near-infrared transmission spectrum of the hot Jupiter XO-1b obtained with HST/NICMOS. The best-fit atmospheric model contains significant amount of H_2O, CH_4, and CO_2. *Figure from Tinetti et al. (2010).*

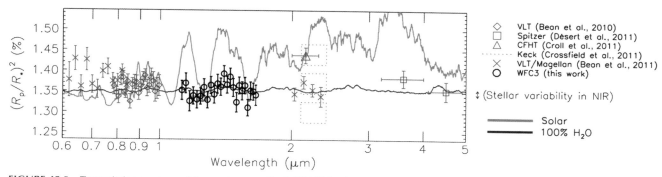

FIGURE 45.9 Transmission spectrum of the nearby super-Earth GJ 1214b, observed in the optical (0.6–1.0 μm) and infrared (>1 μm). The data are far better described by an atmospheric model rich in H_2O (blue line) than by a hydrogen-rich model (red line). NIR, near infrared. *Figure from Berta et al. (2012).*

More recently, similar techniques were used to probe the atmospheres of nearby super-Earths. Figure 45.9 shows the transmission spectrum of the small planet GJ 1214b measured with various instruments covering optical and infrared wavelengths. The very flat spectrum is best explained by an atmosphere rich (>40%) in H_2O, while a solar composition model (i.e. hydrogen rich) is clearly ruled out by the data. In the not so far future, we will be able to use these observational techniques to study small transiting planets and even search for so-called biosignatures in their atmospheres.

4. THE *KEPLER* MISSION

In March 2009, the NASA *Kepler* spacecraft was launched into its own orbit around the Sun. The primary scientific goal of *Kepler* is the determination of the frequency of Earth-size planets near and inside the habitable zone of Sun-like stars. The spacecraft consists of a wide-angle telescope with a mirror of 1 m diameter (Figure 45.10) and 42 CCD detectors to collect the light and perform very precise photometry. *Kepler* is using the transit method to detect extrasolar planets.

An Earth-size planet around a star of the same radius as the Sun produces a transit with a depth of only 84 ppm. Such a small transit signal is completely washed out by the variability of Earth's atmosphere, which is the reason why a transit telescope looking for Earths has to be in space. To achieve the goal to estimate the frequency of Earths, *Kepler* continuously monitors nearly 160,000 stars in a field of 100 square degrees (located roughly between the constellations Lyra and Cygnus).

A large part of the *Kepler* mission consists of the ground-based follow-up program to validate and confirm the *Kepler* planetary candidates. As in the case of ground-based transit searches, one cannot rely on the photometry alone to claim a planet detection. The *Kepler* follow-up observations are not only performed to rule out false positives but also to characterize the host star and, in the best cases, to measure the mass of the transiting planet by using precise radial velocity.

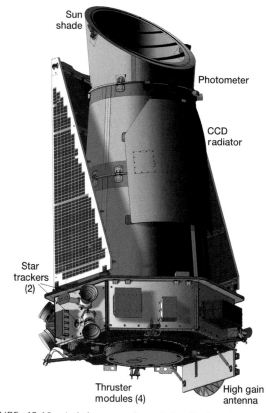

FIGURE 45.10 Artist's conception of the Kepler photometer and spacecraft. A 1-m Schmidt telescope that monitors nearly 160,000 stars in the Cygnus region for planetary transits since March 2009. Results from this mission revolutionized our current understanding of the structure and evolution of planetary systems. *Image courtesy NASA Kepler project.*

Of course, *Kepler* produces candidates much faster than confirmed planet detections. A look at the global properties of these candidates already teaches us a lot about the underlying planet population that creates these candidates. Some fraction of these candidates will be false positives but the vast majority of them are true planets. Over the first 3 years of the mission, some very interesting trends appeared in the parameters of the *Kepler* candidates. These trends truly revolutionized our picture of the extrasolar planet landscape. *Kepler* discovered a large population of small-radii planets with a radius of less than 4 Earth radii, which clearly outnumber giant planets at least for orbital separations less than 1 AU (Figure 45.11). Moreover, it found several surprisingly compact multiplanet systems which resemble more the resonant architecture of the Galilean satellite system than our own solar system. *Kepler* also found circumbinary planets. These are planets orbiting an inner stellar binary, where the stars orbit each other with periods much smaller than the orbital period of the planetary companions. As most stars exist in binary or multiple systems, this tells us that planet formation is quite robust. Planets can also form in the protoplanetary disk around a stellar binary despite the dynamical perturbations caused by the two stars. This also means that the total number of planetary systems in the galaxy may be much larger than previously estimated.

The planets in some of the multiplanetary systems that Kepler discovered are sufficiently close to interact gravitationally with each other. This leads to variations in the orbital parameters of these planets that appear in the transit photometry as changes in the times of the transits of these planets. This effect is called transit timing variations (TTVs). Once TTVs are detected for a system, one can construct a dynamical model, including the masses of the planets, that reconstructs these interactions and the observed TTVs. This method was successfully applied to the compact Kepler-11 six-planet system to constrain the masses (and thus densities) of the planets, without the need of precise radial velocities.

A new data analysis method was developed for the *Kepler* mission, the so-called BLENDER algorithm. BLENDER is used to validate *Kepler* planet candidates in those cases where a mass determination or planet confirmation by other means is not possible. BLENDER constructs light curves for a large variety of possible false-positive scenarios, such as background eclipsing binaries. It then compares these false-positive models with the actual light curve and uses all available information on the field around the target star to estimate the probability that the light curve is produced by a transiting planet orbiting the target star as compared to a false-positive scenario. If the probability for a false detection is very low, the planet is regarded as validated.

Another advantage that the *Kepler* mission has is that the precise *Kepler* light curves can also be used to perform an asteroseismological study of the planet host star. The stellar parameters derived from seismology are more precise than spectroscopically obtained values. And these improved stellar parameters (e.g. the radius of the star) subsequently improve the planetary parameters as well (since the measured depth of the transit is proportional to the radius ratio of the bodies).

Some of the most interesting individual systems that were found by *Kepler* include: the Kepler-9 system, the first system of transiting planets that exhibits TTVs; Kepler-10b, a very hot super-Earth with a radius of 1.4 Earth radii and a high density of 8.8 g/cm^3 (similar to CoRoT-7b); and the Kepler-11 system of six transiting planets, one of the flattest and most compact systems that we know. The planet in the

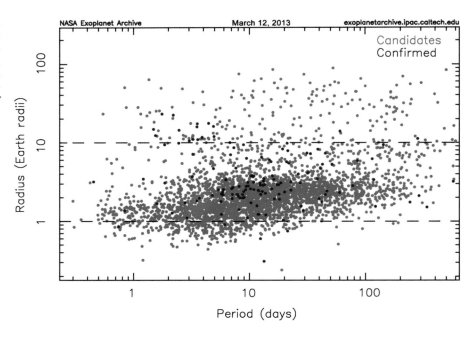

FIGURE 45.11 The distribution of the nearly 3000 Kepler planet candidates (red) in radius versus orbital period. The majority of candidates are smaller than 4 Earth radii. The blue dots show the confirmed Kepler planets. *Image courtesy of NASA Exoplanet Archive: http://exoplanetarchive.ipac.caltech.edu/.*

Kepler-16 system was the first circumbinary transiting planet. The inner binary consists of a K- and an M-type star that orbit each other every 41 days, while the Saturn-sized planet has an orbital period of 229 days. In summer 2012, *Kepler* announced the Kepler-47 system, the first circumbinary multiplanet system with at least two planets, one even in the habitable zone (Figure 45.12). This is further evidence that planet formation appears to be a robust mechanism and that also multiple star systems can host systems with several planetary companions.

Kepler also found small transiting planets inside the habitable zone of their (single) parent star. The first one found, Kepler-22b has a radius of only 2.4 Earth radii and a period of 290 days. Precise RV measurements cannot measure a mass in this case, but they set an upper mass limit, at the 1σ level, of 36 Earth masses. Two planets in the five-planet Kepler-62 system have orbital periods that place them in the habitable zone and measured radii of 1.6 and 1.4 Earth radii.

Despite the fact that the nearly 4000 planet candidates from *Kepler* are not all bona fide planets (a certain percentage are false positives), we can now derive first global statistical properties of the planet population that produces these candidates. From the total numbers of planet candidates of a given size we see immediately that smaller planets, in particular planets smaller than ~ 3 Earth radii, dominate the planet sample. A new study by the team that developed the BLENDER method estimates that the global false-alarm rate for Kepler candidates is $\sim 10\%$. So 90% of the 4000 candidates are very likely to be real planets! The same study presented planet occurrence rates for orbital periods less than 85 days based on the current *Kepler* data and corrected for the false-alarm rate in the different planet size bins. The result is shown in Figure 45.13. We see that there is a sudden jump in the frequency of planets around 3 Earth radii and a relatively flat distribution down to 1 Earth-radius. For planets smaller than that and orbital periods longer than 85 days, the *Kepler* results are still too incomplete to estimate planet occurrence rates. Nevertheless, based on these first estimates, we can say that $\sim 17\%$ of the stars in the *Kepler* field have at least one Earth-sized planet with a period of less than 85 days!

Another surprising result that emerged from the global analysis of the *Kepler* photometric data was the realization that using the photometric variability of our Sun as a benchmark, and as an expectation for the average variability of the stars in the *Kepler* field, underestimated the true variability of the *Kepler* target stars. The average *Kepler* star is $\sim 50\%$ more variable than our Sun. This makes very shallow transit signals more difficult to detect, due to the higher noise level in the light curve. In order to find the transits of true Earth-size planets, several more transits need to be combined to get a significant detection. It thus became necessary to extend the mission by several years to assure sensitivity for Earth-size planets in the

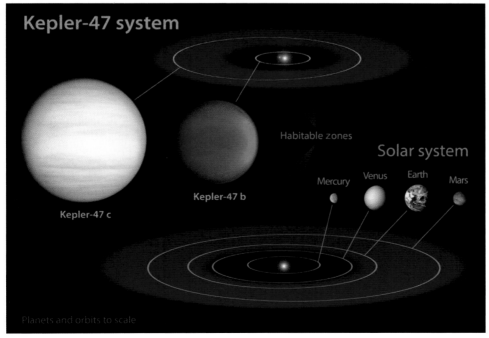

FIGURE 45.12 The Kepler-47 system of circumbinary planets compared to our solar system. The binary star consists of a solartype star and a small red dwarf star that orbit each other every 7.5 days. The two planets have orbital periods of 50 and 303 days, respectively. The inner planet is 3 times larger than the Earth and the outer one has a radius of 4.6 Earth radii and orbits within the habitable zone of this system. *Image courtesy of NASA Kepler mission.*

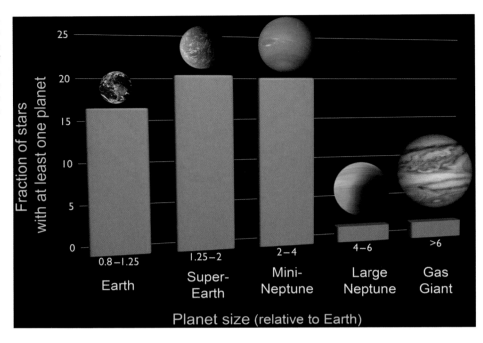

FIGURE 45.13 The planet occurrence rate as measured by Kepler for orbital periods less than 85 days. The primary goal of the extended mission of Kepler is the determination of the frequency of Earth-size planets in or near the habitable zone of Sun-like stars, so for periods close to 1 year. *Figure from Fressin et al. (2013).*

habitable zone (as these planets transit only about once per year). In 2012, NASA decided to extend the mission until 2016, but this extended mission ended prematurely in early 2013 due to a spacecraft malfunction. The final analysis of the entire 4 years of highly precise *Kepler* photometry will likely allow a first estimate of the frequency of Earth-size habitable zone planets, albeit with a larger uncertainty.

5. SUMMARY AND OUTLOOK

We truly live in the golden age of studying extrasolar planets. Detection programs using ground-based telescopes will continue to improve their sensitivity and to extend their search spaces, while space telescopes will play a more important and likely even dominant role in this field in the coming years.

The extrasolar planetary systems discovered so far demonstrate that a surprising variety of planetary systems exists in our galaxy. Although we now know that other stars also have planetary companions, and that planet formation is not unique to our star, many of them have characteristics different from the planets in our solar system. But this is primarily a result of the observational selection effects that are inherent to the techniques used to find these planets. A new kind of planet, the so-called super-Earths, was found to exist in other star systems.

The radial velocity technique was the first successful method that revealed a large number of planetary systems with giant planets on orbits unlike our solar system. These systems have had a much more dynamic past than our planetary system. The gas giant planets found at small orbital separations are probably a result of orbital migration, while the quasirandom distribution of orbital eccentricities might be caused by more violent and frequent interactions between planets in the early evolutionary stages. The fact that the overall mass function of extrasolar planets detected by RV surveys was steeply rising toward lower masses was an early indication that smaller and less massive planets are abundant in the galaxy. We also know today that the metallicity of the star- and planet-forming nebula has an impact on the structure of planetary systems. Most of the radial velocity planets are found around stars that are richer in heavier elements than the Sun. This is strong support for the core accretion model of giant planet formation. The lowest mass planets found around other solartype stars using very precise RV data already approach the mass range of the terrestrial planets.

The results from ground-based transit searches revealed the physical properties of a large number of hot Jupiters (and a few hot Neptunes). For the first time, one was able to determine the bulk density of these planets. Transiting planets helped to move the field from pure detection to a more detailed characterization of exoplanets. By using transmission spectroscopy we probed the atmospheres of these planets.

Microlensing surveys discovered several interesting planetary systems and the global results from these programs deliver important statistics on the galactic population of planets.

FIGURE 45.14 Artist conception of the future Giant Magellan Telescope. It consists of seven 8.4-m mirrors that have an equivalent light-collecting area of a 24.5-m telescope. With these extremely large telescopes like the GMT, the Thirty Meter Telescope, and the European Extremely Large Telescope, scientists will be able to detect and study nearby Earth-like planets and possibly even search for chemical signatures of life. *Image courtesy GMT project.*

The first exoplanets around nearby stars have been directly imaged. These are massive giant planets at large orbital separations that are currently difficult to explain by standard models of planet formation.

Space missions play an increasingly important role in exoplanet research. The European mission CoRoT was the first mission devoted to exoplanets. It found the first transiting super-Earth, CoRoT-7b.

In recent years, NASA's *Kepler* mission really helped to lift the field to the next level. It demonstrated that smaller planets dominate the population of planets. Despite the premature end of the extended mission, a very first estimate of the frequency of Earth-size planets in the habitable zone of Sun-like stars will probably arise from the *Kepler* data. This will be a truly historical result that will resonate not only in scientific circles but also in the general public as well.

But science never stops and *Kepler* is just the beginning of a new era of exoplanet research. There are several new missions and telescopes and instruments currently in preparation and planning that will greatly increase our knowledge of extrasolar planets. NASA is planning to launch the Transiting Exoplanet Survey Satellite (TESS) that will perform an all-sky survey for transiting planets. It can be expected that TESS will find some of the nearest small transiting planets that will be amendable for detailed follow-up studies. NASA's James Webb Space Telescope (JWST) will be an ideal instrument to study extrasolar planets in high detail in infrared wavelengths. The next generation of extremely large aperture ground-based telescopes with mirrors in the 30–40 m range, like the Giant Magellan Telescope (Figure 45.14), the Thirty Meter Telescope, and the European Extremely Large Telescope will allow astronomers in the coming decades to study small, possibly terrestrial, planets and their atmospheres around nearby stars. With JWST and these large telescopes on the ground, it might become possible to search for "biosignatures", chemical atmospheric tracers that are produced by an active biosphere. We are not very far away from obtaining a nearly complete census of the planet population in our galaxy, which will constrain our models of planet formation in general, and to even discover if any of these planets might be habitable.

BIBLIOGRAPHY

Berta, et al. (2012). *Astrophysical Journal, 747,* 35.
Borucki, W. J., Koch, D., Basri, G., Batalha, N., Brown, T. M., Bryson, S. T., et al. (2011). Characteristics of Kepler planetary candidates based on the first data set. *Astrophysical Journal, 728,* 117.
Charbonneau, et al. (2009). *Nature, 462,* 891.
Fressin, et al. (2013). *Astrophysical Journal, 766,* 81.
Haghighipour, N. (2013). The formation and dynamics of super-Earth planets. *Annual Review of Earth and Planetary Sciences, 41,* 469.
Howard, A. W., Marcy, G. W., Johnson, J. A., Fischer, D. A., Wright, J. T., Isaacson, H., et al (2010). The occurrence and mass Distribution of close-in super-Earths, Neptunes, and Jupiters. *Science, 330,* 653.
Kalas, et al. (2008). *Science, 322,* 1345.
Mayor, M., & Queloz, D. (1995). A Jupiter-mass companion to a solar-type star. *Nature, 378,* 355.
The NASA Exoplanet Archive: http://exoplanetarchive.ipac.caltech.edu.
The *Kepler* Mission: http://kepler.nasa.gov/.
The Extrasolar Planet Encyclopedia: http://exoplanet.eu/.
The Exoplanet Data Explorer: http://www.exoplanets.org/.
Tinetti, et al. (2010). *Astrophysical Journal, 712,* L139.

Part IX

Exploring the Solar System

Chapter 46

Strategies of Modern Solar System Exploration

Berndt Feuerbacher
German Aerospace Center, Cologne, Germany (ret.)

Bernhard Hufenbach
European Space Agency, Noordwijk, The Netherlands

Chapter Outline

1. Expanding Human Frontiers into Space — 981
    1.1. Why Should We Explore Space? — 982
    1.2. Exploration Extents into the Universe — 983
        1.2.1. The Human Action Radius is Limited — 984
        1.2.2. The Special Role of Mars — 984
        1.2.3. Exploration Beyond Mars — 985
        1.2.4. Exploration Beyond the Solar System — 986
    1.3. Future Space Exploration Relies on Significant Progress — 986
        1.3.1. Technological Developments — 986
        1.3.2. International Cooperation in Space Exploration — 987
        1.3.3. Human-Robotic Partnership — 987
        1.3.4. The Need for Sustainability — 987
    1.4. The Challenge of Human Exploration — 987
        1.4.1. Radiation Exposure is Most Critical — 988
        1.4.2. Is Shielding the Solution? — 988
2. A Case Study: Human Exploration of Mars — 989
    2.1. Cooperation Models — 990
    2.2. Defining and Communicating the Benefits — 990
    2.3. Focusing on Common Strategies and Intermediate Destinations — 991
    2.4. Mapping the Way to Mars — 992
        2.4.1. Strategic Mission Scenarios — 992
        2.4.2. Design Reference Missions — 992
        2.4.3. Global Exploration Road Map — 992
3. Space Exploration and Society — 993
    3.1. Many Nations are Interested in Space — 993
    3.2. Can Nations Afford Space Exploration? — 993
    3.3. Involving Private Capital — 995
    3.4. Avoiding Contamination of Earth and Celestial Bodies — 995
    3.5. Global Cooperation Becomes Reality — 996
4. Conclusions and Outlook — 996
Bibliography — 996

1. EXPANDING HUMAN FRONTIERS INTO SPACE

Curiosity is a fundamental characteristic of the human species, which is the natural incentive for exploration. Discovering new worlds by exploring the unknown has been the stimulus to develop mankind into its present state. Starting from their origins in Africa, humans have continuously explored their near and far environment and thus settled all habitable ranges of the Earth. Mankind has explored all areas of the globe, from pole to pole, from deep in the sea to the highest mountain tops and to the uppermost levels of the atmosphere.

The challenge of today and of tomorrow is the exploration of space. Like the exploration of the seas, which followed breakthroughs in shipbuilding and navigation, distinct technological developments are a precondition for this step. The advances in rocket propulsion in the middle of the twentieth century, initiated by military research efforts in Germany, were the starting point of a rapid evolution, accelerated by the cold war in the "space race". This led to a solid technological basis for space flight, stimulating research, advanced technology, and international cooperation.

There are various definitions of the term space exploration. In the present context, we define it as follows: "**Space exploration** is the extension of human reach beyond the Earth's atmosphere using spacecraft to access unknown terrains and environments, and to acquire knowledge about space, planets, stars, or other celestial bodies by human and robotic means."

This definition does not include astronomical observations from Earth, which has given mankind insight into space and the celestial bodies since ancient times, as part of space exploration. It does, however, encompass remote observation of any object from space.

The destinations of space exploration are unlimited; however, some restrictions are imposed by the human life span, the energy available for propulsion, and the distances involved. In foreseeable future, our reach will be confined within the planetary system. Distinct constraints will be encountered for human exploration, where the major destination will be the surface of Mars, our neighbor planet.

1.1. Why Should We Explore Space?

Space exploration has started just 50 years ago, and it is a continuous process that will extend over the next centuries. So far, substantial accomplishments have been achieved through space flight that changed paradigms of our world view and improved our daily life. Numerous satellites observe the Earth's surface for a variety of purposes. They range from agricultural support to the mitigation of natural or man-made disaster events. Important evidence is derived from their data concerning the human influence on a global change of climate. A permanent human presence in **low Earth orbit** (LEO) has been established through the International Space Station (ISS), where the microgravity environment allowed for discoveries in medicine and materials science that have led to improved health care and industrial processes on Earth. Communication relies on spacecraft in geostationary orbit, and novel navigation services have become a matter of course worldwide. Industrial structures have been established on Earth through space exploration, with considerable economic impact. Our knowledge about space and the objects in our planetary system has grown substantially, and the understanding of our universe has evolved in a revolutionary way. Such progress for humanity was not expected at the early days of space flight, which were driven by military aspects and the cold war environment.

There is no limit to the exploration of space, and we may expect in future similar progress as in the past. While discoveries and novel technologies cannot be predicted, a number of benefits will undoubtedly emerge from space exploration. Among those are the following:

- *Improvement of the quality of life on Earth* is a prime objective of space exploration. This applies to progress for individuals in areas like safety, food, and health, and even more to collective advances for humanity. It will help to ensure a viable Earth for our grandchildren by mitigating global challenges. This includes the conservation of the Earth's biosphere and climate. Both are global phenomena, so global tools in space are required to monitor changes and find means of reducing negative effects.

- *Human progress and knowledge* will rely on contributions derived from space exploration. There are fundamental questions of mankind, such as

    Where do we come from? What is the origin of life in the universe?
    Are we alone? Does life exist anywhere else than here on Earth?
    Where are we going? How are humanity, the Earth, and the universe developing?

  Definitely, space exploration will deliver important contributions toward answering such questions. Moreover, it will yield a wealth of scientific opportunities that help to understand the solar system in relation to our home planet Earth and the universe as a whole.

- *Technology and economy* will advance in future through progress in space exploration. The technological demands of space are extreme, calling for new approaches in miniaturization, environmental tolerance, reliability, and energy conservation. Space developments will influence industrial processes and strengthen the competitive position of manufacturers on the international markets. Private ventures will increasingly participate in exploration, as opportunities like space tourism and planetary resource extraction emerge.

  Space exploration is an engine of progress, stimulating increased employment in the high-quality job sector. The past rapid evolution of the communication market will progress with new services from space, with the accompanying increase in novel instrumentation and consumer products. In addition, derived services on the ground making use of space assets will rapidly grow.

- *Cooperation* is a necessity in future space exploration. Major space ventures, in particular involving human space travel to distant destinations like Mars, are beyond the capabilities of any single nation. While the early beginning of space flight was dominated by national competition, culminating in the cold war between the two superpowers, in recent times the aspect of cooperation has become more and more relevant. A current example is the ISS, a cooperative venture between 14 nations that is stable since 1998. It has survived political turnover in several partner nations, severe technical and financial problems, and the world economic crisis. There is no doubt that the cooperation in a peaceful technological project of the dimension of the ISS has contributed to improved relations between the participating nations. Scientific and technical cooperation on a global scale, a precondition for

human space exploration, will contribute to peace and a better understanding between nations.
- *Inspiration* is an important feature of space exploration. The fascination of space is prompted by the mere difficulty of reaching and exploring it, and by the rewards in terms of knowledge and discovery of exotic new worlds. Human presence in space dramatically enhances public interest, a fact of utmost significance as it is this very public that carries the load of funding in its role as taxpayer. Space exploration catches the attention of the young generation and is a stimulus to aspire careers in science and technology.
- *Sustained human life beyond Earth* could open a variety of perspectives for future human development. While none of the bodies in our planetary system is directly suitable for human life, numerous planets orbiting stars are presently being identified by searches in the Milky Way (*see* Exoplanets). Some of those could be habitable or be made habitable to be destinations for expansion of the human race. Enormous technology steps are required to support human transportation to even nearby stellar systems. Alternatives such as the transfer of genetic information might be considered.

1.2. Exploration Extents into the Universe

Space is huge, and it contains a vast amount of features and objects worth exploring. Not all of them are as obvious as, for example, a planet. Of interest is "empty" space itself, where the physical properties, magnetic and electric fields, radiation environment, and stellar winds or dust particles may be the object of investigation. In fact, the first scientific satellite, Explorer 1(*see* Planetary Exploration Missions), discovered in 1958 the van Allen belts, a radiation zone around the Earth composed of solar wind and cosmic ray particles. Later this led to the discovery of the magnetosphere, which protects the Earth from hazardous radiation and particles.

Conspicuous potential destinations of exploration are the large and visible objects in nearby space, like the sun and its planets, the Moon, and various natural satellites orbiting planets. There is a vast population of smaller objects, like the **asteroids** accumulated between Mars and Jupiter, and **comets** roving around in the solar system. The latter originate from the **Kuiper belt** or the **Oort cloud**, areas crowded with small bodies, extending from Neptune's orbit to the limits of the solar system.

Most large bodies in our vicinity have been discovered in ancient times, long before the space age. New discoveries through space probes therefore are limited to several moons of the giant gas planets.

Exploration of planetary bodies is usually performed in steps. First, a flyby takes close images and measures selected physical properties. A more thorough study continues with a satellite in orbit around the celestial body, mapping most of the surface and taking detailed remote measurements. Exploration continues with landing a probe on the surface (whenever existing) to make in situ measurements, and possibly releasing a mobile element to investigate the near environment. Returning samples to the Earth for detailed analysis is the following step, before a human landing on a body in the planetary system might be considered. The final step is utilization, which could be scientific, public, or commercial.

A summary of the exploration steps of the large objects in the solar system is presented in Figure 46.1. Apparently all planets and some smaller bodies have been visited by

Objects in the solar system													
Exploration step	Moon	Mercury	Venus	Mars	Mars moons	Comets, Asteroids	Jupiter	Jovian moons	Saturn	Saturn's moons	Uranus	Neptune	Kuiper belt
Discovery	Discovered before the Space Age						Galileo		Cassini				
Flyby	Luna 1 1959	Mariner 10 1974	Mariner 2 1962	Mariner 4 1965	Viking 1 1976	Giotto 1986	Pioneer 10 1973	Galileo 1996	Pioneer 11 1979	Cassini 2004	Voyager 2 1986	Voyager 2 1986	New Horizons
Orbiting	Luna 10 1966	Messenger 2008	Venera 9 1975	Mariner 9 1971		NEAR 2000	Galileo 1995	JUICE	Cassini 2004				
Landing	Luna 9 1966		Venera 4 1967	Viking 1 1976		NEAR 2001				Huygens 2005			
Sample return	Apollo 11 1969			MSR	Phobos Grunt	Hayabusa 2005							
Human exploration	Apollo 11 1969												
Utilization													

FIGURE 46.1 Summary of the exploration of solar system objects. The light blue color marks the exploration steps (left column) successfully performed with the first event marked together with the year of accomplishment. Presently planned exploration steps are indicated in dark blue. The hatched fields mark bodies without an accessible solid surface. NEAR, Near-Earth Asteroid Rendezvous; JUICE, Jupiter Icy Moon Explorer; MSR, Mars Sample Return *Images: NASA and ESA.*

spacecraft (see Planetary Exploration Missions), but only the Moon has been explored including all steps and is now awaiting utilization.

1.2.1. The Human Action Radius is Limited

While robotic spacecraft have investigated nearly the entire solar system, present-day space technology sets strict limitations on the range humans can travel in space. Crew on board is heavy; it requires facilities and consumables. Radiation in space is a hazard, with the dose increasing with exposure time. The safety demands are considerably higher than for robotic probes, and the safe return of the crew is, with the exception of a few debatable projects, a precondition.

A schematic presentation of the **human action radius** is given in Figure 46.2. Man can travel in the vicinity of Earth on various orbits and has established permanent presence in LEO by means of the ISS. The Moon has been visited as part of the Apollo project several times in the late 1980s. Using present-day technology, humans can travel toward the Sun up to about the Venus orbit. Closer to the Sun thermal conditions become very demanding. In the other direction, away from the Sun, planet Mars is the most exciting destination for human exploration. Using appropriate road maps and way stations, and particular by suitable means of **human-robotic partnership**, the human range could be extended beyond Mars and its moons, up to the inner parts of the asteroid belt.

The current human action radius appears quite limited, compared to the size of the solar system and the dimensions of the universe. However, man can be a factor to answering the main quest of exploration within his action radius. In particular, contributions to the fundamental search for the origin of life can be expected. Not all destinations are of equal value in this respect. The Moon offers a special potential as an archive of the history of the planetary system (see The Moon and Lunar Exploration). Without atmospheric erosion and limited geologic activity, the Moon preserves a record of almost its entire impact history of 4.5 billion years, offering important information on the processes in the early evolution of the solar system. Venus, with its hot surface and the extreme atmospheric pressure probably has not preserved much features relating far back in time. Interesting objects are small bodies like asteroids and comets. Comets, distinguished from asteroids by their coma (or tail), originate from the outer regions of the solar system. At the low temperatures prevailing there, volatiles and possible organic molecules have been preserved from the **protoplanetary nebula**, the material from which our solar system has formed. If such a body is deviated into regions close to Earth, it is a source of information about matter essentially unchanged since 4.5 billion years. Different motives stimulate human interest in asteroids, which might be a hazard to Earth. Some of them, approaching the vicinity of Earth as a "**near Earth object**", might develop into targets for commercial ventures mining precious materials for use on Earth.

1.2.2. The Special Role of Mars

Planet Mars has always inspired human fantasy. This is one of our two close neighbor planets, and it exhibits similarities to Earth. Through a telescope, even in early times, features could be identified, which change periodically with seasons. This stimulated the idea of vegetation existing on Mars. The discovery of surface structures, erroneously identified as engineered "channels" (actually due to an incorrect translation from Schiaparellis labeling as "canali", which means riverbeds) led to speculations about intelligent life on the planet.

Early exploration missions to Mars therefore had stimulated significant interest. The first images received from a spacecraft were taken by Mariner 4 in 1965. Even though the quality of the 22 pictures returned to Earth was disappointing from today's standards, the result was a sensation, but at the same time disillusioning; they showed a desolate landscape with no signs of vegetation or water. The landing of the Viking spacecraft in 1976 confirmed this impression, and its search for traces of life was rather negative, however, by no means conclusive.

Nevertheless, Mars is an intriguing planet. It is at the outer edge of the **habitable zone** of our solar system, where the existence of liquid water in principle is possible. In the early state of planetary development Mars could have been quite similar to Earth, temporarily with abundant water and a dense atmosphere (See Mars Atmosphere: History and Surface Interactions). It is intriguing to note that this was at

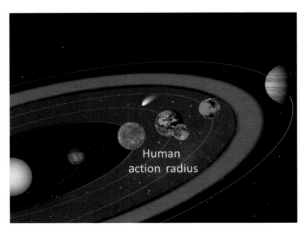

FIGURE 46.2 The human action radius in the solar system extends from the orbit of Venus to the inner part of the asteroid belt, including the Moon, Mars, near-Earth asteroids, and some comets.

a time when life developed on Earth, about 3.5 billion years ago. Due to its smaller size, Mars has developed much faster than Earth (*see* Mars Interior). Today it is a desert planet with a tenuous and toxic atmosphere. At the Mars surface, pressure and temperature conditions are below the triple point of water, so water cannot exist in liquid form for extended periods of time. The atmosphere, consisting mainly of carbon dioxide, is highly oxidizing. Together with the radiation environment this will prevent the persistence of organic material near the surface of Mars. In spite of this hostile environment, traces of extinct or extant life could potentially be found in subsurface layers of Mars, where water exists adsorbed on grains or in the form of ice, protected from the atmosphere and radiation.

It is therefore not surprising that Mars is one of the most popular destinations for space exploration, only second to the Moon, which is so much closer. Since 1960, 42 missions have been launched to the red planet (not including swing-by maneuvers). Of these, only 22 missions have been successful, which demonstrates the difficulty in reaching our neighbor planet. A total of eight landing probes have successfully reached the surface of Mars, and four rovers have been deployed to investigate the environment.

The flight to Mars is quite challenging. The minimum energy trajectory to transport a spacecraft to the planet follows a **Hohmann transfer orbit** (Figure 46.3). This is an elliptical trajectory that has common tangents to circular orbits both at the body of departure and the destination body, with pulses of injection thrust applied at both ends. It makes use of the orbital velocity of the Earth around the Sun. The launch window is defined by the respective positions of the planets at the beginning and the end of the transfer. The Earth—Mars trajectory has a launch window every 780 days, which is the **synodic period** of Mars with respect to Earth. Due to the orbital eccentricities, optimum launch windows exist approximately every 16 years, with the last opportunity in 2003 and the next in 2018. The actual transfer flight takes about 260 days. After a landing maneuver, the window for the return trip opens about 15 months after arrival, so the total time required for a journey to Mars and back with minimum energy requirement is about 2 years and 9 months or 1000 days.

1.2.3. Exploration Beyond Mars

For destinations beyond Mars, the difficulties of spaceflight increase substantially. The distances become very large, which gives rise to transfer times in the order of decades rather than years. Simultaneously, fuel and propulsion requirements increase as the spacecraft moves further out in the solar gravitational field. Solar power becomes increasingly dilute at these distances, imposing demands on spacecraft energy supply and thermal design. The radiation exposure is high due to the long travel time, and it can become massive when approaching the radiation belts of the large planets. All these difficulties increase dramatically if transport of crew is considered.

The giant gas planets Jupiter, Saturn, Uranus, and Neptune have thick atmospheres of mainly hydrogen and helium that gradually increases in density, with possibly a rocky core inside (*see* Interiors of the Giant Planets). In contrast to the terrestrial planets, there is no solid surface at their outer boundaries, so landing a probe does not make sense. Therefore, flyby and orbital missions might possibly be complemented by balloon-like probes in the atmosphere. All the outer planets, however, have numerous moons that have solid surfaces and one of them, Saturn's largest moon Titan, even has an atmosphere. Titan has been visited by the Cassini mission and the landing probe Huygens has successfully descended through the atmosphere to land on the moon's surface.

Icy moons of gas giants have recently attracted increased attention as they may contain large reservoirs of liquid water, where primitive forms of life might exist. The three outer Galilean moons of Jupiter, namely, Europa, Ganymede, and Calisto, hold liquid water under thick surface layers of ice, kept fluid by pressure and tidal heating. It is in particular Europa (*see* Europa) that has stimulated interest due to the observations that hint at a huge ocean of salty water, buried under an ice layer with a thickness of a few kilometers up to several tens of kilometers. The potential existence of microbial life has been speculated there, similar to the hydrothermal vents ("black smokers") observed deep in the oceans of the Earth. Another candidate for the existence of liquid water is Saturn's moon Enceladus, which emits plumes of water vapor in a process of **cryovolcanism**. Different conditions exist on Titan, where lakes of liquid methane were detected. Conditions similar to those on the primordial Earth led to

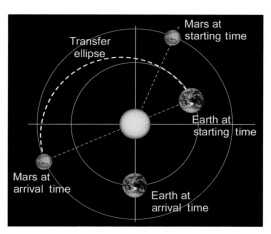

FIGURE 46.3 A minimum-energy flight to Mars relies on the Hohmann transfer ellipse, making use of the orbital velocity of Earth.

the speculation of possible prebiotic compounds or even primitive microbial life forms to exist there. The Huygens lander was not equipped to search for traces of life, so no evidence exists today. It is interesting to note that these observations put the notion of a **habitable zone** around a star in perspective. This is presently defined as a circumstellar zone where planets with a sufficiently dense atmosphere can maintain liquid water at the surface and such can host life.

1.2.4. Exploration Beyond the Solar System

At the present time, four spacecraft are on the way to leave the solar system (Figure 46.4). These are Pioneer 10 and 11, launched to explore the outer solar system in 1972 and 1973, respectively, and Voyager 1 and 2, both launched in 1977 to take advantage of a favorable alignment of the planets. Voyager 1 presently travels through the heliosheath, which separates the solar system from interstellar space. In 2022, Voyager 1 will pass the **heliopause** and become the first man-made object to leave the solar system. In contrast to the Pioneer spacecraft, both Voyager probes are still partly operational and transmit data about electric and magnetic fields, solar wind, and hydrogen distribution.

Traveling at a speed of 17 km/s after four gravity assist maneuvers it will take Voyager 1 about 50 years to leave the solar system. This is an indication of the enormous distances and flight times inherent in interstellar travel. A trip to the closest star Proxima Centauri at this speed would take roughly 75,000 years. Obviously, exploration beyond the solar system requires major progress in propulsion. Novel systems including nuclear energy and continuous propulsion through, for example, electric drives will be necessary. Interstellar spaceships carrying crew are out of scope with present day technology. The problems are not only propulsion, but include the life support, which has to operate with a closed cycle, radiation protection, reliability of all systems and subsystems, and not the least the psychological problems inherent in such ventures. On the other hand, science discovers at a rapid rate **extrasolar planets**. Very likely planets with conditions similar to our home planet will be found, and some of them maybe in nearby stellar systems. It will be just a matter of time to develop technology to explore those potential future habitats for humans.

1.3. Future Space Exploration Relies on Significant Progress

Future space exploration, by its very nature, will have destinations more demanding than previous ventures. Most of them, in particular if a crew will be involved, will therefore require advanced technical developments or even breakthroughs, novel ways of project preparation and long-term planning, enhanced coordination and cooperation between nations on a global scale, and political consensus within and between all participating nations.

1.3.1. Technological Developments

Heavy launch capabilities are a precondition for ambitious exploration tasks. The present development status of chemical rockets has approached its limits. In modern launch vehicles, the structural coefficient, i.e. the ratio of structural mass to total launch mass, has reached a value of 5%. There is not much room for engineering improvements in this field.

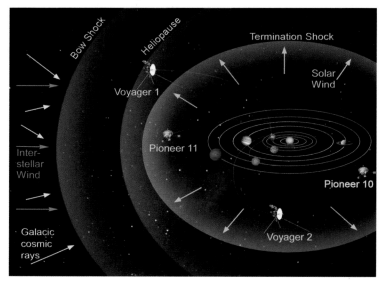

FIGURE 46.4 Schematic of the heliosphere (not to scale), the region where solar fields and particle flows dominate interstellar space. The termination shock was successfully detected by Voyager 1 in 2004 and Voyager 2 in 2007. Voyager 1, launched in 1977, is the farthest man-made object from Earth and is still active.

Various chemical propellants are in use, optimized for solid or liquid rocket motors. Adaptation to specific tasks, including exploration missions, therefore relies on scaling and staging optimization.

Major improvements in the efficiency of transport from the Earth's gravity well into space require new approaches. Several systems using air breathing engines in the first stage are presently under development. Among these are commercial ventures like the suborbital space plane "SpaceshipOne" or the "Skylone" project. Alternatives are projectiles, accelerated by chemical means or by electromagnetic rail guns. Futuristic ideas such as a space elevator will have to await advances in materials development.

For in-space propulsion, presently used systems rely on reactive thrusters, where a mass is emitted at high velocity to propel the spacecraft. Chemical systems could in future be replaced by electromagnetic or nuclear powered engines. Nonreactive propulsion systems like solar sail, magnetic sail, or electromagnetic tether depend on photon or particle flux from the Sun, or on interplanetary fields. The latter are most efficient in the vicinity of the Sun or near strong local magnetic fields like in the Jupiter system.

Novel development is also demanded in spacecraft design. Depending on the exploration destination and the strategy chosen various spacecraft for crew and cargo transport will be required, in addition to habitats, descent- and ascent vehicles, and service and support systems. All this will have to be embedded into extensive research programs on advanced materials, novel manufacturing processes, system management and simulation, and accompanied by parallel advances in instrument technologies.

1.3.2. International Cooperation in Space Exploration

Most space ventures to date are single-nation efforts or bilateral agreements. Notable exceptions are the European Space Agency (ESA), which brings together the knowledge, industrial power, and resources of 20 European nations, and the ISS, a joint effort of five partners including 14 nations. A challenging exploration effort should encompass as many nations as possible, each contributing according to its national interest, capabilities, and resources. This requires a strong and long-term coordination effort to ensure compatibility, efficiency, and to avoid duplication. A framework has to be established of interested nations, open for additional participants. This has to agree on overarching principles such as a governance system, common goals and objectives, basic principles of cooperation including work sharing, interdependency and competition, and the cooperation with commercial partners. From there, mission scenarios and technological road maps will then be developed.

1.3.3. Human-Robotic Partnership

The definition of space exploration hinged to the extension of human reach in space does not imply a priority for human spaceflight. In fact, **human-robotic partnership** will be indispensable to any exploration venture. Any human mission to a body in space will be preceded by robotic flights, investigating the environment and preparing it for human presence. Robotic precursor missions are important sources of scientific knowledge. The synergy between human and robotic action in space will not be limited to preparatory measures. Robotic components in a human space program will be necessary to reduce the risk, to ensure human safety, health, and to enhance the working efficiency. Humans may play a role for future servicing of high-value assets near Earth or Moon (following the example of the Hubble Space Telescope). A balanced use of human and robotic capabilities will be necessary to maximize the return of investment in any space exploration mission.

1.3.4. The Need for Sustainability

The Apollo program was a bold pioneering venture and very successful. It was, however, by no means sustainable. While the scientific results are still highly valuable, much of the technical and operational expertise is lost today. Apollo is neither a precursor nor a model for a future exploration program.

Sustainability refers to the need to make steady progress and continuous delivery of benefits within an affordable budget environment. In the end, the perceived benefits and investments need to be balanced. A sustainable exploration program cannot be understood as a sequence of several independent visits to various destinations, but rather as a continuous, stepwise process that finally leads to self-sufficient presence of humans in space, supported by robotic systems, and exploiting local resources. Therefore, the capability of developing **in situ resource utilization** will be an important component of exploration.

But sustainability is necessary not only in the technical domain. As a multinational effort, the international cooperation agreement signed as the basis for space exploration has to be sufficiently robust to sustain political changes, economic problems, and a temporary decline of public interest. An important ingredient of such an effort is the involvement of private industry, both in the development of an efficient space infrastructure, and in the exploitation of extraterrestrial resources. In the long run, having a sustained and self-sufficient presence in space will also allow humanity to maintain off-world repositories of knowledge and history.

1.4. The Challenge of Human Exploration

With the establishment and operation of the ISS mankind is at home in LEO. The ISS has been inhabited continuously since the year 2000. Some astronauts have logged an extremely long stay in space, with cosmonaut Valeri Polyakov holding the record of 437 days. There are no reports about permanent impact of spaceflight on the health status of astronauts. Apparently, the effects of microgravity on the human physiology can be managed by appropriate countermeasures. These facts appear to allow the conclusion that humans can adapt to the space environment. This, however, is true only for the LEO, where the crew remains within the shield of the Earth's magnetosphere and at least a third of the cosmic radiation is blocked out by the large body of nearby Earth.

The challenges of human spaceflight beyond LEO are numerous. As the distances get large, the travel time increases due to present day limitations in space propulsion. Intermediate stations on the way to distant targets may reduce some difficulties. For destinations beyond the Moon, opportunities for emergency return within a few days diminish. This imposes stringent requirements on safety and reliability of all subsystems, as well as repair and in situ manufacturing of spare parts. In addition, the impact on human psychology should not be neglected. Medical care requires novel approaches. Provisions for in situ medical treatment have to be envisaged, supported by advanced telemedicine as developed in the course of ISS operations.

A crew member requires about 5 kg of consumables per day in the form of food, water, and oxygen. It creates an equal amount of waste in solid, liquid, or gaseous form. All systems used presently for long-time life support, e.g. on Apollo or ISS, rely on resupply from Earth. For long-duration space flight, it is a prerequisite to recycle at least part of the consumables. Ultimately, closed cycle systems are necessary. Environmental control and life support systems provide breathing air, remove carbon dioxide and contaminants, and regulate pressure, temperature, and humidity for human survival and comfort. Only partial reuse is present state of the art. The life support system of the ISS is equipped with a water recovery system that processes waste water from hygiene systems (showers, sinks etc.), urine water, and water vapor collected from the atmosphere. This water is partly recycled for drinking, and part of it is electrolyzed to produce oxygen for air revitalization. Huge development efforts are necessary to produce closed cycle life support systems suitable for long-distance space flights. There are research attempts toward in situ food production, which are still far from a closed cycle production.

1.4.1. Radiation Exposure is Most Critical

The most severe limitation for human spaceflight is the radiation environment. As a spacecraft leaves the protective shield of the Earth magnetosphere, it is exposed to the space radiation environment. This is composed of galactic cosmic rays and energetic solar particles. **Galactic cosmic radiation** (GCR) consists of about 85% high-energy protons, 15% helium (alpha particles), and highly ionized heavy nuclei (HZE particles). **Solar energetic particles** originate from solar flare eruptions, which inject energetic electrons, protons, and alpha particles into space.

The health impact of space radiation environment on the human body can be both acute and chronic. In particular heavy bursts of solar particle events (SPEs) may be lethal to unprotected crew. These events, however, are relatively rare with less than 10 per solar cycle and last only for days. The crew can be protected for these periods in a small, specially shielded shelter. GCR flux is more continuous, its intensity being inversely correlated with the solar cycle. It causes health hazard like DNA damage, increased cancer risk, and may impact the human central nervous system. An illustration of the risk of cancer death for various exposures is presented in Figure 46.5.

1.4.2. Is Shielding the Solution?

Shielding reduces the radiation flux, but its effect has limitations, mainly due to the amount of protective material that has to be transported during the space flight. Present human spacecraft uses the aluminum skin of the crew module as radiation shield. The skin of ISS reduces the radiation flux inside by about 50%. Light atoms like hydrogen are more efficient in shielding. Therefore, plastic material like polyethylene, which is rich in hydrogen, might in future be used for spacecraft skin. This material is light and can be produced with strength higher than aluminum. Liquid hydrogen, which would serve as fuel for propulsion, can be used as a temporary shield. Water carried on the spacecraft is an alternative. Especially for crew protection during SPEs, which are of limited time duration and allow for a warning time of a few hours, small sheltered areas can be created in this way.

An alternative to mass shielding is the active deflection of charged particles by magnetic fields, similar to the magnetosphere that protects Earth. A "magnetic Faraday cage" for a manned flight to Mars has been proposed by Samuel Ting, which, however, requires a total coil mass of more than 5 tons and up to gigawatts of energy.

The effects hazardous to the human body scale with the absorbed dose, which is exposure multiplied by time. This leads to typical mission radiation doses as illustrated in Figure 46.6, based on present-day space transport technology. The data reveal that a human Mars mission is at the limit of the radiation dose acceptable for an astronaut's career. Human space flight beyond Mars, exemplified for a mission to Jupiter's moon Calisto in the diagram, would far exceed this dose. As passive shielding has a heavy mass penalty, the alternative is to reduce travel time. This requires

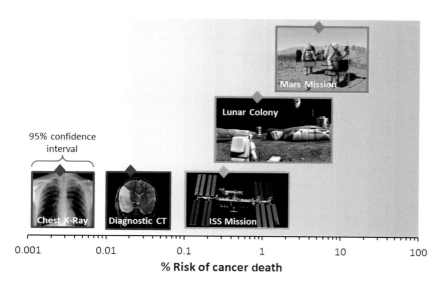

FIGURE 46.5 Risk of cancer death for various space missions, as compared to the risk from medical diagnostic tools on Earth. *Courtesy: Marco Durante.*

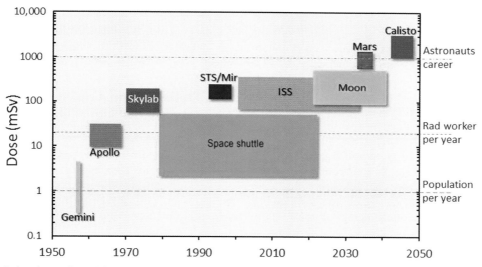

FIGURE 46.6 Radiation dose estimated for various space missions. The dashed horizontal lines indicate the limit levels for the public (lower line, corresponds approximately to the radiation background on Earth), for designated nuclear energy workers (middle), and the maximum level for an astronaut accumulated over his career. *Courtesy: Marco Durante.*

breakthroughs in propulsion technology, including continuous thrust rather than short pulses as foreseen in Hohmann transfer orbits.

2. A CASE STUDY: HUMAN EXPLORATION OF MARS

Human in situ exploration of the Mars surface represents an achievable challenge for the twenty-first century. Sending humans to the surface of Mars, working there efficiently for advancing the knowledge of the Mars system, and returning the crew safely will require long-term planning and preparation over multiple decades. Studies by the International Academy of Astronautics (IAA) (Huntress et al., 2006) and the Committee on Space Research (COSPAR) (Ehrenfreund et al., 2012) have investigated this endeavor. The capabilities required for implementing such a mission as well as knowledge gaps that have to be closed through terrestrial and space-based research are being considered in detail. The resources required for realizing a mission over such an extended period of time exceed the capabilities of any single nation. Broad international cooperation is therefore a mandatory precondition for achieving a human mission to Mars. Affordability considerations will constrain the ability to develop in parallel the large number of new technologies and systems required and therefore call for a stepwise approach. The risk areas are well identified, and significant effort is required to adapt them to the current safety standards related to human space travel. An incremental approach will help to achieve intermediate successes and results, a necessity in a program that is depending on public support for periods of several decades.

A long-term cooperative program therefore requires in preparation a careful procedure that includes

- establishment of a cooperation framework including all nations interested,
- definition of potential benefits in science, economy, and culture,
- agreement on common strategies,
- investigation of various mission scenarios,
- definition of a global road map.

2.1. Cooperation Models

Early space exploration activities have been based on national pride and international competitiveness, driven by the Cold War. With increasing complexity of space projects, nations joined into cooperative missions on a bilateral or multilateral level to share resources and capabilities. Usually one nation is leading in such collaboration and the contributions from partners are of complementary nature.

A special kind of cooperation was created in the ESA, where several nations decided to tackle complex and costly space projects on the basis of equal partnership, each nation pursuing priorities to benefit their citizens. With 20 nations pooling their resources in an organization active for more than 40 years, ESA is a success story of cooperation.

The ISS is the first intercontinental cooperation on a single, long-term space project, which relies on public funding and coordinates national interest on the basis of complementarity and interdependency. It is based on an international treaty signed in 1998, called the ISS Intergovernmental Agreement. This provides the framework for design, development, operation, and utilization of the ISS. The agreement has governed the cooperation successful over several decades, overcoming technical problems, economic turmoil, and changing governments.

While the structure of the ISS partnership has been designed for the specific purpose of developing and operating the ISS, it represents a good basis and reference for the type of partnership required for taking on the challenge of expanding human presence to Mars. The functional needs and characteristics of the cooperation required for the latter differ from the ISS cooperation. Sustainable human exploration of Mars is a venture of several decades, with timescales extending to a century. A more flexible framework is therefore required that can evolve in line with the stepwise implementation of the mission and adapt to new or changing partners along the way. The agreement has to focus on a long-term alignment of national plans of the partners, which depend on public interest, industrial capabilities and experience, resource levels, and specific needs and priorities of their citizens. The cooperation has to include complementarity, interdependency, and competitive elements. Typical characteristics of an international cooperation agreement for human exploration of Mars therefore should be

- a definition of common goals and objectives,
- long-term committing but nonbinding agreements between partners,
- incorporation of national plans and priorities, including planned near-term investments,
- a stimulus for aligning future developments in a direction advancing the common goals,
- identification of strategic knowledge gaps and specific engineering challenges,
- a stepwise approach including intermediate results of high visibility.

In addition, a collaboration agreement on the programmatic and technical side has to be complemented by a coordination framework on the political level and should be paralleled by industrial cooperation on an international basis.

2.2. Defining and Communicating the Benefits

Institutional funding plays to date a dominant role in advancing research and development in space. Private sector investment becomes increasingly more important for the development and provision of space services. However, activities in space driven by the private sector generally exploit knowledge and technologies developed earlier through institutional funding. The bulk of investments required toward realizing human missions to Mars will have to be provided by governments. Gaining and maintaining public support is therefore a critical factor for advancing toward human Mars missions. It is essential to understand and communicate how expanding the sphere of human activity within the Earth–Moon–Mars environment is relevant to society and citizens on Earth. Public sector investments need to ultimately improve the quality of life of citizens. Delivering tangible benefits and reporting back to the investors on benefits achieved will therefore be an important aspect of managing the process toward realizing a human Mars mission. Those benefits will concentrate in the areas of innovation, culture and inspiration, and peaceful global cooperation.

It is imperative for the international consortium to communicate efficiently with political decision makers and with the public regarding socioeconomic benefits resulting from space exploration. This includes the establishment of a common benefits-related framework and vocabulary

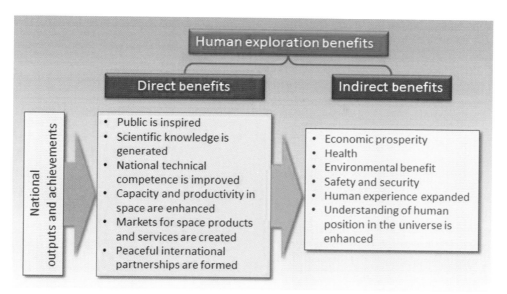

FIGURE 46.7 Logical model for describing how space exploration delivers benefit to society derived from work performed by the International Space Exploration Coordination Group. *Copyright: ISECG.*

among the partners. In this context it is helpful to distinguish between direct benefits, which emerge immediately from exploration outputs of space agencies and industry, with supporting contributions from international partners, and indirect benefits resulting over time from direct benefits (see Figure 46.7).

2.3. Focusing on Common Strategies and Intermediate Destinations

The mission of extending human presence to the Mars surface is of enormous complexity and extends over several decades. It therefore requires long-term common guidance and agreed goals and objectives. A set of guiding principles will have to be developed and agreed upon by the cooperating partners, such as

- a stepwise approach in phases based on capabilities and objectives,
- well-defined intermediate destinations generating results and stimulating public interest,
- robustness to ensure resilience to technical challenges and new developments,
- affordability based on realistic extrapolations of budget developments of the partners,
- flexibility to accommodate changes in partnership or in partner's priorities,
- emphasis on human-robotic partnership.

For the case studied here, the final destination is defined as a sustainable human presence on the surface of planet Mars. In any strategy, various intermediate way stations offer risk reduction, accumulation of experience, and valuable intermediate results. On the way to Mars the following destinations could be considered:

- **ISS**: As an existing asset in LEO, ISS can serve as a laboratory to advance capabilities in specific technologies, for gaining experience in operating critical systems over extended periods of time, for simulating deep space mission operation scenarios and conducting research for mitigating risks to human health and performance in space.
- **LEO**: Institutional and private sector funds have been invested in the development of crew and cargo transportation systems for accessing LEO. These relatively easy accessible orbits represent destinations where the economic feasibility of sustained human presence will be demonstrated first, possibly serving as model for similar activities applied later at other exploration destinations in the Earth—Moon—Mars system.
- **Cis-lunar space**: This embraces destinations beyond LEO, extending to the Moon and including the Earth—Moon **Lagrangian points** as well as high lunar orbits above 100 km.
- **Moon**: Low lunar orbits and the surface of the Moon are important destinations to gain experience and to develop capabilities required for a human mission to Mars. The Moon offers reduced risk due to easy communication to Earth with signal delays in the order of seconds, and crew emergency return to Earth in a 3—5 day time frame.
- **Near-Earth asteroids**: Those are objects from the asteroid belt with trajectories partially or fully within the Mars orbit, some of them crossing the Earth orbit. As intermediate destinations, they are of high scientific

interest and can reduce the risk for deep space exploration. Several commercial institutions have expressed interest in exploiting asteroid resources.
- **Mars system**: The Mars moons Phobos and Deimos, as well as Mars orbits, can serve as intermediate stations preceding human surface exploration of the Mars surface.

2.4. Mapping the Way to Mars

The development of a human space exploration strategy is a stepwise process. Starting on the basis of strategic mission scenarios, suitable selection processes are used to narrow down to a small number of design reference missions, which finally are used to detail a mission road map. In the following, this process is described using examples from the results of the International Space Exploration Coordination Group (ISECG).

2.4.1. Strategic Mission Scenarios

The strategies and intermediate way stations agreed jointly in the cooperating partnership serve as a basis for the development of a set of mission scenarios. These explore potential pathways to the final destination over a time scale of several decades. The missions are considered compatible with the existing capabilities and global development plans, using way stations in various configurations. They are based on consensus between the partners on principles such as affordability and value to stakeholders (see Figure 46.8). Necessarily, assumptions will have to be used in this process. Several scenarios are introduced, accompanied by studies into a suitable level of detail, which allows judgment with respect to the criteria defined by the stakeholders. In this way the number of mission scenarios is reduced.

2.4.2. Design Reference Missions

Within a selected strategic mission scenario, one or several **design reference missions** are defined. A design reference mission is a top-level description of the mission sequence and the capabilities needed to execute it. While usually being destination focused, the reference missions will include capabilities that are reused or evolved from capabilities used at other destinations. It will allow the partners to gain insight in what is required to explore the various destinations, including the capabilities needed and the basic operational concept. An example for a design reference mission developed for a human mission to the lunar surface is shown in Figure 46.9.

2.4.3. Global Exploration Road Map

Derived from the strategic mission scenarios and the design reference missions, a **global exploration road map** emerges. The road map reflects the international effort to collaboratively define technically feasible and programmatically implementable exploration mission scenarios with a common goal, agreed as sending humans to the surface of Mars and returning them safely. Figure 46.10 shows an example of a stepwise road map, starting from the ISS, advancing with readiness of the international partners for conducting human missions to Mars in the post-2030 time frame. It contains a long-term vision based on the national capabilities and priorities of the partners and serves as a nonbinding reference for the nations to inform near-term decisions related to exploration preparatory activities. A global road map will indicate the intended work sharing along the way to Mars, with existing and future capabilities of the participating nations forming a flexible, sustainable, robust, and affordable long-term plan. It

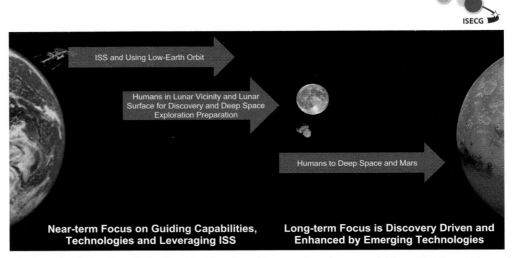

FIGURE 46.8 Early Strategic Mission Scenario developed by ISECG considers two alternatives to reach Mars with different intermediate way stations. *Copyright: ISECG.*

Chapter | 46 Strategies of Modern Solar System Exploration

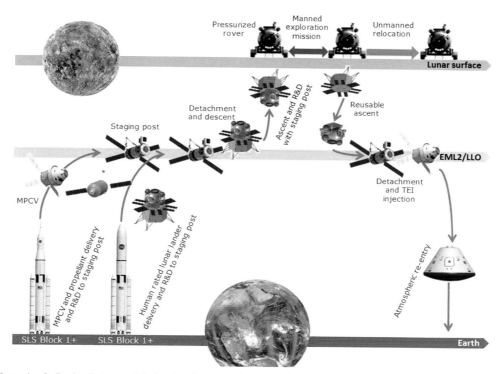

FIGURE 46.9 Example of a Design Reference Mission showing a staged approach for sending humans to the surface of the Moon to live and work in a predeployed pressurized rover. SLS, Space Launch System; MPCV, Multipurpose Crew Vehicle; R&D, Rendezvous and Docking; EML2, Earth−Moon Lagrangian Point 2; TEI, Trans Earth Injection; LLO, Low Lunar Orbit. *Copyright: ISECG.*

contains both robotic and human mission elements in a balanced way and ensures intermediate results of scientific or technical nature. It defines industrial niches for partner nations and allows for bilateral or multilateral partnerships between the participants.

3. SPACE EXPLORATION AND SOCIETY

An ambitious endeavor like the long-term vision of human exploration of our solar system, which relies on the cooperation of many nations of the world, requires careful preparation over an extended time span. A flexible approach in the cooperation framework is a necessity, as individual nations have different priorities and advance their technology with different paces. Diverse views will also prevail on the important question of affordability. In this context it is rewarding the see that a core of a cooperation framework has already been formed and is initiating first steps toward a global exploration plan.

3.1. Many Nations are Interested in Space

Space activities have been started by the two superpowers in the Cold War. With the proliferation of space technologies, other nations developed their own capabilities, but for quite some time space flight appeared to be the domain of rich, highly developed countries. It is only since the late 1990s that more and more nations discovered space as a tool to improve the living conditions of their citizens. This includes numerous developing or emerging economies worldwide. It is significant that 74 of a total of 193 member states of the United Nations (UN) have joined the UN Committee on Peaceful Uses of Outer Space (COPUOS), which has grown into one of the largest Committees of the United Nations. While the number of states having the technical capability to launch humans into space is still restricted to three, there are 11 capable of launching payloads into orbit. Through a network of bilateral and multilateral cooperation, at least 43 nations operate satellites in space, and there are many more that make use of data or results of space payloads for the benefit of their citizens (see Table 46.1). While most of these activities are presently concentrated on space assets looking at Earth or on communication and navigation systems, an increasing number of nations are beginning to extent their interest into space exploration. This is the basis for international cooperation in the preparation of a global effort of space exploration.

3.2. Can Nations Afford Space Exploration?

The concept of affordability is always relative. It depends on current priorities and the perception of values, which are subject to change over time. This is imminent for programs targeting far into the future, like a human mission to Mars

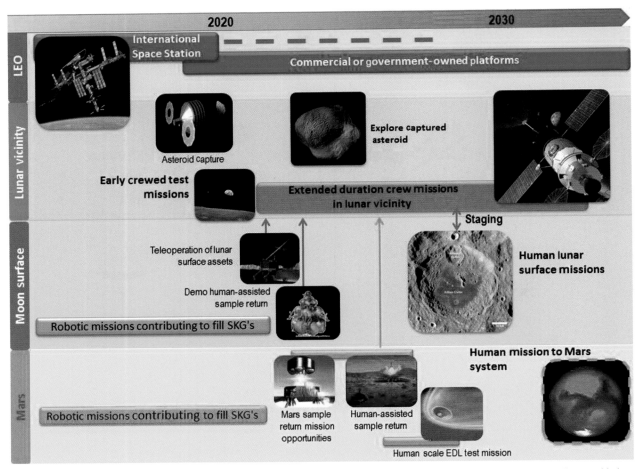

FIGURE 46.10 Overview of the international exploration mission scenario for the next two decades, driven by the long-term goals to enable human missions to Mars and showing operational (International Space Station), planned (early crewed test missions, asteroid capture), and conceptual missions on the pathway to Mars. SKG, strategic knowledge gap; EDL, entry, descent, landing. *Copyright: ISECG.*

TABLE 46.1 Overview of Nations with Vested Interest in Space at Various Levels	
Nations participating in UN COPUOS	74
National Space Agencies or Offices	68
Nations operating satellites	45
Nations with human space programs	26
Nations with orbital launch capabilities	11
Nations with human space transportation capabilities	3
UN COPUOS: United Nations Committee on Peaceful Use of Outer Space.	

extending over half a century. By any means space exploration is expensive. It requires numerous costly technological developments over a very long time scale. In all considerations, however, it should be kept in mind that these funds are spent on Earth on highly qualified jobs and cutting-edge technology.

Calculation of the cost of a human flight to Mars is difficult and depends on the complexity and duration of the mission considered. Most exploration studies refrain from stating potential cost figures. There is a diverging range of estimates and speculations, some of them not very well substantiated. A commercial venture claims cost of 6 billion dollars for bringing, within 10 years, a crew of four to the Mars surface to initiate a settlement, i.e. without considering a return. The National Aeronautics and Space Administration (NASA) "90 Day Study on Human Exploration of the Moon and Mars", presented to President Bush in 1989 did not give explicit cost figures. Cost estimates on the basis of this study arrived at

TABLE 46.2 Assessment of the Affordability of a Human Mission to Mars by Comparison to the Apollo Program

	Reference Year	Billion $	Ratio to GNP %
Apollo cost	1973	24.5	—
US GNP	1973	1.190	2.06
World GNP	1973	5.400	0.45
Mars mission cost	2011	450	—
US GNP	2011	14.991	3.00
World GNP	2011	70.020	0.64

The ratio of the total program cost (over several years) to the gross national product (GNP) in a single base year is used to judge affordability.

258 billion dollars for a three-decade program. This amounts to about 450 billion dollars in 2013 value. Assuming leveled expenditure over the years results in an annual spending figure of 15 billion dollars over 30 years. This is close to the total annual NASA budget of nearly 18 billion dollars, showing clearly that no single nation is able to afford a human exploration program on its own. The situation looks different if compared to the total global annual government spending on civil space programs, which amounts to 73 billion dollars in 2012. From this point of view, an exploration effort appears feasible and affordable on a global scale.

An alternative way to assess affordability is a comparison to the Apollo program. This is summarized in Table 46.2. The total cost of the Apollo project extending over the years 1961–1972 was reported as 25.4 billion dollars in 1973. In order to relate this value to the different economic conditions in 1973 and today, the ratio to the gross national product (GNP) in the relevant year is used. The ratio of the cost of a human mission to Mars in 2011 conditions to the US GNP would be about 50% higher than that of the Apollo program. On the other hand, the impact of the Mars mission cost on the world GNP is only a third compared to that of Apollo on the US GNP. Considering the fact that the Apollo project did not have a negative effect on the US economic development but in contrast contributed substantially to its growth, a global effort to extend human presence to Mars can not only be regarded as affordable, but may even stimulate global economy.

3.3. Involving Private Capital

Space exploration does by no means have to rely exclusively on government funds. In contrast, a global effort should enlarge the basis of stakeholders by including capital and capabilities of commercial entities. Private capital can be used more flexibly, but it does not necessarily reduce overall cost as it always is expected to generate revenues.

NASA has successfully initiated a commercialization of space transport to LEO in its Commercial Orbital Transportation Services program (COTS), which has been extended to crew transport in the Commercial Crew and Cargo Office (C3PO). There are several commercial initiatives aiming at tourist services in suborbital flights. Similar activities are planned for orbital flights and potentially to the Moon.

Two commercial projects have been announced that target destinations beyond the Moon. The enterprise Planetary Resources, founded in April 2012, is aiming at "bringing the natural resources of space within humanity's economic sphere of influence" by mining asteroids for precious minerals or using them as fuel depots in space. Similar intentions have been declared by the company "Deep Space Industries Inc." in January 2013.

In February 2013, the first space tourist Dennis Tito revealed his plans for a privately funded mission to Mars, sending a man and a woman on a flyby trajectory without landing, to be launched in 2018. The SpaceX Company, which successfully performed the first commercial cargo transport to ISS, proposed the "Red Dragon" landing device for Mars. The founder of SpaceX, Elon Musk, announced his intention to establish a colony on Mars on a commercial basis. With a ticket price of half a million dollars, he expects to expand in future to 80,000 colonists. A commercial venture called "MarsOne" plans to send humans to Mars and start a settlement there, to be funded on the basis of a television reality show.

3.4. Avoiding Contamination of Earth and Celestial Bodies

A potential problem in any kind of space exploration is that of contamination, usually summarized under the term **planetary protection**. It pertains to both the avoidance of contamination of celestial bodies by human action and the inverse problem of bringing potentially hazardous extraterrestrial material to Earth. There is concern about the contamination of the Moon with the increasing number of missions planned. Even more severe is the problem on Mars as this is the only planet in our solar system that holds promise of the potential detection of past or present life. This is the main scientific driver for the exploration of Mars. Any activity on Mars, be it human or robotic, carries the danger of contaminating the planet or introducing microbial life forms. COSPAR has elaborated rules for planetary protection (Rummel et al., 2002), which should be carefully observed in any mission to Mars. An early colonization, as foreseen in some concepts of commercial human flight to Mars, is difficult to design compatible with the COSPAR rules.

3.5. Global Cooperation Becomes Reality

In 2006, 14 space agencies from different nations came together to discuss possibilities of jointly developing a vision for peaceful robotic and human space exploration, focusing on destinations within the solar system where humans may one day live and work. Considering common objectives and plans, a joint document was released in May 2007 entitled "The Global Exploration Strategy: The Framework for Coordination". The key finding in this work was the demand to establish a cooperation framework between the agencies to prepare future progress. Consequently, the ISECG was formed between the cooperating agencies as a voluntary, nonbinding international coordination mechanism to exchange information and coordinate plans. The intention is to work together on means of strengthening both individual exploration programs as well as the collective effort.

The ISECG team, consisting of representatives of the presently 14 space agencies involved, was established in November 2007 and meets since at regular intervals. It works on the key principle of an "open and inclusive" approach. The membership is open to any interested space agency. Numerous products have been achieved and published by ISECG. Various reports have been issued and papers are presented in the name of the group at international space congresses. The most important product so far is the "Global Exploration Road map", the first iteration of which has been issued September 2011. This road map reflects the international effort to define feasible and sustainable exploration pathways to the Moon, near-Earth asteroids, and Mars. Beginning with the ISS, it examines possible pathways for space exploration in the next 25 years with the aim to arrive at a single, consolidated plan.

A global coordination process on a long-term vision like space exploration can work only if it involves the political leaders worldwide. A first step in this direction has also been initiated. ESA convened high-level political representatives in Prague, 2009, stimulating an exchange to shape Europe's position in the global space exploration context. This was followed by a similar meeting in Brussels (2010). In 2011, the third International Conference on Space Exploration in Lucca convened government representatives from 28 nations, which committed to begin an open structured high-level policy dialogue on space exploration. A "High-level International Space Exploration Platform" was founded that issued a declaration recognizing the benefit of a continuing dialogue on a worldwide scale to help identify potential areas for international cooperation. The subsequent meeting was hosted by the United States in Washington, DC, in early 2014. It is expected that further discussions will cover joint missions and collaboration on research, and could lead to greater cooperation in areas such as access to space, innovation and space technologies, the use of current and future low-orbit infrastructures, and future human and robotic presence further out in space.

4. CONCLUSIONS AND OUTLOOK

Exploration of space started only some 50 years ago and it has already had profound impact on the development of society: economically, scientifically, and culturally. Progress in human exploration has been politically motivated and accomplished in major steps enabled by the development of new transportation systems. Robotic missions are and have been primarily science-driven. Future space exploration plans promote strongly the concept of human-robotic partnership and a further integration of human and robotic capabilities in space seems likely.

A human mission to Mars has been declared by various space agencies to be the driving goal for exploration within this century. Such an endeavor is technically feasible and programmatically implementable, provided it is carefully planned, executed in a stepwise manner, and realized within a broad international partnership, including members from both the institutional and private sectors. While Mars mission concepts have been studied since the days of Wernher von Braun, there are signs today that such an endeavor could actually become reality: (1) New transportation systems, under development in the United States and Russia are being designed such that they have the potential to evolve toward the capabilities required for human Mars exploration. (2) International coordination at agency level and international dialogue at political level demonstrate the commitment to cooperate toward achieving common exploration goals. Most exploration missions, robotic and human, are conducted presently within international partnerships. (3) Various initiatives in the domain of exploration demonstrate the readiness of the private entities to invest in this domain. It is therefore likely that within the next 30–50 years humans will set foot on Mars, representing what partnership, possibly between different political and economic systems, can achieve for the benefit and future prosperity of the global society.

BIBLIOGRAPHY

Culbert, C., Mongrard, O., Satoh, N., Goodliff, K., Seaman, C., Troutman, P., et al. (2011). *ISECG mission scenarios and their role in informing next steps for human exploration beyond low earth orbit.* IAC-11–D3.1.2.

Durante, M., & Cucinotta, F. A. (2008). Heavy ion carcinogenesis and human space exploration. *Nature Reviews Cancer, 8,* 465.

Ehrenfreund, P., McKay, C., Rummel, J. D., Foing, B. H., Neal, C. R., Masson-Zwaan, T., et al. (2012). Toward a global space exploration program: a stepping stone approach. *Advances in Space Research, 49,* 2–48.

The global space exploration strategy — A framework for coordination (May 2007). Washington, DC: NASA. www.globalspaceexploration.org.

International Space Exploration Coordination Group (September 2011). *The global exploration roadmap*. Washington DC: NASA. www.globalspaceexploration.org.

Hufenbach, B., Laurini, K., Satoh, N., Piedboeuf, J.-C., Lange, C., Martinez, R., et al. (2013). *The 2nd iteration of the global space exploration roadmap*. IAC-13, B3.1.

Huntress, W., Stetson, D., Farquhar, R., Zimmerman, J., Clarke, B., O'Neil, W., et al. (2006). The next steps in exploring deep space—a cosmic study by the International Academy of Astronautics. *Acta Astronautica, 58*, 304.

Rummel, J. D., Stabekis, P. D., Devincenzi, D. L., & Barengoltz, J. B. (2002). COSPAR's planetary protection policy: a consolidated draft. *Advances in Space Research, 30*, 1567–1571.

Chapter 47

A History of Solar System Studies

David Leverington

Stoke Lacy, Herefordshire, United Kingdom

Chapter Outline

1. Babylonians and Greeks — 999
2. Copernicus and Tycho — 1001
3. Kepler and Galileo — 1001
4. Second Half of the Seventeenth Century — 1003
   4.1. The Moon — 1003
   4.2. Saturn — 1004
   4.3. Newton — 1005
5. The Eighteenth Century — 1006
   5.1. Halley's Comet — 1006
   5.2. The 1761 and 1769 Transits of Venus — 1006
   5.3. The Discovery of Uranus — 1006
   5.4. Origin of the Solar System — 1006
   5.5. The First Asteroids — 1007
6. The Nineteenth Century — 1007
   6.1. The Sun — 1007
   6.2. Vulcan — 1007
   6.3. Mercury — 1008
   6.4. Venus — 1008
   6.5. The Moon — 1008
   6.6. The Earth — 1008
   6.7. Mars — 1009
   6.8. Jupiter — 1009
   6.9. Saturn — 1010
   6.10. Uranus — 1010
   6.11. The Discovery of Neptune — 1010
   6.12. Asteroids — 1011
   6.13. Comets — 1011
   6.14. Meteor Showers — 1011
7. The Twentieth Century Prior to the Space Age — 1012
   7.1. The Sun — 1012
   7.2. Mercury — 1013
   7.3. Venus — 1013
   7.4. The Moon — 1013
   7.5. The Earth — 1014
   7.6. Mars — 1014
   7.7. Internal Structures of the Giant Planets — 1015
   7.8. Atmospheres of the Giant Planets — 1015
   7.9. Jupiter — 1015
   7.10. Saturn — 1015
   7.11. Uranus and Neptune — 1016
   7.12. The Discovery of Pluto — 1016
   7.13. Asteroids — 1017
   7.14. Comets — 1017
   7.15. The Origin of the Solar System — 1017
Bibliography — 1017

This chapter gives a brief overview of the history of solar system research from the earliest times up to the start of the space age.

1. BABYLONIANS AND GREEKS

Many early civilizations studied the heavens, but it was the Babylonians of the first millennium B.C. who first used mathematics to try to predict the positions of the Sun, Moon, and visible planets (Mercury, Venus, Mars, Jupiter, and Saturn) in the sky. In this they differed from the Greeks, as the Babylonians were priests trying to predict the movement of the heavenly bodies for religious purposes, whereas the Greeks were philosophers trying to understand why they moved in the way they did. The Babylonians were fascinated by numbers, whereas the Greeks were more interested in geometrical figures.

The accuracy of the Babylonian predictions in the second century B.C. is remarkable. For example, their estimate of the length of the **sidereal** year was within 6 min of its true value, and that of the average **anomalistic month** was within 3 s. In addition, Jupiter's sidereal and **synodic periods** were within 0.01% of their correct values.

Pythagoras (c. 580–500 B.C.) was a highly influential early Greek philosopher who set up a school of philosophers, now known as the Pythagoreans. None of Pythagoras' original writings survive, but later evidence suggests that the Pythagoreans were probably the first to believe that the Earth is spherical, and that the planets all move in separate orbits inclined to the celestial equator. But the

Pythagorean spherical Earth did not spin and was surrounded by a series of concentric crystalline spheres supporting the Sun, Moon, and individual planets. Each had its own sphere, which revolved around the Earth at different speeds, producing a musical sound, the "music of the spheres", as they went past each other.

Hicetus of Syracuse (fl. fifth century B.C.) was the first person to specifically suggest that the Earth spun on its axis, at the center of the universe. This model was further developed by Heracleides who proposed that Mercury and Venus orbited the Sun as it orbited the Earth. Then Aristarchus (c. 310–230 B.C.), who was one of the last of the Pythagoreans, went one step further and proposed a heliocentric (i.e. Sun-centered) universe in which the planets orbit the Sun in the (correct) order of Mercury, Venus, Earth, Mars, Jupiter, and Saturn, with the Moon orbiting a spinning Earth. This was 1700 years before Copernicus came up with the same idea. Aristarchus was also the first to produce a realistic estimate for the Earth––Moon distance, although his estimate of the Earth–Sun distance was an order of magnitude too low.

While the Pythagoreans were developing their ideas, Plato (c. 427–347 B.C.) was developing a completely different school of thought. Plato, who was a highly respected philosopher, was not too successful with his geocentric (i.e. Earth-centered) model of the universe. His main legacy to astronomy was his teaching that all heavenly bodies must be spherical, as that is the perfect shape, and that they must move in uniform circular orbits, for the same reason. Aristotle (384–322 B.C.), a follower of Plato, was one of the greatest of Greek philosophers. His ideas were to hold sway in Europe until well into the Middle Ages. However, his geocentric model of the universe was highly complex, requiring a total of 56 spheres to explain the motions of the Sun, Moon, and planets. Unfortunately, many of its predictions were wrong, and it soon fell into disuse.

Hipparchus (c. 185–120 B.C.), who was the first person to quantify the **precession of the equinoxes**, was aware that the Sun's velocity along the ecliptic was not linear. This was known to the Babylonians and to Callippus of Cyzicus, but they did not seek an explanation. Hipparchus, on the other hand, in adopting Plato's philosophy of uniform circular motion in a geocentric universe, realized that this phenomenon could only be explained if the Sun was orbiting an off-center Earth. However, his estimate of the off-center amount was far too large, although his **apogee** position was in error by only 35′.

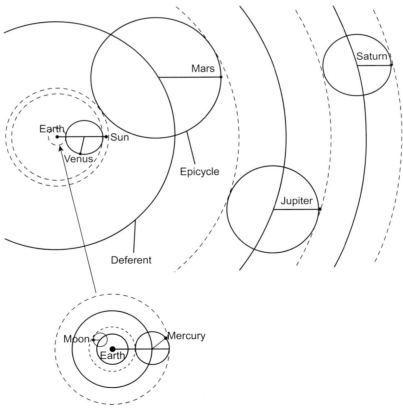

FIGURE 47.1 Ptolemy's model of the universe in which all bodies, except the Sun (and stars), describe epicycles, the centers of which orbit the Earth in deferents. He assumed that there were no gaps between the circle enclosing the furthest distance of one planet, and that just touching the epicycle of the next planet out from the Earth.

Chapter | 47 A History of Solar System Studies

The mathematician Apollonius of Perga (c. 265—190 B.C.) appears to have been the first to examine the properties of epicycles. These were later adopted by Ptolemy (c. A.D. 100—170) in his geocentric model of the universe. In Ptolemy's scheme (Figure 47.1), the Moon, Sun, and planets each describe a circular orbit called an epicycle, the center of which goes in a circle, called a deferent, around a nonspinning Earth. Because the inferior planets, Mercury and Venus, each appear almost symmetrically on both sides of the Sun at maximum **elongation**, he assumed that the centers of their epicycles were always on a line joining the Earth and Sun. For the superior planets he assumed that the lines linking these with the center of their epicycles were always parallel to the Earth—Sun line. Unfortunately, this simple system did not provide accurate enough position estimates, and so Ptolemy introduced a number of modifications. In the case of the Moon, he made the center of the Moon's deferent describe a circle whose center was the Earth. For the planets he introduced the concept of an equant, which was a point in space equidistant with the Earth from the center of the deferent (Figure 47.2). The equant was the point about which the planet's angular velocity appeared to be uniform. Other modifications were also required, but by the time he had finished, he was able to make accurate position estimates for all but the Moon and Mercury. In addition, assuming that there were no gaps between the furthest part of one epicycle and the nearest part of the next, he was able to produce an estimate for the size of the solar system of about 20,000 times the radius of the Earth (or about 120 million km). Although this was a gross underestimate, it gave, for the first time, an idea of how large the solar system really was.

2. COPERNICUS AND TYCHO

There was virtually no progress in astronomy over the next 1000 years, and during this time many of the Greek texts had been lost in Europe. But in the twelfth century Arab translations found their way to Europe, mainly via Islamic Spain. Then in the fourteenth century Ibn al-Shātir (1304—1375), working in Damascus, improved Ptolemy's model by modifying his epicycles and deleting his equant. Interestingly, al-Shātir's system was very much like Copernicus' later system, but with the Earth, not the Sun, at the center.

Copernicus' heliocentric theory of the universe (Figure 47.3) was published in his *De Revolutionibus Orbium Coelestium* in 1543, the year of his death. Interestingly, in the light of Galileo's later problems with the Church, the book was well received. This is probably because of the Foreword, which had been written by the theologian Andreas Osiander and explained that the book described a mathematical model of the universe, rather than the universe itself.

Copernicus (1473—1543) acknowledged that his idea of a spinning Earth in a heliocentric universe was not new, having been proposed by Aristarchus. In addition, Copernicus' theory was based on circular motion and still depended on epicycles, although he deleted the equant. But he had resurrected the heliocentric theory, which had not been seriously considered for almost 2000 years, at the height of the Renaissance, which was eager for new ideas.

In the Middle Ages, Aristotle's ideas were taught at all the European universities. But now Copernicus had broken with the Aristotelian concept of a nonspinning Earth at the center of the universe. Then in 1577 Tycho Brahe (1546—1601) disproved another of Aristotle's ideas. Aristotle had believed that comets are in the Earth's atmosphere, but Tycho was unable to measure any clear parallax for the comet of that year. Finally, Tycho, in his book of 1588, rejected another of Aristotle's ideas, that the heavenly bodies are carried in their orbits on crystalline spheres. This is because, in Tycho's new model of the universe, all the planets, except the Earth, orbit the Sun as the Sun orbits the Earth. This meant that the sphere that carried Mars around the Sun would intercept that which carried the Sun around the Earth, which was clearly impossible if they were crystalline.

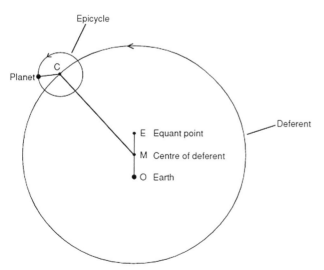

FIGURE 47.2 Ptolemy modified his epicycle theory for the superior planets by moving the Earth O from the center M of the deferent, and by defining an equant point E such that the distance $EM = MO$. He then assumed that the angular velocity of C, the center of the epicycle, is uniform about the equant point E, rather than about the center M of the deferent.

3. KEPLER AND GALILEO

Johannes Kepler (1571—1630) looked at the universe in an entirely different way than his predecessors. The Babylonians

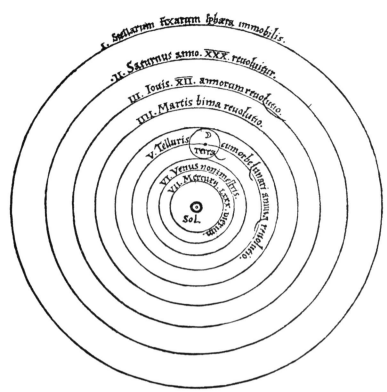

FIGURE 47.3 Copernicus' heliocentric universe, as described in his *De Revolutionibus*, in which the planets orbit the Sun (Sol) and the Moon orbits the Earth (Terra).

had examined it arithmetically, and the Greeks and later astronomers had considered it in geometrical terms. Kepler, on the other hand, tried to understand the structure of the solar system by considering physical forces.

Kepler conceived of a force emanating from the Sun that pushed the planets around their orbit of the Sun such that planetary movement would stop if the force stopped. The magnitude of his force, and hence the linear velocity of the planets, decreased linearly with distance. This should have resulted in the period of the planets varying as their distance squared, but Kepler made a mathematical error and came up with another relationship. Fortuitously, however, his analysis produced remarkably accurate results.

Although Kepler was having some success with this and other theories, he thought he could improve them if he had access to Tycho Brahe's accurate observational data. So Kepler went to see Tycho; a visit that ended with him joining Tycho and eventually succeeding him after his death.

Tycho had initially asked Kepler to analyze Mars' orbit, a task that he continued well after Tycho's death. Kepler published his results in 1609 in his book *Astronomia Nova*, in which he reintroduced the equant, previously deleted by Copernicus. In Kepler's model, all the planets orbited the Sun in a circle, with the Sun off-center, but he could not find a suitable circle to match Mars' observations, even with an equant. So he decided to reexamine the Earth's orbit, as the Earth was the platform from which the observations had been made.

Copernicus had proposed that the Earth moved around the Sun in a circle at a uniform speed, with the Sun off-center. So there had been no need for an equant. But Kepler found that an equant was required to explain the Earth's orbit. However, even adding this, he could not fit a circle, or even a flattened circle to Mars' orbit. And so in desperation he tried an ellipse, with the Sun at one focus, and, much to his surprise, it worked.

Kepler now considered what type of force was driving the planets in their orbits, and concluded that the basic circular motion was produced by vortices generated by a rotating Sun. Magnetic forces then made the orbits elliptical. So Kepler thought that the Sun rotated on its axis, and that the planets and Sun were magnetic.

Initially, Kepler had only shown that Mars moved in an ellipse, but in his *Epitome* of 1618−1621 he showed that this was the case for all the planets, as well as the Moon and the satellites of Jupiter. He also stated what we now know as his third law, that the square of the periods of the planets are proportional to the cubes of their mean distances from the Sun. Finally, in his *Rudolphine Tables*, he listed detailed predictions for planetary positions and predicted the transits of Mercury and Venus across the Sun's disc.

Galileo Galilei (1564–1642) made his first telescopes in 1609 and started his first telescopic observations of the Moon in November of that year. He noticed that the **terminator** had a very irregular shape and concluded that this was because the Moon had mountains and valleys. It was quite unlike the pure spherical body of Aristotle's cosmology.

Galileo undertook a series of observations of Jupiter in January 1610 and found that it had four moons that changed their positions from night to night (Figure 47.4). Galileo presented his early Moon and Jupiter observations in his *Sidereus Nuncius* published in March 1610. By 1612, he had determined the periods of Jupiter's moons to within a few minutes.

Galileo's *Sidereus Nuncius* created quite a stir, with many people suggesting that Galileo's images of Jupiter's moons were an illusion. Kepler, who was in communication with Galileo, first saw the moons himself in August 1610 and supported Galileo against his doubters. The month before, Galileo had also seen what he took to be two moons on either side of Saturn, but for some reason they did not move. Finally in late 1610 he observed the phases of Venus, finally proving that Ptolemy's structure of the solar system was incorrect. As a result, Galileo settled on the Copernican heliocentric system.

Sunspots had been seen from time to time in antiquity, but most people took them to be something between the Earth and Sun. Although Thomas Harriot and Galileo had both seen sunspots telescopically in 1610, it was Johann Fabricius who first published his results in June 1611. He concluded that they were on the surface of the Sun, and that their movement indicated that the Sun was rotating. This was completely against Aristotle's teachings that the Sun was a perfect body.

In the meantime, Galileo had visited the Jesuits of the Roman College to get their support for his work and, in particular, their support for Copernicus' heliocentric cosmology. His reception was very warm, and he was even received in audience by the pope. But, although the Roman Catholic Church did not argue with his observations, outlined above, there was considerable unease at his interpretation. Initially, the Church was prepared to tolerate Galileo's support of the Copernican cosmology, provided he presented this cosmology as a working hypothesis, rather than as a universal truth. But Galileo was stubborn and tried to take on the Church in its interpretation of theology. In this he could not win, of course, and the Church put him on trial, where he was treated very well. Nevertheless, he was forced in 1633 to recant his views and was then placed under house arrest for the remaining 9 years of his life.

4. SECOND HALF OF THE SEVENTEENTH CENTURY

4.1. The Moon

Thomas Harriot (1560–1621) was the first astronomer to record what we now know as the libration in latitude of the Moon, which has a period of 1 month. This occurs because the Moon's spin axis is not perpendicular to its orbit. A little later Galileo detected a libration in longitude, which he thought had a period of 1 day. In fact, it has a period of 1 month and is caused by the eccentricity of the Moon's orbit.

Although Galileo thought that the Moon has an atmosphere, he concluded that there was very little water on the surface as there were no clouds. His early telescopes were not sufficiently powerful, however, to show much surface detail. But over the next few decades, maps of the Moon were produced by a number of astronomers. The most definitive of which were published in 1647 by Johannes Hevelius (1611–1687). They were the first to show the effect of libration.

By mid-century, it was clear that there were numerous craters on the Moon, and in 1665 Robert Hooke (1635–1703) speculated on their cause in his *Micrographia*. He undertook laboratory-like experiments and noted that if round objects were dropped into a mixture of clay and water, features that resemble lunar craters were produced. But he could not think of the source of large objects hitting the Moon. However, he also found that he

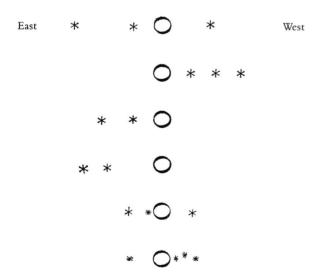

FIGURE 47.4 Galileo's observations of the moons of Jupiter on consecutive nights from 7 to 13 January (excluding 9 January) 1610, as shown in his book *Sidereus Nuncius*.

TABLE 47.1 Key Solar System Discoveries and Observations, 1630–1700

Sun–Earth distance	
1672	Richer, Cassini, and Picard deduce a solar parallax of 9.5 min of arc from observations of the parallax of Mars. John Flamsteed independently deduces a similar value. This implied a Sun–Earth distance of about 22,000 earth radii, or 140 million km.
Moon	
See main text	
Mercury	
1631	First observation of a transit of Mercury by Gassendi, Remus, and Cysat—all independently. It occurred on the date predicted by Kepler.
1639	Phases of Mercury first observed by Zupus.
Venus	
1639	First observation of a transit of Venus by Horrocks and Crabtree.
1646	Fontana observes that Venus' terminator is uneven, attributing the cause to high mountains. (This is now known to be incorrect; Venus is covered in dense clouds.)
1667	Cassini deduces a rotation period of about 24 h (This is now known to be incorrect).
Mars	
1659	Huygens observes Syrtis Major and deduces a planetary rotation period of about 24 h.
1672	Huygens first unambiguously records the south polar cap.
Jupiter	
c. 1630	Fontana, Torricelli, and Zucchi independently observe the main belts.
1643	Riccioli observes the shadows of the Galilean satellites on Jupiter's disc.
1663	Cassini deduces a Jupiter rotation period of 9 h 56 min.
1665	Cassini observes a prominent spot that may be an early appearance of the Great Red Spot.
1690	Cassini observes the differential rotation of Jupiter.
1691	Cassini observes Jupiter's polar flattening, which he estimates to be about 7%.
Saturn	
See main text	

could produce crater-like features if he boiled dry alabaster powder in a container. As a result, he concluded that lunar craters are produced by the collapsed blisters of warm viscous lava.

4.2. Saturn

Christiaan Huygens (1629–1695) and his brother Constantyn finished building a state-of-the-art telescope in early 1655. Shortly afterward Christiaan discovered Saturn's first moon, Titan, which he announced in his *De Saturni* of 1656. The next four moons of Saturn were discovered by Gian Domenico Cassini (1625–1712); Iapetus in 1671, Rhea in 1672, and both Tethys and Dione in 1684.

Huygens had also mentioned in *De Saturni* that he had solved the problem of Saturn's two "moons" observed by Galileo. In fact, the behavior of these moons had been very odd, as they had both completely disappeared in November 1612, reappearing again in mid-1613. Since then, their shape had gradually changed. In 1650, Francesco Grimaldi discovered Saturn's polar flattening, but still the behavior of the moons, then called ansae, was unexplained. Finally, Huygens announced, in his *Systema Saturnium* of 1659, that the ansae were actually a thin, flat solid ring, which was inclined to the ecliptic, and so changed its appearance with time. Then in 1675 Cassini noticed that Saturn's ring was divided in two by a dark line, now called the Cassini Division, going all the way around the planet. Cassini speculated that the two rings were not solid but composed of swarms of small satellites.

Other major observational discoveries of this period are listed in Table 47.1.

4.3. Newton

Kepler had thought that the planets were being pushed around their orbits by a vortex emanating from the Sun but attributed the tides on Earth to the combined attraction of the Sun and Moon by a gravitational force. It seems strange to us that he did not think of this attractive force as having some effect on the orbits of the planets.

René Descartes (1596−1650) also developed a vortex theory to explain the motion of the planets. In his theory, the vortices are in the ether, which is a frictionless fluid filling the universe. In his *Principia* of 1644, Descartes stated that each planet had two "tendencies": one tangential to its orbit and one away from the orbit's center. It is the pressure in the vortex that counterbalances the latter and keeps the planet in its orbit.

In 1664, Isaac Newton (1642−1727) started to consider the motion of a body in a circle. In the following year, he proved that the force acting radially on such a body is proportional to its mass multiplied by its velocity squared, and divided by the radius of the circle (i.e. mv^2/r). From this, he was able to prove that the force on a planet moving in a circular orbit is inversely proportional to the square of its distance from the center. Newton realized that this outward centrifugal force on a planet must be counterbalanced by an equal and opposite centripetal force, but it was not obvious at that time that this force was gravity.

At this time, it was known that gravity acted on objects on the Earth's surface, but it was not known how far from Earth gravity extended. To get a better understanding of this, Newton devised his so-called Moon test. In this test, he compared the force acting on the Moon, because of its motion in a circle, with the force of the Earth's gravity at the Moon's orbit and found that they were not the same. The difference was not large, but it was sufficient to cause Newton to stop work on gravity. In fact, at that time, Newton appears to have thought that the centripetal force was a mixture of the gravitational force and the force created by vortices in the ether, so he may not have been too surprised by his result.

Newton was finally prompted to return to the subject of gravity by an exchange of letters with Robert Hooke in 1679. In the following year, Newton proved that, assuming an inverse square law of attraction, planets and moons will orbit a central body in an ellipse, with the central body at one focus. Then in 1684 he finally rejected the idea of ethereal vortices and started to develop his theory of universal gravitation.

It was during this period that the comet of 1680 appeared. At that time, most astronomers, including Newton, believed that comets described rectilinear orbits. John Flamsteed (1646−1719), on the other hand, believed that comets described closed orbits, and he suggested, in a letter to Edmond Halley (1656−1742), that the 1680 comet had passed in front of the Sun. Newton, who had been sent a copy of this letter, thought, like a number of astronomers, that there had been two comets, one approaching the Sun and one retreating. Further communications between Flamsteed and Newton in 1681 did not resolve their disagreements, causing Newton to drop the subject of cometary orbits. Eventually, Newton returned to the subject, and by 1686 he had changed his position entirely, as he proved that cometary orbits are highly elliptical or parabolic, to a first approximation. So the 1680 comet had been one comet after all. Newton now felt, having solved the problem of cometary orbits, that he could complete his *Principia*, which was published in 1687.

Newton developed his universal theory of gravitation in his *Principia*, which ran to three editions. For example, he used Venus to "weigh" the Sun, and planetary moons to weigh their parent planets, and by the third edition he had deduced the masses and densities for the Earth, Jupiter, and Saturn relative to the Sun (Table 47.2).

Newton realized that if gravity was really universal, then not only would the Sun's gravity affect the orbit of a planet, and the planet's gravity affect the orbit of its moons, but the Sun would also affect the orbits of the moons, and one planet would affect the orbits of other planets. In particular, Newton calculated that Jupiter, at its closest approach to Saturn, would have about 1/217 times the gravitational attraction of the Sun. So he was delighted when Flamsteed told him that Saturn's orbit did not seem to fit exactly the orbit that it should if it was only influenced by the Sun. Gravity really did appear to be universal.

Richer, Cassini, and Picard had found evidence in 1672 that the Earth had an equatorial bulge. Newton was able to use his new gravitational theory to calculate a theoretical value for this **oblateness** of 1/230 (modern value 1/298). He then considered the gravitational attraction of the Moon and Sun on the oblate Earth and calculated that the Earth's spin axis should precess at a rate of about $50''.0$ per annum (modern value $50''.3$). This explained the precession of the equinoxes.

TABLE 47.2 A Comparison of Newton's Results (Relative to the Sun) with Modern Values

	Mass		Density	
	Principia	Modern Value	Principia	Modern Value
Sun	1	1	100	100
Earth	1/169,282	1/332,980	400	392
Jupiter	1/1067	1/1047	94.5	94.2
Saturn	1/3021	1/3498	67	49

5. THE EIGHTEENTH CENTURY

5.1. Halley's Comet

Halley used Newton's methodology to determine the orbits of 24 comets that had been observed between 1337 and 1698. None of them appeared to be hyperbolic, and so the comets were all clearly permanent members of the solar system. Halley also concluded that the comets of 1531, 1607, and 1682 were successive appearances of the same comet as their orbital elements were very similar. But the time intervals between successive perihelia were not the same; a fact he attributed to the perturbing effect of Jupiter. Taking this into account, he predicted in 1717 that the comet would return in late 1758 or early 1759.

Shortly before the expected return of this comet, which we now called Halley's comet, Alexis Clairaut (1713–1765) attempted to produce a more accurate prediction of its **perihelion** date. He used a new approximate solution to the three-body problem that allowed him to take account of planetary perturbations. This showed that the return would be delayed by 518 days due to Jupiter and 100 days due to Saturn. As a result, he predicted that Halley's comet would reach perihelion on about 15 April 1759 ± 1 month. It did so on 13 March 1759, so Clairaut was just 33 days out with his estimate.

5.2. The 1761 and 1769 Transits of Venus

James Gregory (1638–1675) had suggested in 1663 that observations of a transit of Mercury could be used to determine the **solar parallax**, and hence the distance of the Sun from Earth. Such a determination required observations from at least two different places on Earth, separated by as large a distance as possible. In 1677, Edmond Halley observed such a transit when he was on St. Helena observing the southern sky. But, when he returned, he found that Jean Gallet in Avignon seemed to have been the only other person who had recorded the transit. Unfortunately, there were too many problems in comparing their results, which resulted in a highly inaccurate solar parallax.

In 1678, Halley reviewed possible methods of measuring the solar parallax and suggested that transits of Venus would produce the most accurate results. The problem was, however, that these occur in pairs, 8 years apart, only every 120 years. The next pair were due almost 100 years later, in 1761 and 1769.

Joseph Delisle (1688–1768) took up Halley's suggestion and tried to motivate the astronomical community to undertake coordinated observations of the 1761 transit. After much discussion, the French Academy of Sciences sent observers to Vienna, Siberia, India, and an island in the Indian Ocean, while other countries sent observers to St. Helena, Indonesia, Newfoundland, and Norway. Unfortunately, precise timing of the planetary contacts proved much more difficult than expected, resulting in solar parallaxes ranging from $8''.3$ to $10''.6$. Interestingly, several observers noticed that Venus appeared to be surrounded by a luminous ring when the planet was partially on the Sun. Mikhail Lomonsov (1711–1765) correctly concluded that this showed that Venus was surrounded by an extensive atmosphere.

The lessons learned from the 1761 transit were invaluable in observing the next transit in 1769. This was undertaken from over 70 different sites, and analysis of all the results eventually yielded a best estimate of $8''.6$ (modern value $8''.79$) for the solar parallax.

5.3. The Discovery of Uranus

On 13 March 1781, William Herschel (1738–1822), while looking for double stars, noticed what he thought was a comet. Four days later, when he next saw the object, it had clearly moved, confirming Herschel's suspicion that it was a comet. He then wrote to Nevil Maskelyne (1732–1811), the Astronomer Royal, notifying him of his discovery. As a result, Maskelyne observed the object on a number of occasions, but he was unsure as to whether it was a comet or a new planet.

Over the next few weeks a number of astronomers observed the object and calculated its orbit, which was found to be essentially circular. So it was a planet, now called Uranus. It was the first planet to be discovered since ancient times, and its discovery had a profound effect on the astronomical community, indicating that there may yet be more undiscovered planets in the solar system.

A few years later Herschel discovered the first two of Uranus' satellites, now called Titania and Oberon, with orbits at a considerable angle to the ecliptic.

5.4. Origin of the Solar System

Immanuel Kant (1724–1804) outlined his theory of the origin of the solar system in his *Universal Natural History* of 1755. In this he suggested that the solar system had condensed out of a nebulous mass of gas, which had developed into a flat rotating disc as it contracted. As it continued to contract, it spun faster and faster, throwing off masses of gas that cooled to form the planets. However, Kant had difficulty in explaining how a nebula with random internal motions could start rotating when it started to contract.

Forty years later, Laplace (1749–1827) independently produced a similar but more detailed theory. In his theory, the mass of gas was rotating before it started contracting. As it contracted, it spun faster, progressively throwing from its outer edge rings of material that condensed to form the planets. Laplace suggested that the planetary satellites formed in a similar way from condensing rings of material around each of the protoplanets. Saturn's rings did not condense to form a satellite because they were too close to the planet. At face value, the theory seemed plausible, but it became clear in the nineteenth century that the original solar nebula did not have enough angular momentum to spin off the required material.

5.5. The First Asteroids

A number of astronomers had wondered why there was such a large gap in the solar system between the orbits of Mars and Jupiter. Then in 1766 Johann Titius (1729–1796) produced a numerical series that indicated that there should be an object orbiting the Sun with an orbital radius of 2.8 **astronomical units** (AUs). Johann Elert Bode (1747–1826) was convinced that this was correct and mentioned it in his book of 1772. However, what is now known as the Titius–Bode series was not considered of any particular significance, until Uranus was found with an orbital radius of 18.9 AU. This was very close to the 19.6 AU required by the series.

In 1800, a group of astronomers, who came to be known as the Celestial Police, agreed to undertake a search for the missing planet. But before they could start Giuseppe Piazzi (1746–1826) found a likely candidate by accident in January 1801. Unfortunately, although he observed the object for about 6 weeks, he was unable to fit an orbit, and wondered if it was a comet. But Karl Gauss (1777–1855) had derived a new method of determining orbits from a limited amount of information, and in November of that year he was able to fit an orbit. It was clearly a planet, now called Ceres, at almost exactly the expected distance from the Sun. But it was much smaller than any other planet. Then in March 1802 Heinrich Olbers (1758–1840) found another similar object, now called Pallas, at a similar distance from the Sun. At first Olbers thought that these two objects may be the remnants of an exploded planet. But he dropped the idea after the discovery of the fourth such asteroid, as they are now called, in 1807, because its orbit was inconsistent with his theory.

6. THE NINETEENTH CENTURY

6.1. The Sun

Sunspots were still an enigma in the nineteenth century. Many astronomers thought that they were holes in the photosphere, but because the Sun was presumably hotter beneath the photosphere, the Sunspots should appear bright rather than dark. Then in 1872 Angelo Secchi suggested that matter was ejected from the surface of the Sun at the edges of a sunspot. This matter then cooled and fell back into the center of the spot, so producing its dark central region.

In 1843, Heinrich Schwabe found that the number of sunspots varied with a period of about 10 years. A little later Rudolf Wolf analyzed historical records that showed periods ranging from 7 to 17 years, with an average of 11.1 years. Then in 1852, Sabine, Wolf, and Gautier independently concluded that there was a correlation between sunspots and disturbances in the Earth's magnetic field. There were also various unsuccessful attempts to link the sunspot cycle to the Earth's weather. But toward the end of the century, Walter Maunder pointed out that there had been a lack of sunspots between about 1645 and 1715. He suggested that this period, now called the Maunder Minimum, could have had a more profound effect on the Earth's weather than the 11-year solar cycle.

In 1858, Richard Carrington discovered that the latitude of sunspots changed over the solar cycle. In the following year, he found that sunspots near the solar equator moved faster than those at higher latitudes, showing that the Sun did not rotate as a rigid body. This so-called differential rotation of the Sun was interpreted by Secchi as indicating that the Sun was gaseous. In the same year, Carrington and Hodgson independently observed two white light solar flares moving over the surface of a large sunspot. About 36 h later, this was followed by a major geomagnetic storm.

Astronomy was revolutionized in the nineteenth century by Kirchoff's and Bunsen's development of spectroscopy in the early 1860s, which, for the first time, enabled astronomers to determine the chemical composition of celestial objects. Kirchoff measured thousands of dark Fraunhofer lines in the solar spectrum and recognized the lines of sodium and iron. By the end of the century, about 40 different elements had been discovered on the Sun.

Solar prominences had been observed during a total solar eclipse in 1733, but it was not until 1860 that they were proved to be connected with the Sun rather than the Moon. Spectroscopic observations during and after the 1868 total eclipse showed that prominences were composed of hydrogen and an element that produced a bright yellow line. This was initially attributed to sodium, but Norman Lockyer suggested that it was caused by a new element that he called helium. This was confirmed when helium was found on Earth in 1895.

6.2. Vulcan

Newton's gravitational theory had been remarkably accurate in explaining the movement of the planets, but by the nineteenth century there appeared to be something wrong with the orbit of Mercury. In 1858, Le Verrier analyzed data from a number of transits and concluded that the perihelion of Mercury's orbit was precessing at about $565''$/century, which was $38''$/century more than could be accounted for using Newton's theory. As a result, Le Verrier suggested that there was an unknown planet called Vulcan, inside the orbit of Mercury, causing the extra precession. A number of astronomers reported seeing such a planet, but none of the observations stood up to detailed scrutiny, and the idea was eventually dropped.

Einstein finally solved the problem of Mercury's perihelion precession in 1915 with his general theory of relativity. No extra planets were required.

6.3. Mercury

There was considerable disagreement among astronomers in the nineteenth century on what could be seen on Mercury. Some thought that they could see an atmosphere around the planet, but others could not. Hermann Vogel detected water vapor lines in its spectrum, and Angelo Secchi saw clouds in its atmosphere. However, Friedrich Zöllner concluded, from his photometer measurements, that Mercury was more like the Moon with, at most, a very thin atmosphere.

A number of astronomers detected markings on Mercury's disc in the middle of the nineteenth century and concluded that the planet's period is about 24 h. On the other hand, Daniel Kirkwood maintained that it should have a **synchronous rotation** period because of tidal effects of the Sun on its crust. In the 1880s, Giovanni Schiaparelli confirmed this synchronous rotation observationally, and in 1897 Percival Lowell came to the same conclusion. So at the end of the century, synchronous rotation was thought to be the most likely.

6.4. Venus

In the eighteenth century, Venus was thought to have an axial rotation rate of about 24 h. In fact, a 24-h period was generally accepted until in 1890 Schiaparelli and others concluded that it, like Mercury, has a synchronous rotation period.

Spectroscopic observations of Venus yielded conflicting results in the nineteenth century. A number of astronomers detected oxygen and water vapor lines in its atmosphere; however, W. W. Campbell, who used the powerful Lick telescopes, could find no such lines.

6.5. The Moon

The impact theory for the formation of lunar craters was resurrected at the start of the nineteenth century, after the discovery of the first asteroids and a number of meteorites. There now seemed to be a ready source of impacting bodies, which Hooke had been unaware of when he had abandoned his impact hypothesis. But both the impact and volcanic theories still had problems. Most meteorites would not hit the lunar surface vertically, and so the craters should be elliptical, but they were mostly circular. Also, as Grove K. Gilbert pointed out, the floors of lunar craters are generally below the height of their surrounding area, whereas on Earth the floors of volcanic craters are generally higher than their surroundings.

Edmond Halley had discovered in 1693 that the Moon's position in the sky was in advance of where it should be based on ancient eclipse records. This so-called secular acceleration of the Moon could be because the Moon was accelerating in its orbit, and/or because the Earth's spin rate was slowing down. In 1787, Laplace had shown that the observed effect, which was about $10''/\text{century}^2$, could be completely explained by planetary perturbations. But in 1853, John Couch Adams included some of Laplace's second-order terms, which Laplace had omitted, so reducing the calculated figure from $10''/\text{century}^2$ to just $6''/\text{century}^2$. Charles Delaunay suggested that the missing amount was probably because of tidal friction, but it was impossible at that time to produce a reasonably accurate estimate of the effect. In the early twentieth century, Taylor and Jeffreys produced the necessary calculations, showing that Delaunay was correct.

In 1879, George Darwin developed a theory of the origin of the Moon. In this the proto-Earth had gradually contracted and increased its spin rate as it cooled. Then, when the spin rate had reached about 3 h per revolution, it had broken into two unequal parts: the Earth and the Moon. After breakup, tidal forces had caused the Earth's spin rate to slow down and the Moon's orbit to gradually increase in size.

A major problem with this theory was that the Earth would have had a tendency to break up the Moon shortly after separation. It was not clear whether the Moon could have passed through the danger zone before this could have happened.

6.6. The Earth

Karl Friedrich Küstner undertook precise position measurements of a number of stars in 1884 and 1885 from the Berlin Observatory. When he analyzed his results, however, he found that the latitude of the observatory had apparently decreased by about $0.20''$ in a year. Intrigued, the International Commission for Geodesy (ICG) decided to organize a series of observations around the world to define the effect more precisely. These results indicated that the Earth's spin axis was moving, relative to its surface, with a period of about 12 or 13 months.

Seth Chandler had also noticed slight variations in the latitude of the Harvard College Observatory, at about the same time as Küstner was making his measurements, but Chandler had not taken the matter further. Galvanized by Küstner's and the ICG's results, however, he undertook a

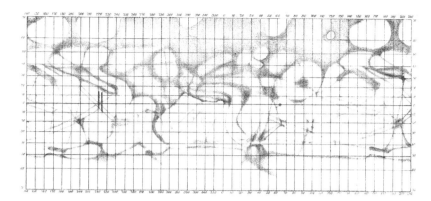

FIGURE 47.5 Schiaparelli's map of Mars produced following the 1881 opposition. A large number of *canali* are seen, many of them double. *From Ball, R. (1897)*.

thorough review of all available data. As a result, he concluded that the observed effect had two components. One had a period of 14 months, and was because of the nonrigid Earth not spinning around its shortest diameter. The other, which had a period of a year, was because of the seasonal movement of water and air from one hemisphere to the other and back.

6.7. Mars

The first systematic investigation of Mars' polar caps had been undertaken in the eighteenth century by Giacomo Maraldi, who found that the south polar cap had completely disappeared in late 1719, only to reappear later. William Herschel suggested that this was because it consisted of ice and snow that melted in the southern summer.

At the end of the eighteenth century, most astronomers thought that the reddish color of Mars was because of its atmosphere. But in 1830, John Herschel suggested that it was the true color of its surface. Camille Flammarion, on the other hand, hypothesized that it was the color of its vegetation.

It was generally believed by astronomers in the mid-nineteenth century that there must be some form of life on Mars, even if it was only plant life, because the planet clearly had an atmosphere and a surface that exhibited seasonal effects. The polar caps were apparently made of ice or snow, and there were dark areas on the surface that may be seas.

Schiaparelli produced a map of Mars, following its 1877 **opposition**, that showed a network of linear features that he called *canali*. This was translated incorrectly into English as canals, which implied that they had been built by intelligent beings. Schiaparelli and others saw more *canali* in subsequent years (Figure 47.5), but other equally competent observers could not see them at all. Percival Lowell then went further than Schiaparelli in not only observing many canali, but interpreting them to be a network of artificial irrigation channels. At the end of the century, the debate as to whether these *canali* really existed was still in full swing.

Spectroscopic observations of Mars in the late nineteenth century yielded conflicting results. Some astronomers detected oxygen and water vapor lines, whereas Campbell at the Lick Observatory could find none. There was also a problem with the polar caps: Calculations showed that the average temperature of Mars should be about $-34\ °C$, yet both polar caps clearly melted substantially in summer, which they should not have done if they had been made of water ice or snow. In 1898, Ranyard and Stoney suggested that the caps could be made of frozen carbon dioxide. But there appeared to be a melt band at the edge of the caps in spring, yet carbon dioxide should sublimate directly into gas on Mars.

Two satellites of Mars, now called Phobos and Deimos, were discovered by Asaph Hall in 1877. Their orbits were extremely close to the planet, and the satellites were both very small. As a result, they were thought to be captured asteroids.

6.8. Jupiter

The Great Red Spot (GRS) was first clearly observed in the 1870s. Then in 1880 an unusually bright, white equatorial spot appeared; it rotated around Jupiter over 5 min faster than the GRS. This gave a differential velocity of about 400 km/h. But the rotation rates of both the white spot and the GRS were not constant, indicating that neither could be surface features as some astronomers had supposed.

White and dark spots were continuously appearing and disappearing on Jupiter, suggesting that they were probably clouds. But the GRS was completely different because, although it changed its appearance and size over time, it was still there at the end of the century. This longevity led astronomers to wonder if it could really be a cloud system.

In 1778, Leclerc, Compte de Buffon, had suggested that rapid changes in Jupiter's appearance showed that it had not completely cooled down since its formation. In the nineteenth century, Jupiter's differential rotation and low density, which were both similar in nature to those of the Sun, caused some astronomers to go even further and wonder if Jupiter was self-luminous. Although this was considered unlikely, the idea had not been completely ruled out by the end of the century.

William Herschel had concluded in 1797 that the axial rotation rates of the four Galilean satellites were synchronous. However, it was not until the 1870s that Engelmann and Burton independently confirmed this for Callisto and the 1890s that Pickering and Douglass confirmed it for Ganymede. The rotation rates of Io and Europa were still unclear.

In 1892, Edward Barnard discovered Jupiter's fifth satellite, now called Amalthea, very close to the planet, when he was observing Jupiter visually through the 36-in. Lick refractor. Amalthea was very small compared with the four Galilean satellites. It was the last satellite of any planet to be discovered visually.

6.9. Saturn

In 1837, Johann Encke found that the A ring was divided into two by a clear gap, now called the Encke Division. Then in 1850 W. C. and G. P. Bond discovered a third ring, now called the C ring, inside the B ring. The new ring was very dark (Figure 47.6) and partly transparent. In 1867, Kirkwood pointed out that any particles in the Cassini Division would have periods of about one-half that of Mimas, one-third that of Enceladus, one-quarter that of Tethys, and one-sixth that of Dione. He concluded that these resonances had created the Cassini Division, which would be clear of particles.

The true nature of Saturn's rings had been a complete mystery in the eighteenth century. Cassini had thought that they may be composed of many small satellites, and Laplace had suggested that they were made of a number of thin solid rings. Others thought that they may be liquid. But in 1857, James Clerk Maxwell proved mathematically that they could not be solid or liquid. Instead, he concluded that they were composed of an indefinite number of small particles.

Two new satellites were found in the nineteenth century: Hyperion by G. P. Bond in 1848 and Phoebe by William Pickering 50 years later. Phoebe was the first satellite in the solar system to be discovered photographically. It was some 13 million km from Saturn, in a highly eccentric, **retrograde** orbit. So it appeared to be a captured object.

6.10. Uranus

Little was known about Uranus in the nineteenth century. William Herschel had noticed that Uranus had a polar flattening, its orientation indicating that its axis of rotation was perpendicular to the plane of its satellites. But observations of apparent surface features produced very different orientations. Uranus' spectrum appeared to be clearly different from those of Jupiter and Saturn, but it was very difficult to interpret. There was even confusion about the discovery of new satellites. It was not until 1851 that William Lassell could be sure that he had discovered two new satellites, now called Ariel and Umbriel within the orbit of Titania. He had, in fact, seen them both some years before, but his earlier observations had been too infrequent to produce clear orbits.

6.11. The Discovery of Neptune

In 1821, Alexis Bouvard tried to produce an orbit for Uranus using both prediscovery and postdiscovery observations. But he could not find a single orbit to fit them. The best he could manage was an orbit based on only the postdiscovery observations; he published the result but admitted that it was less than ideal. However, it did not take long for Uranus to deviate more and more from even this orbit. One possible explanation was that Uranus was being disturbed by yet another planet, and if the Titius-Bode series was correct it would be about 38.8 AU from the Sun.

In 1843, the Englishman John Couch Adams set out to try to calculate the orbit of the planet that seemed to be disturbing the orbit of Uranus. By September 1845, he had calculated its orbital elements and its expected position in the sky, and over the next year, he progressively updated this prediction. Unfortunately, these predictions varied wildly, making it impossible to use them for a telescopic search of the real planet. In parallel, and unknown to both men, Urbain Le Verrier, a French astronomer, undertook the same task. He published his final results in August 1846 and asked Johann Galle of the Berlin Observatory if he would undertake a telescope search for it. Galle and his assistant d'Arrest found the planet within an hour of

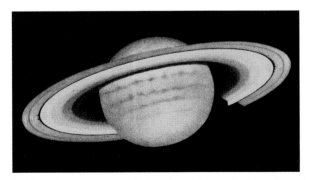

FIGURE 47.6 Trouvelot's 1874 Drawing of Saturn. It clearly shows the dark C ring extending from the inner edge of the B ring to about half-way to the planet. *From Ledger, E. (1882).*

starting the search on 23 September 1846. There then followed a monumental argument between the English and French astronomical establishments on the priority for the orbital predictions. But much of the evidence on the English side was never published, and an "official line" was agreed. That evidence has recently come to light, however, and it is currently being analyzed to establish the exact sequence of events. What is clear, however, is that when Neptune's real orbit was calculated, it turned out to be quite different from either of the orbits predicted by Le Verrier or Adams. So its discovery had been somewhat fortuitous.

Less than a month after Neptune's discovery, William Lassell observed an object close to Neptune, which he thought may be a satellite. It was not until the following July that he was able to confirm his discovery of Neptune's first satellite, now called Triton. Triton was later found to have a retrograde orbit inclined at approximately 30° to the ecliptic.

6.12. Asteroids

The fourth asteroid, Vesta, had been discovered in 1807, but it was not until 1845 that the fifth asteroid was found. Then the discovery rate increased rapidly so that nearly 500 asteroids were known by the end of 1900. As the number of asteroids increased, Kirkwood noticed that there were none with certain fractional periods of Jupiter's orbital period. This he attributed to resonance interactions with Jupiter.

All the early asteroids had orbits between those of Mars and Jupiter, and even as late as 1898 astronomers had discovered only one that had part of its orbit inside that of Mars. But in 1898, Eros was found with an orbit that came very close to that of the Earth, with the next closest approach expected in 1931. This could be used to provide an accurate estimate of solar parallax.

In 1906, two asteroids were found at the Lagrangian points, 60° in front of and behind Jupiter in its orbit. They were the first of the so-called Trojan asteroids to be discovered.

6.13. Comets

Charles Messier discovered a comet that passed very close to the Earth in 1770. Anders Lexell was the first to fit an orbit to it, showing that it had a period of just 5.6 years. With such a short period it should have been seen a number of times before, but it had not. As Lexell explained, this comet had not been seen because it had passed very close to Jupiter in 1767, which had radically changed its orbit. In the late nineteenth century, Hubert Newton examined the effect of such planetary perturbations on the orbits of comets and found that, for a random selection of comets, they were remarkably inefficient. Lexell's comet appeared to be an exception.

Jean Louis Pons in 1818 discovered a comet that, on further investigation, proved to have been seen near previous perihelia. In the following year, Johann Encke showed that the comet, which now bears his name, has an orbit that takes it inside the orbit of Mercury. When the comet returned in 1822, Encke noticed that it was a few hours early and suggested that it was being affected by some sort of resistive medium close to the Sun. In 1882, however, a comet passed even closer to the Sun and showed no effect of Encke's medium. Then in 1933, Wolf's comet was late, rather than early. The problem of these cometary orbits was finally solved in 1950 when Fred Whipple showed that the change in period was caused by jetlike vaporization emissions from the rotating cometary nucleus.

The first successful observation of a cometary spectrum was made by Giovanni Donati in 1864. When the comet was near the Sun, it had three faint luminous bands, indicating that it was self-luminous. Then 4 years later, William Huggins found that the bands were similar to those emitted by hydrocarbon compounds in the laboratory.

Quite a number of cometary spectra were recorded over the next 20 years. When they were first found, they generally exhibited a broad continuous spectrum like that of the Sun indicating that they were scattering sunlight. As they got closer to the Sun, however, the hydrocarbon bands appeared. Then in 1882 Wells' comet approached very close to the Sun. Near perihelion its bandlike structure disappeared to be replaced by a bright double sodium line. In the second comet of 1882, this double sodium line was also accompanied by several iron lines when the comet was very near the Sun. As the comet receded, these lines faded and the hydrocarbon bands returned.

6.14. Meteor Showers

A spectacular display of shooting stars was seen in November 1799, and again in November 1833. They seemed to originate in the constellation Leo. In the following year, Denison Olmsted pointed out the similarities between these two meteor showers and a less intense one in 1832. These so-called Leonid meteors seemed to be an annual event occurring on or about 12 November. Olmsted explained that the radiant in Leo was because of a perspective effect (Figure 47.7). A similar effect was then observed for a meteor shower on 8 August 1834, which appeared to have a radiant in Perseus. Shortly afterward, Lambert Quetelet showed that these were also an annual event.

In 1839, Adolf Erman suggested that both the Leonid and Perseid meteor showers were produced by the Earth passing through swarms of small particles that were orbiting the Sun and spread out along Earth's orbit. But it was still unclear as to the size of the orbit. In 1864, Hubert Newton found that the node of the Leonids' orbit was

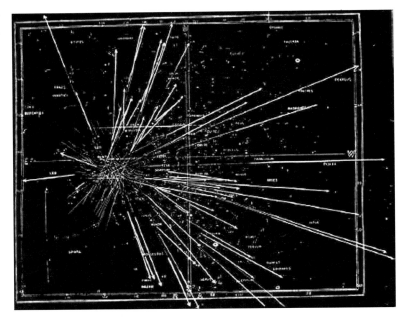

FIGURE 47.7 Paths of the Leonid meteors showing their apparent origin from a common radiant due to parallax. *From Newcomb, S. (1898).*

precessing at about 52″/year. John Couch Adams then showed that only a particle in a 33.25-year orbit would have this nodal precession. So the Leonids were orbiting the Sun in a diffuse cloud every 33.25 years, which explained why the most intense showers occurred with this frequency. The stragglers all around the orbit explained why we saw the Leonids on an annual basis. In 1867, Carl Peters recognized that the source of the Leonid meteor stream was a periodic comet called Tempel–Tuttle. This was just after Schiaparelli had linked the Perseids to another periodic comet, Swift–Tuttle.

7. THE TWENTIETH CENTURY PRIOR TO THE SPACE AGE

7.1. The Sun

In the nineteenth century, most physicists had thought that heat was transported from the interior to the exterior of the Sun by convection. But in 1894, R. A. Sampson suggested that the primary mechanism was radiation. Then, 30 years later, Arthur Eddington used the concept of radiative equilibrium to calculate the temperature at the center of the Sun and found it to be about 39 million K. At about the same time, Cecilia Payne showed that hydrogen and helium were the most abundant elements in the stars. Although this idea was initially rejected, it was soon accepted for both the Sun and stars. As a result, in 1935 Eddington reduced his temperature estimate for the center of the Sun to 19 million K.

However, Eddington's calculations made no assumption on how the Sun's heat was produced, which was still unknown at the time. Earlier, in 1920, Eddington himself had proposed two alternative mechanisms. The heat could be produced either by the mutual annihilation of protons and electrons or by the fusion of hydrogen atoms into helium atoms in some unknown manner. There were other mechanisms suggested by other physicists, but the issue could not be resolved at the time because nuclear physics was still in its infancy. The breakthrough came in 1938 when Charles Critchfield explained how energy could be produced at high temperatures by a chain reaction starting with proton–proton collisions and ending with the synthesis of helium nuclei. Hans Bethe then collaborated with Critchfield to develop this idea. But Bethe also examined an alternative mechanism that relied on carbon as a catalyst to produce helium from hydrogen, in the so-called carbon cycle. Carl von Weizsäcker independently developed this same scheme. Which mechanism was predominant in the Sun depended crucially on temperature, and it was not until the 1950s that it became clear that the proton–proton chain is dominant in the Sun.

In the nineteenth century, the **corona** had been found to have a faint continuous spectrum crossed by Fraunhofer absorption lines, but the conditions in the corona were unclear. Of particular interest was a bright green emission line in the coronal spectrum; Young and Harkness found it in 1869 and originally attributed it to iron. In 1898, however, it was found to have a slightly different wavelength than the iron line. Because no known element generated the required line, it was attributed to a new element called coronium.

At that time, it was assumed that the temperature of the Sun and its corona gradually reduced from the center moving outward. But in the early part of the twentieth century, competing theories were put forward, one for a low-temperature corona and another for a high-temperature one. In 1934, Walter Grotrian analyzed the coronal spectrum and concluded that the temperature was an astonishing 350,000 K. A few years later Bengt Edlén, in a seminal paper, showed that coronal lines are produced by highly ionized iron, calcium, and nickel at a temperature of at least 2 million K. The "coronium" line, in particular, was the product of highly ionized iron. How the temperature of the corona could be so high, when the photosphere temperature is only of the order of 6000 K, was a mystery, which has not been completely resolved even today.

Charles Young discovered in 1894 that, at very high dispersions, many absorption lines in sunspot spectra appeared to have a sharp bright line in their centers. In 1908, George Ellery Hale and Walter Adams found that photographs of the Sun taken in the light of the 656.3-nm hydrogen line showed patterns that looked like iron filings in a magnetic field. This caused Hale to examine sunspot spectra in detail. He found that the Young effect was actually caused by Zeeman splitting of spectral lines in a magnetic field, which was of the order of 3000 G. So sunspots were the home of very high magnetic fields.

Hale then started to examine the polarities of sunspots, and found that spots generally occur in pairs, with the polarity of the lead spot, as they crossed the disc, being different in the two hemispheres. This pattern was well established by 1912 when the polarities were found to be reversed at the solar minimum. They reversed yet again at the next solar minimum in 1923. So the solar cycle was really 22 years, not 11.

Walter Maunder found in 1913 that large magnetic storms on Earth start about 30 h after a large sunspot crosses the center of the solar disc. Later work showed that the most intense storms were often associated with solar flares. In 1927, Chree and Stagg found that smaller storms, which did not seem to be associated with sunspots, tended to recur at the Sun's synodic period of 27 days. Julius Bartels called the invisible source on the Sun of these smaller storms, M regions. Both the so-called flare storms and the M storms were assumed to be caused by particles ejected from the Sun. In 1951, Ludwig Biermann suggested that, to explain the behavior of cometary ion tails, there must be a continuous stream of charged particles emitted by the Sun. Then in 1957, Eugene Parker proposed his theory of the solar wind, which was later confirmed by early spacecraft.

Marconi noticed in 1927 that interference with radio signals in September and October of that year coincided with the appearance of large sunspots and intense aurorae. In the late 1930s, Howard Dellinger carried out a detailed examination of the timing of shortwave radio fade-outs, at numerous receiving stations, and solar flares. He found a reasonable but by no means perfect correlation. The fade-outs seemed to start almost instantaneously after the flare was seen, and they only occurred when the receiving station was in daylight. So Dellinger concluded that they were caused by some form of electromagnetic radiation from the Sun, rather than particles.

7.2. Mercury

The synchronous rotation period of Mercury was gradually accepted as a fact in the twentieth century. But in 1962, W. E. Howard found that Mercury's dark side seemed to be warmer than it should be if it were permanently in shadow. Then 3 years later, Dyce and Pettengill found, using radar, that Mercury's rotation period was not synchronous, but represented two-thirds of its orbital rotation period.

7.3. Venus

There was considerable confusion in the first half of the twentieth century about Venus' rotation period. All sorts of periods were proposed between about 24 h and synchronous (225 days). Then in 1957 Charles Boyer found a distinctive V-shaped pattern of Venus' clouds that had a 4-day period. In 1962, however, Carpenter and Goldstein deduced a period of about 250 days retrograde using radar, which was modified to 243 days in 1965 for the rotation period of Venus' surface. So Venus has a 243-day period, while its clouds have a period of about 4 days, both periods being retrograde.

In 1932, Adams and Dunham concluded that there was no oxygen or water vapor on Venus, but carbon dioxide was clearly present. A few years later, Rupert Wildt calculated that the greenhouse heating of the latter could produce a surface temperature as high as 400 K. Then in 1956, Mayer, McCullough, and Sloanaker deduced a surface temperature of about 600 K by analyzing Venus' thermal radio emissions. The suggestion that Venus' surface temperature could be so high was naturally treated with caution. Shortly afterward, Carl Sagan estimated that the surface atmospheric pressure was an equally incredible 100 bar.

7.4. The Moon

The idea that there may be life on the Moon had fascinated people for centuries. Even respected astronomers like William Herschel had thought that there would be "lunarians" as he called them. But by the start of the twentieth century, it was thought that the most complex life-forms would be some sort of plant life. However, by the 1960s, when the Americans were planning their lunar landings, even this concept had been rejected. Nevertheless, it was thought that there may be some sort of very elemental life, like bacteria, on the Moon.

Bernard Lyot had concluded in 1929, from polarization measurements, that the Moon was probably covered by volcanic ash. Then in the 1950s, Thomas Gold suggested that the Moon may be covered with dust up to a few meters deep. If this was so, it would have provided a major problem for the manned *Apollo* missions.

At the end of the nineteenth century, the key objection to the impact theory for the formation of lunar craters had been that the craters were generally circular, when they should have been elliptical, because most of the impacts would not be vertical. However, after the First World War it was realized that the shape of the lunar craters resembled shell craters. The shell craters were formed by the shock wave of the impact or explosion, so a nonvertical impact could still produce a circular crater. Nevertheless, not all lunar craters have the same general appearance. So, by the start of the space age it was still unclear if they had been produced by volcanic action, meteorite impact, or both.

7.5. The Earth

It was known in the nineteenth century that temperatures in deep mines on Earth increased with depth. That, together with the existence of volcanoes, clearly indicated that the Earth has a molten interior. Calculations indicated that the rocks would be molten at a depth of only about 40 km.

In 1897, Emil Wiechert suggested that the Earth has a dense metallic core, mostly of iron, surrounded by a lighter rocky layer, now called the mantle. A little later, Richard Oldham found clear evidence for the existence of the core from earthquake data. Then in 1914, Beno Gutenberg showed that the interface between the mantle and the core, now called the Wiechert−Gutenberg discontinuity, is at about $0.545r$ from the center of the Earth (where r is its radius).

A little earlier, Andrija Mohorovičić had discovered the boundary between the crust and mantle, now called the Mohorovičić discontinuity, by analyzing records of the Croatian earthquake of 1909. The depth of this discontinuity was later found to vary from about 70 km under some mountains to only about 5 km under the deep oceans.

A number of theories were proposed to try to define and explain the internal structure of the Earth. In particular, Harold Jeffreys produced a theory that assumed that all the **terrestrial planets** and the Moon have a core of liquid metals, mostly iron, and a silicate mantle. But it could not explain how those planets with the smallest cores could have retained a higher percentage of lighter material in their mantles. In 1948, William Ramsey solved this problem when he proposed that the whole of the interior of the terrestrial planets consists of silicates, with the internal pressure in the largest planets causing the silicates near the center to become metallic. Unfortunately this idea became unviable when Eugene Rabe found in 1950 that Mercury's density was much higher than originally thought. It was even higher than that of Venus and Mars, which were much larger planets.

In the mid-twentieth century, most astronomers believed that the planets had been hot when first formed from the solar nebula, but in 1949 Harold Urey suggested that the nebula had been cold. According to Urey, the Earth had been heating up since it was formed because of radioactive decay. Internal convection had then started as iron had gradually settled into the core. Urey believed that the Moon was homogenous because it was relatively small.

At the turn of the nineteenth century, it was thought that radio waves generally traveled in a straight line. So it was a great surprise when Marconi showed in 1901 that radio waves could be successfully transmitted across the Atlantic. Refraction could have caused them to bend to a limited degree, but not enough to cross the ocean. In the following year, Heaviside and Kennelly independently suggested that the waves were being reflected off an electrically conducting layer in the upper atmosphere.

The structure of what we now call the E or Heaviside layer, and of other layers in the ionosphere, was gradually clarified over the next 20 years or so. The 80 km high D layer was found to largely disappear at night, and the higher E layer was found to maintain its reflectivity for only 4 or 5 h after sunset. In addition, it was found that solar flares can cause a major disruption to the ionosphere (see Section 7.1). However, it was not until after the Second World War that the cause of these effects could be examined in detail by first sounding rockets and then by spacecraft. The first major discovery was made by Herbert Friedman in 1949 when he showed that the Sun emits X-rays, which have a major effect on the Earth's ionosphere.

7.6. Mars

There was a great deal of uncertainty about the surface of Mars in the first half of the twentieth century. It was thought unlikely that the linear markings called *canali* really existed, but they were still recorded from time to time by respected observers. In addition, some astronomers thought that the bluish green areas on Mars were vegetation, while others thought that they were volcanic lava.

There was also considerable uncertainty about the spectroscopic observations of Mars. Some observers recognized water vapor and oxygen lines, whereas others found none. But in 1947 Gerard Kuiper clearly found evidence for a small amount of carbon dioxide, and in 1963 Audouin Dollfus found a trace amount of water vapor. Estimates of the surface atmospheric pressure varied from about 25 to 120 mbars. Then in 1963, shortly before the first spacecraft reached Mars, a figure of 25 ± 15 mbars was estimated by Kaplan, Münch, and Spinrad.

It seemed clear that the yellow clouds seen on Mars were dust. In 1909, Fournier and Antoniadi found that they appeared to cover the whole planet for a while. Later Antoniadi found that they tended to occur around perihelion when the solar heating is greatest, and so appeared to be produced by thermally generated winds. Thirty years later, De Vaucouleurs measured the wind velocities as being typically in the range of 60–90 km/h when the clouds first formed.

7.7. Internal Structures of the Giant Planets

It was known in the nineteenth century that the densities of Jupiter, Saturn, Uranus, and Neptune were similar to that of the Sun, and were much less than that of the terrestrial planets. At that time, it was thought that Jupiter, and probably Saturn, had not yet fully cooled down since their formation. As a result, they were probably emitting more energy than they received from the Sun.

In 1923, Donald Menzel found that the cloud top temperatures of Jupiter and Saturn were about 160 K. This compares with temperatures of 120 and 90 K for Jupiter and Saturn, respectively, that would be maintained solely by incident solar radiation. Three years later, Menzel produced modified observed temperatures of 140, 120, and 100 K, for Jupiter, Saturn, and Uranus. So any internally generated heat would be rather low.

In 1923, Harold Jeffreys pointed out that the ratio of the densities of Io and Europa, the innermost of Jupiter's large satellites, to that of Jupiter, was about the same as the ratio of the density of Titan, Saturn's largest satellite, to that of Saturn. He then assumed that the density of the cores of Jupiter and Saturn were the same as these their large satellites. In that case, the thickness of the planetary atmospheres would be about 20% of their radii.

In the following year, Jeffreys included consideration of the moments of inertia of Jupiter and Saturn in his analysis and concluded that their atmospheres would have depths of $0.09 R_J$ and $0.23 R_S$, respectively (where R_J and R_S are the radii of Jupiter and Saturn, respectively). He assumed that beneath their atmospheres there was a layer of ice and solid carbon dioxide, which in turn surrounded a rocky core.

Various schemes were then produced by a number of physicists, of which those of Rupert Wildt in 1938 and William Ramsey in 1951 were probably the most significant. Wildt, who was particularly interested in internal pressures, wanted to find out if matter at the core of the large planets was **degenerate**. His calculations indicated that it was not. Ramsey, on the other hand, developed his theory assuming that the giant planets were made of hydrogen. He then added helium and other ingredients until their densities and moments of inertia were correct. On this basis, he concluded that Jupiter and Saturn were composed of 76% and 62% hydrogen, by mass, respectively, with central pressures of 32 and 6×10^6 bar. At these pressures, most of the hydrogen would be metallic.

The structures of Uranus and Neptune were a problem in Ramsey's analysis because the heavier planet, Neptune, was the smaller. So their constituents could not be the same. Then in 1961 William Porter produced a model that seemed to fit; in this model, Neptune had 74% ammonia and 26% heavier elements, whereas Uranus had less heavy elements and a small amount of hydrogen.

7.8. Atmospheres of the Giant Planets

Vesto Slipher undertook a detailed investigation of the spectra of Jupiter, Saturn, Uranus, and Neptune in the early decades of the twentieth century. He recorded numerous bands for all the planets but had trouble interpreting them. In 1932, Rupert Wildt deduced that a number of the bands in all four planets were due to ammonia and methane. However, subsequent work by Mecke, Dunham, Adel, and Slipher showed that some of the lines had been misattributed, so there was no ammonia in the atmospheres of Uranus and Neptune. This was, presumably, because it had been frozen out at their lower temperatures. Adel and Slipher also concluded that the methane concentration reduced in going from Neptune to Uranus to Saturn to Jupiter.

7.9. Jupiter

In 1955, Burke and Franklin made the unexpected discovery that Jupiter was emitting radio waves at 22.2 MHz. Subsequently, it was found that Jupiter emitted energy at many radio frequencies. Some of it was thermal energy, with an effective temperature of 145 K, but some was clearly nonthermal. The latter was taken to indicate that Jupiter had an intense magnetic field, with radiation belts similar to those that had, by then, been found around the Earth.

Our knowledge of Jupiter's Galilean satellites changed little in the twentieth century before the space age. In 1900, Barnard had observed that the poles of Io appeared to be reddish in color. Then in 1914 Paul Guthnik showed that all four Galilean satellites exhibited synchronous rotation. In the nineteenth century, it was thought that all four satellites probably had atmospheres, but this was considered more and more unlikely as the twentieth century progressed.

7.10. Saturn

A prominent white equatorial spot had been observed on Saturn in 1876. Then in 1903 Edward Barnard discovered another temporary prominent white spot at about 36° N, but its rotation period around Saturn was some 25 min slower. Another equatorial spot that had a similar period to the

1876 equatorial spot appeared in 1933, and another spot that had a similar period to the 1903 spot was observed at about 60 °N in 1960. The velocities of these spots showed that there was an equatorial current on Saturn, similar to that on Jupiter. But the one on Saturn had a velocity of about 1400 km/h, compared with just 400 km/h for Jupiter. It was unclear why Saturn, which is farther from the Sun, and so receives less heat than Jupiter, should have a much faster equatorial current.

Markings on Saturn's rings were seen by a number of observers in the late nineteenth and early twentieth centuries, including the respected observers Etienne Trouvelot and Eugiéne Antoniadi. In 1955, Guido Ruggieri noticed clear radial streaks at both ansae of the A ring, but after further investigation he concluded that they were an optical illusion. It is unclear whether any of these observations were early observations of spokes, of the sort discovered by the *Voyager* spacecraft on the B ring, or not.

In the winter of 1943–1944, Gerard Kuiper photographed the spectrum of the 10 largest satellites of the solar system and found evidence for an atmosphere on Titan and possibly Triton. He could find no such evidence for the Galilean satellites of Jupiter, however.

7.11. Uranus and Neptune

In the nineteenth century, Triton had been found to orbit Neptune in a retrograde sense, and it was unclear at the time whether Neptune's spin was also retrograde. But in 1928 Moore and Menzel found, by observing the **Doppler shift** of its spectral lines, that Neptune's spin was **direct** or **prograde**. So Neptune's largest satellite was orbiting the planet in the opposite sense to the planet's spin. This phenomenon had not been observed before in the solar system for a major satellite.

Kuiper discovered Uranus' fifth satellite, now called Miranda, in 1948. It was orbiting the planet in an approximately circular orbit inside that of the other four satellites. Then in the following year he discovered Neptune's second satellite, now called Nereid, orbiting Neptune in the opposite sense to Triton. Nereid was in a highly elliptical orbit well outside the orbit of Triton. So Nereid was the "normal" satellite in orbiting Neptune direct or prograde, whereas the larger Triton, which was nearer to Neptune in an almost circular orbit, appeared to be the abnormal one.

7.12. The Discovery of Pluto

The discoveries of Uranus and Neptune made astronomers realize that there may well be planets even farther out from the Sun. As Neptune had only been discovered in 1846, and as it was moving very slowly, its orbit was not very well known in the second half of the nineteenth century. However astronomers had much better information on Uranus' orbit, and so they reexamined it to see if there were any unexplained deviations that might indicate the whereabouts of a new planet. Such deviations were soon found, and a number of possible locations for the new planet proposed by various astronomers, including Percival Lowell. A photographic search for the new planet was started at Lowell's observatory, but this was abandoned when Lowell died in 1916.

In 1929, Vesto Slipher, the new director of Lowell's observatory, recruited Clyde Tombaugh to undertake a search for the new planet using a photographic refractor that had been specifically purchased for the task. Tombaugh photographed the whole of the zodiac, and used a blink comparator to find objects that had moved over time. The task was very tedious, but he discovered Pluto in February 1930 after working for 10 months. However, although the planet's orbit was very similar to that predicted by Lowell (Figure 47.8), it was far too small to have perturbed Uranus in the way that Lowell had estimated.

Over the years, the estimated mass of Pluto has gradually reduced from 6.6 M_E (M_E is the mass of the Earth) predicted by Lowell, to 0.7 M_E (maximum) at the time of its discovery, to 0.002 M_E now. Its orbit is highly eccentric, and it has the largest inclination of the traditional planets.

In 1955, Walker and Hardie deduced a rotation period of 6 day 9 h 17 min from regular fluctuations in Pluto's intensity. Little more was known about the planet when the space age started.

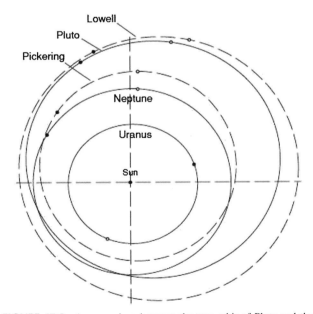

FIGURE 47.8 A comparison between the true orbit of Pluto and that predicted by Lowell and Pickering. Although Lowell's orbit was reasonably close to that of Pluto, the agreement was fortuitous. (The open circles show the positions of the planets in 1900, and the closed circles represent those in 1930.)

7.13. Asteroids

In 1918, Kiyotsugu Hirayama identified families of asteroids based on their orbital radius, eccentricity, and inclination. Initially, he identified three families, Themis (22 members), Eos (21 members), and Koronis (13 members). Hirayama suggested that the three families were each the remnants of a larger asteroid that had fractured. This resurrected, in modified form, the theories of Thomas Wright and Wilhelm Olbers, in the eighteenth and nineteenth centuries. They both believed that there had been a planet between the orbits of Mars and Jupiter that had broken up.

In the nineteenth century, Eros had been discovered with a perihelion of 1.13 AU. In 1932, another asteroid, now called Amor, was found that had an orbit that came even closer to that of the Earth than Eros. Then, just 6 weeks later, the first asteroid, now called Apollo, whose orbit crossed that of the Earth, was discovered. The names of Amor and Apollo have now been given to families of asteroids with similar orbital characteristics.

7.14. Comets

Huggins had shown in the nineteenth century that there were hydrocarbon compounds in the heads of comets, but he was not able to specify exactly which hydrocarbons were involved. Molecular carbon, C_2, was first identified in the head of a comet just after the turn of the century, and by the mid-1950s C_3, CH, CN, OH, NH, and NH_2 had been found in the heads of comets.

Molecular bands were observed in the tail of Daniel's comet by Deslandres, Bernard, and Evershed in 1907 and in the tail of Morehouse's comet by Deslandres and Bernard the following year. These bands were later identified by Alfred Fowler was those of ionized carbon monoxide, (CO^+) and N_2^+. Later CO_2^+ was also found in the tail of a comet.

In the 1930s, Karl Wurm observed that many of the molecules found in comets were chemically very active, and so they cannot have been present there for very long. He suggested, instead, that they had come from the more stable so-called parent molecules $(CN)_2$, H_2O, and CH_4 (methane). In 1948, Pol Swings, in his study of Encke's comet, concluded that the parent molecules were water, methane, ammonia (NH_3), nitrogen, carbon monoxide, and carbon dioxide, all of which had been in the form of ice before being heated by the Sun.

In 1950 and 1951, Fred Whipple proposed his icy-conglomerate model (better known as his dirty snowball theory) in which the nucleus is composed of ices, such as methane, with meteoric material embedded within it. Unfortunately, some of the parent molecules were highly volatile. But in 1952 Delsemme and Swings suggested that these highly volatile elements would be able to resist solar heating better if they were trapped within the crystalline structure of water ice, in what are known as clathrate hydrates.

It was difficult to determine the orbits of long-period comets because they were only observed for the fraction of their orbit when they were close to the Sun. However, a survey of about 400 cometary orbits observed up to 1910 showed that only a tiny minority appeared to be hyperbolic. Strömgren and Fayet then showed that none of these comets had hyperbolic orbits before they passed Saturn or Jupiter on their approach to the Sun. So the long-period comets appeared to be members of the solar system.

In 1932, Ernst Öpik concluded, from an analysis of stellar perturbations, that comets could remain bound to the Sun at distances of up to 10^6 AU. Some years later, Adrianus Van Woerkom showed that there must be a continuous source of new near-parabolic comets to explain the relative numbers observed. Then in 1950 Jan Oort showed that the orbits of 10 comets, with near-parabolic orbits, had an average **aphelion** distance of about 100,000 AU. As a result, he suggested that all long-period comets originate in what is now called the Oort cloud about 50,000 to 150,000 AU from the Sun.

7.15. The Origin of the Solar System

In the early decades of the twentieth century, theories of the origin of the solar system generally focused on the effect of collisions, and close encounters of another star to the Sun. But all the theories were found to have significant problems, so Laplace's theory of a condensing nebula was reconsidered.

Laplace's theory had been rejected in the nineteenth century because the original solar nebula did not appear to have had enough angular momentum. However, in the 1930s, McCrea showed that this would not be a problem if the original nebula had been turbulent.

In 1943, Carl von Weizsacker produced a theory where cells of circulating convection currents, or vortices, formed in the solar nebula after the Sun had condensed. These vortices produced planetesimals that grew to form planets by accretion. Unfortunately, as Chandrasekhar and Kuiper showed, the vortices would not be stable enough to allow condensation to take place. Kuiper then produced his own theory, as did Safronov and others, with the common theme of planetesimals merging to form planets, but none was fully satisfactory.

BIBLIOGRAPHY

Ball, R. (1897). *The story of the heavens.* Plate XVIII.
Hoskin, M. (Ed.), (1997). *Cambridge illustrated history of astronomy.* Cambridge, England: Cambridge Univ. Press.

Hufbauer, K. (1993). *Exploring the sun; solar science since Galileo.* Baltimore: Johns Hopkins Univ. Press.

Koestler, A. (1990). *The sleepwalkers: A history of man's changing vision of the universe.* Penguin Books.

Ledger, E. (1882). *The Sun: Its planets and their satellites.* Plate IX.

Leverington, D. (2003). *Babylon to Voyager and beyond: A history of planetary astronomy.* Cambridge, England: Cambridge Univ. Press.

Newcomb, S. (1898). *Popular astronomy.* p. 403.

North, J. (1995). *The Norton history of astronomy and cosmology.* New York: Norton.

Pannekoek, A. (1961). *A history of astronomy.* New York: Interscience (Dover reprint 1989).

Taton, R., & Wilson, C. (Eds.), (1989). *The General History of Astronomy: Vol. 2. Planetary Astronomy from the Renaissance to the Rise of Astrophysics: Part A, Tycho Brahe to Newton.* Cambridge, England: Cambridge Univ. Press.

Taton, R., & Wilson, C. (Eds.), (1995). *The General History of Astronomy: Vol. 2. Planetary Astronomy from the Renaissance to the Rise of Astrophysics: Part B, The Eighteenth and Nineteenth Centuries.* Cambridge, England: Cambridge Univ. Press.

Chapter 48

X-rays in the Solar System

Anil Bhardwaj
Space Physics Laboratory, Vikram Sarabhai Space Centre, Trivandrum, Kerala, India

Carey M. Lisse
Applied Physics Laboratory, Johns Hopkins University, Laurel, Maryland

Konrad Dennerl
Max-Planck-Institut für extraterrestrische Physik, Garching, Germany

Chapter Outline

1. Introduction — 1019
2. Earth — 1020
   2.1. Auroral Emissions — 1020
   2.2. Nonauroral Emissions — 1022
   2.3. Geocoronal Emissions — 1022
3. The Moon — 1023
4. Mercury — 1025
5. Venus — 1025
6. Mars — 1027
7. Jupiter — 1029
   7.1. Auroral Emission — 1029
   7.2. Nonauroral (Disk) Emission — 1031
8. Galilean Satellites — 1031
9. Io Plasma Torus — 1032
10. Saturn — 1032
11. Rings of Saturn — 1033
12. Comets — 1034
    12.1. Spatial Morphology — 1035
    12.2. X-ray Luminosity — 1037
    12.3. Temporal Variation — 1037
    12.4. Energy Spectrum — 1038
    12.5. Summary — 1038
13. Asteroids — 1039
14. Heliosphere — 1040
15. Summary — 1041
Acknowledgments — 1045
Bibliography — 1045

1. INTRODUCTION

The usually defined range of X-ray photons spans $\sim 0.1-100$ keV. Photons in the lower (<5 keV) end of this energy range are termed soft X-rays. In space, X-ray emission is generally associated with high-temperature phenomena, such as hot plasmas of 1 million to 100 million K and above in stellar coronae, accretion disks, and supernova shocks. However, in the solar system, X-rays have been observed from bodies that are much colder, $T < 1000$ K. This makes the field of planetary X-rays a very interesting discipline, where X-rays are produced from a wide variety of objects under a broad range of conditions.

The first planetary X-rays detected were terrestrial X-rays, discovered in the 1950s. The first attempt to detect X-rays from the Moon in 1962 failed, but it discovered the first extrasolar source, Scorpius X-1, which resulted in the birth of the field of X-ray astronomy. In the early 1970s, the *Apollo* 15 and 16 missions studied fluorescently scattered X-rays from the Moon. Such X-rays originate when energetic photons or particles remove an inner electron from atoms in the irradiated material. When the atom relaxes by filling the resulting gap with an outer shell electron, an X-ray photon with a characteristic energy is emitted. At low X-ray energies, this photon is usually produced by the transition of an $n = 2$ to the $n = 1$ shell electron, and is termed a Kα transition. A Kβ transition would be for an $n = 3$ to $n = 1$ electron, while an Lα photon would be produced by an $n = 3$ to $n = 2$ transition, etc.

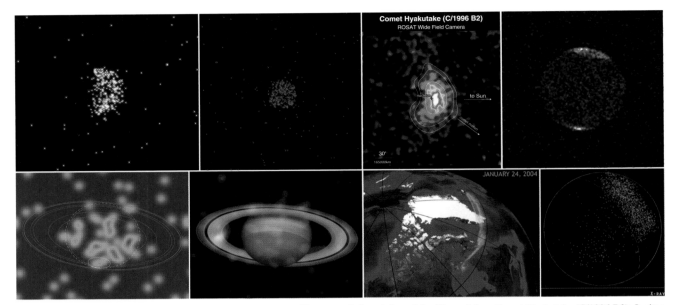

FIGURE 48.1 *Chandra* montage of solar system X-ray sources. Upper panel, from left to right: Venus, Mars, Comet Hyakutake (C/1996 B2), Jupiter (aurora + reflected disk emissions). Bottom panel, from left to right: Saturn, Saturn rings, Earth (auroral emission), Moon (sunlit side emission).

When Kα X-rays are produced in an oxygen atom, it is called oxygen Kα or O-Kα photon.

Launch of the first X-ray satellite *Uhuru* in 1970 marked the beginning of satellite-based X-ray astronomy. The subsequently launched X-ray observatory *Einstein* discovered, after a long search, X-rays from Jupiter in 1979. Before 1990, the three objects known to emit X-rays (in addition to the Sun) were Earth, Moon, and Jupiter. In 1996, *ROSAT* (*Röntgensatellit*) made an important contribution to the field of solar system X-rays by discovering X-ray emissions from comets. This discovery revolutionized the field of solar system X-rays and highlighted the importance of the solar wind charge exchange (SWCX) mechanism in the production of X-rays in the solar system, which will be discussed in this chapter in various sections.

Today, the field of solar system X-rays is very dynamic and in the forefront of new research. During the past one decade or so, our knowledge about the X-ray emission from bodies within the solar system has significantly improved. The advent of higher resolution X-ray spectroscopy with the *Chandra* and X-Ray Multi Mirror Mission (*XMM*)-*Newton* X-ray observatories, followed by the next-generation *SWIFT* and *Suzaku* observatories has been of great benefit in advancing the field of planetary X-ray astronomy. Several new solar system objects are now known to shine in the X-ray (Figure 48.1). At Venus, Earth, the Moon, Mars, Jupiter, Saturn (including its rings), and asteroids, scattered solar X-rays have been observed. The first soft X-ray observation of Earth's aurora by *Chandra* showed that it is highly variable, and the Jovian aurora is a fascinating puzzle that is just beginning to yield its secrets. The X-ray emission from comets, the exospheres of Venus, Earth, and Mars, and the heliosphere are all largely driven by charge exchange between highly charged minor (heavy) ions in the solar wind and gaseous neutral species.

This chapter surveys the current understanding of X-ray emission from the solar system bodies. We start our survey locally, at the Earth, move to the Moon and the nearby terrestrial planets, and then venture out to the giant planets and their moons and rings. Next, we move to the small bodies, comets and asteroids, found between the planets, and finally we study the X-ray emission from the heliosphere surrounding the whole solar system and possibilities of X-rays from extrasolar systems. An overview is provided on the main source mechanisms of X-ray production from each object. For further details, readers are referred to the bibliography provided at the end of the chapter and references therein.

2. EARTH

2.1. Auroral Emissions

Precipitation of energetic charged particles from the magnetosphere into Earth's auroral upper atmosphere leads to ionization, excitation, dissociation, and heating of the neutral atmospheric gas. Deceleration, or braking of precipitating particles during their interaction with atom and molecules in the atmosphere, results in the production of a continuous spectrum of X-ray photons, called **bremsstrahlung**. This is the main X-ray production mechanism in the Earth's auroral zones, for energies above ~ 3 keV; therefore, the X-ray spectrum of the aurora has been found to be very useful in studying the characteristics of energetic electron precipitation. In addition, particles precipitating into the Earth's upper atmosphere give rise to discrete atomic emission lines in the X-ray range. The characteristic inner-shell line emissions for the main species of the Earth's atmosphere are all in the low-energy range (nitrogen Kα at 0.393 keV, oxygen Kα at 0.524 keV, argon Kα at 2.958 keV, and Kβ at 3.191 keV). Very few X-ray observations have been made at energies at which these lines are emitted.

While charged particles spiral around and travel along the magnetic field lines of the Earth, the majority of the X-ray photons in Earth's aurora are directed normal to the field, with a preferential direction toward the Earth at higher energies. Downward-propagating X-rays cause additional ionization and excitation in the atmosphere below the altitude where the precipitating particles have their peak energy deposition. The fraction of the X-ray emission that is moving away from the ground can be studied using satellite-based imagers (e.g. AXIS on *Upper Atmosphere Research Satellite (UARS)* and PIXIE on the *POLAR* spacecraft).

Auroral X-ray bremsstrahlung has been observed from balloons and rockets since the 1960s and from spacecraft since the 1970s. Because of absorption of the low-energy X-rays propagating from the production altitude ($\sim$100 km) down to balloon altitudes (35–40 km), such measurements were limited to >20 keV X-rays. Nevertheless, these early omnidirectional measurements of X-rays revealed detailed information of temporal structures from slowly varying bay events to fast pulsations and microbursts.

The PIXIE instrument aboard *POLAR* is the first X-ray detector that provides true two-dimensional global X-ray images at energies >3 keV. In Figure 48.2 two images taken by PIXIE in two different energy bands are presented. The auroral X-ray zone can be clearly seen. Data from the PIXIE camera have shown that the X-ray bremsstrahlung intensity statistically peaks at midnight, is significant in the morning sector, and has a minimum in the early dusk sector. During solar substorms, X-ray imaging shows that the energetic electron precipitation brightens up in the midnight sector and has a prolonged and delayed maximum in the morning sector due to the scattering of magnetic-drifting electrons and shows an evolution significantly different than that when viewed in the ultraviolet (UV) emissions.

During the onset/expansion phase of a typical substorm, the electron energy deposition power is about 60–90 GW, which produces 10–30 MW of bremsstrahlung X-rays. By combining the results of PIXIE with the UV imager aboard *POLAR*, it has been possible to derive the energy distribution of precipitating electrons in the 0.1–100 keV range with a time resolution of about 5 min (see Figure 48.2). Because these energy spectra cover the entire energy range important for the electrodynamics of the ionosphere, important parameters like the Hall and Pedersen conductivities and the amount of **Joule heating** can be determined on a global scale with larger certainties than parameterized models can do. Electron energy deposition estimated from global X-ray imaging also gives valuable information on how the constituents of the upper atmosphere, like NO_x, is modified by energetic electron precipitation.

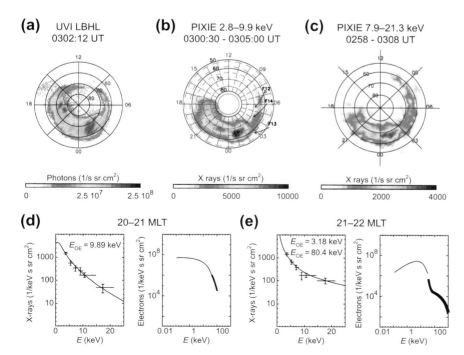

FIGURE 48.2 Earth's aurora. *POLAR* satellite observation on July 31, 1987. (a) UVI and (b, c) PIXIE images in two different energy bands. (d) Left: Measured X-ray energy spectrum. An estimated X-ray spectrum produced by a single exponential electron spectrum with e-folding energy 9.89 keV is shown to be the best fit to the measurements. Right: The electron spectrum derived from UVI (thin line) and PIXIE (thick line). Both plots are averages within a box within 20–21 magnetic local time (MLT) and 64–70° magnetic latitude. (e) Same as (d) but within 21–22 MLT, where a double exponential electron spectrum is shown to be the best fit to the X-ray measurements. *From Østgaard et al. (2001).*

Limb scans of the nighttime Earth at low- to midlatitude by the X-ray astronomy satellite High Energy Astronomy Observatory-1 (*HEAO-1*) in 1977, in the energy range 0.15–3 keV, showed clear evidence of the Kα lines for nitrogen and oxygen sitting on top of the bremsstrahlung spectrum. The High-Resolution Camera (HRC-I) aboard the *Chandra* X-ray Observatory imaged the northern auroral regions of the Earth in the 0.1- to 10-keV X-ray range at 10 epochs (each ~20 min duration) between December 2003 and April 2004. These first soft X-ray observations of Earth's aurora (see Figure 48.3) showed that it is highly variable (intense arcs, multiple arcs, diffuse patches, and at times absent). Also, one of the observations showed an isolated blob of emission near the expected cusp location. Modeling of the observed soft X-ray emissions suggests that it is a combination of bremsstrahlung and characteristic K-shell line emissions of nitrogen and oxygen in the atmosphere produced by electrons. In the soft X-ray energy range of 0.1–2 keV, these line emissions are ~5 times more intense than the X-ray bremsstrahlung.

2.2. Nonauroral Emissions

The nonauroral X-ray emission above 2 keV from the Earth is almost completely negligible except for brief periods during major solar flares (Figure 48.4). However, at energies below 2 keV, soft X-rays from the sunlit Earth's atmosphere have been observed even during quiet (nonflaring) Sun conditions. The two primary mechanisms for the production of X-rays from the sunlit atmosphere are: (1) Thomson (coherent) scattering of solar X-rays from the electrons in the atomic and molecular constituents of the atmosphere and (2) the absorption of incident solar X-rays followed by the resonance fluorescence emission of characteristic K lines of nitrogen, oxygen, and argon. During flares, solar X-rays light up the sunlit side of the Earth by Thomson and fluorescent scattering (Figure 48.4); the X-ray brightness can be comparable to that of a moderate aurora.

Around 1994, the Compton Gamma Ray Observatory (*CGRO*) satellite detected a new type of X-ray source from the Earth. These are very short-lived (1 ms) X-ray and γ-ray bursts (~25 keV–1 MeV) from the atmosphere above thunderstorms, whose occurrence is also supported by the more recent Reuven Ramaty High Energy Solar Spectroscopic Imager (*RHESSI*) observations. It has been suggested that these emissions are bremsstrahlung from upward-propagating, relativistic (megaelectronvolt) electrons generated in a runaway electron discharge process above thunderclouds by the transient electric field following a positive cloud-to-ground lightning event.

2.3. Geocoronal Emissions

In the Earth's exosphere (geocorona), SWCX with neutrals can produce X-rays. This process is now understood as a contribution to the soft X-ray background and to its long-term enhancements (LTEs) seen in the *ROSAT* all-sky survey. The LTEs in the *ROSAT* all-sky survey data are well correlated with the solar wind proton flux, suggesting that SWCX with H in the geocorona is the source of the LTEs. *Chandra* observations of the Moon are particularly interesting because they have cleanly separated

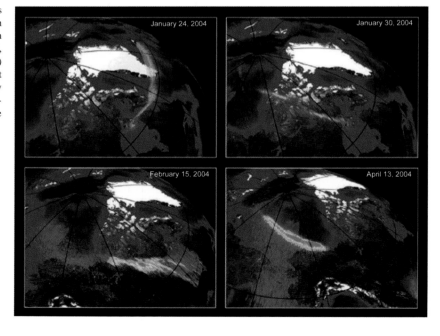

FIGURE 48.3 Four Earth's aurora X-ray images (shown on the same brightness scale) of the north polar regions obtained by *Chandra* HRC-I on different days (marked at the top of each image), showing large variability in soft (0.1–10 keV) X-ray emissions from Earth's aurora. The bright arcs in these *Chandra* images show low-energy X-rays generated during auroral activity. The images are superimposed on a simulated image of the Earth. *From Bhardwaj et al. (2007b).*

Chapter | 48 X-rays in the Solar System

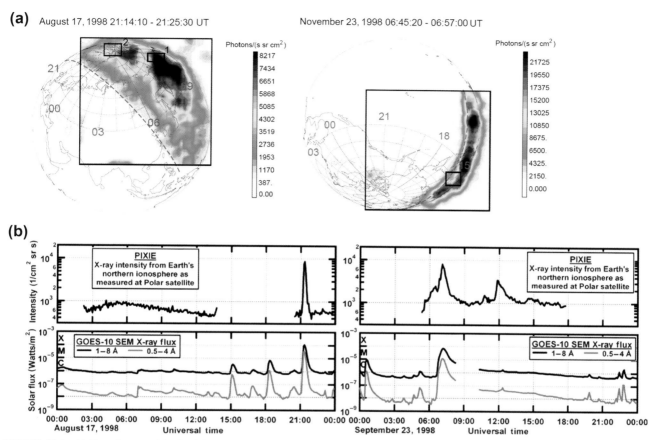

FIGURE 48.4 (a) X-ray images of the Earth from the *POLAR* PIXIE instrument for the energy range 2.9–10.1 keV obtained on August 17, 1998, (left) and November 23, 1998, (right), showing the dayside X-rays during a solar X-ray flare. The grid in the picture is in corrected geomagnetic coordinates, and the numbers shown in red are magnetic local times. The terminator at the surface of the Earth is shown as a red dashed line. *From Petrinec et al. (2000).* (b) Left: a sudden increase and subsequent decrease in X-ray intensity were observed on August 17, 1998, shortly after 21 UT while the PIXIE instrument was observing the Earth's northern hemisphere. This spike in X-ray intensity coincided with an X1 solar X-ray flare, as measured by *GOES-10*. Right: the occurrence of an M7 solar flare at ~07 Universal Time (UT) resulted in a sudden increase and decrease in X-ray flux from the Earth on September 23, 1998. A later increase in the X-ray emissions from the Earth's ionosphere (shortly before 12 UT) was due to increased auroral activity. *From Bhardwaj et al. (2009).*

geocoronal from heliospheric charge exchange emissions and all other contributors to the soft X-ray background and have shown correlation of geocoronal X-rays with the flux of highly charged oxygen ions in the solar wind. *XMM-Newton* and *Suzaku* have observed correlations of the X-ray intensity from the Earth's vicinity with the solar wind flux on timescales of about half a day, and also short-term (~10 min) variations. The correlated variability of X-ray intensity with solar wind, and the lack of correlation with solar X-rays, suggests that the production of SWCX-induced X-rays is taking place in the Earth's magnetosheath.

3. THE MOON

X-ray emissions from the Earth's nearest planetary body, the Moon (see also "The Moon"), have been studied in two ways: close-up from lunar orbiters (e.g. *Apollo 15* and *16*, *Clementine*, *SMART-1*, *Kaguya*, and *Chandrayaan-1*), and more distantly from Earth-orbiting X-ray astronomy telescopes (e.g. *ROSAT* and *Chandra*). Lunar X-rays result mainly from fluorescence of solar X-rays from the surface, in addition to a low level of scattered solar radiation. Thus, X-ray fluorescence studies provide an excellent way to determine the elemental composition of the lunar surface by remote sensing, since at X-ray wavelengths the optical properties of the surface are dominated by its elemental abundances. Elemental abundance maps produced by X-ray spectrometers (XRSs) on the *Apollo 15* and *16* orbiters were limited to the equatorial regions but succeeded in finding geochemically interesting variations in the relative abundances of Al, Mg, and Si, such as the enhancement of Al/Si in the lunar highlands relative to the mare. The D-Compact X-ray Spectrometer (D-CIXS) instrument on

SMART-1 has obtained abundances of Al, Si, Fe, and even Ca at 50-km resolution from a 300-km altitude orbit about the Moon. XRSs capable of higher spatial and spectral resolution have been flown on *Kaguya* and *Chandrayaan-1*. The X-ray charge-coupled devices on *Kaguya* suffered severe radiation damage en route to the Moon and could not perform well. C1XS on *Chandrayaan-1* provided new data sets on composition of Mg, Al, Si, Ca, and Fe at scales of 50 km, which indicates regions that may be richer in the sodic variety of plagioclase than previously thought. These recent experiments also suggest that particle-induced X-ray fluorescence could also contribute to the signal. Future X-ray experiments such as CLASS on *Chandrayaan-2* aim at global elemental mapping with better spatial resolution (~12–25 km) and wider elemental overage (including direct detection of sodium) and will use refined

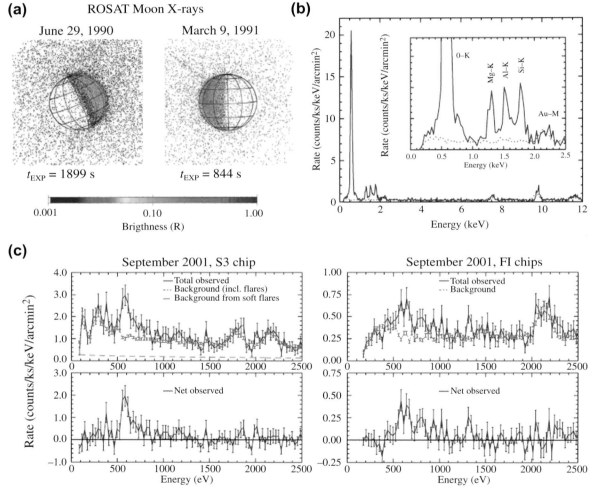

FIGURE 48.5 The Moon. (a) *ROSAT* soft X-ray (0.1–2 keV) images of the Moon at first (left side) and last (right side) quarter. The dayside lunar emissions are thought to be primarily reflected and fluoresced sunlight, while the faint nightside emissions are foreground due to charge exchange of solar wind heavy ions with H atoms in the Earth's exosphere. The brightness scale in R assumes an average effective area of 100 cm^2 for the ROSAT Position Sensitive Proportional Counter (PSPC) over the lunar spectrum. *From Bhardwaj et al. (2007a).* (b) *Chandra* spectrum of the bright side of the Moon. The green dotted curve is the detector background. K-shell fluorescence lines from O, Mg, Al, and Si are shifted up by 50 eV from their true values because of residual optical leak effects. Features at 2.2, 7.5, and 9.7 keV are intrinsic to the detector. *From Wargelin et al. (2004).* (c) Observed and background-subtracted spectra from the September 2001 *Chandra* observation of the dark side of the Moon, with 29-eV binning. The left panel is from the back-illuminated Advanced CCD Imaging Spectrometer - Spectroscopic Array 3 (ACIS-S3) CCD, while the right panel shows the spectrum obtained with the front-illuminated (FI) ACIS-I Advanced CCD Imaging Spectrometer - Imaging Array (ACIS-I) CCDs, which have higher spectral resolution, but lower sensitivity. Oxygen emission from charge exchange is clearly seen in both spectra, and the energy resolution in the FI chips is sufficient that emission from M-L shell transitions of O^{7+} ions is largely resolved from L-K transitions of O^{6+} ions. Emission lines from transitions of what is likely Mg-Kα around 1340 eV, Al-Kα at about 1550 keV, and Si-Kα at approximately 1780 keV are also apparent in the FI spectrum. *See also Ewing et al. (2013).*

methodologies to take into account signals from PIXE, enabling observations even during lunar night. These global maps will be complemented by rovers on the planned missions *Selene-2* as well as *Chandrayaan-2* which are expected to carry alpha particle spectrometers to study local surface chemistry through in situ quantitative analysis.

Early lunar X-ray observations from Earth orbit were made with *ROSAT*. A marginal detection by the Advanced Satellite for Cosmology and Astrophysics (*ASCA*) is also reported. Figure 48.5(a) shows the *ROSAT* images of the Moon; the right image is data from a lunar occultation of the bright X-ray source GX5-1. The power of the reflected and fluoresced X-rays observed by *ROSAT* in the 0.1–2 keV range coming from the sunlit surface was determined to be only 73 kW. The faint but distinct lunar nightside emissions (100 times less bright than the dayside emissions) were originally interpreted as being produced by bremsstrahlung of solar wind electrons of several hundred electronvolts impacting the nightside of the Moon on its leading hemisphere. However, this was before the GX5-1 data were acquired, which clearly show lunar nightside X-rays from the early trailing hemisphere as well. A new, much better and accepted explanation is that the heavy ions in the solar wind charge exchanges (SCWX) with geocoronal and interstellar H atoms that lie between the Earth and Moon result in foreground X-ray emissions between *ROSAT* and the Moon's dark side. This was confirmed by *Chandra* Advanced CCD Imaging Spectrometer (ACIS) observations in 2001 (see Figure 48.5(c)).

The July 2001 *Chandra* observations also provide the first remote measurements that clearly resolve discrete K-shell fluorescence lines of O, Mg, Al, and Si on the sunlit side of the Moon (see Figure 48.5(b)). The observed oxygen Kα (O–Kα for short; which means K-shell emissions from oxygen (O) atoms) line photons correspond to a flux of 3.8×10^{-5} photons/s/cm^2/arcmin^2 (3.2×10^{-14} erg/s/cm^2/arcmin^2). The Mg–Kα, Al–Kα, and Si–Kα lines each had roughly 10% as many counts and 3% as much flux as O–Kα line, but statistics were inadequate to draw any conclusions regarding differences in element abundance ratios between highlands and maria. Later *Chandra* observations of the Moon used the photon counting, high spatial resolution HRC-I imager to look for **albedo** variations due to elemental composition differences between **highlands** and **maria**. The observed albedo contrast was noticeable, but very slight, making remote elemental mapping difficult.

4. MERCURY

Being too close to the Sun, Mercury (see also "Mercury") cannot be observed by any X-ray observatory orbiting around the Earth. Launched on August 3, 2004, the *MESSENGER* (MErcury Surface, Space ENvironment, GEochemistry, and Ranging) spacecraft conducted its first flyby of Mercury on January 14, 2008, followed by two subsequent encounters on October 6, 2008, and September 29, 2009, prior to Mercury orbit insertion. On March 18, 2011, it became the first probe to orbit the planet Mercury. The XRS onboard *MESSENGER* has observed X-ray fluorescence emission from Mercury—thereby providing important information on the elemental composition of its surface. The XRS spectra have revealed Mercury's surface to differ in composition from those of other terrestrial planets and the Moon.

The XRS has also observed electron-induced X-ray emission from the dark side of Mercury. These X-ray fluorescence emissions are produced by interaction of $\sim$1- to 10-keV electrons with Mercury's surface and the abundance results derived from this technique for Mg, Si, and Al are found to be consistent with those derived from solar-induced X-ray fluorescence. In fact, electron-induced X-ray fluorescence has been used to derive the spectrum of electrons precipitating onto the Mercury surface, which is in agreement with the *MESSENGER*'s Energetic Particle Spectrometer (EPS)-measured spectrum. This showed that both the dayside and the nightside of an airless planetary body can be mapped by X-ray fluorescence—while on the dayside solar photons produce fluorescence and on the nightside its electrons. The XRS has also been used to study astrophysical objects during *MESSENGER*'s flybys of Mercury.

In future, the Mercury Imaging X-ray Spectrometer (MIXS) aboard the European Space Agency–Japan Aerospace Exploration Agency *BepiColombo* mission will help further investigate Mercury's X-rays and look for even solar wind-induced charge exchange emission, since the strength of solar wind flux is very high at Mercury.

5. VENUS

The first X-ray observation of Venus was obtained by *Chandra* in January 2001. It was expected that Venus would be an X-ray source due to two processes: (1) charge exchange interactions between highly charged ions in the solar wind and the Venusian exosphere and (2) scattering of solar X-rays in the Venusian atmosphere. The predicted X-ray luminosities were $\sim$0.1–1.5 MW for the first process, and $\sim$35 MW for the second one, with an uncertainty factor of about 2. The *Chandra* observation of 2001 consisted of two parts: grating spectroscopy with Low Energy Transmission Grating (LETG)/Advanced CCD Imaging Spectrometer - Spectroscopic Array (ACIS-S) and direct imaging with ACIS-I. This combination yielded data of high spatial, spectral, and temporal resolution. Venus was clearly detected as a half-lit crescent, exhibiting

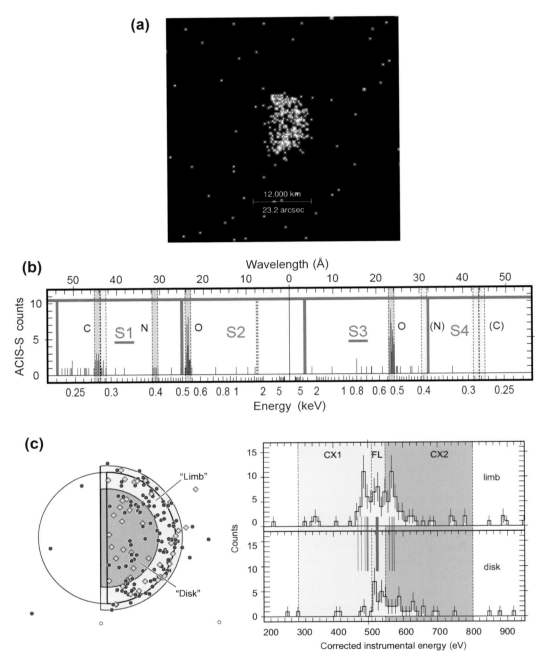

FIGURE 48.6 Venus. (a) First X-ray image of Venus, obtained with *Chandra* ACIS-I on January 13, 2001. The X-rays result mainly from fluorescent scattering of solar X-rays on C and O in the upper Venus atmosphere, at heights of 120–140 km. In contrast to the Moon, the X-ray image of Venus shows evidence for brightening on the sunward limb. This is caused by the fact that scattering takes place on an atmosphere and not on a solid surface. (b) Spectral scan. Scales are given in kiloelectronvolts and angstroms. The observed C, N, and O fluorescent emission lines are enclosed by dashed lines; the width of these intervals matches the size of the Venus crescent (22.8″). *From Dennerl, Burwitz, Englhauser, Lisse, and Wolk (2002).* (c) Spatial and spectral distribution of X-ray photons from Venus, obtained on March 27, 2006 with Chandra. Left: Distribution of the X-ray photons in the energy range 0.3–0.8 keV. Photons in the fluorescence band FL are marked with bright diamonds, and those in the charge exchange bands CX1 and CX2 with dark circles. The extraction regions for the limb and disk spectra are superimposed on light and dark gray, respectively; the circle indicates the geometric size of Venus. Right: X-ray spectra for the limb (top) and the disk region (bottom), with the energy bands CX1, FL, and CX2 marked. The spectrum of the limb region is dominated by two emission lines in the CX1 and CX2 bands. These lines are almost absent in the disk spectrum, which is dominated by emission in the FL band. The energies of characteristic lines expected for charge exchange and fluorescence are marked by vertical lines. *From Dennerl (2008).*

considerable brightening on the sunward limb (Figure 48.6); the LETG/ACIS-S data showed that the spectrum was dominated by K-shell fluorescence of oxygen and carbon (O—Kα and C—Kα), and both instruments indicated temporal variability of the X-ray flux. An average luminosity of 55 MW was found, which agreed well with the theoretical predictions for scattered solar X-rays. In addition to the C—Kα and O—Kα emission at 0.28 and 0.53 keV, respectively, the LETG/ACIS-S spectrum also showed evidence for N—Kα emission at 0.40 keV. An additional emission line was indicated at 0.29 keV, which might be the signature of the C 1s → π* transition in CO_2. The observational results are consistent with fluorescent scattering of solar X-rays by the majority species in the Venusian atmosphere, and no evidence of the 30 times weaker charge exchange interactions was found. Simulations showed that fluorescent scattering of solar X-rays is most efficient in the Venusian upper atmosphere at heights of ~120 km, where an optical depth of 1 is reached for incident X-rays with energy 0.2—0.9 keV.

The second *Chandra* observation of Venus in March 2006 showed clear signature of charge-exchanged X-rays from the exosphere (Figure 48.6), which was marginally detected again in the October 2007 *Chandra* observation. The bright emission feature is O^{6+} emission near 565 eV, which is also an important feature in X-ray spectra of comets and the Martian exosphere.

The appearance of Venus is different in optical light and in X-rays. The reason for this is that the optical light is reflected from clouds at a height of 50—70 km, while scattering of X-rays takes place at higher regions extending into the tenuous, optically thin parts of the thermosphere and exosphere. As a result, the Venusian sunlit hemisphere appears surrounded by an almost transparent luminous shell in X-rays, and Venus looks brightest at the limb because more luminous material is there. Because X-ray brightening depends sensitively on the density and chemical composition of the Venusian atmosphere, its precise measurement will provide direct information about the atmospheric structure in the thermosphere and exosphere. This opens up the possibility of using X-ray observations for monitoring the properties of these regions that are difficult to investigate by other means, as well as their response to solar activity.

6. MARS

The first X-rays from Mars were detected on July 4, 2001, with the ACIS-I detector onboard *Chandra*. In the *Chandra* observation, Mars showed up as an almost fully illuminated disk (Figure 48.7). An indication of limb brightening on the sunward side, accompanied by some fading on the opposite side, was observed. The observed morphology and X-ray luminosity of ~4 MW, about 10 times less than at Venus, was consistent with fluorescent scattering of solar X-rays in the upper Mars atmosphere. A single narrow emission line caused by K-shell fluorescence emission of oxygen dominated the X-ray spectrum.

Simulations suggest that scattering of solar X-rays is most efficient between 110 km (along the subsolar direction) and 136 km (along the terminator) above the Martian surface. This behavior is similar to that seen on Venus. No evidence for temporal variability or dust-related emission was found, which is in agreement with fluorescent scattering of solar X-rays as the dominant process responsible for Martian X-rays. A gradual decrease in the X-ray surface brightness between 1 and ~3 Mars radii is observed (see Figure 48.7). Within the limited statistical quality of the low flux observations, the spectrum of this region (halo) resembled that of comets, suggesting that they are caused by charge exchange interactions between highly charged heavy ions in the solar wind and neutrals in the Martian exosphere (corona). For the X-ray halo observed within 3 Mars radii, excluding Mars itself, the *Chandra* observation yielded a flux of about 1×10^{-14} erg/cm^2/s in the energy range 0.5—1.2 keV, corresponding to a luminosity of 0.5 ± 0.2 MW for isotropic emission, which agrees well with that expected theoretically for the SWCX mechanism.

The first *XMM-Newton* observation of Mars in November 2003 confirmed the presence of the Martian X-ray halo and allowed a detailed analysis of its spectral, spatial, and temporal properties to be made. High-resolution spectroscopy of the halo with *XMM-Newton* Reflection Grating Spectrometer (RGS) revealed the presence of numerous (~12) emission lines at the positions expected for deexcitation of highly ionized C, N, O, and Ne atoms (Figure 48.8). The three Kα lines of O VII emission were resolved and found to be dominated by a long-lived forbidden triple to singlet transition, as expected for charge exchange but not fluorescence or collisionally induced X-ray emission. This was the first definite detection of charge exchange-induced X-ray emission from the exosphere of another planet.

The *XMM-Newton* observation confirmed that the fluorescent scattering of solar X-rays from the Martian disk is clearly concentrated on the planet, and is directly correlated with the solar X-ray flux levels. On the other hand, the Martian X-ray halo was found to extend out to ~8 Mars radii, with pronounced morphological differences between individual ions and ionization states (Figure 48.8). The halo emission exhibited pronounced variability, but, as expected for solar wind interactions, the variability of the halo did not show any correlation with the solar X-ray flux. Mars was found to be dimmer in observations made during solar minimum, suggesting direct correlation with solar activity. *Suzaku* could not even detect Mars in X-rays in an observation in April 2008.

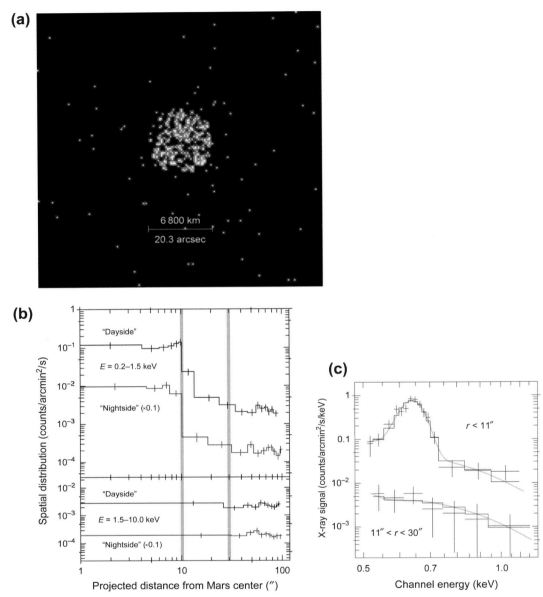

FIGURE 48.7 Mars. (a) First X-ray image of Mars, obtained with *Chandra* ACIS-I. The X-rays result mainly from fluorescent scattering of solar X-rays on C and O in the upper Mars atmosphere, at heights of 110–130 km, similar to Venus. The X-ray glow of the Martian exosphere is too faint to be directly visible in this image. *From Dennerl (2002).* (b) Spatial distribution of the photons around Mars in the soft ($E = 0.2-1.5$ keV) and hard ($E = 1.5-10.0$ keV) energy range, in terms of surface brightness along radial rings around Mars, separately for the dayside (offset along projected solar direction >0) and the nightside (offset <0); note, however, that the phase angle was only $18.2°$. For better clarity, the nightside histograms were shifted by one decade downward. The bin size was adaptively determined so that each bin contains at least 28 counts. The thick vertical lines enclose the region between 1 and 3 Mars radii. (c) X-ray spectra of Mars (top) and its X-ray halo (bottom). Crosses with 1σ error bars show the observed values; the model spectra, convolved with the detector response, are indicated by gray curves (unbinned) and by histograms (binned as the observed spectra). The spectrum of Mars itself is characterized by a single narrow emission line (this is most likely the O-Kα fluorescence line at 0.53 keV; the apparent displacement of the line energy is due to optical loading). At higher energies, the presence of an additional spectral component is indicated. The spectral shape of this component can be well modeled by the same 0.2-keV thermal bremsstrahlung emission which can be used as a "technical" proxy for characterizing the basic spectral properties of the X-ray halo in an instrument-independent way. *From Dennerl (2002).*

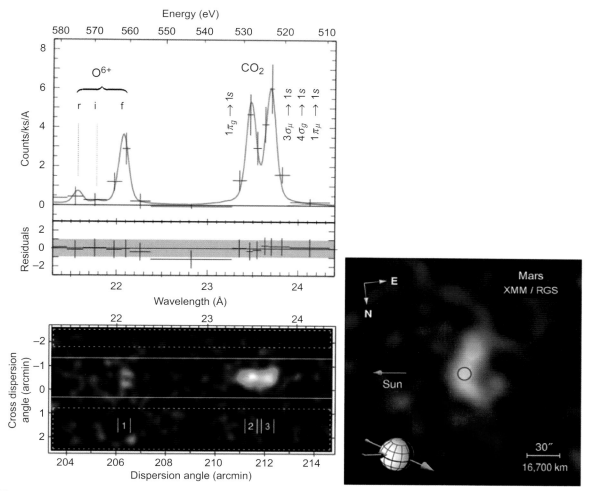

FIGURE 48.8 Mars. Imaging spectroscopy of Mars with *XMM-Newton's* RGS in November 2003 showing images of Mars and its halo in the individual emission lines of ionized oxygen and fluorescence of CO_2. (Top) RGS spectrum obtained from a 100″-wide area along the cross-dispersion direction, showing the region around the CO_2 doublet and the O VII multiplet. (Bottom) dispersed images in the same wavelength range as above. The apparent diameter of Mars during this observation was 12.2″. (Right) X-ray image of Mars in November 2003 with *XMM-Newton*/RGS in the emission lines of charge exchange (green-blue) and fluorescence of solar X-rays (orange). The black circle indicates the size of planet. *From Dennerl et al. (2006).*

7. JUPITER

7.1. Auroral Emission

Like the Earth, Jupiter emits X-rays both from its aurora and its sunlit disk. Jupiter's UV auroral emissions were first observed by the *International Ultraviolet Explorer* (IUE) and soon confirmed by the Voyager 1 ultraviolet spectrometer as it flew through the Jupiter system in 1979 (see Bhardwaj and Gladstone (2000) for review). The first detection of the X-ray emission from Jupiter was also made in 1979; the *Einstein* satellite detected X-rays in the 0.2–3.0 keV energy range from both poles of Jupiter, due to the aurora. Analogous to the processes on Earth, it was expected that Jupiter's X-rays might originate as bremsstrahlung by precipitating electrons. However, the power requirement for producing the observed emission with this mechanism (10^{15}–10^{16} W) is more than two orders of magnitude larger than the input auroral power available as derived from *Voyager* and *IUE* observations of the UV aurora. (The strong Jovian magnetic field excludes the bulk of the solar wind from penetrating close to Jupiter, and the solar wind at Jupiter at 5.2 AU is 27 times less dense than at the Earth at 1 AU.) Precipitating energetic sulfur and oxygen ions from the inner magnetosphere, with energies in the 0.3–4.0 MeV/nucleon range, were suggested as the source mechanism responsible for the production of X-rays on Jupiter. The heavy ions are thought to start as neutral SO and SO_2 molecules emitted by the volcanoes on Io into the Jovian magnetosphere, where they are dissociated and ionized by solar UV radiation, and then swept up into the huge dynamo created by Jupiter's rotating magnetic field (see also "Jupiter" and "Planetary Magnetospheres").

The ions eventually become channeled onto magnetic field lines terminating at Jupiter's poles, where they emit X-rays by first charge stripping to a highly ionized state, followed by charge exchange and excitation through collisions with H_2.

ROSAT's observations of Jupiter X-ray emissions supported this suggestion. The spatial resolution of these early observations was not adequate to distinguish whether the emissions were linked to source regions near the Io torus of Jupiter's magnetosphere (inner magnetosphere) or at larger radial distances from the planet. The advent of *the Chandra* and *XMM-Newton* X-ray observatories revolutionized our understanding of Jupiter's X-ray aurora. High-spatial-resolution (<1 arcsec) observations of Jupiter with *Chandra* in December 2000 (see Figure 48.9) revealed that most of Jupiter's northern auroral X-rays come from a "hot spot" located significantly poleward of the UV auroral zones (20–30 Jupiter's radius, R_J), and not at latitudes connected to the Io plasma torus (IPT) (inner magnetosphere). The hot spot is fixed at 60°–70° magnetic latitude and 160°–180° longitude (in "system III" coordinates rotating with the Jovian magnetosphere) and occurs in a region where anomalous infrared and UV emissions (the so-called flares) have also been observed. On the other hand, auroral X-rays from 70° to 80° S latitude spread almost halfway across the planet (starting at ~300° and spreading through 0° out to 120° longitude). The location of the auroral X-rays connects along magnetic field lines to regions in the Jovian magnetosphere well in excess of 30 Jovian radii from the planet, a region where there are insufficient S and O ions to account for the X-ray emission. Acceleration of energetic ions was invoked to increase the phase space distribution, but now the question was whether the acceleration involved outer magnetosphere heavy ions or solar wind heavy ions.

Surprisingly, *Chandra* observations also showed that X-rays from the Jovian aurora pulsate with a periodicity that is quite systematic (approximately 45-min period) at times (in December 2000) and irregular (20–70 min range) at other times (in February 2003). The 45-min periodicity is highly reminiscent of a class of Jupiter high-latitude radio emissions known as quasiperiodic radio bursts, which had been observed by Ulysses in conjunction with energetic electron acceleration in Jupiter's outer magnetosphere. During the 2003 *Chandra* observation of Jupiter, the Ulysses radio data did not show any strong 45-min quasi-periodic oscillations, although variability on timescales similar to that in X-rays was present. *Chandra* also found that X-rays from the northern and southern auroral regions are neither in phase nor in antiphase, but that the peaks in the south are shifted from those in the north by about 120° (i.e. one-third of a planetary rotation).

The *Chandra* and *XMM-Newton* spectral and spatial observations have now established that X-rays from Jupiter's aurora are basically of two types: (1) soft (~0.1–2 keV) and (2) hard (>2 keV) X-rays (Figure 48.9).

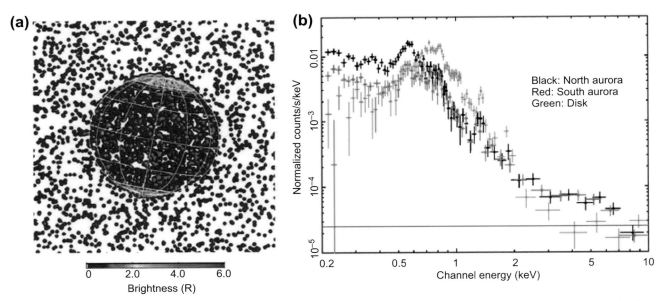

FIGURE 48.9 Jupiter. (a) Detailed X-ray morphology first obtained with *Chandra* HRC-I on December 18, 2000, showing bright X-ray emission from the polar "auroral" regions, indicating the high-latitude position of the emissions, and a uniform distribution from the low-latitude "disk" regions. *From Gladstone et al. (2002).* (b) Combined *XMM-Newton* European Photon Imaging Camera (EPIC) spectra from the November 2003 observation of Jupiter. Data points for the north and south aurorae are in black and red, respectively. In green is the spectrum of the low-latitude disk emission. Differences in spectral shape between auroral and disk spectra are clear. The presence of a high-energy component in the spectra of the aurorae is very evident, with a substantial excess relative to the disk emission extending to 7 keV. The horizontal blue line shows the estimated level of the EPIC particle background. *From Branduardi-Raymont et al. (2007).*

The soft X-rays are line emissions, which are consistent with high-charge states of precipitating heavy O and S ions from the Jovian magnetospheric that are accelerated to attain energies of >1 MeV/nucleon before impacting the Jovian upper atmosphere. There is no evidence for the role of solar wind ions due to the absence of signatures of carbon ions in the X-ray spectrum. Modeling studies which include a source of gaseous neutral atoms (Io's SiO_x and SO_x for Jupiter, Enceladus' H_2O for Saturn), solar UV photolysis and ionization (to S^+, Si^+, and O^+ for Jupiter and H^+ and O^+ for Saturn), followed by pickup, acceleration, and polar precipitation plus charge exchange of the ions at 0–2 keV energies also supports this. The higher (>2 keV) energy X-rays are basically bremsstrahlung from precipitating magnetospheric energetic electrons (see Figure 48.9) and originate from locations that spatially coincide with the Far Ultra Violet (FUV) auroral oval, suggesting that the source of both, X-rays and UV emissions, is the same. The variability on timescales of days suggests a link to changes in the energy distribution of the precipitating magnetospheric electrons and may be related to the solar activity at the time of observation.

7.2. Nonauroral (Disk) Emission

The existence of low-latitude "disk" X-ray emission from Jupiter was first recognized in *ROSAT* observations made in 1994. As for the inner planets, it was suggested that elastic scattering of solar X-rays by atmospheric neutrals (H_2 and He for Jupiter) and fluorescent scattering by the dominant multielectron atom, carbon, via carbon K-shell X-rays (mainly from methane (CH_4) molecules located below the Jovian homopause) were the sources of the disk X-rays.

A general decrease in the overall X-ray brightness of Jupiter observed by *ROSAT* over the years 1994–1996 was found to be coincident with a similar decay in solar activity index (solar 10.7 cm flux). A similar trend is seen in the data obtained by *Chandra* in 2000 and 2003; Jupiter's disk was about 50% dimmer in 2003 compared to that in 2000, which is consistent with variation in the solar activity. A *Chandra* observation in early 2008 during the *New Horizon* flyby also showed a dimmer Jupiter, consistent with a decrease in solar activity.

First direct evidence for temporal correlation between Jovian disk X-rays and solar X-rays was provided by *XMM-Newton* observations in November 2003, which demonstrated that day-to-day variations in disk X-rays of Jupiter are synchronized with variations in the solar X-ray flux, including a solar flare that had a matching feature in the Jovian disk X-ray light curve. *Chandra* observations of December 2000 and February 2003 also support this association between light curves of solar and planetary X-rays. However, there is an indication of higher X-ray counts from regions of low surface magnetic field in the *Chandra* data, suggesting the presence of some particle precipitation.

The higher spatial resolution observation by *Chandra* has shown that nonauroral disk X-rays are relatively more spatially uniform than the auroral X-rays (Figure 48.9). Unlike the $\sim 40 \pm 20$-min quasiperiodic oscillations seen in auroral X-ray emission, the disk emission does not show any systematic pulsations. There is a clear difference between the X-ray spectra from the disk and from the auroral region on Jupiter; the disk spectrum peaks at higher energies (0.7–0.8 keV) than that of the aurora emission (0.5–0.6 keV) and lacks the high-energy component (above ~ 3 keV) present in the latter (see Figure 48.9).

8. GALILEAN SATELLITES

The Jovian *Chandra* observations on November 25–26, 1999, and December 18, 2000, discovered the X-ray emission from the Galilean satellites (Figure 48.10). These satellites are very faint when observed from the Earth's orbit (by *Chandra*), and the detections of Io and Europa, although statistically very significant, were based on ~ 10 photons each! The energies of the detected X-ray events ranged between 300 and 1890 eV and appeared to show a clustering between 500 and 700 eV, suggestive of oxygen K-shell fluorescent emission. The estimated power of the X-ray emission was 2 MW for Io and 3 MW for Europa. There were also indications of X-ray emission from Ganymede. X-ray emission from Callisto seems likely at levels not too far below the *Chandra* sensitivity limit because the heavy ion fluxes of the Jovian magnetosphere are an order of magnitude lower than at Ganymede and Europa, respectively. Emissions from Io were also seen in a February 2003 *Chandra* observation, although at weaker levels.

The most plausible proposed emission mechanism is inner (K-shell) ionization of surface and near-surface atoms by incoming magnetospheric ions followed by prompt X-ray emission. Oxygen should be the dominant emitting atom either on a SiO_x (silicate) or SO_x (sulfur oxides) surface (Io) or on an icy one (the outer Galilean satellites). It is also the most common heavy ion in the Jovian magnetosphere. The extremely tenuous atmospheres of the satellites are transparent to X-ray photons with these energies, as well as to much of the energy range of the incoming ions. However, oxygen absorption in the soft X-ray is strong enough that the X-rays must originate within the top 10 μm of the surface in order to escape. Simple estimates suggest that excitation by incoming ions dominates over electrons and that the X-ray flux produced is within a factor of three of the measured flux. The detection of X-ray emission from the Galilean satellites thus provides a direct measure of the interactions of the

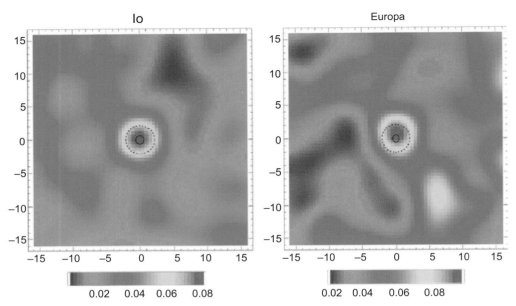

FIGURE 48.10 Galilean Moons. *Chandra* X-ray images of Io and Europa (0.25 keV < E < 2.0 keV) from November, 1999 observations. The axes are labeled in arcsec (1 arcsec ~ 3000 km) and the scale bar is in units of smoothed counts per image pixel (0.492 by 0.492 arcsec). The solid circle shows the size of the satellite (the radii of Io and Europa are 1821 km and 1560 km, respectively), and the dotted circle shows the size of the detect cell. *From Elsner et al. (2002).*

magnetosphere of Jupiter with the satellite surfaces. An intriguing possibility is placement of an imaging XRS onboard a mission to the Jupiter system.

9. IO PLASMA TORUS

The IPT is known to emit at extreme ultraviolet (EUV) energies and below, but it was a surprise when *Chandra* discovered that it was also a soft X-ray source. The 1999 Jovian *Chandra* observations detected a faint diffuse source of soft X-rays from the region of the IPT. The 2000 *Chandra* image, obtained with the HRC-I camera (Figure 48.11), exhibited a dawn-to-dusk asymmetry similar to that seen in the EUV. Figure 48.11 shows the background-subtracted *Chandra*/ACIS-S IPT spectrum for November 25–26, 1999. This spectrum shows evidence for line emission centered on 574 eV (very near a strong O VII line), together with a very steep continuum spectrum at the softest X-ray energies. Although formed from the same source, the spectrum is different from that of the Jovian aurora because the energies, charge states, and velocities of the ions in the torus are much lower—the bulk ions have not yet been highly accelerated. There could be contributions from other charge states because current plasma torus models consist mostly of ions with low charge states, consistent with photoionization and ion-neutral charge exchange in a low-density plasma and neutral gas environment. The 250- to 1000-eV energy flux at the telescope aperture was 2.4×10^{-14} erg/cm^2/s, corresponding to a luminosity of 0.12 GW. Although bremsstrahlung from nonthermal electrons might account for a significant fraction of the continuum X-rays, the physical origin of the observed IPT X-ray emission is not yet fully understood. The 2003 Jovian *Chandra* observations also detected X-ray emission from the IPT, although at a fainter level than in 1999 or 2000. The morphology exhibited the familiar dawn-to-dusk asymmetry.

10. SATURN

The production of X-rays at Saturn (see also "Saturn") was expected because, like the Earth and Jupiter, Saturn was known to possess a magnetosphere with energetic electrons and ions within it; however, early attempts to detect X-ray emission from Saturn with Einstein in December 1979 and with ROSAT in April 1992 were negative and marginal, respectively. Saturnian X-rays were unambiguously observed by *XMM-Newton* in October 2002 and by the *Chandra* X-ray Observatory in April 2003. In January 2004, Saturn was again observed with *Chandra* ACIS-S in two exposures, one on 20 January and the other on 26–27 January, with each observation lasting for about one full Saturn rotation. The X-ray power emitted from Saturn's disk is roughly one-fourth of that from Jupiter's disk, which is consistent with Saturn being twice as far from the Sun and the Earth as Jupiter.

Chapter | 48 X-rays in the Solar System

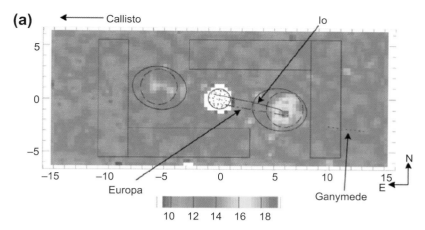
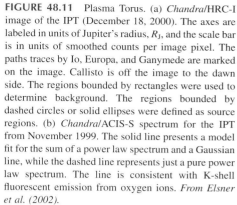
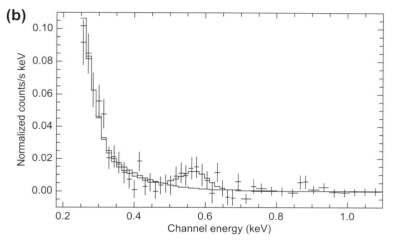

FIGURE 48.11 Plasma Torus. (a) *Chandra*/HRC-I image of the IPT (December 18, 2000). The axes are labeled in units of Jupiter's radius, R_J, and the scale bar is in units of smoothed counts per image pixel. The paths traces by Io, Europa, and Ganymede are marked on the image. Callisto is off the image to the dawn side. The regions bounded by rectangles were used to determine background. The regions bounded by dashed circles or solid ellipses were defined as source regions. (b) *Chandra*/ACIS-S spectrum for the IPT from November 1999. The solid line presents a model fit for the sum of a power law spectrum and a Gaussian line, while the dashed line represents just a pure power law spectrum. The line is consistent with K-shell fluorescent emission from oxygen ions. *From Elsner et al. (2002).*

The January 2004, *Chandra* observation showed (Figure 48.12) that X-rays from Saturn are highly variable—a factor of two–four variability in brightness over 1 week. The bright X-rays from Saturn's south polar cap on January 20 (see Figure 48.12, left panel), which are not evident in the January 26 observation (see Figure 48.12, right panel) and in earlier *Chandra* observations are an extension of the disk X-ray emission of Saturn. No evidence of auroral X-rays from Saturn has been found so far, which could be due to the limited sensitivity of current X-ray observatories.

As is the case for Jupiter's disk, X-ray emission from Saturn is due to the scattering of the incident solar X-ray flux. An X-ray flare has been detected from the non-auroral disk of Saturn during the *Chandra* observation on January 20, 2004, which, taking light travel time into account, coincided with an M6-class flare emanating from a sunspot that was clearly visible from both Saturn and Earth. This was the first direct evidence suggesting that Saturn's disk X-ray emission is principally controlled by processes happening on the Sun. Further, a good correlation has been observed between Saturn X-rays and F10.7 solar activity (see also "The Sun") index, suggesting a solar connection.

The spectrum of X-rays from Saturn's disk is very similar to that from Jupiter's disk (Figure 48.12).

11. RINGS OF SATURN

The rings of Saturn (see also "Planetary Rings"), known to be made of mostly water (H_2O) ice, are one of the most fascinating objects in our solar system. The discovery of X-rays from the rings of Saturn was made with *Chandra* ACIS-S observations in January 2004 and April 2003. X-rays from the rings are dominated by emission in a narrow (~ 130 eV wide) energy band of 0.49–0.62 keV (Figure 48.13). This band is centered on the oxygen Kα fluorescence line at 0.53 keV, suggesting that fluorescent scattering of solar X-rays from oxygen atoms in the surface of H_2O icy ring material is the likely source mechanism for ring X-rays. The X-ray power emitted by the rings in the 0.49- to 0.62-keV band on January 20,

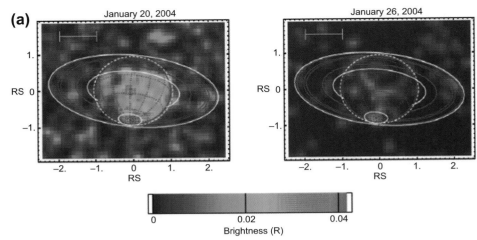
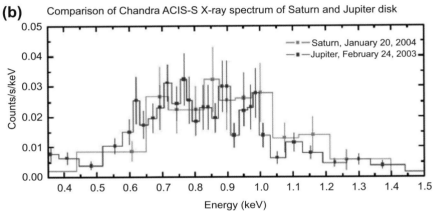

FIGURE 48.12 Saturn. (a) *Chandra* ACIS X-ray 0.24- to 2.0-keV images of Saturn on January 20 and 26, 2004. Each continuous observation lasted for about one full Saturn rotation. The horizontal and vertical axes are in units of Saturn's equatorial radius. The white scale bar in the upper left of each panel represents 10 arcsec. The two images, taken a week apart and shown on the same color scale, indicate substantial variability in Saturn's X-ray emission. *From Bhardwaj et al. (2005a).* (b) Disk X-ray spectrum of Saturn (red curve) and Jupiter (blue curve). Values for the Saturn spectrum are plotted after multiplying by a factor of 5. *From Bhardwaj (2006).*

2004, is 84 MW, which is about one-third of that emitted from the Saturn disk in the 0.24- to 2.0-keV band. The projected rings have about half the surface area of the Saturn disk, consistent with this ratio. The X-ray power emitted by the rings in the 0.49- to 0.62-keV band could vary from 30 to 150 MW depending on the observation period.

Figure 48.13 shows the X-ray image of the Saturnian system in January 2004 in the 0.49- to 0.62-keV band, the energy range where X-rays from the rings were unambiguously detected. The observations of January 2004 also suggested that, similar to Saturn's X-ray emission, the ring X-rays are highly variable—a factor of two to three variability in brightness over 1 week. There is an apparent asymmetry in X-ray emission from the east (morning) and west (evening) ansae (the apparent extremities of the rings, looking like two handles; see Figure 48.13(a)). However, when the *Chandra* ACIS-S data sets of January 2004 and April 2003 are combined, the evidence for asymmetry is not that strong. Recent study by *XMM-Newton* suggests no direct relationship between Saturn disk X-rays and ring X-rays: while disk X-rays follow solar activity, the X-rays from rings do not, suggesting the role of other processes in their production, like meteoric impact-induced spokes or lightning-induced electron beams.

12. COMETS

The *ROSAT* discovery of X-ray emission in 1996 from C/1996 B2 (Hyakutake) created a new class of cold, 10^2-10^3 K X-ray-emitting objects. Observations since 1996 have shown that the very soft ($E < 1$ keV) emission is due to an interaction between the solar wind and the comet's atmosphere (see also "Physics and Chemistry of Comets"), and that X-ray emission is a fundamental property of comets. Theoretical and observational work in the two decades since the discovery has demonstrated that charge exchange collision of highly charged heavy solar wind ions with cometary neutral species is the best explanation for the emission. In fact, of the solar system bodies with associated X-ray emission, comets are the best example of a nearly pure charge exchange emitting system, as their gravitationally unbound atmospheres (or **comae**), are tenuous and highly extended (typically 10^5-10^6 km in radius), unable to scatter

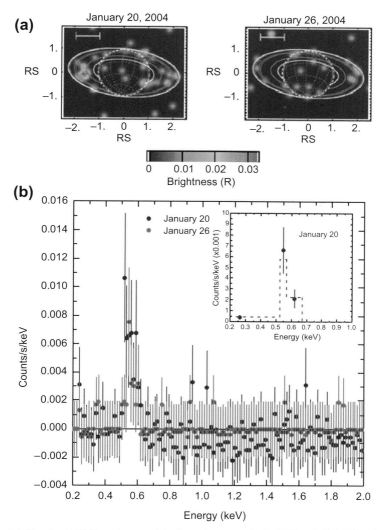

FIGURE 48.13 Saturn's rings. (a) *Chandra* ACIS X-ray images of the Saturnian system in the 0.49- to 0.62-keV band on January 20 and 26–27, 2004. The X-ray emission from the rings is clearly present in these restricted energy band images; the emission from the planet is relatively weak in this band (see Figure 48.12 (a) for an X-ray image of the Saturnian system in the 0.24- to 2.0-keV band). (b) Background-subtracted *Chandra* ACIS-S3-observed X-ray energy spectrum for Saturn's rings in the 0.2–2.0 keV range on January 20 and 26–27, 2004. The cluster of X-ray photons in the ∼0.49- to 0.62-keV band suggests the presence of the oxygen Kα line emission at 0.53 keV in the X-ray emission from the rings. The inset shows a Gaussian fit (peak energy = 0.55 keV, σ = 140 eV), indicated by the dashed line, to the ACIS-observed rings' spectrum on January 20 suggesting that X-ray emissions from the rings are predominantly oxygen Kα photons. *From Bhardwaj et al. (2005b).*

many solar X-rays but highly capable of intercepting a large amount of solar wind ions as they stream away from the Sun.

Recently, X-ray emission has also been detected from a comet (C/2011 W3) flying through the lower solar corona, with the solar X-ray imager X-Ray Telescope (XRT) onboard *Hinode*. This emission is thought to arise from the ionization of cometary material in the corona.

The observed characteristics of the emission can be organized into the following four categories: (1) spatial morphology, (2) X-ray luminosity, (3) temporal variation, and (4) energy spectrum. Each of the observed characteristics depends on the nature of the comet's coma and the solar wind it interacts with. We discuss the typical nature of each of these characteristics next.

12.1. Spatial Morphology

X-ray and EUV images of C/1996 B2 (Hyakutake) made by the *ROSAT* High Resolution Imager (HRI) and Wide Field Camera (WFC) look very similar (Figure 48.14). Except for C/1990 N1, 2P/Encke, 73P/SW-3B, and C/2011 W3 (see below), all EUV and X-ray images of comets have exhibited similar spatial morphologies. The emission is largely confined to the sunward side of the cometary

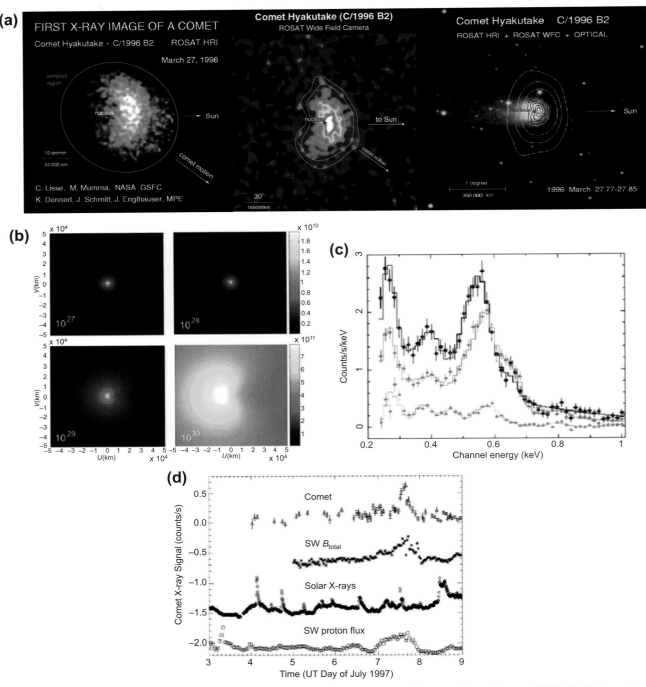

FIGURE 48.14 The rich behavior of X-ray emission seen from comets. (a) Cometary X-ray emission morphology. Images of C/1996 B2 (Hyakutake) on March 26–28, 1996 UT: *ROSAT* HRI 0.1–2.0 keV X-ray, *ROSAT* WFC 0.09–0.2 keV EUV, and visible light, showing a coma and tail, with the X-ray emission contours superimposed. The Sun is toward the right, the plus signs mark the position of the nucleus, and the orbital motion of the comet is toward the lower left in each image. *From Lisse et al. (1996).* (b) Morphology as a function of comet gas production rate (given in terms of molecules/s in the lower right of each panel). Note the decreasing concentration of model source function and the increasing importance of diffuse halo emission in the extended coma as the gas production rate increases. *From Lisse et al. (2005).* (c) *Chandra* ACIS spectra of the X-ray emission from three comets. All curves show ACIS-S3 measurements of the 0.2- to 1.0-keV pulse height spectrum, with $\pm 1\sigma$ error bars and the best-fit emission line + thermal bremsstrahlung model convolved with the ACIS-S instrument response as a histogram. Pronounced emission due to O^{7+} and O^{6+} is evident at 560 and 660 eV, and for C^{5+}, C^{4+}, and N^{5+} emission lines at 200–500 eV. Best-fit model lines at 284, 380, 466, 552, 590, 648, 796, and 985 eV are close to those predicted for charge exchange between solar wind C^{5+}, C^{6+}, C^{6+}/N^{6+}, O^{7+}, O^{8+}, and Ne^{9+} ions and neutral gases in the comet's coma. (Black) ACIS spectra of C/LINEAR 1999 S4 (circles), from Lisse et al. (2001). (Red) Comet McNaught-Hartley spectra (squares), after Krasnopolsky et al. (2003).

coma; almost no emission is found in the extended tails of dust or plasma. The X-ray brightness gradually decreases with increasing cometocentric distance r with a dependence of about r^{-1}. The emission morphology and range depend on the amount of neutral gas pouring out of the comet's nucleus. The brightness merges with the soft X-ray background emission at distances as small as 10^4 km for weakly active comets, but the most actively outgassing comets can be X-ray luminous out to 10^6 km. For the least active comets, the X-ray emission tracks the regions of densest coma gas, usually at the nucleus or along jets and shells; for the highly outgassing comets, the coma is collisionally thick to charge exchange and the region of peak emission is crescent shaped with a brightness peak displaced toward the Sun from the nucleus (Figure 48.14). The distance of this peak from the nucleus appears to increase with increasing gas production rate; for Hyakutake, it was located at $\sim 2 \times 10^4$ km.

Numerical simulations of the solar wind interaction with Hyakutake including charge exchange have been used to generate X-ray images. A global magnetohydrodynamic model and a hydrodynamic model were used to predict solar wind speeds and densities in addition to the X-ray emission around a comet. The simulated X-ray images are similar to the observed images. In the collisionally thick case, the gas production rate can be directly determined from the observed X-ray morphology, as it was demonstrated for four comets observed by *ROSAT* and *XMM-Newton*. It is also possible to deduce the location of the cometary bow shock by a tomographic analysis of the X-ray morphology. This technique was successfully applied to *XMM-Newton* data of comet C/2000 WM1 (LINEAR).

A completely different morphology—a tail-like structure—was observed from Comet Lovejoy (C/2011 W3) with XRT on *Hinode* 30 min after perihelion, when the comet was only $\sim$320,000 km above the solar photosphere. The X-ray emission was similar to the EUV emission, but offset along the local magnetic field lines. This emission, however, is unlikely to result from charge exchange interactions but rather from direct ionization of the cometary material.

12.2. X-ray Luminosity

The observed X-ray luminosity, L_x, of comets is mainly determined by the gas production rate and the heavy ion flux in the solar wind, and also by instrumental parameters like the energy bandpass and the observational aperture. Typical values of L_x range from 0.01 to 1 GW. This is of the order of 10^{-4} of the total luminosity of a comet. For weakly active comets, with $L_x < 0.1$ GW, a roughly linear correlation between optical and X-ray luminosities is observed. For the brightest X-ray comets, a plateau or asymptote in the X-ray production is seen at a maximum value of $\sim 10^{16}$ erg/s. Particularly dusty comets, like C/1995 O1 (Hale–Bopp), 103P/Hartley 2, or 73P/Schwassmann-Wachmann 3 appear to have less X-ray emission than would be expected from their overall optical luminosity L_{opt}. This is most likely a consequence of the fact that the optical luminosity of a comet is dominated by the amount of dust, while the X-ray luminosity is controlled by the amount of gas. The peak X-ray surface brightness decreases with the inverse square of the heliocentric distance r, independent of the gas production rate.

12.3. Temporal Variation

Photometric light curves of the X-ray and EUV emission typically show a long-term baseline level with superimposed impulsive spikes of a few hours' duration, with positive excursions typically three to four times that of the baseline emission level. Figure 48.14 demonstrates the strong correlation found between the time histories of the solar wind proton flux (a proxy for the solar wind minor ion flux), the solar wind magnetic field intensity, and a comet's X-ray emission, for the case of comet 2P/Encke observed almost continuously over the course of 2 weeks in 1997 by *ROSAT* and Extreme Ultraviolet Explorer (*EUVE*) (Figure 48.14(d)). Another long-term study was conducted for comet 9P/Tempel 1 in 2005, supporting the *Deep Impact* mission. Comparison of comet luminosities with time histories of the solar wind proton flux, oxygen ion flux, and solar X-ray flux shows a strong correlation between the cometary emission and the

---

(Green) 2P/Encke spectrum taken on November 24, 2003, multiplied by a factor of 2. The C/1999 S4 (LINEAR) and C/McNaught-Hartley 2001 observations had an average count rate on the order 20 times as large, even though Encke was closer to *Chandra* and the Earth when the observations were being made. Note the 560-eV complex to 400-eV complex ratio of 2–3 in the two bright, highly active comets, and the ratio of approximately 1 for the faint, low-activity comet Encke. *From Lisse et al., op. cit (2005)*. (d) Temporal trends of the cometary X-ray emission. Light curve, solar wind magnetic field strength, solar wind proton flux, and solar X-ray emission for 2P/Encke 1997 on July 4–9, 1997, UT. All error bars are $\pm 1\sigma$. D, *ROSAT* HRI light curve, July 4–8, 1997. ◇, *EUVE* scanner Lexan B light curve July 6–8, 1997, UT, taken contemporaneously with the HRI observations, and scaled by a factor of 1.2. Also plotted are the Comprehensive Solar Wind Laboratory for Long-Term Solar Wind Measurements (WIND) total magnetic field B_{total} (*), the Solar and Heliospheric Observatory *(SOHO)* Charge, Element, and Isotope Analysis System (CELIAS)/Solar Extreme Ultraviolet Monitor (SEM) 1.0–500 Å solar X-ray flux (◇), and the *SOHO* CELIAS solar wind proton flux (boxes). There is a strong correlation between the solar wind magnetic field/density and the comet's emission. There is no direct correlation between outbursts of solar X-rays and the comet's outbursts. *From Lisse et al. (1997)*.

solar wind oxygen ion flux, a good correlation between the comet's emission and the solar wind proton flux, but no correlation between the cometary emission and the solar X-ray flux.

Up until 2005, the temporal variation of the solar wind dominated the observed behavior on all but the longest timescales of weeks to months. A "new" form of temporal variation was demonstrated in the *Chandra* observations of comet 2P/Encke 2003, wherein the observed X-ray emission is modulated at the 11.1-h period of the nucleus rotation. Rotational modulation of the signal should be possible only in collisionally thin (to SWCX) comae with weak cometary activity, where a change in the coma neutral gas density can directly affect the cometary X-ray flux. Imaging of the X-ray emission of comet 103P/Hartley 2, compared to optical ground-based images of the comet obtained during the *Deep Impact Extended* mission flyby of the comet, also seem to show correlated rotational modulation of the comet's X-ray emission.

12.4. Energy Spectrum

Until 2001, all published cometary X-ray spectra had very low spectral energy resolution ($\Delta E/E \sim 1$ at 300–600 eV), and the best spectra were those obtained by ROSAT for C/1990 K1 (Levy) and C/1990 N1 (Tsuchiya-Kiuchi), and by *Beppo*SAX for C/1995 O1 (Hale–Bopp). These observations were capable of showing that the spectrum was very soft. However, due to the limited spectral resolution, continuum emission could not be distinguished from a multiline spectrum, as it would result from the SWCX mechanism. It was found that thermal bremsstrahlung was a good "technical" proxy for characterizing the basic spectral properties in an instrument-independent way, given the limited spectral resolution and the lack of a more realistic model spectrum. All these spectra were consistent with bremsstrahlung temperatures kT between 0.2 and 0.3 keV. Nondetections of comets C/Hyakutake, C/Tabur, C/Hale–Bopp, and 55P/Tempel–Tuttle using the X-Ray Timing Explorer Proportional Counter Array (XTE PCA) (2–30 keV) and ASCA Solid-state Imaging Spectrometers (SIS) (0.6–4 keV) imaging spectrometers were consistent with an extremely soft spectrum.

In 2001, the first high-resolution cometary X-ray spectrum was obtained, using *Chandra* X-ray observatory measurements of the emission from comet C/1999 S4 (LINEAR) as it passed close by the Earth. Discrete line emission signatures due to highly ionized oxygen and carbon were immediately apparent. Higher resolution spectra of cometary X-ray emission are now common. Eight comets were studied with the *Chandra* X-ray observatory spectroscopy in the period 2000 to 2006, covering the transition from solar maximum to solar minimum. Figure 48.15(a) shows the (background-subtracted raw) spectra for eight of the comets, and Figure 48.15(b) and (c) show at which ecliptic latitude and phase in the solar cycle the spectra were observed. It is immediately obvious that there are spectral differences. In Figure 48.15(a), three spectral bands are indicated, dominated by emission from (1) C V, C VI, N VI ("C + N"); (2) by O VII; and (3) by O VIII ions, and the spectra are arranged so that, from top to bottom, flux is systematically shifted from lower to higher energy bands. The quantitative results of the spectral fits clearly show that the flux in the C + N band is anti-correlated to that in the O VIII band (Figure 48.15(d)), indicating that the comets were exposed to different solar wind conditions.

As can be seen in Figure 48.15(b) and (c), all the comets which were observed at high latitudes happened to be there during solar maximum, when the equatorial solar wind had expanded into these regions. This implies that, until 2006, *Chandra* had not observed any comet exposed to the polar wind. This situation changed in October 2007, during solar minimum, when the nucleus of comet 17P/Holmes experienced a spectacular outburst, which increased its dust and gas outflow and optical brightness by almost a million times within hours, from under 17 mag to 3 mag, making it by far the optically brightest comet observable by *Chandra* since its launch. At the time, comet 17P/Holmes was located at a sufficiently high heliographic latitude (19°) to be exposed to the polar wind at solar minimum. It was thus expected that this comet would exhibit considerably different X-ray properties, and in fact this was observed: 17P/Holmes became the first comet where *Chandra* did not detect any significant X-ray emission at all. The most likely explanation for this dramatic X-ray faintness is that the polar wind was so diluted and its ionization so low that only very little X-ray flux was generated by charge exchange at energies above ~ 300 eV. An instrumental effect, i.e. a loss of sensitivity, can definitively be ruled out, because only two months later, another comet, 8P/Tuttle, was observed with *Chandra*, and this comet, at low latitude (3°), was clearly detected in X-rays.

12.5. Summary

Driven by the solar wind, cometary X-rays provide an observable link between the solar corona, where the solar wind originates, and the solar wind where the comet resides. They are the cleanest example of charge exchange-driven X-ray emission, and should prove to be quite valuable in understanding other astrophysical charge exchange systems found wherever cold neutral and hot ionized gases meet—e.g. in the entire heliosphere, in stellar winds, in massive star-forming regions, in the expanding

Chapter | 48 X-rays in the Solar System

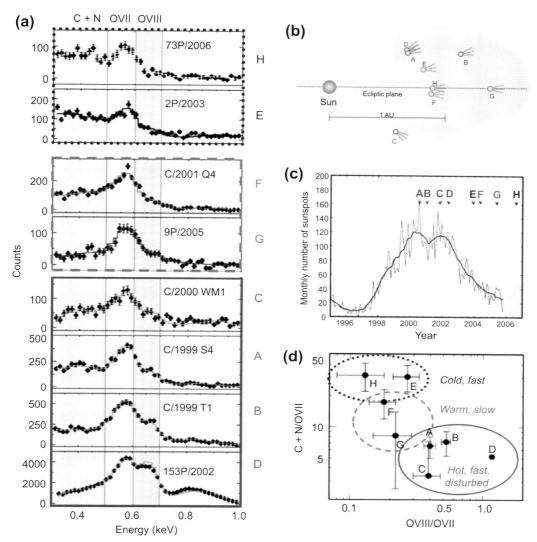

FIGURE 48.15 Summary of the spectral results obtained with *Chandra* for all the comets (denoted by A–H) which were observed from 2000 to 2006. (a) The 0.3- to 1.0-keV pulse height distributions, (b) the ecliptic latitudes, (c) phases in the solar cycle of the observed comets, and (d) the deduced information about the solar wind heavy ion content. Two comets were observed interacting with low ionization temperature but fast winds arising from the bottom of the solar corona (E, H); at least two with high ionization temperature but slow winds arising from the top of the Sun's corona (F, G, and possibly A and C); and two comets (B & D) appeared to have interacted with disturbed solar winds found during flares, coronal mass ejections, or solar sector boundary crossings. *Adapted from Bodewits et al. (2007).*

shells of supernova remnants, in active galaxies, or in clusters of galaxies. In our own solar system, once we have understood the SWCX mechanism's behavior in cometary comae in sufficient detail, we will be able to use comets as probes to measure the solar wind throughout the inner heliosphere. This will be especially useful in monitoring the solar wind in places hard to reach with spacecraft—such as over the solar poles, at large distances above and below the ecliptic plane, and at heliocentric distances greater than a few astronomical units. For example, about one-third of the observed soft X-ray emission is found in the 530- to 700-eV oxygen O^{7+} and O^{6+} lines; observing photons of this energy will allow studies of the oxygen ion charge ratio of the solar wind, which is predicted to vary significantly between the slow and fast solar winds at low and high solar latitudes, respectively.

13. ASTEROIDS

X-rays from asteroids have been studied by experiments on two in situ missions, the X-ray/gamma-ray spectrometer (XGRS) on the Near Earth Asteroid Rendezvous (*NEAR*)–*Shoemaker* mission to asteroid 433 Eros, and the XRS on the *Hayabusa* mission to asteroid 25143 Itokawa (see also "Near Earth Asteroids"). The only attempt to detect X-rays from an asteroid remotely was a 10-ks, observation by *Chandra* on December 11, 2001, of 1998 WT24, but it was unsuccessful. The results of the in situ observations show X-ray emission due to fluorescence and scattering of incident solar X-rays, similar to the emission seen from the surface of the airless Moon. In fact, the best measurements were obtained during a strong solar flare, when the incident solar X-rays were highly amplified. As for the Moon, X-ray spectroscopy of resonantly scattered solar X-rays can be used to map the elemental composition of the surface.

NEAR-Shoemaker entered an Eros orbit on February 14, 2000, and completed a 1-year-long mission around it. Eros at $33 \times 13 \times 13$ km in size is the second largest near-Earth asteroid, and its "day" is 5.27 h long. Eros exhibits a heavily cratered surface with one side dominated by a huge, scallop-rimmed gouge; a conspicuous sharp, raised rimmed crater occupies the other side. The XRS part of the XGRS detected X-rays in the $1-10$ keV energy range to determine the major elemental composition of Eros' surface. The XRS observed the asteroid in low orbit (<50 km) during May 2, 2000, to August 12, 2000, and again during December 12, 2000 to February 2, 2001. These observations suggest that elemental ratios for Mg/Si, Al/Si, Ca/Si, and Fe/Si on Eros are most consistent with a primitive chondrite and give no evidence of global differentiation. The S/Si ratio is considerably lower than that for a chondrite and is most likely due to surface volatilization ("space weathering"). The overall conclusion is that Eros is broadly "primitive" in its chemical composition and has not experienced global differentiation into a core, mantle, and crust, and that surface effects cause the observed departures from chondritic S/Si and Fe/Si.

Hayabusa reached the asteroid 25143 Itokawa on September 12, 2005. The first touchdown occurred on November 19, 2005. The observations made during the touchdown, a period of relatively enhanced solar X-ray flux, returned an average elemental mass ratio of Mg/Si $= 0.78 \pm 0.07$ and Al/Si $= 0.07 \pm 0.03$. These early results suggest that, like Eros, asteroid Itokawa's composition can be described as an ordinary chondrite, although occurrence of some differentiation cannot be ruled out.

The composition and structure of the rocks and minerals in asteroids provide critical clues to their origin and evolution and are a fundamental line of inquiry in understanding the asteroids, of which more than 20,000 have been detected and cataloged. It is interesting to note that for both Eros and Itokawa the compositions derived by remote X-ray observations using spacecraft in close proximity to the asteroid seem consistent with those found using Earth-based optical and infrared spectroscopy.

14. HELIOSPHERE

The solar wind (see also "The Solar Wind") flow starts out slowly in the corona but becomes supersonic at a distance of few solar radii. The gas cools as it expands, falling from $\sim 10^6$ K down to about 10^5 K at 1 AU. The average properties of the solar wind at 1 AU are proton number density $\sim 7/\text{cm}^3$, speed ~ 450 km/s, temperature $\sim 10^5$ K, magnetic field strength ~ 5 nT, and Mach number ~ 8. However, the composition and charge state distribution far from the Sun are "frozen-in" at coronal values due to the low collision frequency outside the corona. The solar wind contains structure, such as slow (300 km/s) and fast (700 km/s) streams, which can be mapped back to the Sun. The solar wind "terminates" in a shock called the heliopause, where the ram pressure of the streaming solar wind has fallen to that of the interstellar medium (ISM) gas. The region of space containing plasma of solar origin, from the corona to the heliopause at ~ 100 AU, is called the heliosphere. A very small part of the solar wind interacts with the planets and comets; the bulk of the wind interacts with neutral ISM gas in the heliosphere and neutral and ionized ISM at the heliopause.

X-ray emission from the heliosphere has also been predicted from the interaction of the solar wind with the interstellar neutral gas (mainly HI and HeI) that streams into the solar system. It has been demonstrated that roughly half of the observed 0.25-keV X-ray diffuse background can be attributed to this process (see Figure 48.16). Solar and Heliospheric Observatory (SOHO) observations of neutral hydrogen Lyman-alpha emission show a clear asymmetry in the ISM flow direction, with a clear deficit of neutral hydrogen in the downstream direction of the incoming neutral ISM gas, most likely created by SWCX ionization of the ISM. The analogous process applied to other stars has been suggested as a means of detecting stellar winds. Also a strong correlation between the solar wind flux density and the *ROSAT* "LTEs," systematic variations in the soft X-ray background of the *ROSAT* X-ray detectors, has been shown. Photometric imaging observations of the lunar nightside by *Chandra* made in September 2001 does not show any lunar nightside emission above an SWCX background. The soft X-ray emission detected from the dark side of the Moon, using *ROSAT*, would appear to be attributable not to electrons spiraling from the sunward to the dark hemisphere, as proposed earlier, but to SWCX in the geocorona and the column of heliosphere between the Earth and the Moon (Section 3).

Just as charge exchange-driven X-rays are emitted throughout the heliosphere, similar emission must occur within the astrospheres of other stars with highly ionized stellar winds that are located within interstellar gas clouds that are at least partially neutral. Although very weak, in

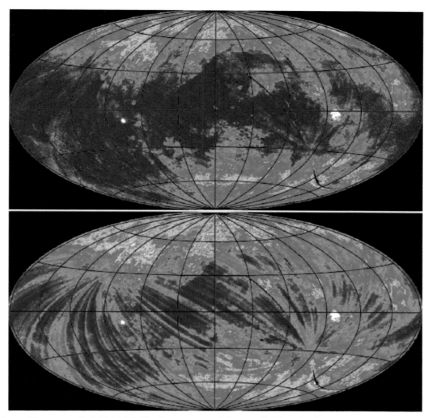

FIGURE 48.16 Heliosphere. (Upper panel) *ROSAT* All-Sky Survey map of the cosmic X-ray background at 1/4 keV. The data are displayed using an Aitoff projection in galactic coordinates centered on the galactic center with longitude increasing to the left and latitude increasing upward. Low intensity is indicated by purple and blue while red indicates higher intensity. (Lower panel) same as above except the contaminating LTEs (SWCX emission) were not removed. The striping is due to the survey geometry where great circles on the sky crossing at the ecliptic poles were scanned precessing at $\sim 1°$/day. *From Snowden et al. (1997).*

principle, this emission offers the opportunity to measure mass loss rates and directly image the winds and astrospheres of other main sequence late-type stars. Imaging would provide information on the geometry of the stellar wind, such as whether outflows are primarily polar, azimuthal, or isotropic and whether or not other stars have analogs of the slow (more ionized) and fast (less ionized) solar wind streams.

15. SUMMARY

Table 48.1 summarizes our current knowledge of the X-ray emissions from the planetary bodies that have been observed to produce soft X-rays. Several other solar system bodies, including Titan, Uranus, Neptune, and inner icy satellites of Saturn, are also expected to be X-ray sources, but they are yet to be detected. X-rays are expected from these bodies due to scattering of solar X-rays as well as SWCX and/or magnetospheric ion precipitation and electron bremsstrahlung. However, due to larger distance from Sun, and hence much reduced solar radiation and solar wind flux, the X-rays produced at these objects would be at level much lower than the detection capability of current X-ray observatories. X-rays would also be produced in extrasolar planets—through processes similar to those in our solar system. However, detecting X-rays from exoplanets would be a challenging task since the flux would be very weak due to large distances involved.

Table 48.2 lists the spacecraft missions and satellite-based observatories mentioned in the text, which have contributed to the growth of planetary X-rays and our current understanding of the processes of X-ray production. Upcoming X-ray observatories, like *Astro-H*, *eROSITA/SRG*, and *Athena*, and planetary missions (*BepiColombo*, *Selene-2*, *Chandra*yaan-2) carrying experiments are also listed, which would have better sensitivity, effective area, and resolution, thus providing better tools to significantly advance the field of solar system X-rays.

TABLE 48.1 Summary of the Characteristics of Soft X-ray Emission from Solar System Bodies

Object	Emitting Region	Power Emitted[1]	Special Characteristics	Possible Production Mechanism
Earth	Auroral atmosphere	10–30 MW	Correlated with magnetic storm and substorm activity	EB + characteristic line emission from atmospheric neutrals due to electron impact
	Nonauroral atmosphere	40 MW	Correlated with solar X-ray flux	FS by atmosphere
Jupiter	Auroral atmosphere	0.4–1 GW	Pulsating (~20–60 min) X-ray hot spot in north polar region	Energetic ion precipitation from magnetosphere and/or solar wind + EB
	Nonauroral atmosphere	0.5–2 GW	Relatively uniform over disk	RS + possible ion precipitation from radiation belts
Moon	Dayside surface	0.07 MW	Correlated with solar X-rays	FS by the surface elements on dayside
	Nightside (geocoronal)		Nightside emissions are ~1% of the dayside	SWCX with geocorona
Comets	Sunward-side coma	0.2–1 GW	Intensity peaks in sunward direction, ~10^5–10^6 km ahead of cometary nucleus and is correlated with solar wind parameters	SWCX with cometary neutrals
Comets in the solar corona	Coma plus tail		Offset from the EUV emission along magnetic field	Ionization of cometary material
Venus	Sunlit atmosphere	50 MW	Emissions from ~120 to 140 km above the surface	FS by C and O atoms in the atmosphere
	Exosphere		Emissions from region 1.2 times Venus radius	SWCX with Venus exospheric neutrals
Mars	Sunlit atmosphere	1–4 MW	Emissions from upper atmosphere at heights of 110–130 km	FS by C and O atoms in the upper atmosphere
	Exosphere	1–10 MW	Emissions extend out to ~8 Mars radii	SWCX with Martian corona
Io	Surface	2 MW	Emissions from upper few micrometers of the surface	Energetic Jovian magnetospheric ions impact on the surface
Europa	Surface	3 MW	Emissions from upper few micrometers of the surface	Energetic Jovian magnetospheric ions impact on the surface
IPT	Plasma torus	0.1 GW	Dawn-dusk asymmetry observed	EB + ?
Saturn	Sunlit disk	0.1–0.4 GW	Varies with solar X-rays	RS + FS by atmosphere + EB
Rings of Saturn	Surface	80 MW	Emissions confined to a narrow energy band around at 0.53 keV.	FS by atomic oxygen in H_2O ice + ?
Asteroid	Sunlit surface		Emissions vary with solar X-ray flux	FS by elements on the surface
Mercury	Dayside		Emissions vary with solar flux	FS of solar X-rays by elements on the surface
	Nightside		Depends on precipitating electron spectrum	Electron-induced FS by elements on the surface
Heliosphere	Entire heliosphere	10^{16} W	Emissions vary with solar wind	SWCX with heliospheric neutrals

SWCX, solar wind charge exchange is the charge exchange of heavy, highly ionized solar wind ions with neutrals; FS, fluorescent scattering of solar X-rays; RS, resonant scattering of solar X-rays by atmospheric constituents; EB, bremsstrahlung from precipitating energetic electrons.
The question mark (?) refer to some other process(es) at work not clearly known as of now.
[1]The values quoted are values at the time of observation. X-rays from all bodies are expected to vary with time. For comparison, the total X-ray luminosity from the Sun is 10^{20} W.

TABLE 48.2 Satellites and Spacecraft Mentioned in the Text, Chronologically Sorted According to Launch Date

Instrument, Mission	Description	Category	Date	Agency, Country
Uhuru	First X-ray satellite	X-ray Astronomy	1970–1973	NASA (USA)
Apollo 15	Fourth manned lunar landing	Moon	July–August 1971	NASA (USA)
Apollo 16	Fifth manned lunar landing	Moon	April 1972	NASA (USA)
HEAO-1	High Energy Astronomy Observatory-1	X-ray Astronomy	1977–1979	NASA (USA)
UV spectrometer Voyager 1	Outer solar system	Jupiter, Saturn	1977–present	NASA (USA)
IUE	International Ultraviolet Explorer	UV Astronomy	1978–1996	NASA, ESA, Science and Engineering Research Council (SERC)
Einstein (HEAO-2)	High Energy Astronomy Observatory 2	X-ray Astronomy	1978–1981	NASA (USA)
PSPC, HRI, WFC ROSAT	Röntgensatellit	X-ray Astronomy	1990–1999	Germany, USA, UK
Ulysses	Solar Wind Observatory	Solar Wind	1990–2009	ESA, NASA
CGRO	Compton Gamma-Ray Observatory	Gamma-ray Astronomy	1991–2000	NASA (USA)
AXIS UARS	Upper Atmosphere Research Satellite	Earth	1991–2011	NASA (USA)
Lexan B EUVE	Extreme Ultraviolet Explorer	EUV Astronomy	1992–2001	NASA (USA)
SIS ASCA	Advanced Satellite for Cosmology and Astrophysics	X-ray Astronomy	1993–2001	ISAS (Japan)
Clementine	Lunar Mission	Moon	1994	Strategic Defense Initiative Organization (SDIO), NASA (USA)
WIND	Solar Wind Observatory	Solar Wind	1994–present	NASA, ESA, ISAS
CELIAS/SEM SOHO	Solar and Heliospheric Observatory	Sun	1995–present	ESA, NASA
PCA RXTE	Rossi X-ray Timing Explorer	X-ray Astronomy	1995–2012	NASA (USA)
NEAR-Shoemaker	Near Earth Asteroid Rendezvous	Asteroid 433 Eros	1996–2001	NASA (USA)
PIXIE POLAR	Earth Magnetosphere	Earth	1996–2008	NASA (USA)
BeppoSAX	Giuseppe Occhialini (Beppo) Satellite per Astronomia a raggi X	X-ray Astronomy	1996–2003	Italy, Netherlands
GOES-10	Weather satellite	Earth	1997–2009	NOAA/NASA (USA)
ACIS, HRC-I Chandra	Chandrasekhar X-ray Observatory	X-ray Astronomy	1999–present	NASA (USA)
EPIC, RGS XMM-Newton	X-ray Multi Mirror Mission	X-ray Astronomy	1999–present	ESA

(Continued)

TABLE 48.2 Satellites and Spacecraft Mentioned in the Text, Chronologically Sorted According to Launch Date—cont'd

Instrument, Mission	Description	Category	Date	Agency, Country
RHESSI	Reuven Ramaty High Energy Solar Spectroscopic Imager	Sun	2002–present	NASA (USA)
XRS Hayabusa	Asteroid Sample Return Mission	Asteroid 25143 Itokawa	2003–2010	JAXA (Japan)
D-CIXS SMART-1	Lunar Mission	Moon	2003–2006	ESA
EPS, XRS MESSENGER	Mercury Surface, Space Environment, Geochemistry and Ranging	Mercury	2004–present	NASA (USA)
SWIFT	Gamma-Ray Burst Mission	Gamma-ray Astronomy	2004–present	NASA, with international participation
Deep Impact (Extended)	Comet Mission	Comets Tempel 1 and Hartley 2	2005–2007, 2008–2013	NASA (USA)
Suzaku	Astro E2	X-ray Astronomy	2005–present	ISAS/JAXA (Japan)
New Horizon	Outer Solar System	Pluto	2006–present	NASA (USA)
XRT Hinode	Solar Observations	Sun	2006–present	JAXA (Japan), with international Participation
X-Ray CCDs Kaguya	Lunar Mission	Moon	2007–2009	JAXA (Japan)
X-Ray CCDs Chandrayaan-1	Lunar Mission	Moon	2008–2009	ISRO (India)
Astro-H	Planned	X-ray Astronomy	~2014	JAXA, NASA
MIXS BepiColombo	Planned	Mercury	~2015	ESA/JAXA
eROSITA SRG	Planned	X-ray Astronomy	~2016	Germany, Russia
Rover Selene-2	Planned	Moon	~2017	JAXA (Japan)
CLASS, orbiter Chandrayaan-2	Planned	Moon	~2016–2017	ISRO (India)
Athena	Planned	X-ray Astronomy	~2028	ESA

CCD, charge-coupled device; ESA, European Space Agency; ISAS; ISRO, Indian Space Research Organisation; JAXA, Japan Aerospace Exploration Agency; NASA, National Aeronautics and Space Administration.

ACKNOWLEDGMENTS

A large part of this chapter is based on the review article by Bhardwaj et al. (2007a), which is a collective effort of several authors, and we deeply acknowledge all the authors of that paper. We also thank the entire solar system X-ray community whose works have led to this review.

BIBLIOGRAPHY

Bhardwaj, A. (2006). *X-ray emission from Jupiter, Saturn, and Earth: a short review.* In *Advances in Geosciences* (Vol. 3); (pp. 215−230). Singapore: World Scientific.

Bhardwaj, A., Elsner, R. F., Waite, J. H., Jr., Gladstone, G. R., Cravens, T. E., & Ford, P. G. (2005a). Chandra Observation of an X-ray Flare at Saturn: Evidence for Direct Solar Control on Saturn's Disk X-ray Emissions. *Astrophysical Journal Letters, 624,* L121−L124.

Bhardwaj, A., Elsner, R. F., Waite, J. H., Jr., Gladstone, G. R., Cravens, T. E., & Ford, P. G. (2005b). The Discovery of Oxygen Kα X-ray Emission from the Rings of Saturn. *Astrophysical Journal Letters, 627,* L73−L76.

Bhardwaj, A., Elsner, R. F., Gladstone, G. R., Cravens, T. E., Lisse, C. M., Dennerl, K., et al. (2007a). X-rays from solar system objects. *Planetary and Space Science, 55,* 1135−1189.

Bhardwaj, A., Gladstone, G. R., Elsner, R. F., Østgaard, N., Waite, J. H., Jr., Cravens, T. E., Chang, S.-W., Majeed, T., & Metzger, A. E. (2007b). First Terrestrial Soft X-ray Auroral Observation by the Chandra X-ray Observatory. *Journal of Atmospheric and Solar Terrestrial Physics, 67,* 179. http://chandra.harvard.edu/press/05_releases/press_122805.html.

Bhardwaj, A., Elsner, R. F., Gladstone, G. R., Branduardi-Raymont, G., Dennerl, K., Lisse, C. M., Cravens, T. E., et al. (2009). X-ray Emission from Planets and comets: Relationship with Solar X-rays and Solar Wind. *Advances in Geosciences, 15,* 229−244.

Bhardwaj, A., & Gladstone, G. R. (2000). Auroral emissions of the giant planets. *Reviews of Geophysics, 38,* 295−353.

Bodewits, D., et al. (2007). Spectral Analysis of the Chandra Comet Survey. *Astronomy and Astrophysics, 469,* 1183−1195.

Branduardi-Raymont, G., Bhardwaj, A., Elsner, R. F., Gladstone, G. R., Ramsay, G., Rodriguez, P., Soria, R., Waite, J. H., Jr., & Cravens, T. E. (2007). *Astronomy and Astrophysics, 463,* 761−774.

Branduardi-Raymont, G., Bhardwaj, A., Elsner, R. F., & Rodriguez, P. (2010). X-rays from Saturn: a study with XMM-Newton and Chandra over the years 2002−05. *Astronomy and Astrophysics, 510,* A73.

Christian, D. J., Bodewits, D., Lisse, C. M., Dennerl, K., Wolk, S. J., Hsieh, H., et al. (2010). Chandra observations of comets 8P/Tuttle and 17P/Holmes during solar minimum. *The Astrophysical Journal Supplement Series, 187,* 447−459.

Dennerl, K. (2002). *Astronomy and Astrophysics, 394,* 1119−1128.

Dennerl, K. (2008). *Planetary Space Science, 56,* 1414.

Dennerl, K. (2010). Charge transfer reactions. *Space Science Reviews, 157,* 57−91.

Dennerl, K., Burwitz, V., Englhauser, J., Lisse, C., & Wolk, S. (2002). Discovery of X-rays from Venus with *Chandra*. *Astronomy and Astrophysics, 386,* 319−330.

Dennerl, K., Lisse, C. M., Bhardwaj, A., Burwitz, V., Englhauser, J., Gunell, H., et al. (2006). First observation of Mars with *XMM-Newton*. High resolution X-ray spectroscopy with RGS. *Astronomy and Astrophysics, 451*(2), 709−722.

Dennerl, K., Lisse, C. M., Bhardwaj, A., Christian, D. J., Wolk, S. J., Bodewits, D., et al. (2012). Solar system X-rays from charge exchange processes. *Astronomische Nachrichten, 333*(4), 324−334.

Elsner, R. F., Gladstone, G. R., Waite, J. H., Crary, F. J., Howell, R. R., Johnson, R. E., et al. (2002). *Astrophysical Journal, 572,* 1077−1082.

Ewing, et al. (2013). *Astrophysical Journal, 763.* article id. 66.

Gladstone, G. R., Waite, J. H., Grodent, D., Lewis, W. S., Crary, F. J., Elsner, R. F., et al. (2002). *Nature, 415,* 1000.

Kharchenko, V., Bhardwaj, A., Dalgarno, A., Schultz, D., & Stancil, P. (2008). Modeling spectra of the North and South Jovian X-ray auroras. *Journal of Geophysical Research, 113,* A08229.

Krasnopolsky, V. A., Greenwood, J. B., & Stancil, P. C. (2004). X-ray and extreme ultraviolet emission from comets. *Space Science Review, 113,* 271−374.

Lisse, C. M., et al. (1996). *Science, 292,* 1343−1348.

Lisse, C. M., et al. (1997). *Icarus, 141,* 316−330.

Lisse, C. M., et al. (2005). Chandra observations of Comet 2P/Encke 2003: first detection of a collisionally thin, fast solar wind charge exchange system. *Astrophysical Journal, 635,* 1329−1347.

Lisse, C. M., et al. (2007). Chandra observations of Comet 9P/Tempel 1 during the deep impact campaign. *Icarus, 190,* 391−405.

Lisse, C. M., Cravens, T. E., & Dennerl, K. (2004). X-ray and extreme ultraviolet emission from comets. In M. C. Festou, H. U. Keller, & H. A. Weaver (Eds.), *Comet II* (pp. 631−643). Tucson: Univ. Arizona Press.

McCauley, P. I., Saar, S. H., Raymond, J. C., Ko, Y.-K., & Saint-Hilaire, P. (2013). Extreme-ultraviolet and X-ray observations of comet Lovejoy (C/2011 W3) in the lower corona. *Astrophysical Journal, 768,* 161−172.

Østgaard, N., Stadsnes, J., Bjordal, J., Germany, G. A., Vondrak, R. R., Parks, G. K., et al. (2001). *Journal of Geophysical Research, 106,* 26081.

Petrinec, S. M., McKenzie, D. L., Imhof, W. L., Mobilia, J., Chenette, D. L., et al. (2000). *Journal of Atmospheric and Solar Terrestrial Physics, 62,* 875−888.

Snowden, S. L., Egger, R., Freyberg, M. J., McCammon, D., Plucinsky, P. P., Sanders, W. T., et al. (1997). *Astrophysical Journal, 485,* 125−135.

Wargelin, B. J., Markevitch, M., Juda, M., Kharchenko, V., Edgar, R., Dalgarno, A., et al. (2004). *Astrophysical Journal, 607,* 596−610.

Wegmann, R., & Dennerl, K. (2005). X-ray tomography of a cometary bow shock. *Astronomy and Astrophysics, 430,* L33−L36.

Wolk, S. J., Lisse, C. M., Bodewits, D., Christian, D. J., & Dennerl, K. (2009). Chandra's close encounter with the disintegrating comets 73P/2006 (Schwassmann-Wachmann 3) fragment B and C/1999 S4 (LINEAR). *Astrophysical Journal, 694,* 1293−1308.

Chapter 49

The Solar System at Ultraviolet Wavelengths

Amanda R. Hendrix, Robert M. Nelson and Deborah L. Domingue
Planetary Science Institute, Tucson, AZ, USA

Chapter Outline

1. A Brief History of UV Astronomy — 1047
2. Nature of Solar System Astronomical Observations — 1049
3. Observations of Planetary Atmospheres — 1049
   - 3.1. Mercury and the Moon — 1050
   - 3.2. Venus — 1050
   - 3.3. Mars — 1051
   - 3.4. Jupiter — 1052
   - 3.5. Saturn — 1055
   - 3.6. Uranus — 1056
   - 3.7. Neptune — 1057
   - 3.8. Pluto — 1057
   - 3.9. Galilean Satellites — 1058
   - 3.10. Titan and Triton — 1059
4. Observations of Solid Surfaces — 1060
   - 4.1. Galilean Satellites — 1061
   - 4.2. Saturnian Satellites — 1063
   - 4.3. Enceladus — 1065
   - 4.4. Uranian Satellites — 1066
   - 4.5. A Comparison of Icy Satellite Systems — 1067
   - 4.6. Pluto and Charon — 1068
   - 4.7. Asteroids and Comets — 1068
   - 4.8. The Moon and Mercury — 1069
   - 4.9. Planetary Rings — 1069
5. Conclusions — 1070
BIBLIOGRAPHY — 1070

Ultraviolet (UV) imaging and spectroscopy are powerful tools for probing planetary atmospheres and surfaces. In this chapter, we review the significant contributions to our understanding of the solar system that have been made by UV observing methods. We cover results from the mid-ultraviolet (MUV, 300–400 nm), near-ultraviolet (NUV, 200–300 nm), far-ultraviolet (FUV, 122–200 nm), and extreme ultraviolet (EUV, 10–121 nm) wavelength ranges. These are shorter than visible and near-infrared (IR) wavelengths and involve photons of increasingly higher energy. UV observations therefore provide unique insight into planetary processes involving more energetic processes that cannot be studied using photons of longer wavelengths. Many of the solar system observations in the UV wavelengths have been performed by Earth-orbiting telescopes, such as the *International Ultraviolet Explorer* (*IUE*) and the *Hubble Space Telescope* (*HST*). We also review results from UV instruments on *Voyager, Galileo, Cassini,* and other spacecraft. Each planet in the solar system has been observed using UV instrumentation on Earth-orbiting telescopes and/or visiting spacecraft. Many of the larger planetary satellites, selected asteroids, and comets have also been observed. These data provide important information regarding the atmospheres and surfaces of solar system objects and the processes shaping their compositions.

1. A BRIEF HISTORY OF UV ASTRONOMY

The UV spectral region is significant to the entire community of astronomers, from those who study nearby objects such as Earth's Moon to those who probe the edge of the observable universe. From the perspective of a planetary astronomer, the UV spectral information is important for determining the composition and understanding the physical processes that are occurring on the surfaces and in the atmospheres of solar system objects.

Prior to the dawn of the space age, spectrophotometry of solar system objects at wavelengths shorter than ∼300 nm had long been desired in order to complement observations made by ground-based telescopes at longer wavelengths.

However, the presence in the Earth's atmosphere of ozone, a strong absorber of UV light between 200 and 300 nm, and molecular oxygen (O_2), which is the dominant UV absorber below 200 nm, prevented astronomers of the 1950s and earlier from observing the universe in this important spectral region.

The UV wavelengths of astronomical sources became observable midway through the past century when instruments could be deployed above the Earth's atmosphere. A rocket or spacecraft provides a platform from which astronomical observations can be made where the light being collected has not been subjected to absorption from the Earth's atmospheric gases. Thus, the space revolution dramatically enhanced the ability of astronomers to access the full spectrum of electromagnetic radiation emitted by celestial objects.

In the 1950s, a series of rocket-flown instruments began to slowly reveal the secrets of the UV universe. The first photometers and spectrometers were flown on unstabilized Aerobee rockets. They remained above the ozone layer for several tens of minutes while they scanned the sky at UV wavelengths. By the early 1960s, spectrometers on three-axis-stabilized platforms launched by rockets on suborbital trajectories were able to undertake observations with sufficient resolution such that individual spectral lines could be resolved in the target bodies.

Shortly thereafter, the military spacecraft designated 1964-83C carried an ultraviolet spectrometer (UVS) into the Earth's orbit. This was followed closely by NASA's launch of the first Orbiting Astronomical Observatory (OAO) satellite in 1966. These space platforms permitted long-duration observations compared to what was possible from a rocket launch on a suborbital trajectory. By 1972, the third spacecraft of the OAO series was launched. It was designated the *Copernicus* spacecraft and was an outstanding success.

In Europe, a parallel pattern of development for exploring the UV sky was under way using sounding rockets followed by orbiting spacecraft. In 1972, the European Space Research Organization launched an Earth-orbiting spacecraft (TD-1A) dedicated to UV stellar astronomy. Such developments set the stage in the 1970s for a joint US–European collaboration, the IUE satellite.

The IUE spacecraft was launched in 1978 into a geosynchronous orbit over the Atlantic Ocean. From there it could be controlled from ground stations in Greenbelt Maryland in the United States or in Villafranca, Spain, by engineers from NASA or European Space Agency. It functioned continuously from launch until it was terminated in 1996 and its capabilities were taken over by instruments on the HST. IUE spectra were recorded in two wavelength ranges of 115–195 nm and 190–320 nm, at either high or low spectral resolution. The IUE had no imaging capability, although spatial discrimination was possible within the largest (10×20 arcsec oval) spectrograph entrance aperture.

Additional Earth-orbiting satellites with UV observing capabilities were launched in the early 1990s. These include NASA's *Extreme Ultraviolet Explorer* satellite (*EUVE*) and the joint US–European HST. HST is in a low-Earth orbit, allowing upgrades to the facilities by astronauts. However, the low orbit reduced the observational duty cycle to 50–60% of that of the IUE in high orbit. UV spectroscopy with HST has been performed with the Goddard High-Resolution Spectrograph (GHRS), the Faint Object Spectrograph (FOS), the Space Telescope Imaging Spectrograph (STIS), the Advanced Camera for Surveys (ACS), and the Cosmic Origins Spectrograph. In 1990 and 1995, the *Hopkins Ultraviolet Telescope* (*HUT*) was flown aboard the US space shuttle as part of the Astro Observatory. The *Far-Ultraviolet Spectroscopic Explorer* (*FUSE*) was launched in 1999 and has spectroscopic capabilities in the 90–120 nm wavelength range.

Many interplanetary spacecraft missions have included UV instruments in their payloads. *Pioneers 10* and *11*, which were launched in 1970 and 1973, respectively, included UV photometers among their scientific instruments. (*See* Planetary Exploration Missions.) These two spacecraft were the first to safely pass through the asteroid belt and fly by Jupiter and Saturn. *Mariner 6* and *7*, Mars flyby missions launched in 1969, and *Mariner 9*, the first spacecraft to orbit Mars, launched in 1971, all carried UV spectrometers. *Mariner 10*, which flew by Mercury three times in 1974 and 1975, carried two EUV spectrometers (an **airglow** spectrometer and an occultation spectrometer) to measure the planet's exospheric composition. *Mariner 10* also made measurements of Earth's Moon after launch. *Pioneer Venus*, which was launched in 1978, was the first US mission dedicated to the exploration of the planet Venus. It included a UVS in its instrument package. Soviet spacecraft missions *Vega 1* and *Vega 2*, launched in 1985, dropped two descent probes into Venus' atmosphere, which included a French–Russian UV spectroscopy experiment. The *Voyager* project sent two spacecraft that included UVS in their instrument payloads, both launched in 1977, to the outer solar system. *Voyager 2* was the first spacecraft to fly by all four of the Jovian planets (Jupiter, Saturn, Uranus, and Neptune). In 1989, the *Galileo* spacecraft was launched. This spacecraft was the first dedicated mission to the Jupiter system, and it included within its scientific instrument payload two UV spectrometers: the EUV (EUV spectrometer, which operated between 50 and 140 nm) and the UVS (covering the 115–430 nm wavelength range). En route to Jupiter, the *Galileo* spacecraft collected UV spectra as it flew by Venus, the Moon, and the asteroids Gaspra and Ida. The Cassini mission, launched in 1997, includes the Ultraviolet Imaging Spectrograph (UVIS) and arrived at Saturn in June 2004. The *Nozomi* spacecraft, launched in 1998, carried two UV instruments; measurements were made of the Moon en route to Mars (Nozomi unfortunately failed to enter Mars' orbit). *Mars Express,* in orbit since December 2003, has an UV instrument called Spectroscopy for Investigation of Characteristics of the Atmosphere of Mars (SPICAM) as part of its payload. In 2011, the *MErcury Surface, Space ENvironment, GEochemistry, and Ranging* (*MESSENGER*) mission entered orbit about Mercury carrying a UV–visible (115–600 nm) spectrometer whose purpose was to observe both the surface and the exosphere of the planet. Also in 2004, Rosetta with its Alice UV instrument (covering 70–205 nm) was launched en route to comet 67P/Churyumov–Gerasimenko, set to enter orbit in 2014. The *New Horizons* mission to Pluto, launched in January 2006, carries the next-generation version Alice instrument, covering 52–187 nm. *Venus Express* has been in orbit at Venus since April 2006 and employs the Spectroscopy for Investigation of Characteristics of the Atmosphere of Venus (SPICAV)/Solar Occultation at Infrared (SOIR) UV and IR instrument for atmospheric studies. The *Lunar Reconnaissance Orbiter* (*LRO*), launched in June 2009, carries an FUV spectrometer, the Lyman Alpha Mapping Project (LAMP). LAMP is designed to peer into the Moon's polar permanently shadowed regions (PSRs) to search for water ice, using interplanetary and interstellar hydrogen emission (Lyman-alpha) as an illumination source.

2. NATURE OF SOLAR SYSTEM ASTRONOMICAL OBSERVATIONS

Most astronomers observe objects that have their own intrinsic energy source, such as stars and galaxies. However, the majority of the observations undertaken by planetary astronomers are of targets that do not emit their own radiation but are observable principally because they reflect the sunlight that falls on them or emit energy as a result of various physical processes. The measured spectrum of a body can thus reveal significant information on the composition of, and processes occurring within, planetary surfaces and atmospheres. The measured spectrum includes absorption features that can determine or constrain the composition of a surface or atmosphere, or emission features that suggest excitation processes in a gas or thermal emission from solids. The measured spectrum often displays solar features (either emission features or spectral continuum). To study the spectrum of the body itself, the solar spectrum is divided out, resulting in what is known as the spectral reflectance. The variation of the reflectance or **geometric albedo** (the reflectance at zero phase angle) as a function of wavelength is used to measure the strength of absorption features, from which the abundance of spectrally active species can be estimated.

Measuring reflected light at UV wavelengths can pose some interesting problems for instrument designers. First, the instrument's spectral sensitivity becomes weaker with decreasing wavelength and so does the Sun's energy output. The energy output of the Sun changes by a factor of 10^3 between the EUV and NUV spectral ranges, which until recently exceeded the dynamic range of UV detectors.

Second, in order to obtain the spectral reflectance of a body, a solar spectrum must be measured, which is no easy task in the UV range. Furthermore, below 180 nm, the spectrum of the Sun is variable. Therefore, a simultaneous spectrum of the Sun (or the reflection spectrum from an object whose spectrum is well understood) must be gathered at the same time that any UV observations are undertaken.

Last, particularly when performing measurements from an Earth-orbiting observatory such as the IUE or HST, solar system objects change positions against the background of stars during the course of an individual observation. In most cases, special tracking rates must be calculated prior to each observing run in order to know the change of the position of the target with time. Inaccurately calculated tracking rates can cause the observed target to drift from the instrument's field of view, thus adding noise and uncertainty to a measurement.

3. OBSERVATIONS OF PLANETARY ATMOSPHERES

With the exception of the innermost planet Mercury (which possesses a surface-bounded exosphere, similar to that observed on the Moon), all the planets in the solar system (and a few planetary satellites) are surrounded by detectable atmospheres. All the planets with atmospheres absorb UV light, and as a result UV observations provide information on the composition of, and processes that are occurring in, the object's atmosphere.

In general, the atmospheres of the terrestrial planets (Mercury, Venus, Earth, and Mars) are considered secondary atmospheres because they evolved after the primordial atmospheres were lost. However, the atmospheres of the four Jovian planets (Jupiter, Saturn, Uranus, and Neptune), because of their strong gravitational attraction and comparatively low temperatures, retained the primordial elements, particularly hydrogen and helium. From ground-based observations, methane (CH_4) and ammonia (NH_3) were identified in the atmospheres of the giant planets and therefore atmospheric processes were suspected of producing a host of daughter products that can be detected at UV wavelengths. (*See* Atmospheres Of The Giant Planets.)

Sunlight entering a planetary atmosphere can experience or initiate a wide variety of processes that contribute to the total energy emitted by the object and observed by an astronomical facility. The objects described previously all possess atmospheres that contribute significantly to their spectral behavior. Astronomical observations of such bodies search for and measure the depths of absorption bands in the spectrum or emission bands due to atmospheric interactions with both solar photons and energetic particles that originate from the solar wind or the planet's magnetosphere. These bands are unique to specific gases; thus, it is possible to identify or eliminate particular gases as candidate materials in the atmospheres of these objects. The interpretation of an UV spectrum can be an arduous task, given that the bands and lines observed in the spectrum may arise from a combination of processes. These include:

1. Single and multiple scattering of photons by aerosols such as haze and dust (**Mie scattering**) and gas (**Rayleigh scattering/Raman scattering**) in the planetary atmosphere.
2. Absorption of the incident UV solar light by atmospheric species.
3. Stimulation of an atmospheric gas by incident sunlight and emission by **fluorescence**, chemiluminescence, or resonant scattering.
4. Photoionization and photodissociation reactions that produce a reaction product in an excited state.
5. Excitation of gas by precipitation of magnetospheric particles.

Each of these processes is associated with a well-understood physical mechanism, the details of which are beyond the scope of this chapter. The reader is referred to the bibliography and other chapters in this book.

Two significant methods of studying atmospheres at UV wavelengths are stellar occultations (observing a star as a body passes in front of the star and measuring the stellar flux as it is diminished) and reflection/airglow measurements (measurements of the backscatter of the solar continuum either by Rayleigh–Raman atomic/molecular scattering or by Mie scattering from atmospheric aerosols). The atmospheric species and density can be constrained by studying the occulted stellar spectrum as it passes through the atmosphere. Limited wavelength facilities can identify some but not all the constituents present and processes ongoing in a planetary atmosphere. The UV data from Earth-orbiting satellites have been used in combination with ground-based observations at other wavelengths and with observations by other spacecraft (including flyby missions) to develop an understanding of the atmospheres of planetary objects. The following discussion summarizes the results of those bodies in the solar system that possess atmospheres.

3.1. Mercury and the Moon

Both Mercury and the Moon have very tenuous atmospheres that are often referred to as surface-bounded exospheres. The atoms in these atmospheres do not collide with each other; rather, they bounce from place to place on the surface. The *Mariner 10* UV airglow experiment detected hydrogen, helium, and oxygen atoms as constituents of Mercury's exosphere. No molecules were detected. The pressure of Mercury's atmosphere was determined to be about 10^{-12} bar (compared to the 1 bar atmospheric pressure at sea level on Earth). Ground-based telescopic observations in the visible region have identified resonant scattering emission features attributed to sodium and potassium and calcium as well. Observations of Mercury's exospheric sodium demonstrate that it is spatially and temporally variable and the variability is not solely related to interactions with the solar environment. Sources for the known exospheric species include impact vaporization, ion sputtering, thermal- and photon-stimulated desorption, crustal outgassing, and neutralization of solar wind ions. The relative importance of these production mechanisms has been debated, and they will only be resolved by careful study of the spatial and temporal variations of the exospheric species. Many of these production mechanisms predict the existence of several species (such as Ar, Si, Al, Mg, Fe, S, and OH) that have not been detected in ground-based observations. The MESSENGER spacecraft carries a UVS as part of the Mercury Atmosphere and Surface Composition Spectrometer (MASCS) instrument package. This spectrometer operates from 115 to 600 nm and its goal is to map the constituents of the atmosphere and provide information to relate them to specific source and production mechanisms. MASCS has detected sodium, potassium and calcium in addition to discovering magnesium (Mg) in Mercury's exosphere. While in orbit about Mercury, this instrument has been mapping the spatial and temporal distribution of these species in addition to searching for other predicted species.

The lunar atmosphere is extremely tenuous and was first detected using instrumentation on *Apollo 17*. The known lunar atmospheric species present in detectable abundances are Ar, He, Na, and K. So far, only upper limits on other species have been set using UV wavelengths. The *Apollo 17* UVS provided upper limits on the number density of H, H_2, O, C, N, and CO. More recently, HST FOS NUV observations of the region away from the surface of the Moon resulted in upper limits on OH, Al, Si, and Mg abundances. LRO LAMP measurements have provided new, lower upper limits on many species and have detected He.

3.2. Venus

For more than half a century, the very dense Venus atmosphere (surface pressure of roughly 90 bar) has been known to be composed principally of carbon dioxide (CO_2) based on the existence of strong spectral absorption features in the near-IR spectrum. Several layers of clouds many kilometers thick composed of sulfuric acid completely cloak the surface. Although these clouds obscure the surface at visual and UV wavelengths, the *Magellan* spacecraft used radar to construct maps of the volcanic surface. Atmospheric measurements in the UV region have been performed by sounding rockets and spacecraft, including the IUE, the *Pioneer Venus* orbiter, and the HUT. A UV image from Pioneer Venus is shown in Figure 49.1. The UV images of Venus' atmosphere show distinctive cloud patterns; in particular, a horizontal "Y"-shaped cloud feature (discovered by Mariner 10 Venus scientists in 1974) is visible near the equator. This feature may suggest atmospheric waves, analogous to high- and low-pressure cells on the Earth. Bright clouds toward Venus' poles appear to follow latitude lines. The polar regions are bright, possibly showing a haze of small particles overlying the main clouds. The dark regions show the location of enhanced sulfur dioxide near the cloud tops.

Within a few years of launch, the IUE identified several important trace constituents, including nitric oxide (NO), and confirmed the presence of several others, such as sulfur dioxide (SO_2). The Vega 1 and 2 probes measured local UV absorptions due primarily to SO_2 and aerosols. UV reflectance spectra obtained during two sounding rocket observations in 1988 and 1991 found that SO_2 is the primary spectral absorber between 190 and 230 nm and that sulfur monoxide (SO) is also present in Venus' atmosphere. The EUV instrument aboard the Galileo spacecraft observed Venus in the EUV wavelength range (55–125 nm) during its flyby. It detected emissions due to helium, ionized oxygen, atomic hydrogen, and an atomic hydrogen–atomic oxygen blend. In 1994, an extreme ultraviolet spectrograph (EUVS) was launched aboard a sounding rocket to observe the Venusian atmosphere from 82.5 to 11 nm. The EUVS identified several species, including N I, N II, N_2, H I, O I, and O II. The results of the EUVS measurements are consistent with earlier observations by the IUE, Pioneer Venus, *Venera 11* and *12*, and the Galileo EUV spectrometer. The EUVE provided the first full EUV (7–76 nm) spectrum of Venus in 1998 and made brightness measurements on the He I (58.4 nm) and O II (53.9 nm) lines. The FUV spectrum of Venus is dominated by the CO fourth-positive band system, as well as by neutral oxygen and carbon features, and has been measured by HUT (82–184 nm) in 1995 and by Cassini UVIS.

The IUE spectra of the Venus dayside and nightside obtained while Venus was near elongation displayed SO_2 absorptions at 208–218 nm, which when combined with the **column densities** reported by the Pioneer Venus orbiter and with ground-based observations are a measure of the SO_2 **mixing ratio** with altitude and its variation at the top of the cloud deck. This provides information on its variation in spatial distribution and permits models to be constructed of the planet's atmospheric dynamics. Observations of the Venus nightside with Pioneer Venus orbiter and the IUE detected the Venus nightglow, which is caused by the emission bands of nitric oxide (NO). Because of the short lifetime of NO on the nightside, this finding implies the rapid dayside–nightside transport of material in the Venus atmosphere. Observations of the Venus dayside have led to the discovery that the dayglow emission is carbon monoxide fluorescence, probably due to fluorescent scattering of solar Lyman-alpha radiation.

UV measurements of the Venus atmosphere using the Venus Express SPICAV have found the presence of a thin ozone layer at Venus, at a mean altitude of 100 km. SPICAV has also characterized the hydrogen exospheric corona and is monitoring SO_2 and the UV dayglow at Venus.

3.3. Mars

The atmosphere of Mars, like that of Venus, is dominated by carbon dioxide, and also consists of small amounts of N_2, H_2O, and their photochemical products. Mars' atmosphere is much less dense than Venus's atmosphere (approximately 14,500 times lower surface pressure) and is relatively transparent at most wavelengths. Therefore, UV to IR observations of Mars reveal information about both its atmosphere and its surface. The observations of Mars by the UVS instruments on Mariner 6 and 7 were the first to reveal the UV dayglow of this planet; later observations by Mariner 9 confirmed and extended these results. NUV

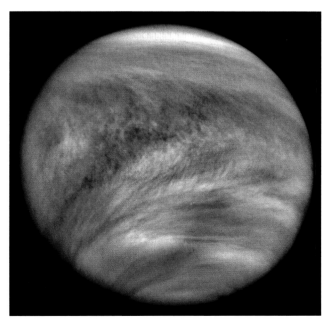

FIGURE 49.1 Pioneer Venus Orbiter cloud photopolarimeter image of Venus at 365 nm.

spectra (Figure 49.2(a)) revealed the presence of CO (a-X) Cameron bands, CO^+ (B-X), CO_2^+ (A-X), and CO_2^+ (B-X) features. FUV spectra displayed oxygen features at 130.4 and 135.6 nm, neutral carbon emission features at 156.1 and 165.7 nm, and the CO (A-X) at fourth-positive bands. All these UV features of the Martian airglow are products of processes involving Mars' CO_2 atmosphere. The CO_2^+ band systems are the result of a combination of photoionization excitation of CO_2 and fluorescent scattering of CO_2^+, and the CO (a-X) and (A-X) bands are due to photon or electron dissociative excitation of CO_2. The presence of escaping hydrogen (Ly-α at 121.6 nm), suggested atomic hydrogen within the atmosphere and also suggested the accumulation of oxygen and loss of water.

More recent (1995) HUT observations (82–184 nm) confirmed these early Mariner results (Figure 49.2(b)), and FUSE measurements detected molecular hydrogen (H_2) emission features at 107 and 116.7 nm for the first time. EUVE provided the first measurements of helium 58.4 nm within the Martian atmosphere. These helium observations have been used to set constraints on outgassing processes. Recent UV observations from the Mars Express spacecraft have made the first detection of the Martian nightglow, revealing nitric oxide emission features similar to those seen on Venus, auroral emissions associated with crustal magnetic field features, and high-altitude CO_2 ice clouds.

Mariner 6 and 7 UVS observations revealed the presence of the Hartley band of ozone (O_3), centered around 255 nm. The feature was detected at the south polar cap, through ratios of south polar spectra to low-latitude spectra (Figure 49.2(c)). Further observations with the UVS on the Mariner 9 orbiter revealed that the north and south polar ozone amount varied with season. Ozone densities were highest in winter and lowest in summer, anticorrelating with atmospheric water vapor content. The correlation between higher amounts of ozone with a cold, clean, dry atmosphere led to the conclusion that ozone is formed through the combination of atomic and molecular oxygen, both of which are more readily present when less water is available. Subsequent HST observations, and particularly the global coverage obtained by Mars Express SPICAM, have studied the seasonal variation of atmospheric ozone and have linked low-latitude ozone abundance variations across the Martian perihelion–aphelion cycle with the large annual water vapor variation due to the eccentricity of Mars' orbit. These observations have also been used to study atmospheric aerosol (dust and cloud) opacities, demonstrating the critical function that is performed by the small amounts of H_2O in the Martian atmosphere, which control the buildup of CO and O_2 and sustain the stability of CO_2. It has also been found that chemical reactions occurring on ice clouds are critical to understanding the observed Martian ozone levels. Ozone is readily destroyed by hydrogen radicals and is therefore a sensitive tracer of the chemistry that regulates the atmosphere of Mars.

3.4. Jupiter

Jupiter, the target of numerous Earth-based observations as well as spacecraft flybys, is composed of 90% hydrogen and 10% helium, with small amounts of ammonia and methane. The uppermost layers of the atmosphere are observable in the UV wavelength range and display products of photochemical processes.

The first FUV spectra of Jupiter were measured in sounding rocket experiments in the late 1960s and the early 1970s. These early measurements displayed H_2 Lyman and Werner band emissions and hinted at the presence of absorption features due to C_2H_2, C_2H_4, and NH_3. Stellar occultation observations by the Voyager UVS were significant in providing measurements of upper atmospheric temperatures. Early IUE observations confirmed that C_2H_2 absorption bands are the dominant features in the 165–185 nm wavelength region. Figure 49.3(a) shows the geometric albedo of Jupiter in the FUV wavelength range, displaying C_2H_2 and NH_3 features, derived using a composite spectrum of Jupiter from the 1978–1980 time frame.

The spectral geometric albedo of Jupiter as measured by the IUE at NUV wavelengths is shown in Figure 49.3(b). Most of this spectral behavior is attributable to hazes that are high above the cloud deck. The best-fit model (solid line in the figure) to the data occurs for a Jovian cloud deck with a geometric albedo of 0.25 and for a haze composed of particles with a single-scattering albedo of 0.42. Although such a result may not be able to provide an unambiguous identification of the materials that compose the haze, it can constrain the eligible candidate materials that are suggested by other observations.

IUE observations also permitted the ammonia–hydrogen mixing ratio to be calculated, and it was found to be 5×10^{-7}. The fact that the IUE was able to observe the absorption features of these species indicates that they are above the Jovian tropopause, where the clouds create an opaque barrier to light emitted from the material underneath and hence make spectral identification of the underlying material impossible. In July 1994, the comet Shoemaker-Levy 9 (SL-9) collided with Jupiter. It was not until the impact of the fragmented comet that studies of this underlying material became possible. The EUVE satellite observed Jupiter before, during, and after this event. EUVE found that 2–4 h after the impact of several of the larger fragments, the amount of neutral helium temporarily increased by a factor of ~ 10. This transient increase is attributed to the interaction of sunlight with the widespread high-altitude remnants of the plumes from the larger impacts. HST also observed this event with the GHRS and the Faint Object Camera (FOC). The UV spectra obtained

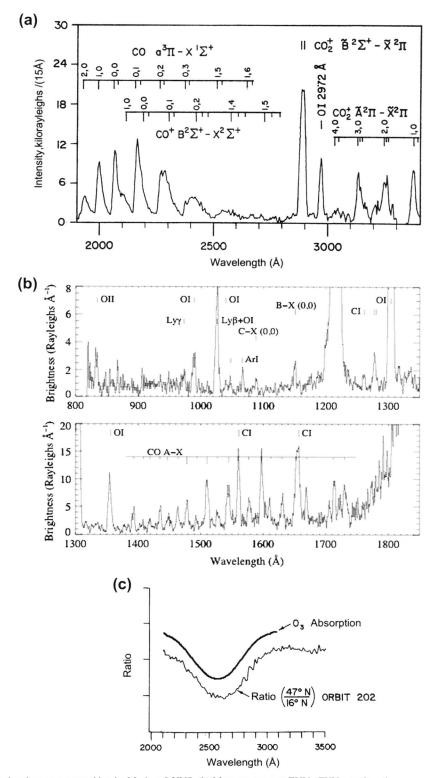

FIGURE 49.2 (a) Mars dayglow as measured by the Mariner 9 UVS. (b) Mars spectrum at EUV–FUV wavelengths as measured by HUT. (c) Ozone at high latitudes on Mars as measured by Mariner 9 UVS. A high-latitude spectrum is shown ratioed to a low-latitude spectrum, compared with a laboratory spectrum of the Hartley band of O_3. *Figures (a) and (c) reproduced with permission from Elsevier. Figure (b) reproduced with permission from AAS Publications.*

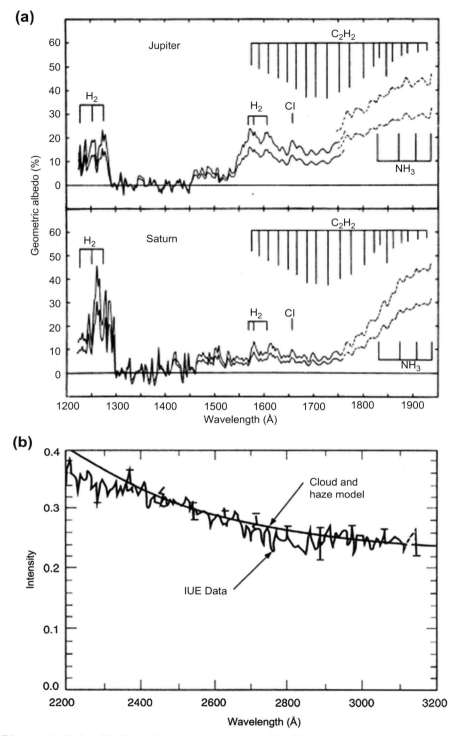

FIGURE 49.3 (a) FUV geometric albedos of Jupiter and Saturn as measured by the IUE. The albedos are derived from composite spectra of the planets between 1978 and 1980. The upper and lower curves for each planet correspond to assumptions of uniform and cosine-limb-darkened disks, respectively. The dashed lines represent data that are uncertain in magnitude owing to the subtraction of scattered light and are regarded as upper limits. (b) The spectral geometric albedo of Jupiter as measured by the IUE. The smooth solid line is the best fit from a model that assumes a layer of haze particles with single-scattering albedo of 0.42 that overlie a cloud deck with geometric albedo of 0.25. *Figure (a) reproduced with permission from AAS Publications. Figure (b) reproduced with permission from Elsevier.*

by HST of Jupiter after the collision of SL-9 identified approximately 10 species of molecules and atoms in the perturbed atmosphere, many of which had never been detected before in Jupiter's atmosphere. Among these were S_2, CS_2, CS, H_2S, and S^+, which are believed to be derived from a sulfur-bearing parent molecule native to Jupiter. The observations also detected stratospheric ammonia (NH_3). Neutral and ionized metals, including Mg II, Mg I, Si I, Fe I, and Fe II, were also observed in emission and are believed to be from the SL-9 comet fragments. The surprising observation was the absence of absorptions due to oxygen-containing molecules.

A major focus of study at UV wavelengths is the polar regions of Jupiter and their impressive exhibit of auroral activity. Jupiter's auroral displays are the most energetic in the solar system. FUV measurements were first made using the Voyager UVS, and subsequent observations were performed by the IUE. The FUV emissions are dominated by the hydrogen Lyman-alpha and the H_2 Lyman and Werner system bands. Synoptic observations of these UV emissions using the IUE have shown that they vary with Jupiter's magnetic (not planetary) longitude, and hence these emissions are magnetospheric phenomena. IUE observations have been used to construct a spatial map of the Lyman-alpha emission and the data indicate that the emitting material is upwelling at about 50 m/s relative to the surrounding material. More intensive UV observations with FUSE and HST instruments GHRS, FOC, STIS, and ACS have measured the temporal variability within the aurora and temperature variations within the auroral ovals seen at both poles. These variations are reflections of possible distortions in the magnetic field of Jupiter. HST measured the first detection of reversed Lyman-alpha emissions, which are linked to variable atomic hydrogen. Estimates of vertical column densities ($1-5 \times 10^{16}$ cm^{-2}) of atomic hydrogen above the auroral source have been made. HST has also detected UV emission from a superthermal hydrogen population. The Galileo spacecraft EUV spectrometer and UVS also observed Jupiter's aurora. These observations have placed constraints on the vertical distribution of methane (CH_4) in Jupiter's atmosphere. Slant methane column abundances are estimated to be 2×10^{16} cm^{-2} in the north and 5×10^{16} cm^{-2} in the south based on the Galileo observations. Cassini UVIS measurements showed that Jupiter's aurora responded strongly to the compression events produced when large solar coronal mass ejections reached Jupiter's magnetosphere. Figure 49.4 displays Jupiter's UV aurorae as imaged by *HST* in 1998. Evident in Figure 49.4 is the auroral "footprint" of Io, where the field line intersecting Io connects to the planet, revealing the magnetospheric relationship between the planet and its moon. Magnetic footprints of the other moons also exist but are less obvious in this image.

Bright H Ly-α emissions have been observed from Jupiter's equatorial region, and the source of this "equatorial bulge" has been debated. The source is likely a combination of charged particle excitation and solar resonance scattering and fluorescence. The emission has been shown to be consistent with resonant scattering of solar Ly-α with a large planetary line width, requiring a fractional ($\sim 1\%$) suprathermal population of fast H atoms in the uppermost atmosphere. The fast atoms are likely due to dissociative excitation of molecular hydrogen (H_2).

3.5. Saturn

Like Jupiter, Saturn's atmosphere is dominated by hydrogen and helium, with traces of water, ammonia, and methane. The FUV spectrum of Saturn was first measured in sounding rocket experiments in 1978. Absorption features in the UV spectrum of Saturn that have been associated with acetylene (C_2H_2) in the upper atmosphere were discovered using early IUE measurements. Figure 49.3(a) displays the FUV geometric albedo of Saturn derived using a composite of IUE Saturn spectra from the 1978–1980 time period. The mixing ratio of the acetylene is about 1×10^{-7}. Although acetylene is a well-known strong absorber of UV radiation, it alone cannot explain the low UV spectral geometric albedo of Saturn that has been reported by the IUE. Other UV-absorbing materials must be present. Comparisons of laboratory spectra of C_2H_2, PH_3, AsH_3, and GeH_4 with the IUE observations show that the best fit model for Saturn's atmospheric UV spectrum includes absorptions by C_2H_2, H_2O, CH_4, C_2H_6, PH_3, and GeH_4. The distribution of PH_3 and GeH_4 decreases with increasing altitude in these models, suggesting that UV photolysis is an important process occurring at higher altitudes.

Pole-to-pole mapping studies of the hydrogen Lyman-alpha emission across Saturn's disk led to the discovery of pronounced spatial asymmetries in the emission. Other observations of hydrogen do not find a variation in intensity with rotational period as with Jupiter. There is no rotational bulge in the Lyman-alpha emission as seen on Jupiter. This is probably due to the fact that Saturn's magnetic pole is coincident with the rotational pole, whereas in Jupiter's case, the poles are offset.

Like Jupiter, Saturn displays auroral activity. On both planets, this auroral activity also creates aerosols that are detectable in the UV as dark-absorbing regions. HST FOC UV observations discovered a dark oval encircling the north magnetic pole that is spatially coincident with the aurora detected by the Voyager UVS. Voyager 2 UV Photopolarimeter Subsystem (PPS) measurements also demonstrate a geographical correlation between the auroral zones of Jupiter and Saturn with UV-dark polar regions.

FIGURE 49.4 HST image of Jupiter aurora. The magnetic "footprint" of Io, marking the location where magnetic field lines joining the moon and Jupiter connect with the planet, is also seen as a bright spot with a tail outside the main auroral oval. WFPC2 is the Wide Field Planetary Camera 2 on HST. *Image credit: J. Clarke, NASA.*

Additional UV observations with the HST FOC of Saturn's northern UV aurora and polar haze support the hypothesis that the polar haze particles are composed of hydrocarbon aerosols produced during H_2^+ auroral activity. More recent HST UV imaging of Saturn's aurorae (Figure 49.5) shows that they behave differently from Jupiter's aurorae, varying in brightness and shifting in latitude. Cassini UVIS has also studied Saturn's UV auroral emissions and has discovered an auroral footprint connected to the moon Enceladus, similar to that of Io in Jupiter's auroral region.

3.6. Uranus

Uranus presents a unique observational circumstance to the inner solar system observer because of the fact that its pole is inclined 89° to the ecliptic and that at the present position in its 84-year orbit about the Sun it presents its pole to the Earth. This unusual inclination, combined with its great distance from the Earth, makes it impossible to use an Earth-based instrument to undertake pole-to-pole comparisons as was done with Jupiter and Saturn. Uranus has a geometric albedo at NUV wavelengths of about 0.5, more than twice that of Jupiter and Saturn. This suggests that additional absorbers are present in the Jovian and Saturnian atmospheres that are not present in the atmosphere of Uranus. Both Uranus and Neptune possess hot thermospheres and stratospheres that are substantially clear of hydrocarbons and other heavy constituents, making the UV albedos higher than those of Jupiter and Saturn. A sharp increase in measured reflectance intensity at wavelengths longward of 150 nm is indicative of acetylene (C_2H_2) present in the atmosphere of Uranus.

Voyager 2 spacecraft observations of Uranus found a very small internal heat source compared to the large internal heat sources found in Jupiter and Saturn. This suggests that there is very little atmospheric mixing driven by heating and buoyancy in the Uranian atmosphere. Thus, UV observations are able to sense a deeper region of the atmosphere.

The UV emissions from Uranus' atmosphere have been measured by the IUE and the Voyager UVS. To increase the signal-to-noise ratio, the IUE observers used principally low-resolution observations and binned broad-wavelength

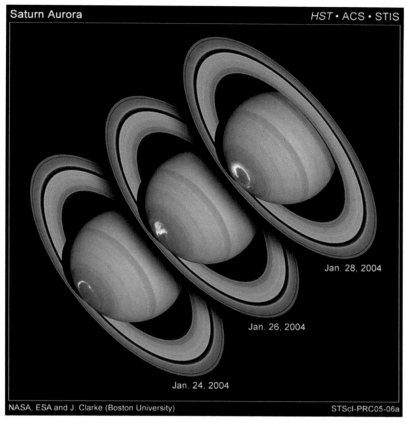

FIGURE 49.5 HST images of Saturn's varying aurorae. *Image credit: J. Clarke, NASA.*

regions together to search for broadband absorbers at UV wavelengths. Analysis of the IUE observations detected acetylene absorptions, which were also detected on Jupiter and Saturn. Based on these observations, the mixing ratio of acetylene is estimated to be 3×10^{-8}. Analysis of the Voyager UVS observations of H_2 band UV airglow emissions shows aurora at both magnetic poles, which are offset from the rotational poles by $\sim 60°$. The auroral emissions on Uranus are very localized in magnetic longitude and do not form complete auroral ovals as are seen on Jupiter and Saturn.

3.7. Neptune

Neptune is so distant that only broadband UV measurements are possible from the Earth's orbit. The geometric albedo of Neptune measured by the IUE is 0.5, which, like that of Uranus, is twice that of Jupiter and Saturn. Below 150 nm, Neptune's albedo is reduced by the higher hydrocarbon abundance carried into its stratosphere by its more vigorous vertical transport. Most of the important data for Neptune at UV wavelengths have come from the UVS onboard the Voyager 2 spacecraft and from HST. CH_4 and C_2H_6 abundances inferred from the Voyager UVS solar occultation experiment are between 0.0006 and 0.005 mole fraction for CH_4 in the lower stratosphere (with a mixing ratio of $5-100 \times 10^{-5}$) and the density of C_2H_6 is estimated to reach 3×10^9 cm^{-3}.

In 1994, HST imaged Neptune in six broadband filters, one of which was in the UV. The goal of these observations was to study the cloud structure on Neptune and compare the measurements with the observations made by Voyager 2. The HST images showed that the Great Dark Spot seen by Voyager no longer existed, but a new large dark feature of comparable size had appeared in the northern latitudes. HST measurements also detected weak carbon monoxide lines at 199.2 and 206.3 nm, suggesting a mixing ratio of $\sim 3 \pm 2 \times 10^{-6}$ in the upper troposphere.

Voyager 2 UVS measurements tentatively identified weak auroral emissions at Neptune's South Pole, interpreted as H_2 emissions. The North Pole was not observed by Voyager, so it is unknown whether this hemisphere displays aurora.

3.8. Pluto

For the dwarf planet Pluto and its large satellite Charon, UV spectroscopy is a potentially rich source of information due to the atmospheric chemistry that is likely occurring in this distant part of the solar system. Pluto's atmosphere is known to contain N_2 predominantly. The additional presence of methane in the atmosphere suggests that photochemical products should include hydrocarbons and nitriles, detectable at UV wavelengths. The IUE obtained a few spectra of Pluto and Charon and observed that the UV albedos vary with rotation. The amplitude of the rotational variation as measured at UV wavelengths by the IUE is greater than the rotational variation measured at longer wavelengths by Earth-based observers, consistent with the presence of an absorbing material that is spectrally active in the 320–480 nm wavelength range; the geometric albedo of Pluto in the NUV is spectrally flat. The composition of the absorbing material is unknown. (See Pluto.)

Observations made with the FOS on HST were used to determine upper limits on Pluto's predicted atmospheric species C_4H_2, C_6H_2, HC_3N, and C_4N_2 of 1.6×10^{16}, 1.8×10^{16}, 2.7×10^{16}, and 4×10^{16} cm^{-2}, respectively. The UV spectrum of Pluto's satellite, Charon, was also measured by the HST FOS and was found to have a spectrally flat geometric albedo in the NUV; Charon's spectrum does not exhibit any absorption or emission features that provide compositional clues.

The Pluto–Charon system is the target of the *New Horizons* flyby mission, which includes the Alice FUV imaging spectrograph (which operates from 52 to 187 nm) to probe the atmospheres and surfaces of these distant worlds.

3.9. Galilean Satellites

Jupiter's moon Io has one of the most unique atmospheres in the solar system: The primary sources of the atmosphere are volcanic emissions and sublimation of SO_2 frost on the surface. The result is a tenuous, patchy atmosphere made up of SO_2, SO, S_2, S, and O; trace species include Na, K, Cl, NaCl, and H. Gaseous SO_2 can be deposited onto the surface at night or during eclipse by Jupiter. Material is also lost to the torus as the ionized material sweeps by Io. Gaseous SO_2 was discovered at Io by the IRIS instrument on Voyager; since this discovery, numerous studies of Io's atmospheric processes have been made at UV wavelengths. IUE, HST, and the Galileo UVS have made measurements of Io at NUV wavelengths (200–350 nm). FUV observations from HST and Cassini UVIS identified emissions from neutral oxygen and sulfur and have been used in mapping the distribution of the SO_2 atmosphere.

Associated with Io is a plasma torus, or a donut-shaped ion cloud, centered at Io's orbital radius. This torus has been studied by Pioneer, Voyager, IUE, HST, EUVE, HUT, FUSE, Galileo, and Cassini. Oxygen, sulfur, and sodium ions are the major constituents of the torus, and protons are present at ~10% abundance; chlorine ions have also been detected. The torus is not uniform, and the density of ions shows various asymmetries depending on Io's position and dawn–dusk timings, in addition to temporal variations.

Intriguing auroral features are a consequence of Io's SO_2 atmosphere, resulting from electron impact excitation of atomic oxygen and sulfur as well as electron dissociation and excitation of SO_2, and have been observed at visible and FUV wavelengths. The Io flux tubes (IFT) and the Io plasma torus are the two primary sources of electrons in the Io environment. Due to the 10° tilt of Jupiter's magnetic field, Io is alternately above and below the magnetic equator (depending on Io's System III Jovian magnetic longitude, λ_{III}), the primary region of the torus electrons. Furthermore, the tangent points between field-aligned electrons and Io's atmosphere change as Jupiter rotates. The interaction of the torus electrons and the IFT electrons with Io's atmosphere has been detected at FUV wavelengths. Equatorial spots (Figure 49.6) have been observed to wobble up and down, reflecting the changing location of the IFT tangent points in time. The equatorial spot on the anti-Jovian hemisphere has been measured to be brighter than the spot on the sub-Jovian hemisphere, likely due to the motion of electrons through Io's atmosphere by the Hall effect, with hotter electrons on the anti-Jovian side. (See Io: The Volcanic Moon.)

Observations with the HST GHRS detected atomic oxygen emissions at 130.4 and 135.6 nm from Jupiter's satellite Europa, which have been interpreted as evidence for a tenuous O_2 atmosphere about this satellite. The source of this oxygen atmosphere is likely sputtering of the icy surface by corotating magnetospheric particles. These

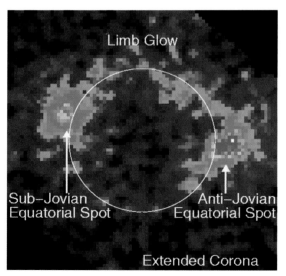

FIGURE 49.6 Io imaged at 135.6 nm by HST STIS. Equatorial "spots" are the result of interaction between electrons flowing along the IFT with Io's SO_2 atmosphere. *Figure courtesy of K. Retherford.*

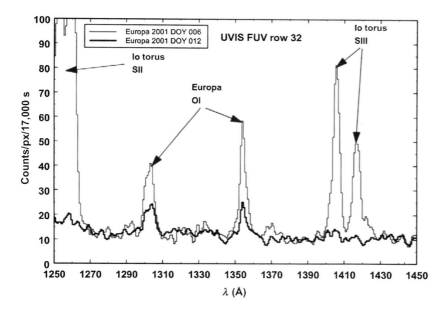

FIGURE 49.7 The FUV spectrum of Europa as measured by Cassini UVIS on January 6, 2001, and January 12, 2001. Neutral oxygen emission features appear at 130.4 and 135.6 nm in both observations. Io torus emission features from ionized sulfur also appear. The presence of the O I features is due to electron dissociation and excitation of a tenuous O_2 atmosphere at Europa. *Figure reproduced with permission from Elsevier.*

emission features were also measured by the Cassini UVIS instrument during the Jupiter system flyby in 1999–2000 (Figure 49.7). Similar oxygen emission features were detected by HST at Ganymede, although it was found that Ganymede's emissions are restricted primarily to the polar regions. The emissions are auroral features produced by dissociative excitation of O_2 by electrons traveling along the field lines of Ganymede's own magnetosphere. The HST imaged Ganymede's auroral emissions at 135.6 nm and found them to be temporally and longitudinally variable (Figure 49.8). *Galileo's* UVS detected hydrogen escaping from Ganymede, possibly due to sputtering of Ganymede's surface by charged particles. Callisto, in contrast to Ganymede and Europa, exhibits only weak oxygen emission features. An analysis of HST measurements found that the oxygen and CO emission features, expected after the discovery of CO_2 gas, are so faint that Callisto's interaction with the magnetosphere is like that of a unipolar inductor and that another species such as O_2 is likely abundant, enhancing the ionosphere and its conductivity. Callisto's ionosphere is apparently of sufficient conductivity to reduce the flow of plasma into its atmosphere and inhibits oxygen emission features in contrast to Ganymede and Europa.

3.10. Titan and Triton

Saturn's satellite Titan and Neptune's satellite Triton are among the largest satellites in the solar system. In addition, they are far from the Sun; therefore, the reduced solar energy allows the atmospheric gases to remain cold enough that they cannot easily escape by thermal processes (*See* Titan; Triton). Both Titan and Triton have nitrogen atmospheres that have been studied using UV techniques.

Titan is a solar system curiosity due to its very thick (~1.5 bar) nitrogen atmosphere, which prevents UV observations of the surface. The nitrogen atmosphere was discovered using a UV solar occultation by the Voyager spacecraft. Ground-based and Voyager spacecraft observations have identified methane (CH_4) as a significant constituent of Titan's atmosphere. Analyses of IUE and HST observations of Titan at NUV wavelengths have placed constraints on the properties of Titan's high-altitude haze and the abundances of simple organic compounds such as acetylene (C_2H_2). At FUV wavelengths, observations by Cassini UVIS demonstrate the presence of molecular and atomic nitrogen based on emission features due to electron dissociation and excitation (Figure 49.9). UVIS measurements are also being used to study the nature of the aerosols in Titan's atmosphere.

Triton's surface contains N_2, CO, and CH_4 frosts, which are highly volatile and in a continual state of exchange between the atmosphere and surface. IUE observations of distant Triton likely tested the limits of IUE's sensitivity. The photopolarimeter on Voyager 2 measured an albedo of 0.59 on all sides of Triton. HST FOS observations in 1993 detected broad apparent absorption features centered around 275 nm and between 200 and 210 nm. The FOS analysis also led to mixing ratio upper limits for atmospheric constituents of OH, NO, and CO of 3×10^{-6}, 8×10^{-5}, and 1.5×10^{-2}, respectively. HST STIS observations from August to September 1999 showed that Triton's albedo in the 250–320 nm range was 15–30% brighter and also spectrally redder than measured by the

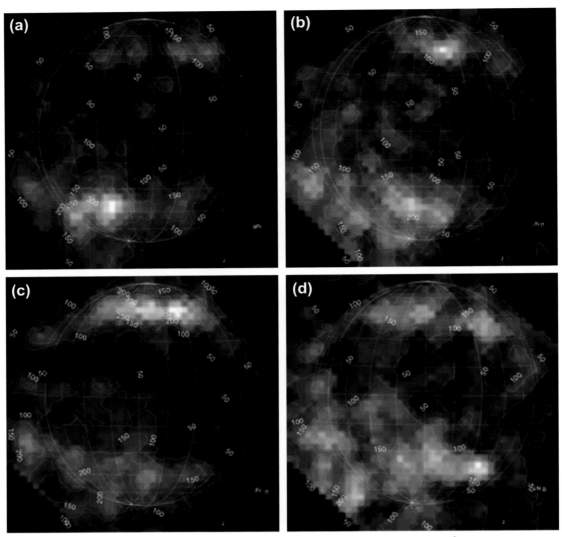

FIGURE 49.8 (A–D) HST STIS images of Ganymede's UV aurora. The images represent neutral oxygen at 1356 Å and reveal brightening in the polar regions that is variable over the four HST orbits during which the imaging took place. *Figure reproduced with permission from AAS Publications.*

HST FOS in 1993, suggesting that Triton's NUV albedo undergoes changes on timescales shorter than the seasonal cycle. Such changes may be due to bright frost deposition or to the emplacement of a relatively dark UV material. At this point, the source of the dramatic, short-timescale changes in UV brightness on Triton remains unknown.

4. OBSERVATIONS OF SOLID SURFACES

Many solid-state materials that make up the surfaces of solar system objects exhibit spectral absorption features, and thus it is possible to identify or constrain the abundance of solid components on the surfaces of these objects. This is accomplished by comparing the spectral geometric albedo of the object with the reflection spectrum of the solid-state materials as measured in the laboratory. The following discussion focuses on UV observations of solid surfaces throughout the solar system. Light is reflected from particulate surfaces by volume scattering and surface scattering. At longer visible and near-IR wavelengths, volume scattering dominates in most materials. At shorter NUV and FUV wavelengths, surface scattering dominates. Absorptions in the NUV are generally due to charge transfer and result in rather broad absorption features, compared with absorption features in the near-IR wavelengths, usually due to weaker electronic transitions. Many nonice materials absorb in the NUV wavelength, including organics, sulfur compounds, and many refractory materials. There is usually an absorption edge between 300 and 370 nm. In contrast, ices such as H_2O and CO_2 are not very absorbing at visible and NUV wavelengths (water and CO_2 ice exhibit strong absorption features near 165 nm). Atmosphereless bodies in

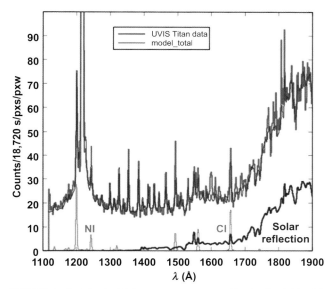

FIGURE 49.9 Titan's FUV spectrum as measured by the Cassini UVIS. The spectrum includes an overall continuum due to reflected solar light, in addition to emission features due to electron dissociation and excitation of molecular and atomic nitrogen. *Figure courtesy of D. Shemansky.*

the solar system experience weathering, whereby their surfaces are affected by bombardment of micrometeorites as well as charged particles, either from the solar wind in the case of the Moon and asteroids or from the magnetosphere of the parent planet in the case of the icy satellites of the outer solar system. Particularly in the case of icy surfaces, radiation-aged surfaces tend to be darker at UV wavelengths. Thus, UV observations of icy surfaces can be used to indicate ice "freshness," or the amount of contamination.

Of importance in studying the UV characteristics of the surfaces of many icy satellites of the outer solar system is the fact that most of these moons orbit synchronously. Their phase-locked nature means that one hemisphere (centered on 0° W) faces the planet at all times. The leading hemisphere is the side that faces the direction of motion of the satellite in its orbit and is centered on 90° W longitude, while the trailing hemisphere has a central longitude of 270° W. Ratios of spectra from the leading and trailing hemispheres are useful tools for studying hemispheric compositional variations, because processes such as plasma and micrometeoroid bombardment tend to affect primarily either the leading or the trailing hemispheres. In the Saturn system, many of the moons are affected primarily on one hemisphere by E-ring grain bombardment. These processes can induce absorptions as a result of ice regolith alteration, chemical and/or structural.

4.1. Galilean Satellites

The first in-depth UV studies of the Galilean satellites (Io, Europa, Ganymede, and Callisto) were accomplished with the use of the IUE satellite. Subsequent disk-integrated observations with the HST supported the initial findings of the IUE, in addition to adding to our knowledge of the composition of the surfaces of these satellites. Galileo studies contributed to disk-resolved studies of these bodies.

The high spatial resolution provided by Voyager and Galileo visible images shows that the Galilean satellites, particularly Io, are variegated in color on continental scales. Compositional information may be derived from high-spectral-resolution studies from Earth or near-Earth orbit because the satellite's synchronous rotation permits any given full-disk observation of a satellite to be associated with a uniquely defined hemisphere of that particular object. The extension of the available spectral range to shorter wavelengths with UV telescopes enhances this data set by permitting the identification of more absorption features, thereby providing further constraints on the compositional models that have been developed. (*See* Planetary Satellites.)

The corotating charged particles of Jupiter's magnetosphere have orbital speeds greater than those of the moons, so that the plasma sweeps by the moons impacting primarily the trailing sides. An in-depth study of several hundred IUE spectra of the Galilean satellites revealed significant hemispheric UV spectral asymmetries that are indicative of compositional variations.

Table 49.1 lists the geometric albedos of the Galilean satellites in three NUV wavelength bands, for the leading and trailing hemispheres. This table displays the significant hemispheric differences in brightness exhibited by these

TABLE 49.1 Ultraviolet Geometric Albedos of the Galilean Satellites

	260–270 nm	280–300 nm	300–320 nm
Io (leading)	0.015 ± 0.001	0.017 ± 0.001	0.042 ± 0.001
Io (trailing)	0.028 ± 0.002	0.030 ± 0.005	0.038 ± 0.003
Europa (leading)	0.213 ± 0.004	0.347 ± 0.010	0.407 ± 0.020
Europa (trailing)	0.118 ± 0.002	0.164 ± 0.004	0.222 ± 0.006
Ganymede (leading)	0.15 ± 0.007	0.190 ± 0.009	0.200 ± 0.001
Ganymede (trailing)	0.050 ± 0.003	0.060 ± 0.004	0.080 ± 0.008
Callisto (leading)	0.040 ± 0.008	0.049 ± 0.001	0.066 ± 0.002
Callisto (trailing)	0.056 ± 0.002	0.064 ± 0.002	0.105 ± 0.008

moons. Io's leading hemisphere is brighter than the trailing hemisphere only in the longest NUV wavelength band. Shortward of 300 nm, Io's trailing side has a higher albedo than its leading side, just the opposite of what is seen when Io is observed at visible wavelengths. (IUE's precursor, OAO-2, measured Io's albedo at 259 nm to be just 3%, in marked contrast to its 70% albedo at visible wavelength. This result was so unusual that it remained in doubt until IUE confirmed it by measuring Io's spectrum in this spectral range.) Io's reversal in brightness associated with orbital phase is more pronounced than for any other object in the solar system and proved to be important in efforts to determine the surface composition variation in longitude across Io's surface. It can be directly inferred from the Io data that there is a longitudinally asymmetric distribution of a spectrally active surface component on Io's surface. The material was determined to be sulfur dioxide (SO_2) frost, which strongly absorbs shortward of ~ 320 nm and is very reflective longward of that wavelength, as a result of IUE observations. Sulfur dioxide frost is in greatest abundance on the leading hemisphere of Io and is in least abundance on the trailing hemisphere. Figure 49.10 shows a Galileo UVS spectrum that displays Io's dramatic increase in albedo at wavelengths longer than 300 nm.

Europa and Ganymede exhibit a variation in brightness at NUV wavelengths with orbital phase that is in the same sense as the variation reported at the visible wavelengths; at all NUV wavelengths, these objects are brighter on their leading sides than on their trailing sides. A gradual decrease in albedo toward shorter wavelengths occurs on both hemispheres of both objects. The ratio of IUE spectra of Europa's trailing hemisphere to its leading hemisphere led to the discovery of an absorption feature present primarily on the trailing hemisphere centered around 280 nm.

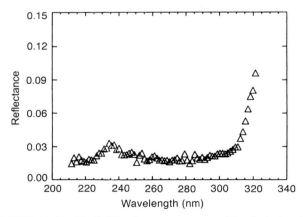

FIGURE 49.10 The NUV reflectance spectrum of Io, displaying the dramatic increase in brightness with wavelength starting at ~ 320 nm, due to the presence of SO_2 frost on the surface. This spectrum was measured by the Galileo UVS. The original discovery of SO_2 frost on the surface of Io in NUV spectra was made using IUE spectra. *Figure reproduced with permission from AAS Publications.*

The absorption feature, displayed in Figure 49.11(a), was attributed to an S—O bond and was suggested to be due to implantation of sulfur ions into the ice lattice on the trailing hemisphere. HST measurements confirmed the absorption feature, and it was suggested that the feature was similar to laboratory spectra of SO_2 frost layered on water ice. Subsequent disk-resolved Galileo UVS measurements showed that the 280-nm absorption feature is strongest in regions associated with visibly dark terrain. These locations have also been found to have relatively high concentrations of nonice material, interpreted to be hydrated sulfuric acid or hydrated salt minerals. An additional Galileo discovery was the presence of hydrogen peroxide (H_2O_2) on Europa, primarily in regions of lower nonice concentrations, such as on the leading hemisphere (Figure 49.11(b)). A study of Europa spectra from the early IUE era (1978–1984) compared to the late IUE era (1995–1996) suggested a temporal variation in Europa's leading and anti-Jovian hemisphere spectra that may be linked with variations in H_2O_2 abundances as a result of temporal variability in the space environment.

The IUE spectra of Ganymede's trailing hemisphere ratioed to the leading hemisphere revealed the presence of a possible absorption feature centered close to 260 nm, although the signal was approaching the IUE detection limits. It was suggested that ozone (O_3) in the ice could explain the apparent absorption feature. Subsequent HST measurements confirmed the presence of the O_3 absorption feature in the ice lattice on the trailing hemisphere. Disk-resolved observations of Ganymede from Galileo showed that the O_3 feature was strongest in the polar regions, and at large solar zenith angles, suggesting a connection with the magnetic field lines, or with photolysis or ice temperatures. Figure 49.11(c) displays the O_3 absorption feature as measured by the Galileo UVS.

Ground-based observations of Callisto have found that its albedo varies with **orbital phase angle** in the opposite sense to that of Europa and Ganymede (i.e. its trailing side has a higher albedo than its leading side). This is also true at NUV wavelengths. The albedo of Callisto decreases shortward of 550 nm and continues to decrease throughout the NUV. Its albedo at all wavelengths is lower than the albedos of Europa and Ganymede. Analysis of many of the IUE spectra, in addition to HST spectra, shows a broad, weak absorption at 280 nm similar to that seen on Europa (Figure 49.11(a)). These observations suggest the presence of SO_2 in a few leading hemisphere regions. The source of this SO_2 is not well understood and may be linked with implantation of neutral sulfur flowing outward from Io; disk-resolved measurements of Callisto from Galileo UVS do not exhibit the 280-nm absorption strongly. Rather, the Galileo observations indicate a different type of absorption at high latitudes, interpreted to be due to some type of organic species that is weathered away (carbonized) at

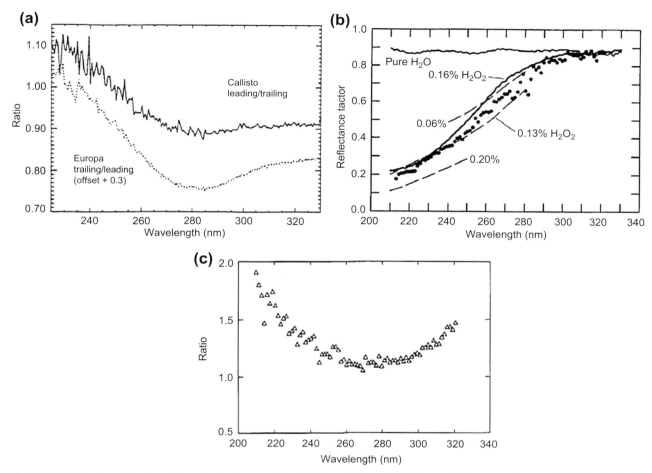

FIGURE 49.11 Spectra of significant absorption features on the icy Galilean satellites. (a) SO_2 absorption features on Europa and Callisto obtained by taking the ratio of the spectra of the trailing to leading hemisphere (Europa) and the spectra of the leading to trailing hemisphere (Callisto). (b) Hydrogen peroxide (H_2O_2) absorption feature as measured on Europa by the Galileo UVS. Also shown are mixture models for varying amounts of H_2O_2 in a water ice mixture, and the spectrum of pure H_2O ice. (c) Ozone (O_3) absorption feature as measured on Ganymede by Galileo UVS. Shown is the ratio of a spectrum from the north polar region to a region on the leading hemisphere. The broad absorption feature mimics O_3 in water ice. *Figures (a) and (c) reproduced with permission from AAS Publications. Figure (b) reproduced with permission from AAAS/Science.*

lower latitudes. The presence of CO_2 in the surface and in the atmosphere of Callisto and the dark nature of the surface suggest that carbon-based species are present across the surface associated with either endogenic or delivered organics. These organics experience chemical modification by UV radiation and are mixed into the regolith by meteoritic bombardment.

4.2. Saturnian Satellites

All of Saturn's large and medium-sized satellites, like the Galilean satellites and Earth's Moon, are in synchronous rotation (with the exception of Phoebe and Hyperion), and exogenic processes lead to hemispheric albedo dichotomies and absorptions in many wavelength ranges. At visual wavelengths, Tethys, Dione, and Rhea all have leading-side albedos that are 10–20% higher than their trailing-side albedos. The albedos of the satellites and their longitudinal asymmetries have been linked to the tenuous E-ring in which they all orbit, made up of fine-grained H_2O ice particles which tend to brighten the surfaces. IR observations of the large satellites of Saturn identified water ice as the principal absorbing species of the optically active surface of Mimas, Enceladus, Tethys, Dione, and Rhea and the trailing hemisphere of Iapetus.

Initial UV observations of the Saturnian satellites were made with IUE for Tethys, Dione, Rhea, and Iapetus. Mimas and Enceladus are too close to Saturn to obtain useful spectra from the Earth: the amount of scattered light from Saturn is too great to be overcome by IUE. Later UV observations of the satellites, including Enceladus, were made by the HST.

More recently, the Cassini UVIS has made numerous observations of the Saturnian satellites, providing clues to their surfaces and processes. In particular, UVIS has performed critical stellar and solar occultations of the Enceladus plume, contributing to the discovery of the amazing activity, and monitoring its vapor output.

The UV geometric albedos for the Saturnian satellites were calculated for three near-UV bandpasses using a combination of IUE and HST data and for a far-UV bandpass using Cassini UVIS data. These are shown in Table 49.2. The NUV albedos were determined using photometric corrections considering Voyager geometric albedos at 350 nm.

The near-UV albedos of the leading hemisphere of Enceladus and Tethys are the highest of the Saturnian satellites and are comparable to the high visual albedo reported by Voyager and ground-based visual observations. Tethys' trailing hemisphere is more absorbing with decreasing wavelength than the leading hemisphere throughout the UV region, whereas little difference was seen at visible wavelengths in the Voyager data set. The leading side of Dione is nearly two times brighter than its trailing side, a greater difference than the brightness variation reported from ground-based and Voyager visual wavelength observations. These hemispheric differences in UV albedo signal the presence of a UV absorber focused on the trailing hemisphere of Tethys and Dione. The leading side of Rhea is $\sim$30% brighter than its trailing side at NUV wavelengths, similar to what is measured at visual wavelengths. All these satellites exhibit a strong absorption in the NUV wavelength, particularly in the region between $\sim$190 and $\sim$250 nm, a spectral region that has not been well studied. Ammonia, which absorbs strongly near 210 nm, could contribute to this absorption, particularly on Enceladus; H_2O_2 (Figure 49.11(b)) is another good candidate. Cassini measurements have not yet zeroed in on the precise nature of this nonice species; other candidates include some type of tholin or carbonaceous material, or perhaps a small amount of iron and/or hematite. Because Dione and Rhea have lower visual albedos than the other satellites, their NUV absorption is not as strong and could signal the presence of one of these latter species, which are also relatively dark in the visible.

The hemispheric albedo asymmetry of Iapetus at visual wavelengths is extremely large (the trailing side is brighter by a factor of five). The leading (dark) hemisphere of Iapetus does not show spectral features consistent with water ice and has an IR spectrum that is nearly featureless. In the case of the dark hemisphere of Iapetus, the source of the darkening material is primarily the moon Phoebe, dust from which forms a giant, tenuous ring surrounding the Saturn system. In the UV wavelength, as in the visual, the leading side of Iapetus is extremely absorbing, and the trailing side albedo is comparable to that of other Saturnian satellites. The leading-side albedo is almost six times lower than the trailing side albedo at the IUE wavelengths, somewhat more than the five times darker than reported at visual wavelengths. The spectral absorber that darkens the leading hemisphere of Iapetus is more absorbing toward shorter wavelengths. Thermal segregation in addition to the Phoebe dust is a critical component in the production of Iapetus' dramatic hemispheric albedo dichotomy.

HST observations of Dione and Rhea (Figures 49.12(a) and 49.12(b)) detected a broad absorption centered around 260 nm. Like Ganymede, the feature has been attributed to the presence of ozone on both satellites. Ozone on these satellites is likely a product of radiolysis, from magnetospheric plasma bombardment.

With the arrival of the Cassini spacecraft at the Saturn system, FUV measurements of the icy satellites have been made with the UVIS. The FUV spectra of the icy satellites are dominated by the strong water absorption feature

TABLE 49.2 Ultraviolet Geometric Albedos of the Saturnian Satellites

	175–185 nm	240–270 nm	280–300 nm	300–320 nm
Enceladus	0.35 ± 0.05	0.95 ± 0.05	1.0 ± 0.05	1.05 ± 0.05
Tethys (leading)	0.29 ± 0.02	0.58 ± 0.02	0.6 ± 0.05	0.67 ± 0.05
Tethys (trailing)	0.18 ± 0.02	0.42 ± 0.02	0.48 ± 0.03	0.55 ± 0.05
Dione (leading)	0.21 ± 0.01	0.46 ± 0.05	0.49 ± 0.03	0.52 ± 0.03
Dione (trailing)	0.13 ± 0.01	0.22 ± 0.05	0.25 ± 0.05	0.29 ± 0.04
Rhea (leading)	0.21 ± 0.01	0.41 ± 0.05	0.44 ± 0.03	0.48 ± 0.03
Rhea (trailing)	0.15 ± 0.05	0.30 ± 0.05	0.34 ± 0.04	0.38 ± 0.04
Iapetus (leading)	0.04 ± 0.02	0.040 ± 0.02	0.04 ± 0.01	0.041 ± 0.01
Iapetus (trailing)	0.20 ± 0.02	0.30 ± 0.03	0.31 ± 0.02	0.32 ± 0.02

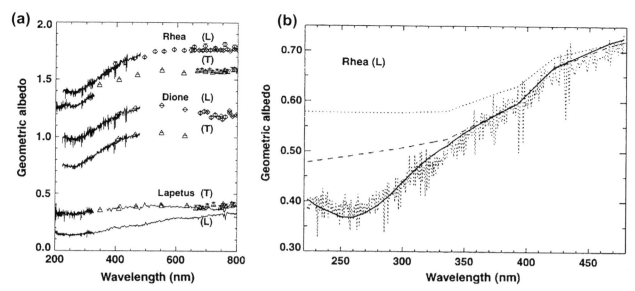

FIGURE 49.12 (a) The NUV geometric albedos of the leading (L) and trailing (T) hemispheres of Rhea, Dione, and Iapetus as measured by HST FOS (thick solid lines). Also shown are longer wavelength albedos (thin lines and discrete points) from other sources. The Rhea albedos are offset by 1.0 and the Dione albedos are offset by 0.5. The spectrum of Iapetus' leading hemisphere has been scaled by a factor of 2.5. (b) Rhea's leading-hemisphere albedo (a) with a model including O_3. *Figure (a) reproduced with permission from Nature Publishing Group. Figure (b) reproduced with permission from Nature Publishing Group.*

at ~165 nm. At wavelengths shortward of ~165 nm, the icy satellites are extremely dark due to the presence of water ice, and in fact are darker than the surrounding interplanetary hydrogen (IPH). Cassini UVIS images of the icy satellites therefore reveal both the dayside and nightside of the icy satellites. Figure 49.13(a) displays the FUV reflectance spectrum of Phoebe, one of the outermost satellites of Saturn and Figure 49.13(b) shows the FUV image of Phoebe; the visible wavelength image is shown in Figure 49.13(c) for comparison.

Cassini observations of Mimas have revealed a "lens"-shaped region at the low latitudes of the leading hemisphere which is bright at NUV wavelengths (338 nm, as measured by the imaging system). This region also exhibits different thermal inertia than the surrounding regions, likely due to bombardment by energetic electrons in this region. The energetic electrons do not appear to affect the FUV spectrum as much as the NUV spectrum, perhaps because of their penetration depth; UVIS measures only a hint of their effects. Instead, UVIS detects the seasonal effects of photolysis—at Mimas and Tethys—which likely produces H_2O_2 at the uppermost layers of the icy regoliths. Hydrogen peroxide is readily produced via photolysis, which is consistent with the UVIS-measured FUV darkening of the summer hemispheres of these moons.

4.3. Enceladus

Enceladus, not easily observed from the Earth's orbit due to its proximity to Saturn and the rings, has recently been the target of key UV measurements from the Cassini spacecraft in orbit around Saturn. Enceladus has long intrigued scientists because it is the brightest object in the solar system; Voyager images revealed vast regions that were evidently crater free, suggesting recent resurfacing by geologic activity. Furthermore, its orbit at the densest part of the broad, tenuous E-ring has suggested that Enceladus could somehow be the source of the E-ring ice particles.

UV measurements from the HST detected the hydroxyl radical, OH, in emission (308.5 nm) in the Saturn system, primarily near the orbit of Tethys. Similarly, Cassini UVIS measured neutral oxygen (at 130.4 nm), in varying amounts, with the greatest abundances near the orbit of Enceladus.

The presence of OH and O suggested that H_2O is produced by erosion of the inner icy satellites of Saturn by micrometeoroid bombardment and is then broken down by photodissociation to produce the neutral species. However, the amounts of H_2O necessary to produce the observed OH and O abundances were not consistent with sputtering rates; an additional source of H_2O was needed—and this remained a mystery until Cassini observations of Enceladus in 2005.

The Cassini spacecraft, through synergistic multi-instrument observations, discovered active water plumes on Enceladus. A stellar occultation by Cassini UVIS measured the presence of water vapor above the limb of the South Pole. Similar UV occultations of other regions of the moon had found no evidence of any gases, indicating that the vapor was locally confined to the South Pole. Surface temperatures, measured by the far-IR

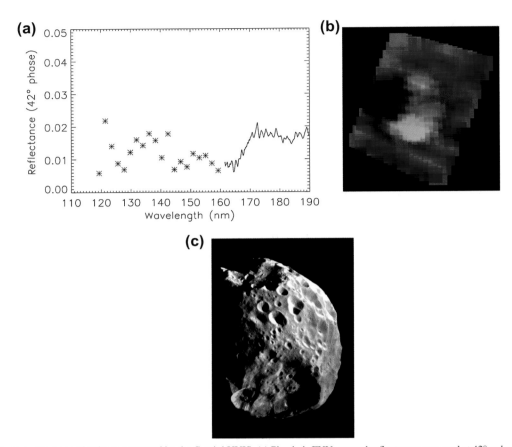

FIGURE 49.13 Saturn's moon Phoebe as measured by the Cassini UVIS. (a) Phoebe's FUV spectral reflectance measured at 42° *solar phase angle*. (b) FUV wavelength image of Phoebe. Red represents 1216 Å where IPH is bright throughout the solar system. Phoebe's water ice makes it darker than the background IPH on both the illuminated and the dark hemispheres so that the entire disk is visible. Blue colors represent longer FUV wavelengths and show that the brightness varies across Phoebe's surface due to solar incidence, topography, and compositional variations. (c) Visible wavelength image of Phoebe from Cassini. *Figures (a), (b), and (c) reproduced with permission from AAAS/Science; ISS image courtesy of NASA.*

spectrometer, were found to be anomalously high; measurements by the magnetometer, mass spectrometer, dust detector, and near-IR spectrometer on Cassini confirmed the presence of gaseous species and expulsion of ice grains from Enceladus' south polar hot spot. It is surprising that such a small, icy body is currently geologically active! The cause of the south polar hot spot and associated plumes is under investigation, and Enceladus remains a primary observational target of the Cassini mission.

4.4. Uranian Satellites

The five major satellites of Uranus—Miranda, Ariel, Umbriel, Titania, and Oberon—are a suite of icy satellites that are situated at about the limit at which the IUE was able to confidently return spectral information. They are so faint that it is not even possible to divide the IUE wavelength range into several bands, as was done with the Jovian and Saturnian satellites. All the spectral information is integrated into one wavelength range, and a geometric albedo can be determined.

The Uranian satellites are in an orbital plane that is parallel to the Uranian equator, and the pole of Uranus' orbit is tilted such that, at the present time, it is pointed toward the Earth. Therefore, only the poles of one hemisphere of the satellites of Uranus are observable with the IUE, and hence it is not possible to construct orbital phase curves and leading/trailing side ratio spectra.

IUE was able to observe Oberon, Uranus's brightest satellite. The IUE result proved to be an important and independent confirmation of results from the Voyager 2 photopolarimeter experiment. The UV geometric albedo of Oberon was found to be 0.19–0.025, an excellent confirmation of the earlier Voyager 2 PPS result of 0.17.

Spectra from 220 to 480 nm were obtained with the HST FOC for the Uranian satellites Ariel, Titania, and Oberon. The inner Uranian satellites Miranda and Puck were also observed from 250 to 800 nm with the HST FOC. The geometric albedos for Ariel, Titania, and Oberon display a broad, weak absorption at 280 nm, similar to the feature seen on Europa. Although this absorption feature on the Galilean satellites has been attributed to SO_2, it has been attributed to OH on the Uranian satellites. Both SO_2 and OH produce an absorption feature near 280 nm; however, the molecule OH (a by-product of the photolysis and radiolysis of water) is unstable at the surface temperatures of the Galilean satellites but is stable at the colder surface temperatures of the Uranian satellites. The 260-nm ozone feature seen on Ganymede, Dione, and Rhea has not been detected in any of the Uranian satellite spectra.

4.5. A Comparison of Icy Satellite Systems

The UV observations of planetary satellites can be integrated with the results of observations at longer wavelengths to provide a comparative assessment of the families of large planetary satellites in the solar system. The IUE- and HST-determined photometric properties of the larger planetary satellites of Jupiter, Saturn, and Uranus are shown in Figure 49.14 as a plot of UV-to-IR color ratio versus UV geometric albedo. The geometric albedos of the Saturnian satellites indicate that in this system there is a wide variation in UV geometric albedo. The Galilean and Uranian satellites have photometric properties that are common within each group. This is consistent with the hypothesis that the surface modification processes that have occurred are similar within the Galilean and Uranian satellite systems, but the two systems have surface modification processes that are distinct from each other. The diverse nature of the photometric properties of the Saturnian satellites suggests that multiple surface modification processes are altering the surfaces of the satellites. Certainly coating/bombardment by E-ring grains, an ongoing process, effectively keeps the optical surface of the Saturnian satellites young and unweathered; this is particularly important for Mimas, Enceladus, and Tethys.

In general, the NUV spectra of icy satellites are dominated by weathering products. Radiolysis and photolysis are extremely important processes at the surfaces of these satellites, and products of these processes are apparent at NUV wavelengths. This is evidenced by the presence of SO_2, O_3, and H_2O_2 in the surfaces of the icy Galilean satellites (and O_3 in the surfaces of some of the icy Saturnian satellites). The icy Galilean satellites are all relatively dark at UV wavelengths. However, water ice, the primary constituent of these surfaces, is bright in the NUV wavelengths. Therefore, another material must be responsible for the UV absorption of the icy Galilean satellites. The most likely darkening agents are elemental sulfur and sulfur-bearing compounds originating from the very young and active surface of Io, which are transported as ions outward from Io's orbit by Jovian magnetospheric processes. These energetic ions and neutrals interact with the icy surfaces of Europa, Ganymede, and Callisto and cause the ices to become darkened at UV wavelengths. This process competes with other processes of surface modification such as infall of interplanetary debris.

In contrast to Jupiter's UV-dark icy satellites, Saturn's icy satellites, particularly those orbiting closer to Saturn, are relatively bright. This is likely related to the presence of the large and tenuous E-ring. Mimas, Enceladus, Tethys, Dione, and Rhea all orbit Saturn within this broad ring of tiny icy grains. The relative velocities between the E-ring particles and the icy satellites explains the overall visual

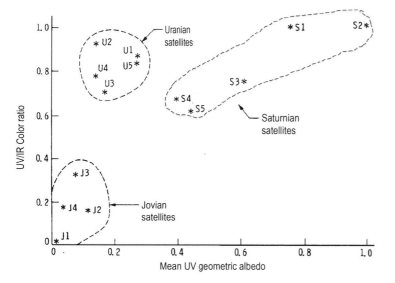

FIGURE 49.14 Comparisons of geometric albedos of the Galilean, Saturnian, and Uranian satellites. (Satellites are numbered such that the closest to the planet is #1; e.g. Io = J_1.) For the Galilean satellites and for Enceladus, Dione, Rhea, and Tethys, the UV albedos are from the IUE and HST. For Mimas, the UV albedo is from Voyager images (350 nm). The IR albedos of the Galilean and Saturnian satellites are from ground-based spectrophotometry. All the Uranian satellite data are from the Voyager photopolarimeter. With the exception of the slight difference in wavelength for Mimas noted previously, all the wavelength ranges are similar. Jovian and Uranian satellites each have distinct color ratios that distinguish the two groups of satellites from each other. The Saturnian satellites have little albedo similarity among themselves. This is consistent with the hypothesis that the surfaces of all the icy Galilean satellites are being modified by a common process. Likewise, the Uranian satellite surfaces may also have a common process of surface modification that is different from the Jovian system. The Saturnian satellite system may have multiple surface modification processes occurring.

brightness, as well as the large-scale longitudinal albedo patterns of the icy satellites. Mimas and Enceladus are both slightly darker on the leading hemispheres than on the trailing hemispheres, likely because the E-ring particles sweep by the trailing hemispheres, brightening them. The leading hemispheres of Tethys and Dione are brighter than the trailing hemispheres because their leading hemispheres sweep by the E-ring particles. Furthermore, Saturn's magnetosphere is different from Jupiter's and appears to be dominated by neutrals rather than by electrons and other charged particles. This difference may have an effect on the ice chemistry that occurs within the surfaces of the icy satellites because much of the ice chemistry occurring on the icy Galilean satellites is the result of charged particle bombardment.

4.6. Pluto and Charon

The first spatially resolvable images of Pluto and its satellite Charon were obtained by the HST FOC at visible (410 nm) and UV (278 nm) bandpasses. The image resolution is sufficient to show the presence of large, longitudinally asymmetric polar cap regions in addition to a variety of albedo markings. The combination of UV and visible images were used to look for regions of clean ice and nonclean (contaminated) ice—either radiation-darkened sites or sites where atmospheric chemistry products were deposited. No positive identification of solids on the surfaces of Pluto and Charon has been made at UV wavelengths. The cleanest ice (bright in UV and VIS bandpasses) was found in a location at the equator, although overall the equator was found to be heavier in contaminated ice than midlatitudes. The north polar region was found to have the cleanest ice.

4.7. Asteroids and Comets

At UV wavelengths, asteroids have been studied by the IUE and HST from Earth-based orbit. The Galileo spacecraft obtained NUV spectra of Ida and Gaspra during its travels through the asteroid belt, and the *Rosetta* Alice instrument studied Lutetia and Steins in the FUV wavelengths. Finally, the Mariner 9 UVS obtained spectra of Phobos and Deimos, the Martian moons that are likely captured asteroids.

The IUE satellite obtained UV observations of ~45 main belt asteroids in the wavelength range between 230 and 325 nm. The geometric albedos for these objects are consistently low, and three major asteroid taxonomic classes seen in the visible persist into the UV. Analysis of the IUE asteroid data shows that the asteroids observed have UV albedos that range from 0.02 for C-class asteroids to 0.08 for M-class asteroids; albedos of S-class asteroids are intermediate. Analysis of a set of IUE Vesta observations covering more than one rotation of Vesta indicates that this unusual asteroid displays UV albedo variations across the surface such that the UV light curve of Vesta is opposite that of the visible light curve. Such a spectral reversal is consistent with a hemispheric dichotomy in composition and/or a variation in geologic age (due to space weathering) across the surface. A study of IUE measurements of S-class asteroids has found that the UV spectral slope may be an important indicator of space weathering and ultimately exposure age; the strong decrease in albedo that is typical in silicates at NUV wavelengths decreases in strength with exposure, consistent with production of nanophase iron via weathering. Measurements at visible—IR wavelengths suggested the presence of hydrated minerals on Ceres; a search for OH (emissions at ~308.5 nm) in IUE spectra, however, found none. HST FOC, and more recently ACS, images of Ceres at UV wavelengths were the first well-resolved images of this largest asteroid. Albedo variations are detectable across the surface. The brightness of the surface in the FUV may rule out the presence of a large amount of water ice; a deep absorption band with a central wavelength of 280 nm appears be present but its cause is unidentified. Analysis of Mariner 9 UVS spectra of Phobos and Deimos show these bodies to be spectrally similar to carbonaceous chondrites. However, analysis of HST FOS data of these moons at UV—visible wavelengths, compared with FOS spectra of a C-type and a D-type asteroid, showed the Martian moons to be more similar to the D-type asteroid than to the C-type asteroid.

Both HST and IUE observed the Centaur asteroid 2060 Chiron, a possible former resident of the Kuiper Belt. Neither instrument detected emission from gaseous species at UV wavelengths, in contrast to CN emissions that have been reported at visible wavelengths. The UV albedo of Chiron is similar to that of some of the Saturnian and Uranian satellites. In particular, Chiron's UV/IR color and UV albedo are very similar to those of Dione. (*See* Kuiper Belt Objects: Physical Studies.)

Observations of comets at UV wavelengths are extremely useful for measuring fluorescence of solar photons by important atomic and molecular species, and thereby studying relative abundances of the vaporizing species and probing the photochemical and physical processes acting in the densest regions of the coma. UV observations of comets were first accomplished by sounding rockets and the *OAO* satellite prior to the launch of the IUE. These observations established the emission of hydroxyl ion at near the limit of ground-based observations, 308.5 nm. This is consistent with a cometary composition dominated by water ice; the hydroxyl ion is a product of exposure to solar radiation.

The IUE observed more than 50 comets (~400 individual spectra). IUE's photometric constancy provided the ability to compare observations of comets that appeared several years apart. Those observed range from short-period comets with aphelion near Jupiter to long-period comets that may be first-time visitors to our solar system.

All the comets observed by the IUE have displayed the 308.5-nm hydroxyl line, which is consistent with water ice being a major part of comet composition. Although all comets appear to have similar principal compositional components (water), each has different amounts of trace components, including carbon dioxide, ammonia, and methane, detected by the IUE and HST. Gas production rates have been derived for species such as H_2O, CS_2, and NH_3. Several comets that were observed over a long time exhibited differences in their dust-to-gas ratios from one observation to the next, consistent with a variation as a function of heliocentric distance.

The first detection of diatomic elemental sulfur in a comet was seen in comet IRAS-Araki-Alcock. The lifetime of the diatomic sulfur in the cometary atmosphere is quite short (~ 500 s). This makes sulfur a useful tracer of the dynamics of the tenuous cometary atmosphere, which appears during the short time that the comet is near the Sun. Analysis of the S I triplet emission band near 181.4 nm in cometary comae spectra taken with the IUE and the HST FOS shows that cometary sulfur, which is present and stored in a variety of volatile species, is depleted in abundance compared to solar abundances. The detection of CS at UV wavelengths in comae is attributed to the presence of CS_2 in the comet. Sulfur detected in the comae in excess of the sulfur attributable to CS_2 is assumed to originate from H_2S and nuclear atomic sulfur in the comet. Using this assumption, models have been used to measure total sulfur versus water abundances, which range between ~ 0.001 and ~ 0.01. (*See* Physics And Chemistry Of Comets.)

UV observations using the IUE, HST, and FUSE have also detected UV CO Cameron band emissions from comets, which is useful for measuring the CO_2 production rate. This rate derived from IUE observations of comet 1P/Halley agrees with the rate measured in situ by the spacecraft *Giotto*. These HST and IUE observations suggest that the level of activity of a comet may be linked to its CO abundance; however, this is based on a small sample of the comet population. FUSE measurements of C/2001 A2 (LINEAR) displayed H_2 emission lines of the Lyman system at 107.2, 111.9, and 116.7 nm, in addition to CO features that suggested both a hot and a cold component of CO, the hot component likely being due to excitation of CO_2 and the cold component being attributed to fluorescent scattering of CO or to electron impact excitation of CO.

4.8. The Moon and Mercury

The first UV observations of the Moon were made at FUV wavelengths using the instrument aboard the Apollo 17 orbiter. It was noted in these measurements that the lunar maria regions, darker than the highlands at visible wavelengths, are brighter than the highlands in the FUV. This was the first indication of the so-called spectral reversal, which was also detected at EUV wavelengths using measurements by the EUVE. This phenomenon is attributed to the concept that FUV measurements probe just the outer layers of the grains (surface scattering, as opposed to volume scattering measured at longer wavelengths), and that space weathering processes may cause the lunar grains to be covered with a fine coating. Lunar samples measured in the laboratory support this idea: lunar soils (presumably more weathered) show spectral reversal, while ground-up lunar rocks (presumably less weathered) do not. Galileo UVS measurements in the NUV showed that the maria are darker than the highlands and that the spectral reversal must occur at a wavelength shorter than ~ 220 nm. The HUT measurement of the lunar surface (a region near Flammarion-C, a border area between mare and highlands) at FUV wavelengths indicated an albedo of $\sim 4\%$ with a slight increase in brightness toward shorter wavelengths. Because of the different spectral behavior at UV versus visible wavelengths, ratio images of UV to visible color images and visible reflectance spectra are used to map spectral trends related to opaque mineral abundance and the combined effects of FeO content and soil maturity. From the *Apollo* samples, it is known that the dominant opaque mineral is ilmenite, which is high in Ti content. Thus, UV/visible ratio images have been used to map Ti content variation in the lunar mare basalts.

More recently, the LRO LAMP has been mapping the Moon at FUV wavelengths. Hints of water frost in the polar PSRs have been found, as has dayside hydration, which varies in abundance with latitude and throughout the day. The hydration appears to be consistent with solar wind-produced H_2O which diffuses through the lunar regolith at warmer temperatures.

The Mariner 10 spacecraft carried a color imager that included a NUV filter (355 nm). Mercury image ratios (UV/visible) have been used to map spectral trends associated with geologic features, using similar methods as used on lunar images. A lower UV/visible ratio suggests more FeO, or more mature soil. Spectrally neutral opaque minerals (such as ilmenite) tend to lead to a higher UV/visible ratio. Mercurian regions believed to be volcanic in origin have been found to have FeO amounts slightly less than average, consistent with ancient lava flows.

4.9. Planetary Rings

The rings of Saturn were successfully observed by the IUE in a series of observations between 1982 and 1985. The spectrum of the rings in the 160–310 nm range is dominated by the water ice absorption edge at ~ 165 nm. More recently, Cassini UVIS has made higher resolution

FIGURE 49.15 Saturn's rings as imaged by Cassini UVIS. This false color two-dimensional representation of Saturn's Cassini Division and A ring was generated from UVIS data obtained during a radial scan of the rings immediately after Saturn Orbit Insertion as Cassini flew over the rings. To generate the image, azimuthal symmetry was assumed. Although there are azimuthal variations in the structure of the rings, they are smaller than the 100-km resolution of this image. Red represents Lyman-alpha emission from IPH (121.6 nm) and shines through gaps and optically thin parts of the ring. Green and blue represent reflection of solar UV light longward of the water ice absorption edge near 165 nm. *Figure courtesy J. Colwell.*

observations of Saturn's rings; an image is shown in Figure 49.15. This image shows a combination of the UV reflectance and transmission of the ring system. The red-colored region at the left is the Cassini Division with a mean opacity of about 0.1, and the thin bright band near the outer edge of the rings is the 300-km-wide Encke gap. Brighter blue—green regions indicate cleaner water ice (less absorption by nonice species). The A-ring material is cleaner than the Cassini Division and the abundance of water ice is seen to increase near the outer edge of the A ring.

Cassini UVIS also utilizes UV stellar occultations to provide high-resolution profiles of the structure of Saturn's rings in three dimensions. (*See* Planetary rings.)

5. CONCLUSIONS

The importance of UV solar system science has been exhibited through discoveries and continuing studies spanning the topics of atmospheric and auroral science, surface composition and space weathering. UV observations of solar system surfaces and atmospheres have been made possible by the IUE and HST orbiting telescopes, along with FUSE, HUT, and EUVE, and have been substantially complemented by interplanetary missions such as Voyager, Mariner, Galileo, Cassini, LRO and New Horizons. The IUE spacecraft provided the astronomical community with the first stable long-term (spanning nearly two decades) observing platform in space, from which astronomers have been able to study regions of the spectrum that are inaccessible from telescopes on the Earth's surface. This foundation, with the support of UV spectrometers incorporated into the payloads of deep space missions, filled an observational void that had existed since the dawn of astronomy. These observations have led to important new discoveries and have provided tests of physical models that have been developed based on ground-based observations. IUE's observing capability was surpassed by HST, which has provided the astronomical community with the opportunity to look at fainter and more distant solar system objects and has led to new discoveries in the UV spectrum. The future of Earth-based orbiting UV telescopes is unclear, but such UV instruments, with ever-improving spectroscopic and imaging capabilities, are vital to understanding solar system objects and complementing longer wavelength observations.

BIBLIOGRAPHY

Barth, C. A. (1985). The photochemistry of the atmosphere of Mars. In *The photochemistry of atmospheres*. San Diego: Academic Press.

Chamberlain, J. W., & Hunten, D. M. (1987). *Theory of planetary atmospheres: An introduction to their physics and chemistry*. New York: Academic Press.

Nelson, R. M., & Lane, A. L. (1987). In Y. Kondo (Ed.), *Exploring the universe with the IUE satellite*. Dordrecht, The Netherlands: D. Reidel.

Nelson, R. M., Lane, A. L., Matson, D. L., Fanale, F. P., Nash, D. B., & Johnson, T. V. (1980). Io: longitudinal distribution of SO_2 frost. *Science, 210*, 784—786.

Carlson, R. W., et al. (1999). Hydrogen peroxide on the surface of Europa. *Science, 283*, 2062—2064.

Clancy, R. T., Wolff, M. J., & James, P. B. (1999). Minimal aerosol loading and global increases in atmospheric ozone during the 1996—1997 Martian northern spring season. *Icarus, 138*, 49—63.

Clarke, J. T., et al. (2009). Response of Jupiter's and Saturn's auroral activity to the solar wind. *Journal of Geophysical Research, 114*(A5). CiteID A05210.

Feldman, P. D., et al. (2000). HST/STIS ultraviolet imaging of polar aurora on Ganymede. *Astrophysical Journal, 535*, 1085—1090.

Feldman, P. D., et al. (2000). Far-ultraviolet spectroscopy of Venus and Mars at 4 Å resolution with the Hopkins ultraviolet telescope on Astro-2. *Astrophysical Journal, 538*, 395—400.

Hansen, C. J., et al. (2008). Water vapour jets inside the plume of gas leaving Enceladus. *Nature, 456*, 477—479.

Hendrix, A. R., et al. (2012). The lunar far-UV Albedo: indicator of hydration and weathering. *Journal of Geophysical Research, 117*, E12001. http://dx.doi.org/10.1029/2012JE004252.

Hendrix, A. R., Cassidy, T. A., Johnson, R. E., & Paranicas, C. (2011). Europa's disk-resolved ultraviolet spectra: relationships with plasma flux and surface terrains. *Icarus, 212*, 736—743.

Hendrix, A. R., Domingue, D. L., & Noll, K. S. (2012). UV properties of planetary ices. In *Solar system ices*. Springer.

Hendrix, A. R., Hansen, C. J., & Holsclaw, G. M. (2010). The ultraviolet reflectance of Enceladus: implications for surface composition. *Icarus, 206*, 608—617.

Hendrix, A. R., & Vilas, F. (2006). The effects of space weathering at UV wavelengths: s-class asteroids. *Astronomical Journal, 132*, 1396—1404.

Lane, A. L., Barth, C. A., Hord, C. W., & Stewart, A. I. F. (1973). Mariner 9 ultraviolet spectrometer experiment: observations of ozone on Mars. *Icarus, 18*, 102–108.

Lefevre, F., Bertaux, J.-L., Clancy, R. T., Encrenaz, T., Fast, K., Forget, F., et al. (2008). Heterogenous chemistry in the atmosphere of Mars. *Nature, 454*, 971–975.

McGrath, M. A., Hansen, C. J., & Hendrix, A. R. (2009). Observations of Europa's tenuous atmosphere. In *Europa*. Univ. Arizona Press.

Noll, K. S., Johnson, R. E., Lane, A. L., Domingue, D. L., & Weaver, H. A. (1996). Detection of ozone on Ganymede. *Science, 273*, 341–343.

Retherford, K. D., Moos, H. W., & Strobel, D. F. (2003). Io's auroral limb glow: Hubble space telescope FUV observations. *JGR, 108*(A8), 1333. http://dx.doi.org/10.1029/2002JA009710.

Stern, S. A. (1999). The lunar atmosphere: History, status, current problems, and context. *Reviews of Geophysics, 37*, 453–492.

Chapter 50

Infrared Views of the Solar System from Space

Mark V. Sykes
Planetary Science Institute, Tucson, AZ, USA

Chapter Outline
1. Introduction   1073
2. The Zodiacal Dust Cloud and Its Sources   1074
3. A Ring of Dust Around the Earth's Orbit   1078
4. Comets and Their Nature   1079
5. Asteroid Physical Properties   1083
6. Pluto and Beyond   1085
7. An Exciting Future   1087
   Information Web Sites   1087

Since 1983, a series of telescopes operating in the thermal infrared have been launched into Earth orbit and now heliocentric orbit. The images and other data returned have resulted in the discovery of new phenomena in the solar system and a new perspective on the processes within it. These observations have focused on comets, asteroids, and interplanetary dust, because until the launch of European Space Agency's (ESA's) Herschel Space Observatory, the major planets and Earth's Moon were too bright to be observed, overwhelming the detectors.

1. INTRODUCTION

At night we see objects in the solar system by the sunlight they reflect. The Moon, planets, comets and (with the help of telescopes) asteroids, and distant Kuiper Belt Objects are visible to the extent that they efficiently reflect that light, coupled with their apparent size. Small particles and dust are basically invisible with the exception of the zodiacal light, seen before sunrise and after sunset at certain times of the year, and the interplanetary particles that give off light as they burn up as meteors in the Earth's upper atmosphere. At thermal wavelengths, the sky is dramatically different (Figure 50.1). The otherwise invisible dust now dominates the view and "familiar" phenomena like comets have a very different appearance that has changed our understanding of their nature.

At thermal wavelengths, we are looking at the objects themselves as sources of light, instead of reflected light from another source like the Sun. All objects in the universe radiate heat. The energy distribution of this radiation with wavelength is a function of the temperature of the source. The Sun, at a temperature of more than 5000 K, radiates primarily at visual wavelengths and appears yellow. Colder sources radiate at longer wavelengths. Thus, the heating element of an electric stove appears orange-red.

Objects in space (e.g. asteroids, comets, and planets) also radiate, but at wavelengths much longer than can be detected by the human eye. This region of the spectrum (generally beyond 5 μm to the submillimeter) is referred to as the thermal infrared. Analysis of the thermal radiation from an object can tell us much about its composition and other physical properties including thermal inertia and grain size distribution and characteristics.

Observing this radiation from ground-based telescopes is complicated by thermal emission from the telescope itself and the atmosphere, both of which are much brighter than the astrophysical sources being observed. This has been compared to observing a star in the daytime with the telescope on fire.

Techniques that allowed objects within tiny patches of sky to be observed by ground-based and aircraft-borne telescopes were developed. Somewhat larger strips of the sky were observed by small-aperture, rocket-borne

FIGURE 50.1 At wavelengths visible to the human eye, the night sky (above) is dominated by black space and pointlike stars. From space, in the thermal infrared the same area of sky (below) is dominated by clouds of interstellar dust and extended solar system structures. Both images span $30° \times 20°$ near the First Point of Ares. The false-color thermal image was constructed from scans made by the Infrared Astronomical Satellite. Interstellar dust, known as "cirrus" is cold (indicated by red). Warm (blue) interplanetary dust reveals rings of dust around the solar system arising from asteroid collisions (one of which is seen as the broad band extending diagonally across the top of the image), and long contrail-like structures consisting of cometary debris (one is seen below the band). *Source (above): A. Mellinger.*

telescopes with tantalizing results. Only by getting above the atmosphere with a cooled telescope would it be possible to study the sky on a large scale at these wavelengths. This was achieved on January 26, 1983, with the launch of the Infrared Astronomical Satellite (IRAS), a joint project of the United States, the United Kingdom, and the Netherlands. It was the first in a series of space-based infrared telescopes, the latest of which include the National Aeronautics and Space Administration (NASA) Spitzer Space Telescope and NASA Wide-field Infrared Survey Explorer (WISE) (Figure 50.2).

These telescopes and their detectors were cryogenically cooled to minimize the noise introduced by the telescope and detectors themselves. In general, their operating temperatures need to be well below that of the sources they wish to observe. For solar system objects, this means well below 20 K to study the Kuiper Belt and beyond (Figure 50.3). This is accomplished by carrying a reservoir of liquid helium (having a temperature between a fraction of a degree and several degrees Kelvin) or solid hydrogen, both of which have a finite lifetime before it is expended. At that point, the telescope and detectors warm up and loose their sensitivity, although more recent facilities, Spitzer and WISE, have continued to provide useful data in their postcryogenic states.

Since the launch of IRAS, infrared detectors have become increasingly sensitive, thus able to study fainter and fainter sources. At the same time, the different spacecraft have operated in different modes in order to focus on different science questions. IRAS was primarily a survey instrument, mapping out the complete celestial sphere almost three times. Since the sky had not been mapped at thermal wavelengths, this was a mission of discovery. The NASA Cosmic Background Explorer (COBE) was also a survey instrument, with the primary goal of understanding the distribution of the cosmic background radiation from the Big Bang. One of its instruments, the Diffuse Infrared Background Experiment operated at thermal and near-infrared wavelengths at lower spatial resolution than IRAS. The ESA Infrared Space Observatory (ISO) and US Ballistic Missile Defense Organization Midcourse Space Experiment (MSX) were primarily pointing instruments, designed to measure specific targets or map out small regions of the sky in detail. Spitzer is also primarily a pointing and small-area mapping instrument. The Japan Aerospace Exploration Agency's (JAXA) Akari spent a portion of its mission generating the first thermal map of the sky since IRAS and COBE, and spent the remainder of its time conducting pointed observations. ESA's Herschel was primarily a pointing instrument, functioning at longer, submillimeter wavelengths. Finally, the WISE telescope was the first telescope to fundamentally reproduce the original IRAS survey for point sources. Even though all these missions were designed to address primarily astrophysical questions, they have been a great boon to our understanding of solar system phenomena.

2. THE ZODIACAL DUST CLOUD AND ITS SOURCES

When we think of the solar system, the image that often comes to mind is the textbook picture of planets orbiting the Sun on concentric orbits, asteroids between Mars and Jupiter, and the occasional comet flying by. However, in the inner solar system we are immersed in a cloud of dust that we see sometimes on the horizon as the zodiacal light (Figure 50.4) and sometimes in the direction opposite the Sun as the gegenschein. (*See* Solar System Dust.)

The zodiacal light is caused by the scattering of sunlight off of small particles near the Earth's orbit viewed when the

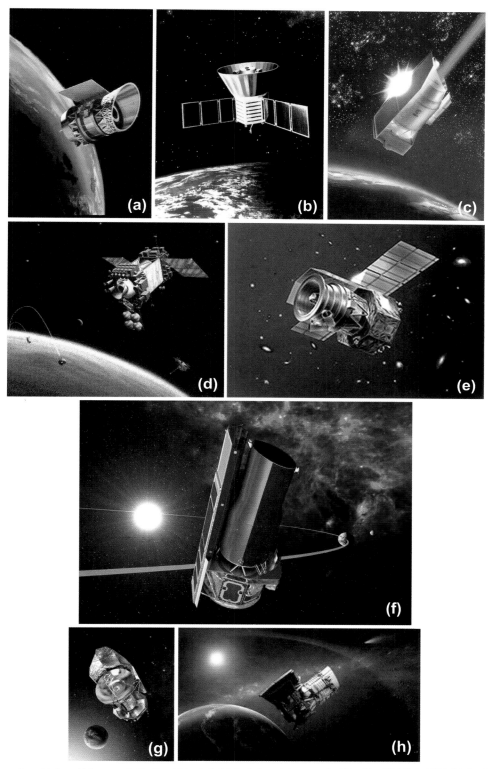

FIGURE 50.2 Space-based infrared telescopes in Earth orbit include (a) the Infrared Astronomical Satellite (IRAS) in 1983, (b) the Cosmic Background Explorer, 1989–1990, (c) the European Infrared Space Observatory (ISO), 1995–1998, (d) the US Air Force Midcourse Space Experiment (MSX), 1996–1997, (e) the Japanese Akari spacecraft, 2006–2011, (f) the Spitzer Space Telescope, which was launched into a heliocentric orbit in 2003, expected to operate 5 years, but has continued to operate in a postcryogenic phase since 2009, (g) the Herschel Space Telescope, 2009–2013, and (h) the Wide-field Infrared Survey Explorer (WISE), which launched in 2009, operated in postcryogen 2010–2011, at which time the spacecraft was put in hibernation. In 2013, WISE was reactivated for a 3 years postcryo phase.

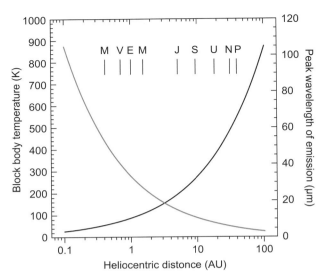

FIGURE 50.3 Bodies decrease in temperature with increasing distance from the Sun as shown by the red curve (which assumes that a body absorbs all incident sunlight). The corresponding wavelength at which the thermal emission spectrum peaks is shown by the black curve. At a given heliocentric distance, bodies that reflect increasing amounts of sunlight have lower equilibrium temperatures and will emit an increasing fraction of their thermal energy at longer wavelengths. For reference, distances of the planets are denoted by their first letters.

geometry is optimal. Comets were long thought to be the origin of the **zodiacal cloud**. However, early estimates of dust production by short-period comets fell far short of that needed to maintain the cloud in steady state against losses from particles spiraling into the Sun. This mechanism, where the absorption and reemission of solar radiation continually decreases particle velocity, is called Poynting–Robertson drag.

A cloud populated by comet emissions would have to be replenished by the occasional capture of "new", highly active comets into short-period orbits. Comet Encke was suggested as one such possible source in the past. Asteroid collisions have also been considered to be a source of interplanetary dust, but there were few observational constraints on estimates of their relative contribution to the cloud as a whole, until the advent of space-based telescopes operating in the thermal infrared (Table 50.1).

At thermal wavelengths, interplanetary dust is seen around the sky, peaking about the **ecliptic** plane (Figure 50.5). It appears brighter as we look closer to the Sun (where it is warmer and denser, hence giving off more thermal radiation). Within this broad band of dust there are structures related to dust-producing processes not seen before the advent of space-based infrared telescopes. The most prominent of these structures are the **dust bands**—parallel rings of dust straddling the plane of the ecliptic (Figure 50.5). These bands arise from collisions in the asteroid belt. When asteroids collide, the resultant

FIGURE 50.4 The zodiacal light from Mauna Kea, Hawaii. It is seen most prominently after sunset in the spring and before dawn in autumn at northern latitudes. *Source: Courtesy M. Ishiguro, ISAS.*

fragments are ejected with velocities that are small compared to the orbital velocity of the original asteroid. Consequently, the orbits of the fragments are close to each other, forming a "family" of smaller asteroids (Figure 50.6). All asteroid orbits precess like tops because of the gravitational influence of Jupiter. Small differences in the semimajor axes of the debris orbits cause them to precess at slightly different rates, so that over time, while their semimajor axes, orbital inclinations, and eccentricities remain roughly the same, their nodes (orbit orientations) become randomized. They are still identifiable as families, but the volume of space they fill is a torus.

These fragments continue to experience collisions and generate smaller and smaller pieces that fill the torus, whose cross-section is shown in Figure 50.7, with peaks in number density in its corners. A torus of asteroid dust, observed from Earth's orbit, would have the appearance of parallel bands of dust, straddling the **ecliptic** (the bands closer to the Sun overlapping those further from the Sun along our line of sight).

Chapter | 50 Infrared Views of the Solar System from Space

TABLE 50.1 Space-Based Telescopes Operating in the Thermal Infrared

Spacecraft	Launch Date	End of Cryogenic Mission	Aperture (cm)	Wavelength Coverage (μm)
IRAS	January 1983	November 1983	57	12–100
COBE (DIRBE)	November 1989	September 1990	19	1.25–240
ISO	November 1995	May 1998	60	2.5–240
MSX	April 1996	September 1997	33	8.3–21.3
Spitzer[1]	August 2003	May 2009	85	3.6–106
Akari	April 2006	November 2011	68.5	1.7–180
Herschel	May 2009	April 2013	350	55–672
WISE[2]	June 2010	September 2010	40	3.4–22

DIRBE, Diffuse Infrared Background Experiment.
[1] Spitzer continues to operate in postcryogenic mode.
[2] WISE continued to operate in postcryogenic mode until February 2011. A second postcryogenic phase commenced in September 2013.

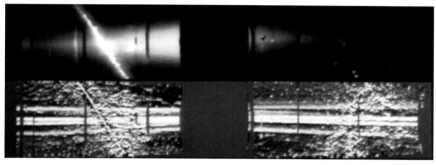

FIGURE 50.5 The zodiacal cloud (top) is seen extending from 0° to 360° in ecliptic latitude from right to left, constructed from scans of the ecliptic plane by IRAS. Ecliptic latitudes between 30° and −30° are shown. The diagonal structure crossing the ecliptic plane near 90° and 270° longitude is the galactic plane. Where the cloud is bright and wide (in latitude), the sky is being scanned at lower solar elongations, picking up the brighter thermal emissions of the warmer dust that lies closer to the Sun. As the satellite scans further away from the Sun at higher solar elongations, it is looking through less dust near the Earth and seeing a greater fraction of colder fainter dust. When filtered to remove its broad component (bottom), the zodiacal cloud reveals dust bands, located out in the asteroid belt and surrounding the inner solar system. Parallax results in their separation being smaller at lower solar elongations, where they are seen at a greater distance. Other solar system structures include dust trails.

Dust production from collisions is continuous down to sizes at which they are finally removed from the production region by radiation forces. When the fragments are around 1 μm in size they are immediately ejected from the solar system along hyperbolic orbits. These are known as **β-meteoroids**. For larger particles the solar radiation field and solar wind act as a friction to their orbital motion (**Poynting–Robertson drag**) and they will slowly spiral past the orbit of the Earth into the Sun. It is thought that the dust ultimately vaporizes and is incorporated into the Sun or recondenses into small particles that are then lost to the solar system as β-meteoroids.

Poynting–Robertson drag stretches out the small particle component of the torus (Figure 50.8), which retains its number density peak near its greatest distance from the ecliptic plane at a given heliocentric distance. When viewed in the thermal infrared from Earth's orbit, it still results in the appearance of distinct parallel bands straddling the ecliptic over all longitudes.

Initially, the **dust bands** were thought to be associated with the principal **Hirayama asteroid families** because of the proximity of their apparent latitudes with the orbital inclinations of those groups. These families are thought to have arisen from the catastrophic disruption of asteroids 100–250 km in diameter more than a billion years ago. If the asteroid belt as a whole was grinding down and generating dust, then it would follow that the most dust would be generated in the regions of greatest asteroid concentration—the largest asteroid families. Assuming this dust to be the main source of the **zodiacal cloud**, the cloud itself would be something expected to change slowly over much of the age of the solar system. An alternative hypothesis proposed that the dust bands arose from more recent collisions of smaller asteroids and that the zodiacal

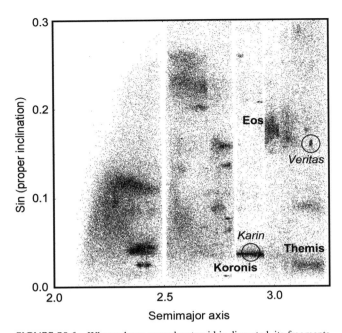

FIGURE 50.6 When a large enough asteroid is disrupted, its fragments are identified as other asteroids having similar orbital elements. The distribution of proper elements of asteroids in the main asteroid belt reveals many of these groupings referred to as families. The principal asteroid families, first identified by Kiyotsugu Hirayama in 1914, are Themis, Koronis, and Eos. The Karin (within Koronis) and Veritas families arose from the disruption of smaller asteroids within the past 10 million years and were identified as the sources of the two most prominent pairs of dust bands by D. Nesvorny.

cloud was highly variable over time. David Nesvorny and colleagues identified the sources of the two most prominent pairs of dust bands as the Karin and Veritas families and determined that the collisions forming these families occurred within the past 10 million years. This demonstrates that the zodiacal cloud, once assumed to be in relative steady state, may vary substantially over time as dust production in a given family slowly declines as more and more of its mass is ground up and removed by radiation forces and a new random collision creates family of debris that generates more dust. A faint inner pair of bands was originally associated with the very ancient Themis family, the largest asteroid family, which was formed by the catastrophic disruption of a 240 km diameter asteroid billions of years ago. Nesvorny and colleagues subsequently identified the origin of these bands as the Beagle cluster within the Themis family, arising from the disruption of a ~20-km-diameter asteroid less than 10 million years ago. Thus, the asteroidal contribution to the zodiacal dust complex appears to be dominated by recent catastrophic disruptions of small asteroids.

The mystery of whether asteroid collisions or short-period comets (or some combination) gives rise to the zodiacal dust cloud has been resolved in favor of Jupiter-family comets. Detailed simulations of the collisional and dynamical evolution of dust from the disintegration of these comets by Nesvorny and colleagues are shown to reproduce to good precision the broad thermal emission observed by IRAS (Figure 50.9).

3. A RING OF DUST AROUND THE EARTH'S ORBIT

As the sky was being mapped for the first time in the thermal infrared by IRAS, something odd was noticed; the sky always seemed to be slightly brighter in the direction opposite the Earth's motion about the Sun than in the direction of the Earth's motion. Since the satellite orbited above the terminator of the Earth and was looking out at different parts of the sky as the Earth orbited the Sun, if it was a difference in the actual sky brightness, would eventually result in a reversal of the direction of enhanced brightness. That did not happen. The "trailing" sky was always brighter. It made no sense that the Earth could be tracked by a large orbiting cloud—such a cloud would not be stable and disperse. Unable to come up with a satisfactory explanation, it was thought to be a strange calibration problem.

In 1993, a graduate student, Sumita Jayaraman, calculated that particles evolving from the asteroid belt past the Earth under **Poynting–Robertson drag** would have that orbital decay interrupted as a consequence of **resonance** interactions with the Earth (where the ratio of particle and

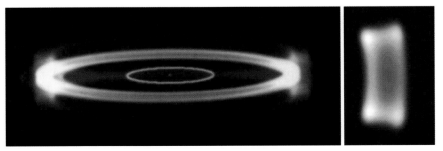

FIGURE 50.7 When nodes of particle orbits that are initially tightly clustered in semimajor axis, inclination, and eccentricity are randomized, they fill a torus. (Left) Tori associated with the principal Hirayama asteroid families would appear as parallel rings when viewed from Earth's orbit. (Right) Viewed in cross-section, particle number densities are maximum near the outer surface and are highest near the corners.

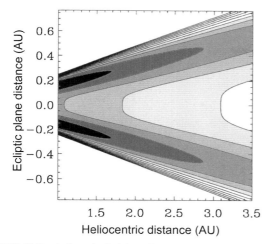

FIGURE 50.8 As hypothetical interplanetary dust particles originating in the Eos family torus migrate in toward the Sun, they contribute to the overall zodiacal cloud. The density contours of their contribution are shown, with darker regions corresponding to increasing particle number densities. The x-axis is heliocentric distance in astronomical units, and the y-axis is roughly the distance above the ecliptic plane. As the particles evolve to smaller heliocentric distances, the number density increases, and the extrema near the upper and lower edges of the cloud component is maintained. Viewed from the Earth, as we scan from the pole to the ecliptic, the column density of particles increases as we approach an angle near their average orbital inclination resulting in the appearance of a pair of parallel dust bands. *Source: Courtesy of W. Reach.*

Earth orbital periods is a ratio of integers). This dust would pile up for a while before continuing its sunward spiral, forming a ring around the Earth's orbit. The resonant ring has a clump (about 10% enhancement in density over the background **zodiacal cloud**) always trailing the Earth by about 0.2 AU in its orbit (Figure 50.10). This resonant structure represents a volume through which particles are circulating around the Sun, to be distinguished from a cloud of self-attracting particles. This explained the IRAS mystery. The existence of the resonance ring structure was confirmed by COBE observations.

The Spitzer Space Telescope is in a heliocentric orbit slowly trailing further and further behind the Earth. Spitzer traveled completely through the trailing cloud, allowing its three-dimensional structure to be probed in detail. It confirmed the predicted gap in dust within 0.1 AU behind the Earth and found the cloud to be azimuthally symmetric, centered about 0.2 AU behind the Earth with a width of 0.08 AU along Earth's orbit. Evidence for substructures was also detected. These observations constrain the production and evolution of particles from the asteroid belt in the size range sensitive to **resonance** with Earth's motion, and provide insights into how such structures in dust disks about other stars may provide details about planets imbedded in those disks. (*See* Extra-Solar Planets.)

4. COMETS AND THEIR NATURE

Comets are members of the solar system that have been known since ancient times. At visible wavelengths they are characterized by distinctive tails of micron-sized dust particles ejected from the nucleus and pushed away under radiation pressure, a coma of gas, ice, and dust surrounding the nucleus, and sometimes an ion tail of gas molecules carried away by the solar wind (its typically blue color arising from electron recombination events) (Figure 50.11). In the 1950s, Fred Whipple developed the standard model of comet nuclei as bodies largely of ice with a mixture of some dust—a "dirty snowball", which actively sublimates on the sunlit side as it moves close to the Sun, explaining the nongravitational components of their motion. He also linked their activity to the maintenance of the **zodiacal cloud**, which required constant replenishment as its constituent particles spiraled into the Sun under **Poynting–Robertson drag**. (*See* Physics and Chemistry of Comets.)

Comets are known to eject large particles from their association with many meteor streams. These particles spread over a comet's orbit and are scattered within its plane. If the comet orbit happens to extend inside the Earth's, these particles will be seen as meteors as the Earth passes through their orbital plane. Because they are striking the Earth's atmosphere from the same direction, meteor streams seem to come from a particular location in the sky.

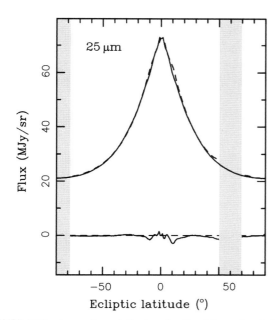

FIGURE 50.9 A scan through the broad zodiacal thermal emission by IRAS (above) is compared to a model of thermal emission from evolved debris of Jupiter-family comets (above). The residual is due to dust from recent asteroid collisions and not collisional activity within the main asteroid belt as a whole (below). *Source: Courtesy of D. Nesvorny.*

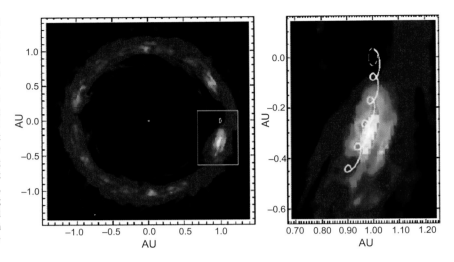

FIGURE 50.10 A simulated image of the Earth's resonant ring (left) showing a close-up of the cloud of dust that trails the Earth in its orbit through the year. The ellipse, in the inset box left and dashed at the top right, shows the orbit of the Earth in a rotating reference frame. The resolution of the image is 0.01 AU in the x and y directions. The cloud is modeled using 12 μm spherical particles of astrophysical silicate. The proximity of the clump of dust behind the Earth near its orbit explains why thermal scans of the sky behind the Earth along its orbit are always brighter than scans in front of the Earth. Over the course of its mission, Spitzer will travel through the dust cloud trailing Earth (right). The "loops" of Spitzer's orbital path and the oval of Earth's motion arises from the small eccentricity of the orbits. *Source: Figure courtesy of S. Jayaraman.*

FIGURE 50.11 Comet West on March 9, 1976, less than half an AU from the Sun after perihelion, exhibiting classic dust (white) and ion (blue) tails. *Source: Image by J. Laborde.*

This is called the radiant. Analysis of meteors as they burn up in the Earth's atmosphere indicates that particles associated with known comets have low (<1 g/cm^3) to modest (<2.5 g/cm^3) mass densities.

Viewed from space in the thermal infrared, many short-period comets were found to have very extensive, narrow trails consisting of millimeter- to centimeter-sized particles extending degrees to tens of degrees across the sky (Figure 50.12). The narrowness of the trails is due to the low velocities with which their constituent particles are ejected. They retain a record of comet emission history over a period of years to centuries. For comets having perihelia interior to the Earth's orbit, trails represent the birth of a meteor stream. The number density of particles within them are such that were the Earth to pass through one, there would be a "meteor storm" equal to or exceeding the famous Leonid storms of 1833 and 1966. First discovered by IRAS, it was inferred that **dust trails** were common to short-period comets. Surveys by Spitzer and WISE suggest this is the case (Figure 50.13).

Space-based infrared observations revealed that comets possessed far more dust than had been thought. Classical "gassy" comets such as P/Encke were found to possess both a significant large particle dust coma and trail (Figure 50.14). Encke's trail was found to extend over 80° of its orbit. It was determined that the ejection of large particles into trails was the principal mechanism by which comets lose mass. These particles quickly devolatilize after leaving the comet nucleus; this means that most of the comet's mass loss is in refractory particles.

The discovery of cometary dust trails has changed the picture of comet nuclei from being primarily icy bodies to objects more akin to "frozen mud balls", because of their much higher than expected fraction of refractory dust. The

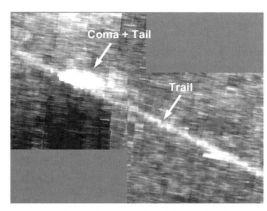

FIGURE 50.12 The most prominent dust trail in 1983 was associated with the short-period comet Tempel 2. The dust coma and tail appear as the fish to the dust trail's stream. Trails are characteristically narrow (as a consequence of the small relative velocities of the constituent dust relative to the nucleus of the parent comet) and trace out a portion of the comet's orbit. The particles ahead of the comet (to the left) are preferentially larger than those following the comet.

FIGURE 50.13 Spitzer has detected the first new dust trails in the infrared since IRAS. Shown are P/Johnson (top) and P/Shoemaker-Levy 3 (bottom). Spitzer is confirming the commonality of such large particle emissions across the short-period comet population. *Source: Figure courtesy of W. Reach.*

fraction of dust-to-gas in comet nuclei provides important information about where they formed and how they evolve, once captured into short-period orbits.

Dust-to-gas mass ratios corresponding to the canonical dirty snowball model range between 0.1 and 1. If we were to compress comet nucleus material so that refractories have a density of 3 g/cm^3 and volatiles had a density of 1 g/cm^3, this would give us a canonical nucleus in which 3−33% of the volume consisted of refractory material.

This picture is based largely on ground-based observations of dust at visual wavelengths, sensitive to particles within a decade or so of 1 μm in size. These observations underestimate the mass fluence of dust from comets.

Most of cometary mass loss appears to be in much larger (and dark) particles, which are difficult to detect at visual wavelengths. This conclusion was also reached after the European Giotto spacecraft was struck by a small number of large particles as it flew by comets Halley and Grigg-Skellerup.

Analysis of the IRAS observations of eight trails indicates that short-period comets lose their mass primarily in refractory particles in the millimeter to centimeter diameter size range. An average dust-to-gas mass ratio of 3 was calculated (Figure 50.15). This was the upper limit inferred for Halley by Giotto (with a nominal value of 2). Assuming the same densities for refractories and volatiles as previously, this corresponds to a comet nucleus that is 75% refractive by mass and 50% by volume (Figure 50.16). Mixing equal volumes of dirt and water in a backyard experiment demonstrates the apt description of such a mixture as a mud ball.

These dust-to-gas ratios also provide insight into the formation location of short-period comets. Dynamical considerations have lead investigators to focus on the proto-Uranus and proto-Neptune regions as that location. Significant amounts of ice have long suggested the outer solar system as the source of short-period comets. Consideration has also been given to their formation beyond the solar system, for instance, in molecular clouds. Models of comets forming in such interstellar locations yield comets dominated by their volatile components, contradicting inferences drawn from space-based thermal infrared observations. On the other hand, it is very interesting that both Pluto and Triton have effective dust-to-gas mass ratios that are identical to the average comet values determined from IRAS and Giotto (Figure 50.15). This is not unexpected if Pluto and Triton accumulated from protocomets in the vicinity of Neptune's orbit.

The existence of **dust trails** indicates that short-period comets are losing mass more rapidly than previously thought. Hence, their lifetime against sublimation may be shorter. A greater fraction of refractory material, however, would allow for the rapid formation of a nonvolatile mantle that is difficult to blow off, progressively choking off cometary activity. Such a mantle was apparent in the Giotto images of the Halley nucleus, which was near **perihelion** at the time. When activity is choked off, the comet would look like an asteroid until such time as sufficient pressure built up from subsurface ices to break through the crust in a burst of resumed activity. The discovery in August 1992 that asteroid 4015 was actually comet P/Wilson-Harrington (last seen in outburst in 1949) provided the first hard evidence of such "dormant" comets in the inner solar system.

FIGURE 50.14 Comet Encke (left) is considered a classic "gassy" comet based on visible wavelength observations showing only a gas coma and no dust tail. *(Image courtesy of J. Scotti.)* An ISO map (right, to scale) of P/Encke and its trail at 11.5 μm, evidencing anisotropic emission and requiring the spin axis of the nucleus to lie nearly in the orbital plane. The inferred dust-to-gas mass ratio of 10–30 is even higher than that inferred from IRAS observations. *Source: Figure courtesy of W. Reach.*

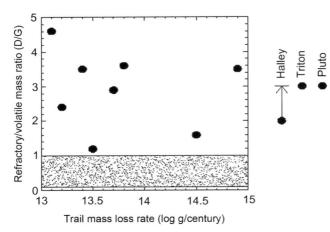

FIGURE 50.15 Dust-to-gas mass ratios are shown for comets having detected dust trails by IRAS. For comparison, values are shown for Halley, Triton, and Pluto. The shaded area spans the "canonical" ratios between 0.1 and 1.

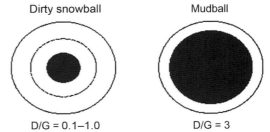

FIGURE 50.16 The canonical "dirty-snowball" model of comets, inferred from ground-based observations at visible wavelengths, is compared to the "frozen mud ball" or "frozen mud-ball" model inferred from space-based observations in the thermal infrared. All refractories are collected at the center in both cases.

In addition to trails associated with known short-period comets, IRAS also detected trails having no known source (Figure 50.17). Unfortunately, since these were discovered in the data long after the mission had ended, it was not possible to follow up the IRAS observations with observations from the ground in order to determine their orbits. So these objects are now lost. However, assuming a cometary origin, the numbers of these **"orphan trails"** suggest that there may be twice as many short-period comets as previously recorded, with the majority of them being less active and hence more difficult to detect by traditional means. The serendipitous detection of orphans requires a space-based thermal infrared survey of the sky having sufficient spatial resolution. More orphans may be waiting in the surveys obtained by the Japanese Akari mission and the WISE mission.

Space-based thermal infrared telescopes have also provided direct compositional information about comets through spectroscopic observations raising questions about the conditions in the early solar system when they formed. ISO made the first detections of crystalline silicates and CO_2 in its observations of the long-period comet Hale-Bopp and Jupiter-family comet P/Hartley 2. Crystalline silicates are formed at high temperatures not associated with the outer solar system. Spitzer measured the spectra of pristine material excavated by the Deep Impact event on P/Tempel 1, finding materials never before seen in comets such as carbonates and clay (which form in the presence of liquid water) as well metal sulfides, polycyclic aromatic

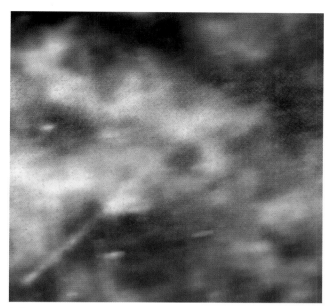

FIGURE 50.17 The brightest of the "orphan trails" detected by IRAS, seen against a background of interstellar clouds. Blue elongated sources are stars distorted by the rectangular shape of the detectors. Orphan trails are probably associated with comets never before detected.

hydrocarbons, and crystalline silicates. Liquid water is also not expected to be present in the outer solar system. These observations suggest perhaps substantial radial mixing of materials forming in hot and cold environments before the comets begin to accrete.

Yan Fernández and colleagues conducted Spitzer observations of 89 Jupiter-family comet nuclei and determined that their **albedos** are uniformly dark, with a cumulative size distribution with slope of $b = -1.9$ that may be enhanced in the 3- to 6-km-diameter range. This same survey showed 24% to be active at large distances from the Sun where water is inactive. In sum, Spitzer indicates that Jupiter-family comets, thought to derive from the Kuiper Belt, have a common origin, but a complex evolutionary history in their journey to the inner solar system. Ultimately, these comets may have been the source of Earth's oceans, as Herschel measurement of the D/H ratio in the Jupiter-family comet P/Hartley 2 found a value closely matching the ocean value.

5. ASTEROID PHYSICAL PROPERTIES

Over a million asteroids with diameters greater than 1 km reside primarily in a belt between Mars and Jupiter. (*See* Main-Belt Asteroids.) The asteroid population extends interior to the orbit of the Earth and beyond Jupiter. These objects are the scattered and disrupted remains of an early population of protoplanets whose continued growth was interrupted early on by mutual gravitational stirring, or the growth and migration of a massive Jupiter, or possibly both. The size distribution of asteroids provides insight into their origins and evolution through collisions and dynamical processes. Asteroid sizes are also important to determine the hazard of near-Earth objects. Asteroids are almost all unresolved to telescopes, and their brightnesses are insufficient for determining their sizes unless their **albedos** are known. By combining visible and thermal infrared observations of an asteroid, its diameter and albedo can be simultaneously determined. The difficulty of making radiometric (thermal) observations of asteroids from ground-based telescopes made for slow growth in the number for which physical properties could be determined. Space-based surveys in the thermal infrared greatly increased the numbers of asteroids for which albedos and diameters were available, providing new information about the composition of the asteroid belt.

Because most space-based infrared telescopes have operated in a pointed mode, targeting specific objects or locations for observations, few have engaged in asteroid surveys. The first significant catalog of asteroid albedos and diameters were derived from the IRAS survey in 1983. This survey resulted in 8210 observations of 2004 asteroids. A more limited survey by MSX in 1996 resulted in observations of 168 asteroids. By and large, such surveys rely upon the detection of asteroids in known orbits, since they are unable to provide sufficient astrometry to determine the orbits of newly discovered asteroids.

During the IRAS mission, there were tens of thousands of asteroids having known orbits. Three decades later that number has increased by an order of magnitude. Greatly increased detector sensitivity, has allowed even nonsurvey satellites to produce an even larger catalog of asteroid diameters and albedos. Over the nearly 6 year nominal mission of Spitzer, about 25,000 serendipitous asteroid measurements were made. It was the WISE mission, however, launched more than 27 years after IRAS, that has produced survey observations of more than 140,000 asteroids, providing the most comprehensive database of their thermal properties.

Known asteroid diameters inferred from WISE are shown in Figure 50.18 as a function of heliocentric distance. The absence of small asteroids with increasing heliocentric distance is a consequence of the limits of detector sensitivity.

Albedo provides insight into composition. Meteorite studies show that very dark surfaces arise from largely carbonaceous materials, while high-albedo surfaces are associated with silicic compositions lacking such carbonaceous material. IRAS confirmed that most C-type asteroids (thought to be carbonaceous) are indeed dark compared to the "stony" S-type asteroids, and that there is a trend toward darker asteroid surfaces with increasing heliocentric distance. This trend is seen in the WISE

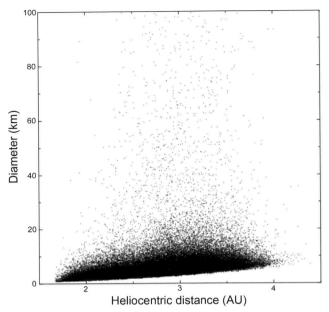

FIGURE 50.18 Main-belt asteroid diameters versus heliocentric distance of detection. The lower limit of detected asteroid sizes reflects the sensitivity limits of WISE detections. Asteroids with diameters less than 18 km dominate.

observations (Figure 50.19). This is consistent with the view that there is not only a primordial composition gradient through the asteroid belt, but that inner belt asteroids (predominantly S-type) were significantly processed by heating in the early solar system, while the outermost asteroids have experienced little heating and have retained a more "primitive" mineralogy.

New observations often result in as many new questions as new answers, and the WISE asteroid observations are no exception. Prior to IRAS, ground-based thermal observations had been preferentially made of the largest asteroids. It was noticed that there was a bimodal distribution in the inferred albedos, which was consistent with the main-belt asteroid population being dominated by dark C-type and bright S-type asteroids. IRAS added large numbers of observations of smaller asteroids and it was found that they had an albedo distribution quite different from the larger asteroids. WISE observations confirm this. Small asteroids have a fairly unimodal distribution spanning a far broader range of albedos than those of larger asteroids (Figure 50.20). Since the small asteroids are fragments of larger asteroids this might imply that surface mineralogies are not representative of interior mineralogies or that significant "space weathering" may affect the surface spectra of the larger bodies.

There has been some question as to whether all asteroid families mark the site of a past catastrophic disruption or whether in some cases asteroids might be clumped together due to dynamical forces such as gravitational perturbations

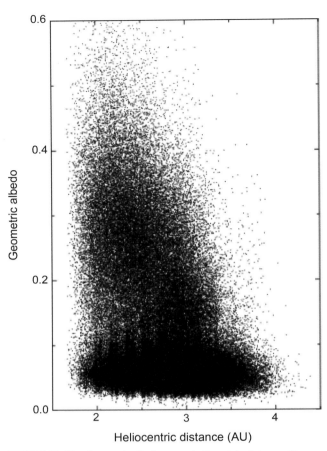

FIGURE 50.19 Geometric albedo versus heliocentric distance of known main-belt asteroids detected by WISE. The high-albedo asteroids are located predominantly in the inner portion of the main asteroid belt. Vertical structures are an artifact of asteroids whose orbits are not well determined.

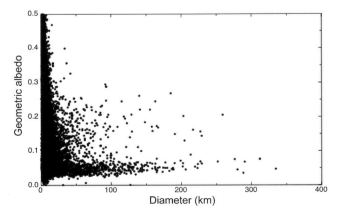

FIGURE 50.20 The geometric albedo of main-belt asteroid detected by WISE displays a bimodal distribution for asteroids larger than ∼50 km, while smaller asteroids display a much greater, roughly unimodal, distribution of albedos.

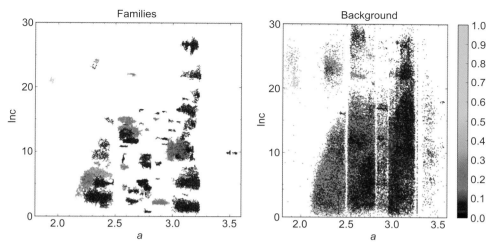

FIGURE 50.21 Proper inclination (inc, in degrees) versus proper semimajor axis (*a*, in AU) for 76 asteroid family members (left) and all nonfamily background objects (right). Color indicates albedo. Family members tend to have uniform albedos compared to corresponding background objects. *Source: Courtesy of J. Masiero.*

on their orbits by Jupiter. Albedo distributions can also provide clues to the origin of some asteroid families. Assuming the parent to have been compositionally homogeneous, the fragments should exhibit similar spectral properties. On the other hand, members of purely dynamical clusters would not be expected to have similar compositions. WISE scanned numerous members of 76 asteroid families and demonstrated that family members had albedos more similar to each other than to the background asteroids nearby, giving support, in those cases, to the asteroid breakup hypothesis (Figure 50.21).

Thermal spectra can provide information including thermal inertia and composition. Spitzer observations of Jupiter Trojan asteroids have yielded results suggesting fine-grained silicates in a relatively low-density, perhaps "fairy-castle" matrix. Does this indicate a possible cometary origin? Evidence for complex carbon compounds on primitive asteroids has also been detected. A Spitzer survey of M-class asteroids, spectrally similar at visible wavelengths, indicates major differences in thermal inertia among them, which could indicate significant differences in surface densities (due to relative age and collisional processing) or conductivity (some M-types are metallic, some may be stony). Thermal infrared observations of asteroids from space-based facilities are revealing an increasingly diverse population of objects.

6. PLUTO AND BEYOND

Thermal radiation from Pluto and its moon Charon was first detected by IRAS (Figure 50.22). The thermal flux of the system was consistent with that of a rapidly rotating **gray body** having an equatorial temperature of ~ 60 K. This information in combination with ground-based spectroscopic measurements and **albedo** maps derived from the mutual eclipses between Pluto and its moon between 1984 and 1990 has provided important insights into the nature and dynamics of the surface of Pluto. (*See* Pluto.)

When methane was first detected in visible wavelengths on the surface of these objects, it was thought that Pluto must be completely covered by the frozen ice, and would be isothermal because of the transport of heat as highly insolated locations would be cooled by sublimation and less insolated locations would be warmed by the condensation of atmospheric methane. Charon was thought to be a less likely location of such a coating of methane frost because of its lower gravity, from which methane would be expected to escape over time.

The detection of an extended atmosphere from a stellar occultation in 1988 and the subsequent detection of nitrogen ice on Pluto's surface required that the volatile surface ices be dominated by nitrogen with a small fraction of the more spectroscopically active methane, and that these surface ices must be very cold, ~ 35 K.

The spectroscopic- and IRAS-derived temperatures appear to contradict each other. Nitrogen ice at the warmer radiometric value would produce an enormous atmosphere that would have been evident in the occultation observations. The surface albedo maps, however, show that Pluto's surface is segregated into bright and dark regions with bright ices generally at higher latitudes. A high-albedo surface is bright at visible wavelengths (reflected light) and faint at thermal wavelengths, while a dark surface is faint at visible wavelengths and bright in the thermal. On Pluto, visible wavelength spectroscopy samples primarily the bright icy polar regions of the planet while space-based thermal observations are dominated by the dark equatorial region.

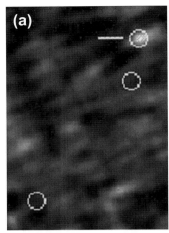
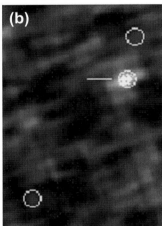

FIGURE 50.22 Pluto−Charon were detected moving across the infrared sky by IRAS. These images were constructed from 60 μm scans for (a) July 13, (b) July 23−24, and (c) August 16 in 1983. The predicted positions of Pluto−Charon at each of these times are indicated by circles. The August 16 position is the lower left circle in (a) and (b).

IRAS tells us that the volatile nitrogen ice—from which the atmosphere derives—is segregated on the surface of Pluto, away from the warmer regions giving rise to the thermal emission detected by the satellite. These warmer regions are probably a mixture of water ice and carbonaceous residue resulting from the radiation processing of methane ice over the age of the solar system. The dark regions are not contributing significantly to the atmospheric gases, which is consistent with a water/organic composition that would have negligible vapor pressure at the temperature inferred.

More sensitive Spitzer observations of Pluto−Charon have provided further insights into the properties of their surfaces, indicating high surface porosity and (on Pluto) the effect of what appears to be significant seasonal migration of surface ices. These observations will complement the spatially resolved studies, in reflected light, of the New Horizons mission as it passes by Pluto and Charon in 2015.

Spitzer and Herschel have been trained on trans-Neptunian objects (TNOs) beyond Pluto, finding that their **albedos** vary considerably. Short-period comets, which tend to have low albedos ∼0.04, are thought to derive from this region. The largest objects (in the dwarf planet category) have been determined to have very high albedos (>0.70). Nominally, an undifferentiated primitive surface exposed to galactic cosmic radiation over the age of the solar system would be expected to be covered with dark complex organics. For large bodies, bright surfaces suggest substantial evolution that allows volatile ices such as nitrogen to migrate to the surface—perhaps evidence of collisional activity or heating from early decay of radioisotopes resulting in differentiation.

Jupiter-family comets are thought to derive from the TNO region, yet Spitzer and Herschel have determined that many TNOs have significantly higher albedos than their cometary descendants. Comet nuclei are covered with dark refractory material. Their corresponding dust trails are likewise dark and refractory. Some smaller TNOs may be bright because their interiors are exposed by collisional activity. In that case, the bright ice and dark dust might not be in an intimate mixture, which would be necessarily dark. In fact, the 0.1−1 cm **dust trail** particles may be the principal size of the refractory component of comets. Thus, infrared studies of these distant bodies may be providing an important clue to the interior physical composition of comets and how they formed.

Beyond the Kuiper Belt lies the question of what additional parts of the solar system wait to be discovered. The sky is filled with cirruslike structures (Figure 50.1) and begs the question of whether any of them are local to the solar system.

IRAS surveyed 96% of the sky twice and 75% of the sky a third time. Images from these surveys were used to conduct the first-ever parallactic survey of the sky. Images of the same location taken weeks to several months apart were compared to search for reflex motion of extended sources, but in the end, nothing was identified in a volume extending 100 AU in some directions and 1000 AU in others—within the sensitivity of IRAS. With the greater sensitivity of current radiometric detectors and the larger apertures of space-based telescopes, the question asked with IRAS can be asked again and again, perhaps one day with a positive result or a definitive answer.

7. AN EXCITING FUTURE

At the time this article is being written, the age of infrared surveys of the sky from space may be drawing down for the foreseeable future. After more than two and a half years in hibernation, WISE is being restarted to continuing scanning the sky again to find more near-Earth asteroids. Its second postcryogen phase will last 3 years. Spitzer continues to operate in its postcryogen phase. In 2018, NASA will launch the massive James Webb Space Telescope (JWST). JWST has a 6.5-m-diameter mirror, and will operate from the near-infrared out to 28 μm, keeping the system cold through a combination of shielding and long-lived cryocoolers. While concentrating on the early universe and far astrophysical sources, JWST will have the capability to conduct detailed studies of near-Earth objects to the faint, small TNOs, bringing new understanding to the most distant regions of our solar system.

INFORMATION WEB SITES

Infrared Astronomical Satellite (IRAS) http://irsa.ipac.caltech.edu/IRASdocs/iras.html.
Infrared Space Observatory (ISO) http://www.iso.vilspa.esa.int.
Midcourse Space Explorer (MSX) http://irsa.ipac.caltech.edu/Missions/msx.html.
Akari (Astro-F) http://www.isas.ac.jp/e/enterp/missions/astro-f/.
Cosmic Background Explorer (COBE) http://lambda.gsfc.nasa.gov/product/cobe/.
Spitzer Space Telescope http://ssc.spitzer.caltech.edu.
Herschel Space Telescope http://sci.esa.int/herschel/.
Wide-field Infrared Survey Explorer (WISE) http://www.jpl.nasa.gov/wise/.

Chapter 51

New Generation Ground-Based Optical/Infrared Telescopes

Alan T. Tokunaga
Institute for Astronomy, University of Hawaii, Honolulu, HI, USA

Chapter Outline

1. Introduction — 1089
2. Advances in the Construction of Large Telescopes and in Image Quality — 1090
3. Advances with Detector Arrays — 1098
4. Advances in Adaptive Optics (AO) — 1099
5. Sky Survey Telescopes — 1100
6. Concluding Remarks — 1104
   Bibliography — 1105

1. INTRODUCTION

The telescope has played a critical role in planetary science from the moment of its use by Galileo in 1608. His observations of the craters on our Moon and the moons of Jupiter were the first astronomical discoveries made with a telescope. The development of larger refracting and reflecting telescopes led to the seminal discoveries of the rings of Saturn, asteroids, the outer planets Uranus and Neptune, new satellites of Mars and the outer planets, and Pluto by 1930.

Although spacecraft missions have revolutionized our understanding of the solar system (there are many examples in this encyclopedia), ground-based telescopes continue to play a very important role in making new discoveries, and this is the focus of this chapter. Figures 51.1–51.4 show a small sample of observations that illustrate the advantages of ground-based observations.

Solar system astronomers typically use telescopes built for other fields of astronomy. However, during the 1970s, NASA constructed ground-based telescopes to support its planetary missions. NASA funded the construction of the 1.7-m Kuiper telescope, the 2.7-m McDonald telescope, the University of Hawaii 2.2-m telescope, and the 3.0-m NASA Infrared Telescope Facility (IRTF) to provide mission support, but currently only the IRTF continues to be funded by NASA for that purpose. NASA also provides funding for searches for near-Earth objects (NEOs) as part of a Congressional directive.

Telescopes are designed to collect and focus starlight onto a detector. Larger telescopes not only allow more light to be collected and put onto the detector but also allow sharper images to be obtained at the **diffraction** limit of the telescope. While building larger telescopes may seem conceptually simple, ground-based observers have to contend with limitations imposed by physics, the atmosphere, technology, and funding. First, the collecting area of a telescope is limited in size. The largest optical telescope in the world presently has an equivalent collecting area of an 11.8-m diameter mirror. Although larger telescopes could be built, there are serious technical and financial difficulties to overcome. Second, the atmosphere limits observations to specific observing "windows" where the atmosphere is transparent; the wavelength range 25–350 μm is largely inaccessible to ground-based observers because of water absorption bands. Third, for infrared observations, the thermal emission of the atmosphere at wavelengths longer than 2.5 μm greatly reduces the sensitivity of observations. To overcome the problems of atmospheric absorption and thermal emission, it is necessary to go to high-mountain sites such as Mauna Kea in Hawaii and Atacama in Chile, or to use balloons, aircraft, or spacecraft. Fourth, atmospheric **seeing** typically limits the sharpness of images to 0.25–0.5 arcsecond at the best high-altitude sites. To achieve diffraction-limited imaging, one must employ special techniques that actively reduce the effects of atmospheric turbulence many times per

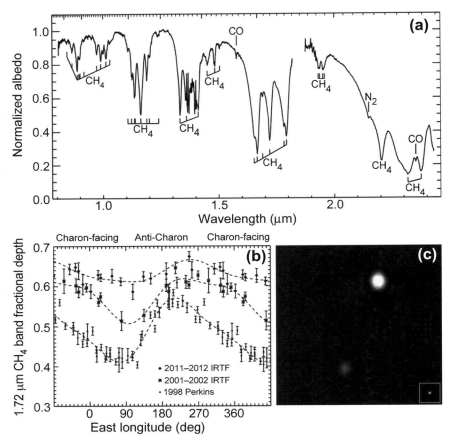

FIGURE 51.1 (a) Near-infrared spectrum of Pluto taken with the NASA Infrared Telescope Facility (IRTF). Absorption bands of methane (CH_4), carbon monoxide (CO), and nitrogen (N_2) ices on the surface of Pluto are shown. (b) Variation in the methane band depth with time and longitude on Pluto. These observations were taken over a period of 14 years and show surprisingly short timescale variations of composition on the surface of Pluto. (c) High angular resolution image of Pluto and its satellite Charon taken with the Gemini North telescope using a speckle camera. The angular resolution is 20 milli-arcseconds at a wavelength of 692 nm. The angular resolution of an unresolved star is shown in the insert in the lower right corner, and so Pluto and Charon are clearly resolved. The angular separation between Pluto and Charon is 0.81 arcsecond. These results show the long time base of observations and the extremely high angular resolution that can be obtained only with ground-based telescopes. *Courtesy of W. Grundy and S. Howell.*

second. One such technique, called **adaptive optics** (AO), is discussed later in Section 4.

Very large and low-noise visible and infrared detector arrays have been developed in the past decade, and this advance has been as significant as the improvement of telescope construction in providing greater observing capability. An important result of large-format detector arrays has been large sky surveys. Key objectives of these sky surveys are to detect asteroids that may present an impact hazard to Earth and to complete the reconnaissance of Kuiper Belt objects (KBOs). The major challenges of these survey projects are obtaining large enough detector arrays to provide the field of view required, and analyzing and storing the tremendous amounts of data that they generate.

In this chapter, we discuss very large telescopes that have been developed in the past 20 years to maximize collecting area, optimize image quality, and achieve diffraction-limited imaging with techniques to reduce the atmospheric turbulence. We also discuss sky survey telescopes that take advantage of the large-format detectors for the detection of solar system objects.

2. ADVANCES IN THE CONSTRUCTION OF LARGE TELESCOPES AND IN IMAGE QUALITY

The Hale 5.1-m telescope went into operation in 1949. It represented the culmination of continual telescope design improvements since the invention of the reflecting telescope by Newton in 1668. The basic approach was to scale up and improve design approaches that were used previously. Figure 51.5 shows the increase in telescope aperture with time. After the completion of the Hale telescope, astronomers recognized that building larger telescopes would require completely new approaches. Simple scaling of the classical techniques would lead to primary mirrors that would be too massive and an observatory (including the telescope enclosure) that would be too costly to build. Since the 1990s, a number of groundbreaking approaches have been tried, and the barrier imposed by classical telescope design has been broken. Table 51.1 shows a list of telescopes with apertures greater than 5 m. The increase in telescope area is a result of advances in telescope construction technology and the willingness of society to bear

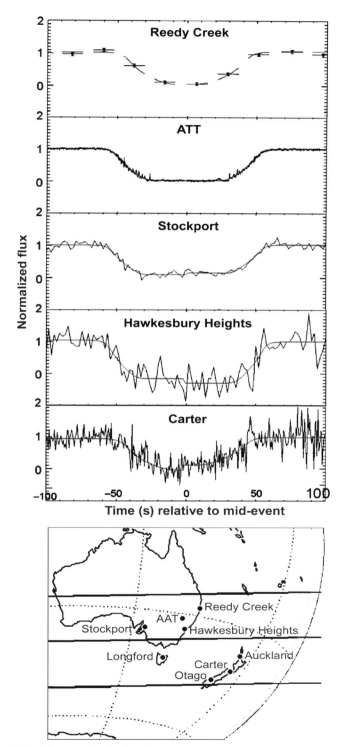

FIGURE 51.2 Occultation of a star by Pluto observed by a network of telescopes ranging in aperture from 0.2 to 4.0 m. The light curves are shown in the upper panel. The lower panel shows the locations of the telescopes. The three lines indicate the midpoint of the occultation track and the upper and lower limits of the track that are determined by the size of Pluto. This type of occultation observation provides a way to measure the size and shape of small solar system bodies that cannot be achieved in any other way. *Courtesy of L.A. Young.*

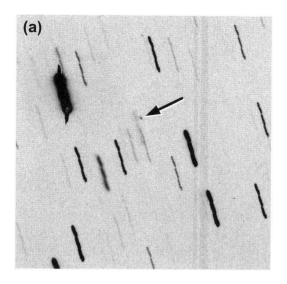

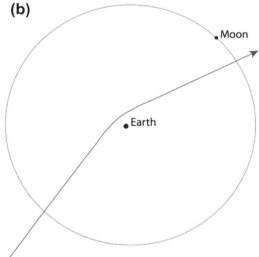

FIGURE 51.3 (a) Image of the asteroid 99942 Apophis. When it was discovered during its last close approach to Earth in 2004, it had a significant probability of striking Earth in the future. Subsequent observations show that it will pass within 5.6 Earth radii of the Earth in 2029 (see panel (b)). However, the future trajectory of the asteroid cannot be predicted well, and the asteroid will have to be carefully monitored with ground-based telescopes. The diameter of the asteroid is about 325 m. Close passages by an asteroid of this size are estimated to occur about once in 1300 years. The danger posed by such close-approaching asteroids has led to funding of asteroid surveys as discussed in Section 5. *Courtesy of R. Tucker, D. Tholen, and F. Bernardi.*

the costs. How much longer can this increase in telescope area continue on the ground?

Several groups in the United States are proposing the next leap in technology to a telescope in the 20–30 m class, and the engineering studies have started. One proposal is the 30-Meter Telescope (TMT), an international consortium consisting of the California Institute of Technology, University of California, Canada, Japan, China, and

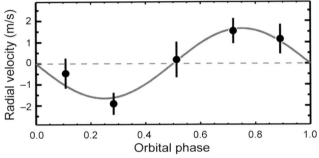

FIGURE 51.4 (Top) Artist conception of Kepler 78 and its planet, Kepler 78b. Kepler 78b was discovered as a transiting object with a radius of 1.2 times that of Earth by the Kepler telescope. With an orbital period of 8.5 h and a distance of only 0.01 AU from the star, the surface temperature of Kepler 78b is estimated to be 2000–2800 °C. (Botttom) The radial velocity variation of Kepler 78 caused by the planet measured with the Keck Telescope. From these measurements, the mass of the planet is estimated to be 1.69 times that of Earth. The mean density of the planet is similar to that of the Earth. It was recently estimated that there might be as many as 40 billion planets that are one to two times the size of Earth in our Galaxy. Ground-based observations such as shown here are essential for obtaining masses of transiting planets. *Courtesy of A. Howard.*

India (http://www.tmt.org/). This project proposes to build a telescope, similar in concept to the Keck telescopes, that will have a primary mirror composed of more than 492 hexagonal segments. As the name implies, the collecting area is equivalent to a circular mirror 30 m in diameter. The other project is the Giant Magellan Telescope, which is supported by a group of public and private institutions in the United States, Australia, and Korea (http://www.gmto.org/). This telescope concept consists of seven 8.4-m mirrors to create a single telescope with the collecting area equivalent to a 22-m circular mirror. In addition to these efforts the European Southern Observatory (ESO) is planning an even larger telescope, the European Extremely Large Telescope (E-ELT), with an equivalent mirror diameter of 39 m using 798 hexagonal mirrors (http://www.eso.org/public/teles-instr/e-elt/). Thus it seems inevitable that a ground-based telescope larger than 10 m will be built.

Five major technical advances have led to the development of large telescopes:

1. Advances in computer-controlled hardware allow correction for flexure of the primary mirror. This has permitted thinner mirrors to be used, reducing the mass of the mirror and the total mass of the telescope. For example, the mass of the ESO Very Large Telescope 8.2-m primary mirror is 23 tons with an aspect ratio (mirror diameter to mirror thickness ratio) of 46. This is a very thin mirror compared with the 5.1-m Hale telescope, which has a weight of 14.5 tons and an aspect ratio of 9.
2. Altitude-azimuth (alt-az) mounts reduce the size of the required telescope enclosure. An 8-m alt-az telescope can fit into the same size enclosure as a 4-m equatorial telescope. An alt-az telescope requires computer-controlled pointing and tracking on two axes, whereas the traditional equatorial mount requires tracking on only a single axis. The Hale telescope is the largest equatorial telescope ever built. All larger telescopes use alt-az mountings. Figure 51.6 illustrates the basic types of telescope mounts, and Figure 51.7 shows examples of telescopes with equatorial and alt-az mounts.
3. Advances in mirror casting and computer-controlled mirror polishing allow the production of larger primary mirrors with shorter focal lengths. A shorter focal length allows the telescope structure to be smaller, thus lowering the weight and cost of the telescope. It also greatly reduces the cost of the dome enclosure. The state of the art in short focal length primary mirrors is a focal length to diameter ratio of 1.14 used in the Large Binocular Telescope (LBT). For comparison, the Hale telescope primary mirror has a focal length to diameter ratio of 3.3. The smaller telescope structure with reduced mass requires less time to reach thermal equilibrium, and its lower mass makes it easier to move. This is extremely important in achieving the best image quality and to efficiently reposition in the telescope.
4. Advances in reducing dome **seeing** led to significant improvement in image quality. Dome seeing is caused by temperature differences within the dome, especially differences between the mirror and the surrounding air. To reduce dome seeing, it is necessary to flush the dome with outside air at night, refrigerate it during the daytime, and cool the primary mirror to about 0.5 °C below the ambient air temperature. Dome seeing is so important that large telescope projects use wind tunnel experiments to determine what type of dome design to employ. Careful attention to dome design is critical in

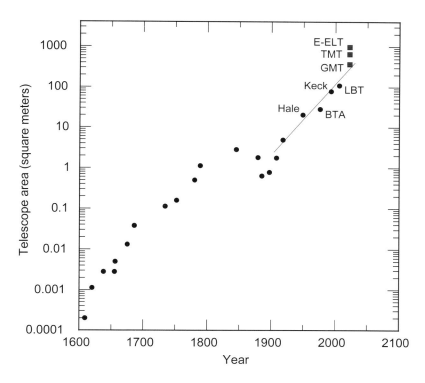

FIGURE 51.5 Increase in telescope area with time. Only the area of the largest telescopes at each time period is shown, so this indicates the envelope of maximum telescope area as a function of year. The time for the telescope area to double is about 26 years from the invention of the telescope in 1608 to the present. However, the doubling time has decreased from about 1900 to the present. The solid line shows a doubling of telescope aperture about every 19 years. The next jump in aperture size will be in the range of 22–40 m with an even shorter doubling time. See Table 51.1 and the text for a discussion of large telescopes.

eliminating dome seeing and achieving the very best seeing at the observatory site. Figure 51.7(b) shows an innovative approach to providing dome flushing by providing slits in the dome.

5. Advances in telescope construction have led to novel methods of reducing the cost of building extremely large telescopes. For example, the 10-m Keck telescopes have mirror segments that make up the primary mirror (Figure 51.7(c)). Although this technique had been used to build radio telescopes, the difficulty of making the segments and the high-precision alignment required at visible wavelengths presented formidable obstacles. Fortunately, the problems of fabricating segmented mirrors and aligning them were solved. The hexagonal mirror segments have a thickness of 75 mm, and so the aspect ratio of the 10-m primary is 133 and the total weight of the glass required is 14.4 tons, about the same weight as the 5-m Hale telescope. Another novel approach uses two 8.4-m primary mirrors on a single structure as in the LBT (Figure 51.7(d)). A third approach involves building a telescope with a fixed vertical elevation. Stars move past the prime focus and are tracked for a limited time. This approach has limitations, but it is much less expensive to build. Two projects (the Hobby–Eberly Telescope and the South African Large Telescope) have adopted this design to achieve 9-m class telescopes at about 15–20% of the cost of an equivalent alt-az telescope.

Large telescopes generally employ one of three different types of primary mirror fabrication: (1) Segmented mirrors. Each segment is figured appropriately, and all segments are aligned so as to act as a single mirror. (2) Thin meniscus mirror using low-expansion glass. Such mirrors are made as thin as possible to be lightweight and to have a short thermal time constant (thus coming into equilibrium with the atmospheric temperature quickly). (3) Thick honeycomb mirror using borosilicate glass. The advantage of using borosilicate glass instead of low-expansion glass is that the former is much cheaper. The disadvantage of borosilicate glass is that the mirror temperature needs to be controlled more carefully. All of these types of primary mirror fabrication approaches have been proven successful. Column (7) in Table 51.1 shows the type of mirror used for each telescope.

All large telescopes use **active optics** to control the shape of the primary mirror. Active optics is the slow adjustment of a mirror to correct aberrations in the image. These adjustments are not fast enough to correct for the atmospheric turbulence, but they can correct for flexure in the telescope structure and for temperature changes that will cause the telescope structure to expand and contract. The process for doing this is illustrated in Figure 51.8. A star is required for the active optics system to be able to compute the deformations on the primary mirror that are needed to correct the image. Although Figure 51.8 illustrates the case for a single mirror, a similar approach is

TABLE 51.1 Telescopes with Apertures Greater than 5 m

(1)	(2)	(3)	(4)	(5)	(6)	(7)	(8)	(9)	(10)
Aperture (m)	Circular Aperture Equiv. (m)	Telescope Name	Location	Date of Operation	Primary f/no	Mirror Type	Mirror Aspect Ratio	Mounting Type	Ref.
2 × 8.4	11.8	Large Binocular Telescope (LBT)	Mt. Graham, Arizona	2008	1.14	Honeycomb	9.4	Alt-Az	1
11.4 × 9.9 Hexagon	10.4	Gran Telescopio Canarias (GTC)	La Palma, Canary Islands	2009	1.65	Segmented	125	Alt-Az	2
11.0 × 9.4 Hexagon	10.0	Keck I	Mauna Kea, Hawaii	1993	1.75	Segmented	133	Alt-Az	3
11.0 × 9.4 Hexagon	10.0	Keck II	Mauna Kea, Hawaii	1996	1.75	Segmented	133	Alt-Az	3
11 × 9.8 Hexagon	9.2	Hobby–Eberley Telescope (HET)	Mt. Fowlkes, Texas	1997	1.4	Segmented	200	Azimuth only	4
11 × 9.8 Hexagon	9.2	Southern African Large Telescope (SALT)	Sutherland, South Africa	2011	1.4	Segmented	200	Azimuth only	5
8.2	8.2	Subaru	Mauna Kea, Hawaii	1999	1.8	Meniscus	41	Alt-Az	6
8.2	8.2	Very Large Telescope (VLT) UT1 Antu	Cerro Paranal, Chile	1998	1.75	Meniscus	46	Alt-Az	7
8.2	8.2	Very Large Telescope (VLT) UT2 Kueyen	Cerro Paranal, Chile	1999	1.75	Meniscus	46	Alt-Az	7

8.2	Very Large Telescope (VLT) UT3 Melipal	Cerro Paranal, Chile	2000	1.75	Meniscus	46	Alt-Az	7
8.2	Very Large Telescope (VLT) UT4 Yepun	Cerro Paranal, Chile	2000	1.75	Meniscus	46	Alt-Az	7
8.0	Gemini North	Mauna Kea, Hawaii	1998	1.8	Meniscus	40	Alt-Az	8
8.0	Gemini South	Cerro Pachon, Chile	2000	1.8	Meniscus	40	Alt-Az	8
6.5	MMT Observatory	Mt. Hopkins, Arizona	1999	1.25	Honeycomb	9	Alt-Az	9
6.5	Magellan I—Walter Baade	Cerro Manqui, Chile	2000	1.25	Honeycomb	9	Alt-Az	10
6.5	Magellan II—Landon Clay	Cerro Manqui, Chile	2002	1.25	Honeycomb	9	Alt-Az	10
6.0	Bol'shoi Teleskop Azimultal'nyi (BTA)	Mt. Pastukhova, Russia	1977	4	Solid	6	Alt-Az	11
5.1	Hale	Mt. Palomar, California	1949	3.3	Honeycomb	8	Equatorial	12

References: (1) http://www.lbto.org/, (2) http://www.eso.org/, (3) http://www.gtc.iac.es/, (4) http://www.keckobservatory.org/, (5) http://www.as.utexas.edu/mcdonald/het/het.html, (6) http://www.salt.ac.za/, (7) http://www.eso.org/, (8) http://www.gemini.edu/, (9) http://www.mmto.org/, (10) http://www.lco.cl/telescopes-information/magellan, (11) http://www.sao.ru/Doc-en/Telescopes/bta/descrip.html, (12) http://www.astro.caltech.edu/palomar/hale.html.

Column (1): The aperture is the diameter of the primary that can collect light. Unless specified, the number given is the diameter of a circular aperture. The LBT consists of two 8.4-m mirrors that are on a single mount, and the light from both mirrors is combined to form a single image. Two-mirror operations started in 2008. The Keck, HET, and SALT telescopes have primary mirrors that are made from hexagonal segments. The primary mirror has a hexagonal shape, and the largest and smallest widths of the hexagon are given.

Column (2): This is the diameter of the equivalent circular aperture equal to the total light collecting area of the telescope. For the HET and SALT telescopes, this is the maximum equivalent circular aperture that is accepted by the prime focus optics. The LBT, Keck, and VLT observatories can combine light from the mirrors for use as an interferometer. This mode of observations is not considered in this table for the purpose of determining the equivalent circular aperture.

Column (5): Year that science operations started.

Column (6): Primary mirror f/no, which is equal to the focal length of the telescope divided by the primary mirror diameter.

Column (7): Honeycomb: Primary mirror that is lightened with a honeycomb structure on the back. Segmented: Primary mirror is made out of hexagonal segments. Meniscus: Single thin concave mirror. Solid: Thick mirror with no light weighting.

Column (8): The aspect ratio is the primary mirror diameter divided by the mirror (or segment) thickness.

Column (9): The azimuth-only telescope mounts conduct observations by tracking objects in the focal plane of the telescope. For such telescopes, the telescope is fixed but the instrumentation tracks the object.

This table is adapted from J.M. Hill's web site: http://abell.as.arizona.edu/~hill/list/bigtel99.htm and http://en.wikipedia.org/wiki/List_of_largest_optical_reflecting_telescopes.

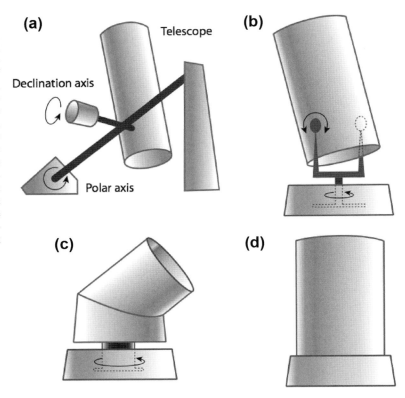

FIGURE 51.6 Schematic of different telescope mounts. (a) Equatorial, (b) altitude-azimuth (alt-az), (c) azimuth only, and (d) fixed. The Hale 5.1-m telescope was the last large telescope to be built with an equatorial mount. The equatorial mount has one axis aligned to the rotation axis of the Earth. (Note: there are many types of equatorial mounts. The Hale telescope uses a type known as the horseshoe equatorial mount.) All fully steerable large telescopes utilize the alt-az mount, such as the Keck, Gemini, Very Large Telescope, and Subaru telescopes (see Table 51.1). In the alt-az mount, the azimuth axis points to the zenith with a perpendicular altitude axis. Two large telescopes built specially for spectroscopy use the azimuth-only mount—the Hobby–Eberly and the South African Large Telescope. These telescopes move only in azimuth and is fixed in altitude. The fixed mount (the telescope points only to the zenith) is used for very specialized applications, such as a liquid mercury mirror telescope or in radio astronomy.

employed for correcting the surface figure of a segmented primary mirror, although the details are quite different.

Efforts to escape the harmful effects of Earth's atmosphere have led to telescopic observations using balloons, aircraft, and rockets. Although we do not discuss space observatories in this chapter, we note here that the Stratospheric Observatory for Infrared Astronomy (SOFIA), a major facility funded by NASA and the German Aerospace Center (DLR) to fly a 2.5-m telescope in the stratosphere, has been recently been completed (http://www.sofia.usra.edu/Sofia/sofia.html). At this high altitude, it is possible to observe throughout the 25–350 μm wavelength range that is inaccessible from the ground. This facility will provide long-term access to a critical wavelength range that otherwise could be exploited only infrequently with spacecraft.

We do not know what ultimately will be the largest ground-based telescope to be built (see Figure 51.5). The limitations arise from the need to be **diffraction** limited, the difficulty of building a suitable enclosure, and the cost. To be competitive with space observatories, all large telescopes must work at the diffraction limit using AO. But the need to be diffraction limited will ultimately cause AO systems to be too complex on an extremely large telescope. An enclosure is necessary to keep the disturbance by wind to acceptable levels, and the cost to build and operate the telescope will be enormous. At some point, it may be more cost-effective to go into space, where gravity and the weather are not factors driving the design.

The drive to build ever-larger telescopes is motivated by the need to collect as much light as possible and thereby increase the signal-to-noise (S/N) ratio of observations. One can derive that for a diffraction-limited telescope and a detector that is background limited, the S/N in a given integration time is proportional to

$$S/N \approx (A * \eta/\varepsilon)^{0.5}/(\text{FWHM}), \quad (51.1)$$

where A is the area of the telescope, η is the total transmission of the optics and the detector quantum efficiency, ε is the background emission, and FWHM is the full width at half maximum of a stellar image. η takes into account all of the light losses that occur from the reflection of the mirrors and transmission losses of lenses as light propagates from the telescope to the detector. To minimize these losses, it is necessary to utilize high reflection coatings on mirrors and lenses as well as to minimize the number of lenses. The detector quantum efficiency is the fraction of light that is absorbed by the detector material. This is near the theoretical maximum of 1.0 at visual wavelengths and about 0.8–0.9 for the 1–15 μm wavelength range. The background emission, ε, arises from the sky emission lines at visual wavelengths and thermal background from the telescope and sky at

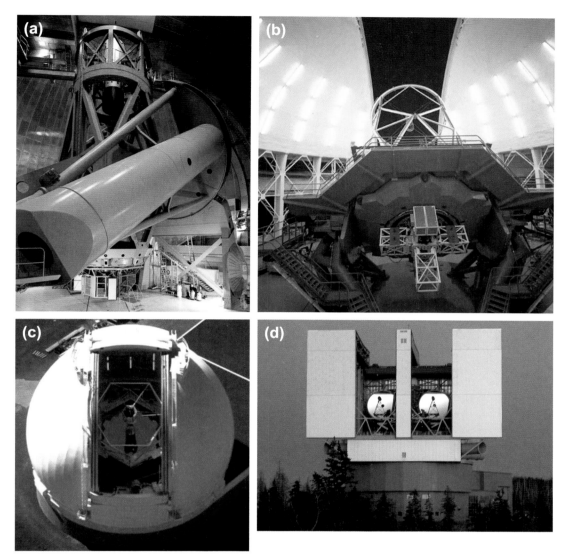

FIGURE 51.7 (a) Hale 5.1-m telescope. The largest telescope to be built in the "classical style" with an equatorial mount, a culmination of about 280 years of development of the reflecting telescope. *(© 2005 Gigapxl Project.)* (b) The 8-m Gemini South telescope. Instruments are mounted on the back of the telescope. These instruments are on the telescope all of the time so that instrument changes can be made very quickly. The dome has vents to allow flushing of the dome by the night air. This allows the telescope and dome to quickly reach equilibrium with the air temperature. *(Courtesy of Gemini Observatory/AURA.)* (c) The 10-m Keck telescope. This image shows one of the two Keck telescopes. The primary mirror consists of 36 hexagonal segments that are aligned to optical precision. The instruments are located on a platform on two sides of the telescope facing the declination bearings. The segmented mirror design approach was first used with the Keck telescopes. The Gran Telescopio Canarias was based on this design, and two of the next-generation large telescopes, the TMT and the E-ELT will use segmented mirrors. *(Courtesy of R. Wainscoat.)* (d) LBT consisting of two 8.4-m primary mirrors. Two mirror science operations started in 2008. The light-gathering power of the two primary mirrors combined is equivalent to an 11.8-m telescope. Both mirrors are on a single structure, and the light from both mirrors can be combined for imaging, spectroscopy, and interferometry. The combined light from the two mirrors will have the angular resolution of a 22.8-m telescope when the LBT is used as an interferometer. *Courtesy of the Large Binocular Telescope Observatory.*

wavelengths longer than 2 μm. To reduce the thermal emission from the telescope, it is necessary to have the highest reflectivity mirrors available and to reduce or eliminate the thermal emission from the secondary mirror. The latter is often accomplished by forming an image of the secondary within the instrument and then blocking it with a cooled metal plate. Then the infrared detector will sense only the thermal emission from the sky and the object being observed.

After maximizing η and reducing ε as much as possible, one can only increase the telescope area and reduce the FWHM to further increase the S/N. Reducing the image FWHM requires decreasing the dome **seeing** to the absolute minimum, building on sites that have good atmospheric

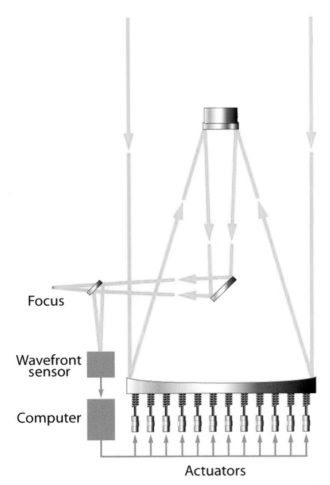

FIGURE 51.8 Schematic of an active optics system. Starlight from the telescope is sent to a beam splitter just in front of the focus that simultaneously sends light to the focus and to a wavefront sensor. The computer analyzes the output of the wavefront sensor and sends control signals to the primary mirror to correct any aberrations in the image. *Courtesy of C. Barbieri.*

seeing, and working at the diffraction limit of the telescope. Astronomical sites in Hawaii, Chile, and La Palma are prime locations for large telescopes because of the good seeing and the good weather they offer.

Figure 51.9 shows the advances in image quality that have been achieved. The development of AO has led to the ability to work at the diffraction limit in the near infrared and to achieve improvements in S/N given by Eqn (51.1). The principles underlying AO are discussed in Section 4.

3. ADVANCES WITH DETECTOR ARRAYS

Initial observations with telescopes were conducted solely with the human eye, but the advantages of using photographic plates to record and archive observations of the sky were quickly exploited beginning in the 1850s. Photographic plates were eventually supplemented with electronic devices like the photomultiplier tube, which amplified the signal from stars by about one million. At infrared wavelengths, there were specialized detectors that employed bolometers, photovoltaic devices, and photoconductive devices. However, photographic plates were a necessity for recording high-resolution images of large areas of sky and recording spectra with a wide wavelength range.

Images recorded by photographic plates depend on the chemical reaction that is induced by a photon of light. Although their efficiency in converting a photon to an image is only a few percent, photographic plates allowed quantitative measurements of the brightness of stars and the strength of spectral lines to be made, and enabled scientists to archive the information for future use.

The next technological revolution came with the invention of the **charge-coupled** device (CCD) in 1973. CCDs are composed of millions of picture elements or pixels. Each pixel is a single detector and is capable of converting photons to electrons. The accumulated electrons can then be sent to an amplifier to be "readout" and recorded by a computer. CCD technology is employed in digital cameras, and just as digital photography has replaced film photography, a similar transformation has taken place in astronomy.

The impact of the CCD on astronomy was immediately apparent. CCDs have two major advantages over the photographic plate: the capability to directly record photons with an efficiency of 80–90% and to store data electronically. The stored data can then be processed with a computer. Until recently, the main deficiency of the CCD relative to the photographic plate was the relatively small amount of sky that could be covered. However, the recent development of very large CCD mosaics now permits larger areas of sky to be covered by a CCD than by a photographic plate. The rapid development of computing power and disk storage has made it practical to use large CCD mosaics. While astronomers have worked hard to develop CCD technology that is optimized for astronomy, the consumer market has driven the development of the necessary computing power and storage. Figure 51.10 shows an example of a state-of-the-art large-format CCD.

There has been a similar revolution in the development of infrared arrays. The first infrared arrays for astronomy were used in the early 1980s. While initially very modest in size (32×32 pixels), infrared arrays now typically contain more than a million pixels. There are several significant differences between CCDs and infrared arrays. One is that a CCD has a single readout amplifier, while an infrared array has one readout amplifier per pixel. The electrons in a CCD are transferred to a single readout amplifier (hence the origin of the term "charge transfer"). Only a single readout amplifier is needed since the readout electronics and the detector material are made out of the same semiconductor

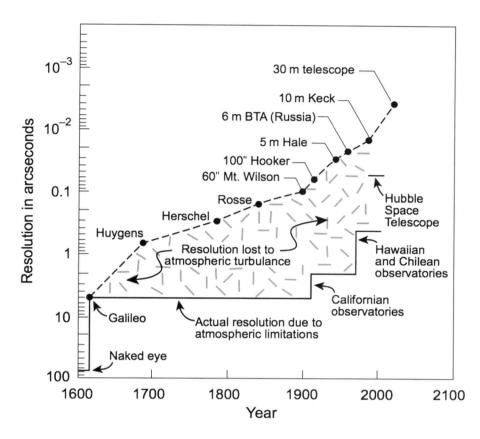

FIGURE 51.9 Improvement in angular resolution at optical wavelengths. The development of adaptive optics (AO) has permitted diffraction-limited observations from ground-based observatories since 1990, largely eliminating the effects of the atmosphere. The dashed line shows the theoretical diffraction-limited resolution for the telescope that can be achieved with AO. The solid line shows the seeing limit imposed by the atmosphere where improvements were obtained by going to sites with very good seeing. The resolution of the Hubble Space Telescope is shown, from which it is evident that the frontier for high angular resolution is being advanced with ground-based observatories. *From P. Bely, 2003.*

material. In an infrared array, the detector material and the readout amplifier have to be made out of different materials, so each pixel must have a separate amplifier. A second difference is that the infrared arrays must be cooled to much lower temperatures. CCDs can operate effectively at about -30 to $-40\,°C$. Infrared arrays must be cooled to liquid nitrogen ($-196\,°C$) or liquid helium ($-269\,°C$) temperatures.

The development of large-format CCDs and infrared arrays has enabled astronomers to undertake large-scale digital sky surveys at visible and infrared wavelengths (see Section 5), just as the use of large photographic plates enabled astronomers to make the first deep sky surveys over 60 years ago.

Figure 51.11 shows an image of Saturn at a wavelength of 18 μm taken with an infrared array. At these wavelengths, we are observing the thermal emission (heat) from the planet. Thus the temperatures in Saturn's atmosphere and the dust particles in the rings can be measured.

4. ADVANCES IN ADAPTIVE OPTICS (AO)

AO is a technique that removes the atmospheric disturbance and allows a telescope to achieve **diffraction**-limited imaging from the ground. This is critical in achieving the maximum S/N given in Eqn (51.1). The basic idea of AO is to first measure the amount of atmospheric disturbance, then correct for it before the light reaches the camera. A schematic of how this can be done is shown in Figure 51.12.

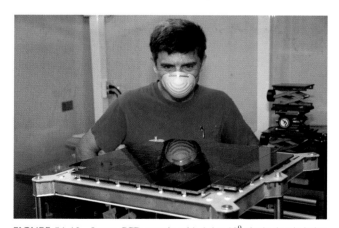

FIGURE 51.10 Large CCD mosaic with 1.4×10^9 pixels that is being used with the Pan-STARRS 1 telescope. This CCD mosaic consists of 60 individual CCDs and is the largest CCD mosaic in use for astronomy. A larger CCD mosaic with more than twice as many pixels is planned for use with the Large Synoptic Survey Telescope. Large CCD mosaics have led to many types of sky surveys and are a driving force for a virtual observatory where large collections of imaging and spectroscopic data are made available for research programs. *Courtesy of the Pan-STARRS project.*

FIGURE 51.11 Image of Saturn and its rings obtained in 2004 with the 10-m Keck I telescope at a wavelength of 17.6 μm. This is a false color image, where higher signal levels are lighter. At these wavelengths, we are seeing the heat radiated by the atmosphere and the rings of Saturn. The South Pole has an elevated temperature ($-182\,°C$) compared with its surrounding. This is likely because the South Pole was illuminated by the Sun for 15 years. *Courtesy of G. Orton, JPL.*

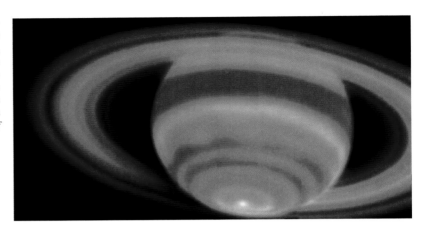

AO requires a star or another object that is bright enough for rapidly and accurately measuring the incoming wavefront. If the object of interest is not bright enough, then it is necessary to use a nearby bright star. This limits the sky coverage, since not every region of the sky will have a bright enough star nearby. If there is no nearby bright star, then it is necessary to use a laser guide star. A laser is pointed in the same direction as the telescope and is used to excite a thin layer of sodium atoms in Earth's ionosphere (at an altitude of 90 km). This provides a point source that acts as an artificial star for the AO system.

Figure 51.13 shows a dramatic photo of laser guide stars in use at the Keck and Subaru telescopes on Mauna Kea. All large telescopes now have AO and laser guide stars. The effect of using AO is striking. It is like placing the telescope into space. An impressive example of how AO can improve image quality is shown in Figure 51.14. A further example of the high angular resolution provided by AO is shown in Figure 51.15. Here the volcanoes on the satellite Io can be individually resolved. Such high angular resolution is impressive and rivals the images from the Voyager and Galileo spacecraft.

We can look forward to the exploration of other solar systems. Figure 51.16 shows an image of an exoplanetary system. The physical characteristics of other solar systems and exoplanets can be studied for the first time, and this is certainly a scientific area where large telescopes with AO are necessary.

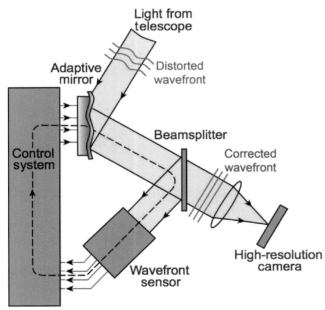

FIGURE 51.12 Simplified diagram of an AO system. Light from the telescope is collimated and sent to an adaptive or deformable mirror. If there were no atmospheric turbulence, the wavefront of the light would be perfectly straight and parallel. The light is then reflected to a beam splitter, where part of the light is reflected to the wavefront sensor. The wavefront sensor measures the distortion of the wavefront and sends a correction signal to the adaptive mirror. The adaptive mirror is capable of changing its shape to remove the deformations in the light wave caused by the atmospheric turbulence. In this way, the light with a corrected wavefront reaches the high-resolution camera, where a diffraction-limited image is formed. *Courtesy of C. Max.*

5. SKY SURVEY TELESCOPES

Although large telescope projects tend to get a lot of attention, recently there has been a corresponding quantum jump in the construction of visible and infrared survey telescopes. This has been made possible by the availability of large-format CCDs and infrared arrays. In addition, the discovery of the Kuiper Belt has led to fundamental advances in our understanding of how our solar system formed. There is a great need to continue the survey of the Kuiper Belt because detailed knowledge of the size and orbit distributions of these objects will allow us to test theories of the orbital migration of the outer planets (Jupiter, Saturn, Uranus, Neptune), the origin of the

Chapter | 51 New Generation Ground-Based Optical/Infrared Telescopes

FIGURE 51.13 Sodium laser guide stars in simultaneous use at the Keck and Subaru telescopes. The laser operates at a wavelength of 5890 Å (0.589 μm), and the laser light is propagated through a smaller telescope attached to the telescopes. It excites sodium atoms at an altitude of 90 km in Earth's atmosphere. The sodium atoms emit light at the same wavelength as the laser, and this is viewed as an artificial star by the telescope. All large telescopes employ laser guide stars to achieve diffraction-limited imaging. This is a long exposure photograph. The laser guide star is barely visible with the naked eye from this angle. *Courtesy of Dan Birchall.*

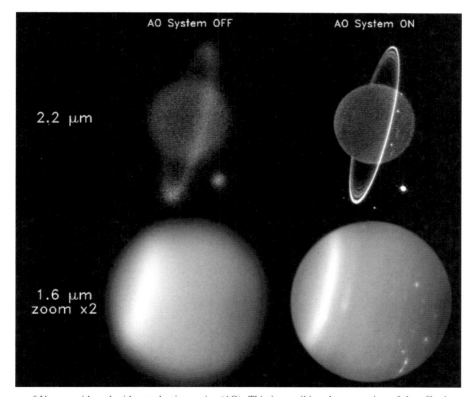

FIGURE 51.14 Images of Uranus with and without adaptive optics (AO). This is a striking demonstration of the effectiveness of AO in removing atmospheric turbulence. One can also see that the signal-to-noise ratio is greatly enhanced because light is concentrated into a diffraction-limited image with AO, thus greatly increasing the ability to detect faint spots and cloud structure. At a wavelength of 1.6 μm, we are seeing reflected light from low-altitude clouds, while at 2.2 μm the high-altitude clouds are revealed. The planet is much darker at 2.2 μm because of absorption of methane gas in the atmosphere. This allows a much longer exposure and for the rings to be seen clearly. The point-like cloud features at 2.2 μm show that in certain places turbulence is very strong, and it is pushing material from lower altitudes into the stratosphere. *Courtesy of H. B. Hammel, I. de Pater, and the W. M. Keck Observatory.*

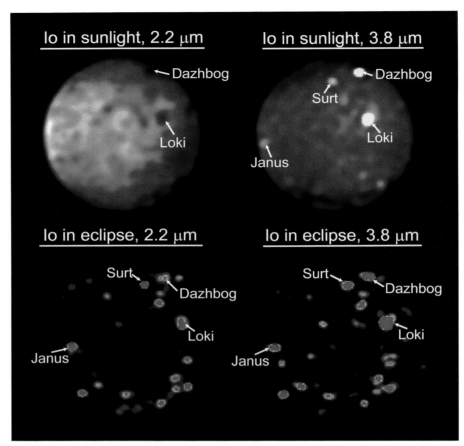

FIGURE 51.15 Adaptive optics images of Jupiter's satellite Io obtained with the Keck Observatory. The angular size of Io is 1.2 arcseconds. (Top): Io in sunlight. At 2.2 μm the reflected light from the Sun dominates, while at 3.8 μm thermal emission (heat) from the volcanoes can be observed. Some volcanoes (Loki, Dazhbog) show up as hot spots at 3.8 μm, but as low-albedo features at 2.2 μm. (Bottom): Io in eclipse. Images of Io taken 2 h later, after the satellite had entered Jupiter's shadow. Without sunlight reflecting off the satellite, even the faintest hot spots can be discerned. The difference in brightness between the two wavelengths gives an indication of the temperature of the spot. Dazhbog, very bright at 3.8 μm, is a low-temperature (500 K) hot spot. Janus, on the other hand, is very bright at both 3.8 and 2.2 μm, indicative of higher temperatures (800 K). *Courtesy of I. de Pater.*

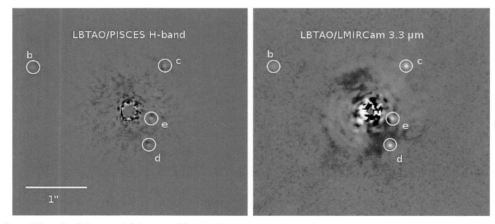

FIGURE 51.16 Large Binocular Telescope AO images of the HR 8799 planetary system in the H-band (1.65 μm) and 3.3 μm. The objects labeled b, c, d, and e are thought to be planets that are less than 13 Jupiter masses, although this depends on the assumed age of the host star, HR 8799. The data presented along with spectra obtained of the objects indicate the atmospheres of the objects are cloudy and have a mixture of temperatures of 400 and 1100 °C. *Courtesy of A.J. Skemer.*

short-period comets, and the cause of the late heavy bombardment of the inner solar system.

There is also an increased awareness that it is important to identify asteroids and comets that could collide with Earth (see Figure 51.3). In 1998 the Congress of the United States directed NASA to identify within 10 years at least 90% of NEOs larger than 1 km that may collide with Earth, and this has been achieved. There are a number of scientific benefits that arise from the NEO surveys, including determining the origin of NEOs, identifying interesting NEOs that could be visited by spacecraft, improving our knowledge of the numbers and sizes of the asteroids in the main asteroid belt, and the discovery of new comets.

The discovery of all NEOs larger than 1 km is important because if such an object collides with Earth, the consequences will be catastrophic. If it is possible to predict that there will be a collision, it may be possible to divert the asteroid so that it misses Earth. The earlier such a prediction can be made, the more likely it is that the diversion is possible. This is a case in which there is a practical use for astronomy.

Table 51.2 shows a list of asteroid search programs that are currently in progress or planned to find asteroids that present a hazard to the Earth. Current productivity of various programs is shown in Figure 51.17, which shows all NEOs discovered irrespective of size. While the NASA directive is aimed at identifying NEOs larger than 1 km in diameter, many NEOs smaller than 1 km are also discovered because of the sensitivity of the search programs and because small objects that come very close to Earth may be bright enough to be detected. For example in 2012, an NEO named 2012 KT42 with an estimated diameter of only 7 m was discovered and it came to within 16,000 km of Earth's surface.

The number of known NEOs has been increasing because of the larger number of funded survey programs and advances in detector arrays that have allowed much larger areas of sky to be covered in a single exposure. The number of NEOs discovered as a function of time is shown in Figure 51.17. Note that while the total number of asteroids that approach Earth are still being discovered at a rapid rate, the number of new asteroids larger than 1 km discovered each year is decreasing, since there are fewer large asteroids left to find. As can be seen in Figure 51.17 most NEOs are discovered by two dedicated asteroid survey programs: the Catalina Sky Survey in Arizona and Pan-STARRS in Hawaii.

There are three major ground-based sky surveys currently under development or study (see Table 51.2). Pan-STARRS 2 is a 1.8-m telescope similar to the Pan-STARRS 1 telescope. The construction work is funded and the telescope and camera should be ready in 2014. The Asteroid Terrestrial-impact Last Alert System (ATLAS) is a pair of small telescopes designed to find asteroids that about to impact with Earth. This project is funded, and

TABLE 51.2 Ground-Based Asteroid Search Telescopes

Survey	Status	Primary Mirror Diameter (m)	Collecting Area[1] (m^2)	f/no	Field-of-View[2] (deg^2)	Magnitude Limit[3]	Entendu[4] ($m^2\ deg^2$)	Ref.
Spacewatch (Arizona)	Operational	0.9	0.6	3.0	2.9	21.5	1.7	1
Catalina Sky Survey (Arizona)	Operational	0.68	0.31	1.8	19.4	19.5	6.0	2
Catalina Sky Survey (Arizona)	Operational	1.5	1.7	1.6	5.0	21.5	8.5	2
Pan-STARRS 1 (Hawaii)	Operational	1.8	1.7	4.4	7.0	22.0	11.9	3
Pan-STARRS 2 (Hawaii)	Commissioning	1.8	1.7	4.4	7.0	22.0	11.9	3
ATLAS 1 & 2 (Hawaii)	In development	0.65	0.16	2.0	29.0	20.0	9.4	4
Large Synoptic Survey Telescope (Chile)	In development	8.4	35.8	1.25	7.0	24.0	251	5

References: (1) http://spacewatch.lpl.arizona.edu/, (2) http://www.lpl.arizona.edu/css/, (3) http://pan-starrs.ifa.hawaii.edu/public/, (4) http://www.fallingstar.com/, (5) http://www.lsst.org/lsst/.
[1]Collecting area is the area of the telescope corrected for any obscuration by the secondary mirror.
[2]Field of view is the area of sky covered in a single exposure.
[3]Magnitude limit is the faintest star recorded at visible wavelengths.
[4]Etendu is a figure of merit that is the collecting area of the telescope multiplied by the field of view. This is a measure of how fast a sky survey can be achieved assuming the cameras operate at the same wavelength and efficiency.

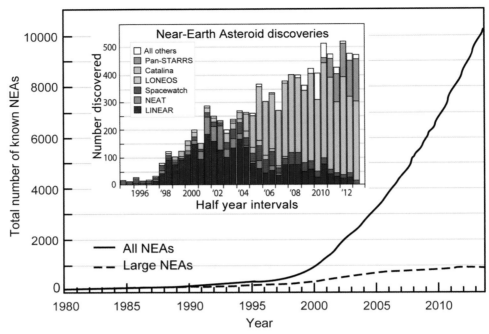

FIGURE 51.17 Cumulative discoveries of near-Earth asteroids. The total number of large near-Earth asteroids (larger than 1 km) is increasing at a slower rate since most of the easy-to-detect near-Earth asteroids have already been discovered. Most near-Earth asteroid discoveries are now made with the Catalina Sky Survey and the Pan-STARRS telescopes. *Courtesy of Alan Chamberlin.*

observations should start in 2015—2016. The proposed Large Synoptic Survey Telescope is currently under development and is envisioned to be a monolithic 8.4-m wide-field telescope (with a collecting area equal to a 6.7-m telescope). It is planned to be completed by 2025 in the Southern Hemisphere.

6. CONCLUDING REMARKS

Space does not allow coverage of all of the relevant subjects related to the vibrant topics of novel telescope construction, optical fabrication techniques, advances in mirror figure control, AO, and detector improvements at visible and infrared wavelengths. The topics covered in this chapter can only hint at the tremendous advances that have taken place in recent years and that carry on unabated. Since the invention of the refractive and reflective telescopes by Galileo and Newton, the construction of ground-based telescopes has continued to challenge the very best minds in physics and engineering. There are now strong scientific drivers to build larger telescopes in the 20—40 m range, and three such projects are underway.

Solar system astronomy is driven by the need to have large telescopes to study very faint objects in the Kuiper Belt and very faint NEOs that may present a hazard to Earth. It is also necessary to have the highest spatial resolution possible by working at the **diffraction** limit of large telescopes. This will enable researchers to study the surface and atmospheric features of the outer planets, dwarf planets, and their satellites. Large telescopes also allow the study of exoplanets, and thus bring about a merging of studies of our solar system with those around distant stars.

Another driver of solar system astronomy is the need to detect and characterize NEOs that may present an impact hazard to Earth. Asteroid search programs are underway to detect these NEOs. These survey programs will play a significant role in greatly expanding our knowledge of the building blocks of our solar system—the asteroids and comets from the inner to the outer reaches of the solar system. These studies are profoundly affecting our understanding of the formation of our solar system and of life itself.

Although we have not emphasized it in this chapter, there is also strong interest in the use of small telescopes for solar system astronomy. Networks of small telescopes are now routinely employed for the study of stellar occultations by small bodies such as asteroids and KBOs (see Figure 51.2). A new type of observatory, the Las Cumbres Observatory Global Telescope Network, provides observing time on small telescopes distributed around the world. The Taiwan American Occultation Survey is a cluster of small telescopes searching for stellar occultations by KBOs. These efforts are likely to increase in importance in the future.

We anticipate continuing growth in telescope and instrument development for many generations to come. This is indeed a period of great innovation—a renaissance in telescope building and instrumentation—that we are fortunate to be able to witness and to participate in.

BIBLIOGRAPHY

Bely, P. Y. (Ed.). (2003). *The design and construction of large optical telescopes*. New York: Springer-Verlag.

Kitchin, C. R. (2003). *Telescopes and techniques*. London: Springer-Verlag.

McLean, I. (1997). *Electronic imaging in astronomy*. Chichester: John Wiley & Sons.

Racine, R. (2004). *The historical growth of telescope aperture* (Vol. 116). Pub. Astron. Soc. Pacific. p. 77.

Silva, D., & McLean, I. S. (2013). *Introduction to Telescopes, in Planets, Stars, and Stellar Systems. Vol. 1: Telescopes and Instrumentation*. In T. D. Oswalt, & I. S. McLean (Eds.). Springer Science + Business Media Dordrecht. See other relevant chapters in this volume.

Tyson, R. K. (2000). *Introduction to adaptive optics*. Bellingham, Washington: Soc. Photo-Optical Instrumentation Eng. NEO web site http://neo.jpl.nasa.gov/stats/.

Zirker, J. B. (2005). *An acre of glass: A history and forecast of the telescope*. Baltimore: The Johns Hopkins University Press.

Chapter 52

The Solar System at Radio Wavelengths

Imke de Pater
Astronomy Department, University of California, Berkeley, CA, USA; Faculty of Aerospace Engineering, Delft University of Technology, Delft, NL; SRON Netherlands Institute for Space Research, Utrecht, The Netherlands

William S. Kurth
Department of Physics and Astronomy, University of Iowa, Iowa City, IA, USA

Chapter Outline

1. Introduction — 1107
   1.1. Instrumentation — 1108
2. Thermal Emission from Planetary Bodies — 1109
   2.1. Thermal or Blackbody Radiation — 1109
   2.2. Radio Emission from a Planet's (Sub)surface — 1109
   2.3. Radio Emission from a Planet's Atmosphere — 1109
   2.4. Terrestrial Planets and the Moon — 1110
      2.4.1. The Moon — 1110
      2.4.2. Mercury — 1110
      2.4.3. Venus and Mars — 1111
   2.5. Giant Planets — 1112
      2.5.1. Radio Spectra — 1112
      2.5.2. Jupiter — 1112
      2.5.3. Saturn — 1114
      2.5.4. Uranus and Neptune — 1116
   2.6. Major Satellites and Small Bodies — 1117
      2.6.1. Galilean Satellites — 1117
      2.6.2. Titan — 1117
      2.6.3. Asteroids and Trans-Neptunian Objects — 1118
   2.7. Comets — 1119
3. Nonthermal Radiation — 1120
   3.1. Low-Frequency Emissions — 1121
      3.1.1. Cyclotron Maser Emissions — 1121
      3.1.2. Other Types of Low-Frequency Radio Emissions — 1123
      3.1.3. Atmospheric Lightning — 1123
   3.2. Synchrotron Radiation — 1123
   3.3. Earth — 1123
   3.4. Jupiter's Synchrotron Radiation — 1124
   3.5. Jupiter at Low Frequencies — 1126
      3.5.1. Decametric and Hectometric Radio Emissions — 1126
      3.5.2. Kilometric Radio Emissions — 1127
      3.5.3. Very Low-Frequency Emissions — 1128
      3.5.4. Ganymede — 1128
   3.6. Saturn — 1128
      3.6.1. Saturn Kilometric Radio Emissions — 1128
      3.6.2. Very Low Frequency Emissions — 1130
      3.6.3. Saturn Electrostatic Discharges — 1130
   3.7. Uranus and Neptune — 1131
4. Future of Radio Astronomy for Solar System Research — 1131
Bibliography — 1131

1. INTRODUCTION

Ground- and space-based radio astronomical observations of planetary objects provide information that is complementary to that obtained at other (visual, infrared (IR), ultraviolet (UV)) wavelengths. We distinguish between thermal and nonthermal emissions. *Thermal radio emission* originates from a body's surface (or more appropriately subsurface) and/or atmosphere, and *nonthermal radio emissions* are produced by charged particles in a planet's magnetosphere. The thermal emission can be used to deduce the structure and composition of a planet's atmosphere and surface layers; the nonthermal radiation provides information about its magnetic field and charged particle distributions therein. Ground-based radio astronomy is essentially limited to frequencies above about 10 MHz because of the shielding effects of the Earth's ionosphere at lower frequencies. Space-based measurements extend the frequency range of solar system radio astronomy as low as a few kilohertz. In this chapter, we discuss radio emissions from a few kilohertz up to $\gtrsim 500$ GHz. Since we cover over nine orders of magnitude in frequency, we can include only brief summaries of a select number of topics.

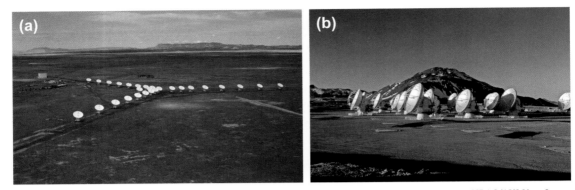

FIGURE 52.1 (a) Aerial photograph of the very large array of radio telescopes in New Mexico. *(Image courtesy: NRAO/AUI Very Large Array is a facility of the National Radio Astronomy Observatory, operated by the Associated Universities, Inc. (AUI), under contract with the National Science Foundation.)* (b) This picture of the Atacama Large Millimeter Array (ALMA) radio antennas on the Chajnantor Plateau in Chile, 16,500 ft above sea level, was taken a few days before the start of ALMA Early Science operations, at which time there were 19 antennas on the plateau. *Image courtesy: ALMA/W. Garnier, ESO/NAOJ/NRAO.*

1.1. Instrumentation

A radio telescope consists of an antenna and a receiver. The antenna can be a simple monopole, dipole, or parabolic dish (Figure 52.1). The sensitivity of the antenna depends on many factors, but the most important are the effective aperture and system temperature. The effective aperture depends on the size of the dish and the aperture efficiency. The sensitivity of the telescope increases when the effective aperture increases and/or the system temperature decreases.

The response of an antenna as a function of direction is given by its antenna pattern, which consists of a "main" lobe and a number of smaller "side" lobes, as depicted in Figure 52.2(a). The resolution of the telescope depends on the angular size of the main lobe. It is common to express the main lobe width as the angle between the directions for which the power is half that at lobe maximum; this is referred to as the half power beam width. This angle depends on the size of the dish and the observing wavelength: For a uniform illumination, the beam width is approximately λ/D radians, with D the dish diameter in the same

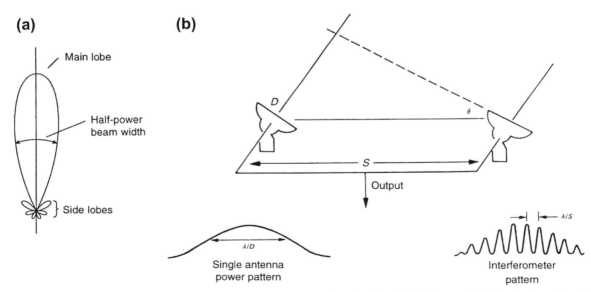

FIGURE 52.2 (a) A generic antenna pattern consists of a "main" lobe and a number of smaller "side" lobes, as depicted in the figure. The half power beam width is the full width at half power. *(After Kraus (1986).)* (b) Top: Geometry of a two-element interferometer. Bottom: Antenna response for a single element of the interferometer (left) and response of the interferometer (right) to an unresolved radio source. *Gulkis, S. & de Pater, I. (2002).*

units as the wavelength λ. Space-borne radio telescopes at low frequencies usually are composed of one or more long cylindrical elements since dish antennas are prohibitive in terms of mass.

The resolution of a radio telescope can be improved by connecting the outputs of two antennas that are separated by a distance S at the input of a radio receiver. The Very Large Array (VLA) in Socorro, New Mexico, consists of a Y-shaped track, with nine antennas along each of the arms (Figure 52.1(a)). This telescope thus provides 351 individual interferometer pairs, each of which has its own instantaneous resolution along its projected (on the sky) baseline S'. Such an array of antennas is needed to construct an image that shows both the large- and small-scale structures of a radio source. At short spacings, the entire object can be "seen", but details on the planet are washed out due to the low resolution of such baselines. At longer baselines, details on the planet can be distinguished, but the large-scale structure of the object gets resolved out, and hence would be invisible on the image unless short spacing data are included as well. Hence, arrays of antennas are crucial to image an object.

2. THERMAL EMISSION FROM PLANETARY BODIES

2.1. Thermal or Blackbody Radiation

Any object with a temperature above absolute zero emits a continuous spectrum of electromagnetic radiation at all frequencies, which is its thermal emission, usually modeled as "*blackbody*" radiation.

Blackbody radiation can be described by Planck's radiation law, which, at radio wavelengths, can usually be approximated by the Rayleigh–Jeans law:

$$B_\nu(T) = \frac{2\nu^2}{c^2} kT \quad (52.1)$$

where $B_\nu(T)$ is the brightness (W/m²/Hz/sr), ν is the frequency (Hz), T is the temperature (K), k is the Boltzmann's constant (1.38×10^{-23} J/K), and c is the velocity of light (3×10^8 m/s). With a radio telescope, one measures the *flux density* emitted by the object. A common unit is the flux unit or Jansky, where 1 Jy = 10^{-26} W/m²/Hz. This flux density can be related to the temperature of the object:

$$S = \frac{abT}{4.9 \times 10^6 \lambda^2} \text{ Jy} \quad (52.2)$$

where λ is the observing wavelength (in meters), $2a$ and $2b$ are the equatorial and polar diameters (in arc seconds), and T is the temperature (in Kelvin). Usually, planets do not behave like a blackbody, and the temperature T in Eqn (52.2) is called the brightness temperature, defined as the temperature of an equivalent blackbody of the same brightness.

2.2. Radio Emission from a Planet's (Sub) surface

Radio observations can be used to extract information about the (sub)surface layers of planetary bodies. The temperature structure of the (sub)surface layers of airless bodies depends on a balance between solar insolation, heat transport within the crust, and reradiation outward. The fraction of the solar flux absorbed by the surface depends on the object's albedo, A, while the energy radiated by the surface (at a given temperature) depends on its emissivity, e (which is 1 for a blackbody). The parameters e and A are related through Kirchoff's law:

$$e = 1 - A \quad (52.3)$$

During the day, a planet's surface heats up and reaches its peak temperature at noon or early afternoon (the exact time depends on the body's thermal inertia, a body's ability to store energy); at night the object cools off. Its lowest temperature is reached just before sunrise. Because it takes time for the heat to be carried downward, there will be a phase lag in the diurnal heating pattern of the subsurface layers with respect to that at the surface, and the amplitude of the variation will be suppressed. At night, heat is carried upward and radiated away from the surface. Hence, while during the day the surface is hotter than the subsurface layers, at night the opposite is true.

Radio waves typically probe ∼ 10 wavelengths into the crust, as determined by the electrical skin depth, L_r, which is equivalent to the depth at which unit optical depth is reached:

$$L_r = \lambda/(2\pi\sqrt{\varepsilon_r}\tan\Delta) \quad (52.4)$$

with λ the wavelength, ε_r the real part of the dielectric constant, and $\tan\Delta$ is the "loss tangent" (or absorptivity) of the material—the ratio of the imaginary to the real part of the dielectric constant. Hence, by observing at different wavelengths, one can determine the diurnal heating pattern of the Sun in the subsurface layers. Such observations can be used to constrain thermal and electrical properties of the crustal layers. The thermal properties relate to the physical state of the crust (e.g. rock versus dust), while the electrical properties are related to the mineralogy of the surface layers (e.g. metallicity).

2.3. Radio Emission from a Planet's Atmosphere

Radio spectra of a planet's atmosphere can be interpreted by comparing observed spectra with synthetic spectra, which are obtained by integrating the equation of radiative transfer through a model atmosphere:

$$B_\nu(T_D) = 2 \int_0^1 \int_0^\infty B_\nu(T) e^{(-\tau/\mu)} d(\tau/\mu) d\mu \quad (52.5)$$

where $B_\nu(T_D)$ can be compared to the observed disk-averaged *brightness temperature*. The brightness $B_\nu(T)$ is given by the Planck function and the optical depth $\tau_\nu(z)$ is the integral of the total absorption coefficient over the altitude range z at frequency ν. The parameter μ is the cosine of the angle between the line of sight and the local vertical. By integrating over μ, one obtains the disk-averaged brightness temperature, to be compared to the observed brightness temperature.

Before the integration in Eqn (52.5) can be carried out, the atmospheric structure, as composition and temperature–pressure profile, needs to be defined. Usually, above and just below the tropopause the temperature–pressure profile as derived from spacecraft data (e.g. occultation data) or from ground-based mid-IR observations is adopted. At deeper layers, the temperature structure is commonly approximated by an *adiabatic lapse rate*, appropriately changed if species condense. The temperature, pressure, and density of an atmosphere are related to one another through the ideal gas law. The shape of absorption/emission lines depends on the temperature and pressure of the environment, and may vary from relatively narrow lines (e.g. Mars, Venus, Titan, Io) to broad quasi-continuum spectra (e.g. giant planets).

2.4. Terrestrial Planets and the Moon

2.4.1. The Moon

Lunar radio astronomy dates back to the mid-1940s, well before the first *Apollo* landing on the Moon. Since the mid-1970s, after a decade of "neglect", there was renewed interest in lunar radio astronomy since radio receivers had improved substantially and laboratory measurements of *Apollo* samples provided a ground truth for several sites on the Moon. By using lunar core samples, one could determine a density profile of the soil with depth near the landing sites, as well as the complex dielectric constant of a variety of rocks and powders, both useful in modeling radio observations.

A microwave image of the full Moon reveals that the maria are ~ 5 K warmer than the highlands. This may result from a difference in albedo (the maria are darker than the highlands), radio emissivity, and/or the microwave opacity. Lunar samples suggest that the microwave opacity in the highlands is somewhat (factor of ~ 2) lower than that in the maria, so that deeper cooler layers are probed in the lunar highlands compared to the maria during full moon (as observed); at new moon the temperature contrast should be reversed, since the temperature increases with depth at night. No observations have yet been reported, however.

2.4.2. Mercury

Radio images of Mercury show a brightness variation across the disk, which displays the history of solar insolation. Figure 52.3(a) shows a radio image at 3.6 cm, probing

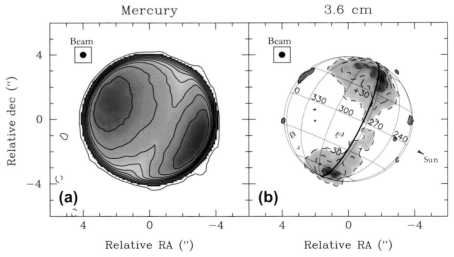

FIGURE 52.3 (a) The 3.6-cm thermal emission from Mercury observed with the VLA. Contours are at 42-K intervals (10% of maximum), except for the lowest contour, which is at 8 K (dashed contours are negative). The beam size is 0.4″ or 1/10 of Mercurian radius. Note the two "hot" regions, discussed in the text. (b) A residual map after subtracting a model image from the map in panel a. Mercury's geometry is indicated, including the direction to the Sun and the morning terminator (heavy line). The hot regions have been modeled well, since they do not show up in this residual map. Instead, large negative (blue) temperatures near the poles and along the morning side of the terminator are visible, likely caused by shadows on the surface resulting from local topography. Contour intervals are in steps of 10 K, which is roughly three times the root mean square noise in the image. *Mitchell, D. L. & de Pater, I. (1994).*

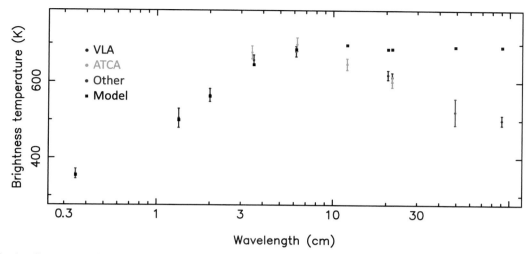

FIGURE 52.4 A radio spectrum of Venus. At short wavelengths, the radio emissions arise approximately from altitudes near Venus' cloud layers. The brightness temperature increases when deeper warmer layers are probed and decreases again at wavelengths long enough to probe down into Venus' surface. VLA = Very Large Array; ATCA = Australia Telescope Compact Array. *Butler, B. J. & Sault, R. J. (2003).*

a depth of ~70 cm. Two "hot" regions are visible, one of them almost opposite to the direction of the Sun. This hot–cold pattern results from Mercury's 3/2 spin–orbit resonance: Mercury rotates three times around its axis for every two revolutions around the Sun. This, combined with the planet's large orbital eccentricity, causes Mercury's peak (noon) surface temperature to vary between 700 K for longitudes facing the Sun at perihelion to 570 K 90° away. While the surface temperature responds almost instantaneously to changes in illumination, the subsurface layers do not, and this variation in solar insolation remains imprinted at depths well below the surface, giving rise to the "hot" regions displayed in Figure 52.3(a).

These hot regions are modeled well, as exemplified by the map in Figure 52.3(b), which shows the difference between the observed map and a thermal model. The viewing geometry is superimposed on the latter image. The negative temperatures near the poles and along the terminator are indicative of areas colder than predicted in the model. This is likely caused by surface topography, which causes permanent shadowing in craters at high latitudes and transient effects in the equatorial regions, where crater floors and hillsides are alternately in shadow and sunlight as the day progresses.

Radio spectra and images, together with *Mariner 10* IR data show that Mercury's surface properties are quite similar to those of the Moon, except for the microwave opacity, which is ~2–3 smaller than that of most lunar samples. This suggests a low ilmenite ($FeTiO_3$) content, the mineral that is largely responsible for the dark appearance of the lunar maria. The absence of minerals bearing iron (Fe) and titanium (Ti) from Mercury's surface has been confirmed via X-ray spectroscopy from the MESSENGER spacecraft, and suggests the absence of extensive basaltic volcanism from Mercury's deep interior.

2.4.3. Venus and Mars

Venus and Mars have atmospheres which consist of over 95% carbon dioxide gas (CO_2). Other than having a similar composition, the atmospheres are very different. The surface pressure on Venus is approximately 90 times that on the Earth, while on Mars it is ~140 times smaller. The shear amount of CO_2 gas on Venus provides so much opacity that Venus' surface can only be probed at wavelengths longward of ~6 cm (Figure 52.4), whereas Mars' atmosphere is essentially transparent at most radio wavelengths. On both planets, CO_2 gas is *photodissociated* (molecules are broken up) by sunlight at high altitudes into carbon monoxide (CO) and oxygen (O). CO gas has strong rotational transitions at millimeter wavelengths, which can be utilized to determine the atmospheric temperature profile and the CO abundance on Venus and Mars in the altitude regions probed.

Since CO is formed in the upper part of the atmosphere, the line is seen in absorption against the warm continuum background on both Venus and Mars (Figure 52.5). On Venus, one probes the so-called *mesosphere* in these transitions, a region between the massive lower atmosphere (altitudes $\lesssim 70$ km), in which the radiative time constant is much greater than a solar day, and the upper atmosphere (altitudes $\gtrsim 120$ km), which has a low heat capacity. In contrast to the lower atmosphere, a strong day-to-night gradient in temperature exists above the mesosphere, which leads to strong winds from the day- to the nightside. This is very different from the retrograde zonal winds observed in the visible cloud

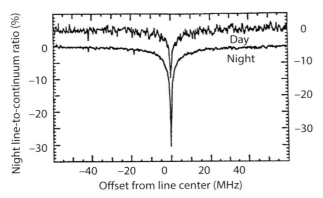

FIGURE 52.5 Spectra of Venus in the $J=1-0$ line. The upper curve is for the dayside hemisphere (when Venus is near superior conjunction), and the lower curve is for the nightside hemisphere (when Venus is near inferior conjunction). *Schloerb, F. P. (1985)*.

layers. These mesospheric winds carry CO, formed on the dayside through photodissociation of CO_2, to the nightside of the planet. The spectra in Figure 52.5, therefore, show the nightside line to be much deeper and narrower than the dayside line.

2.5. Giant Planets

2.5.1. Radio Spectra

At millimeter to centimeter wavelengths, one typically probes altitudes in the giant planet atmospheres from within to well below (pressure of tens of bars) the cloud layers. Representative microwave spectra are shown in Figure 52.6(a) (Jupiter) and (b) (Uranus). They generally show an increase in brightness temperature with increasing wavelength beyond 1.3 cm, due to the combined effect of a decrease in opacity at longer wavelengths, and an increase in temperature at increasing depth in the planet. The main source of opacity is ammonia (NH_3) gas, which has a broad absorption band at 1.3 cm. At longer wavelengths (typically >10 cm) absorption by water vapor and droplets may affect the spectra, while at short millimeter wavelengths the contribution of collision-induced absorption by molecular hydrogen becomes noticeable. On Uranus and Neptune, there is additional absorption by hydrogen sulfide (H_2S) and (perhaps) phosphine (PH_3) gas.

The composition of all four giant planets is dominated by H_2 and He gases, while the condensable gases CH_4, NH_3, H_2S, and H_2O constitute only a small fraction of the total. These gases, however, determine much of the "weather" on these planets. Although only cloud tops are seen "visually", thermochemical equilibrium calculations reveal the presence of a number of cloud layers deeper in the atmosphere, as depicted in Figure 52.6(c) and (d): an aqueous ammonia solution cloud, water ice, a cloud of ammonium hydrosulfide particles ($NH_3 + H_2S \rightarrow NH_4SH$ around 250 K), ammonia and/or hydrogen sulfide ice, and methane ice. The "visible" cloud layers on Jupiter and Saturn are composed of ammonia ice, while Uranus and Neptune are cold enough to allow condensation of methane gas.

To first approximation, the spectra of both Jupiter and Saturn resemble those expected for a solar composition atmosphere, while the spectra of Uranus and Neptune indicate a depletion of ammonia gas compared to the solar value by ~ 2 orders of magnitude. As shown in Figure 52.6(d), this depletion has been explained via formation of an extensive NH_4SH cloud, which essentially removes all NH_3 from the upper atmosphere.

The thermal emission from all four giant planets has been imaged with the VLA. To construct high sign-to-noise images, the observations are integrated over several hours, so that the maps are smeared in longitude and only reveal brightness variations in latitude. The observed variations have typically been attributed to spatial variations in opacity (NH_3, H_2S gases), as caused by a combination of atmospheric dynamics and condensation at higher altitudes. Below we briefly discuss findings for each planet individually.

2.5.2. Jupiter

In situ observations by the *Galileo* probe revealed that the NH_3 and H_2S abundances in Jupiter's deep atmosphere ($P \gtrsim 8$ bar) are three to four times solar, while radio spectra (Figure 52.6) show a subsolar abundance of NH_3 gas at pressures $P < 2$ bar. The apparent decrease in the NH_3 abundance at higher altitudes may be caused by dynamical processes, but the jury is still out on this.

Radio images of Jupiter clearly show bright zonal bands across the disk, co-located with the bright 5-μm bands; these all coincide with the brown belts seen at visible wavelengths (Figure 52.7). The radio and 5-μm bright bands have a higher brightness temperature due to a lower opacity in the belts relative to the zones, so deeper warmer layers are probed in the belts. This phenomenon is suggestive of gas rising in the zones; when the temperature drops below ~ 140 K, ammonia gas condenses out. In the belts the air, now depleted in ammonia gas (i.e. dry air), descends. The low NH_3 abundance enables deep layers to be probed at radio wavelengths, while the absence of clouds allows one to see deep at 5 μm.

Earlier this century, an algorithm has been developed to construct longitude-resolved images of Jupiter, and these maps reveal, for the first time, that the 5-μm hot spots are also hot at radio wavelengths (Figure 52.7(d)). At radio wavelengths, the hot spots indicate a relative absence of NH_3 gas, whereas they suggest a lack of cloud particles in the IR. Models show that ammonia must be depleted down

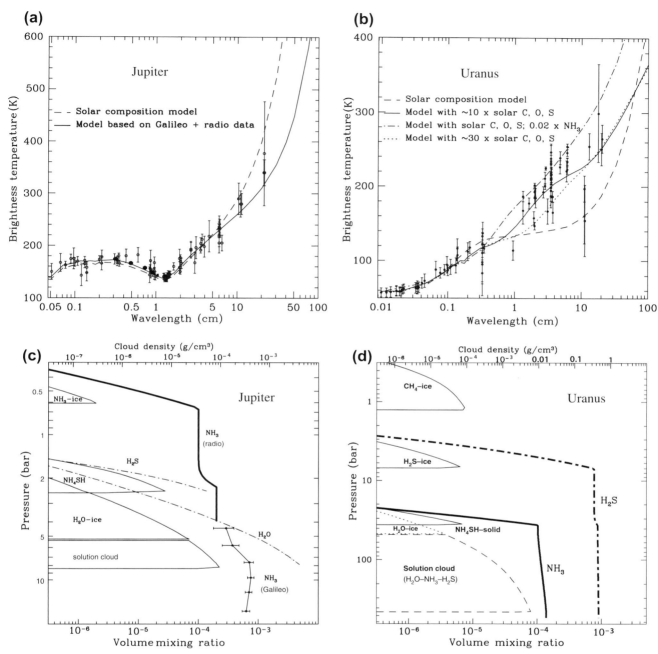

FIGURE 52.6 (a) Microwave spectrum of Jupiter, with superposed models for a solar composition atmosphere (dashed line), and one in which the altitude profiles for the condensable gases were based on the *Galileo* probe data at $P > 8$ bar, decreasing to $\sim 1/2$ the solar value at $P < 2$ bar (solid line). (b) Microwave spectrum of Uranus, with superposed models for a solar composition atmosphere (dashed line), and one in which H_2S and H_2O gases are enhanced by a factor ~ 10 above solar values. In these models, ammonia gas is significantly depleted at higher altitudes in Uranus atmosphere through formation of NH_4SH, so that deeper warmer levels are probed. (c) Cloud structure in Jupiter's atmosphere as calculated assuming thermochemical equilibrium. The altitude profile of ammonia gas, based on *Galileo* and ground-based radio data (used in panel a), is superposed. (d) Cloud structure in Uranus' atmosphere as calculated assuming thermochemical equilibrium and CH_4, H_2O, and H_2S abundances 30 times solar. The altitude profiles for H_2S and NH_3 gas are indicated. *de Pater, I. & Lissauer, J. J. (2010).*

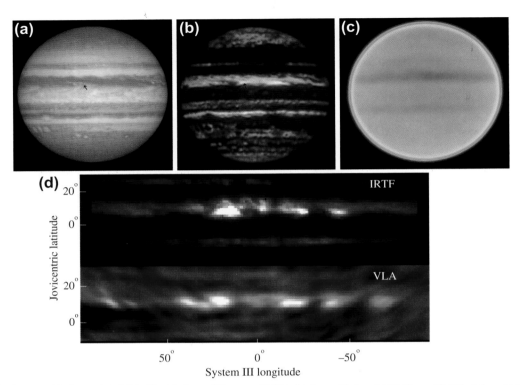

FIGURE 52.7 Images of Jupiter at (a) visible (Hubble Space Telescope (HST)), (b) thermal infrared (IR) (5 μm, NASA's Infrared telescope Facility, IRTF) and (c) radio wavelengths (2 cm, VLA). At visible wavelengths, the planet is seen in reflected sunlight, and its zones and belts show up as white and brown stripes, respectively. At 5 μm and radio wavelengths we receive thermal radiation from the planet. The belts, as well as other brown regions, are warmer than the zones, attributed to a lower opacity in the belts, so deeper warmer layers are probed in belts than zones. The IR and visible light images were taken in October 1995, just 2 months before the Galileo probe entry. The radio image was obtained in January 1996, about 1.5 months after probe entry. The position of the probe entry is indicated by an arrow on the visible light and IR images; the radio image is integrated over ~6–7 h, so that any longitudinal structure is smeared out. *(HST: Courtesy R. Beebe and HST/NASA; IRTF: Courtesy G. Orton; VLA: de Pater, I., et al. (2001)).* (d) A comparison of the North Equatorial Belt of Jupiter at 5 μm (like in panel b) and the radio image in panel c. This longitude-resolved image was constructed by using a novel data reduction technique. *Sault, R. J., et al. (2004).*

to pressure levels of over ~5 bar in the hot spots, the approximate altitude of the water cloud.

2.5.3. Saturn

Images of Saturn's microwave emission at different viewing geometries are shown in Figure 52.8. The planet itself is visible through its thermal emission, while the emission from the planet's rings is dominated by Saturn's thermal radiation reflected off the ring particles. Only a small fraction of the radiation at centimeter wavelengths is thermal emission from the rings themselves.

Like on Jupiter, radio spectra of the atmospheric emission can be interpreted in terms of its ammonia abundance and local variations therein with altitude and latitude. The ammonia and hydrogen sulfide abundances on Saturn are likely ~3–5 times the solar values. The latitudinal structure on Saturn's disk, presumably caused by latitudinal variations in microwave opacity, changes considerably over time.

Figure 52.9 shows a radio map of the planet at 2 cm constructed from Cassini radiometer observations. The deviation in brightness temperature is plotted relative to that expected for a model in which the atmosphere is fully saturated. In many areas the atmosphere appears to be subsaturated (in NH_3 gas, the main source of Saturn's microwave opacity), indicative of subsiding dry air. Note the large extended bright region near 40° N latitude. This is Saturn's large Northern Storm that was visible at that time. The radio-bright areas are suggestive of dry subsiding air, down to ~2–3 bar pressure level. Cassini observations at 2 cm and 5 μm show similar anticorrelations as observed for Jupiter (Figure 52.7), although not all radio-bright areas correspond exactly to 5-μm bright regions.

The classical A, B, and C rings, with the Cassini Division, are clearly visible on Figure 52.8. The inner B ring is brightest, with a brightness temperature of ~10 K. At 1–3 mm the temperature rises to ~20–25 K. In front of the planet, the rings block out part of Saturn's radio

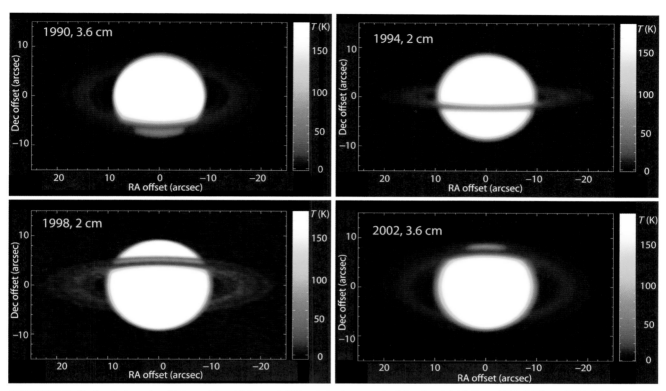

FIGURE 52.8 Radio photographs of Saturn at 2 and 3.6 cm, at different viewing aspects of the planet. (a) 3.6 cm, 1990; (b) 2 cm, 1994; (c) 2 cm, 1998; (d) 3.6 cm, 2002. *Dunn, D. E., de Pater, I., & Molnar, L. A. (2006).*

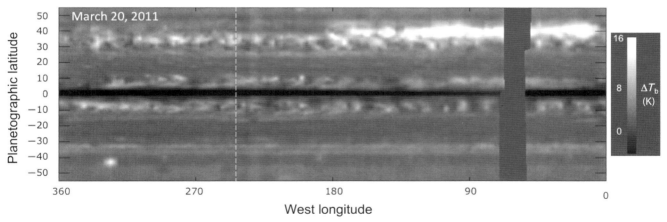

FIGURE 52.9 Cylindrical map of Saturn's 2-cm brightness temperature constructed from Cassini radiometer observations taken in March 2011. The value plotted is the residual brightness relative to a model for a fully saturated atmosphere. The black stripe across the equator is Saturn's ring, blocking the planet's thermal radiation from behind. The planet was mapped by continuous pole-to-pole scans during 14 h when the spacecraft was near periapse. Periapse (indicated by the dashed line) was at 3.72 Saturn radii, where the resolution was best (1.6° in latitude). The resolution degrades linearly with spacecraft distance, out to 5.64 Saturn radii. The spacecraft motion along its trajectory combined with the planet's rotation-combined sweep through the longitude range depicted. *Janssen, M. A., Ingersoll, A., Allison, M. D., Gulkis, S., Laraia, A., Baines, K., et al. (2013).*

emission, resulting in an absorption feature. The west (right) side of the ring is usually somewhat brighter than the east side, which has been attributed to the presence of gravitational "wakes", which are 10- to 100-m-sized density enhancements behind large ring particles which, because of Keplerian shear, travel at an angle to the big particle's orbit. Similar asymmetries have been seen in the A ring in front of the planet.

A combination of radio and radar data show that the ring particles have sizes from ~ 1 cm up to $\sim 5-10$ m, where the number of particles, N, at a given size, R, varies approximately as $N \sim R^{-3}$. Such a particle size distribution would be expected from a collisionally evolved population of particles.

2.5.4. Uranus and Neptune

Uranus is unique among the planets in having its rotation axis closely aligned with the plane in which the planet orbits the Sun. With its orbital period of 84 years, the seasons on Uranus last 21 years. During the Voyager encounter, in 1986, Uranus south pole was facing the Sun (and us). Since that time, this pole has slowly moved out of sight, while the north pole has come into view. Figure 52.10(a) shows a radio (2 cm) image taken with the VLA in the summer of 2003, along with an image at near-IR wavelengths (1.6 μm, panel b) taken with the adaptive optics system on the Keck telescope. The VLA image shows a bright (i.e. hot) south pole, as well as some enhanced brightness in the far north (to the right on the image). At near-IR wavelengths Uranus is visible in reflected sunlight. The bright regions are clouds (presumably CH_4 ice) at high (upper troposphere) altitudes. The bright band around the south pole is at the lower edge of the radio-bright south polar region. Air in this infrared-bright band likely rises, with condensibles forming clouds and the dry air descending over the pole. At radio wavelengths, this dry air allows one to probe deeper warmer layers in Uranus' atmosphere.

Also Neptune's south pole is hot, as shown in Figure 52.10(c). This particular image was obtained with the VLA after its sensitivity was increased by an order of magnitude. For comparison, an image at near-IR wavelengths is shown in Figure 52.10(a). A comparison of the Uranus and Neptune images suggests very similar dynamics in these planets' atmospheres, with air rising at midlatitudes and descending over the poles and the equator.

Prominent emission lines of CO and HCN were detected on Neptune in the early 1990s, with abundances ~ 1000 times higher than predicted from thermochemical

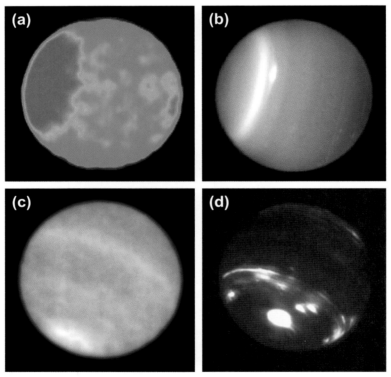

FIGURE 52.10 (a) VLA image of Uranus at 2 cm wavelength taken in the summer of 2003. Note the hot (red) poles. *(Hofstadter, M. D. & Butler, B. J. (2003).* (b) IR (1.6 μm) image of Uranus taken with the Keck adaptive optics system in October 2003. The polar collar around the south pole (left in figure) lines up with the edge of the hot pole seen at radio wavelengths. Several cloud features are visible in the IR image, and the thin line near the right is Uranus' ring system. *(Hammel, H. B., de Pater, I., Gibbard, S., Lockwood, G. W., & Rages, K. (2005).* (c) False-color image of Neptune at a wavelength of 0.9 cm. White is brightest, red dimmer, and black is dimmest. The brightest feature (about 20 K above its surroundings) is centered on the South Pole of the planet, and extends down to $\sim 60°$ S latitude. The equatorial region is marked by a band about 10 K brighter than its surroundings. There is also a hint of a band at $\sim 40°$ S latitude. Observations were made with the upgraded VLA, on August 6 and 9, 2011 *(Butler, B.J., Hofstadter, M., Gurwell, M., Orton, G., & Norwood, J. (2012).* (d) Near-IR (1.6 μm) image of Neptune taken with the Keck adaptive optics system on July 14, 2009 (by Imke de Pater and Heidi B. Hammel). The bright regions are clouds in Neptune's atmosphere.

equilibrium models. Recent work with the Combined Array for Research in Millimeter-wave Astronomy (CARMA) shows that the CO 2−1 and 1−0 lines are consistent with about 1 ppm CO in Neptune's stratosphere and no CO in the troposphere. CO appears to be uniformly distributed over Neptune's disk. Both findings suggest an external source, such as cometary impacts.

2.6. Major Satellites and Small Bodies

2.6.1. Galilean Satellites

Radio spectra of the Galilean satellites are diverse. The brightness temperature at thermal IR wavelengths can be related directly to the satellite's albedo, and hence Callisto, with its relatively low albedo ($A = 0.13$) is warmer than Io and Europa. The brightness temperature at radio wavelengths is determined by the physical temperature and radio emissivity, e, of the subsurface through Kirchoff's laws (Eqn (52.3)), with A being the radar geometric albedo. As expected from the high radar albedos (0.33 for Ganymede and 0.65 for Europa), the observed radio brightness temperatures for Ganymede and Europa are well below the expected physical temperature of the subsurface layers. These high albedos and consequently low emissivities and radio brightness temperatures are likely caused by coherent backscattering in fractured ice.

Since the detection of an ionosphere around Io by the *Pioneer 10* spacecraft in 1973, this satellite is known to posses a tenuous atmosphere. The first detection of a global atmosphere was obtained in 1990, where a rotational line of sulfur dioxide (SO_2) gas was measured at 222 GHz. Io is the only object with an atmosphere dominated by SO_2 gas. Part of the gas is of direct volcanic origin, and part is driven by subliming SO_2 frost, which itself is a product of volcanic eruptions. Several SO_2, as well as SO, lines have been observed. Analysis of the spectra suggests a surface pressure of a few, perhaps up to 40 nbar, covering 5−20% of the surface. The atmosphere may be relatively hot (500−600 K) on the trailing and cooler (250−300 K) on the leading hemisphere.

SO_2 maps have been obtained with the Smithsonian Submillimeter Array (SMA) at ∼346 GHz, with a synthesized beam size slightly smaller than the disk of Io itself. These maps confirm differences in the SO_2 abundance between the leading and trailing hemispheres and have been interpreted to favor an atmosphere supported primarily via sublimation of SO_2 ice.

2.6.2. Titan

Of all solar system bodies, Titan's atmosphere is most similar to that of the Earth, in being dominated by nitrogen gas and with a surface pressure 1.5 times that on the Earth. Methane gas, with an abundance of a few percent, has a profound effect on the atmosphere (*see* Titan). Photolysis and subsequent chemical reactions lead to the formation of hydrocarbons and nitriles. Because CO and the nitriles HCN, HC_3N (cyanoacetylene), and CH_3CN (acetonitrile) have several transitions at (sub)millimeter wavelengths (Figure 52.11), radio observations can be used to constrain the vertical distributions of these species. As expected from

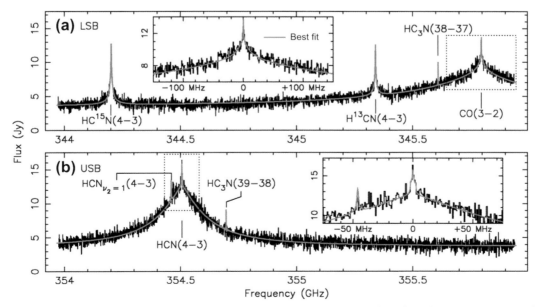

FIGURE 52.11 Submillimeter whole-disk spectrum of Titan obtained with the SMA, with an overlaid best-fit model spectrum. (a) Lower sideband (LSB) containing CO (3−2), $H^{13}CN$ (4−3), $HC^{15}N$ (4−3), and HC_3N (38−37) rotational transitions. (b) Upper side band (USB) with both the ground and vibrationally excited HCN (4−3) $v_2 = 0,1$ transitions and the HC_3N (39−38) transition. *Gurwell, M. A. (2004).*

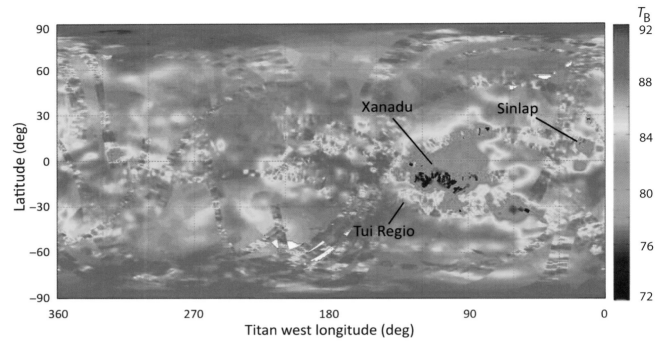

FIGURE 52.12 Map of Titan's absolute brightness temperature derived from Cassini radiometer data at a wavelength of 2 cm. A few regions discussed in the text are indicated. *Janssen, M. & Le Gall, A. (2011).*

photochemical models, their abundances increase with altitude and are highest in the stratosphere.

Disk-resolved spectra, such as obtained with the Submillimeter Array (SMA) and Institute for Radio Astronomy in the Millimeter Range (IRAM) Plateau de Bure Interferometer, also contain information on the zonal wind profile. Although 12-μm spectroscopic measurements had already suggested the winds to be prograde at ~100−300 km altitude, the radio data confirmed the direction of the winds and reported more precise values for the wind speeds in the upper stratosphere and lower mesosphere. At lower altitudes, the winds were determined by measuring the Doppler shift in the radio (communication) signal transmitted from the *Huygens* probe to the *Cassini* orbiter. This signal was recorded by the *very long baseline interferometry network* (VLBI). These measurements revealed weak prograde winds near the surface, rising to ~100 m/s at 100−150 km altitude, with a substantial drop (down to a few m/s) near 60−80 km altitude.

The isotopic carbon and nitrogen ratios were first determined from ground-based radio data (such as in Figure 52.11), and subsequently confirmed/improved by instruments on board the *Cassini* spacecraft and *Huygens* probe. The $^{12}C/^{13}C$ isotope ratio on Titan is very similar to that on the Earth (89), while $^{14}N/^{15}N$ was measured to be several times less than the terrestrial value of 272. This has been explained by a large loss of Titan's primitive atmosphere over time, which would lead to an isotopic fractionation in nitrogen. In contrast, the similar-to-Earth value of $^{12}C/^{13}C$ hints at a continuous or periodic replenishment of methane gas into Titan's atmosphere, such as could happen, for example, through *cryovolcanism* or a methane cycle akin to the hydrology cycle on the Earth.

Radiometry maps of Titan obtained with the *Cassini* spacecraft at 2 cm wavelength can be used with radar and IR measurements to better constrain the surface composition and compactness. Observations show variations in brightness temperature up to ~10 K (Figure 52.12), which are more or less anticorrelated with IR brightness, i.e. the IR/optically bright areas have a low radio brightness temperature.

The effective dielectric constant of Titan's surface is low, averaging about 1.6. There are only a few places where the dielectric constant is close to the value of 3.1 for pure water ice—near the Sinlap crater and the region separating Tui Regio from Xanadu. The overall low value is consistent with a fluffy surface of organic material. In most cases the radar reflectivity and radio emissivity are anticorrelated, as expected from Kirchoff's laws (Eqn (52.3)). In Xanadu, however, the reflectivities are anomalously high, indicative of coherent scattering on scales of a few wavelengths.

2.6.3. Asteroids and Trans-Neptunian Objects

In analogy with the terrestrial planets, a comparison of multiwavelength radio data of small airless bodies with thermophysical models provides information on the (sub)surface properties of the material, as composition and compactness. Radio spectra of several main belt asteroids suggest that these bodies are typically overlain by a layer of fluffy (highly porous) dust a few centimeters thick.

The Microwave Instrument for the Rosetta Orbiter (MIRO) on the European Space Agency's (ESA) Rosetta spacecraft observed thermal emissions from the main belt asteroids 2867 Steins and 21 Lutetia at submillimeter (0.6 mm) to millimeter (1.6 mm) wavelengths during the Rosetta spacecraft encounters with these asteroids in 2008 and 2010, respectively. The MIRO instrument consists of a 30-cm-diameter telescope coupled to a dual-frequency heterodyne radiometer/spectrometer. Steins is a small-diameter ($\sim$5 km) E-type asteroid; Lutetia is 100-km-diameter C- or M-class asteroid. The observed continuum thermal emission from both asteroids allowed their regoliths to be probed to depths of 2 mm–2 cm below their surfaces and their thermal inertias to be measured. Lutetia is covered by a very low thermal inertia ($<$20 J/K/m^2/s$^{0.5}$) and high-emissivity ($>$0.9) fine powder or dust, similar to the lunar regolith. The thermal inertia of Steins is between 450 and 850 J/K/m^2/s$^{0.5}$ and its emissivity is $\sim$0.8. These properties are indicative of a more rocklike surface material, as expected for smaller asteroids. Asteroids covered by powder or dust will have large dayside temperature gradients, $\sim$50 K/cm, and hence observers at radio wavelengths will sense lower temperatures than found on the surface.

It has been challenging to observe trans-Neptunian or Kuiper Belt objects (KBO) at radio wavelengths, including Pluto, due to their small angular extent and low surface temperature. For an object in radiative equilibrium with an albedo of $\sim$0.6, the surface temperature should be $\sim$50 K, consistent with the 53–59 K temperatures for Pluto as measured by the IRAS satellite at 60 and 100 μm. Observations of Pluto with the 30-m IRAM telescope at millimeter wavelengths revealed a brightness temperature of $\sim$30 K, 10 K less than expected $\sim$10 wavelengths below its surface, indicative of a low radio emissivity ($e \approx 0.6 - 0.7$), similar to that seen on Ganymede. Such a low emissivity can be reconciled with a surface composed of icy grains, and hence a relatively high porosity.

Radio measurements of KBOs have been used to determine the size and albedo of several of the largest objects (Eris, Quaoar, Ixion, Varuna, 2002 AW197), in concert with optical measurements and the so-called Standard Thermal Model (STM) to interpret the data. Although one has to be aware of the assumptions made in the STM, which can lead to over- or underestimates of the size and albedo, such measurements are usually our only means to get a reasonable size estimate for these objects.

2.7. Comets

Continuum measurements at radio wavelengths measure the thermal emission from a cometary nucleus and from large dust grains in its coma, while spectroscopic observations provide information on the "parent" molecules in a comet's coma. Upper limits to the radio continuum emission of a few comet nuclei suggest that the temperature gradient in the nucleus may be very steep, or, alternatively, that the emission is substantially suppressed by subsurface scattering.

The most significant advances in cometary radio research have been obtained from spectroscopic studies. The cometary nucleus consists primarily of water ice, which sublimates off the surface when the comet approaches the Sun. H_2O dissociates into OH and H (lifetime $\sim$1 day). Since the early 1970s, the 18-cm OH line has been observed and monitored in many comets, providing indirect information on the production rate and time variability therein of water, a molecule that remains difficult to observe on a routine basis.

The OH line is sometimes seen in emission, and at other times in absorption against the galactic background. The OH emission is maser emission, i.e. stimulated emission from molecules in which the population of the various energy levels is inverted, so that the higher energy level is overpopulated compared to the lower energy level. This population inversion is caused by absorption of solar photons at UV wavelengths. However, this excitation process depends on the comet's velocity with respect to the Sun (heliocentric velocity), referred to as the Swings effect, attributed to the astrophysicist P. Swings. If the heliocentric velocity is such that solar Fraunhofer (absorption) lines are Doppler shifted into the OH excitation frequency, the molecule is not excited. In that case, OH will absorb 18-cm photons from the galactic background and be seen in absorption against the galactic background. If the line is excited, background radiation at the same wavelength (18 cm) will trigger its de-excitation, and the line is seen in emission (maser or stimulated emission). With radio interferometers, the OH emission can be imaged. Such images have, for example, revealed the so-called quenching region directly, a region around the nucleus where collisions between particles thermalize the energy levels of OH molecules, so they no longer produce maser emission (Figure 52.13).

One of the strengths of radio astronomy is the detection of "parent" molecules in a cometary coma, molecules that evolve directly from its icy surface. Such observations are crucial for our understanding of a comet's composition, and, indirectly, of the conditions in the early solar nebula from which our planetary system formed (cometary nuclei have not been altered by excessive heating or high pressures). A growing number of molecular species have been detected at radio wavelengths. Figure 52.14 shows the time evolution of observed production rates for a large number of gases, subliming from comet Hale–Bopp (C/1995 O2). Only the most volatile materials sublime at heliocentric

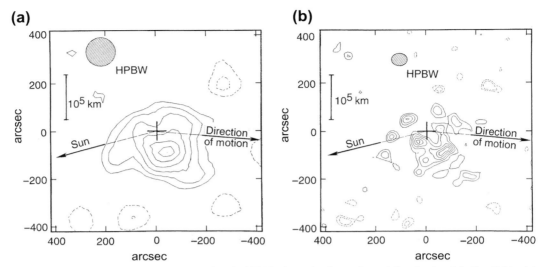

FIGURE 52.13 Contour plots of comet Halley, November 13–16, 1985. The image is taken at the peak flux density of the line (0.0 km/s in the reference frame of the comet). Panel a) shows a low-resolution image (HPBW=3′), and panel b) shows a high-resolution image (HPBW=1′), after the data for both dates were combined. Contour levels for the low-resolution image are 4.9, 7.8, 10.8, 13.7, 16.7, and 18.6 mJy/beam. For the high-resolution image, they are 4.4, 4.4, 6.0, 7.7, 9.3, and 10.4 mJy/beam. Dashed contours indicate negative values. The half power beam width (HPBW), a linear scale, the direction of motion, and the direction to the Sun are indicated in the figures. The cross indicates the position of the nucleus at the time of the observations. *de Pater, I., Palmer, P., & Snyder, L.E. (1986).*

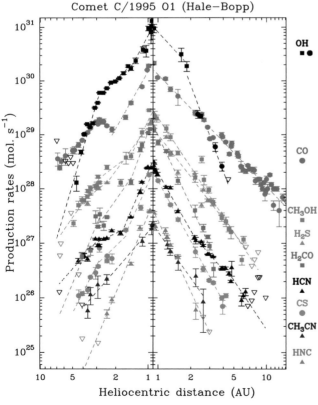

FIGURE 52.14 Time evolution of the production rates of 9 different molecules in C/Hale-Bopp as observed at (sub)mm-wavelengths. Power law fits (dashed lines) are superposed on the data. *Biver, N., et al., 2002.*

distances $r \gtrsim 5$ AU, while OH (from H_2O) becomes dominant at $r < 3$ AU.

Whereas most molecules sublime directly off the cometary nucleus, some gases, such as CO and formaldehyde (H_2CO), are also released from dust grains in a comet's coma (Figure 52.15).

3. NONTHERMAL RADIATION

Nonthermal planetary radio emissions are usually produced by electrons spiraling around magnetic field lines. Until the era of spacecraft missions, we had only received *nonthermal radio emissions* from the planet Jupiter, and these were limited to frequencies $\gtrsim 10$ MHz, since radiation at lower frequencies is blocked by Earth's ionosphere. Strong radio bursts at frequencies below 40 MHz were attributed to emissions via the cyclotron maser instability in which *auroral* electrons with energies of a few to several kiloelectronvolt power the emission, while radiation at frequencies $\gtrsim 100$ MHz was interpreted as synchrotron radiation, emitted by high-energy (megaelectronvolt range) electrons trapped in Jupiter's radiation belts, a region in Jupiter's magnetic field analogous to the Earth's *Van Allen belts*. Like Earth, the magnetic fields of the four giant planets resemble to first approximation a dipole magnetic field and are strong radio sources at low frequencies (kilometric wavelengths). Jupiter's moon Ganymede is also a source of nonthermal radio emissions. The strongest planetary radio emissions usually originate near the auroral regions and are intimately related to auroral processes.

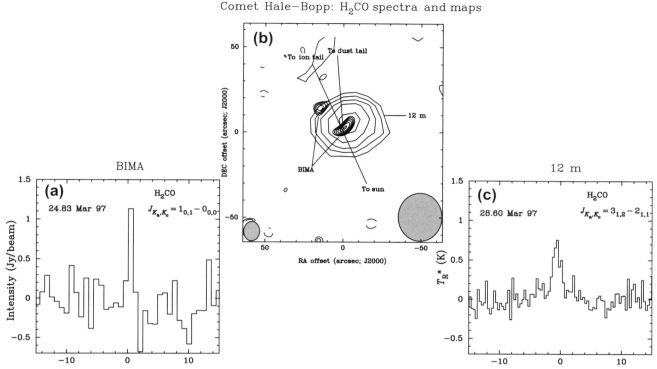

FIGURE 52.15 Images and spectra of H$_2$CO in comet C/1995 O1 (Hale–Bopp) taken with Arizona Radio Observatory (ARO) 12-m telescope and the Berkeley–Illinois–Maryland Association (BIMA) array. Panels (a and c) show spectra at different frequencies. Panel (b) shows a contour map from BIMA at 72.8 GHz (in bold; from spectrum in panel (a)), superposed on the ARO 225.7-GHz image (from spectrum in panel (c)). The synthesized beam for BIMA is shown in the lower left, and that of ARO appears in the lower right. These observations show that formaldehyde originates both from the nucleus and in the coma, where the coma source appears dominated by a single fragment. *Milam, S. N., et al. (2006)*.

A graph of the average normalized spectra of the auroral radio emissions from the four giant planets and the Earth is displayed in Figure 52.16(a). Jupiter is the strongest low-frequency radio source, followed by Saturn, Earth, Uranus, and Neptune. In Sections 3.3–3.7, we discuss the emissions from each planet.

3.1. Low-Frequency Emissions

3.1.1. Cyclotron Maser Emissions

Electron cyclotron maser radiation is emitted at the frequency at which electrons spiral around the local magnetic field lines (the cyclotron or Larmor frequency):

$$\nu_L = \frac{qB}{2\pi m_e c} \quad (52.6)$$

where q is the elemental charge, B is the magnetic field strength, m_e is the electron mass, and c is the speed of light. Propagation of the radiation depends on the interaction of the radiation with the local plasma, or charged particle population. The oscillation of these particles, as caused by the electromagnetic properties of the plasma, leads to a complex interaction between the propagating radiation (the electromagnetic waves) and the local plasma. For example, the radiation can escape its region of origin only if the local cyclotron frequency is larger than the electron plasma frequency:

$$\nu_e = \left(\frac{4\pi N_e q^2}{m_e}\right)^{1/2} \quad (52.7)$$

with N_e being the electron density. Hence, the plasma frequency is the frequency at which electrons oscillate about their equilibrium positions in the absence of a magnetic field. This similarly sets the limit for propagation through Earth's ionosphere at ~ 10 MHz. If the local cyclotron frequency is less than the electron plasma frequency, the waves are locally trapped and amplified, until they reach a region from where they can escape. The cyclotron maser instability also requires a large ratio of ν_L/ν_e. The auroral regions in planetary *magnetospheres* are characterized by such conditions. The mode of propagation (or polarization) of auroral radio emissions is predominantly in the so-called extraordinary (X) sense, and the polarization (direction of the electric vector of the radiation) depends upon the direction of the magnetic field. The emission is right-handed (RH) circularly polarized if the field at the source is

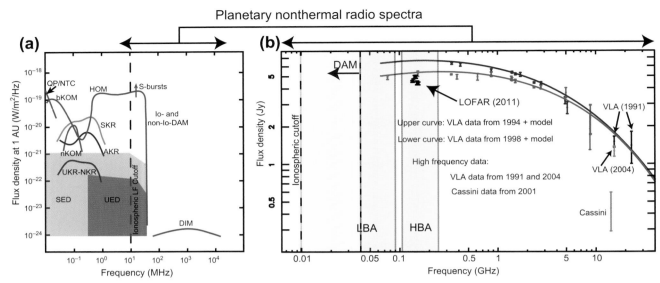

FIGURE 52.16 (a) A comparison of the peak flux density spectrum of the kilometric continuum radio emissions of the four giant planets and the Earth. All emissions were scaled such that the planets appear to be at a distance of 1 AU. Jovian emissions shown include quasi-periodic (QP) bursts, nonthermal continuum (NTC), broadband and narrowband kilometric radiation (bKOM, nKOM), hectometric radiation (HOM), decametric radiation (DAM), and decimetric radiation (DIM). Saturn's kilometric radiation is designated SKR, and its electrostatic discharge emissions are labeled SED. Terrestrial auroral kilometric radiation is designated AKR. UKR and NKR refer to kilometric radiation from Uranus and Neptune, respectively. Uranus' electrostatic discharges are labeled UED. *(Adapted from Zarka, P. & Kurth, W. S. (2005).* (b) Jupiter's synchrotron radio spectrum covers a fraction of the frequency range in the panel (a), as indicated by the arrows above the panels. The spectra were taken in September 1998 (lower curve) and June 1994 (upper curve), augmented with higher frequency VLA data from 1991 to 2004, and an anomalously low measurement taken by the Cassini spacecraft en route to Saturn. Superposed are model calculations that match the data. Preliminary results from Low-Frequency Array (LOFAR) commissioning data taken in November 2011 are indicated in black. The ionospheric cutoff frequency, DAM frequency range, and LOFAR low band (LBA) and high band (HBA) frequency coverage are indicated. *Figure adapted from Kloosterman, J. L., Butler, B., & de Pater, I. (2008). LOFAR data from: Girard, J. N., Zarka, P., Tasse, C., Hess, S., & the LOFAR Collaboration (2012).*

directed toward the observer and left-handed (LH) circularly polarized if the field points away from the observer.[1]

Cyclotron radiation is emitted in a dipole pattern, where the lobes are bent in the forward direction. The resulting emission is like the hollow cone pattern shown in Figure 52.17. The radiation intensity is zero along the axis of the cone, in the direction of the particle's parallel motion, and reaches a maximum at an angle Ψ. Theoretical calculations show that Ψ is very close to 90°. Observed opening angles, however, can be much smaller, down to ∼50°, which has been attributed to refraction of the electromagnetic waves as they depart from the source region.

The cyclotron maser instability derives energy from a few kiloelectronvolt electrons, which have distribution

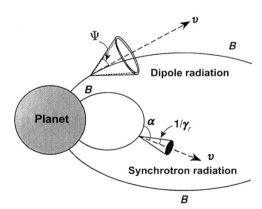

FIGURE 52.17 Radiation patterns in a magnetic field. Indicated is the hollow cone pattern caused by cyclotron (dipole) radiation from nonrelativistic electrons in the auroral zone. The electrons move outward along the planet's magnetic field lines. The hollow cone opening half-angle is given by Ψ. At low magnetic latitudes, in the Van Allen belts, the instantaneous cone of emission of a single relativistic electron is indicated. The angle between the particle's direction of motion and the magnetic field, commonly referred to as the particle's pitch angle, α, is indicated on the sketch. The emission is radiated into a narrow cone with a half width of $1/\gamma$. *de Pater, I. & Lissauer, J. J. (2001).*

---

1 Circular polarization is in the right-hand (RH) sense when the electric vector of the radiation in a plane perpendicular to the magnetic field direction rotates in the same sense as an RH screw advancing in the direction of propagation. Thus, rotation is counterclockwise when propagation is toward and viewed by the observer. RH polarization is defined as positive; left-hand polarization (LH), as negative. In some cases, the radio emissions propagate in the ordinary (O) magneto-ionic mode. In this mode the polarization is reversed. The theory of the cyclotron maser instability does admit the possibility of emission in the ordinary mode. However, it is less common.

functions with a positive slope in the direction perpendicular to the magnetic field. Recent observations in the source of Earth's auroral kilometric radiation (AKR) reveal "horseshoe"-shaped electron distributions that provide a highly efficient (of order 1%) source of free energy for the generation of the radio waves. This distribution is thought to result from parallel electric fields in the auroral acceleration region, the loss of small pitch-angle electrons to the planetary atmosphere, and trapping of reflected electrons. Radio emissions generated in planetary magnetospheres by this mechanism often display a bewildering array of structure on a frequency–time spectrogram including narrowband tones that rise or fall in frequency, sharp cutoffs, and more continuum-like emissions. While it is generally accepted that emissions that rise or fall in frequency are related to tiny sources moving down or up the magnetic field line (hence, to regions with higher or lower cyclotron frequencies), there is no generally accepted theoretical explanation for the fine structure.

3.1.2. Other Types of Low-Frequency Radio Emissions

While the radio emissions generated by the cyclotron maser instability are, by far, the most intense in any planetary magnetosphere, other types of radio emissions do occur that are of interest. Perhaps the most ubiquitous of these is the so-called nonthermal continuum radiation that arises from the conversion of wave energy in electrostatic waves near the source plasma frequency to radio waves, usually propagating in the ordinary mode. There are arguments for both linear and nonlinear conversion mechanisms. The term "continuum" was originally assigned to this class of emissions because they can be generated at very low frequencies (VLFs) and can be trapped in low-density cavities in the outer portions of the magnetosphere when the surrounding solar wind density is higher. The mixture of multiple sources at different frequencies and multiple reflections off the moving walls of the magnetosphere tend to homogenize the spectrum. However, at higher frequencies, these emissions are often created as narrowband emissions from electrostatic bands at the upper hybrid resonance frequency on density gradients in the inner magnetosphere and can propagate directly away from the source, yielding a complex narrowband spectrum. These emissions were first discovered at Earth and have been found at all the magnetized planets. Furthermore, emissions of this nature are also produced by Ganymede's magnetosphere.

Another type of planetary radio emission is closely related to a common solar emission mechanism, the conversion of Langmuir waves to radio emissions at either the fundamental plasma frequency (fp) or its harmonic (2fp). The Langmuir waves are common features of the solar wind upstream of a planetary bow shock, which arises from the interaction of the supersonic flow of solar wind plasma past the planets. This mechanism is a nonlinear mechanism involving three waves: the Langmuir wave, the radio wave, and a low-frequency wave in the case of emission near the plasma frequency or a second Langmuir wave in the case of harmonic emission. The resulting emissions are weak, narrowband emissions.

3.1.3. Atmospheric Lightning

Radio emissions from planets are sometimes associated with atmospheric lightning. The lightning discharge, in addition to producing the visible flash, also produces broad, impulsive radio emissions. If the spectrum of this impulse extends above the ionospheric plasma frequency and if absorption in the atmosphere is not too great, a remote observer can detect the high-frequency end of the spectrum. The "interference" detected with an AM radio on Earth during a thunderstorm is the same phenomenon.

3.2. Synchrotron Radiation

Synchrotron radiation is emitted by relativistic electrons gyrating around magnetic field lines. In essence, this emission consists of photons emitted by the acceleration of electrons as they execute their helical trajectories about magnetic field lines. The emission is strongly beamed in the forward direction (see Figure 52.17) within a cone $1/\gamma$:

$$\frac{1}{\gamma} = \sqrt{1 - \frac{v^2}{c^2}} \quad (52.8)$$

where v is the particle's velocity and c is the speed of light. The relativistic beaming factor $\gamma = 2E$, where E is the energy in mega-electronvolt. The radiation is emitted over a wide range of frequencies, but shows a maximum at $0.29\,\nu_c$, with the critical frequency, ν_c, in megahertz:

$$\nu_c = 16.08 E^2 B \quad (52.9)$$

where the energy E is in mega-electronvolts and the field strength B is in Gauss. The emission is polarized, where the direction of the electric vector depends on the direction of the local magnetic field. Jupiter is the only planet for which this type of emission has been observed. It has been mapped by ground-based radio telescopes and by *Cassini* to provide some of the most comprehensive, although indirect, information about Jupiter's intense radiation belts.

3.3. Earth

The terrestrial version of the cyclotron maser emission, commonly referred to as Auroral Kilometric Radiation

(AKR), has been studied both at close range and larger distances by many Earth-orbiting satellites. The radiation is very intense; the total power is 10^7 W, sometimes up to 10^9 W. The intensity is highly correlated with geomagnetic substorms, thus it is indirectly modulated by the solar wind. It originates in the nightside auroral regions and in the dayside polar cusps at low altitudes and high frequencies and spreads to higher altitudes and lower frequencies. Typical frequencies are between 100 and 600 kHz. Since AKR is generated by auroral electrons, it can be used as a proxy for auroral activity. And, since numerous in situ studies of the terrestrial auroral electron populations and the resulting radio emissions have been carried out, we can apply our understanding of this emission process to similar emissions at other planets where in situ studies have not yet been carried out.

Earth is also a source of nonthermal continuum radiation. Below the solar wind plasma frequency this radiation is trapped within the magnetosphere. The spectrum is relatively smooth down to the local plasma frequency, typically in the range of a few kilohertz, where the emission cuts off at the ordinary mode cutoff. A few observations of this emission also show an extraordinary mode cutoff. Hence, the emission is either generated in both polarizations or some of the initially dominant ordinary mode is converted into the extraordinary mode via reflections or other interactions with the magnetospheric medium. Above the solar wind plasma frequency, typically at a few tens of kilohertz, the "continuum" radiation spectrum exhibits a plethora of narrowband emissions; some of these extend well into the range of a few hundred kilohertz.

While not as important as the auroral radio emissions from an energetics point of view, the low-frequency limit of the continuum radiation at the plasma frequency provides an accurate measure of the plasma density, an often-difficult measurement for a plasma instrument because of spacecraft charging effects.

3.4. Jupiter's Synchrotron Radiation

Jupiter is the only planet from which we receive synchrotron radiation. The variation in total intensity and polarization characteristics during one Jovian rotation (the so-called beaming curves) indicate that Jupiter's magnetic field is approximately dipolar in shape, offset from the planet by roughly one-tenth of a planetary radius toward a longitude of 140°, and inclined by $\sim 10°$ with respect to the rotation axis. Most electrons are confined to the magnetic equatorial plane. The magnetic north pole is in the northern hemisphere, tipped toward a longitude of 200°. The total flux density of the planet varies significantly over time (Figure 52.18). The smoothly varying component seems to be correlated with solar wind parameters, in particular the solar wind ram pressure, suggesting that the solar wind may influence the

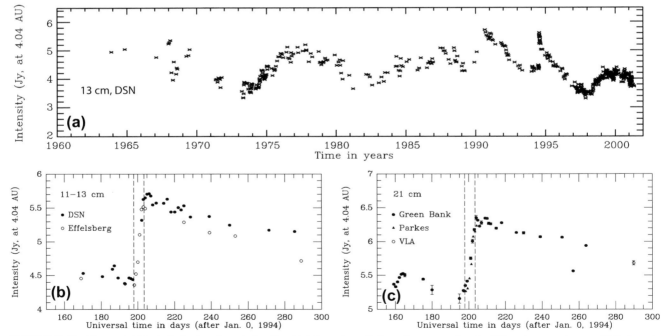

FIGURE 52.18 Time variability in Jupiter's radio emission. Panel (a) shows the radio intensity at a wavelength of 13 cm between the years 1963 through early 2001, as compiled by M.J. Klein. Panels (b) and (c) show Jupiter's radio intensity at 11–13 and 21 cm, respectively, during 1994 up to the summer of 1995. The impact of comet D/Shoemaker–Levy 9 with Jupiter occurred in July of 1994 (indicated by the vertical dashed lines). *Figure from de Pater and Lissauer (2010).*

supply and/or loss of electrons into Jupiter's inner magnetosphere. In addition to variations in the total flux density, the radio spectrum changes as well (Figure 52.16).

An image of Jupiter's synchrotron radiation at a wavelength of 20 cm, obtained with the VLA in 1994, is shown in Figure 52.19(a). Since Jupiter's synchrotron radiation is optically thin, one can use tomography to extract the three-dimensional distribution of the radio emissivity from data obtained over a full Jovian rotation. The example in Figure 52.19(b) shows that most of the synchrotron radiation is concentrated near the magnetic equator, which, due to the higher order moments in Jupiter's field, is warped like a potato chip. The secondary emission regions, apparent at high latitudes in Figure 52.19(a), show up as rings of emission north and south of the main ring. These emissions are produced by electrons at their mirror points and reveal the presence of a rather large number of electrons bouncing up and down field lines that thread the magnetic equator at ∼2.5 Jovian radii. This emission may be "directed" by the moon Amalthea. A fraction of the electrons near Amalthea's orbit undergoes a change in their direction of motion, caused perhaps by interactions with low-frequency plasma waves near Amalthea (such plasma noise was detected by the *Galileo* spacecraft when it crossed Amalthea's orbit), and through interactions with dust in Jupiter's rings, while regular synchrotron radiation losses also lead to small changes in an electron's direction of motion.

Figure 52.16(b) shows radio spectra of Jupiter's synchrotron radiation from 74 MHz up to ≳20 GHz. These spectra show that the electrons in Jupiter's radiation belts do not follow a simple $N(E) \propto E^{-a}$ power law. Well outside the synchrotron radiation region, beyond Io's orbit at 6 Jovian radii, the electron energy spectrum appears to follow a double power law, $N(E) \propto E^{-0.5}(1 + E/100)^{-3}$, consistent with in situ measurements by the *Pioneer* spacecraft. Processes such as radial diffusion, pitch angle scattering, synchrotron radiation losses, and absorption by moons and rings change the electron spectrum. The radio spectra superposed on the data were derived from such models.

Sometime between 1930 and 1970, Jupiter captured a comet, now known as comet D/Shoemaker–Levy 9 (SL9). During a close encounter with the planet, this comet was ripped apart by Jupiter's strong tidal force into over 20 pieces. These comet fragments, all in orbit about Jupiter, were discovered by the Shoemaker–Levy comet hunting team in May 1993. About a year later, from July 16 to 22 (1994), all comet fragments hit Jupiter. These events were widely observed, at wavelengths across the entire electromagnetic spectrum. At IR wavelength these impacts were incredibly bright, while at optical wavelengths the impact sites were visible as dark spots with even the smallest telescopes. This collision also triggered large temporary changes in Jupiter's synchrotron radiation. The total flux density increased by ∼20% (Figure 52.18), the radio spectrum flattened toward shorter wavelengths, and the spatial brightness distribution changed considerably (Figure 52.20(a) and (b)). These changes were brought about by a complex interaction of the radiating particles with shocks and electromagnetic waves induced in the magnetosphere by the series of cometary impacts. Results from models simulating the effects are shown in Figure 52.20(c–f).

Although impacts like those by SL9 were expected to be once in a lifetime events, only 15 years later, in July 2009, amateur astronomer Anthony Wesley in Australia

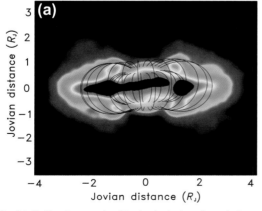 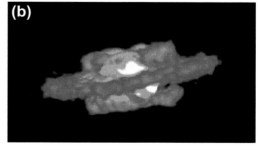

FIGURE 52.19 (a) Radio photograph of Jupiter's decimetric emission at a wavelength of 20 cm, and a central meridian longitude of $\lambda_{cml} \sim 312°$. Magnetic field lines at equatorial distances of 1.5 and 2.5 Jupiter radii are superposed. Field lines are shown every 15°, between $\lambda_{cml} - 90°$ and $\lambda_{cml} + 90°$. The image was taken with the VLA in June 1994. The resolution is 0.3 Jupiter radii, roughly the size of the high-latitude emission regions. *(de Pater, I., et al. (1997)).* (b) Three-dimensional reconstruction of Jupiter's nonthermal radio emissivity, from VLA data taken in June 1994, as seen from the Earth at $\lambda_{cml} = 140°$ ($D_E = -3°$). The planet is added as a white sphere in this visualization. *Adapted from de Pater, I. & Sault, R. J. (1998).*

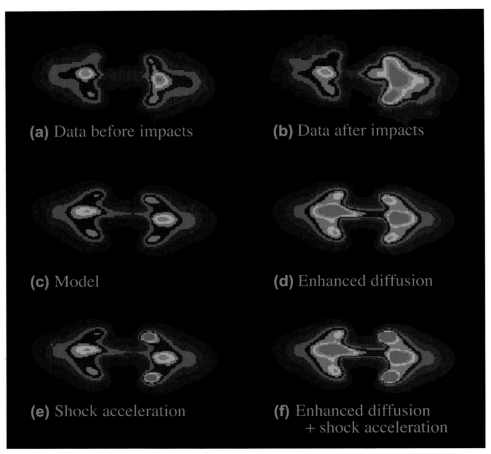

FIGURE 52.20 Real and synthetic false-color images at a wavelength of 20 cm (1.5 GHz) of Jupiter following the impacts of comet D/Shoemaker–Levy 9 with the planet. (a and b) Observations of the synchrotron radiation before (June 1994) and after several impacts (July 19, 1994), respectively. (c) Theoretical emission based on a model of the ambient relativistic electron distribution within a multipole magnetic field configuration. (d) Theoretical synchrotron radiation after an enhancement in the radial diffusion coefficient by a factor of a few million. (e) Enhancement in the theoretical synchrotron radiation, as produced from just shock acceleration. (f) Theoretical synchrotron radiation using the shock model and radial diffusion combined. *Brecht, S. H., de Pater, I., Larson, D. J., & Pesses, M. E. (2001).*

announced the discovery of a brown "spot" on Jupiter, which he attributed to an impact. Indeed, follow-up observations confirmed that Jupiter was hit yet again by an object several hundreds of meters across, this time of asteroidal rather than cometary nature. By chance, Jupiter had been observed with the VLA both before and after the impacts took place, and as in 1994 the radiation belts were again affected. As only one object hit the planet, enhancements in emission were only seen in the main radiation belts, in agreement with expectations (like the model in Figure 52.20(d)).

3.5. Jupiter at Low Frequencies

Jupiter has the most complex low-frequency radio spectrum of all the planets. Examples of most of these are shown in Figure 52.21 and are discussed in this section.

3.5.1. Decametric and Hectometric Radio Emissions

From the ground, Jupiter's decametric (DAM) emission, confined to frequencies below 40 MHz, has routinely been observed since its discovery in the early 1950s, occasionally down to frequencies of 4 MHz. The upper frequency cutoff is determined by the local magnetic field strength in the auroral regions: 40 MHz for RH circularly polarized emissions translates into ∼14 Gs in the north polar region, and 20 MHz for LH into ∼7 Gs in the south.

The dynamic spectra in the frequency–time domain are extremely complex, but well ordered.

On timescales of minutes, the emission displays a series of arcs, like open or closed parentheses (Figure 52.21). Within one storm, the arcs are all oriented the same way. The emissions have been interpreted as coherent cyclotron emissions. The satellite Io appears to modulate some of the

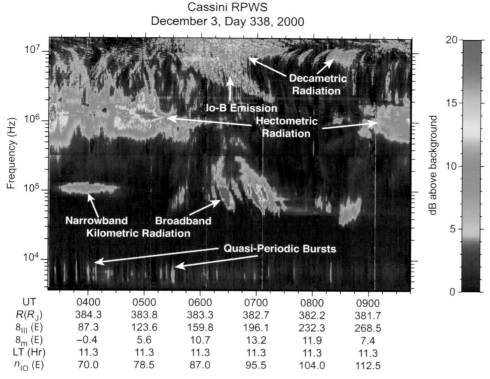

FIGURE 52.21 A representative dynamic spectrum of several of Jupiter's low-frequency radio emissions. The color bar is used to relate the color to the intensity of the emission. The emission is plotted as a function of frequency (along the *y*-axis) and time (along the *x*-axis). *Lecacheux, A. (2001)*.

emissions: Both the intensity and the probability of the occurrence of bursts increase when Io is at certain locations in its orbit with respect to Jupiter and the observer. The non-Io emission originates near Jupiter's aurora, and is produced by electrons that travel along magnetic field lines from the middle to outer magnetosphere toward Jupiter's ionosphere. Particles that enter the atmosphere are "lost". These may locally excite atoms and molecules through collisions, which upon de-excitation are visible as aurora at UV and IR wavelengths. Other electrons are reflected back along the field lines, and produce DAM, where their motion along the field line is reflected in the form of arcs in the radio emission (i.e. a drift with frequency). The Io-dependent emissions are produced at or near the footprints of the magnetic flux tube passing through Io (similar, but much weaker, emissions originate along the flux tubes passing through Ganymede, and perhaps Callisto).

Hectometric (HOM) emissions are, in many ways, indistinguishable from DAM except that they are found at lower frequencies, from a few hundred kilohertz to a few megahertz, with a local maximum near 1 MHz. The source region of HOM must be further from Jupiter than the DAM source. Otherwise, like DAM, HOM is predominantly emitted in the extraordinary mode and is likely generated by the cyclotron maser instability.

Because the dipole moment of Jupiter is tilted by $\sim 10°$ from the rotational axis, most Jovian radio emissions exhibit a strong rotational modulation. Given that Jupiter is a gas giant, this modulation is thought to be the best indicator of the rotation of the deep interior of the planet. The rotation period of the interior is important, for example, because this provides a rotating coordinate system against which the atmospheric winds can be measured. Because these radio observations have been recorded over many decades of time, analysis of these data lead to an extremely accurate determination of Jupiter's rotation period, $9^h\,55^m\,29^s.6854$.

3.5.2. Kilometric Radio Emissions

Between a few kilohertz up to 1 MHz various spacecraft detected both broadband kilometric (bKOM) and narrowband kilometric (nKOM) radiation from Jupiter (Figure 52.21). The lower frequency cutoff for bKOM, ~ 20 kHz (sometimes down to ~ 5 kHz) is likely set by propagation of the radiation through the Io plasma torus. The source of these emissions is at high magnetic latitudes and appears fixed in local time. The forward lobe near the north magnetic pole is of opposite polarization than a "back lobe" of the same source. The nKOM emissions last longer (up to a few hours) than bKOM; are confined to a smaller frequency range, 50–180 kHz; and show a smooth rise and fall in intensity. The recurrence period for nKOM events suggests that the source lags behind Jupiter's rotation by 3–5%, which was the first indication that this emission, in contrast to any other low-frequency emissions, is produced by distinct sources near the outer edge of the plasma torus. *Galileo* and *Ulysses* studies have shown that these emissions occur as a part of an apparently global magnetospheric dynamic event. There is a sudden onset of these emissions, they are visible for a few to several planetary rotations, and finally, they fade away.

3.5.3. Very Low-Frequency Emissions

The *Voyager* spacecraft detected continuum radiation in Jupiter's magnetosphere at frequencies below 20 kHz, both in its escaping and trapped form. As discussed in Section 3.1, radiation can be trapped inside the magnetic cavity if it cannot propagate through the high-plasma-density *magnetosheath*. This trapped emission has been observed from a few hundred hertz up to ~ 5 kHz. Occasionally, it has been detected up to 25 kHz, suggesting a compression of the magnetosphere caused by an increased solar wind ram pressure. Outside the magnetosphere the lower frequency cutoff of the freely propagating radiation corresponds to the plasma frequency in the magnetosheath and appears to be well correlated with the solar wind ram pressure. This escaping component is characterized by a complex narrowband spectrum, attributed to a linear or nonlinear conversion of electrostatic waves near the plasma frequency into freely propagating electromagnetic emissions. The linear mechanism favors ordinary mode radiation, but the trapped emission appears to be a mix of both ordinary and extraordinary radiation, perhaps from the multiple reflections off high-density regions in the magnetosphere and at the magnetopause.

The quasi-periodic (QP), or Jovian type III emissions (in analogy to solar type III bursts, because of their similar dispersive spectral shape) often occur at intervals of 15 and 40 min as observed by *Ulysses*, but neither *Galileo* nor *Cassini* found particularly dominant periodicities at these or other intervals (see Figure 52.21). The emission likely originates near the poles. Simultaneous measurements by the Galileo *and* Cassini spacecraft, both in the solar wind but at different locations, observed similar QP characteristics, suggestive of a strobe light pattern rather than a searchlight rotating with the planet. Within the magnetosphere, the QP bursts can then appear as enhancements of the continuum emission. At the magnetosheath, the lower frequency components of the bursts are dispersed by the higher density plasma, which produces the characteristic type III spectral shape. The 40-min QP bursts were correlated with energetic (~ 1 MeV) electrons observed by *Ulysses*. *Chandra* detected similar periods in X-rays from the auroral region, although not directly correlated with QP bursts themselves. Such observations suggest that the QP bursts are related to an important particle acceleration process, but the details of the relationship and the details of the process remain elusive.

3.5.4. Ganymede

Jupiter's satellite Ganymede has its own magnetosphere embedded within Jupiter's magnetic field. It presents a rich plasma wave spectrum, similar to that expected from a planetary magnetosphere. It is also the source of nonthermal narrowband radio emissions at 15–50 kHz, very similar to the escaping continuum emissions from Jupiter. The more intense cyclotron maser emission, seen from the auroral regions of all giant planets and the Earth, is absent, however. This is almost certainly because the electron plasma frequency is greater than the cyclotron frequency; hence, the cyclotron maser instability does not operate.

3.6. Saturn

Saturn's nonthermal radio spectrum consists of several components, as displayed in Figure 52.21 and discussed in the following section.

3.6.1. Saturn Kilometric Radio Emissions

Saturn's kilometric radiation (SKR) is characterized by a broadband of emission, 100% circularly polarized, covering the frequency range from 20 kHz up to several hundred kilohertz. When displayed in the frequency–time domain, it is sometimes organized in arclike structures, reminiscent of Jupiter's DAM arcs (see Figure 52.22(a)). *Cassini* has revealed some fine structure characteristic of cyclotron maser emissions (Figure 52.22(b)). As on the Earth, the SKR source appears to be fixed at high latitudes primarily in the local morning to noon sector, but it also appears at other local times. The SKR intensity is strongly correlated with the solar wind ram pressure, perhaps suggesting a continuous transfer of the solar wind into Saturn's low-altitude polar cusps. In fact, a detailed comparison between high-resolution *Hubble Space Telescope* images of Saturn's aurora with SKR suggests a strong correlation between the intensity of UV auroral spots and SKR.

Cassini has flown through the top of an SKR source on at least one occasion. The observed SKR was at and just below the electron cyclotron frequency, consistent with generation by the cyclotron maser instability. The source was in an upward current region, consistent with a region of precipitating electrons. Low-energy plasma observations are suggestive of an unstable shell-like distribution. Hence, the SKR source region appears to be similar to the source of AKR at the Earth.

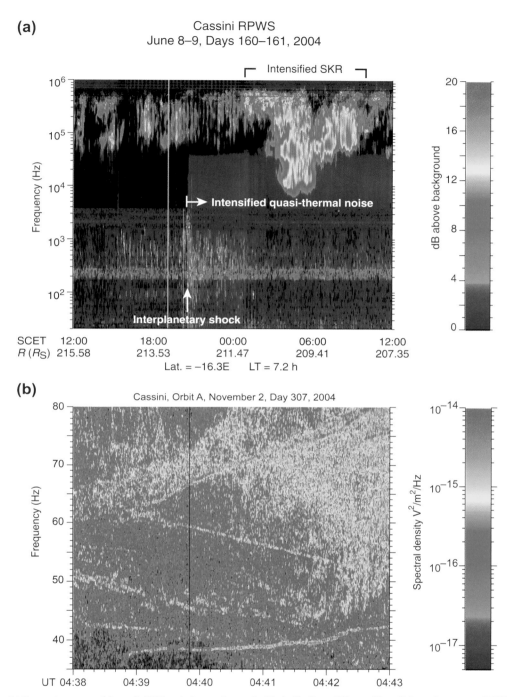

FIGURE 52.22 (a) Dynamic spectra of Saturn's SKR emission as observed with the Radio and Plasma Wave Science Instrument (RPWS) on the Cassini spacecraft. Frequency is plotted against the spacecraft event time (SCET) and distance, R, from the planet in planetary radii (R_S). The local time, LT, is listed below the figure. This illustrates a dramatic intensification of the SKR in response to an interplanetary shock that passed *Cassini* at about 20:30 on June 8, 2004. (b) A high temporal and spectral resolution record of SKR obtained by *Cassini*. Frequency is plotted against universal time (UT). This spectrogram illustrates the complex structure and variations in the SKR spectrum, which is also typical of cyclotron maser emissions at Jupiter and Earth. *After Kurth, et al. (2005).*

Even though the emission is highly variable over time, a clear periodicity at $10^h\ 39^m\ 24^s \pm 7^s$ was derived from the *Voyager* data, which was adopted as the planet's rotation period. Because the emission is tied to Saturn's magnetic field, which is axisymmetric, the cause of the modulation remains a mystery, although it may be indirect evidence of higher order moments in Saturn's magnetic field. Even more mysterious, however, is that the SKR modulation period measured by *Ulysses* and *Cassini* varies by 1% or more (several minutes) on timescales of a few years or less. Clearly, this change in period cannot represent a change in the planet's rotation itself, but there is no commonly accepted explanation.

Cassini has not only confirmed this variation in SKR period but also discovered that, in fact, two variable periods exist with one in the northern hemisphere and one in the south, as displayed in Figure 52.23. A number of other magnetospheric measurements have also suggested this dichotomy of rotational modulation periods. Further, near the Saturnian equinox the two periods appeared to merge or exchange several months afterward. A number of models have been proposed to explain these curious rotational periods, but it seems most likely that they are due to variations in the upper level winds in the atmosphere.

The existence of rotational modulation in Saturn's magnetosphere remains an enigma; however, it is likely that the axisymmetric nature of the magnetic field has uncovered a range of periodic phenomena not unlike the so-called system IV period at Jupiter exhibited by nKOM radiation.

3.6.2. Very Low Frequency Emissions

While the spacecraft was within Saturn's magnetosphere, it detected low-level continuum radiation (trapped radiation) at frequencies below 2–3 kHz (VLF). At higher frequencies, the emission can escape and appears to be concentrated in narrow frequency bands. One of the most predominant narrowband emissions is typically observed around 5 kHz. It is generated by mode conversion from Z-mode emissions near 3 Saturn radii at auroral latitudes. Higher frequency narrowband emissions may be generated by mode conversion from electrostatic upper hybrid bands, again at high latitudes.

3.6.3. Saturn Electrostatic Discharges

Saturn electrostatic discharges (SEDs) are strong, impulsive events, which last for a few tens of milliseconds from a few hundred kilohertz to the upper frequency limit of the *Voyager* planetary radio astronomy experiment (40.2 MHz), and are also detected by the *Cassini* spacecraft. Structure in individual bursts can be seen down to the *Voyager* time resolution limit of 140 μs, which suggests a source size less than 40 km. During the *Voyager* era, episodes of SED emissions occurred approximately every $10^h 10^m$, distinctly different from the periodicity in SKR. Cassini has been able to link SED episodes directly to cloud systems near 35° south planetocentric latitude observed in Saturn's atmosphere by the *Cassini* spacecraft. *Cassini*, however, has found SEDs to be much less common, generally speaking, than *Voyager*. Cassini can go months without seeing the discharges. Perhaps it may be a seasonal effect or related to

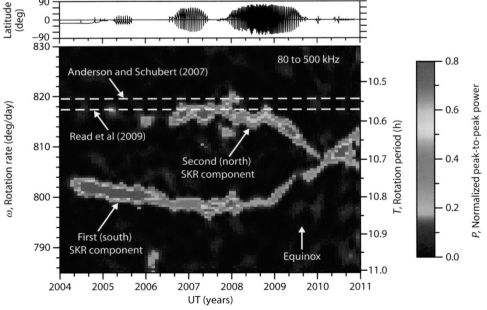

FIGURE 52.23 Top: Latitude of the Cassini spacecraft. Bottom: Frequency–time spectrogram of the normalized peak-to-peak power of the SKR modulation as a function of time and the rotational modulation rate, ω. The white dashed horizontal lines are the internal rotation rates of Saturn inferred by different authors. All rotational modulation rates are less than the inferred internal rotation rate which implies that the magnetosphere is slipping with respect to the rotation of Saturn's interior. *Gurnett, D. A., Groene, J. B., Averkamp, T. F., Kurth, W. S., Ye, S.-Y., & Fischer, G. (2011).*

the extent of ring shadowing on the atmosphere (or ionosphere, if propagation is an issue).

In December 2010 Cassini observed the onset of a very intense lightning storm commensurate with the onset of a planetwide convective storm in the northern hemisphere, likely similar to "Great White Spot" storms which occur approximately once per Saturn year (see also Figure 52.9). This storm quickly encircled the planet at planetocentric latitudes near and around 35°. All previous lightning observed by Cassini had originated near 35° S, hence, this great storm apparently reflects the early spring season in the northern hemisphere.

3.7. Uranus and Neptune

Like Saturn's radio emissions, both smooth and bursty components are apparent in the radio emissions from Uranus and Neptune, and these emissions probably originate in the southern auroral regions of the planets. Note, though, that the magnetic fields of these planets are inclined by large angles (47° for Uranus, 59° for Neptune) with respect to their rotational axes, and hence the auroral regions are not near the rotation poles. The periodicity of the emissions leads to the determination of the rotation periods of both planets, 17.24 ± 0.01 h for Uranus and 16.11 ± 0.02 h for Neptune. The upper bound to the frequency of the emissions is determined by (and indicative of) the planets' surface magnetic field strength.

From Uranus, we have also received impulsive bursts, similar to the SED events of Saturn, which are referred to as Uranus electrostatic discharge events. They were fewer in number and less intensive than the SEDs. If these emissions are caused by lightning, the lower frequency cutoff suggests peak ionospheric electron densities on the dayside of $\sim 6 \times 10^5$ cm^{-3}. In addition to the broadband emissions, both planets also emit trapped continuum and narrowband radiation.

4. FUTURE OF RADIO ASTRONOMY FOR SOLAR SYSTEM RESEARCH

This chapter highlighted the value of radio observations for planetary atmospheres (composition, dynamics), surface composition and structure, comets (parent molecules, source of material, outgassing), and magnetospheres (magnetic field configurations, particle distributions). While writing this review chapter, three new large arrays of telescopes are being commissioned. At centimeter wavelengths, the VLA has been upgraded, and sensitivities have improved by an order of magnitude. The image of Neptune in Figure 52.10(c) was an early result from this upgraded array. The next leap in mapping sources at centimeter wavelengths will be with the Square Kilometer Array, an array of telescopes spread out over 3000 km with a collecting area of 1 km^2. The array will be built in South Africa and Australia, and is expected to be finished by 2024.

At millimeter wavelengths the ALMA is being built in Chile, jointly by the United States, Canada, Europe, and Chile. ALMA will enable detection of, e.g. hundreds of asteroids, "bare" cometary nuclei, emissions from molecular "jets" from comets at high spatial and time resolution, Io's volcanic plumes, Titan's hydrocarbon chemistry, and "proto-Jupiters" in nearby stellar systems.

At long wavelengths, operations of the Low Frequency Array in the Netherlands has just commenced. This array will, for the first time, enable mapping of Jupiter's DAM emissions, and pinpoint its sources with high accuracy.

ESA's *Rosetta* spacecraft, currently on its way to comet 67P/Churyumov–Gerasimenko, carries a microwave instrument, MIRO, with receivers centered at 190 and 562 GHz. Starting in mid-2014, Rosetta will move with the comet from a *heliocentric distance* of ~ 3.5 AU down to perihelion near 1.3 AU. MIRO will measure near-surface temperatures and temperature gradients in the comet's nucleus using broadband channels on both receivers. Particularly exciting is its spectrometer, sensitive to several major volatile species (H_2O, CO, CH_3OH, and NH_3) at extreme high spatial and spectral resolution. These measurements will provide unprecedented information on the outgassing of the comet as a function of heliocentric distance.

Juno, a New Frontiers Jupiter mission, expected to arrive at Jupiter in 2016, will measure the planet's composition down to ~ 100 bar using six microwave radiometers sensitive to wavelengths between 1.3 and 50 cm. It also carries a radio and plasma wave sensor to make in situ measurements of the radio emissions and plasma in the auroral regions.

ESA expects to launch the Jupiter Icy Moon Explorer (JUICE) which is also expected to carry radio and plasma wave instruments. Orbit insertion is currently planned for 2030.

BIBLIOGRAPHY

Biver, N., et al. (1999). Post-perihelion observations of the distant gaseous activity of comet C/1995 O1 (Hale–Bopp) with the Swedish–ESO Submillimeter Telescope (SEST). *Asteroids, Comets and Meteors*.

Biver, N., et al. (2002). The 1995–2002 long-term monitoring of Comet C/1995 O1 (HALE–BOPP) at radio wavelengths. *Earth, Moon and Planets, 90*, 5–14.

Brecht, S. H., de Pater, I., Larson, D. J., & Pesses, M. E. (2001). Modification of the jovian radiation belts by Shoemaker–Levy 9: An Explanation of the data. *Icarus, 151*, 25–38.

Butler, B. J., & Sault, R. J. (2003). Long wavelength observations of the surface of Venus. *IAUSS, 1E*, 17B.

Butler, B. J., Campbell, D. B., de Pater, I., & Gary, D. E. (2004). Solar system science with SKA. *New Astronomy Reviews, 48*(11–12), 1511–1535.

Carr, T. D., Desch, M. D., & Alexander, J. K. (1983). Phenomenology of magnetospheric radio emissions. In A. J. Dessler (Ed.), *Physics of the Jovian magnetosphere* (pp. 226–284). Cambridge, United Kingdom: Cambridge University Press.

de Pater, I., et al. (1986). The brightness distribution of OH around comet Halley. *Astrophysical Journal Letters, 304*, L33–L36.

de Pater, I., & Lissauer, J. J. (2001). *Planetary Sciences*. Cambridge University Press.

de Pater, I., & Lissauer, J. J. (2010). *Planetary sciences* (2nd ed.). Cambridge University Press.

de Pater, I., Kassim, N., & Rucker, H. (Eds.). (2004). *LOFAR planetary and space science*. special issue.

de Pater, I., & Sault, R. J. (1998). An intercomparison of 3-D reconstruction techniques using data and models of Jupiter's synchrotron radiation. *Journal of Geophysical Research Planets, 103*(E9), 19973–19984.

Desch, M. D., Kaiser, M. L., Zarka, P., Lecacheux, A., LeBlanc, Y., Aubier, M., et al. (1991). Uranus as a radio source. In J. T. Bergstrahl, A. D. Miner, & M. S. Matthews (Eds.), *Uranus* (pp. 894–925). Tucson: University of Arizona Press.

Dunn, D. E., de Pater, I., & Molnar, L. A. (2006). Examining the wake structure in Saturn's rings from microwave observations over varying ring opening angles and wavelengths. *Icarus, 192*, 56–76.

Girard, J. N., Zarka, P., Tasse, C., Hess, S., & the LOFAR Collaboration (2012). Jupiter's synchrotron imaging with LOFAR. SF2A, In S. Boissier, P. de Laverny, N. Nardetto, R. Samadi, D. Valls-Gabaud, & H. Wozniak (Eds.).

Gulkis, S., & de Pater, I. (2002). *Radio astronomy, planetary* (3rd ed.). In *Encyclopedia of physical science and technology* (Vol. 13). Academic Press, 687–712.

Gurnett, D. A., Groene, J. B., Averkamp, T. F., Kurth, W. S., Ye, S.-Y., & Fischer, G. (2011). An SLS4 longitude system based on a tracking filter analysis of the rotational modulation of Saturn kilometric radiation. In *Proceedings of the 7th international workshop on planetary, solar and heliospheric radio emissions (PRE VII)* (pp. 51–64), held at Graz, Austria, September 15–17, 2010.

Gurwell, M. A. (2004). Submillimeter observations of Titan: global measures of stratospheric temperature, CO, HCN, HC_3N, and the isotopic ratios $^{12}C/^{13}C$ and $^{14}N/^{15}N$. *The Astrophysical Journal, 616*, L7–L10.

Harrington, J., de Pater, I., Brecht, S. H., Deming, D., Meadows, V. S., Zahnle, K., et al. (2004). Lessons from Shoemaker–Levy 9 about Jupiter and planetary impacts. In F. Bagenal, T. E. Dowling, & W. McKinnon (Eds.), *Jupiter: Planet, satellites & magnetosphere* (pp. 158–184). Cambridge, United Kingdom: Cambridge University Press.

Janssen, M., & Le Gall, A. (2011). Combining active and passive microwave observations of Titan to learn about its surface. *EPSC-DPS*, 2011–276.

Janssen, M. A., Ingersoll, A., Allison, M. D., Gulkis, S., Laraia, A., Baines, K., et al. (2013). Saturn's thermal emission at 2.2-cm wavelength as imaged by the Cassini RADAR radiometer. *Icarus, 226*, 522–535.

Kaiser, M. L., Desch, M. D., Kurth, W. S., Lecacheux, A., Genova, F., Pederson, B. M., et al. (1984). Saturn as a radio source. In T. Gehrels, & M. S. Matthews (Eds.), *Saturn* (pp. 378–415). Tucson: University of Arizona Press.

Kloosterman, J. L., Butler, B., & de Pater, I. (2008). VLA observations of Jupiter's synchrotron radiation at 15 GHz. *Icarus, 193*, 644–648.

Kraus, J. D. (1986). *Radio astronomy*. Powell, Ohio: Cygnus Quasar Books.

Kurth, W. S., et al. (2005). High spectral and temporal resolution observations of Saturn kilometric radiation. *Geophysical Research Letters, 32*, L20S07. doi: 1029/2005 GL022648.

Kurth, W. S., Bunce, E. J., Clarke, J. T., Crary, F. J., Grodent, D. C., Ingersoll, A. P., et al. (2009). Auroral processes. In M. K. Dougherty, L. W. Esposito, & S. M. Krimigis (Eds.), *Saturn from Cassini-Huygens* (p. 333). Springer Science+Business Media B.V. ISBN: 978-1-4020-9216-9.

Lecacheux, A. (2001). Radio observations during the Cassini flyby of Jupiter. In H. O. Rucker, M. L. Kaiser, & Y. Leblanc (Eds.), *Planetary radio emissions V* (pp. 1–13). Vienna: Austrian Academy of Sciences Press.

Milam, S. N., et al. Formaldehyde in comets C/1995 O1 (Hale–Bopp), C/2002 T7 (LINEAR), and C/2001 Q4 (NEAT): investigating the cometary origin of H_2CO. The Astrophysical Journal, 649, 1169–1177.

Mitchell, D. L., & de Pater, I. (1994). Microwave imaging of Mercury's thermal emission: observations and models. *Icarus, 110*, 2–32.

Sault, R. J., et al. (2004). Longitude-resolved imaging of Jupiter at $\lambda = 2$ cm. *Icarus, 168*, 336–343.

Schloerb, F. P. (1985). Millimeter-wave spectroscopy of solar system objects: present and future. In P. A. Shaver, & K. Kjar (Eds.), *Proceedings of the ESO–IRAM–Onsala workshop on (sub)millimeter astronomy* (pp. 603–616). Aspenas, Sweden, June 17–20, 1985.

Thompson, A. R., Moran, J. M., & Swenson, G. W., Jr. (1986). *Interferometry and synthesis in radio astronomy*. New York: John Wiley and Sons.

Zarka, P. (1998). Auroral radio emissions at the outer planets: observations and theories. *Journal of Geophysical Research, 103*, 20159–20194.

Zarka, P., Pederson, B. M., Lecacheux, A., Kaiser, M. L., Desch, M. D., Farrell, W. M., et al. (1995). Radio emissions from Neptune. In D. Cruikshank (Ed.), *Neptune* (pp. 341–388). Tucson: University of Arizona Press.

Chapter 53

Planetary Radar

Catherine D. Neish
Department of Physics and Space Sciences, Florida Institute of Technology, Melbourne, FL, USA

Lynn M. Carter
Planetary Geodynamics Laboratory, NASA Goddard Space Flight Center, Greenbelt, MD, USA

Chapter Outline

1. Introduction 1133
   1.1. Scientific Value of Radar 1133
   1.2. History 1134
2. Techniques 1136
   2.1. Radar Imagers 1136
   2.2. Radar Sounding 1138
   2.3. Radar Topography 1139
3. Target Properties 1141
   3.1. Radar Albedo 1141
   3.2. Radar Polarimetry 1143
4. Radar Measurements of Planetary Bodies 1144
   4.1. Mercury 1144
      4.1.1. Polar Deposits 1144
      4.1.2. Rayed Craters 1147
   4.2. Venus 1147
      4.2.1. Magellan Reveals the Surface 1147
   4.2.2. Post-Magellan Radar Imaging and the Search for Surface Changes 1148
   4.3. Moon 1149
      4.3.1. Polar Environment 1149
      4.3.2. Lunar Geology 1150
   4.4. Mars 1150
      4.4.1. Radar Heterogeneity 1150
      4.4.2. Subsurface Structure 1151
   4.5. Asteroids 1153
      4.5.1. Imaging and Shape Reconstruction 1153
      4.5.2. Orbit Determination and the Yarkovsky Effect 1153
   4.6. Jupiter and Saturn Systems 1155
      4.6.1. Icy Satellites 1155
      4.6.2. Titan 1156
5. The Future of Planetary Radar 1157
Acknowledgments 1158
Bibliography 1158

1. INTRODUCTION

1.1. Scientific Value of Radar

Radar describes the act of transmitting a radio signal towards a target and receiving its echo. Planetary radar refers specifically to the study of the solar system's planets, moons, and small bodies. The principles of planetary radar are similar to those used in other radar systems, such as aircraft navigation and weather forecasting, except that the distances involved are generally orders of magnitude larger, requiring the world's most powerful radar systems.

Unlike most forms of remote sensing, which rely on light sources for which the user has no control (passive astronomy), radar astronomy uses its own transmitting system to illuminate a surface (active astronomy). This gives the user complete control over the transmitted signal, making radar one of the few forms of "experimental astronomy". The transmitters and receivers used in a radar system can be housed in Earth-based telescopes or on orbital spacecraft, and can be colocated (monostatic) or separated in space (**bistatic**).

There are several advantages to observing planetary objects at radio wavelengths, in addition to visual and infrared (VIR) wavelengths. Firstly, radar represents the best way to observe the surface of planets with optically opaque atmospheres, such as Venus and Titan, leading to a better understanding of their surface morphology. Secondly, radar provides a wealth of information on the physical properties of the surface being imaged, on scales not accessible with VIR images. Unlike VIR images, which typically probe the chemical composition of the uppermost surface layer, radar images are sensitive to physical and electrical properties, highlighting differences in the slope, roughness, and dielectric constant of the surface and near subsurface (to depths of centimeters to meters). These properties make radar data very complementary to optical data, since features that are difficult to perceive in optical data are often easily seen in radar data, and vice versa.

Radar also has the advantage of utilizing a range of wavelengths easily accessible from the surface of the Earth. Although the Earth's atmosphere is generally quite transparent at radio wavelengths, there are some frequency limitations when transmitting with Earth-based telescopes. Molecular absorptions in the Earth's atmosphere lead to losses for frequencies >50 GHz (<1 cm) and the ionosphere reflects frequencies <10 MHz (>30 m). Other worlds have different limitations: on Venus, the dense atmosphere leads to absorption at all frequencies above a few gigahertz (<10 cm), and on Titan, the atmosphere limits operation to frequencies <30 GHz (>1 cm). Of course, there are no such limitations when observing airless worlds from orbital platforms. See Table 53.1 for typical frequencies used in planetary radar.

1.2. History

Radio-based detection and tracking was first developed during World War II for air defense and other military purposes (see History of Solar System Exploration). During this time, the US Navy coined the acronym RADAR (RAdio Detection And Ranging). These early radars generated relatively long wavelength (~1 m) radiation, but the development of the cavity magnetron in 1940 allowed for the creation of narrow-beam, higher frequency radar systems.

The birth of planetary radar did not come until 1946, when the US Army Signal Corps station at Fort Monmouth, NJ, USA, obtained the first radar echoes from the surface of the Moon. A Hungarian group led by Zoltan Bay made an independent measurement only a few weeks later. Equipped with a less powerful transmitter than the Americans, Bay had devised a technique in which he sent repeated signals to the Moon and summed the received signals.

Radar imaging progressed rapidly after the war, and synthetic aperture radar (SAR) was developed in the 1950s. As we discuss in Section 2, SAR has the benefit of increasing the along-track resolution of radar images and making the resolution independent of the distance to the target. During this time, radars were also used to map the Moon at a variety of wavelengths in support of the Apollo program, and the first echoes were received from the terrestrial planets. Radar observations allowed for the first accurate determination of the rotation period of Venus and Mercury. These were difficult to determine from optical data alone, due to Venus's obscuring atmosphere and Mercury's proximity to the Sun.

Radar experiments were carried aboard the Apollo spacecraft, the first radars to orbit another planetary object. **Bistatic** radar measurements were made during the Apollo 14, 15, and 16 missions by transmitting a radio signal from the spacecraft and receiving the echoes on the Earth. At short wavelengths, these experiments found that the lunar regolith could be approximated by a surface made of a material with a dielectric constant of ~3, while at long wavelengths, evidence of volume scattering was present. The first SAR to orbit another planet was carried on board the Apollo 17 mission. The Apollo 17 Lunar Sounder Experiment operated at 5 MHz (60 m), 15 MHz (20 m), and 150 MHz (2 m). At short wavelengths, it was used to determine the topography of the lunar surface, while at 60 m, it was capable of penetrating as much as 2 km in dry lunar rocks, providing evidence of subsurface layering.

TABLE 53.1 Frequency Bands Typically Used in Planetary Radar

Band Name	Frequency	Wavelength	Planetary Usage	Notes
Ka-band	27–40 GHz	0.75–1.1 cm		Above K-band
K-band	18–27 GHz	1.1–1.7 cm		From German "*kurz*", for short
Ku-band	12–18 GHz	1.7–2.5 cm	Cassini RADAR	Under K-band
X-band	8–12 GHz	2.5–3.75 cm	Haystack, Goldstone	Used in missile guidance — "X" marks the spot
C-band	4–8 GHz	3.75–7.5 cm	Mini-RF	A compromise between X- and S-band
S-band	2–4 GHz	7.5–15 cm	Arecibo, Goldstone, Apollo 14/15/16, Venera 15/16, Magellan, Mini-RF	Short wave
L-band	1–2 GHz	15–30 cm	Millstone-Hill, Pioneer-Venus	Long wave
P-band	250–500 MHz	60–120 cm	Arecibo, Apollo 14/15/16	Previous—applied to early radar systems
HF-band	3–30 MHz	10–100 m	Apollo 17, MARSIS, SHARAD, Kaguya Lunar Radar Sounder	High frequency

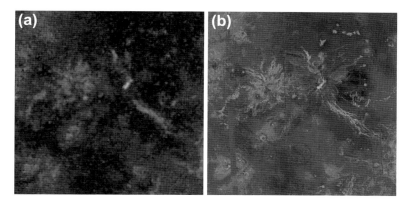

FIGURE 53.1 Our understanding of surface processes on Venus improved greatly between 1978 and 1990, as the resolution of orbital radars improved from ∼25 km (Pioneer-Venus) to ∼100 m (Magellan). Note the differences in this image of Venus, centered between Sif Mons (22° N, 352° E) and Gula Mons (22° N, 359° E), when viewed at (a) a lower resolution at 17 cm (Pioneer-Venus) and (b) a higher resolution at 12.6 cm (Magellan).

The first spaceborne imaging radar used for scientific studies was Seasat, which was launched in 1978 and collected data on the properties of the Earth's ocean and land surfaces. That same year, Pioneer Venus used a radar altimeter to measure the global surface topography of Venus and to create a 25-km resolution map of its equatorial region. Since then, radar images of the surface of Venus have improved from 1 km resolution of the northern hemisphere by Venera 15 and 16 in 1982 to ∼100 m resolution of 98% of the planet by Magellan in the early 1990s (Figure 53.1). Ground-based radar systems have detected echoes from objects as far away as Saturn. The first echoes from Jupiter's moons and Saturn's rings were received in the mid-1970s. These observations showed that Saturn's rings are made of relatively large particles and that icy satellites backscatter radar waves in an unusual way, unlike that seen on rocky objects.

Radar has also revolutionized our knowledge of small bodies in the solar system. The first echo from an asteroid was detected in 1968, and the first echo from a comet was detected in 1980. An upgrade in the Arecibo radio telescope in 1997 provided an order of magnitude improvement in sensitivity, causing the number of asteroids detected and characterized by radar to rise exponentially (Figure 53.2). As of September 2013, radar has detected 16 comets, 397 near-Earth asteroids (NEAs), and 134 main belt asteroids (MBAs). A knowledge of the orbits of these objects has proved to be extremely important for hazard mitigation on the Earth from impacting bodies.

A recent explosion of orbiting radars has increased our knowledge of many of the objects in the solar system. Cassini RADAR (2004 to present) has provided the first high-resolution views of Titan. Mars Advanced Radar for Subsurface and Ionosphere Sounding (MARSIS) (2005 to

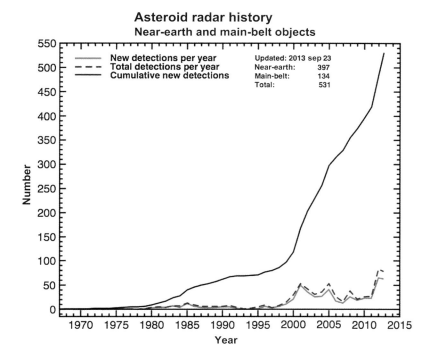

FIGURE 53.2 Improvements in ground-based radar systems have increased our understanding of the surface properties and orbits of many of the small bodies in the solar system. *Image from http://echo.jpl.nasa.gov.*

present) and Mars SHAllow RADar sounder (SHARAD) (2006 to present) have helped to quantify the volatile inventory on Mars. The Miniature Radio-Frequency instrument (Mini-RF) (2009 to present) has provided the first radar images of large portions of the far side of the Moon, and the first **bistatic** radar images of another planetary body at large beta angles. In addition, the MESSENGER mission has provided the first optical views of the non-imaged hemisphere of Mercury, adding contextual information to the ground-based radar images of that world.

2. TECHNIQUES

Radar systems are classified in one of several categories depending on the type of measurement needed. Radar imagers acquire two-dimensional images of planetary surfaces (Section 2.1), radar sounders probe an object's subsurface structure along a satellite track (Section 2.2), and altimeters and interferometric radars acquire accurate height measurements of a planetary surface (Section 2.3). Scatterometry is another type of radar technique, which measures the reflectivity of a surface by sending pulses and measuring the backscattered power in the radar beam. Although not discussed in detail here, scatterometry measurements are crucial for determining the average scattering properties of large areas of a surface and are used extensively for the interpretation of high-resolution radar images.

2.1. Radar Imagers

Radar imagers, whether ground based or orbital (Figure 53.3), produce two-dimensional images of a planetary surface by observing changes in the **Doppler shift** (motion relative to observer) and time delay (distance) of an object. Thus, to obtain high-resolution images, it is necessary to have high-resolution measurements of time, t, and frequency, ν.

For example, if there are two point targets separated by a distance dr, the echoes received from these points will be separated by a time $dt = 2dr/c \sin\theta$, where c is the speed of light and θ is the look angle of the radar. To achieve this range resolution, it is necessary to pulse the transmitted radio wave into discrete time intervals. However, smaller pulses (resulting in images with higher range resolution) contain less energy than longer pulses. Therefore, to achieve both a small dt and a large pulse energy, radar signals are *modulated* over a longer time interval τ. This is often done by phase modulation, wherein a monochromatic signal has its phase shifted by $0°$ or $180°$ at time intervals $dt \ll \tau$, in essence producing a binary code (Figure 53.4). The received signal is cross-correlated with the known transmitted signal, and the returning echo is identified when the correlation between transmitted and received signals is high. At all other times, the correlation is near zero. Note that in many radars, the transmitted pulses are separated by a time T, yielding a pulse repetition frequency $PRF = 1/T$. Between pulses, the radar can switch from transmit to receive mode, to collect the signals from previously transmitted pulses.

Once identified, the pulses can be stacked to produce a time series of echoes at each individual range. The Fourier transform of that time series yields information about the Doppler shift of each point, providing the second dimension of the radar image (Figure 53.4). The resolution can be adjusted after the data is collected by adjusting the length of the Fourier transform, again emphasizing the experimental nature of planetary radar. In essence, a delay-Doppler image maps a three-dimensional (3D) object into a two-dimensional representation (Figures 53.5 and 53.6). This introduces ambiguities into the resultant image, since there exist two points north and south of the subradar point with identical time delay and Doppler shift. However, if the radar beam is smaller than the target or is offset from the rotational equator, these ambiguities can be minimized, and the

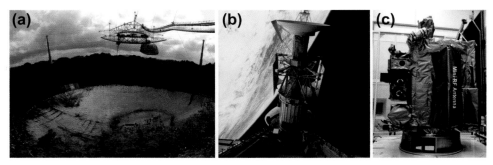

FIGURE 53.3 Typical ground-based and orbital radar systems. (a) The Arecibo radio telescope in Puerto Rico has a 305-m-diameter primary reflector. The S-band transmitter and receiver are housed in the Gregorian dome suspended above the reflector. (b) The Magellan spacecraft weighed 1035 kg, and supported one instrument, the Radar System (RDRS). This instrument used a 3.7-m antenna, which functioned as a radar imager, altimeter, and radiometer. (c) One recent orbital radar—Mini-RF—is orders of magnitude less massive (14 kg) than Magellan, acting as just one of the seven instruments on the Lunar Reconnaissance Orbiter. *Image credit: NASA/GSFC.*

Chapter | 53 Planetary Radar

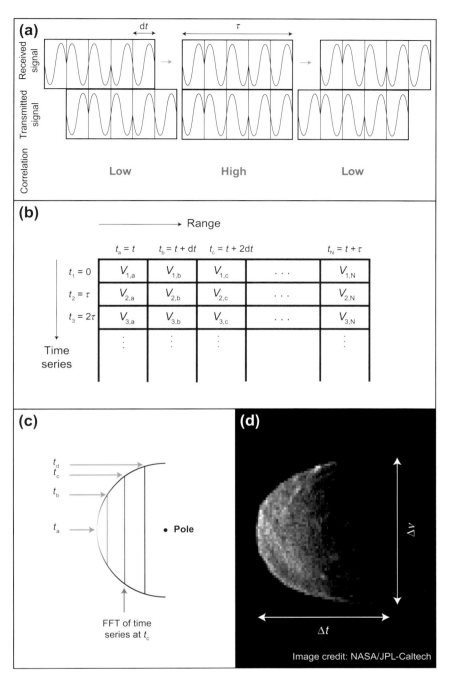

FIGURE 53.4 (a) In radar imaging, the incoming radar pulse is compared with the transmitted signal. The start of the received signal is identified where the correlation is high. (b) The received pulses are then stacked to produce a time series of echoes at each range (delay). The Fourier transform of the time series yields information about the Doppler shift at each point. (c) The radar target viewed from its spin vector. Regions near the limb will have larger time delay and correspondingly smaller radar backscatter. (d) Delay-Doppler image of asteroid 2005 YU55. For reasonably spherical objects, the spread in time delay (Δt) can be used to determine the size of the asteroid, and the spread in Doppler shift ($\Delta \nu$) can be used to determine its rotation rate. *Image credit for part (d): NASA/JPL-Caltech.*

delay-Doppler images can later be transformed into map coordinates.

Orbital radars acquire data in the Doppler direction in a slightly different way than what is described above for ground-based radars. Here, the Doppler (or "along-track") resolution corresponds to the two nearest separable points along a constant delay line. For real aperture radars, this is equal to the width of the antenna footprint, which is generally quite large. For this reason, SARs were developed. SAR exploits the fact that a particular target P is illuminated by the radar from numerous locations along the orbital path. These echoes can be combined in a processor to synthesize a much larger antenna of length D, with a correspondingly smaller beam width, λ/D (Figure 53.7). In SAR systems, the along track resolution is also independent of the distance between the sensor and the target being imaged, h, and depends only on the size of the physical antenna, L. This surprising result can be understood in the

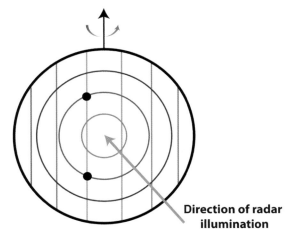

FIGURE 53.5 Delay-Doppler radar images map a 3D object into a two-dimensional image. Circles represent constant delay (*t*) and lines represent constant Doppler shift (ν). Unless the radar beam is smaller than the target or offset from the center of the object, ambiguities are introduced since there exist two points north and south of the subradar point with identical time delay and Doppler shift (black dots).

following way: the farther the sensor is from the target, the larger the footprint is on the ground, leading to a longer synthetic array with a correspondingly finer resolution. The only disadvantage of using larger orbital altitudes is that more power will be required to acquire a detectable echo. Note that unlike ground-based radars, in which the target is generally comparable in size to the radar beamwidth, SAR illuminates a discrete swath on only one side of the subspacecraft track. This eliminates any right—left ambiguity in the image. If the surface topography is known, geometric distortions can be corrected, allowing for registration with optical imagery and other data sets.

2.2. Radar Sounding

Radar sounding instruments are similar to the imaging SAR systems described above, but they utilize long wavelengths (meters to hundreds of meters) and have been optimized for detecting subsurface structure. At shorter wavelengths (meters), sounding radars can be used to probe the upper tens of meters of ice sheets, while at longer wavelengths they are capable of probing hundreds of meters to kilometers through rock and ice. Prior planetary sounding radars spanned a wide range of frequencies: from longest to shortest wavelength, these instruments are the MARSIS radar (0.1–5.5 MHz, or 50–3000 m) on Mars Express, the Lunar Radar Sounder (5 MHz, or 60 m) on the Kaguya spacecraft, the Apollo Lunar Sounder Experiment (5–150 MHz, or 2–60 m) on Apollo 17, and the SHARAD radar (20 MHz, or 15 m) on the Mars Reconnaissance Orbiter.

Sounding radar data are usually displayed as radargrams, with time delay on the *y*-axis and along-track distance along

FIGURE 53.6 (a) The difference between angle—angle space as perceived by the human eye and delay-Doppler space as perceived by radar is well illustrated by this S-band image of Saturn's rings. (b) The outermost "A" ring returns the earliest echo in the delay direction, but orbits more slowly than the "B" ring, making it narrower in the Doppler direction. This leads to four "cross-over" regions, where signal from the two rings add together. *Image modified from Nicholson, P. D., et al. (2005).*

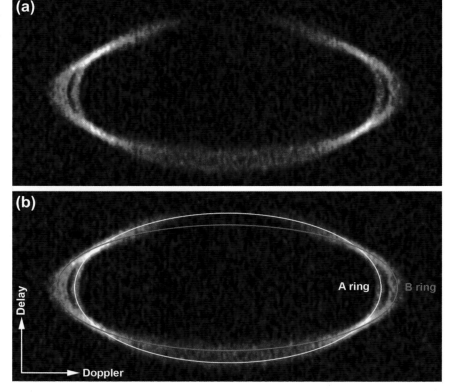

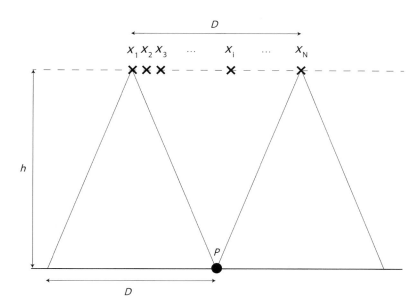

FIGURE 53.7 A SAR is capable of observing the same point P from many different positions, x_i. This improves the along-track or Doppler resolution of the image by increasing the size of the antenna from its physical size L to its synthetic size D. The resolution of the radar is independent of the orbital altitude, h.

the x-axis. Sounding radars have large spatial footprints (hundreds of kilometers), and so are not typically used to make images. However, low-resolution maps of radar reflectivity can be obtained if multiple orbital tracks are mosaicked. During the radargram formation, synthetic aperture processing is applied along track to improve the spatial resolution and signal to noise. SAR "focusing" also combines power from a feature observed at multiple distances and Doppler shifts into a single radargram element, eliminating parabola-shaped time-delay "tails" that are caused by the radar moving toward and away from a topographic feature.

Sounding radars are nadir-pointing instruments, but radargrams can contain "clutter", or power from off-nadir surface elements. These occur at later time delay values than the surface echo and therefore appear to be subsurface features. To verify that apparent subsurface features in the radargrams are not clutter, simulated radargrams are generated using topographic data (Figure 53.8). Clutter is more problematic in rough terrains, at shorter radar wavelengths where there are increased reflections from wavelength-scale surface roughness, and in regions that lack topographic data with sufficient resolution to accurately model the expected radar echo power.

Long-wavelength radars are also sensitive to the ionosphere, which distorts the phases of radar waves passing through. The SHARAD and MARSIS radars must both account for these phase distortions in processing, especially for daytime orbital tracks where electron densities are high due to sunlight interacting with atoms in the atmosphere. The phase distortion is greater at longer wavelengths, and the MARSIS radar has ionospheric modes that are used to probe the structure of the ionosphere and its temporal variability.

Sounding radar is particularly well suited to measurements of dielectric properties. If a subsurface interface is detected, then the round-trip light travel time (Δt) can be measured from the radargram. If an estimate of the thickness or depth of the deposit can be made using separate topographic information (h), the permittivity (ε') can be computed from:

$$\varepsilon' = \left(\frac{c\,\Delta t}{2\,h}\right)^2$$

The loss tangent can also be calculated by measuring how the reflected power from a subsurface interface decreases with greater depth. These dielectric values can then be compared to laboratory-derived values for terrestrial rocks in order to constrain the density and composition of the rocks.

2.3. Radar Topography

Radar systems can be used to acquire topography of a planetary surface using one of a number of different techniques. One of the most straightforward methods is the use of radar altimeters. These instruments provide a measure of the distance between the sensor and the surface at the nadir point by measuring the time delay between the transmission of the signal and the receipt of the echo. The accuracy of the time difference depends mainly on the "sharpness" of the pulse ($dt = 1/B$, where B is the bandwidth of the pulse) and any pulse dispersion resulting from surface roughness. Rough surfaces will return energy from various heights on the surface, leading to a gradual rise in the echo leading edge and a decrease in the echo strength. To reduce these uncertainties, a narrow beam covering a

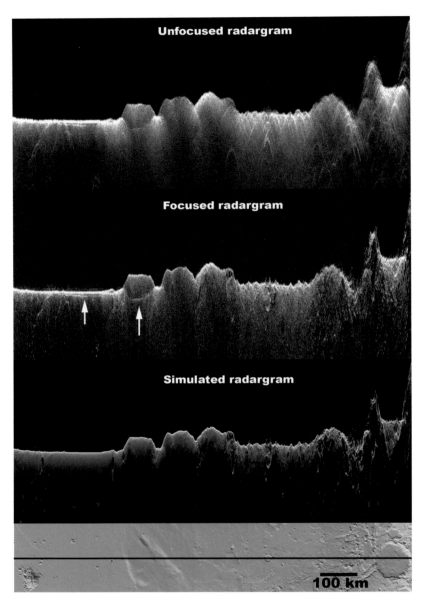

FIGURE 53.8 SHARAD sounding radar data over a portion of the Medusae Fossae Formation (MFF) on Mars. The top two images illustrate the effect of synthetic aperture focusing on the data. The third panel is a simulation of the focused data product that uses Mars Orbital Laser Altimeter data to predict the echoes produced by surface reflections. Subsurface features (arrows) are apparent in the data beneath hills of MFF material that are ∼600 m above the surrounding plains. The lowest panel shows the orbit track and topography. *Source: Carter, L. M., et al. (2009).*

smaller surface area is generally desirable. Conventional altimeters will acquire a single topographic profile along the flight line, but two-dimensional topography can be acquired with a multibeam altimeter. Radar altimeters were flown on the Pioneer Venus, Venera 15/16, Magellan, and Cassini missions.

Solid surface topography can also be measured by stereo imaging (see Stereophotogrammetry). It requires acquiring two images of the same surface area with two different viewing geometries. The principles behind radar stereo imaging are similar to those for the more familiar method of optical stereo imaging, but take into account the geometric principles by which SAR images are formed (i.e. delay-Doppler format rather than angle–angle format). Two-dimensional digital terrain models are produced using highly adjustable image matching algorithms. These match points can be edited interactively to account for noise in the images and any differences in illumination direction. The image matching algorithms require a sensor model, a type of software that converts image pixel location into a ground location in latitude, longitude, and elevation. This sensor model will differ for images acquired with different spacecraft. Stereo topography has been derived with radar images from the Magellan, Cassini, and Lunar Reconnaissance Orbiter (LRO) missions.

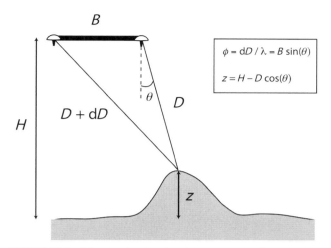

FIGURE 53.9 Diagram showing the relevant geometry for radar interferometric measurements of topography.

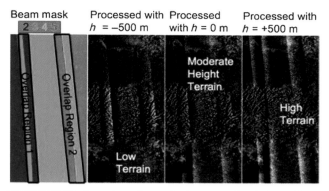

FIGURE 53.10 In regions where radar beams overlap, it is possible to correlate the received power to the antenna gain pattern to estimate surface heights. Three examples from Cassini RADAR are shown here, processed at three different heights. Dark to light banding indicates that the height was underestimated, while light to dark banding indicates that the height was overestimated. *Figure from Stiles, B. W., et al. (2009).*

Radar interferometry, or "InSAR", is another technique by which high-resolution topography can be obtained over large areas using a spacecraft. InSAR works in one of two ways. The first involves a spacecraft flying at an altitude of H that carries a boom with two radar antennas at either end. Radar waves are transmitted to the ground and later received by the antennas on the spacecraft. The phase difference between the waves received at either end of the boom, ϕ, can be measured. Knowing the boom length, B, and the wavelength of the radar waves, λ, one can calculate the incidence angle, θ, using simple geometry:

$$\phi = dD/\lambda = B \sin(\theta)$$

Knowing the incidence angle, θ, the altitude, H, and the direct distance from the antenna to the ground, D (the speed of light times the measured return time divided by 2), you can then solve for the topography:

$$z = H - D \cos(\theta)$$

A schematic of the relevant geometry for radar interferometric measurements is given in Figure 53.9. Alternatively, a spacecraft can make two nearly identical passes over a region, and the phase difference from the two passes can be used to determine the topography. InSAR has been used to determine the topography of the Earth (e.g. the Shuttle Radar Topography Mission and TerraSAR-X), but has yet to be successfully implemented on any planetary mission. However, ground-based telescope interferometry has been used to derive the topography of the Moon, using two telescopes separated by a baseline B to receive the signal.

Finally, a technique known as "SARTopo" has recently been developed that estimates surface heights by comparing the calibration of overlapping radar beams. This technique has been particularly useful for determining the topography of Saturn's moon Titan with the Cassini RADAR instrument, a world with only a limited amount of altimetry and stereo topography. The Cassini RADAR instrument has five different antenna feeds (or beams), and data from overlapping beams are acquired nearly simultaneously. Surface heights are computed by maximizing the correlation between the received power and the antenna gain pattern at each point along track in the overlap region (Figure 53.10). This technique is capable of estimating surface heights for most of the radar-imaged surface of Titan with ~10 km horizontal resolution and a vertical resolution of tens of meters. It has extended the area over which colocated topography and radar imagery is available on Titan by at least an order of magnitude.

3. TARGET PROPERTIES

Radar images can appear superficially quite similar to images acquired at optical wavelengths. However, the information contained in the images—namely, its albedo and polarimetric properties—is very different. While optical images are primarily sensitive to the absorption of light due to varying chemical compositions, radar images are primarily sensitive to the physical and electrical properties of the surface. Electromagnetic waves impinging on **dielectric materials** will excite atoms in the target, such that they become small electromagnetic oscillators and radiate energy. How the energy is radiated will depend on the slope, roughness, dielectric constant, conductivity, etc. of the surface, which in turn affect the albedo (Section 3.1) and polarimetry (Section 3.2) of the radar image.

3.1. Radar Albedo

The intrinsic radar albedo of a surface is characterized by its backscatter cross-section, $\sigma(i)$. This parameter is a function of the incidence angle, i, of a radar, its polarization and wavelength, and the properties of the surface, especially its roughness (Figure 53.11). For example, if a surface is perfectly flat, it will scatter energy away from the radar receiver, at an angle equal to the incidence angle (specular reflection). If a surface is rough at the scale of the radar wavelength, it will scatter energy in every direction, including back toward the radar receiver (diffuse scattering). Most surfaces are some combination of both: small-scale roughness superimposed on large, flat facets. As natural surfaces do not typically vary by more than $\sim 30°$, at small angles the return is dominated by the quasispecular return from the large facets. At larger angles, the return is dominated by the effect of the diffuse scattering from the small-scale roughness.

Radar backscatter functions like those shown in Figure 53.11(b) can be determined by scatterometers measuring the average scattering properties of large areas of a planetary surface as a function of frequency and illumination direction. It is then possible to fit these data to a variety of models that have been developed to approximate the scattering behavior of natural surfaces. These are generally a superposition of the facet model, and the diffuse scattering, or Bragg, model. For example, a facet model commonly used in planetary radar applications is the Hagfors model:

$$\sigma(i) = \frac{C\rho}{2}\left(\cos^4 i + C \sin^2 i\right)^{-3/2}$$

where $C^{-1/2}$ is the root mean square slope of the surface in radians and ρ is the Fresnel reflectivity, related to the dielectric constant by $\rho = (|1-\sqrt{\varepsilon}|/|1 + \sqrt{\varepsilon}|)^2$. For planetary surfaces, the dielectric constant, $\varepsilon = \varepsilon' + i\,\varepsilon''$, will be most sensitive to the density of the rock, and the presence of metals. Higher densities and high concentrations of metals will increase ε from a typical value of 3 for dry rocks. The presence of liquid water will also produce significant effects in radar images, given its extremely high dielectric constant of $\varepsilon \sim 80$. Generally, though, dielectric constant is a secondary effect in radar images compared to surface roughness variations.

A Bragg-type model commonly used in planetary radar applications is the Muhleman model:

$$\sigma(i) = \frac{\beta \cos i}{(\sin i + \alpha \cos i)^3}$$

where α and β are empirical constants. This model was used for Magellan SAR data processing.

Note that scattering occurs not only at the surface but also from the near subsurface. Radars have relatively long penetration depths, and will scatter wherever there is an interface separating two regions of different electromagnetic properties, such as bedrock buried under regolith. The imaginary part of the dielectric constant, ε'', is known as the loss factor, and it determines the ability of the target material to absorb the radar energy and convert it into another form of energy,

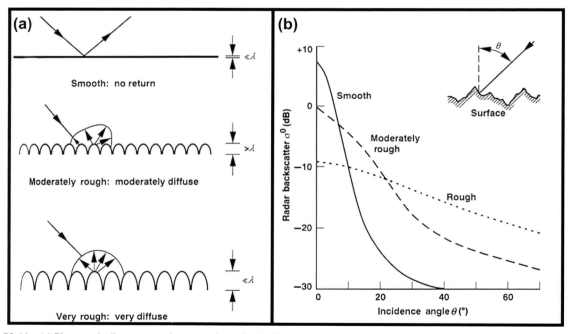

FIGURE 53.11 (a) Planetary bodies scatter radar energy in particular directions depending on the wavelength-scale roughness of its surface. Smooth surfaces scatter energy in the specular direction, away from the radar receiver, and rough surfaces scatter energy in every direction, including back toward the radar receiver. (b) Radar backscatter will also vary depending on the incidence angle of the radar beam. Smooth surfaces have brighter returns when the incidence angle is close to specular, but quickly drop to zero at angles greater than $\sim 30°$. *Figure adapted from Farr, T. G. (1993).*

such as heat. For most materials, this factor decreases with wavelength, such that longer wavelengths penetrate further than shorter wavelengths into a particular material. The penetration depth, or the distance in the subsurface at which the radar power has decreased by 1/e, can be approximated by $L \sim \lambda/(2\pi\varepsilon'^{1/2} \tan \delta)$, where the loss tangent, $\tan \delta = \varepsilon''/\varepsilon'$. For dry low-loss soils, this typically gives $L \sim 10\lambda$.

3.2. Radar Polarimetry

Additional information about target properties can be found in the orientation of the returned radar signal, known as its polarization. In particular, radars that are capable of receiving an echo in two or more known polarizations contain more information about the surface properties than backscatter measurements acquired in a single polarization. Radar signals can be horizontally polarized (perpendicular to the plane of incidence, H) or vertically polarized (in the plane of incidence, V). Alternatively, they can be circularly polarized, a case in which the electric field magnitude remains constant but its direction is constantly rotating. These polarizations can be right handed (clockwise rotation, R) or left-handed (counterclockwise rotation, L). Often, they are simply referred to in terms of the same sense that was transmitted (SC) and the opposite sense that was transmitted (OC). Examples of radars capable of polarimetric measurements include ground-based radars such as Arecibo, which transmits in one circular polarization and receives in two orthogonal circular polarizations, and the recent Mini-RF radars on the Chandrayaan-1 and LRO spacecraft. These radars transmit in one circular polarization, and receive in both horizontal and vertical polarizations, a technique known as "hybrid polarity".

The polarization properties of a surface may be expressed by the Stokes vector. The first Stokes parameter (S_1) is a measure of the total average power in the echo, S_2 and S_3 describe the linearly polarized state of the wave, and S_4 gives the direction and magnitude of the circularly polarized power. For any radar that receives in two orthogonal polarizations, the Stokes vector can be given by:

$$S_1 = \langle |E_H|^2 + |E_V|^2 \rangle = \langle |E_L|^2 + |E_R|^2 \rangle$$
$$S_2 = \langle |E_H|^2 - |E_V|^2 \rangle = 2Re\langle E_L E_R^* \rangle$$
$$S_3 = 2Re\langle E_H E_V^* \rangle = 2Im\langle E_L E_R^* \rangle$$
$$S_4 = 2Im\langle E_H E_V^* \rangle = \langle |E_L|^2 - |E_R|^2 \rangle$$

Here, E represents the complex received voltage in the subscripted polarization (H, V, L, or R), * denotes the complex conjugate, $\langle \ldots \rangle$ denotes averaging, and Re and Im refer to the real or imaginary part of the complex amplitude. Two particularly useful products can be derived from the Stokes vector: the circular polarization ratio (CPR) and the degree of linear polarization (DLP). These are defined as:

$$CPR = \frac{S_1 - S_4}{S_1 + S_4} = \frac{SC}{OC} \text{ and } DLP = \frac{\sqrt{S_2^2 + S_3^2}}{S_1}$$

The CPR is a useful indicator of surface roughness (Figure 53.12). This is because when a circularly polarized radar wave is backscattered off an interface, the polarization state of the wave changes. Thus flat, mirrorlike surfaces, dominated by single-bounce reflections, tend to have high OC returns and low CPR values. Rough surfaces, dominated by multiple-bounce reflections, tend to have approximately equal OC and SC returns, with CPR values approaching unity. Incidence angle also affects the CPR. As we discussed in Section 2.1, low incidence angles are dominated by quasispecular scattering off smooth facets, leading to a larger OC return, while high incidence angles are dominated by diffuse scattering, increasing the SC/OC ratio and hence the CPR.

Rarely, CPR values can exceed unity. CPRs greater than 1 have been observed in rocky areas such as Maxwell Montes on Venus, blocky lava flows on the Earth, lava flows on Mars, and fresh ejecta blankets on the Moon. CPRs up to

Smooth: low CPR (≪1)
Single bounce backscattering flips polarization (OC > SC)

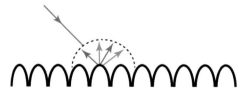

Rough: moderate CPR (~ 0.5–1)
Multiple bounce backscattering randomizes polarization (OC ~ SC)

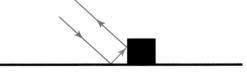

Blocky: high CPR (>1)
Corner reflectors cause double bounce backscattering (OC ≪ SC)

FIGURE 53.12 The roughness of a planetary surface affects the CPR. Smooth surfaces are dominated by single-bounce backscattering, which tends to flip the polarization (OC > SC). Rough surfaces are dominated by multiple-bounce backscattering, which randomizes the polarization (OC ~ SC). Blocky surfaces act as corner reflectors, causing double-bounce or **dihedral scattering** (OC < SC).

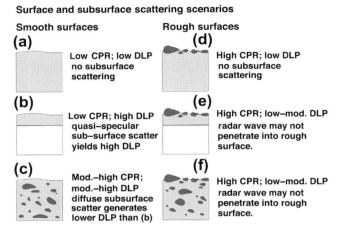

FIGURE 53.13 The CPR and DLP can be used to infer differences in the surface and subsurface structure. Note that smooth surfaces (depicted in a, b, and c) and low-loss tangents will increase the relative amount of subsurface scattering compared to rough surfaces (depicted in d, e, and f). *Figure from Carter, L. M., et al. (2011).*

2 can be produced via multiple reflections from rock edges and cracks, while scattering from natural corner reflectors (**dihedral scattering**) can produce an average CPR up to 3–4. High CPRs can also be caused by the presence of water ice, a phenomena we discuss in Section 4.1. Note, however, that ice cannot account for CPRs greater than 2.

The DLP can be used to infer the presence of subsurface scattering. A circularly polarized wave can be thought of as a combination of vertically and horizontally polarized waves. These two components have different power transmission coefficients, so if the wave penetrates the surface, the reflected wave will change from circular to elliptical. This introduces a linear component to the signal that was not present before, increasing the DLP. Thus, echoes from subsurface interfaces will have a larger DLP than those from most types of surface interfaces. Combined with the CPR, these products can be used to infer differences in the surface and subsurface structure of the planetary body (Figure 53.13). For example, a surface made of a fine-grained material overlying bedrock will have low CPR (from scattering off a smooth surface), but a high DLP (from subsurface scattering off of the bedrock) (Figure 53.14).

4. RADAR MEASUREMENTS OF PLANETARY BODIES

There have been countless studies of planetary bodies using radar, covering a variety of geologic topics. In this chapter, we touch on a selected set of highlights from that body of work and invite the reader to explore the references below for a more complete picture of planetary radar.

4.1. Mercury

Mercury has been studied by ground-based radars for many decades (see Mercury). Indeed, one of the earliest discoveries made by planetary radar was that Mercury was in a 3:2 spin–orbit resonance and not synchronously rotating about the Sun. Radar imaging of the planet began in earnest in the 1990s, producing the best images of the non-Mariner-imaged hemisphere of Mercury prior to the arrival of the MESSENGER spacecraft. In these images, two types of features were most prominent: radar-bright deposits near the north and south poles and rayed craters near the equator and midlatitudes. New optical images from the MESSENGER spacecraft have recently allowed for more detailed study of these features.

4.1.1. Polar Deposits

Radar-bright features were first discovered near Mercury's north pole using Goldstone and the Very Large Array (VLA) in a **bistatic** configuration. These results were soon confirmed by observations at Arecibo, and similar deposits near the south pole were also discovered. These deposits have large same-sense radar reflectivities and CPRs that

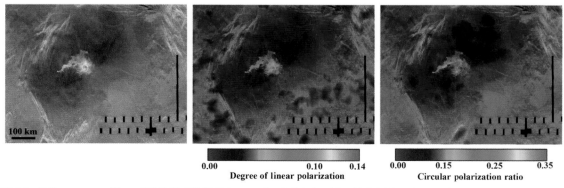

FIGURE 53.14 Galina crater on Venus (47.6° N, 307.1° E) has a radar dark halo with a high DLP and a low CPR. This suggests the presence of a smooth, fine-grained deposit over a buried interface. *Figure from Carter, L. M., et al. (2011).*

exceed unity. Similar radar properties have been observed on the Galilean satellites and Mars' polar ice caps.

These properties are relatively unusual compared to most targets in the solar system, and are thought to be indicative of the coherent backscatter effect in cold water ice (Table 53.2, Figure 53.15). When an incident circularly polarized radar wave is backscattered off an interface, the polarization state of the wave changes. For most surfaces, this leads to a return with more of an "opposite sense" (OC) polarization than a "same sense" (SC) polarization, so the CPR (CPR = SC/OC) remains less than 1. However, in weakly absorbing media (such as water ice), the radar signal can undergo a series of forward scattering events off small imperfections in the material, each of which preserves the polarization properties of the signal. Those portions of the wave front that are scattered along the same path but in opposite directions combine coherently to produce an increase in the SC radar backscatter. This coherent backscatter effect leads to large returns in the same sense (SC) polarization and values for CPR that can exceed unity.

The resolution of the radar images of Mercury improved over time, and many of these radar-bright features were observed to lie within impact craters imaged by Mariner 10. However, many more features could not be mapped in this way, since Mariner 10 only imaged roughly half of the planet during its two encounters in 1974 and 1975. In March 2011, MESSENGER became the first spacecraft to orbit Mercury. Subsequent analysis of the images collected during its 1-year prime mission was able to identify regions of persistent shadow near the north and south poles. For both poles, there is a remarkable correlation between regions of persistent shadow and radar-bright deposits (Figure 53.16(a)). Indeed, within 10° of the north pole, almost all craters larger than 10 km in diameter host radar-bright deposits. This work is an excellent example of the complementary nature of optical and radar imagery in understanding planetary surfaces.

Some craters located at lower latitudes (as far south as 66° N) or smaller than 10 km are also correlated with radar-bright deposits. This is odd because thermal models of Mercury's polar craters suggest that temperatures within such regions are likely to be too high to maintain water ice over geologic timescales, even if it is insulated by a thin (a few decimeters) layer of regolith. Understanding the thermal environment in these craters will be

TABLE 53.2 Radar Properties of the Icy Galilean and Saturnian Satellites Compared to Mars' Polar Caps and the Putative Ice Deposits Near the Poles of Mercury

Planet	SC Radar Albedo	CPR	Wavelength (cm)	References
Europa	1.58 ± 0.15	1.53 ± 0.03	12.6	Ostro et al. (1992)
Ganymede	0.83 ± 0.09	1.43 ± 0.06	12.6	Ostro et al. (1992)
Callisto	0.37 ± 0.03	1.17 ± 0.04	12.6	Ostro et al. (1992)
Enceladus	0.86 ± 0.20	0.83 ± 0.25	12.6	Black, Campbell, and Carter (2007)
Tethys	0.79 ± 0.09	1.22 ± 0.21	12.6	Black et al. (2007)
Dione	0.32 ± 0.07	0.81 ± 0.21	12.6	Black et al. (2007)
Rhea	0.71 ± 0.04	1.17 ± 0.09	12.6	Black et al. (2007)
Mars residual south polar ice cap	0.716 ± 0.009	$>1 - \sim 2.3$	3.5	Butler and Muhleman (1994)
Mercury north polar deposits (>75° N)	0.60–1.36	1.30–1.46	12.6	Harmon, Slade, and Rice (2011)
Mercury south polar deposits (<75° S)	0.52–3.49	0.75–1.67	12.6	Harmon et al. (2011)
Average Mercury	0.0045	0.1	12.6	Harmon (1997), Ostro et al. (2006)

[1] Ostro, S.J., D.B. Campbell, R.A. Simpson, R.S. Hudson, J.F. Chandler, K.D. Rosema, I.I. Shapiro, E.M. Standish, R. Winkler, and D.K. Yeomans (1992), Europa, Ganymede, and Callisto: New radar results from Arecibo and Goldstone, *Journal of Geophysical Research* **97**, 18,227–18,244.
[2] Black, G. J., Campbell, D. B., and Carter, L. M. (2007). Arecibo radar observations of Rhea, Dione, Tethys, and Enceladus. *Icarus* **191**, 702–711.
[3] Butler, B. J., Muhleman, D. O., and Slade, M. A. (1994). Martian Polar Regions: 35-cm Radar Images. *Abstracts of the 25th Lunar and Planetary Science Conference* **25**, 211.
[4] Harmon, J. K., Slade, M. A., and Rice, M. S. (2011). Radar imagery of Mercury's putative polar ice: 1999 - 2005 Arecibo results. *Icarus* **211**, 37–50.
[5] Harmon, J. (1997). Mercury radar studies and lunar comparisons. *Advances in Space Research* **19**, 1487–1496.
[6] Ostro, S.J. (2006). Planetary radar. In "Encyclopedia of the Solar System" (McFadden, L.A., Weissman, P., Johnson, T., Eds.), pp. 735–764. Academic Press.

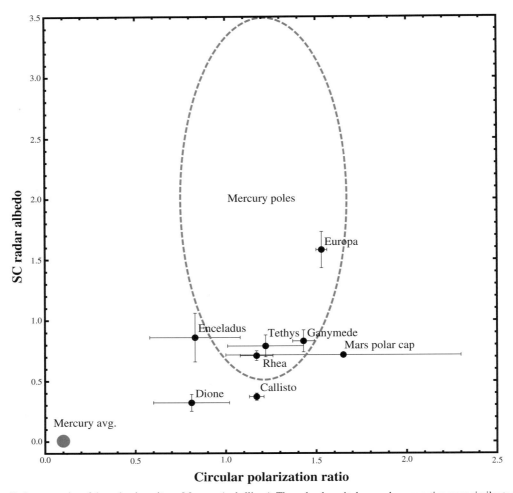

FIGURE 53.15 Radar properties of the polar deposits on Mercury (red ellipse). The polar deposits have radar properties more similar to the icy satellites of Jupiter and Saturn (black dots) than the average rocky surface of Mercury (red dot).

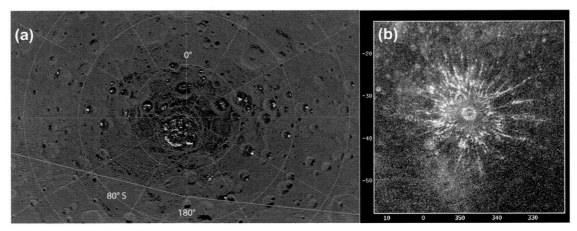

FIGURE 53.16 (a) The radar-bright deposits near Mercury's south pole (white regions) correlate well with regions of permanent shadow as observed by the MESSENGER spacecraft (black regions). *(Image from Chabot, N. L., et al. (2012), Journal of Geophysical Research 118, http://dx.doi.org/10.1029/2012JE004172.)* (b) Debussy crater (85 km diameter, 348° W, 34° S), formerly "Radar A", is one of the most spectacular rayed craters in the solar system. Rough ejecta appears radar bright in this SC radar image. *From Harmon, J. K., et al. (2007).*

key for constraining the timing of the emplacement of the radar-bright deposits. The presence of ice within such craters may suggest that it was emplaced relatively recently. Models of the burial of ice on Mercury through regolith emplacement estimate emplacement times of <50 million years for relatively pure ice covered by 20 cm of regolith.

There had been suggestions that the radar-bright deposits were a material other than water ice, such as sulfur, which would be more stable at the temperatures expected within Mercury's craters. However, recent measurements by the neutron spectrometer on the MESSENGER spacecraft show decreases in the flux of epithermal and fast neutrons that are consistent with the presence of water ice. The radar-bright deposits also have an infrared reflectance consistent with water ice and persist in areas where water ice is predicted to be stable at the surface. In regions where water ice is predicted to be stable only in the near subsurface, the infrared reflectance is lower, possibly indicative of a sublimation lag deposit.

4.1.2. Rayed Craters

In general, the radar reflectivities of the equatorial and midlatitude regions of Mercury are similar to those measured for the Moon. These low reflectivities suggest that the surface of Mercury, like that of the Moon, is covered in a loose regolith. However, unique to Mercury is the presence of several large features with enhanced "same-sense" (SC) backscatter. Large returns in SC radar are indicative of rough surfaces with wavelength-scale scatterers, and high-resolution radar imaging identified these features as fresh impact craters, some with extensive ray systems (Figure 53.16(b)).

These images provided our first look at the craters on the non-Mariner-imaged hemisphere of Mercury. Four craters identified there—originally dubbed "Radar A", "Radar B", "Radar C", and the "ghost feature"—have recently been observed by the MESSENGER spacecraft, and renamed Debussy, Hokusai, Xiao Zhao, and Amaral, respectively. Many of these craters have prominent ray systems, made up of a dense collection of narrow rays not typical of the Moon. On the Moon, radar rays tend to be more sparsely distributed, and extend to longer distances, if they are observed at all. The difference in radar ray appearance on the Moon and Mercury is not well understood, but could be related to differences in gravity, impact velocities, and/or crater maturation rates.

Optical rays are also prominent features on Mercury, but only a small fraction of these have corresponding radar rays. This suggests that crater maturation advances more quickly for radar brightness than for optical brightness. Given that craters on Mercury are observed to weather more quickly at optical wavelengths compared to the Moon, it seems likely that the radar-rayed craters on Mercury are very young.

4.2. Venus

Thick cloud cover has made radar the primary means of studying the surface of Venus (see Venus: Surface and Interior; Venus: Atmosphere; Solar System Dynamics: Rotation of the Planets). In the early 1960s, observations by the Goldstone Tracking Station were used to identify regions of anomalous reflectivity, and the rotation rate was computed for the first time. In the 1970s, delay-Doppler imaging from the Earth and the Pioneer Venus orbiter provided radar maps of portions of the surface at kilometer-scale resolution. Unfortunately, the same side of Venus always faces the Earth during inferior conjunction due to a resonance, so ground-based imaging is limited to one side of the planet. The highest resolution and most comprehensive imaging of Venus was completed in the early 1990s with the Magellan spacecraft. Magellan was not capable of polarimetry and used a horizontal linear polarization for transmitting and receiving, although the spacecraft was rotated for a few orbit tracks to produce vertical polarization image strips. Ground-based radar observations continue to provide new data (at ~3–15 km resolution), including polarimetric mapping and long time baseline imaging to search for possible surface changes.

4.2.1. Magellan Reveals the Surface

The Magellan mission produced a new, high-resolution map of surface features that had been identified in prior data sets but were still enigmatic. It also produced complementary topography and emissivity data. One notable mystery is the provenance of the high-reflectivity, low-radiothermal emissivity material on Venus mountaintops (Figure 53.17(a)). At altitudes above 3 km near the equator (e.g. Beta Regio) and above 5 km at the poles (e.g. Maxwell Montes), the Fresnel reflectivity increases and the emissivity drops from a plains average 0.85 to values as low as 0.3. At the highest altitudes on Maxwell Montes, the emissivity shifts back to normal values. The sharp onset of this reflectivity change with altitude suggests that a temperature- or pressure-dependent process is responsible for the changes. **Bistatic** observations of Maxwell using Magellan and the Deep Space Network yielded dielectric constant values with a large imaginary component suggestive of an electrically conducting or semiconducting surface coating. It is still not certain what material condenses at the surface, but possibilities include minerals with lead or bismuth, or ferroelectrics, magnetite, or metal halides.

Magellan also provided an improved view of the 900 or so Venusian impact craters. The Venus atmosphere does not allow small impactors to reach the surface, so there are few

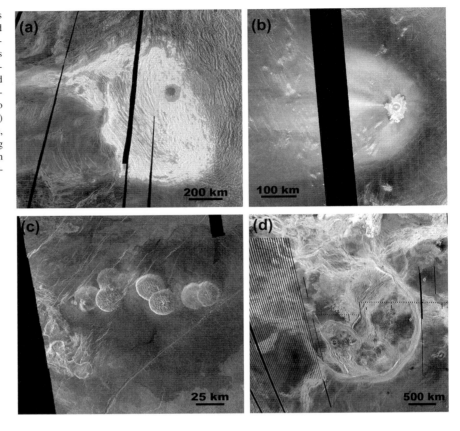

FIGURE 53.17 (a) Maxwell Montes is the highest mountain on Venus (11 km) and displays the unusual high-reflectivity, low-emissivity behavior common at high altitudes on Venus. (b) Adivar crater has a radar bright-parabola created when winds transported impact debris lofted during the impact explosion. (c) Pancake domes east of Alpha Regio are ∼25 km in diameter and ∼750 m high. (d) Artemis Corona is the largest of the coronae, with a diameter of 2100 km. The circular ring fractures encompass a trough with up to 2.5 km of relief, and the interior has numerous fractures, lava flows, and volcanoes.

small-diameter (<5 km) craters. Many craters have long impact melt flows, and some craters have dark haloes or parabola-shaped features that are likely generated as impact ejecta is deposited near the crater or blown by the wind (Figure 53.17(b)). The small number of craters relative to the Moon and Mars, and their statistically even distribution across the planet, suggests that the surface is young (∼500 My).

Dune fields, wind streaks, and **yardang** features are all present in Magellan images, indicating that there is fine-grained material on the surface that is transported by the wind. The radar images also revealed a wide range of volcanic features, from groups of small shield volcanoes on the plains to large lava flow complexes associated with major volcanoes such as Beta Regio, Sif and Gula Mons, and Atla Regio (see Planetary Volcanism). Some volcanic features are unique to Venus, such as the flat-topped pancake volcanoes (Figure 53.17(c)) and the ring-shaped 60 to 2100 km-wide coronae (Figure 53.17(d)). The former are thought to have formed from viscous lavas (having either a high silica content or cooler temperatures), while the latter have a complex relationship between tectonic annuli, lava flows, and interior topography that suggest they are formed by mantle plumes.

4.2.2. Post-Magellan Radar Imaging and the Search for Surface Changes

Magellan provided nearly full-coverage imaging of Venus, but it lacked polarimetry information that could provide more details about the nature of the surface. Comparisons of 1988 Arecibo data with Magellan and terrestrial analog terrains showed that most volcanic flows on Venus have CPR values less than 0.3, similar to terrestrial pahoehoe flows or smoother a'a flows. Imaging of Venus between 1999 and 2012 provided improved signal-to-noise polarimetry due to enhancements to the Arecibo radar system. Fine-grained **pyroclastic deposits** are typically radar dark with low CPR values due to their smooth surface and lack of embedded rocks, and polarimetry can be used to search for possible mantling deposits. A few volcanic deposits have DLP and CPR values that are consistent with fine-grained mantling, including flows near Sif Mons (Figure 53.18). Polarimetry has also shown that some craters have mantling deposits that likely represent deposition of material during impacts (Figure 53.13).

Decades of ground-based radar imaging using the radar system at Arecibo have also enabled a search for surface changes. This is particularly important in the light of recent

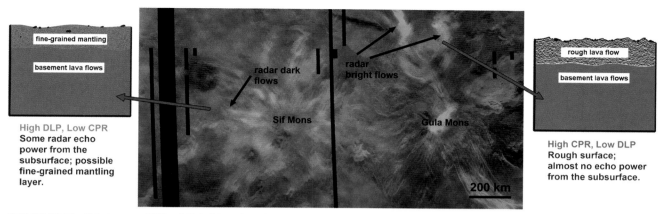

FIGURE 53.18 Polarimetry of Sif and Gula Mons (see also Figure 53.1) illustrate how radar polarimetry can reveal the complexity and variety of different surfaces present in volcanic areas. This image shows the DLP (red) and CPR (green) overlaid on a Magellan radar image. Radar-bright flows have a high CPR values consistent with rough lava flows. Radar-dark flows have higher than average DLP values and low CPRs, which indicates that the radar wave is penetrating into fine-grained material such as **pyroclastics**. *Source: Carter, L. M., Campbell, D. B., & Campbell, B. A. (2011).*

results from the Venus Express infrared instrument that revealed regions near some volcanoes that have a different infrared emissivity compared to surrounding areas. These areas have been interpreted as being less weathered, and hence young (<2.5 Million years), volcanic deposits. If active volcanism occurred in geologically recent times, it could still be ongoing, and if flows of sufficient size were produced, they might be visible in Arecibo images. When comparing radar imaging, it is best to use very similar geometries, and the 1988 and 2012 observing geometries are nearly identical. To date, no temporal changes in radar properties have been found, but studies will continue since the topic of active volcanism is one of the key unanswered questions about Venus.

4.3. Moon

The Moon was the first planetary object to be detected by radar, with the ground-breaking experiment conducted by the US Army Signal Corps in 1946 (see The Moon; Lunar Interior; Exploration of the Moon). It could be argued that it is the most extensively studied object in the solar system (other than the Earth), having been imaged by ground-based radars at a variety of wavelengths and polarizations. Still, given the synchronous rotation of the Moon about the Earth, the far side of the Moon remained unseen at radar wavelengths until SARs were launched in 2008 and 2009 on the Chandrayaan-1 and LRO spacecraft. These spacecraft have allowed for the whole Moon to be imaged at a consistent look angle.

4.3.1. Polar Environment

Like Mercury, the Moon has a reasonably small obliquity (1.5° from the ecliptic), leading to large regions of permanently shadowed areas in depressions near the poles. Given radar's sensitivity to the presence of thick deposits of water ice, and its ability to "see in the dark", it is uniquely qualified to assess the presence of ice in these permanently shadowed regions. Unfortunately, the region of permanent shadow that can be imaged from the Earth is much smaller than the total area of shadow, making the question of polar ice deposits difficult to assess from the Earth. The first attempt to use an orbital spacecraft to search for ice on the Moon was conducted during the Clementine mission. The spacecraft transmitted a 13.2-cm signal from its high-gain antenna toward the lunar south pole, and received the signal on the Earth at **bistatic** angles ranging between 0° and 5°. Targets that exhibit the coherent backscatter opposition effect, such as water ice, will have an opposition surge at $\beta = 0°$, with the signal decreasing sharply at angles greater than $\sim 1°$. A return of this nature was reported over Shackleton crater, but the inference of ice was not supported by other workers in their interpretation of subsequent, higher resolution polarimetric imaging of shadowed crater floors.

The launch of the Mini-SAR instrument on Chandrayaan-1 in 2008 and Mini-RF on LRO in 2009 allowed for high-resolution, monostatic imaging of the lunar poles. Of the permanently shadowed regions identified there, some do appear to have enhanced CPRs (Figure 53.19), but many others do not. Complicating matters, the presence of high CPRs can be explained through other mechanisms, such as **dihedral scattering**, that do not require the presence of water ice. New **bistatic** observations by the Mini-RF instrument on LRO should help to discriminate between high CPRs caused by ice and those caused by rough surfaces. Note that these tests will only be effective if the ice is present as a thick deposit (tens of wavelengths) of nearly pure water ice. If water is present

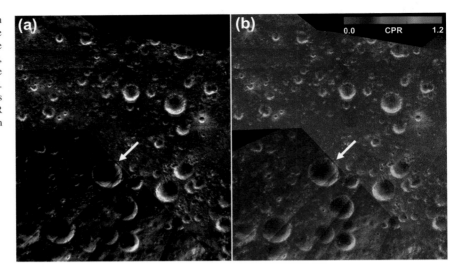

FIGURE 53.19 Several small craters within Peary crater, near the lunar north pole, have high CPRs within their crater rims, but lack the bright ejecta blankets associated with young, fresh craters. This is consistent either with the presence of ice, or rough, blocky crater walls. Shown here is (a) total radar backscatter as observed by Mini-RF on LRO and (b) CPR overlaid on total radar backscatter, scaled from 0 (purple) to 1.2 (red).

in the form of small grains of ice mixed into the regolith, it will be difficult to distinguish from a rocky surface, since the change in dielectric constant caused by the presence of a small concentration of water ice would not produce large scattering differences.

4.3.2. Lunar Geology

Radar has also been an extremely useful tool for studying the geology of the lunar surface. Nearly complete radar maps of the lunar nearside exist for 3.8 cm, 12.6 cm, 70 cm, and 7.5 m wavelengths (in addition, the Mini-RF instrument on LRO has imaged 70% of the lunar far side at 12.6 cm). The radar scattering behavior for the Moon mimics the **dichotomy** between the maria and highlands. The highlands have higher radar backscatter, likely due to a lower loss tangent in the regolith, which allows for more scattering from subsurface rocks. This difference is most pronounced at longer wavelengths, which have larger penetration depths. The dominant control on the microwave loss tangent of the regolith appears to be related to the mineral ilmenite, which contains most of the titanium in **mare** basalts.

Radar is also a particularly powerful tool for studying impact craters on the Moon. Radar exposes rough ejecta blankets, impact melts, and secondary craters not easily seen by optical instruments alone. In particular, radar can be used to date lunar craters. The blocky ejecta around young, fresh craters will break down over time, reducing their roughness and hence their radar backscatter. This decline in radar backscatter is most pronounced for smaller wavelengths, since smaller ejecta blocks will be removed from the population most quickly. The oldest craters will show little to no radar enhancement, as the crater approaches the same stable regolith as the surroundings. Radar is also proving to be extremely useful for identifying and characterizing impact melt deposits around lunar craters. These features appear to be some of the roughest materials on the Moon and can be identified even if partially buried by regolith (Figure 53.20).

4.4. Mars

The ubiquitous dust cover on Mars makes it a prime target for radar instruments that can see below these mantling layers and reveal buried terrain (see Mars: Surface and Interior; Mars Interior). However, Mars is a difficult target for ground-based radar systems because the planet rotates rapidly, which causes echoes that are overspread in frequency and leads to **aliasing** when images are formed. Early radar experiments used continuous-wave observations to generate **power spectra** at different longitudes, or used **radio interferometric synthesis** imaging with a radar transmitter to avoid the radar imaging issues. In 1990, a long code technique was developed at Arecibo Observatory that avoids the overspreading problem, and the first radar images were generated at 12.6 cm wavelength. More recently, two sounding radars, MARSIS (operational in 2005) and SHARAD (operational in 2006) have provided unprecedented information about the subsurface of Mars.

4.4.1. Radar Heterogeneity

Radar imaging from ground-based telescopes has shown that the upper surface of Mars has variable textures and dielectric properties. The most obvious radar feature, first detected in the 3.5-cm-wavelength synthesis imaging data, is an extremely low-reflectivity region near the equator that was dubbed "Stealth". Its anomalously low reflectivity is indicative of a low-density, low-permittivity surface with few embedded wavelength-sized rocks. Much of the Stealth region corresponds to mapped exposures of the Medusae

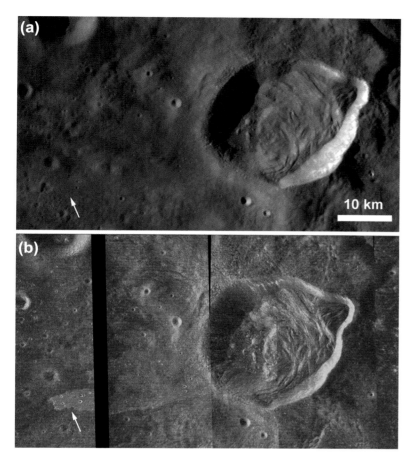

FIGURE 53.20 Image of the crater Gerasimovich D, located on the far side of the Moon, as seen (a) at optical wavelengths by the LRO wide-angle camera, and (b) at radar wavelengths by Mini-RF on LRO. Due to its sensitivity to rough surfaces, radar is able to highlight a previously unrecognized impact melt flow (indicated by an arrow).

Fossae Formation, an extensive equatorial deposit stretching from 140 to 240° E longitude. Optical images show that this deposit has numerous **yardangs**, slope streaks, and dunes and has been heavily wind eroded, consistent with a very fine-grained low-density material. The most likely origin for this formation is **pyroclastic volcanic deposits**, but other hypotheses include **aeolian deposits** or relic polar deposits that may contain ice.

Radar observations of Mars using Arecibo Observatory at S-band (12.6 cm) during the 2005–2012 oppositions produced 3-km-resolution dual-polarization images (Figure 53.21). Radar-bright lava flows surround the Tharsis volcanoes, Olympus Mons, and Elysium Mons. The flows have CPR values approaching and exceeding 1.0, indicating that rough, blocky (at the centimeter to meter scale) lava flows must be common on Mars. Changes in radar backscatter along the flows reveal channels and changes in flow texture. Radar-dark fan-shaped deposits are visible west of the Tharsis volcanoes and correspond to deposits thought to have a glacial origin. The low reflectivity suggests that these deposits have a uniform covering of fine-grained material such as dust or ash with few embedded centimeter-sized or larger rocks. The Marte Vallis channel, which is thought to be a water-carved channel subsequently filled with lava, is very radar bright and links the Cerberus plains with Amazonis Planitia. Both these plains regions, which are dust covered in optical imagery, have detailed lava flow features in the radar imagery.

4.4.2. Subsurface Structure

Sounding radar observations from the MARSIS and SHARAD radars have dramatically improved our understanding of the Martian subsurface, particularly for the polar ice deposits. MARSIS has a longer wavelength (~60 m vs. 15 m) and can therefore detect deeper interfaces, but SHARAD has higher time resolution and can detect surface layering in the upper hundreds of meters. Both MARSIS and SHARAD penetrate through the north polar layered deposits of Mars and reveal remarkable internal stratigraphy (Figure 53.22). The north pole has four packets of finely layered reflectors that are continuous throughout the polar layered deposits separated by featureless intervals that likely indicate changes in Martian climate. The maximum deflection of the subsurface due to ice loading at the north pole is no more than 100 m, which means that either Mars has a lithosphere greater than 300 km thick or the response to the ice loading is currently in a transient state.

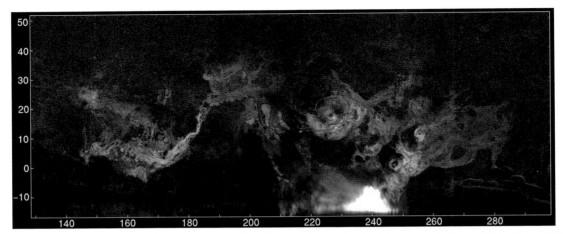

FIGURE 53.21 Radar image of Mars at S-band (12.6 cm wavelength) acquired using Arecibo Observatory during the 2005–2012 oppositions. The radar-bright lava flows associated with (left to right) Elysium, Olympus, and the Tharsis Montes are clearly visible, as are radar-dark fan-shaped deposits west of the Tharsis volcanoes and the equatorial "Stealth" areas south and west of Olympus Mons. Marte Vallis is a very bright sinuous feature near 180° E. *Figure from Harmon, J. K., et al., (2012).*

At the south pole, MARSIS penetrates to the base of the polar layered deposits but SHARAD usually does not. However, SHARAD data show reflection-free zones with no internal layering in the upper part of some radargrams that have dielectric properties and a surface morphology consistent with CO_2 ice. These unusual areas cover a large portion of the south polar layered deposits and have been interpreted as an extensive reservoir of CO_2 capable of increasing the atmospheric mass by 80% if released. As more data is collected, a complete view of layering in both poles is steadily building up and full 3D modeling of the internal structure will be possible in at least some locations. In addition to polar ice deposits, SHARAD has detected subsurface interfaces beneath many of the midlatitude lobate debris aprons, thought to be relic debris-covered glaciers, and confirmed that these deposits have dielectric properties consistent with ice.

SHARAD and MARSIS have also produced interesting results in volcanic and plains regions. Both radars see through exposures of the Medusae Fossae Formation. The permittivity measured from both radars is ∼3, and the MARSIS-derived loss tangent is 0.005, which are values consistent with either low-density volcanic ash or ice. Neither radar instrument detects interfaces within the outcrops, despite evidence of layering in optical imagery. SHARAD penetrates through tens-of-meters-thick lava

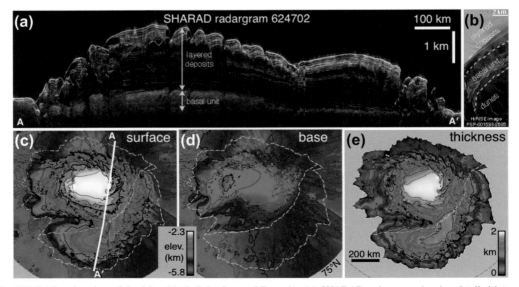

FIGURE 53.22 SHARAD radar view of the Mars North Polar Layered Deposits. (a) SHARAD radargram, showing detailed internal layering. (b) Optical image showing the layered deposits on the basal unit layer. (c) Surface topography with the SHARAD orbital track indicated. (d) The base of the polar layered deposits as measured from SHARAD data. (e) The thickness of the deposits measured from topography and radar data. *NASA/JPL-Caltech/ University of Rome/Southwest Research Institute/University of Arizona.*

flows, and flows northwest of Ascraeus Mons have permittivities and loss tangents that are consistent with low-iron and low-titanium lunar and terrestrial basalts. In Amazonis Planitia, a set of dipping kilometers-long subsurface interfaces have loss tangent values consistent with sedimentary packages and were interpreted as the base of the Vastitas Borealis Formation. In Elysium Planitia, SHARAD detects the buried surface of the original water-carved channels of Marte Valles and mapping of the channels suggests that Cerberus Fossae was the source of the Marte Vallis flows.

4.5. Asteroids

Radar is a powerful tool for post-discovery physical and dynamical characterization of NEAs (see Near-Earth Objects; Main Belt Asteroids). Radar provides high-resolution imagery of these asteroids, which can be inverted to yield shape models, allows for long-term orbit refinement, and is well suited to detect the presence of satellites.

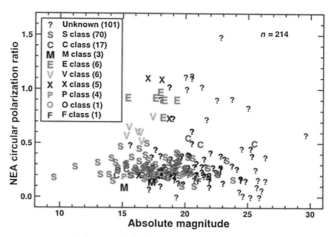

FIGURE 53.23 The CPR of NEAs is correlated with their spectral class. E-, V-, and X-class asteroids have larger CPRs than the more common S-class asteroids, suggesting differences in their mechanical properties or collisional histories. *Figure from Benner, L., et al. (2008).*

4.5.1. Imaging and Shape Reconstruction

Delay-Doppler imaging of asteroids has provided views of these small worlds at resolutions as fine as 4 m. These images provide direct size estimates and apparent rotation rates, as well as information about the asteroid's surface properties, such as the topographic relief, metal concentration, and surface roughness. Surface roughness can be characterized by the CPR of an object, and CPRs for NEAs span the range from near zero (very smooth) to more than unity (very rough). A recent survey of 214 NEAs revealed a correlation between CPR and the visual-infrared spectral class, which is determined by mineralogy (Figure 53.23). This correlation could be a reflection of the intrinsic mechanical properties of the asteroids and the response of different mineral assemblages to impact bombardment, or it may reflect differing formation ages and collisional histories. The highest ratios, and hence the roughest objects, are associated with groups connected to igneous rocky meteorites, such as the V-class, derived from MBA 4 Vesta.

Radar images of asteroids also allow for accurate shape determination. Given the north–south ambiguity present in delay-Doppler images, it is not possible to determine shape from one image along. However, if the radar is not in the target's equatorial plane, then as the object rotates, it will provide unique views of the asteroid in delay-Doppler space. A set of images like this can be inverted to reconstruct the target's 3D shape and spin state. For example, radar observations of (4179) Toutatis in 1992 and 1996 provided high-resolution delay-Doppler images that were used to produce the shape model shown in Figure 53.24. The resulting shape suggests that Toutatis might consist of two separate objects joined in a gentle collision.

These images also allowed for a determination of the asteroid's spin state. It was found to be rotating in a long-axis non–principal axis spin state, with periods of 5.4 days (long-axis rotation) and 7.4 days (long-axis precession).

Radar imaging also provided the first undeniable evidence for NEA binary systems, and has since discovered 26 binary and two trinary systems, representing two-thirds of all known NEAs with companions (see, for example, Figure 53.25). These observations are particularly important, because they provide a constraint on the density of the asteroids, one of the few methods for determining this value. Given the orbital period, T, and semimajor axis, a, of the secondary, it is possible to determine the total mass of the system from Kepler's third law:

$$T^2 = a^3 \left(\frac{4\pi^2}{G(M_1 + M_2)} \right)$$

Then, if one assumes that both asteroids have the same density and are roughly spherical, one can determine the density, ρ, given the radii of the primary and secondary, r_1 and r_2:

$$(M_1 + M_2) = \rho(V_1 + V_2) = \frac{4\pi}{3}\rho(r_1^3 + r_2^3)$$

The density was determined this way for the binary asteroid (5381) Sekhmet, giving $\rho = 1.98 \pm 0.65$ g/cm^2, consistent with its spectral classification as an S-type asteroid.

4.5.2. Orbit Determination and the Yarkovsky Effect

Radar observations of NEAs can be used to reduce the uncertainty in an asteroid's orbital parameters by many

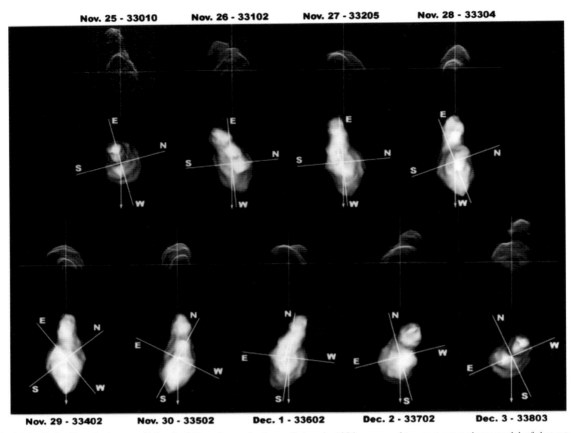

FIGURE 53.24 Delay-Doppler images (3.5 cm) of Toutatis acquired over 9 days in 1996 were used to construct a shape model of that asteroid. The corresponding plane-of-sky appearance is shown beneath the radar image. The illumination direction is from the top for the radar images, and colocated with the viewer for the model images. The arrow indicates the instantaneous spin vector. *Figure from Ostro, S. J., et al. (1999).*

orders of magnitude. The line-of-sight astrometry provided by radar is orthogonal to the plane-of-sight optical astrometry, making the combination a powerful tool for refining orbits. A single radar measurement can often extend the useful lifetime of an orbital solution by ~200 years. This prevents objects from being lost, and allows careful tracking of potentially hazardous close-Earth approaches.

The dominant source of uncertainty in the long-term tracking of asteroids involves the Yarkovsky effect (see Solar System Dynamics; Regular and Chaotic Motion). This is a nongravitational acceleration caused by the anisotropic thermal emission of absorbed sunlight, leading to changes in an object's semimajor axis over time. The magnitude of the Yarkovsky effect is related to an object's mass, size, spin properties, and surface thermal characteristics. This effect can be detected for NEAs up to a few kilometers in size, given precise radar astrometry spanning a decade or more. The first detection came in 2003 for

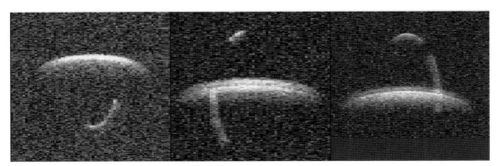

FIGURE 53.25 Hours-long delay-Doppler time exposures of binary asteroid 1999 KW4. Gaps in the secondary orbit are due to breaks in data collection. The primary looks much wider than the secondary because it is a few times larger and rotating much faster. Information about the size of the asteroids and the secondary's orbital parameters can be used to deduce the asteroid's density. *Figure from Chapman, C. R. (2004).*

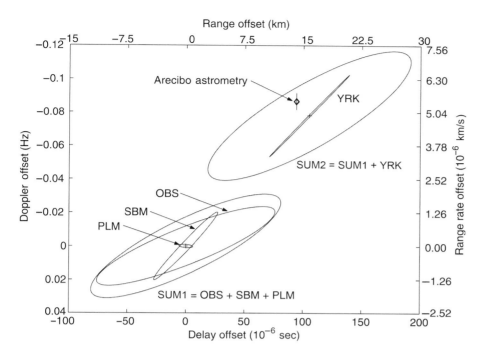

FIGURE 53.26 The observed offset of the asteroid (6489) Golevka's orbit in May 2003 is most consistent with the Yarkovsky effect (YRK), as opposed to uncertainties associated with astrometric measurements (OBS), small body masses (SBM), and planetary masses (PLM). The SUM1 ellipse depicts the 90% confidence region for a non-Yarkovsky prediction, and the SUM2 ellipse depicts the 90% confidence region for the prediction with Yarkovsky accelerations. *Figure from Chesley, S. R., et al. (2003).*

the asteroid (6489) Golevka, using radar observations from 1991 to 2003 (Figure 53.26). This asteroid has a Yarkovsky drift of $(-6.39 \pm 0.44) \times 10^{-4}$ AU/per million years. More recently, semimajor axis drifts were found in 54 NEAs, 42 of which agree with numerical estimates of Yarkovsky drifts. Typical rates are 0.001 AU/per million years for kilometer-sized asteroids. Due to its dependence of asteroid mass and size, the Yarkovsky effect provides one of the few constraints on asteroid density for nonbinary objects. For example, the asteroid Golevka has a density of 2.7 (+0.4/−0.6) g/cm^3 for reasonable values of thermal conductivity.

4.6. Jupiter and Saturn Systems

4.6.1. Icy Satellites

The first echoes from Jupiter's moons were received in the mid-1970s. These observations showed that icy satellites backscatter radar waves in an unusual way, unlike that seen on rocky objects (see Planetary Satellites; Europa; Ganymede and Callisto; Titan; Enceladus). Since then, additional ground-based and space-based observations have refined our understanding of the scattering mechanisms that control the radar properties of these icy worlds, as well as their close cousins at Saturn.

As we discussed in Section 4(a), icy satellites generally have high radar reflectivities and CPRs (see Figure 53.15). This indicates that the radar scattering is dominated by subsurface volume scattering, not single scattering from the vacuum—surface interface, as is typical for rocky bodies. This scattering behavior has been modeled as a coherent backscatter effect resulting from scatterers embedded in the weakly absorbing water ice surfaces of these moons. The observation that the radar cross-sections of the Galilean satellites at 3.5 and 12.6 cm are much higher than those observed 70 cm can be reproduced by a model with scatterers that follow a fairly steep power-law distribution and are no larger than 0.5−1 m in size. Efficient scatterers, like voids, are required to reproduce the observations. Partially absorbing scatterers, such as silicate rocks, cannot reproduce the high radar cross-sections observed.

Saturn's icy satellites also appear to be dominated by subsurface volume scattering, although their radar reflectivities are somewhat lower than those of the Galilean satellites at 13 cm. This suggests that the subsurface scattering is operating less efficiently than on the Galilean satellites, most likely due to a difference in composition. A likely candidate is ammonia, since water ice's absorption length is extremely sensitive to ammonia concentration. At 2.2 cm, the radar albedo of the Saturnian satellites is much higher, although average radar albedos decrease as you move outward from Enceladus and Tethys to Iapetus and Phoebe. This is likely a function of increasing contamination of near-surface water ice.

Two interesting case studies are Enceladus and Iapetus. At 13 cm, Enceladus has higher radar albedo on the leading side than the trailing side, yet is uniformly bright at 2.2 cm. This suggests that the regolith on the trailing side is too fresh for meteoroid bombardment to have developed the large-scale heterogeneities needed to elevate the 13-cm radar albedo, but all of Enceladus is mature enough at small scales to enhance the 2-cm albedo. Iapetus also has a hemispheric asymmetry, but only in the 2-cm data, mimicking the albedo asymmetry that is so obvious at optical wavelengths. The fact that the 13-cm data shows no asymmetry suggests that the leading side's optical contaminant is present to a depth of at least a decimeter, but no further.

4.6.2. Titan

Prior to the Cassini mission, the surface geology of Titan was unknown (see Titan). The Voyager imagers, operating only in the visible, were barely able to penetrate the haze layer than shrouds this world, while near-infrared observations from the Earth had too low a spatial resolution (>50 km) to permit geological interpretation. It was not until 2004 that Cassini began collecting the first high-resolution images of Titan's surface, using SAR at 2.2 cm. The images obtained from these encounters (along with images from the Huygens probe in 2005) provided the first look at the surface of this hazy moon. Through these observations, Titan has been revealed to be an active world, surprisingly Earth-like in appearance. Large dune fields extend around the equatorial region, stream channels carve the landscape, and the poles are covered in lakes of liquid hydrocarbons.

Observations by Cassini RADAR, along with complementary observations by the Visual and Infrared Mapping Spectrometer (VIMS) and Imaging Science Subsystem (ISS), have revealed that the main processes occurring on Titan are aeolian and fluvial modification, tectonics, cryovolcanism, and impact cratering (Figure 53.27). Dunes are one of the most ubiquitous features on Titan, and indeed, the fraction of Titan's surface covered by dunes is larger than on any terrestrial planet. Given its high resolution (~ 350 m), data from Cassini RADAR has been used to map thousands of dunes on Titan (Figure 53.27(b)). The dunes mostly lie in the equatorial regions, between $30°$S and $30°$N, where there is apparently the requisite dry, transportable sediment and winds to move the material. The dunes are principally linear, and their direction indicates that sand transport operates from west to east, suggesting a modestly changing wind regime, with winds that are predominantly westerly. Unlike silicate sand on the Earth, these dunes appear to be composed of sand-sized organic particles, likely sintered from Titan's abundant haze material.

Fluvial features are also common on Titan, and appear remarkably similar to the lakes and streams seen on the Earth. The lakes are concentrated exclusively near the

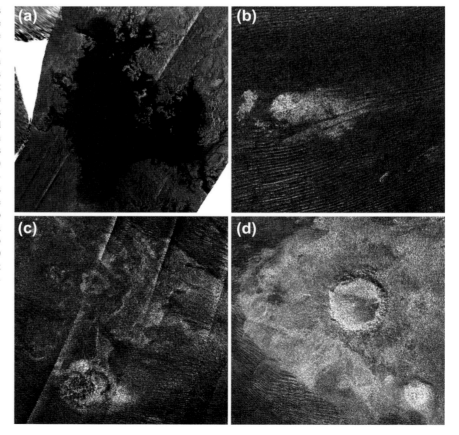

FIGURE 53.27 Cassini RADAR has revealed Titan to be an intriguingly Earth-like world. (a) Lakes of liquid hydrocarbons are common near Titan's north and south poles, including Ligeia Mare, which is larger than Lake Superior on the Earth. Titan's lakes appear dark to radar because they are smooth at radar wavelengths; as of yet, no waves have been detected on Titan. (b) Long, linear dunes of sand-sized organic material extend around Titan's equator and dominate the region between $30°$ S and $30°$ N. Specular reflections of dune slip faces are seen at lower left. (c) Sotra Facula is the best candidate for a cryovolcano on Titan. Here, flowlike features extend to the northeast of two peaks more than 1000 m tall and multiple pits as deep as 1500 m. (d) Like the Earth, Titan has a dearth of recognizable impact craters. Sinlap (~ 80 km in diameter) is one of only ~ 60 craters discovered thus far. The lack of impact craters indicates that Titan's surface is relatively young.

poles, and appear to be composed of a mixture of ethane and methane. Although first observed at near-infrared wavelengths at the south pole, RADAR was able to provide strong supporting evidence for the lake interpretation. The radar return for these features is at the noise floor of the data, consistent with an extremely smooth surface at 2.2 cm (Figure 53.27(a)). RADAR also made the first discovery of lakes at the north pole of Titan, which were in darkness during the first part of the Cassini mission, when Titan was in northern winter. Ground-based radar has also found intriguing evidence for specular reflections near the equator, indicating large surfaces smooth at a wavelength of 12.6 cm. No obvious lakes have been observed here, so these observations may be suggestive of smooth surfaces saturated in liquid hydrocarbons. Streams and valley networks have been observed at all latitudes, from the equator to the poles. Remarkably, the conditions required for sediment transport by liquid methane on Titan are within a factor of a few of those required for transport by liquid water flow on Mars and the Earth, resulting in remarkably similar features given the different surface conditions. Some channels are radar bright, suggesting that they are filled with rough boulders, while others are radar dark and thought to contain either liquid or deposits of smooth sediments.

In addition to the exogenic processes discussed above, there is some evidence for internal geologic activity occurring on Titan. Tectonic terrain has been observed by RADAR in the form of linear chains of radar-bright mountains. The heights of these mountains (as determined from radarclinometry) range from <300 to 2000 m, with slopes approaching 45° in some cases. Radar evidence of cryovolcanism is controversial, but if it is present, discrete cryovolcanoes are not ubiquitous. The best candidate cryovolcano yet observed is Sotra Facula (Figure 53.27(c)). Stereo topography from overlapping RADAR swaths revealed this feature to be a mountain more than 1000 m tall with multiple pits as deep as 1500 m, with flows perhaps 100 m thick extending away from the complex. Note that effusive volcanism, if present on Titan, is likely to consist of liquid ammonia−water lavas.

Unlike most worlds in the solar system, Titan has a relatively small number of confirmed impact craters, indicating that most parts of its surface are relatively young. Many craters on Titan show evidence of erosion, suggesting that (as on Earth) there are processes at work that can modify and erase craters (Figure 53.27(d)). A recent survey of the depths of Titan's craters show that they are on an average several hundred meters shallower than comparably sized craters on Ganymede, supporting this idea. Of the regions on Titan viewed by RADAR, the polar regions appear to have the lowest concentration of impact craters, and the bright equatorial region Xanadu appears to have the high concentration of impact craters, suggesting that it is a remnant of an older surface.

5. THE FUTURE OF PLANETARY RADAR

Many worlds have already been surveyed with radar, but many more remain unexplored. Even those worlds that have been observed at radar wavelengths would benefit from higher resolution data at new wavelengths with dual- or quad-polarization. Orbital radars could be imagined for any solid surface in the solar system, but the two worlds that would benefit especially from new radar observations are Mars and Europa.

Two sounding radars, SHARAD and MARSIS, are currently investigating the Martian subsurface. However, an imaging radar is one of the few instruments not yet flown on any Mars mission. Currently, our best radar images of Mars come from ground-based observatories and offer relatively poor spatial resolutions (see Section 4.4). Imaging radars are a particularly important instrument for characterizing the Martian surface. Mars has been extensively modified by aeolian processes, so that a widespread layer of dust covers large portions of the planet. Orbital SARs allow for the investigation of landscapes buried beneath this dust layer, revealing features unseen in optical images, which probe only the top surficial layer, or by long wavelength radar sounders, which are dedicated to deep (several hundred meters) sounding with low spatial resolution. Without an imaging radar, geological structures such as fluvial channels, ancient shorelines, lava flows, and icy polygonal terrain still await discovery under Mars' dusty surface. Note that imaging radars have been used on the Earth for similar purposes, demonstrating the ease with which one can study ancient fluvial networks in sand-covered regions (Figure 53.28). In addition, an orbital radar would be sensitive to the presence of ice and water, and could therefore help to constrain Mars' volatile inventory. Liquid water's extremely high dielectric constant would make it particularly easy to spot in radar images. Finally, there is the possibility that the radar could be used interferometrically to track changes in Martian surface, such as that caused by the subsidence or swelling of volcanoes.

Orbital radars for Mars have been proposed by several groups. The key system design parameters include the sensitivity of the radar, its wavelength, and its polarization. A radar with a VV polarization is recommended to minimize transmission losses and maximize penetration depth. A wavelength between 30 and 60 cm (L-band) is also recommended, as it allows for a large penetration depth while also being sensitive to differences in wavelength-scale roughness. For most natural surfaces, the wavelength-scale roughness declines with wavelength.

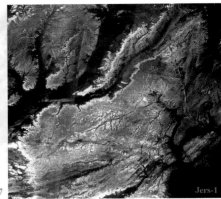

FIGURE 53.28 (left) Optical image and (right) L-band SAR image of northern Sudan, showing an ancient river valley obscured beneath several meters of sand. *Image from Paillou, P., et al. (2006).*

Other workers recommend a slightly longer wavelength—70 cm or P-band—to compensate for the higher iron content on Mars, which may limit penetration depth at shorter wavelengths. Such a radar would be less sensitive to roughness differences in the upper few meters of the subsurface.

Europa is another high-priority target for radar, in this case, an ice-penetrating radar. Europa likely has a saltwater ocean under an ice shell that is several kilometers to several tens of kilometers thick. Since life as we know it depends on a long-lived liquid water environment, this makes Europa a prime candidate in the search for habitable environments in the solar system. Therefore, it is critical to understand the existence and extent of its subsurface ocean, and to determine the presence of any water within the ice shell. Over the past decade, the National Aeronautics and Space Administration has been considering several mission options for exploring Europa. Due to the constrained budget environment, recent studies have focused on reduced-scope options in order to find cost reductions for this mission. One of these proposed missions is a multiple flyby concept, which includes exploring Europa's ice shell for evidence of liquid water within or beneath it. Included in the proposed payload is an ice-penetrating radar. This radar would have two modes: a higher frequency band to study the ice shell above 3 km depth at 10-m vertical resolution (nominally 60 MHz with a 10 MHz bandwidth), and a low-frequency band to search for the ice–ocean interface at a depth of up to 30 km at 100-m vertical resolution (nominally 9 MHz with a 1 MHz bandwidth). Since this radar is heavy, requires significant operating power, and generates large volumes of data, a multiple flyby mission is ideal for this instrument. A multiple flyby mission design allows for high-data-rate science collection followed by days of playback time, and reduces the amount of shielding needed to lower the radiation dose.

In addition to advances in space-based radars, it is expected that ground-based radars will continue to characterize an increasing number of NEAs. Such discoveries will bring with them new scientific information for our understanding of small bodies in the solar system, and may someday protect us from a devastating impact.

ACKNOWLEDGMENTS

The authors would like to dedicate this chapter to Steve Ostro, whose work in radar astronomy provided much of the inspiration and background for this chapter. This work was supported by an appointment to the NASA Postdoctoral Program at the Goddard Space Flight Center, administered by Oak Ridge Associated Universities through a contract with NASA to C.N.

BIBLIOGRAPHY

Benner, L., Ostro, S. J., Magri, C., Nolan, M. C., Howell, E. S., Giorgini, J. D., et al. (2008). Near-Earth asteroid surface roughness depends on compositional class. *Icarus, 198*, 294–304.

Black, G., Campbell, D. B., & Nicholson, P. D. (2001). Icy Galilean satellites: modeling radar reflectivities as a coherent backscatter effect. *Icarus, 151*, 167–180.

Campbell, B. A. (2012). High circular polarization ratios in radar scattering from geologic targets. *Journal of Geophysical Research, 117*, E06008.

Carter, L. M., Campbell, D. B., & Campbell, B. A. (2011). Geologic studies of planetary surfaces using radar polarimetric imaging. *Proceedings of the IEEE, 99*, 770–782.

Carter, L. M., Campbell, B. A., Watters, T. R., Phillips, R. J., Putzig, N. E., Safaeinili, A., et al. (2009). Shallow radar (SHARAD) sounding observations of the medusae fossae formation, Mars. *Icarus 199*, doi: 10.1016/j.icarus.2008.10.007, 2009.

Chapman, C. R. (2004). The hazard of near-Earth asteroid impacts on earth. *Earth and Planetary Science Letters, 222*, 1–15.

Chesley, S. R., Ostro, S. J., Vokrouhlický, D., Capek, D., Giorgini, J. D., Nolan, M. C., et al. (2003). Direct detection of the Yarkovsky effect by radar ranging to asteroid 6489 Golevka. *Science, 302*, 1739–1742.

Farr, T. G. (1993). Radar interactions with geologic surfaces. In: *Guide to Magellan Image Interpretation, JPL Publication, 93–24*, 148.

Harmon, J. K., Nolan, M. C., Husmann, D. I., Campbell, B. A., et al. (2012). Arecibo radar imagery of Mars: the major volcanic provinces. *Icarus, 220*, 990–1030.

Nicholson, P. D., French, R. G., Campbell, D. B., Margot, J.-L., Nolan, M. C., Black, G. J., et al. (2005). *Icarus, 177*, 32–62.

Ostro, S. J., Hudson, R. S., Benner, L. A. M., Giorgini, J. D., Magri, C., Margot, J.-L., et al. (2002). Asteroid radar astronomy. In W. Bottke, A. Cellino, P. Paolicchi, & R. P. Binzel (Eds.), *Asteroids III* (pp. 151–168). Tucson: University of Arizona Press.

Ostro, S. J., Hudson, R. S., Rosema, K. D., Giorgini, J. D., Jurgens, R. F., Yeomans, D. K., et al. (1999). Asteroid 4179 Toutatis: 1996 radar observations. *Icarus, 137*, 122–139.

Paillou, P., Lasne, Y., Heggy, E., Malézieux, J.-M., Ruffié, G., et al. (2006). A study of P-band synthetic aperture radar applicability and performance for Mars exploration: imaging subsurface geology and detecting shallow moisture. *Journal of Geophysical Research, 111*, E06S11.

Pettengill, G. H., Campbell, B. A., Campbell, D. B., & Simpson, R. A. (1997). Surface scattering and dielectric properties. In P. Hunten, L. Colin, T. Donahue, & V. Moroz (Eds.), *Venus II*. Tucson: University of Arizona Press.

Pettengill, G. H., & Dyce, B. R. (1965). A radar determination of the rotation of planet Mercury. *Nature, 206*, 1240.

Phillips, R. J., Davis, B. J., Tanaka, K. L., Byrne, S., Mellon, M. T., Putzig, N. E., et al. (2011). Massive CO_2 ice deposits sequestered in the south polar layered deposits of Mars. *Science, 211*, 838–841.

Slade, M. A., Butler, B. J., & Muhleman, D. O. (1992). Mercury radar imaging - evidence for polar ice. *Science, 258*, 635–640.

Stiles, B. W., Hensley, S., Gim, Y., Bates, D. M., Kirk, R. L., Hayes, A., et al. (2009). Determining Titan surface topography from Cassini SAR data. *Icarus, 202*, 584–598.

Stofan, E. R., et al. (2007). The lakes of Titan. *Nature, 445*, 61–64.

Chapter 54

Remote Sensing of Chemical Elements Using Nuclear Spectroscopy

Thomas H. Prettyman
Planetary Science Institute, Tucson, AZ, USA

Chapter Outline

1. Introduction 1161
2. Origin of Gamma Rays and Neutrons 1162
   2.1. Galactic Cosmic Rays 1162
   2.2. Fundamentals of Neutron Moderation 1163
   2.3. Gamma Ray Production and Transport 1166
3. Detection of Gamma Rays and Neutrons 1167
   3.1. Counting Rate Models 1167
   3.2. Gamma Ray and Neutron Detection 1168
   3.3. Spatial Resolution 1171
4. Missions 1171
5. Science 1175
   5.1. Lunar Prospector 1175
   5.2. Mars Odyssey 1177
6. Future Prospects 1181
Bibliography 1183

1. INTRODUCTION

Nuclear spectroscopy is used to determine the elemental composition of planetary surfaces and atmospheres. Radiation, including **gamma rays** and **neutrons**, is produced steadily by cosmic ray bombardment of the surfaces and atmospheres of planetary bodies. Gamma rays are also produced by the decay of naturally occurring radioelements. The leakage **flux** of gamma rays and neutrons contains information about the abundance of major elements, selected trace elements, and light elements such as H and C. Gamma rays and neutrons can be measured from high altitudes (less than a planetary radius), enabling global mapping of elemental composition by an orbiting spacecraft. Radiation that escapes into space originates from shallow depths (<1 m within the solid surface of airless bodies and those with thin atmospheres through which galactic cosmic rays can penetrate). Consequently, nuclear spectroscopy is complementary to other surface mapping techniques, such as reflectance spectroscopy, which is used to determine the mineralogy of planetary surfaces.

The main benefit of gamma ray and neutron spectroscopy is the ability to reliably identify elements important to planetary geochemistry and to accurately determine their abundance. This information can be combined with other remote sensing data, including surface thermal inertia and mineralogy, to investigate many aspects of planetary science. This article provides an overview of this burgeoning area of remote sensing. The origin of gamma rays and neutrons, their information content, measurement techniques, and scientific results from the Lunar Prospector and Mars Odyssey missions are highlighted.

Cosmogenic nuclear reactions and radioactive decay result in the emission of gamma rays with discrete energies, which provide a fingerprint that can uniquely identify specific elements present within the surface. Depending on the composition of the surface, the abundance of major rock-forming elements, such as O, Mg, Al, Si, Cl, Ca, Ti, Fe, and radioactive elements, such as K, Th, and U can be determined from measurements of the gamma ray spectrum. The geochemical data provided by nuclear spectroscopy can be used to investigate a wide range of topics including the following:

- Determining bulk composition for comparative studies of planetary geochemistry and the investigation of theories of planetary origins and evolution;
- Constraining planetary structure and differentiation processes by measuring large-scale stratigraphic variations within impact basins that probe the crust and mantle;
- Characterization of regional-scale geological units, such as lunar mare and highlands;

- Estimating the global heat balance by measuring the abundance of radioelements such as K, Th, and U;
- Measuring the ratio of the volatile element K to the refractory element Th to determine the depletion of volatile elements in the source material from which planets were accreted and to estimate the volatile inventory of the terrestrial planets.

Neutrons are produced by cosmic ray interactions and are sensitive to the presence of light elements within planetary surfaces and atmospheres, including H, C, and N, which are the major constituents of ices, as well as elements such as Fe, Ti, Gd and Sm, which are strong neutron absorbers. In addition, alpha particles are produced by radioactive decay of heavy elements such as U and Th and have been used to identify radon emissions from the lunar surface, possibly associated with tectonic activity.

Close proximity to the planetary body is needed to measure neutrons and gamma rays because their production rate is relatively low. Unlike optical techniques, distances closer than a few hundred kilometers are needed in order to obtain a strong signal. In addition, sensors that detect neutrons and gamma rays are generally insensitive to their incident direction. Consequently, spatial resolution depends on orbital altitude, and higher resolution can only be achieved by moving closer to the planet. Regional-scale measurements are generally made using nuclear spectroscopy, in contrast to the meter to kilometer scales sensed by optical methods.

Measurements of the solid surface are not possible for planets with thick atmospheres, including the Earth and Venus. Variations in atmospheric composition can be measured and, for example, have important implications to understanding seasonal weather patterns on Mars. Gamma ray and neutron spectroscopy can be applied to investigate the surfaces of planets with little or no atmosphere, such as Mercury, Mars, the Moon, comets, asteroids, and dwarf planets. In principle, the satellites of Jupiter and Saturn could be investigated using nuclear spectroscopy; however, the intense radiation environment within the magnetospheres of these planets may be a limiting factor.

X-ray spectroscopy can also be used to determine elemental composition and is complementary to nuclear spectroscopy. Solar X-rays cause planetary surfaces to fluoresce. The spectrum of fluorescence X-rays can be analyzed to determine the abundance of rock-forming elements such as Fe and Mg. In contrast to nuclear spectroscopy, surface coverage may be limited, especially when solar activity is low; however, high statistical precision for elemental abundances can be achieved during flares. Since the intensity of solar X-rays diminishes with increasing heliocentric distance, the technique is limited to inner solar system bodies. The depth sensitivity of X-ray and nuclear spectroscopy is very different. X-rays are produced close to the surface (typically to depths less than a 100 μm), whereas gamma rays and neutrons sample bulk surface materials to depths of few tens of decimeters. Missions that have used X-ray spectroscopy include Apollo, Near Earth Asteroid Rendezvous (NEAR) (see Near-Earth Objects), and Small Missions for Advanced Research in Technology 1 (SMART-1). The MErcury Surface, Space ENvironment, GEochemistry, and Ranging (MESSENGER) mission used both X-ray and nuclear spectrometers to determine the elemental composition of Mercury, and an X-ray spectrometer was on the payload of Chandrayaan-1, the Indian Space Research Organization's first mission to the Moon.

2. ORIGIN OF GAMMA RAYS AND NEUTRONS

Neutrons and gamma rays are produced by the interaction of energetic particles and cosmic rays with planetary surfaces and atmospheres. While solar energetic particle events can produce copious gamma rays and neutrons, we will focus our attention on **galactic cosmic rays**, which are higher in energy, penetrate more deeply into the surface, and have a constant flux over relatively long periods of time. Gamma rays are also produced steadily by the decay of radioactive elements such as K, Th, and U. A diagram of production and transport processes for neutrons and gamma rays is shown in Figure 54.1.

2.1. Galactic Cosmic Rays

Galactic cosmic rays consist primarily of protons with an average flux of about 4 protons/cm^2/s and a wide distribution of energies extending to many gigaelectronvolts (Figure 54.2; inset). The flux and energy distribution of galactic protons reaching a planetary surface is modulated by the solar cycle (see The Sun). Sunspot counts are a measure of solar activity (Figure 54.2). Higher fluxes of galactic protons are observed during periods of low solar activity. A larger portion of protons penetrate the heliosphere during solar minimum, resulting in a shift in the population toward lower energies. The flux and energy distribution of the cosmic rays are controlling factors in the production rate, energy distribution, and depth of production of neutrons and gamma rays. For example, the neutron counting rates at McMurdo Station in Antarctica are modulated by the solar cycle as shown in Figure 54.2.

The gigaelectronvolt-scale kinetic energy of galactic protons can be compared to the relatively small binding energy of protons and neutrons in the nucleus (for example, 11.2 MeV is required to separate a neutron from a tightly bound ^{56}Fe nucleus). High-energy interactions with nuclei can be modeled as an intranuclear cascade, in which the energy of the incident particle is transferred to the nucleons, resulting in the emission of secondary particles by spallation, followed by evaporation, and subsequent

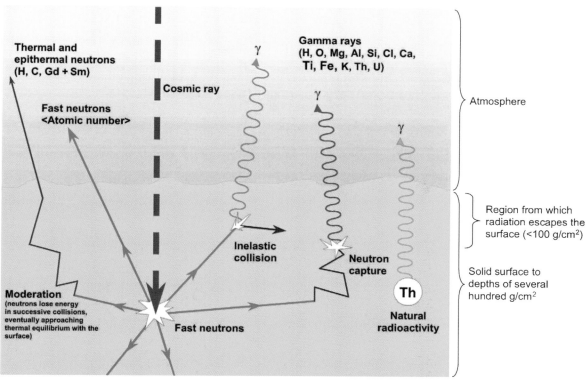

FIGURE 54.1 Overview of the production of gamma rays and neutrons by cosmic ray interactions and radioactive decay. Fast neutrons produced by high-energy cosmic ray interactions undergo inelastic collisions, resulting in the production of characteristic gamma rays that can be measured from orbit. Neutrons lose energy through successive collisions with nuclei and approach thermal equilibrium with the surface. Thermal and epithermal neutrons provide information about the abundance of light elements, such as H and C, and strong thermal neutron absorbers, such as Fe, Ti, Gd and Sm. Fast neutrons are sensitive to the average atomic mass of the surface. Gamma rays produced by neutron capture and inelastic scattering can be used to measure the abundance of rock-forming elements, such as O and Fe. Gamma rays are also produced by the decay of long-lived radioisotopes, including K, Th, and U. While cosmic rays can penetrate deep into the surface, the radiation escaping the surface originates from shallow depths, generally less than 100 g/cm².

deexcitation of the residual nuclei. The secondary particles, which include neutrons and protons, undergo additional reactions with nuclei until the initial energy of the cosmic ray is absorbed. Since most of the gamma ray production is caused by reactions with neutrons, we will focus our attention on how neutrons slow down in matter.

2.2. Fundamentals of Neutron Moderation

Neutrons transfer their energy through successive interactions with nuclei and are eventually absorbed within the surface or atmosphere or escape into space. The process of slowing down via repeated collisions is known as "moderation". There are three general interaction categories that are important in the context of planetary science: (1) nonelastic reactions, in which the incident neutron is absorbed, forming a compound nucleus, which decays by emitting one or more neutrons followed by the emission of gamma rays; (2) elastic scattering, a process that can be compared to billiard ball collisions for which kinetic energy is conserved; (3) neutron radiative capture, in which the neutron is absorbed and gamma rays are emitted.

The probability that a neutron will interact can be expressed in terms of the **microscopic cross-section** of the target nucleus, denoted σ, which is an effective, cross-sectional area with units of barns (1 barn = 10^{-24} cm²). The magnitude of the microscopic cross-section depends on the reaction type, target nucleus, and the energy (E) of the incident neutron. Microscopic cross-sections for natural Fe are shown, for example, in Figure 54.3, for radiative capture, elastic scattering, and inelastic scattering. Inelastic scattering occurs above a threshold determined by the energy required to produce the first excited state of the compound nucleus. The elastic scattering cross-section is constant over a wide range of energies. The cross-section for radiative capture usually varies as $E^{-1/2}$. Consequently, radiative capture is important at low energies. The sharp peaks that appear at high energy (greater than 100 eV) are resonances associated with the nuclear structure of the Fe isotopes. Neutron inelastic scattering is an important energy loss mechanism at high energies (greater than about 0.5 MeV for most isotopes of interest to planetary science).

Under the steady bombardment of cosmic rays, the population of neutrons slowing down in the regolith is, on

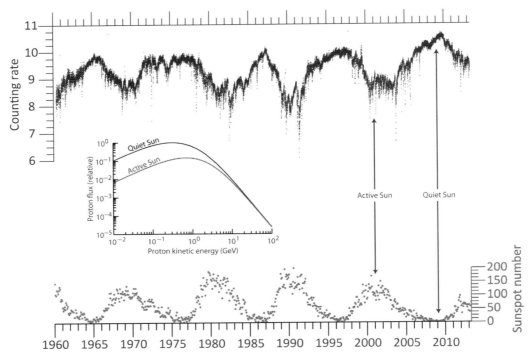

FIGURE 54.2 The variation of neutron counting rates (with units of 10^5 counts per hour) measured at McMurdo Station in Antarctica as a function of time (neutron monitors of the Bartol Research Institute are supported by NSF grant ATM-0000315). Monthly sunspot counts are shown for comparison *(courtesy of SIDC, RWC Belgium, World Data Center for the Sunspot Index, Royal Observatory of Belgium, 1960–2013)*. During periods of low solar activity (low sunspot counts), low-energy galactic cosmic rays penetrate the heliosphere, which results in relatively high neutron production rates. During periods of high solar activity (high sunspot counts), the low-energy galactic cosmic rays are cut off, resulting in lower neutron counting rates. The variation in neutron counting rates is about 20% over the solar cycle. Theoretical galactic proton energy spectra within the heliosphere, representative of quiet and active solar years, are shown (inset).

average, constant with time. The steady state neutron energy, angle, and spatial distributions depend on the composition and structure of the surface and atmosphere. An important property of the neutron population is the scalar flux (φ), which varies with depth beneath the surface and is given by the product of the speed of the neutrons, v (centimeters per second), and the number density of neutrons slowing down in the planetary surface (n neutrons per cubic centimeter); $\varphi = nv$, with units of neutrons per square centimeter per second. The rate at which neutrons interact with nuclei is given by the product of the flux of neutrons, the density of the target nuclei (N nuclei per cubic centimeter), and the microscopic cross-section for the nuclear reaction of interest: $R = \varphi N \sigma$ (interactions per cubic centimeter per second).

Cosmic ray showers can be modeled using Monte Carlo methods, in which the random processes of particle production and transport are simulated. The number of times something interesting happens, such as a particle crossing a surface, is tallied. Statistical averages of these interesting events are used to determine different aspects of the particle population such as fluxes and **currents**. Monte Carlo transport simulations generally provide for the following: a description of the cosmic ray source and the target medium (including geometry, composition, and density of the planetary body); detailed physical models of interaction mechanisms and transport processes (including tabulated data for interaction cross-sections); and a system of tallies.

The general purpose code Monte Carlo N-Particle eXtended (MCNPX) developed by Los Alamos National Laboratory provides a detailed model of cosmic ray showers, including the intranuclear cascade and subsequent interactions of particles within the surface and atmosphere.

For example, a model of the Martian surface used to calculate neutron leakage spectra is shown in Figure 54.4(a). The model includes several layers, representative of the high latitude surface, which is seasonally covered by CO_2 ice due to condensation of atmospheric CO_2 in the polar night, and whose frost-free surface consists of a dry lag deposit covering ice-rich soil. The curvature of Mars was included in the MCNPX calculations along with details of the incident galactic proton energy distribution. The goal was to determine the effect of surface parameters on neutron output, including the **column abundance** (grams per square centimeter) of the layers, their water abundance, and major element composition. The variation of the density of the atmosphere with altitude (the scale height is roughly 11 km) was also modeled.

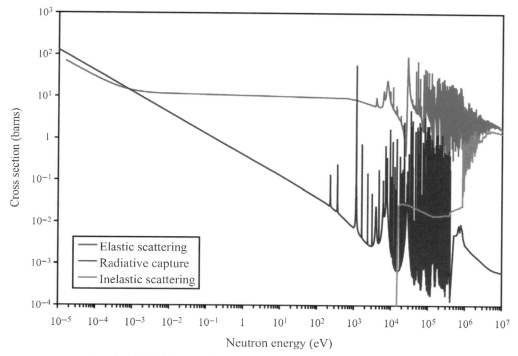

FIGURE 54.3 Neutron **microscopic cross-sections** for natural Fe (see text).

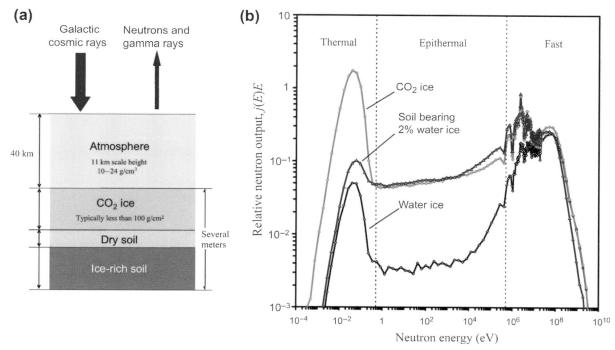

FIGURE 54.4 (a) Model of the Martian surface at high latitudes; (b) the current of neutrons leaking away from Mars for three different solid surface compositions. Neutron energy ranges are indicated. (See text for details.)

The population of neutrons escaping the solid surface or atmosphere can be represented as a current, J, which is the ratio of the number of neutrons escaping into space per galactic cosmic ray incident on the planet. The energy distribution of leakage neutrons is given by the current density $j(E)$, which is the number of escaping particles per unit energy per incident cosmic ray, such that $J = \int_0^\infty j(E)dE$. The current density of neutrons leaking away from Mars was calculated by MCNPX for homogeneous solid surfaces consisting of water ice, which is representative of the north polar residual cap; relatively dry soil bearing 2% water ice, which is representative of dry equatorial regions; and CO_2 ice, which is representative of the seasonal polar caps. The relative neutron output, given by the product of the current density and neutron energy is shown in Figure 54.4(b) for each of these materials. Integrating the current density over all energies gives 5, 3, and 1 for the total number of neutrons escaping the surface per incident cosmic ray proton for the CO_2 ice, dry soil, and water ice surfaces, respectively.

The neutron current density spans 14 decades of energy and can be divided into three broad ranges (Figure 54.4(b)), representing different physical processes: (1) thermal neutrons, which have undergone many collisions, have energies less than about 0.5 eV, and are approaching thermal equilibrium with the surface; (2) epithermal neutrons, which have energies greater than about 0.5 eV and are in the process of slowing down from higher energies; and (3) fast neutrons, with energies greater than the threshold for inelastic scattering (about 0.5 MeV). Absorption and leakage result in a hardened energy distribution for the thermal portion of the spectrum. The most probable neutron energy in the thermal range is slightly higher than would be predicted if the neutrons were in thermal equilibrium with the regolith.

Elastic scattering is the most important loss mechanism for planetary neutron spectroscopy because it provides strong differentiation between H and other more massive nuclei. For elastic scattering, the energy loss per collision varies systematically with atomic mass. The maximum energy that a neutron can lose in a collision is given by fE where $f = 1 - [(A - 1)/(A + 1)]^2$, E is the energy of the neutron before the collision, and A is the atomic mass of the target nucleus. Thus, a neutron could lose all its energy in a single collision with hydrogen ($A = 1$), which has roughly the same mass as a neutron. This fact is readily verified by observing head-on collisions in a game of billiards. In contrast, the maximum energy loss in a collision with C, which is the next most massive nucleus of interest in planetary science, is 28%. For Fe, the maximum energy loss per collision is 7%. The mean energy loss per collision follows a similar trend. Consequently, for materials that are rich in H, such as water ice, energy loss by elastic collisions is high and neutrons slow down more quickly than for materials that do not contain H.

For H-rich materials, the population of neutrons that are slowing down is strongly suppressed relative to materials without H. For example, the epithermal current density for the simulated water ice surface in Figure 54.4(b) is considerably lower than either the soil or CO_2 surfaces. The current density of fast neutrons, which have undergone relatively few collisions following their production, are influenced not only by elastic scattering, but also by variations in neutron production, which depend on the average atomic mass of the medium.

Absorption of neutrons by radiative capture significantly influences the population of thermal neutrons. Elements such as H, Cl, Fe, and Ti have relatively high absorption cross-sections and can suppress the thermal neutron flux. C and O have very low absorption cross-sections compared to H. Consequently, the thermal neutron output for the water ice in Figure 54.4(b) is suppressed relative to the surfaces containing CO_2 ice and soil.

2.3. Gamma Ray Production and Transport

For galactic cosmic ray interactions, gamma rays are produced by neutron inelastic scattering and radiative capture. Deexcitation of residual nuclei produced by these reactions results in the emission of gamma rays with discrete energies. The energies and intensities of the gamma rays provide a characteristic fingerprint that can be used to identify the residual nucleus. Since, in most cases, a residual nucleus can only be made by a reaction with a specific target isotope, gamma rays provide direct information about the elemental composition of the surface.

For example, neutron inelastic scattering with ^{56}Fe frequently leaves the residual ^{56}Fe nucleus in its first excited state, which transitions promptly to ground state by the emission of an 847 keV gamma ray. The presence of a peak at 847 keV in a planetary gamma ray spectrum indicates that the surface contains Fe. The intensity of the peak is related directly to the abundance of elemental Fe in the surface.

Gamma rays produced by the decay of short-lived neutron activation products and long-lived (primordial) radioisotopes also provide useful information about elemental abundance. Radioactive elements such as K, Th, and U can be detected when present in trace quantities. Most notably, the Th decay chain produces a prominent gamma ray at 2.6 MeV, which can be measured when Th is present in the surface at low levels (>1 ppm).

To illustrate a typical gamma ray leakage spectrum, a Monte Carlo simulation of the lunar gamma ray leakage current induced by galactic cosmic ray protons is shown in Figure 54.5. The composition of the surface was assumed to be the mean soil composition from the Apollo 11 landing

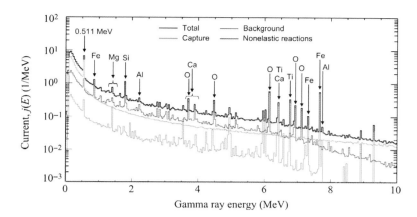

FIGURE 54.5 Simulated current of gamma rays leaking away from the Moon for a composition representative of the Apollo 11 landing site. Gamma rays from radioactive decay are not included in the simulation.

site. Contributions from nonelastic reactions and capture are plotted separately. A background continuum produced by various processes including pion decay is also shown. The peaks correspond to gamma rays that escape into space without interacting with the surface material. The peaks are superimposed on a continuum, which results from the scattering of gamma rays in the surface. The total number of gamma rays escaping the surface per incident cosmic ray proton was 2.7, which is within the range of values for the number of neutrons escaping the Martian surface, presented in Section 2.1.

Gamma ray peaks associated with neutron interactions with major elements are labeled with the target element in Figure 54.5. The intensity (or area) of each peak is proportional to the product of the abundance of the target element and the number density of neutrons slowing down within the solid surface or atmosphere. Specifically, the measured intensity (I) of a gamma ray peak with energy E for a selected reaction can be modeled as the product of three terms: $I \alpha g y R$, where g accounts for attenuation of gamma rays by intervening surface materials and the variation of detection efficiency with gamma ray energy; y is the number of gamma rays of energy E produced per reaction; and $R = \varphi N \sigma$ is the **reaction rate**, the product of the neutron flux, cross-section, and atom density of the target element.

Because gamma rays are produced by neutron interactions, the absolute atom density or, equivalently, the weight fraction of the target element cannot be determined unless the neutron flux is known. Thus, neutron spectroscopy plays an important role in the analysis of gamma ray data. Relative abundances can be determined without knowledge of the magnitude of the neutron flux. For example, the ratio of Fe to Si abundances can be determined from the ratio of the intensities of the prominent Fe doublet (at 7.65 and 7.63 MeV) and the Si gamma ray at 4.93 MeV. Because attenuation of gamma rays by surface materials depends on gamma ray energy and the distribution of gamma ray production with depth, models of the depth profile of the neutron flux are needed in order to analyze gamma ray data.

For homogeneous surfaces, accurate results can be obtained for absolute and relative abundances; however, layered surfaces present a difficult challenge for analyzing nuclear spectroscopy data. Compositional layering of major elements on a submeter scale is widespread on Mars as shown, for example, by the Spirit and Opportunity rovers and the Mars Science Laboratory (MSL) (*see* Mars: Landing Site Geology, Minerology, and Geochemistry). In some cases, geophysical assumptions can be made that simplify the analysis and allow quantitative results to be obtained; however, it is often the case that insufficient information is available. In these cases, it is sometimes possible to establish bounds on composition that are useful for geochemical analysis. Development of accurate algorithms for determining elemental abundances, absolute or relative, requires careful synthesis of nuclear physics with constraints from geology, geophysics, and geochemistry.

3. DETECTION OF GAMMA RAYS AND NEUTRONS

In this section, a simple model of the counting rate observed by orbiting neutron and gamma ray spectrometers (GRSs) is presented along with an overview of radiation detection concepts for planetary science applications.

3.1. Counting Rate Models

The flux of radiation reaching an orbiting spectrometer varies in proportion to the solid angle subtended by the planet at the detector, which depends on orbital altitude. The fractional solid angle of a spherical body is given by

$$\Omega(h) = 1 - \sqrt{1 - R^2/(R+h)^2}, \quad (54.1)$$

where h is the orbital altitude and R is the radius. The fractional solid angle varies from 1 at the surface (for $h = 0$) to 0 far away from the planet. For galactic cosmic ray interactions, the flux of gamma rays or neutrons at the orbiting spectrometer is approximately

$$\phi(h) = 1/2 \Phi J \Omega(h), \qquad (54.2)$$

where Φ is the flux of galactic cosmic ray protons far from the planet (about 4 protons/cm^2/s, depending on the solar cycle and location within the heliosphere), and J is the leakage current. Because alpha particles and heavier nuclei of galactic origin contribute to neutron and gamma ray production, Eqn (54.2) must be multiplied by a factor, approximately 1.4, in order to estimate the total leakage flux.

Equation (54.2) can be used, for example, to calculate the flux of neutrons incident on the *Mars Odyssey* neutron spectrometer (NS). The orbital altitude for *Mars Odyssey* is 400 km, and the volumetric mean radius of Mars is 3390 km. The fractional solid angle, given by Eqn (54.1), is 0.55. The total leakage current for a surface consisting of thick CO_2 ice, representative of the polar seasonal caps during winter, was $J = 5$ (from Section 2.2). Consequently, from Eqn (54.2), the total flux of neutrons at Odyssey's orbit from thick CO_2 deposits is approximately 8 neutrons/cm^2/s. For a surface that is 100% water, which is representative of the north polar residual cap, J was 1, and the total flux at orbital altitude is expected to be about 2 neutrons/cm^2/s.

Radiation detectors, such as the gamma ray and neutron spectrometers on *Mars Odyssey*, count particle interactions and bin them into energy or pulse height spectra, for example, with units of counts per second per unit energy. For both gamma rays and neutrons, the net counting rate (with units of counts per second) for selected peaks in the spectrum is needed in order to determine elemental abundances.

The flux of particles (gamma rays or neutrons) incident on a spectrometer can be converted to counting rate (C), given the intrinsic efficiency (ε) and projected (cross-sectional) area (A) of the spectrometer in the direction of the incident particles:

$$C = \varphi(h) \varepsilon A. \qquad (54.3)$$

The intrinsic efficiency is the probability that an incident particle will interact with the spectrometer to produce an event that is counted. Because particles can pass through the spectrometer without interacting, the intrinsic efficiency is always less than or equal to 1. For example, $\varepsilon \cdot A$ is on the order of 10 cm^2 for the *Mars Odyssey* epithermal neutron detector, which has a maximum projected area (A) of about 100 cm^2. The efficiency-area product ($\varepsilon \cdot A$) varies with the energy and angle of incidence of the particles. So, the value for $\varepsilon \cdot A$ used in Eqn (54.3) must be appropriately averaged over neutron energy and direction.

One of the main sources of uncertainty in measured counting rates is statistical fluctuations due to the random nature of the production, transport, and detection of radiation. While a detailed discussion of error propagation is beyond the scope of this article, the most important result is given here. The statistical uncertainty (precision) in the counting rate is given by $\sigma = \sqrt{C/t}$ where t is the measurement time and C is the mean counting rate. For example, to achieve a precision of 1% ($\sigma/C = 0.01$) when $C = 10$ counts/s, which is typical of the epithermal and thermal counting rates measured by the *Mars Odyssey* NS, a counting time of 1000 s is required. Longer counting times are needed when background contributions are subtracted, for example, to determine counting rates for peaks in gamma ray and neutron spectra. Uncertainties in the counting rate due to random fluctuations propagate to the uncertainties in elemental abundance and other parameters determined in the analysis of spectroscopy data. Long counting times are desired to minimize statistical contributions. Improved precision can be achieved by increasing the counting rate, which can be accomplished through instrument design, by maximizing the efficiency-area product and/or by making measurements at low altitude.

3.2. Gamma Ray and Neutron Detection

Radiation spectrometers measure ionization produced by the interaction of particles within the sensitive volume of a detector. Gamma ray interactions produce swift primary electrons that cause ionization as they slow down in the sensitive volume. Neutrons undergo reactions that produce energetic ions and gamma rays. The recoil proton from neutron elastic scattering with hydrogen can produce measurable ionization. The charge liberated by these interactions can be measured using a wide variety of techniques, two of which are illustrated here.

Semiconductor radiation detectors typically consist of a semiconductor dielectric material sandwiched between two electrodes. An electric field is established in the dielectric by applying high voltage across the electrodes. Gamma ray interactions produce free electron–hole pairs, which drift in opposite directions in the electric field. As they drift, they induce charge on the electrodes, which is measured using a charge-sensitive preamplifier. The amplitude of the charge pulse, or pulse height, is proportional to the energy deposited by the gamma ray. Consequently, a histogram of pulse heights, known as a pulse height spectrum, provides information about the energy distribution of the incident gamma rays.

For example, a diagram of a high-purity germanium (HPGe) detector is shown in Figure 54.6(a) along with a photograph of an HPGe crystal in Figure 54.6(b). The closed-end coaxial geometry is designed to minimize

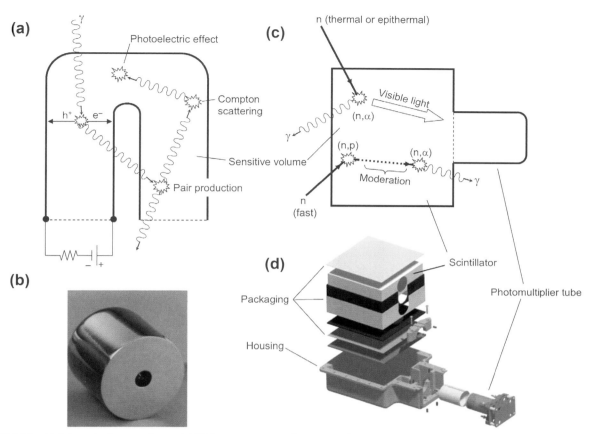

FIGURE 54.6 (a) Schematic diagram of a coaxial HPGe spectrometer and gamma ray interactions; (b) photograph of a HPGe crystal; (c) diagram of a scintillation-based spectrometer with neutron interactions; and (d) assembly diagram for a boron-loaded plastic scintillator for Dawn's Gamma Ray and Neutron Detector, including the housing and packaging designed to withstand the vibrational environment during launch. *Part (b) courtesy of AMETEK, Advanced Measurement Technology, Inc., ORTEC Product Line, 801 South Illinois Avenue, Oak Ridge, TN 37830.*

trapping of carriers as they drift to the electrodes. To minimize noise due to leakage current, the HPGe must be operated at cryogenic temperatures. The requirement for cooling adds to the mass and complexity of the design for space applications.

A hypothetical gamma ray interaction is superimposed on the diagram in Figure 54.6(a). Gamma rays undergo three types of interactions: pair production, Compton scattering, and photoelectric effect. High-energy gamma rays (greater than 1.022 MeV) can undergo pair production, in which the gamma ray disappears and an electron–positron pair is produced. The kinetic energy of the electron and positron is absorbed by the medium. When the positron is annihilated by an electron, two back-to-back (511 keV) gamma rays are produced, which can undergo additional interactions. In Compton scattering, a portion of the energy of the gamma ray is transferred to an electron. The energy lost by the gamma ray depends on the scattering angle. At low energies, the gamma ray can be absorbed by an electron via the photoelectric effect. All these interactions vary strongly with the atomic number (Z) and density of the detector material. High Z, high density, and a large sensitive volume are desired to maximize the probability that all the energy of the incident gamma ray is absorbed in the detector.

A pulse height spectrum for a large volume (slightly larger than the crystal flown on *Mars Odyssey*), coaxial HPGe detector is shown in Figure 54.7. The gamma rays were produced by moderated neutrons, with an energy distribution similar to the lunar leakage spectrum, incident on an Fe slab. Well-defined peaks corresponding to neutron capture and inelastic scattering with Fe appear in the spectrum. For example, the doublet labeled Fe(1) corresponds to gamma rays (7646 and 7631 keV) produced by neutron capture with Fe. The peaks labeled Fe(2) are shifted 511 keV lower in energy and correspond to the escape of one of the 511 keV gamma rays produced by pair production in the spectrometer. The continuum that underlies the peaks is caused by external Compton scattering and the escape of gamma rays that scattered in the spectrometer. Gamma rays from neutron capture with H and the radioactive decay of K and Th are also visible.

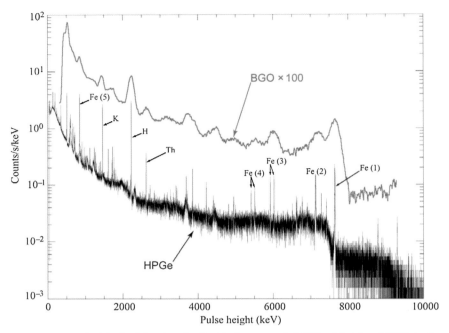

FIGURE 54.7 Gamma ray spectra acquired in the laboratory by liquid nitrogen-cooled HPGe (black) and BGO (red) spectrometers. The BGO spectrometer is part of Dawn's Gamma Ray and Neutron Detector (GRaND). To improve visualization, the spectrum for BGO has been multiplied by 100. The source was moderated neutrons, with energy distribution similar to a planetary leakage spectrum, incident on an iron slab. Gamma rays from natural radioactivity in the environment are also visible (from K at 1461 keV and Th at 2615 keV). A gamma ray at 2223 keV from neutron capture with H (from polyethylene in the moderator) is a prominent feature in the HPGe and BGO spectra. Major gamma rays from neutron interactions with Fe that are resolved by the HPGe spectrometer are labeled: (1) 7646 and 7631 keV doublet from neutron capture; (2) their single escape peaks; (3) 6019 and 5921 keV gamma rays from neutron capture; (4) their single escape peaks; and (5) 846.7 keV gamma ray from neutron inelastic scattering. *HPGe spectrum courtesy of S. Garner, J. Shergur, and D. Mercer of Los Alamos National Laboratory.*

Scintillators provide an alternative method of detecting ionizing radiation, which can be used for gamma ray and neutron spectroscopy. Scintillators consist of a transparent material that emits detectable light when ionized. The light is measured by a photomultiplier tube or photodiode, which is optically coupled to the scintillator. The amount of light produced and the amplitude of the corresponding charge pulse from the photomultiplier tube and pulse processing circuit is proportional to the energy deposited by the radiation interaction.

A diagram of a boron-loaded, plastic scintillation detector is shown in Figure 54.6(c) along with an assembly diagram of flight sensor (Figure 54.6(d)). Thermal and epithermal neutrons are detected by the ^{10}B(n, $\alpha\gamma$) ^{7}Li reaction (see glossary for an explanation of **reaction notation**). The recoiling reaction products (alpha particle and ^{7}Li ion) produce ionization equivalent to a 93 keV electron, which makes a well-defined peak in the pulse height spectrum. The area of the peak depends on the flux of incident thermal and epithermal neutrons. Thermal neutrons can be filtered out by wrapping the scintillator in a Cd foil, which strongly absorbs neutrons with energies below about 0.5 eV. Thus, the combination of a bare and Cd-covered scintillator can be used to separately measure contributions from thermal and epithermal neutrons. Above about 500 keV, light is produced by recoiling protons from neutron elastic scattering with hydrogen in the scintillator. Fast neutrons (greater than about 500 keV) can be detected by a prompt pulse from proton recoils followed a short time later by a second pulse, corresponding to neutron capture of the moderated neutron by ^{10}B. This characteristic, double-pulse time signature can be used to identify, and separately measure, fast neutron events.

Scintillators are also used routinely for gamma ray spectroscopy. For example, a pulse height spectrum acquired by a bismuth germanate (BGO) scintillator is shown in Figure 54.7. The source was exactly the same as measured by the HPGe spectrometer, and the two spectra share similar peak features. Note, however, that the peaks measured by BGO are considerably broader than those measured by HPGe. For BGO, the width of the peaks is controlled by statistical variations in the number of scintillation photons produced by full-energy gamma ray interactions. Similar dispersion occurs for charge carriers (electrons and holes) produced in the HPGe crystal; however, the effect is far less pronounced. Pulse height resolution is determined by the full width at half maximum (FWHM) of the gamma ray peaks, which is often quoted for a standard gamma ray (e.g. the 1332 keV gamma ray from ^{60}Co). Spectrometers that produce narrower peaks have higher resolution and, in principle, contain more detailed information about the incident gamma ray spectrum and backgrounds.

The ability of the HPGe technology to resolve individual peaks is coveted by the planetary spectroscopist; however, the added cost and complexity of HPGe relative to scintillation technology has made scintillators competitive for some missions. HPGe has been successfully flown on three planetary missions (Table 54.1). A comparison between whole Moon spectra acquired by Lunar Prospector (BGO), Chang'E-2 ($LaBr_3$), and Kaguya (HPGe) is shown in Figure 54.8.

Other technologies that have been flown for gamma ray and neutron detection include ^{3}He ionization chambers (for thermal and epithermal neutron detection on Lunar Prospector) and various scintillators, including Tl-doped NaI on NEAR and Apollo, Tl-doped CsI on Phobos and Chang'E-1 and -2, and Ce-doped $LaBr_3$ on Chang'E-2. Lanthanum halides such as $LaBr_3$ are very bright scintillators and thus have excellent pulse height resolution; however, the presence of radioactive contaminants within these scintillators (e.g. ^{138}La and Th decay products) produce backgrounds that interfere with measurement of radioelements in planetary regoliths. The Dawn mission deployed a large-volume BGO scintillator as the primary GRS and demonstrated a new compound semiconductor technology (CdZnTe). CdZnTe-based spectrometers have significantly better pulse height resolution compared to BGO and, in contrast to HPGe, can be operated at ambient temperatures.

3.3. Spatial Resolution

The spatial resolution that can be achieved by a spectrometer depends on the angular distribution of radiation emitted from the surface, the angular response of the spectrometer, and the altitude of the orbit. The angular response of most spectrometers is roughly isotropic or weakly dependent on incident direction. Consequently, the spectrometer is sensitive to radiation emitted from locations from underneath the spectrometer all the way out to the limb. Due to their increased area, off-nadir regions contribute more to the counting rate than regions directly beneath the spacecraft.

When the spectrometer passes over a point source of radiation on the surface, the counting rate as a function of distance along the orbital path has an approximately Gaussian shape, with the peak occurring when the spacecraft passes over the source. Consequently, the ability of the spectrometer to resolve spatial regions with different compositions depends on the FWHM of the Gaussian, which as a general rule is approximately 1.5 times the orbital altitude. For example, the lowest orbital altitude of *Lunar Prospector* was 30 km for which the FWHM was 45 km or 1.5° of arc length. For *Mars Odyssey*, the orbital altitude was 400 km, and the FWHM was approximately 600 km or 10° of arc length.

The broad spatial response of gamma ray and neutron spectrometers must be considered in the analysis and interpretation of data, especially where comparisons to high resolution data (for example, from optical spectroscopy) are concerned. It may be possible to increase the resolution of a spectrometer by the addition of a collimator, which would add mass to the instrument and also reduce the precision of the measurements. Alternatively, spatial deconvolution and instrument modeling techniques can sometimes be employed to study regions that are smaller in scale than the spatial resolution of the spectrometer.

4. MISSIONS

Since the dawn of space flight, nuclear spectroscopy has been used for a wide variety of applications, from astrophysics to solar astronomy. Planetary science missions with gamma ray and/or neutron spectrometers on the payload are listed in Table 54.1. While nuclear spectroscopy was used on earlier missions to the Moon, Mars, and the surface of Venus, the first major success was the Apollo Gamma Ray Experiment, which flew on the Apollo 15 and 16 missions, providing global context for lunar samples. Phobos II traveled to Mars and provided a glimpse of the regional composition of the western hemisphere, which includes Tharsis and Valles Marineris. Due to the small size of Eros and high orbital altitudes, the GRS on NEAR provided little useful information about Eros until the NEAR landed on the asteroid. Once on the surface, the NEAR GRS acquired data with sufficient precision to determine the abundance of O, Mg, Si, Fe, and K. NEAR also had an X-ray spectrometer that provided complementary information about surface elemental composition. The first intended use of neutron spectroscopy for global mapping was on Mars Observer, which was lost before reaching Mars.

Lunar Prospector was the first mission to combine gamma ray and neutron spectroscopy to provide accurate, high-precision global composition maps of a planetary body. Missions that followed, including 2001 Mars Odyssey, MESSENGER, and Dawn also included neutron and GRSs on their payloads. The results of Lunar Prospector and Mars Odyssey are detailed in Section 5.

Recent missions to the Moon, including the Japanese Kaguya (also known as the SELenological and ENgineering Explorer (SELENE)) mission and the Chinese Chang'E missions used gamma ray spectroscopy to confirm many of the initial findings of Lunar Prospector regarding the geochemistry of the Moon. The SELENE mission, with its high-resolution HPGe spectrometer, directly measured the concentration of U and the U/Th ratio, supplementing measurements of K and Th by Lunar Prospector. A NS

TABLE 54.1 Summary of Planetary Science and Exploration Missions with Gamma Ray and/or Neutron Spectrometers. Missions prior to Apollo, including Luna and Ranger, are not listed (for information about these missions, see Bibliography).

Mission	Country/Program	Launch Date(s)	Status	Planet or Minor Body	Orbit	Mapping Duration[1]	Gamma Ray Spectrometer	Neutron Spectrometer	Results and/or Objectives[2]
Apollo 15 and 16	US	July 26, 1971 April 16, 1972	Completed	Moon	Equatorial orbit covering 20% of the lunar surface	10.5 days (Apollo 15 and 16 combined)	NaI(Tl) with plastic anticoincidence shield	None	Maps of major and radioactive elements, including Fe, Ti, and Th
Venera 8	USSR	March 27, 1972	Completed	Venus	Not applicable (N/A); descent module/lander	42 min of data acquisition on the surface	NaI(Tl)	None	Abundances of K, Th, and U; found K/Th ratio similar to that of Earth rocks
Phobos II[3]	USSR	July 12, 1988	Lost during Phobos encounter	Mars and Phobos	Elliptical, equatorial orbit, 900 km periapsis, 80,000 km apoapsis	2 orbits analyzed	CsI(Tl)	None	Abundances of O, Si, Fe, K and Th in two equatorial regions in the western hemisphere
Mars Observer	US, NASA Mars Exploration Program	September 25, 1992	Lost prior to orbital insertion	Mars	400 km altitude circular polar mapping orbit[2]	1 Mars year[2]	HPGe, passively cooled	Boron-loaded plastic scintillators	Global maps of major elements and water-equivalent hydrogen (objectives not achieved)
Near Earth Asteroid Rendezvous (NEAR)	US, NASA Discovery Program	February 17, 1996	Completed	Eros	Useful data acquired following successful landing on Eros	7 days on the surface	NaI(Tl) with BGO anticoincidence shield	None	Abundances of O, Mg, Si, Fe, and K
Lunar Prospector	US, NASA Discovery Program	January 6, 1998	Completed mission by planned impact in a south polar crater	Moon	High- and low-altitude circular polar mapping orbits (100 km and 30 km, respectively)	300 days at high altitude; 220 days at low altitude	BGO with boron-loaded plastic anticoincidence shield	^{3}He gas proportional counters and boron-loaded plastic scintillator	Discovery of enhanced water-equivalent hydrogen associated with polar cold traps; global maps of major and radioactive elements
2001 Mars Odyssey	US, NASA Mars Exploration Program	April 7, 2001	Completed primary mission; ongoing extended mission	Mars	400 km altitude circular polar mapping orbit	Over 6 Mars years completed; gamma ray spectra available through September of 2009	HPGe, passively cooled	Boron-loaded plastic scintillators (NS); Stilbene and ^{3}He tubes (HEND[4])	Distribution and layering of water-equivalent hydrogen; seasonal variations in CO_2 ice and noncondensable gasses; and global maps of major and radioactive elements
MESSENGER	US, NASA Discovery Program	August 3, 2004	Primary mission completed; extended mission underway	Mercury	Elliptical polar mapping orbit with periapsis at 400 km altitude at >60° N latitude, 15,000 km apoapsis	1 year primary mission began in March of 2011	HPGe, actively cooled with boron-loaded plastic anticoincidence shield	^{6}Li-loaded glass and boron-loaded plastic scintillator	Mapped elemental composition of the northern hemisphere; detected large amounts of hydrogen, consistent with deposits of ice in permanently shadowed polar craters

Chapter | 54 Remote Sensing of Chemical Elements Using Nuclear Spectroscopy

Mission	Agency	Launch Date	Status	Target	Orbit	Duration	Gamma-Ray Spectrometer	Neutron Spectrometer	Key Results/Objectives[2]
Dawn	US, NASA Discovery Program	September 27, 2007	Completed Vesta encounter; en route to Ceres	Main-belt bodies 4 Vesta and 1 Ceres	Survey, high-, and low-altitude circular polar mapping orbits	5 months data acquisition at low altitude (210 km) around Vesta; Ceres planning underway	CdZnTe and BGO	^{6}Li-loaded glass and boron-loaded plastic scintillators	Global elemental ratios confirm Vesta as the HED parent body; discovered extensive hydrogen deposits on Vesta, consistent with exogenic delivery of hydrated minerals by carbonaceous chondrites; search for water ice and evidence for aqueous alteration at Ceres
SELENE (Kaguya)	Japan	September 14, 2007	Completed mission by planned impact into lunar surface (June 11, 2009)	Moon	High- and low-altitude polar mapping orbits (100 km and 30 km × 50 km, respectively).	1.5 years duration (70 days accumulation of data at low altitude)	Actively cooled HPGe with BGO anticoincidence shield	None	Global maps of U, Th, K, and Ca. Maps of Fe, Ti, Si, Mg, Al, and H in preparation
Lunar Reconnaissance Orbiter (LRO)	US, NASA Lunar Exploration Program	June 18, 2009	Primary mission completed; extended mission underway	Moon	~50 km circular polar mapping orbit	1 year primary mission	None	Eight ^{3}He tubes, four with passive shielding intended for collimation of epithermal neutrons (LEND[4])	Confirmed the presence of enhanced hydrogen at the lunar poles; efficacy of collimation debated
Chang'E-1 and -2	China National Space Administration	October 24, 2007 and October 1, 2010	Completed	Moon	200 and 100 km circular polar mapping orbits for Chang'E-1 and -2, respectively	1 year, ~85 days accumulation time (Chang'E-1); 178 days accumulation (Chang'E-2)	CsI(Tl) (Chang'E-1) and LaBr$_3$ (Chang'E-2), both with CsI(Tl) anticoincidence shields	None	Distribution of radioelements, confirming results of Lunar Prospector and SELENE
Mars Science Laboratory (MSL)	US, NASA Mars Exploration Program	November 26, 2011	Deployed within Gale crater; first drive August 22, 2012; Data acquired within the 400 m traverse completed in the first 90 sols	Mars	N/A	1 Mars year (5 to 20 km traverse)	None	^{3}He tubes with active neutron interrogation with a pulsed, 14 MeV, D-T neutron generator (DAN[4])	Preliminary results presented for WEH abundance, H-layering, and the abundance of neutron-absorbing elements
BepiColombo	European Space Agency	2015	Planning for launch	Mercury	Elliptical, polar mapping orbit (400 × 1500 km)	1 year	LaBr$_3$ scintillator[4]	Design based on HEND[4]	Global maps of elemental abundances

[1] Refers to the time periods during which gamma ray and/or neutron data were acquired.
[2] Objectives are listed for Mars Observer, Dawn at Ceres, and BepiColumbo.
[3] Neutron and gamma ray spectrometers were flown on Phobos 1, which was launched on 7 July, 1988; however, Phobos 1 was lost during the cruise phase of the mission. The Mars 4 and 5 missions (USSR, 1973) flew identical sodium iodide gamma ray spectrometers. A few gamma ray spectra were acquired by Mars 5 while in an elliptical orbit around Mars (apoapsis 32,560 km, periapsis 1760 km, and inclination 35° to the equator).
[4] The High Energy Neutron Detector (HEND) on Odyssey, the Lunar Exploration Neutron Detector (LEND) on LRO, the Dynamic Albedo of Neutron (DAN) instrument on MSL, and the gamma and neutron spectrometers on BepiColumbo were provided by the Russian Federation.

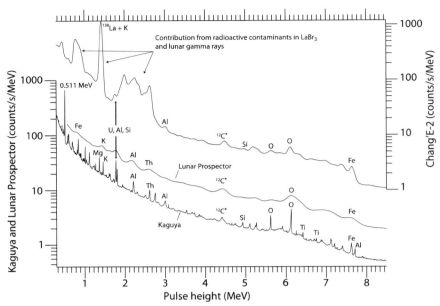

FIGURE 54.8 Gamma ray spectra acquired by Lunar Prospector (BGO), Kaguya (HPGe) and Chang'E-2 (LaBr$_3$) missions averaged over the whole Moon. The data were acquired in their respective low-altitude orbits (see Table 54.1). Prominent peaks, mostly of lunar origin, are labeled. For Chang'E-2, strong contributions from lunar K were determined by subtracting the background produced by radioactive contaminants within LaBr$_3$. The Kaguya spectrum contains some peaks (e.g. reactions with Al) that have strong background contributions from nearby structural materials. The relatively low mass of housing materials and boom deployment minimized background contributions for the Lunar Prospector Gamma Ray Spectrometer. The spectrum acquired by Kaguya has broader peaks than that of the spectrum shown in Figure 54.7, which was acquired in the laboratory. The mechanical (Stirling) cryocooler used in flight operations does not cool the HPGe crystal as effectively as liquid nitrogen and is also a source of microphonic noise. In addition, low-energy tailing due to radiation damage accrued during flight is evident. *Lunar Prospector data are from the NASA Planetary Data System; Kaguya data are courtesy of N. Hasabe, Research Institute for Science and Engineering, Waseda University, Japan; Chang'E-2 data are courtesy of M. Zhu, Space Science Institute, Macau University of Science and Technology, China.*

provided by the Russian Federation was flown on NASA's Lunar Reconnaisance Orbiter (LRO). LRO's Lunar Exploration Neutron Detector (LEND) included a massive collimator intended to provide high spatial-resolution measurements of hydrogen deposits within permanently shadowed craters. The efficacy of the collimator is the subject of controversy among scientists. Nevertheless, LEND's uncollimated neutron detectors confirmed the presence of enhanced hydrogen deposits at the lunar poles discovered by Lunar Prospector.

Data acquired by MESSENGER's GRS were analyzed to map the abundance of K, Th, U, Na, S, Ca, Al, and Fe on Mercury at high northern latitudes. Measurements of the K/Th ratio indicate that the material from which Mercury formed had similar volatile-to-refractory content as other terrestrial planets, in contrast to the Moon, which is volatile depleted. High S and Na contents show that the Mercury's surface was not depleted in volatiles, which implies that present models of Mercury's formation are inadequate. Despite Mercury's high bulk Fe content, its surface is Fe-poor, which remains a puzzle. Neutron spectroscopy revealed high concentrations of H near the north pole, consistent with thick deposits of water ice trapped within permanently shadowed polar craters.

The NASA Dawn mission successfully completed operations at the giant, main-belt asteroid 4 Vesta and is now in interplanetary cruise for an encounter with the dwarf planet Ceres in 2015. Dawn's Gamma Ray and Neutron Detector (GRaND) found global Fe/O and Fe/Si ratios consistent with that of the howardite, eucrite, and diogenite meteorites, providing chemical evidence that Vesta is the parent body of these achondrites. Neutron spectroscopy revealed relatively high abundances of H (at least 400 μg/g) in low-albedo regions near Vesta's equator where water-ice is not stable. The hydrogen observed by GRaND is thought to be exogenous, in the form of hydrated minerals emplaced within Vesta's otherwise basaltic regolith by the gradual infall of carbonaceous chondrites. High-velocity impacts into this volatile-rich material may have formed pitted terrain seen in relatively young craters on Vesta.

The Dynamic Albedo of Neutrons (DAN) experiment on the NASA MSL is the first rover-based NS. DAN features active neutron interrogation with a 14 MeV pulsed deuterium–tritium (D–T) neutron generator as well as the ability to acquire passive neutron counting data. With MSL operations underway, DAN is expected to provide data on the abundance and layering of **water-equivalent hydrogen (WEH)** as well as information about neutron-absorbing elements such as Cl and Fe as the rover traverses Gale crater.

Finally, the launch of BepiColumbo, a European-led mission to Mercury, is anticipated in 2015. BepiColumbo's proposed orbit should enable the acquisition of global compositional maps using nuclear spectroscopy.

5. SCIENCE

Lunar Prospector and *Mars Odyssey* acquired high-precision gamma ray and neutron data sets for the Moon and Mars, respectively. Highlights of the science carried out on these missions are presented along with a description of their instrumentation. The Moon and Mars are very different, both in their origin and composition. With the possible exception of polar water ice, the Moon is depleted in volatile elements and has no atmosphere. The lunar surface has been extensively modified by cratering and basaltic volcanism. Mars has a tenuous atmosphere, extensive water ice deposits, and seasonal CO_2 caps. Volcanic, aqueous, and eolian processes have continued to shape the surface of Mars long past the primordial formation of the crust. The differences between these two bodies will provide the reader with insights into the wealth of information provided by nuclear spectroscopy and the challenges faced in the analysis of the data. For *Lunar Prospector*, emphasis is placed on the combined analysis of neutron and gamma ray data to determine the abundance of major and trace radioactive elements. For *Mars Odyssey*, results from the NS for global water abundance and the seasonal caps are presented.

5.1. Lunar Prospector

Lunar Prospector was a spin-stabilized spacecraft, with the spin axis perpendicular to the plane of the ecliptic. The instruments were deployed on booms to minimize backgrounds from the spacecraft (Figure 54.9(a)). The payload included a large-volume BGO GRS, which was surrounded by a boron-loaded plastic anticoincidence shield (Figure 54.9(b)). The shield served two purposes: (1) to suppress the Compton continuum caused by gamma rays escaping the BGO crystal and to reject energetic particle events; and (2) to measure the spectrum of fast neutrons from the lunar surface using the double-pulse technique described in Section 3.2. Sn- and Cd-covered ^{3}He gas proportional counters were used to detect and separately measure thermal and epithermal neutrons. Gamma ray and neutron spectroscopy data were acquired for long periods of time (Table 54.1), providing full coverage of the Moon at 100 and 30 km altitude.

The data were analyzed to determine global maps of surface elemental composition. The resulting abundance data were mapped on different spatial scales, depending on the precision of the data and the altitude of the spectrometer. Results of the analysis were submitted to the NASA Planetary Data System and include the following data sets: the abundance of hydrogen from neutron spectroscopy

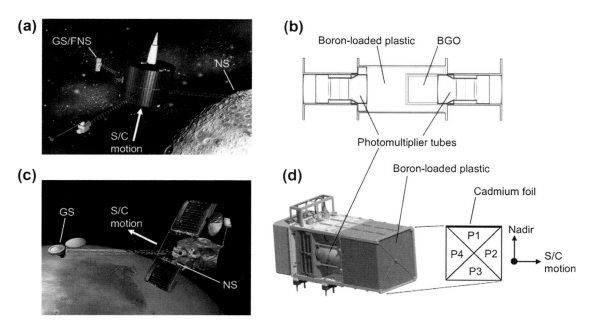

FIGURE 54.9 (a) Annotated artist's conception of Lunar Prospector; (b) Cross-sectional view of the gamma ray and fast neutron spectrometer (GS/FNS); (c) annotated artist's conception of 2001 Mars Odyssey; (d) engineering drawing of the neutron spectrometer on Odyssey cut away to show the boron-loaded plastic scintillators. A schematic diagram of the arrangement of scintillators and their orientation relative to spacecraft motion and nadir is also shown. *Parts (a) and (c) courtesy of NASA.*

(0.5° **equal angle map**); the average atomic mass from fast neutron spectroscopy (2° **equal-area maps**); the abundance of major oxides, including MgO, Al_2O_3, SiO_2, CaO, TiO_2, and FeO, and trace incompatible elements K, Th, and U (2°, 5°, and 20° equal-area maps) using a combination of gamma ray and neutron spectroscopy; and the abundance of the rare-earth elements (Gd + Sm) from neutron spectroscopy (0.5° equal angle map).

Perhaps the most significant result of *Lunar Prospector* was the discovery of enhanced hydrogen at the poles in association with craters in permanent shadow, which are thought to be cold traps for water ice. If present in ample quantities, water ice could be an important resource for manned exploration. Consequently, the polar cold traps are a prime target for future missions. The presence of cold-trapped volatiles within Cabeus crater near the lunar south pole was recently confirmed by the NASA Lunar CRater Observation and Sensing Satellite (LCROSS) mission, in which the ejecta plume generated by the deliberate impact of the launch vehicle's upper stage with the crater floor was analyzed. Data acquired by *Lunar Prospector* and the *Lunar Reconnaissance Orbiter* were used in the selection of the target crater. Geochemical results from the analysis of neutron and gamma ray spectra fully reveal the dichotomy in the composition of the Moon, with a near side that is enriched in incompatible elements and mafic minerals and a thick far-side crust primarily consisting of plagioclase feldspar.

Global geochemical trends observed by *Lunar Prospector* are not significantly different from trends observed in the sample and meteoritic data; however, there are some notable discrepancies that point to the existence of unique lithologies that are not well represented by the lunar sample data. Interpretation and analysis of the data is ongoing with emphasis on regional studies. For example, the impact that formed the South Pole Aitken (SPA) basin could have excavated into the mantle. Analysis of the composition of the basin floor may reveal information about the bulk composition of the mantle and lower crust.

The analysis of major and radioactive elements was carried out using a combination of gamma ray and neutron spectroscopy data. A typical gamma ray spectrum is shown in Figure 54.10 for a 20° equal-area pixel in the western mare. Two intense, well-resolved peaks, labeled in Figure 54.10, were analyzed to determine the abundance of Fe and Th. In addition, a spectral unmixing algorithm similar to those used to analyze spectral reflectance data was developed to simultaneously determine the abundance of all major and radioactive elements from the gamma ray spectrum.

Lunar gamma ray spectra can be modeled as a linear mixture of elemental spectral shapes. The magnitude of the spectral components must be adjusted to account for the nonlinear coupling of gamma ray production to the neutron number density (for neutron capture reactions) and the flux of fast neutrons (for nonelastic reactions). Once the adjustment is made, a linear least squares problem can be

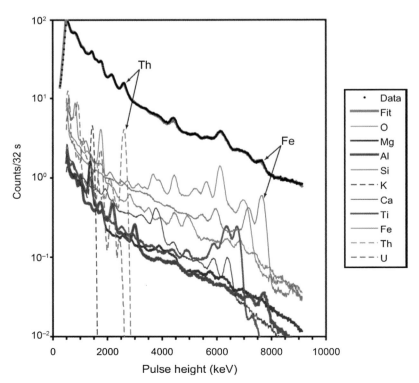

FIGURE 54.10 Lunar Prospector gamma ray spectrum for a 20° equal-area pixel in the western mare is compared to the fitted spectrum and elemental spectral components.

solved to determine elemental weight fractions. Fitted elemental spectral shapes are shown, for example, in Figure 54.10.

Abundance maps for selected elements determined by *Lunar Prospector* are shown in Figure 54.11. To provide context for the elemental abundance maps, a map of topography determined by *Clementine*, superimposed on a shaded relief image, is shown in Figure 54.11(a). The far side includes the feldspathic highlands and the SPA basin. The near side consists of major basins, including Procellarum and Imbrium, which contain mare basalts. The mare basalts are rich in Fe, with the highest concentrations occurring in western Procellarum (Figure 54.11(b)). The low abundance of FeO in the highlands, which are rich in plagioclase feldspar, reflects a significant lunar geochemical trend in which mafic silicate minerals are displaced by plagioclase, which is Fe-poor.

A large portion of the western near side is enriched in radioactive elements such as Th (Figure 54.11(c)). K, Th, and U are incompatible with major lunar minerals and were likely concentrated in the residual melt during lunar differentiation. Consequently, their distribution on the surface and with depth has important implications to lunar evolution. The association of high Th concentrations with the mare suggests that heating by radioactive elements may have significantly influenced lunar thermal evolution and mare volcanism.

The distribution of TiO_2 is shown in Figure 54.11(d) as a $5°$ equal-area map. The low spatial resolution of the TiO_2 map compared to FeO and Th is a consequence of the relatively low intensity of the Ti gamma rays and their position in the gamma ray spectrum near strong peaks from O and Fe (Figure 54.10). The abundance of TiO_2 can be used to classify mare basalts. Strong spatial variations in the abundance of TiO_2, for example, indicate that different source regions and processes were involved in creating the basalts that comprise the mare. The highest concentrations of TiO_2 are found in Tranquillitatis as shown in Figure 54.10(d); however, high concentrations are also found in western Procellarum. The abundances of Fe and Ti observed in western Procellarum suggest that this region has a unique composition that is not well represented by the lunar samples.

5.2. Mars Odyssey

As of this writing, 2001 *Mars Odyssey* is in an extended mission having successfully completed over six Mars years of mapping (each Mars year is 687 days). *Odyssey* is in a circular polar mapping orbit around Mars at an altitude of approximately 400 km (Table 54.1). The nuclear spectroscopy payload consists of a GRS, a NS, and a Russian-supplied high-energy neutron detector. Gamma ray and neutron spectroscopy data acquired by *Mars Odyssey* provide constraints on geochemistry, the water cycle, climate history, and atmospheric processes, including atmospheric dynamics and atmosphere–surface interactions (*see* Mars Atmosphere: History and Surface Interactions).

Since the discovery of abundant subsurface WEH at high latitudes, *Odyssey*'s gamma ray and neutron spectrometers have continued to provide a wealth of new information about Mars, including the global distribution of near-surface WEH, the elemental composition of the surface, seasonal variations in the composition of the atmosphere at high latitudes, and the column abundance of CO_2 ice in the seasonal caps. This information has contributed to our understanding of the recent history of Mars. The climate is driven strongly by short-term variations in orbital parameters, principally the obliquity, and the surface distribution of surface water ice is controlled by atmosphere–surface interactions. The discovery of anomalously large amounts of WEH at low latitudes, where water ice is not stable, has stirred considerable debate about the mineral composition of the surface and climate change.

The GRS on *Odyssey* is boom-mounted, passively cooled, HPGe spectrometer, similar in design to the instrument flown on *Mars Observer* (Figure 54.9(c)). The NS is a deck-mounted instrument that consists of a boron-loaded plastic block (roughly 10 cm on a side), which has been diagonally segmented into four prisms and read out by separate photomultiplier tubes (Figure 54.9(d)). The orientation of the spacecraft is constant such that one of the prisms faces nadir (P1), one faces zenith (P3), one faces in the direction of spacecraft motion (P2), and one faces opposite the spacecraft motion (P4). P1 is covered with a Cd foil that prevents thermal neutrons from entering the prism. Consequently, P1 is sensitive to epithermal and fast neutrons originating from the surface and atmosphere.

Neutrons with energy less than the gravitational binding energy of Mars, approximately 0.13 eV, corresponding to an escape speed of about 5000 m/s, travel on parabolic trajectories and return to Mars unless they decay by beta emission. The mean lifetime of a neutron is approximately 900 s. The most probable energy for neutrons in thermal equilibrium with the surface of Mars (for the mean Martian temperature of 210 K) is 0.018 eV, which corresponds to a neutron speed of 1860 m/s. Consequently, a significant portion of the thermal neutron population travels on ballistic trajectories and arrive at the spectrometer from above and below. Neutrons that leave the atmosphere with energies less than about 0.014 eV, just below the most probable energy, cannot reach the 400 km orbital altitude of *Odyssey*. Consequently, gravitational binding has a significant effect on the flux and energy distribution of neutrons at *Odyssey*'s orbital altitude, and, in contrast to the simplified discussion in Section 3.1, gravitational effects must be accounted for in models of the flux and instrument response.

To separate thermal and epithermal neutrons, the NS makes use of the orbital speed of the spacecraft, which is

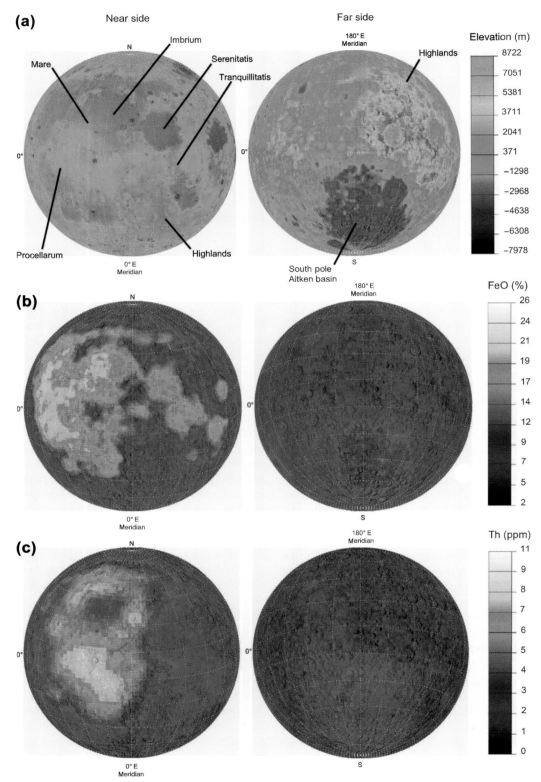

FIGURE 54.11 Orthographic projections of the lunar near and far sides: (a) elevation; (b–d) abundance (as percent or parts per million (ppm) g/g) of selected elements. The map data are superimposed on a shaded relief image. *FeO data courtesy of NASA Planetary Data System; image courtesy of United States Geological Survey.*

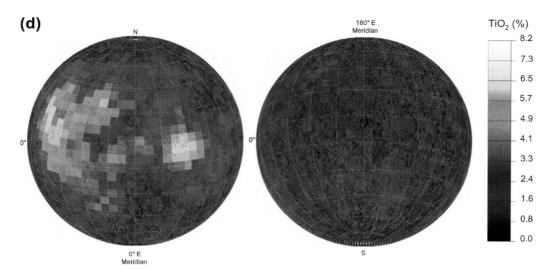

FIGURE 54.11 (continued).

approximately 3400 m/s, the same speed as a 0.06 eV neutron. Neutrons arriving at orbital altitudes with velocities less than that of the spacecraft (most of the thermal neutron population) cannot enter P4. So, P4 is primarily sensitive to epithermal neutrons. In contrast, P2 "rams" into thermal neutrons that arrive at the orbital altitude ahead of the spacecraft. P2 has roughly the same sensitivity as P4 for epithermal neutrons. Consequently, the thermal flux is given by the difference between the counting rates for P2 and P4.

Thermal, epithermal, and fast neutrons are sensitive to surface and atmospheric parameters, including the abundance and layering of hydrogen in the surface, the presence of strong neutron absorbers such as Cl and Fe in the Martian rocks and soil, the presence of CO_2 ice on the surface, the column abundance of the atmosphere, and the enrichment and depletion of noncondensable gasses, N_2 and Ar, as CO_2 is cycled through the seasonal caps (Table 54.2). The effect of these parameters on the neutron counting rate can be explored using a simple physical model of the surface and atmosphere as described in Section 2.2 (Figure 54.4(a)). Models of the counting rate are then used to develop algorithms to determine parameters from observations.

For example, the variation of thermal, epithermal, and fast neutron counting rates as a function of water abundance in a homogeneous surface is shown in Figure 54.12(a). Epithermal and fast neutrons are sensitive to hydrogen (as described in Section 2.2) and their counting rates decrease monotonically with water abundance. Both are insensitive to the abundance of elements in the surface other than hydrogen, as illustrated in Figure 54.12(a) by changing the abundance of Cl, which is a strong thermal neutron absorber. In contrast, thermal neutrons are sensitive to variations in major-element composition and relatively insensitive to hydrogen when the abundance of WEH is less than about 10%. Epithermal neutrons are a good choice for determining the WEH abundance because of their high counting rate and relative insensitivity to other parameters.

TABLE 54.2 Sensitivity of Neutron Energy Ranges to Mars Surface and Atmospheric Parameters

Type	Energy Range	Major Interactions	CO_2-Free Surface Parameters[1]	Atmospheric/Seasonal Parameters
Fast	>0.5 MeV	Inelastic scattering, elastic scattering	WEH abundance and layering, Average atomic mass	Atmospheric mass, CO_2 ice column abundance <100 g/cm^2
Epithermal	0.5 eV (**Cd cutoff**) to 0.5 MeV	Elastic scattering	WEH abundance and layering	Atmospheric mass, CO_2 ice column abundance up to about 150 g/cm^2
Thermal	<0.5 eV (hardened Maxwellian energy distribution)	Elastic scattering, capture (absorption)	WEH abundance, absorption by Fe, Cl, Ti. Layering of WEH and absorbers	CO_2 ice column abundance up to about 1000 g/cm^2, absorption by N_2 and Ar

[1] The surface in the northern or southern hemisphere during summer following the recession of the seasonal cap.

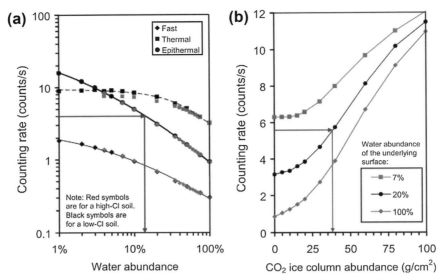

FIGURE 54.12 (a) Variation of thermal, epithermal, and fast neutron counting rates as a function of water abundance for a soil composition with low Cl abundance (black symbols). The red symbols correspond to a soil with higher Cl abundance, similar to the average composition of soils at the Pathfinder landing site. Note that the epithermal and fast neutron counting rates are unaffected by the change in Cl abundance. Because Cl is a strong absorber of thermal neutrons, the thermal neutron counting rate is sensitive to Cl abundance. (b) Variation of epithermal counting rate as a function of CO_2 ice column abundance covering homogeneous surfaces containing 7%, 20%, and 100% water ice (mixed with dry soil). Observed counting rates can be converted directly to water-equivalent hydrogen abundance or CO_2 ice column abundance using the model results in parts (a) and (b) as indicated by the arrows. The counting rate during the summer, which is a measure of the water abundance of the underlying surface, must be known in order to select the correct trend for CO_2 ice column abundance.

Measured epithermal counting rates can be converted directly to WEH as indicated by the arrows in Figure 54.12(a).

A map of WEH determined from measured epithermal counting rates is shown in Figure 54.13. In order to avoid contributions from the seasonal CO_2 ice, the northern and southern high latitudes only included counting rates measured during their respective summers. The algorithm for determining WEH included corrections for minor variations in the counting rate due to changes in the atmospheric column abundance with topography. The map gives a lower bound on WEH. Higher WEH abundances could be present if the surface is stratified, for example, with a dry top layer covering a water-rich medium.

The minimum WEH abundance on Mars ranges from 2% in equatorial and midlatitude regions to nearly 100% for the north polar water ice cap. Low abundances of WEH are found in regions such as northern Argyre Planitia, the midlatitude, southern highlands, Solis Planum, and the eastern flanks of the Tharsis Montes. Correlations between WEH and topography suggest that some aspects of the surface distribution of WEH can be explained by regional and global weather patterns. Moderate WEH abundances (8–10%) can be found in large equatorial regions, for example, in Arabia Terra. Ice stability models predict that water ice is not stable at equatorial latitudes on Mars under present climate conditions. Consequently, the moderate abundances of WEH may be in the form of hydrated minerals, possibly as magnesium sulfate hydrate. High abundances of WEH are found at high northern and southern latitudes (poleward of 60°). A detailed analysis of neutron and gamma ray counting rates suggests that the high-latitude surface outside of the residual caps consists of soil rich in water ice covered by a thin layer of dessicated material (soil and rocks). This result is consistent with models that predict that water ice is stable at shallow depths at high latitudes. Similar terrestrial conditions are observed in the Dry Valleys of Antarctica, where ice is stable beneath a dry soil layer that provides thermal and diffusive isolation of the ice from the atmosphere.

Seasonal variations on Mars are driven by its obliquity relative to the orbital plane, which is similar to that of Earth. In the polar night in the northern and southern hemispheres, atmospheric CO_2 condenses to form ice on the surface. Approximately 25% of the Martian atmosphere is cycled into and out of the northern and southern seasonal caps. Consequently, the seasonal caps play a major role in atmospheric circulation. The main questions about the seasonal caps that remain unanswered concern the local energy balance, polar atmospheric dynamics, and CO_2 condensation mechanisms. Seasonal parameters constrained by neutron spectroscopy include the column abundance of CO_2 ice on the surface and the column abundance of noncondensable gasses (N_2 and Ar) in the atmosphere. For example, analyses of gamma ray and neutron spectroscopy data reveal that the southern

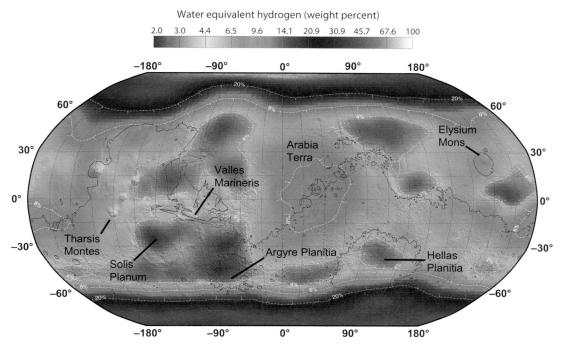

FIGURE 54.13 Global map of the abundance of WEH within the surface of Mars. The map gives a lower bound on the abundance of WEH. Contours for 4%, 8%, and 20% WEH are shown as dashed white lines. The black contour line corresponds to 0 km elevation. The map data are superimposed on a shaded relief image. *Elevation data and shaded relief image courtesy of the NASA Mars Orbiter Laser Altimeter Science Team.*

atmosphere is strongly enriched in N_2 and Ar during cap growth. The observed enrichment may be caused by the formation of a strong polar vortex accompanying the condensation flow, which inhibits **meridional mixing** of the polar atmosphere with lower latitudes.

Based on simulations, the epithermal neutron counting rate generally increases with the column abundance of CO_2 ice on the surface; however, the trend depends on the abundance of water ice in the underlying surface as is shown in Figure 54.12(b). The sensitivity of epithermal neutrons to CO_2 ice is higher for surfaces that contain more water ice. At high latitudes, the column abundance of CO_2 can be determined from seasonal epithermal counting rates, given the counting rate during summer, when no CO_2 ice is present.

Maps of epithermal counting rates are shown in Figure 54.14(a). The extent of the seasonal caps can be seen by comparing maps of the northern and southern hemispheres during the two time periods shown in Figure 54.14(a). For example, during late southern winter, low counting rates are observed in the northern high latitudes, corresponding to the summertime CO_2 frost-free surface, which contains abundant water ice. In early northern spring, elevated epithermal counting rates are observed in the northern hemisphere, corresponding to CO_2 ice on the surface.

During their respective winters, the counting rate at high latitudes increases toward the poles, which indicates that the CO_2 ice column abundance increases with latitude. The observed spatial variation is expected since the polar night lasts longer at higher latitudes and frost has more time to accumulate. The time variation in epithermal counting rates for the north and south poles (poleward of 85°), shown in Figure 54.14(b), reveals the cyclic behavior of the seasonal caps during two Mars years. The total inventory of CO_2 in the seasonal caps determined from epithermal counting data is similar to that predicted by general circulation models (GCMs) (for example, see Figure 54.14(c)). The ability to measure the thickness of the CO_2 caps in the polar night is unique to gamma ray and neutron spectroscopy. Local ice column abundances determined by nuclear spectroscopy can be compared to GCM predictions, providing information needed to improve physical models of the seasonal caps and the polar energy balance.

6. FUTURE PROSPECTS

The first two decades of the twenty-first century will likely be remembered as a golden age of planetary exploration, when new discoveries were made throughout the solar system from data acquired by an international armada of robotic spacecraft. Ongoing missions continue to provide scientists and the general public a close-up view of alien worlds, which hitherto were fuzzy points of light in the viewfinder of a telescope. Nuclear spectroscopy has played an important role in this process, providing unique chemical information inaccessible by telescopic observations.

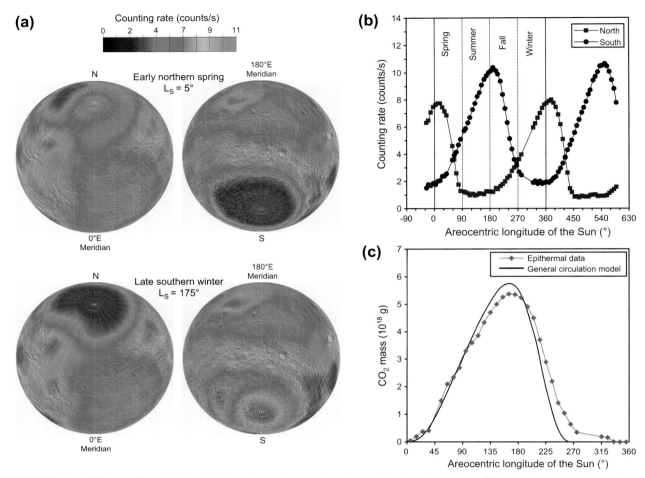

FIGURE 54.14 (a) Orthographic projections of epithermal counting rates in northern and southern hemispheres of Mars during early northern spring and late southern winter; (b) Epithermal counting rate as a function of time (**Areocentric longitude**) at the north and south poles (poleward of 85° latitude); (c) total mass of CO_2 in the southern seasonal cap poleward of 60° S from a general circulation model is compared to that determined from the epithermal counting data. *General circulation model results courtesy of NASA Ames Research Center and the New Mexico State University Department of Astronomy.*

Space exploration, whether by robots or humans, is not merely a source of wonder—it has the potential to provide us with a better understanding of the world in which we live and is a source of inspiration for future engineers and scientists. The development and application of nuclear instrumentation in support of these endeavors is expected to continue in the foreseeable future.

In the short term the planetary science community is focused on completing missions currently underway, such as MSL, MESSENGER, LRO, and Dawn. For example, the 2015 rendezvous of the Dawn spacecraft with the dwarf planet Ceres promises to provide the first spectacular view of an enigmatic, icy body, which does not appear to be represented in the meteorite collection. Surface expressions of a briny, liquid-water ocean possibly hidden beneath the crust of Ceres could be revealed by nuclear spectroscopy. Dawn's GRaND is sensitive to Cl, which is a tracer for aqueous processes, as well as volatiles such as ammonia ice, which, if present, may indicate recent cryovolcanism. A high-latitude, subsurface ice table, if detected by GRaND, could provide further clues needed to determine how water has shaped Ceres.

In the decades to come, nuclear spectroscopy will continue to be used in the exploration of the Moon, Mars, and asteroids, including preliminary investigations by robots in advance of human exploration and possible commercial endeavors. Missions to Jupiter's icy moons and the Martian satellite Phobos are also envisioned. On the Moon, nuclear spectrometers may be used on rovers or incorporated into borehole logging tools to search for and characterize water ice deposits in polar craters. On Mars, they may be included on future rovers (e.g. the 2020 follow-on to MSL), landers, weather stations, and drilling systems, for example, to search for water deep within the Martian cryosphere. In addition, there may be opportunities for low-altitude, high-spatial resolution measurements of portions of the Martian surface from an airplane or balloon.

Finally, continued effort is needed to analyze and interpret nuclear spectroscopy data that has already been acquired by missions to Mercury, Mars, the Moon, and asteroids. Mission science and instrument teams lack the resources to harvest all the information from these data sets. Thus, careful archiving of data in a stable repository accessible to the scientific community, such as NASA's Planetary Data System, is an invaluable legacy of many planetary science missions. Students of nuclear spectroscopy are sure to make new discoveries and will find the data useful for planning future missions far beyond the Moon and Mars.

BIBLIOGRAPHY

Arnold, J. R., Metzger, A. E., Anderson, E. C., & Van Dilla, M. A. (1962). Gamma rays in space, Ranger 3. *Journal of Geophysical Research, 67*(12), 4878–4880.

Bielefeld, M. J., Reedy, R. C., Metzger, A. E., Trombka, J. I., & Arnold, J. R. (1976). Surface chemistry of selected lunar regions. In *Proceedings of the 7th lunar and planetary science conference* (pp. 2661–2676).

Boynton, W. V., Feldman, W. C., Mitrofanov, I. G., Evans, L. G., Reedy, R. C., Squyres, S. W., et al. (2004). The Mars Odyssey gamma-ray spectrometer instrument suite. *Space Science Reviews, 110*(1–2), 3783.

Duderstadt, J. J., & Hamilton, L. J. (1967). *Nuclear reactor analysis*. John Wiley & Sons.

Elphic, R. C., Lawrence, D. J., Feldman, W. C., Barraclough, B. L., Gasnault, O. M., Maurice, S., et al. (2002). Lunar Prospector neutron spectrometer constraints on TiO_2. *Journal of Geophysical Research, 107*(E4), 5024. http://dx.doi.org/10.1029/2000JE001460.

Feldman, W. C., Maurice, S., Lawrence, D. J., Little, R. C., Lawson, S. L., Gasnault, O., et al. (2001). Evidence for water ice near the lunar poles. *Journal of Geophysical Research, 106*, 23231–23251.

Feldman, W. C., Ahola, K., Barraclough, B. L., Belian, R. D., Black, R. K., Elphic, R. C., et al. (2004). Gamma-ray, neutron, and alpha particle spectrometers for the Lunar Prospector mission. *Journal of Geophysical Research Planets, 109*(E7), E07S06.

Feldman, W. C., Prettyman, T. H., Maurice, S., Plaut, J. J., Bish, D. L., Vaniman, D. T., et al. (2004). Global distribution of near-surface hydrogen on Mars. *Journal Geophysical Research Planets, 109*(E9), E09006.

Knoll, G. F. (1989). *Radiation detection and measurement* (2nd ed.). John Wiley & Sons.

Lawrence, D. J., Feldman, W. C., Barraclough, B. L., Binder, A. B., Elphic, R. C., Maurice, S., et al. (2000). Thorium abundances on the lunar surface. *Journal of Geophysical Research, 105*(E8), 20307–20331.

Lawrence, D. J., Feldman, W. C., Elphic, R. C., Little, R. C., Prettyman, T. H., Maurice, S., et al. (2002). Iron abundances on the lunar surface as measured by the Lunar Prospector gamma-ray and neutron spectrometers. *Journal of Geophysical Research, 107*(E12), 5130. http://dx.doi.org/10.1029/2001JE001530.

Lawrence, D. J., Feldman, W. C., Goldsten, J. O., Maurice, S., Peplowski, P. N., Anderson, B. J., et al. (2013). Evidence for water ice near Mercury's north pole from MESSENGER neutron spectrometer measurements. *Science, 339*(6117), 292–296.

Metzger, A. E., Trombka, J. I., Peterson, L. E., Reedy, R. C., & Arnold, J. R. (1973). Lunar surface radioactivity: preliminary results of the Apollo 15 and Apollo 16 gamma-ray spectrometer experiments. *Science, 179*(4075), 800–803.

Miller, R. S., Nerurkar, G., & Lawrence, D. J. (2012). Enhanced hydrogen at the lunar poles: new insights from the detection of epithermal and fast neutron signatures. *Journal of Geophysical Research, 117*, E11007. http://dx.doi.org/10.1029/2012JE004112.

Mitrofanov, I. G., Litvak, M. L., Varenikov, A. B., Barmakov, Y. N., Behar, A., Bobrovnitsky, Y. I., et al. (2012). Dynamic albedo of neutrons (DAN) experiment onboard NASA's Mars science laboratory. *Space Science Reviews, 170*, 559–582.

Peplowski, P. N., Evans, L. G., Hauck II, S. A., McCoy, T. J., Boynton, W. V., Gillis-Davis, J. J., et al. (2011). Radioactive elements on Mercury's surface from MESSENGER: implications for the planet's formation and evolution. *Science, 333*(6051), 1850–1852.

Prettyman, T. H., Feldman, W. C., Mellon, M. T., McKinney, G. W., Boynton, W. V., Karunatillake, S., et al. (2004). Composition and structure of the Martian surface at high southern latitudes from neutron spectroscopy. *Journal of Geophysical Research, 109*, E05001. http://dx.doi.org/10.1029/2003JE002139.

Prettyman, T. H., Hagerty, J. J., Elphic, R. C., Feldman, W. C., Lawrence, D. J., McKinney, G. W., et al. (2006). Elemental composition of the lunar surface: analysis of gamma ray spectroscopy data from Lunar Prospector. *Journal of Geophysical Reserach, 111*, E12007. http://dx.doi.org/10.1029/2005JE002656.

Prettyman, T. H., Feldman, W. C., & Titus, T. N. (2009). Characterization of Mars' seasonal caps using neutron spectroscopy. *Journal of Geophysical Research, 114*, E08005. http://dx.doi.org/10.1029/2008JE003275.

Prettyman, T. H., Feldman, W. C., McSween, H. Y., Jr., Dingler, R. D., Enemark, D. C., Patrick, D. E., et al. (2011). Dawn's gamma ray and neutron detector. *Space Science Reviews, 163*, 371–459. http://dx.doi.org/10.1007/s11214-011-9862-0.

Prettyman, T. H., Mittlefehldt, D. W., Yamashita, N., Lawrence, D. J., Beck, A. W., Feldman, W. C., et al. (2012). Elemental mapping by Dawn reveals exogenic H in Vesta's regolith. *Science, 338*(6104), 242–246.

Reedy, R. C. (1978). Planetary gamma-ray spectroscopy. In *Proceedings of the 9th lunar and planetary science conference* (pp. 2961–2984).

Surkov, Yu. A., Barsukov, V. L., Moskaleva, L. P., Kharyukova, V. P., Zaitseva, S. Ye., et al. (1989). Determination of the elemental composition of Martian rocks from Phobos 2. *Nature, 341*, 595–598.

Vinogradov, A. P., Surkov, Yu. A., Chernov, G. M., Kirnozov, F. F., & Nazarkina, G. B. (1968). Gamma investigation of the Moon and composition of the lunar rocks. In A. Dollfus (Ed.), *Moon and planets II* (pp. 77–90).

Vinogradov, A. P., Surkov, Yu. A., & Kimozov, F. F. (1973). The content of uranium, thorium, and potassium in the rocks of Venus as measured by Venera 8. *Icarus, 20*, 253–259.

Yamashita, N., Hasebe, N., Reedy, R. C., Kobayashi, S., Karouji, Y., Hareyama, M., et al. (2010). Uranium on the Moon: global distribution and U/Th ratio. *Geophysical Research Letters, 37*, L10201.

Zhu, M.-H., Chang, J., Ma, T., Ip, W.-H., Fa, W., Wu, J., et al. (2013). Potassium map from Chang'E-2 constraints the impact of Crisium and Orientale basin on the moon. *Scientific Reports, 3*. http://dx.doi.org/10.1038/srep01611.

Chapter 55

Probing the Interiors of Planets with Geophysical Tools

W. Bruce Banerdt
Jet Propulsion Laboratory, California Institute of Technology, Pasadena, CA, USA

Véronique Dehant
Royal Observatory of Belgium, Brussels, Belgium

Robert Grimm
Southwest Research Institute, Boulder, CO, USA

Matthias Grott
German Aerospace Center (DLR), Institute of Planetary Research, Berlin, Germany

Philippe Lognonné
Institut de Physique du Globe de Paris, Paris, France

Suzanne E. Smrekar
Jet Propulsion Laboratory, California Institute of Technology, Pasadena, CA, USA

Chapter Outline

1. Introduction — 1185
2. Investigating Planetary Interiors Using Seismology — 1186
   2.1. Terrestrial Seismology — 1186
   2.2. Lunar Seismology — 1187
   2.3. Mars Seismology — 1190
   2.4. Jovian Seismology — 1190
3. Investigating Planetary Interiors Using Gravity and Dynamics — 1190
4. Investigating Planetary Interiors Using Heat Flow Measurements — 1192
   4.1. Measurement Techniques — 1192
   4.2. The Heat Flow of the Earth — 1193
   4.3. Lunar Heat Flow — 1194
   4.4. Future Investigations — 1196
5. Investigating Planetary Interiors Using EM Sounding — 1198
   5.1. Natural Sources — 1198
   5.2. Measurement Techniques — 1199
   5.3. Terrestrial Exploration — 1199
   5.4. Magnetic Sounding of the Moon — 1200
   5.5. Magnetic Sounding of the Galilean Satellites — 1201
   5.6. Future Exploration — 1202
6. Summary — 1203
Bibliography — 1203

1. INTRODUCTION

Understanding the structure and processes of the deep interior of the planets is crucial for understanding their origin and evolution, as well as the surface history and processes. Planetary interiors record evidence of conditions of planetary **accretion** and **differentiation**. The crust of a planet is generally thought to form initially through **fractionation** of an early magma ocean, with later addition through partial melting of the mantle and resulting volcanism. Thus the volume (thickness) and structure of the crust can place strong constraints on the depth and evolution of a putative magma ocean. Interior processes also exert significant control on surface environments. The structure of a planetary interior and its dynamics control heat transfer within a planet through **advected** mantle material, heat conducted through the lithosphere, and volcanism.

Volcanism in particular controls the timing of volatile release, and influences the availability of water and carbon. The existence and strength of any planetary magnetic field depends in part on the size and state of the core.

Even the comprehension of how life developed and evolved on the Earth requires knowledge of the Earth's thermal and volatile evolution and how mantle and crustal heat transfer, coupled with volatile release, affected habitability at and near the planet's surface. Although geophysics can provide information about past processes and states required to reach this understanding, it primarily provides a "snapshot" at the present time of how the Earth behaves. This "boundary condition" is a powerful constraint on all models that describe the history of the Earth and attempt to place the evolution of life in this framework, as such models must evolve to this present state.

However, this realm of the deep planetary interior is inaccessible to direct observation, with tens to thousands of kilometers of rock blocking our "view"; note that the deepest hole ever drilled is the Kola SG-3 in Russia, which reached 12.262 km in 1989. Whereas some deep material is inevitably transported to the surface (for example, through volcanism, impact excavation, and outgassing), these materials typically undergo complex changes on their journey that are difficult to unravel and are subject to sampling biases, and thus provide limited insight.

The most effective way to explore the interior of a planet is through geophysics, as it provides tools that allow for "remote sensing" of the interior. Seismology, together with surface heat flow, magnetic and gravity field measurements, and electromagnetic (EM) techniques, have revealed the basic internal workings of the Earth: its thermal structure, its compositional **stratification**, as well as significant lateral variations in these quantities. For example, seismological, magnetic, and **paleomagnetic** measurements revealed the basic components of seafloor spreading and subduction, and seismology and EM measurements have mapped the structure of the core, compositional and phase changes in the mantle, three-dimensional velocity anomalies in the mantle related to **subsolidus convection**, and lateral variations in lithospheric structure. Additionally, seismic information placed strong constraints on Earth's interior temperature distribution and the mechanisms of **geodynamo** operation (see Magnetic Field Generation).

This chapter describes the four major classes of geophysical techniques that are used to probe deep planetary interiors: seismology, geodesy (which includes both gravimetry and planetary dynamics), heat flow, and electromagnetism. We choose to exclude techniques that have limited penetration depths (such as gamma-ray scattering, which samples to about a meter; microwave/radar techniques, which penetrate a few meters to a few kilometers; and active/exploration seismology, which typically concerns itself with depths less than 10 km), as well as a few exotic techniques (such as neutrino and muon tomography) whose use in planetary exploration is likely beyond any practical horizon.

2. INVESTIGATING PLANETARY INTERIORS USING SEISMOLOGY

2.1. Terrestrial Seismology

Seismology is the most powerful tool available for probing the interior of planets. Seismic waves (for terrestrial planets) or acoustic waves (for giant gas planets or the Sun) propagate at **seismic velocities**, or the speed of sound. These velocities, as well as the amplitude, polarization, and propagation direction of seismic waves, are directly related to quantities such as the temperature, pressure, and/or composition of the material through which the waves travel.

Modern seismology started in 1889, with the detection in Germany by Ernst von Rebeur-Paschwitz of ground displacements associated with waves generated by an earthquake in Japan. In quick succession, compositional (as manifested in seismic velocity) discontinuities between the mantle and core (the core—mantle boundary, or CMB, discovered by Richard Dixon Oldham in 1906) and the crust and mantle (the Moho discontinuity, discovered by Andrija Mohorovičić in 1909) were detected through the analysis of seismograms, as was the solid inner core, somewhat later, by Inge Lehmann in 1936 (Figure 55.1).

Waves are either reflected or **refracted** by these discontinuities, and they propagate in ways very similar to optical rays. Two types of **body waves** can propagate inside a planet: compressional waves (or P-waves), which generate pressure changes and displacements along the propagation direction, and shear waves (or S-waves), which do not generate a pressure change but rather displacements perpendicular to the propagation direction. The latter cannot exist in fluids and have velocities about 1.7 times smaller than the P velocities, and thus arrive after the P-waves. S-waves are fully reflected by solid—fluid boundaries, and these reflections are key to detecting underground fluids, such as the liquid iron of the outer core or underground oil and natural gas reservoirs. Because of its power and flexibility, seismology has been used extensively for both basic research and resource exploration, and now enables three-dimensional imaging of deep crustal structure or of the convection patterns in the Earth's mantle, through the small changes in seismic velocities generated by both temperature differences and the preferred orientation of the component crystals in mantle rocks in the direction of convective motion.

In addition to the crust—mantle and core—mantle boundaries, several additional global discontinuities have been found in the mantle and in the core by seismology. In the mantle, these discontinuities mark the transition between an olivine upper mantle (at 600 km depth for one

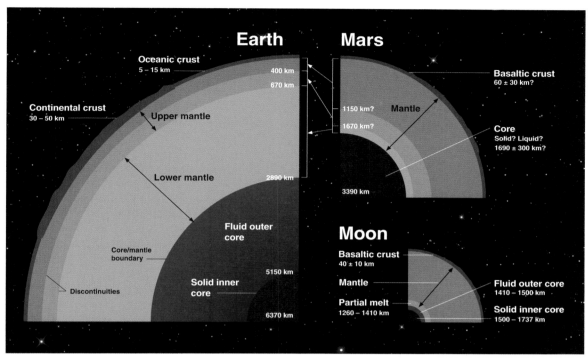

FIGURE 55.1 Interior structure of the Earth and Moon, as seen by seismology. The expected structure of Mars is shown in comparison. Shown are the depths to the major interior boundaries of the planets, between the crust, mantle, outer core, and inner core, as well as discontinuities at the two major phase boundaries. On the Moon there also appears to be a partially-molten zone at the base of the mantle. For the Earth the range of observed crust thicknesses are given, whereas for the Moon and Mars the uncertainty in the average crust thickness is shown. A smaller planet with weaker gravity has a slower pressure increase with depth; consequently, most of the Martian mantle is in a pressure state similar to the Earth's upper mantle, while most of the Moon's mantle corresponds to the conditions in only the first 300 km of the Earth.

of the most widely accepted average Earth models, called PREM, or Preliminary Reference Earth Model) and a perovskite lower mantle (at 770 km depth). In the region between these two discontinuities, called the transition zone, the dominant mineral, olivine, is in a spinel crystal structure. These two discontinuities may play a major role in controlling the convection of the mantle and therefore its geodynamic evolution. Much of modern seismology focuses on the details of these discontinuities, such as their topography or the thin transition zones just above the CMB (perhaps associated with a post-perovskite phase in the mantle) and outer-inner core boundary (due to the accumulation of lighter elements in a slushy layer just above the solid inner core).

Two other major discoveries have marked terrestrial seismology. The first was the measurement of global oscillations of the Earth, which are bell-like global vibrations with discrete periods directly related to the Earth's structure and composition. This was first done following the 1960 9.5 magnitude earthquake in Chile, which remains the largest earthquake ever observed with seismometers. These data, along with those subsequently recorded for quakes with magnitudes larger than 7–7.5, have revealed not only the natural resonant frequencies of the Earth (called eigenfrequencies, with periods ranging between 52 min and about 1 min) but also a fine structure of these tones, mathematically comparable to the fine structure of atoms in quantum mechanics.

The second major discovery was made in the early twenty-first century. Much of the ground noise measured during the quiet periods (when no coherent seismic waves from quakes are noticeable) is in fact the superposition of seismic waves excited by ocean waves and atmospheric winds and turbulence. It was shown that analysis of these vibrations can be used to measure not only the seismic speeds between two locations but also very small time variations of these seismic speeds, associated, for example, with changes in the structure of volcanoes prior, during and after an eruption or in the crustal structure after an earthquake. These findings opened the area of time-dependent seismology, which aims to detect temporal changes in the Earth's structure.

2.2. Lunar Seismology

In a sense, the long-term goal of planetary seismology is to repeat this geophysical success story on other planets. Planetary seismology began almost at the same time as space exploration itself; one of the instruments carried by the Ranger 3 mission, launched toward the Moon on January

26, 1962, was a seismometer. This first attempt, as well as those of Rangers 4 and 5 (also launched in 1962), all failed, but the first Apollo mission, Apollo 11, successfully deployed the first planetary seismometer. This instrument was powered by solar panels and did not survive the end of the second lunar day. The next five Apollo missions, however, each had a long-lived nuclear-powered seismometer called a Passive Seismic Experiment (PSE; Figure 55.2), able to record both long-period and high-frequency ground displacements. With the exception of Apollo 13, they were all successfully installed and worked until 1977, forming a network of four stations. This regional network (stations 12 and 14 were only about 180 km apart and formed one corner of an approximately equilateral triangle, with stations 15 and 16 at the other corners, each about 1100 km away), was complemented by local seismic arrays at the Apollo 15, 16, and 17 sites for active seismic studies of the shallow lunar subsurface. A gravimeter, deployed at the Apollo 17 site, was also able to detect seismic waves with a much lower sensitivity than the Apollo PSE seismometers. These instruments remain the only seismometers operated on the Moon, and this is likely to remain the case until at least 2020, as none of the confirmed missions of this decade plan to deploy new seismometers on the Moon.

Due to their extraordinary sensitivity, the Apollo PSE instruments revolutionized our understanding of the lunar interior. Even more surprising is that their data are still being used today to perform original research, more than 35 years after they were turned off.

The principle of a seismometer is to detect the small vibrations of the ground associated with passing seismic waves. Although technology development since the 1960s has allowed higher and higher sensitivities, the fundamental design on which virtually all seismometers are based has remained the same for more than 120 years: a mass, called the proof mass, supported by a spring. When the ground (and the instrument which is fixed to it) moves, the inertia of the mass causes it to remain stationary, which is physically equivalent to a motion with respect to the frame of the instrument. The displacement or velocity of the mass is detected by a transducer. The Apollo seismometers were able to detect ground displacement with amplitudes of about 0.5 Å, which is of the order of the radius of a hydrogen atom. These measurements were performed simultaneously by a three-axis (vertical and two mutually perpendicular horizontal directions) long-period instrument, focused on periods around 2 s, and by a single-axis (vertical) short-period instrument, focused on vibrations with periods of around 0.125 s.

These instruments discovered a seismically active Moon and detected about 12,500 moonquakes during more than 7 years of operation. The Moon is still only weakly active compared to the Earth; all these events have very small magnitudes with respect to terrestrial standards, and the largest moonquake has a **moment magnitude** (M) of only about 4, as compared to $M \sim 9.5$ of the largest quake detected on the Earth (the 1960 Chile earthquake). This is a factor of 10^8 smaller in terms of long-period amplitudes.

Three types of seismic signals were detected (Figure 55.3): about 30 were tectonic quakes, including the largest ones. Like most earthquakes, they occurred in the upper part of the planet, at depths no greater than about 200 km. Much more numerous (almost 9000 events) were the deep moonquakes, occurring at depths between 800 and 1200 km. These events were found to occur with the periodicity of the lunar orbit around the Earth, and occurred at a limited number of epicenters (less than 300), where quakes repeated regularly during the 7 years of network operations. These quakes appear to be associated with the solid-body tidal stresses caused by the Earth on the Moon, which are modulated by small variations of distance between the two bodies. Most of these quakes are very small, most with M smaller than 2 and only a few reaching magnitude 3. The last class of events is meteoroid impacts, with almost 2000 impacts detected. In contrast to the Earth, no atmosphere

FIGURE 55.2 On the left is an Apollo PSE seismometer on the lunar surface. The image on the right shows the Viking seismometer, on the deck of the Viking Lander (the shorter box near the center of the picture).

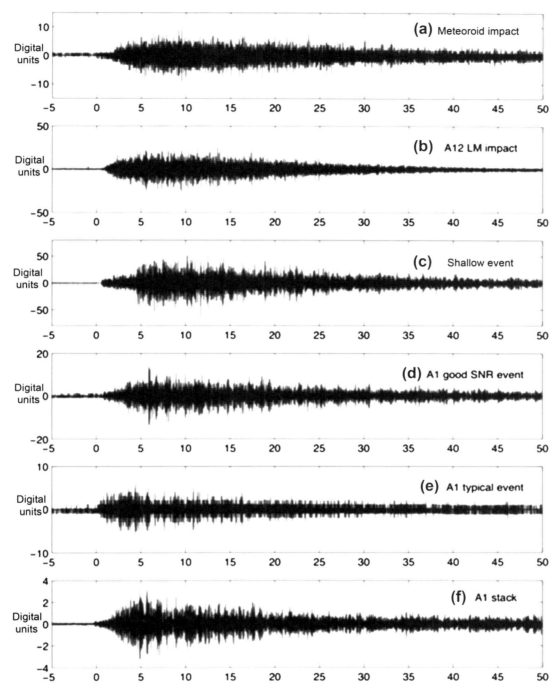

FIGURE 55.3 Examples of lunar seismograms for (a) natural impacts, (b) artificial impacts, (c) tectonic quakes, and (d–f) deep moonquakes. All seismograms are recorded on long period channels by the Apollo Passive Seismic Experiment; one digital unit corresponds to 0.5 Å of ground displacement at 0.5 Hz; *x*-axis is time in minutes. The tectonic quake in this example is one of the largest magnitude shallow events and has a comparable amplitude to the chosen artificial impact. Note the low amplitude and poor data quality of typical deep moonquakes in (e). However, the repeated similar waveforms of deep moonquakes from distinct source regions allows stacking of many events to improve the signal-to-noise ratio (f). Note: LM is Lunar Module, SNR is Signal-to-Noise Ratio. *From Lognonne and Johnson (2007).*

protects the Moon from **meteoroids**, and they can therefore generate seismic waves when they hit the lunar surface with mean impact velocities of about 20 km/s. All these seismic events were characterized with seismograms very different from those typically seen on the Earth. Due to the lack of atmosphere, the surface and subsurface lunar materials are extremely dry. Moreover, the billions of years of impacts have generated a highly fractured, low-seismic-velocity zone near the lunar surface, which is very effective in trapping and scattering waves. These conditions result in extremely long duration (sometimes hours in length) seismic signals, even for very small quakes. Such long reverberations are never observed on the Earth.

Thanks to these seismic sources, both the location of many of these events and the associated travel time of the seismic waves between the **epicenter** and the seismic stations have been determined, enabling scientists to model the interior of the Moon and to identify a crust, a mantle, and a core, like on the Earth. The crustal thickness has a mean value of 40 ± 5 km. From there down to about 1400 km, a relatively homogeneous mantle was seen, possibly relatively hot and **attenuating** at depths larger than 1100–1200 km. And finally, a 350-km-radius liquid core was discovered in early 2012, more than 40 years after the transmission of the Apollo data, thanks to new seismic processing techniques and to the capability of the modern computers to perform calculations beyond the reach of the 1970s' Apollo seismologists. Seismology has also been able, through the measurement of seismic velocities in the mantle, to constrain the temperature at the bottom of the mantle. Because of the different melting temperatures for iron and iron alloys, this suggests that the lunar core is composed of iron mixed with some light element rather than pure iron. Evidence for seismic discontinuities in the lunar mantle remains weak. A possible discontinuity at a depth of 500 km, proposed by the Apollo seismic experiment team, is still controversial, as it might in fact be related to the poor resolution of the Apollo network at this depth. Thus much more remains to be done on the Moon in terms of seismology, and several prospective projects plan to redeploy seismometers on our natural satellite.

2.3. Mars Seismology

The seismic exploration of Mars was the logical next step after the Moon, and this motivated the inclusion of seismometers in the payload of the two Viking landers, launched by National Aeronautical and Space Administration (NASA) to Mars in 1976 (Figure 55.2). Unfortunately, due to cost and technical limitations, the seismometers were about 10,000 times less sensitive than the Apollo seismometers and were mounted on the lander deck, not placed in direct contact with the ground. In addition, the seismometer on Viking 1 failed to uncage (the seismometers were locked, or "caged", during flight to avoid damage) due to a technical failure. The seismometer onboard Viking 2 lasted for 19 months on Mars, but due to operational limitations it accumulated only a little more than 3 months of usable data and did not detect any marsquakes. During the day, the data were dominated by wind-generated lander vibrations, while during the night, its quality was limited by the seismometer's poor sensitivity and coupling to the ground. Given the various limitations, this lack of marsquake detection remains compatible with a Mars activity comparable to that of the Earth away from tectonic plate boundaries (where most of the activity is located). Thus it is generally accepted that Mars likely has a level of seismic activity between this upper limit and that of the Moon. Constraining this level of activity will be the first goal of the Interior exploration using Seismic Investigations, Geodesy and Heat Transport (InSight) project, which is currently under development by NASA. This mission aims to deploy a seismometer on Mars in September 2016, together with other geophysical sensors (heat flow, geodetic, and magnetic measurements, in addition to some atmospheric sensors).

2.4. Jovian Seismology

Mars' seismic activity will therefore remain hidden for a few more years. But the next planet, Jupiter, has finally revealed its continuous vibration after almost 30 years of attempted observations. In this case, data are obviously not coming from seismometers, but from remote observation through telescopes on the Earth. The technique consists of measuring the vertical velocity of the Jovian optical surface with spectroscopic techniques originally developed for solar and stellar seismology. Thanks to a steady increase in sensitivity, the observations have finally detected continuous oscillations with an amplitude of about 0.5 m/s, putting Jupiter third on the list of the bodies of the solar system for which global oscillations have been detected (after the Earth and the Sun). It is likely that a significant improvement in these observations will be made in the next 5 years, and we can therefore reasonably expect that by the end of this decade, comparative seismology will be possible among three terrestrial bodies (Earth, Moon, and Mars) and between two fluid ones (the Sun and Jupiter).

3. INVESTIGATING PLANETARY INTERIORS USING GRAVITY AND DYNAMICS

Historically, the most common method for exploring planetary interiors has been through observed variations in their gravitational fields, orientation, and rotation, which can be collectively described as geodesy. The data can be obtained as a by-product of the communication and navigation systems of planetary probes flying by, in orbit around, or landed on the planets. Gravity and rotation provide information primarily on the distribution of mass within a planet, including density differences due to overall structure (crust, mantle, and core) and from processes such as volcanism and tectonic deformation.

The major disadvantage of these techniques is their inherent ambiguity. It can be shown mathematically (**Gauss' theorem**) that for any gravity field measured outside a body, there are an infinite number of distinct mass distributions inside the body that could generate that field (a similar result holds for rotational dynamics). While this seems daunting, in practice, there is usually other information that can be used to constrain the possible (or likely) distributions. Thus, geodetic methods are most powerful when used in combination with other techniques.

Spatial variations in the gravity field of planetary bodies are obtained by monitoring the trajectory of passing or orbiting spacecraft. For example, if an orbiter passes over a buried body with higher density than its surroundings, it will experience a higher gravitational attraction than otherwise and will be pulled slightly downward in its orbit. This deviation is determined via Doppler (which gives velocity) and ranging (which gives distance) measurements on radio links between the Earth and the spacecraft. These measurements use the precision transmitters and receivers at large deep space antennas such as the ESTRACK stations (European Space Agency (ESA) TRACKing stations) at Perth and Madrid and the NASA DSN (Deep Space Network) antennas at Goldstone, Madrid, and Canberra (see Figure 55.4). The precision possible with these systems is impressive, with resolution of velocity changes of fractions of a millimeter per second and absolute locations in inertial space of a few centimeters now possible.

For the analysis of these radio science data and for theoretical simulations of planetary interiors with which to compare them, sophisticated numerical codes are used to compute accurate orbits of spacecraft from radio tracking data. Spatial and temporal variations in the gravity field can be used to determine physical properties of the interior and atmosphere of the planet. Since the beginning of the space age, the large-scale structure of the gravity field of planets and moons has been successfully used to determine the moment of inertia, which is a measure of the global density distribution and an important constraint on the interior structure.

FIGURE 55.4 Deep Space Network 70-m station at Goldstone, California.

In addition to global mass distribution, regional or local properties above target surface features can also be measured from the deviation of spacecraft trajectories. As an example, the Mars radio science experiment (MaRS) aboard Mars Express acquired gravity data during a number orbits above the Tharsis volcanoes, which form the largest volcanic region in the solar system.

The data analysis shows that the overall density of the volcanoes is higher than the average density of the Martian crust, in agreement with the basaltic composition of many Martian meteorites probably originating in the Tharsis area. One volcano, Ascraeus Mons, differs from the others in being of lower density in its upper part, although its overall density remains high. If the Tharsis Montes were built in succession by a single moving mantle plume, this suggests that Ascraeus Mons formed as the last of the Tharsis Montes. These data also show that Olympus Mons, the highest mountain in the solar system, lacks a low-density root, which indicates that it was built on a lithosphere of high rigidity, whereas the other volcanoes partly sank within a less rigid lithosphere.

Solid-body tides, which can be observed through their time-variable effect on the gravity field, can also provide information on the deep interior. These are particularly sensitive to global fluid layers such as a liquid iron core in terrestrial planets or an internal subsurface ocean in icy satellites. From the latest available data on the **moment of inertia** and the tidal amplitude of Mars, the best estimates yet obtained have been determined for the core size and composition of Mars. These show that the core size is expected to be between 1715 and 1850 km and that the weight fraction of sulfur in the core is between 13% and 18%. The addition of sulfur decreases the melting temperature of iron, and for the current estimates of the internal temperature of Mars, this high sulfur estimate implies that the core of Mars is entirely liquid and contains no solid inner part, in contrast to the Earth.

Constraints on planetary interiors can also be obtained from rotation variations. Three broad classes of rotation variations are usually considered: rotation rate (also called length-of-day) variations, orientation changes of the rotational axis with respect to inertial space (**precession** and **nutation**), and orientation changes of the planet's surface with respect to the rotation axis (polar motion and polar wander). These are due to both internal (angular momentum exchanges between solid and liquid layers) and external (gravitational torques) causes. As the rotational response depends on the planet's structure and composition, insight into the planetary interior can also be obtained. This is particularly so for the rotational variations due to well-determined external gravitational causes, such as for the nutations of Mars and the **librations** of Mercury and natural satellites.

Observation of the rotation may be performed using orbital measurements or, more precisely, by direct tracking of landers on the surface of a planet. The latter will be the case with Rotation and Interior Structure Experiment (RISE) onboard the InSight Lander. Precise Doppler tracking from the Earth of a location on the Martian surface over extended periods (months to years) can be used to obtain Mars' rotation behavior. More specifically, measuring the relative position of the lander on the surface of Mars with respect to the terrestrial ground stations allows the reconstruction of Mars' time-varying orientation and rotation in space. Precession (long-term changes in the rotational orientation occurring over many tens of thousands of years) and nutations (periodic changes in the rotational orientation occurring on subannual timescales) as well as polar motion (motion of the planet's surface with respect to its rotation axis) are determined from this experiment and are used to obtain information about Mars' interior. Precession measurements improve the determination of the moment of inertia of the whole planet, which is particularly sensitive to the radius of the core. Using geochemical constraints to specify a plausible range of possible compositions, the core radius is expected to be determined with a precision of a few tens of kilometers (compared to ∼150 km currently). A precise measurement of variations in the orientation of Mars' spin axis also enables an independent (and more precise) determination of the size of the core via the core resonance in the nutation amplitudes. The amplification of this resonance depends on the size, moment of inertia, and flattening of the core. For a large core, the amplification can be very large, ensuring the detection of the free core nutation and determination of the core moment of inertia.

At the same time, measurement of variations in Mars' rotation rate reveals variations of the angular momentum due to seasonal mass transfer between the atmosphere and polar caps and zonal winds.

Investigation of the dependence of rotation variations, gravity field, and tidal variations on interior and atmosphere properties and orbital motion characteristics is essential to understand the interior and evolution of terrestrial planets. These studies include the development of advanced models of rotation, the construction of detailed models for the structure and dynamics of solid and fluid layers of the planets, the investigation of the dynamical response of these models to both internal and external forcing, the modeling of the orbital motion of large bodies of our solar system, and the inclusion of general relativistic effects into the data analysis.

4. INVESTIGATING PLANETARY INTERIORS USING HEAT FLOW MEASUREMENTS

The amount of heat escaping from a planet's interior is diagnostic of the subsurface temperature structure, which in turn is connected to the activity we observe on the surface. Essentially all geologic activity, except for meteorite impacts and wind and water erosion (which are powered by the sun), are driven by the internal heat engine of a planet. In addition, many material parameters of crustal and mantle rocks are temperature dependent, and subsurface temperatures influence the velocity of seismic waves, the **rheology** of crustal and mantle rocks, as well as the capability of crustal rocks to record **remanent magnetic fields** through the magnetic minerals' **Curie temperatures**. In addition, subsurface temperatures drive hydrothermal circulation in the crust, and, apart from pressure, temperature is the main factor controlling melting in the crust and upper mantle. Temperature differences also drive solid-state convection in the mantle and determine tectonic surface deformation to a considerable degree.

Heat emanating from a planetary interior derives in unknown proportions from primordial heat accumulated during accretion and core formation, and heat released by the decay of radioactive elements. Apart from the short-lived isotopes ^{26}Al and ^{60}Fe that have considerable influence on the earliest evolution of a planetary body, decay of ^{40}K, ^{232}Th, ^{235}U, and ^{238}U is the main source of heat in planetary interiors. Heat is transported to the surface through mantle convection and thermal diffusion, before it is finally radiated to space. Therefore, planets act like heat engines, and planetary cooling from initial temperatures after accretion and core formation is governed by the efficiency of the different heat transport mechanisms.

In order to quantify the heat flow from the Earth's interior, the first terrestrial heat flow measurements were conducted in the late 1930s by Bullard and Benfield, and tens of thousands of measurements have been performed since. It was found that heat flow varies substantially with tectonic setting, composition, and the age of rocks, and the average amount of heat lost from the Earth's interior was estimated to be of the order of 40 TW. It is estimated that only about 35% of the Earth's heat loss is balanced by present-day heat generation in the deep interior, while the bulk of the radiated energy is drawn from primordial heat accumulated during accretion and differentiation and previous radioactive decay. This implies that the Earth's interior is cooling more efficiently than a purely conductive sphere, a process likely driven by the plate tectonics cycle and the associated transport of hot mantle material to the surface at midocean ridges. However, the fraction of the heat released in the interior to that which is radiated to space, known as the Urey ratio, could have been significantly larger in the past due to greater radiogenic heat production and possibly less efficient cooling in previous epochs of the Earth's evolution.

4.1. Measurement Techniques

A measurement of the planetary heat flow requires knowledge of the subsurface temperature gradient as well as the thermal conductivity of the material in which the measurement is made. The amount of heat passing perpendicularly through a surface of unit area is then given by Fourier's Law, $F = k\, dT/dz$, where k is the thermal conductivity (in units of W/m/K), dT/dz is the thermal gradient (in units of K/m), and F is heat flow, or, more precisely, the heat flux density, in units of W/m^2.

Apart from the heat originating in the planetary interior, temperatures at shallow crustal depths can be significantly affected by **diurnal**, annual, or even climatic changes of the surface temperature. While daily temperature fluctuations penetrate crustal rocks to depths of only a few meters, annual temperature variations have noticeable effects down to a few tens of meters, while glaciation cycles, which act on 100,000-year timescales, can affect subsurface temperatures to depths of a few kilometers. In order to measure the planetary heat flow, measurements must be conducted below the influence of these perturbations, or, alternatively, their influence must be taken into account by modeling.

Therefore, heat flow measurements on the Earth's continents are generally performed in boreholes whose depths extend from a few tens to a few hundreds of meters. The thermal gradient in the boreholes is determined by logging temperatures for extended periods of time, and the thermal conductivity of the host rock is generally determined from drill cores which are investigated in the laboratory. In contrast, because of its extremely stable surface temperature, measurements in the oceanic crust can be performed at much shallower depth, and probes derived from the original design by Bullard (see Figure 55.5) and having lengths of several meters are usually employed. This reduction in measurement depth is made possible by the almost isothermal conditions at the bottom of the oceans. In ocean bottom measurements, thermal conductivity of the sediments is usually determined from drill cores using needle probes in the laboratory, but can also be estimated from the cooling curves of the probes in situ. Integrated gradient and thermal conductivity probes using active heating techniques to determine thermal conductivity in situ were proposed by Christoffel and Calhaem in the late 1960s, and have been widely applied since then.

First attempts to measure extraterrestrial heat flows were made using ground-based radio observations in the early 1960s, but reliable results were only later obtained during the Apollo missions, where Bullard-like probes were emplaced in the lunar subsurface by the Apollo astronauts. Remote sensing techniques have also been proposed for extraterrestrial heat flow measurements, and microwave emissions, which are sensitive to temperatures at different depths for different wavelengths, could in principle be used to determine the thermal gradient in the lunar subsurface. Lacking insolation, surface temperatures in permanently shadowed polar craters are strongly

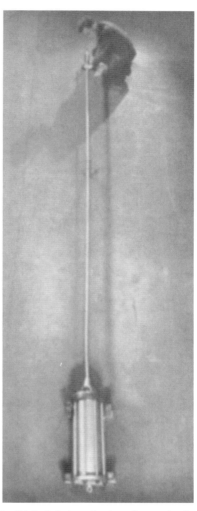

FIGURE 55.5 Original design of a heat flow probe for marine measurements by *Bullard (1954)*. The approximately 5-m-long and 3-cm-thick probe is equipped with temperature sensors to record the thermal gradient as well as the cooling of the probe after insertion into the sediments. Data logging equipment is located in the pressure-proof cylinder at the top of the probe (bottom of the image). *From Bullard (1954).*

influenced by interior heat flow, and infrared emissions from such regions could be used to estimate heat flow near the lunar poles from orbital measurements. However, there are large uncertainties associated with the interpretation of such data, so in situ measurements using borehole logging techniques remain the most reliable for surface heat flow determination to date.

4.2. The Heat Flow of the Earth

The heat flow from the Earth's interior has been determined at more than 20,000 field sites in both continental and oceanic terrains, and a representation of the data is shown in Figure 55.6. Surface heat flow is largest near divergent plate boundaries where hot mantle material is advected

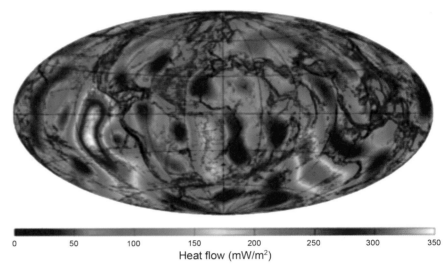

FIGURE 55.6 Color-coded map of the terrestrial heat flow derived from 24,774 measurements at 20,201 sites (*Pollack et al. (1993)*), plotted over a shaded relief topographic map (Hammer projection). Heat flow is at a maximum at midocean ridges where hot mantle material is advected toward the surface by mantle convection. Minimum heat flow is observed on the continental shields.

toward the surface, and heat flow in the oceanic lithosphere averages 101 mW/m^2. In contrast, continental heat flow is significantly less and measures only 65 mW/m^2 on an average. This difference is a direct consequence of the difference in heat transport efficiency, which is partly advective in the oceanic crust, but purely conductive in the continents. Heat flow on the continents is dominated by the heat released through radioactive decay in the crustal rocks, and is at a minimum in old continental regions. Taken together, the average heat loss from the surface of the Earth is 87 mW/m^2, corresponding to a total heat loss of 44.2 TW. Using numerical simulations, it has been shown that this heat loss corresponds to a present-day cooling rate of the Earth's mantle of 50 ± 25 K/Gyr.

The heat flow pattern observed on the Earth's surface reflects the action of the plate tectonics cycle, and surface heat flow is thus indicative of interior processes. Plate tectonics (see Evolution of Planetary Interiors and Earth: Surface and Interior) appear to be absent on planets other than the Earth, which is usually attributed to the lack of surface water on these bodies. Water acts to weaken the crustal rocks and facilitates breakup and finally subduction of the lithospheric plates.

The identification of large-scale heat flow variations on planets other than the Earth would further our understanding of heat transport mechanisms in these bodies, ultimately shedding light on the question why plate tectonics as observed on the Earth is unique in the solar system. Through analysis of the global heat flow data set it was found that the surface heat flow field correlates well with the seismic shear wave velocity in the upper mantle, thus establishing a link between seismic and heat flow investigations. For the Earth, heat flow data is collected and archived by the International Heat Flow Commission, and the global data set can be accessed through their website.

4.3. Lunar Heat Flow

In contrast to the Earth, structures like subduction zones, back-arc volcanoes, or midocean ridges are absent on the Moon, indicating the absence of plate tectonics. Instead, the Moon's interior can be viewed as an immobile outer shell underlain by a convecting mantle, a style of convection termed "stagnant lid convection". In fact, there is debate as to whether the Moon continues to convect. In any case, since no hot material is being transported directly to the surface, the heat flow pattern on the Moon only weakly reflects any mantle convection pattern, thus making the lunar surface similar to Earth's continental shields from a heat flow perspective.

Therefore, lunar heat flow mainly reflects the distribution and abundance of heat-producing elements in the lunar crust. Surface abundances of radioactive elements can be determined from orbit using gamma-ray spectroscopy, as gamma-ray data are sensitive to elements in the upper few decimeters of the crustal material. In this way, the Lunar Prospector mission between 1998 and 1999 determined elemental abundances on the lunar surface. It was found that K and Th vary widely across the surface, being significantly enriched in a region in the Oceanus Procellarum. This unique geochemical province was subsequently termed the Procellarum KREEP (for potassium (K), rare earth elements (REE), and phosphorus (P)) Terrane (PKT). A map of thorium on the Moon is shown in Figure 55.7. Thorium in the PKT is enriched by a factor of up to 10 with respect to the surrounding highlands, and although slightly elevated Th concentrations are observed

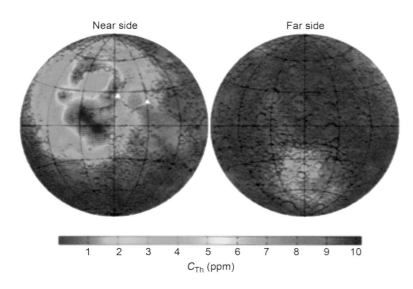

FIGURE 55.7 Surface abundance of Th as determined from Lunar Prospector gamma-ray data. The Procellarum KREEP terrane exhibits Th abundances which are elevated by up to a factor of 10 with respect to the lunar highlands. The Apollo 15 and 17 landing sites are indicated by triangles.

in the South Pole—Aitken Basin on the lunar farside, such large concentrations are absent anywhere else on the surface. However, from remote sensing data alone, the depth extent of the deposits cannot be determined, and in situ heat flow experiments are necessary to address this question.

While deep boreholes are needed to reliably determine the heat flow from the interior of the Earth, drilling deep holes on extraterrestrial bodies is inhibited by the mass of the necessary equipment. For the foreseeable future, the subsurface of the Moon will only be directly accessible to a depth of a few meters, and the deepest holes drilled by the Apollo astronauts extended to 2.4 m depth (Figure 55.8). Given that surface temperature fluctuations on the Moon are large, deeper boreholes might seem to be required for heat flow measurements. But the shallow depth of the boreholes is compensated by the low thermal conductivity of the lunar regolith, which rapidly attenuates surface temperature fluctuations. Thus nearly isothermal conditions are attained at a depth of only about 1 m, and temperature gradients can be reliably determined from data collected at shallow depth. Similar approaches would be feasible on Mercury or Mars. The latter environment is more challenging due to the presence of the Martian atmosphere, which acts to increase thermal conductivity and thus the penetration depth of temperature perturbations.

The history of extraterrestrial heat flow measurements starts with the ill-fated Apollo 13 mission, and the first successful measurement was not performed until Apollo 15. Technical problems during the emplacement of the probes on Apollo 16 frustrated a successful measurement during that mission, and only one further successful measurement was performed during Apollo 17 before the Apollo program ended. Material including images, videos, and radio communication transcriptions connected to the

FIGURE 55.8 The heat flow experiment during emplacement on Apollo 15. The drill rack is in the foreground and the Apollo lunar surface drill is on the surface behind the borestem, which is emplaced in the ground. The dark, two-segment rod in the left hand of the crewman is the heat probe. The white probe emplacement tool is in his right hand (*Apollo image AS-15-92-12407*).

Apollo missions is accessible through the Apollo Lunar Surface Journal Web sites maintained by NASA.

For Apollo 15, thermal gradients were determined to depths of up to 1.4 m, while the Apollo 17 probes were emplaced to a final depth of 2.4 m (Figure 55.9). Thermal

FIGURE 55.9 The Apollo 17 heat flow probe after emplacement in the borehole. The probe is connected to the Apollo Lunar Surface Experiment Package central station in the background. *Apollo image AS17-134−20496.*

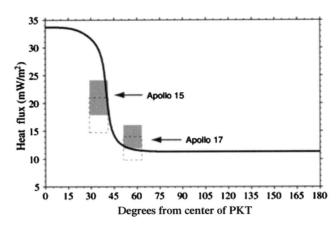

FIGURE 55.10 Modeled present-day lunar heat flow as a function of distance from the center of the Procellarum KREEP terrane (PKT). Initial heat flow estimates derived for the Apollo 15 and 17 measurements are indicated in gray, and subsequent corrections accounting for lateral heat transport in the lunar crust are shown as dashed rectangles. The sizes of the rectangles indicate the measurement uncertainty. *Figure from Wieczorek & Phillips (2000).*

conductivity was determined in situ from a line heat source method, and thermal diffusivity κ was independently estimated from the attenuation of the annual temperature wave. Together with measurements of density ρ and heat capacity c_p on drill core samples returned to the Earth, thermal conductivity $k = \kappa/\rho c_p$ could be independently estimated from the thermal diffusivity measurement. Results of the latter method were finally given preference over the line heat source data, as it was concluded that the direct heating method likely overestimated thermal conductivity due to regolith compaction during probe emplacement.

From these measurements, Langseth and colleagues estimated heat flows of 21 and 16 mW/m² at the Apollo 15 and 17 sites, respectively, and the average lunar heat flow was estimated to be 18 mW/m². Accounting for effects of lateral heat transport, these values were later revised downward and an average heat flow of 12 mW/m² was determined. This heat flow corresponds to a bulk lunar Uranium concentration of 20 ppb, a value close to that derived for the silicate fraction of the Earth, lending support to the hypothesis that the Moon reaccreted from the debris after a giant impact of a Mars-sized planetary body with the proto-Earth. However, it should be kept in mind that the wealth of new data returned by recent lunar missions (SELENE, Gravity Recovery and Interior Laboratory (GRAIL)) suggests that the average crustal thickness is less than previously thought and that it maintains significant porosity to depths of tens of kilometers. Therefore, a fresh look at the implications of the Apollo measurements is warranted.

Models of the heat production in the lunar interior indicate that the observed concentration of heat-producing elements in the Procellarum region must be the surface expression of a larger reservoir at depth. The Apollo heat flow data are compatible with the local presence of a KREEP layer up to 10 km thick at the base of the lunar crust (see Figure 55.10), which would be in line with the persistent volcanic activity observed in the region. Youngest mare basalts are only slightly over 1 Gyr old, and it has been speculated that the concentration of heat-producing elements in the PKT could have been caused by large-scale mantle flow following the gravitational overturn of a freezing magma ocean that likely covered the Moon following accretion. However, further heat flow measurements would be needed to corroborate this view and, ideally, future measurements should be conducted in the center of the PKT and/or far away in the lunar highlands. The latter would be unaffected by the surface concentrations of heat-producing elements and be more representative of the global average.

4.4. Future Investigations

Future lunar heat flow measurements are currently being studied in the frame of an international collaboration termed the International Lunar Network, but definite plans for mission implementation do not exist at present. However, the recently selected Discovery-class mission InSight, which is due to launch in 2016, will place a geophysical lander carrying a heat flow probe in the southern Elysium region of Mars in September 2016. The heat flow probe, termed the Heat flow and Physical Properties Package, or HP³ for short, is built to access the Martian regolith to a depth of up to 5 m by means of a hammering mechanism, emplacing a suite of temperature sensors into the subsurface. The overall measurement approach is similar to that taken by Bullard or Langseth, and a depth-resolved measurement of the subsurface temperatures will be used to determine the thermal gradient. Active heating elements inside HP³ will be used to determine the thermal conductivity in situ, and the attenuation of the annual temperature wave will be used to independently estimate thermal diffusivity and to provide a consistency check for the thermal conductivity value determined from active heating.

While a single heat flow measurement is hardly enough to confidently constrain the average heat flow from a planet, the InSight measurement will provide an important baseline. In addition, the seismic experiment on InSight will provide an estimate of crustal thickness, which can be used to validate thickness models derived from gravity data. The thickness of the crust is a key constraint needed to interpret local heat flow measurements in terms of the global average. Furthermore, the heat flow pattern on the Martian surface is expected to be much simpler than that of either the Earth or the Moon for two reasons: First, Mars currently lacks a plate tectonics cycle (although it may have possessed one during its earliest evolution), and second, Mars does not show any geochemically anomalous regions like the PKT of the Moon (compare Figure 55.11). Therefore, a first global estimate of the average heat flow can be derived from the InSight data, but further measurements in different locations are clearly desirable.

Another device built to measure the energy balance at the surface of an extraterrestrial body is the Rosetta MUPUS instrument (MUlti PUrpose sensor for Subsurface observation), currently on its way to comet 67p/Churyumov–Gerasimenko. Its goal is to measure the heat flow into the comet, the largest unknown contribution to the surface energy balance of the comet. The instrument will be delivered to the cometary surface onboard the Rosetta Philae Lander and is shown in Figure 55.12. MUPUS consists of a 35-cm-long rod equipped with temperature sensors and heaters, and a hammering mechanism is mounted on top of the rod to emplace it into the ground. The instrument will then determine the surface temperature, subsurface thermal gradient, as well as the thermophysical properties of the cometary regolith, thus quantifying the heat flow into the comet and help to detail the solar energy input responsible for driving activity and liberating gas and dust to the cometary coma and tail.

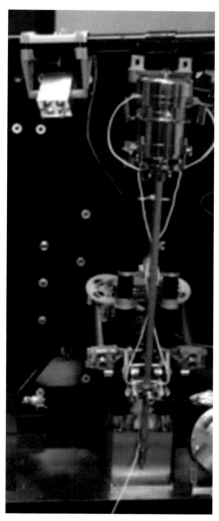

FIGURE 55.12 The Multi Purpose Sensor (MUPUS) onboard the Rosetta Lander Philae, currently on its way to rendezvous the comet 67p/Churyumov–Gerasimenko in 2014. The MUPUS probe on the right-hand side of the picture houses 16 temperature sensors, which can simultaneously be operated as heaters for thermal conductivity determination. MUPUS will be deployed away from the lander using an extendable boom. The golden cap houses a hammering mechanism which will drive the probe into the cometary surface. *From Spohn et al. (2007).*

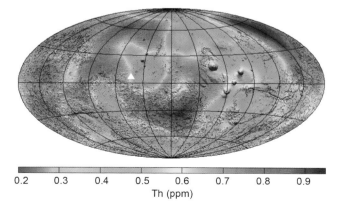

FIGURE 55.11 Abundance of Th on the Martian surface as determined from gamma-ray spectroscopic measurements of the Mars Odyssey spacecraft.

Other measurement approaches which have been proposed to determine heat flow on extraterrestrial bodies include the so-called flux plates, which can measure surface heat flow in environments with constant surface temperatures. Flux plates are placed on the planetary surface from which the heat flow is to be measured, and the temperature difference across a layer of known thermal conductivity is recorded. Such a device would be suited to determine the heat flow on Venus, whose rocky surface precludes drilling to any significant depth. The constant cloud cover and very dense atmosphere of Venus result in very stable surface temperatures, such that a flux plate measurement could be successfully executed there. Such a measurement would provide a very important constraint on how Earth-sized planets without plate tectonics lose their heat. Many models of the interior evolution of Venus speculate that it should currently be heating up, resulting in a present-day Urey ratio >1.

5. INVESTIGATING PLANETARY INTERIORS USING EM SOUNDING

EM sounding encompasses a wide variety of methods used to sense subsurface structure from depths of less than a few meters to a thousand kilometers or more. EM sounding is distinct from surface-penetrating radar (or its seismological cousins) in being inductive rather than wavelike: the diffusive transfer of energy is akin to that in heat flow, groundwater, and electrical circuits. This transition to induction occurs at low frequency, in the kilohertz to megahertz range depending on ground electrical conductivity. Thus a vast part of the EM spectrum, from below 1 µHz to perhaps 1 MHz, is within the realm of EM sounding. In this section we will be focusing on the so-called passive techniques—those that use natural EM energy in planetary magnetospheres, ionospheres, and atmospheres as the source.

Electrical conduction in the familiar sphere of Earth's upper crust is largely electrolytic, i.e. by movement of ions in water—conductivity increases with porosity and fluid content. In frozen H_2O systems, conduction through ice and hydrates is protonic, through charge defects resulting from lattice substitutions of salts, acids, and bases. Analogous substitutions by iron, aluminum, oxygen, and hydrogen in silicates result in electronic semiconduction, both from electrons and holes. Conductivity is also strongly temperature dependent. In contrast, the outer portions of all other silicate or icy bodies in the solar system are likely to be much more resistive than the Earth, for the simple reason that free water is lacking. This opens up these regions to EM sounding at higher frequencies (see below), which in turn provides better measurements and more convenient signal integration.

The effective penetration depth for EM waves is a strong function of frequency and resistivity. The EM skin depth (in meters) is given by $\delta = 500/\sigma^{1/2}f^{1/2}$, where σ is the conductivity (S/m) and f is the frequency (Hz). This relation indicates what frequencies are necessary to probe to a specified depth at a particular average conductivity. For example, terrestrial exploration geophysics seeking targets at several tens of meters to several kilometers depth utilizes frequencies from ~10 mHz to 10 kHz, but mantle studies require 10 µHz or lower. Conversely, the lunar lower mantle has been adequately sounded in the millihertz range, but better characterization of the more resistive upper mantle and crust will require frequencies up to 1 Hz or higher.

In practice, the true conductivity vs depth is determined from the apparent conductivity (or other measurement parameters; see below) vs frequency using classical inverse methods. Note that unlike the potential field methods discussed above, EM inversions are not formally nonunique. The inversion can be thought of as assigning the conductivity at the highest measured frequency to the uppermost layer, and progressively solving for the conductivity at greater depths by comparing the apparent conductivity at the next depth to the true conductivity of the overlying material.

5.1. Natural Sources

Within the low-frequency branch of diffusive EM, the primary division of techniques is between those that use an artificial source (a transmitter) and those that rely on natural sources. The former enjoy considerable flexibility and high signal to noise but require significant additional resources and are generally limited to investigation depths of several kilometers. On Earth, abundant energy exists at less than 1 Hz from magnetospheric pulsations and the interaction of the magnetosphere with diurnal heating of the ionosphere. Above 1 Hz, the ground–ionosphere waveguide allows lightning energy to be recorded globally as the low-frequency Schumann resonances and regionally at higher frequencies. These signals are collectively known as spherics. These sources of energy for EM sounding will likely be present on other bodies with magnetospheres, ionospheres, and chargeable atmospheres. In interplanetary space, temporal variations in the plasma density of the solar wind provide signals that have already been exploited for lunar sounding. Finally, special circumstances around specific planets can provide unique sources; in particular, the inclination of satellite orbits with respect to the static magnetic field of Jupiter yields an apparent time variation at the orbital period. Overall, some kind of ambient energy is likely present at most bodies that would enable EM sounding.

TABLE 55.1 Approaches to Natural Source EM Sounding

Method	Measurements	Number of Stations	Comments
Transfer function (TF)	Magnetic field B	2	Determine source field by distant second station unaffected by target
Transfer function-1 (TF-1)	B	1	Special cases using prior knowledge of source field *or* target is "perfect" conductor
Geomagnetic depth sounding (GDS)	B	3 or more	Compute impedance from vertical field and horizontal gradients
Geomagnetic depth sounding (GDS-1)	B	1	Special case where horizontal gradient can be computed using known periods and length scales
Magnetotellurics (MT)	B + Electric field	1	General single-station method
Wave tilt (WT)	Electric field	1	General single-station method

5.2. Measurement Techniques

The fundamental quantity that must be derived in any sounding is the frequency-dependent EM impedance Z, and it is the variety of approaches to Z that lead to more individual techniques in EM than in any other geophysical method. A commonly used parameter, the apparent conductivity, is simply related to the impedance as $\sigma_a = \mu\omega/Z^2$, where μ is the permeability and ω is the angular frequency. Apparent conductivity is useful because it is dimensionally identical to true conductivity but instead represents an aggregate response of the target.

Two known quantities are necessary to determine the impedance (see Table 55.1). One of these quantities is nearly always the magnetic field B near the target, i.e. the sum of source and induced magnetic fields. Note that measurement need not be made at the surface, as long as it is at an altitude substantially less than one skin depth at the highest frequency of interest. This can include spacecraft if the desired exploration depth is sufficiently large. Active methods can measure either electric or magnetic fields; the second known quantity is the transmitted signal.

The magnetic transfer function (TF) method is straightforward, measuring three components of the magnetic field at a location on or near a planetary body and comparing that to a measurement distant from the body. This relates the sum of source and induced fields to the source field alone, from which the induced field and internal conductivity structure can be derived. The TF method was used for Apollo-era lunar soundings. In a few special cases (TF-1) of accurately characterized natural signals—the Earth's ring current or the time variation introduced by the motion of the Galilean satellites in Jupiter's main field—the source can be specified a priori, so a single platform is sufficient. Single-magnetometer characterization can also be performed where the target can be approximated as a perfect conductor, like the Moon's core.

Geomagnetic depth sounding (GDS) uses surface arrays of magnetometers to determine impedance from the ratio of vertical B to the magnitude of the horizontal gradient of B. Because the wavelength in the ground $\lambda = 2\pi\delta$, GDS requires array spacing comparable to the skin depth in order to resolve the relevant horizontal wave structure. This calls for dense arrays to resolve the outer tens to hundreds of kilometers of planetary bodies.

In the magnetotelluric (MT) method, measurement of the electric field E supplies the required second piece of information and enables complete EM soundings from a single station. Finally, the wave-tilt (WT) method is preferable for aerial surveys because the quadrature horizontal E (containing most of the inductive signal) can be readily determined by comparison to the vertical E.

Only magnetic methods have been used heretofore in planetary exploration because of their simplicity: electric field measurements are more challenging at low frequencies and noninductive contributions to E must be identified. Nonetheless, those methods using E have the significant advantage of complete soundings from a single vehicle that do not require a priori knowledge or special conditions.

5.3. Terrestrial Exploration

Regional EM soundings of the crust and upper mantle of the Earth began in the 1960s using magnetometer arrays (GDS) and have since been supplemented by MT. The deepest soundings in the mantle use very long period signals from reasonably well-known sources. Two examples here illustrate how EM sounding can elucidate regional

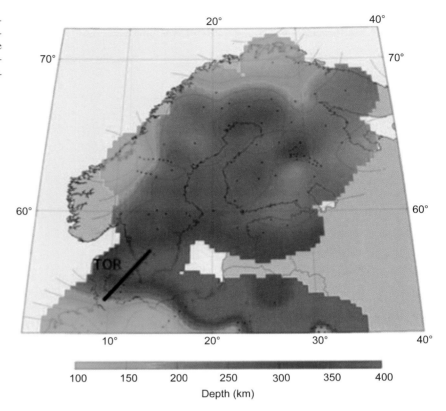

FIGURE 55.13 Thickness of the thermal lithosphere in northern Europe compiled from magnetotelluric soundings. Profile "TOR" crosses the Trans-European Suture Zone, a region of thickening caused by multiple plate tectonic collisions. *Korja (2007)*.

variations in upper mantle structure (Figure 55.13) or deep mantle structure (Figure 55.14). At present, fully three-dimensional images of the Earth are emerging, using decades of observations from hundreds of geomagnetic observatories.

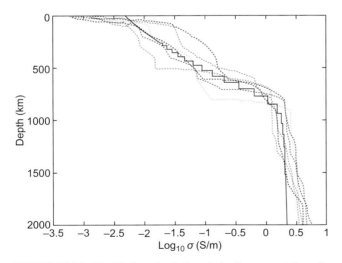

FIGURE 55.14 Electrical conductivities at six European stations (in various colors) and global average (black) from MT and GDS methods (*Semenov & Kuvshinov (2012)*). Much current work in interpreting these kinds of profiles using laboratory data focuses on the tradeoff of temperature (including partial melting) vs dopants such as trace amounts of water.

Variations in electrical conductivity revealed by EM sounding are interpreted in terms of temperature and composition. The former dominate Figure 55.13 where the thickness of cold lithosphere is doubled in the Trans-European Suture Zone. Both may affect the deep profiles of Figure 55.14. The effect of trace quantities of water (hundreds of ppm) in the Earth's mantle is hotly debated.

5.4. Magnetic Sounding of the Moon

During the Apollo Program, a variety of lunar surface experiments were performed in geophysics, space physics, and astrophysics. Simultaneously, the robotic Explorer program was reaching full stride. The Apollo 12 surface magnetometer performed continuous dayside measurements for several months; throughout this time the Explorer 35 spacecraft, also including a magnetometer, orbited the Moon at distances up to several thousand kilometers. The two instruments jointly recorded fluctuations in the Earth's magnetotail and the solar-wind-embedded interplanetary magnetic field from 10^{-5} to 3×10^{-2} Hz. EM sounding was performed using the frequency-dependent magnetic TFs. Due to the fact that the outer portions of the Moon are so electrically resistive (because it is both cold and dry), measured EM waves at even the highest observed frequency penetrate a few hundred kilometers before being able to induce measurable eddy

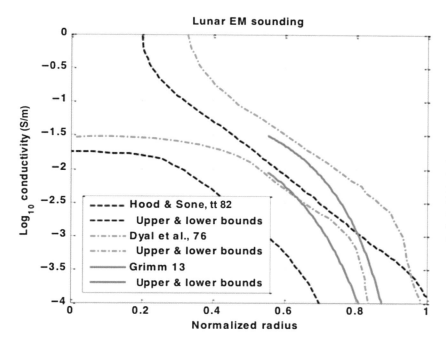

FIGURE 55.15 Electrical conductivity of the Moon, derived by comparing the time-varying magnetic field at the lunar surface with that of a distant satellite (transfer function method). Conductivity increases with depth due to increasing temperature. The large uncertainty at depth is because the volume fraction is so small. Results are also poorly constrained at depths less than a few hundred kilometers (normalized radius >0.8) because this region is so resistive it is transparent to the millihertz signals used for the soundings. Conductivities in the midmantle (normalized radius = 0.7, 500 km depth) indicate higher than expected temperatures (1600 K), perhaps due to higher radionuclide abundance or a thermally insulating megaregolith. Alternatively, impurities such as Fe^{3+}, Al^{3+}, or H^+ can raise the mantle conductivity.

currents. Hence there are large uncertainties in the electrical properties of the lunar crust and upper mantle. Conversely, the lowest frequencies ($\ll 10^{-4}$ Hz) fully penetrate the Moon but are relatively insensitive to the innermost 20–30% of the radius (Figure 55.15).

Although the Apollo Explorer TFs established an upper limit to the size of a lunar core, better bounds were determined many years later during the Lunar Prospector mission. External magnetic field lines are nearly parallel when the Moon is passing through Earth's magnetotail. If the permeability–conductivity product of the core is very large—effectively equivalent to infinite conductivity—the induced fields will completely exclude the source field, so the induced dipole produces a perturbation to the parallel external field lines that depends only on the core size (Figure 55.16). The resulting bounds 340 ± 90 km are consistent with seismology and laser ranging. It is also worth noting that magnetic **hysteresis** measured with Apollo Explorer indicates 4–14 wt% free iron, depending on assumptions of bulk mineralogy.

5.5. Magnetic Sounding of the Galilean Satellites

Among the most significant results of the Galileo mission was the discovery of EM induction signatures in the magnetometer records for flybys of Io, Europa, and Callisto, and perhaps for Ganymede as well. Because the conductivity of solid ice or rock is small, salty subsurface oceans on the icy satellites and a global partial melt layer on Io are the most likely causes of these signatures. Jupiter's massive magnetic moment produces field amplitudes of 4–1700 nT at the Galilean satellites and because this field is not orthogonal to the plane of the satellites, it varies periodically with **synodic periods** of 10–13 h as viewed from the satellites. The result is a source field variation of $\sim 10^{-6}$ Hz with known amplitude and direction, so the TF can be evaluated using the Galileo spacecraft alone (TF-1). The depth to and conductivity of the

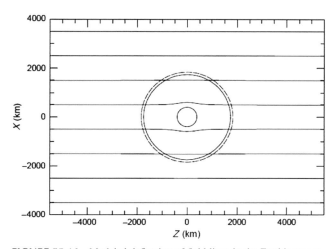

FIGURE 55.16 Modeled deflection of field lines in the Earth's magnetotail by a perfectly conducting lunar core of radius 400 km (inner circle). This is a feature of time-varying signals: eddy currents in the conducting core exclude the primary magnetic field, whereas a static field would be focused into a magnetically permeable core. A range of core radii 250–430 km was found to match small perturbations observed by the Lunar Prospector spacecraft at the outermost circle. *Hood et al. (1999).*

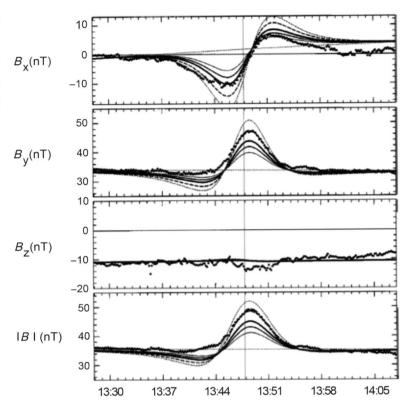

FIGURE 55.17 Galileo magnetometer data (dots) at Callisto flyby on June 25, 1997, compared to various induction models for internal oceans (lines). The temporal variation in Jupiter's static field due to its inclination with respect to Callisto's orbit causes inductive eddy currents to flow in the satellite's interior, forming an induced dipole. This dipole is evident in the raw data above and because its orientation always opposes the main field, it cannot have a static origin. *Zimmer et al. (2000).*

internal water or magma oceans can be inferred to be tens of kilometers and ~1 S/m, respectively. Skin effects allow these fields to penetrate through much of the satellite, but ocean thicknesses are harder to estimate because EM sounding is sensitive only to the conductance (the product of conductivity and thickness) when trying to assess a highly conductive layer. Nonetheless, magma and water ocean thicknesses of 100–200 km have been derived (Figures 55.17 and 55.18).

5.6. Future Exploration

Many missions can exploit EM sounding of planetary interiors using natural sources and simple sensor configurations. A significant magnetosphere makes Mercury the most Earth-like in terms of classical natural EM sources, yet frequent **reconnection** events define an even more dynamic environment. Its core will also be a prime EM target and the conductivity profile in the mantle may show whether it is convecting or whether the lithosphere comprises the entire mantle. The iron content of the mantle may also be recovered.

With a shielding ionosphere and no defined external sources, EM soundings of Venus must be performed from the surface or atmosphere, but both ionospheric disturbances and lightning are known to exist. A balloon in the benign middle atmosphere can perform long-duration remote sensing of the subsurface using MT or WT. The dearth of water on Venus implies that these signals will penetrate tens of kilometers into the ground enabling the mean thickness and lateral variations of the **thermal lithosphere** to be determined.

FIGURE 55.18 Internal structure of Io revealed by EM sounding (TF-1 method). Red layer ~40 to >90 km deep is the global partial melt. *Khurana et al. (2011).*

FIGURE 55.19 Concept for a lunar geophysical station including a magnetotelluric sounder. Three spring-launched orthogonal wires have electric field sensors at their tips, whereas boom at left supports fluxgate and search-coil magnetometers. Other instruments include seismometer, heat flow probes (2), and retroreflector. *Courtesy Jet Propulsion Laboratory.*

EM soundings of the Moon can be improved by expanding the frequency range: shallower skin depths at higher frequencies can resolve the outer resistive portions, and stable long-term measurements may detect the core and distinguish molten silicate from iron. Low-altitude orbital coverage or a surface network (Figure 55.19) would allow geographic variations to be resolved, for example, if the Moon's most volcanically active regions are also distinct in the upper mantle.

The cold surface and lower heat flow of Mars result in a cryosphere many kilometers thick, putting liquid groundwater and potential stable microbial habitats out of reach of conventional planetary remote or in situ sensing. This conductivity contrast would be comparable to that observed for the Galilean satellites, and its depth would provide a ready estimate of geothermal gradient.

The extremely low frequency time variations developed at the satellites of Jupiter by their inclined orbits with respect to the main field will also be present at Uranus and Neptune. This will enable single-magnetometer global TF soundings for close flybys of satellites of the ice giants. Saturn's satellites have negligible inclination with respect to the main field, but for all the giant planets, the diversity of magnetospheric phenomena may provide a rich spectrum of EM signals. Titan is of course different because of its atmosphere; like Venus, MT or WT measurements from landed or aerial assets could exploit ionospheric disturbances or lightning.

6. SUMMARY

For the past 40 years, the geophysical exploration of the deep interiors of the terrestrial planets had lagged far behind the exploration of their surfaces. This is partly due to the difficulty in making the necessary precise measurements, which often requires multiple long-lived surface platforms for the best results, and partly due to the difficulty in generating the same level of public excitement as a surface mission, with its easily communicated imagery and sense of exploration. However, the scientific value of these measurements is unquestioned, and we now stand at the beginning of a new era of geophysical missions. The GRAIL mission has recently completed its mapping of the gravity field of the Moon at unprecedented resolution, and MESSENGER (MErcury, Surface, Space ENvironment, GEochemistry and Ranging) is currently exploring the interior of Mercury with gravity and magnetic measurements from orbit. Also, the InSight mission to Mars will, starting in late 2016, obtain surface seismic, heat flow, and magnetic measurements which will allow us to address many of the fundamental questions regarding the structure, origin, and evolution of Mars. Looking farther ahead, ESA is studying INSPIRE (Integrated Network of Seismic Probes to Invigorate Robotic Exploration), a Mars geophysical network mission to use multiple landers to sharpen the seismic resolution of that planet's interior, and there are plans within NASA to explore the possibility of a habitat/environment on Europa using geophysical techniques to probe the extent of its putative subcrustal ocean. Thus the next decade promises to be an exciting one for planetary interior exploration.

BIBLIOGRAPHY

Bullard, E. (1954). The flow of heat through the floor of the Atlantic Ocean. *Proc. Roy. Soc. London A, 222*, 408–429.

Grimm, R. (2002). Low-frequency electromagnetic exploration for groundwater on Mars. *Journal of Geophysical Research, 107*, 1–29.

Hood, L., et al. (1999). Initial measurements of the lunar induced magnetic dipole moment using Lunar Prospector Magnetometer data. *Geophysical Research Letters, 26*, 2327–2730.

Jaupart, C., & Mareschal, J.-C. (2007). *Heat flow and thermal structure of the lithosphere*. In G. Schubert (Ed.), *Treatise on geophysics* (Vol. 10); (pp. 217–252). Oxford: Elsevier Ltd.

Khurana, K., et al. (2011). Evidence of a Global Magma Ocean in Io's Interior. *Science, 332*, 1186.

Korja, T. (2007). How is the European lithosphere imaged by magnetotellurics. *Surveys in Geophysics, 28*, 239–272.

Lognonné, P., & Johnson, C. (2007). *Planetary seismology*. In G. Schubert (Ed.), *Treatise on geophysics* (Vol. 10); (pp. 69–122). Oxford: Elsevier Ltd.

Pollack, H. N., et al. (1993). Heat flow from the Earth's interior: Analysis of the global data set. *Reviews of Geophysics, 31*(3), 267–280.

Semenov, A., & Kuvshinov, A. (2012). Global 3-D imaging of mantle electrical conductivity based on inversion of observatory C-responses - II. Data analysis and results. *Geophysical Journal International, 191*, 965–992.

Spohn, T., Seiferlin, K., Hagermann, A., Kollenberg, J., Ball, A. J., Banaszkiewicz, M., et al. (2007). MUPUS – a thermal and mechanical properties probe for the Rosetta Lander Philae. *Space Science Reviews, 128*, 339–362.

Wieczorek, M. (2007). *The gravity and topography of the terrestrial planets*. In G. Schubert (Ed.), *Treatise on geophysics* (Vol. 10); (pp. 165–206). Oxford: Elsevier Ltd.

Wieczorek, & Phillips. (2000). The Procellarum KREEP Terrane: Implications for mare volcanism and lunar evolution. *Journal of Geophysical Research, 105*, 20417–20430.

Zimmer, C., et al. (2000). Subsurface oceans on Europa and Callisto: Constraints from Galileo magnetometer observations. *Icarus, 147*, 329–347.

Chapter 56

Planetary Exploration Missions

James D. Burke
The Planetary Society, Pasadena, CA

Chapter Outline

1. **Introduction** — 1206
2. **Program Evolution** — 1206
   - 2.1. Launch Services — 1206
   - 2.2. Tracking and Data Acquisition — 1207
   - 2.3. Spacecraft — 1208
   - 2.4. Operations — 1209
   - 2.5. Reliability and Quality Assurance — 1209
   - 2.6. Management — 1209
3. **Sun and Heliosphere** — 1209
   - 3.1. Helios — 1210
   - 3.2. Solar Maximum Mission — 1210
   - 3.3. Ulysses — 1210
   - 3.4. SOHO — 1210
   - 3.5. Advanced Composition Explorer — 1210
   - 3.6. TRACE — 1211
   - 3.7. Genesis — 1211
   - 3.8. Solar Dynamics Observatory — 1211
4. **Mercury** — 1211
   - 4.1. Mariner 10 — 1211
   - 4.2. MESSENGER — 1211
5. **Venus** — 1211
   - 5.1. Mariner 2 — 1211
   - 5.2. Veneras 4 Through 16 and Vega — 1211
   - 5.3. Mariner 5 — 1211
   - 5.4. Mariner 10 — 1212
   - 5.5. Pioneer Venus — 1212
   - 5.6. Vega 1 and 2 — 1212
   - 5.7. Magellan — 1212
   - 5.8. Venus Flyby — 1212
   - 5.9. Venus Express — 1212
6. **Earth** — 1212
   - 6.1. Resurs — 1213
   - 6.2. Galileo Earth Flybys — 1213
   - 6.3. Terra — 1213
   - 6.4. TOPEX/Poseidon and Jason-1 and Jason-2 — 1213
   - 6.5. GRACE — 1213
   - 6.6. Envisat — 1213
   - 6.7. Aqua — 1213
   - 6.8. Aura — 1213
7. **Moon** — 1213
   - 7.1. Luna 1, 2, and 3 — 1214
   - 7.2. Ranger 7, 8, and 9 — 1214
   - 7.3. Zond 3 — 1214
   - 7.4. Luna 9 and 13 — 1214
   - 7.5. Luna 10, 11, 12, and 14 — 1214
   - 7.6. Lunar Orbiter 1–5 — 1214
   - 7.7. Surveyor 1, 3, 5, 6, and 7 — 1214
   - 7.8. Zond 5, 6, 7, and 8 — 1214
   - 7.9. Apollo 8 — 1214
   - 7.10. Apollo 10 — 1215
   - 7.11. Apollo 11 — 1215
   - 7.12. Apollo 12 — 1215
   - 7.13. Apollo 13 — 1215
   - 7.14. Apollo 14 — 1215
   - 7.15. Apollo 15, 16, and 17 — 1215
   - 7.16. Luna 16, 17, 20, 21, and 24 — 1215
   - 7.17. Clementine — 1216
   - 7.18. Lunar Prospector — 1216
   - 7.19. SMART-1 — 1216
   - 7.20. Kaguya/SELENE — 1216
   - 7.21. Chang'e 1 — 1216
   - 7.22. Chandrayaan-1 — 1216
   - 7.23. Lunar Reconnaissance Orbiter — 1216
   - 7.24. LCROSS — 1216
   - 7.25. ARTEMIS — 1216
   - 7.26. Chang'e 2 — 1216
   - 7.27. GRAIL — 1216
   - 7.28. LADEE — 1216
   - 7.29. Chang-E 3 and Yutu — 1217
8. **Mars** — 1217
   - 8.1. Mariner 4 — 1217
   - 8.2. Mariner 6 and 7 — 1217
   - 8.3. Mars 2 and 3 — 1217
   - 8.4. Mariner 9 — 1218
   - 8.5. Mars 4, 5, 6, and 7 — 1218
   - 8.6. Viking 1 and 2 — 1218
   - 8.7. Phobos 1 and 2 — 1218
   - 8.8. Mars Pathfinder and Mars Global Surveyor — 1218
   - 8.9. Failures of the 1990s — 1218
   - 8.10. Mars Odyssey — 1218
   - 8.11. Spirit and Opportunity — 1218
   - 8.12. Mars Express — 1218
   - 8.13. Mars Reconnaissance Orbiter — 1219

8.14. Phoenix	1219	9.8. Rosetta	1220
8.15. Curiosity	1219	9.9. Deep Impact	1220
8.16. Mangalyaan	1219	9.10. Dawn	1220
9. Small Bodies	**1219**	**10. Outer Planets and Moons**	**1221**
9.1. International Cometary Explorer	1219	10.1. Pioneers 10 and 11	1221
9.2. The Halley Armada	1219	10.2. Voyagers 1 and 2	1221
9.3. Galileo En Route Encounters	1219	10.3. Galileo	1221
9.4. NEAR-Shoemaker	1220	10.4. Cassini–Huygens	1221
9.5. Deep Space 1	1220	10.5. New Horizons	1221
9.6. Stardust	1220	10.6. JUNO	1222
9.7. Hayabusa (Muses-C)	1220	**11. Conclusion**	**1222**

1. INTRODUCTION

Immediately upon the launching of *Sputnik* in 1957, it was clear that technical and political conditions would soon permit humans to realize a dream of centuries—exploring the Moon and planets. With large military rockets plus advanced radio techniques and the dawning skills of robotics, it would be possible eventually to send spacecraft throughout the solar system.

At first, however, the effort mostly failed. Driven by Cold War desires to show superiority in both military and civil endeavor, the Soviet and US governments sponsored hectic attempts to penetrate deep space, using strategic-weapon boosters, cobbled-together upper rocket stages, and hastily prepared robotic messengers. In time, as the equipment became more reliable and the management more capable, successes came—but in-flight failures have continued for decades to afflict all deep space programs. Lunar and planetary exploration is barely achievable even with the finest skills.

Here, where our purpose is to trace the development of flight missions, we do not dwell on the failures. We thus list only those missions that yielded some data in accord with their objectives (See Table 4 of the Appendix for a more complete listing).

In the early years, the Soviet Union garnered all the main firsts: the first escape from Earth's gravity, the first man and first woman in orbit, the first lunar impact, the first lunar landing, and the first lunar orbit. But the US program came from behind and scored the first data from a planet, Venus, and ultimately the grand prize, the first human exploration of the Moon.

Although Cold War rivalry provided emotional stimulus and government support, both programs were scientific right from the start. The earliest satellites were launched in support of the International Geophysical Year (IGY). Every mission carried some instruments to elucidate the character of its target body or region, and this largely continued as more nations and agencies joined the program. As a result, there is now a huge body of data, some of it still unexamined, from flight missions complementing an important archive of ground-based and Earth-orbiting telescopic observations of the Moon, planets, and small bodies in the solar system.

In what follows, recent and established near-future missions will be emphasized and only the most important earlier missions will be described. More complete lists are given in cited references.

Exploration of the Sun's domain by robots and of the Moon by humans has now placed us in a position to build strong hypotheses about the origin and evolution of the solar system and life and also to begin the study of other such systems as they are discovered. The missions that made this possible are an unprecedented expression, on a grand international scale, of peaceful human values and achievement.

2. PROGRAM EVOLUTION

2.1. Launch Services

Sputnik, which orbited on October 4, 1957, galvanized a huge response from the United States. Less than 12 years later, two astronauts walked on the Moon. In both USSR and the United States, an existing military rocket legacy enabled launch of the first satellites in 1957 and 1958. In time, space mission developers in other nations, driven primarily by a desire to have assured, independent access to space and also by a desire for their own organic technology advancement, began to provide their own launch services, at first for low Earth orbit (LEO) missions and later for missions beyond LEO, including geosynchronous Earth orbit (GEO), lunar, interplanetary, and planetary ventures. The required upper rocket stages were, and remain, an exasperating cause of failures.

The search for lower cost launch services, regarded as a key to future space development, has led over decades to the spending of resources equaling billions of dollars in studies and aborted vehicle developments, with as yet no promising result. However, work continues on a variety of approaches including air launch, hybrid air-breathing and rocket propulsion, and alternatively just extreme simplification in booster design.

Commercial launch services, now being encouraged by partial government subsidy, may offer another route toward lower costs.

Even without a radical launch cost reduction, a human breakout into the solar system is conceivable through the use of extraterrestrial resources. With energy and especially materials collected off the Earth, in a manner that has come to be called in situ resource utilization, great savings are possible in the mass that must be lifted from the Earth. This technique has yet to be demonstrated at a large enough scale for its true potential and its real comparative costs to be known.

2.2. Tracking and Data Acquisition

Without some way of delivering robotic mission results to the Earth, it does not matter what else works or does not work. In the time before the invention of radio, space science fiction authors assumed that signaling with light beams would be used. In a way they were right: Optical communications using lasers may yet become the method of choice in certain applications. Meanwhile, however, telemetry, tracking and orbit determination, command, and science in deep space entirely depend on radio technique.

For the first satellites, tracking stations were improvised based on previous military communications systems. For missions to the Moon and beyond, however, it was necessary to adapt methods used by strategic defense radar developers and radio astronomers. Huge antennas, supersensitive receivers, transmitters with enormous power output, and advanced data recording and processing all were needed.

From the outset, a difference in philosophy guided Soviet and American deep space engineers. In the then secretive USSR, the initial plan was to have spacecraft turn on their transmitters only when over Soviet territory, thus requiring ground stations in only the eastern hemisphere (In response to that, an American deep space signals intercept site was built in Eritrea). In the United States, on the other hand, the policy called for continuous contact, meaning that stations would have to be located worldwide, with of course a worldwide ground and space communications system for command, control, and data acquisition. That led to the

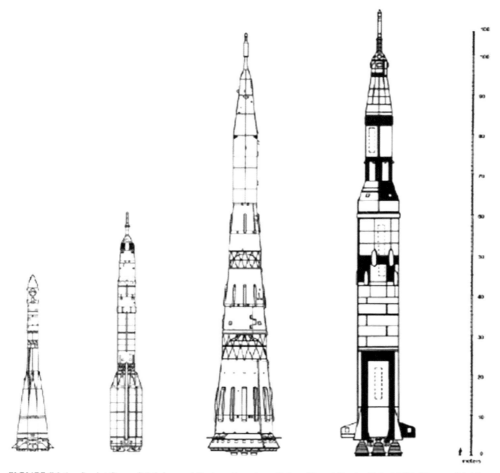

FIGURE 56.1 Soviet Soyuz/Molniya and Proton, American Saturn V, and Soviet N-1. *NASA History Division.*

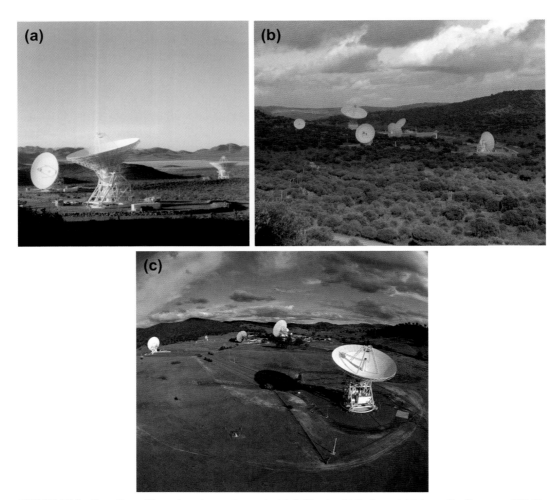

FIGURE 56.2 Deep Space Network stations today are in (a) California, (b) Spain, and (c) Australia. *Courtesy of NASA.*

creation of the Deep Space Network (DSN) whose stations today are in California, Spain, and Australia. For *Apollo*, a dedicated network was built, and it was backed up by the DSN plus a few specially equipped radio astronomy sites (Figure 56.1).

Meanwhile, the Soviet system evolved. At first located only in the Crimea, the Soviet network expanded to include sites in the Far East and in the central USSR, plus a fleet of tracking ships offshore. As additional nations joined in exploring the solar system and the cosmos beyond, many more stations were built for both tracking and radio science. Figure 56.2 shows three typical examples of the modern deep space stations that now exist in several countries.

2.3. Spacecraft

From a modest beginning, robotic spacecraft in LEO and GEO have evolved into the thousands of diverse science and applications machines that have been sent into orbit. Among these are large, multifunction craft devoted to observing the Earth as a planet, such as the European *Envisat* and the American *Terra* and *Aqua*.

Meanwhile, spacecraft designed to explore the solar system beyond the Earth underwent a similar evolution. The most important early mission was that of *Luna 3* in 1959, ending centuries of speculation by returning the first images of the Moon's far side. Soon after that, spacecraft design began to elaborate on the features that are essential in interplanetary space: attitude stabilization for pointing cameras and high-gain antennas, capable onboard data handling systems, long-duration power supplies, and long-surviving electronic equipment.

Three interplanetary technologies, whose mathematical foundations have been known for a century or more, are now starting to be applied. They are solar electric propulsion, weak stability boundary flight paths, and solar sailing. All may supplement or supplant chemical spacecraft propulsion with much increased efficiency in exchange for much longer transit times.

For human spaceflight, the earliest craft were mainly just capsules capable of sustaining life and returning safely to the Earth. But as space stations in LEO and flight beyond LEO became program objectives, more functions became the responsibility of human pilots and other crew members. The *Apollo* and space shuttle designs took full advantage of human capacities, while Soviet missions continued to make more use of teleoperation and onboard automation, as shown by the routine automated dockings of *Soyuz* and *Progress* servicing craft with the ISS.

2.4. Operations

In even the earliest lunar and planetary missions, it was necessary to keep track of the spacecraft's trajectory and issue commands for onboard functions both engineering and scientific. Gradually a humans-and-machines art developed, represented today by large rooms full of people and displays backed by buildings full of computers and data systems. Initially centered in main theaters, as missions have become more complex, these facilities have become dispersed, providing work spaces for the many specialized flight management and scientific teams working during a mission. With the Internet and other modern communications available, scientists can now reside at their home institutions and participate in missions in real time.

The latest trend is toward increasing onboard autonomy, holding the promise of reducing the large staffing needed round the clock to control missions. Some degree of autonomy, for example, stabilization with Sun and star references, is needed anyway in deep space, simply because of the round-trip signal times to distant spacecraft, tens of minutes for Mars and Venus and many hours in the outer solar system.

Operations have become more and more dependent on software whose design and verification now constitute one of the main cost items in each new mission's budget. With the maturing of the operations art have come numerous stories of remarkable rescues when a distant robot (or, as in *Apollo 13*, a human crew) got into trouble, but there are also instances where a mistake on the Earth sent a mission to oblivion.

2.5. Reliability and Quality Assurance

A vital part of the deep space exploration art is the creation of systems having but a small chance of disabling failures, plus an ability to work around failures when they do occur. One reason for the high cost of lunar and planetary missions is the need for multiple levels of checking, testing, reviewing, and documentation at every stage from the manufacturing of thousands of tiny components through assembly into subsystems and systems for both ground and flight, organization of human teams capable of imagining and analyzing failure scenarios and designing around them, and finally launching and controlling a mission during its years or decades of activity.

These costs are aggravated by the nature of deep space exploration as a work of building very complicated systems (hardware, software, and human—machine complexes) in ones and twos, as distinct from the repetitive manufacture of highly reliable items such as cars or computers whose teething troubles can be eliminated in early prototype testing. In a sense, every lunar or planetary mission is a first effort.

2.6. Management

In the twentieth century, as cold and hot warfare became more and more technological, a suite of skills, traditions, and managerial methods grew and created the capability of planning and executing large complicated projects. Many disciplines were involved, ranging from what became known as systems engineering all the way to new ways of organizing academic institutions, industries, and government agencies. The sometimes maligned worldwide military—industrial complex is a product of those developments, and it was the seedbed of the world's deep space programs.

The great lunar contest of the midtwentieth century highlighted some stark differences between American and Soviet management methods and organizations. At the outset, both used existing military hardware and existing military ways of working, but over time, the programs evolved along different paths. With their head start, the Soviets garnered all the early prizes in robotic lunar exploration, but when planning began for human lunar exploration the Soviet system fell into both technical and managerial disarray.

3. SUN AND HELIOSPHERE

The emphasis in this chapter is on missions to the Moon and planets. Now that star—planet aggregates are at last being observed as a class of known objects in the cosmos, it is essential for us to include at least a part of the story of missions devoted to our own star as host of a planetary system.

Our tale begins with the IGY. Centuries of ground-based investigations of sunspots and solar and terrestrial magnetism, plus decades of ionospheric and auroral research, had led by the midtwentieth century to a drive by scientists for a worldwide campaign of coordinated measurements resembling previous efforts such as international polar years. The new element now was the knowledge that rockets could take instruments beyond Earth's atmosphere and even into orbit. In both the USSR and the United States, satellite experiments were planned and announced in support of this goal, and in 1957 and 1958, it was achieved.

Explorer I found an excess of radiation saturating its detector. *Explorer IV* showed that this radiation is due to energetic particles trapped in the Earth's magnetic field, the Van Allen belts. Then, in 1962, an instrument aboard *Mariner II*, en route to Venus, confirmed predictions of a fast outward flow of plasma from the Sun—the solar wind, now known to bathe the entire solar system out to the boundary of the heliosphere, where it meets the oncoming, tenuous interstellar medium. *Voyager 1* and *2* are now entering that interaction region, more than 100 AU from the Sun. Over the next 5—10 years, they are expected to continue to yield information on phenomena at the outer limits of the Sun's domain.

Meanwhile, over the past five decades, many spacecraft have journeyed into interplanetary space, investigating the particles and fields environment of the solar system or the Sun itself. The first international solar mission, *Helios*, a US—German cooperative mission, with interplanetary spacecraft observing the solar wind and radiation, was launched in the mid-1970s.

Now the Sun is continuously observed from space. In the aggregate, as described in the chapter on the heliosphere, these investigations have shown a common portrait, with variations, of what happens as the Sun's streaming plasma, coronal mass ejections, and electromagnetic radiations interact with the magnetic fields, ionospheres, and atmospheres and surfaces of solar system bodies. These effects are most dramatic when they result in spectacular comet plasma tails, but they are also important in causing magnetic storms and driving the evolution of atmospheres due to dissociation of molecules and ionization and sweeping away of atoms.

Voyager 1 has reached the limit of the heliosphere and is entering interstellar space, 110 AU from the Sun.

Study of these interactions as they are imagined to have happened in the ancient past, for example, when our star is thought to have gone through a huge energetic T Tauri phase, enables not only the analyses of early planetary history here but also productive reasoning about what may be observed in other star—planet systems as they are found.

3.1. Helios

Two German spacecraft, launched by National Aeronautics and Space Administration (NASA) *Titan-Centaurs* in 1975 and 1976, explored solar phenomena between the Earth's orbit and as close as 0.29 AU from the Sun. An arrangement of mirrors and radiators enabled the spinning spacecraft to survive the consequent extreme heating.

3.2. Solar Maximum Mission

Launched in 1980, *Solar Maximum Mission (SMM)* carried a suite of instruments investigating the Sun at the height of the sunspot cycle. Ultraviolet, X-ray, gamma-ray, and visible light observations combined to give a picture of the Sun's total radiation and its variations due to flares. The spacecraft failed and was dramatically rescued by a shuttle crew in 1984, whence it continued until atmospheric reentry in 1989.

3.3. Ulysses

Launched in 1990, European Space Agency's (ESA's) *Ulysses* used Jupiter's gravity to kick its orbit out of the plane of the ecliptic and send the spacecraft back inward, passing over the Sun's poles to survey a region never before explored. Now the craft goes out to the distance of Jupiter's orbit and back to the Sun every 5 years; its functions were switched off in 2009.

3.4. SOHO

The ESA/NASA Solar and Heliospheric Observatory (SOHO), launched by the American *Atlas-Centaur* in 1995, orbits about the L1 Lagrangian libration point 1.5 million km sunward from the Earth, where its 14 instruments continuously observe phenomena relevant to understanding the solar interior, the solar atmosphere, and the solar wind (Figure 56.3). SOHO's observations are immediately fed to users via the Internet at umbra.nascom.nasa.gov. The mission has already made observations through most of an 11-year solar cycle, and it is expected to continue for several more years. It too survived a massive onboard failure with a dramatic rescue—this time by remote control from the Earth.

3.5. Advanced Composition Explorer

The *Advanced Composition Explorer (ACE)*, a NASA mission with nine instruments and an international team of

FIGURE 56.3 The ESA/NASA Solar and Heliospheric Observatory.

20 investigators, was launched in 1997. Like SOHO, it orbits in the L1 region where it continuously surveys the isotopic and elemental composition of particles from the solar corona, the interplanetary medium, and the interstellar space. In 1998, the ACE data system began providing public, real-time observations that can give warning of solar events that cause geomagnetic storms.

3.6. TRACE

A small *Explorer* satellite launched in 1998, TRACE provides nearly continuous solar coronal observations with high spatial and temporal resolution, complementing the data from SOHO.

3.7. Genesis

In an audacious venture using gravity assist at Earth and libration orbiting for 2 years near L1, the *Genesis* mission, launched in 2001, in 2004 returned a capsule to the Earth bearing actual samples of the solar wind and interplanetary medium embedded in ultraclean collector plates. Due to a failure to signal its parachute to open, the capsule crashed in the Utah desert, but not all was lost: A number of the collector units survived in condition good enough for the recovery of isotopic information and other science data.

3.8. Solar Dynamics Observatory

The 3-ton Solar Dynamics Observatory spacecraft, launched in 2010, is in an inclined geosynchronous orbit near the 100° West meridian. Its three instruments continuously observe the solar atmosphere driven by changing magnetic fields and ultimately causing the solar variations that drive space weather.

4. MERCURY

4.1. Mariner 10

Flight to the innermost planet began with *Mariner 10*, launched on November 3, 1973, (Figure 56.4). It was the first mission to use gravity assist, flying by Venus on February 5, 1974, en route to Mercury, where it arrived on 29 March. Then using Mercury gravity assist, it flew by again on September 21, 1974, and March 16, 1975.

4.2. MESSENGER

The *MESSENGER* spacecraft, launched on August 2, 2004, entered obit about Mercury in 2011 after an Earth gravity assist in 2005, Venus gravity assists in 2006 and 2007, then three Mercury assists in 2008 and 2009. The spacecraft carries a suite of instruments to investigate Mercury's

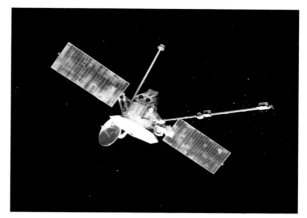

FIGURE 56.4 Flight to the innermost planet, Mercury, began with *Mariner 10*, launched on November 3, 1973. *NASA.*

surface and interior composition, its gravity and magnetic fields, its particles and radiation environment, and the polar regions where it confirmed Earth-based radar observations showing the presence of ices in permanently shadowed craters.

5. VENUS

5.1. Mariner 2

The first mission to return data from another planet, *Mariner 2* in 1962, had amazing escapes from disaster. During ascent, its *Atlas* went into uncontrolled rolling and miraculously stopped in an orientation such that the *Agena* upper stage could deliver the spacecraft onto a trajectory toward Venus. En route, the spacecraft survived a series of mortal threats, and shortly after flying by Venus, it succumbed to overheating. But during the flyby, as described in the chapters on Venus of this book, it produced proof of the planet's hellish greenhouse.

5.2. Veneras 4 Through 16 and Vega

First to enter another atmosphere, *Venera 4* in 1967 carried the emblem of USSR to Venus. It began the Soviets' most successful interplanetary program. *Venera* missions of increasing complexity and scientific yield continued to be launched at nearly every celestial mechanics opportunity until 1983, and then in 1985, the two Vega spacecraft, en route to Halley's Comet, delivered balloons into the Venus atmosphere. Scientific results of this decades-long exploration are described in the chapter Venus.

5.3. Mariner 5

Launched 2 days after *Venera 4* in 1967, *Mariner 5* made flyby observations, including ultraviolet cloud imaging,

FIGURE 56.5 Vega 1, a large Soviet spacecraft, flew by Venus in 1985 en route to Halley's Comet.

which revealed the rapid rotation and spiraling equator-to-pole circulation of the Venusian atmosphere.

5.4. Mariner 10

During its gravity assist flyby of Venus in 1974 en route to Mercury, *Mariner 10* made observations of the Venusian atmosphere and ionosphere, confirming the equator-to-pole circulation and absence of a magnetosphere.

5.5. Pioneer Venus

The two *Pioneer Venus* spacecraft, launched in 1978, had complementary objectives. *Pioneer Venus 1* went into orbit with a radar altimeter to survey the surface through the planet's permanent cloud cover. *Pioneer Venus 2* delivered four probes into the atmosphere to measure its character and composition down to the surface.

5.6. Vega 1 and 2

Two large Soviet spacecraft *Vega 1* and *2* (Figure 56.5) flew by Venus in 1985 en route to close encounters with Halley's Comet. Their spherical entry capsules released balloons that were inflated and floated in the Venus atmosphere, returning data for several days.

5.7. Magellan

The ubiquitous clouds of Venus forever hide the planet's surface from outside visual examination. *Venera* landers in 1975–1981 gave close-up surface panoramas and in 1983 radars on the *Venera 15* and *16* orbiters mapped most of the northern hemisphere. Long delayed through years of attempts to gain government approval, *Magellan* (Figure 56.6) was finally launched in 1989 into a series of orbits enabling it

FIGURE 56.6 *Magellan* was launched in 1989 into a series of orbits enabling it to map the entire planet using synthetic-aperture radar.

to map the entire planet using synthetic-aperture radar. Once the radar mission was complete, the spacecraft was moved into a lower orbit to map the Venusian gravity field and to test aerobraking techniques.

5.8. Venus Flyby

En route to Jupiter, the *Galileo* spacecraft performed a gravity assist flyby at Venus in February 1990. Spacecraft observations included infrared imaging of the planet's cloud layers and even surface features, through infrared "windows" in the atmosphere and clouds. MESSENGER and Cassini–Huygens also flew by Venus for gravity assists en route to their destinations.

5.9. Venus Express

By modifying the design to cater to the hot environment near Venus, but otherwise using many proven components and operational techniques, ESA was able to mount a low-cost mission to place in a 24-hour Venus orbit a spacecraft based on the successful *Mars Express* described later below. Launched in 2005, the mission has delivered unique images of Venus' north polar cloud vortex.

6. EARTH

Among the thousands of spacecraft launched to date, at least hundreds have made some contributions to the study of our Earth as a planet. Here, we make no attempt at a catalog of all those ventures. Instead, we highlight a few recent and representative missions that illustrate the state of humans' ongoing endeavor to understand the Earth's interior, its oceans and lands, its atmosphere, its evolution, and its fate, including that of its biosphere.

Observations by ISS astronauts, and their descriptions for onlookers, constitute a rich legacy that goes beyond science to give us all an appreciation of our precious and beautiful planet.

Scientific and applications satellites not only look down upon the Earth but also observe our planet's electromagnetic and particle environments, including auroral kilometric radio emissions and X-ray and even gamma-ray outbursts from thunderstorms. These energetic upward-going phenomena are characteristic of other planets too: auroras are seen on Jupiter and Saturn and must exist on some extrasolar planets. Thus, the study of our own fascinating home has implications throughout the cosmos.

6.1. Resurs

Soviet and Russian film-return photoreconnaissance satellites have operated over many years for Earth observation. Civil uses have been publicized since 1979, with increasingly capable camera systems used for both applications and science.

Corresponding US imagery was mostly kept classified until 1995, when much previously secret overhead reconnaissance information was released for public use in historical and scientific studies.

6.2. Galileo Earth Flybys

En route to Jupiter (see Outer Planets section below) the *Galileo* spacecraft made gravity assist passes at the Earth in 1990 and 1992. Spectrometric observations were made to simulate a search for evidence of life on an unknown planet, and the data did show an out-of-equilibrium, oxygen-rich atmosphere.

6.3. Terra

Launched in 1999, NASA's *Terra* spacecraft carries five advanced international radiometric and spectrometric instruments observing global phenomena of land, oceans, and atmosphere. Measuring Earth's radiation budget, its carbon cycle, and evolution of its climate and biosphere are the main mission goals.

6.4. TOPEX/Poseidon and Jason-1 and Jason-2

Launched in 1992, 2001, and 2008, respectively, as parts of a collaboration between NASA and the French national space agency CNES, *TOPEX/Poseidon, Jason-1, and Jason-2* use radar altimetry and very precise orbit determination to determine ocean topography, aiding studies of currents, winds, and climate effects including El Niño.

6.5. GRACE

In a collaboration among NASA, the German space agency DLR, and other partners, two small satellites, *GRACE*, launched in 2002 use very precise measurements of the distance between them to gain knowledge of the bumps and hollows in Earth's gravity field, leading to information on the exchanges of mass, momentum, and energy between oceans and atmosphere. The measurements are so precise that they have revealed changes in groundwater masses over large regions.

6.6. Envisat

ESA's 8200-kg Earth observing satellite, *Envisat*, launched in 2002, carries 10 large instruments including a synthetic-aperture radar, a radar altimeter, and a suite of radiometers and spectrometers recording atmospheric, ocean, ice, land, and biosphere data, spanning the spectrum from ultraviolet to microwave frequencies. Its polar orbit gives global coverage.

6.7. Aqua

NASA's *Aqua* satellite, launched in 2002, carries six radiometric and spectrometric instruments surveying Earth's water cycle, sea and land ice, atmospheric temperature, aerosols and trace gases, and soil moisture, so as to increase the understanding of climate and Earth's radiation balance, with both physical and biological influences.

6.8. Aura

Launched in 2004, the *Aura* satellite's four instruments complement those of *Terra* and *Aqua* by measuring atmospheric chemistry, including the formation and dissipation of polar ozone holes and the distribution of greenhouse gases.

7. MOON

After centuries of careful naked-eye and telescopic observation from Earth, the Moon has at last become a body to be investigated by robots, visited by human explorers, and perhaps ultimately inhabited by the people of a first outward wave of civilization. At its beginning, scientific lunar exploration was caught up in the great twentieth-century struggle between the United States and the USSR. With the end of the USSR, the program fell victim to low priority and languished for decades, but now a lively international revival is in progress. Here, we list the most important robotic missions of the past, then briefly mention the grand *Apollo* venture and its failed Soviet competitor, and finally remark on the new missions now established in a widening group of countries.

7.1. Luna 1, 2, and 3

The *Luna* Soviet missions in 1959 yielded the first escape from Earth's gravity, the first lunar impact, and the first farside images ending centuries of speculation.

7.2. Ranger 7, 8, and 9

After two nonlunar tests and three failed attempts to deliver seismometers to the lunar surface, the NASA *Ranger* missions, launched by *Atlas-Agenas* in 1964 and 1965, yielded thousands of high-resolution television (TV) images of the lunar surface showing that all features are mantled by the impact-generated regolith.

7.3. Zond 3

A Soviet planetary spacecraft, *Zond 3*, launched on a test flight including a lunar flyby, this mission in 1965 returned improved imagery of parts of the Moon's far side.

7.4. Luna 9 and 13

After many Soviet lunar failures in 1960–1965, *Luna 9 and 13* in 1966 (Figure 56.7) achieved history's first and third successful lunar touchdowns, delivering image panoramas showing fine surface details.

FIGURE 56.7 *Luna 9*, a Soviet spacecraft, achieved history's first successful lunar touchdown, delivering image panoramas showing fine surface details.

7.5. Luna 10, 11, 12, and 14

These Soviet missions, *Luna*, in 1966 and 1968 achieved the first entry into lunar orbit and made some measurements of lunar gravity and geochemistry.

7.6. Lunar Orbiter 1–5

Designed to image landing sites on the Moon in support of Apollo, the first three of the *Atlas-Agena*-launched *Lunar Orbiter* NASA photographic missions were so successful that the last two were given the expanded task of mapping the entire Moon.

7.7. Surveyor 1, 3, 5, 6, and 7

NASA's *Surveyor 1*, launched by Atlas-Centaur, achieved the first lunar soft landing and returned TV mosaics of its surroundings. In addition to imagery, the *Surveyors* in 1966 and 1967 yielded information on the mechanical and chemical properties of the regolith.

7.8. Zond 5, 6, 7, and 8

The *Zond* Soviet spacecraft, launched from 1968 to 1970 by large Proton vehicles, flew on circumlunar trajectories, returning to the Earth after passing over the Moon's farside. They were test flights for a never-completed human lunar flight program. Payloads consisted of environmental instrumentation and biological specimens including tortoises. The later flights demonstrated an ingenious skip reentry, dipping briefly into the atmosphere over the Indian Ocean and then traveling on to land in central Asia.

7.9. Apollo 8

When in 1961 US President John F. Kennedy called for starting *Apollo*, he had asked his advisors to describe a program in which "we can win" in competition with the USSR. Observation of Soviet lunar launch preparations and test flights led to a decision to send a human crew to the Moon as soon as possible. The risky *Apollo 8* mission in 1968 was the result. It went into lunar orbit with only the Command and Service Modules (CSMs) because the lunar landing module (LM) was not yet available. Thus there was no prospect of saving the mission in "LM Lifeboat" mode as had to be done in *Apollo 13* (see below). The *Apollo 8* crew broadcast TV images and a Christmas voice reading of the first chapter of Genesis in the King James Bible from lunar orbit, took photos, made visual observations, and returned safely to splashdown in the Pacific Ocean.

7.10. Apollo 10

In the final rehearsal for a lunar landing in 1969 (after *Apollo 9's* successful Earth-orbiting test of the LM), the *Apollo 10* crew exercised all LM functions in low lunar orbit, rendezvoused with the CSM, and returned safely to the Earth.

7.11. Apollo 11

Apollo 11, the mission that won the greatest peaceful international contest placed, on July 20, 1969, the first human footprints on the Moon. The LM crew gathered rock and soil samples and installed a set of long-lived instruments on the surface. Meanwhile, a photographic survey from the orbiting CSM covered landing sites for future missions.

7.12. Apollo 12

An outstanding achievement in 1969 by the *Apollo 12* ground and flight crews is shown in Figure 56.8. Navigating to a landing within 170 m of *Surveyor 3*, which had been sitting on the Moon for 31 months, the LM crew walked over to the *Surveyor*, cut off its camera and soil sampler claw, and returned them to the Earth. The mission also brought back a new harvest of rocks, soils, orbital and surface imagery, and other science data.

7.13. Apollo 13

When the *Apollo 13* spacecraft was en route to the Moon in 1970, an oxygen tank in the service module exploded. The dramatic rescue of the mission during the following week is an epic tale of devotion and ingenuity by the ground and flight crews. Moving out of the crippled CSM into the LM, the crew used the LM descent engine to adjust their trajectory to a circumlunar return to the Earth. In the midst of the emergency, they even managed to obtain some lunar farside photographs.

7.14. Apollo 14

Continuing to expand *Apollo*'s scientific capabilities, the 1971 Apollo 14 mission's surface exploration included a hand-drawn cart for carrying instruments.

7.15. Apollo 15, 16, and 17

During three *Apollo* missions in 1971 and 1972, human lunar scientific exploration showed its real potential. With augmented geological training of astronauts, plus one crew member a professional geologist, plus a rover to carry the LM crew on extended surface traverses, plus a suite of remote sensing instruments on the CSM, these missions yielded a cornucopia of information that is described in the chapter on Moon.

7.16. Luna 16, 17, 20, 21, and 24

During the *Apollo* years the USSR had three lunar programs. The first was the robotic science program that began in 1959 and continued with increasing capabilities until 1976. The second was the Proton-launched circumlunar *Zond* (a name meaning sounder) human precursor tests. The third was the human lunar landing effort based on the giant N-1 vehicle that failed in four launch attempts.

Lunas 16 through *24* were emissaries of the first program. The Proton-launched *Luna 16, 20,* and *24* (Figure 56.9) drilled into the regolith, encapsulated small soil samples, and returned them to the Earth. *Luna 17* and *21* delivered Lunokhod rovers to the Moon's surface.

FIGURE 56.8 Astronaut Pete Conrad examines Surveyor 3's camera and soil sampler claw in 1969.

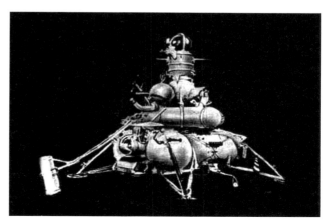

FIGURE 56.9 The *Luna 16* spacecraft.

7.17. Clementine

The mission that revived lunar exploration in 1994 after its decades of stasis, *Clementine,* had an innovative management and technical plan. Proposed as a test of instrument technologies for the American Strategic Defense Initiative, it was sponsored by the Ballistic Missile Defense Organization and NASA, managed by the Naval Research Laboratory, and launched from the Pacific Missile Range on a Titan II-G.

During 2 months in lunar orbit, it mapped the entire Moon at many wavelengths and hinted at the presence of theoretically predicted excess volatiles, possibly a signature of cold-trapped water ice near the lunar poles.

7.18. Lunar Prospector

Launched in 1998 by an Athena solid-fueled vehicle, the NASA *Lunar Prospector* continued the trend toward small, highly capable lunar spacecraft and relatively low mission costs. With neutron, gamma-ray, and alpha-particle spectrometers plus measurements of lunar magnetic and gravity fields, the mission yielded data on the Moon's surface composition and its geochemical and geophysical properties. By sending convincing evidence of excess hydrogen near both poles, it added confidence to the *Clementine* findings of possible polar ices.

7.19. SMART-1

ESA's first lunar mission, *SMART-1,* was launched in 2003 with a small, highly advanced spacecraft demonstrating solar electric propulsion, onboard autonomy, and several new instrument technologies. Spiraling slowly outward from the Earth and then inward toward the Moon, the craft was captured by the Moon's gravity late in 2004 and began science operations in lunar orbit in 2005, whence it delivered a fine harvest of imaging and other remote sensing data until its planned crash into the Moon on September 3, 2006.

7.20. Kaguya/SELENE

Japan and China both launched lunar polar remote sensing orbiters in 2007, returning a rich harvest of multispectral observations as well as gravity and selenodesy data obtained from precise tracking. Kaguya/SELENE carried 14 science instruments and two small subsatellites.

7.21. Chang'e 1

China's first lunar mission gave excellent data on lunar surface and interior properties, including a global thermal map from microwave radiometry.

7.22. Chandrayaan-1

Chandrayaan-1, India's first lunar venture, a remote sensing polar orbiter launched in 2008 and ended in 2009, among other instruments carried an American infrared spectrometer, the Moon Mineralogy Mapper, that confirmed the existence of varying amounts of water at the lunar surface (see Moon chapter).

7.23. Lunar Reconnaissance Orbiter

This US mission, launched in 2009, continues to provide huge amounts of data including imaging of such high quality as to show landed spacecraft from previous missions.

7.24. LCROSS

This mission, opportunistically taking advantage of a large available payload excess, was launched with Lunar Reconnaissance Orbiter. The launch vehicle's spent upper stage was targeted to crash in a shadowed polar crater while a following spacecraft observed the impact debris cloud, illuminated by sunlight, with an infrared spectrometer, yielding confirmation of water ice.

7.25. ARTEMIS

Another ingenious mission of opportunity, this venture uses two of the five THEMIS spacecraft, moved from their original orbits, observing auroras and other magnetospheric phenomena, into new libration orbits inward and outward from the Moon.

7.26. Chang'e 2

The second Chinese lunar orbiter, after completing its mapping mission in 2010, was sent off toward asteroid Toutatis.

7.27. GRAIL

Launched in 2011 and using the same measurement principle as the GRACE mission at the Earth, the twin lunar orbiters, with the varying distance between them recorded at extreme precision, have yielded a new high-resolution map of the Moon's gravity. In addition, one spacecraft carries a camera, dedicated to educational outreach, targeted by school children.

7.28. LADEE

Delivered into Lunar orbit in 2013, this mission is to investigate the Moon's tenuous atmosphere and the

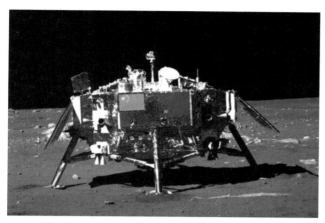

FIGURE 56.10 *Chang-E 3* lunar lander (CNSA)

FIGURE 56.11 *Yutu* (Jade Rabbit) lunar rover (CNSA).

levitated dust phenomena observed by robotic landers and Apollo astronauts. It also carries an ultra-broadband laser communications technology demonstration experiment.

7.29. Chang-E 3 and Yutu

In December 2013 China's *Chang-E 3* landed in northern Mare Imbrium and deployed the rover Yutu (Jade Rabbit). See Figures 56.10 and 56.11. Both spacecraft began observing the lunar surface, subsurface and sky with suites of instruments including ground-penetrating radar and ultraviolet telescopes. After landing both spacecraft survived their first lunar night, but after the second night Yutu was found to be partly inoperative due to temperature extremes after a solar panel mechanism fault. Yutu continued to operate as a fixed observer.

8. MARS

With nineteenth-century telescopic observation showing polar caps and other indications of an atmosphere and changing surface features, Mars became the planet of choice for speculation about other life in the cosmos and about human travel to other worlds.

These pervasive ideas have since driven planetary program priorities with the result that huge resources have been devoted to Martian robotic exploration and to studies of the prospect of human ventures to Mars. But Mars has proved to be a difficult destination: Failure has been an ever present hazard—not only in flight missions but also in the councils where budget decisions are made.

In what follows, we concentrate on successes, but those must be seen as just the most visible parts of a remarkable, decades-long striving toward a possible breakout of humanity beyond the bounds of Earth.

8.1. Mariner 4

Mars launch opportunities occur about every 26 months. In both the United States and the USSR, the October 1960 window was the favored first chance. The Soviets did launch, with two upper stage vehicle failures. During the 1962 window, the Soviets tried three launches, one of which sent *Mars 1* toward the planet. That spacecraft failed en route. In 1964, NASA launched two *Atlas-Agenas* with one success. *Mariner 4* flew by Mars and returned 22 images of the cratered southern highlands, leading to the impression of a Moon-like Mars, proved false by later missions.

8.2. Mariner 6 and 7

Two Mars flyby missions, *Mariner 6* and *7*, launched by *Atlas-Centaurs* in 1969, demonstrated the rapid advance of deep space data acquisition technology. Their imaging was greatly improved over that of *Mariner 4* in both quality and quantity, and in addition, infrared spectrometry gave some first indications of Martian surface compositions. They still covered mainly southern, including polar, ancient landforms, omitting the vast volcanoes and canyons discovered by *Mariner 9*.

8.3. Mars 2 and 3

During the 1971 Mars window, the USSR and United States each launched two missions. The Soviet *Mars 2* and *3* orbiter/landers both arrived successfully into orbit at the planet; *Mars 2* returned some orbital science data but its lander crashed. *Mars 3,* in addition to its orbital operations, delivered its lander with a small tethered mobile platform. But the transmissions from the lander ceased only 20 s after touchdown.

8.4. Mariner 9

The *Atlas-Centaur* carrying *Mariner 8* failed but *Mariner 9* became the most rewarding Mars mission up to its time, waiting out a global dust storm in orbit and then sending imagery of most of the Martian surface until its mission ended in 1972, revealing enormous volcanoes, canyons, apparent river channel networks, sapping collapse features and clouds, as well as images of the two small moons, Phobos and Deimos.

8.5. Mars 4, 5, 6, and 7

At the 1973 opportunity, the Soviets made an all-out effort to upstage the American *Viking* missions planned for 1975. They launched four large spacecraft, all of which arrived in the vicinity of Mars but each of which ultimately failed for a different reason. *Mars 4* failed to brake into orbit but did return some flyby data; *Mars 5* entered orbit, sent some images, and failed after 22 days; *Mars 6* released a lander that failed during descent; and *Mars 7*'s lander missed the planet.

8.6. Viking 1 and 2

In 1975, two large NASA orbiter/landers, *Viking 1* and *2*, were launched by powerful *Titan-Centaurs*. Arriving in June and July 1976, they entered orbit and began surveying for landing sites. The *Viking 1* lander set down in Chryse Planitia on 20 July and the *Viking 2* lander descended to Utopia Planitia on 3 September on the opposite side of Mars. While the orbiters began imaging the whole planet and making spectrometric remote sensing observations, during descent the landers measured atmospheric composition. Then the landers began to operate a suite of instruments for imaging their surroundings and determining meteorological, geological, and biological properties. At first, microbial activity was suspected, but eventually most scientists concluded that no life did or could exist in the soil samples. The Viking 1 landing site was renamed the Thomas A. Mutch Memorial Station.

8.7. Phobos 1 and 2

After a long pause in Martian exploration, in 1988, two large and complex Soviet spacecraft, *Phobos 1* and *2*, were launched by *Proton* vehicles toward the vicinity of Mars. *Phobos 1* was lost en route due to a human error in ground control. *Phobos 2* arrived and began phasing orbits for a rendezvous with the little moon, where it was to make close-up observations and deposit two small landing packages, one of them a hopping rover. Imagery and some other data of Phobos and Mars were obtained, but the spacecraft failed before the landings could occur.

8.8. Mars Pathfinder and Mars Global Surveyor

The 1996 launch window saw the revival of detailed American exploration of Mars. NASA's *Pathfinder* delivered a lander, named the Carl Sagan Memorial Station, and a small rover, named Sojourner, which explored nearby surroundings in the Ares Vallis outwash plain. The *Global Surveyor* spacecraft entered an eccentric orbit and was delicately aerobraked down into circular mapping orbit over a period of months, the long period being needed due to structural failure of the attachment of one solar panel. The mission has yielded a continuing stream of imaging and other data, revolutionizing scientists' knowledge and modeling of Martian geology and atmospheric processes.

8.9. Failures of the 1990s

The years 1992, 1998, and 1999 saw three US missions fail during arrival at the planet: *Mars Observer, Mars Climate Orbiter,* and *Mars Polar Lander.* An elaborate international Russian mission's launch, *Mars 96,* failed in 1996—a series of events that led in the United States to a management overhaul and in Russia to the end of Mars exploration for the time being.

8.10. Mars Odyssey

Mars Odyssey, a NASA orbiter launched in 2001, is instrumented for measurements complementing those of the *Global Surveyor.* With infrared/visible, gamma-ray, and particle spectrometers, it produces thermal imaging enabling evaluation of surface physical properties, subsurface elemental chemistry, and the planet's radiation environment. *Odyssey's* findings have greatly stimulated interpretations of many of Mars' landforms as resulting from the action of subsurface briny water, ice, and carbon dioxide.

8.11. Spirit and Opportunity

In an intense 3-year effort, two NASA Mars rover missions, *Spirit* and *Opportunity,* were prepared for the 2003 launch opportunity. Both succeeded, and the two rovers continue to make discoveries in Meridiani Planum and Gusev Crater, on opposite sides of the planet, reinforcing the orbiters' findings of a history dominated by the effects of water.

Spirit finally became mired and its mission has ended, but Opportunity made its way to the large crater Endeavour where it goes on adding to the history of the planet's watery past.

8.12. Mars Express

ESA's *Mars Express* orbiter, launched in 2003, delivered the small British *Beagle-2* lander, which failed, and has then gone on to yield excellent three-dimensional imaging, plus spectrometric measurements indicating, among other findings, that there is a correlation between regions of enhanced water vapor and methane concentrations in the atmosphere. Mars Express also carries a ground-penetrating radar for detecting the signatures of subsurface brines and ices.

8.13. Mars Reconnaissance Orbiter

Launched in 2005 and delivered in 2006 into aerobraking orbit at Mars, *Mars Reconnaissance Orbiter* has increased by orders of magnitude the quantity and quality of remote sensing data from Mars, because of its powerful radio system and advanced on-board instruments and system software. An example of MRO's amazing imaging capacity is given in Figure 56.12.

8.14. Phoenix

A reflight of the failed *Mars Polar Lander*, in 2007, the *Phoenix* spacecraft, during its 5-month surface mission, confirmed the existence of high-latitude ice within centimeters of the Martian surface. Droplets of brine were observed on parts of the lander.

8.15. Curiosity

Delivered into the huge ancient crater Gale in 2012 by an elaborate entry, descent, and landing system including a rocket sky crane, at the time of this writing, the 1-ton rover is in fine working condition and has begun traveling toward its goal of exploring the layered deposits of Mount Sharp (officially Aeolis Mons). Some of its chemical analyses have revealed all elements necessary to support life (Figure 56.13).

8.16. Mangalyaan

Launched in 2013 to arrive at Mars in September 2014, this first Indian planetary mission is to investigate the Martian atmosphere, including the methane detected by previous orbiters; image the surface; and demonstrate technologies for exploring Mars.

9. SMALL BODIES

As scientists have come to realize that comets and asteroids contain clues to the ancient history of the solar system—clues largely obliterated by geologic processes in planets and moons—missions to small bodies have increased in importance. Also, studies of cratering and meteorite records show that near-Earth asteroids present both a threat and an opportunity. The threat is that of devastating impacts and the opportunity is that of useful resources not found in the Moon.

9.1. International Cometary Explorer

After completing its solar mission as *ISEE-3* (see Sun and Heliosphere sections), the spacecraft was retargeted and renamed *International Cometary Explorer*. It flew through the tail of Comet Giacobini–Zinner in 1985, and then continued on in heliocentric orbit where it sent low-rate data for the next several years. It is now returning toward the vicinity of Earth.

9.2. The Halley Armada

As Halley's comet arrived near the Sun in 1986 on its 76-year orbit, it was met by spacecraft from Japan, Europe, and the USSR. Comet enthusiasts lamented the absence of the United States from this once-in-a-lifetime opportunity. Japan's *Suisei* and *Sakigake* made distant observations of the ultraviolet coma. ESA's *Giotto* passed within 600 km of the nucleus collecting imaging, spectra, and detailed chemical data. The Soviet *Vega 1* and *2* flew by at intermediate distances after their productive en route encounters with Venus (see Venus section).

9.3. Galileo En Route Encounters

While en route to Jupiter on its long journey with gravity assists at Venus and Earth, the NASA *Galileo* spacecraft flew by two Asteroids, 951 Gaspra in 1991 and 243 Ida in 1993, and obtained close-up imagery, spectra, and other

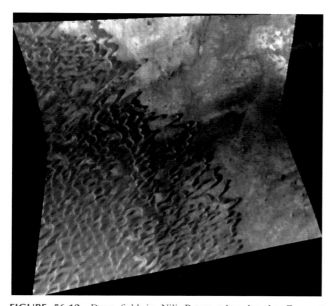

FIGURE 56.12 Dune field in Nili Patera taken by the Compact Reconnaissance Imaging Spectrometer for Mars (CRISM). The image was taken in 544 colors in the wavelength range between 0.36 and 3.92 μm and shows features as small as 20 m across. © *NASA/JPL/JHUAPL/USGS.*

FIGURE 56.13 With an unprecedented landing technique NASA's Mars Science Laboratory was lowered with a sky crane to the surface on August 4, 2012. After crossing the Gale crater plane, rover *Curiosity* will climb 5500 m elevation Aeolis Mons. © *NASA/JPL/JHUAPL/USGS*.

measurements. A highlight of the Ida encounter was the discovery of the tiny moon Dactyl orbiting Ida.

9.4. NEAR-Shoemaker

Launched in 1996, the NASA spacecraft *NEAR-Shoemaker* entered orbit about Asteroid 433 Eros in 2000, delivered imagery, spectrometric data, and gravitometric data. After 1 year in orbit, it was commanded to a gentle touchdown, which it survived, even though not designed for landing.

9.5. Deep Space 1

The NASA craft *Deep Space 1* launched in 1998 to demonstrate solar electric propulsion, autonomous navigation, and other new technologies encountered Asteroid 9969 Braille in 1999, although it only returned a few distant, low-resolution images. Its mission was extended to a close flyby of Comet Borrelly on September 22, 2001, and successfully imaged the nucleus at visible and infrared wavelengths.

9.6. Stardust

With the goal of collecting samples of cometary dust and returning them to the Earth, NASA's *Stardust* mission, launched in 1999, flew by Asteroid 5535 Annefrank in 2002, and encountered Comet Wild 2 in 2004, returning imaging data. The sample return capsule successfully parachuted to Earth on January 4, 2006, with its precious cargo of thousands of cometary (and interstellar) dust particles.

9.7. Hayabusa (Muses-C)

JAXA mission with assistance from NASA, *Hayabusa* was launched in 2003 and used solar electric propulsion to rendezvous with Asteroid 25143 Itokawa in September 2005. It returned multispectral imaging and gravity data and collected surface samples for return to the Earth. Despite problems causing long delays and calling for heroic action by mission planners and controllers, the returning payload was recovered in Australia in June 2010.

9.8. Rosetta

Rosetta, an ESA mission launched in March 2004 with the *Ariane 5,* is scheduled to arrive at Comet Churyumov−Gerasimenko in 2014 after three Earth gravity assists and one at Mars. It was also targeted to flyby Asteroids 2867 Steins in 2008 and 21 Lutetia in 2010.

9.9. Deep Impact

With the goal of determining the physical and chemical makeup of a cometary nucleus, NASA's *Deep Impact* mission, launched in January 2005, successfully delivered a 370-kg projectile to a 10 km/s collision with Comet Tempel 1 on July 4, 2005. Imaging from the impactor and the flyby spacecraft returned the highest resolution pictures of a comet to date and documented the impact event, which provided new insights into the nature of cometary nuclei.

9.10. Dawn

The ongoing Dawn mission is beginning to deliver the full benefits of solar electric propulsion. Launched in 2007, it orbited and mapped Asteroid Vesta until 2012 and is now en route to Asteroid Ceres, potentially yielding comparative planetological information on the two largest, and notably different, asteroids (Figures 56.14 and 56.15).

FIGURE 56.14 With its xenon-ion thrusters, Dawn will be the first space probe orbiting two different planetary bodies beyond Earth: Vesta (December 2001), and Ceres (2015, artist's rendition). *NASA*.

FIGURE 56.15 The Hayabusa sample container after landing in Australia. *JAXA*.

10. OUTER PLANETS AND MOONS

In 1610, when Galileo observed four bright specks moving near Jupiter, he set in motion a quest that culminated in the twentieth century with history's greatest robotic exploration program, giving never-to-be-repeated first close looks at the giant outer planets and their retinue of moons and rings.

10.1. Pioneers 10 and 11

Two NASA missions launched the first two of five human artifacts to escape forever from the Sun's domain. Leaving Earth in 1972, *Pioneer 10* flew by Jupiter in 1973 with imaging and magnetospheric measurements. Its signal continued to be detected at the Earth until 2003. After launch in 1973, *Pioneer 11*'s flyby trajectory was adjusted so that, at its encounter in 1974, Jupiter's gravity would fling it onward toward Saturn, where it flew by in 1979. Each spacecraft carried a golden plaque illustrating humans and encoded information on where and when in the cosmos the flight had originated.

10.2. Voyagers 1 and 2

Launched in 1977 by Titan-Centaurs and still operating today, the NASA missions *Voyagers 1* and *2* are a mighty achievement. *Voyager 1* flew by Jupiter in 1979 and Saturn in 1980, whence it headed toward the heliopause, the boundary between the Sun's realm and that of interstellar space. En route it delivered a family portrait of the solar system showing Earth as a pale blue dot. *Voyager 2* was targeted to a Jupiter flyby and then to Saturn, where Saturn's gravity would send it on to Uranus and Neptune, taking advantage of a planetary alignment that happens at intervals of 173 years. *Voyager 2* passed Uranus in 1986 and Neptune in 1989. The two *Voyagers* returned a vast harvest of imagery, geochemical and geophysical data on the giant planets and their moons and rings, and magnetospheric information. Each one carried a golden phonograph and video record showing characteristics of our planet, its inhabitants, and human civilization. In 2005, *Voyager 1* detected the heliopause, and in 2006, it passed 100 astronomical units from the Sun. At the time of this writing, it has detected passage from the heliosphere into interstellar space.

10.3. Galileo

NASA's *Galileo* mission was launched by the space shuttle plus the *Inertial Upper Stage* in 1989 after a fraught history of replanning and delays including one due to the *Challenger* disaster. *Galileo* entered Jovian orbit in 1995, having made one Venus and two Earth gravity assist flybys en route. During the flybys, some science data were collected, including multispectral observations of the Earth and Moon. Galileo performed the first two asteroid flybys, targeted to Gaspra and Ida, and was in position to image the Comet Shoemaker—Levy 9 impacts on Jupiter. At arrival in the Jovian system, the spacecraft delivered a probe into the huge planet's atmosphere, reaching a level of 55 bars and 152 °C. Despite the failure of the orbiter's high-gain antenna to deploy, the mission returned a large volume of imaging, spectra, and other data on the planet and its moons. In 2003, the craft was commanded to a Jupiter impact, with destruction in Jupiter's atmosphere to keep it from becoming a contamination risk toward any possible biology in the putative subsurface oceans of Europa and Ganymede.

10.4. Cassini—Huygens

Launched in 1997 by a *Titan-Centaur*, NASA Saturn orbiter spacecraft *Cassini-Huygens* carried ESA's *Huygens* probe designed to enter the dense atmosphere of the huge moon Titan (Figure 56.13). With a 1998 Venus gravity assist and a 2000 Jupiter flyby with some scientific observations en route, the combination entered Saturn's orbit in 2004. The probe descended to Titan in 2005, delivered remarkable images, and survived landing on the surface for more than 2 h. Both spacecraft returned unique new observations that will cause active scientific analysis and argument for years to come (Figures 56.16 and 56.17).

10.5. New Horizons

Launched in 2006 at such a high speed that it passed Jupiter in less than 13 months from Earth departure, the *New Horizons* spacecraft's purpose is to investigate the surfaces and atmospheres of Pluto and its large moon Charon during a flyby in 2015. After that it is expected to continue

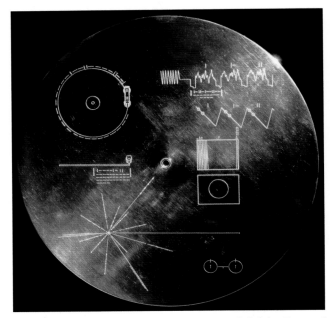

FIGURE 56.16 At the end of 2012, scientists concluded that *Voyager 1* had crossed the heliopause and, as the first man-made object, had been entering the interstellar space. On board is the famous Golden Record, inspired by the late Carl Sagan. *NASA*.

functioning for several more years, exploring the mysteries of the Kuiper Belt, that far-out region of the solar system where the first representatives of a likely multitude of small, icy objects have already been discovered by telescopic observation from the Earth and from the Hubble Space Telescope.

10.6. JUNO

Launched in 2011, another grand mission to Jupiter and its moons is in progress. After a gravity assist flyby of the

FIGURE 56.17 *Cassini* was launched in 1997. *NASA*.

Earth in 2013 the spacecraft is expected to enter Jovian orbit in 2016.

11. CONCLUSION

Thus continues the first, magnificent phase of investigation throughout the Sun's domain. Meanwhile, spaceflight in the inner solar system is reinvigorated as robotic missions to the Moon and Mars take on the purpose, in addition to science, of searching for extraterrestrial energy and material resources and acting as precursors to renewed human exploration and perhaps ultimately settlement of communities off the Earth. And discoveries of planets (hundreds confirmed, thousands of candidates) orbiting other stars tell us that exploration of star and planet systems has a limitless future.

Chapter 57

Exploration and Analysis of Planetary Shape and Topography Using Stereophotogrammetry

Jürgen Oberst, Klaus Gwinner and Frank Preusker
Planetary Geodesy Department, German Aerospace Center, Institute of Planetary Research, Berlin, Germany

Chapter Outline

1. Introduction — 1223
2. Missions and Cameras — 1224
   2.1. Framing Cameras — 1224
   2.2. Scanners — 1224
   2.3. Cameras on Surface Landers and Rovers — 1224
   2.4. Built-in Stereo Capability — 1225
   2.5. Acquisition of Ancillary Orbit and Pointing Data — 1225
   2.6. Camera Calibration — 1225
3. Coordinate Systems — 1225
4. Photogrammetric Processing — 1225
   4.1. Image Matching — 1226
   4.2. Triangulation and Bundle Block Adjustment Techniques — 1226
   4.3. DTM Interpolation — 1227
   4.4. Orthorectification — 1227
5. Quality Assessment — 1227
6. Examples — 1228
7. Summary and Conclusions — 1231
Acknowledgments — 1232
Bibliography — 1233

1. INTRODUCTION

Acquiring information on the surface morphology is among the foremost goals in the exploration of the solar system **planets, satellites**, and minor bodies. In addition to its key role in understanding geologic structures and geophysical processes, morphology is also a critical factor in the coregistration of images for studies of spectrometry, photometry, and temporal change. Geodetic reference systems, maps, and appropriate data visualizations are essential contributions to modern planetary remote sensing missions. With the availability of modern digital camera systems and improved processing techniques, the resolution and quality of digital terrain models (DTMs) covering planetary surfaces has improved considerably in the past 20 years.

DTMs are of great interest for studies of surface features over a variety of scales, from giant volcanoes to small sand dunes. Useful physical parameters can be obtained, such as excavation depths and ejecta volumes for craters of various sizes. Slope measurements in DTMs are used to assess mass wasting effects. Precise topographic data are also used to study illumination conditions and surface shading effects. From precise topography, we can predict areas to be in daylight or in shadow at any given time. Also, terrain models are used to rectify images (i.e. to correct images for local displacements that are caused by surface topography), which is essential for the production of geometrically accurate maps or the coregistration of data sets. In combination with gravity models, topography is essential to assess crustal thickness.

Stereophotogrammetry is a well-established technique in terrestrial mapping that exploits perspective effects in images acquired with different viewing angles to derive information on object positions in three dimensions. The discipline name "photogrammetry" was introduced more than a century ago to express that images were used for precise positional measurements. The term "stereogrammetry" is an abridgment often found in the American literature. An analogous term, "radargrammetry", has been coined to describe the similar geometric analysis and topographic mapping capability of side-looking radar images.

Stereophotogrammetry has a unique performance concerning the geometric resolution of three-dimensional (3-D) data products and does not require assumptions regarding the physical properties of the surface to be analyzed. In contrast, one of the alternative techniques, photoclinometry or "shape from shading", uses brightness information to constrain slopes and topography, and requires knowledge of the radiometric properties of the reflecting surface and any intervening atmosphere, which is typically limited. Other more direct observations of shadows or surface contours against the dark sky (limb observations) have been used to estimate topographic profiles or to determine heights of individual features ("shape from silhouette"), when adequate stereo image coverage was not available. The latter techniques fall within the definition of photogrammetry but in this paper, we focus on the stereophotogrammetric techniques alone.

Alternative topographic mapping techniques, such as laser altimetry or the above-mentioned radargrammetry are not discussed in this paper. However, at the end, we will discuss synergies of stereo mapping with laser altimetry, as the combination of these two techniques has shown valuable benefits in modern planetary missions.

2. MISSIONS AND CAMERAS

The coordinates of a 3-D point are determined by computing viewing vectors from two or more different observation positions and by computing the intersection point of the rays defined by these viewing vectors. The exterior orientation of the camera, i.e. the positions of the viewing posts at the time of exposure and the camera pointing (more specifically, the rotation matrix relating the target coordinate system to image coordinates) must be precisely known or inferred to determine the viewing vectors. For convenience, positions and viewing vectors are computed in the coordinate system fixed to the target body. In the case that the target-fixed coordinate system (e.g. a planet's rotational model) is not sufficiently well known, this must be modeled in the process.

Viewing vectors are computed from the exterior orientation parameters of the camera in combination with the measured image coordinates of features and of calibration data describing the imaging system itself. The latter include focal length, image distortion parameters, and mounting offsets of the sensor system typically determined from calibration campaigns carried out either before flight on ground or in-flight, e.g. by using observations of star fields.

As navigation and calibration data are typically not as well known as one might wish, sophisticated techniques for reconstructing orbit and/or pointing data in combination with parameters of the imaging system have been developed. Depending on the properties of the available data and the desired characteristics of the surface model, these steps play an important role in establishing a reliable geometric framework for the reconstruction of topography.

2.1. Framing Cameras

Framing cameras are equipped with frame **Charge-Coupled Device (CCD)** or **Active Pixel Sensor (APS)** array sensors or even (in the early mapping of the Earth and Moon) photographic film. Here, mathematically simple geometric conversions exist that relate positions of a feature on the sensor surface to viewing vectors for each sensor **pixel**. The image geometry is very stable, i.e. the relative positions of the pixels are precisely known. Stereo images are acquired by pointing the camera from different body-fixed positions to the surface feature of interest. Alternatively, for small objects as asteroids, the camera may remain at some inertially fixed position, while the asteroid rotates beneath the camera position.

2.2. Scanners

For spacecraft in low planetary orbits, it is advantageous to use line scanners rather than framing cameras for topographic mapping. Line scanners are equipped with linear CCD or APS sensors, which are operated at high rates. Images are obtained along the spacecraft ground track, while the spacecraft moves ("pushbroom mode"). The stereo processing of scanner images or the combination of several such scans requires that the spacecraft trajectory and attitude of the sensor are known or modeled on a quasi-continuous basis, according to the high sampling rate of the imager. By using multiple line sensors side by side, large image blocks can be obtained. For example, the High-Resolution Imaging Science Experiment (HiRISE) camera on Mars Reconnaissance Orbiter has 14 separate line sensors on its large focal plane. Sensors equipped with filters can provide multispectral images. One of the advantages of scanners over framing cameras is that they can produce quickly very large-format images, limited by available onboard data storage only.

2.3. Cameras on Surface Landers and Rovers

A special but increasingly important case of planetary photogrammetry is the recovery of topography from stereo images obtained by cameras on surface landers or rovers. Such spacecraft are typically equipped with pairs of framing cameras mounted on a mast, on the rover body, or both. Dual lenses, separated horizontally by a fixed and precisely calibrated base length are used to obtain the perspective effects. The stereo reconstruction process is challenged by the fact that ranges to features of interest and corresponding perspective effects vary greatly. Perspective effects are large near the camera, and approach zero at the horizon, depending on the separation and angular field of view of the dual cameras. Also, due to the specific viewing conditions ("frog perspective") and a possibly rough surface, large parts of the image may be hidden from the camera, thus preventing the construction of a contiguous terrain model from only one pair of stereo images.

2.4. Built-in Stereo Capability

Using multiple sensors pointed at different along-track viewing directions, stereo images can be obtained simultaneously during a single orbital pass. For spacecraft in orbits dedicated to surface mapping, it may be worthwhile carrying cameras with such a built-in stereo capability. As the first camera of this kind, the high-resolution stereo camera (HRSC) line scanner on Mars Express is equipped with nine sensors, five of which are pointed at different viewing directions and are dedicated to stereo imaging. While the spacecraft moves, the stereo sensors will scan an area several times under different viewing angles. Similar camera systems featuring dual or triple line arrays for terrain mapping have flown on the Japanese Kaguya, the Indian Chandrayaan-1, and the two Chinese Chang'e missions. Alternatively, dual-frame cameras pointed forward and backward can be used. The built-in stereo capability does not require any dedicated pointing or spacecraft maneuvers. The stereo images are typically obtained within minutes under near-identical illumination conditions, which facilitates the data analysis.

2.5. Acquisition of Ancillary Orbit and Pointing Data

For the stereo processing excellent navigation data are required. Spacecraft orbits are usually obtained from radio range and Doppler measurements. Long tracking arcs of several hours are used to determine the trajectory of the spacecraft at any time. The attitude of the spacecraft and, hence, pointing of the cameras, is obtained by star sensors in combination with gyroscopes mounted on the craft.

2.6. Camera Calibration

For the processing, also good knowledge of the camera geometric parameters is required. This includes the focal length and principle point of the camera as well as alignment parameters with the spacecraft coordinate system. Parameters of image distortion can be relevant as well, usually expressed in polynomial functions relating the image coordinates at which a feature is observed to the "ideal" coordinates for a distortionless camera. Camera geometric properties are normally determined on the ground before flight. However, after launch, updates of camera calibration parameters may be required. Typically, dedicated imaging campaigns of star fields (in which positions of stars are accurately known) are used to verify the camera calibration parameters.

3. COORDINATE SYSTEMS

Established coordinate systems are essential for the geometrically correct reconstruction of planetary topography, where images taken from different positions and at different times are analyzed jointly. The definitions of coordinate systems for planets and their satellites, the asteroids, and the comets, enjoy common standards established by the International Astronomical Union working group on Cartographic Coordinates and Rotational Elements. The coordinate systems are defined by their rotational states with the z-axis pointing into the direction of the rotational axis, and the equator constraining the plane of the x- and y-axes. The origin is located at the bodies' center of mass. The choice of the prime meridian (i.e. where $y = 0$) is not unique for planets and satellites. If bodies are in locked rotation (i.e. it is always the same hemisphere that is facing the central body), the mean subplanet point typically defines the prime meridian. Alternatively, the principle axes of the **moment of inertia** are used to define this meridian. For example, two such body-fixed coordinate systems, the Mean Earth and the Principle Axis system are currently in use for the Moon. As a practical matter, once sufficiently detailed images are available, longitudes are usually defined relative to a surface feature, e.g. a prominent crater. The longitude assigned to the reference feature can be chosen to provide continuity with earlier coordinate definitions for the same body.

The orientation of the rotational axis is specified by the intersection of this axis with the celestial sphere. For some planetary bodies (Moon, Mars), sufficient observational data are available to describe a secular motion of the rotational axis on the celestial sphere. Objects in tidal locks, like Moon, Mercury, and small satellites are subject to tidal interaction and show physical librations (slight oscillations about their equilibrium rotation state). Photogrammetric measurements of librational amplitude and phase represent important constraints for planetary interior models.

4. PHOTOGRAMMETRIC PROCESSING

Significant advances in the application of photogrammetric techniques in planetary exploration (as well as earth observation) were made with the availability of digital image sensors, increased computing power, and development of improved techniques and algorithms for the full exploitation of the sensor data. Despite the apparent ease of capturing 3-D views with human eyes, the derivation of geometrically accurate gridded DTMs covering large contiguous areas, and suitable for quantitative terrain analysis, is a complex and computationally intensive process.

4.1. Image Matching

Stereoscopic measurements are based on registering the different positions of a common feature in the image planes of stereo images. The process of identifying corresponding image points (called tiepoints or conjugate points) and of accurately measuring the image coordinates of these points is commonly denoted as "image matching".

Once the position and attitude of the sensor and the geometric properties of the camera are determined through the analysis described in Section 4.2, the image coordinates of a dense set of points (as required to generate a contiguous terrain model) can be converted to absolute coordinates including the distance of the object from the camera. Sophisticated matching techniques are required to produce a high density of corresponding points, as the goal is a contiguous model of surface topography with highest possible spatial resolution.

So-called area-based image matching methods have proven to perform well for this task. Coordinates of tiepoints are sought via comparisons of gray values in small image patches (e.g. 11×11 pixels, or larger). By means of least-squares techniques, positions of the tiepoints can be established with subpixel precision using minimization techniques. The currently available matching techniques applied to planetary surfaces differ in a number of aspects. For example, different strategies are used in how similarity of patches is measured (using minimum deviation in image brightness or maximum mutual information content), how the search process is initialized (with the help of an approximate 3-D model or random seed points) and how the search is organized (using grid search, hierarchical multiresolution, or region growing approaches). The available matching algorithms also differ in their capability to avoid false correlations, and by their different approaches to subpixel matching.

High-performance matching tools regularly apply a range of control procedures for avoiding miscorrelations. These can include quality thresholds of the correlation process or consistency checks obtained from 3-D analysis of the resulting corresponding points. The use of multiple (i.e. more than two) stereo images is valuable, because the redundancy introduced allows for verifying and improving overall precision.

Image matching techniques generally have to cope with various challenges associated with "real life". For example, differing solar illumination conditions in stereo pairs can limit the performance of image matching. For cameras with built-in stereo capabilities, this effect is minimized because stereo images are obtained near-simultaneously.

Occasionally, DTMs are constructed from large numbers of overlapping images, where appropriate matching partners must be identified beforehand. For large blocks of images, this may become a challenging sorting task. At the end, the image matching is carried out individually for each "stereo model", where we may take advantage of parallel computing, if available.

4.2. Triangulation and Bundle Block Adjustment Techniques

If camera position and pointing were accurately known, accurate absolute ground coordinates of points could be derived from the successfully matched points without much computational effort. However, the typical errors in the orientation data make complex block adjustments necessary.

The main goal of the photogrammetric bundle block adjustment is to obtain an improved model of the positions and orientations of the sensor during the image acquisition. The adjustment has to warrant that the 3-D surface point positions resulting from the joint analysis of many individual images form a geometrically consistent and rigid model of the surface. Also, every point of the model must be positioned at its correct position in the planet-fixed coordinate system.

The mathematical backbone of this adjustment are the so-called collinearity equations, which define the relationship between the coordinates of points in the images via the orientation data to the corresponding surface points for one image. From multiple observations of large numbers of surface features in large numbers of images, large systems of the collinearity equations are assembled, which are simultaneously solved in several iterative steps.

Sufficiently large numbers of conjugate points (or "tiepoints") must be determined in overlapping areas of the images, as input for the adjustment. This is usually accomplished by automated techniques, such as the image matching, described above. Block adjustments can become computationally challenging, when several thousands of images are involved (see examples below), in which case sophisticated matrix inversion schemes are required to reduce computing time and memory allocations.

The tiepoints are the input (observations) to the block adjustment, whereas the 3-D coordinates of the points on the surface (object points) and the orientation data of each image are the unknowns to be determined. The nominal position and pointing data are used as starting values to begin the iterative process. If appropriate observations are available, it is also possible to improve further unknown parameters, e.g. planet rotation models, spacecraft orbit parameters, or camera constants, by varying the relevant rotation parameters until the minimum of error totals is reached.

For small isolated image blocks, systematic errors in the orientation data may remain uncorrected after the adjustment. Even though coordinates of ground points are internally consistent, the resulting DTMs may have vertical and lateral offsets or small tilt with respect to the surface-fixed coordinate systems at the end. This is caused by the fact that these offsets or tilts are not well constrained by the limited surface coverage of the images. The issue may be resolved when ground control points with specified absolute coordinates (e.g. from laser altimetry) are included in the calculations. The absolute positioning of the DTMs is typically well constrained if large regional blocks of images or even closed global image coverage are available (see examples below).

The results of the block adjustment are improved values for the orientation data and the coordinates of the related surface points. Even without subsequent DTM processing, the improved navigation data are of great interest, as the data enables the construction of geometrically accurate image mosaics.

4.3. DTM Interpolation

The coordinates of matched points are used to compute 3-D object (i.e. surface) points by means of forward ray intersection. The adjusted position and pointing data are used to define the viewing rays for each image. As a result, a large number of object points represented in body-fixed coordinates as well as information on ray-intersection accuracy for each combination of corresponding rays are obtained. The accuracy estimates provide verification for image matching and the block adjustment.

In the case that DTMs are constructed from multiple sets of stereo partner images, point clouds from each single stereo model are merged. This resulting point cloud will generally show variations in point density. It is therefore desirable to convert the irregular point cloud into a contiguous surface model. This is achieved either by point triangulation, which will partly preserve the pattern of the point distribution. More commonly, particularly in cases of billions of 3-D points, the points are integrated into a regular grid of height values in a map coordinate system. The 3-D points therefore are first transformed to geographic latitude/longitude/height coordinates, and finally converted to line/sample coordinates in map space using some appropriate map projection, e.g. sinusoidal, stereographic, etc.

The chosen grid spacing of the raster DTM must properly relate to the spacing of the previously derived object points. If several object points are located within a DTM pixel, these are combined and represented as one average value. DTM pixels without any object point information should represent a subordinate fraction of the grid and, where present, are filled using neighborhood information. The grid values of a DTM represent height values above some chosen reference body (usually a sphere having the planet's mean radius, but in some cases an ellipsoid of revolution or a gravitational equipotential surface defined by gravity field measurements).

4.4. Orthorectification

Production of DTMs is seldom the final photogrammetric step of mapping. Images (including those of the stereopair used to create the DTM, but possibly others as well) are usually projected onto the detailed topographic surface in accordance with the 3-D coordinates of each pixel. The horizontal map coordinates are then used to locate the feature in map projection. Provided the DTM is sufficiently detailed and the orientation data of the images to be rectified are precisely known, this process of orthorectification removes the parallax distortions in the original images and provides an undistorted "overhead" view. It is therefore worth noting that even when a sufficiently detailed DTM is available from other sources such as altimetry, the photogrammetric adjustment processes are essential for registering different observations so they can be compared at full resolution to assess subtle differences such as temporal change, spectral features, and photometric effects.

5. QUALITY ASSESSMENT

While ground control points are widely available for the land surface of the Earth, the quality of surface models for planetary bodies must be assessed almost entirely by analysis of internal consistency and comparisons with complementary remote sensing data.

Parameters derived directly from the stereo image processing represent a most obvious quality criteria. Among these, the 3-D ray intersection error, i.e. the Root Mean Square (RMS) deviation of the minimum distance between rays defining a 3-D point from an ideal intersection point, is the most powerful measure. However, simple intersections obtained from only two rays can be biased by certain conditions of projective geometry as well as certain illumination conditions. In contrast, triple and higher fold intersections have been shown to provide very reliable estimates of point precision using terrestrial airborne images for example.

Topographic profiles obtained from orbital laser altimetry are reliable reference data for stereophotogrammetric DTMs (Figure 57.1), as the individual laser measurements typically have a height accuracy of few meters or better. Laser altimeter data also show a high degree of geometric consistency on regional and global scale, which is useful, because of the occasional height offsets and model tilts of DTMs, as discussed above (Section 4.2).

Comparisons are particular useful where the laser footprint on the surface and the sampling distance along the

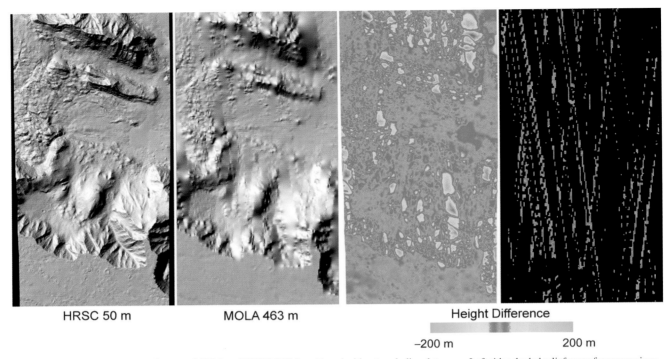

FIGURE 57.1 Difference maps between MOLA and HRSC DTMs, with and without excluding data gaps. Left side: shaded relief maps for comparison. Note that the persisting height deviation observed when measurement gaps on both sides are excluded can be often associated with prominent relief elements such as narrow valleys and ridges, etc. Ophir Labes, Valles Marineris, south is up. *Source: ESA/DLR/FU Berlin (G. Neukum).*

altimeter track are similar to the resolution of the stereo data. This is approximately the case for HRSC and Mars Orbiter Laser Altimeter data. Here, laser altimeter data may be used to assess effective resolutions of the stereo DTMs. The altimeter tracks are typically merged to gridded DTMs, where areas between the tracks are filled by interpolation. When comparisons are made, care must be taken to cope with possible interpolation errors.

Quality assessments can also be made if several DTMs are available for the same area. The comparison can also be performed in the spectral domain using Fourier analysis of the topography. Fourier spectra are also sometimes used to separate "natural" spectral components from a higher frequency noise. These limits are subject to interpretation, however, as the "true" frequency distribution related to the surface alone is basically unknown.

Another class of quality checks makes use of qualitative comparisons of reconstructed surface shapes and corresponding images to study visibility of well-known features, e.g. craters. A crater detection curve showing the fraction of craters of a given diameter range seen in the DTM, typically sheds light on the quality and resolution limits of the DTM.

6. EXAMPLES

With availability of digital image data, planetary photogrammetry has made enormous advances in the past 20 years (Table 57.1). The two Viking Orbiters delivered a large volume of stereo images, for which computer-aided production of topographic models of a planetary surface was carried out for the first time. Beginning in 1990, a sophisticated stereo camera HRSC and a related photogrammetric software system were developed for the Mars '94 mission. However, after the failed launch of this mission in 1996, it was the large volumes of digital image data provided by the Clementine mission as well as by an airborne version of HRSC, from which the first stereo terrain models could be produced using this software (Figure 57.2).

Ever since, stereo data are obtained on a regular basis during planetary missions using various stereo viewing schemes or using dedicated cameras with built-in stereo capability. The Galileo spacecraft arrived in Jupiter **orbit** in 1995. During its operation until 2003, the spacecraft engaged in multiple flybys of the Galilean satellites. Many stereo images were obtained, notably for Ganymede. These revealed the unusual character of Ganymede's surface, attesting to the satellite's early tectonic activity (Figure 57.3).

In 2003, the Mars Express mission was launched, carrying the HRSC camera, originally intended for flight on Mars '94. The spacecraft, moving in its elliptic Mars orbit, is still operating up to the present day. Images are typically obtained during pericenter passes, where spatial resolutions

TABLE 57.1 Missions during Which Substantial Volumes of Stereotopographic Data were Obtained

Mission	Orbit Insertion	Target	Camera(s)	Camera Type	Strategy
Clementine	February 1994	Moon	Ultraviolet–visible (UVVIS)	Frame	1,5
Galileo	December 1995	Jupiter/Satellites	Solid State Imager (SSI)	Frame	2
Cassini	July 2004	Saturn/Satellites	Imaging Science Subsystem (ISS), Narrow-Angle Camera (NAC), and Wide-Angle Camera (WAC)	Frame	2
Mars Express	December 2003	Mars/Phobos	HRSC	Multiline scanner	3
Mars Global Surveyor	September 1997	Mars	Mars Orbiter Camera (MOC)	Scanner	2
Mars Reconnaissance Orbiter	March 2006	Mars	HiRISE, CTX	Scanner	2
NEAR	February 2000	Eros	(CCD camera)	Frame	2
Rosetta		Lutetia	Osiris, NAC, and WAC	Frame	2
MESSENGER	March 2011	Mercury	MDIS, NAC, and WAC	Frame	2
DAWN	July 2011	Vesta, Ceres	Dawn framing camera	Frame	2
SELENE (Kaguya)	October 2007	Moon	Terrain Camera (TC)	Dual line scanner	4
Chandrayaan-1	November 2008	Moon	TMC	Dual Line Scanner	4
Lunar Reconnaissance Orbiter	June 2009	Moon	Lunar Reconnaissance Orbiter Camera (LROC) NAC	Scanner	1
			LROC WAC	Pushframe	5

[1] Spacecraft is tilted across track for stereo viewing.
[2] Spacecraft and body-fixed camera specifically pointed for stereo observations.
[3] Built-in stereo capability; five panchromatic sensors looking forward, downward, and backward, all of which may be used for stereo reconstructions.
[4] Built-in stereo capability; two or three sensors looking forward and backward, respectively.
[5] Stereo effects in overlapping WAC images.

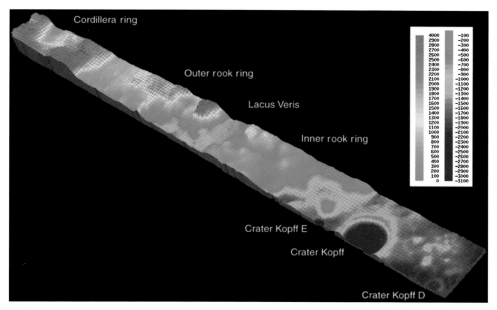

FIGURE 57.2 Oblique view of Mare Orientale digital elevation model derived from Clementine data in the early 1990s. *Source: NASA/DLR.*

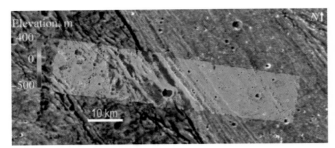

FIGURE 57.3 Stereotopographic model for Ganymede in a transition area between bright and dark surface material. *Source: NASA/JPL/DLR.*

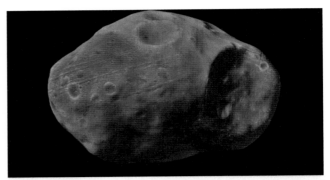

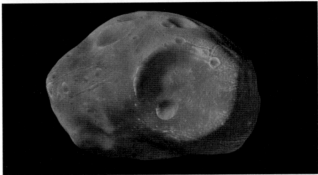

FIGURE 57.5 Phobos synthetic views, computed from a global shape model and overlaid orthoimages, derived from the superresolution channel of the HRSC on Mars Express. *Source: ESA/DLR/FU Berlin (G. Neukum).*

of up to 10 m are achieved. Based on its built-in stereo capability (Table 57.1), stereo images have been acquired systematically for almost the entire surface of Mars and are used to reconstruct terrain models, which are released to the public as HRSC standard products with typical grid spacing of 50–75 m (Figure 57.4). Likewise, stereo images have been obtained during multiple flybys of the Martian satellite Phobos, from which full shape models are being obtained (Figure 57.5). Several other missions to Mars have obtained stereo images of selected targets, including Mars Reconnaissance Orbiter with its powerful HiRISE (0.3 m/pixel) (Figure 57.6) and the Context Camera (CTX) (6 m/pixel), both of which obtain stereo coverage by rolling the spacecraft to image the same target from different angles on multiple flybys.

The Japanese spacecraft Kaguya (former name, SELENE), launched in September 2007, carried a dedicated Terrain Camera (TC) which had two line sensors, looking forward and backward. Stereo images at spatial resolution of 10 m were obtained for most of the Lunar surface, from which large regional terrain models were produced. The Indian Chandrayaan-1, launched 1 year after, had an even higher resolution (5 m) stereo scanner TMC, used for delivery of regional DTMs. The Lunar Reconnaissance Orbiter, launched on June 18, 2009, and still in operation, returns large volumes of images on a regular basis. The **Narrow-Angle Camera** (NAC) is equipped with two linear CCD sensors mounted side by side, which create high-resolution swaths across the lunar surface, typically 5 km wide, at a resolution of 0.5 m/pixel. By combination with overlapping images from adjacent orbits (as for HiRISE and CTX), local high-resolution stereo models can be produced. In contrast, the **Wide-Angle Camera** (WAC) is equipped with an array sensor, operated in the push-frame mode. Using stereo effects in

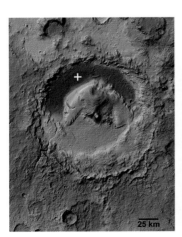
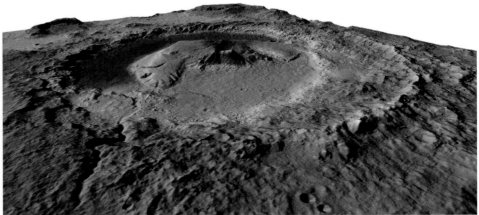

FIGURE 57.4 Regional 50 m DTM of the Mars Science Laboratory (MSL) landing site in Gale Crater derived from HRSC data. *Source: ESA/DLR/FU Berlin (G. Neukum).*

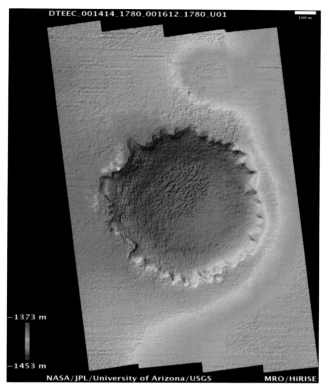

FIGURE 57.6 HiRISE DTM of Victoria crater, Mars. This terrain model with 1 m grid spacing was one of the first DTMs produced from the submeter resolution stereo images of the HiRISE instrument. It covers an area of Meridiani Planum explored by the Mars Exploration Rover "Opportunity". Planetary Data System (PDS) data set DTEEC_001414_1780_001612_1780_U01. *Image credit: NASA/JPL/ University of Arizona/USGS.*

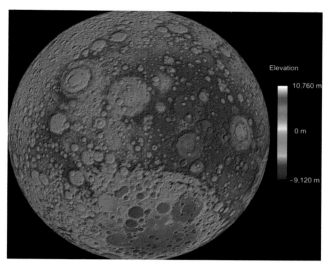

FIGURE 57.7 Far side of the Moon. Elevations are given above 1737.4 km lunar sphere. The model is a subset of the "GLD100" which was published in 2012. *Source: NASA/GSFC/Arizona State University/ DLR.*

overlapping consecutive images a near-global topographic model at a resolution of 100 m/pixel ("GLD100") was generated (Figure 57.7), which complements the global topographic model by LRO's Laser Altimeter Lunar Orbiter Laser Altimeter (LOLA).

MESSENGER (Mercury Surface, Space Environment, Geochemistry and Ranging) was launched in 2004 and arrived in Mercury orbit in 2011. The spacecraft is equipped with the MESSENGER Dual Imaging System, which includes a wide- and a narrow-angle framing camera. While some stereo images were obtained for selected regions during early planet flybys in 2008 and 2009, the large volumes of stereo images, currently being obtained during the orbital mission aim at global coverage (Figure 57.8). Full coverage of the planet was initially obtained with the camera system looking nadir. On the second month, the imaging sequence was repeated under similar lighting conditions, with the cameras being tilted along the orbit track to obtain corresponding stereo viewing. The harsh thermal environment of Mercury requires sophisticated models for calibrations of focal length and distortion of the camera as well as associated software adaptations.

The Dawn spacecraft was launched in September 27, 2007, and reached **asteroid** Vesta in July, 2011. Dawn has completed three mapping orbit phases: the Survey Orbit, the High Altitude Mapping Orbit (HAMO), and the Low Altitude Mapping Orbit (LAMO). The Survey and HAMO orbits were designed to map the entire illuminated surface stereoscopically using the onboard framing camera. From the stereo images, a global shape model of Vesta could be obtained, which includes a rotational pole solution (Figure 57.9). In 2015, Dawn will arrive at asteroid Ceres, the largest asteroid (recently reclassified as a **dwarf planet**) in our solar system, to complete the mission. Several other spacecraft have engaged in flybys of smaller asteroids (e.g. Galileo and Rosetta) and **comet** nuclei (Deep Space 1, STARDUST, Deep Impact) or have carried out orbital missions (NEAR spacecraft, Hayabusa-1), during which stereo images were obtained and processed to derive hemispheric terrain or global shape models.

7. SUMMARY AND CONCLUSIONS

In the past years, production of DTMs from planetary images has become one of the most important tools for planetary exploration, and many missions carrying cameras have provided useful 3-D data products. While the production of large and geometrically accurate stereo models requires considerable know-how and powerful software tools, several teams worldwide have specialized in this processing and analysis of stereo images.

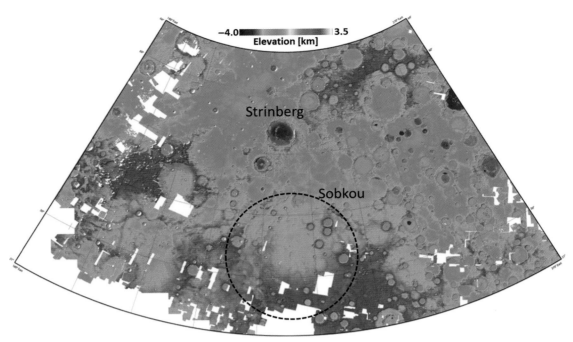

FIGURE 57.8 A DTM (hill-shaded color-coded heights) of Mercury H03 quadrangle (Shakespeare), computed from MESSENGER stereo images. The model was computed from 9500 stereo pairs by the Narrow- and Wide-Angle camera images and contains 2.3 billion object points. The raster grid is 250 m/pixel. The DTM is produced in a Lambert (conformal conic) projection centered at 225°E. Gaps are due to missing data to be filled within the remaining mission time. *Source: NASA/Johns Hopkins University Applied Physics Laboratory/Carnegie Institution of Washington/USGS/DLR.*

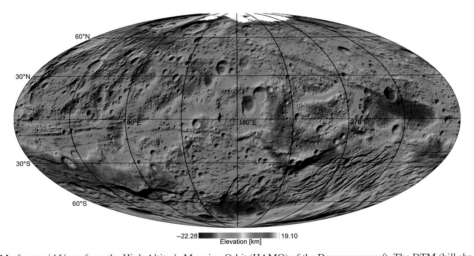

FIGURE 57.9 DTM of asteroid Vesta from the High Altitude Mapping Orbit (HAMO) of the Dawn spacecraft. The DTM (hill shaded and color coded) has a lateral spacing of 100 m and is produced in Mollweide projection (equal-area) centered at 180°. Heights are given with respect to ellipsoid (285/285/229 km). This DTM consists of about 25,000 stereo pairs from which 14.6 billion object points were generated with a mean object point error of 7.5 m. White areas are gaps in the stereo coverage. *Source: NASA/JPL-Caltech/UCLA/MPS/DLR/IDA.*

In the coming years, we will see further planetary exploration missions with cameras, which will have dedicated stereo capability. This includes the Bepi Colombo mission to Mercury, or the Jupiter Icy Moon Explorer (JUICE) mission which will orbit Jupiter's satellite Ganymede. Also, asteroid sample return missions are being planned, where stereo imaging will play a crucial role for landing site selection and assessment.

ACKNOWLEDGMENTS

We wish to thank our colleagues Thomas Roatsch, Klaus-Dieter Matz, Ralf Jaumann, Bernd Giese, Stephan Elgner, and Marita Wählisch (all at German Aerospace Center, DLR), who contributed data and support to this paper. In particular, we wish to thank Frank Scholten (DLR) and Randy Kirk (US Geological Survey) who provided thorough reviews to an earlier version of this manuscript.

BIBLIOGRAPHY

Gwinner, K., Scholten, F., Preusker, F., Elgner, S., Roatsch, T., Spiegel, M., et al. (2010). Topography of Mars from global mapping by HRSC high resolution digital terrain models and orthoimages: characteristics and performance. *Earth and Planetary Science Letters, 294*(3–4), 506–519.

Heipke, C., Oberst, J., Albertz, J., Attwenger, M., Dorninger, P., Dorrer, E., et al. (2007). Evaluating planetary digital terrain models — The HRSC DTM test. *Planetary and Space Science, 55*(14), 2173–2191.

Kirk, R. L., Howington-Kraus, E., Hare, T., Dorrer, E., Cook, D., Becker, K., et al. (1999). Mapping the Pathfinder landing site in three dimensions. *Journal Geophysical Research, 104*(E4), 8869–8887.

Kirk, R. L., Howington-Kraus, E., Rosiek, M. R., Anderson, J. A., Archinal, B. A., Becker, K. J., et al. (2008). Ultrahigh resolution topographic mapping of Mars with MRO HiRISE stereo images: meter-scale slopes of candidate Phoenix landing sites. *Journal of Geophysical Research, 113*(E12). CiteID E00A24.

Planetary photogrammetry. In J. C. McGlone, E. M. Mikhail and J. S. Bethel (Eds), *Manual of photogrammetry* (5th ed.), ISBN-10: 1570830711.

Preusker, F., Oberst, J., Head, J. W., Watters, T. R., Robinson, M. S., Zuber, M. T., et al. (2011). Stereo topographic models of Mercury after three MESSENGER flybys. *Planetary and Space Science, 59*(15), 1910–1917.

Scholten, F., Oberst, J., Matz, K.-D., Roatsch, T., Wählisch, M., Speyerer, E. J., et al. (2012). GLD100: the near-global lunar 100 m raster DTM from LROC WAC stereo image data. *Journal of Geophysical Research, 117*(E3). http://dx.doi.org/10.1029/2011JE003926. American Geophysical Union.

Thornhill, G. D., Rothery, D. A., Murray, J. B., Cook, A. C., Day, T., Muller, J. P., et al. (1993). Topography of Apollinaris Patera and Ma'adim Vallis: automated extraction of digital elevation models. *Journal of Geophysical Research*. ISSN: 0148-0227, *98*(E12), 23,581–23,587.

Willner, K., Oberst, J., Hussmann, H., Giese, B., Hoffmann, H., Matz, K.-D., et al. (2010). Phobos control point network, rotation, and shape. *Earth and Planetary Science Letters, 294*, 547–553.

Appendix

TABLE 1: SELECTED ASTRONOMICAL CONSTANTS

Astronomical unit, AU	$1.4959787066 \times 10^{11}$ m
Speed of light, c	2.99792458×10^{8} m/s
AU in light time	499.00478353 s
Gravitational constant, G	6.67259×10^{-11} m^3/kg/s^2
Mass of the Sun, M_S	1.9891×10^{30} kg
Gaussian gravitational constant $= \sqrt{G \cdot M_s}$	0.01720209895 AU$^{3/2}$/day/solar mass$^{1/2}$
Solar constant	1368 W/m^2
Sun–Jupiter mass ratio	1047.83
Earth–Moon mass ratio	81.3484
Equatorial radius of the Earth, R_e	6378.137 km
Obliquity of the ecliptic (J2010)	23°26′ 21.406″
Earth sidereal day	23 h 56 min 4.0916 s
Sidereal year (J2000)	365.25636 days
Semimajor axis of the Earth's orbit	1.00000105726665 AU
Parsec, pc	206,264.806 AU
Age of the solar system	4.568×10^{9} years
Age of the galaxy	$\sim 13 \times 10^{9}$ years

TABLE 2: PHYSICAL AND ORBITAL PROPERTIES OF THE SUN, PLANETS AND DWARF PLANETS

Name	Mass (kg)	Equatorial Radius (km)	Density (g/m^3)	Rotation Period	Obliquity (degrees)	Escape Velocity (km/s)	Semimajor Axis (AU)	Eccentricity	Inclination (degrees)	Period (years)
Sun	1.989×10^{30}	695,500	1.41	25.38 days	7.25[1]	617.7	—	—	—	—
Mercury	3.301×10^{23}	2440	5.43	58.646 days	0	4.25	0.38710	0.205631	7.0049	0.2408
Venus	4.868×10^{24}	6052	5.24	−243.020 days	177.33	10.36	0.72333	0.006773	3.3947	0.6152

(Continued)

—cont'd

Name	Mass (kg)	Equatorial Radius (km)	Density (g/m^3)	Rotation Period	Obliquity (degrees)	Escape Velocity (km/s)	Semimajor Axis (AU)	Eccentricity	Inclination (degrees)	Period (years)
Earth	5.973×10^{24}	6378	5.51	23.935 h	23.44	11.19	1.00000	0.016710	0.0000	1.0000
Mars	6.417×10^{23}	3396	3.93	24.622 h	25.19	5.03	1.52366	0.093412	1.8506	1.8808
Ceres[2]	8.7×10^{20}	473	2.0	9.075 h	—	0.52	2.76668	0.078375	10.5834	4.6
Jupiter	1.898×10^{27}	71,492	1.33	9.925 h	3.13	59.54	5.20336	0.048393	1.3053	11.862
Saturn	5.684×10^{26}	60,268	0.69	10.656 h	26.73	35.49	9.58256	0.055723	2.4845	29.457
Uranus	8.682×10^{25}	25,559	1.27	−17.24 h	97.77	21.29	19.2021	0.044405	0.772	84.011
Neptune	1.024×10^{26}	24,764	1.64	16.11 h	28.32	23.5	30.0476	0.001121	1.7692	164.79
Pluto[2]	1.314×10^{22}	1195	1.83	6.387 days	122.53	1.23	39.4817	0.250249	17.0890	247.68
Haumea[2]	?	620	?	3.915 h	?	?	43.17	0.19245	28.19	283.61
Makemake[2]	?	715	?	22.48 h	?	?	45.67	0.15709	29.00	308.62
Eris[2]	1.67×10^{22}	1.163	2.52	25.9 h	—	1.38	67.958	0.437083	43.8853	560.23

Orbital data for January 1, 2000.
[1]Solar obliquity relative to the ecliptic plane 0.055723219.
[2]Dwarf planet.
http://nssdc.gsfc.nasa.gov; http://ssd.jpl.nasa.gov; http://solarsystem.nasa.gov Sicardi et al., 2011, Size, density, albedo, and atmosphere limit of dwarf planet Eris from a stellar occultation, EPSC Abstracts Vol. 6, EPSC-DPS2011-137-8.

TABLE 3: PHYSICAL AND ORBITAL PROPERTIES OF THE SATELLITES OF PLANETS AND DWARF PLANETS

EARTH

Name	Semimajor Axis (10^3 km)	Orbital Inclination (degrees)	Orbital Eccentricity	Orbital Period (days)	Mean Radius (km)	Mass (10^{20} kg)	Density (10^3 kg/m^3)	Year of Discovery
Moon	384.40	0.0554	5.16	27.322	1737.4	734.8	3.34	—

MARS

Name	Semimajor Axis (10^3 km)	Orbital Inclination (degrees)	Orbital Eccentricity	Orbital Period (days)	Mean Radius (km)	Mass (10^{20} kg)	Density (g/cm^3)	Year of Discovery
Phobos	9.38	0.0151	1.08	0.319	11.1	1.1×10^{-4}	1.872	1877
Deimos	23.46	0.0002	1.79	1.262	6.2	1.5×10^{-5}	1.471	1877

Appendix

JUPITER

Jupiter—Small Inner Satellites

Name		Semimajor Axis (10^3 km)	Orbital Inclination (degrees)	Orbital Eccentricity	Orbital Period (days)	Mean Radius (km)	Mass (10^{20} kg)	Density (10^3 kg/m^3)	Year of Discovery
J16	Metis	128.0	0.0012	0.02	0.295	22	—	—	1979
J15	Adrastea	129.0	0.0018	0.05	0.298	8	—	—	1979
J5	Amalthea	181.4	0.0032	0.38	0.498	84	0.15	—	1892
J14	Thebe	221.9	0.0176	1.08	0.675	49	—	—	1979

Jupiter—Themisto Group of Irregular Satellites

Name		Semimajor Axis (10^3 km)	Orbital Inclination (degrees)	Orbital Eccentricity	Orbital Period (days)	Mean Radius (km)	Mass (10^{20} kg)	Density (10^3 kg/m^3)	Year of Discovery
J18	Themisto	7507	43.08	0.242	130.0	4.5	—	—	2000

Jupiter—Himalia Group of Irregular Satellites

Name		Semimajor Axis (10^3 km)	Orbital Inclination (degrees)	Orbital Eccentricity	Orbital Period (days)	Mean Radius (km)	Mass (10^{20} kg)	Density (10^3 kg/m^3)	Year of Discovery
J13	Leda	11,165	27.46	0.164	240.9	9	—	—	1974
J6	Himalia	11,461	27.50	0.162	250.6	80	—	—	1904
J10	Lysithea	11,717	28.30	0.112	259.2	19	—	—	1938
J7	Elara	11,741	26.63	0.217	259.6	39	—	—	1904
	S/2000 J11	12,555	28.30	0.248	287.0	2			2000

Jupiter—Galilean Satellites

Name		Semimajor Axis (10^3 km)	Orbital Inclination (degrees)	Orbital Eccentricity	Orbital Period (days)	Mean Radius (km)	Mass (10^{20} kg)	Density (10^3 kg/m^3)	Year of Discovery
J1	Io	421.8	0.036	0.0041	1.769	1821.5	893.2	3.528	1610
J2	Europa	671.1	0.466	0.0094	3.551	1562.1	480.0	3.013	1610
J3	Ganymede	1070.4	0.177	0.0013	7.155	2632.3	1481.9	1.942	1610
J4	Callisto	1882.7	0.192	0.0074	16.69	2409.3	1075.9	1.834	1610

Jupiter—Ananke Group of Irregular Satellites

Name		Semimajor Axis (10^3 km)	Orbital Inclination (degrees)	Orbital Eccentricity	Orbital Period (days)	Mean Radius (km)	Mass (10^{20} kg)	Density (10^3 kg/m^3)	Year of Discovery
J34	Euporie	19,302	145.8	0.144	550.7	1	—	—	2001
J35	Orthosie	20,721	145.9	0.281	622.6	1	—	—	2001

(Continued)

—cont'd

Jupiter—Ananke Group of Irregular Satellites

Name		Semimajor Axis (10³ km)	Orbital Inclination (degrees)	Orbital Eccentricity	Orbital Period (days)	Mean Radius (km)	Mass (10²⁰ kg)	Density (10³ kg/m³)	Year of Discovery
J33	Euanthe	20,799	148.9	0.232	620.6	1.5	—	—	2001
J29	Thyone	20,940	148.5	0.229	627.3	2	—	—	2001
J40	Mneme	21,069	148.6	0.227	620.0	1	—	—	2003
J22	Harpalyke	21,105	148.6	0.226	623.3	2	—	—	2000
J30	Hermippe	21,131	150.7	0.210	633.9	2	—	—	2001
J27	Praxidike	21,147	149.0	0.230	625.3	3.5	—	—	2000
J42	Thelxinoe	21,162	151.4	0.221	628.1	1	—	—	2003
J24	Iocaste	21,269	149.4	0.216	631.5	2.5	—	—	2000
J12	Ananke	21,276	148.9	0.244	610.5	14	—	—	1951

Jupiter—Carme Group of Irregular Satellites

Name		Semimajor Axis (10³ km)	Orbital Inclination (degrees)	Orbital Eccentricity	Orbital Period (days)	Mean Radius (km)	Mass (10²⁰ kg)	Density (10³ kg/m³)	Year of Discovery
J43	Arche	22,931	165.0	0.259	723.9	1.5	—	—	2002
J38	Pasithee	23,096	165.1	0.267	719.5	1	—	—	2001
J50	Herse	23,097	164.2	0.200	715.4	1	—	—	2003
J21	Chaldene	23,179	165.2	0.251	723.8	2	—	—	2000
J37	Kale	23,217	165.0	0.260	729.5	1	—	—	2001
J26	Isonoe	23,217	165.2	0.246	725.5	2	—	—	2000
J31	Aitne	23,231	165.1	0.264	730.2	1.5	—	—	2001
J25	Erinome	23,279	164.9	0.266	728.3	1.5	—	—	2000
J20	Taygete	23,360	165.2	0.252	732.2	2.5	—	—	2000
J11	Carme	23,404	164.9	0.253	702.3	23	—	—	1938
J23	Kalyke	23,583	165.2	0.245	743.0	2.5	—	—	2000
J47	Eukelade	23,661	165.5	0.272	746.4	2	—	—	2003
J44	Kallichore	24,043	165.5	0.264	764.7	1	—	—	2003

Jupiter—Pasiphae Group of Irregular Satellites

Name		Semimajor Axis (10³ km)	Orbital Inclination (degrees)	Orbital Eccentricity	Orbital Period (days)	Mean Radius (km)	Mass (10²⁰ kg)	Density (10³ kg/m³)	Year of Discovery
J45	Helike	21,263	154.8	0.156	634.8	2	—	—	2003
J32	Eurydome	22,865	150.3	0.276	717.3	1.5	—	—	2001

—cont'd

Jupiter—Pasiphae Group of Irregular Satellites

Name		Semimajor Axis (10^3 km)	Orbital Inclination (degrees)	Orbital Eccentricity	Orbital Period (days)	Mean Radius (km)	Mass (10^{20} kg)	Density (10^3 kg/m^3)	Year of Discovery
J28	Autonoe	23,039	152.9	0.334	762.7	2	—	—	2001
J36	Sponde	23,487	151.0	0.312	748.3	1	—	—	2001
J8	Pasiphae	23,624	151.4	0.409	708.0	29	—	—	1908
J19	Megaclite	23,806	152.8	0.421	752.8	3	—	—	2000
J9	Sinope	23,939	158.1	0.250	724.5	19	—	—	1914
J39	Hegemone	23,947	155.2	0.328	739.6	1.5	—	—	2003
J41	Aoede	23,981	158.3	0.432	761.5	2	—	—	2003
J17	Callirrhoe	24,102	147.1	0.283	758.8	3.5	—	—	1999
J48	Cyllene	24,349	149.3	0.319	737.8	1	—	—	2003
J49	Kore	24,543	145.0	0.325	779.2	1	—	—	2003

Jupiter—Satellites yet to be Named

Name	Semimajor Axis (10^3 km)	Orbital Inclination (degrees)	Orbital Eccentricity	Orbital Period (days)	Mean Radius (km)	Mass (10^{20} kg)	Density (10^3 kg/m^3)	Year of Discovery
S/2003 J2	2857.041	151.8	0.380	982.5	1	—	—	2003
S/2003 J3	18339.885	143.7	0.241	504.0	1	—	—	2003
S/2003 J4	23257.920	144.9	0.204	723.2	1	—	—	2003
S/2003 J5	24084.180	165.0	0.210	759.7	2	—	—	2003
S/2003 J9	22441.680	164.5	0.269	683.0	0.5	—	—	2003
S/2003 J10	24249.600	164.1	0.214	767.0	1	—	—	2003
S/2003 J12	19002.480	145.8	0.376	533.3	0.5	—	—	2003
S/2003 J15	22000.000	140.8	0.110	668.4	1	—	—	2003
S/2003 J16	21000.000	148.6	0.270	595.4	1	—	—	2003
S/2003 J18	20700.000	146.5	0.119	606.3	1	—	—	2003
S/2003 J19	22800.000	162.9	0.334	701.3	1	—	—	2003
S/2003 J23	24055.500	149.2	0.309	759.7	1	—	—	2003
S/2010 J1	23314.335	163.2	0.320	723.2	1	—	—	2010
S/2010 J2	20307.150	150.4	0.307	588.1	0.5	—	—	2010
S/2011 J1	20155.290	162.8	0.296	580.7	0.5	—	—	2011
S/2011 J2	23329.710	151.9	0.387	726.8	0.5	—	—	2011

SATURN

Saturn—Ring Satellites

Name		Semimajor Axis (10^3 km)	Orbital Inclination (degrees)	Orbital Eccentricity	Orbital Period (days)	Mean Radius (km)	Mass (10^{20} kg)	Density (10^3 kg/m^3)	Year of Discovery
	S/2009 S1	117	0.000	0.000	0.000	0.2	—	—	2009
S18	Pan	133.6	0.000	0.000	0.575	10	—	—	1981
S35	Daphnis	136.5	0.000	0.000	0.594	3.5	—	—	2005
S15	Atlas	137.7	0.000	0.000	0.602	16	—	—	1980
S16	Prometheus	139.4	0.000	0.002	0.613	46.8	—	—	1980
S17	Pandora	141.7	0.000	0.004	0.629	40.6	—	—	1980
S11	Epimetheus	151.4	0.335	0.021	0.69	58.3	—	—	1980
S10	Janus	151.5	0.165	0.007	0.70	90.4	—	—	1980
S53	Aegaeon	167.5	0.001	0.000	0.808	0.2	—	—	2008
S1	Mimas	185.6	1.566	0.021	0.94	198.2	0.38	1.152	1789
S32	Methone	194	0.000	0.000	1.01	1.5	—	—	2004
S49	Anthe	197.7	0.100	0.001	1.04	0.5	—	—	2004
S33	Pallene	211	0.000	0.000	1.14	2	—	—	2004
S2	Enceladus	238.1	0.010	0.000	1.37	252.3	1.1	1.606	1789
S3	Tethys	294.7	1.091	0.000	1.89	536.3	6.2	0.956	1684
S13	Telesto	294.7	1.118	0.001	1.89	12	—	—	1980
S14	Calypso	294.7	1.500	0.001	1.89	9.5	—	—	1980
S34	Polydeuces	377.2	0.175	0.019	2.74	2	—	—	2004
S4	Dione	377.4	0.028	0.002	2.74	562.5	11	1.469	1684
S12	Helene	377.4	0.213	0.000	2.74	16	—	—	1980

Saturn—Regular Satellites

Name		Semimajor Axis (10^3 km)	Orbital Inclination (degrees)	Orbital Eccentricity	Orbital Period (days)	Mean Radius (km)	Mass (10^{20} kg)	Density (10^3 kg/m^3)	Year of Discovery
S5	Rhea	527.07	0.33	0.0010	4.518	764.5	23.1	1.233	1672
S6	Titan	1221.87	0.31	0.0288	15.95	2575.5	1346	1.880	1655
S7	Hyperion	1500.93	0.62	0.0274	21.28	133	5.6×10^{-2}	0.569	1848
S8	Iapetus	3560.85	8.30	0.0283	79.33	734.5	18.1	1.088	1671

Saturn—Irregular Satellites

Name		Semimajor Axis (10^3 km)	Orbital Inclination (degrees)	Orbital Eccentricity	Orbital Period (days)	Mean Radius (km)	Mass (10^{20} kg)	Density (10^3 kg/m^3)	Year of Discovery
S24	Kiviuq	11111	45.71	0.334	449.2	8	—	—	2000
S22	Ijiraq	11124	46.44	0.316	451.4	6	—	—	2000

—cont'd

Saturn—Irregular Satellites

Name		Semimajor Axis (10³ km)	Orbital Inclination (degrees)	Orbital Eccentricity	Orbital Period (days)	Mean Radius (km)	Mass (10²⁰ kg)	Density (10³ kg/m³)	Year of Discovery
S9	Phoebe	12944.3	174.8	0.164	548.2	120	8.3×10^{-2}	1.633	1898
S20	Paaliaq	15200	45.13	0.364	686.9	11	–	–	2000
S27	Skathi	15541	152.6	0.270	728.2	4	–	–	2000
S26	Albiorix	16182	33.98	0.478	783.5	16	–	–	2000
S37	Bebhionn	17119	35.01	0.469	834.8	3	–	–	2004
S38	Erriapus	17343	34.62	0.474	871.2	5	–	–	2000
S39	Siarnaq	17531	45.56	0.295	895.6	20	–	–	2000
S47	Skoll	17665	161.2	0.464	878.3	3	–	–	2006
S21	Tarvos	17983	33.82	0.531	926.2	7.5	–	–	2000
S52	Tarqeq	18009	46.09	0.160	887.5	3.5	–	–	2007
S51	Greip	18206	179.8	0.326	921.2	3	–	–	2006
S44	Hyrrokkin	18437	151.4	0.333	931.8	4	–	–	2004
S35	Mundilfari	18685	167.3	0.210	952.6	3.5	–	–	2000
S50	Jarnsaxa	18811	163.3	0.216	964.7	3	–	–	2006
S31	Narvi	19007	145.8	0.431	1003.9	3.5	–	–	2003
S38	Bergelmir	19338	158.5	0.142	1005.9	3	–	–	2004
S23	Suttungr	19459	175.8	0.114	1016.7	3.5	–	–	2000
S43	Hati	19856	165.8	0.372	1038.7	3	–	–	2004
S39	Bestla	20129	145.2	0.521	1083.6	3.5	–	–	2004
S40	Farbauti	20390	156.4	0.206	1086.1	2.5	–	–	2004
S30	Thrymr	20474	176.0	0.470	1094.3	3.5	–	–	2000
S36	Aegir	20735	166.7	0.252	1116.5	3	–	–	2004
S45	Kari	22118	156.3	0.478	1233.6	3.5	–	–	2006
S41	Fenrir	22453	164.9	0.136	1260.3	2	–	–	2004
S48	Surtur	22707	177.5	0.451	1297.7	3	–	–	2006
S19	Ymir	23040	173.1	0.335	1315.4	9	–	–	2000
S46	Loge	23065	167.9	0.187	1312.0	3	–	–	2006
S42	Fornjot	25108	170.4	0.206	1490.9	3	–	–	2004

Saturn—Satellites yet to be Named

Name	Semimajor Axis (10³ km)	Orbital Inclination (degrees)	Orbital Eccentricity	Orbital Period (days)	Mean Radius (km)	Mass (10²⁰ kg)	Density (10³ kg/m³)	Year of Discovery
S/2004 S071	9800	165.1	0.580	1103	3	–	–	2004
S/2004 S12	19650	164.0	0.401	1048	2.5	–	–	2004
S/2004 S13	18450	167.4	0.273	906	3	–	–	2004

(Continued)

―cont'd

Saturn―Satellites yet to be Named

Name	Semimajor Axis (10³ km)	Orbital Inclination (degrees)	Orbital Eccentricity	Orbital Period (days)	Mean Radius (km)	Mass (10²⁰ kg)	Density (10³ kg/m³)	Year of Discovery
S/2004 S17	18600	166.6	0.259	986	2	—	—	2004
S/2006 S1	18981	154.2	0.130	970	3	—	—	2006
S/2006 S3	21132	150.8	0.471	1142	3	—	—	2006
S/2007 S2	16560	176.7	0.218	800	3	—	—	2007
S/2007 S3	20519	177.2	0.130	1100	2.5	—	—	2007

URANUS

Uranus―Regular Satellites

Name		Semimajor Axis (10³ km)	Orbital Inclination (degrees)	Orbital Eccentricity	Orbital Period (days)	Mean Radius (km)	Mass (10²⁰ kg)	Density (10³ kg/m³)	Year of Discovery
U6	Cordelia	49.8	0.085	0.000	0.335	20	—	—	1986
U7	Ophelia	53.8	0.104	0.010	0.376	21	—	—	1986
U8	Bianca	59.2	0.193	0.001	0.435	26	—	—	1986
U9	Cressida	61.8	0.006	0.000	0.464	40	—	—	1986
U10	Desdemona	62.7	0.113	0.000	0.474	32	—	—	1986
U11	Juliet	64.4	0.065	0.001	0.493	47	—	—	1986
U12	Portia	66.1	0.059	0.000	0.513	68	—	—	1986
U13	Rosalind	69.9	0.279	0.000	0.558	36	—	—	1986
U27	Cupid	74.4	0.099	0.001	0.613	5	—	—	2003
U14	Belinda	75.3	0.031	0.000	0.624	40	—	—	1986
U25	Perdita	76.4	0.470	0.012	0.638	10	—	—	1986
U15	Puck	86.0	0.319	0.000	0.762	81	—	—	1985
U26	Mab	97.7	0.134	0.002	0.923	4	—	—	2003
U5	Miranda	129.9	4.338	0.001	1.410	235.8	0.66	1.201	1948
U1	Ariel	190.9	0.041	0.001	2.520	578.9	13.5	1.665	1851
U2	Umbriel	266.0	0.128	0.004	4.140	584.7	11.7	1.400	1851
U3	Titania	436.3	0.079	0.001	8.710	788.9	35.3	1.715	1787
U4	Oberon	583.5	0.068	0.001	13.46	761.4	30.1	1.630	1787

Uranus—Irregular Satellites

Name		Semimajor Axis (10³ km)	Orbital Inclination (degrees)	Orbital Eccentricity	Orbital Period (days)	Mean Radius (km)	Mass (10²⁰ kg)	Density (10³ kg/m³)	Year of Discovery
U22	Francisco	4276	145.2	0.146	266.6	11	—	—	2001
U16	Caliban	7231	140.9	0.159	579.7	36	—	—	1997
U20	Stephano	8004	144.1	0.229	677.4	16	—	—	1999
U21	Trinculo	8504	167.1	0.220	759.0	9	—	—	2001
U17	Sycorax	12,179	159.4	0.522	1288.3	75	—	—	1997
U23	Margaret	14,345	56.6	0.661	1694.8	10	—	—	2003
U18	Prospero	16,256	152.0	0.445	1977.3	25	—	—	1999
U19	Setebos	17,418	158.2	0.591	2234.8	24	—	—	1999
U24	Ferdinand	20,901	169.8	0.368	2823.4	11	—	—	2003

NEPTUNE

Neptune—Regular Satellites

Name		Semimajor Axis (10³ km)	Orbital Inclination (degrees)	Orbital Eccentricity	Orbital Period (days)	Mean Radius (km)	Mass (10²⁰ kg)	Density (10³ kg/m³)	Year of Discovery
N3	Naiad	48.2	4.74	0.000	0.294	29	—	—	1989
N4	Thalassa	50.1	0.205	0.000	0.311	40	—	—	1989
N5	Despina	52.5	0.065	0.000	0.335	74	—	—	1989
N6	Galatea	62.0	0.054	0.000	0.429	79	—	—	1989
N7	Larissa	73.5	0.201	0.001	0.555	86	—	—	1989
	S/2004 N1	105.3	0.000	0.000	0.950	9	—	—	2013
N8	Proteus	117.6	0.039	0.000	1.122	208	0.5	1.2	1989

Neptune—Irregular Satellites

Name		Semimajor Axis (10³ km)	Orbital Inclination (degrees)	Orbital Eccentricity	Orbital Period (days)	Mean Radius (km)	Mass (10²⁰ kg)	Density (10³ kg/m³)	Year of Discovery
N1	Triton	354.8	156.8	0.000	5.88	1352.6	214.0	2.061	1846
N2	Nereid	5513.4	7.23	0.751	360.1	170	0.3	1.5	1949
N9	Halimede	15728.0	134.1	0.571	1879.7	30.5	—	—	2002
N11	Sao	22422.0	48.51	0.293	2914.1	20	—	—	2002
N12	Laomedeia	23571.0	34.74	0.424	3167.9	20	—	—	2002
N10	Psamathe	46695.0	137.4	0.450	9115.9	19	—	—	2003
N13	Neso	48387.0	132.6	0.495	9374.0	30	—	—	2002

PLUTO

Name		Semimajor Axis (10³ km)	Orbital Inclination (degrees)	Orbital Eccentricity	Orbital Period (days)	Mean Radius (km)	Mass (10²¹ kg)	Density (10³ kg/m³)	Year of Discovery
P1	Charon	19.60	0.0	0.00005	6.3872	603.6	1.59	1.7	1978
P5	Styx	42.41	0.0	0.00001	20.162	4–14			2012
P2	Nix	48.69	0.0	0.00000	24.855	23–70	4.50×10^{-5}	—	2005
P4	Kerberos	57.75	0.4	0.00000	32.168	7–22	1.65×10^{-5}		2011
P3	Hydra	64.72	0.3	0.00554	38.202	29–86	4.79×10^{-5}	—	2005

Brozović, M. et al. (2014) The orbits and masses of satellites of Pluto. *Icarus* (doi:10.1016/j.icarus.2014.03.015).

HAUMEA

Name	Semimajor Axis (10³ km)	Orbital Inklination (Degrees)	Orbital Eccentricity	Orbital Period (Days)	Mean Radius (km)	Mass (10²¹ kg)	Density (10³ kg/m³)	Year of Discovery
Namaka								2005
Hi'aka				49	~310			2005

ERIS

Name	Semimajor Axis (10³ km)	Orbital Inclination (degrees)	Orbital Eccentricity	Orbital Period (days)	Mean Radius (km)	Mass (10²⁰ kg)	Density (10³ kg/m³)	Year of Discovery
Dysnomia	33	—	—	~14	200	—	—	2005

This table summarizes the physical and dynamical parameters of the satellites of planets and dwarf planets of the solar system. The satellites are ordered by their primary and the distance from the primary. Major satellites of the giant planets have been marked by shading.
http://home.dtm.ciw.edu/users/sheppard/satellites/, http://solarsystem.nasa.gov, http://ssd.jpl.nasa.gov, Hussmann et al. (2009), Basic data of planetary bodies. In W. Martinsen (Editor-in-chief), Landolt–Börnstein: Numerical Data and Functional Relationships in Science and Technology, New Series. Group VI, Vol. 4: Astronomy, Astrophysics and Cosmology; Subvolume B: Solar System (p. 163–181). Heidelberg: Springer. Brozović, M. et al. (2014) The orbits and masses of satellites of Pluto. *Icarus* (doi:10.1016/j.icarus.2014.03.015).

TABLE 4: SOLAR SYSTEM EXPLORATION MISSIONS

Prepared by Susanne Pieth, DLR German Aerospace Center, Institute of Planetary Research, Berlin.

Introduction

This is a compilation of robotic and human exploration missions to the bodies of the solar system except for Earth observation missions. Missions that failed to launch in the early time of space flight are not listed as are missions under development.

Missions to the Sun

Mission	Launch Date	Description
Pioneer 5	11 Mar 1960	NASA mission, solar and interplanetary research, solar orbit between the orbits of Venus and Earth, successful first testing of digital data transmission, end of mission: 26 Jun 1960
Pioneer 6	16 Dec 1965	NASA mission, study of the solar wind, the interplanetary electron density, solar and galactic cosmic rays, and the interplanetary magnetic field from a solar orbit between the orbits of Venus and Earth, last tracking: 8 Dec 2000
Pioneer 7	17 Aug 1966	NASA mission, study of the solar wind, interplanetary electron density, solar and galactic cosmic rays, and interplanetary magnetic field from a solar orbit at 1.1 AU, last tracking: Mar 1995
Pioneer 8	13 Dec 1967	NASA mission, study of the solar wind, the interplanetary electron density, solar and galactic cosmic rays, and interplanetary magnetic field, cosmic dust, and electric fields from a solar orbit at 1.1 AU, last tracking: 22 Aug 1996
Pioneer 9	08 Nov 1968	NASA mission, study of the solar wind, the interplanetary electron density, solar and galactic cosmic rays, and interplanetary magnetic field, cosmic dust, and electric fields from a solar orbit at 1.1 AU, contact until May 1983, unsuccessful attempt to contact the spacecraft in 1987
Skylab	26 May 1973	NASA's first manned space station (171 days), 150,000 images of the Sun with the Apollo Telescope Mount
Explorer 49 (Radioastronomy Explorer B)	10 Jun 1973	NASA mission, second of a pair of Radioastronomy Explorer satellites, study low-frequency radio emissions from the planets, the Sun, and the galaxy from lunar orbit, last contact: Aug 1977
Helios 1 (Helios A)	10 Dec 1974	German–American mission, study of the solar wind, magnetic and electric fields, cosmic rays, and interplanetary dust from a elliptic solar orbit, closest approach to the Sun: 47 million km, loss of contact: 16 Mar 1986
Helios 2 (Helios B)	16 Jan 1976	German–American mission, study of the solar plasma, the solar wind, magnetic and electric fields, cosmic rays, and cosmic dust from an elliptic solar orbit, closest approach to the Sun: 43.5 million km, end of mission: Dec 1981
Solar Maximum Mission	14 Feb 1980	NASA mission, coordinated monitoring of solar activity especially solar eruptions during a period of maximum solar activity, successful repair mission on STS-41C, data collection until 24 Nov 1989, Earth atmosphere reentry: 2 Dec 1989
Hinotori (Astro-A)	21 Feb 1981	Japanese mission, monitoring of solar eruptions during a period of maximum solar activity from Earth's orbit, Earth atmosphere reentry: 11 Jul 1991
Ulysses	06 Oct 1990	European–American mission, study of the Sun poles, corona, solar wind, solar magnetic field, plasma waves, and cosmic radiation from heliocentric orbit with high inclination after flyby at Jupiter on 8 Feb 1992, end of mission: 29 Jun 2009
Yohkoh	31 Aug 1991	Japanese–American–British mission, study of high-energy radiation during solar eruptions from Earth's orbit, successor to Hinotori, end of mission: 14 Dec 2001, Earth atmosphere reentry: 12 Sep 2005
SAMPEX	03 Jul 1992	"Solar Anomalous and Magnetospheric Particle Explorer", NASA mission, study of the anomalous components of cosmic rays, of solar energetic particles emission from the Sun, and of magnetospheric relativistic electrons from Earth's orbit, Earth atmosphere reentry: 13 Nov 2012
Koronas-I	02 Mar 1994	Russian–Ukrainian mission, study of the Sun in ultraviolet light and X-ray from Earth's orbit, end of mission: 4 Mar 2001
Wind	01 Nov 1994	NASA mission, study of the incoming solar wind, magnetic fields, and particles from a halo orbit around L1 Lagrangian point, still active
SOHO	12 Dec 1995	"Solar and Heliospheric Observatory", European–American mission, study of the inner structure and the physical processes that form the solar corona from a halo orbit around L1 Lagrangian point, still active

(*Continued*)

—cont'd

Mission	Launch Date	Description
ACE	25 Aug 1997	"Advanced Composition Explorer", NASA mission, measurements of the solar wind between Sun and Earth to allow "geomagnetic storm warnings" with a 1-h lead time from the orbit at L1 Lagrangian point, still active
TRACE	02 Apr 1998	"Transition Region and Coronal Explorer", NASA mission, study of solar eruptions, the photosphere, the geometry and dynamics of the upper solar atmosphere from a Sun-synchronous Earth orbit, end of mission: 22 Jun 2010
Koronas-F	31 Jul 2001	Russian–Ukrainian solar observatory, study of the Sun in ultraviolet light and X-ray from Earth's orbit
Genesis	08 Aug 2001	NASA mission, collection of samples of solar wind at the L1 Langrangian point and its return to Earth after two years, parachute of the sample return capsule failed to deploy on reentry on 8 Sep 2004
RHESSI	05 Feb 2002	"Reuven–Ramaty High-Energy Solar Spectroscopic Imager", NASA mission, study of the particle acceleration and energy release during solar eruptions from Earth's orbit, still active
SORCE	25 Jan 2003	"Solar Radiation Climate Experiment", NASA mission, precise measurements of solar radiation in different wavelengths from X-ray to near infrared from Earth's orbit, still active
STEREO	26 Sep 2006	"Solar Terrestrial Relations Observatory", NASA mission, consists of two probes, three-dimensional study of the structure and the evolution of solar storms on its way through space, heliocentric orbit ahead, and behind the Earth, still active
Hinode (SOLAR-B)	23 Sep 2006	Japanese mission, study of the interactions between the magnetic field and corona from a sun-synchronous orbit about Earth, still active
Koronas-Foton	30 Jan 2009	Russian mission, study of processes during solar flares, acceleration mechanisms, propagation, and interaction of fast particles in the solar atmosphere from Earth's orbit, end of mission due to a failed power system in Jan 2010
SDO	11 Feb 2010	"Solar Dynamics Observatory", NASA mission, study of the solar atmosphere in different wavelengths, solar activity and space weather, measurements of solar interior, plasma of solar corona, and radiation from Earth's orbit, still active
Picard	15 Jun 2010	French microsatellite, study of the solar irradiance, the diameter and solar shape and the Sun's interior through helioseismology from Earth's orbit, still active
IRIS	28 Jun 2013	"Interface Region Imaging Spectrograph", NASA mission, study of the outer solar atmosphere from Earth orbit, still active

Missions to Mercury

Mission	Launch Date	Description
Mariner 10	03 Nov 1973	NASA mission, first mission to two planets, Venus flyby and three Mercury flybys, study of the environment, atmosphere, surface, and figure parameters of Mercury and Venus, >10,000 images, 57% of Mercury covered, closest approach: 327 km, end of mission: 24 Mar 1975
MESSENGER	03 Aug 2004	"Mercury Surface, Space Environment, Geochemistry and Ranging", NASA mission, study of the chemical composition of the surface, geology, magnetic field, core, poles, exosphere and magnetosphere of the planet from orbit, orbit entry on 18 Mar 2011 after three flybys, still active

Appendix

Missions to Venus

Mission	Launch Date	Description
Venera 1	12 Feb 1961	Soviet mission, first flyby of a satellite at Venus, closest approach: 99,800 km, radio contact lost at 7 million km distance on 4 Mar 1961
Mariner 1	22 Jul 1962	NASA mission, intended to perform a Venus flyby, lost 293 s after launch due to a failure in the guidance commands
Mariner 2	26 Aug 1962	NASA mission, backup for Mariner 1, flyby at Venus, study of the atmosphere, magnetic field, charged particle environment, and mass, closest approach: 34,773 km, last transmission: 3 Jan 1963
Zond 1	22 Apr 1964	Soviet mission, loss of radio contact soon after 14 May 1964, Venus flyby at a 100,000-km distance on 14 Jul 1964, solar orbit
Venera 2	12 Nov 1965	Soviet mission, television (TV) system, and scientific instruments, Venus flyby, closest approach: 23,950 km, because of radio interference no data transmission possible, solar orbit
Venera 3	16 Nov 1965	Soviet mission, landing on the surface, atmosphere entry, communication system failed at an altitude of 32 km, impact on the planet
Venera 4	12 Jun 1967	Soviet mission, atmosphere entry, release of instruments, descent on the night side, transmission of the atmosphere and surface data for 96 min until it reached an altitude of 24.96 km
Mariner 5	14 Jun 1967	NASA mission, refurbished backup of Mariner 4, study of the structure of the atmosphere, its radiation and magnetic field environment, closest approach: 4094 km, contact lost on 4 Dec 1974
Venera 5	05 Jan 1969	Soviet mission, study of the atmosphere, release of a capsule into the atmosphere, data transmission on 16 May 1969 for 53 min
Venera 6	10 Jan 1969	Soviet mission, study of the atmosphere, release of a capsule into the atmosphere, data transmission on 17 May 1969 for 51 min
Venera 7	17 Aug 1970	Soviet mission, study of the atmosphere and other phenomena of the planet, release of a landing capsule on 15 Dec 1970, transmission of data for 23 min after landing
Venera 8	27 Mar 1972	Soviet probe and lander, successful landing on 22 Jul 1972, transmission of data from the surface for 50 min
Mariner 10	03 Nov 1973	NASA mission, first mission to two planets, study of the environment, atmosphere, surface, and figure parameters of Mercury and Venus, Venus flyby, and three Mercury flybys, closest approach to Venus: 5768 km, first images from Venus, end of mission: 24 Mar 1975
Venera 9	08 Jun 1975	Soviet mission, orbiter and lander, successful landing on 22 Oct 1975, operated for 53 min after landing, first images from the surface
Venera 10	14 Jun 1975	Soviet mission, orbiter and lander, successful landing on 25 Oct 1975, operated for 65 min after landing, images from the surface
Pioneer Venus 1	20 May 1978	NASA mission, study of the atmosphere, ionosphere, and surface from the orbit, end of mission: Aug 1992, spacecraft burnt up in the Venusian atmosphere on 8 Oct 1992
Pioneer Venus 2	08 Aug 1978	NASA mission, multiprobe spacecraft (five atmospheric probes), release of the probes on 16 Nov 1978 and 20 Nov 1978, transmission of data from the surface for 67 min from one of the probes
Venera 11	09 Sep 1978	Soviet mission, study of the chemical composition, the nature of the clouds and the thermal balance of the atmosphere, descent into the atmosphere on 25 Dec 1978, soft landing, transmission of data from the surface for 95 min
Venera 12	14 Sep 1978	Soviet mission, study of the chemical composition, the nature of the clouds and the thermal balance of the atmosphere, descent into the atmosphere on 21 Dec 1978, soft landing, transmission of data from the surface for 110 min

(Continued)

—cont'd

Mission	Launch Date	Description
Venera 13	29 Oct 1981	Soviet mission, soft landing on 1 Mar 1982, first panoramic images through several filters, examination of soil samples, operated for 127 min on the surface
Venera 14	01 Nov 1981	Soviet mission, soft landing on 3 Mar 1982, panoramic images through several filters, examination of soil samples, operated for 57 min on the surface
Venera 15	02 Jun 1983	Soviet mission, study of the surface properties with Synthetic Aperture Radar from orbit, radar images of the northern hemisphere, resolution: 1–2 km, end of mission: 10 Jul 1984
Venera 16	07 Jun 1983	Soviet mission, study of the surface properties with Synthetic Aperture Radar from orbit, radar images of the northern hemisphere, resolution: 1–2 km, end of mission: Oct 1984
Vega 1	15 Dec 1984	Soviet mission, Venus flyby in Jun 1985 and comet Halley flyby, release of a lander and balloon for studies of the central cloud cover, balloon operated in the atmosphere for 46.5 h
Vega 2	21 Dec 1984	Soviet mission, Venus flyby in Jun 1985 and comet Halley flyby, release of a lander and balloon for studies of the central cloud cover, balloon operated in the atmosphere for ~47 h
Magellan	04 May 1989	NASA mission, study of landforms and tectonics, impact processes, erosion, deposition, chemical processes, and data for modeling the interior radar mapping of 95% of the surface with Synthetic Aperture Radar, highest resolution: 75 m/pixel, contact lost on 12 Oct 1994
Galileo	18 Oct 1989	NASA mission, flyby at Venus on its way to Jupiter on 10 Feb 1990, images of the cloud cover
Cassini	15 Oct 1997	NASA–ESA mission, two flybys at Venus on its way to the Saturnian system on 26 Apr 1998 and 24 Jun 1999
MESSENGER	03 Aug 2004	"Mercury surface, space environment, geochemistry and ranging", NASA mission, images from Venus during two flybys on its way to Mercury on 24 Oct 2006 and 5 Jun 2007
Venus Express	09 Nov 2005	European Space Agency (ESA) mission, study of the complex dynamics, and chemistry of the planet and the interaction between the atmosphere and surface from orbit, orbit entry on 11 Apr 2006, still active
Akatsuki (Planet-C)	20 May 2010	Japanese Aerospace Exploration Agency (JAXA) mission, study the dynamics of the atmosphere from orbit, orbit entry failed and satellite flew by Venus, next possibility to enter orbit in 2016/2017

Missions to the Moon

Mission	Launch Date	Description
Luna 1	02 Jan 1959	Soviet mission, study of interplanetary space, flyby, closest approach: 5995 km, data transmission for 62 h after launch
Pioneer 4	03 Mar 1959	American mission, flyby of the Moon at a distance of 60,000 km
Luna 2	12 Sep 1959	Soviet mission, first impact on the Moon on 13 Sep 1959
Luna 3	04 Oct 1959	Soviet mission, 29 images from the near and the far side for the first time, after flight around the Moon spacecraft burnt up in the Earth's atmosphere on Apr 1960
Ranger 3	04 Oct 1959	NASA mission, study of the surface prior to impact, rough landing of a seismometer capsule, to collect gamma-ray data in flight and study radar reflectivity of the surface, missed lunar orbit due to malfunctions

—cont'd

Mission	Launch Date	Description
Ranger 4	23 Apr 1962	NASA mission, study of the surface prior to impact, rough landing of a seismometer capsule, to collect gamma-ray data in flight and study radar reflectivity of the surface, impact on 26 Apr 1962, loss of radio communication on launch day
Ranger 5	18 Oct 1962	NASA mission, study of the surface prior to impact, rough landing of a seismometer capsule, to collect gamma-ray data in flight and study radar reflectivity of the surface, malfunction during injection on the Moon trajectory, passed within 725 km of the Moon
Luna 4	02 Apr 1963	Soviet mission, astronavigation system failed, missed the Moon by about 8400 km
Ranger 6	30 Jan 1964	NASA mission, study the lunar surface in high-resolution prior impact, impact on the edge of Mare Tranquillitatis on 2 Feb 1964, due to a failure of the camera system no images were transmitted
Ranger 7	28 Jul 1964	NASA mission, study the lunar surface in high-resolution prior impact, impact in Mare Nubium on 31 Jul 1964, 4308 images during approach
Ranger 8	17 Feb 1965	NASA mission, study the lunar surface at high-resolution prior impact, impact in Mare Tranquillitatis on 20 Feb 1964, >7000 images during approach
Ranger 9	21 Mar 1965	NASA mission, study the lunar surface at high-resolution prior impact, impact in the Alphonsus crater on 24 Mar 1965, >5800 images
Luna 5	09 May 1965	Soviet mission, testing of a lunar soft landing, impact on the Moon on 12 May 1965 due to malfunctions of the main retrorocket system
Luna 6	08 Jun 1965	Soviet mission, attempted soft landing, midcourse correction failed, missed the Moon by 161,000 km on 11 Jun 1965
Zond 3	18 Jul 1965	Soviet mission, flyby at a distance of 9200 km on 20 Jul 1965, 25 images in very good quality from the far side, transmission ended in Mar 1966
Luna 7	04 Oct 1965	Soviet mission, system tests for landing, due to the loss of attitude control impact in Oceanus Procellarum on 7 Oct 1965
Luna 8	03 Dec 1965	Soviet mission, system tests for landing, soft landing failed, impact in Oceanus Procellarum on 6 Dec 1965
Luna 9	31 Jan 1966	Soviet mission, first soft landing on the Moon on 3 Feb 1966, panoramic images of the surface, end of mission: 6 Feb 1966
Luna 10	31 Mar 1966	Soviet mission, first artificial satellite of the Moon, operated for 56 days in orbit, end of mission: 30 May 1966
Surveyor 1	30 May 1966	NASA mission, soft landing after direct injection into a lunar impact trajectory, 10,338 images, 1000 using red, green, and blue filters on the first lunar day, 812 images on the second lunar day, end of mission: 13 Jul 1966
Lunar Orbiter 1	10 Aug 1966	NASA mission, study of the surface for potential Surveyor and Apollo landing sites from orbit, photographic coverage of ~5 million km^2, transmission of 229 images, mission ended with impact on 29 Oct 1966
Luna 11	24 Aug 1966	Soviet mission, study of lunar gamma- and X-ray emission and lunar gravitational anomalies from orbit, battery failed on 1 Oct 1966
Surveyor 2	20 Sep 1966	NASA mission, attempted soft landing, after unsuccessful path correction out of control, impact south of Copernicus crater on 23 Sep 1966
Luna 12	22 Oct 1966	Soviet mission, images of the surface from orbit, data transmission ended on 19 Jan 1967
Lunar Orbiter 2	06 Nov 1966	NASA mission, study of the surface for potential Surveyor and Apollo landing sites from orbit, transmission of 817 images, mission ended with impact on 11 Oct 1967

(*Continued*)

—cont'd

Mission	Launch Date	Description
Luna 13	21 Dec 1966	Soviet mission, soft landing near Seleucus crater, panoramic images at different sun angles, loss of contact: 30 Dec 1966
Lunar Orbiter 3	05 Feb 1967	NASA mission, study of the surface for potential Surveyor and Apollo landing sites from orbit, transmission of only 626 images due to failure in the image transport system, mission ended with impact on 9 Oct 1967
Surveyor 3	17 Apr 1967	NASA mission, soft landing in the eastern part of Oceanus Procellarum, transmission of 6326 images, last transmission of 4 May 1967
Lunar Orbiter 4	04 May 1967	NASA mission, broad systematic photographic survey of the surface, transmission of 546 images, 99% of the near side covered, mission ended with impact on 31 Oct 1967
Surveyor 4	14 Jul 1967	NASA mission, soft landing failed, impact on the Moon on 17 Jul 1967
Explorer 35	19 Jul 1967	NASA mission, study of interplanetary plasma, magnetic field, energetic particles, and solar X-rays from lunar orbit, end of mission on 24 Jun 1973 after successful operation for 6 years
Lunar Orbiter 5	01 Aug 1967	NASA mission, broad systematic photographic survey of the surface and imaging of additional potential Surveyor and Apollo landing sites from orbit, 844 images, mission ended with impact on 18 Aug 1967
Surveyor 5	08 Sep 1967	NASA mission, soft landing in Mare Tranquillitatis on 11 Sep 1967, 18,006 images and soil analyses during the first lunar day, 19,118 images in total over four lunar days, loss of contact: 17 Dec 1967
Surveyor 6	07 Nov 1967	NASA mission, soft landing in Sinus Medii, 14,500 images and $\sim$55 soil analyses; 15,000 images from a new position (stereoscopic coverage), last contact: 14 Dec 1967
Surveyor 7	07 Jan 1968	NASA mission, soft landing 40 km north of the rim of Tycho crater; 21,038 images for the first time using a polarizing filter, soil analyses, transmission ended on 21 Feb 1968
Luna 14	07 Apr 1968	Soviet mission, study of the gravitational interaction of the Earth and lunar masses, lunar gravitational field, solar charged particles and cosmic rays, and the motion of the Moon, mission success is not proved
Zond 5	14 Sep 1968	Soviet mission, test of the return of a space probe with biological payload after lunar flyby, flight around the Moon and back to the Earth, closest approach to the Moon: 1950 km, first successful Soviet circumlunar Earth-return mission, mission ended on 21 Sep 1968
Zond 6	10 Nov 1968	Soviet mission, aerodynamic return of biological payload, closest approach: 2420 km, photographs of the lunar near and far side from distances of $\sim$11,000 and 3300 km, controlled reentry on 17 Nov 1968
Apollo 8	21 Dec 1968 to 27 Dec 1968	NASA mission, first manned space flight, photographic exploration of the planned Apollo landing site and other areas, test of command module systems, 10 orbits around the Moon, extensive photography of the lunar surface and six live TV transmissions, safe return to Earth on 27 Dec 1968
Apollo 10	18 May 1969 to 26 May 1969	NASA mission, manned landing simulation in the lunar orbit, closest approach: 15,185 m, extensive photography of the lunar surface and TV images, safe return to Earth on 26 May 1969
Luna 15	13 Jul 1969	Soviet mission, study of the circumlunar space, the lunar gravitational field, and the chemical composition of lunar rocks, lunar surface photography, mission ended with impact on 21 Jul 1969
Apollo 11	16 Jul 1969 to 24 Jul 1969	NASA mission, first manned Moon landing, landing in Mare Tranquillitatis, return of soil, rock samples, and photographs, safe return to Earth on 24 Jul 1969

Appendix

—cont'd

Mission	Launch Date	Description
Zond 7	07 Aug 1969 to 14 Aug 1969	Soviet mission, color photography of the Moon and Earth from varying distances, closest approach: 2000 km, aerodynamic return, soft landing on Earth on 14 Aug 1969
Apollo 12	14 Nov 1969 to 24 Nov 1969	NASA mission, second manned Moon landing, landing in Oceanus Procellarum, return of soil, rock samples, and photographs, safe return to Earth on 24 Nov 1969
Apollo 13	11 Apr 1970 to 17 Apr 1970	NASA mission, attempted manned Moon landing, abortion after explosion of an oxygen tank in the service module, successful return to Earth on 17 Apr 1970
Luna 16	12 Sep 1970	Soviet mission, first return of soil samples with an unmanned remote controlled semiautomatic probe, safe landing in Mare Foecunditatis, liftoff of the return capsule 26 h later back to Earth, capsule landed on 24 Sep 1970
Zond 8	20 Oct 1970	Soviet mission, closest approach: 1120 km, return trajectory over the northern hemisphere of Earth, color and black/white images from the Moon and Earth, splashdown in the Indian Ocean on 27 Oct 1970
Luna 17	10 Nov 1970	Soviet mission, soft landing in Mare Imbrium, remote controlled semiautomatic lunar rover "Lunokhod 1", >200 panoramic images, 20,000 other images, >500 soil analyses, operation ceased officially on 4 Oct 1971
Apollo 14	31 Jan 1971 to 09 Feb 1971	NASA mission, third manned Moon landing, landing near Fra Mauro crater in eastern Oceanus Procellarum, return of soil, rock samples, and photographs, safe return to Earth on 9 Feb 1971
Apollo 15	26 Jul 1971 to 07 Aug 1971	NASA mission, fourth manned Moon landing, landing in the Rima Hadley area, first manned lunar rover, return of soil, rock samples, and photographs, safe return to Earth on 7 Aug 1971
Luna 18	02 Sep 1971	Soviet mission, impact in Mare Foecunditatis after 54 orbits, signals ceased at the moment of impact
Luna 19	28 Sep 1971	Soviet mission, study of lunar gravitational field and location of mascons from lunar orbit, high-resolution images of the lunar surface, end of mission: 20 Oct 1972
Luna 20	14 Feb 1972	Soviet mission, landing at the northeastern edge of Mare Foecunditatis, return of samples, return of the capsule on 25 Feb 1972
Apollo 16	16 Apr 1972 to 27 Apr 1972	NASA mission, fifth manned Moon landing, landing in Cayley plateau near Descartes crater, manned lunar rover, return of soil, rock samples, and photographs, safe return to Earth on 27 Apr 1972
Apollo 17	07 Dec 1972 to 19 Dec 1972	NASA mission, sixth and last manned Moon landing, landing in Taurus–Littrow, manned lunar rover, return of 113 kg of lunar samples and photographs, safe return to Earth on 19 Dec 1972
Luna 21	08 Jan 1973	Soviet mission, soft landing in Le Monnier crater, remote controlled semiautomatic lunar rover "Lunokhod 2", 86 panoramic images, >80,000 TV pictures, end of mission: 4 Jun 1974
Explorer 49 (Radioastronomy Explorer B)	10 Jun 1973	NASA mission, second of a pair of Radio Astronomy Explorer satellites, lunar orbit, study low-frequency radioemissions from the planets, the Sun, and the galaxy, last contact: Aug 1977
Luna 22	29 May 1974	Soviet mission, study of the magnetic field, surface gamma-ray emissions, composition of lunar surface rocks and gravitational field from orbit at varying altitudes, end of mission: Nov 1975
Luna 23	28 Oct 1974	Soviet mission, attempted sample return, after lunar orbit landing in Mare Crisium, sample collection failed
Luna 24	14 Aug 1976	Soviet mission, landing at the southeastern edge of Mare Crisium, return of 170 g of lunar samples, return of the capsule on 22 Aug 1976

(Continued)

—cont'd

Mission	Launch Date	Description
Galileo	18 Oct 1989	NASA mission, flybys at the Moon on its way to Jupiter in Dec 1990 and Dec 1992, multispectral images
Hiten (Muses-A)	24 Jan 1990	Japanese mission, test and verify technologies for future missions, subsatellite "Hagaromo" released to lunar orbit but data transmission failed, several flybys at the Moon of Hiten before lunar orbit and impact on the Moon on 10 Apr 1993
Clementine	25 Jan 1994	American mission, test of sensors and spacecraft components, study of the surface mineralogy of the Moon from orbit, multispectral mapping of the whole Moon with resolutions of 125–250 m/pixel, planned flyby at asteroid 1620 Geographos failed, end of mission: Jun 1994
Lunar Prospector	07 Jan 1998	NASA mission, study of the surface composition and possible deposits of polar ice, measurements of magnetic and gravity fields and lunar outgassing events from polar orbit, end of mission with impact on 31 Jul 1999
SMART-1	27 Sep 2003	ESA mission, test of a solar powered ion engine, study of the geology, morphology, topography, mineralogy, geochemistry, and exospheric environment from orbit, end of mission with impact on 3 Sep 2005
Kaguya (SELENE)	14 Sep 2007	JAXA mission, two subsatellites "Ouna" and "Okina", study of mineralogy, topography, geography, and gravity field from orbit, end of mission with impact on 10 Jun 2009
Chang'e 1	24 Oct 2007	Chinese mission, test of technologies for future missions, mapping of lunar surface chemistry, stereo images from lunar orbit, end of mission with impact on 1 Mar 2009
Chandrayaan 1	22 Oct 2008	Indian mission with Indian and international payload, test of space technologies, global high-resolution mapping with 5m resolution, mineralogical mapping, laser ranging, release of a Moon Impact Probe on 14 Nov 2008, loss of contact: 28 Aug 2009
Lunar Reconnaissance Orbiter (LRO)	18 Jun 2009	NASA mission, mapping of the surface, characterization of future landing sites regarding terrain roughness, usable resources, and radiation environment from orbit, still active
Lunar Crater Observation and Sensing Satellite	18 Jun 2009	NASA mission, impactor, launched along with LRO, consists of a Shepherding Spacecraft (S—S/C) attached to the Centaur upper stage, search for water ice by directing the Centaur upper stage to impact the surface while observing the collision from behind, impacted on the lunar surface on 9 Oct 2009
Chang'e 2	01 Oct 2010	Chinese mission, modified backup satellite of Chang'e 1, high-resolution images of the surface, search for landing sites from orbit, left the Moon orbit to L2 Lagrangian point on 9 Jun 2011 for measurements of the solar wind, left the L2 point to asteroid 4179 Toutatis, flyby on 13 Dec 2012 at the closest distance of 3.2 km, presumably still active
GRAIL	10 Sep 2011	"Gravity Recovery and Interior Laboratory", NASA mission, consisted of two satellites launched together, high resolution and accuracy mapping of the lunar gravity field to study the structure of the crust and lithosphere, the asymmetric thermal evolution of the Moon, the subsurface structure of basins and of mascons, orbit entry in Dec 2011 and Jan 2012, end of mission with impacts of the satellites on the Moon on 17 Dec 2012
LADEE	07 Sep 2013	"Lunar Atmosphere and Dust Environment Explorer", NASA mission to study the exosphere and dust in the vicinity of the Moon with a neutral mass spectrometer, ultraviolet–visible spectrometer, and lunar dust experiment, technical demonstration of a laser communication terminal, still active
Chang'e 3	01 Dec 2013	Chinese mission, soft landing in northern Mare Imbrium on 14 Dec 2014, release of rover Yutu, still active

Appendix

Missions to Mars

Mission	Launch Date	Description
Marsnik 1 (Mars 1960A)	10 Oct 1960	Soviet mission, first soviet planetary probe, failure of the third stage, parking orbit not reached, highest altitude: 120 km, reentry
Marsnik 2 (Mars 1960B)	14 Oct 1960	Soviet mission, failure of the third stage, parking orbit not reached, highest altitude: 120 km, reentry
Sputnik 22 (Mars 1962A)	24 Oct 1962	Soviet mission, attempted Mars flyby, failure, upper stage exploded in the Earth orbit
Mars 1	01 Nov 1962	Soviet mission, study of the interplanetary space between Earth and Mars, flyby at Mars at a distance of 193,000 km on 19 Jun 1963, loss of contact: 21 Mar 1963
Sputnik 24 (Mars 1962B)	04 Nov 1962	Soviet mission, upper stage failed fragmenting the spacecraft, could not leave the Earth orbit, reentry on 19 Jan 1963
Mariner 3	05 Nov 1964	NASA mission, study the vicinity of Mars and obtain photographs, failure of shell separation, no Mars flyby possible, loss of contact
Mariner 4	28 Nov 1964	NASA mission, first successful flyby at Mars on 14 Jul 1965, closest approach: 9844 km, 22 images of the Martian surface, communication ceased on 21 Dec 1967
Zond 2	30 Nov 1964	Soviet mission, flyby at Mars on 6 Aug 1965, closest approach: 1500 km, failure of the communication system in early May 1965
Zond 3	18 Jul 1965	Soviet mission, flyby at the Moon at a distance of 9200 km on 20 Jul 1965, transmission ended in Mar 1966, flight to Mars orbit
Mariner 6	25 Feb 1969	NASA mission, study of the surface and atmosphere during close flyby on 31 Jul 1969 at a distance of 3431 km, 75 images returned
Mariner 7	27 Mar 1969	NASA mission, spacecraft identical to Mariner 6, study of the surface and atmosphere during close flyby at 5 Aug 1969 at a distance of 3430 km, 126 images returned
Mars 1969A	27 Mar 1969	Soviet mission, attempted Mars orbiter, explosion of the third stage 7 min after launch
Mars 1969B	02 Apr 1969	Soviet mission, attempted Mars orbiter, failure of the first stage promptly after launch
Mariner 8	08 May 1971	NASA mission, attempted Mars orbiter, malfunction of the Centaur upper stage, reentry
Cosmos 419	10 May 1971	Soviet mission, attempted Mars orbiter, reached parking orbit around the Earth, could not leave parking orbit due to upper stage failure, reentry on 12 May 1971
Mars 2	19 May 1971	Soviet mission, orbiter and lander, study of the surface, atmosphere, temperature, topography, and physical properties, reached Mars orbit on 21 Nov 1971, first release of landing capsule, soft landing failed, end of mission: 22 Aug 1972
Mars 3	28 May 1971	Soviet mission, identical to Mars 2, study of the surface, atmosphere, temperature, topography, and physical properties, reached Mars orbit on 2 Dec 1971, release of landing capsule, soft landing, lander instruments worked for 20 s only, end of mission: 22 Aug 1972
Mariner 9	30 May 1971	NASA mission, mapping of the surface and study of temporal changes of the atmosphere and surface of Mars, reached Mars orbit on 14 Nov 1971, first artificial satellite of a planet, >7000 images of the surface, maximum resolution: 100 m/pixel, end of mission: 27 Oct 1972
Mars 4	21 Jul 1973	Soviet mission, attempted orbiter, could not reach Mars orbit due to technical malfunctions, flyby at a distance of 2200 km on 10 Feb 1974
Mars 5	25 Jul 1973	Soviet mission, study the composition, structure, and properties of the Martian atmosphere and surface, reached Mars orbit on 12 Feb 1974, orbiter worked a few days only, transmission of data from the atmosphere and images of a small part of the southern hemisphere
Mars 6	05 Aug 1973	Soviet mission, flyby bus and descent module, reached Mars on 12 Mar 1974, landing in Margaritifer Sinus, contact lost

(*Continued*)

—cont'd

Mission	Launch Date	Description
Mars 7	09 Aug 1973	Soviet mission, flyby bus, and descent module, reached Mars on 9 Mar 1974, landing capsule missed its target
Viking 1	20 Aug 1975	NASA mission, orbiter and lander, high-resolution images of the surface, study of the structure and composition of the atmosphere and surface, search for evidence of life, reached Mars orbit on 19 Jul 1976, soft landing on 20 Jul 1976 in Chryse Planitia, both orbiter and landers (Viking 1 and 2) transmitted panoramic images and other data, all in all 55,000 images (including of the moons), coverage of the whole surface with resolutions from 100 to 200 m/pixel, regionally up to 30 m, some images up to 8 m/pixel, orbiter shut down: 7 Aug 1980, lander operated in the automated mode until Nov 1982
Viking 2	09 Sep 1975	NASA mission, orbiter and lander, identical to Viking 1, high-resolution images of the surface, study of the structure and composition of the atmosphere and surface, search for evidence of life, reached Mars orbit on 7 Aug 1976, soft landing on 3 Sep 1976 in Utopia Planitia, end of orbiter mission: 25 Jul 1978, lander operated until 11 Apr 1980
Phobos 1	07 Jul 1988	Soviet mission, attempted Mars orbiter and flyby at Phobos with release of small landers, loss of contact due to a fault signal on 2 Sep 1988, recovery of the probe not possible
Phobos 2	12 Jul 1988	Soviet mission, Mars orbiter and flyby at Phobos with release of a small lander, reached Mars orbit on 29 Jan 1989, thermal images of a nearly 1500-km broad strip at the equator, resolution: $\sim$ 2 km/pixel, 9 images of Phobos, loss of communication on 27 Mar 1989 before release of probe
Mars Observer	25 Sep 1992	NASA mission, attempted Mars orbiter, study of the surface and climate, loss of communication on 21 Aug 1993 three days before orbit insertion at Mars, the most likely cause: Explosion of the engine during injection maneuver
Mars Global Surveyor	07 Nov 1996	NASA mission, replacement for Mars Observer, Mars orbit insertion started on 12 Sep 1997, one year longer aerobraking to mapping orbit due to a malfunction of one of the solar panels, >243,000 images and high-resolution global topographic map using laser altimetry, mapping of the surface magnetization, loss of contact: 2 Nov 2006
Mars-96	16 Nov 1996	Russian mission with international participation, instable Earth orbit due to a malfunction of the fourth rocket stage, loss of probe and fourth stage, reentry on 17 Nov 1996
Mars Pathfinder	04 Dec 1996	NASA mission, soft landing on 4 Jul 1997 in Ares Vallis, rover "Sojourner" was released on 6 Jul 1997, lander and rover worked until loss of contact on 27 Sep 1997, >17,000 images
Nozomi (Planet B)	03 Jul 1998	Japanese mission, study of the atmosphere from orbit, Mars orbit insertion failed, mission abandoned on 9 Dec 2003
Mars Climate Orbiter	11 Dec 1998	NASA mission, study of weather and climate, water and carbon dioxide budget from orbit, loss of probe during orbit insertion on 23 Sep 1999
Mars Polar Lander	03 Jan 1999	NASA mission, study of weather and climate, water and carbon dioxide budget, landing failed on 3 Dec 1999
Deep Space 2	03 Jan 1999	NASA mission, part of the New Millennium Program, consisted of two micropenetrators to explore the sub-surface of Mars near the South Pole, were attached to the Mars Polar Lander, landing failed on 3 Dec 1999
2001 Mars Odyssey	07 Apr 2001	NASA mission, detailed mineralogical mapping of the surface and study of the radiation environment from orbit, also communication relay for future landing missions, still active
Mars Express	02 Jun 2003	ESA mission, orbiter and lander "Beagle 2", lander separation on 19 Dec 2003, landing failed, orbit insertion on 25 Dec 2003, global high-resolution imaging, mineralogical mapping, study of atmospheric composition and processes, still active

Appendix

—cont'd

Mission	Launch Date	Description
Spirit (Mars exploration Rover A)	10 Jun 2003	NASA mission, landing in Gusev crater on 4 Jan 2004, rover with scientific instrumentation and a daily range of 100 m, explore the habitability of the area, study of climate and geology, loss of contact: 22 Mar 2010
Opportunity (Mars exploration Rover B)	08 Jul 2003	NASA mission, landing in Meridiani Planum on 25 Jan 2004, rover with scientific instrumentation and a daily range of 100 m, explore the habitability of the area, study of climate and geology, still active
Mars Reconnaissance Orbiter	12 Aug 2005	NASA mission, study of the current climate, observation of the surface from orbit using a high-resolution camera and search for landing sites, orbit entry on 10 Mar 2006, still active
Phoenix	04 Aug 2007	NASA mission, small stationary lander, study of the surface at high latitudes, observation of the polar climate and weather, composition of the lower atmosphere, geomorphology, and role of water, soft landing in the north polar region at 68.15° N and 125.9° W on 25 May 2008, end of mission: 2 Nov 2008
Phobos Grunt	08 Nov 2011	Russian lander mission to Phobos, exploration of the landing site, soil sampling from the surface and return to Earth, could not reach trajectory to Mars due to a failure in the propulsive unit, loss of the spacecraft, Earth atmosphere reentry on 15 Jan 2012
Yinghuo-1	08 Nov 2011	Chinese mission, "lightning bug", tandem flight with Phobos Grunt, measurements of the gravity field for one year from orbit around Mars, loss of the satellite, Earth atmosphere reentry on 15 Jan 2012
Mars Science Laboratory	25 Nov 2011	NASA mission, rover "Curiosity", exploration of the mineralogy, geochemistry and habitability, successful landing in Gale crater on 6 Aug 2012, nominal mission duration: one Martian year (687 days), still active
Mangalyaan (Mars Orbiter mission)	05 Nov 2013	Indian mission, study of surface features, morphology, mineralogy, and atmosphere, orbit insertion planned for 24 Sep 2014, still active
MAVEN	18 Nov 2013	"Mars Atmosphere and Volatile EvolutioN", NASA mission, study the upper atmosphere, ionosphere, and interactions with the solar wind, scheduled to reach Mars on 22 Sep 2014, still active

Missions to Asteroids

Mission	Launch Date	Description
Galileo	18 Oct 1989	NASA mission, flyby at 951 Gaspra (Oct 1991) and 243 Ida (Aug 1993) on its way to Jupiter
NEAR Shoemaker	17 Feb 1996	"Near Earth Asteroid Rendezvous", NASA mission, study asteroid 433 Eros from close orbit, flyby at 253 Mathilde on 27 Jun 1997, in orbit around 433 Eros from Feb 2000 to Feb 2001, afterward controlled descent to Eros on 12 Feb 2001, contact lost: 28 Feb 2001
Cassini	15 Oct 1997	NASA mission, flyby at 2685 Masursky (Jan 2000) on its way to the Saturnian system
Deep Space 1	24 Oct 1998	NASA mission, test of new technologies (ion propulsion) for use in space, flyby at asteroid 9969 Braille (29 Jul 1999) and comet Borrelly (22 Sep 2001), end of mission: 18 Dec 2001

(Continued)

—cont'd

Mission	Launch Date	Description
Hayabusa (Muses-C)	09 May 2003	Japanese mission, orbiter and lander "Minerva" with sample return from the surface of asteroid 25143 Itokawa and technology demonstration, reached Itokawa in Sep 2005, lander was deployed on 12 Nov 2005, landing of the orbiter on 25 Nov 2005, sample acquisition and restart to Earth, reentry capsule with samples entered Earth's atmosphere on 13 Jul 2010
Rosetta	02 Mar 2004	ESA mission, orbiter and lander "Philae", flyby at 2867 Šteins (5 Sep 2008) and 21 Lutetia (10 Jul 2010) on its way to comet Churyumov–Gerasimenko, landing planned for Nov 11, 2014
Dawn	27 Sep 2007	NASA mission, rendezvous and orbit of asteroid 4 Vesta and dwarf planet 1 Ceres, arrival at Vesta and in orbit from 14 Aug 2011 to 5 Sep 2012, will reach 1 Ceres on 1 Feb 2015, still active

Missions to Jupiter

Mission	Launch Date	Description
Pioneer 10	03 Mar 1972	NASA mission, first spacecraft to the outer solar system, study of the planet Jupiter, flyby at Jupiter at a distance of 200,000 km on 4 Dec 1972, numerous images of the equatorial region and a few satellites, first probe which left the solar system, last contact on 23 Jan 2003
Pioneer 11	06 Apr 1973	NASA mission, second spacecraft to explore Jupiter and the outer solar system and first to explore Saturn and its main rings, flyby at Jupiter at a distance of 42,000 km on 3 Dec 1974, numerous images of Jupiter and a few satellites, last contact at a distance of 44.7 AU at the end of 1995
Voyager 2	20 Apr 1977	NASA mission, study of the outer planets, flyby at Jupiter on 9 Jul 1979, closest approach: 722,000 km, 18,000 images of Jupiter and its moons, still active on its way out of the solar system
Voyager 1	05 Sep 1977	NASA mission, study of the outer planets, flyby at Jupiter on 5 Mar 1979, closest approach: 349,000 km, 18,000 images of Jupiter and its moons, entered interstellar space on 25 Aug 2012
Ulysses	06 Oct 1990	European–American mission, study of the Sun, flyby at Jupiter on 8 Feb 1992, end of mission: 29 Jun 2009
Galileo	18 Oct 1989	NASA mission with international participation, first spacecraft with a complex trajectory with gravitational assists, orbiter, and atmospheric probe, study of Jupiter's atmosphere and magnetosphere, and the Galilean satellites, arrival at Jupiter on 7 Dec 1995, end of mission with impact into Jupiter on 21 Sep 2003
Cassini–Huygens	15 Oct 1997	NASA–ESA mission, study of the Saturnian system, flyby at Jupiter in Dec 2000 on its way to the Saturnian system
New Horizons	19 Jan 2006	NASA mission, study of Pluto and its moon Charon and continue to the Kuiper Belt, flyby at Jupiter on 28 Feb 2007 on its way to the Pluto–Charon system
Juno	05 Aug 2011	NASA mission, study of the atmosphere, magnetic field, gravity field, and magnetosphere of Jupiter from polar orbit, after Earth flyby on 9 Oct 2013 arrival at Jupiter on 5 Jul 2016, still active

Appendix

Missions to Saturn

Mission	Launch Date	Description
Pioneer 11	06 Apr 1973	NASA mission, second spacecraft to explore Jupiter and the outer solar system and first to explore Saturn and its main rings, flyby at Saturn at a distance of 20,800 km on 1 Sep 1979, 440 images of Saturn and a few satellites, last contact at a distance of 44.7 AU at the end of 1995
Voyager 2	20 Aug 1977	NASA mission, study of the outer planets, flyby at Saturn on 26 Aug 1981 at a distance of 102,000 km, 16,000 images of Saturn and its moons, still active on its way out of the solar system
Voyager 1	05 Sep 1977	NASA mission, study of the outer planets, flyby at Saturn on 12 Nov 1980 at a distance of 124,000 km, about 16,000 images of Saturn and its moons, entered interstellar space on 25 Aug 2012
Cassini–Huygens	15 Oct 1997	NASA–ESA mission, study of Saturn, its rings and satellites from orbit, release of atmospheric probe "Huygens" into Titan's atmosphere in Dec 2004, still active

Missions to Uranus

Mission	Launch Date	Description
Voyager 2	20 Aug 1977	NASA mission, study of the outer planets, flyby at Uranus on 24 Jan 1986 at a distance of 107,000 km, 7000 images of Uranus and its moons, still active on its way out of the solar system

Missions to Neptune

Mission	Launch Date	Description
Voyager 2	20 Aug 1977	NASA mission, study of the outer planets, flyby at Neptune on 25 Aug 1989 at a distance of 29,240 km, 9000 images of Neptune and its moons, still active on its way out of the solar system

Missions to Dwarf Planets

Mission	Launch Date	Description
New Horizons	19 Jan 2006	NASA mission, study of Pluto and its moon Charon during flyby and continue to the Kuiper Belt, arrival at the Pluto—Charon system in Jul 2015, still active
Dawn	27 Sep 2007	NASA mission, rendezvous and orbit of the asteroid 4 Vesta and dwarf planet 1 Ceres, arrival at Vesta and in orbit from 14 Aug 2011 to 5 Sep 2012, will reach 1 Ceres on 1 Feb 2015, still active

Missions to Comets

Mission	Launch Date	Description
ISEE 3/ICE	12 Aug 1978	"International Sun Earth Explorer 3", NASA mission, renamed "International Cometary Explorer", study of the interaction between the solar wind and a cometary atmosphere, flight through the plasma tail of comet Giacobini—Zinner on 11 Sep 1985
Vega 1	15 Dec 1984	Soviet mission, flyby at comet Halley on 6 Mar 1986 after Venus flyby in Jun 1985
Vega 2	21 Dec 1984	Soviet mission, flyby at comet Halley on 9 Mar 1986 after Venus flyby in Jun 1985
Sakigake	08 Jan 1985	Japanese mission, flyby at comet Halley on 11 Mar 1986, loss of contact: 15 Nov 1995
Giotto	02 Jul 1985	ESA mission, study of comet Halley during flyby on 13 Mar 1986 at a distance of 605 km, images of the nucleus, flyby at comet Grigg—Skellerup on 10 Jul 1992 at a distance of 200 km, end of mission: 23 Jul 1992
Suisei (Planet-A)	18 Aug 1985	Japanese mission, flyby at comet Halley on 8 Mar 1986 at a distance of 151,000 km, end of mission: 22 Feb 1991
Galileo	18 Oct 1989	NASA mission with international participation, first spacecraft with a complex trajectory with gravitational assists, images from traces after impact of Shoemaker-Levy 9 fragments on Jupiter, 17—22 Apr 1994
NEAR	17 Feb 1996	"Near Earth Asteroid Rendezvous", NASA mission, study asteroid 433 Eros from close orbit, flyby at comet Hyakutake on 24 Mar 1996 on its way to asteroid 433 Eros
Deep Space 1	24 Jan 1998	NASA mission, test of new technologies (ion propulsion) for use in space, flyby at asteroid 9969 Braille (29 Jul 1999) and comet Borrelly (22 Sep 2001), end of mission: 18 Dec 2001
Stardust	07 Feb 1999	NASA mission, study of comet P/Wild 2 and collect samples of dust and volatile matter from the comet's coma during flyby on 31 Dec 2003, safe sample return to Earth on 15 Jan 2006; mission extension as Stardust/NEXT ("New exploration of Tempel 1"), study of comet Tempel 1 during flyby on 15 Feb 2011 at a distance of ∼200 km, loss of contact: 25 Mar 2011
CONTOUR	03 Jul 2002	"Comet Nucleus Tour", failed NASA mission, close flybys at comets Encke and Schwassmann-Wachmann-3 and possibly comet d'Arrest, could not leave Earth orbit, loss of contact: 15 Aug 2002
Rosetta	26 Feb 2004	ESA mission, orbiter and lander "Philae", study of comet 67 P/Churyumov-Gerasimenko from "orbit" and soft landing on the nucleus, landing planned for Nov. 2014, still active
Deep Impact	12 Jan 2005	NASA mission, study of comet Tempel 1 during flyby on 3 Jul 2005, release of an impactor to the nucleus on 4 Jul 2005, observation of the impact; mission extension as EPOXI ("Extrasolar Planet Observation/Xtended Investigation of Comets"), observation of comet Hartley 2 on 4 Nov 2010 at a distance of 700 km, end of mission: 19 Sep 2013

Appendix

Missions to the Kuiper Belt

Mission	Launch Date	Description
New Horizons	19 Jan 2006	NASA mission, study of Pluto and its moon Charon during flyby and continue to the Kuiper Belt, arrival at the Pluto–Charon system in Jul 2015, still active

Bibliography

Table 4

Chronology of Lunar and Planetary Exploration: http://nssdc.gsfc.nasa.gov/planetary/chrono.html.

Engelhardt, W. (2005). *Galileo Cassini Giotto: Raumsonden erforschen unser Planetensystem*. Frankfurt am Main: Verlag Harri Deutsch, 551 pp.

Huntress, W. T., & Marov, M. Y. (2011). *Soviet Robots in the solar system: Mission technologies and discoveries*. Springer-Praxis, 453 pp.

Harvey, B., & Zakutnyaya, O. (2011). *Russian space probes: Scientific discoveries and future missions*. Springer-Praxis, 514 pp.

Leitenberger, B.: Articles about spaceflight and space (in German) http://www.bernd-leitenberger.de/raumfahrt.shtml.

Mission home pages of NASA and ESA (http://nasa.org and http://esa.int).

Morrison, D. (1982). *Voyages to Saturn*. NASA, 227 pp., NASA SP-451.

Morrison, D., & Samz, J. (1980). *Voyage to Jupiter*. NASA, 199 pp., NASA SP-439.

NSSDC Master Catalog Information: information on planetary missions, experiments and data sets held at NSSDc from the NSSDC Master Catalog: http://nssdc.gsfc.nasa.gov/planetary/projectdata.html.

Pieth, S., & Köhler, U. (2013). *Our solar system: A short introduction to the bodies of our solar system* (3rd, and enhanced ed). DLR Institute of Planetary Research, 96 pp.

Stooke, P. J. (2012), 1953–2003. *The international Atlas of Mars exploration: The first five decades* (Vol. 1). Cambridge University Press, 359 pp.

Ulivi, P., & Harland, D. M. (2007). *Robotic exploration of the solar system. Part 1: The Golden Age 1957–1982*. Springer-Praxis, 534 pp.

Glossary

Aberration The shift in direction of incoming light or material flow produced by the motion of the observer. In the solar system, the aberration angle of the solar wind is approximately the tangent of V_{perp}/w, where V_{perp} is the observer's speed perpendicular to the direction from the Sun and w is the solar wind speed. This effect causes comet plasma tails and the magnetic tails of Earth and Jupiter to lag behind (in the sense of motion) the radial direction.

Ablation Removal of material. Meteors ablate during their passage through the atmosphere.

Absolute magnitude *(H)* A measure of the brightness of an object. It is defined as the brightness if the object were at 1 AU each from the Sun and the Earth, and viewed at 0 degrees phase angle. For a given albedo, smaller absolute magnitudes correspond to larger objects. A difference of 5 absolute magnitudes corresponds to a factor of 10 in radius for objects with the same albedo.

Accretion The process of building larger bodies from smaller ones through low velocity collisions where the particles stick to one another, or where the gravity of a relatively large body draws the smaller bodies to it.

Achondrite A meteorite from a body that has melted and differentiated.

Actinic flux The solar flux used in calculating photodissociation rates, corresponding to the mean intensity at a given point in the atmosphere.

Active optics The controlled deformation or displacement of optics to compensate for slowly varying effects such as flexure or temperature changes. Typical timescale for updates is longer than one second.

Active Pixel Sensor (APS) A image sensor made of an integrated circuit containing an array of pixel sensors, each of which contains a photodetector and an amplifier. APS are commonly used in digital consumer cameras.

Adaptive optics An observational technique where the phase perturbations induced by the Earth's atmospheric turbulence, responsible for the blur in the images obtained, is corrected in real-time on the incident wavefront reaching the telescope. These perturbations are measured by a wavefront sensor. Opposite phase corrections are then applied using a thin deformable mirror in the pupil plane. The timescale for updates is typically about 1/1000 of a s.

Adiabatic A process occurring without exchange of heat with the surroundings.

Adiabatically *See* adiabatic.

Adiabatic compression Compression of a gas without exchange of heat. Expansion or compression of rising or sinking air masses in planetary atmospheres is commonly assumed to be driven by adiabatic processes.

Adiabatic temperature gradient In a convective fluid where there is no heat transfer between the rising and sinking parcels with the environment, the temperature profile follows a so-called adiabat. The dry adiabatic gradient (with altitude) of the Earth's atmosphere is roughly -7 K/km. That of the Earth's mantle (with depth) about 0.3 K/km.

Adiabatic temperature lapse rate (or temperature gradient) *See* Adiabatic temperature gradient.

Adsorption The formation of a thin layer of gas, liquid, or solid on the surface of a solid or, more rarely, a liquid. There are two types. A single layer of molecules, atoms, or ions can be attached to a surface by chemical bonds. Alternatively, molecules can be held onto a surface by weaker physical forces.

Advected Transported by advection. Advection is the transport of a quantity, typically heat, by the bulk motion of material.

Aeolian deposits Materials deposited by the wind, such as sand dunes.

Aerosol In atmospheric physics, aerosol is a generic name for any particle (cloud, dust, haze) suspended in the air, although in the Earth science community the term is usually restricted to apply to haze rather than cloud particles.

Aerogel Synthetic porous ultralight solid material with extremely low density, most commonly made of silica gel (also called "frozen smoke"). The Stardust mission was equipped with dedicated aerogel collectors that successfully collected dust particles in the coma of comet Wild 2 and interstellar particles along the spacecraft"s interplanetary trajectory and returned them to Earth.

Agglomerate Similar to aggregate.

Age Time elapsed since some event at a discrete time, t_0.

Aggregate A collection of separate particles or units gathered into a mass or whole, e.g. a dust ball is a loose aggregate.

Agglutinate Common particle in the lunar soil, usually about 60 m in size, consisting of rock, mineral, and glass fragments bonded together by glass (that also contains submicron metal droplets) produced by meteorite impact.

Airglow The emission of light by an atmosphere. Airglow may result from resonant scattering, fluorescence, impact by charged particles, or radiative decay of atoms, ions, or molecules left in an excited state by some chemical reaction.

Albedo (p) A ratio of scattered to incident electromagnetic radiation power, most commonly light. It is a unitless measure of a surface or body's reflectivity. The geometric albedo of an astronomical body is the ratio of its total brightness at zero phase angle to that of an idealized fully reflecting, diffusively scattering (Lambertian) disk with the same cross section. The visual geometric albedo refers to this quantity when taking into account only electromagnetic radiation in the visual range. The bond albedo is the fraction of total power in the electromagnetic incident radiation that is scattered back out into space, taking into account all wavelengths. The bond albedo (A) is related to the geometric albedo (p) by the expression $A = pq$, where q is the phase integral.

Alfvén point Heliocentric distance at which the solar wind flow speed exceeds the local Alfvén speed. Usually thought to occur at a heliocentric distance of ~ 0.04 AU.

Alfvén speed The speed of propagation of disturbances in a magnetized plasma that bend a magnetic field without changing its magnitude.

Alfvén waves A wave propagating in a magnetized plasma in which the magnetic field oscillates transverse to the propagation direction. The propagation speed is given by the Alfvén speed $v_A = B/\sqrt{4\pi\rho}$, where B is the magnetic field strength and ρ is the mass density.

Algae Any of a large group of mostly aquatic organisms that contain chlorophyll and other pigments and can carry on photosynthesis, but that lack true roots, stems, or leaves; they range from microscopic unicellular organisms to very large multicellular structures.

Aliasing Overlapping of radar echos at different frequencies or at different time delays.

Alkali plagioclase An aluminum silicate mineral rich in sodium (*See* Feldspar).

Allochthonous Describes a rock unit that has been moved into its present location.

Alpha particle Helium nucleus having mass four times and charge twice that of a proton.

Amino acid Any organic compound containing an amino acid (-NH$_2$) and a carboxyl (-COOH) group; specifically, one of the so-called building blocks of life, a group of 20 such compounds from which proteins are synthesized during ribosomal translation of messenger RNA.

Amorphous Having no crystalline form.

Angrites A group of twelve basaltic achondrites with $\sim$4.56 billion year ages.

Angle of repose The maximum slope at which loose material does not fall downhill.

Angular momentum Property of orbiting or rotating objects, usually expressed as mvr, where m is the mass, v is the velocity, and r is the distance from the center of rotation. The Earth and Moon have orbital angular momentum on account of their revolution around the Sun and spin angular momentum because of axial rotation. Angular momentum is conserved unless forces act to change it.

Anhydrous silicates Silicates lacking in water content.

Anomalistic month The time between successive passages of the Moon through perigee.

Anorthosite An igneous rock formed almost exclusively of plagioclase. It forms the outer layer of the Moon.

Antenna gain Ratio of an antenna's sensitivity in the direction toward which it is pointed to its average sensitivity in all directions.

Antipodes The opposite points on the surface of a sphere, given by a line through the center of the sphere.

Aphelion The point in the elliptical orbit of a planet, comet, or asteroid farthest from the Sun.

Aphelion distance (Q) The farthest distance from the Sun of an object in an elliptical orbit, given by $Q = a(1 + e)$, where a is the object's semimajor axis and e is its eccentricity.

Apoapse Point on an orbit farthest from the center of gravity, called aphelion for orbits about the Sun and apogee for orbits about the Earth.

Apogee The point in the orbit of the Moon or an artificial satellite, furthest away from the Earth.

Apojove The point in an orbit around Jupiter, farthest from the planet.

Apse Informal synonym for *Longitude of perihelion*.

Arc seconds, arcsec, second of arc An angle equal to 1/3600 of a degree, or 1/60 of an arc min. The Sun subtends an angle of $\sim$1919 a s on average when viewed from the Earth.

Arc minute, arcmin, minute of arc An angle equal to 1/60th of a degree, or 60 a s.

Areocentric Sun longitude (L_S) An angular measure of the Martian year. $L_S = 0°$ corresponds to the vernal equinox, marking the beginning of Spring in the northern hemisphere. $L_S = 90°, 180°,$ and $270°$ correspond to the summer solstice, autumnal equinox, and winter solstice, respectively.

Argument of perihelion (w) In an orbit around the Sun, the angle between the ascending node and the perihelion point, measured in the body's orbital plane and along its direction of motion.

Asteroid A rocky, carbonaceous, or metallic body, smaller than a planet and orbiting the Sun. Most asteroids are in semistable orbits between Mars and Jupiter, but others are thrown onto orbits crossing those of the major planets. Also called a **minor planet.**

Asteroid family Population of asteroids that share similar orbital elements, such as semi-major axis, eccentricity, and orbital inclination. The family members are thought to be fragments of asteroid collisions. About one-third of asteroids in the main belt are members of an asteroid family.

Aspect ratio The ratio between two dimensions of something; for a disk, the ratio between its vertical thickness and its diameter.

Asthenosphere A low viscosity zone that lies between the lithosphere and the mantle.

Astrology A belief system in which the future of individuals is predicted based upon the date and location of their birth and the positions of the moon and planets relative to the Sun or Earth at specific times.

Astrometry Precise measurement of an object's position in space.

Astronomical unit (AU) Commonly thought of as the mean distance of the Earth from the Sun. It is more formally the distance at which a massless particle in a circular orbit would have an orbital period of 1 Gaussian year, equal to 365.256898326 days. It is equal to 149.59787066×10^6 km, or about 92.955807×10^6 miles.

Astrophysical disk A rotating system of countless orbiting constituents in which random and/or vertical velocities are small compared to the orbital velocities (equivalently, in which all orbital **inclinations** are small).

Aurora Atmospheric emissions excited by the precipitation of energetic magnetospheric and solar particles, most frequently at high latitudes.

Autochthonous Describes a rock unit that has been formed in place.

Autotrophy Literally, self-feeding; the capacity of an organism to obtain its essential nutrients by synthesizing nonorganic materials from the environment, rather than by consuming organic materials; photosynthetic green plants and chemosynthetic bacteria are examples of autotrophic organisms.

(B-R) color A color scale for astronomical objects. Light from astronomical objects consists of all the colors of the rainbow: red, orange, yellow, green, blue, indigo, and violet (roygbiv). The (B-R) color measures the proportion of red to blue Light from an object. A red astronomical object has (B-R) $\sim$ 2, the Sun has (B-R) = 1.03, and a blue astronomical object has (B-R) $\sim$ 0.

Bar Unit of pressure, equal to 10^6 dyn/cm^2 or 10^5 Pa; the standard sea level pressure of the Earth's atmosphere is 1.013 bar. Typical planetary interior pressures are measured in megabars (Mbar) or 10^6 bar.

Baroclinic instability A 3-dimensional process common in the midlatitude troposphere wherein cold polar air pushes underneath hot low-latitude air, which transports heat toward the poles and produces complex circulation patterns that generate much of the winter rainfall in the midlatitudes.

Baroclinic, barotropic Barotropic is a region of uniform temperature distribution; a lack of fronts. Everyday being similar (hot and humid with no cold fronts to cool things off) would be a barotropic type of atmosphere, such as we find at tropical latitudes. In a baroclinic region, distinct air mass regions exist. Fronts separate warmer from colder air. In a synoptic scale baroclinic environment, you will find the polar jet, troughs of low pressure (mid-latitude cyclones), and frontal boundaries. There are clear density gradients in a baroclinic environment caused by the fronts. Mid-latitude cyclones are found in a baroclinic environment.

Barycenter The center of mass of a system of two or more gravitationally bound (orbiting) bodies.

Basalt An igneous rock primarily composed of plagioclase and pyroxene. On Earth, oceanic crust is primarily basaltic in composition.

Beta meteoroid Small meteoroid for which the solar radiation pressure force is comparable to solar gravitational attraction and hence leaves the solar system on an unbound orbit.

Biologically available nitrogen Chemical forms of nitrogen such as NH_3 that can be used by organism as nutrients.

Birch–Murnaghan equation A complex semi-empirical equation of state for planetary mantles that relates pressure to density and temperature.

Bistatic Describes an observation that involves the use of a transmitter and a receiver at separate locations. The angle between the transmitter and receiver is known as the bistatic angle.

Blackbody A blackbody is defined as an object that absorbs all radiation that falls on it at all frequencies and all angles of incidence; none of the radiation is reflected.

Blackbody radiation Continuous spectrum of electromagnetic radiation emitted by an object that absorbs all radiation incident on it.

Body wave A seismic wave, or vibration, that travels through the interior of a planet. There are two types of body waves: P-waves, or compressional waves, have vibrational motion in the same direction as the wave is traveling, and are equivalent to sound waves in air; S-waves, or shear waves, have vibrational motion perpendicular to the direction of travel, and move somewhat more slowly than P-waves.

Bolide A meteoric fireball.

Bolometric albedo Fraction of the incident energy (at all wavelengths) reflected by the surface.

Bond albedo Ratio of the total radiation reflected in all directions from a solar system object to the total incidence flux.

Bound orbit Circular or elliptical orbit about a central body (e.g., Sun, planet).

Breccia Rock composed of fragments derived from previous generations of rocks.

Bremsstrahlung Electromagnetic radiation that is emitted when an energetic electron is deflected by an ion. It is also called free-free emission because both the electron and ion are free in an ionized plasma. The term is borrowed from German and means "braking radiation" because the deflected electron loses energy (by the emitted photon) and is slowed down.

Brightness temperature The temperature a body would have if it were a blackbody producing the same brightness as the observed object at the same wavelength. It can also be defined as the radiant intensity scaled to units of temperature by $X^2/2 k$ where X is wavelength, and k is Boltzmann's constant.

Brittle deformation Irreversible deformation of rocks and solid materials occurring at relatively high stresses and low temperatures and resulting in the generation of cracks and faults or in frictional movement along existing fractures.

CAI: calcium-aluminum rich inclusion Minor component of primitive meteorites composed of refractory minerals.

Caloris Basin The largest recognized impact basin on Mercury, approximately 1550 km in diameter (960 miles).

Caldera Large volcanic crater, usually greater than 1 km in diameter. A caldera is many times the size of any associated vent(s). Calderas are formed either by collapse (most often) or explosion.

Carbonaceous (C-type) material Carbon-silicate material rich in simple organic compounds, such as that found in carbonaceous meteorites, which are believed to be among the most primitive (unaltered since their formation in the solar nebula) objects found in the solar system. They contain complex carbon compounds (hydrocarbons, amino acids), made mostly from the elements C, H, O, and N. C-type material is low albedo, spectrally flat and exists on the surfaces of several outer planet moons.

Catalytic cycle Series of chemical reactions facilitated by a substance that remains unchanged.

Cd-cutoff The neutron capture cross section for natural cadmium (Cd) is very high for thermal neutrons, but drops sharply for energies greater than about 0.5 eV, which is sometimes referred to as the "Cd cutoff" energy. Consequently, Cd is an excellent filter that absorbs thermal neutrons, but allows epithermal and fast neutrons to pass through.

Centaur A small body in a heliocentric orbit whose average distance from the Sun lies between the orbits of Jupiter and Neptune, and that has a Tisserand parameter with respect to Jupiter greater than three. Typically, the orbits of Centaurs also cross one or more of the orbits of the other giant planets. Centaurs are part of the population of ecliptic comets. They are most likely derived from the Kuiper belt and the Scattered disk. Eventually, some Centaurs may evolve into the terrestrial planets zone and become short-period comets.

Chaotic motion A dynamical situation in which the error in the prediction of the long-term motion of a body grows exponentially with time. This exponential growth leads to an inability to predict the location of the body. Chaotic motion can be confined to fairly narrow regions of space so that the orbit of the body will not change much over time. This is the case for the planets in our solar system. However, for most comets and some asteroids that we observe, chaotic motion leads to sudden and drastic changes in the orbit of the body.

Chaotic terrain Areas of the martian surface where the ground has collapsed to form a surface of jostled blocks standing 1–2 km below the surrounding terrain.

Charge-coupled device (CCD) A solid-state device used to record light electronically. A typical CCD has thousands to millions of tiny detectors arranged in a grid pattern. Each detector element is called a pixel. These devices record images electronically. CCDs have completely replaced photographic plates in astronomy due to their very high efficiency in capturing light.

Chasmata Term used in planetary geology to refer to long, relatively narrow, steep-sided troughs.

Chemoautotrophy The capacity of an autotrophic (self-feeding) organism to derive the energy required for its growth from certain chemical reactions (e.g., methanogenesis) rather than from photosynthesis; some bacterial forms are chemoautotrophic organisms.

Chiron-type comet A Centaur that displays cometary activity.

Chlorofluorocarbons (CFCs) Various compounds made with the halogens chlorine and fluorine. Their stability made them favored refrigerants until it was discovered that this also makes them efficient atmospheric ozone destroyers.

Chondrite A meteorite from an asteroid that never grew hot enough to melt and differentiate. It usually contains chondrules and their fragments.

Chondrule Millimetre-sized bead of rock found in large numbers in primitive meteorites.

Chromophore Any coloring material.

Chromosphere Lower atmosphere of the Sun, above the photosphere and beneath the transition region, with a vertical height extent of about 2000 km and a temperature range of 6000–20,000 K.

Circular polarization ratio Ratio of radar echo power received in the same sense of circular polarization as transmitted (the SC sense) to that received in the opposite (OC) sense.

Circular restricted three-body problem A special case of the problem of calculating the gravity-controlled motion of three bodies. In the circular restricted three-body problem, the two massive bodies follow circular orbits about each other, and the mass of the third body is negligible. The motion of comets and asteroids can often be approximated with the circular restricted three-body problem with the Sun and Jupiter, or sometimes the Sun and Neptune, as the two massive bodies.

Clathrate Clathrate compound, or cage compound **A.** chemical substance consisting of a crystalline lattice of one type of molecule trapping and containing a second type of molecule. A clathrate therefore is a material that is a weak composite, with molecules of suitable size captured in spaces left by the other crystalline molecule. Water ice often forms clathrates with more volatile molecules.

Co-latitude The co-latitude is the complementary angle of the latitude, i.e. the difference between 90° and the latitude. It increases from zero at the North Pole to 90° at the equator and 180° at the South Pole.

Column abundance The product of density (g/cm^3) and geometric thickness (cm). Measures the mass per unit area of an atmospheric or surface layer.

Column density The number of molecules above a column of unit area in an atmosphere.

Coma The freely outflowing atmosphere of gas and dust around the nucleus of a comet. The nucleus and coma of a comet together are often called the head.

Comet A body containing a significant fraction of ices, smaller than a planet or dwarf planet and orbiting the Sun, usually in a highly eccentric orbit. Most comets are stored beyond the planetary system in two large reservoirs: the Kuiper belt beyond the orbit of Neptune and the Oort cloud at near-interstellar distances. Comets become "active" when their ices sublimate and carry gas and dust into the coma.

Comet dust trail A contrail-like structure extending behind a comet close to its orbit, and sometimes a short distance ahead of the comet, consisting of large particles (p $< \sim 10^{-3}$) emitted at low velocities from the nucleus. Trails are distinguished from comet tails, which consist of much smaller particles, more sensitive to solar radiation pressure.

Cometary mantle or crust A layer of refractory material covering some or all of a comet nucleus' surface. When thick enough, a cometary mantle will choke off outgassing over that area.

Conjunction Occurs when two or more planetary bodies appear in the same area of the sky.

Conservation of angular momentum Fundamental physical law requiring that the quantity of angular momentum, p, be conserved (constant) for objects in orbit around a primary body: $p = mvr$, were m is mass, v is velocity, and r is the distance from the primary body, and for rotating objects. Angular momentum is not conserved in the presence of modifying torques.

Contact surface In the vicinity of a comet, the surface that separates outflowing cometary plasma from the slowed solar wind that is approaching the comet.

Continents Part of the Earth's crust of silicic composition that lies mostly above sea level and is less dense than both the mantle and the oceanic crust.

Convection Transport of energy by mass motion. In turbulent regions of planetary atmospheres and interiors, rising parcels of hot air (or rock) and sinking parcels of cool air (or rock) transport energy outward from the interior.

Core The central part of a differentiated planet, satellite or asteroid. Terrestrial planets have iron-rich cores. Jovian planets have rocky-iron cores.

Core accretion model A model for the formation of giant planets beginning with the growth of a solid core, followed by accretion of gas from the surrounding protoplanetary disk.

Coriolis acceleration Component of the acceleration on a rotating planet that acts perpendicular to the motion and balances the horizontal pressure gradient in an atmosphere or ocean. It causes circulation around high- and low-pressure centers. It is strongest in the polar regions and weakest in the tropics.

Corona Upper atmosphere of the Sun, extending above the transition region out into the heliosphere, with a dominant temperature of 1–2 million K in the lowest 100,000 km. The corona is visible during solar eclipses, and extends outward many solar radii.

Coronae Circular to oval feature surrounded by concentric ridges and fractures.

Coronal mass ejection (CME) Magnetic instabilities in the solar corona that lead to eruption of filaments, prominences, and magnetic flux ropes, which propagate as ejected mass out into the heliosphere, often accompanied by flare phenomenon. Known as an interplanetary coronal mass ejection (ICME) when observed in the solar wind far from the Sun.

Cosmochemistry The field of chemistry that concerns itself with the chemistry of extraterrestrial matter.

Coulomb interaction or collision The interaction of charged particles at large distances through the Coulomb force. The interaction is sometimes called a Coulomb collision because it produces a change of particle momentum similar to that in a conventional collision.

Cowling's theorem A mathematical theorem proved by Cowling that states that an axisymmetric magnetic field cannot be maintained by dynamo action.

Crater rays Diffuse, elongate geologic feature arrayed radially around relatively young, fresh impact craters on planetary surface. The rays form where dispersed cluster of impacting ejecta fragments form chains of secondary craters. The secondaries in turn excavate relatively bright regolith from the subsurface that has been less darkened by space weathering. Crater rays are most distinctive with the Sun overhead to accentuate albedo differences.

Cratering flow-field Penetration of the impactor, followed by movement of target materials in an impact event in response to the passage of the shock and rarefaction, or decompression, waves.

Cratering flow-field Movement of target materials in an impact event in response to the passage of the shock and rarefaction, or decompression, waves.

Critical point Temperature and pressure for a given material above which there is no distinction between the liquid and gas phases.

Crust The chemically distinct, less dense, outer shell of a planet or satellite formed by melting of the interior.

Cryovolcanism Volcanism where the volcanic materials are melted ices, such as water, ammonia, and methane, as distinguished from the common high-melting point volcanic materials of the terrestrial planets, such as basalt and rhyolite. The melted ices freeze on the surface, forming "lava flows" composed of ice.

Cryovolcano An icy volcano. See cryovolcanism.

Cumulate Plutonic igneous rock composed of crystals accumulated by floating or sinking in the silicate melt, or magma.

Curie temperature The temperature below which permanent magnetization can be maintained in a ferromagnetic material. Below the Curie temperature the individual atomic-level magnetic moments are "frozen" in place, whereas above this temperature they are largely free to realign.

Current Strictly, the product of the velocity and number density of particles. The dot product of the current and the unit normal vector of a surface yields the net number of particles crossing the surface per unit area. In the chapter "Remote Chemical Sensing Using Nuclear Spectroscopy," current is taken to be the net number of particles crossing out of a planetary surface per cosmic ray, which is dimensionless.

Cusps Features of planetary magnetospheres near the "magnetic poles", where inward closure of magnetic field lines allows plasma from the Sun direct access to the planet's surface or upper atmosphere.

Cycloidal Motion of a cycloid, which is the curve defined by a fixed point on a wheel as it rolls.

Cyclostrophic balance On Earth, the surface rotation typically surpasses the zonal winds, and the pressure gradient force generated by the unequal solar fluxes at low and high latitudes is balanced by the Coriolis force, in what is called a geostrophic balance. On Titan (and Venus), the opposite is true, and pressure gradients are balanced by strong centrifugal forces arising from the rapid rotation of the atmosphere. This balance, typical of cyclones, is called cyclostrophic.

Debris disk A disk of dust and debris in orbit around a star. Contrary to the dense protoplanetary disks, debris disks are not primordial. Rather they are secondary disks formed from debris released from planetesimals like asteroids and comets. Debris disks are considered the massive counterparts of the Zodiacal and **Kuiper** belt dust disks in our solar system.

Design reference mission Conceptual definition of a mission scenario responding to a set of mission objectives, including as a minimum the overall mission architecture, major architecture elements, and concept of operations.

Degenerate matter Matter at very high pressures where the normal atomic structure is destroyed.

Deterministic system Dynamical system in which the individual bodies move according to fixed laws described mathematically in the form of equations of motion. A deterministic system can still give rise to chaotic, unpredictable motion because of the finite precision with which any physical measurement or numerical computation can be made.

Deuterium/deuterated Heavy form of the hydrogen atom, consisting of one proton and one neutron. A deuterated molecule, such as CH_3D, deuterated methane, has one or more deuterium atoms in place of hydrogen.

Diapir A body of rock or ice that has moved upward due to buoyancy, attaining an inverted teardrop or pear shape, and piercing and displacing the overlying layers.

Diaplectic glass Glass phase produced from minerals by the destruction of internal structural order, without melting, by the passage of a shock wave.

Dichotomy A division into two opposite groups. Often used to describe the difference between the nearside mare terrain and the farside highland terrain on the Moon. Also used to describe the difference between the northern lowlands and the cratered southern highlands on Mars.

Dielectric materials Materials with few free charges, where current flows by small shifts in electrons bound to constituent atoms (as opposed to conductors, which have a large number of free electrons). Dielectric materials include most rocks, regolith, and ice.

Differentiated A body that has melted and separated into layers of different composition and density.

Differentiation Melting and fractionation of a planet, moon or asteroid into multiple layers or zones of different chemical composition; e.g., core, mantle and crust. High-density materials sink and low-density materials float.

Diffraction A physical process in which light from different parts of a mirror or lens interfere with each other. As a result of diffraction, the image of a star is not absolutely sharp; instead the full width at half maximum (FWHM) of the image is given by $\theta = 0.252*\lambda/D$ a s, where the wavelength is given by λ in micrometers and D is the telescope diameter in meters. In the absence of atmospheric seeing, the image of a star would be diffraction-limited if the optics were polished sufficiently well.

Diffraction grating An object with finely and uniformly spaced grooves, commonly seen in a physics or optics laboratory; because it scatters different wavelengths of light at different angles, a diffraction grating separates light into its **spectrum** (i.e. like a prism, it creates a rainbow).

Dihedral scattering Scattering from two surfaces that form a right angle. Dihedral surfaces are commonly referred to as corner reflectors.

Dipole magnetic field The shape of the field lines around a short bar magnet. This field can be visualized by sprinkling iron filings on a piece of paper on top of a bar magnet.

Direct or prograde motion Orbital or axial motion of a body in the solar system that is counterclockwise as seen from north of the ecliptic.

Disk instability model A model for giant planet formation in which such planets form by the gravitational collapse of a portion of a protoplanetary disk.

Diurnal Occurring every day. A diurnal cycle is one that repeats daily.

Diurnal stresses Twice-daily forces exerted on a planet, which are derived from gravitational forces between the planet and a satellite.

Doppler shift Difference between the frequencies of the radar echo and the transmission, caused by the relative velocity of the target with respect to the radar. Also, the shift in frequency of spectral lines due to the motion of a light source toward or away from an observer.

Drainage basin Geomorphic entity that contains a drainage network. Typically a bowl-shaped catchment in humid areas, drainage basins in arid regions can be quite flat. Drainage patterns typically reflect the topography of the drainage basins that contain them.

D-type material Primordial, low-albedo material thought to be rich in organic compounds. It is redder than C-type material.

Ductile A ductile material is deforming under tensile stress. Ductility is an element of plasticity.

Dust bands Apparent parallel bands of dust arising from comminuted asteroid collisional debris filling a roughly Sun-centered torus and extending past the Earth as a consequence of Poynting–Robertson drag.

Dust tail The broad, relatively featureless tail of a comet consisting of micron-sized dust particles being driven away from the nucleus and coma by solar radiation pressure. The dust particles are on independent orbits around the Sun under reduced gravity. The dust tail appears whitish or yellowish from sunlight scattered by the dust particles.

Dwarf planet A new term created by the IAU in 2006 to describe bodies orbiting the Sun that are round (in hydrostatic equilibrium) but are not massive enough to have cleared their zones. Dwarf planets include Ceres, Pluto, and Eris (2003 UB313). Other main-belt asteroids and Kuiper belt objects are potential candidates for this classification as more is learned about them.

Dynamical inertia The increase in the radiative time constant due to mixing of more massive, deeper layers of an atmosphere. The thin atmosphere at high altitudes would be expected to respond rapidly to changes in sunlight, but if there is substantial vertical circulation, the changes will occur more slowly as the mixing increases the effective mass of the layer under consideration.

Dynamo The conversion of mechanical energy into electromagnetic energy through electromagnetic induction. In a planetary dynamo, motions in an electrically conducting fluid amplify magnetic field efficiently enough to maintain it faster than it can decay through Ohmic dissipation.

Dynamo action Mechanism for the generation of magnetic fields in stars and planetary interiors resulting from convection of a rotating and electrically conducting fluid.

α-effect The process of generating magnetic field through helical motions.

ω-effect The process of generating magnetic field through large scale zonal shear flows.

Eccentricity A measure of the departure of an orbit from circular. For an elliptical orbit, the eccentricity, e, is equal to $(1-b^2/a^2)^{1/2}$, where a and b are the semimajor and semiminor axes of the ellipse, respectively. Circular orbits have $e=0$; elliptical orbits have $0 < e < 1$; radial and parabolic orbits have $e=1$; and hyperbolic orbits have $e > 1$.

Echo bandwidth Dispersion in Doppler frequency of a radar echo, i.e., the width of the echo power spectrum.

Ecliptic, ecliptic plane The plane of the Earth's orbit around the Sun. The planets, most asteroids, short-period comets, Kuiper belt objects and Scattered-Disk objects follow orbits with small or moderate inclinations (or tilts) relative to the ecliptic.

Ecliptic comet A comet with a Tisserand parameter with respect to Jupiter greater than 2. They generally have a small or moderate inclination to the ecliptic. The designation of such objects as comets does not necessarily imply visible cometary activity because comets typically do not become active until they pass well within Jupiter's orbit. The term "comet" assumes that the body contains a substantial fraction of water ice. Ecliptic comets include Centaurs and Jupiter-family comets.

Eddy-driven jet The jet stream occurring in the mid-latitudes, driven by eddies generated by baroclinic instabilities.

Ejecta Debris excavated from depth by an impact and deposited onto the surface outside the rim of the impact crater. The lateral extent of the ejecta depends on the gravity of the body and the velocity of the impactor.

Ekman layer Idealized model of the planetary boundary layer in which the mean flow (as in an atmosphere or ocean) is modified near the ground by friction (either laminar or turbulent). The resulting variation of flow speed and direction with height is described by the Ekman spiral.

Electromagnetic induction The generation of an electromotive force from a changing magnetic flux, governed by Faraday's Law.

Elevation angle The angle between a flat surface (such as Saturn's rings) and the observer's line of sight.

El Nino-Southern Oscillation *See* ENSO.

Elongation The angular distance between the Sun and a planet or other solar system body as viewed from the Earth.

Embays To form a protective barrier.

Embryo Differentiated protoplanet resulting from the accretion of km-sized planetesimals.

Emissivity The ratio of radiant energy flux from a material to that from a blackbody at the same temperature. A blackbody is an ideal material that absorbs all radiant energy incident upon it and emits radiant energy at the maximum possible rate per unit area at each wavelength for a given temperature. A blackbody has an emissivity of 1 across the entire spectrum. Real materials have an emissivity between 0 and 1 for a given wavelength.

Encke-type comet An ecliptic comet whose entire orbit is interior to the orbit of Jupiter and which has a Tisserand parameter with respect to Jupiter greater than 3.

Endogenic Forming from within.

ENSO El Nino, "the child," and the Southern Oscillation. El Nino is the episodic appearance of warm water off the coast of South America, often at Christmas time, that devastates Peruvian fishing (usually one fifth of the world's catch), causes drought conditions in Australia, and weakens the monsoon in India. The Southern Oscillation is the historical name for the global (not just southern) atmosphere-ocean oscillation for which El Nino years are the extreme.

Entropy Broadly, the degree of disorder, or randomness in a system; in thermodynamics, a measure of the amount of heat energy in a closed system that is not available to do work. In a condition of low entropy (high efficiency), the system will convert to energy a large portion of the heat transferred to it from an external source (no actual system can utilize 100% of the heat it receives).

Enzymes Proteins that catalyze, or accelerate, chemical reactions.

Epicenter The point on a planet's surface that is directly above the hypocenter or focus of a seismic event (or quake), the point within the planet where a quake originates.

Equal angle map For mapping, the surface of a planet is subdivided into spatial elements called pixels on which quantities such as counting rates and elemental abundances are specified. In an equal area map, all of the pixels span the same angle in latitude and longitude. Consequently, a parallel near a pole is divided into the same number of pixels as a parallel at the equator and the area of the pixels varies with latitude.

Equal area map The longitude-range for pixels at different latitudes is adjusted so that all of the pixels have approximately the same area. Consequently, a parallel near the pole is divided into fewer pixels than at the equator. The span of pixels in latitude and longitude at the equator is used to specify the map.

Equation of state Equation relating the pressure of a given material to its temperature and density, typically derived from experimental and theoretical considerations.

Equinox The passing of the Sun through a planet's equatorial plane, so that the northern and southern hemispheres receive sunlight in equal measure; associated with the seasons of spring and autumn (cf. **solstice**).

Equilibrium vapor pressure The ambient pressure of the gas phase over a condensed phase when the gas and condensed phase are in thermodynamic equilibrium (i.e., when the rate of condensation from gas to ice equals the rate of sublimation from ice to gas). In effect, vapor pressure is a measure of the amount of gas an ice or liquid layer at a specified temperature will evolve in a closed container (or planetary atmosphere). Vapor pressures are extremely sensitive functions of temperature and are also related to the composition and structure of a given ice or liquid.

Escape velocity Minimum velocity required to escape from the surface of a body to infinity.

Europium anomaly and Eu* Because the Moon is highly reduced, europium is divalent on the Moon and hence is mostly separated from the other smaller trivalent rare earth elements because it is concentrated in plagioclase feldspar. The degree of enrichment or depletion is given by Eu/Eu*, where Eu is the measured abundance and Eu* is the abundance expected if Eu had the same relative concentration as the neighboring rare earth elements, samarium and gadolinium.

Exosphere The outermost part of an atmosphere, characterized by very low densities and very long mean-free paths, and usually isothermal.

Extended scattered disk Collection of objects with orbits with semimajor axis >50 AU, large eccentricity, and perihelion distance large enough to avoid destabilizing encounters with Neptune. The apparent similarity with the orbits of objects in the scattered disk suggests that the latter extended further in perihelion distance in the past, due to a different orbital architecture of the planets or of the environment of the solar system. The most prominent members of the extended scattered disk population are 2000 CR105 and 90377 Sedna.

External comet A returning comet with a semimajor axis greater than $\sim$ 34.2 AU. External comets have Tisserand parameters with respect to Jupiter less than 2. Also known as a **long-period comet.**

Extrasolar planet A planetary companion to a star other than the Sun.

Fault A planar fracture or discontinuity in a rock with significant displacement.

Ferrel cells The overturning circulation in the midlatitudes, poleward of the Hadley cells. Unlike the Hadley cell, the Ferrel cell is not thermally driven but rather is mechanically driven by baroclinic instabilities and wave generation in this latitude range.

Feldspar A common group of aluminum silicate minerals.

Field of view The region visible in a single image taken by a camera or telescope.

Filaments Near-horizontal magnetic field lines on the Sun suspended above magnetic inversion lines that are filled with cool and dense chromospheric mass, seen on the solar disk.

Flexural response Behavior of the elastic, near-surface portion of a planet when a load is applied.

Flares A magnetic instability in the solar corona that impulsively releases large energies that go into heating of coronal and chromospheric plasma, as well as into acceleration of high-energy particles. A flare is usually accompanied by impulsive emission in gamma rays, hard x-rays, soft x-rays, EUV, and radio emission.

Fluctus (pl., flucti) Term meaning (on Io) a volcanic flow field.

Fluorescence Photons emitted immediately after electron decay. The electron had been elevated to a higher energy state by external stimulation of their parent atoms, ions, and molecules. In planetary atmospheres, the external stimulation is usually sunlight or electrons.

Flux The flux of particles (denoted p with units of $cm^{-2}s^{-1}$) given by the product of the speed of the neutrons, v (cm/s), and the number density (particles per cm^3).

Flux density Power per unit area and per unit frequency interval received from an object. The units of flux density are Janskies: $1 \text{ Jy} = 10^{-26} \text{ W/m}^2/\text{Hz}$.

Flux transfer event A localized spatial region in which magnetic reconnection links the solar wind magnetic field to a planetary magnetic field producing a configuration that transports flux from the day side to the night side of the planet.

Fractionation Separation of elements or isotopes based on their masses or chemistry.

Fraunhofer lines Dark lines in a radiance spectrum, produced when chemical elements in a cold gas absorb light of specific energies from a broad-spectrum photon source. In the solar spectrum, Fraunhofer lines originate from gas in the outer regions of the Sun, which are too cold to directly produce emission lines of the elements they represent.

Frequency-time spectrogram A graph of the emission intensity as a function of frequency and time. Usually the intensity is shown on a gray scale ranging from black to white or one of several color schemes, with frequency plotted along the vertical y-axis and time along the horizontal x-axis.

Galactic cosmic rays Energetic particles, including photons, electrons, protons, and heavy ions, that originate outside the heliosphere.

Galilean satellites The four major satellites of Jupiter: Io, Europa, Ganymede and Callisto, discovered by Galileo in 1610.

Gamma ray A high energy quantum of electromagnetic radiation (photon) emitted by nuclear transitions. Gamma rays originate from nuclear processes, such as radioactive decay and the de-excitation of residual nuclei produced by nuclear reactions.

Gamma-ray spectroscopy In planetary science, the estimation of abundances of elements from their emissions of gamma rays. The rays come from the natural radioactivity of uranium, thorium, or potassium, or emissions from other elements when their nuclei have been excited by cosmic rays.

Gas chromatography A chemical technique for separating gas mixtures, in which the gas is passed through a long column containing a fixed absorbent phase that separates the gas into its component parts.

Gas drag Drag force experienced by a solid object when it moves through a surrounding gas.

Gauss coefficients Radial coefficients of the spherical harmonic expansion of the scalar potential for a planetary magnetic field in an insulator.

Gauss' Theorem Gauss' theorem (also known as the divergence theorem) states that the outward flux of a vector field through a closed surface is equal to the volume integral of the divergence over the region inside the surface. Intuitively, it states that the sum of all sources minus the sum of all sinks gives the net flow out of a region. When applied to gravity, it can be restated as Gauss' law of gravitation: "The integrated gravitational flux through any closed surface is proportional to the enclosed mass."

Gaussian year The orbital period of a massless particle in a circular orbit with a semimajor axis of 1 AU, equal to 365.256898326... days. Formally, the Gaussian year is defined as $2\pi/k$, where k is the Gaussian gravitational constant, 0.01720209895.

Geochemistry The study of the chemical components of the lithosphere of the Earth and other planets, chemical processes and reactions that produce and modify rocks and soils, and the cycles of matter and energy that transport chemical components in space and time.

Geodesy The measurement and representation of Earth's topography, its gravitational field and geodynamic phenomena (e.g., polar motion, tides, and crustal motion) in 3-dimensional, time-varying space.

Geodynamo The self-sustaining process responsible for maintaining a planet's magnetic field in which the kinetic energy of convective motion of the planet's liquid core is converted into magnetic energy.

Geologic unit In planetary geology, a volume of rock (or ice) having a consistent age range, that is identifiable by distinctive morphology, surface texture, composition, or other measurable physical property that relates to the rock's origin.

Geomagnetic activity Disturbances in the magnetized plasma of a magnetosphere associated with fluctuations of the surface field, auroral activity, reconfiguration and changing flows within the magnetosphere, strong ionospheric currents, and particle precipitation into the ionosphere.

Geomagnetic storm The response of the Earth to the arrival of an interplanetary medium disturbance, usually associated with a CME.

Geomagnetism The Earth's magnetic field, which is approximately a magnetic dipole, with the magnetic poles offset from the corresponding geographic poles by approximately $11.3°$, and extending several tens of thousands of kilometers into space.

Geometric albedo *See* Albedo.

Geomorphology Science of landscape analysis. Geomorphic investigations deal with the processes and timescales of landscape formation and degradation.

Geospace The Earth's magnetosphere and upper atmosphere, including the ionosphere.

Geostationary orbit An orbit around the Earth in which a satellite has an orbital period equal to the Earth's rotation period. A geostationary satellite remains above the same point on the Earth's surface at an altitude of about 36,000 km.

Geostrophic A geostrophic flow (ocean) or wind (atmosphere) results from the balance between the Coriolis effect and the pressure gradient force.

Giant Planets The giant planets in the solar system are Jupiter and Saturn. Jupiter is 300 times more massive than the Earth and about 10 times as large. Saturn is about 9 as large as Earth but only about 100 times as massive. Compare the planetary data in table 2 of the appendix. Uranus and Neptune are smaller and are usually called the sub-giants.

Global contraction Reduction in the radius of a planetary body due to cooling of its interior, placing the near-surface region under compressional stress. Global contraction is thought to be an important driver of tectonic deformation on Mercury.

Global exploration roadmap Defines technical feasible and programmatically implementable pathways toward human missions to Mars, reflecting policies and plans of space agency's and defining future mission concepts as well as required preparatory activities.

Graben A pair of conjugated normal faults that create a narrow, linear trough as the lithosphere is pulled apart under extension.

Granite Light-colored intrusive rock containing more than 50% silica. On Earth, continents are largely granite and other high silica rocks.

Grain-size Dimension of individual crystals in a rock that strongly influences its rheological properties.

Grand Tack model Model describing a hypothetical episode during the formation of the planets in which Jupiter migrated inwards and then outwards across the asteroid belt, scattering away many of the planetary building blocks from the asteroid belt and the region now occupied by Mars.

Gravitational focusing The tendency of an object's trajectory to curve toward a massive body due to gravitational attraction.

Gravitational instability Spontaneous collapse of a portion of a protoplanetary disk due to mutual gravitational attraction. This can refer to either the solid or gaseous component of the disk.

Graybody A sphere that absorbs a fraction of all light incident on it, and reemits that energy in all directions as thermal radiation. A gray body which absorbs all light incident on it is considered a "black body".

Greenhouse effect A radiative effect that occurs when the atmosphere is more transparent in the visible than infrared (IR) radiation. Visible sunlight easily penetrates through the atmosphere and is absorbed at deep levels. Because the atmosphere is relatively opaque in the IR, the atmosphere radiates IR radiation both up (to space) and down (to the surface). Since the surface receives not only sunlight but IR energy from the atmosphere, its temperature becomes higher than would occur in absence of an atmosphere. In steady state, energy balance is achieved, with escaping heat energy almost exactly equalling absorbed solar energy; in contrast to popular myth, heat is not "trapped."

Gyro radius Radius of the orbit of a charged particle gyrating in a magnetic field.

Gyrofrequency The frequency of the circular motion of a charged particle perpendicular to a magnetic field.

Habitable zone The region of space around a star in which a geologically active, rocky planet can maintain liquid water on its surface.

Hadley cell *See* Hadley circulation.

Hadley circulation A major component of atmospheric circulation driven directly by latitude-averaged heat sources and sinks. Warm air rises in regions near the equator, flows poleward at higher altitudes, and loses heat in the colder, higher latitude regions. The cooler, denser air then descends and has a flow component near the surface back toward the low-latitude heat source, which completes a circulation cell. The near-surface and high-altitude branches of the flow have eastward ("trade wind") and westward components, respectively, arising from Coriolis forces. When the heat source is located on the equator, the Hadley circulation tends to be symmetric about the equator, but the Hadley circulation is asymmetric about the equator if the heat source is located off the equator, as occurs during solstice seasons on Earth and Mars.

Half-life The time in radioactive decay in which half of the isotope has decayed to its daughter isotope.

Hall conductivity The electric conductivity in the ionosphere perpendicular to the electrical and the magnetic field.

Halley-type comet A returning comet with a semimajor axis less than ~ 34.2 AU. Halley-type comets have Tisserand parameters with respect to Jupiter less than 2.

Heat flow Heat emitted (or received) at the surface of a body that is ultimately radiated to (or absorbed from) space.

Heavy elements In astrophysics, all elements other than hydrogen and helium.

HED meteorites The "howardite–eucrite–diogenite" achondrite meteorites believed to originate from the asteroid Vesta.

Heliocentric A Sun-centered coordinate system.

Heliopause Interface between the heliosphere and the interstellar plasma; the outer boundary of the heliosphere.

Heliosphere The region of space, surrounding the Sun that is carved out in the interstellar medium by the solar wind. This region extends to about 100 AU in radial distance from the Sun.

Heliospheric current sheet The surface in interplanetary space separating solar wind flows of opposite magnetic polarity; the interplanetary extension of the solar magnetic equator.

Heliospheric magnetic field Remnant of the solar magnetic field dragged into interplanetary space by the solar wind.

Helium rain The precipitation of helium droplets due to the dissociation of helium from hydrogen that occurs at the molecular-metallic hydrogen transition in Jupiter and Saturn.

Heterotrophy Literally, other-feeding; the condition of an organism that is not able to obtain nutrients by synthesizing nonorganic materials from the environment, and that therefore must consume other life forms to obtain the organic products necessary for life; e.g., animals, fungi, most bacteria.

Highlands A topographic region that stands above its surroundings. A prominent example is the lunar highlands.

High-reflectance plains (HRP) Material that dominates some of Mercury's largest deposits of smooth plains, notably in Caloris basin and in the North Polar region. HRP is higher in albedo and more red than average Mercury surface material.

Hill sphere Region around a secondary in which the secondary's gravity is more influential for the motion of a particle about the secondary than is the tidal influence of the primary.

Hilly and lineated terrain The broken-up surface of Mercury at the antipode of the Caloris impact basin.

Hirayama asteroid families Clusters of asteroids having orbits with similar semimajor axes, eccentricities, and inclinations. The most prominent of these were discovered by Kiyotsugu Hirayama in 1914.

Hohmann transfer orbit Elliptical trajectory to achieve low-energy transfer of spacecraft between two different circular orbits, involving orbital maneuvers consisting of thrust impulses tangential to the orbits at the beginning and end of the transfer ellipse.

Homopause Level in an atmosphere, above the stratosphere, at which gases cease being uniformly mixed and separate by diffusion, with the lighter elements diffusing upward.

Horseshoe orbits Librating orbits encircling the L_3, L_4, and L_5 Lagrangian points in the circular restricted three-body problem. These orbits appear to be shaped like horseshoes in the frame rotating with the mean motion of the system.

Hot Jupiter An extrasolar gas giant planet at a very small orbital separation of 0.03–0.05 AU from its host star and with an orbital period of a few days. The proximity of the discovered hot Jupiters to their host stars is probably a result of inward orbital migration.

Hot poles The alternating perihelion subsolar points on Mercury at the 0° and 180° meridians.

Hot spots Regions of enhanced thermal emission on Io, a marker of volcanic activity. The term does not imply a particular eruption mechanism.

Hugoniot elastic limit Stress at which a rock or mineral's response to shock changes from elastic to plastic. Stresses above the Hugoniot elastic limit cause the rock or mineral to deform plastically.

Human action radius The range around the Earth where humans can travel by spacecraft within the present state of technology. It includes the Moon, Venus, Mars, and near Earth objects like asteroids or comets.

Human–robotic partnership Complementarity of robotic and human capabilities in space for advancing exploration goals. It refers to the role of robotic missions in preparing human missions as well as mission concepts, which combine robotic and human capabilities in space.

Hydrated A mineral in which water molecules or hydroxyl radicals are attached to the crystalline structure.

Hydrodynamic escape A limiting case of atmospheric escape that occurs when the escape rate is so rapid that the atmosphere at high altitudes reaches an outward velocity comparable to the speed of sound. This occurs if the thermal energy of the gas molecules becomes comparable to the gravitational binding energy. Hydrodynamic escape allows the upper atmosphere of a planet to escape wholesale, as opposed to the usually slower processes of Jeans-type thermal leakage or solar wind ion pickup.

Hydrogen cloud The huge cloud of atomic hydrogen surrounding most comets. The hydrogen cloud is produced by the dissociation of water and the hydroxyl molecule (OH).

Hydrostatic equation Relationship that says pressure is equal to the weight of gas or liquid above the level of interest.

Hyperbolic orbit Unbound orbit in which the object escapes the gravitational attraction of the central body. Examples are orbits of beta meteoroids and interstellar grains.

Hypsometry Geodetic observations of terrain elevations with respect to sea level.

Hysteresis The dependence of a system not only on its current environment but also on its past environment. This is typically manifested as the lagging of an effect behind its cause, as when the change in magnetism of a body lags behind changes in the magnetic field to which it is subjected.

Ice dwarf The term given to the planetesimals believed to have been created in large numbers during the formation of the giant planets and later scattered to the Oort cloud or ejected from the solar system by close encounters with the forming giant planets. Pluto and Triton are thought to be among the largest remnants of this population.

Ice Mixture of water, ammonia, methane, and other volatile compounds in the interiors of jovian planets, not literally in the form of condensed "ice."

IDP Interplanetary dust particle, collected by aircraft in the stratosphere.

Imaging spectrometer An instrument that can take pictures of a scene at many wavelengths simultaneously, sampling a significant portion of the **spectrum** at each pixel.

Impact melt Melt of target rocks resulting from the waste heat generated in an impact event. When solidified, it can be either glassy or crystalline and contain clasts of rock and mineral debris from unmelted portions of the target.

Inclination The angle between the plane of the orbit of a planet, comet, or asteroid and the ecliptic plane, or between a satellite's orbit plane and the equatorial plane of its primary. Inclination takes on values between 0° and 180°.

Inclination (orbital) For asteroids this is the tilt of an object's orbit with respect to the solar system ecliptic.

Incompatible elements Element that upon silicate melting tends to be largely concentrated in the liquid phase.

Inertia matrix In space, inertia is defined by a symmetric 3×3 inertia matrix where the diagonal elements couple torque around one axis to acceleration about the same axis, and where the off-diagonal terms called products of inertia couple torque around one axis to acceleration about another axis.

Insolation The flux of sunlight at all wavelengths falling on a body. For the Earth this amounts to a flux of 1.368×10^6 ergs 21 cm s.

Integral of the motion Any function of the position and velocity coordinates of an object that remains constant with time along all orbits. In the circular restricted three-body problem, the Jacobi constant is an integral of the motion. The Jacobi constant can be approximated by the Tisserand parameter.

Infrared light Electromagnetic radiation with wavelengths ranging from 900 nm to about a millimeter on the **spectrum**; because it is emitted by objects at temperatures commonly experienced on Earth and other planets, infrared is commonly thought of as "heat radiation".

In-situ resource utilization (ISRU) Use of resources found or manufactured on a celestial body to advance a space mission. This could be life support (oxygen, food and water), energy, propellant, construction or other materials.

Intercrater plains The oldest plains on Mercury that occur in the highlands and formed during the period of late heavy bombardment.

Intraplate volcanism Volcanic events not associated with phenomena occurring at plate margins and generally thought to be related to magmatism generated by mantle plumes.

Intrusion Geological structure of igneous material that forces its way into an existing formation. Examples are dykes, sheet-like intrusions of magma that later solidified or diapirs, bodies of (partial) melt that intruded driven by their bouyancy relative to the solid surrounding rock.

Invariable plane The plane passing through the center of mass of the solar system, which is perpendicular to its total angular momentum vector. The invariable plane is inclined $0.5°$ to the orbital plane of Jupiter and $1.6°$ to the ecliptic.

Ionopause The surface separating ionospheric plasma and the solar wind in the vicinity of an unmagnetized planet.

Ionosphere Outer portion of an atmosphere where charged particles are abundant.

Isolation mass The mass of a planetary embryo if it sweeps up all the accessible solid material in its vicinity.

Isostasy A condition of gravitational balance (similar to floating) in which crustal rocks are buoyantly supported from below by denser mantle rocks. The crustal rocks above subside into the mantle, or the mantle rises, until the mass in each column of rock is equalized.

Isotopes variant of a given chemical element characterized by the same number of protons but a different number of neutrons.

Jacobi constant An integral of the motion in the circular restricted three-body problem. It is proportional to the total orbital energy of the small body in a reference frame rotating with the two massive bodies.

Jeans escape The process by which fast (energetic or hot) molecules of an atmosphere escape into space. The energy distribution of a gas at a given temperature has a hot tail—a few atoms moving faster than the rest. If, at an altitude where collisions between molecules are rare, the molecules in the hot tail move faster than the local escape velocity, they can escape to space. This process is fastest for hot atmospheres of light gases (hydrogen, helium) on bodies with low gravity.

Jets The observed, collimated emission of gas and dust that occurs in restricted areas on the surface of a cometary nucleus. Jets are usually active on the sunlit side of the nucleus.

Joule heating Heating that occurs when a current flows through a resistive medium. In the high atmospheres of the giant planets, it may be an important process in heating the atmosphere to high temperature as currents of charged particles driven by magnetospheric electric fields collide with the neutral atmosphere atoms, which provide resistance.

Jovian planet A planet like Jupiter, which is composed mostly of hydrogen, with helium and other gases, but possibly with a silicate/iron core. Also called a gaseous or a giant planet. The jovian planets are Jupiter, Saturn, Uranus, and Neptune.

Jupiter-family comet An ecliptic comet with a Tisserand parameter between 2 and 3. It is typically on a low to moderate inclination orbit, with a semimajor axis less than that of Jupiter's orbit. Most Jupiter-family comets are in orbits that cross or closely approach Jupiter's orbit.

Ka-Band Ka-band is the part of the microwave band of the electromagnetic spectrum that ranges from 26.5 to 40 GHz. The corresponding range of wavelengths is 0.75 to 2.5 cm. In space missions the reserved frequency band is around 34 GHz (34.2 to 34.7 GHz) for the uplink (Earth to Space) and around 32 GHz (31.8 to 32.3 GHz) for the downlink (Space to Earth).

K or kelvin Unit of absolute temperature. The freezing and boiling points of water are 273.16 K and 373.16 K, respectively. The temperature in Kelvin is calculated from the temperature in °C by adding 273,15 K.

Kepler mission A four year NASA mission using a 1 m space telescope to monitor a large target field of 160,000 stars with the transit technique. *Kepler* has the sensitivity to detect planets as small as Earth.

Keplerian orbit The path that a body would follow if it were subject only to the gravitational attraction of its primary, e.g., a planet orbiting the Sun, a satellite orbiting a planet.

Keplerian velocity The speed with which a solid body moves on a circular orbit about a larger body.

Kepler's laws Three rules that describe the unperturbed motion of planets about the Sun (and of moons about planets): (1) Planets move on elliptical paths with the Sun at one focus. (2) An imaginary line from the Sun to a planet sweeps out area at a constant rate. (3) The square of a planet's orbital period varies as the cube of the semimajor axis of its orbit.

Kirkwood gaps Zones in the asteroid belt that have been depleted of objects due to mean-motion orbital resonances with Jupiter.

Klystron Vacuum-tube amplifier used in planetary radar transmitters.

Kozai resonance A resonance where an object's nodal precession rate is equal in magnitude and direction to its periapse precession rate. Objects within a Kozai resonance undergo oscillations in eccentricity and inclination that are out of phase (i.e., when one increases, the other decreases). Kozai resonances affect the motion of Pluto and some comets and asteroids in the solar system.

Kuiper belt Generally used to refer to the population of trans-Neptunian bodies, i.e., those with semimajor axes >30 AU. In a more detailed classification, which partitions the trans-Neptunian population into the Kuiper belt, the scattered disk and the extended scattered disk, the name "Kuiper belt" is associated with a collection of bodies on essentially stable, low inclination, low eccentricity orbits. Almost all Kuiper belt objects discovered so far have semimajor axes <50 AU, which argues for the Kuiper belt having an outer edge at approximately that location. *See also* chapters 42, 43 and 44 about Pluto, Kuiper belt dynamics and about the physical properties of Kuiper belt objects.

Lagrangian points The five locations in the circular restricted three-body problem at which the net gravitational and centrifugal forces in the frame rotating with the massive bodies is zero. The first three Lagrangian points, L_1, L_2, and L_3, lie on the line connecting the massive bodies; all three colinear Lagrangian points are unstable. The L_4 and L_5 Lagrangian points each make equilateral triangles with the two massive bodies; orbits about the triangular Lagrangian points are stable to small perturbations provided the ratio between the masses of the two bodies is ≥ 27.

Larmor radius The radius (also known as the gyroradius or cyclotron radius) of motion of a charged particle in a magnetic field. It depends on the mass of the particle, the magnetic field strength, and the speed of the particle perpendicular to the magnetic field lines.

Landforms Natural physical features of a planet's surface.

Langmuir probe Instrument used to measure electron and ion densities. The external sensor is usually a stiff wire and the current is measured as different voltages are applied.

Laplace resonance Occurs when three or more orbiting bodies have a simple integer ratio between their orbital periods.

Last universal common ancestor The hypothetical latest living organism from which all currently living organisms descend.

Late heavy bombardment (LHB) An apparent event 4.1 to 3.8 billion years ago, during which a large number of impactors collided with inner solar system planetary bodies. The record of this event exists in the crater populations of Mercury, the Moon, and Mars. The LHB happened "late" in the solar system's accretion period, when Earth and the other rocky planets formed and gained most of their mass (although this period is still early in the history of the solar system as a whole). The LHB may have happened when solar system dynamics caused the orbits of many main-belt asteroids to become elliptical and intersect inner solar system planets.

Latent heat Heat that is released or absorbed during a phase change, i.e., vapor or liquid or ice to ice or liquid or vapor. Latent heat contributes to heating and cooling the atmosphere in regions where ice and liquid clouds form and dissipate. It also contributes to the heat capacity of a parcel of gas/cloud and therefore influences the adiabatic temperature gradient.

Lava flow A body of silicate rock on a planetary surface, both while it is flowing into place and after it has come to rest.

Leading hemisphere When a moon is locked in synchronous rotation, the leading hemisphere faces in the direction of orbital motion.

Libration A small oscillation around an equilibrium configuration, such as the angular change in the face that a synchronously rotating satellite presents toward the focus of its orbit.

Lightcurve A graph of an object's brightness versus time. Since asteroids and cometary nuclei are usually not perfect spheres, the observed projected area of the object varies as the object rotates. The time difference between the peaks of the lightcurve provide a measure of the object's rotation rate and the shape of the lightcurve can be statistically modeled to derive the object's shape.

Limb-darkening The darkening of the observed edges a planetary disk or a star. This may be due to the scattering properties of the surface (if, for example, it is a strongly backscattering surface, like an icy one) or more usually to the presence of an optically thick atmosphere. It is often characterized by an exponent k, the Minnaert exponent, for a scattering law of the form $I = I_0 \mu^k / \mu_0^{k-1}$, where μ and μ_0 are the cosines of the angle between the normal at a given point and the observer and sun respectively and I_0 is the brightness of the center of the disk. $k = 0.5$ corresponds to a flat disk (rather like the moon), while $k = 1$ is a Lambertian disk with strong limb-darkening. $k < 0.5$ corresponds to limb-brightening, typical of a scattering but optically thin region above an absorbing (dark) region in the atmosphere.

Linear stability Mathematical technique used to describe the stability of a dynamical system subject to small perturbations.

Liquidus Temperature above which all minerals present in a rock are molten.

Lithophile Material made of elements that are commonly found in rocks, such as Si, O, Al, Ca, and Fe; derived from the Greek, meaning "rock-loving."

Lithosphere The rigid outer shell of a planetary body, generally including a chemically distinct crust and part of the upper mantle; For a more detailed discussion of planetary lithospheres see chapter 9 on the thermal evoluton of terrestrial planets. *See also* Thermal lithosphere.

Lobate scarp A long sinuous cliff (*See* Thrust fault)

Longitude of perihelion (ϖ) The sum of the longitude of the ascending node and the argument of perihelion.

Longitude of the ascending node (Ω) The nodes of an orbit are the points where the orbit crosses some reference plane, usually the ecliptic. The ascending node is where the orbit crosses the reference plane from south to north. The longitude of the ascending node is the angle between the location of the ascending node and some standard direction in the reference frame, usually the direction of the vernal equinox.

Long-period comet A comet with an orbital period of more than 200 years. Some long-period comets have orbital periods of millions of years.

Lorentz force Force exerted by a magnetic field on a moving charged particle. This force is always perpendicular to the motion of the particle.

Low Earth orbit (LEO) Spacecraft orbit with altitudes between 160 km (below this limit atmospheric drag leads to rapid orbital decay) and 2000 km above the Earth surface. All human spaceflight except the Apollo missions has taken place in LEO.

Low-reflectance material (LRM) Material that occurs on Mercury within and surrounding some craters and impact basins that have exposed material from several or more kilometers below the surface. LRM is lower in albedo and less red than average Mercury surface material.

Lyapunov exponent Measure of the rate of divergence of two nearby trajectories in a system. A positive Lyapunov exponent is associated with chaotic motion, and its inverse gives an estimate of the timescale for exponential separation of nearby orbits.

Macroscopic cross section The product of the number density (number per cm^3) of the target nuclei and the microscopic cross section. The macroscopic cross section has units of cm^{-1} and gives the probability per unit path length that a particle will undergo an interaction, for example, in a planetary surface.

Mafic Dense, Fe- and Mg-rich silicate minerals, such as those that dominate the mantle; usually refers to basalts and other refractory igneous rock types.

Magma ocean Partially or completely molten silicate mantle of a terrestrial body resulting from extreme heating due to large impacts or radioactive decay.

Magnetic flux rope A magnetic flux tube characterized by axial fields near its center and by helical fields with increasing pitch angles away from the axis of the rope near the outer edge.

Magnetic induction equation A source-diffusion equation for magnetic field generation derivable from Maxwell's equations and Ohm's law.

Magnetic reconnection A magnetic instability that can be triggered in the solar corona, where the topology and connectivity of magnetic field lines change; believed to be the primary cause of flares and coronal mass ejections.

Magnetic Reynolds number A dimensionless group of parameters that describes the ratio of magnetic induction to magnetic diffusion. A critical value must be attained for a self-sustained dynamo to occur.

Magnetic storm A prolonged interval of intense geomagnetic activity often lasting for days.

Magnetic wind Zonal flows driven by Lorenz forces.

Magnetopause The outer boundary of a magnetosphere between the solar wind region and a planets' magnetic field region, where a strong thin current generally flows.

Magnetosheath The region between a planetary bow shock and magnetopause in which the shocked solar wind plasma flows around the magnetosphere.

Magneto-rotational instability Coupling of a protoplanetary disk's magnetic field to the gas's orbital motion leading to an inward flow of mass and outward transport of angular momentum.

Magnetic secular variation Time variations of a planetary magnetic field.

Magnetic wind Zonal flows driven by Lorenz forces.

Magnetohydrodynamic limit An approximation made to Maxwell's equations and Ohm's law used in the study of dynamics of electrically conducting fluids when velocities are much smaller than the speed of light.

Magnetosphere The region of space around a planet or satellite dominated by its intrinsic magnetic field and associated charged particles.

Magnetostrophic balance It is achieved in the momentum equation when the Lorenz and Coriolis forces are dominant and balance each other.

Magnitude A logarithmic unit of brightness. Large magnitude values correspond to faint objects. The Sun, the faintest star visible with the unaided eye, and the faintest Kuiper belt object seen with the *Hubble Space Telescope* have magnitude values of $-26.74, +6$, and $+28$, respectively. For every change by five magnitudes, the brightness changes by a factor of 100. One magnitude equals a factor of $100^{1/5}$ or ~ 2.5119 in brightness. All magnitudes are scaled to the flux of Alpha Lyrae, also named Vega, which is designated as magnitude 0.

Magnetotail The part of a planetary magnetosphere that is pushed antisunward into a tail-like feature by the pressure of the solar wind.

Main sequence When stars are plotted on a graph of their luminosity versus their surface temperature (or color), most stars fall along a line extending from high-luminosity, high-surface temperature stars, to low- luminosity, low-surface temperature stars. This plot is known as the Hertzsprung-Russell diagram, and the line is known as the main sequence. Stars spend the majority of their lifetimes on the main sequence, during which they produce energy by hydrogen fusion within their cores.

Mantle The interior of a generic Earthlike (terrestrial) planet consists of an iron-rich core surrounded by a rock mantle underneath a rock crust. The crust forms by partial melting and extrusion of mantle rock. Mantle and crustal rock is solid but may flow over geologic timescales of hundreds of million years.

Mantle convection Movement of material within the Earth's mantle occurs because the density of constituent rock is related to temperature. Buoyant hot material thus tends to rise and cooler material tends to sink over time. In our experience, convection is most commonly associated with liquids or gases, however, deep within the Earth, plastic and even solid rock under pressure can convect, and thus transport heat away from the core, through the mantle, and ultimately toward the surface.

Mare (pl., maria) Latin word for "sea," used first by Galileo to refer to the dark patches on the lunar surface, now known to be basaltic lava flows.

Mascons Regions of the Moon of excess mass concentrations per unit area, identified by positive gravity anomalies and associated with basalt-filled multiring basins.

Mass wasting The downslope movement of rock, regolith, and soil under the influence of gravity.

Maxwellian distribution The distribution of particle velocities for a gas in thermal equilibrium.

Mean anomaly The mean anomaly is an angle that defines the position of a fictive body moving with a uniform velocity along an orbit and with the same orbital period as the true body.

Mean motion The mean motion is the mean value of the angular velocity of a planet or a satellite while progressing around its elliptical orbit.

Mean-motion resonance An orbital resonance in which the orbital periods of the bodies involved are in a simple integer ratio. For example, Pluto is in a 2:3 mean-motion resonance with Neptune; it completes two orbits around the Sun for every three of Neptune.

Meridional circulation Motions of the atmosphere in the plane defined by the vertical and latitudinal coordinates. Atmospheric motions in the vertical and north-south directions participate in the meridional circulation.

Meridional mixing Mixing of the atmosphere along meridians (lines of constant longitude), for example, between polar regions and mid-latitudes.

Mesosphere The layer of the atmosphere overlying the stratosphere and below the mesopause. In the mesosphere and stratosphere - collectively called the middle atmosphere — radiative heat transfer dominates over convection.

Metallic hydrogen High-pressure (≥ 1.4 Mbar) metallic form of hydrogen found in the interiors of Jupiter and Saturn.

Meteor Light phenomenon that results from the entry of a meteoroid from space into Earth's atmosphere.

Meteorite Meteoroid that has reached the surface of Earth or another planet without being completely vaporized.

Meteoroid A small fragment of an asteroid or comet that is in interplanetary space. When a meteoroid enters a planetary atmosphere and begins to glow from friction with the atmosphere, it is called a **meteor**. A fragment that survives atmospheric entry and can be recovered on the ground is called a **meteorite**.

Methane clathrate A clathrate consisting of methane ice. *See* clathrate.

Micrometeoroid Meteoroid smaller than about 0.1 mm in size.

Micron, micrometer, or μm One millionth of a meter.

Microscopic cross section An effective area that gives the probability that a particle (for example, a neutron or gamma ray) will undergo a reaction with a target nucleus (or atom). The microscopic cross section has units of barns per nucleus. One barn is 10^{-24} cm^2.

Mid-ocean ridges Active undersea volcanic chain at which the Earth's oceanic crust is generated by magma solidification.

Mie scattering The scattering of sunlight by atmospheric particles such as aerosols.

Miller-Urey experiments Laboratory experiments in which mixtures of gases representing the composition of planetary atmospheres or the Earth's early atmosphere were placed in sealed vessels and exposed to various forms of energy such as UV and energetic particles. In general the experiments tended to produce more complex combinations of molecules from the initial gases.

Mineral Naturally-occurring substance of specified chemical composition and physical properties having a characteristic atomic structure and/or crystalline form.

Minor planet Another term for an asteroid.

Mixing ratio Fractional mass of a particular component of an intimate mixture.

Moho Short form for Mohorovicic discontinuity. A discontinuity in seismic velocity (and density) that characterizes the boundary between the Earth's cust and the mantle. Named after discoverer Andrija Mohorovicic (1857–1936).

Molecular cloud Cold, dense, region of the interstellar medium containing molecular hydrogen:H_2, often the site of star formation.

Moment magnitude A measure of the size of earthquakes in terms of the energy released, on a logarithmic scale. The magnitude is based on the seismic moment of the earthquake, which is equal to the rigidity of the Earth multiplied by the average amount of slip on the fault and the size of the area that slipped. The moment magnitude scale is similar to the more famous Richter scale, but is preferred by seismologists due to its direct relationship to the physics of an earthquake.

Moment of inertia The moment of inertia is the inertia for rotational motion. For a larger moment of inertia of a given principal axis of inertia a larger torque along that axis is needed for a given rotational response.

Moon A body in orbit around another larger body, known as the primary, such as a planet, dwarf planet, or asteroid. Also called a **satellite**.

Morphology Study of the shape of landforms on a planetary surface.

Multiringed basin The largest category of impact features, characterized by multiple, concentric rings of massifs in their interiors. Smaller features have a single peak ring; still smaller ones have a single central peak near the center of the crater; and the smallest craters are "simple" and lack an elevated central region.

Near-Earth asteroid *See* Near Earth object.

Near-Earth object (NEO) Any object, such as an asteroid or comet, orbiting the Sun with a perihelion distance less than 1.3 AU.

Nearly isotropic comet (NIC) A population of comets with orbits that are randomly inclined to the ecliptic plane. By definition, the Tisserand parameter of NICs is less than 2. Also known as a **long-period comet**.

Neutron A neutral particle with mass similar to that of the proton. Neutrons and protons are the primary constituents of the atomic nucleus. Neutrons liberated from the nucleus by nuclear reactions decay by beta emission with a mean lifetime of 900 s.

New comet A nearly isotropic or long-period comet that is entering the planetary region for the first time since it was placed in the Oort cloud. Dynamically new comets are usually taken to be nearly isotropic comets with original semimajor axes greater than 10,000 AU.

Newton's laws Three laws of motion and one of gravity that describe aspects of the physical world: (1) A body remains at rest or in uniform motion unless it is acted upon by an external force. (2) The acceleration of a body is directly proportional to the force acting upon it and inversely proportional to its mass. (3) For every action, there exists an equal and opposite reaction. (Gravity) The gravitational attraction between any two spherically symmetric objects is proportional to the product of their masses and inversely proportional to the square of the distance between their centers.

Nice model A model designed to explain the orbital architecture of the outer Solar System based on early migration of the orbits of the giant planets as they gravitationally scattered numerous smaller planetesimals. *See* chapter 2 on the Origin of the Solar System for more detail.

Near Infrared Mapping Spectrometer (NIMS) A near infrared mapping spectrometer on the Galileo Mission to make multi-spectral images for atmospheric and moon surface chemical analysis. The NIMS instrument was sensitive to 0.7–5.2 micrometer wavelength IR light. The telescope associated with NIMS with an aperture of 229 mm used only mirrors and no lenses. The spectrometer of NIMS used a grating to disperse the light collected by the telescope. The dispersed spectrum of light was focused on detectors of indium antmonide and silicon.

Node One of the two points where a body's orbit crosses a reference plane, such as the ecliptic.

Nongravitational force A force not due to gravity that acts on comets and asteroids and that can significantly alter their orbits. The most important nongravitational force for comets is the reaction force due to the outgassing of volatile materials from the day side of the nucleus.

Nonsynchronous rotation The state of a satellite whose rotation period is not equal to its orbital period.

Nonthermal escape Atmospheric escape of gases in processes that do not depend on the temperature of the bulk upper atmosphere. Nonthermal escape can occur when a neutral species is photo-ionized by solar extreme ultraviolet radiation and recombines with an electron to form a fast neutral atom. This is called photochemical escape. Alternatively, a fast ion can impart its charge to a neutral atom through collision or charge exchange, and become a fast neutral atom with escape velocity. Today, photochemical escape is important for the loss of carbon, oxygen, and nitrogen from Mars.

Nonthermal radio emission Radio emission produced by processes other than those which produce thermal emission. In particular, in planetary science we are concerned with cyclotron and synchrotron radiation. Cyclotron radiation is emitted by (nonrelativistic) electrons, often in the auroral (near-polar) regions of a planet's magnetic field at the frequency of gyration around the magnetic field lines (cyclotron frequency). The emission is like a hollow cone pattern. Synchrotron radiation is produced by relativistic (i.e., particle velocity approaches the speed of light) electrons. This radiation is strongly beamed in the direction in which the particle is moving. There are other types of nonthermal radio emissions that involve coupling from various plasma wave modes to radio waves.

Normal modes A particular type of oscillation in a system that maintains a constant amplitude and frequency.

Nucleosynthesis The creation of stable and unstable isotopes in stars.

Nucleus The central cometary body that is the source of all the other cometary features. The nucleus is composed of volatile materials, primarily water ice, and dust particles composed of both silicates and organics. Sublimation produces molecular gases and releases the dust particles. The nucleus has a typical radius of 1–10 km.

Nutation Nutation is the periodic change in the orientation of a planet due to gravitational torques exerted by other solar system bodies. The tip of the rotation axis describes a periodic motion, superimposed on the precession motion, with periods related to the periods of the orbital elements of the bodies involved in the gravitational torque. Nutation causes the motion in the direction of precession to be faster or slower (called nutation in longitude) and also periodically changes the obliquity (called nutation in obliquity). *See* chapter 8 on Planetary Rotation for more detail.

Oblateness A measure of the amount to which the shape of a planet or other body differs from a perfect sphere.

Obliquity The angle between a planet's equator and its orbital plane. Earth's current obliquity of 23.5° is sufficient to cause seasons.

Observational bias The effect that some astronomical objects are easier to discover and observe than others, generally because they are brighter. The observed sample of comets suffers from severe bias because, in general, only those comets that pass well within the orbit of Jupiter become bright enough to be observable.

Occultation Obscuration of a body brought about by the passage of another body in front of it. Occultations of stars by planets or by planetary rings allow the observer to probe the atmospheric structure of the planet or the structure of the rings. Occultation of stars by asteroids allow the observer to determine the size of the asteroid.

Ohmic dissipation Dissipation of energy due to electrical resistance.

Oersted Unit of magnetic intensity in the centimeter-gram-second system, equivalent to the gauss.

Oligarchic growth Self-regulated stage of planet formation that follows runaway growth.

Olivine A mineral with chemical formula $(Mg,Fe)_2SiO_4$ that is abundant in the Earth's upper mantle (to 670 km depth). Its properties are often used to characterize the upper mantle.

Oort cloud A large reservoir of several times 10^{12} cometary nuclei surrounding the planetary system and extending from a few thousand AU to about 100,000 AU from the Sun. The outer Oort cloud, beyond ∼10,000 AU, is roughly spherical; the inner Oort cloud is flattened toward the ecliptic plane. The existence of the Oort cloud is inferred from the semimajor axis distribution of long-period comets.

Opacity The ability of an atmosphere to absorb (or sometimes scatter) radiation. Also called optical depth. A beam of monochromatic radiation passing through an atmosphere with an optical depth of one will have its intensity reduced by a factor of $e (= 2.718)$, while an optical depth of four absorbs or scatters 98% of the radiation. Opacity is a function of wavelength as well as of the pressure, temperature, and composition of the region of the atmosphere under consideration.

Opposition effect Nonlinear surge in brightness as a celestial object approaches being viewed at zero phase angle.

Opposition The position of a superior planet, a comet or an asteroid when it is opposite the Sun in the sky, i.e., when its elongation approaches 180°.

Optical depth Measure of the integrated extinction of light along a path through a medium, such as an atmosphere or the disk of particles forming a ring. Normal optical depth refers to the extinction along a path perpendicular to the ring plane.

Orbit The path of a planet, asteroid, or comet around the Sun, or of a satellite around its primary. Most bodies are in closed elliptical orbits. Some comets and asteroids are thrown on to hyperbolic orbits, which are not closed and which will escape the solar system.

Orbital elements The six parameters that uniquely specify an object's orbit and it location within the orbit. Two parameters, semimajor axis and eccentricity, enumerate the size and shape of the orbit. Three angles, inclination, longitude of the ascending node, and argument of perihelion, describe the orbit's orientation in space. Finally, the mean anomaly specifies the position of the object along the orbit.

Orbital phase angle Angular position of a satellite in orbit about its primary object, measured counterclockwise when viewed from the north.

Orbital inclination (of a solar system body) The angle between the plane of the body's orbit and the ecliptic plane.

Orbit-orbit resonance Condition in which two objects have orbital periods in the ratio of small integers. Orbit-orbit resonances are commonly found between Jupiter and minor planets in the asteroid belt, between Neptune and bodies in the Kuiper belt, and in the satellite systems of Jupiter and Saturn. Also known as a **mean-motion resonance**.

Orogenic, orogeny Process of mountain building, with uplift generally occurring as a result of tectonic plate collisions.

Orphan trail A dust trail that does not appear to be connected to any cometary source. This might arise as a consequence of planetary perturbations causing a shift in a comet's orbit and disconnecting it from a more distant portion of its dust trail.

Outflow channels Large channels that start full size and have few if any tributaries. They may be up to several tens of kilometers across and thousands of kilometers long and are believed to have been formed by large floods.

Oxidants Chemical compound that readily transfers oxygen atoms.

Paleomagnetism The study of the record of a planet's magnetic field through the remanent magnetism in rocks.

Palimpsest Flattened, circular bright patches on Ganymede and Callisto that are believed to be the remnants of ancient large impact structures.

Panspermia A theory by which life spreads through the solar system and the galaxy by spores carried on dust grains or small particles.

Parallax The apparent change in the position of a nearby star on the celestial sphere when measured from opposite sides of the Earth's orbit, usually given in seconds of arc.

Parautochthonous Describes a rock unit that has been moved only slightly into its present location.

Partial melting Melting of one or more minerals composing a rock.

Particle-size distribution The function that specifies the number of particles of different sizes in a given population.

Parsec The distance at which a star would have a parallax of 1 a s, equal to 206,264.8 AU, or 3.261631 lightyears, abbreviated: pc. One thousand parsecs are equal to a kiloparsec, abbreviated: kpc.

Patera (pl., paterae) A collective term for a variety of unusual, saucer-shaped, shallow volcanic constructs that often have a central crater or caldera.

Peak ring A ring of mountains or massifs near the center of a small impact basin. Smaller impact features have a single central peak near the center of the crater; still smaller craters are "simple" and lack an elevated central region. Larger impact features have multiple, concentric rings of massifs in their interiors.

Periapse Point on an orbit closest to the central body, called **perihelion** for orbits about the Sun and **perigee** for orbits about the Earth.

Perigee The closest point to the Earth of the elliptical orbit of the Moon or an artificial satellite.

Perihelion distance (q) The closest distance to the Sun an object reaches in its orbit, given by $q = a(1-e)$, where a is the object's semimajor axis and e is its eccentricity.

Perihelion Point in a heliocentric orbit when it is closest to the Sun.

Perijove The point in an orbit around Jupiter when the object is closest to the planet.

Periodic comet Traditionally, a comet with an orbital period of less than 200 years; also known as a short-period comet.

Permafrost zone Near-surface zone within which temperatures are always below 0° C. It may or may not contain ground ice.

Petrology The study of the nature and history of mineralogic phases and chemical compositions of rocks, and conclusions regarding their origins. One aim of mineralogy and petrology is to decipher the history of igneous and metamorphic rocks.

Phase angle Angle between the illumination source (the Sun), a given object, and the observer with the object at the vertex.

Phase function The curve describing the change in brightness of a body as a function of the phase angle, the angle between the observer, the body, and the Sun. Usually expressed in astronomical magnitudes per degree.

Phase integral Integrated value of the function that describes the directional scattering properties of a surface.

Phase space Multidimensional space in which the coordinates are, for example, the positions and the velocities.

Photoautotrophy The capacity of an autotrophic (self-feeding) organism to derive the energy required for its growth from sunlight by means of photosynthesis; green plants are photoautotrophic.

Photodissociation A chemical reaction in which a chemical compound is broken down by photons.

Photolysis Process that occurs when a molecule absorbs light of sufficiently high energy (usually ultraviolet light) and breaks apart.

Photosphere A thin, 300 km thick layer above the solar surface from where most of the optical emission (white light) is irradiated, with a temperature of ~ 6000 K.

Photon-stimulated desorption A process that liberates atoms from a planetary surface, whereby a photon breaks bonds holding the surface molecules together, and ejects one or more atoms.

Phototactic/phototaxis The movement of an organism in response to light, either toward or away from the source; e.g., certain microorganisms are phototactic and will migrate in the direction of sunlight.

Phylogenetic Refers to organisms that are related to each other through evolution.

Physiographic Referring to the physical appearance of the landscape.

Pixel A "picture element." One element in a CCD or infrared detector array.

Plagioclase A mineral of the Feldspar group with chemical formula $NaAlSi_3O_8 - CaAl_2Si_2O_8$ that has a relatively low melting temperature and is thus typically found in crustal rock.

Planar deformation features Planar, micrometer-sized bands of intense deformation or glass that occur in minerals due to the passage of a shock wave.

Planetary protection The guiding principle in space missions to celestial bodies ensuring avoidance of their harmful contamination and also adverse changes in the environment of the Earth resulting from the introduction of extraterrestrial matter. Planetary Protection is included in the United Nations Outer Space Treaty of 1967, signed by 101 nations.

Planet According to the new IAU definition passed in 2006, a planet must have three qualities: (1) it must be round, indicating its interior is in hydrostatic equilibrium; (2) it must orbit the Sun; and (3) it must have gravitationally cleared its zone of other debris. According to the definition, our solar system has eight planets: Mercury, Venus, Earth, Mars, Jupiter, Saturn, Uranus and Neptune. Although most astronomers have accepted the new definition, some are campaigning to have it changed. See The Solar System and Its Place in the Galaxy and Pluto and Charon for more discussion.

Planetary embryo Large solid body formed by runaway and oligarchic growth.

Planetesimal A small solid body formed in the early solar system by accretion of dust and ice (if present) near the central plane of the solar nebula. The terrestrial planets, asteroids, comets, and cores of the giant planets are generally thought to have formed through the accretion and aggregation of planetesimals.

Planetary rings An **astrophysical disk** adorning a planetary body.

Plasma Ionized medium in which electrons have been stripped from neutral matter to make a gas of charged ions and electrons.

Plasma beta Ratio of gas pressure to magnetic field pressure within a plasma.

Plasma tail The narrow, highly structured tail consisting of molecular ions (and electrons) confined to magnetic field lines wrapped around the head of the comet. The plasma tail is normally attached to the head region. The exception is when disconnection events occur. The orientation of the tail, approximately anti-sunward, is produced by the solar wind interaction. The plasma tail appears blue because of resonance scattering of sunlight from ionized carbon monoxide molecules.

Plasmoid A region within a magnetosphere in which plasma is confined by a magnetic structure that is not directly linked to the planet.

Plate tectonics Theory describing the motion of rigid plates over the Earth's surface driven by subduction of oceanic lithosphere and, to a lesser extent, by spreading of mid-ocean ridges.

Plume A hot blob or column of material that rises through the mantle, typically causing uplift of the surface and volcanism.

Poloidal field Component of a solenoidal field that can be written as the curl of a curl of a function in the radial direction. The term was coined by Elsasser to refer to magnetic field in the direction "toward the poles".

Polar wandering Changes in the direction of the magnetic pole relative to its orientation in space.

Polymict A rock unit consisting of fragments of various pre-existing rock units.

Polymorph Crystal form of a mineral that has a different crystal structure from that of the original mineral.

Polysaccharide Any of a group of carbohydrates consisting of long chains of simple sugars; e.g., starch, glycogen.

Power spectra Graphs of the radar backscatter power versus frequency. Measuring the received power in each frequency bin allows for one to derive the mean radial velocity and albedo of an object.

Poynting-Robertson effect, P-R drag Drag on interplanetary particles caused by their interaction with solar radiation, which causes the particles to lose orbital angular momentum and to spiral in towards the Sun.

Precession Precession is the slow change in the orientation of a planet due to gravitational torques exerted by other solar system bodies in which the tip of the rotation axis describes a large circle on the celestial sphere around the normal to the orbital plane. *See* chapter 8 Rotation of Planets for more detail.

Precession of the equinoxes The slow rotation of the equinoxes with respect to the stars. It has a period of about 26,000 years for the Earth.

Primary body Celestial body (usually the Sun or a planet) around which a planet or a moon, respectively, or secondary body, orbits.

Primitive meteorite *See* Chondrite.

Primordial The original form, e.g. the primordial solar nebula would be the original, unaltered state.

Principle axis of inertia Each (celestial) body has a set of mutually perpendicular axes, called principal axes of inertia, for which the off-diagonal terms of the **inertia matrix** are zero. A torque along a principal axis only affects the rotational acceleration about that axis.

Prominences Cool and dense mass structures suspended above the chromosphere, observed above the solar limb, which are called filaments when seen on the solar disk.

Protoplanetary nebula A disk of gas and dust that surrounds a newborn star, from which the planets, asteroids, and comets are thought to form.

Protostar A star in the process of formation, which is luminous due to the release of gravitational potential energy from the infall of nebula material.

P-wave velocity Seismic body wave velocity associated with particle motion (alternating compression and expansion) in the direction of wave propagation.

Pyroclastic materials Fragmented materials ejected during an explosive volcanic eruption, including ash, pumice, and rock fragments.

Pyroclastic vent A typically rimless, scalloped-rimmed depression formed by the explosive eruption of pyroclastic volcanic materials.

Pyroxene A mineral of chemical formula $(Mg, Fe, Ca)SiO_3$ that is typically found in mantle rock.

Radar albedo Ratio of a target's radar cross section in a specified polarization to its projected area, hence a measure of the target's radar reflectivity.

Radar cross section Most common measure of a target's scattering efficiency, equal to the projected area of that perfect metal sphere that would give the same echo power as the target if observed at the target's location.

Radial velocity technique Observational method used to detect stellar reflex motions by measuring the line-of-sight component of a star's space velocity vector. If the radial velocity can be measured with a precision of a few m/s then the reflex motion due to planetary companions become detectable. Today, this is the most successful method for finding extrasolar planets.

Radiation belts Toroidal zones containing charged particles that are magnetically trapped in a planetary dipole field. The Van Allen belts around the Earth include ions and electrons with energies from hundreds of keV to tens of MeV.

Radiation pressure The force per unit area an object feels due to solar photons hitting its surface.

Radio interferometric synthesis The addition of signals from a set of smaller telescopes, used to produce an image with the same angular resolution of a much larger telescope.

Radio spectrum A graph of the brightness temperature as a function of wavelength or frequency.

Radiolysis The dissociation of molecules by high-energy radiation.

Radiometric modeling Interpretation of the measured thermal emission of an asteroid by modeling to provide an estimate of the asteroid's surface temperature distribution and albedo. A dark asteroid, for instance, would absorb more of the visible sunlight since it has a low albedo, but would radiate that additional energy in thermal wavelengths, showing a warmer surface temperature. Combining data on thermal "temperature" and visible reflectance can thus provide both the albedo of an object and an estimate of its size. The accuracy of the model depends on various assumptions made about the nature of the surface.

Radiometric modeling Measures the thermal emission of an asteroid to provide an estimate of the asteroid's surface temperature and albedo. A dark asteroid, for example, would absorb more of the visible sunlight because it has a low albedo, but it would radiate that additional energy at thermal wavelengths, showing a warmer surface temperature. Combined data on thermal "temperature," and visible reflectance can provide the albedo of an object and an estimate of its size.

Radio occultation The passing of a radio beam through a planet's atmosphere. Attenuation and refraction (bending) of the beam, generally by phase delay, can be used to measure the density of electrons in the planet's ionosphere and the density of the gas in its atmosphere. The abrupt cutoff of the signal can also be used to make a precise measurement of the planet's radius.

Raman scattering Inelastic scattering of sunlight by gas molecules in an atmosphere, such that the scattered photon is shifted in frequency.

Random walk A series of movements in which the direction and size of each move is randomly determined.

Rayed crater *See* Crater rays.

Rayleigh number Non-dimensional parameter characterizing the vigor of convection in a convecting fluid. It can be interpreted as the ratio between the bouyancy forces driving the flow to the viscous forces that retard the flow and the loss of bouyancy through diffusion.

Rayleigh scattering The scattering of sunlight by gas molecules in an atmosphere.

Reaction notation The notation for nuclear reactions given by T (i, p) R, where T is the target nucleus, i is the incident particle, p indicates the particles(s) produced by the reaction, and R is the product nucleus. For example, the notation for neutron inelastic scattering with ^{56}Fe is $^{56}Fe(n,n'y)$, where n' is the scattered neutron and y is the associated gamma ray. Similarly, neutron capture with ^{56}Fe is denoted $^{56}Fe(n,y)^{57}Fe$, where the product isotope has been appended.

Reaction rate The rate (R) at which gamma rays or neutrons interact with nuclei is given by the product of the flux (0), the number density of the target nuclei (N nuclei per cm^3) and the microscopic cross section (0) for the selected reaction: $R = yNa$ (interactions per unit time).

Reconnection A process in which the magnetic configuration changes as if two field lines were broken and reconnected in a new configuration. This can occur when two plasmas containing oppositely directed magnetic fields flow toward each other.

Red slope A increase in albedo with increasing wavelength that is common in the visible to near-infrared portion of the spectrum of sunlight reflected from silicate planetary surfaces.

Reductants Compounds or catalysts that result in the loss of an electron.

Reflectance spectroscopy The study of the physical and mineralogical properties of materials over the wavelength range of reflected electromagnetic radiation. Light interacts with the atoms and crystal structure of materials producing a diagnostic set of absorptions and reflectances.

Refracted The manner in which the path of a wave is bent when it passes from one medium to another in which its speed is different. This is commonly observed in optics as light passes from air into glass or water, but is a general property of all wave phenomena.

Refraction Bending of a light ray as it traverses a boundary, for example, between air and glass or between space and an atmosphere.

Refractories Materials not deformed or damaged by high temperatures. Classic refractories are high-melting oxides, like silica and alumina, but also carbides, nitrides, sulfides, and pure carbon. In our terminology, refractories are materials that are not modified by space conditions (temperature and vacuum) in the inner solar system. The opposite are volatile materials, e.g., ices that rapidly sublimate close to the Sun.

Refractory element In planetary science, an element that has a strong chemical bonds causing it to remain in the solid phase at relatively high temperatures.

Refractory inclusion See **CAI**.

Regolith The outermost unconsolidated fragmental layer on some airless planets, satellites and asteroids that results from the breakup of rocks by repeated impacts of meteoroids.

Regular motion A trajectory that does not display chaos.

Regular satellite A satellite with low orbital eccentricity and inclination.

Remanent magnetization Permanent magnetism in a rock after removal of the magnetizing field.

Remnant magnetic field The field generated by the permanent magnetization left behind in a ferromagnetic material (such as iron) after an external magnetic field is removed.

Rendezvous mission A space mission in which the spacecraft visits the target object, e.g. an asteroid, and may enter orbit around it, "hover" close to it, or touch down on its surface.

Resonance A situation in which two orbiting bodies have orbital frequencies (related to the time they take to complete their orbits or for their orbits to precess) that are in a simple integer ratio. Objects in resonance exert a regular gravitational influence on each other. Depending upon the particular resonance involved, resonance can either stabilize an orbit, as in the case of Pluto, or destabilize an orbit as near the Kirkwood gaps in the asteroid belt. Strong satellite resonances open gaps at particular locations in broad planetary rings; weaker ones drive radial and vertical wave trains.

Resonant emission A quantum process in which photons of specific energies or wavelengths are absorbed by an atom or ion and then reemitted at the same wavelength.

Resurfacing The process of renewing a planetary surface by volcanism, cratering, erosion and sedimentation.

Retrograde, retrograde motion Orbital or rotational motion in the solar system that is clockwise as seen from north of the ecliptic. Nearly isotropic comets with inclinations greater than 90° have retrograde orbits. Triton is in a retrograde orbit around Neptune. Venus is in retrograde rotation.

Returning comet A nearly isotropic or long-period comet that is returning to the planetary region for at least the second time. Returning comets are usually taken to be nearly isotropic comets with semimajor axes less than 10,000 AU.

Reynolds number Dimensionless number that governs the conditions for the occurrence of turbulence in fluids.

Reynolds stress Transfer of momentum to the axisymmetric zonal velocity field from the correlation of small-scale zonal and radial velocity fields through the nonlinear inertial force.

Rheology The characteristics of the response of material in response to an applied force, under conditions in which they respond with plastic or viscous flow rather than deforming elastically. For example, a commonly used rheological parameter is viscosity.

Ring current A current carried by energetic particles that flows at radial distances beyond a few planetary radii in the near-equatorial regions of a planetary magnetosphere.

Roche limit, Roche zone The distance from a planet or the Sun, within which another body will be disrupted because tidal forces from the planet exceed the self-gravity of the smaller body, unless the material strength of the body is strong enough to hold it together. For nonrotating bodies of equal density and zero strength, the Roche limit is about 2.2 planetary radii.

Rock A rock is a solid usually microscopically heterogeneous mixture of minerals, glass, and rock fragments and may contain remnants of organisms. The crust and the mantle of terrestrial (Earthlike) planets are composed of rock. Rock understood as a mixture of iron, silicon, magnesium and other refractory elements may also be found in the deep interiors of giant planets.

Rotational lightcurve A graph depicting the variation in brightness of an object verus time as it rotates on its axis. This variation can be caused by nonsphericity (i.e., shape effects) or albedo markings; for objects as large as Pluto and Charon, albedo markings usually dominate. *See also* lightcurve.

Runaway growth Stage of planetary growth in which the largest planetesimals grow rapidly while most others remain small.

Satellite A body in orbit around a planet, dwarf planet, or an asteroid. Also called a **moon**.

Scale height The vertical distance over which atmospheric pressure or density falls by $1/e = 0.368$; equal to kT/mg, where k is Boltzmann's constant, T is temperature, m is the mean mass of the gas, and g is the acceleration of gravity.

Scattered disk A collection of $\sim 10^9$ icy planetesimals in high-eccentricity, low-to-moderate inclination orbits beyond Neptune. Scattered disk objects typically have semimajor axes of order 50–100 AU. The scattered disk is probably the primary source of the Jupiter-family comets. Scattered disk objects may have escaped the Kuiper belt billions of years ago, and/or may be scattered Uranus-Neptune planetesimals.

Scattering law Function giving the dependence of a surface element's radar cross section on viewing angle.

S-band S-band is the part of the microwave band of the electromagnetic spectrum that ranges from 1.55 to 5.2 GHz. The corresponding range of wavelengths is 7.5 to 15 cm. In space missions the reserved frequency band is around 2.1 GHz (2.0 to 2.15 GHz) for the uplink (Earth to Space) and around 2.3 GHz (2.2 to 2.45 GHz) for the downlink (Space to Earth).

Scintillator A Transparent material that coverts the kinetic energy of charged particles, such as electrons produced by gamma ray interactions or alpha particles and recoil protons produced by neutron reactions, into flashes of light detectable by a photomultiplier tube or photodiode. A wide variety of organic and inorganic materials scintillate and can be used for radiation detection and spectroscopy.

Secondary crater A crater produced by the impact of blocks of ejecta from a primary impact by a comet or asteroid.

Secular Continuing or changing over a long period of time.

Secular cooling Loss of heat from the interior of a planetary body occurring over geological timescales.

Secular perturbations Long-term changes to the orbit of a body caused by the distant gravitational perturbations of the planets and other bodies.

Secular resonance Near-commensurability among the frequencies associated with the precessions of the line of nodes and/or apsides.

Seeing Blurring of the image of an astronomical object caused by turbulence in the Earth's atmosphere Atmospheric seeing at the very best observatory sites, such as Mauna Kea, is about 0.5 arcsec and can be as good as 0.25 arcsec. Seeing can be improved using adaptive optics.

Seismic normal mode As a consequence of a strong earthquake free oscillation of the Earth can be induced. The oscillation resembles the resonant oscillation of a bell. The normal mode oscillation can be recorded by a seismometer and used to measure the elastic properties of the Earth's interior.

Seismic tomography Technique that uses information on the travel times of elastic waves generated by earthquakes to infer the three-dimensional structure of the Earth's interior.

Seismic velocities The velocities by which seismic (acoustic) waves propagate through the Earth and other planetary interiors. Because the seismic velocities are related to the density and the elastic moduli, the seismic velocities can be used to calculate the three-dimensional structure of the Earth's interior. The seismic velocities are inversely related to the travel times of seismic waves.

Seismometer The instrument by which seismic events (Earthquakes) are recorded. A seismometer in its simplest form is a pendulum that records the motion of the ground.

Semiconductor Semiconductors, such as germanium, silicon, and CdZnTe, can be used to detect gamma rays. Swift electrons produced by Compton and photoelectric interactions ionize the semiconductor, producing electron-hole pairs. The electrons and holes drift under the influence of an applied electric field to electrical contacts. As they drift, the electrons and holes induce charge on contacts, which can be measured by a charge-sensitive preamplifier. The amplitude of the charge pulse is proportional to the energy deposited by the gamma ray, which enables semiconductors to be used for spectroscopy.

Semimajor axis (a) Commonly thought of as the mean distance of the orbit of a body from its primary. More formally, it is one half of the longer of the two axes of an ellipse describing the orbit, that passes through both foci of the ellipse.

Semiminor axis (b) One half of the minor axis (short diameter) of an elliptical orbit.

Sensitivity to initial conditions A situation in which a tiny change in an object's initial state (position and/or velocity) will make a big change in its final trajectory. Sensitivity to initial conditions is a necessary condition for chaos.

Separatrix Boundary of a resonance, separating resonant or librating motion inside the resonance from nonresonant or circulating motion outside.

Shield volcano Broad volcano with a large summit pit formed by collapse and gently sloping flanks, built mainly from overlapping, fluid, basaltic lava flows.

Shock A discontinuous, nonlinear change in pressure commonly associated with supersonic motion in a gas, plasma, or solid.

Shock metamorphism Permanent physical, chemical, and mineralogical changes in rocks, resulting from the passage of a shock wave.

Shock wave Compressional wave, resulting from an impact or explosion, which travels at supersonic velocities.

Short-lived isotopes Radioactive isotopes with half-lives much shorter than the age of the solar system.

Short-period comet A comet with an orbital period <200 years. Short-period comets include **Jupiter-family** and **Halley-type** comets.

Sidereal Relative to the fixed stars. A sidereal period is the orbital period of a planet around the Sun relative to the stars. A sidereal year is the orbital period of the Earth around the Sun relative to the stars.

Siderophile element Element that tends to join with iron and is predominantly found in a planet's core.

Silicate A compound containing silicon and oxygen.

Smooth plains The youngest plains on Mercury with a relatively low impact crater abundance.

SNC meteorites Group of meteorites (Shergotty-Nakhla-Chassigny) believed to be derived from Mars because of their young ages, basaltic composition, and inclusion of gases with the same composition as the Martian atmosphere.

Solar activity cycle Cycle of ~11 year duration characterized by waxing and waning of various forms of solar activity such as sunspots, flares, and coronal mass ejections.

Solar corona The hot, tenuous outer atmosphere of the Sun from which the solar wind originates.

Solar energetic particles Ions (usually protons) and electrons generated in solar flares and in the corona and solar wind by shock waves, with energies above hundreds of keV for electrons and above an MeV per nucleon for ions.

Solar flare A disturbance in the solar atmosphere characterized by a sudden, localized enhancement in electromagnetic emission from visible to x-ray wavelengths.

Solar nebula The cloud of dust and gas out of which the Sun and planetary system formed.

Solar phase angle See **phase angle**.

Solar flux The radiant energy received per unit area per second from the Sun.

Solar wind A magnetized, highly ionized plasma that flows radially out from the solar corona at supersonic and super-Alfvenic speed.

Solstice The passing of the Sun to its farthest latitude away from a planet's equatorial plane, so that the northern or southern hemisphere (as the case may be) receives maximum sunlight; associated with the seasons of summer and winter (cf. **equinox**).

Solidus Line or surface in a phase diagram below which the system is completely solid.

Solidus Temperature above which the mineral with the lowest melting temperature among those that form a rock melts.

Space debris Man-made particulates littered in space.

Space exploration The extension of human reach beyond the Earth's atmosphere using spacecraft to access unknown terrains and environments, and to acquire knowledge about space, planets, stars, or other celestial bodies by human and robotic means.

Space ultraviolet That part of the ultraviolet electromagnetic spectrum that can only be observed from space because the Earth's atmosphere is opaque at those wavelengths; commonly thought of as the region below wavelengths of 3000 Angstrom.

Glossary

Space weather The variable level of geomagnetic activity controlled by the conditions in the solar wind.

Space weathering A process acting on the surface of planetary and asteroidal bodies that changes their surface optical properties over time.

Spectral absorption A particular wavelength of light that is selectively absorbed by a particular material. Patterns of absorptions can serve as "fingerprints" to remotely identify surface or atmospheric materials.

Spectrometer An instrument that measures the profile of light scattered or emitted by an object as a function of wavelength (i.e. the intensity of light at different points of the **spectrum**).

Spectrum The range of wavelengths (equivalently, colors, though the concept extends beyond **visible light**) that characterize electromagnetic radiation (i.e. light); sometimes, the profile of brightnesses at each of those wavelengths.

Spheroid A spheroid is a body that is approximately spherical but not perfectly round. More strictly it can be defined as an ellipsoid of revolution (or biaxial ellipsoids). Fast rotating planets can be well approximated by an oblate spheroid, which is flattened at the poles, and can be generated by revolving an ellipse around its major axis (the equator is a perfect circle but any meridian circle passing through the poles is an ellipse flattened at the equator). A better representation of the shape of a planet is that of a triaxial ellipsoid, where also the equator is an ellipse.

Spiral bending wave A transverse wave propagating through an **astrophysical disk**, generated by a **resonance** that pumps up the orbital **inclinations** of individual disk particles.

Spiral density wave A compression wave propagating through an **astrophysical disk**, generated by a **resonance** that pumps up the orbital **eccentricities** of individual disk particles.

Spin-orbit resonance Simple numerical relationship between the spin period of a planet or satellite and its orbital period. Most natural satellites in the solar system are in the 1:1 spin-orbit resonance, also called the synchronous spin state.

Sputtering An atmospheric loss process that occurs when ions that have been picked up by the magnetic field embedded in the solar wind impact a planetary atmosphere and undergo charge exchange. Charge exchange neutralizes the ions, which can impart their large energies to surrounding particles by collision. Upward-directed energetic particles can then escape. This process may have been important on Mars after it lost its magnetic field and its upper atmosphere was no longer shielded from the solar wind. Sputtering can also occur when energetic particles from the solar wind or a planetary magnetosphere strike the surface of an airless planet or satellite and cause atoms of the surface materials to escape.

Stagnant lid Immobile and stiff layer comprising the crust and the uppermost mantle of all terrestrial planets other than Earth and of the Moon.

Stellar metallicity The amount of chemical elements, heavier than hydrogen and helium, contained in a star. Observations indicate that stellar metallicity is a critical factor in the efficiency of the formation and/or orbital migration of extrasolar planets.

Stellar occultation When a planet or asteroid passes in front of a star and the star is briefly hidden from view. Such events can be used to probe the size and also the atmospheric structure of the planet (or asteroid) doing the occulting, or the structure of rings around a planet.

Stellar reflex motion The movement of a star along its orbit around the barycenter of a star/companion system. If a star has no companions (stars or planets), the barycenter coincides with the star's own center of mass, and no reflex motion exists.

Strain Measure of the deformation representing the relative displacement between particles in a stressed body.

Stratification A layered configuration in which each layer has a distinct composition or physical property, such as density.

Stratigraphy Study of rock layers.

Stratosphere Region in an atmosphere overlying the troposphere that is strongly stabilized against convection by heating because of the absorption of ultraviolet radiation from the Sun. *Stratum* is Latin for "layer."

Stream structure Pattern of alternating flows of low- and high-speed solar wind.

Streaming instability A positive feedback in which small particles drifting inwards in a protoplanetary disk begin to accumulate at a particular location, slowing their drift rate as a result, allowing more particles to catch up and further enhancing their number.

Stress The forces acting within a body of material as a result of external forces tending to change the shape or volume of the body.

Stromatolite A geological feature formed by the conversion of loose, unconsolidated sediment into a coherent layer, as a result of the growth, movement, or activity of microorganisms; e.g., blue-green algae. Microfossils associated with stromatolite formation are an important form of evidence for early life on Earth, and thus a search for stromatolites could undertaken on other planets in sites where liquid water might have accumulated.

Subaerial Referring to landscapes, such as islands or continents, that are exposed to the air.

Subduction The process by which convective forces cause lithospheric plates to be carried into the mantle.

Sublimation The phase change of a solid directly to gas, as in the conversion of ice directly into vapor.

Subsolidus convection The convection of material at temperatures below its freezing point. This can occur because a solid material can often act as a fluid over long time scales, particularly at temperatures near its melting point.

Substorm The elementary disturbance of the magnetosphere that produces geomagnetic activity.

Subtropical jet The jet stream occurring at the poleward edges of the Hadley cell.

Sulfur allotropes Sulfur cooled rapidly from different temperatures, resulting in different colors.

Super-Earth A newly discovered class of planets. Typically, a super-Earth has a larger radius and higher mass as Earth, but is smaller and less massive as icy giants like Uranus and Neptune.

Superior geocentric conjunction The point in a planetary satellite's orbit where it is directly opposite Earth, such that the satellite lies on a straight line connecting Earth, the planet, and the satellite.

Surface of section Means of studying the regular or chaotic nature of an orbit by plotting a sequence of points in two dimensions that can represent all or part of the coordinates of the orbit in phase space.

Synchronous orbit An orbit whose period is equal to the rotation period of the primary.

Synchronous orbit distance The average distance of a body to its primary on a synchronous orbit.

Synchronous rotation Dynamical state caused by tidal interactions in which a satellite presents the same face toward the primary, because the satellite's rotation period is equal to its orbital period.

Synodic period The time required for a body in the solar system to return to the same position relative to the Sun (i.e. elongation) as seen by an observer on the Earth.

Synodic rotation period Apparent rotation period of a target that is moving relative to the observer (who may also be moving), to be distinguished from the sidereal rotation period measured with respect to the fixed stars.

Tadpole orbits Orbits that librate about the stable L_4 or L_5 triangular Lagrangian points in the restricted three-body problem. These orbits appear to be shaped like tadpoles in the frame rotating with the mean-motion of the massive bodies.

Taxon Grouping of organisms or bodies with similar characteristics.

Taylor–Proudman Theorem Mathematical theorem stating that for a fluid dominated by Coriolis and gradient forces, the fluid velocity will be uniform in the direction parallel to the rotation axis.

Tectonic framework The global or large-scale pattern of fractures and folds formed by crustal deformation.

Tensile stresses A stress defined (*See* stress) as the internal forces in a continuum resulting from external forces can be compressional, tensional and shearing. The latter is termed the shear stress while the former two are termed normal stresses, compressional and tensile.

Tephra Generic term for all volcanic fragments that are explosively ejected from a volcano.

Tension fractures Fractures that propagate near-vertically into previously unfractured brittle surface material in response to tensile stress.

Termination shock A discontinuity in the outer heliosphere where the solar wind slows from supersonic to subsonic as it interacts with the interstellar plasma.

Terminator The boundary between the illuminated and non-illuminated parts of a planet, satellite, asteroid, or cometary nuclei.

Terrane A particular type of terrain. Generally used to denote the kind of terrain dominated or formed by a particular geomorphic process regime, such as a volcanic terrane or an aeolian terrane.

Terrestrial planet A planet like the Earth with an iron core and a silicate mantle and crust. The terrestrial planets are Mercury, Venus, Earth, and Mars.

Tessera (pl. tesserae) Tile-like, polygonal terrain interpreted to be intensely deformed and cut by at least two directions of ridges and/or grooves.

Thermal boundary layer The layer next to the surface confining a convecting fluid across which heat is transferred by conduction and across which the temperature gradient is steep. Examples are the so-called planetary boundary layer in the atmosphere, the air layer near the ground, and the thermal lithosphere in a planetary mantle.

Thermal desorption Process of heating to drive off volatile gases.

Thermal lithosphere The outermost layer of a planet that is cooled by conduction, in contrast to the hotter, inner layers that are cooled by convection.

Thermal diffusion Transport of heat resulting from a temperature gradient and caused by microscopic transfer of internal energy.

Thermal emission Electromagnetic radiation emitted by a body, typically at infrared wavelengths, due to its temperature.

Thermal inertia A material property that is a measure of the time it takes for the material to respond to temperature changes. Thermal inertia is mathematically defined in terms of the physical properties of the material as $(kpC)^{1/2}$, where k is the thermal conductivity, p is the bulk density, and C is the specific heat capacity.

Thermal radio emission Continuous radio emission from an object that results from the object's temperature. Blackbody radiation is a form of thermal radio emission.

Thermal wind A wind shear developed in one direction due to a temperature gradient in an orthogonal direction.

Thermosphere, exosphere Outer parts of an atmosphere, heated by ionizing radiation and cooled by conduction. The exosphere is essentially isothermal and is also characterized by very long mean free paths.

Tholeiitic Referring to basaltic rocks generally found on the ocean floor, erupted from oceanic ridge zones or from shield volcanoes. Such rocks are considered in the mafic family.

Tholus (pl., tholi) Dome or shield. Small tholi are scattered across Io.

Three-body problem Problem of the motion of three bodies moving under their mutual gravitational attraction. In the restricted three-body problem, the third body is considered to have negligible mass such that it does not affect the motion of the other two bodies.

Thrust fault A fault where the block on one side of the fault plane has been thrust up and over the opposite block by horizontal compressive forces.

Tidal despinning The reduction in rotation rate of a satellite or planet due to energy dissipation by friction in the tidal bulge raised by the parent planet or star. Ultimately the despinning body becomes tidally locked, with on side always facing the parent body.

Tidal heating Energy deposited in a satellite due to the dissipation of energy from tidal deformation.

Time delay Time between transmission of a radar signal and reception of the echo.

Tisserand parameter A nearly-conserved quantity in the circular restricted three-body problem. The Tisserand parameter for comets with respect to Jupiter is used to recognize returning comets even if their orbits were changed by a close approach to Jupiter, and to classify their orbits.

TNT An explosive that releases approximately 4.184 GJ of energy per ton. The explosive energy of nuclear weapons is normally expressed in terms of kilotons or megatons of TNT. These units are also used to express the energy released in fireball events and asteroid impacts.

Toroidal field Component of a solenoidal field that can be written as the curl of a function in the radial direction and hence has no radial component.

Trailing hemisphere When a moon is locked in synchronous rotation, the trailing hemisphere faces away from the direction of orbital motion.

Transit photometry technique An observational method to find extrasolar planets by the periodic dimming of the host star when the planet crosses in front of it and blocks some of the light of the star. A radius for the planet can be calculated from the amount of the observed dimming.

Transit An event in which a smaller object passes in front of a larger object, blocking some of the light from the larger object.

Transient crater The crater excavated in a hyper-velocity impact, prior to the collapse of the surrounding crater walls.

Transition region The vertical zone in the Sun where the temperature climbs from 20,000 K above the cool chromosphere to 1 million K in the hot corona.

Triaxial ellipsoid A 3-dimensional surface defined by three axes that are elliptical in cross section and used to describe the shape of a body.

Troposphere Lowest level of an atmosphere dominated by vertical mixing and often containing clouds, where temperature falls off with height at close to the neutrally stable (adiabatic) lapse rate. Earth's troposphere contains 80% of the mass of its atmosphere and most of the water vapor, and consequently most of the weather. Terminated at the top by the **tropopause.** On Earth, the troposphere extends to 14 km (equatorial) and 9 km (polar); on Venus, to 65 km. *Tropos* is Greek for "turning."

True anomaly The true anomaly is an angle that defines the actual position of a body moving along an orbit. It is the angle between the direction of periapsis and the position of the body, as seen from the main focus of the ellipse.

Turbulent concentration The concentration of large numbers of similarly sized particles in stagnant regions in a turbulent gas.

Type I migration Gradual inward spiraling of a planet as it loses angular momentum via gravitational interactions with nebular gas. This affects planetary embryos and rocky planets.

Type II migration Change in the size of a planet's orbit when the planet is massive enough to clear a gap in the disk. Migration is typically inward.

Ultraviolet light Electromagnetic radiation with wavelengths between 10 and 300 nm on the **spectrum**, emitted by energetic objects such as stars or fluorescing gas.

Urey ratio Non-dimensional number indicating the ratio between internal heat production and surface heat loss in a planetary body.

Van Allen belts Region in the Earth's magnetic field, inside of ~ 4 Earth's radii, filled with energetic particles. Other magnetized planets have similar radiation belts.

Vernal equinox The direction of the Sun as viewed from the Earth as it crosses the celestial equator moving northward. On Earth, the vernal equinox denotes the beginning of spring in the northern hemisphere.

Very long baseline interferometry (VLBI) A type of astronomical interferometry in which a signal from an astronomical radio source is collected at multiple radio telescopes on Earth. This allows observations of an object that are made simultaneously by many radio telescopes to be combined, emulating a telescope with a size equal to the maximum separation between the telescopes.

Viscoelastic heating Heating (such as tidal heating) produced by nonrecoverable (permanent) deformation in the viscous portion of a viscoelastic body, i.e., a body that behaves with both viscous and elastic components of deformation.

Viscosity Property of a fluid that resists flow; fluid dynamic stiffness or, in a sense, internal friction. For lava flows that typically have remarkably little excess energy above their solidus, viscosity can be the determining factor for the magnitude and morphology of lava flow fields associated with volcanoes and is often exponentially dependent on the core temperature of the flow.

Viscous overstability A phenomenon that can occur under certain conditions in **planetary rings**, in which collisions among the ring particles cause small density fluctuations to grow into large, oscillation density variations.

Viscous relaxation Process whereby topographic features become subdued over time due to the flow of the surrounding geologic material.

Visible light Electromagnetic radiation with wavelengths between about 300 and 900 nm on the **spectrum**, emitted prolifically by our Sun and readily detected by human eyes.

Volatile Any substance that outgasses or produces a significant vapor pressure at a given temperature. Ice is a volatile on Earth (T = 270–300 K), but involatile in the outer solar system (T < 100 K). By contrast, the ices of CH_4, CO, and N_2 are volatile throughout the planetary region wherever $T > 30$ K. Also, chemical compounds or elements contained in magmas that are generally released as gases to the atmosphere during a volcanic eruption.

Volatile element In planetary science, an element that has weak chemical bonds causing it to enter a gas phase at relatively low temperature.

Volcano A positive topographic feature on a planetary surface consisting of rocks that have been erupted from the interior and accumulated around the eruption site.

Vorticity A measure of the circulation of a region of the atmosphere. Spots having cyclonic vorticity rotate counterclockwise in the northern hemisphere and clockwise in the southern hemisphere. Terrestrial hurricanes have cyclonic vorticity. The large stable spots on the giant planets are anticyclones.

Warm poles The alternating aphelion subsolar points on the surface of Mercury at the 90° and 270° meridians.

Water ice The primary volatile constituent of comets. Water ice comes in three forms: amorphous, cubic, and hexagonal. Amorphous ice, believed to be the form in the deep interior of cometary nuclei, is characteristic of ices formed at very low temperatures. It has no crystalline structure. At higher temperatures, typically 100–150 K, energy is available to convert the ice to the lower energy cubic form; this transition releases energy. A similar transition from cubic ice to hexagonal ice also releases energy at ~ 180 K. Cubic ice and hexagonal ice are collectively known as crystalline ice. The water ice nearest to the surface of the cometary nucleus is thought to be hexagonal.

Water-equivalent hydrogen (WEH) Gamma ray and neutron spectrometers are sensitive only to the abundance of hydrogen, which is sometimes expressed as the equivalent weight fraction of water. If all of the hydrogen is in the form of H_2O, then the relationship between the weight fraction of hydrogen (w_H) and the weight fraction of water (w_{water}) is $w_{water} = 9 w_H$.

Wavelength, wavenumber Wavenumber is the inverse of wavelength, having units of inverse length. In spectroscopy, the wavenumber ν of electromagnetic radiation is defined as $\nu = 1/\lambda$, where λ is the wavelength in vacuum. This quantity is commonly specified in cm^{-1}, called a reciprocal centimeter, or inverse centimeter.

Western boundary current Strong ocean current that runs along the western edge of an ocean basin as a result of the much slower eastward group velocity of Rossby waves (planetary waves) relative to the westward group velocity. The Gulf Stream is a well-known example.

Window A spectral region in a planetary atmosphere that is relatively transparent between two regions that have higher opacity. A window region can be important for remote sensing of a planetary surface and for limiting the extent of a greenhouse effect.

Wrinkle ridge A low, sinuous ridge formed on lava plains that can extend for up to several hundred kilometers. Wrinkle ridges are tectonic features thought to have formed when basaltic lava cooled and contracted. They commonly occur overlying ring structures in lava-filled impact basins.

X-Band X-band is the part of the microwave band of the electromagnetic spectrum that ranges from 5.2 to 10.9 GHz (one fines sometimes up to 12 GHz). The corresponding range of wavelengths is 2.4 to 3.75 cm. In space missions the reserved frequency band is around 7.2 GHz (7.1 to 7.25 GHz) for the uplink (Earth to Space) and around 8.4 GHz (8.4 to 8.5 GHz) for the downlink (Space to Earth).

X-ray spectroscopy In planetary science, the estimation of abundances of elements from their emissions of X-rays. The rays come from the fluorescence of atomic nuclei in low-energy X-rays, when the nuclei have been stimulated by high-energy X-rays such as from the Sun.

Yardang A streamlined hill carved from rock through wind abrasion.

Yarkovsky effect A nongravitational force that arises from the asymmetric thermal reradiation of incident sunlight on the surface of a rotating body, that can lead to significant orbital evolution of kilometer-sized and smaller objects.

Yield stress The yield stress is the stress beyond which a material begins to deform plastically instead of elastically. An elastic deformation will be completely reversed after the stress is removed. A plastic deformation is (in part) irreversible.

Zodiacal cloud The cloud of interplanetary dust in the solar system, lying close to the ecliptic plane. The dust in the zodiacal cloud comes from both comets and asteroids.

Zodiacal light Diffuse glow seen on the Earth in the west after twilight and in the east before dawn, that appears wedge-shaped and lies along the ecliptic. It is widest near the horizon and is caused by the reflection of sunlight from the myriads of interplanetary dust particles concentrated in the ecliptic plane.

Index

Note: Page Numbers followed by f indicate figures; t, tables.

A

Aberration angle, 697
Absolute magnitude, 930
Accretion, 186, 1185−1186
ACE. *See* Advanced Composition Explorer
Achondrites, 34−37
ACRs. *See* Anomalous cosmic rays
Active optics, 1093−1096
Adams, John Couch, 862−863, 1008, 1010−1011
Adams ring, 899f, 897f
Adaptive optics (AO), 1089−1090
   advances for ground-based telescopes, 1099−1100, 1100f−1102f
   system schematic, 1098f
Adiabatic gradient, 189−190
Adiabatic lapse rate, 1110
Adrastea, 761t, 770, 770f
   relative size of, 760f
Advanced Composition Explorer (ACE), Sun exploration by, 1210−1211
Advanced Spaceborne Thermal Emission and Reflection (ASTER), 446−447
Advected mantle material, 1185−1186
Advected material, 1193−1194
Aegaeon, 761t
   relative size of, 760f
Aerogel, 665
Aerosols
   giant planets
      atmosphere, 728−733, 729f−730f, 730t, 731f−733f
      meridional circulation, 731
      thermochemical equilibrium theory, 732
   Neptune and Uranus, stratospheric, 731, 731f
Akari spacecraft, 1075f, 1077t
Albedo, 437−438, 909, 932
   asteroid, 1084f
   bolometric, 854
   insight into composition from, 1083−1084
   Jupiter-family comet, 1083
   Moon, 1025
   NEOs, 614−616, 615f−616f, 618f
   radar, 1141−1143, 1142f
   Saturnian satellites, 1145t
   Trans-Neptunian objects, 1086
Alfvén point, 270−271
Alfvén speed, 138−139
Alfvén waves, 247

Aliasing, 1150
ALMA. *See* Atacama Large Millimeter Array
Alpha Capricornids, 661t
Alpha particle spectrometer, Apollo science experiments with, 561t
Alpha Particle X-ray Spectrometer (APXS), 399t
Alpha particles, 261
Alpha Proton X-ray Spectrometer (APXS), 399t
ALSEP. *See* Apollo Lunar Surface Experiments Package
Amalthea, 761t, 770, 770f
   relative size of, 760f
Amino acids, 210, 210f−211f, 218t
Ammonia
   Ganymede and Callisto with, 818
   giant planet atmospheres with, 726t, 727
Amorphous interstellar silicates, 663
Ananke, 761t
   relative size of, 760f
Angular momentum, 674−675, 883, 925−926
Anomalistic month, 999
Anomalous cosmic rays (ACRs), 139−140
Anomaly, 175
Anthe, 761t
   relative size of, 760f
AO. *See* Adaptive optics
Aphelion, 68, 1017
Aphelion distances, 675, 709
Apoapse, 855, 892
Apogee, 1000
Apollo Lunar Surface Experiments Package (ALSEP), 541−542
Apollo program, 518−533, 559−566
   Apollo 8, 1214
   Apollo 10, 1215
   Apollo 11, 518f−519f, 519−520, 983f, 1215
   Apollo 12, 520−522, 520f−521f, 1215, 1215f
   Apollo 13, 1215
   Apollo 14, 522−525, 522f−523f, 1215
   Apollo 15, 524f, 525, 1215
   Apollo 16, 525−526, 525f−526f, 1215
   Apollo 17, 526−527, 527f−528f, 1215
   benefits of human exploration, 565−566
   CSM of, 559, 560f, 561t
   landing sites of, 505f, 559f
   location of, 504f
   LRV of, 559, 559f

   lunar exploration by, 557t
   orbital remote sensing, 562−563
   samples and analysis, 559−560, 563f, 564t
   surface experiments, 560−562, 561t, 565f
April Lyrids, 661t
Apse, 708
APXS. *See* Alpha Particle X-ray Spectrometer; Alpha Proton X-ray Spectrometer
Aqua, Earth exploration by, 1213
Aquarids, 661t
Arago ring, 897f
Arc second, 910
Arend-Roland comet, 695, 696f
Ares Vallis, 398t, 402f
Ariel, 761t, 774−775, 774f−775f
   relative size of, 760f
ARTEMIS, Moon exploration by, 1216
Aspect ratios, 883
ASP/SRP. *See* Soil Mechanics experiment
ASTER. *See* Advanced Spaceborne Thermal Emission and Reflection
Asteroid belt, 48−49
Asteroid families, 663
Asteroidal dust, 663−664
Asteroids, 4, 103, 583−602, 983
   astrobiological potential for, 228
   classes, 595−599, 596t, 597f
      A-class, 596−597
      B-class, 596
      C-class, 596, 598
      D-class, 596
      E-class, 597
      F-class, 596
      G-class, 596
      M-class, 597
      P-class, 596
      Q-class, 598
      R-class, 596−597
      S-class, 597−598
      V-class, 596−597
   defined, 583−585, 584f
   density of, 586−587, 588f
   discoveries, 585
   geometric albedo of, 1084f
   imaged by spacecraft, 584f
   inclination of, 1085f
   locations, 587−593, 588f
   near-Earth, 14f

Asteroids (*Continued*)
  19th century studies on, 1011
  numbering and naming conventions, 622—623
  numbers, and names of, 585
  orbits, 587—593
    asteroid families, 591—592
    evolution of, 68—70, 68f—70f, 591
    meteorites and, 592—593, 592f
    special classes, 590—591
    zones, orbits, and distributions, 587—590, 588f—589f
  physical characteristics and composition, 593—599
    asteroid satellites, 594—595, 594f
    infrared views, 1083—1085, 1084f—1085f
    surfaces, 593—594, 593f—594f
    taxonomy, and distribution of classes, 595—599, 596t, 597f
    telescopic observations of, 595
  porosity of, 586—587, 588f
  puzzles and promise with, 599—601
    dinosaur extinction, 599
    origins of asteroids, 599—600
    spacecraft missions to asteroids, 600—601, 601t
    telescopic searches and exploration, 599
  radar imaging for, 1153—1155
    imaging and shape reconstruction, 1153, 1153f—1154f
    orbit determination, 1153—1155, 1155f
    Yarkovsky effect, 1153—1155, 1155f
  radio wavelengths thermal emission from, 1118—1119
  rotation rates for, 586—587, 588f
  sizes and shapes of, 585—586, 586f
  solar system studies on first, 1007
  solar winds interactions, 142—144
  solid surfaces observations with UV for, 1068—1069
  20th century studies, 1017
  volcanic features of differentiated, 108
  X-ray emissions with, 1039—1040
Asthenosphere, 470—471
Astrobiology, 209—232
  ecological requirements for, 212t
  ecology of, liquid water, 211—212, 212t—213t
  history of life on earth and, 211f, 214—217, 215f—216f
  life defined for, 210—214, 210f—211f
  life in solar system, 219—228
    asteroids, 228
    comets, 228
    Enceladus, 228
    Europa, 227
    giant planets, 226—228, 226f
    Mars, 220—226, 221f, 221t, 223f, 225f
    Mercury and Moon, 220
    Titan, 227—228
    Venus, 220
  limits to life with, 219, 219t
  Mars in, 220—226
    early history of, 222—224, 223f, 225f
    meteorites from, 226
    subsurface life on, 224—226, 225f
    Viking landers' results for, 220—222, 221f, 221t
  microbial life and, 216
  molecules of life and, 210f
  origin of life, 217—219, 217f, 218t
  other stars in, 229—230, 229t
  properties of life for, 214t
  search for life on Mars, Europa, or Enceladus, 228—229
Astrometry, 858
  extrasolar planets detected with, 958—959
Astronomical units (AU), 5, 261, 264—265, 909, 1007
Astrophysical disks, 883
Atacama Large Millimeter Array (ALMA), 1108f
Atlas, 761t, 773, 774f
  relative size of, 760f
Atmosphere
  Earth, 423—444
    atmosphere-ocean system, 441—444
    circulation, 428—433
    climate, 436—441
    vertical structure of, 425—428, 426f
  Moon (Earth), 537—538
  Venus, 305—322
    circulation and dynamics of, 316—319
    clouds and hazes in, 314—316
    composition of, 307t, 311—314, 311f
    evolution of atmosphere and climate, 319—322
    observations, 306—308
    temperatures, 308—311, 309f
Atmospheric seeing, 1089—1090
Attenuating mantle, 1190
AU. *See* Astronomical units
Aura, Earth exploration by, 1213
Aurigids, 661t
Auroral electrons, 1120
Azimuthal clumps and kinks, 900—901, 900f—901f

B

Babylons, 999—1001, 1000f—1001f
Ballerina skirt model, 265, 265f
Barnard, Edward, 1015—1016
Baroclinic instability, 429
Barycenter, 911, 958
Basalts, 101, 457, 471—472
Belinda, 761t
  relative size of, 760f
Beta Pictoris, 8, 9f
Bethe, Hans, 238, 1012
Bianca, 761t
  relative size of, 760f
Biased sample, 709—710
Biermann, Ludwig, 1013
Bigg, E. K., 781
Binder, A. P., 781
Biologically available nitrogen, 856, 859
Bistatic radar, Apollo science experiments with, 561t

Blackbody radiation, 1109
BLENDER algorithm, 974
Bode, J. E., 1007
Bode's law, 7, 8t
Body waves, 1186
Bolometric albedo, 854
Bond, G. P., 1010
Bond ringlet, 896f
Bonneville crater, 406f
Born, Ignaz Edler von, 625—626
Borrelly comet, 698f
Bottke, William, 610
Bouguer gravity anomaly map, 543f
Bound orbits, 667—669
Breccias, 638
Bremsstrahlung, 254, 1020—1021
Brittle deformation, 193
Brown, Michael, 941—942, 945

C

C1XS. *See* Chandrayaan-1 X-ray Spectrometer
Calcium-aluminum-rich inclusions (CAIs), 34—37, 35f, 665, 666f
Caldera, 101—102
Caliban, 761t
  relative size of, 760f
Callisto, 813—830
  albedo, 1061t, 1145t
  ammonia on, 818
  density of, 540t, 815
  exploration of, 813—816
    astronomical observations, 813—815, 814f
    discovery, 813
    spacecraft exploration, 815—816
  formation and evolution of, 819—820
  future exploration of, 828—829
  Galileo spacecraft image of, 814f
  impact craters, 823—825
    distribution, 824—825
    structure, 823—824, 824f
    surface ages, 824—825
  interior structures of, 816, 817f
  internal oceans, 816—818, 817f
  magnetic fields on, 818—819
    induction fields and oceans, 819
    intrinsic fields, 818—819
  magnetospheres interactions with, 154—156, 154t
  mass of, 815
  NIMS of, 814—815, 820, 821f
  planetary dynamos with, 134
  properties of, 761t
  radius of, 540t
  relative size of, 760f
  satellite-to-parent mass ratio of, 540t
  spacecraft exploration of, 815—816
    Cassini, 815
    Galileo, 815—816
    New Horizons, 815
    Pioneer 10 and 11, 815
    Ulysses, 815
    Voyager 1 and 2, 815

Index

surface materials of, 820–823
  atmosphere interactions, 821–822
  composition, 820–821, 821f
  regolith, 822–823, 822f
tectonism and volcanism, 825–828
  bright terrain, 825–827, 826f–827f
  dark terrain, 827–828, 828f
  grooved terrain, 826–827, 827f
Caloris Basin, 284
Calypso, 761t, 774f
  relative size of, 760f
Campbell, W. W., 1008
Canada France Hawaii Telescope (CFHT), 833
Canyons, Mars, 367–368, 367f
Carbon, giant planet atmospheres with, 726–727, 726t
Carbon dioxide
  Mars atmosphere with, 352–353
  Venus atmosphere with, 311
Carbon monoxide, Venus atmosphere with, 313
Carbonaceous chondrites, 34–37, 638–639, 639f
  composition of, 638–639
  organic constituents, 639
Carbonaceous material, 657
Carme, 761t
  relative size of, 760f
Carrington, Richard, 1007
Cassini, Giovanni Domenico, 1004
Cassini Composite Infrared Spectrometer (CIRS), 856
Cassini spacecraft
  exploration by, 983f
  Ganymede and Callisto exploration by, 815
  interplanetary dust detection by, 667, 668t, 669
  magnetic field highlights from, 122t
  Titan surface viewed by, 843–845, 844f–845f
  Venus space missions by, 307–308
Cassini-Huygens spacecraft
  flyby missions to satellites by, 764t
  Titan, 833–834, 834f
Catalytic cycle, 311–312
CCD. See Charge-coupled device
Centaur, 712
  albedo, 942
  databases of known, 943, 943t
  dynamical classes, 943–944
  light curves, 946–947
Centaur objects, 942, 949, 950f
  discovery of, 942, 943f
Ceres, 12–14
CFHT. See Canada France Hawaii Telescope
CGRO. See Compton Gamma-Ray Observatory
Champ, magnetic field highlights from, 122t
Chandler, Seth, 1008–1009
Chandrayaan space missions
  Chandrayaan-1, 568–569, 569f
  lunar exploration by, 557t
  Moon exploration by, 1216
  Moon space missions by, 493–494

Chandrayaan-1 X-ray Spectrometer (C1XS), 569
Chang'e space missions
  Chang'e 1, 569
    lunar exploration by, 557t
    Moon exploration by, 1216
    Moon space missions by, 493–494
  Chang'e 2, 569
    lunar exploration by, 557t
    Moon exploration by, 1216
  Chang'e 3 Lunar Lander and Rover, 572
Chaos, Europa landforms and, 801–806, 803f–805f
Chaotic motion, 63–68
  concept of, 63
  three-body problem as paradigm for, 63–68, 63f
Chaotic rotation, 76–78
  Hyperion, 77, 77f–78f
  other satellites, 77–78
  spin-orbit resonance, 76–77, 76f
Chapman, S., 262
Charge-coupled device (CCD), 760, 1098
Charged particle lunar environment experiment (CPLEE), Apollo science experiments with, 561t
"Charming" ringlet, 897f
Charon, 13, 13f
  barycenter, 911
  density of, 913–914
  diameter of, 942f
  discovery of, 910–911, 910f
  internal structure of, 918f
  magnitudes, albedos, and diameters of, 944t
  orbit of, 912–913, 913t
  Pluto comparison, 920–921, 921f, 921t
  radii of, 913–914
  size comparison to United States, 942f
  solid surfaces observations with UV for, 1068
  spectra of, 921f
  system mass of, 912–913
Chasmata, Venus, 339, 339f
ChemCam, 399t
ChemMin, 399t
Chiron-type comets, 710f, 712
Chondrites, 192–193, 635–641
  breccias, 638
  carbonaceous chondrites, 34–37, 638–639, 639f
    composition of, 638–639
    organic constituents, 639
  petrographic properties, 635–638, 636f, 637t, 638f
  shock, 639–640, 640t
  weathering, 641, 641f
Chondrules, 34–37, 34f–35f, 635–641, 647–648
  carbonaceous, 34–37
  components of, 647–649
    Ca- and Al-Rich Inclusions, 648
    interstellar grains, 648–649, 649f
  enstatite, 34–37
  ordinary, 34–37

Chromosphere, 241–243
  chromospheric dynamic phenomena, 242–243, 243f
  materials, giant planets atmospheres with, 730–731, 730t
  physical properties, 241–242, 242f
Chromospheric dynamic phenomena, 242–243, 243f
Chryse Planitia, 398t, 402f
Circular restricted three-body problem, 710–711
Circular velocity, 58
Circumplanetary disks, 52
CIRS. See Cassini Composite Infrared Spectrometer
Cis-lunar space, 991–992
Clementine, 566, 566f
  lunar exploration by, 557t, 1216
Clouds
  giant planets, 728–733, 729f–730f, 730t, 731f–733f
  molecular, 31–32, 31f
  Titan, 840–841, 840f
  Venus atmosphere, 314–316
    appearance and motions, 314
    cloud chemistry, 316
    global variability, 315, 315f
    lightning, 316
    vertical layering, 314–315, 315f
CMB. See Core-mantle boundary
CMEs. See Coronal mass ejections
COBE. See Cosmic Background Explorer
Colatitude, 165
Cold cathode gage experiment, Apollo science experiments with, 561t
Collisions
  cosmic dust dynamics and, 677, 677f
  Coulomb, 147–148
  hazards of NEOs
    collision frequency, 621
    collision magnitude, 620–621
  marginally collisional plasma, 273–274
Column abundance, 1164
Coma, 665, 693–695, 693f–694f, 694t
Comet dust, 662t, 665–667, 666f
Comet trails, 660
Cometary activity, 660
Comets, 4, 15, 103, 983
  apse node of, 708
  Arend-Roland, 695, 696f
  astrobiological potential for, 228
  Borrelly, 698f
  chemistry of, 683–704, 700f
  Chiron-type, 710f, 712
  coma of, 665, 693–695, 693f–694f, 694t
  cometary classification flowchart, 710f
  "dirty-snowball" model of, 1082f
  dust-to-gas mass ratios for, 1082f
  Encke, 1082f
  external, 710f, 712
  formation and ultimate fate of, 700–702, 702f
  Giacobini-Zinner, 693f
  Hale-Bopp, 684f, 693–695, 694f, 696f–697f
  Halley, 685t, 686–687, 686f, 691f, 700f

Comets (*Continued*)
  Halley-type, 688–689, 710f, 712
  history of studies on, 685–689, 689f
  Hyakutake, 698, 699f
  hydrogen cloud of, 693–695, 693f–694f, 694t
  infrared views of, 1079–1083, 1080f–1083f
  Jupiter-family, 710f, 712–713
  layered structure of nucleus of, 690, 690f
  LINEAR, 690–691, 692f
  long-period, 688–689
  McNaught, 696, 697f
  NEO relationship and Tisserand parameter, 609–610, 609f
  19th century studies on, 1011
  node of, 708
  orbits
    chaotic, 706–707
    dynamics of, 706–709, 706f, 709f
    ecliptic comets, 711–712, 712f
    evolution of, 70–71
    Kozai resonance in, 708, 708f
    nearly isotropic comets, 711–712, 711f
    orbital distribution, 712–713
    regular, 706–707
    retrograde, 708
    taxonomy of, 709–713, 710f–711f
  orphan trails of, 1082, 1083f
  periodic, 709
  physics of, 683–704
    nucleus, 689–693, 690f–692f
  populations and dynamics of, 705–720
  radio wavelengths thermal emission from, 1119–1120, 1120f–1121f
  reservoirs, 713–719
    Oort cloud, 713–714, 714f
    Oort cloud and scattered disk formation, 717–719, 717f–718f, 719f
    scattered disk, 713–717, 715f–716f
  rotation of nuclei of, 690, 691f
  Shoemaker-Levy, 690–691, 692f
  solar winds interactions, 142–144
  solid surfaces observations with UV for, 1068–1069
  space missions to, 683–685, 684f, 685t, 686f, 688f
    Deep Impact mission, 684–685, 685t, 687f
    Deep Space 1, 684, 685t
    Giotto spacecraft, 683, 685t
    Stardust Wild 2 mission, 684, 685t
    Stardust-NExT, 684–685, 685t
    VEGA 1 and 2, 683, 685t
  tails of, 695–698, 696f–699f
  Tempel 1, 685t, 687f–688f
  20th century studies, 1017
  ultraviolet spectra of, 698–699, 700f
  West, 1080f
  X-ray emissions with, 1034–1039, 1039f
    energy spectrum, 1038
    spatial morphology, 1035–1037, 1036f–1037f
    temporal variation, 1037–1038
    X-ray luminosity, 1037

Command/Service Module (CSM), 559, 560f, 561t
Compositional convection, 127
Compton Gamma-Ray Observatory (CGRO), 254
Continents, 189–190
Contrast reversal, 885
Convection, 101, 185, 239–240, 240f
Convective flows, 127, 128f
Coordinated Compact Reconnaissance Imaging Spectrometer for Mars (CRISM), 97–98
Copernicus, solar system studies with, 1001, 1002f
Cordelia, 761t, 775–776
  relative size of, 760f
Core, 162–163, 185
  Earth
    inner core, 473
    outer core, 472
  Mars, 380f, 382–383
  Moon (Earth), 551–552, 552f
Core accretion, 49
Core-mantle boundary (CMB), 162–163, 463
Core-mantle boundary heat flow, 202–204, 203f–204f
Core-mantle differentiation, 186
Coriolis acceleration, 431
Coriolis force, 428–429
Corona, 236, 243–248, 1012
  active regions, 243, 243f
  coronal heating, 248, 249f
  coronal holes, 244
  coronal magnetic field, 245–246, 246f
  dynamics of, 244–245
  hydrostatics of coronal loops, 244, 244f
  MHD oscillations of coronal loops, 246–247, 247f
  MHD waves in, 247–248, 247f–248f
  quiet-sun regions, 243–244
  transition region of, 245
Coronae, 106
  Venus, 339–340, 339f
Coronal heating, 248, 249f
Coronal holes, 244
Coronal loops
  hydrostatics of, 244, 244f
  MHD oscillations of, 246–247, 247f
Coronal magnetic field, 245–246, 246f
Coronal mass ejections (CMEs), 239, 248–258, 258f, 259f, 479, 480f, 483–484, 483f. *See also* Interplanetary coronal mass ejections
  transient solar wind and, 268–271, 268f
    characteristics of ICMEs, 270, 270t
    field line draping about fast ICMEs, 271
    heliospheric disturbances driven by fast, 269, 269f–270f
    magnetic field topology of ICMEs, 270–271, 271f
    magnetic flux balance problem, 270–271, 271f

other forms of solar activity, 268–269
Cosmic Background Explorer (COBE), 663, 1075f, 1077t
Cosmic dust, 657–682
  dynamics and evolution of, 675–678
    collisions, 677, 677f
    dust in interplanetary space, 678, 679f, 679t–680t
    gravity and Keplerian orbits, 675–676
    interaction with magnetic field, 677–678, 677f
    radiation pressure, 676, 676f
  future studies on, 678–682
  infrared views of, 1074–1078, 1076f–1077f, 1077t, 1078f–1079f
  manifestations of, 659–675, 659f
    comet dust, 662t, 665–667, 666f
    extrasolar debris disks, 674–675
    interplanetary dust particles, 660–663, 662t
    lunar microcraters, 664–665, 664f
    meteors, 660, 661t, 662f
    zodiacal light, 663–664
  spacecraft measurements of, 667–674, 668t, 670f
    interplanetary dust, 667–669, 670f
    interstellar dust in heliosphere, 673–674, 674f–675f
    planetary dust streams, 669–670, 670f–671f
    tenuous dusty planetary rings, 670–673, 672f
Cosmic ray detector experiment (CRD/LSCRE), Apollo science experiments with, 561t
Cosmic rays
  anomalous, 139–140
  galactic, 1162
Cosmic-ray exposure (CRE), meteorites chronometry with, 650–652, 651f–652f
Cosmochemical fractionations, meteorites with, 644–645, 645f
Cosmochemistry, 30
Coulomb collisions, 147–148
Coulomb interactions, 273–274
CPLEE. *See* Charged particle lunar environment experiment
Crater rays, 293–294
Crater size-frequency distribution measurements (CSFDs), 515–516, 516f
Craters
  Bonneville, 406f
  complex, 84, 85f
  dimensions, 88–89, 88t
  Eagle, 406f
  Endurance, 375f
  formation of, 89–92, 90f
  Gale, 376f, 398t, 402–404, 404f
  Gusev, 398t, 401–402, 403f, 405–406, 406f, 412–413, 413f
  impact, 83–89
  Jackson, moon, 85f

Jezero, 350f
lunar microcraters, 664–665, 664f
Mars geology, 405f–406f, 412, 412f
planetary evolution with, 94–97
  biosphere evolution, 95–97
  early crustal evolution, 95
  impact origin of earth's moon, 94–95
as planetary probes, 97–99
  morphologic and geologic, 99
  spectral composition, 98–99
  water and ices, 97–98, 98f
ring basins of, 84, 86f
shape, 83–88, 84f–87f
simple, 83–84
Triton geology with, 872f
Winslow, Mars, 84f
CRD/LSCRE. See Cosmic ray detector experiment
CRE. See Cosmic-ray exposure
Cressida, 761t
  relative size of, 760f
Cretaceous-Tertiary Event, 599, 603
CRISM. See Coordinated Compact Reconnaissance Imaging Spectrometer for Mars
Critical core mass, 50
Crust, 101
  Earth, 469–470
  Mars, 380f, 384–385, 384f
  Moon (Earth), 550–551, 551f
Crust formation, 199–201, 200f
  Mars, 395–396
Crustal plateaus, Venus, 338–339, 338f
Crusts, 185
Cryovolcanic designation, 868
Cryo-volcanism, 107–108, 985–986
Crystallization age, meteorites chronometry with, 652–654, 653t, 654f
CSFDs. See Crater size-frequency distribution measurements
CSM. See Command/Service Module
Cupid, 761t
  relative size of, 760f
Curie temperatures, 1192
Curiosity, Mars exploration by, 1219. See also MSL curiosity
Currents, 1164
Cyclotron maser emissions, 1121–1123

D

DAN. See Dynamic Albedo of Neutrons experiment
Daphnis, 761t
  relative size of, 760f
Darwin, George, 1008
DAWN spacecraft, asteroids mission, 600, 601t
Debris disks, 31, 670f
Deep Impact, space missions to comet by, 684–685, 685t, 687f
Deep Space 1, space missions to comet by, 684, 685t
Degenerate, 1014

Deimos
  Mars Express flyby missions to, 764t
  properties of, 761t
  relative size of, 760f
Delaunay, Charles, 1008
Delisle, Joseph, 1006
Dellinger, Howard, 1013
Demos, satellites of, 760f, 769–770, 769f
Dense narrow rings, 896–897
Deoxyribonucleic acid. See DNA
Depletion, 313
DEs. See Disconnection events
Descent Imager Spectral Radiometer (DISR), 837
Desdemona, 761t
  relative size of, 760f
Design reference missions, Mars, 992, 993f
Despina, 761t, 776
  relative size of, 760f
Deterministic chaos, 56
Deuterated hydrogen, 726
Deuterium, 726
Deuterium to hydrogen (D/H) ratio, Venus atmosphere with, 312
D/H ratio. See Deuterium to hydrogen ratio
Diapirism, 871
Dielectric materials, 1141
Differentiated asteroids, volcanic features, 108
Differentiation, 102, 186, 1185–1186
Diffraction, 884–885, 1096
Diffraction grating, 892
Diffraction limit, 1089–1090
Diffraction-limited imaging, 1099
Digital terrain models (DTMs), 1223
Dihedral scattering, 1143–1144
Dione, 761t, 770–772, 771f–772f
  albedo, 1064t, 1145t
  relative size of, 760f
Direct imaging, extrasolar planets detected with, 962–963, 962f
Direct spin, 1016
Disconnection events (DEs), 697
Disk instability, 49
Disks, planetary, dense broad, 890–896, 890f–891f
  structures formed by distant moons, 895–896
    confinement of edges and ringlets, 895–896
    spiral waves, 895–896, 896f
  structures formed by nearby moons, 892–895
    gap edges and moonlet wakes, 892, 893f
    propellers, 892–895, 894f–895f
  structures formed spontaneously, 891–892
    overstabilities, 892
    self-gravity wakes, 891–892, 891f–892f
DISR. See Descent Imager Spectral Radiometer
Dissipative forces, 72–75
  gas drag, 73–74
  Poynting-Robertson drag, 73
  radiation force, 72–73
  tidal evolution and resonance, 75
  tidal interactions, 74–75
  Yarkovsky effect, 73

Diurnal climatic changes, 1193
DNA (Deoxyribonucleic acid), 211
Dome seeing, 1092–1093
Donati, Giovanni, 1011
Doppler shift, 1016, 1136
DTM interpolation, 1227
DTMs. See Digital terrain models
DTREM. See Dust detector experiment
Dust bands, 1076–1078
Dust detector experiment (DTREM), Apollo science experiments with, 561t
Dust trails, 1081, 1086
Dusty rings, 897–901, 897f
  azimuthal clumps and kinks, 900–901, 900f–901f
  dust production and destruction, 897–898
  nongravitational forces, 898
  resonantly confined material, 898–900, 899f
Dwarf planets, 3, 5
  orbits of, 6t
  physical parameters for, 10t
Dynamic Albedo of Neutrons (DAN) experiment, 1174
Dynamical friction, 43–44
Dynamo, 125–126, 472. See also Planetary dynamo
Dynamo action, 185
Dynamo region geometry, 129

E

Eagle crater, 406f
Earth
  accretion of, 48f
  bulk composition of, 32f
  characteristics, overview, 424–425
  day length, 424–425
  density of, 380f
  El Niño Southern Oscillation, 425
  EM sounding for investigating, 1199–1200, 1200f
  exploration missions, 1212–1213
    Aqua, 1213
    Aura, 1213
    Envisat, 1213
    Galileo Earth Flybys, 1213
    GRACE, 1213
    Resurs, 1213
    Terra, 1213
    TOPEX/Poseidon and Jason-1 and Jason-2, 1213
  heat flow measurements for, 1193–1194, 1194f
  historical events of, 215f
  magnetic field highlights from missions to, 122t
  magnetospheres with, 141t
  19th century studies on, 1008–1009
  nonthermal radio emissions of, 1123–1124
  orbit ring of dust, infrared view, 1078–1079, 1080f
  planetary dynamos with, 131–133, 132f
  as Rosetta stone, 477
  rotation unique to, 183–184

Earth (*Continued*)
 seismology, 1186—1187, 1187f
 solar wind properties with, 141t
 Super-Earths, 958, 966—967
 temperature structure for, 426f
 20th century studies on, 1014
 volcanic features of, 101—103, 102f
 Western boundary currents, 424
 X-ray emissions, 1020—1023, 1020f
  auroral emissions, 1020—1022, 1021f—1022f
  geocoronal emissions, 1022—1023
  nonauroral emissions, 1022, 1023f
Earth, atmosphere, 423—444
 circulation, 428—433
  global-scale, observed, 429—432, 430f—431f
  Hadley cell, 429—430
  insights from other atmospheres, 432—433
  processes, 428
  rotation influences, 428—429
 climate, 436—441
  feedbacks, 437—438
  greenhouse effect, 437
  ice ages, 439—441, 440f
  recent times, 438—439, 438f—439f
  volatile inventories, 441
 vertical structure of, 425—428, 426f
  exosphere, 427—428
  ionosphere, 427—428
  mesosphere, 427
  stratosphere, 425—427
  thermosphere, 427
  troposphere, 425—426
Earth, atmosphere-ocean system, 441—444
 interplanetary evidence for life, 441—444, 442f
  oxygen and methane, 443—444
  radio emissions, 442—443
  surface features, 443
Earth, oceans, 433—436
 atmosphere-ocean interactions, 436
 circulation, 434—435
 oceans on other worlds, 436
 salinity, 435—436
 structure, 433—434
Earth, surface and interior, 445—478
 landform types, 450—456
  subaerial landforms, 453—456, 454f, 455t
  submarine landforms, 450—453, 451f
 other planets guided by, 445—447, 446f
 physiographic provinces of, 447—456
  basic divisions, 447—450, 448f—449f
  landform types, 450—456
  terrestrial geomorphological features, 455t
  terrestrial vs planetary landscapes, 456
 processes, 456—462, 457f
  constructive, landscape, 456—457, 458f—459f
  destructive, geomorphic, 457—462, 460f—461f
 radial structure, 469—473, 470f
  crust, 469—470
  inner core, 473
  lower mantle, 471—472
  outer core, 472
  transition zone, 471
  upper mantle, 470—471
 seismic sources, 466—469, 467f—468f
 in three dimensions, 473—477, 474f—477f
 tools for study, 462—466, 462f, 464f—466f
  global seismic tomography, 473, 474f
  three-component seismograph system, 463, 463f
Earth Observing System (EOS), 446
Earthquake, seismic sources, 466—469, 467f—468f
Eccentricity, 6, 58, 159—160, 858, 896, 926—927, 943—944
Eclipses, lunar, 495—496
Ecliptic, 5, 56—57, 943—944
Ecliptic comets, 711—712, 712f, 933—934
 surface number density, 934f
Ecliptic plane, 663, 1076
Eddy-driven jet, 431
Edgeworth, Kenneth, 926
Edlén, Bengt, 1012—1013
E-ELT. *See* European Extremely Large Telescope
Einstein's general theory of relativity, 55—56
Ejecta, 284, 888
El Niño Southern Oscillation, 425
Elara, 761t
 relative size of, 760f
Elliptical motion, 56—57, 57f
Elliptical orbits, 58
Elongation, 1001
EM sounding
 future exploration with, 1202—1203, 1202f, 1203f
 Galilean satellites, 1201—1202, 1202f
 investigating planetary interiors using, 1198—1203
 measurement techniques, 1199, 1199t
 Moon, 1200—1201, 1201f
 natural sources, 1198
 terrestrial exploration, 1199—1200, 1200f
Enceladus, 851—860
 albedo, 1064t, 1145t
 astrobiological potential for, 228
 density of, 852t
 evolution of, 858—859
 future analysis of data for, 859
 gravity, 852, 853f
 history, 851—852, 851f, 852t
 mass of, 852t
 properties of, 761t
 radius of, 852t
 relative size of, 760f
 shape, 852, 853f
 solid surfaces observations with UV for, 1065—1066
 south polar region of, 854—856
  jets, 854—856, 855f
  regional geology, 854, 854f
  thermal signature, 855f, 856
  vapor component of plume, 856
 structure of, 856—858, 857f
 surface composition of, 852
 surface geology and tectonics of, 852—854, 853f
 topography, 852, 853f
Encke comets, 1082f
Encke Gap, 892, 893f
Encke Gap ringlets, 897f, 901f
Endurance crater, 375f
Energetic particles, 148—150, 149t
 solar wind and, 276—277
Energy, 58
Enstatite chondrules, 34—37
Envisat, Earth exploration by, 1213
Eolian processes, Mars, 418
EOS. *See* Earth Observing System; Equations of state model
Epicenter, 1190
Epimetheus, 761t, 773, 774f
 relative size of, 760f
EPOXI spacecraft, space missions to comet by, 685t
Epstein drag, 73—74
Equations of state (EOS) model, giant planets interiors in, 747—750, 748f
 helium, 749
 hydrogen, 748—749, 749f
 ices, 749
 mixtures, 749—750
 rock, 749
Equinox, 892, 903
 precession of, 1000
Eris, 13
 magnitudes, albedos, and diameters of, 944t
Erman, Adolf, 1011—1012
Erosion
 Mars, 368
 Moon, 512—513
ESA. *See* European Space Agency
Escape velocity, 58
ESO. *See* European Southern Observatory
Eta Aquarids, 661t
Europa, 793—812
 albedo, 1061t, 1145t
 astrobiological potential for, 227, 810
 atmospheres observations with UV for, 1058—1059, 1059f
 density of, 540t
 exploration history, 793—794, 794f
  Galileo spacecraft, 793—794, 794f
  Pioneer 10 and 11 spacecrafts, 793
 formational and compositional models for, 794—795
 future exploration for, 810—811
 internal structure of, 795, 795f
 Laplace resonance, 796f
 magnetospheres interactions with, 154—156, 154t
 properties of, 761t, 794t
 radius of, 540t
 relative size of, 760f
 rocky mantle of, 795, 795f
 satellite-to-parent mass ratio of, 540t

tectonic patterns, global for, 795–797
  diurnal stressing, 796–797, 797f
  nonsynchronous rotation, 797
tides of, 795–797
  tidal evolution, 795–796, 796f
  tidal heating, 796
Europa, landforms, 797–806
  lenticulae and chaos, 801–806
    chaos, 803–805, 803f–805f
    impact structures, 805–806, 805f–806f
    lenticulae, 801–803, 802f–803f
  ridges, troughs, bands, 797–801
    cycloidal structures, 798–799, 800f
    double ridges, 797f, 798, 799f
    folds, 800–801, 802f
    pull-apart bands, 799–800, 801f
    ridge complexes, 797f, 798
    triple bands, 799, 800f
    troughs, 797f, 798
  surface age and evolution, 808–810
    geological evolution, 809–810, 809f
    surface age estimated, 808–809
    surface history, 809–810, 809f
  surface composition, 806–807, 806f–807f
  surface physical processes, 807–808, 808f
  thermal state, 806–807, 806f–807f
European Extremely Large Telescope (E-ELT), 1091–1092
European Southern Observatory (ESO), 1091–1092
European Space Agency (ESA), 683
EUV. *See* Extreme ultraviolet
Evolution
  of asteroids orbits, 591
  cosmic dust, 675–678
    collisions, 677, 677f
    dust in interplanetary space, 678, 679f, 679t–680t
    gravity and Keplerian orbits, 675–676
    interaction with magnetic field, 677–678, 677f
    radiation pressure, 676, 676f
  Enceladus, 858–859
  Europa landforms
    geological evolution, 809–810, 809f
    surface age and, 808–810
    surface age estimated, 808–809
    surface history, 809–810, 809f
  Ganymede and Callisto formation and, 819–820
  lunar formation and, 551–552, 553f–554f
  of lunar magma, 499f
  Mars, 393–396, 417–419
    chemical evolution, 395–396, 417
    crust formation, 395–396
    early dynamo, 395
    eolian processes, 418
    extraction of water, 396
    geologic evolution of landing sites, 418–419
    implications for habitable world, 419
    origin of igneous rocks, 417
    surface water, 417

thermal evolution, 393–395, 394f
    weathering on mars, 417–418
  Moon, thermal, 502–503
  planetary satellites formation, 766–767
  tidal on Europa, 795–796, 796f
  Titan, 841
  Venus, atmosphere and climate, 319–322
    atmospheric gases sources, 319–320
    escape processes, 320–321
    evolutionary climate models, 321–322, 322f
    surface pressure and CO2 budget, 320
    volcanism, 319–320
    water loss, 320–321
Evolution of interior, 205–206, 206f–207f
Exoplanets. *See* Extrasolar planets
Exosphere, 305–306
  Earth, 427–428
  of mercury, 288–292, 289f–290f
  Mercury, 288–292, 289f–290f
Exploration missions, 1205–1222. *See also* Lunar exploration
  Earth, 1212–1213
    Aqua, 1213
    Aura, 1213
    Envisat, 1213
    Galileo Earth Flybys, 1213
    GRACE, 1213
    Resurs, 1213
    Terra, 1213
    TOPEX/Poseidon and Jason-1 and Jason-2, 1213
  Mars, 1217–1219
    Curiosity, 1219
    failures of 1990s, 1218
    Mangalyaan, 1217f, 1219
    Mariner 4, 360, 360t, 1217
    Mariner 6 and 7, 360, 360t, 1217
    Mariner 9, 360, 360t, 1218
    Mars 2 and 3, 1217
    Mars 2 spacecraft, 4-7, 360t
    Mars 4, 5, 6, and 7, 1218
    Mars Express, 345, 360t, 1218–1219
    Mars Observer, 360t
    Mars Odyssey, 1218
    Mars Pathfinder and Mars Global Surveyor, 1218
    Mars Reconnaissance Orbiter, 345, 351–352, 360t, 1219
    Opportunity and Opportunity rover, 345, 360t
    Phobos 1 and 2, 1218
    Phobos spacecraft, 360t
    Phoenix, 1219
    Spirit and Opportunity, 1218
    Spirit Rover, 345, 360t
    surface and interior, 359–361, 360t
    Viking 1 and 2, 1218
    Viking mission, 360t, 405, 405f
  Mercury, 1211
    Mariner 10, 283, 1211, 1211f
    MESSENGER spacecraft, 283–284, 285f, 286t, 1211

Moon, 1213–1217
    Apollo 8, 1214
    Apollo 10, 1215
    Apollo 11, 1215
    Apollo 12, 1215, 1215f
    Apollo 13, 1215
    Apollo 14, 1215
    Apollo 15, 16, and 17, 1215
    ARTEMIS, 1216
    Chandrayaan-1, 1216
    Chang'e 1, 1216
    Chang'e 2, 1216
    Clementine, 557t, 566, 566f, 1216
    GRAIL, 1216
    Kaguya/SELENE, 1216
    LADEE, 1216–1217
    LCROSS, 1216
    Luna 1, 2, and 3, 1214
    Luna 9 and 13, 1214, 1214f
    Luna 10, 11, 12, and 14, 1214
    Luna 16, 17, 20, 21, and 24, 1215, 1215f
    Lunar Orbiter 1-5, 1214
    Lunar Prospector, 1216
    Lunar Reconnaissance Orbiter, 1216
    Ranger 7, 8, and 9, 1214
    SMART-1, 1216
    Surveyor 1, 3, 5, 6, and 7, 1214
    Zond 3, 1214
    Zond 5, 6, 7, and 8, 1214
  outer planets and moons, 1221–1222
    Cassini-Huygens, 1220f–1221f, 1221
    Galileo, 1221
    JUNO, 1222
    New Horizons, 1221–1222
    Pioneers 10 and 11, 1221
    Voyagers 1 and 2, 1221
  program evolution, 1206–1209
    launch services, 1206–1207
    management, 1209
    operations, 1209
    reliability and quality assurance, 1209
    spacecraft, 1208–1209
    tracking and data acquisition, 1207–1208, 1207f–1208f
  small bodies, 1219–1220
    Dawn, 1219f–1220f, 1220
    Deep Impact, 1220
    Deep Space 1, 1220
    Galileo En Route Encounters, 1219
    Halley Armada, 1219
    Hayabusa (Muses-C), 1220
    International Cometary Explorer, 1219
    NEAR-Shoemaker, 1217f, 1220
    Rosetta, 1220
    Stardust ., 1220
  Sun and heliosphere, 1209–1211
    Advanced Composition Explorer (ACE), 1210–1211
    Genesis, 1211
    Helios, 1210
    SOHO, 1210
    Solar Dynamics Observatory, 1211

Exploration missions (*Continued*)
  Solar Maximum Mission (SMM), 1210, 1210f
  TRACE, 1211
  Ulysses, 1210
  Venus, 1211–1212
    Magellan, 1212, 1212f
    Mariner 2, 307–308, 1211
    Mariner 5, 307–308, 1211–1212
    Mariner 10, 307–308, 1212
    Pioneer Venus, 1212
    Vega 1 and 2, 1212, 1212f
    Vega spacecraft (1 and 2), 307–308
    Veneras 4 Through 16, 307–308
    Veneras 4 Through 16 and Vega, 1211
    Venus Express, 1212
    Venus Flyby, 1212
Explosion-dominated eruptions, 786f, 787–788
External comet, 710f, 712
Extinct radioactives, meteorites chronometry with, 654–655, 654t
Extrasolar debris disks, 674–675
Extrasolar planets, 52–53, 53f, 957–978, 986
  atmosphere, 972–973, 972f–973f
  atmospheres of, 741–742
  defined, 957
  detection techniques, 958–963
    astrometry, 958–959
    direct imaging, 962–963, 962f
    microlensing, 961
    radial velocity method, 959–960
    timing method, 961–962
    transit photometry, 960–961
  future outlook for, 976–977
  giant planets, interior modeling for, 755–756, 756f
  Kepler mission, 973–976, 974f–976f
  observations, 963–973
    hot Neptunes & super-Earths, 966–967
    planets around sun-like stars, 964–967, 964f, 965f
    pulsar planets, 963–964, 963t
  planetary dynamos with, 135
  planetary rings of, 890
  radial velocity planets, 964–967, 964f, 965t
  transit searches for, 967–971, 967f, 969f
    direct images of extrasolar planets, 970–971, 971f
    microlensing survey results, 969–970, 970f
Extreme ultraviolet (EUV), 479, 480f, 481, 1047
  giant planets with, 738

F

F ring, 897f
Far-ultraviolet (FUV), 1047
Fast rotating model (FRM), 615
Feeding zone, 44
Feldspathic Highlands Terrane (FHT), 503
Ferrel cells, 431
FHT. *See* Feldspathic Highlands Terrane

Field of view, 885
Filaments, 242–243, 249, 250f
Flare plasma dynamics, 252–253, 253f
Flares, 236
Flexural response, 852
Flow-dominated eruptions, 788, 788f
Flux, 1161
Flux transfer events (FTEs), 151–152
Forced nutation, 181
Fowler, Albert, 1017
Fractionation, 1185–1186
Fracture belts, Venus, 339, 339f
Fraunhofer lines, 289
Free core nutation, 181
FRM. *See* Fast rotating model
Fronts, 429
FTEs. *See* Flux transfer events
FUV. *See* Far-ultraviolet

G

Galactic cosmic radiation (GCR), 988
Galactic cosmic rays, 262, 1162
Galatea, 761t, 776
  relative size of, 760f
Gale crater, 376f, 398t, 402–404, 404f
Galilean satellites
  EM sounding for investigating, 1201–1202, 1202f
  magnetic field highlights from missions to, 122t
  radio wavelengths thermal emission from, 1117
  X-ray emissions, 1031–1032, 1032f
Galileo Earth Flybys, Earth exploration by, 1213
Galileo Galilei
  Moon drawings, 556, 556f
  solar system studies with, 1001–1003, 1003f
Galileo spacecraft, 983f
  asteroids flyby, 600, 601t
  flyby missions to satellites by, 764t
  Ganymede and Callisto exploration by, 815–816
  Ganymede and Callisto image from, 814f
  interplanetary dust detection by, 667, 668t, 669
  magnetic field highlights from, 122t
  planetary satellites observation with, 769
  Venus space missions by, 307–308
Gamma ray spectrometer, Apollo science experiments with, 561t
Gamma rays, 1162
  detection of, 1167–1171
    counting rate models, 1167–1168
    high-purity germanium detector for, 1168–1169, 1169f
    radiation spectrometers for, 1168
    scintillators for, 1170–1171
    semiconductor radiation detectors for, 1168
    spatial resolution, 1171
    techniques and interments for, 1168–1171, 1169f–1170f

emission, 254–255, 256f
origin of, 1162–1167, 1163f
  galactic cosmic rays, 1162–1163, 1164f
  gamma ray production and transport, 1166–1167, 1167f
  neutron moderation, 1163–1166, 1165f
spectroscopy, 300
Ganymede, 813–830
  albedo, 1061t, 1145t
  ammonia on, 818
  atmospheres observations with UV for, 1058–1059, 1060f
  density of, 540t, 815
  exploration of, 813–816
    astronomical observations, 813–815, 814f
    discovery, 813
    spacecraft exploration, 815–816
  formation and evolution of, 819–820
  future exploration of, 828–829
  Galileo spacecraft image of, 814f
  impact craters, 823–825
    distribution, 824–825
    structure, 823–824, 824f
    surface ages, 824–825
  interior structures of, 816, 817f
  internal oceans, 816–818, 817f
  Laplace resonance, 796f
  magnetic fields on, 818–819
    induction fields and oceans, 819
    intrinsic fields, 818–819
  magnetospheres interactions with, 154–156, 154t, 156f
  mass of, 815
  NIMS of, 814–815, 820, 821f
  nonthermal radio emissions of, 1128
  planetary dynamos with, 134
  properties of, 761t
  radius of, 540t
  relative size of, 760f
  satellite-to-parent mass ratio of, 540t
  spacecraft exploration of, 815–816
    Cassini, 815
    Galileo, 815–816
    New Horizons, 815
    Pioneer 10 and 11, 815
    Ulysses, 815
    Voyager 1 and 2, 815
  surface materials of, 820–823
    atmosphere interactions, 821–822
    composition, 820–821, 821f
    regolith, 822–823, 822f
  tectonism and volcanism, 825–828
    bright terrain, 825–827, 826f–827f
    Callisto, 828, 828f
    dark terrain, 827–828, 828f
    grooved terrain, 826–827, 827f
Gap edges, 892, 893f
Gas chromatograph, 221
Gas Chromatograph Mass Spectrometer (GCMS), 399t, 837
Gas drag, 73–74
Gas exchange (GEx) experiment, 221
Gas giant planets. *See* Giant planets

Gas retention age, meteorites chronometry with, 652, 653f
Gas-starved disks, 52
Gauss, Karl, 1007
Gauss' theorem, 1190–1191
GCMs. See Global Circulation models
GCMS. See Gas Chromatograph Mass Spectrometer
GCR. See Galactic cosmic radiation
GDS. See Geomagnetic depth sounding; Great Dark Spot
GEM. See Giotto Extended Mission
Geminids, 661t
Genesis, Sun exploration by, 1211
Geodynamo, 1186
Geologic units, 284
Geomagnetic activity, 150
Geomagnetic depth sounding (GDS), 1199, 1199t
Geophysical tools, 1185–1204
   EM sounding, 1198–1203
      future exploration with, 1202–1203, 1202f, 1203f
      Galilean satellites, 1201–1202, 1202f
      measurement techniques, 1199, 1199t
      Moon, 1200–1201, 1201f
      natural sources, 1198
      terrestrial exploration, 1199–1200, 1200f
   gravity and dynamics, 1190–1192, 1191f
   heat flow measurements, 1192–1198
      Earth's heat flow, 1193–1194, 1194f
      future investigations, 1196–1198, 1197f
      lunar heat flow, 1194–1196, 1195f–1196f
      measurement techniques, 1192–1193, 1193f, 1195f
   seismology, 1186–1190
      Jovian, 1190
      lunar, 1187–1190, 1188f–1189f
      Mars, 1190
      terrestrial, 1186–1187, 1187f
Geostrophy, 428–429, 434
GEx. See Gas exchange experiment
GI. See Gravitational instability
Giacobini-Zinner comet, 693f
Giant planets, 4, 12, 12f
   astrobiological potential, 226–228, 226f
   gas, 49–51
   Hubble Space Telescope view of, 733–734, 733f
   ice, 49–51
   radio wavelengths thermal emission from, 1112–1117
   relative sizes of, 729f
   ring systems of, 19f
   Voyager images of, 733–734
Giant planets, atmospheres, 723–742
   chemical composition, 724–728, 725f, 726t, 728f–729f
      ammonia, 726t, 727
      carbon, 726–727, 726t
      helium, 725–726, 726t
      hydrogen, 725–726, 726t
      hydrogen cyanide, 726t, 728
      methane, 726t, 727, 728f
      nitrogen, 726–727, 726t
      oxygen, 726–727, 726t
      phosphine, 726t, 727
      sulfur, 726–727, 726t
      water, 726t, 727
   chromosphere materials in, 730–731, 730t
   clouds and aerosols, 728–733, 729f–730f, 730t, 731f–733f
   dynamical meteorology of troposphere of stratosphere, 733–738, 735f–736f
   energetic processes in high, 738–741, 739t, 740f–741f
   extreme UV with, 738
   magnetic fields with, 739, 739t, 740f
   physical properties of, 724t
   20th century studies on, 1015
   zonal flow among, 734–735, 735f
Giant planets, interiors, 743–758
   constraints, 745–747
      atmosphere, 746–747
      gravitational field, 745–746, 746f
      magnetic field, 747
   equations of state model for, 747–750, 748f
      helium, 749
      hydrogen, 748–749, 749f
      ices, 749
      mixtures, 749–750
      rock, 749
   interior modeling, 750–756, 751f–752f
      extrasolar giant planets, 755–756, 756f
      Jupiter, 752–753, 753f
      Neptune, 754–755, 755f
      Saturn, 753f, 754
      Uranus, 754–755, 755f
   20th century studies on, 1015
Gilbert, Grove K., 862–863
Giotto Extended Mission (GEM), 685t
Giotto spacecraft, 983f
   space missions to comet by, 683, 685t
Global Circulation models (GCMs), 174
Global contraction, 297
Global exploration road map, Mars, 992–993, 994f
Global Seismic Network (GSN), 463, 463f
Global seismic tomography (GST), 473, 474f
Grabens, 103, 284, 292–293
GRACE, Earth exploration by, 1213
GRAIL, Moon exploration by, 1216
Grain size, 194
Grand Tack model, 45
Granulation, 239–240, 240f
Gravimeter, Apollo science experiments with, 561t
Gravitational focusing factor, 43
Gravitational harmonics, 751f
Gravitational instability (GI), 43
Gravity
   cosmic dust dynamics and, 675–676
   Enceladus, 852, 853f
   investigating planetary interiors using, 1190–1192, 1191f
   Moon (Earth), 500–501, 501f
      measurements, 542–543, 543f
   rings used to probe planet's, 903
Gravity Recovery and Interior Laboratory (GRAIL), 542, 543f, 571–572
   lunar exploration by, 557t
   Moon space missions by, 493–494
Gray body, 1085
Great Dark Spot (GDS), 733
Great Red Spot (GRS), 727, 733, 1009
Greeks, 999–1001, 1000f–1001f
Greeley, R., 783–785
Greenhouse effect, 307
   Earth, 437
   Mars
      carbon dioxide, 352–353
      hydrogen-aided, 353–354
      methane-aided, 353
      sulfur dioxide, 353
Ground-based rotational observation, 181, 182f
GRS. See Great Red Spot
GSN. See Global Seismic Network
GST. See Global seismic tomography
Gusev Crater, 398t, 401–402, 403f, 405–406, 406f, 412–413, 413f
Gyro radius, 274, 291–292

H

Habitable zone, 984–986
Hadley cell, 429–430
Hale, George Ellery, 1013
Hale-Bopp Comet, 684f, 693–694, 697f
   hydrogen cloud of, 693–694, 694f
   neck-line structure of, 695, 696f
   striae in dust tail of, 695, 696f
Hall, Angeline Stickney, 769–770
Hall, Asaph, 769, 1009
Halley, Edmund, 862–863, 1005, 1008
Halley Multicolor Camera (HMC), 683
Halley's comet, 685t, 686–687, 686f
   rotating nucleus of, 691f
   solar system studies on, 1006
   ultraviolet spectra of, 700f
Halley-type comets (HTCs), 688–689, 710f, 712
Hard x-ray emission, 254
Harrington, R. S., 941
Harriot, Thomas, 1003
HASI. See Huygens Atmospheric Structure Instrument
Hawaiian volcanic eruptions, 114–115, 114f–115f
Hayabusa spacecraft, 618–619, 983f
HCS. See Heliospheric current sheet
Heat flow
   core-mantle boundary, 202–204, 203f–204f
   Io, 780, 780t, 788–790
   Moon (Earth), 545–546
Heat flow experiment (HFE), Apollo science experiments with, 561t
Heat flow measurements, 1192–1198
   Earth's heat flow, 1193–1194, 1194f
   future investigations, 1196–1198, 1197f

Heat flow measurements (*Continued*)
  lunar heat flow, 1194–1196, 1195f–1196f
  measurement techniques, 1192–1193, 1193f, 1195f
Heavy ion content, solar wind and, 276, 276t
Hectometric (HOM) emissions, 1127
HED. *See* Howardite, eucrite, and diogenite meteorites
Helene, 761t, 774f
  relative size of, 760f
Helin, Eleanor, 604–605
Heliopause, 272–273
Helios spacecraft
  Helios 1 and /2, interplanetary dust detection by, 668t, 669
  Sun exploration by, 1210
Helioseismology, 238–239, 239f
Heliosphere, 245, 261
  solar wind and, 20–22
  X-ray emissions with, 1040–1041
Heliospheric current sheet (HCS), 697
Heliospheric magnetic field, 263–264, 264f
  ballerina skirt model, 265, 265f
Helium
  giant planet atmospheres with, 725–726, 726t
  giant planets interiors in EOS model with, 749
HEOS 2, interplanetary dust detection by, 667, 668t
Herschel, John, 1009
Herschel, William, 1006, 1010, 1013
Herschel ringlet, 896f
Herschel Space Telescope, 1075f, 1077t
Hevelius, Johannes, 1003
HFE. *See* Heat flow experiment
High Resolution Imaging Science Experiment(HiRISE), 97–98
Highland rocks, 506, 506f
Highlands, 1025
High-purity germanium (HPGe) detector, 1168–1169, 1169f
High-reflectance plains (HRP), 294
Hill sphere, 60–61, 61f
Himalia, 761t
  relative size of, 760f
Hipparcos, 959
Hirayama, Kiyotsugu, 1017
Hirayama asteroid families, 1077–1078, 1078f
HiRISE. *See* High Resolution Imaging Science Experiment
Hiten (MUSES-A), lunar exploration by, 557t
Hiten spacecraft, interplanetary dust detection by, 668t
HMC. *See* Halley Multicolor Camera
Hohmann transfer orbit, 985, 985f
HOM. *See* Hectometric emissions
Homopause level, 724
Horseshoe orbits, 60
Hot Jupiters, 53, 964
Hot Neptunes, 53, 966–967
Hot poles, 286
Howardite, eucrite, and diogenite (HED) meteorites, 34–37, 633, 640
HPGe. *See* High-purity germanium detector

HRP. *See* High-reflectance plains
HST. *See* Hubble Space Telescope
HTCs. *See* Halley-type comets
Hubble Space Telescope (HST), 13f, 683, 833, 959
  Neptune and Uranus as seen by, 733, 733f
Human action radius, 984, 984f
Human-robotic partnership, 984, 987
Huygens, Christiaan, 1004
Huygens Atmospheric Structure Instrument (HASI), 834–835
Huygens ringlet, 896f
Huygens spacecraft, exploration by, 983f
Hyakutake comet, 698, 699f
Hydra, discovery of, 910f, 911
Hydrodynamic escape, 46, 346, 920
Hydrogen
  deuterated, 726
  giant planet atmospheres with, 725–726, 726t
  giant planets interiors in EOS model with, 748–749, 749f
  ortho, 724–725
  para, 724–725
Hydrogen cyanide, giant planet atmospheres with, 726t, 728
Hydrogen-aided greenhouse, Mars atmosphere with, 353–354
Hydrostatic equation, 310–311
Hydrostatic equilibrium models, 236–237
  on Mars, 385
Hyperbolic orbits, 58
Hyperion
  chaotic rotation with, 77, 77f–78f
  properties of, 761t
  relative size of, 760f

I

Iapetus, 761t, 770–772, 771f–772f
  albedo, 1064t
  planetary rings of, 889–890
  relative size of, 760f
IAU. *See* International Astronomical Union
ICB. *See* Inner core boundary
Ice
  giant planets interiors in EOS model with, 749
  Mars, 372, 372f
ICE. *See* International Cometary Explorer
Ice dwarf planet, 923
Ice line, 33
Ice-giant planets, 49–51
ICME. *See* Interplanetary coronal mass ejections
IDP. *See* Interplanetary dust particle
IKAROS spacecraft, interplanetary dust detection by, 668t
IMAGE spacecraft, 150–151, 151f
Imaging spectrometer, 885
IMF. *See* Interplanetary magnetic field
Impact basins, Mercury, 293–294, 293f–294f
Impact breccias, 507–508
Impact craters, 83–89
  crater dimensions, 88–89, 88t
  crater shape, 83–88, 84f–87f
  Ganymede and Callisto, 823–825
    distribution, 824–825
    structure, 823–824, 824f
    surface ages, 824–825
  Mars, 364–365
    crater morphology, 365, 365f
    cratering rates, 364–365
    Martian timescale, 364–365
  Mercury, 293–294, 293f–294f
  planetary evolution with, 94–97
    biosphere evolution, 95–97
    early crustal evolution, 95
    impact origin of earth's moon, 94–95
  as planetary probes, 97–99
    morphologic and geologic, 99
    spectral composition, 98–99
    water and ices, 97–98, 98f
  Venus, 326–329, 327f–328f
Impact hazards, NEOs, 619–622, 619f–620f
  collision frequency, 621
  collision magnitude, 620–621
  mitigation measures, 621–622, 622f
Impact heating, Mars atmosphere with, 353
Impact processes, 89–94
  crater formation, 89–92, 90f
  Moon, 508–509, 508f, 574
  target rock changes in
    melting, 94
    solid effects, 92–94, 93f
In situ resource utilization, 987
Inclination, 6, 663, 896, 927
  asteroid, 1085f
Incompatible elements, 199
Infrared Astronomical Satellite (IRAS), 13, 660, 662f, 1073–1074, 1075f, 1077t
  observations of comets by, 702, 702f
Infrared excess, 30–31
Infrared light, 885
Infrared scanning radiometer, Apollo science experiments with, 561t
Infrared Space Observatory (ISO), 833, 1075f, 1077t
Infrared spectrometer (IRIS), 832
Infrared Telescope Facility (IRTF), 1089, 1090f
Infrared views, 1073–1088
  asteroid properties, 1083–1085, 1084f–1085f
  comets, 1079–1083, 1080f–1082f
  Earth's orbit ring of dust, 1078–1079, 1080f
  future for, 1087
  Pluto and beyond, 1085–1087, 1086f
  zodiacal dust cloud, 1074–1078
INMS. *See* Neutral Mass Spectrometer
Inner core boundary (ICB), 472
Insolation, 911
Integral of motion, 710–711
Intercrater plains, 106, 284, 292–293
International Astronomical Union (IAU), 4
  planets definition by, 5
  web address for, 943t
International Cometary Explorer (ICE), space missions to comet by, 685t
International Space Station (ISS), 982

Index

LEO established with, 987−988, 991−992
Interplanetary coronal mass ejections (ICME)
  characteristics of, 270, 270t
  field line draping about fast, 271
  magnetic field topology of, 270−271, 271f
Interplanetary dust, spacecraft measurements of, 667−669, 670f
Interplanetary dust particle (IDP), 20, 20f, 33−38, 660−663, 662t
  geochemistry of, 33, 34f
Interplanetary magnetic field (IMF), 290
Interstellar dust in heliosphere, spacecraft measurements of, 673−674, 674f−675f
Intrapatera eruptions, 788, 789f
Intraplate volcanism, 200
Intrusion, 101−102
Io, 779−792
  albedos, 1061t
  atmosphere of, 790−791
  atmospheres observations with UV for, 1058−1059, 1058f
  aurora of, 790, 791f
  density of, 540t
  exploration of, 780−783, 782f
  future exploration for, 791−792
  Galilei's discovery of, 779
  heat flow of, 780, 780t, 788−790
  images, Galileo spacecraft, 779f, 784f
  Laplace resonance, 780−781, 796f
  magnetospheres interactions with, 154−156, 154t
  physical properties of, 780t
  Pioneer 10 flyby, 781
  "pizza moon" nickname, 782
  properties of, 761t
  radius of, 540t
  relative size of, 760f
  satellite-to-parent mass ratio of, 540t
  sulfur dioxide on, 780
  surface of, 783−787, 784f−786f
  torus, 790−791
  volcanic eruptions on, 787−788
    Lokian eruptions, 788, 789f
    Pillanian eruptions, 786f, 787−788
    Promethean eruptions, 788, 788f
  volcanic features of, 106−107, 107f
  volcanoes
    Loki plume, 782, 782f
    Pele plume, 782, 782f
    Prometheus plume, 788f
    Ra Patera, 783−785, 785f
    Thor, 784f
    Zamama, 784f
  Voyager 1 spacecraft flyby, 782, 782f
Io plasma torus, X-ray emissions, 1032, 1033f
Ionosphere
  Earth, 427−428
  Venus, 313−314, 314f
IRAS. See Infrared Astronomical Satellite
IRIS. See Infrared spectrometer
Irons, 34−37, 36f
Irradiance spectrum of sun, 236, 237f
Irregular satellites, 52

IRTF. See Infrared Telescope Facility
ISO. See Infrared Space Observatory
Isolation mass, 44
Isotopes, Venus atmosphere with, 313
ISS. See International Space Station

J

Jackson crater, 85f
Jacobi constant, 710−711
Janus, 761t, 773, 774f
  relative size of, 760f
Jason-1 and Jason-2, Earth exploration by, 1213
Jayaraman, Sumita, 1078−1079
Jeans, James, 831
Jeans escape, 320−321
Jeffreys, Harold, 463, 1014
Jeffreys ringlet, 896f
Jet Propulsion Laboratory (JPL), web address for, 943t
Jewitt, David, 915, 941−942
Jezero Crater, 350f
Joule heating, 739, 1021
Jovian planet evolution, 756−757
  Kelvin-Helmholtz model in, 756−757
Jovian seismology, 1190
JPL. See Jet Propulsion Laboratory
Juliet, 761t
  relative size of, 760f
June Lyrids, 661t
Jupiter
  atmospheres observations with UV for, 1052−1055, 1054f
  cloud layers on, 730f
  colors on, 730
  cross section of interior structure of, 743−744, 744f
  diagram for hydrogen of, 749f
  equatorial radius of, 724t
  Galileo discovery of satellites of, 759
  GRS of, 727, 733, 1009
  hot Jupiters, 964
  interior modeling for, 752−753, 753f
  Kelvin-Helmholtz model in, 756−757
  length of day of, 724t
  magnetic field highlights from missions to, 122t
  magnetospheres with, 141t
  mass of, 724t
  19th century studies on, 1009−1010, 1010f
  nonthermal radio emissions of
    decametric and hectometric radio emissions, 1126−1127
    Ganymede, 1128
    kilometric radio emissions, 1127−1128
    low frequencies, 1126−1128, 1127f
    synchrotron radiation, 1124−1126, 1124f−1126f
    very low-frequency emissions, 1128
  physical properties of, 724t
  planetary dynamos with, 133

  planetary rings of, 671−672, 672f, 888, 888f−889f
  radar imaging for, 1155−1157
  radio wavelengths thermal emission from, 1112−1114, 1113f−1114f
  ring-moon systems of, 884f
  size relative to other planets, 729f
  small satellites of, 770, 770f
  solar wind properties with, 141t
  surface gravity of, 724t
  temperature of, 724t
  20th century studies on, 1015
  X-ray emissions, 1020f, 1029−1031
    auroral emissions, 1029−1031, 1030f
    nonauroral emissions, 1030f, 1031
Jupiter-family comets, 710f, 712−713, 925
  albedos, 1083

K

Ka-band, 182
Kaguya, 568, 568f
  lunar exploration by, 557t
Kaguya/SELENE, Moon exploration by, 1216
KBO. See Kuiper belt object
Keeler Gap, 892, 893f
Kelvin-Helmholtz model
  Jovian planet evolution in, 756−757
  Jupiter and Saturn in, 756−757
  Neptune and Uranus in, 757
Kepler, Johannes
  planetary motion laws of, 56, 56f, 159−160, 160f
  solar system studies with, 1001−1003
Kepler mission, 958
  BLENDER algorithm, 974
  extrasolar planets in, 973−976, 974f−976f
  transit timing variations in, 974
Keplerian motion, 55−57
  elliptical motion, 56−57, 57f
  orbit in space, 56−57, 57f
  orbital elements, 56−57, 57f
  perturbed, 59
  planetary motion laws, 56, 56f
Keplerian orbits, cosmic dust dynamics and, 675−676
Keplerian velocity, 42
Kepler's third law, 892
Kirkwood, Daniel, 1008
Kirkwood gaps, 69−70, 588−589
Kivelson, Margaret, 818
Kolmogorov-Arnol'd-Moser theory, 63
Kozai resonance, 708, 708f
KREEP (Rocks rich in K, P, and REEs), 497−498, 498f
Kuiper belt, 5, 346, 715, 925−942, 983
  dust from, 669
  ecliptic comets, 933−934, 934f
  historical perspective, 925−926
  mass of, 954
  orbital dynamics, 926−930, 927f, 929f
  properties, physical and orbital, 930−931, 931f
  trans-Neptunian objects

Kuiper belt (*Continued*)
  orbital and dynamical structure, 926–930, 927f, 929f
  primordial sculpting, 934–938, 936f, 938f
  size distribution and total mass, 931–933, 932f
Kuiper belt object (KBO), 5, 8–9, 700–701, 941–956
  albedo, 942, 944t, 946
  binaries, 952–954, 953t
    mutual events, 953–954
    origins, 954
    system mass, 952, 954f
  brightness, 944–945
    absolute magnitude, 944t, 945
    apparent magnitude, 943f, 944–945, 944t
    luminosity function, 944t, 945, 945f
  brightness variation, 946–949, 946f
    amplitude, 946f, 947
    density, 948
    period of rotation, 946–947, 946f, 947t
    porosity, 948–949
    shape, 947–948, 947f–948f
  classical, 928
  color of, 930–931
  composition, 949–952
    spectroscopy, 951–952, 951f–952f
    surface color, 949–951
  databases of known, 943, 943t
  diameters of, 944t, 945–946
  discovery of, 942, 943f
  distribution of, 927f, 936f
  dynamical classes, 943–944
  formation process of, 932–933
  future work on, 955
  inclination of, 931f
  light curves, 946–947
  naming of, 943
  New Horizons spacecraft flyby of, 954–955
  position of, 942f
  surface color of, 949–951
    Centaur objects, 949, 950f
    classical KBOs, 949, 950f
    reasons for color patterns, 950–951, 950f
    scattered disk objects, 949, 950f
Küstner, Karl Friedrich, 1008

L

Labeled release (LR) experiment, 222
LACE. *See* Lunar atmosphere composition experiment
Lade spacecraft, interplanetary dust detection by, 668t
LADEE, Moon exploration by, 1216–1217
Lafayette Martian meteorite, 627f
Lagrangian points, 60–61, 60f, 991–992
Land Surface Data Acquisition and Analysis Center (LPDAAC), 446
Langmuir probe, Venus space missions by, 308
Laplace, Pierre, 1006–1007
Laplace resonance, on Io, 780–781, 796f
Laplace ringlet, 896f

Large Binocular Telescope (LBT), 1092–1093
Larissa, 761t, 776
  relative size of, 760f
Laser altimeter, Apollo science experiments with, 561t
Laser ranging, 543–544, 544f
Laser Ranging (LLR) experiment, 543–544, 544f
Laser ranging retroreflector, Apollo science experiments with, 561t
Last universal common ancestor, 214–215
Late Heavy Bombardment (LHB), 293, 937
Late veneer, 46
Laws of motion, 57
LBT. *See* Large Binocular Telescope
LCROSS, Moon exploration by, 1216
Le Verrier, Urbain, 1010–1011
Le Verrier ring, 897f, 899f
LEAG. *See* Lunar Exploration Analysis Group
Leakage flux, 1161
LEAM. *See* Lunar ejecta and meteorites experiment
Leda, 761t
  relative size of, 760f
Leibacher, J., 238–239
Leighton, R., 238–239
Lenticulae, Europa landforms and, 801–806, 802f–803f
LEO. *See* Low Earth orbit
Leonids, 661t, 662f
Leverrier, U.J.J., 6
Levison, H., 7, 688–689
Lewis, J. S., 869
Lewis, John, 816–817
Lexell, Anders, 1011
LHB. *See* Late Heavy Bombardment
Librations, 175–177, 175f–176f, 852, 1191
Life, 209–210
  defined, 210–214, 210f–211f
  ecological requirements for, 212t
  ecology of, liquid water, 211–212, 212t–213t
  history on earth of, 211f, 214–217, 215f–216f
  limits to, 219, 219t
  microbial, 216
  molecules of, 210f
  origin of, 217–219, 217f, 218t
  about other stars, 229–230, 229t
  properties of, 214t
  search for on Mars, Europa, or Enceladus, 228–229
  in solar system, 219–228
    asteroids, 228
    comets, 228
    Enceladus, 228
    Europa, 227
    giant planets, 226–228, 226f
    Mars, 220–226, 221f, 221t, 223f, 225f
    Mercury and Moon, 220
    Titan, 227–228
    Venus, 220
  theories for, 212–214, 214t
Light curves, Centaur, 946–947

Limb darkening, 831
Lincoln Near-Earth Asteroid Research program, 610–611
LINEAR comet, 690–691, 692f
Linear stability, 191–192
Liquidus curves, 186–187
Lithosphere
  Earth, 470–471, 475
  Mercury, 297–298
Lithosphere surface deformation, 204–205, 205f
LLR. *See* Laser Ranging experiment; Lunar Laser Ranging
LNPE. *See* Neutron probe experiment
Lobate scarps, 284
LOD variations, 171–175, 172f–174f
Loki plume, 782, 782f
Lokian eruptions, 788, 789f
LOLA. *See* Lunar Orbital Laser Altimeter
Long-period comets (LPCs), 688–689
Long-term spin evolution, 160–164, 162f
Lorentz force, 162–163
Low Earth orbit (LEO), 982
  ISS and establishment of, 987–988, 991–992
Lowell, Percival, 1008, 1016
Lowell Observatory
  Near-Earth Object Search of, 610–611
  web address for, 943t
Low-reflectance material (LRM), 294
LPCs. *See* Long-period comets
LPDAAC. *See* Land Surface Data Acquisition and Analysis Center
LR. *See* Labeled release experiment
LRM. *See* Low-reflectance material
LRO. *See* Lunar Reconnaissance Orbiter
LRV. *See* Lunar Roving Vehicle
LSPE. *See* Seismic profiling experiment
Luna 17/Lunokhod 1, 531, 531f–532f
Luna landing sites, 505f
Luna spacecraft
  Luna 1, 983f
    Moon exploration by, 1214
  Luna 2, Moon exploration by, 1214
  Luna 3, Moon exploration by, 1214
  Luna 9, 983f, 1214, 1214f
  Luna 10, 983f, 1214
  Luna 11, Moon exploration by, 1214
  Luna 12, Moon exploration by, 1214
  Luna 13, Moon exploration by, 1214, 1214f
  Luna 14, Moon exploration by, 1214
  Luna 16, 527–529, 529f, 531f, 1215, 1215f
  Luna 17, Moon exploration by, 1215, 1215f
  Luna 20, 529, 530f–531f, 1215, 1215f
  Luna 21, Moon exploration by, 1215, 1215f
  Luna 24, 529–530, 530f–531f, 1215, 1215f
  lunar exploration by, 557t
  planetary satellites observation with, 768
Luna21/Lunokhod 2, 532–533, 533f–534f
Lunar Atmosphere and Dust Environment Explorer, 572
Lunar atmosphere composition experiment (LACE), Apollo science experiments with, 561t

Lunar Crater Observation and Sensing Satellite (LCROSS), 302, 569–571
  lunar exploration by, 557t
  Moon space missions by, 493–494
Lunar ejecta and meteorites experiment (LEAM), Apollo science experiments with, 561t
Lunar exploration, 555–580
  Apollo program, 518–533, 559–566
    Apollo 8, 1214
    Apollo 10, 1215
    Apollo 11, 518f–519f, 519–520, 983f, 1215
    Apollo 12, 520–522, 520f–521f, 1215, 1215f
    Apollo 13, 1215
    Apollo 14, 522–525, 522f–523f, 1215
    Apollo 15, 524f, 525, 1215
    Apollo 16, 525–526, 525f–526f, 1215
    Apollo 17, 526–527, 527f–528f, 1215
    benefits of human exploration, 565–566
    CSM of, 559, 560f, 561t
    landing sites of, 505f, 559f
    location of, 504f
    LRV of, 559, 559f
    orbital remote sensing, 562–563
    samples and analysis, 559–560, 563f, 564t
    surface experiments, 560–562, 561t, 565f
  early space age, 557–559, 557t, 558f–559f
  meteorites' importance for, 572–573
  objectives for science, 573–578
    lunar resources, 576–578
    science from moon: lunar-based astronomy, 576
    science on moon, 575–576
    as stepping stone for exploring other destinations, 576–578
  objectives for studies of Moon, 573–575
    atmospheric and dust environment, 575
    bombardment history of inner solar system, 573
    crustal rocks diversity, 574
    impact processes, 574
    regolith processes, 574–575, 575f
    structure and composition of lunar interior, 573–574
    volatiles at lunar poles, 574
    volcanism, 574
  post-Apollo, 566–572
    Chandrayaan-1, 568–569, 569f
    Chang'e 1 and 2, 569
    Chang'e 3 Lunar Lander and Rover, 572
    Clementine, 557t, 566, 566f, 1216
    Gravity Recovery and Interior Laboratory, 571–572
    Kaguya, 568, 568f
    LCROSS, 571
    Lunar Atmosphere and Dust Environment Explorer, 572
    Lunar Prospector, 566–567, 567f
    Lunar Reconnaissance Orbiter, 569–571, 570f–571f
    SMART-1, 567
  telescopic exploration, 556–557
  Galileo's Moon drawings, 556, 556f
Lunar Exploration Analysis Group (LEAG), 573
Lunar formation and evolution, 551–552, 553f–554f
Lunar Laser Ranging (LLR), 182–183
Lunar meteorites, 33–34, 644
Lunar microcraters, 664–665, 664f
Lunar Orbital Laser Altimeter (LOLA), 569–570
Lunar Orbiter 1-5, Moon exploration by, 1214
Lunar orbiters, lunar exploration by, 557t
Lunar phases, 495–496
Lunar Prospector, 566–567, 567f
  gamma ray and neutron data from, 1175–1177, 1175f–1176f
  lunar exploration by, 557t
  magnetic field highlights from, 122t
  Moon exploration by, 1216
Lunar Prospector spacecraft, 302
Lunar Reconnaissance Orbiter (LRO), 302, 569–571, 570f–571f
  lunar exploration by, 569–570, 557t
  Moon exploration by, 1216
  Moon space missions by, 493–494
Lunar rocks, 505–508
  highland rocks, 506, 506f
  impact breccias, 507–508
  magnesian suite, 507
  mare basalts, 507, 507f
Lunar Roving Vehicle (LRV), 559, 559f
Lunar seismology, 1187–1190, 1188f–1189f
Lunar stratigraphy, 513–516, 514f
  age determination, 515, 515f
  age record of lunar maria, 515–516, 516f
  late heavy bombardment, 513–515
Lunar volatiles, 536–537, 537f
Lunokhod 1, 2, lunar exploration by, 557t
Lunokhod landing sites, 505f
Luu, Jane, 915, 941–942
Lysithea, 761t
  relative size of, 760f

M

Mab, 761t
  relative size of, 760f
Magellan
  Venus exploration with, 1212, 1212f
  Venus space missions by, 307–308
Magma oceans, 186–187
  Moon, 497–499, 498f–499f
Magnesian suite rocks, 507
Magnetic fields
  coronal, 245–246, 246f
  cosmic dust interaction with, 677–678, 677f
  dynamo mechanism for planetary, 125–126
    conditions for, 126
    defined, 125–126
    generation regions, 125t, 126
  Ganymede and Callisto, 818–819
    induction fields and oceans, 819
    intrinsic fields, 818–819
  generation, 121–136
  giant planets, 739, 739t, 740f
  heliospheric, 263–264, 264f
  history, 202–204, 203f–204f
  ICME and topology of, 270–271, 271f
  Mercury, 286–288
  Moon, 501–502, 502f
  observations of planetary, 121–125, 122t
    sources of, 121–122
    spatial characteristics of, 122–125, 123f–125f, 125t
    temporal characteristics of, 125
  photospheric, 240
  planetary dynamo, standard, 127–129
    driving forces in, 127
    fluid motions in dynamo regions with, 127, 128f
    generation mechanisms with, 127–129, 128f
    beyond standard dynamo, 129
  planetary dynamos, 131–135
    ancient Mars, 134–135
    ancient Moon, 134
    Earth, 131–133, 132f
    extrasolar planets, 135
    Ganymede, 134
    Jupiter, 133
    Mercury, 133
    Neptune, 134
    planetary bodies lacking dynamos, 135
    Saturn, 133–134
    small bodies, 135
    Uranus, 134
  radiation patterns in, 1122f
  remanent, 1192
  simulations and experiments with, 129–131
    dynamo experiments, 131, 132f
    numerical dynamo simulations, 129–131
  solar wind in, 482, 482f
  Sun, 265, 265f
  turbulence and solar wind with, 277–278, 277f–278f
Magnetic flux rope, 269
Magnetic reconnection, 146–147, 242–243, 262
  solar flare with, 249, 250f
Magnetic techniques, Moon (Earth), 544–545, 544f
Magnetohydrodynamics (MHD), 235
  oscillations of coronal loops, 246–247, 247f
  waves, 247–248, 247f–248f
Magnetopause, 137, 138f, 290
Magnetorotational instability (MRI), 32
Magnetospheres, 4, 137–158, 885, 1121–1122
  blunt body shape of, 484f
  circulation, 486f, 488f
  defined, 137–138, 138f
  Dungey's cartoon of, 485f, 488f
  dynamics, 150–153, 150f–152f, 154f
  geomagnetic activity, 150

Magnetospheres (*Continued*)
    interactions with moons, 154–156, 154t, 156f
    of mercury, 288–292, 291f–292f
    planetary magnetic fields, 144–146, 145t, 146f–147f, 148t
    plasmas, 146–150
        energetic particles, 148–150, 149t
        sources, 146–148, 149f
    rings used to probe planet's, 903, 903f–904f
    types of, 138–144
        heliosphere, 138–140, 139f–140f
        magnetized planets, 144
        solar winds interactions, 142–144, 143f
        unmagnetized planets, 140–142, 141f, 141t, 142f
Magnetotail, 290
Magnetotellurics (MT), 1199, 1199t
Magsat, magnetic field highlights from, 122t
MAHLI. *See* Mars Hand Lens Imager
Main sequence star, 5
Mangala Valles, 351f
Mangalyaan, Mars exploration by, 1217f, 1219
Mantle, 101, 162–163, 185
    advected mantle material, 1185–1186
    Earth
        lower mantle, 471–472
        upper mantle, 470–471
    Europa, 795, 795f
    Moon (Earth), 551
Mantle heat budget, 201–202, 202f
Mantle waves, 465, 465f
Mare basalts, 507, 507f
Marginally collisional plasma, 273–274
Maria, 1025
Mariner spacecraft
    magnetosphere observed by, 153
    Mariner 2, 983f
        Venus exploration with, 307–308, 1211
    Mariner 4, 983f
        Mars exploration by, 360, 360t, 1217
    Mariner 5, Venus exploration with, 307–308, 1211–1212
    Mariner 6 and 7, Mars exploration by, 360, 360t, 1217
    Mariner 9, 983f
        flyby missions to satellites by, 764t
        Mars exploration by, 360, 360t, 1218
    Mariner 10, 983f
        magnetic field highlights from, 122t
        Mercury exploration by, 283, 1211, 1211f
        Venus exploration by, 307–308, 1212
Mars
    astrobiological potential for, 220–226
        early history of, 222–224, 223f, 225f
        meteorites from, 226
        subsurface life on, 224–226, 225f
        Viking landers' results for, 220–222, 221f, 221t
    Atmosphere, 343–358
        Martian geologic timescale, 357f
        mechanisms for producing wetter environments, 352–354
        Milankovitch cycles, 354, 355f
        past climates, 348–356, 350f–352f
        volatile inventories and their history, 344–348
    atmospheres observations with UV for, 1051–1052, 1053f
    exploration missions, 1217–1219
        Curiosity, 1219
        failures of 1990s, 1218
        Mangalyaan, 1217f, 1219
        Mariner 4, 360, 360t, 1217
        Mariner 6 and 7, 360, 360t, 1217
        Mariner 9, 360, 360t, 1218
        Mars 2 and 3, 1217
        Mars 4, 5, 6, and 7, 1218
        Mars Express, 1218–1219
        Mars Odyssey, 1218
        Mars Pathfinder and Mars Global Surveyor, 1218
        Mars Reconnaissance Orbiter, 1219
        Phobos 1 and 2, 1218
        Phoenix, 1219
        Spirit and Opportunity, 1218
        Viking 1 and 2, 1218
    general characteristics of, 381t
    historical events of, 215f
    human exploration case study, 989–993
        cooperation models, 990
        defining and communicating benefits, 990–991, 991f
        design reference missions, 992, 993f
        destinations and strategies, 991–992
        global exploration road map, 992–993, 994f
        mapping way to Mars, 992–993
        strategic mission scenarios, 992, 992f
    magnetic field highlights from missions to, 122t
    magnetospheres with, 141t
    19th century studies on, 1009, 1009f
    planetary dynamos with ancient, 134–135
    planetary rings of, 888–889
    radar imaging for, 1150–1153
        radar heterogeneity, 1150–1151, 1152f
        subsurface structure, 1151–1153, 1152f
    radio wavelengths thermal emission from, 1111–1112
    role in solar system exploration, 984–985, 985f
    satellites of, 760f, 769–770, 769f
    seismology, 1190
    solar wind properties with, 141t
    space missions
        Mars 2, 4-7, 360t
        Mars Express, 345, 360t
        Mars Observer, 360t
        MER opportunity, 398t, 405–406, 406f
        MER spirit (SPI), 398t, 405–406, 406f
        MPF, 398t, 405, 405f
        MRO, 345, 351–352, 360t
        MSL curiosity, 398t, 406, 407f
        Opportunity rover, 345, 360t
        Phobos1 & 2, 360t
        Phoenix, 398t, 406, 406f
        Spirit Rover, 345, 360t
        Viking 1 & 2, 360t, 405, 405f
    synodic period of, 985
    20th century studies on, 1014–1015
    volcanic features of, 104–105, 104f
    Winslow crater on, 84f
    X-ray emissions, 1020f, 1027, 1028f–1029f
Mars, interior structure and evolution, 379–396
    core of, 380f, 382–383
    crust of, 380f, 384–385, 384f
    density of, 380, 380f
    evolution of Mars, 393–396
        chemical evolution, 395–396
        crust formation, 395–396
        early dynamo, 395
        extraction of water, 396
        thermal evolution, 393–395, 394f
    formation and differentiation in, 381–382
    global interior structure, 390–393
        global geodesy data, 390–392
        model results, 392–393, 392f
    mantle of, 380f, 383–384, 383f
    principles of, 385–390
        basic equilibrium equations, 385, 386f
        heat sources, 385–387, 387f
        heat transport, 387–390
            conduction, 387–388, 388f
            convection, 388–390, 389f
Mars, landing sites, 397–420
    Ares Vallis, 398t, 402f
    Bonneville crater, 406f
    Chryse Planitia, 398t, 402f
    Eagle crater, 406f
    evolutionary implications from, 417–419
        chemical evolution, 417
        eolian processes, 418
        geologic evolution of landing sites, 418–419
        implications for habitable world, 419
        origin of igneous rocks, 417
        surface water, 417
        weathering on mars, 417–418
    Gale Crater, 398t, 402–404, 404f
    geology, 409–412
        craters, 405f–406f, 412, 412f
        eolian deposits, 411–412, 411f
        outcrops, 409f–410f, 410
        rocks, 406f, 409, 410f
        soils, 410
    Gusev Crater, 398t, 401–402, 403f, 405–406, 406f, 412–413, 413f
    instruments used to analyze, 399t
    Meridiani Planum, 398t, 402, 403f
    mineralogy and geochemistry, 412–417
        rocks, 405f–406f, 412–416, 412f–416f
        soils, 411f, 416–417
    in remotely sensed data, 404–409
        global compositional units, 407–409, 407f–409f
        surface physical properties, 404–407, 405f–407f
    Utopia Planitia, 398t
    Vastitas Borealis, 398t

Mars, surface and interior, 359—378
  canyons of, 367—368, 367f
  dunes of, 373f
  general characteristics of, 361—364
    global structure, 362—363
    global topography, 363—364, 364f
    orbital and rotational constants, 361—362, 361t
    physiography, 363—364
    planet formation, 362—363
    surface conditions, 362
  ice of, 372, 372f
  impact cratering of, 364—365
    crater morphology, 365, 365f
    cratering rates, 364—365
    Martian timescale, 364—365
  Mars exploration, 359—361, 360t
  poles of, 373
  surface view of, 373—376
    Burns Cliff, 375f
    Columbia Hills, 374f
    Endurance crater, 375f
    Gale crater, 376f
    Meridiani Planum, 374, 375f
    plains of Gusev, 374f
  tectonics of, 367
  volcanism of, 365—367, 366f
  water of, 368—372
    dark streaks, 371—372, 372f
    erosion and weathering, 368
    gullies, 371, 371f
    outflow channels, 370—371, 370f—371f
    valley networks, lakes, deltas, 368—370, 369f—370f
  wind of, 372—373, 373f
Mars 2 and 3, Mars exploration by, 1217
Mars 2 spacecraft, 4-7, Mars exploration with, 360t
Mars 4, 5, 6, and 7, Mars exploration by, 1218
Mars atmosphere, 343—358
  Martian geologic timescale, 357f
  mechanisms for producing wetter environments, 352—354
    carbon dioxide greenhouse, 352—353
    hydrogen-aided greenhouse, 353—354
    impact heating, 353
    methane-aided greenhouse, 353
    producing fluvial features in cold climates, 354
    sulfur dioxide greenhouse, 353
  Milankovitch cycles, 354, 355f
  past climates, 348—356, 350f—352f
  present climate, 348—356
    properties of, 344t
  volatile inventories and their history, 344—348
    sources and losses of volatiles, 346—348, 348f
    volatile abundances, 344—346, 344t—345t, 346f
  wind modification of surface, 355—356
Mars Express
  Mars exploration by, 1218—1219
  Mars exploration with, 345, 360t

Mars Global Surveyor
  flyby missions to satellites by, 764t
  magnetic field highlights from, 122t
  Mars exploration by, 1218
Mars Hand Lens Imager (MAHLI), 399t
Mars Observer, Mars exploration with, 360t
Mars Odyssey
  gamma ray and neutron data from, 1177—1181, 1179t, 1180f
  Mars exploration by, 1218
Mars Orbiter Laser Altimeter (MOLA), 348f, 401f
Mars pathfinder (MPF), 398t, 405, 405f
  Mars exploration by, 1218
Mars Reconnaissance Orbiter (MRO)
  flyby missions to satellites by, 764t
  Mars exploration by, 1219
  Mars exploration with, 345, 351—352, 360t
Mars Science Laboratory (MSL), 400
MARSIS, 1135—1136, 1139, 1151
Martian meteorites, 644
Maskelyne, Nevil, 1006
Mass spectrometer, Apollo science experiments with, 561t
Mass wasting, 293
Mass-independent fractionation (MIF), 348
Mast Camera (MASTCAM), 399t
MASTCAM. See Mast Camera
Matson, D., 789
Maunder, Walter, 1007, 1013
Maxwell, James Clerk, 1010
Maxwell model, 385
Maxwell ringlet, 896f
MB. See Mössbauer Spectrometer
McNaught comet, 696, 697f
Mean motion, 167
Mean-motion resonances, 712, 927
MECA. See Microscopy, Electrochemistry, and Conductivity Analyzer
Meddies, 435
Menzel, Donald, 1015
MER opportunity (OPP), 398t, 405—406, 406f
MER spirit (SPI), 398t, 405—406, 406f
Mercury, 283—304
  astrobiological potential for, 220
  atmospheres observations with UV for, 1050
  characteristics of, 284, 287f
  density of, 380f
  exosphere of, 288—292, 289f—290f
  exploration of, 283—284, 285f, 286t, 1211
    Mariner 10, 283, 1211, 1211f
    MESSENGER spacecraft, 283—284, 285f, 286t, 1211
  geologic features of, 292—301, 293f
    impact craters and basins, 293—294, 293f—294f
    surface composition, 299—301, 299f—300f
    tectonics and topography, 297—299, 297f—299f
    volcanic plains and vents, 294—296, 295f—296f

history of, 303—304
  geologic history, 303
  origin, 303—304
  thermal history, 303
internal structure of, 286—288
magnetic field highlights from missions to, 122t
magnetic field of, 286—288
magnetosphere of, 141t, 288—292, 291f—292f
motion of, 284—286
19th century studies on, 1008
planetary dynamos with, 133
radar imaging for, 1144—1147
  polar deposits, 1144—1147, 1145t, 1146f
  rayed craters, 1147
radio wavelengths thermal emission from, 1110—1111, 1110f
recent surface features of, 301—303
  hollows, 302—303, 302f
  radar-bright polar deposits, 301—302, 301f
regolith, 284, 293
solar wind properties with, 141t
solid surfaces observations with UV for, 1069
temperature of, 284—286
20th century studies on, 1013
volcanic features of, 106, 106f
X-ray emissions, 1025
Meridiani Planum, 398t, 402, 403f
Meridional circulation, 731
Meridional flows, 127
Mesosphere, 305—306
  Earth, 427
MESSENGER spacecraft
  exploration by, 983f
  magnetic field highlights from, 122t
  Mercury exploration by, 283—284, 285f, 286t, 1211
Metabolic pathways, 213t
Metallicities, 53, 53f
Meteor showers, 661t
  19th century studies on, 1011—1012, 1012f
Meteorites, 33—38, 625—657
  achondrites, 34—37
  asteroid derived, 641—643
    changes in meteorite populations, 642—643
    meteorite-asteroid connection, 641—642, 642f
  asteroids and, 592—593, 592f
  chemical and isotopic constituents of, 644—647
    cosmochemical fractionations, 644—645, 645f
    noble gases, 646—647, 647f
    non-noble gas elements, 644—646
  chondrules, 34—37, 34f—35f, 635—641, 647—648
    Ca- and Al-Rich Inclusions, 648
    components of, 647—649
    interstellar grains, 648—649, 649f
  chronometry, 649—655, 649f
    cosmic-ray exposure ages, 650—652, 651f—652f
    crystallization age, 652—654, 653t, 654f

Meteorites (*Continued*)
    extinct radioactives, 654–655, 654t
    gas retention age, 652, 653f
    terrestrial ages, 650, 650f
  classification, 628–641
    breccias, 638
    carbonaceous chondrites, 638–639, 639f
    chondrites, 635–641
    general, 628–632, 630f–631f
    oxygen isotopes and interpretation, 633–635, 635f
    petrographic properties, 635–638, 636f, 637t, 638f
    shock, 639–640, 640t
    specific classes, 631f–633f, 632–633, 634t
    weathering, 641, 641f
  genesis of, 628f
  geochemistry of, 33, 34f
  HED, 34–37, 633, 640
  historical background, 643–644, 643f
  irons, 34–37, 36f
  Lafayette Martian, 627f
  from larger bodies, 643–644
  lunar, 33–34, 644
  from Mars, 226
  Martian, 644
  mass spectrometric measurements on, 37
  minerals common in, 627t
  Noblesville H chondrite, 627f
  number of falls, 626t
  orbital evolution of, 70
  parent body to Earth, 626–628, 629f
  SNC, 33–34, 633
  stony-iron, 34–37
Meteoritic organics, 666
β-meteoroids, 1077
Meteoroids, 7, 657, 903, 1188–1190
Meteors, 657, 660, 661t, 662f
Meter-sized barrier, 42–43
Methane, giant planet atmospheres with, 726t, 727, 728f
Methane-aided greenhouse, Mars atmosphere with, 353
Methone, 761t, 774f
  relative size of, 760f
Metis, 761t, 770, 770f
  relative size of, 760f
MHD. *See* Magnetohydrodynamics
MI. *See* Microscopic Imager
Microlensing, extrasolar planets detected with, 961
Micrometeoroids, 670f
Microscopic Imager (MI), 399, 399t
Microscopy, Electrochemistry, and Conductivity Analyzer (MECA), 399t
Midcourse Space Experiment (MSX), 1075f, 1077t
Midocean ridges, 185
Midultraviolet (MUV), 1047
MIF. *See* Mass-independent fractionation
Milankovitch cycles, 354, 355f
Milky Way galaxy, 24–27, 24f–25f
Miller-Urey experiments, 217

Mimas, 761t, 770–772, 771f
  relative size of, 760f
Mineral grains, 665
Mineralogical composition, 665
Miniature Synthetic Aperture Radar (Mini-SAR), 569
Miniature Thermal Emission Spectrometer (Mini-TES), 399, 399t
Minimum-mass solar nebula (MMSN), 32–33
Mini-SAR. *See* Miniature Synthetic Aperture Radar
Mini-TES. *See* Miniature Thermal Emission Spectrometer
Miranda, 761t, 774–775, 774f–775f
  relative size of, 760f
Mixtures, giant planets interiors in EOS model with, 749–750
MMSN. *See* Minimum-mass solar nebula
Models, 249–252, 251f–252f
Moderately volatile elements, 35f, 38, 38f
Modes of convection, 193–196, 194f, 194t, 195f–196f
Modulated signals, 1136
MOI. *See* Moon's moment of inertia
MOLA. *See* Mars Orbiter Laser Altimeter
Molecular clouds, 31–32, 31f
Moment magnitude, 1188
Moment of inertia, 1191
Moon (Earth), 493–538, 991–992. *See also* Lunar exploration
  albedo, 1025
  Apollo and Luna landing sites, 518–533
    Apollo 8, 1214
    Apollo 10, 1215
    Apollo 11, 518f–519f, 519–520, 983f, 1215
    Apollo 12, 520–522, 520f–521f, 1215, 1215f
    Apollo 13, 1215
    Apollo 14, 522–525, 522f–523f, 1215
    Apollo 15, 524f, 525, 1215
    Apollo 16, 525–526, 525f–526f, 1215
    Apollo 17, 526–527, 527f–528f, 1215
    location of, 504f
    Luna 1, 2, and 3, 1214
    Luna 9 and 13, 1214, 1214f
    Luna 10, 11, 12, and 14, 1214
    Luna 16, 527–529, 529f, 531f, 1215, 1215f
    Luna 17, 21, and 24, 1215, 1215f
    Luna 17/Lunokhod 1, 531, 531f–532f
    Luna 20, 529, 530f–531f, 1215, 1215f
    Luna 24, 529–530, 530f–531f
    Luna21/Lunokhod 2, 532–533, 533f–534f
    overview, 504f, 518–519
  astrobiological potential for, 220
  atmospheres observations with UV for, 1050
  EM sounding for investigating, 1200–1201, 1201f
  exploration missions, 1213–1217
    ARTEMIS, 1216
    Chandrayaan-1, 1216
    Chang'e 1, 1216
    Chang'e 2, 1216
    Clementine, 557t, 566, 566f, 1216

    GRAIL, 1216
    Kaguya/SELENE, 1216
    LADEE, 1216–1217
    LCROSS, 1216
    Lunar Orbiter 1-5, 1214
    Lunar Prospector, 1216
    Lunar Reconnaissance Orbiter, 1216
    Ranger 7, 8, and 9, 1214
    SMART-1, 1216
    Surveyor 1, 3, 5, 6, and 7, 1214
    Zond 3, 1214
    Zond 5, 6, 7, and 8, 1214
  heat flow measurements for, 1194–1196, 1195f–1196f
  interior of, 499–503
    cross-section, 499f
    gravity, 500–501, 501f
    magnetic field, 501–502, 502f
    seismicity, 499f, 500
    thermal evolution, 502–503
  Jackson crater, 85f
  Luna landing sites, 504f
  lunar atmosphere and environment, 537–538
  lunar crust and lunar terranes, 503, 504f–505f
  lunar rocks, 505–508
    highland rocks, 506, 506f
    impact breccias, 507–508
    magnesian suite, 507
    mare basalts, 507, 507f
  lunar volatiles, 536–537, 537f
  Lunokhod landing sites, 504f
  magma ocean of, 497–499, 498f–499f
  magnetic field highlights from missions to, 122t
  19th century studies on, 1008
  orbit of, 495–496
    general characteristics, 495, 495f
    librations and tidal forces, 496
    lunar phases and eclipses, 495–496
  origin of, 496–497, 497f
  physical properties of, 496
  planetary dynamos with ancient, 134
  radar imaging for, 1149–1150
    lunar geology, 1150, 1151f
    polar environment, 1149–1150, 1150f
  radio wavelengths thermal emission from, 1110
  regolith, 517–518, 517f, 549–550, 550f
    properties and sources, 518
    thickness, 517–518
  seismology, 1187f
  significance of landing sites for, 533–536, 535f
  solid surfaces observations with UV for, 1069
  space missions, 493–494
  stratigraphy and surface ages, 513–516
    age determination, 515, 515f
    age record of lunar maria, 515–516, 516f
    late heavy bombardment, 513–515
    lunar stratigraphy, 513, 514f
  surface of, 508–513
    erosion, and degradation, 512–513
    impact processes, 508–509, 508f

space weathering, 512–513
tectonic processes, 511–512, 512f
volcanic processes, 509–511, 509f–511f
20th century studies on, 1013–1014
volcanic features of, 103–104, 103f
X-ray emissions, 1020f, 1023–1025, 1024f
Moon (Earth), interior, 539–554
bulk lunar properties, 539–541, 540t, 541f
density of, 540t
lunar formation and evolution, 551–552, 553f–554f
methods used to probe, 541–549
Apollo core samples, 542, 542f
compositional studies, 546
gravity measurements, 542–543, 543f
heat flow, 545–546
laser ranging, 543–544, 544f
magnetic techniques, 544–545, 544f
seismology, 546–549, 546f–548f, 548t, 549f
radius of, 540t
satellite-to-parent mass ratio of, 540t
schematic cross-section of, 540f
structure, 540f, 549–552
core, 551–552, 552f
crust, 550–551, 551f
mantle, 551
regolith, 517–518, 517f, 549–550, 550f
Moonlet wakes, 892, 893f
Moon's moment of inertia (MOI), 540–541
Mössbauer Spectrometer (MB), 399, 399t
MPF. See Mars pathfinder
MRI. See Magnetorotational instability
MRO. See Mars Reconnaissance Orbiter
MSL. See Mars Science Laboratory; MSL curiosity
MSL curiosity (MSL), 398t, 406, 407f
MSX. See Midcourse Space Experiment
MT. See Magnetotellurics
Muhleman model, 1142
Multi Purpose Sensor (MUPUS), 1197, 1197f
MUPUS. See Multi Purpose Sensor
MUV. See Midultraviolet

N

Naiad, 761t
relative size of, 760f
Nanedi Vallis, 350f
NASA. See National Aeronautics and Space Administration
National Aeronautics and Space Administration (NASA), 446
web address for, 943t
National Solar Observatory (NSO), 247
N-body problem, 71–72
NEAR. See Near-Earth Asteroid Rendezvous spacecraft
Near Earth object, 984
NEAR Shoemaker spacecraft, 618, 618f
Near-Earth Asteroid Rendezvous spacecraft (NEAR)
asteroids flyby, 600, 601t
exploration by, 983f

Near-Earth asteroid thermal model (NEATM), 615
Near-Earth Asteroid Tracking system, 610–611
Near-Earth asteroids (NEAs), 933, 991–992
Near-Earth object (NEO), 603–624
comets and, 609–610
evidence for extinct comets, 610
Tisserand parameter, 609–610, 609f
impact hazards, 619–622, 619f–620f
collision frequency, 621
collision magnitude, 620–621
mitigation measures, 621–622, 622f
in-situ studies, 618–619
Hayabusa, 618–619
NEAR Shoemaker, 618, 618f
numbering and naming conventions, 622–623
origins, 606f, 607–610
dynamical history, 607–608, 609f
meteor shower associations, 608–609
relationship to comets, 609–610
relationship to main belt asteroids, 607
relationship to meteorites, 608, 608f
physical properties, 612–617, 613f
brightness, 613
color and taxonomy, 616–617
density, 616
mineralogy, 617, 617f
rotation rates, 614, 615f
shape, 613, 613f–614f
size and albedo, 614–616, 615f–616f, 618f
population, 610–612
search programs and techniques, 610–611, 610f
size, number of NEOs, 611–612, 612f
significance, 606–607
early solar system remnants, 606, 606f
exploration destinations, 606–607
hazard assessment, 606
resource potential, 606–607
subgroups of
Amors, 604–605, 604f–605f
Apollos, 604, 604f–605f
Atens, 604, 604f–605f
Atiras, 604, 604f–605f
total of discovered, 604f
Near-infrared mapping spectrometer (NIMS), Ganymede and Callisto data with, 814–815, 820, 821f
Nearly isotropic comets (NICs), 711–712, 711f
Near-ultraviolet (NUV), 1047
NEAs. See Near-Earth asteroids
NEATM. See Near-Earth asteroid thermal model
Nelson, R., 781
NEO. See Near-Earth object
NEOShield project, 622, 622f
Neptune
atmospheres observations with UV for, 1057
axial inclination of, 724t
cloud layers on, 730f

discovery of, 19th century studies on, 1010–1011
equatorial radius of, 724t
GDS of, 733
hot Neptunes, 966–967
Hubble Space Telescope image of, 733, 733f
interior modeling for, 754–755, 755f
Kelvin-Helmholtz model in, 757
length of day of, 724t
magnetic field highlights from missions to, 122t
magnetospheres with, 141t
mass of, 724t
nonthermal radio emissions of, 1118
orbit of, 665
orbital period of, 724t
photochemicals in stratosphere of, 729f
physical properties of, 724t
planetary dynamos with, 134
planetary rings of, 887–888, 888f
position of, 942f
radio wavelengths thermal emission from, 1116–1117
satellites of, 776–777, 862f
inner moons, 776–777, 776f
outer irregular moons, 777
size relative to other planets, 729f
solar wind properties with, 141t
stratospheric aerosols on, 731
surface gravity of, 724t
temperature of, 724t
20th century studies on, 1016
Voyager 2's trajectory through, 865f
Nereid, 761t, 776
relative size of, 760f
Nesvorny, David, 1077–1078
Neugebauer, M., 262
Neutral Mass Spectrometer (INMS), 836
Neutrinos, 238
Neutron probe experiment (LNPE), Apollo science experiments with, 561t
Neutrons, 1162
detection of, 1167–1171
counting rate models, 1167–1168
high-purity germanium detector for, 1168–1169, 1169f
radiation spectrometers for, 1168
scintillators for, 1170–1171
semiconductor radiation detectors for, 1168
spatial resolution, 1171
techniques and intermnents for, 1168–1171, 1169f–1170f
origin of, 1162–1167, 1163f
galactic cosmic rays, 1162–1163, 1164f
gamma ray production and transport, 1166–1167, 1167f
neutron moderation, 1163–1166, 1165f
semiconductor radiation detectors for, 1168
New Horizons spacecraft
exploration by, 983f
Ganymede and Callisto exploration by, 815
interplanetary dust detection by, 668t, 669
Kuiper belt object flyby by, 954–955

Newton, Hubert, 1011
Newton, Isaac, 1005
  solar system studies with, 1005, 1005t
Newton's Laws, 55–57
  second law of motion equation, 57
  universal law of gravity equation, 57
Nice model, 51
NICs. See Nearly isotropic comets
NIMS. See Near-infrared mapping spectrometer
Nitrogen
  biologically available, 856, 859
  giant planets atmospheres with, 726–727, 726t
  Triton, liquid seas of, 863–864, 864f
  Venus atmosphere with, 311
Nix, discovery of, 910f, 911
Noble gases
  meteorites with, 646–647, 647f
  Venus atmosphere with, 313
Noblesville H chondrite, 627f
Node, 708
Nongravitational forces, 707
Nonthermal escape, 347
Nonthermal radio emissions, 1107, 1120
Nozomi spacecraft, interplanetary dust detection by, 668t
NSO. See National Solar Observatory
Nuclear spectroscopy, 1161–1184
  future prospects for, 1181–1183
  gamma rays and neutrons detection, 1167–1171
    counting rate models, 1167–1168
    spatial resolution, 1171
    techniques and interments for, 1168–1171, 1169f–1170f
  gamma rays and neutrons origin, 1162–1167, 1163f
    galactic cosmic rays, 1162–1163, 1164f
    gamma ray production and transport, 1166–1167, 1167f
    neutron moderation, 1163–1166, 1165f
  missions, 1171–1175, 1172t, 1174f
  science, 1175–1181
    Lunar Prospector, 1175–1177, 1175f–1176f
    Mars Odyssey, 1177–1181, 1179t, 1180f
Nucleosynthesis, 38–41, 39f–40f, 41t
Nutation, 169–171, 170f, 171t, 1191
  free core, 181
NUV. See Near-ultraviolet

O

Oberon, 761t, 774–775, 774f–775f
  relative size of, 760f
Oblateness, 1005
Obliquities, 160, 160t, 164, 286, 909–910
Occultations, 885
Oceans
  Earth, 433–436
    atmosphere-ocean interactions, 436
    atmosphere-ocean system, 441–444
    circulation, 434–435
    interplanetary evidence for life in, 441–444, 442f
      oxygen and methane, 443–444
      radio emissions, 442–443
      surface features, 443
    salinity, 435–436
    structure, 433–434
  Ganymede and Callisto, 816–818, 817f
    induction fields with, 819
  on other worlds, 436
Odin-type plains, 296f
Olbers, Heinrich, 1007
Olbers, Wilhelm, 1017
Oldhaim, Richard, 1014
Oligarchic growth, 44–45
Olmsted, Denison, 1011
Oort, J. H., 8
Oort cloud, 8, 713–714, 714f, 929–930, 983
  formation of, 717–719, 717f–718f, 719f
Ophelia, 761t
  relative size of, 760f
Öpik, Ernst, 1017
OPP. See MER opportunity
Opportunity and Opportunity rover, Mars exploration by, 345, 360t, 1218
Opposition, 1009
OPR. See Ortho- to para-hydrogen ratio
Optical depth, 314, 885
Orbit
  inclinations, 671–672
  Neptune, 665
  small bodies, 58–63
  small bodies, dissipative forces and, 72–75
  in space, 56–57, 57f
  stability in, 71–72
Orbital constants, Mars, 361–362, 361t
Orbital elements, 56–58, 57f
Orbital evolution, 68–71
  asteroids, 68–70, 68f–70f
  comets, 70–71
  meteorites, 70
  small satellites/rings, 71
Orbital resonances, 48
Orbit-orbit resonance, 75
Orcus, magnitudes, albedos, and diameters of, 944t
Ordinary chondrules, 34–37
Orionids, 661t
Orphan trails, 1082, 1083f
Orsted, magnetic field highlights from, 122t
Ortho- to para-hydrogen ratio (OPR), 699
Ortho-hydrogen, 724–725
Outcrops, Mars, 409f–410f, 410
Oxygen
  Earth, 443–444
  giant planet atmospheres with, 726–727, 726t
  meteorites, isotopes and interpretation, 633–635, 635f
  Venus atmosphere with, 311–312
Ozone, Venus atmosphere with, 311–312

P

PAH. See Polyaromatic hydrocarbons
Paleomagnetic measurements, 1186
Pallasites, 34–37, 37f
Pallene, 761t, 774f
  relative size of, 760f
Pan, 761t, 773, 774f
  relative size of, 760f
Pancam. See Panoramic Camera
Pandora, 761t, 773, 774f
  relative size of, 760f
Panoramic Camera (Pancam), 399, 399t
Panoramic Survey Telescope and Rapid Response System, 610–611
Panspermia theories, 217
Parabolic orbits, 58
Para-hydrogen, 724–725
Parameter of ellipse, 159–160
Parker, E. N., 20–21
Parker Spiral, 482, 482f
Partial melt, 194
Particle acceleration and kinematics, 253–254, 255f
Particle-size distribution, 885
Pasiphae, 761t
  relative size of, 760f
Pele plume, 782, 782f
Perdita, 761t
  relative size of, 760f
Periapse, 56–57, 892
Perihelion, 68, 675, 1006, 1081
Periodic comet, 709
Perseids, 661t
Phase angles, 884–885
Phase functions, 885
Phase space, 63
PHO. See Potentially hazardous object
Phobos spacecraft
  flyby missions to satellites by, 764t
  Mars exploration by, 360t, 1218
  Mars Express flyby missions to, 764t
  properties of, 761t
  relative size of, 760f
  satellites of, 760f, 769–770, 769f
Phoebe, 773, 773f
  properties of, 761t
  relative size of, 760f
Phoenix (PHX), 398t, 406, 406f
  Mars exploration by, 1219
Pholus, rotation period of, 947t
Phosphine, giant planet atmospheres with, 726t, 727
Photodissociated gas, 1111
Photometry, planetary satellites observation with, 767
Photon-stimulated desorption (PSD), 288
Photosphere, 239–241
  defined, 239
  granulation and convection, 239–240, 240f
  photospheric magnetic field, 240
  sunspots, 240–241, 241f
Photospheric magnetic field, 240
Phototactic microorganisms, 215–216

Phreato-Magmatic activity, 116—117
PHX. See Phoenix
Phylogenetic tree, 211f, 216
Pickering, William, 1010
Pilcher, Carl, 915
Pillanian eruptions, 786f, 787—788
Pioneer spacecraft
　Pioneer 8 and 9, interplanetary dust detection by, 667, 668t
　Pioneer 10, 983f
　　flyby missions to satellites by, 764t
　　Ganymede and Callisto exploration by, 815
　　interplanetary dust detection by, 668t, 669
　　Io flyby, 781
　　magnetic field highlights from, 122t
　Pioneer 11, 983f
　　flyby missions to satellites by, 764t
　　Ganymede and Callisto exploration by, 815
　　interplanetary dust detection by, 668t, 669
　　magnetic field highlights from, 122t
　　planetary satellites observation with, 769
Pioneer Venus Multiprobe and Orbiter, Venus space missions by, 307—308
Pioneer Venus probes, 307—308, 1212
PKT. See Procellarum KREEP Terrane
Plains fractures, Venus, 340—341
Planetary cores, 127—128
Planetary dust streams, spacecraft measurements of, 669—670, 670f—671f
Planetary dynamos, 131—135
　ancient Mars, 134—135
　ancient Moon, 134
　driving forces in, 127
　　compositional convection, 127
　　radiogenic heat sources, 127
　Earth, 131—133, 132f
　extrasolar planets, 135
　fluid motions in dynamo regions with, 127, 128f
　　convective flows, 127, 128f
　　meridional flows, 127
　　zonal flows, 127
　Ganymede, 134
　generation mechanisms with, 127—129, 128f
　Jupiter, 133
　mechanism for, 125—126
　　conditions, 126
　　generation regions, 125t, 126
　Mercury, 133
　Neptune, 134
　planetary bodies lacking dynamos, 135
　Saturn, 133—134
　small bodies, 135
　standard, 127—129
　beyond standard dynamo, 129
　　alternative driving mechanisms, 129
　　dynamo region geometry, 129
　　influence of external fields, 129
　　laterally-varying boundary conditions, 129
　　radially-varying physical properties, 129
　　stably stratified layers, 129
　Uranus, 134
Planetary embryos, 43—44, 186

Planetary impacts, 83—100
　impact craters, 83—89
　　crater dimensions, 88—89, 88t
　　crater shape, 83—88, 84f—87f
　impact processes, 89—94
　　crater formation, 89—92, 90f
　　target rock changes in, 92—94, 93f
　planetary evolution with, 94—97
　　biosphere evolution, 95—97
　　early crustal evolution, 95
　　impact origin of earth's moon, 94—95
　as planetary probes, 97—99
　　morphologic and geologic, 99
　　spectral composition, 98—99
　　water and ices, 97—98, 98f
Planetary interior modeling, 750—756, 751f—752f
　extrasolar giant planets, 755—756, 756f
　Jupiter, 752—753, 753f
　Neptune, 754—755, 755f
　Saturn, 753f, 754
　Uranus, 754—755, 755f
Planetary interiors, 126—127, 129
Planetary interiors, evolution of, 185—208
　constraints on and models for, 198—206
　　core-mantle boundary heat flow, 202—204, 203f—204f
　　crust formation, 199—201, 200f
　　evolution of interior, 205—206, 206f—207f
　　lithosphere surface deformation, 204—205, 205f
　　magnetic field history, 202—204, 203f—204f
　　mantle heat budget, 201—202, 202f
　　plate tectonics, 205—206, 206f—207f
　　surface heat flow, 201—202, 202f
　　thermal evolution, 204—205, 205f
　　volcanic history, 199—201, 200f
　formation and early evolution, 186—189, 186f—187f, 187t, 188f
　modeling interior dynamics with, 197—198, 198f—199f
　modes of convection with, 193—196, 194f, 194t, 195f—196f
　rock rheology with, 193—196, 194f, 194t, 195f—196f
　subsolidus convection with, 189—193, 190f—192f, 193t
Planetary interiors using geophysical tools, 1185—1204
　EM sounding, 1198—1203
　　future exploration with, 1202—1203, 1202f, 1206f
　　Galilean satellites, 1201—1202, 1202f
　　measurement techniques, 1199, 1199t
　　Moon, 1200—1201, 1201f
　　natural sources, 1198
　　terrestrial exploration, 1199—1200, 1200f
　gravity and dynamics, 1190—1192, 1191f
　heat flow measurements, 1192—1198
　　Earth's heat flow, 1193—1194, 1194f
　　future investigations, 1196—1198, 1197f
　　lunar heat flow, 1194—1196, 1195f—1196f

　　measurement techniques, 1192—1193, 1193f, 1195f
　seismology, 1186—1190
　　Jovian, 1190
　　lunar, 1187—1190, 1188f—1189f
　　Mars, 1190
　　terrestrial, 1186—1187, 1187f
Planetary magnetic fields
　dynamo mechanism for, 125—126
　　conditions for, 126
　　defined, 125—126
　　generation regions, 125t, 126
　generation of, 121—136
　magnetospheres with, 144—146, 145t, 146f—147f, 148t
　observations of, 121—125, 122t
　　sources of, 121—122
　　spatial characteristics of, 122—125, 123f—125f, 125t
　　temporal characteristics of, 125
　planetary dynamo, standard, 127—129, 128f
　planetary dynamos in planets, 131—135, 132f
Planetary motion laws, 56, 56f
Planetary perturbations, 58—63
　Keplerian motion and resonances, 59
Planetary protection, 995—996
Planetary rings. See Rings, planetary
Planetary shape, stereophotogrammetry in analysis of, 1223—1234
Planetesimals, 8, 43, 705, 858, 904
Planets
　definition of, 4—5
　dwarf, 3, 5, 6t
　giant, 4, 12, 12f, 19f
　　astrobiological potential, 226—228, 226f
　ice dwarf, 923
　orbits of, 6t
　physical parameters for, 10t
　pulsar, 963—964, 963t
　rotation of, 159—184
　terrestrial, 5, 11f, 1014
　triaxial ellipsoid, 175
Planetshine, 885
Plasma, 261, 290
　magnetospheric, 146—150
　　energetic particles, 148—150, 149t
　　sources of, 146—148, 149f
　solar wind, kinetic properties with, 273—276
　　marginally collisional plasma, 273—274
　　solar wind electrons, 274—276, 275f
　　solar wind ions, 274, 275f
Plasmoid, 151—152
Plate tectonics, 101, 185, 205—206, 206f—207f
Plinian activity, 115—116, 116f
Plumes, 450
Pluto, 13, 13f, 909—924
　albedo, 909, 914, 921t
　atmosphere, 918—920
　　composition, 918—919
　　escape, 920
　　structure, 919—920, 919f
　atmospheres observations with UV for, 1057—1058

Pluto (*Continued*)
   Charon comparison, 920−921, 921f, 921t
   density of, 913−914, 921t
   diameter of, 942f
   discovery, 909−910
   discovery of, 20th century studies on, 1016, 1016f
   discovery of Pluto's satellites, 910−911, 910f
   historical background, 909−911, 910t
   infrared views, 1085−1087, 1086f
   interior, 917−918
     bulk composition, 917−918
     density, 917
     internal structure, 917−918, 918f
   magnetospheres with, 141t
   magnitudes, albedos, and diameters of, 944t
   mutual events, Pluto and Charon, 913−914
   orbit and spin of, 910t, 911−913
     Charon's orbit and system mass, 912−913, 913t
     heliocentric orbit, 911
     lightcurve, 911−912, 912f, 921t
   origin pf, 922−923
   planetary rings of, 888−889
   pole direction of, 911−912
   position of, 942f
   radius of, 913−914, 921t
   rotation period of, 911−912, 921t
   satellite system origin, 922−924
   size comparison to United States, 942f
   solar system context for, 923−924
   solar wind properties with, 141t
   solar winds interactions, 142−144
   solid surfaces observations with UV for, 1068
   spectra of, 921f
   surface properties and appearance, 914−917
     albedo and color, 914
     solar phase curve, 914
     surface appearance and markings, 912f, 916−917, 916f−917f
     surface composition, 914−915
     surface temperature, 915
Pluto-Charon barycenter, 911
P-mode waves, 238−239, 239f
Poincare, Henri, 63
Polar motion, 178−181, 179f−180f
Polarimetry, planetary satellites observation with, 768
Poles
   hot, 286
   Mars, 373
   Moon, volatiles at, 574
   warm, 286
Polyaromatic hydrocarbons (PAH), 639
Polydeuces, 761t
   relative size of, 760f
Pons, Jean Louis, 1011
Porous flow, 199
Porter, William, 1015
Portia, 761t
   relative size of, 760f
Potentially hazardous object (PHO), 603
Power spectra, 1150

Poynting-Robertson drag, 72−73, 898, 1077−1079
Poynting-Robertson effect, 667, 676, 676f
PR. *See* Pyrolytic release experiment
Prebiotic organics, 218t
Precession, 166−169, 167f, 896, 1191
Precession of equinoxes, 1000
Preliminary Earth model (PREM), 469, 470f
Presolar grains, 34−37
Principal axes of inertia, 168
Procellarum KREEP Terrane (PKT), 500, 567
Prograde spin, 1016
Promethean eruptions, 788, 788f
Prometheus, 761t, 773, 774f
   relative size of, 760f
Prometheus plume, 788f
Prominences, 249, 250f
Propellers, 892−895, 894f−895f
Proplyds. *See* Protoplanetary disks
Proteus, 761t, 776−777, 776f
   relative size of, 760f
Protoplanetary disks (Proplyds), 30−33, 30f
   flared, 33
Protoplanetary nebula, 705, 984
PSD. *See* Photon-stimulated desorption
Ptolemy, 1000f−1001f, 1001
Puck, 761t, 775−776
   relative size of, 760f
Pulsar planets, 963−964, 963t
P-waves, 462−463, 462f
Pyroclast dispersal (into vacuum), 117
Pyroclastic deposits, 102, 1148
Pyroclastic vents, 295−296, 296f
Pyrolytic release (PR) experiment, 220−221
Pythagoras, 999−1000

Q

Quadrantids, 661t
Quiet-sun regions, 243−244

R

Ra Patera, 783−785, 785f
Radar, planetary, 1133−1160
   frequencies used in, 1134t
   future of, 1157−1158, 1158f
   history, 1134−1136, 1135f
   planetary measurements using, 1144−1157
     asteroids, 1153−1155
     asteroids - imaging and shape reconstruction, 1153, 1153f−1154f
     asteroids - orbit determination, 1153−1155, 1155f
     asteroids - Yarkovsky effect, 1153−1155, 1155f
     icy satellites, 1155−1156
     Jupiter, 1155−1157
     Mars, 1150−1153
     Mars - radar heterogeneity, 1150−1151, 1152f
     Mars - subsurface structure, 1151−1153, 1152f

     Mercury, 1144−1147
     Mercury - polar deposits, 1144−1147, 1145t, 1146f
     Mercury - rayed craters, 1147
     Moon, 1149−1150
     Moon - lunar geology, 1150, 1151f
     Moon - polar environment, 1149−1150, 1150f
     Saturn, 1155−1157
     Titan, 1156−1157, 1156f
     Venus, 1147−1149
     Venus - Magellan surface imaging, 1147−1148, 1148f
     Venus - post-Magellan radar imaging, 1148−1149
   satellites observation with, 768
   scientific value of, 1133−1134
   target properties, 1141−1144
     radar albedo, 1141−1143, 1142f
     radar polarimetry, 1143−1144, 1143f−1144f
   techniques, 1136−1141
     radar imagers, 1136−1138, 1136f−1139f
     radar sounding, 1138−1139, 1140f
     radar topography, 1139−1141, 1141f
Radar albedo, 1141−1143, 1142f
Radar polarimetry, 1143−1144, 1143f−1144f
Radar sounder, Apollo science experiments with, 561t
Radial drift, 42
Radial velocity method, 957−958
   extrasolar planets detected with, 959−960
   planets around sun-like stars with, 964−967, 964f, 965t
Radiation force, 72−73
Radiation pressure, 898
Radiation spectrometers, 1168
Radio astronomy, future of research with, 1118
Radio emission, 256, 257f
Radio interferometric synthesis, 1150
Radio occultations, 885
Radio spectra, 1112, 1113f
Radio telescopes
   Atacama Large Millimeter Array, 1108f
   Very Large Array, New Mexico, 1108f
Radio wavelengths, 1107−1132
   instrumentation for, 1108−1109, 1108f
   nonthermal low-frequency emissions, 1121−1123
     atmospheric lightning, 1123
     cyclotron maser emissions, 1121−1123
     other low-frequency emissions, 1123
   nonthermal radiation, 1107, 1120−1131, 1122f
     Earth, 1123−1124
     Neptune, 1118
     Uranus, 1118
   nonthermal radiation by Jupiter
     decametric and hectometric radio emissions, 1126−1127
     Ganymede, 1128
     kilometric radio emissions, 1127−1128
     low frequencies, 1126−1128, 1127f

synchrotron radiation, 1124–1126, 1124f–1126f
very low-frequency emissions, 1128
nonthermal radiation by Saturn, 1128–1131
electrostatic discharges, 1130–1131
kilometric radio emissions, 1128–1130, 1129f–1130f
very low frequency emissions, 1130
nonthermal synchrotron radiation, 1122f, 1123
thermal emission from planetary bodies, 1107, 1109–1120
asteroids, 1118–1119
comets, 1119–1120, 1120f–1121f
Galilean satellites, 1117
giant planets, 1112–1117
Jupiter, 1112–1114, 1113f–1114f
major satellites and small bodies, 1117–1119
Mars, 1111–1112
Mercury, 1110–1111, 1110f
Moon, 1110
Neptune, 1116–1117
planet's atmosphere, 1109–1110
planet's (sub)surface, 1109
radio spectra, 1112, 1113f
Saturn, 1114–1116, 1115f
terrestrial planets and Moon, 1110–1112
thermal or blackbody radiation, 1109
Titan, 1117–1118, 1117f
trans-Neptunian objects, 1118–1119
Uranus, 1116–1117, 1116f
Venus, 1111–1112, 1111f–1112f
Radioactive isotopes, 186–187
Radiogenic heat sources, 127
Radiometry, planetary satellites observation with, 767–768
Ramaty High-Energy Spectroscopic Solar Imager (RHESSI), 254–255
Ramsey, William, 1014
Random walk, 708–709
Ranger spacecraft
planetary satellites observation with, 768
Ranger 7, 8, and 9, Moon exploration by, 1214
Ranger probes, lunar exploration by, 557t
Rare earth element (REE), 497–498, 498f
RAT. See Rock Abrasion Tool
Ravi Vallis, 351f
Ray paths, 462–463, 462f
Rayleigh number, 191–192
Reaction rate, 1167
Reconnection events, 290, 1202
Recurring Slope Lineae (RSL), 352, 352f
Red slope, 300
REE. See Rare earth element
Reflectance spectroscopy, 299
Reflected waves, 1186
Refraction, 308
Refractory elements, 289
Refractory rocky, 657
Refractory surfaces, 914
Regolith
Ganymede and Callisto, 822–823, 822f
Mercury, 284, 293

Moon, 517–518, 517f, 549–550, 550f
processes, 574–575, 575f
properties and sources, 518
thickness, 517–518
Regular satellites, 52
Remanent magnetic fields, 1192
Resonances, 59, 852, 888, 1078–1079
examples of
Hill sphere, 60–61, 61f
horseshoe orbits, 60
Lagrangian points, 60–61, 60f
ring particles, 61–63
shepherding, 61–63
tadpole orbits, 60
orbit-orbit, 75
secular, 70
spin-orbit, 162–164, 175
3:2 spin-orbit, 162–164
tidal evolution and, 74–75
Resonant emission, 288
Resurfacing history, Venus, 326–329, 327f–328f
Resurs, Earth exploration by, 1213
Retrograde, 305–306, 1010
Retrograde orbit, 708
Revolution periods, 160, 160t
Reynolds number, 73–74
Rhea, 761t, 770–772, 771f
albedo, 1064t, 1145t
planetary rings of, 889–890
relative size of, 760f
Rheology, 856, 1192
RHESSI. See Ramaty High-Energy Spectroscopic Solar Imager
Ribonucleic acid. See RNA
Ridge belts, Venus, 340, 340f
Ridges, troughs, bands, Europa, 797–801
cycloidal structures, 798–799, 800f
double ridges, 797f, 798, 799f
folds, 800–801, 802f
pull-apart bands, 799–800, 801f
ridge complexes, 797f, 798
triple bands, 799, 800f
troughs, 797f, 798
Ring current, 148–149
Ring particles, 61–63
Rings, planetary, 15–20, 883–906
age and origins of, 904–905
astrophysical disks compared to, 883
defined, 883
dense broad disks, 890–896, 890f–891f
structures formed by distant moons, 895–896
confinement of edges and ringlets, 896
spiral waves, 895–896, 896f
structures formed by nearby moons, 892–895
gap edges and moonlet wakes, 892, 893f
propellers, 892–895, 894f–895f
structures formed spontaneously, 891–892
overstabilities, 892
self-gravity wakes, 891–892, 891f–892f

dense narrow rings, 896–897
dusty rings, 897–901, 897f
azimuthal clumps and kinks, 900–901, 900f–901f
dust production and destruction, 897–898
nongravitational forces, 898
resonantly confined material, 898–900, 899f
giant planets, 19f
Jupiter, 671–672, 672f
learning about rings, 884–885, 884f, 886f
by location, 885–890
exoplanets, 890
Jupiter, 888, 888f–889f
Mars and Pluto, 888–889
Neptune, 887–888, 888f
Rhea and Iapetus, 889–890
Saturn, 885–887, 886f
Uranus, 887, 887f
orbital evolution of, 71
Saturn, 672–673
B ring spokes, 672
E ring, 672–673, 672f
X-ray emissions, 1033–1034, 1035f
solar system probing using rings, 901–903
interplanetary meteoroids, 902, 902f–903f
planet's gravity, 903
planet's magnetosphere, 903, 903f–904f
solid surfaces observations with UV for, 1069–1070, 1070f
tenuous dusty, 670–673, 672f
RM. See Rossiter-McLaughlin effect
RNA (Ribonucleic acid), 211
Roche limit, 883
Rock Abrasion Tool (RAT), 399, 399t
Rock rheology, 193–196, 194f, 194t, 195f–196f
Rocks
giant planets interiors in EOS model with, 749
highland, 506, 506f
impact processes with, 92–94, 93f
lunar, 505–508, 506f
impact breccias, 507–508
magnesian suite, 507
mare basalts, 507, 507f
mantle of Europa, 795, 795f
Mars
evolutionary implications of igneous rocks, 417
geology, 409, 410f
igneous, 417
mineralogy and geochemistry, 405f–406f, 412–416, 412f–416f
refractory, 657
volcanic, 101
Rocks rich in K, P, and REEs. See KREEP
Romer, Olaus, 780
Rosalind, 761t
relative size of, 760f
Rosetta Philae Lander, 1197, 1197f
Rossiter-McLaughlin (RM) effect, 968
Rotation, 159–184
Earth atmosphere influenced by, 428–429
libration, 175–177, 175f–176f

Rotation (*Continued*)
  LOD variations, 171–175, 172f–174f
  long-term evolution of orientation with, 164–165, 164f
  long-term spin evolution with, 160–164
  nutation, 169–171, 170f, 171t
  observation of planet, 181–184
    Earth's particular case, 183–184
    ground-based, 181, 182f
    spacecraft based, 182–183, 182f–183f
  observed rotation state of planets, 159–160
  origin of, 160–164
  precession with, 166–169, 167f
  rotational flattening of planets with, 165–166
  tidal dissipation with, 161–164
    effects on orbit of, 163–164, 164f
    long-term spin evolution, 161–163, 162f
    tidal torque, 161, 161f
  wobbles and planet interiors, 177–181, 177f–178f
    forced nutation, 181
    free core nutation, 181
    polar motion, 178–181, 179f–180f
Rotation axis, 160
Rotation periods, 160, 160t
Rotation rates, NEOs, 614, 615f
Rotational flattening of planets, 165–166
Rotational lightcurve, 914
R-process isotopes, 39
RSL. *See* Recurring Slope Lineae
Ruggieri, Guido, 1016
Runaway growth, 186

S

SAC-c, magnetic field highlights from, 122t
Sakigake spacecraft, space missions to comet by, 685t
Samarium-neodymium isochron diagram, 654f
Sampling System (SA/SPaH), 399t
SAR. *See* Synthetic aperture radar
SA/SPaH. *See* Sampling System
Satellites, 5, 15–20, 17f, 51–52
  chaotic rotation with, 77–78
  Galilean, atmospheres observations with UV for, 1058–1059, 1058f–1060f
  Galilean, solid surfaces observations with UV for, 1061–1063, 1061t, 1062f–1063f
  icy, 107–108
  icy, radar imaging for, 1155–1156
  icy satellite systems comparison with UV, 1067–1068
  irregular, 52
  origin in Pluto of, 922–924
  planetary, 51–52
  Pluto, 910–911, 910f
  properties of, 761t
  radio wavelengths thermal emission from, 1117–1119
  regular, 52
  Saturnian, albedo, 1145t
  Saturnian, solid surfaces observations with UV for, 1063–1065, 1064t, 1065f–1066f
  tidal interactions with, 74–75
  Uranian, solid surfaces observations with UV for, 1066–1067, 1067f
Satellites, planetary, 759–778
  characteristics, 759–765
    discovery, 759–764, 761t, 764t
    properties, physical and dynamical for, 764–765
  flyby missions to, 764t
  formation of, 765–767
    evolution, 766–767
    theoretical models, 765–766
  individual, 769–777
    Jupiter's small satellites, 770, 770f
    Mars: Phobos and Demos, 760f, 769–770, 769f
    Neptune's inner moons, 776–777, 776f
    Neptune's outer irregular moons, 777
    Saturnian system, 770–774
    Saturn's medium-sized icy satellites, 770–772, 771f–772f
    Saturn's small satellites, 773–774, 773f–774f
    Uranus' medium-sized icy satellites, 774–775, 774f–775f
    Uranus' small satellites, 775–776
  observation of, 767–769
    photometry, 767
    polarimetry, 768
    radar, 768
    radiometry, 767–768
    spacecraft exploration, 768–769
    spectroscopy, 767
    telescopic, 767–768
  relative sizes of, 760f
Satellites, small, orbital evolution of, 71
Saturn
  atmospheres observations with UV for, 1054f, 1055–1056, 1056f
  axial inclination of, 724t
  cloud layers on, 730f
  diagram for hydrogen of, 749f
  equatorial radius of, 724t
  interior modeling for, 753f, 754
  Kelvin-Helmholtz model in, 756–757
  length of day of, 724t
  magnetic field highlights from missions to, 122t
  magnetospheres with, 141t
  mass of, 724t
  19th century studies on, 1010
  nonthermal radio emissions of, 1128–1131
    electrostatic discharges, 1130–1131
    kilometric radio emissions, 1128–1130, 1129f–1130f
    very low frequency emissions, 1130
  orbital period of, 724t
  physical properties of, 724t
  planetary dynamos with, 133–134
  planetary rings of, 672–673, 885–887, 886f
    B ring spokes, 672
    E ring, 672–673, 672f
    X-ray emissions, 1033–1034, 1035f
    radar imaging for, 1155–1157
    radio wavelengths thermal emission from, 1114–1116, 1115f
  satellites of, 770–774
    medium-sized icy satellites, 770–772, 771f–772f
    small satellites, 773–774, 773f–774f
  size relative to other planets, 729f
  solar wind properties with, 141t
  surface gravity of, 724t
  temperature of, 724t
  20th century studies on, 1015–1016
  X-ray emissions, 1032–1033, 1034f
Saturn electrostatic discharges (SEDs), 1130–1131
Saturn's kilometric radiation (SKR), 1128–1130, 1129f–1130f
S-band, 182
S-band transponder, Apollo science experiments with, 561t
Scale heights, 310–311
Scattered disk, 8–9, 51, 713–717, 715f–716f, 926–927
  formation of, 717–719, 717f–718f, 719f
  objects, 949, 950f
Schiaparelli, Giovanni, 1008
Scientific Instrumentation Module (SIM), 560f
Scintillators, 1170–1171
SDO. *See* Solar Dynamics Observatory
Secchi, Angelo, 1008
Secondary craters, 293–294
Secular cooling, 189
Secular resonances, 70, 930
Sedna, magnitudes, albedos, and diameters of, 944t
SEDs. *See* Saturn electrostatic discharges
Seismic profiling experiment (LSPE), Apollo science experiments with, 561t
Seismic tomography, 185
Seismic velocities, 1186
Seismicity, Moon, 499f, 500
Seismology, 1186–1190
  Jovian, 1190
  lunar, 1187–1190, 1188f–1189f
  Mars, 1190
  Moon (Earth), 546–549, 546f–548f, 548t, 549f
  terrestrial, 1186–1187, 1187f
SELENE (Kaguya), Moon space missions by, 493–494
Semiconductor, 1168
Semimajor axis, 925
Semiminor axis, 159–160
SEP. *See* Surface electrical properties
Separatrices, 67–68, 67f
SHARAD, 1135–1136, 1139, 1140f, 1151
Shepherding, 61–63
Shergottite-Nakhlite-Chassignite (SNC) meteorites, 33–34, 633
Shield volcanoes, 101
Shock, chondrites, 639–640, 640t

Shoemaker, Gene, 618
Shoemaker-Levy comet, 690–691, 692f
Short-lived isotopes, 38–41, 39f–40f, 41t
Sidereal, 305–306
Sidereal year, 999
Silicate compounds, 186
SIM. *See* Scientific Instrumentation Module
Sinope, 761t
   relative size of, 760f
SKR. *See* Saturn's kilometric radiation
Sky survey telescopes, 1100–1104, 1103t, 1104f
SMA. *See* Smithsonian Submillimeter Array
Small missions for advanced research in technology-1 (SMART-1), 493–494, 567
   lunar exploration by, 557t, 1216
Smithsonian Submillimeter Array (SMA), 1117
SMM. *See* Solar Maximum Mission
Smooth plains, 284, 292–293
SNC. *See* Shergottite-Nakhlite-Chassignite meteorites
Soft X-ray Telescope (SXT), 248
SoHO. *See* Solar and Heliospheric Observatory
Soil Mechanics experiment (ASP/SRP), Apollo science experiments with, 561t
Soils, Mars
   geology, 410
   mineralogy and geochemistry, 411f, 416–417
Solar activity cycle, 262
Solar and Heliospheric Observatory (SoHO), 246–247
   Sun exploration by, 1210
Solar corona, 236, 261
Solar Dynamics Observatory (SDO), 248
   Sun exploration by, 1211
Solar dynamo, 239
Solar energetic particles, 988
Solar flares, 248–258, 261–262
   coronal mass ejections, 256–258, 258f, 259f
   filaments and prominences, 249, 250f
   flare plasma dynamics, 252–253, 253f
   gamma-ray emission, 254–255, 256f
   hard x-ray emission, 254
   magnetic reconnection, 249, 250f
   models, 249–252, 251f–252f
   particle acceleration and kinematics, 253–254, 255f
   radio emission, 256, 257f
Solar latitude effects, 265, 266f
Solar Maximum Mission (SMM), Sun exploration by, 1210, 1210f
Solar nebula, 8, 31–32, 186
Solar parallax, 1006
Solar system
   architecture of, 5–22
   fate of, 27–28
   habitable zone of, 984–986
   nature and composition of, 9–15, 10t
   origin of
      18th century studies, 1006–1007
      20th century studies on, 1017
   place in galaxy, 3–28, 24f–25f
   Pluto's context in, 923–924
   rings used to probe, 901–903
   stability of, 72
Solar system, origin, 22–24, 23f, 29–54
   asteroid belt in, 48–49
   extrasolar planets in, 52–53, 53f
   gas and ice giant planets in, 49–51
   meteorites in, 33–38
      achondrites, 34–37
      chondrules, 34–37, 34f–35f
      irons, 34–37, 36f
   nucleosynthesis in, 38–41, 39f–40f, 41t
   planetary growth, early stages in, 41–43, 42f
   planetary satellites in, 51–52
   proplyds in, 30–33, 30f
   short-lived isotopes with, 38–41, 39f–40f, 41t
   star formation in, 30–33, 30f
   terrestrial planets formation in, 43–48, 44f–45f, 47f–48f
Solar system dynamics, 5–9, 55–80
   chaotic motion in, 63–68
      concept of, 63
      three-body problem as paradigm for, 63–68, 63f
   chaotic rotation in, 76–78
      Hyperion, 77, 77f–78f
      other satellites, 77–78
      spin-orbit resonance, 76–77, 76f
   dissipative forces in, 72–75
      gas drag, 73–74
      Poynting-Robertson drag, 73
      radiation force, 72–73
      tidal evolution and resonance, 75
      tidal interactions, 74–75
      Yarkovsky effect, 73
   Keplerian motion in, 55–57
      elliptical motion, 56–57, 57f
      orbit in space, 56–57, 57f
      orbital elements, 56–57, 57f
      planetary motion laws, 56, 56f
   orbital evolution of minor bodies in, 68–71
      asteroids, 68–70, 68f–70f
      comets, 70–71
      meteorites, 70
      small satellites/rings, 71
   planetary orbits stability in, 71–72
      n-body problem for, 71–72
   planetary perturbations in, 58–63
      Keplerian motion and resonances, 59
      resonance examples, 60–61
   two-body problem in, 57–58
      energy, circular velocity, escape velocity, 58
      laws of motion and universal gravitation, 57
      orbital elements, 58
      reduction to one-body case, 57–58
Solar system exploration, 981–998
   challenge of, 987–989
      radiation exposure, 988, 989f
      shielding as solution, 988–989, 989f
   expanding human frontiers with, 981–989
   exploration extents, 983–986, 983f
      exploration beyond Mars, 985–986
      exploration beyond solar system, 986, 986f
      human action radius is limited, 984, 984f
      special role of Mars, 984–985, 985f
   Hohmann transfer orbit, 985, 985f
   human exploration of Mars case study, 989–993
      cooperation models, 990
      defining and communicating benefits, 990–991, 991f
      design reference missions, 992, 993f
      destinations and strategies, 991–992
      global exploration road map, 992–993, 994f
      mapping way to Mars, 992–993
      strategic mission scenarios, 992, 992f
   outlook for, 996
   progress needed for future of, 986–987
      human-robotic partnership, 987
      international cooperation, 987
      need for sustainability, 987
      technological developments, 986–987
   reason to explore space, 982–983
   society and, 993–996
      contamination avoidance, 995–996
      global cooperation, 996
      many nations interested, 993, 994t
      nations finance, 993–995, 995t
      private capital financing, 995
Solar system studies, 999–1018
   Babylons and Greeks, 999–1001, 1000f–1001f
   Copernicus and Tycho, 1001, 1002f
   18th century, 1006–1007
      discovery of Uranus, 1006
      first asteroids, 1007
      Halley's comet, 1006
      origin of solar system, 1006–1007
      1761 and 1769 transits of Venus, 1006
   Kepler and Galileo, 1001–1003, 1003f
   19th century, 1007–1012
      asteroids, 1011
      comets, 1011
      Earth, 1008–1009
      Jupiter, 1009–1010, 1010f
      Mars, 1009, 1009f
      Mercury, 1008
      meteor showers, 1011–1012, 1012f
      Moon, 1008
      Neptune's discovery, 1010–1011
      Saturn, 1010
      Sun, 1007
      Uranus, 1010
      Venus, 1008
      Vulcan, 1007–1008
   17th century, second half, 1003–1005, 1004t
      Moon, 1003–1004
      Newton, 1005, 1005t
      Saturn, 1004

Solar system studies (*Continued*)
  20th century, pre-space age, 1012–1017
    asteroids, 1017
    comets, 1017
    Earth, 1014
    giant planets' atmospheres, 1015
    giant planets' internal structures, 1015
    Jupiter, 1015
    Mars, 1014–1015
    Mercury, 1013
    Moon, 1013–1014
    origin of solar system, 1017
    Pluto's discovery, 1016, 1016f
    Saturn, 1015–1016
    Sun, 1012–1013
    Uranus and Neptune, 1016
    Venus, 1013
Solar tide, 319
Solar wind, 5, 261–280, 288
  CME and transient, 268–271
    characteristics of ICMEs, 270, 270t
    coronal mass ejections, 268, 268f
    field line draping about fast ICMEs, 271
    heliospheric disturbances driven by fast, 269, 269f–270f
    magnetic field topology of ICMEs, 270–271, 271f
    magnetic flux balance problem, 270–271, 271f
    other forms of solar activity, 268–269
  discovery, 261–262
    early indirect observations, 261–262
    first direct observations in, 262
    Parker's model in, 262
  elemental abundances in, 276t
  energetic particles with, 276–277
  heavy ion content with, 276, 276t
  heliosphere with, 20–22
  heliospheric current sheet, 264–265
    ballerina skirt model, 265, 265f
    Sun's large-scale magnetic field, 265, 265f
  heliospheric magnetic field, 263–264, 264f
  kinetic properties of plasma, 273–276
    marginally collisional plasma, 273–274
    solar wind electrons, 274–276, 275f
    solar wind ions, 274, 275f
  magnetospheres with, 141t
  properties of, statistical in ecliptic plane at 1 AU, 262–263, 263t
  solar latitude effects, 265, 266f
  space weather with, 482, 482f–483f
  termination of, 272–273, 273f
  turbulence and magnetic field with, 277–278, 277f–278f
  variation with distance from Sun, 271–272, 272f
  velocity fluctuations with, 277–278, 277f–278f
Solar wind composition experiment (SWC), Apollo science experiments with, 561t
Solar wind electrons, 274–276, 275f
Solar wind ions, 274, 275f

Solar wind stream
  evolution, 266–268
    kinematic stream steepening, 266
    shock formation, 266–267, 267f
    two and three dimensions, 267–268, 267f
  structure of, 264–265, 264f
Solidus, 109
Solidus curves, 186–187
Solstice, 903
Space debris, 659
Space exploration, 981
Space ultraviolet, 914
Space weather, 152–153, 479–492
  astrophysics implications from, 491
  atmospheric effects of, 487–489, 489f
  geospace role in, 484–487, 485f–488f
  practical aspects of, 489–491
  solar and heliospheric roles in, 481–484, 484f
    CMEs, 479, 480f, 483–484, 483f
    Parker Spiral, 482, 482f–483f
    solar spectrum, 481, 481f–482f
    solar wind, 482, 482f–483f
  Sun-Earth connection, 479, 480f
Space weathering, 293–294
  Moon (Earth), 512–513
Space-based infrared telescopes
  Akari spacecraft, 1075f, 1077t
  Cosmic Background Explorer, 663, 1075f
  Herschel Space Telescope, 1075f, 1077t
  Infrared Astronomical Satellite, 13, 660, 662f, 702, 702f, 1073–1074, 1075f, 1077t
  Infrared Space Observatory, 833, 1075f
  Midcourse Space Experiment, 1075f, 1077t
  Spitzer Space Telescope, 30–31, 675, 1073–1074, 1075f
  Wide-field Infrared Survey Explorer, 1075f, 1077t
Spacecraft based rotational observation, 182–183, 182f–183f
Spacecraft exploration. *See also* Exploration missions
  planetary satellites observation with, 768–769
Spectrometer, 885
Spectroscopy, planetary satellites observation with, 767
Spectrum, 885
Spherically symmetric Earth model (SSEM), 469
Spheroid, 175
SPI. *See* MER spirit
Spin axis, 160
Spin-orbit resonance, 76–77, 76f, 162–164, 175
  chaotic rotation with, 76–77, 76f
Spiral bending wave, 896
Spiral density wave, 896, 896f
Spiral waves, 895–896, 896f
Spirit, Mars exploration by, 1218
Spirit Rover, Mars exploration with, 345, 360t
Spitzer Space Telescope, 30–31, 675, 1073–1074, 1075f, 1077t
Sputtering, 288, 347

SSEM. *See* Spherically symmetric Earth model
SSI. *See* Surface Stereo Imager
Stably stratified layers, 129
Stagnant lid, 185
Standard thermal model (STM), 615
Star, main sequence, 5
Star formation, 30–33, 30f
Stardust Wild 2 mission, space missions to comet by, 684, 685t
Stardust-NExT, space missions to comet by, 684–685, 685t
Stellar metallicity, 966
Stellar nucleosynthesis, 38–41, 39f–40f, 41t
Stellar occultations, 885, 913
Stereophotogrammetry, 1223–1234
  coordinate systems, 1225
  examples, 1228–1231, 1229f, 1229t, 1230f–1232f
  missions and cameras, 1224–1225
    acquisition of ancillary orbit and pointing data, 1225
    built-in stereo capability, 1225
    camera calibration, 1225
    cameras on surface landers and rovers, 1224–1225
    framing cameras, 1224
    scanners, 1224
  photogrammetric processing, 1225–1227
    DTM interpolation, 1227
    image matching, 1226
    orthorectification, 1227
    triangulation and bundle block adjustment techniques, 1226–1227
  quality assessment, 1227–1228, 1228f
STM. *See* Standard thermal model
Stokes' drag, 73–74
Stony-iron meteorites, 34–37
Strain, 193
Strain rate, 109
"Strange" ringlet, 896f
Strategic mission scenarios, Mars, 992, 992f
Stratification, 1186
Stratigraphy, Moon (Earth), 513–516, 514f
  age determination, 515, 515f
  age record of lunar maria, 515–516, 516f
  late heavy bombardment, 513–515
Stratosphere, 305–306
  Earth, 425–427
  giant planets, 733–738, 735f–736f
  interplanetary dust particles in, 658f
Stresses, 193
Stromatolites, 215–216, 216f
Strombolian activity, 111–112, 112f
Subduction, 195–196, 447, 456, 472, 475
Sub-satellites Magnetometer, Apollo science experiments with, 561t
Subsolidus convection, 189–193, 190f–192f, 193t, 1186
Substorms, 150, 153
Subtropical jet, 431
Suisei spacecraft, space missions to comet by, 685t

Index

Sulfur, giant planet atmospheres with, 726–727, 726t
Sulfur dioxide
  Mars atmosphere with, 353
  Venus atmosphere with, 312–313
Sun, 235–260
  chemical composition of, 235–236
  chromosphere and transition region, 241–243
    chromospheric dynamic phenomena, 242–243, 243f
    physical properties, 241–242, 242f
  corona, 243–248
    active regions, 243, 243f
    coronal heating, 248, 249f
    coronal holes, 244
    coronal magnetic field, 245–246, 246f
    dynamics of, 244–245
    hydrostatics of coronal loops, 244, 244f
    MHD oscillations of coronal loops, 246–247, 247f
    MHD waves in, 247–248, 247f–248f
    quiet-sun regions, 243–244
    transition region of, 245
  exploration missions, 1209–1211
    Advanced Composition Explorer, 1210–1211
    Genesis, 1211
    Helios, 1210
    SOHO, 1210
    Solar Dynamics Observatory, 1211
    Solar Maximum Mission, 1210, 1210f
    TRACE, 1211
    Ulysses, 1210
  interior, 236–239
    helioseismology, 238–239, 239f
    models for, 236–237
    neutrinos, 238
    solar dynamo, 239
    thermonuclear energy source, 238
  irradiance spectrum of, 236, 237f
  large-scale magnetic field of, 265, 265f
  models for
    hydrostatic equilibrium models, 236–237
    time-dependent numerical simulations, 236–237
  19th century studies on, 1007
  photosphere, 239–241
    defined, 239
    granulation and convection, 239–240, 240f
    photospheric magnetic field, 240
    sunspots, 240–241, 241f
  physical parameters for, 10t
  physical properties of, 236t
  solar flares and CMEs, 248–258
    coronal mass ejections, 256–258, 258f, 259f
    filaments and prominences, 249, 250f
    flare plasma dynamics, 252–253, 253f
    gamma-ray emission, 254–255, 256f
    hard x-ray emission, 254
    magnetic reconnection, 249, 250f
    models, 249–252, 251f–252f
    particle acceleration and kinematics, 253–254, 255f
    radio emission, 256, 257f
  spectroscopic determination of composition of, 34f
  structure of, 235–236, 236f
  20th century studies on, 1012–1013
Sunspots, 240–241, 241f
Super-Earths, 958, 966–967
SuperWASP, 968
Surface composition, Mercury, 299–301, 299f–300f
Surface electrical properties (SEP), Apollo science experiments with, 561t
Surface heat flow, 201–202, 202f
Surface Stereo Imager (SSI), 399t
Survey and Spacewatch programs, 610–611
Surveyor 1, 3, 5, 6, and 7, Moon exploration by, 1214
Surveyor landers, lunar exploration by, 557t
SWC. See Solar wind composition experiment
SXT. See Soft X-ray Telescope
Sycorax, 761t
  relative size of, 760f
Synchronous orbit, 903
  orbit-orbit resonance, 75
Synchronous orbit distance, 890–891
Synchrotron radiation, 1122f, 1123
Synodic periods, 985, 999, 1201–1202
Synthetic aperture radar (SAR), 1134, 1139f

T

T Tauri stars, 32
Tadpole orbits, 60
Taurids, 661t
Taylor-Proudman effect, 429
Taylor-Proudman theorem, 127
TDS. See Thermal degradation sample experiment
Tectonic stresses, 106
Tectonics
  Europa, 795–797
    diurnal stressing, 796–797, 797f
    nonsynchronous rotation, 797
  Mars, 367
  Mercury, 297–299, 297f–299f
  Moon (Earth), 511–512, 512f
  Venus, 337–341
    chasmata, 339, 339f
    coronae, 339–340, 339f
    crustal plateaus, 338–339, 338f
    fracture belts, 339, 339f
    plains fractures, grids, and polygons, 340–341
    ridge belts, 340, 340f
    tessera terrains, 338–339, 338f
    wrinkle ridges, 340
Tectonism, Ganymede and Callisto, 825–828
TEGA. See Thermal and Evolved Gas Analyzer

Telescopes
  ground-based, 1089–1106
    active optics system schematic, 1098f
    adaptive optics advances, 1099–1100, 1100f–1102f
    advantages of, 1089, 1090f–1092f
    with apertures greater than 5 meters, 1094t
    asteroid search with, 1103t
    construction advances, 1090–1098, 1093f, 1094t, 1096f–1099f
    detector arrays advances, 1098–1099, 1099f–1100f
    with equatorial and alt-az mounts, 1097f
    importance of, 1089–1090, 1090f–1092f
    sky survey telescopes, 1100–1104, 1103t, 1104f
    types of telescope mounts, 1096f
  planetary satellites observation with, 767–768
Telesto, 761t, 774f
  relative size of, 760f
Tempel 1 comet, 685t, 687f–688f
Temperature
  Mars, principles of
    conduction, 387–388, 388f
    convection, 388–390, 389f
    heat transport, 387–390
  Mercury, 284–286
  Venus, atmosphere, 308–311, 309f
    lower atmosphere, 309–310, 309f
    middle atmosphere, 309f, 310
    surface, 308–309
    upper atmosphere, 310–311
Termination shock, 272–273
Terminator, 1003
Terra, Earth exploration by, 1213
Terrestrial ages, meteorites chronometry with, 650, 650f
Terrestrial planets, 5, 11f, 186–187, 1014
Terrestrial planets formation, 43–48, 44f–45f, 47f–48f
Terrestrial seismology, 1186–1187, 1187f
TES. See Thermal Emission Spectrometer
Tessera terrains, Venus, 338–339, 338f
Tethys, 761t, 770–772, 771f
  albedo, 1064t, 1145t
  relative size of, 760f
TF. See Transfer function
Thalassa, 761t
  relative size of, 760f
Thebe, 761t, 770, 770f
  relative size of, 760f
Theia, 51
THEMIS. See Thermal Emission Imaging System
Thermal and Evolved Gas Analyzer (TEGA), 399t
Thermal boundary layer, 189–190
Thermal degradation sample experiment (TDS), Apollo science experiments with, 561t
Thermal desorption, 288
Thermal diffusion, 189–190

Thermal Emission Imaging System (THEMIS), 99
Thermal Emission Spectrometer (TES), 402
Thermal evolution, 204–205, 205f
Thermal lithosphere, 1202
Thermal radio emission, 1107
Thermochemical equilibrium theory, 732
Thermonuclear energy source, 238
Thermosphere, 305–306
　Earth, 427
　Triton, 868
Thor, 784f
3:2 Spin-orbit resonance, 162–164
Three-body problem
　chaotic orbits in, 65–67, 65f–67f
　location of regular and chaotic regions in, 67–68, 67f
　as paradigm for chaotic motion, 63–68, 63f
　regular orbits in, 64–65, 64f–65f
Three-component seismograph system, 463, 463f
Thrust faults, 292–293
Tidal despinning, 297
Tidal dissipation, 161–164
　effects on orbit of, 163–164, 164f
　long-term spin evolution, 161–163, 162f
　tidal torque, 161, 161f
Tidal evolution, 75
Tidal forces, Moon's orbit and, 496
Tidal heating, 286
Tidal torque, 161, 161f
Tides of Europa, 795–797
　tidal evolution, 795–796, 796f
　tidal heating, 796
Time-dependent numerical simulations, 236–237
Timing method, extrasolar planets detected with, 961–962
Tisserand parameter, 710–711
　comet classification system based on, 688–689
　comet-NEO relationship, 609–610, 609f
Titan, 831–850
　astrobiological potential for, 227–228
　atmosphere, 834–841
　　chemical composition, 833t, 835–837, 836f
　　clouds, 840–841, 840f
　　dynamical processes, 837–839
　　haze, 839–840
　　lateral and temporal variations, 836f, 837–838
　　observations with UV for, 1059–1060, 1061f
　　thermal structure, 834–835, 835f
　　three-dimensional view and waves, 838–839
　　zonal circulation, 837
　Cassini-Huygens mission, 833–834, 834f
　density of, 540t
　discovery of, 760, 831–832
　exchange processes, 847–848, 848f
　exploration, 832–834, 832f, 832t–833t, 834f
　future for data and exploration of, 848–849
　interior of, 847–848, 848f
　limb darkening on, 831
　magnetic field highlights from missions to, 122t
　magnetospheres interactions with, 154–156, 154t
　orbital and body parameters of, 832f
　origin and evolution of, 841
　properties of, 761t
　radar imaging for, 1156–1157, 1156f
　radio wavelengths thermal emission from, 1117–1118, 1117f
　radius of, 540t
　relative size of, 760f
　satellite-to-parent mass ratio of, 540t
　surface of, 841–847
　　Cassini orbiter view, 843–845, 844f–845f
　　in situ data, 845–847, 846f
　　Voyager and other pre-Cassini missions, 842, 843f
　Voyager spacecraft observation of, 768
Titan ringlet, 896f
Titania, 761t, 774–775, 774f–775f
　relative size of, 760f
Titius, Johann, 1007
TNOs. See Trans-Neptunian objects
Tombaugh, Clyde, 941
Toomre stability criterion, 49
TOPEX/Poseidon, Earth exploration by, 1213
Topography
　Mars, 363–364, 364f
　Mercury, 297–299, 297f–299f
　stereophotogrammetry in analysis of, 1223–1234
TRACE, Sun exploration by, 1211
Trans-Atlantic Exoplanet Survey (TrES), 968
Transfer function (TF), 1199, 1199t
Transit of star, 890
Transit photometry
　extrasolar planets detected with, 960–961
　technique, 957–958
Transit timing variations (TTVs), 974
Transition region, 241–243
　chromospheric dynamic phenomena, 242–243, 243f
　of corona, 245
　physical properties, 241–242, 242f
Transition zone, Earth, 471
Trans-Neptunian objects (TNOs)
　albedos, 1086
　orbital and dynamical structure, 926–930, 927f, 929f
　primordial sculpting, 934–938, 936f, 938f
　radio wavelengths thermal emission from, 1118–1119
　size distribution and total mass, 931–933, 932f
Tremaine, Scott, 941–942
TrES. See Trans-Atlantic Exoplanet Survey
Triaxial ellipsoid planet, 175
Triggered collapse, 32
Triton, 776, 861–882
　atmosphere and surface, 872–878
　　atmosphere, 872–874, 874f–875f
　　Hili Plume, Mahilani Plume, 873, 875f
　　plume models, 874–877
　　polar cap and climate, 877–878
　atmospheres observations with UV for, 1059–1060
　characteristics, 866–868, 866f, 866t, 867f
　cryovolcanic designation, 868
　discovery and orbit, 862–863
　geology, 868–872
　　bright polar terrains, 871–872
　　cantaloupe terrain, ridges, and fissures, 870f–871f, 871
　　craters distribution, 872f
　　diapirism, 871
　　history, 872, 872f
　　smooth plains and zoned maculae, 869f, 870
　　undulating, high plains, 868–869
　　walled and terraced plains, 869–870, 869f
　internal structure model for, 866f
　magnitudes, albedos, and diameters of, 944t
　near-infrared telescopic reflectance spectra of, 867f
　origin and evolution of, 878–881, 879f–880f
　　exchange capture model, 879, 879f
　plume models, 874–877
　　eruption velocity and temperature, 876
　　plumes as jets, 875–876
　　subsurface energy transport, 876–877, 877f
　pre-Voyager astronomy, 863–865
　　liquid nitrogen seas, 863–864, 864f
　　radius, mass, and spectra, 863
　　similarities with Pluto, 864–865
　properties of, 761t, 866t
　relative size of, 760f
　thermosphere, 868
　troposphere, 868
　volcanic region on, 869f
　Voyager 2 mission encounter, 865, 865f
Trojans, 15
Troposphere, 305–306
　Earth, 425–426
　giant planets, 733–738, 735f–736f
　Triton, 868
Trouvelot, Etienne, 1016
TTVs. See Transit timing variations
Turbulent concentration, 43
Two-body problem, 57–58
　energy, circular velocity, escape velocity, 58
　laws of motion and universal gravitation, 57
　orbital elements, 58
　reduction to one-body case, 57–58
Tycho, solar system studies with, 1001
Type-I migration, 44–45
Type-II migration, 50–51

U

Ulrich, R., 238–239
Ultraviolet astronomy, history of, 1047–1049

Ultraviolet light (UV), 479, 480f, 481, 885
  space, 914
Ultraviolet wavelengths, 1047−1072
  planetary atmospheres observations with, 1049−1060
    Europa, 1058−1059, 1059f
    Galilean satellites, 1058−1059, 1058f−1060f
    Ganymede, 1058−1059, 1060f
    Io, 1058−1059, 1058f
    Jupiter, 1052−1055, 1054f
    Mars, 1051−1052, 1053f
    Mercury, 1050
    Moon, 1050
    Neptune, 1057
    Pluto, 1057−1058
    Saturn, 1054f, 1055−1056, 1056f
    Titan, 1059−1060, 1061f
    Triton, 1059−1060
    Uranus, 1056−1057
    Venus, 1050−1051, 1051f
  solar system astronomical observations with, 1049
  solid surfaces observations with, 1060−1070
    asteroids, 1068−1069
    Charon, 1068
    comets, 1068−1069
    Enceladus, 1065−1066
    Galilean satellites, 1061−1063, 1061t, 1062f−1063f
    icy satellite systems comparison, 1067−1068
    Mercury, 1069
    Moon, 1069
    planetary rings, 1069−1070, 1070f
    Pluto, 1068
    Saturnian satellites, 1063−1065, 1064t, 1065f−1066f
    Uranian satellites, 1066−1067, 1067f
Ulysses spacecraft
  Ganymede and Callisto exploration by, 815
  interplanetary dust detection by, 667, 668t, 669
  Sun exploration by, 1210
Umbriel, 761t, 774−775, 774f−775f
  relative size of, 760f
Universal gravitation law, 57
Ur. See Urey ratio
Uranus
  atmospheres observations with UV for, 1056−1057
  axial inclination of, 724t
  cloud layers on, 730f
  discovery of, 1006
  equatorial radius of, 724t
  Hubble Space Telescope image of, 733, 733f
  interior modeling for, 754−755, 755f
  Kelvin-Helmholtz model in, 757
  length of day of, 724t
  magnetic field highlights from missions to, 122t
  magnetospheres with, 141t
  mass of, 724t
  19th century studies on, 1010
  nonthermal radio emissions of, 1118
  orbital period of, 724t
  physical properties of, 724t
  planetary dynamos with, 134
  planetary rings of, 887, 887f
  radio wavelengths thermal emission from, 1116−1117, 1116f
  satellites of, 774−776
    medium-sized icy satellites, 774−775, 774f−775f
    small satellites, 775−776
  size relative to other planets, 729f
  solar wind properties with, 141t
  stratospheric aerosols on, 731, 731f
  surface gravity of, 724t
  temperature of, 724t
  20th century studies on, 1016
Urey ratio (Ur), 201
Ursid, 661t
Utopia Planitia, 398t
UV. See Ultraviolet light
UV spectrometer, Apollo science experiments with, 561t

V

Varuna, rotation period of, 947t
Vastitas Borealis, 398t
Vega spacecraft (1 and 2)
  space missions to comet by, 683, 685t
  Venus exploration by, 307−308, 1211−1212, 1212f
Venera 4, 983f
Venera 9, 983f
Veneras 4 Through 16, Venus exploration by, 307−308, 1211
Venus
  astrobiological potential for, 220
  atmospheres observations with UV for, 1050−1051, 1051f
  density of, 380f
  exploration missions, 1211−1212
    Magellan, 1212, 1212f
    Mariner 2, 307−308, 1211
    Mariner 5, 307−308, 1211−1212
    Mariner 10, 307−308, 1212
    Pioneer Venus, 1212
    Vega 1 and 2, 1212, 1212f
    Veneras 4 Through 16 and Vega, 1211
    Venus Express, 1212
    Venus Flyby, 1212
  history of exploration of, 324−325
  magnetic field highlights from missions to, 122t
  magnetospheres with, 141t
  19th century studies on, 1008
  radar imaging for, 1147−1149
    Magellan surface imaging, 1147−1148, 1148f
    post-Magellan radar imaging, 1148−1149
  radio wavelengths thermal emission from, 1111−1112, 1111f−1112f
  1761 and 1769 transits of, 1006
  solar wind properties with, 141t
  20th century studies on, 1013
  volcanic features of, 105−106, 105f
  X-ray emissions, 1020f, 1025−1027, 1026f
Venus, atmosphere, 305−322
  circulation and dynamics of, 316−319
    meridional wind field, 317
    meteorology, 318
    polar vortex, 318−319, 318f
    surface and lower atmosphere wind profiles, 316−317, 317f
    tides, 319
    upper atmosphere, 319
    zonal winds and superrotation, 317
  clouds and hazes in, 314−316
    appearance and motions, 314
    cloud chemistry, 316
    global variability, 315, 315f
    lightning, 316
    vertical layering, 314−315, 315f
  composition of, 307t, 311−314, 311f
    carbon dioxide, 311
    carbon monoxide, 313
    D/H ratio, 312
    ionosphere, 313−314, 314f
    isotopes, 313
    nitrogen, 311
    noble gases, 313
    oxygen and ozone, 311−312
    sulfur dioxide, 312−313
    water vapor, 312, 313t
  evolution of atmosphere and climate, 319−322
    atmospheric gases sources, 319−320
    escape processes, 320−321
    evolutionary climate models, 321−322, 322f
    surface pressure and CO2 budget, 320
    volcanism, 319−320
    water loss, 320−321
  observations, 306−308
    earth-based, 306−307, 307t
    near-IR images, 306, 307f
    space missions, 307−308
    UV images of clouds, 306f
  temperatures, 308−311, 309f
    lower atmosphere, 309−310, 309f
    middle atmosphere, 309f, 310
    surface, 308−309
    upper atmosphere, 310−311
Venus, surface and interior, 323−342
  composition, 332−335
    global implications, 332−334, 332t−333t, 334f
    surface weathering, 334−335
  general characteristics, 325−326
    orbital rotations and motions, 325
    physiography, 325−326
    radius, 325−326
    surface conditions, 326
    topography, 325−326, 326f−327f
    views of surface, 326, 327f
  impact craters and resurfacing history, 326−329, 327f−328f
  interior processes, 329−332, 331f−332f

Venus, surface and interior (*Continued*)
  radar for improved understanding of, 1135f
  tectonics, 337–341
    chasmata, 339, 339f
    coronae, 339–340, 339f
    crustal plateaus, 338–339, 338f
    fracture belts, 339, 339f
    plains fractures, grids, and polygons, 340–341
    ridge belts, 340, 340f
    tessera terrains, 338–339, 338f
    wrinkle ridges, 340
  volcanism, 335–337, 335f–337f
Venus Express
  Venus exploration with, 1212
  Venus space missions by, 307–308
Venus Flyby, Venus exploration with, 1212
Venus space missions
  Akatsuki, 325
  Cassini, 307–308
  Galileo, 307–308, 325
  Langmuir probe, 308
  Magellan, 307–308
  Mariner 2, 5, and 10, 307–308, 325
    flyby by Mariner 2, 324
    Mariner 10 on way to Mercury., 324
  Pioneer Venus Multiprobe and Orbiter, 307–308
  Pioneer Venus probes, 307–308, 324
  Vega 1 and 2, 307–308, 325
  Venera 4-16, 307–308, 324
  Venus Express, 307–308, 325
Very Large Array, New Mexico, 1108f
Very Large Telescope (VLT), 833
Viking mission
  flyby missions to satellites by, 764t
  Mars exploration with, 360t, 405, 405f
  Mars reconnaissance by, 220–222, 221f, 221t
  Viking 1, 983f
  Viking 1 and 2, Mars exploration by, 1218
VIRTIS. *See* Visible and infrared thermal imaging spectrometer
Viscosity, 190–191, 457, 471, 896
Viscous accretion, 32
Viscous overstability, 892
Viscous relaxation, 190–191
Visible and infrared thermal imaging spectrometer (VIRTIS), 325
Visible light, 885
VLT. *See* Very Large Telescope
Vogel, Hermann, 1008
Volatile element, 289
Volcanic eruptions
  classification of processes in, 108–109
  effusive, 109–111
  explosive, 111–117
    basic considerations, 111
    Hawaiian activity, 114–115, 114f–115f
    Phreato-Magmatic activity, 116–117
    Plinian activity, 115–116, 116f
    pyroclast dispersal (into vacuum), 117
    Strombolian activity, 111–112, 112f
    Vulcanian activity, 112–114, 113f
  on Io, 787–788
    Lokian eruptions, 788, 789f
    Pillanian eruptions, 786f, 787–788
    Promethean eruptions, 788, 788f
  lava flows with, 109–111
Volcanic features, 101–108
  differentiated asteroids, 108
  Earth, 101–103, 102f
  icy satellites: cryo-volcanism, 107–108
  Io, 106–107, 107f
  Mars, 104–105, 104f
  Mercury, 106, 106f
  Moon, 103–104, 103f
  Venus, 105–106, 105f
Volcanic history, 199–201, 200f
Volcanic plains, Mercury, 294–296, 295f–296f
Volcanic rocks, 101
Volcanic vents, Mercury, 294–296, 295f–296f
Volcanism, 101–120
  Ganymede and Callisto, 825–828
  intraplate, 200
  Mars, 365–367, 366f
  Moon, 574
  Moon (Earth), 509–511, 509f–511f
  planetary interiors inferred from, 117–118
  Venus, 335–337, 335f–337f
  Venus atmosphere with, 319–320
Volcanoes, 101
  Io
    Loki plume, 782, 782f
    Pele plume, 782, 782f
    Prometheus plume, 788f
    Ra Patera, 783–785, 785f
    Thor, 784f
Voyager spacecraft (1 and 2), 983f
  flyby missions to satellites by, 764t
  Ganymede and Callisto exploration by, 815
  giant planets viewed by, 733–734
  Io flyby, 782, 782f
  magnetic field highlights from, 122t
  planetary satellites observation with, 768
  Titan surface viewed by, 842, 843f
  Triton encounter with, 865, 865f
Vulcan, 19th century studies on, 1007–1008
Vulcanian activity, 112–114, 113f

W

Waldmeier, Max, 244
Warm poles, 286
WASP. *See* Wide Angle Search for Planets
Water
  astrobiology, ecology of, 211–212, 212t–213t
  giant planet atmospheres with, 726t, 727
  impact crater with, 97–98, 98f
  Mars
    dark streaks, 371–372, 372f
    erosion and weathering, 368
    evolutionary implications with, 417
    gullies, 371, 371f
    outflow channels, 370–371, 370f–371f
    valley networks, lakes, deltas, 368–370, 369f–370f
  Venus, atmosphere and loss of, 320–321
  Venus atmosphere with vapor, 312, 313t
Wave tilt (WT), 1199, 1199t
Wavelength, 896
Weathering
  chondrites, 641, 641f
  Mars, 368, 417–418
  meteorites, 641, 641f
  Moon, 512–513
  space, 293–294
  Venus, surface, 334–335
Weizsäcker, Carl von, 1017
Western boundary currents, 424
Wetherill, George, 47–48, 609
Whipple, Fred, 1017
Whipple model, 687
Wide Angle Search for Planets (WASP), 968
Wide-field Infrared Survey Explorer (WISE), 1073–1074, 1075f, 1077t
Wide-field Infrared Survey Explorer's NEOWISE project, 610–611
Widmanstätten pattern, 632
Wiechert, Emil, 1014
Wildt, Rupert, 1013
Williams, D., 783–785
Winslow crater, 84f
WISE. *See* Wide-field Infrared Survey Explorer
Witteborn, F., 781
Wobbles, 177–181, 177f–178f
  forced nutation, 181
  free core nutation, 181
  polar motion, 178–181, 179f–180f
Wolf, Rudolph, 1007
Wolfe, C., 238–239
Wright, Thomas, 1017
Wrinkle ridges, 297, 297f
  Venus, 340
WT. *See* Wave tilt
Wurm, Karl, 1017

X

X-band, 182
X-ray astronomy, birth of field of, 1019–1020
X-ray emissions, 1019–1046
  asteroids, 1039–1040
  comets, 1034–1039, 1039f
    energy spectrum, 1038
    spatial morphology, 1035–1037, 1036f–1037f
    temporal variation, 1037–1038
    X-ray luminosity, 1037
  Earth, 1020–1023, 1020f
    auroral emissions, 1020–1022, 1021f–1022f
    geocoronal emissions, 1022–1023
    nonauroral emissions, 1022, 1023f
  Galilean satellites, 1031–1032, 1032f
  heliosphere, 1040–1041
  Io plasma torus, 1032, 1033f

Index

Jupiter, 1020f, 1029–1031
    auroral emissions, 1029–1031, 1030f
    nonauroral emissions, 1030f, 1031
Mars, 1020f, 1027, 1028f–1029f
Mercury, 1025
Moon, 1020f, 1023–1025, 1024f
Saturn, 1032–1033, 1034f
Saturn's rings, 1033–1034, 1035f
Venus, 1020f, 1025–1027, 1026f
X-ray Fluorescence Spectrometer (XRFS), 399t
    Apollo science experiments with, 561t
X-ray light, 479, 480f, 481
X-ray luminosity, 1037
X-ray spectroscopy, 300
XRFS. *See* X-ray Fluorescence Spectrometer

Y

Yardang features, 1148
Yarkovsky, I. O., 591
Yarkovsky effect, 70, 72–73
    orbit determination for asteroids, 1153–1155, 1155f
Yield stress, 193

Z

Z Machine, 747–748, 748f
Zodiacal cloud, 20, 1077–1079, 1077f
Zodiacal dust cloud, infrared views of, 1074–1078, 1076f–1077f, 1077t, 1078f–1079f
Zodiacal light, 657, 663–664
Zöllner, Friedrich, 1008
Zonal flows, 127
Zond 3, Moon exploration by, 1214
Zond 5, 6, 7, and 8, Moon exploration by, 1214

Jupiter, 1020f, 1029–1031
    auroral emissions, 1029–1031, 1030f
    nonauroral emissions, 1030f, 1031
Mars, 1020f, 1027, 1028f–1029f
Mercury, 1025
Moon, 1020f, 1023–1025, 1024f
Saturn, 1032–1033, 1034f
Saturn's rings, 1033–1034, 1035f
Venus, 1020f, 1025–1027, 1026f
X-ray Fluorescence Spectrometer (XRFS), 399t
    Apollo science experiments with, 561t
X-ray light, 479, 480f, 481
X-ray luminosity, 1037
X-ray spectroscopy, 300
XRFS. *See* X-ray Fluorescence Spectrometer

Y

Yardang features, 1148
Yarkovsky, I. O., 591
Yarkovsky effect, 70, 72–73
    orbit determination for asteroids, 1153–1155, 1155f
Yield stress, 193

Z

Z Machine, 747–748, 748f
Zodiacal cloud, 20, 1077–1079, 1077f
Zodiacal dust cloud, infrared views of, 1074–1078, 1076f–1077f, 1077t, 1078f–1079f
Zodiacal light, 657, 663–664
Zöllner, Friedrich, 1008
Zonal flows, 127
Zond 3, Moon exploration by, 1214
Zond 5, 6, 7, and 8, Moon exploration by, 1214